T0093709

Eponym Dictionary of Fishes

Bo Beolens, Michael Grayson and Michael Watkins

Whittles Publishing

Published by
Whittles Publishing Ltd.,
Dunbeath,
Caithness, KW6 6EG,
Scotland, UK

www.whittlespublishing.com

ISBN 978-184995-498-3

Printed and bound by CPI Group (UK) Ltd, Croydon, CR0 4YY

Contents

Introduction

In 2012 Dr Peter Uetz, who is one of the brains behind that invaluable resource 'The Reptile Database', wrote a review of our publication, *The Eponym Dictionary of Reptiles,* finishing with the following words: "....Beolens and co-authors have produced a great book that is fun to read. Notably, they have already published similar books on birds and mammals (Beolens and Watkins 2003; Beolens et al. 2009) and reportedly have a companion volume on amphibians in press. If they live long enough to work through the 30,000 species of fish, a future eponym dictionary of vertebrates may keep saving biologists from buying *People* magazine for years to come."

After *The Eponym Dictionary of Reptiles* was published, we did indeed publish *The Eponym Dictionary of Amphibians* (2011) and followed that with *The Eponym Dictionary of Birds* (2014). We decided that we would try and fulfil his expectations, if only in part, as the *Eponym Dictionary of Fishes* only covers the names of people after whom fish have been named. However, this does mean that, spread over all our published volumes, we have covered everyone after whom a vertebrate has been named.

We are well aware that this is a never-ending task as every day someone somewhere finds a new fish and has it described and, as underwater is an environment which is hostile to most land-based vertebrates, we cannot just dive under the waves and count how many more species are waiting to be discovered! The use of modern technologies, especially manned submersibles has, if anything, reinforced the sense of impossibility of ever achieving 100% coverage.

Sadly, a few months into starting on fishes Mike Watkins was diagnosed with cancer. After surgery could not remove the tumour the prognosis was poor. Mike never gave up the fight and was determined to make a solid contribution to this book. After a year of fighting the disease he had winkled out the vast majority of fish eponyms and identified a large proportion of those who were being honoured. Sadly, in a few short weeks after his health deteriorated and he sadly passed. We dedicate this book to his memory.

How it is done

We are the authors, but this volume reflects an incredible collaboration from far too many people to fully acknowledge (although we try to cover the most prolific contributors in the acknowledgements section). Some people have given many hours freely and generously, others answered a query or two and very many people mentioned herein have corrected their own entries, and those of their friends and families.

Much of the research has used online resources but this has had to be supplemented and expanded from library research and contact with many experts around the world. For example, researching the entry for Zander was a great example of what it sometimes takes to identify eponyms. Tracking down the specific name *zanderi* required an international effort! We approached Christopher Scharpf (of the wonderful ETYFish Project), who, in turn asked his German friend and colleague Erwin Schraml if he could research Zander's identity. Erwin asked Hans-J Paepke, retired Curator of Ichthyology at the Berlin Museum of Natural History and a scholar of ichthyological history. Dr Paepke forwarded our query to Dr Natalia Chernova, an ichthyologist at the Zoological Institute of the Russian Academy of Sciences in St. Petersburg. Dr Chernova spent a considerable amount of time digging through library and museum archives to compile everything she could find about Zander and we are grateful to them all for the facts for that entry. It just goes to show how wonderfully cooperative scientists and academics, both professional and amateur, are across national divides!

Who is it for?

Vernacular names of animals often contain a person's name (such names are called 'eponyms'). Furthermore, many scientific names contain the Latinised name of its discoverer, or some other person thought worthy of the honour, whether it be the second part of a binomial for a species. Indeed, some genera names are also eponyms. So, this book is for the amateur ichthyologist, the student of zoology or anyone else interested in taxonomy, nomenclature or fishes.

How to use this book

The following abbreviations may be used in the lists of species and the general texts.

Alt.	Alternative Common Name
S	Synonym
JS	Junior Synonym of
Orig.	Original Scientific Name

AMNH	American Museum of Natural History, New York
BM	British Museum
BMNH	British Museum of Natural History, London
CSIRO	Commonwealth Scientific and Industrial Research Organisation
IRD	Institut de Recherche pour le Développement
MNHN	Muséum Nationale d'Histoire Naturelle, Pars
MRAC	Musée Royale de l' Afrique Centrale, Tervuren, Belgium
NOAA	National Oceanic and Atmospheric Administration, Washington D.C.
ORSTOM	Office de la recherche scientifique et technique outre-mer
RMCA	Royal Museum for Central Africa, Tervuren, Belgium
RMNH	Rijksmuseum van Natuurlijke Histoire, Leiden, Netherlands
ROM	Royal Ontario Museum

USNM	Smithsonian Institution (National Museum of Natural History)
UN	United Nations
WW1	First World War 1914-1918
WW2	Second World War 1939-1945

This book is arranged alphabetically by the names of the people after whom fishes have been named. Generally, the easiest way to find what you are looking for is to look it up under the name of the person that is apparently embedded in the animal's common or scientific name. We say 'apparently', as things are rarely as simple as they seem. In some names, for example, the apostrophe implying ownership is a transcription error; others may have been named after places. So, we have included any names where we think confusion might arise, but we do not promise to have been completely comprehensive in that respect. You should also beware of spelling. Surf the net and you may well find animals' names spelt in a number of different ways – that greatest resource is also full of inaccuracies and misinformation. We have tried to include entries on those alternatives that we have come across. Where a surname begins with an accented letter (such as 'Ö') which is 'Latinised' using several letters (i.e. Oe) we list the name alphabetically under the proper accented name, but we try to also list the 'Latinised version' alphabetically. For example, Öser will appear correctly spelt alphabetically but the Latinised Oeser will also appear with '(See **Öser**)'. We don't bother to do this when the entry would be in the same place anyway.

Each biography tries to follow a standard format: first, you will find the name of the person honoured; next, there follows a list of fishes named after that person, arranged in order of the year in which they were described. (This list gives common names, scientific names, names of the people who first described each species and the date of the original descriptions – in that sequence). Alternative English names follow in parentheses and are each preceded by the abbreviation Alt.; different scientific names (where taxonomists have disagreed) are preceded by the abbreviation S. (synonym) etc; finally, there is a brief biography of that individual. We try to include at least one publication that the subject wrote or of which he or she is a co-author; space restrictions mean that we cannot have exhaustive lists and we think that one example is enough. The reader who is curious for more will always be able to Google their way to a bibliography, using the example quoted as a starting point. In a work of this size we cannot add all the sources to each entry and feel a bibliography would also be ridiculously long and not necessarily helpful. This is an eponym dictionary NOT a scientific paper. If facts are dubious or uncheckable we say so, and where we have had to speculate we acknowledge that too. To assist you in your search, we have cross-referenced the entries by adding (in bold) the person's alternative names or names of their relatives who also appear elsewhere in the book as some fishes are named in different ways after the same person.

A person's fame does not get them a correspondingly long entry – in fact, often the opposite. Very famous people, like Charles Darwin, have fairly brief write-ups. He is so well known, and has so much written about him, that it is necessary for us just to indicate that he is commemorated.

Sometimes animals are named in the vernacular after the finder, the person who wrote the scientific description or some other person of the latter's choice. Thus: Beolens Fish *Piscatorius graysoni* B Beolens, 2019 would appear under entries for both Beolens and

Grayson. When more than one person has thought a species new, the fish may get more than one set of names, so it can warrant an entry in several places! We cannot hope to cover all redundant or junior synonyms, but try to include those which have appeared very prominently or over a long time or where the jury may still be out.

There is a very great number of recent namings of fossil fish. As the rate at which fossil remains are discovered and described seems to be increasing exponentially and the disagreement among the palaeontologists seems epidemic, we decided that we would ignore any prehistoric extinctions; in simplistic terms, before Columbus discovered America.

In the last 100 years many cities and countries have changed names. We have normally put in the name by which the subject would have known it, putting in brackets the name by which it is now known e.g. Salisbury (Harare) Rhodesia (Zimbabwe) or sometimes in reverse: Sri Lanka (formerly Ceylon).

What's in a name?

Tracking down the provenance of eponymous fish names, and finding out about the individuals responsible for them, can prove to be fraught with difficulties. The final list includes a few where the same species has been named after two people. The names include some which sound like people's names but in fact are not, plus indigenous peoples, fictional characters, Biblical references and references to mythology. Additionally, there are entries for a *very* few names of people whom we have been unable to identify!

Common names

Many species of fish currently have no accepted common name. In such cases, we have resisted the temptation to invent names. Thus the reader will often encounter the use of the term 'sp.' For example, Bagrid Catfish sp. means that this is a species of catfish of the family Bagridae, but one that lacks a common name (at least in English).

Describers and namers

New species are usually first brought to the notice of the scientific community in a formal, published description of a type specimen, essentially a dead example of the species, which will eventually be lodged in a scientific collection. The person who describes the species will give it its scientific name, usually in Latin but sometimes in Latinised ancient Greek! Sometimes the 'new' animal is later reclassified and then the scientific name may be changed. This frequently applies to generic names (the first part of a binomial name), but specific scientific names (the second part of a binomial), once proposed, usually cannot be amended or replaced – there are precise and complicated rules governing any such changes.

The scientific names used in this book are largely those used in the ETYFish database and we must here acknowledge what a wonderful help it has been in providing basic leads to where we could find more extensive particulars. Especially we thank Christopher Scharpf of ETYFish. We have benefitted from a very enjoyable collaboration with him, swapping sources and helping each other with identifications.

We may have missed a few recently published taxonomic changes (although we have tried to be comprehensive up until the final proof of the book), but we have put the name of the original describer (authors) after every entry. Because alterations to taxonomy have been so radical, and so swiftly changing, we decided never to put brackets round changed entries.

Although we have used current scientific names as far as possible, these are not always as universal as the casual observer might suppose. There is no single 'world authority' on such matters.

There are no agreed conventions for English names and indeed the choice of vernacular names is often controversial. Often the person who coined the scientific name will also have given it a vernacular name, which may not be an English name if the describer was not an English speaker. On the other hand, vernacular names have often been added afterwards, frequently by people other than the authors of the description. In this book, therefore, when we refer to an animal having been NAMED by someone, we mean that that person gave it the ENGLISH name in question. We refer to someone as the AUTHOR (or DESCRIBER) when they were responsible for the original description of the species and hence for its scientific name. As we have said above, it is the author of the scientific description's name, which is given after the scientific name in the biographies. It is also worth noting that some species have been identified and even named well before a scientific description is published. Often authors acknowledge provenance in the name, sometimes not, and sometimes the collector, namer or supplier of the specimen is mentioned in the paper or book but not honoured in the name they publish.

BIOPAT is a German society that encourages people, organisations and companies to donate money for taxonomic research. They are rewarded by having a species named after themselves or someone they nominate. The effect is that the describers often have little or no knowledge of the people after whom they name species!

Animals named after more than one person

Throughout the text you may come across several different names for the same species. In some cases, these names are honorifics; for example, among amphibians, *Chari's Bush Frog Philautus charius Rao, 1937 is the same as Seshachar's Bush Frog.* This peculiarity has sometimes come about through simple mistakes or misunderstandings – such as believing juveniles or females to be a different species from the adult male. In some cases, the same animal was found at about the same time in two different places and only later has it emerged that this is the same animal named twice. Some of these duplications persist even today, with the same species being called something different in different places or by different people. However, sometimes the apparent duplication is that a different name may be used for different subspecies to distinguish them from the nominal form. Even more confusingly, a vernacular name given in one place ends up in common usage, despite an alternative being designated by the scientific authors.

Male or female

In some cases, we know that an animal is named after a man, even though its scientific name is in the feminine. This mostly occurs when a name ends in the letter 'a'. Presumably, the reason for this is that many singular Latin nouns ending in 'a' are feminine – for example, 'mensa' means 'table' (nothing very feminine about that), and the possessive/genitive case is 'mensae', not 'mensai'. There are a number of masculine Latin nouns, e.g. 'agricola' that means 'farmer', but they are declined as though they are feminine, the convention being that the feminine form is adopted in such cases. This convention has been falling into disuse in recent years. It is quite striking how many modern namings ignore it, and the masculine ending is becoming more widespread. For instance, two birds described in the 1920s named after Cervera (a wren and a rail) both have the binomial *cerverai*. There are also cases where female eponyms are wrongly 'Latinised' as masculine. In recent times the laborious process of 'correcting' this has been followed, but more often the 'incorrect' ending stays. The same can be said for species named after a group or several people, for example a family. This should mean the ending is '*orum*' so one named after my wife and I would be *beolensorum*. However, it is often not the case and the author chooses to honour it as if after just one person when it comes to the Latinised name.

Red herrings

Further confusion arises from a number of animals which appear to be named after people, but – upon closer examination – turn out to be named after a place that was itself named after a person. We have included these with an appropriate note, as other sources of reference will not necessarily help the enquirer. For example, 'Victoria's' fish, bird or whatever, could be named directly after Queen Victoria, the Australian State of Victoria or after Lake Victoria each of the latter being themselves eponymous.

Academic qualifications and job titles

We know that there are considerable differences between the levels of expertise and knowledge that students must demonstrate in order to achieve any qualification in different countries and academic disciplines. We found it very difficult to devise any way of comparing between these demands and expectations except at the most basic comparison level. We have, therefore, as we are brought up in the English academic tradition, resorted to using the expression bachelor's and master's degrees and doctorates and have to apologize to our readers from other traditions that these expressions may not match their own experiences. This extends to abbreviations so we use BSc to indicate a degree of bachelor of science even although the persons degree from an English-speaking university might be BS in their country.

This rather simplistic approach means that we have had to sacrifice some of the finer points of higher education. As examples, we have ignored the higher achievement of 'habilitation' which is used as a post-doctoral qualification in some European, Central

Asian, North African and Latin American countries nor have we succeeded in finding any intermediate English degree to equate to the 'licentiate' status that we see in some countries, probably because its meaning varies enormously from country to country.

Incidentally we do not regard 'Professor' as a qualification, but as a job title. 'Doctor' may be a qualification or a title and when someone has a medical qualification we try and make it clear. Similarly if a person has an honorary doctorate and has not earned the title through solid academic work and research, we will highlight that. 'Professor' can become 'Prof' and 'Doctor' can be shortened to 'Doc' and often these are semi-humorous titles of affection or respect or just nick-names.

Military, naval and clerical titles are also just job descriptions, even though the persons attaining such ranks may have actually achieved something to get them! Our policy is to always give the most senior rank or designation that the individual reached as it is thereby much easier to find them if the reader wants to research that person's life further. Looking for 'Lieutenant-General Smith' is always easier than 'Lieutenant Smith', let alone "Private Smith' even if they were honoured in a name when they held a lower rank. Similarly, looking for an obscure curate is made easier if he later became a bishop.

We have tended to use original clerical titles, or directly translated them from the language they appear in. For example, Père David is used partly as that in itself is an eponym but also as it is well known, whereas other clerics of the same rank, religion and nation might appear as 'Father', and so forth. Catholic titles are used for Catholics, Islamic ones for followers of Islam, etc. We follow the same principle when mentioning political offices.

Deliberate misspellings in scientific names

Some Latin rendering of names may confuse because of slight alphabet differences. For example, any name beginning 'Mc' may, in a binomial, be rendered 'mac'. It is wise, therefore, to search for both spellings. Confusion may also arise as other alphabets (such as Cyrillic) have fewer or larger numbers of letters, some of which are often interchangeable when written in the modern English version of the Roman alphabet. Examples are V and W, and J and Y, and letter combinations such as Cz and Ts for Czar and Tsar.

By convention, diacritical marks, such as accents in French and the tilda used in Spanish and Portuguese, have to be ignored in scientific names and the phonetic sense of them expressed in other ways. The Scandinavian letters å, ä, ö and ø are normally expressed as aa, ae, oe and oe and the German ö and ü as oe and ue. In the English name, either spelling is acceptable. We have tried to use correct accents in people's names or their book titles etc.

There are differences in how diacritical marks can alter alphabetical orders. Orthography is the set of conventions that govern how a particular language is written down. Some languages are regulated by various academies and two examples of them are French and Swedish. In French, the order in which letters modified by the use of a diacritical mark appear after the original letter mean that a French sequence will look like "....c d e ê è ê f g..." The Swedes, however have three extra vowels which are regarded as entirely separate letters and form the last three letters of the Swedish alphabet, thus "... x y z å ä õ". We have endeavoured to respect the various traditions, as well as formal rules, when deciding the order of entries.

Many other languages (including English) have no rules formally set down but just develop in a *de facto* manner. Furthermore, there is no universal way of recording 'given' names and 'family' names. In the UK 'Bo Beolens' would indicate that my family name is 'Beolens' and my given name 'Bo'. Were I Portuguese there would, perhaps, follow another name (often not used) indicating my male parent's line. Incidentally 'Bo' is not my given name but a nickname coined in my youth which is used by everyone except my late parents. If I had an entry in this book my name would appear thus: Richard 'Bo' Beolens (actually I would have other forenames and a hyphenated surname too, rather than my *nom de plume* 'Bo Beolens').

Chinese names are often written with the family name first followed by the (usually hyphenated) given name(s). This can be very confusing to westerners. Russians most often have a given name, a patronymic and a family name. Wherever possible we try to render names in the 'right' way but subject to usage. That is if someone is generally listed, recorded and known by a name that, in their own land and language is inaccurate, we stick with the error so that those wanting to know more about them will finds them in the literature and online.

Acknowledgements:

Major Contributors

Michelle Denise Freeborn (q.v.) for the use of her wonderful illustration of a John Dory

Steffen Heller (q.v.) has been very helpful identifying, and giving some detailed biographical information on aquarists in general and in particular, German killifish enthusiasts that appear in this work.

Dr Jean Henri Huber (q.v.) has been extremely helpful on confirming Cyprinodont eponyms and supplying details. His website is a terrific resource too: http://www.killi-data.org/

Gary William Lange is a complete fanatic regarding American Rainbow fish and founded the Rainbowfish Study Group of North America. He has helped us with many eponyms pertaining to rainbowfish and their enthusiasts worldwide.

Christopher Scharpf created and continues to extend his database of all fish etymology: http://www.etyfish.org/ he has personally supplied, corrected or enhanced a great many entries.

Dr Jevgeni Shergalin has been indispensable as a supplier, confirmer or expander of information on a great many Russians after whom fish has been named. We first corresponded with him when he started to help us with our Eponym Dictionary of Birds.

Thanks also go to…
…the very many people who appear in this work and who have helped with their own entries or those of relatives or colleagues.

A

Abadie

Djebba Mormyrid *Marcusenius abadii* GA Boulenger, 1901

Captain George Howard Fanshawe Abadie (1873–1904) was an officer in the 16th Lancers who was working as a political officer and administrator in Northern Nigeria (1897–1904), where he died suddenly from a 'malignant fever'. He was a Fellow of both the Royal Geographical Society and the Zoological Society of London. He collected fish along the River Niger and presented the mormyrid holotype to the BMNH.

Abbott, JF

Oriental False Gudgeon genus *Abbottina* DS Jordan & HW Fowler, 1903

Dr James Francis Abbott (1876–1926) was an American zoologist who also became a diplomat and a special adviser on Japanese affairs. He graduated from Stanford (1899) and went to Japan where he taught English at the Imperial University of Tokyo and at the Japanese Naval Academy. He returned to America and taught zoology at Washington (1904–1917), during which period he was one of the co-founders of the St Louis Zoological Society and the St Louis Zoo (1910). He returned to Japan (1919) as commercial attaché at the US Embassy. He was one of the US representatives at the Washington Naval Conference (1921). He wrote on ichthyology in papers such as: *The Marine Fishes of Peru* (1899) and on Japanese policies in: *Japanese expansion and American policies* (1916).

Abbott, W

Veracruz Whiff *Citharichthys abbotti* CE Dawson, 1969

Dr Walter Abbott worked with Charles Dawson at the Gulf Coast Research Laboratory. Dawson said in his etymology: "*…I take pleasure in naming this species after my colleague, Dr. Walter Abbott, who assisted in the collection of the type series.*"

Abbott. WL

Deep-sea Spiny Eel sp. *Notacanthus abbotti* HW Fowler, 1934

Dr William Louis Abbott (1860–1936) was a naturalist and collector. Initially qualifying as a physician at the University of Pennsylvania and working as a surgeon at Guy's Hospital in London, he decided not to pursue medicine but to use his private wealth to engage in scientific exploration. As a student (1880) he had collected in Iowa and Dakota, and in Cuba and San Domingo (1883), in the company of Joseph Krider, son of the taxidermist John Krider. He went to East Africa (1887), spending two years there. He studied the wildlife of the Indo-Malayan region (1891), using his Singapore-based ship 'Terrapin', and made large collections of mammals from Southeast Asia for the USNM in Washington DC. He moved on to Thailand (1897) and spent 10 years exploring and collecting in and around the China Sea. He provided much of the Kenya material in the USNM and was the author of *Ethnological Collections in the United States National Museum from Kilima-Njaro, East Africa* (1890–1891). He returned to Haiti and San Domingo (1917), exploring the interior and discovering more new species. He retired to Maryland but continued the study of birds all his life. Amongst other taxa 18 birds, two amphibians, three mammals and two reptiles are named after him.

Abdoli

Stone Loach sp. *Paraschistura abdolii* J Freyhof, G Sayyadzadeh, HR Esmaelli & MF Geiger, 2015

Dr Asghar Abdoli is an Iranian fish ecologist who collected this species together with Jörg Freyhof (2007). He is currently Associate Professor at Shahid Beheshti University Department of Biodiversity and Ecosystem Management. The Gorgan University of Agricultural Sciences and Natural Resources, Gorgan, Iran awarded his BSc, Department of Fisheries and Environment, Faculty of Natural Resources, University of Tehran his MSc and Laboratoire d'ecologie des hydrosystèmes fluviaux, University Lyon 1, Villeurbanne, France his PhD. He was a researcher at the Iranian Fisheries Research Organization and Research Scientist at the Ecological Academy of the Caspian Sea (1991–1994) then became a lecturer at Gorgan University of Agricultural Sciences & Natural Resources – Fisheries and Environment (1994–2005) then head of department Biodiversity and Ecosystems Management at the Environmental Sciences Research Institute, Shahid Beheshti University, Tehran (2005–2008) before his current appointment. His published papers include: *The role of temperature and daytime as indicators for the spawning migration of the Caspian lamprey* Caspiomyzon wagneri *Kessler, 1870* (2017).

Abdul Kalam

Airbreathing Catfish sp. *Horaglanis abdulkalami* Subhash Babu Kallikadavil, 2012

Dr Avul Pakir Jainulabdeen Abdul Kalam (1931–2015) was a scientist in physics and aerospace engineering, who became a statesman, as the 11th President of India (2002–2007).

Abdurakhmanov

Abdurakhmanov's Tadpole-Goby *Benthophilus abdurahmanovi* DAO Ragimov, 1978

Yusif A Abdurakhmanov (b.1912) was one of the leading ichthyologists of Azerbaijan. He wrote: *Freshwater Fishes of Azerbaijan* (1962).

Abe, G

Eelpout sp. *Japonolycodes abei* K Matsubara, 1936

Genkiti Abe of Nisiura, Aiti Prefecture, Japan, collected the eelpout type. He is also recorded as having 'cordially supplied material' from his area to Ituo Kubo for his study of shrimps.

Abe, K

Abe's Mangrove Goby *Mugilogobius abei* DS Jordan & JO Snyder, 1901

Kakichi Abe, of Tokyo, was a former student of the senior author at Stanford University. He accompanied the authors in their travels throughout southern Japan, to the 'great advantage' of their work.

Abe, T

Shining Tubeshoulder *Sagamichthys abei* AE Parr, 1953
Sea Toad sp. *Chaunax abei* Y Le Danois, 1978
Abe's Flying Fish *Cheilopogon abei* NV Parin, 1996
Spikefish sp. *Paratriacanthodes abei* JC Tyler, 1997
Abe's Puffer *Pao abei* TR Roberts, 1998
[Alt. Brown-spot Puffer]
Yellowfin Sea Bream *Dentex abei* Y Iwatsuki, M Akazaki & N Taniguchi, 2007

Professor Tokiharu Abe (1911–1996) was a Japanese ichthyologist at University Museum, University of Tokyo. He was particularly well-known for his taxonomic studies of Asian pufferfish (*Tetraodontidae*), describing a number of species. He was an honorary foreign member of the American Society of Ichthyologists and Herpetologists. He died from a cerebral haemorrhage in a hospital in Tokyo.

Abe, Y

Abe's Angelfish *Centropyge abei* GR Allen, F Young & PL Colin, 2006

Dr Yoshitaka Abe (b.1940) became (2000) Executive Director of the Aquamarine Fukushima Aquarium, Japan. He is also a founder committee member of the International Aquarium Forum. After graduating from Tokyo University he joined the Ueno Zoo. Later (1968–1969) he was a researcher at the Kuwait Institute for Scientific Research, then worked at the Tama Zoo and Tokyo Sea Life Park. Among his publications is the co-written: *Fishes of Kuwait* (1972). The etymology states: "*...The species is named* abei *in honour of Dr. Yoshitaka Abe, without whose faith, guidance and support, none of the work would have been possible. Dr. Abe is the director of Aquamarine Fukushima, a world class public aquarium in Fukushima Prefecture, Japan. He is largely responsible for many innovations in aquarium science and design including display of large tunas up to 100kg, jellyfish keeping, and the first public display of large hammerhead sharks. Aquamarine Fukushima and Dr. Abe provided the entire budget and material support for the deepwater operations that resulted in the collection of the new angelfish.*"

Abeele

Lake Tanganyika Cichlid sp. *Greenwoodochromis abeelei* M Poll, 1949
[Syn. *Limnochromis abeelei*]

Marcel Henri Joseph Van Den Abeele (1898–1980) was the General Administrator of the Belgian Colonies. He was also an agronomist who was Director General of the National Institute for Agronomic Studies of the Belgian Congo. The etymology says: "*Espèce dédiée à M. M. Van den Abeele, Administrateur général des Colonies, en hommage de reconnaissance pour les encouragements multiples et les nombreuses marques d'intérêt témoignés à la mission.*" This is a reference to the Belgian Hydrobiological Mission to Lake Tanganyika (1946–1947), during which the holotype was collected.

Abel

Robber Tetra sp. *Brycinus abeli* HW Fowler, 1936

Henri Abel was the Administrator at Fort Sibut, Central African Republic, which is where the tetra holotype originated. The etymology states that he 'developed native interest and materially assisted the expedition' that collected the holotype.

Abendanon

Sailfin Silversides sp. *Telmatherina abendanoni* MCW Weber, 1913

Eduard Cornelius Abendanon (1878–1962) was a Dutch malacologist and mining engineer. He was born in Pati, Indonesia. He was educated at Delft Technical College and took study tours in Europe. He was employed by the Indian Government (1900–1901), then the Dutch East Indies Mining Service (1901–1906). During this time (1903–1905), at his own expense he undertook an exploration tour in China and a voyage round the world. He was part of the Central Celebes Expedition, including studying the results (1907–1918). He also (1915–1920) conducted comprehensive geologic surveys of Sulawesi and was the first to describe the basic rocks of this region, as well as in Mainland China. He was Extraordinary Professor at Amsterdam University (1921–1925). During that time Weber was Professor of Zoology there. Among his publications is: *The Red River Geology of Sichuan Province (China)* (1906). He is also remembered in the name of at least one mollusc, *Tylomelania abendanoni*.

Abernethy

Lanternshark sp. *Etmopterus abernethyi* JAF Garrick, 1957
[Treated as a junior synonym of *E. lucifer*, but taxonomy of this genus requires further work]

Fred Abernethy (d.1995) was chief engineer on board the 'Holmwood' when it was captured and sunk by German raiders operating around New Zealand and in the Pacific (late 1940). He was captive on board one of the raiders until released on to Emirau Island (east of New Guinea). He wrote of his experiences in: *A Captive's Diary* (1985). He later worked on the research vessel 'M T Thomas Currell' and contributed greatly to the collection of New Zealand elasmobranchs. He was on the New Zealand Chatham Islands 1954 Expedition. Garrick named the shark when commercial fisherman Richard Baxter (q.v.) (1956) caught it.

Abhoya

Garra sp. *Garra abhoyai* SL Hora, 1921

No etymology is given by Hora, but perhaps named after Baboos Abhoya Churn Chowdry (fl. 1899), who was a scientific illustrator at the Indian Museum in Calcutta.

Abilhoa

Armoured Catfish sp. *Ancistrus abilhoai* AG Bifi, CS Pavanelli & CH Zawadzki, 2009

Dr Vinícius Abilhoa is a zoologist and curator of fishes and head of the laboratory, Museu de História Natural do Capão de Imbuia, Curitiba, Paraná, Brazil. He is also the curator of the fish collection of the Capão da Imbuia Natural History Museum. The Pontifical

Catholic University of Paraná awarded his bachelor's degree (1989) and the Federal University of Paraná his master's (1998) and doctorate (2004). He did post-doctoral work at the Japan Biodiversity Centre (2006), and later (2011) trained in Statistics for Ecology and Conservation at the Smithsonian Conservation Biology Institute. He co-wrote: *Fishes of the Atlantic Rain Forest Streams: Ecological Patterns and Conservation* (2011). He collected some of the type series, and was thanked for his great assistance to the describers.

Abramov

Roughy sp. *Hoplostethus abramovi* AN Kotlyar, 1986

Alexey Aleksandrovich Abramov is a Russian ichthyologist whose colleague and friend, Kotlyar, honoured him for their "*...many years of working together.*" He works at the Russian Federal Research Institute of Fishery and Oceanography. He wrote: *Species composition and distribution of Epigonus* (Epigonidae) *in the world ocean* (1992).

Abrau

Abrau Sprat *Clupeonella abrau* SM Maliatsky, 1930

This is a toponym; it refers to Lake Abrau, near Novorossiysk, Russia, the only known distribution of the species.

Abrishamchian

Crested Loach sp. *Paracobitis abrishamchiani* H Mousavi-Sabet, S Vatandoust, MF Geiger & J Freyhof, 2019

Dr Mir-Jafar Abrishamchian (1930–2018) and Ali Abrishamchian (1954–2007). The species is named in honour of Mir-Jafar Abrishamchian and his son Ali Abrishamchian (1954–2007), described as "great benefactors" in Guilan Province (Iran), to respect their developmental services in support of the University of Guilan and its students. (NB. As more than one person is honoured, the binomial should be in the plural form = *abrishamchianorum*)

Ab'sáber

Armoured Catfish sp. *Harttia absaberi* OT Oyakawa, I Fichberg & F Langeani-Neto, 2013

Aziz Nacib Ab'Sáber (1924–2012) was a geographer, geologist and environmentalist who was Emeritus Professor of the University of São Paulo. The etymology states that his "*... contributions represent a landmark in the knowledge of geography, ecology and geomorphology of the Brazilian territory*". He published over 480 works including *Ecossistemas do Brasil* (2006).

Abuelorum

Grandparents Clingfish *Tomicodon abuelorum* WA Szelitowski, 1990

Abuelorum is a Latinized derivation from the Spanish word *abuel,* which means grandparent, and here refers to the author's grandparents: Irma Drescher and Martha and Telford Walker.

Abyss

Deep Blue Chromis *Chromis abyssus* RL Pyle, JL Earle & BD Greene, 2008

Not a true eponym but named for the documentary film *Pacific Abyss*, produced by the British Broadcasting Corporation (BBC), which funded the expedition during which the holotype was collected.

Accorsi

Neotropical Rivuline sp. *Austrolebias accorsii* DTB Nielsen & D Pillet, 2015

Bruno Accorsi is an environmentalist who helped Dalton Nielsen and Didier Pillet collect the type in Bolivia. He also helped collect at least one other holotype.

Acero

Hagfish sp. *Eptatretus aceroi* AP Fernandez & B Fernholm, 2014

Blue-spotted Barnacle Blenny *Acanthemblemaria aceroi* PA Hastings, RI Eytan & AP Summers, 2020

Dr Arturo Acero Pizarro (b.1954) is Associate Professor (1986–2018) at the Universidad Nacional de Colombia. His bachelor's degree was awarded by Universidad Jorge Tadeo Lozano (1977), his master's by the University of Miami (1983) and his doctorate by the University of Arizona (2004). He was an instructor at Universidad Jorge Tadeo Lozano (1977–1978) and a Marine Biologist at Instituto de Investigaciones Marinas de Punta de Betín, INVEMAR (1981–1986). His published papers include: *Anotaciones ecológicas y sistemáticas sobre los peces de la familia Pomacentridae en el Caribe colombiano* (1978) and *Real identity of the northern Colombian endemic sea catfish* Galeichthys bonillai *Miles, 1945 (Siluriformes: Ariidae)* (2006). (See also **Arturo**)

Achilles

Achilles Tang *Acanthurus achilles* G Shaw, 1803

Achilles was the mythical Greek hero of the Trojan War. His only vulnerable spot was his heel (hence 'Achilles' heel'). The tang has a (roughly) heel-like mark on its body.

Ackley

Ackley's Ray *Rostroraja ackleyi* S Garman, 1881

[Alt. Ocellate Ray]

Lieutenant (later Rear Admiral) Seth Mitchell Ackley (1845–1908) was the officer commanding the Coast Survey steamer 'Blake' (1877). He joined the navy (1862) entering the US Naval Academy. After graduating (1866) he was commissioned as a Second Lieutenant. His first command (1876–1877) was 'RS Wyoming'. He was stationed at Olongapo and Cavite and also served as a member of the Naval War College at Newport and the General Board of the US Navy. Garman thanked him "for much valuable assistance".

Acosta

Suckermouth Catfish sp. *Astroblepus acostai* CA Ardila Rodríguez, 2011

Dr Eduardo Acosta Bendek (1930–2014) was a physician and Director, for more than 35 years, of the Universidad Metropolitana de Barranquilla, Colombia, where the author worked. The latter thanked Acosta "for cooperation with his research both in and outside of the institution."

ACSI

Lake Tanganyika Catfish sp. *Chrysichthys acsiorum* M Hardman, 2008

Driftwood Catfish sp. *Spinipterus acsi* A Akama & CJ Ferraris, 2011

ACSI stands for All Catfish Species Inventory. The *Chrysichthys acsiorum* binomial is to honour all those who are involved in achieving ACSI's objectives.

Acuen

Armoured Catfish sp. *Hisonotus acuen* GSC Silva, FF Roxo & C Oliveira, 2014

The Xavante are an indigenous ethnic group who live in the region inhabiting region between Rio das Mortes and Rio Culuene, Mato Grosso, Brazil. They are known in anthropological literature as *acuen*. (See also **Xavante**)

Adamastor

Adamastor's Freshwater Stingray *Potamotrygon adamastor* JP Fontenelle & MR De Carvalho, 2017

Adamastor is a mythic character created by the Portuguese poet, Luís Vaz de Camões, who gave his creation a 'backstory' as one of the Giants in Greek mythology, although the name does not actually appear in Greek myth.

Adams

Twilight Goby *Varicus adamsi* RG Gilmore, JL Van Tassell & L Tornabene, 2016

Michael L Adams (1934–1993) was the late 'famed' research submersible pilot at Johnson Sea Link, Harbor Branch Oceanographic Institution, who had been a US Navy Diver during the Korean war. He 'painstakingly' captured Bahaman specimens of the goby during a 30 to 45-minute chase (1987) using a 26-ton submarine in 'simultaneous multiple thrust, multidirectional mode'. He was one of the five original research submersible pilots in the United States.

Adamson, JC

Adamson's Grunter *Hephaestus adamsoni* E Trewavas, 1940

Charles Thomas Johnston Adamson (1901–1978) emigrated from England to Australia (1923). He worked as a sheep shearer and in the sugarcane fields until moving to Papua (1926) to prospect for gold. He was on the Archbold Papua Expedition (1933–1934) and was co-leader of a second expedition (1936). The etymology says: "*Early in 1938 there was received at the British Museum (Natural History) a collection of Papuan fishes from Mr. J. C. Adamson, a member of the government party sent to open up a region in the upper part of the basin of the Kikori River.*" He was a patrol and police officer (1935–1939) and served in the Royal Australian Navy (WW2) in the North Atlantic, the Indian Ocean and finally off

Papua. He owned and ran a plantation in Papua (1945–1964), retiring to live in Cooktown, Queensland (1964–1978). When his health failed, he shot himself. A bird is also named after him.

Adamson, TA

Adamson's Goby *Coryogalops adamsoni* M Goren, 1985

Thomas A Adamson, at the time of the discovery, worked in the Section of Fishes, the Natural History Museum of Los Angeles County. He was honoured for his 'genuine interest in Indo-West Pacific fishes' and his 'valuable contributions to ichthyology'. Among his papers are the co-authored: *A Review of the Monotypic Indo-Malayan Labrid Fish Genus Xenojulis* (1982) and *Batesian Mimicry between a Cardinalfish (Apogonidae) and a Venomous Scorpionfish (Scorpaenidae) from the Philippine Islands* (1983).

Adan

Papaloapan Killifish *Profundulus adani* SE Dominguez–Cisneros, E Velázquez-Velázquez, CD McMahan & WA Matamoros, 2021

Adán Enrique Gómez González (1981–2018) was a Mexican ichthyologist who worked at the Museo de Zoología, Instituto de Ciencias Biológicas, Universidad de Ciencias y Artes de Chiapas. He was honoured for his contributions to the study and conservation of freshwater fishes in southern México. He was working to protect the butterflies of a reserve where illegal logging and other activities were taking place, probably by a criminal cartel. He was murdered in an apparent robbery attempt in January 2018, but it might have been to stop his conservation campaign.

Addison

Ornate Sleeper Ray *Electrolux addisoni* L Compagno & PC Heemstra, 2007

Mark Ramsay Addison (b.1967) is the South African owner/operator of an underwater filming and expedition company 'Blue Wilderness' founded in 1997. He and his company filmed some of the most amazing sequences, such as the Sardine Run, in the BBC's 'Blue Planet' for which they won an Emmy. He started out to become a lawyer and studied at Rand Afrikaans University (1988–1990). His father, Brent Addison, was a marine scientist so he was exposed to the marine life off southern Africa from an early age. He collected the ray holotype. He has been described as part man, part shark, part comedian(!). After graduating, Mark started what became South Africa's largest dive charter boating company (1990–1996). He said of the ray's naming, '...They just ran out of ideas and named it after me'. Interestingly, the sleeper ray's genus is named after the Electrolux™ Vacuum Cleaner company because of the well-developed electrogenic properties that the ray has, and its vigorous sucking action when feeding.

Adelaide

Checkerboard Zipper Loach *Paracanthocobitis adelaideae* RA Singer & LM Page, 2015

Adelaide Singer is the daughter of the first author, Randal Singer.

Adele (Fielde)

Swatow Thryssa *Thryssa adelae* CL Rutter, 1897

Adele Marion Fielde (1839–1916) was an American Baptist missionary. She was engaged to a Baptist missionary candidate to Siam (Thailand) and became a Baptist herself to join him there, where they intended to marry. When she arrived (1865) she learned that he had died months earlier during her long sea voyage. She took over his ministry but did not fit in with the Baptist missionary community. Her dancing, card playing, and associations with the diplomatic community resulted in her dismissal from the mission. She returned home but was later reinstated and reassigned to China. She settled in the city of Swatow, Guangdong, where she started training local women as evangelists and Bible teachers. During her 20 years in China she established schools and trained c.500 'Bible women'. After her retirement, she returned to the USA and became involved in scientific research. She sent a 'considerable collection' of fishes from the port of Swatow, China, including the type of the eponymous species, to the University of Indiana (1885).

Adelscott

Celebration Whiptail *Spicomacrurus adelscotti* TN Iwamoto & N Merrett, 1997

This may look like an eponym but is in fact the name of a beer! The authors wrote: "*The name comes from a notably fine French ale with which we celebrated the discovery of [this] new species.*"

Adler

Waryfish sp. *Scopelosaurus adleri* VV Fedorov, 1967

'Adler' was the name of the fishery research trawler that collected the holotype.

Adloff

Banded Pearlfish *Austrolebias adloffi* CGE Ahl, 1922

Alfred Hubert Adloff (b.1874) was a German-born Brazilian (1913) artist, sculptor and fish culturist. He discovered the holotype (1921). His mouldings and sculptures adorn many important buildings in the city of Port Alegre. He fell on hard times (1937) and moved in with his son, after which nothing more is recorded of him. He discovered a number of fish species, including the pearlfish somewhere in Rio Grande do Sul, Brazil.

Adolf

Adolf's Eelpout *Lycodes adolfi* JG Nielsen & SA Fosså, 1993

Professor Adolf Severin Jensen (1866–1953) was a Danish zoologist who specialised in Arctic fish. He worked at the Zoological Museum, Copenhagen and was Professor at Copenhagen University. He carried out extensive research on the fisheries of West Greenland and was a member of the committee of the Three-year Expedition there (1931–1934). Adolf S. Jensen Land is a peninsula in the southern limit of King Frederick VIII Land, northeastern Greenland that is also named after him.

Adolf Friedrich

Lake Kivu Cichlid sp. *Haplochromis adolphifrederici* GA Boulenger, 1914

Adolf Friedrich Albrecht Heinrich, Duke of Mecklenburg-Schwerin (1873–1969) was a German explorer in Africa as well as a colonial politician. He led scientific expeditions to Central Africa traversing east to west (1907–1908) – during which the type was collected – and to Lake Chad (1910–1911). He was the last governor of Togoland (1912–1914). Nine birds, two mammals, two reptiles and an amphibian are also named after him.

Adolfo

Adolfo's Cory *Corydoras adolfoi* WE Burgess, 1982

Adolfo Schwartz was an aquarium-fish collector and exporter, Turkys Aquarium Manaus, Brazil. He discovered this species (1982) in the Rio Negro.

Adonis

Adonis Tetra *Lepidarchus adonis* TR Roberts, 1966
Polka Dot Lyretail Pleco *Acanthicus adonis* IJH Isbrücker & H Nijssen, 1988
Adonis Shrimp-Goby *Myersina adonis* K Shibukawa & U Satapoomin, 2006

In Greek mythology, Adonis was a god of vegetation and rebirth, as well as being an ideal of masculine beauty.

Adriaens

Squeaker Catfish sp. *Atopodontus adriaensi* JP Friel & TR Vigliotta, 2008

Dr Dominique Adriaens (b.1970) is a Belgian biologist. Ghent University awarded his biology MSc (1992) and PhD (1998). He became an Associate (2001) then Full Professor (2011) at the Department of Biology, Ghent University, where he is also the head of the Research group: Evolutionary Morphology of Vertebrates. He is also a Research Associate at AMNH. Most of his studies focus on functional and evolutionary morphology of the feeding and locomotor apparatus of fish, but also on aspects of ecomorphology and skeletal deformities. He has published more than 150 papers, of which most are on fish (catfishes, syngnathid fishes, anguilliform fishes, cichlids, sparids, etc.) such as the co-written *Functional consequences of extreme morphologies in the craniate trophic system* (2009). He was honoured in the name because he brought the existence of this species to the authors' attention.

Adriani

Sulawesi Ricefish genus A*drianichthys* MCW Weber, 1913

Nicolaus Adriani (1865–1926) was a Dutch Christian missionary for the Nederlandsch Bijbelgenootschap. He studied linguistics at Leiden, being awarded his PhD there (1893). He was sent by the Dutch Bible Society to Poso, Sulawesi, as a linguist. While there he collected natural history specimens, including fish. He wrote a number of papers and booklets on linguistics and translations of parts of the bible into local languages.

Aelianus

Adriatic Grayling *Thymallus aeliani* A Valenciennes, 1848

Claudius Aelianus (ca.175–ca.235) was a Roman naturalist and author, who wrote about this species (or one very like it) in his: *De Natura Animalium.*

Aelsbroeck

Elizabethville Mormyrid *Cyphomyrus aelsbroecki* M Poll, 1945

R P van Aelsbroeck (fl.1912–fl.1938) collected in the Belgian Congo for the Musée du Congo Belge (now RMCA), Tervuren, Belgium in the 1930s and 1940s. He collected the holotype (1937) in the River Busira (a tributary of the Zaire River), and a second specimen near Elizabethville; which Poll found surprising in that a new species was discovered so close to a big city.

Aeolos

Oriental Sillago *Sillago aeolus* DS Jordan & BW Evermann, 1902
[Alt. Oriental Trumpeter Whiting]

In Greek mythology, Aeolos was the keeper of the winds and king of the mythical floating island of Aiolia. There is no etymology for the sillago, and the fish might be named after the *meaning* of the name Aeolos: 'nimble'.

Aesculapius

Panther Danio *Danio aesculapii* SO Kullander & F Fang, 2009
[Syn. *Brachydanio aesculapii*]

In Greek and Roman mythology Aesculapius (or Asclepius) was the god of medicine. The staff of Asclepius, a rod with one or two snakes wrapped around it, is a symbol of medicine. Before the danio was formally named, it was known in the aquarium trade as Danio sp. 'Snakeskin'.

Afele

Afele's Dwarf-Goby *Eviota afelei* DS Jordan & A Seale, 1906

Afele was a Samoan boy who collected the holotype from coral heads in Pago Pago. Nothing more was recorded about him.

Afife

Loach sp. *Cobitis afifeae* J Freyhof, E Bayçelebi & M Geiger, 2018

Afife Jale (1902–1941) was a Turkish stage actress. She is best known as the first Muslim theatre actress in Turkey.

Afuer

Peruvian Butterfly Ray *Gymnura afuerae* SF Hildebrand, 1946

Not an eponym but a toponym; from the type locality – Lobos de Afuera Island, Peru.

Afzelius

Small-toothed Pellonula ssp. *Pellonula leonensis afzeliusi* A Johnels, 1954

Dr Björn Arvid Afzelius (1925–2008) was a biologist who specialized in spermatology. He studied zoology at the Wenner-Gren Institute, University of Stockholm, which awarded his doctorate (1957) and went on to become Assistant Professor there (1968). The Karolinska Institute gave him an honorary doctorate (1981) before he became Professor there (1982) until retirement, when he became Professor Emeritus. He was also Assistant Professor of biophysics at Johns Hopkins University, USA. He wrote over 300 scientific papers and the book: *Cellen i text och bild* (1967). He and the describer, Alf Johnels, were both members of the expedition that collected the holotype in the Gambia River (1950). Half a dozen plants and a crab are also named after him.

Agafonova

Ridgehead sp. *Poromitra agafonovae* AN Kotlyar, 2009

Tatyana Borisovna Agafonova (1950–2004) worked at the Russian Federal Research Institute of Fishery and Oceanography where she was both a friend and a research colleague of the author. She wrote: *Systematics and distribution of* Cubiceps *(Nomeidae) of the world ocean* (1994).

Agaléga

Viviparous Brotula sp. *Majungaichthys agalegae* W Schwarzhans & PR Møller, 2011

A toponym not an eponym; named after the Agaléga Islands, Mauritius, Indian Ocean.

Agassiz, AER

Snailfish sp. *Liparis agassizii* FW Putnam, 1874
Bumblebee Catfish sp. *Lophiosilurus alexandri* F Steindachner, 1876
Agassiz's Slickhead *Alepocephalus agassizii* GB Goode & TH Bean, 1883
Creole Damsel *Pomacentrus agassizii* R Bliss Jr, 1883
Spotfin Dragonet *Foetorepus agassizii* GB Goode & TH Bean, 1888
Cusk Eel sp. *Monomitopus agassizii* GB Goode & TH Bean, 1896
Longfin Scorpionfish *Scorpaena agassizii* Goode & TH Bean, 1896
Grideye Spiderfish *Ipnops agassizii* S Garman, 1899
Agassiz's Smooth-head *Leptochilichthys agassizii* S Garman, 1899
Lanternfish sp. *Diaphus agassizii* CH Gilbert, 1908

Alexander Emanuel Rodolphe Agassiz (1835–1910) was born in Switzerland but emigrated to the USA with his eminent palaeontologist father, Louis Agassiz (see next entry), and made a fortune out of copper mining. He graduated from Harvard (1855) and took a second degree (BSc.) (1857) after studying engineering and chemistry. He joined the US Coast Survey (1859) as an assistant, becoming a specialist marine ichthyologist. He worked (1860–1866) as an assistant at the Museum of Natural History that his father had founded at Harvard. He had become involved as an investor in a copper mining venture in Michigan and (1866) became the treasurer of the enterprise. After a struggle, he made the company prosperous, merged and acquired other companies, and expanded the conglomerate of which he became president (1871–1910). He returned to Harvard (1870s) to pursue his interests in natural history, giving $500,000 for the Museum of Comparative Zoology there and being its Curator (1874–1885). He visited Peru and Chile (1875) to look at the copper mines and to survey Lake Titicaca. He helped in the examination and classification of the specimens collected by Wyville Thomson on the *Challenger* Expedition and took part in three dredging expeditions (1877–1880) on the Coast Survey's vessel *Blake*. He published much on marine zoology in bulletins and two books: *Seaside Studies in Natural History*, co-written with his stepmother Elizabeth Cary Agassiz (1865) and *Marine Animals of Massachusetts Bay* (1871). Bliss, who named the damsel, was one of his students. He died at sea aboard *Adriatic*.

Agassiz, JLR

Shortnose Greeneye *Chlorophthalmus agassizi* CL Bonaparte, 1840
Rio Skate *Rioraja agassizii* JP Müller & FGJ Henle, 1841
Carache Negro (pupfish) *Orestias agassii* A Valenciennes, 1846
Arapaima sp. *Arapaima agassizii* A Valenciennes, 1847
Agassiz's Glassfish *Ambassis agassizii* F Steindachner, 1866
[Alt. Agassiz's Perchlet]
Spring Cavefish *Forbesichthys agassizii* FW Putnam, 1872
Agassiz's Dwarf Cichlid *Apistogramma agassizii* F Steindachner, 1875
Spotted Cory *Corydoras agassizii* F Steindachner, 1876
Headstander Characin sp. *Leporinus agassizii* F Steindachner, 1876
White Salema *Xenichthys agassizii* F Steindachner, 1876
Graery Threadfin Seabass *Cratinus agassizii* F Steindachner, 1878
Characin sp. *Aphyocharax agassizii* F Steindachner, 1882
Silver Trout *Salvelinus agassizii* S Garman, 1885 (Extinct)
Sea Catfish sp. *Cathorops agassizii* CH Eigenmann & RS Eigenmann, 1888

Jean Louis Rudolphe Agassiz (1807–1873) was a Swiss-American geologist, glaciologist and zoologist whose speciality was ichthyology. He studied at Zurich, Heidelberg and Munich, where he qualified as a physician (1830), and in Paris under Cuvier (1831). While still a student he was tasked with working on the Spix and Martius Brazilian freshwater fish collection. He became Professor of Natural History at the Lyceum de Neuchâtel (1832). He was the first person to propose scientifically that the Earth had been subject to an ice age and to study ice as a subject, having lived in a special hut built on a glacier in the Alps (1837). He went to the USA (1846) to study American natural history and geology and to deliver a course of zoology lectures. He visited again (1848) and remained there for the rest of his life, becoming Professor of Zoology and Geology at Harvard, where he founded and directed the Museum of Comparative Zoology (1859–1873). Latterly he took up studies of Brazilian fishes again and led the Thayer expedition to Brazil (1865). He established the Marine Biological Laboratory (1873). Three reptiles are also named after him, as well as at least four insects, three marine invertebrates and four mountains!

Agnes

Characin sp. *Moenkhausia agnesae* J Géry, 1965

Agnes Frobenius was honoured in the fish's binomial at the request of Harald Schultz (who collected the holotype), but no other information was given about her. There was an Agnes Susanne Schulz who was employed at the Frobenius Institut (1929–1959), an artist (scientific draughtsman) on the Frobenius expedition to Natal. Harald Schultz (1909–1966) was also a guest scientist at the institute (1961). Perhaps she was in some way related to the expedition leader Leo Frobenius (1873–1938)?

Agostinho

Armoured Catfish sp. *Ancistrus agostinhoi* AG Bifi, CS Pavanelli & CH Zawadzki, 2009

Dr. Ângelo Antônio Agostinho is a Brazilian zoologist, limnologist and ichthyologist. The State University of Londrina awarded his bachelor's degree (1975), the Federal University of Paraná his master's (1979) and the Federal University of São Carlos his doctorate (1985) in Ecology and Natural Resources. Presently he is professor at the State University of Maringá and a member of the advisory council of the National Institute of Limnology. He is also researcher 1a of the National Council of Scientific and Technological Development. He was specifically honoured in the name of this species "in recognition of his myriad contributions to our knowledge of ecology of Neotropical fishes, and his great participation in the establishment of the Núcleo de Pesquisas em Limnologia, Ictiologia e Aquicultura, Nupélia, one of the most important centers of research in ecology of fishes of the Latin America."

Aguana

Long-whiskered Catfish sp. *Exallodontus aguanai* JG Lundberg, F Mago-Leccia & P Nass, 1991

Leonidas Aguana is a technician at Instituto da Zoologia Tropical, Universidad Central de Venezuela. The etymology states that his "…friendship and intrepid collaboration in the field have contributed greatly to the authors' research and to Venezuelan ichthyology."

Aguaruna

Long-whiskered Catfish genus *Aguarunichthys* DJ Stewart, 1986
Tetra sp. *Hemigrammus aguaruna* FCT Lima, EV Cortes Roldán & RP Ota, 2015

The Awajun people, better known by the name Aguaruna, the second-largest native population in the Peruvian Amazon, occupy a portion of the Río Morona basin (Departamento Loreto, Peru), where these fish occur.

Aguirre Pequeno

Soto la Marina Shiner *Notropis aguirrepequenoi* S Contreras-Balderas & R Ivera-Tiellery, 1973

Dr Eduardo Aguirre Pequeño (1904–1988) was a Mexican biologist, author and humanist who founded the faculties of biology and agronomy at the Universidad Autónoma de Nuevo León, Monterrey. He studied medicine at the Monterrey Medical School (1926–1932) and after graduating studied parasitic diseases, such as amoebiasis and malaria, affecting people who live in the tropics. He wrote: *Cultivation techniques in the diagnosis of amoebiasis* (1946).

Ahl, CGE

Waryfish genus *Ahliesaurus* E Bertelsen, G Krefft & NB Marshall, 1976
Burmese Flying Barb *Esomus ahli* SL Hora & DD Mukerji, 1928
Claroteid Catfish sp. *Parauchenoglanis ahli* M Holly, 1930
African Rivuline sp. *Aphyosemion ahli* GS Myers, 1933
Electric Blue Hap (cichlid) *Sciaenochromis ahli* E Trewavas, 1935

Dr Christoph Gustav Ernst Ahl (1898–1945) was an ichthyologist, herpetologist and aquarist. He served in the artillery in the First World War (1916). He studied natural science at Humboldt University, Berlin (1919–1921), where he was awarded his doctorate. He was at the Department of Ichthyology and Herpetology, Zoological Museum (1921–1941), becoming Curator of Herpetology (1923) and later Director. He was Editor-in-Chief of the magazine *Das Aquarium* (1927–1934). Having joined the Nazi Party (1930s) to keep his job, he was expelled for indiscipline (1939). He was sacked by the Museum (1941), probably as his scientific work was 'superficial and careless and his knowledge of the literature poor', rather than because he had been recalled to the Wehrmacht (1939). He fought in Poland and North Africa and was reported as missing in action in Herzegovina (1945). He wrote 170 papers on fish and amphibians, but many of the names he coined are no longer considered valid. Two reptiles and seven amphibians are named after him.

Ahl, JN

Worm Eel genus *Ahlia* DS Jordan & BM Davis, 1891

Dr Jonas Nicholaus Ahl (1765–1817) was a physician and one of Linnaeus' students. He published his thesis: *De Muraena et Ophichth* (1789) which according to Jordan's dedication "*furnishes the beginning of our systematic arrangement of the eels.*"

Ahlander

Danio sp. *Devario ahlanderi* S Kullander & M Norén, 2022

Erik Åhlander is a long-time Senior Assistant in the ichthyology and herpetology collection of the Swedish Museum of Natural History, where the two describers work.

Ahlstrom

Lanternfish sp. *Metelectrona ahlstromi* RL Wisner, 1963
Paddletail Onejaw *Monognathus ahlstromi* SN Raju, 1974
Sabretooth Fish sp. *Evermannella ahlstromi* RK Johnson & GS Glodek, 1975
Ahlstrom's Waryfish *Scopelosaurus ahlstromi* EOG Bertelsen, G Krefft & NB Marshall, 1976
Pencil Smelt sp. *Nansenia ahlstromi* K Kawaguchi & JL Butler, 1984

Dr Elbert Halvor Ahlstrom (1910–1979) was an ichthyologist and biologist who spent more than four decades working for the National Fisheries Service. He worked for the Fish and Wildlife Service and the Commercial Fisheries Office, then became senior scientist at, and first Director (1959) of, the Southwest Fisheries Center, National Marine Fisheries Service, La Jolla, California. Among his publications, was: *Sardine Eggs and Larvae and Other Fish Larvae* (1959).

Ahmet

Stone Loach sp. *Seminemacheilus ahmeti* S Sungur, P Jalili, S Eagderi & E Çiçek, 2018

Ahmet Sungur (1990–2017) was the brother of the senior author, Turkish ichthyologist Sevil Sungur. He passed away in an accident when just 27 years old.

Ai Nonaka

Ai's Bandfish *Owstonia ainonaka* WF Smith-Vaniz & GD Johnson, 2016

Ai Nonaka is the wife of GD Johnson. The name is a combination of the first (Ai) and last name (Nonaka) of the second author's wife in appreciation for her valuable assistance with this study. (Dave Johnson is an ichthyologist and research zoologist who is curator of marine larval fishes at the NMNH. (Also see **Johnson, GD**)

Ajuricaba

Characin sp. *Astyanax ajuricaba* MMF Marinho & FCT Lima, 2009

Ajuricaba was paramount chief of the Manau Indians. They formerly inhabited the Rio Negro area of Brazil, where this species occurs. The Manau went to war against Portuguese slavers (1727) but were unsuccessful in resisting the assaults of the Portuguese. Ajuricaba was arrested by slavers but avoided any trial as a rebel by drowning himself in the Rio Negro while shackled; an act of bravery that became a symbol of Indian resistance against Portuguese oppression. The Portuguese also acknowledged his act of heroism.

Akaje

Whip Stingray *Hemitrygon akajei* JP Müller & FGJ Henle, 1841
[Alt. Japanese Red Stingray]

The original description has no etymology but, though the binomial looks as if it could be an eponym, it is merely a translation: *aka* is Japanese for red, referring to its bright orange-red underside and *jei* is Japanese for ray/skate.

Akama

Banjo Catfish sp. *Micromyzon akamai* JP Friel & JG Lundberg, 1996
Driftwood Catfish sp. *Ageneiosus akamai* FRV Ribeiro, LH Rapp Py-Daniel & SJ Walsh, 2017

Dr Alberto Akama is a Brazilian ichthyologist at the Museu Paraense Emílio Goeldi (Belém, Pará, Brazil). The University of São Paulo awarded his BSc (1993), his masters (1999) and his PhD (2004). He worked as a professor at the Federal University of Tocantins (2008–2013) before joining the museum. He was honoured in the banjo catfish's name for his 'enthusiastic help' in collecting the type series, and for the driftwood catfish "*…for his many contributions to the systematics of neotropical catfishes.*"

Akawaio

Bluntnose Knifefish genus *Akawaio* JA Maldonado-Ocampo et al., 2013

The Akawaio Amerindians who live in the region of the upper Mazaruni River, Guyana, were thanked for their valuable help while the describers studied the fishes of their territory.

Akhtar

Stone Loach sp. *Triplophysa akhtari* MA Vijayalakshmanan, 1950
Sisorid Catfish sp. *Glyptosternon akhtari* EG Silas, 1952

Kazmi Sayed Ali Akhtar (b.1899) was an Afghan botanist attached to the Faculty of Medicine in Kabul, Afghanistan. He collected for both Kew and the Zoological Survey of India to whom he gave a 'very valuable and interesting' collection of fishes from Afghanistan, including the holotype of the loach.

Akihito

Goby genus *Akihito* RE Watson, P Keith & G Marquet, 2007
Imperial Goby *Platygobiopsis akihito* VG Springer & JE Randall, 1992
Akihito's Goby *Exyrias akihito* GR Allen & JE Randall, 2005
Emperor Reefgoby *Priolepis akihitoi* DF Hoese & HK Larson, 2010

Emperor Akihito of Japan (b.1933) is noted for his many contributions to goby systematics and phylogenetic research. In extension of his father's interest in marine biology, the Emperor is a published ichthyological researcher, and has specialised in studies within the taxonomy of the family Gobiidae. He has written papers for scholarly journals such as: *Gene* and the *Japanese Journal of Ichthyology*. In addition, many of the type specimens were supplied by the Biological Laboratory of the Imperial Household in Tokyo. (See also **Imperator**)

Akiko

Clingfish sp. *Lepadichthys akiko* GR Allen & MV Erdmann, 2012

Akiko Shiraki Dynner (d.<2007) had, according to the etymology, a "*...lifelong attachment to the ocean.*" She and her husband Alan, Chairman of the New England Aquarium, were avid scuba divers. He established a fund in her name (Akiko Shiraki Dynner Fund for Ocean Exploration and Conservation) and was a strong supporter of marine conservation initiatives in the Bird's Head Seascape, which encompasses Cenderawasih Bay, West Papua, Indonesia, where the clingfish was first found.

Akil

Beysehir Bleak *Alburnus akili* F Battalgil, 1942 (Extinct)

Dr Akil Muhtar Özden (1877–1949) was Professor of Pharmacodynamics and Clinical Therapy at Istanbul University Medical School and an internal medicine expert. He was at Askeri Tibbiye Idadisi (military pre-medical college), and after graduation he attended Mekteb-i Tibbiye-i Askeriye (military medical school) for a year then went to Geneva where he continued his education. He eventually prepared his doctoral dissertation there before returning to Istanbul at the invitation of the government, where he established the discipline of pharmacodynamics and founded a modern internal medicine department at Istanbul University. He retired (1943) and was elected to parliament (1946). He is best remembered for inventing the 'liver function test' still used in modern medicine.

Akiri

Claroteid Catfish sp. *Notoglanidium akiri* L Risch, 1987

Pamela Jeanne Akiri (b. 1944) is an American biologist who moved to Nigeria with her husband and worked at Rivers State University of Science and Technology, Port Harcourt, Nigeria. She collected the catfish holotype.

Alama

Soldierfish sp. *Ostichthys alamai* M Matsunuma, Y Fukui & H Motomura, 2018

Ulysses B Alama is on the staff of the Museum of Natural Sciences, College of Fisheries and Ocean Sciences, University of the Philippines, Visayas. "*The specific name* alamai *is in recognition of Mr. Ulysses Alama, Museum of Natural Sciences, College of Fisheries and Ocean Sciences, University of the Philippines Visayas, for his great contributions to the authors' and other collaborators' surveys at Iloilo [Panay Island] during 2013–2017, when the new species was collected. These surveys resulted in the field guide 'Commercial and Bycatch Market Fishes of Panay Island, Republic of the Philippines.*"

Alastair

Australian Hagfish sp. *Eptatretus alastairi* MM Mincarone & B Fernholm, 2010

Alastair Graham (b.1964) was honoured for help and hospitality offered to the second author. (See **Graham**, also see **Al Graham**)

Albatross

Grenadier genus *Albatrossia* DS Jordan & CH Gilbert, 1898

Greeneye sp. *Chlorophthalmus albatrossis* DS Jordan & EC Starks, 1904

'Albatross' was a steam-driven research vessel owned by the US Fish Commission. It was used for many cruises and expeditions. The greeneye species was collected when the 'Albatross' dredged off the coast of Japan.

Albert JS

Armoured Catfish sp. *Hisonotus alberti FF* **R**oxo, GSC Silva, BT Waltz & JE Garcia-Melo, 2016

Bluntnose Knifefish sp. *Brachyhypopomus alberti* GR Crampton, CDCM de Santana, JC Waddell & NR Lovejoy, 2017

Dr James Spurling Albert (b.1964) is a biologist and ichthyologist who is Professor in the Department of Biology (2015) at the University of Louisiana, Lafayette. The University of California, Berkeley awarded his bachelor's degree (1988) and the University of Michigan, Ann Arbor, both his master's (1991) and his doctorate (1995). He then became Assistant Professor, Nippon Medical School, Tokyo (1995–1999), Assistant Professor, Department of Biology, University of Florida (1999–2004), Assistant Professor, Department of Biology, Graduate Faculty, University Louisiana at Lafayette (2005–2009), the Associate Professor (2010–2014) there before taking his current position. He has published c.100 scientific articles in the areas of Evolutionary Biology, Biogeography, Systematic Ichthyology, Tropical Aquatic Biodiversity, and Conservation Biology such as the co-written: *Historical Biogeography of Neotropical Freshwater Fishes* (2011). He serves on the editorial boards of several scientific journals, has been principle investigator on five multi-investigator projects documenting aquatic Amazonian biodiversity, and has edited three books on the diversity and evolution of Amazonian fishes. He was honoured "*...for his dedication and contributions to the studies of neo-tropical freshwater fishes.*"

Albert (King)

Albert's Synodontis *Synodontis alberti* L Schilthuis, 1891
[Alt. Bigeye Squeaker Catfish, High-fin Synodontis]
African Barb sp. *Enteromius alberti* M Poll, 1939

King Albert I (1875–1934) was King of the Belgians (1909–1934). Although the patronym is not identified in either of the original descriptions of these fish, the holotypes were taken in the former Belgian Congo (now DRC).

Albert (Parr)

Gulf Chimaera *Hydrolagus alberti* HB Bigelow & WC Schroeder, 1951

Dr Albert Eide Parr (1890–1991) was an oceanographer and marine biologist who was born and raised in Norway, where he took his first degree at the University of Oslo. He was in the Norwegian Merchant Marine and undertook postgraduate research at Bergen Museum. Yale awarded his doctorate. He worked at the New York Aquarium (1926) from where he was recruited to curate Harry Payne Bingham's fish collection, which was donated to Yale – and Parr went along with it. At Yale, he became Director of their oceanographic laboratory (1930s) and rose through Yale's academic ranks, becoming Professor of Oceanography (1938) and Director of their Peabody Museum (1938–1942). He became Director of the AMNH (1942–1959) and reorganised many departments. He was editor of the authors' *Fishes of the Western North Atlantic* monographs and was honoured for his many contributions to ichthyology.

Alberto

Ghost Knifefish sp. *Apteronotus albertoi* LAW Peixoto, GM Dutra, A Datovo, NA Menezes & CD de Santana, 2021

Alberto Carvalho is manager of the Laboratório Multiusuário de Processamento de Imagens de Microtomografia Computadorizada de Alta Resolução do Museu de Zoologia da Universidade de São Paulo (MZUSP). He was honoured for his support in generating μCT scan images for researchers, especially for the project "Diversity and Evolution of the Gymnotiformes".

Alcock

Cusk–eel genus *Alcockia* GB Goode & TH Bean, 1896
Alcock's Tongue-Sole *Cynoglossus carpenteri* AW Alcock, 1889
Conger Eel sp. *Acromycter alcocki* CH Gilbert & F Cramer, 1897
Searobin sp. *Lepidotrigla alcocki* CT Regan, 1908
Pale-spot Whipray *Himantura alcockii* N Annandale, 1909
Arabian Catshark *Bythaelurus alcockii* S Garman, 1913
Tripodfish *Halimochirurgus alcocki* MCW Weber, 1913
Alcock's Croaker *Atrobucca alcocki* PK Talwar, 1980
Cusk-eel sp. *Neobythites alcocki* JG Nielsen, 2002

Lieutenant-Colonel Dr Alfred William Alcock (1859–1933) was a British physician, herpetologist, carcinologist and naturalist. His education was interrupted when his father lost his money (1876) and Alfred was forced to leave school. He went to India to a coffee plantation in Malabar and worked as an agent recruiting unskilled labourers (1878–1880). He became an assistant master in a European boys' school in Darjeeling (1880) and was able to return to England to start his medical training (1881). He graduated as a physician from Aberdeen University (1885), joined the Indian Medical Service and served (1886–1888). He transferred to the Indian Marine Survey and served at Surgeon-Naturalist on board the survey ship 'Investigator' (1888–1892). He became Superintendent of the Indian Museum in Calcutta (1893) but resigned (1906) after a disagreement with the Viceroy, Lord Curzon, and returned to London where he worked on tropical medicine at the School of Tropical Medicine. He wrote: *A Naturalist in Indian Seas* (1902) and *Entomology for Medical Officers* (1911). The catshark was named after him because he was the first (1896) to note that it was a distinct species. Two reptiles are also named after him.

Aldrich

Mullet genus *Aldrichetta* GP Whitley, 1945

Fred C Aldrich was Chief Inspector of Fisheries and Game, Perth, Western Australia. We have failed to locate further information about him.

Aldrovandi

Halosaur genus *Aldrovandia* GB Goode & TH Bean, 1896

Dr Ulisse Aldrovandi (1522–1605) was an Italian naturalist. He studied law and medicine, graduating as a physician at Bologna (1553). He was accused of heresy before the Inquisition (1549) but was able to clear himself. He was appointed Professor of Philosophy and Lecturer on Botany at the University of Bologna (1551), becoming Professor of Natural History (1561). He became the first Director of Bologna's botanical garden (1568). He was instrumental in the founding of Bologna's public museum. He willed his huge collection of natural history specimens to the Senate of Bologna, but these were gradually distributed among a variety of institutions. A bird and a reptile are named after him.

Alessandrini

Bumphead Sunfish *Mola alexandrini* C Ranzani, 1839
[Alt. Southern Sunfish, Ramsay's Sunfish]

Antonio Alessandrini (1786–1861) was an Italian physician and anatomist, (Ranzini latinized his name). He taught comparative anatomy and veterinary science at the University of Bologna. He was also president of the Accademia delle Scienze of the city and of the Zoology Division of the Ottava Riunione degli Italiani Scienziati in Genoa (1846). He wrote many articles on zoology in general and parasitology in particular e.g. *Observazioni antomiche intorno a diverse specie di entozoarii de genere Filaria* (1838). Alessandrini had been at working on a detailed anatomical study of the sunfish's gills, which he later published.

Alexander (The Great)

Turkish toothcarp sp. *Paraphanius alexandri* F Akşiray, 1948

The binomial means 'of Alexander', but the etymology is not explained. It may be named after Alexandria, the historical name of Iskenderun, Turkey the type locality, or in honour of Alexander the Great (356–323 BC) who founded the settlement that eventually bore his name.

Alexander, AM

Alexander's Damsel *Pomacentrus alexanderae* BW Evermann & A Seale, 1907

Annie Montague Alexander (1867–1950) was an American philanthropist and fossil collector who established the fossil collection at the University of California Museum of Paleontology, Museum of Vertebrate Zoology and sponsored collecting expeditions. The etymology says that the species was "*…named for Miss Annie M Alexander, of Oakland, California, in recognition of her interest and work in zoology.*" She shared her life with Louise Kellogg for forty-two years. Lake Alexander in Alaska is named after her as are at least seventeen plants species (others are named after Kellogg).

Alexander (Agassiz)

Pac-Man Catfish *Lophiosilurus alexandri* F Steindachner, 1876

Alexander Emanuel Rodolphe Agassiz (1835–1910) (See **Agassiz, AER**)

Alexander (Santos)

Neotropical rivuline sp. *Austrolebias alexandri* HP Castello & RB Lopez, 1974

Alejandro Fernandez Santos (1960–1961) was the son of Jorge Osvaldo Fernandez Santos, a dealer in aquarium fishes and collector of the type specimen of this species. He died tragically young.

Alexandra (Bazikalova)

Baikal Sculpin sp. *Cottocomephorus alexandrae* DN Taliev, 1935

Aleksandra Yakovlevna Bazikalova was the wife of Dmitrii Nikolaevich Taliev (q.v.) and, like him, worked at the Limnological Research Station at Baikal on systematisation of Amphipoda. She published papers including descriptions of new taxa, such as: *Concerning the systematics of the Baikalian amphipods* (1935) and: *The feeding habits of benthophagous fish in the Maloye* More (1959). She has a number of other species such as marine worms named after her.

Alexandra (Potanina)

Stone Loach sp. *Triplophysa alexandrae* AM Prokofiev, 2001

Alexandra Viktorovna Potanina née Lavrovskaya (1843–1893) was a Russian explorer and geographer who was also the wife and companion of Grigory Nikolayaevich Potanin (1835–1920) (q.v.) with whom she explored Central Asia. They made four expeditions; Mongolia (1876–1877), Zaisan (1879–1880), northern China, eastern Tibet and central

Mongolia (1884–1886) and finally Tibet and the Gobi Desert (1892–1893) during which she became ill and died. She and her husband wrote: *The Buryats* (1891) for which the Russian Geographical Society, of which she was the first female member, awarded them its gold medal. (See also **Potanin**)

Alexandre

Brazilian Snapper *Lutjanus alexandrei* RL Moura & KC Lindeman, 2007

Alexandre Rodrigues Ferreira (1756–1815) was an early Brazilian naturalist. He made an extensive journey across the interior of the Amazon Basin to Mato Grosso (1783–1792), during which he collected and described the local flora and fauna as well as writing about the indigenous people. He became known as the 'Brazilian Humboldt'. The etymology reads: "*The specific name honors the pioneer Brazilian naturalist Alexandre Rodrigues Ferreira (1756–1815), whose many years of field work in Brazil during the late 18th Century remain underrecognized due to the confiscation of his and others' collections at Lisbon's Museu da Ajuda in 1808.*" His *Diary of a Philosophical Voyage* was published long after his death (1887). A reptile is also named after him.

Alexandrini

Bumphead Sunfish *Mola alexandrini* C Ranzani, 1839
[Alt. Southern Sunfish, Ramsay's Sunfish]

(See **Alessandrini**)

Alexandrine

Alexandrine Torpedo *Torpedo alexandrinsis* FM Mazhar, 1987

A toponym; this electric ray is named after the city of Alexandria in Egypt, not a person.

Alfaro

Knife Livebearer genus *Alfaro* SE Meek, 1912
Pastel Cichlid *Cribroheros alfari* SE Meek, 1907

Don Anastasio Alfaro (1865–1951) was an archaeologist, geologist, ethnologist, zoologist and famous Costa Rican writer. From a young age he collected birds, insects, minerals and plants. He took his first degree at the University of Santo Tomás (1883). He urged the President to create a National Museum (1885) and then he dedicated much of his life to it, becoming Director shortly after it was established (1887). He spent his life teaching and exploring as well as continuing to collect, discovering a number of new taxa that carry his name. He wrote a number of books, including one on Costa Rican mammals, and also wrote poetry. He was much admired throughout Europe and the Americas and corresponded with all the leading naturalists of his day. He provided a small collection of fishes from Costa Rica to Meek, including the cichlid type. Among other taxa, three mammals, two birds and an amphibian are named after him.

Alfian

Alfian's Flasher Wrasse *Paracheilinus alfiani* GR Allen, MV Erdmann & NLA Yusmalinda, 2016

Rahmad 'Yann' Alfian of Lombok was the authors' dive guide and the collector of the holotype in Lembata Island, Indonesia.

Alfie

Alfie's Goby *Gobiosoma alfiei* JC Joyeux & RM Macieira, 2015

Erasmo Alfredo 'Alfie' Amaral de Carvalho-Filho (b.1950) is a Brazilian ichthyologist at the University of São Paulo Zoological Museum. His BSc is in biology with his main interests being the taxonomy and systematics of marine ichthyology. He has an MBA in marketing. He began studying fish in 1970. He has described five new fish, written 37 papers, chapters and books including the book *Fish: Coastal Brazil* (1992). He has also published over 200 articles in fishing magazines. The authors describe him as a "*...self-made ichthyologist in his spare time*", and honour him for "*...his contribution to the advancement in the diversity and taxonomy of Brazilian marine fishes and his friendship.*" He also owns a research and marketing company called 'Fish'. He is currently (2018) working on the description of nine new fish, several papers on marine fish behaviour, a reef guide and a new book about all (c.2,300) Brazilian marine fish. He is also a photographer and has made around 2,600 dives.

Alfonso

Scrapetooth Characin sp. *Parodon alfonsoi* A Londoño-Burbano, C Román-Valencia & DC Taphorn Baechie, 2011

Alfonso Londoño Orozco is the senior author's father and was thanked for "*...his support and personal inspiration through the years*".

Alfred (Prince)

Reef Manta Ray *Manta alfredi* JLG Krefft, 1868

Prince Alfred, Duke of Edinburgh (1844–1900) was a son of Queen Victoria who ruled as Duke of Saxe-Coburg and Gotha (1893) in the Prussian Empire. He earlier served in the Royal Navy and commanded 'HMS Galatea' in a circumnavigation (1867–1868). During this voyage he visited Australia twice, surviving an assassination attempt (1868), when he was shot in the back (the would-be assassin was promptly, arrested, tried and hanged). The 'Sydney Illustrated News' in which Krefft published his article, stated that Prince Alfred had been presented with photographs of the manta.

Alfred (Duvaucel)

Rohtee *Osteobrama alfredianus* A Valenciennes, 1844

Alfred Duvaucel (1796–1824) was a French naturalist who was Cuvier's (q.v.) stepson. He was sent (1817) by Cuvier with Diard (q.v.) to collect in India for the MNHN, Paris. They established a botanical garden in Chanannagar (1818–1819), and collected in Sumatra under contract to Stamford Raffles, but were dismissed when Raffles discovered that they were sending most of their collections back to Paris rather than to him. Duvaucel died in India. Among other taxa, four birds, two reptiles and a mammal are named after him.

Alfred, ER

Asian Stream Catfish sp. *Pseudobagarius alfredi* HH Ng & M Kottelat, 1998
Licorice Gourami sp. *Parosphromenus alfredi* M Kottelat & PKL Ng, 2005

Eric Ronald Alfred (1931–2019) was an ichthyologist and Curator of the Raffles Museum, Singapore (1957–1967). He became Director (1967–1972) following a year's attachment to the British Museum and the Rijksmuseum van Natuurlijke Histoire, Leiden. He collected fish species from the Malay Peninsula. He went with the museum's zoological collection when it was moved to the Maritime Museum, Sentosa. A bird and a reptile are also named after him.

Alfred (MItchell)

Atlantic Flashlightfish *Kryptophanaron alfredi* CF Silvester & HW Fowler, 1926

Alfred Mitchell (1832–1911) was a wealthy American business man – having married the heiress to the Tiffany fortune! They lived in a house of more than 60 rooms, built to look like a Roman villa, in Jamaica. The house was far in advance of its time as it had electric light, running water, a sauna and an indoor swimming pool which was filled with sea water – upon the surface of which, Ulrich Dahlgren (1870–1946), the Princeton biologist, found the flashlightfish holotype floating. Mitchell imported the second car to arrive in Jamaica – a Rolls Royce – and kept a menagerie of peafowl and monkeys. The house was sold after 1914 and the new owners abandoned it. It is now a ruin, but MItchell's real legacy is to be found on a nearby uninhabited island, now populated by the descendants of his monkeys – so the local name of the island is now Monkey Island.

Alfreda

Characin sp. *Bryconamericus alfredae* CH Eigenmann, 1927

Alfreda Bingham née Mitchell (1874–1967) was the heiress to the Tiffany jewellery fortune and wife of explorer-politician Hiram Bingham. She divorced him (1937) and married Henry Gregor (1886–1964). She was interested in Peruvian natural history and gave "*... material assistance in making possible an expedition to the Urubamba.*" (See also **Bingham**)

Al Graham

Al Graham's Sandperch *Parapercis algrahami* JW Johnson & J Worthington Wilmer, 2018
Blind Cusk-eel sp. *Barathronus algrahami* JG Nielsen, JJ Pogonoski & SA Appleyard, 2019

Alastair (Al) Graham (b.1964), Collection Manager at CSIRO Marine Research, Hobart, was honoured in the binomial for his longstanding efforts in building and maintaining Australian ichthyological collections and his helpful cooperation with taxonomical research. (See **Graham** + also see **Alastair**)

Ali

Pamphylian Spring Minnow *Pseudophoxinus alii* F Küçük, 2007

Ali Küçük is the father of the author, Turkish ichthyologist Fahrettin Küçük.

Alia

Smalleye Pygmy Shark *Squaliolus aliae* H-T Teng, 1959

Huang A-li is the name of Teng's wife. He named the shark, which was first caught inTaiwanese waters, after her "*...for her continuous encouragement and assistance over the past 20-some years*".

Alice (Birtwistle)

Tropical Sand Goby sp. *Favonigobius aliciae* AWCT Herre, 1936

Alice Birtwistle was married to William Birtwistle (1890–1953) who was the officer in charge of the Fisheries Department of Straits Settlements and Federated Malay States. She hosted the author during his field trip to Singapore. The etymology says: "*I take pleasure in naming this species in honor of Mrs. Alice Birtwistle, who, with her husband, Mr. W. Birtwistle, Officer-in-Charge of the Fisheries Department, S.S. and F.M.S., was most hospitable during my stay in Singapore.*" (Also see **Birtwhistle**)

Alice (Boring)

Lanternfish sp. *Diaphus aliciae* HW Fowler, 1934

Dr Alice Middleton Boring (1883–1955) was a zoologist, biologist, geneticist and cytologist. Her bachelor's degree (1904) and her doctorate (1910) were both awarded by Bryn Mawr College, Pennsylvania. She taught zoology at the University of Maine (1910–1918). She worked at Yenching University, China (1923–1950), though she spent the Second World War in an internment camp before being repatriated to America. She returned to America permanently (1950) and taught at Smith College (1950–1955). She co-wrote: *Handbook of North China Amphibia and Reptiles* (1932). Two amphibians are named after her.

Alice (Ferreira)

Dwarf Basslet sp. *Serranus aliceae* A Carvalho Filho & CEL Ferreira, 2013

Alice Ferreira is the daughter of the second author.

Alice (Holland)

Alice's Argentine *Argentina aliceae* DM Cohen & SP Atsaides, 1969

Alice Holland was a former Secretary, Bureau of Commercial Fisheries Systematics Laboratory, USNM. She was honoured for her "devoted" services to ichthyology.

Alice (Jouy)

Southern Leatherside Chub *Lepidomeda aliciae* PL Jouy, 1881

Alice Elizabeth Jouy née Craig (1853–1880) was the describer's wife. She accompanied her husband on his collecting expeditions in China, Japan and Korea, where he was a Chinese Custom's official for several years and attached to the US Legation, and latterly to Mexico and Arizona.

Alicia

African Barb sp. *Enteromius aliciae* R Bigorne & C Lévêque, 1993
[Syn. *Barbus aliciae*]

Alice Lévêque is the daughter of the junior author, Christian Lévêque.

Ali Daei

Sisorid catfish sp. *Glyptothorax alidaeii* H Mousavi-Sabet, S Eagderi, S Vatandoust & J Freyhof, 2021

Ali Daei (b. 1969) is an Iranian former professional footballer and football manager. He was captain of the Iranian national team (2000–2006) and played in the German Bundesliga for Arminia Bielefeld, Bayern Munich and Hertha Berlin. He is regarded as one of the best Asian footballers of all time. He was honoured in the binomial for his humanitarian activities after the 2018 Sarpol-e Zahab earthquake, where the Seimareh River (type locality of the catfish) is situated.

Alikunhi

Alikunhi's Blind Catfish *Horaglanis alikunhii* Subhash Babu Kallikadavel & CKG Nayar, 2004

Dr Kolliyil Hameed Alikunhi (1918–2010) was an Indian aquaculturist. He held a master's degree from the University of Madras (Chennai) and a honorary doctorate awarded by the University of Bombay (Mumbai). This species was named after him "*...for his contributions to fishery science in general and Indian fisheries in particular.*"

Alipio

Armoured Catfish sp. *Isbrueckerichthys alipionis* WA Gosline III, 1947

Darter Tetra sp. *Characidium alipioi* H-P Travassos, 1955

Dr Alipio de Miranda-Ribeiro (1874–1939) was a Brazilian zoologist, herpetologist and ichthyologist who did much, with Adolpho and Bertha Lutz, to enhance the collections at the Brazilian National Natural History Museum, Rio de Janeiro. He initially studied medicine but joined the National Museum (1894) before completing the course. He was Assistant Naturalist for the Museum (1897) and Secretary, Department of Zoology (1899), Deputy Head & Professor, Zoology Department (1910–1929). He created the Inspectorate of Fisheries (1911), the first South American oceanographic service and was its first Director (1911-1912). He wrote the 5-volume: *Fauna Brasiliensis* (1907–1915). Many taxa, including seven amphibians and two birds, are named in his honour. (See also **Ribeiro**)

Alis

Escolar sp. *Rexea alisae* CD Roberts & AL Stewart, 1997

Alis Velvet Skate *Notoraja alisae* B Séret & PR Last, 2012

'Alis' is the name of a ship used by the Institut de Recherche pour la Dévelopement to research off the coast of New Caledonia. The ship is, in turn, named after a local wind.

Alison

Alison's Blue Devil *Paraplesiops alisonae* DF Hoese & R Kuiter, 1984

Alison Kuiter is the junior author's wife. In a news article (2004) he explained that "...She was the one who saw it [the fish] first".

Aliye

Loach sp. *Cobitis aliyeae* J Freyhof, E Bayçelebi & M Geiger, 2018

Fatma Aliye Topuz (1862–1936) was an important Turkish novelist, columnist, essayist and women's rights activist and a humanitarian.

Allan

Licorice Gourami sp. *Parosphromenus allani* B Brown, 1987

Allan J Brown (1911–2009) was the husband (1939–2009) of author Barbara Brown née Demaree. After an economics degree from Berkeley (1933) and post graduate studies in agronomy he worked for the Soil Conservation Service (1937–1945) then working until retirement for the Oakland Park Department as the Horticultural Superintendent (1946–1973). They made at least three collecting trips to Sarawak. (Also see **Brown, B & A**)

Allard

Allard's Clownfish *Amphiprion allardi* W Klausewitz, 1970

Jacques Allard had a fish collecting station in Kenya which the aquarist Herbert Axelrod visited. Wolfgang Klausewitz named the clownfish after him as Allard had supplied many adult and sub-adult specimens and gave him a lot of support during his expedition to East Africa.

Allardice

Allardice's Moray *Anarchias allardicei* DS Jordan & EC Starks, 1906

Robert Edgar Allardice (1862–1928) was a mathematician who was educated and taught at Edinburgh University, which he left (1892) to become Professor of Mathematics at Stanford University. There he was a friend and colleague of the senior author David Starr Jordan (1851–1931) (q.v.). He retired as Emeritus Professor (1927). He helped Jordan to collect the holotype.

Allaud

Allaud's Haplo (cichlid) *Astatoreochromis alluaudi* J Pellegrin, 1904

Charles A Alluaud (1861–1949) was an entomologist and explorer who collected the type. The common name is merely a common misspelling. (See **Alluaud**)

Allen, EJ

Viviparous Brotula sp. *Cataetyx alleni* LW Byrne, 1906

Dr Edgar Johnson Allen (1866–1942) was a distinguished British marine biologist. He was the Director of the Marine Biological Association (1894–1936). He was elected a Fellow of the Royal Society (1914) and won the Gold Medal of the Linnean Society (1926) and the Royal Society's Darwin Medal (1936). The citation for his Darwin Medal reads: "*In recognition of his long continued work for the advancement of marine biology, not only by his own researches but by the great influence he has exerted on very numerous investigations at Plymouth.*" In his description of the brotula, Byrne wrote: "*...which I propose to name in honour of my friend Dr. E. J. Allen, the Director of the Association.*"

Allen, G

Toadfish genus *Allenbatrachus* DW Greenfield, 1997

Dr George Allen (1923–2011) was a fisheries biologist at Humboldt State University, Arcata, California which he joined (1956) and where he taught for about 30 years and retired (1985) as Emeritus Professor. He served in the US military for three years in WW2 and completed his bachelor's degree at the University of Wyoming after being de-mobilised.

Allen, GR

Anglerfish genus *Allenichthys* TW Pietsch, 1984
Allen's Perchlet *Plectranthias alleni* JE Randall, 1980
Allen's Chromis *Chromis alleni* JE Randall, H Ida & JT Moyer, 1981
Allen's Tubelip *Labropsis alleni* JE Randall, 1981
Andaman Damsel *Pomacentrus alleni* Burgess, 1981
Allen's Skate *Pavoraja alleni* JD McEachran & JD Fechhelm, 1982
Allen's River Garfish *Zenarchopterus alleni* BB Collette, 1982
Allen's Sandburrower *Creedia alleni* JS Nelson, 1983
Little Rainbow Wrasse *Dotalabrus alleni* BC Russell, 1988
Allen's Combtooth Blenny *Ecsenius alleni* VG Springer, 1988
Goby sp. *Stenogobius alleni* RE Watson, 1991
Allen's Slender Blenny *Tanyemblemaria alleni* PA Hastings, 1992
Allen's Cardinalfish *Cheilodipterus alleni* O Gon, 1993
Kimberley Blenny *Cirripectes alleni* JT Williams, 1993
Glassy Perchlet sp. *Parambassis alleni* NC Datta & BL Chaudhuri, 1993
Allen's Glider-Goby *Valenciennea alleni* DF Hoese & HK Larson, 1994
Allen's Cling-Goby *Stiphodon allen* RE Watson, 1996
[Probably a synonym of *Stiphodon semoni*]
Allen's Rainbowfish *Chilatherina alleni* DS Price, 1997
Allen's Shrimp-Goby *Tomiyamichthys alleni* A Iwata, N Ohnishi & T Hirata, 2000
Sabah Dottyback *Manonichthys alleni* AC Gill, 2004
Abrolhos Jawfish *Opistognathus alleni* WF Smith-Vaniz, 2004
Allen's Coralbrotula *Diancistrus alleni* W Schwarzhans, PR Møller & JG Nielsen, 2005
Allen's Sole *Leptachirus alleni* JE Randall, 2007
Snake-eel sp. *Ophichthus alleni* JE McCosker, 2010
Goby sp. *Schismatogobius alleni* P Keith, C Lord & HK Larson, 2017
Allen's Clingfish *Flabellicauda alleni* K Fujiwara, KW Conway & H Motomura, 2021

Dr Gerald Robert 'Gerry' Allen (b.1942) is an American-born Australian ichthyologist. The University of Hawaii awarded his PhD (1971), after which he began work (1972) as an ichthyologist at the Australian Museum, Sydney. He then moved to be Curator of Fishes at the Department of Ichthyology, Western Australian Museum, Perth (1974–1997). He worked (1997–2003) for Conservation International, preparing distribution maps for all known reef fishes and has been full time consultant there ever since (2003). He continues as a Research Associate of the Western Australian Museum. He has a particular interest in freshwater fish of New Guinea and Northern Australia. He has taken part in many collecting trips, such as the Western Australian Museum expedition to the Kimberley Coast

(1991) aboard 'North Star IV'. During these trips, he has logged over 7,500 dives. Among 32 books and over 300 papers, he co-wrote *The Marine Fishes of North-Western Australia: A Field Guide for Anglers and Divers* (1988). He wrote the three-volume: *Reef Fishes of the East Indies* (2013) and won the Bleeker Award (2013), NOGI award (2017) among others. His other interests include underwater photography and pursuits outside the marine environment include bird watching, rock climbing, mountaineering and bicycle racing, for which he has been eight-time state veteran champion. He was President of the Australian Society for Fish Biology (1979–1981). To date he has described 13 new genera and over 575 species. (Also see **GR Allen**)

Allen, JA

Characin sp. *Ctenobrycon alleni* CH Eigenmann & WL McAtee, 1907

Joel Asaph Allen (1838–1921) was a zoologist chiefly interested in mammals and birds. He studied under Louis Agassiz and accompanied him to Brazil (1865). He made a number of field trips in North America, and led an expedition for the Northern Pacific Railroad (1873). He was an Assistant in Ornithology at the Museum of Comparative Zoology, Harvard (1870), and was Curator of the Department of Mammals and Birds, American Museum of Natural History, New York (1885–1921). In addition to naming many species, he made important studies on geographic variation relative to climate. Allen's recognition of "variation within populations and intergradation across geographic gradients" helped to overturn the typological species concept current in the mid-1800s, setting out the principle that intergrading populations should be treated as subspecies instead of separate species. This idea led to the widespread adoption of trinomials by American zoologists, a practice that Allen helped to spread through his editorship of the *Auk* and through the American Ornithologists' Union code of nomenclature. He wrote many scientific papers and edited *Auk* and the *Bulletin of the Nuttall Ornithological Club*. Eight birds, ten mammals and a reptile are named after him.

Allen, WF

Stripefin Ronquil *Rathbunella alleni* CH Gilbert, 1904

W F Allen. The author says that a "...*Mr. W. F. Allen discovered the species at Monterey Bay, California*". Two specimens had been 'taken on long lines', though it is unclear whether Allen was himself the long-line fisherman or whether he had somehow acquired the fish.

Allen, WR

Pencil Catfish sp. *Apomatoceros alleni* CH Eigenmann, 1922
Allen's Anchovy *Anchoviella alleni* GS Myers, 1940
Characin sp. *Prodontocharax alleni* JE Böhlke, 1953

Dr William Ray Allen (1885–1955) was a zoologist at the University of Kentucky where he was Professor of Zoology (1922–1948). He was recognised for his contributions to the knowledge of South American fishes and collected the catfish and characin holotypes. He collected in Peru and crossed the Andes (1918–1919) with Eigenmann (q.v.), under the auspices of the National Academy of Sciences. He co-wrote: *Fishes of Western South America (*1942).

Allis

Sculpin sp. *Porocottus allisi* DS Jordan & EC Starks, 1904

Edward Phelps Allis (1851–1947) was one of the foremost comparative anatomists and evolutionary morphologists in the early twentieth century. Recently, many unpublished illustrations and colour drawings (such as of serial sections of primitive actinopterygian fishes) of his were discovered at the Museum of Comparative Zoology, Harvard University. Among his published papers was: *The cranial anatomy of the mail-cheeked fishes* (1909) and *Anatomy of fishes. Collected papers* (1913). The authors say the sculpin was "*...Named for Edward Phelps Allis, of Milwaukee*."

Allison

Tailspot Tetra sp. *Bryconops allisoni* C Silva-Oliviera, ALC Canto & FRV Ribeiro, 2019

Dr Antonio Machado-Allison (b.1945) is a Venezuelan ichthyologist. He is Professor Emeritus of the Institute of Zoology and Tropical Ecology at the Central University of Venezuela where he gained his BSc in Biology (1971) His PhD was awarded by the George Washington University (1982). He began teaching at CUV (1971), eventually becoming full professor (1991). He has written more than 100 papers, articles and books including: *The Caribbean fish of Venezuela* (1996) and *The fish of the plains of Venezuela* (2005). He was honoured "*...in recognition of his contributions to the knowledge of the taxonomy of Bryconops*."

Alluaud

Alluaud's Haplo (cichlid) *Astatoreochromis alluaudi* J Pellegrin, 1904
Alluaud's Catfish *Clarias alluaudi* GA Boulenger, 1906
African Barbel sp. *Labeobarbus alluaudi* J Pellegrin, 1909
Labeo sp. *Labeo alluaudi* J Pellegrin, 1933

Charles A Alluaud (1861–1949) was a French entomologist, botanist and naturalist who came from a wealthy family, living in a chateau where the painter Corot was a frequent visitor. His father was President of the Royal Porcelain Factory, Limoges. When his parents died, he inherited enough wealth to travel extensively, including on scientific expeditions to Ivory Coast, Tunisia, Morocco, Sudan, Niger, the Canary Islands, the Seychelles and Madagascar (1887–1930). He collected throughout, mostly insects, which he sent back to the MNHN. He wrote 165 entomological papers. Four reptiles and three Odonata are named after him. (Also see **Allaud**)

Almaça

Portuguese cyprinid sp. *Iberochondrostoma almacai* MM Coelho, NP Mesquita & MJ Collares-Pereira, 2005

Dr Carlos Alberto da Silva Almaça (1934–2010) was a biologist and zoologist. His bachelor's degree (1957) and doctorate (1968) were both awarded by the University of Lisbon where he went on to be Professor of Zoology at their Faculty of Science (1979–2004) and was Director of the Bocage Museum of Natural History in Lisbon (1984–2004). He wrote c.250 papers including: *Fish species and varieties introduced into inland waters* (1995).

Aloy

African Barb sp. *Enteromius aloyi* B Román, 1971

Hermano Isidro Aloy (b.1925) was a Spanish biologist and mathematics teacher, and a member of the teaching order, the Christian Brothers. He was a missionary and teacher in West Africa for over 45 years. He was honoured in the binomial for "…his assistance at all times."

Alport

Gurnard Perch sp. *Helicolenus alporti* FL de Castelnau, 1873

Moreton Allport (1830–1878) was an English-born Australian colonial naturalist, whose family moved to Tasmania when he was a baby (1831). He followed his father into law, being admitted as a solicitor of the Supreme Court of Tasmania (1852), but was also a keen and accomplished naturalist. He added greatly to zoological knowledge - particularly ichthyology and botany - of Tasmania, as well as to anthropology. He became an expert on Tasmanian fish which he collected, catalogued, described and illustrated sending a great many new fish specimens back to the BMNH and elsewhere. He was also instrumental in introducing trout and other fish into Tasmanian waters. He was a Vice-President of the Royal Society of Tasmania. Castelnau wrote (misspelling his surname): "*I have dedicated this sort to Mr. Moreton Alport, of Hobart Town, who has done so much for the cause of acclimatisation.*"

Alta

Driftwood Catfish sp. *Duringlanis altae* HW Fowler, 1945
[Syn. *Centromochlus altae*]

Alta Dunn was the wife of herpetologist Emmett Reid Dunn, who first brought this species to Fowler's attention. (Also see **Dunn**)

Altuve

Piranha sp. *Serrasalmus altuvei* MV Ramírez, 1965

Néstor Altuve González was a sylvicuturist and forestry expert. He joined the Ministry of Agriculture, Venezuela (1960) and became Director of Natural Resources at that Ministry (1964). He was working (1968) in Rome for the UN Food Programme (FAO) when he was engaged by Costa Rica to help devise laws to prevent over-logging of forests.

Álvarez, J

Alvarez's Silverside *Atherinella alvarezi* E Díaz-Pardo, 1972
[Alt. Gulf Silverside]
Yellowfin Gambusia *Gambusia alvarezi* C Hubbs & VG Springer, 1957
Chiapas Swordtail *Xiphophorus alvarezi* DE Rosen, 1960
Potosi Pupfish *Cyprinodon alvarezi* RR Miller, 1976
Tepehuan Shiner *Cyprinella alvarezdelvillari* S Contreras-Balderas & M de Lourdes Lozano-Vilano, 1994

Dr José Álvarez del Villar (1903–1986) was a Mexican biologist and ichthyologist who is regarded as the founder of modern Mexican ichthyology. His bachelor's degree and doctorate were awarded by the National School of Biological Sciences and his two master's degrees from the universities of Alabama and Michigan (1945). He described 35 new species, including the Mexican Brook Lamprey *Tetrapleurodon geminis* (1964) and founded the National Collection of Mexican Freshwater Fishes of the National Polytechnic Institute. Among his writings are the co-written: *Cuatro especies nuevas de peces dulceacuicolas del sureste de Mexico* (1952) and *Los peces del valle de México* (1957). The Mexican Ichthyology Society awards the 'Dr José Álvarez del Villar' prize in his memory. He collected a number of the type specimens.

Alvarez, M

African Airbreathing Catfish sp. *Channallabes alvarezi* B Román, 1971

Mario Álvarez was honoured in the binomial "*...in gratitude and friendship*" (translation), but otherwise remains mysterious.

Alvey

Makira Rainbowfish *Bedotia alveyi* CC Jones, WL Smith & JS Sparks, 2010

Dr Mark Alvey (b.1955) is an American museums communications administrator. He graduated from the University of Iowa (1978) with a BA in Film Studies, received his MA at the University of Texas, Austin (1985), and completed his PhD there (1995) with the dissertation: *The Semi-Anthology and Series Drama: Sixties Television in Transition*. He became a Departmental Assistant in the Photography Archives (1990) at the Field Museum of Natural History in Chicago. Over the next three decades he held a number of posts at the Field Museum, chiefly providing administrative and communications support for five consecutive Vice Presidents of Science/Collections. He has also served as a content specialist on a number of Field Museum exhibitions. He has published on American television and film, Field Museum history, and taxidermy.

Alvheim

St. Brandon's Sandy *Novaculops alvheimi* JE Randall, 2013

Alvheim's Pufferfish *Chelonodontops alvheimi* PN Psomadakis, K Matsuura & H Thein, 2018

Alvheim's Goby *Thorogobius alvheimi* M Sauberer, T Iwamoto & H Ahnelt, 2018

Oddgeir Berg Alvheim (b.1944) of the Institute of Marine Research in Norway. He has sailed on the RV 'Dr. Fridtjof Nansen" for twenty-six years (from 1986) accumulating a vast knowledge of tropical fish species, and over the years he has established a large photo database of fish and other marine organisms from tropical and subtropical waters. An example of his published papers is: *Investigations on capelin larvae off northern Norway and in the Barents Sea in 1981–84* (1985). He was honoured in the sandy's name because it was he who suggested that the two fishes might be a new species, one female and one male. In the goby etymology he was honoured for his many photographic contributions to the FAO Species Identification Guides and "*...for his assistance and advice to the second author during three surveys aboard the R/V Dr. Fridtjof Nansen, during which type was collected in 2005.*" In fact he has taken part in c.70 surveys on three generations of the RV Dr. Fridtjof Nansen.

Ama

Headstander Characin sp. *Leporinus amae* MP de Godoy, 1980

This species is named after AMA (Assessoria Para Meio Ambiente), an agency of Electrosul, a Brazilian energy company. It was set up to encourage health and environmental initiatives in southern Brazil.

Amad

Sulawesi Freshwater Goby sp. *Mugilogobius amadi* MWC Weber 1913

Raden Mas Amad was a Javanese surgeon who helped Dutch mining engineer and geologist Edward C Abendanon collect fishes and molluscs in Sulawesi, Indonesia (presumably including the type of this goby). Amad also sketched fishes and recorded their native names.

Amanda (Bleher)

Ember Tetra *Hyphessobrycon amandae* J Géry & A Uj 1987

Amanda Flora HIlda Bleher née Kiel (1910–1991) was a German limnologist and adventurer, the mother of explorer and ornamental fish wholesaler-supplier Heiko Bleher, who collected the holotype. The tetra is named after her to acknowledge her interest in and knowledge of the freshwater fauna and flora of Brazil. She took her four children to Brazil and lived with native peoples in the Mato Grosso (1951–1953). She returned to West Germany but was imprisoned (1955–1958), branded as a spy for having traded fishes in East Germany so she decided to settle finally in Brazil (1959). She wrote: *Iténez – River of Hope*, published posthumously (2005).

Amanda (Gimeno)

Freshwater Stingray sp. *Potamotrygon amandae* TS Loboda & MR de Carvalho, 2013

Amanda Lucas Gimeno (1984–2006) was a Brazilian biologist who graduated (2005) from the University of São Paulo, where the senior author had been one of her undergraduate colleagues. She was killed when there was a collapse of an external awning of an amphitheatre at the State University of Londrina during the 26th Brazilian Congress of Zoology.

Amanda Jane

Amanda Jane's Cory *Corydoras amandajanea* DD Sands, 1995

Mrs Amanda Jane Sands is the author's wife, whom he thanked, for "*…unending help and assistance*" during his research.

Amaoka

Lefteye Flounder sp. *Parabothus amaokai* NV Parin, 1983
Gurnard sp. *Pterygotrigla amaokai* WJ Richards, T Yato & PR Last, 2003
Amaoka's Dreamer *Oneirodes amaokai* H-C Ho & T Kawai, 2016

Dr Kunio Amaoka is a Japanese ichthyologist who is Professor Emeritus of Hokkaido University. He has written numerous papers including a number with Kawai such as: *A new righteye flounder,* Poecilopsetta pectoralis *(Pleuronectiformes: Poecilopsettidae), from New Caledonia* (2006). The etymology for the gurnard reads: "*Named after the eminent and recently retired Japanese scientist, Kunio Amaoka, (formerly at Hokkaido University), for his many contributions to ichthyology.*"

Amar

Snakehead sp. *Channa amari* A Dey, BR Chowdhury, R Nur, D Sarkar, L Kosygin & S Barat, 2019

[Syn. *Channa brunnea* – the two names, apparently applying to the same species, were published within weeks of each other]

Amar Chandra Dey (b.1951) is a retired schoolteacher. He is the father of the first author, Arpita Dey, who works at the Aquaculture and Limnology Research Unit, Department of Zoology, University of North Bengal, Darjeeling. Amar gave Arpita five of these fish. The etymology says it was named "*…to honour Amar Dey who collects this new species and donating of the live specimens.*"

Amaral

Banjo Catfish genus *Amaralia* HW Fowler, 1954

Dr Afranio Pompilio Bastos do Amaral (1894–1982) was a physician, zoologist, and herpetologist and was Director of Instituto Butantan (1919–1921 and 1928–1938). *Time* magazine described him on his arrival in Manhattan (1929) as 'the soft-voiced suave herpetologist'. He originally trained as a physician at the Medical School of Bahia. His Ph.D. was awarded by Harvard, where he also taught. He was Director of the Antivenin Institute of America (1927) and conducted the first major study of the incidence of snakebite in Texas. He published, with Barbour: *Notes on Some Central American Snakes* (1924). Nine reptiles are named after him.

Ambrosetti

Armoured Catfish sp. *Pterygoplichthys ambrosettii* EL Holmberg, 1893

Dr Juan Bautista Ambrosetti (1865–1917) was an Argentinian archaeologist, anthropologist and naturalist. His first expedition was to Chaco province (1885). He became Director of Zoology at the Entre Rios Province Museum, Parana, and (1903) Professor of Archaeology at the University of Buenos Aires where he established the Museum of Ethnography (1904). He discovered (1908) the previously lost ruins of the Omaguaca civilisation (10th century AD) in Jujuy province. A bird is named after him.

Amelia

Armoured Catfish sp. *Corymbophanes ameliae* NK Lujan, JW Armbruster, DC Werneke, TF Teixeira & NR Lovejoy, 2019

Amelia was a Patamona Amerindian girl who disappeared near Amaila Falls, Guyana, in the late 19th century. The falls are named after her although her name was spelled incoreectly. The species occurs near these falls.

Amemiya

Bagrid Catfish sp. *Hemibagrus amemiyai* S Kimura, 1934

Dr Ikusaku Amemiya (b.1889) was a marine biologist who was Professor, Tokyo Imperial University, Chairman of its Agricultural and Fishery Section and Director (1936) of the Oceanology Laboratory there. He was a graduate of the University and spent his career there. His many papers include: *Notes on Experiments on the Early Developmental Stages of the Portuguese, American and English Native Oysters, with Special Reference to the Effect of Varying Salinity* (1926). Kimura wrote: "*…The author thanks Prof. I. Amemiya of Tokyo Imperial University for his constant guidance and valuable advice.*"

Amiet

Amiet's Lyretail *Fundulopanchax amieti* AC Radda, 1976

Professor Jean-Louis Amiet (b.1936) is a French zoologist, herpetologist, entomologist, ecologist, and ichthyologist, formerly at Université de Yaoundé, Cameroon, from where he retired to Europe after 29 years. The University of Lille awarded his doctorate (1963). He wrote: *Faune du Cameroun* (1987). Five amphibians and a reptile are named after him.

Amikam

Goby sp. *Callogobius amikami* M Goren, A Miroz & A Baranes, 1991

Amikam Gorovitch was killed in a diving accident in Eilat, Israel; the goby's type locality. We have no other information about him.

Amirhossein

Garra sp. *Garra amirhosseini* HR Esmaeili, G Sayyadzadeh, BW Coad & S Eagderi, 2016

Amirhossein is the son of the first author.

Amir Kabir

Amirkabir's Bleak *Alburnus amirkabiri* H Mousavi-Sabet, S Vatandoust, S Khataminejad, S Eagderi, K Abbasi, M Nasri, A Jouladeh & ED Vasil'eva, 2015

Mirza Taghi Khan Farahani aka Amir Kabir (1807–1852) is regarded as the first Iranian modernizer and reformer, who served as Chief Minister (1848–1852) under the Shah Naser al-Din Shah Qajar. A power struggle in the Iranian government finally resulted in his arrest and expulsion from the capital, followed by an order for his execution. The type specimen of this bleak was collected in Markazi province, in which Amir Kabir was born.

Ammer

Ammer's Rainbowfish *Melanotaenia ammeri* GR Allen, PJ Unmack & RK Hadiaty, 2008

Raja Ampat Dottyback *Pseudochromis ammeri* AC Gill, GR Allen & M Erdmann, 2012

Maximilian Johannes 'Max' Ammer (b.1961) of Sorong, West Papua, is a Dutch diver and conservationist. He had a technical education (as a toolmaker), then served in the Dutch Army as Para/Commando. Later he restored classical Harley-Davidson and Indian motorcycles as a profession, before founding Papua Diving and the Raja Ampat Research &

Conservation Centre. He went to the region in the 1990s following his passion to search for submerged World War II aircraft. His other interests include all aviation, Christianity and making exploration trips into the heart of Papua.

AMS

Dottyback genus *Amsichthys* AC Gill & AJ Edwards, 1999

AMS is institutional code in ichthyology for the Australian Museum (Sydney). It was honoured for the "...generous help, encouragement and friendship given by staff of the Australian Museum's Ichthyology Section to the first author during this study and throughout his career and training."

Ana

Characin sp. *Astyanax anai* A Angulo, AC Santos, M López, F Langeani & CD McMahan, 2018

Ana Rosa Ramírez-Coghi was (2000–2018) assistant collection manager of the Universidad de Costa Rica fish collection. She was honoured in the name "...*for her dedicated service, since the year 2000, as the assistant collection manager of the Universidad de Costa Rica fish collection.*" (See also **Anai**)

Anabel

Loach sp. *Cobitis anabelae* J Freyhof, E Bayçelebi & M Geiger, 2018

Anabel Perdices is an ichthyologist, biogeographer and evolutionary biologist at the Museo Nacional de Ciencias Naturales, Madrid. She was honoured because she "...*dedicated parts of her scientific life to the research on the diversity and phylogeny of the genus* Cobitis."

Anai

Characin sp. *Astyanax anai* A Angulo, AC Santos, M López, F Langeani & CD McMahan, 2018

Asociación ANAI is is a Costa Rican non-profit association "dedicated to helping people put into practice community and landscape-level initiatives that integrate nature conservation and the well-being of the people who conserve and sustainably use nature." It was honoured in the fish's binomial for "...*exceptional work in support of the knowledge and conservation of aquatic environments and in the promotion and execution of sustainable development initiatives in the Talamanca region of eastern Costa Rica-western Panama*" (where this species occurs). The name also represents a tribute to Ana R. Ramírez Coghi (see **Ana** above).

Anais

Pencil Catfish sp. *Trichomycterus anaisae* AM Katz & WJEM Costa, 2021

Maria Anais Barbosa Segadas Vianna, Brazilian zoologist and ichthyologist (see **Barbosa, MA**).

Anak

Oyster Pompano *Trachinotus anak* JD Ogilby, 1909
[Alt. Giant Oystercracker]

Anak is the name of a biblical giant. The 'sons of Anak' are first mentioned in *Numbers 13*, when Moses sends twelve spies to scout out the land of Canaan. The spies report that the sons of Anak live in that region, and that they (the spies) felt like mere grasshoppers in the presence of these giants.

Anauls

Salmonid sp. *Coregonus anaulorum* IA Chereshnev, 1996

This name was first coined by Kaganowsky (1933) but only became available for use (1996) and the etymology was unexplained. It presumably refers to the Anauls, a sub-tribe of the Yukaghir people of northern Siberia (17th century), who lived in the area where this fish is found.

Ancieta

Sea Chub sp. *Medialuna ancietae* N Chirichigno F., 1987

Dr Felipe Ancieta Calderón (1922–2001) was the Director of Fisheries Investigations for Inland Waters, Peru. He wrote a number of papers such as: "*Outline of the Principal Needs of Peru in Fisheries and Related Investigations*" (1981). He was trained (1951) in the USA at the University of Michigan.

Ancon

Hagfish sp. *Eptatretus ancon* HK Mok, LM Saavedra-Diaz & A Acero Pizarro, 2001

This refers to the research vessel 'Ancon' from which the holotype was collected in the Caribbean.

Anders

Spiketail Platyfish *Xiphophorus andersi* MK Meyer & M Schartl, 1980

Professor Dr Fritz Wilhelm Anders (1919–1999) was a German geneticist. He served in WW2 and was a prisoner of war in Russia (1943–1948). After returning to Germany he studied Biology at the University of Mainz. He was appointed (1964) as founder and Director of the Institute of Genetics at the University of Giessen. He studied the development of cancer via research using *Xiphophorus* fish (easily propagated through many generations) as experimental models.

Andersen

Andersen's Lantern Fish *Diaphus anderseni* AV Tåning, 1932

Dr N C Andersen was the ship's doctor on board the Danish research ship 'Dana' on a number of cruises. Tåning named this fish after him as a memorial to his late friend.

Anderson, A

Catface Grouper *Epinephelus andersoni* GA Boulenger, 1903
[Alt. Brown-spotted Rockcod]

Alexander Anderson of Durban, South Africa, was a collector and 'dealer in natural history specimens.' Boulenger said in his etymology that Anderson: "...*takes a great interest in the fishes in which he deals, and who brought the fishes over with him on a recent visit to England*".

Anderson, AJ

Anderson's Shortnose Pipefish *Micrognathus andersonii* P Bleeker, 1858

Dr A J Anderson was a physician on the Cocos Islands who collected fish specimens. He wrote a number of papers including: *A few remarks concerning a parasitic fish, found in the Holothuria of the Cocos Islands* (1860). He and G. Clunies-Ross who was 'Superintendant' of the islands, sent a collection of 120 specimens to Bleeker. Bleeker writes: "Ik heb deze verzamelingen te danken aan de welwillendheid van den heer Dr A. J. Anderson, geneesheer op de Kokos-eilanden en J. G. C. Ross, tegenwoordigen beheerder dier eilanden."

Anderson, C

Slickhead sp. *Alepocephalus andersoni* HW Fowler, 1934

Charles Anderson (1876–1944) was a mineralogist and palaeontologist who became Director of the Australian Museum (1921–1940). The University of Edinburgh awarded his BA and MA (1989), BSc (1900) and doctorate (1908). He joined the Australian Museum (1901) where he stayed for the rest of his career. Fowler wrote that he: "…*contributed much to my delightful stay in Sydney.*"

Anderson, CJ

Three-spotted Tilapia *Oreochromis andersonii* FL de Castelnau, 1861

Charles John Andersson (1827–1867) aka Karl Johan Andersson explored Southwest Africa (Namibia) in the 19th century. He was a Swedish explorer, hunter, trader and amateur naturalist. Castelnau named a whole series of fish after European explorers of Africa, without giving detailed etymologies. In this case, it seems likely he intended this person, despite Andersson's name having a double-s: Castelnau seems to have mis-spelt it.

Anderson, CW

Armoured Catfish sp. *Corymbophanes andersoni* CH Eigenmann, 1909

Charles Wilgress Anderson (b.1867) was a geologist and the Government Surveyor in the Department of Lands and Mines, British Guiana (Guyana). He was one of the commissioners who dealt with the Guyana-Venezuela border dispute (1905). He co-wrote *The Geology of the Goldfields of British Guiana* (1908). He also collected botanical specimens, which he sent to Kew. Presumably he must in some way have given help to Eigenmann during the latter's expedition to Guyana, during which the holotype was collected.

Anderson, J

African catfish genus *Andersonia* GA Boulenger, 1900
Tibetan catfish sp. *Glaridoglanis andersonii* F Day, 1870
Chinese cyprinid sp. *Anabarilius andersoni* CT Regan, 1904

John Anderson (1833–1900) was a qualified physician who became Professor of Comparative Anatomy at the Medical School in Calcutta and Director of the Indian Museum there (1865). He joined an expedition to Burma and Yunnan in southwest China as naturalist (1868). A second expedition (1875) only collected in Burma. Boulenger wrote of Anderson: "…*to whose exertions during the latter years of his life Science is indebted for much progress in the zoology of the Nile region*." He was elected a Fellow of the Royal Society (1879). Three amphibians, four birds, three mammals, a dragonfly and eight reptiles are also named after him.

Anderson, JME

Anderson's Toadfish *Torquigener andersonae* G Hardy, 1983

Dr Jennifer M E Anderson was Hardy's colleague at the University of New South Wales. In the etymology he says: "*It is named for Dr Jennifer M E Anderson, a very friendly and pleasant colleague, with whom I shared working facilities whilst at the University of New South Wales*."

Anderson, ME

Thermal Vent Eelpout sp. *Thermarces andersoni* RH Rosenblatt & DM Cohen, 1986
Bearded Seadevil sp. *Linophryne andersoni* O Gon, 1992
Anderson's Eelpout *Pachycara andersoni* PR Møller, 2003
Anderson's Mudbrotula *Dermatopsoides andersoni* PR Møller & W Schwarzhans, 2006
Viviparous Brotula sp. *Microbrotula andersoni* W Schwarzhans & JG Nielsen, 2011

Dr Michael Eric Anderson is an American ichthyologist. His bachelor's degree was awarded by California State University, Hayward (1973), Moss Landing Marine Laboratories, California State University and Colleges, Moss Landing his master's (1977) and Virginia Institute of Marine Sciences, College of William and Mary, Gloucester Pt., Virginia, his doctorate (1984). He worked for various organisations, being based at Moss Landing Marine Laboratories (1973–1981) and at the California Academy of Sciences, San Francisco (1983–1989) and as Senior Ichthyologist, South African Institute for Aquatic Biodiversity, J L B Smith Institute of Ichthyology. He is particularly noted for his work on systematics of deep-sea fishes.

Anderson, MP

Torrent Catfish sp. *Liobagrus andersoni* CT Regan, 1908

Malcolm Playfair Anderson (1879–1919) was an American zoologist educated at secondary level in Germany, returning to the USA to study zoology and graduate at Stanford University (1904). From age 15 he took part in collecting expeditions to Arizona, Alaska and California. He joined the Cooper Ornithological Club (1901) and wrote a number of articles on ornithology, but did not confine himself to that subject. He was chosen to conduct the Duke of Bedford's Exploration of Eastern Asia for the Zoological Society of London (1904). He took photographs and extensive notes on the collections and wrote several short stories about the people with whom he lived and worked in the Orient. He was again in western China (1909 and 1910). He was in Peru with Osgood (q.v.) (1912). He died after falling from scaffolding at the shipyards in Oakland, California. Four mammals and a bird are named after him.

Anderson, RC

Sole sp. *Aseraggodes andersoni* JE Randall & SV Bogorodsky, 2013

Dr R Charles Anderson is a marine ecologist who was one of the collectors of the sole holotype. He assisted the first author during the latter's visits to the Republic of Maldives, and is the author of: *Reef Fishes of the Maldives* (2007).

Anderson, WD

Bucktoothed Slopefish *Symphysanodon andersoni* A Kotthaus, 1974
Forgetful Snake-eel *Lethogoleos andersoni* JE McCosker & JE Böhlke, 1982

Professor Dr William D Anderson Jr (b.1933) was a marine biologist and ichthyologist at Grice Marine Biological Laboratory, College of Charleston, South Carolina. The University of South Carolina awarded his PhD (1960). He is now Professor Emeritus, College of Charleston. He is a described as "*...a friend and ichthyologist, who made specimens available to authors.*" His current interests are: Systematics of fishes: Symphysanodontidae (slopefishes), worldwide; Anthiinae (anthiine seabasses, family Serranidae), Atlantic and eastern Pacific; Callanthiidae (splendid perches), worldwide; Lutjanidae (snappers), worldwide and the History of biology; particularly the history of natural history investigations in South Carolina and the history of the evolution-creationism controversy. Among his many publications are the co-written: *Natural history investigations in South Carolina from colonial times to the present* (1999) and *Review of Atlantic and eastern Pacific anthiine fishes (Teleostei: Perciformes: Serranidae), with descriptions of two new genera* (2012).

Anderson, WeW

Pufferfish sp. *Sphoeroides andersonianus* JE Morrow Jr, 1957

Wendell W Anderson Sr., (1901–1959) was an investment banker and yachtsman. He loved sport fishing and science and satisfied both by funding two marine expeditions for Yale University's Bingham Oceanographic Laboratory and Peabody Museum to New Zealand (1948) and Ecuador and Peru (1953). His son 'Jack' shared his father's interests and participated in the expeditions. Also present were ichthyologists, oceanographers and technicians.

Anderson, WiW

Florida Torpedo *Torpedo andersoni* HR Bullis, 1962

William Wyatt Anderson (1909–1993) was an American ichthyologist who was a Fisheries Research Biologist at the State Game and Fish Commission, Coastal Fisheries Division, Brunswick, Georgia, USA, and a close friend of the describer. He was honoured as a "*... colleague and mentor, whose labors have contributed immeasurably to our knowledge of the marine fauna of the southeastern United States.*" They co-wrote a number of papers such as: *Searching the sea bed by sub* (1970).

Andersson, CJ

(See **Anderson, CJ**)

Andersson, JG

Icefish sp. *Neosalanx anderssoni* CH Rendahl, 1923

Johan Gunnar Andersson (1874–1960) was a Swedish archaeologist, geologist and palaeontologist who was Director of Sweden's National Geological Survey. He was on the

Swedish Antarctic Expedition (1901–1903). He made a number of important archaeological and paleontological finds in China, including the first remains of 'Peking Man'. He obtained the holotype of this species. A bird is also named after him. (Also see **Gunnar**)

Andersson, KA

Rhombic Lanternfish *Krefftichthys anderssoni* E Lönnberg, 1905

Karl Andreas Andersson (1875–1968) was the zoologist of the Swedish Antarctic Expedition led by Otto Nordenskjöld (1902–1903). He collected the holotype of this species. He wrote: *Die Pterobranchier der schwedischen Südpolarexpedition 1901–1903 nebst Bemerkungen über* Rhabdopleura normani *Allmann* (1908). Andersson Peak in Antarctica is named after him.

Anderton

Painted Latchet *Pterygotrigla andertoni* Waite, 1910

Thomas Anderton worked at the Port Chalmers fish hatchery, New Zealand. The original text says of the holotype: "*Length 294 mm. Bay of Plenty; trawled by Mr. Thomas Anderton, of the Portobello Fish-hatchery, Port Chalmers.*" We have been unable to find more about him.

Andhra

Andhra Anchovy *Stolephorus andhraensis* Babu Rao, 1966

This is a toponym; it refers to the state of Andhra Pradesh, India, the type locality.

Andover

Andover Lionfish *Pterois andover* GR Allen & M van N Erdmann, 2008

Named after the Andover group of companies: "*...dedicated to promoting greater public appreciation of the oceans and marine conservation in Asia.*"

Andre

Estuarine Frillfin Goby *Bathygobius andrei* HE Sauvage, 1880

Édouard François André (1840–1911) was a French horticulturalist, landscape architect and designer, famously designing city parks in Monte Carlo and Montevideo before which he was part of the redesign of Paris open spaces. He won a competition to design Sefton Park, Liverpool (1866), and went on to design around 100 public parks all over Europe and the Russian Empire. He also edited (1870) *L'Illustration Horticole*. He collected plants in the Andes (1875–1876), including Ecuador, and from this, many plants were introduced into Europe. He wrote: *Bromeliaceae Andreanae. Description et Histoire des Bromeliacées récoltées dans la Colombie, l'Ecuador et la Venezuela* (1889). In the original description of the goby, it states only that 'André' collected the holotype, but we believe this to be the person honoured.

Andreas

Ande's Lanternfish *Centrobranchus andreae* CF Lütken,1892

Capt. A F Andréas, who is described as a 'tireless collector', collected this and other marine fishes for the University of Copenhagen Zoological Museum. He did not confine himself to ichthyology, but collected other taxa and also has a reptile named after him.

Andres

Characacin sp. *Hemibrycon andresoi* C Román-Valencia, 2003
[Syn. *Bryconamericus andresoi*]

Andrés Córdoba B is a Colombian biologist. He collected the holotype and provided ecological data and observations.

Andrew (Rao)

Slaty Leaf-fish *Nandus andrewi* HH Ng & Z Jaafar, 2008
Dwarf Snakehead sp. *Channa andrao* R Britz, 2013
Tropical Barb sp. *Oreichthys andrewi* JDM Knight, 2014
Stone Loach sp. *Schistura andrewi* B Solo, Lal Ramliana, S Lalronunga & Lalanun Tiunga Vanchhawng, 2014
Chameleonfish sp. *Badis andrewraoi* S Valdesalici & S Van der Voort, 2015
Slender Ricefish sp. *Oryzias andrewi* TR Roberts, P Chakraborty, K Yardi & P Mukherjee, 2021

Andrew Arunava Rao is an ornamental-fish collector and breeder who owns and runs 'Malabar Tropicals'. He lists some of the problems of collecting fish in the wild including '...hoodlums and security forces and a few times from wild elephants.' He co-wrote: *Ornamental Aquarium Fish of India* (1999). He collected the barb type, and was honoured for his "*...enthusiasm and support to ichthyology around the world*" and for the snakehead for his "*support of the ichthyological exploration of the freshwater fish fauna of India.*"

Andrew (Smith)

Cape Whitefish *Pseudobarbus andrewi* KH Barnard, 1937

Dr Sir Andrew Smith (1797–1872) was a Scottish surgeon, explorer and zoologist (see **Smith, Andrew** for his biography).

Andrews

Lefteye Flounder sp. *Arnoglossus andrewsi* DE Kurth, 1954

Ernest E 'Dick' Andrews was Secretary of the Tasmanian Licensed Fishermen's Association (having been a commercial fisherman all his working life) when he was appointed (1950) Chief Inspector of Fisheries, Tasmania. Kurth says of him and the eponymous fish: "*Mr. E. E. Andrews, Chief Inspector of Fisheries for Tasmania, whilst examining material brought up in the dredges noticed a small flat fish of unusual appearance which he preserved and sent to the author for identification. It was suspected that the specimen belonged to an undescribed species and as the fish was immature, efforts were made to obtain more. Several larger specimens were taken subsequently, but they were extensively damaged by the dredge and it was not until the following year that the series of seven fish used in the following description was obtained.*"

Andriashev

Eelpout genus *Andriashevia* VV Fedorov & AV Neyelov, 1978
Sculpin genus *Andriashevicottus* VV Fedorov, 1990
Andriashev's Spiny Pimpled Lumpsucker *Eumicrotremus andriashevi* Perminov, 1936
Chukot Char *Salvelinus andriashevi* LS Berg, 1948

Andriashev's Lanternfish *Protomyctophum andriashevi* VE Becker, 1963
Pineapple Rattail *Idiolophorhynchus andriashevi* YI Sazonov, 1981
Little-eyed Skate *Bathyraja andriashevi* VN Dolganov, 1985
Deep-sea Smelt sp. *Bathylagus andriashevi* SG Kobyliansky, 1986
Flabby Whalefish sp. *Gyrinomimus andriashevi* VV Fedorov, AV Balushkin & IA Trunov, 1987
Tripodfish sp. *Bathypterois andriashevi* KJ Sulak & YN Shcherbachev, 1988
Batfish sp. *Halieutopsis andriashevi* MG Bradbury, 1988
Snailfish sp. *Paraliparis andriashevi* DL Stein & LS Tompkins, 1989
Conger Eel sp. *Gnathophis andriashevi* ES Karmovskaya, 1990
Stonefish sp. *Minous andriashevi* SA Mandritsa, 1990
Snailfish sp. *Osteodiscus andriashevi* DL Pitruk & VV Fedorov, 1990
Lightfish sp. *Polymetme andriashevi* NV Parin & OD Borodulina, 1990
Morid Cod sp. *Physiculus andriashevi* YN Shcherbachev, 1993
Spiny Plunderfish sp. *Harpagifer andriashevi* VP Prirodina, 2000
Andriashev's Dwarf Snailfish *Psednos andriashevi* N Chernova, 2001
Eel Cod sp. *Muraenolepis andriashevi* AV Balushkin & VP Prirodina, 2005
Eelpout sp. *Zoarces andriashevi* NV Parin, SS Grigoryev & ES Karmovskaya, 2005
Barreleye sp. *Dolichopteryx andriashevi* NV Parin, TN Belyanina & SA Evseenko, 2009
Manefish sp. *Platyberyx andriashevi* E Kukuev, NV Parin & IA Trunov, 2012

Professor Dr Anatoly Petrovich Andriyashev (1910–2009) was a Russian marine biologist. Leningrad State University awarded his bachelor's degree (1933), his doctorate (1937) and his further doctorate of science (1951). He became a researcher at the Zoological Institute of the Russian Academy of Sciences (1943) and a professor there (1970). He wrote: Fishes of the northern seas of the USSR (1964).

Andromakhe

Characin genus *Andromakhe* GE Terán, MF Benitez & JM Mirande, 2020

Andromakhe (or Andromache) is a character in Greek mythology: the wife of Hector, Prince of Troy, and the mother of Astyanax. The name was chosen when the authors placed several species formerly in the genus *Astyanax* into this new genus.

Anduze

Thorny Catfish genus *Anduzedoras* A Fernández-Yépez, 1968
Banjo Catfish sp. *Ernstichthys anduzei* A Fernández-Yépez, 1953
Driftwood Catfish sp. *Trachelyopterichthys anduzei* CJ Ferraris Jr & JM Fernandez, 1987
Anduze's Pencilfish *Nannostomus anduzei* JM Fernandez & SH Weitzman, 1987

Dr Pablo Jose Anduze (1902–1989) was a Venezuelan entomologist. He was schooled at the Maison de Melle, Ghent, Belgium and the Saint George's College, Weybridge, England. On his return to Venezuela he became head of the Department of Entomology of Service Prophylaxis of Yellow Fever (1940), then Head of Entomology, National Institute of Hygiene (1941). He was member of the National Commission for the study of onchocerciasis and Chief Medical Entomology Creole Petroleum Corporation (1946–1951). The Academy of Physics, Mathematics and Natural History made him a fellow (1947). He became head of the Commission of Zoology (1950) – the Franco-Venezuelan Expedition that determined the location of the sources of the Orinoco. He wrote up his experience as: *Shailili-ko*: the story of his search for the sources of the Orinoco, his scientific observations and the most comprehensive study ever done on the Yanomami people.

Andruzzi

Somalian Cavefish *Phreatichthys andruzzii* D Vinciguerra, 1924

Major (later Colonel) Dr Alcibiade Andruzzi was an Italian military surgeon and physician practicing in Italian Somaliland (now part of Somalia) and the Director of the local Colonial Health Service. The collectors gave the holotype to him and he passed it to the author at the Museo Civivo di Storie Naturale di Genova. He co-wrote: *Osservazioni sulle fripanosi del bestiame nella Somalia Italiana* (1924).

Andy Sabin

Malagasy Blue-spotted Guitarfish *Acroteriobatus andysabini* S Weigmann, DA Ebert & B Séret, 2021

Andrew 'Andy' Sabin is the President of Sabin Metal Corporation, and a well-known New York philanthropist committed to species conservation. In particular he has supported many projects in Madagascar, including research on lemurs, tortoises and frogs. Despite his conviction that climate change is real and caused by human activities, and his funding of the Sabin Center for Climate Change Law at Columbia University, Sabin has described himself as a 'staunch supporter' of Donald Trump. He was honoured in the guitarfish's binomial for his support of the 'Lost Sharks' project at the Pacific Shark Research Center in California.

Angel

Stone Loach sp. *Triplophysa angeli* PW Fang, 1941

Fernand Angel (1881–1950) was a French zoologist and herpetologist. He joined the *MNHN*, Paris (1905) as an Assistant Taxidermist for Leon Vaillant and then François Mocquard (q.v.), working there until he died. The only break in his career was his French Army service (1914–1918). Nine reptiles and three amphibians are named after him.

Angela

Ângela's Catshark *Parmaturus angelae* KD de Araújo Soares, MR de Carvalho, PR Schwinger & OBF Gadig, 2019
[Alt. Brazilian Filetail Catshark]

The etymology says the fish was "… *dedicated to the last author's granddaughter, Ângela.*"

Angela (Watters)

African Killifish sp. *Nothobranchius angelae* BR Watters, B Nagy & DU Bellstedt, 2019

Angela Watters is the wife of the first author, Brian Watters (q.v.), who honoured her for 'unwavering support' for his many field trips to Africa to study killifish. (Also see **Watters**)

Angelo

Angelo's Eelpout *Pachycara angeloi* R Thiel, T Knebelsberger, T Kihara & K Gerdes, 2021

Ângelo Miguel de Oliveira Mendonca is the husband of the third author, marine biologist Terue Cristina Kihara.

Anhaguapitã

Armoured Catfish sp. *Rineloricaria anhaguapitan* MS Ghazzi, 2008

Anhaguapitã in Tupí legend is a devil who clashed with St. Peter. In the legend, the demon became rain and small stones and so created the Uruguay River of southern Brazil.

Anisits

Bloodfin Tetra *Aphyocharax anisitsi* CH Eigenmann & CH Kennedy, 1903
Paraná Sailfin Catfish *Pterygoplichthys anisitsi* CH Eigenmann & CH Kennedy, 1903
Pencil Catfish sp. *Homodiaetus anisitsi* CH Eigenmann & DP Ward, 1907
Buenos Aires Tetra *Hyphessobrycon anisitsi* CH Eigenmann, 1907

Juan Daniel Anisits (1856–1911) was born in Hungary and became a Paraguayan citizen. He worked at the the National University of Paraguay and collected many specimens that he sent to Eigenmann for study.

Anita

Armoured Catfish sp. *Rineloricaria anitae* MS Ghazzi, 2008

Ana Maria de Jesus Ribeiro di Garibaldi aka Anita Garibaldi (1821–1849) was the Brazilian wife and comrade-in-arms of Italian revolutionary Giuseppe Garibaldi, a key figure in the Ragamuffin War (Revolução Farroupilha), a failed war of secession from the Brazilian Empire (1835–1845). She is regarded in Brazil as a symbol of Brazilian republicanism and as a national heroine. She followed Garibaldi to Italy (1848) when they fought against the Austrian Empire, but had to flee after Rome fell (1849) to the assault of French troops. She had malaria and died near Ravenna during the retreat of the Garibaldian Legion.

Ann Natalia

Danio sp. *Devario annnataliae* S Batuwita, M De Silva & S Udugampala, 2017

Natalie Ann Ratnaweera (1990–2012) was a Sri Lankan wildlife enthusiast.

Anna

Armoured Catfish sp. *Squaliforma annae* F Steindachner, 1881
[Now regarded as a synonym of *Aphanotorulus emarginatus*]

Steindachner is notorious for failing to give useful etymologies, and this is one of them. The identity of Anna remains a mystery.

Anna (Weber)

Anna's Dottyback *Pseudoplesiops annae* M Weber, 1913

Anna Antoinette Weber van Bosse (1852–1942) was a Dutch botanist. With her husband M. C. W. Weber (d.1937), she collected in the East Indies (1888–1890 and 1899–1900). When Anna died their estate at Eerbeek was bequeathed to the Gelders Landschap foundation. Their house is now an adult education centre. Their library and scientific correspondence were bequeathed to the Artis Library, University of Amsterdam, and to the Zoological Museum of the University of Amsterdam (Institute of Taxonomic Zoology). A bird is also named after her. There is no etymology, but it seems very likely that the fish is named after Weber's wife.

Anna (Weigmann)

Anna's Sixgill Sawshark *Pliotrema annae* S Weigmann, O Gon, RH Leeney & AJ Temple, 2020

Anna Weigmann Huerta (b.2017) is the niece of the first author. The species was named in her honour "…to express its relationship to Pliotrema kajae, named after the first author's daughter Kaja Magdalena Weigmann."

Anna Maria

Mormyrid sp. *Marcusenius annamariae* P Parenzan, 1939

We have been unable to establish whom Pietro Parenzan (1902–1992) had in mind when he named this species.

Annamar

Gurnard sp. *Lepidotrigla annamarae* L del Cerro & D Lloris, 1997

Annamar del Cerro (b.1990) is the daughter of the senior author. "*This species is named for the seventh birthday of Annamar, the older daughter of Lluis del Cerro.*"

Annandale

Anandale's Giant Danio *Devario annandalei* BL Chaudhuri, 1908
Annandale Loach *Lepidocephalichthys annandalei* BL Chaudhuri, 1912
Annandale's Skate *Rajella annandalei* MCW Weber, 1913
Scaly-nape Goby *Pleurosicya annandalei* J Hornell & HW Fowler, 1922
Sisorid Catfish sp. *Glyptothorax annandalei* SL Hora, 1923
Annandale's Guitarfish *Rhinobatos annandalei* JR Norman, 1926
Annandale's Garra *Garra annandalei* SL Hora, 1921
Sole sp. *Zebrias annandalei* PK Talwar & Chakrapany, 1967

Dr 'Thomas' Nelson Annandale (1876–1924) was a zoologist (primarily entomologist and herpetologist) and anthropologist who became Superintendent of the Indian Museum, Calcutta which still houses his insect and spider collection. He was instrumental in establishing a purely zoological survey, not combined with anthropology, undertaking several expeditions, (from 1899) most notably the Annandale-Robinson expedition that collected in Malaya (1901–1902). He went to India (1904) as Deputy Superintendent at the museum, becoming Director (1907). He also became the first Director of the Zoological Survey of India (1916–1924). He was also noted for his work on the biology and anthropology of the Faroe Islands and Iceland. He wrote or co-wrote a number of scientific papers (1903–1921), including: *The Aquatic and Amphibious Molluscs of Manipu* (1921). He was honoured in the name of the guitarfish for his contributions to Indian ichthyology, which included an account of this species (1909) and in the goby name in 'slight recognition for his work on Indian fishes'. Six amphibians, five odonata, four reptiles and a mammal are also named after him.

Anne Patrice

Basslet sp. *Tosanoides annepatrice* RL Pyle, BD Greene, JM Copus & JE Randall, 2018

Anne Patrice Greene is the mother of Brian David Greene (q.v.) of the Association for Marine Exploration, one of the authors, who collected all known specimens of this species. She was honoured "*...in recognition of the support and encouragement she has consistently provided to Brian's exploration of the deep coral reefs of Micronesia.*" (Also see **Greene, BD**)

Annie (Keith)

Goby sp. *Stiphodon annieae* P Keith & RK Hadiaty, 2015

Mrs Annie Keith is the senior author's wife. He named this species after her for her patience and unfailing support during field trips in the Pacific islands.

Annie (Leveque)

African Barb sp. *Enteromius anniae* C Lévêque, 1983

Annie Lévêque née Roux is the wife of the describer, Christian Lévêque.

Ansorge

Hingemouth *Phractolaemus ansorgii* GA Boulenger, 1901
West African Cichlid sp. *Thysochromis ansorgii* GA Boulenger, 1901
Lutefish sp. *Citharidium ansorgii* GA Boulenger, 1902
African Whiptailed Catfish *Phractura ansorgii* GA Boulenger, 1902
Agberi Mormyrid *Petrocephalus ansorgii* GA Boulenger, 1903
African Barb sp. *Enteromius ansorgii* Boulenger, 1904
Slender Stonebasher *Hippopotamyrus ansorgii* GA Boulenger, 1905
Spiny-eel sp. *Mastacembelus ansorgii* GA Boulenger, 1905
Angolan Cyprinid sp. *Labeobarbus ansorgii* GA Boulenger, 1906
[Syn. *Varicorhinus ansorgii*]
Cunene Labeo *Labeo ansorgii* GA Boulenger, 1907
African Characin sp. *Alestes ansorgii* GA Boulenger, 1910
Claroteid Catfish sp. *Chrysichthys ansorgii* GA Boulenger, 1910
African Freshwater Pipefish *Enneacampus ansorgii* GA Boulenger, 1910
Shellear sp. *Kneria ansorgii* GA Boulenger, 1910
African Tetra sp. *Nannopetersius ansorgii* GA Boulenger, 1910
Ansorge's Fangtooth Pellonuline *Odaxothrissa ansorgii* GA Boulenger, 1910
Guinean Bichir *Polypterus ansorgii* GA Boulenger, 1910
African Cyprinid sp. *Raiamas ansorgii* GA Boulenger, 1910
African Rivuline sp. *Epiplatys ansorgii* GA Boulenger, 1911
Characin sp. *Nannocharax ansorgii* GA Boulenger, 1911
Squeaker Catfish sp. *Synodontis ansorgii* GA Boulenger, 1911

Ornate Ctenopoma *Microctenopoma ansorgii* GA Boulenger, 1912
Ansorge's Neolebias *Neolebias ansorgii* GA Boulenger, 1912

Dr William John Ansorge (1850–1913) was an English explorer and collector who was active in Africa in the second half of the 19th century. He wrote: *Under the African Sun* (1899). He collected a number of new species of fish from the Niger delta including the type specimens of the *Kneria* and the eponymous pellonuline. Four mammals, one amphibian, three reptiles and 19 birds are named after him. (Also see **Gulielm**)

ANSP

Academy Eel *Apterichtus ansp* JE Böhlke, 1968

This is an acronym for the Academy of Natural Sciences of Philadelphia, where Böhlke worked.

Antenor

Neotropical Rivuline sp. *Hypsolebias antenori* J Tulipano, 1973

Dr Antenor L de Carvalho (1910–1985). (See **Carvalho**)

Antigone

Boarfish genus *Antigonia* RT Lowe, 1843

Lowe did not explain his reason for coining this name, but presumably it is derived from mythology as FISHBASE suggests: Antigone was the daughter of Oedipus and Jocasta, who defied her uncle, King Creon, by performing funeral rites over her brother Polynices and was condemned to be immured in a cave. These are deep-water fish so the cave is not an obvious connection.

Antinori

Black Lampeye *Micropanchax antinorii* D Vinciguerra, 1883
Tunisian Barb *Luciobarbus antinorii* GA Boulenger, 1911

Marchese Orazio Antinori (1811–1882) was an Italian zoologist. He travelled and collected in Ethiopia (1876–1882). He was head of a scientific station in Shoa (1876–1881), which was set up by the Royal Geographic Society of Italy of which he was a founding member. He was a college 'drop-out': he studied the classics but left (1828) without getting a diploma. His family was noble but not well off, so he needed to earn a living. He spent the next ten years pursuing ornithology and taxidermy, moving to Rome (1837) to work for Bonaparte mounting skins, etc. He undertook various writing and curating tasks until, in a time of political turmoil (1848), he was shot in the right arm which nearly had to be amputated. He then taught himself to write and draw with his left hand. He continued to make a living as a taxidermist, travelling to Greece and Turkey (1850s) and eventually to Syria and Egypt. He explored in Sudan until 'the continuous rains, the fevers, the dysentery and lack of drinking water threatened to bury us all'. He continued to travel, notably to Sardinia with Salvadori (1863) and to Tunisia (1866). When he arrived in Ethiopia (1876) he wrote to his friend Doria telling him it was the most wonderful place he had seen and did not leave with the rest of the expedition, preferring to continue his work there for the rest of his life. A mammal and two birds are named after him.

Antipa

Danube Delta Gudgeon *Romanogobio antipai* PM Bănărescu, 1953

Grigore Antipa (1867–1944) was a Romanian zoologist, biologist, ichthyologist, limnologist and oceanographer who studied the fauna of the Danube Delta and the Black Sea. He was Director (1892–1944) of the Bucharest Natural History Museum (now named after him) and totally re-organised it, including a move to new, purpose-built premises. He founded the School of Hydrobiology and Ichthyology in Romania and was Professor there.

Antipholus

Longnose Cusk-eel *Ophidion antipholus* RN Lea & CR Robins, 2003

The brothers Antipholus are characters in Shakespeare's *The Comedy of Errors.*

Anton (Greshoff)

African Characin sp. *Distichodus antonii* L Schilthuis, 1891

Anton (sometimes Antoine) Greshoff (1856–1905) was a trader who arrived in the Congo (1877) and was there for more than 20 years, running the company's business at Boma. He was also Dutch Consul in Leopoldville. He was an active collector, particularly of entomological specimens. He supplied ethnographic material to the Ethnographical Museum, Leiden, where he presented a large percentage of the spears now on display. This particular species of fish he presented to the Zoological Museum of Utrecht University. An amphibian is also named after him.

Anton, I

Sanggau Betta *Betta antoni* HH Tan & PKL Ng, 2006

Irwan Anton is based in Pontianak, West Kalimantan, Indonesia. The authors honoured him "in recognition of his generous help and gift of specimens."

Anton Bruun

Lanternfish sp. *Diaphus antonbruuni* BG Nafpaktitis, 1978
Croaker sp. *Atrobucca antonbruun* K Sasaki, 1995
Snailfish sp. *Notoliparis antonbruuni* DL Stein, 2005

Anton Frederik Bruun (1901–1961) (see also **Bruun**). The etymology for the lanternfish states that it also commemorates the research vessel, 'Anton Bruun', named after him and from which the holotype was collected. The croaker is named after the vessel due to its "… contributions to the biology of the Indian Ocean fishes."

Antoncich

Japanese Flyingfish sp. *Cheilopogon antoncichi* LP Woods & LP Schultz, 1953

Michael Antoncich was a commercial fisherman from Monterey, California. He was on the 'Operation Crossroads' project which was the scientific study of the effects of the Bikini Atoll nuclear tests in the mid-1940s. The US Navy asked the Smithsonian to send experts (which included Schultz) to the Marshall Islands, part of a team of botanists, zoologists, geologists, and oceanographers from universities, oceanographic institutes, and government research bureaus. Professional fishermen were recruited to help the scientists and Michael Antoncich was one such. Schultz said at the time: "*For some reason I do not fear the Atom Bomb era that is with the world. There are and will be I am confident, still enough good men and women in this world to control properly the various advancements of men.*"

Antonia

Sailfin Silverside sp. *Telmatherina antoniae* M Kottelat, 1991

Antoinette Kottelat-Kloetzli is the wife of the author, honoured "…in acknowledgment of her help at all stages of this and many other research projects." (Also see **Kloetzli**)

Antonio

Glass Knifefish sp. *Eigenmannia antonioi* LAW Peixoto, GM Dutra & WB Wosiacki, 2015

Antônio da Silva Wanderley was the first author's grandfather.

Anu

Ghost Knifefish sp. *Apteronotus anu* CDCM de Santana & RP Vari, 2013

The Añu indigenous people lived along the shores of Lake Maracaibo, Venezuela, in traditional houses which they built above the lake. This reminded early European explorers of Venice, and it has been suggested that may have been the basis for calling the region Venezuela.

Anudarin

Asian Gudgeon sp. *Microphysogobio anudarini* J Holcík & K Pivnicka, 1969

Anudarin Dashidorzhi is a Mongolian ichthyologist who was Professor of Zoology, State University, Ulan Bator. Among other works he co-wrote: *Mongolian Grayling* Thymallus brevirostris *Kessler from the Dzabakhan River Basin* (1972).

Anzueto

Central American Killifish sp. *Pseudoxiphophorus anzuetoi* DE Rosen & RM Bailey, 1979

[Syn. *Heterandria anzuetoi*]

Roderico Anzueto is a Guatemalan naturalist, who works for El Zayab S.A., an organisation involved in conservation. He wrote *Guatemalan Beaded Lizard (Heloderma horridum charlesbogerti) on the Pacific Versant of Guatemala* (2010). He was a friend and field companion of the senior author. He also has a reptile named after him.

Aoki

Mizaki Worm Eel *Scolecenchelys aoki* DS Jordan & JO Snyder, 1901

Kumakichi Aoki was a highly skilled fisherman who was assistant to Kakichi Mitsukuri at the Marine Laboratory at Misaki diuring the late 19th and early 20th centuries. He was known as 'Kuma-San' and was an expert at traditional Japanese long-line fishing, being described as "…one of the best [fish] collectors in Japan".

Aoyagi

Coral Goby sp. *Gobiodon aoyagii* K Shibukawa, T Suzuki & M Aizawa, 2013

Dr Hyoji Aoyagi (1912–1971) was a Japanese ichthyologist. He described a number of new species such as the blenny *Ecsenius yaeyamaensis* (1954). Among his published papers is: *Studies on the coral fishes of the Riu-Kiu Islands* (1949). He was the first to provide accounts of this goby species from Japan (as *G. rivulatus*) including an excellent illustration.

Apache

Apache Trout *Oncorhynchus apache* RR Miller, 1972

This species is named after the White Mountain Apache tribe which is concerned for the welfare of this endangered species.

Apai

Neotropical Rivuline sp. *Austrolebias apaii* WJEM Costa, PG Laurino, R Recuero & HM Salvia, 2006

Dr Arnoldo Apai is a Uruguayan biologist, aquarium hobbyist, and collector of fish from the rivers and streams of Uruguay.

Apang

Hillstream Catfish sp. *Amblyceps apangi* P Nath & SC Dey, 1989

Sri Gegong Apang (b.1949) is an Indian politician. He was Chief Minister (ex-head of fisheries) of Arunachal Pradesh, India (1980–1999 & 2003–2007). He was previously head of the fisheries department. He was arrested (2010) for alleged corruption.

Apiaka

Corydoras sp. *Corydoras apiaka* VC Espíndola, MRS Soares, LR Rosa & MR Britto, 2014
Characin sp. *Aphyodite apiaka* ALH Esguícero & RM Castro, 2017

Apiaká are an indigenous people who originally occupied the middle and lower Rio Arinos, Mato Grosso, Brazil. This tribe is known for facial tattoos, bravery in battles, and anthropophagic rites after fights.

Apolinar

Scrapetooth Characin sp. *Parodon apolinari* GS Myers, 1930

Brother Apolinar María (1867–1949) was a missionary monk and ornithologist who collected in Colombia. He was Director of the Institute La Salle in Bogota (1914). (See also **Maria**)

Apollo

Characin sp. *Leporinus apollo* BL Sidlauskas, JHA Mol & RP Vari, 2011

Apollo was the god of sun, music and healing in Greek and Roman mythology, referring to the fish's extremely slender form that is "reminiscent of the arrow that was Apollo's favoured weapon and predominant symbol, and the yellow cast of body and fins and rounded shape of the lateral markings, which evoke the sun that was one of Apollo's primary aspects."

Appel

Prickly Footballfish *Himantolophus appelii* FE Clarke, 1878

Mr. Appel of Hokitika, New Zealand. In a transcript of a paper read before the Westland Institute (1877), Clarke thanks "...*Mr. Appel, V.S., of this town, to whom this last fish was sent by its collector, and who has kindly allowed me to figure and describe it.*" We know nothing more about him.

Ara

Roughtail Catshark *Galeus arae* JT Nichols, 1927
[Alt. Marbled Catshark]

This species is named after the yacht 'Ara', owned by William K Vanderbilt, which collected the first two specimens.

Arachas

Armoured Catfish sp. *Microlepidogaster arachas* F de O Martins, BB Calegari & F Langeani Neto, 2013

The Arachás were a native people who once lived in the area drained by the Rio Araguari – where this fish is found – in Minas Gerais, Brazil. They were exterminated by the Caiapós in the 1750s.

Araga

Basslet sp. *Liopropoma aragai* JE Randall & L Taylor, 1988

Chuichi Araga is a Japanese ichthyologist who was Curator at the Seto Marine Biological Laboratory of Kyoto University. He co-wrote: *Coastal fishes of southern Japan* (1975) and papers such as: *A New Species of Deep-dwelling Razorfish from Japan* (1986). The etymology says: "Named in honor of Chuichi Araga of the Seto Marine Biological Laboratory of Kyoto University, who made the specimen and photograph available to us for study."

Arambourg

Ethiopian Barb sp. *Enteromius arambourgi* J Pellegrin, 1935

Camille Louis Joseph Arambourg (1885–1969) was a French vertebrate palaeontologist and palaeo-ichthyologist, with a particular interest in fish fauna from the Cretaceous and Cenozoic. He graduated in agronomy at the Institut National Agrionomique (1908) and settled near Oran, Algeria. After serving in the French army in WW1 and fighting in the Dardanelles campaign (1915), in Serbia (1915) and Macedonia (1916–1918), he became Professor of Geology at the Agricultural Institute of Algeria (1920–1930). He moved to Paris (1930) and later was Professor of Palaeontology at the MNHN in Paris (1936–1955) when Pellegrin (q.v.) was Director. Most importantly he organised an expedition, the *Mission Scientifique de l'Omo*, to the Omo River valley in Ethiopia (1932 to 1933) where he collected fossils and explored along parts of Lake Rudolf (now Lake Turkana). He was also interested in palaeoanthropology and having excavated remains of Cro-magnons, he disputed that the Neanderthals had been simian and brutish. He was elected a member of the French

Academy of Sciences (1961). There is no etymology, but there is no real doubt that this fish is named after Camille Arambourg.

Aramburu

Characin sp. *Astyanax aramburui* LC Protogino, AM Miquelarena & HL López, 2006

Raúl Horacio Arámburu (1924–2004) was a researcher and professor of the Museo de La Plata, Buenos Aires. He was the first Professor of Ichthyology in Argentina. This species of fish is endemic to Argentina.

Aramis

Stone Loach sp. *Schistura aramis* M Kottelat, 2000

Aramis is a fictional character, one of *The Three Musketeers* – see various novels by Alexandre Dumas. Two other loach species were named *athos* and *porthos* (q.v.).

Aranda

Armoured Catfish sp. *Parotocinclus arandai* LM Sarmento-Soares, P Lehmann A & RF Martins-Pinheiro, 2009

Arion Túlio Aranda is an ichthyological colleague of the authors. He works for the Oswaldo Cruz Institute, which awarded his masters. He was the Manager & Curator of their biological collections (2007–2013), having worked there (2004–2007) in a more junior capacity. Previously (1999–2004) he was Collection Manager at the Ichthyology Collection of the National Museum of Brazil. The authors honoured him "*...for his talent for catching fish and knowledge of their behaviour.*" He helped collect the type series.

Arcas

Iberian Cyprind sp. *Achondrostoma arcasii* F Steindachner, 1866

Laureano Pérez Arcas (1824–1894) was a Spanish entomologist, malacologist and ichthyologist who started out intending to become a lawyer. He qualified at the University of Madrid (1843) and then switched to zoology. He was awarded a doctorate in zoology (1846) and there became a professor (1847). His most important work was: *Treatise on Zoology* (1861), which became a textbook in Spanish schools and in other Spanish speaking countries. He and Steindachner were friends and used to share specimens.

Arctowski

Dark-mouthed Skate *Bathyraja arctowskii* L Dollo, 1904

Henryk Arctowski (né Artzt) (1871–1958) was a Polish scientist, meteorologist and explorer who took part in the Belgian Antarctic Expedition that took place over the course of three years (1897, 1898 & 1899) with the voyages of the 'Belgica' under the command of Adrien de Gerlache de Gomery. During the expedition studies were made of the Antarctic regions and Southern oceans. The results generated over seventy scientific reports over the course of the next five decades. Arctowski with others including Roald Amundsen were among the first to overwinter in the Antarctic. He studied maths, physics and astronomy at the University of Liège and chemistry and geology at the Sorbonne (1888–1893). He was appointed to the expedition (1895) and on its completion he lived in Brussels analysing its results. On a lecture tour in London he met the American actress and opera singer Arian Jane Addy and they later (1909) married and they moved to New York where he headed the science division of NY Public Library (1911–1919). Having taken Belgian nationality (1909), he became an American citizen (1915). He lived in exile for many years but was instrumental in restoring Polish independence after WW1 and returned to the newly independent Poland (1920–1939) where he turned down a ministerial position to become Professor of geophysics and meteorology at Jan Kazimierz University. When he and his wife went to America for a conference (1939) Poland was invaded by USSR and Germany and he lost his home and possessions and could not return. Thereafter he worked as a research associate at the Smithsonian until his death, although he had had to retire earlier (1950) because of ill-health. Several geographical features, an Antarctic station and a medal are also named after him. Louis Dollo (q.v.) was one of those who studied the findings of the expedition and reported on them.

Ardila, CA

Naked-back Knifefish sp. *Gymnotus ardilai* JA Maldonado-Ocampo & JS Albert, 2004

Characin sp. *Lebiasina ardilai* AL Netto-Ferreira, H López-Fernández, DC Taphorn & EA Liverpool, 2013

Carlos Arturo Ardila Rodríguez is an ichthyologist who teaches at Universidad Metropolitana, Barranquilla, Colombia. He has made numerous contributions to the systematics of pencilfishes.

Ardila, CJ

Suckermouth Catfish sp. *Astroblepus ardilai* CA Ardila Rodríguez, 2012

Suckermouth Catfish sp. *Astroblepus ardiladuartei* CA Ardila Rodríguez, 2015

Carlos Julio Ardila Duarte is a Colombian biologist and is the describer's son. *Astroblepus ardilai* was named for his scientific illustrations of the fishes of Bolivar, Colombia. He collected the holotype of *A. ardiladuartei.*

Ardila, R

Characin sp. *Creagrutus ardilai* CA Ardila Rodriguez, 2021

Rodolfo Ardila Rodriguez is the author's brother. He helped to collect the holotype of this species (2001).

Ardila Duarte

Suckermouth Catfish sp. *Astroblepus ardiladuartei* CA Ardila Rodríguez, 2015

Carlos Julio Ardila Duarte (See **Ardila, CJ**).

Aredvi

Loach sp. *Paraschistura aredvii* J Freyhof, G Sayyadzadeh, HR Esmaeili & M Geiger, 2015

Aredvi Sura Anahita is the Avestan language name of an Indo-Iranian cosmological figure venerated as the 'Divinity of the Waters' and associated with fertility, healing and wisdom.

Arena

Messina Scorpionfish *Scorpaenodes arenai* M Torchio, 1962

Giuseppe Arena of Ganzirri was an Italian collector for the Museo Civico di Storia Naturale di Milano. He collected type specimen.

Arguelles

Goby sp. *Mugilogobius arguellesi* HA Roxas & GL Ablan, 1940

Angel S Arguelles (1888–1952) was a soil chemist. He was Director, Philippine Bureau of Science.

Arguni

Cardinalfish sp. *Glossamia arguni* RK Hadiaty & GR Allen, 2011
Arguni Rainbowfish *Melanotaenia arguni* Kadarusman, RK Hadiaty & L Pouyaud, 2012
Sleeper Gudgeon sp. *Mogurnda arguni* GR Allen & RK Hadiaty, 2014

These are toponyms, referring to Arguni Bay, Papua Barat, Indonesia.

Argus

Spotted Scat *Scatophagus argus* Linnaeus, 1766
Peacock Hind *Cephalopholis argus* JG Schneider, 1801
Argus Wrasse *Halichoeres argus* ME Bloch & JO Snyder, 1801
Spotted Pipefish *Stigmatopora argus* J Richardson, 1840
Snakehead sp. *Channa argus* TE Cantor, 1842
Argus Moray *Muraena argus* F Steindachner, 1870
Long-finned Gurnard *Lepidotrigla argus* JD Ogilby, 1910
Eye-spot Grenadier *Coelorinchus argus* MCW Weber, 1913
Peacock Flounder *Pseudorhombus argus* MCW Weber, 1913
Armoured Catfish sp. *Hypostomus argus* HW Fowler, 1943
Goby sp. *Gnatholepis argus* HK Larson & D Buckle, 2005

Argus (or Argos) was a 100-eyed watchman in Greek mythology. The god Hermes killed him, after which the goddess Hera placed Argus's eyes into the peacock's train. Species named after him – including an amphibian, two birds and four reptiles – tend to be marked with eye-like spots.

Ariane

Tetra sp. *Hyphessobrycon arianae* A Uj & J Géry, 1989

Ariane Devore is a colleague of the senior author who thanked her for her encouragement during the course of his research.

Ariel

Characin sp. *Priocharax ariel* SH Weitzman & RP Vari, 1987

Named for Ariel (an airy spirit), "*...in reference to the tiny and translucent nature of this fish in its natural habitat*."

Arif

Stone Loach sp. *Schistura arifi* MR Mirza & PM Bănărescu, 1981

It is known that M Arif collected the holotype, but that is the extent of what we know.

Arimaspi

Characin sp. *Leporinus arimaspi* MD Burns, BW Frable & BL Sidlauskas, 2014

The Arimaspi in Greek legend were a mythical people of northern Scythia. The name was conferred due to the large black spot on the fish's body, reminiscent of the single, centrally located eye that the Arimaspians were said to have.

Aristone

Snakehead sp. *Channa aristonei* J Praveenraj, T Thackeray, SG Singh, A Uma, N Moulitharan & BK Mukhim, 2020

Aristone M 'Bah' Ryndongsngi (b.1998) is Project Field Coordinator at FXB India, which supports marginalised communities. St Anthony's College, Shillong awarded his BSc (2019). He describes himself as a fish supplier and collects from the wild in northeast India. The etymology says "*This species is named after Aristone M. Ryndongsngi from Meghalaya, in recognition of his discovery of this new species and assistance to the authors during the field work*."

Aristotle

Aristotle's Catfish *Silurus aristotelis* S Garman, 1890
Tuzla Chub *Squalius aristotelis* M Özuluğ & J Freyhof, 2011

Aristotle (384–322 BC) was a Greek philosopher who lived (348–345 BC) in Assos, Turkey, which is where the chub holotype was taken.

Arleo

Pencil Catfish sp. *Trichomycterus arleoi* A Fernández-Yépez, 1972

Octavio Arleo PIgnatoro (1920–2005) was a collector and taxidermist at the Museo de Ciencias Naturales de Caracas, Venezuela. He and the author collected this catfish together (1949).

Armbruster

Sea Catfish sp. *Notarius armbrusteri* R Betancur-Rodríguez & A Acero Pizarro, 2006
Thorny Catfish sp. *Rhinodoras armbrusteri* MH Sabaj Pérez, 2008
Armoured Catfish sp. *Panaque armbrusteri* NK Lujan, M Hidalgo & DJ Stewart, 2010

Dr Jonathan William Armbruster (b.1969) is Professor and Curator of Fishes at Auburn University, Alabama. The University of Illinois awarded both his bachelor's degree (1991) and his doctorate (1997). He wrote: *Standardized measurements, landmarks, and meristic counts for cypriniform fishes* (2012).

Armitage, AE

Armitage Angelfish *Apolemichthys armitagei* JLB Smith, 1955
[Probably a hybrid between *Apolemichthys trimaculatus* and *A. xanthurus*]

Colonel A E Armitage of the Seychelles, who, according to Smith's etymology, "...was deeply interested in the fishes of that area".

Armitage, D

Rasbora sp. *Rasbora armitagei* A Silva, K Maduwage & R Pethiyagoda, 2010

David Armitage is a retired British civil servant who was a principal scientific officer with the Ministry of Agriculture, Fisheries and Food (1971–2005) where he worked on multi-disciplinary grain storage technology, He is better known as an aquarist and expert on labyrinthine fish and a leading member of the Anabantoid Association of Great Britain whose journal he has edited since 1981. He has travelled and collected in Asia (China, Kalimantan, Malaysia, Myanmar, Sri Lanka and Thailand) and Africa (Congo, Gabon, Ghana, Madagascar, South Africa, Uganda and Zambia).

Armstrong

Lefteye Flounder sp. *Arnoglossus armstrongi* EOG Scott, 1975

Philip Armstrong – about whom we know nothing other than the etymology's statement that the holotype was caught at Bridport, northern Tasmania "*...by Mr Philip Armstrong, in whose honour the species is named.*"

Arnaz

Arnaz's Damselfish *Chrysiptera arnazae* GR Allen, MV Erdmann & PH Barber, 2010
Cat's-Eye Cardinalfish *Siphamia arnazae* GR Allen & MV Erdmann, 2019

Arnaz Erdmann née Mehta is the wife of the second author and his 'best friend' who, according to the damselfish etymology "*...has selflessly supported his extensive field time in the Bird's Head region (northern Papua New Guinea and eastern Indonesia), where this species occurs.*" She discovered the cardinalfish while diving at Sideia Island, Milne Bay Province, PNG (December 2016).

Arnegard

Mormyrid sp. *Petrocephalus arnegardi* S Lavoué & JP Sullivan, 2014

Dr Matthew Arnegard (b.1967) is an evolutionary biologist who was a member of the 'Mintotom Team' of researchers in the laboratory of Professor Carl Hopkins, Cornell University (N.B. Mintotom is the plural form of the word for mormyrid in the Fang language of western Central Africa). These researchers have conducted field studies on African electric fishes for more than 15 years and Dr Arnegard filmed the fish by becoming one of a group of hunting mormyrids; his co-written paper: *Electric organ discharge patterns during group hunting by a mormyrid fish* (2005) is about this experience. He is now a scientist at the Fred Hutchinson Cancer Research Center in Seattle, Washington and among his other co-written works is: *Genetics of ecological divergence during speciation* (2013).

Arnold, A

Pacific Leaping Blenny *Alticus arnoldorum* A Curtiss, 1938

Augusta Arnold née Foote (1844–1903) was an American amateur naturalist; in fact America's first woman to write about the nature of the shoreline. Her magnum opus, and life's work, was the classic: *The Sea-Beach at Ebb-Tide: A Guide to the Study of the Seaweeds and the Lower Animal Life Found Between Tidemarks* (1901). The eponym is not identified in Anthony Curtiss' privately printed book on the fauna of Tahiti, but we believe Augusta Arnold to be one of the persons honoured because her book is mentioned by Curtiss and may well have inspired him to follow in her path. However, *arnoldorum* is a plural form ('of the Arnolds') and the identity of the other 'Arnold(s)' remains unknown.

Arnold, JP

African Tetra genus *Arnoldichthys* GS Myers, 1926
Arnold's Killi *Fundulopanchax arnoldi* GA Boulenger, 1908
Black-base Tetra *Pseudocheirodon arnoldi* GA Boulenger, 1909
Armoured Catfish sp. *Otocinclus arnoldi* CT Regan, 1909
Splash Tetra *Copella arnoldi* CT Regan, 1912
Characin sp. *Myloplus arnoldi* CGE Ahl, 1936

Johann Paul Arnold (1869–1952) was a German aquarist in Hamburg. He sent specimens to Boulenger for study and identification. He co-wrote: *The alien freshwater fish* (1936).

Arnoldi

Stone Loach sp. *Triplophysa arnoldii* AM Prokofiev, 2006

Dr Lev Vladimirovich Arnoldi (1903–1980) was a Russian entomologist and hydrobiologist, in which latter discipline he graduated from Moscow University (1927) and completed post-graduate studies (1927). He was both awarded a doctorate and admitted to the Zoological Institute of the Russian Academy of Sciences (1944) and was promoted to Professor (1954). He worked at hydrobiology stations at Lake Sevan (Armenia) (1927–1932), in Batumi (1932–1934) and Sebastopol (1934–1941). Forced to evacuate in the face of the German invasion of Russia (1941) he worked as an entomologist in southern Kazakhstan (1941–1942) and then moved to Dushambe (Tajikstan) (11942–1944) where the Sebastopol hydrobiology unit had been re-located. During his career, he produced c.120 scientific publications on hydrobiology and entomology.

Arnoult

Sand Darter sp. *Gobitrichinotus arnoulti* A Kiener, 1963
Upside-down Catfish sp. *Synodontis arnoulti* B Roman, 1966
Madagascan Killifish sp. *Pachypanchax arnoulti* PV Loiselle, 2006

Jacques Arnoult (1914–1995) was a French ichthyologist and herpetologist. He graduated in biology, agriculture, and hydrobiology at Université de Toulouse. He was in charge of zoological research at the Institute of Scientific Research, Madagascar (1951), and became an Assistant in the Department of Reptiles and Fish, Muséum National d'Histoire Naturelle, Paris (1954) during which time Kiener was a colleague. He collected frequently in Africa and was notably successful in getting his live specimens to breed in captivity. He was Director of the Aquarium, Monaco (1968–1981). A reptile is also named after him.

Aron

Aron's Blenny *Ecsenius aroni* VG Springer, 1971

Dr William I Aron (b.1930) was Director of the Oceanography and Limnology Program at the Smithsonian Institution. He was (1971) also Director of NOAA's Office of Ecology and Environmental Conservation, where he was responsible for full consideration of environmental protection matters and for liaison with governmental and other organizations involved in conservation and ecology. Among his publications is *Midwater Trawling Studies in the North Pacific* (1959). He was honoured by Springer "*...in appreciation for his making possible my field work in the Red Sea.*"

Arraya

Bluntnose Knifefish sp. *Brachyhypopomus arrayae* WGR Crampton, CDCM de Santana, JC Waddell & NR Lovejoy, 2017

Mariana Arraya is a Bolivian biologist at Unidad de Limnologia y Recursos Acuaticos, Universidad Mayor de San Simon. She was thanked in the etymology "*...for her assistance in collecting the type series in Bolivia.*"

Arrietty

Short-horn Triangular Batfish *Malthopsis arrietty* H-C Ho, 2020

The fish is named after the miniature character of the Japanese animated fantasy film 'Arrietty the Borrower' (Japan) or 'The Secret World of Arrietty' (North America), in reference to the miniature size of the species. These are loosely based on Mary Norton's classic novel The Borrowers, about a 14-year-old girl named Arrietty and her parents, who are Borrowers and thus only inches tall.

Arrigo

Júcar Nase *Parachondrostoma arrigonis* F Steindachner, 1866

Arrigo (d.1865) was a Professor in Valencia. Steindachner says he was a "dear friend" who died of cholera. However, he did not provide any first name(s) and we have not been able to identify him.

Ars-Cuttoli

Goby sp. *Schismatogobius arscuttoli* P Keith, C Lord & N Hubert, 2017

The Ars-Cuttoli Foundation funded the authors' research in Indonesia.

Artaxerxes

Lake Victoria Cichlid sp. *Haplochromis artaxerxes* PH Greenwood, 1962

Artaxerxes (d.424BC) was a King of Persia (465BC–424BC). He was also known as Longimanus ('long handed'), and the cichlid was named with reference to its long pectoral fins.

Artedi

Sculpin genus *Artedius* CF Girard, 1856
Sculpin genus *Artediellus* DS Jordan, 1885
Sculpin genus *Artedielloides* VK Soldatov, 1922
Sculpin genus *Artediellina* AY Taranetz, 1937
Sculpin genus *Artediellichthys* AY Taranetz, 1941
Northern Cisco *Coregonus artedi* CA Lesueur, 1818
Bluntnose Knifefish sp. *Hypopomus artedi* JJ Kaup, 1856

Peter Artedi (1705–1735) was a Swedish ichthyologist, sometimes regarded as 'the father of ichthyology". He began studying fishes in Uppsala before Linnaeus, with whom he became good friends. Artedi left Uppsala for London (1834), then moved on to Amsterdam (1835), where he could study a large private fish collection. However, he fell into one of the canals and drowned. Linnaeus finished his fish monograph manuscript and published it. Artedi had arranged his fishes in genera, the genera in 'maniples', the 'maniples' in orders and the orders in a class; a system which Linnaeus more or less took over.

Arturo

Malpelo Clingfish *Acyrtus arturo* J Tavera, S Rojas-Velez & E Londoño-Cruz, 2021

Dr Arturo Acero of the Universidad Nacional de Colombia was honoured "for his contributions to the knowledge of Neotropical fishes and for mentoring many generations of ichthyologists, including the first author..." (See **Acero**)

Arunachalam

Garra sp. *Horalabiosa arunachalami* JA Johnson & R Soranam, 2001
[Syn. *Garra arunachalami*]

Dr Muthukumarasamy Arunachalam (b.1955) is an Indian ecologist and ichthyologist who was a Research Fellow at the University of Kerala (1977–1989). He worked as Research Director for a Non-Governmental Organisation (1990–1993) and joined Manonmaniam Sundaranar University as a lecturer (1994–1997), Reader (1997–2005) and since 2005 has been Professor and head of Sri Paramakalyani Centre for Environmental Sciences, Manonmaniam Sundaranar University, Tamilnadu, India. He has describer 13 new species of fish. He co-wrote: *A New Record of the Marine Puffer Fish Genus,* Chelonodon *(Tetraodoniformes, Tetraodonitae) from Freshwater Habitat of Western Ghats, India* (1999).

Arup

Garra sp. *Garra arupi* K Nebeshwar Sharma, W Vishwanath & DN Das, 2009

Dr Arup Kumar Das (b.1952) is a Professor in the Botany Department, Rajiv Gandhi University,

Itanagar Arunchal Pradesh, India where he is also the co-ordinator of the Centre of Excellence in Biodiversity, which gave support to the authors' study of north-eastern Indian fishes.

Arx

Arrowtooth Eel sp. *Ilyophis arx* CH Robins, 1976

Arx is Latin for castle, alluding to Dr Peter Henry John Castle (1934–1999), who "*laid the foundations of modern work on synaphobranchid eels*". (See **Castle**)

Asaeda

Asaeda's Flounder *Monolene asaedae* HW Clark, 1936

Toshio Asaeda was described in the etymology as "*...the clever and accomplished artist of the expedition.*" This being the Templeton Crocker Expeditions of the California Academy of Sciences (1932–1933).

Asakusa

Snake-eel sp. *Ophichthus asakusae* DS Jordan & JO Snyder, 1901

Not a true eponym, as it is named after the Asakusa Aquarium in Tokyo, from which the holotype was obtained.

Asano

Crested Flounder sp. *Samariscus asanoi* A Ochiai & K Amaoka, 1962

Conger Eel sp. *Gnathophis asanoi* ES Karmovskaya, 2004

Dr Hirotoshi Asano was (1987) a member of the Department of Fisheries, Faculty of Agriculture, Kinki University, Japan. The conger species was named after him to commemorate his "*...significant contributions to the study of congrid eels of the Japanese Archipelago.*"

Ascanius

Yarrell's Blenny *Chirolophis ascanii* JJ Walbaum, 1792

Peter Ascanius (1723–1803) was a Norwegian biologist who was one of Linnaeus's students. He taught zoology and mineralogy in Copenhagen (1759–1771) and later worked as a supervisor at the mines in Kongsberg and elsewhere in Norway. His best-known work was the five-volume illustrated *Icones rerum naturalium* (1767). Ascanius discovered the giant oarfish (1772).

Ashby-Smith

Ctenopoma sp. *Ctenopoma ashbysmithi* KE Banister & RG Bailey, 1979

Lieutenant Adrian Ashby-Smith (1952–1976) took part in The Zaire River Expedition (1974–1975), during which five new species of fish were identified. He was quartermaster of the team that explored Mount Sangay, Ecuador (1976). During their assent the volcano erupted, sending a mass of hot rocks into the air that rained down on the party and everyone was hit and dislodged from where they were on the snow. Although everyone was injured, four survived but two - including Ashby-Smith - died of their injuries. The etymology says: "*This species is named in the memory of 2nd Lt. Adrian Ashby-Smith who helped the fish team for part of its stay in Zaire. He was killed in 1976 whilst exploring a volcano in Ecuador.*"

Aschemeier

Deepwater Flathead sp. *Brachybembras aschemeieri* HW Fowler, 1938

Charles R W Aschemeier (1892–1973) was a taxidermist at the US National Museum. His field notebook, for when he served as a collector for the museum during the Collins-Garner Expedition to the French Congo (1917–1918), is held by the Department of Botany there.

ASFRAC

Central American Cichlid sp. *Tomocichla asfraci* R Allgayer, 2002

This name is an acronym coined from the initial letters of l'Association France Cichlid, for its promotion of the family Cichlidae.

Ashley

Neotropical Rivuline sp. *Papiliolebias ashleyae* DTB Nielsen & R Brousseau, 2014

Ashley Kimberly Brousseau is the daughter of the second author Dr Roger Brousseau, who collected the first specimen of the species. (Also see **Brousseau**)

Ashmead

Redtail Barb *Discherodontus ashmeadi* HW Fowler, 1937

Charles C Ashmead lived in Philadelphia and collected locally in Pennsylvania. He was an early local contributor to the Philadelphia Academy of Natural Sciences' fish collection.

Asmuss

Russian Bitterling *Acheilognathus asmussii* B Dybowski, 1872
[Alt. Spiny Bitterling; Syn. *Acanthorhodeus asmussii*]

Dr Eduard Philibert Assmuss (1838–1882) was a German entomologist. He wrote: *Parasiten der Honigbiene* (1865). The type collections of Dybowski and Asmuss are held in the Museum of Geology, Tartu, Estonia.

Asoka

Asoka Barb *Systomus asoka* M Kottelat & R Pethiyagoda, 1989

Asoka Mivanpalana was a Sri Lankan aquarist who discovered this fish in the 1950s.

Assasi

Arabian Picasso Triggerfish *Rhinecanthus assasi* P Forsskål, 1775

Not an eponym but reported by Forsskål as being a local Arabic name for the fish – 'Azzazi'.

Assessor

Devilfish (Plesiopid) genus *Assessor* GP Whitley, 1935

Meaning 'assistant' or 'helper', the allusion is not explained, but might perhaps refer to Francis (Frank) A McNeill (1896–1969), Whitley's colleague at the Australian Museum, who, as carcinologist, probably helped Whitley with his ichthylogical collections (and for whom *A. macneilli* is named). (See **Macneil**)

Astakhov

Soldierfish sp. *Myripristis astakhovi* AN Kotlyar, 1997
Velvetfish sp. *Cocotropus astakhovi* AM Prokofiev, 2010

Dr Dmitry Alekseevich Astakhov is an ichthyologist at the Laboratory of Oceanic Ichthyofauna, P P Shirshov Institute of Oceanology, Russian Academy of Sciences, Moscow. He co-wrote: *Preliminary annotated list of species of the family Chaetodontidae (Actinopterygii) from Ly Son Islands (South China Sea, Central Vietnam)* (2016). He collected and provided holocentrid fishes from Vietnam.

Astorqui

Black Midas Cichlid *Amphilophus astorquii* JR Stauffer, JK McCrary & KE Black, 2008

Father Ignacio Astorqui (1923–1994) was a Spanish Jesuit priest, teacher and naturalist who researched the freshwater fish of Nicaragua where this species is found. He entered the priesthood at nineteen and after ordination took a masters in biology studying Nicaragua's freshwater fish. He went on to become a professor at the Central American College in both Granada and in Managua. He was also a parish priest in San Blas for the rest of his life. His published works include: *Fish from the basin of the great lakes of Nicaragua* (1974).

Astyanax

Characin genus *Astyanax* SF Baird & CF Girard, 1854

Astyanax was the son of Hector in Greek mythology. See Homer's *Iliad* for details. The reasoning for its use for a genus of characins is not explained.

Asurini

Armoured Catfish sp. *Pseudancistrus asurini* GSC Silva, FF Roxo & C de Oliveira, 2015

The Asurini are an indigenous people who inhabit the right margin and middle portions of Rio Xingu, near the town of Altamira, Pará, Brazil.

Atahualpa

Characin sp. *Moenkhausia atahualpiana* HW Fowler, 1907
Dwarf Cichlid sp. *Apistogramma atahualpa* U Römer, 1997
Whale Catfish sp. *Paracetopsis atahualpa* RP Vari, CJ Ferraris Jr & MCC de Pinna, 2005

Atahualpa (d.1533) was the last ruler of the Incas of Peru. He became emperor (1532) after defeating his half-brother Huáscar in a civil war. When the Spanish arrived, Francisco Pizarro captured Atahualpa and eventually had him executed by strangulation. According to Römer's etymology of the cichlid, this murder is a "*...perfect metaphor for the continuing destruction of the cultures of the indigenous peoples of South America and destruction of their environment by 'modern' man.*" The asteroid 4721 Atahualpa and an amphibian are also named after him.

Atavai

Polynesian Wrasse *Pseudojuloides atavai* JE Randall & HA Randall, 1981

Not an eponym but from the Tahitian word for 'pretty'.

Athena

Athena Chromis *Chromis athena* GR Allen & MV Erdmann, 2008

Not named directly after the Greek goddess, but with reference to the sailing yacht *Athena* which the authors say: "…served as our base of operations during the cruise upon which the new species was first discovered."

Athos

Stone Loach sp. *Schistura athos* M Kottelat, 2000

Athos is a fictional character, one of The Three Musketeers – see various novels by Alexandre Dumas. It joins two other *Schistura* species named *aramis* and *porthos* (q.v.).

Atil

Lake Beyşehir Loach *Oxynoemacheilus atili* F Erk'akan, 2012

Nothing more is known of Mr Ahmet Tuncay Atil other than him being honoured in this fish's binomial.

Atkinson, EFT

Burmese Carplet *Amblypharyngodon atkinsonii* E Blyth, 1860

Edwin Felix Thomas Atkinson (1840–1890) was an Irish entomologist who was also a lawyer, a specialist in Indian law. He joined the Indian Civil Service (1862) and collected widely in India and South-east Asia. He wrote: *Catalogue of the Insecta. Order Rhynchota. Suborder Hemiptera-Heteroptera. Family Capsidae* (1890) and many books and articles on the history, fauna and flora of the Kumaon HIlls. He died in Calcutta (Kolkata) from Bright's disease.

Atkinson, EL

Crocodile Icefish sp. *Cryodraco atkinsoni* CT Regan, 1914

Dr Edward Leicester Atkinson (1881–1929) was a Royal Navy surgeon, parasitologist and Antarctic explorer. He grew up in Trinidad and qualified as a physician (1906). He was a member of the scientific staff on the 'Terra Nova' Expedition (1910–1913) which collected the icefish holotype. He was at the expedition's Cape Evans base for nearly a year (1912) acting as physician looking after patients. He also led the party that found the tent with the bodies of Scott, Bowers and Wilson. He served with distinction in WW1 being wounded several times and decorated twice. Some Antarctic cliffs are also named after him.

Atkinson, M

Dash-dot Barb *Enteromius atkinsoni* RG Bailey, 1969

Maurice Atkinson was a Fisheries Officer in the Lake Victoria and Tanzanian Fisheries Service (1964). The etymology describes him as one "*…who had a wide interest in the biology and correct identification of East African fishes, and whose contributions in the realm of fisheries development and training, will long be valued by his colleagues and students alike*".

Atkinson, WS

Pacific Yellowtail Emperor *Lethrinus atkinsoni* A Seale, 1910

William Sackston Atkinson (1864–c.1925) was a zoology graduate of Stanford University and went on a number of expeditions principally as illustrator. He made paintings of fish on a number of occasions and is mentioned in at least one paper by Alvin Seale and DS Jordan. Seale gives no apparent etymology, but we speculate that this could well have been the intended person.

Atlantis

Atlantic Legskate *Cruriraja atlantis* HB Bigelow & WC Schroeder, 1948

This is named after the Woods Hole Oceanographic Institute research vessel 'Atlantis', which collected three new species of skates along the coasts of Cuba, including this one.

Atropos/Atropus

Trevally genus *Atropus* L Oken, 1817

Cleftbelly Trevally *Atropus atropos* ME Bloch & JG Schneider, 1801

In Greek mythology, Atropos - known as the 'inflexible' or 'inevitable' - was the oldest of the Three Fates.

Attalus

Bakir Shemaya *Alburnus attalus* M Özuluğ & J Freyhof, 2007

Attalus I (269–197 BC) ruled Pergamon, a Hellenistic city-state in Asia Minor (now Anatolia, Turkey). The fish is endemic to that area.

Attems

African Airbreathing Catfish sp. *Clariallabes attemsi* M Holly, 1927

Carl August Graf Attems-Petzenstein (1868–1952) was an Austrian invertebrate zoologist and myriapodologist at the Natural History Museum, Vienna, where he was the author's colleague. He joined the staff of the Vienna Museum (1905). We cannot be sure this allocation is correct, as the etymology does not identify any person, but it seems the most probable answer. He visited the zoological station, Naples (1899) and Crete (1900). Attems served in the Austro-Hungarian army in WW1. He described around 1,800 new myriapod taxa. He wrote *The Myriopoda of South Africa* (1928). He died whilst following his normal route from home to the museum.

Attenborough

African Killifish sp. *Nothobranchius attenboroughi* B Nagy, BR Watters & DU Bellstedt, 2020

Sir David Frederick Attenborough (b.1926) is famous as a maker of wildlife television programmes. He studied natural sciences at Cambridge and joined a firm of publishers (1950), where he did not stay long before joining the BBC in the early days of its post-war television service. He has been associated with the BBC, first as an employee and later as a freelance journalist, virtually ever since. He rose high in the organisation's ranks, becoming controller of BBC2 and responsible for introducing colour television to Britain, yet his first love was not administration but photojournalism. He has made some of the most stunning series of nature programmes and produced excellent books to accompany them, such as *The Life of Birds* (1998). He was honoured in the killifish name in recognition of his dedicated efforts to promote biophilia: raising awareness of the wonders and beauties of nature for so many people worldwide, promoting awareness of the importance of biodiversity conservation, and above all, inspiring so many researchers in the field of natural history, including the authors of this paper. A bird, two mammals and three reptiles, among other taxa are also named after him.

Attiti

Loach Goby sp. *Protogobius attiti* RR Watson & C Pöllabauer, 1998

Chief Attiti is one of the Melanesian chiefs of the Goro tribe, New Caledonia. He was honoured because he was "*...quick to recognize the new goby, but stated there was no name for it among the Melanesians*".

Atukorale

Glowlight Carplet *Horadandia atukorali* PEP Deraniyagala, 1943

Mr Vicky 'Athu' Atukorale (fl. 1965) was a Sri Lankan naturalist who first drew Deraniyagala's attention to this fish. He wrote: *Notes on Birds Collected at Kumana Sanctuary on 17th May, 1949 for the Dehiwela Zoo* (1949). An amphibian is named after him.

Atururi

Aturiri's Pygmy-goby *Trimma aturirii* R Winterbottom, M van N Erdmann & NKD Cahyani, 2015

Brigadier Abraham Octavanianus Atururi (b.1950) was Governor of West Papua province (2003–2005, 2006–2011, 2012–2017) during the time the goby was collected having previously been Vice-Governor (1996–2000) and Sorong Regent (1992–1997). He was honoured for his "*...energetic efforts to ensure the conservation and wise, sustainable use of the hugely diverse coral reefs of the Bird's Head Seascape [where this goby occurs] for the benefit of the Papuan people*"

NB: since the authors consistently misspelled the Governor's name as "Aturiri" it does not represent a typo or printer's error, and therefore cannot be corrected.

Atz

Atz's Numbfish *Narcine atzi* MR de Carvalho & JE Randall, 2003

Dr James Wade Atz (1915–2013) was Curator Emeritus Vertebrate Zoology and Dean Bibliographer, Department of Ichthyology at AMNH, which he joined (1964), having been Associate Curator of the New York City Aquarium (c.1959–1964). He was Curator at the AMNH (1970–1981). He was also Adjunct Professor of Biology, Graduate School of Arts and Science on New York University where he had earned his Master's degree (1952) and his Doctorate. Among his many publications are several books on aquarium and tropical fish, whilst for the benefit of the American Jewish community he produced: '*a definitive lists of kosher and non-kosher fishes*'. He was reported to have a wicked sense of humour and enjoyed winding up the well-known chat-show host, Johnny Carson, on just how fish have sex and how some change sex to adjust the male/female sex ratio. He was honoured "*...for his many contributions to different aspects of ichthyology, and for his unparalleled enthusiasm for the study of fishes.*"

Aubyne

Characin sp. *Parapristella aubynei* CH Eigenmann, 1909

William St Aubyn (1855–1914) ran a sugar estate in Guiana. The St Aubyn family were early settlers in British Guiana (Guyana) and appear to have had involvement in sugar estates in the Demerara region. We cannot be certain which St Aubyn is involved but believe it to be the above, who 'became an overseer on the sugar estates in Demerara, British Guiana'. Whoever he was, he hosted Eigenmann during his stay and "*...did everything in his power to further the interests*" (of Eigenmann's expedition).

Audley

Collared Large-eye Bream *Gymnocranius audleyi* JD Ogilby, 1916

Audley Raymond Jones was a friend of the author and an amateur ichthyologist. Ogilby wrote: "*...Named after my friend Audley Raymond Jones, to whom I am indebted for much interesting information regarding the habits of our fishes.*"

August

Black Moray *Muraena augusti* JJ Kaup, 1856

Auguste Henri André Duméril. (1812–1870) was a friend of Kaup (see **Dumeril, AHA**).

Aurora

Reef Surfperch *Micrometrus aurora* DS Jordan & CH Gilbert, 1880
Aurora Rockfish *Sebastes aurora* CH Gilbert, 1890
Yellowmargin Basslet *Liopropoma aurora* DS Jordan & BW Evermann, 1903
Lake Malawi Cichlid sp. *Maylandia aurora* WE Burgess, 1976
Pinkbar Goby *Amblyeleotris aurora* NVC Polunin & HR Lubbock, 1977
Dawn Threadfin Bream *Nemipterus aurora* BC Russell, 1993
Pearl Darter *Percina aurora* RD Suttkus & BA Thompson, 1994

Aurora was the Roman goddess of the dawn. In the case of the goby, the name refers to the pattern on its caudal fin being '...reminiscent of the rising sun'. The name is often applied to species with pink/yellow colouration.

Austin

Austin's Guitarfish *Rhinobatos austini* DA Ebert & O Gon, 2017

Austin Ebert is a 'flat shark' enthusiast, who was honoured on the occasion of his graduation from the University of Southern California. He is the senior author's nephew.

Autran

Brazilian Cichlid sp. *Australoheros autrani* FP Ottoni & W Costa, 2008

Felipe Tavares Autran was a student in Laboratório de Sistemática e Evolução de Peixes Teleósteos, Universidade Federal do Rio de Janeiro, Brazil, (1990s). He was the first to study this species. (Also see **Tavares**)

Avá-canoeiros

Armoured Catfish sp. *Lamontichthys avacanoeiro* A de Carvalho Paixão & M Toledo-Piza Ragazzo, 2009

The Avá-canoeiros are an indigenous Brazilian people. Historically they lived in the upper Rio Tocantins basin, Goiás, where this catfish is found.

Avaiki

Star-spangled Goby *Kelloggella avaiki* L Tornabene, B Deis & MV Erdmann, 2017

Avaiki is not an eponym but a toponym – a Polynesian word referring to the sacred homeland of ancestors, and specifically on the island of Nuie (where this goby is endemic) referring to an area of tide pools and coastal caverns near the type locality that are revered as the sacred bathing pools of kings.

Avi

Sarawak Loach sp. *Homalopteroides avii* ZS Randall & LM Page, 2014

Lawrence 'Avi' Greenberg (1982–2011) (See **Greenberg)**

Avicenna

Zagros Spined Loach *Cobitis avicennae* H Mousavi-Sabet, S Vatandoust, HR Esmaeili, MF Geiger & J Freyhof, 2015

Abū 'Alī al-Husayn ibn 'Abd Allāh ibn Al-Hasan ibn Ali ibn Sīnā (aka Ibn-Sīnā or, in the Latinized version, Avicenna) (ca. 980–1037) was a Persian polymath, regarded as one of the greatest of the thinkers and writers of the Islamic Golden Age. He is known to have written about 450 books and, remarkably, some 240 have survived.

Avidor

Avidor's Pygmy-Goby *Trimma avidori* M Goren, 1978

A Avidor was an ichthyologist at Tel Aviv University, Israel. He co-wrote: *Biosociology and ecology of pomacentrid fishes around the Sinai Peninsula (northern Red Sea)* (1974). He collected the type.

Awa

Lumpfish sp. *Lethotremus awae* DS Jordan & JO Snyder, 1902

This is a toponym. The etymology says: "*The species is known from specimens about 300 millimeters in length, from Kominato, in the province of Awa, at the mouth of Tokyo Bay.*"

Awl

Maracaibo Mojarra *Eugerres awlae* LP Schultz, 1949

Aime Rebecca Awl née Motter (1887–1973) was an artist at the US National Museum. She was a student at Johns Hopkins Medical Art Department then became a 'scientific delineator' at the Smithsonian. Her work appeared in a wide range of scientific publications as well as the Encyclopaedia Britannica. According to the etymology she had "*...willingly and expertly drawn for [Schultz] very numerous figures of new fishes over a period of years*" (including type of this species).

Axel

Wolftrap Seadevil sp. *Thaumatichthys axeli* AF Bruun, 1953

Prince Axel Christian Georg of Denmark (1888–1964) served in the Navy and gained the rank of Admiral. He was a keen sailor and fisherman. He was also a member of the International Olympic Committee, was on the board of the SAS airline, and was a pioneer of motorsport.

Axelrod

Characin genus *Axelrodia* J Géry, 1965
Cardinal Tetra *Paracheirodon axelrodi* LP Schultz, 1956
Calypso Tetra *Hyphessobrycon axelrodi* HP Travassos, 1959
Axelrod's Cory *Corydoras axelrodi* F Rössel, 1962
Axelrod's Tetra *Neolebias axelrodi* M Poll & JP Gosse, 1963
Characin sp. *Schultzites axelrodi* J Géry, 1964
Characin sp. *Brittanichthys axelrodi* J Géry, 1965
Lake Malawi Cichlid sp. *Cynotilapia axelrodi* WE Burgess, 1976
Danio sp. *Sundadanio axelrodi* MR Brittan, 1976
Axelrod's Rainbowfish *Chilatherina axelrodi* GR Allen, 1979
Axelrod's Clown Blenny *Ecsenius axelrodi* VG Springer, 1988
Axelrod's Reef-Bass *Pseudogramma axelrodi* GR Allen & DR Robertson, 1995
Gabon Cichlid sp. *Parananochromis axelrodi* A Lamboj & ML Stiassny, 2003
African Cyprinid sp. *Labeobarbus axelrodi* A Getahun, MLJ Stiassny & GG Teugels, 2004
Axelrod's Tube-snouted Ghost Knifefish *Sternarchorhynchus axelrodi* CDCM de Santana & RP Vari, 2010

Dr Herbert Richard Axelrod (1927–2017) was an expert on tropical fishes, an entrepreneur and a publisher of pet books. He was in the US Army in Korea, after which he attended New York University which awarded his doctorate in Education. He worked at the aquariums at the AMNH, New York. He introduced many new species to the aquarium-fish market, including some that were new to science. He was also a collector of rare musical instruments and donated 4 Stradivarius stringed instruments to the Smithsonian (1988). He donated his collection of fossil fish to the University of Guelph, Canada (1989). He got into trouble over valuations of assets he sold, including the Stradivarius instruments, which may not have been exactly as represented. He was indicted for tax fraud (2004), went on the run to Cuba and beyond, being eventually arrested in Berlin and extradited to the USA where he was convicted (1905) and sentenced to 18 months imprisonment. He wrote: *Handbook of Tropical Aquarium Fishes* (1955) whilst serving in the US armed forces in Korea. According to Burgess his "*encouragement and help (including a trip to Lake Malawi to collect these fishes) has made [Burgess'] study of mbuna possible*" (See also **Herbert Axelrod**, also see **Evelyn (Axelrod)**)

Ayliffe

Indian Mimic Goatfish *Mulloidichthys ayliffe* F Uiblein, 2011

Neville Ayliffe was a dive operator at Sodwana. The etymology thanks him as one "*…who has assisted the South African Institute of Aquatic Biodiversity in acquiring important fish collections during many years. He collected the holotype and two paratypes with a speargun from shallow reefs in Sodwana Bay, KwaZulu-Natal, South Africa.*"

Ayling

Crimson Cleaner Wrasse *Suezichthys aylingi* BC Russell, 1985

Dr Anthony Michael Ayling (b.1947) is a well-known Australian marine biologist. He is known for his ichthyological studies and diving exploits and popular series of books and booklets. His BSc in zoology (1968) and PhD (1976) were awarded by the University of New Zealand. He was a post-doctoral student there (1976–1979) before becoming a consultant in the field of marine biological resource and impact assessment and management (1980–2002). The etymology says: "*This species is named in honor of Dr. A. M. Ayling who first recognized this fish as a new species and collected most of the type-specimens.*" Among his published works are: *Collins Guide to the Sea Fishes of New Zealand* (1982), *Sharks & Rays* (2006) and the co-authored: *Sea Critters* (2006) and *Marine Life* (2009).

Aymonier

Chinese Algae-Eater *Gyrinocheilus aymonieri* G Tirant, 1883

Étienne François Aymonier (1844–1929) was a French explorer and linguist who was the first archaeologist to make a systematic survey of the ruins of the Khmer empire in what is now parts of Cambodia, Thailand, Laos and Vietnam. Originally a corporal in an infantry regiment, he succeeded in educating himself and passing the qualifying examination to go to Saint-Cyr from where he was commissioned as a lieutenant (1868) and was posted to the colonial forces in French Indochina. Here he was variously employed (1869–1875, 1876–1885 & 1886–1889), including acting as the French representative (1879–1881) in what was then the French protectorate of Cambodia. He returned to France and was the first headmaster of l'École Colonial (1889) until he left the army and colonial administration with the honorary rank of colonel (2005). He wrote: *Le Cambodge* (3 vols. 1900–1904). He helped collect the holotype.

Ayraud

Ocean Leatherjacket *Nelusetta ayraud* JRC Quoy & JP Gaimard, 1824

Jean Jacques Victor Ayraud (1789–1821) was a naval surgeon who died in the Maritime Hospital at Fort-Royal, Martinique. On this basis we assume this to be the person the fish was named after. The author wrote: "*The name given to this fish recalls one of the many victims of yellow fever among naval medics. It was in the last epidemic that M. Ayraud died in Martinique, after seeing one of his comrades, an ensign, succumb to the same disease."* (translation).

Ayres

Western River Lamprey *Lampetra ayresii* A Günther, 1870

Dr William Orville Ayres (1817–1887) was an American ichthyologist, ornithologist and physician. He graduated from Yale (1837) and spent the next 15 years teaching. He then returned to Yale to study medicine, being awarded his MD (1854), after which he spent 20 years in medical practice in California. He also served as Professor of the Theory and Practice of Medicine in the Toland Medical College, San Francisco and also became Curator of Ichthyology at the California Academy of Sciences. Having moved to Chicago and lost financially there, he returned to New Haven (1878) practicing medicine but also lecturing on Diseases of the Nervous System in the Yale Medical School. Throughout he collected natural history specimens, especially fish and birds, lodging them with the Academy.

Ayson

Blind Electric Ray *Typhlonarke aysoni* A Hamilton, 1902

Lake Falconer Ayson (1855–1927) gave the holotype to its describer after the vessel 'Doto' trawled it up on a research cruise, of which Ayson was the leader. Ayson had been a farm labourer, a rabbit inspector, an acclimatisation officer and finally Chief Inspector of Fisheries.

Azaghal

Plated Catfish sp. *Aspidoras azaghal* LFC Tencatt, J Muriel-Cunha, J Zuanon, MFC Ferreira & MR Britto, 2020

Azaghâl was a king of the Dwarves in Tolkien's Middle Earth stories. The name references Terra do Meio (Pará, Brazil, type locality), loosely translated as 'Middle Earth' in English; and also, the fact that this catfish occurs in a mountainous region and reaches a relatively small size: typical features of Tolkien's fictional dwarves.

Azpelicueta

Armoured Catfish *sp. Farlowella azpelicuetae* GE Terán, GA Ballen, F Alonso, G Aguilera & JM Mirande, 2019

Dr María de las Mercedes Azpelicueta is Principal Investigator in the Vertebrate Zoology department of the University of La Plata, having been a researcher there for many years (1976–2015). She has written or co-written at least 75 papers from *Una nueva cita y ampliacion de la distribucion de dos especies para la ictiofauna argentina* (1980) to *Morphology and molecular evidence support the validity of* Pogonias courbina *(Lacepède, 1803) (Teleostei: Sciaenidae), with a redescription and neotype designation* (2019). She was honoured "*...in recognition of her prominent contributions to ichthyology, especially the systematics of Argentinian fishes. She described numerous species and was essential to the formation of subsequent generations pf freshwater fish systematists in Argentina."*

B

Babaran

Panay Anchovy *Stolephorus babarani* H Hata, S Lavoué & H Motomura, 2020

Professor Dr Ricardo P Babaran is the 10th Chancellor (2017) of the University of the Philippines Visayas, and Professor in the College of Fisheries and Ocean Sciences there. HIs BSc was awardedc by the University of Washington and his PhD by Kagoshima University. He was honoured for his 'great contributions' to surveys by the authors and other collaborators at Iloilo (Philippines), when the new anchovy species was collected. These surveys (2013–2017) resulted in the field guide: *Commercial and Bycatch Market Fishes of Panay Island, Republic of the Philippines* (Motomura *et al.* 2017). Dr Babaran has authored more than sixty-five papers such as *Payao fishing and its impacts to tuna stocks: A preliminary analysis* (2006).

Babault

Lake Tanganyika Cichlid sp. *Pseudosimochromis babaulti* J Pellegrin, 1927

Guy Babault (1888–1963) was a French traveller, naturalist, conservationist and collector. He was collecting in British East Africa (1912–1920) and is known to have been in India and Ceylon (now Sri Lanka). He wrote about his extensive collecting missions, as in: *Chasses et recherches zoologiques en Afrique orientale anglaise* (1917) and *Recherches zoologiques dans les provinces centrales de l'Inde et dans les régions occidentales de l'Himalaya* (1922). Many of the animals he collected can be seen in the Muséum de Bourges, gifted (1927) on his return from another trip to East Africa. At least one book was written about his journeys; *Voyage de M. Guy Babault dans l'Afrique Orientale Anglaise 1912–1913*. There were further specimens gifted to the museum by his wife (1932). A mammal and two birds are also named after him.

Babcock

Redbanded Rockfish *Sebastes babcocki* WF Thompson, 1915

John Pease Babcock (1855–1936) was a fisheries expert, civil servant, conservationist and author, who helped to establish the scientific and institutional underpinnings of fisheries conservation in British Columbia although he had no formal education in fisheries science. He was a Fisheries administrator in California (1891–1901 & 1910–1912) and British Columbia fisheries commissioner (1901–1910 & 1912–1933). Much of his work was to preserve or reestablish salmon runs in rivers. He wrote a handbook: *The game fishes of British Columbia* (1910) and also wrote short stories.

Babka

Caspian Goby genus *Babka* BS Iljin, 1927

Russian slang for an old woman or grandmother, and a local name for gobies.

Bacallado

Canary Moray *Gymnothorax bacalladoi* EB Böhlke & A Brito, 1987

Dr Juan José Bacallado Aránega (b.1939) is the Director of the Museum of Natural Science, Tenerife, Canary Islands and Professor of Animal Biology at the University of La Laguna. The University of Madrid awarded his bachelor's degree (1969) and the University of La Laguna his doctorate (1973). He was honoured '*...for contributions to study of marine fauna of Canary Islands.*'

Bacchet

Anthiine sp. *Dactylanthias baccheti* JE Randall, 2007

Philippe Bacchet (b.1960) is a French photographer in French Polynesia. He studied engineering, did military service and was a professional musician for three years. He went to Tahiti as an electronics engineer, but was developing his skills as a photographer and a diver. He took many photographs of nature but was increasingly drawn to the marine environment. He has written articles and longer works such as *îles et lumières* (1999) and being lead author on *Guide des poissons de Tahiti et ses îles* (2007). The etymology states: "*This species is named in honor of Philippe Bacchet who provided the holotype and its color photograph.*"

Bacchus

Red Codling *Pseudophycis bachus* JR Forster, 1801

Bacchus was the Roman god of wine. Forster used a variant spelling of the name.

Băcescu

Clingfish sp. *Apletodon bacescui* A Antoniou-Murgoci, 1940
Snailfish sp. *Menziesichthys bacescui* TT Nalbant & RF Mayer, 1971
Dragonet sp. *Neosynchiropus bacescui* TT Nalbant, 1979
Fanfin Anglerfish sp. *Caulophryne bacescui* A Mihai-Bardan, 1982

Professor Dr Mihai C Băcescu (1908–1999) was a Romanian zoologist, oceanologist and curator. After graduating (1933) at Iasi Univeristy he studied for his doctorate, which was awarded (1938) by the Constanta Bio-Oceonographic Institute. He won a scholarship to France in 1939 and worked at the National Museum of Natural History in Paris, the Oceanographic Museum of Monaco, at the Marine Biological Stations at Banyuls-sur-Mer and at Roscoff. He then became Head of Animal Morphology Department at Iasi. He went on to become head of department at the Museum of Natural History in Bucharest and head of the laboratory at the Fisheries Research Institute with the rank of university professor (1939–1954). He became (1954) the Director of the Fisheries Research Institute, and (1958) the head of the Oceanology Department of the Academy and Director of the National Museum of Natural History 'Grigore Antipa' where he stayed for three decades. He was 80 when he retired (1988). He participated in several scientific expeditions to the coasts of Peru and Chile (1965), Mauritania (1975), Arabia (1977) and Tanzania (1973–1974). He wrote an incredible 480 papers on a wide range of fauna, describing more than 300 new taxa including several new families. He also published a number of books particularly on

Romanian fauna and the ecology of the Black Sea. Around 70 species bear his name. He was the first to catch and preserve the snailfish. The etymology for the dragonet states that it is: "*Dedicated to Mihai Băcescu…on the occasion of his 70th anniversary.*"

Bach, J

Armoured Catfish sp. *Peckoltichthys bachi* GA Boulenger, 1898

Dr José Bach was a physician working at La Plata on the Rio Juruá of Brazil in the late 1800s, He was also the Director of the museum in Manaus and is recorded as having collected rodents for the museums in Para and in Rio de Janeiro. He collected the holotype of this species.

Bach, JS

Bach's Catshark *Bythaelurus bachi* S Weigmann, DA Ebert, PJ Clerkin, MFW Stehmann & GJ Naylor, 2016

Johann Sebastian Bach (1685–1750) was described in the etymology as: "*…a musical genius and one of the greatest composers of all time.*" There is no further need here for a lengthy biography.

Bachmann

Eelpout sp. *Lycenchelys bachmanni* AE Gosztonyi, 1977

Dr Axel O Bachmann (1927-2017) was Professor in the Faculty of Natural Sciences at Buenos Aires University (which awarded his chemistry degree and his biology doctorate) and the Natural History Museum there. Among his c.150 published papers is: *A catalog of the types of Staphylinidae (Insecta, Coleoptera) deposited in the Museo Argentino de Ciencias Naturales, Buenos Aires* (2017). The etymology says: "*The specific name honours Lic. Axel O Bachmann from Buenos Aires University Biology Dept., who very kindly helped the author in his early days in the biological sciences.*"

Bacon

Sheepshead Minnow ssp. *Cyprinodon variegatus baconi* CM Breder, 1932

Daniel Bacon was Charles M Breder Jr's host at Long Cay, Bahamas, during a collecting trip (1930) during which this subspecies was discovered. As well as Bacon and his wife offering accommodation, Daniel often accompanied Breder aboard the 'SS Munarga' and collected fish with him.

Bader

Corydoras sp. *Corydoras baderi* R Geisler, 1969

Herbert Bader of Hannover, Germany is described in the etymology as a "superb aquarist and travel companion". We can find nothing more about him.

Baduel

Leporinus sp. *Leporinus badueli* J Puyo, 1948

This is a toponym reflecting the fact that the holotype was taken near Mont Baduel, French Guiana.

Baensch, H-A

Golden Livebearer *Poeciliopsis baenschi* MK Meyer, AC Radda, R Riehl & W Feichtinger, 1986
Baensch's Damsel *Pomacentrus baenschi* GR Allen, 1991
Dwarf Cichlid sp. *Apistogramma baenschi* U Römer, I Hahn, E Römer, DP Soares & M Wöhler, 2004

Hans-Albrecht Baensch (1941–2016) was a German aquarist and publisher. Initially Hans worked for his father's (q.v.) company 'Tetra Werke' (founded 1951) and travelled for them worldwide. He also took part in three Amazon expeditions, during which he was involved in the discovery of several new fish species. He founded his own company Mergus Verlag (1977) and co-wrote and published the modern classic of the aquarium hobby: the first of 26 volumes in the series *Aquarium Atlas* (1982). He published the book in which the description of the damsel appeared and, together with his father, wrote *Beginner's Aquarium Digest: The Magic of Aquariums* (1975). The etymology for the livebearer merely says it is dedicated to Hans A Baensch. (Also see **Rosita** & **Hans Baensch**)

Baensch, U

Lake Malawi Cichlid sp. *Aulonocara baenschi* MK Meyer & R Riehl, 1985

Dr Ulrich Baensch was a biologist who invented (1955) TetraMin, the first universal dry food suitable as the staple food for a multitude of popular aquarium fishes. He was the father of Hans-Albrecht Baensch (q.v.). Among other works he wrote: *Blooming Bromeliads* (1996) and co-wrote: *Tropical Aquarium Fish* (1991) and, together with his son, wrote *Beginner's Aquarium Digest: The Magic of Aquariums* (1975).

Baenziger

Stone Loach sp. *Tuberoschistura baenzigeri* M Kottelat, 1983

(See under **Bänziger**)

Baer

Volkhov Whitefish *Coregonus baerii* K Kessler, 1864
Siberian Sturgeon *Acipenser baerii* JF Brandt, 1869
Baer Pugolovka *Benthophilus baeri* K Kessler, 1877

Karl Ernst von Baer (Karl Maksimovich) (1792–1876) was a versatile and well-travelled Estonian of German extraction, a naturalist and explorer of Siberia, Novaya Zemlya and the Caspian Sea region. He graduated (1814) as a physician and later took further training in anatomy, eventually joining the staff of Königsberg University (1817) and becoming Professor of Zoology (1821). He was the Director of the Zoological Museum of Königsberg, which he himself had established, then Professor of Anatomy (1826) within which he began his embryo research. He is known as the 'father of Estonian science and world embryology', not only discovering the egg cell, but also that embryos have similar developmental stages in virtually all animals - now known as the Baer Laws. He was also one of the cofounders of the Russian Geographical Society and edited a number of publications on geography. He worked in Austria and Germany before settling in Russia as Head of the Anthropological

Museum of the St Petersburg Academy of Sciences and the Director of the Department of Foreign Literature of the Library there. He was active in St Petersburg researching in geography, ichthyology, ethnography, anthropology and craniology. He worked briefly in the Ministry of Public Education (1862–1867). He is further remembered in 'Baer's Rule', which is about how riverbanks are symmetrical because of the rotation of the earth. He was a contemporary of Darwin (q.v.), with whom he corresponded, and spent the last years of his life writing critiques of Darwin's theories on evolution. In addition to a bird, seven different geographical objects on different continents bear the name of Baer, and there is a street named after him in Tartu and his portrait graced the Estonian 2-Kroon banknote (since superseded when Estonia adopted the Euro).

Bagan

Bagan Anchovy *Stolephorus baganensis* JDF Hardenberg, 1931

This is a toponym; it refers to the type locality, Bagansiapiapi, Indonesia.

Bahai

Stone Loach sp. *Turcinoemacheilus bahaii* HR Esmaeili, G Sayyadzadeh, M Özuluğ, MF Geiger & J Freyhof, 2014

Bahā' al-Dīn Muhammad ibn Husayn al-'Āmilī (aka Shaykh-i Bahā'ī) (1547–1621) was a Persian scholar, architect, mathematician, astronomer, philosopher and poet. He wrote more than 100 books and monographs and was one of the first astronomers to suggest the movement of the Earth, ante-dating Copernicus. In Iran, his birthday is now celebrated as National Architect Day.

Bahamonde

Bahamonde's Sole *Aseraggodes bahamondei* JE Randall & RC Meléndez, 1987

Professor Nibaldo Bahamonde Navarro (b. 1924) is a Chilean marine biologist. He graduated (1943) and then took a teaching qualification. He was (1946) Professor of Biological & Chemical Sciences at the University of Chile. During the 1950s he carried out research on marine fauna in Europe including at the University of Bergen, Norway. During the 1960s he continued to teach, with his interests turning to ecology and conservation. He went to the Juan Fernandez Islands (1965) to investigate a rumour, which fortunately turned out to be true, that the Juan Fernandez Fur Seal (*Arctocephalus philippii*), then thought extirpated in the middle of the 19th century, was in fact still extant. He has published over 200 books, articles and papers including, with G R Pequeño in 1975, *Peces de Chile, Lista sistematica*. He became Professor Emeritus of the University of Chile, Santiago (2004), where he had long worked. A mammal and three amphibians are also named after him.

Bailey, RM

Lake Tanganyika Cichlid genus *Baileychromis* M Poll, 1986
Rough Shiner *Notropis baileyi* RD Suttkus & EC Raney, 1955
Black Sculpin *Cottus baileyi* CR Robins, 1961
Smoky Madtom *Noturus baileyi* WR Taylor, 1969
Characin sp. *Astyanax baileyi* DE Rosen, 1972

Oriental Cyprinid sp. *Chagunius baileyi* WJ Rainboth, 1986
Hairy Puffer *Pao baileyi* S Sontirat, 1989
Armoured Catfish sp. *Hypoptopoma baileyi* AE Aquino & SA Schaefer, 2010
Spiny Dwarf Catfish sp. *Scoloplax baileyi* MS Rocha, H Lazzarotto de Almeida & LH Rapp Py-Daniel, 2012

Dr Reeve Maclaren Bailey (1911–2011) was an American ichthyologist. The University of Michigan awarded both his bachelor's degree (1933) and doctorate (1938). He started his doctorate studies in herpetology but switched to ichthyology and worked for two summers on the New York State Biological Survey (1935–1937). He taught at Iowa State College (now Iowa State University) (1938–1944) and was also Head of the Iowa Fisheries Research Unit. He returned to the University of Michigan's Museum of Zoology where he worked as Assistant Curator (1945–1948) then Curator until retiring as Curator Emeritus of Fishes and Professor Emeritus of Zoology. He wrote on American freshwater fishes, such as: *Fishes of South Dakota* (1962).

Bailey, VO

White River Springfish *Crenichthys baileyi* CH Gilbert, 1893

Vernon Orlando Bailey (1864–1942) was an American naturalist and ethnographer. As a young Minnesota farmer, he sent many natural history specimens to C Hart Merriam (q.v.), the then head of the US Biological Survey. Bailey joined the Survey (1887), soon becoming its Chief (1889–1902). He retired in 1933. During his time with the Survey he undertook many field trips, some with his wife, including six to Texas. He married Florence Augusta Bailey (née Merriam) (1899) who also worked for the Survey. He wrote: *The Mammals and Life Zones of Oregon* (1936) and also a number of papers including, for example, his field notes for his (1909) trip to Lincoln Co., Coos Co., and Curry Co., Oregon. He is also honoured in the names of six mammals and a reptile.

Baillon

Small Spotted Dart *Trachinotus baillonii* BG de Lacepède, 1801
Baillon's Wrasse *Symphodus bailloni* A Valenciennes, 1839

Louis Antoine François Baillon (1778–1851) was a French naturalist and collector. He worked as an assistant at the MNHN, Paris (1798–1799). He wrote: *Catalogue des Mammifères, Oiseaux, Reptiles, Poissons et Mollusques Testacés Marins Observés dans l'Arrondissement d'Abbeville* (1833). Threee birds are also named after him. His father, Jean François Emmanuel Baillon (1742–1801) was also an amateur naturalist.

Bains

Rocky Kurper *Sandelia bainsii* FL Castelnau, 1861

Andrew Geddes Bain (1797–1864) was a noted Scotish-born South African (emigrated 1816) geologist, engineer, palaeontologist and explorer. He started exploring and collecting zoological specimens, writing about his journeys and illustrating his articles. He tried farming and surveying for military roads under the Corps of Royal Engineers (1838). He became engineering inspector for the Cape Roads Board (1845) and constructed several passes. The construction work led to an interest in geology and fossil collecting. As this is a South African fish, named for the "savant géologue M. Bains" we believe that Castelnau made a slight error and put an incorrect 's' on the end of Bain's name. He also has a South African whiskey named after him.

Baird, IG

Stone Loach sp. *Schistura bairdi* M Kottelat, 2000

Dr Ian George Baird (b.1966) is a Canadian geographer who lived and worked in Laos and Thailand for about 20 years (mid-1980s to the late 2000s). The University of Victoria, in British Columbia, awarded his master's degree (2003) and the University of British Columbia his doctorate (2008). He is now Associate Professor of Geography at the University of Wisconsin-Madison. He conducted various wild-capture fisheries studies in southern Laos in the 1990s and has been conducting research regarding wild-capture fisheries and dams in the Mun River Basin in northeastern Thailand since 2014. He does research with ethnic Lao people living in Laos, Thailand and Cambodia, Brao people in Laos and Cambodia and Hmong people in Thailand and Laos. He also has experience working with former Lao and Hmong refugees living in the United States, France, and Canada. His research interests include Mekong fish and fisheries, the impacts of dams in the Mekong Basin, land grabbing and plantation development in mainland Southeast Asia, Lao and Hmong diaspora, identities issues (particularly ethnic and indigenous identities) and Southeast Asian history. He has written or co-written five books including the co-written: *People, Livelihoods and Development in the Xekong River Basin of Laos* (2008) as well as over 100 peer-reviewed articles and chapters in edited books. He collected the holotype of the eponymous species.

Baird, SF

Mottled Sculpin *Cottus bairdii* CF Girard, 1850
Triplespine Deepwater Cardinalfish *Sphyraenops bairdianus* F Poey, 1861
Bumphead Damsel *Microspathodon bairdii* TN Gill, 1862
Marlin-spike Grenadier *Nezumia bairdii* GB Goode & TH Bean, 1877
Baird's Slickhead *Alepocephalus bairdii* GB Goode & TH Bean, 1879
Lancer Dragonet *Callionymus bairdi* DS Jordan, 1888
Red River Shiner *Notropis bairdi* CL Hubbs & AI Ortenburger, 1929

Spencer Fullerton Baird (1823–1887) was an American zoologist and giant of American ornithology. He organised expeditions with the steamer 'Albatross'. Baird was Assistant Secretary and then Secretary (1878) of the Smithsonian. He wrote: *Catalogue of North American Birds* (1858). The young Baird became a friend of John James Audubon and sent him specimens. He is commemorated in the names of fourteen birds, five mammals, two reptiles and an amphibian.

Baissac

Frenchman Seabream *Polysteganus baissaci* MM Smith, 1978

Jean de Boucherville Baissac was an ichthyologist, fisheries officer (1947) and accomplished watercolour artist and illustrator of fish. Much of his work is exhibited at the Blue Penny Museum, Port Louis, Mauritius. Among his books are works on agriculture, soldiering and geology as well as: *Contribution à l'étude des poissons de l'Ile Maurice* (1949). He recognised this species as undescribed (1956) and was honoured in the name for that and for his many years working with Mauritian fishes. The author says: "*Through his energy, collecting and publications he has produced a comprehensive list of the fishes of that area, so useful to subsequent workers.*"

Baka

Dwarf Moon Tetra *Bathyaethiops baka* T Moritz & UK Schliewen, 2016

This species is named after the Baka people, a native hunter-gatherer tribe in Cameroon.

Baker, F

Needlefish sp. *Nomorhamphus bakeri* HW Fowler & BA Bean, 1922

Frederick 'Fred' Baker (1854–1938) was an American physician, naturalist and amateur malacologist in San Diego, California. He was the prime mover in founding the Marine Biological Institution, which became the Scripps Institution of Oceanography. He was also a co-founder of the Zoological Society of San Diego. He discovered this needlefish.

Baker, H

Malabar Baril *Barilius bakeri* F Day, 1865
Indian Cyprinid sp. *Osteobrama bakeri* F Day, 1873

Rev Henry Baker Jr (1819–1878) was born in Travancore, India, where his parents were missionaries for the Church Missionary Society. He was sent to England (1839) to complete his education and train for the church. He was ordained as a deacon (1842) then as a priest (1843). He returned to India (1843) and worked among the Travancore hill tribes for the rest of his life. He died in Madras. He was clearly interest in natural history as he is mentioned as having in one instance 'obtained several specimens' and, in another, as having collected the holotype.

Bakossi

Lake Bermin Cichlid sp. *Coptodon bakossiorum* MLJ Stiassny, UK Schliewen & WJ Dominey, 1992

Named after the Bakossi people of Cameroon, who regard Lake Bermin as a holy place.

Balay

Claroteid Catfish sp. *Parauchenoglanis balayi* HE Sauvage, 1879
Ogooue River Mormyrid *Petrocephalus balayi* HE Sauvage, 1883

Dr Noel Eugene Balay (1847–1902) was a French explorer and colonial administrator. He qualified as a physician in Paris (1874) then took part in Savorgnan de Brazza's first two expeditions to West Africa (1875 & 1879). He was Lieutenant-Governor of Gabon (1886–1889), Governor of French Guinea (1891–1900) and Governor-General of French West Africa (1900–1902) during a period of a serious epidemic of Yellow Fever. He died from an undiagnosed fever, almost certainly a victim of the same disease. He collected the holotype.

Balbo

Balbo Sabretooth *Evermannella balbo* A Risso, 1820

Prospero Balbo, Count of Vinadio (1762–1837) was an Italian politician and President of the Turin Academy of Sciences (1815–1837) and Rector of the University of Turin (1805–1814). We cannot be sure that this is the right attribution but Risso mentions him many times in the same volume that contains the original description.

Balboa

Anchovy sp. *Anchoviella balboae* DS Jordan & A Seale, 1926

This is a toponym; it refers to the type locality, Balboa, Panama.

Baldasseroni

Labeo sp. *Labeo baldasseronii* L Di Caporiacco, 1948

Vincenzo *Baldasseroni* (1884–1963) was the Director of *La Specola* Natural History Museum, Florence. Among his published works are papers such as: *Nuovo contributo alla conoscenza dei lombrichi italiani* (1912) and a textbook: *General Entomology Course* (1935).

Baldwin, AH

Lantern Bass *Serranus baldwini* BW Evermann & MC Marsh, 1899

Baldwin's Razorfish *Iniistius baldwini* DS Jordan & BW Evermann, 1903

Albertus Hutchinson Baldwin (1865–1935) was an American artist. He studied at Yale University and École des Beaux-Arts, Paris. He was employed by the federal government Departments of Agriculture, Interior and Commerce to illustrate scientific reports. Apart from scientific illustrations he was a fine watercolourist, particularly of waterside scenes. The razorfish etymology says: "*…This species is named for Mr. Albertus H Baldwin in recognition of his paintings of American and Hawaiian fishes.*"

Baldwin, CC

Sea Bass genus *Baldwinella* WD Anderson & PC Heemstra, 2012

Dr Carole C Baldwin is an ichthyologist who is a research zoologist at the Division of Fishes, National Museum of Natural History, Smithsonian Institution. James Madison University awarded her BSc and the College of Charleston her MSc in marine biology. The College of William and mary awarded her PhD in marine science. She has published more than 100 papers such as: *Larvae of Diploprion Bifasciatum, Belonoperca Chabanaudi and Grammistes Sexlineatus (Serranidae: Epinephelinae) with a Comparison Of Known Larvae of other Epinephelines* (1991) and *More new deep-reef basslets (Teleostei, Grammatidae, Lipogramma), with updates on the eco-evolutionary relationships within the genus* (2018). She was honoured in recognition of her contributions to the understanding of the systematics of serranid fishes.

Baldwin, WJ

Baldwin's Pipefish *Dunckerocampus baldwini* ES Herald & JE Randall, 1972

Baldwin's Major *Stegastes baldwini* GR Allen & LP Woods, 1980

Wayne J Baldwin is an American ichthyologist based at the Hawaii Institute of Marine Biology who has described several new fish. Among his published papers are: *A new chaetodont fish, Holacanthus limbaughi from the Eastern Pacific* (1963) and *Stolephorus pacificus, a new species of tropical anchovy (Engraulidae) from the western Pacific Ocean* (1984). The pipefish etymology says: "*...In recognition of his study of Pacific fishes, the redstripe pipefish,* Dunckerocampus baldwini, *is named in honour of Wayne J. Baldwin who with the junior author collected the first specimens of this new species*".

Balguerías

Snailfish sp. *Paraliparis balgueriasi* J Matallanas Garcia, 1999

Dr Eduardo Balguerías is a Spanish marine biologist working at Centro Oceanográfico de Canarias, Instituto Español de Oceanografia (IEO). He started his professional career (1979) as collaborator at the University of La Laguna and as scientific observer onboard fishing vessels working for the IEO. He became Senior Scientist there (1989) and was head of their reasrch programme in Africa (1991–2008). He became Deputy Director (2008–2010) and is now General Director (2010). Among his published papers are: *The origin of the Saharan Bank cephalopod fishery* (2000) and *On the identity of Octopus vulgaris Cuvier, 1797 stocks in the Saharan Bank (Northwest Africa) and their spatio-temporal variations in abundance in relation to some environmental factors* (2002). He was described in the etymology as a "*... pioneer of the Spanish fishing investigations in the Southern Ocean.*"

Balik

Antalya Bleak *Alburnus baliki* NG Bogutskaya, F Küçük & E Ünlü, 2000
Sakarya Barb *Capoeta baliki* D Turan, M Kottelat, FG Ekmekçi & HO İmamoğlu, 2006
Murat Trout *Salmo baliki* D Turan, I Aksu, M Oral, C Kaya & E Bayçelebi, 2021

Dr Süleyman Balik is a Turkish ichthyologist. Ege University in Izmir awarded all his degrees: bachelor's (1969), master's (1972) and doctorate (1974). He became a professor in the Faculty of Fisheries, Ege University (1988) having previously been an assistant professor (1980–1987). He has written widely on Turkish freshwater fishes including: *Freshwater Fish in Anatolia, Turkey* (1995) and co-writing: *Growth features of an endemic population of Chondrostoma holmwoodii* (2010).

Ball

Ball's Pipefish *Cosmocampus balli* HW Fowler, 1925

Dr Stanley Crittenden Ball (1885–1956) was an American teacher, fish researcher and curator at the Peabody Museum. Yale University awarded his PhD (1915). After brief appointments at the Massachussetts Agricultural College in Amherst, Springfield (Massachusetts) College, and the Bernice Pauahi Bishop Museum in Honolulu, Hawaii, he returned permanently to Yale (1926) where he taught biology courses (entomology and ornithology) and curated the zoology collections until his retirement (1954). The etymology says: "*...Named for Stanley C. Ball, of the Bishop Museum, in slight acknowledgment of his interest in the fishes of Oceania.*"

Ballesteros

Pencil Catfish sp. *Trichomycterus ballesterosi* CA Ardila Rodríguez, 2011

Dr Jesús Ballesteros Correa is a biologist at the University of Córdoba, Colombia, which has awarded him two bachelor's degrees, in biology (1986) and chemistry (1992). The Pontificia Universidad Javeriana, Bogota, awarded his master's (2001) and his doctorate (2015). His major interest in the study of mammals and among his publications is the co-written *Bats from the plains of the Tigre and Manzo River of the PNN-paramillo (Córdoba-Colombia)* (2009). He collected the holotype of this species.

Ballieu

Lined Coris *Coris ballieui* L Vaillant & H Sauvage, 1875
Spotfin Scorpionfish *Sebastapistes ballieui* H Sauvage, 1875
Blacktail Wrasse *Thalassoma ballieui* L Vaillant & H Sauvage, 1875

Pierre Étienne Théodore Ballieu (1828–1885) was French consul to the Sandwich Islands (Hawaii) (1869–1878). He was interested in natural history and collected himself, including the type of the Hawaiian bird known as the Palila (*Loxioides bailleui*), which is also named after him (although Oustalet misspelt his name in the binomial as Baillieu).

Balloch

Balloch's Blue-eye *Kiunga ballochi* GR Allen, 1983

Dr David Balloch (b.1950) is an Australian biologist who was a staff biologist at the Ok Tedi mine in Papua New Guinea, during which time he collected the species. The University of Stirling awarded his biology BA and the University of Asron his MSc in biology of water management. He was a Contract Officer and Research Hydrobiologist (1975–1977) then Project Manager and Research Fellow (1977–1980) at the Applied Hydrobiology Research Station in England. He became Senior Environmental Biologist, Ok Tedi Mining Limited (PNG) (1982–1985) before becoming Associate (1985–1987) then Associate Director (1987–1991) with NSR Environmental Consultants operating out of Melbourne. He became Director of his own company, David Balloch & Associates Pty. Ltd., Environmental Consultants (1991–2006) since when he has been Managing Director, EnviroGulf Consulting, Dubai, United Arab Emirates. He has produced a dozen reports such as *Effects of a paper mill effluent on the water quality and biota of the River North Esk, Scotland* (1973).

Balon

Balon's Ruffe *Gymnocephalus baloni* J Holčík & K Hensel, 1974
Tilapia sp. *Tilapia baloni* E Trewavas & DJ Stewart, 1975

Eugene Kornel Balon (1930–2013) was a Canadian ichthyologist and writer. He was born in Czechoslovakia of Polish extraction. The Universitry of Prague awarded his BCs (1952), MSc (1953) and PhD (1962). By the late 1960s he worked at the Research Institute of Fish Culture and Hydrobiology, University of South Bohemia, in České Budějovice, and in the Slovak Academy of Sciences in Bratislava. He was then an ichthyologist and FAO expert at the Central Fisheries Research Institute FAO/UNDP, Chilanga, Zambia (1967–1971) when he was invited to work in the USA, but instead settled in Canada. There he worked as a scientist at the Faculty of Zoology, University of Toronto (1972), then in Guelph at the departments of Zoology (1972–1995) and Integrative Biology (since 1995) of the

University of Guelph and Axelrod Institute of Ichthyology. He helped establish the journal 'Environmental Biology of Fishes' and became its editor-in-chief (1974–2002). He wrote more than 340 scientific works, including nine monographs and two textbooks.

Balston

Balston's Pygmy Perch *Nannatherina balstoni* CT Regan, 1906

William Edward Balston (1848–1918) was a wealthy and successful English businessman who was interested in ornithology in particular, and natural history in general. Two birds and two mammals are also named after him.

Balushkin

Whipnose Angler sp. *Gigantactis balushkini* VE Kharin, 1984
Naked-head Toothfish sp. *Gvozdarus balushkini* OS Voskoboinikova & A Kellermann, 1993
Palemouth Snailfish *Psednos balushkini* DL Stein, NV Chernova & AP Andriashev, 2001
Dreamer sp. *Phyllorhinichthys balushkini* TW Pietsch, 2004
Balushkin's Antarctic Sculpin *Bathylutichthys balushkini* OS Voskoboinikova, 2014
Balushkin's Gargoyle Fish *Coelorinchus balushkini* W Schwarzhans, T Mörs, A Engelbrecht, M Reguero & J Kriwet, 2016
Ridgehead ssp. *Scopeloberyx malayanus balushkini* AN Kotlyar, 2004

Dr Arcady Vladimirovich Balushkin (1948-2021) was an ichthyologist at the Zoological Institute, Russian Academy of Sciences, St. Petersburg, where he was Head of the Laboratory of Ichthyology. His first degree was awarded by Perm University and his PhD in Biology was awarded by the University of Leningrad (1979) and his DSc by St Petersburg University (1997). He started his career as a junior Science Worker at the Leningrad State Institute of River & Lake Fisheries (1975–1977) then at the Zoological Institute USSR Academy of Science in Leningrad (1977–1985), Senior Science Worker (1985–1997) and Leading Science worker (1997–1998) there. He published over 150 papers including the last co-written: *Melanostigma meteori sp. n. (Zoarcidae) – a new Pelagic Eelpout species from the Meteor Bank (Southeastern Atlantic), with Remarks on the Polymerization of the lateral line in the family* (2019) and books such as: *Morphological bases of the systematics and phylogeny of the nototheniid fishes* (1989).

Balzan

Argentine Humphead Cichlid *Gymnogeophagus balzanii* A Perugia, 1891
Neotropical Rivuline sp. *Trigonectes balzanii* A Perugia, 1891

Dr Luigi Balzan (1865–1893) was an Italian naturalist who set out on a grand tour of South America (1890), traveling alone by whatever means he could find, through Argentina, Brazil, Paraguay, Peru, and Bolivia. He collected the cichlid type during that trip. He later became Professor of Natural Sciences in Asunción, Paraguay. He wrote: *Voyage de M E Simon au Venezuela 1887–1888* (1892). A reptile, a bird and an amphibian are named after him.

Bam

Bam's Damsel *Chromis bami* Randall & McCosker, 1992

Foster Bam (b.1927) is an American lawyer and underwater photographer. Yale awarded his BA (1950) and Columbia & Yale his LLB (1953). He is a team member and life trustee at the Bermuda Institute of Ocean Sciences. He is also chairman of the American Fly-fishing Museum, California Academy of Science. The etymology describes him as a "*...friend, photographer, and diving companion*" of the authors.

Bamileke

Bamileke Killi *Aphyosemion bamilekorum* AC Radda, 1971

This is not a true eponym, being named for a people: the Bamileke are an indigenous people in Cameroon.

Ba Nar

Stone Loach sp. *Nemacheilus banar* J Freyhof & DV Serov, 2001

This is not a true eponym as the fish is named after the Ba Nar, an ethnic minority people in the area of Vietnam where this species was collected. Two species of amphibians are also named after them.

Bănărescu

Stone Loach genus *Petruichthys* AGK Menon, 1987
Loach sp. *Parabotia banarescui* TT Nalbant, 1965
Stone Loach sp. *Oxynoemacheilus banarescui* GB Delmastro, 1982
Stone Loach sp. *Mesonoemacheilus petrubanarescui* AGK Menon, 1984
Bănărescu's Barb *Capoeta banarescui* D Turan, M Kottelat, FG Ekmekçi & HO İmamoğlu, 2006
Gudgeon sp. *Squalidus banarescui* IS Chen & YC Chang, 2007
Riffle Minnow sp. *Alburnoides petrubanarescui* NG Bogutskaya & BW Coad, 2009

Dr Petru Mihai Bănărescu (1921–2009) was a Romanian ichthyologist and naturalist. The Faculty of Sciences, Department of Natural Science, University of Clui awarded his bachelor's degree (1944) and his doctorate (1949). He taught zoology and biogeography at the University of Clui (1944–1952) and then was a researcher at the Romanian Institute of Fish Research until he moved to the Institute of Biology in Bucharest (1970). He made many trips abroad in Europe, the Near East and the USA. He was an honorary member of both the European and American Associations of Herpetologists and Ichthyologists and published over 300 scientific papers during his career. (Also see **Petru Banarescu**)

Bancroft

Bancroft's Numbfish *Narcine bancroftii* E Griffith & CH Smith, 1834
[Alt. Brazilian/Lesser Electric Ray]

Dr Edward Nathaniel Bancroft (1772–1842) was a British surgeon at St Georges, London (1805–1811). He was educated at St Johns College, Cambridge where he graduated as Bachelor of Medicine (1794). He became an army physician (1795) and saw service in Portugal, Egypt and the Windward Islands. He returned to study at Cambridge where he obtained his doctorate (1804). He became a physician in Kingston, Jamaica (1811) 'for health reasons', where he married, later becoming the Deputy Inspector General of Army Hospitals in Jamaica. He was the author of several scientific papers such as: *On several fishes of Jamaica* (1831). He was also a double-agent spy during the American Revolution. A fine watercolourist, he not only studied electric fish but also painted them and the scientific description of the numbfish was based on his illustration.

Bandeiras

Armoured Catfish sp. *Neoplecostomus bandeirante* FF Roxo, C de Oliveira & CH Zawadzki, 2012

The early Portuguese settlers of São Paulo explored (16th-18th centuries) the unknown and mapless interior of Brazil in excursions known as *bandeiras*. They were, by our standards, unpleasant people as they were hunting for indigenous people to enslave them, and to search for mineral wealth, such as silver, gold and diamonds. The work they did made the founding of new cities possible and created the geographic demarcation of the Brazilian territory.

Bandula

Bandula Barb *Pethia bandula* M Kottelat & R Pethiyagoda, 1991

Ranjith Bandula, who is a collector of ornamental fishes, discovered the eponymous species near Galapitamada, Sri Lanka (1987) and gave the holotype to Rodney Jonklaas (q.v.), at whose house Rohan Pethiyagoda first saw the fish in an aquarium (1991). Bandula then helped Pethiyagoda to collect the type series. The vernacular name 'Bandula Barb' was coined by Jonklaas and not by the authors.

Baniwa

Ghost Knifefish sp. *Apteronotus baniwa* CDCM de Santana & RP Vari, 2013

The Baniwa are an indigenous people whose home territory is included in the type locality; Río Orinoco basin, Venezuela.

Bankier

Chinese Demoiselle *Neopomacentrus bankieri* J Richardson, 1846

Robert Austin Bautues Bankier (1815–1853) was a naturalist and Royal Naval surgeon on HMS Blenheim (1842) and was Surgeon-in-Charge HMS Alligator and later HMS Minden, both of which were hospital ships. Previously he was Assistant Surgeon (1836) and Surgeon (1843) serving on HMS Herald (1938), HMS Pelorus (1839), HMS Bentinck (1841) and HMS Winchester (1842). He was a founding member of the China Medico-Chirurgical Society (1945). His brother James was also a naval surgeon. The description says: "*The only example we have seen of this species was sent to Haslar Museum from Hong Kong, by Surgeon R A Bankier, R.N.*"

Banks

Saddle Barb *Barbodes banksi* AW Herre, 1940

Edward H 'Bill' Banks (1903–1988) was, as a British colonial administrator, District Officer in Sarawak, a zoologist, naturalist and Curator of Sarawak Museum, Kuching (1925–1945).

He was interned at Batu Lintang during the Japanese occupation (1942–1945) (WW2). He wrote: *A Naturalist in Sarawak* (1949) and retired the following year. A bird is also named after him.

Banneau

Pencil Catfish sp. *Trichomycterus banneaui* CH Eigenmann, 1912

Henri Banneau was described as being a Parisian commercial traveller with wide South American experience. Importantly for Eigenmann, Banneau appears to have had his own steamer on the Magdalena River, Colombia, and set his crew to fishing for specimens for Eigenmann. Banneau himself was also keen on fishing and collecting plus he relieved Eigenmann "entirely of the vexations of handling" his baggage. We have not been able to find other details of him

Banner

Banner's Pipefish *Cosmocampus banneri* ES Herald & JE Randall, 1972

Dr Albert Henry 'Hank' Banner (1914–1985) was Professor of Zoology at the University of Hawaii for more than thirty years. His diverse interests in natural history resulted in a corpus of work dealing not only with his special interest in snapping shrimp (family Alpheidae), but also publications on ciguatera fish poisoning and on the effects of pollution on coral reefs, as well as compilations of animal and plant names from Pacific islands. He worked with his wife Dora May 'Dee' Banner (1916–1985) and produced more than 100 publications, such as: *Preliminary report on marine biology study of Onotoa Atoll, Gilbert Islands* (1952) and co-written with his wife: *The alpheid shrimp of Australia. Part 2: The genus Synalpheus* (1975). The etymology states "*…Named banneri in honor of Dr A H Banner whose welcome field efforts resulted in the capture of the holotype of this species.*"

Bannikov

Ridgehead sp. *Scopeloberyx bannikovi* AN Kotlyar, 2004

Dr Aleksandr Fedorovich Bannikov (b.1954) is an ichthyo-palaeontologist. The University of Moscow awarded his bachelor's degree (1976) and doctorate (1980) He has worked at the Palaeontology Institute, Russian Academy of Sciences, Moscow as a post-doctoral fellow (1976–1979), junior scientist (1980–1986), research scientist (1986–1988) and since 1988, senior scientist. He co-wrote: *Phylogenetic revision of the fish families Luvaridae and Kushlukiidae (Acanthuroidei), with a new genus and two new species of Eocene Luvarids* (1995). The author honoured him because he "*…repeatedly rendered [Kotlyar] invaluable aid in his investigations.*"

Bannwarth

Pipefish sp. *Lissocampus bannwarthi* PGE Duncker, 1915

Dr Emile Bannwarth was a physician who lived in Cairo until the outbreak of WW1 (1914). He and his wife collected much material for the Hamburg Natural History Museum. He discovered this species.

Banon

Sandperch sp. *Parapercis banoni* JE Randall & T Yamakawa, 2006

Rafael Bañón Díaz (b.1961) is a Spanish research scientist and fisheries ecologist focusing on research on fish taxonomy, small scale fisheries, fishers and fisheries anthropology. The University of Santiago de Compostela awarded his degree in marine biology (1988) and the University of Vigo his masters (1997) and doctorate (2016). He worked as a Senior Research Scientist, Instituto Español de Oceanografía IEO, Vigo, Spain (1996–1998), as a Biologist at Xunta de Galicia, Santiago de Compostela, Spain (1999–2015) at the Instituto de Investigaciones Marinas (IIM CSIC) in Vigo (2016–2017) and again at the Xunta de Galicia (2018). His more than eighty published papers include the co-authored: *Commented checklist of marine fishes from the Galicia Bank seamount (NW Spain)* (2016) and *New insights into the systematics of North Atlantic Gaidropsarus (Gadiformes, Gadidae): flagging synonymies and hidden diversity)* (2018). He provided the authors with specimens and colour photographs.

Bänziger

Stone Loach sp. *Tuberoschistura baenzigeri* M Kottelat, 1983

Dr Hans Bänziger (b.1941) is a Swiss entomologist at the Department of Entomology and Plant Pathology, Faculty of Agriculture, Chiang Mai University, Thailand. He has worked there for nearly 40 years, previously as a Principal Researcher. He wrote: *Remarkable New Cases of Moths Drinking Tears in Thailand* (1992) in which he describes the sensations he felt when moths drank his own tears. The author honoured Bänziger for the help he received during his collection trip in Chiang Mai.

Baranes

Threadfin Bream sp. *Parascolopsis baranesi* BC Russell & D Golani, 1993

Moray sp. *Gymnothorax baranesi* DG Smit, E Brokovich & S Einbinder, 2008

Dr Albert 'Avi' Baranes (b.1949) was born in Alexandria, Egypt. He lived in Paris (1956–1970) and the Faculté des Sciences, University of Paris awarded his bachelor's degree. He is both Honorary Consul of France and Director of the Marine Biology Laboratory at Eilat, part of the Hebrew University of Jerusalem, where he was awarded both his master's and doctorate in oceanography (1986). He has published 50 papers on sharks and deep-sea fishes of the Red Sea, such as, with Ben-Tuvia (q.v.): *Two rare carcharhinids,* Hemipristis elongatus *and* Iago omanensis, *from the northern Red Sea* (1979) and *Ichthyofauna of the rocky coastal littoral of the Israeli Mediterranean* (2007) as well as two books on sharks. His main research interests are sharks of the Red Sea and the taxonomy and biology of Red Sea deepwater fish.

Barazer

Perchlet sp. *Chelidoperca barazeri* S-H Lee, M-Y Lee, M Matsunuma & W-J Chen, 2019

Jean-François Barazer was the first Captain of the Research Vessel 'Alis', a French fisheries patrol vessel. It took part in a series of biodiversity expeditions carried out (2007–2017) mainly in the West Pacific, under the Tropical Deep-Sea Benthos (TDSB) program and the cooperative project between Taiwan and France entitled 'Taiwan-France marine diversity exploration and evolution of deep-sea fauna'. The etymology says that the fish "*…is named* barazeri *for honor of Mr. Jean-François Barazer, the captain of R/V Alis. He is an expert in organizing trawling operations, deep-sea biodiversity surveys, and cruise arrangements. Without his support and great efforts, the discovery of new species in many studies including ours carried out through the TDSB program would not be possible.*"

Barbara (Brown)

Licorice Gourami sp. *Parosphromenus barbarae* TH Hui & J Grinang, 2020

Barbara Brown née Demaree is the widow of Allan Brown, who first collected this species in Sarawak. (See also **Allan** and **Brown, B & A**)

Barbara (Williams)

Lake Victoria Cichlid sp. *Haplochromis barbarae* PH Greenwood, 1967

Barbara Williams is a scientific illustrator including such books as Greenwood's *Fishes of Uganda* (1958). The etymology says: "Named in honour of Mrs. Barbara Williams, whose drawings illustrate this and others of my papers."

Barber

Hawaiian Lionfish *Dendrochirus barberi* F Steindachner, 1900

Captain Barber found the type specimen in plankton during a trip from Honolulu to Cape Horn (1896–1897). We have been unable to identify him further.

Barber, P

Barber's Clownfish *Amphiprion barberi* GR Allen, JA Drew & L Kaufman, 2008

Dr Paul Henry Barber is an American evolutionary biologist and ecologist. The University of Arizona awarded his BSc (1991) and the University of California, Berkeley, his PhD (1998). His career has been as a researcher and teacher, with post doctoral work at Harvard (1999–2001), as Assistant Professor of Biology (2002–2008) then Adjunct Professor (2008–2018) at Boston University as well as other posts. He has published widely. The etymology states that: "*…This species is named* Amphiprion barberi *in honour of Dr. Paul Barber of Boston University, USA in recognition of his valuable contributions to our understanding of genetic relationships of Indo-Pacific coral reef organisms.*"

Barbero

Croaking Tetra sp. *Mimagoniates barberi* CT Regan, 1907

Dr Andrés José Camilo Barbero Crosa (1877–1951) was a Paraguayan scientist, physician, pharmacologist and philanthropist. From a wealthy background he qualified in pharmacy (1898) and went on to qualify as a physician (1903), after which he dedicated himself to teaching. With others he founded the Paraguay Scientific Society. He used his wealth to establish libraries, museums and societies. Regan's etymology is minimal, referring to specimens having been collected in Paraguay by Dr. A. Barbero. The Instituto Dr. Andrés Barbero is named after him as is a bird.

Barbosa, GN

Darter Characin sp. *Characidium barbosai* NJ Flausino, FCT Lima, FA Machado & MRS Melo, 2020

Gerson Natalício Barbosa is a District Attorney from Mato Grosso, Brazil. He was honoured for his commitment to enforcing environmental laws, as he was one of the conceivers of the 'Agua para o Futuro' project, which is surveying, protecting and restoring springs in the urban area of Cuiabá, Mato Grosso.

Barbosa, MA

Brazilian Cichlid sp. *Australoheros barbosae* FP Ottoni & WJEM Costa, 2008
Pencil Catfish sp. *Cambeva barbosae* WJEM Costa, CRM Feltrin & AM Katz, 2021

Maria Anais Barbosa Segadas Vianna is a Brazilian zoologist and ichthyologist at the Laboratory of Systematics and Evolution of Fish Teleosteos, Department of Zoology, Federal University of Rio de Janeiro. Among her published works are: *Trichomycterus puriventris (Teleostei: Siluriformes: Trichomycteridae), a new species of catfish from the rio Paraíba do Sul basin, southeastern Brazil* (2012) written with the junior author and *Description of two new species of the catfish genus* Trichomycterus *(Teleostei: Siluriformes: Trichomycteridae) from the coastal river basins, southeastern Brazil* (29013). She was honoured as she had helped the authors with the field and laboratory work.

Barbour, RW

Teardrop Darter *Etheostoma barbouri* RA Kuehne & JW Small, 1971

Professor Dr Roger William Barbour (1919–1993) was a vertebrate zoologist who taught and researched at the University of Kentucky. Morehead State Teachers College awarded his bachelor's degree in biology (1938), Cornell University his master's (1939) and his doctorate (1949). His studies were interrupted by army service (WW2). Most of his career was with the Department of Zoology, University of Kentucky, Lexington as Instructor (1950–1952), Assistant Professor (1952–1956), Associate Professor (1956–1968) and then Professor (1968–1984). He was in Indonesia (1957–1959) as a member of the Kentucky Contract Team at the Institute Teknologi di Bandung. He wrote *The Amphibians & Reptiles of Kentucky* (1971), *Bats of America* (1982) and, with RA Kuehne: *The American Darters* (1983). He was also a keen wildlife photographer.

Barbour, T

Sole genus *Barbourichthys* P Chabanaud, 1934
Velvet Whalefish genus *Barbourisia* AE Parr, 1945
Barbour's Seahorse *Hippocampus barbouri* DS Jordan & RE Richardson, 1908
Ratfish sp. *Hydrolagus barbouri* S Garman, 1908
Characin sp. *Moenkhausia barbouri* CH Eigenmann, 1908
Pencil Catfish sp. *Trichomycterus barbouri* CH Eigenmann, 1911
Cuban Ribbontail Catshark *Eridacnis barbouri* HB Bigelow & WC Schroeder, 1944

Dr Thomas Barbour (1884–1946) was an American zoologist. He graduated from Harvard (1906) and obtained his PhD there (1910). He became an Associate Curator of Reptiles and Amphibians at the Harvard Museum of Comparative Zoology, and was its Director (1927–1946). He became Custodian of the Harvard Biological Station and Botanical Garden Soledad, Cuba (1927). He was Executive Officer in charge of Barro Colorado Island Laboratory, Panama (1923–1945). During his time at the museum he explored in the East Indies, the West Indies, India, Burma, China, Japan, and South and Central America. He was famously jovial good company and would invite all and sundry to eat and converse next door to his office in the 'Eateria' in which his secretary, Helen Robinson, prepared the food for his thousands of guests. Something of an all-rounder, he wrote many articles and books, including *The Birds of Cuba* (1923) and *Naturalist at Large* (1943). He also co-wrote *Checklist of North American Amphibians and Reptiles*. . He was honoured in the cat shark name for the "*...constant assistance*" he gave the authors in their studies of western North Atlantic sharks. His special area of interest was the herpetology of Central America and 24 reptiles are named after him, as are three birds, two mammals and four amphibians.

Barel

Lake Victoria Cichlid sp. *Haplochromis bareli* MJP van Oijen, 1991

Dr Cornelis Dirk 'Kes' Nico Barel (b.1942) is a Dutch zoologist and ichthyologist at the University of Leiden. The etymology reads thus: "*...This species is named in honour of Dr C. D. N. Barel, the initiator of the Haplochromis Ecology Project, who first collected specimens of this species. His stimulating enthusiasm and interest in all aspects of biology engaged many biology students in cichlid research. His research has contributed much to our knowledge of the Lake Victoria haplochromine cichlids.*" Among his many publications is the paper: *Towards a constructional morphology of cichlid fishes (Teleostei, Perciformes)* (1983) and longer works such as: *Vissen in de oudheid* (Fish in Ancient Times) (2002).

Bargibant

Bargibant's Pygmy Seahorse *Hippocampus bargibanti* GP Whitley, 1970

Georges Bargibant is a biologist and diver at the IRD Centre in Nouméa, New Caledonia, considered to be one of the leading specialists in Indo-Pacific coral environments. He was collecting specimens of Muricella (soft corals) for the Nouméa museum aquarium (now Lagoon Aquarium) and was about to dsect one when he noticed a pair of these tiny seahorses. His co-written book on the soft corals is: *Coral Reef Gorgonians of New Caledonia* (2001) and he has published articles and co-written at least one other book: *Sponges of New Caledonia* (1998).

Bario

Hillstream Loach sp. *Gastromyzon bario* HH Tan, 2006

This is not a true eponym; the Bario are the indigenous people who live on the Bario plateau in the Kelabit highlands, northern Sarawak, Borneo.

Barlow

Lake Malawi Cichlid sp. *Maylandia barlowi* KR McKaye & JR Stauffer, 1986
Dwarf Cichlid sp. *Apistogramma barlowi* U Römer & I Hahn, 2008

Dr George Webber Barlow (1929–2007) was Professor of Integrative Biology (1966–1993) Emeritus at UC Berkeley. He entered UCLA to study medicine but graduated with a BA in biology (1951). He then spent two years in the coast guard before returning to graduate school at UCLA. There he was awarded his MA (1955) and PhD (1958) which was followed by a post-doctoral fellowship at NIMH. After two years in Bavaria he took a post as Assistant Professor of Zoology at the University of Illinois (1960–1963) and Associate Professor (1963–1966) the took the same position at UC Bwerkeley and reseach ethologist at the University's Museum of Vertebrate Zoology becoming full professor (1970). He published 163 papers and three books most notably *Cichlid Fishes: nature's Grand Experiment in Evolution* (2000). He said that "*...My earliest memories are of trying to keep sea anemones and hermit crabs alive in a small bowl and fishing from the local dock. Birds, however, were to become my passion. And eventually all my time, and the little money I made from delivering newspapers in the wee hours of the morning, were invested in racing pigeons. I spared enough money, however, to indulge in tropical fishes, including cichlids.*" The etymology for the dwarf cichlid calls him "*... one of the most productive and notably leading behavioural ichthyologists of the last decades*" and among the first to "*seriously discuss consequences of Environmental Sex Determination.*"

Barnard

Sole genus *Barnardichthys* P Chabanaud, 1927
Barnard's Lanternfish *Symbolophorus barnardi* AV Tåning, 1932
Bigthorn Skate *Rajella barnardi* JR Norman, 1935
Barnard's Robber *Hemigrammopetersius barnardi* AW Herre, 1936
Blackback Barb *Enteromius barnardi* RA Jubb, 1965
Barnard's Dentex *Dentex barnardi* J Cadenat, 1970
Barnard's Rock Catfish *Austroglanis barnardi* PH Skelton, 1981
Blackchin Dwarf Snailfish *Psednos barnardi* N Chernova, 2001

Dr Keppel Harcourt Barnard (1887–1964) was a UK-born South African invertebrate zoologist particularly interested in marine crustaceans. He graduated from Cambridge then studied law but, being more interested in science was for a short time honorary naturalist at the Plymouth Marine Biology Laboratory. He worked at the South African Museum, Cape Town (1911–1964), initially as a marine biologist, becoming Assistant Director (1924) and then Director (1946–1956). He undertook many collecting expeditions in South Africa and Mozambique, and in three donkey-trek expeditions in Namibia (1924–1926). A keen mountaineer, he was Secretary of the Mountain Club of South Africa (1918–1945). He wrote on molluscs and entomology as well as fish such as: *A Monograph of the Marine Fishes of South Africa* (1925). Two reptiles are named after him.

Barnes, AT

Barnes' Garden Eel *Gorgasia barnesi* BH Robison & TM Lancraft, 1984

Anthony T Barnes was a colleague and shipmate of the authors aboard the research vessel 'Alpha Helix', from which the holotype was collected. In the 1970s he worked at the Department of Biological Sciences and Marine Science Institute, University of California, Santa Barbara and co-wrote: *Bioluminescence in the mesopelagic copepod Gausia princeps* (T. Scott) (1972).

Barnes, CA

Barnes' Silverside *Hypoatherina barnesi* LP Schultz, 1953

Dr Clifford Adrian Barnes (1905-1995) was an American marine scientist who was Professor of Oceanography at the University of Washington, Seattle (1947–1973) his alma mater and Professor Emeritus after retirement. His publications include: *Circulation near the Washington Coast* (1954) and *An Oceanographic Model of Puget Sound* (1954). He was project officer on 'USS Bowditch' during Operation Crossroads (1946). LP Schultz was also part of the observation team. A research vessel owned by the university is named after him.

Barnes, WR

Barnes's Lanternfish *Gonichthys barnesi* GP Whitley, 1943

William R Barnes (1886–1962) was a taxidermist and the last member of a family that gave 90 years' service to the Australian Museum, Sydney, which he joined as an attendant (1907), transferring to the taxidermy department (1922). He retired (1950) and moved to Santa Monica, California, near where his son and grandson lived. The collections of fishes he made in New South Wales and at Lord Howe Island 'contributed largely' to a reorganisation of the Australian Museum's fish collection. A bird is also named after him.

Barnett

Dragonfish sp. *Photonectes barnetti* CI Klepadlo, 2011

Dr Michael Alvin Barnett (1945–1988) was an American marine biologist whose bachelor's degree was awarded by Yale (1967). The Scripps Institution of Oceanography awarded his doctorate. He worked as an ecologist for an engineering consulting company in Washington D.C. He collected the holotype (1971).

Barra

Black Spiny Ashiura Pleco *Megalancistrus barrae* F Steindachner, 1910

This is a toponym; Barra is a place near the type locality, Rio São Francisco, Brazil.

Barrall

Barrall's Pygmy-goby *Trimma barralli* R Winterbottom, 1995

Glen E Barrall is a dive master and underwater photographer 'par excellance', who provided Winterbottom with many 'superb' slides of species of *Trimma* in their natural habitat, including those used in the description of this one.

Barraway

Katherine River Gudgeon *Hypseleotris barrawayi* HK Larson, 2007

Sandy Barraway was, according to the etymology: "*...the 'traditional' (i.e. aboriginal) owner of the Sleisbeck country (land or region), who had great knowledge of the fauna and stories associated with that country.*"

Barreimi

Garra sp. *Garra barreimiae* HW Fowler & H Steinitz, 1956

This is a toponym; the type locality is Barreimi, Oman.

Barrett

Blue-spotted Basslet *Lipogramma barrettorum* A Nonaka & DR Robertson, 2018

Craig Barrett (b.1939) and his wife Barbara Barrett née McConnell (b.1950). Craig was a US business man who rose to become Chairman of Intel (1998–2009). He was educated at Stanford (1957–1964), being awarded a PhD in material science. After retirement he joined the faculty of Thunderbird School of Global Management, Arizona where Barabara has served as President of the Board. She is an American business woman who is chairman of Aerospace Corporation. Her bachelor's, master's and law degree were all awarded by Arizona State Univeristy. She also served as US Ambassador to Finland (2008–2009). She is also a keen aviator. They gave an endowment of $10 million to Arizona State University. They were honoured for their support of the Smithsonian's Deep Reef Observation Project (DROP), during which the holotype was collected.

Barrie

Barrie's Lanternshark *Etmopterus brosei* DA Ebert, RW Leslie & S Weigmann, 2021

Barrie Rose (1947–2016) was a South African naturalist and tour guide. An accomplished sports fisherman, he joined the Sea Fisheries Research Institute as a research technician (1968). He left Sea Fisheries (1990) to become the manager of I&J's deep-sea trawler fleet, where he worked tirelessly to mitigate the impact that fisheries had on seabirds. He died after slipping from a cliff while angling.

Barriga

Characin sp. *Creagrutus barrigai* RP Vari & AS Harold, 2001

Dr Ramiro E Barriga Salazar is an ichthyologist, ecologist and taxonomist at the Universidad Politecnica, Quito, Ecuador. This fish was named after him in acknowledgement of his many contributions to the knowledge of the freshwater fishes of Ecuador, and for his assistance to the authors. He co-wrote: *Two New Chaetostoma Group (Loricariidae: Hypostominae) Sister Genera from Opposite Sides of the Andes Mountains in Ecuador, with the Description of One New Species* (2015).

Barro

Basslet sp. *Selenanthias barroi* P Fourmanoir, 1982

Monsieur Barro collected the type from the 'R/V Vauban' which is operated by IRD (formely ORSTOM) based in Noumea. The etymology says: "*L'espèce a été dédiée à M. Barro qui l'a récoltée a bord du R/V Vauban*." Unfortunately, no first name is given.

Barrois

Tigris Barb *Capoeta barroisi* LCE Lortet, 1894

Dr Theodore Barrois (1857–1920) was a French biologist, naturalist, physician and politician. He lectured at the Faculty of Medicine and Pharmacy of Lille (1885) where he became Professor of Zoology (1886) and (1894) Professor of Parasitology, a position specially created for him. He travelled and collected widely, including in Lapland (1881) and Palestine and Syria (1890) as well as spending time based at the biological research station at Concarneau. He suddenly abandoned science for politics (1898) and was elected as a deputy to the national assembly in Paris and was re-elected (1902), but no longer spoke in parliament as he was constrained (1902–1920) by important administrative functions in connection with the Pasteur Institute in Lille and the Central Committee that dealt with the French coal industry. He collected the holoytpe in Syria and wrote a monograph on Syrian Lakes that included a description of the fish.

Barron

Oriental Cyprinid sp. *Paralaubuca barroni* HW Fowler, 1934

P A R Barron was a collector, mainly of herpetological specimens, for the BMNH in London and the Raffles Museum, Singapore. He lived in Chiang Mai, Thailand and was a member of the Natural History Society of Siam. He wrote: *On the breeding of the toad Bufo macrotis* (1918). A reptile is also named after him.

Barry

Clingfish genus *Barryichthys* KW Conway, GI Moore & AP Summers, 2019
Viviparous Brotula sp. *Zephyrichthys barryi* W Schwarzhans & PR Møller, 2007

Dr J Barry Hutchins (b.1946). (See **Hutchins**).

Barry Brown

Stellate Scorpionfish *Scorpaenodes barrybrowni* DE Pitassy & CC Baldwin, 2016

Barry Baxter Brown (b.1965) is a freelance photographer with a degree in commercial art who is currently working with the Smithsonian Institution documenting new Caribbean deep-water species and building a one-of-a-kind database. His work has been used by numerous magazines from *National Geographic* to *Vogue* as well as scores of specialist publications. The etymology says: "…Named in honor of Barry Baxter Brown, who worked for Substation Curaçao and free-lance photographer (www.coralreefphotos.com), who has patiently, diligently, and expertly taken photographs of hundreds of fishes and invertebrates captured alive by DROP investigators. He has generously shared his photographs, and they have enhanced numerous scientific and educational publications."

Barsukov

Stubbeard Plunderfish *Pogonophryne barsukovi* AP Andriashev, 1967
Barsukov's Green Notothenia *Gobionotothen barsukovi* AV Balushkin, 1991
Stinger Flathead sp. *Plectrogenium barsukovi* SA Mandritsa, 1992
Manefish sp. *Caristius barsukovi* E Kukuev, NV Parin & IA Trunov, 2013

Vladimir Viktorovich Barsukov (1922–1989) was a Russian Ichthyologist, who was a Senior Research Scientist at the Zoological Institute of the USSR Academy of Science and Director of its Ichthyology Laboratory for seventeen years. He was awarded his zoology degree by Perm University (1949) having had his education disrupted by WW2 during which he served in the military, was wounded twice, and awarded medals for valour. He began at the Zoological Institute with post graduate study (1949) culminating in his docorate (1953); his dissertation was published as *The Fauna of the USSR, The Family Anarhichidadidae*. He worked for two years at Borok Biological Station then returned to the institute where he worked for the rest of his life. He wrote more than 100 papers and his major work was the book: *Rockfishes of the World's Oceans...* (1981). He took part in many field trips including the second Soviet Antarctic Epedition aboard 'Ob'. The etymology for the Green Notothenia says: "*...This is a tribute to the memory of a remarkable person and scientist who made an enormous contribution to the development of Soviet and world ichthyology.*"

Bartels

Bartels' Dragonet *Synchiropus bartelsi* R Fricke, 1981

Harald Bartels is a biologist. Among his published papers is: *Age and growth studies on fish from the Schapenbruchteich in the nature reserve Riddagshausen near Braunschweig* (1993). Fricke wrote: "The new species is named in honour of Mr. Harald Bartels, Braunschweig, for his continued interest in my studies."

Barthem

Driftwood Catfish sp. *Tetranematichthys barthemi* LAW Peixoto & WB Wosiacki, 2010

Dr Ronaldo Borges Barthem is a marine biologist and ichthyologist. The Federal University of Rio de Janeiro awarded his bachelor's degree (1977), the National Institute of Amazonian Research (1981) his Master's Degree in Fresh Water Biology and Inner Fishing (1981) and the State University of Campinas his doctorate in Ecology from the State University of Campinas (1990). At present, he is a researcher at the Museu Paraense Emílio Goeldi, Belém, Brazil. He co-wrote: *Goliath catfish spawning in the far western Amazon confirmed by the distribution of mature adults, drifting larvae and migrating juveniles* (2017). He is regarded as an expert on the fisheries ecology in the Amazon.

Bartlett, HH

Shortjaw Cisco ssp. *Coregonus zenithicus bartletti* W Koelz, 1931

Dr Harley Harris Bartlett (1886–1960) was an American botanist, ethnographer and collector. He graduated from Harvard (1908) and joined the US Plant Industry Bureau in Washington (1909). He became Assistant Professor of Botany, University of Michigan (1915), Professor (1921), Head of the Botany Department (1922–1947) and Director of the Botanical Garden (1919–1955). He collected in Sumatra (1918 and 1926) and the Philippines (1935 and 1940) plus collecting expeditions to Tibet, Malaysia, Formosa (Taiwan), Guatemala, British Honduras (Belize), Panama, Haiti, Argentina, Uruguay and Chile. A bird is named after him.

Bartlett, N & P

Bartletts' Anthias *Pseudanthias bartlettorum* JE Randall & R Lubbock, 1981

Nathan A Bartlett (1927–2014) was an electical engineer, private pilot, navy veteran, avid SCUBA diver and published underwater photographer. He and his wife Patricia B Bartlett moved back to the USA from Kwajalein, Marshall Islands (1979). The etymology states: "*Named* bartlettorum *in honor of Nathan and Patricia Bartlett, formerly of Kwajalein, Marshall Islands, whose underwater photos of this fish first revealed its existence.*"

Barton, FO

Robber Tetra sp. *Brycinus bartoni* JT Nichols & FR La Monte, 1953
[Syn. *Alestes bartoni*]

Frederick Otis Barton (1899–1992) was a wealthy American who was a deep-sea diver, inventor and actor. He and William Beebe made a number of record dives in the oceans, culminating with a descent to 1,372 metres (1949) in the Pacific Ocean. He wrote: *The World Beneath the Sea* (1953). He was closely involved in discovering and collecting fish, including the holotype of this species.

Barton (Bean)

Barton's Cichlid *Nosferatu bartoni* TH Bean, 1892
[Syn. *Herichthys bartoni*]
Alberca Silverside *Chirostoma bartoni* DS Jordan & BW Evermann, 1896 (Extinct)

Barton Appler Bean (1860–1947). (See **Bean, BA**)

Barton (Worthington)

Lake Victoria Cichlid sp. *Haplochromis bartoni* PH Greenwood, 1962

Dr Edgar Barton Worthington (1905–2001) (See **Worthington** & also **Stella**)

Bartsch

Cusk-eel sp. *Luciobrotula bartschi* HM Smith & L Radcliffe, 1913

Paul Bartsch (1871–1960) was born in Poland but educated in the USA. He joined the Smithsonian staff (1896) and stayed until retiring (1942) when he was assistant curator, division of molluscs. He was part of an expedition to the Philippines (1907–1910) aboard the ship *Albatross* among many other expeditions. He taught at a number of universities and was a world authority on molluscs. He also organised Washington's first Boy Scout group and was a keen member of the DC branch of the National Audubon Society. His account of the Philippines expedition was published in *Copeia* (1941). Two reptiles and a bird are named after him.

Basabe

Pete Basabe's Butterflyfish *Prognathodes basabei* RL Pyle & RK Kosaki, 2016
[Alt. Orangemargin Butterflyfish]

Peter K Basabe was a veteran local diver from Kona who, over the years, has assisted with the collection of reef fishes for numerous scientific studies and educational displays. Basabe, an experienced deep diver himself, was instrumental in providing support for the dives that produced the first specimen of the fish that now bears his name. The etymology says: "... *We take great pleasure in naming this species basabei, in honor of Peter K. Basabe, long-time diver, aquarium fish collector and resident of Kona, Hawai'i, both for his role in the collection of the first specimen of this new species in 1998, and more generally for his extensive contributions and assistance to many researchers (especially the authors) in the ichthyological community."* The fish had first been recognised as a new species when observed from a submersible by 'Deetsie' Chave (q.v.).

Bashford Dean

African Characin sp. *Microstomatichthyoborus bashforddeani* JT Nichols & L Griscom, 1917

Dr Bashford Dean (1857–1928) was an American zoologist, ichthyologist and acknowledged expert on mediaeval armour. The College of the City of New York awarded his bachelor's degree (1886) and Columbia University, where he was later Professor of Zoology, his doctorate (1890). He is the only person to have held concurrent positions at two of New York's great museums, the Metropolitan Museum of Art, where he was Honorary Curator of Arms and Armour, and the American Museum of Natural History, where he was a colleague of the senior author. He wrote: *Bibliography of Fishes* (1916) and *Catalogue of European Court Swords and Hunting Swords Including the Ellis, De Dino and Reubell Collections* (1929). He died unexpectedly after undergoing surgery.

Basil / Basileus

Dreamer sp. *Oneirodes basili* TW Pietsch III, 1974

Lanternfish sp. *Diaphus basileusi* VE Becker & VG Prut'ko, 1984

Dr Basil (Vasilis) George Nafpaktitis (1929–2015) (Basileus is a latinization of Basil) was a Greek-born American ichthyologist and marine biologist. The American University, Beirut, Lebanon awarded both his bachelor's (1962) and master's degrees (1963). Harvard awarded his doctorate (1966). His academic career (1966–1998) was at the University of Southern California, Dornsife and also (1982–1998) at The University of Crete where he founded the Institute of Marine Biology of Crete. He retired (1998) as Emeritus Professor of Biological Sciences at both universities. He was honoured for his investigations, particularly of the genus *Diaphus*. He was Pietsch's professor when (1969) the latter took his MA at USC. He died from complications associated with Parkinson's disease.

Basilevsky

Pufferfish sp. *Takifugu basilevskianus* S Basilewsky, 1855

Dr Stepan Ivanovich Basilewsky (aka Vasilevski) was a Russian physician and ichthyologist. He wrote: *Ichthyographia Chinae borealis* (1855). This is an interesting case, as there is no etymology and it appears that the author named the fish after himself. Possibly he used a name that had appeared in script but not as a formal description.

Basim

Haditha Cavefish *Caecocypris basimi* KE Banister & MK Bunni, 1980

Dr Basim M Al-Azzawi was a collector with the Natural History Research Centre, University of Baghdad. He lectured on blindfishes to the Port Macquaries Convention (2015). He collected the holotype (1977).

Baskin

Spiny Dwarf Catfish sp. *Scoloplax baskini* MS Rocha, RR de Oliveira & LH Rapp Py-Daniel, 2008

Dr Jonathan N Baskin (b.1939) is a biologist and ichthyologist who was a member of the faculty at California State Polytechnic University which he joined (1971) and where he is now Professor Emeritus. Harvard awarded his bachelor's degree (1961), the University of Miami, Florida his master's (1965) and the City University of New York in conjunction with the AMNH, his doctorate (1973). He was a Research Assistant, Institute of Marine Science, University of Miami (1961–1964) and occupied the same position at the AMNH (1964–1965). He wass a lecturer in Biology at Queens College, City University of New York (1965–1967 & 1968–1971) with a position as a Training Fellow at the Smithsonian (1967–1968). He is now the Owner of and Principal Scientist at San Marino Environmental Associates, San Marino, California. He co-wrote: *Scoloplax dicra, a minute new catfish from the Bolivian Amazon* (1976).

Bassargin

Fathead sp. *Eurymen bassargini* GU Lindberg, 1930

Vladimir Grigorievich Basargin (1838–1893), Russian geographer, cartographer and Vice Admiral. He was educated at the school of the naval cadet corps, leaving with the rank of marine Guard (1853) and joining the navy as midshipman (1854). He served on the sloop 'Udav' (1857), the corvette 'Rynda' (1858–1860) serving in the far east when he was promted to Lieutenant. He joined an expedition to the Pacific (1861) and appointed Commander of the 'Rynda' (1862). He was promoted to Captain (1872) serving on the frigate 'Prince Pozharsky' (1867–1877). He was appointed commander of the warship 'Peter the Great' (1880) then the cruiser 'Dmitry Donskoy' (1883). He accompanied (1880–1891) the Tsarevich Nicolas Alexandrovich of Russia (later Tsar Nicholas II of Russia) on his journey that took him to Trieste in the Far East and then accompanied the heir to the throne of Russia, out of Vladivostok, he crossed Siberia to reach St. Petersburg and was promoted to Vice Admiral (1892). He was also a member of the Board of the St. Petersburg Society of Homeopathic Physicians. A headland, Barsargin Cape near Vladivostok, is named after his father who was Governor of Astrakhan.

Bastar

Indian Catfish sp. *Clupisoma bastari* AK Datta & AK Karmakar, 1980

This is a toponym referring to the Bastar District of Madhya Pradesh, India, where the holotype was taken.

Bastian

Squeaker Catfish sp. *Synodontis bastiani* J Daget, 1948

M (probably Monsieur) Bastian collected the holotype in the Ivory Coast. We have not been able to find anything more about him.

Bates, GL

West African Cichlid sp. *Benitochromis batesii* GA Boulenger, 1901

Robber Tetra sp. *Brycinus batesii* GA Boulenger 1903
[Syn. *Alestes batesii*]
African Carp sp. *Labeobarbus batesii* GA Boulenger, 1903
African Bumblebee Catfish *Microsynodontis batesii* GA Boulenger, 1903
Squeaker Catfish sp. *Chiloglanis batesii* GA Boulenger, 1904
Kribi Mormyrid *Paramormyrops batesii* GA Boulenger, 1906
Squeaker Catfish sp. *Synodontis batesii* GA Boulenger, 1907
Bumba Mormyrid *Mormyrops batesianus* GA Boulenger, 1909
Bates' Killi *Aphyosemion batesii* GA Boulenger, 1911
Labeo sp. *Labeo batesii* GA Boulenger, 1911
African Barb sp. *Raiamas batesii* GA Boulenger, 1914
West African Cyprinid sp. *Prolabeo batesi* JR Norman, 1932

George Latimer Bates (1863–1940) was an American farmer and amateur ornithologist. He was born in Illinois, and travelled and farmed in Cameroon, West Africa (1895–1931). He wrote: *Handbook of the Birds of West Africa* (1930) and a number of articles, notably: *Birds of the Southern Sahara and Adjoining Countries* (1933). He also left an unpublished manuscript of: *Birds of Arabia*, subsequently utilised by Meinertzhagen for his 1950s work on the subject. He supplied both Boulenger and Norman at the BMNH with specimens, including the holotypes of many of the fish named after him. Twenty birds, several plants, four mammals, three amphibians and a reptile are named after him.

Bates, HW

Bates' Sabretooth Anchovy *Lycengraulis batesii* A Günther, 1868
Bluntnose Knifefish sp. *Brachyhypopomus batesi* WGR Crampton, CDCM de Santana, JC Waddlee & NR Lovejoy, 2017

Henry Walter Bates (1825–1892) was an English explorer and naturalist. He was the first to describe animal mimicry for science. Bates explored the Amazon with Alfred Russel Wallace, where they collected 14,000 specimens; 8,000 of which were new to science (1848–1859). He was largely self-taught, having left school at 12 when apprenticed to a hosier, yet ten years later published his first scientific paper. His most famous publication was: *The Naturalist on the River Amazons*, subtitled *A Record of the Adventures, Habits of Animals, Sketches of Brazilian and Indian Life, and Aspects of Nature under the Equator, during Eleven Years of Travel* (1863). After returning from the expedition he worked as Assistant Secretary of the Royal Geographical Society (1864) and became a Fellow of the Linnean Society and of the Royal Society. A bird and a reptile are also named after him. The description of the anchovy does not mention Bates, but it is extremely likely that he is the right man as he sent specimens to Günther for study, from the right place at the right time. The knifefish was named in honour of Bates's contributions to the natural history of the Amazonian region where it occurs.

Bath

Bath's Goby *Pomatoschistus bathi* PJ Miller, 1982
Bath's Combtooth Blenny *Ecsenius bathi* VG Springer, 1988

Dr Hans Walter Bath (1924–2015) was a German ichthyologist. He focused much of his research on the taxonomy of the family Blenniidae. He was a corresponding member of

Senckenberg Gesellschaft, Frankfurt Natural History Museum (1965). He was honoured as he collected the goby holotype, and for his valuable work on the systematics of Mediterranean gobies. Amongst other published papers he wrote: *Gammogobius steinitzi n. gen. n. sp. aus dem westlichen Mittelmeer (Pisces: Gobioidei: Gobiidae)* (1971).

Batman

Armoured Catfish sp. *Otocinclus batmani* A Lehmann, 2006

Batman, who has a bat shape for a symbol, is a comic book super-hero who first appeared in *Detective Comics* (1939). The etymology refers to the single W or bat-shaped vertical spot on the caudal fin of this species.

Battalgil

Battalgil's Spined Loach *Cobitis battalgili* MC Băcescu, 1962
Beysehir Minnow *Pseudophoxinus battalgilae* NG Bogutskaya, 1997
Eyilikler Gudgeon *Gobio battalgilae* AM Naseka, F Erk'akan & F Küçük, 2006
Gediz Shemaya *Alburnus battalgilae* M Özuluğ & J Freyhof, 2007

Dr Fahire Battalgil (1902–1948) was a Turkish ichthyologist and Turkey's first female professor She used the name Battalgazi after 1943. Her early life was spent in Damascus, Syria (part of the Ottoman Empire until the end of WW1) where she graduated (1924) from Beleb Alem High School. She then attended and grauated from the Faculty of Sciences, University of Istanbul (1926) and was appointed as an assistant dissector at the Faculty of Science and Physiology (1927) but transferred the same year to the Zoological Institute. She studied zoology and Comparative Anatomy in Paris at the Sorbonne (1931–1932). She became an assistant professor at the University of Istanbul (1933) and became full professor (1944). She wrote: *A new and little-known fish in Turkey* (1944). (See also **Fahire**)

Batuna

Batuna's Damsel *Amblyglyphidodon batunaorum* GR Allen, 1995

Hanny and Ineke (misspelled Inneke by Allen) Batuna are the owners of Manado Murex Resort, Indonesia. They provided accommodations, boat transport, and general logistic assistance during Allen's visit to Manado Bay, Sulawesi, the ttype locality (1994). The species was originally spelled *batunai*; but since the name honours more than one person, emendment was necessary.

Bauchot

Rosette Torpedo *Torpedo bauchotae* J Cadenat, C Capapé & M Desoutter, 1978
Bauchot's Goby *Callogobius bauchotae* M Goren, 1979
Perchlet sp. *Plectranthias bauchotae* JE Randall, 1980
Combtooth Blenny sp. *Microlipophrys bauchotae* P Wirtz & H Bath, 1982
Conger Eel sp. *Ariosoma bauchotae* C Karrer, 1983
Stargazer sp. *Uranoscopus bauchotae* R Brüss, 1987
Scorpionfish sp. *Neomerinthe bauchotae* SG Poss & G Duhamel, 1991
Moray sp. *Channomuraena bauchotae* L Saldanha & J-C Quéro, 1994
Goose-billed Croaker *Paranebris bauchotae* LN Chao, P Béarez & DR Robertson, 2001

Dr Marie-Louise Bauchot née Boutin (b.1928) is a French ichthyologist who was Curator of Ichthyology and Assistant Manager at the Muséum National d'Histoire Naturelle, Paris (1952–1991). Amongst other works she wrote a catalogue of the fish type specimens at the MNHN and *La Vie des Poissons* (1967). She collaborated on a number of books and papers, such as: *Les Poissons* (1954) with her husband Roland Bauchot, who was also a biologist (primarily a herpetologist and ichthyologist). She co-sponsored the Marshall Island expedition that collected the goby holotype. In the scorpionfish etymology, the authors say it was "*...named after Marie Louise Bauchot in recognition of her numerous contributions to ichthyology.*"

Baudon

African Barb sp. *Enteromius baudoni* GA Boulenger, 1918
African Loach Catfish sp. *Paramphilius baudoni* J Pellegrin, 1928

Alfred Baudon (1875–1932) was a French colonial administrator whose career was spent in French Equatorial Africa, starting as Harbour Master in Dakar (1899) and culminating with being the French Equatorial Africa's Technical Delegate at the Marseilles Colonial Exhibition (1922). He sent a collection of fishes from the Shari River to the BMNH, and this included the holotype of the eponymous barb.

Baunt

Whitefish (salmonid) sp. *Coregonus baunti* FB Mukhomediyarov, 1948

This is a toponym. It is named after the Baunt Administrative district in Siberia, where this species occurs.

Baxter, R

New Zealand Lantern Shark *Etmopterus baxteri* JAF Garrick, 1957

Richard Baxter was a retired New Zealand fisherman who took part in 'experimental line fishing' to collect shark specimens for the Zoology Department of Victoria University College, Wellington (1955). He 'collected' one from 500 fathoms close inshore in the Kaikoura region. He also caught another 'new' shark - *Etmopterus abernethyi* - that was later revised as an already known species: *Etmopterus lucifer.*

Baxter, RE

Black Fathead *Cubiceps baxteri* AR McCulloch, 1923

R E Baxter was an amateur naturalist on Lord Howe Island. He sent 'an interesting collection' of fish specimens to the Australian Museum, including the fathead holotype.He was still making observations and sending material in the 1950s.

Bayer

Hookjaw Moray *Enchelycore bayeri* LP Schultz,1953

Dr Frederick Merkle Bayer (1921–2007) was a marine biologist, specialising in soft corals, at the Smithsonian (1947–1961 & 1975–1996) and was Professor at the Marine Science School of the University of Miami (1961–1975). He retired as Emeritus Curator of the Smithsonian

(1996). He served in the US Army Air Force in the Pacific in WW2. The University of Miami awarded his bachelor's degree and the George Washington University both his master's (1954) and his doctorate in taxonomy (1958). Whilst at the University of Miami he took part in a number of expeditions to the Caribbean and to West Africa. He was sent to Bikini Atoll (1947) to study the effects of nuclear testing, only two years after the explosions.

Bayon

Lake Victoria Cichlid sp. *Haplochromis bayoni* GA Boulenger, 1909

Dr Enrico Pietro (later Henry Peter) Bayon (1876–1952) was an Italian-born British physician and entomologist. He studied engineering at the University of Genoa before graduating in medicine (1902) at the University of Würzburg. He was an assitant pathologist at the University of Geneva (1902–1905) before practicing medcine in Genoa and London and then aboard various ships. He studied sleeping sickness in Uganda (1907–1910) and himself succumbed to the disease. After serving in Riga (1913) and at Robben Island, South Africa (1913–1914), he volunteered in the Red Cross during WW1 and then was a pathologist at the British War Hospital. His later years were spent studying disease in birds, although he published widely on the history of medicine.

Bê

Oriental Cyprinid sp. *Opsariichthys bea* TT Nguyen, 1987

Mai Dinh Yen is the wife of the author TT Nguyen, Bê is a reference to her. (See **Mai DY**)

Beadle

Lake Nabugabo Cichlid sp. *Haplochromis beadlei* E Trewavas, 1933

Leonard Clayton Beadle OBE (1905–1985) was a pioneering limnologist, biologist and zoologist. His lifelong interest in freshwater biology started when, as an undergraduate, he took part in an expedition to South America (1926). He was the biologist, chemist and zoologist on the Cambridge East African Lakes Expedition (1930–1931), whose mission was to study the biology and geological history of Lakes Baringo and Rudolf in the East African Rift Valley. He made three further African trips; Algeria (1938) to work on oases and saline water, Uganda (1949) where he was head of the Zoology Department of Makerere University and a Trustee of Uganda Parks (1949–1966), and lastly again to Uganda on a Royal Society biological programme to Lake George. He was (1979) Honorary Senior Research Fellow at the University of Newcastle-Upon-Tyne and Emeritus Professor of Zoology and Wellcome Research professor at Makerere University, Uganda. He wrote: *The Art of Science* (1955) and *The Inland Waters of Tropical Africa: An Introduction to Tropical Limnology* (1974). The etymology says: "*Named in honour of Mr. L. C. Beadle, chemist and zoologist on the Expedition*" referring to the Cambridge Expedition to the East African Lakes (1930–1931).

Bean, BA

Scorpionfish sp *Neomerinthe beanorum* BW Evermann & MC Marsh, 1900
Gar Characin sp. *Ctenolucius beani* HW Fowler, 1907
Scaleless Black Dragonfish *Melanostomias bartonbeani* AE Parr, 1927

Barton Appler Bean (1860–1947) was the younger brother of T H Bean. He worked under his brother at the US National Museum (1881), becoming Assistant Curator of Fishes there (1890–1932) until his retirement. During this time, he occasionally worked as an investigator for the United States Fish Commission. He also made trips to New York, Florida and the Bahamas. He wrote more than 60 papers, including: *Fishes of the Bahama Islands* (1905) and, with others, *The Fishes of Maryland* (1929). He died falling from a bridge. The scorpionfish is named after both BA and TH Bean.

Bean, TH

Naked Sand Darter *Ammocrypta beanii* DS Jordan, 1877
Deepwater Dab *Poecilopsetta beanii* GB Goode, 1881
Bean's Sawtooth Eel *Serrivomer beanii* TN Gill & JA Ryder, 1883
Bean's Bigscale *Scopelogadus beanii* A Günther, 1887
Green Guapote (cichlid) *Mayaheros beani* DS Jordan, 1889
Bean's Searobin *Prionotus beanii* GB Goode, 1896
Scorpionfish sp. *Neomerinthe beanorum* BW Evermann & MC Marsh, 1900
Silverside sp. *Atherinella beani* SE Meek & SF Hildebrand, 1923

Tarleton Hoffman Bean (1846–1916) was an American ichthyologist and elder brother of B A Bean who was also an ichthyologist. Colombian (George Washington) University awarded his MD (1876) and Indian University awarded an MS (1883) on the basis of his accomplishments. He was a forester (his first interest was botany), fish culturist, conservationist, editor, administrator and exhibitor. He worked as a volunteer at the Fish Commission laboratory in Connecticut (1874) where he met Baird and George Brown Goode. He spent two decades (1875–1895) working in Washington at the National Museum (1878–1888) and the Fish Commission (1888–1895). After resigning from the Fish Commission, he was Director of the New York Aquarium (1895–1898). He then worked on forestry or fishery exhibits at the World's Fairs in Paris (1900) and St Louis (1905). He was a fish culturist for New York State (1906) until his death in a road traffic accident. He wrote nearly 40 papers with Goode, culminating in *Oceanic Ichthyology* (1896). The scorpionfish is named after both BA and TH Bean.

Beate

Beate's Coralbrotula *Diancistrus beateae* W Schwarzhans, PR Møller & JG Nielsen, 2005

Mrs Beate Schwarzhans is the first author's wife. He wrote of her thanking her for her 'most valuable support' during the many years he was engaged in the study of this genus.

Beau Perry

Beau's Wrasse *Cirrhilabrus beauperryi* GR Allen, J Drew & P Barber, 2008

Named for Beau Perry, son of Claire and Noel Perry, supporters of Conservation International.

Beaufort

Loach genus *Beaufortia* SL Hora, 1932
Beaufort's Mouth Almighty *Glossamia beauforti* MCW Weber, 1907
Beaufort's Goby *Stenogobius beauforti* MCW Weber, 1907

Sole sp. *Aseraggodes beauforti* P Chabanaud, 1930
Beaufort's Loach *Syncrossus beauforti* HM Smith, 1931
Spotlight Rasbora *Rasbora beauforti* JDF Hardenberg, 1937
Crocodile Fish *Cymbacephalus beauforti* LW Knapp, 1973

Professor Lieven Ferdinand de Beaufort (1879–1968) was a Dutch zoologist whose main interests were first fish and then birds. As a student, he participated (1902–1903) in the first scientific expedition to New Guinea, headed by the geographer Professor A Wichman. He undertook a second voyage to the Dutch East Indies (1909–1910). He regularly published (1904–1921) about birds collected in New Guinea by him, his friend (later Professor) Cosquino de Bussy, and others. In his early years, he was an assistant of Max Weber, whom he succeeded as Director of the Zoological Museum of the University of Amsterdam (1922–1949), and was Extraordinary Professor in Zoogeography (1929–1949) there. He was one of the founders (1901) of the *Nederlandsche Ornithologische Vereeniging* (the Dutch Ornithological Society), becoming Secretary (1911–1924) and Chairman (1924–1956). He was also a member of the editorial board of the society's journal *Ardea* (1924–1956). He edited the 11-volume: *Fishes of the IndoAustralian Archipelago* (1911–1962). Two birds, a mammal and a reptile are named after him. (Also see **De Beaufort**)

Beaven

Creek Loach *Schistura beavani* A Günther, 1868

Captain Robert Cecil Beavan (1841–1870) served in India with the Bengal Staff Corps for ten years. He collected birds and their eggs, as well as fish which were presented to the BMNH. His health was poor and he was twice invalided home to England, on the second occasion dying at sea. He wrote: *The Avifauna of the Andaman Islands* (1867) and a series of papers in *Ibis* mostly as 'notes on various Indian birds' (1865–1868). He also (probably) wrote: *Descriptions of two imperfectly known Species of Cyprinoid Fishes from the Punjab, India. By Lieut. Reginald Beavan, F.R.G.S.* Four birds are named after him.

Beccari

Parasitic Catfish sp. *Vandellia beccarii* L Di Caporiacco, 1935

Dr Odoardo Beccari (1843–1920) was an Italian botanist. He explored the Arfak Mountains during extensive zoological exploration with D'Albertis (1872–1873), recorded in *Wanderings in the Great Forests of Borneo.* He also explored and collected in the Celebes (Sulawesi) and Sumatra, where he found the Titan Arum or Corpse Flower *Amorphophallus titanum*, the world's largest flower. Seeds of it were sent to the Royal Botanic Gardens at Kew and were successfully grown, flowering for the first time in cultivation (1889). He also collected in Ethiopia. Four mammals, six reptiles and an amphibian are named after him.

Becker

Lanternfish sp. *Protomyctophum beckeri* RL Wisner, 1971
Morid Cod sp. *Physiculus beckeri* YN Shcherbachev, 1993
Cutlassfish sp. *Aphanopus beckeri* NV Parin, 1994
Gloved Snailfish *Palmoliparis beckeri* AV Balushkin, 1996

Vladimir Eduardovich Becker (1925–1995) was a Russian ichthyologist at the Institute of Oceanology, Moscow. He was the first to recognize the distinctness of the species of lanternfish and was specifically mentioned for his 'extensive and valuable' studies on myctophid fishes collected by various Russian expeditions.

Beckford

Beckford's Pencilfish *Nannostomus beckfordi* A Günther,1872
[Alt. Golden Pencilfish]

F J B Beckford (1842–1920) of Winchester was a naturalist and one of the early bee-keepers in Hampshire. He presented the pencilfish holotype to the BMNH, it being part of a collection he had made on the coast of Demerara, British Guiana (Guyana).

Beddome

Pencil Weed-Whiting *Siphonognathus beddomei* RM Johnston, 1885

Captain Charles Edward Beddome (1839–1898) was a London-born Australian pastoralist and businessman who bought land in Queensland with a bequest from his father. He was also a magistrate and briefly (1873–1874) a Police Inspector. He then (1874) with Frank Jardine, established a pearl shelling station on Naghir (in the Torres Straits). He later seems to have moved to Tasmania, where he was appointed one of 25 Fisheries Commissioners under the 1889 Fisheries Act to manage Tasmania's fisheries. He wrote such articles as: *Description of two new marine shells dredged off Three Hut Point, D'entrecasteaux Channel, Tasmania* (1881). He also collected ferns in Tasmania, and was a member of the Malacological Society (1893).

Bedot

Madagascan Rainbowfish genus *Bedotia* CT Regan, 1903

Maurice Bedot (1859–1927) was a Swiss expert on Hydrozoans. After studying at the University of Geneva and time at the Naples Zoological Station, he devoted his career to the study of marine organisms. He became Director of the Geneva Natural History Museum. He travelled to Malaysia (1890) where he, with Camille Picet, discovered many new taxa some of which bear his name – such as the spider *Dyschiriognatha bedoti*. Picot died (1893) and Bedot later (1897) married his widow.

Beebe

Southern Whitetail Major *Stegastes beebei* JT Nichols, 1924
Wolftrap Seadevil sp. *Lasiognathus beebei* CT Regan & E Trewavas, 1932
Miragoane Gambusia *Gambusia beebei* GS Myers, 1935
Bluntnose Knifefish sp. *Brachyhypopomus beebei* LP Schultz, 1944
Pink Blenny *Paraclinus beebei* C Hubbs, 1952

Dr Charles William Beebe (1877–1962) was an American naturalist, ornithologist, marine biologist, entomologist, explorer and author. He began his working life looking after the birds at the Bronx Zoo (New York) and became Curator of Ornithology, New York Zoological Society (1899–1952), and Director, Department of Tropical Research (1919). He conducted a series of expeditions for the New York Zoological Society including deep dives in the Bathysphere including (1934) a record descent of 923 metres (3,028 feet) off Nonsuch Island, Bermuda. He set up a camp (1942) at Caripito in Venezuela for jungle studies and (1950) bought 92 hectares (228 acres) of land in Trinidad and Tobago, which became the New York Zoological Society's Tropical Research Station (Asa Wright Nature Centre). He married Helen Elswyth Thane Ricker (1900–1981), who wrote romantic novels (pen name Elswyth Thane). He was known for his scientific writing for the public as well as academia. Many of his writings were popular books on his expeditions, such as: *Two bird-lovers in Mexico* (1905) and *Beneath tropic seas* (1928). He made enough money from them to finance his later expeditions. Two amphibians and a bird are named after him.

Beeblebrox

Viviparous Brotula sp. *Bidenichthys beeblebroxi* CD Paulin, 1995

Zaphod Beeblebrox is a character in Douglas Adams' *Hitchhiker's Guide to the Galaxy* series.

Beggin

Blue Panaque *Baryancistrus beggini* NK Lujan, M Arce Hernández & JW Armbruster, 2009

Chris D Beggin, of Nashville, Tennessee. owns a fish and reptile retail outlet called The Aquatic Critter. He was honoured and thanked "*...for his financial support of the authors' research, ethical ornamental-fish business practices, and influence on the professional development of the first author.*"

Behnke

Snake River Fine-spotted Cutthroat Trout *Oncorhynchus clarkii behnkei* MR Montgomery, 1995

Dr Robert J Behnke (1929–2013), known as 'Dr Trout', was a fisheries biologist who was regarded as the world's authority on salmonid fishes. He served in the US Army (1952–1954) during the Koreaan War. The University of Connecticut awarded his bachelor's degree in zoology (1957) and the University of California, Berkeley both his master's and doctorate (1965) in ichthyology. He joined the US Fish and Wildlife Service, based at Colorado State University (1966–1974), and was a Professor at Colorado State University (1975–1999). He retired as Professor Emeritus of Fishery and Wildlife Biology, Colorado State University. He wrote: *Trout and Salmon of North America* (2002).

Behr

Oriental Cyprinid sp. *Incisilabeo behri* HW Fowler, 1937
[Syn. *Bangana behri*]

Otto Frederick Behr (1861–1934) was an amateur naturalist and ornithologist who was active from the 1890s onwards. With his brother Herman, he owned and operated a lumbering and milling business in Lopez, Sullivan County, Pennsylvania. He supplied specimens to the Academy of Natural Sciences of Philadelphia as well as to Fowler, who named the fish in Behr's honour. The brothers were both honorary members of the Delaware Ornithological Club, affiliated to the Academy of Natural Sciences.

Behre

Characin sp. *Brycon behreae* SF Hildebrand, 1938

Dr Ellinor Helene Behre (1886–1982) was an American zoologist and marine biologist who collected the holotype. Radcliffe College awarded her bachelor's degree and the University of Chicago her doctorate. She joined the faculty of Louisiana State University (1920) where she was a Professor until retiring (1957). She co-wrote: *Breeding Behavior, Early Embryology, and Melanophore Development in the Anabantid Fish, Trichogaster trichopterus* (1953).

Bélanger

Belanger's Croaker *Johnius belangerii* G Cuvier, 1830
Burmese Cyprinid sp. *Osteobrama belangeri* A Valenciennes, 1844

Charles Paulus Bélanger (1805–1881) was a French traveller. His voyage is commemorated in his published journal: *Voyage aux IndesOrientales, par le Nord de l'Europe, les Provinces du Caucase, la Géorgie, l'Arménie et la Perse, suivi de Détails Topographiques, Statistiques et Autres sur le Pégou, les Îles de Java, de Maurice et de Bourbon, sur le Cap-de-BonneEspérance et Sainte Hélène, pendant les Années 1825, 1826, 1827, 1828 et 1829*. His trip was part of the French expedition across Europe to India undertaken in order to make a botanical garden at Pondicherry. The expedition collected vast numbers of specimens of dried and living plants and seeds, as well as fish, birds, crustaceans, molluscs and a few mammals. Bélanger became the Director of the Botanical Gardens in Martinique (1853). A mammal, a bird and a reptile are all named after him.

Belayew

Seabream sp. *Sparidentex belayewi* SL Hora & KS Misra, 1943

Dimitry D Belayew was a Specialist in Fisheries at the Directorate General of Agriculture, in Baghdad, Iraq, where the species was collected. The authors based their description on two collections that were received by the Zoological Survey of India (1941 & 1943), sent with photographs by Belayew. Among his published works was: *The geographical position of Iraq and its effect on the life of fish* (1955).

Belcher

Spottail Spiny Turbot *Psettodes belcheri* ET Bennett, 1831
Belcher's Dragonet *Callionymus belcheri* J Richardson, 1844
Rock Pipefish *Phoxocampus belcheri* JJ Kaup, 1856

Admiral Sir Edward Belcher (1799–1877) was a British explorer. After exploring the Pacific coast of America (1825–1828) he was in command of the 'Samarang' and surveyed the coast of Borneo, the Philippine Islands and Formosa (Taiwan) (1843–1846). He also explored the Arctic (1852–1854) searching for Franklin. He was court-martialled (1854) for abandoning three ships during this search but was acquitted. He was the author of: *The Last of the Arctic Voyages; Being a Narrative of the Expedition in HMS Assistance, under the Command of ... in Search of Sir John Franklin, During the Years 1852–53–54 with Notes on the Natural History by Sir John Richardson* (1855). He was promoted to admiral (1872). A reptile and a bird are named after him.

Belding

Paiute Sculpin *Cottus beldingii* CH Eigenmann & RS Eigenmann, 1891

Lyman Belding (1829–1917) was a professional bird collector who wrote a series of articles about his trips, such as: *Collecting in the Cape Region of Lower California, West* (1877) and *A Part of my Experience in Collecting* (1900). Six birds, a mammal and a reptile are also named after him.

Belinda

Bluntnose Knifefish sp. *Brachyhypopomus belindae* WGR Crampton, CDCM de Santana, JC Waddell & NR Lovejoy, 2017

Dr Belinda Siew-Woon Chang is a biologist who became Associate Professor of Ecology and Evolutionary Biology at the University of Toronto, Canada (2003). Before this she taught at Harvard (1996–1999) and was a post-doctoral fellow at Rockefeller University (1999–2002). Princeton University awarded her bachelor's degree (1988) and Harvard her doctorate (1995). She co-wrote: *Recreating ancestral proteins* (2000).

Beling

Northern Whitefin Gudgeon *Romanogobio belingi* EP Slastenenko, 1934

Dr Demeter (Dimitry) Beling was a German ichthyologist and hydrobiologist who was Professor of Zoology, University of Kiev and Director of the Dnieper Biological Station in the 1920s. He was evacuated to Germany in WW2. He was Professor of Zoology at the University of Göttingen (1949). He was an authority on Ukrainian fishes and co-wrote: *Benthophiloides brauneri n. g., n. sp., ein für das Schwarzmeerbassin neuer Vertreter der Familie der Gobiidae* (1925).

Bell

Bell's Flasher Wrasse *Paracheilinus bellae* JE Randall, 1988

Lori Jane Colin née Bell is a marine biologist. The University of California awarded her BA in Biology (1979) and Florida State University her MSc (1991). Her formative years were spent at Miyake-jima, Japan, where she worked on fish behaviour, especially of the clownfish *Amphiprion clarkii*. She has worked in the tropical Pacific since first going to the Marshall Islands (1981). She farmed giant clams in Papua New Guinea and studied the reproduction of the West Indian Topshell for her masters' degree in the Bahamas. She did contract work for the National Cancer Institute for 22 years, returning to the Pacific (1992) where she has been ever since running the Coral Reef Research Foundation lab in Palau. Through this work her interest in sponge taxonomy and biology developed, broadening her marine career to both fish and invertebrate biology. Her published papers include: *Morphological and genetic variation in Japanese populations of the anemonefish, Amphiprion clarkii* (1982) and, co-written with her husband (Patrick L Colin): *Aspects of the spawning of labrid and scarid fishes (Pisces: Labroidei) at Enewetak, Marshall Islands, with notes on other families* (1991). She also co-authored the description of a new species of *Naso* which she discovered at Enewetak, with Jack Randall: *Naso caesius, a new acanthurid fish from the Central Pacific* (1992). More recently she has co-authored a Palau sponge book: *Splendid Sponges of Palau* (2016). The etymology honoured her as "*...one of the collectors of the type specimens, in recognition of her research on Marshall Islands fishes.*"

Bell-Cross

Gorgeous Barb *Clypeobarbus bellcrossi* RA Jubb, 1965
African Barb sp. *Coptostomabarbus bellcrossi* MFL Poll, 1969
Lake Tanganyika Cichlid sp. *Greenwoodochromis bellcrossi* M Poll, 1976

Graham Bell-Cross (1927–1998) was a South African ichthyologist, zoogeographer and naturalist. He became the first full time ichthyologist at the Natural History Museum of Zimbabwe (1971), having previously worked for the Zambia Department of Game and Fisheries. He went on to become Deputy Executive Director there. He wrote on many East African ichthyological subjects including: *The fishes of Zimbabwe* (1988). He collected the type series of the cichlid. (Also see **Neef**))

Bellemans

African Rivuline sp. *Nothobranchius bellemansi* S Valdesalici, 2014

Marc Bellemans (b.1953) is a Belgian ecologist and ichthyologist and keen killifish aquarist. He graduated (1979) as a fisheries biologist and joined the UN Food and Agriculture Organisation. He has been (since 1993) a private consultant travelling extensively, in particular in Africa. Among his publications is: *Nothobranchius nubaensis (Cyprinodontiformes: Nothobranchiidae) a new annual killifish from Sudan and Ethiopia* (2009) written with S Valdesalici as senior author.

Bellido

Neotropical Rivuline sp. *Spectrolebias bellidoi* DTB Nielsen & D Pillet, 2015

Waldo Bellido Villavicencio is a Bolivian veterinarian and environmentalist, and a friend of the junior author.

Bellingshausen

Eelpout sp. *Lycenchelys bellingshauseni* AP Andriashev & Y Permitin, 1968

Admiral Fabian Gottlieb Thaddeus von Bellingshausen (German) or Faddy Faddeyevich Bellinggauzen (Russian) (1778–1852) was a Russian officer of Baltic German descent in the Imperial Russian navy, who rose to become admiral. He was a cartographer and explorer. He took part in the first Russian circumnavigation and later led another circumnavigation that discovered the Antarctic continent, three days before a British ship did the same.

Belloc

Longfin Bonefish *Pterothrissus belloci* J Cadenat, 1937

Gérard Belloc was Directeur du Laboratoire de l'Office Scientifique et Technique des Pêches Maritimes at La Rochelle (1920s & 1930s). He was Directeur à l'institut océanographique d'Alger (1950s). He was the leading research scientist on the cruise aboard the research vessel 'Tench' during which the holotype was acquired. Among other publications, he wrote a monograph on the species of Hake.

Bellotti

Brotula genus *Bellottia* EH Giglioli, 1883
Argentine Pearlfish *Austrolebias bellottii* F Steindachner, 1881
Senegal Seabream *Diplodus bellottii* F Steindachner, 1882
Dash-dot Tetra *Hemigrammus bellottii* F Steindachner, 1882
Red Pandora *Pagellus bellottii* F Steindachner, 1882
Duckbill Eel sp. *Leptocephalus bellottii* U D'Ancona, 1928

Cristoforo Bellotti (1823–1919) was an Italian ichthyologist and palaeontologist, who was also very interested in the breeding of silkworms. He was President of the Italian Society of Natural Sciences (1902–1903). It is not clear if he collected the fish holotypes or just supplied Steindachner with specimens from his collection at Museo Civico di Storia Naturale, Milan, where he was honourary conservator (1857–1904).

Belon

Needlefish genus *Belone* Linnaeus, 1761

Dr Pierre Belon (du Mans) (1517–1564) (pen name Petrus Bellonius Cenomanus) was a French diplomat, writer, traveller and naturalist who was deeply interested in antiquity and extolled the virtues of the Renaissance. He studied medicine in Paris, being awarded the degree of doctor (1542) then became the pupil of Valerius Cordus the botanist with whom he travelled to Germany. After this he undertook a scientific journey through Greece, Crete, Asia Minor, Egypt, Arabia and Palestine (1546–1549). He published an account of his travels as: *Observations* (1553). He undertook a second journey (1557) in northern Italy and its environs. He wrote widely on ichthyology, ornithology, botany, anatomy, architecture and Egyptology and his other works, often illustrated with anatomical drawings, included: *Histoire naturelle des estranges poissons* (1551), *De aquatilibus* (1553), and *L'Histoire de la nature des oyseaux* (1555). He was murdered by thieves on a return journey to Paris.

Beltran

Blackfin Pupfish *Cyprinodon beltrani* J Álvarez del Villar, 1949

Dr Enrique Beltrán Castillo (1903–1994) was a Mexican naturalist, botanist and zoologist and one of Mexico's first conservationists. He was a student of Alfonso Herrera at the UNAM (National Autonomous University of Mexico) which awarded his first degree, masters (1926) and PhD (1933). He was appointed by Herrera to head up two marine commissions (1923 & 1926) that were established to study and improve the use of Mexico's coastal fisheries. He held a number of academic positions and was Professor of Natural Sciences at the National School of Biological Science (1922–1939). He headed the Department of Protozoology at Mexico's Institute of Health and Tropical Diseases (1939–1952), then became founder Director (1952) of the Mexican Institute of Natural Resources. He was Professor Emeritus (1991). He wrote many papers and several (mainly text) books. The naming of the pupfish was conferred to celebrate his 25 years as a biologist.

Belyaev

Snailfish sp. *Pseudoliparis belyaevi* AP Andriashev & DL Pitruk, 1993

Dr Georgi Mihailovich Belyaev (1913–1994) was a Russian zoologist and oceanologist who specialised in Echinodermata, and in the fauna from the very deepest ocean depths. Together with colleagues from the Institute of Oceanology, he got accepted into Soviet science the system of vertical biological zonality of the world's oceans, and in particular of the abyss and ultraabyss zones. His most important work is: *Hadal Bottom Fauna of the World Ocean* (1972) and he had numerous papers published on the subject, such as: *Deep-sea Trenches and their Fauna* (1989).

Belyanina

Deep-sea Tripod Fish sp. *Bathymicrops belyaninae* JG Nielsen & N Merrett, 1992

Dr Tat'yana Nikolaevna Belyanina is an ichthyologist and illustrator. She worked for the Shirshov Institute of Oceanology, Russian Academy of Sciences in the 1980's. Her work focused on the fish families: Opisthoproctidae, Exocoetidae and Bregmacerotidae. She wrote: *The larvae of some rare mesopelagic fishes from the Caribbean and the Gulf of Mexico* (1981). She was thanked in the etymology as "*...colleague and former shipmate, for kindly providing her illustrations of larval Bathymicrops*". (Also see **Parbevs**)

Bemin

Cameroon Cichlid sp. *Coptodon bemini* DFE Thys van den Audenaerde, 1972

Lake Bermin (sometimes spelled Bemin), Cameroon, is where this cichlid is endemic.

Bemis

Pale Ghost Shark *Hydrolagus bemisi* DA Didier Daget, 2002

Dr William E Bemis is a vertebrate anatomist, currently Professor of Ecology and Evolutionary Biology at Cornell University, the university that awarded his first degree (1976). The University of Michigan awarded his masters (1978) and the University of California, Berkeley his doctorate (1983). He is also Emeritus Professor, Biology Department, University of Massachusetts Amherst. His chief interest is in the evolution of bony and cartilaginous fishes. Among his many papers and longer works is the co-written: *Functional Vertebrate Anatomy* (2001). Didier honoured him as her "...longtime mentor and friend, and a leader in ichthyological research".

Benatti

Corydoras sp. *Corydoras benattii* VC Espíndola, LFC Tencatt, FM Pupo, L Villa-Verde & MR Britto, 2018

Laert Benatti was honoured in the etymology thus: "*The specific name,* benattii, *honours the late Laert Benatti for his humanitarian work, providing fresh water from artesian wells to poor communities in Brazil.*"

Benchley

Ninja Lanternshark *Etmopterus benchleyi* VE Vásquez, DA Ebert & DJ Long, 2015

Peter Benchley (1940–2006) was an American author and screenwriter, best known for writing *Jaws* (1974). Later in life, Benchley came to regret writing such sensationalist literature about sharks and became a supporter of marine conservation. His grandfather was Robert Benchley (1889–1945), the famous wit who visited Venice and cabled his editor, saying 'Streets full of water. Please advise'.

Bendire

Malheur Sculpin *Cottus bendirei* TH Bean, 1881

Major Charles Emil Bendire né Karl Emil Bender (1836–1897), born in Germany, was an American oologist, zoologist and army surgeon. He collected birds' eggs (1860s and 1870s) while stationed at frontier posts throughout the Department of Columbia and was famous for the copious notes he made on everything he observed. Fellow officers sent Bendire feathers and eggs from other posts in the West. He became Honorary Curator of Oology at the Smithsonian (1883) and compiled the two-volume *Life Histories of North American Birds* (1892). He personally oversaw the watercolour illustrations to ensure accuracy. He died of Bright's disease. His remarkable collection of 8,000 eggs is housed at the USNM in Washington. Fans of Westerns might like to know that he also once argued Chief Cochise into a truce. A lake, a mountain in Oregon, a mammal and five birds are named after him.

Benedetto

Benedetto's Pipefish *Corythoichthys benedetto* GR Allen & MV Erdmann, 2008

Benedetto 'Bettino' Craxi (1934–2000) was an Italian politician and Prime Minister of Italy (1983–1987). The new species was named *benedetto* to honour the request of Baroness Angela Vanwright von Berger, who successfully bid to support the conservation of this species at the Blue Auction in Monaco (2007) and has given generously to support Conservation International's Bird's Head Seascape marine conservation initiative. Benedetto Craxi was the 'beloved friend' of the Baroness.

Benine

Characin sp. *Moenkhausia beninei* FCT Lima & IM Soares, 2018

Dr Ricardo C Benine is an evolutionary taxonomist, particularly of fish fauna, who is an assistant professor in the Department of Zoology at São Paulo State University, Brazil. He has published more than fifty scientific papers, articles and book chapters such as: *Taxonomic revision and molecular phylogeny of* Gymnocorymbus *Eigenmann, 1908 (Teleostei, Characiformes, Characidae)* (2015). The etymology says: "*...The specific name honors our dear friend Ricardo C. Benine, and is in recognition for his contributions for the knowledge of characid fishes, and particularly those belonging to the genus* Moenkhausia. "

Benjamin, M

Spiny Cockscomb *Alectrias benjamini* DS Jordan & JO Snyder, 1902
Sleeper Catfish *Entomocorus benjamini* CH Eigenmann, 1917

Marcus Benjamin (1857–1932) was a chemist, educated at the School of MInes, Columbia Unversity. He switched to becoming an editor, and from 1896 he edited the Smithsonin's publications. In WW1 he served in naval intelligence.

Benjamin, P

Ringeye Pygmy-goby *Trimma benjamini* R Winterbottom, 1996
[Alt. Redface Dwarfgoby, Redface Goby)

Peter Benjamin owns Benjamin Film Laboratory (Toronto), and is a keen amateur ichthyologist. He provided Winterbottom with free 35mm film and processing over two decades and has participated in several collecting expeditions.

Benjamin (Crampton)

Bluntnose Knifefish sp. *Brachyhypopomus benjamini* WGR Crampton, CD de Santan, JC Waddell & NR Lovejoy, 2017

Benjamin Thomas David Crampton (b.1972) is a British and Irish diplomat and amateur ornithologist. After graduating in Disaster Management at Coventry University (1998) and working as a Balkans research analyst at the British Foreign Office, Ben went to Macedonia as an adviser to the EU Special Representative for the implementation of the (2001) Ohrid Peace Agreement. He went to Kosovo (2004) as special advisor to the former head of the UN mission there, Soeren Jessen-Pedersen. He became (2007) deputy head of the European Union planning team in Pristina, preparing the ground for the international mission to follow the declaration of independence although he left shortly before independence (2008). He was Head of Office Belgrade-Pristina Dialogue (2009–2013), Team Leader for the EU mission EUCAP Nestor in Somalia & Kenya (2013–2014) and is now Chief of Staff to the EU Special Representative for the Horn of Africa, based in Brussels. With his father, Richard Crampton, an Oxford historian he wrote: *Atlas of Eastern Europe in the Twentieth Century* (1996). He collected the holotype of this species while on an expedition to Peru (2003). The senior author is his brother. (Also see **Crampton**)

Benji

Torrent Catfish sp. *Parachiloglanis benjii* RJ Thoni & DB Gurung, 2018

Dasho 'Benji' Paljor Jigme Dorji (b.1943) is an environmentalist, judge and diplomat who has been a long-standing advocate for the protection of nature in Bhutan (where this catfish is endemic). He is the son of a former prime Minister and cousin to the King. He was Magistrate of Paro (1969–1972), High Court Judge (1972–1974) and First Chief Justice of the High Court of Bhutan (1974–1987). Known as the 'godfather of conservation' he established Bhutan's first NGO, the Royal Society for the protection of nature (1987) to conserve the Black-necked Crane. He served in government as Deputy Minister for Social Services (1988–1991) and Deputy Minister to the National Environment Commission (1994–1997). In between (1992–1994) he was Ambassador to the United Nations and European capitals.

Bennett, ET

Bennett's Butterflyfish *Chaetodon bennetti* G Cuvier, 1831
[Alt. Eclipse Butterflyfish, Bluelashed Butterflyfish]
Bennett's Stingray *Hemitrygon bennetti* JP Müller & FGJ Henle, 1841
[Alt. Frill-tailed Stingray; Syn. *Dasyatis bennetti*]
Spiny Turbot *Psettodes bennettii* F Steindachner, 1870

Bennett's Flyingfish *Cheilopogon pinnatibarbatus* ET Bennett, 1831

Edward Turner Bennett (1797–1836) was a British zoologist, elder brother of the botanist John Joseph Bennett. He practiced as a surgeon in London but was chiefly interested in zoology. He tried to establish an entomological society (1822) that became incorporated in the Linnaean Society and eventually the Zoological Society of London of which he was Secretary (1831–1836). He wrote a number of works including: *The Tower Menagerie* (1829), but his most significant ichthyological work was co-contributing the section on fish in: *Zoology of Beechey's Voyage* (1839). His anonymous contributions to batoid literature, which appear in a (1830) memoir on the life of Thomas Stamford Raffles, are cited several times by the authors of the stingray description. The turbot was named after him as Bennett had described the related species *Psettodes belcheri*. Five mammals and a bird are also named after him.

Bennett, JW

Bennett's Sharpnose Puffer *Canthigaster bennetti* P Bleeker, 1854

John Whitchurch Bennett (1790–1853) was a British military officer, official, printer and naturalist. He was in the Royal Marines (1806–1815) then transferred to the army and was posted to Ceylon (Sri Lanka). He served there (1816–1819) and became a civil servant (1819–1827) but left under a clound, having been accused of financial mismanagement. He was a printer in London but suffered bankruptcy (1839) and was sent to Fleet Prison. He was the author of: *A selection from the most remarkable and interesting fishes found on the coast of Ceylon* (1828–1830), whose illustration of this species, Bleeker said, captured its salient features. He named the Great Trevally *Scomber heberi*, in honor of Reginald Heber (q.v.), Bishop of Calcutta, who had supported his ichthyological research. He also wrote: *Ceylon and Its Capabilities: An Account of Its Natural Resources, Indigenous Productions, and Commercial Facilities* (1843).

Bennett, MVL

Bluntnose Knifefish sp. *Brachyhypopomus bennetti* JP Sullivan, JAS Zuanon & C Cox Fernandes, 2013

Dr Michael Vander Laan Bennett (b.1931) is Professor of Neuroscience, Albert Einstein College of Medicine, Bronx, New York. He was an undergraduate at Yale University, which awarded his BSc in zoology (1952). He was a Rhodes Scholar at Oxford where he was awarded his DPhil in physiology (1957), after which he was a research associate (1958–1959), Assistant Professor (1959–1961) & Associate Professor (1961–1966) in neurology at Columbia University, New York. He later moved to the Albert Einstein College where he was Professor of Anatomy (1967–1974) and Professor of Neuroscience (1974) among other posts and is now Distinguished Professor (2005). He carried out pioneering research (1960s) on electric organs and electroreceptors in fish and on gap junctions as one class of electrical synapses. He has published more than 300 papers such as the co-written: *Electrophysiology of electric organ in Gymnotus carapo* (1959) and *Gap junctions and electrical synapses* (2009). He has been an editor of a number of journals including the Journal of Neurobiology and Journal of Neuroscience. He reported studying a knifefish with a monophasic electric organ discharge that is likely to have been this eponymous species.

Bennett, TH

Bennett's Perchlet *Plectranthias bennetti* GR Allen & F Walsh, 2015
Lemon-striped Pygmy Hogfish *Bodianus bennetti* MF Gomon & F Walsh, 2016
Bennett's Fairy Basslet *Tosanoides bennetti* GR Allen1 & F Walsh, 2018

Timothy Horace Bennett (b.1960) is an Australian diver and marine aquarium collector. He began competitive spear-fishing when he was 15, and went on to become a professional fisherman (1986). As well as diving for crayfish, mussels and pearl shells he collects living specimens for the aquarium trade. Nowadays he specialises in collecting deep-water fish using a rebreather. He caught the basslet holotype.

(As an aside, the vernacular name 'Lemon-striped Pygmy Hogfish' refers not only to the unique yellow pattern present in many individuals but also the great assistance generously provided by 'Lemon' Yi-Kai Tea in documenting this species)

Benoit

Benoit's Lanternfish *Hygophum benoiti* A Cocco, 1838

Luigi Benoit (1804–1890) was an Italian naturalist and marine biologist, ornithologist and conchologist who specialised in Sicilian fauna. He was also a political reformer and was sentenced to 18 years imprisonment (1828) but was released (1832). He wrote: *Ornitologia Siciliana* (1840). He was a friend of the author.

Benone

Bluntnose Knifefish sp. *Hypopygus benoneae* LAW Peixoto, GM Dutra, CD de Santana & WB Wosiacki, 2013

Naraiana Loureiro Benone is a biologist and zoologist who has a bachelor's degree (2010), and a master's degree (2012) both awarded by Universidade Federal do Pará, Belem, Brazil, where he is now a PhD student. She co-wrote: *Effect of waterfalls and the flood pulse on the structure of fish assemblages of the middle Xingu River in the eastern Amazon basin* (2015). She collected most of the type series of this species.

Benson

Lanternfish sp. *Lampanyctus bensoni* HW Fowler, 1934

Richard Dale Benson Jr (1876–1949) of Philadelphia, "…t*o whom Fowler was indebted for many collections of American fishes*". Benson was also able to obtain (1939) for the Academy of Natural Sciences of Philadelphia an intact specimen of the Pygmy Sperm Whale, *Kogia breviceps*. Fowler seems to have been successful in organising the intelligentsia of New England into showering him with collections of fishes. It is very noticeable how often he expresses his thanks and, mostly, couched in very similar terms!

Bentinck

Araucanian Herring *Strangomera bentincki* JR Norman, 1936

Victor Frederick William Cavendish-Bentinck (1897–1990) was a British diplomat and businessman. He entered the Diplomatic Service (1919) and served in Paris, Athens (1932) and Santiago de Chile (1933) where he accumulated and donated a collection of Chilean marine fishes to the BMNH. The collection contained the herring holotype. He was then British Ambassador to Poland (1945–1947). He became a businessman after resigning from the Diplomatic Service and on the death of his elder brother (1980) he succeeded to the family title as the 9th Duke of Portland.

Bentley, AC

Many-ray Cusk *Microbrotula bentleyi* ME Anderson, 2005

Andrew Charles Bentley is an ichthyologist. The University of Witwatersrand (1988) and the University of Port Elizabeth, South Africa (1991) both awarded him bachelor's degrees and the University of Port Elizabeth also his master's (1996). He was Collections Manager, Ichthyology, South African Institute for Aquatic Biodiversity (1997–2001) and is now Collections Manager, Ichthyology, University of Kansas Biodiversity Institute, Lawrence. He co-wrote *Sublittoral sand dollar* (Echinodiscus bisperforatus) *communities in two bays on the South African south coast* (1994).

Bentley, WH

Bentley's Mormyrid *Marcusenius bentleyi* GA Boulenger, 1897

Dr William Holman Bentley (1855–1905) was a Baptist missionary in the Congo (Zaire) (1879–1884 & 1886–1900). He published: *Dictionary and Grammar of the Kongo Language* (1887), a work that is still used today, and translated the New Testament into Kikongo (1893). For these and other achievements the University of Glasgow awarded him an honorary doctorate of divinity. He acquired the mormyrid holotype at Stanley Falls.

Bentos

Ornate Tetra *Hyphessobrycon bentosi* M Durbin Ellis, 1908

Colonel Bentos was a volunteer collector on the Thayer Expedition (1865–1866) during which the holotype was collected. We have been unable to find further information.

Ben-Tuvia

Cardinalfish subgenus *Bentuviaichthys* JLB Smith, 1961
Ben-Tuvia's Goby *Didogobius bentuvii* PJ Miller, 1966
Xenia Coral Pipefish *Siokunichthys bentuviai* E Clark, 1966
Twohorn Gurnard *Lepidotrigla bentuviai* WJ Richards & VP Saksena, 1977
Ben-Tuvia's Deepwater Dragonet *Callionymus bentuviai* R Fricke, 1981
Goby sp. *Lobulogobius bentuviai* M Goren, 1984
Eilat Electric Ray *Heteronarce bentuviai* A Baranes & JE Randall, 1989

Dr Adam Ben-Tuvia (1919–1999) was born in Krakow, Poland, but died in Jerusalem, Israel. He initially studied Agriculture at the Yaggiellonian University of Krakow (1937–39), interrupted by WW2. After the war he settled in Palestine, changing his studies to Biology and Zoology at the Hebrew University of Jerusalem (1944–1947), where he was awarded his MSc (1947). He was employed (1949–1964) at the Sea Fisheries Research Station, Haifa (Ministry of Agriculture). Here he specialised in fisheries biology, exploratory fishing and fish systematics. During this time, he finished his PhD at the Hebrew University (1955). He then worked (1964–1968) at the FAO Department of Fisheries, Fishery Resources and Exploitation Division before returning to the Sea Fisheries Research Station in Haifa (1969–1972). When his mentor Professor Heinz Steinitz passed away (1972), he moved to the Zoological Department of the Hebrew University of Jerusalem, where he taught ichthyology and took care of the extensive fish collection. He wrote numerous scientific papers in the field of systematics, biology and zoogeography. He described nine fish species as new to science.

Berbixe

Bumblebee Catfish sp. *Microglanis berbixae* I Tobes, A Falconí-López, J Valdiviezo-Rivera & F Provenzano-Rizzi, 2020

'Berbixe' is the affectionate nickname of María Resurección Sesma Lizari, mother of the senior author Ibon Tobes Sesma. She was honoured "*for its* [sic] *unconditional support and infinite love.*"

Berdmore

Sisorid Catfish sp. *Exostoma berdmorei* E Blyth, 1860
Burmese Loach *Lepidocephalichthys berdmorei* E Blyth, 1860
Sheatfish (catfish) sp. *Pterocryptis berdmorei* E Blyth, 1860
Blyth's Loach *Syncrossus berdmorei* E Blyth, 1860

Major Hugh Thomas Mathew Berdmore (1811–1859) was commissioned into the Bengal Artillery (1830) and served in Burma (Myanmar). He left the army and joined the civil administration as Assistant Commissioner in Northern Tenasserim (Myanmar). At the time of his death he was deputy Commissioner of Martaban (now Mottama). He amassed a large natural history collection which he presented to the Asiatic Society of Bengal (1856), and which Blyth reported on. Two reptiles, an amphibian and two mammals are also named after him.

Berezovsky

Stone Loach sp. *Homatula berezowskii* A Günther, 1896

Mikhail Mikhailovitch Berezowski (or Berezovsky) (1848–1911) was a Russian explorer, geographer and zoologist, who graduated in zoology from the Biology Faculty of St Petersburg University. He was an associate of his more famous contemporary Valentin Bianchi, together with whom he wrote: *Aves Expeditionis Potanini per Provinciam Gansu et Confinia 1884–1887* (1905). He took part in fourteen expeditions as zoologist and botanist; he was in northwest Mongolia (1876–1878), Gansu and Szechuan (Sichuan), China (1884–1887) and again (1892–1894). He was head of expeditions to China and Central Asia (1902–1908) including an expedition to Kuçar, Turkestan (1905–1907). He collected many written items as well as natural history specimens and his collection of 1,876 oriental text items is held in the Institute of Oriental Manuscripts in St Petersburg. Many of his ornithological specimens are now at the Zoological Museum of Moscow University. A reptile, a mammal and two birds are named after him.

Berg, FGC

Characin genus *Bergia* F Steindachner, 1891
Long-whiskered Catfish genus *Bergiaria* CH Eigenmann & AA Norris, 1901
Goosehead Scorpionfish *Scorpaena bergii* BW Evermann & MC Marsh, 1900
Naked Characin *Gymnocharacinus bergii* F Steindachner, 1903
Pufferfish sp. *Pao bergii* CML Popta, 1905
Blotched Sand Skate *Psammobatis bergi* TL Marini, 1932
Castaneta *Nemadactylus bergi* JR Norman, 1937

Dr Frederico Guillermo 'Carlos' (Friedrich Wilhelm Carl) Berg (1843–1902) was a Latvian entomologist and naturalist. After a number of years working in commerce, he became a conservator of entomological specimens at the Riga Museum. He joined the Museo Nacional, Buenos Aires (1873) and went on expeditions to Patagonia (1874) and Chile (1879). He worked at the Museo Nacional de Historia Natural, Montevideo (1890–1892) and was Director, Museo Nacional, Buenos Aires (1892–1901). He was honoured in the scorpionfish's name for "…his excellent work on South American fishes." Popta gives no etymology for the pufferfish, but it seems likely that Frederico Berg is intended. A mammal, an amphibian and three birds are named after him.

Berg, LS

Sculpin sp. *Taurocottus bergii* VK Soldatov & MN Pavlenko, 1915
Issyk-Kul Dace *Leuciscus bergi* DN Kashkarov, 1925
Volga Dwarf Goby *Hyrcanogobius bergi* BS Iljin, 1928
Lumpfish sp. *Cyclopteropsis bergi* AM Popov, 1929
Stone Loach sp. *Oxynoemacheilus bergianus* AN Derjavin, 1934
Bulgarian Barbel *Barbus bergi* G Chichkoff, 1935
Berg's Lake Sculpin *Limnocottus bergianus* DN Taliev, 1935
Bottom Skate *Bathyraja bergi* VN Dolganov, 1983

Lev (Leo) Semionovitch (Semenovich) Berg (1876–1950) was a Russian geographer and zoologist born in Bender, Moldavia. He established the foundations of limnology in Russia with his systematic studies on the physical, chemical, and biological conditions of fresh waters, particularly of lakes. His work in ichthyology was also noted in the palaeontology, anatomy, and embryology of Russian fish. He wrote: *Natural Regions of the USSR* (1950). He was honoured in the name of the skate having been the first to informally describe it (See also **Leo Berg**).

Berlanga

Galapagos Thread Herring *Opisthonema berlangai* FH Berry & I Barrett, 1963

Father Tomás de Berlanga (1487–1551) was the fourth Bishop of Panama (1534). He sailed to Peru to sort out a dispute (1535) but the wind dropped and ocean currents carried the ship to some islands. As he reported his discovery to King Charles V of Spain, he is credited with discovering what are now known as the Galapagos Islands. He returned to Spain (1537) accompanied by his pet caiman.

Bernacchi

Emerald Rockcod *Trematomus bernacchii* GA Boulenger, 1902

Louis Charles 'Bunny' Bernacchi (1876–1942) was a physicist and astronomer who took part in several expeditions to Antarctica. His father was Italian, but he was born in Belgium and as a child (1883) moved to Hobart, Tasmania. He studied astronomy, magnetism, meteorology and physics at Melbourne Observatory. He stood (1910) unsuccessfully, as a Liberal for the UK parliament and served in the Royal Navy during WW1 including in the anti-submarine division and the American destroyer squadron, rising to the rank of Lt. Commander. He was physicist on the (1898) expedition that was the first to endure winter on the Antarctic continent. He was also physicist on Scott's expedition to Antarctica (1901–1904). There Cape Bernacchi, Bernacchi Head and Bernacchi Bay are all named after him. He succeeded Shackleton as editor of the South Polar Times, and imparted his Antarctic findings in scientific writings and by travelling in Tasmania giving presentations. He was one of the organisers of the British Polar Exhibition (1930) in London and founded the Antarctic Club. Among his books are: *To the South Polar Regions* (1901); *The Polar Book* (1930); *A very gallant gentleman* (1933); and *Saga of the Discovery* (1938).

Bernardino

Yaqui Sucker *Catostomus bernardini* CF Girard, 1856

This is a toponym referring to Rio de San Bernardino, Sonora, Mexico.

Bernatzik

Oriental Barb sp. *Opsarius bernatziki* FP Koumans, 1937

Hugo Adolph Bernatzik (1897–1953) was an Austrian anthrioplogist, travel writer and photographer. He served in the Austro-Hungarian army in WW1 and after the war set out to be a physician, but abandoned his studies (1920) to become a businessman. His first wife died (1924) and he thereafter travelled widely taking photographs and became a professional photographer. His travels (1924–1937) encompassed Spain, Mauretania, Egypt, Somalia, Sudan, Romania, Albania, Portuguese Guinea, New Guinea, Solomon Islands, Bali, Lapland, Burma (Myanmar), Thailand, Vietnam, Cambodia and Laos. He had plans to visit China, but the outbreak of WW2 meant he was in Austria for its duration. After WW2, he visited Morocco (1949–1950). He studied ethnology, geography and anthropology at The University of Vienna, which awarded him a doctorate (1932). He applied (1935) for habilation to become a Professor at the University of Graz. The application was accepted (1936) and he was appointed to the Institute of Geography at the University of Graz (1939). Among his writings are: *Südsee. Travels in the South Sea* (1935) and *The Spirits of the Yellow Leaves* (1938). He collected the holotype of this species. He died from a tropical disease

Berndt

Blotcheye Soldierfish *Myripristis berndti* DS Jordan & BW Evermann, 1903
Berndt's Moray Eel *Gymnothorax berndti* JO Snyder, 1904
Pacific Beardfish *Polymixia berndti* CH Gilbert, 1905

E Louis Berndt (b.1851) was born in Germany and emigrated to Hawaii (1883). He was described in the eel's etymology as "*the efficient inspector of fisheries in Honolulu*"; a position to which he was appointed (1900). He supplied a number of fish specimens, which he had found in the Honolulu fish market, to Jordan.

Berney

Berney's Shark Catfish *Neoarius berneyi* GP Whitley, 1941

Frederic Lee Berney (1865–1949) was an Australian naturalist and ornithologist who was an honorary member and President of the RAOU (1933–1934). He was born in Croydon, England and emigrated to Australia about 1890. He worked on various properties in Queensland until enlisting in the army in WW1 and served in Egypt with the 2nd Light Horse Regiment (1916–1917). After returning from the war he bought Barcarolle Station, Queensland where he lived (1919–1939) and then retired to Melbourne. It appears he got bored with city life as he returned to Queensland, taking a book-keeper's job on a station in the west of the state. A bird genus and three bird species are named after him.

Bernhard

Bernhard's Elephant-snout fish *Mormyrus bernhardi* J Pellegrin, 1926

R P Bernhard collected the holotype near Nairobi, Kenya. Our enquiries have yielded nothing further.

Berovides

Cuban Killifish sp. *Rivulus berovidesi* RR Silva, 2015

Professor Dr Vicente Berovides Alvarez (b.1941) is a Cuban biologist, environmentalist and conservationist. He was a genetics specialist, animal behaviourist and evolutionary biologist at the Faculty of Biology, Universidad de la Habana, Cuba. He is a specialist in Population Biology and Conservation of Biotic Resources and has developed more than twenty research projects on Animal Genetics, Ecology and Tourism, and conservation of species. He has also collaborated with 180 national and foreign publications and is the author of nine books on various topics of the environment. He was honoured "*…in recognition of his life-long dedication and contribution to train several generations of new researchers in biological sciences.*"

Berrisford

Onrust Clipfish *Clinus berrisfordi* ML Penrith, 1967

C D Berrisford was a marine biologist who worked at the Natal Regional Laboratory of CSIR, University of Cape Town. He collected the type with Michael Penrith, the husband of the author Mary Louise Penrith of the State Museum of Namibia. He wrote: *Biology and zoogeography of the Vema seamount: A report on the first biological collection made on the summit* (1969) and co-wrote: *The ecology and chemistry of sandy beaches and nearshore submarine sediments of Natal: Pollution criteria for sandy beaches in Natal* (1967) when he was associated with the National Water Research Institute of Fountain Valley, USA. A marine worm, which he also collected, is also named after him.

Berry (Levy)

Whiptail sp. *Pentapodus berryae* GR Allen, MV Erdmann & WM Brooks, 2018

Beryl 'Berry' Rae Levy is the stepdaughter of the third author William M Brooks. The etymology says: "*The new species is named berryae in honor of Beryl ("Berry") Rae Levy, the third author's stepdaughter.*" (Also see **Stuart (Brooks)**, **Jack Brooks** & **Bill (Brooks)**)

Berry, FH

Slope Bass *Symphysanodon berryi* WD Anderson, 1970
Lizardfish sp. *Ahliesaurus berryi* EOG Bertelsen, G Krefft & NB Marshall, 1976
Berry's Grenadier *Mesovagus berryi* CL Hubbs & TN Iwamoto, 1977

Frederick Henry Berry (1927–2001) was an American systematic ichthyologist. He has also been described as a *bon vivant*, wise counsellor, gambler, benefactor, disputant, careful scientist, charmer and friend by Collette and W D Anderson Jr in his obituary. He left high school (1944) and served two terms with the US Navy Reserve (1945–1946 & 1951–1952). Florida University awarded his BS (1954) and MS (1955). He began taxonomic research on fish at the US Bureau of Commercial Fisheries (1956–1969) partly in California at Scripps. He spent several years trying to develop commercial farmed pompano in South Florida, then (1973) worked in marine monitoring. He led a group working on Bluefin Tuna (1975–1979) and then on endangered species protection until retirement (1978–1987). However, he was re-hired by NMFS to work for IOCARIBE working in Colombia and the Caribbean. Even after 1997 when under medical care he continued with his work. He described 11 new species and wrote many papers on fish and sea turtles, from: *Food of the Mudfish* (1955) to *MEXUS-Gulf Sea Turtle Research 1977–85* (1987) and 45 botanical papers. He died of lung cancer.

Bertel

Dreamer genus *Bertella* TW Pietsch, 1973
Onejaw sp. *Monognathus berteli* JG Nielsen & KE Hartel, 1996
Gulper Eel sp. *Saccopharynx berteli* KA Tighe & JG Nielsen, 2000

(See **Bertelsen** below)

Bertelsen

Bertelsen's Lanternfish *Diaphus bertelseni* BG Nafpaktitis, 1966
Barbeled Dragonfish sp. *Eustomias bertelseni* RH Gibbs, TA Clarke & JR Gomon, 1983
Flabby Whalefish sp. *Parataeniophorus bertelseni* TA Shiganova, 1989
Pike Conger Eel sp. *Gavialiceps bertelseni* ES Karmovskaya, 1993
Morid Cod sp. *Physiculus bertelseni* YN Shcherbachev, 1993
Slickhead sp. *Conocara bertelseni* YI Sazonov, 2002

Dr Erik 'Bertel' Bertelsen (1912–1993) was a celebrated Danish fisheries biologist and ichthyologist at the Zoological Museum of the University of Copenhagen. He is regarded as the all-time world authority on ceratioids (anglerfishes) of which he wrote: *The Ceratioid Fishes; Octogeny, Taxonomy, Distribution and Biology* (1951) as well as a number of authorative papers and several beautifully illustrated monographs. He went on a number of expeditions including the cruise of the Russian research vessel 'Vityaz'

(1988). In his etymology Karmovskaya described him as a "*leading ichthyologist and outstanding person.* "

Bertha Lutz

Stingray sp. *Hypanus berthalutzae* FF Petean, GJP Naylor & SMQ Lima, 2020

Bertha Maria Júlia Lutz (1894–1976) was a Brazilian zoologist, politician and pioneering feminist, who founded the Brazilian Federation for Feminine Progress (1922). Her father was a Swiss physician and her mother an English nurse. She studied zoology at the Sorbonne, Paris, and was a member of the Brazilian Parliament for a short period (1936–1937) until the coup d'état by Getúlio Vargas. Her main zoological interest was in amphibians. The stingray was named after her to represent 'female empowerment in Brazil'. Two lizards and three frogs are also named after her.

Berthold

Berthold's Killi *Scriptaphyosemion bertholdi* E Roloff, 1965

Karl Berthold was a German aquarist. He travelled quite widely, managing to collect in some difficult locations (1930s). He was part of Erhard Roloff"s group that collected or were breeders of tropical fish. He was honoured as a "*...diligent Aphyosemion breeder*" (translation). Roloff mentions that Karl Berthold was the only one to maintain *Aphyosemion filamentosum* throughout WW2 but lost all his stock (Christmas Day 1947). Some reports have mentioned this was due to a heating problem...which is true to a point. Roloff stated that Berthold was carrying a tank containg 80 young which slipped out of his hands onto another glass tank containing the brood stock. Both tanks broke up, sending the fish over a heating unit which killed all the fish immediately. This effectively wiped out all of this particular form.

Bertin

Conger Eel sp. *Bathycongrus bertini* M Poll, 1953
Thread Eel *Serrivomer bertini* M-L Bauchot, 1959
Onejaw sp. *Monognathus bertini* EOG Bertelsen & JG Nielsen, 1987

Dr Léon Bertin (1896–1954) was a French zoologist and ichthyologist at the Muséum National d'Histoire Naturelle, Paris and a noted scientific illustator. His BSc (1917), masters (1920) and doctorate were awarded (1925) by the Ecole Normale Superieure, the latter for his thesis on sticklebacks: *Recherches bionomiques, biométriques et systématiques sur les épinoches (Gastérostéidés).* He studied geology under Alfred Lacroix and invertebrate biology under Louis Bouvier at the NMNH and (1938) worked in the herpetology laboratory there, later succeeding Jacques Pellegrin (q.v.). He headed the Zoological Society of France (1949). Bauchot, who described the Thread Eel, was his assistant (1952). He wrote and illustrated: *Catalogue des types de poissons du Muséum national d'histoire naturelle* (part 5) (1950) and *Les Poissons singuliers* (1954) among more than 90 written or co-written papers and longer works. He died in a car accident.

Bertoni

Parasitic Catfish sp. *Paravandellia bertoni* CH Eigenmann, 1917
[Junior Synonym of *Paravandellia oxyptera*]

Arnoldo de Winkelried *Bertoni* (1879–1973) was a Paraguayan zoologist. He was born in Switzerland and arrived in Missiones, Argentina (1884), later moving to Paraguay (1886). He was Professor of Zootechnics and Zoology, School of Agriculture and Model Farm (1903–1906) and (1930s) Professor of Zoology, Animal Science, Entomology and Phytopathology, School of Agriculture. He was also a member of the Scientific Society of Paraguay. He wrote *Fauna paraguaya, catalogos sistematicos de los vertebrados del Paraguay* (1914). His father was Dr Moises Santiago Bertoni (1857–1929) who established the Swiss colony 'William Tell' on the Parana River on the border with Argentina. The site is now called Puerto Bertoni.

Besnard

Polkadot Catshark *Scyliorhinus besnardi* S Springer & V Sadowsky, 1970

Professor Wladimir Besnard (1890–1960) was the founder and first Director of Instituto Oceanografico da Universidade de Sao Paulo. Born in France of Russian origin he studied zoology and comparative anatomy in Kiev. The Governor of Sao Paulo State invited him to Brazil where he organised and headed what became the Instituto. There he researched marine biology and physical oceanography. A Brazilian oceanographic research vessel is named 'Professor Wladimir Besnard' after him and he is honoured in the names of at least eight marine organisms. The type of the catshark was trawled by the institute's research vessel.

Besouro

Glass Knifefish sp. *Eigenmannia besouro* LAW Peixoto & WB Wosiacki, 2016

Manoel Henrique Pereira (1895–1924) was known by his nickname, Besouro Mangangá (The Mangangá Beetle). He came from the Recôncavo region of Bahia, Brazil (where this knifefish occurs), and was a legendary figure in the Afro-Brazilian martial art capoeira.

Beta

Danio genus *Betadevario* PK Pramod, F Fang, K Rema Devi, TY Liao, TJ Indra, KS Jameela Beevi & SO Kullander, 2010

Beta Mahatvara aided the authors, but the genus is only partly named after him as it closely resembles the genus *Devario* and Beta is Greek for second. However, the name is also a reference to Beta Mahatvara, who, in the words of the etymology "*...made great efforts to make the material available for this study*."

Bethan

Corydoras sp. *Corydoras bethanae* RF Bentley, S Grant & LFC Tencatt, 2021

Bethan is the daughter of one of the authors, British aquarist Steven Grant, a specialist in catfish and loaches who has kept fish for twentyfive years.

Beuchey

Neotropical Rivuline sp. *Moema beucheyi* S Valdesalici, DTB Nielsen & D Pillet, 2015

Michael Beuchey is an environmentalist and was one of the collectors of this species along with the authors, Jean Marc Beltramon and Christine Lambert. They have often collected together in the Bolivian Amazon.

Bhanot

Pipefish genus *Bhanotia* Hora, 1926

Kali Das Bhanot. Dr Sunder Lal Hora's etymology says the genus was named after his 'esteemed friend'.

Bhimachar

Stone Loach sp. *Schistura bhimachari* SL Hora, 1937

Dr Brahmananda Srinivasachar Bhimachar (1906–1979) was a lecturer at the Department of Zoology, Intermediate College, University of Mysore, Bangalore, India in the 1930s. He was a specialist in applied zoology and fisheries and was awarded his doctorate by the University of Calcutta (1946). He was Fisheries Officer, Government of Mysore (1940–1947) and worked for the Central Marine Fisheries Research Institute, Barrackpore as Senior Research Officer (1947–1953) and as Director (1954–1966). He wrote: *A Study of the Medulla Oblongata of Cyprinodont Fishes with Special Reference to their Feedng Habits* (1937). He collected the holotype of this species. Among his other interests was astrology.

Bickell

Sixgill Stingray *Hexatrygon bickelli* PC Heemstra & MM Smith, 1980

David 'Dave' Bickell was a South African journalist, formerly the angling correspondent for the 'Eastern Province Herald'. The holotype of this stingray was a near-perfect female specimen washed up on a beach at Port Elizabeth in South Africa (1980) where Bickell discovered it. He became a volunteer worker for the protection of turtles.

Biddulph

Snowtrout sp. *Schizothorax biddulphi* A Günther, 1876

Sir John Biddulph (1840–1921) was an army colonel and naturalist who originally was commissioned into the 19th Lancers (1858) and fought during the Indian Mutiny. He was subsequently in the political department of the government of British India, and was a member of the Second Yarkand Mission across the Himalayas (1873–1874). During this expedition he collected zoological specimens. He was posted to Gilgit, Kashmir (1877–1881) and later served on the staff of the Viceroy of India (1892–1896). He wrote: *The Pirates of Malabar and an Englishwoman in India Two Hundred Years Ago* (1907). Three birds are named after him.

Biden

Brotula genus *Bidenichthys* KH Barnard, 1934

C Leo Biden of Cape Town was described as a *'knowledgeable angler'*, and a person "*…to whom the South African Museum is indebted for many specimens and much information.*" He wrote: *Sea-Angling Fishes of the Cape* (1948).

Biff

Sea Catfish sp. *Notarius biffi* R Betancur-Rodríguez. & A Acero Pizarro, 2004

Dr Eldredge 'Biff' Bermingham is an American biologist and biogeographer. Cornell University awarded his bachelor's degree (1977) and the University of Georgia his doctorate (1988) after which he returned to Cornell as a postdoctoral fellow (1987–1989). He worked for the Smithsonian (1988–2013), first as a Staff Biologist (1988–2001), then as a Senior Scientist (2002–2013) during which time he was Director, Smithsonian Tropical Research Institute (2007–2013). He is now Chief Science Officer, Patricia and Philip Fost Museum of Science, Miami, which he joined (2014).

Bigelow

Barbeled Dragonfish sp. *Eustomias bigelowi* WW Welsh, 1923
Deep-sea Tripod Fish sp. *Bathypterois bigelowi* GW Mead, 1958
Barracudina sp. *Lestidium bigelowi* MJF Graae, 1967
Bigelow's Ray *Rajella bigelowi* M Stehmann, 1978
Blurred Lantern Shark *Etmopterus bigelowi* Shirai Shigeru & H Tachikawa, 1993

Dr Henry Bryant Bigelow (1879–1967) was an American zoologist. After graduating at Harvard (1901) he worked with the ichthyologist Alexander Agassiz (q.v.) and accompanied him on several major marine expeditions including one on 'Albatross' (1907). He was Curator of the Museum of Comparative Zoology (1905) then joined the Department of Invertebrate Zoology, Harvard (1906–1966) and founded WHOI (Woods Hole Oceanographic Institution) (1930) and became its first Director (1931–1940). He published several books and over 100 scientific papers, including co-writing: *Fishes of the Gulf Main* (1953). At least twenty-six species including more than fifteen marine organisms are named after him as is the NOAA research vessel 'Henry B Bigelow'. He was honoured in the name of the Lantern Shark as he had first described it but misidentified it as *E. pusillus.*

Bigerri

Adour Minnow *Phoxinus bigerri* M Kottelat, 2007

This is not a true eponym, but named after a people; the Bigerri, a tribe in ancient Gaul (France) in whose former territory the holotype was acquired.

Bigorne

African Barb sp. *Enteromius bigornei* C Lévêque, G Teugels & DFE Thys van den Audenaerde, 1988

Rémy Bigorne (b.1954) is a French ichthyologist and hydrobiologist at MNHN, Paris. He worked for ORSTOM (1980s) which became IRD where he is an engineer and currently concentrates on the systematics and biogeography of Neotropical freshwater fishes. He co-wrote: *Note sur la systématique des* Petrocephalus *(Teleostei, Mormyridae) d'Afrique de l'Ouest.* (1991).

Bill

Triplefin sp. *Helcogramma billi* PEH Hansen, 1986

Dr William Farr 'Bill' Smith-Vaniz (b.1941) collected all the specimens examined by the author. (See **Smith-Vaniz**)

Bill Brooks

Citron Mudgoby *Priolepis billbrooksi* GR Allen, MV Erdmann & William M Brooks, 2018

William 'Bill' Mathews Brooks is the third author's son. (Also see **Jack Brooks** and **Stuart (Brooks)** & **Berry (Levy)**)

Billsam

Rattail sp. *Hymenocephalus billsam* NB Marshall & T Iwamoto, 1973

This species is named for two ichthyologists: William '**Bill**' H Longley (1881–1937) and Samuel '**Sam**' F Hildebrand (1883–1949). The etymology explains: "*...named for ichthyologists William (Bill) H. Longley (1881–1937) and Samuel (Sam) F. Hildebrand (1883–1949), who noted the existence of this species in 1941 but did not describe it.*" [Iwamoto later amended the spelling to "*billsamorum*", believing that to be the correct grammatical ending for a patronym that honours two men, but since the original spelling did not have the the usual patronymic "*i*", we treat it as a noun in apposition that does not require amendment] (See **Longley** and **Hildebrand**)

Billy Kriete

Kriete's Tonguefish *Symphurus billykrietei* TA Munroe, 1998

William H 'Billy' Kriete Jr (d.1988) was a fisheries scientist at the Division of Science & Services, Virginia Institute of Marine Science. Among his published papers are: *Status of American Shad Stocks in Virginia* (1980) and the co-written *An Overview of The Status of Alosa Stocks in Virginia* (1982).

Bilsel

Great Beysehir Spined Loach *Cobitis bilseli* F Battalgil, 1942

Dr Mehmet Cemil Bilsel (1879–1949) was a Turkish lawyer, academic and politician who graduated from the Istanbul School of Law (1903). He taught in various law schools in different parts of the Ottoman Empire (1899–1925), including at the University of Istanbul (1921–1925). He was the first Dean of the newly established Ankara University Faculty of Law (1925–1934). He served in the Turkish parliament (1943–1949). The University of Toulouse awarded him an honorary Doctorate of Law (1948). He was honoured for his support of and interest in the fauna of Turkey.

Bineesh

Sweeper sp. *Pempheris bineeshi* JE Randall & BC Victor, 2015

Short-tail Whipray *Maculabatis bineeshi* BM Manjaji-Matsumoto & PR Last, 2016

Dr K K Bineesh (b.1981) is an Indian biologist who holds a master's degree and a doctorate awarded by Cochin University of Science and Technology, Kerala (2015). He works at the National Bureau of Fish Genetic Resources, Lucknow. He co-wrote: *Redescription of the rare and endangered Broadfin Shark Lamiopsis temminckii (Müller & Henle, 1839) (Carcharhiniformes: Carcharhinidae) from the northeastern Arabian Sea* (2016).

Bingham H

Characin sp. *Ceratobranchia binghami* CH Eigenmann, 1927

Dr Hiram Bingham III (1875–1956) was an academic, explorer and politician. Yale awarded his bachelors' degree (1898), the University of California, Berkeley a further degree (1900) and Harvard his doctorate (1905). He taught history and politics at Harvard and then Princeton and was a lecturer in South American History, and later professor at Yale. He was a delegate at the First Pan American Scientific Congress in Santiago, Chile (1908) and travelled home via Peru where he visited pre-Colombian cities, which led him to organise an expedition. Thus, he was the Director, Yale Peruvian Expedition (1911), which rediscovered the Incan city of Machu Pichu (the fact for which he is most remembered today). He made several further trips to Peru (1912, 1914 & 1915). He wrote *The Lost City of the Incas* (1948). He was captain of the Connecticut National Guard (1916) and became an aviator (1917) and organised the United States Schools of Military Aeronautics at eight universities to provide ground school training for aviation cadets. He served the Aviation Section, US Signal Corps and the Air Service, attaining the rank of lieutenant colonel. In Issoudun, France, Bingham commanded the Third Aviation Instruction Center and subsequently wrote of his wartime experiences in *An Explorer in the Air Service* (1920). He entered politics (1922) as Lieutenant-Governor of Connecticut and became a Senator for Connecticut in the US Congress (1924–1933). Many consider him to be the inspiration for the character *Indiana Jones*. A bird and an amphibian are also named after him. See also **Alfreda**.

Bingham HP

Barbeled Dragonfish sp. *Eustomias binghami* AE Parr, 1927

Wolftrap Seadevil sp. *Thaumatichthys binghami* AE Parr, 1927

Harry Payne Bingham (1887–1955) of Cleveland, Ohio was an American philanthropist and businessman with an interest in natural history. He financed and led a number of expeditions, using his private yacht as base and including in the party, the author, Albert Eide Parr. He founded the Bingham Oceanographic Collection at the Peabody Museum of Natural History, Yale University.

Binh

False Gudgeon sp. *Abbottina binhi* V H Nguyễn, 2001

Bình Nguyễn helped the author, Nguyễn Văn Hảo, of the Research Institute for Aquaculture, to collect the holotype. We have been unable to throw further light on who this person is and whether they are related to the author.

Bini

Mottled Flounder *Pseudorhombus binii* E Tortonese, 1955

Professor Giorgio Bini (1906–?) was a naturalist, gastronome and birdwatcher who was said to be the foremost authority on Mediterranean fish. Among other books, papers and reports he wrote *Freshwater Fish of Italy* (1962). He carried out a fisheries mission along the coasts of Chile and Peru (1949–1950) on behalf of the Central Fisheries, Fish & Hydrobiology Laboratory of Rome. During this he collected 'valuable fish material' now in the Genoa

Museum of Natural History. Among this material, studied by Enrico Tortonese was the flounder which had not previously been described; he dedicated the fish to his friend. Bini was also despatched to Tunisia by the Italian government in the early 1960s to investigate a maritime dispute in the Gulf of Tunis. Alan Davison, a writer on food, said of Bini the gastronome that he "*...weighed a ton and broke three of our dining room chairs!*" Another food writer said that Bini said that the finest tasting fish of the Mediteranean was the Moray Eel.

Biorn

Combtooth Blenny sp. '*Salarias biorni*' HW Fowler, 1946

[Not currently recognised. In *Praealticus*?]

Major Dr Carl Biorn (1915–2005) was a physician who graduated from the University of Minnesota and became an intern at the San Bernardino County Hospital, San Bernardino, California until WW2 intervened. He then joined the US Army Medical Corps, serving as a major in Honolulu at the 147th General Hospital and at a Surgical Portable Hospital in Tinian, Saipan and Okinawa. After the war he resumed a fellowship at the Mayo Clinic and after four years joined Palo Alto Medical Clinic as a urologist. He was also a clinical professor of surgery at Stanford University School of Medicine. Fowler wrote that Biorn: "*...of Minneapolis, who greatly assisted in making the present collection of fishes possible.*"

Bipul

Snakehead sp. *Channa bipuli* J Praveenraj, A Uma, N Moulitharan & H Bleher, 2018

Bipul Das is the proprietor of 'Wild Caught Ornamental Fish' and lives in Guwahati, Assam. He studied biochemistry at Pragiyotish College there. He collected the type with Heiko Bleher. The etymology says that the: "*...specific epithet 'bipuli' is named to honor Bipul Das, who also found recently this new species, and for his immense support in the collection and donating of the live specimens.*" On his facebook page he sums up himself thus: "*Love fish, born for fish...*"

Birchmann

Birchmann's Swordtail *Xiphophorus birchmanni* P Lechner & AC Radda, 1987

[Alt. Sheepshead Swordtail]

Heinz Birchmann was an Austrian aquarist from Vienna who helped the senior author, P Lechner, collect the type specimen (1987) in Hidalgo State, Mexico, during a botanical expedition. He was the first to import the species to Europe and breed it there.

Bird

Birdi Loach *Botia birdi* BL Chaudhuri, 1909

W J A Bird was a colonial officer in the Indian Sub-continent (1894). He was awarded a medal for developing a method of coagulating latex (1906). He was the Superintending Engineer of the irrigation branch of the Punjab Provincial Civil Service at Rupar, Punjab, India when he collected the type (1908). He also managed tea estates (1904–1912) in Ceylon

(Sri Lanka) and appears to have bought his own estate there (1912) while continuing to manage another.

Birdsong

Birdsong's Mudskipper *Boleophthalmus birdsongi* EO Murdy, 1989
Goby sp. *Stiphodon birdsong* RE Watson, 1996
Fin-joined Goby *Gobulus birdsongi* DF Hoese & S Reader, 2001

Dr Ray S Birdsong (1935–1995) was an American ichthyologist. Florida State University awarded his bachelor's and master's degrees and the University of Miami his doctorate. Between school and college, he worked for two years as a pipefitter and then in the army for three years as a cryptanalyst. He joined (1968) the Department of Biological Sciences at the Old Dominion University where, apart from one year (1970–1971) at the National Science Foundation, he spent his whole career. He was honoured in the mudskipper's name "*...for his contributions towards a better understanding of gobioid osteology*" and in a goby's name as "*a long time friend and colleague of the senior author, who provided valuable assistance and inspiration to* [his] *goby work*." While not prolific, he wrote a number of important papers almost all concerning gobioid fishes. He died from 'complications' following viral pneumonia.

Birindelli

Armoured Catfish sp. *Loricaria birindellii* MR Thomas & MH Sabaj Pérez, 2010
Thorny Catfish sp. *Platydoras birindellii* LM Sousa, MS Chaves, A Akama, J Zuanon & MH Sabaj, 2018

Dr José Luís Olivan Birindelli (b.1979) is a Brazilian ichthyologist. The Federal University of São Carlos awarded his bachelor's degree (2004), and Universidade de São Paulo his master's (2006) and doctorate (2012). He is currently an Adjunct Professor of the Department of Animal and Plant Biology at the State University of Londrina. He is also Treasurer of the Brazilian Society of Ichthyology and Editor of the Brazilian Society of Ichthyology Bulletin. He was President of the International Symposium on Phylogeny and Classification of Neotropical Fishes (Londrina, 2017). He was the leader of the Pipe Expedition (2007) to Serra do Cachimbo, Brazil, during which many undescribed fishes, including the holotype of the *Loricaria* species, were discovered. He focuses his research on the morphology and evolution of thorny catfishes (family Doradidae) and headstanders (family Anostomidae). The etymology for the *Platydoras* species says that the name: "*...honors our colleague and friend José Luís Olivan Birindelli, for enriching our knowledge of doradid catfishes and inspiring the next generation of Neotropical ichthyologists with his enthusiasm and integrity*." He has written around 80 papers, book chapters and articles and his major publications are on doradid catfishes that includes the proposal of a phylogenetic hypothesis for Doradidae and Auchenipteridae (2014), the study of morphology of the gas bladder in catfishes (2012), and the taxonomic revisions of several genera.

Birkhahn

Neotropical Rivuline sp. *Cynodonichthys birkhahni* H-O Berkenkamp & VMF Etzel, 1992

Holger Birkhahn is a German aquarist. According to the etymology he was honored for 'intensive' (translation) collecting work during two trips to Panama, including helping to collect the type of this species.

Birtwistle

Glass Goby sp. *Gobiopterus birtwistlei* AW Herre, 1935
Oriental Barb sp. *Poropuntius birtwistlei* AW Herre, 1940

William Birtwistle (1890–1953) was the Director of the Fisheries Department, Singapore and in charge of all Fisheries in the Federated Malay States (1920s–1945). Before WW2 he frequently visited Japan and learned to speak Japanese. He was greatly respected by Japanese ichthyologists and as a result he was not imprisoned during the Japanese occupation (1942–1945), being one of very few British officials who were ordered to carry on research and ensuring that the museum collections and the botanical gardens were not looted and vandalized. In Birtwistle's case, the Japanese administration received a direct order from a well-known Japnese ichthyologist, Emperor Hirohito. This species was named after him as he had greatly helped Herre in his study of Malayan fishes. (Also see **Alice (Birtwhistle)**)

Biruni

Barb/Scraper sp. *Capoeta birunii* H Zareian & HR Esmaeili, 2017

Abū Rayḥān Muḥammad ibn Aḥmad Al-Bīrūnī aka Abū Rayḥān Bīrūnī (973–1048), (Gregorian Callendar: 1566–1638) known as Al-Biruni in English was an Iranian scholar and polymath. He was from Khwarezm, a region which encompasses modern-day western Uzbekistan and northern Turkmenistan, but spent most of his life in what is now central-eastern Afghanistan. Al-Biruni is regarded as one of the greatest scholars of the medieval Islamic era and was well versed in physics, mathematics, astronomy, and natural sciences, and also distinguished himself as a historian, chronologist and linguist. Bīrūnī's catalogue of his own literary production up to his 65th lunar/63rd solar year (the end of 427/1036) lists 103 titles divided into 12 categories: astronomy, mathematical geography, mathematics, astrological aspects and transits, astronomical instruments, chronology, comets, an untitled category, astrology, anecdotes, religion, and books he no longer possesses.

Bischoff

Hagfish sp. *Eptatretus bischoffii* AF Schneider, 1880

Dr Theodor Ludwig Wilhelm Bischoff (1807–1882) was a German physician and biologist who taught anatomy at Heidelberg (1835–1843). He was Professor of anatomy and physiology at Giessen (1843–1855) and Professor of Anatomy at Munich University (1854). His particular study was embryology and he wrote a number of detailed papers on different mammal ovum (1842–1854). He was a colleague of Schneider's.

Bitter

Neotropical Rivuline sp. *Papiliolebias bitteri* WJEM Costa, 1989
African Rivuline sp. *Aphyosemion bitteri* S Valdesalici & W Eberl, 2016

Bloch's Tongue Sole. *Paraplagusia bleekeri* M Kottelat, 2013

Dr Pieter Bleeker (1819–1878) was an ichthyologist and army surgeon commissioned (1841) by the Dutch East India Company. He was apprenticed to an apothecary (1831–1834) and became interested in anatomy and zoology. He qualified as a surgeon at Haarlem (1840) and went to Paris, working in hospitals while attending Blainville's lectures. He was stationed in the Dutch East Indies (Indonesia) (1842–1860) and acquired 12,000 fish specimens, most of which are today in the RMNH, Leiden. He returned to Holland (1860), taking with him his collection, which was sold after his death. He wrote over 500 scientific papers describing 511 new genera and 1,925 new species! He wrote a monumental 36-volume work: *Atlas ichthyologique des Indies Orientales Neerlandaises* (1862–1877). He was honoured in the shrimpgoby name for his "*significant contributions to Indo-Pacific fish research, including authorship of the Amblyeleotris genus*." It might look as if Bleeker did the unthinkable and named *Paracetopsis bleekeri* after himself, but not so. What seems to have happened in this case is that Bleeker reported the name *Paracetopsis bleekeri* as having been coined by "Guich. (Mus. Paris)" = i.e. Antoine Alphone Guichenot. But, for some reason, Guichenot's use of this name is not recognised today – perhaps it wasn't published in a scientific journal – so Bleeker's own publication of the name (in Atlas ichthyologique des Indes Orientales Néêrlandaises, publié sous les auspices du Gouvernement colonial néêrlandais. Tome II) becomes the first 'official' usage.

Bleher

Silverside genus *Bleheratherina* Aarn & W Ivantsoff, 2009
Bleher's Rainbowfish *Chilatherina bleheri* GR Allen, 1985
Firehead Tetra *Hemigrammus bleheri* J Géry & V Mahnert, 1986
Rainbow Snakehead *Channa bleheri* J Vierke, 1991
Congo cichlid sp. *Steatocranus bleheri* MK Meyer, 1993
African Tetra sp. *Phenacogrammus bleheri* J Géry, 1995
Characin sp. *Leporinus bleheri* J Géry, 1999
African Tetra sp. *Nannaethiops bleheri* J Géry & A Zarske, 2003
Blue-Eye sp. *Kiunga bleheri* GR Allen, 2004

Heiko Bleher (b.1944) is a German explorer and ornamental fish wholesaler and supplier. He lived in Brazil for much of his childhood (see entry for **Amanda** for details) before moving to study at the University of South Florida (1962) and then returned to Rio de Janeiro (1964), opening an aquarium and starting to collect on his own account. He moved his operations centre to Germany (1967). During the 1970s he expanded his operations to include Africa, Asia and Oceania and introduced more than 4,000 aquarium fish species to the market (1965–1997) He provided most of the specimens for Géry's study, including at least one holotype. (See also **Amanda** & **Heiko**)

Bimba

Armoured Catfish sp. *Hypostomus bimbai* CH Zawadzki & I de S Penido, 2021

José Manoel dos Reis Machado (1899–1974), commonly called Mestre Bimba, was a master practitioner of the Brazilian martial art of capoeira. Mestre Bimba created the Regional style of capoeira, characterised by more acrobatic moves and more aggressive punches. He was also responsible for the legalization of capoeira in 1930.

Bloch

Bloch's Topknot *Zeugopterus regius* PJ Bonnaterre, 1788
[Alt. Eckström's Topknot]
Bloch's Gizzard Shad *Nematalosa nasus* ME Bloch, 1795
Bloch Razorbelly Minnow *Salmostoma balookee* ME Bloch, 1795
Snubnose Pompano *Trachinotus blochii* BGE Lacepède, 1801
Winghead Shark *Eusphyra blochii* GLCFD Cuvier, 1817
Hottentot Seabream *Pachymetopon blochii* A Valenciennes, 1830
Ringtail Surgeonfish *Acanthurus blochii* A Valenciennes, 1835
Twoblotch Ponyfish *Nuchequula blochii* A Valenciennes, 1835
Bloch's Catfish *Pimelodus blochii* A Valenciennes, 1840
Bluntnose Guitarfish *Acroteriobatus blochii* JP Müller & FGJ Henle, 1841
Bloch's Tongue Sole *Paraplagusia blochii* P Bleeker, 1851 NCR
Long-tail Tripodfish *Tripodichthys blochii* P Bleeker, 1852
Paeony Bulleye *Priacanthus blochii* P Bleeker, 1853
Spotback Corydoras *Corydoras blochi* H Nijssen, 1971

Marcus Élieser Bloch (1723–1799) was a German physician and naturalist specialising in ichthyology. At first unable to read or write German, he became tutor to a Jewish surgeon's family in Hamburg, where he learned Latin, German and the basics of anatomy. He was subsequently able to move to Berlin and study for his doctorate in natural history and medicine and later practice medicine there (1747). He wrote several books and the twelve–volume magnum opus: *Allgemeine Naturgeschichte der Fische* (The Natural History of Fish) (1782–1795), which included over 400 of his beautifully hand-painted illustrated plates. He then set out to catalogue every fish species. J G T Schneider completed this work after Bloch died as: *Blochii System Ichthyologia ME* (1801), which described 1,519 species. Bloch's collection included around 1,500 specimens including a preserved specimen of the guitarfish that the authors studied. The razorbelly minnow binomial is a replacement name; Bloch's name of 1795 – *Cyprinus clupeoides* – is now viewed as ambiguous, but 'Bloch' has remained attached to the fish's common name.

Block

Green Panchax *Aplocheilus blockii* JP Arnold, 1911

Captain Block (no other information available), collected and imported this species to Germany as an aquarium fish.

Blohm

Driftwood Catfish sp. *Epapterus blohmi* RP Vari, SL Jewett, DC Taphorn Baechle & CR Golbert, 1984

Tomas Blohm (1926–2008) was a Venezuelan cattle rancher, naturalist and conservationist whose ranch, Hato Masagural in the llanos of Venezuela, was kept as close to unspoilt and original as a wildlife sanctuary. Many zoologists are grateful to him, not only for help but also for hospitality. In the case of this species, the etymology is typical of many tributes, saying Blohm "...generously made his ranch available to the authors, which greatly facilitated their research". He used his land as a place for many research programmes, such as the establishment (late 1970s) of a captive breeding programme to ensure the survival of the Orinoco Crocodile *Crocodylus intermedius*.

Blokzeyi

Goby sp. *Stenogobius blokzeyli* P Bleeker, 1860

G H Blokzeyl discovered this species. He may be the same person as A H G Blokzeyl who was (1856) the first Dutch Governor of Bali. He collected fishes and reptiles and sent collecections of both to the Leiden Museum.

Bloyet

Bloyet's Haplo (cichlid) *Astatotilapia bloyeti* H Sauvage, 1883

Captain A Bloyet collected while mapping in Zanzibar, where he was head of Kondoa Station, in the late 19th century. This included many species such as crustacea and insects named after him. He wrote: *De Zanzibar a la station de Kondoa* (1890). The description makes just a passing mention of 'Monsieur Bloyet', but the fact that the holotype came from Kondoa in East Africa makes us sure that it is a reference to the above.

Blomberg

Ecuadorian cichlid sp. *Andinoacara blombergi* N Wijkmark, SO Kullander & R Barriga Salazar, 2012

Rolf David Blomberg (1912–1996) was a Swedish explorer, filmmaker, photographer and author who first went to Ecuador (1934). He was a neutral war correspondent (WW2) and was based in Indonesia where he helped people held in Japanese concentration camps. He returned to Ecuador after the Second World War and made a number of expeditions to contact indigenous tribes, also exploring in Colombia, Brazil and Peru, and made films in Indonesia and Australia. He settled permanently in Ecuador (1968). He wrote: *Såna djur finns* (1951*). He was honoured* for his several expeditions in Ecuador and also has two amphibians named after him.

Blyth

Blyth's Loach *Syncrossus berdmorei* E Blyth, 1860
Stone Cat *Myersglanis blythii* F Day, 1870
Mahseer sp. *Neolissochilus blythii* F Day, 1870
Clouded Archerfish *Toxotes blythii* GA Boulenger, 1892

Edward Blyth (1810–1873) was an English zoologist and author. He was Curator of the Museum of the Asiatic Society of Bengal (1842–1864) and wrote its catalogue. He wrote a series of monographs on cranes, gathered together and published as: *The Natural History of Cranes* (1881). Hume said of him, '*Neither neglect nor harshness could drive, nor wealth nor worldly advantages tempt him, from what he deemed the nobler path. [He was] ill paid and subjected ... to ceaseless humiliations.*' In a similar tribute, Arthur Grote wrote '*Had he been a less imaginative and more practical man, he must have been a prosperous one... All that he knew was at the service of everybody. No one asking him for information asked in vain.*' Twenty-six birds, three mammals, two reptiles and an amphibian are named after him.

Boardman

Australian Sawtail Catshark *Figaro boardmani* GP Whitley, 1928

William Boardman (1906–1963) was a naturalist at the Australian Museum who was a friend and colleague of the describer and who collected the holotype from a trawler. He made regular collecting trips, such as one of ten days on North West Island, Queensland with Melbourne Ward (1929). He wrote a number of scientific papers such as: *Two new species of the genus* Notoscolex *(Oligochaeta) from Ulladulla, New South Wales* (1931) in the Records of the Australian Museum, and another with Whitley: *Marine animals from Low Isles, Queensland (*1929).

Boavista

Characin sp. *Characidium boaevistae* F Steindachner, 1915

A toponym: Boa Vista is the capital of Brazilian state of Roraima, on the western bank of the Rio Branco, which is where the holotype originated. The binomial is often wrongly spelled *boavistae*, without the first *e*.

Bob Miller

San Ignacio Pupfish *Cyprinodon bobmilleri* M de L Lozano-Vilano & S Contreras-Balderas, 1999

Robert Rush Miller (1916–2003) (See **Miller, RR**)

Bob Wisner

Hagfish sp. *Eptatretus bobwisneri* B Fernholm et al., 2013

Robert 'Bob' Lester Wisner (1912–2005) was honoured for his "*...valuable assistance with C.B. McMillan's myxinid research as well as his other contributions to ichthyology.*" (See **Wisner**, also see **Fern**)

Bobo

African Barb sp. *Enteromius boboi* LP Schultz, 1942

Bobo was a local man who helped William H Munn, the director of the Washington DC Zoological Park, to collect fishes in Liberia. No more is recorded of him.

Bocage

Iberian Barbel sp. *Luciobarbus bocagei* F Steindachner, 1864
Angolan Catfish sp. *Schilbe bocagii* ARP Guimarães, 1884
Claroteid Catfish sp. *Chrysichthys bocagii* GA Boulenger, 1910

José Vicente Barboza du Bocage (1823–1907) was a Portuguese zoologist and politician. He studied at the University of Coimbra (1839–1846). He was the Curator of Zoology at the Museu Nacional de Lisboa, and greatly expanded the collection with items brought back from

Portugal's colonies. He also wrote a book about how to collect and prepare such specimens. He retired from most scientific work (1880) except the directorship of the museum and went into politics, becoming the Minister of Navy and Ultramarine Possessions and later Minister of Foreign Affairs (1883–1886). He published numerous works, including over 200 papers, on mammals, herpetology and birds as well as fishes, such as: *Diagnose de algumas espécies inéditas da família Squalidae que frequentam os nossos mares* (1864). Twelve birds, six mammals, ten reptiles and four amphibians are named after him.

Bockmann

Driftwood Catfish sp. *Centromochlus bockmanni* LM Sarmento-Soares & PA Buckup, 2005

Characin sp. *Astyanax bockmanni* RP Vari & RMC Castro, 2007

Armoured Catfish sp. *Hisonotus bockmanni* M de *Carvalho & A Datovo da Silva, 2012*

Three-barbeled Catfish sp. *Pimelodella bockmanni* V Slobodian & MNL Pastana, 2018

Dr Flávio Alicino Bockman works at the Laboratório de Ictiologia de Ribeirão Preto, Departamento de Biologia, Faculdade de Filosofia, Ciências e Letras, Universidade de São Paulo, Brazil. He was awarded his bachelor's degree (1990) by the University of Santa Ursula. He is now a Professor at Universidade de São Paulo, where he was awarded his doctorate (1998), after which he did post-doctoral studies at the Federal University of Rio de Janeiro (1999–2000). He is noted for his contributions to the knowledge of neotropical catfishes and fishes of the upper Rio Paraná basin, Brazil. He co-wrote: *Dorsolateral head muscles of the catfish families Nematogenyidae and Trichomycteridae (comparative anatomy and phylogenetic analysis* (2010).

Bocourt

King Bagrid *Mystus bocourti* P Bleeker, 1864

Basa Catfish *Pangasius bocourti* HE Sauvage, 1880

Chisel-tooth Cichlid *Cincelichthys bocourti* L Vaillant & J Pellegrin, 1902

Marie Firmin Bocourt (1819–1904) was a French zoologist and artist. He followed his father, who engraved copper plates for Muséum National d'Histoire Naturelle, Paris. He became a preparator for Bibron (1834). He was officially designated 'Museum Painter' (1854). He was sent to Siam (Thailand) (1861), where he made an important collection that he took back to Paris. He visited Mexico and Central America (1864–1866). He co-wrote, with Duméril and Mocquard, *Études sur les reptiles et les batraciens* (1870). He collected the cichlid type. Eighteen reptiles and three amphibians are also named after him.

Boddart

Boddart's Goggle-eyed Goby *Boleophthalmus boddarti* PS Pallas, 1770

Pieter Boddaert (1730–1796) was a Dutch physician-naturalist, zoologist, ornithologist, and physiologist who confirmed Pallas' belief that this species, which he first saw in the museum of pharmacist L. Juliaans, represented an undescribed species. He lectured on natural history at the University of Utrecht after he had qualified there as a physician (1764). He wrote *Elenchus Animalium* (1785) and corresponded regularly with Linnaeus.

Bodenhamer

Armoured Catfish sp. *Ancistrus bodenhameri* LP Schultz, 1944

Raymond L Bodenhamer worked for the Lago Petroleum Corporation as a warehouseman at La Salina. He greatly aided the author in regard to transportation while he was in Venezuela.

Boehlke

(See **Böhlke, EB & JE**)

Boeke

St. Maarten Pejerry *Melanorhinus boekei* JM Metzelaar, 1919

Dr Jan Boeke (1874–1956) was Professor of histology and embryology at the University of Utrecht. He graduated in medicine (1896) from the University of Amsterdam continuing to study there (1900) and at the deep-sea institute in Naples focussing on fish embryology. He studied the fisheries of the Netherlands West Indies (1904–1905) and collected the type there. He became Director of the Anatomy Institute at Leiden (1906–1912). He taught in Batavia (now Jakarta) (1913–1914). He spent the rest of his career at Utrecht apart from two other short periods in Indonesia. Among his c.100 publications perhaps the best known is the book: *Problems of Nervous Anatomy* (1940). He died in Indonesia.

Boers

Golden Spadefish *Platax boersii* P Bleeker, 1853

Major W J A W Boers (b.1814) was an Infantry officer (later Lieutenant-Colonel) in Gombong, Java. He was later in Leiden, The Netherlands. He was one of many people who supplied Bleeker with specimens from the Dutch East Indies.

Boeseman

Croaker genus *Boesemania* E Trewavas, 1977
Boesman's Tetra *Hemigrammus boesemani* J Géry, 1959
Corydoras sp. *Corydoras boesemani* H Nijssen & IJH Isbrücker, 1967
Speckled Catshark *Halaelurus boesemani* S Springer & JD D'Aubrey, 1972
Boeseman's Rainbowfish *Melanotaenia boesemani* GR Allen & NJ Cross, 1980
Redfin Dwarf Monocle Bream *Parascolopsis boesemani* DM Rao & KS Rao, 1981
Boeseman's Skate *Okamejei boesemani* H Ishihara, 1987
Trinidad Killifish sp. *Poecilia boesemani* FN Poeser, 2003
Freshwater Stingray sp. *Potamotrygon boesemani* RS Rosa, MR de Carvalho & C de Almeida Wanderley, 2008

Dr Marinus Boeseman (1916–2006) was a Dutch ichthyologist, working in the Department of Zoology, Rijksmuseum van Natuurlijke Histoire, Leiden, becoming Curator of Fishes (1947–1981). Universiteit Leiden awarded his master's degree (1941). He was in the Dutch resistance during WW2 and was arrested (1943) but survived imprisonment at Dachau, though for years his health was so badly affected that he could not work. He collected fishes in El Salvador (1953) and travelled in New Guinea (1954–1955). He caught polio (1957) and suffered from a permanent disability in his right arm, but still took part in collecting expeditions to, inter alia, Surinam and Trinidad. In his etymology of the skate, Ishihara thanks him as otherwise he would not have been aware of problems in the systematics of Japanese *Raja*. The citation for the stingray honours him as the person who "*…contributed substantially to our knowledge of both South American ichthyology (including chondrichthyans) and zoological history.*" He collected the rainbowfish. A reptile and an amphibian are named after him.

Bogardus

Sand Knifefish sp. *Gymnorhamphichthys bogardusae* JG Lundberg, 2005

Mrs Joan Bogardus Spears (1939–2002) was a member of the music staff of St. Catherine of Siena Church, Greenwich, Connecticut for more than 20 years. The etymology says she was "*…a descendant of the earliest Dutch settlers in New York, whose avid interests in life's diversity on Earth taught and inspired her children to support its scientific discovery and documentation.*" Mrs Spears' daughter, Dorothy, provided 'generous support' of Lundberg's work.

Boggiani

Cichlid genus *Boggiania* A Perugia, 1897
[Now usually placed in *Crenicichla*]

Guido Boggiani (1861–1902) was an Italian artist, ethnologist and photographer. He went to Argentina (1887) and subsequently traveled through the interiors of Brazil, Bolivia and Paraguay (1887). He returned to Paraguay a number of times and made expeditions to the Gran Chaco (1888) and again (1901), after which he disappeared until his remains were found (1904). It is probable that he was ritually killed by local native people who took the trouble to bury his camera - at that time cameras were thought of by some peoples as stealing their souls. He wrote a number of ethnological works on Paraguayan indigenous people, including: *Notizie etnografiche sulla tribù dei Ciamacoco, etc.* (1994).

Böhlke, EB & JE

Tetra genus *Boehlkea* J Géry, 1966
False Moray genus *Boehlkenchelys* KA Tighe, 1992
Böhlke 's Clingfish *Tomicodon boehlkei* JC Briggs, 1955
False Penguin Tetra *Thayeria boehlkei* SH Weitzman, 1957
Diamond Blenny *Malacoctenus boehlkei* VG Springer, 1959
Roughhead Triplefin *Enneanectes boehlkei* RH Rosenblatt, 1960
Barcheek Blenny *Coralliozetus boehlkei* JS Stephens, 1963
Characin sp. *Pseudochalceus bohlkei* GE Orcés-Villagómez, 1967
Characin sp. *Characidium boehlkei* J Géry, 1972
Cryptic Bearded Goby *Barbuligobius boehlkei* EA Lachner & JF McKinney, 1974
Thorny Catfish sp. *Rhinodoras boehlkei* GS Glodek, GL Whitmire & GE Orcés-Villagómez, 1976
Characin sp. *Acestrocephalus boehlkei* NA de Menezes, 1977
Corydoras sp. *Corydoras boehlkei* H Nijssen & IJH Isbrücker, 1982

Sand Stargazer sp. *Dactyloscopus boehlkei* CE Dawson, 1982
Neotropical Rivuline sp. *Cynodonichthys boehlkei* JH Huber & J-F Fels, 1985
Carolina Pygmy Sunfish *Elassoma boehlkei* FC Rohde & RG Arndt, 1987
Onejaw sp. *Monognathus boehlkei* EOG Bertelsen & H Nielsen, 1987
Characin sp. *Leporinus boehlkei* JS Garavello, 1988
Characin sp. *Pseudocurimata boehlkei* RP Vari, 1989
Boehlke's Goby *Psilotris boehlkei* DW Greenfield, 1993
Böhlke's Coralbrotula *Ogilbia boehlkei* PR Møller, W Schwarzhans & JG Nielsen, 2005

James Erwin Böhlke (1930–1982) and his wife Eugenia Louisa Böhlke née Brandt (1929–2001) were both ichthyologists who worked at the Academy of Natural Sciences of Philadelphia, where he was Curator of Ichthyology (1954–1982). He was a master's student at Stanford University when he first met his future wife. Academically he was behind her, and she became the breadwinner whilst he finished his studies (1953). Stanford had awarded her bachelor's degree (1949) and master's (1951) before they married (1951). She had three children to look after and her research took a back seat, so it was not until after her husband's death that her own career took off: entirely through her determination and in the face of opposition from the Philadelphia hierarchy, as it was her husband and not her who had the tenure at the Academy. She died of cancer. The tetra genus is named after James; the false moray genus after them both. Although the above species are named after James, other fish are named after Eugenia under her nickname 'Genie' (see **Jacobus Boehike** & **Genie**).

Boitone

Lyre-fin Pearlfish *Simpsonichthys boitonei* AL de Carvalho, 1959

Dr José Boitone was head of Zoological Services at the Zoo in Brasilia. At the new zoo being constructed at the new capital of Brazil, Brasilia, a zoo official named Saturnino Maciel de Carvalho was responsible for collecting live fishes to feed the aquatic birds of the zoo. During one of these collections (1959) he noticed the beautiful appearance of five specimens of a small fish. He brought these to the head of Zoological Services, José Boitone. Sr. Boitone, not being familiar with pisciculture, took it upon himself to send specimens to the renown scientist and ichthyologist, Dr. Antenor de Carvalho, in Rio de Janeiro. Dr. Carvalho described and classified this new species, giving it the name in honour of Sr. Boitone for having sent the specimens for classification.

Böker

Goby sp. *Gobius boekeri* E Ahl, 1931

Dr Hans Böker (1886–1939) was a Mexican born German anatomist and zoologist. He studied medicine at the University of Freiburg (1906), graduating as an anotomist (1911) and physician (1913) having also studied at Kiel and Berlin. He worked at the Anatomical Institute of the University of Freiburg (1912–1932) becoming Associate Professor (1927). He was then Professor of Anatomy (1932–1938) and Director of the anatomical institute in Jena. He took up a new post at the University of Cologne but was only there seven months befor his death. He undertook numerous research trips including to Corsica, the Canaries and North Africa. He co-wrote the two-volume: *Introduction to the comparative biological anatomy of vertebrates* (1935–1937) among other works. He collected the holotype.

Bokermann

Neotropical Rivuline sp. *Pterolebias bokermanni* H Travassos, 1955
[Junior Synonym of *P. longipinnis*]
Neotropical Rivuline sp. *Simpsonichthys bokermanni* AL de Carvalho & CAG da Cruz, 1987

Dr Werner Carlos Augusto Bokermann (1929–1995) was a Brazilian herpetologist and ornithologist. He received his doctorate in zoology at the Bioscience Institute, São Paulo University. He became Head, Bird Section, Fundaçào Parque Zoológico de São Paulo, and stayed there throughout his working life. He described a great number of new species, including at least sixty frogs and toads, and he published a great many scientific papers. Many other taxa, including thirteen amphibians, a mammal and a bird, are named after him.

Boklund

African Rivuline sp. *Nothobranchius boklundi* S Valdesalici, 2010

Jørn Boklund is a Danish aquarist who collected the holotype (2009) from seasonal pools in North Luangwa National Park, Zambia. He is a friend of the author.

Boldingh

Soft-coral Goby *Pleurosicya boldinghi* MCW Weber, 1913

Lieutenant Helenus Johannes Boldingh (1868–1954) was one of three officers aboard the Dutch 'HMS Siboga' expedition (1899–1900) to the East Indies, led by Weber (q.v.), that collected the type. While Weber led the scientific research, much of the hydrographic observations were undertaken by the ship's officers. Boldingh retired from active service (1903) and was later (1933) a teacher at the Dutch Royal Naval Institute.

Bolin

Lanternfish genus *Bolinichthys* JR Paxton, 1972
Sculpin genus *Bolinia* M Yabe, 1991
Bigmouth Sculpin *Hemitripterus bolini* GS Myers, 1934
Bolin's Lanternfish *Protomyctophum bolini* AF Fraser-Brunner, 1949
Grand Lanternfish *Gymnoscopelus bolini* AP Andriashev, 1962
Bolin's Clingfish *Lepadichthys bolini* JC Briggs, 1962
Lanternfish sp. *Notoscopelus bolini* BG Nafpaktitis, 1975
Blind Cusk Eel sp. *Paraphyonus bolini* JG Nielsen, 1974
[Syn. *Aphyonus bolini*]

Dr Rolf Ling Bolin (1901–1973) was an American ichthyologist and marine biologist who was a specialist in lanternfish. The University of Utah awarded his bachelor's degree in graphic arts (1925). He changed course and went to Stanford (1928) to study biology and there was awarded his doctorate (1934). He stayed at Stanford for his entire career, becoming Professor of Marine Biology and Oceanography (1949–1967) and retiring as Professor Emeritus. He was particularly associated with Stanford's Hopkins Marine Station where he was based for over 40 years. He wrote: *A review of the myctophid fishes of the Pacific coast of the United States and of Lower California* (1939).

Bolivar

Characin sp. *Creagrutus bolivari* LP Schultz, 1944

Simón Bolívar (1783–1830) 'The Liberator' was one of the great men of Latin American history and needs no biography here. Many places are named after him.

Bollinger

American Sole sp. *Trinectes hubbsbollinger* RR Duplain, F Chapleau & TA Munroe, 2012

John A Bollinger, Scripps Institution of Oceanography, California, had been a student of Carl L Hubbs (1894–1979). The sole is also named after Hubbs (q.v.), as he and Bollinger first recognized it as an undescribed species. Bollinger co-wrote: *A New Species of Trinected (Pleuronectiformes: Achiridae), with Comments on the other Eastern Pacific Species of the Genus* (2001).

Bollman

Goby genus *Bollmannia* DS Jordan, 1890
Large-tooth Flounder sp. *Hippoglossina bollmani* CH Gilbert, 1890

Charles Harvey Bollman (1868–1889) was an America naturalist principally interested in fish and myriapods, considered by Jordan to be one of the most brilliant and promising naturalists ever known despite being just 20 when he died. He graduated from Indian University (1889) where he was Jordan's student, and was immediately appointed an assistant in the US Fish Commission. He published 13 papers (1887–1889) including in co-authorship with Jordan, and described no less than 65 new species of North American myriapods. The etomology says of him: "*...whose untimely death while engaged in the exploration of the rivers of Georgia, took place while this paper was passing through the press.*" He died of dysentery contracted while collecting fish in the swamps of Waycross, Georgia, USA. The millipede genus, *Bollmania*, is also named after him.

Bollons

Bollons' Rattail *Coelorinchus bollonsi* C McCann & DG McKnight, 1980

Captain John Peter Bollons (1862–1929) was an English born, New Zealand amateur ornithologist, naturalist, ethnographer and mariner who captained government steamers which annually provisioned the subantarctic islands. He went to sea (1876) and first landed in New Zealand as a survivor of the wreck (1881) of the barquentine 'England's Glory' near the entrance of Bluff Harbour. After the wreck, he joined the 'Bluff' pilot cutter and then transferred to a government ketch, engaged in suppressing seal poaching. He became a master mariner (1892) and was employed (1893) by the Marine Department of the New Zealand Government. He liked to name his children after ships that he had commanded, so his son was called Tutanekai and his daughter Hinemoa. He died unexpectedly after a hernia operation. A bird is also named after him.

Bonaparte

Bristlemouth genus *Bonapartia* GB Goode & TH Bean, 1896
Shortfin Spiny Eel *Notacanthus bonaparte* A Risso, 1840

Smallnose Fanskate *Sympterygia bonapartii* JP Müller & FGJ Henle, 1841
Ghost Knifefish sp. *Apteronotus bonapartii* F de L de Castelnau, 1855
Napoleon Snake-eel *Ophichthus bonaparti* JJ Kaup, 1856

Prince Charles Lucien Bonaparte (originally Jules Laurent Lucien) (1803–1857) was a nephew of Napoleon Bonaparte and a renowned biologist and zoologist. He was much travelled and spent many years in the USA cataloguing birds. He has been described as the 'father of systematic ornithology'. He eventually settled in Paris and commenced his: *Conspectus Generum Avium*, a catalogue of all bird species, which was unfinished on his death. Its publication was heralded as a major step forward in achieving a complete list of the world's birds. He also wrote: *American Ornithology* (1825) and *Iconografia della Fauna Italica - Uccelli* (1832). Swainson described Bonaparte as "*...destined by nature to confer unperishable benefits on this noble science*". His treatise on the fauna of Italy introduced several new sharks and rays to science. 20 birds and a mammal are also named after him.

Bond, CE

Sucker sp. *Catostomus bondi* GR Smith, JD Stewart & NE Carpenter, 2013

Professor Dr Carl E Bond (1920–2007) was an American ichthyologist. Oregon State College awarded his bachelor's (1947) and master's (1948) and the University of Michigan awarded his PhD (1963). He began his teaching and research career in Oregon State College's Fisheries and Wildlife Department (1949) as assistant aquatic biologist. Specialising in the study of freshwater fishes, he became one of the world's leading authorities on the sculpins (*Cottidae*). Bond was also involved in research internationally, working in Latin America and Africa, as well as India and Iran on Peace Corps projects (1967–1971). He was also Assistant Dean of the Graduate School (1969–1974). He became Professor Emeritus of Fisheries (1984). He wrote: *Biology of Fishes* (1979).

Bond, FE

Silver Tetra *Gymnocorymbus bondi* HW Fowler, 1911

Francis 'Frank' Edward Bond (1867–1923) was an American banker and stockbroker, as well as an amateur ornithologist. He led the (1911) expedition to the Orinoco River, Venezuela, where he collected the holotype for the Academy of Natural Science of Philadelphia. He was also the father of the ornithologist James Bond (see below), whose name Ian Fleming used for his British spy hero.

Bond, FF

Bond's Cory *Corydoras bondi* WA Gosline III, 1940
Pencil Catfish sp. *Schultzichthys bondi* GS Myers, 1942
Long-finned Glass Tetra *Xenagoniates bondi* GS Myers, 1942
Neotropical Rivuline sp. *Anablepsoides bondi* LP Schultz, 1949

Franklyn F Bond (1897–1946) was an American biologist and ichthyologist at the University of Rochester, New York. He studied at Cornell. He explored the rainforests of South and Central America, discovered new animal and plant species and collected several thousand fish. He researched mosquito-control fishes in Venezuela (1938–1940) and collected the holotypes of these species.

Bond, J

Sharpnose Lizardfish *Synodus bondi* HW Fowler, 1939

James Bond (1900–1989) was an American ornithologist educated in England at Harrow and Cambridge, which latter awarded his bachelor's degree (1922). He worked for a Philadelphia banking firm (1922–1925), resigning to take part in an expedition to the Amazon run by the Philadelphia Academy of Sciences for whom he subsequently worked as an ornithologist and Curator of Birds. His area of expertise was the Caribbean and he wrote: *Birds of the West Indies* (1936). He obtained the lizardfish holotype in Jamaica (1935). Ian Fleming knew of Bond and his work as an ornithologist, and asked him for permission to use his name for Fleming's British Secret Service agent hero. Bond agreed and in the first Bond film, *Dr No*, Sean Connery, as 007, can be seen examining a copy of *Birds of the West Indies*. James Bond the ornithologist suffered many years of cancer before eventually succumbing. Three birds are named after him.

Bond, R

Hispaniola Pupfish *Cyprinodon bondi* GS Myers, 1935

Dr Richard Marshall Bond (1903–1976) was an American zoologist, botanist, ornithologist and limnologist, an Assistant Biologist at Yale (1928–1930), which awarded his first degree (1926) and PhD (1932). He was a fellow at the Bishop Museum (1930–1931). The original text says that "*...Dr. R. M. Bond collected the types during an ecological investigation of the Hispaniolan lakes*" (1933). He published the results such as: *Investigatiosn of some Hispaniolan lakes. II Hydrobiology and Hydrogeography* (1935). He also worked as an Instructor, Biology (1932–1933), National Research Fellow (1933–1934), Teacher Santa Barbara School (1934–1935), Associate Wildlife Technican, National Parks Service (1935–1938) and Regional biologist, Soil Conservation Service, US Dept. Agriculture (1938–1951). He became Director, Land & Water Development Program, Virgin Islands Corporation (1952) and Agriculturist-in-Charge, Virgin Islands Agricultural Program Agricultural Research Service until retirement (1953–1967).

Bonelli

Padanian Goby *Padogobius bonelli* CL Bonaparte, 1846

Franco Andrea Bonelli (1784–1930) was an Italian zoologist, ornithologist and collector. He taught at the University of Turin (1809–1811), then became the Curator of the museum there and totally recatalogued the collection according to scientific principles. The collection grew during his tenure to become one of Europe's greatest assemblages of ornithological specimens. He wrote: *Catalogue des Oiseaux du Piémont* (1811) and three birds are named after him. He named this goby *Gobius fluviatilis* in a manuscript used by Cuvier & Valenciennes (1837), which was preoccupied by *G. fluviatilis* Pallas (1814).

Bonilla

New Granada Sea Catfish *Notarius bonillai* C Miles, 1945

Dr Heliodoro Bonilla Guzmán (1908–1961) was a Colombian veterinarian who qualified (1936) at the National University, where he was later appointed Professor of Zootechnics

(1937). He worked for the Ministry of Economics (1939–1941), was national director for cattle raising (1943–1947), including being the Director of the Department of the Ministry, Bogota, Colombia that dealt with fishes.

Bonne

Bonnes' Ricefish *Oryzias bonneorum* LR Parenti, 2008

Cornelis Bonne (1890–1948) and Dr Johanna Bonne-Wepster née Wepster (1892–1978) were both systematic entomologists who worked throughout Indonesia in the early 20th century and collected fish. Among their papers, they wrote: *Diagnoses of New Mosquitoes from Surinam. With a note on Synonymy* (Diptera, Culicidae) (1919) and she wrote: *New Mosquitos (Dipt.) from the Nethwerlands Indies* (1934) under the auspices of the Institute for Tropical Hygiene, Royal Colonial Institute, Amsterdam.

Bont

Occassional-Shrimp Goby *Drombus bontii* P Bleeker, 1849

Jacob de Bondt, aka Jacobus Bontius (1592–1631), was a Dutch physician and pioneer of tropical medicine. His *Historiae naturalis et medicae Indiae orientalis* (1631) was cited by Bleeker several times in other works. So, although the patronym is not identified in the goby's description, we feel this is very likely whom Bleeker intended.

Bont

Sailfin Silversides sp. *Telmatherina bonti* MCW Weber & LF de Beaufort, 1922

This is not an eponym but based on a local name for these fish, 'bonti-bonti'.

Booth

Booth's Pipefish *Halicampus boothae* GP Whitley, 1964

Julie Booth is an Australian marine naturalist, and a pioneer in discovering the behaviour of marine turtles by habituating them to her presence. She collected the holotype on Lord Howe Island (1962). She was resident on Fairfax Island (1969). Among her published works is: *Behavioural studies on the Green Turtle (*Chelonia mydas*) in the sea* (1972), which included underwater observations of green turtles near Fairfax Island (1972–1974).

Boquilla

Three-barbeled Catfish sp. *Cetopsorhamdia boquillae* CH Eigenmann, 1922

This is a toponym; it refers to a town of the same name on the Río Cauca, Colombia, where the holotype was collected.

Borchgrevink

Bald Notothen *Pagothenia borchgrevinki* GA Boulenger, 1902

Carsten Egeberg Borchgrevink (1864–1934) was an Anglo-Norwegian Polar explorer. He was educated at Gjertsen College, Oslo, and later (1885–1888) at the Royal Saxon Academy of Forestry at Tharandt, Saxony, in Germany. He then spent four years as a government surveyor in Queensland and New South Wales and then spent time teaching languages.

He began exploring (1894) on a Norwegian whaling expedition when he became the first man to set foot on the mainland of Antarctica. He was Commander of the British Southern Cross Antarctic Expedition (1898–1900) that over-wintered on mainland Antarctica, and reached further south than any previous expedition.

Boreham

Boreham's Sole *Aseraggodes borehami* JE Randall, 1996

Roland Stanford 'Bud' Boreham (1924–2006) was an American businessman best known for heading up Baldor Electric. He graduated with bachelor degrees in physics and meteriology from UCLA and served in the Pacific during WW2. He began at Baldor as a sales manager (1961) and rose to CEO (1978), Chairman of the Board (1981) until he retired (2004). He wrote *The Three-Legged Stool* (1999) about how to be successful in business. His father Howard was Randall's school chum. He was honoured in recognition of his support for Randall's ichthyological research.

Borelli

Characin sp. *Characidium borellii* GA Boulenger, 1895
Armoured Catfish sp. *Hypostomus borellii* GA Boulenger, 1897
Pencil Catfish sp. *Trichomycterus borellii* GA Boulenger, 1897
Headstander Characin sp. *Schizodon borellii* GA Boulenger, 1900
Borelli's Dwarf Cichlid *Apistogramma borellii* CT Regan, 1906
[Alt. Umbrella Cichlid]

Dr Alfredo Borelli (1857–1943) was an Italian arachnologist, entomologist and zoologist who worked at the Turin Museum (1881–1913) after completing his studies at the university there. He spent three years (1893–1894, 1895–1896 & 1899) exploring and collecting in Argentina, Bolivia and Paraguay. When he returned to Italy, he found his estates impoverished and then led an almost hermit-like existence at the University. He published many scientific papers. He collected the type of the cichlid. An amphibian, four birds and two reptiles are also named after him as are a number of invertebrates.

Boretz

Rattail sp. *Malacocephalus boretzi* YI Sazonov, 1985

Dr Leonid A Boretz (also spelled Borets) was a Russian ichthyologist at the Pacific Scientific Research Fisheries Centre, Vladivostok. He collected the first specimens of this species and provided them for study. He wrote: *Annotated List of Fishes of Far-Eastern Seas* (2000).

Borisov

East Siberian Cod *Arctogadus borisovi* PA Dryagin, 1932

Pavel Gavrilovich Borisov, also spelled Borissov (b.1889), was a Russian ichthyologist, explorer and collector who was in Sakhalin (1911–1913). He wrote: *A Handbook of catching fish of the USSR* (1964).

Bork

Blue Tetra *Knodus borki* A Zarske, 2008

Dieter Bork is a German ichthyologist and aquarist. He supplied the holotype of this species. He co-wrote: *South American Dwarf Cichlids* (1997).

Borley

Borley's Redfin Hap (cichlid) *Copadichromis borleyi* TD Iles, 1960

Sir Henry John Hawkins Borley was Director of the Game Fish & Tsetse Control Department of Nyasaland (Malawi). He was a member of the 1939 Fishery Survey Team to Lake Malawi, along with Dr CKR Bertram and Dr E Trewavas (q.v.), and they were all co-authors of: *Report on the fish and fisheries of Lake Nyasa* (1942). Borley was knighted (1963).

Borno

Large-spotted Soapfish *Rypticus bornoi* W Beebe & J Tee-Van, 1928

Eustache Antoine Francois Joseph Louis Borno (1865–1942) was a Haitian politician who served as President of Haiti (1922–1930) during the period of the American occupation.

Borodin

Three-barbeled Catfish sp. *Imparfinis borodini* GF Mees & PC Cala, 1989

Nikolai Andreyevich Borodin (1861–1937) was a Russian Cossack ichthyologist and fisheries expert who passed out of military college with distinction and then studied at St Petersburg University (1879–1885). He had radical ideas and was arrested (1886) but released with a caution. He travelled in Europe and North America (1891–1893) to study ichthyological stations. He was Professor and Chief Specialist in Fish Culture at the Petrograd Agricultural College (1899–1917), though during WW1 he was sent to the USA (1915) and was employed by the Government in the development of cold storage for perishable foodstuffs. After the 1917 revolution he worked in the Ministry of Agriculture for the Kolchak government and was sent by it to the USA again (1919) to buy agricultural machinery. Whilst he was in the USA, he learned of the defeat of the Kolchak army by the Bolsheviks and decided not to return to Russia. He worked as an assistant in the Brooklyn Museum of Art and Science (1926–1927) and at the US Museum of Natural Science (1927–1928). Finally, he went to Harvard and was the Curator of Fishes at the Museum of Comparative Zoology (1928–1937). He wrote: *Ideals and Reality* (1930), his memoirs (1879–1919).

Bortayro

Pencil Catfish sp. *Silvinichthys bortayro* LA Fernández & MCCde Pinna, 2005

Gonzalo Padilla Bortayro is an Argentine biologist who is responsible for biodiversity monitoring of the activities of Minera Alumbera, a mining group for which he works, in the Catamarca region. He co-wrote: *Parasitic nematodes of freshwater fishes from the province of Catamarca, Argentina* (2011). He was the first person to collect this species and brought it to the authors' attention.

Boruca

Characin sp. *Lebiasina boruca* WA Bussing, 1967
[Syn. *Piabucina boruca*]

The Boruca are an indigenous people who used to live in southern Costa Rica, where this species is endemic.

Bosc

Striped Blenny *Chasmodes bosquianus* BG de Lacepède, 1800
Naked Goby *Gobiosoma bosc* BG de Lacepède, 1800
Sea Chub sp. *Kyphosus bosquii* BG de Lacepède, 1802
Four-spot Megrim *Lepidorhombus boscii* A Risso, 1810

Louis-Augustin Guillaume Bosc d'Antic (1759–1828) was a French naturalist and botanist. He was President of the French Natural History Society (1790). After his friend Mme. Roland was guillotined (1793), he had to hide in the Forest of Montmorency, returning to Paris after Robespierre's fall. After the coup d'état (1799) he could only support himself by mass-producing articles for scientific periodicals. He became inspector of the gardens of Versailles and publicly-owned nurseries. His manuscript provided the basis of Lacepède's descriptions.

Boschma, H

Bronzestripe Grunt *Haemulon boschmae* J Metzelaar, 1919
Boschma's Frogfish *Lophichthys boschmai* M Boeseman, 1964
Giant Lampfish *Parvilux boschmai* CL Hubbs & RL Wisner, 1964
Three-barbeled Catfish sp. *Pimelodella boschmai* JWB Van der Stigchel, 1964

Professor Dr Hilbrand Boschma (1893–1976) was a Dutch zoologist, herpetologist, and expert on crustaceans. Boschma studied botany and zoology at the University of Amsterdam where his doctoral dissertation (1919) was on the neck skeleton of crocodiles. He went to the former Dutch East Indies (1920–1922), where he studied embryology, functional morphology in reptiles and amphibians, and stony corals. He joined a Danish expedition to the Kai Islands (1922) and sampled and studied corals and wrote papers on several new species of fire corals. He returned to the Netherlands, becoming Chief Assistant at the Zoological Laboratory of the State University at Leiden and started (1925) giving lectures in general zoology for medical students and becoming (1931) Professor of Zoology. He turned his attention to invertebrates at Rijksmuseum van Natuurlijke Histoire, Leiden (1922), where he became Director (1934–1958); still lecturing (1958–1963) and writing (until 1974). Other taxa, including crustaceans, two reptiles and a bird, are named after him. The grunt etymology merely says: "…*Named after a zoologist of the Amsterdam University*", which fits with Boschma's attendance.

Boschung

Slackwater Darter *Etheostoma boschungi* BR Wall & JD Williams, 1974
Mobile Chub *Macrhybopsis boschungi* CR Gilbert & RL Mayden, 2017

Dr Herbert T 'Bo' Boschung (1925–2015) was Professor Emeritus of Biology at the University of Alabama. He was nationally recognized for his work on freshwater fishes in the Southeast and for identifying many new species of fish. He served in the US Army and was awarded a Purple Heart. He received his bachelor's and masters' degrees from the University of Alabama, then joined the teaching staff (1950) rising to Professor of Biology until retirement (1987). He was Director of the Alabama Museum of Natural History (1966–1978) and was curator of fishes for UA's Ichthyological Collection (1966–1987). He was also Director of the Alabama Marine Laboratory (1966–1967). His best known work is *Fishes of Alabama* (2004) and he was senior author of *Audubon Society's Field Guide to North American Fishes* (1983). His interests went beyond ichthyology, as he was deeply concerned about the quality of the environment and conservation.

Boseto

Sleeper sp. *Eleotris bosetoi* MI Mennesson, P Keith, BC Ebner & PJR Gerbeaux, 2016

David Boseto is an ichthyologist and aquatic ecologist and a friend of the senior author. He is Co-Director of Ecological Solutions Solomon Islands. The University of the South Pacific, Fiji, awarded his bachelor's and master's degrees. He has conducted scientific research and field surveys and collections, and collaborated with other scientific expeditions in Solomon Islands, Fiji, Kosrae, Australia, New Caledonia, Japan and USA. He co-authored two chapters of the book *Vulnerability of Tropical Pacific Fisheries and Aquaculture to Climate Change. Secretariat of the Pacific Community, Noumea, New Caledonia* (2011). He has also co-authored a number of scientific papers. He was honoured for his 'extensive and enthusiastic' work on the freshwater fauna of the Solomon Islands. He also helped collect the holotype.

Bosha

Boris' Swift-tail Flying Fish *Cypselurus bosha* IB Shakhovskoy & NV Parin, 2019

Boris Vladimirovich Shakhovskoy is the father of the first author. He provided significant material and technical assistance. 'Bosha' is formed by a combination of the first syllable of his name and surname.

Bospor

Big-eyed Icefish *Channichthys bospori* GA Shandikov, 1995

This is a toponym referring to the Crimean city of Kerch by its former name Bospor – home of the research institute UNGNIRO, which is engaged in Antarctic research.

Bossche

Small-eyed Flathead *Cymbacephalus bosschei* P Bleeker, 1860

Jules Felicien Romain Stanislas van den Bossche (1819–1889) was a Dutch colonial administrator who was Resident van Banka at the time of Bleeker's description. He was taken as an infant to the Dutch East Indies (1822) and stayed for the rest of his life. He served in the army, beginning his administrative career (1846) as an Inspector at Palembang, becoming Controller (1846–1855). He became a Governer of the Dutch Gold Coast (1857–1858) then moved back to the East Indies. He served as Governor of Sumatra's west coast (1862–1869). He was discharged honourably from service at his own request (1871) and was thereafter in commerce.

Bostock

Freshwater Cobbler *Tandanus bostocki* GP Whitley, 1944

Rev George James Bostock (1833–1881) was an Anglican Church minister and naturalist. Cambridge awarded his bachelor's degree (1856) and in the same year he was ordained as a deacon and as priest (1860). He was the Chaplain of Fremantle, Western Australia (1860–1875) during which time he was an active collector of fishes for Castelnau, including the holotype of this species. Having returned to England, he was Vicar of Kirkby Wharfe, Yorkshire (1875–1881).

Boticario

Pencil Catfish sp. *Listrura boticario* MC de Pinna & W Wosiacki, 2002
Neotropical Rivuline sp. *Moema boticarioi* WJEM Costa, 2004
[Syn. *Aphyolebias boticarioi*]

Named after a Brazilian nature conservation institution; Fondação Grupo Boticário de Proteção À Natureza.

Botta

Waspfish sp. *Vespicula bottae* HE Sauvage, 1878
[Probably a Junior Synonym of *V. trachinoides*]

Paul-Emile Botta (1802–1870) was a traveller, diplomat, archaeologist and physician. He spent a year on board the French ship 'Héros' as ship's surgeon and naturalist. The vessel traded on the Californian coast (c.1827) under the command of Captain Auguste Duhaut-Cilly, who recorded expeditions ashore with Botta. He was physician to Mohammed Ali, Pasha of Egypt (1830), and was the French Consul in Alexandria (1833). He wrote *Notes on a Journey in Arabia* and *Account of a Journey in Yemen* (1841). He was appointed Consular Agent for Iraq (1840) and became French Consul at Mosul (1842). He identified Khorsabad as being the site of biblical Nineveh, the ancient capital of Assyria. Botta excavated there (1843–1846) and discovered a dictionary for Class III cuneiform script dating from the seventh century BC. He wrote about his discoveries in the 5-volume *Monuments de Ninivé, Décoverts et Décrits par Botta, Mésurés et Déssinés per E. Flandin* (1849–1850). Three mammals, a bird and a reptile are named after him. There is no etymology in Sauvage's original, but many of his species are named after prominent naturalists of the time and we are in little doubt that Paulo Emilio Botta was intended.

Bottego

Bottego's Ethiopian Minnow *Neobola bottegoi* D Vinciguerra, 1895
Labeo sp. *Labeo bottegi* D Vinciguerra, 1897

Captain Vittorio Bottego (1860–1897) was an Italian explorer. He was an artilleryman and a skilled horseman who wanted adventure and to become a hero, so he arranged a transfer to Eritrea (1887). He set out on a journey of exploration from Berbera with Captain Matteo Grixoni (1892). They reached the upper flow of the Juba River and penetrated to its source (1893). After parting ways with Grixoni, Bottego reached Daua Parma and discovered the Barattieri waterfalls, finally reaching Brava (September 1893). The expedition lost 35 men en route. Bottego set off again (1895) under the auspices of the Italian Geographical Society, with a contingent of 250 local troops. He later tried crossing Ethiopia and was offered a truce, but turned it down and was killed in the fighting. The King of Ethiopia kept his men imprisoned for two years, and only when they were released did word of Bottego's fate reach the Italian colonial regime. A mammal, a bird and two reptiles are named after him.

Bottome

Shorthead Blenny *Emblemariopsis bottomei* J Stephens, 1961

Peter Bottome Deery (1937–2016) was a Venezuelan businessman. There is not much in the way of an etymology. However, Stephens does record that 'Peter Bottome and party' collected the type at Yonqui, Los Roques Archipelago, Venezuela. We believe that this refers to Peter Bottome Deery.

Boucard

Balsas Shiner *Graodus boucardi* A Günther, 1868
[Syn. *Notropis boucardi*]

Adolphe Boucard (1839–1905) was a French naturalist who collected in Mexico and spent c.40 years killing hummingbirds for science and the fashion trade. He moved to London (1890) but passed his later years in his villa near Ryde on the Isle of Wight. He was author of: *The Hummingbird* (1891). He wrote (1894) that '*Nowadays the mania of collecting is spread among all classes of society, and that everyone possesses, either a gallery of pictures, aquarels, drawings, or a fine library, an album of postage stamps, a collection of embroideries, laces ... and such like, a collection of hummingbirds should be the one selected by ladies. It is as beautiful and much more varied than a collection of precious stones and costs much less ...*' Nine birds and a reptile are named after him.

Bouchelle

Crystal Tetra *Roeboides bouchellei* HW Fowler, 1923

Dr Theodore W Bouchelle (d.1935) was the surgeon and metallurgist of the Eden Mining Company, Nicaragua. He was a member of the Philadelphia Academy of Sciences, and being resident in Nicaragua (1904) he suggested that the Academy send an expedition, which he would host as well as taking care of the health of its members (1922). A bird is named after him, as is a gastropod that he collected.

Bouchet

Bouchet's Dragonet *Callionymus boucheti* R Fricke, 2017

Philippe Bouchet (b.1953) is a French zoologist and malacologist. He is (1973–present) Senior Professor at the L'Institut de Systématique, Évolution, Biodiversité, MNHN. His particular interest is the exploration and description of biodiversity, especially marine invertebrates. He has taken field trips to three different oceans and described over 600 new species of molluscs. Among his numerous publications are the co-written: *A New Species of* Lunovula *from the South Pacific (Gastropoda: Ovulidae, Pediculariinae)* (2018) and *Frogs and Tuns and Tritons A Molecular Phylogeny and Revised Family Classification of the Predatory Gastropod Superfamily Tonnoidea (Caenogastropoda)* (2019). The etymology says: "*This new species is named in honour of Philippe Bouchet (MNHN, Paris), appreciating the excellent organisation of numerous expeditions exploring the biodiversity of tropical seas, including the KAVIENG 2014 Expedition to New Ireland.*"

Bouet

Liberian Swamp Eel *Typhlosynbranchus boueti* J Pellegrin, 1922

Dr Georges Théodore Louis Bouët (1869–1957) was an army physician, ornithologist and colonial administrator, who served (1896–1900) in Sudan, Madagascar (1900–1904), Ivory Coast, Niger and Dahomey (Benin) (1906–1930). He was seconded to the Ministry of Foreign Affairs and was the French Chargé d'Affaires and French Consul in Monrovia, Liberia (1917–1927). He was an active collector for MNHN, Paris. He wrote *Oiseaux de l'Afrique tropicale* (1955).

Bougainville

Gulf Gurnard Perch *Neosebastes bougainvillii* G Cuvier, 1829
Bougainville's Anglerfish *Histiophryne bougainvilli* A Valenciennes, 1837
Short-snouted Shovelnose Ray *Aptychotrema bougainvillii* JP Müller & FGJ Henle, 1841 [Junior Synonym of *A. rostrata*]

Admiral Baron Hyacinthe Yves Philippe Potentien de Bouganville (1781–1846) was a French naval officer in command of the corvette 'Espérance'. He was a Midshipman on 'Le Geographie' under Baudin (1800–1803) and later took part in a circumnavigation (1824–1826). Two mammals and a bird are named after him.

Bouillon

African Loach Catfish sp. *Leptoglanis bouilloni* M Poll, 1959

Jean Bouillon (1926–2009) was a Belgian marine biologist and zoologist. He was on the faculty of l'Université Libre de Bruxelles (1955–1991) as a Professor in charge of both the Zoology and Marine Biology laboratories. He founded and was the Director of the King Leopold III biological Station at Laing Island, Madang, Papua New Guinea (1975–1994). He co-wrote: *An introduction to Hydrozoa* (2006). He was a member of the expedition to the River Congo during which the holotype was collected.

Boujard

Armoured Catfish sp. *Paralithoxus boujardi* F Fisch Muller & IJH Isbrücker, 1993

Dr Thierry Boujard (b.1959) is a hydrobiologist, ichthyologist and physiologist. He was at Laboratoire d'Hydrobiologie de Guyane (INRA) when the fish was described (1984–1990). He worked at Guelph University (Ontario Canada) (1990–1991) as well as at the Fish Nutrition Laboratory (INRA St-Pée, France (1991–2002). He is now deputy head of Department of Animal Physiology and Livestock Systems, INRA Paris. He co-wrote *Poissons de Guyane* (1997), was co-editor of *Food Intake in Fish* (2001) and has published more than 200 scientific papers.

Boulenger

Pike-Characin genus *Boulengerella* CH Eigenmann, 1903
Lake Tanganyika Cichlid genus *Boulengerochromis* J Pellegrin, 1904
Mormyrid genus *Boulengeromyrus* L Taverne & J Géry, 1968
African Pike-Characin sp. *Phago boulengeri* L Schilthuis, 1891

Boulenger's Featherfin Tetra *Bryconaethiops boulengeri* J Pellegrin, 1900
Boulenger's Snaggletooth *Astronesthes boulengeri* JDF Gilchrist, 1902
Armoured Catfish sp. *Hypostomus boulengeri* CH Eigenmann & CH Kennedy, 1903
Suckermouth Catfish sp. *Astroblepus boulengeri* CT Regan, 1904
Boulenger's Toadfish *Batrachoides boulengeri* CH Gilbert & Starks, 1904
Short-headed Sculpin *Cottinella boulengeri* LS Berg, 1906
Tetra sp. *Hyphessobrycon boulengeri* CH Eigenmann, 1907
Characin sp. *Pseudocurimata boulengeri* CH Eigenmann, 1907
Lake Tanganyika Cichlid sp. *Lepidiolamprologus boulengeri* F Steindachner, 1909
Labeo sp. *Labeo boulengeri* D Vinciguerra, 1912
Thorny Catfish sp. *Opsodoras boulengeri* F Steindachner, 1915
African Airbreathing Catfish sp. *Heterobranchus boulengeri* J Pellegrin, 1922
Chinese Gudgeon sp. *Xenophysogobio boulengeri* TL Tchang, 1929
Squeaker Catfish sp. *Euchilichthys boulengeri* JT Nichols & JR La Monte, 1934
Lake Tanganyika Cichlid sp. *Xenotilapia boulengeri* M Poll, 1942
Snake-eel sp. *Dalophis boulengeri* J Blache, J Cadenat & A Stauch, 1970
Boulenger's Anthias *Sacura boulengeri* PC Heemstra, 1973
African Barb sp. *Labeobarbus boulengeri* EJ Vreven, T Musschoot, J Snoeks & UK Schliewen, 2016
Mango Tilapia ssp. *Sarotherodon galilaeus boulengeri* J Pellegrin, 1903

George Albert Boulenger (1858–1937) was a Belgian-British zoologist at the British Museum, London. He graduated from the Free University in Brussels (1876) and worked at the Muséum des Sciences Naturelles, Brussels as assistant naturalist studying amphibians, reptiles and fishes, until moving to London (1880) and taking British nationality (1882). He began work at the BMNH cataloguing the amphibian collection. He became First-class Assistant there (1882), a position he held until retiring (1920). His output was prodigious: nearly 2,600 species described including 1,096 fishes, and 877 scientific papers aided by his almost photographic memory. He was also a violinist and polyglot. He retired to grow and study roses, producing 34 papers and two books on botanical subjects. He was a member of the American Society of Ichthyologists and Herpetologists and was elected its first honorary member (1935). Belgium conferred (1937) on him the Order of Léopold, the highest honour awarded to a civilian. 41 amphibians and 73 reptiles are named after him.

Bounthob

Oriental Cyprinid sp. *Metzia bounthobi* K Shibukawa, P Phousavanh, K Phongsa & A Iwata, 2012

Dr Bounthob Praxaysombath is a zoologist, entomologist and arachnologist at the National University of Laos, Vientiane, where he is Professor and Head of the Deapartment of Biology, Faculty of Science. His published papers are mainly on spiders and freshwater fish. He co-wrote: *Selenops muehlmannorum spec. nov. from Southern Laos (Araneae: Selenopidae)* (2011).

Bourdarie

African Barb sp. *Enteromius bourdariei* J Pellegrin, 1928

Paul Bourdarie (1864–1950) was a French ethologist and sociologist. After visiting the French Congo (1893), he was Secretary General of the African Society of France (1894–1897) and went on a number of tours (1896–1898) during which he lectured on the domestication of elephants in Africa. He was Secretary General of the African Society of France (1894–1897). He promoted the planting of cotton in Africa and was Professor at the Free College of Social Sciences (1908–1914) and lectured on the history and sociology of French Equatorial Africa. He was Managing Director of the Colonial Cotton Association (1917–1921) and a member of the Board of Colonies (1920). He was instrumental in the creation of the Academy of Overseas Sciences and was elected its Permanent Secretary (1922) and was a member of the committee that studied the feasibily of a Trans-Saharan railway (1929). He wrote: *L'éléphant d'Afrique* (1910).

Bourée

Longbarb Dragonfish *Flagellostomias boureei* E Zugmayer, 1913

Henri Jean Alfred Bourée (b.1873) was a French naval officer. He entered the navy (1873) serving on a number of vessels ncluding a submarine, and was promoted to Liuetenant (1904). After a four-year leave of absence in the USA, he was seconded to the service of the Prince of Monaco (1911). Zugmayer honoured him thus: "*Je prie M. H. Bourée, aide de camp de S. A. S. le Prince.*" (This being Prince Albert I of Monaco (1848–1922), who devoted much of his life to oceanography, with whom Zugmayer spent time aboard his Research Vessel named Pricess Alice after his wife). Bourée was a talented photographer and early film-maker. He wrote: *L'océanographie vulgarisée: De la surface aux abîmes* (1912) with Prince Albert. There is a record (1933) of him being president of a photography society.

Bourget

Characin sp. *Astyanax bourgeti* CH Eigenmann, 1908

D Bourget was a French naturalist who was resident in Rio de Janeiro. He was a member of the Thayer Expedition to Brazil (1865–1866), engaged by Agassiz as a collector and preparator. He collected the holotype.

Bourguy

Armoured Catfish sp. *Microlepidogaster bourguyi* A de Miranda Ribeiro, 1911

Dr Hermillo Bourguy Macedo de Mendonça was a Professor (1901) and became (1924) Curator of Zoology at the National Museum of Brazil, Rio de Janeiro. Ribeiro studied under him (c.1907) and later became his deputy at the museum.

Bourret

Garra sp. *Garra bourreti* J Pellegrin, 1928

René Leon Bourret (1884–1957) was a French zoologist. He first arrived in Indochina as a very young soldier (1900) and later worked as a surveyor (1907) and undertook a number of geological surveys (1919–1925). He also undertook a comprehensive herpetological survey of Vietnam and studied Indochinese fauna (1922–1942). He became a Professor at the École Supérieure des Sciences, Université Indochinoise, Hanoi (1925). Among his publications is: *Les Tortues de l'Indochine* (1941), the first detailed monograph to deal with all the chelonians of South-East Asia. Among other taxa, seven reptiles, two amphibians, a bird and a mammal are named after him.

Boutan

Boutan's Sillago *Sillago boutani* J Pellegrin, 1905

Louis Boutan (1859–1934) was a French zoologist and underwater photographer. He studied biology and natural history at the University of Paris before being assigned to organise the French exhibit at Melbourne International Exhibition (1880). He stayed in Australia for 18 months, travelling and studying the animals. He took an appointment at Lille University (1886) also learning to dive. He was appointed professor at Laboratoire Arago (1893) and, with his brother, developed equipment for underwater photography, including lighting. He wrote: *La Photographie sous-marine et les progrès de la photographie* (1898). He led a scientific mission in Indochina to investigate improvements to rice and the culture of pearl oysters (1904–1908). After his return to France he became (1910) Professor of zoology and animal physiology at the University of Bordeaux. Boutan and his brother worked on a diving suit for the French army (1914–1916). After the war, he began research into the artificial production of pearls, one of the first people to investigate this subject. He became (1921) director of the Station Biologique d'Arcachon. Later (1924) he became chair of general zoology of the faculty of science at the University of Algiers and Director of the Station d'Aquaculture et de Pêches de Castiglione and inspector for the Algerian fisheries. He retired to Algeria (1929).

Boutchanga

Claroteid Catfish sp. *Notoglanidium boutchangai* DFE Thys van den Audenaerde, 1965

Honoré Boutchanga was a Gabonese technical assistant, Eaux et Forêts (Waters and Forests). He collected the holotype.

Bouvier

Yellowstone Cutthroat Trout *Oncorhynchus clarkii bouvieri* DS Jordan & CH Gilbert, 1883

Charles E Bendire (1836–1897) in an unpublished manuscript suggested the name *Salmo bouvieri* for this fish in honour of 'Captain Bouvier'. Bendire did not identify Bouvier except to say he was a friend. Both Bouvier and Bendire served as officers in the Union army in the American Civil War and we think that it could be John Vernou Bouvier Sr (1843–1926) who was Second Lieutenant, 20th New York State Militia. He was severely wounded (1862) and captured by Confederate forces. He was part of a prisoner exchange (1862) and re-joined the Union Army (1863) as First Lieutenant, 80th New York infantry and was promoted to Captain (1865). Among his descendants is Jacqueline Lee Bouvier who was first married to John Fitzgerald Kennedy, President of the USA and later to Aristotle Onassis, Greek shipping magnate and billionaire.

Bovalius

Characin sp. *Poecilocharax bovaliorum* CH Eigenmann, 1909

Edward and Dr Edwin Bovalius were members of the Essequibo Exploration Company in British Guiana (Guyana). They were a great help to Eigenmann in providing boats and guides, without which he would probably not have reached the Tumatumari and Kaieteur regions.

Bovalli

Armoured Catfish sp. *Paralithoxus bovallii* CT Regan, 1906

Carl Erik Alexander Bovallius (or Bowallius) (1849–1907) was a Swedish biologist. He was Associate Professor of Zoology at Uppsala Universitet, but gave up teaching and travelled in South America (1897–1900), founding a rubber plantation in Trinidad (1901). In 1904, he began further explorations in Brazil and the Guianas. He died in Georgetown, British Guiana (Guyana). He wrote on crustaceans, in papers such as: *A New Isopod from the Swedish Arctic Expedition of 1883 Described* (1885) A mammal, a reptile and a crustacean genus are named after him.

Bove

African Loach Catfish sp. *Phractura bovei* A Perugia, 1892

Lieutenant Giacomo Bove (1852–1887) was an Italian explorer. He was trained as a naval officer and served as a midshipman on the 'Governolo' ethnographical scientific expedition to the Far East (1872–1873) which mapped the coast of Borneo and visited Malaya (Malaysia), the Philippines, China and Japan. He was chosen to represent Italy on board the 'Vega' – the expedition led by Adolf Erik Nordenskiöld to search for the north-east passage. He had great responsibilities put on him as sailing master, was in charge of the ship's chronometers and made the astronomical observations to enable the ship's position to be fixed. He explored Tierra del Fuego (1881–1882), during which his ship 'San Jose' was driven ashore in a storm. They were rescued by a cutter sent by the English missionaries at Ushuaia. He made a second expedition to Argentina (1883–1884), sailing up the Parana River and exploring as far as the falls at Iguazu. He explored the Congo River (1885–1886) but was not impressed at italy's chances of involvement in the Congo. He contracted fever, returned to Italy and resigned from the navy (1886). Shortly afterwards, he committed suicide at the age of 35 (1887). He wrote: *Notes of a journey in the Missions and high Parana* (1885). He collected the holotype of this species.

Bové

Mormyrid sp. *Petrocephalus bovei* A Valenciennes, 1847

Nicolas Bové (1802–1842) was a Luxembourg gardener, botanist and collector. He worked (1823–1826) in the garden of Chateau de Preisch, before returning to his home town (1826–1827) to study ancient languages. He moved to Paris to work at the Jardin des Plantes of the Muséum National d'Histoire Naturelle, Paris, meanwhile studying natural history under the tutors at the Museum. He was recommended and appointed Director of the Cairo gardens of Ibrahim Pasha and started collecting plants in Egypt, Saudi Arabia and the Yemen in his first year, then in Sinai, Palestine and Syria (1831–1832). The latter trip was primarily on behalf of the Egyptian government, who charged Bové with reporting on these neighbouring regions and the plants that could be introduced to Egypt from there. He returned to Paris (1833) and was appointed as Director of a large garden of acclimatisation at Birkadem, Algeria and collected flora there too. He supplied the type material as well as a great deal of other botanical and zoological specimens to the MNHN. He also collected in other parts of North Africa (Mauritania & Sudan), Madagascar and other Middle Eastern areas (Turkey, Palestine, Israel, Yemen, Saudi Arabia, Syria & Lebanon and the Red Sea) (1830–1841) as well as in France. His particular interest was in algae and spermatophytes.

Bowen, WW

African Minnow sp. *Opsaridium boweni* HW Fowler, 1930

Professor Wilfrid Wedgwood Bowen (1899–1987) was an ornithologist, entomologist who was born in Barbados, educated at Cambridge which awarded his bachelor's degree, and became Professor of Zoology and Director of the Museum of Natural History, Dartmouth College, New Hampshire (1932–1966), which also awarded him an honorary master's degree. He served in the British Army in WW1 and was in the Sudan (1922–1931) where he was Deputy Curator of the Sudan Government Museum, Khartoum. He was with the Gray African Expedition (1929), during which he collected the holotype. He wrote: *Catalogue of Sudan Birds* (1926). Two birds are named after him.

Bowers, GM

Snipe Eel sp. *Avocettina bowersii* S Garman, 1899
Rattail sp. *Bathygadus bowersi* CH Gilbert, 1905
Cheat Minnow *Pararhinichthys bowersi* EL Goldsborough & HW Clark, 1908
Bowers' Parrotfish *Chlorurus bowersi* JO Snyder, 1909

George Meade Bowers (1863–1925) was a Republican politician in the House of Representatives (1916–1923). He was the United States Commissioner of Fish and Fisheries (1898–1903) and Director of the US Bureau of Fisheries (1903–1913).

Bowers

Bowers Bank Snailfish *Careproctus bowersianus* CH Gilbert & CV Burke, 1912

This is a toponym referring to Bowers Bank, a marine region off Kamchatka.

Bowes

Rhomboid Chromis *Chromis bowesi* BG Arango, HT Pinheiro, C Rocha, BD Greene, RL Pylem JM Copus, B Shepherd & LA Rocha, 2019

William Ketcham 'Bill' Bowes Jr. (1926–2016) was an American venture capitalist, co-founder of US Venture Partners, and visionary Bay Area philanthropist who was devoted to advancing science spanning biotech, medicine and other disciplines, usually through the foundation he established (1991). Stanford University awarded his BA in economics (1950 after it was interrupted by WW2) and Harvard his MBA (1952). He served in the infantry in combat landings in Japan and the Philippines. He was honoured as the lead donor in the Hope for Reefs initiative from the Califiornia Acdemy of Science.

Bowman

Bowman's Rainbowfish *Melanotaenia bowmani* G R Allen, PJ Unmack & RK Hadiaty, 2016

Ron Bowman (b.1924) is an Australian aquarist and breeder of rainbowfishes. He is a life member and past president of the Australia and New Guinea Fishes Association. The etymology says: "*The new species is named bowmani in honour of Ron Bowman, a widely respected Australian aquarist, in recognition of his many years of rainbowfish breeding expertise and knowledge sharing as well as countless contributions and exemplary leadership in connection with the Australia New Guinea Fishes Association (ANGFA) and its journal* 'Fishes of Sahul'. The choice of Bowman was suggested to Dr Allen by ANGFA.

Boyd Walker

Professor Brotula *Ogilbia boydwalkeri* PR Moller, W Schwarzhans & JG Nielsen, 2005

Professor Dr Boyd W Walker (1917–2001) The etymology says: "*Named in honour of the late Dr. Boyd Walker, who studied* Ogilbia *for many years.*" (See **Walker, BW**)

Boyer

Boyer's Sand Smelt *Atherina boyeri* A Risso, 1810

Guillaume Boyer de Nice was a poet. The species was described in a book about the ichthyology of Nice. Risso chose to name the smelt after a historic son of that city; the poet who was a "*…troubadour distingué du troisième siècle.*"

Boyle, C

Peppermint Angelfish *Centropyge boylei* RL Pyle & J Randall, 1992

Charles 'Chip' J Boyle is a diver and fish collector, He is the owner of 'Cook Islands Aquarium Fish Ltd'.

Boyle, HS

Pencil Catfish sp. *Trichomycterus boylei* JT Nichols, 1956

Howarth Stanley Boyle (1894–1951) was collecting in North America and publishing the results (1900s). He collected in Central America (1911) and spent two years (1914–1915) undertaking fieldwork in a number of countries in South America (Colombia, Bolivia and Argentina) as assistant to Miller for the AMNH. He left AMNH (1917) to join the navy at the Naval Base Hospital. He published widely (1915–1956), both alone and with a number of co-authors. A reptile and a bird are named after him.

Boyle, W

Broad-banded Pipefish *Dunckerocampus boylei* RH Kuiter, 1998

Bill Boyle, an underwater fish photographer, brought this species to the author's attention.

Brachet

African Rivuline sp. *Micropanchax bracheti* HO Berkenkamp, 1983

Heinrich 'Heinz' Brachet (1919–2011) was an aquarist of Wiesbaden, West Germany, who was Chairman of the Wiesbadener Aquarienverein for 25 years. He had 80 tanks and many breeding successes, especially of guppies. He made collecting trips in southeast Asia, Africa and some islands in the Indian and Pacific Oceans including three collecting trips to Togo (1978–1980).

Bradbury

Dreamer sp. *Oneirodes bradburyae* MG Grey, 1956
Batfish sp. *Coelophrys bradburyae* H Endo & G Shinohara, 1999
Bradbury's Triangular Batfish *Malthopsis bradburyae* HC Ho, 2013

Dr Margaret G 'Maggie' Bradbury (1927–2010) was an artist and zoologist. She became (1947) Staff Artist of the Department of Zoology at the Chicago Natural History Museum (now Field Museum of Natural History, FMNH). She was awarded her Zoology BSc (1955) by the University of Chicago. That summer she made her first overseas collecting trip to the Bahamas. She then went on to the Stanford University graduate school, working part time as a teacher and technical assistant. Stanford awarded her PhD (1963) by which time she had already become Assistant Professor of Biology at MacMurray College. Her dissertation was on batfish which she continued to study for the next four decades. She returned (1963) to California to join the faculty of San Francisco State College (now University, SFSU) as Assistant Professor of Biological Science, becoming Associate Professor (1967) and full Professor (1971) until retiring (1994). She taught a wide variety of courses including Ichthyology, Biology of Fishes, Comparative Vertebrate Anatomy, History of Biology, Evolution, and Introductory Biology. She took part in a number of cruises to the Indian Ocean, Pacific etc. She published widely particularly on batfish such as: *Batfishes of the Galapagos Islands with descriptions of two new species of* Dibranchus *(Teleostei: Ogcocephalidae)* (1999). (Also see **Margaret (Bradbury)**)

Brahm

Brahm's Dwarfgoby *Eviota brahmi* DW Greenfield & L Tornabene, 2014

Brahm Kai Erdmann is the son of Canadian marine biologist Mark V Erdmann, Curator Emeritus ROM. Brahm was honoured because he pointed out that fishes similar to *E. nigriventris* captured at Raja Ampat did not match the photos of that species in his father's book, *Reef Fishes of the East Indies* (2012).

Brahmachary

Dwarf Snakehead sp. *Channa brahmacharyi* P Chakraborty, K Yardi & P Mukherjee, 2020

Professor Dr Ratan Lal Brahmachary (1932–2018) was an Indian biochemist at the Indian Statistical Institute although he initially studied astrophysics switching to molecular biology (1957). He studied animal behaviour (from 1970) particularly big cats and visited Africa fourteen times as well as South America, Borneo and Indonesia then (1979) concentrating on mammal pheromones. He once said: "*Biology is as fascinating as probing the mysteries of the physical universe. The inner universe of an organism or of an ecosystem is as challenging as the outer Universe of the expanding cosmos.*" He wrote numerous papers and several books including *My Tryst With Big Cats* (2013) and *The Neurobiology of Chemical Communication*

(2014). He was awarded a DSc by the University of Calcutta (2008). He was honoured as a "*distinguished ethologist, biochemist and pioneer in tiger pheromone research.*" He died of pneumonia and left his body to science.

Brahui

Stone Loach sp. *Triplophysa brahui* E Zugmayer, 1913

This is not a true eponym; the Brahui are an ethnic group of people in Baluchistan, Pakistan and live in the locality where the holotype was acquired.

Brandt

Sculpin sp. *Myoxocephalus brandtii* F Steindachner, 1867
Pacific Redfin *Tribolodon brandtii* B Dybowski, 1872
White PIranha *Serrasalmus brandtii* CF Lütken, 1875
Kura Loach *Oxynoemacheilus brandtii* KF Kessler, 1877

Johann Friedrich (Fedor Fedorovich) von Brandt (1802–1879) was a German zoologist, surgeon, pharmacologist and botanist who moved to Russia (1831). He explored Siberia and was founding Director of the Zoological Museum of the Academy of Science, St Petersburg. He described several birds from western North America, including an eponymous cormorant. He produced works on systematics, zoogeography, comparative anatomy and the palaeontology of mammals. He co-wrote the two-volume: *Medical Zoology* (1829–1833) and solely wrote: *Descriptiones et Icones Animalium Rossicorum Novorum vel minus Rite Cognitorum* (1836) dealing in particular with Russia. Other taxa such as insects, two birds, five mammals and a reptile are named after him.

Branicki

Sand Grunt *Pomadasys branickii* F Steindachner, 1879
Armoured Catfish sp. *Chaetostoma branickii* F Steindachner, 1881

It is not known which member of the Branicki family is honoured in the names of these species but it is probably Hieronim Florian Radziwill Konstanty Count Branicki (1824–1884). The Branicki family are so important and influential in Polish zoology that it is worth a little deviation to explain who they were and what they achieved. The first generation of the family were Alexander, Count Branicki (1821–1877), Hieronim Florian Radziwill Konstanty Count Branicki (1824–1884) and his wife Jadwiga (Hedwiga) Potocka, Countess Branicka (1827–1916) and the youngest brother Count Wladyslaw (Ladislas) Michael Branicki (1848–1914). The next generation is represented by the son of Hieronim and Jadwiga, Xavier, Count Branicki (1864–1926) and Countess Anna Branicka (1876–1953), who was a daughter of Count Wladyslaw (Ladislas) Michael Branicki. The wealthy aristocratic Polish Branicki family was very interested in hunting and Alexander and Konstanty started hunting abroad (1863), visiting Egypt, Sudan, Palestine and Algeria. Zoologists such as Taczanowski (q.v.) and Waga accompanied them, and they were persuaded to take a more scientific approach to their favourite pastime (1867). They employed Konstanty Jelski (q.v.), amongst others, to collect for them in South America. This resulted in the creation of the Branicki Museum in Poland. Poland was then part of the Imperial Russian Empire and they feared that the museum in St Petersburg might be covetous and transfer the collections, so

Konstanty decided to protect it by establishing a private museum. This was achieved (1887) after his death by his nephew Wladyslaw and his own son Xavier. Following the Treaty of Versailles (1919), Poland became independent and the collections were transferred to the state. Eight birds and a mammal are named after various members of the family.

Branner

Branner's Livebearer *Micropoecilia branneri* CH Eigenmann, 1894
Three-barbeled Catfish sp. *Rhamdia branneri* JD Haseman, 1911

Dr John Casper Branner (1850–1922) was an American geologist who graduated (1882) from Cornell and was awarded his doctorate (1885) from the University of Indiana. He was at Stanford University as Professor of Geology (1891–1913), and as President of the University (1913–1915). He was an expert on Brazilian geology and visited the country several times, notably as leader of the Branner-Agassiz Expedition (1899). He wrote *Notes on the fauna of the islands of Fernando de Noronha.* (1888). A mineral, Brannerite, is also named after him, as is an amphibian.

Bransford

Characin sp. *Astyanax bransfordii* TN Gill, 1877

Dr John Francis Bransford (1846–1911) was an assistant US naval surgeon (1872–1890) on the Nicaragua and Panama Canal surveys (1872–1888). He made three separate herpetological collections: in Nicaragua (1875 and 1885) and in Panama (1875). He was recalled to the colours (1898) for the Spanish-American War, and he retired again (1901) with the rank of Surgeon. He published a number of books such as: *Archæological researches in Nicaragua* (1881). An amphibian and a reptile are also named after him.

Brashnikow

Braschnikow's Shad *Alosa braschnikowi* NA Borodin, 1904

(See **Bražnikov** below)

Brauer

Brauer's Lanternfish *Gymnoscopelus braueri* E Lönnberg, 1905
Barbeled Dragonfish sp. *Eustomias braueri* E Zugmayer, 1911
Wormline Peckoltia *Peckoltia braueri* CH Eigenmann, 1912
Snaketooth sp. *Chiasmodon braueri* MCW Weber, 1913
Brauer's Dragonfish *Photonectes braueri* E Zugmayer, 1913
Shovel-nose Grenadier *Coelorinchus braueri* KH Barnard, 1925
Bristlemouth sp. *Cyclothone braueri* PC Jespersen & AV Tåning, 1926

Professor Dr August Bernhard Brauer (1863–1917) was a German herpetologist and ichthyologist. He graduated from Humboldt-Universität, Berlin, in natural sciences (1885) and took his doctorate there (1892). He collected in the Seychelles (1897) and was on the *Valdivia* expedition (1898), describing (1908) the fishes they collected. He became Professor at Berlin University (1905) and Director of the university's Zoological Museum (1906). He was appointed Professor of the Zoological University, Berlin (1914). A mammal, an amphibian and a reptile are named after him.

Braun, A

African Cyprinid sp. *Labeobarbus brauni* J Pellegrin, 1935
[Syn. *Varicorhinus brauni*]

André Braun de Ter Meeren (1907–1982) was the local district administrator of the Belgian authorities in the Congo. The entomologist Louis Burgeon wrote of Braun making useful collections in the region of Lake Kivu (the type locality) (January 1934) and in the same paragraph reports that Babault had given him some interesting specimens. It is notable that Burgeon named a beetle *Promegalonychus brauneanus* (1933), presumably after the same Braun who wrote: *Etude sur les migrations des divers clans composant actuellement la population des Bahavu* (1934). The only reference in the etymology is that Braun collected the type with the explorer Guy Babault, but we believe André to be the Braun in question.

Braun, CD

Braun's Whale Shark *Rhincodon typus* A Smith, 1829
[Alt. Whale Shark]

Camrin D Braun is an American ichthyologist. The College of Idaho awarded his bachelor's degree (2011) and King Abdullah University of Science and Technology, Saudi Arabia, his master's (2013). His doctorate (2018) was earned at the Woods Hole Oceanographic Institution in USA, and he is now a Postdoctoral Research Associate at the Applied Physics Laboratory at the University of Washington, Seattle. He wrote *Movements of juvenile whale sharks* (Rhincodon typus) *in the Red Sea* (2010) and it is only subsequent to the publication of that thesis that his name has been occasionally attached to this fish, though generally it is simply known as the Whale Shark. Braun is also a National Geographic Young Explorer and an elected member of the Explorer's Club.

Braun, J

Blue-lined Hulafish *Trachinops brauni* GR Allen, 1977
Braun's Pughead Pipefish *Bulbonaricus brauni* CE Dawson & GR Allen, 1978
Braun's Wrasse *Pictilabrus brauni* JB Hutchins & SM Morrison, 1996

John Braun of Perth. The paper describing the pipefish does not help further identify the honouree, saying only that J Braun caught the pipefish holotype off Western Australia (November 1976). In the hulafish etymology he is honoured for his much-appreciated assistance in the field.

Brauner

Goby sp. *Benthophiloides brauneri* DE Beling & BS Iljin, 1927

Aleksandr Aleksandrovich Brauner (1857–1941) was a Ukrainian zoologist, taxonomist and archaeologist who was Professor of Zoology at Odessa (1915). He collected the goby holotype. A bird is also named after him.

Brausch

Blood-throat Cichlid *Thoracochromis brauschi* M Poll & DF Thys van den Audenaerde, 1965

Georges Brausch was a Belgian government official in the (then) Belgian Congo. He wrote: *Belgian Administration in the Congo* (1961). He collected the holotype.

Bravo

Bravo's Barbelgoby *Gobiopsis bravoi* AW Herre, 1940

Pablo Bravo was a scientific illustrator including for reports of the US Fish & Wildlife Service: e.g. *Juvenile Forms of Neothunnus macropterus, Katsuwonus pelamis and Euthynnus yaito from Philippines Seas* (1950). He illustrated many of Herre's (q.v.) papers, such as: *Gobies of the Philippines and the China Sea* (1927).

Bray

Viviparous Brotula sp. *Saccogaster brayae* JG Nielsen, W Schwarzhans & DM Cohen, 2012

Dianne J Bray is Senior Collections Manager, Vertebrate Zoology, Museums Victoria, Australia which she joined (1999) having previously worked at the Australian Museum, Sydney for 25 years. She has a master's degree awarded by Macquarie University. She wrote *Why do we depend on estuaries? An environmental educational resource* (1993). She was honoured for her support with material for the authors' revision of the genus.

Brayton

Tamaulipas Shiner *Notropis braytoni* DS Jordan & BW Evermann, 1896

Dr Alembert Winthrop Brayton (1848–1926) was a physician and a naturalist. After studying at Cornell (1871–1872) he moved to Butler College, which awarded his bachelor's degree (1878) and a master's (1880). In the meantime, he qualified as a physician at Indiana Medical College (1879) and in the same year was appointed Professor of Chemistry, Toxicology and Medical Jurisprudence at the College of Physicians of Indianapolis. He was appointed Professor of these same subjects at the Medical College of Indiana (1881), transferring (1885) to become Professor of Physiology and ultimately to the chair of Pathology, Clinical Medicine and Dermatology. He was editor of the *Indiana Medical Journal* (1892–1911). In the etymology of the Tamaulipas Shiner, Jordan wrote "*...with pleasant memories of our explorations in Georgia and the Carolinas*" and Brayton was part of Jordan's expedition (1877) under the auspices of the US Fish Commissioner. Brayton wrote: *A catalogue of the birds of Indiana, with keys and descriptions of the groups of greatest interest to the horticulturist* (1879) and co-wrote with Jordan: *Contributions to North American Ichthyology [microform]: based primarily on the collections of the United States National Museum* (1878)

Bražnikov

Caspian Marine Shad *Alosa braschnikowi* NA Borodin, 1904
Lumpfish sp. *Cyclopteropsis brashnikowi* PJ Schmidt, 1904
Brashnikov's Catfish *Tachysurus brashnikowi* LS Berg, 1907
Hairhead Sculpin *Trichocottus brashnikovi* VK Soldatov & MN Pavlenko, 1915
Eelpout sp. *Gymnelopsis brashnikovi* VK Soldatov, 1922
Snailfish sp. *Liparis brashnikovi* VK Soldatov, 1930

Vladimir Konstantinovich Bražnikov (or Braschnikow) (1870–1921) was a Russian zoologist, ichthyologist and entomologist with an interest in Diptera who lived in Tokyo for many years. He was a chief of fisheries who organised several expeditions in the Amur River basin. He led an expedition (1912) and organised the first Russian School of Fisheries, in Moscow (1913). He noted (1898) that the shad was a distinct variety of *A. saposchnikowii.*

Brazza

African Barb sp. *Enteromius brazzai* J Pellegrin, 1901

Jacques (or Giacomo) C Savorgnan de Brazza (1859–1888) trained as a naturalist but dreamed of following his elder brother as an explorer. He studied geology in the Alps and became well-known as a mountaineer. He was appointed (1883) the Director of Le Mission de l'Ouest Africain. Jacques amassed a fish collection, mainly from the Ogowe River that was sent to MNHN, Paris (c.1885) and which Pellegrin subsequently studied. Jecques often accompanied his elder brother, Count Pierre Paul François Camille Savorgnan de Brazza (1852–1905), who was a distinguished French explorer of the Congo and later became Governor of the French Congo. The city of Brazzaville is named after him. It is not clear whether one or other of the brothers collected the specimen, although it seems likely. His health never recovered from his time in the Congo and he died in Roma of scarlet fever. Two birds are also named after him.

Breder

Confused Podge *Pseudogramma brederi* SF Hildebrand, 1940
Halfbeak sp. *Hyporhamphus brederi* A Fernández-Yépez, 1948

Dr Charles Marcus Breder Jr (1897–1983) was an experimental and behavioural ichthyologist whose doctorate was honorary, awarded by the University of Newark (1938). After leaving high school he worked for the US Bureau of Fisheries (1919–1921) and for the New York Aquarium (1921–1944) and became acting director (1937) and director (1940). He was Curator and Chairman of the Department of Fishes at the AMNH (1944–1965) and was an interim director of the Mote Marine Laboratory (1967). He was a leader in the study of cavefishes and was a dominant figure in this area of research in the 1940s and 1950s. He led the expedition known as 'The Aquarium Cave Expedition to Mexico' (1940). He died from Alzheimer's disease.

Brédo

Mweru Elephantfish *Campylomormyrus bredoi* M Poll, 1945
African False Sardine (cyprinid) sp. *Engraulicypris bredoi* M Poll, 1945
[Syn. *Mesobola bredoi*]

Hans Joseph Anna Erich Richard Brédo (1903–1991) was a Belgian government entomologist who collected in the Congo (1930s). He was sent to the Congo to advise the Belgian government on all locust matters (1930) and later was appointed Director of the International Red Locust Team (1949–1951). His major achievement was a scientific explanation for locusts swarming (i.e. a succession of environmental conditions affecting the behaviour of these usually non-gregarious insects). He wrote a number of scientific papers such as: *La lutte biologique et son importance econimique au Congo Belge* (1934). Two birds are also named after him.

Bree

Bree's Cory *Corydoras breei* IJH Isbrücker & H Nijssen, 1992

Dr Peter Johannes Hendrikus van Bree (1927–2011) was a marine mammalogist. He was Curator of Mammals (1960–1992) at the Zoological Museum of Amsterdam, University of Amsterdam, where he had been awarded a bachelor's degree (1955) and a doctorate (1958). He officially retired (1992) but was asked to stay on as an honorary collaborator for several years. He wrote *On Globicephala sieboldii Gray, 1846, and other species of pilot whales* (1971). This catfish species was named after him to mark his retirement.

Breeden

Blackcheek Moray *Gymnothorax breedeni* JE McCosker & JE Randall, 1977

Victor Ellis Breeden (1927–1998) was president of the Charline H Breeden Foundation, which gave financial support to the expedition off the Comoros Islands, which collected the moray holotype and studied other Comoran fishes. He had been married (1955) to Charline Breeden nee Humphreys (1929–1972).

Breidohr

Angostura Cichlid *Vieja breidohri* U Werner & R Stawikowski, 1987

Hans-Günther Breidohr (1938–2015) was a German cichlid aquarist from Wuppertal who was Uwe Werner's friend and collecting partner. Together they made about 30 trips to Central and South America, starting with a trip to Costa Rica (1980). He was responsible for the discovery and first import of many New World cichlids.

Breitenstein

Asian Stream Catfish genus *Breitensteinia* F Steindachner, 1881

Dr Heinrich Breitenstein (1848–1930) was a German physician who served with the Dutch East Indies army for 21 years. While in Borneo he collected herpetofauna that Steindachner purchased for the Naturhistorisches Museum Wien, Austria. He published his memoirs as *21 Jahre in Indien; Aus dem Tagebuchen eines Militärarztes* (1899). A reptile is named after him.

Brembach

Viviparous Halfbeak sp. *Nomorhamphus brembachi* D Vogt, 1978

Dr Manfred Brembach (b.1949) is a German aquarist, passionate photographer and musician. He taught, and also wrote children's books as well as a volume about his home river, the Trave. He wrote a book about his two research trips to Malaysia, Indonesia and Singapore where he went to remote locations looking for live-bearing fish, of which he said "I discovered and described six new species of fish. That was the basis for my doctoral thesis." He has been in six bands and made a number of CDs all recorded at his studio, known as the 'ornamental fish studio' as it is in what was his father's aquarium shop. He was honoured for his many contributions to the knowledge of Hemiramphidae (the halfbeak's presumed family at the time).

Brett

Goby sp. *Coryogalops bretti* M Goren, 1991

Gregg Brett was an ichthyologist at the East London Museum, South Africa who is now retired. He helped collect the type specimen.

Breunig

Goby sp. *Gymnogobius breunigii* F Steindachner, 1879

Professor Dr Ferdinand Breunig was Professor of Natural History, Imperial Gymnasium at Schotten (Vienna) and Curator of their natural history collection. He was honoured as a token of Steindachner's 'special veneration' (translation) of him.

Breuseghem

Rectangle-spot Moon Tetra *Bathyaethiops breuseghemi* M Poll, 1945

Dr Raymond Van Breuseghem was a mycologist at the Belgian Institute of Tropical Medicine, Antwerp. He collected the holotype.

Brevoort

Menhaden genus *Brevoortia* TN Gill, 1861
Tropical Pomfret *Eumegistus brevorti* F Poey, 1860
Hairfin Lookdown *Selene brevoortii* TN Gill, 1863
Lined Rockskipper *Salarias brevoorti* HW Fowler, 1946
[Now regarded as a Junior Synonym of *Blenniella bilitonensis*]

James Carson Brevoort (1818–1887) was an American collector of coins and rare books. He was private secretary to Washington Irving, then US Ambassador to Spain (1838–1839) and afterwards travelled in Europe (1939–1843). After returning to New York from Europe he served in a number of capacities in institutions like the Astor Library and the Academy of Natural Sciences and developed his collections to include both books on and specimens of entomology and ichthyology. He wrote a number of short books such as: *Notes on some figures of Japanese fish : taken from recent specimens by the artists of the US Japan expedition* (written 1856) and *Verrazno the Navigator.*

NB Regarding the pomfret, Poey first uses the spelling with one o (*brevorti*), which therefore must stand. But, in the index of the same work, he uses the spelling Brevoorti – which he presumably intended all along.

Brian Greene

Latigo Fairy Wrasse *Cirrhilabris briangreenei* Y Tea, RL Pyle & LA Rocha, 2020

The etymology says: We name this species *Cirrhilabrus briangreenei* in honor of Brian D. Greene, who in addition to collecting the type specimens, has contributed extensively towards the study and exploration of coral-reef diversity (particularly on MCEs) through deep technical diving. (See Greene, BD)

Brianne

Batangas Groppo *Grammatonotus brianne* WD Anderson, BD Greene & LA Rocha, 2016

Brianne M Greene née Atwood was the second author's wife. (See above)

Brichard

Blind Spiny Eel *Mastacembelus brichardi* M Poll, 1958
African Barb sp. *Enteromius brichardi* M Poll & JG Lambert, 1959
Brichard's Synodontis *Synodontis brichardi* M Poll, 1959
Congo Cichlid sp. *Teleogramma brichardi* M Poll, 1959
Cherry Red Congo Tetra *Alestopetersius brichardi* M Poll, 1967
Brichard's Lampeye *Poropanchax brichardi* M Poll, 1971
Lake Tanganyika Cichlid sp. *Chalinochromis brichardi* M Poll, 1974
Fairy Cichlid *Neolamprologus brichardi* M Poll, 1974
Blue-eyed Tropheus (cichlid) *Tropheus brichardi* MHJ Nelissen & DFE Thys van den Audenaerde, 1975
Lake Tanganyika Catfish sp. *Phyllonemus brichardi* LM Risch, 1987
Lake Tanganyika Cichlid sp. *Telmatochromis brichardi* P Louisy, 1989

Jean-Pierre Brichard (1921–1990) was a Belgian explorer, and a collector and exporter of fish for the aquarium trade. He fought in the Belgian army (1940) and after the defeat of Belgium, was a member of the Belgian underground resistance movement. He was bored by post-war Belgium and the family emigrated to the Belgian Congo (Zaire) where he developed a passion for tropical fish and started exporting. When the Congo became independent and descended into civil war, because of the terrible atrocities his family returned to Belgium, but he stayed behind. He continued his business activity, albeit with a machine gun, mounted and manned 24 hours a day at the top of the stairs. In the 1970s he was fortunate in finding a buyer for his business and assets and the family moved to Burundi and settled at Bujumbara on Lake Tangyanika. They founded a new business, 'The Fishes of Burundi' that is still run by the family. He wrote: *Fishes of Lake Tanganyika* (1980). He collected a number of the eponymous species including the spiny eel.

Brien

Mormyrid genus *Brienomyrus* L Taverne, 1971
African Rivuline sp. *Nothobranchius brieni* M Poll, 1938
African Loach Catfish sp. *Belonoglanis brieni* M Poll, 1959
African Loach Catfish sp. *Dolichamphilius brieni* M Poll, 1959
Marbled Lungfish ssp. *Protopterus annectens brieni* M Poll, 1961
Lake Tanganyika Cichlid sp. *Paracyprichromis brieni* M Poll, 1981

Paul Louis Philippe Brien (1894–1975) was a Belgian biologist, zoologist and politician. He was at the Free University Brussels as a student (1918–1922), becoming Doctor of Natural Sciences, after which he became a research assistant there. He rose to become a lecturer in animal physiology (1928) and then to professor until retirement (1930–1964). He collected fishes in the Belgian Congo (1937). He was passionately anti-Fascist, being President of the Popular Front (1935–1936), and was elected local senator for the Communist Party in Brabant (1936) but resigned after a few months. During WW2, his political opinions did not find favour with the occupiers of Belgium and he was arrested, interrogated and imprisoned by the Gestapo (1942). Among his 350 publications were: *Elements of Zoology and Notions of Comparative Anatomy* (1938) and *Biology of animal reproduction - Blastogenesis, Gametogenesis, Sexualization* (1966). An amphibian is also named after him.

Briggs, EA

Briggs' Crested Pipefish. *Histiogamphelus briggsii* AR McCulloch, 1914

Dr Edward Alfred Briggs (1890–1969) was a zoologist at the Australian Museum (1912–1918). He collected the holotype off Tasmania (1914). He was Lecturer in Zoology at the Sydney Technical College (1916–1932), at the University of Sydney (1933–1935) going on to be Reader in Zoology there (1944–1945). He also led New Guinea expeditions (1924–1927) and was the editor (1928–1932) of reports on the Australian Antarctic Expedition (1911–1914).

Briggs, JC

Clingfish genus *Briggsia* MT Craig, 2009
Clingfish sp. *Aspasmodes briggsi* JLB Smith, 1957
Clingfish sp. *Propherallodus briggsi* M Shiogaki & Y Dotsu, 1983
Clingfish sp. *Tomicodon briggsi* JT Williams & JC Tyler, 2003

John C 'Jack' Briggs (1920–2018) was a prominent scholar and a pioneer in the field of biogeography where his contributions are widely recognised. He started out as a forestry major at Oregon State University but switched to fish and wildlife. While in college he earned a private pilot's license, which came in handy when he joined the Army Air Corp's Air Transport division (1943). After the war he studied biology under the GI Bill obtaining his MA (1947) and his PhD (1952) from Stanford. Over the years he held teaching, research and administrative positions at various universities in the USA and Canada concluding with a 30-year career at the Marine Science Department at University of South Florida, latterly as professor emeritus. He continued his writing after moving away from Florida, first in association with University of Georgia and then back to Oregon State University. His research interest evolved from fish life history and systematics in the early days to historic and contemporary distributions and biodiversity of living things on land and sea. He remained active in research until his final days, still publishing papers (2017) and working on three more (2018). In addition to over 150 scientific articles and six books and monographs he also wrote a science-fiction book for his grandchildren *A Mesozoic Adventure* (2007) and an autobiography *A Professorial Life* (2009). One etymology says: "*The species is named in honor of John C. Briggs, Georgia Museum of Natural History, in recognition of his pioneering monograph on clingfish systematics.*"

Brigitte

Mosquito Rasbora *Boraras brigittae* D Vogt, 1978

Brigitte Vogt is the wife of the author Dieter Vogt (b.1933).

Brousseau

Neotropical Rivuline sp. *Spectrolebias brousseaui* DTB Nielsen, 2013

Dr Roger D Brousseau is a Californian aquarist and killifish hobbyist. He has made many collecting trips to Bolivia and discovered this species there (2008). He wrote: *A Hobbyist's Guide to South American Annual Killifish* (1994). He described another fish with Nielsen - *Spectrolebias pilleti*. (Also see **Ashley**)

Broussonnet

Violet Goby *Gobioides broussonnetii* BG de Lacepède, 1800
Striped Drum *Umbrina broussonnetii* G Cuvier, 1830
Broussonnet's Mullet *Mugil broussonnetii* A Valenciennes, 1836

Dr Pierre Marie Auguste Broussonet (1761–1807) was a physician-naturalist. He was honoured for his contributions to natural history, particularly his studies of the flora and fauna of Morocco. Lacepède, Cuvier and Valenciennes all apparently misspelled Broussonet's name, with an extra 'n'; since this spelling is in prevailing usage, amendment is not likely.

Brown

Nigerian Tongue Sole *Cynoglossus browni* P Chabanaud, 1949

Dr Brown was Director of the Fisheries Researches Institute of British West Africa. We can find no further biographical detail.

Brown, B & A

Browns' Betta *Betta brownorum* KE Witte & J Schmidt, 1992

Barbara Brown née Demaree and Allan J Brown (1911–2009) were the first to collect this species in Sarawak. (See also **Allan** & **Barbara (Brown)**)

Brown, CB

Characin sp. *Moenkhausia browni* CH Eigenmann, 1909

Charles Barrington Brown (1839–1917) was a British geologist. He was a government surveyor in British Guiana (Guyana) (1867–1873). He discovered the Kaieteur Falls (1870) and explored the Amazon River and its tributaries (1873–1875). He was in British Guiana and Surinam, examining possibilities for mining for gold (1887–1891). He also worked in Burma (Myanmar) in relation to the country's ruby mines, and mined for gemstones in Ceylon (Sri Lanka), the USA and Australia. He wrote: *Canoe and Camp life in British Guiana* (1876).

Brown, JC

Danio sp. *Devario browni* CT Regan, 1907

Dr John Coggin Brown (1884–1962) was a geologist who was a superintendent of the Geological Survey of India. The Mining, Geological and Metallurgical Institute of India awards the 'Dr J Coggin Brown Memorial Gold Medal' in his honour. He wrote extensivlely on Asian geology in general and on India in particular, such as: *India's Mineral Wealth* (1936). He collected the holotype. (See also **Coggin**)

Brown, M

Lake Victoria Cichlid sp. *Haplochromis brownae* PH Greenwood, 1962

Dr Margaret 'Peggy' Varley née Brown (1918–2009) was an influential British fish ecologist who was born in India. Her PhD was awarded (1945) when she defended her dissertation that described the growth physiology of Brown Trout *Salma trutta*, a seminal work that is still cited today. She married (1955) fellow ecologist George C Varley and published under her husband's name, such as *Physiology of Fishes* (1957), which is said to have effectively created the field of ecophysiology. She worked with Greenwood as a visiting scientist (c.1950–1951) at the East African Freshwater Fisheries Research Organization in Jinja, Uganda, on the shore of Lake Victoria, where he was a Research Officer (1950–1957). There she worked with tropical-fish ecologist Rosemary Lowe-McConnell (1921–2014), who was also in Uganda at the time and with whom she later studied fish together in the Brazilian Amazon. She was also one of the founders (1969) of the Open University.

Brown, R

Spiny-tailed Leatherjacket *Acanthaluteres brownii* J Richardson, 1846

Dr Robert Brown (1773–1858) was a botanist, educated at Aberdeen and Edinburgh, who studied the flora of Scotland (1791). He was an army official in Ireland (1795) and naturalist to Captain Flinders's (q.v.) Australasian expedition (1801–1805) on board *Investigator*. He was librarian to the Linnean Society and to Sir Joseph Banks (q.v.) and wrote: *Prodromus Florae Novae Hollandiae et insula VanDiemen* (1810). Richardson wrote: "…*During Captain Flinder's voyage of discovery round Australia, Mr. Ferdinand Bauer made highly finished coloured drawings of fish which are now in the possession of Dr. Robert Brown, and which this gentleman has kindly permitted me repeatedly to examine*." Three birds are also named after him.

Browne, CE

Stone Loach sp. *Physoschistura brunneana* TN Annandale, 1918

Charles Edward Browne (b.1861) was political adviser at Yawnghwe (Nyaung Shwe), Southern Shan State, Burma (Myanmar). He was born in Darjeeling, India, where his father was serving in the British Army. He was educated in England and joined the Burma Civil Service (1880). He held ministerial appointments in the Chin Hills and the Southern Shan State (1886). He stayed on in Burma after he retired (pre-1927) and lived at Kalaw.

Browne, P

Caribbean Moonfish *Selene brownii* G Cuvier, 1816

Dr Patrick Browne (1720–1790) was an Irish physician, botanist and historian. He studied medicine in Paris and Leiden and graduated from the University of Rheims (1742). After a short period of further study, he took up a post at St Thomas's Hospital, London. After this he lived for many years in the Caribbean, in Antigua, to which he had made an earlier visit

(1737), Jamaica, Saint Croix and Montserrat, but when he retired (1771) it was to his native County Mayo. He wrote: *Civil and Natural History of Jamaica* (1756), which included new names for 104, mostly botanical, genera. A species of hutia, a Jamaican endemic rodent, is also named after him.

Brownfield

Brownfield's Wrasse *Halichoeres brownfieldi* GP Whitley, 1945

Edward John Brownfield was the then Acting Chief Inspector of Fisheries and Game, Perth, Australia. He began his public service career as a junior clerk (1921) and his appointment as Deputy Inspector was cancelled (1953).

Brownrigg

Surge Damselfish *Chrysiptera brownriggii* JW Bennett, 1828

Sir Robert Brownrigg (1759–1833) was an army officer and colonial administrator. He was commissioned (1775) and became Military Secretary to the Duke of York (1795–1803). He was made Quartermaster-General to the Forces (1803–1811) and made colonel (1804–1833). He was appointed Governor of Ceylon (1812–1820) and (1815) acquired the last part of Ceylon not under British rule. In recognition of this he was made a baronet (1816). He fought against the Great Rebellion (1817–1818), managing to defeat the Ceylonese insurgents with re-inforcements sent from India. He was promoted to full General (1819) but left Ceylon the next year (1820). There is no etymology given in the original description, in Bennett's *A selection from the most remarkable and interesting fishes found on the coast of Ceylon*. We think this may be because the author believed it would be self-evident, due to Brownrigg's fame.

Bruce, AJ

Pipefish sp. *Bulbonaricus brucei* CE Dawson, 1984

Dr Alexander James 'Sandy' Bruce (b.1929) studied as a medical doctor, but left that career to follow his true passion: the taxonomy of shrimps. He settled in Australia (1970s) to become the Director of Heron Island Research Station on the southern end of the Great Barrier Reef. From there he moved to Darwin to become Curator of Crustacea and head of the Natural Science program in the Northern Territory Museum. On retirement he moved to Brisbane and an honorary research position at the Queensland Museum. He collected the holotype of this pipefish at Pangani, Tanzania.

Bruce, CA

Gray's Stone Loach *Balitora brucei* JE Gray, 1830

Charles Alexander Bruce (1793–1871) was an explorer, author and soldier in the Honourable East India Company's Army best known as the father of the India tea industry. His elder brother, Robert Bruce (d.1824), a trader and explorer, learned of a native tribe in Assam that was growing a variety of tea unknown at that time to the outside world. He carefully negotiated their permission to acquire plants and seeds, but died soon afterwards, but had told his young brother about it. Charles fought in the First Anglo-Burmese War (1824) when he was posted to Sadiya as Commandant of gunboats. The East India Company assigned him the task to start tea plantations (1835). He resigned his commission to start (1836) a tea plantation of indigenous plants in Assam and was soon (1839) in full production. He wrote: *An Account of the Manufacture of the Black Tea, As Now Practised at Suddeya in Upper Assam, By The Chinamen Sent Thither For That Purpose* (1838). He lived out his life in Assam. It is not certain that the fish is named after Charles but his fame, dates and the East India Company connection (JE Gray worked for them) makes him most likely.

Bruce, GE

Ogun Mormyrid *Marcusenius brucii* GA Boulenger, 1910
Waziristan Snowtrout *Schizocypris brucei* CT Regan, 1914

Major George Evans Bruce (1867–1949) was a colonial administrator in Waziristan (now divided between Afghanistan and Pakistan) who made a small collection of fishes in southern Waziristan and presented it to the BMNH. It included the holotype of the snowtrout. A 'Major G. E. Bruce', presumably the same man, presented the mormyrid holotype to the British Museum. He wrote: *The tribes of Waziristan: notes on Mahsuds, Wazirs, Daurs etc* (1929)

Bruce, WS

Cusk-eel sp. *Holcomycteronus brucei* L Dollo, 1906

Dr William Speirs Bruce (1867–1921) was a Scottish oceanographer and zoologist. He was training at Edinburgh University to become a physician but gave up his studies to go as a scientific assistant on the Dundee Whaling Expedition to Antarctica (1892–1893). He went on expeditions during the 1890s to Novaya Zemla, Spitsbergen, Franz Josef Land. He fell out with the Royal Geographical Society over an appointment to Scott's Discovery Expedition and organised his own Scottish National Antarctic Expedition (1902–1904), during which the cusk-eel holotype was collected. He made a number of expeditions to the Arctic (1907–1920) but after his death was almost entirely forgotten. HIs doctorate was honorary and awarded by the University of Aberdeen. He was a co-founder of Edinburgh Zoo.

Bruce (Colette)

Bruce's Argentine *Argentina brucei* DM Cohen & SP Atsaides, 1969
Dr Bruce Baden Collette (b.1934) (See **Collette**)

Bruce (Turner)

Santa Catarina Sabrefin *Campellolebias brucei* R Vaz-Ferreira & B Sierra de Soriano, 1974
[Alt. Swordfin Killifish]

Dr Bruce Jay Turner (b.1945) is an American ichthyologist who supplied the first specimens of this killifish to the authors. He is Associate Professor Emeritus, Departmernt of Biological Sciences, College of Science, Virginia Tech. His bachelor's degree was awarded by the Brooklyn College, City University New York (1966) and his master's (1967) and doctorate (1972) were awarded by Department of Zoology, UCLA, after which he did post-doctoral work at the Neuropsychiatric Institute there (1972–1974), Rockefeller

University (1974–1976) and at the Museum of Zoology, University of Michigan (1976–1978). He joined the Deptartment of Biological Sciences at Virginia Tech (1978) and taught courses in Evolutionary Biology, Genetics, Ichthyology and Evolutionary Genetics until his retirement (2010). He co-wrote: *Enjoy Your Killifish* (1967), edited *Evolutionary Genetics of Fishes* (1984) and is the author or co-author of more than 80 papers and book chapters such as: *The evolutionary genetics of a unisexual fish, Poecilia formosa* (1982). His work has involved the evolutionary genetics of pupfish (*Cyprinodon*) populations in both the western American deserts and the Caribbean islands, the elucidation of ecological polymorphisms and chromosomal divergence in goodeid fishes of the genus *Ilyodon* from central Mexico.

Bruce Thompson

Ouachita Darter *Percina brucethompsoni* HW Robison, RC Cashner, ME Raley & TJ Near, 2014

Dr Bruce Alan Thompson (1946–2007) was, for nearly three decades, an Associate Professor at the Louisiana State University and researcher at the Coastal Fisheries Institute. Cornell awarded his BSc (1968) and Tulane University, New Orleans his PhD (1977). He was a "*significant figure in North American ichthyology and fisheries biology, especially to the southeast Gulf Coast region.*" According to one obituary "*His research interests not only spanned both freshwater and marine fishes of the southeastern U.S., but extended to several cosmopolitan deep-sea fish families (e.g., Aulopidae, Bramidae, and Percophidae), and to problems in Europe and the Southern Hemisphere as well. He was also known for his passionate conservationism and his eagerness to bring appreciation of biodiversity, habitat protection.*" He died from a very aggressive kidney cancer just three weeks after diagnosis.

Bruggemann

Sweeper sp. *Pempheris bruggemanni* JE Randall & BC Victor, 2015

Professor J Henrich Bruggemann is Director of the Laboratoire d'Ecologie Marine, Université de la Réunion. He collected the paratype (off Mauritius) and the holotype (off La Réunion). Among his published papers is the co-wrtten: *Evolutionary Dynamics in the Southwest Indian Ocean Marine Biodiversity Hotspot: A Perspective from the Rocky Shore Gastropod Genus Nerita* (2014).

Brum

Freshwater Stingray sp. *Potamotrygon brumi* GJ Devincenzi & GW Teague, 1942
[Junior Synonym of *Potamotrygon brachyura*]

Baltasar Brum Rodríguez (1883–1933) was a Uruguyan politician who was President of Uruguay (1919–1923). During a period when he was Minister of Public Instruction, he was, according to the senior author, the only politician in charge of education who showed any interest in the Museum of Natural Sciences in Montevideo. Brum was also interested in Uruguayan history and folklore. He committed suicide by publicly shooting himself as a protest at the dictatorial rule of one of his successors.

Brummer

Ghost Eel *Pseudechidna brummeri* P Bleeker, 1858

[Alt. White Ribbon Eel]

Lieutenant Colonel Dr O Brummer was a military surgeon in the Royal Dutch East Indies Army and collected in Timor while stationed in Java. He collected the eel holotype. Bleeker (q.v.) was also an army surgeon in the East Indies (from 1842).

Brunei

Loach sp. *Neogastromyzon brunei* HH Tan, 2006

This is not an eponym; it is named after both the Sultanate of Brunei itself and the indigenous Brunei people of Borneo.

Brunelli

Sea Catfish sp. *Arius brunellii* G Zolezzi, 1939

The original text does not help, but the likelihood is that this species is named after Gustavo Brunelli (1881–1960), who was a zoologist, biologist and limnologist. He was Director, Central Laboratory for Hydrology, Rome and head of the fisheries service, part of the MInistry of Agriculture (1919). He wrote: *Theories about the origin and evolution of life* (1923). He led the expedition during which the holotype was collected.

Brüning

Brüning's Killi *Scriptaphyosemion brueningi* E Roloff, 1971

Christian Brüning (d.1943) was a German aquarist. He pioneered the import of fish – especially *Aphyosemlon* species – from the West African coast before the war, and was known for numerous essays on newly introduced aquarium fish. Roloff also honoured Professor Ladiges, "*...whom I would like to thank in return for his kind support in examining the new species*.' He received many valuable suggestions from Christian Brüning at a young age. Brüning died as the result of an air raid on Hamburg.

Bruno (Alvares)

Trindade Scaled-Blenny *Malacoctenus brunoi* RZP Guimarães, GW Nunan & JL Gasparini, 2010

Dr Bruno Álvares da Silva Lobo (1884–1945) was a physician and scientist; a physiologist who became Director (1915–1923) of the National Museum of Brazil particularly interested in entomology. He studied medicine at Rio de Janeiro Medical School where he later taught. He was also involved in politics. The etymology says: "*We honor Mr. Bruno Álvares da Silva Lobo, from 1915 to 1923 director of Museu Nacional, Rio de Janeiro, Brazil, who organized and participated in the pioneering Barroso Expedition to Trindade Island*."

Bruno (Bove)

Neotropical Rivuline sp. *Hypsolebias brunoi* WJEM Costa, 2003

Pencil Catfish sp. *Trichomycterus brunoi* MA Barbosa & WJEM Costa, 2010

Bruno Bove de Costa is the son of author WJEM Costa (q.v.). He is a biologist and herpetologist who graduated with a bachelor's degree from University of Veiga de Almeida (2013). He is a student of amphibians. He was thanked for "...*valuable help in collecting Trichomycterus and observations in the field*." (Also see **Claudia (Bove)** & **Costa, WJEM**)

Brunswig

Cusk-eel sp. *Lamprogrammus brunswigi* AB Brauer, 1906

H Brunswig was first officer aboard the research vessel 'Valdivia', the vessel employed on the first German expedition to explore deep seas. The holotype of this species was collected during the Valdivia expedition (1898–1899). This identification is tentative, as the circumstances and evidence point towards him, but there is no proof.

Bruun, AF

Onejaw sp. *Monognathus bruuni* L Bertin, 1936
Flabby Whalefish sp. *Gyrinomimus bruuni* RR Rofen, 1959
Viviparous Brotula sp. *Cataetyx bruuni* JG Nielsen & O Nybelin, 1963
Bruun's Cutthroat Eel *Histiobranchus bruuni* PHJ Castle, 1964
Halfbeak sp. *Hemiramphus bruuni* NV Parin, BB Collette & YN Shcherbachev, 1980

Dr Anton Frederik Bruun (1901–1961) was a Danish oceanographer and ichthyologist. The University of Copenhagen awarded his doctorate (1927). He was elected as the very first chairman of the Intergovernmental Ocanographic Commission, participating in many oceanographic circumnavigations including taking part in the 'Dana' cruises from which the type was collected. Among his c.100 publications he wrote: *Flying-fishes (Exocoetidae) of the Atlantic: systematic and biological studies* (1935), *Cephalopoda (Zoology of Iceland)* (1945) and *The Galathea Deep Sea Expedition, 1950–1952: described by members of the expedition* (1956). The former American presidential yacht 'USS Williamsburg' is now called 'R/V Anton Bruun' after him.

Bruun (Ship)

Blind Cusk Eel sp. *Barathronus bruuni* JG Nielsen, 1969
Lanternfish sp. *Hygophum bruuni* RL Wisner, 1971
Hatchetfish sp. *Polyipnus bruuni* AS Harold, 1994

The research vessel 'Anton Bruun', from which the holotypes of these species were collected, is named after the Danish oceanographer and marine biologist Dr Anton Frederik Bruun (above). The vessel was formerly the US Presidential yacht 'USS Williamsburg'. (See **Anton Bruun**)

Bruynis

Goby sp. *Schismatogobius bruynisi* LF de Beaufort, 1912

Lieutenant J L Bruynis commandeding the military post at Honitetu, Western Ceram, Indonesia. He wrote: *Twee landschapen op Timor (Amarassi en Zuid-Beloe* (1919). He helped de Beaufort "in every possible way" (e.g., providing quarters).

Bryan

Bryan's Sand Dart *Kraemeria bryani* LP Schultz, 1941

Edward Horace Bryan, Jr. (1898–1985) was Curator of Collections, Bernice P. Bishop Museum, Honolulu (1919–1968). He was a 'good friend' of the author. He went to Hawaii at his uncle's invitation to attend college (1916) where he studied entomology, later teaching the course there. He wrote a number of books including *Ancient Hawaiian Life* (1938) and *Panala'au Memoirs* (1974). He wrote that W A Bryan, a colleague at the museum, on meeting him said: 'If we are to masquerade around here under the same surname we might as well get acquainted.' They never found common ancestors but became good friends. A shearwater named after him was collected on Midway Atoll (1963) but lay misidentified in a museum drawer until Pyle of the Institute for Bird Populations noted differences in measurement and it was confirmed a new species (*Puffinus bryani*) through DNA analysis. It was presumed extinct but later rediscovered (2012).

Bucchich

Bucchich's Goby *Gobius bucchichi* F Steindachner, 1870

Gregorio Marijo Bucchich, aka Grgo Bucˇič (1829–1911), was a Croatian naturalist. Steindachner honoured him as his "...*highly esteemed friend*" who had collected the goby holotype.

Buchanan, AR

Sisorid Catfish sp. *Glyptothorax buchanani* HM Smith, 1945

A R Buchanan was working as a clerk for the SIngapore docks board (1906) and next appears as an employee of the Borneo Cmmpany, collecting fishes from the River Mechem, northern Thailand (1935). Perhaps as a result of that he was elected a member of the Natural History Society of Siam (1936).

Buchanan, F

Schilbeid Catfish sp. *Proeutropiichthys buchanani* A Valenciennes, 1840
Bluetail Mullet *Crenimugil buchanani* P Bleeker, 1853
Burmese Eel-Goby *Taenioides buchanani* F Day, 1873

Dr Francis Hamilton Buchanan (1762–1829) was a Scottish physician. He was Assistant Surgeon on board a man-of-war but had to retire through ill health. His health stayed poor for some years, but (1794) he took a job as Surgeon for the East India Company in Bengal. The voyage to India seemed to restore his health fully and, once there, he started collecting plants in Pegu, Ava and the Andaman Islands. He presented these specimens, with drawings, to Sir Joseph Banks. He also studied fish. He worked for the Board of Trade in Calcutta (1798), then (1800–1807) continued to study plants and animals in Mysore until his return to Britain. Buchanan collected in both India and Nepal. He returned to Calcutta to become Superintendent of the botanical gardens there, and later became Surgeon to the Governor-General of India. He returned to Britain (1815) following the death of his elder brother, being heir to the family seat at Leney, Perthshire, following which he changed his name to Hamilton, his mother's maiden name. He spent the rest of his life at Leney, 'improving' the gardens by planting many exotics there. He wrote: *Travels in the Mysore* and *A History of Nepal*. Two birds are named after him. There is no etymology for the catfish, so the allocation to Francis Hamilton Buchanan must remain uncertain.

Buchanan, JL

Ghost Shiner *Notropis buchanani* SE Meek, 1896

John Lee Buchanan (1831–1922) was president of the Arkansas Industrial University (now University of Arkansas) (1894–1902), where Meek was teaching when he described this species. Buchanan was awared a bachelor's degree (1856) and a master's (1858) by Emory and Henry College, Virginia, which in the same year elected him professor to teach ancient languages. During the American Civil War, he worked in the ordnance department of the Confederate government then resumed his teaching career (1865). He was Superinedent of Public Schools in Virginia (1885–1894).

Bucher

Cuban Mosquitofish sp. *Gambusia bucheri* LR Rivas, 1944

George C Bucher assisted Rivas who collected the type. We have been unable to add any biographical information.

Buchholz

Freshwater Butterflyfish *Pantodon buchholzi* WKH Peters, 1876
African Barb sp. *Raiamas buchholzi* WKH Peters, 1876

Dr Reinhold Wilhelm Buchholz (1837–1876) was a German physician, zoologist and explorer. He was a member of the German polar expeditions to Spitzbergen (Svalbard) (1868) and to East Greenland in the 'Hansa' (1869–1870). He was in Africa with Reichenow and collected plants in West Cameroon (1872–1875). He collected the holotype of the butterflyfish.

Buckley

Scrapetooth Characin sp. *Parodon buckleyi* GA Boulenger, 1887
Three-barbeled Catfish sp. *Pimelodella buckleyi* GA Boulenger, 1887

Clarence Buckley (d.1880s) was a (probably English) collector of natural history specimens in Ecuador and Bolivia and possibly Colombia (1860s–1870s). Nothing seems to be recorded of his early life, so that even his nationality is uncertain. He remained in South America until c.1880, but details of his death appear to be unrecorded. Buckley acquired specimens for a number of museums, collectors and ornithologists, with c.5,000 of his butterflies going to the creationist William C Hewitson, an amateur English lepidopterist who produced many fine watercolours of them. The BMNH houses a large fish collection made by Buckley, as well as over 80 species of birds. Six birds, four amphibians and a reptile are named after him, as well as butterflies such as *Asterope buckleyi*. By the time Boulenger described the above two fish species, he was referring to 'the late' Clarence Buckley.

Buckup

Neotropical Livebearer sp. *Phalloceros buckupi* PHF Lucinda, 2008

Dr Paulo Andreas Buckup (b.1959) is an ichthyologist (1994–present) who is a Professor in the Vertebrates Department of the Brazilian National Museum at the Federal University of Rio de Janeiro (UFRJ) and Curator of the fish collection. The Federal University of Rio

Grande do Sul awarded his BSc in zoology (1981) and his masters from the University of Rio Grande (FURG) (1984) in biological oceanography. The University of Michigan awarded his PhD in biological sciences (1991). He was a postdoctoral fellow at the Field Museum (1991–1992) and Philadelphia Academy of Science (1992–1994). Among his c.80 publications, he wrote catalogues of both the of the marine and the freshwater species of fish of Brazil. He has authored or co-authored the descriptions of numerous species of fish, including catfish and characiform fishes. The geographic scope of his work ranges from southern Brazil to the Andean and northern portions of South America. His work includes major contributions to the composition, evolution and phylogenetic position of the Crenuchidae. Recently he has been working on molecular diversity and evolution of neotropical fishes. He was honoured "*...in recognition of his many contributions to Neotropical ichthyology*."

Budgett

Squeaker Catfish sp. *Synodontis budgetti* GA Boulenger, 1911

John Samuel Budgett (1872–1904) was an embryologist, ichthyologist, naturalist, artist and explorer, employed by Trinity College, Cambridge. He made a minimum of four West African expeditions, making a very extensive collection of African fishes. He suffered so badly from malaria [eventually fatally] that he was not passed as fit for service in the Boer War, so he went exploring again. He succeeded in documenting the embryonic development of *Polypterus,* an African freshwater fish genus. A mammal and two amphibians are named after him.

Budker

Lefteye Flounder sp. *Parabothus budkeri* P Chabanaud, 1943

Paul Budker (1900–1992) was a French ichthyologist specialising in sharks, but also interested in cetaceans. He founded and directed a Whaling Research Centre, which was run in connection with National Museum of Natural History and the Laboratory of Colonial Fisheries (Paris), where he was Deputy Director. He published more than 70 works on tropical fishing, sharks and whales, including *Summary report of a mission in the Red Sea and the French Somali Coast (*1939), *Whales and Whaling* (1958) and *The Life of Sharks* (1972).

Buen

Goby genus *Buenia* BS Iljin, 1930

Don Fernando de Buen y Lozano (1895–1962) (See **DeBuen** & also see **Odón de Buen**)

Buffe

Three-striped African Glass Catfish *Pareutropius buffei* R Gras, 1961

J Buffe was Director, Eaux et Forêts (Waters and Forests) Service, Dahomey (Benin). He wrote: *Les Pêcheries en branchage „acadja" des lagunes du Bas-Dahomey* (1958). He was one of the party which collected the catfish holotype (1957).

Buffon

Buffon's River Garfish *Zenarchopterus buffonis* A Valenciennes, 1847

Count Georges Louis Leclerc de Buffon (1707–1788) is one of the giants of zoology and widely written about, hence the brevity of this entry. He is most famous for having developed the species concept, the basis of all taxonomy. Louis XV appointed him to the Academy of Sciences (1734) and later Director of the Jardin du Roi (1739), which Buffon transformed into the present Jardin des Plantes, Paris. He envisaged that his *Histoire Naturelle Générale et Particulière* (1749–1804) would encompass 50 volumes, but 'only' 36 had been produced by the time of his death, and a further 8 posthumously. Two mammals and eight birds are named after him.

Buhse

Namak Scraper (barb) sp. *Capoeta buhsei* KF Kessler, 1877

Friedrich Alexander Buhse (1821–1898) was a Latvian (of German extraction) who was an explorer and botanist. He studied at Dorpat, Estonia and collected in Armenia, Azerbaijan and Persia (Iran) (1847–1849). He co-wrote: *Enumeration of the collected on a journey through Transcaucasia and Persia plants* (1860). He collected the holotype.

Buitendijk

Fintail Serpent Eel *Neenchelys buitendijki* MCW Weber & LF de Beaufort, 1916

Pieter Buitendijk (1870–1932) was a ship's surgeon, mainly on steamers trading between Amsterdam and Java. He collected fish and marine invertebrates for the Leiden museum. He was the surgeon aboard the 'Siboga' expedition, which collected the eel holotype in the Java Sea.

Buller

Slender Thread Herring *Opisthonema bulleri* CT Regan, 1904
Sarabia Cichlid *Paraneetroplus bulleri* CT Regan, 1905

Dr Audley Cecil Buller (1853–1894) was an American naturalist and collector of mammals, reptiles and fish. He collected the holotypes of several taxa including the Thread Herring. When collecting for the AMNH he travelled one thousand miles across the Sierra de Nayarit and ranges of the Sierra Madre to Zacatecas, then the least known area of Mexico, collecting many vertebrate specimens. Two mammals and a reptile are also named after him.

Bullis

Pugnose Bass genus *Bullisichthys* LR Rivas, 1971
Deepwater Squirrelfish *Sargocentron bullisi* LP Woods, 1955
Lined Lantern Shark *Etmopterus bullisi* HB Bigelow & WC Schroeder, 1957
Bullis Skate *Dipturus bullisi* HB Bigelow & WC Schroeder, 1962
Barracudina sp. *Stemonosudis bullisi* RR Rofen, 1963
Eelpout sp. *Lycenchelys bullisi* DM Cohen, 1964
Bullish Conger *Bathycongrus bullisi* Smith DG & RH Kanazawa, 1977

Harvey Raymond Bullis Jr (1924–1992) was an American malacologist who was head of the Bureau of Commercial Fisheries Exploratory Base at Pascagoula, Mississippi and Chief, Gulf Fisheries Exploration & Gear Research, US Fish and Wildlife Service. He wrote a number of pamphlets and scientific papers, including: *Gulf of Mexico Shrimp Trawl Designs* (1951). At least eight fish and other marine organisms are named after him. US Fish and Wildlife Service vessels under his control collected many fish during cruises off the coasts of Florida, and in the Gulf of Mexico, Caribbean and along the South American coast.

Bullock, DS

Pencil Catfish genus *Bullockia* MGE Arratia Fuentes, A Chang Garrido, S Menu-Marque & G Rojas M,1978

Red Jollytail (smelt) *Brachygalaxias bullocki* CT Regan, 1908

Professor Dr Dillman Samuel Bullock (1878–1971) was an agriculturalist, naturalist and Methodist missionary. He graduated from Michigan State University (1902) and worked for the American Missionary Society, Bunster Agricultural School in Angol, Chile (until 1912). He then returned and taught in the USA (1912–1921), before returning to South America. For a year he worked as Agricultural Commissioner, US Department of Agriculture at the US embassy, Buenos Aires. He returned to teach at the School in Chile (1923–1958). He collected the holotype of the smelt (1907). In connection with a specimen of a toad that Bullock had discovered, the author of the description of that amphibian, Schmidt, wrote: '…We are indebted to Dr D. S. Bullock of El Vergel, Angol, Chile, for the two specimens of this frog, collected in the Nahuelbuta forest… …it is pleasant to name the new form for Dr Bullock, in recognition of his services to Chilean natural history and of his long-continued friendly cooperation with Chicago Natural History Museum', showing how much Bullock's efforts were appreciated. A bird and that amphibian are named after him.

Bullock, TH

Bluntnose Knifefish sp. *Brachyhypopomus bullocki* JP Sullivan & CD Hopkins, 2009

Dr Theodore Holmes Bullock (1915–2005) was a comparative neurobiologist, interested in both invertebrates and vertebrates. The University of California, Berkeley awarded his bachelor's degree (1936) and his doctorate (1940). He was a post-doctoral fellow and taught at Yale (1940–1944), taught at the University of Missouri (1944–1946) and at the University of California, Los Angeles (1946–1966). He then joined the University of California, San Diego (1966–1982), retiring as Professor Emeritus. He co-wrote: *Structure and Function in the Nervous System of Invertebrates* (1965). Among his discoveries are electroreceptors in weakly electric fish and the pit organ in pit vipers.

Bulmer

Bulmer's Goby *Glossogobius bulmeri* GP Whitley, 1959

Ralph Neville Hermon Bulmer (1928–1988) was an ethnobiologist at the National University, Canberra, Australia. He was also a social anthropologist who was a member, University of Auckland. He worked and collected in the Papua New Guinea highlands (1955–1976). Among other works he co-wrote: *Karam classification of frogs* (1971). He collected and presented the goby type to the author. An amphibian is also named after him.

Burcham

Dusky Sculpin *Icelinus burchami* BW Evermann & EL Goldsborough, 1907

James S Burcham (d.1905). The etymology says: "*This interesting species is named for Mr. James S. Burcham, a young naturalist of great promise, who lost his life at Lake McDonald, November 12, 1905, while in the employ of the Bureau of Fisheries.*"

Burchell

Burchell's Redfin *Pseudobarbus burchelli* A Smith, 1841

William John Burchell (1781–1863) was an English explorer-naturalist who lived on St Helena (1805–1810) as a merchant and then as the local schoolmaster and official botanist. He went to the Cape of Good Hope (1810), undertaking a major exploration of the South African interior (1811–1815) during which he travelled more than 7,000km through largely unexplored country. He collected over 50,000 specimens and documented this adventure in the two-volume: *Travels in the Interior of Southern Africa* (1822 & 1824). He was renowned as a meticulous collector, botanist and artist. He returned to London (1815) to work on his collections. He spent two months in Lisbon (1825) and then went to Brazil, where he collected extensively before again returning home (1830). He became increasingly reclusive and in his last two years suffered serious mental illness, eventually committing suicide. Among other taxa, a reptile and a mammal are named after him.

Burek

Mardi Gras Wrasse *Halichoeres burekae* DC Weaver & LA Rocha, 2007

Joyce Burek and her husband Frank are wildlife photographers whose speciality is marine subjects. Since being certified as divers (1980) they have spent four decades logging over 2000 dives taking 100,000 images! Their photos have appeared in aquariums, calendars, brochures, books, magazines, newspapers and posters with numerous publishers including Audubon Society, National Geographic Society and the National Wildlife Federation. According to the etymology she was honoured because "*...she first photographed the terminal phase male of this species, and in appreciation of the talents and generosity of Joyce and Frank Burek for donating their underwater photographs to the FGBNMS and their ongoing support of Sanctuary research and education. The Bureks photographed both initial and terminal stages of the fish and brought the species to the attention of researchers for identification.*"

Buresch

Buresch's Loach *Oxynoemacheilus bureschi* P Drensky, 1928

Dr Ivan Yosipov Buresch (1885–1980) was a Bulgarian zoologist and entomologist, Director of the Royal (later National) Museum of Natural History and Professor at the Institute of Zoology, Sofia (1914–1959). Of Czech descent, he graduated in natural science at Charles University, Prague (1909), and undertook postgraduate study at Munich University. An amphibian and a bird are also named after him.

Burg

Berg River Redfin *Pseudobarbus burgi* GA Boulenger, 1911

This is not an eponym but a toponym as, despite the spelling of the vernacular name, this species is named after the Burg River.

Bürger, H

Blackspotted Catshark *Halaelurus buergeri* JP Müller & FGJ Henle, 1838
Deepsea Jewfish *Glaucosoma buergeri* J Richardson, 1845
Ribbed Gunnel *Dictyosoma burgeri* J van der Hoeven, 1855
Inshore Hagfish *Eptatretus burgeri* CF Girard, 1855
Goldfish ssp. *Carassius auratus burgeri* H Temminck & H Schlegel, 1846

Dr Heinrich Bürger (1804/06*-1858) was a German physicist and biologist employed by the Dutch government. He studied mathematics and astronomy at Göttingen (1821–1822) and called himself 'doctor' but there is no evidence of his obtaining a PhD. He first travelled to Batavia (Jakarta), Java (1824) and gained a third-class degree of apothecary (1825). The Dutch Government appointed him (1825) assistant to the German physicist P F von Siebold. He and Siebold collected natural history specimens for RMNH, Leiden (1820s). Bürger and Siebold collected together throughout the Dutch East Indies (Indonesia), then in Japan (1837–1838). Bürger was honourably discharged from government service (1843) and set up a rice and sugar business as well as mining and insurance. Two amphibians are named after him.

* *His definitive birth date is unknown: the oldest birth date is the one given by Bürger himself, but it appears likely that he used it to appear two years older than he really was.*

Burger, R

Characin sp. *Astyanax burgerai* AM Zanata & P Camelier de Assis Cardosa, 2009

Rafael Burger was described as a student who *'enthusiastically'* collected the holotype in Bahia State, Brazil.

Burgess, GH

Broad-snouted Lantern Shark *Etmopterus burgessi* JA Schaaf-Da Silva & DA Ebert, 2006

George H Burgess (b.1949) is an American ichthyologist. The University of Rhode Island awarded his bachelor's degree and the University of Florida awarded his master's (1978). He is Director of the Florida Program for Shark Research of the Florida Museum of Natural History. His research interests are the Life history, ecology, systematics, fishery management and conservation of elasmobranchs; shark attack; systematics and biogeography of fishes; management and conservation of aquatic ecosystems. His fieldwork includes the survey of the marine ichthyofauna of southwestern Florida, particularly the Florida Keys and Straits of Florida and the movements of reef-dwelling elasmobranchs through reef passages in Belize. He is Vice-Chair of the IUCN/SSC Shark Specialist Group. Among his many publications are: *Description of six new species of lantern-sharks of the genus* Etmopterus *(Squaloidea: Etmopteridae) from the Australasian region* (2002) and *Management of sharks and their relatives* (2000). He was honoured in the name of the lantern shark for his contributions to the systematics of *Etmopterus*.

Burgess, WE

Burgess' Butterflyfish *Chaetodon burgessi* GR Allen & WA Starck, 1973
Burgess' Cory *Corydoras burgessi* HR Axelrod, 1987

Dr Warren E Burgess is an American ichthyologist and research scientist who co-wrote: *Dr Burgess's Atlas of Marine Aquarium Fishes* (1988) and the *Colored Atlas of Miniature Catfish* (1992). He was honoured "*…for contributions to the study of catfishes, particularly Corydoras.*"

Burke

Snailfish sp. *Liparis burkei* DS Jordan & WF Thompson, 1914

Dr Charles Victor Burke (1882–1958) was an American ichthyologist. Stanford University awarded his BA (1907), MA (1908) and PhD in zoology (1912). He taught at Washington State University. He kept a journal *Cruise of the Albatross* during the Pacific research expedition from Alaska to Hawaii and Japan (1906). He particularly studied the Cyclogasteridae (snailfish) as evidenced by his co-written: *A new genus and six new species of fishes of the family Cyclogasteridae* (1913).

Burmeister

Pencil Catfish sp. *Trichomycterus burmeisteri* LS Berg, 1895
[Junior Synonym of *Hatcheria macraei*]

Professor Karl Hermann Konrad Burmeister (1807–1892) was an ornithologist who was Director, Institute of Zoology, Martin Luther University, Halle Wittenberg, Germany (1837–1861). He made large collections during two expeditions to Brazil (1850–1852) and the La Plata region, Argentina (1857–1860). He lived in Argentina (1861–1892), being founding Director, Museo Nacional, Buenos Aires until retirement (1880). He was in the Prussian civil service but won his release by using the very inventive excuse that a persistent stomach complaint was caused by arsenic emissions in the museum, and by the drinking water in Halle that had high sulphate content! He wrote: *Reise nach Brasilien* (1853). Three birds, two amphibians, a reptile and a mammal are named after him.

Burns

Characin sp. *Lepidocharax burnsi* KM Ferreira, NA Menezes & I Quagio-Grassiotto, 2011

Dr John Robert Burns is Professor Emeritus of Zoology at George Washington University, Washington, DC. Brooklyn College, New York, awarded his bachelor's degree (1968) and the University of Massachusetts, Amherst his master's (1972) and doctorate (1974). He worked in El Salvador (1974–1978) of which period over half was as part of the Peace Corps. He co-wrote: *Relationships among major lineages of characid fishes (Teleostei: Ostariophysi: Characiformes), based on molecular sequence data* (2010).

Burr

Brook Darter *Etheostoma burri* PA Ceas & LM Page, 1997

Dr Brooks M Burr (b.1950) is an American ichthyologist who is Professor Emeritus and Curator of Fishes at Southern Illinois University (1977–2010). Greenville College awarded his BA in biology (1971) and the University of Illinois awarded both his MSc in Zoology (1974) and PhD in Zoology (1977). He retired from teaching, advising, and administration. Working now as private contractor for PG Environmental, LLC. He is also a past Secretary of the American Society of Ichthyologists and Herpetologists. Perhaps his most well-known publication is the *Field Guide to Freshwater Fishes: North America, North of Mexico* (1991) and as one of the editors of *Freshwater Fishes of North America: Volume 1* (2014). The paper describes him as an 'esteemed colleague' of the authors and an accomplished ichthyologist, who first brought the darter to their attention.

Burrage

Rattail sp. *Nezumia burragei* CH Gilbert, 1905

Sharpchin Slickhead *Bajacalifornia burragei* CH Townsend & JT Nichols, 1925

Vice-Admiral Guy Hamilton Burrage (1867–1954) was Commander-in-Chief, United States Navy. He commanded United States Naval Forces in Europe (1926–1928) and brought back to the USA Charles Lindbergh and his aeroplane 'Spirit of St. Louis'. Whilst he was only a lieutenant he was navigating and executive officer of the *Albatross,* which collected the holotypes. He was Commander of the 'Albatross' (1910–1912).

Burrell

Slender Platanna-Klipfish *Cancelloxus burrelli* JLB Smith, 1961

Cyril J Burrell. According to Smith he "*...has constantly provided valuable aid in my researches*". We have been unable to find any biographical detail.

Burridge

Mary's Pygmygoby *Trimma burridgeae* R Winterbottom, 2016

Dr Mary Elizabeth Burridge is an ichthyologist who is an Assistant Curator at the Department of Natural History, Royal Ontario Museum, Toronto, Canada. She was honoured for her extensive work in revising the genus *Priolepis,* her ongoing work on barcoding *Trimma* spp., her field work collecting fishes in the Philippines and Viêt Nam, and her extensive contributions to the maintenance of the fish and frozen tissue collections at the Royal Ontario Museum. Among her publications are *Systematics of the Acanthophthalmus kuhlii complex (Teleostei: Cobitidae), with the description of a new species from Sarawak and Brunei* (1992) and the co-written ROM Fieldguide: *Freshwater Fishes of Ontario* (2009).

Burrough

Burrough's Damselfish *Pomacentrus burroughi* HW Fowler, 1918

Dr Marmaduke Burrough (1797–1844) was a physician and diplomat. He obtained fishes (presumably including the type of this one) from Manila, Philippines, which made their way to the Academy of Natural Sciences of Philadelphia (where Fowler worked). The etymology reads: "*For Dr. Marmaduke Burrough, 1798?–1844, who obtained fishes at Manila, which found their way to the Academy collection.*"

Burrows

Canterbury Mudfish *Neochanna burrowsius* WJ Phillipps, 1926

Alfred Burrows was an amateur fish enthusiast of West Oxford in the South Island of New Zealand. "*He sent two specimens of this fish to the Dominion Museum. They were packed alive in a tin box together with a quantity of damp earth, sent by parcel-post on a journey lasting over thirty hours, and arrived alive and extremely active.*"

Burt

Gulf Butterfish *Peprilus burti* HW Fowler, 1944

Professor Dr Charles Earle Burt (1904–1963) was a herpetologist. He took his bachelor's degree at Kansas State Agricultural College, and his master's degree (1927) and doctorate (1930) at the University of Michigan. He worked at the American Museum of Natural History (1929–1930). He taught at Trinity College, Waxahachie, Texas (1930–1931) and at Southwestern College, Winfield, Kansas (1932–1944). He then became owner and manager of Quivira Specialties Co. of Topeka, Kansas (suppliers of such useful items as live toads as food for hog-nosed snakes) and taught at Kansas State College. His wife, May Danheim Burt, was a teacher of home economics but was just as interested as Charles in herpetology, and they co-wrote several articles and papers (see Danheim). Burt wrote: *A Key to the Lizards of the US and Canada* (1936). He died of cancer. A reptile is also named after him.

Burt Jones

Burt's Damselfish *Chrysiptera burtjonesi* GR Allen, MV Erdmann & NKD Cahyani, 2017

Burt Jones is a photographer and underwater guide "*par excellence*". He and his partner Maurine Shimlock were pioneers for the promotion of dive tourism at the Solomon Islands (where this species appears to be endemic) and, more recently, "*…have been instrumental in the tremendous popularity of the West Papuan region by means of their excellent underwater guidebook to the area and creation of the highly informative Bird's Head Seascape website (birdsheadseascape.com).*"

Burton

Lake Tanganyika Cichlid sp. *Xenotilapia burtoni* M Poll, 1951

There is no explicit etymology in Poll's original, but the text indicates that this is a toponym referring to Burton Bay, Lake Tanganyika, where the species is endemic. The bay is named after Sir Richard Francis Burton (q.v.).

Burton, EM

Lanternfish sp. *Diaphus burtoni* HW Fowler, 1934
Blotchside Darter *Percina burtoni* HW Fowler, 1945

Edward Milby Burton (1898–1977) was originally a Charleston insurance broker who was Director of the Charleston Museum, South Carolina (1932–1972). He collected local fishes for the museum and invited Fowler to study them in order to get his own identifications authenticated. He wrote: *The Siege of Charleston, 1861–1865* (1982).

Burton, RF

Frillgoby sp. *Bathygobius burtoni* AWE O'Shaughnessy, 1875

Lake Tanganyika Cichlid sp. *Astatotilapia burtoni* A Günther, 1894

Sir Richard Francis Burton (1821–1890) was a noted British explorer, linguist, author and devotee of erotica. He began his career as an army officer but is best remembered as an explorer who, among much more, searched for the source of the Nile. He was appointed British Consul in Trieste (at that time part of the AustroHungarian Empire) (1872) and lived there for the rest of his life. He was fluent in over 20 languages and devoted his time to literature, translating the *Kama Sutra* (1883), *The Arabian Nights* (1885) and *The Perfumed Garden* (1886) into English. Immediately after his death his widow burned all his papers - an action that was condemned as one of the greatest acts of literary vandalism of all time. With John Hanning Speke, he discovered Lake Tanganyika, the cichlid type locality. A bird, a mammal and a reptile are named after him.

Burton, RW

Indian Barb sp. *Poropuntius burtoni* DD Mukerji, 1933

Lieutenant-Colonel Richard Watkins Burton (1868–1963) was originally commissioned in to the Lancashire Fusiliers (1889) and served in the Madras Regiment of the Indian Army. He was also an explorer and big game hunter and a noted conservationist; he wrote a pamphlet (1950) urging the Indian Government to conserve Indian wildlife and this led to the formation of the Indian Wildlife Board (1952). He wrote: *A history of Shikar in India* (1952). He collected the holotype of this species.

Busakhin

Busakhin's Beardfish *Polymixia busakhini* AN Kotlyar, 1992

Sergey Vasilevich Busakhin (b.1946) was an ichthyologist who abandoned ichthyology and became a professional artist. He contributed greatly to beryciform systematics with his revision of the family Berycidae (at the time, Polymixiidae was placed in the order Beryciformes): *Systematics and distribution of the family Berycidae (Osteichthyes) in the world ocean* (1982). Among his other publications is: *Trachichthodes druzhinini Busakhin, a new species of the family Berycidae (Osteichthyes) from the Indian Ocean* (1981). He is also Kotlyar's friend.

Büscher

Lake Tanganyika Cichlid sp. *Neolamprologus buescheri* W Staeck, 1983

Dr Heinz H Büscher (b.1942) is a German ichthyologist and aquarist living in Switzerland who discovered this cichlid (1982) and collected the type. He worked as a Lab Technician in Basel (1962–2002). He has made a number of collecting trips to Lake Tanganyika. He has written quite extensively about cichlids. His c.45 papers include: *Tropheus moorii — Beobachtungen im natürlichen Lebensraum* (1981) to *Polypterus - an archaic member of the Actinopterygii in Lake Tanganyika* (2018). He was awarded an honourary doctorate by the University of Basel (2013).

Busck

Goby sp. *Sicydium buscki* BW Evermann & HW Clark, 1906

Dr August Busck (1870–1944) was a Danish-American entomologist who worked for the Bureau of Entomology, US Department of Agricukture, USNM. Both his master's degree and his doctorate were awarded (1893) by the Royal University, Copenhagen. He wrote: *Descriptions of tineoid moths (Microlepidoptera) from South America* (1911). He collected the holotype of this species.

Bussarawit

Arrowtooth Eel sp. *Dysomma bussarawiti* AM Prokofiev, 2019

Dr Somchai Bussarawit is the Chief of Museum and Aquarium at Phuket Marine Biological Center, Thailand. His published papers include the co-written: *Phylogenetic analysis of Thai oyster (Ostreidae) based on partial sequences of the mitochondrial 16S DNA gene* (2006) and *Larvae of commercial and other oyster species in Thailand (Andaman Sea and Gulf of Thailand)* (2012). The species was dedicated to him as he had made the material available for Prokofiev.

Busse

Busse's Stickleback *Pungitius bussei* NA Warpachowski, 1888
[Alt. Amur Stickleback]

Fyodor Fyodorovich Busse (1838–1897) was a Russian geographer and archaeologist who was Chairman of the Society for the Study of the Amur Region. He excavated in the Amur River basin where this species was found and published (1888) on his work.

Bussing

Bussing's Stargazer *Platygillellus bussingi* CE Dawson, 1974
Bussing's Drum *Umbrina bussingi* MI López, 1980
Slickhead sp. *Talismania bussingi* YI Sazonov, 1989
Neotropical Cichlid sp. *Cribroheros bussingi* PV Loiselle, 1997
Bussing's Mudbrotula *Gunterichthys bussingi* PR Møller, W Schwarzhans & JG Nielsen, 2004
Characin sp. *Roeboides bussingi* WA Matamoros Ortega, P Chakranarty, A Angul Sibaja, CA Garita-Alvarado & CD McMahan, 2013
Tetra sp. *Hyphessobrycon bussingi* RR Ota, FR Carvalho & CS Pavanelli, 2020

William 'Bill' A. Bussing (1933–2014) was an American ichthyologist at the School of Biology, Universidad de Costa Rica where he taught for over four decades, retiring as Professor Emeritus (1991). Military service in Korea interrupted his education but the University of Southern California awarded his master's in biology (1965). He described more than 50 new species. He wrote widely, alone or with others, including around 70 scientific papers and seven books, such as: *Peces costeros del caribe de Centro América meridional* (2010). (Also see **Eric**)

Busson

Goby sp. *Schismatogobius bussoni* P Keith, N Hubert, GV Limmon & H Darhuddin, 2017

Frédéric Busson is an ichthyologist at MNHN, Paris. He is recognized for "...*all his work to improve our knowledge on Indonesian freshwater fishes.*"

Bustamante

Mexican Dace *Evarra bustamantei* GL Navarro, 1955 (Extinct)

Miguel Francisco Nepomuceno María José Ignacio Bustamante Praxedis Septién (1790–1844) was the first Mexican to scientifically describe a Mexican fish, *Cyprinus viviparus* – now *Girardinichthys viviparus* (1837). He was Professor of Botany at the Royal School of Mines and in charge of the Royal Botanical Gardens, Mexico City (1826–1844).

Bustamente

Goby sp. *Awaous bustamantei* R Greeff, 1882
Goby sp. *Sicydium bustamantei* R Greeff, 1884

Gabriel de Bustamenté was a Brazilian slave trader on São Tomé Island, Gulf of Guinea. His occupation did not stop Greef describing him as the 'hospitable and intelligent' (translation) owner of a farm situated on São Tomé.

Butcher

Southern Black Bream *Acanthopagrus butcheri* ISR Munro, 1949

Alfred Dunbavin Butcher (1915–1990) was an Australian zoologist. He studied agricultural science at Geelong College, then Science at the University of Melbourne which awared his BSc (1939) and MSc (1943). They also conferred an honourary DSc (1986). He worked at the Fisheries and Game Branch of the Victorian Government (1941) travelling and consulting fishweremen as well as producing articles and pamphlets such as *Conservation of the bream fishery* (1945). He was promoted (1947) to Inspector of Fisheries then (1949) Director of Fisheries and Game. He was again promoted to Deputy Director of the newly established Ministry of Conservation (1973) until retiring from public service (1978). He was Appointed an ex officio member of the Zoological Board of Victoria (1947) becoming chairman (1962–1985), and was on the board of many conservation bodies. He became the President of the Royal Society of Victoria (1971) and was a Trustee of the World Wildlife Fund Australia (1979–84). Munro honoured Butcher "...*who has made an extensive study of its economic biology in the Gippsland Lakes.*"

Buth

Spadenosed Guitarfish *Pseudobatos buthi* KM Rutledge, 2019

Dr Donald George Buth (b.1949) is an ichthyologist who is a professor in the Department of Ecology and Evolutionary Biology at the University of California, Los Angeles. The University of Illinois awarded his BSc (1971), his BA (1972), his MSc (1974) and his PhD (1978). He was a teaching assistant at the University of Illinois (1971–1978) and a post-doctoral researcher at UCLA (1978–1979) before becoming Lecturer in Biology there (1980) and Assistant Professor of Biology that same year. His research foci are systematics and evolution, particularly of reptiles and fish. Among his publications are: *Gene expression in the desert tortoise, Gopherus agassizi: tissue sources and buffer optima for the gene products of 73 loci* (1999) and *Should mitochondrial DNA sequences be used in phylogenetic studies?* (2010). The etymology reads: "*Named in honor of my mentor, UCLA ichthyologist Donald Buth, who provided me with the opportunity to describe this new species and whose support and guidance has been instrumental in my scientific career.*" He is also honoured in the name of a parasitic fish leech.

Butler, AW

Pacific Molly *Poecilia butleri* DS Jordan, 1889

Amos William Butler (1860–1937) was an American naturalist. He was a founding member of the Brookville (Indiana) Society of Natural History (1881). His bachelor's degree (1894) and master's (1900) were awarded by Indiana University. The State of Indiana employed him, first in the Department of Geology and Resources as an ornithologist (1896–1997) and then as Secretary to the Board of State Charities (1897–1923). He was a friend of David Starr Jordan, who described the molly. A reptile is named after him.

Butler, G

Butler's Frogfish *Tathicarpus butleri* JD Ogilby, 1907

Dr Arthur Graham Butler (1872–1949) was a physician and medical historian. He was in general practice in Gladstone (1902–1907) c100 km from the type locality. He went on to win the DSO at Gallipoli (1915). We cannot be certain this was the person honoured, due to the small amount of information given by Ogilby, who says only that 'Dr A Graham Butler' discovered the frogfish near Port Curtis, Queensland.

Butler, WH

Sharpnose Grunter *Syncomistes butleri* RP Vari, 1978
Sailfin Catfish *Paraplotosus butleri* GR Allen, 1998

Dr William Henry 'Harry' Butler (1930–2015) was born in Perth and trained as a teacher. He began working for corporate and government bodies as an environmental consultant and collector (1963) and undertook a major study of Western Australian animals. He collected more than 2,000 examples of mammals, 14 new to science. A passionate conservationist, he presented the popular ABC television series 'In the Wild' (from 1976). He received the Australian of the Year award (1979) and was awarded an honorary Doctorate of Science by the Edith Cowan University in Perth (2003). One mammal, a bird and six reptiles are named after him.

Büttiker

Garra sp. *Garra buettikeri* F Krupp, 1983

Dr William Büttiker-Otto (1921–2009) was a Swiss medical entomologist, parasitologist and environmental scientist. His master's degree in biology (1945) and his doctorate in zoology (1947) were both awarded by the Federal Institute of Technology, Zurich. He worked for the pharmaceutical company Ciba-Geigy, Basel (1948 & 1959–1975) and for the same company in Riyadh, Saudi Arabia (1975–1981). In between he worked for an English pest control company (1949–1956) in England, Southern Rhodesia (Zimbabwe), Sudan, South Africa and Mauritius and for the WHO, Geneva (1958–1958). After retiring (1981) he became Environmental Science Expert for the Saudi Meteorology and Environmental Protection Administration, Jedah (1982–1985). He was Honorary Collaborator, Natural History Museum, Basel (1986) and was a visiting professor at the Faculty of Science, University of Kuwait (1988). He collected the holotype. He wrote: *Memories of a Scientist: The Carp Expedition to the Save River in Zimbabwe and Mozambique* (2008).

Büttikofer

Zebra Tilapia *Heterotilapia buttikoferi* AAW Hubrecht, 1881
African Airbreathing Catfish sp. *Clarias buettikoferi* F Steindachner, 1894
West African Cichlid sp. *Pelmatochromis buettikoferi* F Steindachner, 1894
Claroteid Catfish sp. *Parauchenoglanis buettikoferi* CML Popta, 1913
Büttikofer's Bichir *Polypterus palmas buettikoferi* F Steindachner, 1891

Johan Büttikofer (1850–1927) was a Swiss zoologist. He made two collecting trips to Liberia (1879–1882 & 1886–1887) but curtailed the second due to ill health. He did, however, make one more trip there (1888). He accompanied Nieuwenhuis to Borneo (1893–1894) where they explored and collected. He was then Director of the Rotterdam Zoo (1897–1924). He wrote: *Zoological researches in Liberia. A list of birds, collected by the author and Mr F. X. Stampfli during their last sojourn in Liberia*. A number of birds, three mammals and two reptiles are also named after him.

Buys

Namib Hap (cichlid) *Thoracochromis buysi* M Penrith, 1970

Peter J Buys is a South African taxidermist, herpetologist and ornithologist who was Curator, National Museum of Namibia (1957–1987). He co-wrote: *Snakes of Namibia* (1983). He collected the first specimen of the hap.

Buytaert

African Rivuline sp. *Aphyosemion buytaerti* AC Radda & J-H Huber, 1978

John Buytaert (1944–2012) was a Belgian aquarist. He was the co-discoverer of this species in the Congo Republic.

Bwathondi

Lake Victoria Cichlid sp. *Haplochromis bwathondii* P Niemantsverdriet & F Witte, 2010

Professor Dr Philip O J Bwathondi was Director General of the Tanzania Fisheries Research Institute (TAFIRI) (1983–2006). During this period, he supported the research of the Haplochromis Ecology Survey Team (HEST) in many ways.

Byblis

Byblis Goby *Knipowitschia byblisia* H Ahnelt, 2011

Byblis was a mythical figure, the twin sister of Caunos, who founded the ancient city Caunos on the southwest Anatolian coast. The ruins of this city are located close to Lake Köycegiz, Turkey, the type locality.

Byers

Notchtail Stargazer *Dactyloscopus byersi* CE Dawson, 1969

Major and Mrs Joseph Byers The etymology says: "*I take pleasure in naming this species after Major and Mrs. Joseph Byers in recognition of their interest and support of my ichthyological studies.*" In which case, as more than one person is honoured, the binomial should correctly be *byersorum*.

Bynni

Niger Barb *Labeobarbus bynni* P Forsskål, 1775

Not an eponym, but, according to Forsskål, the local Arabic name for the fish.

Bynoe

Bynoe's Goby *Amblygobius bynoensis* J Richardson, 1844

Dr Benjamin Bynoe (1804–1865) was a British naval surgeon. He was appointed as Assistant Surgeon on HMS *Beagle* (1831). His superior, Robert McCormick, was so annoyed that Darwin, instead of himself, was treated as the ship's naturalist that he resigned from the expedition (1832), returning to England. Consequently, Bynoe was promoted to Surgeon and served for the rest of that voyage (1836). He was given the same position on the *Beagle*'s third voyage (1837–1843). He was a great success as a naturalist and collector of both mammals and birds. He wrote the first description of the birth of marsupials. Bynoe Harbour in Australia was named after him (1839), as are a bird, a mammal and a reptile. Using the form *bynoensis* makes this look more like a toponym than an eponym, but Richardson wrote: "*The specific name has been bestowed in honour of Benj. Bynoe, Esq., Surgeon in the Royal Navy.*"

Byrne

Ordinary Eel *Ethadophis byrnei* RH Rosenblatt & JE McCosker, 1970

John Byrne was the San Diego resident who had the enormous good fortune to see an eel stick its head up above the sand as he walked along a beach at low tide. He grabbed it and took it to the Scripps Institution of Oceanography, where the experts identified it as a new species. No other specimen was recorded until 2019.

C

Caballero

Characin sp. *Astyanax caballeroi* S Contreras-Balderas & R Rivera-Teillery, 1985

Dr Eduardo Caballero y Caballero (1904–1974) was a Mexican biologist and helminthologist. The National Autonomous University of Mexico awarded his bachelor's degree (1928), his master's (1934) and his doctorate (1938). He was the first Professor of Zoology at the Instituto Tecnológico de Mexico (1947) and became Head of the Parasitology Department at La Escuela Nacional de Ciencias Biológicas del Instituto Politécnico de la S.E.P (1958).

Cable

Cable's Goby *Eleotrica cableae* I Ginsburg, 1933

Louella E Cable (1900–1986) was a U.S. government biologist and skilled scientific illustrator. The University of South Dakota awarded her BA (1926) and her Fisheries MA (1927) and much later (1959) the University of Michigan awarded her PhD. She became a Junior Aquatic Biologist (1927) for the US Bureau of Fisheries and retired from the US Fish & Wildlife Service (1970). She published widely. She illustrated this goby for Ginsburg and alerted him to its ventral fins not being united.

Cabra

African Cichlid sp. *Pelmatolapia cabrae* GA Boulenger, 1899

Lieutenant-General Alphonse Cabra (1862–1932) was a Belgian army officer (1878–1924) and surveyor and explorer in the Congo. He settled the limits of the Portuguese enclave of Cabinda (Angola) (1897–1899), the boundaries with the rest of Angola (1901–1902) and the French Congo (1903). He carried out an 'inspection' (1905–1906) from Mombasa to Boma, accompanied by his wife, who became the first European woman to traverse Africa from East to West. He was invalided home to Belgium (1906), serving in various military posts there including commanding the fortress of Namur (WW1), eventually retiring as Commander of the 2nd Army Corps. He collected the cichlid type when serving as a Captain.

Cabrillo

Channel Islands Clingfish *Rimicola cabrilloi* JC Briggs, 2002

Juan Rodriguez Cabrillo (1499–1543) was a Spanish marine navigator who explored the west coast of North America for the Spanish Empire; the first European to sail the Californian coast, the type locality.

Cacah

Characin sp. *Characidium cacah* AM Zanata, TC Ribeiro, FA Araújo-Porto, TC Pessali & L Oliveira-Silva, 2020

Carlos Bernado Mascarenhas Alves of the Universidade Federal de Minas Gerais, Brazil, is known to his friends as 'Cacá.' He was honoured for his "...*great contribution to the knowledge of the ichthyofauna of the rio das Velhas basin*" (where this fish is endemic) and for being one of the characin's first collectors. Among his published works is: *Evaluation of fish passage through the Igarapé Dam fish ladder (rio Paraopeba, Brazil), using marking and recapture* (2007).

Cacique

Characin sp. *Creagrutus cacique* N Flausino Jr & F C T Lima, 2019

Nilso Estevão da Silva. The etymology says that the name 'cacique' "...derives from the Taino word kasike, and is used in both Portuguese and Spanish to designate an Amerindian chief. It honors our friend Nilso Estevão da Silva, nicknamed Cacique, a technician of the Universidade Federal do Mato Grosso at Cuiabá, who has participated on ichthyological expeditions from the middle 1980's to the present, and contributed to the collection of an enormous amount of fishes from across Mato Grosso state, Brazil. Nilso's grandfather is a Bororo Amerindian."

Cadenat

Cadenat's Rockfish *Scorpaena loppei* J Cadenat, 1943
Ghanaian Tongue Sole *Cynoglossus cadenati* P Chabanaud, 1947
Guinean Sole *Synaptura cadenati* P Chabanaud, 1948
Cadenat's Chromis *Chromis cadenati* GP Whitley, 1951
Guinean Flagfin *Aulopus cadenati* M Poll, 1953
Cadenat's Sole *Pegusa cadenati* P Chabanaud, 1954
Clingfish sp. *Opeatogenys cadenati* JC Briggs, 1957
Short-tail Eel sp. *Coloconger cadenati* RH Kanazawa, 1961
African Barb sp. *Enteromius cadenati* J Daget, 1962
Barracudina sp. *Lestidiops cadenati* G Maul, 1962
Broadfoot Legskate *Cruriraja cadenati* HB Bigelow & WC Schroeder, 1962
Longfin Sawtail Catshark *Galeus cadenati* S Springer, 1966
West African Rockhopper *Entomacrodus cadenati* VG Springer, 1967
Moroccan White Seabream *Diplodus cadenati* R de la Paz, M-L Bauchot & J Daget, 1974
Black Slimehead *Hoplostethus cadenati* J-C Quéro, 1974
Duckbill sp. *Bembrops cadenati* MK Das & JS Nelson, 1996
Benguela Hake ssp. *Merluccius polli cadenati* MP Doutre, 1960

Dr Jean Cadenat (1908–1992) of the University of Dakar, was a French researcher and ichthyologist at the 'Office de la recherche scientifique et technique outre-mer', where he was Director of the Marine Biological Section of the Institute Français d'Afrique Noire, Gorée, Senegal. He is honoured in the names of many other taxa, including flatworms and copepods. Much of his work concerned the marine fauna of the West African coast, from Mauritania to Nigeria, and his publications include: *Poissons de mer du Sénégal (1950).*

Cadman

Scalebreast Gurnard *Lepidotrigla cadmani* CT Regan, 1915

J Cadman collected the type and presented it to the BMNH. The etymology reveals that the holotype was part of a collection of fish trawled off Lagos, Nigeria, and presented to the British Museum by "…*Mr. J. Cadman of the Western Fisheries Ltd.*" The collection included other taxa including several undescribed crabs.

Cadwalader

African Tetra sp. *Brachypetersius cadwaladeri* HW Fowler, 1930

Charles Meigs Biddle Cadwalader (1885–1959) was an American philanthropist who was Director of the Academy of Natural Sciences, Philadelphia (1937–1951). The mineral Cadwaladerite is named after him.

Cahn

Slender Chub *Erimystax cahni* CL Hubbs & WR Crowe, 1956

Dr Alvin Robert Cahn (1892–1971) was an American biologist, naturalist and, by accident, an archaeologist. Cornell awarded his bachelor's degree (1913), the University of Wisconsin his master's (1915) and the University of Illinois his doctorate (1924). He taught at the University of Wisconsin (1913–1918), at the Agricultural and Mechanical College of Texas (now Texas A & M University (1919–1922) and at the University of Illinois (1922–1935). He was Chief of the Wildlife Development Division, Tennessee valley Authority (1936–1940). After the attack on Pearl Harbor (1941), as a naval reservist he was called to duty and was stationed at Dutch Harbor, Aleutian Islands (1942–1945). When the base was expanded, he discovered remains of earlier inhabitants. He did much collecting and excavating and saved a great deal that would otherwise have been bulldozed as construction to increase the size of the naval air base progressed. After WW2, he did a lot of work on Japanese crustaceans and molluscs, including a study of oyster cultivation whilst he was part of the US Occupation forces. In this period, he became a great friend and manager of the Japanese world flyweight-boxing champion, Yoshio Shirai. Cahn stayed on in Japan and in old age suffered from dementia. Shirai and his family cared for Cahn until his death, and as Cahn had no family he made Shirai his sole heir and it was to Shirai that his ashes were given after his cremation. Among his publications was: *The effect of carp on a small lake: the carp as a dominant* (1929). He collected the chub holotype.

Cahual

Pencil Catfish sp. *Tridentopsis cahuali* M de las Mercedes Azpelicueta, 1990

Cahual is the name both of an aboriginal Araucanian chief and of a private protected area in which the type series was collected.

Caiapo

Whale Catfish sp. *Cetopsis caiapo* RP Vari, CJ Ferraris Jr & MCC de Pinna, 2005

Named for the Caiapo Amerindian tribe that used to inhabit the area of the Rio Tocantins drainage system.

Cailliet

Philippine Angelshark *Squatina caillieti* JH Walsh, DA Ebert & L Compagno, 2011

Dr Gregor Michel Cailliet (b.1943) is an ichthyologist and marine biologist who was Professor at Moss Landing Marine Laboratories (1972–2009) and Professor Emeritus since then. He started (2001) the Pacific Shark Research Center at MLML, being still Co-Director with Dr. David A. Ebert. He also has been (since 1997) Consultant, Regional Water Quality Control Board, California Energy Commission, and Aspen Environmental Group. The University of California, Santa Barbara awarded both his bachelor's degree (1966) and doctorate (1972). Among his publications is: *Fishes: A Field and Laboratory Manual on Their Structure, Identification and Natural History* (1986). He has been honoured for his contributions to ichthyology, especially chondrichthyan age and growth. Gregor Caillet told us that at a lunch with Ray Troll (q.v.) and Nancy Burnett (q.v.), they decided to form a very select club called 'The Chondronomes' - as they all had cartilaginous fishes named after them!

Caira

Borneo Sand Skate *Okamejei cairae* PL Last, Fahmi & H Ishihara, 2010

Dr Janine N Caira (b.1957) is a Canadian parasitologist who is Professor of Ecology and Evolutionary Biology at the University of Connectictut. She has written a number of papers, mostly on parasitism in general and tapeworms of fish in particular, such as: *A new species of Phoreiobothrium (Cestoidea: Tetraphyllidia) from the Great Hammerhead shark Sphyrna mokarran and its implications for the evolution of the onchobothriid scolex* (1996). She also co-wrote: *Sharks and Rays of Borneo* (2010).

Caitlin

Cenderawasih Dottyback *Pictichromis caitlinae* GR Allen, AC Gill & MV Erdmann, 2008

Ms. Caitlin Elizabeth Samuel (b.1993) is the daughter of a supporter of a conservation initiative. The etymology says: "*The new species is named* caitlinae *for Ms. Caitlin Elizabeth Samuel, a young Canadian leader, on the occasion of her sixteenth birthday. This name honours the request of her mother, Kim Samuel Johnson, who successfully bid to conserve this species at the Blue Auction in Monaco on 20 September 2007 in support of Conservation International's Bird's Head Seascape marine conservation initiative.*"

Cajorarori

Hyperostotic Gurnard *Pterygotrigla cajorarori* WJ Richards & T Yato, 2012

The binomial name is formed from the first two letters of the senior author's five granddaughters - **CA**rolyn, **JO**sephine, **RA**chel, **RO**semary, and **RI**ley.

Cala

Characin sp. *Creagrutus calai* RP Vari & AS Harold, 2001
Catfish sp. *Trichomycterus calai* CA Ardila Rodriguez, 2019

Dr Plutarco Cala Cala (b.1938) is an ichthyologist who is Professor Emeritus at Universidad Nacional de Colombia where he was first awarded a bachelor's degree (1962). His doctorate was awarded by the University of Lund, Sweden (1975), where he had been a research assistant (1965–1967) and has since been a visiting professor. He wrote: *On the ecology of the ide* Idus idus *(L.) in the River Kvlingen, South Sweden* (1970). (See also **Plutarco**)

Caldas

Characin sp. *Bryconamericus caldasi* C Román-Valencia, RI Ruiz-C, DC Taphorn & CA Garcia-Alzate, 2014

Francisco José de Caldas (1768–1816) was a Colombian geographer, lawyer and self-taught naturalist. He took part in a number of expeditions, including one with Alexander von Humboldt. He lived in what was then called New Granada (now mostly part of Colombia) and pushed hard for independence, rebelling against the Spanish régime. He was executed by order of Pablo Morillo, Count of Cartagena. When Caldas was about to be executed, some people present at the site appealed for the life of the scientist. Morillo reportedly replied: "*Spain does not need savants*".

Calderón

Northern Iberian Spined Loach *Cobitis calderoni* M Băcescu, 1962

Enrique C Calderón was the chief engineer at the Central Station of Hydrobiology, Madrid. He wrote: *Raising brown and rainbow trout in very warm waters* (1966). He collected the holotype.

Caldwell, DK

Caribbean Blenny *Emblemaria caldwelli* JS Stephens, 1970
Pipefish sp. *Syngnathus caldwelli* ES Herald & JE Randall, 1972
[Junior Synonym of *Cosmocampus howensis*]

Dr David Keller Caldwell (1928–1990) was the Curator of Ichthyology at the Los Angeles County Museum of Natural History and Research Associate, Florida State Museum. Among his c.80 publications are the paper: *Observations on the distribution, coloration, behaviour and audible sound production of the spotted dolphin,* Stenella plagiodon *(Cope)* (1966) co-written with his wife Melba C Caldwell née Wilson (1921–1991), and the book *Marine and freshwater fishes of Jamaica* (1966).

Caldwell, HR

Oriental Cyprinid sp. *Spinibarbus caldwelli* JT Nichols, 1925
Chinese Loach sp. *Vanmanenia caldwelli* JT Nichols, 1925

Harry Russell Caldwell (1876–1971) was an American Methodist missionary at Yenping, Fukien, China (1900). He was a keen hunter and amateur naturalist, who took part in the AMNH's Central Asiatic Expeditions (1916–1925). He wrote: *Blue Tiger – Strange Adventures of a Missionary in China* (1924), which includes his attempts to trap a blue-morph tiger, as well as: *South China Birds* (1931). A mammal and an amphibian are also named after him.

Calmon

Dusky Piranha *Pristobrycon calmoni* F Steindachner, 1908

Miguel Calmon du Pin e Almedia (1879–1935) was a Brazilian engineer and politician who was Minister and Secretary of State for Industry, Transportation and Public Works (1906–1909) and, later, Minister of Agriculture, Commerce and Industry (1922–1926). Though Steindachner did not provide an etymology, he did name a catfish after this man [see **Cameron**] as a "*...token of my respect and gratitude*", and it seems highly likely the piranha's binomial also honours this person.

Calvert

Characin sp. *Phenacogaster calverti* HW Fowler, 1941

Dr Philip Powell Calvert (1871–1961) was an American entomologist, one of the true giants of odonatology. University of Philadelphia awarded a certificate in biology (1892) and his PhD (1895). He was also a teacher there as Assistant Instructor (1892–1897), Instructor (1897–1907), Assistant Professor (1907–1912) and full Professor (1912–1939). He was at University of Berlin (1895–1896) and took a sabbatical year in Costa Rica (1909–1910). He was associated with the American Entomological Society for many years serving on its council, then as Vice-President (1894–1898) and President (1900–1915). He was Associate Editor (1893–1910) and Editor (1911–1943) of *Entomological News*. He was also an accomplished illustrator. Throughout he published widely and was recognised as an authority on Odonata: thirteen of which are named after him.

Camelia

Snailfish sp. *Crystallichthys cameliae* TT Nalbant, 1965

Danube Delta Dwarf Goby *Knipowitschia cameliae* TT Nalbant & V Oțel, 1995

Camelia Iliana Nalbant is the wife of the senior author, Romanian ichthyologist Teodor T Nalbant (1933–2011).

Cameron

Armoured Catfish sp. *Pareiorhaphis cameroni* F Steindachner, 1907

Miguel Calmon du Pin e Almedia (1879–1935) was a Brazilian engineer and politician (see **Calmon**, above). Steindachner managed to badly misspell the binomial, using *cameroni* when he intended *calmoni*. He corrected the spelling in 1908, and some taxonomists have starting using *calmoni* despite the general rule that an initial usage should be retained.

Campbell, GG

Cardinalfish sp. *Apogon campbelli* JLB Smith, 1949

Blackspot Skate *Dipturus campbelli* JH Wallace, 1967

Dr George Gordon Campbell (1893–1977) was a South African physician and naturalist who inspired the Foundation for Marine Biological Research and was President of it, for which he was honoured in the name of the skate. He went to study medicine at Edinburgh University (1914) but only qualified after the end of WW1, during which he was an early member of the Royal Flying Corps in the days when dog-fighting meant firing a revolver at someone in an aeroplane somewhere close by! He was President of the Royal Society of South Africa (1968) and Chancellor of the University of Natal.

Campbell, DH

Campbell's Goby *Istigobius campbelli* DS Jordan & JO Snyder, 1901

Dr Douglas Houghton Campbell (1859–1953) was a botanist and a Professor at Stanford University. The University of Michigan awarded his masters (1882) and he taught high school botany until completing his PhD (1886). His post-doctoral research took him to Gwerman and he returned to take up the post of professor at Indiana University (1888–1891) during which time he wrote the text book: *Elements of Structural and Systematic Botany*. He left there to found the botany department at Stanford, where he stayed for the rest of his career (1891–1925). During which he wrote the standard works: *The Structure and Development of Mosses and Ferns* (1895) and *University Textbook of Botany* (1902). He travelled throughout the Pacific region resulting in his *Outline of Plant Geography* (1926). His death has been described as "…the end of an era of a group of great plant morphologists." He was honoured for his interest in the flora of Japan and 'in all things Japanese'.

Campbell, S

New Guinea Tiger Perch *Datnioides campbelli* GP Whitley, 1939

Flight-Lieutenant Stuart Alexander Caird Campbell (1903–1988) was an Australian air force officer, aviator, andministrator and businessman. Sydney University awarded his BE in mechanical and electrical engineering (1926) then he immediately enlisted. Part of his initial duties was surveying the Great barrier Reef. He served (1929) in the seaplane-carrier HMAS *Albatross* before being becoming senior pilot to the British, Australian and New Zealand Antarctic Research Expeditions (1929–1930 & 1930–1931). He pioneered flying in Antarctica in a Gipsy Moth seaplane. He was promoted to Flight Lieutenant (1930) and was awarded the Polar medal (1934). He worked his passage to England joining an air transport company returning to Australia as its field manager undertaking aerial surveys there and in New Guinea. After WW2 he worked for the Department of Civil Aviation becoming Director (1948). He ran a successful business in New Guinea before retiring to Queensland. He was honoured as he collected the type series.

Campbell

Campbell's Whiptail *Coelorinchus kaiyomaru* T Arai & T Iwamoto, 1979

We think this use of 'Campbell' is as a toponym, after Campbell Island in the New Zealand subantarctic. It is known that the 'Kaiyo Maru' cruised in that area (1970–1971). (See also **Kaiyo Maru**)

Campello

Neotropical Killifish genus *Campellolebias* R Vaz Ferreira & B Siera de Soriano, 1974

African Rivuline sp. *Kryptolebias campelloi* WJEM Costa, 1990

Gilberto Campello Brasil, a Brazilian aquarist. (See **Gilberto**)

Campos

Pencil Catfish sp. *Listrura camposae* P de Miranda Ribeiro, 1957

Antonia Amaral Campos was a Brazilian ichthyologist at the Department of Zoology, Secretariat of Agriculture, State of São Paulo. She wrote *Sôbre os Caracídios do Rio Mogi-guaçu (Estado de São Paulo)* (1945). The binomial was originally spelt *camposi* but repeated usage appears to have meant that the (correct) feminine form *camposae* has become acceptable. She collected the holotype of this species.

Campos-da-Paz

Ghost Knifefish sp. *Apteronotus camposdapazi* CD de Santana & PC Lehmann-Albornoz, 2006

Glass Knifefish sp. *Eigenmannia camposi* EE Herrera-Collazos, AM Galindo-Cuervo, JA Maldonado-Ocampo & M Rincón-Sandoval, 2020

Dr Ricardo Campos-da-Paz is a Brazilian biologist and zoologist. The Federal University of Rio de Janeiro awarded his bachelor's degree (1990), the University of São Paulo his master's (1992) and his doctorate (1997). His post-doctoral studies were all at the Federal University of Rio de Janeiro (1997–1998, 1999–2000 & 2003–2005). Since 2006 he has been an Associate Professor at the Federal University of the State of Rio de Janeiro where he is head of the Department of Ecology and Marine Resources. He wrote: *Previously undescribed dental arrangement among electric knifefishes, with comments on the taxonomic and conservation status of* Tembeassu marauna *Triques (Otophysi: Gymnotiformes: Apteronotidae)* (2005). He discovered this species.

Camps

Highlands Rainbowfish *Chilatherina campsi* GP Whitley, 1957

Norman John Ernest Camps was an assistant preparator at the Australian Museum (1949–1955). He was one of the collectors of the original specimen, on his exploratory trip to Papua New Guinea for the museum under the leadership of the Curator of Mammals, Ellis Troughton (1954), which collected over a thousand species including 220 birds. His job was to collect, skin and preserve specimens. The fish were later examined by Whitley, his colleague at the Australian Museum. He had also been part of a collecting expedition to northwest Australia (1952); the largest of its kind by the museum at the time. He resigned from the museum (1955) and became a coffee plantation owner in the Mount Hagen area of New Guinea where he stayed for many years before retiring to Australia.

Candidus, E

Black-stripe Dwarf Cichlid *Taeniacara candidi* GS Myers, 1935

Edward Candidus was an American aquarist. The etymology says that the holotype was "*... received from Mr. Ed. Candidus of Morsemere, New Jersey, for whom the species is named*" and "*...whose aquarium collection is famed for the ichthyological rarities it contains.*"

Candidus (Carvalho)

Pencil Catfish sp. *Trichomycterus candidus* P de Miranda Ribeiro, 1949

Dr José Cândido de Melo Carvalho (1914–1994) was a Brazilian entomologist who was a world expert on Hemiptera. The Universidade Federal de Viçosa awarded his bachelor's degree (1929), the University of Nebraska his master's (1940) and the University of Iowa his doctorate (1942). He was Director of Museu Paraense Emílio Goeldi, Belém (1955–1960) and Professor and Director of Museu Nacional do Rio de Janeiro. He published a huge number of papers and articles, being best remembered as the chief editor of the *Atlas da Fauna Brasileira* (1978). He was also an athlete, as he was part of the Brazilian team that competed at the Berlin Olympics (1936). He collected the holotype of this species.

Candolle

Connemara Clingfish *Lepadogaster candolii* A Risso, 1810

Augustin Pyramus de Candolle (1778–1841) was a Swiss botanist. He spent four years at the Geneva Academy, studying science and law according to his father's wishes, but moved to Paris (1798) after Geneva had been annexed to the French Republic. He published his: *Essai sur les propriétés médicales des plantes* (1804) and was granted a doctor of medicine degree by the medical faculty of Paris. He documented hundreds of plant families and created a new plant classification system. Although his main focus was botany, he also contributed to related fields such as phytogeography, agronomy, paleontology, medical botany and economic botany. He was appointed (1807) as Professor of Botany in the medical faculty of the University of Montpellier and took the botany chair there (1810). He wrote his *Théorie élémentaire de la botanique* (1813) and spent the rest of his life trying to complete his natural system of botanical classification. He would have been well known in French scientific circles at the time of Risso's description so, though there is no etymology, it seems likely this is the man Risso intended. Also, Risso is known to have honoured other botanists. Further, it seems there have been attempts made, at times, to 'correct' the binomial to such spellings as *candollei* or *candollii*.

Canes

Neotropical Rivuline sp. *Melanorivulus canesi* DTB Nielsen, 2017

Paulo José Ferreira Canes was a Brazilian environmentalist and aquarist. The etymology says the fish was "*…named in honour of the late aquarium hobbyist and environmentalist Paulo José Ferreira Canes.*"

Canestrini

Canestrini's Goby *Pomatoschistus canestrinii* AP Ninni, 1883

Giovanni Canestrini (1835–1900) was an Italian naturalist and biologist who was Ninni's friend. He studied natural sciences at the University of Vienna. He was a lecturer at the University of Modena (1862–1869) prior to becoming Professor of Zoology and comparative anatomy at the University of Padua, where he established a bacteriology laboratory. He was an early convert to Darwinism and translated Darwin's work into Italian. He wrote some 200 scientific papers and other publications including: *Origine dell'uomo* (1866) and the three-volume: *Compendio di zoologia e anatomia comparata* (1869–1871). He described the similar-looking *Gobius pusillus* (=*P. knerii*) (1862).

Caneva

Blenny sp. *Microlipophrys canevae* D Vinciguerra, 1880

Giorgio Caneva. Vinciguerra wrote: "*The careful and patient research of a young friend of mine, Mr. Giorgio Caneva, put me in mind, last summer, to examine an overwhelming quantity of Blennius...*" (Translation). We are unable to find any biographical information on him.

Canosa

Argentine Croaker *Umbrina canosai* C Berg, 1895

Don Sabas Canosa was a 'preparater' and later conservator at the National Museum of Montevideo, Uruguay. He was responsible for taxidermy and other forms of preservation of specimens. He also collected for the museum.

Cantera

Pacu (characin) sp. *Colossoma canterai* GJ Devincenzi, 1942
[Junior Synonym of *Piaractus mesopotamicus*]

Procopio Cantera was an Assistant and the librarian at the Museo de Historia Natural, Montevideo. He appears to have worked at the museum for 65 years.

Canto

Tetra sp. *Hyphessobrycon cantoi* TC Faria, KLA Guimarães, LRR Rodrigues, C Oliveira & FCT Lima, 2021

André Luiz Colares Canto is Curator of the fish collection of the Universidade Federal do Oeste do Pará. He was honoured for his contribution to the knowledge of fishes from the rio Tapajós basin (Pará, Brazil), where this tetra occurs.

Cantor

Cardinalfish sp. *Apogonichthyoides cantoris* P Bleeker, 1851
Gourami sp. *Trichopodus cantoris* A Günther, 1861
Goby sp. *Scartelaos cantoris* F Day, 1871
Croaker sp. *Johnius cantori* P Bleeker, 1874
Striped Codlet *Bregmaceros cantori* DM Milliken & ED Houde, 1984

Dr Theodore Edvard Cantor (1809–1860) was a Danish amateur zoologist and Superintendent Physician of the European Asylum, Bhowanipur, Calcutta. This was part of the Honourable East India Company's Bengal Medical Service. He collected in Penang and Malacca. He was interested in tropical fish, and (c.1840) the King of Siam gave him some Bettas, commonly known as fighting fish. He published an article about them that led to 'Betta fever', the popular craze in Victorian England for keeping such fish. He wrote: *Catalogue of Malayan fishes* (1850). Two mammals, eleven reptiles and a bird are named after him.

Canut

Hoary Catshark *Apristurus canutus* S Springer & PC Heemstra, 1979
Grey Skate *Dipturus canutus* PR Last, 2008

Canutus is Latin for grey or ash-coloured and nothing to do with the tide-defying King Canute.

Capart

Lake Tanganyika Cichlid sp. *Trematocara caparti* M Poll, 1948

Dr André Capart (1914–1991) was an oceanographer who worked at the Royal Museum of Natural History in Belgium. He was an Assistant Naturalist (1938) and completed his PhD (1941) while there. He was on several missions to Central Africa, including being a member of the Océanographe de la Mission Hydrobiologique to Lake Tanganika (1946–1947) during which he collected the holotype. He also took part in expeditions to Antarctica, Papua and several oceans. He became Professor of Oceanography at the Catholic University of Louvain.

Capellen

North Pacific Crestfish *Lophotus capellei* CJ Temminck & H Schlegel, 1845

Godert Alexander Gerard Philip, Baron van der Capellen (1778–1848), was a Dutch statesman who was Governor-General, Dutch East Indies (Indonesia) (1815–1825). He appears not to have been a great success as he annoyed everyone, sparking the bloody Java War (1825–1830). He arranged a loan from Great Britain, thereby mortgaging the East Indies in the process. It was a step too far for the authorities in the Netherlands and he was sacked.

Capurro

Leaping African Mullet *Mugil capurrii* A Perugia, 1892

Captain Guiseppe Capurro collected fish in the Antilles and Senegal. Perugia wrote that he named the mullet in honour of Capt. Guiseppe Capurro, who collected type in Senegal, and to whom the Museo Civico di Genova "owes many interesting animals collected by him during his travels" (translation).

Carberry

Threadfin Anthias *Nemanthias carberryi* JLB Smith, 1954

Smith's etymology says only that the fish was: "...*Named for J E Carberry of Malindi, Kenya who greatly assisted my work.*" Since Smith gives no further details, we cannot be entirely certain, but believe this to be John Evans Carberry (1892–1970) who ran a business in Malindi at this time. He was born John Freke Evans, 10th Lord Carbery, but changed his name by deedpoll (1921). He outraged the colony (1938) by preoposing a toast at the Muthainge Club of: "*To hell with England, long live Germany*". He also spent a year in gaol for 'irregular currency transactions', which, characteristically he claimed was the "*best year of my life*".

Cardona

Twinhorn Blenny *Coralliozetus cardonae* BW Evermann & MC Marsh, 1899

This is a toponym referring to a little islet off Playa de Ponce, Puerto Rico, on the reef of which the type was collected.

Cardoso, AR

Banjo Catfish sp. *Hoplomyzon cardosoi* TP Carvalho, RE Reis & JP Friel, 2017

Dr Alexandre Rodrigues Cardoso graduated from the Pontifical Catholic University of Rio Grande do Sul (1997), also being awarded his masters (2003) and PhD (2008) there. He is described in the etymology as a "*…dear colleague who prematurely passed away … for his humbleness, positive attitude, and dedicated friendship.*" The fish is dedicated in his honour and in memory of him and his contribution to the taxonomy of neotropical fishes.

Cardoso, J

Yellowfish sp. *Labeobarbus cardozoi* GA Boulenger, 1912

José Cardoso (also spelled Cardozo) was, according to the etymology, the Governor-General of Angola. He gave Ansorge (q.v.), who collected the holotype, help with his arrangements for his collecting expedition. We cannot find that anyone of that name was Governor-General of Angola at that time and wonder if either Ansorge or Boulenger promoted someone beyond his rank? However, there was a Dr António José Cardoso de Barros who was Secretary General of Angola at the time Ansorge was collecting. Among his duties was the oversight of education, health and religious institutions, so we speculate that this might have been the man honoured.

Carine

Squeaker Catfish sp. *Synodontis carineae* EJ Vreven & A Ibala Zamba, 2011

Dr Carine Plancke is a Belgian anthropologist. The University of Leuven awarded her master's degree; her doctorate (2010) was awarded by them and the Ecole des hautes études en sciences sociales. She is a postdoctoral researcher at the Centre for Research on Culture and Gender, University of Ghent and also is a Research Associate at the Laboratoire d'anthropologie sociale, Collège de France, Paris. She did postdoctoral studies at the centre for Dance Research, University of Roehampton, England. She worked with the Punu people in the Nyanga area, Congo Republic and wrote numerous articles in peer-reviewed journals such as: *On Dancing and Fishing: Joy and the Celebration of Fertility among the Punu of Congo-Brazzaville* (2010). She particularly likes the colour yellow, referring to the vivid yellow shine of this species in life.

Carl (Eigenmann)

Characin genus *Carlana* E Strand, 1928

Characin genus *Carlastyanax* J Géry 1972

Professor Dr Carl Henry Eigenmann (1863–1927). (See **Eigenmann, CH**)

Carl (Hubbs)

Threeline Prickleback *Esselenichthys carli* WI Follett & ME Anderson, 1990

Carl Leavitt Hubbs (1894–1979) (see **Hubbs, CL**)

Carla (Lindenaar-Sparrius)

Carla's Cory *Corydoras carlae* H Nijssen & IJH Isbrücker, 1983

Mrs Carla C Lindenaar-Sparrius was in charge of administration for 11 years at the Department of Ichthyology, Zoölogisch Museum, Amsterdam, where the authors were curators of fishes. She also occasionally collected for the museum.

Carla (Pavanelli)

Bumblebee Catfish sp. *Microglanis carlae* HS Vera Alcaraz, WJ da Graça & OA Shibatta, 2008

Dr Carla Simone Pavanelli (b.1967) is a Brazilian biologist and curator at the Center for Research in Limnology, Ichthyology and Aquaculture, at the University of Maringá (UEM), for the last thirty years. She is the leader of the Neotropical Fish Systematics Research Group there, and acts as a reviewer of several national and foreign periodicals, besides being since 2015 the Editor-in-chief of the *Neotropical Ichthyology* journal (official journal of the Brazilian Society of Ichthyology). The State University of Maringá awarded her degree in biological science (1987); the Federal University of Rio de Janeiro awarded her masters' (1994) and the Federal University of São Carlos her PhD (1999). She was honoured for her contributions to Neotropical ichthyology, where she has published more than 70 articles, three books, several book chapters, among which are the co-written: *Neotropical Siluriformes as a model for insights on determining biodiversity of animal groups* (2015), *Revision of the trans-Andean scrapetooths genus* Saccodon *(Ostariophysi: Characiformes: Parodontidae)* (2015) and *Ten years after "Peixes da planície de inundação do alto rio Paraná e áreas adjacentes": revised, annotated and updated* (2018). She has also advised many undergraduate and postgraduate students. She is interested in conserving the biodiversity of fish.

Carlarius

Sea Catfish genus *Carlarius* AP Marceniuk & NA Menezes, 2007

Dr Carl James Ferraris, Jr. (See **Ferraris**)

Carl Bond

Snailfish sp. *Paraliparis carlbondi* DL Stein, 2005

Professor Dr Carl E Bond (1920–2007) was an American ichthyologist. Oregon State College awarded his bachelor's (1947) and master's (1948) and the University of Michigan awarded his PhD (1963). He began his teaching and research career in Oregon State College's Fisheries and Wildlife Department (1949) as Assistant Aquatic Biologist. Specialising in the study of freshwater fish, he became one of the world's leading authorities on sculpins (Cottidae). Bond was also involved in research internationally, working in Latin America and Africa, as well as India and Iran in Peace Corps projects (1967–1971). He was also Assistant Dean of the Graduate School (1969 to 1974). He became Professor Emeritus of Fisheries (1984). He wrote: *Biology of Fishes* (1979).

Carleton

Stone Loach sp. *Schistura carletoni* HW Fowler, 1924

Rev Marcus Manard Carleton (1826–1898) was an American Presbyterian Missionary and commercial fruit grower. He was at Amherst College, Massachusetts (1848) and graduated

(1851) before studying theology at Columbia Seminary and East Windsor (now Hartford) Seminary where he graduated and was was ordained (1854). He married (Celestia Bradford d.1881) that year and was immediately sent as a missionary to Ambala, northern India (1854–1898). During that period he moved from Kullu to what is now the city of Ani to grow fruit (1870s), settling native farmers who converted to Christianity on land there. (One of his daughters was called Annie and the town was named after her). He also established a leper colony at Sabathu. He remarried the daughter of a missionary, Eliza Calhoun (1884). Of his six children, three graduated at the same college, two became doctors and a daughter became a medical missionary in India. He collected many Indian freshwater fishes, including the holotype of the eponymous loach. He lived out his life in India, never even returning to the USA on leave, and died of heart disease.

Carletto

Neotropical Rivuline sp. *Hypsolebias carlettoi* WJEM Costa & DTB Nielsen, 2004

André Carletto (b.1972) is a Brazilian aquarist who helped collect the type. The etymology says: "*Named in honour of André Carletto, in recognition to his valuable help on several collecting trips.*"

Carl Hubbs

Giant Hagfish *Eptatretus carlhubbsi* CB McMIllan & RL Wisner, 1984
Southern Scythe Butterflyfish *Prognathodes carlhubbsi* T Nalbant, 1995

Professor Carl Levitt (Leavitt) Hubbs (1894–1979). (See **Hubbs**, and also see **Carl**)

Carlotta

Rainbow Bream *Sargochromis carlottae* GA Boulenger, 1905

Charlotte Sclater was the wife of the zoologist William Lutley Sclater (1863–1944) (q.v.) who was Director of the South African Museum, and who presented the type to the British Museum (Natural History).

Carlsberg

Hatchetfish sp. *Valenciennellus carlsbergi* AF Bruun, 1931
Carlsberg's Lanternfish *Electrona carlsbergi* ÅV Tåning, 1932
Dreamer sp. *Oneirodes carlsbergi* CT Regan & E Trewavas, 1932
Amarsipa (marine glassfish sp.) *Amarsipus carlsbergi* RL Haedrich, 1969

The Carlsberg Foundation, which owns the famous Danish brewery, financed the 'Dana' (q.v) fishery research cruises that collected the holotypes of *Valenciennellus carlsbergi* and *Valenciennellus carlsbergi* which is named after the Carlsberg Laboratory, Copenhagen, the research arm of the Carlsberg Foundation.

Carmabi

Candy Basslet *Liopropoma carmabi* JE Randall, 1963

This is an acronym made up of the initial letters of an institution: **Car**aibisch **Ma**rien-**Bi**ologisch Instituut. *Randall wrote in his description: "Remarks — C. carmabi is one of the* *most complexly and beautifully colored of West Indian fishes, and it is with pleasure that I name it in honor of the Caribbean Marine Biological Institute ("Caraibisch Marien-Biologisch Instituut", Carmabi) of Curaçao and its most cooperative staff headed by Ingvar Kristensen."*

Carmen

Eelpout sp. *Santelmoa carmenae* J Matallanas, 2010

Dr Carmen Benito González is the manager of the radiology lab at the Biology Faculty, University of Barcelona, Spain.

Carlos

Characin sp. *Bryconamericus carlosi* C Román-Valencia, 2003

Carlos Román-Valencia is the author's son.

Carlson

Carlson's Damsel *Neoglyphidodon carlsoni* GR Allen, 1975

Dr Bruce Carlson is an ichthyologist-aquarist. He collected the damselfish type, provided detailed notes on habitat, and contributed many new pomacentrid records for the Fiji Islands, according to the author. He began keeping freshwater fish (1956) but switched to saltwater (1962) and says of himself: "*As a teenager in Michigan there were many cold wintry nights when I dreamed of diving on a coral reef and seeing reef fish in their natural environment. In 1967 I took up SCUBA diving, and a year later I bought my first camera-in-box that could take underwater photos. I finally fulfilled my dream of seeing a coral reef in 1972 when I joined the Peace Corps and was sent to Fiji where I resided for nearly four years.*" The University of Michigan awarded his zoology degree (1971) after which he joined the Peace Corps. During this time, he met Gerry Allen for the first time. He was awarded his PhD by the University of Hawaii (1976) and started work at Waikiki Aquarium (1976–2002). He moved to Georgia (2002) as part of the design team for the Georgia Aquarium, working there until retirement (2011) when he moved back to Hawaii but continues his research work.

Carnegie, A

Neotropical Killifish sp *Cnesterodon carnegiei* JD Haseman, 1911

Armoured Catfish sp. *Dolichancistrus carnegiei* CH Eigenmann, 1916

Andrew Carnegie (1835–1919) was a great industrialist and philanthropist. Among his many gifts was the foundation of the four Carnegie Museums in Pittsburg, Pennsylvania, which come under the umbrella of the Carnegie Institute. The Carnegie Museum of Natural History supported and published many of Eigenmann's studies.

Carnegie, MC

Characin genus *Carnegiella* CH Eigenmann, 1909

Margaret Cameron Carnegie (1897–1990) was the only child of the famous industrialist and philanthropist Andrew Carnegie (1835–1919) (see previous entry). She married Roswell Miller Jr. (1919) and had four children. The Carnegie Museum co-sponsored Eigenmann's 1908 expedition to British Guiana (Guyana) and published his report.

Carol

Carol's Gurnard *Lepidotrigla carolae* WJ Richards, 1968

Carol is the wife of author, Dr William J 'Bill' Richards. (Also see **Richards**)

Carol

Fissi (Lake Barombi Mbo cichlid sp.) *Sarotherodon caroli* M Holly, 1930

One of those annoying cases where the author did not deign to provide an etymology. The fish was originally described (in German) under the name *Tilapia caroli* but without any information about 'Carol'. Possibly 'Carol' = Karl, and refers to the German natural history collector Karl Albert Haberer (q.v.), who supplied a collection of fishes from Cameroon to Naturhistorisches Staatsmuseum (Vienna).

Carol (Youmans)

Driftwood Catfish sp. *Balroglanis carolae* RP Vari & CJ Ferraris Jr, 2013
[Syn. *Centromochlus carolae*]

Carol Youmans is Management Support Specialist, Invertebrate Zoology Department, the Smithsonian. She worked in the National Museum of Natural History (1994–2018) since which she works in the Office of the Provost. The etymology thanks her: "*...for 'invaluable' assistance to both authors over the years, particularly the senior author.*" She informed us: "*I worked with Rich Vari for several years as the Management Support Specialist for four of the science departments at the National Museum of Natural History including Vertebrate Zoology. In that position, I handled all administrative aspects, such as his finances, procurement, travel, etc. He was a good person and I sincerely miss him.*"

Caroline (Ajootian)

Snailfish sp. *Psednos carolinae* DL Stein, 2005

Caroline Ajootian (b.1952). The etymology says: "*Named in honor of Caroline Ajootian, for her unfailing support and encouragement of snailfish research.*"

Caroline (Henry)

Banded Sculpin *Cottus carolinae* TN Gill, 1861

Caroline Henry (1839–1920) was the daughter of Professor Joseph Henry (q.v.) who took Gill under his wing when he was Secretary of the Smithsonian at the time when Gill was beginning his career. It was noted that Gill delighted in female company but never married. Gill's etymology says: "*I have given myself the pleasure of dedicating this fine species to my estimable young friend, Miss Caroline Henry.*"

Caroline (Lynch)

Freshwater Goby sp. *Lentipes caroline* DB Lynch, P Keith & FL Pezold III, 2013

Caroline Lynch (DNF) is the first author's daughter. The fish is also named after the: "*beautiful islands of Micronesia for which she was named,*" i.e., the Caroline Islands, where it occurs.

Caroline (Paugy)

Characin sp. *Brycinus carolinae* D Paugy & C Lévêque, 1981

Mrs Caroline Paugy was the wife of the senior author, Didier Paugy.

Caroliterti

Northern Iberian Chub *Squalius carolitertii* I Doadrio Villarejo, 1988

Carlos III (1716–1788) was King of Spain (1759–1788). The binomial is a Latinized version of 'Carlos the third'. He founded the Museo Nacional de Ciencias Naturales, Madrid (1777) where the holotype of this species is housed.

Carotta

Carotta's Barbel *Barbus carottae* PG Bianco, 1998
[Syn. *Messinobarbus carottae*]

Arianna Martha Maria Carotta of Bolzano, Italy helped collect the holotype.

Carpentaria

Gulf of Carpentaria Anchovy *Stolephorus carpentariae* CW De Vis, 1882

This is a toponym; the Gulf of Carpentaria drainage is in Queensland, Australia, and the type locality was the Norman River.

Carpenter, A

Deepwater Cardinalfish sp. *Brephostoma carpenteri* AW Alcock, 1889
Hooked Tongue Sole *Cynoglossus carpenteri* AW Alcock, 1889

Captain Alfred Carpenter RN DSO was the Commander of H M Indian Marine Survey Steamer 'Investigator'. He surveyed the exact delimitation of the Bay of Bengal. Alcock, who was Surgeon-Naturalist to the Survey, calls him "…the pioneer of scientific hydrography in India" and says of him that he was a "…highly scientific officer" to whom he was "… indebted for much more than the facts alone." His son, Captain Alfred Francis Blakeney Carpenter won the VC when commanding 'HMS Vindictive' in WW1.

Carpenter, KE

Carpenter's Flasher Wrasse *Paracheilinus carpenteri* JE Randall & R Lubbock, 1981
Carpenter's Yellowtop Jewelfish *Meganthias carpenteri* WD Anderson, 2006

Dr Kent E Carpenter is Professor of Biological Sciences at Old Dominion University, Virginia, USA. Florida Institute of Technology awarded his BSc in marine biology (1975), the University of Hawaii awarded his PhD. His research emphasis is in the systematics and evolution of marine fishes. He is also a long-term collaborator with the Food and Agriculture Organization of the United Nations Species Identification and Data Programme for Fisheries, producing identification guides for regions such as the western Pacific and the western and eastern Atlantic Oceans. He has done fieldwork in the Caribbean, West Africa, and the Philippines. In addition to research and teaching responsibilities, he is also the coordinator for the IUCN Global Marine Species Assessment, completing the first global review of every marine vertebrate species, and of selected marine invertebrates and marine plants, to determine conservation status and possible extinction risk for about 20,000 marine species. He was the principal collector of the type specimens.

Carpenter, KJ

Carpenter's Chimaera *Chimaera lignaria* DA Didier Dagit, 2002

Kevin J Dagit (See **Dagit**). The binomial means 'of wood', referring to Dagit's profession as a woodworker.

Carpenter, M

Slope Soapfish *Rypticus carpenteri* CC Baldwin & LA Weigt, 2012

Michael Carpenter was station manager for the Smithsonian's research station at CarrieBow Cay, Belize, for over three decades. The authors mention his "*...dedication to maintaining this remote station benefited a multitude of marine scientists (and marine science). We thank him for his good-natured support in the field, and the first author is grateful for his enduring friendship.*"

Carpenter, WD

Goby sp. *Tukugobius carpenteri* A Seale, 1910
[Syn. *Rhinogobius carpenteri*]

William Dorr Carpenter (1879–1958) was an American naturalist. He helped to collect the holotype of this species. Two birds are named after him.

Carpenter, WK

Guinean Amberjack *Seriola carpenteri* FJ Mather, 1971

William K Carpenter is a big game fisherman who is a friend of the author. The etymology says: "*This species is named for my friend William K. Carpenter of Fort Lauderdale, Florida. Mr Carpenter, an outstanding big game fisherman, has long been the President and leading sponsor of the International Game Fish Association.*"

Carr, AF

Seminole Goby *Microgobius carri* FW Fowler, 1945
Stipple Wormfish *Microdesmus carri* CR Gilbert, 1966

Professor Dr Archibald 'Archie' Fairly Carr Jr (1909–1987) was an American ecologist and conservationist. He was Professor of Zoology, University of Florida, where he studied zoology before settling on a career as a herpetologist, becoming one of the world's leading experts on sea turtles. The Dr Archie Carr Wildlife Refuge, Costa Rica, was established and named in his honour. He wrote: *Ulendo: Travels of a Naturalist in and out of Africa* (1954). Two reptiles and two amphibians are also named after him.

Carr, JK

Dwarf Sicklefin Chimaera *Neoharriotta carri* HR Bullis & JS Carpenter, 1966

James K Carr (1914–1980) was Under Secretary at the US Department of the Interior (1961–1964). He was honoured in the name of the chimaera for: "*...his great personal interest and counsel in the (Bureau of Commercial Fisheries') exploratory fishing programs.*"

Carrie

Goby genus *Carrigobius* JL Van Tassell, L Tornabene & RG Gilmore, 2016

Not an eponym but a toponym referring to Carrie-Bow Cay, Belize, home of the Smithsonian Institution's field station, where many specimens of *C. amblyrhynchus* were collected.

Carriker

Carriker's Anchovy *Anchoviella carrikeri* HW Fowler, 1940
Scrapetooth Characin sp. *Parodon carrikeri* HW Fowler, 1940

Melbourne Armstrong Carriker Jr (1879–1965) was one of the great early naturalists (primarily an ornithologist and entomologist) of Central and northern South America. His was the first modern systematic publication of the birds of Costa Rica (1910), in which he listed 713 species for the country. He greatly enhanced the bird collections of the Carnegie Museum (Pittsburgh) and the USNM. He is the subject of a biography written by his son Professor Melbourne Romaine Carriker. Ten birds, a mammal and an amphibian are named after him. He collected the holotype of the characin.

Carrillo

Characin sp. *Hemibrycon carrilloi* G Dahl, 1960

Dr Jorge Carrillo was Director, Fisheries Department, Colombian Ministry of Agriculture. He was noted for his: "*...enthusiastic work in defence of the Colombian fauna.*"

Carrington

Bonneville Speckled Dace *Rhinichthys osculus carringtonii* ED Cope, 1872

Edward Campbell Carrington (1851–1917) was a zoologist and naturalist with the US Geological Survey. He graduated from Columbia college and became a criminal lawyer. However, he joined the Geological Survey as a 'genial and well-liked' zoologist. He collected the dace holotype in Wyoming during the expedition led by Dr F V Hayden (1870–1871) to what is now Yellowstone National Park, wherein Carrington Island was named after him.

Carrión

Armoured Catfish sp. *Chaetostoma carrioni* JR Norman, 1935
[Syn. *Lipopterichthys carrioni*]

Professor Clodoveo Carrión Mora (1883–1957) was the Ecuadorian natural scientist of the 20th century. He was a palaeontologist and naturalist who came from a literary family. Recognizing that he had aptitude for the sciences but none for letters, he travelled to England and studied at universities in Manchester and London, emerging after 10 years as an engineer. Having returned to Ecuador he became Professor of Natural Sciences, Colegio Bernardo Valdevisio. He collected the holotypes of the two reptiles that are also named after him.

Carrow

Blenny sp. *Malacoctenus carrowi* P Wirtz, 2014

Frank Carrow was an American conservationist and philanthropist. Wirtz's etymology says the blenny was: "*...named in honor of Frank Carrow, whose interest in marine conservation led to his creation and funding of the Carrow Foundation, a charitable organization that supports a broad range of marine conservation activities.*"

Carter, GS

Characin sp. *Phenacogaster carteri* JR Norman, 1934

Dr George Stuart Carter (1893–1969) was a zoologist at Cambridge University where he had gained his doctorate and lectured (1938–1960). During WW1, he served in the British Army in the infantry (1914–1917) and the Royal Engineers (1917–1919). He worked at the Stazione Zoologica, Naples (1922–1923) and as a lecturer at Glasgow University (1923–1930). He led the expedition to British Guiana (Guyana) during which the holotype was collected. He co-wrote: *Notes on the Habits and Development of Lepidosiren paradoxa* (1930).

Carter (Gilbert)

Cylindrical Lantern Shark *Etmopterus carteri* S Springer & GH Burgess, 1985

Dr Carter Rowell Gilbert (1930-2022). (See **Gilbert, CR**)

Cartier

Snake-eel sp. *Benthenchelys cartieri* HW Fowler, 1934

Oscar Cartier was a German herpetologist and ichthyologist who was at the University of Wurzburg, which published his thesis (1872). He studied Philippine fishes and wrote: *Beschreibungen neuer Pharyngognathen – Ein Beitrag zur Kenntniss der Fische des philippinischen Archipels* (1874).

Cartledge

Characin sp. *Leporellus cartledgei* HW Fowler, 1941

Franklin Fisher Cartledge (1869–1949) was a Professor in Philadelphia. He supplied the author with many local fishes.

Carvalho, AL

Armoured Catfish sp. *Hypostomus carvalhoi* A Miranda Ribeiro, 1937
Armoured Catfish sp. *Harttia carvalhoi* A Miranda Ribeiro, 1939
Neotropical Rivuline sp. *Austrolebias carvalhoi* GS Myers, 1947
Plated Catfish sp. *Aspidoras carvalhoi* H Nijssen & IJH Isbrücker, 1976

Dr Antenor Leitao de Carvalho (1910–1985) was a Brazilian herpetologist and ichthyologist, specializing in frogs. He became a pilot in the merchant marine (1927–1932), and whenever in port (Rio) he volunteered to help out at the museum. He became Field Collector for the Museu Nacional, Rio de Janeiro (1933), undertaking a number of expeditions in Brazil. He became Curator of Herpetology (1941) and, eventually, the museum's Vice Director. He collected specimens of many different taxa from all over Brazil (1930s and 1940s). He wrote: *A Preliminary Synopsis of the Genera of American Microhylid Frogs* (1954). Nine amphibians and five reptiles are also named after him. (See also **Leitao**)

Carvalho, MR

Characin sp. *Tetragonopterus carvalhoi* BF Melo, RC Benine, TC Mariguela & C de Oliveira, 2011

Dr Marcelo Rodrigues de Carvalho works at the Departamento de Zoologia, Instituto de Biociências, Universidade de São Paulo where he is a professor. He studied at Universidade Santa Úrsula, USU, Brasil for his first degree (1987–1991). He did a doctorate through a programme on evolutionary biology run jointly (1992–1999) by New York City University and AMNH, where he did subsequently undertake post-doctoral research (1999–2001).

Cary

Rainbow Seaperch *Hypsurus caryi* L Agassiz, 1853

Thomas Cary (1824–1888) of San Francisco was a businessman, amateur naturalist, and the brother-in-law of Louis Agassiz, who described the species. Cary was honored for procuring specimens that confirmed Jackson's claims (see *Embiotoca jacksoni*) of viviparity.

Casale

Somali Mormyrid *Mormyrus casalis* D Vinciguerra, 1922

Captain Ugo Casale was an Italian army officer who was part of the Italian occupation forces in Itallian Somaliland (now part of Somalia). He collected fish and other taxa and supplied Italian museums with many specimens (1910–1915), including the holotype of this fish (1910) which he sent to the Museo Civico di Storia Naturale di Genova.

Casatti

Croaker sp. *Plagioscion casattii* O Aguilera & DR de Aguilera, 2001

Lilian Casatti is a Brazilian ichthyologist, biologist and ecologist. She is (2009–present) Associate Professor at São José do Rio Preto, Universidade Estadual Paulista. Paulo State University awarded her BSc (1993). Paulista State University Júlio de Mesquita Filho awarded her masters' (1996) and zoology doctorate (2000). Among her published papers are the co-written: *More of the Same: High Functional Redundancy in Stream Fish Assemblages from Tropical Agroecosystems* (2015) and *Intraspecific and interspecific trait variability in tadpole meta-communities from the Brazilian Atlantic rainforest* (2019). The authors named the croaker after her "*...for her valuable contribution to the study of freshwater Sciaenidae*."

Cashner

West African killifish sp. *Epiplatys cashneri* F Pezold, K Ford & RC Schmidt, 2021

Robert Charles Cashner (1942–2018) was an American fish biologist whose career was spent at the University of New Orleans, Louisiana. He began as an Instructor in the Department of Biological Sciences and rose through the ranks, becoming a Full Professor (1987), a Research Professor (1993), and serving as Chair of Biological Sciences (1993–1996). He became Dean of the Graduate School (1996) and retained this position even after also taking on the role of Vice-Chancellor for Research and Sponsored Programs (2001). In 2005, he was part of the administrative team that fought to save the university during the Hurricane Katrina disaster, after which he stayed on in both roles to further help the long recovery until he retired (2008). Most of his research focused on freshwater fish. He was honoured in the killifish binomial in part for being a mentor and teacher to the first author.

Cashibo

Armoured Catfish sp. *Loricariichthys cashibo* CH Eigenmann & WR Allen, 1942

The Cashibos are an indigenous tribe for whom the site where the holotype was collected, Lake Cashiboya in Peru, was named.

Cashner

West African killifish sp. Epiplatys cashneri F Pezold, K Ford & RC Schmidt, 2021

Robert C Cashner (1942-2018) University of New Orleans (Louisiana, USA), a mentor and teacher to the first author and a friend to many; "A recognized authority on North American freshwater fishes, his legacy also includes descriptions of two killifish species from North America."

Cassius

Lake Victoria Cichlid sp. *Haplochromis cassius* PH Greenwood & KDN Barel, 1978

Derived from Shakespeare's Julius Caesar (Act I, Scene II), "*Yond Cassius has a lean and hungry look*"; which look is shared by the cichlid.

Castello

Castello's Apron Numbfish *Discopyge castelloi* RC Menni, G Rincón & ML Garcia, 2008

Dr Hugo Patricio Castello is an Argentine zoologist, marine biologist and ichthyologist who gave the holotype of this electric ray to the describers, suggesting it might be new to science. He worked at Universidad del Salvador, Buenos Aires and later as *Chief of the Marine Mammal Laboratory, Argentine Museum of Natural Sciences, Buenos Aires.*

Castelnau

Amazon Pellona *Pellona castelnaeana* A Valenciennes, 1847
Castelnau's Jawfish *Opistognathus castelnaui* P Bleeker, 1859
Castelnau's Herring *Herklotsichthys castelnaui* JD Ogilby, 1897
Spotback Skate *Atlantoraja castelnaui* A Miranda-Ribeiro, 1907
Dwarf Stonebasher *Pollimyrus castelnaui* GA Boulenger, 1911

Francis(co) Louis Nompar de Caumont, Comte de Laporte de Castelnau (1810–1880) was a career diplomat and naturalist who was born in London, studied natural science in Paris, and then led a French scientific expedition to study the lakes of Canada, the USA and Mexico (1837–1841). He led the first expedition (1843–1847) to cross South America from Peru to Brazil, following the watershed between the Amazon and the Río de la Plata systems. Following this he took several diplomatic posts. He lived in Melbourne (1864–1880), being Consul-General (1862) and then French Consul (1864–1877). Despite becoming almost completely blind when returning to Paris, he published 15 volumes on geography, botany and zoology of South America (1850s) and several papers on Australian fishes in the 1870s. He sent the holotype of the Amazon Pellona to Valenciennes. Five birds and a reptile are named after him.

Castex

Vermiculate Freshwater Stingray *Potamotrygon castexi* HP Castello & DR Yagolkowski, 1969
[Junior Synonym of *Potamotrygon falkneri*]

Dr Mariano Narciso Antonio José Castex Ocampo (b.1932) is an Argentine ichthyologist and physician who was Professor of Biology at the Zoology Department, Universidad del Salvador, Buenos Aires (1960s). He was ordained as a Jesuit priest but even though he left the church (1970s) he received a PhD in Canonical Law. After leaving the priesthood, he married, practised as a psychiatrist and was a consultant on mental problems in criminal cases. He was honoured in the name of the ray as the person "*...whose studies in the last years have cast many and new lights on this difficult group of elasmobranchs*".

Castillo

Thorny Catfish sp. *Rhynchodoras castilloi* JLO Birindelli, MHS Sabaj Pérez & DC Taphorn Baechle, 2007

Otto E Castillo G is a Venezuelan biologist at the Museum of Zoology, National Experimental University of the Western Llanos Ezequiel Zamora, Venezuela. He co-wrote: *Two New Species of Thicklip Thornycats, Rhinodoras Genus* (Teleostei: Siluriformes: Doradidae) (2008). The eymology says he: "*...collected much of the type material, for his lifelong dedication to the study and stewardship of his country's rich diversity of freshwater fishes.*"

Castle

Conger Eel genus *Castleichthys* DG Smith, 2004
Castle's Moray *Gymnothorax castlei* EB Böhlke & JE Randall, 1999
Duckbill Eel sp. *Facciolella castlei* NV Parin & ES Karmovskaya, 1985
Castle's Conger *Gnathophis castlei* ES Karmovskaya & JR Paxton, 2000
Deepwater Big-eyed Worm Eel *Scolecenchelys castlei* JE McCosker, 2006
Conger Eel sp. *Bathycongrus castlei* DG Smith & H-C Ho, 2018

Dr Peter Henry John Castle (1934–1999) was an ichthyologist at the School of Biological Sciences, Victoria University, Wellington, New Zealand, which awarded all his degrees; bachelor's (1955) master's (1958), doctorate (1964) and DSc. (1998). He was a member of the faculty there (1958–1998). He was an expert on eel systematics and all the new taxa he described are eels. He took part in many research cruises in the Pacific as well as in New Zealand waters, and in his career published nearly 100 papers including: *A reassessment of the eels of the genus* Bathycongrus *in the Indo-west Pacific* (2005). (Also see **Arx**)

Castor

Longsnout Scorpionfish *Pontinus castor* F Poey, 1860
Lokundi Mormyrid *Hippopotamyrus castor* P Pappenheim, 1906

In Roman mythology, Castor and his brother Pollux were the heavenly twins. However, in Latin *Castor* also means 'beaver', and the mormyrid was so named because of the teeth which protrude from its lower jaw, like a beaver's incisors.

Castro, M

Galapagos Barnacle Blenny *Acanthemblemaria castroi* JS Stephens & ES Hobson, 1966

Miguel Castro was an Ecuadorian naturalist resident on the Galapagos Islands. He became (1960) the first conservation officer at the Charles Darwin Research Station, Academy Bay, Santa Cruz Island, Galapagos.

Castro, RMC

Armoured Catfish sp. *Rineloricaria castroi* IJH Isbrücker & H Nijssen, 1984
Pencil Catfish sp. *Cambeva castroi* MCC de Pinna, 1992
Glass Knifefish sp. *Sternopygus castroi* ML Triques, 2000

Ricardo Macedo Corrêa e Castro is a Brazilian zoologist and ichthyologist who is a full professor, Department of Biology, Universidade de São Paulo. The Federal University of Rio de Janeiro awarded his bachelor's degree (1977) and the University of São Paulo awarded his master's (1984) and his doctorates in Biology and Vertebrate Zoology (both 1990). He was a member of the BIOTASP/FAFESP Coordination group (1996–1999). He co-wrote: *Ichthyofauna diversity of the Upper Paraná River: present composition and future perspectives* (2007). He collected the holotypes and made them available for study.

Castro Aguirre

Veracruz White Hamlet *Hypoplectrus castroaguirrei* LF Del Moral Flores, JL Tello-Musi & JA Martínez-Pérez, 2012
Mojarra sp. *Eugerres castroaguirrei* AF González-Acosta & M de L Rodiles-Hernández, 2013
[Junior Synonym of *Eugerres mexicanus*]

José Luis Castro Aguirre (1943–2011) was a Mexican ichthyologist. The National School of Biological Sciences at the National Polytechnic Institute (ENCB-IPN) awarded his master's (1974) and PhD (1986). He worked at the National Fisheries Institute and the Food and Agriculture Organization in the 1960s, and later was a professor and a researcher at the ENCB-IPN, Universidad Autónoma Metropolitana (1979–1987), the Interdisciplinary Center of Marine Science (1976–1979 & 1994–2011) and Northeast Center of Biological Research (1987–1994). He wrote: *Catálogo sistemático de los peces marinos que penetran en aguas continentales de México, con aspectos zoogeográficos y ecológicos* (1978) the first catalogue of the fish of mexico's estuaries and around 150 other papers. He described about a dozen new fish species.

Catala

Comoro Mullet *Agonostomus catalai* J Pellegrin, 1932
Cardinalfish sp. *Jaydia catalai* P Fourmanoir, 1973

Dr René Catala (1901–1988) was a French biologist and entomologist who lived and worked in Madagascar (1920–1937), after which he returned to Paris to complete his doctorate.

Immediately after the end of WW2 he was involved (1946) in setting up the French Oceania Institute, which evolved to become ORSTOM (1964) and IRD (1998). He was Director of the biological station at Nouméa, New Caledonia where he also founded the Noumea Aquarium (1956). He collected the holotype of the cardinalfish.

Catania

Thorny Catfish sp. *Leptodoras cataniai* MH Sabaj Pérez, 2005

Dr David Catania is Senior Collection Manager, Department of Ichthyology, California Academy of Sciences where he has worked since 1985. He wrote: *X-ray Ichthyology: The Structure of Fishes* (2002).

Catarina

Catarina Allotoca *Allotoca catarinae* F de Buen, 1942

This is a toponym referring to the town of Laguna de Santa Catarina, Mexico

Catchpole

Blind Spiny Eel *Mastacembelus catchpolei* HW Fowler, 1936

Captain Geoffrey Catchpole of Njiana farm, Bunia, Congo, collected the holotype. We have found nothing more about him.

Catesby

False Moray genus *Catesbya* JE Böhlke & DG Smith, 1968

Mark Catesby (1683–1749) was an English naturalist, artist and traveller. He made two journeys to the Americas (1712–1719 & 1722–1726). He refers to the colonies as the Carolinas as they were then known. He predates Linnaeus, Bartram and Audubon, all of whom were influenced by him. Lewis and Clark consulted Catesby's work on their cross-country expedition (1804–1806). This work was: *The Natural History of Carolina, Florida and the Bahama Islands: Containing the Figures of Birds, Beasts, Fishes, Serpents, Insects and Plants* (1731–1743). During his travels, Catesby observed that birds migrate. This discovery was entirely contrary to the then prevailing view that birds hibernated in caves or under ponds in the winter. He published these observations in an essay (1747) *On the passage of birds*. He observed the similarity in the features of the Native Americans and peoples of Asiatic origin and was the first person to hypothesise the existence in the distant past of a landbridge between Asia and the Americas. He also mentioned the trends in animal size and species diversity as he progressed farther south. He used to ship his snake specimens back to England in jars of rum, which sailors sometimes drank, ruining his specimens! Many plant and animal taxa, including five birds, an amphibian and two reptiles, are also named after him.

Catherine (de Beaufort)

Waigeo Rainbowfish *Melanotaenia catherinae* LF de Beaufort, 1910

Catherine de Beaufort was the wife of the author Lieven Ferdinand de Beaufort (1879–1968) (See also de **Beaufort**)

Catherine (Ozouf-Costa)

Snailfish sp. *Careproctus catherinae* AP Andriashev & DL Stein, 1998

Dr Catherine Ozouf-Costa (b.1951) is an ichthyologist at the Laboratoire d'Ichthyologie, MNHN, Paris. Among her publications is: *Cytogenetics of the Antarctic icefish* Champsocephalus gunnari *Lönnberg, 1905 (Channichthidae, Notothenioidei)* (1997) and she edited *Techniques of Fish Cytogenetics* (2015).

Catherine (Robins)

False Moray sp. *Robinsia catherinae* JE Böhlke & DG Smith, 1967

Catherine Robins née Hale was honoured for her contributions to the knowledge of this 'fascinating group of eels' (the generic name honors her husband) (see also **Robins CH & CR**)

Caudan

Rattail sp. *Coelorinchus caudani* JBFR Köhler, 1896

'Caudan' was a French vessel, from which the holotype was collected.

Caunos

Caunos Goby *Knipowitschia caunosi* H Ahnelt, 2011

Caunos was a mythological figure, the twin brother of Byblis (q.v.) and founder of the ancient city Caunos; the ruins of this city are located close to Lake Köycegiz, Turkey, the type locality.

Causse

Causse's Scorpionfish *Pteroidichthys caussei* H Motomura & Y Kanade, 2015

Sleeper sp. *Giuris caussei* P Keith, M Mennesson & C Lord, 2020

Romain Causse is manager of the fish collection at MNHN whose research foci are systematics and ecology of Southern, Antarctic and Indo-Pacific fish species. He made a collection of fish around the Kerguelen Islands (2007). He and Hyroyuki Motomura are friends, and he has published jointly with Motomura on the subject of scorpionfish, such as: *Re-description of the Indo-Pacific scorpionfish* Scorpaenodes guamensis *(Quoy & Gaimard1824) (Scorpaenidae), a senior synonym of seven nominal species* (2016). He co-wrote: *Biogeographic patterns of fish* (2014). He was honoured as a friend but also because he made specimens available to the author.

Cauvet

Kindia Killi *Scriptaphyosemion cauveti* R Romand & C Ozouf-Costaz, 1995

Christian Cauvet is a French aquarist and fish collector who has taken part in a number of overseas trips, such as to Sierra Leone (1984). He also takes photographs which are often used on tropical fish websites. He wrote: *Discovering Killies* (2014).

Cavery

Cauvery Rasbora *Rasbora caverii* TC Jerdon, 1849

This is a toponym referring to the Kaveri (anglicised as Cauvery or Cavery) River in southern India, where the species occurs.

Caxarar

Armoured Catfish sp. *Otocinclus caxarari* SA Schaefer, 1997

The Caxarar were an indigenous tribe of people who used to inhabit lowland regions of the Rio Guapore, southwest of Porto Velho, Brazil.

Caycedo

Colombian Blenny *Emblemaria caycedoi* A Acero P., 1984

Ivan Enrique Caycedo Lara (d.1978) was a Colombian marine biologist. The etymology says: "*The species is named in honor of the late Ivan Enrique Caycedo Lara, the best of the Colombian young marine biologists, killed through ignorance.*" (He died SCUBA diving while making observations for his work.) The document centre of the library of the Colombian Marine Research Institute is also named after him.

Cea

Cea's Scorpionfish *Rhinopias cea* JE Randall & LH DiSalvo Chalfant, 1997

Dr Alfredo Cea Egaña (1934–2016) was a Chilean physician, ichthyologist, wreck diver, photographer, film maker and spearfishing hobbyist. He founded (1967) the Submarine Research Center at the Universidad Católica del Norte after having spent some time in Spain studying underwater medicine. He spent time working at a hospital in the Easter Islands and also studying the local marine cultural history and natural history. Ill health forced him to give up diving, and in his later years he wrote of his experiences and scientific perspectives and also on naval battles. He was honoured "*…in recognition of his contribution to the knowledge of the fishes of Easter Island, the documentation of their native names, and for his dedication to the people of Rapa Nui as their physician for many years.*"

Cecilia

Snake-eel sp. *Bascanichthys ceciliae* J Blache & J Cadenat, 1971

The original description has no etymology so we have not been able to make a positive identification. Cecilia may be a female relation of one of the authors.

Cecilia (Contreras Lozano)

Villa Lopez Pupfish *Cyprinodon ceciliae* M de L Lozano-Vilano & S Contreras-Balderas, 1993 (Extinct)

Cecilia Contreras Lozano is the senior author's daughter and the junior author's niece. She helped on the trip (1988) when the species was collected. The Mexican spring in which it lived has dried up (1991), causing the pupfish's extinction. (Also see **Veronica**)

Celeste

Pikeblenny sp. *Chaenopsis celeste* J Tavera, 2021

Celeste (b. 2014) is the daughter of the author, Colombian biologist José Tavera, and his "main source of inspiration."

Celia

Celia's Aphysemion *Aphyosemion celiae* JJ Scheel, 1971

Celia Epie was the eldest daughter of John Epie, Manager of Meanja Rubber Estate, Cameroon. Scheel collected the type specimens from a stream near Epie's home.

Celia (Bueno)

Long-whiskered Catfish sp. *Hypophthalmus celiae* MW Littmann, JG Lundberg & MS Rocha, 2021

Celia Bueno (b.1984) is a Swiss biologist. She became (2014) the vertebrates and molluscs assistant curator at the Museum d'Histoire Naturelle de Neuchâtel, Switzerland, where she started work as a 'cultural mediator'. The University of Neuchâtel awarded her bachelor's degree (2005) and master's in biology, ethology, ecology and evolution (2008). During the latter she did a month exchange at the University of Bielefeld, Germany, and five months at the University of Alberta, Canada. She co-wrote: *Mammals collected by Johann Jakob von Tschudi in Peru during 1838–1842 for the Muséum d'Histoire Naturelle de Neuchâtel, Switzerland* (2020). She provided material that helped the authors to "*…correct the long-confused taxonomy of the genus*" [*Hypophthalmus*].

Celsa

Pencil Catfish sp. *Trichomycterus celsae* CA Lasso-Alcalá & F Provenzano Rizzi, 2003

Dr Josefa Celsa 'Celsi' Señaris (b.1965) is a biologist and herpetologist at Fundación La Salle de Ciencia Naturales, Caracas, Venezuela, which she joined (1986) as Assistant, becoming Assistant Director (1994) and Director (2004), Natural History Museum La Salle. The Universidad Central de Venezuela awarded her baccalaureate (1990) and the University of Santiago de Compostela, Spain her doctorate (2001). She is a representative member in Venezuela for 'Conservation International'. Her nickname is 'Celsi' and so it is acknowledged in the Glass Frog genus *Celsiella* that is also named after her.

Cemal

Stone Loach sp. *Oxynoemacheilus cemali* D Turan, C Kaya, G Kalayci, E Bayçelebi & I Aksu, 2019

Dr Cemal Turan is a Turkish professor (2015–present) in the faculty of marine sciences and technology at Iskenderun Technical University, Iskenderun. His first degree was awarded by Ankara University (1992) and his masters in molecular genetics by the University of Wales (1995). The University of Hull awarded his PhD (1997). He began his career at Mustafa Kemal University as a researcher (1993–1998) then assistant lecturer (1998–2000), lecturer (2000–2006) and professor (2006–2015). He has written or co-written c.160 papers such as: *Microsatellite DNA reveals genetically different populations of Atlantic bonito* Sarda sarda *in the Mediterranean Basin* (2015), and a number of books such as: *Jellyfish of the Black Sea and Eastern Mediterranean Waters* (2018). He was honoured for "*...his contributions to the molecular exploration of Turkish freshwater fishes.*"

Cépède

Broadnose Sevengill Shark *Notorynchus cepedianus* F Péron, 1807
American Gizzard Shad *Dorosoma cepedianum* CA Lesueur, 1818
Bernard Germaine Etienne de la Ville, Comte de Lacépède (1756–1825). (See **LaCépède**)

Cerberus

Eelpout sp. *Thermarces cerberus* RH Rosenblatt & DM Cohen, 1986
Collared Wriggler sp. *Paraxenisthmus cerberusi* R Winterbottom & A Gill, 2006

Cerberus was the three-headed dog who guarded the entrance to Hades in Greek mythology. Three reptiles are also named after Cerberus. The wriggler was rather fancifully named "*...in allusion to the relatively toothy attributes of the genus, and to the black juveniles and red and black adults, colors which are often associated with the darkness and flames of the Christian concept of the Underworld.*"

Cervigón

Finspot Ray *Rostroraja cervigoni* HB Bigelow & WC Schroeder, 1964
Spotfin Porgy *Calamus cervigoni* JE Randall & DK Caldwell, 1966
West African Catshark *Scyliorhinus cervigoni* C Maurin & M Bonnet, 1970
Stardrum sp. *Stellifer cervigoni* NL Chao, A Carvalho-Filho & J de Andrade Santos, 2021

Dr Fernando Cervigón Marcos (1930–2017) was a Spanish-born Venezuelan ichthyologist, who founded (1983) and was president of the Museo del Mar, Margarita Island, Venezuela. The University of Barcelona awarded his doctorate. He was Scientific Director of the marine research station on Margarita Island (1961–1970) and Professor at Universidad del Oriente (1970–1980). He wrote the five-volume: *Marine fishes of Venezuela* (1991–2000). He was the first (1960) to recognise the catshark as a distinct species. He was honoured in the ray name for giving its authors the opportunity to describe Venezuelan specimens.

Cesar Pinto

Driftwood Catfish sp. *Glanidium cesarpintoi* R Ihering, 1928
Armoured Catfish sp. *Parotocinclus cesarpintoi* P Miranda Ribeiro, 1939

Dr César Ferreira Pinto (1896–1964) was an entomologist and helminthologist at Instituto Oswaldo Cruz, Rio de Janeiro. He also worked for the Brazilian Malaria Service. He wrote: *Zooparasitos de interesse médico e veterinário* (1938). He collected or supplied the holotypes and photographs of these two species.

Cetti

Mediterranean Trout *Salmo cettii* CS Rafinesque, 1810

Fr Francesco Cetti (1726–1778) was an Italian Jesuit priest, zoologist and mathematician who wrote: *Storia Naturale di Sardegna* (1776). A bird is named after him.

Chabanaud

Arrowhead Soapfish *Belonoperca chabanaudi* HW Fowler & BA Bean, 1930
Squeaker Catfish sp. *Atopochilus chabanaudi* J Pellegrin, 1938
Chabanaud's Tonguefish *Symphurus chabanaudi* MN Mahadeva & TA Munroe, 1990

Dr Paul Chabanaud (1876–1959) was a French ichthyologist and herpetologist. He took his first degree at Poitiers (1897). He volunteered (1915) his services to the Muséum National d'Histoire Naturelle, Paris under Louis Roule, who asked him to identify herpetological specimens and sent him on a scientific expedition to French West Africa (1919). He travelled to Senegal and Guinea before walking 1,200 km through southern Guinea and Liberia, returning to France (1920), when he became a Preparator of Fishes at the Museum with a special interest in flatfish. He took his doctorate at the Sorbonne (1936). He wrote 40 papers on herpetology (1915–1954). In dedicating the tonguefish to him, the authors wrote: "*...Named in honour of Paul Chabanaud who contributed greatly to our knowledge of flatfish, especially the genus* Symphurus." Three reptiles and an amphibian are named after him.

Chabaud

Moustache Grouper *Epinephelus chabaudi* FL Castelnau, 1861

Castleneau merely tells us that the holotype was provided from Algoa Bay by 'M. Chabaud' (presumably = Monsieur Chabaud). As no first name given, we believe the mostly likely person is one or other of two brothers: Gustavus Henry Pullen Chabaud (1826–1877), or Louis Antoine Chabaud (1831–1901). They were the sons of John Chabaud, after whom a number of plants are named. They both lived in South Africa where Louis is described as being Inspector of Native Territories.

Chabert

Cave-dwelling Pencil Catfish sp. *Trichomycterus chaberti* JP Durand, 1968

Jacques Chabert is a French speleologist and cave explorer. He is President of the Paris Speleological Club. He helped collect the holotype in the Bolivian Andes.

Chagres

Three-barbeled Catfish sp. *Pimelodella chagresi* F Steindachner, 1876
Armoured Catfish sp. *Ancistrus chagresi* CH Eigenmann & RS Eigenmann, 1889
Neotropical Silverside sp. *Atherinella chagresi* SE Meek & SF Hildebrand, 1914

This is a toponym referring to a river, the Rio Chagres, Panama.

Chaignon

Spring Minnow sp. *Pseudophoxinus chaignoni* LL Vaillant, 1904

Vicomte Henri de Chaignon (1833–1917) was a French naturalist who is recorded in the annals of the Société d'Histoire Naturelle d'Autun, of which he was Vice President, as being interested in ornithology, geology and mineralogy. He spent 6 winter-spring months of each year (1902, 1903 & 1904) in Tunisia and wrote: *Contributions À L'histoire Naturelle De La Tunisie* (1904). He collected the holotype.

Chaim

Goby sp. *Silhouettia chaimi* M Goren, 1978

Chaim Goren was the grandfather of the author, Menachem Goren, who honoured him "... *for his encouragement in my studies*."

Chale

African Loach Catfish sp. *Amphilius chalei* L Seegers, 2008

Dr Francis Markus Mpendakulya Chale (b.1947) was awarded his BSc (1971) by the University of Dar el Salaam and his MSc (1977) and PhD (1982) by the University of Michigan. He was employed as a Research officer by the Tanzania Fisheries Research Institute (1977–1985) and by the University of Dar es Salaam, Tanzania as a lecturer (1985–1991). He was Senior Lecturer at the University of Guyana (1991–1996) and Senior Researcher at the Lake Tanganyika Biodiversity Project (1997–1999). He was Senior Lecturer at the Open University of Tanzania (2001–2011), Associate Professor at Ruaha University College (2011–2014) and is now Associate Professor at Teofilo Kisanji University (2017–present). He wrote, among other papers: *Preliminary studies on the ecology of Mbasa* (Opsaridium microlepis *(Gunther)) in Lake Nyasa around the Ruhuhu River* (2011). His current research interest is on the *Opsaridium* (African minnow) species of the Lower Ruhuhu river which is the largest inflow into Lake Nyasa. For many years Chale assisted Seegers in the exportation of live and preserved fishes from Tanzania.

Chalias

Ruby-headed Fairy Wrasse *Cirrhilabrus chaliasi* Y-K Tea, GR Allen & M Dailami, 2021

Vincent Chalias (b.1956) is a skilled underwater photographer, field biologist, aquarist and proponent of coral and fish aquaculture in Bali. He has a master's degree in Marine Aquaculture. He "greatly assisted" in the description of this wrasse species through his excellent underwater photographs and detailed field observations. He helped to set up the first Indonesian Coral mariculture farms at the beginning of this century and spent over 20 years in Indonesia running them. He has been extensively diving and documenting corals all over the Indo-Pacific and has written at least 100 articles on reefs, corals and aquariums.

Challenger (FRV)

Challenger Skate *Rajella challengeri* PR Last & M Stehmann, 2008

This species is named after the Tasmanian fisheries research vessel 'Challenger' (named after its famous antecedent, see below), which was used to survey deep-water fish resources off Tasmania (1980s).

Challenger (HMS)

Longnose Tapirfish *Polyacanthonotus challengeri* LL Vaillant, 1888
Snailfish sp. *Paraliparis challengeri* AP Andriashev, 1993

'HMS Challenger' was a deep-sea research vessel, famous for the Challenger Expedition

(1872–1876) which contributed hugely to oceanography and led to the discovery of over 4,000 previously unknown species.

Chalmers (Mitchell)

Whale Catfish sp. *Cetopsis chalmersi* JR Norman, 1926
[Junior Synonym of *Cetopsis gobioides*]

Sir Peter Chalmers Mitchell (1864–1945) was a Scottish zoologist. His early career was as a lecturer in Oxford and in London. As Secretary of the Zoological Society of London (1903–1935) he was responsible for many of the developments and improvements at the London Zoo and created the open zoological park known as Whipsnade Zoo. He died following an accident in which he was struck by a taxi after stepping off a bus.

Chalmers (Salsbury)

Freshwater Sleeper sp. *Microdous chalmersi* JT Nichols & CH Pope, 1927
[Syn. *Sineleotris chalmersi*]

Chalmers Salsbury was the son of Clarence Grant Salsbury (q.v.) of the American Presbyterian Mission of Hainan, China. He was honored for his interest and aid in Nichols' work.

Chalumna

Coelacanth *Latimeria chalumnae* JLB Smith, 1939

This is a toponym referring to the Chalumna River. It is ironic that such a famous marine fish is named after a river! The first known specimen was taken (1938) near the mouth of the Chalumna River, South Africa, by Captain Hendrik Goosen.

Chame

Chame Point Anchovy *Anchoa chamensis* SF Hildebrand, 1943

This is a toponym; Chame Point, Panama is the type locality.

Chamula

Mexican Blue-eyed Livebearer *Priapella chamulae* M Schartl, MK Meyer & B Wilde, 2006

Not a true eponym but named for a people; the native population of the Chamula, who live in central Chiapas and on the Tabasco border, Mexico. An amphibian is also named after them.

Chan

Blenny sp. *Medusablennius chani* VG Springer, 1966

William L Chan was the first to notice the distinctness of this species (in a note in the jar containing the type specimens collected in 1955) which Springer examined (1963) at Stanford University's George Vanderbilt Foundation collection. We believe this to be Dr William L Chan of the Marine Science Laboratory, Chinese University of Hong Kong and Fisheries Research Station, (established 1952) who was later Senior Small-Scale Fisheries

Development Adviser at FAO/UNDP-SCSP, Manila, Philippines. He co-wrote *Carapus homei Commensal in the Mantle Cavity of Tridacna sp. in the South China Sea* (1972).

Chance

Short-stripe Goby *Elacatinus chancei* W Beebe & G Hollister, 1933

Colonel Edwin M Chance (d.1954) was President of United Engineers & Constructors Inc. (which built power plants) and a philanthropist. According to the etymology, his 'interest and generosity' made the West Indian expedition (during which the type was collected) possible.

Chang E

Stone Loach sp. *Homatula change* M Endruweit, 2015

In Chinese mythology Chang E is the lunar goddess, said to have incredible beauty.

Chang, FI

Characin sp. *Creagrutus changae* RP Vari & AS Harold, 2001
Iquitos Tiger Pleco *Panaqolus changae* BR Chockley & JW Armbruster, 2002
Armoured Catfish sp. *Chaetostoma changae* NJ Salcedo-Maúrtua, 2006
Tetra sp. *Hemigrammus changae* RP Ota, FCT Lima & MH Hidalgo, 2019

Fonchii Ingrid Chang Matzunaga (1963–1999) was a Peruvian ichthyologist of both Japanese and Chinese ancestry. She worked at the Museo de Historia Natural, Lima until her untimely death. She drowned in a boating accident near Lake Rimachi after her rubber boots filled with water, preventing her from rising to the surface; additionally, she was shocked by an electric eel and knocked unconscious. Her boatman also drowned in the same accident. (See also **Fonchi**)

Chantre

Antakya Barbel *Carasobarbus chantrei* HE Sauvage, 1882
Sheatfish (catfish) sp. *Silurus chantrei* HE Sauvage, 1882
Chantre's Pupfish *Aphanius chantrei* C Gaillard, 1895
[Probably a Junior Synonym of *Aphanius danfordii*]

Ernest Chantre (1843–1924) was a French naturalist, geologist, archaeologist and anthropologist in Asia Minor undertaking many trips accompanied by his wife (1890–1894). He was also an early photographer. He became Deputy Director, Museum of Lyon (1877–1910) and was Professor of Anthropology (1908–1924). He is most well known for having excavated the first Sumerian cuneiform tablets in Turkey (1892–1893). Among his publications (1887–1912) are: *Mission in Cappadocia* (1899) and *Anthropological research in East Africa: Egypt* (1904). He collected the barbel holotype. A bird is also named after him.

Chao

Stardrum sp. *Stellifer chaoi* O Aguilera, OD Solano & J Valdez, 1983
Chao's Tube-snouted Ghost Knifefish *Sternarchorhynchus chaoi* CDCM de Santana & RP Vari, 2010

Dr Ning Labbish Chao is an ichthyologist, fisheries scientist and biologist whose bachelor's degree was awarded by Tunghai University, Taiwan (1968), his master's by Northeastern University (1972), and doctorate by Virginia Institute of Marine Science (1976). He is currently a professor at the Federal University of Amazonas and also a research associate at the Smithsonian. He co-wrote: *Three fishes in one: Cryptic species in an Amazonian floodplain forest specialist* (2011).

Chapal

Smallmouth Silverside *Chirostoma chapalae* DS Jordan & JO Snyder, 1899
Chapala Chub *Yuriria chapalae* DS Jordan & JO Snyder, 1899

This is a toponym; referring to Laguna de Chapala, Mexico.

Chapare

Three-barbeled Catfish sp. *Pimelodella chaparae* HW Fowler, 1940
Characin sp. *Astyanax chaparae* HW Fowler, 1943

A toponym referring to the Río Chaparé, Bolivia.

Chaper

Chaper's Characin *Brycinus chaperi* HE Sauvage, 1882
African Killifish sp. *Epiplatys chaperi* HE Sauvage, 1882

Maurice Armand Chaper (1834–1896) was a botanist and a conchologist, who was educated at l'Ecole polytechnique and in geology at l'Ecole des Mines de Paris; he appears to have left both establishments without any diploma or qualification. He served in the National Guard in the Franco-Prussian War (1870–1871) and was arrested by the Commune. He travelled very widely, visiting and prospecting (1874–1896), in Venezuela, the Rocky Mountains, Borneo, the Cape of Good Hope, the Urals, Tuscany, India, Panama (as part of the Canal study group), Senegal and Transylvania. He collected the holotypes of these fish.

Chapin

Squeaker Catfish sp. *Acanthocleithron chapini* JT Nichols & L Griscom, 1917
Barred Seabass *Centrarchops chapini* HW Fowler, 1923

Dr James Paul Chapin (1889–1964) was an American ornithologist. He was joint leader of the LangChapin Expedition, which made the first comprehensive biological survey of the Belgian Congo (1909–1915). He was Ornithology Curator for the AMNH and President of the Explorers' Club (1949–1950). He wrote: *Birds of the Belgian Congo* (1932), which largely earned him the award of the Daniel Giraud Elliot Gold Medal that year. He famously discovered the Congo Peafowl *Afropavo congensis* after he kept a puzzling feather from a native headdress and was able to match it, a quarter of a century later, with the plumes on two dusty specimens in a museum in Belgium that a curator had labelled as juvenile domestic peacocks. Sixteen birds, two mammals, one amphibian and four reptiles are named after him.

Chapleau

Madang Sole *Aseraggodes chapleaui* JE Randall & M Desoutter-Meniger, 2007

François Chapleau (b.1955) is a Professor in the Department of Biology at the University of Ottawa. The Universidty of Montreal awarded both his BSc and MSc and Queen's University, Ontario awarded his PhD (1986). He then undertook post-doctoral research at the Australian Museum (Sydney, Australia) and at the Canadian Museum of Nature in Ottawa. He gained tenure in the Department of Biology of the University of Ottawa (1989), later becoming full-professor (2000). He has published nearly 50 referred papers and book chapters, mostly dealing with flatfish taxonomy, phylogeny and evolution. His most important publication is *Pleuronectiform relationships: a cladistic reassessment* (1993). He was named University of Ottawa "professor of the year" in 2013. He was honoured "... *in recognition of his research on soleid fishes*.". He also does research on the fish communities and populations of Canada's National Capital region.

Chaplin

Papillose Blenny *Acanthemblemaria chaplini* JE Böhlke, 1957

Charles Clifford Gordon Chaplin (1906–1991) was a British ichthyologist born in India. He married (1937) an American (Lousie Davis Catherwood 1906–1983) in Philadelphia and moved there, serving in the British Consulate during WW2 and later taking US citizenship. He was a Research Associate in Fishes at the Academy of Natural Sciences of Philadelphia and worked (1940s) in Nassau, Bahamas for 15 years collecting over 500 fish species: sixty-five of which were new to science. His colleague was Dr James E Böhlke and together they wrote: *Fishes of the Bahamas and Adjacent Tropical Waters* (1968). They pioneered the use of SCUBA when collecting. He compiled *A Fishwatchers Guide to West Atlantic Coral Reefs* illustrated by Sir Peter Scott with a pioneering waterproof edition for divers to use underwater. He, with others, founded Exuma Cays Land and Sea Park, one of the world's first underwater marine reserves. He died of an aortic aneurysm.

Chapman, FM

Suckermouth Catfish sp. *Astroblepus chapmani* CH Eigenmann, 1912
Pencil Catfish sp. *Trichomycterus chapmani* CH Eigenmann, 1912

Frank Michler Chapman (1864–1945) was Curator of Ornithology for the AMNH, New York (1908–1942). He photographed and collected data on North American birds for c.50 years and did much to popularise birdwatching in the USA in the 20th century. He began publishing *Bird Lore* (1899) magazine, which became a unifying national forum for the Audubon movement. He had been an enthusiastic collector but became a leading light in the conservation movement. His interest in protection can be traced to a walk in New York City (1886) where he observed that three quarters of ladies wearing hats had feathers in them. He wrote: *Handbook of Birds of Eastern North America* (1903), *The Distribution of Bird Life in Colombia* (1917) and *The Distribution of Bird Life in Ecuador* (1926). Two mammals are named after him. In his etymology, Eigenmann did not make it clear who the 'Chapman' he was honouring was. However, thirty years later he mentioned a 'Dr F. M. Chapman' who had been a traveling companion in South America, and it is most likely that he had this ornithologist in mind.

Chapman, JW

Garden Eel sp. *Heteroconger chapmani* AW Herre, 1923

Dr James Wittenmyer Chapman (1880–1964) was an American Presbyterian missionary and myrmecologist. Harvard awarded his doctorate after which he worked for a time at the Boston Department of Public Parks before he, his wife (Ethel Robinson d.1958) and two children left for Dumagete, on the island of Negros Oriental in the Philippines. He became Professor of Biology and Zoology, Silliman Institute there, where he taught biology (1916–1947). After the Japanese invaded, he and his wife refused to surrender and lived a peripatetic life in the jungle (1942–1943). To save his collection of ants, he buried it and, miraculously, it survived and he was able to recover it after the war. They were eventually captured by the Japanese and interned in Manila (1944–1945). After recuperating in the USA, they returned to Silliman to help rebuild the education system; they finally retired to California (1947). There is now a research foundation named in his honour at the university. He wrote a number of papers on ants and he and his wife co-wrote: *Escape to the Hills* (1947) an account of their time evading the Japanese occupation forces and of their internment.

Chapman, WM

Glowtail Pipefish *Dunckerocampus chapmani* ES Herald, 1953
Collared Wriggler sp. *Xenisthmus chapmani* LP Schultz, 1966
Chapman's Blenny *Entomacrodus chapmani* VG Springer, 1967

Dr Wilbert Mcleod 'Wib' Chapman (1910–1970) was an American ichthyologist. His biology degree was awarded by the University of Washington. He worked (1933–1942) as a biologist at the International Fisheries Commission of the Washington State Department of Fisheries, and for the US Wildlife Service. He became (1942–1947) Director of the Aquarium, and Curator of the Department of Ichthyology of the California Academy of Sciences. He became (1947–1948) Director of the University of Washington School of Fisheries. He was then a Special Assistant to the Secretary of State for Fisheries and Nature (1948–1951). He became (1951–1959) Research Director at the American Tunaboat Association, San Diego. For most of the rest of his career he was director of research or resources for commercial companies. Among his many published works are books such as: *Fishing in troubled waters* (1949) and *Comments on Fishery Oceanography* (1962).

Chappuis

African Loach Catfish sp. *Doumea chappuisi* J Pellegrin, 1933

Dr Pierre-Alfred Chappuis (1891–1960) was a French-born Swiss zoologist, entomologist and bio-speleologist who specialized in isopods. He led the Mission Scientifique de l'Omo, Ethiopia (1933), during which the holotype was collected. A toad is also named after him.

Chapra

Indian River Shad *Gudusia chapra* F Hamilton, 1822

This is a toponym; Chapra District (also known as Saran), Bihar, India is the type locality.

Charar

Least Silverside *Chirostoma charari* F de Buen, 1945

Not an eponym but derived from the local Mexican vernacular *charal* or *charari*, meaning minnow.

Charcot

Charcot's Dragonfish *Parachaenichthys charcoti* LL Vaillant, 1906
Snailfish sp. *Paraliparis charcoti* G Duhamel, 1992

Jean-Baptiste Auguste Étienne Charcot (1867–1936) was a French scientist, physician, sportsman and leader of the French Antarctic Expedition (1904–1907). He explored Rockall (1921) and Eastern Greenland and Svalbardfrom (1925–1936). He won two silver medals in sailing at the Summer Olympics (1900). He died when his vessel, *Pourquoi-Pas?* was wrecked in a storm off the coast of Iceland (1936).

Chardewall

Pearlfish sp. *Encheliophis chardewalli* E Parmentier, 2004

The binomial is a combination of the surnames of Michel Chardon (see below) and Pierre Vandewalle (q.v.)

Chardon

Pearlfish sp. *Encheliophis chardewalli* E Parmentier, 2004

Dr Michel Chardon is a Belgian ichthyologist at the Laboratory of Functional Morphology, Institute of Zoology, University of Liège, Belgium where he is now an Emeritus Professor. He co-wrote the catchily-titled: *Osteology and Myology of the Cephalic Region and Pectoral Girdle of Bunocephalus knerii, and a Discussion On the Phylogenetic Relationships of the Aspredinidae (Teleostei: Siluriformes)* (2001). (See also **Chardewall**)

Charlene

Charlene's Anthias *Pseudanthias charleneae* GR Allen & MV Erdmann, 2008

Charlene, Princess of Monaco, née Charlene Lynette Wittstock (b.1978) was born in Bulawayo, Rhodesia (Zimbabwe), but the family relocated to South Africa (1989). As a swimmer she won three gold and one silver medals at the All-Africa Games (1999) and represented South Africa at two Commonwealth Games and the 2000 Sydney Olympics. She met the prince (2000) at a swimming competition. They later married (2011). She retired from competitive swimming in 2007). The name was given to honour the request of HSH Prince Albert II of Monaco.

Charpin

Sleeper sp. *Giuris charpini* P Keith & M Mennesson, 2020

Nicolas Charpin is an ichthyologist, hydrobiologist and diver who is President (2017) of the NGO 'Vies d'Ô douce', in New Caledonia, which he founded. He has previously worked as a diver and the captain of the first electric fishing boat in Europe. He co-wrote: *Non-native species led to marked shifts in functional diversity of the world freshwater fish faunas* (2018). He was honoured in the binomial for "...*all his work for the improvement of knowledge and the protection of the freshwater fauna of New Caledonia*," where the sleeper occurs. He collected the holotype (2019).

Charrua

Neotropical Rivuline sp. *Austrolebias charrua* WJE Costa & MM Cheffe, 2001

Armoured Catfish sp. *Hisonotus charrua* AE Almirón, M de las M Azpelicueta, JR Casciotta & T Litz, 2006

Brazilian cichlid sp. *Australoheros charrua* O Říčan & SO Kullander, 2008

The Charrua were aborigines who lived along the Uruguayan coast of the Rio de la Plata.

Charusin

Common Bleak ssp. *Alburnus alburnus charusini* SM Herzenstein, 1889

Alexey Nikolayevich Charusin (1864–1933) (See **Kharuzin**)

Chaseling

Snake-eel sp. *Yirrkala chaselingi* GP Whitley, 1940

Rev Wilbur Selwyn Chaseling (1910–1989) was a Methodist missionary who established Yirrkala Mission in Arnhem Land, Australia, and was its first Superintendant (1934). He was very active in encouraging the local tribe to paint on bark and established a trade in their art as a local industry. Although this was not intended to serve as a record of the people' history, culture and traditions, it was recorded as such by anthropologists. He wrote a number of books about the people of Arnhem Land including: *Yulengor, nomads of Arnhem Land* (1957). He provided the snake-eel holotype to Whitley.

Chatham

Snailfish sp. *Psednos chathami* DL Stein, 2012

This is a toponym, referring to the type locality: the Chatham Rise area of ocean floor east of New Zealand.

Chauche

Chauche's Aphyosemion *Aphyosemion chauchei* J-H Huber & JJ Scheel, 1981

Maurice Chauche (d.2012) was a French aquarist and fish photographer. He was honored for his 'complete devotion' (translation) to the study of aquarium fishes.

Chaudhuri

Sisorid Catfish sp. *Exostoma chaudhurii* SL Hora, 1923

Dr B L Chaudhuri (d.1931) was an Indian ichthyologist who was Assistant Superintendent of the Indian Museum, Kolkata. He held both a BA and BSc as well as a doctorate. He identified this catfish as *E. vinciguerrae* (1919).

Chaves, D

Nicaraguan Gizzard Shad *Dorosoma chavesi* SE Meek, 1907

Dioclesiano Chaves (1844–1936) was a Nicaraguan taxidermist and naturalist who founded the first museum in Nicaragua (1897). It was later (1922) designated to be the National Museum of Nicaragua, with collections covering zoology, entomology, botany and archaeology. Meek visited Nicaragua (1907) and he and Chaves became friends and worked together. Meek wrote that the species was named after Chaves: "*...for assistance in collecting in Lakes Tiscapa and Managua*."

Chaves, FA

Chaves' Lanternfish *Lampadena chavesi* R Collett, 1905

Colonel Francisco Alfonso Chaves e Melo (1857–1926) was a naturalist, meteorologist and geophycist. He was also a keen photographer whose collection of about 6,000 photographs is housed at the Museum at Ponta Delgada, Azores, Portugal of which he was the Director. He wrote: *Zoological bibliography of the Azores* (1906). He supplied the holotype of this species.

Chaytor

African Rivuline sp. *Scriptaphyosemion chaytori* E Roloff, 1971

Professor Dr Daniel Emanuel Babatunji Chaytor is a Sierra Leonean ichthyologist. He was a professor for two decades at the Institute of Marine Biology and Oceanography (IMBO) then became (1993) Vice Chancellor of the University of Sierra Leone. The University of Aberdeen awarded his Zoology BSc (1958) and the University of London awarded his PhD (1961). He then returned to Sierra Leone to become a Lecturer in Zoology, Senior Lecturer (1956–1969) at Fourah Bay College and later was Associate Professor at Njala University. He was (1969–1971) visiting Professor at MIT, USA returning to Fourah Bay as Professor of Zoology (1971) then being appointed (1973) as Professor and Director of IMBO. He organised and participated in biannual scientific cruises (1976–1990). He published very widely, with many ichthyology or fisheries papers, and he initiated the Bulletin of IMBO (1976).

Cheeseman

Cheeseman's Puffer *Lagocephalus cheesemanii* FE Clarke, 1897

Thomas Frederick Cheeseman (1846–1923) was a New Zealand botanist and naturalist. Born in Hull, England, he and his parents moved to New Zealand (1854). His first botanical publication was (1872) about the plant life in the Waitakere Ranges. He became Curator of the newly founded Auckland Museum (1874) and added to its collections – including many bird specimens shot by his younger brother, William. Cheeseman published papers every year for the rest of his life. He wrote: *The Manual of the New Zealand Flora* (1906). The type specimen was given to Clarke by Cheeseman.

Cheeya

Rasbora sp. *Brevibora cheeya* TT Liao & HH Tan, 2011

Named for a minor oriental deity: Cheeya, along with Beiya, hunts ghosts for the death-god Yama.

Cheffe

Neotropical killifish sp. *Austrolebias cheffei* MV Volcan, C Barbosa, LJ Robe & LE Krause Lanés, 2021

Morevy Moreira Cheffe is a Brazilian ichthyologist who works in the Ichthyology Sector, Fauna Division, Special Group for the Study and Protection of the Aquatic Environment of Rio Grande do Sul and has been associated with the Federal University of Rio de Janeiro. Among his publications are, co-written with Wilson Costa (q.v.): Austrolebias univentripinnis *sp nov (Teleostei Cyprinodontiformes: Rivulidae): a new annual killifish from the Mirim Lagoon basin, southern Brazil* (2005) and *Two new seasonal killifishes of the* Austrolebias adloffi *group from the Lagoa dos Patos basin, southern Brazil (Cyprinodontiformes: Aplocheilidae)* (2017). He was honoured for "…his contribution to the knowledge of annual fish in southern Brazil".

Chefoo

Chefoo Thryssa *Thryssa chefuensis* A Günther, 1874

This is a toponym; Chefoo, Shantung Province, China, was the type locality.

Chekopa

Lake Malawi Cichlid sp. *Mylochromis chekopae* GF Turner & JD Howarth, 2001

This is a toponym referring to the type locality; Chekopa trawl station on the southeastern arm of Lake Malawi.

Chemnitz

Snub-nosed Spiny Eel *Notacanthus chemnitzii* ME Bloch, 1788

Johann Hieronymus Chemnitz (1730–1800) was a German naturalist, conchologist and theologian. He was chaplain to the Royal Danish Embassy in Vienna (1757–1768) and was subsequently the chaplain attached to the garrison at Helsingør (1969–1771) and then at Copenhagen (1771). After the death of Friedrich Heinrich Wilhelm Martini (1778), who had written: *The new systematic Conch Cabinet,* Chemnitz published eight volumes (1779–1795) to add to the three published by Martini in his lifetime. Chemnitz supplied Bloch with specimens of both Arctic and North Atlantic fishes, including the holotype of this eponymous one.

Chen, D-M

Hairy Whipnose Anglerfish *Gigantactis cheni* H-C Ho & K-T Shao, 2019

Din-Moo Chen is described in the etymology as an excellent fisherman who collected (off northeast Taiwan) most samples for the authors' studies, including the type series of both new species in the publication which included the eponymous species and *Oneirodes formosanus.*

Chen, I-S

Face-stripe Pygmy Goby *Trimma cheni* R Winterbottom, 2011

Dr I-Shiung Chen is Director of the Institute of Marine Biology at the National Taiwan Ocean University. He completed his PhD at the University of Bristol, UK. He was honoured for his numerous (90+) publications on Indo-Pacific reef fishes in general, and gobies in particular. His many papers include the co-written: *New record of a brackish water goby (Perciformes: Gobiidae: Acentrogobius) from Taiwan* (2015).

Chen, J-P

Cardinalfish sp. *Ostorhinchus cheni* M Hayashi, 1990
Triplefin sp. *Enneapterygius cheni* S-C Wang, K-T Shao & S-H Shen, 1996
Sole sp. *Aseraggodes cheni* JE Randall & H Senou, 2007

Dr Jen-Ping Chen is an ichthyologist who works at the National Taiwan University, Taipei, where he is a Full Professor and at the Institute of Zoology of Academia Sinica, National Museum of Marine Biology and Aquarium, Taiwan. He co-wrote: *A new species of the lizardfish genus* Synodus *(Aulopiformes: Synodontidae) from the western Pacific Ocean* (2016).

Chen, JTF

Chinese Hillstream Loach *Pseudogastromyzon cheni* Y-S Liang, 1942
Gudgeon sp. *Gobiobotia cheni* PM Bănărescu & TT Nalbant, 1966
Hagfish sp. *Eptatretus cheni* SC Shen & HJ Tao, 1975

Dr Johnson (Jian-shan) T F Chen (1898–1988) was a Taiwanese vertebrate zoologist who graduated as a teacher in Peking (Beijing) (c.1920) and then studied in France and the United Kingdom. He went to Taiwan to help with reconstruction after the defeat of the Japanese and became the Director of the Taiwan Museum, Taipei (1945–1965) and a Professor at the National Taiwan University and at Donghai University (1945–1972). He was out of favour with the Taiwanese authorities, which removed him from his post with the Museum and did not offer him any alternative official employment, so he left Taiwan and lived in the USA (1972–1982). He visited Mainland China (1978) and twice lectured at the Institute of Oceanography on China Sea in Guangzhou (1980 & 1981). Finally, he decided to return to China for good and settled in Shanghai (1982) and worked at the Shanghai Museum of Natural History. Amongst other works he wrote: *A review of the sharks of Taiwan* (1963) and *Synopsis of the Vertebrates of Taiwan* (1969).

Chen, T-P

Chen's Worm Eel *Neenchelys cheni* JTF Chen & HT-C Weng, 1967

Tung-Pai Chen was an ichthyologist who was Chief, Section of Fisheries, Joint Sino-American Commission on Rural Reconstruction (JCCR). He was honoured for 'financial support and kind encouragement'. He wrote: *The culture of Tilapia in race paddies in Taiwan* (1954) and *Aquaculture Practices in Taiwan* (1976).

Chen Ti-Ti

Goby sp. *Rhinogobius cheni* JT Nichols, 1931

Ch'en Ti-Ti was a tiger hunter who was "...widely known under the name 'Da-Da'". He collected the holotype.

Chen, W-J

Sleeper Shark sp. *Somniosus cheni* H-H Hsu, C-Y Lin & S-J Joung, 2020

Wen-Jong Chen, of the Taitung Xin Gang District Fisherman's Association, was honoured for his contributions of shark samples and his assistance with Taiwanese and international shark research for over 30 years.

Chen, YY

Loach sp. *Formosania chenyiyui* C-Y Zheng, 1991

Chen Yi-Yu (b.1944) is an ichthyologist, taxonomic zoologist and biologist who was the first to recognise that *Formosania chenyiyui* is a distinct species. Xiamen University awarded his degree in biology (1964). He became Professor at the Institute of Hydrobiology, Chinese Academy of Sciences, Wuhan (1989) and served as its Director (1991–1995). He was Vice President of the Chinese Academy of Sciences (1995–2003). He has published in excess of 100 scientific papers, articles and monographs.

Cheng

Chinese Cyprinid sp. *Schizopygopsis chengi* P-W Fang, 1936
[Syn. *Schizopygopsis malacanthus chengi*]

Professor Cheng Wan-Chun (1908–1987) was a botanist who started out as a plant collector (late 1920s). He became one of one of the world's leading authorities on the taxonomy of gymnosperms. He was assistant botanist at the Herbarium of the Biological Laboratory of the Science Society of China (1936). He worked at the National Central University, Nanking (1944) and was instrumental in identifying the Dawn Redwood *Metasequoia glyptostroboides* that had previously only been known from fossils. He collected the holotype of the fish named after him.

Chereshnev

Chereshnev's Horsefish *Zanclorhynchus chereshnevi* AV Balushkin & MY Zhukov, 2016

Dr Igor Aleksandrovich Chereshnev (1948–2013) was a biologist at, and Director (2005–2013) of, the Institute of Biological Problems of the North of the Far East branch of the Russian Academy of Science. He wrote, or co-wrote, about 40 papers such as: *Phylogenetic and Taxonomic Relationships of Northern Far Eastern Phoxinin Minnows, Phoxinus and Rhynchocypris (Pisces, Cyprinidae), as Inferred from Allozyme and Mitochondrial 16S rRNA Sequence Analyses* (2006) and *Speciation and genetic divergence of three species of charr from ancient Lake El'gygytgyn (Chukotka) and their phylogenetic relationships with other representatives of the genus Salvelinus* (2015).

Chermock

Vermilion Darter *Etheostoma chermocki* HT Boschung, RL Mayden & JR Tomelleri, 1992

Ralph Lucien Chermock (1918–1977) was primarily a lepidopoterist. He was Professor of Biology at Alabama University (1947–1966) and Director, Alabama Natural History Museum (1960–1966). He was the founder of the University of Alabama's Ichthyological Collection. An amphibian is also named after him.

Chernoff

Tetra sp. *Bryconops chernoffi* C Silva-Oliveira, FCT Lima & JD Bogata-Gregory, 2018

Dr Barry Chernoff (b.1951) is an American ichthyologist, a professor of biology who holds the Chair of Environmental Studies and is (2003–present) Director of the College of the Environment at the Wesleyan University, Middletown, USA. He teaches courses in Environmental Studies, Tropical Ecology, Aquatic Ecosystem Conservation, and Quantitative Analysis for the departments of Biology and Earth and Environmental Sciences. His research centres on the freshwater fishes of the Neotropical region, primarily those in the Amazon region. His research includes ecology, evolutionary biology and conservation. He has made more than thirty expeditions to twelve countries. He formerly held professorial and curatorial positions at the Field Museum, University of Chicago, University of Illinois at Chicago, University of Pennsylvania and the Academy of Natural Sciences of Philadelphia. He holds a visiting position at Universidad Central de Venezuela. He has authored and co-authored more than 87 scientific papers, articles and books including *Geographic and environmental variation in Bryconops sp. cf. melanurus (Ostariophysi: Characidae) from the Brazilian Pantanal* (2006). He also co-wrote the script for a short documentary film entitled *Understanding Biodiversity* that was awarded finalist status at a number of film festivals, including Cannes, Toronto and Sundance. The etymology sauys: "*...the specific epithet honors Barry Chernoff, and is in recognition for his contributions to the taxonomy of* Bryconops, *as well as for ichthyology as a whole.*"

Chesnon

Warty Poacher genus *Chesnonia* T Iredale & GP Whitley, 1969

Cygnus G Chesnon (dates unknown) published a rare book on the natural history of Normandy, France (1835), in which he proposed the name *Occa* (now = *Somateria*) for a genus of eider ducks. Noticing that the original name for the Warty Poacher genus – *Occa* Jordan & Evermann 1898 – was thus preoccupied by *Occa* Chesnon 1835, Iredale & Whitley renamed the fish genus in honour of Chesnon. Unfortunately no details of Chesnon's life seem to be recorded.

Chester

Longfin Hake *Phycis chesteri* GB Goode & TH Bean, 1878

Captain Hubbard C Chester (1836–1886) made a number of whaling voyages before being appointed sailing master of the Arctic exploring steamer *Polaris* (1871). He was one of those who survived the ill-fated *Polaris* expedition (an attempt to reach the North Pole). He was later employed by the US Fish Commission and engaged in deep-sea work. He collected the holotype of this species.

Chevalier

African Rivuline sp. *Epiplatys chevalieri* J Pellegrin, 1904

Auguste Jean Baptiste Chevalier (1873–1956) was a French botanist, taxonomist and explorer in Africa, particularly Côte d'Ivoire. His degree (1896) and PhD (1901) were awarded by the Universit of Lille where he worked as assistant to C E Bertrand. He took part in a mission to French Sudan (1899–1900) and (1905) established the botanical garden in Dalaba, French Guinea. He also explored and collected in the Neotropics and Asia (1913–1919). Later (1929) he became a professor in Paris. He published throughout his career, such as: *Sur l'existence probable d'une mer récente dans la région de Tombouctou* (1901) and *La forêt du Brésil* (1929). He discovered the rivuline species.

Chevey

Sheatfish (catfish) sp. *Kryptopterus cheveyi* J-D Durand, 1940
[Syn. *Micronema cheveyi*]

Dr Pierre Chevey (1900–1942) was a French ichthyologist who graduated at the exceptionally young age of nineteen and became a teacher in Clermont-Ferrand in order to continue his studies. The Sorbonne awarded his doctorate (1925). He went to Indochina whilst doing his military service, which he combined with collecting for the Oceanography Institute of Monaco. He stayed on as an Assistant under Armand Krempf (q.v.), eventually becoming Interim Director of the Oceanography Institute (1931) and then Director (1935). He wrote: *Rapport sur les pêcheries ou bouchots de la baie du.*

Ch'i

Bitterling ssp. *Tanakia himantegus chii* C-P Miao, 1934

Chen-Ju Ch'i was the Director of the Bureau of Education of Honan (now Henan) Province, China, and was thanked by the author for his 'kind support' of the author's study of Kiangsu fishes.

Chico (Filho)

Characin sp. *Astyanax chico* JR Casciotta & AE Almirón, 2004

Chico Mendes né Francisco Alves Mendes Filho (1944–1988), who was murdered by a rancher, was a Brazilian rubber tapper, trade union leader, environmentalist and rainforest activist. He was born at a time when schools were prohibited on rubber plantations (for fear that the peasants might learn to read and do arithmetic!) and he only learned to read when he was 18. The Chico Mendes Institute for Conservation of Biodiversity is named in his honour.

Chico (Teixeira)

Characin sp. *Characidium chicoi* WJ da Graça, RR Ota & WM Domingues, 2019

Francisco 'Chico' Alves Teixeira is a retired laboratory assistant formerly of Universidade Estadual de Maringá, Núcleo de Pesquisas em Limnologia, Ictiologia e Aquicultura. He was honoured for his over thirty years' experience in biological material samples and because he was an 'excellent partner' in numerous field trips in Brazil, including the one when this species was collected.

Chihiroe

Fringehead sp. *Neoclinus chihiroe* R Fukao, 1987

Chihiro Okazaki is the the wife of Dr Toshio Okazaki (q.v.), whom Fukao credits with leading to the discovery of *Neoclinus chihiroe* sister taxon *N. okazakii*. Interestingly, 'Chihiroe' also means 'thousand fathoms'. (Also see **Okazaki**)

Chilca (Bay)

Pacific Menhaden ssp. *Ethmidium maculatum chilcae* SF Hildebrand, 1946

This is a toponym; Chilca Bay, Peru, was the type locality.

Children

Indian Catfish sp. *Silonia childreni* WH Sykes, 1839

John George Children (1777–1852) was a British entomologist, who was also very interested in electricity and published notes on it (1808–1813). He became a Fellow of the Royal Society, London (1807) and later Secretary. He visited Pennsylvania (1802) and was in Spain and Portugal (1808–1809) and his diary records include details of the Peninsular War. He worked in the British Museum (1816–1840). Audubon collected and described a bird in his: *Ornithological Biography or an Account of the Habits of the Birds of the United States of America* (1831), writing that he named it after Children "*...as a tribute of sincere gratitude for the unremitted kindness which he has shewn me*." The mineral childrenite is named after him, as is a bird and reptile.

Chilola

Lake Malawi Cichlid sp. *Placidochromis chilolae* M Hanssens, 2004

This is a toponym referring to the type locality; Chilola Bay, Lake Malawi.

Chilton

Longsnout Flathead *Thysanophrys chiltonae* LP Schultz, 1966

There is nothing in the original description that helps, but we feel reasonably sure that this is named after the USS *Chilton*, which hosted the survey team led by Schultz looking at reef fish after tha Bikini Atoll (Marshall Islands) nuclear tests.

Chilton, MA

Pencil Catfish sp. *Trichomycterus chiltoni* CH Eigenmann, 1928

Colonel M A Chilton was the military attaché of the American Embassy in Santiago de Chile. He accompanied Eigenmann on his tour of the 'Switzerland of Chile' (i.e., Chilean Lake District, southern Chile).

Chin, P-K

Betta sp. *Betta chini* PKL Ng, 1993
Barb sp. *Osteochilus chini* J Karnasuta, 1993
Borneo Loach sp. *Neogastromyzon chini* H-H Tan, 2006

Datuk (title) Chin Phui-Kong (b.1923–fl.2016) is a Malaysian ichthyologist. He was sent (1941) to Nationalist China for his secondary education and was stranded there when WW2 broke out. He joined the British forces, becoming a lieutenant specialising in demolition and explosives in Force 136 of the resistance movement against the Japanese occupying forces and trained in India and Ceylon (Sri Lanka) before service in Malaya. He returned to China to complete his education, graduating BSc Marine Biology from Amoy (now Xiamen) University. He worked at the Fisheries Department, Sabah (1950–1978) including being the Director (1970–1978). He wrote: *Marine Food Fishes and Fisheries of Sabah* (1998) and co-wrote: *The fresh-water fishes of North Borneo* (1962). As a past President of the Sabah Society, he was honoured with the title 'Datuk' and also 'for his contributions to the ichthyology of Sabah and Sarawak.' The barb species is said to have been named after Chin Khuikong 'for his efforts in collecting type', we believe this to be the same person.

Chiou

Snake-eel sp. *Xyrias chioui* JE McCosker, W-L Chen & H-M Chen, 2009

Captain Jiun-Shiun Chiou captured a number of eel specimens that he gave to the laboratory of the National Taiwan Ocean University, Keelung.

Chip

Peruvian cichlid sp. *Tahuantinsuyoa chipi* SO Kullander, 1991

Not an eponym but a word in the Shipibo language of Amazonian Peru meaning 'sister', with reference to its close relationship with *T. macantzatza*.

Chiping

Gudgeon ssp. *Gobio gobio chipingi* PM Bănărescu & TT Nalbant, 1964

Dr Chih Ping (1886–1965). (See **Ping C**)

Chipoka

Lake Malawi Cichlid sp. *Melanochromis chipokae* DS Johnson, 1975

This is a toponym referring to Chipoka Island, Lake Malawi, Malawi, the type locality.

Chirio

African Rivuline sp. *Pronothobranchius chirioi* S Valdesalici, 2013

Laurent Vincent Chirio (b.1957) is a French/Italian herpetologist who has spent most of his career teaching and counselling in Africa and working as a freelance herpetological expert. The University of Orléans awarded bachelor's degree (1977) and his master's in Animal Biology (1979) and Natural Sciences (1980). He taught in various schools in France (1981–1984), in Algeria (1984–1986), in Niger (1986–1990 & 2004–2008), in the Central

African Republic (1990–1999), in Cameroon (1996–2001), in Réunion (2001–2004), in Senegal (2008–2011), in Gabon (2001–2014), combined with counselling at the RIyadh International French School in Saudi Arabia. He co-wrote: *Biogeography of the reptiles of the Central African Republic* (2006). A toad is also named after him.

Chisumulu

Lake Malawi Cichlid sp. *Labidochromis chisumulae* DSC Lewis, 1982

This is a toponym referring to Chisumulu Island, east-central Lake Malawi, where it is endemic.

Chitamwebwa

Lake Tanganyika Cichlid sp. *Neolamprologus chitamwebwai* P Verburg & R Bills, 2007

Deonatus Chitamwebwa is a Director of Tanzania Fisheries Research Institute (TAFIRI) and Officer-in-Charge of their Kigoma research station. He was honoured in gratitude for his excellent hospitality and assistance…and because of his elongated shape and small body depth to length ratio that makes him stand out, as it does with this cichlid species!

Chittenden

Mexican Flounder *Cyclopsetta chittendeni* BA Bean, 1895

Dr John Franks Chittenden (1843–1895) was an English physician who moved to the West Indies (1871). He was Government Medical Officer for Montserrat (1872) and District Medical officer in Trinidad until retirement (1888) as well as a Justice of the Peace there. He was inaugural Vice-President of the Trinidad Medical Association (1891) at Port of Spain, Trinidad.

Chloe (Expeditions)

Goby sp. *Smilosicyopus chloe* RE Watson, P Keith & G Marquet, 2001

This species is named after the Chloé Expéditions I and II to New Caledonia, during which the holotype was collected.

Chloe (Starks)

Goby subgenus *Chloea* DS Jordan & JO Snyder, 1901

Chloe Lesley Starks (1866–1952) was an artist and naturalist at Hopkins Biological Laboratory, Stanford University. She provided many of the illustrations in the authors' monograph of Japanese gobies.

Chlupaty

Chameleon Sand Tilefish *Hoplolatilus chlupatyi* W Klausewitz, JE McCosker, JE Randall & H Zetzsche, 1978

Peter Chlupaty was a German marine aquarist and author. He wrote books and articles on keeping marine fish such as the book: *Guide to Marine Aquarium* Keeping (1965) and the article: *The harlequin bass* (1977). He was honoured in the name for "…sending the first imported fish of the genus *Hoplolatilus* to the Seckenberg Museum for scientific processing." (Translation)

Choat

Choat's Wrasse *Macropharyngodon choati* JE Randall, 1978
West African Parrotfish *Sparisoma choati* LA Rocha, A Brito & DR Robertson, 2012

Dr John Howard Choat is Professor Emeritus of Marine Biology at the School of Marine and Tropical Biology, James Cook University, Queensland, Australia. He has done extensive scientific work on parrotfishes. His more than 100 publications include: *A functional analysis of grazing in parrotfishes (family Scaridae): the ecological implications* (2004) and *Giant coral reef fishes display markedly different susceptibility to night spearfishing* (2018).

Chochamanda

African Rivuline sp. *Nothobranchius chochamandai* B Nagy, 2014

Professor Auguste Chocha Manda is an ichthyologist at thge University of Lubumbashi, Katanga, DR Congo. His PhD thesis at the Université de Namur was on the evolutionary genetics of a catfish (2010). Among his published papers is: *On some chubbyhead minnows of the Upper Lualaba (Upper Congo basin: DR Congo): the case of E. motebensis (Cypriniformes: Cyprinidae) and the populations of the Kundelungu highland plateau* (2018). He was honoured for "… his dedication in researching the fishes of his country." He also collected the type with Nagy.

Chocó

Cichlid genus *Chocoheros* O Říčan & L Piálek, 2016

Named after the Embera-Wounaan indigenous tribe, also known as the Chocó in their language, and after the Chocó biogeographic area and area of endemism encompassing the distribution of the genus. Combined with *Heros*, neotropical cichlid name meaning 'hero', to mean 'hero of the Chocó'.

Choi

Choi's Spiny Loach *Cobitis choii* I-S Kim & Y-M Son, 1984

Ki-Chul Choi (1910–2002) was an ichthyologist who was Professor of Zoology, College of Education, Seoul National University, South Korea and at the Korean Institute of Freshwater Biology, Seoul. He investigated the distribution patterns of freshwater fishes in Korea and published much on that subject, including, as co-author: *A Revision of the Eleotrid Goby Genus in Japan, Korea and China* (1985). (See also **Kichulchoi**)

Chong Ling-Chun

Chinese Cyprinid sp. *Poropuntius chonglingchungi* TL Tchang, 1938

Chong Ling-Chung collected 'some interesting fishes' from Yunnan, China. He was a collector for the Fan Memorial Institute of Biology, Peking (Beijing).

Chong, LT

Snowtrout sp. *Schizothorax chongi* PW Fang, 1936

Chong L T was an ornithologist at Academia Sinica, Peking (Beijing) active in the 1930s. He helped the author collect specimens, including the holotype of this species. He wrote: *Contributions to the ornithology of Kwangsi I Dicruridae* (1932) and the co-written *Notes additionnelles sur l'avifaune du Kwangsi* (1937).

Chopra

Glowlight Danio *Celestichthys choprae* SL Hora, 1928
Stone Loach sp. *Triplophysa choprai* SL Hora, 1934
[Syn. *Indotriplophysa choprai*]

Dr Bashambhar Nath Chopra (1898–1966) was an Indian zoologist and ichthyologist whose doctorate was awarded by Panjab University, Lahore (now in Pakistan). He was Assistant Superintendent Zoological Survey of India, Calcutta (Kolkata) (1923–1944) and then its Director (1944–1947). He was Fisheries Development Adviser, Ministry of Agriculture, to the Government of India (1947–1958). He was President of the Zoology and Entomology Section of the Indian Science Congress (1943) and a Founder Fellow of the Zoological Society of India. He edited: *Handbook of Indian Fisheries* (1951) and wrote: *Fisheries and Fishing Industry in India* (1958). He led a number of expeditions and during two of them he collected the holotypes.

Ch'orti

Cichlid genus *Chortihero* O Říčan & K Dragová, 2016

Named after the Ch'orti' people, an indigenous Maya people of southeastern Guatemala, northwestern Honduras and northern El Salvador, and for whom the Chortis Block (one of the main geological components of Middle America) is named and to whose northern part this genus is the oldest and most isolated endemic lineage. Combined with *Heros*, neotropical cichlid name meaning 'hero', to mean 'hero of the Chortis'.

Christiane

Christiane's Pygmygoby *Trimma christianeae* GR Allen, 2019

Eva Christiane Waldrich (b.1957) is an underwater photographer and diver who is co-owner of the dive resort 'Villa Markisa' in Bali. She had the opportunity (1999) to learn SCUBA diving in Tulamben, Bali, Indonesia. This changed her life completely and she immediately became interested in marine life in general and nudibranchs and sea slugs in particular. She moved to Bali (2001) and started building (2006) the dive resort. She has logged well over 4300 dives, mostly in Indonesia. She is particularly interested in nudibranchs and has taken pictures of over 1200 different species. Many of them are illustrated in: *Indo-pacific nudibranchs and sea slugs* (2008), *Nudibranch & sea slug identification* (2015). The author wrote: "*The new species is named in honor of Christiane Waldrich, who photographed this fish while scuba diving and brought it to my attention.*"

Christine (Huber)

Neotropical Rivuline sp. *Anablepsoides christinae* JH Huber, 1992

Christine Huber is the wife of the author. He honoured her for her patience and assistance while editing his *Rivulus* review.

Christine (Karrer)

European Dwarf Snailfish *Psednos christinae* AP Andriashev, 1992

Dr Christine Karrer is a German ichthyologist at the Hamburg University Zoological Museum. (See **Karrer**)

Christy

Christy's Lyretail *Aphyosemion christyi* GA Boulenger, 1915
Squeaker Catfish sp. *Atopochilus christyi* GA Boulenger, 1920
Christy's Elephantfish *Campylomormyrus christyi* GA Boulenger, 1920
Squeaker Catfish sp. *Microsynodontis christyi* GA Boulenger, 1920
Coppernose Barb *Raiamas christyi* GA Boulenger, 1920
[Syn. *Opsaridium christyi*]
Lindi Mormyrid *Petrocephalus christyi* GA Boulenger, 1920
Lake Malawi Cichlid sp. *Lethrinops christyi* E Trewavas, 1931
Lake Malawi Cichlid sp. *Aristochromis christyi* E Trewavas, 1935
Lake Tanganyika Cichlid sp. *Neolamprologus christyi* E Trewavas & M Poll, 1952
Lake Tanganyika Cichlid sp. *Greenwoodochromis christyi* E Trewavas, 1953

Dr Cuthbert Christy (1863–1932) was a physician and zoologist who was particularly known for his work on sleeping sickness. He qualified (1892) as a physician at Edinburgh University having won a bursary there. He travelled in the West Indies and South America (1892–1895), subsequently joining the army as Medical Officer. He became Senior Medical in northern Nigeria (1898–1900) then medical officer on plague duty in Bombay (1900–1902) and later was in Uganda (1902) and the Congo (1903) as part of a three-man mission investigating trypanosomiasis. He was involved in similar work in Ceylon (Sri Lanka) (1906), East Africa (1906–1909), then Nigeria and Wset Africa (1909–1910). He was in the Congo again (1911–1914) for the Belgian Government, latterly exploring the forest and Rwenzori Mountains. At this time, he wrote: *The African rubber industry and Funtumia elastica* (1911). He served in Africa and Mesopotamia (Iraq) (WW1). After the war, he explored in the Sudan (1920–1923), Nyasaland (Malawi) and Tanganyika (Tanzania) (1925–1928) for BMNH and was a member of a League of Nations commission enquiring into slavery and forced labour in Liberia, known as the 'Christy Commission'. He was Director of the Congo Museum, Tervuren, Belgium. He wrote the report on the commission: *Report of the International commission of inquiry into the existence of slavery and forced labor in the republic of Liberia. Monrovia, Liberia, September 8, 1930* (1931). He was on a zoological expedition to the Congo (1932) when he was gored by a buffalo that he had fired at, and died later from the wounds. He collected a number of the fish types including those of the cichlids. A mammal, a bird, three reptiles and two amphibians are also named after him.

Chrysi Cristina

Diyarbakir Loach *Schistura chrysicristinae* TT Nalbant, 1998

Cristina Ana 'Chrysi' or 'Crina' Hoinic (1967–1997) was a Romanian entomologist. She initially intended to study medicine, but having enrolled (1987) for biology at Bucherest University she graduated (1991) then completed her MA in ethology (1992). That year she obtained a post at the Department of Entomology, 'Grigore Antipa' National Museum of Natural History, Bucharest, Romania working on the systematics of leaf-beetles. She published her first paper on *Macroplea* (1994), which she illustrated beautifully herself. She tragically died as the result of a medical procedure to see if she was a suitable kidney donor for her twin sister. Nalbant wrote an obituary of her in which he said she was a: "... *wonderful friend and colleague. My debts to her are beyond the words*". He also said: "*She died smiling after twelve days in a coma, the day before her thirtieth birthday*".

Chu, X-L

Loach sp. *Yunnanilus chui* J-X Yang, 1991
Sisorid Catfish sp. *Pareuchiloglanis chui* X Li, W Dao & W Zhou, 2020

Xin-Luo Chu (sometimes given as Chu Xin-Luo) is a Chinese ichthyologist of the Kunming Institute of Zoology. He co-authored the seminal two volume monograph *The Fishes of Yunnan, China* (1989 & 1990).

Chu, Y-T

Carp sp. *Gymnocypris chui* TL Tchang, TH Yueh & HC Hwang, 1964
Chu's Croaker *Nibea chui* E Trewavas, 1971

Professor Yuan-Ting Chu (Yuanding Zhu) (1896–1986) was Director of the Shanghai Fisheries Institute, Shanghai, China (1958). He wrote: *Contributions to the ichthyology of China* (1930). We cannot be sure about the carp, as the original text is unhelpful, so it must remain only a best guess that *Gymnocypris chui* is named after him.

Chubb

Dusky Rubberlip *Plectorhinchus chubbi* CT Regan, 1919

Ernest Charles Chubb (1884–1972) was an ornithologist who became Curator of the Museum in Durban, South Africa. His father Charles (1851–1924), a fellow ornithologist, was a Curator at the British Museum but was knocked down and killed by a car as he left the premises (1924). Ernest became President of the Southern Africa Association for Advancement of Science (1945). He wrote a paper entitled *Record of nesting of skimmer at St Lucia* (1943). Five birds are also named after him.

Chula

Yellow-stripe Mangrove Goby *Mugilogobius chulae* HM Smith, 1932

Luang Chula Cachanagupta was Director of the Department of Fisheries of Siam (now Thailand).

Chulabhorn

Carplet sp. *Amblypharyngodon chulabhornae* C Vidthayanon & M Kottelat, 1990
Chaofa Chulabhorn Loach *Physoschistura chulabhornae* A Suvarnaraksha, 2013

Dr HRH Princess Chulabhorn Mahidol (b.1957) of Thailand has a bachelor's degree in chemistry awarded by Kasetsart University (1979) and a doctorate of science degree from Mahidol University (1985). She is well known for her energetic promotion of scientific research. She is President of the Chulabhorn Research Institute, is an Honorary Fellow of the Royal Society of Chemistry in London and was awarded UNESCO's Einstein Medal (1986) for her efforts in promoting scientific collaboration.

Chun

Chun's Telescopefish *Gigantura chuni* A Brauer, 1901

Carl Chun (1852–1914) was a German marine biologist and Professor of Zoology at the University of Leipzig (1892–1914). He led the German Deep-Sea Expedition (1898–1899) to the subantarctic seas. Many other marine taxa are named after him as well as a bird.

Churamani

Sisorid Catfish sp. *Glyptothorax churamanii* Rameshori Yumnan & W Vishwanath, 2012

Churamani (aka Lalchharliana) gave the authors 'immense help' in the collection of this species. We can find no other details about him.

Chuvasov

Snaggletooth sp. *Astronesthes tchuvasovi* NV Parin & OD Borodulina, 1996

Vladimir Mikhailovich Chuvasov was the leading technician of the Laboratory of Oceanic Fauna, P P Shirshov Institute of Oceanology, Moscow. He was the authors' companion on many research cruises.

Ci

Marmara Chub *Squalius cii* J Richardson, 1857

This is a toponym not an eponym, referring to the type locality: the River Gemlik (northwest Turkey) that, according to the author, was "*...anciently named Cius*"

CIAD

Cortez Flounder *Etropus ciadi* AM van der Heiden & HG Plascencia González, 2005

Not an eponym but an acronym that stands for **C**entro de **I**nvestigacion en **A**limentacion y **D**esarrollo. Both the authors have held research positions in that Mexican institution.

Cibele

Bumblebee Catfish sp. *Microglanis cibelae* LR Malabarba & JKF Mahler Jr, 1998

Cibele Barros Indrusiak has a bachelor's degree in biology from the Federal University of Rio Grande do Sul (1991) and master's degree in Wildlife Management awarded by the National University of Córdoba (1997). She is an Environmental Analyst, Brazilian Institute for the Environment and Renewable Natural Resources, Brasilia. She co-wrote: *Ecology and Conservation of the Jaguar* (Panthera onca) *in Iguaçu National Park, Brazil* (2004).

Çiçek

Loach sp. *Oxynoemacheilus ciceki* S Sungur, P Jalili & S Eagderi, 2017

Dr Erdoğan Çiçek is a Turkish ichthyologist who became an assoiate professor (2010) and full professor (2015) in the Department of Biology at Nevşehir Hacı Bektaş Veli University, Turkey. Istanbul University awarded his Associate Degree (1992) and Çukurova University his other degrees, culminating in his PhD (2006). He is honoured for his 'valuable contribution to the knowledge of freshwater fishes of Turkey'.

Cifuentes

Olive Grouper *Epinephelus cifuentesi* RJ Lavenberg & J Grove, 1993

Miguel Cifuentes was a park administrator, though we only know what the etymology tells us: "*Sr. Miguel Cifuentes, former lntendente of the Galapagos National Park in appreciation of his assistance in the field work at the Galapagos Islands.*"

CIMAR

Chilean Roundray *Urotrygon cimar* S López & WA Bussing, 1998

[Alt. Denticled Roundray]

This is an abbreviation for Centro de Investigación en Ciencias del Mar y Limnologia, a research centre of the Universidad de Costa Rica (where both authors work) in honour of its 20th anniversary.

Cincotta

Diamond Darter *Crystallaria cincotta* SA Welsh & RM Wood, 2008

Daniel Anthony 'Dan' Cincotta (b.1952) is Adjunct Assistant Professor of Ichthyology at West Virginia University (2005–2018) (where Stuart Welsh is an assistant professor), and the Senior Ichthyologist with the Wildlife Diversity Unit of West Virginia Division of Natural Resources (1995–present) and non-game fish biologist for the West Virginia Natural Heritage Program. York College, Pennsylvania awarded his BSc in Biology (1975) and Frostburg State College his MSc in fisheries management (1980). He was at the WV Division of Natural Resources as a research biologist (1978–1980) and planning and environmental coordination biologist (1980–1995). He was honoured for his commitment toward the preservation, conservation and management of studies of West Virginia fishes, including efforts toward conservation of the Elk River drainage and its diverse ichthyofauna. Among his more than 50 published papers on fishes are his: *Fish Responses to Temperature to Assess Effects of Thermal Discharge on Biological Integrity* (1982) and a popular article written with Dr Welsh on *The Moundbuilders* (2005).

Cinderella

Brazilian cichlid sp. *Teleocichla cinderella* SO Kullander, 1988

Not an eponym but referring to the cichlid's black and grey skin giving it an 'ashy' appearance, so using the same root as the fairy story character.

Citerni

Genale Mormyrid *Mormyrops citernii* D Vinciguerra, 1912

Captain Carlo Citerni (1873–1918) was an Italian explorer. He took part in Bottego's second expedition (1895) to Lake Rudolph. He led the expedition to mark the border between Italian Somaliland and Ethiopia (1910–1911). He co-wrote: *L'Omo. Viaggio d'esplorazione nell'Africa Oriental* (1899). A reptile was named after him, while many Italian cities and towns have a 'Via Carlo Citerni' named in his memory.

Claea

Stone Loach genus *Claea* M Kottelat, 2010

In Greek mythology, Claea was a nymph of a sacred cave on Mount Kalathion in Messenia in the Peloponnese. Kottelat chose this nymph's name to replace another 'nymph' name – *Oreias* Sauvage, 1874 – which name was pre-occupied.

Claire

Claire's Fairy Wrasse *Cirrhilabrus claire* JE Randall & RL Pyle, 2001

Claire T Michihara is the wife of Chip Boyle, the collector of the type specimens in Rarotonga. They co-own Export Aquarium Ltd at Matavara in the Cook Islands.

Clark, E

Flap-headed Goby sp. *Callogobius clarkae* M Goren, 1978
Dusky Crawler *Sticharium clarkae* A George & VG Springer, 1980
Barred Triplefin *Enneapterygius clarkae* W Holleman, 1982
Wrasse sp. *Pteragogus clarkae* JE Randall, 2013
Genie's Dogfish *Squalus clarkae* MO Pfleger, RD Grubbs, CF Cotton & TS Daly-Engel, 2018

Dr Eugenie Clark (1922–2015), sometimes known as 'The Shark Lady', was a Japanese-American ichthyologist. Hunter College awarded her bachelor's degree (1942) and New York University her master's (1946) and doctorate (1950). She was the founding Director (1955–1967) of the Mote Marine Laboratory, Sarasota, Florida. She joined the faculty at the University of Maryland College Park (1968) retiring as Professor Emeritus of Zoology. Known for her study of sharks, she was also a pioneer of Scuba diving for research purposes. She wrote: *Lady with a Spear* (1951) and was happy to use her fame to promote marine conservation. She was honoured in the goby species for her 'outstanding contributions to the knowledge of the fishes of the Red Sea' (where this goby occurs, which species was originally and often incorrectly spelled *clarki*). The authors of the dogfish species say: "*This species is named in honor of Dr. Eugenie Clark, a pioneer in the field of marine science broadly, and elasmobranch biology in the Gulf of Mexico specifically. Among her many accomplishments, Dr. Clark completed multiple submersible dives, led hundreds of research expeditions, and founded Mote Marine Laboratory in Southwest Florida which continues to conduct research in the Gulf of Mexico and abroad. Dr. Clark's dedication to elasmobranch biology and marine conservation has served as a source of inspiration for countless scientists, including these authors. Her history of deep sea research and passion for fauna of the Gulf of Mexico inspired the etymology presented herein.*" (Also see **Genie (Clark)** & **Genia** & **Niki**)

Clark, FN

Pacific Bearded Brotula *Brotula clarkae* C Hubbs, 1944

Dr Frances Naomi Clark (1894–1987) was the only woman in the male-dominated field of marine biology. She worked for the California Department of Fish and Game (1924–1956) including 17 years as Director, California State Fisheries Laboratory. She wrote: *California marine fisheries investigations, 1914–1939* (1982).

Clark, HC

Sleeper sp. *Erotelis clarki* SF Hildebrand, 1938

Dr Herbert Charles Clark (1877–1960) was the first Director of Gorgas Memorial Laboratory, Panama (1929–1954). United Fruit Company previously employed him as Director of Laboratories and Preventive Medicine. He also organized an annual census of the snake population of Panama (1929–1953). An amphibian and five reptiles are named after him.

Clark, HW

Wormfish genus *Clarkichthys* JLB Smith, 1958
Goby sp. *Ctenogobius clarki* BW Evermann & TH Shaw, 1927

H Walton Clark (1870–1941) was Assistant Curator of fishes at the California Academy of Sciences. He was honoured for his 'valuable' studies of the flora and fauna of Lake Maxinkuckee, Indiana, USA, although this is far from where the (Chinese) goby occurs. He described the type species of the wormfish genus, *Clarkichthys bilineatus* (1936).

Clark, J

Clark's Clownfish *Amphiprion clarkii* JW Bennett, 1830
[Alt. Yellowtail Clownfish]

John Clark was an engraver who engraved plates for JW Bennett and aquatinted the results to produce superb illustrations. The author wrote that the clownfish was "*...named after the engraver, Mr. John Clark, who has not only done ample justice to, but laid the author under very great obligations to him for his able assistance in the present work.*"

Clark, JH

Desert Sucker *Catostomus clarkii* SF Baird & CF Girard, 1854

Lieutenant John Henry Clark (1830–1885) was an American surveyor, naturalist and collector. He was a student of Spencer Fullerton Baird (q.v.) at Dickinson College (c.1844). He was a zoologist on the US/Mexican Border Survey (1850–1855), during which period he collected the holotypes of about 100 new vertebrate species. Under the auspices of the USNM he conducted the Texas Boundary Survey (1860). Two reptiles, a mammal, a bird and an amphibian are named after him.

Clark, K

Marbled Knifefish *Adontosternarchus clarkae* F Mago-Leccia, JG Lundberg & JN Baskin, 1985

Kate Rodriguez-Clark ran a research station in Venezuela with her husband. She collected the holotype of this species.

Clark, RS

Clark's Fingerskate *Dactylobatus clarkii* HB Bigelow & WC Schroeder, 1958

Dr Robert Selbie Clark (1882–1950) was a British marine zoologist and explorer who was a member of Shackleton's Imperial Transantarctic Expedition (1914–1917), being one of those left on Elephant Island when Shackleton sailed in an open boat to South Georgia to get help. Aberdeen University awarded his master's degree (1908). He gained a DSc degree (1925). He was zoologist to the Scottish Oceanographical Laboratory (1911–1913) and naturalist to the Plymouth Marine Biological Association (1913–1914 & 1919–1925), having served in minesweepers in the Royal Navy (1917–1919). He was Director of the Fisheries Research Laboratory, Aberdeen (1925–1934) and Superintendent of Scientific Investigations for the Fishery Board for Scotland (1934–1948). He was honoured in the name of the skate for his: *Rays and skates, A revision of the European species* (1926).

Clark, W

Cutthroat Trout *Oncorhynchus clarkii* J Richardson, 1837

Captain William Clark (1770–1838) was an American explorer and soldier who later became Governor of Missouri Territory. In 1803, Meriwether Lewis recruited Clark to share command of the newly formed Corps of Discovery, whose mission was to explore the territory of the Louisiana Purchase and establish trade with Native Americans. The Lewis & Clark Expedition (1804–1806) crossed the American continent to the Pacific, and during this expedition many new species were collected.

Clarke, MA

Angolan Cyprinid sp. *Labeobarbus clarkeae* KE Banister, 1984
[Syn. *Varicorhinus clarkeae*]

Margaret Anne Clarke worked at Department of Zoology, BMNH, London. Banister wrote that she: "*…gave so much assistance during the course*" of his work on the genus of the species named after her. She co-wrote: *A revision of the large Barbus (Pisces, Cyprinidae) of Lake Malawi with a reconstruction of the history of the southern African Rift Valley lakes* (1980).

Clarke, MR

Dreamer sp. *Oneirodes clarkei* GN Swinney & TW Pietsch III, 1988

Professor Dr Malcolm Roy Clarke FRS (1930–2015) of the Marine Biological Association, UK, was a malachologist known for his work with cephalopods. He was formerly at the National Institute of Oceanography. Among his published papers is *Vertical Distribution of Cephlapods at 30°N 23°W* (1974).

Clarke, RD

Clingfish sp. *Tomicodon clarkei* JT Williams & JC Tyler, 2003

Professor Dr Raymond D Clarke was (1972–2012) Professor of Biology, and is now professor emeritus, at Sarah Lawrence College, Bronxville, New York. He continues to research in ecology, evolutionary biology and marine biology. He has published at least seventeen papers, such as: *Water flow controls distribution and feeding behavior of two co-occurring coral reef fishes: II. Laboratory experiments* (2009). He collected the holotype. The etymology reads: "*The species is named in honor of Raymond D. Clarke, Professor of Biology at Sarah Lawrence College, Bronxville, New York, who collected the holotype and only known specimen during his studies of the behavioral ecology of chaenopsid blennies at Carrie Bow Cay, Belize.*"

Clarke, TA

Lanternfish sp. *Scopelengys clarkei* JL Butler & EH Ahlstrom, 1976

Dr Thomas Arthur Clarke (1940–2013) was an American ichthyologist. The University of Chicago awarded his bachelor's degree (1962) and the Scripps Institute of Oceanography, La Jolla, California his doctorate (1968). He worked at the Hawaii Institute of Marine Biology and at the Department of Oceanography, University of Hawaii; he was Professor at both institutions (1967–2000). He wrote: *Pelagic fishes of the genus* Eustomias, *subgenus* Dinematochirus (Stomiidae), *in the Indo-Pacific with the description of twelve new species* (2001).

Clarke (Island)

Clarke's Triplefin *Trinorfolkia clarkei* A Morton, 1888

Despite the common name and binomial, this is actually a toponym referring to Clarke Island, off northeast Tasmania, which was the type locality.

Clark Hubbs

Texas Silverside *Menidia clarkhubbsi* AA Echelle & DT Mosier, 1982
Signal Triplefin *Lepidonectes clarkhubbsi* WA Bussing, 1991
San Felipe Gambusia *Gambusia clarkhubbsi* GP Garrett & RJ Edwards, 2003

Dr Clark Hubbs (1921–2008) (See **Hubbs**)

Claude

Reef-cave Brotula *Grammonus claudei* C de la Torre y Huerta, 1930

Georges Claude (1870–1960) was a French engineer and inventor. Among his inventions is a system of generating energy by pumping cold water up from the depths of the oceans, and while pumping up water in Matanzas Bay (Cuba) accidentally discovered this fish. He also invented neon lighting. He was an active collaborator of the Germans in WW2. He was arrested (1944), tried, convicted, stripped of all honours and distinctions and sentenced to life imprisonment (1945) but released (1950). He wrote: *L'Électricité à la portée de tout le monde* (1901).

Claudia (Bove)

Neotropical Rivuline sp. *Moema claudiae* WJEM Costa, 2003
Pencil Catfish sp. *Trichomycterus claudiae* MSR Barbosa & WJEM Costa, 2010

Dr Claudia Petean Bove (b.1961) is a botanist specialising in aquatic plants who is titular professor at the Federal University of Rio de Janeiro and also (since 2002) a researcher at the National Museum of Brazil. She is also married to WJEM Costa. The Federal University of Rio de Janeiro awarded both her bachelor's degree (1983) and master's (1990). The University of São Paulo awarded her doctorate (1996). She was a visiting scholar at Kew (2012) and MNHN, Paris (2014). She wrote more than 80 papers and three books. Her research is focused on the systematics, phylogeny and biogeography of the river weeds, the biology of aquatic vascular plants, and freshwater ecosystems. The etymology of the catfish thanks her for her help and companionship during the trip that collected the holotype. She was the co-discoverer of the rivuline. (Also see **Bruno (Bove)** & **Costa, WJEM**)

Claudia (Fricke)

Claudia's Dragonet *Synchiropus claudiae* R Fricke, 1990

Claudia Fricke is the author's sister. She was honoured "*...for her continued interest in and support of my studies on callionymid fishes.*" (Also see **Fricke**)

Claudia (Rocha)

Claudia's Wrasse *Halichoeres claudia* JE Randall & LA Rocha, 2009

Claudia Rocha is the wife of the second author, Luiz Rocha. Her BSc was awarded by the Federal University of Paraiba (1996) and her MSc by the Federal University of Pernambuco (1999). She is a marine biologist who is (since 2015) a Research & Curatorial Associate at the California Academy of Science where she was previously (2011–2015) a researcher. She has also been a Research Associate at the University of Texas (2009–2011), the University of Hawaii (2005–2008) and the Smithsonian Tropical Research Institute (2003–2005). She has published papers with Luiz such as: *Roa rumsfeldi, a new butterflyfish (Teleostei, Chaetodontidae) from mesophotic coral ecosystems of the Philippines Launched to accelerate biodiversity research* (2017). She was thanked in the etymology for her continued support and help with lab work.

Claudine

African Barb sp. *Labeobarbus claudinae* L De Vos & DFE Thys van den Audenaerde, 1990

Claudine Mauel (d.1985) was a resident of Gisenyi, Rwanda (near to where the holotype was acquired). She died in a road accident.

Clausen, CHS

African Loach Catfish sp. *Phractura clauseni* J Daget & A Stauch, 1963
African Rivuline sp. *Fundulopanchax clauseni* JJ Scheel, 1975
[Syn. *Fundulopanchax gardneri clauseni*]
African Barb sp. *Enteromius clauseni* DFE Thys van den Audenaerde, 1976

Claus Herluf Stenholt Clausen (1921–2002) was a California-born Danish ichthyologist and biologist. He collected in Liberia (1965). He wrote: *Tropical Old World Cyprinodonts* (1967). He collected the holotype.

Clausen, CP

Characin sp. *Hemibrycon clausen* CAA Rodriguez, 2020

Dane Christian Peter Clausen founded (1888) the Clausen Brewery in Floridablanca, Colombia. He was a Danish immigrant to Colombia (1882) who started the modern Colombian beer industry, passing the brewery on to his son (1917). He also established a hardware store there (1892) and other businesses.

Clayton

Mexican Goby *Ctenogobius claytonii* SE Meek, 1902

Powell Clayton (1833–1914) was US Minister (ambassador) to Mexico. He was honoured for his 'many courtesies' during Meek's field work in that country (1901), which he extended (1903).

Clea

Clea's Triplefin *Enneapterygius clea* R Fricke, 1997

Clea Fricke (b.1990) is the author's daughter. She was seven years old at the time of the fish's description. (Also see **Claudia (Fricke)** & **Fricke**)

Cleaver

Tasmanian Mudfish *Neochanna cleaveri* EOG Scott, 1934

F Cleaver of West Ulverstone, Tasmania, discovered the holotype (1933) burrowed inside the root of a eucalyptus tree, where it had been aestivating. We have not been able to discover more about him.

Cleisthenes

Flounder genus *Cleisthenes* DS Jordan & EC Starks, 1904

The original text reads: "*…Cleisthenes, the effeminate, an Athenian noted by Aristophanes*". No reason is given for why the name is appropriate for a flatfish. This is not the same person as the Cleisthenes noted for being 'the father of Athenian democracy', but another character of that name who seems to have been the butt of Aristophanes' jokes.

Clemence

Yellow Swordtail *Xiphophorus clemenciae* J Álvarez, 1959

Clemence Álvarez was the wife of the author Dr José Álvarez del Villar (1903–1986).

Clemens, HB

Mottled Scorpionfish *Pontinus clemensi* JE Fitch, 1955

Harold B Clemens was Senior Marine Biologist at the California Department of Fish & Game (c.1950s-c.1970s). The author wrote: "*…The single specimen… …was one of several fish species taken with hook and line in 300 feet of water by H B Clemens, May 3, 1954… … Clemens, a guest aboard the tuna clipper 'Mayflower' was biologist in charge on an official tagging trip for the California Department of Fish & Game.*" He wrote a number of articles appearing in their 'Fish Bulletin', such as: *The migration, age and growth of Pacific Albacore (Thunnus alalunga) 1951–1958* (1961).

Clemens, J

Philippine Barb sp. *Barbodes clemensi* AW Herre, 1924

Joseph Clemens (1862–1936) was born in to an English mining family, which, when he was a child, emigrated from Cornwall to Pennsylvania. He graduated (1894) from Dickinson College, Pennsylvania, as a minister in the American Methodist Episcopalian Church. He he had taken a very wide range of subjects including five different foreign languages, physics, chemistry and psychology. He was a pastor in Pennsylvania (1894–1901). He joined the US military as a chaplain and served in Hawaii and the Philippines (1901–1917) and in France at the end of WW1 (1918) after which he retired having been wounded. He returned to the Philippines as a missionary (1922–1931) and moved to Borneo (1931–1936). He and his wife, Mary, were keen botanists and had made their first joint study at Mount Kinabalu, North Borneo (1915), during which in a six-week period they discovered and collected 101 new plant species. According to the etymology they later: '*...made the first scientific collections around Lake Lanao, Mindanao, Philippines*.' After returning to Borneo (1931) they were commissioned by the BMNH in London to collect plants. He died in New Guinea after contracting food poisoning caused by the contaminated meat of a wild boar.

Clemens, WA

Longfin Gunnel *Pholis clemensi* RH Rosenblatt, 1964

Dr Wilbert Amie Clemens (1887–1964) was a Canadian zoologist and educator. The University of Toronto awarded his BA in Biology (1912) and his MA (1913). Cornell awarded his PhD (1915). He was an instructor in the Department of Zoology at the University of Maine (1915–1916) and a lecturer and Assistant Professor in Biology and Limnobiology at the University of Toronto (1916–1924). He then became the Director of the Pacific Biological Station of the Biological Board of Canada (1924–1940), after which he was Head of the Department of Zoology at the University of British Colombia (1940–1952) and professor emeritus until his death. He undertook various investigations of Canadian lakes and held a number of ancilliary posts and society offices. He wrote: *An ecological study of the mayfly Chirotenetes* (1917).

Clementina

Armoured Catfish sp. *Ancistrus clementinae* H Rendahl, 1937

This is a toponym referring to the type locality, the Río Clementina system in Ecuador.

Cleopatra

Stone Loach sp. *Nemacheilus cleopatra* J Freyhof & DV Serov, 2001

Cleopatra (69–30 BC) was Queen of Egypt (52–30 BC), probably best remembered today for her role in Shakespeare's: *Antony and Cleopatra.* She was a legendary beauty, leading the authors to name this 'elegant' fish after her.

Cléva

Broadheaded Catshark *Bythaelurus clevai* B Séret, 1987

Dr Régis Cléva is a zoologist at Muséum Nationale d'Histoire Naturelle, Paris. He specialises in the study of crustaceans and co-wrote: *Report on some caridean shrimps (Crustacea: Decapoda) from Mayotte, southwest Indian Ocean* (2012). He collected the holotype.

Cleveland

Arrow Goby genus *Clevelandia* CH Eigenmann & RS Eigenmann, 1888

Daniel Cleveland (1838–1929) was a founding partner and President of the San Diego Society of Natural History. He was honoured because he had: "*...done much towards making known the fauna and flora of Southern California*".

Cliff

Sandy Ridgefin Eel *Callechelys cliffi* JE Böhlke & JC Briggs, 1954

Dr Frank Samuel Cliff (1928–2000) was an American biologist and herpetologist who collected the holotype whilst he was a graduate student at Stanford University, which awarded his doctorate (1951). He was an Associate Professor at the Department of Zoology, Colgate University, Hamilton, New York (1958–1959). He taught in the Biology Department, California State University, Chico (1959–1991).

Clifford Pope

Goby sp. *Rhinogobius cliffordpopei* JT Nichols, 1925

Clifford Hillhouse Pope (1899–1974) was an American herpetologist. He spent several years in China (1920s) as a member of expeditions organised by the AMNH, New York. He became a fluent speaker of Chinese. He was Assistant Curator of Herpetology, American Museum (1928–1935) but was sacked after a disagreement with the Director, Gladwyn Kingsley Noble. To support himself, unemployed in the Great Depression, he wrote a number of popular books on herpetology. He was (1941–1954) Curator of Reptiles, Field Museum, Chicago. He took early retirement and went to live in California where he continued to write. He wrote: *Reptiles of China* (1935). His wife, Sarah H Pope was also a herpetologist. Two amphibians and four reptiles are named after him.

Climax

Pearleye sp. *Scopelarchoides climax* RK Johnson, 1974

This species is named after the Climax Expeditions to the central Pacific Ocean, during which the holotype was collected. The etymology makes it clear that the dedication is to include the scientists and crews of the expedition and, in particular the leader of the expeditions, Dr John A McGowan (q.v.) who was then an associate professor of oceanography at the University of California, San Diego and is now Professor Emeritus of Oceanography, Scripps institute of Oceanography, University of California, San Diego. In WW2, he served in the US Navy in the Pacific. Oregon State University awarded his bachelor's and master's degrees and the Scripps Institute his doctorate. (See also **McGowan**)

Clinton

Beaded Darter *Etheostoma clinton* RL Mayden & SR Layman, 2012

President William Jefferson 'Bill' Clinton né Blyth III was the 42nd President of the USA. He was honoured for his lasting environmental accomplishments.

Clistenes

Characin sp. *Characidium clistenesi* MRS de Melo & VC Espíndola, 2016

Dr Alexandre Clistenes de Alcântara Santos is a Brazilian ichthyologist at Feira de Santana State University where he is a Full Professor. and a colleague of the authors The Federal Rural University of Rio de Janeiro awarded his bachelor's degree (1986), his master's (1996) and his doctorate (2003) after which he undertook post-doctoral studies at the same university (2008–2009). He specialises in the natural history of the fishes of Chapada Diamantina, Bahia, where this species occurs. He co-wrote: *Assessing patterns of ichthyofauna by an artisanal fishery in the Bay of All Saints, Brazil* (2012).

Clive R

Rattail sp. *Nezumia cliveri* T Iwamoto & N Merrett, 1997

Dr Clive Douglas Roberts (b.1952) collected the holotype. (See **Roberts, CD**)

Coad

Coad's Riffle Minnow *Alburnoides coadi* H Mousavi-Sabet, S Vatandoust & I Doadrio Villarejo, 2015

Scraper (barb) sp. *Capoeta coadi* NH Alwan, H Zareian & HR Esmaeili, 2016

Dr Brian William Coad (b.1946) is a British/Canadian ichthyologist. The University of Manchester awarded his BSc in Biology (1970) and the University of Waterloo, Canada his MSc in Zoology (1972). The University of Ottawa awarded his doctorate (1976). He was an associate professor at the Department of Biology, Pahlavi University, Shiraz, Iran (1976–1979), after which he was at the Ichthyology Section, National Museum of Natural Sciences (now Canadian Museum of Nature), Ottawa, as Research Associate (1979–1981), Associate Curator (1981–1986) and Curator of Fishes (1986–1991). He was Research Scientist at the Canadian Museum of Nature, Ottawa, Canada (1991–2016) and, since retiring (2017) a Research Associate there. Among his huge output (1972–2017) of scientific papers (380 and counting) and books are: *Guide to the Marine Sport Fishes of Atlantic Canada and New* England (1992), *Encyclopedia of Canadian Fishes* (1995), *Expedition Field Techniques: Fishes* (1998), *Fishes of Tehran Province and adjacent areas* (2008), *Freshwater Fishes of Iraq* (2010), *Fishes of Afghanistan* (2014) and the upcoming *Marine Fishes of Arctic Canada* (2017). See: http://www.briancoad.com/

Coates, CW

Naked-back Knifefish sp. *Gymnotus coatesi* ER La Monte, 1935

Christopher William Coates (1899–1974) served in the US Army in WW1 - he admitted to lying a little about his age in order to be allowed to enlist! After WW1, he became a produce broker, but had a great interest in ichthyology so that when the New York Aquarium was given $10,000 (1930) for a tropical fish exhibit he was put in charge and served as Curator (1930–1956), Director (1956–1964) and retired as Curator Emeritus. He was the first person to use electric eels to power light bulbs, demonstrating their biologically generated electricity.

Coates, D

Eeltail Catfish sp. *Neosilurus coatesi* GR Allen, 1985
Coate's Goby *Glossogobius coatesi* DF Hoese & GR Allen, 1990
Coates' Catfish *Neoarius coatesi* PJ Kailola, 1990

Dr David Coates is a marine biologist and limnologist whose bachelor's and master's degrees were awarded by the University of Newcastle-upon-Tyne, England. His doctorate was gained through the Open University, United Kingdom (1986). He worked in Sudan and the upper reaches of the Nile (late 1970s), then moved to Fisheries Research Laboratory of the Papua New Guinea Department of Primary Industry, becoming Chief Scientist. He worked for the field programme of the United Nations' FAO (1986–1997), for the fisheries programme of the Mekong River Commission (1997–2001) and worked as a team co-ordinator involving environmental considerations in the fisheries and water resources in the lower Ganges basin (2001–2003). He now works for the Secretariat of the Convention on Biological Diversity, Montreal as a programme officer - Inland waterways and ecosystems. He helped to collect all the types.

Coates, G

Coates' Shark *Carcharhinus coatesi* GP Whitley, 1939
[Alt. White-cheek Shark, Whitecheek Whaler, Widemouth Blackspot Shark]

George Coates (d.1980) of Townsville, Queensland was an artist and illustrator who collected the holotype. He supplied many fish to Whitley. He wrote: *Fishing on the Barrier Reef and Inshore* (1943). A training restaurant at the Barrier Reef Institute of TAFE (Training and Further Education) and, appropriately, a fishing boat are named after him.

Coatlicue

Mexican Cichlid sp. *Vieja coatlicue* LF Del Moral-Flores, E López-Segovia & T Hernández-Arellano, 2018

This is named for the Aztec goddess Coatlicue; the "*goddess of life and death, sovereign of the land and fertility*" (translation).

Coats, A

Seaperch sp. *Lepidoperca coatsii* CT Regan, 1913

Major Andrew Coats (1852–1930) was not only a sponsor of the Scottish National Antarctic Expedition (see next entry) but he himself was a polar explorer. He led the British Sport-hunting and Scientific Expedition to the Arctic (1898); primarily a private expedition to shoot bears and walruses. He later was a major in the sixth battalion of the Imperial Yeomanry during the Boer War (1900–1901) for which he was awarded the DSO.

Coats, J

Antarctic Jonasfish *Notolepis coatsorum* L Dollo, 1908

Sir James Coats, Jr (1834–1913) 1st Baronet and his brother Major Andrew Coats together donated £30,000 towards the Scottish National Antarctic Expedition (1902–1904) during which the holotype was collected by the expedition's scientific leader William Speirs Bruce. They were members of the family of Paisley cotton thread manufacturers, a company that after over 250 years is still in business today. Coats Land in Antarctica is also named after them.

Cocco

Sabretooth fish genus *Coccorella* L Roule, 1929
Scorpionfish genus *Coccotropsis* KH Barnard, 1927
Cocco's Lanternfish *Gonichthys cocco* A Cocco, 1829
Deepwater Cardinalfish sp. *Microichthys coccoi* WPES Rüppell, 1852

Anastasio Cocco (1799–1854) was an Italian naturalist, marine biologist and pharmacist. It may seem that Cocco named *Gonichthys cocco* after himself, but it is in fact: "*...dedicated to the memory of my dear father, who died very prematurely, and whose loss will never stop bringing me to tears.*" He wrote: *Su di alcuni Salmonidi del mare di Messina; lettera al Ch. D. Carle Luciano Bonaparte* (1838).

Cocha

Characin sp. *Grundulus cochae* C Román-Valencia, HJ Paepke & F Pantoja, 2003

A toponym, being named after La Cocha Lake, southern Colombia, where it is endemic.

Cochu

Characin sp. *Tyttocharax cochui* W Ladiges, 1949
Barredtail Cory *Corydoras cochui* GS Myers & SH Weitzman, 1954
Characin sp. *Creagrutus cochui* J Géry, 1964

Ferdinand 'Fred' Cochu, a partner at Paramount Aquarium, is an American tropical-fish importer. (See also **Fredcochu**)

Codrington

Upper Zambezi Yellowfish *Labeobarbus codringtonii* GA Boulenger, 1908
Green Hap (cichlid) *Sargochromis codringtonii* GA Boulenger, 1908

Thomas Codrington (1829–1918) was an engineer and antiquarian who was an inspector for local government. He wrote on such matters as how to maintain roads, but is now remembered for: *Roman Roads in Britain* (1903) which was the first attempt by anyone to try and fully catalogue the remains of the Roman transport network. He visited (1908) his son, Sir Robert Edward Codrington (1869–1908), who was the colonial administrator of northwest Rhodesia (Zambia). Whilst living in Livingstone he made a collection of fishes which he later presented to the BMNH in London. He wrote: *Some Notes on the Neighbourhood of the Victoria Falls (Rhodesia)* (1909).

Coffea

West African Cichlid sp. *Coptodon coffea* DFE Thys van den Audenaerde, 1970

This is a toponym referring to Mount Coffee Dam Lake, St. Paul River, Liberia, the type locality.

Coggin

Chinese cyprinid sp. *Poropuntius cogginii* BL Chaudhuri, 1911

Dr John Coggin Brown (1884–1962) (See **Brown JC**).

Cohen

Cortez Pikeblenny *Chaenopsis coheni* JE Böhlke, 1957
Pearlfish sp. *Echiodon coheni* JT Williams, 1984
Short-tail Eel sp. *Thalassenchelys coheni* PHJ Castle & SN Raju, 1975
Gulf Snailfish *Liparis coheni* KW Able, 1976
Cohen's Puffer *Pelagocephalus coheni* JC Tyler & JR Paxton, 1979
Morid Cod sp. *Physiculus coheni* CD Paulin, 1989
Cohen's Whiptail *Nezumia coheni* T Iwamoto & N Merrett, 1997
Cohen's Mudbrotula *Gunterichthys coheni* PR Møller, W Schwarzhans & JG Nielsen, 2004
Cusk-eel sp. *Luciobrotula coheni* JG Nielsen, 2009

Dr Daniel Morris Cohen (1930–2017) was an ichthyologist who was a research associate at the California Academy of Sciences. Stanford University awarded his bachelor's degree (1952), master's (1954) and doctorate (1958). He was an Assistant Professor of biology and Curator of Fishes at the University of Florida, Gainsville (1957–1958). He worked for the US Bureau of Commercial Fisheries Ichthyological Laboratory, based in the Smithsonian (1959–1982), being Director (1960–1982) of the National Systematics Laboratory. He became Chief Curator of Life Sciences and Deputy Director of Research and collections at the Los Angeles County Museum of Natural History (1982–1995). He went on a number of collecting cruises including the famous cruise of the Russian research vessel 'Vitjaz' to the Western Indian Ocean (1988). (Also see **Daniel**)

Coillot

Sicklefin Devil Ray *Mobula coilloti* J Cadenat & P Rancurel, 1960
[Junior Synonym of *Mobula tarapacana*]

M Coillot was the chief engineer of the oceanographic research vessel 'Reine Pokou'. He harpooned and captured the holotype.

COLAS

Sleeper sp. *Oxyeleotris colasi* L Pouyaud, Kadarusman & RK Hadiaty, 2013

The COLAS Group of companies in Indonesia, which co-sponsored the Lengguru-Kaimana expedition (2010), during which the holotype was collected.

Colclough

Colclough's Shark *Brachaelurus colcloughi* JD Ogilby, 1908
[Alt. Bluegrey Carpetshark]

John Colclough was a friend of the author. He was a keen fisherman, member of the Amateur Fisherman's Association of Queensland and collector, and later sent a considerable collection to Ogilby from the Aru Islands and the Northern Territory of Australia.

Cole, HI

False Bandit Dottyback *Pseudochromis colei* AW Herre, 1933
Cole's Rockskipper *Istiblennius colei* AW Herre, 1934

Howard I Cole (1892–1966) was Chief Chemist for the Philippine Health Service at the leper colony on Culion Island, Philippines, the type locality. His "*enthusiastic cooperation and generous aid alone made possible* [Herre's] *large and interesting collection*" of fishes from that island.

Cole, LJ

Golden Silverside *Menidia colei* CL Hubbs, 1936

Dr Leon Jacob Cole (1877–1948) was an American geneticist and ornithologist. He graduated from the University of Michigan (1901) and was awarded his PhD (1906) by Harvard. As an undergraduate he took part on the Harriman expeition to Alaska (1899). He conducted an early fish survey of Yucatán (where this species is endemic) and published his collections jointly with Thomas Barbour (1906) having spent time collecting birds. He was in charge of the Division of Animal Breeding and Pathology at the Rhode Island Experiment Station, as well as an instructor in zoology at Yale University (1906–1910). He joined the University of Wisconsin (1910) to initiate the Department of Experimental Breeding, a forerunner of the university's Department of Genetics, becoming a professor (1914), and was Professor of Genetics (1918–1947).

Cole, WW

Cole's Char *Salvelinus colii* A Günther, 1863

William Willoughby Cole, 3rd Earl of Enniskillen (1807–1866) was an Irish palaeontologist and politician. From a young age, he started collecting fossil fishes and amassed a large collection which was subsequently acquired by the British Museum. He made strenuous efforts helping Günther to obtain this species for study.

Coleman, HW

Thai Barb sp. *Discherodontus colemani* HW Fowler, 1937
[Syn. *Barbodes colemani*]

Dr Henry Waldburg Coleman (1847–1907) was a physician who graduated (1868) from College of Physicians and Surgeons at Columbia University, New York, and practiced at Trenton, New Jersey (1882). Fowler described him as being an early contributor to the fish collection at the Academy of Natural Sciences of Philadelphia.

Coleman, N

Sandperch sp. *Parapercis colemani* JE Randall & MP Francis, 1993
Coleman's Pygmy Seahorse *Hippocampus colemani* RH Kuiter, 2003

Neville Coleman (1938–2012) was an Australian award-winning environmental photographer, underwater explorer and conservationist with an interest and expertise in nudibranchs, sea grass, marine sponges, and various types of coral. He started Scuba diving in Sydney (1963) and led (1969) a four-year Australian Coastal Marine Expedition, the first underwater photographic fauna survey ever attempted around an entire continent. This work led to his: *Australian Marine Fishes In Colour* (1974) the first of over 60 books on marine life (as well as numberous papers, articles and on-line guides) ending with his last book, the *Nudibranchs Encyclopedia* (2008). He discovered around 450 taxa including the pygmy seahorse species and at least two eponymous invertebrates.

Coles

Lightnose Skate *Breviraja colesi* HB Bigelow & WC Schroeder, 1948
[Alt. Bahama Skate]

No etymology is given but we believe the skate to be named after Russell Jordan Coles (b.1865). He was a tobacco trader in Virginia and a keen amateur fisherman and 'shark watcher' who spent three or four months each year anywhere between Florida and Newfoundland in search of the biggest fishes, which he sometimes collected for the AMNH. Among his friends and fishing companions was Theodore Roosevelt (he gave a memorial address about him to the Explorers Club). He was known as 'Doctor' as he had studied medicine, but never qualified as a physician. However, at Roosevelt's suggestion, Trinity University, Hartford, Connecticut awarded him an honorary doctorate (1918). He wrote, among other scientific articles: *Natural history notes on the devil-fish, Manta birostris (Walbaum) and Mobula oflfersi (Müller)* (1916).

Colin

Colin's Angelfish *Centropyge colini* WF Smith-Vaniz and JE Randal, 1974
Colin's Damsel *Pomacentrus colini* GR Allen, 1991
Tiny New Guinea Longtail Dragonet *Callionymus colini* R Fricke, 1993
Colin's Fairygoby *Tryssogobius colini* HK Larson & DF Hoese, 2001
Belize Sponge Goby *Elacatinus colini* JE Randall & PS Lobel, 2009
Colin's Pygmygoby *Trimma corerefum* R Winterbottom, 2016

Patrick Lynn 'Pat' Colin (b.1946) is a coral-reef biologist at the Coral Reef Research Foundation. His BSc (1968), MSc (1970) and PhD (1973) in Marine Sciences were awarded by the University of Miami Rosensteil School of Marine and Atmospheric Sciences. He taught (1974) at the Department of Marice Sciences at the University of Puerto Rico. He became (1979) Senior Scientist for the University of Hawaii's Mid-Pacific Research Laboratory at Enewetak Atoll, Marshall Islands. He and his wife Lori relocated (1983) to the Motupore Island Research Station of the University of Papua New Guinea where they ran the activities of the research and educational facility. Then (1987) he worked at the Caribbean Marine Research Center, Bahamas, as Senior Scientist to study the fast disappearing Nassau grouper spawning aggregations of the Atlantic. He co-founded (1991) the Coral Reef Research Foundation (CRRF), where he worked (from 1992) in Federated States of Micronesia and Palau and where he remains President. His books include: *Neon Gobies* (1975), *Caribbean Reef Invertebrates and Plants* (1978), *Marine Environments of Palau* (2009) and the co-written *Tropical Pacific Invertebrates* (1995). Work by CRRF for the US National Cancer Institute resulted in the collection of nearly 200 new species of invertebrates and algae which have been described by many different taxonomists. He collected the fairygoby and damsel types. Randall honoured him for his exceptional doctoral thesis on the comparative biology of the genus *Elacatinus* and for his help (providing color photographs and guidance) with their research. He is interested in deep coral reefs, their fauna and oceanography, as well as general coral reef ecology. (See also ***Corerefum***)

Collart

African Barb sp. *Enteromius collarti* M Poll, 1945

Albert Désiré Clément Hubert Collart (1899–1993) was a Belgian entomologist at the Royal Institute of Natural Sciences of Belgium with a passion for bookplates. He was collecting actively in the late 1920s through 1930s and much of his collections are held at the Essig Museum of Entomology. He went to the Congo (1923) as a health worker with little spare time, but he still managed to collect insects, birds, local artefacts and fish including the barb type. He returned home for good (1930) because of ill health, and spent the rest of his career at the Royal Museum. He was Research Associate from (1932–1935), Assistant Naturalist intern (1935–1936), Assistant Naturalist (1936–1941), Assistant Curator (1941–1950), Curator (1950–1952), Laboratory Director (1952–1964) and Scientific Collaborator (1965). He published papers in the Bulletin of the Society of Entomology in Belgium. A species of odonata is also named after him.

Collett

Longtail Catfish sp. *Olyra collettii* F Steindachner, 1881
Characin sp. *Moenkhausia collettii* F Steindachner, 1882
Alaska Snailfish *Careproctus colletti* CH Gilbert, 1896
Rattail sp. *Gadomus colletti* DS Jordan & CH Gilbert, 1904

Dr Robert Collett (1842–1913) was a Norwegian zoologist and ichthyologist. He worked at Christiania (Oslo) Museum (1871–1913), first as an Assistant Curator, then Curator (1874) and Director (1892) and was Professor of Zoology at the university from 1884. He wrote: *Bemserkninger til Norges Pattedyrfauna* (1876). A mammal, an amphibian, six birds and two reptiles are named after him.

Collette

Toadfish genus *Colletteichthys* DW Greenfield, 2006
Parrotfish sp. *Nicholsina collettei* LP Schultz, 1968
Creole Darter *Etheostoma collettei* RS Birdsong & LW Knapp, 1969
Collette's Coralblenny *Ecsenius collettei* VG Springer, 1972
Collette's Herring *Herklotsichthys collettei* T Wongratana, 1987

Collette's Basslet *Liopropoma collettei* JE Randall & L Taylor, 1988
Halfbeak sp. *Dermogenys collettei* AD Meisner, 2001
Tetra sp. *Bryconops collettei* B Chernoff & A Machado-Allison, 2005
Collette's Viviparous Brotula *Ungusurculus collettei* W Schwarzhans & PR Møller, 2007
Halfbeak sp. *Hyporhamphus collettei* HM Banford, 2010
Cusk-eel sp. *Lepophidium collettei* CR Robins, RH Robins & ME Brown, 2012
Stardrum sp. *Stellifer collettei* NL Chao, A Carvalho-Filho & J de Andrade Santos, 2021

Dr Bruce Baden Collette (b.1934) is an American ichthyologist. Cornell University awarded his bachelor's degree (1956) and his doctorate (1960). He joined the National Marine Fisheries Service Systematics Laboratory (1963), later becoming its Director. He is now a Senior Scientist at the National Marine Fisheries Service Systematics Laboratory in the Smithsonian. As at the date of writing, Dr Collette has described 4 genera, 42 species and 6 subspecies of fishes and produced a prodigious output of papers, articles, chapters, books etc., including: *Fishes of Bermuda : history, zoogeography, annotated checklist, and identification keys* (1999) and *Systematics of the tunas and mackerels (Scombridae)* (2001).

Collie

Spotted Ratfish *Hydrolagus colliei* GT Lay & ET Bennett, 1839

Lieutenant Dr Alexander Collie (1793–1835) was the naval surgeon and naturalist on an expedition (1825–1828) led by Captain Frederick Beechey on 'HMS Blossom'. This made some significant zoological discoveries during the voyage from Chile to Alaska. Collie collected many specimens that did not survive the return journey to England in good condition, but he made some coloured drawings of taxa he thought were new and also took extensive notes. Collie also collected many live birds that went on to be exhibited in London Zoo. Dr Collie went to Perth as a colonial administrator where he died. When aboard 'HMS Sulphur' he discovered what is now the Collie River in Western Australia. A town in Australia, a mammal, a bird and a reptile are named after him.

Collier

Neotropical killifish sp. *Rivulus collieri* JH Huber, 2021

Glen E Collier, of the University of Tulsa (Oklahoma, USA), was honoured for his pioneering and long-term (25 years) genetic and molecular contributions to killifish knowledge, including Rivulus, his "preferred" genus.

Collignon

Sailfin Weever *Trachinus collignoni* C Roux, 1957

Jean Collignon was an oceanographer and marine biologist who was Head of Research at ORSTOM. He was a colleague of Charles Roux and they were co-authors of: *Mollusques, crustacés, poissons marins des côtes d'A E F en collection au Centre d'Océanographie de l'Institut d'Etudes Centrafricaines de Pointe-Noire* (1957); they also co-authored a Key to the most important marine fish off the coast. *Clef Pour la Détermination des Prinipaux poisons Marins*. He wrote many other papers such as: *La systématique des Sciaenidés de l'Atlantique oriental* (1959).

Collingwood

Borneo Barb sp. *Barbonymus collingwoodii* A Günther, 1868

Dr Cuthbert Collingwood (1826–1908) was a surgeon and physician who became a physician in order to pursue a career as a naturalist. Oxford University awarded both his bachelor's degree (1849) and his master's (1852). He became a Bachelor of Medicine (1854) and a Member of the Royal College of Physicians (1859). He also studied botany in Paris (1851). He was at sea as an unpaid surgeon naturalist on 'HMS Serpent' and others of the Royal Navy's surveying vessels in the Far East (1866–1867). He was a Fellow of the Linnaean Society and was a member of the audience on the famous occasion when Darwin's and Wallace's papers on natural selection were first read. He abandoned (1872) natural history for the study of religion, joined the Swedenborgian church and became convinced that Darwin, with whom he had corresponded and was on good terms, was wrong. He left England and lived in Paris (1901–1907). He wrote: *Rambles of a Naturalist on the Shores and Waters of the China Sea* (1868). He presented the holotype to the BMNH in London.

Collins

Armoured Catfish sp. *Parotocinclus collinsae* RE Schmidt & CJ Ferraris Jr, 1985

Dr Margaret James Strickland Collins (1922–1996) was an African-American biologist, zoologist and entomologist who specialised in the study of termites. West Virginia State University awarded her bachelor's degree (1943) and master's. The University of Chicago awarded her doctorate (1950), becoming only the third black woman zoologist in the USA. She taught at Florida A&M University (early 1960s), at Federal City College, Washington (1969–1976) and at Howard University (1963–1969 & 1977–1983). She was active in the civil rights movement to the extent that after she spoke at a white university on genetics and molecular biology, her department at Florida A&M, where she was the dean, received a bomb threat. After retiring as Professor of Zoology she worked as a Research Associate at the Smithsonian (1983–1996). She worked extensively in the field and when at Alfred Emerson Field Station, Kartabo, Guyana, made it possible for the senior author to collect fishes in Guyana. She co-wrote: *Science and the Question of Human Equality* (1981). She died in the Cayman Islands while on a research trip.

Commerson

Commerson's Frogfish *Antennarius commerson* BG de Lacépède, 1798
Narrow-barred Spanish Mackerel *Scomberomorus commerson* BG de Lacépède, 1800
Commerson's Freshwater Goby *Awaous commersoni* JG Schneider, 1801
Small-spotted Grunter *Pomadasys commersonnii* BG de Lacépède, 1801
Talang Queenfish *Scomberoides commersonnianus* BG de Lacépède, 1801
Commerson's Sole *Synaptura commersonnii* BG de Lacépède, 1802
White Sucker *Catostomus commersonii* BG de Lacépède, 1803
Commerson's Anchovy *Stolephorus commersonnii* BG de Lacepède, 1803
Armoured Catfish sp. *Hypostomus commersoni* A Valenciennes, 1836
Blue-spotted Cornetfish *Fistularia commersonii* E Rüppell, 1838

Dr Philibert Commerson (1727–1773) was known as 'doctor, botanist and naturalist of the King'. He accompanied the French explorer Louis Antoine de Bougainville on his round-the-world expedition (1766–1769) on board 'La Boudeuse' and 'L'Etoile'. He was primarily a botanist, but he has a wide diversity of animal species named after him, including two birds and two mammals. He also discovered the vine *Bougainvillea* (1760s), naming it after the expedition leader.

Compagno

Lantern Shark sp. *Etmopterus compagnoi* R Fricke & I Koch, 1990
Tigertail Skate *Leucoraja compagnoi* M Stehmann, 1995

Dr Leonard Joseph Victor Compagno (b.1943) is an American ichthyologist, internationally recognized as authority on shark taxonomy. His early career was in the USA as a research assistant in chemistry (1964) and ornithology (1965–1966) before becoming a Curatorial Assistant of Fishes at Stanford (1966–1967). He worked in various jobs within and without the university while studying for his PhD. Stanford University awarded his doctorate (1979). Since 1977 he has been Research consultant and contract writer for Fisheries Division of Food and Agriculture Organization, United Nations and the Save Our Seas Foundation, Geneva. He was Adjunct Professor at San Francisco State University (1979–1985) and Curator of Fishes, Iziko Museum, Cape Town, South Africa then (1987) became Director of the Shark Research Centre there until retiring (2008). He is currently (2014) Director of the international Shark Research Institute at Princeton and Extraordinary Professor, Department of Zoology & Botany, Stellenbosch University, South Africa. He has written over 1000 scientific papers and longer works including: *Sharks of the World* (2002) and *A Field Guide to the Sharks of the World* (2005). He has described or co-described 52 species, genera or families of *Elasmobranchii*. A claim to fame is that he is mentioned in the credits for the film *Jaws* (1975). He has awards and honours too numerous to mention and belongs to many societies and organisations, including in particular the American Society of Ichthyologists and Herpetologists, the American Elasmobranch Society, the Royal Society of South Africa and Oceania Chondrichthyan Society, Australia. He divides his time between living in South Africa, New Jersey and Oregon, USA. He was honoured in the lantern shark name for his research on South African sharks.

Compine

Yellowfish sp. *Labeobarbus compiniei* HE Sauvage, 1879

Though there is no etymology in the original text, the species is probably named after Louis Alphonse Henri Victor du Pont, Marquis de Compiègne (1846–1877). After qualifying brilliantly with a bachelor's degree and a degree in law, he was sent to Algeria to explore (1867–1868). Back in Paris he partied and ran up great debts. His family rallied round and cleared his debts after making it clear that they were not pleased! He then went to the USA and explored the Florida Everglades (1869). The outbreak of the Franco-Prussian War (1870) brought him back to France to join the army. He was captured and held as a prisoner-of-war (1870–1871) and on his release found himself involved in the fighting associated with the civil war known as the Paris Commune (1871). Once the fighting was over, he left for South America and visited Lake Maracaibo, Panama and Nicaragua. He explored the Ogooue River, Gabon (where the holotype was taken) (1872–1874), but was forced to turn back after cannibals attacked the expedition slaughtering most of the native porters. His health was severely undermined by malaria and he was not fit for further exploring so he accepted the post of Secretary General of the Khedive's (Ruler of Egypt) Geographical Society in Cairo (1875–1877). He got into an argument at a ball with a German who so insulted Compiègne's companion that a duel became inevitable. He was shot and wounded, the wound became infected and he died.

Condé

Small-scale Glass Tetra *Charax condei* J Géry & Knöppel, 1976
African Barb sp. *Enteromius condei* V Mahnert & J Géry, 1982
Armoured Catfish sp. *Rhadinoloricaria condei* IJH Isbrücker & H Nijssen, 1986
Condé's Wrasse *Cirrhilabrus condei* GR Allen & JR Randall, 1996

Dr Bruno Condé (1920–2004) was a French zoologist and speleologist, whose doctorate in natural science was awarded by the Faculty of Sciences, University of Nancy (1952), where he had started work as a botany collector (1943). He became Deputy Director of the university's Museum of Zoology (1955), its Director (1960) and a professor at the University (1962). He was the Founding President of the Nancy Speleological Society (1961). He retired (1989) but continued private research.

Conklin

Freckled Cardinalfish *Phaeoptyx conklini* CF Silvester, 1915

Dr Edwin Grant Conklin (1863–1952) was an American biologist, zoologist and embryologist. He graduated from Ohio Weslayan University and from Johns Hopkins University, Baltimore, where he was awarded his doctorate. He was Professor of Biology, Ohio Wesleyan University (1891–1894), Professor of Zoology, Northwestern University (1894–1896), at the University of Pennsylvaia (1896–1908) and at Princeton University (1908–1933), where he remained after retiring as an independent researcher and lecturer. He wrote: *Direction of Human Evolution* (1921).

Connell

Harlequin Goldie *Pseudanthias connelli* PC Heemstra & JE Randall, 1986
Sweeper sp. *Pempheris connelli* JE Randall & BC Victor, 2015

Dr Allan D Connell (1943–2016) was a South African marine biologist whose initial interest, including his PhD, was in entomology. He worked for Natural Resources and Industrial Research (NRE) until retirement (2004) and remained an Honorary Research Associate of SAIAB. Much of his career was spent monitoring the impact of marine pollution outfalls on coastal environments for the CSIR. However, according to Professor Alan Whitfield who knew Dr Connell as a colleague for many years, "his real skills and passion as a naturalist were evident with his knowledge and contributions to marine larval fish research. His backyard hobby of raising a wide variety of fish larvae from the egg to the early juvenile stage was accompanied by detailed developmental photographs of each specimen and associated meticulous documentation, a record that he made freely available via the web to ichthyologists in South Africa and abroad." In recent years, as a volunteer he contributed genetic material from an amazing variety of southern African marine fish species to the Barcode of Life Database (BoLD). According to the etymology he was honoured as his "*... untiring efforts to document the fishes of KwaZulu-Natal have resulted in the discovery of many new species. He collected several of the recent lots of the types of this species, as well as numerous specimens, photographs, and tissue samples of other species of fishes from southern Africa.*" He sadly died while diving.

Connie

Wrasse genus *Conniella* GR Allen, 1983
Popondetta Blue-eye *Pseudomugil connieae* GR Allen, 1981
Connie's Wrasse *Conniella apterygia* GR Allen, 1983
[Alt. Mutant Wrasse, Rowley Shoals Wrasse]

Connie Lagos Allen is the wife of the author, Dr Gerald Ray Allen (b.1942). (Also see **Allen, GR** & **GR Allen**)

Conrad

Pencil Catfish sp. *Trichomycterus conradi* CH Eigenmann, 1912

Bernard S Conrad, Georgetown, Washington "*...greatly assisted the expedition* (that collected the type) *with advice and guidance*". We can find nothing more about him.

Constancia (White)

Feather-fin Pearlfish *Simpsonichthys constanciae* GS Myers, 1942

Constance Millicent White née Rowe (1903–1980) was a British artist, angler and ichthyologist who was the second wife (1938) of US General Thomas Dresser White (q.v.). The were both avid anglers, and she often accompanied him on outings and painted in watercolour the fishes he caught while serving in Brazil, and the landscapes of the places he fished. She was decorated by the Brazalian government for her contributions to the science of ichthyology. Myers named two species of pearlfish that they found, one after Constance and one after her husband. (Also see **White, TD**)

Constanza

Deepwater Cardinalfish sp. *Epigonus constanciae* EH Giglioli, 1880

Constanza Giglioli née Casella (1849–1940) was a writer of educational books. She married the author (1871) Professor Dr Enrico Hillyer Giglioli (1845–1909) and he refers to her in the etymology as his 'beloved companion' (translation). They had four children including, Odoardo, who was a writer on art. (Also see **Giglioli**)

Contreras-Balderas

Ajijic Silverside *Chirostoma contrerasi* CD Barbour, 2002

Dr Salvador Contreras-Balderas (1936–2009) was a Mexican ichthyologist and biologist. The University of Nuevo León awarded his bachelor's degree (1961) and Tulane University, New Orleans his master's (1966) and doctorate (1975). Nearly all his professional life was devoted to teaching and research at the School of Biological Sciences,

Universidad Autónoma de Nuevo León where he founded not only the ichthyological collection, but also its herpetological and ornithological collections. Among the academic positions he held was Director General of Scientific Research, Universidad Autónoma de Nuevo León and among his honours was the award of the title 'Distinguished Professor' (1993).

Conway, DB

Cape Knifejaw *Oplegnathus conwayi* J Richardson, 1840

David Barry Conway (d.1832) was a Royal navy surgeon who was medical superintendant on two different convict ships to Australia. He was appointed Assistant-Surgeon Royal Navy (1811) and was then employed as Hospital Mate at the Royal Hospital Haslar (1815). He was promoted to Surgeon (1822) and appointed to the 'Harrier' (1824). He became Surgeon Superintendant (1827) aboard 'Manlius' which sailed for New South Wales and kept a diary of that and subsequent voyages. He died 'after just a few hours of illness' in Chatham, UK where he was described as 'surgeon in ordinary of that fort'. The original text refers to the type having come from "*Mr Conway, formerly medical superintendent of a convict ship, and since deceased.*"

Conway, K

Conway's Clingfish *Lepadichthys conwayi* K Fujiwara & H Motomura, 2020

Dr Kevin Conway is an Associate Professor of Ichthyology and Curator of Fishes at Texas A&M University. The University of Glasgow awarded his BSc and Imperial College London & the BMNH awarded his MSc. Saint Louis University awarded his PhD. He has a special interest in clingfish. Among his recent publications is *A new species of the livebearing fish genus Poeciliopsis from northern Mexico (Cyprinodontiformes: Poeciliidae)* (2019).

Cook, J

Cook's Cardinalfish *Ostorhinchus cookii* WJ Macleay, 1881
Cook's Rattail *Coelorinchus cookianus* C McCann & DG McKnight, 1980

Captain James Cook (1728–1779) was one of the most famed explorers of all time, so there are many biographies to consult. He commanded 'HMS Endeavour' on his first expedition, the achievements of which included the discovery of eastern Australia (1770) and much of New Zealand. He commanded 'HMS Resolution' on his last two expeditions. He was killed in a skirmish with natives in Hawaii. There is no etymology for the rattail, but it seems likely that Captain Cook is intended.

Cook, SF

Cook's Swellshark *Cephaloscyllium cooki* PR Last, B Séret & WT White, 2008
Clown Wedgefish *Rhynchobatus cooki* PR Last, PM Kyne & LJV Compagno, 2016

Sidney 'Sid' F Cook (1953–1997) was a shark fishery conservationist and biologist. The swellshark describers say his: "*…energy, dedication and contribution to shark conservation is sadly missed.*" He wrote: *Cooks Book Guide To the Handling & Eating of sharks and skates* (1985).

Cooke, CM

Bigeye genus *Cookeolus* HW Fowler, 1928
Cooke's Sandburrower *Crystallodytes cookei* HW Fowler, 1923
Prickly Shark *Echinorhinus cookei* V Pietschmann, 1928
Worm Eel sp. *Scolecenchelys cookei* HW Fowler, 1928

Dr Charles Montague Cooke Jr (1874–1948) was a malacologist at the Bishop Museum, Hawaii and Curator of the snail collection (1902–1948). He was born into a wealthy family in Honolulu and his resources enabled him to acquire some extensive collections. Yale awarded his bachelor's degree (1897) and his doctorate (1901). He took part in a number of expeditions to the South Pacific and led the Museum's Mangarevan Expedition (1934). Fowler honoured him for the: "*...encouragement of his steadfast friendship*".

Cooke, RG

False Bronze Sea-catfish *Notarius cookei* P Acero & R Betancur-Rodriguez, 2002

Dr Richard George Cooke (b.1946) is a British archaeologist who is a staff scientist at the Smithsonian Tropical Research Institute, Panama. The University of Bristol, England, awarded his bachelor's degree (1968) and London University his doctorate (1972). He co-wrote: *Pre-Colombian use of freshwater fish in the Santa Maria biogeographical province, Panama* (2008). He provided the type series of this species.

Coomans

Cooman's Grenadier Anchovy *Coilia coomansi* JDF Hardenberg, 1934

Dr Louis Coomans de Ruiter (1898–1972) was a Dutch ornithologist, botanist, collector and entomologist. He was an administrator in the East Indies (Indonesia) (1921–1948); Palembang, Sumatra (1926–1928), Western Borneo (1928–1934), Palembang again (1936–1938), Manado, Celebes (1938–1941) and Makassar (1941–1942). He was then at the RMNH (1949). He collected the holotype of this species. A bird is also named after him.

Cooper, BJ

African Rivuline sp. *Nothobranchius cooperi* B Nagy, BR Watters & DU Bellstedt, 2017

Barry J Cooper is an American aquarist who collects and breeds killifish. He has written articles with Brian Watters. The authors said of him that he is a "*...renowned collector and breeder of killifish, for his significant contributions to the field study of* Nothobranchius *and to the killifish hobby in general.*"

Cooper, CF

Red-bar Anthias *Pseudanthias cooperi* CT Regan, 1902
Cooper's Dragonet *Callionymus cooperi* CT Regan, 1908
Flathead sp. *Seychelliceps cooperi* CT Regan, 1908
[Syn. *Suggrundus cooperi*]

Clive Forster Cooper, FRS (1880–1947) was an English palaeontologist. He was educated at Trinity College, Cambridge. He joined J Stanley Gardiner's expedition to the Maldive and Laccadive islands (1900) to make collections and study the formation of coral reefs, then joined the staff of the North Sea Fisheries Commission for Scientific Investigation (1902–1903). He returned to Cambridge and went on an expedition to the Seychelles with Gardiner on 'HMS Sealark' (1906). He met Dr C W Andrews at the BMNH, who introduced him to palaeontology. He went with Andrews to Egypt (1907). He spent a year (1909–1910) at the AMNH, New York, before returning to Cambridge. He was appointed Director, Cambridge University Museum of Zoology (1914). He worked (1914–1918) on malaria at the School of Tropical Medicine, Liverpool. He became Cambridge University Reader in Vertebrates, Fellow and Bursar of Trinity Hall, and also became a Fellow of the Royal Society (1936). He was Director of the BMNH (1938–1947), actually living in the Museum, ensuring the survival of the collections despite severe bomb damage (1940 & 1945). He was knighted (1946), but died two months before he was due to retire (1947). He was described as modest and shy, appreciated paintings, and was a watercolorist and draughtsman of some skill. An amphibian is also named after him.

Cooper, JM

Short-maned Sand-eel *Phaenomonas cooperae* G Palmer, 1970

Mrs Jane M Cooper of the Department of Cooperative Society, Suva, Fiji, collected the holotype while living at Tarawa Atoll, Betio, in the Gilbert Islands (Kiribati, Western Pacific). She also collected botanical specimens. She wrote: *A List of Gilbertese Fish Names* (1962), *Poisonous Fish in the Gilbert Islands* (1962) and *Ciguatera and Other Marine Poisoning in the Gilbert Islands* (1964).

Cope

Characin genus *Copeina* HW Fowler, 1906
Characin genus *Copella* GS Myers, 1956
Northern Leatherside Chub *Lepidomeda copei* DS Jordan & CH Gilbert, 1881
Characin sp. *Brachychalcinus copei* F Steindachner, 1882
Characin sp. *Moenkhausia copei* F Steindachner, 1882
Bluntsnout Smooth-head *Xenodermichthys copei* TN Gill, 1884
Blacksnout Seasnail *Paraliparis copei* GB Goode & TH Bean, 1896
Thorny Catfish sp. *Leptodoras copei* A Fernández-Yépez, 1968
Cope's Corydoras *Corydoras copei* H Nijssen & IJH Isbrücker, 1986

Edward Drinker Cope (1840–1897) was an American palaeontologist, anatomist, herpetologist and ichthyologist. He studied under Baird at the Smithsonian (1859), at the British Museum, London, and the Jardin des Plantes, Paris (1863–1867). He was Professor of Comparative Zoology and Botany, Haverford College, Pennsylvania (1864–1867), and was appointed Curator, Philadelphia Academy of Natural Sciences (1865). He was the palaeontologist on the Wheeler Survey (1874–1877) west of the 100th meridian in New Mexico, Oregon, Texas and Montana. He was a professor at the University of Pennsylvania: of Geology and Mineralogy (1889–1895) and Zoology and Comparative Anatomy (1895–1897). He was senior naturalist (1878) on the periodical 'American Naturalist', which he co-owned. He wrote many articles including: *On the fishes of the Ambylacu River* (1872). In his will, he donated his body to science. His cause of death was listed as uremic poisoning. It was rumoured that he died of syphilis, but when (1995) permission was granted for his skeleton to be medically examined, no such evidence was found. 59 reptiles and 19 amphibians are also named after him.

Copeland, E

Characin sp. *Leporinus copelandii* F Steindachner, 1875
Copeland's Tetra *Hyphessobrycon copelandi* M Durbin Ellis, 1908

Edward Copeland was a volunteer on the Thayer Brazilian Expedition (1865) that collected the type specimens. The expedition leader Louis Agassiz (q.v.) described him as an extremely able and enthusiastic volunteer. Durbin says the tetra was named after Herbert Copeland who was a volunteer on the Thayer Expedition. However, there is no other evidence of anyone of that name being on the expedition and we believe he may have confused 'Herbert' for Edward.

Copeland, EB

Snake-eel sp. *Pisodonophis copelandi* AW Herre, 1953

Edwin Bingham Copeland (1873–1964) was an American botanist, tropical agriculturalist, physiologist, teacher and administrator. He was at the Universoty of Wisconsin (1891–1894) but graduated from Stanford as part of its first graduating class (1895). He then studied at Leipzig (1895) and Halle which awarded his doctorate (1896). He spent several years in a number of positions studying physiology; University of Indiana (1897–1898), California State Normal School (1899), University of West Virginia (1899–1900) and Stanford (1901–1903) before going to the Philippines (1903) to take up the post of Systematic Botanist of the Bureau of Government Laboratories (later Bureau of Science), Manila. He founded the Philippines College of Agriculture (1909) (now the University of the Philippines, Los Baños). He was the Dean and Professor of Plant Physiology there (1909–1917). He returned to the USA and grew rice and was for a time Associate Curator at the University of California, Berkeley (1927–1931). He was also Director of the Botanical Gardens back in Manila (1932–1935) after which he retired from the service and returned to America. He wrote more than 120 books and papers including the book: *Elements of Philippine Agriculture* (1910), *The Coconut* (1914), *Rice* (1924) and the two-volume *Fern Flora of the Philippines* (1958). He described 35 new genera and over 600 species of ferns with a personal herbarium of c.25,000 species! He and the author Herre were friends. He died of pneumonia.

Copeland, HE

Channel Darter *Percina copelandi* DS Jordan, 1877

Professor Herbert Edson Copeland (1849–1876) was a zoologist and father of Edwin Bingham Copeland (see above). Cornell University awarded his bachelor's degree (1872) and his master's (1875). He went to Indiana University (1875) as a pupil of David Starr Jordan (q.v.) with whom he wrote: *Johnny Darters* (1876). Jordan erected a genus *Copelandia*, 1877 in his memory after his early and unexpected death, but this genus is no longer recognised as valid.

Copley

Copley's Wormfish *Gunnellichthys copleyi* JLB Smith, 1951

Hugh Copley (d.1959) was Fish Warden of Kenya, previously being Acting Game Warden in the 1940s and 1950s. He wrote nature guides such as: *Common Freshwater Fish of Kenya* (1958) and *Small Mammals of Kenya* (1950) and articles such as: *The Tilapias of Kenya Colony* (1952). He collected 'valuable ichthyological material', including the type of the eponymous wormfish.

Coppinger

Swallowtail Dart *Trachinotus coppingeri* A Günther, 1884

Dr Richard William Coppinger (1847–1910) was a naval surgeon and naturalist. Born in Dublin, he studied medicine at Queen's University and entered the medical department of the navy. He served as a naturalist on 'HMS Alert' (1878–1882) during a voyage of exploration that took the ship to Patagonia, Polynesia and the Mascarene Islands. He was Inspector-General, Hospitals and Fleets (1901–1904). An amphibian and two birds are also named after him.

Coquenan

Armoured Catfish sp. *Pseudancistrus coquenani* F Steindachner, 1915

This is a toponym referring to where the species occurs in the Kukenan River basin, Venezuela.

Coquette

Conger Eel sp. *Ariosoma coquettei* DG Smith & RH Kanazawa, 1977

Named after the US Bureau of Commercial Fisheries research vessel 'Coquette', from which the holotype was collected.

Cordemad

Peruvian cichlid sp. *Bujurquina cordemadi* SO Kullander, 1986

The name is based on an acronym, referring to the Corporación Departamental de Desarrollo de Madre de Dios, which "*...greatly facilitated the collecting around Puerto Maldonado in 1983 that led to the discovery of this species.*"

Corerefum

Colin's Pygmygoby *Trimma corerefum* R Winterbottom, 2016

Not an eponym but an arbitrary combination of letters (pronounced '*core-ref-um*') reflecting the Coral Reef Research Foundation in Palau. The etymology states: "*This organization, established and operated by Pat and Lori Colin, has not only spearheaded marine research in Palau through the efforts of its founders, but has also provided laboratory and research facilities while acting as a home away from home for innumerable scientists working on marine organisms of the Palauan I slands for almost a quarter of a century.*"

Cornel

Cornel's Drummer *Kyphosus cornelii* GP Whitley, 1944
[Alt. Western Buffalo Bream]

Jeronimus Cornelisz (1598–1629) was a Dutch apothecary and East India Company merchant. He led (1629) one of the bloodiest mutinies in history after the merchant ship 'Batavia' was wrecked in the coral islands off Australia. The etymology says: "*I name this fish after the 'villain' of the Batavia mutiny which, more than 300 years ago, occurred by the place where it was caught – Cornelius.*"

Cornelia

Cornelia's Rockskipper *Entomacrodus corneliae* HW Fowler, 1932

Cornelia Elizabeth Pinchot née Bryce (1881–1960) was a politician, political activist, and wife (1914) of Gifford Pinchot, conservationist and governor of Pennsylvania; so was the 'first lady of Pennsylvania'. She was a prohibitionist and supporter of the poorly paid and ran unsucesfully for congress (1928 & 1932).

Coriat

Corydoras sp. *Corydoras coriatae* WE Burgess, 1997

Mrs Nery Coriat, according to the etymology, is a "*...supplier of aquarium fishes from Peru who has worked for the past 25 years in the Peruvian fish business and has contributed a great deal to the industry.*"

Cornet

Stanley Pool Mormyrid *Stomatorhinus corneti* GA Boulenger, 1899

Dr Jules Cornet (1865–1929) was a physician, geologist, explorer and naturalist, known as the 'father of Geology in the Congo'. He studied medicine at Ghent University (1883), which awarded his doctorate in natural sciences (1890). He became a 'dissector' at the Institute of Zoology and Comparative Anatomy there. He was a member of the Bia-Franqui expedition (1891–1893) to Katanga where he prospected, discovered and mapped the major copper deposits of the area. He returned to the Congo (1895) to study the proposed route of a railway. He became Professor of Mineralogy, Geology and Palaeontology at the School of Mines, Mons, Belgium (1897–1929). A mineral from Katanga is named 'cornetite' in his honour.

Corrigan

Catshark sp. *Galeus corriganae* WT White, RR Mana & GJP Naylor, 2016

Dr Shannon Leah Corrigan (b.1982) is an Australian molecular geneticist. Macquarie University Sydney awarded her BSc (2004) and PhD (2010), and here she worked as a demonstrator (2004–2008) and laboratory manager (2006–2009) and associate lecturer (2009–2010). She was a molecular technician at the Australian National Fish Collection (2010–2011). Since then she has been a post-doctoral researcher at the South Australia Research & Development Institute (2011) and Hollings Marine Laboratory, College of Charleston (2011) under Professor Gavin Naylor. Among her published papers is: *Genetic and reproductive evidence for two species of ornate wobbegong shark on the Australian East Coast* (2008). She was honoured for her "*...extensive molecular population and phylogenetic work on sharks has contributed toward an improved understanding of sharks species relationships.*"

Cortes

Freckled Splitfin *Ilyodon cortesae* J Paulo-Maya & P Trujillo-Jiménez, 2000

Maria Teresa Cortés was a Mexican ichthyologist. The etymology says: "*This species is named after Maria Teresa Cortés, who played an important role in the developement of the Mexican Ichthyology.*"

Cortez

Bigfin Eelpout *Lycodes cortezianus* CH Gilbert, 1890
Cortez Rockfish *Sebastes cortezi* W Beebe & J Tee-Van, 1938
Cortez Ray *Beringraja cortezensis* JD McEachran & Miyake, 1988

The etymologies do not help, but we believe these species are named after the Sea of Cortez where the holotypes were caught, or in the case of the eelpout, Cortez Bank off San Diego, California.

Cortez, H

Delicate Swordtail *Xiphophorus cortezi* DE Rosen, 1960

Hernán Cortés de Monroy y Pizarro (1485–1547) was a Spanish conquistador whose expedition to Mexico (1519–1521) brought about the collapse of the Aztec empire.

Corti

Headstander Characin sp. *Schizodon corti* LP Schultz, 1944

Not an eponym: *corti* is the common name for this fish in the area of Lake Maracaibo, Venezuela.

Coskuncelebi

Spirlin sp. *Alburnoides coskuncelebii* D Turan, C Kaya, I Aksu, E Bayçelebi & Y Bektas, 2019

Dr Kamil Coskuncelebi is a Turkish biologist and botanist. He is a Professor (1992–present) at the Department of Biology at Karadeniz Technical University, Turkey. He is the author of around 70 papers. One of his areas of study is the effect of plant extracts on fish pathogens. The etymology reads thus: "*This species is named in the honour of Prof. Dr. Kamil Coskuncelebi who is a specialist for flowering plants of Turkey and a well-known Turkish plant taxonomist in Karadeniz Technical University, Department of Biology, Turkey.*"

Costa, L

Neotropical Rivuline sp. *Spectrolebias costai* KJ Lazara, 1991

Luis de Camargo Costa is an ornamental fish trader in Arunã, Goias State, Brazil. He discovered the rivuline with aquarist Rosario LaCorte (1982).

Costa, M

Costa's Crested Flounder *Samaris costae* J-C Quéro, DA Hensley & AL Maugé, 1989

Maria-José Costa is a marine biologist. She retired as Professor at the Faculty of Sciences, University of Lisbon and is currently a researcher at the Centre for Marine and Environmental Sciences. She was Vice-President of the National Fisheries Research Institute and President of the Animal Biology Department of the Faculty of Sciences of the University of Lisbon, Director of the Oceanography Center, Member of the Scientific Council of Marine Sciences and Environment of the Foundation for Science and the Technology and President of the Portuguese Society of Natural Sciences. She has over 150 published papers, book chapters and a book *Fish of Portugal* (2018) as well as articles in magazines and newspapers. She is the president of AMONET, the Women's Scientists Association and was one of 100 scientists honored by Living Science on Women's Day.

Costa, OG

Goldblotch Grouper *Epinephelus costae* F Steindachner, 1878
Characin sp. *Moenkhausia costae* F Steindachner, 1907

Dr Oronzio Gabriele Costa (1787–1867) was an Italian physician and zoologist. and it is most likely, but not proven, that these fish are named after him. He taught zoology at the University of Naples. He was president of the Accademia Pontaniana in Naples (1846). He wrote 126 papers, mainly on entomology and also wrote: *Fauna Vesuviana* (1827) and *Fauna del regno di Napoli* (1829) which Steindachner mentions in his description of the grouper.

Costa, PAS

Lanternbelly sp. *Verilus costai* WW Schwarzhans, M Mincarone & BT Villarins, 2020

Dr Paulo Alberto Silva da Costa is a marine biologist who is Associate Professor in the Department of Ecology and Marine Resources at UNIRIO, Brazil. He served as coordinator of the population dynamics and stock assessment area of the REVIZEE Program on the central coast (1996–2007). There he coordinated two commissions of the French Oceanographic Ship Thalassa in Brazil, surveying fishing resources on the Brazilian continental slope. He is the author of more than fifty scientific articles and book chapters, and was editor of two collections of works on fishing and fish ecology in deep sea regions of the Brazilian coast. He was honoured in recognition of his contribution to the knowledge of deep-sea fishes from Brazil.

Costa, WJEM

Pencil Catfish sp. *Listrura costai* L Villa-Verde, H Lazzarotto de Almieda & SMQ Lima, 2012
Characin sp. *Nematocharax costai* PH Bragança, MA Barbosa & JL Mattos, 2013
Corydoras sp. *Corydoras costai* FP Ottoni, MA Barbosa & AM Katz, 2016

Dr Wilson Jose Eduardo Moreira da Costa is a Brazilian biologist and ichthyologist. The Federal University of Rio de Janeiro awarded his bachelor's degree (1983) and the University of São Paulo his doctorate in Zoology (1989). He is presently a Full Professor of the Department of Zoology of the Biology Institute of the Federal University of Rio de Janeiro. He worked for six months in the fish collections of the Royal Museum for Central Africa, Tervuren, Belgium (2016). He created (1990) the Laboratory of General and Applied Ichthyology as a unit of the Biology Institute of the Federal University of Rio de Janeiro. By reuniting (1993) the old fish collections of the Department of Zoology, Federal University of Rio de Janeiro, also creating the Special Collection of Fish Teleosteos, he achieved the formation of the largest collections of the order Cyprinodontiformes. He wrote: *Inferring Evolution of Habitat Usage and Body Size in Endangered, Seasonal Cynopoeciline Killifishes from the South American Atlantic Forest through an Integrative Approach (Cyprinodontiformes: Rivulidae)* (2016). (Also see **Bruno (Bove)** & **Claudia (Bove)**)

Cotinho

Characin sp. *Moenkhausia cotinho* CH Eigenmann, 1908

Major J M S Cotinho of the Brazilian army was described as the 'Brazilian attaché' of the Thayer Expedition (1865–1866). The holotype was collected during this expedition.

Cotronei

Garden Eel sp. *Gorgasia cotroneii* U D'Ancona, 1928

Giulio Cotronei (1885–1962) was an Italian zoologist who graduated from the University of Naples and became Director at the Institute of Comparative Anatomy, Università di Roma. D'Ancona, who described this species, was also based there. Except for service as a cavalry officer in WW1 (1915–1918) he spent his life in scientific research and study.

Couard

Whitefin Hammerhead *Sphyrna couardi* J Cadenat, 1951

Monsieur Couard was Director of shark fisheries off the coast of Senegal, the type locality. Unfortunately we can find nothing more about him.

Couch, DN

Monterrey Platyfish *Xiphophorus couchianus* CF Girard, 1859

General Darius Nash Couch (1822–1897) was a soldier and administrator. He was also an explorer, taking leave of absence to lead a zoological expedition in Mexico. Geisler wrote about a garter snake that Couch collected in 1853, "... *named in honor of its indefatigable discoverer, Lt. D. N. Couch, who, at his own risk and cost, undertook a journey into northern Mexico, when the country was swarming with bands of marauders, and made large collections in all branches of zoology.*" He was commissioned into the US Army (1846) and sent to Mexico, where he fought at the Battle of Buena Vista (1847). He returned to Washington DC (1854) and in the next year he resigned his army commission and became a merchant and manufacturer in New York and Massachusetts (1855–1861). At the outbreak of the American Civil War he rejoined the army as a colonel, then became brigadier-general of volunteers. He offered to resign on the grounds of ill health (1863) but was persuaded to stay by being promoted to major general! He was posted to Pennsylvania and put in charge of all the ceremonies associated with the consecration of the National Cemetery at Gettysburg (1865), the occasion of Abraham Lincoln's Gettysburg Address. After the Civil War he appears to have again resigned from the army and was Collector of the Port of Boston (1866–1867), President of a Virginia mining and manufacturing concern (1867–1877), and an administrator in the state of Connecticut (1877–1884). Two birds, two reptiles and an amphibian are also named after him.

Couch, J

Couch's Goby *Gobius couchi* PJ Miller & MY El-Tawil, 1974

Dr Jonathan Couch (1789–1870) was an English naturalist. After schooling and some years pupillage with two local physicians he attended Guy's and St Thomas's Hospitals to finish his medical traing. He returned (c.1809) to his home village of Polperro for the rest of his life, practicing medicine there for 60 years. He was fascinated by the natural world and trained some local fisherman to make observations and take specimens for him. He was described (1836) by Yarrell (to whose *Brirtish Fishes* he contributed greatly) as 'that indefatigable ichthyologist of Cornwall'. He wrote: *A History of the Fishes of the British Islands* (1862–1867) as well as numberous other books and papers on fish and a variety of other topics from local history, the Cornish language, such as *Glossary of Words in Use in Cornwall* (1880), and customs and folklore, including *The Folklore of a Cornish Village* (1855 & 1857).

Coues

Lake Chub genus *Couesius* DS Jordan, 1878

Triplewart Seadevil *Cryptopsaras couesii* TN Gill, 1883

Dr Elliott Ladd Coues (pronounced 'cows') (1842–1899) was a US Army Surgeon who graduated from Columbia University (1861) then being appointed Assistant-Surgeon in the army (1864). He was one of the founders of the American Ornithologists' Union and wrote: *Key to North American Birds* (1872). He was appointed surgeon and naturalist to the US Northern Boundary Commission (1876). He lectured in anatomy at Columbia's medical school (1877–1882) during which (1881) he resigned his commission to devote himself to science, becoming professor there (1882–1887). Amongst other works he wrote: *Handbook of Field and General Ornithology* (1890), in which he set out in meticulous detail how to 'collect' and preserve birds, *A Checklist of North American Birds* (1873) and the 5-volume *Key to North American Birds* (1872–1903). He wrote many papers and books on birds, mammals and other taxa and edited the journals of the Lewis and Clark expedition (1893). Three mammals and five birds are also named after him. He died following surgery to his throat.

Coulon

Toothed Goby *Deltentosteus collonianus* A Risso, 1820

Paul-Louis-Auguste Coulon (1777–1855) was a businessman and civic leader who was interested in science and nature and helped found the local museum and endowed the local library. The etymology is not explained by Risso but, according to Valenciennes (1837), was in honour of Monsieur Coulon of Neufchâtel, Switzerland, "*an enlightened amateur of natural history*" (translation) who we belive to be the above. (NB Risso provided a different spelling, *colonianus*, in 1827)

Coulter, GW

Cunene Hap (cichlid) *Sargochromis coulteri* G Bell-Cross, 1975

Dr George W Coulter was formerly the Senior Fisheries Research Officer, Department of Wildlife, Fisheries and National Parks of the Government of Zambia and later Deputy

Director, Forum Fisheries Agency. Around the time of the cichlid's description, he was writing such articles as: *The biology of Lates species (Nile perch) in Lake Tanganyika and the status of the pelagic fishery for Lates species and Luciolates stappersi* (1976).

Coulter, JM

Pygmy Whitefish *Prosopium coulterii* CH Eigenmann & RS Eigenmann, 1892

Dr John Merle Coulter (1851–1928) was an American botanist. Hanover College, Indiana awarded his bachelor's degree (1870), his master's (1873) and his doctorate (1883) and there he was Professor of Natural Sciences (1871–1879). He was Professor of Botany, Wabash College (1879–1891), during which period he received an honorary doctorate (1884) from Indiana University. He was President and Professor of Botany, Indiana University (1891–1893), President of Lake Forest College (1893–1896), Professor and Head, Department of Botany, University of Chicago (1896–1925) and Dean. Boyce Thompson Institute for Plant Research, Yonkers, New York (1925–1928). He was botanist to the US Geological Survey in the Rocky Mountains (1872–1875) which led to his writing: *Manual of the Botany of the Rocky Mountain Region* (1885), where this species occurs. He and his family survived (1909) the loss of the White Star liner 'Republic', that sank after a collision with 'Florida' - the first time that Marconi radio was used for a distress call at sea, resulting the loss of six lives but the saving of some 1,500.

Courtenay

Socorran Soapfish *Rypticus courtenayi* LV McCarthy, 1979

Dr Walter Rowe Courtenay, Jr (1933–2014) was an American ichthyologist and oceanographer. Vanderbilt University awarded his BA (1956) and the University of Miami his MSc (1961) and PhD (1965). He worked at Duke University (1963–1965) and Boston University (1965–1967) before becoming a professor in the Department of Biological Sciences at Florida Atlantic University until retirement (1967–1999). Among his many published papers are: *Atlantic fishes of the genus* Haemulon *(Pomadasyidae): systematic status and juvenile pigmentation* (1961) and *Atlantic fishes of the genus* Rypticus *(Grammistidae): systematics and osteology* (1967).

Courtet

Squeaker Catfish sp. *Synodontis courteti* J Pellegrin, 1906

Henri Courtet (1858–1912) was a French soldier who served mainly in French colonial possessions. He joined the French army (1874) and served in New Caledonia (1882 & 1888), Tahiti (1882–1885), Cochinchina (Vietnam) (1889–1890), Dahomey (1891–1893), Benin (1894–1896), Madagascar (1897–1898), Senegal (1899–1900) and Chari (Chad) (1900–1902). He was later in France as Administration Officer to the Chief Engineer in Cherbourg (1904) and then posted to Paris. He was described as a being from the Colonial Artillery when a member of French 1902–1903 mission 'Mission Chari-Lac Chad' to study the region between Ubangi River and Lake Chad, during which the holotype was collected. He wrote: *Observations geologiques recuellies par la mission Chari-lac Tchad* (1905).

Cousins

Eelpout sp. *Pachycara cousinsi* PR Møller & N King, 2007

Michael Cousins is the partner of the second author, Nicola King.

Cousseau

Cousseau's Skate *Bathyraja cousseauae* JM Díaz de Astarloa & E Mabragaña, 2004 [Alt. Joined-fins Skate]

Dr Maria Berta Cousseau is an Argentine ichthyologist who is Professor Emeritus at the Department of Sciences, Universidad Nacional de Mar del Plata, Argentina. She wrote: *Peces marinos de Argentina: Biologia, distribucion, pesca* (2000) and co-wrote *Ictiología. La vida de los peces sudamericanos* (2010). She was honoured for her contributions to knowledge of the marine fishes of Argentina.

Cowley

Cowley's Torpedo Ray *Tetronarce cowleyi* DA Ebert, DL Hass & MR de Carvalho, 2015

Dr Paul D Cowley is principle scientist at the South African Institute for Aquatic Biodiversity (since 2000) and honorary professor and Research Associate at the Department of Ichthyology and Fisheries Science at Rhodes University. He has published around 100 papers such as the co-authored: *Potential effects of artificial light associated with anthropogenic infrastructure on the abundance and foraging behaviour of estuary associated fishes* (2013). The ray was named "*...in recognition of his contributions to the study of fishes in southern Africa.*"

Cox, H

Danio sp. *Devario coxi* SO Kullander, MM Rahman, M Noren & AR Mollah, 2017

Captain Hiram Cox (1760–1799) was employed by the Honourable East India Company in Superintendent of Palongkee on the border of Bengal (Bangladesh) and Burma (Myanmar) and the town of Cox's Bazaar is named after him. He was a cultured man, a member of the Asiatic Society and famous for the Cox-Forbes theory that chess originated as a game for four players. The etymology is rather strange, in having three ways in which it can be understood: "*The specific name is a noun in the genitive case, referring to the type locality area (Cox's Bazar), but could also be understood as referring to the gene fragment used to identify the species (cytochrome c oxidase subunit 1, often shortened to COXI); or to Hiram Cox (1760–1799), after whom Cox's Bazar was named.*"

Cox, JC

Cox's Gudgeon *Gobiomorphus coxii* G Krefft, 1864

Krefft does not explain the eponym, but it probably honours James Charles Cox (1834–1912), who also collected a snake that Krefft described in a companion paper. James Cox was an Australian physician and naturalist – in particular, a conchologist. He was first president of the New South Wales Board of Fisheries, and a Trustee of Sydney Museum.

Coxey

Characin sp. *Brycon coxeyi* HW Fowler, 1943

W Judson Coxey of the Academy of Natural Sciences of Philadelphia was an entomologist. He made at least three expeditions to Ecuador and obtained the holotype.

Coxiponé

Armoured Catfish sp. *Curculionichthys coxipone* FF Roxo, GSC Silva, LE Ocha & C de Oliveira, 2015

The Coxiponé indigenous people inhabit the margins of Rio Cuiabá, Mato Grosso, Brazil, where this catfish occurs.

Craddock

Waryfish sp. *Scopelosaurus craddocki* E Bertelsen, G Krefft & NB Marshall, 1976

Dr James Edward 'Jim' Craddock (1937–2009) was a naturalist and ichthyologist who was Oceanographer Emeritus when he retired. The University of Louisville, Kentucky, awarded his bachelor's degree (1958) and his doctorate (1965). He first joined the Woods Hole Oceanographic Institution, Massachusetts (1964) as a post-doctoral fellow. He was an associate at the Museum of Comparative Zoology, Harvard during the 1980s and 1990s. He wrote: *Food habits of Atlantic White-sided Dolphins (*Lagenorhyncus acutus*) off the Coast of New England* (2009, published posthumously). He was honoured "*for his many contributions to our knowledge of deep-sea fishes.*" He died from complications of pneumonia. (Also see **Jim Craddock**)

Cragin

Arkansas Darter *Etheostoma cragini* CH Gilbert, 1885

Professor Francis Whittemore Cragin (1858–1937) was a herpetologist, palaeontologist, librarian and curator. He graduated from Harvard (1882) then travelled across the USA collecting natural history specimens. He established (1883) the Washburn Biological Survey of Kansas. He was librarian (1884–1887) and curator at the Kansas Academy of Science. The etymology states: "*The types of this species were obtained by Prof. F. W. Cragin...*"

Craig

Five-bar Grouper *Epinephelus craigi* BW Frable, SJ Tucker & HJ Walker, 2018

Dr Matthew Thomas Craig (b.1976) is a fisheries scientist at National Oceanic and Atmospheric Administration. Occidental College, Los Angeles awarded his Biology BA (1998) and his MA (2000), while the University of California awarded his PhD (2005). He was a post-doctoral researcher at Hawaii Institute of Marine Biology at the University of Hawaii at Mānoa (2005–2009) when he was appointed as Associate Professor in the Department of Marine Science at the University of Puerto Rico (2009–211). He was then Adjunct Assistant Professor & Research Associate in the Department of Marine Science and Environmental Studies at the University of San Diego (2012–2014) and in their Department Environmental and Ocean Science (2012–2015), followed by a post as Research geneticist at NOAA (2015–present). He was co-author of *Groupers of the World* (2011). Dr. Craig's research broadly focuses on the systematics, conservation, and management of marine fishes.

Cramer

Darkblotched Rockfish *Sebastes crameri* DS Jordan, 1897
Oregon Chub *Oregonichthys crameri* JO Snyder, 1908

Frank Cramer (1861–1948) was an ichthyologist and biologist at Stanford University. Lawrence College, Wisconsin awarded his bachelor's degree (1886). He taught biology at Olivet College, Michigan (1886–1889), then moved to Palo Alto, California to study at Stanford, which awarded his master's degree (1893). David Starr Jordan was a great influence on Cramer, whom he persuaded to open a Preparatory School for Boys (1891); this school was renamed Manzanita Hall (1892). He wrote: *On the cranial characters of the genus Sebastodes (rock-fish)* (1895). He helped to collect the holotype of the eponymous chub.

Crampton

Crampton's Tube-snouted Ghost Knifefish *Sternarchorhynchus cramptoni* CDCM de Santana & RP Vari, 2010
Toothless Characin sp. *Cyphocharax cramptoni GC Bortolo & FCT de Lima, 2020*

Dr William Gareth Richard 'Will' Crampton (b.1969) is a British ichthyologist. He was awarded his BA in Zoology (1991) and PhD (1996) by the University of Oxford. His doctoral thesis was entitled *The Electric Fishes of the Upper Amazon: Ecology and Signal Diversity*. He then served as a staff researcher and Director of Aquatic Biodiversity Research at the Instituto Mamirauá / Ministério de Ciência e Tecnologia, Brazil (1997–2001). This was followed by post-doctoral fellowships at the University of Florida, Florida Museum of Natural History (2001–2005) and the University of Toronto, Department of Biological Sciences (2004–2005). He was first employed by the University of Central Florida (2006), where he is now (2019) an Associate Professor in the Department of Biology. His research program focuses on the sensory ecology and taxonomy of Neotropical electric fishes (order Gymnotiformes) and the biodiversity of Amazonian fish. During his career, he has undertaken field research across tropical and sub-tropical South and Central America, to date resulting in over 90 authored or co-authored peer-reviewed journal articles and book chapters. He has described or co-described over 50 taxa of gymnotiform electric fish. (Also see **Benjamin (Crampton)**)

Cranbrook

Hillstream Loach sp. *Gastromyzon cranbrooki* HH Tan & ZH Sulaiman, 2006

Dr Gathorne Gathorne-Hardy, 5th Earl of Cranbrook, (b.1933) is a mammalogist, ornithologist, zooarchaeologist and environmental biologist. His bachelor's degree (1956) and master's (1960) were both awarded by Cambridge and his doctorate by Birmingham University (1960). He worked at the Sarawak Museum, Kuching (1956–1958) as a collector and to study cave swiftlets. He was in Indonesia (1960–1961) and on the staff of the Zoology Department, University of Malaya, Malaysia (1961–70), where he was instrumental is setting up the University's Field Studies Centre at Ulu Gombak, Selangor. He returned to England (1970) and succeeded to the earldom (1978) and was active in promoting environmental issues in Parliament. He wrote: *Mammals of Borneo* (1965) and co-wrote: *Belalong: A tropical rainforest* (1994). He was honoured 'for his contributions to the study of biodiversity in Southeast Asia.'

Cranch

Kokuni Catfish *Chrysichthys cranchii* WE Leach, 1818

John Cranch (1785–1816) was a British explorer of tropical Africa and also an accomplished natural historian. He took part in an expedition (1816) led by Captain J. K. Tuckey to discover the source of the Congo River. A bird is also named after him.

Crandell

Characin sp. *Characidium crandellii* F Steindachner, 1915

Roderic Crandall (b.1885), an American geologist who worked for the Brazilian Geological Survey, is probably the intended person, despite Steindachner using the spelling 'Crandell'. He lived in Boa Vista, capital of the state of Roraima, near the Rio Branco, which is one of the localities where this species can be found, so he seems a likely candidate.

Crane

Flabby Whalefish sp. *Cetomimus craneae* RR Harry, 1952

Dr Jocelyn Crane (1909–1998) was a carcinologist at Department of Tropical Research, New York Zoological Society (now called the Wildlife Conservation Society). Smith College, Northampton, Massachusetts awarded her bachelor's degree (1930) and an honorary master's degree (1947). She worked for the New York Zoological Society (1930–1971), rising from being a laboratory assistant (1930) to Director (1962–1966). She wrote: *Notes on the Biology and Ecology of Giant Tuna,* Thunnus thynnus *Linnaeus, observed at Portland, Maine* (1936). After retiring (1971) she studied Art History and was awarded a doctorate by the Institute of Fine Arts, New York University (1991).

Cranwell

New Zealand Sand-diver *Tewara cranwellae* LT Griffin, 1933

Lucy May Cranwell (1907–2000) was a New Zealand botanist. The University of Auckland awarded her BA (1928) and masters in botany (1929). She was recruited aged just 21 as Curator of Botany at Auckland Museum just a few weeks after graduating. Louis Griffin was Assistant Director. She studied plants in many of North Island New Zealand's forests and islands, collecting over 4000 plants for the museum's herbarium in her fourteen years there. She is well remembered for having prepared a booklet for servicemen during WW2 *Food is where you find it: A Guide to Emergency Foods in the Western Pacific* (1943). That year she married a US airforce officer and moved to the USA (1944) where, after working at Harvard, she became a Research Affiliate in palynology at the University of Arizona. "*The type was brought to me by Miss Lucy Cranwell, after whom I have specifically named this fish.*" The type was taken at Smugglers' Bay, Whangarei Heads, (1931).

Crawford

Titicaca Pupfish sp. *Orestias crawfordi* V Tchernavin, 1944

George Ivor Crawford CBE (1910–2011) was a British malacologist and held various offices in the Malacological Society of London and Unitas Malacologica. After graduating from Trinity College, Cambridge and working at the Marine Biological Laborarory, Plymouth, he worked as Curator of Molluscs at the BMNH (1935–1946). He was Deputy Leader of the Percy Sladen Trust Titicaca Expedition (1937) and in charge of all its collections. He provisionally sorted around 20,000 specimens and deposited them with the BMNH. He was also Assistant Secretary, Department of Education & Science (1971). He died peacefully in his sleep aged 100. A number of Amphipoda are named after him.

Creaser

Roughcheek Sculpin *Ruscarius creaseri* CL Hubbs, 1926

Dr Charles William Creaser (1897–1965) was an ichthyologist who was at the Museum of Zoology, University of Michigan (1923) and where he was awarded his Doctorate (1926). He was a Professor at Wayne University, Michigan (1940). He wrote: *The skate, Raja erinacea Mitchill, a laboratory manual* (1927).

Creed

Sandburrower genus *Creedia* JD Ogilby, 1898

John Mildred Creed (1842–1930), a friend of Ogilby's, was an English-born Australian physician and politician. University College London awarded his MRCS (1866) and Edinburgh University his LRCP (1866). He went to Australia as a surgeon aboard 'Anglesey' and went on a Northern territory expedition as Medical Officer (1867). He was in medical practice (1868–1882) apart from a brief stint in the legislative Assembly (1872–1874), before being renominated and serving the rest of his life (1885–1930). He wrote: *My Recollections of Australia and Elsewhere, 1842-1914* (1916). Ogilby, in naming the fish after him says of him: "*…to whose unfailing kindness and support my present position in science is mostly due.*"

Crimmen

Corydoras sp. *Corydoras crimmeni* S Grant, 1997

Oliver Crimmen (b.1954) joined the staff of the Fish Section of the BMNH (1973) as an Assistant Scientific Officer and is now Senior Curator, Fish, Life Sciences Department, Vertebrates Division. The University of London awarded his bachelor's degree (1992). Among his experiences in the preservation of fishes, he collaborated with the artist, Damien HIrst, in preserving a tiger shark as 'an artwork'. He co-wrote: *Two erroneous, commonly cited examples of 'swordfish' piercing wooden ships* (1996). (Also see **Ollie**)

Crinog

Tetra sp. *Pristella crinogi* FCT Lima, RA Caires, CC Conde-Saldaña, JM Mirande & FR Carvalho, 2021

Cristiano de Campos Nogueira ('Crinog') is a Brazilian herpetologist who is a post doctoral researcher at the Universidade de São Paulo, which awarded his BSc (1998), his MSc (2001) and PhD (2006). He had previously been awarded a degree in International Relations by the University of Brasilia. He researched 'Biogeography, biodiversity and conservation of cisandean Squamate Reptiles' at the Sao Paulo Research Foundation (2016–2020). Among his published material are: *New records of squamate reptiles in the central Brazilian cerrado: Emas National Park region* (2001) and *Contrasting patterns of phylogenetic turnover in amphibians and reptiles are driven by environment and geography in Neotropical savannas* (2021). He was thanked for his "...enthusiastic help during the fish survey at the Estação Ecológica Serra Geral do Tocantins in 2008, when the new species was discovered."

Cristian

Characin sp. *Bryconamericus cristiani* C Román-Valencia, 1999

Cristian Román-Valencia is one of the author's twin sons. He showed an early knowledge and enthusiasm for ichthyology. He is now studying biology at Universdad del Valle, Colombia. He co-wrote: *Trophic and Reproductive Ecology of a Neotropical Characid Fish* Hemibrycon brevispini *(Teleostei: Characiformes) (*2014).

Cristina

Thorny Catfish sp. *Nemadoras cristinae* MH Sabaj Pérez, H Arce, LM Sousa & JLO Birindelli, 2014

Mrs Maria Cristina Sabaj Pérez, is a teacher at Friends' Central School, Wynnewood, Pennsylvania. She is also the senior author's wife who thanked her not only for her contributions to the collection of the type series but also for looking to his well-being.

Crocker

Mote Sculpin *Normanichthys crockeri* HW Clark, 1937
Browncheek Blenny *Acanthemblemaria crockeri* W Beebe & J Tee-Van, 1938
California Flashlightfish *Protomyctophum crockeri* RL Bolin, 1939

Charles Templeton Crocker (1885–1948) was a member of a California family that made its money from railways, having invested in the first transcontinental American railroad. He was more interested in exploring the South Pacific than in the South Pacific Railroad Company and had a beautiful yacht built, the 'Zaca', that he used as a floating base for a number of 'Templeton Crocker expeditions'. These started (1930) with a voyage to Fiji, then (1932) to the Galapagos Islands and (1933) to the Solomon Islands. He wrote: *The Cruise of the Zaca* (1933), and made a film called: *People and Dances of Oceania*. A reptile is named after him.

Croil

Goby genus *Croilia* JLB Smith, 1955

Dr John Frederick Croil Morgans (See **Morgans**). According to the etymology, he "*... observed these agile small creatures while diving, and it was only with great difficulty that he caught them, for they live in burrows in the bottom, to which they retire when startled.*"

Cromer

Shellear genus *Cromeria* GA Boulenger, 1901

Lord Cromer Evelyn Baring (1841–1917), 1st Earl of Cromer, was a soldier (1858–1877) and a particularly oppressive and racist colonial administrator. He was British Controller-General in Egypt (1879–1883) and Consul-General of Egypt (1883–1907) setting back education, and the country's financial independence. At home, he opposed women's suffrage. He wrote: *Modern Egypt* (1908).

Crook

Butterfly Ray sp. *Gymnura crooki* HW Fowler, 1934

Alfred Herbert Crook (b.1873) was a biologist and Headmaster of Queen's College, Hong Kong (1925–1930), who also lectured on biology at the Hong Kong College of Medicine. He was amateur naturalist who contributed articles to 'The Hong Kong Naturalist' of which he was co-editor. He wrote: *The Flowering Plants of Hong Kong: Ranunculaceae to Meliaceae: with Thirty Diagrams, Comprising Drawings of Parts of Over 100 Different Species* (1959).

Crosnier

Spoon-nose Eel *Mystriophis crosnieri* J Blache, 1971
Long-tailed Groppo *Grammatonotus crosnieri* P Fourmanoir, 1981
Madagascar Skate *Dipturus crosnieri* B Séret, 1989
Cusk-eel sp. *Neobythites crosnieri* H Nielsen, 1995

Dr Alain Georges Paul Crosnier (1930-2021) was a French biologist and carcinologist at the MNHN, Paris, who published much on crustaceans. He initiated the deep trawling surveys off Madagascar (1970s).

Cross

Cross' Damsel *Neoglyphidodon crossi* GR Allen, 1991

Norbert J Cross is an Honorary Associate of the Western Australian Museum (Perth), where Allen works, and with whom he has co-authored papers such as: *Description of Five New Rainbowfishes (Melanotaeniidae) From New Guinea* (1980) and the book: *Rainbowfishes of Australia and Papua New Guinea* (1982).

Crozier

Indian Pale-edged Stingray *Telatrygon crozieri* E Blyth, 1860

Dr William Crozier (1816–1862) was an anatomy and physiology professor. He qualified MRCS at Barts (1839) and AS (1842). He won a medal during the first Sikh or Sutlej war (1845–1846) while serving in the Indian Medical Corps. He was aboard the 'SS Simla' in the Red Sea when he died. The patronym is not identified by Blyth but is probably in honour of him as he was his colleague at (and finance chair of) the Asiatic Society of Bengal.

Cruls

Armoured Catfish sp. *Hypostomus crulsi* YFF Soares, P de PU Aquino, JC Bagley, F Langeani & GR Colli, 2021

Luiz Ferdinando Cruls (1848–1908) was a Belgian (naturalized Brazilian) engineer and astronomer who served as director of the Imperial Observatory of Rio de Janeiro. He led the Central Plateau Exploration Commission of Brazil, which was responsible for demarcating an area for the installation of Brasília, the future capital of Brazil.

Cruxent

Characin genus *Cruxentina* A Fernández-Yépez, 1948
[Now usually included in *Steindachnerina*]
Three-barbeled Catfish sp. *Pimelodella cruxenti* A Fernández-Yépez, 1950

José Maria Cruxent (1911–2005) was a Spanish-born Venezuelan anthropologist and archaeologist. He left Spain for Venezuela after the end of the Spanish Civil War. He was Director, Museo de Ciencias de Caracas (1949–1952), at which museum his fish collection from his Orinoco expedition (1951) is housed. He wrote: *Venezuelan Archaeology* (1963). He collected the holotype of *Pimelodella cruxenti.*

Cruz, CAG da

Neotropical Rivuline sp. *Notholebias cruzi* WJEM Costa, 1988

Dr Carlos Alberto Gonçalves da Cruz (b.1944) is a Brazilian herpetologist who was at the Universidade Federal Rural do Rio de Janeiro (1992), and is now a Research Associate at the Museu Nacional, Rio de Janeiro. He co-wrote: *Phyllomedusa: posição taxonômica, hábitos e biologia (Amphibia, Anura, Hylidae)* (2002). Four frogs are named after him.

Cruz, J

Dwarf Cichlid sp. *Apistogramma cruzi* SO Kullander, 1986

José Cruz Rodriguez was, as the etymology states: "... *motorista and skilled volunteer co-collector on visits both to Pebas in 1981 and Mazán in 1984*", who was honoured "...*in recognition of his tireless help on those trips which added considerably to the inventory of the Peruvian ichthyofauna*."

CSIRO

Loweye Snailfish *Paraliparis csiroi* DL Stein, NV Chernova & AP Andriashev, 2001

CSIRO stands for Commonwealth Scientific and Industrial Research Organisation.

Cuffy

Armoured Catfish sp. *Loricaria cuffyi* A Londoño-Burbano, A Urbano-Bonilla & MR Thomas, 2020

Cuffy (also spelled Coffy, Kofi and Koffi) was a West African man taken into slavery to work in the Dutch colony of Berbice (in present-day Guyana). He led a revolt (1763) of more than 2500 slaves and declared himself Governor of Berbice, but the insurrection was defeated, partly due to divisions within the rebel forces. Today, Cuffy is considered Guyana's first national hero.

Culebra

Panamanian Worm Blenny *Stathmonotus culebrai* A Seale, 1940

This is a toponym, referring to the type locality, Port Culebra, Costa Rica.

Cuming

Atoyac Chub *Notropis cumingii* A Günther, 1868
Cuming's Barb *Pethia cumingii* A Günther, 1868

Hugh Cuming (1791–1865) was an English naturalist and conchologist, once described as the 'Prince of Collectors'. At thirteen he was apprenticed to a sail maker, meeting seafarers that sparked his imagination. He shipped out to Valparaiso, Chile (1819). There he collected (1822–1826) and shipped specimens to England, establishing a successful business and using the money he saved to buy a ship, 'The Discoverer', specifically built for collecting and storing natural history specimens (1826). He collected (mainly shells and plants) in the Neotropics mostly along the Pacific coast from Chile to Mexico (1828–1830), Polynesia (1827–1828) and the East Indies (1836–1840). There he often recruited schoolchildren to collect along the shoires and forest and he amassed the greated collection made by an individual: 130,000 plants, 30,000 shells and large nmbers of birds, reptiles, mammals and insects as well as numerous living orchids. Much of his material was sent to the BMNH and living plants to English botanical gardens. He preceded Darwin in having collected in the Galápagos (1829). His shell collection is housed in the Linnean Library in London. A mammal, three birds and seven reptiles are named after him, as are at least three orchids.

Cummings

Dusky Shiner *Notropis cummingsae* GS Myers, 1925

Mrs J H Cummings (b.1885) and her husband (b.1863) are referred to as amateur naturalists but in practice they were professionals. They had acquired (1921) the business in Wilmington, North Carolina, of the local aquarist and wildlife dealer, Mr T P Lovering, and were operating very profitably from their houseboat at Smith's Creek, Wilmington, selling flora and fauna to all parts of the USA. They had George S Myers (q.v.) as a guest on their houseboat and supplied him with such a large number of herpetological specimens that he had no need to go collecting for himself in that part of the state. They seem to have collaborated with Myers over a long period, as his papers include correspondence from them (1925–1938). They also supplied a number of specimens of bats to the Smithsonian. Myers wrote in his etymology: "*...The species is named for Mrs. J. H. Cummings, in recognition of her investigation of the Wilmington fauna and flora... ...to whose kind help the success of my trip is entirely due*". We have been unable to find anything more about them.

Cunningham

Cunningham's Triplefin *Helcogrammoides cunninghami* FA Smitt, 1898

Robert Oliver Cunningham (1841–1918) was a Scottish naturalist. He was appointed (1866) Professor of Natural History in the Royal Agricultural College, Cirencester, but resigned after six months as he had been appointed by the Admiralty to collect plants as Naturalist on board 'HMS Nassau'. This ship's voyage was to survey the Straits of Magellan and the west coast of Patagonia. Cunningham collected this triplefin species during the voyage (1868) but could not identify it.

Cunnington

Lake Tanganyika Cichlid genus *Cunningtonia* GA Boulenger, 1906
Lake Tanganyika Catfish sp. *Dinotopterus cunningtoni* GA Boulenger, 1906
Lake Tanganyika Cichlid sp. *Lepidiolamprologus cunningtoni* GA Boulenger, 1906
Spiny Eel sp. *Mastacembelus cunningtoni* GA Boulenger, 1906

Dr William Alfred Cunnington (1877–1958) was a British zoologist, naturalist and anthropologist who was employed as a Demonstrator in Zoology, Cambridge. He led the third Tanganyika (Tanzania) expedition (1904–1905) on behalf of the ZSL, during which the holotypes of these species were collected. He also visited and collected in Brazil and Paraguay (1926–1927). He mainly published on crustacea.

Cupid

Cupid Cichlid *Biotodoma cupido* JJ Heckel, 1840
[Alt. Green-streaked Eartheater]

The etymology is not explained, but perhaps named for Cupid, the Roman god of love and desire; or generically meaning 'desire' or 'longing', perhaps alluding to the desirable nature of its attractive appearance in life.

Curasub

Yellow-spotted Sand Goby *Coryphopterus curasub* CC Baldwin & DR Robertson, 2015

Not an eponym but named for the manned submersible 'Curasub', which collected the type. It is owned and operated by Substation Curaçao and was honoured "*...for its contributions to increasing our knowledge of the Caribbean deep-reef fish fauna.*"

Curran

Pacific Flagfin Mojarra *Eucinostomus currani* B Zahuranec, 1980

Howard Wesley Curran is a biologist whose PhD dissertation: *A systematic revision of the gerrid fishes referred to the genus* Eucinostomus, *with a discussion of their distribution and speciation* (1942) concerned the same genus of fish. The mojarra was originally named in a MSc thesis of 1967; *The gerreid fishes of the genus Eucinostomus in the Eastern Pacific* (Zahuranec, B. *MSc Thesis*, La Jolla, CA: Univ. of California). However, it was not mentioned in a recognised publication until 1980, and this has to be taken as the date of the citation. That publication was *Taxonomía, ecología y estructura de las comunidades de peces en lagunas costeras con bocas efímeras del Pacífico de México*. Publicaciones Especiales, Centro de Ciencas del Mar y Limnología, Universidad Nacional Autónoma de México. No. 2: 1–306, which made no mention of the person honoured in the etymology!

Currie

Labeo sp. *Labeo curriei* HW Fowler, 1919

Rolla Patteson Currie (1875–1960) was an entomologist and priest. He graduated from the University of North Dakota (1893) and joined the staff of the Smithsonian for whom he took part in many field trips including to British Columbia and Liberia (1897). He joined the Bureau of Entomology, Department of Agriculture and worked there as an entomologist (1901–1944), specialising in odonata. He enrolled as a student at the Virginia Theological Seminary, Alexandria, Virginia (1944) and was ordained a priest (1946). He collected the holotype.

Curupira

Eartheater Cichlid sp. *Satanoperca curupira* RR Ota, SO Kullander, GC Depra, WJD da Graça & CS Pavanelli, 2018

Curupira refers to a mythological creature of Brazilian folklore that protects the forest and its inhabitants, punishing those who hunt for pleasure or who kill breeding females or defenseless juveniles; the legend reveals the relationship between indigenous people and the forest, showing respect for life.

Cutter

Cutter's Cichlid *Cryptoheros cutteri* HW Fowler, 1932

Victor Macomber Cutter (1881–1952) was President of the United Fruit Company. He was honoured for supporting members of the Academy of Natural Sciences in Central America, or, as the etymology puts it, for his "*...deep interest in neotropical zoology, and his cordial assistance to the Academy [of Natural Sciences of Philadelphia] on numerous occasions.*"

Cuvier

Smalleye Squaretail *Tetragonurus cuvieri* A Risso, 1810
Tiger Shark *Galeocerdo cuvier* F Péron & CA Lesueur, 1822
Pearl Wrasse *Anampses cuvier* JRC Quoy & JP Gaimard, 1824
Pike-Characin sp. *Boulengerella cuvieri* JB von Spix & L Agassiz, 1829
Cuvier's Clown Wrasse *Coris cuvieri* ET Bennett, 1831
Longsnout Stinger *Inimicus cuvieri* JE Gray, 1835
Bartail Jawfish *Opistognathus cuvierii* A Valenciennes, 1836
Japanese Crucian Carp *Carassius cuvieri* CJ Temminck & H Schlegel, 1846
Titicaca Pupfish sp. *Orestias cuvieri* A Valenciennes, 1846
Noodlefish *Salanx cuvieri* A Valenciennes, 1850
Dorado Characin sp. *Salminus cuvieri* A Valenciennes, 1850
[Junior Synonym of *Salminus brasiliensis*]
Kafue Pike Characin *Hepsetus cuvieri* F de L de Castelnau, 1861
Lesser Tigertooth Croaker *Otolithes cuvieri* E Trewavas, 1974

Georges Léopold Chrétien Frédéric Dagobert Baron Cuvier (1769–1832), better known by his pen name Georges Cuvier, was a French naturalist and one of the scientific giants of his age. He supported the geological school of thought termed 'catastrophism', according to which paleontological discontinuities are evidence of sudden and widespread catastrophes, and extinctions took place suddenly. The harshness of his criticism towards scientific opponents, and the strength of his reputation, discouraged other naturalists from speculating about the gradual transmutation of species, right up until Darwin's time. Cuvier is also famed for having stayed in a top government post, as permanent secretary at the Academy of Sciences, through three regimes, including Napoleon's. Cuvier's research on fish was begun in 1801, and culminated in a work co-written with Achille Valenciennes: *Histoire naturelle des poisons* (1828–1831), which described c.5,000 species of fishes. Seven birds, six reptiles, three mammals and an amphibian are named after him.

Cuyuní

Bandit Cichlid sp. *Guianacara cuyunii* H López-Fernández, DC Taphorn Baechle & SO Kullander, 2006

This is a toponym referring to the Cuyuní River, Bolívar, Venezuela; the type locality.

Cwynar

Barbeled Dragonfish sp. *Bathophilus cwyanorum* MA Barnett & RH Gibbs, 1968

Edward A Cwynar (b.1942) and Shigeru Yano. The etymology states that this name encompasses: "*...the surnames of two very capable colleagues who were largely responsible for the success of the midwater trawling program.*" Edward A Cwynar (b.1942) was brought to Hawaii as a 3-year old child. He attended Kailua High School, Hawaii and that is all we can find about him. Shigeru Yano is a long-line fisherman from Honolulu, Hawaii. The etymology describes him as a "*...friend and fellow fisherman,*" whose maintenance and operation of nets and associated equipment contributed greatly to the success of the authors' trawling expedition. (See **Yano, S**)

Cyrano

Cyrano Spurdog *Squalus rancureli* P Fourmanoir & J Rivaton, 1979
Rattail sp. *Nezumia cyrano* NB Marshall & T Iwamoto, 1973
Garra sp. *Garra cyrano* M Kottelat, 2000

Hercule-Savinien de Cyrano de Bergerac (1619–1655) was a French dramatist, poet and duelist. Edmond Rostand wrote (1897) a play: *Cyrano de Bergerac*, which was based loosely on his life. It revolved around the fact that Cyrano had a very prominent nose: as do the fish named after him.

Czekanowski

Czekanowski's Minnow *Rhynchocypris czekanowskii* B Dybowski, 1869

Dr Aleksander PIotr Czekanowski (1833–1876) was a Polish geologist who explored Siberia. He qualified as a physician and geologist at Kiev and Dorpat universities. He took part in the January Uprising (1863) and was exiled to Siberia where he took part in and led several expeditions in Eastern Siberia where a mountain range is named after him. Being in exile did not deter the Russian Geographical Society from awarding him a gold medal (1870). Having been released from exile, he became custodian of the Mineralogy Museum of the Imperial Academy of Sciences (1876). In the absence of an etymology we believe this to be the man honoured.

Czerski

Cherski's Sculpin *Cottus czerskii* LS Berg, 1913
Chewrskii's Thicklip Gudgeon *Sarcocheilichthys czerskii* LS Berg, 1914
Cherski's Char *Salvelinus czerskii* PA Dryagin, 1932

Alexander Ivanovich Chersky (Czerski) aka Tschersky (1879–1921) was a Russian of Polish descent who was an ornithologist, naturalist and researcher of the Russian Far east. He collected at least one of the holotypes. His father was the famous geologist, explorer and naturalist, Jan Czerski (1845–1892). He was born in the back of a cart when his mother was fleeing a great fire in the city of Irkutsk. His father undertook (1890) a three year-long expedition in the extreme northeast of Russia and his son went too, taking part in the scientific work. After his father's death he graduated in physics and maths from St Petersburg University (1904) and followed his father's footsteps into the natural sciences and spent his life exploring remote regions. He became (1908) a conservator in the museum of the Society for the Study of the Amur Region in Vladivostok. For seven yrears he made annual expeditions and trips to various regions to study the flora and fauna of Primorsky Krai. He collected over two thousand bird specimens. He left the museum (1915) and went to the Commander Islands as a caretaker for fish and fur industries. For three years he lived on these islands, studying the biology of game animals – fur seals, sea otters, foxes – along the way collecting fish specimens. He published throughout his life. He committed suicide, leaving a note that his belongings should go to his staff. A bird is also named after him as is a cape near Vladivostok.

D

Dabra

Dabra Goby *Feia dabra* R Winterbottom, 2005

This is an arbitrary combination of letters combining the first few letters of the given names of Winterbottom's son, David, and of Bradley Hubley who is entomology collection manger in the Department of Natural History at the Royal Ontario Museum. Having studied both entomology and invertebrate zoology at the University of Toronto he joined the museum as a summer student (1984), then joined full time (1985) and stayed ever since. He has participated in more than two dozen field trips around the world from East Asia, the Neotropics and Russia. Both contributed "*...immeasurably to the success and wellbeing of the Palau biodiversity expedition team*" that collected the type.

Dabry

Dabry's Sturgeon *Acipenser dabryanus* AHA Duméril, 1869
Humpback *Chanodichthys dabryi* P Bleeker, 1871
Chinese Lizard Gudgeon *Saurogobio dabryi* P Bleeker, 1871
Large-scale Loach *Paramisgurnus dabryanus* CP Dabry de Thiersant, 1872
Stone Loach sp. *Claea dabryi* HE Sauvage, 1874
Freshwater Sleeper sp. *Micropercops dabryi* HW Fowler & TH Bean, 1920

Claude-Philibert Dabry de Thiersant (1826–1898) was French consul at Hankow (one of the three cities whose merging formed modern-day Wuhan, China) and an amateur naturalist. He co-wrote: *La Médicine chez les Chinois* (1863). He was a fish culturist and interested in ichthyology. He collected the holotype of *Paramisgurnus dabryanus* and described the fish using a museum name coined by Guichenot (q.v.). However, by writing the description, he became the 'official' author of a name that honours himself. A bird is named after him.

Dadiburjor

Dadio (Indian barb) *Laubuka dadiburjori* AGK Menon, 1952

Sam J Dadiburjor was an aquarist in Bombay (Mumbai). He acquired this fish from Cochin and, having a pair, decided to breed it in captivity. He sent specimens to the Laboratories of the Indian Zoological Survey. Menon thought it new to science and asked for more specimens, which Dadiburjor supplied (five) (1950). He also gave a detailed description of their breeding habits, etc.

Daedalus

New Guinea Worm Eel *Neenchelys daedalus* JE McCosker, 1982

In Greek mythology Daedalus was a craftsman and inventor. When he and his son Icarus were imprisoned by King Minos, they escaped by building wings out of wax. The etymology refers to "...*the Greek artisan who escaped from his Earth-bound prison and ascended into heaven.*" This is a rather fanciful allusion to the eel's midwater habitat, having left the bottom substrate typical for most of its relatives and risen to dwell in midwater.

Daemel

Saddletail Grouper *Epinephelus daemelii* A Günther, 1876

Eduard C F Dämle, also known as Dämel or Daemel (1821–1900), was a German entomologist, and a collector and dealer in natural history specimens. He was in Australia (1871–1874), collecting for the Godeffroy Museum in Hamburg (to which he sent the holotype of the grouper). After returning to Germany he revived his own private natural history business. A frog and a snake are also named after him.

Daget

Redchin Panchax *Epiplatys dageti* M Poll, 1953
West African Cichlid sp. *Coptodon dageti* DFE Thys van den Audenaerde, 1971
Claroteid Catfish sp. *Chrysichthys dageti* LM Risch, 1992
African Characin sp. *Nannocharax dageti* FC Jerep, RP Vari & EJ Vreven, 2014

Dr Jacques Daget (1919–2009) was a French ichthyologist who was a professor at MNHN, Paris and Director of Ichthyology (1975–1984). His education was interrupted by WW2. He was in the French Air Force and later (1941) enrolled at the Sorbonne and at the MNHN to study botany and zoology. He was arrested and deported as a slave labourer to a textile mill in Czechoslovakia (1944–1945). After returning to Paris to MNHM, he was asked to undertake a study mission in Mali (1946–1947). He was in Mali again (1950–1960) creating and managing a new laboratory of Hydrobiology. He then worked for ORSTOM in Mali (1960–1963) and subsequently (1963–1964) for IRD in Chad, where a research laboratory on Lake Chad was named in his honour (1969). His publications exceed 200 including: *Fish Fouta Dialon and Lower Guinea* (1962).

Dagit

Carpenter's Chimaera *Chimaera lignaria* DA Didier Dagit, 2002

Kevin J Dagit was, at the time of the description, the describer's husband. He is a woodworker, carpenter and "...*supporter of research on chimaeroid fishes in his spare time*". The second part of the binomial is the sort of play on words much loved by taxonomists: '*lignaria*' means 'of wood', a reference to Dagit's occupation.

Dagmar

Dagmar's Dragonet *Foetorepus dagmarae* R Fricke, 1985

Dagmar Hansen of Mönchengladbach, Germany, was presumably a friend of the author who wrote of her: "...to whom I am indebted for encouragement in various ways."

Dahl, C

Dahl's Seahorse Seahorse *Hippocampus dahli* JD Ogilby, 1908
[Alt Lowcrown Seahorse}

Christian Dahl sent the type to Ogilby at the Australian Museum.He discovered it in Moreton Bay attached to some seaweed which was curled around a boat's anchor line. Ogilby placed it in the collection of the Queensland Amateur Fisherman's Association so we presume Mr Dahl was an amateur angler.

Dahl, G

Characin sp. *Eretmobrycon dahli* C Román-Valencia, 2000
[Syn. *Bryconamericus dahli*]

Professor George Dahl (1905–1979) was a Swedish biologist and ichthyologist who visited Colombia (1936–1939). He returned there in 1948, settling in Sincelejo, Sucre Province, where he worked at Liceo Bolivar and where a foundation is now named after him at Universidad de Sucre. He was at Institute de Ciencias Naturales, Bogotá, Colombia (1961). A turtle, which he collected the holotype of, is also named after him. (Also see **Martha (Dahl)**)

Dahl, K

Toothless Catfish *Anodontiglanis dahli* H Rendahl, 1922
Dahl's Frogfish *Batrachomoeus dahli* H Rendahl, 1922

Professor Dr Knut Dahl (1871–1953) was a Norwegian naturalist, explorer and collector. He was in Australia (1894–1896), collecting specimens for the Zoological Museum, Norwegian University. He wrote: *In Savage Australia: An account of a hunting and collecting expedition to Arnhem Land and Dampier Land* (English publication, 1926). Two mammals and an amphibian are named after him.

Dahlgren

Lanternfish sp. *Diaphus dahlgreni* HW Fowler, 1934

Dr Ulric Dahlgren (1870–1946) was a zoologist and histologist, He spent his entire career at Princeton University, New Jersey, from which he graduated with a bachelor's degree (1894) and a master's (1896). He was also Director of Princeton's Mount Desert Island Biological Laboratory, Bar Harbor and was Assistant Director, Marine Biological Laboratory, Woods Hole (1898–1906). He wrote: *The Maxillary and Mandibular Breathing Valves of Teleost Fishes* (1898). The etymology states he was honoured for his work on 'luminous animals'.

Dainelli

Snowtrout sp. *Schizopyge dainellii* D Vinciguerra, 1916
Yellowfish sp. *Labeobarbus dainellii* G Bini, 1940

Giotto Dainelli (1878–1968) was an Italian geographer and geologist. He graduated in natural sciences at the University of Florence (1900) and later studied at the University of Vienna. He was a lecturer at the University of Florence (1903–1914), was Professor of Geography at the University of Pisa (1914–1921), taught geology at the University of Naples (1921–1923) and was Professor of Geography and Geology at the University of

Florence (1924–1953). He explored in Eritrea (1905–1906), was a member of De Filippi's expedition to the Karakorum region of Asia (1913–1914) and led his own expedition to Tibet (1930), during which he discovered the source of the Yarkand River. Later, he led another expedition to Lake Tana, Ethiopia (1936–1937). He wrote in excess of 600 scientific and popular publications including: *My trip in Western Tibet* (1932). He led the expedition during which the holotype of *Labeobarbus dainellii* was collected.

Dalerocheila

Hotlips Pygmygoby *Trimma dalerocheila* R Winterbottom, 1984

Holly Arnold. This is hardly an obvious eponym, as *daleros* means hot or fiery, and *cheilos* means lip, referring to the goby's red snout and lips. However, the name also alludes to Ms. Arnold, who helped collect the holotype, and who, according to Richard Winterbottom, in the 'male dominated' society of the Chagos Archipelago (the type locality), was called a 'scarcely appropriate nickname' = Holly Hotlips.

Dalgleish

East Coast Flounder *Arnoglossus dalgleishi* C von Bonde, 1922
Spotted Tinselfish *Xenolepidichthys dalgleishi* JDF Gilchrist, 1922

Commodore James Dalgleish (1891–1964) was a Scottish-born chief of the Seaward Defense Force (later the South African Navy) (1941–1946). He served in the Royal Navy in WW1 and later transferred to the South African Naval Service for whom, as a LIeutenant, he commanded a number of survey vessels including 'Pickle', from which the holotype of the tinselfish was collected. He retired (1946) and returned to Scotland.

Dali

Pencil Catfish sp. *Trichomycterus dali* PP Rizzato, EPD Costa, E Trajano & ME Bichuette, 2011

Salvador Dali (1904–1989) was a Spanish surrealist artist, noted for his long moustache. This species has very long barbels.

Dall

Mudminnow genus *Dallia* TH Bean, 1880
Blue-banded Goby *Lythrypnus dalli* CH Gilbert, 1890
Calico Rockfish *Sebastes dallii* CH Eigenmann & CH Beeson, 1894

William Healey Dall (1845–1927) was an America naturalist. His interests were wide-ranging, including anthropology, meteorology and oceanography as well as palaeontology, zoology, and in particular malacology. He was a pupil of Louis Agassiz of Harvard's Museum of Comparative Zoology. He undertook field work in Alaska and along the coasts of the USA, initially as an assistant to Kennicott. His findings were published as: *Alaska and Its Resources*. He was appointed as Acting Assistant to the US Coast Survey (1870). He continued to collect and send his specimens to Agassiz at Harvard. He wrote: *Tribes of the Extreme Northwest* (1887). He published over 1,600 papers, reviews, and commentaries; describing 5,302 species, many of them molluscs. He dredged one of the Blue-banded Goby type specimens off Catalina Harbor, California, USA. Four mammals are named after him.

Dalwigk

Black Codling *Physiculus dalwigki* JJ Kaup, 1858

Reinhard Carl Friedrich von Dalwigk (1802–1888) was a German politician in the State of Hesse. He was anti-Prussian but after the Austro-Prussian War (1866), in which Hesse was an ally of Austria, he resigned (1871) and retired into private life. Kaup says that he was a 'dear friend'.

Daly

Amirante Dwarfgoby *Eviota dalyi* DW Greenfield & L Gordon, 2019

Ryan Daly is a South African zoologist and oceanographer who is Research Director (2017–present) at 'Save Our Oceans' D'Arros Research Centre. Cape Town University awarded his BSc in zoology and oceanology and Rhodes University his master's in marine biology and his PhD on bull sharks. According to the etymology he was honoured as it was he: "… *who photographed and collected the holotype and who has played a major role in surveying the fishes of the Seychelles Islands*" He co-wrote (with G Stevens & K Daly): *Rapid marine biodiversity assessment records:16 new marine fish species for Seychelles, West Indian Ocean* (2018). His wife and fellow researcher Clare Keating Daly is Programme Director at SOS.

Dam

Madagascan Cichlid sp. *Paretroplus damii* P Bleeker, 1868

Douwe Casparus van Dam (1827–1898) was a Dutch explorer, avid hunter and naturalist. He became friends through hunting with Dutch naturalist and merchant François Pollen (1842–1888) (q.v.). Together they made an expedition to Madagascar (1863) for the Rijksmuseum of Natural History. They visited Réunion (1864), Madagascar and returned (1866) via Mauritius. During this trip they collected the type. With Pollen he co-wrote: *Recherches sur la Faune de Madagascar et de ses dependances* (1868). He revisited Madagascar (1868–1873) without Pollen, although the latter financed the expedition. He became a wine merchant (1873) but was given a post as manager of the newly separated Rijksmuseum of Geology & Minerology (1878). He was accused of stealing silver and gold from the museum and dismissed (1891). A bird and a butterfly are also named after him.

Damas

Flagfin sp. *Leptaulopus damasi* S Tanaka, 1915

Not an eponym, but apparently derived from its Japanese name, *Eso-damashi.*

Damas, HFC

Ctenopoma sp. *Microctenopoma damasi* M Poll & H Damas, 1939
Shellear sp. *Parakneria damasi* M Poll, 1965

Dr Hubert François Constant Damas (1910–1964) was a Belgian ichthyologist who succeeded his father Philippe-Désiré Damas (1877–1959) as Professor of Zoology, University of Liège, which had awarded his doctorate (1933). He was in the Congo (1934–1936) and made a number of other expeditions to Africa, including to Ruanda-Urundi and Katanga (1957). He collected the holotypes. He died in a car accident.

Damasceno (Nogueira)

Armoured Catfish sp. *Ancistrus damasceni* F Steindachner, 1907

Colonel José Damasceno Nogueira acted as Steindachner's host in Filomeno, Brazil (1903), near where the holotype was acquired. The original text does not help, but it is most likely that the catfish's binomial honours the colonel.

Damasceno (Soares)

Neotropical Rivuline sp. *Stenolebias damascenoi* WJEM Costa, 1991

Joao Damasceno Soares (d.c.2001) was a Brazilian aquarist and collector of tropical fish who discovered the species in the Pantanal near his home in Mato Grosso. He turned his hobby into his job by collecting fish for sale to fellow aquarists. He sent five specimens to Dr Wilson Costa, and two proved to be of this new species.

Damla

Dalaman Spined Loach *Cobitis damlae* F Erk'akan & FS Özdemir, 2014

Sedef Damla Erk'akan (b.1989) is the daughter-in-law of the first author, Dr Füsun Erkakan who is Professor at the Faculty of Science Department of Biology, Section of Hydrobiology, Hacettepe University, Ankara, Turkey. She is married to Dr Erkakan's son Evren (q.v.) and graduated from METU Chemical Engineering Department (2012) then took her Master Science Degree from Hacettepe University Chemical Engineering Department (2015). Now (2017) she is studying as a PhD student in Ankara University, Chemistry Department. (Also see **Erdal** & **Evren**)

Dammerman

Cardinalfish sp. *Nectamia dammermani* M Weber & LF de Beaufort, 1929
[Syn. *Apogon dammermani*]

Dr Karel Willem Dammerman (1888–1951) was a Dutch field zoologist, botanist and collector who worked in the East Indies. He took his PhD (1910), then was appointed Assistant Entomologist of the Botanical Laboratories of the Botanic Gardens, Buitenzorg, Bogor, Java (1910). He became Chief of the Zoological Museum and Laboratory (1919) and ultimately (1932) Director of the Botanical Gardens. He was Chairman of the Society of Nature Protection, Dutch East Indies (1919–1932). On retirement (1939) he returned to Holland and joined the Museum of Natural History, Leiden (1942). He was also Editor of the Botanical Gardens journal *Treubia*. He wrote: *Preservation of Wildlife and Nature Reserves in the Netherlands Indies* (1929) and *The Agricultural Zoology of the Malay Archipelago* (1929), as well as an article on the Orang Pendek, a small 'yeti' of Sumatra, still unknown to science. He also wrote (1948) an extensive book on the fauna of Krakatau. He was known for long periods of silence, which caused one of his assistants to name a newly discovered snail *Thiara carolitaciturni*, meaning 'Karel the Silent'; a parody on 'William the Silent', a nickname of William I of OrangeNassau. He collected the holotype of this species. Five birds, an amphibian and a mammal are named after him.

Dana

Marine Hatchetfish genus *Danaphos* AF Bruun, 1931
Dreamer Anglerfish genus *Danaphryne* EOG Bertelsen, 1951
Whalefish genus *Danacetichthys* JR Paxton, 1989
Pencil Smelt sp. *Xenophthalmichthys danae* CT Regan, 1925
Dreamer sp. *Dolopichthys danae* CT Regan, 1926
Dana Viperfish *Chauliodus danae* CT Regan & E Trewavas, 1929
Dana Duckbill Eel *Nessorhamphus danae* EJ Schmidt, 1931
Dana Lanternfish *Diaphus danae* AV Tåning, 1932
Footballfish sp. *Himantolophus danae* CT Regan & E Trewavas, 1932
Barracudina sp. *Macroparalepis danae* V Ege, 1933
Barbeled Dragonfish sp. *Stomias danae* V Ege, 1933
Lefteye Flounder sp. *Monolene danae* AF Bruun, 1937
Slickhead sp. *Rouleina danae* AE Parr, 1951
Big-scale Fish *Melamphaes danae* AW Ebeling, 1962
Fangtooth Smooth-head *Bathyprion danae* NB Marshall, 1966
Dana Pearleye *Scopelarchoides danae* RK Johnson, 1974
Hatchetfish sp. *Polyipnus danae* AS Harold, 1990
Barbeled Dragonfish sp. *Eustomias danae* TA Clarke, 2001

The succession of Danish research vessels named 'Dana' seem to have been greatly favoured by describers of fish, and often the etymology includes: "*...from which the holotype was collected*." Dana IV entered service in 1981. Her predecessor is famous for a circumnavigation of the world in the third Dana expedition (1928–1930). In one case (*Polyipnus danae*) the acknowledgement is also for the contributions of the Carlsberg Foundation Dana Expeditions (1928–1930) to deep-sea ichthyology (q.v. **Carlsberg**).

Dan Cohen

Morid Cod sp. *Gadella dancoheni* YI Sazonov & YN Shcherbachev, 2000

Dr Daniel Morris Cohen (b.1930). (See **Cohen** & **Daniel**)

Dandara

Black Jacundá *Crenicichla dandara* HR Varella & PMM Ito, 2018

Dandara was an Afro-Brazilian warrior of Brazil's colonial period. According to legend, Dandara and her husband Zumbi fiercely defended the community of Palmares, a safe haven for escaped slaves in the coastal state of Alagoas, Brazil. Nowadays, she has become a symbol of the struggle against racism and the exploitation of black women. The day of Zumbi's death, 20 November 1695, is celebrated as the Dia da Consciência Negra [Black Awareness Day] throughout Brazil.

Danford

Danford's Killifish *Anatolichthys danfordii* GA Boulenger, 1890
[Syn. *Aphanius danfordii*]
Carpathian Lamprey *Eudontomyzon danfordi* CT Regan, 1911

Charles George Danford (1843–1928) was a Scottish geologist, palaeontologist, zoologist, artist, traveller and explorer. He was in Asia Minor (Turkey) (1875–1876 & 1879). The Danford Iris was named after his wife, who introduced it to England. A reptile and a bird are also named after him.

Dani

Armoured Catfish sp. *Parotocinclus dani FF* Roxo, GSC Silva & C de Oliveira, 2016

Daniela Fernandes Roxo is the first author's sister.

Daniel

Argentine sp. *Glossanodon danieli* NV Parin & YN Shcherbachev, 1982

Dr Daniel Morris Cohen (1930-2017) (see **Cohen**).

Daniela

Daniela's Soldierfish *Ostichthys daniela* WD Greenfield, Je Randall & PN Psomadakis, 2017

Daniela Basili is the third author's wife.

Daniels

Daniel's Catfish *Cochlefelis danielsi* CT Regan, 1908

Major William Cooke Daniels (1871–1918) was a wealthy American who was a partner in a department store, 'Daniels and Fishers', Denver, Colorado. He financed and led the Cooke-Daniels ethnographic expedition to British New Guinea (1903–1904) during which the holotype was collected and presented to the BMNH. His military rank dates from his service as a volunteer officer in the Spanish-American War in Cuba (1898). He died whilst on a visit to Argentina.

Danilevski

Danilevskii's Dace *Leuciscus danilewskii* KF Kessler, 1877

Bojak (Sevan Trout ssp.) *Salmo ischchan danilewskii* M Gulelmi, 1888

Nikolai Yakovlevich Danilevski (sometimes transcribed as Nikolay Danilewsky) (1822–1885) was Russian naturalist, economist, ethnologist, philosopher and historian who was a great proponent of the pan-Slavic ideal. He passed the examinations for a master's degree at the University of St Petersburg, but was arrested before being able to defend his thesis (1849), for belonging to the Petrashevsky Circle (regarded as very dangerous liberals!) of which Dostoyevsky was also a member. He was imprisoned and then exiled to northwest Russia to work in local administration. He was part of von Baer's expedition to assess fisheries on the River Volga and the Caspian Sea (1852–1856), after which he was at the state's Agricultural Department of the Property Ministry and was responsible for the organisation of many expeditions to the White and Black Seas, the Caspian and the Arctic Ocean and all points in between. He was head of a commission, which set the rules for using running water in the Crimea (1872–1879) and Director of the Nikitsky Botanical Gardens, near Yalta. He wrote: *Examination of Fishery Conditions in Russia* (1872).

Danner

Riedling (Austrian salmonid sp.) *Coregonus danneri* D Vogt, 1908

H Danner of Linz, Austria was a fisheries inspector. He collected the holotype of this species. We can find no further biographical data.

Dannevig

Cusk-eel genus *Dannevigia* GP Whitley, 1941

Harold Christian Dannevig (1860–1914) was Director of Fisheries for Australia, and collected the holotype. He was lost at sea when his fisheries research vessel 'Endeavour' disappeared without a trace.

Dao Van Tien

Vietnamese Cyprinid sp. *Ancherythroculter daovantieni* PM Bănărescu, 1967

Professor Dao Van Tien (1920–1995) was a Vietnamese teacher and Professor of Biology at Viet Minh University and National University of Hanoi (1946–1986). He was educated in Hanoi under the French colonial administration and graduated from Université de l'Indochine (1942) where he was also awarded his master's degree (1944). He was a primatologist and is widely acknowledged as the father of his field in Vietnam, although he is probably best known for asserting his belief in the existence of 'Forest Man', a supposed primitive hominid reported from remote parts of Asia. A mammal and a reptile are also named after him.

da Paz

Neotropical Rivuline sp. *Melanorivulus dapazi* WJEM Costa, 2005

Dr Ricardo Campos da Paz (b.1963) is a Brazilian evolutionary biologist, zoologist and ichthyologist. The Federal University of Rio de Janeiro awarded his BSc (1990) and the University of São Paulo awarded his zoology MSc (1992) and PhD (1997). He served as Visiting Professor of the Institute of Biology (Department of Zoology) of UFRJ. He is Associate Professor at the Federal University of the State of Rio de Janeiro (UNIRIO), Coordinator of the Laboratory of Neotropical Ichthyology (LABIN), and Head of the Department of Ecology and Marine Resources (DERM), as well as Substitute Director of the Institute of Biosciences IBIO) of UNIRIO. He is interested in the systematics and taxonomy of groups of electric fish of the Neotropical Region (Gymnotiformes), as well as Biogeography, Evolution, the History of Biology, Systematics and Taxonomy. His publications range from: *Phylogenetic systematics of American knifefishes: A review of the available data* (1998) to *A new species of* Eigenmannia *Jordan and Evermann (Gymnotiformes: Sternopygidae) from the upper rio Paraguai basin* (2017). He discovered this species.

Dardenne

Lake Tanganyika Latticed Cichlid *Limnotilapia dardennii* GA Boulenger, 1899

Léon-Louis Dardenne (1865–1912) was a Belgian artist described as 'peintre de la mission scientifique du Katanga' (1898–1900). His artwork from the expedition was published by MRAC (1965). In *On the fishes obtained by the Congo Free State Expedition under Lieut.*

Lemaire in 1898, Boulenger is not very informative, as he just refers to the coloured sketches made "*by M. Dardenne, the excellent artist attached to the expedition.*"

Darge

M'bam Killifish *Aphyosemion dargei* J-L Amiet, 1987

Philippe Darge (b.1933) was a French entomologist. He wrote a large number of papers, from *Description d'une forme nouvelle de Colias electo L.: Colias electo mamengoubensis subsp* (1968) to *Données complementaires sur le genre Antistathmoptera Tams, 1935, avec description d'une espèce nouvelle pro venant d'une forêt côtiere de Tanzanie. (Lepidoptera, Saturniidae, Urotini)* (2015). He was also Publisher of the 'Saturnafrica' journal which carried many of his articles. He has authored over 238 taxon names. He was honoured as he encouraged Amiet to research and write.

Darling, J

Zambezi Hap (cichlid) *Pharyngochromis darlingi* GA Boulenger, 1911

Captain James Johnston ffolliott Darling (1859–1929) was an Irish naturalist, zoological collector and prospector in Rhodesia (Zimbabwe) at the end of the nineteenth century and beginning of the twentieth. Having failed his final medical exam in his native Dublin, and argued with his father, he opted for a new life in South Africa (1883) joining the Cape Mounted Rifles as a medical orderly, rising to Sergeant in the Cape Field Artillery before purchasing his release (1886). He briefly ran a pharmacist's shop (1888) but gave this up to prospect for gold and to become a settler. He collected natural history specimens, sending many to museums in London and Dublin (1894–1897). He wrote accounts of hunting trips and *Six Months in Mashonaland* (unpublished) in which, among other things, he described treatments for snake-bite. Boulenger wrote *A list of the fishes, batrachians and reptiles collected by Mr. J. ffolliott Darling in Mashonaland, with descriptions of new species* (1902). There are records of collections of insects and arachnids being donated to the South Africa Museum (1896 & 1897). He farmed and studied local flora and fauna but eventually retired and returned to Ireland (c.1900) and was a 'gentleman farmer' there. He is also commemorated in the scientific name of an arachnid, two mammals and a frog.

Darom

Darom's Goatfish *Upeneus davidaromi* D Golani, 2001

Dr David Darom (b. 1943) is an Indian-born Israeli marine biologist and nature photographer. He spent 35 years as head of the Department of Scientific Photography at the Hebrew University of Jerusalem. He has published on both fauna and flora, including being co-author with Daniel Golani of: *Handbook of the Fishes of Israel* (1997).

D'Arros

D'Arros Pipefish *Cosmocampus darrosanus* CE Dawson & JE Randall, 1975

This is a toponym referring to D'Arros Island in the Seychelles, where there is a research station.

Dartevelle

African Barb sp. *Labeobarbus dartevellei* M Poll, 1945

Dr Edmond Dartevelle (1907–1956) was a Belgian palaeontologist, geologist and explorer in Africa. He graduated with a doctorate from the Free University of Brussels (1932). He joined the Museum of the Belgian Congo at Tervuren (MRAC) (1936) and was in the Belgian Congo (Zaire) (1937–1938 & 1946–1949) for the museum. He was Curator of Invertebrates MRAC (1949–1956). He collected the holotype of this species. An amphibian is also named after him.

Darvell

Sweeper sp. *Pempheris darvelli* JE Randall & BC Victor, 2014

Dr Brian William Darvell (b.1948) is a chemist, academic and marine conservationist who is honorary professor at the School of Dentistry at Birmingham University, UK. His BSc was awarded by the University of Wales, Cardiff (1969) as was his MSc (1972). His PhD was awarded by the University of Birmingham (1975) as was his DSc (2006). After post-doctoral work at Melbourne University (1978–1980), he was Senior Lecturer, then Reader, in the faculty of dentistry at the University of Hong Kong (1980–2009), then professor at Kuwait University (2009–2014). He was also founding Chairman of the Hong Kong Marine Conservation Society and a founding member of Reef Check Hong Kong, and one of the first to pioneer anti-shark-finning campaigns. Among his publications is a text-book: *Materials Science for Dentistry* (10th ed., 2018). His research interests include test methods for materials that are interpretable in service and solubility of calcium phosphates, while his personal interests include diving, books and playing guitar. He was honoured for his fieldwork in Oman resulting in the specimens and photographs of the sweeper species. Darvell accompanied Keith D P Wilson on an expedition to Oman at the authors' request to photograph and collect *Pempheris*. He is also honoured in the names of three marine invertebrates.

Darwin

Galápagos Sheephead Wrasse *Semicossyphus darwini* L Jenyns, 1842
Darwin's Slimehead *Gephyroberyx darwinii* JY Johnson, 1866
Darwin's Toadfish *Marilyna darwinii* F de L de Castelnau, 1873
Remo Flounder *Oncopterus darwinii* F Steindachner, 1874
Galápagos Batfish *Ogcocephalus darwini* CL Hubbs, 1958
Snailfish sp. *Paraliparis darwini* DL Stein & NV Chernova, 2002
Darwin's Mudskipper *Periophthalmus darwini* HK Larson & T Takita, 2004
Darwin's Sanddab *Citharichthys darwini* BC Victor & GM Wellington, 2013
Electric Knifefish sp. *Gymnotus darwini* R Campos-da-Paz & CD de Santana, 2019

Charles Robert Darwin (1809–1882) was the prime advocate, together with Wallace (q.v.), of natural selection as the driver of speciation. To quote from his seminal work: *On the Origin of Species by Means of Natural Selection* (1859): "*I have called this principle, by which each slight variation, if useful, is preserved, by the term Natural Selection.*" Darwin was naturalist on 'HMS Beagle' on her scientific circumnavigation (1831–1836). In South America, he found fossils of extinct animals that were similar to extant species. On the Galápagos Islands, he noticed many variations among plants and animals of the same general type as those in South America. Darwin collected specimens for further study everywhere he went. On his return to London he conducted thorough research of his notes and specimens. Out of this study grew several related theories: evolution did occur; evolutionary change was gradual taking thousands or millions of years; the primary mechanism of evolution was 'natural selection'; and the millions of species alive today arose from a single original life form. However, Darwin held back on publication for many years, not wanting to offend Christians, especially his wife. The sanddab etymology says "*...the description serves as a somewhat belated recognition of the 150th anniversary of the publication of* The Origin of Species *in London in October 1860, mitigated to some small degree by the knowledge that the dilatory nature of the endeavor would not be particularly foreign to Darwin's sensibilities.*" Darwin is remembered in the names of numerous other taxa, including 24 birds, eight reptiles, four amphibians and four mammals.

Darwin, (City)

Darwin Sole *Leptachirus darwinensis* JE Randall, 2007

This is a toponym, named after the Australian city.

Das

Goby sp. *Oligolepis dasi* PK Talwar, TK Chatterjee & MK Dev Roy, 1982

Apurba Kumar Das was the Officer-in-Charge, Zoological Survey of India, Andaman & Nicobar Regional Station. He made specimens of the goby available for study. He co-authored *Glimpses of Animal Life of Andaman & Nicobar Islands* (1985).

Daubenton

Stone Loach sp. *Schistura daubentoni* M Kottelat, 1990

François d"Aubenton-Carafa (1923–2008) was a French ichthyologist and naturalist who unusually had no university education. After being demobilized after WW2 he went to work for L'École Pratique des hautes Études at the MNHN, Paris. Finding this boring, he spent as much time as possible in the ichthyological laboratories instead. He was part of a Hydrobiology Mission to Mali (1954). He worked closely with Maurice Blanc in Paris and on a mission to the Mekong River system in Cambodia (1965). He retired (1989) to help his son run an antique shop. He collected the holotype.

Dauguet

Sea Catfish sp. *Cryptarius daugueti* P Chevey, 1932

Paul Daughet (b.1883) was a French merchant seaman and commander (1925–1939) of 'de Lanessan,' a research vessel under the control of Institut Océanographique de l'Indochine, from which the holotype was collected. He was at sea in the period when sail was giving way to steam (1901–1916), and then in the French Navy in WW1 (1916–1919). In WW2, he and the 'de Lanessan' were taken over by the French Navy and assigned to dredge the port of Saigon, but Daughet became ill and was demobilised on grounds of ill-health (1939). Chevey seems to have made a mistake in the spelling of Daughet's name in the binomial.

Daul

Molly (livebearer) sp. *Poecilia dauli* MK Meyer & AC Radda, 2000

Günter Daul was a German aquarist from Berlin. He collected widely, including lungfish from Africa (1969). He collected swordtails in Quintana Roo, Mexico while on holiday there (1975) and imported them to Germany where they were established as aquarium fishes. A group of aquarists collected the type in Venezuela.

David (Acero)

David's Angel Shark *Squatina david* A Acero, JJ Tavera, R Anguila & L Hernández, 2016

David [pronounced 'dahveed'] Acero (d. 2011) was the son of the first author, Arturo Acero.

David Darom

Darom's Goatfish sp. *Upeneus davidaromi* D Golani, 2001

See under **Darom**.

David, Père JPA

Chinese Cyprinid sp. *Xenocypris davidi* P Bleeker, 1871
Sisorid Catfish sp. *Chimarrichthys davidi* HE Sauvage, 1874
Goby sp. *Rhinogobius davidi* HE Sauvage & CP Dabry de Thiersant, 1874
Loach sp. *Formosania davidi* HE Sauvage, 1878
Snowtrout sp. *Schizothorax davidi* HE Sauvage, 1880
Gudgeon sp. *Sarcocheilichthys davidi* HE Sauvage, 1883

Fr Jean Pierre Armand David (1826–1900), often known as Père David, was a French Lazarist priest as well as a fine zoologist. Ordained in 1851, he was shortly afterwards sent to Peking (Beijing). As a missionary to China, he became the first Westerner to observe many animals, including the Giant Panda *Ailuropoda melanoleuca* and the deer *Elaphurus davidianus* famously named after him. He co-wrote: *Les Oiseaux de Chine* (1877). He collected the holotypes of several of the above fish species. The French naturalist Alphonse MilneEdwards classified many of the specimens David collected. He collected thousands of specimens and had many taxa named after him, including 15 birds, five mammals, three odonata, an amphibian and a reptile.

David Sands

Sands' Cory *Corydoras davidsandsi* BK Black, 1987

Dr David Dean Sands (b.1951) is an English aquarist who is an established researcher in human and animal psychology with a general zoology background. He took up the hobby of fishkeeping in the late 1970s, taking his first collecting trip to Brazil (1979) and since taking many others to Peru, Guyana, Venezuela, etc. He went on to undertake a PhD at Liverpool University in ethology (1995). Among his published works are books such as: *Bumper Guide to Tropical Cichlids* (1999) and a number on pet behaviour. He also is a human-companion-animal practitioner, based in Lancashire, and specialises in the treatment of behavioural conditions in dogs, cats, birds, horses and exotic animals as well as being an internationally established author and photographer. He has been featured in several UK television documentaries and programmes such as 'Fish People', 'Absolutely Animals', 'Pet Rescue', 'Potty about Pets' and BBC's '999' and consulted on and appeared in 'To The Ends Of the Earth' (Amazon) filmed for Channel 4 and Anglia Survival. He also regularly contributes to UK TV and Radio news programs on related human and pet companionship subjects, including being a consultant for the 2005 Animal Planet & Channel 5 series, 'Britain's Worst Pet', and a documentary 'Suicide Bridge Dogs' about dogs that have died leaping off a gothic bridge in Scotland. He was honoured for his: "*underestimated contribution to the popularisation of catfishes*", despite being called an: "*amateur Corydoras taxonomist*."

David Shen

Jawfish sp. *Stalix davidsheni* W Klausewitz, 1985

David Shen discovered this jawfish in the Red Sea and photographed it. He has sponsored field studies in the Red Sea and other ichthyological research, and is also, according to the etymology, an 'excellent' diver and underwater photographer. (Also see **Shen, D**)

David Smith

Smith's Brotula *Ogilbia davidsmithi* PR Møller, W Schwarzhans & JG Nielsen, 2005
[Alt. Cortez Brotula]
Flores Mud Moiray *Gymnothorax davidsmithi* JE McCosker & JE Randall, 2008

Dr David G Smith (see **Smith, DG**)

Davidson

Xantic Sargo *Anisotremus davidsonii* F Steindachner, 1875

George Davidson (1825–1911) was a geodesist, astronomer, geographer, surveyor, engineer and revered colleague of the author who became the President of the California Academy of Sciences (1871–1887). He spent many yeares surveying and mapping in California, Alaska and Panama as well as setting up his own observatory and became the first Professor of Geography at the University of California, Berkeley (1898–1905). He wrote numerous papers and longer works including *Origin and Meaning of the Name California* (1910). He was honoured for his contributions to the research of natural history in California, where this species was described.

Davies

Davies' Stingaree *Plesiobatis daviesi* JH Wallace, 1967
[Alt. Deepwater Stingray, Giant Stingaree]

Dr David Herbert Davies (1922–1965) was a South African ichthyologist and oceanographer. The University of Cape Town awarded his BSc (1942). He then served in the South African Air Force, rising to Lieutenant with active service in Italy and North Africa. After the war, he became Senior Demonstrator in Zoology at Cape Town University (1945) and then (1946) a researcher with the Division of Sea Fisheries, quickly rising to its Chief Biologist (during this time he also took his MSc and PhD). He took part in a Danish circumnavigation (1951). He left Cape Town (1957) to join the staff of the Institute of Marine Resources in California. He then became the first Director (1958) of the Oceanographic Research Institute, a division of the South African Association for Marine Biological Research of South Africa. Its research vessel is also named after him, as is its library. He was also Research Professor at the University of Natal (1960). He wrote many papers (1947–1965) but was probably best known for: *About Sharks and Shark Attack* (1964). He was killed in an accident whilst attending a scientific meeting. He was honoured as the person "*...who was responsible for the initiation of research on the batoid fishes of the east coast of Southern Africa.*"

Davis, WB

Scrapetooth Characin sp. *Apareiodon davisi* HW Fowler, 1941

William Baldwin Davis lived in Philadelphia. Fowler mentions that he was: "*...indebted for many American fishes*" from his large circle of friends and acquaintances. W B Davis was one of that circle, but we can find nothing more about him.

Davis, WG

Pencil Catfish sp. *Cambeva davisi* JD Haseman, 1911
[Syn. *Trichomycterus davisi*]

The etymology says it was named after 'Dr Davis', which is less than helpful. A strong possibility is that it is intended for Walter Gould Davis (1851–1919) an American engineer and meteorologist who was responsible for the Argentine Meteorological Service which he joined (1876) and was Director of (1885–1915). Haseman also acknowledges a Dr Davis from Corumbá for helping him during his voyage, without giving further information about this person. An American newspaper report (1914) mentions an American dentist named Dr Davis living in Corumbá, so we are still left with guesses!

Davis Singh

Sisorid Catfish sp. *Glyptothorax davissinghi* A Manimekalan & HS Das, 1998

Davis Franc Singh was Senior Scientist, Sálim Ali Center for Ornithology and Natural History. The survey that collected the holotype was instigated by him. He was also mentioned for 10+ years of fish and fish-habitat conservation work in the Western Ghats of India.

Davison

Choctawhatchee Darter *Etheostoma davisoni* OP Hay, 1885

D M Davison. The etymology simply says: "*Named in honor of Mr. D.M. Davison, one of its collectors.*" Davison was one of two persons who supplied a small collection of fish specimens from Florida, but we know nothing more about him.

Dawkins

Barb genus *Dawkinsia* R Pethiyagoda, M Meegaskumbura & K Maduwage, 2012

Dr Clinton Richard Dawkins (b.1941) is an ethologist and evolutionary biologist who was Professor for Public Understanding of Science (1995–2008) at Oxford, which had awarded both his bachelor's degree (1962) and his doctorate (1966) and where he had taught as a lecturer (1970–1989) and a reader in zoology (1990–1994). Previously he had been Assistant Professor of Zoology at the University of California Berkeley (1967–1969) where he became very involved in anti-war demonstrations. His most famous and influential book is: *The Selfish Gene* (1976). He was honoured '*...for his contribution to the public understanding of science.*' He is, perhaps, the most famous present day 'evangelistic' atheist.

Dawn Arnall

Dawn's Shrimpgoby *Vanderhorstia dawnarnallae* GR Allen, 2019

Dawn L Arnall is a businesswoman and philanthropist who was the wife (2000–2008) of the late Roland E Arnall (1939–2008) the billionaire US Ambassador to the Netherlands (2006–2008), they were co-directors of Ameriquest Capital Corporation a sub-prime mortgage lender. The etymology states that: "*The new species is named* dawnarnallae *in honour of Dawn Arnall, who both funded the expedition that led to the discovery of this species and has provided critical support and advice to the Bird's Head Seascape marine conservation initiative that now protects the habitat of this new species. It is a pleasure and an honor to name this beautiful shrimp goby in recognition of her invaluable support.*"

Dawson, CE

Dawson's Pipefish *Syngnathus dawsoni* ES Herald, 1969
Chain Pearlfish *Echiodon dawsoni* JT Williams & RL Shipp, 1982
Reefgoby sp. *Priolepis dawsoni* DW Greenfield, 1989

Charles Eric 'Chuck' Dawson (1922–1993) was a Canadian-born American ichthyologist. He fought in the Canadian Army in WW2 as an infantryman, surviving the Dieppe raid in which he lost an eye and collected a legful of shrapnel (1942). He graduated from the University of Miami (1954), after which he joined the University of Texas Institute of Marine Science, Port Aransas as research scientist. He worked at the Gulf Coast Research Laboratory, Ocean Springs, Mississippi (1958–1985). He wrote: *Atlantic sand stargazers (Pisces: Dactyloscopidae), with description of one new genus and seven new species* (1982). He also wrote: *A new subspecies of the Gulf pipefish, Syngnathus scovelli maki (Pisces: Syngnathidae)* (1972) with Herald. Greenfield wrote that it was Dawson "*...who first recognized that populations of* Priolepis *from Brazil might differ taxonomically, and whose extensive collecting activity in South America has done much to further our understanding of the Brazilian Fish Province.*"

Dawson,EW

New Zealand Catshark *Bythaelurus dawsoni* S Springer, 1971

Elliot Watson Dawson (1930-2020) was a biologist with the New Zealand Oceanographic Institute of the New Zealand Department of Scientific & Industrial Research (1955–1990), principally involved in carrying out surveys of the marine benthic fauna of the New Zealand sub-Antarctic region. He led a joint New Zealand/United States expedition to the Ross Sea (1965), making the first bathymetric charts of the Balleny Islands. The highlight of his career was leading the multidiscipline Royal Society Cook Bicentenary Expedition to the South Pacific (1969). He now is an Honorary Research Associate of the Museum of New Zealand Te Papa Tongarewa, Wellington, where he works on decapod crustaceans of the Southwest Pacific islands. As well as various papers on crustaceans, molluscs, brachiopods and sub-fossil birds, his publications include a comprehensive bibliography of King Crabs of the

World and their fisheries. He has also written accounts of the little-known German Transit of Venus expedition to the Auckland islands (1874) and a comprehensive bibliography of the human and natural history of these sub-Antarctic islands. He is a graduate of the University of New Zealand and of Cambridge University where he was the first holder of the John Stanley Gardiner Studentship in Zoology. Springer named this species after him as he had brought the first specimens to Springer's attention.

Dawydoff

Slender Moray sp. *Strophidon dawydoffi* AM Prokofiev, 2020

Konstantin Nikolaevich Davydov (1877–1960) was a Russian zoologist at the Institute of Oceanography in Nha Trang, Viet Nâm (near where this eel was collected). His bachelor's degree was awarded by the University of St Petersburg (1901) where he studied further before becoming a professor at the University of Perm. He travelled through Syria, Palestine, Indonesia, and the north of European Russia (Olonets Krai). He worked in France (1923) and Indochina and became (1949) a corresponding member of the Academy of Sciences in Paris. He wrote a number of text books and many papers. He was honoured for his contribution to the knowledge of various groups of marine invertebrates.

Day

Round Herring genus *Dayella* Talwar & Whitehead, 1971
Day's Baril *Barilius evezardi* F Day, 1872
Day's Mystus *Mystus bleekeri* F Day, 1877
Day's Glassy Perchlet *Parambassis dayi* P Bleeker, 1874
Characin sp. *Roeboides dayi* F Steindachner, 1878
Day's Catfish *Nedystoma dayi* EP Ramsay & JD Ogilby, 1886
Bagrid Catfish sp. *Batasio dayi* D Vinciguerra, 1890
Ceylon Killifish *Aplocheilus dayi* F Steindachner, 1892
Brown Spike-tailed Paradise Fish *Pseudosphromenus dayi* W Köhler, 1908
Black-spotted Spiny Eel *Mastacembelus dayi* GA Boulenger, 1912
Day's Sardinella *Sardinella dayi* CT Regan, 1917
Blenny sp. *Praealticus dayi* GP Whitley, 1929
Loach sp. *Botia dayi* SL Hora, 1932
[Has been treated as a Junior Synonym of both *B. almorhae* and *B. rostrata*]
Loach sp. *Schistura dayi* SL Hora, 1935
Indian Airbreathing Catfish sp. *Clarias dayi* SL Hora, 1936
Snow Trout sp. *Osteobrama dayi* SL Hora & KS Misra, 1940
Day's Goby *Acentrogobius dayi* FP Koumans, 1941
Day's Pellona *Pellona dayi* T Wongratana, 1983
Day's Thryssa *Thryssa dayi* T Wongratana, 1983

Dr Francis Day CIE (1829–1889) was Inspector-General of Fisheries in India (1871–1877) and Burma and an ichthyologist. He became the medical officer in the Madras Presidency East India Company services (1852). He wrote the two volumes on fish in: *The Fauna of British India, including Ceylon & Burma* (1875–1878) in which he described 1400 species. He also wrote: *Fishes of Malabar* (1865). Later Günther questioned much of his work. Day loathed Günther to the extent that he arranged for his own collections to be sold to the Australian Museum for £200 instead of them going to the BMNH. He retired (1877) to Cheltenham where he died of stomach cancer. Two mammals and a reptile are also named after him.

NB Steindachner gives no etymology for the characin, so this allocation is uncertain. Day is associated with fishes of the Indian region, whereas the characin is Neotropical.

Deacon

Las Vegas Dace *Rhinichthys deaconi* RR Miller, 1984 (Extinct)

Dr James 'Jim' Everett Deacon (1934–2015) was an American ichthyologist, limnologist and environmentalist. The University of Kansas awarded his doctorate (1960). He was a member of the faculty (1960–1968) and full Professor (1968–1988) at the University of Nevada, Las Vegas. After retirement, he became Distinguished Professor Emeritus in the Departments of Environmental Studies and Biological Sciences. His main research focus was the biology and conservation of desert fishes and later sustainable water use. To study such fish, he often scuba-dived in water-filled subterranean caverns. Among his published works are the co-written: *Fishes of the Wakarusa River in Kansas* (1961) and *Daily and yearly movement of the Devil's Hole pupfish,* Cyprinodon diabolis *Wales in Devil's Hole, Nevada* (1983). The etymology states that he was honoured for the fact that his: "*...concern about the conservation status of many fishes from the Southwest has aroused interest on their behalf and whose ecological studies have provided the necessary biological information needed to aid their survival.*"

Dean

Longnose Dogfish genus *Deania* DS Jordan & JO Snyder, 1902
Black Hagfish *Eptatretus deani* BW Evermann & EL Goldsborough, 1907
Philippine Chimaera *Hydrolagus deani* HM Smith & L Radcliffe, 1912
Prickly Snailfish *Paraliparis deani* CV Burke, 1912

Dr Bashford Dean (1857–1928) was an American zoologist, ichthyologist and acknowledged expert on mediaeval armour. The College of the City of New York awarded his bachelor's degree (1886) and Columbia University, where he later was Professor of Zoology, his doctorate (1890). He is the only person to have held concurrent positions at two of New York's great museums, the Metropolitan Museum of Art, where he was Honorary Curator of Arms and Armour, and the American Museum of Natural History. He wrote: *Bibliography of Fishes* (1916) and *Catalogue of European Court Swords and Hunting Swords Including the Ellis, De Dino and Reubell Collections* (1929). He was honoured in the name of the hagfish for his work on the embryology of *E. stoutii.*

Deansmart

Smart's Blind Cave Loach *Schistura deansmarti* C Vidthayanon & M Kottelat, 2003

Dean Smart (See **Smart, D**)

Dearborn

Dearborn's Pupfish *Cyprinodon dearborni* SE Meek, 1909

Ned Dearborn (1865–1948) was an American ornithologist. Dartmouth University awarded his bachelor's degree (1891) and New Hampshire State College his master's (1898). He was Assistant Curator of Birds at the Field Museum, Chicago (1901–1909), then worked with the US Biological Survey (1909–1920). He started and ran the Dearborn Fur Farm, Sackets Harbour, New York (1920–1948). He wrote *Catalogue of a collection of birds from Guatemala* (1907). In the spring of 1908, Dearborn and a Mr John F Ferry made a short visit to the islands of Aruba, Curaçao and Bonaire to collect zoological material for the Field Museum. During this trip, Dearborn collected the pupfish holotype on Curaçao. The etymology simply states: "*Collected by Dr N Dearborn, for whom the species is named.*" Two birds are also named after him.

Deason

Searobin sp. *Lepidotrigla deasoni* AW Herre & DE Kauffman, 1952

Dr Hilary J Deason was Chief of the Office of Foreign Activities, US Fish and Wildlife Service. He joined the (then) Bureau of Fisheries (1928) after graduating with a BA, MSc and MA, going on to complete his PhD (1936) all at the Michigan University. He took a tour of Asian fisheries (1949), spending three weeks in the Philippines and time in Siam, India and Pakistan. He then went to London to discuss fisheries and attend the Whaling Commission. He was later (c.1959) Director of Science Library program of the American Association for the Advancement of Science. He was honoured as someone "*…who has taken great interest in Philippine fish and fisheries and has zealously promoted their study.*"

De Bauw

African Glass Catfish *Pareutropius debauwi* GA Boulenger, 1900

Major Guillaume Clément Adolphe De Bauw (1865–1914) was a Belgian Army officer commissioned as a 2nd Lieutenant in the 2nd regiment of line infantry (1886). He embarked for the Congo (1897), where he served with various periods of leave in between until finally returning to Belgium (1913) He collected the holotype of this species and is also recorded as collecting botanical specimens in the Congo (1906). He wrote: *La zone Uere-Bomu* (1901).

De Beaufort

De Beaufort's Flathead *Cymbacephalus beauforti* LW Knapp, 1973
[Alt. Crocodile Fish]

Professor Lieven Ferdinand de Beaufort (1879–1968) (See **Beaufort**)

Debora

Blue Velvet Angelfish *Centropyge deborae* KN Shen, HC Ho & CW Chang, 2012

Deborah Smith is the wife of Walt Smith, who collected the type series and provided the underwater photograph and detailed collecting data of the new species.

De Buen

Chignahuapan Silverside *Poblana ferdebueni* A Solórzano Preciado & Y López-Guerrero, 1965
Snailfish sp. *Paraliparis debueni* AP Andriashev, 1986
Hagfish sp. *Myxine debueni* RL Wisner & CB McMillan, 1995
De Buen's Coris (wrasse) *Coris debueni* JE Randall, 1999

Don Fernando de Buen y Lozano (1895–1962) was a Spanish ichthyologist and oceanographer. He and his brother Rafael were the first curators of the ichthyological collection at the Malaga laboratory (1915–1926). They enhanced the collection with many specimens collected by Spanish trawlermen off Morocco, which Fernando catalogued. He worked and lived in Mexico, Uruguay and Chile. He was Director of the Department of Oceanography and Fisheries Service in Uruguay and Professor of Hydrobiology and Protozoology in the Faculty of Arts and Sciences. He was honoured in the name of the hagfish for his 'extensive work' on South American fishes. (Also see **Odón de Buen & Buen**)

Decary

Goby sp. *Acentrogobius decaryi* J Pellegrin, 1932

Raymond Decary (1891–1973) was a colonial administrator in Madagascar (1916–1944). He was a naturalist, botanist, geologist and ethnographer, and was interested in everything to do with Madagascar, contributing over 40,000 specimens of Malagasy flora to the Paris herbarium. He qualified in law (1912) before army service, when he was seriously wounded at the Battle of the Marne (1914) and unable to resume active service. He went to Madagascar (1916) as an officer in the Reserve, thus releasing a fully fit officer for active service. He trained as a colonial administrator (1921), returning to Madagascar (1922). He undertook seven scientific expeditions in the island (1923–1930) and became Director of Scientific Research there (1937). He was again in the French Army in Madagascar (1939–1944), returning to France after the Liberation. Demobilised (1945) he retired to private life, continuing his research. He wrote: *Malagasy Fauna* (1950). Three reptiles, a bird and an amphibian are named after him.

Decken

Pangani Suckermouth (catfish) *Chiloglanis deckenii* W Peters, 1868

Baron Karl Klaus von der Decken (1833–1865) was a German explorer who died in Somalia. He explored in East Africa and was the first European to try to climb Mount Kilimanjaro. Decken explored the region of Lake Nyasa on his first expedition (1860). He ascended Kilimanjaro (1862) to 13,780 feet, seeing its permanent snowcap and also establishing its height as about 20,000 feet. Another expedition (1863) took him to Madagascar, the Comoro Islands and the Mascarene Islands. Then, in Somalia (1865), he sailed the Jubba River, where his ship 'Welf' foundered in the rapids above Bardera, where local Somalis killed him and three other Europeans. He sent a considerable quantity of specimens from Somalia to the Museum in Hamburg. His letters were edited and published (1869) in book form under the title: *Reisen in Ost-Afrika*. The giant lobelia *Lobelia deckenii*, two mammals and a bird are named after him.

Deckert, K

Cameroon cichlid sp. *Coptodon deckerti* DFE Thys van den Audenaerde, 1967

Professor Dr Kurt Deckert (1907–1987) was (1946–1973) Curator of Fishes, Museum für Naturkunde der Humboldt-Universität zu Berlin, having initially started working there as a trainee and assistant in the herpetology department (1939). His whole career was spent there, apart from time in the military and as a prisoner of war (1940–1946). He had been a zoology and botany student there (1928) and was awarded his doctorate (1937). During his military deployment in North Africa, he took the opportunity to study the local fauna and flora. During this time, he not only collected for the Zoological Museum, but also sent living reptiles to Berlin. Among his published works were *Miracle World of the Deep Sea* (1950) and co-authored with his wife zoologist Giselda Haagen, *How Animals Behave* (1974). He helped the author during his visit to the Museum, and afterwards "*...spent many hours providing necessary and useful data for his cichlid research.*"

Deckert, GD

Golden Mojarra genus *Deckertichthys* FJ Vergara-Solana, FJ GarcIa-Rodriguez, JJ Tavera, E De Luna & J De La Cruz-AgUero, 2014

Dr Gary Dennis Deckert is an American ichthyologist who was Professor at the University of Northern Illinois. He has published numerous papers. He was honoured as it was he "*...who first recognized the distinctiveness of this new taxon, and for his contribution to the study of Gerreidae.*

Decorse

African Rivuline sp. *Aphyosemion decorsei* J Pellegrin, 1904

Spiny Eel sp. *Mastacembelus decorsei* J Pellegrin, 1919

Dr Gaston-Jules Decorse (1873–1907) was an army physician interested in botany, natural history, ethnography and linguistics. He travelled in Madagascar, where he collected botanical specimens (1898–1902) and joined a French expedition to Lake Chad (1902–1904) where he also collected much material that his friend Pellegrin studied. He co-wrote: *Rabah et les Arabes du Chari* (1905). Four reptiles and a bird are named after him.

Deetsie

Deetsie's Cardinalfish *Apogon deetsie* JE Randall, 1998

Dr Edith 'Deetsie' Chave née Hunter was an ichthyologist and marine biologist at the University of Hawaii and Associate Director of the Honolulu Aquarium. She married (1969) fellow marine zoologist Dr *Keith Ernest Chave (1928–1995)*, Department of Oceanography, University of Hawaii. She wrote: Ecological Requirements of *Six Species of Cardinal Fishes Genus* Apogon *in Small Geographical Areas of Hawaii* (1971) and co-wrote: *Hawaiian Reef Animals* (1973) and *In deeper waters – Photographic Studies of Hawaiian Deep-Sea Habitats and Life-Forms* (1998). She helped collect the holotype.

De Forges

Chesterfield Island Stingaree *Urolophus deforgesi* B Séret & PR Last, 2003

Dr Bertrand Richer-de-Forges (b.1948) is a French carcinologist and marine biologist who (since 1984) has been based at ORSTOM, Noumea, New Caledonia. He studied at the Pierre-et-Marie-Curie University and at Museum National d'Histoire Naturelle, Paris. He has organized many dredging expeditions in New Caledonia and the southwest Pacific Ocean and has collected numerous novelties in every group of marine animals. He co-wrote: *Lagons et récifs de Nouvelle-Calédonie* (2004). A large part of the zoological results from his expeditions are published in a series of the Muséum National d'Histoire Naturelle called: *Tropical Deep Sea Benthos*. He was honoured for promoting the exploration of the bathyal fauna off New Caledonia and for collecting valuable fish specimens from cruise surveys.

Degen

Degen's Leatherjacket *Thamnaconus degeni* CT Regan, 1903
Lake Victoria Catfish sp. *Bagrus degeni GA* Boulenger, 1906
Lake Victoria Cichlid sp. *Platytaeniodus degeni* GA Boulenger, 1906
[Syn. *Haplochromis degeni*]
Katonga River Mormyrid *Petrocephalus degeni* GA Boulenger, 1906
Labeo sp. *Labeo degeni GA* Boulenger, 1920

Edward J E Degen (1852–1922) was a Swiss naturalist and ornithologist who lived and worked in Australia, Africa and England; he died in London. He was educated in Basle and Paris. He was a member of the Field Naturalists Club of Victoria, Australia (1894) and was working for the National Museum in Melbourne (1894–1895). He collected reptiles, mammals and fish in East Africa (1895–1905) for the BMNH, but is particularly noted for his interest in birds and especially their moult, on which he wrote two important papers (1894 & 1895) and the extensive paper: *Ecdysis, as Morphological Evidence of the Original Tetradactyle Feathering of the Bird's Fore-limb, based especially on the Perennial Moult in Gymnorhina tibicen* (1902). After leaving Africa he worked as an articulator/taxidermist at the BMNH. He helped to collect the holotype of the labeo and made a watercolour painting of it, from which Boulenger wrote the official description. A reptile, a bird and an amphibian are named after him.

Degreef

Guatemalan Rivuline sp. *Cynodonichthys degreefi* GE Collier, 2016

Jaap-Jan De Greef (b.1957) was born on Sumatra and grew up in Mozambique, Tanzania, Suriname, Ecuador and Curacao. He then moved to Florida, where he now lives, to study marine biology at Florida State University, which awarded his BSc. He began collecting fish when just three years old. In Ecuador, aged 16, he collected fish for the Ecuadorian National Fish Institute in Guayaquil. In Curacao, he volunteered for a marine biological research station diving, collecting specimens, and assisting with other research projects. He works in the family business managing a tomato packing house, but whenever he can he travels the world collecting fish for researchers and fish farmers, and leads aquarium fish collecting eco-tours mostly in the Amazon. He continues to be an obsessive hobby aquarist with an 80–aquarium fish house, several ponds and 25 vats of fish.

DeGruy

DeGruy's Chromis *Chromis degruyi* R Pyle, JL Earle & BD Greene, 2008

Michael V deGruy (1951–2012) was a world-renowned award-winning nature cinematographer and marine scientist from Alabama. He is noted for his work on documentaries such as *Life in the Freezer, Trials of Life, Pacific Abyss* and *The Blue Planet*. He died in a helicopter crash soon after take-off from an airport at Nowra, near Sydney in Australia along with the pilot Andrew Wight who was a writer and TV producer. The etymology says the fish was named: "*...to honor Michael V. DeGruy, in recognition of the sincere enthusiasm and determination he demonstrated while attempting to collect the first adult specimen of this species.*"

Deguide

African Barb sp. *Enteromius deguidei* H Matthes, 1964

R Deguide was a collector for MRAC in the Ikela region of the Belgian Congo (DRC) (1955–1956). Matthes honoured him as a way of thanking him for his assistance during Matthes' researches in that region.

De Haven

Lanternfish sp. *Diaphus dehaveni* HW Fowler, 1934

Isaac Norris De Haven (1847–1924) of Philadelphia, Pennsylvania was a birder and sportsman and another of those many friends and associates for whom Fowler was 'indebted for many local fishes'.

Deheyn

African Tetra sp. *Phenacogrammus deheyni* M Poll, 1945

Jean Jacques Louis Joseph Deheyn (1914-2009) was a diplomat, educationalist, agronomist and naturalist. He qualified at the Agricultural Institute, Hainault, Belgium (1933), the Institut des Hautes Études, Brussels (1937) and the University of Brussels (1947). He was an Agricultural Officer, Leopoldville, Belgian Congo (1937–1947), Head of Agricultural Education there (1947–1954) and Director, Technical Education, Leopoldville (1955–1960). He then worked variously for the International Bank for Reconstruction and Development (1961–1962) and UNESCO. He collected the holotype.

Deignan

Oriental Bitterling sp. *Acheilognathus deignani* HM Smith, 1945

Herbert 'Bert' Girton Deignan (1906–1968) was a fellow of the John Simon Guggenheim Memorial Foundation (1952) and worked for the USNM (1938–1962), where he was an Associate Curator of Birds. He graduated from Princeton (1928) with a degree in European languages, then (1928–1932) taught English at a school at Chiang Mai, Siam (Thailand). He was an associate of Alexander Wetmore, who helped him get a temporary job at the USNM (1933). Deignan then worked at the Library of Congress (1934–1935) before returning to his old job in Thailand (1935–1937) and combining teaching with collecting birds for Wetmore. He returned to the USA and worked at the USNM (1938–1944) before going back to southern Asia as an agent of the Office of Strategic Services (the forerunner to the CIA). He returned to the USNM again (1946) until retiring (1962) to Switzerland. He wrote: *The Birds of Northern Thailand* (1945), *Type Specimens of Birds in the United States National Museum* (1961) and *Checklist of the Birds of Thailand* (1963). He collected the holotype of this species. He is also remembered in the name of a reptile and the names of nine birds.

Deissner

Deissner's Licorice Gourami *Parosphromenus deissneri* P Bleeker, 1859

F H Deissner was a local Officer of Health in Indonesia. He made a collection of fish from the island of Banka (Bangka) which he sent to Bleeker. There appears to be nothing more recorded about him.

De Jong

Cuban Basslet *Gramma dejongi* BC Victor & JE Randall, 2010

Arie De Jong (b.1954) is owner-operator of De Jong Marinelife, a marine aquarium-fish supplier based in the village of Spijk, The Netherlands. His parents started the business (1958), which Arie later took over, travelling around the tropics looking for fishes, and sometimes uncovering new species. He first recognized the basslet as a new species and provided the type specimens.

Dekeyser

Armoured Catfish genus *Dekeyseria* Rapp Py-Daniel, 1985

Dr Pierre Louis Dekeyser (1914–1984) was a zoologist and ethnologist. In the 1950s he worked at L'Institut Français d'Afrique Noire in Dakar (Senegal). He wrote a number of papers on both of his subjects: *Les mammiferes de l'Afrique noire francaise* (1955) and, jointly with A Villiers: *Contribution a l'etude du peuplement de la Mauritanie. Notations ecologiques et biogeographiques sur la faune de l'Adrar* (1956). He also published several articles on the avifauna of Brazil and seems to have worked in Sao Paulo in the early 1980s. A mammal is named after him.

De Kimpe

Roundbelly Pellonuline *Laeviscutella dekimpei* M Poll, PJP Whitehead & AJ Hopson, 1965

Squeaker Catfish sp. *Synodontis dekimpei* D Paugy, 1987

Paul De Kimpe (b.1927) was the Fisheries Officer at Cotonou, Dahomey (now Republic of Benin). He collected not only the above two species but many others for the Musée de l'Afrique Centrale in Tervuren, Belgium (MRAC). He was Project Manager, Regional Fish-culture Training and Research, a programme sponsored by the UN's FAO (1969) based in Abidjan (Ivory Coast) and covering also the Central African Republic and Cameroon.

Delacour

Sisorid Catfish sp. *Oreoglanis delacouri* J Pellegrin, 1936

Dr Jean Theodore Delacour (1890–1985) was a French-American ornithologist renowned for discovering and rearing some of the rarest birds in the world. He was born in Paris and died in Los Angeles. In France (1919–1920) he created the zoological gardens at Clères and donated them to the French Natural History Museum in Paris (1967). He undertook a number of collecting expeditions to Indochina, particularly in search of rare pheasants. He wrote: *Birds of Malaysia* (1947) and co-wrote: *Birds of the Philippines* (1946) among many other ornithological books, as well as his memoirs: *The Living Air: The Memoirs of an Ornithologist*. A number of taxa are named after him including eighteen birds, an amphibian, a reptile and three mammals.

Delais

Black-faced Blenny *Tripterygion delaisi* J Cadenat & J Blache, 1970

Michel Delais worked at the Institute Française d'Afrique Noire at Gorée in Senegal under Dr J Cadenat. He captured virtually all the specimens that the authors studied, including the blenny holotype off Senegal (1950).

Delalande

Delalande's Blenny *Malacoctenus delalandii* A Valenciennes, 1836
Brazilian Sharpnose Shark *Rhizoprionodon lalandii* A Valenciennes, 1839

Pierre Antoine Delalande (1787–1823) was a French naturalist and explorer who worked for Muséum National d'Histoire Naturelle, Paris. He collected in the region around Rio de Janeiro (1816) with Auguste de Saint-Hilaire, and in the African Cape with his nephew, Jules Verreaux, and Andrew Smith (1818). Later Geoffroy Saint-Hilaire employed him as a taxidermist. He collected the holotypes of the two eponymous fishes. Three birds and three reptiles are named after him.

Deland

Deland's Dragonet *Synchiropus delandi* HW Fowler, 1943

Dr Judson de Land of Philadelphia, "to whom I am indebted for American fishes", seems to have been one of Fowler's large circle of specimen-providers. By the time of the description, Fowler refers to him as 'the late' Dr Judson de Land.

Delfin

Kelpfish sp. *Chironemus delfini* C Porter, 1914
Large-tooth Flounder sp. *Paralichthys delfini* G Pequeño Reyes & R Plaza, 1987

Dr Federico T Delfin was a Chilean ship's surgeon and an ichthyologist who was a research associate at the Museo de Valparaiso, Chile. Among his publications were: *Nuevo pez para la fauna de Chile* (1899) and *Catálogo de los peces de Chile* (1901). He was honoured in the flounder's name "*…just over 80 years after his death*" (translation), for his contributions to Chilean ichthyology.

Delhez

Claroteid Catfish sp. *Chrysichthys delhezi* GA Boulenger, 1899
Barred Bichir *Polypterus delhezi* GA Boulenger, 1899

Henri Paul Delhez (1870–1900) was a Belgian artist and naturalist who made collections of fishes in the Congo, and whose field observations Boulenger found most useful in the preparation of his book on Congo fishes.

de Lima

Armoured Catfish sp. *Hypostomus delimai* CH Zawadzki, RR de Oliveira & T Debona, 2013

Characin sp. *Charax delimai* NA de Menezes & CAS de Lucena, 2014

Tetra sp. *Gymnocorymbus flaviolimai* RC Benine, BD de Melo, RMC Castro & C de Oliveira, 2015

Tetra sp. *Hyphessobrycon delimai* TF Teixeira, AL Nett-Ferreira, JLO Brindelli & LM Sousa, 2016

Dr Flávio César Thadeo de Lima (b.1974) is a Brazilian zoologist and ichthyologist who is a specialist in the systematics of freshwater Neotropical fish of the order Characiformes. The Universidade de São Paulo awarded his bachelor's degree (1998), his master's (2001) and his doctorate (2006). His post-doctoral studies were at the same university (2007–2010). He is (since 2011) an associate researcher at Universidade Estadual de Campinas and is Curator of the ichthyology collection in the university's Museum of Zoology. He has over 170 publications to his name including the co-written: *New catfish of the genus* Aspidoras *(Siluriformes: Callichthyidae) from the upper rio Paraguai system in Brazil* (2001) and *Hyphessobrycon vanzolinii, a new species from Rio Tapajós, Amazon basin, Brazil (Characiformes: Characidae)* (2016).

Dell

Clingfish genus *Dellichthys* JC Briggs, 1955

Richard 'Dick' Kenneth Dell (1920–2002) was Director, Dominion Museum (now Te Papa). After study at Auckland University College he took a teacher's course but then joined the New Zealand Artillery during WW2, serving in the Solomon Islands, Egypt and Italy. After the war he was a malacologist at the Dominion Museum, building up the collection to 30,000 specimens. While in that post he studied for his masters at Victoria University of Wellington. He took part in the Chatham Islands Expedition (1954) and his publication of the results *The Archibenthal Mollusca of New Zealand* (1956) earned him his doctorate. He worked on the Antarctic collections, publishing a monograph on the Antarctic's bivalves, chitons and scaphopods (1964). He was promoted to Assistant Director (1961) then Director (1966) until retirement (1980), thereafter concentrating on writing. He published his standard work *Antarctic Mollusca with special reference to the Fauna of the Ross Sea* (1990) and a total of 150 papers. He was honoured in the name for his interest in the shore fishes of New Zealand and his generosity in providing Briggs with 'material of great value' for his monograph. He died after a long illness.

Deloach

Two-bar Triplefin *Enneanectes deloachorum* BC Victor, 2013

Spotfin Fangblenny *Adelotremus deloachi* WF Smith-Vaniz, 2017

Ned DeLoach (b.1944) graduated from Texas Tech University (1967) and moved to Florida to take up diving and underwater photography. Along with Paul Humann, he has published a series of field guides on coral reef fauna, such as *Reef Fish Identification: Baja to Panama* (2004). The fangblenny etymology states that DeLoach's publications "*...have encouraged numerous divers and fish watchers to become more aware of the importance of protecting the threatened marine environment and fauna*". The triplefin is named after Ned and his wife Anna, who is also a scuba diver, underwater naturalist and founder of the blog *BlennyWatcher* (2011).

Delsman

Asian Bonytongue sub-genus *Delsmania* HW Fowler, 1934
Delsman's Flounder *Etropus delsmani* P Chabanaud, 1940

Hendricus Christoffel Delsman (1886–1969) was a fisheries biologist. He was Director of the Fishery station in Batavia (Jakarta), Dutch East Indies (Indonesia). He co-wrote: *De indische zeevisschen en zeevisscherij* (1934).

Del Solar

Trident Grenadier *Coryphaenoides delsolari* N Chirichigno F & T Iwamoto, 1977

Enrique del Solar Cáceda (1911–1990) was a Peruvian marine biologist, who acted as technical and scientific advisor to the Sociedad Nacional de Pesqueria (National Fisheries Society) from its foundation (1951) to 1968. He represented Peru at various international conferences relating to fisheries and marine resources. The crab genus *Delsolaria* is also named after him.

De Luca

Neotropical Rivuline sp. *Hypsolebias delucai* WJEM Costa, 2003
[Syn. *Simpsonichthys delucai*]

André C De Luca is a Brazilian ichthyologist and ornithologist at the Federal University of Rio de Janeiro. He was the first collector of this species. He has described at least nine taxa, all with Wilson Costa, such as: *Rivulus unaensis, a new aplocheiloid killifish of the subgenus Atlantirivulus from eastern Brazil (Cyprinodontiformes: Rivulidae)* (2009) and *Rivularus cajariensis, a new killifish from the Guiana Shield of Brazil, eastern Amazon (Cyprinodontiformes: Rivulidae)* (2011).

Delvari

Delvari's Loach *Paraschistura delvarii* H Mousavi-Sabet & S Eagderi, 2015

Rais Ali Delvari (1882–1915) was an independence fighter and activist against British colonialism. He is now remembered as the national hero of Iran, who organised popular resistance against British troops, which had invaded Iran in 1915.

Delyamure

Chornaya Gudgeon *Gobio delyamurei* J Freyhof & AM Naseka, 2005

Dr Semion Lyudvigovich Delyamure was a Soviet zoologist, parasitologist and helmintholgist at Taurida National V I Vernadsky University, Simferopol in the Crimea where he was Professor of Zoology (1949–1980) and established and maintained an internationally famous centre for the study of the helminths of marine mammals. He also founded the university's zoological museum (1965). He wrote several books and scientific papers on Crimean fishes, but most of his work (1950s through 1980s) concerned parasites, such as: *Key to Parasitic Nematodes. Vol III: Strongylata* (1961) and *Helminthofauna of marine mammals (ecology and phylogeny).* (1968).

Demason

Lake Malawi Cichlid sp. *Chindongo demasoni* AF Konings, 1994

Laif DeMason of Homestead, Florida, USA is an aquarist and cichlid expert. He is the author of many articles in Cichlid News Magazine as well as books such as: *A Guide To The Tanzanian Cichlids of Lake Malawi* (1995). The etymology says of DeMason "*... without whom our preliminary survey could not have been completed. Mr. DeMason, importer, exporter, and breeder of cichlids, has greatly stimulated the keeping of cichlids in the USA.*"

Dementjev

Minnow sp. *Rhynchocypris dementjevi* FA Turdakov & KV Piskarev, 1954

Petr Petrovich Dementiev (recently deceased at time of the fish's description) was an ichthyologist who worked in Kyrgystan, which is the type locality.

Demerara

Driftwood Catfish sp. *Auchenipterus demerarae* CH Eigenmann, 1912

This is a toponym, referring to the Demerara River in Guyana.

Demeuse

Congo cichlid sp. *Haplochromis demeusii* GA Boulenger, 1899
[Syn. *Thoracochromis demeusii*]

Dr Fernand Alexandre Robert Demeuse (1863–1915) was a Belgian scientist and photographer who was a professor of natural science. After studying at Liège University, he joined the botanist Auguste Linden on an expedition to Central Africa (1886) and then continued exploring with a trader (1887–1889). He explored, collected and photographed in the upper Congo. He again explored (1892–1893) and on his return was made an honorary member of the Royal Anthropological Society. The etymology says that 'J De Meuse' collected the type. However, we believe this is likely a mistaken initial or transcription error.

Demidoff

Common Percarina *Percarina demidoffii* A von Nordmann, 1840

Count Anatoly Nikolaievich Demidov (Demidoff) (1813–1870) was a Russian diplomat and patron of the arts and sciences. He organised (1837–1838) a scientific expedition of 22 scholars, writers and artists to southern Russia and the Crimea, at a cost of 500,000 francs. The results were published as *Voyage dans la Russie méridionale et la Crimée* (4 vol., 1840–1842). The description of the percarina was published in the 3rd volume.

Demir

Eastern Aegean Bleak *Alburnus demiri* M Özuluğ & J Freyhof, 2007

Dr Muzaffer Demir is a Turkish marine biologist who was a Professor at the University of Istanbul and Director of the Hydro-biological Station. His publications (1950s-2000s) mainly related to marine organisms, including the co-written: *Biological and hydrological factors controlling the migration of mackerel from the Black Sea to the Sea of Marmara* (1955). He was honoured "*…for great contributions to the knowledge of Turkish benthic invertebrates and marine fishes.*" A shrimp is also named after him.

Dempster

Oblique-banded Stingfish *Minous dempsterae* WN Eschmeyer, LE Hallacher & KV Rama Rao, 1979

Lillian J Dempster (1905–1992) worked for the California Academy of Sciences. The University of Washington awarded her degree in languages; knowledge she put to good use in Washington DC during WW2, as she knew Russian, German, French, Spanish, Italian, Portuguese and Swedish! She joined the California Academy of Sciences as a secretary (1946) but was soon curating and administering the ichthyology department, becoming research assistant then (1955) Assistant Curator and (1961) Associate Curator working with Curator Bill Follett for 40 years. Among her published papers is the co-written *List of Fishes of California* (1979). The etymology reads thus: "*The species is named for Lillian J. Dempster, a friend and colleague, in recognition for her assistance in the preparation of this and other papers on scorpionfish.*"

Denise

Denise's Pygmy Seahorse *Hippocampus denise* SA Lourie & JE Randall, 2003

Denise Tackett née Nielsen (1947–2015) was a well-respected American writer and photographer of marine life in general and reef life in particular. She travelled the tropics for many years, primarily in the Indo-Pacific, photographing the beauty and biodiversity of coral reefs. She lived in twelve different countries and for ten years owned an international travel and photography business. She and her husband Larry also spent 13 years collecting sponges and other organisms in a search for disease-treating natural compounds being sought by the Cancer Research Institute of Arizona State University. She was the author of three books including: *Reef Life: Natural History and Behaviors of Marine Fishes and Invertebrates* (2002). She discovered the eponymous species.

Denison

Denison's Barb *Sahyadria denisonii* Day F, 1865
Stone Loach sp. *Schistura denisoni* Day F, 1867

Sir William Thomas Denison (1804–1871) became Lieutenant Governor of Van Diemen's Land (Tasmania) (1846) and Governor-General of New South Wales, Van Diemen's

Land, Victoria, South Australia, and Western Australia (1854). He was later Governor of Madras (1861–1866), before returning to England via the newly opened Suez Canal. His last public appointment sounds contemporary: he was Chairman of an enquiry into the pollution of rivers in Britain. He was a conchologist with a collection of 8,000 species of Australian shells. Port Denison, Western Australia, is named after him as is a genus of reptiles.

Dennis Yong

Betta sp. *Betta dennisyongi* HH Tan, 2013

Dennis Yong Ghong Chong is a Malaysian naturalist. The etymology says he is: "*...a distinguished and knowledgeable naturalist well experienced in many facets of tropical Southeast Asian fauna and flora with an avid interest in labyrinth fishes. He has accompanied the author on many trips and shared many interesting stories, tips and gastronomic delights.*"

Denny

Marbled Snailfish *Liparis dennyi* DS Jordan & EC Starks, 1895

Charles Latimer Denny (1861–1919) was a wealthy real estate agent in Seattle (the city that his father, Arthur Armstrong Denny, politician, banker and industrialist, founded). The authors honoured him "*...in recognition of his active and intelligent interest in the natural history of Washington.*"

Denoncourt

Golden Darter *Etheostoma denoncourti* JR Stauffer & ES van Snik, 1997

Dr Robert F 'Bob' Denoncourt is an American ichthyologist. He was Associate Professor of the Department of Biological Sciences at York College of Pennsylvania (1970s) and affiliated to Loyola University Maryland. Among his published papers are: *A systematic study of the gilt darter* Percina evides *(Jordan & Copeland) (Pisces, Percidae)* (1969) and *A Freshwater Population of the Mummichog,* Fundulus heteroclitus, *from the Susquehanna River Drainage in Pennsylvania* (1978)

Denton

Rattail sp. *Kumba dentoni* NB Marshall, 1973

Dr Sir Eric James Denton (1923–2007) was a marine biologist who was originally a physicist. He taught at Aberdeen University (1948–1955), during which time Aberdeen University awarded his doctorate (1952). He joined the Marine Biological Association Laboratory, Plymouth (1956) and worked there until retirement (1987) as Emeritus Professor, Bristol University. He was the Director (1974–1987).

De Pauw

Squeaker Catfish sp. *Synodontis depauwi* GA Boulenger, 1899

Louis François De Pauw (1876–1943) was head taxidermist and preparator at the Royal Belgian Institute of Natural Sciences, where he supervised the first reconstruction of an Iguanodon in the St. George Chapel, Brussels (1882). Later he became curator of collections at l'Université libre de Bruxelles and installed there an exhibition of Congo fishes (1897), for which Boulenger honoured him.

Depierre

Claroteid Catfish sp. *Notoglanidium depierrei* J Daget, 1980

Daniel Depierre is the Chief Engineer of Rural Engineering and Water and Forests. He was previously at the National Superior School of Agronomy, Yaoundé, Cameroon. He wrote *Wild Mammals of Cameroon* (1992). He collected fish and his collection, including the holotype of this species, was published by Daget.

de Pinna

Depinna's Catfish *Aspidoras depinnai* M Ribeiro de Britto, 2000
Pencil Catfish sp. *Listrura depinnai* L Villa-Verde, J Ferrer dos Santos & LR Malabarba, 2014

Dr Mário Cesar Cardoso de Pinna is a Brazilian ichthyologist and evolutionary biologist who is the deputy director of the Museum of Zoology, University of São Paulo, Ipiranga. He graduated in biological sciences at Universidade Federal do Rio de Janeiro (1988) and was awarded his doctorate in evolutionary biology by the City University of New York (1992). He did post-doctoral work at both the Smithsonian (1993–1994) and the Field Museum, Chicago (1994–1995). He wrote: *Higher-level phylogeny of Siluriformes, with a new classification of the order (Teleostei, Ostariophysi)* (1993).

Deppe

Nautla Cichlid *Herichthys deppii* J Heckel, 1840

Ferdinand Deppe (1794–1861) was a horticulturist, naturalist, explorer, collector and artist. He arrived in Mexico (1824) with Count von Sack, an irresolute 'expedition leader' who soon returned to Germany while Deppe stayed in Mexico (1827). He made a brief visit home, returning to Mexico with botanist Wilhelm Schiede (until 1836). Many of the specimens he collected went to the Berlin Zoology Museum. He collected the cichlid holotype. Four reptiles, two birds and a mammal are also named after him.

Deraniyagala

Indian Long-tailed Sand-Eel *Bascanichthys deraniyagalai* AGK Menon, 1961
Redneck Goby *Schismatogobius deraniyagalai* M Kottelat & R Pethiyagoda, 1989

Professor Paulus Edward Peiris Deraniyagala (1900–1973) was a palaeontologist and zoologist, particularly a herpetologist, as well as being an accomplished artist. He was Director of all National Museums, Ceylon (Sri Lanka) (1939–1963) and Vice Chancellor of the University of Ceylon. He was also the Dean of the Faculty of Arts at the Vidyodaya University (now Sri Jayewardenepura University) (1961–1964). He wrote at least 93 papers and at least nine books, including: *Some Vertebrate Animals of Ceylon* (1949), *The Pleistocene of Ceylon* (1956) and *A general guide to the Colombo National Museum* (1961). He took a number of study trips to China. He had a wide range of interests, including early humans, reptiles and elephants. In his youth, he was a notable flyweight boxer who defeated the champion of the British Empire's Armed Forces (1923). Six species of reptile are named after him.

Derby, CF

Armoured Catfish sp. *Loricariichthys derbyi* HW Fowler, 1915

C F Derby collected the holotype from the Rio Jaguaribe at Barro Alto, Brazil (1913). We have been unable to find anything more about him.

Derby, OA

Armoured Catfish sp. *Hypostomus derbyi* JD Haseman, 1911

Dr Orville Adalbert Derby (1851–1915) was an American geologist. Cornell University awarded his bachelor's degree (1873) and his doctorate (1874). He was an assistant professor at Cornell (1873–1875). He was nominated as assistant to the Geographical Commission of the Empire of Brazil (1875–1877) and then joined the National Museum in Rio de Janeiro and served also as Director, Geographic and Geological Commission of São Paulo (1886–1904). He founded the first botanical gardens in São Paulo. Haseman commented that Derby: "*…has spent thirty-five years in the cause of science in Brazil, and who rendered me more assistance than any other man in South America*". He wrote: *O picos altos do Brasil* (1889). He led a solitary life, living in hotel rooms. He committed suicide a few months after becoming a Brazilian citizen.

De Rham

Pencilfish genus *Derhamia* J Géry & A Zarske, 2002
African Characin sp. *Brycinus derhami* J Géry & V Mahnert, 1977
Neotropical Rivuline sp. *Anablepsoides derhami* J-F Fels & JH Huber, 1985
Madagascar Rainbowfish sp. *Rheocles derhami* MLJ Stiassny & DM Rodríguez, 2001
Characin sp. *Cyphocharax derhami* RP Vari & F Chang, 2006

Dr Patrick Henri de Rham (b.1936) is a Swiss zoologist, botanist, ichthyologist and aquarist in Lausanne, and was Expert Ecologist to the Coopération Technique Suisse. He is a Director of the Aquatic Conservation Network. (Also see **Patrick** and **Rham**)

Derjavin

Sculpin sp. *Radulinopsis derjavini* VK Soldatov & GU Lindberg, 1930
Spined Loach sp. *Cobitis derzhavini* ED Vasil'eva, EN Solovyeva, BA Levin, & VP Vasil'ev, 2020

Aleksandr Nikolayeviç Derjavin (Derzhavin) (1878–1963) was a Russian biologist. He graduated from the Department of Natural Sciences at Kazan University (1902) after which he began to research. He participated in the Kamchatka Expedition organised by the Russian Geographical Society and engaged in fish research in the rivers in the southern part of the peninsula (1908–1910). He was at Astrakhan Echo-laboratorial Laboratory (1910–1912), then became the head of the Department of Agriculture at Baku (1912–1927). While working at the Institute of Zoology, he developed scientific bases for designing fish farms in connection with increasing stocks of sturgeon. He wrote *Restoration of Caspian*

Sea Reservoir Resources (1941), *Restoration of Sturgeon's Reserve Resources* (1947) and *Azerbaijan's Sweet Fish* (1949).

Derjugin

Georgian Shemaya *Alburnus derjugini* LS Berg, 1923
Leather-fin Lumpsucker *Eumicrotremus derjugini* AM Popov, 1926
Ronquil sp. *Bathymaster derjugini* GU Lindberg, 1930
Eelpout sp. *Lycodapus derjugini* AP Andriashev, 1935
Derjugin's Snailfish *Careproctus derjugini* NV Chernova, 2005

Professor Dr Konstantin Michailovich Derjugin (1878–1938) was an oceanographer and marine zoologist at Leningrad State University, and manager of the Oceanic Division of the State Hydrological Institute in Leningrad. He conducted major research on classifying Arctic aquatic taxa and re-established the Kola meridian water-sampling station. He also helped to organise over fifty expeditions in the White Sea. He led the Russian Pacific Expedition on the research vessel 'Gagara' (1932–1935). A reptile and a bird are also named after him, as is Deryugina Bay in the Sea of Okhotsk which was surveyed during the 'Gagara' voyage.

de Roy

Deroy's Cusk-eel *Ogilbia deroyi* M Poll & JJ van Mol, 1966

André de Roy moved (1955) with his wife Jacqueline, daughter Tui and son Gil to Academy Bay, Santa Cruz Island, Galápagos Islands, which is where he helped collect the holotype. He was also a commercial fisherman, and helped the authors by collecting and/or providing type material. He and his wife were avid collectors of Galapagos marine molluscs. He co-wrote: *Shelling in the Galapagos* (1967).

Derzhavin

(see **Derjavin** above)

de Santana

Glass Knifefish sp. *Eigenmannia desantanai* LAW Peixoto, GM Dutra & WB Wosiacki, 2015

Dr Carlos David de Santana is a Brazilian ichthyologist who is presently a post-doctoral fellow at the Smithsonian. The Universidade Federal de Pernambuco awarded his bachelor's degree, the UNuiversidaade Catolica de Pernambuco his master's and Instituto Nacional de Pesquisas da Amazonia his doctorate. He co-wrote: *A New Species of the Glass Electric Knifefish Genus* Eigenmannia *Jordan and Evermann (Teleostei: Gymnotiformes: Sternopygidae) from Río Tuíra Basin, Panama* (2017).

Descamps

Descamps' Strange-tooth Cichlid *Ectodus descampsii* GA Boulenger, 1898

Captain Georges Descamps was a Belgian officer who was prominent in the fight against slave traders in East Africa. He was captain and commandant of the Belgian antislavery movement at Lake Tanganyika. He obtained the type and sent it to Boulenger.

Desfontaines

Tunisian Blue-lip Cichlid *Astatotilapia desfontainii* BG de Lacepède, 1802

René Louiche Desfontaines (1750–1833) was a French botanist and collector in Tunisia and Algeria (1783–1785). He became Professor of Botany at Le Jardin des Plantes, Paris (1786) and later Director, MNHN and was a friend and colleague of Lacepède. He found the cichlid in the thermal waters of Tunisia. A bird is also named after him.

De Silva

Lesser Swamp Eel *Monopterus desilvai* RM Bailey & C Gans, 1998

Dr Pilippu Hewa Don Hemasiri de Silva (1927-2020) was a zoologist and herpetologist who was Director of the National Museum of Ceylon (Sri Lanka) (1965–1981). Among his many publications are *Museums in Ceylon* (1966) and *One hundred years of museum publications, 1877–1977* (1977). The etymology says: "*We take pleasure in naming this species for Dr. P. H. S. H. de Silva, former director of the National Museums of Ceylon, herpetologist and zoologist. This acknowledges his personal hospitality and support to C.G. during field work on the island as well as much professional advice on local conditions and natural history.*" We think the different initial may be an error. He also has a reptile named after him.

Desio

Pupfish sp. *Aphanius desioi* L Gianferrari, 1933
[Junior Synonym of *A. fasciatus*]

Count Ardito Desio (1897–2001) was an Italian mountaineer, geologist, cartographer and explorer. The University of Florence awarded his degree in Natural Science (1920). He was an assistant at that university (1921–1923) and at the University of Pavia (1923–1924) and the University of Milan (1924–1927). He was lecturer in Physical Geography, Geology and Palaeontology (1928–1931), Professor of Geology at the University of Milan, and Applied Geology at the Engineering School of Milan (1932–1972). Concurrently, he was a consultant geologist for the Edison Company for hydroelectric plants in Italy, Spain, Switzerland, Greece, Turkey and Brazil; and in the same capacity for the Public Power Corporation of Greece. He became Professor Emeritus at the University of Milan (1973). He explored in the Alps and Apennines (1920) and Libya (1926 & 1930–1933, 1935 & 1936, 1936–1940). He also explored in Iran, northern Pakistan, Afghanistan, the Philippines, Tibet and Antarctica.

Desjardin

Indian Sailfin Surgeonfish *Zebrasoma desjardinii* ET Bennett, 1836

Julien François Desjardins (1799–1840) was a French zoologist who spent several years on Mauritius. He was one of the founders of the *Société d'Histoire Naturelle de l'Île Maurice* (1829) and the Society's first secretary. He left Mauritius for France (1839) with a large collection of natural history specimens, but died not long after his return.

de Soto

Gulf Sturgeon *Acipenser oxyrinchus desotoi* VD Vladykov, 1955

Hernando de Soto (1496–1542) was a conquistador and an explorer in the Gulf of Mexico. His reputation was of a good fighter, horseman and tactician but he was notorious for his brutality. He was in Nicaragua and led an expedition up the coast of the Yucatan peninsula to seek a (non-existent) passage between the Atlantic and Pacific Oceans (1530). He became one of Pizzaro's commanders in the destruction of the Incan Empire. He led an expedition from Florida through what are now the states of Georgia, Alabama, Arkansas and Louisiana. He died on the banks of the Mississippi River, which he was one of the first Europeans to see and cross.

Desoutter

Crested Flounder sp. *Samariscus desoutterae* J-C Quéro, DA Hensley & AL Maugé, 1989

Martine Desoutter-Méniger has spent her entire career at Museum National d'Histoire Nationale, Paris, where she studied the collections and made a special study of Acanthomorphs.

Despax

Headstander Characin sp. *Hypomasticus despaxi* J Puyo, 1943

Raymond Justin Marie Despax (1886–1950) was a French zoologist, entomologist, and herpetologist. He wrote: *Sur trois collections de reptiles et de batraciens provenant de l'archipel Malais* (1912). An amphibian and three reptiles are also named after him.

Desta

Triplefin sp. *Enneapterygius destai* E Clark, 1980

Prince Alexander (also Iskinder and Eskander) Desta (1934–1974) was Admiral of the former Ethiopian Imperial Navy; he was one of 60 imperial officials who were executed when the Derg took over the country (where this blenny occurs in the Red Sea).

Devaney

Deepwater Cardinalfish sp. *Epigonus devaneyi* O Gon, 1985

Dr Dennis M Devaney (1938–1983) was a prominent echinoderm researcher. His BA was awarded by the Occidental College, California (1960), his MA by the University of California at Los Angeles and his PhD by the University of Hawaii (1968) following which he was a Postdoctoral Fellow at the Smithsonian (1968–1969). He was an Invertebrate Zoologist at the Bishop Museum, Hawaii (1967) rising to Chairman of Invertebrate Zoology there. His publications include: *Shallow-water echinoderms from British Honduras, with a description of a new species of* Ophiocoma *(Ophiuroidea)* (1974). He died in a diving accident off the island of Hawaii.

Dev Dev

Stone Loach sp. *Schistura devdevi* SL Hora, 1935
Oriental Cyprinid sp. *Bangana devdevi* SL Hora, 1936

Dev Dev Mukerji (1903–1937) was an Indian zoologist and ichthyologist. He studied at St Xavier's College, Calcutta (1919) and was awarded his Zoology BSc (1923), then attended Calcutta University which awarded his MSc (1925). He joined the Zoological Survey of

India as a technical assistant (1926) where he became Hora's assistant. He published his first paper (1927) on two 'pug-headed' catfish specimens. Among his publications is the co-written: *European Species of Fish from the Tavoy Coast, Burma* (1936). He worked on a number of collections and in the field. He died in Calcutta after a brief illness; at the time he was working on a bulletin on Indian freshwater fishes for the Malarial Survey of India. Hora wrote his obituary.

De Vecchi

Somalian Blind Barb *Barbopsis devecchii* L Di Caporiacco, 1926

Cesare Maria De Vecchi, Count of Val Cismon (1884–1959), was a colonial administrator and a Fascist politician who was elected to the Chamber of Deputies (1921). He was governor of Italian Somaliland (now part of Somalia) (1923–1928). He was the first Italian ambassador to the Vatican after the Concordat (1929) and was Minister of Education (1935–1936). He served (1936–1940) as Governor of the Aegean Dodecanese Islands, which were part of the Kingdom of Italy (1912–1943). He voted in favour of the deposing of Mussolini (1943) and was sentenced to death in absentia. He acquired a Paraguayan passport and escaped to Argentina, only returning to Italy later (1949). He refused to take an active part in post-war Italian politics or governments and lived quietly in Rome.

De Venanzi

Ghost Knifefish sp. *Adontosternarchus devenanzii* F Mago-Leccia, JG Lundberg & JN Baskin, 1985

Dr Francisco De Venanzi (1917–1987) was a physician and scientist who became the first Rector, Universidad Central de Venezuela, Caracas (1959–1963). He qualified as a physician (1942) at the Central University of Venezuela and completed a master's degree in biochemistry at Yale (1945. He became a Professor in the Faculty of Medicine, Central University of Venezuela, resigning (1951) in protest at the actions of the military junta, then in charge. He was honoured for having encouraged the senior author to study fishes.

Devi

Devi's Loach *Mesonoemacheilus remadevii* CP Shaji, 2002

Dr Karunakaran Rema Devi (See **Rema Devi**)

Devidé

Armoured Catfish sp. *Hisonotus devidei* FF Roxo, GSC Silva & BF Melo, 2018

Renato Devidé is a 'dear friend' of the authors and collector of the catfish holotype. He was honoured "...for his immeasurable contribution during more than 30 years as an academic technician in the LBP [Laboratório de Biologia e Genética de Peixes, Universidade Estadual Paulista, Botucatu, Brazil] fish collection, assisting and coordinating expeditions that resulted in numerous scientific publications, theses and dissertations in the fields of ecology, cytogenetics, population genetics, taxonomy, systematics and evolution of Neotropical fishes."

De Ville

Characin sp. *Brycon devillei* F de L de Castelnau, 1855

Émile de Ville (1824–1853) was a French naturalist and taxidermist who collected in South America (1843–1847). King Louis-Philippe of France ordered Castelnau (q.v.), accompanied by Deville (among others), to explore Brazil and Peru during this period. Nine birds and an amphibian are named after him.

Devincenzi

Armoured Catfish sp. *Pseudohemiodon devincenzii* JS Señorans, 1950

Dr Garibaldi José Devincenzi (1882–1943) was a Uruguayan naturalist and ichthyologist, who was qualified as a physician (1909). He was Director, Uruguay Museum, Montevideo (1912–1942). He produced the first systematic catalogue of Uruguayan fish, reptiles, mammals and birds, and wrote: Peces del Uruguay (1924). An amphibian is named after him.

De Vis

Devis' Anchovy *Encrasicholina devisi* GP Whitley, 1940
Australian Slender Wrasse *Suezichthys devisi* GP Whitley, 1941

Charles Walter De Vis (1829–1915) was a zoologist and clergyman. Magdalene College, Cambridge awarded his BA (1849) and he became a deacon (1852) and then a Rector in Somerset (1855–1859). He gave up his work as a clergyman (1862) to become Curator, Queens Park Museum, Manchester. He emigrated from England to Australia (1870), where he became Librarian, School of Arts, Rockhampton, Queensland. He published many popular articles under the pen name of 'Thickthorn', which brought him to the attention of the Trustees of the Queensland Museum. They recruited him to be the museum's first Director (1882–1905); he remained a consultant (until 1912). He was a founding member of the Royal Society of Queensland (1884) and its President (1888–1889), and a founder and first Vice President of the Australasian Ornithologists' Union (1901). He wrote c.50 papers on herpetological subjects (1881–1911). Two mammals, five reptiles and two birds are named after him.

De Vos

Tanzanian Cichlid sp. *Neolamprologus devosi* R Schelly, MLJ Stiassny & L Seegers, 2003
Wenje Mormyrid *Marcusenius devosi* BJ Kramer, PH Skelton, HF van der Bank & Wink, 2007
African Rivuline sp. *Fenerbahce devosi* R Sonnenberg, T Woeltjes & JR Van der Zee, 2011
African Barb sp. *Enteromius devosi* G Banyankimbona, EJ Vreven & J Snoeks, 2012
Squeaker Catfish sp. *Chiloglanis devosi* RC Schmidt, HL Bart & DW Nyingi, 2015

Dr Luc De Vos (1957–2003) was a Belgian ichthyologist who was Curator of Fishes at the Nairobi Museum, Kenya. The Catholic University of Louvain awarded his doctorate (1984). After he had spent five years at the Africa Museum, Tervuren (MRAC) (1979–1984), he lived and researched in Rwanda (1984–1987). He was professor at the Kisangani University, Congo (1988–1990) where he made extensive collections for the museum. Due to political problems, he returned to Belgium (1990). He taught at the University of Leuven (Louvain) until he could go back to Africa again, returning again to Belgium (1996). He joined the Flemish Technical Cooperation and was posted to Nairobi, Kenya, to become the head of the newly formed Ichthyology Department at the National Museums of Kenya. He collected in virtually every river in Kenya and re-discovered the Giant Pancake-headed Catfish *Pardiglanis tarabinii*. He died from kidney failure. One etymology says of him that he: "*…dedicated so much of his career to expanding our knowledge of the fishes of East and Central Africa. His sudden and untimely death [from a kidney blockage] is a great loss to our community.*"

Devries

Snailfish sp. *Paraliparis devriesi* AP Andriashev, 1980

Dr Arthur L DeVries (b.1938) was Professor of Animal Biology at the University of Illinois 1976–2011) and is now Professor Emeritus. The University of Montana awarded his zoology degree (1960), following which he took a research position with Stanford based for nearly two years at McMurdo Sound, Antarctica culminating in his PhD thesIs (1968). He was a post-doctoral fellow at the University of California then was research physiologist at Scripps Institution of Oceanography (1971–1976), making another five trips to McMurdo Sound. Among his key interests are how fish survive in cold waters as evidenced by many of his extensive published papers including: *Chemical and Physical Properties of Freezing Point-depressing Glycoproteins from Antarctic Fishes* (1969) and *Structure of a peptide antifreeze and mechanism of adsorption to ice* (1977). He continues to visit the Antarctic. He caught the type of the snailfish.

Dewa

Dartfish sp. *Navigobius dewa* DF Hoese & H Motomura, 2009

Shin-ichi Dewa is an amateur ichthyologist from Kagoshima, Japan, the type locality. He has co-authored at least seven papers including: *Global warming and comparison of fish fauna in southern Japan* (2014). He collected the type. (See also **DewaPyle**)

de Waal

Checked Goby *Redigobius dewaali* MCW Weber, 1897

B H de Waal was appointed (1892) to be the Consul-General of the Netherlands in Cape Town, South Africa. During the Boer War he was active in support of pro-Boer organisations based in the Netherlands.

Dewapyle

Yellow-striped Hogfish *Terelabrus dewapyle* Y Fukui & H Motomura, 2015

The binomial combines the surnames of the collectors of the type specimens: Shin-ichi Dewa (q.v.) and Richard L Pyle (q.v.)

Deweger

Vieja (sea bass) *Paralabrax dewegeri* J Metzelaar, 1919

Mr De Weger (d.1910) worked for the Royal West Indian Mail. He collected in Trinidad, Haiti, Venezuela, Suriname and Curaçao (1906–1908) as well as purchasing specimens during his travels. The sea bass description mentions that: "*Two specimens of 18 cm. from*

Guanta, Venezuela where it was discovered by Mr. De Weger, in August—September, 1906" And elsewhere: "*Special mention should be made of some scores of very fine fishes collected by Mr. De Weger, officer of the Royal West Indian Mail Cy. from Trinidad, Venezuela, Haïti and other Islands of the West Indian Archipelago in 1907. This excellent collector was drowned 3 years afterwards.*"

Dewindt

Lake Tanganyika Cichlid sp. *Aulonocranus dewindti* GA Boulenger, 1899

Dr Jean Charles Louis De Windt (1876–1898) was a Belgian geologist on the Congo Free State Expedition led by Charles Lemaire. He and the expedition artist both drowned. He qualified at Ghent University. The etymology states: "*This species is named in memory of the distinguished young geologist, Dr De Windt, attached to Lieut. Lemaire's expedition, who was accidentally drowned in Lake Tanganyika.*"

DeWitt, HH

Crocodile Icefish sp. *Chionobathyscus dewitti* AP Andriashev & AV Neyelov, 1978
DeWitt's Plunderfish *Pogonophryne dewitti* RR Eakin, 1988
Rockcod (notothen) sp. *Paranotothenia dewitti* AV Balushkin, 1990
Antarctic Dragonfish sp. *Acanthodraco dewitti* KE Skóra, 1995
Brown Ribbed Snailfish *Paraliparis dewitti* DL Stein, NV Chernova & AP Andriashev, 2001

Dr Hugh Hamilton DeWitt (1933–1995) was an American ichthyologist and herpetologist. Stanford University awarded his BA (1955), his MA (1960) and PhD (1966). He was Assistant Professor of Marine Science at the University of South Florida (1967–1969); Assistant Professor of Zoology at the University of Maine, Orono (1969–1970) and Assistant Professor of Oceanography, Walpole, Maine (1970–1972). The rest of his career was spent there as Associate Professor (1972–1979), Professor (1979–1981), Chairman (1976–1979) and eventually Professor of Zoology (1981). He led a number of field trips and expeditions for the National Science Foundation to the Southern Ocean and Antarctic (1962–1963, 1964, 1966, 1967, 1975 & 1976). The etymology for the plunderfish reads: "*I take pleasure in naming this species for Hugh H. DeWitt who not only provided the specimen but who has for many years contributed immeasurably to our knowledge of Antarctic fishes. He has inspired and guided me in my ichthyological research since my graduate study in his laboratory at the University of Maine.*" A mountain is also named after him. (Also see **Joanne**)

Dewy Sea

Ruby Seadragon *Phyllopteryx dewysea* J Stiller, NG Wilson & GW Rouse, 2015

Mary 'Dewy' Lowe was honoured "for her love of the sea and her support of seadragon conservation and research". She is the co-founder of the Lowe Family Foundation. The binomial apparently combines her nickname with the English word 'sea'.

Dgebuadze

Loach sp. *Barbatula dgebuadzei* AM Prokofiev, 2003

Dr Yuri Yulianovich Dgebuadze is an ichthyologist and biologist who spent his entire career at the Severtsov Institute of Ecology and Evolution of the Russian Academy of Sciences, Moscow, where he was a professor, Deputy Director and an Academician (2011). His bachelor's degree was awarded by Moscow State University (1971) and his doctorate in biology by the A N Severtsov Institute (1975). He also holds a DSc degree (1998). Among his publications, he co-wrote: *Four fish species new to the Omo-Turkana basin, with comments on the distribution of* Nemacheilus abyssinicus *(Cypriniformes: Balitoridae) in Ethiopia* (1994) and the book: *Ecological Patterns of Growth Variability* (2001). He collected the holotype of the eponymous loach.

Dhanis

Lake Tanganyika Cichlid sp. *Tangachromis dhanisi* M Poll, 1949

Not a direct eponym but named after the 'Baron Dhanis', the boat that transported Max Poll and his team around Lake Tanganyika during the expedition that collected the holotype. (Also see **Elavia**)

Dhanze

Spiny Eel sp. *Macrognathus dhanzei* LA Kumar, 2020

Professor Dr J R Dhanze (b.1949) was Professor and Head of the Department of Fisheries Resource Management, at the College of Fisheries, Central Agricultural University, Lembucherra, Agartala, Tripura (2006–2014). Previously, he was Head of Fisheries department at CSKHPKV, Palampur, Himachal Pradesh (1988–2006) and an Assistant Zoologist, Zoological Survey of India (1970–1988). He is currently working as consultant in the Centre of Excellence programme of Department of Biotechnology, supported on Fishery and Aquaculture there. He has 45 years' professional experience in teaching, research, extension, administration and institutional building.

Dhont

Squeaker Catfish sp. *Synodontis dhonti* GA Boulenger, 1917
Lake Tanganyika Cichlid sp. *Telmatochromis dhonti* GA Boulenger, 1919
African Airbreathing Catfish sp. *Clarias dhonti* GA Boulenger, 1920
Labeo sp. *Labeo dhonti* GA Boulenger, 1920

G Dhont-De Bie was a member of the Corps Expéditionnaire Belge en Afrique de l'Est. He was stationed at Albertville in the Belgian Congo (DRC) and accompanied Stappers' (q.v.) expedition to Lakes Tanganyika and Moero (1911–1913). He collected the holotypes of these eponymous species.

Diallo, A

Blenny sp. *Parablennius dialloi* H Bath, 1990

Amadou Diallo of the Musée de la Mer, Gorée, Senegal, provided specimens and "helpfully supported" (translation) Bath's research in Senegal.

Diallo, S

Squeaker Catfish sp. *Chiloglanis dialloi* RC Schmidt & F Pezold, 2017

Samba Diallo is a Guinean fisheries biologist who is (2002) Director of the National Center for Fisheries Science of Boussoura, Guinea. His master's was awarded by Conakry University (2001). He provided logistical support and assisted in the field during two expeditions (2003 & 2013). The etymology says that "*...his efforts are largely responsible for the success of the expeditions and the subsequent descriptions of different new species.*"

Diamouangana

African Barb sp. *Enteromius diamouanganai* G Teugels & V Mamonekene, 1992

Dr Jean Diamouangana is an ecologist who was the UNESCO National Project Director in Mayombe, Congo, which supported the authors' work. He is also Director of 'Groupement pour l'Etude et la Conservation de la Biodiversité pour le Dévelopment', Brazzaville, Republic of Congo. Among his publications are: *La Réserve de la Biosphère de Dimonika (Congo)* (1995) and the co-written: *Attitudes Towards Forest Elephant Conservation Around a Protected Area in Northern Congo* (2017).

Diana

Diana's Hogfish *Bodianus diana* BGE Lacepède, 1801

Diana, the Roman goddess of the chase and the moon; an allusion to the beautiful coloration and form of the species.

Diane

Orangeflag Blenny *Emblemariopsis dianae* JC Tyler & PA Hastings, 2004

Diane M Tyler was managing editor of the Scientific Contributions Series of the Smithsonian Institution Press. Among her publications is (co-written with the senior author Dr James Chase Tyler (q.v.) to whom she is married): *Natural history of the sea fan blenny,* Emblemariopsis pricei *(Teleostei:Chaenopsidae), in the western Caribbean* (1999). The etymology by Tyler & Hastings (2004) reads: "*The species is named in honour of Diane M. Tyler of the Smithsonian Institution Press, in recognition of her studies of the behavioural ecology of chaenopsids at Carrie Bow Cay; she is the co-collector of most of the type specimens of this new species, and her dedicated collecting efforts over the years in and around Carrie Bow Cay have procured many important materials.*"

Dianne

Fiji Sponge-Goby *Bryaninops dianneae* HK Larson, 1985

Dianne 'Di' J Bray is an Australian ichthyologist at Museum Victoria (since 1999). Macquarie University awarded her MSc where her thesis was about the systematics of Zeid fishes. Previously she was Senior Collections Manager at the Australian Museum, Victoria, where she managed the ichthyology and herpetology collections (c.1975–1999). Among her published papers is the co-written: *Family Ateleopodidae* (2006) and she was co-editor of Fishes of Australia's Southern Coast (2008). She has also undertaken field trips such as aboard 'RV Lewia' during fieldwork in Vanuatu (1996). She collected the goby holotype.

Diard

Loach sp. *Sewellia diardi* TR Roberts, 1998

Pierre-Medard Diard (1794–1863) was a French explorer and naturalist. Encouraged by his stepfather, Georges Cuvier, he left Europe (1817) for India to collect for the MNHN, Paris. With Alfred Duvaucel, he was employed as a naturalist by Sir Thomas Stamford Raffles to collect for him in Sumatra (1818–1821). However, they were summarily dismissed after Raffles discovered that they had sent most of the material they had collected to MNHN, rather than to him. Diard then went to Indochina, visiting Annam (becoming one of the first Europeans to visit Angkor Wat) and Vietnam (c.1821) where he was probably the first person to collect freshwater fishes for science. He next visited Malaya, then joined the Dutch East India Company in Batavia (Jakarta) and collected in the East Indies (Indonesia) (1827–1848). He created the Buitenzorg Botanical Gardens in Java, where he was largely responsible for the introduction of sugar cane and the breeding of the silkworm moth. A mammal, four birds and a reptile are also named after him.

Diaz, J

Patzcuaro Allotoca *Allotoca diazi* SE Meek, 1902

José de la Cruz Porfirion Diaz Mori (1830–1915) was President of Mexico (1876–1880 & 1884–1911).

Diaz, M

Bluntnose Knifefish sp. *Brachyhypopomus diazae* A Fernández-Yépez, 1972

María Isabel 'Betty' Diaz was the author's secretary.

Dick

Dick's Damsel *Plectroglyphidodon dickii* F Liénard, 1839
[Alt. Blackbar Devil]

The Honorable George F Dick was Colonial Secretary of Mauritius and the President of the Natural History Society of Mauritius (1836–1850). He retired to his native Ireland.

Dickfeld

Dickfeld's Slender Cichlid *Julidochromis dickfeldi* W Staeck, 1975

Alf Dickfeld was a German aquarist. It was his idea to undertake the fish-collecting expedition to Zambia during which the holotype was collected.

Didi

Barb sp. *Pethia didi* SO Kullander & F Fang, 2005

Johan Bernard Didi Kullander (b.1999) is the authors' son. His parents stated he: "*had to repeatedly suffer their parents' absence searching for these and other fish in faraway places.*" 'Didi' in Mandarin Chinese means 'little brother'. (Also see **Kullander** & **Tiantian**)

Didier

Falkor Chimaera *Chimaera didierae* PJ Clerkin, DA Ebert & LM Kemper, 2017

Dr Dominique Ann Didier (b.1965) is an American marine biologist and ichthyologist. The Illinois Wesleyan University awarded her bachelor's degree in biology (1987) and University of Massachusetts, Amherst her doctorate (1991). As an undergraduate, she initially wanted to study bats, but was advised that many people were studying bats whereas nobody was looking at chimaeras (ratfish). Initially sceptical, she is now a leading world expert on these fishes. As an Associate Professor of Biology at Millersville University of Pennsylvania, her research interests focus on the systematics and evolution of Chondrichthyan fishes with a particular focus on the chimaeroids. Among her notable publications is a monograph on the: *Phylogenetic Systematics of Extant Chimaeroid Fishes (Holocephali, Chimaeroidei).* More recently her publications have focused on taxonomic descriptions of new species of chimaeroid fishes and she and her colleagues have synthesized much of this work in the second edition of: *Biology of Sharks and Their Relatives*, (*Phylogeny, Biology, and Classification of Extant Holocephalans* by Didier, Kemper, and Ebert) (2012).

Dido

Goby genus *Didogobius* PJ Miller, 1966

Dido was a princess of Tyre (in modern-day Lebanon), who sailed the eastern Mediterranean (where *D. bentuvii* occurs) to found the city of Carthage and become its first queen. She is particularly noted for her role in Virgil's *Aeneid*.

Dieffenbach

New Zealand Longfin Eel *Anguilla dieffenbachii* JE Gray, 1842

Dr Johann Karl Ernst Dieffenbach (1811–1855) was a German physician, naturalist, explorer, linguist and writer. He began studying medicine at Giessen University (1828) but had to flee the country having been agitating for political reform. He again began studying medicine at Zurich, but was again in trouble and was imprisoned for two months for various offences including duelling. He was later expelled from the country (1836), fortunately already having been awarded his MD. He arrived in London (1837) and built a reputation as a scientist, becoming acquainted with the likes of Charles Darwin, Charles Lyell and Richard Owen. He was appointed as naturalist to the New Zealand Company (1839) and travelled widely in New Zealand (1839–1841), assessing settlement areas, surveying, mapping and collecting natural history specimens on behalf of BMNH. Dieffenbach and James Heberley became the first Europeans to climb Mount Taranaki (1839) and he also visited the Chatham Islands. He was unimpressed with the settlers and their attitude toward the Maoris and tried, unsuccessfully, to extend his contract there. Back in England he tried to plea for the Maori people and culture to be protected from further settlement. He wrote: *Travels in New Zealand* (1843). He made the first translation into German of Darwin's: *The Voyage of the Beagle* (1844). He established a career as a translator and tried in vain to raise enough funds for passage back to New Zealand. He returned to Giessen (1848) and become Professor of Geology and Director of the Geological Museum at the University there. A bird and the plant genus *Dieffenbachia* are also named after him.

Diehl

Thorny-tail Skate *Dipturus diehli* JMR Soto & MM Mincarone, 2001

Professor Fernando Luiz Diehl (b.1959) is a Brazilian oceanographer. His bachelor's degree in oceanography was awarded (1984) and his master's in geography (1997). He was Director of the School of Marine Sciences of Universidade do Vale do Itajaí - UNIVALI (1992–2004), and was President of the Brazilian Association of Oceanography - AOCEANO six times (1990–2009) and President of the Latin American Association of Researchers in Marine Sciences - ALICMAR (2007/2009 and 2011/2013). He was honoured in recognition of his: "*...extensive work and tireless dedication to oceanography in Brazil.*"

Dieuzeide

Spiny Gurnard *Lepidotrigla dieuzeidei* M Blanc & J-C Hureau, 1973

Dr Jean René Dieuzeide (b.1900) was a French marine biologist. He wrote: *Catalogue des poissons des côtes algériennes* (1955). We believe this to be the person honoured in the binomial, who is mentioned just as 'Jean Dieuzeide' in the etymology. A number of marine invertebrates are also named after him.

Diguet

Banded Cleaner Goby *Tigrigobius digueti* J Pellegrin, 1901
Pale Garden Eel *Heteroconger digueti* J Pellegrin, 1923
Blenny sp. *Hypsoblennius digueti* P Chabanaud, 1943

Léon Diguet (1859–1926) was a French chemical engineer and geologist who was employed at a copper mine at Santa Rosalia, Baja California (1889–1892). He became interested in local natural history and collected specimens for MNHM, Paris including the holotype of the eponymous species. The management at the MNHN was so impressed that they sent him back to Mexico and employed him as a full-time explorer and collector. He explored Baja California (1893–1894), identifying important early rock paintings and reporting on them in *L'Anthropologie* (1895). He revisited Mexico several times before WW1. He wrote the posthumously published: *Les cactacées utiles du Mexique* (1928). The Barrel Cactus *Ferocactus diguetii* is also named after him.

Dimi Pavlov

Pavlov's Goatfish *Upeneus dimipavlov* F Uiblein & H Motomura, 2021

Dr. Dimitri Alexandrovich 'Dimi' Pavlov is (1974–1991 and currently) Principal Scientist in the biology department at Lomonosov Moscow State University. He has also worked at the Institute of Marine Research Norway (1991–1995) and the Russian-Vietnamese Tropical Research and Technology Centre (1999). He is a fish biologist with over 40 years of professional experience who conducted investigations in Russia (the regions of the former Soviet Union and the White Sea), Norway, and Vietnam. His professional interests are connected with different aspects of fish biology, reproduction, and ontogeny including aquaculture. The etymology honours him "*...for collecting, photographing and donating mullid specimens from Vietnam (including the holotype of Upeneus dimipavlov) for taxonomic research.*"

Discovery

Deepsea Tripodfish genus *Discoverichthys* N Merrett & JG Nielsen, 1987
Snailfish sp. *Careproctus discoveryae* G Duhamel & NJ King, 2007

'Discovery' (built 1962) was a Royal Research Ship from which the holotypes were caught. The present day 'RRS Discovery' was built in 2013.

Disi

Slopefish sp. *Symphysanodon disii* MA Khalaf & F Krupp, 2008

Dr Ahmad Mohammad Disi (b.1942) is Professor of Zoology at the University of Jordan. His PhD was given by the University of Wisconsin, Madison (1976). He was honoured in the fish's binomial "*...in recognition of his contributions to our knowledge of the vertebrate fauna of Jordan.*" He is co-author of: *Fishes of the Gulf of Aqaba* (2002).

Disney

Squeaker Catfish sp. *Chiloglanis disneyi* E Trewavas, 1974

Ronald Henry Lambert Disney (b.1938) is a British medical entomologist. He is a Senior Research Associate, University Museum of Zoology, Cambridge. He wrote: *Tasmanian Phoridae (Diptera) and some additional Australasian species* (2003). He collected the holotype of this species.

Djaja

Priapiumfish sp, *Neostethus djajaorum* LR Parenti & KDY Louie, 1998

The Djaja family - Rachmat, Jootje and their three children (Ike, Yuni and Andi) – were honoured "*...for their kindness and extraordinary support of* [the authors'] *fieldwork in Sulawesi.*" Nothing more was recorded.

Dlouhy

Armoured Catfish sp. *Hypostomus dlouhyi* C Weber, 1985

Carlo Dlouhy is a biologist and ichthyologist of Czech origin who worked at Muséum d'histoire naturelle de Genève and since 1975 has lived in Paraguay. He makes regular yearly visits to Geneva but otherwise acts for the museum as a guide and collaborator for all the museum's researchers who go to Paraguay. He co-wrote: *Ultimo Paraguay : Expéditions et aventures du Muséum d'histoire naturelle de Genève au Paraguay* (2009). He collected the holotype of this species.

Do Tu

Cave Loach sp. *Homatula dotui* DT Nguyen, H Wu, L Cao & E Zhang, 2021

Do Van Tu is a Vietnamese malacologist, and a member of the Institute of Ecology and Biological Resources, Vietnam Academy of Science and Technology. He caught the type specimens of this loach.

Doadrio

Splitfin sp. *Xenotoca doadrioi* O Domínguez-Domínguez, DM Bernal-Zuñiga & KR Piller, 2016

Dr Ignacio Doadrio (b.1957) is a Spanish ichthyologist at the National Museum of Natural Science. The etymology says: "*The name of the species is derived from the name of the prestigious ichthyologist Dr. Ignacio Doadrio, Museo Nacional de Ciencias Naturales, Spain, who has strongly contributed to the study and knowledge of Mesoamerican fish diversity.*"

Doak

Arrow-tooth Lizardfish *Synodus doaki* BC Russell & RF Cressey Jr, 1979

Wade Thomas Doak (1940–2019) was a pioneering New Zealand diver and underwater naturalist, who discovered this species at Poor Knights Islands, off eastern Northland, New Zealand. He wrote: *Sharks and Other Ancestors* (1976).

Dobson

Krishna Barb *Hypselobarbus dobsoni* F Day, 1876

Colonel Andrew Francis Dobson (1848–1921), an army surgeon, was commissioned into the Royal Army Medical Corps (1872). He was attached to the Madras Medical Service, Bangalore (1876) as Brigade Surgeon when he provided the author with a collection of c.170 fishes he had made in the Deccan.

Dockins

Sandperch sp. *Parapercis dockinsi* JE McCosker, 1971

Dr Donald Martin Dockins (1930–1975) was a zoologist at the Scripps Institution of Oceanography. He was described as "*the intrepid captor of many of the type specimens*" in the Juan Fernández Islands, Chile, (1965).

Döderlein, LHP

Blackthroat Seaperch genus *Doederleinia* F Steindachner, 1883
Flyingfish sp. *Cheilopogon doederleinii* F Steindachner, 1887
Döderlein's Cardinalfish *Ostorhinchus doederleini* DS Jordan & FO Snyder, 1901
Striped Jewfish *Stereolepis doederleini* GU Lindberg & ZV Krasyukova, 1969

Dr Ludwig Heinrich Philipp Döderlein (1855–1936) was a zoologist and palaeontologist. He studied at Erlangen and Munich universities, and his doctorate was awarded by Strasbourg (1877). He taught in Japan (1879–1881), making a collection of Japanese fauna (1880–1881). He became Curator and Director, Zoological Collections, Strasbourg (1882), and was appointed as Assistant Professor (1883) at Université de Strasbourg, becoming Professor of Zoology (1891). He was sacked and expelled from the city (1919) because he was a German; Alsace and Lorraine had been restored to France by the Treaty of Versailles. He was head of the Zoological Collections, München Staatliches Museum für Naturkunde, with the title of Honorary Professor Emeritus of Taxonomy (1923–1927). A reptile is named after him.

Doderlein, P

Wrasse sp. *Symphodus doderleini* DS Jordan, 1890

Pietro Doderlein (1809–1895) was an Italian zoologist and geologist. He was Professor of Zoology at the University of Modena (1839–1862) and then Professor of Zoology and Comparative Anatomy at the University of Palermo (1862–1894). He founded the Zoological Museum there (1862) and was Director (1863–1894). It is now named after him, as the Museo di zoologia Pietro Doderlein.

Doello-Jurado

Southern Thorny Skate *Amblyraja doellojuradoi* AJ Pozzi, 1935

Professor Martin Doello-Jurado (1884–1948) was a marine biologist and malacologist, who was also interested in palaeontology. He directed several oceanic explorations of the Magellan Region and elsewhere (1914–1925 & 1921) aboard 'A R A Patria' while teaching at the University of Buenos Aires. He was a Professor of Natural Sciences and Director, Museo Argentino de Ciencias Naturales, 'Bernardino Rivadavia', Buenos Aires (1924–1946). A reptile is named after him.

Doflein

Jellynose Fish sp. *Ijimaia dofleini* H Sauter, 1905
Doflein's Lanternfish *Lobianchia dofleini* E Zugmayer, 1911

Dr Franz Theodor Doflein (1873–1924) was a German ichthyologist and herpetologist. He studied natural sciences at the universities of Strasbourg and Munich (1893–1898). He joined the Bavarian State Collection of Zoology, Munich (1901), and became Professor, Department of Zoology, University of Freiburg (1912) and at the University of Breslau (Wroclaw, Poland) (1918). He made a number of collecting expeditions, to Central America and USA (1898), China, Japan (where he made a notable collection of marine specimens in Sagami Bay) and Ceylon (Sri Lanka) (1904–1905). He also collected in Macedonia (1917–1918), when he was attached to the German Army. He wrote: *Lehrbuch der Protozoenkunde* (1906). Sauter remarked that it was thanks to Doflein's encouragement that he studied marine life. An amphibian and a bird are named after him.

Dogar Singh

Manipur Baril (cyprinid) *Opsarius dogarsinghi* SL Hora, 1921
[Syn. *Barilius dogarsinghi*]

Sardar Dogar Singh was State Overseer, Manipur, India when Hora visited Manipur (1920). He gave Hora 'material assistance' in the collection of specimens, including the holotype of this species, and helped arrange survey tours for him.

Dolganov

Fathead sp. *Gilbertidia dolganovi* SA Mandrytsa, 1993

Dr Vladimir Nikolaevich Dolganov (b.1949) is a Russian ichthyologist with a particular interest in sharks and rays. He was at the Institute of Marine Biology, Russian Academy of

Sciences, Vladivostok (2008). He wrote: *Manual for identification of cartilaginous fishes of Far East seas of USSR and adjacent waters* (1983).

Dollfus

Tonguefish sp. *Cynoglossus dollfusi* P Chabanaud, 1931
Goby sp. *Vanneaugobius dollfusi* CL Brownell, 1978
Dollfus' Stargazer *Uranoscopus dollfusi* R Brüss, 1987

Robert Philippe F Dollfus (1887–1976) was a French zoologist, ichthyologist and parasitologist, professor at the MNHN, Paris and Director of the National Centre for Scientific Research. He co-organised the 'Vanneau' expeditions (1923–1926) along Morocco's Atlantic coastline, during which the goby holotype was collected. He organised or participated in other expeditions, including one he led to Egypt (1927–1929). He published over 400 papers and longer works such as the monograph: *Les Echinides de la Mer Rouge* (1981). His checklist of Moroccan Atlantic fishes (1955) was described by Brownell as "… *very helpful to subsequent workers in the field*."

Dollo

Spotted African Lungfish *Protopterus dolloi* GA Boulenger, 1900
[Alt. Slender Lungfish]
Tilapia ssp. *Sarotherodon nigripinnis dolloi* GA Boulenger, 1899

Dr Louis Antoine Marie Joséph Dollo (1857–1931) was a French-born Belgian palaeontologist, known for his work on fossil reptiles. He qualified with a degree in civil engineering at the École Centrale de Lille (1877) then worked in mining for five years, all the while developing his interest in palaeontology. He supervised the excavation of Iguanodon fossils found at Bernissart, Belgium (1878–1882) and moved to Belgium (1879) becoming Assistant-Naturalist, Royal Museum of Natural History, Brussels until retirement (1882–1925). He also taught palaeontology at the Université Libre de Bruxelles (1909–1912). He proposed what is now known as 'Dollo's Law' (1893), which states that 'complex organs, once lost, can never be regained in exactly the same form' or that evolution cannot be reversed. He appraised lungfish (1895) and hypothesized that they evolved from Devonian 'crossopterygians' (primitive lobe-finned bony fishes believed to be the forerunner to four-legged vertebrates). Among his publications are: *Première note sur les dinosauriens de Bernissart* (1882) and *La Paléontologie éthologique* (1910). An amphibian is also named after him.

Dolly Varden

Dolly Varden Trout *Salvelinus malma* JJ Walbaum, 1792

Dolly Varden is a character in Charles Dickens' *Barnaby Rudge* (1841). We have found a letter from a lady giving an account of the origin of this name. It is well worth quoting in full! *My grandmother's family operated a summer resort at Upper Soda Springs on the Sacramento River just north of the present town of Dunsmuir, California. She lived there all her life and related to us in her later years her story about the naming of the Dolly Varden trout. She said that some fishermen were standing on the lawn at Upper Soda Springs looking at a catch of the large trout from the McCloud River that were called 'calico trout' because of their spotted, colorful markings. They were saying that the trout should have a better name. My grandmother, then a young girl of 15 or 16, had been reading Charles Dickens'* Barnaby Rudge *in which there appears a character named Dolly Varden; also the vogue in fashion for women at that time (middle 1870s) was called 'Dolly Varden', a dress of sheer figured muslin worn over a bright-colored petticoat. My grandmother had just gotten a new dress in that style and the red-spotted trout reminded her of her printed dress. She suggested to the men looking down at the trout, "Why not call them 'Dolly Varden'?" They thought it a very appropriate name and the guests that summer returned to their homes (many in the San Francisco Bay area) calling the trout by this new name. David Starr Jordan, while at Stanford University, included an account of this naming of the Dolly Varden Trout in one of his books.*

Dolomieu

Smallmouth Bass *Micropterus dolomieu* BGE Lacepède, 1802

Déodat Gratet de Dolomieu (1750–1801) was a French geologist. He was at first a strong supporter of the French Revolution, but the murder of his friend the Duc de la Rochefoucauld and the beheading of several of his relatives made him re-evaluate: he became a supporter of Napoleon Bonaparte. Bernard Lacepède (q.v.) named the bass after his friend expressing joy and relief that Dolomieu had been released from solitary confinement after twenty-one months. He had been locked up by Italy during a border dispute with France. He was ill when imprisoned and held in such bad conditions that his health deteriorated further. Sadly, by the time the description appeared in print, he had died. The mineral dolomite is also named after him.

Domira

Lake Malawi Cichlid sp. *Placidochromis domirae* M Hanssens, 2004

A toponym referring to the type locality; Domira Bay, Lake Malawi.

Donascimiento

Colombian Cave Catfish sp. *Trichomycterus donascimientoi* CA Castellanos-Morales, 2018

Long-whiskered Catfish sp. *Hypophthalmus donascimientoi* MW Littmann, JG Lundberg & MS Rocha, 2021

Dr Carlos Luis DoNascimiento Montoya (b.1973) is a Venezuelan ichthyologist. The University of Central Venezuela awarded his zoology degree (2001) and his PhD (2013). He was a researcher at the Caracas Natural History Museum (2002–2004) and an ichthyology Research Assistant at Drexel University (2004–2005). He became Curator of the freshwater fish collection at Instituto de Investigación de Recursos Biológicos Alexander von Humboldt (2014–2019 and is (2006–present) Associate Professor at the University of Carabobo, Colombia. Among his published papers is: *A new catfish species of the genus* Trichomycterus *(Siluriformes: Trichomycteridae) from the río Orinoco versant of Páramo de Cruz Verde, Eastern Cordillera of Colombia* (2015). He was honoured in the name of the cave catfish for his 'invaluable orientation' in the author's research on the genus *Trichomycterus*.

Donny

Mormyrid sp. *Genyomyrus donnyi* GA Boulenger, 1898

General Baron Albert-Ernest Donny (1841–1923) was Vice-President of the Société d'Études Colonials (1894–1912) and President (1912–1923). This society sponsored the first major collection of fishes from the Congo. He was a Belgian army officer; Second Lieutenant (1860–1901) then Lieutenant-General (1901–1923). He wrote: *Manuel du voyageur et du résident au Congo* (1900).

Dooley

Bankslope Tilefish *Caulolatilus dooleyi* FH Berry, 1978

James K Dooley is Professor of Biology at the Adelphi University, New York, where he has spent his career (since 1973) and a passionate conservationist. The University of Miami awarded his BSc (1964) and the University of South Florida his MA (1969). His PhD was awarded by the University of North Carolina (1974). Among his published work are: *An annotated check-list of the fishes of the Canary Islands* (1985) and *A new species of deepwater tilefish (Percoidea:Branchiostegidae) from the Philippines, with a brief discussion of tilefish systematics* (2012). He and FH Berry have collaborated.

Dor

Flap-headed Goby sp. *Callogobius dori* M Goren, 1980
Sole sp. *Soleichthys dori* JE Randall & TA Munroe, 2008

Dr Menachem Dor (1901–1998) was a zoologist and Hebrew scholar. (Dor was an adopted surname made up from an acronym of his parents' names, David and Rachel Klugaft). He was strongly influenced by anarchist theories and was a vegetarian although he continued to follow Jewish traditions. He emigrated from Austria to Israel in the 1930s, having first visited aged 19. He attended the Sorbonne (1935) to study zoology and returned to Israel joining a kibbutz and also establishing a research centre on field mice. He was one of the founders of Beit Berl College and built a nature institute there where he taught. He published a number of studies using his zoological and Hebrew knowledge to identify species mentioned in the Old Testament. He was honoured for his contribution to the knowledge of fishes of the Red Sea (where the goby and sole occur). Indeed, he published a book entitled: *Checklist of the Fishes of the Red Sea* (1984).

Dora

Chobe Sand Catlet *Zaireichthys dorae* M Poll, 1967

Dora Machado, who collected the holotype, was the wife of zoologist António de Barros Machado (1912–2002).

Dora Demir

Loach sp. *Cobitis dorademiri* F Erk'akan, F Özdemir & SC Özeren, 2017

Dora Demir Özdemir is the son of the second author, Dr Filiz Özdemir.

Dorbigny

Cloudy Doradid *Rhinodoras dorbignyi* R Kner, 1855

Alcide Charles Victor Dessalines d'Orbigny (1802–1857) (See **Orbigny**)

Doria

Iranian Bleak sp. *Alburnus doriae* F De Filippi, 1865
Bumblebee Goby *Brachygobius doriae* A Günther, 1868
Gilded Goatfish *Upeneus doriae* A Günther, 1869
Dragonfin Tetra *Pseudocorynopoma doriae* A Perugia, 1891
Golden Eel-Loach *Pangio doriae* A Perugia, 1892
Banjo Catfish sp. *Bunocephalus doriae* GA Boulenger, 1902
Bagrid Catfish sp. *Leiocassis doriae* CT Regan, 1913

Marchese Giacomo Doria (1840–1913) was an Italian zoologist and ornithologist who collected in Persia (Iran) (1862–1863) and in Borneo with Odoardo Beccari (1865–1866). He was the founder and first Director of the Natural History Museum in Turin (1867–1913). He visited other European museums, building up relationships with other naturalists and exchanging specimens. Six mammals, three amphibians, two birds and eight reptiles, among other taxa, are named after him.

Dorion

Characin sp. *Astyanax dorioni* DE Rosen, 1970

Robert Charles Dorion (b.1926) is an entrepreneur and investor. He was awarded a bachelor's degree in Naval Sciences by Dartmouth College (1946). He has worked in a number of commercial enterprises and since 1952 has lived in Guatemala, showing a keen interest in natural history. The etymology thanks him for: "*...continuing assistance for our field efforts in Guatemala since 1963, and whose companionship and hard work during several field trips have always been greatly appreciated.*"

Döring

Swamp Eel sp. *Synbranchus doeringii* H Weyenbergh, 1877

[Junior Synonym of *Synbranchus marmoratus*]

Oscar Döring (1844–1917) and Adolf Döring (1848–1925) were brothers, both members of the so called 'Córdoba group' of German scientists invited to Argentina by Burmeister to the University of Córdoba. Dr Oscar Döring was a mathematician and an agricultural meteorologist. He became Professor of Mathematics at the University of Córdoba (1874–1880). Dr Adolf Döring was a zoologist, chemist and geologist who lived in Argentina (1872–1925) and became Professor of Organic Chemistry, University of Cordoba (1875) and also Professor of Zoology (1892–1916). Weyenbergh did not specify which person he was honouring in the swamp eel's binomial, though Adolf is most likely given his zoological interests.

Dority

Dority's Rainbowfish *Glossolepis dorityi* GR Allen, 2001

Dan Dority is an American missionary in New Guinea. He is also an aquarist; a rainbowfish enthusiast who collected the type (2000) with David Price (a New Zealand missionary) from a small lake in West Papua. With Gerry Allen they caught some more a few months later and photographed them in Dan's home aquarium. The etymology says: "*This species is named* dorityi *in honour of Dan Dority for his efforts in collecting the type specimens.*"

Dorothea

Dorothea's Wriggler *Allomicrodesmus dorotheae* LP Schultz, 1966

Dorothea Bowers Schultz was the wife of the author Dr Leonard Peter Schultz (1901–1986). She illustrated many of the new species in his monograph (though not this wriggler).

Dorothy

Dorothy's Sculpin *Triglops dorothy* TW Pietsch III & JW Orr, 2006

Dorothy Thomlinson Gilbert (1929–2008) was the wife of William W Gilbert, the late grandson of Charles Henry Gilbert, and a noted philanthropist. The UW School of Fisheries (now the School of Aquatic and Fishery Sciences) of Stanford University was the recipient (1998) of the 'Dorothy T. Gilbert Endowed Ichthyology Research Fund', established by her. In 2008, the fund Endowed Professorship was established in the UW College of Ocean and Fisheries Science (now the College of the Environment) with the initial occupant of that position being the distinguished UW ichthyologist, Theodore Wells Pietsch III (q.v.). The etymology states that: "*The species is named in honor of Dorothy Thomlinson Gilbert, great granddaughter-in-law of the eminent ichthyologist and fisheries biologist Charles Henry Gilbert, for her generous and steadfast support to graduate students in ichthyology at the University of Washington, Seattle, in establishing the William W. and Dorothy T. Gilbert Ichthyology Research Fund.*"

Dorsey

Three-barbeled Catfish sp. *Pimelodella dorseyi* HW Fowler, 1941

Lewis M Dorsey Jr. appears to have been one of Fowler's many contacts in the Philadelphia area, to whom Fowler stated he was 'indebted for local fishes.' He may have been an architect or draftsman (fl. 1904–1936).

Dotsu/Dotu

Goby genus *Dotsugobius* K Shibukawa, T Suzuki & H Senou, 2014
Goby sp. *Silhouettea dotui* K Takagi, 1957
Dartfish sp. *Parioglossus dotui* I Tomiyama 1958

Yoshie Dotsu is a Japanese ichthyologist. His name was given as Yosie Dôtu in his earlier publications. He worked at the Fisheries Laboratory, Kyushu University, and at the Faculty of Fisheries, Nagasaki University. Among his publications is: *The Life History of the Eleotrid Fish,* Mogurnda obscura *Temminck Et Schlegel* (1964). He collected the types of both species named after him.

Do Tu

Cave Loach sp. *Homatula dotui* DT Nguyen, H Wu, L Cao & E Zhang, 2021

Do Van Tu is a Vietnamese malacologist, and a member of the Institute of Ecology and Biological Resources, Vietnam Academy of Science and Technology. He caught the type specimens of this loach.

Douglas, NH

Tuskaloossa Darter *Etheostoma douglasi* RM Wood & RL Mayden, 1993

Neil H *Douglas* was (1962) Professor of Biology (now emeritus) at Northwestern Louisiana State University. He received the Outstanding Achievement Award of the Southern Division of the American Fisheries Society (2012) for his half-century of contributions to southern ichthyology, as a field biologist, researcher, mentor, teacher and curator. He wrote: *Freshwater Fishes of Louisiana* (1974). His fish collection at ULM is now recognised as the largest regional fish collection in North America, and the third largest university-based fish collection in the world.

Douglas, RAM

Grey Morwong *Nemadactylus douglasii* J Hector, 1875
[Alt. Porae, Blue Morwong]

Captain Sir Robert Andrews Mackenzie Douglas, 3rd Baronet of Glenbervie (1837–1884), was an army officer, farmer and politician. He was commissioned (1854) and served in the Crimea War, Aden and the Indian Mutiny after which he was sent to New Zealand. There he led a company of the 57th Regiment (1857) but retired by sale of his commission (1867). He bought an 1,800–acre farm near Whangarei, was a JP and served in the House of Representatives (until 1879). The etymology states: "*I have named this fine species in honour of Sir Robert Douglas, Bart., to whose kind hospitality I was indebted for a pleasant fishing excursion at Ngunguru, which afforded me many novelties.*"

Doutre

Violet Skate *Dipturus doutrei* J Cadenat, 1960
[Alt. Javelin Skate]

Michel-Pierre Doutre was a French veterinary surgeon at Laboratoire National de l'Elevage et de Recherches Vétérinaires, Dakar, Senegal (off the coast of which this skate occurs). He was chief fisheries officer for Senegal. He wrote: *Les merlus de Senegal. Mise en evidence d'une nouvelle espèce* (1960), naming it *Merluccius polli cadenati* after Cadenat (1960).

Dov

Tailspot Cardinalfish *Apogon dovii* A Günther, 1862
Spotted-head Sargo *Genyatremus dovii* A Günther, 1864
Guapote (cichlid) *Parachromis dovii* A Günther, 1864
White-eye *Oxyzygonectes dovii* A Günther, 1866
Dove's Longfin Herring *Opisthopterus dovii* A Günther, 1868
Fine-spotted Moray *Gymnothorax dovii* A Günther, 1870

John Melmoth Dow (1827–1892) (see **Dow**). As a purist, when Günther latinized Dow's name for his binomials he used a v instead of w, because there is no w in Latin. This led to such errors as someone assuming the longfin herring was named after a person called 'Dove'.

Dowe

Pacific Four-eyed Fish *Anableps dowei* TN Gill, 1861

Another version of Dow (see next entry).

Dow

Dow's Mojarra *Eucinostomus dowii* TN Gill, 1863
Flapnose Sea Catfish *Sciades dowii* TN Gill, 1863
Dow's Toadfish *Daector dowi* DS Jordan & CH Gilbert, 1887

Captain John Melmoth Dow (1827–1892) was an American naturalist and explorer, in addition to being a ship's master and shipping agent. He first went to Central America (1851) and remained involved in Central and South America coastal trading (1851–1876). He was appointed captain of the 'SS Constitution' (1853) and opened the Central American service of the Panama Railroad Company whilst in command of 'SS Colombus'. After leaving the sea he became a shipping agent in Colón, Panama (at that time part of New Granada). He collected plants and animals in Central America, notably Costa Rica. A Zoological Society of London paper by Albert Günther (1869) is entitled: *An account of the fishes of the states of Central America, based on collections made by Capt. J. M. Dow, F. Godman, Esq., and O. Salvin, Esq.*" He worked for the American Packet Service and sent plants to Britain. There are a number of plants (particularly orchids) named after him.

Downing

Downing's Shrimp-goby *Amblyeleotris downingi* JE Randall, 1994

Dr Nigel Downing (b.1951) is a South African-born British marine biologist. His BA (1973), MA (1977) and PhD (1979) were all awarded by Cambridge University. He initially researched in the Casamance (Senegal and The Gambia) (1974–1975) for his PhD on sawfish, but later switched to studying the peach-potato aphid. He worked at the Kuwait Institute for Scientific Research (1980–1988) where he specialised on the coral reefs and islands off Kuwait. There he hosted Jack Randall, providing logistical support on a couple of his trips to collect in the Gulf, during which they came across the goby. He also discovered a new species of coral there, which was named after him *Acropora downingi*. He took some time out (1988–1990) to travel back overland, towing a caravan with his family through Iraq, Jordan, Israel, Syria, Turkey, Greece and western Europe to the UK. He has been a businessman (since 1990) but has also twice (1991 & 1992) returned to Kuwait's reefs looking at the potential impact of oil release during the Gulf War, which happily showed there was none. Since being in business he spent c.9 years (1999–2008) working on a project on the coral atoll, Aldabra, southern Seychelles to assess the long-term recovery of the fish and coral communities, forming the Aldabra Marine Programme. Contacted (2015) by a fellow researcher on Gambia's sawfish, he returned (2017) sponsored by 'Save Our Seas Foundation', to The Gambia after 42 years. The sawfish are now feared to be almost extinct in the region. In private life he is interested in sports and holds the record of being the oldest Briton (64) to swim the Straits of Gibraltar from Spain to Morocco!

Drach

Yellowfin Soapfish *Diploprion drachi* R Roux-Estève, 1955
Conger Eel sp. *Uroconger drachi* J Blache & M-L Bauchot, 1976

Pierre Drach (1906–1998) was a French marine biologist. He was a professor at the Zoological Laboratory, Faculty of Sciences, University of Paris (1956). He was Deputy Director of the National Centre for Scientific Research (1962) and Director, Station Zoologique, campus of the Laboratoire Arago at Banyuls (1964–1976).

Drachenfels

West African Cichlid sp. *Pelvicachromis drachenfelsi* A Lamboj, D Bartel & E Dell'Ampio, 2014

Ernst-Otto von Drachenfels is a German biologist, businessman and aquarist since childhood. He regularly collects overseas, such as in French Guiana (2013) and has over 60 aquariums and a water-flea breeding pool in his garden. He is a friend of the authors and a promotor of their research. They say in the paper that "*...without his help this work would not have been possible.*"

Drachmann

Drachmann's Lanternfish *Diaphus drachmanni* AV Tåning, 1932

Dr Anders Bjørn Drachmann (1860–1935) was a Danish classical philologist who became a Professor at the University of Copenhagen (1892) where his doctorate had been awarded (1891). He was also president of the Carlsberg Foundation (q.v.) (1926–1933), which financed the *Dana* expedition (q.v.) that collected the holotype.

Dracula

Dracula Shrimp-goby *Stonogobiops dracula* NVC Polunin & HR Lubbock, 1977
Dracula Fish *Danionella dracula* R Britz, KW Conway & L Rüber, 2009

This is not a true eponym as the fish are named after a fictional character; Count Dracula, the vampire creation of Bram Stoker in his Gothic novel of that name (1897). The male of the *Danionella* has long fangs, and the *Stonogobiops* is endowed with sharp white teeth.

Drechsel

Barbeled Dragonfish sp. *Eustomias drechseli* CT Regan & E Trewavas, 1930

Commodore Christian Frederik 'Frits' Drechsel (1854–1927) served in the Danish navy (1870–1906). He had various commands (1874–1887) in the Caribbean, the Faeroes and Iceland. He had a great interest in fisheries, and was in charge of the fisheries inspection at Skagen (1883–1886) and instrumental in the establishment of a Danish biology station (1889). He was President of the Dana Committee for the Study of the Sea (1908–1925), which managed the *Dana* Expedition that collected the type of this species. (See also **Dana**).

Drewes

Sea Bass sp. *Serranus drewesi* T Iwamoto, 2018

Dr Robert Clifton Drewes (b.1942) is an American herpetologist at the California Academy of Sciences. He took his bachelor's degree at San Francisco State University (1969) and his doctorate (1981) at the University of California, Los Angeles. He is particularly noted for his work on the island of São Tomé, and the etymology honours him "*...for his dedicated efforts in leading 12 separate scientific and educational expeditions to São Tomé e Principe to explore and document the diverse fauna and flora of that country and to inspire and educate the country's citizens as to the biological wealth and uniqueness of where they live.*" A reptile and two frogs are named after him, as is a species of fungus found on São Tomé (*Phallus drewesii*).

Drjagin

Dryagin's Char *Salvelinus drjagini* MV Logashev, 1940

Striped Bystranka ssp. *Alburnoides taeniatus drjagini* FA Turdakov & KV Piskarev, 1955

Pavel Amphilokhievich Drjagin (1893–1977) was a Russian ichthyologist, hydrobiologist and fisheries specialist. He pioneered fisheries research in Russia and the management of inland waters. He wrote: *Les groupes bioecologiques des poissons, leur origine* (1936). (In the absence of etymologies, there is no definite proof, but we believe these fish are almost certainly named after the above)

Dromio

Shorthead Cusk-eel *Ophidion dromio* RN Lea & CR Robins, 2003

In Shakespeare's play: *The Comedy of Errors*, the identities of the brothers Dromio were confused throughout the play. The allusion is referring to how this species had been widely and incorrectly reported as *O. beani* (a junior synonym of *O. holbrooki*). Three Shakespearian characters lend their names to cusk-eels (see also **Antipholus** & **Puck**)

DROP

Four-fin Blenny *Haptoclinus dropi* C Baldwin & R Robertson, 2013

Not an eponym but named for the Smithsonian Institution's Deep Reef Observation Project (DROP).

Drummond

Pearlfish sp. *Echiodon drummondii* W Thompson, 1837

Dr James Lawson Drummond (1783–1853) was an Irish physician, botanist and naturalist. He had surgical training at Belfast Academy and became a naval surgeon in the Royal Navy (1807–1813). He enrolled at the University of Edinburgh and graduated there as a physician (1814). He was physician to the Belfast Dispensary (18114–1818) and Professor of Anatomy and Physiology, Belfast Academy (1818–1835) and Professor of Botany (1835–1836). He was the first President, Faculty of Medicine, of which he was one of the co-founders. He discovered the holotype dead on a beach.

Drummond Hay

Speckled Hind *Epinephelus drummondhayi* GB Goode & TH Bean, 1878

Colonel Henry Maurice Drummond Hay (1814–1896), was a noted illustrator, botanist, ichthyologist, and ornithologist. The etymology says the fish is dedicated to "*...Colonel H M Drummond Hay, C.M.Z.S., of Leggieden, Perth, Scotland, formerly of the British Army, by whom the species was first discovered at the Bermudas in 1851.*"

Druzhinin

Redfish (Berycid) sp. *Centroberyx druzhinini* SV Busakhin, 1981
Roughy sp. *Hoplostethus druzhinini* AN Kotlyar, 1986

Dr Anatoly Dmitrievich Druzhinin (1926–1979) was a marine biologist and fisheries scientist at the All-Russian Research Institute of Fisheries and Oceanography (VNIRO), where he was Professor and head of the pelagic fish laboratory. He wrote: *Distribution, biology and fisheries of drums (or croakers) – Sciaenidae family – throughout the world's oceans* (1974). He led a number of expeditions to the southeast Pacific where a seamount is named after him.

Dryagin

Dryagin's Char *Salvelinus drjagini* MV Logashev, 1940

Petr Amphilokhovich Drjagin (see under **Drjagin**).

Duarte

Ghost Knifefish sp. *Adontosternarchus duartei* CDCM de Santana & RP Vari, 2012

Cleber Duarte of Instituto Nacional de Pesquisas da Amazônia, Manaus, collected the majority of the specimens on which the description of this species was based. He co-wrote: *Comparison of the relative efficiency of two fishing gears in sandy beaches in lower Purus river, Amazonas, Brazil* (2013).

Dubois, A

Dubois' Panchax *Epiplatys duboisi* M Poll, 1952
[Syn. *Aphyoplatys duboisi*]
Dubois' Freshwater Puffer *Tetraodon duboisi* M Poll, 1959

A Dubois was a Belgian pharmacist who was also an aquarist. He collected the types of these species in the Congo. Unfortunately we can find nothing more about him.

Dubois, J

Lake Tanganyika White-spotted Cichlid *Tropheus duboisi* G Marlier, 1959

Jean Dubois was a limnologist and a colleague of the author, Georges Marlier. This *might* refer to J Thomas Dubois who worked at the Centre de Recherches du Tanganika, part of the Institut pour la Recherche Scientifique en Afrique Centrale. He wrote: *Evolution de la température, de l'oxygène dissous et de la transparence dans la Baie Nord du lac Tanganika* (1958).

Duboulay

Scribbled Angelfish *Chaetodontoplus duboulayi* A Günther, 1867

Duboulay's Rainbowfish *Melanotaenia duboulayi* F de L de Castelnau, 1878

Francis Houssemayne du Boulay (1837–1914) was a collector, entomologist and natural history artist, sending specimens to British and Australian museums. He is best known for sending Coleoptera to BMNH. Born in Kent, England he shipped to Perth, WA (c.1858). He was located at Minnannooka Station near Geraldton in WA with his brother and started collecting insects there. He was in Victoria (c.1869) and collected around Cooktown and Rockhampton (1870), eventually settling in Beechworth, Victoria until his death. He was also an accomplished artist and presented plates of Coleoptera to the National Museum, Melbourne. The rainbowfish etymology records that a man called Du Boulay collected the original specimen in the Richmond River, northern New South Wales (during 1870s), and we believe this to be the same person.

DUCCIS

Ghost Knifefish sp. *Sternarchella duccis* JG Lundberg, C Cox Fernandes & JS Albert, 1996
[Syn. *Magosternarchus duccis*]

DUCCIS is an acronym of the ichthyological society, the Duke University Center for Creative Ichthyological Studies. It is pronounced 'doo-sis'.

Duc Huu Nguyen

Vietnamese Cyprinid sp. *Opsariichthys duchuunguyeni* TQ Huynh & IS Chen, 2014

Professor Dr Nguyen Huu Duc (See **Nguyen, HD**)

Duellman

Pencil Catfish sp. *Trichomycterus duellmani* MGE Arratia Fuentes & SA Menu-Marque, 1984

Dr William Edward Duellman (1930-2022) was a herpetologist regarded as the world authority on Neotropical frogs. He became Curator of Herpetology, University of Kansas (1959), and retired (1997) as Curator Emeritus and Professor Emeritus, Department of Ecology and Biological Evolution. His wife, Linda Trueb, is also a herpetologist and Curator of Herpetology, University of Kansas. They co-wrote *Biology of Amphibians* (1986). Nine reptiles and seven amphibians are named after him.

Dugès

Lerma Catfish *Ictalurus dugesii* TH Bean, 1880
Opal Allotoca *Allotoca dugesii* TH Bean, 1887

Dr Alfredo Augusto Delsescautz Dugès (1826–1910) was a French physician who qualified in medicine in Paris. He emigrated to Mexico where he became Professor of Natural History, Universidad de Guanajuato, Mexico. He is regarded as being the father of Mexican herpetology, as he was the first to define Mexican herpetofauna in Linnaean terms. Two birds and six reptiles are named after him. He collected the holotypes of these two fish species.

Duhamel

Snailfish sp. *Paraliparis duhameli* AP Andriashev, 1994

Dreamer sp. *Spiniphryne duhameli* TW Pietsch III & ZH Baldwin, 2006

Dr Guy Duhamel (b.1953) is a professor (1992) at the Department of Life Adaptations at MNHN, Paris & Curator of Marine Fish, Head of the Department of Aquatic Sciences (2008–2016) at MNHN. He is a professor of ichthyology there (1992) and was in charge of the vertebrate collection (2003–2006). Previously he was a research scientist at TAAF (1979–1981), FIOM and the SFI (1981–1982) and CNRS (1982–1988) then Assistant professor at MNHN (1988–1992). The University of Paris awarded his PhD (1987). He is a specialist in the fish family Liparidae with description of new species. Among his publications, he has produced a fish guide of the Kerguelen and Crozet Islands *Poissons des îles Kerguelen et Crozet, guide régional de l'océan Austral* (2005). He is involved in conservation and management of marine exploited Antarctic fish (especially Patagonian Toothfish and Mackerel Icefish) with many contributions to the CCAMLR (Commission for the Conservation of Antarctic Marine Living Resources). He organised the first Symposium on the Kerguelen Plateau (Duhamel & Welsford, 2011).

Duida

Centipede Knifefish *Steatogenys duidae* FR La Monte, 1929

This is a toponym referring to Mount Duida, Venezuela, where the holotype was acquired.

Duka

Mahseer sp. *Neolissochilus dukai* F Day, 1878

Lieutenant-Colonel Dr Theodore Duka (1825–1908) was a Hungarian lawyer and physician. He qualified as a lawyer, but was involved in the abortive Hungarian Revolution and, as he was on the losing side at the Battle of Schwechat (1848) and in danger of arrest and execution, he went into exile in Paris (1849). He went from there to London (1850) where he enrolled as a medical student. He qualified at St Andrews (1853) and joined the Bengal Presidency of the Honourable East India Company as a medical officer. He served in India (1854–1864 & 1866–1874). In retirement, he wrote a biography of Alexander Csoma de Koros, the Hungarian linguist and explorer of Tibet, was honoured by the Austro-Hungarian Emperor Franz Joseph, and the University of Budapest awarded him an honorary doctorate of medicine (1899). He sent examples of this species to the author.

Dulkeit

Snailfish sp. *Liparis dulkeiti* VK Soldatov, 1930

Dr Georgy Dzheymsovich Dulkeit (1896–1988) was a Russian zoologist. He was at the Tomsk Institute of Technology saddlery department but left for military service (1916). He worked as a hunter, fisherman and teacher of geography before heading a major expedition to 'Dalryba' on Shantar Islands. He was awarded the equivalent of a doctorate (1938) and headed a scientific department at the Altai State Reserve (1940–1951), but continued as a member of the Scientific Council and coordinator of zoological studies of the reserve. Then (1952) until retirement (1965) he worked at the Stolby Nature Reserve.

He was awarded a further doctorate (1971). Among his many publications he wrote: *Ichthyofauna of Teletskoe Lake and the Biya River* (1949) and *Materials on Fish of the Shantar Sea* (1988). He was awarded a Ministry of Agriculture diploma for his book: *The hunting fauna, questions and methods for evaluating the performance of hunting areas of the Altai-Sayan mountain taiga.*

Dumas

Omba Rainbowfish *Melanotaenia dumasi* MCW Weber, 1907

Joannes Maximiliaan Dumas (1856–1931) was a feather merchant, surveyor and self-taught naturalist who collected in New Guinea (1899–1917), gaining a reputation as an indefatigable and intrepid explorer. He also collected in the Moluccas for Everett (q.v.), Stresemann (q.v.) and others. He was part of a scientific expedition of Sentani Lake and surrounding areas sponsored by the Treub Company and Royal Dutch Geographical Society (1903). He spent seven years (1907–1915) in the interior as naturalist and surveyor on an extended military mapping and collecting expedition. Two birds are also named after him.

Duméril, AHA

Lanternfish sp. *Diaphus dumerilii* P Bleeker, 1856
Longtail Sole *Apionichthys dumerili* JJ Kaup, 1858
Chinese Longsnout Catfish *Tachysurus dumerili* P Bleeker, 1864
African Airbreathing Catfish sp. *Clarias dumerilii* F Steindachner, 1866
Suco Croaker *Paralonchurus dumerilii* F Bocourt, 1869
Grooved Mullet *Chelon dumerili* F Steindachner, 1870
Chinese Gudgeon sp. *Saurogobio dumerili* P Bleeker, 1871
Southern Fiddler Ray *Trygonorrhina dumerilii* FL Castelnau, 1873

Auguste Henri André Duméril (1812–1870), was a physician and zoologist like his father (see next entry) and followed much in his father's footsteps. He was an ichthyologist and herpetologist at MNHN, Paris. He produced the two-volume: *Histoire naturelle des poissons, ou Ichtyologie générale* (1865–1870). He died during the Siege of Paris (which event led to defeat for France in the Franco-Prussian War).

Duméril, AMC

Greater Amberjack *Seriola dumerili* A Risso, 1810
Atlantic Angel Shark *Squatina dumeril* CA Lesueur, 1818
Yelloweye Filefish *Cantherhines dumerilii* H Hollard, 1854
[Alt. Barred Filefish, Barred Leatherjacket]

Dr André Marie Constant Duméril (1774–1860) was a French zoologist who qualified as a physician (1793). He was Professor of Anatomy, Muséum National d'Histoire Naturelle, Paris (1801–1812), during which time he published his *Zoologie analytique* (1806). He later became Professor of Herpetology and Ichthyology (1813–1857). He built up the largest herpetological collection of the time. Toward the end of his career his son assisted him. He retired entirely (1857) and his son took over his professorship. A very large number of taxa are named after the two men, including 6 amphibians, 2 birds and 21 reptiles.

Hollard's description of the filefish refers to the 'illustrious scholar' (translation) who put the specimen at Hollard's disposal, and might refer to either Duméril.

Duncker

Pipefish genus *Dunckerocampus* GP Whitley, 1933
Priapiumfish sp, *Phallostethus dunckeri* CT Regan, 1913
Dunker's Pugnose Pipefish *Bryx dunckeri* JM Metzelaar, 1919
Duncker's River Garfish *Zenarchopterus dunckeri* E Mohr, 1926
Duncker's Pipehorse *Solegnathus dunckeri* GP Whitley, 1927
Bigspot Barb *Barbodes dunckeri* E Ahl, 1929
Duncker's Pipefish *Halicampus dunckeri* P Chabanaud, 1929
Freshwater Pipefish sp. *Microphis dunckeri* B Prashad & DD Mukerji, 1929

Dr Paul Georg Egmont Duncker (1870–1953) was a German zoologist and ichthyologist. He studied at the universities of Berlin, Freiburg and Kiel where he was awarded his doctorate (1895). He was peripatetic (1895–1900), working in Karlsruhe, Plymouth, Naples, Long Island (New York) and Würzburg before joining an expedition to the South Seas (1900). He left the expedition when in Malaya (1901) and was Curator of the Selangor State Museum, Kuala Lumpur (1901–1902). He returned to Europe and worked for a year in Naples but eventually joined the staff of the Hamburg Zoological Museum (1907) and took part in the first year (1908–1909) of the Hamburg Südsee-Expedition (1908–1910). He returned to the museum (1909) and became Curator and Professor (1928), and retired (1934).

Dunn

Driftwood Catfish sp. *Tatia dunni* HW Fowler, 1945

Dr Emmett Reid Dunn (1894–1956) was a leading American herpetologist of his time. He took both his bachelor's degree (1915) and his master's (1916) at Haverford College, Pennsylvania, where he later became Professor of Biology (1934). He took his doctorate (1921) at Harvard, where he worked at the Museum of Comparative Zoology. He was a Zoology Assistant, Smith College (1916–1928). He visited London, Paris and Berlin to study their museum collections (1928). He was Secretary of the *Journal of American Society of Ichthyologists and Herpetologists* (1924–1929) and was President of the American Society of Ichthyologists and Herpetologists (1930–1931). From 1937 he was closely associated with the Philadelphia Academy of Natural Sciences, becoming Curator of Herpetology (1944). He tried to become an army officer (WW1) but was rejected, as it was considered that his weekend pursuit of salamanders and snakes was unbecoming in an officer and a gentleman! Instead he was an ensign in the US Navy (1917–1918). Eighteen reptiles and twelve amphibians are named after him. (Also see **Alta**)

Dunsire

Dunsire's Cave Garra *Garra dunsirei* KE Banister, 1987

Andy Dunsire is an ex-RAF radio engineer and speleologist who was employed (from early 1970s) in Oman to help keep the fighter planes of the Sultan of Oman's Air Force in the air. He is also famous as an explorer of Arabian caves, and collected the first specimens of the eponymous fish (1980).

Duperrey

Saddle Wrasse *Thalassoma duperrey* JRC Quoy & JP Gaimard, 1824

Captain Louis Isidore Duperrey (1786–1865) was a French naval officer (1802). He was second in command and marine hydrologist on board *L'Uranie* during its circumnavigation (1817–1820), then was appointed (1821) to command *La Coquille* for its circumnavigation (1822–1825). A reptile and a bird are also named after him.

Dupont

Central African cichlid sp. *Chilochromis duponti* GA Boulenger, 1902

Dr Édouard-François Dupont (1841–1911) was a Belgian naturalist, geologist and palaeontologist. The University of Louvain awarded his doctorate and he went on to be Director of the RNHM (1868–1909), where he added to and re-organised the entire collections. The etymology says: "*...I have much pleasure in naming the species in honour of the eminent Director of the Brussels Museum, one of the pioneers in the geological exploration of the Congo Basin.*"

Dupouy

Banjo Catfish genus *Dupouyichthys* LP Schultz, 1944
Armoured Catfish sp. *Chaetostoma dupouii* A Fernández-Yépez, 1945

Walter Dupouy (1906–1978) was a Venezuelan anthropologist-biologist, who also published on museology, history, speleology, ornithology, geography, folklore, journalism, drama and novels, despite having been self-taught. He was Director, National Museum of Natural Sciences, Caracas (1940–1948). He encouraged and supported Fernández-Yépez' study of catfishes.

Duquesne

Black Redhorse *Moxostoma duquesnei* CA Lesueur, 1817

This is not a true eponym, but is named after Fort Duquesne (now Pittsburgh, Pennsylvania, USA) so actually a toponym. The fort was named after Marquis Duquesne, the governor-general of New France (i.e. the area of North America colonised by the French).

Duranton

Three-barbeled Catfish sp. *Mastiglanis durantoni* M De Pinna & P Keith, 2018

Michel Duranton is an entomologist and ichthyologist from French Guiana. His publications include the co-written: *Anthribidae de Guyane française: plantes nourricières de trois espèces* (1996). According to the etymology, the species name was dedicated to: "... *Michel Duranton, who collected with the second author the first specimens of the species, in recognition of his fascinating work on the fauna of French Guiana.*"

Durbin

Characin sp. *Bryconops durbini* CH Eigenmann, 1908
Tetra sp. *Hemigrammus durbinae* RP Ota, FCT Lima & CS Pavanelli, 2015

Marion Lee Ellis née Durbin (1887–1972) was an American ichthyologist, entomologist, limnologist and environmental toxicologist. She worked at the University of Missouri, Columbia and made a highly detailed study of the genus *Hemigrammus*. She wrote: *On the species of Hasemania, Hyphessobrycon and Hemigrammus collected by JD Haseman for the Carnegie Museum* (1911). (See also **Marion** and **Ellis**)

Durin

Driftwood Catfish genus *Duringlanis* S Grant, 2015

Durin the Deathless, eldest of the Seven Fathers of the Dwarves in Tolkien's *Lord of the Rings*. The allusion is to the small size of species in this subgenus.

Durini

Schilbid Catfish sp. *Schilbe durinii* L Gianferrari, 1932

Countess Maria Teresa Camozzi Durini di Monza née Camozzi (1892–1943) was the wife of Count Hercules Luchino Durini di Monza (1876–1968) and she collected the type at Lake Tanganyika. He was co-leader of the Beragiola-Durini expedition (1930) with Giuseppe Carlo Odoardo Angelo Maria Baragiola (b.1890). Both men were members of the Italian parliament and their mission was not purely exploratory, but also political - to investigate any chances for Italy to acquire a colony in West Africa. They set out in motorcars to drive across Africa from west to east. There is doubt about the true type locality of this species, as no other specimens have been found at Lake Tanganyika. The patronym is not identified and, although the countess collected the type, the count led the expedition and it is most probable he was honpoured in the name.

Durrell

Don Tadpole-Goby *Benthophilus durrelli* VS Boldyrev & NG Bogutskaya, 2004

Gerald 'Gerry' Malcolm Durrell (1925–1995) is best known for the Durrell Wildlife Preservation Trust, Jersey, Channel Islands. He was born in India and first went to England (1928) upon his father's death. The family lived a Bohemian existence on Corfu (1935–1939). His first expedition (1947), to the British Cameroons (Cameroon), was financed by his inheritance from his father. He sold the animals he brought back and so financed further expeditions to British Guiana (Guyana), but decided to set up his own zoo. He founded his zoo in Jersey (1958) with the help of his first wife, Jacqueline Sonia Wolfenden, from whom he was later divorced (1979). He married Lee McGeorge Wilson (1979), a naturalist, zookeeper, and author from Tennessee. She has carried on the work that Gerald started. He wrote: *My Family and Other Animals* (1956), which was a financial success and provided funding for more expeditions. A mammal, a reptile and an amphibian are also named after him. The etymology says the goby's name is in honour of Jerald (sic) Durrell the 'famous English animal writer.'

Dursun Avşar

Stone Loach sp. *Seminemacheilus dursunavsari* E Çiçek, 2020

Professor Dr Dursun Avşar is a Turkish biologist and marine ecologist who is head of the laboratory in the Faculty of Fishes at the Cukurova University in Adana where he has worked all of his career (1980–now). The Middle East Technical University Marine Science Institute awarded (1987) his MSc and his PhD (1993). He wrote: *A new species record for the Central and Eastern Mediterranean; Sphoeroides cutaneus (Günther, 1870)(Pisces: Tetraodontidae)* (1999) and *Balıkçılık Biyolojisi ve Popülasyon Dinamiği* (2016) among numerous scientific papers. The etymology states: "*The new species is named after Professor Dr Dursun Avsar... ...for his supports as my supervisor.*"

Durville

Bourbon Chromis *Chromis durvillei* J-C Quéro, J Spitz & J Vayne, 2010

Patrick Durville (b.1963) is an ichthyologist on the island of Réunion, where he helped set up the Aquarium de La Réunion. He made good use of an underwater eruption (2007) that brought to the surface many new fish. He was co-author of: *Checklist of Reunion Island Fishes* (2004). The etymology says: "*Les auteurs ont dédié cette espèce à Patrick Durville, ichtyologiste, avec Thierry Mulochau, de l'aquarium de Saint-Gilles, île de La Réunion. Coauteurs du catalogue des poissons de l'île (Letourneur et al., 2004), ils ont tout de suite perçu le grand intérêt que présentaient, à la suite de l'éruption, ces poissons flottant à la surface et ont largement contribué à la collecte et à la mise en collection de ces spécimens.*"

Dury

Black Darter *Etheostoma duryi* JA Henshall, 1889

Charles Dury was an amateur entomologist. The etymology states: "*Mr. Charles Dury, recently, while collecting insects in East Tennessee, collected a small lot of fishes, consisting of five species, two of which seem to be new to science.*"

Dusén

Armoured Catfish sp. *Isbrueckerichthys duseni* A Miranda Ribeiro, 1907

Dr Per Karl Hjalmar Dusén (1855–1926) was a Swedish naturalist, botanist, cartographer, explorer and bryologist. His first overseas collecting expedition was to Cameroon (1890). He was in Argentina on the Princeton expeditions to Patagonia (1896–1899). He was on board the 'Antarctic', responsible for the cartography, as a member of Nathorst's expedition to Spitzbergen (1899). Princeton awarded him an honorary doctorate (1904). He collected the holotype of this species. An amphibian and a reptile are also named after him.

Dussumier

Rainbow Sardine genus *Dussumieria* A Valenciennes, 1847
Malabar Glassy Perchlet *Ambassis dussumieri* G Cuvier, 1828
Sin Croaker *Johnius dussumieri* G Cuvier, 1830
Lesser Bream *Brama dussumieri* G Cuvier, 1831
Eyestripe Surgeonfish *Acanthurus dussumieri* A Valenciennes, 1835
Dussumier's Ponyfish *Karalla dussumieri* A Valenciennes, 1835
Lance Blenny *Aspidontus dussumieri* A Valenciennes, 1836
Streaky Rockskipper *Istiblennius dussumieri* A Valenciennes, 1836
Dussumier's Mudskipper *Boleophthalmus dussumieri* A Valenciennes, 1837

Flat Toadfish *Colletteichthys dussumieri* A Valenciennes, 1837
Whitecheek Shark *Carcharhinus dussumieri* JP Müller & FGJ Henle, 1839
Indian Airbreathing Catfish sp. *Clarias dussumieri* A Valenciennes, 1840
Blacktip Sea Catfish *Plicofollis dussumieri* A Valenciennes, 1840
Labeo sp. *Labeo dussumieri* A Valenciennes, 1842
Dussumier's Halfbeak *Hyporhamphus dussumieri* A Valenciennes, 1847
Gold-spotted Grenadier Anchovy *Coilia dussumieri* A Valenciennes, 1848
Dussumier's Thryssa *Thryssa dussumieri* A Valenciennes, 1848

Jean-Jacques Dussumier (1792–1883) was a French merchant, collector, traveller and ship owner. He was most active (1816–1840) in South-eastern Asia and around the Indian Ocean. He collected molluscs and fish, a large number being named after him. He was also interested in cetaceans and reported on sightings he had made whilst at sea, and harpooned specimens of dolphins and porpoises. He corresponded on the subject with Georges Cuvier (1769–1832), who wrote a number of the formal scientific descriptions. A mammal, four birds and five reptiles are also named after him.

Dutoit

Dutoit's Dottyback *Pseudochromis dutoiti* JLB Smith, 1955

Petrus Johann du Toit (1888–1967) was a veterinary scientist and former president of the South African Council for Scientific and Industrial Research, which provided financial assistance to Smith.

Dutra

Glass Knifefish sp. *Eigenmannia dutrai* LAW Peixoto, MNL Pastana & GA Ballen, 2020

Dr Guilherme Moreira Dutra is a Brazilian zoologist who is a post doctoral fellow at the Museu de Zoologia da Universidade de São Paulo, with a special interest in the taxonomy and systematics of Neotropical fish. His BSc was awarded by the Pontifical Catholic University of Minas Gerais (2008), his MSc (2011) and PhD (2015) by the Federal University of Pará. Among his publications is: *Rapid assessment of the ichthyofauna of the southern Guiana Shield tributaries of the Amazonas River in Pará, Brazil* (2020). He was honoured "*…for his contribution to the field of ichthyology, in particular to the taxonomy of Eigenmannia.*"

Dutton

Claroteid Catfish sp. *Chrysichthys duttoni* GA Boulenger, 1905

Dr Joseph Everett Dutton (1874–1905) was a British parasitologist and co-leader of the Dutton-Todd expedition to the Congo that collected the holotype (1903–1905). He graduated as a physician (1897) and was appointed to Liverpool Infirmary as a house surgeon and then as a house physician. He joined an expedition to Nigeria (1900) and went on a solo expedition to Gambia (1901). He was investigating tick disease in the Congo Free State by means of performing post mortems, caught the disease himself and died from it. He is famous for discovering one of the trypanosomes that cause sleeping sickness. It is worthy of note that it took two months for the news of his death to reach the closest telegraph station

Dybowski, BT

Naked Osman *Gymnodiptychus dybowskii* KF Kessler, 1874
Korean Sandlance *Hypoptychus dybowskii* F Steindachner, 1880
Prickleback sp. *Pholidapus dybowskii* F Steindachner, 1880
Little Baikal Oilfish *Comephorus dybowskii* AA Korotneff, 1904

Benedykt (Benoit) Tadeusz Dybowski (1833–1930) was a Polish biologist who was born in what is now Belarus. He was an ardent proponent of Darwin's theory of evolution. He was appointed Adjunct Professor of Zoology in Warsaw (1862), but after the failure of the (1863) Uprising (against the Russian Empire) he was banished and spent time as a political exile in Siberia. Here, support from the Zoological Cabinet at Warsaw allowed him to undertake investigations into the natural history of Lake Baikal and other parts of the Soviet Far East. He was pardoned (1877) and went to Kamchatka as a physician (1878). He was appointed (1883–1906) to the Chair of Zoology, University of Lemburg, Poland (Lviv, Ukraine) until his retirement. A mammal, three birds and two amphibians are also named after him.

Dybowski, J-T

Squeaker Catfish sp. *Euchilichthys dybowskii* L Vaillant, 1892

Jean-Thadée Emmanuel Dybowski (1856–1928) was a French (with Polish parents) agronomist, naturalist and explorer of (especially equatorial) Africa. He led a Congo expedition (1891) and wrote accounts of his travels, to Chad (1893): *La Route du Tchad*, and the Congo (1912): *Le Congo Méconnu*. Dybowski established new gardens and plantations in Tunisia and organised schools of agriculture. Later (c.1908) he became French Inspector General of colonial agriculture. The Dybowski family comprised many scientists, including an outstanding arachnologist, so it can be difficult to track down quite what is named after whom. Dybowski was largely responsible for the isolation and introduction of the psychotropic drug ibogaïne. Dybowski and Landrin isolated the alkaloid, which they named ibogaïne from the bark of the root (1901) and showed it to have the same psychoactive properties as the root itself.

Dydymov

Dydymov's Hookear Sculpin *Artediellus dydymovi* VK Soldatov, 1915

Named after the Research vessel 'Lieutenant Dydymov' from which the type specimens of this fish (and many others) was caught. The vessel is in turn named after Lieutenant Akim Grigorevitch Dydymov, a Russian naval officer who served in the Far East.

E

Eakin

Eakin's Plunderfish *Pogonophryne eakini* AV Balushkin, 1999

Dr Richard Reynolds Eakin was a biologist at the Department of Life Sciences, University of New England, Maine. His primary focus has been on marine species of Antarctic waters. He has worked on the genus *Pogonophryne*, sometimes in association with Balushkin, and they co-authored papers such as: *A new species of* Pogonophryne *from East Antarctica* (2015).

Eapen

Malabar Swamp Eel *Monopterus eapeni* PK Talwar, 1991

Dr K C Eapen was a marine biologist who worked for the Marine Biological Laboratory at Trivandrum, part of the Department of Marine Biology and Oceanography, University of Kerala. He described the eel (1963) but used a preoccupied name (*Monopterus indicus*) in his paper: *A new species of* Monopterus *from South India* (1963).

Earle

Earle's Splitfin *Luzonichthys earlei* JE Randall, 1981
Orange-striped Wrasse *Cirrhilabrus earlei* JE Randall & RL Pyle, 2001
Earle's Soldierfish *Myripristis earlei* JE Randall, GR Allen & DR Robertson, 2003
Sea-Whip Goby sp. *Bryaninops earlei* T Suzuki & JE Randall, 2014

John L Earle is a founder member of the Association for Marine Exploration. Princeton University awarded his bachelor's degree. He served in the US Navy as a pilot during the Vietnam War, and later became an airline plot but is now retired. He is a Research Associate in Zoology, Bishop Museum, Hawaii and (since 1994) he has been a rebreather diver engaged in pioneering deep reef exploration. He co-wrote: *Xanthichthys greenei, a new species of triggerfish (Balistidae) from the Line Islands* (2013). He collected a specimen of the soldierfish in the Marquesas, and provided an underwater photograph used in the description. He discovered the goby and collected the type with the author. He also collected the splitfin from Fanning Island for the Bishop Museum.

Earll

Carolina Hake *Urophycis earllii* TH Bean, 1880

Robert Edward Earll (1853–1896) was an ichthyologist and an honorary museum curator at USNM. He graduated from Northwestern University (1877) and worked for the US Fish Commission as a fish culturist (1878). He worked for the 10th US Census (1879–1982) collecting data on fisheries. He became Chief of Division, Fisheries, US Fish Commission

(1883). He wrote: *Materials for a history of the mackerel fishery* (1881). He used to scour a fish market in Charleston, South Carolina, for new holotype specimens.

Earnshaw

Earnshaw's Hawkfish *Amblycirrhitus earnshawi* R Lubbock, 1978
[Alt. White Hawkfish]

George Earnshaw was formerly a resident of Ascension Island and a member of their Historical Society. He provided the author with specimens of a number of local fish species including the hawkfish that he first collected (1975). He also sent other marine specimens to other researchers.

Eastman, CR

Gecko Catshark *Galeus eastmani* DS Jordan & JO Snyder, 1904

Dr Charles Rochester Eastman (1868–1918) was an American palaeontologist, particularly interested in fossil fish. He graduated Harvard (1891) and was awarded his MSc by Johns Hopkins University (1892) after which he undertook his PhD awarded by Munich (1894). He taught geology and palaeontology at Harvard and Radcliffe while further studying fossil fish under Agassiz (q.v.). He became Curator of Vertebrate Palaeontology at the Museum of Comparative Zoology and then Curator at Carnegie Museum Pittsburgh (1910). He was also an assistant geologist for the US Geological Survey's New England Division. He wrote more than one hundred scientific papers. Moreover, he was the foremost authority on the literature of fishes and produced a massive bibliography. His body was found drowned at Long Beach where he had been walking while recuperating from Spanish Flu. It was supposed that he collapsed and fell into the water.

Eastman, JT

Thickskin Snailfish *Paraliparis eastmani* DL Stein, NV Chernova & AP Andriashev, 2001
Antarctic Eelpout sp. *Ophthalmolycus eastmani* J Matallanas, 2011

Dr Joseph Thornton Eastman (b.1944) is Professor Emeritus of Anatomy, Ohio University. The University of Minnesota awarded his BA (1966), MS (1968) and PhD (1970). He was Instructor in Anatomical Sciences at the University of Oklahoma Medical Center (1970–1971) and Assistant Professor of Anatomical Sciences there (1971–1973) before moving to Brown University as Assistant Professor of Anatomy (1973–1979). He moved to Ohio University as an associate professor (1979–1989) before becoming full Professor there (1989–2014). His focus is on the nature of fish diversity in the Antarctic marine ecosystem, with emphasis on the notothenids. He has written a book entitled *Antarctic Fish Biology: evolution in a unique environment* (1993). He was also a contributor to *Fishes of the Southern Ocean* (1990) and authored about 100 other papers on Antarctic fishes. He serves on the editorial boards of *Antarctic Science* (1999–present) and *Polar Biology* (2013–present). He now lives in Minneapolis, Minnesota. The snailfish etymology states that: "*The new species is named after Joseph T Eastman in honour of his valuable studies on the natural history, physiology, and origins of the Antarctic fish fauna.*"

Eastward

Glass Knifefish sp. *Rhabdolichops eastwardi* JG Lundberg & F Mago-Leccia, 1986

'Eastward' is the name of a Research Vessel that at one time was owned by the Duke University Oceanographic Program. It supported two ichthyological expeditions to the lower Orinoco.

Eaton, AE

Eaton's Skate *Bathyraja eatonii* A Günther, 1876

Reverend Alfred Edmund Eaton (1845–1929) was an English explorer, entomologist and naturalist who published many scientific papers (1860s–1920). He collected in the Kerguelen Islands with the Transit of Venus Expedition (1874–1875) and collected the holotype of the skate in Royal Sound (now Golfe du Morbihan), Kerguelen. Two birds are named after him.

Eaton, P

Durban Sweeper *Pempheris eatoni* JE Randall & BC Victor, 2014

Patrick Eaton, about whom we have failed to find more information than is contained in the original text. The etymology states: "*…This species is named* Pempheris eatoni in honor of Patrick Eaton who collected the holotype and five other adult specimens for this study." Elsewhere in the paper, the authors write: "*We are most grateful to Dr. Allan D. Connell for his collections and photographs of* Pempheris from KwaZulu-Natal; his *cousin Patrick Eaton for the special effort to collect fish from the difficult exposed rocky shore of South Africa…*"

Ebeling

Ebeling's Fangjaw *Sigmops ebelingi* MG Grey, 1960

Bigscale sp. *Melamphaes ebelingi* MJ Keene, 1973

Alfred W Ebeling (b.1931) is an American ichthyologist who was at Scripps Institution of Oceanography. Grey honoured Ebeling for his 'interest and assistance' during the course of Grey's preliminary review of the family. He co-wrote: *Fishes: A Field and Laboratory Manual on Their Structure, Identification, and Natural History* (1985).

Ebisu

Fishgod Blenny *Malacoctenus ebisui* VG Springer, 1959

Ebisu is a Japanese god who is the patron of fishermen.

Ebrardt

Orange-finned Halfbeak *Nomorhamphus ebrardtii* CML Popta, 1912

Friedrich Clemens Ebrardt aka Ebrard (1850–1935) studied history and economics at Erlangen University, then moved to Göttingen University and went on to complete his doctorate (1872) at the University of Tübingen, after which he spent four years in the army. He became (1876) Assistant Librarian at the University of Strasbourg and Curator a year later, then Librarian (1878). He moved to become senior Librarian for the city of Frankfurt am Main and later Director and was given Prussian citizenship. He also attained high office in the church. The etymology just refers to 'Dr Ebrardt' with no details of other names, but described him as a Privy Councillor to the state of Hesse-Nassau (then controlled by Prussia). He was also an honorary member of the Frankfurt Geographical and Statistical Association that mounted the 1911 expedition to the Lesser Sunda Islands, led by Elbert. Although there is no further information given on Dr Ebrardt we believe this to be the person honoured.

Eccles

Lake Malawi Cichlid sp. *Diplotaxodon ecclesi* WE Burgess & HR Axelrod, 1973
Lake Malawi Cichlid sp. *Placidochromis ecclesi* M Hanssens, 2004

Dr David Henry Eccles (1932-2021) was a British ichthyologist, limnologist and naturalist who was Senior Fisheries Research Officer of Malawi. He worked for about two decades on the biology, evolution and taxonomy of Lake Malawi cichlids. He described around forty species. Among his published papers is: *An outline of the physical limnology of Lake Malawi (Lake Nyasa)* (1974). The etymology for the *Diplotaxodon* reads: "*...Named for Dr. David H. Eccles, whose work is helping to clarify the taxonomic confusion prevailing in the Lake Malawi cichlids.*"

Echeagaray

Maya Gambusia *Gambusia echeagarayi* J Álvarez del Villar, 1952

Luis Echeagaray Bablot (d.1984) was the manager of water resources in the southeast of Mexico. He took a great interest in the work of the author and facilitated his collecting.

Echidna

Moray Eel genus *Echidna* JR Forster, 1788

In Greek mythology Echidna was a female monster, and mother of many other monsters including the Hydra. Hesiod described her as "half a nymph with glancing eyes and fair cheeks, and half again a huge snake, great and awful, with speckled skin."

Ecklon

Clingfish genus * Eckloniaichthys* JLB Smith, 1943
Kelp Weedfish *Heteroclinus eckloniae* RJ McKay, 1970

This is not a direct eponym as the species and genus are named after the kelp *Ecklonia* in which they tend to live. The kelp was named after Christian Friedrich Ecklon (1795–1868) a Danish botanical collector.

Ecklon, FL

Naked Carp sp. *Gymnocypris eckloni* S Herzenstein, 1891

F L Ecklon was an assistant to Przewalski (q.v.) on his second expedition into Tibet (1883–1885). Przewalski wrote that his services were 'invaluable', but very little seems to be recorded of him.

Eckström

Eckström's Topknot *Zeugopterus regius* PJ Bonnaterre, 1788
[Alt. Bloch's Topknot]

Carl Ulrik Ekström (1781–1859) was a Swedish priest and naturalist who studied medicine at Uppsala (1804) but was later ordained (1807). He was a parish priest, but got to know a number of scientists and undertook his own research and collected his observations as 'from thirty years of hunting'. He wrote papers on migratory birds (1826, 1827 & 1829) and became a member of the Swedish Academy of Science (1830). He was an 'inspector' of the Swedish Natural History Museum (1831). He was given a parish in Gothenburg so he could continue his zoological studies there. He was one of the authors of *A History of Scandinavian Fishes* and wrote a practical thesis on suitable ways to fish for herring, cod, ling, mackerel, lobsters and oysters (1845). We speculate that at some point he came to have his name attached to the fish's vernacular name.

Eclancher

Harlequin Wrasse *Bodianus eclancheri* A Valenciennes, 1846

Charles René Augustin Leclancher (1804–1857) was a French naval surgeon. He assisted Gaimard (q.v.) in making zoological collections during an expedition aboard the corvette *Recherche* (1836). He took part in the circumnavigation of *La Vénus* (1836–1839), and in a voyage to the Far East in *La Favorite* (1841–1844). Two birds are named after him. Valenciennes seems to have taken his name as L'Eclancher and dropped the initial L.

Economidis

Goby genus *Economidichthys* PG Bianco, AM Bullock, PJ Miller & FR Roubal, 1987
Lake Trichonis Blenny *Salaria economidisi* M Kottelat, 2004

Professor Dr Panos Stavros Economidis was formerly Director of the Ichthyology and Zoology Laboratories in the School of Biology, Aristotle University of Thessaloniki and was the main contributor to its fish collection. He is now Emeritus Professor. He is an expert in fish taxonomy, distribution and conservation, and in fisheries management with more than four decades of experience. He has written or co-written many papers including: *Fish Fauna of Greece in a Protected Greek Lake: Fish biodiversity, impact of introduced fish species on the ecosystem* (2008). He was honoured in the genus' name for his 'contributions to Greek ichthyology', while Kottelat called him "...a unique historical, gastronomic and oenological cicerone." (Also see **Panos**)

Economou

Riffle Minnow sp. *Alburnoides economoui* R Barbieri, J Vukić, R Šanda, Y Kapakos & S Zogaris, 2017

Dr Alcibiades N Economou is Research Director, Institute of Inland Waters, Hellenic Centre for Marine Research. He co-wrote: *Osteological abnormalities in laboratory reared sea-bass (*Dicentrarchus labrax*) fingerlings* (1991). He was honoured *"...for 'significant' contributions to the biogeography and ecology of Greek fishes."*

Edds

Sisorid Catfish sp. *Pseudecheneis eddsi* HH Ng, 2006
Stone Loach sp. *Balitora eddsi* KW Conway & RL Mayden, 2010

Dr David Ray Edds (b.1954) is an ichthyologist and aquatic ecologist. The University of Kansas, Lawrence, awarded his bachelor's degree (1977), Oklahoma State University, Stillwater, his master's (1984) and his doctorate (1989). He joined the Department of Biological Sciences, Emporia State University, Emporia, Kansas (1989) and is now Roe R Cross Distinguished Professor there. His main research interests and expertise include the community structure of aquatic environments, fish ecology, anthropogenic impacts on rivers and streams, and fishes of Nepal, but he has also studied the ecology of freshwater mussels, aquatic turtles, and invasive zebra mussels. He is a specialist in Nepalese fishes, having volunteered there (1977–1979) and studied there (1984–1986, 1996, 2015). Among his published writings is: *Fishes in Nepal: ichthyofaunal surveys in seven nature reserves* (2007). He is also a co-author of the book *Kansas Fishes* (2014). He collected the type series of *Balitora eddsi.*

Edelmann

Morid Cod sp. *Gadella edelmanni* AB Brauer, 1906

J Edelmann was a machinist on the *Valdivia* Expedition (1888–1899) during which the holotype was collected.

Eden

Snowtrout sp. *Schizothorax edeniana* J McClelland, 1842

George Eden, 1st Earl of Auckland (1784–1849), was Governor-General of India (1836–1842). Although not absolutely certain, it is likely that the binomial refers to him as the etymology (rather typical of Victorian manners) states that it is named "*…in honour of a Nobleman to whom Science is indebted for the opportunities afforded Mr. Griffith of extending his Botanical Researches from the Straits of Malacca into Central Asia*". The Griffith referred to is the botanist and physician William Griffith (1810–1845) who was in the Madras Medical Service (1832–1845) and collected the holotype. He worked with John McClelland on a mission to examine tea cultivation in northeast India. All sorts of things were named after Eden, including a mountain and a town.

Edith

Batik Betta *Betta edithae* J Vierke, 1984

Edith Korthaus (1932–1987) was a German amateur aquarist and tropical fish hobbyist. Together with Dr Walter Foersch (q.v.), she collected this and many other fish in swamps in Kalimantan (South Borneo) (1979).

Edmondson

Blenny sp. *Enchelyurus edmondsoni* HW Fowler, 1923
[Junior Synonym of *Enchelyurus brunneolus*]
Edmondson's Pipefish *Halicampus edmondsoni* V Pietschmann, 1928

Charles Howard Edmondson (1876–1970) was Hawaii's first marine biologist (arriving there 1920). He taught at the College of Hawaii (1920–1942), the Director of the Cooke Marine Laboratory and then full-time Curator at Bishop Museum (1942–1962). Among his published papers are: *The Protozoa of Iowa* (1906) and *Resistance of woods to marine borers in Hawaiian waters* (1955).

Edmunds

Edmund's Spurdog *Squalus edmundsi* PR Last, WT White & JD Stevens, 2007

Dr Matthew J 'Matt' Edmunds is an Australian ecologist. His bachelor's degree (1990) and his doctorate in zoology (1995) were both awarded by the University of Tasmania, Hobart. He was employed as a marine biologist by Consulting Environmental Engineers Pty Ltd (1990–1999) and became (1999) Director, Principal Ecologist at Australian Marine Ecology Pty Ltd, a company that provides scientific consulting services. He was honoured for his "*...high-quality, preliminary research*" on Australian *Squalus* during a summer vacation scholarship at CSIRO Marine Laboratories in the early 1990s.

Eduard

Lake Edward cichlid sp. *Schubotzia eduardianus* GA Boulenger, 1914
Lake Edward cichlid sp. *Haplochromis eduardii* CT Regan, 1921

Not an eponym but a toponym, referring to the type locality where these fish are found: Lake Edward (DRC/Uganda). The use of a *u* rather than a *w* in Eduard/Edward reflects the fact that there is no w in Latin, so 'purists' avoid it when coining Latinized binomials.

Edward

Mombasa Pencil Wrasse *Pseudojuloides edwardi* BC Victor & JE Randall, 2014

Jason Edward is the owner of Greenwich Aquaria, New York, a commercial company specialising in customised aquariums and their maintenance as well as supplying fish. He was instrumental in obtaining the male type specimens (Kenyan coast) and generously supplying them to the authors.

Edward Raney

Fluvial Shiner *Notropis edwardraneyi* RD Suttkus & GH Clemmer, 1968

Dr Edward Cowden Raney (1909–1984), American ichthyologist (see **Raney**).

Edwards, CL

Spaghetti Eel *Moringua edwardsi* DS Jordan & CH Bollman, 1889

Dr Charles Lincoln Edwards (1863–1937) was an American zoologist. Indiana University awarded his bachelor's degree (1886) and his master's (1887). He was a graduate student at Johns Hopkins Biological Laboratory when he collected the holotype. He completed his doctorate at the University of Leipzig (1890) and then taught at Clark University (1890–1892) and at the University of Texas (1892–1894). He was Professor of Biology at the University of Cincinnati (1894–1900) and Professor of Natural History at Trinity College, Hartford, Connecticut (1900–1910). He was appointed (1912) as Naturalist of the Park Department of the City of Los Angeles to plan a zoo and aquarium for the city. He was also Professor of Embryology and Histology at the medical department of the University of South California. He wrote: *The floating laboratory of marine biology of Trinity College* (1905) and *Life in the Sea* (1935).

Edwards, CR

Slipmouth (Ponyfish) sp. *Leiognathus edwardsi* Evermann & Seale, 1907
[Junior Synonym of *Equulites leuciscus*]

Brigadier General Clarence Ransom Edwards (1859–1931) was a US Army officer (1883–1922) who was commander of the 26th Division during WW1 and the first Chief of the Bureau of Insular Affairs, US War Department. During part of his career he was on detached service as Professor of Military Science & Tactics at St. John's College (now Fordham University), from which he received an honorary degree. He acquired a collection of Filipino fishes, including type of this one, during the Philippine-American War (1899–1902).

Edwards, G

Puffadder Shyshark *Haploblepharus edwardsii* HR Schinz, 1822

George Edwards (1694–1773) was an illustrator, naturalist, and ornithologist. He was Librarian, Royal College of Physicians, London (1733–1764), and corresponded regularly with Linnaeus. Edwards wrote the four-volume: *A Natural History of Birds* (1743–1751). The first known reference to the Puffadder Shyshark was in some drawings by Edwards, of specimens that were later lost. Cuvier described the species (1817) as 'Scyllium D'Edwards' based on Edwards' drawings, and although he was not assigning a scientific name it was translated by the German zoologist Schinz (1822) as *Scyllium edwardsii*. Four birds, a reptile and mammal are named after him.

Edwards (Milne-)

Snailfish sp. *Paraliparis edwardsi* L Vaillant, 1888

Sir Alphonse MilneEdwards (1835–1900) was a French zoologist and palaeontologist. He was Professor of Zoology at the MNHN, Paris, (1876–1892), becoming Director (1892). His interest in fossil birds led to the publication of: *Recherches Anatomiques et Paleontologiques pour servir a l'Histoire des Oiseaux Fossiles de la France* (1867 and 1872). He had a close working relationship with Prince Albert I of Monaco and may have been influential in the prince establishing the Oceanographic Museum in Monte Carlo. The 'Prix Alphonse MilneEdwards' was created (1903) in his memory. Seven mammals, two birds and a reptile are named after him. Alphonse's father, Henri Milne-Edwards (1800–1885) was born in Belgium, the 27th son of a reproductively prolific Englishman, and went on to become a renowned French naturalist. He became Professor of Hygiene and Natural History at the Collège Central des Arts et Manufactures (1832). He wrote works on crustaceans, molluscs and corals. Many marine organisms were named after him.

Edwin

Brown Darter *Etheostoma edwini* CL Hubbs & MD Cannon, 1935

Dr Edwin Phillip Creaser (1907–1981) was a biologist at the University of Michigan interested in crustaceans, particularly decapods. Among his written works was: *Descriptions of some new and poorly known species of North American crayfishes* (1933) and he co-wrote with Hubbs and others: *The Cenotes of Yucatan : a zoological and hydrographic survey* (1936). He provided the authors with specimens for description. A reptile is also named after him. (Also see **Creaser, CW**)

Eetveide

African Characin sp. *Eugnathichthys eetveldii* GA Boulenger, 1898

Stanislas Marie Léon Edmond van Eetvelde (1852–1925) was a Belgian diplomat. He went to China (1871) to explore trade openings but he decided (1873) to join the Chinese Customs Service, not returning to Belgium until 1877. He then spent time as Consul General in India (1878–1884) before becoming General Secretary of the Congo Free State (where the holotype was collected).

Eeyore

Tasmanian Codling genus *Eeyorius* CD Paulin, 1986

This can only be named after Eeyore, the very gloomy donkey in A A Milne's: *Winnie-the-Pooh* books. As Eeyore was meant to live in a wood, and fish, generally, live under water, we cannot see much of a connection: although the author says in the etymology "… *Named for Eeyore, a literary character who lived in damp places.*"

Eggers

Eggers' Killifish *Nothobranchius eggersi* L Seegers, 1982

Gerhard (Gerd) Eggers, a German aquarist, was co-discoverer of this species. The author honoured him for "…*his camaraderie during two trips to Tanzania, which led to the discovery of this species; he also collected plants and animals other than fish, which he generally made available.*"

Eggleston

Goby genus *Egglestonichthys* PJ Miller & P Wongrat, 1979

Eggleston's Bumblebee Goby *Egglestonichthys bombylios* HK Larson & DF Hoese, 1997

Dr David Eggleston was Fisheries Research Officer at the Fisheries Research Station, Hong Kong. It was he who collected the type of *E. patriciae* and provided Miller with an "*invaluable store of gobioid material taken during fisheries research in Hong Kong*". (Also see **Patricia (Eggleston)**)

Eggvin

Eggvin's Lumpsucker *Eumicrotremus eggvinii* E Koefoed, 1956

[Now thought to be a Junior Synonym of *E. spinosus*]

Dr Jens Konrad Eggvin (1899–1989) led the oceanographical cruise on the research vessel G O *Sars* during which the Eggvin bank was discovered. The species was caught with a dredge on that bank. So, this could be either a toponym or an eponym, but the wording of the etymology doesn't make it entirely clear which: "*The fish was caught with a dredge on the Eggvin bank, discovered by the research ship m/s 'G. O. Sars' during an oceanographical cruise under the leadership of Dr. phil. Jens Eggvin.*"

Ehrenberg

Porgy sp. *Evynnis ehrenbergii* A Valenciennes, 1830

Blackspot Snapper *Lutjanus ehrenbergii* W Peters, 1869

Dr Christian Gottfried Ehrenberg (1795–1876) was a German naturalist, comparative anatomist and microscopist; one of the foremost scientists of his time. He started studying theology at Leipzig (1815) but changed direction (1817) and went to Berlin to study medicine. He worked on fungi (1820) and lectured at the University of Berlin, where he became Professor of Medicine (1827). He travelled, mainly with his friend Hemprich (1820–1825), in northeast Africa and the Middle East. He travelled with Humboldt to Asia (1829). He met Darwin at Oxford (1847). He is regarded as the founder of micro-palaeontology. On his death his collection of specimens was deposited at the Museum für Naturkunde, Berlin. He was the first person to establish that phosphorescence in the sea is caused by the presence of plankton-like micro-organisms. A mammal, an amphibian and two birds are also named after him.

Ehrhardt

Ehrhardt's Cory *Corydoras ehrhardti* F Steindachner, 1910

Wilhelm Ehrhardt (1860–1936) was a German collector and taxidermist who was born in Guyana. He collected professionally in Brazil (1897–1935), with a gap (1920–1927) when he lived in Hamburg, kept a shop with 'zoological supplies' and traded in natural history specimens. His herpetological collection is in the Museum für Naturkunde, Berlin, which was one of his clients. The place and date of his death are not known but are conjectured as being Brazil (1936). An amphibian is named after him. The cory (catfish) etymology does not identify the person intended, but it is most probably named after Wilhelm.

Ehrhorn

Lanternfish sp. *Diaphus ehrhorni* HW Fowler, 1934

Edward Macfarlane Ehrhorn (1862–1941) was an entomologist and horticulturist. He was educated in Germany, Switzerland, England and Stanford University and employed in California (1892–1904) in connection with horticultural quarantine and insect control. He was entomologist and Chief of the Division of Plant Inspection, Honolulu, Hawaii, where he first arrived (1909).

Ehrich

African Killifish sp. *Micropanchax ehrichi* HO Berkenkamp & VMF Etzel, 1994

Christian Ehrich was a German landscaper and agronomist and one of Etzel's friends who had invited him to west Sumatra (early 1970s). This led to them making several fish-collecting trips in a number of tropical counties. He was a team leader in the Kangra and Dhauladar project in India (1976–1984) and worked for technical cooperation in Indonesia, India, Bolivia, Mali and Brazil before returning to Berlin. In the 1990s he was working on integrated land use planning.

Eibl

Eibl's Angelfish *Centropyge eibli* W Klausewitz, 1963
[Alt. Blacktail Angelfish, Red-stripe Angelfish]

Dr Irenäus Eible-Eibesfeldt (1928–2018) was a German biologist who collected the type in the Nicobar Islands. He studied zoology at the University of Vienna (1945–1949), gaining his doctorate. His first posts were as Research Associate at the local Biological Station at Wilhelminenberg, then at the Institute for Comparative Behaviour Studies in Altenberg. He then joined the Max-Planck-Institute for Behavioral Physiology (1951–1969). During that time, he took part in or led expeditions to the Caribbean and Galapagos (1953–1954 & 1957) and to the Maldives and Nicobars under Hans Hass (1957–1958). He became Honorary Scientific Director of the International Institute for Submarine Research in Vaduz, Liechtenstein (1957–1970). During this time he was visiting professor at Chicago University (1961) and University of Minnesota (1967), and then became Professor of Zoology at Munich (1970). He was Head of a Research Group for Human Ethology which became an independent Research Institute for Human Ethology of the Max-Planck-Society (1970–1996) and Professor Emeritus there (1996–2014). After twenty years of research in Animal Behaviour and Marine Biology, in the 1960s Irenaeus Eibl-Eibesfeldt began with research studies in Human Ethology. Through his studies of expressive behaviour of those born deaf and blind, and his cross-cultural documentation programme of human behaviour, he contributed significantly to establish the discipline of Human Ethology including writing a textbook on it.

Eichhorn

Pencil Catfish sp. *Ituglanis eichorniarum* A Miranda Ribeiro, 1912

Not a true eponym, as it is named after the botanical genus *Eichhornia* (water hyacinth). The holotype was caught amongst this type of plant. The genus is named after Johann Albrecht Friedrich von Eichhorn (1779–1856), a Prussian statesman.

Eichler

Eichler's Dottyback *Pseudochromis eichleri* AC Gill, GR Allen & M Erdmann, 2012

Dieter Eichler is a German underwater photographer who was the first to photograph this species. He wrote: *Tropical marine life: the identification handbook for divers and snorkelers* (1995) and *Gefährliche Meerestiere erkennen: Gefahren, richtiges Verhalten, Erste Hilfe* (2010).

Eichwald

Goby subgenus *Eichwaldiella* GP Whitley, 1930
Kura Chub *Alburnoides eichwaldii* F De Filippi, 1863

Professor Dr Karl Eduard von Eichwald (1795–1876) was a Baltic German from Lithuania (in his day under Russian control). He took a degree in medicine (1819) at Vilnius University after studying in Berlin (1814–1817) and Paris (1818). He was a member of various expeditions to the Caspian Sea, Azerbaijan and the Caucasus. He was Professor of Zoology and Comparative Anatomy at Vilnius University (1827–1838), then moved to St Petersburg, where he was a Member of the Russian Academy of Science. He produced the three–volume: *Zoologia specialis, quam expositis animalibus tum vivis, tum fossilibus potissimuni rossiae in universum, et poloniae in specie, in usum lectionum publicarum in Universitate Caesarea Vilnensi. Zawadski, Vilnae* (1829–1831). *Alburnoides eichwaldii* was previous reported by him as a variety of *Alburnus alburnus.* An amphibian is also named after him.

Eidi

Cave Loach genus *Eidinemacheilus* IH Segherloo, N Ghaedrahmati & JA Freyhof, 2016

Eidi Heidari, Lorestan Bureau of Environment, Iran, is the ranger who protects the spring in which species of this genus are found.

Eigenmann, CH

Glass Knifefish genus *Eigenmannia* DS Jordan & BW Evermann, 1896
Clinid sp. *Ribeiroclinus eigenmanni* DS Jordan, 1888
Slender Clingfish *Rimicola eigenmanni* CH Gilbert, 1890
Three-barbeled Catfish sp. *Pimelodella eigenmanni* GA Boulenger, 1891
Characin sp. *Astyanax eigenmanniorum* ED Cope, 1894
Plateau Chub *Evarra eigenmanni* AJ Woolman, 1894
Thorny Catfish sp. *Ossancora eigenmanni* GA Boulenger, 1895
Shelf Goby *Bollmannia eigenmanni* S Garman, 1896
[Binomial sometimes amended to *eigenmanniorum*]
Suckermouth Catfish sp. *Astroblepus eigenmanni* CT Regan, 1904
Characin sp. *Bryconamericus eigenmanni* BW Evermann & WC Kendall, 1906
Eigenmann's Whiptail Catfish *Rineloricaria eigenmanni* J Pellegrin, 1908
Onesided Livebearer sp. *Jenynsia eigenmanni* JD Haseman, 1911
Three-barbeled Catfish sp. *Pimelodella eigenmanniorum* A Miranda Ribeiro, 1911
Characin sp. *Carlana eigenmanni* SE Meek, 1912
Splashing Tetra sp. *Copella eigenmanni* CT Regan, 1912
Needlefish sp. *Potamorrhaphis eigenmanni* A Miranda-Ribeiro, 1915
Eigenmann's Anchovy *Anchoa eigenmannia* SE Meek & SF Hildebrand, 1923
Eigenmann's Doradid *Orinocodoras eigenmanni* GS Myers, 1927
Characin sp. *Othonocheirodus eigenmanni* GS Myers, 1927
Eigenmann's Piranha *Serrasalmus eigenmanni* JR Norman, 1929
Cavefish sp. *Typhlichthys eigenmanni* HH Charlton, 1933
[Junior Synonym of *Typhlichthys subterraneus*]
Pencil Catfish sp. *Pygidianops eigenmanni* GS Myers, 1944
Long-whiskered Catfish sp. *Propimelodus eigenmanni* JWB Van der Stigche, 1946
Characin sp. *Moenkhausia eigenmanni* J Géry, 1964

Professor Dr Carl Henry Eigenmann (1863–1927) was a German-born American ichthyologist. He graduated from Indiana University with a bachelor's degree (1886) and was awarded a doctorate by the same university (1889). He was Curator, San Diego Natural History Society (1888), then became Professor of Zoology, Indiana University (1891). He travelled in much of the Americas, his last expedition being to Peru, Bolivia, and Chile (1918–1919). He died after suffering a stroke (1927). Among his most notable works are: *A revision of the South American Nematognathi or Cat-Fishes* (1890) and *The American Characidae* (1917–1929). Three of the species named above also appear in the list for his wife (see next entry) as they were named after them both. It is not clear after which Eigenmann *Anchoa eigenmannia* is named; it may be meant to commemorate both of them.

Eigenmann, R

Characin sp. *Astyanax eigenmanniorum* ED Cope, 1894
Shelf Goby *Bollmannia eigenmannorum* S Garman, 1896
[Original spelling of binomial = *eigenmanni*, but name honours two persons]
Three-barbeled Catfish sp. *Pimelodella eigenmanniorum* A Miranda Ribeiro, 1911

Rosa Eigenmann née Smith (1858–1947), was married (1887) to Carl (above) and was also a noted ichthyologist. She studied under David Starr Jordan for two years at the University of Indiana but had to return home due to illness in her family. She published her first paper (1880). After their marriage (1887) they spent time at Harvard studying Agassiz's collection (1887). Rosa was the first woman to be allowed to attend graduate-level classes at Harvard. They collaborated over research and publishing their finds, but her contribution grew less and less as she had five children to look after. Even after his death she remained scientifically inactive. In honouring them, Garman said they: "*…have added so much to our knowledge of the American Gobiidae*", but neglected to give the correct plural form of the binomial.

Eilperin

Sleeper sp. *Calumia eilperinae* GR Allen & M van N Erdmann, 2010

Juilet Eilperin (b.1971) is a Pulitzer Prize-winning environmental journalist who works for the Washington Post (1998–>2020) as Senior National Affairs Correspondent and the on-line newspaper the Huffington Post. She graduated BA (1992) magna cum laude in Politics and Latin American Studies from Princeton University. After graduating she went to Seoul, South Korea, on a Scholarship which allowed her to cover politics and economics for an English-language magazine. Returning to Washington, she wrote for Louisiana and Florida papers at States News Service and then joined Roll Call newspaper (1994). She joined The Washington Post (1998) as its House of Representatives reporter. She has covered (since 2004) the environment for the national desk, reporting on science, policy and politics in areas including climate change, oceans, and air quality. She served (2005) as the McGraw Professor of Journalism at Princeton University, teaching political reporting to a group of undergraduate and graduate students. She was honoured as the person "*…who has continued to expose* [promote?] *and support the Bird's Head Seascape marine conservation program [encompassing Cenderawasih Bay, West Papua, Indonesia, type locality] through her excellent reportage of the initiative*."

Einar

Smooth-head genus *Einara* AE Parr, *1951*

Einar Laurentius Koefoed (1875–1963) (See **Koefoed**)

Einthoven

Brilliant Rasbora *Rasbora einthovenii* P Bleeker, 1851

Dr Jacob Einthoven (1825–1866) was a surgeon who qualified at Groningen and spent his career in the Dutch East Indian Army and later as a public health civil servant at Semarang, Java (1846–1866). He either collected the type or supplied Bleeker with it. He died from a stroke. (One of his sons, Willem Einthoven invented the electrocardiogram for which he was awarded the Nobel Prize in 1924).

Eisen

Red-tail Splitfin *Xenotoca eiseni* CL Rutter, 1896

Dr Gustav Eisen (1847–1940) was a Swedish-born American polymath with a particular interest in marine invertebrates and earthworms (a genus of the latter, *Eisenia*, being named after him). He was a graduate (1873) of Uppsala University. He became a member (1874), and later Life Member (1883), of the California Academy of Science. He went on to become (1893) the 'Curator of Archaeology, Ethnology, and Lower Animals' at the Academy, later changed to 'Curator of Marine Invertebrates'. He was appointed (1938) as an 'Honorary Member', which is considered the highest honour from the Academy. His diverse interests included art and art history, archaeology, anthropology, agronomy, horticulture, history of science, geography, cartography, cytology, and protozoology, as well as marine invertebrate zoology. He also studied malaria-vector mosquitoes, founded a vineyard in Fresno, introduced avocados and Smyrna figs (he wrote a book on the history of figs) to California, campaigned to save the giant sequoias, and wrote a multivolume book about the Holy Grail! He made a large collection of dragonflies from the Cape region of Baja California (1893–1894). Mt. Eisen, in the Sierra Nevada in California, was also named after him (1941), as were numerous worms, odonata, some algae and plants. His ashes were later interred at Redwood Meadow near the foot of the eponymous peak. He collected the splitfin type.

Eisenhardt

Galapagos Grey Skate *Rajella eisenhardti* DJ Long & JE McCosker, 1999

Emil Roy Eisenhardt (b.1939) is Director Emeritus of the California Academy of Sciences. He attended Dartmouth College and served in the US Marine Corps before going to law school. He practiced law for twelve years (1966–1978) in San Francisco and taught at University of California, Berkeley's Boalt Hall School of Law, from which he had graduated (1965). He was President of Oakland Athletics (1982–1989). He was an Executive Director of the California Academy of Sciences (1989–1995), since when he has been in private consultancy. He was honoured in the skate's name for his generous assistance to the authors and their colleagues.

Eisentraut

Lake Barombi-Mbo Cichlid sp. *Konia eisentrauti* E Trewavas, 1962

Professor Dr Martin Eisentraut (1902–1994) was a German zoologist and collector. He was on the staff of the Berlin Zoological Museum, working on bat migration and on the physiology of hibernation, when he went on his first overseas trip, to West Africa (1938). He left Berlin to become Curator of Mammals at the Stuttgart Museum (1950–1957). He then became Director of the Alexander Koenig Museum in Bonn, where he lived for the rest of his life. He made six trips to Bioko and Cameroon (1954–1973). Some of the material collected on these trips is still being studied, but the fish were passed to Trewavas (q.v.) to study. Eisentraut published many scientific papers and three books, including: *Notes on the Birds of Fernando Pó Island, Spanish Equatorial Africa* (1968). He also published a slim volume of poems. Five mammals, two birds and a reptile are also named after him.

Ekmekçi

Grusinian Scraper *Capoeta ekmekciae* D Turan, M Kottelat, SG Kirankaya & S Engin, 2006

Stone Loach sp. *Seminemacheilus ekmekciae* B Yoğurtçuoğlu, C Kaya, MF Geiger & J Freyhof, 2020

Dr Fitnat Güler Atalay née Ekmekçi is an ichthyologist at the Department of Biology, Faculty of Sciences, Hacettepe University, Ankara, Turkey where she has been full professor since 1983. Her interests include fish ecology, freshwater fish systematics and taxonomy. She has co-written a number of articles on threatened fishes such as: *Threatened Fishes of the World, Cobitis puncticulata* (1998).

Eksa

Velvetfish sp. *Cocotropus eksae* AM Prokofiev, 2010

Dr Eugenia K Sytchevskaya is Curator of Fossil Fishes at the Palaeontological Institute of the Russian Academy of Science. Among her published papers is: *Freshwater fish fauna from the Triassic of Northern Asia* (1999).

Elaine

Triplefin sp. *Enneapterygius elaine* W Holleman, 2005
Snake-eel sp. *Luthulenchelys heemstraorum* JE McCosker, 2007
Perchlet sp. *Plectranthias elaine* PC Heemstra & JE Randall, 2009
Yellowtail Blenny *Cirripectes heemstraorum* JT Williams, 2010
Greater Sweeper *Pempheris heemstraorum* JE Randall & BC Victor, 2015
Heemstras' Frogmouth *Chaunax heemstraorum* HC Ho & WC Ma, 2016
Rough Skate sp. *Leucoraja elaineae* DA Ebert & RW Leslie, 2019
Elaine's Threadfin Bream *Nemipterus elaine* BC Russell & G Gouws, 2020

Elaine Margaret Heemstra née Lawrence formerly Grant is Scientific Illustrator Emeritus at the South Africa Institute for Aquatic Biodiversity. She was the second wife (1991) of the senior author of the perchlet, Dr Phillip Clarence Heemstra (q.v.). She was honoured in the triplefin name by Holleman "*...in recognition of her considerable and excellent contribution to the illustration of Indo-Pacific fishes, including several in this paper.*" The snake-eel, goatfish, sweeper and frogmouth are named after both Phillip and Elaine. (Also see **Heemstra**)

Elat

Eilat Sand-Goby *Hazeus elati* M Goren, 1984

This is a toponym named after Elat (also spelled Eliat), Gulf of Aqaba, northern Red Sea, Israel; the type locality.

Elavia

Lake Tanganyika Cichlid sp. *Plecodus elaviae* M Poll, 1949

Captain Elavia Nairman was the commander of the steamer 'Baron Dhanis' at Lake Tanganyika during most of the scientific cruises undertaken by the Belgian Hydrobiological Mission (1946–1947), during which the type was collected. The cichlid was: "*dédicatoire au Capitaine Elavia, Commandant du Baron Dhanis sur le lac Tanganyika*". (Also see **Dhanis**)

Elbert

Red-barred African Killifish *Aphyosemion elberti* CGE Ahl, 1924

Dr Johannes Eugen Wilhelm Elbert (1878–1915) was a Danish geologist, geographer and naturalist. The University of Greifswald awarded his PhD. He explored and collected with his wife Hetta in Indonesia first as a member of the Selenka Expedition (1907) then as leader of the Sunda Expedition (1909–1910) on behalf of the Frankfurt Society for Geography and Statistics. He explored in the Nusa Tenggara islands in Indonesia e.g. Bali, Lombok, Sumbawa, Salayer, Tukang Besi, Flores and Wetar; and also on the islands off Sulawesi, such as Muna and Buton, exploring the geographical relationship between the Asian and Australian fauna. He was later in Cameroon (1913–1914) where the fish is found. At the outbreak of WW1, he fled to Spain, where he died of heart failure after having suffered from sleeping sickness. Two amphibians and three birds are also named after him. While Ahl says only that Dr Elbert was the original collector, we believe it to be the above.

Eldon

Eldon's Galaxias *Galaxias eldoni* B McDowall, 1997

G Anthony 'Tony' Eldon worked for the New Zealand Ministry of Agriculture and Fisheries for over thirty years until retirement (1994). He was honoured then for his 'enthusiastic commitment' to the study and conservation of New Zealand's native freshwater fishes. He wrote: *Observations on Growth and Behaviour of Galaxiidae in Aquariums* (1969) and a number of articles with McDowall.

Eleanor

Characin sp. *Pyrrhulina eleanorae* HW Fowler, 1940

Mrs Eleanor Morrow was the wife of William Penn Morrow, who led the Peruvian expedition that collected the holotype.

Eleonora

Armoured Catfish sp. *Nannoplecostomus eleonorae* AC Ribeira, FCT Lima & EHL Pereira, 2012

Eleonora Trajano is a Brazilian zoologist, biologist and biospeleologist. The Institute of Biosciences, University of São Paulo, awarded her bachelor's degree (1977), her master's (1981), and her doctorate (1987). She was Professor of the Institute of Bioseceinces, Department of Zoology (1981–2006), Full Professor (2006–2012) and, although retired ranked as Senior Professor (2012–2015). She taught Vertebrate Zoology, Speleology, Underground Biology and Ethics and Conservation. She is now Professor Collaborator,

Department of Ecology and Evolutionary Biology, Centro Universitário São Camilo. The etymology specifically mentions "her key contributions to the knowledge of the diversity of Brazilian troglobitic fishes, including fishes of the karst area of São Domingos (where this catfish occurs)". She co-wrote: *Subterranean biodiversity in the Serra da Bodoquena karst area, Paraguay River basin, Mato Grosso do Sul, Southwestern Brazil* (2014)

Elera

Twinspot Chromis *Chromis elerae* HW Fowler & TH Bean, 1928

Castro de Elera (b.1852) was a Dominican friar and zoologist who was the second Director of the University of Santo Tomas Natural History Museum. He added to the collections by collecting all over the Philippines. He was author of the three-volume: *Catalago sistematico de toda la fauna de Filipinas* (1895–1896), which is cited several times by the authors.

Elias (Freyhof)

Loach sp. *Oxynoemacheilus eliasi* B Yoğurtçuoğlu, C Kaya & J Freyhof, 2022

Elias Freyhof is the son of the third author, Dr Jörg Freyhof (q.v.). The etymology says that Elias "…always suffered from the absence of his father being in the field to search for loaches".

Elias

Characin sp. *Chrysobrycon eliasi* JA Vanegas-Ríos, M de las Mercedes Azpelicueta & H Ortega, 2011

Elias Vanegas G was the senior author's father.

Elizabeth (Agassiz)

Dwarf Cichlid sp. *Apistogramma elizabethae* SO Kullander, 1980

Mrs Elizabeth Cabot Cary Agassiz (1822–1907) was the second wife of J L R Agassiz (q.v.), participant of the Thayer Expedition (1865–1866), and principal author of a book about that journey that they collaborated on. (In addition, the name reflects this cichlid's similarity to *A. agassizii*.)

Elizabeth (Fowler)

Deepwater Anthias sp. *Odontanthias elizabethae* HW Fowler, 1923

Elizabeth K Fowler was the wife of the author Henry Weed Fowler.

Engelbrecht

Loach Catfish sp. *Amphilius engelbrechti* DN Mazungula & A Chakona, 2021

Johan Engelbrecht worked for the Mpumalanga Parks Board, South Africa. 'The late' zoologist and environmental scientist was honoured for his contributions to ichthyological research and conservation of freshwater fishes in the Mpumalanga and Limpopo Provinces of South Africa.

Elizaveta

Sultan Sazlığı Minnow *Pseudophoxinus elizavetae* NG Bogutskaya, F Küçük & MA Atalay, 2006

Elizaveta Bogutskaya (b.1981) is the daughter of the senior author Dr Nina G Bogutskaya. She is a physician and lives in Poland.

Ellen

Ellen's Damselfish *Chrysiptera ellenae* GR Allen, M van Nydeck Erdmann & NKD Cahyani, 2015

Dr Ellen R Gritz is professor and chairman of the Department of Behavioural Science and Frank T McGraw Memorial Chair in the Study of Cancer at The University of Texas MD Anderson Cancer Center. She is a leading expert in smoking and cancer. She was honoured for "*...her valued friendship, illustrious career in cancer prevention research, and generous support of the authors' East Indian reef fish investigations.*"

Elliot

Flag-in Cardinalfish *Jaydia ellioti* F Day, 1875
Elliot's Triggerfish *Balistes ellioti* F Day, 1889
Red-eye Threefin *Helcogramma ellioti* AWCT Herre, 1944

Sir Walter Elliot (1803–1887) was a career civil servant in the Indian Civil Service, Honourable East India Company, Madras (1821–1860). He was Commissioner for the Administration of the Northern Circars (1845–1854) and a member of the Council of the Governor of Madras (1854–1860). A distinguished Orientalist, his interests included botany, zoology, Indian languages, numismatics, and archaeology. He was a regular correspondent of Charles Darwin (q.v.). His Indian herbarium was given to the Edinburgh Botanic Garden. He retired to Scotland and, despite blindness, worked on local natural history projects. Two mammals, two reptiles and a bird are also named after him.

Ellis, M

Ghost Knifefish sp. *Apteronotus ellisi* AS Alonso de Arámburu, 1957

Max Mapes Ellis (1887–1953) was a zoologist and biologist. He graduated from Vincennes University (1907) and was awarded his doctorate (1909) by Indiana University. He became (1909) Assistant Professor, Zoology Department, University of Colorado, Boulder. He was the leader of the Gimbel Expedition (1911) to the headwaters of the Amazon and published about its discoveries in: *Gymnotid Eels of Tropical America* (1913). He worked closely with the US Bureau of Fisheries (1925–1942), researching on propagating mussels. He concluded that the overall health of the Mississippi River basin, where he was working, was the determinant and published a co-written report: *Determination of Water Quality* (1946). His wife, with whom he often worked, was Marion Lee Ellis née Durbin (see next entry) who was an ichthyologist, entomologist, limnologist and environmental toxicologist.

Ellis, ML

Characin sp. *Bryconacidnus ellisae* NE Pearson, 1924
Corydoras sp. *Corydoras ellisae* WA Gosline III, 1940

Marion Lee Ellis née Durbin (1887–1972). (See **Durbin**).

Ellis, PV

Broad-bodied Toadfish *Riekertia ellisi* JLB Smith, 1952

P V Ellis of Bizana, Eastern Cape, South Africa 'caught this fish'. We know nothing more about him.

Elochin

Elochin's Sculpin *Abyssocottus elochini* DN Taliev, 1955

This is a toponym referring to the type locality; Cape Erokhin (Elochin), Lake Baikal.

Elomione

Freshwater Goby sp. *Sicyopterus elomionearum* C Lord, P Keith, R Causse & P Amick, 2020

Eloïse and Hermione Lord, are the daughters of the first author, Clara Lord (see **Lord, C**). Lord explains that the binomial is a combination of their names and "*...wishes to salute their great interest in their mother's work*".

Elsa

Loach sp. *Oxynoemacheilus elsae* S Eagderi, P Jalili & E Çiçek, 2018

Elsa Eagderi is the daughter of the first author.

Elsk

Lefteye Flounder sp. *Chascanopsetta elski* VP Foroshchuk, 1991

This is named after the scientific Research Vessel 'Elsk' which is an adapted refrigerated trawler.

Elsman

Elsman's Whipnose Anglerfish *Gigantactis elsmani* E Bertelsen, TW Pietsch III & RJ Lavenberg, 1981

Kai L Elsman was a scientific illustrator, particularly of ichthyological and entomological subjects. The etymology says: "*This species is named for the late Kai L. Elsman, whose superb illustrations have added immeasurably to this revision.*"

Eltanin

Snailfish sp. *Paraliparis eltanini* DL Stein & LS Tompkins, 1989
Snailfish sp. *Careproctus eltaninae* AP Andriashev & DL Stein, 1998
Eltanin Handfish *Pezichthys eltanani* PR Last & DC Gledhill, 2009

These species are named after the 'USS Eltanin' of the US Antarctic Expedition.

Elvira

Eelpout sp. *Santelmoa elvirae* J Matallanas, 2011

Elvira Matallanas is the wife of the author Jesus Matallanus.

Emanuel

Pencil Catfish sp. *Trichomycterus emanueli* LP Schultz, 1944

Juan F Emanuel was governor of the Goajira district, Venezuela. He acted as the author's guide when he was collecting in the Maracaibo Basin lowlands.

Emanuel (D'Oliveira)

Cape Verdes Basslet *Liopropoma emanueli* P Wirtz & UK Schliewen, 2012

Emanuel Charles D'Oliveira (b.1958) is a PE teacher and professional diver as well as a recreational diving instructor. He graduated with a physical education degree and masters in Sports Science from Moscow University. He later took diving courses in Florida. He has worked with underwater archaeological teams. He made the documentary *Cabo Verde: uma história submerse* and wrote: *Cabo Verde: na rota dos* naufrágios (2005). The etymology says of him: "...*whose knowledge of the marine fauna of Santiago Island has been a great help to the first author during many dives.*"

Emery, AR

Emery's Gregory *Stegastes emeryi* GR Allen & JE Randall, 1974
Emery's Pygmygoby *Trimma emeryi* R Winterbottom, 1985

Dr Alan Roy Emery (b.1939) is a Trinidad-born Canadian ichthyologist. The University of Toronto awarded his BSc (1962), McGill University his MSc (1964) and the University of Miami his PhD (1968). He was research & teaching assistant at Toronto and Montreal (1959–1965) then Research Assistant at the Institute of Marine Sciences, Miami, Florida (1965–1968). He became a Research Scientist at the Ontario Ministry of Natural Resources (1968–1972) but was also Research Associate to Royal Ontario Museum, Toronto becoming Associate Curator there (1969–1980). He was then Curator, Ichthyology and Herpetology (1980–1983) as well as Associate Professor at the University of Toronto (1976–1983) when he became President of the Canada Museum of Nature (1983–1996) and President of Kivu Nature (1997–2006). He was honoured as a "...*friend, colleague, and diving buddy, in memory of our expeditions to the Indo-Pacific.*"

Emery, JB

Double Whiptail *Pentapodus emeryii* J Richardson, 1843

James Barker Emery (c.1794–1889) was a gifted artist and amateur naturalist who joined the Royal Navy (1808) and was First Lieutenant aboard the *'Beagle'* during a survey of the Australian coast (1837–1841). He recorded, in a series of vivid watercolours, numerous fish of the continent's western and northern coasts. Emery identified some of the fish by their scientific names, and where not he recorded as much useful data as possible. His paintings are mostly unpublished, although six were reproduced in *Icones piscium* or *Plates of Rare*

Fishes by Sir John Richardson in London (1843). Point Emery, in northern Australia, was named during the voyage in honour of Emery's successful searches for water there.

Emil

Congo Cichlid sp. *Thysochromis emili* G Walsh, A Lamboj & MLJ Stiassny, 2019

Emil Woolf Kentridge-Young is the grandson of South African artist and animator of c.17 films (1989–2015), William Kentridge (b.1955) and his wife Dr Anne Stanwix, a rheumatologist at the Wits University Donald Gordon Medical Centre. The authors named it after him in recognition of their support of their research in Africa.

Emily (Howsman)

Pugnose Minnow *Opsopoeodus emiliae* OP Hay, 1881

Mrs Mary Emily 'Mollie' Hay née Howsman (1849–1931) was the author's wife. Unusually for the time she went to college at Eureka, Woodford County, Illinois. Eureka College was only the third college in the USA to admit women on an equal basis with men. Before her marriage (1870) she was a schoolteacher.

Emily (Irving)

Emily's Shrimpgoby *Tomiyamichthys emilyae* GR Allen, MV Erdmann & IV Utama, 2019

Emily Irving is a Canadian diver, photographer and conservationist. She was a sponsor of the authors' latest collecting trip. The etymology says: "*It is a pleasure to name this magnificently ornamented new species in honour of Canadian diver Emily Irving, who has accompanied and assisted the authors on numerous ichthyological expeditions and is a dedicated supporter of marine conservation and exploration efforts worldwide*".

Emine

Beyazsu Chub *Alburnoides emineae* D Turan, C Kaya, FG Atalay-Ekmekçi & EB Doğan, 2014

Mrs Emine Turan is the senior author's mother.

Emma (Karmovskaya)

Moray sp. *Gymnothorax emmae* AM Prokofiev, 2010
Conger Eel sp. *Ariosoma emmae* DG Smith & H-C Ho, 2018

Dr Emma Stanislavovna Karmovskaya (b.1937) is a Senior Researcher at the Laboratory of Oceanic Ichthyofauna of the Institute of Oceanology, P P Shirshov RAS. She graduated in zoology from Odessa State University (1960) and began working as a research fellow at the herring laboratory of the Polar Institute of Fisheries (PINRO, Murmansk). She moved to Moscow (1961) and worked at the Department of Invertebrate, Zoology Moscow State University (1965–1968), and subsequently at the Institute of Fisheries and Oceanography (CNIITEIRH) (1969–1972), during which period, as a volunteer at the same Institute's laboratory of Oceanography, she began to study mesopelagic eels and the morphology and systematics of anguilliform fishes. She got a permanent job as a researcher at the same lab (now called the Laboratory of Oceanic Ichthyofauna) and became a Senior Researcher there (2009). She has been involved in several marine cruises and international conferences and symposia. As an expert on eels she was repeatedly invited to work on the collections of various museums around the world including BMNH in London, ZMUC in Copenhagen, AMS in Sydney and MNHN in Paris. Her present research interests include the study of species composition, geographic and bathymetric distribution of eels occurring over the continental slope and underwater rises. She has published 47 scientific articles, and has (as at 2015) described 41 new taxa of eels.

Emma (Psomadakis)

Emma's Basslet *Pseudanthias emma* AC Gill & PN Psomadakis, 2018

Emma Psomadakis is the daughter of the second author, marine biologist and ecologist Dr Peter Nick Psomadakis.

Emre

Loach sp. *Cobitis emrei* J Freyhof, E Bayçelebi & M Geiger, 2018

Yunus Emre (c.1238–1320) was a Turkish folk poet, philosopher and sufi mystic who was the pioneer of Turkish poetry in Anatolia.

Endeavour

Furry Coffinfish *Chaunax endeavouri* GP Whitley, 1929
Endeavour Skate *Dentiraja endeavouri* PR Last, 2008

These species are named after the fisheries investigation steamship 'Endeavour' that was lost with all hands off the coast of Australia (1914). It was responsible for collecting the first specimens of the skate species, and many of Australia's continental shelf fish species, in the early years of the 20th century.

Endler

Endler's Livebearer *Poecilia wingei* FN Poeser, M Kempkes & IJH Isbrücker, 2005

Dr John Arthur Endler (b.1947) is an ethologist and evolutionary biologist noted for his work on the adaptation of vertebrates to their unique perceptual environments and the ways colour patterns evolve. The University of Edinburgh awarded his PhD and he then worked at the University of California and James Cook University, North Queensland. He was appointed as an Anniversary Professor of Animal Behaviour in the School of Psychology at the University of Exeter, England (2006). He is currently (2009–2018) working at the Centre for Integrative Biology at Deakin University, Victoria, Australia. He has been elected fellow of both the US (2007) and Australian (2012) Academies of Science. He has published widely including books such as: *Geographic Variation, Speciation, and Clines* (1977) and *Natural Selection in the Wild* (1986), and articles such as the co-written: *Ornament colour selection, visual contrast and the shape of colour preference functions in great bowerbirds* (2006). The livebearer was first collected from Laguna de Patos in Venezuela by Franklyn F Bond (1937) and later rediscovered by Endler (1975). The latter were the first examples of this fish to make it to the aquarium trade. (Also see **Winge**)

Endlicher

Saddled Bichir *Polypterus endlicherii* JJ Heckel, 1847

Stephan Ladislaus Endlicher aka István László Endlicher (1804–1849) was an Austrian botanist and numismatist who was appointed Professor of the University of Vienna and was Director of the university's botanical gardens (1840). He was born in Bratislava (now in Slovakia but part of the Austro-Hungarian Empire in his day). He started out by reading theology but was appointed to reorganise the manuscript collection at the Austrian National Library (1836). He resigned when he was not made President of the Imperial Academy of Science, gave his library and herbarium to the state, and for the rest of his life enjoyed several hours of conversation every week with the Emperor Ferdinand. He must have had many idle hours in which to browse the collections as he reportedly discovered the bichir holotype in the fish collection at the Natural History Museum in Vienna.

Endo

Anglerfish sp. *Lophiodes endoi* H-C Ho & K-T Shao, 2008
Grenadier sp. *Kuronezumia endoi* N Nakayama, 2020

Dr Hiromitsu Endo (b.1964) of the Laboratory of Marine Biology is a Professor (since 1996) in the Faculty of Science, Kochi University, and Curator there. Kochi University awarded his BSc (1988) and MSc (1990), and his PhD was awarded by Hokkaido University (1995). He was a Research Fellow, Tohoku National Fisheries Research Institute (1995–1996), then Assistant Professor (1996–1999), Lecturer (1999–2003), Associate Professor (2004–2010) and Professor (2010–present) all at Kochi University. He has also been Curator and collection manager of BSKU fish collection there (1996–present). His interests include taxonomy and systematics of marine bottom-dwelling fishes, early life history of rattail fishes and the evolution of gadiforme fishes. Among his more than 50 publications are the co-written: *A new batfish, Coelophrys bradburyae (Lophiiformes: Ogcocephaidae) from Japan, with comments on the evolutionary relationships of the genus* (1999) and *Phylogeny of the order Gadiformes (Teleostei, Gadiformes)* (2002). He was honoured in the anglerfish binomial "*...in recognition of his excellent work in ichthyology, his friendship, and for supplying specimens for this study.*"

Engel

Kanda (mullet) *Osteomugil engeli* P Bleeker, 1858

Chris Engel was a prolific scientific illustrator who provided many of the illustrations in Bleeker's works, including his monumental 36-volume *Atlas Ichthyologique des Orientales Neerlandaises* (1862–1878).

Florence (de Rapleye Foerderer)

Characin sp. *Bryconops florenceae* C Silva-Oliveira, RP Ota, MH Sabaj & LH Rapp Py-Daniel, 2021

Florence de Rapleye Foerderer (1926–1999) granted $7 million each in her will to the Academy of Natural Sciences of Philadelphia, the Philadelphia Zoo and Gallaudet University (a liberal arts college for deaf and hard-of-hearing students). "Florence held a great love for animals and her generous bequest continued the Foerderer family's long history of civic involvement and philanthropy in the Philadelphia area."

Engelhard

Orangebar Anthias *Pseudanthias engelhardi* GR Allen & WA Starck, 1982

Charles W Engelhard Jr (1917–1971) was an American businessman in mining, metals and horse racing, and also a philanthropist. The Charles Engelhard Foundation, headed by his wife after his death and by their children following her death (2004), provides funding to a wide range of causes including education, medical research, cultural institutions, and wildlife and conservation organisations. The etymology says: "*The species is named in honor of the late Charles Engelhard ans* [sic] *his family in recognition of their numerous and generous philanthropic contributions.*" And further "*The extensive ongoing program at Escape Reef* [on the northern Great Barrier Reef] *is generously supported by the Charles Engelhard Foundation. We would like to particularly acknowledge the interest and support of Susan Engelhard O'Connor in making this work possible.*"

Engelsen

African Airbreathing Catfish sp. *Clarias engelseni* S Johnsen, 1926

Harald Engelsen (1883–1954) was a Norwegian naval surgeon and researcher into tropical and industrial diseases. He reported cases of lead poisoning among shipyard workers (1931–1933), and was awarded the University of Oslo's gold medal and the Order of St Olav for his work in improving seamen's health and welfare. He was head of medical services in the Royal Norwegian Navy (1937–1940). He collected the holotype of this species.

Englert

Englert's Scorpionfish *Scorpaenodes englerti* WN Eschmeyer & GR Allen, 1971

Anton Franz Englert aka Father Sebastian (1888–1969) was a Capuchin friar, Roman Catholic priest, missionary, linguist, ethnologist and naturalist. He grew up in Bavaria and entered (1907) the novitiate Order of Friars Minor Capuchin where he received his religious name of Sabastian and was ordained (1912). He was an army chaplain during WW1 and then worked as a parish priest in Munich for five years. He asked to be a missionary and was sent (1922) to Chile. There he conducted ethnological and linguistic research into Mapuche culture and the Mapudungun language. He published studies in Araucanian literature (1934–1938). He worked as a missionary priest on Rapa Nui (1935) for the rest of his life. He was perhaps the only non-Rapa Nui to have mastered their language and, although he celebrated Mass in Latin, he preached, heard confessions and catechized the faithful in the Rapa Nui language. He wrote: *La tierra de Hotu Matu'a* (1948) a study of their history, language and social customs. The etymology reads thus: "*This species is named in honor of Father Sebastian Englert, who lived on Easter Island for over 30 years and was an avid student of the archaeology and natural history of the island. He died in New Orleans, Louisiana, while on a tour of the United States to raise funds for the restoration of archaeological sites on Easter Island. He was buried in the courtyard of his small church on January 18, 1969, the same day that Randall and Allen arrived on the island.*" The Easter Island anthropological museum is named after him.

Enis

Blue-band Betta *Betta enisae* M Kottelat, 1995

Enis H Widjanarti is DOI-ITAP Regional Coordinator for Indonesia Protected Areas. Prior to joining the DOI-ITAP/Indonesia Project, she was a Communications, General Affairs and HR Manager at the IAR Indonesia Foundation (2012–2014). Before this she worked at the UNESCO Office in Jakarta as Programme Assistant (1998–2012); Wetlands International Indonesia Programme as Wetlands/Freshwater Fish Ecologist (1990–1995) and Freshwater Fish Ecologist/Technical Secretary (1997–1998). She has been a co-author with Kottelat such as: *The fishes of Danau Sentarum National Park and the Kapuas Lakes area, Kalimantan Barat, Indonesia* (2005) and in her own right such as: *A checklist of freshwater fishes of Danau Sentarym National Park and adjacent areas, Kapaus Hulu, West Kalimantan* (1996).

Enoch

Three-barbeled Catfish sp. *Pimelodella enochi* HW Fowler, 1941

Dr George F Enoch appears to have been one of Fowler's large number of friends and acquaintances in the Philadelphia area to whom he was "*...indebted for various local fishes.*" He was (1938) one of three physicians at the Philadelphia county prison.

Enya

Characin sp. *Leporinus enyae* MD Burns, M Chatfield, JLO Birindelli & BL Sidlauskas, 2017

Eithne Pádraigín Ní Bhraonáin (anglicised as Enya Patricia Brennan) is known to the world as 'Enya' (b.1961), an Irish singer, songwriter and musician. She was in the band 'Clannad' as a backing vocalist and keyboard player (1980) but left (1982) to pursue a solo career. She has sold over 75 million albums worldwide. According to the authors her "*...beautiful song 'Orinoco Flow' celebrates the flow of the mighty Orinoco River, which the new species inhabits.*" This was her greatest hit.

Eos

Northern Redbelly Dace *Chrosomus eos* ED Cope, 1861
Rosy Clingfish *Tomicodon eos* DS Jordan & CH Gilbert, 1882
Bigeye Seabass *Pronotogrammus eos* CH Gilbert, 1890
Pink Rockfish *Sebastes eos* CH Eigenmann & RS Eigenmann, 1890
Boarfish sp. *Antigonia eos* CH Gilbert, 1905
Parazen sp. *Stethopristes eos* CH Gilbert, 1905
Dawn Tetra *Hyphessobrycon eos* M Durbin Ellis, 1909
Sun Loach *Yasuhikotakia eos* Y Taki, 1972
Snailfish sp. *Liparis eos* ZV Krasyukova, 1984
Hagfish sp. *Rubicundus eos* B Fernholm, 1991
[Syn. *Eptatretus eos*]

Eos in Greek mythology was the goddess of the dawn. In zoology, her name is generally applied to species that have 'rosy' coloration, sometimes combined with orange or yellowish hues that might be reminiscent of a sunrise.

EPA

Acara (cichlid) sp. *Aequidens epae* SO Kullander, 1995

Whale Catfish sp. *Denticetopsis epa* RP Vari, CJ Ferraris & MC de Pinna, 2005

The Expedição Permanente de Amazônia – EPA – collected large series of scientifically valuable fishes.

Ephes

Ephesus Goby *Knipowitschia ephesi* H Ahnelt, 1995

This is a toponym for Ephesus, western Anatolia, Turkey (and an ancient Greek city and later capital of the Roman province of Asia), in whose vicinity this goby was found.

Episcopa

Roundnose Minnow *Dionda episcopa* CF Girard, 1856

Major-General John Pope (1822–1892) was an American Unionist general in the American Civil War (the binomial is a Latinisation of his surname). He graduated from West Point (1842) and was commissioned into the Corps of Topographical Engineers. He served in the Mexican-American War (1846–1848) and spent most of his career until the outbreak of the Civil War (1861–1865) as a surveyor and topographical engineer in Florida, Minnesota and New Mexico. After a successful campaign in the western theatre, he was brought east to command the Army of Virginia. He was decisively defeated by Generals Robert E Lee and Stonewall Jackson at the Second Battle of Bull Run (1862), and was subsequently employed in wars against native Americans in the Dakota War (1862) and (post 1865) against the Sioux and the Apache. He led the party that collected the minnow holotype.

Eppley

Armoured Catfish sp. *Parotocinclus eppleyi* SA Schaefer & F Provenzano, 1993

Captain Dr Marion Eppley (1883–1960) was a physical chemist whose master's degree (1912) and doctorate (1919) were both awarded by Princeton. He served in the US Navy in WW1 (1917–1918) as a lieutenant-commander and in WW2 as a captain (1941–1945). He established the Eppley Laboratory, Newport, Rhode Island (1917) to experiment in manufacturing cells, previously imported from Germany. He then established The Eppley Foundation for Research (1947) and by his will on his death, the Eppley Charitable Trust was established (1960), the income from which goes to the Eppley Foundation. This foundation supported the authors' research in Venezuela, which resulted in the discovery of this species.

Epting

Armoured Catfish sp. *Hypostomus eptingi* HW Fowler, 1941

William J Epting was a citizen of Somerset County, Maine, "*...who secured many American fishes for the Academy of Natural Sciences of Philadelphia.*" It is very noticeable that Fowler uses very similar words in his etymologies of what must have been a very large circle of friends and acquaintance who collected for him and the Academy. We cannot find anything more about Epting, but we believe he must have been one of Fowler's circle.

Erber

Neotropical Rivuline sp. *Anablepsoides erberi* H-O Berkenkamp, 1989

Hans Joachim Erber is a German aquarist. He was honoured for his "*...longtime work breeding killifish and assisting the author for many years.*"

Erdal

Stone Loach sp. *Oxynoemacheilus erdali* F Erk'akan, TT Nalbant & SC Özeren, 2007

Erdal Erk'akan (b.1950) is the husband of the senior author Dr Füsun Erk'akan, who is a zoologist and ecologist and a professor at Hacettepe University, Ankara, Turkey. She has also named a fish species after their son Evren and daughter-in-law Damla. Erdal is a well-known underwater photographer. He was a civil engineer, graduating from METU Civil Engineering Department (1975) becoming the managing partner of Yüksel Proje (1978). He trained as a photographer (1977) and in painting (1990–1995). He began taking underwater photographs (1992) and is a founding member of the Underwater Research Society. (Also see **Evren** & **Damla**)

Erdman

Greygreen Blenny *Malacoctenus erdmani* CL Smith, 1957

Donald S Erdman was an American ichthyologist, fishery biologist and angler. He was a Scientific Aid in the Division of Fishes, USNM (1947–1950). He participated (1948) in a fisheries survey of the Persian Gulf and Red Sea under the auspices of the Arabian American Oil Company, on which he collected nearly 5,000 fishes for the USNM. He became Fishery Biologist, Department of Agriculture & Commerce, Puerto Rico. He wrote many papers and a number of books including: *Recent Fish Records from Puerto Rico* (1956), *Inland Game Fishes of Puerto Rico* (1972) and *Nombres vulgares de peces en Puerto Rico* (*Common names of fishes in Puerto Rico)* 1974. The etymology says: "*I take pleasure in naming this species for Donald S. Erdman, who in recent years has added much to our knowledge of the fishes of Puerto Rico.*"

Erdmann

Clingfish genus *Erdmannichthys* KW Conway, K Fujiwara, H Motomura & AP Summers, 2021
Triton Tilefish *Hoplolatilus erdmanni* GR Allen, 2007
Erdmann's Wrasse *Halichoeres erdmanni* JE Randall & GR Allen, 2010
Deep-reef Cardinalfish *Apogonichthyoides erdmanni* TH Fraser & GR Allen, 2011
Erdmann's Fangblenny *Meiacanthus erdmanni* WF Smith-Vaniz & GR Allen, 2011
Erdmann's Dottyback *Pseudochromis erdmanni* AC Gill & GR Allen, 2011
Erdmann's Pygmygoby *Trimma erdmanni* R Winterbottom, 2011
Spotted-belly Catshark *Atelomycterus erdmanni* Fahmi & WT White, 2015
Erdmann's Dwarf-goby *Eviota erdmanni* L Tornabene & DW Greenfield, 2016

Dr Mark van Nydeck Erdmann (b.1968) is an American reef fish expert and marine senior advisor with Conservation International Indonesia. Duke University awarded his bachelor's degree in biology (1990) and the University of California, Berkeley his doctorate (1997). He wrote: *The ecology, distribution and bioindicator potential of Indonesian coral reef stomatopod communities* (1997). In honouring him in the name of the dwarf-goby, the authors said he: "*...has tirelessly photographed and collected numerous individuals of* Eviota, *many of which are new to science, including this species.*" The authors of the pygmygoby named it after him for his "*...deep interest in* Trimma *(and other fishes, of course), his enthusiastic collection and documentation of specimens of this genus for the present author's research program, his friendship, and for the superb job he does for Conservation International's Indonesian Marine Program.*"

Erebus

Australian River Gizzard Shad *Nematalosa erebi* A Günther, 1868

'HMS Erebus' was a British warship used in surveying and exploring and from which the holotype was collected.

Erhan

Ceyhan Scraper *Capoeta erhani* D Turan, M Kottelat & FG Ekmekçi, 2008

Dr Erhan Ünlü (b.1957) is a biologist and aquatic toxicologist at Dicle University in southeastern Anatolia, Turkey. He graduated as a primary school teacher (1975) and entered Dicle University (1976) to study zoology and graduated (1980). He was awarded his doctorate (1989) and in the same year became Assistant Professor. He became a Lecturer in Hydrobiology in the Department of Zoology (1993) and Professor (1994). He became Professor of Zoology (1999) and on the creation of a separate Department of Hydrobiology (2005) also the department's Professor. He co-wrote: *Heavy metals in mullet, Liza abu, and catfish, Silurus triostegus, from the Atatürk Dam Lake (Euphrates), Turkey* (2004). He was honoured for his contribution to the authors' research on the fishes of Anatolia.

Eric

Eric's Goby *Evermannia erici* WA Bussing, 1983

Eric A Bussing is the son of the author William A Bussing. He said it was his son: "*who first called attention to these inconspicuous fishes and captured several by hand to convince me they were not really the young of other goby species stranded by the outgoing tide.*" (Also see **Bussing**)

Eric Anderson

Eelpout genus *Ericandersonia* G Shinohara & H Sakurai, 2006

Dr Michael Eric Anderson (see **Anderson, ME**).

Eric Roberts

Suswa Rainbowfish *Melanotaenia ericrobertsi* GR Allen, PJ Unmack & RK Hadiaty, 2014

Eric Roberts is a pilot with Associated Mission Aviation, West Papua, New Guinea and an aquarium enthusiast and collector. The etymology says: "*The new species is named* ericrobertsi *in honour of Eric Roberts, a pilot with Associated Mission Aviation (AMA), Papua Province, Indonesia. Eric is an aquarium fish enthusiast who collected live specimens and is responsible for the introduction of this species to the aquarium hobby.*"

Erica

Armoured Catfish sp. *Hypostomus ericae* P Hollanda Carvalho & C Weber, 2005
Tetra sp. *Hyphessobrycon ericae* CLR Moreira & FCT Lima, 2017

Dr Erica Maria Pellegrini *Caramaschi is an ichthyologist and a professor at* the Departamento de Ecologia, Universidade Federal do Rio de Janeiro, Brazil. Her bachelor's degree was awarded (1975) by the Faculdade de Ciências Médicas e Biológicas de Botucatu and both her master's (1980) and doctorate (1986) by the Universidade Federal de São Carlos. An amphibian is named after her. She helped to collect the holotype of *Hypostomus ericae*, and presented the authors with the first specimens of the tetra species. The etymology apologises that she had to wait nearly 20 years for the description to be completed and published!

Eriksson

Blind Cusk Eel sp. *Nybelinella erikssoni* O Nybelin, 1957

John Eriksson was the surgeon aboard the Swedish ship 'Albatross' during its circumnavigation (1947–1948), during which the holotype was collected. He wrote: *Deep Sea and Volcanic islands: A Naturalist's Diary from Albatross Worldwide* (1953).

Erk'akan

Loach sp. *Cobitis erkakanae* J Freyhof, E Bayçelebi & M Geiger, 2018

Dr Füsun Erk'akan is an evolutionary biologist and zoologist who is a Professor at the Department of Biology, Faculty of Science, Hacettepe University, Ankara, Turkey. She was honoured for "*...her contributions to the exploration of the species diversity of Cobitis.*" Among her published papers is one she co-authored with Freyhof, Perdices (q.v.) and others: *An overview of the western Palaearctic loach genus Oxynoemacheilus (Teleosei: Nemacheilidae). Ichthyological Exploration of Freshwaters* (2012).

Ernest Magnus

Madagascan Cichlid sp. *Ptychochromis ernestmagnusi* JS Sparks & MLJ Stiassny, 2010

Ernest Magnus (1908–1983) was honoured at the request of the family of Rudolf G Arndt, a German-American marine biologist and ichthyologist, whose 'generous gift' supported the authors' research (Magnus was Arndt's uncle and was instrumental in helping Arndt's family survive in Berlin after World War II and then immigrate to New York City (1950), providing "*...food, clothing, shelter, love, many kindnesses and moral support.*" NB the authors misspelled Arndt's first name as 'Rudolph'.

Ernst

Banjo Catfish genus *Ernstichthys* AFernández-Yépez, 1953

Dr Adolfo (also spelled Adolf) Ernst (1832–1899) was a biologist, born in Prussia and a graduate of the University of Berlin. He emigrated to Venezuela (1861). He taught at the Central University of Venezuela, where he became Professor of Natural Science (1874) and where he was awarded a doctorate of philosophy (1889). He founded the National Museum (1874) and became Director, National Library (1876). He wrote: *Enumeración sistemática de las especies de moluscos hallados hasta ahora en los alrededores de Caracas y demás partes de la República* (1876).

Erondina

Banjo Catfish sp. *Bunocephalus erondinae* AR Cardoso, 2010

Erondina Rodrigues Cardoso was the author's mother.

Erume

Indian Halibut *Psettodes erumei* ME Bloch & JG Schneider, 1801

This is not an eponym but based on the fish's local name in Tamil.

Erwan

Slanted Pygmygoby *Trimma erwani* J Viviani, JT Williams & S Planes, 2016

Dr Erwan Delrieu-Trottin is an evolutionary biologist and diver. He is currently at Universidad Austral de Chile, Instituto de Ciencias Ambientales y Evolutivas in Chile. He has written at least sixteen published papers including: *Shore fishes of the Marquesas Islands, an updated checklist with new records and new percentage of endemic species* (2015). He was honoured for having helped collect the goby holotype in the Marquesas (as well as being on other French Polynesia expeditions), and because he worked late into the night with Williams taking tissue samples and processing specimens.

Esaki

Snout-spot Goby *Amblygobius esakiae* AW Herre, 1939

Professor Dr Teizo (sometimes Teiso) Esaki (1899–1957) was a Japanese entomologist. He was Chairman of the Entomological Laboratory of the Faculty of Agriculture and was appointed as Dean of the Faculty of Agriculture (1948), while concurrently holding the laboratory chairmanship at Kyushu Imperial University (then Kyushu University). A born linguist he spoke English, German, French, Italian and Hungarian as well as Esperanto. He was an influential and erudite teacher with a wide knowledge of entomology, zoology and the history of biology, although his specialty was taxonomy of Hemiptera and he particularly loved butterflies. He collected in Micronesia (1936–1939), including obtaining the goby holotype. Most of the specimens he collected were deposited in the Kyusyu University. He was one of the International Commissioners on Zoological Nomenclature. He published a great many papers on zoography, entomology history as well as about many insect taxa and has a dragonfly named after him.

Esau

Hairyfish *Mirapinna esau* EOG Bertelsen & NB Marshall, 1956

Esau, in the Book of Genesis in the Old Testament of the Bible, is described as being 'a hairy man'. The allusion is not explained but presumably refers to the "*...dense pile of hair-like outgrowths*" covering nearly the entire body.

Escherich

Caucasian Bleak *Alburnus escherichii* F Steindachner, 1897
Ankara Barbel *Luciobarbus escherichii* F Steindachner, 1897
Escherich's Killi *Aphyosemion escherichi* CGE Ahl, 1924

Dr Karl Escherich (1871–1951) was a German physician, entomologist and collector who graduated in medicine (1893) and took a doctorate in zoology (1896). He became Professor of Forest Zoology at Tharandt Forest School (1907). He was Professor of Applied Zoology at Ludwig-Maximilians University, Munich (1914) and Rector there (1933–1936). He visited the USA (1911) and was in the Cameroons (1912). He was an early supporter of Adolph Hitler and took part in the abortive putsch (1923), but after that appears to have tried to distance himself – although still a member of the Nazi party. He collected the bleak and barbel while in Turkey and the killifish in Cameroon. A bird is also named after him.

Eschmeyer

Cape Rockfish *Trachyscorpia eschmeyeri* GP Whitley, 1970
Eschmeyer's Scorpionfish *Rhinopias eschmeyeri* B Condé, 1977
Scorpionfish sp. *Phenacoscorpius eschmeyeri* NV Parin & SA Mandritsa, 1992
Ghost Knifefish sp. *Apteronotus eschmeyeri* CDCM de Santana, JA Maldonado, W Severi & GN Mendes, 2004
Scorpionfish sp. *Scorpaenopsis eschmeyeri* JE Randall & DW Greenfield, 2004

Dr William Neil 'Bill' Eschmeyer (b.1939) is an American ichthyologist and taxonomist who is now Curator Emeritus, California Academy of Sciences, San Francisco and a Research Associate, Florida Museum of Natural History, Gainesville. The University of Miami awarded his PhD. He co-wrote: *Xenaploactis, a new genus for Prosopodasys asperrimus* Günther (Pisces: Aploactinidae), *with descriptions of two new species* (1980) and is also famous as the founder and developer of the database *Catalog of Fishes* that is hosted by the California Academy of Sciences.

Eschricht

Bulbous Dreamer *Oneirodes eschrichtii* CF Lütken, 1871

Professor Dr Daniel Frederik Eschricht (1798–1863) was a Danish physician, physiologist and naturalist. He studied medicine and surgery at Frederiks Hospital in Copenhagen. After he had qualified (1822) he became a general practitioner on the island of Bornholm. He was appointed as Reader in Physiology and Obstetrics (1829), Assistant Professor (1830) and Professor of Anatomy and Physiology at the University of Copenhagen (1836). (Lütken was Professor of Zoology there, so they were colleagues and fellow naturalists.) He was a famous authority on whales, on which subject he wrote a number of papers. A whale genus is also named after him.

Eschwartz

Tetra sp. *Hyphessobrycon eschwartzae* A García-Alzate, C Román-Valencia & H Ortega, 2013

Armoured Catfish sp. *Andeancistrus eschwartzae* NK Lujan, SV Meza-Vargas & RE Barriga-Salazar, 2015

Eugenie 'Ersy' Chavannee Schwartz (1951–2015) was a New Orleans-based artist and benefactor – a surrealist sculptress – who financially supported the expedition that collected the holotype of *Hyphessobrycon eschwartzae* and whose support through the Coypu Foundation enabled the authors' research into catfish in general and *Andeancistrus eschwartzae* in particular.

Esguicero

Serrasalmid characin sp. *Utiaritichthys esguiceroi* TNA Pereira & RMC Castro, 2014

Dr André Luiz Henríques Esguícero (b.1980) is a Brazilian ichthyologist. He graduated in Biology at Centro Universitário Barão de Mauá (2002) and was awarded his doctorate by the Faculdade de Filosofia, Ciências e Letras de Ribeirão Preto of Universidade de São Paulo (2010). He was then part of a post-doctoral project at the Laboratório de Ictiologia de Ribeirão Preto (LIRP), University of São Paulo, entitled 'Taxonomic review and phylogeny of the subfamily Aphyoditeinae (Characiformes: Characidae)' (2012–2015). He is engaged (2018) in a post-doctoral project at the Laboratório de Ictiologia de Ribeirão Preto, University of São Paulo which is due to run for four years (2022). Among his publications he co-wrote: *Fragmentation of a Neotropical migratory fish population by a century-old dam* (2010) and *Taxonomic revision of the genus* Atopomesus *Myers,1927 (Characiformes: Characidae), with comments on its phylogenetic relationships* (2016). He collected the holotype.

Esmaeili

Middle Eastern Toothcarp genus *Esmaeilius* J Freyhof & B Yoğurtçuoğlu, 2020

Nase sp. *Chondrostoma esmaeilii* S Eagderi, A Jouladeh-Roudbar, SS Birecikligil, E Çiçek & BW Coad, 2017

Dr Hamid Reza Esmaeili is an Iranian biologist and ichthyologist at Shiraz University, Shiraz where he is full Professor of Ichthyology. He is Editor-in-Chief of the Iranian *Journal of Ichthyology*. He co-wrote: *Geographic Variation, Distribution, and Habitat of* Natrix tessellata *in Iran* (2011). (Also see **Ghazal**)

Esmark

Norway Pout *Trisopterus esmarkii* S Nilsson, 1855

Greater Eelpout *Lycodes esmarkii* R Collett, 1875

Lauritz Martin Esmark (1806–1884) was a Norwegian zoologist. He was Conservator of the Zoological Museum of the University of Christiana (Oslo), and the first person to observe that the pout represented a separate species. He wrote: *Carcinologiske Bidrag til den skandinavske Fauna* (1865).

Espe

Barred Pencilfish *Nannostomus espei* H Meinken, 1956
Espe's Rasbora *Trigonostigma espei* H Meinken, 1967

Heinrich Espe was a dealer in ornamental fish. He acquired the first example of the rasbora species (1959) and later (1965) imported a number to his base in Bremen. Subsequently he sent specimens to Meinken to investigate whether they were a distinct species or a colour morph of *T. heteromorpha.*

Esperanza

Corydoras sp. *Corydoras esperanzae* DM Castro, 1987

Esperanza Rocha is the author's wife. He thanked her "*...for her help during the 'elaboration' of his paper on Colombian Corydoras.*"

Espinoza

Grunt sp. *Anisotremus espinozai* EA Acevedo-Álvarez, G Ruiz-Campos & O Domínguez-Domínguez, 2021

Eduardo Espinoza is a marine biologist, conservationist and resource manager at the Galapagos National Park Marine Reserve, Ecuador, where he heads the Marine Ecosystem Monitoring programme. He has worked in the Galapagos Islands for three decades (1991–2021), initially as a research officer for the Darwin Foundation (1991–1998) before completing his master's at the University of Tokyo. He and his team were responsible for the discovery of a hammerhead shark nursery in the Galapagos Islands. He has contributed to around 50 scientific papers such as: *Assessing the Efficacy of a Marine Reserve to Protect Sharks with Differential Habitat Use (2020). The authors say he has: "...strongly contributed to the conservation and knowledge of Galapagos Archipelago fish diversity."*

Esquivel

Bulldog Eelgoby *Taenioides esquivel* JLB Smith, 1947

The etymology is not explained, and has not proved explainable. Theories include that 'Esquivel' was possibly a member of Smith's staff, or a man from Portuguese East Africa who assisted Smith during the 1946 expedition in which the holotype was collected. (Smith named several new species in the same paper in honour of individuals in his party, but 'Esquivel' is not included in his acknowledgments)

ESSI

Goby sp. *Schismatogobius essi* P Keith, C Lord & HK Larson, 2017

Named after an NGO: ESSI (Ecological Solutions, Solomon Islands), which tries to improve taxonomic and ecological knowledge of species and ecosystems throughout the Solomon Islands.

Esther

Red Zebra Cichlid *Maylandia estherae* A Konings, 1995

Esther Grant of Salima, Malawi was the wife of cichlid exporter Stuart Malcolm Mariot Grant (1937–2007). They hosted Konings when he visited. She "*...was instrumental in the renewed opening up of the Mozambique waters for the collection of ornamental fishes.*"

E T

African Cichlid genus *Etia* UK Schliewen & ML Stiassny, 2003

Ethelwynn 'E T' Trewavas (1900–1993), British Museum (Natural History). Her "*...ground-breaking work on cichlid biology spanned some 60 years,*" and her "*creativity, humility and kindness are legendary.*" (See **Ethelwynn** & **Trewavas**)

Ethelwynne

Garra sp. *Garra ethelwynnae* AGK Menon, 1958
Goby-like Cichlid sp. *Gobiocichla ethelwynnae* TR Roberts, 1982
Chitande Cichlid *Aulonocara ethelwynnae* MK Meyer, R Riehl & H Zetzsche, 1987

Dr Ethelwynn Trewavas (1900–1993). (See **Trewavas**, also see **ET**). According to the etymology for the goby-like cichlid she was, at the time: "*...in her eighty-second year and still studying Cichlidae from all parts of Africa including Cameroon.*"

Etheridge

Notch-headed Marblefish *Aplodactylus etheridgii* JD Ogilby, 1889

Robert Etheridge, Jr. (1846–1920) collected two specimens of this fish on the Admiralty Islets near Lord Howe Island. He was primarily a palaeontologist and took part, as assistant field geologist, in the geological Survey of Victoria (Australia) (1866–1869). After this he tried gold mining, but returned to England (1871) where he married then spent time as an underground manager in a coalmine in Wales (1873). He became palaeontologist to the Geological Survey of Scotland, and then (1874–1886) an assistant in the geology department of the British Museum (natural history) under his father. He returned to Australia (1887), becoming palaeontologist to the Geological Survey of New South Wales and to the Australian Museum, Sydney, exploring and collecting such as on Lord Howe Island. He co-wrote the monumental *Geology and Palaeontology of Queensland and New Guinea* (1892). He was acting curator (1893) and then Director of the Australian Museum (1895–1917). He wrote over 355 papers and monographs and collaborated on another 60. Numerous invertebrates and fossils are named after him, as were a goldfield, a river, a glacier and a range of hills.

Etnier

Cherry Darter *Etheostoma etnieri* RW Bouchard, 1977
Coosa Chub *Macrhybopsis etnieri* CR Gilbert & RL Mayden, 2017

Dr David A Etnier (b.1937) is Emeritus Professor of Zoology at the University of Tennessee which houses the 'Etnier Ichthyological Collection' of over 40,000 specimens that he began (1960s). He is also an ardent conservationist. The University of Minnesota awarded his PhD (1966). His research covers systematics, biogeography, and biology of freshwater fishes of the southeast USA; also the systematics, distribution, and larval taxonomy of Trichoptera (caddisflies) of central and eastern North America. His many published papers and books include: *Fishes of Tennessee* (1994). According to one etymology he has "*...made many contributions to southeastern ichthyology and aquatic biology, including co-authorship of the definitive book on the fishes of Tennessee.*"

NB He is known to the public as a man who challenged the might of government when he tried to stop the Tennessee Valley Authority (TVA) from building a dam that would wipe out the only known colony of a fish he discovered; the Snail Darter *Percina tanasi*. Eventually the so-called God Committee decided the fish should be protected, but the dam went ahead when a bill was passed exempting the TVA from federal law! Ironically Dr Etnier discovered another colony of the darter later, as did several others.

Etzel

African Rivuline sp. *Epiplatys etzeli* HO Berkenkamp, 1975
African Rivuline sp. *Scriptaphyosemion etzeli* HO Berkenkamp, 1979
Characin sp. *Characidium etzeli* A Zarske & J Géry, 2001

Dr Vollrad Max Friedrich Etzel (1944–2012) was a German veterinary surgeon and ichthyologist at the Institute for Fish and Fishery Products, Cuxhaven (1980–2010). He was noted for his love of travel, visiting West Africa, India, Thailand and New Guinea. He co-wrote: *Description of* Cnesterodon raddei *sp. n. from a swamp near Resistencia, Rio Paraná basin, Argentina (Teleostei: Cyprinodontiformes: Poeciliidae)* (2001). He helped collect the rivuline holotypes.

Eugenia (Böhlke)

Characin sp. *Aulixidens eugeniae* JE Böhlke, 1952

Eugenia Louisa Böhlke née Brandt (1929–2001). (See **Böhlke EB**)

Eugenia (Sytchevskaya)

Stone Loach sp. *Indotriplophysa eugeniae* AM Prokofiev, 2002

Dr Eugenia K Sytchevskaya is a palaeontologist and ichthyologist who is Curator of Fossil Fishes at the Paleontological Institute, Russian Academy of Sciences, Moscow. She wrote: *Palaeogene freshwater fish fauna of the USSR and Mongolia* (1986). Prokofiev named this fish after her in recognition of her help and support and for critically reading his manuscript. A beetle is also named after her.

Euxin

Black Sea Garfish *Belone euxini* A Günther, 1866
[Syn. *Belone belone euxini*]

This is a toponym referring to Pontus Euxinus, an old name for the Black Sea.

Evans, D

Blackmouth Lantern Shark *Etmopterus evansi* PR Last, GH Burgess & B Séret, 2002

David Evans is an Australian fisheries scientist who worked for CSIRO at the Marmion Marine Laboratories, Western Australia. The etymology honours: "*...David Evans, who* *over the last decade has meticulously selected and donated valuable taxonomic specimens (including several excellent* Etmopterus *specimens of which two have been designated as holotypes) collected by commercial trawlers from the tropical deepwater of Western Australia.*" He co-wrote: *Crustaceans from the deepwater trawl fisheries of Western Australia* (1991).

Evans, ERGR

Antarctic Dragonfish sp. *Prionodraco evansii* CT Regan, 1914

Admiral Edward Ratcliffe Garth Russell Evans RN CB, 1st Baron Mountevans (1880–1957), known as 'Teddy' Evans was a Royal Navy Officer and Antarctic explorer. He was seconded to the 'Discovery' Antarctica Expedition (1901–1904), serving on the relief ship but planning his own expedition putting iot to one side when he was offered second-in-command of Scott's ill-fated expedition to the South Pole and as captain of the Terra Nova (1910–1913). He accompanied Scott to within 150 miles of the Pole, but was sent back in command of the last supporting party. On the return he became seriously ill with scurvy and only narrowly survived. (There is no etymology in the original paper, but eponymous fishes in Regan's article were all named after members of the Terra Nova expedition and Regan later (2016) collected that information together and reveals who he named fish after. Regan named the fish after the, then, Commander.) Teddy Evans had an illustrious career, commanding a destroyer WW1, in the Home Fisheries Protection Squad then in command of the Battlecruiser 'HMS Repulse'. He retired from the navy (1941) but had a civil defence role WW2. After which, he was given a peerage and sat in the Lords as a Labour Party member.

Evans, FV

Yellowback Anthias *Pseudanthias evansi* JLB Smith, 1954

Frank V Evans. Smith's etymology reads: "*...Named in honour of Frank V Evans esq., of Durban, who has greatly assisted my work*." We have failed to find any biographical information and assume this to be an amateur collector.

Evans, JR

African Barb sp. *Enteromius evansi* HW Fowler, 1930

J R Evans worked for Companhia de Diamantes de Angola. He joined the (1929) Prentiss N Gray African Expedition in Angola for two weeks as guide and interpreter. In the report of the expedition W Wedgwood Bowen wrote of him: "*...without the help and unrivalled knowledge of the country which Mr J R Evans placed so freely at our disposal, the expedition would have met many difficulties...*" He was an American who was resident in Angola but we have not been able to find anything more about him. The holotype was collected during this expedition.

Evans, JW

Armoured Catfish sp. *Spatuloricaria evansii* GA Boulenger, 1892

Dr John William Evans (1857–1930) was a British geologist. He was President of the Geological Society of London (1924–1929). He took part in an expedition (1891–1892) to the Matto Grosso region of Brazil, during which he obtained the holotype of this species, and wrote: *The Geology of Matto Grosso (particularly the Region drained by the Upper Paraguay)* (1894).

Evelyn (Axelrod)

Evelyn's Cory *Corydoras evelynae* F Rössel, 1963
Spot-tail Curimata *Curimatopsis evelynae* J Géry, 1964
Lake Malawi Cichlid sp. *Docimodus evelynae* DH Eccles & DSC Lewis, 1976

Evelyn Theresa Axelrod née Miller (1928–2020) was the wife of the philanthropist, the late Herbert Richard Axelrod (1927–2017) (q.v.). The cichlid etymology states that she and her husband *"have done so much to promote interest in the fishes of Lake Malawi."*

Evelyn (Gordon)

Puebla Platyfish *Xiphophorus evelynae* DE Rosen, 1960

Evelyn Gordon was instrumental in collecting the first known samples (1939) when accompanying her husband Dr Myron Gordon on his collecting trip to Mexico. Rosen's review of the genus *Xiphophorus* is dedicated to the memory of Dr Myron Gordon (1899–1959) for his quarter of a century of contributions to the biology of this and other groups of fishes. He was primarily a geneticist who used these fish for his cancer research. (Also see **Gordon**)

Evelyn (McCutcheon)

Shark-nose Goby *Elacatinus evelynae* JE Böhlke & CR Robins, 1968

Evelyn McCutcheon née Shaw (1894–1977) was a writer, psychologist and ichthyologist. She wrote a number of books including: *Our Island Kingdom* (1922) and *The Island Songbook* (1927) a collection of ballads, ditties, dirges, lullabies and anthems of the Bahama Islands. This included *The John B Sails* which her husband introduced to Carl Sandburg, and which is better known as the Beach Boys favourite *The Sloop John B.* She also wrote scientific papers such as *Minimal light intensity and the dispersal of schooling fish* (1961). She was married to John Tinny McCutcheon (1870–1949) the Pulitzer prize winning political cartoonist, known as the 'Dean of American Cartoonists'. His family owned Blue Lagoon Island (1916–1979). The authors described her as the "*…gracious mistress of Treasure Island*" (Salt Cay in the Bahamas), "*…where the senior author and Mr Charles C G Chaplin have spent many pleasant hours observing and collecting her fishes and enjoying her fine hospitality.*"

Everett

Clown Barb *Barbodes everetti* GA Boulenger, 1894
Borneo Cyprinid sp. *Nematabramis everetti* GA Boulenger, 1894
Sleeper sp. *Hypseleotris everetti* GA Boulenger, 1895
Rasbora sp. *Rasbora everetti* GA Boulenger, 1895

Alfred Hart Everett (1848–1898) was a British civil servant who worked as an administrator in North Borneo. He was also a naturalist who collected for wealthy patrons such as the

Marquis of Tweeddale and Walter Rothschild, who in turn named various species after Everett. He was engaged by the Royal Society (1878–1879) to explore 'the Caves of Borneo' in search of remains of ancient humans. It is believed that a jawbone from an orang-utan, which he found in a cave, may have been used in the 'Piltdown Man' hoax. Everett was interested in all aspects of natural history and anthropology. His death made the front page of the *Sarawak Gazette*. 29 birds, six mammals, four amphibians and three reptiles are also named after him. He collected the holotypes of these fish species.

Evermann

Goby genus *Evermannia* DS Jordan, 1895
Sabretooth genus *Evermannella* HW Fowler, 1901
Goby genus *Evermannichthys* JM Metzelaar, 1919
Inotted Lizardfish *Synodus evermanni* DS Jordan & CH Bollman, 1890
Evermann's Conger *Congrosoma evermanni* S Garman, 1899
Jawfish sp. *Opistognathus evermanni* DS Jordan & JO Snyder, 1902
Silverside sp. *Alepidomus evermanni* CH Eigenmann, 1903
Kamchatka Flounder *Atheresthes evermanni* DS Jordan & EC Starks, 1904
Evermann's Cardinalfish *Zapogon evermanni* DS Jordan & JO Snyder, 1904
Evermann's Lanternfish *Symbolophorus evermanni* CH Gilbert, 1905
Umpqua Dace *Rhinichthys evermanni* JO Snyder, 1908

Dr Barton Warren Evermann (1853–1932) was a schoolteacher (1876–1886) and a student at Indiana University, where he was awarded his bachelor's degree (1886), master's (1888) and doctorate (1891). He surveyed the North American Pacific Northwest fish fauna and worked for the Bureau of Fishes in Washington (1891–1914) in various roles including being Chief Scientist that he combined with lecturing on zoology at Cornell (1900–1903), Yale (1903–1906), and later Stanford, after he became Director of the Museum, California Academy of Sciences (1914). He was Jordan's student and later scientific associate; Metzelaar honoured him for his 'kind assistance'. A bird and a reptile are also named after him.

Evers

Ricefish sp. *Oryzias eversi* F Herder, RK Hadiaty & AW Nolte, 2012
Corydoras sp. *Corydoras eversi* LFC Tencatt & MR Britto, 2016

Hans-Georg Evers (b.1964) is a German aquarist based in Hamburg. He was chief editor of the aquarium magazine 'Amazonas' (2005–2018). He travels widely, particularly in South America and southern Asia, to discover new freshwater fish - including the two species named after him. He has written books including: *Fische im Gesellschaftsaquarium* (1999) and a wide variety of articles including the co-written: *A new species of* Corydoras *Lacépède, 1803 (Siluriformes: Callichthyidae) from the* río Madre de Dios basin, Peru (2016).

Evezard

Day's Baril *Barilius evezardi* F Day, 1872
Stone Loach sp. *Indoreonectes evezardi* F Day, 1872

General George Charles Evezard (1826–1901) was a soldier who joined the Bombay Army of the Honourable East India Company as an Ensign (1843). He served in various regiments of Native Infantry up to the time of the Indian Mutiny (1857), after which he was Cantonment Magistrate, Poona (Pune) (1858) before being posted to the Bombay Staff Corps. He assisted in the gathering of the holotypes of both species as well of many other natural history specimens, as he was a conchologist who collected in India for nearly 30 years. The Smithsonian eventually acquired his collection of some 10,000 specimens (1922). He settled in retirement at Khotagiri, Tamil Nadu and was recorded as being alive (1899), it being remarked upon that he had never, since 1863, taken home leave to England.

Evliya

Lycian Spring Minnow *Pseudophoxinus evliyae* J Freyhof & M Özuluğ, 2010

Evliya Çelebi (1611–1683), also known as Mehmed Zilli, was an Ottoman Turk who travelled through the Ottoman Empire and its surrounding lands (1630–1672). He visited Vienna (1865–1866) and noted the similarity between some words in Persian and German; this was a very early recognition of the genetic relationship between two Indo-European languages. He made notes of what he saw and wrote a 10–volume work: *Seyahatname,* which translates as: *Book of Travels.*

Evren

Ceyhan Sportive Loach *Oxynoemacheilus evreni* F Erk'akan, TT Nalbant & SC Özeren, 2007

Ceyhan Spined Loach *Cobitis evreni* F Erk'akan, SC Özeren & TT Nalbant, 2008

Evren Erk'akan (b.1987) is the son of the senior author Professor Dr Füsun Erk'akan and Erdal Erk'akan (q.v.). He graduated from Atılım University, Civil Engineering Department (2008) and took his MSc at METU Civil Engineering Department (2014). He won the examination of the Conservatory of the Başkent University (2016) and is now (2017) studying there as a scholarship post-graduate student (2017). (Also see **Damla** & **Erdal**)

Evseenko

Eel Cod sp. *Muraenolepis evseenkoi* AV Balushkin & VP Prirodina, 2010

Dr Sergei Afanasievich Evseenko (b.1949) is a Russian ichthyologist who was Senior, then Leading, Research Scientist and now Chief of the Laboratory of Oceanic Ichthyofauna, P P Shirshov Institute of Oceanology, Russian Academy of Sciences (1983–present). Moscow State University awarded his ichthyology degree (1972), the All-Union Research Institute of Marine Fisheries and Oceanography (VNIRO), Moscow his PhD (1979) and the P P Shirshov Institute of Oceanology, Russian Academy of Sciences his Dr Sci (1998). Before his current position he was (1972–1983) Junior, then Senior Research Scientist at the All-Union Research Institute of Marine Fisheries and Oceanography, Moscow, Laboratory of Oceanic Fishes. He has published more than 100 papers including *Larval Engyophrys sentus Ginsburg, 1933 (Pisces, Bothidae) from the American Mediterranean Sea* (1977), *Early life history stages of peacock flounder Bothus lunatus (Bothidae) from the western and central tropical Atlantic* (2008) and *Fishes of Russian seas. Annotated catalogue* (2014). He was honoured for his considerable contribution to the study of Antarctic fishes. (Also see **Parbevs**)

Eydoux

Pelagic Porcupinefish *Diodon eydouxii* CNF Brisout de Barneville, 1846
Searobin sp. *Lepidotrigla eydouxii* HE Sauvage, 1878

Joseph Fortuné Théodore Eydoux (1802–1841) was a French naturalist who became a naval surgeon (1821). He was aboard *La Favorite* in the East Indies (1830–1832) and was also was a member of the crew of *La Bonite*, which circumnavigated the globe (1836–1837). He co-wrote: *Voyage autour du monde exécuté pendant les années 1836 et 1837 sur la corvette La Bonite* (1841). He died in Martinique. A reptile is also named after him.

Eyre

Eyre's Dwarf-goby *Eviota eyreae* DW Greenfield & JE Randall, 2016
Shrimp-goby sp. *Tomiyamichthys eyreae* GR Allen, MV Erdmann & MU Mongdong, 2020

Janet Van Sickle Eyre (b.1955) is an American diver and volunteer at the Reef Environmental Education Foundation. Her BA in geology was awarded (1976) by Harvard. She spent 25 years working in the finance industry equipment leasing in San Francisco, USA. She certified in SCUBA (1994) and discovered a passion for marine fauna. When she retired from corporate life (2001) she entered a second career as a volunteer citizen scientist conducting fish surveys for REEF as part of its mission 'to protect biodiversity and ocean life by actively engaging and inspiring the public through citizen science, education, and partnerships with the scientific community'. She told us that she "*...loves reading fish identification books, finding as many fish as she can on every dive, and bugging the world's leading ichthyologists with photographs of fish she can't positively identify. This led to cooperative relationships with some ichthyologists wherein she provides them with photographs, range information, and occasionally fish samples, to further their research.*" Ms Eyre said she "*...feels naked in the water without a slate and camera and hopes to find many undescribed species of fish in her lifetime while conducting surveys.*" When at home in San Francisco, she is a volunteer in the Ichthyology Department at the California Academy of Sciences. She collected and photographed this goby in Fiji, and was also 'of great assistance in the authors' studies of the genus'.

Ezenam

Kezenoi-am Trout *Salmo ezenami* LS Berg, 1948

This is a toponym; referring to Lake Ezenam, Daghestan, Caucasus, Russia which is the type locality.

F

Fabricio

Cascarudo Catfish sp. *Callichthys fabricioi* C Román-Valencia, PC Lehmann-Albornoz, & AM Muñoz, 1999

Fabricio Lehmann Gonzalez (b.1936) is a self-taught Colombian naturalist and sport fisherman. He was an enthusiastic supporter of the authors' expedition to the Popayán region (Colombia).

Fabricius

Black Dogfish *Centroscyllium fabricii* JCH Reinhardt, 1825
Slender Eel-Blenny *Lumpenus fabricii* JCH Reinhardt, 1836
Gelatinous Snailfish *Liparis fabricii* HN Krøyer, 1847

Bishop Otto Fabricius (1744–1822) was a Danish missionary, explorer and naturalist in Greenland (1768–1773). His zoological observations enabled him to publish *Fauna Groenlandica* (1780), written in Latin after his return to Denmark. He was the first to study many species of Greenland fishes, giving their local Inuit vernacular names and details of how the Inuit caught them.

Facciolà

Witch-Eel genus *Facciolella* GP Whitley, 1938
Facciola's Sorcerer *Facciolella oxyrhyncha* C Bellotti, 1883

Dr Luigi Facciolà (1851–1943) was an Italian physician and ichthyologist who was the first to recognise and describe the Witch-Eel genus (1911) but in doing so used a preoccupied name (*Nettastomella*). He made a particular study of fishes to be found in the Straits of Messina.

Fagen

Horseface Unicornfish *Naso fageni* JE Morrow, 1954

Captain R W Fagen. We can say no more about him than is contained in James Morrow's (q.v.) etymology, which states the fish is named after: "*...Captain R W Fagen of Miami, Florida, friend and guide on several ichthyological expeditions*".

Fahire

Tefenni Minnow *Chondrostoma fahirae* W Ladiges, 1960
Küçük Menderes Spined Loach *Cobitis fahirae* F Erk'akan, FG Atalay-Ekmekçi & TT Nalbant, 1998

Dr Fahire Battalgil (1902–1948) (See **Battalgil**).

Fahrettin

Pisidian Spring Minnow *Pseudophoxinus fahrettini* J Freyhof & M Özuluğ, 2010
Gudgeon sp. *Gobio fahrettini* D Turan, C Kaya, E Bayçelebi, I Aksu & Y Bektas, 2018
Trout sp. *Salmo fahrettini* D Turan, Y Bektas, C Kaya & E Bayçelebi, 2020

Fahrettin Küçük is a Turkish zoologist and biologist who is a professor and department head at Egirdir Fisheries Faculty, Fisheries Basic Sciences, Süleyman Demirel University, Turkey. Ankara University awarded his bachelor's degree in biology (1983), the Mediterranean University, Sciences Institute, his master's (1991) and the SDI institute of Fisheries Engineering Science his doctorate (1996). He co-wrote: *Alburnus baliki, a new species of cyprinid fish from the Manavgat River system, Turkey* (2000). He was honoured for his contribution to the knowledge of Turkish, particularly Central Anatolian, fishes.

Fainzilber

Lake Malawi Cichlid sp. *Pseudotropheus fainzilberi* W Staeck, 1976
[Syn. *Maylandia fainzilberi*]

Misha Fainzilber was an exporter of tropical fish, based in Tanzania. He facilitated Staeck's access to Lake Malawi.

Fairchild

New Zealand Torpedo *Tetronarce fairchildi* FW Hutton, 1872

Captain John Fairchild (1834–1898) was a New Zealander who was master on government steamers, and supply vessels to sub-Antarctic islands. He was in command of the 'Luna' when he discovered the holotype stranded on the mud inside Napier Harbour (1868). Fairchild died in an accident whilst superintending the loading of some iron rails and was struck on the head by a chain. A bird is named after him.

Fairweather

Picotee Livebearer *Phallichthys fairweatheri* DE Rosen & RM Bailey, 1959

Rev Dr Gerald Fairweather was on the (1949) New York Aquarium expedition to British Honduras. The etymology says: "*...This species is named in honor of the Rev. Gerald Fairweather in acknowledgment of his participation in obtaining extensive scientific collections of fishes in British Honduras.*"

Fajardo

Spotted Filefish *Thamnaconus fajardoi* JLB Smith, 1953

Mussolini P Fajardo collected the type in Mozambique. We can find no further biographical details.

Falkner

Swamp Eel genus *Falconeria* DA Larrañaga, 1923 NCR
[Junior Synonym of *Synbranchus*]

Large-spot River Stingray *Potamotrygon falkneri* MN Castex & I Maciel, 1963

Thomas Falkner (sometimes Falconer) (1702–1784) was born in Manchester, England. He studied medicine and natural sciences in London, as a student of Mead and Newton. The Royal Society of London commissioned Falkner to go to South America to study the medicinal properties of herbs. His passage to Buenos Aires was as a doctor on a slave ship. Soon after arriving he became very ill. He was aided in his quest by a Jesuit and converted to Catholicism and (1732) entered the Society of Jesus, spending the next ten years studying in Cordoba. He was sent (1742) as a missionary explorer along the Salado and then south. He helped found a mission (1746) and studied the local flora and fauna until a general uprising of the Indians destroyed the missions. Sent to another mission in Areco he continued his studies before returning to Córdoba (1752), where he founded the Department of Mathematics. He was arrested as a spy and, with 40 other Jesuits, was deported from Argentina to Spain (1776) and then returned to England. Castex (also a priest) honoured him for his apostolic and scientific work in 18th-century Argentina.

Fang, F

Borneo Cyprinid genus *Fangfangia* R Britz, M Kottelat & HH Tan, 2012
Danio sp. *Devario fangfangae* M Kottelat, 2000
Osum Riffle Minnow *Alburnoides fangfangae* NG Bogutskaya, P Zupančič & AM Naseka, 2010
Danio sp. *Devario fangae* SO Kullander, 2017

Dr Fang Kullander née Fang (1962–2010) was a Chinese ichthyologist. Zhanjiang Fisheries College, Guangdong, China awarded her bachelor's degree (1984) and Hebei University her master's (1987). She first went to Sweden (1992) as a visiting scientist and stayed on to be awarded her doctorate by Stockholm University (2001). She worked as a Research Associate, Division of Ichthyology, Institute of Zoology, Chinese Academy of Sciences, Beijing (1987–1993). She travelled and researched (1996–2001) in many countries including Myanmar, Brazil, French Guiana, Paraguay and India and worked for ECOCARP, an organisation wishing to discover new species for aquaculture in China (2001–2003). From 2003 she was the curator of the Swedish FishBase team. In private life she was married to Sven Oscar Kullander (q.v.) with whom she described a number of species. She was also interested in music and very keen on singing, becoming one of the founders of the Stockholm Chinese Choir. She died from cancer.

Fang, L-S

Fang's Pygmy-goby *Trimma fangi* R Winterbottom & I-S Chen, 2004

Professor Lee-Shing Fang (b.1951) is a Chinese ichthyologist. He was head of the National Museum of Marine Biology and Aquarium, Taiwan. The University of California, San Diego, awarded his PhD (1982). He has been affiliated for many years to Cheng Shui University, Kaohsiuing, Taiwan. He has been Chairman of the Taiwan Coral Reef Society (1998–1999); Director General of the National Museum Marine Biology and Aquarium, Ping Tung, Taiwan (2000–2006); Director of the Institute of Marine Biodiversity and Evolution, National Dong-Hwa University, Hawlien, Taiwan (2004–2006) and General Manager of Aquala Consultant Company Ltd., Hong Kong, (since 2007). He is a world-renowned expert in marine life conservation and researcher in stream fish ecology, whose many endeavours in protecting fauna and conserving natural resources have earned him international respect parallel to that of Dr Jane Goodall in her effort to protect chimpanzees. He was honoured for his "*enthusiastic support of the second author's current research program.*"

Fang, P-W

Chinese Cyprinid sp. *Xenocypris fangi* TL Tchang, 1930
Loach sp. *Pseudogastromyzon fangi* JT Nichols, 1931
Bitterling sp. *Rhodeus fangi* CP Miao, 1934
Gunnel sp. *Pholis fangi* KF Wang & S-C Wang, 1935
Pufferfish sp. *Pao fangi* J Pellegrin & P Chevey, 1940
Chinese Cyprinid sp. *Onychostoma fangi* M Kottelat, 2000

Fang Ping-Wen (sometimes transcribed as Bingwen) (1903–1944) was a Chinese ichthyologist at the Metropolitan Museum of Natural History, and Biological Laboratory of the Science Society of China, Academia Sinica, Nanking (Nanjing). Among his publications is the co-written:

Faour

Neotropical Rivuline sp. *Hypsolebias faouri* Britzke, DTB Nielsen & C. de Oliveira, 2016

Amer Faour Martín is an ichthyologist and environmentalist. He co-authored the paper: *A new species of Simpsonichthys (Cyprinodontiformes: Rivulidae) from the Rio São Francisco basin, northeastern Brazil* (2010) and has written with Dalton Nielsen such as: *Hypsolebias trifasciatus, new species* (2014). He discovered this species.

Farfan

Titicaca Pupfish sp. *Orestias farfani* LR Parenti, 1984
[Junior Synonym of *O. luteus*]

L Edgar Farfan Vizcardo was the Director of the Laboratorio de Puno at Instituto del Mar del Peru. The author honoured him in the name "*…to commemorate my appreciation for his generosity and support given to research on Lago Titicaca.*"

Faridpak

Faridpak's Spine Loach *Cobitis faridpaki* H Mousavi-Sabet, ED Vasil'eva, S Vatandoust & VP Vasil'ev, 2011

Farhad Faridpak (1910–1995) was an Iranian ichthyologist and fishery scientist who worked on the fish of the Caspian Sea. Most notably, he was the founder of the Iranian Northern Fisheries and was a member of the board of the Iran Fishery Organisation.

Farlow

Whiptail Catfish genus *Farlowella* CH Eigenmann & RS Eigenmann, 1889

Dr William Gilson Farlow (1844–1919) was an American physician and botanist who specialised in algae. Harvard awarded his bachelor's degree (1866) and his medical qualification (1970). He taught at Harvard as adjunct professor of botany (1874–1879) and as Professor of Cryptogamic Botany from 1879. He received honorary degrees from three universities: Harvard, Glasgow and Wisconsin-Madison. He was also President of several learned societies: American Society of Naturalist (1899), National Academy of Sciences (1904) and Botanical Society of America (1911). He wrote: *Marine Algae of New England* (1881).

Farwell

Stone Loach sp. *Triplophysa farwelli* SL Hora, 1935

Brigadier Arthur Evelyn 'Tiny' Farwell (1898–1976) was Military Attaché to the British Legation at Kabul, Afghanistan in the 1930s. He was a lieutenant in the 15th Sikhs, Indian Army (1916), and retired from the Army (1948) having achieved the rank of Brigadier-General in WW2. When a Major he sent a collection of fishes (including the holotype of this species) to the Bombay Natural History Society.

Note: Anyone whose nickname is 'Tiny' has to be very tall – and he was!

Fasolt

African Barb sp. *Labeobarbus fasolt* P Pappenheim, 1914

This is not a true eponym in that Fasolt is a fictional giant in 'Das Rheingold', the first opera in Wagner's cycle 'Der Ring des Nibelungen'.

Fassl

Characin sp. *Astyanax fasslii* F Steindachner, 1915
Pencil Catfish sp. *Trichomycterus fassli* F Steindachner, 1915

Anton Heinrich Hermann Fassl (1876–1922) was a German commercial collector, particularly of Lepidoptera and Coleoptera. He maintained a dealership in Berlin and Bohemia. He travelled in Colombia (1907–1909), in Peru (1913–1914) and Brazil (1920, 1921 & 1922) where he died – probably at the upper Madre de Dios River. Among others he supplied Hartert (q.v.) & Jordan. He wrote a number of papers including: *Einige kritische Bemerkungen zu J. Röbers „Mimikry und verwandte Erscheinungen bei Schmetterlingen“* (1922). (See also **Fuessi**).

Fatio

Whitefish (salmonid) sp. *Coregonus fatioi* M Kottelat, 1997

Dr Victor Fatio de Beaumont (1838–1906) was a Swiss zoologist and physiologist. The University of Leipzig awarded his doctorate. He contracted typhoid (1861), causing him to forget nearly all his physiological knowledge. He went to Paris and studied zoology under Henri MilneEdwards at MNHN (1862) and became an expert on phylloxera, a disastrous infection of grapevines, writing: *État de la question phylloxérique en Europe en 1877* (1878). His major work was the 6-volume *Faune des Vertébrés de la Suisse* (1869–1904). A bird is named after him.

Faughn

Faughn's Mackerel *Rastrelliger faughni* T Matsui, 1967
[Alt. Island Mackerel]

James Lawrence Faughn (1910–1985) worked at the Scripps Institution of Oceanography (1947–1974). He served in the US Navy (1927–1942) as a warrant officer spending a great deal of time at sea. Discharged because of a collapsed lung, he later served as an engineer and captain in the Merchant Marine and the Army Transport Corps. When Scripps were able to expand their oceanographic fleet, Faughn was recommended to take charge of it. He was employed as Marine Superintendent (Engineering), in which capacity he directed the conversion into research vessels 'Horizon', 'Spencer F. Baird', 'Crest', and 'Paolina-T'. Except for a leave of a few months (1953), he worked on several nuclear bomb tests and (1956) headed the scientific group for a resurvey of Eniwetok Atoll, requested by the U.S. Navy. When the Southeast Asia Project was under consideration (1958) he was designated as Captain of the SIO Research Vessel 'Stranger' from which he led the Naga expedition (1959–1961). He wrote (1963) its summary report and compiled its first full volume (1974). After his return from Thailand (1961), he carried out some sea tests for the Advanced Oceanography Group of SIO. He captained other vessels and took various roles until his retirement (1974).

Faulkner

Yoknapatawpha Darter *Etheostoma faulkneri* KA Sterling & ML Warren, 2020

William Cuthbert Faulkner (1897–1962) was an American author, poet and screen writer from Oxford, Lafayette County, Mississippi who is primarily known for his stories set in the fictional Yoknapatawpha County, based on the nearly identical Lafayette County. He was also an avid hunter and angler. He wrote 13 novels and many short stories and his best-known works are probably the novels *The Sound and the Fury* (1929) and *As I Lay Dying* (1930) and for co-writing scripts for *To Have and Have Not* (1944) and *The Big Sleep* (1946). He was awarded the Nobel Prize for Literature (1949) for "...his powerful and artistically unique contribution to the modern American novel". He fell from his horse, developed a thrombosis and died of a heart attack three weeks later.

Faure

Scalybreast Gurnard *Lepidotrigla faurei* JDF Gilchrist & WW Thompson, 1914

Named after a boat. The type series was "...procured by the Cape government trawler 'P Faure' off the Natal coast."

Faustino

Lanternfish sp. *Diaphus faustinoi* HW Fowler, 1934

Dr Leopoldo Alcarez Faustino (1892–1935) was a Filipino zoologist. He attended the University of the Philippines but left without graduating (1912). He went to Columbus, Ohio, and worked as a messenger boy for Western Union, saving enough money to enrol in Ohio State University, where he graduated in engineering (1917). He returned to Manila (1918) and was employed as an assayer by the Bureau of Science and became lecturer in metallurgy, University of the Philippines (1920). Sponsored by the government he returned to Ohio State University (1921–1922) for degrees in mining engineering, moving to Stanford University for a master's degree (1922) and a doctorate (1924). He underwent field training with the US Geological Survey (1921–1924) and returned to the Philippines (1924). He became a lecturer in geology at the University of the Philippines (1926–1930).

He was Acting Director of the National Museum (1930), then Director (1933) and chief of the division in mineral resources, Department of Agriculture and Commerce (1934). He wrote a number of short works such as: *Coral reefs of the Philippine Islands* (1931). A bird is also named after him. Fowler's etymology mentions *"with pleasant memories of our trip to Krakatau."* He died of cancer.

Fay

Goby sp. *Sicydium fayae* VE Brock, 1942

Rosemary Fay Brock was the wife of the author, Dr Vernon Brock (1912–1971) (Also see **Brock**).

Fea

Snow Trout sp. *Osteobrama feae* D Vinciguerra, 1890
Sisorid Catfish sp. *Pareuchiloglanis feae* D Vinciguerra, 1890

Leonardo Fea (1852–1903) was an Italian explorer, painter and naturalist. He was Assistant at the Natural History Museum in Genoa and liked exploring far-off, little-known countries. He visited Burma (Myanmar), islands in the Gulf of Guinea, and the Cape Verde Islands. Two mammals, six birds, four amphibians, five reptiles and a dragonfly are named after him.

Feconat

Dwarf Cichlid sp. *Apistogramma feconat* U Römer, DP Soares, CRG Dávila, F Duponchelle, J-F Renno & I Hahn, 2015

This is an acronym referring to the Federation of the Native Communities of the Tigre (FECONAT). The name was given in recognition of the Federation's struggle to protect the environment of tribal lands in the Rio Tigre drainage of Peru (where this cichlid occurs) from the impacts of modern society; the "*...ongoing battle for Indian civil rights by FECONAT against powerful industrial opponents has repeatedly brought severe environmental problems to public awareness.*"

Fedorov

Seaperch sp. *Helicolenus fedorovi* VV Barsukov, 1973
Fedorov's Catshark *Apristurus fedorovi* VN Dolganov, 1985
Fedorov's Skate *Bathyraja fedorovi* VN Dolganov, 1985
[Alt. Cinnamon Skate]
Roughy sp. *Hoplostethus fedorovi* AN Kotlyar, 1986
Fedorov's Lumpsucker *Microancathus fedorovi* SA Mandritsa, 1991
Morid Cod sp. *Physiculus fedorovi* YN Shcherbachev, 1993
Snaggletooth sp. *Astronesthes fedorovi* NV Parin & OD Borodulina, 1994
Snailfish sp. *Careproctus fedorovi* AP Andriashev & DL Stein, 1998
Eelpout sp. *Zoarces fedorovi* LA Chereshnev, MV Nazarkin & DA Chegodaeva, 2007

Dr Vladimir Vladimirovich Fedorov (1939–2011) was a Russian ichthyologist. His doctorate was awarded by Leningrad State University (1979). He worked at the Pacific Research

Institute of Fisheries and Oceanography, Valdivostock (1961–1975) and at the Zoological Institute Academy of Sciences, Leningrad (later St Petersburg) (1975). He collected the holotype of the skate.

Fedtschenko

Syr Darya Sturgeon *Pseudoscaphirhynchus fedtschenkoi* KF Kessler, 1872

Alexei Pavlovich Fedtschenko (or Fedchenko) (1844–1873), a graduate in zoology and geology from Moscow University, was a naturalist and explorer of Central Asia. The Fedchenko Glacier in the Pamirs is named after him, as is the asteroid 3195 Fedchenko. He died while climbing on Mont Blanc. After his death, the Russian government published accounts of his discoveries and explorations. Two reptiles are also named after him

Feegrade

Yoma Danio *Brachydanio feegradei* SL Hora, 1937

Lieutenant Dr Egbert Stanley Feegrade (b.1884) of the Indian Medical Department was a physician and Special Malaria Officer with the Public Health Department of Burma (1905–>1933) and an amateur naturalist. In WW1, he served with a Field Ambulance Unit. He made an important collection of mollusc shells from around Hsipaw and Lashio in the Northern Shan States (1926) and at least one such is named after him. He described a new fruit fly (1928). He was promoted (1929) to first class surgeon after returning to India. He undertook surveys of malarial outbreaks and took measures such as spraying to eradicate the mosquitos in those areas. He collected the danio holotype.

Fehlmann

Harlequin Catshark *Ctenacis fehlmanni* S Springer, 1968
Labrisomid Blenny sp. *Paraclinus fehlmanni* VG Springer & R Trist, 1969
Robbermask Goby *Stenogobius fehlmanni* RE Watson, 1991
Goby sp. *Smilosicyopus fehlmanni* LR Parenti & JA Maciolek, 1993
Wormfish sp. *Paragunnellichthys fehlmanni* CE Dawson, 1969
Flathead sp. *Rogadius fehlmanni* LW Knapp, 2012

Dr Herman Adair Fehlmann (1917–2005) of George Vanderbilt Foundation, Bangkok, and the Smithsonian was primarily an ichthyologist but also an herpetologist. His herpetological specimens were deposited in the Stanford University Museum. He wrote: *Ecological distribution of fishes in a stream drainage in the Palau Islands* (1960). He was honoured in the catshark's name for "*…setting high standards for field treatment of shark specimens collected for study*." He collected the type of the wormfish. A reptile is named after him.

Feldberg

Armoured Catfish sp. *Ancistomus feldbergae* RR de Oliveira, LH Rapp Py-Daniel, JAS Zuanon & MS Rocha, 2012
[Syn. *Peckoltia feldbergae*]

Dr Eliana Feldberg is a Brazilian biologist and geneticist at Instituto Nacional de Pesquisas da Amazônia, Manaus, Brazil. The Federal University of São Carlos awarded her bachelor's degree (1979) and master's (1983). Her doctorate was awarded by Instituto Nacional de Pesquisas da Amazônia (1990), where she is a researcher and head of the Animal Genetics Laboratory. She co-wrote: *Chromosome mapping of repetitive sequences in four Serrasalmidae species (Characiformes)* (2014).

Feldmann

River Tongue Sole *Cynoglossus feldmanni* P Bleeker, 1854

Surgeon-Major A Feldmann was a Dutch Colonial Army doctor stationed in Samarang, East Borneo, about whom we have failed to find more information.

Felippone

Armoured Catfish sp. *Rineloricaria felipponei* HW Fowler, 1943

Dr Florentino Silvestre Feliponne Bentos (1852–1939) was a Uruguayan naturalist, biologist, bryologist and physician, who qualified (1882) in Montevideo and studied and qualified as a chemist in Paris (1885). He was the first Director, Faculty of Medicine, Montevideo, Professor of Chemistry, University and Athenaeum of Montevideo and deputy director, Museum of Natural History, Montevideo. He was very interested in all Uruguayan flora and fauna and sent specimens to all the leading experts. He collected the holotype of the catfish.

Felix

Felix's Mormyrid *Mormyrus felixi* J Pellegrin, 1939

Henri Jacques-Félix (1907–2008) was a French botanist who specialised in tropical African flora. He explored in West Africa and collected particularly in Gabon, Cameroon, Guinea and the Ivory Coast (1930–1966) and worked at MNHN, Paris and as Director ORSTOM. Among his published books and papers are: *La vie et la mort du Lac Tchad: rapports avec l'agriculture et l'élevage* (1947), *Flora of Gabon: Melastomataceae* (1963) and the co-written: *Grasses of tropical Africa: Poaceae.* (1962). He collected the holotype in Cameroon. He is also commemorated in the name of at least one plant, *Jacquesfelixia dinteri*.

Fellmann

African Rivuline sp. *Aphyosemion fellmanni* JR Van der Zee & R Sonnenberg, 2018

Emmanuel Fellmann is a French aquarist who collected this species in the Congo. He made four expeditions to the Republic of Congo to study killifish and to get a better idea of the distribution of *Aphyosemion* and *Epiplatys* species in the southern part of the country. He wrote: *Guide pratique des voyages aquariophiles* (2014) about collecting in west Africa.

Fellowes

Aegean Chub *Squalius fellowesii* A Günther, 1868

Sir Charles Fellowes (1799–1860) was the British archaeologist who discovered (1838) the ruins of cities of ancient Lycia (now in southwestern Turkey). He wrote: *A Journal Written During an Excursion in Asia Minor* (1839). He made two subsequent visits to Asia Minor (Anatolia) (1839 & 1841) at the behest of the British Museum and collected a considerable amount of Lycian works of art. He was a fine artist himself and made sketches of all the ancient sites he explored. His work was used to illustrate Byron's works including his famous epic poem: *Childe Harold*. He presented the holotype of the eponymous species to the BMNH.

Feltrin

Neotropical Rivuline sp. *Cynopoecilus feltrini* WJEM Costa, PF de Amorim & JL de Oliveira Mattos, 2016

Caio Feltrin is a Brazilian biologist working at the Atlantic Forest Wildlife Sanctuary. His focus has been river systematics, ecology and ichthyology. The etymology records that the fish is named after: "...*Caio Feltrin, in recognition of his dedication in inventorying the fish fauna of southern Brazil.*"

FEN Mutis

Jawfish sp. *Opistognathus fenmutis* A Acero Pizarro & R Franke-Ante, 1993

FEN + Mutis is an arbitrary combination of letters in honour of the José Celestino Mutis Fund of the National Electric Finance Company (FEN, Bogotá), a patron of natural sciences in Colombia which financed the senior author's studies.

Ferdebuen

Chignahuapan Silverside *Poblana ferdebueni* A Solórzano & Y López, 1965

Don Fernando de Buen y Lozano (1895–1962) (see **de Buen**)

Ferdowsi

Scraper (barb) sp. *Capoeta ferdowsii* A Jouladeh-Roudbar, S Eagderi, L Murillo-Ramos, HR Ghanavi & I Doadrio, 2017

Abu'l-Qāsim Ferdowsi Tusi (c.940–1020) was a Persian poet and the author of the epic poem Shahnameh ('Book of Kings'), which is the national epic of Iran. The Shahnameh was originally composed by Ferdowsi for the princes of the Samanid dynasty, who were responsible for a revival of Persian cultural traditions after the Arab invasion of the seventh century. The poem chronicles the legendary history of the pre-Islamic kings of Iran.

Fern

Lanternfish sp. *Nannobrachium fernae* RL Wisner, 1971

Fern Wisner (d.1990) was the wife of the author, Robert Wisner (q.v). He wrote of her: "... *who deserves far more than this small token of my gratitude for her long sufferance of the preoccupation and associated neglectful acts so prevalent among workers in ichthyology.*" (Also see **Wisner**)

Fernandez, AA

Headstander Characin sp. *Laemolyta fernandezi* GS Myers, 1950

Dr Augustín Antonio Fernández-Yépez (1916–1977) was a Venezuelan architect and ichthyologist. (See **Yepez**)

Fernandez, AJ

Tetra sp. *Hyphessobrycon fernandezi* AA Fernández-Yépez, 1972

Dr Alberto José Fernández-Yepez was a Venezuelan ornithologist, and the author's brother. The etymology states that Alberto dedicated his life to natural science, and "*departed too soon*" (translation). A biological station is also named after him.

Fernández-Yépez

Panda Uaru *Uaru fernandezyepezi* R Stawikowski, 1989

Dr Augustín Antonio Fernández-Yépez (1916–1977) (See **Fernandez, AA and Yepez**)

Fernholm

Hagfish sp. *Paramyxine fernholmi* CH Kuo, KF Huang & HK Mok, 1994
Hagfish sp. *Myxine fernholmi* Wisner & CB McMillan, 1995
Hagfish sp. *Eptatretus fernholmi* CB McMillan & RL Wisner, 2004

Bo Fernholm (b.1941) is a Swedish zoologist. He is Professor Emeritus of Vertebrate Zoology at the Swedish Museum of Natural History. His specialist area is hagfish systematics. He also served 20 years as Swedish commissioner to the International Whaling Commissioner and as Swedish scientific representative to the Convention for the Conservation of Antarctic Marine Living Resources (CCAMLR). He was honoured for his many contributions to hagfish knowledge.

Ferrant

African Rivuline sp. *Aphyosemion ferranti* GA Boulenger, 1910

Viktor Ferrant (1856–1942) was a naturalist and entomologist who was born and died in Luxembourg. He studied trade and commerce in Paris to follow his father into the family business, but his real interest was in natural history. He joined the Luxembourg 'Service Agricole' (1890) rising to head (1902). He became a conservator of the natural history section of the State Museum (1892) until retirement (1924): but remained until his death since no replacement could be found. He was co-founder (1890) of the Société des Naturalistes Luxembourgeois. He was also a member of the Belgian Entomological Society and the Natural History Museum in Paris. He donated the type specimens for study.

Ferraris

Driftwood Catfish subgenus *Ferrarissoaresia* S Grant, 2015
Ferraris' Goby *Exyrias ferrarisi* EO Murdy, 1985
Armoured Catfish sp. *Niobichthys ferrarisi* SA Schaefer & F Provenzano Rizzo, 1998
Chameleonfish sp. *Badis ferrarisi* SO Kullander & R Britz, 2002
Ghost Knifefish sp. *Apteronotus ferrarisi* CDCM de Santana & RP Vari, 2013
Driftwood Catfish sp. *Centromochlus ferrarisi* JLO Birindelli, LM Sarmento-Soares & FCT Lima, 2015

Dr Carl James Ferraris Jr (b.1950) is an American ichthyologist and fish taxonomist who is a recognised authority on catfish. He was awarded his bachelor's degree from Cornell University (1972), a master's degree from Oklahoma State University (1976), and a PhD from the City University of New York (1989). He was at the department of ichthyology, American Museum of Natural History (1981–1991), the California Academy of Sciences (1991–2000), where he is currently a research associate, and worked for the Florida Museum of Natural History (2003–2009). He has written nearly 100 peer-reviewed taxonomic publications which included reports of the discovery of 67 previously unnamed species and nine genera of fishes. The subgenus *Ferrarissoaresia* is named after both Ferraris and Luisa Maria Sarmento-Soares (q.v. under **Soares**). He collected the goby when with Murdy in the Philippines. (See also **Carlarius**)

Ferreira, AR

Black Arowana *Osteoglossum ferreirai* RH Kanazawa, 1966

Alexandre Rodrigues Ferreira (1756–1815) was the first Brazilian naturalist to explore the Amazon and Pantanal biomes in the states of Pará and Mato Grosso (1783–1792), sponsored by the Portuguese government; he became known as the Brazilian Humboldt. He studied law and then natural philosophy and mathematics at the University of Coimbra, Portugal, being awarded his baccalaureate (1777) then studied natural history there and was awarded his doctorate (1779). He catalogued specimens and wrote scientific papers at the Ajuda Museum, Lisbon (1778–1783) and became a corresponding member of the Lisbon Academy of Science. He left for Brazil (1783) and there followed the course of the Amazon and its tributaries, studying not only the fauna and flora, but also the indigenous people and their languages and customs (1783–1792). He returned to Lisbon (1793) and became Director of the Natural History Museum and of the Botanical Gardens (1793–1815). A mammal, an amphibian and a reptile are also named after him, as well as various insects and marine invertebrates.

Ferreira, EJG

Whale Catfish sp. *Cetopsidium ferreirai* RP Vari, C J Ferraris Jr & MCC de Pinna, 2005
Glass Knifefish sp. *Archolaemus ferreirai* RP Vari, CDCM de Santana & WB Wosiacki, 2012

Dr Efrem Jorge Gondim Ferreira (b.1954) is a Brazilian ichthyologist, who is a Senior Researcher III at Instituto Nacional de Pesquisas da Amazônia, Manaus, and Professor of the Postgraduate Programme in Fresh Water Biology and Inland Fisheries. His bachelor's degree was awarded by the Fisheries Engineering Department (1976) and both his master's (1981) and doctorate (1992) by Instituto Nacional de Pesquisas da Amazônia. He wrote: *Composition and ecological aspects of the fish fauna of a section of the Tronbetas River, in the* área *influenced by the future UHE Cachoeira Porteira* (1993); and co-wrote: *Rio Negro: rich life in poor water* (1988); *The Smithsonian Atlas of the Amazon* (2003); *A review of the South American cichlid genus* Cichla*, with description of nine new species (Teleostei:Cichlid)* (2006), and *Rio Branco: fish, ecology and conservation of Roraima* (2007). He was honoured in the catfish's name for having collected all the known specimens of this species and for his contributions to our knowledge of the fishes of the Brazilian Amazon.

Ferrer (Aledo)

Ferrer's Goby *Pseudaphya ferreri* O de Buen & J-L Fage, 1908

Jaume Ferrer Aledo (1854–1956) was a pharmacist and an amateur ichthyologist from Menorca. His pharmacy degree was awarded (1877) by the University of Barcelona. He opened a pharmacy on Menorca but, around the turn of the century, sold it and divided his time between work as an administrator and marine studies. He published his first article and his important *Catalogue of Fish of Menorca* (1906) in a local journal. He published (1906–1923) his *Fauna de Menorca* chapter by chapter in the Menorca Magazine. He studied the fishes of the Balearic Islands and sent specimens to the authors, including the type of the eponymous goby. He corresponded often with Odón de Buen (q.v.) sending him drafts of his papers as well as specimens.

Ferrer (Castellanos)

Pencil Catfish sp. *Trichomycterus ferreri* CA Ardila Rodríguez, 2018

Jorge de Jesus Ferrer Castellanos is a Colombian zoologist and botanist, whom the etymology describes as an *'eminent scientist'* (translation) – though he seems yet to have made his mark on an international level. He helped to collect the catfish holotype.

Ferrier

Rattail sp. *Coryphaenoides ferrieri* CT Regan, 1913

James G Ferrier was Secretary of the Scotia Committee; 'Scotia' is ship from which the holotype was collected during the Scottish National Antarctic Expedition (1902–1904).

Festa

Suckermouth Catfish sp. *Astroblepus festae* GA Boulenger, 1898
Characin sp. *Astyanax festae* GA Boulenger, 1898
Sea Catfish sp. *Cathorops festae* GA Boulenger, 1898
Neotropical Livebearer sp. *Pseudopoecilia festae* GA Boulenger, 1898
Characin sp. *Lebiasina festae* GA Boulenger, 1899
Guayas Cichlid *Mesoheros festae* GA Boulenger, 1899
Garra sp. *Garra festai* E Tortonese, 1939

Dr Enrico Luigi Festa (1868–1939) was an Italian naturalist who worked for the Zoological Museum of the University of Turin (1899–1923), first as Deputy Assistant Professor and retiring as Honorary Vice Director. His wealth enabled him to travel extensively and collect natural history specimens. He visited Panama and Ecuador (1895–1898), Palestine (1905) and made an expedition (early 1920s) to Cyrenaica in Libya. He also visited and collected in Egypt, Syria, Jordan, Rhodes and Sardinia. Six reptiles, two birds and three amphibians are also named after him.

Fibreno

Fibreno Trout *Salmo fibreni* S Zerunian & G Gandolfi, 1990

This is a toponym referring to the Fibreno river basin, including Lake Posta Fibreno, Italy, where this species is endemic.

Field

Two-tone Chromis *Chromis fieldi* JE Randall & JD DiBattista, 2013

Richard Field is an underwater photographer, at first free diving (1964) but latterly (early 1990s) using SCUBA. He was the first to suspect that the Indian Ocean population of *C. dimidiata* might represent a different species; he provided photographs taken in the Red Sea in the vicinity of Jeddah, Saudi Arabia, that gave the authors the opportunity to note colour variation. He also collected four specimens of the new species. He says that this "... *was a Field family affair, my wife Mary and son Francis helped in collecting specimens of Red Sea Chromis dimidiata to demonstrate that it is a different species to that found in the Indian Ocean.*" He wrote the ebook: *Reef Fishes of Oman* (2016) and *Reef Fishes of the Red Sea* (1998).

Figaro

Catshark genus *Figaro* GP Whitley, 1928

Barber Goby *Elacatinus figaro* I Sazima, RL Moura & R de S Rosa, 1997

Figaro is the eponymous: *Barber of Seville* (opera by Rossini) and of Mozart's: *The Marriage of Figaro.* The goby was named after the barber for its cleaning behaviour. No explanation was given by Whitley as to why he thought the name was appropriate for a genus of sharks.

Figueiredo

Characin sp. *Creagrutus figueiredoi* RP Vari & AS Harold, 2001

Southern Atlantic Sharpnose Puffer *Canthigaster figueiredoi* RL Moura & RMC Castro, 2002

Characin sp. *Kolpotocheirodon figueiredoi* LR Malabarba, FCT Lima & SH Weitzman, 2004

Anchovy sp. *Lycengraulis figueiredoi* MV Loeb & AV Alcântara, 2013

Characin sp. *Knodus figueiredoi* ALH Esguícero & RMC Castro, 2014

Dr José Lima de Figueiredo (b.1943) aka 'Ze Lima' is a Brazilian ichthyologist who is a former researcher and Curator of Fishes at the Zoological Museum of the University of São Paulo. He is a professor at the University of São Paulo, which awarded his bachelor's degree (1969) and doctorate (1981). He co-wrote: *The northernmost record of Bassanago albescens and comments on the occurrence of Rhynchoconger guppyi (Teleostei: Anguilliformes: Congridae) along the Brazilian coast* (2011). (Also see **Lima**)

Figueroa

Three-barbeled Catfish sp. *Pimelodella figueroai* G Dahl, 1961

Adalberto Figueroa Potes was a Colombian zoologist and agronomist at the Facultad de Agronomía, Universidad Nacional de Colombia, Palmira, where he was Professor Emeritus. He wrote: *Initial catalogue of Cochinillas del Valle del Cauca. (Homoptera-Coccoidea)* (1946). He was honoured "*...for his incessant work and important contributions to the knowledge of the fauna of Colombia.*"

Filewood

Sailback Houndshark *Gogolia filewoodi* LJV Compagno, 1973

Lionel Winston Charles Filewood (b.1936) is an Australian biologist who worked (1960s-1980s) for the Department of Agriculture, Stock and Fisheries, Papua *New Guinea. The University of Sydney awarded his bachelor's degree (1958). He co-wrote: Scientific names used in Birds of New Guinea and tropical Australia* (1976). He was honoured "*...for his work on the poorly known elasmobranch fauna of New Guinea.*" An Australian court sentenced him to 9 months' imprisonment for child pornography offences (2010). A bird is named after him.

Filippi

Kura Bleak *Alburnus filippii* KF Kessler, 1877

Dr Filippo de Filippi (1814–1867) was an Italian physician, traveller and zoologist. He visited, among other places, Alaska, Persia, Mongolia and Turkestan. He succeeded Carlo Gené as Professor of Zoology at the Museum of Natural History in Turin. His efforts to disseminate knowledge of Darwin's evolutionary theory included the presentation of a seminal lecture (1864) 'L'uomo e le scimmie' (Man and the Apes). Filippi was the scientist who accompanied the Duke of Abruzzi's expedition to Alaska, and he also led an expedition (1862) to explore Persia (now Iran). He set out on a government-sponsored scientific voyage aboard 'Magenta', being replaced by Giglioli who published the results. Two reptiles and three birds are also named after him.

Finley

Cameroon Cichlid sp. *Benitochromis finleyi* E Trewavas, 1974

Lee Finley has been an American aquarium hobbyist for over fifty years. He has regularly written on various aquaria topics over the past 40 years and has over 345 published articles to his credit. For 12 years he wrote a monthly book review column 'Aquarist's Library' in Aquarium Fish Magazine, and for almost ten years (over two separate runs) a monthly catfish column 'Catfish Corner' in Tropical Fish Hobbyist Magazine. He wrote the book: *Catfishes* (2009) and co-wrote *Tropical Fish* (2003). Now 'retired' from being a columnist, his writing interests remain. He also ran a predominately mail-order book business 'Finley Aquatic Books' for a decade, dealing in both new and used literature covering all aspects of the aquarium hobby and aquatic natural history. He succeeded in breeding this cichlid in an aquarium. He 'generously' provided a photograph of the living fish for Trewavas's article, which is why he was honoured in the name.

Fiolent

Fiolent's Smooth-head *Conocara fiolenti* YI Sazonov & AN Ivanov, 1979

Rover sp. *Plagiogeneion fiolenti* NV Parin, 1991

Named after the research vessel 'R/V Fiolent', from which the holotypes were collected.

Firat

Euphrates Spring Minnow *Pseudophoxinus firati* NG Bogutskaya, F Küçük & MA Atalay, 2006

Firat Nehri is the Turkish name for the Euphrates River.

Firestone

African Loach Catfish sp. *Paramphilius firestonei* LP Schultz, 1942

This species was named to recognise the Smithsonian-Firestone Expedition to Liberia (1940) during which the holotype was collected.

Fischer, EN

Fanfin Skate *Pseudoraja fischeri* HB Bigelow & WC Schroeder, 1954

E N Fischer made skilful drawings for the describers' publications on fishes from the Gulf of Maine, so this skate was named after him. We have been unable to find out anything more about him.

Fischer, GA

Fischer's Victoria Squeaker *Synodontis afrofischeri* FM Hilgendorf, 1888
Lake Victoria Cichlid sp. *Haplochromis fischeri* L Seegers, 2008

Dr Gustav Adolf Fischer (1848–1886) was a German physician, naturalist and explorer of East and Central Africa. He was an army physician who (1876) joined the expedition of the Denhardt brothers (Clemens and Gustav) to Wituland (coastal northern Kenya) and explored there again the following year. He explored the Tana River (1878), again together with the Denhardt brothers who were very influential in the region. He stayed on until (1890) the sultanate of Wituland came under British influence, after which the brothers lost all their property without compensation. Fischer stayed in Zanzibar (1878–1882) as a physician. He made (1882) an expedition into the interior of East Africa (Masai lands), sponsored by the Hamburger Geographischen Gesellschaft, then returned to Germany (1883). He undertook his last mission (1885) to meet Emin Pasha, who was living in southern Sudan but who fled to Uganda because of a local uprising. Fischer reached Nyanza on Lake Victoria's east coast, but was unable to go further, so returned to Zanzibar. He died in Berlin the year after of a tropical fever. Eighteen birds and a mammal are also named after him.

Fischer, W

Armoured Catfish sp. *Chaetostoma fischeri* F Steindachner, 1879

W Fischer, described by Steindachner as a 'dear friend', sent a collection of river fishes from Panama which included the holotype of this species. We have not been able to identify Fischer further.

Fischer, W

Longfin Crevalle Jack *Caranx fischeri* WF Smith-Vaniz & KE Carpenter, 2007

Dr Walter Fischer worked for the Marine Resources Department of the UN Food and Agriculture Organisation. He inspired experts worldwide with his work to create sets of species identification sheets (1973), synopses and catalogues (1976) leading to the start of a comprehensive data programme (1989). The etymology says: "*We take great pleasure in naming this new species* Caranx fischeri *in honor of our friend and colleague Dr. Walter Fischer (retired) for his vision and dedication in initiating the Species Identification and Data*

Programme of the Food and Agriculture Organization of the United Nations (Fischer, 1989). In numerous ways this program has been an invaluable resource for marine fisheries biologists and ichthyologists generally."

Fishelson

Snailfish sp. *Liparis fishelsoni* JLB Smith, 1967
Frillgoby sp. *Bathygobius fishelsoni* M Goren, 1978
Fishelson's Pygmy-goby *Trimma fishelsoni* M Goren, 1985

Dr Lev Fishelson (1923–2013) was an ecologist and marine biologist at Tel-Aviv University. He was honoured for his 'well-known' and 'tremendous' contributions to the knowledge of the Red Sea, where these eponymous species are found.

Fisher, CG

Driftwood Catfish sp. *Trachelyopterus fisheri* CH Eigenmann, 1916

Carl Graham Fisher (1874–1939) was an entrepreneur in the fields of automobiles and motor racing in Indianapolis, Indiana and real estate in Miami Beach, Florida. He was worth an estimated $100 million (1926), but lost the lot in the Stock Market Crash (1929), after which he lived in a small cottage on Miami Beach. Fisher: "*...helped to make possible a second expedition to the type locality [Columbia] of this species*" and was probably a first cousin of Eigenmann's student, Homer Glenn Fisher (1888–1918), who died of pneumonia whilst in the US Medical Reserve Corps.

Fisher, JC

Garra sp. *Garra fisheri* HW Fowler, 1937
[Probably a Junior Synonym of *Garra fuliginosa*]

Dr J C Fisher was described by Fowler as: "an early contributor to the natural history collections of the Academy" (i.e. Academy of Natural Sciences of Philadelphia). He made donations to that academy (1850s-1860s) of both animal and mineral specimens, but we have not been able to identify him further.

Fisher, WK

Fisher's Seahorse *Hippocampus fisheri* DS Jordan & BW Evermann, 1903
Fisher's Angelfish *Centropyge fisheri* JO Snyder, 1904
Green Labeo *Labeo fisheri* DS Jordan & EC Starks, 1917
Translucent Goby *Chriolepis fisheri* AW Herre, 1942

Dr Walter Kenrick Fisher (1878–1953) was a botanist and zoologist. His entire career was spent at Stanford University, which he entered as a student (1897), receiving his bachelor's degree (1901), his master's (1903) and his doctorate (1906) and continuing to teach there as Assistant in Zoology (1902–1905), Instructor (1905–1909), Assistant Professor (1909–1925) and Professor (1925–1943); retiring then as Professor Emeritus Zoology. He was also Director of the Hopkins Marine Station at Stanford. He was a member of two (1902 & 1904) of the cruises of the US Fish Commission's research vessel 'Albatross'. He was Curator of the California Academy of Sciences (1916–1932). After retiring he continued his research into starfishes and, until his death, was at work on collections from the Smithsonian of which he was a Research Associate. His many publications include: *Starfishes of the Hawaiian Islands* (1906) and: *A new genus of sea stars (Plazaster) from Japan, with a note on the genus Parasterina* (1941). He collected the holotype of the Green Labeo in Ceylon (Sri Lanka) by using a cast-net. While the etymology for the seahorse clearly states it is named in honour of Walter **V** Fisher of Stanford University, we believe the initial 'V' may be a transcription error.

Fisk

Unicorn Crestfish *Eumecichthys fiski* A Günther, 1890

Rev George Henry Redmore Fisk (1829–1911) was a collector of South African zoological curiosities, including the holotype of this species which he sent to the BMNH. He became an Archdeacon in the Anglican Church of the Province of South Africa (1890) and secretary of the Diocesan Finance Commission. He was a member of the South African Philosophical Society (1877–191) and became one of only two honorary members when the Philosophical Society became the Royal Society of South Africa (1908).

Fison

Fison's Lefteye Flounder *Arnoglossus fisoni* JD Ogilby, 1898

Cecil Shuttleworth Fison (1840–1899) was a British-born Australian Inspector of Fisheries for Queensland (1897). Previously he had been Examiner of seamanship and navigation (1882), Shipwright Surveyor (1887) and Inspector of Oyster Fisheries (1892). Ogilby wrote: "*I have much pleasure in naming this pretty little species for my friend Mr. Cecil S Fison, Inspector of Fisheries for Queensland, from whom I have received much kindness and useful information during my recent visit to Brisbane.*"

Fissunov

Lanternfish sp. *Myctophum fissunovi* VE Becker & OD Borodulina, 1971
Fairygoby sp. *Tryssogobius fisunovi* AM Prokofiev, 2017

Georgy Kasyanovich Fissunov (or Fisunov) was a senior technician-oceanologist aboard the research vessel 'Vityaz' (q.v.). He is described as an "*…enthusiastic and unsurpassed master in the art of fishing with a cast net,*" and to whom the P P Shirnov Institute of Oceanology owes a debt for creating: "*…one of the world's largest collections of epipelagic fishes numbering many thousands of specimens.*" Although the fairygoby was described in 2017 it was collected by Fisunov in 1973. We have been unable to find any further details about him.

Fitch

Pacific Scabbardfish *Lepidopus fitchi* RH Rosenblatt & RR Wilson, 1987
Deepsea Bigeye *Priacanthus fitchi* WC Starnes, 1988
Morid Cod sp. *Notophycis fitchi* YI Sazonov, 2001

John Edgar Fitch (1918–1982) was a marine biologist and ichthyologist who was Director of the US Fisheries Laboratory, San Diego. He wrote: *Offshore fishes of California* (1958).

Fitzroy (River)

Creek Whaler (shark) *Carcharhinus fitzroyensis* GP Whitley, 1943

This is a toponym; the holotype was collected in the estuary of the Fitzroy River, Western Australia.

Fitzroy

Livebearer genus *Fitzroyia* A Günther, 1866 NCR
[Junior Synonym of *Jenynsia*]

Admiral Robert Fitzroy (1805–1865) was a hydrographer and meteorologist who invented the Fitzroy Barometer. He is probably best remembered for having been in command of 'HMS Beagle' (1828–1836), during the last five years of which Charles Darwin was on board as naturalist. He became Member of Parliament for Durham (1841) and was Governor of New Zealand (1843–1845); he was dismissed for taking the view that Maori land claims were just as valid as those of the settlers. He founded and became first Director of the Meteorological Office in London (1854). He was a creationist who felt guilty that the voyage of the 'Beagle' was used to undermine the scriptures, and appealed to Darwin to recant. Fitzroy committed suicide during a bout of severe depression. A mammal is also named after him.

Fitzsimons

Circular Seabat *Halieutaea fitzsimonsi* JDF Gilchrist & WW Thompson, 1916

Frederick William FitzSimons (1870–1951) was an Irish-born South African naturalist and herpetologist particularly interested in snakes and their venom. He became Director, Port Elizabeth Museum and Snake Park (1906–1936). He was a dynamic personality, appointed to run a 'sleepy' museum, and quickly energised it and the local inhabitants. He wrote: *Snakes* (1932). A reptile and a bird are also named after him.

Flavio Lima

Tetra sp. *Gymnocorymbus flaviolimai* RC Benine, BF de Melo, RMC Castro & C de Oliveira, 2015

Dr Flávio César Thadeo de Lima. (See **De Lima**)

Flavius Joseph

Jordan Mouthbrooder *Astatotilapia flaviijosephi* LCE Lortet, 1883

Titus Flavius Josephus (37–c.100), a Romano-Jewish scholar, historian and hagiographer, is mentioned several times in Lortet's study of Lake Tiberias (Sea of Galilee in Israel); Flavius reported a thriving fishing industry on the lake and believed the occurrence of a catfish (*Clarias gariepinus*) in the lake was due to underground connections to the Nile.

Fleming

Highfin Dragonfish *Bathophilus flemingi* W Aron & PA McCrery, 1958

Dr Richard Howell Fleming (1909–1989) was a Canadian-born oceanographer. The University of British Columbia awarded both his bachelor's and master's degrees. He joined (1931) the Scripps Institution of Oceanography, University of California, which awarded his doctorate (1935), and where he taught until 1946. he was chief oceanographer, US Navy Hydrographic Office, Washington. D.C. (1946–1951). He was Chairman Department of Oceanography, University of Washington, Seattle (1951–1967) and Professor of Oceanography until retiring as Professor Emeritus (1980). He co-wrote: *The Oceans* (1942).

Flerx

Hardhead Snailfish *Lopholiparis flerxi* JW Orr, 2004

William C Flerx is an American fisheries biologist who worked at the National Marine Fisheries Service. He has regularly led collecting expeditions on research vessels such as on 'R/V David Starr Jordan' (1986) and 'R/V Pacific Knight' (1994). He collected the holotype and recognised it as a distinct new species. According to the etymology he also "*…has taken extraordinary care to preserve many other significant specimens captured along the west coast from Alaska to California.*" Among his published papers is the co-written: *Demersal fish surveys off central California… …1987–1989.* (1989)

Fleurieu

Flower Cardinalfish *Ostorhinchus fleurieu* BGE Lacepède, 1802

Charles Pierre Claret, Comte de Fleurieu (1738–1810), was a French explorer and hydrographer. He fought in several battles during the Seven Years' War (1756–1763). He was Minister of the Navy under King Louis XVI and was active in planning naval operations against England during France's involvement in the American War of Independence (1778–1783). He wrote: *Découvertes des Français dans le Sud Est de la Nouvelle-Guinée en 1768 et 1769* (1790).

Flinders, M

Eastern School Whiting *Sillago flindersi* RJ McKay, 1985

Matthew Flinders (1774–1814) was an English explorer and navigator who joined the British Navy and trained as a navigator, having wanted to be a sailor and explorer ever since reading *Robinson Crusoe*. He sailed to Australia on 'HMS Reliance' (1795) as a midshipman, and with George Bass, the ship's physician, explored south of Sydney in a tiny boat called *Tom Thumb*. As a Lieutenant (1798), Flinders was given command of the *Norfolk* and discovered the passage between the Australian mainland and Tasmania named Bass Strait after his friend, while its largest island would later be named Flinders Island. He returned to England (1800) and married, but was sent exploring again (1802). He circumnavigated the mainland of Australia aboard the *Investigator* (1802–1803). The French later captured him in Mauritius, treating him as a spy and holding him captive (1803–1810). He then returned to England a broken man in ill health, but still managed to write the story of his circumnavigation under the title: *A Voyage to Terra Australis* (1814) published the day he died. Two birds are also named after him.

Flinders

Pygmy Thornback Skate *Dentiraja flindersi* PR Last & DC Gledhill, 2008
[Syn. *Dipturus flindersi*]

This species is named for the Flindersian Province; the name given to the western warm-temperate biogeographic region of Australian coastal waters, which is where this species occurs.

Flora

Tidepool Snailfish *Liparis florae* DS Jordan & EC Starks, 1895

Mrs Flora Hartley Greene (1865–1948) was (at the time of the fish being named) Assistant Curator of the Museum of Stanford University. She was awarded her BSc by Indiana State Normal School, her BA by Stanford University (1895) and her MSc by the University of Missouri (1909). She married Dr Charles Wilson Greene (1895) a fellow student at Stanford. (Just days before the opening (by David Starr Jordan) of the Hopkins Seaside Laboratory, Charles Wilson Greene, Flora Hartley together with Bradley Davis of Ann Arbor and Ora Boring of California, were four enthusiastic young graduate students of Stanford. They swept out the shavings, arranged tables, stocked aquaria, and set the minor mechanism of that laboratory into operation before it opened (June 27, 1892). During its first session Greene served as one of two assistants, with Flora Hartley, at the laboratory for the summer.) In later life she served as Special Agent in US Children's Bureau, Counsellor of the American Home Economics Association, on the Missouri State Council of Defence, the Executive Board of League of Women Voters, the Red Cross Board, the Missouri Society for Crippled Children, and the Eugene Field Foundation. She held many offices in the Missouri Federation of Women's Clubs, such as Chairman of Home Economics, Child Welfare, Public Welfare, and State Crippled Children's Campaign. (See also **Greene, Charles W**)

Florence

Deepwater Cardinalfish genus *Florenciella* GW Mead, 1965

As Giles Mead did not include an etymology (in contrast to the other genus named in the same paper) we must assume the omission was deliberate. We can only speculate. There was more than one Florence in his life; for example, his father was married to a woman named Florence Ford – but she *wasn't* GW Mead's mother and she died (1921) before Giles was born – his father later re-marrying a woman named Elise. GW Mead seems to have had a sister named Florence Mead Mackay about whom we can find nothing.

Floros

Floros' Goatfish *Upeneus floros* F Uiblein & G Gouws, 2020

Dr Camilla Floros is a South African marine biologist who is Project Leader (2020) at TRAFFIC, the wildlife trade monitoring network. She is also a Research Associate (2018) of the South African Association for Marine Biological Research, Oceanographic Research Institute (Durban, South Africa). She was also a consultant (2019–2020) at OceanAfrica, having been a Coral Reef Ecologist at the Oceanographic Research Institute for the previous decade (2007–2017). Her PhD was awarded by UKZN (2010) as was her MSc (2002).

Rhodes University awarded her marine biology degree (1999). She organised the collection of the goatfish holotype and two paratypes and provided photographs of the types, as well as of other specimens seen at the type locality: Sodwana Bay, South Africa.

Foa

Lake Tanganyika Cichlid sp. *Cyathopharynx foae* LL Vaillant, 1899

Édouard Foà (1862–1901) was a French explorer, geographer and big-game hunter in Africa. In a crossing of that continent, Foa is said to have shot about 500 animals; mainly to supply specimens to the Paris Natural History Museum. He travelled over 7,200 miles, mostly on foot, from the Zambezi delta in the east to the Congo River mouth in the west. He became a Fellow of the Royal Geographical Society (1894). He wrote: *After Big Game in Central Africa* (1899) about his exploits. His death at such a young age was blamed on "*germs...contracted during his African journeys*". He collected the type. A mammal is also named after him

NB Vaillant later attempted to correct the spelling ("*foai*") to reflect gender, but since the name was published before the first edition of the Code (1906), the original spelling stands.

Fodor

African Tetra sp. *Micralestes fodori* H Matthes, 1965

Dr István Fodor was a Hungarian philologist, linguist, interpreter and Arabist who led an expedition to Wadi Haifa in the Sudan (1965) in search of the Magyarabs: reputedly a mixed Hungarian and Nubian tribe descended from a detachment of Hungarian soldiers sent to garrison Wadi Haifa when Sultan Selim I conquered the area (1517). He visited the Belgian Congo (Democratic Republic of Congo), where the holotype originated, and studied a number of Congolese languages. He wrote: *The problems in the classification of the African Languages* (1969). The etymology just refers to 'Dr Fodor' as the collector of the holotype, but we are confident this is the person referred to.

Foersch

African Rivuline genus *Foerschichthys* JJ Scheel & R Romand, 1981
Foersch's Betta *Betta foerschi* J Vierke, 1979
African Rivuline sp. *Nothobranchius foerschi* RH Wildekamp & HO Berkenkamp, 1979
Foersch's Fire Barb *Desmopuntius foerschi* M Kottelat, 1982

Dr Walter Foersch (1932–1993) of Munich, Germany, was a physician and a well-known aquarist. He is often quoted as a prime example of the 'citizen scientist', someone untrained in a discipline whose amateur study uncovers previously unknown scientific facts. He co-wrote: *Aquarium* (1971). He helped collect the barb holotype, and was the first to breed *Nothobranchius foerschi* in captivity.

FOIRN

Brazilian cichlid sp. *Dicrossus foirni* U Römer, IJ Hahn & PM Vergara, 2010

This is an acronym referring to Federacão das Organizações Indígenas do Rio Negro, officially abbreviated as FOIRN. The authors acknowledge FOIRN for giving repeated permission to travel on the tribal land of the village communities of different indigenous groups in the middle and upper Rio Negro, as well as for carrying out observations on wildlife, especially insects and fish, in these areas. The name is also intended to highlight the fact that the basic human rights of indigenous peoples, who depend on large functional ecosystems for all their resources, are still in question in most parts of Amazonia when business projects (such as logging, mining, or the building of hydroelectric dams) are planned in the neotropical rainforests.

Folgor

Folgor's Scorpionfish *Neomerinthe folgori* E Postel & C Roux, 1964

This is not a true eponym, in that it encompasses the captain and crew of the lobster fishing boat 'Folgor'.

Follett

Follett's Pipefish *Syngnathus folletti* ES Herald, 1942
Oxeye Oreo *Allocyttus folletti* GS Myers, 1960
Northern California Brook Lamprey *Entosphenus folletti* VD Vladykov & E Kott, 1976
Eelpout sp. *Lycenchelys folletti* ME Anderson, 1995

Wilbur Irving 'Bill' Follett (1901–1992) was a lawyer by profession but an ichthyologist by choice. He was Curator, Department of Ichthyology, California Academy of Sciences (1947–1969).

Fonchi

Armoured Catfish genus *Fonchiiichthys* IJH Isbrücker & JP Michels, 2001
Armoured Catfish genus *Fonchiiloricaria* MS Rodriguez, H Ortega & R Covain, 2011
Armoured Catfish sp. *Hypostomus fonchii* C Weber & JI Montoya-Burgos, 2002

Fonchi Ingrid Chang Matzunaga, Peruvian ichthyologist (1963–1999). **(See Chang)**

Fontane

Stechlin Cisco *Coregonus fontanae* LP Schulz & J Freyhof, 2003

Theodor Fontane (1819–1898) was a German poet. He published several books extolling his love for Brandenburg. He dedicated his last book to the people and landscape around Lake Stechlin, Germany, where this species is endemic.

Fontanes

Giant Prawn-goby *Amblyeleotris fontanesii* P Bleeker, 1853

Surgeon Major H R F Fontanes was a Dutch East Indian Army physician in Macassar, Celebes. He provided a collection of fishes from Bulucumba, Sulawesi, Indonesia, presumably including the type of the goby.

Fontoynont

Dusky Glass Perch *Ambassis fontoynonti* J Pellegrin, 1932

Dr Antoine Maurice Fontoynont (1869–1948) was a pathologist who was the President of the Malagasy Academy and Director of the Antananarivo School of Medicine. He lived in Madagascar (1898–1948), where the glassfish is endemic.

Forbes

North American Cavefish genus *Forbesichthys* DS Jordan, 1929
Galapagos Grunt *Orthopristis forbesi* DS Jordan & EC Starks, 1897
Barrens Darter *Etheostoma forbesi* LM Page & PA Ceas, 1992

Stephen Alfred Forbes (1844–1930) was an entomologist, self-taught naturalist, farmer and ecologist. He was the first Director of the Illinois State Laboratory of Natural History (1877) combining this with the rôle of State Entomologist (1882). He joined the faculty of the Illinois Industrial University (University of Illinois) (1885). His roles were combined into the Illinois Natural History Survey with him as the first Chief (1917–1930). He is regarded as "the founder of the science of ecology in the United States". He wrote: *The lake as a microcosm* (1887).

Ford

Armoured Catfish sp. *Pseudacanthicus fordii* A Günther, 1868

George Henry Ford (1809–1876) was a South African-born artist. Sir Andrew Smith employed him to make drawings and paintings of specimens he collected, and Ford was also employed by the Cape Town Museum (1825). Ford followed Smith when he returned to London (1837). He was employed as an artist at the British Museum, where he stayed for the rest of his life. Günther only says, *"I have named it after Mr. Ford, whose merits in herpetology are well known by his truly artistical drawings"*, without identifying him further, but logic dictates this to be the man Günther honoured. A reptile, also described by Günther, is named after him.

Fordice

Driftwood Catfish sp. *Auchenipterus fordicei* CH Eigenmann & RS Eigenmann, 1888

Morton William Fordice (1864–1939) was a farmer and politician and, according to the etymology, 'a student of American fishes.' He graduated with a bachelor's degree from Indiana University (1866). He wrote: *A review of the American species of* Stromateidea (1884).

Forest

Short-maned Sand-Eel *Phaenomonas foresti* J Cadenat & C Roux, 1964
Perchlet sp. *Plectranthias foresti* P Fourmanoir, 1977

Professor Jacques Forest (1920–2012) was a French carcinologist and field biologist. He was a soldier for a year in WW2 and after release from the army studied at the University of Lille. After graduating, he worked for the Office Scientifique et Technique des Pêches Maritimes for some years. He worked at MNHN (1949–1989), where he was involved with some scientific journals: *Bulletin du Muséum national d'Histoire naturelle* and *Crustaceana*. In retirement, he stayed involved in the latter (1989–2003). He was a member of several expeditions on the oceanic research vessel 'Calypso' including its cruise of the Cape Verde Islands (1959) that he led, during which the sand-eel holotype was collected.

Foresti

Characin sp. *Moenkhausis forestii* RC Benine, TC Mariguela & C de Oliveira, 2009
Armoured Catfish sp. *Microplecostomus forestii* GSC Silva, FF Roxo, LE Ochoa & C de Oliveira, 2016

Dr Fausto Foresti is a Brazilian geneticist and ichthyologist who was Professor of Cellular Biology, Universidade Estadual Paulista Júlio de Mesquita Filho, where he was awarded his doctorate and where he joined the faculty and taught (1969–2016), retiring as Professor Emeritus. He co-wrote: *Chromosome formulae of Neotropical freshwater fishes* (1988).

Forget

Lake Titicaca Pupfish sp. *Orestias forgeti* L Lauzanne, 1981

J M Forget was a friend of the author, as the original text says: "*Cette espèce est dèdièe à mon ami J.M. Forget.*" Unfortunately, no further biographical information is given.

Fornasini

Thornback Cowfish *Lactoria fornasini* GG Bianconi, 1846

Cavaliere Carlo Antonio Fornasini (1805–1865) was a collector who operated in the area around Inhambane, Mozambique (1839). He left Italy for unknown reasons (probably political) for Portugal, proceeding thence to Mozambique. He mainly collected spiders and botanical specimens, including the first example of the cycad that Bertolini named *Encephalartos ferox* (1851). A reptile and an amphibian are also named after him.

Forsskål

Elongate Tiger Fish *Hydrocynus forskahlii* G Cuvier, 1819
Labeo sp. *Labeo forskalii* E Rüppell, 1835
Red Sea Hardyhead Silverside *Atherinomorus forskalii* E Rüppell, 1838
Red Sea Goatfish *Parupeneus forsskali* P Fourmanoir & P Guéze, 1976

Dr Peter Forsskål (1732–1763) was a Swedish explorer, orientalist, naturalist and a student of Linnaeus. After a chequered academic career, he eventually graduated from Uppsala (1751). He studied oriental languages and philosophy at Göttingen (1753–1756), achieving his doctorate. He returned to Uppsala and published his dissertation on civil freedom for which he received a warning from the crown and the government censored it. However, he was appointed (1760) by King Frederick V of Denmark to join an expedition to Egypt (1761) and then on to Arabia (1762). He collected both botanical and zoological specimens but fell ill with malaria and died. A bird and a reptile are also named after him.

Forsten

Forsten's Parrotfish *Scarus forsteni* P Bleeker, 1861

Eltio Alegondas Forsten (1811–1843) collected in the East Indies (1838–1843). He was primarily a botanist and was interested in the pharmaceutical properties of plants, on which he wrote at least one scientific paper: *Dissertatio botanicpharmaceuticomedica inauguralis de cedrela febrifuga* (1836) published at Leiden. We assume that he was Dutch. Six birds and three reptiles are also named after him. Bleeker provides no etymology, but the holotype comes from Celebes where he collected.

Forster, JR

Forster's Hawkfish *Paracirrhites forsteri* JG Schneider, 1801
Bigeye Barracuda *Sphyraena forsteri* G Cuvier, 1829
Yellow-eye Mullet *Aldrichetta forsteri* A Valenciennes, 1836
Bastard Trumpeter *Latridopsis forsteri* FL Castelnau, 1872
Beaked Salmon sp. *Gonorynchus forsteri* JD Ogilby, 1911
[Alt. Beaked Sandfish]

Johann Reinhold Forster (1729–1798) was a German clergyman in Danzig (now Gdansk, Poland). He became a naturalist and accompanied James Cook aboard 'HMS Resolution' on his second voyage around the world (1772–1773) that extended further into Antarctic waters than anyone had before. He discovered five new species of penguin. However, he gained a reputation as a constant complainer and troublemaker. His complaints about Cook continued after his return and became public, destroying Forster's career in England. He went to Germany and became a Professor of History and Mineralogy. Unpleasant and troublesome to the end, Forster refused to relinquish his notes of the voyage. They were not found and published until c.50 years after his death. His son Johann George Adam was also on Cook's voyage as an artist. He observed the mullet in New Zealand (1769); while he identified it as *Mugil albula* (=*cephalus*), his notes provided sufficient descriptive data for Valenciennes to determine it was a separate and new species. A mammal and eleven birds are also named after him.

Forster, W

Australian Lungfish *Neoceratodus forsteri* G Krefft, 1870

William Forster (1818–1882) was an Australian farmer, journalist, magistrate, poet and politician (1856–1876) who was Premier of New South Wales (1859–1860) once described by a fellow politician as: "*disagreeable in opposition, insufferable as a supporter, and fatal as a colleague*." It was while he was Minister of Lands, New South Wales (1868–1870) that he presented two specimens of this 'great amphibian' to the Australian Museum. Science had not by then identified it as a fish! A town in NSW is also named after him.

Forsyth

Redfin Needlefish ssp. *Strongylura notata forsythia* CM Breder Jr, 1932

This is a toponym referring to Lake Forsyth, Andros Islands, Bahamas.

Fortuita

Cubango Kneria *Parakneria fortuita* MJ Penrith, 1973

This is not an eponym, but an interesting name nonetheless. It reflects the 'fortuitous circumstances' (on a Friday the 13th) that were involved in finding this species, when the collectors pitched camp on the wrong river and were forced to collect there because a burned-down bridge and a washed-out road prevented access to the correct river, the Okavango.

Foster

New Zealand Torrentfish *Cheimarrichthys fosteri* J Haast, 1874

Mrs J C Foster was a New Zealand naturalist. The etymology says the fish is named after: "… *Mrs. J. C. Foster, of Sumner… …who made a collection of fishes… …in the Otira, where that alpine torrent leaves its picturesque gorge*". The binomial should really take the feminine form, *fosterae*.

Fourcroy

Guyanan Croaker *Pachypops fourcroi* BGE Lacepède, 1802

Antoine François, Comte de Fourcroy (1755–1809) was a noted French chemist who qualified as a physician in Paris (1780). He taught chemistry (1783–1787) at the Veterinary School of Alfort. He was a lecturer in chemistry at the college of the Jardin du Roi, (as was Lacepède) where his lectures attained great popularity. His *Entomologia Parisiensis, sive, Catalogus insectorum quae in agro Parisiensi reperiuntur* (1785), co-written with Étienne Louis Geoffroy, was a major contribution to systematic entomology. He and Lacepède were colleagues as members of the national Assembly. Cape Fourcroy on Bathurst Island, Northern Territory, Australia is also named after him.

Fourmanoir

Goby sp. *Tomiyamichthys fourmanoiri* JLB Smith, 1956
Yellow-spotted Tilefish *Hoplolatilus fourmanoiri* JLB Smith, 1964
Black-striped Combtooth Blenny *Ecsenius fourmanoiri* VG Springer, 1972
Doublespot Perchlet *Plectranthias fourmanoiri* JE Randall, 1980
Dentex sp. *Dentex fourmanoiri* M Akazaki & B Séret, 1999

Pierre Fourmanoir (1924–2007) was a French ichthyologist who mainly worked in New Caledonia for the Paris Museum's ichthyology laboratory. He began working for ORMSTOM (1948) being assigned (1950–1963) to Madagascar. He took part in two missions to Vietnam (1963–1964) and Guyana (1966–1967), after which he was assigned to the ORSTOM Centre of Nouméa (1969–1981) describing many reef fish. He described 61 new fish species including two sharks and wrote around 200 papers.

Fowler, EL

Characin sp. *Brycon fowleri* G Dahl, 1955

Rev Ernest L Fowler (1907–1966) was an American Baptist missionary in Colombia (1934–1966). He was murdered by a band of thieves posing as policemen. Dahl, who was his friend, wrote in the etymology of Fowler's: "…*generous help and encouragement.*"

Fowler, HW

Cardinalfish genus *Fowleria* DS Jordan & BW Evermann, 1903

Characin genus *Fowlerina* CH Eigenmann, 1907
Frogfish genus *Fowlerichthys* CD Barbour, 1941
Fowler's Danio *Devario regina* HW Fowler, 1934
Fowler's Snake-eel *Ophichthus fowleri* DS Jordan & BW Evermann, 1903
Halftooth Characin sp. *Bivibranchia fowleri* F Steindachner, 1908
Armoured Catfish sp. *Harttia fowleri* J Pellegrin, 1908
Fowler's Bumblebee Catfish *Cephalosilurus fowleri* JD Haseman, 1911
Philippines Dottyback *Pseudochromis fowleri* AW Herre 1934
Pygmy Scorpionfish *Sebastapistes fowleri* V Pietschmann, 1934
Jellynose Fish sp. *Ijimaia fowleri* L Howell Rivero, 1935
Fowler's Rockskipper *Litobranchus fowleri* AW Herre, 1936
Mekong Cyprinid sp. *Lobocheilos fowleri* J Pellegrin & P Chevey, 1936
[Junior Synonym of *Lobocheilos rhabdoura*]
Shellskin Alfonsino sp. *Ostracoberyx fowleri* K Matsubara, 1939
Loach sp. *Schistura fowleriana* HM Smith, 1945
Fowler's Cory *Corydoras fowleri* JE Böhlke, 1950
Fowler's Surgeonfish *Acanthurus fowleri* LF de Beaufort, 1951
Mud Eel sp. *Panturichthys fowleri* A Ben-Tuvia, 1953
Fowler's Pearlfish *Onuxodon fowleri* JLB Smith, 1955
Fowler's Large-toothed Conger *Bathyuroconger fowleri* DG Smith, H-C Ho & F Tashiro, 2018

Henry Weed Fowler (1878–1965) was an American herpetologist, ornithologist, ichthyologist and artist. He was self-taught, although he briefly studied at Stamford under DS Jordan (q.v.). He joined the Academy of Natural Sciences in Philadelphia and worked as an assistant (1903–1922), Associate Curator of Vertebrates (1922–1934) and Curator of Fish and Reptiles (1934–1940) and finally Curator of Fish (1940–1965). He was a co-founder of the American Society of Ichthyologists and Herpetologists (1927) and its first treasurer. He travelled throughout the USA and also visited China. He was a prolific ichthyologist and published (1899 >) very widely across many taxa (666 profusely illustrated publications). Steindachner does not identify the patronym of the *Bivibranchia* species, but it seems likely that H.W. Fowler is intended. Similarly there is no explicit etymology for the conger, but this clearly honours H.W. Fowler as he reported the eel as *Silvesterina (=Bathyuroconger) parvibranchialis* (1934). Two amphibians are also named after him.

Fowler, SL

Borneo River Shark *Glyphis fowlerae* LJV Compagno, WT White & RD Cavanagh, 2010

Dr Sarah Louise Fowler (b.1958) joined Naturebureau (1989) as Managing Director. She is still (2014) associated with it as a non-executive director and consultant. She has spent her lifetime working for the conservation of sharks. Her bachelor's degree was awarded by the University College of North Wales, Bangor (1979) and her master's by University College, London (1981). She is a founding trustee of the Shark Trust. She also led the study of the family found off Sabah (1996), leading to the discovery of three new species for which she is honoured in the name. She co-wrote: *Collins Field Guide to Sharks of the World* (2005)

Frances

Golden Skiffia *Skiffia francesae* DI Kingston, 1978

Frances Voorhees Hubbs 'Fran' Miller (1919–1987) was an American ichthyologist – which seems hardly surprising as both her parents and her younger brother were ichthyologists, as was her future husband. Her zoology degree was awarded by the University of Michigan (1940). She worked at the lab of the University Hospital (1940–1944) and married Robert Rush Miller (q.v.) She made numerous collecting trips with her parents and husband to Japan, Honduras, Nicaragua and Mexico. She became a Research Associate in Ichthyology (1980) at the Michigan Museum of Zoology and she assisted her husband throughout his career. Together they spent many years studying the freshwater fishes of Mexico, culminating in a book. According to the etymology she was honoured in recognition of her help in furthering our understanding of Mexican fishes. She died of cancer a year after diagnosis. (Also see **Hubbs, Miler, RR**)

Francesca

Neotropical Rivuline sp. *Papiliolebias francescae* S Valdesalici & RD Brousseau, 2014

Francesca Fontana is the wife of the senior author.

Franchetti

Ethiopian Cichlid sp. *Danakilia franchettii* D Vinciguerra, 1931

Baron Raimondo Franchetti (1889–1935) was a wealthy Italian explorer and adventurer. He organised the expedition that collected the cichlid holotype at his own expense. He was killed in a plane crash in the Egyptian desert.

Francis

Francis' Goatfish *Upeneus francisi* JE Randall & P Guézé, 1992
Blacktip Morwong *Cheilodactylus francisi* CP Burridge, 2004

Dr Malcolm Philip Francis (b.1954) is a New Zealand fisheries scientist and marine ecologist who has spent over thirty-nine years studying the population biology of coastal and pelagic fishes. The University of Auckland awarded his BSc (1974), MSc (1976) and PhD (1993). Additionally, he has a resource management MSc awarded by Canterbury University. He was Biology Lecturer, Atenisi Institute University, Nuku'alofa, Tonga (1975–1976) and Biology Tutor, Mander Portman and Woodward Tutorial College, London (1980). He became Scientist and Group Leader, Fisheries Research Centre, Ministry of Agriculture and Fisheries, Wellington (1981–1995) before taking the post of Project Director at New Zealand's National Institute of Water and Atmospheric Research (1995–2000). He is now Principal Scientist there (2000–present), where he specialises in the biology, behaviour and bycatch of sharks. He has published over 100 scientific papers and 200 research reports and also written or edited three books and his publications include: *Checklist of the coastal fishes of Lord Howe, Norfolk, and Kermadec Islands, southwest Pacific Ocean* (1993, on-line 2019) and *Coastal fishes of New Zealand: identification, biology, behaviour* (2012). He is a keen scuba diver and underwater photographer and has made multiple diving expeditions to New Zealand's Three Kings Islands and Kermadec Islands and subtropical Norfolk and Lord

Howe islands to study the fishes and other marine life. He was honoured "*…in recognition of his research on the fishes of New Zealand, Lord Howe Island, and the Kermadec Islands.*" (Also see **Malcolm (Francis)**)

Francis

Horn Shark *Heterodontus francisci* CF Girard, 1855

The original description has no etymology but the specific name probably derives from the city of San Francisco (although the holotype was collected from Monterey Bay rather than San Francisco Bay).

Francisco

Armoured Catfish sp. *Hypostomus francisci* CF Lütken, 1874

This is a toponym; the holotype was collected from the Rio São Francisco, Brazil.

Franco Rocha

Armoured Catfish sp. *Hisonotus francirochai* R Ihering, 1928

Francisco Franco da Rocha (1864–1933) was a pioneering Brazilian psychiatrist. He was founder of the Hospital Psiquiátrico do Juqueri (Região Metropolitana de São Paulo).

Franke

Ucayali Tetra *Hyphessobrycon frankei* A Zarske & J Géry, 1997

Hanns-Joachim Franke (1925–1995) was an aquarist, who, along with ichthyologist-aquarist Patrick de Rham (q.v.), first collected this species in Peru (1979). He wrote: *Handbuch der Welskunde* (1985).

Franoux

Freshwater Goby sp. *Sicyopterus franouxi* J Pellegrin, 1935

Roger Franoux (d.1947) was a friend and collaborator of René Catala (q.v.); together they collected the holotype in Madagascar. He was murdered during the Malagasy uprising.

Frans Seda

Frans' Cardinalfish *Ostorhinchus franssedai* GR Allen, RH Kuiter & JE Randall, 1994

Franciscus Xavierus 'Frans' Seda (1926–2009) was an Indonesian politician who served as a minister (1964–1973) in the governments of President Sukarno and President Suharto. He was also owner of Sao Wisata Resort, Flores, Indonesia. He encouraged the study of Maumere Bay fishes and provided logistical support during the authors' visits.

Frans Vermeulen

Neotropical Rivuline sp. *Anablepsoides fransvermeuleni* S Valdesalici, 2015

Neotropical Rivuline sp. *Rachovia fransvermeuleni* HO Berkenkamp, 2020

Frans B M Vermeulen of Aruba, Dutch West Indies, is a friend of the first author and collected the *Anablepsoides* holotype. He is an amateur ichthyologist who calls himself a hobbyist not a scientist. However, he has specialised in killifish for half a century. He has authored many papers such as: *Rivulus tomasi (Teleostei: Cyprinodontiformes: Rivulidae), a new killifish from Tobogán de la Selva, middle Orinoco river drainage in the Amazonas Territory, southwest Venezuela* (2013) and a series of books such as his four-volume series *New World Killifish.*

Franz

Snailfish sp. *Liparis franzi* T Abe, 1950
[Syn. *Liparis punctulatus franzi*]

Dr Victor Franz (1883–1950) was a German anatomist, malacologist and zoologist who became a professor at Jena. He published his first scientific paper aged 19. He first described this species but thought it to be the very similar *Liparis liparis,* found in European waters, in *Die japanischen Knochenfische der Sammlungen Haberer und Doflein* (1910).

Franz (Uiblein)

Franz's Cusk *Neobythites franzi* JG Nielsen, 2002

Dr Franz Uiblein is a Norwegian/South African ichthyologist and marine biologist, whose doctorate was awarded by the University of Vienna (1989). He was Curator of Fishes, Research Institute Senckenberg, Frankfurt (1992–1993) and has lectured at the University of Salzburg since 1997 and as Guest Professor (2002–2003). He is an Honorary Research Associate, South African Institute of Aquatic Biodiversity, Grahamstown, South Africa since 2010 and at Vietnam National Museum of Nature, Hanoi, Vietnam since 2015. He is presently (since 2003) Principal Scientist, IMR, Bergen, Norway. He co-wrote: *Review of the Indo-West Pacific ophidiid genera Sirembo* and *Spottobrotula* (*Ophidiiformes, Ophidiidae), with description of three new species* (2015.) (Also see **Gloria (Eschevari)**)

Franz Werner

Goby Killi *Aphyosemion franzwerneri* JJ Scheel, 1971

Franz Werner was an American aquarist and killifish keeper from Detroit. He was described in the etymology as the 'late' Franz Werner, 'modern amateur student of rivulins' and an 'eminent American killifish fancier'.

Fraser, AGL

Fraser's Danio *Devario fraseri* SL Hora, 1935
Dharna Barb *Puntius fraseri* SL Hora & KS MIsra, 1938

Dr Albert Glen Leslie Fraser (b.1887) was in the service of the Medical Department of India (1915–1942). He was an amateur herpetologist and zoologist. He wrote the three-part: *The Snakes of Deolali* (1936–1937) and *Fish of Poona* (1942). He collected the holotype of the Dhana Barb as part of the collection of 4,463 specimens he sent to the Zoological Survey of India at their request, following the snakes he had previously sent.

Fraser, C

Marine Hatchetfish sp. *Polyipnus fraseri* HW Fowler, 1934

Dr Charles McLean Fraser (1872–1946) was a marine biologist who is mainly remembered for his work on hydroids. His doctorate was awarded by the University of Iowa (1911). He taught at a high school in Nelson, British Columbia (1903) and, having previously worked there (1908), he became the second curator of the Pacific Biological Station, Departure Bay (1912–1924). He was Head of Department of Zoology, University of British Columbia, Vancouver (1920–1940). He took part in the Allan Hancock Pacific Expeditions (1930s) which collected from Southern California to Peru an also in the Caribbean. He wrote: *Hydroids of the Pacific coast of Canada and the United States* (1937).

Fraser, H

Fraser's Lanternfish *Gymnoscopelus fraseri* A Fraser-Brunner, 1931

Captain Hugh Fraser collected the holotype with a net of his own construction. We have not been able to further identify this person.

Fraser, M & V

Goatfish sp. *Parupeneus fraserorum* JE Randall & DR King, 2009

Michael Dennis Fraser (b. 1955) and Valda Jean Fraser (b. 1957) are a married (1978) couple. He obtained a BA LLB degree (1978) from UKZN and was admitted as an attorney (1984). She qualified with a MA in isiZulu from UKZN (1996) and worked as a lecturer in isiZulu at UKZN (until 2006). She has discovered a number of new sea slugs, and is co-author (with Dennis King) of *The Reef Guide: fishes, corals, nudibranchs & other invertebrates East & South Coasts of Southern Africa* (2014) and *More Reef Fishes & Nudibranchs: East and South Coast of Southern Africa* (2002). She is currently working on publication of a book on the nudibranchs of the east coast of southern Africa with a team of citizen scientists under the guidance of Dr Terence Gosliner. Mike and Valda have retired to Pumula near Port Shepstone on the KZN south coast. They qualified as scuba divers (1988) and rebreather divers (2010). Both Mike and Valda are avid underwater photographers, citizen scientists and ocean explorers. The etymology says: "*We are pleased to name this species collectively for Michael D. Fraser and Valda J. Fraser, he for collecting the type specimens, and she for her underwater photograph of Plate 1D, our first awareness of the species.*" At least one sea slug is named after her – *Bornella valdae* (2009) – and a hermit crab is named after both of them: *Pagurus fraserorum* (2018).

Fraser, TH

Blenny sp. *Meiacanthus fraseri* WF Smith-Vaniz, 1976
Blenny sp. *Dodekablennos fraseri* VG Springer & AE Spreitzer, 1978
Cardinalfish sp. *Siphamia fraseri* O Gon & GR Allen, 2012

Dr Thomas Henry Fraser is an ichthyologist specialising in cardinalfish, who works at Florida Museum of Natural History, University of Florida, Gainesville and at Mote Marine Laboratory, Sarasota, Florida. He wrote: *Comparative osteology of the shallow water cardinal fishes [Perciformes: Apogonidae] with reference to the systematics and evolution of the family* (1972).

Fraser-Brunner

Greeneye sp. *Parasudis fraserbrunneri* M Poll, 1953
Spikefish sp. *Mephisto fraserbrunneri* JC Tyler, 1966

Alec Frederick Fraser-Brunner (1906–1986) was a British ichthyologist at the BMNH. His career also included being curator of the Van Kleef Aquarium in Singapore and the (now closed) aquarium at Edinburgh Zoo. He was also an artist, who 'made models and artwork for new Fish Gallery' (BMNH 1931). He is further noted for having designed Singapore's national symbol – the Merlion – for the Singapore Tourism Board.

Frechkop

Frechkop's Sole *Microchirus frechkopi* P Chabanaud, 1952

Dr Serge Isaacovich Frechkop (1894–1987) was a Russian-born Belgian mammologist. He began university in Moscow (1913) but this was interrupted by WW1, and he was eventually awarded his doctorate (1924). He then left Russia and settled in Belgium. He worked part-time in industry but also at the University of Brussels zoological laboratory. He then joined (1930) the Royal Belgian Institute of Natural Science.

Fred Cochu

Cochu's Blue Tetra *Boehlkea fredcochui* J Géry, 1966

Ferdinand 'Fred' Cochu (See **Cochu**)

Freddy

Needletooth Cusk *Epetriodus freddyi* DM Cohen & JG Nielsen, 1978

Norman Bertram 'Freddy' Marshall (1915–1996) was an ichthyologist and marine biologist at the BMNH (1947–1972). He became Professor of Zoology, Queen Mary College, University of London (1972–1977), retiring as Professor Emeritus. Cambridge awarded his bachelor's degree (1937). After graduating he worked on plankton research at the Department of Oceanography, University of Hull (1937–1941), after which he was in army operations research. He was based in Antarctica (1944–1946) as part of Operation Tabarin, to establish a permanent base at Hope Bay. He wrote: *Aspects of Deep Sea Biology* (1954). Marshall Peak in Antarctica is named after him.

Frederick

Sorong Rainbowfish *Melanotaenia fredericki* HW Fowler, 1939
[Syn. *Charisella fredericki*]

Dr Frederick E Crockett (1907–1978) was a physician and explorer. He took part as a dog driver in the Antarctic Expedition (1928–1930) where Mount Crocket was named after him. He took part in a film of the expedition *With Byrd at the South Pole* (1930) which won an Oscar for cinematography. With his wife, the anthropologist Charis Crockett née Denison, he organised the Denison-Crockett South Pacific Expedition (1937–1938) aboard their schooner 'Chiva' with S Dillon Ripley (q.v.) as the zoologist. There they lived in a house on stilts near Sorong, New Guinea. He wrote about that experience in *The House in the Rainforest* (1942).

Fredrod

African Rivuline sp. *Scriptaphyosemion fredrodi* J-P Vandersmisson, VMF Etzel & HO Berkenkamp, 1980

Fred Wright and Rod Roberts are British aquarists. They are both members of the British Killifish Association and were the original collectors of the type and of many other fish in West Africa.

Freeborn

Snailfish sp. *Paraliparis freeborni* DL Stein, 2012

MIchelle Denise Freeborn (b.1969) (m. Moore) is a London born artist and scientific illustrator. She was awarded her BTEC by Richmond upon Thames College, then spent a year studying fine art at Bath College of Education. Both her father and grandfather worked in the film industry creating special effects and make-up and her first job (aged 17) was working in a make-up effects department. She spent the next fifteen years working on films in the prosthetics/make-up department and costume fabrication. She and her husband moved to New Zealand where she obtained work (2001) at the Te Papa Museum of New Zealand as she has a passion for caring for and investigating the natural environment. She produced scientific illustrations for several New Zealand institutions including Te Papa, The National institute of Water and Atmospheric research (NIWA), Department of Conservation (DOC) and the Parliamentary Commissioner for the Environment, also working with numerous visiting scientists from overseas institutions like the Smithsonian and Oregon State University. For Te Papa she Illustrated over 1500 species of fishes primarily for the book *The Fishes of New Zealand* (2015), which received the 2016 Whitley Medal for outstanding publication in Australasian zoology. She says that working with David Stein on the *Liparidae* was one of her most challenging and ultimately rewarding tasks, as most of what they worked on was so new to science. She now concentrates on oil painting and her 'two lovely daughters.' Though she much appreciated having *P. freeborni* named after her, Stein should really have used the feminine form *freebornae*.

Freeman, BJ

Freeman's Tube-snouted Ghost Knifefish *Sternarchorhynchus freemani* CDCM de Santana & RP Vari, 2010

Etowah Bridled Darter *Percina freemanorum* TJ Near & GR Dinkins, 2021

Dr Byron 'Bud' J Freeman (b.1950) is a zoologist whose bachelor's degree was awarded by the University of Georgia (1972). Samford University, Birmingham, Alabama awarded his master's (1974) and the University of Georgia, Athens his doctorate (1980). He joined the staff of the Department of Zoology, University of Georgia (1974) and has worked there in a number of capacities. Since 2004 he has been Director and Curator of Zoology, Georgia Museum of Natural History. He co-wrote: *Investigating hydrologic alteration as a mechanism of fish assemblage shifts in urbanizing streams* (2005). He was thanked for 'invaluable' assistance to the senior author. The darter is named after both Byron Freeman and his wife, Mary Freeman (below).

Freeman, HW

Poacher genus *Freemanichthys* T Kanayama, 1991

Slender Wenchman *Pristipomoides freemani* WD Anderson, 1966

Dr Harry Wyman Freeman (1923–2012) was Professor of Biology at College of Charleston, his alma mater. During WW2 he served in the US Navy. The University of South Carolina awarded his master's and Stanford his PhD in biology specialising in ichthyology. He taught at the University of South Carolina for nine years (1951–1960) before joining the staff at College of Charleston for 29 years (1960–1989) including 18 as the biology department chair. After retirement he continued to research marine biology.

Freeman, M

Etowah Bridled Darter *Percina freemanorum* TJ Near & GR Dinkins, 2021

Dr Mary C Freeman (b.1958) is an American ecologist, whose degrees were awarded by the University of Georgia: BSc in biology (1979), MSc in entomology (1982) and PhD in forest resources (1990). She works (1992–now) as a research ecologist for the US Department of Interior, addressing the management, ecology and conservation of river systems and biota. The darter is named after both Mary and Byron Freeman (above).

Freiberg

Freiberg's Peacock (cichlid) *Aulonocara jacobfreibergi* DS Johnson, 1974

[Alt. Malawi Butterfly, Eureka Red Peacock, Fairy Cichlid]

Freiberg's Mbuna (cichlid) *Labidochromis freibergi* DS Johnson, 1974

Jacob 'Jack' Freiberg was an American importer of tropical fish, based in New Jersey. He brought the *Labidochromis* to Johnson's attention and first made it available to aquarium hobbyists in the USA. (See **Jacob Freiberg**)

Freire

Armoured Catfish sp. *Hypostomus freirei* I de S Penido, TC Pessali & CH Zawadzki, 2021

Paulo Reglus Neves Freire (1921–1997) was a Brazilian educator and philosopher, He is best known for his influential work *Pedagogy of the Oppressed* (1968), which is generally considered one of the foundational texts of the critical pedagogy movement. He was honoured for his pioneering role in contemporary humanistic pedagogy, promoting "substantial advances in the theory of literacy and methods of teaching," which in turn has "generated great influence in several areas of knowledge such as philosophy, theology, anthropology, social work and sociology."

Frembly

Blue-striped Butterflyfish *Chaetodon fremblii* ET Bennett, 1828

Lieutenant John Frembly RN was a marine surveyor and geologist who was Assistant Maritime Surveyor of HM Surveying Brig 'Investigator' (1823–1824). He was Mate aboard HMS 'Blossom' which, among other things, surveyed Belize harbour (1929). He was also a corresponding member of the Zoological Society and wrote: *A description of several new species of Chitones, found on the coast of Chile, in 1825: with a few remarks on the method of taking and preserving them* (1828). The butterflyfish holotype was part of a collection of fish "*presented to the Zoological Society by John Frembly, Esq., R.N., who accompanied the late expedition to the Pacific Ocean, under the command of Lord Byron.*" We assume this to be Byron's mission on HM Frigate 'Blonde'.

Freminville

Bullnose Eagle Ray *Myliobatis freminvillei* CA Lesueur, 1824
Galápagos Sea Chub *Girella freminvillii* A Valenciennes, 1846

Christophe/Chrétien-Paulin de la Poix Chevalier de Fréminville (1787–1848) was a French naval officer and naturalist. He was on board 'La Syrène', which attempted to discover the Northwest Passage (1806). He was an expert of the history and archaeology of the late Middle Ages and on the history of Brittany, and of the Templars in particular. Toward the end of his life an old episode affected him and he became deranged. He was in command of the French frigate 'La Néréide' in the West Indies (1822). He fell from some rocks and was lucky enough to be rescued from drowning and nursed back to health by a beautiful local girl, Caroline. He had to sail to Martinique but when he returned, he found she had drowned herself, thinking that he had deserted her. He took away some of her dresses as keepsakes (1842) and spent the last six years of his life wearing her old clothes. A reptile is named after him.

French

Foxfish *Bodianus frenchii* CB Klunzinger, 1879
[Alt. Fox Wrasse]

This is a puzzle, as there is no obvious etymology in the original text. Klunzinger clearly meant this to be eponymous as he used a capital on the binomial = *Frenchii*, as was usual at that time when naming animals after persons. The only clue is that Klunzinger named several fish species after Baron Dr Ferdinand Jacob Heinrich von Müller (1825–1896), and mentions 'Herrn French, Assistent des Freiherrn Dr. v. Müller'.

Frerichs

Thickbody Skate *Amblyraja frerichsi* G Krefft, 1968

Thomas Frerichs was captain of the research vessel 'Walther Herwig' from which the holotype was collected. He was honoured for: "*...his keen interest in deepsea catches, which assured us many precious discoveries*".

Frey

Bearded Silverside ssp. *Atherion elymus freyi* LP Schultz, 1953

Dr David Grover Frey (1915–1992) was a limnologist who was Professor of Zoology at the Department of Biology, Indiana University (1951–1986). The University of Wisconsin awarded his bachelor's degree (1936), his master's (1938) and PhD (1940). He joined (1940) the US Fish and Wildlife Service and worked on salmon in the Columbia River in Washington and oysters in Chesapeake Bay. Much of his published work reflects his focus on water fleas, for example: *Comparative Morphology and Biology of Three Species of Eurycercus (Chydoridae, Cladocera) with a Description of Eurycercus macrocanthus sp. nov.* (1973) and *A New Species of the Chydorus sphaericus Group (Cladocera, Chydoridae) from Western Montana* (1985). He accumulated a laboratory of over 10,000 specimens, now at the Smithsonian. After naval service in WW2 he was hired as Associate Professor (1950) then Professor of Zoology (1955–1986). He was a specialist in limnology (aquatic ecology) and already an authority on the *Cladocera*. He was active in several national and international limnological organisations, which included serving as President of the American Society of Limnology and Vice President of the International Association of Limnology. He travelled extensively to attend conferences and to conduct research on lakes around the world, visiting forty-four countries across six continents. He collected the holotype.

Freycinet

Indian Speckled Carpetshark *Hemiscyllium freycineti* JRC Quoy & JP Gaimard, 1824
Six-spined Leatherjacket *Meuschenia freycineti* JRC Quoy & JP Gaimard, 1824
Pug-headed Mudskipper *Periophthalmodon freycineti* JRC Quoy & JP Gaimard, 1824

Captain Louis Claude de Saules de Freycinet (1779–1841) was a French navigator who was involved in mapping the Western Australian coast (1803). He later explored in the Pacific (1817–1820) on 'L'Uranie', making extensive collections of natural history specimens. After *L'Uranie* was wrecked during the return voyage, he bought 'La Physicienne'. Freycinet was admitted into the French Academy of Sciences (1825). An amphibian and two birds are named after him.

Freyhof

Riffle Minnow sp. *Alburnoides freyhofi* D Turan, C Kaya, E Bayçelebi, Y Bektaş & FG Ekmekçi, 2017

Dr Jörg Arthur Freyhof (b.1964) is a German biologist at the Leibniz-Institute of Freshwater Ecology and Inland Fisheries, Berlin. The University of Bonn awarded his zoology degree (1993) and his PhD (1997) for his thesis: *Habitat structures and fish communities in the Rivers Sieg, Germany*. He was a Research Assistant at the Zoological Research Institute and Museum Alexander Koenig, Bonn, and Collection management assistant and head of research projects (1993–2000). He then joined the Leibniz-Institute as a research scientist (2000–2014). He became (2014–2016) Executive Director of the Group of Earth Observations - Biodiversity Observation Network at the German Center for Integrative Biodiversity Research before returning to the Leibniz-Institute (2016–2019). He was honoured "*…for his contribution to the knowledge of the ichthyofauna of the Middle East*."

Fricke

Viviparous Brotula sp. *Lapitaichthys frickei* W Schwarzhans & PR Møller, 2007
Tetra sp. *Hyphessobrycon frickei* EC Guimarães, PS de Brito, PHN Bragança, AM Katz & FP Ottoni, 2020

Dr Ronald Fricke (b.1959) is a fish taxonomist. He was at the Technische Universität Braunschweig (1980–1986) and in the same period acted as Curator of Ichthyology and Herpetology at the Staatliches Naturhistorisches Museum, Braunschweig, with an interim period (1983–1984) at King's College, London. He was at the University of Hamburg

(1986–1987) and at Freiburg (1988), where his doctorate was awarded (1989). He has been Curator of Ichthyology, Staatliches Museum für Naturkunde, Stuttgart, Germany since 1988 and since 2000 has been an independent expert for the EU Commission (diversity, distribution and conservation of NATURA 2000 species in the Atlantic, Continental and Mediterranean regions). He co-wrote: *Raja pita, a new species of skate from the Arabian/ Persian Gulf (Elasmobranchii: Rajiformes)* (1995) and: *The coastal fishes of Madeira Island - new records and an annotated checklist* (2008). (Also see **Claudia (Fricke)** and **Clea**)

Friderici

Threespot Leporinus *Leporinus friderici* ME Bloch, 1794

Jurriaan François de Friderici (1751–1812) was Governor General of Suriname (1792–1801); initially under Dutch rule, but he was allowed to retain his position when the British conquered Suriname (1799). After the restitution of the colony to The Netherlands, Friderici was quickly replaced. He sent fish specimens to Bloch, including this species.

Fridman

Orchid Dottyback *Pseudochromis fridmani* W Klausewitz, 1968

David Fridman was a reef biologist and one of the founders, and first Curator of Coral World (1975) (now Underwater Observatory Marine Park) in Eilat, Israel. He wrote *Guide, Coral World Eilat* (1980) and co-wrote *Wonders of the Red Sea* (1990). He collected the type.

Friedrichsthal

Yellowjacket Cichlid *Parachromis friedrichsthalii* J Heckel, 1840

Baron Emanuel von Friedrichsthal (1809–1842) was an Austrian traveller, botanist, amateur archaeologist and early photographer. He was educated in Vienna and at the Theresian Military Academy, then entered government service but leaving after a short time to begin his scientific travels. He travelled throughout the Balkans (1830s) documenting his findings in two books which included many botanical descriptions: *Reise in die südlichen Theile von Griechenland* (*Journey to the Southern Parts of Greece*, 1838) and *Serbiens Neuzeit in geschichtlicher, politischer, topographischer, statistischer und naturhistorischer Hinsicht* (*Modern Serbia in Historical, Political, Topographical, Statistical, and Natural-Historical Respects*, 1840). He was posted to Central America (1840) as First Secretary to the Austrian Legation to Mexico. It is thought he fell ill with Malaria and returned to Europe (1841) but died the following year. In Mexico he spent much time studying Mayan ruins in Yucatan and Chiapas including Chichen Itza and was the first person to take daguerreotypes of the Mayan ruins. The text mentions 'Herr Baron von Friedrichsthal' and his interest in natural history so we are sure this is the intended recipient, especially as he travelled in Central America where this cichlid is found and sent many natural history specimens from Central America to the Vienna Museum.

Friel

Luapula River Mormyrid *Petrocephalus frieli* S Lavoué, 2012

African Loach Catfish sp. *Amphilius frieli* AW Thomson & LM Page, 2015

Dr John Patrick Friel is an American biologist and ichthyologist who is Curator of Fishes, Amphibians and Reptiles, Cornell University Museum of Vertebrates. His doctorate was awarded by Duke University, Durham, North Carolina (1995), after which he was a post-doctoral researcher at the Department of Biological Sciences, Florida State University, Tallahassee (1995–1998). He co-wrote: *Three new species of African suckermouth catfishes, genus* Chiloglanis *(Siluriformes: Mochokidae), from the lower Malagarasi and Luiche rivers of western Tanzania* (2011). He helped to collect the mormyrid holotype. The catfish was named after him "*...in recognition of his excellent contributions to the study of African fishes.*"

Fries

Black Sea Roach *Rutilus frisii* A von Nordmann, 1840
Fries's Goby *Lesueurigobius friesii* AW Malm, 1874

Dr Bengt Fredrik Fries (1799–1839) was a Swedish biologist, entomologist and ichthyologist. He began studying law at the University of Lund, but switched to natural history, graduating with a master's degree (1823) and then medicine (1827). He was Associate Professor of Natural History (1824), then of Anatomy (1828). He became Professor and Curator at Naturhistoriska Museet, Stockholm (1831). He was co-author of: *Skandinaviens Fiskar, malade efter lefvande exemplar och ritade l' a sten* (Series of volumes 1836–1857, continued after his death by Ekström and Sundevall), which work Nordmann cites in his description of the roach species (despite omitting an *e* from Fries's name in the binomial). Fries reported the (now) eponymous goby as *Gobius gracilis* (=*Pomatoschistus minutus*) (1838).

Fritel

Alima Mormyrid *Marcusenius friteli* J Pellegrin, 1904

Paul-Honoré Fritel (1867–1927) was a paleo-biologist at MNHN, Paris, a scientific illustrator and a colleague of the describer. He wrote a number of papers including: *Paléobotanique (plantes fossiles)* (1903), *Paléobotanique (animaux fossiles)* (1903) and *Histoire naturelle de la France Partie 23, Géologie* (1906) as well as illustrating others.

Fritsch

North African Barbel sp. *Carasobarbus fritschii* A Günther, 1874

Dr Karl Wilhelm Georg Freiherrn von Fritsch (1838–1906) was a German geologist, palaeontologist, biologist and natural history collector. He studied natural sciences at the University of Göttingen where he graduated (1862) and was awarded his doctorate by the University of Zurich (1863), whilst working as a lecturer. He was at the Senckenberg Nature Research Society, Frankfurt-am-Main (1867–1873) as a mineralogist and geologist and was appointed Professor of Geology at the University of Halle (1873) and Full Professor (1874) and occupied himself with paleo-botanical research. He visited Madeira and the Canary Islands (1862) and wrote: *Reisebilder von den Canarischen inseln* (1867). He collected the holotype of this species during a visit to Morocco (1872).

Fritz

Guadaloupe Hagfish *Eptatretus fritzi* RL Wisner & CB McMillan, 1990

Frithjof 'Fritz' Frockney Ohre (1910–2003) was a Californian farmer and amateur ichthyologist. He took part in a number of Scripps Institute of Oceanography expeditions (1967–1973) on which he made a photographic record of the voyages and specimens taken. He was honoured as: "...*friend, willing, eager, and industrious volunteer*" who helped the authors collect specimens.

Froehlich

Corydoras sp. *Corydoras froehlichi* LFC Tencatt, MR Britto & CS Pavanelli, 2016

Armoured Catfish sp. *Hypostomus froehlichi* CH Zawadzki, G Nardi & LFC Tencatt, 2021

Dr Otávio Froehlich (1958–2015) was a Brazilian zoologist whose bachelor's degree was awarded by the University of São Paulo (1984) whist the Universidade Federal de Mato Grosso do Sul awarded both his master's (2003) and doctorate (2010) and where he was an Adjunct Professor. It is clear that this species was named after him by his friends, as a memorial to him shortly after his death.

Froggatt

Froggatt's Catfish *Cinetodus froggatti* EP Ramsay & JD Ogilby, 1886

[Alt. Small-mouthed Salmon Catfish]

Walter Wilson Froggatt (1858–1937) was an Australian entomologist He was entomologist and assistant zoologist on the Royal Geographical Society of New South Wales' 'Bonito' expedition to New Guinea (1885). He was employed by William Macleay as a collector and at the Macleay Museum (1886–1889), broken by a visit to England (1888). He was assistant and collector at the Sydney Technical Museum (1889–1896), then was government entomologist to the agricultural department, New South Wales government (1896–1923). Before his final retirement, he acted as forestry entomologist for the Forestry Department (1923–1927). He also undertook assignments to the Solomon Islands (1909) and New Hebrides (Vanuatu) (1913), and lectured at the University of Sydney (1911–1921). He wrote: *Australia Insects* (1907).

Fromm

Neotropical Rivuline sp. *Cynodonichthys frommi* HO Berkenkamp & V Etzel, 1993

Daniel W Fromm is an American aquarist from New Jersey. According to the etymology he was honoured for "...*active field work and fish collections in Costa Rica and Panama during various trips, and for breeding the fish he collected and publishing the result.*"

Fryer, G

Lake Malawi Cichlid sp. *Sciaenochromis fryeri* AF Konings, 1993

Dr Geoffrey Fryer (b.1927) was a Fisheries Research Officer, Joint Fisheries Research Organisation of Northern Rhodesia and Nyasaland. The etymology says he was: "...*former Fisheries Research Officer of the Joint Fisheries Research Organization stationed in Nkhata Bay, who systematically observed the fishes of Lake Malawi in their natural habitat.*"

Fryer, JCF

False Moray sp. *Xenoconger fryeri* CT Regan, 1912

Sir John Claud Fortescue Fryer (1866–1948) was the first Entomologist at the Ministry of Agriculture and Fisheries (1914–1920) and the first Director of the Plant Pathology Laboratory at Harpenden, England (1920–1944). He then became Secretary of the Agricultural Research Council (1944). He collected in the Seychelles and on Aldabra Island (1908–1909), and in Ceylon (Sri Lanka) (1911). He was President of the Royal Entomological Society (1938–1939) and was a Fellow (1948) of the Royal Society. He collected the holotype of this eel. He died suddenly of Pneumonia.

Fu

Bagrid Catfish sp. *Tachysurus fui* CP Miao, 1934
[Syn. *Pseudobagrus fui*]

Fu Tung-sheng was at Honan (Henan) Museum, China. We know nothing more about him.

Fuchigami

Striped Loach ssp. *Cobitis striata fuchigamii* J Nakajima, 2012

Nobuyoshi Fuchigami is the Onga River Environment Conservation Monitor. He discovered this subspecies in the Onga River system, Kyushu, Japan.

Fuentes

Fuentes' Wrasse *Pseudolabrus fuentesi* CT Regan, 1913

Professor Maturana Francisco Fuentes (1876–1934) was a botanist at the University of Chile and National Museum. He made a collection of fish, including this wrasse, at Easter Island (1911).

Fuerth - see under Fürth

Fuessl

Armoured Catfish sp. *Dolichancistrus fuesslii* F Steindachner, 1911

This is believed to be an error by the printer, with Steindachner's spelling *füsslii being intended to read fasslii (see* **Fassl**). Fassl may have collected the holotype of this species, as he was certainly in the right region at the right period - he was collecting butterflies in Colombia. Steindachner made sure he got Fassl's name right when he named other taxa after him later (1915).

Fuges

Diminuitive Worm Eel *Pseudomyrophis fugesae* JE McCosker, EB Böhlke & JE Böhlke, 1989

Mrs Mary Fuges née Hitchcock (1916–2010) was the scientific illustrator for the Academy of Natural Sciences, Philadelphia. She was well known, to quote from the etymology, for her

"*...artistic ability ... meticulous attention to detail ... and her patience and encouragement throughout the preparation*" of the eel volume of the series: *Fishes of the Western North Atlantic*.

Fugler

Armoured Catfish sp. *Hemiancistrus fugleri* MM Ovchynnyk, 1971
[Junior Synonym of *Hypostomus annectens*]

Professor Dr Charles M Fugler (1929–1999) was an American herpetologist and explorer. He explored in Ecuador for many years and collected a wide number of herpetological and ichthyological specimens, many of which he presented to the University of Illinois Museum of Natural History. He was on the staff of Auburn University, Alabama and also taught biology at the University of North Carolina (1990). Louisiana State University, Museum of Natural Science, where he was a graduate student (1958), has a Charles M. Fugler Fellowship in Tropical Vertebrate Biology. His publications include: *Noteworthy Snakes from Puebla and Veracruz, Mexico* (1958) and *Biological Notes on Rana tigrina in Bangladesh and Preliminary Bibliography* (1984). A snake is also named after him.

Fukui

Cardinalfish sp. *Ostorhinchus fukuii* M Hayashi, 1990

Syojiro Fukui of the Japan Ichthyological Society collected the holotype of this species.

Fukushima

Freshwater Goby sp. *Rhinogobius fukushimai* T Mori 1934

Tsunekichi Fukushima was one of Mori's 'military guards' who was honoured for his 'most faithful services' rendered during Mori's expedition to Jehol (now called Chengde), Hebei Province, China, the type locality.

Fukuzaki

Tapertail Ribbonfish *Trachipterus fukuzakii* JE Fitch, 1964

Ben Fukuzaki was a Japanese American who lived on Terminal Island, San Pedro, California. His great interest in: "*...the creatures of the sea has led him to save and donate to science most of the animals he captures that are either unknown to him or which he recognizes as rare or unusual*," including the first two known specimens of this species. He was an amateur cameraman and left behind silent colour footage of life in the Japanese American community in California from before and just after WW2. This fascinating footage is preserved in the Japanese American National Museum where it was gifted (2002).

Fülleborn

Fülleborn's Mouthbrooder *Haplochromis fuelleborni* FM Hilgendorf & P Pappenheim, 1903
Fülleborn's Labeo *Labeo fuelleborni* FM Hilgendorf & P Pappenheim, 1903
Fülleborn's Squeaker Catfish *Synodontis fuelleborni* FM Hilgendorf & P Pappenheim, 1903

Lake Rukwa Lampeye *Micropanchax fuelleborni* E Ahl, 1924
Blue Mbuna (cichlid) *Labeotropheus fuelleborni* E Ahl, 1926

Dr Friedrich Georg Hans Heinrich Fülleborn (1866–1933) was a German physician, parasitologist and army officer. He served as a military doctor with the German Army in East Africa (1896–1910). He was an expert on tropical diseases and parasitology and became Director of the Hamburg Institute for Marine and Tropical Diseases (1930). Ten birds, an amphibian, a reptile and dragonfly are also named after him. He collected the holotypes of most of these species.

Fuller

Corydoras (catfish) sp. *Corydoras fulleri* LFC Tencatt, SA dos Santos, H-G Evers & MR Britto, 2021

Ian Alexander McDonald Fuller (b.1946) is a British aquarist and ichthyologist who specialises in Corydoras catfish. He runs the 'Corydoras World' website and is, as he puts it: "A toolmaker by trade, and self confessed Corydoras nut by admission." He started keeping fish half a century ago (1970) and breeding Corydoras shortly after (1974). He often speaks at conventions and is a past President of the Catfish Study Group. Among his publications is *Breeding Corydoradine Fishes* (2001) and the co-written book: *Identifying Corydoradine Catfish* (2005).

Funkner

Neotropical Rivuline sp. *Moema funkneri* S Valdesalici, 2019

George Funkner is a Scottish killifish breeder, collector, researcher, dealer and photographer, who collected the type series in Bolivia. He has a fish house in Glasgow with over 260 tanks.

Furic

Deepwater Flathead sp. *Bembradium furici* P Fourmanoir & J Rivaton, 1979

Pierre Furic was master of the Research Vessel 'Vauban' (1969–1987), which worked in the Philippines, off Madagascar and New Caledonia. He then commanded the NO 'Alis' until he retired (1987–1990). He was noted for his dredging skill, hard work and constant courtesy.

Furneaux

Furneaux Scorpionfish *Scorpaenopsis furneauxi* GP Whitley, 1959

Captain Tobias Furneaux (1735–1781) was an English navigator and Royal Navy officer who accompanied James Cook on his second voyage of exploration. He was one of the first people ever to circumnavigate the globe in both directions and he went on to command a British vessel during the American Revolutionary War. He served as second lieutenant of HMS 'Dolphin' under Captain Samuel Wallis on the latter's voyage round the globe (1766–1768) during which he was the first to set foot on Tahiti, hoisting a pennant, turning a turf, and taking possession of the land in the name of His Majesty (1767). He was given command (1771) of 'Adventure', which accompanied James Cook (in 'Resolution') on his second voyage, during which he was twice separated from his leader. On the former occasion he explored a great part of the south and east coasts of Van Diemen's Land (now

Tasmania), and made the earliest British chart there. Most of his names survive; Cook, visiting the shore-line on his third voyage, confirmed Furneaux's account and named after him the Furneaux Group at the eastern entrance to Bass Strait. After 'Adventure' was finally separated from 'Resolution' off New Zealand (1773), Furneaux returned home alone, bringing with him Omai of Ulaietea (Raiatea), the first South Sea Islander to travel to Great Britain. Furneaux was made a navigator (1775). During the American Revolutionary War, he commanded 'Syren' in the British attack upon Charleston, South Carolina. Whitley says of Furneaux in his etymology that "*...a biography has recently been given by F S Blight*": a reference to Blight's *Captain Tobias Furneaux, R.N., of Swilly* (1952).

Furness

Sheatfish (catfish) sp. *Pterocryptis furnessi* HW Fowler, 1905

Dr William Henry Furness III (1866–1920) MD, FRGS was an American anthropologist and explorer most famed for a paper he wrote about time spent with head-hunters in Borneo: *Home Life of Borneo Head Hunters* (1901). He also wrote: *The island of stone money: Uap of the Carolines* (1919) and various articles such as: *Observations on the Mentality of Chimpanzee and Orangutan* (1916). A mammal is named after him.

Furnier

Whitemouth Croaker *Micropogonias furnieri* AG Desmarest, 1823

Marcellin Fournier supplied specimens from Cuba. For some reason, Desmarest chose to omit the *o* from Fournier's name in the scientific binomial *furnieri*. We have been unable to unearth any further information.

Fürth

Pacific Ilisha *Ilisha fuerthii* F Steindachner, 1875

Congo Sea Catfish *Cathorops fuerthii* F Steindachner, 1876

White Stardrum *Stellifer fuerthii* F Steindachner, 1876

Ignác (Ignatius) Fürth was the Austrian Consul (1871) at Panama. He donated many fish specimens to the Vienna Museum. (Despite the name 'Congo' Sea Catfish, the species is found in Central America!)

Furuno

Duckbill sp. *Chrionema furunoi* O Okamura & F Yamachi, 1982

The authors did not provide an etymology. Given that. elsewhere in the book they did give etymologies when species were named in someone's honour, we speculate that this does not honour an individual. We suspect this has something to do with the Furuno Japanese electronics company, whose main products are marine electronics such as fish finders and radar systems, so they may have offered equipment or helped the collectors in some other way.

Furzer

Turquoise Killifish *Nothobranchius furzeri* RA Jubb, 1971

Richard E Furzer was an American hobbyist aquarist and aviculturist who discovered this species in Zimbabwe (1968) where he grew up and spent 25 years. He greatly promoted it to other hobbyists when he returned to the USA. The author notes: "*...through whose efforts this beautiful fish was introduced to Nothobranchius enthusiasts.*" One of the country's largest exotic-bird importers, he pleaded guilty to five felony counts (1993) when accused of using false documentation to illegally ship into the United States nearly 2,400 African Grey Parrots (between 1998 & 1990), worth up to $1,000 each. He was sentenced to 18 months in jail and ordered to pay $75,000 restitution. In a TV interview (1997) he said: *I was aware that the birds did not come from the Ivory Coast or Guinea, and to my mind, it didn't make any difference. They don't have border laws in west Africa like we do here, and they don't care where the birds came from, and I really didn't think that it was a crime. I realized that I was lying to the government, telling them the birds came from one country instead of another, but morally, it was fine by me.*

Fylla

Round Ray *Rajella fyllae* CF Lütken, 1887

This species is named after the Danish vessel 'Fylla' which was used for expeditions to Greenland (1884 & 1886), on one of which the holotype was collected.

G

Gabardini

Manduba (catfish) *Ageneiosus gabardinii* FJJ Risso & ENP de Risso, 1964
[Junior Synonym of *Ageneiosus inermis*]

Mario Gabardini was an Argentinean line fisherman and conservationist. He donated the holotype of this species.

Gabiru

Thorny Catfish sp. *Hassar gabiru* JLO Birindelli, DF Fayal & WB Wosiacki, 2011

Dr Leandro Melo de Sousa is known to his friends as 'Gabiru'. He is an ecologist and a zoologist. The University of São Paulo awarded his bachelor's degree (2000) and doctorate (2010), whilst his master's was awarded (2004) by Instituto Nacional de Pesquisas da Amazônia, where he also did post-doctoral work (2011). He is currently an Adjunct Professor, Universidade Federal do Pará, Altamira. He co-wrote: *Seasonal changes in the assembly mechanisms structuring tropical fish communities* (2017).

Gabriel

Pencil Catfish sp. *Trichomycterus gabrieli* GS Myers, 1926

This is a toponym referring to the type locality; the Sao Gabriel rapids, Rio Negro, Brazil.

Gabriel, J

Frosted Snake-Blenny *Ophiclinus gabrieli* ER Waite, 1906

Joseph Gabriel (b.c.1847–1922) was a Welsh pharmacist who moved to Australia and set up a street chemist shop (1870s), going on to own three shops (1888). He was an honorary collector for the National Museum of Victoria. He was also an early member of the Field Naturalists' Club of Victoria. His reports on marine expeditions to the Bass Strait islands were a major part of the twenty-one papers he published in the 'Victorian Naturalist'. Waite wrote: "*While in Melbourne I met Mr. Joseph Gabriel, who is interested in Mollusca, and he has since kindly sent to the Trustees some small fishes, taken by means of the dredge.*"

Gabriel (Wosiacki)

Plated Catfish sp. *Aspidoras gabrieli* WB Wosiacki, TG Graças Pereira & RE Reis, 2014

Gabriel P Wosiacki is the senior author's son. The species was named "*...as an encouragement of his growing interest in zoology*". From that we infer that Gabriel is still quite young!

Gabriele

Three-barbeled Catfish sp. *Rhamdia gabrielae* RC Angrizani & LR Malabarba, 2018

Gabriele Volkmer is the wife of the senior author, Rafael Angrizani of the Laboratório de Ictiologia, Departamento de Zoologia, Universidade Federal do Rio Grande do Sul (UFRGS), Brasil.

Gabriella

Gabriella's Grouper *Epinephelus gabriellae* JE Randall & PC Heemstra, 1991
[Alt. Multispotted Grouper]

Dr Gabriella Bianchi is a marine biologist who has worked at the Norwegian Institute of Marine Research and was Coordinator, Marine and Inland Fisheries Service, at the Fisheries and Aquaculture Department of the Food & Agriculture Organization of the United Nations in Rome. Among her published papers are: *Relative merits of using numbers and biomass in fish community studies* (1992) and *Impact of Fishing on Size Composition and Diversity of Demersal Fish Communities* (2000). She photographed specimens in Oman and provided additional data to the authors.

Gage

Southern Brook Lamprey *Ichthyomyzon gagei* CL Hubbs & MB Trautman, 1937

Simon Henry Gage (1851–1944) was a histologist and embryologist at Cornell University, which awarded his first degree (1877). He became Assistant Professor of Physiology and Lecturer in Microscopical Technology (1881–1889), then Associate Professor (1889–1893). He was one of the most remarkable, influential and important figures in the history of American microscopy. The author described him as: 'one of the foremost students of the lampreys', who brought this 'interesting and distinct species' to his attention. He wrote what is now considered a classic textbook: *The Microscope* (1923).

Gaige

Big Bend Gambusia *Gambusia gaigei* CL Hubbs, 1929

Professor Frederick McMahon Gaige (1890–1976) was an American entomologist, herpetologist and botanist who was Director of the Zoological Museum, University of Michigan. His wife, Helen Beulah Thompson Gaige, was a well-known herpetologist. In their honour the American Society of Ichthyologists and Herpetologists makes an annual award to a graduate student of herpetology. A bird and an odonate are named after him.

Gaimard

African Coris (wrasse) *Coris gaimard* JRC Quoy & JP Gaimard, 1824
Redeye Mullet *Mugil gaimardianus* AG Desmarest, 1831
[Type lost & description ambiguous: name suppressed & replaced by *M. rubrioculus*]

Joseph (or Jean, according to some sources) Paul Gaimard (1796–1858) was a French naval surgeon, explorer and naturalist. He made a voyage to Australia and the Pacific (1817–1819) aboard the 'Uranie', during which time he kept a journal, published as: *Journal du voyage de circumnavigation, tenu par Mr Gaimard, chirurgien à bord de la corvette l'Uranie.* Though he continued with his journal, further entries were lost when the ship was wrecked off the Falklands and he had to continue his journey on board the 'Physicienne', the ship that had rescued the expedition and then been purchased as a replacement. He was aboard the 'Astrolabe', under the command of Dumont d'Urville, when it visited New Zealand (1826). He led an expedition (1838–1840) aboard the 'Récherche' to northern Europe, visiting Iceland, the Faeroe Islands, northern Norway, Archangel and Spitsbergen. His contemporary, the zoologist Henrik Krøyer, who went with Gaimard to Spitsbergen (1838), described him thus: '*He was of medium build, with curly black hair and a rather unattractive face, but with a charming and agreeable manner*.' He was something of a dandy and, when visiting Iceland, handed out sketches of himself. Two birds, a mammal and a reptile are named after him.

Gairdner

Columbia River Redband Trout *Oncorhynchus mykiss gairdnerii* J Richardson, 1836

Dr Meredith Gairdner (1809–1837) was a British naturalist. He received a medical degree from Edinburgh University, Scotland. He left for Canada (1832) where he was employed as a physician by the Hudson's Bay Company and stationed on the Columbia River for about two years. There he studied local natural history, collecting materials to prepare a monograph of the vast riparian area. He also witnessed an eruption of Mount St Helens. Unfortunately, he contracted tuberculosis when he was sent to treat an outbreak. The disease left him weak and he moved to Hawaii, hoping the climate would help; instead it proved fatal. He left bequests to help poor Hawaiian children. He wrote: *Observations During a Voyage from England to Fort Vancouver, on the Northwest Coast of America* (1834).

Galadriel

Lago Izabal Pipefish *Pseudophallus galadrielae* CIA Dallevo-Gomes, GMT Mattox & M Toledo-Piza, 2020

Galadriel is an elf-queen in Tolkien's 'Lord of the Rings' trilogy. She is the bearer of the ring *Nenya*, also known as the ring of water. The authors say the name "*…is used herein in reference to the additional bony rings diagnostic of the new species and its association with freshwater habitats*."

Galán

Armoured Catfish sp. *Ancistrus galani* A Pérez & AL Viloria Petit, 1994

Carlos Alberto Galán (b.1949) is a Basque (Spanish) speleologist and biologist whose base is the Bio-speleology Laboratory, Aranzadi Science Society, Alto de Zorroaga, San Sebastián, Spain. He went to Venezuela at the age of five and went through the Venezuelan education system until graduating BSc (1966). He lived in Spain (1966–1970) and Argentina (1970–1977) studying at Barcelona, Navara and La Plata University which awarded a BSC in biological science. HIs bachelor's degree involved study in Navarre, Barcelona and the University of the Basque Country as well as Argentina where it was eventually awarded. He also studied at the University of the Basque Country. He wrote: *The Underground River of Ekain, the fauna of caves and the origin of the caves (Izarraitz massif, Gipuzkoa, Basque Country)* (1992). He collected the holotype of this cave-dwelling species. Among his many other publications are: *The Protection of the Caroni River Basin: environmental studies*

(1984) and *Underground rivers and karstic aquifers of Venezuela: Inventory, situation and conservation* (2017).

Galathea

Galathea Sculpin *Antipodocottus galatheae* RL Bolin, 1952
Galathea Sand Dart *Kraemeria galatheaensis* RR Rofen, 1958
Blind Cusk Eel sp. *Sciadonus galatheae* JG Nielsen, 1969
Galathea Gizzard Shad *Nematalosa galatheae* JG Nelson & N Rothman, 1973
Cusk Eel sp. *Abyssobrotula galatheae* JG Nielsen, 1977
Batfish sp. *Halieutopsis galatea* MG Bradbury, 1988
Flabby Whalefish sp. *Danacetichthys galathenus* JR Paxton, 1989
Galathea Assfish *Bassozetus galatheae* JG Nielsen & N Merrett, 2000
Viviparous Brotula sp. *Bellottia galatheae* JG Nielsen & PR Møller, 2008

This is not a true eponym as it is named after a ship. The British sloop 'HMS Leith' was bought by the Danish Expedition Foundation and renamed 'HDMS Galathea" mimicking the name of the vessel which undertook a scientific voyage known as the Danish Galathea Expedition (1845–1847). The new ship undertook the second Danish Galathea Expedition (1950–1952), more or less replicating the route of the first, then a third expedition (1969), and a fourth (2006–2007) which involved a circumnavigation of the world. The first known specimens of the above fish were collected during 'Galathea' expeditions.

Gale, A

Gale's Carp-Gudgeon *Hypseleotris galii* JD Ogilby, 1898
[Alt. Firetail Gudgeon]

Albert Gale, of the Royal Zoological Society of New South Wales, was an aquarist. He discovered the gudgeon in a 'stone tank' at the Royal Botanic Garden (Sydney) and bred it in the aquarium. He was honoured as finder of the species, but also as a friend of the author.

Gale, CF

Blue-lined Leatherjacket *Meuschenia galii* ER Waite, 1905
Gale's Pipefish *Campichthys galei* G Duncker, 1909

Charles Frederick Gale (1860–1928) was an Australian civil servant. Having tried (and presumably failed at) prospecting, he went on to be a stock inspector (1893–1897), Inspector of Pearl-Shell Fisheries at Shark Bay (1897–1899) and then Chief Inspector of Fisheries at Perth (1899–1915). During the latter period he was also Chief Protector of Aboriginals (1908). After marrying (1914) he took a long leave to Japan. He died of pneumonia. The etymology for the leatherjacket says" "*...At the request of Mr. Woodward, this fish is named after Mr. C. F. Gale, Chief Inspector of Fisheries, Western Australia.*"

Gale, J

Cenderwasih Epaulette Shark *Hemiscyllium galei* GR Allen & MV Erdmann, 2008

Jeffrey Gale of Las Vegas, Nevada, USA, is an avid underwater photographer, shark enthusiast and benefactor of the marine realm. He successfully bid to support the conservation of this species at the Blue Auction in Monaco (2007) and has given generously to support Conservation International's Birds Head Seascape marine conservation initiative.

Gallagher, FR

Gallagher's Thicklip Thornycat *Rhinodoras gallagheri* MH Sabaj Pérez, DC Taphorn & G Castillo, 2008

Francis Richard Gallagher (b.1935) was the mailroom supervisor, Academy of Natural Sciences of Philadelphia (1967–2003). In the etymology, he was noted "*...for dedicated service to the global community of taxonomists and systematists via the shipping and receiving of countless loans of biological specimens.*"

Gallagher, MD

Omani Garra sp. *Garra gallagheri* F Krupp, 1988

Major Michael Desmond Gallagher (1921–2014) was a zoologist and soldier who fought in North Africa, Sicily, Italy, France, Belgium and Germany. He retired from the army (1976) and then lived in the Sultanate of Oman (1977–1998) as a member of the staff of the Office for Conservation of the Environment there, during which period he set up (1985), and was Curator of, the Oman Natural History Museum. He took part in the Zaire River expedition (1974–1975). He was a member of Royal Geographical Society, London. He wrote: *The Birds of Oman* (1980) and *Snakes of the Arabian Gulf and Oman* (1993). Twenty-seven wildlife species, including a mammal and a reptile, have been named after him. He collected the holotype of the eponymous fish.

Galusda

Characin sp. *Cheirodon galusdae* CH Eigenmann, 1928

Piedro Galusda was the superintendent of the state hatchery at Lautaro, Chile, where he planned the collecting of fish during Eigenmann's expedition. As part of his work at the hatchery, he succeeded in introducing several forms of trout into Chilean rivers.

Galvis

Characin sp. *Hemibrycon galvisi* C Román-Valencia, 2000

Ghost Knifefish sp. *Apteronotus galvisi* CDCM de Santana, JA Maldonado-Ocampo & WGR Crampton, 2007

Germán Galvis Vergara is a Colombian biologist whose bachelor's degree was awarded by the University of Strasbourg (1967). He works at the Department of Biology, Universidad Nacional, Santafé de Bogotá, Colombia where he became an Associate Professor (1975). He wrote: *Freshwater fauna of the Tayrona Park* (1986). Román-Valencia named this species after Galvis to thank him for his provision of funding and comparative material for study.

Galzin

Speckled Garden-Eel *Gorgasia galzini* PHJ Castle & JE Randall, 1999

Dartfish sp. *Parioglossus galzini* JT Williams & D Lecchini, 2004
Galzin's Podge *Pseudogramma galzini* JT Williams & J Viviani, 2016

Dr René Galzin (b.1950) is Professor and Director of Studies at École Practique des Hautes Études at the Sorbonne, France. His particular research focus is on tropical and Mediterranean fish. The University of Montpellier awarded his doctorate (1985). He was Director of the Laboratoire d'Ichtyoécologie Tropicale et Méditerranéenne, Perpignan, France and Centre de Recherches Insulaires et Observatoire de l'Environnement at Moorea, French Polynesia. Among his publications he co-wrote: *Biodiversity research on coral reef and island ecosystems: scientific cooperation in the pacific region* (2009). He assisted Williams in the collection of all specimens of the dartfish that were known at the time.

Gama

Neotropical Rivuline sp. *Anablepsoides gamae* WJEM Costa, PHN de Bragança & PF de Amorim, 2013

Cecile de Souza Gama is a Brazilian ichthyologist researching at the Institute of Scientific and Technical Research of Amapa State, Brazilian amazon. Her BSc was awarded by the Federal University of Rio de Janeiro (1997), her master's by the Federal University of Juiz de Fora (2000) and her PhD from the Federal University of Paraiba (2013). Among her published works is the co-written: *An inventory of coastal freshwater fishes from Amapá highlighting the occurrence of eight new records for Brazil* (2016).

Gamero

Gamero's Woodcat *Entomocorus gameroi* F Mago-Leccia, 1984

Dr Alonso Gamero Reyes (1923–1980) was Dean of the Faculty of Science, School of Biology, Universidad Central de Venezuela (1958–1980), where the central university library is named after him. He graduated in biology from the National Pedagogical Institute, Caracas, Venezuela and then did postgraduate studies at the University of Michigan where his doctorate was awarded. He was Director, Laboratory of Biology, National Institute (until 1958). He co-wrote: *temporary Immobilization of Salamander Larvae by means of Electric Shock* (1951). He was honoured as it was Gamero "*…who guided the author's introduction to ichthyology*". He died from heart disease.

Gammon

Goby genus *Gammogobius* H Bath, 1971

R Gammon of Occidental College, Los Angeles, collected the type. We have been unable to unearth any biographical information.

Gandhi

Characin sp. *Astyanax gandhiae* RI Ruiz-Calderon, C Román-Valencia, DC Taphorn, PA Buckup & H Ortega, 2018

Maria Gandhi Calderon was the mother of the first author.

Ganymede

Grenadier sp. *Coelorinchus ganymedes* AM Prokofiev, 2021

Ganymede was a young man in Greek mythology, taken to Olympus for his beauty to become the cupbearer of the gods. The name refers to this fish's "graceful appearance and elegant colour" (translation).

Garavello

Armoured Catfish sp. *Harttia garavelloi* OT Oyakawa, 1993
Bumblebee Catfish sp. *Microglanis garavelloi* OA Shibatta & RC Benine, 2005
Armoured Catfish sp. *Rhinolekos garavelloi* F de Oliveira Martins & F Langeani, 2011

Dr Julio Cesar Garavello is a Brazilian zoologist and ichthyologist. The University of São Paulo awarded his bachelor's degree (1968), his master's (1972) and his doctorate (1980). He is presently a professor at the Universidade Federal de São Carlos. Among his publications he co-wrote: *Longitudinal distribution of the ichthyofauna in a tributary of Tietê River with sources on the Basaltic Cuestas of São Paulo, Southeastern Brazil* (2016). (Also see **Julio**)

Garbe

Garbei Cory *Corydoras garbei* R Ihering, 1911
Armoured Catfish sp. *Pareiorhaphis garbei* R Ihering, 1911

Father Ernesto Wilhelm Garbe (1853–1925) was a German-born Brazilian zoologist collecting in Brazil at the turn of the 19th/20th centuries. He worked at the Paulista Museum of the University of São Paulo with Hermann von Ihering (q.v.) and was one of many collaborators of Ihering on his *Revista do Museu Paulista*. An amphibian and a bird are named after him.

Garcia-Barriga

Long-whiskered Catfish sp. *Pimelodus garciabarrigai* G Dahl, 1961

Dr Hernando García-Barriga (1913–2005) was a botanist. The School of San Bartolomé awarded his bachelor's degree and additionally he obtained a bachelor's degree in botanical biology (1939) from the School of Agronomy and Veterinary Medicine, Bogota (now the Faculty of Veterinary and Zootechnics, National University of Colombia), where he was already a professor. He founded (1938) the Botanical Institute, now National Institute of Natural Sciences. He founded (1939) both the first Botanical Gardens in Colombia and the Museum of Natural History. He was a co-founder (1950) of the District University Francisco José de Caldas and its Faculty of Forest Engineering which awarded him an honorary doctorate (1968). He edited: *Medicinal flora of Colombia: Medical botany* (1974). He was a member of the expedition that collected the holotype.

Garcia Marquez

Pencil Catfish sp. *Trichomycterus garciamarquezi* CA Ardila Rodriguez, 2016

Gabriel Garcia Márquez (1927–2014) was a Colombian novelist, screenwriter and journalist. He won the Nobel Prize in Literature (1982). Two of his best-known works are *One Hundred*

Years of Solitude (1967) and *Love in the Time of Cholera* (1985). Many of his works explore the theme of solitude.

Gardiner

Cardinalfish sp. *Apogonichthyoides gardineri* CT Regan, 1908
Longtail Dragonet *Callionymus gardineri* CT Regan, 1908
Perchlet sp. *Plectranthias gardineri* CT Regan, 1908
Gardiner's Butterflyfish *Chaetodon gardineri* JR Norman, 1939

John Stanley Gardiner (1872–1946) was a zoologist and oceanographer. He was Professor of Zoology and Comparative Anatomy, Cambridge University (1909–1937). He travelled in the Indo-Pacific, visiting the Maldives (1899) and Fiji some time earlier. He wrote: *The Natives of Rotuma* (1898). Many of his other writings concerned corals and coral reefs. At the time when the butterflyfish was described, he was Secretary to the Committee of the 'John Murray' Expedition (1933–1934). Two reptiles and an amphibian are named after him.

Gardner, G

Neotropical Rivuline sp. *Hypsolebias gardneri* WJEM Costa, PF Amorim & JLO Mattos, 2018

George Gardner (1810–1849) was a Scottish physician, botanist and naturalist. He was awarded his MD by the University of Glasgow (1829). He published *Musci Britannici, or Pocket Herbarium of British Mosses arranged and named according to Hooker's "British Flora"* (1836) a work which impressed the Duke of Bedford sufficiently for him to sponsor Gardner's travels. He travelled and collected in Brazil (1836–1841). On his return he was elected a fellow of the Linnean Society (1842). He was appointed (1843) Superintendent of the botanic gardens at Peradeniya, Ceylon (now Sri Lanka) and island botanist. While there he wrote: *Travels in the Interior of Brazil, principally through the Northern Provinces and the Gold Districts, during the years 1836–41* (1846). He had made extensive collections towards a complete *Flora Zeylanica*, but his early death meant it was never published. The etymology says that the species is named in: "*...honour of Scottish naturalist George Gardner, who was in the Caatinga during his trip to Brazil between 1836 and 1841, making rich natural history collections. His reports on the region, and the numerous plant species and Cretaceous fossil fish collected by him represent important landmarks of our knowledge about Caatinga biodiversity*." Several plants were also named after him.

Gardner, RD

Gardner's Killi *Fundulopanchax gardneri* GA Boulenger, 1911
[Alt. Blue Lyretail, Steel-blue Killifish; Syn. *Fundulus gardneri*]

Captain R D Gardner collected the types at Okwoga on a small tributary of the Cross River in Nigeria. We have been unable to find any further biographical data.

Garman

Yucatan Flagfish genus *Garmanella* CL Hubbs, 1936
Gibbous Shiner *Cyprinella garmani* DS Jordan, 1885
American Sole sp. *Catathyridium garmani* DS Jordan, 1889

Parras Characodon *Characodon garmani* DS Jordan & BW Evermann, 1898
Hagfish sp. *Myxine garmani* DS Jordan & JO Snyder, 1901
Armoured Catfish sp. *Hypostomus garmani* CT Regan, 1904
Sagami Grenadier *Ventrifossa garmani* DS Jordan & CH Gilbert, 1904
Garman's Lanternfish *Diaphus garmani* CH Gilbert, 1906
Pouty Snailfish *Paraliparis garmani* CV Burke, 1912
Cusk-eel sp. *Monomitopus garmani* HM Smith & L Radcliffe, 1913
Natal Electric Ray *Heteronarce garmani* CT Regan, 1921
Characin sp. *Megaleporinus garmani* NA Borodin, 1929
Headstander Characin sp. *Laemolyta garmani* NA Borodin, 1931
Brown-spotted Catshark *Scyliorhinus garmani* HW Fowler, 1934
Rosette Skate *Leucoraja garmani* GP Whitley, 1939
Sawtooth Eel sp. *Serrivomer garmani* L Bertin, 1944
Freshwater Stingray sp. *Potamotrygon garmani* JP Fontenelle & M De Carvalho, 2017

Dr Samuel Trevor Walton Garman (1843–1927) was an American naturalist, most noted as an ichthyologist and herpetologist. He graduated in Illinois (1870), became a schoolteacher and was Professor of Natural Science at a seminary in Illinois (1871–1872). He became Louis Agassiz's special student (1872) and (1873) became Assistant Director of the Herpetology and Ichthyology Section, Museum of Comparative Zoology, Harvard. He was in South America with Alexander Agassiz (1874) and surveyed Lake Titicaca. Harvard awarded him two honorary degrees, a BS (1898) and an AM (1899). He wrote a number of books as well as many papers including: *The Plagiostomia (Sharks, skates and rays)* (1913). A bird, an amphibian, four reptiles and at least four other fish are also named after him.

Garnier

Garnier's Limia *Limia garnieri* LR Rivas, 1980

Emmanuel Garnier was the Director of the Fisheries Service in Haiti. He was of great assistance to the author during his visit to Haiti (1951).

Garnot

Yellowhead Wrasse *Halichoeres garnoti* A Valenciennes, 1839

Prosper Garnot (1794–1838) was a French naval surgeon, naturalist and collector who worked closely with Lesson (q.v.). They were both on board *La Coquille* during its circumnavigation of the world (1822–1825). Lesson and Garnot co-authored the zoological section of the voyage's report: *Voyage Autour du Monde Exécuté par Ordre du Roi sur la Corvette La Coquille Pendant les Années 1822–1825*, which was published in Paris (1828–1832). A grass genus is named after him as well as a bird and a reptile.

Garrett, A

Hawaiian Filefish *Thamnaconus garrettii* HW Fowler, 1928
[Syn. *Pseudomonacanthus garrettii*]

Andrew Garrett (1823–1887) was an American explorer, naturalist and illustrator who specialised in malacology and ichthyology. He spent much of his life in the South Seas.

He was employed as a collector by the Museum Godeffroy, Hamburg (1861–1885) and also illustrated the work of other zoologists. He wrote a number of papers on molluscs including: *The terrestrial Mollusca inhabiting the Society Islands* (1884). He collected the type. A bird is also named after him.

Garrett, LM

Rattail sp. *Bathygadus garretti* CH Gilbert & CL Hubbs, 1916

Lieutenant-Commander LeRoy Mason Garrett (1857–1906) was an US Navy officer who had served on the fisheries steamer 'Albatross' (1883–1885) and later was master and commander of the vessel (1904–1906), from which the holotype was collected. The expedition to the Northwest Pacific was judged to have been a success, but Garrett was lost overboard during a storm on the return voyage from Japan.

Garrick

San Blas Skate *Dipturus garricki* HB Bigelow & WC Schroeder, 1958
Longnose Houndshark *Iago garricki* P Fourmanoir & J Rivaton, 1979
Azores Dogfish *Scymnodalatias garricki* YI Kukuev & II Konovalenko, 1988
Eelpout sp. *Pachycara garricki* ME Anderson, 1990
Northern River Shark *Glyphis garricki* LJV Compagno, WT White & PR Last, 2008
Garrick's Catshark *Apristurus garricki* K Sato, AL Stewart & K Nakaya, 2013

Dr John Andrew Frank (Jack) Garrick (1928–2018) was a New Zealand ichthyologist with a specialization in elasmobranchs. He worked at Victoria University, Wellington (1950–1990) and became Professor of Zoology there (1971). He carried out the first exploratory deep-sea sampling using specially adapted cone nets, baited traps and longlines, regularly to depths greater than 2000 m. Many new and rare species were obtained by use of these innovative techniques. He was responsible for the notable discovery of the first New Zealand specimens of Orange Roughy (1957), which subsequently formed the basis of a multi-million-dollar fishery. In the etymology of the Northern River Shark, White says that he: "*…discovered this species in the form of two newborn males from Papua New Guinea and supplied radiographs, morphometrics, drawings and other details of these specimens (since lost) to the senior author.*"

Garth

Dragonet sp. *Foetorepus garthi* A Seale, 1940

Dr John Shrader Garth (1909–1993) was an American zoologist, an expert in crabs and Lepidoptera. While still an undergraduate, he first accompanied Captain C Allan Hancock aboard his yacht, 'Velero III', on the first of four expeditions (1931–1935) to the Galapagos Islands and the coasts of Central and South America. His master's was awarded (1945) after which he studied at Cornell (1937) and the University of Pennsylvania (1940), with his PhD being awarded (1941). His later expeditions included voyages to the Gulf of California (1936, 1937 & 1940), along the coasts of Peru and Ecuador (1938), and to Colombia, Venezuela, and Trinidad (1939) during which he amassed collections. He served as a civilian instructor in maps and charts at the Santa Ana Army Air Base, California, (1942–1944) and obtained a direct commission in the Sanitary Corps of the Army Medical Department. He then joined the biology department at USC, becoming Associate Professor (1952) and full professor

(1967) until retirement (1975). He wrote numerous publications, the best known being a two-volume series on the Majid crabs of the eastern Pacific.

Gary Lange

Gary Lange's Rainbowfish *Melanotaenia garylangei* JA Graf, F Herder & RK Hadiaty, 2015

[Alt. Golden Rainbowfish]

Gary William Lange (b.1954) is an American rainbowfish enthusiast in St Louis, USA. He has kept fish since the age of ten. He currently maintains around 100 tanks and keeps about 30 varieties of rainbowfish and blue-eyes. The University of Missouri awarded his BA in Biology (1976), after which he worked as a research biochemist for Monsanto, Searle, Pharmacia and Pfizer – all of those companies pretty much without changing his desk as the companies were taken over. His specialty was characterizing proteins especially with MALDI mass spectrometry. He founded (1987) the Rainbowfish Study Group of North America. He retired (2009) using his time to collect and give talks on rainbowfish. He writes articles and has had many of his fish photographs illustrate magazines. He has made four collecting trips to Australia (since 1993) and six to Papua (2005 on). He was the first person to discover this species near the Brazza River Basin in West Papua when collecting with Graf and Dan Dority (q.v.). He has also had a hand in naming a number of fish sent to Dr Gerry R Allen, such as *Melanotaenia ericrobertsi, M. bowmani* and *M. grunwaldi.*

Gascoyne

Coral Sea Gregory *Stegastes gascoynei* GP Whitley, 1964

HMAS 'Gascoyne', is a vessel from which the type was collected during a survey of islands in the Coral Sea.

Gasparini

Pencil Catfish sp. *Trichomycterus gasparinii* MA Barbosa, 2013

João Luis Rosetti Gasparini is a Brazilian zoologist and ichthyologist, who first collected this species (2001). His bachelor's degree was awarded by Universidade Federal do Espirito Santo (2004) and he now works there in the Oceanography and Ecology Department. He also acts as an independent environmental consultant. He wrote: *Ilha da Trindade e Arquipélago Martin Vaz Pedaços de Vitória no Azul Atlântico* (2004).

G A Steven

Steven's Goby *Gobius gasteveni* PJ Miller, 1974

George Alexander Steven (1901–1958) was a British zoologist with interests ranging from seals to seabirds, but primarily fish. He worked at the Plymouth Laboratory. He was honoured as he had worked extensively on the fishes of the western English Channel and was one of the first to recognize the eponymous goby as a species new to the area.

Gates

Burmese Golden Rasbora *Microdevario gatesi* AW Herre, 1939

Dr Gordon Enoch Gates (1897–1987) was a biologist, zoologist and lumbricologist who graduated from Colby College, Waterville, Maine with a bachelor's degree (1919). Harvard awarded his master's degree (1920) and a doctorate (1934). He was head of the Biology Department, Judson College, Rangoon, Burma (Yangon, Myanmar) (1921–1941) and the Professor of Biology, Rangoon College (1941–1946), but was evacuated to India (1942) as the Imperial Japanese army advanced through Burma. He was a Fellow at the Museum of Comparative Zoology (1946–1947) and returned to his alma mater, Colby College, as Professor of Biology (1948–1951). In retirement, he became a research fellow at Tall Timber Research Station (1967) and had at his disposal a small army to dig holes and collect specimens for him. Herre refers to Gates as a 'distinguished lumbricologist' and that his [Herre's] visit to Rangoon 'would have been of little avail' had it not been for Gates' help. A lumbricologist studies earthworms, and Gate wrote a huge amount about earthworms, his first paper being: *Some new earthworms from Rangoon, Burma* (1925) and his last: *Farewell to North American megadriles* (1982).

Gaucher

Neotropical Rivuline sp. *Anablepsoides gaucheri* P Keith, L Nandrin & P-Y Le Bail, 2006

Philippe Gaucher became (2007) Scientific Director of the Centre National de la Recherche Scientifique and Mission Parc de la Guyane, Cayenne, French Guiana. He co-wrote: *The Tailless Amphibians of French Guiana* (2000). He found the first specimens.

Gauguin

Lizardfish sp. *Trachinocephalus gauguini* FA Polanco, A Acero P & R Betancur-R, 2016

Paul Gauguin (1848–1903) was a French artist, largely unappreciated until after his death. He was a successful stockbroker, but when the Paris stock market crashed (1882) his earnings also crashed. He died in the Marquesas islands, where this lizardfish is found.

Gauss

Lanternfish sp. *Lepidophanes gaussi* AB Brauer, 1906

'Gauss' is the name of the ship from which the holotype was collected during the first German South Polar Expedition to the South Pole (1901–1903). It is not entirely clear if this species is named after the vessel or the German mathematician and physicist Carl Friedrich Gauss (1777–1855) after whom the ship is named.

Gautami

Gautama Thryssa *Thryssa gautamiensis* Babu Rao, 1971

This is a toponym referring to the Gautami branch of Godavari Estuary (Andhra Pradesh, India), where the holotype originated.

Gay

Gay's Wrasse *Pseudolabrus gayi* A Valenciennes, 1839

South Pacific Hake *Merluccius gayi* A Guichenot, 1848
Morwong sp. *Nemadactylus gayi* R Kner, 1865

Claude Gay (1800–1873) was a French botanist. He went to Chile (1828–1932), then returned to France before going back to Chile (1834–1840). He was later in the USA (1859–1860) studying American mining techniques. He wrote many works including his 24-volume magnum opus: *Historia Fisica y Politica de Chile* (1843–1851). An amphibian and three birds are named after him.

Geay

Neotropical Rivuline sp. *Laimosemion geayi* L Vaillant, 1899
Bandit Cichlid sp. *Guianacara geayi* J Pellegrin, 1902
Half-banded Pike-Cichlid *Crenicichla geayi* J Pellegrin, 1903
Red-tailed Silverside *Bedotia geayi* J Pellegrin, 1907
Characin sp. *Markiana geayi* J Pellegrin, 1909
Characin sp. *Cheirodontops geayi* LP Schultz, 1944

Martin François Geay (1859–1910) was a pharmacist, natural history collector, and traveller. He worked as a pharmacist for the Panama Canal Company (1886–1887). He led an expedition to Colombia and Venezuela for the Ichthyology and Herpetology Laboratory, Muséum National d'Histoire Naturelle, Paris (1888–1895). He collected plant and animal specimens in Madagascar (1904–1907) and during an expedition to French Guiana (1897–1899) and again in Madagascar (1908–1909). He left Madagascar (1910) for Australia and died in Melbourne (1910). A reptile and an amphibian are also named after him.

Geck

Congo Cichlid sp. *Orthochromis gecki* FDB Schedel, EJWMN Vreven, BK Manda, E Abwe, BK Manda & UK Schliewen, 2018

Jakob Geck is a German ichthyologist and volunteer researcher. He is a "*...passionate, German fish naturalist,*" who was honoured in the name for his "*...dedicated volunteer work and untiring support*" for the ichthyology section of the Bavarian State Collection of Zoology (Munich), and because his "*great experience in keeping rheophilic cichlids contributed to the knowledge of behaviour and ecology of many cichlid taxa, including* O. katumbii *and* O. indermauri."

Geddes

Mexican Cichlid sp. *Cichlasoma geddesi* CT Regan, 1905

Sir Patrick Geddes (1854–1932) was a Scottish biologist, sociologist, geographer, francophile, town planner and philanthropist. He studied mining at the Royal College of Mining (1874–1977) but did not finish his degree. He was then a demonstrator in the Department of Physiology in University College London (1977–1878). He travelled widely and worked in India and Israel. His published works include *The Evolution of Sex* (1889). He is known to have visited Mexico – where he went temporarily blind.

Geerts, G

Congo Blind Barb *Caecobarbus geertsii* GA Boulenger, 1921

G Geerts was an engineer and railway director in the Belgian Congo. He was an amateur speleologist and was one of a party that (1917) explored a cave at Thysville, Congo (now Mbanza-Ngungu, DRC) and discovered this species. After WW1, Geerts took specimens of it to the Musée du Congo Belge. Nothing more is known about him.

Geerts, M

Lake Malawi Cichlid sp. *Copadichromis geertsi* A Konings, 1999

Martin Geerts is an independent researcher, aquarist and leading light of the Dutch Cichlid Association. He wrote: *The Cichlid Companion* (1998) and was one of the compilers of the Catalogue of Cichlids. He was honoured for his "*...knowledge of the scientific aspects of cichlids and for his support of the author during several expeditions to Lakes Malawi and Tanganyika.*"

Gegarkun

Gegharkuni Trout *Salmo ischchan gegarkuni* KF Kessler, 1877

This is a toponym referring to the historical name for Lake Sevan, Armenia (also spelled Gegharkuni), where this trout occurs.

Gegenbaur

Spiny Eel sp. *Tilurus gegenbauri* RA von Kölliker, 1853

Carl (or Karl) Gegenbaur (1826–1903) was a German comparative anatomist who was a strong supporter of Darwin and who demonstrated that comparative anatomy provides convincing evidence supporting the theory of evolution. He graduated from the University of Würzburg (1851) where the eel's describer Kölliker had been one of his teachers. He travelled in Italy and Sicily before returning to Würzburg (1854) as an independent teacher. He was Professor of Anatomy at the University of Jena (1855–1873) and at the University of Heidelberg (1873–1901). He wrote: *Grundriss der vergleichenden Anatomie* (1874).

Geijskes

Sharpsnout Stingray *Fontitrygon geijskesi* M Boeseman, 1948

Dr Dirk Cornelis Geijskes (1907–1985) was a biologist, ethnologist and entomologist. He graduated from the University of Basel, Switzerland (1935). He worked in Suriname (1938–1965), firstly in agricultural research and (1954) as chief biologist and Director, Suriname Museum, Paramaribo. After returning to the Netherlands he worked at the Leiden Museum, becoming Curator (1967). He wrote: *Natuurwetenschappejik onderzoek van Suriname: 1945–1965* (1967). He collected the stingray holotype and supplied Boeseman with many fish specimens from Suriname. Many different taxa including insects and an amphibian are named after him.

Geiser

Largespring Gambusia *Gambusia geiseri* C Hubbs & CL Hubbs, 1957

Dr Samuel Wood Geiser (1890–1983) was a biologist and historian of science. He taught at Guilford College, North Carolina (1914–1916), Upper Iowa University (1917–1919), Washington University in St Louis (1922–1924) and Southern Methodist University (1924–1957). His PhD was awarded by Johns Hopkins University (1922). He taught a number of subjects but is best known for his work on isopods. He wrote papers and contributed chapters as well as editing, and wrote at least one book: *Naturalists of the Frontier* (1937).

Geisler

Characin sp. *Iguanodectes geisleri* J Géry, 1970
Dwarf Cichlid sp. *Apistogramma geisleri* H Meinken, 1971
Characin sp. *Microschemobrycon geisleri* J Géry, 1973
Stone Loach sp. *Mustura geisleri* M Kottelat, 1990
[Syn. *Schistura geisleri*]
Characin sp. *Hemigrammus geisleri* A Zarske & J Géry, 2007

Dr Rolf Geisler (1925–2012) was a German aquarist, limnologist and ichthyologist, a professor at the University of Freiburg. He was also a long-time friend of Meinken and collected the dwarf cichlid type. He researched tropical fish extensively, particularly travelling in Asia and South America. He was also a noted collector and '…one of the most influential aquarists ever'. Among his published papers and books, he wrote: *Aquarium Fish Diseases* (1960) and *Wasserkunde: Für die aquaristische Praxis* (1964). His eyesight diminished greatly in his later years. He died of a stroke.

Gemellaro

Cocco's Lanternfish *Lobianchia gemellarii* A Cocco, 1838

Dr Carlo Gemellaro (1787–1866) was an Italian physician, surgeon, naturalist, volcanologist and geologist, and a friend of the author. His medical qualifications were gained at the University of Catania, Sicily (1808) and for seven years he served as a regimental surgeon in the British Army. He travelled from one end of Europe to the other (1813–1817) collecting rocks, minerals, fossils etc. He returned to Catania (1817) where he was one of the founders of the Academy of Natural Sciences (1824) and became a Professor at the University of Catania (1830–1847) and also Rector of the university (1847). He died from cancer of the throat. He wrote: *A farewell to the greatest volcano in Europe* (1865).

Gené

South European Nase *Protochondrostoma genei* CL Bonaparte, 1839

Professor Carlo Giuseppe Gené (1800–1847) was an Italian author and naturalist. He studied at the University of Pavia and later (1828) became Assistant Lecturer in Natural History there. He collected in Hungary (1829) and made four trips to Sardinia (1833–1838), mostly collecting insects. On Bonelli's death (1830) Gené succeeded him as Professor of Zoology and Director of the Royal Zoological Museum in Turin. An amphibian, two birds and three dragonflies are named after him.

Genia

Croaker sp. *Atrobucca geniae* A Ben-Tuvia & E Trewavas, 1987

Dr Eugenie Clark (1922–2015) (See **Clark, E**)

Genie (Böhlke)

Cleaner Goby *Elacatinus genie* JE Böhlke & CR Robins, 1968
Moray sp. *Uropterygius genie* JE Randall & D Golani, 1995
Snake-eel sp. *Ophichthus genie* JE McCosker, 1999

Eugenia Louisa Böhlke née Brandt (1929–2001) (See **Böhlke**)

Genie (Clark)

Genie's Dogfish *Squalus clarkae* MO Pfleger, RD Grubbs, CF Cotton & TS Daly-Engel 2018

Dr Eugenie Clark (1922–2015) (See **Clark, E**)

Geoffrey

Lake Malawi Cichlid sp. *Alticorpus geoffreyi* J Snoeks & R Walapa, 2004

Dr Geoffrey Fryer (b.1927) is a retired British carcinologist, biologist, limnologist, ecologist and ichthyologist. He worked on several aspects of the fishes of Lake Malawi where the cichlid is endemic, but is most known for 'his pioneering work' on the lake's rock-dwelling cichlid communities. Among his published works are: *Cichlid Fishes of the Great Lakes of Africa* (1972), *A Natural History of the Lakes, Tarns and Streams of the English Lake District* (1991) and *The Freshwater Crustacea of Yorkshire* (1993).

Geoffroy

Corydoras sp. *Corydoras geoffroy* BG de Lacepède, 1803
Geoffroy's Wrasse *Macropharyngodon geoffroy* JRC Quoy & JP Gaimard, 1824

Étienne Geoffroy Saint-Hilaire (1772–1844) was a French naturalist. He trained for the Church but abandoned it to become Professor of Zoology (aged 21!), when the Jardin du Roi was renamed Le Musée National d'Histoire Naturelle. In his *Philosophie Anatomique* (1818–1822) and other works, he expounded the theory that all animals conform to a single plan of structure. This was strongly opposed by Cuvier (q.v.), who had been his friend, and (1830) a widely publicised debate between the two took place. Despite their differences, the two men did not become enemies; they respected each other's research, and Geoffroy gave one of the orations at Cuvier's funeral (1832). Modern developmental biologists have confirmed some of Geoffroy Saint-Hilaire's ideas. Sixteen mammals, seven birds and a reptile are named after him.

George, R

George's Basslet *Pseudanthias georgei* GR Allen, 1976

Dr Ray W George (b.1929) is an Australian carcinologist at the Western Australian Museum (1953–1984) who was Senior Curator of Crustacea (1965–1984) there and an adjunct senior lecturer at the University of Western Australia. He was also the leading authority on Rock Lobsters which he studied in Australia and the Seychelles. Among his many publications spread over five decades are: *The Status of the 'White' Crayfish in Western Australia* (1958) and *Tethys origin and subsequent radiation of the spiny lobsters (Palinuridae)* (2006). He collected the basslet holotype when on board the 'RV Diamantina' off Cape Inscription, Western Australia.

George (Clipper)

Blackbelly Argentine *Argentina georgei* DM Cohen & SP Atsaides, 1969

George E Clipper was a technician in charge of x-raying fishes and reading the resultant radiographs at Bureau of Commercial Fisheries Systematics Laboratory, USNM. The authors expressed their gratitude for his 'efficient assistance'.

George (Dussumier)

Long-billed Halfbeak *Rhynchorhamphus georgii* A Valenciennes, 1847

Georges Dussumier is described in the etymology as a friend of Valenciennes. We have failed to identify him further, but he was presumably a relative of Jean-Jacques Dussumier (1792–1883) (q.v.), a French merchant and collector after whom Valenciennes named several other fish species.

George Gill

Gill's Goby *Hetereleotris georgegilli* AC Gill, 1998

George Burton Gill (1925–1994) was the author's father.

George (Leacock)

Roundscale Spearfish *Tetrapturus georgii* RT Lowe, 1841

George Butler Leacock supplied Lowe with the spearfish holotype and "...*rendered valuable assistance to the cause of ichthyology of this island* [Madeira]... ...*generally as well as in this instance*." Leacock was a businessman in Madeira associated with 'Leacock's Madeira Wine', although by the time of Lowe's description he had sold up and severed links with that company (1835).

George (Myers)

San Marcos Gambusia *Gambusia georgei* C Hubbs & AE Peden, 1969 (Extinct)

Dr George Sprague Myers (1905–1985) (see **Myers, GS**)

George Miller

Plaincheek Puffer *Sphoeroides georgemilleri* RL Shipp, 1972

George C Miller is an American zoologist formerly of the National Marine Fisheries Service, Southeast Fisheries Center, Miami Laboratory. Most of his published papers concern fish taxonomy with his focus being Searobins, but he has also written on commercial fisheries such as: *Commercial Fishery and Biology of the Fresh-water Shrimp*, Macrobrachium, *in the Lower St. Paul River, Liberia, 1952–53* (1971). He aided Shipp in the collection of Central American fishes used in his study.

George

Silver Mullet *Paramugil georgii* JD Ogilby, 1897

This is a toponym referring to the George's River, Australia, in the estuary of which the type specimen was taken (1895).

Georgette

Georgette's Tetra *Hyphessobrycon georgettae* J Géry, 1961

Georgette Géry was the wife of the author Dr Jacques Géry (1917–2007) (q.v.). Although the original text was silent, the author later revealed that this was the lady honoured. (Also see **Georgia**)

Georgia

Characin sp. *Parapristella georgiae* J Géry, 1964
False Rummy-nose Tetra *Petitella georgiae* J Géry & H Boutière, 1964
Characin sp. *Tetragonopterus georgiae* J Géry, 1965
African Rivuline sp. *Aphyosemion georgiae* JG Lambert & J Géry, 1968

Georgette Géry (See **Georgette**)

Georgiana

Antarctic Starry Skate *Amblyraja georgiana* JR Norman, 1938

This is a toponym referring to the fact that the holotype was caught in the coastal waters of South Georgia and was named after that island.

Georgiana

Banded Sweep *Scorpis georgiana* A Valenciennes, 1832

This is a toponym referring to the fact that the type locality was 'Port du Roi-George' (King George Sound), Western Australia.

Gephart

Thorny Catfish sp. *Trachydoras gepharti* MH Sabaj Pérez & M Arce Hernádez, 2017

George W Gephart Jr. is a businessman, financier and not-for-profit leader who was President & CEO of the Academy of Natural Sciences of Drexel University (2010–2017). He is also a keen naturalist, birder and environmentalist. His BA was awarded by Yale (1975) and his master's in business administration by the University of Pennsylvania (1979). He was honoured for his "*...bold, deft and heartfelt leadership of a Glorious Enterprise into its third century*" ['Glorious Enterprise' alludes to the title of a (2012) book about the Academy, which is America's oldest natural history museum]

Gerald Allen

Viviparous Brotula sp. *Microbrotula geraldalleni* W Schwarzhans & JG Nielsen, 2012

Gerald Ray Allen (b.1942) (See **Allen GR**)

Gercilia

Neotropical Cichlid sp. *Aequidens gerciliae* SO Kullander, 1995

Maria Gercília Mota Soares is a fisheries biologist who is Senior Researcher III at the National Research Institute of the Amazon (*Instituto Nacional de Pesquisas da Amazonia (INPA), Manaus, Brazil*), Researcher at the Federal University of Amazonas, Researcher & Coordinator at the Federal University of Amazonas and a Member of the Editorial Board at Hiléia (UEA). The Federal University of Ceará awarded her Agricultural Engineering degree (1973) and the NRIA her master's (1978) and PhD (1993) in freshwater biology and fishing. She has published quite widely such as: *Impact of flood pulse (or floods and ebbing) on living beings in wetland areas of Central Amazonia* (2003). Kullander says that the cichlid was "*...named for Maria Gercília Mota Soares, who first studied the species in 1980.*"

Gerda

Viviparous Brotula sp. *Bythites gerdae* JG Nielsen & DM Cohen, 1973

'Gerda' is the name of a research vessel that belonged to the Mote Marine Laboratory, Miami, from which the holotype was collected.

Gerlach

Oriental Cyprinid sp. *Onychostoma gerlachi* WKH Peters, 1881

Dr Johann Gerhard Heinrich Carl Gerlach (1843–1913) was a German physician who lived for part of his life in Hong Kong from where he sent a collection of fishes, including the holotype of this species, to Senckenbergischen Naturforschenden Gesellschaft, of which he was a member. He qualified as a physician in Prussia (1868). He is known to have been in Hong Kong when he was a signatory to a petition to have the medical profession regulated (1882) and he was one of the original nine doctors licensed to practice for payment when compulsory registration was introduced (1884). He remained in practice in Hong Kong until at least 1891.

Gerlache

Antarctic Dragonfish genus *Gerlachea* L Dollo, 1900

Baron Adrien Victor Joseph de Gerlache de Gomery (1866–1934) was an officer in the Belgian Royal Navy who led the Belgian Antarctic Expedition (1897–1899). He studied engineering at the Free University of Brussels (1885) but quit to join the navy (1886). He graduated from Osten Nautical College and worked on fisheries protection vessels. He signed on as a seaman (1887) on the English vessel 'Cragie Burn' bound for San Francisco, but it failed to get around Cape Horn and was sold for scrap. He spent some time in Uruguay & Argentina before returning to Europe. He tried unsuccessfully to join an expedition to the Congo, then proposed a plan to the Royal Geographical Society for an Antarctic Expedition. He bought the 'Patria', had it refitted and renamed it 'Belgica' staffed with an international crew including Roald Amundsen. They set sail from Antwerp (16th August 1897), but met ill fortune when they were trapped in the ice for over seven months. Several crew members went mad, while others suffered scurvy. He wrote about the expedition as *Quinze Mois dans l'Antarctique* (1902). An Antarctic strait was named in his honour. He took part in several later Antarctic and Arctic expeditions. He died of paratyphoid.

Germain

Germain's Blenny *Omobranchus germaini* HE Sauvage, 1883

Louis Rodolphe Germain (1827–1917) was a veterinary surgeon in the French colonial army, serving in Indochina (Vietnam) (1862–1867), and went to New Caledonia (1975–1878). He made zoological collections in his spare time, donating them to the MNHN. A mammal is named after him as are four birds.

Gernaert

Gernaert's Halfbeak *Hyporhamphus gernaerti* A Valenciennes, 1847

Benoit Gernart (c.1797–1843) was French Consul-General in China based in Canton (1827–1837), and later in West Java and at Macao where he obtained the halfbeak.

Geronimo

Rattail sp. *Coelorinchus geronimo* NB Marshall & TN Iwamoto, 1973

'Geronimo' is a research vessel owned by the U.S. Bureau of Commercial Fisheries (now the National Marine Fisheries Service). The vessel made a large collection of grenadiers from the Gulf of Guinea, but, interestingly, the holotype of this species was collected by her sister ship 'Undaunted'.

Gerrard

Gerrard's Stingray *Maculabatis gerrardi* JE Gray, 1851
[Alt. White-spotted Whipray, Sharpnose Stingray]
Blight Redfish *Centroberyx gerrardi* A Günther, 1887

Edward Gerrard (1810–1910) worked as an attendant in Gray's department at the British Museum (1841–1896). He was Gray's 'right-hand man' and looked after the galleries and storerooms and also helped Grey with shark and ray identification. He also preserved and registered bottled animals and compiled a catalogue of osteological specimens at the British Museum. Two reptiles are named after him.

Gerring

Toadfish sp. *Daector gerringi* CH Rendahl, 1941

Gösta Gerring (1913–1946) of the Swedish Museum of Natural History collected the holotype.

Gerringer

Cusk-eel sp. *Leucicorus gerringerae* WW Schwarzhans, JG Nielsen & BC Mundy, 2022

Dr. Mackenzie Gerringer is an Assistant Professor at the State University of New York at Geneseo. Her PhD in Marine Biology was awarded by the University of Hawaii (2017). Her current research centres on the physiology and ecology of deep-sea animals.

Gertrud (Dudin)

Lake Malawi Cichlid sp. *Aulonocara gertrudae* A Konings, 1995

Dr Gertrud Konings-Dudin née Dudin is a German biologist who became (2007) Assistant Professor of Biology, El Paso Community College. The Free University of Berlin awarded her Biology Degree (1974) and her PhD (1977). She was a research associate at the University of Heidelberg, University of Rotterdam and American University of Beirut (1977–1988). She was then an Instructor at the University of Maryland (1989–1996) afterwards teaching at El Paso (1997–2001) and as a biology consultant at Dynatec Labs (1997–2007). She is also the wife of the author Ad Koning. He honoured her "*...for her moral support, her interest in cichlids, and for her patience*."

Gertrude (Merton)

Spotted Blue-eye *Pseudomugil gertrudae* MCW Weber, 1911

Mrs Gertrude Merton was the wife of the German naturalist Dr Hugo Merton. She accompanied him to the Aru Archipelago (Indonesia) (1907–1908). (Also see **Merton**)

Géry

South American Darter genus *Geryichthys* A Zarske, 1997
Géry's Killi *Scriptaphyosemion geryi* JG Lambert, 1958
Three-barbeled Catfish sp. *Pimelodella geryi* JJ Hoedeman, 1961
Characin sp. *Elachocharax geryi* SH Weitzman & RH Kanazawa, 1978
Geryi Cory *Corydoras geryi* H Nijssen & IJH Isbrücker, 1983
Géry's Piranha *Serrasalmus geryi* M Jégu & GM Santos, 1988
[Alt. Violet-line Piranha]
Characin sp. *Microcharacidium geryi* A Zarske, 1997
Characin sp. *Knodus geryi* FCT Lima, HA Britski & F de A Machado, 2004
Tetra sp. *Hyphessobrycon geryi* EC Guimarães, PS de Brito, PHN Bragança, AM Katz & FP Ottoni, 2020

Dr Jacques Géry (1917- 2007) was a French physician and ichthyologist. He initially worked as a hospital doctor in Strasbourg and during WW2 he cared for wounded English prisoners-of-war in Germany. After the war, he worked as a plastic surgeon at Briey (1947–1960). His interest in fishes was such that he studied and took a second degree (1960), and from then until the end of 2006 he was a full-time ichthyologist. He wrote: *Characoids of the World* (1977). He died of cancer. (Also see **Georgia**)

Gesmone

Neotropical Rivuline sp. *Maratecoara gesmonei* DTB Nielsen, M Martins & R Britzke, 2014

Gesmone Fernando Godoy is an ornamental fish trader in Brazil. He discovered this species.

Gestetner

African Barb sp. *Labeobarbus gestetneri* KE Banister & RG Bailey, 1979

This is not a true eponym in that the species is named after a boat, 'David Gestetner'. It was used by the collectors during their Zäire River expedition.

Ghazal

Ghazal Goby *Silhouettea ghazalae* M Kovačić, R Sadeghi & HR Esmaeili, 2020

Ghazal Esmaeili is the daughter of the third author. (Also see **Esmaeili**)

Ghesquière

Busira Mormyrid *Marcusenius ghesquierei* M Poll, 1945

Jean Hector Paul Auguste Ghesquière (1888–1982) was a botanist and entomologist who collected for the Belgian National Herbarium. He was in Angola, Belgian Congo (DRC), Cameroon, Niger, São Tomé and Principe, and Uganda (1918–1938). He collected almost 100 species of fish at Eala (Congo) (1933–1937).

Ghetaldi

Popovo Minnow *Delminichthys ghetaldii* F Steindachner, 1882

Baron Frano Getaldić-Gundulić aka Francesco Ghetaldi-Gondola (1833 -1899) was a politician and Mayor of Dubrovnik, Kingdom of Dalmatia, Austro-Hungarian Empire (now in Croatia) (1889–1899). He was also an understanding and sympathetic landowner and horticulturist who founded a School of Agronomy at Lapad (a suburb of Dubrovnik), and introduced such exotic vegetables as Brussels sprouts! He introduced electric street lighting in the city of Dubrovnik and promoted tourism. He fought in the Franco-Prussian War (1870–1871), and also founded the Dubrovnik Philatelic Society (1890). Steindachner honoured him for having made possible the collection of the holotype from an underground cave in Herzegovina. He committed suicide one day after the accounts for the treasury were examined.

Ghigi

Rhodes Minnow *Ladigesocypris ghigii* L Gianferrari, 1927
[Syn. *Squalius ghigii*]

Professor Alessandro Ghigi (1875–1970) was Rector of Bologna University (1930–1943) where he had gained his first degree and doctorate (1902) as well as becoming Professor of Zoology there (1922) and Director of its Institute of Zoology. Early in his career he taught at the Agriculture secondary school and at the University of Ferrara. One of his major interests was breeding hybrids of birds. He was a friend of Delacour, who gave him pheasant specimens he had collected on his various expeditions to Vietnam. Ghigi developed pheasant hybrids and mutations in captivity, including one known as Ghigi's Golden Pheasant. Unfortunately, he was also a Fascist, believing in the superiority of white people and the particular inferiority of mixed-race people. On the positive side, he was very keen on nature conservation and wrote hundreds of scientific papers. He collected the holotype of this fish species.

Ghisolfi

Neotropical Rivuline sp. *Hypsolebias ghisolfii* WJEM Costa, ALF Cyrino & DTB Nielsen, 1996

Julio Cezar Ghisolfi is a Brazilian biologist, aquarist and photographer living in Artur Nogueira, São Paulo. He was honoured as a 'fervent' breeder and collector of annual killifishes who is active in the fish-keeping world.

Giacopini

Characin sp. *Bryconops giacopinii* A Fernández-Yépez, 1950

José Antonio Giacopini Zárraga (1915–2005) was a lawyer, a historian and a military adviser, revolutionary and also a personal friend of Hugo Chávez Frías, sometime President of Venezuela. He was Governor, Amazonas State of Venezuela, when he sponsored an archaeological expedition during which the holotype was collected. He wrote: *The Llaneros. Cowboys of Venezuela* (1994).

Gianeti

Armoured Catfish sp. *Farlowella gianetii* GO Ballen Chapporo, MM de L Pastana & LAW Peixoto, 2016

Dr Michel Donato Gianeti is a biologist and oceanographer. Universidade Estadual Paulista, São Paulo, awarded his bachelor's degree (2002), the Federal University of Rio Grande his master's (2005) and the University of São Paulo his doctorate (2011). He is currently collection manager at the ichthyological collection of the Museu de Zoologia da Universidade de São Paulo. The authors acknowledged him for 'kind assistance' provided during visits to the collection and through loan/data request management. He co-wrote: *A hermaphrodite guitarfish,* Rhinobatos horkelii *(Müller & Henle, 1841)* (Rajiformes: Rhinobatidae), *from southern Brazil* (2007)

Giard

Pink Hap (cichlid) *Sargochromis giardi* J Pellegrin, 1903

Dr Alfred Mathieu Giard (1846–1908) was a French zoologist. After studying natural sciences at the École Normale Supérieure he worked as a preparator at a Paris lab. His PhD was awarded (1872) and he became a professor of natural history at Lille University (1873–1882). He founded (1874) a biological station at Wimereux. He lectured at the École Normale Supérieure (1887) then (1888) until his death held the chair of the evolution of living organisms at Paris. He died on his birthday.

Giaretta

Neotropical Rivuline sp. *Melanorivulus giarettai* WJEM Costa, 2008

Pencil Catfish sp. *Trichomycterus giarettai* MA Barbosa & AM Katz, 2016

Dr Ariovaldo Antonio Giaretta (b.1966) is a Brazilian herpetologist who specialises in amphibians. The State University of Campinas awarded his bachelor's degree (1990), master's (1994) and doctorate (1999). He is currently a full Professor, Department of Zoology, Universidade Federal de Uberlândia, Brazil. He co-wrote: *Calls of Paratelmatobius gaigeae (Cochran 1938)* (Anura, Leptodactylidae) (2013) He collected the types of both species.

Gibbons

Kelpfish genus *Gibbonsia* JG Cooper, 1864

Dr William P Gibbons (1812–1897) was a physician and naturalist from Delaware. His medical degree was awarded by the University of New York (1846). He was interested in ichthyology and botany. He moved to San Francisco (1853) where he practiced medicine and pursued his interest in botany across California. He was founder of the California Academy of Sciences, its Corresponding Secretary (1853–1855) and Curator of Geology & Minerology (1855). He moved to Almedia County (1863) where he again studied medicinal plants, serving as chairman of the committee on botany of the California State Medical Society. Cooper wrote of him: "*...whose descriptions of our viviparous fishes, published in 1854, by the Academy, have only of late been awarded the credit they deserve.*" He was referring to his paper: *Description of four new species of viviparous fish* (1854). He described a number of fish such as the Tule Perch *Hysterocarpus traski* (1853).

Gibbs, PE

Gibbs' Pipefish *Festucalex gibbsi* CE Dawson, 1977

Dr Peter Edwin Gibbs (b.1938) collected the type material. He was associated with the Marine Biological Association of the United Kingdom and published numerous articles in their journal such as: *Observations on the populations of Scoloplos armiger at Whitstable* (1968).

Gibbs, RH

Tallapoosa Shiner *Cyprinella gibbsi* WM Howell & JD Williams, 1971
Barbeled Dragonfish sp. *Eustomias gibbsi* RK Johnson & RH Rosenblatt, 1971
Waryfish sp. *Scopelosaurus gibbsi* E Bertelsen, G Krefft & NB Marshall, 1976
Whipnose Angler sp. *Gigantactis gibbsi* E Bertelsen, TW Pietsch III & RJ Lavenberg, 1981
Lizardfish sp. *Synodus gibbsi* RF Cressey Jr, 1981
Ridgehead sp. *Poromitra gibbsi* NV Parin & OD Borodulina, 1989
Snaggletooth sp. *Astronesthes gibbsi* OD Borodulina, 1992
Gibbs' Lanternfish *Nannobrachium gibbsi* BJ Zahuranec, 2000

Robert Henry Gibbs Jr. (1929–1988) was an American ichthyologist and conservationist. He was a long-standing curator at the Smithsonian and an expert on deep-sea and pelagic fishes. He was a member of the American Society of Ichthyologists and Herpetologists, which honoured him posthumously with the establishment of the Robert H Gibbs Jr. Memorial Award for Excellence in Systematic Ichthyology. He was also a keen collector of odonata across many US States, Canada and Mexico - his meticulous field notebooks are in the Smithsonian. He died of lung cancer.

Giddings

Galápagos Catshark *Bythaelurus giddingsi* JE McCosker, DJ Long & CC Baldwin, 2012

Al Giddings is an underwater filmmaker and naturalist. He has also been active in the commercial film business as the cameraman for: *The Deep* (1977) and as co-producer of: *Titanic* (1997). Moreover, he is a friend of the senior author.

Gigas

Arapaima *Arapaima gigas* HR Schinz, 1822
Giant Seabass *Stereolepis gigas* WO Ayres, 1859
Giant Blenny *Scartichthys gigas* F Steindachner, 1876
Giant Sea Catfish *Arius gigas* GA Boulenger, 1911
Lanternfish sp. *Diaphus gigas* CH Gilbert, 1913
Giant Sawbelly *Hoplostethus gigas* AR McCulloch, 1914
Marine Hatchetfish sp. *Argyropelecus gigas* JR Norman, 1930
Mekong Giant Catfish *Pangasianodon gigas* P Chevey, 1931
Pale Deepsea Lizardfish *Bathysauroides gigas* T Kamohara, 1952
Spotted Stingaree *Urolophus gigas* TD Scott, 1954
Giant Skate *Dipturus gigas* R Ishiyama, 1958
Sonora Blenny *Malacoctenus gigas* VG Springer, 1959
Goby sp. *Scartelaos gigas* Y-T Chu & H-W Wu, 1963
Lake Malawi Cichlid sp. *Labidochromis gigas* DSC Lewis, 1982
Goby sp. *Rhinogobius gigas* Y Aonuma & I-S Chen, 1996
Lake Victoria Cichlid sp. *Haplochromis gigas* O Seehausen & E Lippitsch, 1998
Chinese Catfish sp. *Xiurenbagrus gigas* Y Zhao, J Lan & C Zhang, 2004
Giant Triangular Batfish *Malthopsis gigas* HC Ho & KT Shao, 2010

Gigas was a giant in Greek mythology, the child of Uranus and Gaea. The name is applied to taxa that are giants of their kind.

Giglioli

Tanzanian Cichlid sp. *Haplochromis gigliolii* GJ Pfeffer, 1896

Professor Dr Enrico Hillyer Giglioli (1845–1909) was an Italian zoologist, anthropologist, photographer and ornithologist who graduated from the University of Pisa (1864). He started teaching zoology at the University of Florence (1869) and was Director, Royal Zoological Museum, Florence. He succeeded Filippo de Filippi on the *Magenta* expedition (a government-sponsored circumnavigation) after de Filippi's death from cholera in Hong Kong (1867). Giglioli also wrote up the expedition's results and report. Two birds, an amphibian and a mammal are also named after him. (Also see **Constanza**)

Gilbert, CH

Fathead genus *Gilbertidia* C Berg, 1898
Dogtooth Characin genus *Gilbertolus* CH Eigenmann, 1907
Gilbert's Large Lanternfish *Dianthus adenomus* CH Gilbert, 1905
Rockpool Blenny *Hypsoblennius gilberti* DS Jordan, 1882
Bigmouth Sanddab *Citharichthys gilberti* OP Jenkins & BW Evermann, 1889
Cheekspot Goby *Ilypnus gilberti* CH Eigenmann & RS Eigenmann, 1889
Orangefin Madtom *Noturus gilberti* DS Jordan & BW Evermann, 1889

Corvina Drum *Cilus gilberti* JF Abbott, 1899
Landia Silverside *Membras gilberti* DS Jordan & CH Bollman, 1890
Gilbert's Garden Eel *Ariosoma gilberti* JD Ogilby, 1898
Dogface Witch-eel *Facciolella gilbertii* S Garman, 1899
Gnomefish sp. *Scombrops gilberti* DS Jordan & JO Snyder, 1901
Galapagos Blue-banded Goby *Lythrypnus gilberti* E Heller & RE Snodgrass, 1903
Gilbert's Irish Lord *Hemilepidotus gilberti* DS Jordan & EC Starks, 1904
Gilbert's Flyingfish *Hirundichthys gilberti* JO Snyder, 1904
Gilbert's Hairfin Anchovy *Setipinna gilberti* DS Jordan & EC Starks, 1905
Gilbert's Cardinalfish *Zoramia gilberti* DS Jordan & A Seale, 1905
Gilbert's Spiny Flathead *Hoplichthys gilberti* DS Jordan & RE Richardson, 1908
Flathead Helmet Gurnard *Dactyloptena gilberti* JO Snyder, 1909
Small-disk Snailfish *Careproctus gilberti* CV Burke, 1912
Armoured Gurnard sp. *Scalicus gilberti* DS Jordan, 1921
Rattail sp. *Coelorinchus gilberti* DS Jordan & CL Hubbs, 1925
Large-eye Toadfish *Batrachoides gilberti* SE Meek & SF Hildebrand, 1928
Scaled Sculpin sp. *Icelus gilberti* AY Taranetz, 1936
Cortez Bonefish *Albula gilberti* E Pfeiler, AM van der Heiden, RS Ruboyianes & TD Watts, 2011
Kern River Rainbow Trout *Oncorhynchus mykiss gilberti* DS Jordan, 1894

Dr Charles Henry Gilbert (1859–1928) was an ichthyologist and fishery biologist, whose main area of study was Pacific salmon. He received his bachelor's degree (1879) from Butler University, Indiana, but moved to take his master's (1882) and doctorate (1883) at Indiana University, the first-ever doctorate awarded by that university. Baird asked David Starr Jordan (q.v.) to do a survey of the US west coast fisheries (1879), and Gilbert went as Jordan's assistant on an expedition from British Columbia to Southern California. This expedition lasted a year and was the start of a 50-year-long study of Pacific fishes by Gilbert and Jordan. Gilbert taught at Indiana University (1880–1884 & 1889), at the University of Cincinnati (1885–1888), and at the newly founded Stanford (1890–1925), retiring as Emeritus Professor. He served as naturalist-in-charge on cruises of the US Fish Commission's vessel 'Albatross' in Alaskan waters (1880s & 1890s), Hawaii (1902) and Japan (1906). The Gilbert Fisheries Society was established (1931) at the College of Fisheries, University of Washington, later becoming (1989) the Gilbert Ichthyology Society. Alone or jointly he described 117 new genera and 620 species of fish, writing 172 papers on them. Three reptiles are also named after him.

Gilbert, CR

Yellow Jawfish *Opistognathus gilberti* JE Böhlke, 1967
Gilbert's Blenny *Cirripectes gilberti* JT Williams, 1988
Goby sp. *Sicydium gilberti* RE Watson, 2000
Gulf Coast Pygmy Sunfish *Elassoma gilberti* FF Snelson, TJ Krabbenhoft & JM Quattro, 2009
Carolina Hammerhead Sphyrna gilberti JM Quattro, WB Driggers, JM Grady, GF Ulrich & MA Roberts, 2013

Dr Carter Rowell Gilbert (1930-2022) was an American zoologist and ichthyologist. He was first to notice the variation in vertebrae among what were thought to be scalloped hammerheads, and led to the description of *Sphyrna gilberti.* Ohio State University awarded his bachelor's degree (1951) and his master's (1953), and the University of Michigan his doctorate (1960). He worked for Florida State Museum and the University of Florida, Gainesville (1961–1998), retiring as Curator Emeritus. Like many naturalists he was a keen philatelist, a hobby he took up at the age of six. He wrote among other papers: *A revision of the Hammerhead Sharks (family Sphyrnidae),* (1967). He wafs also the friend and former colleague of JE Böhlke and the collector of the jawfish type. (See also **Carter**)

Gilbert, J

Gilbert's Grunter *Pingalla gilberti* GP Whitley, 1955

John Gilbert (c.1812–1845) was an English naturalist and explorer in south-western Western Australia. He was the principal, and assiduous, collector of birds for Gould (1840–1842) who, despite high expectations, served Gilbert badly, often leaving him with insufficient funds and equipment and barely acknowledging his huge contribution of specimens, descriptions and detailed observations. Gilbert's specimens were sent back to Europe on board *Beagle* and *Napoleon.* Gilbert once stayed with a settler, Mrs Brockman, who wrote: "*He used to go out after breakfast... ...and we seldom saw him until late afternoon, when he would come in with several birds and set busily to work to skin and fill them out before dark... He was an enthusiast at his business, never spared himself, and often came in quite tired out from a long day's tramp after some particular bird, but as pleased as a child if he succeeded in shooting it.*" His employer, the Zoological Society of London trained him as a taxidermist. He was Curator of the Shropshire and North Wales Natural History Society in Shrewsbury (until 1837). He left for Australia (1838) becoming naturalist on Ludwig Leichhardt's expedition to Port Essington (1844–1845). At the Gulf of Carpentaria, Aborigines speared him to death (1845). Two of the expedition, aboriginals themselves, probably caused the attack, having treated the people they met very badly including raping a local aboriginal woman. Leichhardt's account says Gilbert heard the noise of the attack in the night and rushed from his tent with his gun, only to receive a spear in the chest. The only words he spoke were: "*Charlie, take my gun, they have killed me*" before dropping lifeless to the ground. Four birds, a reptile and two mammals are also named after him.

Gilbert, M

Lake Victoria Cichlid sp. *Haplochromis gilberti* PH Greenwood & JM Gee, 1969

Michael Gilbert was an experimental fisheries officer in Kenya at the East African Freshwater Fisheries Research Organization (1965). The etymology says: "*The species is named in honour of Mr. Michael Gilbert, Experimental Fisheries Officer of the East African Freshwater Fisheries Research Organization. Michael Gilbert's enthusiasm and skill have added considerably to our knowledge of the Lake Victoria fishes.*"

Gilbert (surgeon)

Characin sp. *Cyphocharax gilbert* JRC Quoy & JP Gaimard, 1824

M Gilbert was a French naval surgeon who died of yellow fever in the Antilles. Nothing more is known, not even if the M stands for 'Monsieur' or for a first name.

Gilberto

Neotropical Rivuline sp. *Cynolebias gilbertoi* WJEM Costa, 1998

Gilberto Campello Brasil (1945–d.2008?) was a Brazilian aquarist and amateur ichthyologist. He trained as a chemical enginner having a BSc in physics and a PhD in chemical engineering. But his passion, from 1954 was aquaria and he collected a number of new plants and fish, particularly killifish. He took almost annual trips throughout Brazil and beyond to collect and worked closely with Wilson Costa ichthyologist at the Federal University of Rio de Janeiro. He was a technical advisor to the Ministry of the Environment when he disappeared (2008) and is presumed dead. He was seen to leave his apartment alone, stop to refuel at a petrol station, drove away and was never seen again. His cousin donated his collection to the Botanical Garden of Brasilia. (Also see **Campello** and **Gilberto Brasil**)

Gilberto Brasil

Neotropical Rivuline sp. *Hypsolebias gilbertobrasili* WJEM Costa, 2012

Gilberto Campello Brasil (See **Gilberto,** also see **Campello**)

Gilchrist

Round Herring genus *Gilchristella* HW Fowler 1935
Gilchrist's Round Herring *Gilchristella aestuaria* JDF Gilchrist, 1913
Prison Goby *Caffrogobius gilchristi* GA Boulenger, 1898
Grenadier Cod *Tripterophycis gilchristi* GA Boulenger, 1902
Ripple-fin Tongue Sole *Cynoglossus gilchristi* CT Regan, 1920
Gilchrist's Scorpionfish *Scorpaenopsis gilchristi* JLB Smith, 1957

Dr John Dow Fisher Gilchrist (1866–1926) was a Scottish-born South African ichthyologist and naturalist. He studied at Edinburgh for his BSc, and his doctorate (in geology) was awarded by the University of Jena, Germany (1894). He taught zoology at Edinburgh University for a while before taking up a post in South Africa. He was the first marine biologist of the Cape Colony Department of Agriculture (1895). He became Director of Fisheries and of the Marine Biology Survey of South Africa (1896–1907), and Professor (1905–1926) at Cape Town University. He co-wrote: *The Freshwater Fishes of South Africa* (1913). He led (1920) marine survey expeditions in a converted whaler, 'Pickle'. He took early retirement due to ill health (1926) and travelled to Europe to recover, but died soon after he returned to South Africa.

Gildi

Clipperton Cardinal Soldierfish *Myripristis gildi* DW Greenfield, 1965

Mrs Gildi Greenfield is the author's wife, of whom he wrote in the etymology that her "*... efforts in translating numerous foreign publications have added considerably...*" to his revision of the genus.

Giles

Tonguefish sp. *Symphurus gilesii* AW Alcock, 1889

Dr Major George Michael James Giles IMS (1853–1916) was a physician, entomologist and army officer in the Indian Medical Service (1880). He was Surgeon-Naturalist to the marine Survey of India (1884–1888) on board RIMS 'Investigator'. His published papers include: *On the structure and habits of Cyrtophium* (1885) and *Descriptions of seven additional new Indian Amphipods* (1890). He was later sent to Assam to look at Beriberi and compiled a report: *A report of an investigation into the causes of the diseases known in Assam as Kaȉ_la-Azaȉ_r and Beri-Beri* (1890). He was promoted to surgeon major (1890) and retired as Lieutenant-Colonel (1901) to England. He initiated work with mosquitos, publishing (1899) a description of species experimented with in relation to malaria. He subsequently published a handbook on mosquitos: *A handbook of the gnats or mosquitoes giving the anatomy and life history of the Culicidae, together with descriptions of all species noticed up to the present date* (1900). He emigrated to Canada (1912) but returned to England as part of the Canadian Army Medical Corps (1914). He was in charge of a hospital until ill-health forced his retirement (1916), and he died six months later.

Gill, EL

Clanwilliam Catfish *Austroglanis gilli* KH Barnard, 1943

Edwin Leonard Gill (1877–1956) worked in various museums in the United Kingdom before being appointed Director, South African Museum, Cape Town (1925). He implemented and carried through a programme of modernisation and expansion. The museum suffered financially (WW2) and Gill decided to ameliorate the situation by resigning (1942) so that his salary would not be a burden. He was a skilled hobby taxidermist at home, where his sister helped him. An amphibian and two birds are named after him.

Gill, TN

Goby genus *Gillichthys* JG Cooper, 1864
Triplefin genus *Gilloblennius* GP Whitley & WJ Phillips, 1939
Gill's Molly *Poecilia gillii* R Kner, 1863
Dusky Blenny *Malacoctenus gilli* F Steindachner, 1867
Twospot Brotula *Neobythites gilli* GB Goode & TH Bean, 1885
Manacled Sculpin *Synchirus gilli* TH Bean, 1890
Bronze-spotted Rockfish *Sebastes gilli* RS Eigenmann, 1891
Gill's Sand Lance *Ammodytoides gilli* TH Bean, 1895
Flabby Whalefish sp. *Cetomimus gillii* GB Goode & TH Bean, 1895
Spiny Sucker Eel *Lipogenys gilli* GB Goode & TH Bean, 1895
Gill's Cusk-eel *Bassogigas gillii* GB Goode & TH Bean, 1896
Goby sp. *Eutaeniichthys gilli* DS Jordan & JO Snyder, 1901
Characin sp. *Cyphocharax gillii* RS Eigenmann & CH Kennedy, 1903
Eelpout sp. *Zoarces gillii* DS Jordan & EC Starks, 1905
Choelo Halfbeak *Hyporhamphus gilli* SE Meek & SF Hildebrand, 1923
Viviparous Halfbeak *Zenarchopterus gilli* HM Smith, 1945

Professor Theodore Nicholas Gill (1837–1914) was an American ichthyologist, malacologist, mammologist and librarian, a zoologist at George Washington University and associated with the Smithsonian Institution for more than half a century. His father wanted him to enter the Church, but this seemed an unattractive calling to Gill who decided to qualify as a lawyer instead. He was fortunate enough to come to the attention of Spencer Baird (c.1857) whom he met in Washington while en route to the West Indies, where he made an important collection, especially of freshwater fishes from Trinidad. He went to Newfoundland (1859), these two trips being the only extensive fieldwork Gill ever carried out. He was put in charge of the Smithsonian's Library (1862), which was transferred (1866) to the Library of Congress, where Gill served as Senior Assistant Librarian (1866–1874). During his career, he produced over 500 papers of which 388 were on ichthyology. For a time, he also edited the ornithological magazine 'The Osprey'. A marine mammal is also named after him.

Gillbanks

Gillbanks' Globe Fish *Arothron gillbanksii* FE Clarke, 1897
[Junior Synonym of *A. firmamentum*]

Mr Gillbanks was "*...foreman of works at the mole*" (jetty) at Moturoa, Taranaki, New Zealand. He found the fish washed up alive, put it in a bucket of seawater and presented it to Clarke.

Gilliss

Southern Bass sp. *Percilia gillissi* CF Girard, 1855

James Melville Gilliss (1811–1865) was a US Navy astronomer who founded the United States Naval observatory. When just fifteen years old he joined the navy as a midshipman (1827) and made a number of training voyages, passing his first midshipman examinations (1833). He then took a leave of absence to study at the University of Virginia and in Paris (1835), being recalled (1836) to the Depot of Charts and Instruments giving him a chance to practice astronomy. He was made officer in charge (1837) and promoted (1838) to lieutenant. He made observations published as *Astronomical Observations made at the Naval Observatory, Washington* (1846), the first star catalogue published in the USA. He headed a naval astronomical expedition to Chile (1849–1852) which not only made observations but also collected fauna and flora for the Smithsonian. He later led two expeditions to observe the solar eclipses in Peru (1858) and Washington Territory (1860). He was put in charge of the observatory (1861) and was promoted to Captain (1862). He collapsed and died of a stroke aged just 53.

Gill Morlis

Neotropical Cichlid sp. *Crenicichla gillmorlisi* SO Kullander & CAS Lucena, 2013

Walter A Gill Morlis is a Paraguayan ichthyologist who is a fisheries officer of the Itaipú Binacional, Ciudad del Este, Paraguay. He co-authored: *A new species of* Otothyropsis *(Siluriformes: Loricariidae) from the upper Río Paraná basin, Paraguay, with a discussion of the limits between Otothyropsis and Hisonotus* (2017). The etymology says: "*Named for ichthyologist Walter A Gill Morlis A., fisheries officer of the Itaipú Binacional, Ciudad del Este, Paraguay, who contributed considerably to the PROVEPA surveys of fishes in tributaries of the*

Paraná River, and in special recognition of his strong long term engagement in the inventory of the fishes of the río Paraná."

Gilmore

Pepperfin Cusk-eel *Lepophidium gilmorei* CR Robins, RH Robins & ME Brown, 2012

Dr R Grant Gilmore was awarded his doctorate by the Florida Institute of Technology (1988). He is a Senior Scientist with Estuarine, Coastal and Ocean Science, which he founded (2004) after working (1972–2004) with the Harbor Branch Oceanographic Institution, Fort Pierce, Florida and Dynamic Corp., Kennedy Space Center. He wrote: *Hypothermal mortality in marine fishes of south-central Florida* (1978).

Gilson

Lake Titicaca Pupfish sp. *Orestias gilsoni* VV Tchernavin, 1944

Hugh Cary Gilson (1910–2000) was a botanist who was leader of the Percy Sladen Trust Titicaca Expedition (1937). He went to Trinity College, Cambridge to read classics but switched to Natural Sciences, specialising in zoology and anatomy and achieving a double first. When he returned from South America he taught zoology at Cambridge. He became the Director of the Freshwater Biological Association (1946–1973). He wrote: *The British Palmate Orchids* (1930) and *The Nitrogen Cycle* (1937).

Gimbel

Ghost Knifefish sp. *Porotergus gimbeli* MM Ellis, 1912

Jacob 'Jake' Gimbel (1876–1943) was an Indiana philanthropist. He financed the Gimbel Expedition (1911) to British Guiana (Guyana) where the holotype was collected. The author, Ellis (q.v.) was the expedition's leader.

Ginsburg

Goby genus *Ginsburgellus* JE Böhlke & CR Robins, 1968
Seaboard Goby *Gobiosoma ginsburgi* SF Hildebrand & WC Schroeder, 1928
Darkblotch Goby *Parrella ginsburgi* CB Wade, 1946
Ginsburg's Tonguefish. *Symphurus ginsburgi* NA de Menezes & G de Quadros Benvegnú Lé, 1976

Isaac Ginsburg (1886–1975) was a Lithuanian-born goby taxonomist at the US National Museum and for the US Fish & Wildlife Service. He studied ichthyology at Cornell University and, after graduating, spent a short time as an aid in the Division of Fishes, United States National Museum (1917). He worked (1922–1956) with the Bureau of Fisheries until his retirement. His main focus was on the marine fishes of the Gulf of Mexico. He identified many gobies for SF Hildebrand's and WC Schroeder's monograph on fishes of Chesapeake Bay, and called attention to how the *Gobiosoma* species differed from *G. bosc.* Böhlke honoured him in the genus name, describing him as a 'prominent student of American gobies'. He was a colleague of Wade's who honoured him for his 'work with this difficult family'.

Girard, AAA

African Barb sp. *Labeobarbus girardi* GA Boulenger, 1910

Alberto Arthur Alexandre Girard (1860–1914) was a French-Portuguese zoologist at Museu Bocage, Lisbon. Principally interested in marine zoology, he spent time aboard the trawler 'Machado' observing species coming up from the depths. A reptile is also named after him.

Girard, CF

Topminnow genus *Girardinus* F Poey, 1854
Splitfin (goodeid) genus *Girardinichthys* P Bleeker, 1860
Arkansas River Shiner *Notropis girardi* CL Hubbs & AI Ortenburger, 1929
Potomac Sculpin *Cottus girardi* CR Robins, 1961

Dr Charles Frédéric Girard (1822–1895) was a French herpetologist and ichthyologist who was Louis Agassiz's pupil and assistant at Neuchâtel and who moved with Agassiz to the USA. He was in Cambridge, Massachusetts (1847–1850), and worked with Baird (1850–1857), establishing the Smithsonian. He became an American citizen (1854) and while continuing his work at the Smithsonian, studied medicine and graduated MD from Georgetown College (1856). He briefly visited Europe (1860). During the American Civil War, he sided with the Confederacy and supplied the Confederate army with medical and surgical supplies. He left the USA (1864), returned to France and practised medicine there, serving as a physician during the siege of Paris (1870) during the Franco-Prussian War. The etymology for the sculpin says that it was "*...named in honour of Charles Girard, an early student of the genus.*" Two mammals and five reptiles are also named after him.

Gislen

Characin sp. *Astyanax gisleni* G Dahl, 1943

Dr Torsten Richard Emanuel Gislén (1893–1954) was a Swedish zoologist who was interested in herpetology and echinoderms. He studied botany at Uppsala, but his interests turned more towards zoology and his doctorate was awarded (1924) in the latter subject. He was Professor of Zoology at Lund University (1932–1954) and gave Dahl the chance of working at the Zoological Institute of Lund. He wrote: *Den uppgående solens land* (1933).

Giti

Giti Damselfish *Chrysiptera giti* GR Allen & MV Erdmann, 2008

Giti Tire Company of Singapore is ranked the eleventh biggest company by revenue in the world. The name was given at the request of its Chairman, Enki Tan, and his wife, Cherie Nursalim, Vice Chairman of GITI Group, who have given 'generously' to Conservation International's Bird's Head Seascape marine conservation initiative and successfully bid to support the conservation of this species at The Blue Auction, a black-tie charity auction in Monaco (20 Sept. 2007).

Gjellerup

Northern Tandan *Neosilurus gjellerupi* MCW Weber, 1913

Snake-eel sp. *Yirrkala gjellerupi* MCW Weber & LF de Beaufort, 1916

Gjellerup's Mouth Almighty *Glossamia gjellerupi* MCW Weber & LF de Beaufort, 1929

Dr Knud Gjellerup (1876–1950) was a Danish physician and botanist. He was an army surgeon in the Dutch East Indies army in Sumatra, Java and the Celebes (Sulawesi) (1904–1921). He was a member of a detachment charged with exploring in Dutch North New Guinea (1909–1912), during which time he was a member of the Dutch North New Guinea Humboldt Bay Expedition (1909–1910) and the Dutch-German Boundary Delimitation Expedition (1910). He also explored extensively in German North New Guinea (1911–1912). He practiced as an otolaryngologist (ear-nose-and-throat specialist) in central Java (1921–1932). He returned to Denmark, lived in Copenhagen (1932–1936) and Bornholm (1936–1950), and travelled in Europe, North America and eastern Asia. He collected the holotype.

Gladisfen

Prickly Dreamer *Spiniphryne gladisfenae* W Beebe, 1932

'Gladisfen' was a ship built in the 1880s as a New York harbour tugboat. It enjoyed a new lease of life as Beebe's scientific research vessel in Bermuda (early 1930s). The dreamer holotype was taken 'six miles south of Nonsuch Island, Bermuda'.

Gladkov

Spiny Loach sp. *Cobitis gladkovi* VP Vasil'ev & ED Vasil'eva, 2008

Professor Nikolay Alekseyevich Gladkov (1905–1975) was a Russian ornithologist. He joined the Natural Science Department of Moscow University (1926) and was concurrently made chief of the Biological Station in StaroPershino. He became a researcher at the Zoological Museum of Moscow Lomonosov University (1934), eventually becoming Director (1964–1969). During his Red Army service (WW2) he was captured by the Germans and held in a camp in France until liberated by American forces (1944). He returned to the Zoological Museum (1947) and became Chief of the Ornithological Division, leaving (1954) for the Zoogeography Department of Moscow University. He remained as Professor of Zoogeography there, combining this with being Director, Zoological Museum of Moscow Lomonosov University (1964–1969). He co-wrote the multi-volume: *Birds of the Soviet Union* (English translation 1966–1969). Two birds are named after him.

Gladys (Baudon)

African Loach Catfish sp. *Phractura gladysae* J Pellegrin, 1931

Gladys Baudon was the daughter of Alfred Baudon (q.v.), a French colonial administrator in West Africa. Baudon sent several collections of fish specimens to Pellegrin. Gladys apparently helped her father in his fisheries research, but we can locate no further biographical details.

Gladys (de Gonzo)

Gladys' Cory *Corydoras gladysae* PA Calviño & F Alonso, 2010

Gladys Ana María Monasterio de Gonzo is an ichthyologist whose master's degree was awarded by Universidad Nacional del Litoral (1999). She was an Adjunct Professor, Facultad de Ciencias Naturales, Universidad Nacional de Salta (1983–2008), having been a researcher there (1977–1983). She was also director of the university's Museum of Natural Sciences (1993–1994). She wrote: *Peces de los ríos Bermejo, Juramento y cuencas endorreicas de la provincia de Salta* (2003) which is echoed in the etymology that states she was honoured "*...for contributions to the diversity, distribution and biology of fishes in the Salta province of Argentina*" and that she was the first person to collect this species.

Glaphyra

Gyre Flyingfish *Prognichthys glaphyrae* NV Parin, 1999

Glafira Nikiforovna 'Tanya' Pokhil'skaya worked in the Department of Nekton and the Laboratory of Oceanic Ichthyofauna (Russian Academy of Sciences) (1948–1990). Parin has used her 'high-quality' (translation) illustrations in many of his papers.

Glauert

Glauert's Seadragon *Phycodurus eques* A Günther, 1865*
[Alt. Leafy Seadragon]
Glauert's Anglerfish *Allenichthys glauerti* GP Whitley, 1944

Ludwig Glauert (1879–1963) was born in England and trained as a geologist. He moved to Perth, Western Australia (1908), and joined the geological survey as a palaeontologist. He volunteered at the Western Australian Museum (1908–1910), joining the permanent staff (1910) as Scientific Assistant, then Keeper of Geology and Ethnology (1914). He worked on the Margaret River caves (1909–1915), studying remains from the Pleistocene. He served in the Australian Army (1917–1919) and then studied Australian material in the British Museum before returning to Perth (1920) as keeper, Western Australian Museum's biological collections, becoming Curator (1927) and eventually Director (1954). His interests were legion – he was the leading authority on Western Australian reptiles, used his own money to buy books for the museum, and helped with the taxidermy. He retired (1956) but went on working on reptiles and scorpions. Three birds, a reptile and two amphibians are also named after him.

* It may seem odd that a species can have an eponymous common name when the recipient wasn't even born when it was discovered; it is explained by synonymity. Whitley (1939) described the species as *Phycodurus glauerti* believing it to be new to science. Though this was incorrect, Glauert's name became attached to the species in vernacular usage.

Glen

Chinese Sleeper *Perccottus glenii* BN Dybowski 1877
[Alt. Amur Sleeper]

Colonel Nikolay Alexandrovich Glen was honoured because of his efforts that improved the well-being of the Ussuri River area (Ussuriland) of Russia, the type locality.

Glenda

Slender Triplefin *Enneanectes glendae* RH Rosenblatt, EC Miller & PA Hastings, 2013

Glenda G Rosenblatt (d.2014) was the wife of the senior author, Richard Rosenblatt (q.v.).

Glenys

Twoblotch Ponyfish *Nuchequula glenysae* S Kimura, R Kimura & K Ikejima, 2008

Glenys Jones (b.1952) was honoured for her work on the revision of Australian leiognathid fishes (1985) at the Division of Fisheries Research, CSIRO Marine Laboratories, Cronulla, New South Wales. The University of New South Wales awarded her BSc (1975) after which she was a Scientific Officer at NSW State Pollution Control Commission (1975–1977), then joined CSIRO Division of Fisheries and Oceanography (1979–1983). She was publications officer at the Commission for the Conservation of Antarctic Marine Living Resources (1985–1986) and landscape designer at NatureScapes Australia for three years (1986–1989). She then joined the Tasmanian Parks & Wildlife Service as an Assistant Planning Officer, World Heritage Area (1989–2004) and now as a Planner, Policy & Projects - Coordinator Evaluation (2005 on). She is married to marine research scientist Dr Keith John Sainsbury (Also see **Sainsbury**)

Gloerfelt

Pufferfish sp. *Torquigener gloerfelti* GS Hardy, 1984
Tilefish sp. *Branchiostegus gloerfelti* JK Dooley & PJ Kailola, 1988

Thomas Gloerfelt-Tarp (b.1949) is a Danish independent consultant fisheries expert currently based in Australia. The University of Copenhagen awarded his master's in Zoology & Botany (1978) after which he took his teaching certificate (1979). He began his career as a Marine Biologist on the GTZ research vessel in Indonesian waters (1979–1984) becoming head of fisheries there (1984–1990). He was then a Project Leader for the Malawi and German Fisheries and Aquaculture Development (1990–1993) then worked for the Danish International Development Agency (1993–1995) reviewing fisheries in Asia and Africa. He was with the Asian development Bank, Manila (1996–2005) then in Fiji (2005–2009). He began working as a consultant (2010) and has continued this all over the Pacific. He co-wrote: *Trawled Fishes of Southern Indonesia and Northwest Australia* (1984).

Gloria (Arratia)

Chilean Pupfish sp. *Orestias gloriae* I Vila, S Scott, MA Mendez, F Valenzuela, P Iturra & E Poulin, 2012

Dr Gloria Arratia is a Chilean biologist, ichthyologist and ichthyo-palaeontologist whose research focus is fish evolution, particularly of 'ray-finned' fish. She is currently an Associate Research Scientist and a fish curator at the Biodiversity Institute of the University of Kansas. Among her c.150 published papers are the co-written: *Outstanding features of a new Late Jurassic pachycormiform fish from the Kimmeridgian of Brunn, Germany* (2013) and the solely authored: *Morphology, taxonomy, and phylogeny of Triassic pholidophorid fishes (Actinopterygii, Teleostei)* (2013).

Gloria (Echevarria)

Cusk-eel sp. *Neobythites gloriae* F Uiblein & JG Nielsen, 2018

Gloria Jansen Echevarria is the wife of the first author Dr Franz Uiblein (See **Franz**).

Glover

Pufferfish sp. *Lagocephalus gloveri* T Abe & O Tabeta, 1983
[Junior Synonym of *Lagocephalus cheesmanii*]
Glover's Anglerfish *Rhycherus gloveri* TW Pietsch III, 1984
Glover's Hardyhead *Craterocephalus gloveri* LE Crowley & W Ivantsoff, 1990
Goby sp. *Tasmanogobius gloveri* DF Hoese 1991
Dalhousie Goby *Chlamydogobius gloveri* HK Larson, 1995
Dalhousie Catfish *Neosilurus gloveri* GR Allen & MN Feinberg, 1998

Dr Charles John Melville Glover (1935–1992) was ichthyologist at the South Australian Museum, Adelaide (1964–1991) and was Curator of Fishes (1967–1991). He was an expert on Central Australian fishes, and made many expeditions to the central Australian desert (1967–1992) particularly to the area around Dalhousie Springs. He wrote: *Freshwater Fish of South Australia* (1971) and *Adaptations of a central Australian gobiid fish* (1973).

Gmelin

Dagestan Spirlin *Alburnoides gmelini* NG Bogutskaya & BW Coad, 2009

Dr Samuel Georg Gotlieb Gmelin (1744–1774) was a member of the German family that produced several important scientists and naturalists. He qualified as a physician at the University of Leiden (1763) but by inclination was a marine biologist with a passion for algae. He wrote: *Historia Fucorum* (1768), a seminal work as it was the first such book on marine biology dealing exclusively with algae and using the binomial system of nomenclature. He was appointed Professor of Botany at St Petersburg (1766) and was sent (1767) to explore the Volga and Don rivers and the western and eastern coasts of the Caspian Sea (1868–1774). While traveling in the Caucasus he was captured by a Kaitak tribesman, Usmey Khan, and held for ransom. He was ill-treated and died in captivity in Derbent (in present-day Dagestan but then part of Persia). The results of his explorations were published as the four-volume: *Reise durch Russland zur untersuchung der drey natur-reiche* (1770–1784) the last of which was part completed by Güldenstadt (q.v.) and finally edited by Pallas (q.v.) after Güldenstadt's death.

Gnanadoss

Conger Eel sp. *Ariosoma gnanadossi* PK Talwar & P Mukherjee, 1977

D A S Gnanadoss worked (1960s-1970s) at the Central Institute of Fisheries Operatives, Madras Unit, whose fishing trawler collected the holotype. He was in Nigeria (1980s) advising on inshore fisheries development and is now an independent fisheries consultant based in Canada. He wrote the article: *Fisheries education and training in India* (1977).

Gobron

Squeaker Catfish sp. *Synodontis gobroni* J Daget, 1954

M (probably Monsieur) Gobron was a volunteer at Laboratoire de Diafarabé, Mail. He collected the holotype. We have been unable to find any more about him.

Godeffroy

Tailface Sleeper *Calumia godeffroyi* A Günther, 1877
Godeffroy Worm Eel *Scolecenchelys godeffroyi* CT Regan, 1909

Johann Cesar Godeffroy (1813–1885) was a member of a German trading house importing copra from the Pacific and was interested in natural history in general and ornithology in particular. His family was French but had moved to Germany to avoid religious persecution. His fleet of ships traded largely in the Pacific, where he used them as floating collection bases, with paid collectors on board to search for zoological specimens as well as trading commercially. He established around 45 trading posts and bought land and property, laying the foundations of German colonial power. He used his natural history collection to found a museum in Hamburg (1860), naming it after himself. He employed well-known naturalists in Hamburg, such as Otto Finsch and Gustav Hartlaub. Eventually he neglected commerce and the company went bankrupt (1879). The worm-eel is perhaps named after the Godeffroy museum rather than directly after J.C. Godeffroy, as the type was housed there. A reptile and two birds are named after him.

Godlewski

Godlewski's Baikal Sculpin *Limnocottus godlewskii* B Dybowski, 1874

Wiktor Ignacy Aleksandrovich Godlewski (1831–1900) was a Polish collector. He was exiled in Siberia after the failed Polish uprising (1863) in Ukraine. With him were other Polish naturalists who were also exiles, including Benedykt Dybowski who trained him as a collector. He settled in a village by Lake Baikal (1867). He studied the local fauna (1867–1877) thanks to the Zoological Cabinet in Warsaw, as the exiles were refused help by the EasternSiberian Department of Imperial Russian Geographical Society. They sent their material to the Cabinet, and Taczanowski (q.v.) negotiated the sale of part of it to the museums of Western Europe. The funds obtained, together with donations from the Branicki brothers, enabled the exiles' survival and continued study. Most of their research on the lake took place in winter when they studied ice that Godlewski himself chiselled out, thanks to his technical abilities and great physical strength. They also explored the region (1876) around the Angara River source and Kosogol Lake (Khubsugul), Mongolia, leading to the realisation that the Lake Baikal fauna was unique. They were eventually allowed to return to Poland (1876). Two birds are also named after him.

Godman

Southern Checkmark Cichlid *Theraps godmanni* A Günther, 1862
Central American Rivuline sp. *Cynodonichthys godmani* CT Regan, 1907

Dr Frederick DuCane Godman (1834–1919) was a British naturalist (particularly entomologist and ornithologist) who, with his friend Osbert Salvin (and 18 others), co-founded the BOU and compiled the massive *Biologia Centrali Americana* (1888–1904). They presented their joint collection to the BMNH periodically (1885–1870). Godman qualified as a lawyer, but was wealthy and did not need to work, so devoted his life to ornithology. He visited Norway, Russia, the Azores, Madeira, the Canary Islands, India, Egypt, South Africa, Guatemala, British Honduras (Belize) and Jamaica, sometimes with Salvin. The trip to India (1886) was with H J Elwes to whose sister Edith he had been married until she died (1875). He remarried (1891) Sir Francis Chaplin's sister Alice and she accompanied him on his African and West Indies travels. The Godman-Salvin Medal, a prestigious award of the BOU, is named after them. Godman collected the rivuline holotype in Guatemala. Five reptiles, four birds, three mammals and an amphibian are also named after him. [Although Günther credited 'Godman' (one n) with collecting the type, he did not explicitly state that he named the cichlid after him, therefore his apparent Germanized spelling (double n) is retained. Interestingly, CT Regan also used the Germanic spelling (double n) for *Rivulus* (now *Cynodonichthys*) *godmani* (1907), but emended it later that year; a spelling that is in prevailing usage.]

Goebel

Caspian Goby sp. *Ponticola goebelii* K Kessler, 1874
[Syn. *Ponticola ratan goebelii*]

A man named Goebel provided the goby holotype to the Academy of Sciences in Saint Petersburg, Russia. This probably refers to Friedemann Adolph Goebel (1826–1895), a German geologist who was a member of the Academy when Karl Kessler was there. He is known to have made studies in the Caspian Sea region.

Goeldi

Three-barbeled Catfish genus *Goeldiella* CH Eigenmann & A Norris, 1900
Pencil Catfish sp. *Trichomycterus goeldii* GA Boulenger, 1896
Long-whiskered Catfish sp. *Cheirocerus goeldii* F Steindachner, 1908
Halftooth Characin sp. *Hemiodus goeldii* F Steindachner, 1908
Goeldi's Tube-snouted Ghost Knifefish *Sternarchorhynchus goeldii* CDCM de Santana & RP Vari, 2010

Emil August Goeldi (1859–1917) was a Swiss-Brazilian naturalist and zoologist. He studied zoology in Germany with Ernst Haeckel, and (1884) was invited by Ladislau de Souza Mello Netto, the influential Director of the Brazilian Museu Imperial e Nacional, to work at that institution. Later (1894) he reorganised the Pará Museum of Natural History and Ethnography: the institution now (since 1902) bears his name – Museu Paraense Emílio Goeldi. He became well-known for his studies of Brazilian birds and mammals. Due to failing health Goeldi returned to Switzerland (1905) to teach biology and physical geography at the University of Bern until his death. He was reputed to be a racist who did not like Brazilians and worked in a Swiss enclave! He wrote: *Aves do Brasil* (1894). He was also a noted figure in public health and epidemiology, because he studied the mechanism of transmission of yellow fever and advocated combatting mosquitos as the vector of the disease. Two mammals, two amphibians and three birds are named after him as are insects, arthropods and even plants.

Goerner

Corfu Dwarf Goby *Knipowitschia goerneri* H Ahnelt, 1991
See under **Görner**.

Goethe

Characin sp. *Hoplocharax goethei* J Géry, 1966

Charles Matthias Goethe (1875–1966) was an American eugenicist, entrepreneur, conservationist and philanthropist. He founded Sacramento State College (now California State University, Sacramento). The university's arboretum and botanic gardens were named after him, but the university changed the names (2005) as an expression of disgust at his racist views, praise of Nazi Germany and advocacy for eugenics.

Goetzee

Red Velvetfish *Gnathanacanthus goetzeei* P Bleeker, 1855

J W Goetzee, a friend of the author, discovered the species in Hobart, Tasmania. He sent it with others to Bleeker, warning him how poisonous the fish was to eat.

Goheen

Liberian Mormyrid *Mormyrus goheeni* HW Fowler, 1919

Dr Sylvanus McIntyre E Goheen (1813–1851) was a Methodist medical missionary in Liberia (1837–1842). He was co-editor of a newspaper: *Africa's Luminary*, locally produced in Monrovia and first appeared (1839). He was in practice in Lebanon, Illinois and teaching at McKendree College, Lebanon, Illinois (1846). He became infected by gold fever and set out for California, reaching there (1850), set up in practice but died within six months. He was the first person known to have collected fishes in Liberia.

Gökhan

Bleak sp. *Alburnus goekhani* M Özuluğ, MF Geiger & J Freyhof, 2018

Colonel Gökhan Peker (d.2017) was the cousin of the senior author, Turkish ichthyologist Müfit Özuluğ. He died tragically in a helicopter accident.

Gokkyi

Mountain Carp sp. *Psilorhynchus gokkyi* KW Conway & R Britz, 2010

This is a toponym referring to Gokkyi, a small village above the type locality in Myanmar. The name was given to honour the hospitality and help extended to the second author during his collection trip (November 2009).

Golani

Moray sp. *Uropterygius golanii* JE McCosker & DG Smith, 1997
Golani's Lizardfish *Saurida golanii* BC Russell, 2011
Round Herring sp. *Etrumeus golanii* J DiBattista, JE Randall & BW Bowen, 2012
Red Sea Silverside *Hypoatherina golanii* D Sasaki & S Kimura, 2012

Dr Daniel Golani is an ichthyologist and Curator of Life Sciences, Faculty of Science, Hebrew University of Jerusalem, which awarded his doctorate (1988). He became a Researcher and Curator of the Fish Collection (1991) and Coordinator of the Natural History Collections (1995). Among his c.190 published papers he co-wrote: *Review of the Moray Eels (Anguilliformes: Muraenidae) of the Red Sea* (1995) and among his books is: *Checklist of the Mediterranean Fishes of Israel* (2005). He was the first person to bring specimens of the moray species to the authors' attention.

Goldie

Papuan Black Snapper *Lutjanus goldiei* W Macleay, 1882
Goldie River Mullet *Cestraeus goldiei* W Macleay, 1883
Goldie River Rainbowfish *Melanotaenia goldiei* W Macleay, 1883

Andrew Goldie (1840–1891) was born in Scotland and died in Port Moresby, Papua New Guinea. He moved to New Zealand (1862), working there as a nurseryman (c.1873). He then travelled to Britain (1874) and Melbourne (1875) before beginning his explorations in New Guinea (1876). Whilst collecting there (1877) he discovered and named the Goldie River after himself and discovered traces of gold. He then explored the south coast of New Guinea (1878), naming the Blunden River, Milport Harbour and Glasgow Harbour. Goldie returned to Sydney (1878) to have gold samples assayed. He bought land near Hanubada, New Guinea, and established a trading store (1878). He named the Redlick group of islands and discovered Teste Island in Freshwater Bay (1879). He was given £400 as compensation by the government (1886) when they decided to remove European settlement from the Hanubada area, and he used it to purchase 50 suburban acres and three town allotments on which he built Port Moresby's first store (1897). It has been suggested that many of the scientific discoveries claimed by Goldie were actually made by his associate, Carl von Hunstein. Five birds are also named after him. In the (1882) snapper description, Macleay refers to Andrew Goldie but in the (1883) mullet description he refers to 'Alex' Goldie; we assume this is a mere *lapsus calami*. The rainbowfish may be named after the river rather than directly after Goldie, as the original (brief) description is not explicit.

Goldman

Mexican Freshwater Toadfish *Batrachoides goldmani* BW Evermann & EL Goldsborough, 1902

Major Edward Alphonso Goldman (1873–1946) was a field naturalist and mammalogist who was born in Illinois. Edward Nelson (q.v.) hired him (1892) to assist his biological investigations of California and Mexico, and then as Field Naturalist and eventually Senior Biologist with the United States Bureau of Biological Survey. He spent c.14 years collecting all over Mexico. Their biological exploration and collecting expeditions (1892–1906) are said to have been: 'among the most important ever achieved by two workers for any single country'. They investigated in every Mexican state, collecting 17,400 mammals and 12,400 birds, and amassing an enormous fund of information of the country's natural history. The best account of their work is Goldman's: *Biological Investigations in Mexico* (1951). Goldman also had an honorary position with the USNM, as an Associate in Zoology (1928–1946). He was part of the Biological Survey of Panama (1911–1912) during the canal's construction. His results were published in: *The Mammals of Panama* (1920). He was President of the Biological Society of Washington (1927–1929) and of the American Society of Mammologists (1946). He assisted the United States government in negotiating with Mexico to protect migratory birds (1936). Goldman's bibliography includes over 200 titles. He named over 300 forms of mammal, most of them subspecies. Fifteen mammals, eleven birds, an amphibian and a reptile, as well as Goldman Peak in Baja California, bear his name. He helped to collect the holotype of this species.

Goldmann

Goldmann's Goby *Istigobius goldmanni* P Bleeker, 1852
Croaker sp. *Johnius goldmani* P Bleeker, 1855
[Binomial sometimes amended to *goldmanni*]

Carel Frederik Goldmann (1800–1862) was Resident at Timor (1835–1836) and later Governor of the Moluccas (1855) where he hosted Alfred Russell Wallace during his visit there.

Goldshmidt, O

Reefgoby sp. *Priolepis goldshmidtae* M Goren & A Baranes, 1995

Ms Orit Goldshmidt was at the Interuniversity Institute of Marine Sciences, Eilat, Israel. She collected the goby holotype in the northern Gulf of Aqaba.

Goldschmidt, T

Lake Victoria Cichlid sp. *Haplochromis goldschmidti* F Witte, I Westbroek & MP de Zeeuw, 2013

Dr Paul-Tijs Goldschmidt (b.1953) is a Dutch writer and evolutionary biologist. He is currently Writer in Residence of the Artis Bibliotheek, University of Amsterdam. He lived in Tanzania (1981–1986) studying the cichlids of Lake Victoria and was honoured in appreciation for his work on haplochromine cichlids there and for calling the world's attention to their human-induced extinction. He wrote: *Darwin's Dreampond: Drama on Lake Victoria* (1996).

Goliath

Goliath Tigerfish *Hydrocynus goliath* GA Boulenger, 1898
Goliath Hagfish *Eptatretus goliath* M Mincarone & AL Stewart, 2006

Goliath of Gath (about 1,030 BC) was a Philistine warrior of giant size who was killed with a slingshot by David, later King of the Jews (see 1 Samuel XVII.iv, Old Testament of the Bible). His name is sometimes used as a binomial to denote the exceptional size of a species. Four birds, three mammals, an amphibian and a reptile are named after this character.

Gollum

Smooth-hound Shark genus *Gollum* LJV Compagno, 1973
Gollum Galaxias *Galaxias gollumoides* B McDowall & WL Chadderton, 1999
Gollum Snakehead *Aenigmachanna gollum* R Britz, VK Anoop, N Dahanukar & R Raghavan, 2019

The genus and these species are named after the fictional character, Gollum, in J R R Tolkien's *The Hobbit* and *Lord of the Rings*. The galaxias binomial, meaning 'Gollum-like', is justified by the authors as Gollum was "a dark little fellow with big round eyes who sometimes frequents a swamp" – attributes shared by this New Zealand fish. The snakehead is believed to dwell in subterranean aquifers.

Golovan

Whipnose Angler sp. *Gigantactis golovani* E Bertelsen, TW Pietsch III & RJ Lavenberg, 1981

George A Golovan was a Russian Oceanographer and ichthyologist who was, at the time, at the Institute of Oceanography, Academy of Sciences, USSR. He made his large collection available to the authors. Among his published papers are: *Preliminary data on the composition and distribution of the bathyal ichthyofaunal (in the Cap Blanc area)* (1974) and *Rare and first finds of cartilaginous (Chondrichthyes) and bony (Osteichthyes) fishes from the continental slope off West Africa* (1978).

Golubtsov

Loach sp. *Barbatula golubtsovi* AM Prokofiev, 2003

Dr Alexander S Golubtsov is a Russian ichthyologist who is Senior Researcher and acting head of laboratory at the A N Servertsov Institute of Ecology and Evolution, Russian Academy of Sciences, Moscow. Among his more than fifty published papers he co-wrote: *Observations on reproduction of the Lake Tana barbs* (1999). For many years he has focused on Ethiopian fish fauna and wrote: *Fishes of the Ethiopian Rift Valley* (2002). He collected the holotype of this loach.

Gomes, AI

Shrimp Eel *Ophichthus gomesii* FL de Castelnau, 1855

Dr Antônio Ildefonso Gomes de Freitas (1794–1859) was a physician and botanist in Rio de Janeiro whom Castelnau (q.v.) consulted, and with whom he stayed. Gomes cured Castelnau's illness. Before settling in Rio de Janeiro to practice medicine he had already accompanied Augustin François César Prouvençal de Saint-Hilaire on his expedition to northeast Brazil (1816–1822).

Gomes, AL

Armoured Catfish sp. *Aphanotorulus gomesi* HW Fowler, 1942
Characin sp. *Characidium gomesi* HP Travassos, 1956

Dr Alcides Lourenço Gomes (1916–1991) was one of eight technicians at the Experimental Biology and Pisciculture Station at Pirassununga, São Paulo, Brazil (established 1939). He wrote some papers mainly on fish such as: *A small collection of fishes from Rio Grande do Sul, Brazil* (1947). He became interested in odonata while the entomologist Santos worked at the station (1938–1944), and he has a dragonfly named after him.

Gomes, JF

Characin sp. *Leporinus gomesi* JC Garavello & GM dos Santos, 1981

João Gomes da Silva collected the type (1978) and paratypes of this, and another similar species, in the Rio Aripuanã basin, Mato Grosso, Brazil, which he donated to the authors. He was associated with the Universidade Federal se São Carlos as is Julio Garavello who initially described them in his unpublished doctoral thesis, although the name dates to a brief description in a published abstract (1981) and later re-described them (1990).

Gomes, UL

Gomes' Round Ray *Heliotrygon gomesi* MR de Carvalho & NR Lovejoy, 2011

Dr Ulisses Leite Gomes (b.1955) is a Brazilian ichthyologist. The Universidade Santa Úrsula awarded his bachelor's degree (1979) and the Universidade Federal do Rio de Janeiro his master's (1989) and doctorate (2002). He is an associate professor at the Universidade do Estado do Rio de Janeiro, where he has been working on elasmobranch taxonomy since 1988. He was honoured in the name of the ray as a 'pioneer in the study of elasmobranch morphology and systematics in Brazil, and an esteemed colleague and collaborator of the first author'.

Gomez, A

Colombian Goby *Bollmannia gomezi* P Acero, 1981

Alfredo Gómez Gaspar is a Venezuelan marine scientist. He was the author's teacher and friend and 'stimulated his interest in ichthyology'. He has written books and papers such as: *Density of round sardine Sardinella aurita eggs, zooplankton abundance and hydrography in the araya peninsula and south of Margarita Island, Venezuela* (2016).

Gomez, E

Gomez's Shrimpgoby *Tomiyamichthys gomezi* GR Allen & MV Erdmann, 2012

Dr Edgardo Dizon Gomez (1938-2019) was a Filipino biologist who was National Scientist of the Philippines (2014). De La Salle University awarded his bachelor's degree, St Mary's University of Minnesota his master's and the University of California his PhD. He was the founding Director (1973) and later Professor emeritus of the Marine Sciences Institute, at the University of the Philippines. He was been a leading proponent of the conservation of coral reefs and other coastal marine communities.

Gomez, JA

Gomez's Corydoras *Corydoras gomezi* DM Castro, 1986

Juan A Gómez was Director, Centro de Investigaciones Cientificas, Universidad de Bogotá Jorge Tadeo Lozano. He was acknowledged for his 'permanent support' of the author, Castro, and his project regarding the freshwater fishes of Colombia.

Gomojunov

Spinyhook Sculpin *Artediellus gomojunovi* A Taranetz, 1933

Konstantin Azarievich Gomoyunov aka Gomojunov (1889–1955) was a Russian oceanographer and zoologist who led an Arctic expedition (1936–1937) on board the 'Nerpa' which was the first systematic oceanographic survey of the western and central

Kara Sea. He worked variously at the State Far Eastern University, the Vladivostok branch of the Geographical Society, the Pacific Scientific-Production Station, the Institute of the Fish Industry, the Dalgeofizin, the All-Union Arctic Institute. He conducted a series of surveys in arctic waters as well as lecturing in zoology. Among his published papers are: *Hydrological description of Amurskii Bay and Suifun River* (1926) and *Expedition of Far Eastern Geophysical Institute in Amur River estuary* (1932).

Gomon

Cusk-eel sp. *Enchelybrotula gomoni* DM Cohen, 1982
Pale Smiling Whiptail *Ventrifossa gomoni* TN Iwamoto & A Williams, 1999
Squarechin Snailfish *Paraliparis gomoni* DL Stein, NV Chernova & AP Andriashev, 2001
Gomon's Tuskfish *Choerodon gomoni* GR Allen & JE Randall, 2002
Viviparous Brotula sp. *Dactylosurculus gomoni* W Schwarzhans & PR Møller, 2007
Hagfish sp. *Eptatretus gomoni* MM Mincarone & B Fernholm, 2010
Gomon's Gurnard *Pterygotrigla gomoni* PR Last & WJ Richards, 2012
Gomon's Frogmouth *Chaunax gomoni* HC Ho, T Kawai & F Satria, 2015
Gomon's Flathead *Cociella martingomoni* H Imamura & C Aungtonya, 2020

Dr Martin Fellows Gomon (b.1945) is an US/Australian ichthyologist who first went to Australia to Museum Victoria Melbourne (1979) where he is Senior Curator, Ichthyology. Florida State University Tallahassee awarded his bachelor's degree (1967) and the University of Miami his master's (1971) and doctorate (1979). He co-wrote two field guides: *Fishes of Australia's South Coast* (1994) and *Fishes of Australia's Southern Coast* (2008). He has been instrumental in obtaining funding for the website 'Fishes of Australia' designed to give comprehensive information on Australia's 4000+ species of fishes.

Gon

Gon's Cardinalfish *Archamia bleekeri* A Günther, 1859
[*Archamia goni* J-P Chen & K-T Shao, 1993 is a JS]
Eelpout sp. *Pachycara goni* ME Anderson, 1991

Dr Ofer Gon (b.1949) is a South African ichthyologist who is Senior Aquatic Biologist at the JLB Institute of Ichthyology. He co-wrote: *A new species of the cardinalfish genus* Jaydia *(Teleostei: Apogonidae) from the Philippines* (2015). He was honoured in the eelpout's binomial for "…his contributions to knowledge of the cold-water marine fishes of the southern hemisphere." (Also see **Jenny**)

Gonzales

Vanuatu Snapper *Paracaesio gonzalesi* P Fourmanoir & J Rivaton, 1979

Dr Pedro C Gonzales is a Philippine zoologist. He was the Curator of Zoology at the National Museum of Manilla and is now Curator Emeritus. He has specialised in the fauna of Mount Isarog and published on it; such as with S M Goodman *The birds of Mt. Isarog National Park, southern Luzon, Philippines, with particular reference to altitudinal distribution* (1990), and with four other co-authors: *Mammalian diversity on Mt. Isarog, a threatened center of endemism on southern Luzon Island, Philippines* (1999). He co-wrote: *A Guide to the Birds of the Philippines* (2000). He saw this species at a fish market and recognised that it was probably undescribed. The etymology reads thus: "*L'espèce est dédiée à P. Gonzales, curateur de zoologie au Muséum national de Manille, qui a remarqué ce Poisson au marché de Manille.*" A mammal is also named after him.

Gonzalez, M

Thorny Catfish sp. *Amblydoras gonzalezi* A Fernández-Yépez, 1968

Marcelo González Molina (1923–2000) was a Venezuelan civil engineer. He provided access for the author to visit the locality where the holotype was collected.

Gonzalez, PR

Characin sp. *Eretmobrycon gonzalezoi* C Román-Valencia, 2002

Pana Rigoberto Gonzalez is Curator of Fishes, Smithsonian Tropical Research Institute, Panama. He helped Román-Valencia during his stay there. He wrote: *Notes on Geographic Distribution. First record of Gymnotus henni (Albert, Crampton and Maldonado, 2003) in Panama: phylogenetic position and electric signal characterization* (2013)

Good

Goby sp. *Ebomegobius goodi* AWCT Herre, 1946
African Loach Catfish sp. *Paramphilius goodi* RR Harry, 1953

Dr Albert Irwin Good (1884–1975) was an American Presbyterian missionary and naturalist in Cameroon (1909–1949). He translated thirty-three books of the Old Testament into the Bulu language and edited various Bulu, Mabea, and Banok language hymnals as well as writing a number of books, including one on Bulu folktales. He collected natural history specimens, primarily insects and birds, many of which were donated to the Carnegie, Cleveland, and Stanford University Museums. A Fellow of the Royal Geographic Society, he also published articles dealing with African natural history. The etymology describes him as an "*…ardent collector of West African fishes*" who collected the goby holotype. A bird is also named after him.

Goode

Goodeid genus *Goodea* DS Jordan, 1880
Bluefin Killifish *Lucania goodei* DS Jordan, 1880
Quillfish *Ptilichthys goodei* TH Bean, 1881
Southern Eagle Ray *Myliobatis goodei* S Garman, 1885
Chilipepper Rockfish *Sebastes goodei* CH Eigenmann & RS Eigenmann, 1890
Great Pompano *Trachinotus goodei* DS Jordan & BW Evermann, 1896
Goode's Croaker *Paralonchurus goodei* CH Gilbert, 1898

Dr George Brown Goode (1851–1896) was an American ichthyologist and museum administrator at the Smithsonian. He graduated from the Wesleyan University, Middletown, Connecticut (1870) and then studied under Agassiz at Harvard (1870–1871). He returned to Wesleyan as Curator of their new natural history museum (1872), dividing his time between it and the Smithsonian in winters and working in the field with the US Fish Commission. He joined the Smithsonian full time (1877) and became Assistant Secretary

(1887). He had little time for research, but nevertheless produced more than a hundred papers, such as: *Catalog of the Fishes of the Bermudas* (1876) and *Oceanic Ichthyology, A Treatise on the Deep-Sea and Pelagic Fishes of the World, Based Chiefly upon the Collections Made by the Steamers Blake, Albatross, and Fish Hawk in the Northwestern Atlantic* with Tarleton Hoffman Bean. His work on Bermuda fishes later formed the basis for his PhD, granted by Indiana University, Bloomington (1886). He also produced the 7-volume: *The Fisheries and Fishery Industries of the United States* (1884–1887). He died of pneumonia. At least five other taxa are named after him.

Goode(n)bean

Codling sp. *Laemonema goodebeanorum* C Meléndez & DF Markle, 1997
Palefin Dragonet *Synchiropus goodenbeani* T Nakabo & KE Hartel, 1999

Artificial combinations of the names of both George Brown Goode (above) and Tarleton Hoffman Bean (q.v.). (See **Goode** and **Bean, TH**)

Goodlad

Goodlad's Stinkfish *Callionymus goodladi* GP Whitley, 1944

James Goodlad (1902–1984) and his brother Matthew both emigrated to Australia from Shetland (1920) and both became Fisheries Inspectors at Albany, Western Australia. James obtained the holotype of the stinkfish (1943).

Goodyear

Barbeled Dragonfish sp. *Photostomias goodyeari* CP Kenaley & KE Hartel, 2005

Richard Hugo Goodyear (1943–2013) worked at Centre de Ciencias del Mar y Limnologia, Universidad de Panama. He co-wrote: *Study on the biology and ecology of the Chiriqui River estyary, Pedregal* (1981). He was recognised for his contributions to the systematics of stomiid fishes and 'distinguished contributions to ichthyology'.

Gopi

Sisorid Catfish sp. *Glyptothorax gopii* L Kosygin, U Das, P Singh & BR Chowdhury, 2019

Dr K C Gopi is an Indian ichthyologist working at the Fish Division of the Zoological Survey of India. Among his published papers is *Fish fauna of Kozhikode District, Kerala, South India* (2006). The etymology says that the catfish was named after "*…K.C. Gopi (retired scientist) of the Zoological Survey of India, Kolkata, honouring his contribution to the Indian Ichthyology.*"

Gorda

MImic Sanddab *Citharichthys gordae* W Beebe & J Tee-Van, 1938

This is a toponym referring to the Gorda Banks, Lower California.

Gordias

Snake-eel genus *Gordiichthys* DS Jordan & BM Davis, 1891

In Greek mythology Gordias was a farmer who, by fulfilling an oracle, became King of Phrygia (in modern Turkey). His ox-cart was preserved as a relic, tied to a post with an intricate knot. In 333 BC, while wintering in Phrygia, Alexander the Great attempted to untie the knot. When he failed, he sliced it in half with a stroke of his sword, producing the required result (the socalled 'Alexandrian solution' to a problem).

Gordon

Northern Platyfish *Xiphophorus gordoni* RR Miller & WL Minckley, 1963

Dr Myron Gordon (1899–1959) was an American biologist and geneticist. He conducted pioneering cancer research using *Xiphophorus* fish. Cornell University awarded his BSc (1925), master's (1925–1926) and PhD in zoology, limnology and genetics (1929). During his studies he worked at Bronx Zoo (1920) and as game keeper at the State of Maryland Game Farm (1921–1923), then as collector for Cornell's College of Agriculture (1924). He was an instructor at Long Island (1925–1926) and a biologist at the NY State Biological Survey (Fisheries) (1927) and as an Investigator at the Carnegie Institute in Florida (1928) and Woods Hole Marine Laboratory Massachusetts (1929). He worked at the Heckscher Foundation at Cornell (1925–1929 & 1931–1932), in the intervening years between his work with the Heckscher Foundation, Gordon was National Research Fellow at Cornell University, where he was a member of the first expedition to Mexico. He was at Yale (1938), the JS Guggenheim Foundation (1938–1941) then a Research Associate at New York Zoological Society (1941–1943). He became (1944) Assistant Curator of Fishes at the New York Aquarium. During his collection trips, he discovered and named several new species. (Also see **Evelyn (Gordon)**)

Goren

Cardinalfish sp. *Siphamia goreni* O Gon & GR Allen, 2012

Dr Menachem Goren is an Israeli zoologist, fish taxonomist and marine biologist. Tel Aviv University awarded all his degrees; bachelor's (1967), master's (1969) and doctorate (1975). After post-doctoral work at the Senckenberg Museum, Frankfurt-am-Main (1976) he joined the faculty of Tel Aviv University (1978) as a Research Associate in the Department of Zoology and Director of the University's Zoological Museum and is now Principal Research Associate in the ichthyological laboratory there. He has been on marine biology expeditions in the Mediterranean, the Red Sea, the Gulf of Suez and the Indian Ocean including being the Chief Scientist on the Israeli-Egyptian expedition to the Red Sea (1994). Among his writings is: *The freshwater fishes of Israel* (1975).

Gorgas

Garden Eel genus *Gorgasia* SE Meek & SF Hildebrand, 1923

General William Crawford Gorgas (1854–1920) was an epidemiologist and Surgeon General of the US Army. He is known throughout the world as the conqueror of the mosquito and of the malaria and yellow fever it transmits. His pioneering efforts in halting an epidemic of yellow fever by the application of sanitary measures enabled the USA to complete the Panama Canal after earlier attempts had fallen before the onslaught of the insidious insect. The Gorgas Memorial Laboratory was named in his honour. The describers stated that they received a great deal of help from Gorgas' department when they were collecting in Panama. A rodent is also named after him.

Gorgona

Gorgonian Tonguefish *Symphurus gorgonae* P Chabanaud, 1948

This is a toponym; named for Gorgona Island off Colombia, or as the paper puts it: "… *provenant de la côte Pacifique de Colombie, île Gorgona à 30 fathoms…*"

Gorgona

Characin sp. *Compsura gorgonae* BW Evermann & EL Goldsborough, 1909

This is a toponym; Gorgona, in the Panama Canal Zone, is where this species was first collected.

Gorman

Rattail sp. *Coelorinchus gormani* T Iwamoto & KJ Graham, 2008

Terence Brian 'Terry' Gorman was a Senior Biologist, Fishery Scientist and diver. His MSc thesis at Queens University, Belfast (1965) was published as: *Yellow-eyed mullet, Aldrichetta forsteri (Curvier and Valenciennes) in Lake Ellesmere* and *Biological and economic aspects of the elephant fish, Callorhynchus milii Bory, in Pegasus Bay and the Canterbury Bight*. He is credited with having pioneered deepwater fishery research in the 1970s and 1980s with the New South Wales Fisheries Research Vessel 'Kapala'. He was lead author of *The Kapala midwater trawling survey in New South Wales* (1979).

Görner

Corfu Dwarf Goby *Knipowitschia goerneri* H Ahnelt, 1991

Manfred Görner has supported Harold Ahnelt's ichthyological work for many years. Unfortunately, nothing more is said of him.

Gorodinski

Dragonfish sp. *Photonectes gorodinskii* AM Prokofiev, 2015

Andre Aleksandrovich Gorodinski is an amateur lepidopterist, coleopterist, entomologist, insect collector and explorer who is a friend of the author. A number of insects are also named after him.

Gosline, JM

Hagfish sp. *Eptatretus goslinei* MM Mincarone, D Plachetzki, CL McCord, TM Winegard, B Fernholm, CJ Gonzalez & DS Fudge, 2021

John Moffit Gosline (1943–2016) was an American biologist who became Professor Emeritus in the Department of Zoology at the University of British Columbia (UBC), Canada. His work on protein elastomers included studies on slug slime and hagfish slime. He also held a patent on how to transform hagfish slime threads into silk.

Gosline, WA

Biting Blenny *Plagiotremus goslinei* DW Strasburg, 1956
Longarm Brotula *Calamopteryx goslinei* JE Böhlke & DM Cohen, 1966
Hawaiian Spikefish *Hollardia goslinei* JC Tyler, 1968
Cutthroat Eel sp. *Dysomma goslinei* CH Robins & CR Robins, 1976
Banded Allotoca *Allotoca goslinei* ML Smith & RR Miller, 1987

Dr William Alonzo Gosline (1915–2002) was an ichthyologist and botanist at the University of Michigan. Harvard awarded his bachelor's degree (1938) and Stanford his doctorate (1941). After graduating he enlisted in the British Army as an ambulance driver serving in North Africa. He worked in Brazil after the war and at the Zoology Department and Museum of Zoology, University of Michigan (1945–1948) and the University of Hawaii (1949–1972). He returned to the University of Michigan (1972) as an adjunct curator. He wrote c.100 published papers and co-wrote the: *Handbook of Hawaiian Fishes* (1965). In the last five years of his life his eyesight failed. He walked every day by a lake through the 60 acres of woodland at his home in which he and his botanist wife, Alice, catalogued 253 species of flowering plants. He died on one of his morning walks.

Gosse

Smoothbelly Pellonuline *Congothrissa gossei* M Poll, 1964
African Characin sp. *Neolebias gossei* M Poll & JG Lambert, 1964
Pale-spotted Cory *Corydoras gossei* H Nijssen, 1972
Lake Malawi Cichlid sp. *Lethrinops gossei* WE Burgess & HR Axelrod, 1973
Dwarf Cichlid sp. *Apistogramma gossei* SO Kullander, 1982
Borneo Loach sp. *Glaniopsis gossei* TR Roberts, 1982
Characin sp. *Leporinus gossei* J Géry, P Planquette & P-Y Le Bail, 1991

Jean-Pierre Gosse (1924–2001) was a Belgian ichthyologist who was Curator of Vertebrates, Institut Royal des Sciences Naturelles de Belgique. He had explored and collected in Amazonia before collecting in the Congo. He then worked at Musée Royal de l'Afrique Central, Tervuren and wrote: *Les Poissons Du Bassin De L'Ubangi* (1968). He collected the type series of the dwarf cichlid and was honoured for that and for his "*...substantial contribution to South American ichthyology in collecting and taxonomic studies.*"

Gosztonyi

Eelpout genus *Gosztonyia* J Matallanas, 2009
Eelpout sp. *Dieidolycus gosztonyii* ME Anderson & G Pequeño, 1998

Dr Atila Esteban Gosztonyi is an Argentinian ichthyologist and marine biologist at the National Scientific and Technical Research Council, Buenos Aires, Argentina. Among his published works are: *Results of the research cruises of FRV Walther Herwig to South America. XLVIII. Revision of the South American Zoarcidae (Os¬teichthyes, Blennioidei) with the description of three new genera and five new species* (1977) and *A redescription of* Diplomystes mesembrinus (*Siluriformes: Diplomystidae*) (1998). The etymology says the eelpout was: "*...in honour of our friend and colleague Dr. Atila Esteban Gosztonyi, Centro*

Nacional Patagonico, Puerto Madryn, Argentina, for his contributions to temperate South American ichthyology, especially his pioneering work on the Zoarcidae."

Gothe

Galaxias sp. *Brachygalaxias gothei* K Busse, 1983

Karl Heinz Gothe from Taka, Chile, was a friend of the author who was taken by him to the holotype's locality and helped to collect it.

Goto, A

Goto's Red Sculpin *Procottus gotoi* VG Sideleva, 2001

Dr Akira Goto is Professor at the Graduate School of Fisheries Science, Hokkaido University. He started there as Assistant Professor (1977), becoming Associate Professor (1986) and full professor (2006). He is now (2011) Professor of Hokkaido University of Education. Dr Valentina G Sideleva has written papers with Goto, such as: *Description of new species Cottus kolymensis* (2012).

Goto, HE

Goto's Herring *Herklotsichthys gotoi* T Wongratana, 1983

H E Goto was an entomologist who was at the Zoology Department, Imperial College of Science and Technology, University of London, and was the Director of Wongratana's studies in London. Among a great many published papers (1951–1982) he co-wrote: *On some new and disputed synonymy in British Collembola* (1964) and *The Effect of Carbon Dioxide anaesthesia on Collembola* (1971) and among his books is: *Animal Taxonomy (Studies in Biology)* (1982).

Gottwald

Eartheater Cichlid sp. *Geophagus gottwaldi* I Schindler & W Staeck, 2006

Jens Gottwald (b.1967) is a German collector, aquarist and dealer who is an expert in South American pike cichlids. He has kept fish since he was six years old, developing a passion for cichlids when twelve. His company, Aquatarium, imports directly and he also works with another company, Panta Rhei. He collected the holotype (2001) in Venezuela on one of his fifteen trips to South America collecting fish; latterly mostly to Brazil.

Gouan

Clingfish genus *Gouania* A Risso, 1810

Dr Antoine Gouan (1733–1821) was a French naturalist, a pioneer of Linnaean taxonomy in France, who was born and lived in Montpellier. He studied in Toulouse but returned to Montpelier to study medicine and was awarded his doctorate there (1752). He practiced medicine at Saint-Éloi Hospital in Montpellier but soon turned his attention to natural history. He wrote a catalogue of the plants at the Montpelier Botanical Gardens, *Hortus regius monspeliensis* (1762), the first in France to use Linnaeus' binomial system. He was subsequently put in charge of the gardens and classifying its plants. He wrote the treatise *Historia Piscicum* (1770), expanding Linnaean ichthyology. Many plants bear his name such as *Ranunculus gouanii*, Gouan's Buttercup.

Gould, EJ

Deepwater Cardinalfish *Apogon gouldi* WF Smith-Vaniz, 1977

Edwin Jay Gould (1932–1993) was an investor in real estate and wildlife conservationist. He graduated from the University of Virginia and served five years in the US Army, based in Korea. He was a keen angler and sponsored the Bermuda Expedition (1975), during which the holotype of this species was collected. He was a regular supporter of the ichthyological expeditions of the Academy of Natural Sciences of Philadelphia.

Gould, J

Gould's Wrasse *Achoerodus gouldii* J Richardson, 1843
[Alt. Western Blue Groper]

John Gould (1804–1881) was the son of a gardener at Windsor Castle who became an illustrious British ornithologist, artist and taxidermist. Gould was born in Dorset, England, and became acknowledged around the world as 'The Bird Man'. He was employed as a taxidermist by the newly formed Zoological Society of London and travelled widely in Europe, Asia and Australia. He was arguably the greatest and certainly the most prolific publisher and original author of ornithological works in the world. In excess of 46 volumes of reference work were produced by him in colour (1830–1881). He published 41 works on birds, with 2,999 remarkably accurate illustrations by a team of artists, including his wife. His first book, on Himalayan birds, was based on skins shipped to London, but later he travelled to see birds in their natural habitats. Gould and his wife, Elizabeth arrived on board *Parsee* in Australia (1838) to spend 19 months studying and recording the natural history of the continent. By the time they left, Gould had not only recorded most of Australia's known birds, and collected information on nearly 200 new species, but he had also gathered data for a major contribution to the study of Australian mammals. His best-known works include: *The Birds of Europe*, *The Birds of Great Britain*, *The Birds of New Guinea* and *The Birds of Asia*. Gould was commercially minded and pandered to Victorian England's fascination with the exotic, particularly hummingbirds, with which he is particularly associated. His superb paintings and prints of these and other birds were greatly sought after, so much so that he had trouble keeping up with the demand. Forty-seven birds, five mammals, two reptiles and an amphibian are named after him. Richardson refers only to 'Mr Gould', but context strongly suggests this is a reference to the famous ornithologist.

Goulding

Whale Catfish sp. *Helogenes gouldingi* RP Vari & H Ortega, 1986
Characin sp. *Cynopotamus gouldingi* NA de Menezes, 1987
Ghost Candiru *Stauroglanis gouldingi* MCC de Pinna, 1989
Toothless Characin sp. *Cyphocharax gouldingi* RP Vari, 1992
Blue TIger Piranha *Serrasalmus gouldingi* WL Fink & A Machado-Allison, 1992
Characin sp. *Brycon gouldingi* FCT Lima, 2004
Neotropical Killifish sp. *Fluviphylax gouldingi* PHN de Bragança, 2018

Dr Michael Goulding (b.1950) is a conservation ecologist and scientist at the University of Florida and with the Amazon Conservation Alliance with which he has worked for over 30 years. The University of California Los Angeles awarded his doctorate. He wrote: *The Fishes and the Forest: Explorations in Amazonian Natural History* (1981).

Gowers

Lake Victoria Cichlid sp. *Haplochromis gowersii* E Trewavas, 1928

Sir William Frederick Gowers (1875–1954) was Governor of Uganda (1925–1932). He first went to Africa (1899) in commerce but left (1902) to take a post in the colonial service in Nigeria rising to Lieutenant-Governor. He helped fisheries scientist Michael Graham (q.v.) conduct his research on Lake Victoria where this cichlid is endemic.

Goya

Pencil Catfish sp. *Ituglanis goya* A Datovo da Silva, PPU Aquino & F Langeani-Neto, 2016

Characin sp. *Moenkhausia goya* G de C Deprá, VM Azevedo-Santos, OB Vitorino Jr, FCP Dagosta, MMF Marinho & RC Benine, 2018

The Goyá, according to the catfish etymology were "an enigmatic and pacific indigenous group that supposedly inhabited the region of the modern state of Goiás in central Brazil," they were "utterly exterminated by the XVIII century by the first Bandeirantes explorers from southeastern Brazil." (See **Bandeiras**)

G R Allen

Goby genus *Grallenia* K Shibukawa & A Iwata, 2007

Dr Gerald Ray Allen (b.1942) The genus was named to honour "*...his great contribution to our knowledge of the diversity of coral-reef fishes.*" (See **Allen, GR**)

Graeffe

Blue Salmon-Catfish *Neoarius graeffei* R Kner & F Steindachner, 1867

Dr Eduard Gräffe (1833–1916) was born in Switzerland, died in part of the Austro-Hungarian Empire (Slovenia) but regarded himself as an Austrian. He was a physician, author, zoologist and ornithologist. Godeffroy (q.v.) employed him (1860) to organise his personal collection and to start his personal museum. He travelled to the Pacific (1861) as a paid collector for Godeffroy and (1862–1870) visited nearly all the islands between longitudes 170° and 180°E. He returned to Hamburg (1871) with a new collection to add to Godeffroy's museum, and wrote: *Travels in the Interior of Viti Levu*. Godeffroy's business was in a bad state (1874) and he could no longer afford to pay Graeffe, who took the position of an inspector at the Austrian Zoological Station in Trieste. two birds are named after him.

Graells

Ebro Barbel *Luciobarbus graellsii* F Steindachner, 1866

Professor Dr Mariano de la Paz Graëlls y de la Aguera (1809–1898) of Madrid was a botanist, entomologist and malacologist. He qualified in medicine and natural sciences at the University of Barcelona, where he was first an Associate and later full Professor of Physics and Chemistry. There was an epidemic in Barcelona (1835) which he was prominent in combating. He moved to Madrid (1837) as Professor of Zoology at the Museum of Natural Sciences and Director of the Botanical Gardens. He joined a scientific expedition to the Pacific (1845), establishing the Spanish Natural History Society's facilities to allow for the acclimatisation of tropical plants. The phylloxera outbreak virtually wiped out European vineyards in the latter part of the 19th century, and Graells, as a senior person in the Council of Agriculture, had to deal with its effect in Spain. He was a founding member of the Spanish Academy of Exact Sciences and was honoured in Spain and by several foreign governments. He wrote: *Subfamilia felina fauna mastodologica* (1897). A mammal, a bird and a damselfly are named after him.

Graffin

Armoured Catfish sp. *Sturisoma graffini* A Londoño-Burbano, 2018

Greg Walter Graffin (b.1964) is the lead singer of the punk rock band Bad Religion (c1979–c1985, reformed 1986) and a solo artist (1997). As a solo artist he released the albums *American Lesion* (1997), *Cold as the Clay* (2006) and *Millport* (2917). Apart from being a singer-songwriter he is a multi-instrumentalist, college lecturer and author. He has a master's from UCLA and a PhD in Zoology from Cornell University and lectures part time on life sciences, palaeontology and evolution as a professor at University of California, Los Angeles and Cornell. He has written a number of papers and books such as: *Is Belief in God Good, Bad or Irrelevant? A Professor and Punk Rocker Discuss Science, Religion, Naturalism & Christianity* (2006), *Anarchy Evolution : Faith, Science, and Bad Religion in a World Without God* (2010) and *The Population Wars* (2015).

Graham, A

Grahams' Skate *Dipturus grahami* PR Last, 2008

[Binomial sometimes amended to the plural form, *grahamorum*]

Deepwater Scorpionfish sp. *Lythrichthys grahami* H Wada, Y Kai & H Motomura, 2021

Alastair Graham (b.1964) is an Australian Marine Biologist whose bachelor's degree was awarded by Macquarie University. He worked in the Fish Department, the Australian Museum, Sydney, Australia (1988–1990) and since 1990 has been the Fish Collection Manager at the Australian National Fish Collection, CSIRO Marine and Atmospheric Research, Hobart, Australia. He is jointly honoured in the name of the skate with Ken Graham (below) for their "*...very important, but very different contributions to the knowledge of Australian sharks and rays*". (Also see **Alastair** & **Al Graham**)

Graham, DH

New Zealand sleeper-gudgeon genus *Grahamichthys* GP Whitley, 1956

Flabby Whalefish sp. *Gyrinomimus grahami* LR Richardson & J Garrick, 1964

David H Graham was described as a 'veteran' New Zealand ichthyologist and marine biologist. He was resident marine biologist at Otago Marine Station (1926). He was appointed as scientist at the Portobello Marine Station (1930) but was dismissed (1932) as his salary could no longer be financed in the period of the Great Depression. He wrote: *A Treasury of New Zealand Fishes* (1953).

Graham, J

Snowtrout sp. *Schizothorax grahami* CT Regan, 1904
Golden-line Barbel *Sinocyclocheilus grahami* CT Regan, 1904
Stone Loach sp. *Triplophysa grahami* CT Regan, 1906
Sheatfish (catfish) sp. *Silurus grahami* CT Regan, 1907
Graham Minnow *Anabarilius grahami* CT Regan, 1908

Rev Dr John Graham (d.1947) was a British member of the Chinese Inland Mission in Yunnanfu. He collected in Yunnan Province (1900–1920) and sent many natural history specimens to BMNH, including the holotypes of the fishes named after him. He married an American missionary (1930) with whom he went on leave to the USA (1936–1937) and was again in the USA on leave when he died. A reptile and an amphibian are also named after him.

Graham, JD

Rio Grande Darter *Etheostoma grahami* CF Girard, 1859

Colonel James Duncan Graham (1799–1865) was a topographical engineer. He graduated from West Point (1817) and served as an artillery officer (1817–1829) until his particular skills were recognised and he transferred to the Corps of Topographical Engineers as a Captain, being promoted to Major (1838). He was the astronomer on the surveying party that established the boundary between the USA and Texas (then an independent nation) (1839–1840). He resurveyed the famous Mason-Dixon Line (1848–1850). He discovered that there are lunar tides on the Great Lakes (1854). He served on the boundary commission for the United States and Mexico, during which the darter holotype was collected. He was promoted to Lieutenant-Colonel (1861) and Colonel (1863). He died from exposure. Three reptiles are also named after him.

Graham, JW

Lake Magadi Tilapia *Alcolapia grahami* GA Boulenger, 1912

J W Graham (no other information available) collected the type in a hot soda lake in Kenya.

Graham, KJ

Graham's Whiptail *Coryphaenoides grahami* TN Iwamoto & YN Shcherbachev, 1991
Graham's Conger *Gnathophis grahami* ES Karmovskaya & JR Paxton, 2000
Graham's Swallowtail *Odontanthias grahami* JE Randall & PC Heemstra, 2006
Eastern Longnose Spurdog *Squalus grahami* PR Last, WT White & JD Stevens, 2007
Grahams' Skate *Dipturus grahami* PR Last, 2008
[Binomial sometimes amended to the plural form, *grahamorum*]

Kenneth 'Ken' John Graham (b.1947), originally from New Zealand, spent his career as an ichthyologist, biologist and fisheries research scientist in Australia. Canterbury University, Christchurch awarded his bachelor's degree (1968). He worked for New South Wales State Fisheries, Sydney (1972–2008). Since 2005 he has been a Research Associate of the Australian Museum in Sydney. During his career as a sea-going scientist he collected tens of thousands of specimens of fishes and invertebrates, most of which now reside in the Australian Museum's mollusc, fish and marine invertebrate collections. They form the most complete record of the distribution of trawl-caught animals on the continental shelf and slope off southeast Australia. Much of his research concentrated on deep-water commercial fishery trawl surveys and by-catch assessments on board the research vessel 'Kapala'. A number of other marine species are named after him, some with the binomial *kengrahami* including a new genus of shellfish named after the vessel, '*Kapala kengrahami*'. He wrote: *Distribution, population structure and biological aspects of* Squalus *spp. (Chondrichthyes: Squaliformes) from New South Wales and adjacent Australian Waters* (2005) and was senior author of: *Changes in relative abundance of sharks and rays on Australian South East Fishery trawl grounds after twenty years of fishing* (2001).

Note that the skate is also named after Alastair Graham (q.v.).

Graham, M

Graham's Stonebasher *Hippopotamyrus grahami* JR Norman, 1928

Michael Graham (1898–1972) was a fisheries biologist and ichthyologist who worked for the Crown Agents. He collected the holotype in Lake Victoria and wrote: *The Victoria Nyanza and its fisheries – A report on the fishing survey of Lake Victoria 1927–1928* (1929). He was Director of Research at the Fishery Laboratories, Lowestoft, England (1945–1958).

Graham, V

Goby sp. *Gobioides grahamae* G Palmer & AC Wheeler, 1955

Violet Graham (c.1911–1991) was a botanist who presented a large number of fish from British Guiana to the British Museum (Natural History), including the type of this one, along with extensive field notes and colour sketches of most of the fishes collected.

Graham, WM

African Rivuline sp. *Epiplatys grahami* GA Boulenger, 1911

Dr William M Graham worked in the West Africa Medical Service. He collected the holotype along with many other specimens in swamps near Lagos, Nigeria. He also took early photographs of natural history specimens. His main interest seems to have been medical entomology, particularly blood-sucking midges. He also wrote repots such as: *Entomological Observations made in Southern and Central Ashanti* (1907) and on his findings such as: *On the Larval and Pupal Stages of West African Culicidæ* (1910). Some of his collection techniques were novel, for example: "*As the result of his experience in Ashanti, Dr. W. M. Graham states that Midges are best caught in glass tubes, when settled and sucking on a bare arm, by inverting a tube over the insect, and, when the latter is safely inside, slipping a sheet of paper underneath, thus closing the mouth of the tube.*"

Grandidier

Madagascan Cichlid sp. *Ptychochromis grandidieri* HE Sauvage, 1882

Alfred Grandidier (1836–1921) was a French explorer, geographer, herpetologist and ornithologist who made his first visit to Madagascar in 1865. He became fascinated by the island and returned in 1866 and 1868. He wrote *Histoire naturelle des oiseaux de Madagascar* in 1876. In 1866 he recovered bones of what turned out to be *Aepyornis maximus* – the huge extinct Elephant Bird. The mineral Grandidierite, which is found in Madagascar, is named after him. Along with Henri Joseph Léon Humblot (1852–1914) he collected the cichlid type. Eight reptiles, five mammals, a bird and amphibian are also named after him.

Grandperrin

Grandperrin's Giant Sawbelly *Hoplostethus grandperrini* CD Roberts & MF Gomon, 2012

Dr René Grandperrin is an oceanographer and zoologist who is now a retired chief scientist of ORSTOM. He worked at Noumea, New Caledonia, and led deepwater fish explorations off that island. He was honoured for his support for collaborative fieldwork between French and New Zealand scientists. He co-wrote: *Poissons de Nouvelle-Calédonie* (2016).

Gransabana

Neotropical Rivuline sp. *Laimosemion gransabanae* CA Lasso Alcalá, DC Taphorn Baechle & JE Thomerson, 1992

This is a toponym. The 'Gran Sabana' is a high-altitude savannah landscape in Venezuela where the fish is found.

Grant, JA

Lake Victoria Cichlid sp. *Haplochromis granti* GA Boulenger, 1906

Colonel James Augustus Grant (1827–1892) was a Scottish naturalist and explorer. After completing his education at Marischal College, Aberdeen, he joined the British army and served in India during the Sikh Wars (1849) and the Indian Mutiny (1857–1858), during which he was wounded. He spent considerable time (1860–1863) in Africa with John Hanning Speke searching for the source of the Nile. He never saw the source, as he was unable to walk for six months because of debilitating leg ulcers. He kept a record of the journey and published it as *A Walk across Africa* (1864), in which he described '…*the ordinary life and pursuits, the habits and feelings of the natives*'. He served as an intelligence officer in the Abyssinian campaign (1868), retiring with the rank of Lieutenant Colonel. Two mammals and a bird are also named after him.

Grant, W

Grant's Leporinus *Leporinus granti* CH Eigenmann, 1912

William Grant was described by Eigenmann as my '…most efficient Indian guide' when he was in British Guiana (Guyana) (1908), especially when collecting on the Potaro River. Grant became an enthusiastic fisherman and continued to send specimens to Eigenmann after he had returned to the USA.

Grant, WR

Azores Rockling *Gaidropsarus granti* CT Regan, 1903

William Robert Ogilvie-Grant (1863–1924) was a Scottish ornithologist. He was Curator of Birds at the BMNH (1909–1918), having started work there aged 19. He enlisted with the First Battalion of the County of London Regiment at the beginning of WW1 and suffered a stroke whilst helping to build fortifications near London (1916). He is famed for describing a number of well-known species, such as the huge Philippine Eagle *Pithecophaga jefferyi*. He wrote: *A Handbook to the Game Birds* (1895). He acquired a collection of fishes from the Azores, including the holotype of this species. He is also remembered in the names of 21 birds, an amphibian, a reptile and two mammals.

Grassé

Kneria genus *Grasseichthys* J Géry, 1964

Dr Pierre-Paul Grassé (1895–1985) was a French entomologist and biologist. He was studying medicine at the University of Bordeaux when WW1 interrupted his studies; by the end of that war he was a military surgeon. After the war, he studied in Paris and concentrated purely on natural sciences, graduating as a biologist. He was a professor in the zoology department, École Nationale Supérieure Agronomique de Montpellier (1921–1926) and Vice-Director of the École supérieure de sériciculture (1926–1929). He was Professor of Zoology at the Université de Clermont-Ferrand (1929–1935), Assistant Professor at the Sorbonne (1935–1944) and Professor of Zoology and Evolution (1944–1975). He travelled and collected in Africa (1933–1934, 1938–1939, 1945 & 1948). He co-wrote: *Traité de zoologie, anatomie, systématique, biologie* – just known as *le Grassé* (starting 1948), in 52 volumes of which 33 appeared in his lifetime.

Gratzianow

Freshwater Sculpin sp. *Cottus gratzianowi* VG Sideleva, AM Naseka & ZV Zhidkov, 2015

Valerian Ivanovich Gratzianow (Gratsianov) (1876–1932) was a Russian ichthyologist and zoologist. After graduating from Moscow University, he taught ichthyology at the Moscow Agricultural Institute (1902–1907). He was then an assistant to the professor of vertebrate zoology at its zoological museum (1903–1913) and at the N Yu Zograph laboratory (1914). He was the author of the first taxonomic review of Russian fishes: '*Versuch einer Übersicht der Fische des Russischen Reiches in systematischer und geographischer Hinsicht*' (Gratzianow 1907).

Grauer

Lake Tanganyika Cichlid *Bathybates graueri* F Steindachner, 1911
Lake Tanganyika Catfish sp. *Chrysichthys graueri* F Steindachner, 1911
[Syn. *Bathybagrus graueri*]
Lake Kivu Cichlid sp. *Haplochromis graueri* GA Boulenger, 1914

Rudolf Grauer (1870–1927) was an Austrian explorer and zoologist who collected extensively during an expedition to the Belgian Congo (1909) and again (1910–1911) on an expedition paid for by the Austrian Imperial Museum. His research focused on the Albertine Rift. He suffered from actinomycosis contracted in Africa and eventually succumbed to it. He is also commemorated in the names of other taxa including seventeen birds, two mammals, an amphibian and two reptiles.

Gravely

Garra sp. *Garra gravelyi* N Annandale, 1919

Dr Frederic Henry Gravely (1885–1965) was an arachnologist, botanist, entomologist and zoologist. He studied at Manchester University and was a preparator of zoology there (1907–1909). He was Assistant Superintendent, India Museum, Calcutta (Kolkata) and was Superintendent of the Government Museum of Madras (1920–1940) where he revived the Bulletin of the museum, and embarked on the scientific preservation, study and interpretation of the museum's collections as well as enlarging them, especially the invertebrates. His insect and spider collection is in the Indian Museum, Calcutta. He was a prolific writer and among his many works are: *Account of the oriental Passalidae (Coleoptera)* (1914), *Notes on the habits of Indian insects, myriapods and arachnids* (1915) and *Notes on Hindu Images* (posthumously 1977). He collected the holotype of this garra species.

Gravier

Red Sea Mimic Blenny *Ecsenius gravieri* J Pellegrin, 1906

Charles Joseph Gravier (1865–1937) was a French biologist and zoologist. He was a school teacher (1883–1885) before becoming a professor of natural history in Grenoble (1887). He was awarded his PhD (1896) and went on to a post at the MNHN. He was appointed (1917) to the Chair of Zoology at the museum. Numerous marine species are named after him in his area of special interest, sea anemones and corals. Among his publications are: *Madréporaires provenant des campagnes des yachts 'Princesse-Alice' et 'Hirondelle II' (1893–1913)* (1920) and *Hexactinidés provenant des campagnes des yachts 'Princesse-Alice' et 'Hirondelle II' (1893–1913)* (1922).

Gray, JE

Gray's Stone Loach *Balitora brucei* JE Gray, 1830
Gray's Grenadier Anchovy *Coilia grayii* J Richardson, 1845
Gray's Pipefish *Halicampus grayi* JJ Kaup, 1856
Gray's Char *Salvelinus grayi* A Günther, 1862

John Edward Gray (1800–1875) was a British ornithologist and entomologist. He started at the British Museum (1824) with a temporary appointment at 15 shillings a day but rose to Curator of Birds (1840–1874) and then Head of the Department of Zoology. Gray published descriptions of a large number of animal species, including many Australian reptiles and mammals, and was the leading authority on many reptiles. He was also an ardent philatelist and claimed that he was the world's first stamp collector; he wrote: *A Hand Catalogue of Postage Stamps for the use of the Collector* (1862). He worked at the museum with his brother George Robert Gray and together they published: *Catalogue of the Mammalia and Birds of New Guinea in the Collection of the British Museum* (1859). He wrote: *Gleanings from the Menagerie and Aviary at Knowsley Hall* (1846–1850), which was illustrated by Edward Lear. Gray suffered a severe stroke (1869), paralysing his right side, including his writing hand, yet he continued to publish to the end of his life by dictating to his wife, Maria Emma, who had always worked with him as an artist and occasional co-author. He wrote or co-wrote over 500 scientific papers, including many describing new species such as: *Description of twelve new genera of fish, discovered by Gen. Hardwicke, in India, the greater part in the British Museum* (1831). Twenty-three reptiles, nine mammals, three birds and an amphibian are also named after him.

Gray, WB

Blotched Cusk-eel *Ophidion grayi* HW Fowler, 1948

Captain William B Gray was a famous fisherman who was hired (1946) with his boat 'Beau Gregory' to re-stock the Marineland, Florida aquarium after WW2. Along with Fred D. Coppock he founded the Miami Seaquarium (opened 1955). He was the author of *Creatures of the Sea* (1960). He collected the holotype of this species.

Grebnitzki

Prickleback sp. *Pholidapus grebnitskii* TH Bean & BA Bean, 1897
[Junior Synonym of *Pholidapus dybowskii*]
Snailfish sp. *Liparis grebnitzkii* PJ Schmidt, 1904

Dr Nikolay Aleksandrovich Grebnitsky (1848–1908) was a Russian civil servant and naturalist. He wrote: *Commander Islands* (1902), where he was Governor (1877–1907) and was noted for bringing the trapping and hunting for furs under regulation. He lived on Bering Island and was interested in natural history and science, writing on the fauna, geology, and people of the Commander Islands. He collected or otherwise obtained numerous natural history specimens, and among the recipients were the Zoological Museum of the Imperial Academy of Sciences, exiled Polish zoologist B T Dybowski (q.v.), and visiting naturalist L H Stejneger (q.v.) of the US National who he collected with on Bering Island (1885). Grebnitsky participated repeatedly in the diplomatic negotiations with the United States, Japan, and the United Kingdom concerning exploitation of biological resources and protection of Pacific waters. He is also honoured in the names of a bivalve *Mysella grebnitzkii*, a deep-sea crab *Hapalogaster grebnitzkii* and the glass shrimp *Archaeomysis grebnitzkii.*

Greedo

Armoured Catfish sp. *Peckoltia greedoi* JW Armbruster, DC Werneke & M Tan, 2015

Greedo of Rodia is a fictional character from the 1977 film *Star Wars: Episode IV – A New Hope*. He was a bounty hunter killed by Hans Solo in Chalmun's Spaceport Cantina. This species apparently has a 'remarkable resemblance' to Greedo – particularly large, with dark eyes and puckered lips.

Greeley, AW

Sculpin genus *Greeleya* DS Jordan, 1920 NCR
[Now included in *Oligocottus*]
Green Puffer *Sphoeroides greeleyi* CH Gilbert, 1900

Dr Arthur White Greeley (1875–1904) was an American ichthyologist and physiologist. He graduated from Stanford (1898) and spent a post-graduate year in zoology going on a fur-seal expedition in Alaska and to Brazil with the Banner-Agassiz expedition as collector. He taught school for a year before returning as a fellow in physiology at Chicago University, taking his PhD two years later. He was appointed Assistant Professor at the Washington University, St Louis, and spent three summers instructing in physiology at Wood's Hole Marine Laboratory, Massachusetts. He died after an appendectomy. He is also remembered in the names of two marine crustaceans.

Greeley, JR

Mountain Brook Lamprey *Ichthyomyzon greeleyi* CL Hubbs & MB Trautman, 1937

Dr John R Greeley (1904–1964) was Assistant Director of the Institute for Fisheries Research. Cornell awarded his PhD. Hubbs was a colleague of his at the University of Michigan. He collected the holotype. He co-wrote: *Fishes of the Western North Atlantic* (1963).

Green, AH

Lobefin Snailfish *Polypera greeni* DS Jordan & EC Starks, 1895
[Syn. *Liparis greeni*]

Ashdown Henry Green (1840–1927) was a London-born Canadian naturalist and collector. He studied Civil Engineering in England but was advised to move to a cooler, drier climate for his health, so moved to Victoria on Vancouver Island and never went back. He explored Selkirk Mountain range (1865) to locate a government road. He kept a diary of this expedition now in the BC Archives. He received orders (1865) to return to Victoria and settled in the Cowichan Valley on a farm near Somenos Lake where he was a founding member of the Cowichan Lending Library and Institute. He was then appointed (1871–1880) Divisional Engineer for the Canadian Pacific Railway, working on the location of the CPR line across British Columbia. Green set out (1878) to survey reserve allotments as part of the Joint Indian Reserve Commission. He continued as a surveyor for the reserve Commission throughout its existence and (1913) was appointed as technical officer to the McKenna-McBride Royal Commission on Indian Affairs until retirement (1918). He was an avid sport fisherman, publishing fishing articles in the Journal of Natural History on fish species, identifying nine species previously not seen in BC and discovered two species new to science; the Lake Chub and Lobefin Snailfish. The BC Museum currently has a collection of his preserved fish.

Green, HT

Green's Moon Tetra *Bathyaethiops greeni* HW Fowler, 1949

Harold T Green (d.1964) was a taxidermist and artist who was the Curator of Exhibits, Academy of Natural Sciences of Philadelphia, where he worked (1920–1964). He looked after the fish specimens collected by William K Carpenter during his African expeditions (1946–1948). Green was a member of the (1947–1948) expedition to French Equatorial Africa. He went on a number of other expeditions, usually accompanying a patron who wanted to shoot big game, ostensibly for 'educational purposes'.

Green, MJ

Labeo sp. *Labeo greenii* GA Boulenger, 1902
Armoured Catfish sp. *Ancistrus greeni* IJH Isbrücker, 2001

M J Green was an artist. He drew the first illustration of the holotype of *Labeo greenii*. He wrote: *Primary Book A* (1882) and a number of others in the 'Barnes' Popular Drawing Series' including the: *Teacher's Manual*. He illustrated C.T. Regan's: *A monograph of the fishes of the family Loricariidae* (1904).

Green, RGH

Twinbar Goby *Nesogobius greeni* DF Hoese & HK Larson, 2006

Dr Robert 'Bob' Geoffrey Hewett Green (1925–2013) was a Tasmanian farmer. He was very interested in ornithology and photography, becoming a professional wildlife photographer (1953) and honorary ornithologist at the Queen Victoria Museum and Art Gallery, Launceston, Tasmania (1959). He sold his farm (1960) and joined the museum staff, becoming Curator (1962). The University of Tasmania awarded him an honorary doctorate (1987). A reptile is named after him.

Green, T

Sunset Fairy-Wrasse *Cirrhilabrus greeni* GR Allen & MP Hammer, 2017

Tim Green of Monsoon Aquatics, Darwin, Australia, collected the type specimens. He collects fish and corals for the aquarium trade.

Greenberg

Sarawak Loach sp. *Homalopteroides avii* ZS Randall & LM Page, 2014

Lawrence 'Avi' Greenberg (1982–2011) was "*...an inspiration to and missed friend of the first author and many others; the diagnostic lateral cephalic stripe of this species, reminiscent of a smile, is a symbol of Avi's gentle disposition and goodhearted nature.*" It appears that he committed suicide. (See also **Avi**)

Greene, BD

Greene's Triggerfish *Xanthichthys greenei* RL Pyle & JL Earle, 2013

Brian David Greene (b.1980) is an ichthyologist associated with the Bernice Pauahi Bishop Museum and the University of Hawaii where he studied for his zoology BSc (2001) and where he currently studies for his PhD. He began collecting fish aged five and took up SCUBA aged eleven. After thousands of dives across Micronesia, he left Kwajalein (1998) to study Marine Science and Zoology at the University of Hawaii. Since 2001 he has participated in dozens of deep reef exploratory expeditions utilizing mixed gas rebreathers. He specialises in the collection of fish species new to science. His adventures have taken him across the Indo-Pacific with such institutions as Academia Sinica, the Association for Marine Exploration (where he is (2003) on the board of Directors), BBC, Bernice Pauahi Bishop Museum, California Academy of Sciences, Conservation International, National Geographic, NOAA, Paris Museum, and the University of Hawaii. He now lives in Kailua-Kona, HI, and continues to collaborate on deep reef expeditions with Bishop Museum, Cal Academy, and the University of Hawaii and is a contractor with National Geographic. He has written with the senior author such as: *Tosanoides obama, a new basslet (Perciformes, Percoidei, Serranidae) from deep coral reefs in the Northwestern Hawaiian Islands* (2016) and with both authors: *Five New Species Of The Damselfish Genus Chromis (Perciformes: Labroidei: Pomacentridae) From Deep Coral Reefs In The Tropical Western Pacific* (2008). He has dived and collected on reefs with Pyle and others such as off the Philippines (2015). He was a member of the deep diving team that collected the triggerfish holotype (2005). (Also see **Brian Greene**, **Brianne** & **Anne Patrice**)

Greene, Carroll W

Wedgespot Shiner *Notropis greenei* CL Hubbs & AI Ortenburger, 1929

Dr Carroll Willard Greene (b.1901) was an ichthyologist who worked for the New York State Conservation Department. The University of Michigan awarded his bachelor's degree (1925) and doctorate (1934). He was briefly Assistant Curator of Fishes at the University's Museum (1930). Green was Hubbs' student and Hubbs wrote in the etymology for the shiner that Green, "*...is now engaged in making an ichthyological survey of Wisconsin*", and indeed this resulted in: *The distribution of Wisconsin fishes* (1935). Earlier he had written: *The Smelts of Lake Champlain* (1930).

Greene, Charles W

Shoshone Sculpin *Cottus greenei* CH Gilbert & Culver, 1898
Greene's Midshipman *Porichthys greenei* CH Gilbert & EC Starks, 1904

Charles Wilson Greene (1866–1947) was an American professor of physiology and pharmacology from Indiana. He graduated from Leland Stanford (1892) shortly after marrying fellow student Flora Hartley. He was then a physiology instructor (1893–1896) then began his PhD which was awarded (1898) by Johns Hopkins University. He taught school (1889–1891) then at Stanford (1891–1900). He took a post as Professor of Physiology and Pharmacology at the University of Missouri (1900) and established its experimental pharmacology lab. He undertook research for the US Bureau of Fisheries (1901–1911). His researches covered the structure and function of phosphorescent organs in the toadfish, the circulatory system of the hagfish, the physiology of the Chinook Salmon, and the influence of inorganic salts on the cardiac tissues. He described (1899) the sense and phosphorescent organs of the midshipman species in: *The phosphorescent organs in the Toad-fish Porichthys notatus Girard* (1899). Among his longer works were: *Experimental Pharmacology* (1905) and *Textbook of Pharmacology* (1914). (Also see **Flora**)

Greene, R

Loach sp. *Schistura greenei* M Endruweit, 2017

Richard Greene was a Library Technician at the Smithsonian Institution Entomology Library. He was honoured for his "*...persistent support over many years*".

Greenfield

False Papillose Blenny *Acanthemblemaria greenfieldi* WF Smith-Vaniz & FJ Palacio, 1974
White-lined Toadfish *Sanopus greenfieldorum* B Collette, 1983
Soldierfish sp. *Myripristis greenfieldi* JE Randall & T Yamakawa, 1996
Greenfield's Mudbrotula *Dermatopsis greenfieldi* PR Møller & W Schwarzhans, 2006

Viviparous Brotula sp. *Microbrotula greenfieldi* ME Anderson, 2007
Greenfield's Blenny *Starksia greenfieldi* CC Baldwin & CI Castillo, 2011

Dr David Wayne Greenfield (b.1940) is an ichthyologist, marine biologist, zoogeographer and ecologist who is now a Research Associate at the California Academy of Sciences. California State University, Humboldt awarded his bachelor's degree (1962) and the University of Washington his doctorate in fisheries (1966). He was an Assistant Professor at California State University, Fullerton (1966–1970), an Associate Professor (1970–1977) and Professor of Biological Sciences (1977–1984) at North Illinois University, DeKalb. He was Professor of Biology at the University of Colorado, Denver (1984–1987) and Professor of zoology, University of Hawaii, Manoa (1987–2003) of which, in retirement, he is Emeritus Professor. He was involved in Coral-Reef Biology at the Tropical Studies Center, Belize (1972–1981). Among his more than 150 published papers he co-wrote: *Two new dwarf gobies from the Western Pacific (Teleostei: Gobiidae: Eviota)* (2014). The toadfish is named after David and his wife, Teresa Arámbula Greenfield, who was also a professor at the University of Hawaii, Manoa (of Women's Studies and Education) (1991–2000), as both are credited with the collection of the holotype.

Greenway

Mekong Cyprinid sp. *Mystacoleucus greenwayi* J Pellegrin & PW Fang, 1940
Greenway's Livebearer *Scolichthys greenwayi* DE Rosen, 1967
Greenway's Grunter *Hannia greenwayi* RP Vari, 1978

Dr James Cowan Greenway Jr (1903–1989) was an American collector and ornithologist. He was described as an eccentric, shy and sometime reclusive character. His BA was awarded by Yale (1926) and for several years he worked as a newspaper reporter. He accompanied his close friend Delacour (q.v.) on expeditions to Vietnam and Laos (1929–1930 & 1938–1939), a major objective of which was to search for pheasants. They co-wrote: *VIIe Expedition Ornithologique en Indochine française* (1940). He also participated in several collecting expeditions in the Caribbean. Greenway was an author and editor of *Checklist of Birds of the World* (started 1931). He became Assistant Curator of Birds (1932–1952) at Harvard's Museum of Comparative Zoology and then Curator of Birds (1952–1960), although the former was interrupted by his WW2 service on aircraft carriers in the southwest Pacific. He left MCZ for personal reasons, basing himself on the family estate and was both a trustee (1960–1971) and research associate at the AMNH for the rest of his life. Late in life (1978) he participated in a collecting exhibition in New Caledonia. Rosen honoured him in the name of the livebearer as Greenway had supported Rosen's fieldwork in Guatemala. Greenway's 'generous financial support' also made possible the expedition to Western Australia that obtained the grunter holotype. Two reptiles and two birds are also named after him.

Greenwood

Lake Tanganyika Cichlid genus *Greenwoodochromis* M Poll, 1983
African Barb sp. *Enteromius greenwoodi* M Poll, 1967
Greenwood's Happy (cichlid) *Sargochromis greenwoodi* G Bell-Cross, 1975
Lake Malawi Cichlid sp. *Diplotaxodon greenwoodi* J Stauffer Jr & KR McKaye, 1986

Lake Victoria Black Velvet Cichlid *Haplochromis greenwoodi* O Seehausen & N Bouton, 1998
[Syn. *Neochromis greenwoodi*]

Peter Humphry Greenwood (1927–1995) was an English ichthyologist who was brought up in South Africa and spent much of his working life at the BMNH. He served in WW2 as an able seaman in the Royal Navy (1944–1945) and after being demobilised studied zoology at the University of Witwatersrand, South Africa, which awarded his bachelor's degree (1950). He enrolled at London University (1950) and obtained a Colonial Fisheries Research Studentship with the East African Fisheries Research Organisation (1951) based on Lake Victoria, Uganda. He was appointed Research Officer and stayed in Uganda (1951–1957). He joined the BMNH (1959) as a Senior Research Officer, retiring (1989) as Deputy Chief Scientific Officer. After retiring, he moved to Grahamstown, South Africa and continued researching at the J L B Smith Institute of Ichthyology. He visited the BMNH again (1995), borrowed back his old room to do a bit of work, and was found there the next day having suffered a fatal stroke. Seehausen said of him that he: "*…devoted much of his life to the study of the evolution and systematics of the Lake Victoria cichlid species flock and laid the foundation for the modern systematics of these fishes*". (Also see **Humphrey**)

Gregg

Hagfish sp. *Myxine greggi* MM Mincarone, D Plachetzki, CL McCord, TM Winegard, B Fernholm, CJ Gonzalez & DS Fudge, 2021

John Gregg is an American marine geologist who founded (1985) the Gregg Drilling company. He is also the founder and president of the Western Flyer Foundation. He bought (2015) the Western Flyer; the fishing boat known for its use by John Steinbeck and Ed Ricketts in their (1940) expedition to the Gulf of California, the notes from which culminated in their book: *The Log from the Sea of Cortez* (1951).

Gregory, J

Shark Bay Eel-Blenny *Notograptus gregoryi* GP Whitley, 1941

John Gregory was Fisheries Officer at Shark Bay, Western Australia, the type locality. He was honoured as he had 'greatly assisted' Whitley in his professional capacity there.

Gregory, JW

Gregory's Labeo *Labeo gregorii* A Günther, 1894

Professor Dr John Walter Gregory (1864–1932) was a geologist, geographer, stratigrapher, invertebrate palaeontologist, geomorphologist and explorer. He left Stepney Grammar School at 15 to became a clerk at wool sales in the City of London. His growing interest in the natural sciences led him to attend evening classes at the London Mechanics' Institute (Birkbeck College). He matriculated (1886) and gained his BSc. (1891) and DSc (1893). Meanwhile (1887) he was appointed Assistant in the geological department of the BMNH. His first journey outside Europe (1891) was to study the geological evolution of the Rocky Mountains and the Great Basin of western North America. He was seconded as naturalist to a large expedition to British East Africa (1892) and when this collapsed he set out on his own with a party of forty Africans. In five months, he completed scientific observations in fields ranging from structural geology and physical geography to anthropology, and from mountaineering and glacial geology to the malarial parasites. His major success was the study of the 'Great Rift Valley' summarized in two books: *The Great Rift Valley* (1896) and *The Rift Valleys and Geology of East Africa* (1921). Other major scientific expeditions included the first crossing of Spitsbergen (1896), in the West Indies (1899), Libya (1908), southern Angola (1912) and a 1500-mile walk with his son through Burma to southwestern China and Chinese Tibet (1922). He became Professor of Geology and Mineralogy at the University of Melbourne (1899–1904) and was also Director of the Geological Survey of Victoria (1901–1902) and Director of the British National Antarctic Expedition (1901). He became Professor of Geology at Glasgow University until retirement (1904–1929). In all he published twenty books and over 300 papers and received many honours and awards. He joined an expedition to Peru, aged 68, and was drowned when his canoe overturned in the Urubamba River. He collected the labeo holotype. Two odonata are also named after him.

Gregory, WK

Cusk-eel sp. *Dicrolene gregoryi* ES Trotter, 1926
Reef Bass *Pseudogramma gregoryi* CM Breder, 1927
Gregory's Ghost Flathead *Hoplichthys gregoryi* HW Fowler, 1938

Dr William King Gregory (1876–1970) was an American zoologist, primatologist and palaeontologist. Columbia University, New York awarded his bachelor's degree (1900), master's (1905) and doctorate (1910). He was a member of the scientific staff of the AMNH (1911–1944) becoming Curator of Comparative Anatomy, Ichthyology and Anthropology, the faculty of Columbia University (1916–1945), retiring as Professor of Vertebrate Palaeontology. At the time of the bass description he was Breder's 'chief' at the AMNH. He accompanied Beebe on the Arcturus Oceanographic Expedition (1925) to the Sargasso Sea during which the cusk-eel holotype was collected. He also collected in Australia (1921 & 1922), Central Africa (1929) and South Africa (1938). He was originally of the opinion that the Piltdown Man hoax was genuine, and wrote: *The Dawn Man of Piltdown, England* (1914).

Grell

African Rivuline sp. *Aphyosemion grelli* S Valdesalici & W Eberl, 2013

Wolfgang Grell (d.2001) was a member of the DKG (German African Rivuline Association). He was a talented rivuline collector, breeder, and photographer who only one year before his death (May 2001) directed Eberl's attention to the probable existence of an unknown species in the area east of Sindara, Gabon.

Grenfell

Schilbid Catfish sp. *Schilbe grenfelli* GA Boulenger, 1900

George Grenfell (1849–1906) was an explorer and a Baptist missionary in Cameroon (1875–1877) and the Congo Free State (Democratic Republic of the Congo) (1877–1906). He explored very extensively along many of the rivers of the Congo basin and among his discoveries were the pygmy Batwa peoples. He was appointed as a representative for

Belgium to establish the boundary line between Belgian and Portuguese possessions in the area of Luanda (1891). He originally trusted King Leopold and the Belgian administration of the Congo Free State but came to believe that it was evil. He died from Blackwater fever. He collected the holotype of this species.

Gresens

African Rivuline sp. *Fundulopanchax gresensi* HO Berkenkamp, 2003

Horst Gresens is a German aquarist who has collected in Cameroon for more than thirty years (1984–2012) and breeds tropical fish, particularly killifish. He, according to the etymology "*...together with V. Schwoiser, Kumba, H. Gresens, Meckenheim, and colleagues, did much field work with this species" over two decades.*

Greshake

Williams' Mbuna (cichlid) *Maylandia greshakei* MK Meyer & W Förster, 1984

Alfons Greshake is the former Managing Director (1995) of an aquarium fish importer based in Germany.

Greshoff

Greshoff's Mormyrid *Marcusenius greshoffii* L Schilthuis, 1891
Squeaker Catfish sp. *Synodontis greshoffi* L Schilthuis, 1891
Spiny Eel sp. *Mastacembelus greshoffi* GA Boulenger, 1901

Anton (sometimes Antoine) Greshoff (1856–1905) was a Dutch trader who arrived in the Congo (1877) and was there for more than 20 years, running the company's business at Boma. He was also Dutch Consul in Leopoldville. He was an active collector, particularly of entomological specimens. He presented several species from the Congo to the Zoological Museum of Utrecht University and supplied ethnographic material to the Ethnographical Museum, Leiden, where he presented a large percentage of the spears now on display.

Grewingk

Baikal Yellowfin *Cottocomephorus grewingkii* B Dybowski, 1874

Konstantin Ivanovich Grewingk (1819–1887) was a Baltic-German from Estonia who was Professor of Geology and Minerology at the University of Tartu (then Yur'yev) (1864) where he graduated. His PhD was awarded by the University of Jena. He became (1845) the keeper of mineralogical collections of the Academy of Sciences, and librarian (1852) of the Mining Institute. He travelled to Olonetsk and Arkhangel Provinces, to Sweden, Norway and the Urals and was the first person to explore the Kanin Peninsula. His whole career was devoted to the study of geology, mineralogy and archaeology of the Baltic Territory, in the course of which he published about 170 scientific papers and articles. His article *Zur Archäologie des Balticum und Russlands* (1874) summarized everything then known of the prehistoric archaeology of the northern and north-western part of Russia. He also published (1877) a seminal paper on funeral customs and the ship graves of the eastern Baltic coast.

Grey, G

Beaked Sandfish sp. *Gonorynchus greyi* J Richardson, 1845

Sir George Grey (1812–1898) was a soldier, explorer, colonial governor, premier and scholar. He explored Western Australia on government-financed expeditions, to Hanover Bay and Shark Bay (1837–1839). On the first expedition an Aborigine, whom he afterwards shot, speared him; nevertheless, he championed the cause of assimilation, and respect for the aboriginal people. He became Governor of New Zealand (1845), where his greatest success was his management of Maori affairs. He scrupulously observed the terms of the Treaty of Waitangi, assuring Maoris that their land rights were fully recognised. He became Governor of the Cape Colony (1853) and High Commissioner for South Africa. He sought to convert the frontier tribes to Christianity, to 'civilise' them. Grey supported mission schools and built a hospital for African patients. When he returned to New Zealand he was elected to parliament, but he remained a keen naturalist and botanist, and established extensive collections and important libraries at Cape Town and Auckland. He wrote books on Australian aboriginal vocabularies and his Western Australian explorations, and also took a scholarly interest in the Maori language and culture. Three birds, two mammals and a reptile are named after him.

Grey, M

Arrow Stargazer *Gillellus greyae* RH Kanazawa, 1952
Grey's Deep-sea Smelt *Bathylagichthys greyae* DM Cohen, 1958
Roundtail Duckbill *Bembrops greyae* M Poll, 1959
[Binomial originally given as *greyi*]
Alligator Searobin *Peristedion greyae* GC Miller, 1967
Bristlemouth sp. *Manducus greyae* RK Johnson, 1970

Mrs Marion Grey née Griswold (1911–1964) studied at Wellesley College (1929–1931). She started working at the Field Museum, Chicago, as a volunteer (1941), was unpaid in charge of Division of fishes (the regular incumbent, Loren Woods, was away in the US Navy) (1941–1946) and was an Associate, Division of Fishes, Department of Zoology, Field Museum (1943–1964). She was a member of the museum's 1948 expedition to Bermuda. She wrote: *The distribution of fishes found below a depth of 2000 meters* (1956). The bristlemouth was named after her for her contributions to the knowledge of deep-sea fishes. She died as a result of a series of strokes. (See also **Marion**)

Griem

Goldspotted Tetra *Hyphessobrycon griemi* JJ Hoedeman, 1957

Karl Griem (d.1954) was a collector and dealer in ornamental fish. He collected in the Amazon basin (1933–1934) and supplied the New York Aquarium with a large collection of Brazilian fishes (1937).

Griessinger

Spikefin Goby *Discordipinna griessingeri* DF Hoese & P Fourmanoir, 1978

Mr S Griessinger collected one of the paratypes. No other information was recorded and we have been unable to track down further information.

Griffin, LE

Hook-fin Cardinalfish *Ostorhinchus griffini* A Seale, 1910
Three-barbeled Catfish sp. *Pimelodella griffini* CH Eigenmann, 1917

Professor Lawrence Edmonds Griffin (1874–1949) was a herpetologist at Missouri Valley University. He collected in the Philippines early in the 20th century. He was custodian of herpetology at the Carnegie Museum, Pittsburgh (1915–1920), and is mentioned in their Annals. He wrote: *A Check-list and Key of Philippine Snakes* (1911). A reptile is named after him. Eigenmann does not identify the person he named the catfish after, but he and Lawrence Griffin were colleagues at the Carnegie and it seems very likely that Eigenmann intended to honour L E Griffin.

Griffin, LT

Griffin's Spiny Dogfish *Squalus griffini* WJ Phillipps, 1931
[Alt. Northern Spiny Dogfish, Grey Spiny Dogfish]
Griffin's Moray *Gymnothorax obesus* GP Whitley, 1932
[Alt. Speckled Moray]

Louis Thomas Griffin (1870–1935) was a British-born New Zealand ichthyologist and collector. He served time in South Africa as Superintendent of Pretoria's National Zoological Gardens. He arrived in New Zealand (1908) and joined the staff of the Auckland Museum (1908) as a Preparatory and had risen to Assistant Director at the time of his death. He wrote, among other papers and longer works: *Additions to the Fish Fauna of New Zealand* (1923) and *Revision of the Eels of New Zealand* (1937). He described the eponymous moray (1927) as the first record of *Gymnothorax meleagris* (Shaw, 1795) in the Bay of Plenty, New Zealand. Whitney & Phillipps (1939) gave it the name *G griffini* as a new species, saying: "*…We rename the New Zealand species after our late friend Louis T. Griffin, author of several valuable papers on New Zealand fishes.*" Though that name proved to be invalid, Griffin's name has remained attached as a common name for *G obesus* – to which the name *G. griffini* was mis-applied.

Griffis

Griffis' Angelfish *Apolemichthys griffisi* BA Carlson & LR Taylor, 1981

Nixon Griffis (1917–1993) was a conservationist, a Trustee of the New York Zoological Society and Patron of the New York Aquarium. He graduated from Cornell (1940) and served as a lieutenant in the Army Signal Corps in WW2. He then joined the investment banking firm of Hemphill, Noyes & Co., as a partner where his father was a principal. His father acquired control of Brentano's and Nixon served as President and Chairman, expanding the chain and then selling it to Macmillan Publishing (1962). After this he concentrated on his interest in conservation. He travelled the world to gather breeding groups of endangered species. He also sponsored underwater archaeological explorations of Bronze Age and Byzantine wrecks off the coast of Turkey and was a founding director and past president of the American Littoral Society and a member of the Explorers Club. He wrote: *Cooking Conch* (1983) and *The Mariner's Guide To Oceanography* (1987).

Griffith

Stone Loach sp. *Triplophysa griffithi* A Günther, 1868

Edward Griffith (1790–1858) was a zoologist and one of the original members of the Zoological Society of London. His day job was as a solicitor and a Master in the Court of Common Pleas. He wrote: *General and Particular Descriptions of the Vertebrated Animals* (1821) and translated (and added to) Cuvier's *Règne animal* (1827–1835).

Griffiths, J

Griffiths' Razorfish *Iniistius griffithsi* JE Randall, 2007

Jeremy Griffiths is the son of Owen Griffiths who noted that this was a 'new' fish and arranged for a photograph to be taken and a second specimen collected. (Owen Griffiths is an Australian zoologist, malacologist and conservationist who went to Mauritius (1986) and created a Crocodile and Tortoise Reserve.) When informed of the plan to describe the fish in his honour, he asked that it be named for his son Jeremy, one of the two fishermen who caught the type specimens. This is reflected in the etymology which says the holotype was taken on a hand-line by Jeremy Griffiths and Tonio Isidore (2006), off the coast of Mauritius.

Griffiths, JD

Griffiths' Pygmy-goby *Trimma griffithsi* R Winterbottom, 1984

Major John D Griffiths, of the Royal Signals, was the leader of the (1978–1979) British Armed Forces Chagos Expedition. According to the etymology, his "*…leadership and hard work resulted in a highly successful expedition*" during which the goby holotype was collected.

Grigorjev

Shotted Halibut *Eopsetta grigorjewi* SM Herzenstein, 1890
Prickleback sp. *Stichaeus grigorjewi* SM Herzenstein, 1890

Alexander Vasilevich Grigoriev (1848–1908) was a Russian botanist and ethnographer who became Secretary of the Imperial Russian Geographical Society (1883–1903). St Petersburg University awarded his degree (1870) after which (1871) he taught anatomy and plant physiology at the St Petersburg Practical Technical Institute. He made a Geographical Society oceanographic expedition to the White Sea (1876), and he was on the expedition (1879) to the Siberian coast aboard 'SS Nordenskjold', which ran aground off Japan. He had to spend time in Japan and used it to study the Ainu people before returning to Russia (1880). There he passed his collections to the Geographical Society and they elected him Secretary (1883). He collected in the Solovetskie Islands (1886) and Novaya Zemlya (1887), passing the specimens to the museum of the Academy of Science. He helped plan further expeditions culminating in his last great project an expedition to Tibet (1899–1902) and writing an account published posthumously (1918). He retired (1903) but remained on various committees until his death.

Grimaldi

Mirrorbelly *Monacoa grimaldii* EJG Zugmayer, 1911
Barbeled Dragonfish sp. *Aristostomias grimaldii* EJG Zugmayer, 1913
Rattail sp. *Hymenocephalus grimaldii* MWC Weber, 1913

Albert Honoré Charles Grimaldi (1848–1922), Albert I, Prince of Monaco was an oceanographer who founded Monaco's Institut Océanographique. He also (1911) created the Monte Carlo Rally, an automobile race designed to attract tourists to Monaco.

Grimm

Tadpole Goby sp. *Benthophilus grimmi* K Kessler, 1877
Southern Caspian Sprat *Clupeonella grimmi* K Kessler, 1877
Braschnikov's Shad ssp. *Alosa braschnikowi grimmi* NA Borodin, 1904

Dr Oscar Andreevich von Grimm (1845–1921) was a Russian ichthyologist who was the Chief Inspector of Russian fisheries. He was well known internationally as he represented Russia at international fisheries conferences and exhibitions. He collected the holotype of the Southern Caspian Sprat and the goby. Borodin frequently cited Grimm's work: *The Herring of Astrakhan* (1887).

Grinnell, F

Morid Cod sp. *Physiculus grinnelli* DS Jordan & EK Jordan, 1922

Fordyce Grinnell Jr (1882–1943) was a lepidopterist in Southern California. Stanford awarded his bachelor's degree (1918). His elder brother was Joseph Grinnell (see below). He was Assistant Curator of Entomology, Southwest Museum (1916). He was a former student of D S Jordan who wrote that Grinnell provided 'efficient assistance' by visiting Honolulu fish markets daily. He wrote: *A Butterfly of the High Sierra Nevada – Behr's Alpine Sulphur* (1913).

Grinnell, J

Philippines Dragonet *Synchiropus grinnelli* HW Fowler, 1941

Dr Joseph Grinnell (1877–1939) was a field biologist and zoologist who became professor and Director at the Museum of Vertebrate Zoology, University of California. He is noted for introducing a method of recording precise field observations known as the Grinnell System and was instrumental in shaping the philosophy of the United States National Park System. Among his publications is: *A Distributional Summation of the Ornithology of Lower California* (1929) and he was also editor of the journal 'Condor' for over 30 years. Fowler says the dragonet was: "*...named for the late Dr. Joseph Grinnell, of the Museum of Vertebrate Zoology, Berkeley, Calif.*" He died of a heart attack. Nine birds and a reptile are also named after him.

Griselda

Snailfish sp. *Careproctus griseldea* D Lloris, 1982 NCR
[Junior Synonym of *Careproctus albescens*]

Griselda Lloris is the author's 'little daughter'.

Griswold

Borneo Loach sp. *Protomyzon griswoldi* SL Hora & KC Jayaram, 1952

John Augustus 'Gus' Griswold Jr (1912–1991) was an ichthyologist, aviculturist and ornithologist who took part in expeditions to Borneo (1936), Thailand (1937) and Peru (1939). His collection of Gastromyzonid fishes from Borneo was a great help to the authors. He became Curator of Birds, Philadelphia Zoological Gardens (1947). Among his written works is the book: *Up Mount Kinabalu* (1939). A reptile, an amphibian and a bird are also named after him.

Gritsenko

Char sp. *Salvelinus gritzenkoi* ED Vasil'eva & VM Stygar, 2000

Oleg F Gritzenko (also spelled Gritsenko) is a salmon biologist at the Russian Federative Research Institute of Fisheries and Oceanology, Moscow. He organised the 1999 expedition to the North Kurile Islands, Russia, where this species occurs. He wrote: *Aquatic biological resources of the northern Kurile Islands* (2000).

Grixalva

Suckermouth Catfish sp. *Astroblepus grixalvii* A von Humboldt, 1805

Don Mariano Grixalva was a Colombian described by Humboldt as a 'respectable scholar' who "*...disseminated at Popayan [*Colombia, where this catfish occurs*] a taste for the physical sciences, which he himself cultivated with success*". The whole tone of the etymology is incredibly condescending! We have not been able to find anything more about him.

Groeneveld

Stingfish sp. *Minous groeneveldi* M Matsunuma & H Motomura, 2018

Rokus Wessel Groeneveld (b.1966) is a Dutch underwater photographer hobbyist. After studying at the Rijks Hoge School voor Tuin en landschapsinrichting in Boskoop (1986–1990), which awarded his Bachelor of Built Environment degree, he spent 22 years working for an architecture company in Rotterdam. Since then (2012), he has worked as a technical project manager for the Municipality of Gouda. He started diving (1997) in Vietnam and was so captivated that he took a course and qualified as a dive master. A keen photographer, he wanted to take underwater photos too. He has written a number of articles and his photographs have appeared in magazines and books such as *Tropical Marine Wildlife of Australia* (2017) and websites including the one he runs: www.diverosa.com where Mizuki Matsunuma first saw his photos of the stingfish. The etymology says that the name is: "*... in honor of Mr Rokus Groeneveld (Gouda, Netherlands), who provided us with excellent underwater photographs of the new species.*" He and his wife and diving buddy Sanne particularly enjoy diving at night. Rokus did not yet know he had been honoured until we got in touch with him to check this entry!

Grohmann

Lefteye Flounder sp. *Arnoglossus grohmanni* CL Bonaparte, 1837

Francesco (Franz) Saverio Grohmann was an Austrian 'raccoglitore' (collector) who was a commercial dealer in natural history specimens. He is particularly noted for his insects from Sicily. He wrote at least one scientific paper: *Nuova descrizione del camaleonte siculo* (1932).

Gronovius

Man-of-war Fish *Nomeus gronovii* JF Gmelin, 1789

Wolf Characin sp. *Hoplerythrinus gronovii* A Valenciennes, 1847

Laurens Theodorus Gronovius (1730–1777) (sometimes Laurentius Theodorus Gronovius or Laurens Theodore Gronow, or simply 'Laurenti) was a Dutch naturalist, zoologist, botanist and ichthyologist. He amassed one of the most extensive zoological and botanical collections of his day and is noted for having developed a novel way of preserving fish skins. He played a significant part in the classification of fish. Linnaeus acknowledged that Gronovius inspired him. His father was a notable botanist and both his sons were also notable scholars. He wrote: *Museum ichthyologicum* (1754), in which he described over 200 species of fish. A reptile is also named after him.

Groot

Indonesian Leaffish *Pristolepis grootii* P Bleeker, 1852

Cornelis De Groot van Embden (1817–1896) was a Dutch natural historian who worked in Indonesia as a mining engineer. He enlisted in the lancers (1833) rising to Sergeant-Major, but resigned (1840) as he was unable to become an officer. He enrolled in the newly formed Royal Academy (1843), taking his exam three years later (1846) and was appointed as a water management engineer in Indonesia. He spent two years at the Engeland mine leaving (1848) as chief engineer. He wrote: *Herinneringen Aan Blitong* (1887) describing the natural history and ethnography of Belitung Island off Sumatra and its people. Bleeker wrote that he honoured a man: "*...to whom science owes the first knowledge of the freshwater fauna*" [translation] of Belitung.

Gross

Gross' Stinkfish *Callionymus grossi* JD Ogilby, 1910

Major George Gross (d.1909) was an amateur natural historian, at one time Consul for Switzerland and also a schoolmaster for twenty-four years teaching German (1886) at Brisbane Grammar School, Queensland. He led the school cadet corps and was a first class shot. Ogilby wrote in his etymology: "*Named for my friend and colleague, the late Major George Gross, one of the leading conchologists of Queensland.*"

Grosskopf

Long-whiskered Catfish sp. *Pimelodus grosskopfii* F Steindachner, 1879

Th. Grosskopf collected specimens, including the holotype of this species, in Colombia (1876) for the Berlin Museum. We can find no other reference to him, or even his full forename.

Grosvenor

Characin sp. *Bryconamericus grosvenori* CH Eigenmann, 1927
Rockcut Goby *Gobiosoma grosvenori* CR Robins, 1964

Dr Gilbert Hovey Grosvenor (1875–1966) was a Turkish-born American who is considered to have been the father of photo-journalism. Amherst College, Massachusetts, awarded his bachelor's degree (1897). He was the first employee (1899) of the National Geographic Society as fulltime editor of the *National Geographic Magazine* (1899–1954) and became its president (1920–1954). He was also one of the prime movers behind the creation of USA's National Park Service. His wife was Elsie May Bell (1878–1964), the daughter of the inventor of the telephone, Alexander Graham Bell. He also has a sandstone arch in Utah named after him. Eigenmann's etymology thanks Grosvenor "*…whose kindly interest made possible the expedition to Peru.*"

Grouser

Hagfish sp. *Eptatretus grouseri* CB McMillan, 1999

David 'Grouser' Allen McMillan (b.1957). (See **McMillan, DA**)

Grube

Amur Grayling *Thymallus grubii* B Dybowski, 1869

Dr Adolph Eduard Grube (1812–1880) was a Prussian zoologist whose doctorate was awarded by the University of Königsberg, Prussia (Kaliningrad, Russia) (1837). He was Professor of Zoology, University of Dorpat, Estonia (1843–1856) and finally at the University of Breslau (Wroclaw) (1856–1880), He wrote: *Die Familien der Anneliden* (1850), He had been the author's professor.

Grumm(-Grzhimailo), G

Minnow sp. *Phoxinus grumi* LS Berg, 1907
Stone Loach sp. *Hedinichthys grummorum* AM Prokofiev, 2010

Grigory Efimovich Grumm-Grzhimailo (1860–1936) was an explorer, zoogeographer and an entomologist with a particular interest in Coleoptera. He travelled widely in Central Asia (1884–1890), collecting over 32,000 insects. He wrote a number of papers and book chapters including: *Lepidoptera nova vei falciparum cognita regionis palaearcticae* (1899). His younger brother (below) jointly collected and both are credited with collecting the holotype of *Hedinichthys grummorum,* Grigory led the Chinese expedition that collected (1891) the holotype of *Phoxinus grumi* and so it is named after him alone, as is also a bird and a reptile. A third brother, Lieutenant Michael Yefvimovich GrummGrzhimailo (b.1862), accompanied Grigory on his (1889–1890) expedition to Western China during which they collected larger mammals including wild Przewalski's horses and 1,048 birds.

Grumm(-Grzhimailo), V

Stone Loach sp. *Hedinichthys grummorum* AM Prokofiev, 2010

Vladimir Efimovich Grumm-Grzhimailo (1864–1928) was the younger brother of Grigory (above) and was a metallurgist who graduated from the St Petersburg Institute of Mines

(1885). He worked in the Urals after graduating. He taught at the St Petersburg Polytechnic Institute (1907–1918) becoming a Professor (1911). After the Russian Revolution, he was Professor at the Urals Institute of Mines (1918–1924). He created the Bureau of Metallurgical and Heat Engineering Designs, Moscow (1924) and worked there until his death. He jointly collected with Grigory and both are credited with collecting the holotype of the loach.

Grunwald

Grunwald's Rainbowfish *Melanotaenia grunwaldi* GR Allen, PJ Unmack & RK Hadiaty, 2016

Norbert Grunwald was a German aquarist who was editor of the European Rainbowfish group. He devoted much of his life to keeping rainbowfishes and for many years was connected with the Internationale Gesellschaft für Regenbogenfische (IRG) and its journal *Regenbogenfisch*. Gary Lange (q.v.) submitted the fish to Dr Allen and suggested that Grunwald should be honoured in this way.

Gruschka

Triplefin Blenny sp. *Enneapterygius gruschkai* W Holleman, 2005

Dr Victor Gruschka Springer (b.1928) is a marine biologist and zoologist who is Senior Scientist emeritus, Division of Fishes at the Smithsonian, NMNH. Emory University awarded his Biology BA (1948), the University of Miami his MSc (1954) and the University of Texas his PhD in zoology (1957). He is a specialist in the anatomy, classification, and distribution of fishes, with a special interest in tropical marine shorefishes. Holleman described him as the "*…doyen of blennioid systematics*". He wrote or co-wrote, among over 90 other papers: *Revision of the Blenniid Fish Genus Omobranchus with Descriptions of Three New Species and Notes on Other Species of the Tribe Omobranchini* (1975) and *Two new species of the labrid fish genus* Cirrhilabrus *from the Red Sea* (2013) as well as a popular book called: *Sharks in Question, the Smithsonian Answer Book* (1989).

Gruvel

Guinea Flathead *Solitas gruveli* J Pellegrin, 1905
African Barb sp. *Labeobarbus gruveli* J Pellegrin, 1911
Gruvel's Dragonet *Diplogrammus gruveli* JLB Smith, 1963

Jean Abel Gruvel (1870–1941) was a French marine biologist and ichthyologist, who was the first Director of the Institute of Oceanography, Aquarium and Maritime Museum of Natural History at Dinard, France (1935). He was present at the founding (1906) of Port Etienne (now Nouadhibou, Mauretania). He reported upon some of the collections made during the Travailleur and Talisman scientific expeditions (1880–1883). He wrote: *Les pécheries des côtes du Sénégal as des rivières du sud* (1908). He collected the holotype of the barb species.

Gruzov

Stonefish sp. *Inimicus gruzovi* SA Mandrytsa, 1991

Evgeni Nikolaevich Gruzov (1933–2010) was a Russian zoologist and hydrobiologist. He was one of the pioneers of the study of seabed fauna and biocoensis (an association of different organisms forming a closely integrated community) of the Antarctic coastal waters using light diving equipment. He worked at the Echinodermata department of the St Petersburg Zoological Institute (1969–1998), becoming department head (1984).

Guahibo

Armoured Catfish sp. *Hemiancistrus guahiborum* DC Werneke, JW Ambruster, NK Lujan & DC Taphorn, 2005

Characin sp. *Chrysobrycon guahibo* JA Vanegas-Rios, A Urbano-Bonilla & M d M Azpelicueta, 2015

The Guahibo are a tribe of people living in parts of southern Venezuela and western Colombia. The authors of the catfish description thanked members of this tribe for the help they gave in collecting specimens.

Guairaca

Glass Knifefish sp. *Eigenmannia guairaca* LAW Peixoto, GM Dutra & WB Wosiacki, 2015

Guairacá was a semi-legendary chief of the Guaraní people, who sought to protect their land from European colonisers.

Guaitipan

Armoured Catfish sp. *Sturisomatichthys guaitipan* A Londoño-Burbano & RE Reis, 2019

Guaitipan (also known as Gaitana) was cacique (leader) of the Timaná tribe, who occupied the Colombian Andes in the upper Magdalena valley of the Meta department of Colombia (where this catfish occurs). She fought against the Spanish invasion (1539–1540), ending in her defeat due to treason by the cacique Matambo.

Guarani

Characin sp. *Diapoma guarani* V Mahnert & J Géry, 1987

Paraguayan Cichlid sp. *Australoheros guarani* O Říčan & SO Kullander, 2008

Named after the Guarani people of Paraguay. They also have a reptile, a mammal and a dragonfly named after them.

Guaymas

Guaymas Goby *Quietula guaymasiae* OP Jenkins & BW Evermann, 1889

This is a toponym after Guaymas, Sonora, Mexico, where the holotype was acquired.

Guchereau

Glass Knifefish sp. *Distocyclus guchereauae* FJ Meunier, M Jégu & P Keith, 2014

Corinne Guchereau is an administrator at MNHN. She was thanked for facilitating the technical aspects of the authors' work for 15 years.

Gudger

Oriental Cyprinid sp. *Sikukia gudgeri* HM Smith, 1934
Greenback Skate *Dipturus gudgeri* GP Whitley, 1940
[Alt. Bight Skate]

Dr Eugene Willis Gudger (1866–1956) was an associate in ichthyology at the AMNH, New York, which he joined (1919) and later became Curator of Fishes. His doctorate was awarded by Johns Hopkins University, Baltimore (1905). He was a world authority on whale sharks. He was honoured "*...in appreciation of his work on fishes and their bibliography.*"

Gudrun

Barbeled Dragonfish sp. *Astronesthes gudrunae* NV Parin & OD Borodulina, 2002

Mrs Gudrun Schulze is a technician at the fish collection of the Institut für Seefischerei, Hamburg. She was remembered in 'sincere gratitude for all her help' in the authors' study.

Gueldenstädt

Russian Sturgeon *Acipenser gueldenstaedtii* JF von Brandt & JTC Ratzeburg, 1833
[Alt. Diamond Sturgeon]

Professor Johann Anton Gueldenstaedt (or Güldenstädt) (1745–1781) was a Baltic-German, born in Riga (Latvia), then part of the Russian Empire. He was a physician, natural scientist and traveller. He made several expeditions to the Caucasus and TransCaucasus regions (1768–1773) for the Imperial Academy of Science in St Petersburg, where he was a Professor (1771). He was the author of diaries containing extensive geographical, biological and ethnographical material on the Caucasas and Ukraine, as commissioned by the Empress Catherine II. Pallas (q.v.) published his: *Reisen durch Russland und im Caucasischen Gebürge* posthumously (1787–1791). A mammal and a bird are also named after him.

Guenther

See under **Günther, A**

Guerne

Barbeled Dragonfish sp. *Photostomias guernei* R Collett, 1889

Jules Germain Maloteau de Guerne (1855–1931) was a French zoologist and geographer. For three years (1885–1887) he was Prince Albert of Monaco's personal zoologist on the research cruises aboard his yacht *L'Hirondelle.* He wrote: *Excursions Zoologiques Dans Les Iles De Fayal Et De San Miguel: Acores* (1888). (See also **Grimaldi**)

Guézé

Mauritius Gurnard *Pterygotrigla guezei* P Fourmanoir, 1963
Guézé's Butterflyfish *Prognathodes guezei* AL Maugé & R Bauchot, 1976
Reunion Angelfish *Apolemichthys guezei* JE Randall & AL Maugé, 1978

Paul Guézé is a French marine biologist from Réunion. He was associated with the Académie de la Réunion, as well as the Laboratoire d'Ichthyologie at MNHN. He collected marine fauna off Réunion. Among his publications is the co-written (with Randall): *The goatfish Mulloidichthys mimicus n. sp. (Pisces, Mullidae) from Oceania, a mimic of the snapper Lutjanus kasmira (Pisces, Lutjanidae)* (1980) and he has also published with Fourmanoir. He collected at least some of the holotypes.

Guggenheim

Angular Angelshark *Squatina guggenheim* TL Marini, 1936

The John Simon Guggenheim Memorial Foundation funded the studies of the angelshark's describer, Tomas Leandro Marini (1902 - 1984).

Guiart

Lake Victoria Cichlid sp. *Haplochromis guiarti* J Pellegrin, 1904

Dr Jules Guiart (1870–1965) was a parasitologist who was also interested in the history of medicine. He was awarded his medical doctorate (1896) and his doctorate in natural sciences (1901). He began (1894) his career at the University of Paris Faculty of Medicine as a zoology preparer, becoming (1901) Associate Professor of Natural History there. He then (1906) became the chair of teaching of parasitology and natural history of the Faculty of Medicine and Pharmacy of Lyon and also became (1920) Curator of the Museum of the History of Medicine. He became Secretary General of the Société Zoologique de France. He was Pellegrin's friend who described him as a 'distinguished and devoted' General Secretary. Among his many publications was: *Histoire de la Médecine Française* (1947).

Guibe

Goby sp. *Oxyurichthys guibei* JLB Smith, 1959

Guinean Flounder *Bothus guibei* A Stauch, 1966

Dr Jean Marius René Guibé (1910–1999) was a French zoologist and herpetologist at the Muséum National d'Histoire Naturelle, Paris, where he was Professor of Zoology (Reptiles and Fish) (1957–1975). He wrote *Les batraciens de Madagascar* (1978). Nine amphibians and three reptiles are named after him.

Guichenot

Cyprinid sp. *Paracanthobrama guichenoti* P Bleeker, 1864

Gudgeon sp. *Coreius guichenoti* HE Sauvage & CP Dabry de Thiersant, 1874

Cave Hawkfish *Cirrhitichthys guichenoti* H Sauvage, 1880

Antoine Alphone Guichenot (1809–1876) was a French zoologist, primarily interested in herpetology and ichthyology. He taught and researched for the MNHN, Paris and undertook collecting trips for them, most extensively in Algeria. He scaled back to just being an assistant naturalist (1856). Three reptiles are also named after him.

Guignard

African Rivuline sp. *Scriptaphyosemion guignardi* R Romand, 1981

Alain Guignard is a French aquarist belonging to the Killiclub de France, Paris. He made several collecting trips to Africa, including one where he helped collect the type of the eponymous species.

Guilbert

Loach Goby sp. *Rhyacichthys guilberti* G Dingerkus & B Séret, 1992

Eric Guilbert is an entomologist at the Muséum National d'Histoire Naturelle (Paris), who helped collect the type specimen.

Guilcher

Yellowfin Red Snapper *Lutjanus guilcheri* P Fourmanoir, 1959

Dr André Guilcher (1913–1993) was a French geographer, researcher on coral reefs and author of *Coral Reef Geomorphology* (1988), along with nearly 650 papers. He lectured, then was Professor of Geography, at the University of Nancy (1947–1957) then at the Geographical Institute of the Sorbonne (1957–1970) and the University of Western Brittany (1970–1981). He completed his PhD (1948) on the geography of southern Brittany. He travelled very extensively, often addressing conferences and teaching at universities in Europe and North America. He directed the oceanographical work of the research vessel 'Kornog' and dived with Jacques Cousteau on the 'Calypso'.

Guild

African Barb sp. *Enteromius guildi* PV Loiselle, 1973

Paul Douglas Guild (b.1943) was a friend and colleague of the author when they were US Peace Corps volunteers at the Da Na Fisheries Station, Sokode, Togo for three years in the late 1960s. They co-wrote: *Manuel de Pisciculture* (1967), an unpublished report of a Peace Corps project. He was again in Togo in the 1980s as an USAID Project Officer and was a Regional Executive of USAID in Kiev, Ukraine (2007) and, whilst there, was accused of sexually assaulting two teenage boys for whom he was *in loco parentis*. He was found guilty by a US court in January (2008) on one count of sexual abuse of a minor and two counts of assault, one of which was in connection with the 14-year-old student. A jury acquitted Guild on five other counts of sexual abuse, but sentenced him to 4¼ years in prison.

Guillem

Snailfish sp. *Careproctus guillemi* J Matallanas, 1998

Guillem is the author's son. The fish was named for him and his 'inexhaustible scientific curiosity'.

Guillet

Guillet's Goby *Lebetus guilleti* E Le Danois, 1913

Pierre Guillet (1866–1918) was an artist known for his delicate pastel seascapes. The etymology says: "*...This goby's colour invites us to dedicate it to our friend, the excellent artist P. Guillet*" (translation).

Guiral

African Barb sp. *Enteromius guirali* A Thominot, 1886

Léon Guiral (1858–1885) was a former French naval quartermaster, a naturalist and a member of de Brazza's expeditions in the French Congo (arrived 1882). He died, probably of Yellow Fever, on Christmas Day 1885 at Libreville, and his: *Le Congo français, du Gabon à Brazzaville* was published posthumously (1886). A reptile is also named after him. He collected the barb holotype in the Congo (1885).

Guirao

Eastern Iberian Barbel *Luciobarbus guiraonis* F Steindachner, 1866

Dr Angel Guirao y Navarro (1817–1890) was a physician, botanist and naturalist. He graduated in medicine at the University of Valencia (1841). He achieved doctorates at the University of Madrid (1844 in Medicine and 1861 in Science). He represented Murcia as a Senator of the Kingdom (1876–1885). He did not practice as a physician, but was Interim Professor of Natural History in the Provincial Secondary School in Murcia (1842–1846) and became Full Professor (1847). He was a man of independent means and financed the creation of a botanical garden in Murcia in addition to the Museum of Natural Sciences in Murcia, as well as making numerous donations and contributions to the Madrid Museum of Natural History. He was a member of many scientific institutions and societies in Spain and abroad and is inextricably associated with the teaching of natural history in the province of Murcia. Steindachner made a short visit to Murcia and thanked Guirao for his 'kindness and friendship'.

Gulielm

Angolan Barb sp. *Labeobarbus gulielmi* GA Boulenger, 1910

Gulielm is a version of the name William, and here refers to the forename of Dr William John Ansorge (1850–1913). (See **Ansorge**)

Gulliver

Nurseryfish *Kurtus gulliveri* F de L Castelnau, 1878
Giant Glassfish *Parambassis gulliveri* F de L Castelnau, 1878

Thomas Allen Gulliver (1848–1931) was born in England. He and his brother Benjamin seem to have arrived in Australia whilst still children. Thomas became an employee of the Postal & Telegraph Department in Queensland, and eventually Post and Telegraph master at Townsville. He supervised the construction of the telegraph line to Cape York. A suburb of Townsville is named Gulliver after him. He was also a field natural history collector in northern Queensland for the Melbourne Botanical Gardens (1865–1891). Castelnau wrote that "*Mr. Gulliver, who…has done much for the zoology of that remote part of Queensland, has sent me two collections of fishes from this river*" (the Norman River).

Gunaikurnai

Shaw Galaxias *Galaxias gunaikurnai* TA Raadik, 2014

The Gunai/Kurnai is an indigenous nation and the traditional inhabitants of the Gippsland region of Victoria, Australia.

Gunawan (Kasim)

Licorice Gourami sp. *Parosphromenus gunawani* I Schindler & H Linke, 2012

Gunawan 'Thomas' Kasim is an Indonesian aquarium fish collector and exporter. Together with Horst Linke and others, he collected the type specimens of this species in Sumatra. (Also see **Vera (Kasim)**)

Gunawan, T

Tiene's Dwarfgoby *Eviota gunawanae* DW Greenfield, L Tonabene, MV Erdmann & DN Pada, 2019.

Dr Tiene Gunawan is one of Indonesia's foremost marine conservationists. She originally trained as an architect at Universitas Katolik Parahyangan (1982–1988). Texas A&M University awarded her MSc (1995) and the Institute Pertanian Bogor her PhD (2003). She has worked as an independent consultant (2005–2008 & 2014–2016) and for Conservation International (2008–2014) and at Chemonics International (2015). She joined USAID SEA Project (2016) as Deputy Chief of Party. She has, according to the etymology, "*…dedicated the past two decades to expanding the marine protected area network of West Papua and formulating policies to protect the biodiverse marine ecosystems contained therein.*" She also helped to plan and launch the marine biodiversity survey of the Fakfak coastline that led to the discovery of the dwarfgoby.

Gundrizer

Stone Loach sp. *Triplophysa gundriseri* AM Prokofiev, 2002

Aleksey Nikolaevich Gundrizer was a Russian ichthyologist who worked in Siberia. He was a Professor at the Institute of Biology and Biophysics, Tomsk State University (1990). He co-wrote: *Fishes of West Siberia* (1984). Gundrizer had described this species of loach (1962) but used a preoccupied name.

Gunn

Gunn's Leatherjacket *Eubalichthys gunnii* A Günther, 1870

Ronald Campbell Gunn (1808–1881) was a South African-born Australian botanist and politician. He was educated in Scotland and appointed to the Royal Engineers in Barbados, but left (1829) to go to Tasmania where he became (1830) Superintendent of a convict barracks in Hobart. He met local botanist R W Lawrence who encouraged his interest in natural history. He was appointed as a magistrate (1836–1838) and travelled around Tasmania. He became (1839) Private Secretary to Sir John Franklin who was Clerk of the executive and legislative council. He took on the management of an estate (1841), giving him time to study Tasmania's flora and also collected fauna for the BMNH and studied geology. He took other positions but retired (1876) due to ill health.

Gunnar

Mackerel Icefish *Champsocephalus gunnari* E Lönnberg, 1905

Johan Gunnar Andersson (1874–1960). (See **Andersson, JG**).

Gunter

Brotula genus *Gunterichthys* CE Dawson, 1966
Shoal Flounder *Syacium gunteri* I Ginsburg, 1933
Finescale Menhaden *Brevoortia gunteri* SF Hildebrand, 1948

Dr Gordon Gunter (1909–1998) was a marine biologist. Louisiana State Normal College awarded his bachelor's degree and the University of Texas his master's and his doctorate. He was, variously, a marine biologist with the Texas Game, Fish and Oyster Commission and the United States Bureau of Fisheries in shrimp and oyster investigations; Senior Marine Biologist Scripps Institution of Oceanography and also worked with the commercial oyster industry. He was a zoologist for the Louisiana Department of Conservation and was Director of the Gulf Coast Research Laboratory (1955–1971). He supplied the type of the menhaden species. An American research vessel, belonging to NOAA, was named after him (1998).

Günther, A

Pufferfish genus *Guentheridia* CH Gilbert & EC Starks, 1904
Jellynose Fish genus *Guentherus* B Osório, 1917
Günther's Catfish *Horabagrus brachysoma* A Günther, 1884
Mexican Barred Snapper *Hoplopagrus guentherii* TN Gill, 1862
Günther's Wrasse *Pseudolabrus guentheri* P Bleeker, 1862
Channeled Rockfish *Setarches guentheri* JY Johnson, 1862
Malabar Spiny Eel *Macrognathus guentheri* F Day, 1865
Günther's Loach *Mesonoemacheilus guentheri* F Day, 1867
Sweeper sp. *Parapriacanthus guentheri* CB Klunzinger, 1871
New Zealand Brill *Colistium guntheri* FW Hutton, 1873
Rainbow Prigi *Hypseleotris guentheri* P Bleeker, 1875
Searobin sp. *Lepidotrigla guentheri* FM Hilgendorf, 1879
Morid Cod sp. *Lepidion guentheri* EH Giglioli, 1880
Günther's Mouthbrooder *Chromidotilapia guntheri* HE Sauvage, 1882
Crested Bandfish *Lophotus guntheri* RM Johnston, 1883
Suckermouth Catfish sp. *Astroblepus guentheri* GA Boulenger, 1887
Günther's Grenadier *Coryphaenoides guentheri* L Vaillant, 1888
Tribute Spiderfish *Bathypterois guentheri* AW Alcock, 1889
Günther's Waspfish *Snyderina guentheri* GA Boulenger, 1889
Sardine Characin sp. *Steindachnerina guentheri* CH Eigenmann & RS Eigenmann, 1889
Günther's Flounder *Laeops guentheri* AW Alcock, 1890
Characin sp. *Triportheus guentheri* S Garman, 1890
Squeaker Catfish sp. *Euchilichthys guentheri* L Schilthuis, 1891
Gudgeon sp. *Acanthogobio guentheri* SM Herzenstein, 1892
Bordello Slickhead *Rouleina guentheri* AW Alcock, 1892
Red-tail Notho (Zanzibar killifish) *Nothobranchius guentheri* GJ Pfeffer, 1893
Halosaur sp. *Halosaurus guentheri* GB Goode & TH Bean, 1896
Günther's Lanternfish *Lepidophanes guentheri* GB Goode & TH Bean, 1896
Staring Pearleye *Scopelarchus guentheri* AW Alcock, 1896

Cusk-eel sp. *Porogadus guentheri* DS Jordan & HW Fowler, 1902
Barbed Brotula *Selachophidium guentheri* JDF Gilchrist, 1903
Armoured Catfish sp. *Pseudancistrus guentheri* CT Regan, 1904
Armoured Catfish sp. *Sturisoma guentheri* CT Regan, 1904
Günther's Sabre-gill *Champsodon guentheri* CT Regan, 1908
Diamond-back Puffer *Lagocephalus guentheri* A Miranda-Ribeiro, 1915
Lake Malawi Cichlid sp. *Mylochromis guentheri* CT Regan, 1922
Günther's Butterflyfish *Chaetodon guentheri* E Ahl, 1923
Günther's Deepwater Dragonet *Callionymus guentheri* R Fricke, 1981
Lake Malawi Cichlid sp. *Aulonocara guentheri* DH Eccles, 1989

Dr Albert Karl Ludwig Gotthilf Günther (1830–1914) was a German-born British zoologist, ichthyologist and herpetologist. He is most noted for describing more than 340 reptile species and recognising (1867) that the Tuatara is not a lizard, but belongs to an entirely separate order of reptiles. He was educated at the Stuttgart Gymnasium. His family wanted him to train to be a minister of the Lutheran Church, for which he moved to the University of Tübingen. Like his brother before him he switched to medicine, graduating MD (1858), in which year he also published a handbook of zoology for medical students. He joined the British Museum (1857) as an assistant curator, Zoological Department, where his first task was to classify its 2000 snake specimens. He became a naturalised British subject (1874) and changed his second two Christian names to Charles Lewis. On the death of his sponsor John Edward Gray he took over as Curator of Zoology (1875–1895). He became President, Biological Section, British Association for the Advancement of Science (1880) and was President of the Linnean Society (1881–1901). He wrote: *The Reptiles of British India* (1864) and his magnum opus, the eight-volume: *Catalogue of Fishes* (1859–1870). A remarkable three mammals, two birds, twenty-six amphibians and sixty-seven reptiles are named after him.

Gunther, ER

Yellowfin Notothen *Patagonotothen guntheri* JR Norman, 1937

Eustace Rolfe Gunther (1902–1940) was a British junior zoologist on the Discovery Oceanographic Expedition (1925–1927) during which he studied the habits of whales and life in the Antarctic and southern oceans. He also charted the currents off the Peru coast. He was killed on active duty as Second Lieutenant in the Seventy-Second Searchlight Regiment. A marble tablet recording this (featuring a whale on the ocean) commemorates him on the wall of St Mary's Church, Heacham, Norfolk, England.

Günther, F

Armoured Catfish sp. *Schizolecis guntheri* A Miranda Ribeiro, 1918

Francisco Günther (d.1912) collected specimens for the Museu Paulista, São Paulo, including the holotype of this species.

Guppy, PL

Mimic Blenny *Labrisomus guppyi* JR Norman, 1922
[Syn. *Gobioclinus guppyi*]

Conger Eel sp. *Rhynchoconger guppyi* JR Norman, 1925
Reticulated Tilefish *Caulolatilus guppyi* W Beebe & J Tee-Van, 1937

Plantagenet 'Planty' Lechmere 'Jim' Guppy (1871–1934) was a naturalist in Trinidad and Tobago, co-founder of the Trinidad Field naturalists Club. An early advocate of biological controls, he successfully used the sugarcane froghopper to reduce pests. After retiring from Government service (1929) for some years he worked independently as an expert and dealer in tropical fish for aquaria. He moved to Guyana where he lived (1934) collecting for the American Museum of Natural History in New York, and the New York Zoological Gardens, among other organisations. He introduced the 'Guppy' fish – named after his father (below) – to England (1906), Europe and the United States, where it rapidly became the world's most popular aquarium fish, a status it has maintained ever since. He wrote: *Insect Enemies in Tobago* (1922), *A descriptive catalogue of the fishes of Trinidad and Tobago* (1936) and co-edited *Fauna of Trinidad (Vol 1)* (1940). He died in Panama whilst on a world cruise.

Guppy, RJL

Guppy *Poecilia reticulata* W Peters, 1859
[*Girardinus guppii* (Günther, 1866) is a junior synonym]

Robert John Lechmere Guppy (1836–1916) was a British-born civil engineer, schools inspector, civil servant and naturalist in Trinidad & Tobago, who contributed much to the geology, palaeontology and zoology of the West Indies. He was the son of Robert Lechmere Guppy (1808–1894) who was a British lawyer who emigrated to Trinidad and became Mayor of San Fernando, Trinidad. RJL was raised in England by his maternal grandfather and was heir to a castle but had no interest in that role and left England when eighteen. He visited Australia and Tasmania before being ship-wrecked off the New Zealand coast (1856). He lived with the Maoris there for two years and mapped the area before making sail for Trinidad. He was a civil engineer and helped in the construction of the Cipero Railway. He became Trinidad's first Chief Inspector of Schools until his retirement (1891). He served as President of the Scientific Association of Trinidad, as well as of the Royal Victoria Institute Board. He wrote about seventy papers (1863–1913) such as: *On the Tertiary Mollusca of Jamaica* (1866). He is famous for the discovery of the popular aquarium fish, the guppy. He sent specimens of it from Trinidad to Albert Günther at the BMNH (1866). Although first described by Peters from a specimen in Venezuela (1859), and again under a different name by De Filippi from Barbados (1861), the common name 'guppy' stuck.

Gurjanowa

Goby sp. *Pennatuleviota gurjanowae* AM Prokofiev, 2007

Professor Dr Evpraskia Fedorovna Gurjanova (1902–1981) was an eminent Russian zoologist, hydrobiologist and carcinologist at the Zoological Institute, Leningrad (1929) who described a number of new crustaceans. She was researching and publishing even before she graduated from Leningrad University (1924). She began at the institute after taking her higher degree. She collected the goby type (1959). She travelled across Russia, as far as Bering Island and the Sea of Japan and Pacific (1930–1960s) and took part, as deputy chief, on the Kuril-Sakhalin ZIN-TINRO expedition (1946–1949). She was part of the Soviet-Chinese expedition in the Yellow Sea (1956–1960) and led an expedition to Vietnam (1961). She then took part in an expedition to Cuba (1963, 1965 & 1968). She wrote c.200 papers (1920s-1960s) such as *Contribution to the zoogeography of fareastern seas* (1935). Much of her work was about the zoology of far northern species and the Russian far eastern seas. She also has a seaweed, a marine worm and other taxa named after her.

Gurney

Redtail Barb *Enteromius gurneyi* A Günther, 1868

John Henry Gurney (1819–1890) was a banker in Norwich, England, and an amateur ornithologist who worked at the BMNH. Most of his writing was on the birds of his own county, but he also wrote on collections of African birds, as well as editing the works of others, and he had a particular interest in birds of prey. He was able to use his influence so that Günther received a lot of specimens sent to him from Port Natal (now Durban), South Africa. Nine birds are also named after him. His son (1848–1922), who shared the same name, was also an ornithologist.

Gurroby

Blacksaddle Wrasse *Halichoeres gurrobyi* BC Victor, 2016

Chabiraj 'Yam' Gurroby operates Ornamental Marine World Ltd., with his children Mohesh and Meneeka Gurroby; a small company specialised in exporting live tropical marine fishes of Mauritius. They also support reef protection and restoration and keep a 'reef tank' preserving live coral and reef fauna and flora. The etymology says the name was "*…in recognition of his 35 years of efforts in observing and collecting the fishes of Mauritius*".

Gurvich

Dwarf Sculpin *Procottus gurwicii* DN Taliev, 1946

Georgi (Yuri) Semenovich Gurvich (b.1906) was a Russian hydro-biologist and benthologist. He graduated from Leningrad State University (1930) and then worked at the hydro-meteorological station 'Umba' on the White Sea coast that had just been founded. He wrote many papers such as: *The sea (marine) fisheries of the north-eastern part of the Gulf of Kandalaksha* (1934) before the second world war, but none after and his fate is unknown. He does not appear in lists of war victims, Stalin's purges or the holocaust (his surname implies he was Jewish).

Gushiken

Rosy Grubfish *Ryukyupercis gushikeni* T Yoshino, 1975

Soko Gushiken of the Okinawa Development Agency found and photographed a specimen of this new species at Naha Wholesale Market (1971). He provided Tetsuo Yoshino with many specimens. Among his published papers is: *Phylogenetic Relationships of the Perciform Genera of the Family Carangidae* (1988).

Guthrie

Bengal Spaghetti-Eel *Moringua guthriana* J McClelland, 1844

Colonel Charles Seton Guthrie (1808–1875) of the Bengal Engineers, a unit in the army of the Honourable East India Company (1828–1857), was a wealthy man and accumulated an impressive collection of Mughal art and artefacts as well as a famous collection of coins. Some of his collections are now housed in the Victoria and Albert Museum, London. He was honoured for '…service rendered by him to natural history'.

Gutierrez

Characin sp. *Hemibrycon gutierrrezi* CA Ardila Rodriguez, 2020

Carlos Gutierrez & Juanita Gomez and their six children (Emma, Cecilia, Carmen, Lucila, Sor Maria Ines & Father Carlos – none living) were all philanthropists carrying out social welfare projects and establishing schools. The local library is named aftwer Lucila. The etymology says: "*The Gutierrez epithet is a tribute of the author to the Gutierrez Gomez family, declared the most distinguished of the 20th century in Floridablanca, Department of Santander Colombia*" (Translation). The local authoprity created an award for exemplary citizens in their name.

Gutsell

Tuckasegee Darter *Etheostoma gutselli* SF Hildebrand, 1932

Dr James Squier Gutsell (1887–1976) was a fisheries biologist who was associate aquatic biologist with the Bureau of Fisheries. He made a collection of fishes (1930) while studying the effects of discharged trade waste into streams for the Bureau at the request of the North Carolina Department of Conservation. The fish were studied by Hildebrand who discovered that one was a new darter species which he named after Dr Gutsell. He wrote the book: *Danger to Fisheries from Oil and Tar Pollution of Waters* (1923) and among his published papers are: *Fingerling Trout Feeding Experiments, Leetown, 1938* (1939) and *Frozen Fish in Hatchery Diets May Be Dangerous* (1940).

H

Haacke

Southern Sole *Aseraggodes haackeanus* F Steindachner, 1883
Wavy Grubfish *Parapercis haackei* F Steindachner, 1884

Johann Wilhelm Haacke (1855–1912) was a German zoologist. He studied zoology at the University of Jena which awarded his PhD (1878) and worked there as an assistant and at Kiel University. He emigrated to New Zealand (1881), working at the museums in Dunedin and Christchurch. He moved on (1882) to Australia taking a post as Director of the Adelaide Natural History Museum until resigning (1884) after falling out with the board of management. He returned to Germany (1886) and was Director of Franfurt Zoo (1888–1893) and lecturer at Darmstadt University (1893–1897), then working as a private scholar and teacher. Steindachner provides no etymology (not unusual for him) but we believe this is the person intended.

Haas, F

Catalonian Barbel *Barbus haasi* R Mertens, 1925
Sickle Barb *Enteromius haasianus* LR David, 1936

Dr Fritz Haas (1886–1969) was a German zoologist and malacologist specialising in freshwater snails and mussels. He described over three hundred new species. His doctorate was awarded by the University of Heidelberg. He was Curator of invertebrate zoology at the Senckenberg Museum, Frankfurt-am-Main (1911–1936). He travelled extensively in Europe (1910–1919), the Americas and was a member of the Hans Schomburgk expedition to Southern Africa (1931–1932). The Nazis sacked him because he was Jewish so he emigrated to the USA, where he became Curator at the department of lower invertebrates at the Field Museum, Chicago (1938–1959). He wrote what is regarded as his major work: *Superfamilia Unionacea* (1969). He collected the holotype of both species and at least one freshwater snail is also named after him.

Haas, K-H

African Rivuline sp. *Aphyosemion haasi* AC Radda & E Pürzl, 1976
[Syn. *Aphyosemion cameronense haasi*]

Karl-Heinz Haas is a cichlid specialist from Stuttgart and was an old friend of the authors who "*...with his collecting trips in the mid seventies made important contributions to the study of the fish of Gabon.*"

Haast

Bandfish sp. *Cepola haastii* J Hector, 1881

Dr Sir Johann Franz Julius von Haast (1822–1887) was a German-born geologist. He became a naturalised New Zealander, where he worked for the Canterbury provincial government. He was instrumental in founding the Canterbury Museum, becoming its first Director. He was the German Consul in New Zealand (1880) and the first New Zealander to be awarded the Royal Geographical Society's Gold Medal, for his work on moas. The Haast Pass in the Southern Alps, the Haast River, and the town of Haast, all in New Zealand, are named after him as well as three birds.

Habbema

Mountain Grunter *Hephaestus habbemai* M Weber, 1910

Lieutenant D Habbema (b.1880) was a Sumatran-born officer in the Dutch East Indian Army. He was appointed as second-lieutenant (1902) and retired with the rank of Colonel (1929). He was appointed, through his knowledge of the Dayaks, as the leader of the 42-man military escort supporting the Lorentz Expedition (1909–1910) to northern New Guinea, which also consisted of many Dayak bearers and 20 convict labourers. He also took part in collecting, especially botanical specimens. A Papuan lake was also named after him by Lorenz as is an orchid, *Phreatia habbemae*.

Haber

Yellow-banded Basslet *Lipogramma haberorum* CC Baldwin, A Nonaka & DR Robertson, 2016

[Binomial originally given as *haberi* and later amended to the plural form]

Spencer and Tomoko Haber are an American couple who funded, and participated in, a submersible dive by the Smithsonian's Deep Reef Observation Project that resulted in the collection of a paratype (Curaçao, southern Caribbean).

Haberer

Graceful Catshark *Proscyllium habereri* FM Hilgendorf, 1904
Claroteid Catfish sp. *Chrysichthys habereri* F Steindachner, 1912
Claroteid Catfish sp. *Gephyroglanis habereri* F Steindachner, 1912
Yellowfish sp. *Labeobarbus habereri* F Steindachner, 1912

Dr Karl Albert Haberer (1864–1941) was an independent German anthropologist, naturalist, student of politics, and collector in Japan and China. He studied medicine and natural history in Strasbourg, Berlin and Munich (until 1898), then made several expeditions to East Asia (1899–1904). Much of this period was spent in Yokohama, Japan. His voyages were supported by the State of Bavaria in exchange for collected material he donated to Bavarian museums. Besides paleontological and anthropological material, he collected enormous amounts of zoological material in Japan, primarily marine invertebrates, but also birds and mammals. The Boxer Rebellion (1899) forced him to stay in Peking (Beijing) for longer than he planned, but he bought some 'Dragon Bones' (fossilized bones used in traditional Chinese medicine) in a market. Among them was a human-like molar that eventually led to the discovery of 'Peking Man' (*Homo erectus pekinensis*). He wrote: *Schädel und Skeletteile aus Peking* (1902). He also collected in Cameroon during the early years of the twentieth century, where he obtained the holotypes of the catfish species. A bird is named after him.

Habluetzel

Neotropical Rivuline sp. *Papiliolebias habluetzeli* S Valdesalici, DTB Nielsen, RD Brousseau & J Phunkner, 2016

Dr Pascal István Hablützel is a research biologist. His PhD was granted by the University of Zurich (2009). He worked at the Centro de Investigación de Recursos Aquáticos, Universidad Autónoma del Beni 'José Ballivián', Campus universitario 'Hernán Melgar Justiniano., Trinidad, Bolivia (2009–2010); the Laboratory of Biodiversity and Evolutionary Genomics (2010–2015), Belgium and Vlaams Instituut Voor De See, Belgium (2017) where he is Senior Researcher. Among his more than 20 published papers is: *A preliminary survey of the fish fauna in the vicinity of Santa Ana del Yacuma in Bolivia (río Mamoré drainage)* (2012). He was the first to document this species.

Hadhrami

Arabian Barb sp. *Arabibarbus hadhrami* K Borkenhagen, 2014

This is not a true eponym, in as much as the name honours the Hadhrami people of Hadhramaut Province in Yemen; the area where this species occurs.

Hadiaty

Renny's Ricefish *Oryzias hadiatyae* F Herder & S Chapuis, 2010

Pipefish sp. *Choerichthys hadiatyae* GR Allen, MV Erdmann & N Hidayat, 2020

Dr Renny Kurnia Hadiaty (1960–2019) was a zoologist and ichthyologist at the Division of Zoology, Research Center for Biology, Indonesian Institute of Sciences. Cibinong, Indonesia, where she became Curator of Ichthyology (2008). Her doctorate was awarded by The Faculty of Biology, University of General Soedirman, Purwokerto (1985). She took part in six expeditions in and around Indonesia (2003–2011). She co-wrote: *The endemic Sulawesi fish genus* Lagusia *(Teleostei: Terapontidae)* (2012).

Haeckel

Freckled Catshark *Scyliorhinus haeckelii* A Miranda-Ribeiro, 1907

Smallspine Spookfish *Harriotta haeckeli* C Karrer, 1972

Dr Ernst Heinrich Philipp August Haeckel (1834–1919) was an evolutionary biologist, zoologist, philosopher, and artist. He qualified as a physician in Berlin (1857). He studied zoology at Friedrich-Schiller-Universität Jena (1859–1862) and was Professor of Comparative Anatomy (1862–1909). He travelled extensively in the Canary Islands (1866–1867) and in Dalmatia, Egypt, Turkey and Greece (1869–1873). He met Thomas Huxley and Charles Darwin, whose theories he embraced and promoted. He wrote: *The Riddle of the Universe* (1901). An asteroid, 12323 Häckel, is named after him, as are two mountains, one in the USA and the other in New Zealand. A research vessel has been named after him, and the spookfish is named both after the vessel (from which the holotype was taken) and the man himself.

Haedrich

Mocosa Ruff *Schedophilus haedrichi* N Chirichigno F., 1973

Dr Richard Lee Haedrich (1938–2017) was an ichthyologist, marine biologist and oceanographer who was a specialist in this group of fishes and how they relate to the marine environment. Harvard awarded his PhD (1966) and he then spent a year in Denmark as a Fulbright Fellow doing further fish research. Following this, he returned to a position as a research scientist at the Woods Hole Oceanographic Institute. He was leading scientist on numerous research cruises, initially out of Woods Hole, Massachusetts, and later out of the Bedford Institute of Oceanography in Dartmouth, Nova Scotia. He was (2005–2017) Professor Emeritus at Memorial University in St. John's, Newfoundland, where he taught fisheries biology and oceanic biogeography (from 1979) becoming research professor (1999) and played a lead role in directing oceanic research. He was also co-chair of the ocean fish subdivision of Canada's Endangered Species Committee (1999–2004). Among his more than 130 publications is the co-written book: *Deep-sea Demersal Fish and Fisheries* (1997). He enjoyed cycling and playing trombone in the easternmost jazz band in North America.

Haemus

Bulgarian Bullhead sp. *Cottus haemusi* BT Marinov & CI Dikov, 1986

In Greek mythology Haemus was a king of Thrace whom the gods turned into a mountain range; 'Haemus Mons' (the Balkans).

Hafez

Stone Loach sp. *Turcinoemacheilus hafezi* K Golzarianpour, A Abdoli, R Patimar & J Freyhof, 2013

Khwāja Shamsud-Dān Muhammad Hāfez-e Shārizi (1315 or 1317–1390) is best known by his pen name Hāfez. He was, according to the etymology: "...*one of the most famous and influential Persian lyric poets.*" Despite his profound effect on Persian life and culture and his enduring popularity, few definite details of his life are known.

Hafiz

Searobin sp. *Pterygotrigla hafizi* WJ Richards, T Yato & PR Last, 2003

Ahmed Hafiz's career started at the Ministry of Fisheries and Agriculture, Maldives. He worked at the Marine Research Centre from its very beginning and was Director (1999–2004). He became Deputy Minister of Fisheries & Agriculture and is on the Board of the telecommunications company Dhiraagu (2018). Among his published works he co-authrored the three-volume *Common Reef Fishes of the Maldives* (1998).

Hagedorn

Hagedorn's Tube-snouted Ghost Knifefish *Sternarchorhynchus hagedornae* CDCM de Santana & RP Vari, 2010

Dr Mary M Hagedorn is a marine biologist and research physiologist at the Smithsonian Institute (1993–present). Tufts University awarded her bachelor's (1975) and master's (1976) degrees and the Scripps Institute of Oceanography, University of California, San Diego her doctorate (1983). She worked at the University of California (1978–1979) in behavioural research and as a post graduate (1979–1983). She was then at Cornell (1984–1986) and the University of Oregon (1986–1992) before joining the Smithsonian. She is a specialist in coral reefs. She has undertaken field research in Panama (ten field trips 1979–1986), Venezuela (twice 1981–1982), The Gambia (1984), Florida (1985 & 2005), Peru (four times 1990–1999), Hawaii (2002 & 2003), Puerto Rico (2006 & 2009), Singapore and Sulawesi (2010), Belize (2011) and Australia (2011). In addition, she has taught as an Assistant Professor. She co-wrote: *Toward the cryopreservation of Zebrafish embryos: Tolerance to osmotic dehydration* (2016). She collected the type series of this species.

Hagen

Halfbeak sp. *Nomorhamphus hageni* CML Popta, 1912
Hagen's Goby *Sicyopterus hageni* CML Popta, 1921

Dr Bernhard Hagen (1853–1919) was a German physician and amateur naturalist. After studying medicine at Munich University, he was employed by a planting company in Sumatra. Here he made several, mainly zoological, collecting expeditions. He co-wrote the report of the (1911) expedition to the Lesser Sunda Islands, mounted by the Frankfurt Geographical and Statistical Association and led by Elbert. The Astrolabe Company employed him in New Guinea (1893–1895), then he returned to Germany (1895), but revisited New Guinea (1905) with his wife. He was a section head (1897–1904) at the Senckenberg Museum in Frankfurt, founding their Ethnology Department. He published widely on zoology, geography and ethnography. The description of the halfbeak gives no details of who is honoured, but it seems most likely to be this Hagen who travelled and collected in Indonesia where the fish is found and whom Popta does acknowledge in the goby etymology. Two bIrds, two mammals and a reptile are named after him.

Hagey

Scorpionfish sp. *Idiastion hageyi* JE McCosker, 2008

Harry R Hagey (b.1941) is a board member (since 2017) of Nature Conservancy. Northwestern University awarded his MBA. He is the retired CEO and Board Chair of the San Francisco investment firm Dodge & Cox. He served the firm for more than 39 years and was also a Governor of the Investment Counsel Association of America (ICAA) Board. He has served on the boards of St. Luke's Hospital, the Lucille Packard Children's Hospital in Palo Alto and the California Academy of Sciences. He and his wife Shirley, a trustee of TNC's Idaho chapter (they are are restoring several miles of spring creeks on their ranch near Bellevue, Idaho and providing much needed habitat for native fauna), are founding partners of the Science for Nature and People Partnership (SNAPP) and continue to foster its mission. They also serve on TNC's North American Advisory Group.

Hahn

Twig Catfish sp. *Farlowella hahni* H Meinken, 1937
Sand Knifefish sp. *Rhamphichthys hahni* H Meinken, 1937

Carlos Hahn was an Argentine aquarist who collected specimens in the Corrientes area, where he lived. He and the author had a long-standing relationship and correspondence. The holotype of the catfish species came from Hahn's collection.

Haines

Ridged Catfish *Amissidens hainesi* PJ Kailola, 2000

Allan K Haines is a fisheries scientist. He worked in the Fisheries Division, Department of Primary Industry, Konedobu, Papua New Guinea, where he surveyed the country's river systems (1972–1976). He wrote: *An ecological survey of fish of the lower Purari River system, Papua New Guinea* (1979).

Hajomayland

Lake Malawi Cichlid sp. *Maylandia hajomaylandi* MK Meyer & M Schartl, 1984

Hans Joachim Mayland (See **Mayland**)

Halboth

Armoured Catfish sp. *Parotocinclus halbothi* PC Lehmann-Albornoz, H Lazzarotto & RE dos Reis, 2014

Dário Armin Halboth (1965–2003) was a Brazilian ichthyologist, described as an 'excellent field biologist'. He was one of the first researchers to study the effects on fish communities of bauxite tailings deposited in an Amazonian lake He described the ecological features of the fishes living in streams of Amapá State, Brazil.

Hale, A

Oriental Barb sp. *Discherodontus halei* PGE Duncker, 1904

Abraham Hale (1854–1919) lived at Kivala, Selangor in the Straits Settlements (now part of Malaysia). He was also known as R Blake though why he occasionally used a pseudonym is a mystery. He was chairman of the committee that ran the Selangor State Museum when Duncker took over the museum's 'small, ill-kept collection of stuffed fishes'. He was grateful to Hale for 'sympathetic kindness'. Hale joined the service of the Perak government (1884) and was District Officer and Inspector of Mines, Kinta district, Perak (1885). He was interested in anthropology and wrote: *Folklore and the Menangkabau Code in the Negri Sembilan* (1898). He retired to England and worked for the Malayan Information Agency, London.

Hale, HM

Hale's Drombus (goby) *Drombus halei* GP Whitley, 1935
Hale's Wobbegong *Orectolobus halei* GP Whitley, 1940
[Alt. Gulf Wobbegong]

Herbert Mathew Hale (1895–1963) worked at the South Australian Museum, Adelaide (1914–1960). He became the Director's Assistant (1917), Assistant in Zoology (1922), Zoologist (1925) and Director (1928).

Hall

Red Sea Longnose Filefish *Oxymonacanthus halli* NB Marshall, 1952

Major Harold Wesley Hall (1888–1964) was an Australian zoologist, collector, explorer, philanthropist and entrepreneur who promoted 'Hall's fortified wines'. He supported natural history research, having made a fortune through the Mt Morgan goldfields. The etymology states: "*...I have much pleasure in naming this species after Major H W Hall, M.C., the owner of Motor Yacht Manihine.*" It was during an expedition to the Gulf of Aqaba that the type was collected. A bird is also named after him.

Haller, E

African Rivuline sp. *Aphyosemion halleri* AC Radda & E Pürzl, 1976

Ernst Haller of Stuttgart was co-collector (with the authors) of the species in Cameroon (1975). He had flown to Cameroon to make up the third member of the party with Radda and Pürzl. The plan was to hire a car, drive from Ebolowa via Ambam into Gabon, where they would investigate the north of that country. The Ntem is the border between Cameroon and Gabon, but the ferry across it had broken down, so the authors were forced to leave Ernst behind in Ambam in charge of the car. It was there, initially on his own, but then with the authors, that he caught a representative of the *Aphyosemion cameronense* group of killifish near the Ambam Catholic Mission. The specimen differed from the blue specimens of *A. cameronense*, which they had previously found further to the north. Subsequently specimens were brought back to Europe, and bred and photographed by Haller.

Haller, G

Haller's Roundray *Urobatis halleri* JG Cooper, 1863

George Morris Haller (1851–1889) was the young son of an American army officer Granville Owen Haller. Haller Junior was injured in the foot while wading along the shore in San Diego Bay (1862). He was treated by Dr Cooper who suspected that the boy had been hit by a stingray and set out to examine the local forms of stingray. Rightly believing it to be new to science, he described the ray and named it after his patient. George Haller fought at Gettysburg (alongside his father) at the age of 12 (1863). In later life, he became a lawyer. He was drowned when the canoe in which he was traveling capsized.

Hallstrom

Papuan Epaulette Carpetshark *Hemiscyllium hallstromi* GP Whitley, 1967

Sir Edward John Lees Hallstrom (1886–1970) was born in Coonamble, New South Wales, Australia. He was a pioneer of refrigeration, a philanthropist and leading aviculturist. He began work in a furniture factory aged 13, but later opened his own factory to make ice-chests and then wooden cabinets for refrigerators. He eventually designed and manufactured the first popular domestic Australian refrigerator. Hallstrom made generous donations to medical research, children's hospitals and the Taronga Zoo in Sydney, becoming an honorary life Director there. He visited New Guinea (1950). There is a research collection of 1,600 rare books on Asia and the Pacific at the University of New South Wales Library known as the Hallstrom Pacific Collection purchased with funds Hallstrom gave to the Commonwealth government (1948) for the purpose of establishing a library of Pacific affairs and colonial administration. When he was trustee and chairman of Taronga Zoological Park the shark holotype and paratype were kept there alive in captivity. A mammal and three birds are named after him.

Halstead, BW

Halstead's Toadfish *Reicheltia halsteadi* GP Whitley, 1957

Dr Bruce Walter Halstead ne Newton Bruce Mellars (1920–2002) was a physician, marine biologist and toxicologist. University of California awarded his bachelors degree in Zoology and Loma Linda Univeristy his MD. He was founder and Director of the World Life Research Institute at Colton, California, USA. He was consultant to over 40 governmental and international agencies, including WHO; UNESCO; the US Army, Navy, and Air Force and National Institutes of Health of numerous foreign governments, domestic and foreign universities; research institutes; and the pharmaceutical industry. He travelled to over 150 countries and studied marine creatures. He wrote seventeen books including a tropical fish guide and *Dangerous Marine Animals that Bite, Sting or are Non-edible* (1978) and more than 300 scientific publications. He was honoured for his … "studies on poisonous and venomous fishes."

Halstead, RA

Gold-bar Sand-diver *Trichonotus halstead* E Clark & M Pohle, 1996
Halstead's Sandy *Novaculops halsteadi* JE Randall & PS Lobel, 2003

Robert A 'Bob' Halstead (1944–2018) was an early innovator in the development of dive tourism, pioneer diving instructor and underwater photographer. After obtaining his degree in physics from King's College and a post graduate degree in education at Bristol University, he left London to take a teaching post in the Bahamas where he was introduced to Scuba diving. He moved on (1973) to take an educational post with the Australian Government, becoming Head of the Science Department at the Sogen National High School Papua New Guinea. There he met his future wife (1976) Dinah (q.v.) (also a teacher), as his student on a diving course he was teaching. They formed Papua New Guinea's first full time sport-diving business (1977). They started (1986) the first PNG live-aboard boat operation, Telita Cruises. Bob and Dinah explored many underwater regions of coastal Papua New Guinea. He was author of hundreds of articles and eight diving books, such as: *The Coral Reefs of Papua New Guinea* (1998), *Great Barrier Reef* (1999) and *Coral Sea Reef Guide (2000).* The sand-diver is named after both Bob and Dinah. (Also see **Dinah**).

Haludar

Indian Barb genus *Haludaria* R Pethiyagoda, 2013

Haludar (fl. 1807) was a Bengali artist who (ca. 1797) made the illustrations that were later incorporated into Francis Hamilton's: *Gangetic Fishes* (1822).

Hamilton, F

Hamilton's Thryssa *Thryssa hamiltonii* JE Gray, 1835
Burmese Mullet *Sicamugil hamiltonii* F Day, 1870
Mountain Carp sp. *Psilorhynchus hamiltoni* KW Conway, DE Dittmer, LE Jezisek & HH Ng, 2013

Dr Francis Hamilton né Buchanan (1762–1829) was an ichthyologist and botanist who qualified as a physician at Glasgow (1783) and was to have been a ship's surgeon, but ill-health that year prevented him taking up the post. When recovered (1794) he joined the East India Company's Bengal service as an Assistant Surgeon. He collected botanical specimens as he travelled. His botanical drawings were so admired that a number were presented to Joseph Banks, to whom he regularly sent specimens. He studied the fishes of the Ganges and was often employed on survey work on all sorts of subjects, including fisheries. He went on to become Superintendent of the Calcutta Botanical Gardens (1814). His family name at birth was Buchanan, but he dropped it and took the name Hamilton, his mother's maiden name (1815). He signed his name as 'Francis Hamilton' or 'Francis Hamilton (formerly Buchanan)'. He wrote: *Account of the Fishes of the Ganges* (1822). A bird and a reptile are also named after him.

Hamilton, H

Smallscale Waryfish *Scopelosaurus hamiltoni* ER Waite, 1916

Harold Hamilton (1885–1937) was a biologist who graduated from Otago University, New Zealand. He was employed by the New Zealand Geological Survey and was entomological collector for the Dominion Museum, Wellington. He was on board 'Aurora' in the Antarctic (1913–1914). He was zoologist on the Macquarie Expedition (1911–1913) during which the holotype, partly eaten and found on a beach, was collected. He became the first Director of the School of Maori Arts, Rotorua. He wrote: *Biological Diary (1911–1913)*. An amphibian is named after him.

Hamilton, JS

Hamilton's Barb *Enteromius afrohamiltoni* RS Crass, 1960

Lieutenant Colonel James Stevenson-Hamilton (1867–1957) was appointed the first Head Warden of the Kruger National Park (1902) after the Boer War. He was known as 'Skukuza' by his staff at Kruger National Park, a Shangaan name meaning either 'he who sweeps clean' or 'he who turns everything upside down'. The main rest camp's name was changed from Sabie Bridge to Skukuza to honour him (1936). He was employed in the Sudan civil service (1917) until retirement (1946). He wrote: *Animal Life in Africa* (1912) and published a number of maps for parts of southern Africa. Two mammals are also named after him. He collected the holotype of this species.

Hamilton, W

Common Toadfish *Tetractenos hamiltoni* J Richardson, 1846

William Hamilton was the surgeon-superintendent on convict ships which stopped at Port Jackson (c.1825). He joined the service (1797) and was entered into the list of naval medical officers (1814) and made three voyages on convict ships to Australia; on the 'Elizabeth' (1818), the 'Maria' (1829) and the 'Norfolk' (1825). He kept a diary of all three voyages. He applied for a land grant to settle in Australia. Richardson's etymology refers to 'Surgeon Hamilton of the Royal Navy' and says the holotype came from Port Jackson.

Hamilton, WD

Bluntnose Knifefish sp. *Brachyhypopomus hamiltoni* WGR Crampton, CDCM de Santana, JC Waddell & NR Lovejoy, 2017

William Donald 'Bill' Hamilton (1936–2000) was an English evolutionary biologist who was educated at Cambridge. He lectured at Imperial College, London (1964–1977) and became Professor of Evolutionary Biology, University of Michigan (1978–1984), finishing his career as the Royal Society Research Professor in the Department of Zoology, Oxford (1984–2000). He co-wrote: *The evolution of cooperation* (1981).

Hamilton, WJ

Rustyside Sucker *Thoburnia hamiltoni* EC Raney & EA Lachner, 1946

Dr William John 'Wild Bill' Hamilton, Jr. (1902–1990) was a vertebrate zoologist, naturalist and mammologist, whose association with Cornell University started (1920) when he entered as an undergraduate and ended with his death, only being interrupted by service (1942–1945) as a Captain in the army medical corps in WW2. Cornell awarded his BS (1926), MS (1928) and PhD (1930) and he was a graduate assistant (1926–1930). He was then appointed (1930) as an instructor in vertebrate zoology, assistant professor (1937), associate professor (1942) and professor (1947) eventually retiring (1963) as Emeritus Professor. He helped found (1947) the Department of Conservation (now Department of Natural Resources). In his career, he wrote more that 230 scientific articles and papers including the textbook: *Mammals of Eastern North America* (1943). He was president of both the American Society of Mammologists and the Ecological Society of America. He was a great gardener and in retirement transformed the garden at his house into a true botanical garden, including many species previously unknown in New York State.

Hamlin

Poacher sp. *Podothecus hamlini* DS Jordan & CH Gilbert, 1898

Hon. Charles Sumner Hamlin (1861–1938) was an American lawyer who graduated from Harvard (1886). He became US Assistant Secretary of the Treasury (1893–1897 & 1913–1914) and the first Chairman of the Federal Reserve (1914–1916). He wrote pamphlets on statistical and financial subjects, and: *Index Digest of Interstate Commerce Laws* (1907) and the *Index Digest of the Federal Reserve Bulletin* (1921). He twice stood unsuccessfully for Governor of Massachusetts. Jordan says of him: "*under whose auspices the fur seal investigations of 1896 and 1897 were carried on by the United States Fur Seal Commission.*" (The holotype of the poacher was taken during these investigations).

Hamlyn

Purple Eagle Ray *Myliobatis hamlyni* JD Ogilby, 1911

Dr Ronald Hamlyn-Harris (1874–1953) was an English-born entomologist. He studied in Naples, Italy (1901) and was awarded a doctorate by the Eberhard Karl University, Tübingen, Germany, (1902). He went to Australia (1903) and became a schoolmaster at Toowoomba Grammar School, Queensland. He revitalised science teaching, raised funds for a new laboratory and gave popular lectures. He was Director of the Queensland Museum, Brisbane (1910–1917); this came as a great relief to Ogilby (q.v.) who had been trying to combine being an administrator with conducting research. Ogibly and Hamlyn-Harris became life-long friends and used to go fishing together. He then ran a fruit farm (1917–1922). He was put in charge of the Australian Hookworm Campaign's central laboratory in

Brisbane (1922–1924) and was Brisbane's city entomologist (1926–1934), and later a full-time lecturer at the university in Brisbane (1936–1942). A bird is also named after him.

Hammarlund

Armoured Catfish sp. *Hemiancistrus hammarlundi* CH Rendahl, 1937

Carl Theodor Waldemar Hammarlund (1884–1965) was a Swedish botanist and mycologist who taught at Lund University (1924–1932). He collected in Guadeloupe (1933) and in Peru and Bolivia (1934). He collected the holotype of this species.

Hamwi

Stone Loach sp. *Oxynoemacheilus hamwii* F Krupp & W Schneider, 1991

Dr Adel Hamwi was Professor and Director of Zoology, Faculty of Science, University of Damascus and President of the National Committee of Oceanography. He was formerly (1969) the Director of Syria's first marine biology laboratory. He also had an interest in scorpions.

Hana (Mitsukuri)

Threadfin Dartfish *Ptereleotris hanae* DS Jordan & JO Snyder, 1901
[Alt. Thread-tail Dart-Goby]

Hana was the daughter of zoologist Kakichi Mitsukuri (1857–1909) of the Imperial University of Tokyo; he collected the type.

Hana (Raza)

Loach sp. *Oxynoemacheilus hanae* J Freyhof & YS Abdullah, 2017

Hana Ahmad Raza (b.1987) is an Iraqi Kurdish biologist and conservationist who works for Nature Iraq in Sulaymaniyah. The College of Sciences at the University of Sulaimani awarded her BSc in biology, after which (2009) she took up her current post as a Wildlife Conservationist Project Manager at Nature Iraq Organization. She accompanied the senior author during fieldwork in Iraqi Kurdistan. She has undertaken field surveys, co-ordinated training, given talks and presentations and taken part in documentary films on environmental education and public awareness. She is currently Project Manager for the project that is safeguarding the long-term persistence of the Persian Leopard through the establishment of a protected area in Qara Dagh, Sulaimani, in the Kurdistan Region of Iraq. Among her publications she is co-author of the book *Nature Iraq* (2017) and first author of papers such as *First Photographic Record of the Persian Leopard Panthera pardus saxicolor in Kurdistan, northern Iraq* (2012). She won (2017) the 'Future For Nature' award for her work in wildlife conservation in Iraq.

Hancock

Hairy Seabat *Halieutaea hancocki* CT Regan, 1908

Lieutenant Hancock, RN, was First Lieutenant of HMS 'Sealark', the vessel which undertook the Percy Sladen Expedition (1905) headed by J Stanley Gardiner during which the seabat holotype was taken.

Hancock, GA

Hancock's Blenny *Acanthemblemaria hancocki* GS Myers & ED Reid, 1936
Sandtop Goby *Gobulus hancocki* I Ginsburg, 1938

Captain George Allan Hancock (1875–1965) was an oil magnate, banker, businessman and philanthropist. He was also a musician who played cello in the Los Angeles Symphony Orchestra. He donated over seven million dollars to the University of Southern California and founded what later became the Hancock Institute for Marine Studies. He had a motor vessel built, 'Velero III', which he used for private oceanographic research and exploration and then donated it to the university. He led an expedition that collected the goby type. Ginsburg's etymology honoured him "*...in recognition of his interest in the scientific exploration of Pacific waters.*"

Hancock, J

Talking Catfish *Platydoras hancockii* A Valenciennes, 1840

John Hancock (1808–1890) was a British naturalist, landscape architect, ornithologist and taxidermist. He is regarded as the father of modern taxidermy, as he introduced the style of dramatic preparation. He wrote: *Catalogue of the Birds of Northumberland and Durham* (1874). He described (1829) but misidentified this catfish as *Doras* (now *Platydoras*) *costatus*.

Hand

Lanternfish sp. *Diaphus handi* HW Fowler, 1934

Henry Walker Hand of Cape May, New Jersey, was yet another person to whom Fowler (q.v.) was indebted for many fishes, this time from Cape May. The Hand family were among the first to settle in the Cape May area and the earliest known Hand was one of the passengers on the 'Mayflower'.

Handlirsch

Handlirsch's Minnow *Pseudophoxinus handlirschi* V Pietschmann, 1933

Dr Anton Peter Josef Handlirsch (1865–1935) was an Austrian entomologist and palaeontologist who originally trained as a pharmacist, obtaining a master's degree at the University of Vienna. He joined the Department of Entomology at the Natural History Museum, Vienna (1892) eventually retiring as Director (1922). The University of Graz awarded him an honorary doctorate (1923) and he qualified as a Professor at the University of Vienna (1924). He suffered a stroke from which he never completely recovered (1928) but was, nevertheless, appointed an Assistant Professor (1931). His principal work of over 1,400 pages and more than 50 plates was on fossil insects and was published in sections: *Die Fossilen Insekten* (1906–1908). He is regarded as the founder of insect palaeontology.

Haneda

Haneda's Glowbelly *Acropoma hanedai* K Matsubara, 1953
Haneda's Ponyfish *Secutor hanedai* K Mochizuki & M Hayashi, 1989

Dr Yata Haneda (1907–1995) was Director of the Yokosuka City Museum (until 1974) but continued to research there after retirement, studying fireflies. He began studying agriculture but his father wished him to be a doctor so he studied medicine. In his early career he was Director of the Raffles Museum in Singapore during the Japanese occupation, and he was interned when it was liberated by the British. After the war he worked at the Japanese Department of Health Education. He wrote much about bioluminescence in fish including: *The luminescent systems of pony fishes* (1976) and he co-wrote the book *Bioluminescence in Progress* (1966). He was honoured for this study and the Haneda Bioluminescence Collection in Yokosuka Museum.

Hanitsch

Borneo Loach sp. *Glaniopsis hanitschi* GA Boulenger, 1899

Dr Karl Richard Hanitsch (1860–1940) was a German biologist and museum curator. He was Demonstrator, Zoology, University College, Liverpool and then Director, Raffles Library and Museum, Singapore (1895–1919). He wrote: *An expedition to Mt. Kinabalu, British North Borneo* (1900). He collected the holotype of this species. An Australian 50–cent postage stamp, issued by Christmas Island (1977), bears his portrait. A reptile and an amphibian are also named after him.

Hankinson

Brassy Minnow *Hybognathus hankinsoni* CL Hubbs, 1929

Thomas Leroy Hankinson (1876–1935) was an American ichthyologist, herpetologist and ornithologist. His two bachelor's degrees were awarded by Michigan Agricultural College, East Lansing (1898) and Cornell University (1900). He was Professor of Zoology and Physiology at East Illinois State Normal College (1902–1919), ichthyologist at the Roosevelt Wildlife Experimental Station, New York (1919–1921) and Professor of Zoology at Michigan State Normal College (1921–1935) and was a research associate, University of Michigan (1935). He was associated with the University of North Dakota (1922) and surveyed the fish population of that state for which Hubbs honoured him in this species' binomial. He appears to belong to that select group of people who became professors without having achieved a doctorate.

Hannelore

Hannelore's Killi *Aphyosemion hanneloreae* AC Radda & E Pürzl, 1985

Hannelore Pürzl is the wife of the junior author Eduard Pürzl.

Hannerz

Hannerz' Lampeye *Poropanchax hannerzi* JJ Scheel, 1968
[Syn. *Poropanchax luxophthalmus hannerzi*]

Dr Ulf Hannerz (b.1942) is a Swedish anthropologist and amateur aquarist who was a fish hobbyist as a child. He is Emeritus Professor of social anthropology at Stockholm University where he worked (1976–2007) and was awarded his PhD (1969). He is also a member of the Royal Swedish Academy of Sciences. Among his many published papers and nine books are two classic books on urban anthropology: *Exploring the City: Inquiries Toward an Urban Anthropology* (1980) and *Soulside: Inquiries into Ghetto Culture and*

Community (2004). Ulf collected live specimens of the killifish in Nigeria and sent them to Scheel (1961). He gained notability when he appeared on the first episode of the television game show *Kvitt eller dubbelt - 10.000 kronorsfrågan* (literally: *Double or Nothing – The 10,000 Kronor Question*), which was based on the American television show *The $64,000 Question*. In the first episode (1957) when 14-years-old under his nickname *Hajen* (*The Shark*), he was quizzed on the subject 'tropical aquarium fish'. He was asked which of the seven displayed fishes had eyelids; he answered "*hundfisk*" (mudminnow) but the host said the correct answer was "*slamkrypare*" (mudskipper); when viewers informed the show's producers that mudskippers retract their eyes into a dermal cup and technically do not have eyelids, Hannerz was allowed to return to the show and advance in the competition going on to win the prize; since then, "*slamkrypare*" has entered the Swedish language as a term for an incorrectly formulated question in a quiz.

Hans

Moray sp. *Gymnothorax hansi* PC Heemstra, 2004

Hans Fricke (b.1941) is a filmmaker and ethologist. After study in Berlin he completed his zoology PhD in Munich (1968). He was a visiting scientist at the Max Planck Institute for Behavioral Physiology in Seewiesen and visiting professor of the Hebrew University of Jerusalem. He is (1988) Honorary Professor at the Ludwig-Maximilians-University Munich. His research areas are the ecology and social behaviour of marine organisms. In addition to scientific work, Fricke has written articles for the magazines *Geo* and *National Geographic*. For his documentary and television films he has been awarded numerous international prizes, including the Golden Nautilus. He was honoured for his 'pioneering contributions' to the study of fish behaviour and deep demersal communities of the Comoros Islands, Red Sea and Indo-Pacific region. He is also known for the use of the research submarines 'Geo' and 'Jago' for the study of aquatic fauna and his research on the behaviour of the coelacanth.

Hans Baensch

Fort Maguire Aulonocara (cichlid) *Aulonocara hansbaenschi* MK Meyer, R Riehl & H Zetzsche, 1987

[Often regarded as a junior synonym of *Aulonocara stuartgranti*]

Hans-Albrecht Baensch (1941–2016). (See **Baensch, H-A**)

Hansen, G

Hansen's Lanternfish *Hygophum hanseni* AV Tåning, 1932

Captain Georg Hansen was master of the 'Thor', the first Danish research ship equipped for scientific work on the oceans, and later of 'Dana' (q.v.). He had about 30 years' service with the Danish Marine investigations. The holotype was collected by 'Dana'.

Hansen, H

Characin sp. *Hasemania hanseni* HW Fowler, 1949

Henrik Hansen was a breeder and exporter of aquarium fish at Gulf Fish Hatchery, Florida. He acquired living specimens of this species from Brazil and sent them to Fowler for study.

Hanson

Striped Rockcod *Trematomus hansoni* GA Boulenger, 1902

Nicolai Hanson (1870–1899) was a Norwegian zoologist and Antarctic explorer. He graduated with a zoology degree from the University of Oslo. He married just before he took part in the 'Southern Cross' Expedition and his daughter was born after he left for Antarctica. He had been seriously ill during the voyage from England. After having arrived at the expedition's winter camp at Cape Adare he was well enough to carry out parts of the planned scientific activities, but died (14 October 1899) apparently of an intestinal disorder and became the first person ever to be buried in Antarctica at his own request. A small peak, Mount Hanson south of Cape Adare, is also named after him. (Also see **Nicolai**)

Harada

Bitterling sp. *Rhodeus haradai* R Arai, N Suzuki & S-C Shen, 1990

I Harada was the author of: *The freshwater fishes of Hainan Island* (1943), wherein he reported this species under the name *Rhodeus spinalis*. We believe this to be Isokiti Harada, who is mainly remembered as a limnologist, and had some of his work published in German, such as: *Zur Acanthocephalenfauna von Japan* (1935).

Haraguchi

White-spine Butterflyfish *Roa haraguchiae* T Uejo, H Senou & H Motomura, 2020

Mrs Yuriko Haraguchi was a volunteer at the Kagoshima University Museum. She was honoured because she "*...has kindly supported our ichthyological research and fish collection management at the Kagoshima University Museum as a volunteer.*"

Harald

Blue Discus *Symphysodon haraldi* LP Schultz, 1960
[Perhaps a form of *S. aequifasciatus*]
Tetra sp. *Hemigrammus haraldi* J Géry, 1961

Harald Schultz (1909–1966) – see next entry. In his etymology, LP Schultz (no relation) says of Harald that he "*...has collected numerous new and rare South American fishes,*" including discus specimens used in Schultz' review of the genus. (See **Harald Schultz** & **Schultz**)

Harald Schultz

Crystal Red Tetra *Hyphessobrycon haraldschultzi* H Travassos, 1960
Schultz's Cory *Corydoras haraldschultzi* J Knaack, 1962
[Alt. Mosaic Corydoras]

Harald Schultz (1909–1966) was a Brazilian ethnologist, anthropologist and fish collector who studied native South American tribes. He worked at the Museu Paulista, São Paulo (1947–1966). He wrote: *Hombu: Indian Life in the Brazilian Jungle* (1962). As a fish hobbyist, he also made a particular study of piranhas. He is commemorated in the names of other Brazilian taxa including an amphibian. (See also **Harald** & **Schultz**)

Harald Sioli

Neotropical Rivuline sp. *Atlantirivulus haraldsiolii* HO Berkenkamp, 1984

Dr Harald Sioli (1910–2004) was a German biologist and limnologist considered to be the founder of Amazonian ecology. He was Director of the Max Planck Institute of Limnology (1957–1978). He wrote more than 150 papers and longer works, including the book: *The Amazon: Limnology and Landscape Ecology of a Mighty Tropical River and its Basin* (1984). He first went to Brazil (1934) to conduct studies on toads in Brazil's dry northeast. During WW2 he was interned, but after the war ended he began (1945) the first limnological study of the Amazon. He used the local terms 'white water', 'black water' and 'clear water' to develop a scientific classification system for the Amazon's rivers, researching the relationship between water quality and soil chemistry. Much of his work was with the Brazilian Institute Nacional de Pesquisas da Amazonia in Manaus. Together with its director Djalma Batista he founded the magazine 'Amazoniana'. He spent his last years writing his memoires.

Harasti

Harasti's Pipefish *Stigmatopora harastii* G Short & A Trevor-Jones, 2020

Dr David Harasti (b.1975) is an Australian marine biologist, qualified diver and underwater photographer who is a Senior Research Scientist at the Port Stephens Fisheries Institute, NSW. His BAppSc degree was awarded by the University of Canberra (1997) and his PhD by the University of Technology, Sydney (2014). He has written or co-written over sixty-five published papers such as: *Population dynamics and life history of a geographically restricted seahorse, Hippocampus whitei* and a book *The Marine Life of Bootless Bay – Papua New Guinea* (2007) as well as a number of articles on scuba diving, marine species and photography. His research primarily focuses on threatened marine species, particularly Syngnathids (seahorses and pipefish) and he has worked for the past two decades on their conservation in Australian waters.

Harcourt Butler

Burmese Snakehead *Channa harcourtbutleri* N Annandale, 1918

Sir Spencer Harcourt Butler (1869–1938) was educated at Harrow and Bailiol College, Oxford, then almost immediately (1890) became an officer of the Indian Civil Service. He served as Lieutenant Governor of Burma (1915–1917) and later in the same position for the United Provinces of Agra and Oudh (1918–1921), then becoming the first Governor there (1921–1922). He returned to Burma again as Lieutenant Governor (1922–1923) and became Governor there (1923–1927).

Hardenberg, JDF

Hardenburg's Burrfish *Cyclichthys hardenbergi* LF de Beaufort, 1939
Sea Catfish sp. *Aspistor hardenbergi* PJ Kailola, 2000
[Syn. *Hemiarius hardenbergi*]

Dr Johann Dietrich Fran(s/z) Hardenberg (b.1902) was a Dutch marine biologist, botanist, ichthyologist and entomologist who was awarded his doctorate by the University of Utrecht (1927). He worked in the Dutch East Indies (Indonesia) (1920s–1940s) at the Laboratory for the Investigation of the Sea, Batavia (Jakarta) and was Director there (1942) at the time of the Japanese invasion. He resumed work (1946) after the Japanese surrender and stayed on after Indonesian Independence (1949) still being Director of the Laboratory for the Investigation of the Sea (1956). He was back in the Netherlands in the 1960s, reporting on control of birds in relation to agriculture for the Dutch Ministry of Agriculture. He wrote: *The Fishfauna of the Mouth of the Rokan River* (1931). He is commemorated in the names of other aquatic taxa, such as the jellyfish *Acromitus hardenbergi*. Kailola named the catfish after him partly for his contributions to Indo-Australian ichthyology, and partly because he had first recognised that it was a new species. De Beaufort records that Dr Hardenberg sent him the type of the burrfish (1937).

Hardin

Andean Pupfish sp. *Orestias hardini* LR Parenti, 1984

Dr Tim S Hardin is a consultant biologist and instream flow specialist working for the Oregon Department of Fish & Wildlie (since 2006). His BSc was awarded by Knox College and his master's in aquatic biology/limnology by UCSB. Colorado State University awarded his PhD in Fisheries. He collected the type series in northern Peru (1979).

Hardman

Asian Stream Catfish sp. *Pseudobagarius hardmani* HH Ng & MH Sabaj Pérez, 2005

Dr Michael Hardman (b.1978) is a British ichthyologist and herpetologist at the Natural History Museum of Los Angeles County. His doctorate was awarded by the University of Florida, Gainesville (2002). He was a post-doctoral researcher at the Finnish Museum of Natural History (2009–2012). He wrote: *A New species of Catfish Genus* Chrysichthys *from Lake Tanganyika (Siluriformes: Claroteidae)* (2008).

Hardwicke

Hardwicke's Pipefish *Solegnathus hardwickii* JE Gray, 1830
Sixbar Wrasse *Thalassoma hardwicke* JW Bennett, 1830
Finless Sleeper Ray *Temera hardwickii* JE Gray, 1831

Major-General Thomas Hardwicke (1756–1835) served in the Bengal army of the Honourable East India Company. He was an amateur naturalist and collector who was the first to make the Red Panda *Ailurus fulgens* widely known, through a paper that he wrote (1821): *Description of a New Genus…from the Himalaya Chain of Hills between Nepaul and the Snowy Mountains.* Cuvier stole a march on Hardwicke in formally naming the Red Panda because Hardwicke's return to England was delayed. He collected reptiles in India and published on them (1827) with Gray. He also collected the type of the sleeper ray. Five reptiles, a mammal and a bird are named after him.

Harkishore

Indian Loach sp. *Mustura harkishorei* DN Das & A Darshan, 2017
[Syn. *Physoschistura harkishorei*]

Harkishore Das was the father of the senior author. He inspired his son to take up fisheries research as an academic career.

Harmand

Harmand's Sole *Brachirus harmandi* HE Sauvage, 1878
Mekong Carp sp. *Cosmochilus harmandi* HE Sauvage, 1878
Sea Catfish sp. *Hemiarius harmandi* HE Sauvage, 1880
Oriental Cyprinid sp. *Paralaubuca harmandi* HE Sauvage, 1883
Largescale Silver Carp *Hypophthalmichthys harmandi* HE Sauvage, 1884

Dr François-Jules Harmand (1845–1921) was a French Navy Surgeon, and a naturalist and explorer in Indochina (1873–1877). He was Civil Commissioner-General in Tonkin (1883), held consular posts in Thailand (1881), India (1885) and Chile (1890), and was French Ambassador to Japan (1894–1906). Two birds are named after him. He collected the holotypes of these fish species.

Haroldo (Britski)

Dwarf Pleco sp. *Parotocinclus haroldoi* JC Garavello, 1988

Dr Heraldo Antonio Britski. *(See* **Britski***)*

Haroldo (Travassos)

Pike Cichlid sp. *Crenicichla haroldoi* JA Luengo & HA Britski, 1974

Haroldo Pereira Travassos (1922–1977) was a Brazilian ichthyologist and editor of museum journals. He was a graduate of the National Veterinary School of Medicine, but made a career in the Ichthyology Department of the Brazilian National Museum (1944). He was the son of Lauro Pereira Travassos (q.v.).

Harper

Redeye Chub *Notropis harperi* HW Fowler, 1941

Dr Francis Harper (1886–1972) was a naturalist who collected the holotype of this species whilst retracing the footsteps of the 18th century naturalists, William and John Bartram. Cornell University awarded both his bachelor's degree (1914) and his doctorate (1925). He taught at a college briefly but soon left to work for museums and government agencies. His first major trip was as a zoologist for the Geological Survey of Canada to Lake Athabasca (1914). He served in the US Army in France (1917–1919) as a rat-catcher (politely described as a rodent control officer). After the end of WW1, he resumed exploring, returning to Lake Athabasca (1920). He made a number of other trips north, the last being to the Ungava Peninsula (1953) but otherwise concentrated on the Bartrams, father and son, and published extensively on them including annotated editions of John Bartram's: *Diary of a Journey through the Carolinas, Georgia, and Florida 1765–1766* and William Bartram's *Report to Dr John Fothergill 1773–1774.*

Harrer

Maroni Eartheater (cichlid) *Geophagus harreri* J-P Gosse, 1976

Heinrich Harrer (1912–2006) was an Austrian writer, sportsman, explorer and mountaineer. He is best known for being on the four-man climbing team that made the first ascent of the North Face of the Eiger in Switzerland, and for his books *Seven Years in Tibet* (1952) and

The White Spider (1959). He was awarded the title 'professor' by the President of Austria. He was a member of the Nazi party and was pictured with Adolf Hitler, but he said of being a member of the party that it was a mistake made in his youth before he learned to think for himself. When in India he was interred when WW2 broke out. He escaped a number of times. He became (1948) a salaried official of the Tibetan government, translating foreign news and acting as the Court photographer. He met the Dalai Lama and became his tutor in English, geography, and some science. He settled back in Austria (1952) and took up writing, though he continued to travel as a mountaineer. He was also a champion skier and golfer. He travelled in Brazil, Suriname and French Guiana, as is shown in Gosse's etymology of the cichlid: "*Cette nouvelle espèce est cordialement dédiée au Professeur H. Harrer de Kitzbühel, en remerciement pour l'aide qu'il nous a apportée lors de missions ichtyologiques, entre autres au Surinam et en Guyane française où cette espèce fut récoltée.*" He co-collected the type specimen (1969) (the other collector probably being King Leopold III who was part of the collecting party as an amateur entomologist). A film was made of *Seven Years in Tibet* (1997) with Brad Pitt playing Harrer.

Harrington

Reef Silverside *Atherina harringtonensis* GB Goode, 1877

This is a toponym referring to Harrington Sound, Bermuda, where the type was taken.

Harrington, JL

Harrington's Stonebasher (mormyrid) *Hippopotamyrus harringtoni* GA Boulenger, 1905

Lieutenant-Colonel Sir John Lane Harrington (1865–1927) was the British Government's Resident Agent, Consul and later Minister Plenipotentiary, in Addis Ababa (Ethiopia) (1897–1909). He personally led (1903) an expedition to the border region between Ethiopia and Sudan and included in it was Charles Singer (1876–1960), who was a trained zoologist as well as doubling up as the expedition's medical officer. The zoologist Oldfield Thomas wrote in the etymology of a rodent that he named after Harrington: "*in honour of Col. Harrington, the British Resident at Addis Ababa, to whose assistance all British travellers in Abyssinia are so much indebted*". Two reptiles are also named after him.

Harrington, MW

Scalyhead Sculpin *Artedius harringtoni* EC Starks, 1896

Mark Walrod Harrington (1848–1926) was an American botanist, astronomer and meteorologist who became President of the University of Washington. The University of Michigan awarded his MSc amd MA (1868) and he taught there (1868–1876). He then studied at the University of Leipzig, Germany, served as a professor of Mathematics at the University of Peking, China, and was appointed Chief of the Weather Bureau in Washington DC. When he was removed from the Weather Bureau by President Grover Cleveland he was recruited as President of the University of Washington (1895–1897) then returned to the Weather Bureau (1897–1899). He disappeared (1908) until found by his wife in a New Jersey mental hospital; a condition she attributed to him having been struck by lightning when at the Bureau. He never recovered and died in that institution. A bird is also named after him.

Harriott

Chimaera genus *Harriotta* GB Goode & TH Bean, 1895

Thomas Harriott (c.1560–1621) was an English astronomer, mathematician, ethnographer and translator, who published first English work on American natural history (1588). He is sometimes credited with the introduction into Britain of the potato. When in Carolina he learned the Algonquian language. Sir Walter Raleigh hired him as a mathematics tutor.

Harris (Hamlyn-)

Harris's Flathead *Inegocia harrisii* AR McCulloch, 1914

Dr Ronald Hamlyn-Harris (1874–1953) was Director of the Queensland Museum and friend of author, who wrote that he was: "*…indebted for much valuable assistance when working on the collections under his charge.*" A bird is also named after him. (See also **Hamlyn**)

Harris, H & N

Shrimp-Goby sp. *Amblyeleotris harrisorum* MS Mohlmann & JE Randall, 2002

Hamilton and Nancy Harris were friends of the junior author Dr John Ernest Randall Jr. They sponsored the authors' expedition to Kiribati, where this goby is endemic.

Harrison, AC

Harrison's Barb *Osteochilus harrisoni* HW Fowler, 1905

Alfred Craven Harrison Jr (1869–1925) was one of a group of explorers and amateur anthropologists who made five long trips (including two circumnavigations) (1895–1903) to collect ethnographic and natural history specimens for the University of Pennsylvania Museum, Philadelphia. They studied the Dayak people in Borneo, the Nagas in Assam and Ainu in Japan and visited around 20 countries, mainly in Southeast Asia. Harrison was a member of a family that had made its fortune out of sugar and he is recorded as owning a sugar plantation in Cuba (1903), so he appears to have given up exploring and joined the family firm. He collected the holotype of this species.

Harrison, JB

Blackstripe Pencilfish *Nannostomus harrisoni* CH Eigenmann, 1909

Sir John Burchmore Harrison (1856–1928) was the Government Geologist, Georgetown, British Guiana (Guyana) and Director, Department of Science and Agriculture, British Guiana at the time of Eigenmann's 1908 expedition. He graduated in science at Christ's College Cambridge (1878) and researched into agricultural chemicals before being appointed Professor of Chemistry and Agricultural Science in Barbados (1879–1889). He served in British Guiana (1889–1926) and wrote: *The geology of the goldfields of British Guiana (1908).*

Harrison (Williams)

Gulper Eel sp. *Saccopharynx harrisoni* W Beebe, 1932

Harrison Charles Williams (1873–1953) was a multi-millionaire businessman who left his successful bicycle manufacturing business (1903) and took a job with a New York carpet sweeper business. He created American Gas & Electric Co. (1906) then Central States Electric Corp (1912) and went on to make a fortune (the equivalent of c.$10 billion in today's money). He was America's richest man at the time, and often hosted the Prince of Wales (Edward VIII) when he visited the USA. Williams bought the 'Vanadis', then the largest private yacht afloat, renamed her 'Warrior', and refitted her for his own oceanographic and pleasure purposes. He provided financial support for three major natural history expeditions: William Beebe's Galápagos Expedition (1923) and Arcturus Expedition (1925) for the New York Zoological Society; and George P. Putnam's Arctic Greenland Expedition (1926) for the American Museum of Natural History. He was also an interested supporter of Beebe's later Bermuda Oceanographic Expeditions (ca. 1929–1932). Beebe says of him that it was "*...through whose continued interest and support these Bermuda Oceanographic Expeditions have been made possible.*"

Harrisson

Dumb Gulper Shark *Centrophorus harrissoni* AR McCulloch, 1915
Antarctic Dragonfish sp. *Racovitzia harrissoni* ER Waite, 1916

Charles Turnbull Harrisson (1866–1914) was a biologist from Tasmania who took part in the Mawson Antarctic Expedition (1911–1914) and was responsible for collecting many interesting fish species. On his return from Antarctica, Harrisson continued his work as a biologist by conducting research for the Commonwealth Department of Fisheries. In late 1914, he joined the 'Endeavour' voyage to resupply the meteorological station at Macquarie Island, which had been established by Mawson's expedition. Once resupply was completed on 3 December 1914, the 'Endeavour' departed Macquarie Island and was expected to arrive in Hobart a week later. The vessel was never heard of again, nor was any distress signal sent.

(There seems to be disagreement as to when Harrisson was born, as the years 1866, 1867 and 1869 can all be found in the literature)

Harrower

American Coastal Pellona *Pellona harroweri* HW Fowler, 1917

Dr David Elson Harrower (1890–1970) was a member of the Philadelphia Academy of Natural Science, with interests in both anthropology and zoology. He studied forestry at Penn State (1909–1913) and after graduating undertook post graduate studies in botany at the University of Pennsylvania and as a research fellow at the University of Chicago which included his trips to Panama. He visited the Panama Canal Zone (1915 & 1916) and made collections of fishes there. He went to France (June 1917) in the American Red Cross Volunteer Ambulance Corps, then took a commission in the French Army artilliary but transferred to the US army (1918). He was awarded the Croix de Guerre for his bravery as a spotter in a balloon. Post war he returned to take his MA at the Universrity of Pennsylvania and doctorate at Cornell, He collected five hundred artefacts of the Rama, Mosquito and Samu tribes of Nicaragua (1924). He taight at the Woodmere Academy until retirement (1946). He was an associate member of the Delaware Ornithological Club when Fowler was President and wrote a number of articles in their publication.

Harry, GV

Ozark Chub *Erimystax harryi* CL Hubbs & WR Crowe, 1956

George V Harry (d.1979) was a one of Hubbs' graduate students at the University of Michigan when he conducted two extensive surveys of Missouri fishes (1940 & 1941). The work was interrupted by WW2 and was never resumed, even though he is shown as being in the final stage of study for his doctorate there (1947). He co-wrote: *The food and habits of gars (*Lepisosteus *spp.) considered in relation to fish management* (1943). He was a member of the party that collected the holotype.

Harry, RR

Scaly Paperbone *Scopelosaurus harryi* GW Mead, 1953

Dr Robert Rees Harry-Rofen (1925–2015). (See **Rofen**)

Hart

Hart's Rivulus *Anablepsoides hartii* GA Boulenger, 1890

Dr John Hinchley Hart (1847–1911) was a British botanist. After serving eleven years in Jamaica, Hart took up the post of Superintendent of the Trinidad Botanic Gardens (1887–1908). This was at a time when there was a great role for the gardens to play in the economic life of the island; principally in helping decision makers to decide upon appropriate crops and associated planting. Hart immediately began the process of reorganising the department, as he found many specimens were poorly preserved and labelled. He presented the British Museum with specimens of this fish.

Hartel

Hartel's Dwarf Snailfish *Psednos harteli* NV Chernova, 2001
Snaketooth sp. *Chiasmodon harteli* MRS Melo, 2009

Karsten Edward Hartel is an ichthyologist who joined (1975) the Museum of Comparative Zoology, Harvard University. He was Curatorial Associate in Ichthyology there (1975-2018. He has joined many collecting cruises out of Woods Hole Oceanographic Institution.

Hartert

North African Barb sp. *Carasobarbus harterti* A Günther, 1901

Dr Ernst Johann Otto Hartert (1859–1933) was a German ornithologist and oologist. He travelled extensively, often on behalf of his employer Walter (Lord) Rothschild, who wrote his obituary. He was the Ornithological Curator of Rothschild's private museum at Tring (1892–1929), which later became an annexe to the BMNH. He then returned to Germany (1930). He collected the holotype of this species of fish. A reptile and no less than 61 birds are named after him.

Hartt

Armoured Catfish genus *Harttia* F Steindachner, 1877
Armoured Catfish genus *Harttiella* M Boeseman, 1971
Characin sp. *Prochilodus harttii* F Steindachner, 1875

Three-barbeled Catfish sp. *Pimelodella harttii* F Steindachner, 1877
Banjo Catfish sp. *Bunocephalus hartti* TP Carvalho, AR Cardoso, JP Friel & RE Reis, 2015

Charles Frederick Hartt (1840–1878) was a Canadian geologist, palaeontologist and naturalist, who was a member of the Thayer Expedition (1865–1866) to Brazil, a country in which he was a specialist. Acadia College, Wolfville, Nova Scotia awarded his bachelor's degree (1860). He worked as an assistant to Louis Agassiz at the Museum of Comparative Zoology, Harvard (1861–1864). He stayed on in Brazil after the Thayer Expedition until 1868 when he was elected Professor of Natural History, Vassar College, but instead accepted a position at Cornwall University, Ithaca, New York. He was a member of the four Morgan expeditions to Brazil (1870–1878). He died in Rio de Janeiro of yellow fever.

Hartweg

Soconusco Gambusia *Brachyrhaphis hartwegi* DE Rosen & RM Bailey, 1963
Tailbar Cichlid *Vieja hartwegi* JN Taylor & RR Miller, 1980

Dr Norman Edouard 'Kibe' Hartweg (1904–1964) was a herpetologist whose specialty was the distribution and taxonomy of turtles. He worked (1927–1964) at the University of Michigan, where he took his doctorate (1934) and was Assistant Curator, Herpetology (1934) and Curator of Reptiles, then Curator (1947). He had a broad interest in Mexican biology and made valuable fish collections during his field studies in México (where the cichlid is endemic). Two reptiles, two amphibians and a bird are also named after him.

Hartwell

Three-barbeled Catfish sp. *Pimelodella hartwelli* HW Fowler, 1940

Robert Hartwell was one of three volunteers from Cleveland, Ohio who accompanied William C Morrow (q.v.) on the 1937 expedition to Peru during which the holotype was collected.

Hartzfeld

Hartzfeld's Wrasse *Halichoeres hartzfeldii* P Bleeker, 1852
Hartzfeld's Cardinalfish *Ostorhinchus hartzfeldii* P Bleeker, 1852
Hook-nosed Sole *Heteromycteris hartzfeldii* P Bleeker, 1853

Dr Joseph Hartzfeld (1815–1885) was a German physician who was a Principal Medical Officer of the Royal Dutch East Indies Army (1841–1869). He took Dutch nationality (1861) and after retiring was in private practice in The Hague. He sent Bleeker collections of fish from Ambon.

Harvey, EN

Gulf Flashlightfish *Phthanophaneron harveyi* RH Rosenblatt & WL Montgomery, 1976

Dr Edmund Newton Harvey (1887–1959) was an American zoologist who was a leading expert on bioluminescence, probably inspired during an expedition to the South Pacific (1913). The University of Pennsylvania awarded his bachelor's degree and Columbia University, New York, his doctorate. He was a member of the faculty of Princeton where he was a professor (1919) and the Henry Fairfield Osborn Professor (1933). He collaborated in the invention of the centrifuge microscope and was a pioneer in the general field of electro-encephalography. He wrote: *The Nature of Animal Light* (1920).

Harvey, G

Cayman Greenbanded Goby *Tigrigobius harveyi* Victor, 2014

Dr Guy Harvey (b.1955) is a Jamaican marine wildlife artist and conservationist. Aberdeen University awarded his Marine Biology degree (1977) and the University of the West Indies his PhD (1982). He depicted (1985) Ernest Hemingway's fishing story *The Old Man and the Sea* in a series of 44 original pen-and-ink drawings and displayed them at an exhibition in Jamaica. Based on the positive response he received at this show, he began painting full-time and was soon (1988) providing custom artwork for use on a variety of products. His depictions of sealife, especially of sportfish such as marlin, are popular with sportfishermen and have been reproduced in prints, posters, T-shirts, jewellery, clothing and other consumer items. Harvey is also a very vocal and active advocate for marine conservation. He established (1999) the Guy Harvey Research Institute at Nova Southeastern University in Fort Lauderdale (Florida, USA), and the Guy Harvey Ocean Foundation (2008). He was honoured for his 'extensive support' for research and conservation of sharks and gamefishes in the Cayman Islands, where he moved to (1999).

Harvey, W

Licorice Gourami sp. *Parosphromenus harveyi* B Brown, 1987

Willi Harvey (1916–2013) was a German tropical fish breeder, particularly of killifish. He was a glider pilot in WW2 and was captured by the British and held prisoner in Scotland. When released, his family were all in East Germany so he settled in Scotland opening a shop in Edinburgh selling tropical fish he bred. He married and moved to England where he took a job as a professional fish breeder in an established business. He later worked as a patternmaker in an engineering firm, but continuing his aquarist hobby. When he retired to Scotland he kept a fish house with a variety of difficult-to-breed species. He was with Allan Brown when they discovered the gourami species in Malaysia (1984) although Professor Peter Ng made the official description in Singapore when the *Aquarist & Pondkeeper* magazine folded, which was where Brown had announced the name before he could publish the description.

Hasan

Marqīyah Spring Minnow *Pseudophoxinus hasani* F Krupp, 1992

This is a toponym referring to Nab' Hasan, the source of a stream in Syria named Nahr Marqiya where this species is endemic.

Haseman

Tetra genus *Hasemania* MD Ellis, 1911
Tetra sp. *Hyphessobrycon hasemani* HW Fowler, 1913
Duckbill Knifefish *Parapteronotus hasemani* MM Ellis, 1913
Pencil Catfish sp. *Potamoglanis hasemani* CH Eigenmann, 1914

Thorny Catfish sp. *Centrodoras hasemani* F Steindachner, 1915
Characin sp. *Characidium hasemani* F Steindachner, 1915
Three-barbeled Catfish sp. *Imparfinis hasemani* F Steindachner, 1915
Thorny Catfish sp. *Leptodoras hasemani* F Steindachner, 1915
Armoured Catfish sp. *Otocinclus hasemani* F Steindachner, 1915
Driftwood Catfish sp. *Pseudepapterus hasemani* F Steindachner, 1915
Neotropical Livebearer sp. *Pamphorichthys hasemani* AW Henn, 1916
Characin sp. *Apareiodon hasemani* CH Eigenmann, 1916
Twig Catfish sp. *Farlowella hasemani* CH Eigenmann & LE Vance, 1917
Characin sp. *Moenkhausia hasemani* CH Eigenmann, 1917
Three-barbeled Catfish sp. *Pimelodella hasemani* CH Eigenmann, 1917
Armoured Catfish sp. *Rineloricaria hasemani* IJH Isbrücker & H Nijssen, 1979

John Diederich Haseman (1882–1969) was a zoologist and ichthyologist. He graduated from Indiana University (1905) and taught there (1905–1906). As a graduate student, he was sent to Brazil as a last-minute substitute to represent the Carnegie Museum and to collect fishes on a museum-sponsored expedition. (He had previously gone on an expedition to Cuba with Dr Eigenmann, and then under Eigenmann's guidance he went again on his own to Cuba at a later date.) Upon arriving (1907), he found the main expedition about to set out. It was decided that he should run his own solo expedition, which lasted two and a half years and covered large areas of Argentina, Bolivia, Brazil, Paraguay and Uruguay. He never went back to the university but continued to study ichthyology in the field. He wrote: *Some Factors of Geographical Distribution in South America* (1912). He said, of his travels through the wilds of South America: "*After the noises of the day the hush which comes at night-fall causes even the hardened traveler at times to shudder. No man over fifty years of age should attempt to enter this region. A hard heart and cold blood are useful to him who invades it.*" A reptile is named after him, as is a dragonfly.

Hass

Ringed Blenny *Starksia hassi* W Klausewitz, 1958
Spotted Garden-Eel *Heteroconger hassi* W Klausewitz & I Eibl-Eibesfeldt, 1959

Hans Hass (1919–2013) was an Austrian biologist, underwater cinematographer and scuba-diving pioneer. He switched from reading law to zoology (1940), and graduated with a PhD from the University of Berlin (1943) at the Faculty of Biology. His thesis was the first scientific research using an autonomous rebreather diving equipment. He had been excused from serving in the German military because of poor circulation in his feet. Whilst WWII raged, he rented a boat in Greece (1942) and sailed in the Aegean, taking underwater photos and film. His film *Menschen unter Haien* had its world premiere in Zurich (1947) and he went on to make many commercial films; some of them featuring himself and his second wife, Lotte, who was also an expert diver.

Hasselquist

Elephant-snout Fish *Mormyrus hasselquistii* A Valenciennes, 1847

Frederik Hasselquist (1722–1752) was a traveler and naturalist. He was one of Linnaeus' students at Uppsala. Linnaeus often lamented the lack of knowledge of the natural history of Palestine, and this inspired Hasselquist. He reached Smyrna (1749), visited Asia Minor, Egypt and Cyprus in addition to Palestine, and made extensive collections. These eventually reached Sweden after he had died near Smyrna on his way home. LInnaeus edited and published Hasselquist's notes and journals as *Iter Palæstinum, Eller, Resa til Heliga Landet, Förrättad Infrån år 1749 til 1752* (1757). A reptile is also named after him.

Hasselt

Malay Combtail *Belontia hasselti* G Cuvier, 1831
Hasselt's Bonylip Barb *Osteochilus hasseltii* A Valenciennes, 1842
Hasselt's Loach *Lepidocephalichthys hasselti* A Valenciennes, 1846
Hasselt's Flap-headed Goby *Callogobius hasseltii* P Bleeker, 1851
Hasselt's Bamboo Shark *Chiloscyllium hasseltii* P Bleeker, 1852
Sheatfish (catfish) sp. *Silurichthys hasseltii* P Bleeker, 1858

Dr Johan Coenraad van Hasselt (1797–1823) was a Dutch physician, zoologist, botanist and mycologist. He studied medicine at the University of Groningen. He undertook an expedition for the Netherlands Commission for Natural Sciences to Java (1820) with his friend Heinrich Kuhl to study the fauna and flora of the island. They left from Texel, stopping at Madeira, the Cape of Good Hope and Cocos Island en route. Kuhl died after eight months; van Hasselt continued the work for another two years before he too died of disease and exhaustion. However, during their time there they sent the Museum of Leiden 200 skeletons, 200 skins of mammals from 65 species, 2,000 bird skins, 1,400 fishes, 300 reptiles and amphibians, and many insects and crustaceans. Bleeker described the goby from an illustration made by Hasselt. A bird, a mammal and a toad are also named after him, as well as several plants.

Hasson

African Rivuline sp. *Nothobranchius hassoni* S Valdesalici & RH Wildekamp, 2004

Dr Michel Hasson (b.1955) is an ardent lover of nature, active in several associations working to preserve biodiversity in Central Africa including the *Musée Royal de l'Afrique Centrale* and the National Park of Upemba. Among his publications is the co-written book: *Birds of Katanga* (2012). A beetle is also named after him.

Hastings

Clingfish sp. *Briggsia hastingsi* MT Craig & JE Randall, 2009
Cortez Barnacle Blenny *Acanthemblemaria hastingsi* H Lin & GR Galland, 2010

Dr Philip Alan 'Phil' Hastings (b.1951) is Professor of Marine Biology at Scripps Institution of Oceanography, University of California, San Diego and Curator of Marine Vertebrates. His research interests include fish systematics and evolution, biogeography, evolution of behaviour and ecology and marine conservation. The University of South Florida awarded his BA, the University of West Florida his MSc and the University of Arizona his PhD. He has published over 100 papers, and books including: *Fishes: A Guide to Their Diversity* (2015). Some of his papers have described new fish species such as *Gobiesox lanceolatus, a new species of clingfish (Teleostei: Gobiesocidae) from Los Frailes submarine canyon, Gulf of California, Mexico* (2017).

Haswell

Slender Sand-diver *Creedia haswelli* EP Ramsay, 1881
Armoured Flathead *Hoplichthys haswelli* AR McCulloch, 1907

Professor William Aitcheson Haswell (1854–1925) was a zoologist who was born in Scotland and took both his bachelor's and master's degrees at Edinburgh (1878). Soon after, for reasons of health, he left for Australia where he remained for the rest of his life. He became Curator of the Queensland Museum, Brisbane (1879) but moved to Sydney (1880), where he gave a course of public lectures on zoology. He joined HMS 'Alert' (1881) on a surveying cruise of the Great Barrier Reef, studying especially crustaceans. He became Acting Curator, Australian Museum (1882), (where McCulloch worked under him) and taught at the University of Sydney, where he became Professor of Biology (1889–1913). After the post was split between botany and zoology, Haswell became Professor of Zoology. He resigned (1917), becoming Professor Emeritus. He co-wrote *A Text-book of Zoology* (1898), which remained the standard textbook for zoology courses in Australia for many years. He was a Fellow of the Royal Society, London, and of the Royal Society of New South Wales, in addition to being for 33 years a trustee of the Australian Museum.

Hatai

Hatai's Ponyfish *Photopectoralis hataii* T Abe & Y Haneda, 1972

Dr Shinkishi Hatai (1876–1963) was (1924) the first Director of Asamushi Marine Biological Station and the first professor in biology at Tohoku University, Sendai, Japan. He was also founder of the Palau Tropical Marine Biological Laboratory. The etymology says that the name: "*...is in reference to the late Prof. Shinkishi HATAI. Through his courtesy the present authors enjoyed studies at the Palao Tropical Biological Station, Palau Islands, during their twenties. They wish to dedicate this new species to Prof. HATAI.*"

Hatcher

Pencil Catfish genus *Hatcheria* CH Eigenmann, 1909
Neotropical Silverside sp. *Odontesthes hatcheri* CH Eigenmann, 1909

John Bell Hatcher (1861–1904) was an American palaeontologist who is famous for having excavated (1889) the first fossil remains of Torosaurus. He graduated from Yale (1884) having worked as a coal miner to pay for his education. He was an assistant at Yale (1884–1893), then Curator of Vertebrate Palaeontology and Assistant Curator of Geology at Princeton (1893–1900), during which period he made three expeditions to Patagonia. He became Curator of Palaeontology and Osteology at the Carnegie Museum of Natural History (1900) until his death from typhoid. He wrote: *Diplodocus Marsh: Its Osteology, Taxonomy, and Probable Habits, with a Restoration of the Skeleton* (1901) on the *Diplodocus carnegii*, which he described and named after his patron, Andrew Carnegie. A reptile and a bird are also named after him.

Hatooka

Orange-blotched Eel *Apterichtus hatookai* Y Hibino, J Shibata & S Kimura, 2014

Kiyotaka Hatooka is a Japanese aquarist, researcher and ichthyologist. He is the Curator of Ichthyology at the Osaka Museum of Natural History. Among his published papers are descriptions of new fish, such as: *Uropterygius nagoensis, a new muraenid eel from Okinawa (1984) and A new moray eel (Gymnothorax: Muraenidae) from Japan and Hawaii (1992). He also contributed to Fishes of Japan (2002).* The authors honoured him "*...for his contribution to the taxonomy of Japanese anguilliform fishes.*"

Hatta

Worm-Eel sp. *Muraenichthys hattae* DS Jordan & JO Snyder, 1901

Saburo Hatta né Santaro Nakamura (1865–1935) graduated (1891) in zoology at Imperial University, Tokyo. He taught high school German and biology (1891–1893) then zoology at Peer's School, Tokyo (1893–1904) and at the College of Agriculture, which became part of the Imperial University, until resigning (1929) and becoming professor emeritus. His doctorate was awarded (1908) and he became professor during this time, as well as Curator in the University's museum, bringing togather the collection which was scattered among several schools and colleges. He was released for a two-year period (1913–1914) to study museuems in the USA and Europe. His latter years were mostly spent writing and editing. He was honoured for his 'excellent' paper on Japanese lampreys.

Haug

Squeaker Catfish sp. *Synodontis haugi* J Pellegrin, 1906

Pastor Ernest Haug (1871–1915) was a French Protestant missionary (1895–1915) and a correspondent of Muséum National d'Histoire Naturelle, Paris. He established his permanent residence at Ngomo, Ogowe River, Gabon. He made a major collection of the fishes of that area. Moreover, he took note of all their local names, in three different languages. Pellegrin made a study of the collection (1906) and published on them (1914). He wrote: *Le Bas Ogooué: Notice géographiqe et ethnographique* (1903). Because of his outstanding contributions to science, Haug was posthumously awarded the Prix Secques of the Paris Zoological Society (1916). After his death, his brother continued (1918) to send specimens that Ernest had collected. A reptile is named after him.

Haulleville

Mormyrid sp. *Petrocephalus haullevillii* GA Boulenger, 1912

Baron Alphonse de Haulleville (1860–1938) was the Director (1910–1927) of the Musée du Congo Belge at Tervuren (RMCA). As such he would have been well known to Boulenger who was at BMNH, having moved there from Belgium (1890).

Hauxwell

Characin sp. *Ctenobrycon hauxwellianus* ED Cope, 1870

Armoured Catfish sp. *Loricariichthys hauxwelli* HW Fowler, 1915

John Hauxwell (1827–1919) was an English commercial natural history collector mentioned by Henry Walter Bates in: *The Naturalist on the River Amazons* (1864). Bates writes of being shown how to use 'a blow-gun, by Julio, a Juri Indian, then in the employ of Mr. Hauxwell, an English bird-collector'. We can surmise that some of the birds were first collected by this method, better preserving the skin than shooting, but doubt that any fish were. He collected, over nearly 40 years in the upper Amazon basin in Peru. Cope named this species as Hauxwell acquired most of the species, including the holotype of the characin, that are mentioned in Cope's paper on the fishes of the Marañón River in Peru. Seven birds and two dragonflies are also named after him.

Hay

Cypress Minnow *Hybognathus hayi* DS Jordan, 1885

Dr Oliver Parry Hay (1846–1930) was a palaeontologist and zoologist who was ordained as a priest (1870), preached one sermon and then deserted religion for science. He received a medical degree from Indiana Medical College and a doctorate from the University of Indiana (1884 or 1887). He was a Professor of Biology and Geology at Butler University, Indianapolis (1879–1891); hired as they assumed he must be the most religious candidate. However, he did not to support the administration's religious views and eventually resigned due to differences over the teaching of evolution. He then worked at the Field Museum, Chicago (1891–1900), the AMNH, New York (1900–1908) and the Carnegie Institute, Washington DC (1908–1930). After he retired (1926) he devoted himself to the study of languages and was multi-lingual when he died. He collected the holotype of the eponymous minnow.

Hayashi

Hayashi's Cardinalfish *Pseudamia hayashii* JE Randall, EA Lachner & TH Fraser, 1985
Cheek-streaked Goby *Echinogobius hayashii* A Iwata, S Hosoya & Y Niimura, 1998
Four-eyed Pygmygoby *Trimma hayashii* K Hagiwara & R Winterbottom, 2007

Dr Masayoshi Hayashi is Curator, Yokosuka City Museum. Among his many published papers he co-wrote: *New Record of Pseudocalliurichthys pleurostictus (Callionymidae) from Amami O-shima, Japan* (1991). He lent the authors a specimen of the cardinalfish when he learned of their research on the genus. He collected the first specimens of the goby and gave it its Japanese name (Moyoushinobi-haze) and referred (1997) to the pygmygoby as an undescribed species and provided a photograph of the holotype for the description.

Heal

Masked Stargazer *Gillellus healae* CE Dawson, 1982

Elizabeth Heal was Dawson's assistant. We cannot add to the information in the etymology, which says: "*Named after Elizabeth Heal in partial recognition for her years of efficient and willing assistance, without which my ichthyological studies would surely be more difficult.*"

Heald

Heald's Skate *Dentiraja healdi* PR Last, WT White & J Pogonoski, 2008
[Alt. Leyland's Skate)

David I Heald is a zoologist, ichthyologist and botanist who has had a most varied career. The University of Western Australia awarded his bachelor's degree (1969). He worked as a diving marine scientist at Watermans Marine Research Laboratory, Department of

Fisheries and Wildlife, Western Australia (1970–1987). He made a complete career change and worked for a variety of groups as a financial planner and investment consultant (1988–2008). Since leaving the financial business he has worked variously at a call centre, as a gas-meter reader, offering garden services and since 2011 as a team member for a petrol service station. He wrote: *The Commercial Shark Fishery in Temperate Waters of Western Australia* (1987). It was he who discovered the species off Western Australia (early 1980s).

Hearst

Gigantic Worm-Eel *Pylorobranchus hearstorum* JE McCosker, 2014

William Randolph Hearst III (b.1949) and his wife Margaret Hearst are friends of the author, John McCosker (q.v.), and are well known philanthropists. William is the grandson of the famous newspaper magnate and a businessman in his own right, as well as president of the eponymous foundation. They sponsored the (2011) Hearst Philippine biodiversity expedition of the California Academy of Science, which McCosker took part in, that collected the type.

Heber

Blacktip Trevally *Caranx heberi* JW Bennett, 1830

Reginald Heber (1783–1826) was an English cleric, poet, man of letters and hymn writer who became Bishop of Calcutta. He gained fame at the University of Oxford as a poet. After graduation he made an extended tour of Scandinavia, Russia and Central Europe. He was ordained (1807) and took over his father's old parish, Hodnet, Shropshire. He also wrote hymns and general literature, including a study of the works of the 17th-century cleric Jeremy Taylor. He was consecrated Bishop of Calcutta (1823) and travelled widely and worked to improve the spiritual and general living conditions of his flock. Arduous duties, a hostile climate and poor health led to his collapse and death after less than three years in India. Memorials were erected there and in St Paul's Cathedral, London. A collection of his hymns appeared soon after his death. Bennett wrote that he named the fish "*…as a tribute of respect to the memory of departed worth and excellence in the late Right Rev. Bishop Heber*".

Heckel

Taran Roach *Rutilus heckelii* A von Nordmann, 1840
Threadfin Acara *Acarichthys heckelii* J Müller & FH Troschel, 1849
Driftwood Catfish sp. *Centromochlus heckelii* F De Filippi,1853
Thorny Catfish sp. *Scorpiodoras heckelii* R Kner, 1855
Hazar Bleak *Alburnus heckeli* F Battalgil, 1943
Pike Cichlid sp. *Crenicichla heckeli* A Ploeg, 1989

Johann Jakob Heckel (1790–1857) was a self-taught Austrian zoologist and taxidermist with a particular interest in ichthyology. He rose through the ranks to become Director of the Fish Collection at the Vienna Natural History Museum. He undertook little fieldwork, preferring to study and catalogue specimens sent to the museum and he worked with many of the leading ichthyologists of his generation. He wrote more than 60 works, including most notably: *Ichthyolgie* (1843) and the co-written: *Die Süßwasserfische der österreichischen Monarchie, mit Rücksicht auf die angränzenden Länder bearbeitet* that he worked on for 24 years, dying before its eventual publication (1858). He proposed the genus *Crenicichla* and *Acarichthys* (1840) and described the former's first 10 species (nine still valid today) and was the first to seriously study cichlids and revise the family. His probable cause of death was from a bacteriological infection that he caught whilst extracting the skeleton of a dead sperm whale.

Hecq

Lake Tanganyika Cichlid sp. *Lepidiolamprologus hecqui* GA Boulenger, 1899
Lake Tanganyika Cichlid sp. *Xenochromis hecqui* GA Boulenger, 1899

Captain Célestin Hecq (1859–1910) served in the Belgium Forces stationed in the Congo that were fighting the slave trade. He was commander of the fort at Albertville (now Mtoa). At one point he had to flee with his troops to Rwanda (1898). Boulenger refers only to 'Lieutenant Hecq' collecting fishes at Lake Tanganyika. Interstingly, the type specimen of the *Lepidiolamprologus* was found in the mouth of a large catfish (*Auchenoglanis*) collected by Hecq.

Hector, G

Hector's Goby *Koumansetta hectori* JLB Smith, 1957

Gordon Hector (1918–2001) was Chief Secretary to the Government of the Seychelles, in which capacity he was honoured for his 'great assistance' to Smith's work about the archipelago. His history degree was awarded by Lincoln College, Oxford after which he served during WW2 in The King's African Rifles in East Africa and Burma. It was Africa which, in his words, 'got into my blood'. Demobbed he spent a year at Oxford before joining the Colonial Service (1946) as a District Officer, rising to District Commissioner in Kenya. He was then seconded to the Seychelles (1952–1956), latterly as Assistant Governor. He spent ten years in Basututland (Lesotho) through their independence. He retired (1966) becoming Clerk to the Court of Aberdeen University (1966–1976) and other posts until becoming secretary to the assembly council of the Assembly of the Church of Scotland (1980–1985) in Edinburgh. He was a descendant of Sir James Hector (below).

Hector, J

Hector's Clingfish, *Gastroscyphus hectoris* A Günther, 1876
Hector's Lanternfish *Lampanyctodes hectoris* A Günther, 1876

Dr Sir James Hector (1834–1907) was a Scottishborn Canadian geologist who took his medical degree at Edinburgh and, as both geologist and surgeon, was part of the Palliser expedition to western North America (1857–1860). He discovered and named many landmarks in the Rocky Mountains, including Kicking Horse Pass, the route later taken by the Canadian Pacific Railway. He returned to Scotland via the Pacific Coast, the California goldfields, and Mexico. He became the Director of the Geological Survey of New Zealand (1865) and eventually the Curator of the Colonial Museum in Wellington (Museum of New Zealand Te Papa). He wrote: *Outlines of New Zealand Geology* (1886). A bird and two mammals are also named after him.

Hedin

Stone Loach genus *Hedinichthys* CH Rendahl, 1933

Dr Sven Anders Hedin (1865–1952) was the much-honoured foremost explorer of eastern Europe and Asia in his day, as well as a writer, illustrator, geographer and photographer. In addition to being the leader of the Sino-Swedish expedition (1927–1935) to China, he made three famous expeditions (1894–1904). The University of Berlin awarded his original doctorate (1892) and later in life he received no less than 8 honorary doctorates from leading European universities. His life is so well documented that it needs little amplification here beyond stressing that he was not really interested in scientific investigation. His life work, published posthumously was his: *Central Asia Atlas*.

Hedley

Southern Peacock Sole *Pardachirus hedleyi* JD Ogilby, 1916

Charles Hedley (1862–1926) was described by Ogilby as "*...the premier conchologist of Australia*." In poor health, he had to finish schooling at home in Yorkshire. He moved to New Zealand (1881) in the hope the climate would better suit his chronic asthma, but as it did not he moved on to Sydney (1882), then inland in NSW. When this failed he moved to Queensland becoming a fruit farmer, finding that living close to the see kept him reasonably healthy. An accident prevented him pursuing that life and he moved to Brisbane (1888) where he volunteered at the Queensland Museum and was offered a job (1889). He travelled to Papua New Guinea (1890) where he made a collection of land shells but caught Malaria and returned to Australia. He was appointed Assistant Conchologogist at the mueum (1891) becoming Conchologist when his boss retired (1896). He was later Assistant Curator (1908) and then Acting Director (1920), but was pushed sideways after falling out wirth a trustee becoming Principle Keeper of Collections before taking up a post (1924) as Scientific Director of the Great Barrier Reef Committee. He published many papers and the book: *Wild Animals of the World: being a popular guide to Taronga Zoological Park* (1919). Three birds are also named after him.

Heemstra

East African Skate *Okamejei heemstrai* JD McEachran & JD Fechhelm, 1982
Sand-diver sp. *Pteropsaron heemstrai* JS Nelson, 1982
Orangehead Anthias *Pseudanthias heemstrai* H Schuhmacher, F Krupp & JE Randall, 1989
Flathead sp. *Cociella heemstrai* LW Knapp, 1996
Conger Eel sp. *Kenyaconger heemstrai* DG Smith & ES Karmovskaya, 2003
South African Sole sp. *Aseraggodes heemstrai* JE Randall & O Gon, 2006
Manefish sp. *Neocaristius heemstrai* IA Trunov, EI Kukuev & NV Parin, 2006
Snake-eel sp. *Luthulenchelys heemstraorum* JE McCosker, 2007
Viviparous Brotula sp. *Mascarenichthys heemstrai* W Schwarzhans & PR Møller, 2007
Yellowtail Blenny *Cirripectes heemstraorum* JT Williams, 2010
Heemstras' Goatfish *Upeneus heemstra* F Uiblein & G Gouws, 2014
Greater Sweeper *Pempheris heemstraorum* JE Randall & BC Victor, 2015

Heemstras' Frogmouth *Chaunax heemstraorum* HC Ho & WC Ma, 2016
Clingfish sp. *Lepadichthys heemstraorum* K Fujiwara & H Motomura, 2021

Dr Phillip Clarence Heemstra (1941–2019) was an American/South African marine ichthyologist and scientific diver. The University of Illiinois awarded his BSc (1963) and MSc (1968) and the University of Miami awarded his doctorate (1974) and he became a biologist at the Marine Laboratory of the US Department of Natural Resources in Florida (1974–1978). He moved to South Africa (1978) and spent time on board the 'Fridtjof Nansen' research vessel in the Indian Ocean. He was Curator of Sea Fish (1978–2001) and then Curator Emeritus with the South African Institute for Aquatic Biodiversity at Grahamstown. With his second wife (1991), the scientific illustrator Elaine Grant, he has carried out a number of underwater fish surveys. They co-wrote: *Coastal Fishes of Southern Africa* (2004) and he wrote numerous papers and other books. He was a member of the editorial board of *Copeia*. He was reported to be one of the few people on the planet to have tasted coelacanth, though we don't know if he had chips with it. He was honoured in the skate name as he had made available to the authors specimens of the new species and for being 'extremely cooperative' in generally supplying them with South African elasmobranch material. The snake-eel, goatfish, sweeper, frogmouth and clingfish are named after both Phillip and Elaine. (Also see **Elaine**)

Heiko

Characin sp. *Moenkhausia heikoi* J Géry & A Zarske, 2004

Heiko Bleher (b.1944). (See **Bleher** & also **Amanda (Bleher)**)

Hein, G

Characin sp. *Characidium heinianum* A Zarske & J Géry, 2001

Günter Hein is a German freelance ichthyologist. He assisted the senior author's collecting trip to Bolivia in various ways, including helping to collect the holotype. He is the author of: *Hyphessobrycon pando sp. n., a new rosy tetra from northern Bolivia* (*Teleostei, Characiformes, Characidae*) (2008).

Hein, W

Arabian Sea Meagre *Argyrosomus heinii* F Steindachner, 1902
Sole sp. *Solea heinii* F Steindachner, 1903

William Hein (1861–1903) was an Austrian linguist, folklorist, orientalist and ethnographer who worked in the anthropological-ethnographic department of the Vienna Natural History Museum. He founded (1894) the Folklore Association and (1895) the Journal of Austrian Folklore and in the same year the Vienna Museum of Folklore. With his wife, he collected the types of both species in Qishn, Yémen (1901–1902). The ethnographical materials he collected there became the basis of his most important publication on Hadrami texts.

Heinemann, B

Eelpout sp. *Lycodes heinemanni* VK Soldatov, 1916

B A Heinemann collected a number of zoological specimens in the northwest Pacific, including the Sea of Japan (1907). We have failed to identify him further.

Heinemann, H

Heinemann's Killifish *Aphyosemion heinemanni* HO Berkenkamp, 1983

Hendrik Heinemann is a German aquarist. He discovered this killifish species in Cameroon.

Heisei

Ribbon Goby *Oxyurichthys heisei* FL Pezold, 1998

Not an eponym but the Japanese for 'peace succeeds' or 'realised peace'. This is the name of the era of Emperor Akihito. It was chosen to recognise the many contributions to gobioid systematics made by Emperor Akihito and members of the Laboratory of Ichthyology, Akasaka Imperial Palace, working under his direction.

Heiser

Pitcairn Rainbow Wrasse *Thalassoma heiseri* JE Randall & A Edwards, 1984

Dr John B 'J B' Heiser is an American ichthyologist and marine biologist. He wrote his Cornell PhD thesis (1981) on the classification of the genus *Thalassoma*. His biology degree was awarded by Purdue University. He was very well travelled and taught in the field on every continent, having visited seventy countries. He attended Shoals Marine Laboratory in the Gulf of Maine (1967) and went on to become its Director (1979–1994). In the 1980s, he oversaw the launching of its 47-foot vessel, the 'R/V John M. Kingsbury', to transport people and supplies from the mainland and support offshore research and educational cruises. He was one of the authors of *Vertebrate Life* (1979). He is an accomplished SCUBA diver and underwater photographer.

Helen (Holleman)

Helen's Triplefin *Ceratobregma helenae* W Holleman, 1987

Helen Holleman is the wife of the author; Wouter Holleman. His etymology simply says: "The species I have named for my wife, Helen."

Helen (Larson)

Helen's Pygmy Pipehorse *Idiotropiscis larsonae* CE Dawson, 1984

Helen Larson (see **Larson**).

Helen (Newman)

Helen's Pygmygoby *Trimma helenae* R Winterbottom, MV Erdmann & NKD Cahyani, 2014

Helen Newman (d.2014) was a British marine biologist and co-founder of the Sea Sanctuaries Trust which was set up to conserve the 'Coral triangle' of Indonesia. Exeter University awarded her BSc (1980) and the University of Science, Malaysia her MSc in marine biology (1984). She set up (1989) her own environmental consultancy 'Newman Biomarine' and also consulted for Conservation International (2008–2011). She was

Director of Underwater World Singapore (1987–1991). She also organised the installation of mooring buoys around Bali (where she lived) to protect coral reefs from boat anchors, all the while working with local people to establish a marine park in Raja Empat to protect one of the world's remaining areas of pristine natural beauty. She was leader of the survey that led to the discovery of this goby. She was honoured for her "...*tireless conservation efforts over the past decade on behalf of Raja Ampat and its indigenous communities.*"

Helen (Oulton)

Helen's Goby *Didogobius helenae* JL Van Tassell & A Kramer, 2014

Helen Gay Oulton organised local logistics for Earthwatch Expeditions for ten years. She spent an additional ten years organising logistics for the senior author in the Canary Islands (where this goby occurs). She was honoured for her "...*help and dedication, without which none of the authors' research would have been possible, and for her love of the people, fauna and flora of the Canary Islands.*"

Helen (Randall)

Helen's Dartfish *Ptereleotris helenae* JE Randall, 1968
[Alt. Hovering Goby]
Perchlet sp. *Plectranthias helenae* JE Randall, 1980

Helen Randall née Au is a fellow zoologist and wife of the author Dr John 'Jack' Ernest Randall Jr. (Also see **Randall**)

Helen (Zondagh)

Helen's Clipfish *Clinus helenae* JLB Smith, 1946

Helen Evelyn Zondagh (1877–1951) was the author's mother-in-law. There is absolutely nothing in Smith's text to help identify who the 'Helen' was, whom he intended to commemorate in this fish's name, so any identification must be speculative.

Helena

Mediterranean Moray *Muraena helena* C Linnaeus, 1758
Scorpionfish sp. *Pontinus helena* WN Eschmeyer, 1965

According to Eschmeyer's description of the scorpionfish "...*the name is from Greek mythology, Helena sister of Castor*" - a reference to Helen of Troy. Linnaeus gives no etymology but it seems probable that Helen of Troy was intended. Though the eel would not be considered beautiful by most people, Izaak Walton remarked (1653) that the Romans called this eel the 'Helena of their feasts' (see: *The Compleat Angler*). Possibly Linnaeus was inspired by Walton's comment or, given his known sense of humour, perhaps he ironically named an 'ugly' eel after a legendary beauty!

Helena (Ivantsoff)

Drysdale Hardyhead *Craterocephalus helenae* W Ivantsoff, LEL Crowley & GR Allen, 1987

Helena Ivantsoff is the wife of the senior author, Walter Ivantsoff.

Helfrich, C

Helfrich's Membrane-Eye Carp *Rasborichthys helfrichii* P Bleeker, 1857

Lieutenant-Colonel C Helfrich was an army surgeon in the Royal Netherlands East Indies Army. He was stationed on Java and sent the holotype, which he collected in South Borneo, to Bleeker.

Helfrich, P

Helfrich's Dartfish *Nemateleotris helfrichi* JE Randall & GR Allen, 1973

Dr Philip Helfrich (b.1927) is a marine biologist. At the time he was honoured he was Associate Director, Hawaii Institute of Marine Biology (University of Hawaii) (Director (1975) and now Director Emeritus), and Director, Eniwetok Marine Biological Laboratory. He wrote the book: *Fish poisoning in the tropical Pacific* (1961) and among his published papers is: *Ciguatera fish poisoning. 2. General patterns of development in the Pacific* (1968). He was among those first to collect this dartfish.

Helga

False Boarfish *Neocyttus helgae* EW Holt & LW Byrne, 1908

'Helga' was a yacht that belonged to King Edward VII of Great Britain. The holotype of this species was collected from her, off the southwest coast of Ireland, in the course of investigations by the Fisheries Branch of the Department of Agriculture and Technical Instruction for Ireland.

Helker

Dwarf Cichlid sp. *Apistogramma helkeri* I Schindler & W Staeck, 2013

Oliver Helker is a German aquarist, who caught the fish-keeping 'bug' when he was eight years old. He co-wrote the article: *Männerüberschuss-Frauenüberschuss-Paarweise??!! Wie hält man Guppy, Platy & Co.richtig?* (2014). He brought this species of cichlid to the attention of the authors, as well as providing photos and information on the collecting site in Venezuela.

Heller, E

Heller's Barracuda *Sphyraena helleri* OP Jenkins, 1901
Heller's Anchovy *Anchoa helleri* CL Hubbs, 1921
Characin sp. *Hemibrycon helleri* CH Eigenmann, 1927

Edmund Heller (1875–1939) was an American zoologist. Heller collected specimens on the Galapagos Islands during the Hopkins-Stanford Expedition (1898). Stanford University awarded his BA in zoology (1901); during his time there, he collected in the Mojave and Colorado Desserts. He led several expeditions to Africa (1909–1912), which he wrote about in collaboration with Theodore Roosevelt as: *Life-histories of African Game Animals* (1914) and *China* (1916). He took part in a number of other expeditions including to British Columbia (1914) and Peru (1915). He became Curator of Mammals at the Field Museum, Chicago (1926–1928) during which time he took part in other collecting trips. Then he was Director of Washington Park Zoo (1928–1935) and the Fleishhacker Zoo, San Francisco

(1935–1929). He collected the characin holotype. Five birds, three mammals and three reptiles are also named after him.

Heller, KB

Green Swordtail *Xiphophorus hellerii* JJ Heckel, 1848
Yellow Cichlid *Thorichthys helleri* F Steindachner, 1864
Characin sp. *Cyphocharax helleri* F Steindachner, 1910

Karl Bartholomäus Heller (1824–1880) was an Austrian botanist and naturalist who explored in Mexico (1845–1848 and again in 1850) and yet again as a member of the (1864) scientific mission instituted by Napoleon III during which he mapped the state of Tabasco. He was a Professor in Vienna at the Theresianum. He collected cichlid type while exploring México. He wrote: *Reisen in Mexico in den Jahren 1845–48* (1853) and *Darwin und der Darwinismus* (1869). The characin description does not identify who was honoured, but it is likely that this species was named after the above.

Hellner

Neotropical Rivuline sp. *Hypsolebias hellneri* HO Berkenkamp, 1993
Neotropical Rivuline sp. *Moema hellneri* WJEM Costa, 2003

Steffen Hellner (b.1961) is a German aquarist and independent researcher specialising in killifish from an early age, having acquired his first aquarium aged 12. He is also a keen keeper of caudate amphibians (salamanders). After leaving school he served in the Bundeswehr for two years, later becoming an embedded officer in the US Army. His day job is as a freelance journalist working in advertising and marketing for major industrial companies. He has undertaken a number of field trips to Brazil (three times), Gabon, Sierra Leone, Turkey and the USA (eleven times). On his second trip to Brazil (1992) he discovered *Hypsolebias hellneri.* The other species was named in his honour in recognition of his scientific support for Wilson Costa over many years. He wrote: *Killifish: A Complete Pet Owner's Manual* (1990).

Helmer

Characin sp. *Characidium helmeri* AM Zanata, LM Sarmento-Soares & RF Martins-Pinheiro, 2015

Dr José Luis Helmer is a Brazilian ichthyologist who, since 1976, has made a detailed study of this species and who collected part of the type series. He was honoured for his pioneer studies on the natural history of the freshwater fishes of Espírito Santo and Bahia (Brazil).

Helsdingen

Twostripe Goby *Valenciennea helsdingenii* P Bleeker, 1858

W F C van Helsdingen was a Dutch civil servant, who provided Bleeker with a number of well-preserved fishes from the Gorong Archipelago (Indonesia), including the type of this goby.

Heming

Bristlemouth sp. *Triplophos hemingi* AF McArdle, 1901

Captain Thomas Henry Heming RN (b.1856) was (1898–1907) Commander, Royal Indian Marine Survey Steamer 'Investigator' from which the holotype was collected. He carried out soundings in the Humber estuary (1909).

Hemingway

Spinycheek Scorpionfish *Neomerinthe hemingwayi* HW Fowler, 1935

Ernest Miller Hemingway (1899–1961) is too famous a writer to need an extensive biography here. However, most people will know his book: *The Old Man and the Sea* (1952) but may not know he was a very keen sports fisherman, particularly catching marlin off Florida and Cuba.

Hempel

Hempel's Whalefish *Cetomimus hempeli* G Maul, 1969

Dr Gotthilf Hempel (b.1929) is a retired German marine biologist and oceanographer. He studied geology and biology at the Universities of Mainz and Heidelberg, where he was awarded his doctorate (1952). He worked (1952–1967) as a scientific assistant, variously at Wilhelmshaven, Heligoland and Hamburg. He was Professor at the Institute of Oceanography, University of Kiel (1967–1981). He was founder and first Director, Alfred Wegener Institute for Polar and Marine Research, Bremerhaven as well as Director, Institute for Polar Ecology, University of Kiel (1981–1992). He became the first Director, Centre for Marine Tropical Ecology, University of Bremen (1992–1994). He liked to go on expeditions and managed to spend more than 1,000 days on board research vessels. He wrote: *Early Life History of Marine Fish: The Egg Stage* (1980). He was leader of the sixth leg of the voyage of the 'Meteor' during which the holotype was collected.

Hemphill

Blackbelly Blenny *Stathmonotus hemphillii* TH Bean, 1885

Henry Hemphill (1830–1914) was an American biologist and malacologist particularly interested in land and freshwater molluscs. He earned his living as a mason and bricklayer. A Californian resident, he collected extensively along the coast of California, as well as during trips to Florida where he collected specimens of marine fauna at Key West on behalf of the US National Museum. A slug genus, *Hemphillia*, is also named after him.

Hemprich

Red Sea Dwarf Lionfish *Dendrochirus hemprichi* M Matsunuma, H Motomura & SV Bogorodsky, 2017

Dr Wilhelm Friedrich Hemprich (1796–1825) was a German physician, collector, naturalist and explorer. He co-wrote *Natural Historical Journeys in Egypt and Arabia* (1828) with Ehrenberg (q.v.), whom he had met whilst studying medicine in Berlin. They were invited to serve (1820) as naturalists on an expedition to Egypt and they continued to journey and collect in the region, including the Lebanon and the Sinai Peninsula before returning to Egypt and on to Ethiopia. Hemprich died of fever in the Eritrean port of Massawa. He was honoured in the lionfish's binomial for his great contributions to the zoology of the Red Sea. Four birds, a mammal and two reptiles are also named after him.

Henderson, D

Chocolate Toadfish *Chatrabus hendersoni* JLB Smith, 1952

Master David Henderson found the type specimen 'thrown up by a storm' in Algoa Bay, South Africa. As used in the etymology, 'Master' probably refers to a boy under age 12.

Henderson MR

Mahseer sp. *Neolissochilus hendersoni* AW Herre, 1940

Murray Ross Henderson (1899–1982) was a Scottish botanist who graduated at Aberdeen University (1921). He worked in the Straits Settlements, Malaya (1921–1924) and as Curator of the Singapore Herbarium (1924–1941), interrupted by being Director, Penang Botanical Garden (1937–1938). Upon the Japanese capture of Singapore, he escaped by boat to Sumatra, walked across Sumatra and found a boat to Colombo, Ceylon (Sri Lanka) and thence to Durban, South Africa and worked at Kirstenbosch Botanic Gardens, Cape Town until end of WW2. He returned to Singapore (1946) and was Director of the Botanic Gardens there (1949–1954) until retiring to Aberdeen. He wrote: *Common Malayan Wildflowers* (1961). Herre thanked him in his etymology for his hospitality and help in making it easy for him to collect freshwater fishes on Penang.

Henderson, PA

Bluntnose Knifefish sp. *Brachyhypopomus hendersoni* WGR Crampton, CDCM de Santana, JC Waddell & NR Lovejoy, 2017

Dr Peter Alan Henderson (b.1954) is a British fish biologist. Imperial College, London awarded his bachelor's degree (1975) and his doctorate (1979). He was a research scientist Centre Electrical Research Laboratories (1978–1991) and at Oxford (1991–1997). He was Director, Fawley Aquatic Research Laboratories (1991–1995) and has been Director of Pisces Conservation Ltd., Oxford since 1995. He was fisheries consultant for Projeto Mamiraua Amazonia, Brazil (1990–1997). He co-wrote: *Direct evidence that density-dependent regulation underpins the temporal stability of abundant species in a diverse animal community* (2014).

Hendra

Betta (Fighting Fish) sp. *Betta hendra* I Schindler & H Linke, 2013

Hendra Tommy is the Indonesian owner-operator of Kurnia Aquarium and Pet Shop which collects, breeds and sells tropical fish domestically and abroad. The etymology says the name is "*...in honour of Hendra Tommy, (Kurnia Aquarium, Palangkaraya, Kalimantan Tengah, Borneo), who discovered and exported the species.*"

Hendrichs

Mexican Rivuline sp. *Cynodonichthys hendrichsi* J Álvarez de Villar & J Carranza, 1952

Pedro Hendrichs Perez was a Mexican government worker. The etymology says that the fish is named "*...in memory of Pedro Hendrichs Perez, National Institute of Statistics and Geography (Aguascalientes City, México), father of Pedro H. D. Hendrichs, who helped collect type.*"

Hendrickson

Asian Stream Catfish sp. *Akysis hendricksoni* ER Alfred, 1966

Professor Dr John Roscoe Hendrickson (1921–2002) was a zoologist who was an expert on sea turtles and fought tenaciously for their conservation. His bachelor's degree was awarded (1944) by the University of Arizona, after which he spent the rest of the Second World War in the US Navy and was discharged (1946). He went to the University of California, Berkeley, and gained a master's degree (1949) and doctorate (1951). He moved (1959) to Kuala Lumpur as the first Professor of Zoology of the newly created University of Malaya. He was Vice-Chancellor of the University of Hawaii (1963–1969). He was Professor and Director, Marine Biology Programme, University of Arizona (1969). He wanted to be remembered for something worthwhile and arranged that a fund, the John R Hendrickson Scholarship Fund, be set up for the benefit of outstanding young naturalists in Malaysia. He also worked on amphibians and wrote: *Ecology and systematics of the salamanders of the genus Batrachoseps* (1954). An amphibian is named after him.

Hendrik

Eastern Spiny Seahorse *Hippocampus hendriki* RH Kuiter, 2001
[Junior Synonym of *Hippocampus angustus*?]

Hendrik Kuiter is presumably a relative of Rudie Herrman Kuiter (b.1943) the author, who is a Dutch-born Australian underwater photographer, taxonomist and marine biologist. The etymology says it was named for Hendrik Kuiter "*...in recognition of his keen interest in seahorses that he successfully conveyed to classmates and teachers.*" From this, we assume he is Rudie's son.

Hengel

Hengel's Rasbora *Trigonostigma hengeli* H Meinken, 1956
[Alt. Glowlight Rasbora]

J van Hengel was a Dutch ornamental fish importer-exporter in Amsterdam. He used to send specimens to Meinken for identification.

Hengstler

African Rivuline sp. *Nothobranchius hengstleri* S Valdesalici, 2007

Holger Hengstler of Munich is an aquarist who collected the type in northern Mozambique. He co-wrote: *Nothobranchius krammeri n. sp. (Cyprinodontiformes: Nothobran chi ida e): a new annual killifish from the Meronvi River basin, north eastern Mozambique* (2008) with Valdesalici. He has been described by one of his fellow aquarists as 'really crazy' because he went to the Congo during the civil war in search of *Nothobranchius malassai*!

Henle

Bigtooth River Stingray *Potamotrygon henlei* F de L de Castelnau, 1855
Brown Smooth-hound *Mustelus henlei* TN Gill, 1863

Dr Friedrich Gustav Jakob Henle (1809–1885) was a German physician, anatomist and zoologist whose main subjects were ichthyology and human biology. He was Professor of Anatomy in Zurich (1840–1844), at Heidelberg (1844–1852) and at Göttingen (1852–1885). Though he is best remembered today for his role in the development of modern medicine, he also co0produced, with Johann Müller, the first authoritative work on sharks and rays (1839–1841).

Henn

Characin sp. *Brycon henni* CH Eigenmann, 1913
Characin sp. *Phenacobrycon henni* CH Eigenmann, 1914
Sea Catfish sp. *Hexanematichthys henni* HG Fisher & CH Eigenmann, 1922
[Junior Synonym of *Chinchaysuyoa labiata*]
Naked-back Knifefish sp. *Gymnotus henni* JS Albert, WGR Crampton & JA Maldonado-Ocampo, 2003

Arthur Wilbur Henn (1890–1959) was an American herpetologist and ichthyologist. He was Eigenmann's student and also his successor as Curator of Fishes, Carnegie Museum of Natural History. He was the longest serving Treasurer of the American Society of Ichthyologists and Herpetologist (1931–1949). He wrote: *The Voracity of the South American Hoplias* (1916).

Henny Davies

Lake Malawi Cichlid sp. *Placidochromis hennydaviesae* WE Burgess & HR Axelrod, 1973

Henny Davies was the wife of Peter Davies, a former exporter of aquarium fish from Malawi. The couple left Malawi in the 1970s. The authors wrote that this: "*...husband and wife team has made many new aquarium fishes available*". (Also see **Peter Davies**)

Henric

Armourhead Catfish sp. *Cranoglanis henrici* L Vaillant, 1893

Henri Prince d'Orléans (1867–1901) was born in England, but was a French Bourbon prince of the Royal House. His place of birth did not prevent him from being a firm Anglophobe, given to making diatribes against Britain. He was an explorer and geographer who collected in China and Tibet (1889–1890), East Africa (1892) and Indochina (1895). He also discovered the source of the Irrawaddy River. He wrote: *Around Tonkin and Siam* (1894) and *From Tonkin to India* (1897). He died in Saigon (Ho Chi Minh city), Vietnam. Six birds are named after him.

Henrique

Twig Catfish sp. *Farlowella henriquei* A Miranda Ribeiro, 1918

Captain Henrique Silva collected the holotype in Brazil, but that is all we know about him.

Henry, J

Henry's Mormyrid *Isichthys henryi* TN Gill, 1863

Joseph Henry (1797–1878) was the first Secretary of the Smithsonian Institution (1846–1878). He was apprenticed as a watchmaker and silversmith, aged 13. However, he went on to become a scientist and engineer, having been given a free education at the Albany Academy (1819–1824). He was appointed Professor of Mathematics and Natural Philosophy there (1826). He invented prolifically, including the electromechanical relay (1835). He was a friend of the author Theodore Gill. A bird is also named after him.

Henry, JM

Goby sp. *Rhinogobius henryi* AW Herre, 1938

James McClure Henry (1880–1958) was a Presbyterian missionary in China (1909–1919). He joined the staff of Lingnan University, Canton (1919), where he was President (1924–1927) and Provost (1927–1948). He was interned in Canton by the Japanese (1943). He was honoured for his "*...continued interest in, and warm support of*" Herre's studies of Chinese fishes.

Henry, W

Henry's Epaulette Shark *Hemiscyllium henryi* GR Allen & M van N Erdmann, 2008
[Alt. Triton Epaulette Shark, Walking Shark]

Wolcott Henry is a professional underwater photographer who is based in Washington DC. He is a certified Smithsonian Institution science diver and has been contracted to the National Geographic Society since 1997. He co-wrote: *Wild Ocean* (1999). He was honoured for his generous support of Conservation International's marine initiatives, including taxonomy of New Guinea fishes.

Henschel

Neotropical Rivuline sp. *Anablepsoides henschelae* WEJM Costa, PHN de Braganca & PF de Amorim, 2013

Elisabeth Henschel is a researcher at the Laboratory of Systematics and Evolution of Teleost Fishes, Institute of Biology, Federal University of Rio de Janeiro, Brazil. She is currently a master's student of zoology at the Federal University of Rio de Janeiro which awarded her BA (2015). She has undertaken a number of collecting trips in the Amazon. Among her published papers are: *A new catfish species of the* Trichomycterus hasemani *group (Siluriformes: Trichomycteridae), from the Branco river basin, northern Brazil* (2016) and *Position of enigmatic miniature trichomycterid catfishes inferred from molecular data (Siluriformes)* (2017) for which Wilson Costa is a junior author. He honoured her in the name for her "*...valuable help during collecting trips in the Amazon.*"

Hensel

Armoured Catfish sp. *Rineloricaria henselii* F Steindachner, 1907
Characin sp. *Astyanax henseli* F de Melo & PA Buckup, 2006

Reinhold Friedrich Hensel (1826–1881) was a German naturalist, ichthyologist and paleontologist. He taught natural history in Berlin (1850–1860). He travelled to the southern Brazilian province of Rio Grande do Sul (1863–1866) to make a study of fishes on behalf of Berliner Akademie, where he subsequently became Professor of Zoology (1867). He wrote: *Beiträge zur Kenntnis der Wirbelthiere Süd-Brasiliens* (1868). Three amphibians and a reptile are named after him.

Henshall

Alabama Bass *Micropterus henshalli* CL Hubbs & RM Bailey, 1940

Dr James Alexander Henshall (1836–1925) was a physician and writer on angling and flyfishing, known as the 'apostle of the black bass'. He wrote: *Book of the Black Bass – Comprising Its Complete and Scientific and Life History with a Practical Treatise On Angling and Fly Fishing and a Full Description of Tools, Tackle and Implements* (1881), *Camping and Cruising in Florida* (1884), *More about Black Bass* (1889), *Bass, Pike, Perch and other Game Fishes of America* (1903) and combined the information in another title: *Book of the Black Bass* (1904). The etymology states: "*We take pleasure in naming this form for the late James A Henshall, to whom credit is largely due not only for raising the black basses to their position of high esteem in the minds of the sportsmen of the country, but also for determining their proper nomenclature.*"

Henshaw

Henshaw's Snake-Eel *Brachysomophis henshawi* DS Jordan & JO Snyder, 1904

Lahontan Cutthroat Trout *Oncorhynchus clarkii henshawi* TN Gill & DS Jordan, 1878

Henry Wetherbee Henshaw (1850–1930) was a naturalist, ornithologist and ethnologist. He was a keen naturalist as a child and particularly interested in birds. Ill-health prevented him entering Harvard and he was invited on a voyage during which he collected specimens. When his health recovered, he abandoned the idea of Harvard and undertook another collecting trip (1870) in Florida. He was the naturalist on the Wheeler survey of the American West (1872–1879). He worked for the US Bureau of Ethnology (1879–1893) and edited *American Anthropologist* (1888–1893). He visited Hawaii several times (1894–1904). He joined the US Department of Agriculture (1905), working in the Biological Survey, and became (1910–1916) the official in charge. He was a prolific writer and among his longer works are: *Perforated stones from California* (1887) and *Birds of the Hawaiian Islands* (1902). His *Fifty Birds of Farm & Orchard* (1913) sold over 200,000 copies. A reptile and two birds are also named after him.

Hensley

Lefteye Flounder sp. *Engyprosopon hensleyi* K Amaoka & H Imamura, 1990

Euripos Jewelfish *Odontanthias hensleyi* WD Anderson & G García-Moliner, 2012

Dr Dannie Alan Hensley (1944–2008) was an American ichthyologist. San Bernardino Valley College awarded his first degree (1966) and the University of South Florida his PhD (1978). He published his first paper before graduating (1965) and was still publishing four decades later, such as: *Revision of the genus Asterorhombus (Pleuronectiformes: Bothidae) (2005)*. He was an ichthyologist at the Marine Research Laboratory Florida (1974–1978) and then became a post-doctoral research associate at Florida Atlantic University. From here he moved to the University of Puerto Rico for twenty-eight years, becoming Assistant Professor (1980–1984) then Associate Professor (1984–1991) and finally Full Professor (1991). The etymology for the flounder is one of the more humourous we have come across: it says the fish is named after D A Hensley "*...who resembles this species in having a slim body.*"

Hentz

Feather Blenny *Hypsoblennius hentz* CA Lesueur, 1825

Nicholas Marcellus Hentz (1797–1856) was a French-born American entomologist, arachnologist and teacher. His family left France after the downfall of Napoleon and he is known to have worked as a tutor of French and painting for a wealthy family on a plantation on Sullivan's Island near Charleston. He attended Harvard to study medicine but abandoned his studies and taught school, going on to take the chair of modern languages at the University of North Carolina (1927) where he was awarded an honorary MA (1829). He left there (1831) partly because his Catholicism was not liked and French was dropped from the curriculum. He taught female academies with his wife for the next eighteen years. He wrote a number of papers on the spiders of the USA and two French language text books (1822), as well as a novel (1825). His magnum opus, which he also illustrated, was: *The Spiders of the United States* published posthumously (1875). The describer refers only to 'Mr Hentz of Charleston' (South Carolina) but we believe this to be our man.

Heok

Elongated Kerala Catfish *Mystus heoki* M Plamoottil & NP Abraham, 2013

Dr Heok Hee Ng is a biologist and ichthyologist at the Lee Kong Chian Natural History Museum, National University of Singapore. The University of Michigan awarded his doctorate. He co-wrote: *The Glyptothorax of Sundaland: A revisionary study (Teleostei: Sisoridae)* (2016), and was honoured for his contributions to catfish taxonomy.

Heok Hui

Bagrid Catfish sp. *Pseudomystus heokhuii* KKP Lim & HH Ng, 2008

Dr Heok Hui Tan is an ichthyologist who is a lecturer at the National University of Singapore which awarded all his degrees: bachelor's (1996) master's (1999) and Doctorate (2003). He co-wrote: *A new species of glass-perch from Belitung Island, Indonesia (Teleostei: Ambassidae: Gymnochanda)* (2011). He was honoured as he brought this fish to the authors' attention.

Hepburn

Hepburn's Blenny *Parenchelyurus hepburni* JO Snyder, 1908

Lieutenant A J Hepburn was a naval executive officer who was commanding aboard the US Bureau of Fisheries steamer 'Albatross'. He had previously joined the 'Pensacola' (1904).

Hephaestus

Grunter genus *Hephaestus* CW De Vis, 1884

Cory sp. *Corydoras hephaestus* WM Ohara, L Tencatt & MR de Britto, 2016

In Greek mythology, Hephaestus was the god of fire, blacksmiths, metal-workers and other artisans. De Vis gives no reason for his choice of name. The *Corydoras* was so named due to the red (fiery) colour of its fins and parts of its body.

Hera

African Rivuline sp. *Aphyosemion hera* J-H Huber, 1998

In Greek mythology, Hera was queen of the gods. The name was applied to this species because of the female's unusually beautiful colours.

Herald

Pipefish genus *Heraldia* JR Paxton, 1975
Cardinalfish sp. *Jaydia heraldi* AW Herre, 1943
Herald's Angelfish *Centropyge heraldi* LP Woods & LP Schultz, 1953
Sand Stargazer sp. *Dactyloscopus heraldi* CE Dawson, 1975
Pipefish sp. *Cosmocampus heraldi* RA Fritzsche, 1980
Sole sp. *Aseraggodes heraldi* JE Randall & P Bartsch, 2005

Dr Earl Stannard Herald (1914–1973) was an American ichthyologist. The University of California, Los Angeles awarded his bachelor's degree (1937), the University of California, Berkeley his master's (1939) and Stanford University awarded his doctorate (1943). He served in the US Army in WW2 and studied the effects of the atom bomb test on the reef fish of Bikini Atoll (1946). He worked for the US Fish and Wildlife Service in the Philippines (1947–1948). He later presented (1952–1966) the popular science TV show *Science in Action*. He wrote: *Living Fishes of the World* (1961). He was killed in a scuba diving accident off Lower California.

Heraldo

Armoured Catfish sp. *Hypostomus heraldoi* CH Zawadzki, C Weber & CS Pavanelli, 2008

Dr Heraldo Antonio Britski. (See **Britski**)

Herbert

Pencil Catfish sp. *Ituglanis herberti* A Miranda Ribeiro, 1940

Herbert Franzłoni Berla (1912–1985) was a Brazilian ornithologist and entomologist who worked for the National Museum in Rio de Janeiro. He made a collecting trip to Pernambuco (1946) where he catalogued c.160 bird species. He made a particular study of mites which live on birds. Two birds are named after him.

Herbert Axelrod

Black Neon Tetra *Hyphessobrycon herbertaxelrodi* J Géry, 1961
Lake Tebera Rainbowfish *Melanotaenia herbertaxelrodi* GR Allen, 1980
[Alt. Yellow Rainbowfish]

Dr Herbert Richard Axelrod (1927–2017) (See **Axelrod**; also see **Evelyn (Axelrod)**)

Herklots

Herring genus *Herklotsichthys* GP Whitley, 1951
Longfin Catshark *Apristurus herklotsi* HW Fowler, 1934

Dr Geoffrey Alton Craig Herklots (1902–1986) was a British biologist, botanist and ornithologist at the University of Hong Kong (1928–1941). He was interned under the Japanese occupation (1942–1945). On returning to England (1945) he joined the Colonial Service and was Principal of the Imperial College of Tropical Agriculture, Trinidad (1953–1960). He was later Reader in Biology at Hong Kong University (1980). Among other papers and books, he wrote: *Hong Kong Birds* (1953) and *Common marine food-fishes of Hong Kong, 50 species described in English and Chinese (with 16 European and 18 Chinese recipes)* (1961). He was a friend of Henry Fowler (q.v.), whose etymology says of Herklots: "*...with many fond memories of the China Sea and Java.*"

Herlihy

Forktail Blenny *Petroscirtes herlihyi* HW Fowler, 1946
[Junior Synonym of *Meiacanthus atrodorsalis*]

Lieutenant Colonel William J Herlihy was commanding officer of a group who, according to Captain Ernest Tinkham (q.v.), during WWII "*...played a very interesting role in helping to stop the Japanese coming across the Owen Stanley Range from Buna into Port Moresby. He was at Buna and many other places in the fight along the northeastern shores of New Guinea.*" Tinkham collected the blenny and sent the fish to Fowler, suggesting Herlihy be honoured "*...Because of his kindly interest in my efforts he deserves remembrance.*"

Herman

Cape Damsel *Similiparma hermani* F Steindachner, 1887

Lieutenant Herman (forename not given) collected the type in the Cape Verde Islands, either on the way to, or back from, the Congo. We have been unable to find out anything further.

Hermann

Dwarf Sturgeon *Pseudoscaphirhynchus hermanni* KF Kessler, 1877

Kessler does not provide any name other than 'Hermann'. It is believed to refer to a ship's officer who was probably a member of the (1877) Aralo-Caspian Expedition. Hermann apparently gave the holotype to the zoologist and explorer Nikolai Severtsov (1827–1885), who coined the name in a presumably unpublished paper. There is too much conjecture for our liking, but we have not been able to get any further. If anyone knows more or better, please let us know.

Hermann (Ihering HFA)

Armoured Catfish sp. *Hypostomus hermanni* R Ihering, 1905

Hermann von Ihering (1850–1930). (See **Ihering, HFA**)

Hernan

Amazonian Anchovy sp. *Anchoviella hernanni* MV Loeb, HR Varella & NA Menezes, 2018

Hernán Ortega is a Peruvian ichthyologist and marine biologist specialising in aquatic ecosystems. He is associated with the Museo de Historia Natural, Universidad Nacional Mayor de San Marcos. He has written around 50 papers such as the co-written: *Freshwater fishes and aquatic habitats in Peru: Current knowledge and conservation* (2008) and *Estado de conservación y distribución de los peces de agua dulce de los Andes Tropicales* (2016). His best-known longer work is: *Annotated Checklist of the Freshwater Fishes of Peru* (1986). He was honoured "*…in recognition of his contribution to knowledge about the diversity of fishes of Peru and his support for many researchers, either by making material available or by guiding students.*"

Hernandez

Cortez Hake *Merluccius hernandezi* CP Mathews, 1985

Captain Felipe Hernandez Ascencio was a seaman in the Mexican Merchant Navy. This fish was named after him, and for the officers and crew of the research vessel 'Alejandro de Humboldt', from which the holotype was collected, for the all the services and help they gave.

Herre

Herre's Dwarfgoby *Eviota herrei* DS Jordan & A Seale, 1906
Herre's Moray *Gymnothorax herrei* W Beebe & J Tee-Van, 1933
Mindanao Barb sp. *Barbodes herrei* HW Fowler, 1934
Banded Spikefish *Paratriacanthodes herrei* GS Myers, 1934
Herre's Sole *Aseraggodes herrei* A Seale, 1940
Herre's Pipefish *Siokunichthys herrei* ES Herald, 1953
Stone Loach sp. *Mesonoemacheilus herrei* PM Bănărescu & TT Nalbant, 1982

Dr Albert William Christian Theodore Herre (1868–1962) was an ichthyologist, ecologist, botanist and lichenologist. He gained his bachelor's degree in botany and later took his master's (1905) and doctorate (1908), both in ichthyology, at Stanford. He became acting head, Biology Department, University of Nevada (1909–1910), was vice-principal of a high school, Oakland (1910–1912), taught at a school in Washington State (1912–1915), and was then head of the Science Department, Western Washington College of Education (1915–1919). He then went to the Philippines, where he was Chief of Fisheries, Bureau of Science, Manila (1919–1928). He was Curator of Zoology, Natural History Museum, Stanford (1928–1946). After retiring he returned to the Philippines (1947) as a member of the Fishery Program, US Fish and Wildlife Bureau, and then worked in the School of Fisheries, University of Washington (1948–1957). After his second retirement, he researched and collected lichens (1957–1962). He was Jordan's student at the time he was honoured in the goby's name. A reptile and an amphibian are also named after him.

Herring

Onejaw sp. *Monognathus herringi* E Bertelsen & JG Nielsen, 1987
Tubeshoulder sp. *Normichthys herringi* YI Sazonov & NR Merrett, 2001

Dr Peter John Herring is an oceanographer and deep-sea biologist, whose bachelor's degree and doctorate were both awarded by Cambridge. During study for his doctorate he spent 18 months at sea on the International Indian Ocean Expedition. He joined the Institute of Oceanographic Sciences, England, as an oceanic biologist (1966). He is interested in the physiology and ecology of bioluminescence in the deep-sea environment. He has participated, often as Chief Scientist, in more than 60 research cruises on vessels of five different countries. On retirement (2000) he became an Honorary Professor at the Southampton Oceanography Centre. He has written and edited more than 250 scientific papers and books including: *Deep Oceans* (1972), *Bioluminescence in Action* (1978) *Light and Life in the* Sea (1990) and *The Biology of the Deep Ocean* (2001).

Hertz

Banjo Catfish sp. *Bunocephalus hertzi* ALH Esguicero, RM Castro & TNA Pereira, 2020

Dr Hertz Figueiredo dos Santos is a Brazilian Biologist at the Laboratório de Ictiologia de Ribeirão Preto, Universidade de São Paulo (from 1984) and qualified open water diver. His BSc was awarded by the Educational Organization Barão de Mauá - Ribeirão Preto, São Paulo. (1982), and his MSc (1998) and doctorate in Zoology (2005) from the Institute of Biosciences from the University of São Paulo. His publications include: *Structure and composition of the stream ichthyofauna of four tributary rivers of the upper Rio Paraná basin, Brazil* (2004) and *Structure and composition of the rich ichthyofauna of four tributary rivers of the upper Rio Paraná basin, Brazil* (2005). He was co-discoverer of this species and was honoured "*in deep appreciation for his approximately three decades of unfailing contribution*" to the study of Neotropical fishes.

Herwerden

Blackbanded Gauvina *Oxyeleotris herwerdenii* MCW Weber, 1910

Captan J H Hondius van Herwerden served in the Dutch Government Navy. He was honoured for his knowledge of the coast and rivers of New Guinea, where the type was collected.

Herwig

Cape Verde Skate *Raja herwigi* G Krefft, 1965
Waryfish sp. *Scopelosaurus herwigi* E Bertelsen, G Krefft & NB Marshall, 1976
Whipnose Angler sp. *Gigantactis herwigi* E Bertelsen, TW Pietsch & RJ Lavenberg, 1981
Herwig Lanternfish *Metelectrona herwigi* PA Hulley, 1981
Snailfish sp. *Volodichthys herwigi* A Andriashev, 1991

These species are named after the research vessel 'Walther Herwig', from which all the holotypes were caught. The vessel was named after Walther Herwig (1838–1912), who is regarded as the founder of German fisheries science.

Herz

Oriental Striped Shiner *Pungtungia herzi* S Herzenstein, 1892
Korean Perch sp. *Coreoperca herzi* S Herzenstein, 1896

Alfred Otto Herz (1856–1905) was a German entomologist and professional collector in Asia for Dr Otto Staudinger & Andreas Bang-Haas – a dealership based in Dresden. Though he mainly collected beetles and butterflies, he also picked up some fish specimens.

Herzberg

Pemecou Sea Catfish *Sciades herzbergii* ME Bloch, 1794

Bloch did not identify whom he was naming this fish after, but a probable candidate is Count Ewald Friedrich von Herzberg (1725–1795), a distinguished Prussian statesman and savant and one of Bloch's sponsors.

Herzenstein

Chinese Barb genus *Herzensteinia* YT Chu, 1935
Gudgeon sp. *Gnathopogon herzensteini* A Günther, 1896
Yellow-striped Flounder *Pseudopleuronectes herzensteini* DS Jordan & JO Snyder, 1901
Righteye Flounder sp. *Cleisthenes herzensteini* PJ Schmidt, 1904
Sculpin sp. *Gymnocanthus herzensteini* DS Jordan & EC Starks, 1904
Herzenstein's Rough Sculpin *Asprocottus herzensteini* LS Berg, 1906
Herzenstein's Catfish *Tachysurus herzensteini* LS Berg, 1907
Stone Loach sp. *Triplophysa herzensteini* LS Berg, 1909

Solomon Markovich Herzenstein (1854–1894) was a Russian zoologist and ichthyologist who graduated in physics and mathematics at St Petersburg University. He was Curator of the Zoological Museum, Zoological Institute, Russian Academy of Sciences (1880–1894) and taught zoology at the University for Women (1881–1889). He made a number of expeditions to the Kola Peninsula (1880–1887).

Herzog

Herzog's Killi *Aphyosemion herzogi* AC Radda, 1975

Wolfgang Herzog is a German aquarist, and another intrepid member of the DKG (Deutsche Killifisch Gemeinschaft) who makes expeditions to remote areas in search of killifish. He "kindly provided the results of his successful fish collections in Gabon".

Hessfeld

Neotropical Livebearer sp. *Brachyrhaphis hessfeldi* MK Meyer & V Etzel, 2001

Gerhard Hessfeld is a German aquarist based in Hamm. He helped collect the type specimen in Panama.

Heudelot

Smoothmouth Sea Catfish *Carlarius heudelotii* A Valenciennes, 1840
Dotterel Filefish *Aluterus heudelotii* HLGM Hollard, 1855
Blackchin Tilapia ssp. *Sarotherodon melanotheron heudelotii* AHA Duméril, 1861

Jean-Pierre Heudelot (1802–1837) was a French botanist, agriculturalist, plant collector and explorer in Guinea, Senegal, Madagascar and Cape Verde Islands (1825–1837). He was appointed Director of the Royal Plantations in Senegal. He supplied the holotypes of these fish.

Heurn

Mouth Almighty sp. *Glossamia heurni* M Weber & LF de Beaufort, 1929

Willem Cornelis van Heurn (1887–1972) was a Dutch taxonomist, civil engineer, botanist, educationalist and collector who worked for a period at the Natural History Museum, Leiden. He came from a wealthy family but chose to work all his life. He went to Suriname (1911), to Simeulue (off Sumatra) (1913), and to Dutch New Guinea (1920–1921). He then lived in the Dutch East Indies (mostly Java) (1924–1939), where he ran a laboratory for sea research; studied rat control on Java, Timor and Flores; was a schoolteacher; and served as head of the Botany Department at the Netherlands Indies Medical School before returning to Holland. Wherever he travelled or settled he collected natural history specimens, which he meticulously prepared and labelled. Most he sent to the Leiden Museum, where he himself worked as an Assistant Curator for Fossil Mammals (1941–1945). He was a prolific writer, publishing c.100 articles on a wide range of topics, including such gems as: *The safety instinct in chickens* (1927), *Cannibalism in frogs* (1928), *Do tits lay eggs together as the result of a housing shortage?* (1955) and *Wrinkled eggs* (1958). It was said of him in a memorial booklet published by the museum: 'He made natural history collections wherever he went and gave his attention to almost all animal groups. He was an excellent shot, and a competent preparator; his mammal and bird skins are exemplary.' Five birds, a mammal and a reptile are named after him.

Heusinkveld

Lake Victoria Cichlid sp. *Haplochromis heusinkveldi* F Witte & ELM Witte-Maas, 1987

Dr W A Heusinkveld was a lecturer at the Department of Biophysics, Leiden University. He wrote a number of papers such as: *Determination of the differences between the thermodynamic and the Practical Temperature Scale in the range 630 to 1063 °C from radiation measurements* (1966). The authors say: "*In some way, this long, slender fish with its grey suit [body] made us think of him, when we first caught it.*"

Hewett

Hewett's Coris (wrasse) *Coris hewetti* JE Randall, 1999

Jeremy Hewett collected the holotype (with a spear) from the reef at the head of Anaho Bay, Nuku Hiva, Marquesas Islands (1957). We have failed to find out more details about him.

Hext

Rattail sp. *Coryphaenoides hextii* AW Alcock, 1890
Cusk-eel sp. *Tauredophidium hextii* AW Alcock, 1890

Rear-Admiral John Hext (1842–1924) was a British naval officer who served (1857–1889). He was seconded to the Royal Indian Marine (1883) for a five-year appointment as Director, but in fact served in the post for 15 years (1883–1898). His career in the Royal Navy involved service in West Africa, Egypt, the East Indies (Indonesia) and Burma (Myanmar). He was honoured for his generous support of the 'HMS Investigator' expedition to the Arabian Sea, during which the holotypes were collected.

Heyland

Heyland's Suckermouth Catfish *Kronichthys heylandi* GA Boulenger, 1900

Herbert Kyffin Heyland (1849–1944) was a British civil engineer who worked in Brazil where he was involved in setting up a company (1906). He collected the holotype of this species and presented it to the BMNH.

Heyliger

Blenny sp. *Phenablennius heyligeri* P Bleeker, 1859

Raymond V Heyliger sent a collection of fishes from Palembang, south Sumatra. He was a Dutch official in Batavia (Jakarta).

Heyningen

Molo Snake-Eel *Hemerorhinus heyningi* MWC Weber, 1913

Lieutenant Cornelis E Hoorens van Heyningen was an officer aboard the 'Siboga' during the expedition led by Weber (1899–1900) that collected the holotype.

Hialmar

Stone Loach sp. *Triplophysa hialmari* AM Prokofiev, 2001

Dr Carl Hialmar Rendahl (1891–1969). (See **Rendahl**)

Hiatt

Hiatt's Basslet *Acanthoplesiops hiatti* LP Schultz, 1953

Dr Robert Worth Hiatt (1913–1997), Senior Professor of Zoology at the University of Hawaii, was a marine ecologist, conchologist and hydrologist. The University of California, Berkeley awarded his PhD (1941). He joined the University of Hawaii (1943) teaching and researching until retirement (1969), and also founded and directed the Hawaii marine laboratory and other university posts. He was consular officer and secretary at the US Embassy in Tokyo (1970–1973) then became President of the University of Alaska (1973–1977). He was one of the first scientists to use the newly developed SCUBA in his work when he undertook a marine research expedition to Yap in Micronesia (1946). He was part of Schultz' team to the Marshall Islands in connection with the atom-bomb tests of Operation Crossroads (1946); he collected some paratypes and was one of three biologists Schultz praised for their skill in "*swimming, diving, and collecting unusual fishes*". His ashes were scattered in the waters near the Hawaii Institute of Marine Biology which he had established.

Hicks

Hicks' Toadfish *Torquigener hicksi* GS Hardy, 1983

Dr Geoffrey R F Hicks was a New Zealand marine zoologist at the Portobello Marine Laboratory, Otago, New Zealand. He then (1981) became Curator of Crustacea at the National Museum there. He has written a number of papers with Hardy such as *Proposed rempoval of Leiolopisma fasciolare from checklist of New Zealand Scinicidae (Reptilia: lacertilian, and a note of the name Hombronia* (1980) and was one of the two editors of *Awesome Forces: The Natural Hazards That Threaten New Zealand* (1998). The etymology simply says: "*The species is named for Dr. Geoffrey R. F. Hicks, a close friend and colleague.*" At least one Limnoria (gribble worms) that he collected is also named after him.

Hidalgo

Suckermouth Catfish sp. *Astroblepus hidalgoi* CA Ardila Rodríguez, 2013

Three-barbeled Catfish sp. *Cetopsorhamdia hidalgoi* D Faustino-Fuster & LS de Souza, 2021

Max H Hidalgo is a Peruvian biologist and ichthyologist who is an Assistant Professor, Biology Sciences Faculty, Universidad Nacional Mayor de San Marcos de Lima. He co-wrote: *Annotated list of the fish of the continental waters of Peru: Current state of knowledge, distribution, uses and conservation aspects* (2012). He was honoured "*...for his contributions to the study of the freshwater fishes of Peru.*"

Hien

Vietnamese Cyprinid sp. *Opsariichthys hieni* TT Nguyen, 1987

Hiên is Professor Mai Dinh Yen (b.1933) who is an ichthyologist in charge of vertebrate animal studies at the Biology Department, Hanoi University, Vietnam (1956–1986). He was visiting professor at Tiemcen University, Algeria (1986–1989). He is also the author's father, so perhaps 'Hiên' is a nickname used by the family.

Hieronymus

Swamp Eel sp. *Synbranchus hieronymi* H Weyenbergh, 1877

[Junior Synonym of *Synbranchus marmoratus*]

Dr Georg Hans Emmo Wolfgang Hieronymus (1846–1921) was a German botanist whose doctorate was awarded by the University of Halle (1872). He was one of the so-called Córdoba group of German scientists invited to Argentina by Burmeister. He became Director of the botanical museum at the University of Córdoba and Professor of Botany (1874–1883). He returned to Germany and lived in Breslau (1883–1892) and then moved to Berlin, where he was Curator of the botanic gardens and of the botanic garden museum (1892–1921).

Higman

Jawfish sp. *Lonchopisthus higmani* GW Mead, 1959

Smalleye Smooth-hound *Mustelus higmani* S Springer & RH Lowe, 1963

James Booth Higman (1922–2009) was an American fisheries biologist. Western Maryland College awarded his bachelor's degree and University of Miami his master's. He was a Captain in the Chemical Corp, US Army (1942–1946), fighting from Utah Beach, Normandy to the River Elbe. He taught at the Rosenstiel School of Marine and Atmospheric Science, University of Miami for thirty years and was a professor there. He co-wrote: *A review of shrimp capture and culture fisheries of the United States* (1993). He collected the smooth-hound holotype while serving as an observer on an exploratory fishing expedition for the Bureau of Commercial Fisheries, US Department of the Interior. He was honoured for "*... initial interest in the species and his care in the preparation of excellent notes on its natural history.*" He was honoured in the jawfish name for his "*...efforts aboard the 'Coquette', a shrimp trawler, (having) helped secure a collection of fishes off Suriname, including type of this species.*"

Higuchi

Higuchi's Sea Catfish *Cathorops higuchii* AP Marceniuk & R Betancur-Rodriguez, 2008
Thorny Catfish sp. *Doras higuchii* MH Sabaj Pérez & JLO Birindelli, 2008
Higuchi's Tube-snouted Ghost Knifefish *Sternarchorhynchus higuchii* CD de Santana & RP Vari, 2010

Dr Horácio Higuchi is an ichthyologist and evolutionary biologist. The University of São Paulo awarded his bachelor's degree (1979) and his master's (1982) whilst Harvard awarded his doctorate (1991). He is presently an Associate Researcher at Museu Paraense Emílio Goeldi, Belém, Brazil. He wrote: *An updated list of ichthyological collection sites of the Thayer Expedition to Brazil (1865–66)* (1996).

Hilaire

Characin sp. *Brycon hilarii* A Valenciennes, 1850
Dorado Characin sp. *Salminus hilarii* A Valenciennes, 1850
Scrapetooth Characin sp. *Parodon hilarii* JT Reinhardt, 1867

Auguste François César Provençal de SaintHilaire (1779–1853) was a French naturalist and botanist who explored in Brazil (1816–1822). He was President of the French Academy of Sciences (1835). He wrote: *Voyage dans les Provinces de Rio de Janeiro et de Minas Geraes* (1830). A bird is named after him.

Hilda (Fernández-Yépez)

Tetra sp. *Hyphessobrycon hildae* A Fernández-Yépez, 1950

Mrs Hilda Fernández-Yépez née Canelon Fernandez was the author's wife.

Hilda (Jubb)

Hilda's Grunter (catfish) *Amarginops hildae* G Bell-Cross, 1973
[Syn. *Chrysichthys hildae*]

Mrs Hilda M Jubb was the wife of Rex A Jubb, an ichthyologist at Albany Museum, Grahamstown, South Africa. They moved to Albany (1961) after the ichthyology department at Rhodes University became overcrowded. She was also an accomplished artist; as the etymology has it "*...whose excellent fish illustrations of Southern African freshwater fishes [including type of this species] have been admired by all.*"

Hilde

Hilde's Darter Dragonet *Callionymus hildae* R Fricke, 1981

Hildegard Handermann was a German biologist at the Zoologisches Institut, Technische Universität Braunschweig. She co-wrote at least one paper with Fricke (q.v.): *The compatible critical swimming speed: a new measure for the specific swimming performance of fishes* (1987). Fricke's etymology says: "*The species is named in honour of Miss Hildegard Handermann, Braunschweig, for her continued interest in my studies.*"

Hildebrand

Goby sp. *Sicydium hildebrandi* CH Eigenmann, 1918
Neotropical Rivuline sp. *Cynodonichthys hildebrandi* GS Myers, 1927
Dark-tailed Worm-Goby *Microdesmus hildebrandi* ED Reid, 1936
Hildebrand's Goby *Gobiosoma hildebrandi* I Ginsburg, 1939
Characin sp. *Creagrutus hildebrandi* LP Schultz, 1944
Least Madtom *Noturus hildebrandi* RM Bailey & WR Taylor, 1950
Chiapas Killifish *Tlaloc hildebrandi* RR Miller, 1950
Hildebrand's Pipefish *Cosmocampus hildebrandi* ES Herald, 1965
Central American Halfbeak *Hyporhamphus roberti hildebrandi* DS Jordan & BW Evermann, 1927

Dr Samuel Frederick Hildebrand (1883–1949) was born of immigrant parents who never learned to speak English and was brought up as a farm boy in Indiana. He worked for the US Bureau of Fisheries, Washington DC (1910–1949), as a scientific assistant (1910–1914), Director of the US Fisheries Biological Stations, Beaufort, North Carolina (1914–1918 & 1925–1931) and Key West, Florida (1918–1919), Ichthyologist in Washington (1919–1925) and Senior Ichthyologist (1931–1939). He took part in a number of expeditions, including two with Meek to Panama (1910–1911 & 1912), two on his own (1935 & 1937) and one to Central America with Foster (1924). Schultz honoured him for his "extensive contributions on the fish fauna of Panama". (Also see **Bilsam** & **Longley**)

Hildebrandt

Hildebrandt's Elephant-snout *Mormyrus hildebrandti* W Peters, 1882

Johannes Maria Hildebrandt (1847–1881) was a German botanist and explorer. He began work (1869) at the Berlin Botanical Garden, before making a number of expeditions (1872–1881) to Arabia, East Africa, Madagascar and the Comoro Islands (1872–1881). He collected zoological as well as botanical specimens, including the holotype of this fish. He died in Madagascar of yellow fever. His father, who was a painter and entomologist, named a beetle after him. Additionally, three reptiles, two mammals, two birds and an amphibian are also named after him.

Hildegard

African Rivuline sp. *Epiplatys hildegardae* HO Berkenkamp, 1978

Hildegard Bergenkamp is the wife of the author, aquarist and amateur ichthyologist Heinz Otto Berkenkamp.

Hilgendorf

Hilgendorf's Saucord *Helicolenus hilgendorfii* LHP Döderlein, 1884
African Tetra sp. *Alestopetersius hilgendorfi* GA Boulenger, 1899
Japanese Cyprinid sp. *Pungtungia hilgendorfi* DS Jordan & HW Fowler, 1903
Freshwater Pufferfish sp. *Pao hilgendorfii* CML Popta, 1905

Dr Franz Martin Hilgendorf (1839–1904) was a German zoologist and palaeontologist. He entered the University of Berlin (1859) to study philology and moved to Tübingen University (1861) where he was awarded a doctorate (1863) for a thesis on a geological subject. He worked at, and studied in, the Zoological Museum in Berlin (1863–1868). He was Director of the Hamburg Zoological Gardens (1868–1870). He was a private lecturer at the Polytechnic Institute in Dresden (1871–1872) and then became a lecturer at the Imperial Medical Academy in Tokyo (1873–1876). On his return to Germany he became an assistant to Wilhelm Peters (q.v.), and worked in various departments of the Berlin Museum until ill health forced him to retire. A mammal is also named after him.

Hill, CW

Hill's Driftfish *Psenes hillii* JD Ogilby, 1915
[Alt. Hill's Blubber-fish, Hill's Eyebrow-fish]

Charles William Hill ws the Lightkeeper at Cowan Cowan, Moreton Bay, Queensland, Australia. He provided the author with the type.

Hill, G

African Airbreathing Catfish sp. *Clarias hilli* HW Fowler, 1936

Gordon Hill was a volunteer assistant in the Department of Fishes and Reptiles, Academy of Natural Sciences of Philadelphia. We have been unable to find any more about him.

Hill, R

Atlantic Sailfin Flyingfish *Parexocoetus hillianus* PH Gosse, 1851
Caribbean Lantern Shark *Etmopterus hillianus* F Poey, 1861

Richard Hill (1795–1872) was a Jamaican anti-slavery activist, judge, botanist and naturalist who was in England (1827–1830) and was closely involved with men like Wilberforce and other prominent members of the Anti-Slavery Society. He was then in Santo Domingo (1830–1832). He finally returned to Jamaica (1832) and became a stipendiary magistrate (1834–1872). He was Vice-President of the Jamaica Society for the Encouragement of Agriculture and other Arts and Sciences (1844–1849). He corresponded with Poey and Darwin. He was honoured by Poey for his: *Contributions to the Natural History of the Shark* (1850) and other writings on fishes. A bird is also named after him.

Hill, W

Leathery Grunter *Scortum hillii* FL Castelnau, 1878

Walter Hill (1820–1904) was a Scottish botanist who became the first Curator (aka Superintendent) of the Brisbane Botanical Gardens. He started out as an apprentice to his brother, David, then head gardener at Balloch Castle, Dumbartonshire, Scotland. Later he worked at the Royal Botanic Garden Edinburgh, and then (1843) moved to the Royal Botanic Gardens, Kew. Arriving in Sydney (1852) he tried his luck in the goldfields, and later (1855) was botanist to an expedition to North Queensland during which most of the party were killed by indiginous people. Following this he became the Curator of the Brisbane gardens until retirement (1881). He was also made Colonial Botanist of Queensland (1859–1881), making collecting expeditions (Cape York in 1862 and Daintree in 1873). He was respponsible for introducing exotic fruit species such as mango, pawpaw, tamarind, etc. and first cultivated the native nut we know as the Macadamia. Castelnau's test refers to: "*…Mr. Hill, the able Director of the Brisbane Botanical Gardens.*"

Hilomen

Hilomen's Wrasse *Halichoeres hilomeni* JE Randall & GR Allen, 2010

Dr Vincent V Hilomen of the University of the Philippines Los Baños is Project Manager of the Marine Key Biodiversity Areas Project, Philippines. He was honoured in recognition of his efforts in obtaining collection and export permits, and arranging the shipment of the type specimens to the authors.

Hinano

Hinano's Dwarfgoby *Eviota hinanoae* L Tornabene, GN Ahmadia & JT Williams, 2013

Hinano Murphy is President of Te Pu Atitia (Atitia Center) and Associate Director of Administration Outreach at US Berkeley Gump Research Station in Mo'orea, French Polynesia. She is also Cultural Director of the Tetiaroa Society, where her husband is Executive Director. Along with her husband Frank, she was instrumental in facilitating research in Mo'orea, thus resulting in the discovery of this goby. The etymology says: "*Named for Hinano Murphy; where Hinano, a popular name among Tahitian women, is also the vernacular Tahitian name for the male flower of the Pandanus plant as well as the name of a popular beer in Tahiti.*"

Hinde

Garra sp. *Garra hindii* GA Boulenger, 1905

Dr Sidney Langford Hinde (1863–1930) was Medical Officer of the Interior in British East Africa and a Captain in the Congo Free State Forces, as well as a naturalist, ethnographer and collector. He was also a Provincial Commissioner in Kenya and collected there. He wrote: *The Fall of The Congo Arabs* (1897) and co-wrote: *The Last of the Masai* (1901). Eight mammals, two birds and a reptile are also named after him. He collected the holotype of the eponymous species.

Hinds

Hinds' Dragonet *Callionymus hindsii* J Richardson, 1844

Dr Richard Brinsley Hinds (1811–1846) was a British naval surgeon, botanist and malacologist who sailed on the (1835–1842) voyage of HMS 'Sulphur' to explore the Pacific. He studied at St Barts, gaining his honours degree (1830) and admission to the Royal College of Surgeons (1833). He then joined the navy (1835) as Assistant Surgeon, and later the same year was appointed surgeon to the 'Sulphur'. The voyage sailed west and circumnavigated the globe. He edited the natural history reports of that expedition and they were published (1844) as: *The Botany of the Voyage of H.M.S. Sulphur* and *The Zoology of the Voyage of H.M.S. Sulphur*. His health had suffered on the voyages and he was permitted to go to Australia (1845), probably suffering from tuberculosis. He died not long after settling there.

Hine

Quillback ssp. *Carpiodes cyprinus hinei* MB Trautman, 1956

Professor James Stewart Hine (1866–1930) was an American zoologist, ornithologist and entomologist. Ohio State University awarded his BSc (1893) and he spent his entire career there or at the University Museum. He was Assistant in Horticulture (1894), Assistant in Entomology (1895–1899), Assistant Professor of Zoology (1899–1902), Associate Professor of Entomology (1902–1925) and then worked full-time at the Museum (1927–1930). His particular area of interest was the Tabanidae (horse flies & relatives), but he was interested in all aspects of entomology and other areas of natural history. He collected on the US Gulf Coast (1903) and Central America including Guatemala (1905), particularly of water-associated insects. Other trips were to California, Arizona and Mexico (1907), Alaska (1917 & 1919) and Florida and Cuba (1923). He took part in the Katmai Expeditions (1915–1916) under the Auspices of the National Geographical Society as part of the University of Ohio. He also spent a month studying at the BMNH (1925). He was also a keen beekeeper and was President of the Ohio Beekeepers Association. He wrote a number of books and papers over several decades including: *Tabanidae of the western United States and Canada* (1904) and *Robberflies of the genus Erax* (1919). The author, Trautman, and he were friends and used to collect together. He died of a heart attack while preparing for Christmas.

Hing

Stone Loach sp. *Schistura hingi* AW Herre, 1934

Hing Ah was a collector employed in the 1930s by Dr Geoffrey Alton Craig Herklots of Hong Kong University. Herre said his "*...patience and skill enabled* (him) *to get many specimens*".

Hinsby

Girdled Goby *Nesogobius hinsbyi* AR McCulloch & JD Ogilby, 1919
[Alt. Tasmanian Orange-spotted Sandgoby]

George Hinsby of Hobart was a fellow of the Royal Society of Tasmania. The goby was originally named by R.M. Johnston (q.v.) in 1903, but not properly described – which is why McCulloch and Ogilby are credited with the 'official' description. Johnston honoured Hinsby as the latter had 'presented many specimens' to the Tasmanian Museum.

Hinton

Lanternfish genus *Hintonia* AF Fraser-Brunner, 1949

Martin Alister Campbell Hinton (1883–1961) was a zoologist at the BMNH where he worked (1910–1945), being Keeper of Zoology (1936–1945). A trunk, belonging to Hinton, was discovered (1970) at the BMNH. It was found to contain animal bones and teeth, stained and carved in a manner very like the Piltdown finds. It raised the real possibility that Hinton was involved in the long-running deception known as 'The Piltdown Man'. The author honoured him for: "*friendly help and encouragement of the most practical kind.*"

Hipoliti

Rusty Reefgoby *Priolepis hipoliti* JM Metzelaar 1922

The etymology simply says: "*Named after its native collector*", who was presumably a native of Curaçao, Dutch West Indies, named Hipólite/Hipolito

Hippolyta

Dwarf Cichlid sp. *Apistogramma hippolytae* SO Kullander, 1982

In Greek mythology, Hippolyta was queen of the Amazons. The cichlid occurs in the Amazon basin, but apparently the name has no special significance.

Hirai

Flyingfish sp. *Cypselurus hiraii* T Abe, 1953

Masaji Hirai of the Manazuru Branch of Tokaiku Suisan Kenkujo (Kanakawa Prefecture, Japan) had been collecting fishes for Abe for some years. He collected a 'fine specimen of the young' of this species (1952).

Hiramatsu

Goby sp. *Vanderhorstia hiramatsui* A Iwata, K Shibukawa & N Ohnishi, 2007

Wataru Hiramatsu was working at the Laboratory of Marine Biology, Kochi University when he collected the eponymous goby. Among his publications is: *First record of the blenniid fish Laiphognathus multimaculatus from Japan* (1990).

Hirundo

Swallow Grenadier *Sphagemacrurus hirundo* R Collett, 1896

Hirundo is Latin for swallow. The reference is to 'L'Hirondelle', Prince Albert of Monaco's yacht, from which the holotype was collected.

Hittites

Anatolian Gudgeon *Gobio hettitorum* W Ladiges, 1960

Hittitic Spring Minnow *Pseudophoxinus hittitorum* J Freyhof & M Özuluğ, 2010

This is not a true eponym in that the fish are not named after a particular person, but a people. The Hittites were an ancient people and civilisation in Anatolia (in present day Turkey.)

Hjort

Barethroat Slickhead *Asquamiceps hjorti* E Koefoed, 1927

Gulper Eel sp. *Saccopharynx hjorti* L Bertin, 1938

Dr Johan Hjort (1869–1948) was a Norwegian oceanographer, fisheries scientist and marine zoologist and one of the leading men of his time in his field. After commencing studying at the University of Christiana (now Oslo) to become a physician he switched to zoology and studied at the University of Munich, which awarded his doctorate (1892) after he had worked at the Stazione Zoologica, Naples on embryology. He was Curator of the University of Christiana's Zoological Museum (1892–1896). He spent a year (1896–1897) at the University of Jena and then was appointed to the University of Christiana's biological station at Drøbak. He was Director of the Norwegian Institute of Marine Research, Bergen (1900–1916). After disagreement with Norwegian politicians, he resigned (1916) and lived in Denmark and England (Cambridge University) for some years, but returned to the University of Oslo as a Professor (1921). He was one of the founders of the International Council for the Exploration of the Sea (ICES) and Norway's delegate to it (1902–1938) and its President (1938–1948). He was co-leader with Sir John Murray of the Michael Sars Expedition (1910) during which the holotype was collected. The two men co-wrote: *The Depth of the Ocean* (1912), a book that Bertin cited.

Hnilicka

Upper Grijalva Livebearer *Poeciliopsis hnilickai* MK Meyer & D Vogel, 1981

Erich Hnilicka of Puebla is a Mexican aquarist, collector and amateur limnologist who has acted as interpreter to other visiting aquarists and ichthyologists. He collected the holotype of *Xiphophorus meyeri* (named after the senior author) in Coahuila State, Mexico.

Ho, H-C

Worm-Eel sp. *Pylorobranchus hoi* JE McCosker, K Loh, J Lin & H-M Chen, 2012
Ho's Sandperch *Parapercis hoi* JW Johnson & H Motomura, 2017

Dr Hsuan-Ching Ho (b.1978) is a Taiwanese ichthyologist who is Associate Professor at the Institute of Marine Biology at the National Dong Hwa University, Taiwan (2014–present). His BSc was awarded by the National Taiwan Ocean University (2000) which also awarded his MSc (2002) and PhD (2010). He was a Research Assistant, Biodiversity Research Center, Academia Sinica (2000–2010) then Postdoctoral Research Assistant there. He was Assistant Researcher at the National Museum of Marine Biology & Aquarium (2010) before taking his current post. He has already published over 150 papers such as: *Systematics and biodiversity of eels (orders Anguilliformes and Saccopharyngiformes) of Taiwan* (2015) and *Annotated checklist and type catalog of fish genera and species described from Taiwan* (2011). His research foci are fish taxonomy, nomenclature of fishes, the history of fish studies, fish biodiversity of Taiwan. He has described over 80 species.

Ho (Ting)

Bagrid Catfish sp. *Tachysurus hoi* J Pellegrin & PW Fang, 1940

Ho Ting Chieh was a Chinese ichthyologist and geneticist at Wuhan University. He collected the holotype and presented it, along with other Chinese fishes, to MNHN, Paris.

Hoa

Loach sp. *Schistura hoai* VH Nguyen, 2005

This is not an eponym, but derived from *hoa*, the Vietnamese word for flower, referring to the fish's body pattern.

Hoang

Rattail sp. *Coelorinchus hoangi* TN Iwamoto & KJ Graham, 2008

Dr Tuan Hoang, M.D., is a friend and patron of the senior author and was honoured for his "*long and enthusiastic support*" of ichthyology at the California Academy of Sciences, where Iwamoto is Curator of Ichthyology, Emeritus.

Hobart

Palepore Snailfish *Paraliparis hobarti* DL Stein, NV Chernova & AP Andriashev, 2001

This is a toponym referring toTasmania's capital city, Hobart.

Hobelman

Kottelat's Rasbora *Rasbora hobelmani* M Kottelat, 1984

Paul Hobelman is a teacher of English. He went to Thailand (1969) as a member of the Peace Corps, signed up for two years. He and his wife, Nuanphan, also a teacher of English, moved around a great deal, being in Honolulu at the University of Hawaii (1973–1975) where he received a master's degree in language teaching. They were then in Wichita, Kansas (1975–1977), Taif, Saudi Arabia (1977–1978), Kuwait University (1978–1980), Chiang Mai University, Thailand (1980–1986 & 1988–1993), with a year (1987–1988) teaching in Kanazawa, Japan. They returned to Wichita and taught (1993–2010) until ill health forced his retirement. Kottelat thanked him for his "*…hospitality, friendship, help and enduring [Kottelat] for several weeks of fieldwork*" in Thailand.

Hodgart

Indian Spaghetti-eel *Monopterus hodgarti* BL Chaudhuri, 1913

Torrent Catfish *Parachiloglanis hodgarti* SL Hora, 1923

Richard Arthur Hodgart (b.1883) was employed by the Indian Museum, Calcutta, as a Zoological Collector and during the Abor Campaign (1911–1912) was attached to the army as an assistant geologist and anthropologist. Later he was a Zoological Collector, Zoological Survey of India. In WW1, the Zoological Survey of India supplied the manpower for the Anglo-Indian Battery and Hogart, who had enlisted in the Indian Army (1917), was on the strength as a bombardier, stationed at Baghdad (1918), where he collected freshwater sponges for Annandale (q.v.). He surveyed caves in the Meghalaya district of northern India (1922). There is nothing in the catfish's description to identify the person being honoured, but R.A. Hodgart is the obvious candidate.

Hoedeman

Armoured Catfish sp. *Exastilithoxus hoedemani* IJH Isbrücker & H Nijssen, 1985

Hoedeman's Hypopygus (knifefish) *Hypopygus hoedemani* CDCM de Santana & WGR Crampton, 2011

Jacobus Johannes Hoedeman (1917–1982) was a Dutch ichthyologist and aquarist. He wrote: *Rivulid Fishes of the Antilles* (1958), co-wrote: *Naturalists' Guide to Fresh-Water Aquarium Fish* (1974) and at least 34 papers such as: *Studies on cyprinodontiform fishes. 11. A new species of the genus Rivulus from Ecuador with additional records of Rivulus from the upper Amazon and Ucayali rivers* (1962).

Hoedt

Hoedt's Waryfish *Scopelosaurus hoedti* P Bleeker, 1860

Dirk Samuel Hoedt (1813–1893) was a Dutch civil servant and naturalist, being Secretary for the Moluccas (1854–1855), where he also collected (1853–1867). Three birds are named after him.

Hoefler

see under **Höfler**

Hoehne

Armoured Catfish sp. *Rineloricaria hoehnei* A de Miranda Ribeiro, 1912
Three-barbeled Catfish sp. *Phenacorhamdia hoehnei* A de Miranda Ribeiro, 1914
Amazonian Cichlid sp. *Rondonacara hoehnei* A de Miranda Ribeiro, 1918

Frederico Carlos Hoehne (1882–1959) was a self-taught Brazilian botanist and conservationist. He finished his formal education (1899) without the possibility of a university education. He was accomplished as a collector and made money by selling orchids. Despite a lack of scientific training, he was appointed (1907) head gardener at the Scientific Institution, Rio de Janeiro, and (1908) went on the first of many field trips for this and other organisations. He was appointed Botanist (1913) of the (Theodore) Roosevelt-Rondon Scientific Expedition to the Mato Grosso region (1913–1914) during which he collected some of the fish types. He worked (1917–1952) for São Paulo State Botanical Institute, becoming Director. The botanical journal *Hoehnea* was named in his honour, as is a frog.

Ho Enot

Vietnamese Barb sp. *Spinibarbus hoenoti* VH Nguyen, VT Do, THT Nguyen & TDP Nguyen, 2015

Ho Enot is a member of the Van Kieu (Bru) ethnic group and leader of Cu Pua commune of Dakrong commune, Dakrong district, Quang Tri province, Viêt Nam. He was honoured because he collected fishes for the Highland Aquatic Resources Conservation and Sustainable Development project.

Hoese

Hoese's Blenny *Starksia hoesei* RH Rosenblatt & LR Taylor, 1971
Hoese's Shore-Eel *Alabes hoesei* VG Springer & TH Fraser, 1976
Hoese's Goby *Glossogobius hoesei* GR Allen & M Boeseman, 1982
Hoese's Sandgoby *Istigobius hoesei* EO Murdy & JD McEachran, 1982

Forktail Pygmygoby *Trimma hoesei* R Winterbottom, 1984
Goby sp. *Silhouettea hoesei* HK Larson & PJ Miller, 1986
Stream Goby sp. *Stenogobius hoesei* RE Watson, 1991
Sleeper sp. *Allomogurnda hoesei* GR Allen, 2003
Hoese's Dwarf-Goby *Eviota hoesei* AC Gill & SL Jewett, 2004
Hoese's Mudbrotula *Dermatopsis hoesei* PR Møller & W Schwarzhans, 2006
Goby sp. *Schismatogobius hoesei* P Keith, C Lord & HK Larson, 2017
Goby sp. *Koumansetta hoesei* M Kovačić, SV Bogorodsky, AO Mal & TJ Alpermann, 2018
Bandtail Snubnose Goby *Pseudogobius hoesei* HK Larson & MP Hammer, 2021

Dr Douglass Fielding 'Doug' Hoese (b.1942) is Senior Fellow, Ichthyology Collection, Australian Museum, Sydney. The University of Texas awarded his bachelor's degree (1964) and the University of California, San Diego, his doctorate (1971). He was a research assistant at the Scripps Institution of Oceanography (1965–1970) and then emigrated to Australia where he was Assistant Curator of Fishes, Australian Museum (1971–1975) and Curator (1976–1981). After a re-organisation of the museum he was Head of Marine Group (1981–1982), Scientific Officer (1982–1988), Chair of the Vertebrate Zoology Division (1983–1987), Senior Research Scientist (1988–1998), Head of the Division of Vertebrate Zoology and Scientific Services (1989–1998), Chief Scientist (1991–1999) and Head of Science (1999–2004). He retired (2004) and is now a Research Fellow. He specialises in gobioid fishes and co-wrote: *Description Of Three New Species Of* Glossogobius *From Australia And New Guinea* (2009). In his etymology, Winterbottom said that Hoese had "*...generously shared his extensive knowledge of gobioid fishes with Winterbottom during field trips, museum visits, and by correspondence, and whose input has frequently saved Winterbottom from making glaring errors.*"

Hoetmer

Neotropical Rivuline sp. *Anablepsoides hoetmeri* DTB Nielsen, AC Baptista Jr & L van de Berg, 2016

Jan Willem Hoetmer is a Dutch biologist and environmentalist who discovered this species in Brazil. He is also an aquarium hobbyist who breeds killifish and who is President of the Dutch Killifish Association (KFN). Among his published papers is: *The Genus Cynolebias* (1980).

Hoeven

Bagrid Catfish sp. *Hemibagrus hoevenii* P Bleeker, 1846
Banded Mulletgoby *Hemigobius hoevenii* P Bleeker, 1851
Hoeven's Carp *Leptobarbus hoevenii* P Bleeker, 1851
Hoeven's Snake-Eel *Pisodonophis hoeveni* P Bleeker, 1853
Frost-fin Cardinalfish *Ostorhinchus hoevenii* P Bleeker, 1854

Dr Jan van der Hoeven (1801–1868) was a Dutch zoologist and physician who took degrees in physics (1822) and medicine (1824) at Leiden University, then practised as a physician (1824–1826). He became Professor of Zoology and Mineralogy at Leiden University (1826–1868) and, a firm traditionalist, was one of the last professors at Leiden to teach in Latin. A bird is also named after him.

Hofer

Lake Chiemsee Whitefish *Coregonus hoferi* LS Berg, 1932

Dr Bruno Hofer (1861–1916) was a German fisheries scientist who is credited with founding the study of fish pathology. He studied at the University of Königsberg, East Prussia (Kaliningrad, Russia), and then at the University of Munich where he was awarded his doctorate (1887). He was an Assistant at the Zoological Institute of Munich (1887) advancing to University Lecturer (1891). He became a lecturer in ichthyology at the Veterinary University of Munich (1896) and became Associate Professor (1898) for zoology and ichthyology and Full Professor (1904). He wrote: *Die Verbreitung der Thierwelt im Bodensee* (1896).

Hoffmann, P & M

Pencilfish sp. *Derhamia hoffmannorum* J Géry & A Zarske, 2002

Peter Hoffmann of Salzgitter, Germany and Dr Martin Hoffmann of Hanover, Germany, collected the type specimens and, according to the etymology, it was they: " *...who managed to collect and acclimate the species and gave us precious information about its biology.*"

Hoffmann, WE

Stone Loach sp. '*Lefua*' *hoffmanni* AW Herre, 1932
[Uncertain which genus this species really belongs in]
Chinese Cyprinid sp. *Toxabramis hoffmanni* SY Lin, 1934

Dr William Edwin Hoffmann (1896–1986) was an American entomologist and biologist who worked on the staff of both the University of Kansas and the University of Minnesota and received degrees from both of them. He took part in the Minnesota Pacific Expedition (1924), mostly studying marine life. He held the chair of biology at Lingnan University, Canton (Guangzhou), China (1924–1951) and was Curator of the Lingnan Museum and founded (1932) the Natural History Survey and Museum. During his time there he made many collecting expeditions into the province, and later to the Philippines (1927, 1940 & 1941) and Taiwan. He twice took leaves of absence to do research in England and Europe, and travelled extensively visiting 75 countries. He and his wife were caught up in the Sino-Japanese war (1937–1945), living under Japanese supervision until the USA went to war against Japan (1941). He was an internee in Canton (1941–1943) until exchanged and repatriated. His wife was in the Philippines when Pearl Harbor was bombed and she was imprisoned until the liberation of Manila (1945). He was Associate Curator of Insects at the Smithsonian (1944–1947) then returned to China for four years (1947–1951), thereafter working at the University of Kansas until retirement (1962). He wrote about a variety of insects including silkworms and crop pests such as: *An Abridged Catalogue of Certain Scutelleroidea (Plataspidae, Scutelleridae and Pentatomidae) of China, Chosen, Indo-China and Taiwan* (1935) and a paper for the Entomological Society of Washington *Insects as Human Food* (1948). He was editor of the Lingnan University Science Journal. A naturalist at heart, he collected and observed insects, fish, birds, reptiles and plants. Herre remarked that Hoffmann accompanied him on some of his field trips in China and also collected some of the type series. Other taxa are named after him, particularly beetles and a dragonfly.

Höfler

African Sergeant *Abudefduf hoefleri* F Steindachner, 1881
Guinean Parrotfish *Scarus hoefleri* F Steindachner, 1881
Höfler's Butterflyfish *Chaetodon hoefleri* F Steindachner, 1882
[Alt. Four-banded Butterflyfish]

W Höfler was a friend of Steindachner who he describes as his 'dear friend' and his best provider of fish specimens from Africa. He collected the types. We have failed to find further details.

Hofmann

Hofmann's Killi *Aphyosemion hofmanni* AC Radda, 1980

Otto Hofmann is an Austrian aquarist. He discovered the species while collecting with Eduard Pürzl in southern Gabon. He also had collected earlier in Cameroon (1977) and later did so with Radda in DR Congo (1982).

Hofrichter

Hofrichter's Clingfish *Gouania hofrichteri* M Wagner, M Kovačić & S Koblmüller, 2020

Robert Hofrichter (b.1957) is an Austrian zoologist, biologist, environmentalist, author, journalist, tour guide and nature photographer. Born in Bratislava (Slovakia), he now lives in Salzburg (Austria). He has written a number of books including an international bestseller on identifying mushrooms and toadstools *The Secret Lives of Mushrooms* (2019) and *The mysterious world of the seas: a journey into the realm of the deep* (2019). He has travelled the seas for three decades and is President of the sea protection organisation MareMundi. His work on European clingfish is credited with sparking the authors' interest in these fishes.

Hogaboom

Central American Cichlid sp. *Amphilophus hogaboomorum* AF Carr & L Giovannoli, 1950

George Beverly Hogaboom (1929–1950) and Peter Hogaboom (1931–1992) were brothers who both helped collect the type specimen. They were born in Honduras to American (USA) parents. George was a US Air Force pilot and died when his plane crashed in British Columbia. The etymology says: "*The species is named in honor of George and Peter Hogaboom, our companions on many profitable collecting trips in Honduras.*"

Hogan

Sole sp. *Synclidopus hogani* JW Johnson & JE Randall, 2008

Alfred Ernest 'Alf' Hogan (b.1949) is an Australian Fisheries Biologist. The University of Queensland awarded his BSc (1976). He worked as Senior Fish Biologist for the Queensland Department of Primary Industries and Fisheries Research Station (1980–2008) and ran (2008–2015) his own consultancy as Principal Ecologist at Alf Hogan and Associates Fish Ecologists until retiring. He was a research student at University of Queensland Veterninary Department (1968–1976), interrupted by National Service (1971–1972). He later spent several years (1977–1980) as a finance clerk in the Department of Defence. He has led and worked on ecology-based projects for mining companies, the sugar industry, water resource managers, natural resource management groups, interstate and local governments and community groups. He has written articles and scientific papers and numerous reports, TV show segments and leaflets as well as the book: *Field Guide to the Fishes of Southern Gulf of Carpentaria Catchments in Queensland* (2010). He has surveyed most of the freshwater reaches of coastal rivers and wetlands in Queensland. The sole was named after him in recognition of his contribution to the knowledge of fish biodiversity and ecology in Queensland.

Hohenacker

North Caucasian Bleak *Alburnus hohenackeri* KF Kessler, 1877

Dr Rudolph Friedrich Hohenacker (1798–1874) was a Swiss physician, botanist and missionary. He went to a Swabian colony, Kirovabad (Gäncä, Azerbaijan), in South Caucasus (1821). He returned to Switzerland (1841) and then lived in Germany (1842–1874). He collected part of the type series of this fish. A reptile is also named after him.

Hoigne

Dwarf Cichlid sp. *Apistogramma hoignei* H Meinken, 1965

Neotropical Rivuline sp. *Pterolebias hoignei* JE Thomerson, 1974

Emil 'Leo' Hoigne (d.1996) was an Argentinian aquarist and collector who relocated to Valencia, Venezuela. He was the collector of both species, which were named after him because of that and for his contribution to the knowledge of Neotropical Annual Cyprinidontiform fishes – as the etymology put it: "*...It was our privilege to know him and share his delight in discovering and keeping annual killifishes.*" He pointed out the distinctive features of the new species in an article published by the American Killifish Assiciation (1969). (Also see **Leo Hoigne**)

Ho Khanh

Hokhanh's Blind Cavefish *Speolabeo hokhanhi* N Dinh Tao, L Cao, S Deng & E Zhang, 2018

Ho Khanh is a speleologist, local guide and homestay proprietor in Phong Nha-Ke Bang National Park, Vietnam. He discovered the cave called Son Doong (1990), the world's largest cave. "*The specific epithet is named in honor of Mr. Ho Khanh who discovered many caves in Phong Nha-Ke Bang National Park. He was a local guide of the cavefish survey conducted by the first author during 2014 into the cave where the type specimens were collected and provided detailed information about the collection site.*"

Holboell

Krøyer's Deep-Sea Anglerfish *Ceratias holboelli* HN Krøyer, 1845

[Alt. Longray/Northern Seadevil]

Carl Peter Holboell (1795–1856) was a captain in the Danish Royal Navy who served in Greenland and became interested in natural history there. At least one Greenland plant is named after him. Holboell was presumed to have died at sea when the ship on which he was sailing to Greenland from Denmark disappeared without trace. Two birds are also named after him.

Holbrook

Eastern Mosquitofish *Gambusia holbrooki* CF Girard, 1859
Band Cusk-eel *Ophidion holbrookii* FW Putnam, 1874
Spot-tail Seabream *Diplodus holbrookii* TH Bean, 1878

Dr John Edwards Holbrook (1794–1871) was a zoologist and herpetologist who has been described as 'the Father of North American Herpetology'. He qualified as a physician (1818) and went to Edinburgh for postgraduate studies (1819). He visited Paris and became friendly with the great French naturalists of the day, including Cuvier and Duméril. He returned to the USA (1822) and practiced medicine in Charleston, South Carolina. He was Professor of Anatomy, Medical School of South Carolina (1824–1854). During the American Civil War, he was a surgeon in the Confederate army, despite his age. After Union troops captured Charleston and his manuscripts and collections were looted, he gave up all scientific research. He was elected to the National Academy of Sciences (1868). He wrote: *North American Herpetology* (1836–1842). Two amphibians and three reptiles are named after him.

Holčík

Riffle Minnow sp. *Alburnoides holciki* BW Coad & NG Bogutskaya, 2012

Dr Juraj Holčík (1934–2010) was a Czechoslovak (later Slovak) biologist and ichthyologist. He graduated from Comenius University, Bratislava with a bachelor's degree (1958) and then worked in the Provincial Museum, Trnava (1958–1960) and the Fisheries Laboratory, Bratislava (1960–1962). He studied zoology at Charles University, Prague, which awarded his doctorate (1966). He lived and worked in Slovakia (1996–2004) including being the first Director of the Institute of Zoology, Slovak Academy of Sciences (1990–2004). After the Czechoslovak Velvet Revolution (1989) he was a non-party Member of Parliament (1990–1992).

Holcom

Holcom's Reef Sole *Aseraggodes holcomi* JE Randall, 2002

Ronald R Holcom lives in Honolulu, Hawaii where the sole was collected. He dives the reefs and takes underwater photographs of marine species. He co-wrote, with Randall: *Antennatus linearis, a New Indo-Pacific Species of Frogfish (Lophiiformes: Antennariidae)* (2001). According to the etymology, he: "*…collected four of the six type specimens and was the first to recognize this species as different from the other Hawaiian species of Aseraggodes*."

Holder

Island Kelpfish *Alloclinus holderi* F Lauderbach, 1907

Dr Charles Frederick Holder (1851–1915) of Pasadena, California was a naturalist, conservationist, writer and pioneer of big-game fishing. From a wealthy family, he graduated from the United States Naval Academy at Annapolis but did not pursue a naval career. Instead he worked as a curator at AMNH before moving to California (1885) where he was a businessman and philanthropist. He wrote forty books, from *Elements of Zoology* (1885) to *Salt Water Game Fishing* (1914). as well as thousands of articles. He was also interested in archeology and excavated Mayan artefacts in Mexico (1910). He died in a car accident.

Holdridge

Central American Livebearer sp. *Brachyrhaphis holdridgei* WA Bussing, 1967

Dr Leslie Rensselaer Holdridge (1907–1999) was an American botanist and ecologist with a particular interest in forests and forestry. His bachelor's degree in forestry (1931) was awarded by the University of Maine. The University of Michigan, which had awarded his master's, also awarded his doctorate in botany (1947). He worked in Puerto Rico (1935–1941), firstly for the Caribbean National Forest and subsequently as a forester for the US Forestry Service. He divided his time (1941–1949) between working in Haiti and in cinchona research in Guatemala and Colombia. He worked (1949–1960) as forester and ecologist for the Inter-American Institute of Agricultural Sciences, Turrialba, Costa Rica, and founded (1954) La Selva research station. He was a senior staff member of the Associate Colleges of the Midwest's Central American field programme (1963–1967) and was engaged (1967) as an ecologist by the Tropical Science Center, San José, Costa Rica which is mentioned in Bussing's etymology thus: "*The species is named in honor of Leslie R. Holdridge, President, Tropical Science Center, San José, Costa Rica, Dr. Holdridge made possible my study of the fishes of the Río Puerto Viejo, as well as the studies of many students of tropical biology, through his outstanding generosity and friendly advise.*" He was a part-time Professor of Dendrology and Ecology (1978–1982) at the Instituto Tecnológico de Costa Rica. He wrote *Forest Environments in Tropical Life Zones: A Pilot Study* (1971).

Holland

Characin genus *Hollandichthys* CH Eigenmann, 1910
Yellow-spotted Skate *Okamejei hollandi* DS Jordan & RE Richardson, 1909
Three-barbeled Catfish sp. *Imparfinis hollandi* JD Haseman, 1911
Holland's Piranha *Serrasalmus hollandi* CH Eigenmann, 1915
Neotropical Livebearer sp. *Pamphorichthys hollandi* AW Henn, 1916
Taiwanese Cyprinid sp. *Spinibarbus hollandi* M Oshima, 1919
Eelpout sp. *Bothrocara hollandi* DS Jordan & CL Hubbs, 1925

Dr William Jacob Holland (1848–1932) was a Jamaican-born American Presbyterian minister, entomologist and palaeontologist. His primary interest was in Lepidoptera and he wrote: *The Butterfly Book* (1898) and *The Moth Book* (1903). He attended Moravian College and Theological Seminary and Amhurst College where he was awarded his BA (1869) and later (1874) Princeton Theological Seminary, leaving to become pastor of a church in Pittsburgh. He was naturalist for the US Eclipse Expedition (1887), which explored Japan. He was Chancellor of the University of Pittsburgh (1891–1901) where he taught anatomy and zoology, and Director of the Carnegie Museum until retirement (1901–1922) having been hired by his friend Andrew Carnegie. He also donated his own collection of 250,000 specimens to the museum. He appears to have been a difficult man to work with. Given to tantrums, he was somewhat of a sycophant towards his betters (including Carnegie), and seemingly condescending when dealing with employees. His trip to Argentina (1912) to install a replica of a *Diplodocus*, at the behest of Carnegie, is told by him in his travel book: *To the River Plate and Back* (1913). A bird and a dragonfly are also named after him.

Hollard

Spikefish genus *Hollardia* F Poey, 1861
Reticulate Spikefish *Hollardia hollardi* F Poey, 1861

Dr Henri Louis Gabriel Marc Hollard (1801–1866) was a French physician and zoologist. He qualified as a physican in Paris (1824) and as a Doctor of Science (1835). He became Professor of Physical and Natural Sciences in Lausanne and Neuchâtel, Switzerland. He was Professor of Zoology and Anatomy at Montpellier. Among other works he wrote: *De l'homme et des races humaines* (1853).

Holleman

Holleman's Triplefin *Enneapterygius hollemani* JE Randall, 1995
Madagascar Sweeper *Pempheris hollemani* JE Randall & BC Victor, 2015
Holleman's Frogmouth *Chaunax hollemani* HC Ho & WC Ma, 2016
Holleman's Pygmygoby *Trimma hollemani* R Winterbottom, 2016

Wouter Holleman is a Research Associate at the South African Institute for Aquatic Biodiversity. His research foci are Taxonomy of Indo-Pacific Tripterygiidae and the Systematics of Southern African and other Indo-Pacific Clinidae. Among his published papers are: *Three new species and a new genus of tripterygiid fishes (Blennioidei) from the Indo-West Pacific Ocean* (1982) and the co-written: *Contrasting Signals of Genetic Diversity and Historical Demography between Two Recently Diverged Marine and Estuarine Fish Species* (2015). Winterbottom described him as a "*…friend, diving partner, collector extraordinaire, and processor of fishes both small and large on my expeditions for too many decades to detail. He is also a world expert on the systematics of tripterygiid and clinid fishes.*"

Hollis, J

Damselfish sp. *Chrysiptera hollisi* HW Fowler, 1946
[Junior Synonym of *Chrysiptera glauca*]

James J Hollis was a ship's carpenter first class. Under Captain Ernest Tinkham (q.v.) he was in charge of 'native' divers on a collecting expedition off the Ryukyu Islands, Japan (1945). He also helped dive the reef and collect fish from rock pools, amassing a collection of 383 fish of 146 species or subspecies, which the Captain sent to Fowler.

Hollis, R

Viviparous Brotula sp. *Thermichthys hollisi* DM Cohen, RH Rosenblatt & HG Moser, 1990

Ralph Hollis (1932–2013) served in the US Air Force for 20 years. His second career was at the Woods Hole Oceanographic Institute as an electronic technician for the Alvin Group (1975–1989), for whom he trained as the pilot of the submersible craft and became the group's chief pilot. He discovered tube-worms at depths of 9,000 feet near the Galapagos Islands and was the first person to re-discover and see the 'Titanic' (1986).

Hollister

Bermuda Beardfish *Polymixia hollisterae* TC Grande & MVH Wilson, 2021

Gloria Elaine Anable née Hollister (1900–1988) was an American ichthyologist, Red Cross Blood Bank pioneer, and ground-breaking conservationist. Connecticut College for Women awarded her Zoology degree (1924) and Columbia University her MSc (1925). During her time at university she was a keen and talented sportswoman. She was a Research Associate (1928–1941) in the Department of Tropical Research of the New York Zoological Society (now the Wildlife Conservation Society) specialising in fish osteology. She also made record-setting dives in a submersible off the coast of Bermuda (1930s), as a key member of William Beebe's (q.v.) bathysphere expeditions. She spent a number of years (1952–1964) campaigning to create the Mianus River Gorge preserve in New York State.

Holly, M

Holly's Pupfish *Aphanius ginaonis* M Holly, 1929

African Characin sp. *Nannocharax hollyi* HW Fowler, 1936

Dr Maximilian Holly (1901-1969) was an ichthyologist who was Curator of Fishes at National Natural History Museum, Vienna. He wrote a number of papers (1920s & 1930s) such as: *Beiträge zur Kenntnis der Fischfauna Persiens* (1929) and was principle editor of *Pisces 3: Crossopterygii* (1933). He was drafted into military service (1940). (Also see **Maria** & **Sperat**)

Holly Arnold – see **Dalerocheila**

Holm

Characin sp. *Creagrutus holmi* RP Vari & AS Harold, 2001

Erling Holm (b.1950) is a Canadian ichthyologist whose family emigrated from Denmark (1958). The University of Toronto awarded his bachelor's degree (1973). He is Assistant Curator of Ichthyology, Department of Natural History at the Royal Ontario Museum, Canada, which he joined (1977) as a Curatorial Assistant. He is an expert on freshwater fishes, especially those of the province of Ontario.

Holmberg

Neotropical Rivuline sp. *Cynolebias holmbergi* FGC Berg, 1897

[Junior Synonym of *Austrolebias elongatus*]

Armoured Catfish sp. *Loricaria holmbergi* MS Rodríguez & AM Miquelarena, 2005

Eduardo Ladislao Holmberg (1852–1937) was an Argentine zoologist, biologist and novelist. He was Director of the zoological garden, Buenos Aires, and is remembered as the first Argentine-born scientist to publish on fishes. He also published on molluscs but his other claim to fame must be as one of the earliest Latin American writers of science fiction: *Viaje maravilloso del señor Nic-Nac al planeta Marte* appeared (1875).

Holmia

Neotropical Rivuline sp. *Anablepsoides holmiae* CH Eigenmann, 1909

This is a toponym. During Eigenmann's expedition to Guyana he discovered the species (1908) near the village of Holmia, located where the Chenapowu River joins the Potaro River above the Kayeteur Falls.

Holmwood

Izmir Nase *Chondrostoma holmwoodii* GA Boulenger, 1896

Frederic Holmwood (1840–1896) was gazetted as an Ensign in the Kent Rifle Volunteers (1861) and promoted to Captain (1866). He became a consular official and went to Zanzibar (1873) as attaché on the staff of Sir John Kirk. He became Consul (1883) and after Kirk's return to England he was Consul-General (1881–1888). He was then British Consul-General at Smyrna (Izmir), Turkey, (1888–1896). He visited the British Consulate in Singapore (1892). He was a keen explorer and naturalist who was a Fellow of the Zoological Society of London. He explored in Kenya from Mombasa inland to Taveta and collected botanical specimens for Kew (1883). He collected the holotype of this species in Turkey. He wrote: *On the Employment of the Remora by Native Fishermen on the East Coast of Africa* (1884). He died from malaria whilst on leave in England.

Holsworth

Holsworth's Grunter *Syncomistes holsworthi* JJ Shelley, A Delaval & MC le Feuvre, 2017

William Norton 'Bill' Holsworth is an Australian mammologist and philanthropist. He and his wife, Carol, founded the Holsworth Wildlife Research Endowment (1989) which financed the expedition on which this species was discovered.

Holt

Small Lanternfish *Diaphus holti* AV Tåning, 1918

Ernest William Lyons Holt (1864–1922) was a British marine naturalist, ichthyologist and soldier who served in the Nile Campaign (1884–1885) and the Third Burmese War (1886–1887) during which he was invalided home. He studied zoology at St Andrew's University (1888). He worked for the Marine Biological Association in Grimsby (1892–1894) and at the Plymouth Marine Laboratory (1895–1898). He worked in Ireland (1899–1914) for the Department of Agriculture and Technical Instruction for Ireland as scientific adviser and fisheries inspector (1908) and Chief Inspector (1914).

Holub

Cape Stumpnose *Rhabdosargus holubi* F Steindachner, 1881

Dr Emil Holub (1847–1902) was a Bohemian (Czech) naturalist who also studied South African fossils. Like his father, he trained as a physician but was always fascinated by wildlife and foreign lands; his compelling ambition was to follow in the footsteps of David Livingstone. He was a physician, zoologist, botanist, hunter, taxidermist, artist and cartographer, an avid collector of specimens and, above all, a keen observer. He practised medicine to pay his way on his first trip to Africa (1872). He travelled extensively in south-central Africa, gathering varied and valuable natural history material, including c.30,000 specimens! On his return from his first trip he wrote *Seven Years in Africa*. He took another trip (1883) to Africa, which ended in disaster after ten weeks when a number of the party died from malaria and all the equipment was lost. When he returned to Europe he fell upon hard times and was forced to sell much of his collection. He, too, eventually died from malaria, contracted on his second trip. Steindachner refers only to 'Dr. Holub' but we are certain this is who he meant. He also has two birds named after him.

Home

Silver Pearlfish *Encheliophis homei* J Richardson, 1846

Sir Everard Home (1756–1832) was an English physician and naturalist, and an early wombat owner (in the early 19th century it was very fashionable to own exotic pets from Australia). He became Sergeant Surgeon to the King (1808) and Surgeon at Chelsea Hospital (1821). He was made a baronet (of Well Manor in the County of Southampton) in 1813. Richardson referred to Home's 'zeal' in collecting objects of Natural History.

Honckeny

Evileye Puffer *Amblyrhynchotes honckenii* ME Bloch, 1785

Gerhard August Honckeny (aka Honkeny) (1724–1805) was a German naturalist, primarily a botanist. He was also an 'Amtmann'; an official to a landed aristocrat. Bloch says in the text: "*This fish lives in East Indian and Chinese waters, and I owe it to my dear friend, Herr Oberamtmann Honkeny: although this specimen is not larger than the drawing taken from it, it is likely to be found larger*" (Translation). He is best known for the two-volume: *Synopsis Plantarum Germaniæ* (1793). A plant genus is also named after him.

Honda

Seoho Bitterling *Acheilognathus hondae* DS Jordan & CW Metz, 1913
[Syn. *Rhodeus hondae*]

Dr K Honda was Director of the Agricultural Station at Suigen (Suwon, South Korea). He "...*obtained for us a fine collection from the pond at this station.*"

Honda

Armoured Catfish sp. *Hypostomus hondae* CT Regan, 1912

This is a toponym referring to the town of Honda, Colombia, where two specimens were collected and presented to the BMNH.

Honess

Honess' Glass Perchlet *Tetracentrum honessi* LP Schultz, 1945

Captain Ralph F Honess was part of the United States Navy Reserve. He made a collection of twenty-seven species of fish in New Guinea (1944) including the type of this perchlet. He gave them to his friend Lieutenant James R Simon who, in turn, donated them to the AMNH.

Hong

Hong's Tonguefish *Symphurus hongae* M-Y Lee & TA Munroe, 2021

Yu-Syun Hong (1959–2018) was the mother of the senior author, Dr Mao-Ying Lee, a Taiwanese ichthyologist and marine biologist who works at the Marine Fisheries Division of the Fisheries Research Institute and Institute of Oceanography. She died in an automobile accident while his study of *Symphurus* diversity was in progress. The etymology notes her financial and moral support of her son's research.

Hongslo

Dwarf Cichlid sp. *Apistogramma hongsloi* SO Kullander, 1979

Thorbjörn Hongslo is an aquarist and fish disease specialist associated with The National Veterinary Institute (Uppsala, Sweden). He collected the type (and that of three congeners: *iniridae, macmasteri, viejita*), "*…entirely on his own initiative.*"

Honor

Estuarine Hardyhead *Craterocephalus honoriae* JD Ogilby, 1912

Honor Coralie Hamlyn-Harris was one of the 3 daughters of Ronald Hamlyn-Harris (1874–1953), the English entomologist who became Director of the Queensland Museum. Ogilby was Ichthyology Curator there and the two men became friends. Ogilby says the hardyhead was "*…Dedicated to Miss Honor Coralie Hamlyn-Harris.*" (Also see **Harris** and **Hamlyn-Harris**)

Hoosier

Hoosier Cavefish *Amblyopsis hoosieri* ML Niemiller, JA Prejean & P Chakrabarty, 2014

This sounds like an eponym but is in reality a sort of toponym. Hoosier is the nickname given to residents of the state of Indiana (where this cavefish occurs).

Hopkins, CD

Mormyrid sp. *Paramormyrops hopkinsi* L Taverne & DFE Thys van den Audenaerde, 1985

Dr Carl Douglas Hopkins (b.1944) is a neuroethologist with expertise on the behaviour and neurobiology of electric fishes. Bowdoin College, Maine, awarded his BA (1966) and The Rockefeller University in New York City awarded his PhD on Animal Behaviour, Neurobiology (1972). His thesis work was on patterns of electric signaling by gymnotiform fishes of Guyana. After a post-doc with T H Bullock at the University of California studying electroreceptors in wave-discharging gymnotids, he became Assistant Professor of Ecology and Behavioural Biology at the University of Minnesota (1973–1982) where he began working on mormyrids from West and Central Africa. After a sabbatical in France (1981–1982) with T Szabo, he moved to the Department of Neurobiology and Behavior at Cornell University in Ithaca, New York, where he is now Professor Emeritus. He wrote many papers on behaviour of electric fishes mainly studied in the field, and on the neurobiology of electroreception with a focus on temporal coding of electric organ discharges (EODs). He is co-editor of *Poissons d'eaux douces et saumâtres de basse guinée: Afrique central de l'ouest vols. 1 and 2* (2007) that includes his chapter on the Mormyridae. He has carried out fieldwork both in Africa and South America and is author of one new gymnotiform species description and co-author of descriptions of several new mormyrid species and one new genus. His laboratory is currently working to describe new species in the species flock of *Paramormyrops*. He collected the holotype (1975).

Hopkins, MN

Christmas Darter *Etheostoma hopkinsi* HW Fowler, 1945

Milton Newton Hopkins, Jr. (1926–2007) was a farmer in southern Georgia, USA as well as a naturalist, conservationist and author. He was interested in birds from the age of eight and became a member of the Georgia Oorinthological Society (1939) when he was 15, and soon (1942) contributed his first field note to its journal *The Oriole*. After naval service during WWII he studied zoology at the University of Georgia, being awarded his master's (1951). He was known primarily as an ornithologist, and was editor of *The Oriole* (1960–1965) publishing numerous observations of Georgia bird life and the book *The Birdlife of Ben Hill County, Georgia and Adjacent Areas* (1975). He wrote: *In One Place: The Natural History of a Georgia Farmer* (2001). He was always concerned to balance agribusiness with conservation. He collected the type in Georgia (1942). He died of an aneurism aged 80.

Hopkins, T

Crisscross Prickleback *Plagiogrammus hopkinsii* TH Bean, 1894
Squarespot Rockfish *Sebastes hopkinsi* F Cramer, 1895
Jawfish sp. *Opistognathus hopkinsi* DS Jordan & JO Snyder, 1902
Velvetnose Brotula *Petrotyx hopkinsi* E Heller & RE Snodgrass, 1903

Timothy Hopkins (1859–1936) was an American philanthropist who financed several expeditions, including the one to Japan where Jordan collected the jawfish holotype. He was one of the original trustees of Stanford University (1885). The etymology for the prickleback says it is named for "*…Mr. Timothy Hopkins, of Menlo Park, Cal., the founder of the Seaside Laboratory at Pacific Grove, Monterey Bay, in commemoration of his services in behalf of science*."

Hopp

Armoured Catfish sp. *Otocinclus hoppei* A de Miranda Ribeiro, 1939

Werner Hopp (b.1886) was a German civil engineer specialising in hydroelectric power plants, a writer, naturalist and orchid enthusiast. He spent a lot of time in South America; he travelled in the Cauca River Valley, Colombia (1920) and the Venezuelan Andes (1921). Note that the author, when coining the binomial, erroneously added an 'e' to his name. He co-wrote: *Blütenzauber der Orchideen* (1957). He collected the holotype of this species.

Hora

Indian Blind Catfish genus *Horaglanis* AGK Menon, 1950
Indian Cyprinid genus *Horalabiosa* EG Silas, 1954
[Genus often merged into *Garra*]
Indian Catfish genus *Horabagrus* KC Jayaram, 1955
Hora's Loach *Yasuhikotakia morleti* G Tirant, 1885
[Alt. Skunk Loach]
Hora Danio *Devario shanensis* SL Hora, 1928

Longtail Catfish sp. *Olyra horae* B Prashad & DD Mukerji, 1929
Sisorid Catfish sp. *Glyptothorax horai* HW Fowler, 1934
Hora's Razorbelly Minnow *Salmostoma horai* EG Silas, 1951
Stone Loach sp. *Schistura horai* AGK Menon, 1952
Indus Catfish *Mystus horai* KC Jayaram, 1954
Bengal River Catfish sp. *Erethistes horai* KS Misra, 1976
Danio sp. *Devario horai* RP Barman, 1983
Burmese Cyprinid sp. *Gymnostomus horai* PM Bănărescu, 1986

Dr Sunder Lal Hora (1896–1955) was a distinguished Indian ichthyologist and biologist who held a doctorate in science from both the Punjab University (1922) and Edinburgh University (1928), having studied for his master's at the Government College, Lahore. He joined the Zoological Survey of India (1919) and was in charge of herpetology and ichthyology (1921–1947). Mukerji and Misra assisted him there and Misra succeeded him as officer in charge of fishes when Hora was promoted. He was Director of Fisheries, Bengal (1942–1947) and then became Director of the Zoological Survey (1947). He published very widely (1920–1956) no less than 427 papers from: *The Fish of Seistan* (1920) through: *On the Malayan affinities of the freshwater fish fauna of Peninsular India, and its bearing on the probable age of the Garo-Rajmahal Gap* (1944) to: *Some observations of the trout farm and hatchery at Achhabal, Kashmir* (1956). He was giving a paper to a meeting of the Asiatic Society when he collapsed with a coronary thrombosis and died three days later.

Hore

Lake Tanganyika Cichlid sp. *Shuja horei* A Günther, 1894

Captain Edward Coode Hore (1848–1912) was a missionary, explorer, navigator and cartographer. He was appointed as a missionary and arrived in Zanzibar (1877) before getting to the Lake Tanganyika Mission (1878). He explored the area south of the lake (1880) and returned to England (1881) where he married and took his Master mariner exam. He returned to the lake (1883) taking with him (in sections) a lifeboat, the 'Morning Star', and met with the shipment of a steam vessel 'Good News' for which he arranged reconstruction. He left again for England in poor health (1888). He was appointed to visit Australia and he arrived in Melbourne (1890) but there resigned from the London Missionary Society. He became (1894) first officer of the 'SS John Williams', but resigned (1900) and settled in Tasmania for the rest of his life. He collected the cichlid type.

Hori

Lake Tanganyika Cichlid sp. *Benthochromis horii* T Takahashi, 2008
Lake Tanganyika Cichlid sp. *Petrochromis horii* T Takahashi & S Koblmüller, 2014

Professor Dr Michio Hori is a fish ecologist at Kyoto University, Japan. He has published nearly two hundred papers including: *Laterality is Universal Among Fishes but Increasingly Cryptic Among Derived Groups* (2017). He was the first to discover that a fish can obtain food easily even if it shares its living area with other fish species which have similar feeding habits but different ways of catching food, focusing on cichlid fish in Lake Tanganyika. He was the first to identify these fish as undescribed species.

Horkel

Brazilian Guitarfish *Pseudobatos horkelii* JP Müller & FGJ Henle, 1841

Dr Johann Horkel (1769–1846) was a German physician and botanist who had an example of this species, preserved in alcohol, that he provided to the authors. He studied medicine at the University of Halle (1787) where he stayed as a lecturer and associate Professor of Medicine (1804–1810). He was Professor of Plant Physiology, University of Berlin (1810–1846).

Hörmann

African Killifish sp. *Nothobranchius hoermanni* B Nagy, BR Watters & DU Bellstedt, 2020

Alwin Hörmann is, according to the etymology a well-known German aquarist and breeder of *Nothobranchius* fishes, who maintained and propagated the type specimens. "*Without his dedication and expertise, the population from the type locality would not have been available for researchers and others*". This is the second new species described from a collecting trip he made with Holger Hengstler (q.v.).

Horn

Lake Tanganyika Cichlid sp. *Bathybates hornii* F Steindachner, 1911

Adolf and Albin Horn were brothers who explored German East Africa (present day Burundi and Tanzania) collecting specimens for the Vienna Museum, where Steindachner was curator of fishes and reptiles. The patronym is not identified in the original text, but the binomial almost certainly honours one or both of the Horn brothers.

Hornaday

Hornaday's Paradise Fish *Polynemus hornadayi* GS Myers, 1936

William Temple Hornaday (1854–1937) was an American zoologist, taxidermist and conservationist. He worked as a taxidermist for Henry Ward's Natural Science Establishment in Rochester, New York, and travelled in India and Southeast Asia (1877–1878) collecting natural history specimens. He collected the holotype of this fish (1877), but it remained undescribed for nearly 60 years. Hornaday later served as first Director of the Bronx Zoo, and today is best remembered for his efforts to conserve the American Bison.

Horst

Yellowline Goby *Elacatinus horsti* JM Metzelaar, 1922

Dr Cornelis J van der Horst (1889–1951) was a Dutch zoologist. After completing his PhD (on fish brains) he studied scleractinina corals that had been collected on the Siboga and Percy Sladen Trust expeditions and wrote monographs thereon. Having only seen dead corals he was keen to see them in life, so undertook a field trip to Curaçao (1920) and collected the goby type during his stay at the old Quarantine Station there. He wrote up the voyage (1924) and said "*It is a splendid sight to look through the glass bottom of a little box on this mixed variety of creatures, but I only could fully enjoy it, when the Chief of Public Works put a diving suit at my disposal ... Words fail to describe the splendour of such a submarine garden*." He made a large collection which was later studied by many others. He became assistant director of the Central Instutute for Brain Research in Amsterdam (1923). He went to South Africa (1928) as Senior Lecturer in the zoology department of Witwatersrand University, rising (1933) to head of department and Professor of Zoology. He continued his study of marine biology, taking many field trips to the island of Inhaca, Mozambique.

Hortle

Hortle's Whipray *Pateobatis hortlei* PR Last, BM Manjaji-Matsumoto & PJ Kailola, 2006 [Syn. *Himantura hortlei*]

Kent Gregory Hortle is an Australian fisheries and environmental consultant whose bachelor's degree in zoology was awarded by Monash University, Melbourne (1979). He has worked (1980>) as a biologist or environmental scientist in Asia and Australia. He provided the first photographs and fresh specimens of the whipray, which were captured during sampling while he was Environmental Monitoring Superintendent at the Freeport mine in Papua, Indonesia (1996–2001). Since 2001 he has worked primarily on fisheries of the Mekong River basin in Laos, Cambodia, Thailand and Vietnam as an advisor to the Mekong River Mekong River Commission or as a consultant. He has published numerous papers and reports covering environment, fish and fisheries, particularly in the Mekong region.

Hose

Rasbora sp. *Rasbora hosii* GA Boulenger, 1895
Bagrid Catfish sp. *Leiocassis hosii* CT Regan, 1906
Borneo Barb sp. *Leptobarbus hosii* CT Regan, 1908

Dr Charles Hose (1863–1929) was a naturalist who lived in Sarawak and Malaysia (1884–1907). He was also a good cartographer who produced the first reliable map of Sarawak. He successfully investigated the principal cause of the disease beriberi. Hose sent huge collections of zoological, botanical and ethnographic material to many museums and institutions, including the British Museum and at least four British universities, one of which (Cambridge) awarded him an honorary DSc (1900). He was an extremely bulky man, which meant that when he went to visit the local tribes in their long-houses they had to reinforce the floors. He was still remembered (1995) for his extreme size! We are indebted to a member of his family for some reminiscences of him. He successfully put a stop to the headhunting raids among the various villages by the simple expedient of organising a boat-race *à la* the University Boat Race over a similar distance and a similar sinusoidal course (although there were twenty-two dugouts with crews of 70+) (1899). This seemed to satisfied the locals' honour. Another story concerns a journey on the Trans-Siberian Railway when the train stopped near Lake Baikal and he acquired three live Baikal Seals, which he put in the luggage rack! Not surprisingly, they did not live very long. As each expired, he skinned them on the train, to the interest and surprise of his fellow travellers. He returned to Sarawak several times after retirement, very possibly in connection with the development of the oilfields at Miri. He became an expert on the production of acetone (used in the manufacture of cordite) as he ran a factory for it at Kings Lynn (WW1). The raw materials for making acetone were maize and horse chestnuts. Among a great deal more he wrote: *Fifty Years of Romance and Research* (1927) and: *The Field Book of a Jungle Wallah*

(1929). Fort Hose in Sarawak, now a museum, was named after him. Nine mammals, five birds, three amphibians and a dragonfly are also named after him.

Hoskyn

Rattail sp. *Coryphaenoides hoskynii* AW Alcock, 1890
Deep-sea Scalyfin sp. *Bathyclupea hoskynii* AW Alcock, 1891

Commander Richard Frazer Hoskyn (1848–1892) was a British naval officer who was Commander of 'HMS Investigator' from which the holotypes were collected. He was an Admiralty Hydrographer when serving on 'HMS Myrmidon' during a survey of Queensland (1886–1887). He died of a virulent and rapid form of tuberculosis. His wife gave birth to a son, some six months after his death.

Hosokawa

Hosokawa's Coral Blenny *Atrosalarias hosokawai* T Suzuki & H Senou, 1999

Masatomi Hosokawa is a Japanese amateur ichthyologist. He first provided the authors with some specimens of this species. He was co-author of: *Catalogue of the fishes of Hyogo prefecture, based on the specimens collected by Toshiyuki Suzuki* (2000) and *The northernmost and the first Japanese record of a bleniid fish Parablennius thysanius (Perciformes: Blenniidae), collected from Iriomote-jima Island, the Ryukyu Islands, Japan* (2010).

Hosoya

Eelpout sp. *Zoarchias hosoyai* S Kimura & A Sato, 2007

Seiichi Hosoya is Director of Okinawa Branch, IDEA Consultants Inc. (a company that integrates consultancy on infrastructure development and environmental conservation). He has published a number of papers such as: *Echinogobius hayashii, A new genus and species of gobiidae* (1998) and *Paedogobius kimurai, a New Genus and Species of Goby (Teleostei: Gobioidei: Gobiidae) from the West Pacific* (2001). He collected the holotype and paratypes and donated them to the first author.

Hossein Panahi

Sisorid catfish sp. *Glyptothorax hosseinpanahii* H Mousavi-Sabet, S Eagderi, S Vatandoust & J Freyhof, 2021

Hossein Panahi-Dezhkooh (1956–2004) was an Iranian actor and poet. He performed in some TV shows written by himself which were not very successful, but later became popular from his writing and acting in a show called *Two Ducks in Fog*. He died of a heart attack at the age of 47.

Houde

Stellate Codlet *Bregmaceros houdei* VP Saksena & WJ Richards, 1986

Dr Edward D Houde is a fisheries scientist and oceanographer. The University of Massachusetts awarded his bachelor's degree (1963) and Cornell University his master's (1965) and doctorate (1968). He worked for the US Bureau of Commercial Fisheries, Miami (1968–1970), for the University of Miami (1970–1980), for the National Science Foundation,

Washington D.C. (1983–1985) and was Professor at the University of Maryland Center for Environmental and Estuarine Studies, Chesapeake Biological Laboratory, Solomons, Maryland (1980–2016), where he is now Professor Emeritus. He wrote: *Fish early life dynamics and recruitment variability* (1987) The authors thanked him for providing the type specimens and reviewing the authors' manuscript. The authors added a really useful note: "*Dr Houde's name is often mispronounced, so please note that the pronounciation of houdei is hood-eye and not how-dee-eye, whodee-eye or hud-dee-eye*")

Houdemer

Oriental Cyprinid sp. *Toxabramis houdemeri* J Pellegrin, 1932

Lieutenant-Colonel Fernand Édouard Houdemer (b.1881) was a French veterinary surgeon who was very interested in the diseases and parasites that affected animals in French Indo-China (Vietnam), where he was stationed in the 1920s. He wrote: *Recherches de parasitologie comparée indochinoise* (1938). He collected the holotype of this species.

Hoult

Lizardfish sp. *Synodus houlti* AR McCulloch, 1921

Captain Hoult was master of the Queensland Government trawler 'Bar-ea-mul'. He obtained the holotype. Unfortunately his first name(s) seem to be unrecorded.

House

Sisorid Catfish sp. *Glyptothorax housei* AW Herre, 1942

E N House was the manager of the Puthutotam Tea Estate in the Anamallai Hills, southern India. He was generous with both hospitality and assistance to the author during the latter's brief visit to the area. We have not been able to find any further details about him.

Houy

Ctenopoma sp. *Ctenopoma houyi* E Ahl, 1927

Dr Reinhard Houy (1881–1913) was a German naturalist and collector of natural history specimens in Africa (1911–1913). He wrote: *Beiträge zur Kenntnis der Haftscheibe von Echeneis* (1909). An amphibian and two birds are named after him.

Howes, A

Characin sp. *Prodontocharax howesi* HW Fowler, 1940

'Mr. Arthur Howes' is described as being the person after whom this species was named but, in contrast, Gordon B Howes (see next entry) collected the holotype and other fishes during the third Academy of Natural Sciences expedition to Bolivia (1937–1938). Perhaps Arthur was a relative of Gordon. (Also see **Maria (Howes)**)

Howes, GB

Three-barbeled Catfish sp. *Pimelodella howesi* HW Fowler, 1940

Gordon B Howes was an American naturalist who collected fishes (and other taxa) during Carriker's expeditions through South America. He collected the holotype of this species.

Howes, GJ

Lake Victoria Cichlid sp. *Haplochromis howesi* MJP van Oijen, 1992
African Loach Catfish sp. *Congoglanis howesi* RP Vari, CJ Ferraris & PH Skelton, 2012
False Sardine sp. *Engraulicypris howesi* MA Riddin, IR Bills & MH Villet, 2016

Dr Gordon John 'Gordi' Howes (1938–2013) was an ichthyologist and fish systematist at the BMNH, which he joined as a scientific assistant (1968). He wrote: *A review of the Anatomy, Taxonomy, Phylogeny and Biogeography of the African Neoboline Fishes* (1984). Van Oijen honoured him for his "*...many excellent contributions to fish taxonomy*."

Howland

Blacksaddle Grouper *Epinephelus howlandi* A Günther, 1873

This is a toponym, referring to Howland Island in the central Pacific.

Howson

Howson's Dottyback *Pseudochromis howsoni* GR Allen, 1995
Damselfish sp. *Chromis howsoni* GR Allen & M van Nydeck Erdmann, 2014
Howson's Coral-Goby *Gobiodon howsoni* GR Allen, 2021

Craig Howson is a long-time friend of the authors and the founding (1987) Director of North Star cruises and owner of the 'luxurious' Australian cruise ship 'True North'. He provided Gerald Ray Allen with numerous collecting and diving opportunities in the Australia-New Guinea region, resulting in the discovery of several new species, including the damselfish and goby.

Hoy

Bloater *Coregonus hoyi* JW Milner, 1874

Dr Philip Romayne Hoy (1816–1892) was an American physician, explorer and naturalist who qualified as a physician in Ohio in 1840, practised medicine there for some years and then, in 1846, settled in Racine on Lake Michigan. In the 1850s he collected specimens in Wisconsin for Baird and Girard and was Naturalist of the Geological Survey of Wisconsin and Fish Commissioner for Wisconsin. He was so enthusiastic a naturalist that, whenever he was on his rounds calling on his patients, he always took extra equipment like a botany book, a butterfly net, etc. in case he encountered interesting specimens. He wrote: *Catalog of the cold-blooded vertebrates of Wisconsin* (1883). His bird collection is housed in the Racine Heritage Museum. A mammal is named after him. He provided the holotype of this species, which is found in the Great Lakes.

Hsiojen

Triplefin sp. *Enneapterygius hsiojenae* S-C Shen, 1994

Hsiojen Lin Shen is the wife of the author, Dr Shieh-Chieh Shen. (See also **Shen, S-C**)

Hu

Chinese Gudgeon genus *Huigobio* PW Fang, 1938

Dr Hsen Hsu Hu (1894–1968) was a botanist. The University of California, Berkeley awarded his bachelor's degree (1916). He then taught at Nanjing Higher Normal School (1916–1923), after which he went to Harvard, where he was awarded his doctorate in science (1925). He was the first President of National Chung Cheng University (1940–1944) and was Professor at and Director of the Fan Memorial Institute of Biology in Peking (Beijing) (1945). An amphibian is also named after him.

Huachi

Pencil Catfish sp. *Silvinichthys huachi* L Fernández, EA Sanabria, LB Quiroga & RP Vari, 2014

A toponym, referring to the type locality; the Río Huerta de Huachi in Argentina.

Huang Chu-Chien

Chinese Cyprinid sp. *Poropuntius huangchuchieni* TL Tchang, 1962

Dr Huang Chu-Chien was a herpetologist at the Institute of Zoology, Academia Sinica, Beijing, China. He wrote: *A general account of our country's amphibian and reptilian resources* (1978).

Huaorani

Armoured Catfish sp. *Otocinclus huaorani* SA Schaefer, 1997

This species is dedicated to the Ecuadorian Huaorani people, who live in the Department of Orellana, Ecuador, where the holotype was collected.

Huascar

Dwarf Cichlid sp. *Apistogramma huascar* U Römer, P Pretor & IJ Hahn, 2006

Huascar was an Inca prince, brother of the last Inca ruler Atahualpa. The name refers to the "large phenotypical similarity" between the species and one already named for Atahualpa, and to their common origin from the region of the Inca state Tahuantinsuyu, which includes present-day Peru.

Hubbard

Hawkfish sp. *Cirrhitops hubbardi* LP Schultz, 1943

Commander Dr H D Hubbard. Schultz wrote: "*Named hubbardi in appreciation of the cooperation rendered by Dr. H. D. Hubbard, Comdr. (M.C.) U.S.N., while we were on the U.S.S. Bushnell in the South Pacific.*"

Hubbs, C

Redside Blenny *Malacoctenus hubbsi* VG Springer, 1959

Dr Clark Hubbs (1921–2008) was an ichthyologist and biologist who followed in his father's footsteps (see Carl Levitt Hubbs, below). The University of Michigan awarded his bachelor's degree (1942) and Stanford his doctorate (1951). His education was interrupted by service in the US Army (1942–1946). His professional career was at the University of Texas, Austin, which he joined as an Instructor in Zoology (1949), and where he founded the University's

Fish Collection. He eventually became Regents Professor (1988) until retiring as Professor Emeritus (1991). (Also see **Clark Hubbs**)

Hubbs, CL

Splitfin (goodeid) genus *Hubbsina* F de Buen, 1940
Gambusia genus *Carlhubbsia* GP Whitley, 1951
Olympic Mudminnow *Novumbra hubbsi* LP Schultz, 1929
Argentine Hake *Merluccius hubbsi* TL Marini, 1933
Rattail sp. *Coelorinchus hubbsi* K Matsubara, 1936
Lake Eustis Minnow *Cyprinodon hubbsi* A Carr, 1936
Icefish sp. *Neosalanx hubbsi* Y Wakiya & N Takahasi, 1937
Surf Sardine sp. *Notocheirus hubbsi* HW Clark, 1937
Rockfish sp. *Sebastes hubbsi* K Matsubara, 1937
Columbia Sculpin *Cottus hubbsi* RM Bailey & MF Dimick, 1949
Copper Redhorse *Moxostoma hubbsi* V Legendre, 1952
Rasbora sp. *Rasbora hubbsi* MR Brittan, 1954
Eelpout sp. *Lycodes hubbsi* K Matsubara, 1955
Bluegill Bully *Gobiomorphus hubbsi* G Stokell, 1959
Blotchwing Flyingfish *Cheilopogon hubbsi* NV Parin, 1961
Bigscale sp. *Melamphaes hubbsi* AW Ebeling, 1962
Lanternfish sp. *Lampanyctus hubbsi* RL Wisner, 1963
Hubbs' Pearleye *Rosenblattichthys hubbsi* RK Johnson, 1974
Maya Needlefish *Strongylura hubbsi* BB Collette, 1974
Kern Brook Lamprey *Lampetra hubbsi* VD Vladykov & E Kott, 1976 [Syn. *Entosphenus hubbsi*]
Waryfish sp. *Scopelosaurus hubbsi* E Bertelsen, G Krefft & NB Marshall, 1976
Lichen Moray *Gymnothorax hubbsi* JE Böhlke & EB Böhlke, 1977
Panamic Cusk-eel *Lepophidium hubbsi* RH Robins & RN Lea, 1978
Neotropical Silverside sp. *Atherinella hubbsi* WA Bussing, 1979
Whitepatched Splitfin *Allodontichthys hubbsi* RR Miller & Uyeno, 1980
Giant Hagfish *Eptatretus carlhubbsi* CB McMillan & RL Wisner, 1984
Snailfish sp. *Paraliparis hubbsi* A Andriashev, 1986
Delta Silverside *Colpichthys hubbsi* CB Crabtree, 1989
Hagfish sp. *Myxine hubbsi* RL Wisner & CB McMillan, 1995
Ricefish sp. *Oryzias hubbsi* TR Roberts, 1998
American Sole sp. *Trinectes hubbsbollinger* RR Duplain, F Chapleau & TA Munroe, 2012
Ives Lake Cisco *Coregonus artedi hubbsi* WN Koelz, 1929
Bridgelip Sucker ssp. *Catostomus colombianus hubbsi* GR Smith, 1966

Professor Carl Levitt (Leavitt) Hubbs (1894–1979) was a giant of American ichthyology. He was assistant curator of fish, amphibians and reptiles at the Field Museum in Chicago (1917–1920), then curator of fish at the Museum of Zoology, University of Michigan (1920–1944). His next position was as Professor of Biology at the Scripps Institution of Oceanography, California (1944–1969). Many other species carry his name, including the whale *Mesoplodon carlhubbsi*. *Octopus hubbsorum* was named for Carl as well as his wife, Laura Cornelia (q.v.) and their son Clark Hubbs (q.v.). The sole species was named after both Carl Hubbs and John A Bollinger (q.v.). Professor and Mrs Hubbs had three children, all of whom became ichthyologists. Their daughter Frances married yet another ichthyologist, Robert Rush Miller (q.v.). (See also **Laura**; **Hubbs, C**; **Carl; Carl Hubbs**)

Hubbs, LC

Hagfish sp. *Eptatretus laurahubbsae* CB McMillan & RL Wisner, 1984

Laura Cornelia Hubbs née Clark (1893–1988) was an ichthyologist and the wife of Carl Levitt Hubbs (above), as well as a friend and co-worker of the authors, and contributed to the life and works of her husband. Both her bachelor's degree (1915) and her master's (1916) were awarded by Stanford University. She worked part time at the University of Michigan Museum of Zoology (1929–1944). (See also **Laura Hubbs**)

Huber, JH

African Rivuline sp. *Epiplatys huberi* AC Radda & E Pürzl, 1981
[Syn. *Aplocheilus huberi*]

Dr Jean Henri Huber (b.1952) is a French ichthyologist who is particularly interested in Cyprinodontiformes. Bordeaux ENS-PC awarded his Engineering degree (1976), Insead, Fontainebleau awarded his MBA (1978) and Nancy University Zoological Museum awarded his PhD (1978). Now (2018) retired, he worked as a pharmaceutical industry executive and director. He has been studying oviparous Cyprinodonts (Cyprinodontiformes) as an official associate member of the MNHM (since 1980). He has described around seventy species, genera and family-group names and published many papers and books such as: *Review of 'Rivulus'. Ecobiogeography-Relationships* (1992), the various editions of Killi-Data (1994, 1996, 2000, 2006, 2007) also on-line at www.killi-data-org (since 2001). The etymology reads thus: (translation) "*We dedicate this new* Epiplatys *species to our friend Dr. Jean Henri Huber, the discoverer of this beautiful pike-like species. In addition, Dr. Huber first worked on the taxonomy and systematics of the* A. *(*E.*)* multifasciatus *superspecies and also helped to better understand the fish fauna of Gabon in particular.*" (Also see **Jean Huber**)

Huber, W

Huber's Knife Livebearer *Alfaro huberi* HW Fowler, 1923

Wharton Huber (1877–1942) was a collecting naturalist and zoologist who was at the Academy of Natural Sciences of Philadelphia (1920–1940), becoming Curator of Mammals (1923). He studied engineering at Lafayette and then worked on the bird collection at Wister Institute, then spent two years working lumber. His first colledting trip was for six months to New Mexico (1915), where he discovered a new duck. He then worked for the Philadelphia Electric Company (1917). He co-discovered the livebearer when collecting in Nicaragua (1922). He also collected in the USA and Mexico (1927–1934). His private collection of North American birds and mammals amounted to 4000 specimens. Fowler honoured him in reference to his efforts to obtain fishes in a country which 'presents so many difficulties'. He was an excellent natural history photographer and also loved gardening. A bird is also named after him.

Hubrecht

Cusk-eel sp. *Dicrolene hubrechti* MCW Weber, 1913

Dr Ambrosius Arnold Willem Hubrecht (1853–1915) was a Dutch biologist. He studied at Utrecht University (1870–1873) and was awarded his doctorate (1874). He moved to Leiden (1873) and was Curator, Ichthyology and Herpetology, Rijksmuseum van Natuurlijke Histoire, Leiden (1875–1882), then returned to Utrecht as Professor of Zoology. The Hubrecht Laboratory in Holland is named after him. He played a significant role in the formation of the *Siboga* Indonesian expedition (1898–1899), during which the holotype was collected. He wrote: *On a New Genus and Species of Agamidae from Sumatra* (1881). A reptile is named after him.

Hucht

Red-cheeked Fairy Basslet *Pseudanthias huchtii* P Bleeker, 1857

Guillaume Louis Jacques (Willem) van der Hucht (1812–1874) was a Dutch hunter, soldier, tea-planter, taxidermist and dealer. He went to the Dutch East Indies (1844) with him as Captain of the ship, as he was in the Merchant Navy (1836–1844). He owned a plantation (1844–1846) and was co-owner of an even larger one (1844–1857). He was given a collection of fish made for the Governor of the Moluccas Islands in Indonesia. He not only allowed Bleeker to study them, but generously allowed him to take any that he thought were new to science. Hucht's collection ended up with the Zoologisch Museum Amsterdam (ZMA). He returned to the Netherlands (1860) and was elected to the House of Representatives (1866–1871). He also commanded the Haarlem Militia (1860–1874). He wrote about his far-eastern travels in: *Mededeelingen over mijne verrigtingen in Indië, als vertegenwoordiger van de Billiton-Maatschappij* (1863).

Hudson

Hudson's Lanternfish *Diaphus hudsoni* Zurbrigg & WB Scott, 1976

'Hudson' was a vessel belonging to the Canadian Coast Guard. The holotype of this species was collected during the Hudson 70 Cruise around the Americas and named in honour of it.

Hudson, CB

Signal Blenny sp. *Emblemaria hudsoni* BW Evermann & L Radcliffe, 1917

Captain Charles Bradford Hudson (1865–1939) was an American Fisheries Commission reporter, soldier, artist and writer. After completing a degree at Colombian University, he studied art in New York and Paris. He then worked as an illustrator of fiction and scientific publications. Because of the quality of his work and his dedication, Hudson was hired to go on expeditions funded by the Bureau of Fisheries. On trips to the Caribbean, the Great Lakes, the rivers of California and the Pacific Coast, he painted examples of fish as they were caught and identified. He was the author of two books: *The Crimson Conquest* (1907) and *The Royal Outlaw* (1917). He worked closely with David Starr Jordon, whose works on fish he illustrated. He volunteered to serve in the the Spanish-American War (1898). Commissioned as a lieutenant, he was sent to Cuba and participated in the victorious Siege of Santiago. He then returned to the Californian coast where he painted his best known works. The authors wrote: "*We take pleasure in naming this new species for our friend, Capt. Charles Bradford Hudson, artist and author, who has succeeded better than any other in depicting on canvas the life colors of American fishes.*"

Hudson, RL

Hudson's Triplefin *Helcogramma hudsoni* DS Jordan & A Seale, 1906

R L Hudson was a scientific illustrator of many marine species, mainly cephlapods and fish. He was one of the artists who provided illustrations for the paper describing the triplefin. Despite his obvious talent, he never seems to have used his first names - only his initials - when illustrating books.

Hügel

Snowtrout sp. *Schizothorax huegelii* JJ Heckel, 1838

Dalmatian Barbelgudgeon *Aulopyge huegelii* JJ Heckel, 1843

Karl Alexander Anselm Freiherr von Hügel (Baron Charles von Hügel) (1795–1870) was an Austrian army officer, diplomat, botanist and explorer primarily remembered for his travels in northern India (1830s). After studying law, he became an officer (1813) in the Austrian Hussars and fought various campaigns against Napoleon. He travelled in Scandinavia, Russia and both France and Italy where he was stationed. He took up residence in Hietzing, Vienna (1824), where he established a botanical garden through which he introduced plants from Australia into European gardens. He was engaged to a Hungarian countess, but when she broke it off to marry the Austrian Chancellor, Charles began a grand tour of Asia and Australia (1831–1836). On his return, he wrote the four-volume: *Kaschmir und das Reich der Siek*. He also founded the Austrian Imperial Horticultural Society. He married a Scottish woman and retired to England (1867) with his family. A bird is also named after him.

Hugh

Cardamon Garra *Garra hughi* EG Silas, 1955

Hugh M Silas, who collected the holotype in India (1953), was the Assistant Manager, Periakanal Estate; and also the author's brother.

Hugo Wolfeld

Southern King Spine Loach *Iksookimia hugowolfeldi* TT Nalbant, 1993

Hugo Wolfeld of Bucharest, Romania, was an aquarist. He was described as one of the 'most able' of aquarium fish breeders and amateur ichthyologists. The loach was named in his memory.

Huguenin

Oriental Barb sp. *Hypsibarbus huguenini* P Bleeker, 1853

[Syn. *Poropuntius huguenini*]

Huguenin's Dragonet *Repomucenus huguenini* P Bleeker, 1858

Otto Fredrik Ulrich Jacobus Huguenin (1827–1871) was a mining engineer in the Dutch East Indies (Indonesia). He joined the government service (1850) and became head of the

Mining Service (1859). He knew and helped Alfred Russel Wallace when Wallace spent five months (end 1858/early 1859) on the island of Batchian.

Hugueny

African Barb sp. *Enteromius huguenyI* R Bigorne & C Lévêque, 1993

Dr Bernard Hugueny is an ecologist and limnologist at Institut de Recherche pour le Développement (IRD), based at MNHN, Paris. He is a Research Unit leader in the Biofresh Project. He co-wrote: *Habitat fragmentation and extinction rates within freshwater fish communities: a faunal relaxation approach* (2010). He was a friend and colleague of the authors.

Hulley

Barbeled Dragonfish sp. *Eustomias hulleyi* JR Gomon & RH Gibbs, 1985
Boarfish sp. *Antigonia hulleyi* NV Parin & OD Borodulina, 2005
Reunion Brtistle-mouth Fish *Argyripnus hulleyi* J-C Quéro, J Spitz & J-J Vayne, 2009
Roughnose Legskate *Cruriraja hulleyi* N Aschliman, DA Ebert & LJV Compagno, 2010

Dr Percy Alexander 'Butch' Hulley (b.1941) is a South African zoologist and ichthyologist. He is Curator of Fishes and Deputy Director of the Iziko South African Museum. He has described many species new to science. He was honoured for his pioneering work on southern African skates.

Hulot

African Tetra sp. *Brachypetersius huloti* M Poll, 1954
Brown Codling *Physiculus huloti* M Poll, 1953
African Barb sp. *Labeobarbus huloti* KE Banister, 1976

André Hulot, an agronomist and hydrobiologist with Institut National pour l'Etude Agronomique du Congo, collected fishes in the Congo River basin (1946). The etymology for the *Labeobarbus* only mentions 'M. (Monsieur) Hulot', but this is probably the same person.

Hulstaert

Butterfly Barb *Enteromius hulstaerti* M Poll, 1945

Révérend Père Gustaaf Hulstaert (1900–1990) was a missionary, an entomologist and botanist in the Belgian Congo from 1925. He wrote: *Rhopaloceres nouveaux du Musee du Congo Belge* (1926). He collected the holotype of this species.

Human

Human's Whaler Shark *Carcharhinus humani* WT White & S Weigmann, 2014

Dr Brett A Human (1974–2011) was an Australian marine biologist and scientific diver at the Department of Aquatic Zoology, Western Australian Museum, Welshpool. He co-wrote: *Is the Megamouth Shark susceptible to mega-distortion? Investigation of the effects of twenty-two years of fixation and preservation on a large specimen of Megachasma pelagios (Chondrichthyes: Megachasmidae)* (2012). Before working in Western Australia, he had been a researcher at the South African Museum. The etymology for the shark reads: "*Named after the late Dr. Brett Human, for important contributions to shark taxonomy in South Africa and Oman in the western Indian Ocean region, and who is sorely missed by his colleagues.*"

Humann

Snake-eel sp. *Ophichthus humanni* JE McCosker, 2010
Humann's Fairy Wrasse *Cirrhilabrus humanni* GR Allen & MV Erdmann, 2012

Paul H Humann (b.1937) is an underwater photographer whose bachelor's degree in biology was awarded by Wichita State University. He qualified as a lawyer at Washburn Law School (1964) and practiced as a lawyer in Wichita, Kansas (about as far from the sea in the USA as it is possible to go!), but he left his practice (1971) to go to the Cayman Islands and become owner/master of the Caribbean's first live-aboard dive cruiser, which he operated for 8 years. He sold the boat to pursue his underwater photography and publishing career and in the 1980s co-founded New World Publications, which publishes Marine Life Identification books including 14 of which he is the primary photographer and author. He is a co-founder of REEF (Reef Environmental Education Foundation) (1990). McCosker, in the eel's etymology, said that he named it for Humann as he was an: "*...author and friend, who has generously aided ichthyologists with his photographs and observations.*"

Humboldt, FWHA

Shortfin Silverside *Chirostoma humboldtianum* A Valenciennes, 1835
Titicaca Pupfish sp. *Orestias humboldti* A Valenciennes, 1846
[Junior Synonym of *Orestias cuvieri*]
Glass Knifefish sp. *Eigenmannia humboldtii* F Steindachner, 1878

Baron Friedrich Wilhelm Heinrich Alexander von Humboldt (1769–1859) was a Prussian naturalist, explorer and politician. After attending universities at Frankfurt an der Oder and Göttingen (1791) he enrolled at the Freiberg Mining Academy to learn natural history and earth sciences to help him with his intended future travels. To complete his experience, he then worked as an Inspector of Mines in Prussia for five years. After two years of disappointments and delays, he explored (1799–1804) in South America, collecting thousands of specimens, mapping and studying natural phenomena. The trip took in parts of Venezuela, Peru, Ecuador, Colombia and Mexico. He returned via the USA, where Thomas Jefferson entertained him at Monticello. He made a journey (1829) of similarly epic proportions, ranging from the Urals east to Siberia. Humboldt's *Personal Narrative* was inspirational to later travellers in the tropics, notably Darwin (q.v.) and Wallace (q.v.). His most famous writing was the 5–volume work: *The Cosmos* (1845–1862). Humboldt did research in many other fields, including astronomy, forestry and mineralogy. The Humboldt Current that runs south to north just off the Pacific coast of South America was named after him, as are five birds, five mammals and two amphibians.

Humboldt

Humboldt Sucker *Catostomus occidentalis humboldtianus* JO Snyder, 1908

This is not an eponym but a toponym, as the species is named after Humboldt County, California, USA.

Hummel

Chinese Cyprinid sp. *Sinilabeo hummeli* E Zhang, SO Kullander & YY Chen, 2006

Dr David Axelsson Hummel (1893–1984) was a Swedish physician, explorer and resistance leader in WW2. He trained in medicine at the Karolinska Institute in Stockholm where he graduated (1923), following which he held a variety of medical appointments. He joined the Hedin expedition to China (1927–1934) and collected the type specimen there (1930). He also collected plants in Gansu as well as zoological specimens and drew maps. When not exploring, he was a district medical officer on the Norwegian-Swedish border (1923–1956). During the war years he helped Norwegian refugees and helped their resistance agents for which he was later honoured. He later (1953) took part in an archaeological expedition to India and was Ship's Physician on whaling expeditions in the Southern Ocean (1957–1963). Two dragonflies are also named after him.

Hummelinck

Neotropical Rivuline sp. *Rachovia hummelincki* LF de Beaufort, 1940
Pike-Cichlid sp. *Crenicichla hummelincki* A Ploeg, 1991

Dr Pieter Wagenaar Hummelinck (1907–2003) was a Dutch naturalist, zoologist and explorer. He took his first degree at Universiteit Utrecht (1935), starting work there in the Zoological Laboratory (1940). His speciality was the study of the fauna of the Netherlands Antilles, which he frequently visited, as well as other West Indian Islands. He was the founder of the Foundation for Scientific Research in Suriname and the Netherland Antilles; and the cichlid name honoured him for this and on the occasion of his 83rd birthday. He retired (1972), leaving behind his significant zoological collections at the university. When the lab closed down (1988) the collection was transferred to the Zoological Museum, Artis, Amsterdam. A mammal and a reptile are named after him. He collected the rivuline holotype (1937).

Humphrey

African Barb sp. *Labeobarbus humphri* KE Banister, 1976

Peter Humphrey Greenwood (1927–1995). (See **Greenwood**)

Hung

Cardinalfish sp. *Jaydia hungi* P Fourmanoir & Do-Thi Nhu-Nhung, 1965

Nguyen Dinh Hung was Director, Vietnamese Oceanographic Institute. He hired the authors to study the fishes in the Institute's collection, and there they found this species' holotype.

Hunt

Red-banded Perch *Hypoplectrodes huntii* J Hector, 1875

The holotype, from the Chatham Islands, was "Presented by F Hunt, Esq." It is tempting to identify this man as Frederick Alfred Hunt (c.1818–1891), self-declared 'king' of Pitt Island in the Chatham Islands, and author of: *Twenty-Five Years' Experience in New Zealand and the Chatham Islands* (1866). Unfortunately, we have no proof of this.

Hunter, HCV

Lake Chala Tilapia *Oreochromis hunteri* A Günther, 1889

Henry Charles Vicars Hunter (1861–1934) was a Fellow of the Zoological Society of London, a big-game hunter and amateur naturalist. He collected the type and provided notes about its distribution to Günther. Two birds and a mammal are also named after him.

Hunter, JG

Crocodile Icefish sp. *Dacodraco hunteri* ER Waite, 1916

John George Hunter (1888–1964) was an Australian doctor and biologist. He was on the 'Aurora' (1911–1913) as a member of the Cape Denison team of the Australasian Antarctic Expedition (1911–1914) during which the icefish holotype was caught. Sydney University awarded his science degree (1909) and he returned there to take his MB (1915) and was awarded a Ch.M (1946). After graduating, Hunter worked briefly as a resident at the Royal Alexandra Hospital for Children. He was a captain in the Australian Army Medical Corps, Australian Imperial Force, serving on the Western Front with the 9th Field Ambulance (1916–1917) and in England with the 2nd Australian General Hospital (1917–1918). He returned to Sydney, went into general practice, and was honorary assistant-physician (1923–1929) at Sydney Hospital and honorary physician (1927–1929) at Royal South Sydney Hospital. He was appointed full-time medical secretary of the New South Wales branch of the British Medical Association (1929) and general secretary (from 1933) of the BM's Federal Council in Australia. He was mobilised as a militia major serving in Sydney (1941–1943). Strangely there is no etymology in Edgar Waite's text, but it seems very likely that he named the icefish after John George Hunter.

Huntsman

Atlantic Whitefish *Coregonus huntsmani* WB Scott, 1987

Dr Archibald Gowalock Huntsman (1883–1973) was a Canadian marine biologist, oceanographer and physician. He was awarded a bachelor's degree from the University of Toronto (1905 & in medicine, 1907). His doctorate from the same University was honorary (1933). He stayed on at the University of Toronto, joining the Department of Zoology (1907) as a lecturer and became Associate Lecturer (1917) and Professor of Marine Zoology (1927–1954). He was a very active participant in the Canadian Fisheries Expedition (1914–1915), which so influenced him that after it he concentrated on problems of fishery science. Simultaneously with his position as an academic at the University of Toronto, he was Curator, St. Andrews Biological Station, New Brunswick (1911–1918) and its Director (1919–1934), and, additionally, Director Fisheries Experimental Station, Halifax, Nova Scotia (1924–1929). He wrote: *The Maritime Salmon of Canada* (1931).

Hurault

Halftooth Characin sp. *Hemiodus huraulti* J Géry, 1964

Jean-Marcel Hurault (1917–2005) was a French ethnologist and geographer at the Institut Géographique National. His work focussed on French Guiana and Cameroon (1946–2003). He developed a type of geography that is now called anthropo-geography. He was an expert on the forest tribes of French Guiana and wrote: *Français et Indiens en Guyane 1604–1972* (1972). He collected the holotype.

Hureau

Eelpout sp. *Lycenchelys hureaui* AP Andriashev, 1979
Hureau's Flounder *Engyprosopon hureaui* J-C Quéro & D Golani, 1990
Snailfish sp. *Paraliparis hureaui* J Matallanas, 1999

Dr Jean-Claude Hureau (b.1935) is a French ichthyologist who was Professor (1992) and Director (1970) of Marine Icthyology at Muséum National d'Histoire Naturelle, Paris until retirement (2001) and is now professor emeritus. He carried out considerable research in the Kerguelen Islands as well as researching the fishes of Terre Adélie, Antarctica, where he spent over a year (1960–1962) at Dumont d'Urville Research Station.

Hurtado

Crescent Gambusia *Gambusia hurtadoi* C Hubbs & VG Springer, 1957

Leopoldo Hurtado Olin (1888–1971) was a Mexican engineer working at the Department of the Economy in the state of Chihuahua, Mexico. He wrote: *Chihuahua: intergración territorial de estado* (1953) and *Chihuahua turístico: agricultura, ganadería, minería, industria, comercio* (1958). The help he provided the authors included informing them of the location of El Ojo de la Hacienda Dolores, the type locality. He was honoured in appreciation for his aid during the authors' collecting trip in June 1951.

Hüser

Night Aulonocara (Lake Malawi cichlid) *Aulonocara hueseri* MK Meyer, R Riehl & H Zetzsche, 1987

Eberhard Hüser of Hildesheim, Germany, is an aquarist specialising in cichlids.

Hutchins

Earspot Snake-Blenny *Ophiclinops hutchinsi* A George & VG Springer, 1980
Short Boarfish *Parazanclistius hutchinsi* GS Hardy, 1983
Tasmanian Codling *Eeyorius hutchinsi* CD Paulin, 1986
MItchell Gudgeon *Kimberleyeleotris hutchinsi* DF Hoese & GR Allen, 1987
Hutchins' Blenny *Cirripectes hutchinsi* JT Williams, 1988
Posidonia Clingfish *Posidonichthys hutchinsi* JC Briggs, 1993
Brownmargin Flathead *Cociella hutchinsi* LW Knapp, 1996
Hutchins' Toadfish *Halophryne hutchinsi* WD Greenfield, 1998
Hutchins' Anglerfish *Lophiocharon hutchinsi* TW Pietsch, 2004
Hutchins' Mudbrotula *Dipulus hutchinsi* PR Møller & W Schwarzhans, 2006

Western Wobbegong *Orectolobus hutchinsi* PL Last, JA Chidlow & L Compagno, 2006
Brown Rat Clingfish *Barryichthys hutchinsi* KW Conway, GI Moore & AP Summers, 2019

Dr J Barry Hutchins (b.1946) was Curator of Fishes at the Western Australian Museum (1998–2007), which he joined as a Technical Officer (1972) and where he worked until he retired. He is now a Research Associate. The University of New South Wales, Sydney awarded his bachelor's degree (1968). After service in Vietnam as a corporal in the infantry, Australian Army (1969–1970), he worked on prawn and scallop trawlers in Queensland (1970–1972) before joining the Western Australian Museum. Murdoch University awarded his first-class honours degree in ichthyology (1979) and a doctorate (1988). He has written many papers on Australian fishes, including three field guides; his first book was: *The Fishes of Rottnest Island* (1979) and he was senior author of: *The Marine and Estuarine Fishes of South-western Australia* (1983) and *Sea Fishes of Southern Australia* (1986) which was based on his seven-month survey of the near-shore waters of southern Queensland, New South Wales, Victoria, Tasmania and South Australia. Dr Hutchins was the first (1983) person to recognize that the wobbegong was a new species. (See also **Barry**)

Hutereau

Mesh-scaled Topminnow *Micropanchax hutereaui* GA Boulenger, 1913
Uele Mormyrid *Petrocephalus hutereaui* GA Boulenger, 1913

Lieutenant Joseph Armand Oscar Hutereau (1875–1914) was a Belgian army officer who explored in the Congo in the early 1900s under King Leopold. King Albert appointed him to lead an ethnographic expedition to the north part of the Congo (1911–1913). It collected over 10,000 objects for the Musée du Congo Belge (now the Musée royal de l'Afrique centrale) in Tervuren, including the type material of both the fish species. Hutereau returned to active service with the Belgian army on the outbreak of WW1 and was killed in action in November 1914. He prepared: *Histoire des peuplades de l'Uele et de l'Ubangi,* which was published posthumously (1927).

Hutomo

Hutomo's Anthias *Pseudanthias hutomoi* GR Allen & Burhanuddin, 1976

Malikusworo Huromo was Senior Research Scientist at the Center for Oceanographic Research, Indonesian Institute of Sciences (LIPI), Jakarta, Indonesia. He studied at Bogor Agricultural University. Among his published papers is: *Progress in Oceanography of the Indonesian Seas: A Historical Perspective* (2005). The etymology says: "*The species is named* hutomoi *in honour of Mr. MALIKUSWORO HUTOMO of the Lembaga Oseanologi Nasional, Jakarta, who assisted in collecting the type specimens.*"

Hutsebaut

African Characin sp. *Belonophago hutsebouti* LP Giltay, 1929

Father Franz Joseph Hutsebaut (1886–1954) was a Catholic missionary in the Belgian Congo from the 1920s until his death in Katanga. Giltay incorrectly spelt his name as *hutsebout* in the binomial. He was clearly a remarkable man, as he once shot a leopard that had crawled through the window of the mission late one night, and he was also known for rearing Okapis prior to exporting them to zoos outside Africa (1927–1941). An amphibian is named after him.

Hutton

Redfin Bully *Gobiomorphus huttoni* JD Ogilby, 1894
New Zealand Ruffe *Schedophilus huttoni* ER Waite, 1910

Frederick Wollaston Hutton (1836–1905) was an English geologist and zoologist who settled in New Zealand. He served in the Indian Mercantile Marine, and then in the army (1855–1865). He saw service in the Crimean War and the Indian Mutiny. He wrote: *Catalogue of the Birds of New Zealand* (1871). The Royal Society of New Zealand established (1909) the Hutton Memorial Fund in his memory. It awards the Hutton Medal and provides grants for the encouragement of research into the zoology, botany and geology of New Zealand. Four birds are named after him.

Huwald

African Rivuline sp. *Callopanchax huwaldi* HO Berkenkamp & VMF Etzel, 1980

Kurt Huwald was an aquarist from Osnabrück who co-founded the DKG, the German Killifish Association, and was a friend of the authors. He was described by others as an 'outstanding breeder of fish'. Among aquarists a standard expression for tiny fish is '1 Hu'. This refers to the fact that people always wanted newly-bred fish from him straightaway, even though they were so tiny one could not even determine their sex.

Huysman

Huysman's Righteye Flounder *Samariscus huysmani* MCW Weber, 1913

J W Huysmans was the artist on the 'Siboga' expedition of which Weber was the leader. The Siboga expedition was a Dutch zoological and hydrographic expedition to Indonesia (1899–1900). Huysmans was assigned to the expedition from the botanical gardens in Bogor, Java.

Hygom

Bermuda Lanternfish *Hygophum hygomii* CF Lütken, 1892

Captain Vilhelm Johannes Willaius Hygom (b.1818) was a Danish merchant seaman who took his master's ticket (1839). He was also a collector of marine and land specimens (1853–1861) for Steenstrup. He found a Giant Squid *Architeuthis dux* off the Bahamas (1855).

Hylton

Filament Blenny *Emblemaria hyltoni* EK Johnson & DW Greenfield, 1976

Nick Hylton was honoured by the authors because he "*...donated his services as captain and crew of the yacht M/S Miss Sabrina during the Miskito Coast Expedition, aided in field work, saved the expedition at Brus Lagoon...*" [1975].

Hyrtl

Hyrtl's Catfish *Neosilurus hyrtlii* F Steindachner, 1867

[Alt. Glencoe Tandan, Yellowfin Tandan, Yellow-finned Eel-Catfish]

Dr Josef Hyrtl (1810–1894) was an Austrian anatomist who qualified as a physician at the University of Vienna. He was Professor of Anatomy at the University of Prague (1837–1845) and Professor of Anatomy at Vienna (1845–1874). He retired from his professorship because his eyesight was failing. As he was Steindachner's colleague in Vienna, it is likely that this species is named after him, though Steindachner does not make an explicit identification.

I

Iago

Houndshark genus *Iago* LJV Compagno & S Springer, 1971

The name of this genus derives from Iago, the villain in Shakespeare's play *Othello.* The authors say that this genus is "a troublemaker for systematists and hence a kind of villain".

Iara

Ghost Knifefish sp. *Compsaraia iara* MJ Bernt & JS Albert, 2017

Iara, also spelled Uiara or Yara, is a water nymph in Brazilian folklore. She is said to reside in rivers and is often blamed for the disappearance of fishermen.

Iasy

Characin sp. *Jupiaba iasy* AL Netto-Ferreira, AM Zanata, JLO Birindelli & LM Sousa, 2009

îasy is the goddess of the moon in the mythology of the Brazilian Tupi people. The epithet was given "in allusion to the crescent-shaped humeral blotch" of this species.

Iban

Betta sp. *Betta ibanorum* HH Tan & PKL Ng, 2004

The Ibans, also called Sea Dayaks, are an indigenous people of northwestern Borneo, where this species occurs.

Ichihara

Japanese Velvet Dogfish *Zameus ichiharai* K Yano & S Tanaka, 1984

Dr Tadayoshi Ichihara (d.1981) was a scientist at the Whales Research Institute, Tokyo, Japan. At the time of his death he was Professor of Behavioural Ecology of Marine Mammals at Tokai University. The University of Tokyo awarded his PhD. Controversially he declared the Indian Ocean population of Blue Whales to be a separate (smaller) species at a time when whaling of Blue Whales was banned (1960s) which was seen by many as an attempt to get around the ban. He wrote a number of papers, including some with Yano & Tanaka, including: *Notes on a Pacific sleeper shark, Somniosus pacificus, from Suruga bay, Japan* (1982). The describers of the dogfish honoured him in its scientific name as he had suggested to them that they study this shark.

Ichikawa

Neko-gigi (Japanese catfish) *Coreobagrus ichikawai* Y Okada & SS Kubota, 1957

Dr Atsuhiko Ichikawa (1904–1991) was a Japanese platyhelminthologist. He originally graduated from the Imperial University, Tokyo but moved (1930) to Hokkaido University, Sapporo (1930) where he was awarded his doctorate (1940) and from where he retired as Professor Emeritus. He wrote: *Metamorphoses of* Hynobius *larvae following removal of fourth branchial and arterial arches* (1931). The etymology states that he was the junior author's 'benefactor' in college.

Ichthyandr

Greeneye sp. *Chlorophthalmus ichthyandri* AN Kotlyar & NV Parin, 1986

The Russian fishery research vessel 'Ichthyandr' (also spelled 'Ikhtiandr') collected the first specimens of this species.

Ida

Sand-lance sp. *Ammodytoides idai* JE Randall & JL Earle, 2008

Hitoshi Ida is a Japanese marine biologist who at the time of the description was working at the School of Marine Biosciences, Kitasato University. He has co-described other species in this genus with Randall & Earle such as: *Ammodytoides pylei, a New Species of Sand Lance (Ammodytidae) from the Hawaiian Islands* and written much about them such as the co-authored paper: *A taxonomic revision of the fishes of the Genus Bleekeria (Perciformes, Ammodytidae)* (2018). The etymology says that the species was named: "*...in honour of our colleague and friend... ...in recognition of his extensive research on the Ammodytidae.*"

Idabel

Basslet sp. *Lipogramma idabeli* L Tornabene, DR Robertson & CC Baldwin, 2018

The *Idabel* submersible was used to collect the type series in the western Caribbean. The name recognises the efforts of its owner-designer and pilot Karl Stanley and engineer Thomas Trudel who made these and other collections of fishes possible by constructing a fish-catching system that converted *Idabel* from an observation-only vessel to one capable of collecting scientific specimens.

Idei

Betta sp. *Betta ideii* HH Tan & PKL Ng, 2006

Takashige Idei is a Japanese researcher of the Betta genus. who has travelled widely in Africa, India, Southeast Asia, and both North and South America. He has been described as an 'intrepid fish collector' by his friend the senior author. He co-wrote: *Stories from the Mekong, part 2. The Cryptocoryne (Araceae) of Chiang Khan District, Loei Province, Thailand* (2017) and, although not a botanist, he has collected a number of species of these water plants and sent them to experts. He also has such an aquatic plant named after him, *Cryptocoryne ideii.*

Idenburg

Idenburg Tandan *Neosilurus idenburgi* JT Nichols, 1940

This is a toponym; the Idenburg River, West Papua, Indonesia, is the locality where this catfish was first collected.

IFAT

Half-striped Penguin Tetra *Thayeria ifati* J Géry, 1959

IFAT is an acronym that stands for the (Cayenne-based) Institut Français d'Amérique Tropicale (French Institute of Tropical America).

Ignesti

Garra sp. *Garra ignestii* L Gianferrari, 1925

Ugo Ignesti (fl.1924–fl.1953) was an Italian zoologist, philologist and collector in Abyssinia (1924) and British Guiana (Guyana) (1931). He wrote: *La Lingua degli Amharic, Trascritta in Caratteri Latini: Grammatica, Esercizi e Vocabolario* (1937). He collected the holotype of this species in Abyssinia (Ethiopia). A bird is also named after him.

Igobi

Pencil Catfish sp. *Cambeva igobi* WB Wosiacki & MCC de Pinna, 2008

Igobi is a tribal chief in the Tupí-Guaraní legend that tells of the origin of the Iguaçu waterfalls, Paraná, near where this catfish occurs.

Ih

New Zealand Piper *Hyporhamphus ihi* WJ Phillipps, 1932

Not an eponym, but derived from the fish's name in Maori: 'Ihe'.

Ihering HFA

Long-whiskered Catfish genus *Iheringichthys* CH Eigenmann & AA Norris, 1900
Characin sp. *Bryconamericus iheringii* GA Boulenger, 1887
Banjo Catfish sp. *Pseudobunocephalus iheringii* GA Boulenger, 1891

Dr Hermann Friedrich AIbrecht von Ihering (sometimes Jhering) (1850–1930) was a German-Brazilian zoologist, malacologist and geologist. He trained as a physician and served in the German army. He went to Rio Grande do Sul, Brazil (1880) and founded the São Paulo Museum (1894), spending 22 years as its first Director (1894–1916). He returned to Germany (1924) and died there. He co-wrote: *Catálogos da Fauna Brasileira. As Aves do Brazil* (1907) with his son, Rudolpho Teodoro Gaspar Wilhelm von Ihering (below). Three mammals, two reptiles, an amphibian and six birds are named after him. (See also Hermann)

Ihering, RTGW

Armoured Catfish sp. *Hypostomus iheringii* CT Regan, 1908
Pencil Catfish sp. *Cambeva iheringi* CH Eigenmann, 1917
Tetra sp. *Hyphessobrycon iheringi* HW Fowler, 1941
Bumblebee Catfish sp. *Microglanis iheringi* AL Gomes, 1946
Three-barbeled Catfish sp. *Cetopsorhamdia iheringi* O Schubart & AL Gomes, 1959

Rodolpho Teodoro Wilhelm Gaspar von Ihering (1883–1939) was a zoologist, biologist and fish culturist sometimes called the father of fish farming in Brazil. He graduated at São Paulo University (1901) and was appointed (by his father who was Director) as a Deputy

Director of the Museu Paulista. He spent a year (1911) at the Estação Biológica de Nápoles and then worked at the Muséum National d'Histoire Naturelle in Paris. When Brazil entered the war, he resigned to set up a factory where he worked for a decade but continued to collect, study and research. He then went to work in the Laboratory of Parasitology, Faculty of Medicine of São Paulo University (1926–1927) but, after this, began in earnest as an ichthyologist working at the Institute of Biological and Agricultural Defence in São Paulo and then (1931) researched fish farming and headed up (1932–1937) the Technical Committee on Fish Farming in the Northeast. He co-wrote: *Catálogos da fauna brasileira: As aves do Brazil* (1907) with his father, and alone wrote, among others: *Dictionary of the animals from Brazil* (1940). He sent a collection of fishes to the Academy of Natural Sciences, Philadelphia, including the holotype of this species. A reptile is also named after him. (See also Rudolph)

Iijma

Jellynose Fish genus *Ijimaia* H Sauter, 1905
Lefteye Flounder sp. *Psettina iijimae* DS Jordan & EC Starks, 1904
Ijima's Snaggletooth *Astronesthes ijimai* S Tanaka, 1908
Rockfish sp. *Sebastes ijimae* DS Jordan & CW Metz, 1913
Japanese Dragonet *Neosynchiropus ijimae* DS Jordan & WF Thompson, 1914
Taiwanese Gudgeon sp. *Squalidus iijimae* M Oshima, 1919

Professor Dr Isao Ijima (1861–1921) was Professor of Zoology at Science College, Imperial University of Tokyo (1885–1921) from where he had graduated (1881). The University of Leipzig awarded his doctorate (1884). He is regarded as the father of parasitology in Japan and was the first President of the Ornithological Society of Japan. Six birds and a reptile are named after him. The original text for the flounder refers only to 'Dr S Ijima', Professor of Zoology, Imperial University of Tokyo; we believe this is an error in the initial.

Ike (Morgan)

Clarence River Cod *Maccullochella ikei* SJ Rowland, 1986

Isaac (Ike) Morgan Rowland was the author's late grandfather, who he describes as "*…a great admirer and angler of Australian native inland fishes, in particular Murray cod.*"

Ike (Rachmatika)

Goby sp. *Lentipes ikeae* P Keith, N Hubert, F Busson & RK Hadiaty, 2014

Ike Rachmatika is a biologist (1982–2010) at the Cibinong Science Centre LIPI, Indonesia. She was formerly at the ichthyology lab at Museum Zoologicum Bogoriense (Cibinong, Indonesia) and worked at the Division of Zoology, Research and Development Centre for Biology, Indonesian Institute of Science. Her MSc in fisheries and allied aqua-cultures was awarded by Auburn University, USA (1995). She co-wrote: *Life After Logging: Reconciling Wildlife Conservation and Production Forestry in Indonesian Borneo* (2005) and *A first look at the fish species of the middle Malinau: taxonomy, ecology, vulnerability and importance* (2005). She was honoured for her work and passion for the freshwater fishes of Indonesia.

Ikeda, K

Chinese Carp sp. *Acrossocheilus ikedai* I Harada, 1943

Kiyoshi Ikeda was a General in the Japanese Imperial Navy during WW2. Harada was doubtless indulging in a little war-time patriotism.

Ikeda, Y

Ryukyu Ruddertail Dragonet *Callionymus ikedai* T Nakabo, H Senou & M Aizawa, 1998

Yuji Ikeda, about whom we are told very little in the etymology. A specimen of this fish was collected alive on Iriomote Island (1995) and was reared to maturity in an aquarium by Ikeda. We assume this is the same person as the Yuji Ikeda recorded being at the Biological Laboratory of the Imperial Household, Tokyo (2005), who supplied 'most of the type specimens' of the goby *Exyrias akihito*. (See Akihito)

Iksookim

Loach genus *Iksookimia* TT Nalbant, 1993

Dr Ik-Soo Kim is a South Korean ichthyologist who is Professor in the Faculty of Biological Sciences, Chonbuk National University, Chonju. He co-wrote: *Study on the Pigmentation of Albinic Bitterlings Acheilognathus signifer (Pisces; Cyprinidae) Based on Its Entire Body, Appendage and Eye* (2010).

Iles

Lake Malawi Catfish sp. *Bathyclarias ilesi* PBN Jackson, 1959
Lake Malawi Cichlid sp. *Copadichromis ilesi* A Konings, 1999

Thomas Derrick Iles (1927–2017) was a British fishery biologist and ichthyologist. He was a colleague of Jackson's at the Joint Fisheries Research Organization of Northern Rhodesia (Zambia) and Nyasaland (Malawi) when he collected the catfish holotype and drew the author's attention to the species. He was educated at Christchurch College, Cambridge, then was at the Fisheries Laboratory, Ministry of Agriculture and Fisheries, Lowestoft, England (1963–1969). He moved (1969) to New Brunswick, Canada, and worked at the St Andrews Biological Station. He was Chief, Marine Fish Division, Department of Fisheries and Oceans, Bedford Institute of Oceanography, Canada (1982) where he remained. Konings named the cichlid for Iles' "*...1960 paper on the 'Utaka' cichlids of Lake Malawi.*" He co-wrote: *The cichlid fishes of the Great Lakes of Africa* (1972). In his youth, he played rugby at a high level for Cardiff (1947–1950). He died peacefully on Christmas day, which would have been his late wife's birthday.

Iljin

Caspian Goby sp. *Knipowitschia iljini* LS Berg, 1931
Eastern Caspian Bighead Goby *Ponticola iljini* ED Vasil'eva & VP Vasil'ev, 1996
Goby sp. *Biendongella iljini* AM Prokofiev, 2015

Boris Sergeevich Iljin (1889–1958) was a Russian ichthyologist. He wrote: *A new species of herring from the southern Caspian* (1927) and was author of 'classic works' (translation) on the systematics of gobies from the Azov, Black and Caspian Sea basins. As well as being a noted taxonomist he was the first to recognise the Eastern Caspian Bighead Goby as being distinct from *P. kessleri*.

Ilse

Characin sp. *Roeboides ilseae* WA Bussing, 1986

Ilse Marie Bussing (b.1972) was the author's daughter. She accompanied him on many collecting trips and helped in both the collection and the sorting of specimens. William and Mary College, Virginia, awarded her bachelor's degree in comparative literature; the University of Costa Rica awarded her master's in Latin-American literature. Since 2012 she has been a Professor teaching modern languages.

Imai

Imai's Pygmy-goby *Trimma imaii* T Suzuki & H Senou, 2009

Keisuke Imai was a volunteer at the Fish Division, Kanagawa Prefectural Museum of Natural History, Kanagawa, Japan. He collected the type specimen.

Imamura

Eelpout sp. *Lycenchelys imamurai* ME Anderson, 2006

Imamura's Ghost Flathead *Hoplichthys imamurai* Y Nagano, MA McGrouther & M Yabe, 2013

Imamura's Sandperch *Parapercis imamurai* JW Johnson & J Worthington Wilmer, 2018

Dr Hisashi Imamura is a Japanese ichthyologist who has, for some years (since 2009), been Associate Professor of the Fisheries Science faculty at the Hokkaido University Museum, Hakodate, which institution awarded his PhD. He was previously Associate Professor (2005–2009) and Assistant Professor (1999–2005) at their museum. He was also (1996–1999) Domestic Research Fellow of Japan Science and Technology Corporation. He is one of the world's experts on the systematics of the flatheads (Family *Platycephalidae*). He has written a number of papers, and longer works such as *Deep Sea Fishes of Peru* (2010). He has been a visiting researcher at the Australian Museum (2006 & 2012). He was honoured for his valuable contributions to the taxonomy of pinguipedid fishes.

Imelda

Shiner sp. *Notropis imeldae* MT Cortés, 1968

Imelda Martinez was one of the collectors of the holotype (1965), in Oaxaca, Mexico.

Imperiorient

Goby sp. *Stiphodon imperiorientis* RE Watson & IS Chen, 1998

The binomial is a combination of the Latin words *imperator* (emperor) and *orientis* (of the east). This is a reference to Emperor Akihito of Japan (See **Akihito**).

Inayat

Inayat's Goatfish *Parupeneus inayatae* F Uiblein, TA Hoang, U Alama, R Causse, OE Chacate, Fahmi, S Garibay & P Matiku, 2018

Iin Inayat Al Hakim is a marine biodiversity research scientist (1986–present) at the Research Centre for Oceanography, Indonesian Institute of Sciences (LIPI), Jakarta, Indonesia. She is in charge of the marine zoology laboratory. She studied at Bogor Agricultural University, and was awarded her MSc by Jakarta National University. Among her published papers are the co-written: *Polychaeta (Annelida) of the Natuna Islands, South China Sea* (2004) and *Turbo-taxonomy: 21 new species of Myzostomida (Annelida)* (2014). She was honoured "*... for her support of fish taxonomy and invaluable logistic assistance when hosting the senior author at the NCIP fish collection at LIPI Jakarta during research visits.*"

Inca

Lake Titicaca Pupfish sp. *Orestias incae* S Garman, 1895
Incan Cusk-eel *Lepophidium inca* CR Robins & RN Lea, 1978

Not a true eponym, but named after the Inca people of Peru.

Indermaur

Zambian Cichlid sp. *Orthochromis indermauri* FDB Schedel, EJW Vreven, BK Manda, E Abwe, AC Manda & UK Schliewen, 2018

Dr Adrian Indermaur (b.1984) is a Swiss ichthyologist who is (2014) Head of the animal facility at the Waler Salzburger Laboratory. The University of Basel awarded his BSc (2007), his MSc (209) and his PhD (2014). He is responsible for the supervision, organisation and development of the facility, including the execution and permission of animal experiments and building and maintaining their fish specimen collection. He is involved in the preparation and assistance of field-work in Zambia and Cameroon. He was the first to document this new species with underwater photographs, videos, and with aquarium observations. As the etymology puts it: "*...thereby contributing to a large extent to our knowledge of behaviour and ecology of this species.*" He loves to travel and is a snowboarder in his spare time.

Indrambarya

Bengal Hagfish *Eptatretus indrambaryai* T Wongratana, 1984

Professor Dr Boon Indrambarya (1907–1994) was a leading Thai marine biologist. He graduated in fish biology from Cornell University and worked as an assistant with the US Fish Commission (1929), and later at the Department of Fisheries, Ministry of Agriculture, Bangkok, Thailand. He was also Director of the Environmental and Ecological Research Institute, the Applied Scientific Research Corporation of Thailand. He was honoured in the name of the hagfish as being "*...one of the senior-most pioneer fisheries biologists of Thailand.*" A building is named after him at the Faculty of Fisheries, Kasetsart University, Bangkok, where he was the Dean (1944–1962) and from where he retired as Professor Emeritus (1974). The National College of Education, Evanston, Illinois awarded him an honorary doctorate (1963).

Indra Montri

Labeo sp. *Labeo indramontri* HM Smith, 1945

Francis Henry Giles (1869–1951) – Phraya Indra Montri Sri Chandra Kumara was the Thai title and name conferred on him by the King of Siam (Thailand). He joined the British Colonial Service (1887) and served in Burma (Myanmar) as an Assistant District Officer. He was 'loaned' to the Siamese Government (1897) for service in the Ministry of Finance, where he eventually became Director-General of Revenues until failing eyesight meant he had to retire (1930). He stayed on after retirement and lived in Thailand until his death. He was President of the Siam Society for many years and Smith's etymology mentions his "*... untiring labors in extending the knowledge of the history, culture, and natural resources of Thailand.*" He was a brilliant linguist, speaking at least 12 of the languages in use in Burma as well as Siamese. He wrote: *Adversaria of Elephant Hunting* (1930).

Indroyono

Indonesian Houndshark *Hemitriakis indroyonoi* WT White, LJV Compagno & Dharmadi, 2009

Dr Dwisuryo Indroyono Soesilo (b.1955) is an engineer and geologist. He is Secretary & Deputy Senior Minister of the Coordinating Ministry for People's Welfare of the Republic of Indonesia and was Chairman, Marine and Fisheries Research Agency, Jakarta, Indonesia (2009). Institut Teknologi Bandung, Bandung, Indonesia awarded his bachelor's degree (1979), the University of Michigan awarded his master's (1981) and the University of Iowa his doctorate (1987). The shark was: "*...named for Dr Indroyono Soesilo, who has provided a great deal of support for shark research in Indonesia and was a strong advocate for the production of the field guide to sharks and rays of Indonesia.*"

Inger, ML

Batfish sp. *Halieutopsis ingerorum* MG Bradbury, 1988

Mary Lee Inger (1918–1985) was the wife of ichthyologist Robert Inger. She was thanked "*for friendship and wise counsel through the years*". The batfish was named after both Mary and her husband (see next entry).

Inger, RF

Borneo Cyprinid sp. *Parachela ingerkongi* PM Bănărescu, 1969
Batfish sp. *Halieutopsis ingerorum* MG Bradbury, 1988
Goby sp. *Stenogobius ingeri* RE Watson, 1991
Borneo Barb sp. *Osteochilus ingeri* J Karnasuta, 1993
Hillstream Loach sp. *Gastromyzon ingeri* HH Tan, 2006

Dr Robert 'Bob' Frederick Inger (b.1920–2019) was an American herpetologist and ichthyologist. He was Curator Emeritus of Amphibians and Reptiles, Field Museum, Chicago, having started as a University of Chicago student volunteer. From the 1950s his special subject was Southeast Asian (particularly Borneo) herpetology. He co-wrote: *Living Reptiles of the World* (1957). The *Parachela ingerkongi* is named after Inger and Chin Phui Kong (q.v.). Seven reptiles and 11 amphibians are also named after him. The batfish is named after both Bob Inger and his wife Mary Lee Inger. (Also see Inger, ML and Kong, CP)

Inglis

Stone Loach sp. *Nemacheilus inglisi* SL Hora, 1935
[Syn. *Schistura rupecula inglisi*]

Charles McFarlane Inglis (1870–1954) was a Scottish naturalist and planter, who went to India (1888). He was Curator, Darjeeling Museum (1926–1948). He co-wrote: *Birds of an Indian Garden* (1924) and two birds and two Odonata are named after him.

Ingolf

Ingolf Duckbill Eel *Nessorhamphus ingolfianus* EJ Schmidt, 1912

The 'Ingolf' was a Danish ship from which the holotype was collected.

Inhaca

Inhaca Fringelip *Cirrhimuraena inhacae* JLB Smith, 1962

A toponym, referring to Inhaca in Mozambique where the holotype was collected.

Inirida

Dwarf Cichlid sp. *Apistogramma iniridae* SO Kullander, 1979

This is a toponym referring to the Rio Inírida, the major river in the area (Guainia, Colombia) where type material was collected.

Innes

Teardrop Tubeshoulder *Holtbyrnia innesi* HW Fowler, 1934
Neon Tetra *Paracheirodon innesi* GS Myers, 1936

Dr William Thornton Innes (1874–1969) was an American publisher and aquarist. Temple University made him a Doctor of Humane Letters (1951). He founded, published and edited *The Aquarium*, a monthly magazine that ran for 35 years (1932–1967). He sent specimens of the tetra – now one of the world's most popular aquarium fishes – to Myers to have them identified.

Inosima

Morid Cod sp. *Lepidion inosimae* A Günther, 1887

This is a toponym, referring to Inosima, a site in Japan from which H.M.S. *Challenger* obtained the first specimens.

INPA

Characin sp. *Bryconops inpai* H-J Knöppel, WJ Junk & J Géry, 1968
Pike Cichlid sp. *Crenicichla inpa* A Ploeg, 1991
Long-whiskered Catfish sp. *Aguarunichthys inpai* JAS Zuanon, LH Rapp Py-Daniel & M Jégu, 1993
Inpa Tube-snouted Ghost Knifefish *Sternarchorhynchus inpai* CDCM de Santana & RP Vari, 2010

INPA is the acronym for Instituto Nacional de Pesquisas da Amazônia, which helped fund the authors' field work (*Aguarunichthys inpai*) and whose staff supplied Ploeg with 'enormous amounts of material' which served partially as the basis for his revision of the cichlid genus.

INRA

Characin sp. *Moenkhausia inrai* J Géry, 1992

INRA is an acronym for the Institut National de la Récherche Agronomique, Centre Antilles-Guyane (French Guiana).

Intes

Whitelip Moray *Gymnothorax intesi* P Fourmanoir & J Rivaton, 1979

Dr André Intès is a French ichthyologist. He worked in New Caledonia and helped in the collection of several new species of eels there. He was at the Centre de Rechecherche Oceanographiques, Abidjian, Ivory Coast and, later, Directeur de Recherches, ORSTOM, Brest (1995–2001). He wrote: *Pêche profonde aux casiers en nouvelle-calédonie et iles adjacentes: essais préliminaires* (1978) and *La Nacre en Polynésie Française* (1982).

Intha

Inle Carp *Cyprinus intha* N Annandale, 1918

This is not a true eponym: the Intha are an ethnic group who live in the region of Inlé Lake (Myanmar), where the holotype was collected.

Investigator

Rattail sp. *Nezumia investigatoris* AW Alcock, 1889
Armoured Gurnard sp. *Scalicus investigatoris* AW Alcock, 1898
Perchlet sp. *Chelidoperca investigatoris* AW Alcock, 1890
Pufferfish sp. *Canthigaster investigatoris* N Annandale & JT Jenkins, 1910
Investigator Pipefish *Cosmocampus investigatoris* SL Hora, 1926
Broadnose Catshark *Apristurus investigatoris* KS Misra, 1962
Scorpionfish sp. *Scorpaenodes investigatoris* WN Eschmeyer & KV Rama-Rao, 1972

Named after the Royal Indian Marine Survey vessel *Investigator*, which made collections of marine fauna in the period 1884–1926.

Io

Andaman Sea Spiny Dragonet *Callionymus io* R Fricke, 1983

In Greek mythology, Io was one of the mortal lovers of the god Zeus. In an attempt to hide her from his wife, Hera, Zeus transformed Io into a white heifer. Fricke gives no rationale for his choice of name.

IOAN

Barbeled Dragonfish sp. *Eustomias ioani* NV Parin & Pokhil'skaya, 1974

IOAN is an acronym for Institut Okeanologii Akademii Nauk (Institute of Oceanology, Academy of Sciences of the USSR) where the authors worked.

Ionah

Jonah's Icefish *Neopagetopsis ionah* O Nybelin, 1947

(See **Jonah**)

Ios

Whipnose Angler sp. *Gigantactis ios* E Bertelsen, TW Pietsch III & RJ Lavenberg, 1981
White-headed Hagfish *Myxine ios* B Fernholm, 1981

IOS stands for the Institute of Oceanographic Sciences, England, which supplied the type specimen of the hagfish. The describers of the anglerfish use the name "*in recognition of important ichthyological contributions made by our colleagues of that institution*". (Note that *ios* is also a Latin word for arrow, and is used in that context in the name of the Arrow Goby, *Clevelandia ios*)

IPN

Lantern Minnow *Tampichthys ipni* J Alvarez del Villar & Navarro, 1953

This is not a true eponym, but refers to an educational institution. The authors both worked at the Instituto Politecnio Nacional (IPN), México.

Ira

Ira's Snook *Centropomus irae* A Carvalho-Filho, J de Oliveira, C Soares & J Araripe, 2019

Dr Maria Iracilda da Cunha 'Ira' Sampaio is a Brazilian evolutionary biologist and geneticist who is an ichthyological researcher, Associate Professor and post-graduate Director (2012) of the Institute of Coastal Studies, at the Federal University of Pará. She graduated from the Federal Rural University of Amazonia in veterinary medicine (1977) then undertook her MSc in genetic biology at Ribeirão Preto Faculty of Medicine (1984), completing her PhD in biological sciences at the Federal University of Pará (1993). She has had more than 200 papers published, as well as writing or contributing to half a dozen books. The etymology reads: "*The new species is named after Dr. Iracilda Sampaio (Federal University of Pará, Bragança, Pará, Brazil) in recognition of her lifelong contribution to the understanding of the genetic diversity of the fauna of the Amazon region, in particular fish, and her profound dedication to science and teaching.*"

Iracema

Sand Knifefish genus *Iracema* ML Triques, 1996

'Iracema' is the name of an eponymous 1865 Brazilian novel by José de Alencar. The story revolves around the relationship between Iracema (a beautiful indigenous woman) and a Portuguese colonist. It is also a female personal name in Brazil, but why the name was chosen for the knifefish is a mystery.

Irav

Grunter sp. *Mesopristes iravi* T Yoshino, H Yoshigou & H Senou, 2002

Dr Richard Peter Vari (1949–2016). The authors' etymology says that the name: "*...iravi is the reversed name in genitive form of Richard P. Vari (USNM), whose contribution to the taxonomy of the family Terapontidae is greatly appreciated.*" It is unclear why they chose to reverse his name. (See Vari)

Iredale

Coral Worm-Eel *Scolecenchelys iredalei* GP Whitley, 1927

Tom Iredale (1880–1972) was an English artist and naturalist who spent much of his life in Australia. He started work apprenticed to a pharmacist, became a clerk, and later Secretary to Gregory MacAlister Mathews, the Australian ornithologist, after working with him for a number of years at the British Museum. While Mathews is credited as the author of: *Birds of Australia*, Iredale is said to have written much of the text. He collected widely in the Kermadec Islands and Queensland, and accompanied Whitley on a collecting trip to Michaelmas Cay, Great Barrier Reef, the eel's type locality.

Irene

Splitfin sp. *Girardinichthys ireneae* AC Radda & MK Meyer, 2003
[Syn. *Hubbsina ireneae*]

Mrs Irene Radda is the wife of the senior author, Alfred C Radda.

Irina (Shandikova)

Pygmy Icefish *Channichthys irinae* GA Shandikov, 1995

Irina Shandikova is the sister of the author, Gennadiy Shandikov.

Irina (Winterbottom)

Headstander sp. *Pseudanos irinae* R Winterbottom, 1980
Irina's Pygmygoby *Trimma irinae* R Winterbottom, 2014

Irina Winterbottom is the wife of the author, Dr Richard Winterbottom. She was honoured "*...as a small token of my appreciation and gratitude for her patience and forbearance of what she once referred to as my 'magnificent obsession' with coral reef fishes.*"

Iris

Cusk Eel sp. *Barathrites iris* E Zugmayer, 1911
Rainbow Cusk Eel *Ophidion iris* CM Breder, 1936
Lake Victoria Cichlid sp. *Haplochromis iris* RJC Hoogerhoud & F Witte, 1981
Strickland Rainbowfish *Melanotaenia iris* GR Allen, 1987

In Greek mythology Iris, an Oceanid, was the messenger of Hera and goddess of the rainbow. The name is usually applied to 'rainbow coloured' species and has also been applied to six birds, a mammal and an amphibian. However, *Ophidion iris* is said to be named for its unusual iris, in which the upper but not the lower part is black.

Iris

Characin sp. *Attonitus irisae* RP Vari & H Ortega, 2000

Iris Margot Samanez Valer is head of the Department of Limnology, Museum of Natural History, National University of San Marcos, Lima, Peru where she is a Senior Professor. Her master's degree was in Aquatic Resources and she is an expert in phyto- and zoo-plankton taxonomy, mainly from Amazonian and high Andean aquatic ecosystems. She co-wrote: *Geographical distribution of Boeckella and Neoboeckella (Calanoida: Centropagidae) in Peru* (2014).

Irolita

Softnose Skate genus *Irolita* GP Whitley, 1931

Irolita is a beautiful female character in a fairy story written by Marie-Catherine Le Jumel de Barneville, Baroness d'Aulnoy (1650–1705). The reasoning behind this choice of name is not explained by the author.

IRSAC

Spotfin Goby-Cichlid *Tanganicodus irsacae* M Poll, 1950
Lake Tanganyika Catfish sp. *Synodontis irsacae* H Matthes, 1959

Based on an acronym: I.R.S.A.C. = Institut pour la Recherche Scientifique en Afrique Centrale.

Irvine

African Catfish genus *Irvineia* E Trewavas, 1943
Spineback Guitarfish *Rhinobatos irvinei* JR Norman, 1931
West African Cichlid sp. *Paragobiocichla irvinei* E Trewavas, 1943
[Syn .*Steatocranus irvinei*]

Dr Frederick Robert Irvine (1898–1962) was a botanist and science teacher who made a collection of fishes near Accra, Gold Coast (Ghana). He collected the type specimens of the guitarfish and the cichlid. He wrote: *The fishes and fisheries of the Gold Coast* (1947).

Irwin

Southern Bass sp. *Percilia irwini* CH Eigenmann, 1928

William Glanton Irwin (1866–1943) was an Indiana philanthropist who sponsored the Indiana University's 'Irwin Expedition' to Chile and Peru. Eigenmann collected fishes during this expedition. Eigenmann typically gives no etymology, but we are sure this is the person intended.

Isaac

Armoured Catfish sp. *Rineloricaria isaaci* MS Rodriguez & AM Miquelarena, 2008

Dr Isaäc J H Isbrücker (**see Isbruecker**).

Isaacs

Onejaw sp. *Monognathus isaacsi* SN Raju, 1974
Lanternfish sp. *Nannobrachium isaacsi* RL Wisner, 1974
Tubeshoulder sp. *Maulisia isaacsi* T Matsui & RH Rosenblatt, 1987

John Dove Isaacs III (1913–1980) was an oceanographer, naturalist, engineer, fisherman and marine scientist. The University of California, Berkeley, awarded his bachelor's degree (1944). He was Professor of Oceanography at the Scripps Institution of Oceanography (1948–1971) and Director, University of California's Institute of Marine Resources (1971–1980).

Isaline

Smallscale Pike-Characin sp. *Acestrorhynchus isalineae* NA de Menezes & J Géry, 1983

Mrs Isaline Drecq was the wife of Guy van den Bossche, who was a member of the expedition to Brazil during which the holotype was collected. At the time of the description, she was referred to as 'the late' Mrs. Isaline Drecq.

Isarankura

Short-jaw Saury *Saurida isarankurai* S Shindo & U Yamada, 1972

Andhi Praja Isarankura (1935–2006) was a Thai fisheries biologist who graduated from the University of Washington, Seattle (1960). He returned to Thailand and worked at the Marine Fisheries Laboratory, Bangkok. (1963–1970). He worked for the UN Food and Agriculture Organization, Rome (1970–1995) overseeing field projects to establish sustainable fisheries in Asia and the Pacific Islands. He retired to Seattle but was diagnosed with prostate cancer (2003).

Isbrücker

Armoured Catfish genus *Isbrueckerichthys* E Derijst, 1996
Armoured Catfish sp. *Hypostomus isbrueckeri* RE dos Reis, C Weber & LR Malabarba, 1990
Amazonian Cichlid sp. *Crenicichla isbrueckeri* A Ploeg, 1991
Armoured Catfish sp. *Farlowella isbruckeri* ME Retzer & LM Page, 1997
Longnose Cory sp. *Corydoras isbrueckeri* J Knaack, 2004
Isbrücker's Hypopygus (knifefish) *Hypopygus isbruckeri* CDCM de Santana & WGR Crampton, 2011

Dr Isaäc J H Isbrücker (b.1944) is a Dutch ichthyologist who is an expert on the taxonomy of South American armoured catfishes (Callichthyidae, Loricariidae). He was a keeper at the aquarium, Amsterdam Zoo (1960–1985) and was responsible for the fish collection, Zoological Museum Amsterdam (1986–2005). He was Ploeg's colleague, friend and teacher. He is now retired but still active in research. He wrote more than 80 papers including: *A treatise of the Loricariidae Bonaparte, 1831, a family of South American mailed catfishes, with emphasis on the subfamily Loricariinae (Pisces, Siluriformes) and a provisional key to the genera of Loricariidae* (2017).

Ischikawa

Japanese chub genus *Ischikauia* DS Jordan & JO Snyder, 1900

See under Ishikawa.

Iselin

Blind Cusk Eel sp. *Paraphyonus iselini* JG Nielsen, 2015

'Columbus Iselin' is a research vessel from which this species was caught.

Iser(t)

Striped Parrotfish *Scarus iseri* ME Bloch, 1789

[Binomial sometimes amended to *iserti*]

Paul Erdmann Isert (1756–1789) was a German botanist, also known for his opposition to the Danish-Norwegian slave trade. In 1788, he published his most famous book: *Reise nach Guinea und den Caribäischen Inseln in Columbia* (Journey to Guinea and the Caribbean Islands in Columbia), wherein he described his experiences with, and his views on, the slave trade.

Ishihara

Abyssal Skate *Bathyraja ishiharai* M Stehmann, 2005

Dr Hajime Ishihara (b.1950) is a Japanese ichthyologist and Environment Specialist who was connected with the Iraq Sea Line Project under JICA (Japan International Corporation Agency) and is now with W & I Associates Corporation, Fujisawa, Japan. Tokyo University of Fisheries awarded his bachelor's degree (1975) and his master's (1977), and the University of Tokyo his doctorate (1990). He is currently working on the reconstruction of the Iraq Crude Oil Pipeline and has been based in Amman, Jordan since 2009. He has written *First record of a skate,* Anacanthobatis borneensis *from the East China Sea* (1984) and co-wrote, with Matthias Stehmann, *A second record of the deep-water skate* Notoraja subtilispinosa *from the Flores Sea, Indonesia* (1990). In honouring him in the name of the Abyssal Skate, Stehmann described him as his 'skatology' colleague and friend of more than 25 years, who produced important revisions of North Pacific *Bathyraja*.

Ishikawa

Japanese chub genus *Ischikauia* DS Jordan & JO Snyder, 1900

Slender Ribbonfish *Trachipterus ishikawae* DS Jordan & JO Snyder, 1901

Ishikawa's Sculpin *Furcina ishikawae* DS Jordan & EC Starks, 1904

Noodlefish sp. *Neosalangichthys ishikawae* Y Wakiya & N Takahashi, 1913

Korean Taimen *Hucho ishikawae* T Mori, 1928

Japanese salmon ssp. *Oncorhynchus masou ishikawae* DS Jordan & EA McGregor, 1925

Professor Dr Chiyomatsu Ishikawa (1861–1935) was a Japanese zoologist, evolutionary theorist and ichthyologist. After graduating from Tokyo University he spent time studying in Germany under August Weismann. He was at the Naples Zoological Station (1887). He was a zoologist at the College of Agriculture, Imperial University, Tokyo and Curator of the Imperial Museum. He was largely responsible for bringing Darwinism to Japan. At some stage, he was also principal of the Dokkyo Middle School, Tokyo. An amphibian is also named after him.

Ishiyama, N & P

Snake-eel sp. *Ophichthus ishiyamorum* JE McCosker, 2010

Nelson (b.1944) and Patsy (b.1937) Ishiyama are siblings and philanthropists who give generous support to ichthyological research. Nelson gave up being a Stanford-qualified lawyer in favour of building (1991) and operating a fishing lodge in Idaho, as he and his wife are avid fly-fishers. He is President of the Ishiyama Corporation and a trustee of the California Academy of Sciences. Patsy is also a keen fisher and regularly visits her brother in Idaho and is Non-Executive Vice Chairman at National Fish & Wildlife Foundation.

Ishiyama, R

Plain Pygmy Skate *Fenestraja ishiyamai* HB Bigelow & WC Schroeder, 1962

Dr Reizo Ishiyama (1912–2008) was a Japanese ichthyologist who graduated from Tokyo Fisheries College (Tokyo University of Marine Science and Technology). After graduating he developed tuberculosis and worked as an engineer at a small fish-farming factory in Tokorozawa, Saitama prefecture, to be able to live a quiet life whilst recovering from the illness. After his health improved, he became an Assistant at the Faculty of Agriculture, Kyoto University (1947–1951). He worked at Shimonoseki College of Fisheries as Assistant Professor (1951–1953) and as Professor (1953–1967). He then was Professor at Tokyo University of Fisheries (1967–1975). Over his career he wrote many articles and papers including *Studies on the rajid fishes (Rajidae) found in the waters around Japan* (1958) and, jointly with Ishihara, *Five new species of skates in the genus* Bathyraja *from the western north Pacific, with reference to their interspecific relationships* (1977). He was honoured in the skate's name for his work on Japanese batoids.

Ishmael

Lake Victoria Cichlid sp. *Haplochromis ishmaeli* GA Boulenger, 1906

George Ishmael was an Interpreter to the Police Court at Entebbe, to whom Edward Degen (Swiss ornithologist who collected the type) was "*...indebted for valuable assistance during his stay in Uganda.*"

Isidore

Mormyrid sp. *Pollimyrus isidori* A Valenciennes, 1847

Isidore Geoffroy Saint-Hilaire (1805–1861) was a French zoologist. Having studied medicine and natural history he became Assistant to his father, Etienne Geoffroy Saint-Hilaire (q.v.), (1824) at the MNHN, Paris. He lectured on ornithology and taught zoology there (1829–1832). He was particularly interested in deviant forms rather than the norm. He published his work on teratology (i.e. what makes organisms deviate from normal), *Histoire générale et particulière des anomalies de l'organisation chez l'homme et les animaux* (in sections 1832–1837). He became Deputy to his father at the Faculty of Science in Paris (1837) and went to Bordeaux to organise a faculty there. He became Inspector of the Academy of Paris (1840),

Professor of the Museum upon the retirement of his father (1841), Inspector General of the University of Paris (1844), and a member of the Royal Council for Public Instruction (1845). On the death of Henri Marie Ducrotay de Blainville (1850), he succeeded him as Professor of Zoology at the Faculty of Science. He founded the Acclimatization Society of Paris (1854) and was its President. He published a number of scientific papers, essays, and longer works on zoology, palaeontology, anatomy, and the domestication of animals, including: *Essais de zoologie generale* (1841) and *Histoire naturelle générale des règnes organiques* (published in three volumes 1854–1862). The mormyrid holotype came from his collection.

Iskandar

Stone Loach genus *Iskandaria* AM Prokofiev, 2009

Iskandar is the Arabic form of Alexander and here relates to Alexander the Great (BC 356–BC 323).

Isla

Tiger Limia *Limia islai* R Rodriguez-Silva & PF Weaver, 2020

Dominic Isla (d.2020) was an aquarist and dealer who ran CoCo Island Aquatics for 45 years (1975–2020). He also wrote on tropical freshwater fish such as: *Endangered Livebearers at Staten Island Zoo* (1995). He brought the first specimens of the Limia from Lake Miragoane in Haiti, although he did not realise it was a novel species. The etymology states that the fish is named "*...in honour of the late Dominic Isla, one of the first collectors of the novel species and the person credited with introducing the tiger limia to the aquarium hobby.*"

ITA

Suckermouth Catfish sp. *Astroblepus itae* CA Ardila Rodríguez, 2011

ITA is an acronym for Instituto Técnico Agrícola, Cáchira, Norte de Santander, Colombia. The name was given to celebrate the Institute's 55th anniversary.

Itany

Dolphin Cichlid *Krobia itanyi* J Puyo, 1943

This is a toponym referring to Itany, the French name of Litani River, Marowijne (or Maroni) River drainage, part of the boundary between Suriname and French Guiana, the type locality.

Ito

Sweeper sp. *Pempheris itoi* HW Fowler, 1931

Kumataro Ito (c.1860–c.1930) was a natural history painter at the Department of Zoology at Tokyo University. He was artist on board the US Bureau of Fisheries Steamer 'Albatross' during the Philippine Expedition (1907–1910) led by HM Smith. Ito was probably hired by Smith on Kamakichi Kishinouye's recommendation. He was honoured in the name for his many colour sketches of Philippine-East Indian fishes (he made 200 paintings of fish and molluscs).

Itou

Oriental Goatfish *Upeneus itoui* M Yamashita, D Golani & H Motomura, 2011

Mr. Masahide Itou is a Japanese ichthyologist based in Kagoshima, Kyushu, Japan. He collected specimens of this goatfish and made them available to the authors.

Ittoda

Samurai Squirrelfish *Sargocentron ittodai* DS Jordan & HW Fowler, 1902

Not an eponym, but apparently taken from a local name the describers encountered in Okinawa (the type locality). *Itto* means 'the number one' (i.e. the best of many).

Ivanova

African Rivuline sp. *Nothobranchius ivanovae* S Valdesalici, 2012

Iva Ivanova is a Bulgarian aquarist. She discovered and collected this species in western Tanzania. She has undertaken a number of collecting trips with Kiril Kardashev who has co-authored papers with the author.

Ivantsoff

Ivantsoff's Blue-eye *Pseudomugil ivantsoffi* GR Allen & SJ Renyaan, 1999

Dr Walter Ivantsoff is an Australian ichthyologist. He was employed by the Macquarie University, Sydney (1968–2000) and is presently a Senior Research Fellow. He was (until 2016) a Research Associate of the Australian Museum, the (then) Director of which, Dr Frank Talbot, supervised his PhD (1970s). He has collected fish in many parts of Australia as well as teaching a course on fish biology. He has written a number of papers such as: *Review of the Australian fishes of the genus Atherinomorus (Atherinidae)* (1991).

Ivindo

Mormyrid genus *Ivindomyrus* L Taverne & J Géry, 1975

This is a toponym; it refers to the Ivindo River, Gabon, where the holotype of *I. opdenboschi* was collected.

Iwama

Long-tailed River Stingray *Plesiotrygon iwamae* R de S Rosa, HP Castello & TB Thorson, 1987

Satoko Iwama (d.1987) was a Brazilian zoologist who was a student and teacher at the Instituto de Botânica, São Paulo. He wrote: *The Pollen Spectrum of the Honey of Tetragonisca angustula angustula Latreille (Apidae, Meliponinae)* (1975) as part of his master's theses.

Iwamoto

Blackbar Drum *Pareques iwamotoi* GC Miller & LP Woods, 1988
Long-spine Anglerfish *Lophiodes iwamotoi* H-C Ho, B Séret & K-T Shao, 2011
Iwamoto's Whiptail *Hymenocephalus iwamotoi* W Schwarzhans, 2014

Dr Tomio Iwamoto (b.1939) is Curator Emeritus, Ichthyology at the Institute for Biodiversity Science and Sustainability at California Academy of Sciences. After spending six months in active duty as an Army reservist, he began working as a fishery biologist for the then US Bureau of Commercial Fisheries (now National Marine Fisheries Service) at the Exploratory Fishing & Gear Research Station in Pascagoula, Mississippi and later at the field station on St Simons Island, Georgia. He then returned to graduate school at the University of Miami, Rosenstiel School of Marine and Atmospheric Science, which awarded his MSc and PhD. He spent a year at Oregon St University as a lecturer before starting employment with the California Academy of Sciences where he was Curator of Ichthyology for 39 years (1972–2011). He has been visiting researcher at a number of museums such as The Australian Museum (1993 & 2003). His principle research interest is the systematics of grenadiers, a group of more than 400 deep-sea fishes related to the codfish; an interest that began in the early 1960s while he was employed as a fishery biologist working in waters of the tropical western Atlantic. In pursuit of this interest, he has cruised on oceanographic and fishery vessels over most of the tropical Atlantic, the Pacific coast of North America from the Aleutian Islands to southern California, the Hawaiian Islands, the Philippines, and other areas. He also undertook a study of the marine fishes of West Africa, collecting there aboard the Norwegian fishery research vessel 'Dr Fridtjof Nansen' (2005, 2007, 2010 & 2012). Among his most important publications is: *Grenadiers (families Bathygadidae and Macrouridae, Gadiformes, Pisces) of New South Wales, Australia* (2001).

Iwatsuki

Silverside sp. *Doboatherina iwatsukii* D Sasaki & S Kimura, 2019

Yukio Iwatsuki is Professor of the Osakan Laboratory of the Department of Marine Biology and Environmental Sciences at the University of Miyazaki. He has worked at the University since 1983 becoming professor 2000 and his current post 2015. He was honoured because his :*...collections of atherinid species greatly contributed to the authors' study*." He has published more than 130 articles such as the co-authored: *Taxonomic review of the genus Argyrops (Perciformes; Sparidae) with three new species from the Indo-West Pacific* (2018).

Izechsohn

dNeotropical Rivuline sp. *Xenurolebias izecksohni* COG da Cruz, 1983

Dr Eugenio Izecksohn (1932–2013) was a Brazilian herpetologist and a specialist in Neotropical anurans. He worked for the Department of Animal Biology, Universidade Federal Rural do Rio de Janeiro, where he graduated (1953) and was an Emeritus Professor. A mammal and nine amphibians are also named after him.

J

Jaarman

Goby sp. *Oligolepis jaarmani* MWC Weber, 1913

Dr Raden Jaarman Soemintral Zeerban aka Ergban was a Javanese physician. He took part in Hendrikus Lorentz's (q.v.) Second South New Guinea Expedition (1909–1910) as one of two expedition physicians. The type was collected during the expedition, possibly by Jaarman.

Jacad

Bigeye Gurnard *Pterygotrigla jacad* WJ Richards & T Yato, 2014

'Jacad' is an invented portmanteau formed from the names of the senior author's two grandsons, Jacob and Cade.

JACE

Zippered Dottyback *Pseudochromis jace* GR Allen, AC Gill & MV Erdmann, 2008

An acronymic eponym composed of the first letter of **J**onathan, **A**lex, **C**harlie and **E**mily, the children of Lisa and Michael Anderson, who successfully bid to conserve this species at the Blue Auction, a black-tie charity auction in Monaco (20 September 2007) in support of Conservation International's Bird's Head Seascape marine conservation initiative.

Jack Brooks

Slender Mudgoby *Gobiopsis jackbrooksi* GR Allen, MV Erdmann & WM Brooks, 2018

John 'Jack' Moldaw Brooks is the third author's son. (Also see **Berry (Levy)**, **Bill Brooks** & **Stuart (Brooks)**)

Jack Randall

Randall's Coralbrotula *Diancistrus jackrandalli* W Schwarzhans, PR Møller & JG Nielsen, 2005

Dr John 'Jack' Ernest Randall Jr. (b.1924) (See **Randall**)

Jack Roberts

Tetra sp. *Hyphessobrycon jackrobertsi* A Zarske, 2014

Jack Roberts. (See **Roberts, J**)

Jackson

Pygmy Leatherjacket *Brachaluteres jacksonianus* JRC Quoy & JP Gaimard, 1824

This is a toponym, referring to Port Jackson in Australia.

Jackson (do Pandeiro)

Armoured Catfish sp. *Parotocinclus jacksoni* TPA Ramos, SY Lustosa-Costa, L de F Barros-Neto & JEL Barbosa, 2021

Jackson do Pandeiro (1919–1982) was the artistic name of José Gomes Filho (né José Gomes da Silva). He was a Brazilian composer and singer of Forró and Samba, two popular genres of Brazilian music. He was born into poverty in the state of Paraíba, north-east Brazil (where this catfish occurs); so poor he stole fruit from farms and picked through garbage for useable leftovers. The young José had no schooling: he worked as a shoeshine boy, supplementing his income with being a baker's assistant, street sweeper, roofer, painter, etc. He joined musical combos playing the tambourine, moving on to zabumba, drums, bongos and eventually pandeiro. He called himself 'Jack' after the actor Jack Perry who played the good guy in silent westerns. The director of a radio programme that José appeared on didn't like the sound of 'Zé Jack do Pandeiro', as José wanted to be known, and changed it to Jackson do Pandeiro, which stiuck. As singer, composer and drummer, he went on to make over 30 albums in a number of genre from samba to jazz.

Jackson, AC

Black Surfperch *Embiotoca jacksoni* L Agassiz, 1853

A C Jackson, while fishing in San Salita (Sausalito) Bay, California, caught an unusual fish that had living fish inside, which he described as "*...perfect miniatures of the mother*". He sketched an outline of the fish and sent it, along with a letter, to Louis Agassiz (1807–1873). We have not been able to discover further information about Jackson.

Jackson, FJ

Jackson's Barb *Enteromius jacksoni* A Günther, 1889
Marbled Mountain Catfish *Amphilius jacksonii* GA Boulenger, 1912
Victoria Robber (African tetra) *Brycinus jacksonii* GA Boulenger, 1912

Sir Frederick John Jackson (1859–1929) was an English administrator and explorer, but also a naturalist and keen ornithologist. He led an expedition financed by the British East Africa Company (1889) to explore the new Kenya colony, later becoming Lieutenant-Governor of the East African Protectorate (1907–1911). He was also Governor of Uganda (1911–1917), describing the country as 'a hidden Eden, a wonderland for birds'. He wrote: *The Birds of Kenya Colony and the Uganda Protectorate*, published posthumously (1938). Twenty-four birds, seven mammals and four reptiles are also named after him.

Jackson, FN

Stargazer sp. *Gillellus jacksoni* CE Dawson, 1982

Felix N Jackson was a museum technician at the Gulf Coast Research Laboratory Museum, Mississippi. He was honoured "*...in partial recognition for his years of competent and willing performance of myriad ichthyological chores.*"

Jackson, PBN

Lake Malawi Cichlid sp. *Copadichromis jacksoni* TD Iles, 1960

Ghost Stonebasher *Paramormyrops jacksoni* M Poll, 1967

Peter Brian Neville Jackson (1923–2007) was a South African biologist and ichthyologist. He was a pilot in the Royal Air Force in WW2. He started studying biology at Rhodes University, Grahamstown (1946), but switched universities and graduated from the University of Cape Town with a master's degree in biology. He went on to become a Zoological Assistant at Cape Town University. He joined the Joint Fisheries Research Organisation of Rhodesia and Nyasaland (1950), and later returned to Grahamstown to a post at the J L B Smith Institute of Ichthyology (1985). He wrote: *Fishes of Northern Rhodesia* (1961).

Jackson, WT

Bearded Eelgoby *Taenioides jacksoni* JLB Smith, 1943

William T Jackson was a South African amateur naturalist. *"Two specimens, one much shrunken, were obtained by Mr. W. T. Jackson, M.A., an enthusiastic and successful collector, after whom the species is named."* He collected them in a mud burrow in St Lucia estuary, Natal Province, South Africa. He co-wrote: *Causes of injury to flooded tobacco plants* (1954).

Jacob Freiberg

Freiberg's Peacock (cichlid) *Aulonocara jacobfreibergi* DS Johnson, 1974
[Alt. Malawi Butterfly, Eureka Red Peacock, Fairy Cichlid]

Jacob 'Jack' Freiberg of African Fish Import, Verona, New Jersey, USA was an importer of tropical fish. He co-collected the type. (Also see **Freiberg**)

NB In the original description the type locality is given as Makanjila Island, which is probably a name invented by the exporter shipping out to New York. Such practice was then common, as localities were rarely revealed to protect locations and sources. There is no such place as Makanjila Island and the type specimens were not collected near Makanjila Point, in the south-eastern part of the lake. Don S Johnson was probably unaware of this deliberate deception, and noted the locality named by importer Jack Freiberg who had this cichlid named after him.

Jacobson

Rasbora sp. *Rasbora jacobsoni* MCW Weber & LF de Beaufort, 1916

Edward Richard Jacobson (1870–1944) was a Dutch businessman and skilled amateur naturalist. He was manager of a trading company in Java, but he also lived for some years in Sumatra. He made extensive collections for Dutch museums, leaving his business (1910) to devote himself to natural history. His main interest was entomology, but he collected other taxa too. He died in an internment camp during the Japanese occupation. Two reptiles, two birds and an amphibian are also named after him.

Jacobus

Blackbar Soldierfish *Myripristis jacobus* G Cuvier, 1829

Jacobus is a latinization of James, and refers to the fish's local name *Frère-Jacques* (Brother Jim) in Martinique, which is the type locality.

Jacobus Boehike

Oriental Barb sp. *Systomus jacobusboehlkei* HW Fowler, 1958

James Erwin Böhlke (1930–1982) was an ichthyologist who worked at the Academy of Natural Sciences of Philadelphia. (See **Böhlke**)

Jaegar

Neotropical Rivuline sp. *Austrolebias jaegari* WJEM Costa & MM Cheffe, 2002

Norberto Henrique Jaegar (d.<2002) was a Brazilian nature photographer. The etymology says he was honoured for his "*enthusiasm and dedication to the conservation of natural areas.*" He collected the type with the authors and generally helped during collecting trips, as well as taking photographs of specimens.

Jäger

African Barb sp. *Labeobarbus jaegeri* M Holly, 1930
[Syn. *Varicorhinus jaegeri*]

Dr Gustav Jäger (1832–1917) was a German physician, naturalist, entomologist and hygienist. There is no etymology in Holly's description of this species, so any identification is provisional. Dr Jäger taught zoology in Vienna before becoming Professor of Zoology at Hohenheim Academy (1868) and then Professor of Physiology at the Stuttgart veterinary school. He abandoned teaching, started a practice as a physician in Stuttgart (1884), and promoted his views on what clothing to wear to protect health. His ideas inspired the creation of the well-known clothing brand – Jaeger. He wrote: *Die Darwinsche Theorie und ihre Stellung zu Moral und Religion* (1869).

Jagor

Philippine River Pipefish *Oostethus jagorii* WKH Peters, 1868
[Syn. *Microphis jagorii*]

Professor Dr Andreas Fedor Jagor (1816–1900) was a German ethnographer and naturalist who travelled in Asia, particularly the Philippines, in the second half of the 19th century, collecting for the Berlin Museum. He wrote *Reisen in den Philippinen* (1873). He described the country thus: 'Few countries in the world are so little known and so seldom visited as the Philippines, and yet no other land is more pleasant to travel in than this richly endowed island kingdom. Hardly anywhere does the nature lover find a greater fill of boundless treasure.' He also wrote about Indonesia and southern Malaya. Among other taxa named after him are two mammals, two reptiles and a bird.

Jagua

Scaled Herring *Harengula jaguana* F Poey, 1865

This is a toponym; it refers to the Bahía de Jagua, Cuba.

Jaime

Jaime's Tube-snouted Ghost Knifefish *Sternarchorhynchus jaimei* CDCM de Santana & RP Vari, 2010

Dr Jaime Ribeiro Carvalho Júnior is a Brazilian molecular biologist at Centro Jovem de Aquarismo. The Universidade Estadual Vale do Acaraú awarded his bachelor's degree (2004) whilst Universidade Federal do Para awarded his master's degree (2008) and his doctorate (2014). He co-wrote: *Production chain of ornamental fish* (2009).

Jakles

Jakles' Loach *Nemacheilus jaklesii* P Bleeker, 1852

Surgeon Lieutenant Colonel P Jakles was a physician in the Dutch East Indian Army stationed in West Sumatra. He collected the holotype and gave it to his friend and colleague, Bleeker.

Jalla

Nembwe (African cichlid) *Serranochromis jallae* GA Boulenger, 1896
[Syn. *Serranochromis robustus jallae*]

Rev. Luigi (Louis) David Jalla (1860–1943) was an Italian protestant missionary for the *Société des Missions Évangeliques,* Paris. He was ordained (1886) and sent to Africa, being in Northern Rhodesia (Zambia) and Southern Rhodesia (Zimbabwe) (1887–1895, 1898–1905, 1907–1911, 1912–1918 & 1919–1922). Peracca wrote (1886) that Jalla collected reptiles and amphibians "*…along the road from Kazangula to Bulawayo.*" He wrote: *Du Cap de Bonne Espérance au Victoria Nyanza. Notes de voyage* (1905). He collected the type.

Jamal

Jamal's Dottyback *Manonichthys jamali* GR Allen & MV Erdmann, 2007

Jamal was an "*…enthusiastic and hard-working young crew member of the diving vessel M.V. Citra Pelangi, who died as the result of a tragic shipboard accident*" during the authors' 2006 exploratory survey of the Fak Fak-Kaimana region of Barat Province (western New Guinea), Indonesia, where this species occurs.

Jamal (Siddiqui)

Fanged Seabream *Sparidentex jamalensis* SA Amir, PJA Siddiqui & R Masroor, 2014

Dr Pirzada Jamal Ahmed Siddiqui is an ichthyologist who was Professor at, and Director of the Centre of Excellence for Marine Biology at the University of Karachi, Pakistan (2009). He is author or co-author of around 120 papers. He was honoured because his: "*… support and contributions to the work on* [the] *marine fauna of Pakistan is immense and noteworthy.*"

James, W

Characin sp. *Moenkhausia jamesi* CH Eigenmann, 1908
James's Anchovy *Anchoviella jamesi* DS Jordan & SA Seale, 1926
Characin sp. *Leporinus jamesi* S Garman, 1929

Dr William James (1842–1910) was a physician, psychologist and philosopher, regarded as perhaps the USA's greatest philosopher. He qualified as a physician at Harvard (1869) but never practiced. He joined as a volunteer on Louis Agassiz's Thayer Expedition (1865–1866) up the Amazon River in Brazil. His academic teaching career (1873–1907) was spent at Harvard. He became a full professor (1885) and retired as Emeritus Professor of Philosophy. His published works are legion, including his great two-volume: *Principles of Psychology* (1890). Whilst on his trip up the Amazon, he collected some fish holotypes.

James (Schultz)

Spot-tailed Dottyback *Pseudochromis jamesi* LP Schultz, 1943

James Schultz (b.1931) is the son of the author Leonard Peter Schultz (1901–1986). He was 12 years old when this fish was named. He was honoured in the binomial because he "*...is interested in natural history.*"

James Tyler

Goldface Toby *Canthigaster jamestyleri* RL Moura & RMC Castro, 2002

James Chase 'Jim' Tyler (b.1935). (See **Tyler**)

Jameson

Jameson's Seaperch *Hypoplectrodes jamesoni* JD Ogilby, 1908

Jonathan Thompson Jameson was an amateur naturalist, and friend of the author, who collected species near his home in Ipswich, Queensland and on the coast there around Woody Bay and the Moreton Bay area. The etymology says that the seaperch was "*...Named for Mr. Jonathan Thompson Jameson, an enthusiastic collector, who has brought me many interesting zoological specimens.*" Ogilby had shown that a species previously named by William Macleay after Jameson, its collector, was only a synonym. Some twenty-four years later he was pleased to name the seaperch in his friend's honour as he had been sad to take that honour from him by the previous downgrading in status of that eponymous fish.

Jan

Bigscale sp. *Melamphaes janae* AW Ebeling, 1962

Jan Ebeling is the wife of the author, Alfred W Ebeling (q.v.). She was honoured for having sorted the first specimens of this species from collections made during the Scripps Institution of Oceanography's Eastropic Expedition (1955).

Jana

Istrian Chub *Squalius janae* NG Bogutskaya & P Zupančič, 2010

Jana Zupančič is, presumably, related to the junior author but we have not been able to establish how. The etymology says that her 'enormous patience and assistance' made the authors' study possible.

Jan Mol

Armoured Catfish sp. *Harttiella janmoli* R Covain & S Fisch-Muller, 2012

Dr Jan H A Mol is a Dutch ecologist based at Anton de Kom University of Suriname where he is Professor of Aquatic Ecology. His doctorate in Agriculture and Environmental Sciences was awarded by Wageningen University, the Netherlands (1995). *He wrote: The Freshwater Fishes of Suriname (2012). (Also see* **Mol***)*

Jane

Glass Knifefish sp. *Archolaemus janeae* RP Vari, CDCM de Santana & WB Wosiacki, 2012

Jane Mertens of the Faculty of Agriculture and Horticulture, Humboldt Universität zu Berlin was thanked "*...for her assistance to the second author...*"

Janet (Camp)

Cave Goby sp. *Didogobius janetarum* UK Schliewen, P Wirtz & M Kovačić, 2018

Janet Camp (with Janet Van Sickle Eyre (below), both of Reef Environmental Education Foundation), generously supported the authors' goby research.

Janet (Claudy)

Scalycheek Goby *Aulopareia janetae* HM Smith, 1945

Janet Elizabeth Claudy was the daughter of the describer, Dr Hugh McCormick Smith (1865–1941) (q.v.). (Also see **Smith, HM**)

Janet (van Sickle Eyre)

Cave Goby sp. *Didogobius janetarum* UK Schliewen, P Wirtz & M Kovačić, 2018

Janet Van Sickle Eyre (b.1955), along with Janet Camp (both of Reef Environmental Education Foundation), gave generous support to the authors' goby research. (See **Eyre**)

Janpap

African Rivuline sp. *Nothobranchius janpapi* RH Wildekamp, 1977

Jan Pap is a Dutch aquarist. He worked as a development assistant in Tanzania and discovered this species there.

Jansen

Jansen's Goatfish *Parupeneus jansenii* P Bleeker, 1856
Jansen's Wrasse *Thalassoma jansenii* P Bleeker, 1856

Albert Jacques Frédéric Jansen (d.1861) was an administrator in the Dutch East Indies (now Indonesia) and lived on Sulawesi (1853–1859). He was appointed Governor-General of Batavia (now Jakarta) (1848). He also collected botanical specimens, which he sent to the Dutch National herbarium in the 1850s. A reptile is also named after him.

Janss

Janss' Pipefish *Doryrhamphus janssi* ES Herald & JE Randall, 1972
Spotback Goby *Tigrigobius janssi* WA Bussing, 1981

Edwin Janss Jr (1915–1989) was a Californian real estate developer, Chairman of Janss Investment Corporation, and a recreational diver. Previously he had bred thoroughbred horses until he joined the family business (1954–1984). He also collected art and was an anti-war protestor – he said that his 'greatest honour ever' was to be included in Nixon's 'enemies list'. He was honoured for his support and encouragement of research in marine sciences. The type series of the goby was collected from the 'RV Searcher' belonging to the Janss Foundation. He suffered from a debilitating stroke and two weeks later committed suicide by jumping from the 12th storey of a building in Santa Monica. His son said that his greatest enjoyment was underwater photography and that he believed that part of the reason his father took his life may have been because his stroke had severely impaired his vision and hearing, and left him physically unable to continue diving.

Janssens

African Barb sp. *Enteromius janssensi* M Poll, 1976

André Janssens (1906–1954) was an entomologist who was an expert on dung beetles. He worked for L'Institut des Parcs Nationaux du Congo Belge (Democratic Republic of the Congo), and took part in a large-scale faunal survey (1946–1949) of Upemba National Park, which is where the barb holotype was collected. He wrote: *Revision des Onitides* (1937).

Jantje

Armoured Catfish sp. *Lithoxus jantjae* NK Lujan, 2008

Jantje (pronounced *yäntchi*) is the nickname of Lujan's mother - something she took with her when she emigrated from the Netherlands to South America. In the etymology Lujan wrote "*…in deep appreciation for her hard work and material and emotional encouragement that promoted [Lujan's] professional development and made this research possible.*"

Janus

Tanzanian Cichlid sp. *Serranochromis janus* E Trewavas, 1964
Cardinalfish sp. *Gymnapogon janus* TH Fraser, 2016

Janus was a Roman God of beginnings and endings, usually depicted as having two faces. Rather obscurely, the describer of the cardinalfish explains: "Here referring to the rounded caudal fin when scales have been sloughed off for some specimens of *Pseudamia* but with internal characters and preopercle spine of *Gymnapogon*."

Jardine

Gulf Saratoga *Scleropages jardinii* W Saville-Kent, 1892
[Alt. Northern Saratoga, Australian Pearl Arowana]

Francis Lascelles 'Frank' Jardine (1841–1919) was a pastoralist and pioneer associated with the exploration and settlement of the Cape York Peninsula in Queensland, where the Jardine River is named after him. With his brother Alexander, he drove (1864–1865) horses and cattle nearly 2,000 kilometres from Rockhampton to Somerset, for which achievement the brothers were both elected Fellows of the Royal Geographical Society. He settled close to Somerset (1866) and became a police magistrate (1868–1875). He found (1890) Spanish dollars on a reef and had them made into a silver dinner service on which elaborate dinners were served to visiting dignitaries. He collected the holotype of the saratoga. He died of leprosy. A mammal is also named after him.

Jari

Neotropical Rivuline sp. *Anablepsoides jari* WJEM Costa, PHN de Bragança & PF de Amorim, 2013

This is a toponym referring to a Brazilian river.

Jarrov

Dr Henry Crecy Yarrow (1840–1929) (See **Yarrow**)

Jarutanin

Srisawat Blind Cave Loach *Schistura jarutanini* M Kottelat, 1990

Khun Jarutanin Kittipong (b.1958) is a Thai aquarium fish dealer in Bangkok. Despite a lack of formal scientific education, he is regarded as a leading expert on Thailand's freshwater fish. His explorations along the rivers of Southeast Asia led to him being given the nickname 'Indiana Jones Thailand'.

Jauaperi

Neotropical Rivuline sp. *Laimosemion jauaperi* WJEM Costa & PHN de Bragança, 2013

This is a toponym referring to the Jauaperi river drainage, Brazil, where the species is found.

Jayakar

Oman Sea Longsnout Goby *Awaous jayakari* GA Boulenger, 1888
Pacific Barracudina *Lestidiops jayakari* GA Boulenger, 1889
Oman Cownose Ray *Rhinoptera jayakari* GA Boulenger, 1895
Jayakar's Seahorse *Hippocampus jayakari* GA Boulenger, 1900
Indian Golden-barred Butterflyfish *Roa jayakari* JR Norman, 1939

Colonel Dr Atmaram Sadashiv Grandin Jayakar (1844–1911) was an Indian surgeon and entomologist. The Indian Medical Service sent him to Muscat (1878), and during his 21 years in the Oman area (1879–1900) he studied the local wildlife and collected specimens, which he donated to the BMNH, London (1885–1899) including the ray holotype. He spent so long in Muscat that he acquired the nickname 'Muscati'. He wrote/collated a book of proverbs, published later as: *Omani Proverbs* (1987). He has two mammals and three reptiles named after him.

Jayaram

Danio sp. *Inlecypris jayarami* RP Barman, 1984
Indian Barb sp. *Systomus jayarami* W Vishwanath & H Tombi Singh, 1986
Indian catfish sp. *Myersglanis jayarami* W Vishwanath & L Kosygin, 1999
Indian catfish sp. *Glyptothorax jayarami* Y Rameshori & W Vishwanath, 2012

Dr Kottore Chidambaram Jayaram (1926–<2012) was Joint Director, Zoological Survey of India, and according to Barman's etymology "*…one of the prominent workers of the fishes of*

India of the present decade". He wrote, among other works: *The freshwater fishes of the Indian region* (1981).

Jayne

Borneo catfish sp. *Ompok jaynei* HW Fowler, 1905

Dr Horace Fort Jayne (1859–1913) was an American zoologist and educator. The University of Pennsylvania awarded his BA (1879) and MD (1882). He was appointed (1884) Professor of Vertebrate Morphology at the Wistar Institute of Anatomy and Biology, then became Director. He became Professor of Zoology at his alma mater (1894–1905). He wrote a number of books including *Abnormalities Observed in the North American Coleoptera* (1880). Fowler's etymology says of him: "*…formerly Director of the Wistar Institute of Anatomy, of Philadelphia, to whom I am principally indebted for this opportunity of studying the fishes of Borneo.*"

Jean Huber

African Rivuline sp. *Aphyosemion jeanhuberi* S Valdesalici & W Eberl, 2015

Dr Jean-Henri Huber (b. 1952) is a French ichthyologist with a particular interest in killifish. He discovered this species. (See **Huber, JH**)

Jeanne

Mottled Cusk-eel *Lepophidium jeannae* HW Fowler, 1941

Dr Jeanne Schwengel née Sanderson (1889–1961) was a malacologist who was a research associate of the Academy of Natural Sciences, Philadelphia. She collected the cusk-eel holotype while dredging for molluscs.

Jeannel

Omo Lampeye *Lacustricola jeanneli* J Pellegrin, 1935

René Gabriel Jeannel (1879–1965) was a French entomologist and coleopterist. He collected in East Africa (1911–1912) with Charles Alluaud during which the type of an eponymous odonate was taken. He also co-wrote the report on the scientific results of that expedition: *Voyage de Ch. Alluaud et R. Jeannel en African Orientale*. Perhaps his most important work was on the insect fauna of caves in the French Pyrenees and the Romanian Carpathians (as a member of the Romanian Academy). He was Director of the MNHN (1945–1951). Among over 500 works he wrote, his most influential were: *Faune cavernicole de la France* (1940) and *La genèse des faunes terrestres* (1942). Pellegrin wrote up the ichthyological part of the report on the 1911–1912 expedition and studied the fish that had been collected.

Jean Pol

Jean-Pol's Killi *Nimbapanchax jeanpoli* HO Berkenkamp & V Etzel, 1979

Jean-Pol Vandersmissen (b.1947) is a Belgian aquarist and accomplished amateur fish photographer who has been President of the Association Killiphile Francophone de Belgique. He owns 180 aquariums and has been interested in fish ever since early childhood, when his parents would sit him in front of his father's fish tank to calm him down! His wife is Josiane Vandersmissen. (Also see **Josiane**)

Jeb

Viviparous Brotula sp. *Calamopteryx jeb* DM Cohen, 1973

Jeb is a contraction of James Erwin Böhlke (1930–1982). (See **Böhlke**)

Jebb

Jebb's Siphonfish *Siphamia jebbi* GR Allen 1993

Dr Matthew Hilary Peter Jebb (b.1958) is an Irish botanist. Both his bachelor's degree and his doctorate were awarded by Oxford. He was Director, Christensen Research Institute Madang, Papua New Guinea, for five years before taking a two-year post-doctoral position at Trinity College, Dublin. He is currently (since 2010) Director, National Botanic Gardens of Ireland at Glasnevin. He was honoured for giving Allen much help, both physical and financial.

Jeff

Jeff's Snubnose Goby *Pseudogobius jeffi* HK Larson & MP Hammer, 2021

Jeff Larson is the husband of the senior author (see **Larson**), who states that Jeff "... has inadvertently learned much about gobioid fishes over 50-something years, so it is high time that he had a Queensland goby named for him." Jeff is also an avid wildlife and sports photographer.

Jeff Johnson

Johnson's Coralbrotula *Diancistrus jeffjohnsoni* W Schwarzhans, PR Møller & JG Nielsen, 2005

Jeffrey W 'Jeff' Johnson. He collected the holotype of this brotula species. (See **Johnson, JW**)

Jeff Williams

Williams' Coralbrotula *Ogilbia jeffwilliamsi* PR Møller, W Schwarzhans & JG Nielsen, 2005

Snake-eel sp. *Apterichtus jeffwilliamsi* JE McCosker & Y Hibino, 2015

Dr Jeffrey T Williams. (See **Williams**)

Jeffreys

Jeffreys' Goby *Buenia jeffreysii* A Günther, 1867

John Gwyn Jeffreys (1809–1885) was a British conchologist and malacologist. From the age of seventeen, he was an apprentice to one of the principal solicitors of Swansea, before going to London, where he qualified as a barrister (1838). He worked as a solicitor in Swansea until being called to the bar in London (1856). However, his great interest was conchology and he not only collected but undertook detailed study of all aspects of molluscs, through which he was elected a Fellow of the Royal Society (1840). He started

(1861) dredging aboard the yacht 'Osprey', continuing after retiring (1866) mostly around the English Channel, the Irish Sea and Shetland Isles but also off Norway. He collected the goby holotype while dredging for invertebrates. He also took part in the 'Porcupine' expeditions (1868 & 1870) and the 'Valorous' expedition to Greenland (1875). He was made High Sheriff of Hertfordshire (1877). His major publication was the five-volume *British Conchology* (1862–1865).

Jégu

Amazonian Cichlid sp. *Crenicichla jegui* A Ploeg, 1986
Armoured Catfish sp. *Chaetostoma jegui* LH Rapp Py-Daniel, 1991
Glass Knifefish sp. *Rhabdolichops jegui* P Keith & FJ Meunier, 2000

Dr Michel Jégu is an ichthyologist at ORSTOM (Office de la Recherche Scientifique et Technique d'Outre-Mer), presently based at the Marseille office (2017). He was at Instituto Nacional de Pesquisas da Amazonia, Manaus, Brazil (1994). He co-wrote: *A new large species of* Myloplus *(Characiformes, Serrasalmidae) from the Rio Madeira Basin, Brazil* (2016). He collected the cichlid paratypes and most of the other specimens on which Ploeg's paper was based, as well as the knifefish holotype.

Jeitteles

Baikal Red Sculpin *Procottus jeittelesii* B Dybowski, 1874

Ludwig Heinrich Christian Jeitteles (1830–1883) was a zoologist, palaeontologist and geologist who was one of founding fathers of seismological research in the mid-nineteenth century. He studied natural history at the University of Vienna and took the teaching exam (1855), and taught in secondary schools (1856–1870). He then taught at the teacher training institute in Salzburg (1870–1874) and the teacher training institute in Vienna (1874–1883). He wrote about and scientifically analysed an earthquake, in *First macroseismic map with geological background* (1858). He also described the genus *Albumoides* among other taxa, in: *Zoologische Mittheilungen. I. Ueber zwei für die Fauna Ungarns neue Fische, Lucioperca volgensis Cuv. Val. und Alburnus maculatus Kessler* (1861).

Jelski

Armoured Catfish sp. *Ancistrus jelskii* F Steindachner, 1876
Characin sp. *Hemibrycon jelskii* F Steindachner, 1876

Professor Constantin (Konstanty) Roman Jelski (1837–1896) was a Polish naturalist connected with the Zoological Museum of Warsaw. He was born in Minsk and studied medicine in Moscow (1853–1856), then natural history in Kiev (1856–1860). He became Curator in the Zoological Museum of Kiev (1862), but had to flee abroad after the January Revolt (1863) in which he participated. He went (1865) with the French navy to Guyana to become a collector for the Branicki (q.v.) family. Five years later he moved to Peru for reasons of health. He was Curator of the National Museum in Lima (1874–1878), where many of his specimens are still displayed. He collected widely, including in Peru as correspondent for the Museum. He was pardoned by the Russian Government (1878) and returned to Poland, becoming Curator, Krakow Museum (1878–1896). A mammal, an amphibian and ten birds are named after him.

Jenkins, CFH

Western Sooty Grunter *Hephaestus jenkinsi* GP Whitley, 1945

Clee Francis Howard Jenkins (1908–1997) was an Australian entomologist and ornithologist. The University of Western Australia awarded his MA, after which he worked at the Western Australia Museum, Perth (1929–1933). He worked for the Western Australian Department of Agriculture (1933–1973) starting as an assistant entomologist and rising to Chief of the Biological Services Division. He wrote a number of books including: *The Wanderings of an Entomologist* (1988). He collected the holotype.

Jenkins, JT

Jenkins' Whipray *Pateobatis jenkinsii* N Annandale, 1909
[Syn. *Himantura jenkinsii*]

James Travis Jenkins (1876–1959) was an ichthyologist who became Fishery Advisor to the Government of Bengal. He was a close collaborator of Annandale's and helped him collect the holotype. Among his writings are: *The Sea Fisheries* (1920) and *A textbook of oceanography* (1921).

Jenkins, OP

Round Herring genus *Jenkinsia* DS Jordan & BW Evermann, 1896
Snake-eel genus *Jenkinsiella* DS Jordan & BW Evermann, 1905 NCR
[Now in *Cirrhimuraena*]
Saltmarsh Topminnow *Fundulus jenkinsi* BW Evermann, 1892
Mussel Blenny *Hypsoblennius jenkinsi* DS Jordan & BW Evermann, 1896
Jenkins' Blenny *Labrisomus jenkinsi* E Heller & RE Snodgrass, 1903
Spot-nape Cardinalfish *Ostorhinchus jenkinsi* BW Evermann & A Seale, 1907

Dr Oliver Peebles Jenkins (1850–1935) was a physiologist and ichthyologist. His doctorate was awarded by Indiana University. He was Professor of Physiology at Leland Stanford Junior University (1891–1895) and Acting Professor of Physiology, Cooper Medical College (1895–1912) holding this position until Cooper Medical College was absorbed by, and became a department of, Stanford University. He was appointed Professor of Physiology and retired eventually as Professor Emeritus of Physiology and Histology at Stanford University. He described many Hawaiian fishes. (Also see **Oliver (Jenkins)**)

Jenkins, RE

Conasauga Logperch *Percina jenkinsi* BA Thompson, 1985

Dr Robert E Jenkins is an American ichthyologist and fisheries biologist who is Emeritus Professor of Biology, Roanoke College. His interest in fish began when he was taken fishing when six years old and began keeping guppies. He studied for his BSc at Roanoke (1961), his master's at Virginia Tech and his PhD at Cornell while he was working at the Smithsonian. Apart from two years (1970s) at Virginia Commonwealth, he spent his whole career at Roanoke until retirement (2007). He has written numerous papers such as the co-written: *Systematics of the Roanoke Bass, Ambloplites cavifrons* (1982) as well as the co-authored book: *Freshwater Fishes of Virginia* (1993).

Jennings

Blenny sp. *Cirripectes jenningsi* LP Schultz, 1943

Alexander Hutchinson Jennings (c.1896–1958) was the grandson of Eli Hutchinson Jennings who took possession of Swains Island, American Samoa (1856). He managed the island as a plantation. The etymology says: "*Named* jenningsi *after Mr Jennings whose grandfather from Long Island settled and colonized Swains Island in 1856. It was through Mr. Jennings' kindness that my visit to his Island [1939] was made most pleasant*." He was less hospitable when it came to local people, as he once (1953) evicted 56 workers and their families from the island.

Jenny

Sole sp. *Aseraggodes jenny* JE Randall & O Gon, 2006

Jenny Gon is the wife of the junior author, Ofer Gon (q.v.).

Jensen, AS

Jensen's Lanternfish *Diaphus jenseni* AV Tåning, 1932
Eelpout sp. *Lycodes jenseni* A Taranetz & AP Andriashev, 1935
Jensen's Skate *Amblyraja jenseni* HB Bigelow & WC Schroeder, 1950
[Alt. Short-tail Skate]

Professor Adolf Severin Jensen (1866–1953) was a Danish zoologist, ichthyologist and malacologist. He did much work on the fauna of Greenland and made several expeditions there, including the Tjalfe Expedition (1908–1909) which he led. He became Malacological Curator, Zoological Museum, Københavns Universitet (1892) and was Professor of Zoology there (1917–1936). He wrote: *The Fishes of East-Greenland* (1904).

Jensen, K

Sulu Sea Skate *Orbiraja jensenae* PR Last & AKP Lim, 2010

Dr Kirsten Jensen is a cestode parasitologist who is Professor of Ecology and Evolutionary Biology at the University of Kansas and Senior Curator, Biodiversity Institute. She specialises on coevolution and the systematics, diversity, and biogeography of the tapeworms that parasitize elasmobranchs. The University of Connecticut awarded her doctorate (2001). The etymology for this species states that, during an extensive field survey of fish markets of Borneo, Dr Jensen captured digital images of all chondrichthyan specimens sampled and provided images of most species for a field guide to the sharks and rays of Borneo (Last, *et al.*). Along with a close colleague, Dr Janine Caira, she has gained a broad knowledge of the taxonomy of the chondrichthyan fauna, as well as their invertebrate parasites. She has published many papers such as: *Four new genera and five new species of lecanicephalideans (Cestoda: Lecanicephalidea) from elasmobranchs in the Gulf of California, Mexico* (2001) and the co-written: *A Monograph on the Lecanicephalidea (Platyhelminthes: Cestoda)* (2005).

Jentink

West African Cichlid sp. *Tylochromis jentinki* F Steindachner, 1894
Indonesian Barb sp. *Diplocheilichthys jentinkii* CML Popta, 1904

Dr Fredericus Anna Jentink (1844–1913) was the Director of the Dutch National Museum of Zoology at Leiden. He was one of five zoologists chosen by the Third International Congress of Zoology (Leiden, 1895) to form a 'codex' on zoological nomenclature; the basis for today's taxonomic process. He described a large number of animal species, particularly mammals, and wrote: *Catalogue ostéologique des mammifères* (1887) and *Catalogue systématique des mammifères* (1892). He was the editor of the museum's journal, *Notes from the Leyden Museum.*

Jenyns

Neotropical livebearer genus *Jenynsia* A Günther, 1866
Jenyns' Sprat *Ramnogaster arcuata* L Jenyns, 1842
Smalltooth Flounder *Pseudorhombus jenynsii* P Bleeker, 1855
American Sole sp. *Catathyridium jenynsii* A Günther, 1862
Pike Characin sp. *Oligosarcus jenynsii* A Günther, 1864
Three-barbeled Catfish sp. *Rhamdella jenynsii* A Günther, 1864
Characin sp. *Astyanax jenynsii* F Steindachner, 1877
Jenyns' Tonguefish *Symphurus jenynsi* BW Evermann & WC Kendall, 1906
Goby sp. *Ophiogobius jenynsi* DF Hoese, 1976

Reverend Leonard Jenyns (1800–1893) was a clergyman and amateur naturalist. He became vicar of Swaffham Priory, Cambridgeshire (1828). His published natural history papers (1830s) were critically acclaimed. This led to an invitation (1836) from George Darwin for him to document the fish collection made by Darwin whilst on the 'Beagle'. He wrote the four-part: *Fish,* which Darwin edited (1840–1842). He moved (1849) to Swainswick near Bath where he founded the Bath Natural History Society. He also wrote an autobiography (1887). He described the goby (1842) but used a preoccupied name.

Jeon

Korean Gudgeon sp. *Microphysogobio jeoni* IS Kim & H Yang, 1999
Goby sp. *Chaeturichthys jeoni* K Shibukawa & A Iwata, 2013

Dr Sang-Rin Jeon is a biologist who is Professor Emeritus at the Faculty of Natural Science, Sang Myung University, Seoul, South Korea. He co-wrote papers and longer works, such as: *Developmental Characteristics of a Freshwater Goby, Micropercops swinhonis, from Korea* (2001), *Coloured illustrations of the freshwater fishes of Korea* (2002) and *Landlocked populations of an amphidromous goby, Rhinogobius sp.* (2004). The goby's describers honoured him because he: "*…very kindly assisted our research of the fishes in Korea, and gave us the opportunity to examine many specimens of Korean fishes, including several type-specimens of this new species.*"

Jerdon

Garra sp. *Garra jerdoni* F Day, 1867
Dwarf Anchor Catfish *Erethistes jerdoni* F Day, 1870
[Alt. Dwarf Moth Catfish; Syn. *Hara jerdoni*]
Jerdon's Carp *Hypselobarbus jerdoni* F Day, 1870
Brown-banded Cusk-eel *Sirembo jerdoni* F Day, 1888

Thomas Claverhill Jerdon (1811–1872) was a British physician with both zoological and botanical interests. He was born in Durham and educated at the University of Edinburgh. He studied medicine and became an Assistant Surgeon in the East India Company. He wrote: *Birds of India* (1862–1864), which according to Darwin was *the* book on Indian birds. He also wrote: *Illustrations of Indian Ornithology* and *The Game Birds and Wildfowl of India*, as well as *Mammals of India* and writings on ants among other taxa. A mammal, eight reptiles, four amphibians and twenty-three birds are also named after him.

Jespersen, A

Jespersen's Hagfish *Myxine jespersenae* PR Møller, TK Feld, IH Poulsen, PF Thomsen & JG Thormar, 2005

Dr Åse Jespersen (b.1955) is a Danish biologist, Associate Professor Emeritus in the Department of Biology, Marine Biological Section, Helsingør; part of the University of Copenhagen. She started her career as a Research Assistant at the State Serum Institute (1983) and held posts in pharmacology and biology. Her specialism is the reproductive systems of marine invertebrates on which she has published widely, such as the co-written: *Sex, seminal receptacles, and sperm ultrastructure in the commensal bivalve Montacuta phascolionis* (2000). She was honoured for her contributions to our knowledge of the reproductive biology of hagfishes.

Jespersen, PC

Onejaw sp. *Monognathus jesperseni* L Bertin, 1936
Crossthroat Sawpalate *Serrivomer jesperseni* M-L Bauchot-Boutin, 1953

Dr Poul Christian Jespersen (1891–1951) was a Danish oceanographer. After graduating (1917) he worked at the Institute for Fisheries Research. He took part in the cruises, including the circumnavigation (1928–1930), of the research vessel 'Dana' from which the holotypes of the above two fishes were collected. He wrote: *Investigations on the Food of the Herring and the Macroplankton in the Waters round the Faroes* (1944). He was also a keen ornithologist.

Jesse

Onejaw sp. *Monognathus jesse* SN Raju, 1974

Jesse F Raju (b. c.1936) is the wife of the author, Solomon Raju. (See **Raju**)

Jessica

Spiny Basslet sp. *Acanthoplesiops jessicae* GR Allen, MV Erdmann & WM Brooks, 2020

Jessica Jean Levy is the third author's stepdaughter.

Jessica (&) Len

Natal Fingerfin *Chirodactylus jessicalenorum* MM Smith, 1980

Jessica and her husband Len Jones (1935–2017) were anglers, divers and artists. He was author of three books of memoires: *Encounters with Sharks, Dolphins & Big Fish* (2003), *Hooks, Spears & Spanners* (2005) and *Mixed Bag* (2008). He had also been a fireman for a while. He was bitten by a 3–metre white shark (1967) while spearfishing off the KwaZulu-Natal coast, and carrying two fish tied to his belt, receiving wounds to his buttock, thigh and forearm and was pushed along twenty meters and almost out of the water. The etymology says: "*Named for Jessica and Len Jones, who not only procured the specimens for me, but for many years, as records officers of the South African Underwater Union, have skilfully identified fish caught by spear-fishermen.*"

Jessie (Brayton)

Blueside Darter *Etheostoma jessiae* DS Jordan & AW Brayton, 1878

Jessie Brayton née Dewey was the wife of the junior author, Alembert Winthrop Brayton (see **Brayton**).

Jessie (Jordan)

Black-striped Salema *Xenocys jessiae* DS Jordan & CH Bollman, 1890

Jessie Jordan née Knight (1866–1952) was the second wife of the senior author, David Starr Jordan (see **Jordan, DS**).

Jewett

Jewett's Coralbrotula *Ogilbia jewettae* PR Møller, W Schwarzhans & JG Nielsen, 2005
Jewett's Dwarf-goby *Eviota jewettae* DW Greenfield & R Winterbottom, 2012

Susan Lee Jewett (formerly Susan J Karnella) (b.1945) is an American ichthyologist who was Collections Manager, Division of Fishes, Smithsonian Institution. Her bachelor's degree was awarded (1967) by the University of Louisville, Kentucky. After working for the University of Louisville Medical School she moved to the Division of Fishes, USNM, Washington D C (1970–2004), rising from being a museum technician to collection manager. She took part in expeditions to Brazil, Cuba, Peru and Venezuela as well as collecting in the North Atlantic with the Woods Hole Oceanographic Institution. She recognised the goby as new (1978), but did not publish on it. However, during the 1970s and 1980s she described at least 8 new *Eviota* species and laid the foundation for the study of the genus. She wrote, or co-wrote, a number of papers such as: *The significance of the mast cell response to bee venom* (1971) and *Fishes of the genus* Eviota *of the Red Sea with descriptions of three new species (Teleostei, Gobiidae)* (1978). (Also see **Sue** & **Susan (Jewett)**)

Ji

Golden-line Fish sp. *Sinocyclocheilus jii* CG Zhang & DY Dai, 1992

Ji Cun-Shan was a Chinese ichthyologist at the Fishery Institute of Guangxi, Zhuang Autonomous Region. He described another member of this genus, *Sinocyclocheilus guilinensis* (1985).

Jim

Jim's Oto (armoured catfish) *Parotocinclus jimi* JC Garavello, 1977

Dr Jorge Jim (1942–2011) was a herpetologist, professor and Head of the Department of Zoology, Universidade Estadual Paulista Júlio de Mesquita Filho, São Paulo, Brazil, where his private collections are housed. His bachelor's degree on agronomy was awarded (1966) by the Universidade Federal Rural do Rio de Janeiro, and both his master's (1970) and doctorate in zoology (1981) by the Universidade de São Paulo. He has been honoured 'for his contribution to the knowledge of Brazilian herpetology, especially anurans'. Four amphibians are named after him.

Jim Craddock

Barbeled Dragonfish sp. *Eustomias jimcraddocki* TT Sutton & KE Hartel, 2004

James 'Jim' Edward Craddock (1937–2009). (See **Craddock**)

Jim, Joe, Bob

Searobin sp. *Lepidotrigla jimjoebob* WJ Richards, 1992

An 'arbitrary combination' of the names of the author's three sons: James, Joseph and Robert Richards.

Jimenez

Suckermouth Catfish sp. *Astroblepus jimenezae* CA Ardila Rodríguez, 2013

Dr Luz Fernanda Jiménez Segura is a Colombian ichthyologist interested in freshwater fishes and migratory fish ecology. She works at the Institute of Biology, Universidad de Antioquia, Medellin, where she is the Director of the ichthyology laboratory and where her doctorate was awarded. She co-wrote *Regionally nested patterns of fish assemblages in floodplain lakes of the Magdalena river (Colombia)* (2012).

Jimmy Carter

Bluegrass Darter *Etheostoma jimmycarter* SR Layman & RL Mayden, 2012

James Earl 'Jimmy' Carter Jr (b. 1924), the 39th President of the USA and well-known birdwatcher, was honoured for his environmental leadership and accomplishments in the areas of national energy policy and wilderness protection, and his life-long commitment to social justice.

Joan (Stroud)

Characin sp. *Ceratobranchia joanae* B Chernoff & A Machado-Allison, 1990

Mrs Joan Milliken Stroud née Milliken (1922–1985) was a philanthropist and environmentalist. The Stroud Foundation provided some of the funding for the authors' research. She died of cancer.

Joan (Wright)

Scalyjaw Koester *Acanthistius joanae* PC Heemstra, 2010

Joan Wright was a research assistant of the describer, Phil Heemstra. Although he promised to name the fish after her, Joan had retired by the time he got round to doing so.

Joan Allen

Joan's Fairy Wrasse *Cirrhilabrus joanallenae* GR Allen, 2000

Joan Mary Allen of Pensacola, Florida, was the author's mother.

Joan Johnson

Pearl of Likoma (cichlid) *Labidochromis joanjohnsonae* DS Johnson, 1974
[Syn. *Pseudotropheus joanjohnsonae*]

Joan Johnson was editor of *Today's Aquarist*, a magazine in which the description first appeared.

Joanne

Antarctic Dragonfish sp. *Bathydraco joannae* HH DeWitt, 1985

Joanne E DeWitt (1936–1994) was the wife of the author Dr Hugh Hamilton DeWitt (1933–1995). As he says in his etymology: "*This species is named for Joanne DeWitt in recognition of her patience during my absences during trips into the Southern Ocean.*" (Also see **DeWitt**)

Joannis

Long-whiskered Catfish sp. *Pimelodus joannis* FRV Ribeiro, CAS Lucena & PHF Lucinda, 2008

Dr John Graham Lundberg (b.1942). *Joannis* is a Latinized version of John, here referring to "*...John Lundberg, in recognition of his many contributions to catfish systematics*". (See **Lundberg, JG**)

Joao Vitor

Characin sp. *Astyanax joaovitori* CAM Oliveira, CS Pavanelli & VA Bertaco, 2017

João Vitor Kadota Oliveira is the son of the senior author, Carlos Oliveira.

Joaquina

Philippine Barb sp. *Barbodes joaquinae* CE Wood, 1968

Joaquina C Wood according to the etymology "*...first collected this fish during her many trips afield while studying Mindanao fauna*". She co-wrote: *A monograph of the fishes of Lake Lanao : a research* (1963). As she collected with the author, we assume they were a husband and wife team but have failed to identify them further. He being Charles E Wood, a zoologist and freshwater biologist in the Philippines associated with the Institute of Research for Filipino culture, University of Mindanao. He wrote: *Two species of Cyprinidae from North Central Mindanao* (1968).

Jobaert

African Killifish sp. *Hypsopanchax jobaerti* M Poll & JG Lambert, 1965

A J Jobaert of Lulua, Democratic Republic of Congo, was Warden of the Muene Ditu Game Reserve (1950s–1960s) and collected for the Musée Royal de l'Afrique Central.

Johann

Blue-grey Mbuna (cichlid) *Pseudotropheus johannii* DH Eccles, 1973

John Johns was a collector of Lake Malawi fishes for the aquarium trade. 'Johann' is a German variant of John. Unfortunately, we have no further information about him.

Johanna

Red-finned Pike-Cichlid *Crenicichla johanna* JJ Heckel, 1840

Based on a local name for this fish in Mato Grosso, Brazil: 'johanna guensa'.

Johanna (Vickers)

Rasbora sp. *Rasbora johannae* DJ Siebert & S Guiry, 1996

Baroness Joan Helen Vickers (1907–1994) was a British Conservative London Councillor and MP and later chairman of the Anglo-Indonesian Society. She was trained as a Norland Nurse and spent several years working for the Red Cross in the Far East, including 14 months in Indonesia. According to the etymology she was "*...a long-time advocate of Indonesian culture*".

Johanna (Wagner)

Deepwater Cisco *Coregonus johannae* G Wagner, 1910 (Extinct)

Johanna Wagner was the wife of the describer, George Wagner (1873–1954), who was a pharmacologist turned zoologist with interests in ornithology, herpetology and ichthyology. He became Professor of Zoology at the University of Wisconsin (1931–1943). The etymology is not particularly helpful, saying: "*Johanna, a slight token of gratitude for my great indebtedness to my life-companion.*"

Johannis Davis

Travancore Skate *Dipturus johannisdavisi* AW Alcock, 1899

John Davis (1550–1605), an English navigator, is better remembered for having a strait rather than a skate named after him: the Davis Strait between Greenland and Labrador. This 'celebrated Elizabethan navigator and explorer' is too famous a figure to need a biography here, but this is what Alcock says in his original description: '*...John Davis who – though best known for his Arctic voyages – piloted three expeditions to the East Indies and lost his life in Indian seas.*'

John

Croaker genus *Johnius* ME Bloch, 1793

John's Snapper *Lutjanus johnii* ME Bloch, 1792

Rev Dr Christoph Samuel John (1747–1813) was a botanist and herpetologist and a medical missionary (1771–1813) at the Danish trading station of Tranquebar (now Tharangambadi), Tamil Nadu, not far from Madras (Chennai). It was a Danish colony (1620–1845), when Denmark sold its possessions in India (including the Nicobar Islands) to Great Britain. Among John's friends was William Roxburgh, the botanist who lived in Madras in charge of the botanical gardens there. John was awarded an honorary doctorate (1795) for his studies in natural history. A mammal and a reptile are also named after him. The original description refers to the missionary 'Herrn John', who we believe is the above.

John, OH

Armoured Catfish sp. *Hypostomus johnii* F Steindachner, 1877

Orestes Hawley Saint John (1841–1921) was a geologist whose major interest was the fossil fishes of the Palaeozoic. He was a member of Thayer Expedition to Brazil (1865–1866). His tasks included keeping lists of fishes for Louis Agassiz (q.v.) as well as collecting (including the holotype of this species). He was assistant geologist on the geographical Survey of Iowa (1866–1869) and leading geologist on the Hayden Survey (1883) of Idaho, Montana and Wyoming. He subsequently led his own expedition to the Wind River Mountains and the Snake River Valley. Mount Saint John in the Teton Range, Wyoming, is named after him.

John Bob

Snoutscale Whiptail *Ventrifossa johnboborum* T Iwamoto, 1982

This fish is named after two ichthyologists Dr 'John' Richard Paxton (q.v.), and Dr Robert 'Bob' John Lavenberg (q.v.), who first recognised that it was a new species. (See **Lavenberg** & **Paxton**)

John Dory

John Dory *Zeus faber* Linnaeus, 1758
[Alt. St Peter's Fish, St Pierre]

There are many alternative names for this fish and almost as many theories as to the origin of the name. The one we like best is that John Dory was a French ship's captain (likely to have been a pirate) who promised King John II of France (reigned 1350–1364) captive Englishmen in exchange for a pardon. He encountered a ship owned by a Cornishman called Nicholl and, after a battle, John Dory was captured. The origin of this version is a ballad, for which the earliest reference is 1609 but probably predates it and has its origin in the Hundred Years' War.

John Fitch

Black Barracudina *Macroparalepis johnfitchi* RR Rofen, 1960

John Edgar Fitch (1918–1982) was a marine and fisheries biologist and ichthyologist who was Director of the US Fisheries Laboratory, San Diego. He wrote: *Offshore fishes of California* (1958). He provided the author with the holotype of this species.

John McCosker

Snake-Eel sp. *Ophichthus johnmccoskeri* A Mohapatra, D Ray, SR Mohanty & SS Mishra, 2018

Dr John Edward McCosker (b.1945) was honoured for his 'vast' contributions to the taxonomy of ophichthid eels (See **McCosker**).

John Paxton

Paxton's Escolar *Rexichthys johnpaxtoni* NV Parin & DA Astakhov, 1987

John R Paxton was curator of the fish collection at the Australian Museum, Sydney (see **Paxton**, also see **John Bob**).

John Randall

Blacknose Butterflyfish genus *Johnrandallia* TT Nalbant, 1974

John Randall, marine biologist at the Bishop Museum, Hawaii. (See **Randall**, also see **Helen (Randall)**)

John Treadwell

Chinese Gudgeon sp. *Romanogobio johntreadwelli* PM Bănărescu & TT Nalbant, 1973

John Treadwell Nichols (1883–1958) (see **Nichols, JT**).

John Voelcker

Cerise Dottyback *Chlidichthys johnvoelckeri* MM Smith, 1953

John Voelcker (1898–1968) was a businessman and prominent amateur ornithologist. Born in England, he attended the Royal Military Academy at Woolwich. He later moved to South Africa, where he was (1935–1950) President of the South African Ornithological Society. He wrote *Memorandum on Ornithology* (1946). He was honoured in the fish's binomial as he had 'greatly assisted' the describer. A bird is also named after him.

Johnels

Claroteid Catfish sp. *Chrysichthys johnelsi* J Daget, 1959

Dr Alf Gunnar Johnels (1916–2010) was a Swedish zoologist and fisheries biologist. Stockholm University awarded his doctorate (1948). He was in the Gambia (1950) to study lungfish and bichirs. He was Professor at the Department of Vertebrate Zoology, Swedish Museum of Natural History (1958–1982) and President of the Royal Swedish Academy (1981–1983). He was one of the people who discovered mercury poisoning in Sweden (1960s); he analysed bird feathers in museum collections and demonstrated that mercury poisoning in birds coincided with the introduction of the pesticide, methyl mercury. He wrote: *Notes on fishes from the Gambia River* (1954). He observed and reported on the first specimens of this species (1954).

Johnson, A

White-spotted Moray *Gymnothorax johnsoni* JLB Smith, 1962

Alf Johnson of Port Elizabeth, South Africa supplied Smith with "*...many rare and valuable specimens*". We have been unable to find more about him, although he is listed as an individual donor to the Sea Fishes Trust that sponsored a book: *Smiths' Sea Fishes* about the work of J L B Smith (q.v.).

Johnson, DS

Johnson's Wrasse *Cirrhilabrus johnsoni* JE Randall, 1988

David S Johnson was an underwater photographer at Kwajalein in the Marshall Islands. He was the first to observe this wrasse, and assisted in collecting the type specimens. He also photographed many of the local fish, included some of those collected there by Randall.

Johnson, ER

Spikefish genus *Johnsonina* GS Myers, 1934
Pencil Catfish sp. *Potamoglanis johnsoni* HW Fowler, 1932
[Syn. *Trichomycterus johnsoni*]

Eldridge Reeves Fenimore Johnson (1899–1986) was a financier and philanthropist and a Trustee of the Academy of Natural Sciences of Philadelphia. His father had become enormously wealthy as he had invented the technique of recording voices and music on wax discs to play on gramophones. Eldridge not only financially supported, but also took part in, Perfilieff's Mato Grosso Zoological and Ethnographic Expedition (1930–1931). The film that they made of the expedition: *Matto Grosso, the Great Brazilian Wilderness* (1931) is regarded as being the first instance of documentary sound recording in the field. He also sponsored the Johnson-Smithsonian Deep-Sea Expedition (1933), during which the spikefish was discovered.

Johnson, GD

White-cheeked Blenny *Acanthemblemaria johnsoni* GR Almany & CC Baldwin, 1996

Dr G David 'Dave' Johnson (b.1945) is an American ichthyologist who is Research Zoologist and Curator of Marine Larval Fishes at the NMNH Smithsonian Institution (1983–present). His 'knowledge of teleostean anatomy and phylogeny is inspirational' according to the etymology. The University of Texas awarded his BA (1967) and the University of California's Scripps Institution of Oceanography his PhD (1977). He was Chairman of the Department of Vertebrate Zoology at the Smithsonian (1992–1999). He has written or co-written more than 150 papers and articles such as: *Development of Fishes of the Mid-Atlantic Bight: Carangidae Through Ephippidae* (1978) and *A 'living fossil' eel (Anguilliformes: Protanguillidae, fam. nov.) from an undersea cave in Palau* (2011). (Also see **Ai Nonaka**).

Johnson, JS

Cozumel Toadfish *Sanopus johnsoni* BB Collette & WA Starck, 1974

John Seward Johnson I (1895–1983) founded the Harbor Branch Oceanographic Institute, Fort Pierce, Florida. He was honoured for his "*generous and extensive patronage of marine science*." He was immensely rich, leaving over $402 million to his third wife: which led to years of wrangling in the courts as his six children by his first two wives were not happy!

Johnson, JW

Johnson's Gurnard Perch *Neosebastes johnsoni* H Motomura, 2004
Johnson's Coralbrotula *Diancistrus jeffjohnsoni* W Schwarzhans, PR Møller & JG Nielsen, 2005

Jeffrey W 'Jeff' Johnson joined the Queensland Museum, Brisbane as a cadet (1977) and worked as assistant to the Curator of Fishes (1977–1995). He became Collection Manager (1995) and also conducts independent and collaborative ichthyological research. He has a diploma in applied science from Queensland University of Technology. He is a recognised authority on the identification and geographic distribution of Indo-Pacific, Australian and particularly Queensland fishes. the curation and collection management of fishes, field collection of fishes, and underwater visual recording of fish species. He has described new genera and species of fishes of several families, including Acanthuridae (unicornfishes), Haemulidae (sweetlips), Pinguipedidae (sand perches), Soleidae (soles) and Uranoscopidae (stargazers). He has written numerous chapters on fish for popular books and guides, as well as writing papers such as: *Pseudopataecus carnatobarbatus, a new species of velvetfish (Teleostei: Scorpaeniformes: Aploactinidae) from the Kimberley coast of Western Australia* (2012) and *Three new species of Parapercis (Perciformes: Pinguipedidae) and first records of P. muronis (Tanaka, 1918) and P. rubromaculata Ho, Chang & Shao, 2012 from Australia* (2018). (Also see **Jeff Johnson**).

Johnson, JY

Slender Codling *Halargyreus johnsonii* A Günther, 1862
Humpback Anglerfish *Melanocetus johnsonii* A Günther, 1864
Halosaur sp. *Halosaurus johnsonianus* L Vaillant, 1888

James Yate Johnson (1820–1900) was a British naturalist who moved to Madeira (1851) and spent the rest of his life there. He studied and collected, mainly for other naturalists, plants and mosses as well as crustaceans, sponges, sea anemones and fish. Vaillant wrote that Johnson was the first to describe *Halosaurus*, 'this curious genus'.

Johnson, MW

Fringelip Snake-eel *Cirricaecula johnsoni* LP Schultz, 1953

Dr Martin Wiggo Johnson (1893–1984) was an American oceanographer. He served in the US Army in WW1. The University of Washington awarded his bachelor's degree (1923) and later awarded him a doctorate. He was Curator of the biological station at Friday Harbor, Washington State, and a Scientific Assistant at the Passamaquoddy International Fisheries Commission (1931–1932) on the US/Canadian border between Maine and New Brunswick. He was an Associate of the University of Washington (1933–1934) and a Research Associate of the Scripps Institution of Oceanography, La Jolla, California (1934) where he became Professor of Marine Biology before retiring as Professor Emeritus (1961). He wrote many papers including: *Some Observations on the Feeding Habits of the Octopus* (1942). Schultz honoured him as thanks for helping him in his fieldwork (1946).

Johnston

Johnson Island Damsel *Plectroglyphidodon johnstonianus* HW Fowler & SC Ball, 1924

This is a toponym referring to Johnston Island in the central Pacific, the type locality.

Johnston, HH

Johnston's Topminnow *Micropanchax johnstoni* A Günther, 1894
Lake Malawi Cichlid sp. *Placidochromis johnstoni* A Günther, 1894
Lake Malawi Cichlid sp. *Docimodus johnstoni* GA Boulenger, 1897
Yellowfish sp. *Labeobarbus johnstonii* GA Boulenger, 1907

Sir Harry Hamilton Johnston (1858–1927) was a formidable English explorer, botanist, linguist and colonial administrator. He was a larger-than-life character and became known as the 'Tiny Giant', as he was just five feet tall. He was an accomplished painter, photographer, cartographer, linguist, naturalist and writer. He began exploring tropical Africa (1882) and met up with Henry Morton Stanley in the Congo (1883). He was in East Africa (1884) then joined the colonial service (1885), taking various posts across Africa, in Cameroon, Nigeria, Liberia, Mozambique, Tunisia, Zanzibar and Uganda. He also established a British Protectorate in Nyasaland (Malawi). Johnston was Queen Victoria's first Commissioner and Consul General to British Central Africa. A member of the Royal Academy of Art, his paintings of African wildlife are exceptional. He spoke over 30 African languages, as well as Arabic, French, Italian, Portuguese and Spanish. He was knighted (1896) and after retirement (1904) he continued his pursuit of natural history. He discovered more than 100 new birds, reptiles, mammals and insects, the most notable being the Okapi (*Okapia johnstoni*). He wrote more than 60 books including: *The Story of My Life* (1923), and more than 600 monographs and short articles. He made the very first Edison cylinder recordings in Africa, which preserved his squeaky voice for posterity. Six birds, eight mammals, three amphibians and two reptiles are also named after him.

Johnston, RM

Johnston's Weedfish *Heteroclinus johnstoni* W Saville-Kent, 1886
Clarence Galaxias *Galaxias johnstoni* EOG Scott, 1936

Robert Mackenzie Johnston (1843–1918) was a scientist and statistician and, according to the galaxias etymology, was "*...the father of Tasmanian ichthyology, who paid considerable attention to the local Galaxiidae.*" He was born in Scotland and emigrated to Australia (1870) where he worked for a railway company in Tasmania. He then transferred (1872) to government service as a clerk in the Audit Department. He became Registrar-General and Government Statistician and a royal commissioner to report on Tasmanian fisheries. He wrote: *Descriptive Catalogue of Tasmanian Fishes* (1882). He and Saville-Kent were fellow members of the Royal Society of Tasmania.

Jojette

Cryptic Triplefin *Cryptichthys jojettae* GS Hardy, 1987

Johanna Henriette 'Jojette' Drost (b.1960) is a New Zealand biologist who currently a horticulture technician. Her mother coined the nickname by combining the first part of her first name and the last part of her middle name. Waikato University awarded her BSc (Botany) and she took a temporary job as a clerk in the Ornithology Department at the National Museum (New Zealand) (1983–1984), but went on many field trips collecting sub-fossil bird bones then tagging Westland Black Petrels. Because she was a SCUBA diver, she was sent with the museum's Graham Hardy on a fish collecting expedition around Northland, NZ during which she was the collector of the paratypes of this species (1983). She has since been involved in horticultural research, co-wrote a paper on apple irrigation (1986), managed her own orchard (2005–2014) and has been involved (2009–present) in insect pest monitoring and training and kiwi-fruit field data and sample collection. Her voluntary role (2017–2019) in Uretara Estuary Managers Inc involves improving wetland and riparian planting to encourage fish passage. She also teaches piano part-time. According to the etymology she "*...participated with much enthusiasm and effectiveness in a number of recent ichthyology coastal reef fish collecting trips.*"

Joka

West African Cichlid sp. *Coelotilapia joka* DFE Thys van den Audenaerde, 1969

Joris Thys van den Audenaerde is the describer's son. We assume that 'Joka' is a nickname, as the etymology says: "*dedicated to my young son Joris.*"

Jonah

Jonah's Icefish *Neopagetopsis ionah* O Nybelin, 1947

The etymology is not explained. However, as the type was found in the stomach of a whale, otherwise filled with krill, we think it is safe to assume this is an allusion to the biblical Jonah who apparently ended up in the stomach of a 'whale' (though the actual words better translate as 'big fish') for three days and nights.

Jonas

Naked-back Knifefish sp. *Gymnotus jonasi* JS Albert & WGR Crampton, 2001

Jonas Alves de Oliveira is a naturalist and technician at Mamirauá Sustainable Development Reserve, Amazonas, Brazil. He co-wrote: *Chromosomal diversity in three species of electric fish (Apteronotidae, Gymnotiformes) from the Amazon Basin* (2014).

Jonasson

Blind Cusk-Eel sp. *Sciadonus jonassoni* O Nybelin, 1957

Axel Jonasson (b.1903) was the instrument-maker of the Oceanographic Institute, Gothenburg, Sweden. He spent five years (1923–1928) in the USA, working on a ferry between San Francisco and Vallejo. He was chief mechanic and trawling master on board the Swedish ship, 'Albatross', from which the holotype was collected. He co-wrote: *New devices for sediment sampling* (1966).

Jones, JM

Slender Mojarra *Eucinostomus jonesii* A Günther, 1879

John Matthew Jones presented the type specimens to the British Museum. According to Günther, Jones "*...has paid especial attention to the fishes occurring at the Bermudas.*" He was the author of: *The Naturalist in Bermuda* (1859) and *On the Vegetation of the Bermudas* (c.1874).

Jones, S

Travancore Loach *Travancoria jonesi* SL Hora, 1941
Jones' Ponyfish *Eubleekeria jonesi* PSBR James, 1971
Jones' Sardinella *Sardinella jonesi* S Lazarus, 1983

Dr Santhappan Jones (1910–1997) is commemorated in the names of the ponyfish and sardinella. It is recorded that a 'Mr S Jones' sent a collection of fishes from the Travancore Hills, Kerala, India, to the Zoological Survey of India; thus we think the Travancore Loach is likely named after the same person, though this is unproven. Santhappan Jones was an Indian zoologist, entomologist and marine biologist. The Maharaja's College of Science (now University College), Trivandrum, awarded his bachelor's degree (1933). He stayed at the same college as Honorary Researcher (1934) until moving to Madras (Chennai). Here he was awarded a master's degree in zoology (1937) by Madras University for his investigation into the breeding of fishes adapted to live in brackish water. The University of Madras also awarded his doctorate (1952). He joined the Travancore State service as an entomologist (1937–1947). He was the Director of the Central Marine Fisheries Institute, Kochi, India (1957–1970). In his retirement, he was Emeritus Scientist at University College, Trivandrum. He had been afflicted by polio (1963) having a lasting handicap, and so in retirement, under the aegis of the Church of South India, he planned and organised a residential home to care for and rehabilitate children of Trivandrum who had similarly contracted polio.

Jones, TM

Barred Livebearer *Pseudoxiphophorus jonesii* A Günther, 1874
[Syn. *Heterandria jonesii*]

Thomas Manson Rymer Jones (1839–1894) discovered the species in Mexico and presented specimens to the British Museum. He was the son of the English surgeon and zoologist Thomas Rymer Jones (1810–1880). He was appointed (1871) as Assistant Engineer on the Mexican Railway.

Jonklaas

Jonklaas's Loach *Lepidocephalichthys jonklaasi* PEP Deraniyagala, 1956
Lipstick Goby *Sicyopus jonklaasi* HR Axelrod, 1972

Rodney Jonklaas (1925–1989) was a pioneering Sri Lankan diver, underwater photographer and zoo administrator. He wrote: *Collecting Marine Tropicals* (1975). One of his friends was Arthur C Clarke (1917–2008), the author who is probably best remembered for *2001: A Space Odyssey* (1968).

Jordan, CB

Blind Cave Tetra *Astyanax jordani* CL Hubbs & WT Innes, 1936

C Basil Jordan of the Texas Aquaria Fish Company, Dallas, was an aquarium fish dealer who received 75 of the 100 specimens of the Mexican Blind Cave Tetras collected by the San Luis Potosi expedition (1936). He was able to breed the fish in his aquaria, and introduced the species to the hobby.

Jordan, DS

Flagfish genus *Jordanella* GB Goode & TH Bean, 1879
Sculpin genus *Jordania* EC Starks, 1895
Eelpout genus *Davidijordania* AM Popov, 1931
Petrale Sole *Eopsetta jordani* WN Lockington, 1879
Yellow Irish Lord *Hemilepidotus jordani* TH Bean, 1881
Gulf Grouper *Mycteroperca jordani* OP Jenkins & BW Evermann, 1889
Northern Ronquil *Ronquilus jordani* CH Gilbert, 1889
Greenbreast Darter *Etheostoma jordani* CH Gilbert, 1891

Charal *Chirostoma jordani* AJ Woolman, 1894
Sucker sp. *Catostomus jordani* BW Evermann, 1895 NCR
[Now included in *Catostomus platyrhynchus*]
Fanfin Angler *Caulophryne jordani* GB Goode & TH Bean, 1896
Shortbelly Rockfish *Sebastes jordani* CH Gilbert, 1896
Jordan's Damsel *Teixeirichthys jordani* C Rutter, 1897
Jordan's Snapper *Lutjanus jordani* CH Gilbert, 1898
Viviparous Brotula sp. *Diplacanthopoma jordani* S Garman, 1899
Redbelly Triplefin *Enneanectes jordani* BW Evermann & MC Marsh, 1899
Poacher sp. *Agonomalus jordani* DS Jordan & EC Starks, 1904
Flame Wrasse *Cirrhilabrus jordani* JO Snyder, 1904
Jordan's Sculpin *Triglops jordani* DS Jordan & EC Starks, 1904
Jordan's Chimaera *Chimaera jordani* S Tanaka, 1905
Batfish sp. *Malthopsis jordani* CH Gilbert, 1905
Rattail sp. *Coelorinchus jordani* HM Smith & TBE Pope, 1906
Shortjaw Eelpout *Lycenchelys jordani* BW Evermann & EL Goldsborough, 1907
Jordan's Tuskfish *Choerodon jordani* JO Snyder, 1908
Surfperch sp. *Ditrema jordani* V Franz, 1910
Brokenline Lanternfish *Lampanyctus jordani* CH Gilbert, 1913
Cod Icefish sp. *Patagonotothen jordani* WF Thompson, 1916
Smooth Lumpfish *Cyclopteropsis jordani* VK Soldatov, 1929
Cherry Snailfish *Allocareproctus jordani* CV Burke, 1930
Ratmouth Barbel sp. *Ptychidio jordani* GS Myers, 1930
Eelpout sp. *Davidijordania jordaniana* PJ Schmidt, 1936
Icefish sp. *Neosalanx jordani* Y Wakiya & N Takahasi, 1937
Jordan's Cod *Gadella jordani* JE Böhlke & GW Mead, 1951
Pahranagut Roundtail Chub *Gila robusta jordani* VM Tanner, 1950
Southern Red Tabira Bitterling *Acheilognathus tabira jordani* R Arai, H Fujikawa & Y Nagata, 2007

Dr David Starr Jordan (1851–1931) was a leading American ichthyologist, physician, educator, peace activist and believer in eugenics; moreover, he was founding President of Stanford University. Oddly he was educated at a local girl's high school and took his botany degree at Cornell. He studied further at Butler University and the Indiana University School of Medicine. His early career was teaching at several small colleges, before joining the natural history faculty at Indiana Bloomington University (1979). He then became the nation's youngest ever University President at Indiana University (1885), and subsequently (1891–1913) became founding President at Stanford and later Chancellor before retiring (1916). He was President of the World Peace Foundation (1910–1914) opposing US involvement in WW1, though his stance on peace apparently stemmed from his belief that war killed off the strongest people in the gene pool. In his later years, he advocated compulsory sterilization under the auspices of the Human Betterment Foundation. A cloud also hangs over him concerning the death of Jane Stanford, President of the Stanford University board of trustees. She died of strychnine poisoning while on holiday in Hawaii. Jordan sailed there and hired a physician to investigate, declaring that she had died of heart failure. Mrs Stanford was reportedly planning to have Jordan removed from his position at the university. He had the quaint notion of refusing to learn his students' names on the grounds that he would forget the name of a fish for every student's name he learned. Apart from many scientific papers he wrote many longer works including, against war: *War and Waste* (1913), an autobiography: *Days of a Man* (1922) and much on zoology including the 4-volume: *Fishes of North and Middle America* (1896–1900). It may appear that he helped to name two fish after himself (*Agonomalus jordani* and *Triglops jordani*), but in fact he and Starks merely published manuscript names already coined by another researcher. Two amphibians are also named after him.

Jörg Bohlen

Loach sp. *Cobitis joergbohleni* J Freyhof, E Bayçelebi & M Geiger, 2018

Dr Jörg Bohlen (b.1965) is a German biologist specializing in freshwater ecology. He has worked at the Fish Genetics Laboratory of the Academy of Science of the Czech Republic (Liběchov) (since 1999). He was a technical assistant at the Museum of Natural History and Prehistorics, Oldenburg (1991–1993) then a diploma student at the Max-Planck-Institute of Limnology (1993–1994) and a PhD student at the Institute of Freshwater Ecology and Inland Fisheries, Berlin (1996–1999). He has published around 40 papers and has researched the diversity, phylogeny and biology of loaches, especially of the genus *Cobitis*.

Jørgen Nielsen

Nielsen's Mudbrotula *Dermatopsis joergennielseni* PR Møller & W Schwarzhans, 2006

Dr Jørgen G Nielsen (b.1932). (See **Nielsen**)

Jørgen Scheel

African Rivuline sp. *Aphyosemion joergenscheeli* JH Huber & AC Radda, 1977

Jørgen Jacob Scheel (1916–1989) (See **Scheel**)

José Lima

Sultan Pleco *Leporacanthicus joselimai* IJH Isbrücker & H Nijssen, 1989
Golden-spot Pleco *Pterygoplichthys joselimaianus* C Weber, 1991

Dr José Lima de Figueiredo (b.1943), aka 'Ze Lima', is a Brazilian ichthyologist who is a former researcher and curator of fishes at the Zoological Museum of the University of São Paulo. He is a professor at the University of São Paulo, which awarded his bachelor's degree (1969) and doctorate (1981). Among his many publications he co-wrote: *The northernmost record of Bassanago albescens and comments on the occurrence of Rhynchoconger guppyi (Teleostei: Anguilliformes: Congridae) along the Brazilian coast* (2011).

Joseph

Crested Cusk-eel *Ophidion josephi* CF Girard, 1858

This is a toponym referring to St. Joseph Island, Texas, USA, where the holotype was collected.

Joshua

Indian Garra sp. *Horalabiosa joshuai* EG Silas, 1954

[Syn. *Garra joshuai*]

Dr J P Joshua was an entomologist who was Professor of Zoology at Madras Christian College. He was Silas' teacher and he led the entomological survey during which Silas collected the holotype.

Josiane

African Rivuline sp. *Epiplatys josianae* HO Berkenkamp & V Etzel, 1983
[Syn. *Epiplatys fasciolatus josianae*]

Josiane Vandersmissen, wife of Jean-Pol Vandersmissen (b.1947) of the French speaking Belgian African Rivuline Association (Also see **Jean Pol**).

Jothy

Perchlet sp. *Plectranthias jothyi* JE Randall, 1996

Alexander A Jothy is a fisheries and marine ecologist, currently a consultant with GSR Environmental Consultants, Malaysia. He graduated with a BSc in zoology and botany from the University of Singapore and a BSc in aquatic ecology, marine biology and fisheries and has a Fellowship Certificate in Marine Fisheries and Marine Pollution from the Bundesforschungsanstalt für Fischerei, Federal Republic of Germany. He worked for the Fisheries Research Institute of Penang, and the Malaysian Fisheries Department as Senior Fisheries Research Officer, and later as Chief of the Freshwater Fish Research Centre, Batu Berendam in Melaka, Malaysia. He collected the first known specimens of the perchlet (1972). He wrote: *Capture Fisheries and the Mangrove Ecosystem* (1983).

Joubin

Stonefish sp. *Inimicus joubini* P Chevey, 1927
[Junior Synonym of *Inimicus japonicus*]

Professor Louis Marie Adolphe Édouard Joubin (1861–1935) was a French zoologist who worked at the MNHN, Paris. He was director of the laboratories at Banyuls-sur-Mer (1882), Roscoff (1884), and later instructor at the University of Rennes, gaining the Chair of Molluscs at MNHN (1903). He was also a one-time President of the Zoological Society of France (1905). He published widely, including: *Les Némertiens* (1894), *Le Fond de la Mer* (1920) and *Éléments de biologie marine* (1928). Pierre Chevy contributed to at least one of Joubin's publications, *Faune ichthyologique de l'Atlantique nord* (Fishes of the North Atlantic) (1929). A squid genus, *Joubiniteuthis*, is also named after him.

Joung

Shortfin Smooth Lanternshark *Etmopterus joungi* JDS Knuckey, DA Ebert & GH Burgess, 2011

Dr Shoou-Jeng Joung (b.1958) is an ichthyologist at the National Taiwan Ocean University, Keelung, where he is an Associate Professor and Chairman, Department of Environmental Biology and Fisheries Science. He has made a particular study of genetic markers in Whale Sharks. He has published more than 30 papers, including co-writing: *Estimation of life history parameters of the sharpspine skates,* Okamejei acutispina, *in the northeastern waters of*

Taiwan (2011) and *Fisheries, management, and conservation for the whale shark,* Rhincodon typus *in Taiwan* (2012). He was honoured for his contributions to chondrichthyan research in Taiwan and for his assistance and support during field surveys conducted by the second and third authors in Taiwanese fish markets.

Jouy

Japanese Minnow sp. *Rhynchocypris jouyi* DS Jordan & JO Snyder, 1901
[Syn. *Rhynchocypris oxycephalus jouyi*]
Korean Cyprinid sp. *Pseudolaubuca jouyi* DS Jordan & EC Starks, 1905

Pierre Louis Jouy (1856–1894) was an American diplomat, amateur naturalist and ethnographer. He co-wrote with Edward Grey a handbook: *Unique Collection of Ancient and Modern Korean and Chinese Works of Art, Procured in Korea During 1883* (1888). He was collecting in Japan (1881) and Korea (1885–1889), mostly with Dr F C Dale. Later (c.1892) he collected in Arizona and New Mexico. There are a number of articles relating to his travels in the USNM annual reports of the 1880s. Five birds are also named after him.

Joyner

Red Tongue-Sole *Cynoglossus joyneri* A Günther, 1878
Saddled Brown Rockfish *Sebastes joyneri* A Günther, 1878

Henry Batson Joyner (1839–1884) was a British engineer and Fellow of the Royal Geographical Society. He served his pupillage (1856–1860) becoming Assistant Engineer (1860–1861) on a railway project. He worked on the construction of the Cwm-Orthin Railway (1862–1868) and then on the water supply for Tunbridge Wells becoming their Resident Engineer (1868–1870). He then worked for the Japanese Imperial Government (1870–1878), initially on the construction of their first railway then on survey work and training their first meteorologists, writing a pamphlet entitled: *The Progress and ultimate results of Meteorology, specially considered in reference to Japan.* While there he made a collection of Japanese sea-fishes which he later presented to the BMNH and which Günther studied. He left Japan (1877) and went to Brazil as Chief Engineer for the water supply and sewerage system for Sao Paulo. He returned to England (1884) in poor health, and sadly died a few months later.

Juan Lang

Neotropical Rivuline sp. *Austrolebias juanlangi* WJEM Costa, MM Cheffe, HM Salvia & T Litz, 2006

Juan Jorge Reichert Lang (1929–2000). (See **Reichert**)

Juba

African Barb sp. *Labeobarbus jubae* KE Banister, 1984
[Syn. *Varicorhinus jubae*]

This is a toponym, referring to the Juba River in Ethiopia.

Jubb

Southern Deepbody *Hypsopanchax jubbi* M Poll & JG Lambert, 1965
African Barb sp. *Labeobarbus jubbi* M Poll, 1967
African Rivuline sp. *Nothobranchius jubbi* RH Wildekamp & HO Berkenkamp, 1979

Dr Reginald Arthur 'Rex' Jubb (1905–1987) was a South African ichthyologist whose bachelor's degree was awarded by Rhodes University, Grahamstown, which also awarded him a DSc. (1970). He moved to Southern Rhodesia (Zimbabwe) and made one of the first collections of Rhodesian fishes (1933). In WW2, he joined the Southern Rhodesian Air Force and was gazetted as a Flight Lieutenant (1940). He was made Honorary Keeper of Fishes for the Queen Victoria Museum in Salisbury (Harare) and established the national collection of fish (1954) in the National Museum in Bulawayo. He joined the Department of Ichthyology, Rhodes University (1957) and after the Department was downsized he moved, with the fish collection, to the Albany Museum (1961) as honorary, unpaid Curator of Fishes. He stayed there until the mid-1970s. He wrote: *Freshwater Fishes of Southern Africa* (1967).

Jubelin

Sompat Grunt *Pomadasys jubelini* G Cuvier, 1830

Jean Jubelin (1787–1860) was the French governor of Senegal (1828–1829) at the time this fish was discovered there. After his tenure in Senegal, he became governor of French Guiana.

Juelin

Licorice Gourami sp. *Parosphromenus juelinae* W Shi, S Guo, H Haryono, Y Hong & W Zhang, 2021

Juelin Wang is the wife of the senior author Wentian Shi, and helped him to collect the paratypes. Her "inspiration and assistance" helped make the authors' study a success.

Julia

Yoke Darter *Etheostoma juliae* SE Meek, 1891

Julia Ringold Hughes (1849–1916) was the wife of ichthyologist Dr Charles Henry Gilbert (1859–1928). (Also see **Gilbert, CH**).

Julian

Julian Galaxias *Paragalaxias julianus* RM McDowall & W Fulton, 1978

This is a toponym, referring to the Julian Lakes in Tasmania where this species is found.

Julien (Keith)

Rapa Freshwater Goby *Stiphodon julieni* P Keith, RE Watson & G Marquet, 2002

Julien Keith (along with Julien Marquet – see next entry) made an extensive collection of fish in French Polynesia (particularly in the Austral Islands). The authors were careless in not using the genitive plural in the binomial - *julienorum* - as the dedication covers two persons. We assume Julien Keith is related to the senior author.

Julien (Marquet)

Rapa Freshwater Goby *Stiphodon julieni* P Keith, RE Watson & G Marquet, 2002

Julien Marquet (along with Julien Keith – see preceding entry) made an extensive collection of fish in French Polynesia (particularly in the Austral Islands). The authors were careless in not using the genitive plural in the binomial – *julienorum* – as the dedication covers two persons. We assume he is related to the third author.

Julio

Headstander Characin sp. *Hypomasticus julii* GM dos Santos, M Jégu & AC Lima, 1996

Dr Júlio Cesar Garavello. (See **Garavello**)

Julius

Leopard Corydoras *Corydoras julii* F Steindachner, 1906

We cannot be sure who is intended in this patronym, due to Steindachner's typical reluctance to provide etymologies in his descriptions. One theory is that it might be named after a man called Julius Michaelis, who is known to have provided Steindachner with fishes from Brazil. Unfortunately, nothing more seems to be known about this person, even if he *is* the right 'Julius'.

Jullia

Glass Tetra sp. *Phenacogaster julliae* ZMS de Lucena & CAS de Lucena, 2019

Jullia is the author's granddaughter who was born during the description of the species.

Jullien

Jullien's Mud Carp *Cirrhinus jullieni* HE Sauvage, 1878
Jullien's Golden Carp *Probarbus jullieni* HE Sauvage, 1880
[Alt. Isok Barb, Seven-striped Barb]

The etymology only helps to the extent that a 'J Jullien' collected both holotypes. We think that this is likely to be Dr Jules Jullien (1842–1897), a French physician and zoologist and an expert on Bryozoa. He served as ship's doctor on a number of expeditions, including the French scientific expedition to Cape Horn (1882–1883). He was President of the Zoological Society of France (1888). He wrote: *Dragages du Travailleur : Bryozoaires espèces draguées dans l'océan Atlantique en 1881* (1882).

Juniperoserra

Peninsular Clingfish *Gobiesox juniperoserrai* HS Espinosa-Pérez & JL Castro-Aguirre, 1996

Junípero Serra y Ferrer (1713–1784) was a Roman Catholic Spanish priest and friar of the Franciscan Order. He founded a mission in Baja California, along with other missions in modern-day California. He was a keen practitioner of self-flagellation, along with other extreme forms of masochism, though such actions did nothing to prevent (and perhaps assisted) his canonisation (2015).

Junk

Characin sp. *Elachocharax junki* J Géry, 1971
Croaker sp. *Pachyurus junki* L Soares & L Casatti, 2000

Dr Wolfgang Johannes Junk (b.1942) is a zoologist, limnologist, botanist and oceanographer whose doctorate was awarded by the University of Kiel (Christian-Albrechts Universität) (1970). He is a Professor at the University of Hamburg, and retired as Professor Emeritus (2007) from being head of the Working Group of Tropical Ecology at the former Max-Planck-Institute for Limnology at Plön, Germany. After retiring in Germany, he has continued to work in Brazil where for over 40 years he has co-operated with the National Amazon Research Institute, Manaus (see entry for INPA) and for 20 years on the ecology of the Pantanal in collaboration with the Federal University of Mato Grosso, Brazil, where he is a visiting researcher. He co-wrote: *Amazonian Floodplain Forests* (2011).

Juquiá

Armoured Catfish sp. *Otothyris juquiae* JC Garavello, HA Britski & SA Schaefer, 1998
Armoured Catfish sp. *Pseudotocinclus juquiae* AK Takako, C Oliveira & OT Oyakawa, 2005

This is a toponym; the Rio Juquiá basin, São Paulo, Brazil, is where the holotypes were collected.

Jureia

Bluntnose Knifefish sp. *Brachyhypopomus jureiae* ML Triques & DK Khamis, 2003

This is a toponym, referring to Juréia Ecological Station, São Paulo, Brazil, which is where the holotype was caught.

Jürgen Schmidt

Jürgen's Badis *Badis juergenschmidti* I Schindler & H Linke, 2010

Dr Jürgen Schmidt is a German aquarist and biologist. Registered blind, he has to use a magnifying glass to see details. He wrote *Aquarium Plants* (2001), *Bede Atlas for Freshwater Aquarium Fishes* (2002) and *AquaGuide to Catfish* (2003). He was honoured "*...for his valuable contributions on the ethology and taxonomy of Southeast-Asian freshwater fishes.*"

Jurubi

Pike-Cichlid sp. *Crenicichla jurubi* CAS de Lucena & SO Kullander, 1992

This is a Tupí-Guaraní word for 'small mouth', referring to its smaller mouth compared to the otherwise similar *C. igara.*

Juruna

Tetra sp. *Hyphessobrycon juruna* TC Faria, FCT Lima & DA Bastos, 2018

The Juruna are an indigenous group originally living in the region of the mouth of the Rio Xingu, but who have emigrated upstream, first to the region of Altamira, and afterwards reaching the region of the Von den Steinen falls (1916). They currently live in the Parque Indigena do Xingu. The binomial also honours a member of the Xavante people, Mário Dzururã (1950–2002), better known as Mário Juruna, who was the first indigenous deputy in Brazilian history.

Jussie

Carache Amarillo (pupfish) *Orestias jussiei* A Valenciennes, 1846

Joseph de Jussieu (1704–1779) was a French botanist and explorer in South America. He is most remembered for introducing the heliotrope to European gardeners. He spent c.35 years in South America before returning to France (1771).

Jussieu

Mauritian Sardinella *Sardinella jussieu* BG de Lacepède, 1803

Dr Antoine Laurent de Jussieu (1748–1836) was a French botanist and physician who qualified in Paris (1770). He was Professor of Botany at the Jardin des Plantes, Paris (1770–1826). He greatly changed botany by not adopting Linnaeus's system of classification, but did keep Linnaeus' binomial system of naming. Many of today's plant families are still attributed to Jussieu. Lacepède was grateful to Jussieu for having shared with him an unpublished manuscript about fish, written (1770) by Philibert Commerson.

Justa

Characin sp. *Moenkhausia justae* CH Eigenmann, 1908

Dr Justa is stated to have been a Brazilian who collected the type specimen. He is mentioned as someone who collected on the Thayer Expedition to Brazil (1865–1866) and worked closely with Major J M S Coutinho (q.v.).

K

Kaaea

Caledonian Red-nose (Goby) *Lentipes kaaea* RE Watson, P Keith & G Marquet, 2002

Kaaea is a Napwé Tribal Chief, who kindly permitted the collection of freshwater fishes on tribal lands in New Caledonia.

Kabeya

Viper Dogfish *Trigonognathus kabeyai* K Mochizuki & F Ohe, 1990

Hiromichi Kabeya was captain of the trawler 'Seiryo-Maru' from which the holotype was caught (1986). He was still master of the vessel in 2006

Kadlec

African Rivuline sp. *Nothobranchius kadleci* M Reichard, 2010

Jaroslav Kadlec (1951–2006) from Brno, Czech Republic, was a renowned aquarist, recognised worldwide for his articles on the ecology and breeding of killifish.

Kafanov

Snailfish sp. *Genioliparis kafanovi* AV Balushkin & OS Voskoboinikova, 2008

Dr Alexander Ivanovich Kafanov (1947–2007) was a Russian eveolutionary biologist. He began collecting molluscs as a child and as a teenager was already very knowledgeable. He studied at the Biological and Soil Science Faculty of Rostov State University, where he simultaneously specialised in the taxonomy of marine molluscs and parasitology in the Department of Zoology and on biogeochemistry in the Department of Geochemistry. By the time he graduated he had already published several papers. He became (1971) Senior Assistant in the Laboratory of Physiological Ecology in the Institute of Marine Biology of the Far East Science Centre of the USSR Academy of Sciences and registered as a PhD student. He went from being a Senior Laboratory Assistant to a Scientific Secretary and the Head of a Laboratory. After defense of his Doctor of Science thesis (1991), he returned to the Institute of Marine Biology where he worked as a Principal Research Scientist, combining scientific and administrative duties. He published more than 240 papers.

Kaguya

Kaguya Dartfish *Navigobius kaguya* AC Gill, Y-K Tea & H Senou, 2017

The species is named after the Moon Princess Kaguya from the Japanese folk tale *Taketori Monogatari* (*The Tale of the Bamboo Cutter*), alluding to small spots on the first dorsal fin, which resemble the graphics used in moon phase charts. The name also acknowledges that the species occurs in Japanese waters. The name was selected by school students at education workshops associated with University of Sydney performances of *2071: A Performance about Climate Change.*

Kaibara

Japanese Loach sp. *Cobitis kaibarai* J Nakajima, 2012

Ekiken Kaibara (1630–1714) was a philosopher and botanist and "*...the first real naturalist and biologist in Japan*". He was the first to record the distribution of spined loaches from Chikushi (modern-day Fukuoka Prefecture), Kyushu Island, Japan. He wrote: *Medicinal herbs of Japan* (1709).

Kaie

Armoured Catfish sp. *Corymbophanes kaiei* JW Armbruster & MH Sabaj Pérez, 2000

Kaie is a legendary character after whom the Kaieutur Falls, Guyana, is named. Some accounts describe him as a brave chieftain who sacrificed himself by canoeing over the falls in order to save his tribe, whereas another version describes Kaie as a burdensome old man who was sent over the falls by his fellow tribesmen. Either way, say the describers, like Kaie, the genus *Corymbophanes* has "never been successful at traversing the falls".

Kailash

Catfish sp. *Glyptothorax kailashi* L Kosygin, P Singh & S Mitra, 2020

Loach sp. *Aborichthys kailashi* B Shangningam, L Kosygin, B Sinha & SD Gurumayum, 2020

Dr Kailash Chandra is an Indian zoologist and conservationist who is Director of the Zoological Survey of India, Kolkata, where he has worked most of his career (1989–present). His PhD was awarded by Kurukshetra University. He has published a great many papers and a co-authored twenty books including: *Faunal Diversity of Indian Himalaya* (2018) and *Current Status of Freshwater Faunal Diversity in India* (2017). The etymology states that it was named "*...honouring his contribution to the faunal diversity of India.*"

Kailola

Pufferfish sp. *Javichthys kailolae* GS Hardy, 1985

Kailola's Hardyhead *Craterocephalus kailolae* W Ivantsoff, LEL Crowley & GR Allen, 1987

Kailola's Deepwater Dragonet *Callionymus kailolae* R Fricke, 2000

Kailola's Sea Catfish *Cathorops kailolae* AP Marceniuk & R Betancur-R., 2008

Dr Patricia 'Tricia' J Kailola is an Australian biologist, fish taxonomist and fisheries scientist consultant, and honorary fellow at the University of the South Pacific, Suva, Fiji. She was at the Department of Zoology, University of Adelaide (1989) and at the Department of Primary Industries and Energy, Bureau of Resource Sciences, Fisheries Research & Development Corporation, Australia (1993). She is a research associate at the Australian Museum, Sydney. She has written several books on tropical fishes, as well as *Australian fisheries resources* (1993). She has made enormous contributions to the systematics of the catfish family *Ariidae*. (See also **Tricia** & **Patricia (Kailola)**).

Kainji

Lowa Elephantfish *Marcusenius kainjii* DSC Lewis, 1974

This is a toponym; it refers to Lake Kainji, an artifical lake in lower Niger River, Nigeria, where the holotype was collected.

Kaiyo Maru

Campbell Whiptail *Coelorinchus kaiyomaru* T Arai & T Iwamoto, 1979 .

The research vessel 'Kaiyo Maru', owned by the Japanese Fisheries Agency, collected the holotype. (See also **Campbell**).

Kaja

Kaja's sixgill sawshark *Pliotrema kajae* S Weigmann, O Gon, RH Leeney & AJ Temple, 2020

Kaja Magdalena Weigmann (b.2015) is the daughter of the first author. As the etymology states. *She had her first contact with chondrichthyan taxonomy when observing with great interest the examination of Pliotrema specimens for the present study. The name "Kaja" also has the Frisian meaning "warrior", referring to the saw-like rostrum.*

Kakuk

Kakuk's Coralbrotula *Ogilbichthys kakuki* PR Møller, W Schwarzhans & JG Nielsen, 2004

Brian Kakuk was a US Navy Diver (1981–1988) who then moved to the Bahamas, where he is co-owner, Bahamas Underground Cave Diving Facility and Diving Safety Officer, Caribbean Marine Research Center, Lee Stocking Island, Bahamas. The etymology says the brotula is named after him as the person "*...who kindly presented a newly-caught specimen.*"

Kalawatset

Umpqua Chub *Oregonichthys kalawatseti* DF Markle, TN Pearsons & DT Bills, 1991

This is not a true eponym in that the Kalawatset were a native Umpqua people who lived in Oregon, USA.

Kalfatak

Goby sp. *Stiphodon kalfatak* P Keith, G Marquet & RE Watson, 2007

Donna Kalfatak is the Director of Vanuatu's Department of Environmental Protection and Conservation. She was a Senior Biodiversity Officer at the Ministry of Climate Change Adaptation, Meteorology and Geo-Hazards, Environment, Energy and Disaster Management, Port VIla, Vanuatu. Among her published papers is: *Genetic and migratory evidence for sympatric spawning of tropical Pacific eels from Vanuatu* (2015).

Kali

Snaketooth genus *Kali* RE Lloyd, 1909

Kali, also known as *Kālikā* ('the black one') is a Hindu goddess, often depicted as having four arms.

Kalisher

Downy Blenny *Gobioclinus kalisherae* DS Jordan, 1904

The etymology simply says: "Named, at the request of Dr. [Joseph] Thompson, in honor of Miss Kalisher, of San Francisco." We have been unable to further identify her.

Kallman

Swordtail sp. *Xiphophorus kallmani* MK Meyer & M Schartl, 2003

Dr Klaus D Kallman (b.1928) was the German-born American Director of the New York Aquarium; a Research Assoiate of the AMNH, New York and a leading fish geneticist. He retired from the Aquarium after thirty-five years (1958–1993). Queens University awarded his BSc (1952) and he studied for his MSc at New York University, which also awarded his PhD (1959) (and later he was awarded an honorary doctorate by the University of Hamburg (1992). Among his publications are: *A New Look at Sex Determination in Poeciliid Fishes* (1984) & a chapter in *Handbook of Genetics* (1975).

Kalpangi

Garra sp. *Garra kalpangi* K Nebeshwar, K Bagra & DN Das, 2012

A toponym, referring to the River Kalpangi in Arunachal Pradesh, India, where the species was first found.

Kalunga

Characin sp. *Hasemania kalunga* VA Bertaco & FR Carvalho, 2010

The Comunidade Quilombo Kalunga are descendants of African slaves who live in the upper Rio Tocantins basin, Goiás, Brazil, near where the holotype was taken. .

Kamakan

Characin sp. *Characidium kamakan* AM Zanata & P Camelier, 2015

The Kamakã indigenous people originally inhabited the lower portion of Río Pardo basin, Bahia, Brazil, which is where the holotype was taken.

Kamalika

Kami's Barb *Puntius kamalika* A Silva, K Maduwage & R Pethiyagoda, 2008

Dr Kamalika 'Kami' Abeyaratne (1934–2004) was a Sri Lankan pediatrician who became an AIDS activist (1997) after she contracted HIV through a contaminated blood transfusion, administered following a near-fatal traffic accident (1995).

Kamamba

Lake Tanganyika Cichlid sp. *Lepidiolamprologus kamambae* SO Kullander, M Karlsson & M Karlsson, 2012

This is a toponym referring to Kamamba Island, Lake Tanganyika, Tanzania; the type locality.

Kamdem

African Rivuline sp. *Fundulopanchax kamdemi* C Akum, R Sonnenberg, JR van der Zee & RH Wildekamp, 2007

Dr André Kamden Toham is a Cameroonese ichthyologist and landscape ecologist who, since 1999, is the Regional Representative, Gabon, for the World Wildlife Fund Central Africa. His doctorate was awarded by the University of Louvain, Belgium (1997) and he did post-doctoral research at Cornell University (1998). He co-wrote *Diversity patterns of fish assemblages in the Lower Ntem River Basin (Cameroon), with notes on potential effects of deforestation* (1998).

Kami

Perchlet sp. *Plectranthias kamii* JE Randall, 1980

Harry T Kami worked at the Division of Fish and Wildlife on Guam. He donated the type specimen to the Bishop Museum, Honolulu, suspecting that it might represent an undescribed species.

Kamikawa

Arbiter Snailfish *Careproctus kamikawai* JW Orr, 2012

Daniel Joseph Kamikawa (b.1961) is a Research Fisheries Biologist for the national Marine Fisheries Service at the Northwest Fisheries Science Center (Oregon) (since 1995). The University of Wisconsin awarded his BSc (1983), following which he was a Biological Technician at the Wisconsin Department of Natural Resources (1983–1985) then a Resource Conservationist at North Cook County (Illinois) Soil and Water Conservation District (1985–1987). He joined the US Forest Service Klamath National Forest (California) as a Biological Aide (1988) before becoming (1989) a fisheries biologist with the National Marine Fisheries Service stationed at Alaska Fisheries Science Center. Concurrently he was a Biological Technician with the US Army Corps of Engineers in Washington State (1987, 1989 & 1992). He was a Biological Technician for the US Forest Service Kings River/Pineridge Ranger Districts (California) (1991) then a Fisheries Biologist with the US Forest Service Estacada Ranger District (Oregon) (1992) and the National Marine Fisheries Service, Northwest Fisheries Science Center (Washington State) (1992–1995) before taking up his current post. His most recent publication was: *Survey Fishes: An Illustrated List of the Fishes Captured during the Northwest Fisheries Science Center's Fishery Resource Analysis and Monitoring Division's West Coast Surveys* (2017). He caught the snailfish believing it to be another Bering Sea species, newly described by James Orr, but it proved to be new to science and was named after him.

Kamohara

Wide-mouthed Flounder genus *Kamoharaia* K Kuronuma, 1940
Kamohara's Sand-shark *Pseudocarcharias kamoharai* K Matsubara, 1936
[Alt. Crocodile Shark]

Kamohara's Grenadier *Coelorinchus kamoharai* K Matsubara, 1943
Bareskin Dogfish *Centroscyllium kamoharai* T Abe, 1966
Sandperch sp. *Parapercis kamoharai* LP Schultz, 1966
Rattail sp. *Nezumia kamoharai* O Okamura, 1970
Slickhead sp. *Narcetes kamoharai* O Okamura, 1984
Armoured Gurnard sp. *Paraheminodus kamoharai* T Kawai, H Imamura & K Nakaya, 2004
Kamohara's Bandfish *Owstonia kamoharai* H Endo, Y-C Liao & K Matsuura, 2015
Stinger Flathead sp. *Plectrogenium kamoharai* K Uesaka, T Yamakawa, M Matsunuma & H Endo, 2021

Dr Toshiji Kamohara (1901–1972) was an ichthyologist who graduated from Tokyo Imperial University (1926). He was a Professor at a High School in Kochi (1928). He was an artilleryman in the Japanese army (1938–1939), and later a Professor in the Zoology Department, Kochi University (1949–1965), retiring as Kochi University Professor Emeritus. He described 52 new species, of which 45 are still valid. He secured the type specimen of the sand-shark from a fish market and presented it to Matsubara. He was honoured in the dogfish's name for "*...his generosity to all ichthyologists.*"

Kampa

Longnose Catshark *Apristurus kampae* LR Taylor, 1972

Dr Elizabeth Maitland Boden née Kampa (1922–1986) was an oceanographer who was chief scientist on board the research vessel 'Argos' (1970) from which the type specimen was collected. She worked at the Scripps Institution of Oceanography, California (1944–1977). The University of California Los Angeles awarded her doctorate (1950). She joined the University of Hawaii (1977).

Kampen

Kampen's Ilisha *Ilisha kampeni* M Weber & LF de Beaufort, 1913
Sepik River Halfbeak *Zenarchopterus kampeni* M Weber, 1913
Freshwater Snake-eel *Lamnostoma kampeni* M Weber & LF de Beaufort, 1916

Pieter Nicolaas van Kampen (1878–1937) was a Dutch herpetologist and ichthyologist who was Professor of Zoology, Leiden University, until his retirement (1917–1931). He was based in the Dutch East Indies (1905–1911), during which time he collected many zoological specimens. He wrote: *The Amphibia of the Indo-Australian Archipelago* (1923). Two amphibians are also named after him.

Kanak

New Caledonian Catshark *Aulohalaelurus kanakorum* B Séret, 1990
Dragonet sp. *Callionymus kanakorum* R Fricke, 2006

The Kanaky are an indigenous people in New Caledonia, where the shark and dragonet were first taken. As Séret puts it, the shark is "*...dedicated to the Melanesian people of New Caledonia*".

Kanayama

Stinger Flathead sp. *Plectrogenium kanayamai* K Uesaka, T Yamakawa, M Matsunuma & H Endo, 2021

Tsutomu Kanayama is a Japanese ichthyologist at Hokkaido University. He was the first to recognise the uniqueness of this species (in 1982).

Kanazawa, R

Cusk-eel sp. *Dicrolene kanazawai* M Grey, 1958

Dr Robert H Kanazawa (1916–1985) was an American zoologist and ichthyologist at the USNM, Division of Fishes. He wrote: *A revision of the eels of the genus* Conger *with descriptions of four new species* (1958).

Kanazawa, T

Sand-lance sp. *Ammodytoides kanazawai* K Shibukawa & H Ida, 2013

Takeshi Kanazawa of the Tokyo University of Fisheries (now Tokyo University of Marine Science and Technology) provided the authors the opportunity to examine the holotype of this species, which he had collected off the Ogasawara Islands (1995) as part of a biological survey by the University.

Kane

Corydoras sp. *Corydoras kanei* S Grant, 1998

Kane is the son of Steven Grant, whose etymology states that Kane "*…has and still is suffering much due to ill health*".

Kanehara

Goby sp. *Phoxacromion kaneharai* K Shibukawa, T Suzuki & H Senou, 2010

Hiroyuki Kanehara is a diver, guide and photographer working at the Diving Service Amamiensis, Amami Ōshima Island, Ryukyu Islands, Japan. He provided ecological and habitat information about the goby and underwater photographs for the authors.

Kaneko

Velvetfish genus *Kanekonia* S Tanaka, 1915

Ichiro Kaneko provided Tanaka with fish specimens from Nagasaki, Japan, including the holotype of *Kanekonia florida*. We have not been able to find further details about him.

Kaningini

Mormyrid sp. *Marcusenius kaninginii* T Kisekelwa, G Boden, J Snoeks & E Vreven, 2016

Professor Boniface Kaningini Mwenyimali is a Congolese fisheries biologist and limnologist who is Director of UERHA (Unité d'Enseignement et de Recherches en Hydrobiologie Appliquée) and rector of l'Institut Supérieur Pédagogique de Bukavu. He supported, in various ways, the sampling expeditions of Tchalondawa Kisekelwa in the Lowa system, Democratic Republic of Congo.

Kapala

Mottled Conger *Poeciloconger kapala* PHJ Castle, 1990
Kapala Lanternfish *Diaphus kapalae* BG Nafpaktitis, DA Robertson & JR Paxton, 1995
Kapala Whiptail *Nezumia kapala* T Iwamoto & A Williams, 1999
Kapala Stingaree *Urolophus kapalensis* GK Yearsley & PR Last, 2006

The FRV 'Kapala' (formerly of the NSW Fisheries Research Institute, Australia), from which the type specimens were collected, was honoured for the "*...extremely valuable fish collections made by the vessel over almost three decades*".

Kappen

Borneo Barb sp. *Osteochilus kappenii* P Bleeker, 1856

Surgeon Lieutenant-Colonel E F J Van Kappen was a military surgeon and health officer in the Dutch East Indies army. He collected fish on Borneo for Bleeker.

Kapur

Indian Catfish sp. *Pseudolaguvia kapuri* R Tilak & A Husain, 1975

Dr A P Kapur was an Indian entomologist who served as Director of the Zoological Survey of India, Calcutta. He wrote a great number of papers, including *Bionomics of some* Coccinellidae *predaceous on aphids and coccids in north India* (1942).

Karabanov

Loach sp. *Barbatula karabanowi* AM Prokofiev, 2018

Dmitry P Karabanow is (2001 onwards) a Senior Researcher at the Russian Academy of Sciences and is currently researching for his PhD in biological sciences. In the etymology, he was thanked for his 'invaluable' help during Prokofiev's 2008 expedition to Mongolia. He studied at the Yaroslavl State University, Faculty of Biology and Ecology (1996–2001), then undertook a postgraduate course at the I D Papanin Institute for biology of inland waters, Russian Academy of Sciences (2001–2004) where his postgraduate thesis was "*Genetic and biochemical adaptations of kilka Clupeonella cultriventris (Nordmann, 1840) when expanding range*" (2009) at the A N Severtsov Institute of Ecology and Evolution. His scientific interests include hydrobiology, biological invasions, genetics and adaptation. His published papers include: *Invasion of a Holarctic planktonic cladoceran Daphnia galeata Sars (Crustacea: Cladocera) in the Lower Lakes of South Australia* (2018).

Karaja

Neotropical Rivuline sp. *Melanorivulus karaja* WJEM Costa, 2007
Armoured Catfish sp. *Ancistrus karajas* RR de Oliveira, LH Rapp Py-Daniel, CH Zawadzki & J Zuanon, 2016

The Karajás are an indigenous Amerindian tribe, living in the Rio Tocantins basin, Pará, Brazil, which is where these fish occur.

Karaman, MS

Albanian Roach *Rutilus karamani* VD Vladykov & G Petit, 1930 NCR

[Junior Synonym of *Leucos basak*]

Dr Mladen Stanko Karaman (1937–1991) was a Yugoslav (Serbian) zoologist who was an expert on isopods. Originally from Skopje (now in the Former Yugoslav Republic of Macedonia), he lived in Split (now in Croatia) (1940–1950) and in Dubrovnik (now in Croatia) (1950–1953), returning to Skopje (1953) where he graduated from the University (1962). He moved to Pristina (now in Kosovo) (1963) as an assistant professor at the Faculty of Invertebrate Zoology. The University of Ljubljana (now in Slovenia) awarded his doctorate (1964) and he became associate professor (1969–1976). He became full Professor of Comparative Morphology and Invertebrate Taxonomy at what was then known as the Kragujevac Faculty of Sciences, Belgrade (then in Yugoslavia and now in Serbia) and was Dean of the Faculty (1980–1982). He was one of many zoologists and biologists produced by this family over four generations. See the next entry for details of his father.

Karaman, SL

Drin Brook Lamprey *Eudontomyzon stankokaramani* MS Karaman, 1974

Stanko Luka Karaman (1889–1959) was the father of the author (q.v.) and also a biologist and zoologist. He founded and managed both the Museum of Natural Science and the zoo in Skopje, Yugoslavia (1926). He became the first director of the Institute of Biology, Dubrovnik (1950) but returned to Skopje as Curator of the Museum of Natural Science (1953). His wife, Zora (1907–1974), became a Professor in the Faculty of Forestry at the university. His son described him as "*...the greatest explorer of freshwater fish fauna in Yugoslavia.*" (See entry above for his son, with details/descriptions of places that were in the former Yugoslavia).

Kardashev

African Rivuline sp. *Nothobranchius kardashevi* S Valdesalici, 2012

Kiril Kardashev is a Bulgarian aquarist who discovered this species in Tanzania (2011).

Karen

Eelpout sp. *Pachycara karenae* ME Anderson, 2012

Karen Lona Anderson is the wife of the author Michael Anderson.

Karin

Karin's Coralbrotula *Diancistrus karinae* W Schwarzhans, PR Møller & JG Nielsen, 2005

Karin Bloch is the wife of the third author, Jørgen Nielsen, who thanked her for her "*most valuable support*" during the many hours he invested in the study of this genus.

Karipuna

Armoured Catfish sp. *Curculionichthys karipuna* GSC Silva, FF Roxo, BF Melo & C Oliveira, 2016

The Karipuna are an indigenous people who live in the region of the Rio Oiapoque, northern Amapá, Brazil, where this species occurs.

Karnella

Karnella's Rover *Emmelichthys karnellai* PC Heemstra & JE Randall, 1977
[Alt. Pink Opelu]

Charles Karnella was honoured because he was the first to recognise that specimens from Bermuda thought to be *E. ruber* were a distinct species (1971). He also supplied specimens for study to the authors (1974). He works at the National Marine Fisheries Service in Hawaii.

Karrer

Karrer's Whiptail *Coelorinchus karrerae* IA Trunov, 1984
Morid Cod sp. *Physiculus karrerae* CD Paulin, 1989
Conger Eel sp. *Parabathymyrus karrerae* ES Karmovskaya, 1991
Duckbill Eel sp. *Facciolella karreri* W Klausewitz, 1995

Dr Christine Karrer is a German biologist and ichthyologist at the Zoologisches Museum der Universität Hamburg. She previously worked at the Zoologisches Museum an der Humboldt Universität zu Berlin, which had awarded her doctorate (1966) and where she became Curator of Ichthyology (1973). She joined the Institute of Sea Fisheries and the Zoological Museum of the University of Hamburg (1976). She was named as an Associate Professor at MNHN, Paris (1978), researching into a collection, which resulted in her study: *Anguilliformes du Canal de Mozambique* (1983). The etymology for one species reads: "*Named in honor of Dr Christine Karrer, Hamburg, in recognition of her contributions to the knowledge of Anguilliformes of the western Indian Ocean and her studies on deep-sea fishes of the Red Sea of the Senckenberg-Museum.*" (Also see **Christine**).

Karsten

Goby genus *Karsten* EO Murdy, 2002
Dreamer (anglerfish) sp. *Dolopichthys karsteni* SL Leipertz & TW Pietsch, 1987
Snaggletooth sp. *Astronesthes karsteni* NV Parin & OD Borodulina, 2002

Karsten Edward Hartel (b.1944) is an American ichthyologist. During the late 1960s he helped upgrade the ornithology collection at the Children's Museum of Boston and had various posts for the Massachusetts Audubon Society; the University of Massachusetts, Museum of Zoology; Massachusetts State Department of Agriculture; and the US Fish and Wildlife Service. The University of Massachusetts awarded his BSc in Wildlife Biology (1974) and while there he curated their ornithology collection (1972–1974). After this (1975) he became a Curatorial Associate and Collection Manager for the Ichthyology Department at the Museum of Comparative Zoology, Harvard. He has undertaken many fieldwork assignments (1973–2006). He was co-author of the first recent guide to ichthyological collection management (1978) and around sixty papers, as well as co-writing: *Inland fishes of Massachusetts* (2002).

Kasai

Coral Goby sp. *Paragobiodon kasaii* T Suzuki & JE Randall, 2011

Masao Kasai (b.1954) is a diver, guide and photographer. He has a degree in Sociology from Ryukyu University. After graduating he moved to Iriomote Island and became a diving instructor (1980), opening the 'Mr Sakana Diving Service' (1981). He has kept a monthly photo diary for twenty years and has published two photographic books. His photographs have appeared on various magazine covers. He discovered this goby and photographed it underwater. He advocates the 'Tatami one tatami' diving principle, which is to enjoy observation and photography carefully without moving around, keeping disturbance to the minimum. His hobby is 'keep fit' and he also has a sommelier qualification (JSA).

Kasawa

Poacher sp. *Occella kasawae* DS Jordan & CL Hubbs, 1925

Masunosuke Kazawa of Sapporo was a graduate student at Stanford University. According to the etymology, he was "...*engaged in the study of the fishes of the Hokkaido*." We have been unable to identify him further.

Kashkin

Barreleye sp. *Ioichthys kashkini* NV Parin, 2004 .

Nikita Ivanovich Kashkin is a Russian ichthyologist who was interested in the ecological study of mesopelagic species. He took part in many expeditions on vessels of the Acoustics Institute of Oceanography (1960s–1980s). He collected the holotype of this species.

Kasouga

Orange Brotula *Dermatopsoides kasougae* JLB Smith, 1943

A toponym referring to the Kasouga River, South Africa. The holotype was caught near the mouth of this river.

Katanga

Striped Topminnow *Micropanchax katangae* GA Boulenger, 1912
[Syn. *Lacustricola katangae*]
African Characin sp. *Microstomatichthyoborus katangae* L David & M Poll, 1937
Squeaker Catfish sp. *Synodontis katangae* M Poll, 1971
Kneria sp. *Kneria katangae* M Poll, 1976

This is a toponym; the holoypes of these fishes were collected in Katanga Province (formerly part of the Belgian Congo, now the Democratic Republic of Congo).

Katavi

Katavi Mouthbrooder *Haplochromis katavi* L Seegers, 1996

This is a toponym referring to Katavi National Park; the type specimens were caught from the Katuma River (western Tanzania) at the border of this park, and it is probably the only species of *Haplochromis* that occurs within it.

Katayama

Yellowstripe Slopefish *Symphysanodon katayamai* WD Anderson, 1970

Swallowtail Anthias sp. *Odontanthias katayamai* JE Randall, AL Maugé & YB Plessis, 1979

Dr Masao Katayama was a Japanese ichthyologist who was a professor at Yamaguchi University. Among his papers, published across more than four decades, are: *A New Serranid Fish Found in Japan* (1954), *A new genus and species of anthinid fish from Sagami Bay* (1964) and the co-written: *Lates japonicus, a New Centropomid Fish from Japan* (1983). William Anderson refers to Katayama supplying fish specimens for him to study.

Katherine (Bemis)

Katherine's Red Sea Flyingfish *Cheilopogon katherinae* IB Shakhovskoy & NV Parin, 2019

Katherine Elliott Bemis is a master's student at the Virginia Institute of Marine Science. Cornell awarded her BA. She was honoured for her 'kind help' with the authors' work on the review of this genus.

Katherine (Meyer)

Katherine's Fairy Wrasse *Cirrhilabrus katherinae* JE Randall, 1992

Katherine A Meyer was the wife of John W Shepard of the Marine Laboratory at Guam. The etymology says: "*This species is named* katherinae *in honour of Katherine A. Meyer, the late wife of John W. Shepard, at his request.*"

Kathy

Mobile Logperch *Percina kathae* BA Thompson, 1997

Kathy Thompson née Smith was the wife of author Bruce Thompson (1946–2007).

Kato, K

Sandperch sp. *Parapercis katoi* JE Randall, H Senou & T Yoshino, 2008

Kenji Kato of the Tokyo Metropolitan Fisheries Experiment Station caught the holotype (1991) with hook and line in the Ogasawara Islands. He also caught two of the paratypes and provided colour photographs of the holotype. With jack Randall, Richard Pyle and others he co-wrote: *Annotated checklist of the inshore fishes of the Ogasawara Islands* (1997).

Kato, S

Kato's Fairy Wrasse *Cirrhilabrus katoi* H Senou & T Hirata, 2000

Damselfish sp. *Chromis katoi* H Iwatsubo & H Motomura, 2018

Shoichi Kato (b.1957) is a PADI diving instructor who has run Regulus Diving on Hachijo Island, Japan, since he opened it (1992). Underwater photography and diving are also his long-term hobbies. He co-wrote: *Shrimps and Crabs of Hachijo Island* (2001) since which he has continued the series with books on Fishes, Damselfish, Wrasse & Parrotfish and is currently working on another (in publication 2020) on nudibranchs and sea-slugs. He collected the type specimens and made them available to the authors, also kindly providing photographs of them.

Kato, T

Jellynose Fish sp. *Guentherus katoi* H Senou, S Kuwayama & K Hirate, 2008

Tatsuya Kato is a Japanese fisherman who collected the holotype (2006).

Katrine

Katrine's Coralbrotula *Diancistrus katrineae* W Schwarzhans, PR Møller & JG Nielsen, 2005

Katrine Worsaae is the wife of the second author, who thanked her for her 'most valuable support' during the many hours he invested in the study of this genus.

Katumbi

Congo Cichlid sp. *Orthochromis katumbii* FDB Schedel, EJWMN Vreven, BK Manda, E Abwe, A Manda & UK Schliewen, 2018

Moïse Katumbi (b.1964) is a businessman and former Governer of Katanga (Democratic Republic of the Congo). He is a 'great fish enthusiast' who supported part of the (2015) ichthyological research field expedition of the Mbisa Congo project; some specimens of this species were collected on his farm.

Katunzi

Lake Victoria Cichld sp. *Haplochromis katunzii* S ter Huurne & F Witte, 2010

Egid F B Katunzi is Director of the Mwanza Centre of the Tanzania Fisheries Research Institute. He has collaborated with the Haplochromis Ecology Survey Team (HEST) of Leiden University (since 1979) when he did his MSc study on food preferences of four haplochromine species from the Mwanza Gulf. He published several papers on Lake Victoria cichlids and has continuously supported HEST fieldwork for 30 years.

Kaudern

Banggai Cardinalfish *Pterapogon kauderni* FP Koumans, 1933

Dr Walter Alexander Kaudern (1881–1942) was a Swedish ethnographer and zoologist. He was also well versed in botany, geology and geography. The University of Stockholm awarded his Zoology PhD (1910). He was Curator of the geological and mineralogy department of the Gothenburg Museum (1928–1932) and then Director (1932–1942). His ethnographic interest was awakened by his natural history collecting expeditions, notably to Madagascar (1906–1907 & 1911–1912). He wrote many zoological papers and reports on his travels, such as *På Madagaskar* (1913). His ethnography collection is held by the Ethnographical Museum, Stockholm. He also travelled on an expedition to Sulawesi (1916–1921) where he collected widely, including the holotype of this cardinalfish which is now popular in marine aquaria.

Kauffman, DE

Scorpionfish sp. *Neomerinthe kaufmani* AWCT Herre, 1952

Donald 'Don' E Kauffman was an American ichthyologist. He was (1947) aboard the 'Theodore N Gill', an exploratory vessel collecting data for the Philippines Fishery

Programme report (1950). He was author of *Research Report on the Washington State Offshore Troll Fishery* (1951). He was Supervisor of the Research Division of the Department of Fisheries in Washington, initially in the Salmon Division. He also co-authored at least one paper with Albert Herre on Philippine fishes: *New and little-known Philippine triglids* (1952). The etymology says: "*Named for my young colleague, Donald Kaufman, enthusiastic and painstaking student of fishes.*" We think the author made an error in the spelling of his surname by omitting an 'f'.

Kaufman, L

Goby sp. *Psilotris kaufmani* DW Greenfield, LT Findley & RK Johnson, 1993

Dr Les S Kaufman is an evolutionary ecologist specialising in the biology and conservation of aquatic ecosystems. He is Professor of Biology at Boston University. He has special expertise in coral reef biology, the evolution and ecology of tropical great lakes fishes, and ecosystem-based management of marine resources. Johns Hopkins University awarded his BSc (1974) and PhD (1980). He conducted post-doctoral research at the Harvard Museum of Comparative Zoology (1980–1983) where he remains an Associate in Ichthyology. He was an evolutionary ecologist at the New England Aquarium, Boston, Massachusetts, USA (1983–1994), serving in turn as Curator of Education, Curator of Exhibit Research and Development, and Chief Scientist. He has worked at Boston University ever since (1994), first as Associate Professor then full professor (2003). He has (2005>) worked as a senior marine scientist and PI for the Marine Management Areas Science Program and (2012>) as Marine Conservation Fellow in the Betty and Gordon Moore Center for Science. He is occasionally involved in the production of popular articles, television, and radio for venues such as National Geographic, Ranger Rick, and the National Public Broadcasting system. Since 2010, he has also served on the advisory board of Healthy Reefs for Healthy People. He has taken part in diving missions at the Aquarius Underwater Laboratory, where he has also performed underwater surgery on fish to implant electronic acoustic tags inside them. He has published widely such as: *Landscape processes and macroevolutionary patterns in East African cichlid fishes* (1995). He collected this goby and took aquarium photographs of it, which he provided to the authors.

Kaufmann

Amu Darya Sturgeon *Pseudoscaphirhynchus kaufmanni* KF Kessler, 1877

Lieutenant-General Konstantin Petrovich von Kaufmann (1818–1882) was trained as a military engineer and served in the Caucasus (1838). He was the commander of the Russian forces that captured Kars (in Eastern Anatolia, Turkey) after a prolonged siege (1855) during the Crimean War. Later, he was Governor-General of Turkestan (1867–1882), during which time he succeeded in expanding the Russian Empire's territory in Central Asia including absorbing the Emirate of Bokhara and capturing Samarkand. He was interested in natural history and was very progressive in his thinking, as he commissioned the first scientific exploration of the lands he controlled and instituted reforms to improve agriculture. Kessler refers to 'Lord' Kaufmann, and refers to help he gave to zoologist Modest Nikolaevich Bogdanov (who originally described the fish in 1874, but in a way that gave no distinguishing features). There seems little doubt that Konstantin Petrovich von Kaufmann is intended.

Kaup

False Moray genus *Kaupichthys* LP Schultz, 1943
Pipefish genus *Kaupus* GP Whitley, 1951
Snake-eel sp. *Yirrkala kaupii* P Bleeker, 1858
Kaup's Arrowtooth Eel *Synaphobranchus kaupii* JY Johnson, 1862
Pipefish sp. *Enneacampus kaupi* P Bleeker, 1863
Morid Cod sp. *Physiculus kaupi* F Poey, 1865

Johann Jakob von Kaup (1803–1873) was a German zoologist and palaeontologist who became the Director of the Grand Duke's natural history 'cabinet' in Darmstadt. He was a proponent of 'natural philosophy', believing in an innate mathematical order in nature, and he attempted biological classifications based on the Quinarian system. This proposed that all taxa are divisible into five subgroups (and if fewer than five subgroups were known, quinarians believed that a missing subgroup must remain to be found). Kaup wrote: *Classification der säugethiere und vögel* (1844). An amphibian and four birds are also named after him.

Kawamura

Black Kokanee (trout) *Oncorhynchus kawamurae* DS Jordan & EA McGregor, 1925

Dr Tamiji Kawamura (1883–1964) was a Japanese biologist and limnologist. Tokyo Imperial University awarded his bachelor's degree and doctorate in zoology (1908). His post-graduate work (1908–1912) was also at Tokyo Imperial University. He was a member of the faculty of Kyoto Imperial University (1914–1943) as a member of the research staff (1914–1922), Professor (1921) and Director (1922–1943), retiring as Professor Emeritus. He was Professor of Animal Physiology and Ecology, Department of Zoology (1926), when he planned what is now the Kiso Biological Station (opened 1933). He wrote: *Freshwater biology in Japan* (1918). He presented the holotype of this species.

Kayabi

Neotropoical Rivuline sp. *Melanorivulus kayabi* WJEM Costa, 2008
Tetra sp. *Hyphessobrycon kayabi* TF Teixeira, FCT Lima & J Zuanon, 2014
Armoured Catfish sp. *Pseudancistrus kayabi* GSC Silva, FF Roxo & C Oliveira, 2015

The Kayabi are a Tupí-speaking Indian nation that lived in the region of the rivers Arinos, dos Peixes and Teles Pires, Mato Grosso, Brazil. They still survive as a small group in a recently established area at the lower Rio Teles Pires, where the tetra can be found. Another group of the Tupi live as a transplanted population at the Xingu Indigenous Park.

Kaynak

Melid Loach *Oxynoemacheilus kaynaki* F Erk'akan, S Özeren & TT Nalbant, 2008

Hüseyin Kaynak was the father of the senior author.

Kaysone

Laotian Cave Loach *Schistura kaysonei* C Vidthayanon & K Jaruthanin, 2002

Kaysone Phomvihanne (1920–1992) was a Laotian politician who was President of Laos (1991–1992). He had become an active revolutionary against French colonialism during the 1940s while studying law in Hanoi.

Kaznakov

Tibetan Cyprinid sp. *Ptychobarbus kaznakovi* AM Nikolskii, 1903
Stone Loach sp. *Triplophysa kaznakowi* AM Prokofiev, 2004
[Syn. *Labiatophysa kaznakowi*]

Alexander Nikolaevich Kaznakov (1872–1933) was a Russian zoologist and naturalist who became Director, Caucasus Museum. He was on Kozlov's Imperial Russian Geographical Society expedition to Mongolia and Tibet (1899–1901). A reptile and a bird are also named after him.

Kazuko

Mexican False Moray *Chlopsis kazuko* RJ Lavenberg, 1988

Kazuko Nakamura was an archivist at the Los Angeles County Museum. She was honoured for "*...single-handedly archiving the Giles W. Mead ichthyological library at the Los Angeles County Museum of Natural History.*"

Kazumbe

Squeaker Catfish sp. *Chiloglanis kazumbei* JP Friel & TR Vigliotta, 2011

George Kazumbe from Kigoma, Tanzania, is described as an 'expert fisherman and friend'. He has helped the authors and several of their colleagues during fieldwork in Tanzania. He describes himself as a field science technician and underwater ranger and runs cichlid safaris.

Keatinge

Fashoda Mormyrid *Petrocephalus keatingii* GA Boulenger, 1901

Dr Henry Porringer Keatinge (1861–1928) was a British physician and surgeon who was trained at Guy's Hospital in London and Durham University where he graduated (1883). He served in the Egyptian Army as a military surgeon (1884–1890) and took part in the Nile Campaign (1885). He was based at, and became Director of, both the Egyptian Government Medical School, Cairo and the Kasr-el-Aini Hospital (1884–1919).

Keeley

Keeley's Dragonet *Callionymus keeleyi* HW Fowler, 1941

Frank J Keeley (1868–1949) worked at the department of mineralogy, Academy of Natural Sciences of Philadelphia. The University of Pennsylvania awarded his bachelor's degree (1892) and he then joined his father's mining firm. He was a dedicated naturalist and joined the Academy as a Life Member (1894), thereafter holding vatious posts including being Curator of the mineralogy collection. His principle focus was microscopy. A mineral, *keeleyite*, was also named after him (1922) but was later found to be an already-named mineral.

Kee Lin

Earthworm Eel sp. *Chendol keelini* M Kottelat & KKP Lim, 1994

Dr Peter Kee Lin Ng is a freshwater field biologist, particularly of freshwater crab taxonomy and conservation. He leads a team of the Systematics and Ecology Laboratory, as a professor at the National University of Singapore, which awarded his BSc (1983) and PhD (1990). He began his career at NUS as a lecturer (1990–1994), then Senior Lecturer (1994–1998) before becoming Professor. Between the two he studied for his Diploma in Education at the National Institute of Education, Nanyang Technological University. He is also Director of both the Raffles Museum of Biodiversity Research and the Tropical Marine Science Institute of Singapore. The etymology says: "*...in honor of Peter Kee Lin Ng, for his support of the authors and many other researchers of Southeast Asian aquatic biology.*"

Keilhack

African Killifish sp. *Micropanchax keilhacki* E Ahl, 1928

Dr Ludwig Keilhack (1884–1915) was a German zoologist and limnologist with interests in ichthyologist and ornithology. He was particularly interested in freshwater fauna and taxonomy. He was Assistant Zoologist at the Berlin Institute (1913). He wrote: *Bemerkungen zur Fischfauna des nördlichen Njassa-Gebietes: einige neue Arten aus den Gattungen Barbus und Synodontis und Beiträge zur Systematik der Gattung Clarias* (1908) and *Bemerkungen über die Verbreitung einiger Chydoriden innerhalb Deutschlands* (1911).

Keim

Eastern Bigmouth Shiner *Notropis dorsalis keimi* HW Fowler, 1909

Thomas Daniel Keim (1879–1968) was a friend of the author and, eventually his brother-in-law when Fowler married his sister, Elizabeth Keim (1884–1970). He helped in the collection of the holotype (1904) of this subspecies, as well as other fishes for the Academy of Natural Sciences of Philadelphia. He wrote: *Notes on the Fauna about the Headwaters of the Allegheny, Genesee and Susquehanna Rivers in Pennsylvania* (1915).

Keith, HG

Spiny Eel sp. *Macrognathus keithi* AWCT Herre, 1940

Henry George Keith (1899–1982) was born in New Zealand to English parents. In 1925, he was appointed Assistant Conservator of Forests for the government of North Borneo (Sabah), was promoted to Conservator of Forests (1931), and later again to Director of Agriculture and Wildlife. He also collected plants for scientific study. Herre honoured him in the fish's binomial for his hospitality and aid during Herre's trips to Sandakan.

Keith, P

Characin sp. *Jupiaba keithi* J Géry, P Planquette & P Le Bail, 1996

Dr Philippe Keith is an aquatic biologist at MNHN, Paris and at the Sorbonne, at both of which he is a Professor. He co-wrote: *Poissons et crustacés d'eau douce des Iles Marquises* (2016). He helped collect the holotype.

Kelaart

Dwarf Snakehead sp. *Channa kelaartii* A Günther, 1861
[Syn. *Channa gachua kelaartii*].

Lieutenant-Colonel Dr Edward Frederick Kelaart (1819–1860) was a British physician, naturalist and zoologist born in Ceylon (Sri Lanka). He qualified as a doctor at Edinburgh and also attended classes in medicine in Paris. While at Edinburgh he made his first contribution as a naturalist, delivering a paper (1839) on 'The timber trees of Ceylon'. He was in the Ceylon medical service and also served in Gibraltar (1843–1845). Whilst stationed there he studied the plants of the Rock of Gibraltar and wrote 'Flora Calpensis – a contribution to the botany and topography of Gibraltar and its neighbourhood'. He returned to Ceylon (1849–1854) and wrote *Prodromus Fauna Zeylanica* (1852), the first work to give scientific classifications to the mammals of Ceylon. Also, in these five years, he produced publications on geology, mammals, birds, reptiles and the cultivation of cotton. He was appointed Naturalist to the Ceylon Government, which paid £200 per annum (a lot of money in the middle of the 19th century) plus expenses on top of his army pay! One of his tasks as official naturalist was to investigate why the Ceylon pearl fisheries had not produced any profit. He investigated the life history of the pearl oyster and produced no less than four reports on the subject, which were published, some posthumously (1858–1863). When the Governor of Ceylon fell ill, and went home to England on board the 'Nubia' (1860), his health was of such concern that Kelaart, accompanied by his wife and five children, was sent along as the Governor's medical attendant. The Governor died two days before 'Nubia' arrived at Southampton, and Kelaart died the next day and was buried in Southampton soon after the ship docked. Four birds, three mammals, two reptiles and an amphibian are also named after him.

Kelber

Yellow Peacock Bass *Cichla kelberi* SO Kullander & EJG Ferreira, 2006

Dieter Kelber is a recreational angler. He was honoured "*…in recognition of his promotion of* Cichla *as sport fishes, and for supporting our study with information and images*" [of two species].

Kelle

Göksu Spined Loach *Cobitis kellei* F Erk'akan, FG Atalay-Ekmekçi & TT Nalbant, 1998

Dr Ali Kelle was a Professor of Ichthyology at Dicle University, Diyarbakir, where he was awarded his doctorate (1978). He wrote: *Taxonomic and Ecological Studies on Fishes Living in Dicle River Waters* (1978). He both collected and donated the holotype.

Keller

Bristlenose Catfish sp. *Ancistrus kellerae* L De Souza, DC Taphorn & JW Armbruster, 2019

Constance Templeton Keller was Chair of the Board of Trustees for the Field Museum. She has also been Chair of Illinois Nature Conservancy. She, and her husband Dennis, have been honoured with the University of Chicago Medal for their wide-ranging philanthropic support of the University (such as the $10M they gave to establish a new centre there, and the $25M they gave to Princeton in 2008). She is active in a wide-range of educational and environmental causes. The etymology says: "*Named in honor of Connie Keller and in gratitude for her leadership as Chair of the Board of Trustees for the Field Museum, where her unparalleled support of research and conservation work has led to the protection of more than 8 million hectares of South America forests and rivers. The lead author is personally inspired and honored to share a love of fly-fishing, the outdoors and conservation with Connie.*"

Kellogg

Goby genus *Kelloggella* DS Jordan & A Seale, 1905
Kellogg's Seahorse *Hippocampus kelloggi* DS Jordan & JO Snyder, 1901
Kellogg's Scorpionfish *Scorpaenodes kelloggi* OP Jenkins, 1903
Yellowfin Red Seabass *Zalanthias kelloggi* DS Jordan & BW Evermann, 1903

Vernon Lyman Kellogg (1867–1937) was an entomologist, evolutionary biologist and science administrator at Stanford University. He studied at the University of Kansas, Stanford and Leipzig. He was Professor of Entomology at Stanford (1894–1920). His academic career was interrupted by two years (1915/1916) spent in Brussels as Director of Hoover's humanitarian American Commission for Relief in Belgium. Initially a pacifist, Kellogg dined with the officers of the German Supreme Command. He became shocked by the grotesque Social Darwinist motivation for the German war machine – "*the creed of survival of the fittest based on violent and fatal competitive struggle is the Gospel of the German intellectuals.*" Kellogg decided these ideas could only be beaten by force and, using his connections with America's political elite, began to campaign for American intervention in the war. He published an account of his conversations in the book *Headquarters Nights* (1917), as well as a number of books and papers such as *Common injurious insects of Kansas* (1892) and *Human Life as the Biologist Sees it* (1922). After the war, he served as the first permanent secretary of the National Research Council in Washington, DC. A 'Liberty Ship' built during WW2 was also named after him.

Kelso

Armoured Catfish sp. *Pseudolithoxus kelsorum* NK Lujan & JNO Birindelli, 2011

Dr George Leslie Kelso (1929–2015) and Carolyn Kelso née Wierichs (1934–2008). He was a family physician but originally trained as a teacher (1946–1951) and then taught (1952–1953). After service in the US Army (1953–1955) he returned to higher education (1956–1958). He then switched to medicine, qualifying (1962) and setting up in practice (1962–2000). She was a teacher. They made a generous contribution to Texas A&M University and to the Winemiller Aquatic Ecology Lab (Winemiller is their son-in-law) and so facilitated important ichthyological discoveries, including this catfish species.

Kelum

Sri Lankan Barb sp. *Puntius kelumi* R Pethiyagoda, A Silva, K Maduwage & M Meegaskumbura, 2008

Kelum Nalinda Manamendra-Arachchi is a Sri Lankan biologist, herpetologist and zoo-archaeologist at the Wildlife Heritage Trust of Sri Lanka. His specialty is the study of amphibians. He co-wrote: *Conservation and biogeography of Amphibians of eastern Sinharaja* (2012). A reptile is also named after him.

Kemal

Ereğli Minnow *Garra kemali* B Hankó, 1925
[Syn. *Hemigrammocapoeta kemali*]

His Excellency Ghazi Mustafa Kemal (also known as Mustafa Kemal Atatürk) (1881–1938) was the founder of the modern Turkish state.

Kemp

Longtail Catfish sp. *Olyra kempi* BL Chaudhuri, 1912
Loach sp. *Aborichthys kempi* BL Chaudhuri, 1913
Kemp's Garra *Garra kempi* SL Hora, 1921
Longspine African Anglerfish *Lophiodes kempi* JR Norman, 1935

Dr Stanley Wells Kemp (1882–1945) was a zoologist, marine biologist and anthropologist. He joined (1903) the Fisheries Research Section, Department of Agriculture, Dublin, as Assistant Naturalist. He joined the Indian Museum, Calcutta (1911) as Superintendent, Zoological Section. There he worked very closely with the Scottish zoologist Dr Nelson Annandale (in 1925 he would write Annandale's obituary in the 'Records of the Indian Museum'). He was on the Abor Punitive Expedition (1911–1912), during which government scientists made extensive natural history collections, including in the Garo Hills. Kemp later joined the Colonial Office (1924) as Director of Research, Discovery Committee, and led the second Antarctic Discovery Expedition (1924) (relating to Whale Fisheries). He was Director of the Plymouth Marine Laboratory (1936–1945). He lost all his personal possessions, his library, and his unpublished works as the result of a German air raid (1941). Three amphibians and a bird are also named after him and his wife.

Kempkes

Guppy sp. *Poecilia kempkesi* FN Poeser, 2013

Michael Kempkes is a friend of the author, and a guppy enthusiast. He and Poeser were both co-describers of another species in this genus, *P. wingei*. He is the author of: *Guppy, Platy, Molly* (2010).

Kendall, RL

Lake Tanganyika Cichlid sp. *Lepidiolamprologus kendalli* M Poll & DJ Stewart, 1977

Robert L Kendall was Managing Editor (1989) of the North American Journal of Fisheries Management part of the American Fisheries Society; Kendall was acting AFS Executive Director (1999). He was also a fish ecologist at the Department of Zoology, Duke University. Among his published papers are: *An Ecological History of Lake Victoria Basin* (1969), and *Coolwater fishes of North America* (1978). He collected the type.

Kendall, WC

Western Atlantic Finless Eel *Apterichtus kendalli* CH Gilbert, 1891

Slick Puffer *Sphoeroides kendalli* SE Meek & SF Hildebrand, 1928

Dr William Converse Kendall (1861–1939) qualified as a physician at Georgetown University, Washington DC (1885). Bowdoin University gave him an honorary doctorate in Science (1935) to celebrate the 50 years since he qualified. He worked for the US Bureau of Fisheries (1890–1930), retiring as Director. He was a world authority on salmon and trout. He was the naturalist on board the US Fish Commission schooner 'Grampus' from which the eel holotype was collected (1889). The etymology for the puffer says: "*The species is named for Dr William Converse Kendall, of the U. S. Bureau of Fisheries, in recognition of his many valuable contributions to our knowledge of American ichthyology.*"

Kennedy, CH

Characin sp. *Psellogrammus kennedyi* CH Eigenmann, 1903

Dr Clarence Hamilton Kennedy (1879–1952) was an entomologist and ichthyologist. Indiana University awarded his bachelor's (1902) and master's (1903) degrees, and an honorary doctorate (1950). He also gained a master's degree from Stanford (1915) whilst his doctorate was awarded by Cornell University (1919). After a number of teaching posts before his doctorate, he worked at Ohio State University, Columbus (1919–1949) in which latter year he retired having been Professor (1933–1949). He co-wrote: *On a collection of fishes from Paraguay, with a synopsis of the American genera of cichlids* (1903).

Kennedy, CW

Blackblotch Pompano *Trachinotus kennedyi* F Steindachner, 1876

C W Kennedy was Lieutenant Commander on the Hassler Expedition (1871–1872) from Boston to San Francisco via the Straits of Magellan. He 'contributed greatly to the favourable results of the expedition' (translation). Franz Steindachner was a member of the scientific party on the expedition.

Kennedy, JF

Characin sp. *Astyanax kennedyi* J Géry, 1964

John Fitzgerald Kennedy (1917–1963) was the 35th President of the USA and needs no biography here.

Kennerly

Western Lake Chubsucker *Erimyzon sucetta kennerlii* CF Girard, 1856

Dr Caleb Burwell Rowan Kennerly (1829–1861) graduated with a bachelor's degree from Dickinson College, Pennsylvania (1849), and then studied medicine, being awarded his doctorate by the University of Pennsylvania (1852). Spencer Fullerton Baird helped him by getting him appointed as surgeon and naturalist to a number of expeditions sponsored by the government: the Pacific Railroad Survey (1853–1854), United States/Mexican Boundary Survey (1855–1857), and United States/United Kingdom joint Northwestern Boundary Survey (1857–1861). Kennerly kept up a correspondence with Baird, who credited him with many discoveries. He died of a sudden brain disorder while returning to Virginia from California and was buried at sea. A reptile is also named after him.

Kennicott

Stripetail Darter *Etheostoma kennicotti* FW Putnam, 1863

Robert Kennicott (1835–1866) was an American naturalist who founded the Chicago Academy of Sciences, and who explored the American Northwest (1857–1859). At 17 he was sent to study under Dr Jared Potter Kirtland in Cleveland. Kennicott worked for Baird at the USNM, largely helping to classify animals collected on the western frontier by army personnel involved in railroad surveys. Through Baird he went to Canada and met Hudson's Bay's chief trader, Bernard Ross, who became a close friend. One of Baird's biographers described Kennicott thus: '*He became the consummate collector, and when more demanding responsibilities intruded upon his direct involvement in collecting and classifying, he became a collector of collectors. Under his training and guidance virtually all the major natural scientists of the nineteenth century developed their enthusiasms and their professional competence.*' After a period as Curator in Chicago he left to explore 'Russian America' and spent the rest of his life in Alaska. Kennicott suffered a second and fatal heart attack near Nulato, Alaska, at the age of just 30 (1866). A town in Alaska now bears his name, and he is also commemorated in the names of two birds.

Kenoje

Ocellate Spot Skate *Okamejei kenojei* JP Müller & FGJ Henle, 1841

The original description has no etymology, but the binomial is not an eponym: it is derived from the Japanese name for the skate, which is *keno-ei*. (See **Zuge** for a similar case).

Ken Smith

Blunt-nosed Grenadier *Nezumia kensmithi* RR Wilson, 2001

Dr Kenneth L Smith Jr. is an open ocean ecologist, Senior Scientist at the Monterey Bay Aquarium Research Institute (2006–present) and also Adjunct Professor, Department of Ocean Sciences, University of California, Santa Cruz. Southern Illinois University awarded his BA and the University of Georgia his PhD. He was Assistant Scientist, Woods Hole Oceanographic Institution (1971–1975), then Assistant Research Biologist (1976–1979) and Associate Research Biologist (1979–1984), Research Biologist (1984–2006) and since then Research Biologist Emeritus at Scripps Institution of Oceanography. He has over 40 years' experience of going to sea and studying extreme ecosystems ranging from the deep ocean to Antarctic icebergs. The main thrust of his research is to understand the impact of a changing climate on deep-sea and polar ecosystems. He has published c.130 papers (1967–2014).

Kent Allen

Eel-Goby sp. *Taenioides kentalleni* EO Murdy & JE Randall, 2002

Kent 'Trig' Allen (b.1952) is an American zoologist and marine biologist. He was an employee (1981–2002) of ARAMCO; Arabian American Company (now better known as Saudi Aramco, Saudi Arabia's national oil company). The University of Texas, Austin, awarded his bachelor's (1974) and master's (1979) degrees. He was working on an environmental project in Saudi Arabia when he collected (1998) and photographed the holotype.

Kenwood

Neotropical Rivuline sp. *Moema kenwoodi* S Valdesalici, 2016

Kenwood Perkins was the father of Brian Perkins, an aquarist who collected the holotype. The fish was named in Kenwood's memory for support he provided to his son while Brian was studying in Peru.

Kerbert

Snaketooth sp. *Kali kerberti* M Weber, 1913

Dr Coenraad Kerbert (1849–1927) was a Dutch biologist who became Director (1890–1927) of the Amsterdam Zoo (Artis) and Chairman of the Dutch Committee for the Protection of Birds. He studied medicine at the University of Utrecht, graduating (1875) then completing his PhD (1876). He became zoological assistant at the Zoölogisch Laboratorium in Amsterdam (1877) and also taught botany and zoology at a Municipal Training College (1878). Unfortunately, due to inflation and the effects of WW1, his directorship of Amsterdam Zoo ended with the zoo being in a state of some neglect. Kerbert was a close friend of Max Weber, who described the snaketooth.

Kerguenne

African Characin sp. *Neolebias kerguennae* J Daget, 1980

Mrs Kerguenne Daget is the widow of the author, French ichthyologist Dr Jaques Daget (1919–2009).

Kerr, AFG

Kerr's Danio *Danio kerri* HM Smith, 1931

[Alt. Blue Danio; Syn. *Brachydanio kerri*]

Dr Arthur Francis George Kerr (1877–1942) was an Irish physician and botanist. He was particularly noted for his study of the flora of Thailand and a number of plants are named after him. Trinity College Dublin awarded his bachelor's degree in Botany (1897) and he went on to take his MD (1901) before taking a post as physician on a ship to Australia. He was posted to Thailand (1902) as an assistant to a Dr Campbell Highet, and later to the British Legation, Bangkok. He was Principal Officer of Health to the government there (1904–1914). During WW1, he served in the Royal Medical Corps in France (1915–1918) but had to withdraw due to ill health. He returned to Bangkok but soon went into private practice. He was then appointed Director of the Botanical Section of the Ministry of Commerce of Siam (1920–1932). For many years (1920–1929) he took an annual botanical tour. He finally returned to England (1932). Much of his botanical collection is held at Kew. He collected the danio type in Siam. Four dragonflies are also named after him.

Kerr, WE

Purple Emperor Tetra *Inpaichthys kerri* J Géry & WJ Junk, 1977

Dr Warwick Estevam Kerr (1922–2018) was a Brazilian agricultural engineer, geneticist and entomologist. He graduated as an agricultural engineer at Escola Superior de Agricultura

Luiz de Queiroz of the University of São Paulo, at Piracicaba, where he received his doctorate. He did post-doctoral studies at the University of California, Davis (1951) and at Columbia University (1952). He was at State University of São Paulo (1958–1964). He was Full Professor of Genetics at the Faculty of Medicine of Ribeirão Preto, University of São Paulo (1964–1971) during which time he was in Manaus as director of INPA (q.v.) (1975–1979). Immediately after officially retiring from the University of São Paulo, he became Full Professor at Universidade Estadual do Maranhão in São Luís (1981–1988), and then moved (1988) to the Universidade Federal de Uberlândia, in Uberlândia, state of Minas Gerais. He co-wrote: *Little mentioned aspects of the Amazonian biodiversity* (2010).

Kersten

Redspot Barb *Enteromius kerstenii* W Peters, 1868

Otto Kersten (1839–1900) was a German chemist and traveller. He was with Baron von der Decken in the unsuccessful attempt to climb Mount Kilimanjaro (1862). Kersten published six volumes of memoirs (1869–1879). He sent a collection of fishes (including the holotype of this species) to Peters in Berlin. A bird is also named after him.

Kerville

Orontes Minnow *Pseudophoxinus kervillei* J Pellegrin, 1911

Henri Gadeau de Kerville (1858–1940) was a French noble; a zoologist, entomologist, botanist and archeologist as well as a keen photographer. Lycée Pierre Corneille awarded his bachelors degree (1877). He collected around his home in Normandy, discovering a species of beetle new to science when only fifteen. He later travelled in Asia Minor and collected (1912) in Syria. Admitted to the Société des Amis des Sciences Naturelles de Rouen at just twenty-one, he went on to be its Secretary, then President. He created a laboratory to study the biology of caves (1910). During WW1, he was a volunteer nurse in a Rouen hospital. He was made Chevalier de la Légion d'Honneur (1933). His best known published work is: *Les Insectes phosphorescents : notes complémentaires et bibliographie générale (anatomie physiologie et biologie) : avec quatre planches chromolithographiées* (1881). Most of his collections are held in Paris. He collected the holotype of this minnow.

Kessler

Kessler's Goby *Ponticola kessleri* A Günther, 1861
Kessler's Gudgeon *Romanogobio kesslerii* BN Dybowski, 1862
Gillbar Barb *Enteromius kessleri* F Steindachner, 1866
Kessler's Sculpin *Leocottus kesslerii* BN Dybowski, 1874
Sculptured Sea Catfish *Notarius kessleri* F Steindachner, 1876
Red Wolf Fish *Erythrinus kessleri* F Steindachner, 1877
Caspian Anadromous Shad *Alosa kessleri* O von Grimm, 1887
Kessler's Loach *Paraschistura kessleri* A Günther, 1889
Chinese Cyprinid sp. *Schizopygopsis kessleri* SM Herzenstein, 1891
Siberian Brook Lamprey *Lethenteron kessleri* VP Anikin, 1905
Scaldback *Arnoglossus kessleri* PJ Schmidt, 1915
Tadpole Goby sp. *Benthophilus kessleri* LS Berg, 1927

Karl Fedorovich (Theodorovich) Kessler (1815–1881) was a Russian-German zoologist and collector who was one of the founders of the St Petersburg Society of Naturalists (1868), and its President (1868–1879). He studied mathematics and natural science at St Petersburg University (1828) and after graduation taught mathematics. He went on to take his master's (1840) and PhD (1842), presenting dissertations on bird anatomy in both cases. He took the Chair of Zoology at Kiev University (1842), after which he dedicated his life to zoology, particularly ornithology, herpetology and ichthyology. He took the zoology chair at St Petersburg University (1861), and throughout his time at both institutions he undertook fieldwork studying the lakes and rivers of the local area. He was elected Dean of St Petersburg University (1870). This, and his ill health, prevented him taking part in the (1874) Fedtschensko expedition to Turkestan and the Aralo-Caspian Expedition (1877), but all the material was sent to him to study. He wrote about fishes and other vertebrates in European Russia in his reports on the two expeditions. In all he wrote 64 zoological works. He was an authority on fishes of the Volga Delta. He described the *Ponticola* goby (1857) but used a preoccupied name. He also described the *Benthophilus* goby (1877) as a variety of *B. grimmi*. A genus and thirty species and subspecies were named after him, of which as many as 14 may remain valid. Two birds are named after him.

Ketmaier

Minnow sp. *Phoxinus ketmaieri* PG Bianco & S De Bonis, 2015
[Probably a synonym of *Phoxinus lumaireul]*

Valerio Ketmaier is a molecular and evolutionary biologist, zoologist and geneticist with a long history of collaborations with the senior author. He has been associated with the University of Rome, Italy, and latterly the University of Potsdam, Germany where he is an Assistant Professor. He has published well over a hundred papers such as: *Allozymic variability and biogeographic relationships in two Leuciscus species complexes (Cyprinidae) from southern Europe, with the rehabilitation of the genus Telestes Bonaparte* (1998) and the co-written *Genetic structure in the wood mouse and the bank vole: contrasting patterns in a human-modified and highly fragmented landscape* (2018). He says of himself: *My education has been in the broad field of molecular systematics and population genetics. I'm generally more interested in questions rather than organisms, which is why I've studied a variety of invertebrate and vertebrate taxa, from molluscs to arthropods, fish, reptiles, birds and mammals. I'm particularly interested in understanding the evolutionary processes in isolated populations to infer current and historical processes at various temporal and spatial scales.*

Keyvan

Loach sp. *Cobitis keyvani* HM Sabet, SV Yerli, S Vatandoust, SC Özeren & Z Moradkani, 2012
[Junior Synonym of *Cobitis faridpaki*]

Professor Dr Amin Keyvan (1930–2007) was professor in the Department of Fisheries Sciences at the Islamic Azad University, Iran. He is described in the etymology as "*...the greatest Iranian ichthyologist researcher, who dedicated his life to the Iranian fishes.*"

Khajuria

Garo Spineless Eel *Garo khajuriai* PK Talwar, GM Yazdani & DK Kundu, 1977

Dr Hitopdeshak Khajuria (1923–1982) was a zoologist who worked for the Zoological Survey of India. He collected the type during the Indo-German Expedition (1957). He wrote extensively on bats as well as other mammals and co-wrote: *Catalogue mammaliana : an annotated catalogue of the type specimens of mammals in the collections of the Zoological Survey of India* (1977). He also wrote: *Taxonomical and ecological studies on the bats of Jabalpur Dist. Madhyapradesh, India* (1980).

Khamtanh

Stone Loach sp. *Schistura khamtanhi* M Kottelat, 2000

Khamtanh Vatthanatham was a Laotian Fisheries Programme Officer, Mekong River Commission, whose help during a (1999) survey led to the discovery of this species. He returned (2007) to his original employers, the Department of Livestock and Fisheries of the Lao Ministry of Agriculture and Forestry.

Khardina

Tetra sp. *Hyphessobrycon khardinae* A Zarske, 2008

Natasha Khardina (b.1979) is an Uzbek photographer who with Bleher (q.v.) collected the holotype.

Kharin

Eelpout sp. *Melanostigma kharini* AV Balushkin & MV Moganova, 2018

Vladimir Emel'yanovich Kharin (1957–2013) was a Russian ichthyologist and herpetologist with the Institute of Marine Biology, Far East Branch, Russian Academy of Sciences, Vladivostok. He had a particular interest in sea snakes, and described several new species.

Kharuzin

Common Bleak ssp. *Alburnus alburnus charusini* SM Herzenstein, 1889

Alexey Nikolayevich Kharuzin (also transcribed as Charusin) (1864–1933) was a Russian ethnographer, anthropologist and statesman who was born in Reval (now Tallinn, Estonia). He graduated in physics and mathematics at Moscow State University. The Imperial Nature Lovers Society employed him on numerous missions to the Crimea, the Caucasus, Central Asia and the Aegean Sea and he was a Russian Geographical Society envoy to Bosnia and Herzegovina (1889–1891). He was special assistant to the governor of Estonia (1891–1901), Head of the Vilno Governor-General's office (1902–1903), Governor of Bessarabia (1904–1906), and later Director of the Department that dealt with the Religious Affairs of Foreign Faiths. He became Russia's Deputy Minister for Internal Affairs (1911). He did not leave Russia after the 1917 Revolution but worked at an agricultural base he had helped to establish near Tver. He became a gardening teacher at the Agricultural Polytechnic in Moscow (1924). He was arrested (1927) but was released without charge. He was arrested a second time (1931) and charged by the NKVD (secret police) with spreading anti-Soviet propaganda. Tried by a secret police tribunal, he was found guilty (1932) and sentenced to three years in the Gulag. He died in prison within a month. (Also see **Charusin**).

Kheel

Congo Minnow sp. *Raiamas kheeli* MLJ Stiassny, RC Schelly & UK Schliewen, 2006

Theodore Woodrow Kheel (1914–2010) was an American lawyer and skilled mediator in industrial disputes. Cornell University awarded his bachelor's degree (1935) and his law degree (1937). He founded (1991) the Nurture Nature Foundation to help try to resolve the conflict between environmentalists and developers.

Khimaera

Armoured Catfish sp. *Hypostomus khimaera* LFC Tencatt, CH Zawadzki & O Froehlich, 2014

From Greek mythology, the Khimaera (or Chimera) was a creature with a hybrid body formed by three animals (lion, goat, snake). The name was chosen because the catfish possesses features of conspicuously distinct species.

Khopa

Mountain Carp sp. *Psilorhynchus khopai* Lalramliana, B Solo, S Lalronunga & Lalnuntluanga, 2014

This is a toponym; Khopai is a small village near the Tuisi River, Mizoram, India, where the holotype was acquired.

Kiabi

Stone Loach sp. *Oxynoemacheilus kiabii* K Golzarianpour, A Abdoli & J Freyhof, 2011

Dr Bahram Hassanzadeh Kiabi is an Iranian ecologist and zoologist who is an Associate Prfofessor, Department of Marine Biology, Department of Biological Sciences, Shahid Beheshti University, Tehran. Michigan State University awarded his doctorate in Fisheries and Wildlife (1978). He was honoured as he has "*...contributed greatly to the knowledge of Iranian wildlife in general and fishes in particular.*" He co-wrote: *Status of the fish fauna in the South Caspian Basin of Iran* (1999). A reptile is also named after him.

Ki Chul Choi

Loach genus *Kichulchoia* I-S Kim, JY Park & TT Nalbant, 1999

Ki-Chul Choi (1910–2002) (See **Choi KC**)

Kidder

Kidder's Livebearer *Carlhubbsia kidderi* CL Hubbs, 1936
[Alt. Champoton Gambusia]

Alfred Vincent Kidder (1885–1963) was an American archaeologist. Harvard awarded his bachelor's degree (1908) and his PhD in anthropology (1914). He made a series of expeditions to the Southwest, many in northeastern Arizona. These expeditions were sponsored by Harvard's Peabody Museum of Archaeology and Ethnology and the associated Robert S

Peabody Museum of Archaeology at Phillips Academy in Andover, Massachusetts. After this he conducted site excavations (1915–1929) at an abandoned pueblo in the Pecos National Historical Park, excavating levels of human occupation over 2000 years, detailing cultural artifacts, pottery fragments and human remains. He wrote: *Introduction to the Study of Southwestern Archaeology* (1924) as well as other works on his excavations. He was Chairman of the Division of Historical Research of the Carnegie Institution (1929–1950) and spent time excavating Mayan ruins in central Amwerica. He was honoured for his 'broad interest' and making possible Hubbs' work in Yucatán, México.

Kido

Kido's Snailfish *Careproctus kidoi* SW Knudsen & PR Møller, 2008

Dr Kaoru Kido worked at the Ishikawa Fisheries Highschool, Japan. He, together with Dr Marmoru Yabe, first recognised this as a probable new species (1988). Kido wrote or co-wrote several papers including: *Phylogeny of the family Liparididae, with the taxonomy of the species found around Japan* (1988), *Redescription of Paraliparis tremebundus (Liparididae)* (1992) and *Fishes collected by the R/V Shinkai Maru around Greenland* (1995).

Kiener, A

Kotsovato (cichlid) *Paretroplus kieneri* J Arnoult, 1960
Kiener's Silverside *Teramulus kieneri* JLB Smith, 1965

André Kiener was a French fisheries researcher who conducted numerous studies in Madagascar (1950s & 1960s). He wrote: *Poissons, pêche et pisciculture à Madagascar* (1963).

Kiener, LC

Pigfish sp. *Congiopodus kieneri* HE Sauvage, 1878

Louis Charles Kiener (1799–1881) was a French conchologist and zoologist. He was curator of the private collection of the Duke of Rivoli (1834). Four birds are named after him.

Kihn

Central American cichlid genus *Kihnichthys* CD McMahan & WA Matamoros, 2015

Pablo Herman Adolfo Kihn Pineda (aka Herman Adolfo Kihn) is a Guatemalan biologist and ichthyologist. With the Universidad de San Carlos de Guatemala, where he is an ichthyology researcher, he conducted a survey of the fish of Guatemala published as: *Peces de las áreas protegidas Guatemaltecas (Zonas costeras y Humedales de la Vertiente del Pacifico)* (2006). He has been conducting research in the field since the 1960s. The etymology says he has 'spent a lifetime' studying the fishes of Guatemala, making invaluable contributions to our understanding of their diversity and distribution. His name is attached to *ichthys* = fish.

Kikuchi

Kikuchi's Minnow *Aphyocypris kikuchii* M Oshima, 1919

Yonetaro Kikuchi (1869–1921) was a collector for the Taipei Museum in Formosa (Taiwan). He collected examples of the Mikado Pheasant (*Syrmaticus mikado*) for Alan Owston (q.v.). A reptile, a bird and a mammal are also named after him.

Kilburn

Toadfish sp. *Perulibatrachus kilburni* DW Greenfield, 1996 .

Richard Neil 'Dick' Kilburn (1942–2013) was a South African malacologist. He graduated from the University of Natal, Pietermaritzburg (1967), and worked as a malacologist at the East London Museum (1968–1969). He then joined the Natal Museum (1970) where he spent the rest of his career, increasing the museum's mollusc collection from 9,000 to nearly 150,000 specimens. He collected the toadfish holotype from a malacological dredge haul.

Kilian

Characin sp. *Cheirodon kiliani* H Campos, 1982

Ernst F Kilian was a German biologist from the University of Giese who was hired to establish laboratories at Universidad Austral, Chile. He was the founding Director of Instituto de Zoología, Universidad Austral, Chile (1958). He returned to Germany (1965). The institute he founded is now named after him.

Kimluan

Loach sp. *Parabotia kimluani* VH Nguyen, 2005

This is a toponym referring to Kim Luân, a village near the Gâm River, Nà Hang District, Vietnam; the type locality.

Kimura, J

Sand-lance sp. *Ammodytoides kimurai* H Ida & JE Randall, 1993

Johnson Kimura worked at the Ogasawara Fisheries Center on Chichi-jima. The etymology says the name refers to: "*Johnson Kimura who assisted us in many ways in collecting and photographing fishes in the Ogasawara Islands*." Elsewhere in the article, he is referred to as: "*Johnson Kimura of the Ogasawara Fisheries Center on Chichi-jima...*".

Kimura, M

Babyface Goby *Paedogobius kimurai* A Iwata, S Hosoya & HK Larson, 2001

Motofumi Kimura works at the Okinawa Prefectural Fisheries Experiment Station. He co-wrote: *Transient Sex Change in the Immature Malabar Grouper, Epinephelus malabaricus, Androgen Treatment* (2014). He discovered this species in Japan.

Kimura, S

Sole sp. *Aseraggodes kimurai* JE Randall & M Desoutter-Meniger, 2007

Dr Seishi Kimura (b.1953) is a Japanese ichthyologist. Kyoto University awarded his PhD (1987). He is Professor at the Fisheries Research laboratory, Mie University, Japan where he was formerly Asociate Professor (1987–2007) and Assistant Professor (1978–1987). His research foci are: taxonomy and phylogeny of some coastal fishes mainly the family *Atherinidae, Carangidae* and *Leiognathidae*. He has described over 30 new taxa and three new genera. Among his major published papers are: *Revision of the genus Nuchequula with descriptions of three new species (Perciformes: Leiognathidae)* (2008), *The red-fin Decapterus group*

(Perciformes: Carangidae) with the description of a new species (2013) and *Taxonomic review of the genus Hypoatherina Schultz 1948 (Atheriniformes: Atherinidae)* (2014). He was honoured by the authors "*...in appreciation of his assistance in this study.*"

Kincaid

Blackfin Sculpin *Malacocottus kincaidi* CH Gilbert & JC Thompson, 1905
Characin sp. *Cynopotamus kincaidi* LP Schultz, 1950

Trevor Kincaid (1872–1970), zoologist, entomologist and oyster farmer was a Canadian-American Professor at the University of Washington, Seattle, USA (1901–1937), retiring as Professor Emeritus. (He once described himself as an 'omniologist') He visited the Pribilof Islands (1897) and was the entomologist attached to the Harriman Alaska Expedition (1899) during which he collected 8000 insects. The University of Washington awarded his bachelor's degree (1899) and his master's (1901). He wrote: *Development of Oyster Industry of the Pacific* (1928). He also left a typescript of an autobiography: *The adventures of an omniologist* (1962). At least 47 plant and animal species are named after him.

Kinch

Soldierfish sp. *Ostichthys kinchi* R Fricke, 2017

Dr Jeffrey 'Jeff' Paul Kinch (b.1965) is an Australian ichthyologist. The Univesity of Queensland awarded his bachelor's degree (1995) and doctorate (1999). Since 2008 he has been the Principal of the National Fisheries College in Kavieng, Papua New Guinea, and founder of the Nago Island Mariculture and Research Facility. He wrote: *Fisheries in Papua New Guinea* (2007).

King, D

Tiger Angelfish *Apolemichthys kingi* PC Heemstra, 1984

Dennis King, a structural engineer and underwater photographer, was born in the UK and learned to SCUBA dive there, but emigrated to South Africa (1968). He has written several books including the co-written *The Reef Guide* (2014) with fellow diver Valda Fraser. He discovered this species near Durban.

King, HH

African Killifish sp. *Micropanchax kingii* GA Boulenger, 1913

Harold Henry King (1885–1954) was a British government entomologist in Sudan (1906–1932). He was in Northern Rhodesia (Zambia) (1934) to experiment with poison dust as an antidote to swarms of locusts.

King, JE

Barbeled Dragonfish sp. *Bathophilus kingi* MA Barnett & RH Gibbs, 1968

Joseph Edwin King (b.1914) was a fishery research biologist at the U.S. Fish and Wildlife Service. He was honoured in the etymology as his: "*...studies of central Pacific midwater fishes resulted in the first known specimens of this species.*" He co-wrote: *Zooplankton abundance in the Central Pacific* (1963).

King, SG

King's Bullhead Catfish *Liobagrus kingi* T Tchang, 1935

Sohtsu G King (1886–1949), also known as Jin Shaoji, was a malacologist. He went to England in the early 20th century and studied electrical engineering at King's College, London. He was one of the founders of the Peking Laboratory of Natural History and its associated natural history society (1925). He was also a committee member, Fan Memorial Institute of Biology, which published Tchang's study. He co-wrote: *The Shells of Peitaiho* (1928).

King, TL

Cosby Creek Trout *Salvelinus kingi* JR Stauffer Jr., 2020 [Probably a population of Salmo fontinalis]

Dr Timothy Lee King (1958–2016) was an American fishery biologist who gathered extensive information on species of Brook Trout throughout their range. He was co-author of numerous articles (2006–2016) such as: *Spatial Structure of Morphological and Neutral Genetic Variation in Brook Trout* (2015). The author describes him as: "*...a great friend and mentor*".

Kingsley

African Tetra sp. *Brycinus kingsleyae* A Günther, 1896
Tailspot Ctenopoma *Ctenopoma kingsleyae* A Günther, 1896
Old Calabar Mormyrid *Paramormyrops kingsleyae* A Günther, 1896
Gabonese Cichlid sp. *Chromidotilapia kingsleyae* GA Boulenger, 1898
African Pike-Characin sp. *Hepsetus kingsleyae* E Decru, E Vreven & J Snoeks, 2013

Mary Henrietta Kingsley (1862–1900) was an ethnographic and scientific writer and explorer. She came from a literary family – two of her uncles were Charles and Henry Kingsley, both well-known Victorian novelists. Her parents both died (1892) and she received an inheritance of £4,300; a very large sum of money in those days. She decided to go to Africa, where she travelled alone, though it was virtually unheard of for European women to travel unaccompanied. She went from Sierra Leone to Angola (1893), living with the local peoples who taught her skills for surviving in the local environment. She returned to England (1894), where she gathered aid and support from Albert Günther (q.v.) and was signed up by Macmillan to publish her travel accounts. She was in Nigeria at the end of 1894 and later (1895) canoed up the Ogooué River and pioneered a new route when she climbed an active volcano, Mount Cameroon (4,040m). She was a most unusual person, being ready and willing to express her opinions forcefully. She described the issue of women's suffrage as being '...a minor question' and upset the Church of England by criticizing missionaries for trying to convert people and to corrupt their existing religions, stating that African cultures and customs, such as polygamy, needed to be protected and not banned. She went on an extensive lecture tour of the UK (1895–1898). When the second Boer War (1899–1902) started, she travelled to Cape Town and volunteered as a nurse. She was stationed at the Simon's Town hospital (1900) where she treated Boer prisoners-of-war. She died of typhoid and, at her own request, was buried at sea. She wrote: *Travels in West Africa* (1897) and *West African Studies* (1899). During her trip up the Ogooué Rver she collected a number of previously unknown fish species.

Kinzelbach

Orontes Nase *Chondrostoma kinzelbachi* F Krupp, 1985

Dr Ragnar Karl Konrad Kinzelbach (b.1941) is a limnologist and parasitologist, who is Professor Emeritus at the University of Rostock, Germany, where he was Professor of General and Special Zoology (1995–2006). The University of Mainz awarded his doctorate (1967) after which he did post-doctoral research in the USA and the UK (1967–1968). He taught at the University of Mainz as a scientific assistant (1968–1971) and as Professor of Zoology and Ecology (1971–1982); and at the University of Darmstadt as Professor of Zoology and Ecology (1982–1995). He was a member of the DFG Tübingen Atlas of the Middle East and has been involved in the study of the fauna of that region throughout his career. He wrote: *Faunal history of freshwater invertebrates of the Northern Levant (Mollusca, Crustacea)* (1987). He has written many papers and books, especially since 'retiring'. His interests continue to be ornithology, entomology, limnic ecology, zoogeography, historical zoology, traveling for field research and collecting. He is honoured in the names of around thirty different species across a variety of taxa, including gastropods, spiders, scorpions, beetles and other insects. He placed the nase holotype at Krupp's disposal.

Kirchmayer

Dwarf Panchax of Goa *Aplocheilus kirchmayeri* HO Berkenkamp & V Etzel, 1986

Josef Kirchmayer is a German aquarist who collected the type as well as importing and breeding the species.

Kirk, J

Kirk's Blenny *Alticus kirkii* A Günther, 1868
Long-tailed Sand-Eel *Bascanichthys kirkii* A Günther, 1870
Kambuzi (cichlid) *Protomelas kirkii* A Günther, 1894
Labeo sp. *Labeo kirkii* GA Boulenger, 1903

Sir John Kirk (1832–1922) was a Scottish physician, naturalist, explorer and administrator in Africa. He took part, as a physician and naturalist, on David Livingstone's second Zambezi expedition (1858–1863). Kirk contributed much to the eradication of the slave trade. He was Vice-Consul in Zanzibar (1866) and then Consul-General (1873). During his service, he persuaded the Sultan to abolish the slave trade (1873) and also to concede mainland territories to the British East Africa Company (1887). Kirk was a polymath whose interests included botany, geography, history, geology, chemistry and photography, as well as the study of Swahili, Arabic, Spanish, Portuguese and French. He was a Fellow of the Royal Botanical Society and collected and sent specimens to the Royal Botanical Gardens at Kew. He collected the first fish from Lake Nyasa to reach Western science and observed that they were almost all endemic. One amphibian, one reptile, six birds and three mammals are also named after him.

Kirk, RG

Red-fin Nothobranchius *Nothobranchius kirki* RA Jubb, 1969

Dr R G Kirk was a fish biologist working for the Agricultural Research Services, Ministry of Natural Resources, Malawi, where he discovered this species. Among his papers are: *The*

Fishes of Lake Chilwa (1967) and his PhD thesis at the University of London *A Study of Tilapia (Sarotherodon) shirana chilwae Trewavas in Lake Chilwa* (1970).

Kirovsky

Neotropical Rivuline sp. *Laimosemion kirovskyi* WJEM Costa, 2004

Alexandre Lantelme Kirovsky (b.1968) is a Brazilian biologist. His biology BSc was awarded by Universidade *Santa Úrsula, Rio de Janeiro* (1993) and his MSc in Freshwater Biology and Inner Fishing was from the National Institute of Amazonian Research (1999 He was Environmental Analyst of the Pico da Neblina National Park in northern Brazil (2002–2004). He was at the Special Secretariat for Aquaculture and Fisheries of the Presidency of the Republic and then at the Ministry of Fisheries and Aquaculture (2004–2012), where he held the positions of Technical Advisor to the General Coordination of Artisanal Fisheries. He was (2012–2013) Environmental Analyst at the Chico Mendes Institute (Instituto Chico Mendes de Conservação da Biodiversidade) and started a new job there (2017).

Kirschbaum

Glass Knifefish sp. *Japigny kirschbaum* FJ Meunier, C Jégu & P Keith, 2011

Dr Frank Kirschbaum is an honorary professor at Humboldt University, Berlin. He specialises in gymnotiform fishes and has succeeded in breeding several species in the laboratory. He co-wrote: *Species delimitation and phylogenetic relationships in a genus of African weakly-electric fishes (Osteoglossiformes, Mormyridae,* Campylomormyrus*)* (2016).

Kishi

Licorice Gourami sp. *Parosphromenus kishii* W Shi, S Guo, H Haryono, Y Hong & W Zhang, 2021

Hiroyuki Kishi is a Japanese explorer with more than 20 years' experience in Indonesia. He discovered both this species and *P. quindecim* and has "contributed much first-hand field information on this genus over the last decade." He has also made a contribution to environmental protection, having established a private conservation park in Kalimantan for the fish species *Betta antoni*.

Kishimoto

Stargazer sp. *Uranoscopus kishimotoi* R Fricke, 2018

Hirokazu Kishimoto, of Tokai University, is a Japanese ichthyologist noted for his research on stargazers. Among his publications is the co-authored: *Revision of a deep-sea stargazer genus Pleuroscopus* (1988) and a textbook he edited *Laboratory Manual on Fundamental Ichthyology* (2017).

Kishinouye

Mugura Grenadier *Coelorinchus kishinouyei* DS Jordan & JO Snyder, 1900
Lined Javelinfish *Hapalogenys kishinouyei* HM Smith & TEB Pope, 1906
Searobin sp. *Lepidotrigla kishinouyi* JO Snyder, 1911

Sisorid Catfish sp. *Euchiloglanis kishinouyei* S Kimura, 1934
[Syn. *Chimarrichthys kishinouyei*] .

Kamakichi Kishinouye (1867–1929) was a fisheries biologist at Tokyo Imperial University, where he was a professor in the Faculty of Agriculture. He led the party that collected the catfish holotype and very shortly afterwards died in Chengtu, China, of a sudden illness. He wrote: *Contributions to the comparative study of the so-called scombroid fishes* (1923).

Kiss

Shellear sp. *Parakneria kissi* M Poll, 1969

Dr R Kiss was a hydrobiologist and a specialist in crustaceans. At the time when he collected the holotype (1969), he was working as the hydrobiologist of Institut pour la Recherche Scientifique en Afrique Centrale at Lwiro, Democratic Republic of Congo. He wrote, among other papers: *Ostracodes de l'Afrique tropicale* (1959), *Entomostraces de la plaine de la Ruzizi* (1960) and *Les poissons et la pêche dans le lac Ihema (Rwanda, bassin moyen de l'Akagera)* (1977).

Kisselevich

Kisselevich's Herring *Alosa braschnikowi kisselevitschi* GP Bulgakov, 1926
[Alt. Caspian Marine Shad ssp.].

Konstantin Andreevich Kisselevich (also spelled Kisselevitz or Kisselevitsch) (b.1882) was a Russian ichthyologist who became Director of the Astrakhan Ichthyological Laboratory (1930). He was an authority on Caspian-Volgan clupeids and wrote: *The clupeids of the Caspian-Volga District* (1923).

Kistnasamy

Natal Shyshark *Haploblepharus kistnasamyi* BA Human & LJV Compagno, 2006

Nadaraj 'Nat' Kistnasamy (b.1938) is a retired South African shark researcher who worked for 45 years at the Oceanographic Research Institute, Durban. He was the shyshark's original discoverer. He was honoured "*…for outstanding efforts and pioneering work in the systematics and taxonomy of the chondrichthyan fauna of southern Africa.*" He turned to shark research when he failed to find a job as an electrician, for which he was originally trained.

Kita

Scorpionfish sp. *Ursinoscorpaenopsis kitai* T Nakabo & U Yamada, 1996

Mr Tsugiyoshi Kita was the person who collected the holotype (1987). We have failed in our endeavours to find more about him.

Kitahara

Willowy Flounder *Tanakius kitaharae* DS Jordan & EC Starks, 1904
Lefteye Flounder sp. *Laeops kitaharae* HM Smith & TEB Pope, 1906

Dr Tasaku Kitahara (1870–1922) was an oceanographer who was a technical officer at the Imperial Fisheries Bureau, Department of Agriculture and Commerce, Tokyo. He founded a fisheries oceanographical organisation in Japan (1909) and was behind a fundamental oceanographic survey (1910). He wrote: *Marine research and fish migration* (1918).

Kittipong

Kittipong's Stingray *Fluvitrygon kittipongi* C Vidthayanon & TR Roberts, 2005
[Alt. Roughback Whipray; Syn. *Himantura kittipongi*].

Khun Jarutanin Kittipong (b. 1958) is a Thai aquarium fish dealer who provided (2004) the original five specimens that formed the basis for the description. (See **Jarutanin**).

Kiyo

Kiyo's Dragonet *Synchiropus kiyoae* R Fricke & MJ Zaiser, 1983

Kiyoe Tanaka was the widow of Tatsuo Tanaka and donated land and facilities to help set up the Tatsuo Tanaka Memorial Biological Station on Miyake-jima, Japan.

Kiyofuji

Snaggletooth sp. *Astronesthes kiyofujii* N Nakayama, S Ohashi & F Tanaka, 2021

Hidetada Kiyofuji is a fisheries scientist at the Laboratory of Marine Bioresource and Environment Sensing, Graduate School of Fisheries Sciences, Hokkaido University. He encouraged and supported the authors' study as group leader of the Skipjack and Albacore Group, Tuna and Skipjack Resources Division, National Research Institute of Far Seas Fisheries, Shizuoka, Japan. Among his publications is the co-written: *Northward migration dynamics of skipjack tuna* (Katsuwonus pelamis) *associated with the lower thermal limit in the western Pacific Ocean* (2019).

Kiyomatsu

Rockfish sp. *Sebastes kiyomatsui* Y Kai & T Nakabo, 2004

Dr Kiyomatsu Matsubara (1907–1968) (See **Matsubara, K**).

Klatt

Isparta Minnow *Crossocheilus klatti* C Kosswig, 1950

Dr Paul Erich Berthold Klatt (1885–1958) was a German zoologist. He graduated at the University of Berlin (1904) and taught at the Berlin Agricultural University (1908–1914). His career was interrupted by military service (WW1), in which he was wounded. He worked at the Institute of Genetics, Berlin (1918–1923) and was Professor at the University of Hamburg (1923–1928), and at the University of Halle (1928–1933), returning to Hamburg (1934–1954). A bird is also named after him.

Klausewitz

Characin genus *Klausewitzia* J Géry, 1965
Characin sp. *Leporinus klausewitzi* J Géry, 1960
Galapagos Garden Eel *Heteroconger klausewitzi* I Eibl-Eibesfeldt & F Köster, 1983
Klausewitz's Garden Eel *Gorgasia klausewitzi* J-C Quéro & L Saldanha, 1995
Perchlet sp. *Plectranthias klausewitzi* U Zajonz, 2006

Klausewitz's Goby *Lotilia klausewitzi* K Shibukawa, T Suzuki & H Senou, 2012

Dr Wolfgang Klausewitz (1922–2018) was a German zoologist, ichthyologist and marine biologist. He served in the Wehrmacht in WW2, fighting (1941–1945) in North Africa, Italy and France. He worked for the Field Investigations Agency (1946–1947). He studied zoology, botany, anthropology and psychology at the University of Frankfurt (1947–1952). He was put in charge of the Fish Section of the Naturmuseum Senckenberg, Frankfurt (1954) and retired as a Professor (1987). He was interested in the fishes collected in the Red Sea and Indian Ocean and took part in a number of cruises, including that of the research vessel 'Xarifa' under Hans Hass (q.v.) (1957) and a visit to the Galapagos Islands. He investigated the fish population of the lower Main River (a tributary of the Rhine) and later this research proved to be of great importance, as the levels of pollution in some rivers reached the point at which they were biologically dead. In retirement, he continued to be active in research. He described many Indo-Pacific fishes, including the goby genus *Lotilia* and the species *L. graciliosa* from which 'his' goby was split.

Klay

Bicolor Basslet *Lipogramma klayi* JE Randall, 1963

Gerrit Klay was an aquarist and aquarium-fish collector who collected the holotype. He worked at the Cleveland Aquarium in Ohio and was later Director of the SharkQuarium at Marathon, Florida, as well as having worked at the aquarium in Curaçao (Netherlands Antilles).

Klazinga

Sharpsnout Snake-Eel *Apterichtus klazingai* M Weber, 1913

M D Klazinga was the Chief Engineer on board the 'Siboga', during the expedition (1899–1900) on which the holotype was collected. Weber describes him as 'vaillaint' (brave), but fails to give his full names.

Klein

Klein's Butterflyfish *Chaetodon kleinii* ME Bloch, 1790
[Alt. Sunburst Butterflyfish, Blacklip Butterflyfish]
Razorbelly Scad *Alepes kleinii* ME Bloch, 1793
Klein's Sole *Synapturichthys kleinii* A Risso, 1827

Jacob Theodor Klein (1685–1759) was a German jurist, historian, botanist, zoologist and mathematician. His publications on fishes include: *Historiae Piscium Naturalis promovendae Missus primus Gedani* (1744).

Kleinenberg

Morid Cod sp. *Eretmophorus kleinenbergi* EH Giglioli, 1889

Dr Nicolaus Kleinenberg (1842–1897) was a Baltic German zoologist and evolutionary morphologist. The University of Jena awarded his doctorate. He was one of the first assistants at the Stazione Zoologica, Naples. Its founder, Anton Dohrn, was one of Kleinenberg's closest friends, helping him to found the Marine Station at Messina, where Kleinenberg was a professor and became Director, Zoological Institute of Messina University. He wrote: *Hydra - Eine anatomisch-entwicklungsgeschichtliche untersuchung* (1872).

Klepadlo

Dragonfish sp. *Photonectes klepadloae* AM Prokofiev & BW Frable, 2021

Cynthia Klepadlo (1945–2020) was a long-time collection manager at the Scripps Institution of Oceanography, La Jolla, California. California State University, Long Beach awarded her BSc (1974) and MSc (1983) and she was a lecturer there (1981–1988) in the computer science and electrical engineering departments. She was a volunteer at Scripps (1985) and a biological assistant there (1986). Having proving her value, she was hired (1989) as Museum Scientist in the Marine Vertebrate Collection where she remained until retiring (2007). She remained as a volunteer for the rest of her life. She had a particular interest in the genus *Photonectes*, describing or co-describing seven new species of these dragonfish.

Kloetzi

Stone Loach sp. *Schistura kloetzliae* M Kottelat, 2000

Antoinette Kottelat-Kloetzi is the author's wife. He thanked her "*...for her help and support during the author's field work in Laos and on other projects.*" (Also see **Antonia**).

Klunzinger

Klunzinger's Wrasse, *Thalassoma rueppellii* CB Klunzinger, 1871
Brown Sole *Achirus klunzingeri* F Steindachner, 1880
Klunzinger's Mullet *Liza klunzingeri* F Day, 1888
Sleeper sp. *Eleotris klunzingerii* GJ Pfeffer, 1893
Klunzinger's Ponyfish *Equulites klunzingeri* F Steindachner, 1898
Western Carp-Gudgeon *Hypseleotris klunzingeri* JD Ogilby, 1898
Tailed Sole *Leptachirus klunzingeri* M Weber, 1907
Rough Bullseye *Pempheris klunzingeri* AR McCulloch, 1911
Black-headed Chromis *Chromis klunzingeri* GP Whitley, 1929
East African Silverside *Hypoatherina klunzingeri* JLB Smith, 1965

Dr Carl Benjamin Klunzinger (1834–1914) was a German physician and zoologist. He studied medicine at Tübingen and Würzburg. He journeyed to Cairo (1862) where he learnt Arabic and worked as a physician at Kosseir, a Red Sea port (1864) where he amassed a collection of fishes and other marine organisms (1864–1869). He compared his collections with those in Stuttgart, Frankfurt and Berlin (1869), distinguishing many new species. He again collected by the Red Sea (1872–1875) before returning to Stuttgart, where he became Professor of Zoology at the University (1884). He studied the collection of Australasian fishes that Ferdinand von Müller (1825–1896) had accumulated. Klunzinger identified about 50 new species from Australia and New Zealand. Among other works he wrote: *Synopsis der Fische des Rothen Meeres* (1870). He pointed out that Day had incorrectly identified the mullet as *Mugil carinatus* (= *L. carinata*) in 1876.

Knaack

Corydoras sp. *Corydoras knaacki* LFC Tencatt & H Evers, 2016

Joachim Knaack (1933–2012) was a German physician, biologist, amateur ichthyologist and aquarist, who devoted more than 60 years of his life to the study of South American catfishes, especially *Corydoras*. He wrote: *Ein neuer Panzerwels aus Brasilien (*Corydoras guapore*) (Pisces, Teleostei, Callichthyidae)* (1961).

Knapp

Fat-lipped Wormfish *Microdesmus knappi* CE Dawson, 1972
Lanternfish sp. *Diaphus knappi* BG Nafpaktitis, 1978
Flathead sp. *Grammoplites knappi* H Imamura & K Amaoka, 1994
Hagfish sp. *Myxine knappi* RL Wisner & CB McMillan, 1995
Perchlet sp. *Plectranthias knappi* JE Randall, 1996

Dr Leslie William Knapp (1929–2017) was an ichthyologist at the Department of Vertebrate Zoology, National Museum of Natural History, Smithsonian. Cornell awarded his bachelor's degree (1952) and PhD (1964) and the University of Missouri his masters (1958). He was honoured in the name of the hagfish for supplying the authors with study material and, in the wormfish's etymology, for his many 'valuable specimens and personal courtesies'. He has published widely over fifty years, including contributing to: *A checklist of the fishes of the South China Sea* (2000). (Also see **Leslie Knapp**).

Knauer

Cameroon Cichlid sp. *Sarotherodon knauerae* D Neumann, MLJ Stiassny & UK Schliewen, 2011

Barbara Knauer was a technician at the Max Planck Society at Seewiesen, Germany. According to the etymology, she "*...substantially supported* [the third author Dr Ulrich Schliewen] *as a technician and friend*" during his Ph.D. studies.

Kner

Shellear genus *Kneria* F Steindachner 1866
Dalmatian Nase *Chondrostoma knerii* JJ Heckel, 1843
Kner's Goby *Pomatoschistus knerii* F Steindachner, 1861
Thorny Catfish sp. *Oxydoras kneri* P Bleeker, 1862
Stone Sculpin *Paracottus kneri* B Dybowski, 1874
Characin sp. *Schizodon knerii* F Steindachner, 1875
Toothless Characin sp. *Curimata knerii* F Steindachner, 1876
Orange-spotted Therapon *Mesopristes kneri* P Bleeker, 1876
Characin sp. *Galeocharax knerii* F Steindachner, 1879
Silver Dollar (characin) sp. *Myleus knerii* F Steindachner, 1881
Kner's Banjo Catfish *Bunocephalus knerii* F Steindachner, 1882
Whiptail Catfish sp. *Farlowella knerii* F Steindachner 1882
Pencil Catfish sp. *Trichomycterus knerii* F Steindachner, 1882
Big-eye Mandarinfish *Siniperca knerii* S Garman, 1912

Armoured Catfish sp. *Sturisomatichthys kneri* MA Ghazzi, 2005

Dr Rudolf Kner (1810–1869) was an Austrian zoologist specialising in ichthyology. He studied medicine in Linz and Vienna, receiving his degrees in medicine and surgery (1835) from the University of Vienna. He then worked with Heckel (q.v.) and others at the National History Museum, Vienna (1836–1841). He was Professor of Natural Science at the University of Lemburg, Austria (now Lviv, Ukraine) (1841–1849). He returned to the University of Vienna as the first Professor of Zoology in Austria (1849) where he became Steindachner's teacher and friend.

Knezevic

Skadar Rudd *Scardinius knezevici* PG Bianco & M Kottelat, 2005

Borivoj Knezevic (1948–1988) was a Yugoslav (Montenegrin) biologist who dedicated his life to the study and conservation of the freshwater fishes of Montenegro. He co-wrote: *Utilization of filamentous algae by fishes in Skadar Lake, Yugoslavia* (1978).

Knight, R

Bugeye Dottyback *Amsichthys knighti* GR Allen, 1987

Ronald Knight, Sr. of Manus, Papua New Guinea, was honoured *"for his generous hospitality during a collecting visit to the island in 1982."*

Knight (Jordan)

Hawaiian Eyebar Goby *Gnatholepis knighti* DS Jordan & BW Evermann, 1903

Knight Starr Jordan (1888–1947) was the senior author's son. He was honoured as it was he who first noticed this goby in a pond at Moana Hotel at Waikiki Beach, near Honolulu, Hawaii.

Knipowitsch

Goby genus *Knipowitschia* BS Iljin, 1927
Poacher sp. *Sarritor knipowitschi* GU Lindberg & AP Andriashev, 1937
Knipowitsch's Snailfish *Careproctus knipowitschi* NV Chernova, 2005
Anzeli Shad *Alosa caspia knipowitschi* BS Iljin, 1927

Nikolai Mikhailovich Knipovich (also given as Knipowitsch) (1862–1939) was a Russian oceanographer, marine zoologist and ichthyologist. The Saint Petersburg State University awarded his bachelor's degree (1886) and his master's (1892). He became an Assistant Professor at the Saint Petersburg State University (1893) and worked at the Zoological Museum of the St Petersburg Academy of Sciences (1894–1921). He was also Professor of Biology and Zoology at the First Women's Medical Institute (now the St Petersburg State Medical University) (1911–1930). He led many scientific expeditions: to the Russian Arctic (1898, 1900 & 1908), to the Caspian Sea (1886, 1904, 1912–1913, 1914–1915 & 1931–1932), to the Baltic Sea (1902) and to the Black Sea and the Sea of Azov (1922–1927). He wrote: *Identification guide of the fishes of the Black and Azov Seas* (1923).

Knoepffler

Knoepffler's Elephantfish *Boulengeromyrus knoepffleri* L Taverne & J Géry, 1968

Louis-Philippe Knoepffler (1926–1984) was a French biologist, primarily a herpetologist and ichthyologist. He worked at the Biological Mission of Gabon and was honoured by the authors as a 'friend and colleague', and for "*...contributions to the herpetology and ichthyology of Gabon*". He co-wrote: *Vivre et survivre dans la Nature* (1987), published posthumously.

Kobayashi, BN

Kobayashi's Bristlemouth *Cyclothone kobayashii* M Miya, 1994

Dr Bert Nobuo Kobayashi is an ichthyologhist whose doctorate was awarded by the University of California, San Diego. He was the first person to recognise this species as distinct from *C. pseudopallida* in a PhD dissertation: *Systematics, zoogeography and aspects of the biology of the bathypelagic fish genus* Cyclothone *in the Pacific Ocean* (1973).

Kobayashi, M

Kobayashi's Batfish *Malthopsis kobayashii* S Tanaka, 1916

The type was taken by Mr Mansaku Kobayashi of Yokkaichi, Mya Prefecture. We have not been able to find further details about him.

Kobyliansky

Deep-sea Smelt sp. *Bathylagichthys kobylianskyi* O Gon & AL Stewart, 2014
Bigscale sp. *Melamphaes kobylyanskyi* AN Kotlyar, 2015

Dr Stanislav Genrikhovich Kobylyansky is an ichthyologist who is a Senior Researcher at the P P Shirshov Institute of Oceanology, Russian Academy of Sciences, Moscow. He wrote: *Two New Species of Green Eyes of the Genus* Chlorophthalmus *(Chlorophthalmidae, Aulopiformes) from the Continental Slope and Submarine Rises of the Western Tropical Part of the Indian Ocean* (2013). He provided the author with melamphaid fishes collected on cruise 29 of the 'Akademik Ioffee' research vessel.

Koch, R

Goby sp. *Didogobius kochi* JL Van Tassell, 1988

Rudolf Koch ran a dive shop on the south shore of Gran Canaria (Canary Islands). He took the author Van Tassell diving, filled his tanks and repaired equipment problems (all for no charge). He was honoured for this generous help and because his knowledge of underwater habitats led to the discovery of new species and range extensions for gobioid fishes in the Canary Islands.

Koch, HJ

Platanna Klipfish *Xenopoclinus kochi* JLB Smith, 1948

H J Koch from Somerset West, South Africa, was a malacologist who collected the holotype with his wife Anne during their field work in conchology.

Kock

Snailfish sp. *Paraliparis kocki* NV Chernova, 2006

Dr Karl-Hermann Kock participated in the first German Antarctic expedition after World War II and has since taken part in a further 27. He graduated as an external PhD student from Kiel University (1981) with a study on three icefish species. When the Convention on the Conservation of Antarctic Marine Living Resources (CCAMLR) came into force (1982), Kock represented Germany in the Scientific Committee (1984 to 2015).

Koefoed

Koefoed's Searsid *Searsia koefoedi* AE Parr, 1937

Einar Koefoed (1875–1963) was a Danish-born marine biologist, though much of his work was done in Norway. He took part in the 'Michael Sars' (q.v.) expedition of 1910, and was a pioneer researcher on Atlantic species such as the Pollack (1932).

Koelz

Calico Surfperch *Amphistichus koelzi* CL Hubbs, 1933
Pearl Dace ssp. *Margariscus margarita koelzi* CL Hubbs & KF Lagler, 1949

Dr Walter Norman Koelz (1895–1989) was an American zoologist, botanist, anthropologist, fisheries biologist and collector whose doctorate was awarded by the University of Michigan (1920). He was on the McMillan Expedition to the American Arctic (1925). He joined the Himalayan Research Institute and was in India (1930–1932 and 1933). He explored (1934–1941) Persia (Iran), Nepal and parts of India, collecting c.30,000 bird specimens for the Zoological Museum, University of Michigan, and c.30,000 plants for the university's herbarium. Two birds are named after him.

Koepcke

Three-barbeled Catfish sp. *Myoglanis koepckei* F Chang, 1999

Professor Dr Hans-Wilhelm Koepcke (1914–2000) was a German (born in what is today northwest Poland) zoologist, herpetologist and ornithologist. He studied at Kiel University, culminating in a PhD (1947). He took a post at the Javier Prado Museum of Natural History in Lima, Peru, and spent much of his life studying South American – particularly Peruvian – fauna. He was one of the founders of the Biological Station Panguana, which has conducted research in tropical rainforests for over four decades. When he eventually returned to Hamburg, he worked at the Herpetology Department and taught at the Zoological Institute and the Museum of the University of Hamburg. He published widely, with perhaps his most significant work being the two-volume: *Die Lebensformen: Grundlagen zu einer universell gültigen biologischen Theorie* (1971 & 1973). An amphibian and a reptile are named after him.

Koji

Kojis' Perchlet *Plectranthias kojiorum* K Koeda, N Muto & H Wada, 2021

Koji Abe, of Cube International, collected the holotype. Koji Wada, owner of the Blue Harbor Aquarium Factory (aquarist shop) in Osaka, donated that holotype to the describers.

Kokraimoro

Kokraimoro Tube-snouted Ghost Knifefish *Sternarchorhynchus kokraimoro* CD de Santana & RP Vari, 2010

The Kokraimoro is a group within the Kayabo tribe whose ancestral lands included the holotype's locality in the Rio Xingu, Pará, Brazil.

Kolaev

Lanternfish sp. *Protomyctophum kolaevi* AM Prokofiev, 2004

V F Kolaev collected the holotype (1981). We have no other information on Kolaev, though he was probably a worker for TINRO (Pacific Scientific Research Fisheries Centre), Russia's largest scientific fishery.

Koller

African Characin sp. *Distichodus kolleri* M Holly, 1926
Oriental Gudgeon sp. *Gobiobotia kolleri* PM Bănărescu & TT Nalbant, 1966

Dr Otto Koller (1872–1950) of the Vienna Natural History Museum was an Austrian biologist and zoologist who was the first ichthyologist to examine specimens on which the species *Gobiobotia kolleri* is based (1927). He collected herpetofauna in Costa Rica (1930) and mammals in Turkey and southeastern Europe. He wrote: *Fische von der Insel Hai-nan* (1927).

Kolombatovic

Goby sp. *Gobius kolombatovici* M Kovačić & PJ Miller, 2000

Juraj Kolombatovic (1843–1908) was a Croatian Professor of mathematics, a natural scientist, taxonomist and ichthyologist. He collected around the Adriatic, mostly in what used to be Yugoslavia, particularly Dalmatia in Croatia, in the 1880s, 1890s and early 1900s. He wrote notes and scientific papers on mammals, reptiles and amphibians, and described nine new species of fish. He is also remembered for his work on the reforestation project of Marjan hill near the city of Split. He was honoured in the goby's name as the only Croatian naturalist to have worked intensively on small inshore fish. A bat is also named after him.

Kolthoff

Checkered Wolf Eel *Lycenchelys kolthoffi* AS Jensen, 1904

Gustav Isak Kolthoff (1845–1913) was a Swedish zoologist and taxidermist. As a taxidermist he was employed (1864) at Skara University where he prepared more than 400 birds as well as many mammals. He also hunted and collected (1870) and so supplied his already-stuffed specimens to the University. When the position of conservator became vacant at Uppsala he was given the job (1878). He participated in an expedition, as conservator and entomologist, to Greenland (1883) and a few years later to Spitsbergen. He then led an expedition to Svalbard and northeast Greenland (1900). He was given an honorary doctorate (1907).

Komp

Characin sp. *Astyanax kompi* SF Hildebrand, 1938

William H Wood Komp (1893–1955) was a medical entomologist in the United States Public Health Service. Born in Japan to American parents, he came to the USA as a young child (1895). Rutgers University awarded his bachelor's degree (1916) and his master's (1917). He was a Fellow in Agriculture at Cornell (1917). He joined the United States Public Health Service as an Ensign (1918) and became a Captain (1944). He was greatly involved in malaria research (since 1921) and control, was a visiting staff member at Gorgas Memorial Laboratory, Panama (1931–1947) and accompanied the author in investigating the Volcán region, Panama, where this characin was discovered. He was on loan to the Rockefeller Foundation for research on yellow fever in Colombia (1936) and was a consultant to a number of companies, including United Fruit (1924–1930) and Creole Petroleum, Venezuela (1936). He wrote: *Anopheline mosquitoes of the Caribbean region* (1942).

Konem

Mountain Carp sp. *Psilorhynchus konemi* B Shangningam & W Vishwanath, 2016

B D Konem Anal was the father of the first author, Bungdon Shangningam.

Kong, CF

Borneo Cyprinid sp. *Parachela ingerkongi* PM Bănărescu, 1969

Chin Phui Kong (1923–c.2016) was a Malaysian fisheries biologist whose name forms the second part of the binomial (the first part refers to Robert F Inger (q.v.)). Bănărescu said that Kong and Inger were the first biologists to give an 'adequate description' of specimens of *Oxygaster* (now *Parachela*) *oxygastroides* from North Borneo. The two men wrote: *The Fresh-Water Fishes of North Borneo* (1962).

Kong (Urbina)

Austral Cod *Guttigadus kongi* DF Markle & R Meléndez C, 1988

Ismael Kong Urbina (1942–2008) was a zoologist and ichthyologist at the University of Antofagasta, Chile, where he taught Marine Ecology and Aquaculture Engineering and was appointed Professor of Zoology (1980). The University of Chile, Antofagasta, awarded his bachelor's (1968) and his master's (1980) degrees. He collected the type material of this species.

Kongpheng

Stone Loach sp. *Schistura kongphengi* M Kottelat, 1998

Kongpheng Bouakhamvongsa works at the Department of Livestock and Veterinary, Ministry of Agriculture, Laos and is the Laotian national officer, Assessment of Fisheries Component, Mekong River Commission. He assisted Kottelat in the field (1997).

Konings

Lake Malawi Cichlid sp. *Aulonocara koningsi* P Tawil, 2003
Lake Malawi Cichlid sp. *Placidochromis koningsi* M Hanssens, 2004
Lake Tanganyika Cichlid sp. *Hemibates koningsi* FDB Schedel & UK Schliewen, 2017

Dr Adrianus Franciscus Johannes Marinus Maria 'Ad' Konings (b.1956) is a Dutch ichthyologist, cichlid aquarist, photographer, writer and publisher who began collecting cichlids at the age of fourteen. He studied medical biology at the University of Amsterdam (1974), culminating in the award of a PhD (1980). He then became a researcher in molecular biology at the Erasmus University, Rotterdam (1980–1986) before moving to Germany. Here he began writing and breeding cichlids. With his wife he started his own publishers, Cichlid Press (1991). He has since relocated his press in Texas (1996). He occasionally leads expeditions to Lake Malawi. His field work on Lake Malawi cichlids in general, and the genus *Aulonocara* in particular, has contributed considerably to our knowledge of these fishes. Among his publications are the books: *Cichlids from Central America* (1989), and *Malawi Cichlids in Their Natural Habitat* (2001).

Koningsberger

Koningsberger's Herring *Herklotsichthys koningsbergeri* M Weber & LF de Beaufort, 1912

Jacob Christiaan Koningsberger (1867–1951) was a Dutch biologist and politician. He studied biology at Utrecht University (1885) and left (1889) with a doctorate to become the assistant to the Professor of Botany. He worked as a teacher of botany and zoology (1891–1894). He then went to Java as a biologist for the coffee production industry there, focusing on pest insects and later (1897 & 1901) publishing a two-volume study: *De dierlijke vijanden der koffiecultuur op Java*. He was then employed at the Botanical Gardens at Buitzenborg (Bogor), Java. He wrote his magnum opus about the flora and fauna of Java in twelve parts (from 1911): *Java, zoölogisch en biologisch*. He left the gardens in the hands of a superintendant and established an Agricultural School (1903). A Department of Agriculture was established on Java (1905) and he twice acted as Deputy Director (1907 & 1909). All the scientific parts of the department were joined with the Botanical Gardens and he was appointed the first Director (1911). He was appointed to various political posts but was not comfortable with local politics and resigned (1919) to return to the Netherlands. He stayed in politics until finally withdrawing into private life (1929).

Konopicky

Armoured Catfish sp. *Rineloricaria konopickyi* F Steindachner, 1879

Eduard Konopicky (1841–1904) was Steindachner's scientific illustrator. David Starr Jordan described him (1905) as having produced "*the best illustrations of fishes made by any artist.*"

Konstantinov

Fathead sp. *Cottunculus konstantinovi* NA Myagkov, 1991

Konstantinov Konstantin Gavrilovich (1918–1983) was a Russian ichthyologist, writer and poet. After service in WW2, Moscow State University awarded his BSc (1946), after which he held a post-graduate studentship at the Russian Academy of Sciences (1946–1949). He worked at PINRO (1953–1983) and was Head of a laboratory on the Barents Sea (1961–1964). He took part in more than 20 scientific expeditions and was the author of over 150 publications, including the textbook: *Промысловая ихтиология* (Industrial Ichthyology). A fisheries research vessel is also named after him.

Koper

Guppy sp. *Poecilia koperi* FN Poeser, 2003

Michel Koper was described in Poeser's etymology as being a friend of the author "…with whom discussions have helped to keep my thinking flexible."

Kops

Short-headed Tonguesole *Cynoglossus kopsii* P Bleeker, 1851
Singapore Glassy Perchlet *Ambassis kopsii* P Bleeker, 1858

George François de Bruijn (or Bruyn) Kops (1820–1881) was a Dutch naval officer who was a friend of Peter Bleeker and who studied the geology, history and ethnography of the Riau Archipelago.

Korechika

Korechika's Dwarf Goby *Eviota korechika* K Shibukawa & T Suzuki, 2005

Korechika Yano is a Japanese diving instructor and underwater photographer who provided the holotype of this goby. (See **Yano**).

Korjakov

Korjakov's Rough Sculpin *Asprocottus korjakovi* VG Sideleva, 2001

Evgeniy Alekseevich Koryakov was (1940s) an ichthyologist at the Lake Baikal Limnological Station.

Kornelia

Aulonocara Chizumulu (Lake Malawi cichlid) *Aulonocara korneliae* MK Meyer, R Riehl & H Zetzsche, 1987
[Alt. Aulonocara Blue Gold, Blue Orchard Aulonocara]

Kornelia Meyer is the wife of the first author, Manfred Meyer.

Kornilov/Kornilova

Pearlside sp. *Maurolicus kornilovorum* NV Parin & SG Kobyliansky, 1993

Nikolay Pavlovich Kornilov and his wife Galina Nikolayevna Kornilova are fisheries scientists. They were honoured for their help receiving samples, organising research expeditions, and sharing data on the ecology and distribution of deep-sea fishes.

Korotneff

Baikal Smalleye Sculpin *Abyssocottus korotneffi* LS Berg, 1906

Dr Alexei Alexeievich Korotneff (1852–1914) was a Russian marine biologist. The University of Moscow awarded his PhD (1881) and he was a professor at Kiev University (1886–1912).

Korthaus

African Rivuline sp. *Nothobranchius korthausae* H Meinken, 1973

Torpedo Rasbora sp. *Pectenocypris korthausae* M Kottelat, 1982

Mrs Edith Korthaus (1923–1987) was a German aquarist and botanist who collected in Tanzania (1973) and in Kalimantan, Indonesia (1978). She was Editor-in-Chief of the magazine *Das Aquarium* (1972).

Korup

African Loach Catfish sp. *Amphilius korupi* PH Skelton, 2007

A toponym referring to the Korup National Park, Cameroon, where the species is found.

Koshewnikow

(See **Kozhevnikov**).

Kosswig

Scaleless Killifish genus *Kosswigichthys* F Sözer 1942
Kosswig's Barb *Carasobarbus kosswigi* W Ladiges, 1960
[Alt. Kiss-lip Himri]
Kosswig's Loach *Turcinoemacheilus kosswigi* PM Bănărescu & TT Nalbant, 1964
Van Barb *Capoeta kosswigi* MS Karaman, 1969
Kosswig's Barbel *Luciobarbus kosswigi* MS Karaman, 1971
[Junior Synonym of *Barbus lacerta*]
Chub sp. *Squalius kosswigi* MS Karaman, 1972
Riffle Minnow sp. *Alburnoides kosswigi* D Turan, C Kaya, E Bayçelebi, Y Bektas & F G Ekmekçi, 2017

Dr Curt Kosswig (1903–1982) was a German zoologist and geneticist whose doctorate in genetics was awarded by the University of Berlin (1927). He was an Assistant Professor at the University of Münster, Germany (1927–1933) and Professor at Braunschweig University of Technology (1933–1937). He was one of 190 German academics who emigrated, for political reasons, from German to Turkey in the 1930s. He became Professor and Director of the Zoological Institute, Istanbul University (1937–1955). He returned to Germany and worked at the Zoological Institute, University of Hamburg (1955–1969). He wrote: *Zoogeography of the Near East* (1955). He died in Hamburg but was buried in Istanbul. An amphibian and a bird are also named after him.

Kotaka

Kotaka's Argentine *Glossanodon kotakamaru* H Endo & K Nashida, 2010

This species is named after the Research Vessel 'Kotaka-maru' which belonged to the National Research Institute of Fisheries Science, Kochi, Japan. Paratypes of this species and many other scientific specimens from Tosa Bay, Japan, were collected during cruises of this vessel.

Kotlyar

Slickhead sp. *Talismania kotlyari* YI Sazonov & AN Ivanov, 1980
Kotlyar's Cubehead *Cubiceps kotlyari* TB Agafonova, 1988
Kotlyar's Whiptail *Mataeocephalus kotlyari* YI Sazonov, YN Shcherbachev & T Iwamoto, 2003

Deepwater Legskate sp. *Sinobatis kotlyari* MFW Stehmann & S Weigmann, 2016

Dr Alexander Nikolaevich Kotlyar (b.1950) is a biologist and ichthyologist who was an associate curator of ichthyology at the Zoological Museum of Moscow University in the 1990s and is now Chief Scientist, Laboratory of Oceanic Ichthyofauna, P.P. Shirshov Institute of Oceanology, Russian Academy of Sciences. He has more than 500 publications to his name, including: *Beryciformes fishes of the World Ocean* (1996) and *Dictionary of animal names in five languages: Fishes. Latin, Russian, English, German and French* (1989), covering approximately 11,700 names.

Kotschy

Arsuz Bleak *Alburnus kotschyi* F Steindachner, 1863
[Alt. Iskenderun Shah Kuli]
One-spot Seabream *Diplodus kotschyi* F Steindachner, 1876

Karl Georg Theodor Kotschy (1813–1866), an Austrian botanist, explorer, and collector, was the son of a botanically-minded evangelical pedagogical theologian. He visited Cilicia, Syria, Egypt, and the Sudan (1836–1838), as well as Cyprus (1840) and Asia Minor (1842). He traveled in Persia (Iran) (1842–1843) and returned to Vienna via Turkey. He next went to Egypt and Palestine (1855), Cyprus, Asia Minor, and Kurdistan (1859). A reptile is also named after him, as is the plant genus *Kotschya*.

Kottelat

Loach genus *Kottelatlimia* TT Nalbant, 1994
Rasbora genus *Kottelatia* TY Liao, SO Kullander & F Fang, 2009
Kottelat's Rasbora *Rasbora hobelmani* M Kottelat, 1984
Dwarf Mono *Monodactylus kottelati* VR Pethiyagoda, 1991
Rasbora sp. *Rasbora kottelati* KKP Lim, 1995
Sulawesi Halfbeak sp. *Tondanichthys kottelati* BB Collette, 1995
Torrent Loach sp. *Neogastromyzon kottelati* HH Tan, 2006
Menderes Barbel *Luciobarbus kottelati* D Turan, FG Ekmekçi, A Ilhan & S Engin, 2008
Cilician Pike Chub *Squalius kottelati* D Turan, BT Yilmaz & C Kaya, 2009
Antalya Trout *Salmo kottelati* D Turan et al., 2014
Stone Loach sp. *Schistura kottelati* HA Tuan, HN Thao & NX Quang, 2018
Sisorid Catfish sp. *Exostoma kottelati* A Darshan, W Vishwanath, S Abujam & D Narayan Das, 2019

Dr Maurice Kottelat (b.1957) is a Swiss ichthyologist who specializes in Eurasian freshwater fishes and has described more than 440 fish species previously unknown to science. The University of Neuchâtel awarded his Licentiate in Sciences (1987) and the University of Amsterdam his doctorate (1989). Later, the University of Neuchâtel awarded him an honorary degree (2006). He has researched in Thailand (1980) and in Indonesia where he was co-discoverer of *Paedocypris progenetica*, possibly the smallest fish in the world. He has mostly worked as a freelance taxonomist and has not pursued an academic career, paid or unpaid. He was President of the European Ichthyological Society (1997–2007 & 2012). He is also a Commissioner of the International Commission on Zoological Nomenclature. He wrote among much else: *The fishes of inland waters of Southeast Asia: a catalogue and core bibliography of the fishes known to occur in freshwaters, mangroves and estuaries* (2013). (See also **Maurice**)

Kotthaus

Kotthaus' Deepwater Dragonet *Callionymus kotthausi* R Fricke, 1981

Dr Adolf Kotthaus was a Swiss ichthyologist. He was a fisheries biologist at the Biologische Anstalt Helgoland, Hamburg (c.1936 on). He was Head of the Department of Ichtyology there (1964) becoming Director (1972) retiring a year later (1973). He wrote many papers such as: *Flagelloserranus : a new genus of serranid fishes; With the description of 2 new species [Pisces, Percomorphi]* (1970) as well as the book: *Fische des Indischen Ozeans* (1967), which was about the results of the ichthyological investigations during the Indian Ocean Expedition of the Research Vessel 'Meteor' (1964–1965).

Koumans

Goby genus *Koumansetta* GP Whitley, 1940
Goby sp. *Lesueurigobius koumansi* JR Norman, 1935
Goby sp. *Aulopareia koumansi* AWCT Herre, 1937

Dr Frederik Petrus Koumans (1905–1977) was a Dutch ichthyologist who was Curator of Fishes at Rijksmuseum van Natuurlijke Historie (Leiden, Netherlands). Leiden University awarded his MSc (1927) and PhD (1931). He was a goby taxonomist and wrote many papers on them, such as: *A preliminary revision of the genera of the gobioid fishes with united ventral fins* (1931) and *Biological results of the Snellius expedition. XVI. The Pisces and Leptocardii of the Snellius expedition* (1953). He prepared a description of *Koumansetta rainfordi* when he visited the Australian Museum (1938) but did not name it at the time. As Whitley was 'unable, through the exigencies' of WWII to continue his correspondence with Koumans, he named the genus after him, writing that this: "*...will enshrine memories of happier days of our meetings in Leiden and Sydney*". Norman honoured him for his "*...kindness in examining a specimen of this species, and for his opinion as to its probable systematic position.*".

Kovacev

Varna Gudgeon *Gobio kovatschevi* G Chichkoff, 1937

Vasily Kovacev was a Bulgarian ichthyologist. He was honoured for his research on the fishes of his country, and especially for his book: *Freshwater Fish Fauna of Bulgaria* (1922).

Kozhevnikov

Sculpin sp. *Cottus koshewnikowi* VI Gratzianov, 1907

Professor Grigorii Aleksandrovich Kozhevnikov (1866–1933) was a Russian entomologist who became (1904–1930) a Professor at the Imperial Moscow University and Director of their Zoological Museum and Laboratory. After graduating (1888) from Moscow University he became an assistant in the museum (1889). He was particularly involved in the study of bees, and initiated the study of the *Anopheles* genus of mosquito. After being dismissed from his post he was appointed Professor in the palaeontology department at the Geological Institute. Kozhevnikov was one of the foremost proponents of 'zapovedniki', a series of inviolable nature reserves which would serve as a control group in relationship to areas of human inhabitation and allow scientists to test the impact of human activity on the environment. After his dismissal he became pessimistic about conservation being respected.

Kozlov

Snowtrout sp. *Schizothorax kozlovi* AM Nikolskii, 1903

General Petr Kuzmich Kozlov (1863–1935) was a Russian researcher in Central Asia who was one of Przewalski's (q.v.) companions on his fourth and last expedition. He was sent to Tibet to improve relations there and led the Mongolian-Tibetan (1899–1901 & 1923–1926) and Mongolian-Sichuan (1907–1909) expeditions. He stopped on the Silk Road, just inside the presentday Chinese border with Mongolia, when he discovered KharaKhoto, the 'Black City' (1908), which had been described by Marco Polo. It was 'the city of his dreams' and his excavations uncovered many scrolls, which he took back for study, along with geographic and ethnographic materials. The holotype of this species was collected during his 1899–1901 expedition. He wrote: *Mongoliya I Kam*. Five mammals, five birds and a reptile are also named after him.

Kramer, AF

Sand-dart genus *Kraemeria* F Steindachner, 1906

Professor Augustin Friedrich Krämer (1865–1941) was born in Chile to German parents who returned to Germany (1867). He studied medicine (1883–1889) and zoology at the Universities of Tübingen, Kiel and Berlin, but with a year (1884) spent doing his compulsory military service. He went to sea (1889) as a ship's doctor in the Imperial German Navy and made several voyages, mainly to German colonies in the Pacific. This included a stay of 12 months at Apia in Samoa, where he collected the type specimen of *Kraemeria samoensis*, and where he became very interested in Samoan ethnography. He made a collection of ethnographic objects which he sent to the Lindenmuseum für Völkerkunde (Museum of Ethnography) in Stuttgart and the collection at the University of Tübingen. Over a period of years, he visited the Bismarck Archipelago, the Marshall Islands, the Gilbert Islands, the Hawaiian Islands and (1897) Peru and Chile including the Straits of Magellan. He explored in the Bismarck Archipelago and the Caroline Islands (1907) and was leader of the Hamburg Pacific Expedition to the Carolines on board the *Peiho* (1909). He returned to Germany (1911) and in that year he was given a three-year contract as Director at the Stuttgart Museum of Ethnography, after which he lectured on ethnology at Tübingen University. He settled down in Stuttgart and stayed there until his death. A plant and a mammal are also named after him.

Kramer, LB

Kramer's Bulldog Mormyrid *Marcusenius krameri* PA Maake, O Gon & ER Swartz, 2014

Dr Leo Bernd Kramer is a German ichthyologist who was Professor at the Zoological Institute of the University of Regensburg, Germany until he retired (2009). He was honoured "*...for his contributions to the systematics of southern African mormyrids.*" He has researched animal behaviour and behavioural physiology, and is particularly interested in electro-communication, writing on it in: *Electrocommunication in teleost fishes. Behavior and experiments* (1990) and *Electroreception and communication in fishes* (1996).

Kramer, WH

European Mudminnow *Umbra krameri* JJ Walbaum, 1792

Wilhelm Heinrich Kramer (d.1765) (also known by the 'Latinised' version of his name, Guilielmi Henrici Kramer) was an Austrian naturalist who was originally trained in medicine, which he practised in Bruck (close to Vienna). He wrote: *Elenchus Vegetabilium et Animalium per Austriam Inferiorem Observatorum* (1756). This was one of the first works to adopt Linnean binomial nomenclature.

Krammer

African Rivuline sp. *Nothobranchius krammeri* S Valdesalici & H Hengstler, 2008

Werner Krammer of Pöttmes, Germany, is an aquarist who was the first to successfully breed this species.

Krasyukova

Krasyukova's Perch *Caprodon krasyukovae* VE Kharin, 1983

Dr Zoya Valentinovna Krasyukova (c.1928–1991) was a Soviet ichthyologist at the Laboratory of Ichthyology, part of the Zoological Institute of Russian Academy of Sciences having been an assistant at Leningrad University. She undertook several collecting expeditions to the Volga and wrote a number of guides and monographs on freshwater fish. She was also one of the authors of *Fishes of the Sea of Japan and Adjoining Parts of the Yellow Sea* (1989).

Kraus

Krauss' Basketmouth Cichlid *Caquetaia kraussii* F Steindachner, 1878

Dr Christian Ferdinand Friedrich von Krauss (1812–1890) was a German botanist, malacologist, traveller and collector. He was an apothecary's apprentice and started work as a pharmacist, but then studied mineralogy, zoology and chemistry at Tübingen and Heidelberg, where he excelled academically and was awarded a PhD (1836). He went to South Africa (1838), collecting and studying the fauna, flora and geology of the area around Cape Town. He spent the next two years travelling widely in South Africa before returning to Europe (1840). In England he sold many of his plant specimens to the British Museum. He returned to Germany, where he became (1856) Director of the Royal Natural History Cabinet in Stuttgart. Krauss had collected 2,308 species (mostly flowering plants) of which 340 species and 34 genera were new to science. Several were named after him.

Krefft, G

Lanternfish genus *Krefftichthys* PA Hulley, 1981
Krefft's Smooth-head *Herwigia kreffti* JG Nielsen & V Larsen, 1970
Twin-striped Pearleye *Scopelarchoides kreffti* RK Johnson, 1972

Krefft's Dreamer *Oneirodes kreffti* TW Pietsch, 1974
Eelpout sp. *Aiakas kreffti* AE Gosztonyi, 1977
Krefft's Skate *Malacoraja kreffti* MFW Stehmann, 1978
Whipnose Angler sp. *Gigantactis kreffti* E Bertelsen, TW Pietsch & RJ Lavenberg, 1981
Krefft's Lanternfish *Symbolophorus kreffti* PA Hulley, 1981
Hagfish sp. *Nemamyxine kreffti* CB McMillan & RL Wisner, 1982
Barbeled Dragonfish sp. *Eustomias kreffti* RH Gibbs, TA Clarke & JR Gomon, 1983
Snailfish sp. *Paraliparis kreffti* AP Andriashev, 1986
Krefft's Snaggletooth *Astronesthes kreffti* RH Gibbs & JF McKinney, 1988
Flabby Whalefish sp. *Procetichthys kreffti* JR Paxton, 1989
Cod Icefish sp. *Patagonotothen kreffti* AV Balushkin & MFW Stehmann, 1993
Wrinkled Slickhead *Conocara kreffti* YI Sazonov, 1997
Brazilian Blind Electric Ray *Benthobatis kreffti* G Rincón, MFW Stehmann & CM Vooren, 2001
Deepbody Cusk *Diplacanthopoma kreffti* DM Cohen & JG Nielsen, 2002

Dr Gerhard Krefft (1912–1993) was a German ichthyologist and herpetologist whose great uncle was Johann Ludwig Gerard Krefft (below). His doctorate was awarded by Hamburg University (1938). He went to the Canary Islands (1939) but was recalled to Germany after six months and spent WW2 in the German army. He founded the fish collection at Institut für Seefischerei, Hamburg, now held by the Zoological Museum in Hamburg. He was leader of the expeditions of the research vessel 'Walther Herwig' until his retirement (1977). He first collected the Brazilian Blind Electric Ray (1968) and was honoured for his contributions to elasmobranch systematics. He was honoured in the skate's name for his numerous publications on zoogeography and taxonomy of cartilaginous fishes (particularly the *Rajidae*) and Atlantic meso- and bathypelagic bony fishes; for his leadership of the Ichthyology Group of the Institute of Sea Fisheries and its extensive scientific fish reference collection; and for his knowledge and encouragement during more than 10 years of collaboration with Stehmann.

Krefft, JLG

Krefft's Frill-Goby *Bathygobius krefftii* F Steindachner, 1866
Freshwater Long-Tom (needlefish) *Strongylura krefftii* A Günther, 1866
Snubnose Garfish ssp. *Arrhamphus sclerolepis kreffti* F Steindachner, 1867

Johann Ludwig (Louis) Gerard Krefft (1830–1881) was a German-born Australian adventurer, artist, zoologist and palaeontologist. He emigrated from Germany to the USA (1851) and worked as an artist in New York, then sailed for Australia (1852) to join the gold rush. He was a miner (1852–1857) before joining the National Museum in Melbourne as a collector and artist. He seems to have had a temper: he feuded with the Museum trustees and was dismissed (1874). He refused to accept the dismissal and barricaded himself in his office. He was later carried out of the building, still sitting on his chair, deposited in the street, and the door locked behind him. He felt he had been hard done by and so set up a rival 'Office of the Curator of the Australian Museum' and successfully sued the trustees for a substantial sum of money. That proved to be the end of his career and he never worked seriously again, but wrote natural history articles for the Sydney press. He wrote: *The Snakes of Australia* (1869) and *The Mammals of Australia* (1871). Three reptiles, three amphibians, two birds and a mammal are named after him.

Krefft, P

Loach Catfish sp. *Amphilius krefftii* GA Boulenger, 1911

Dr Paul Krefft (1872–1945) was a German physician and herpetologist at the Stuttgart Museum. He made expeditions to German East Africa, where he collected the catfish holotype. He wrote: *Das Terrarium* (1908).

Kreiser

Kreiser's Killifish *Profundulus kreiseri* WA Matamoros, JF Schaefer, CL Hernández & P Chakrabarty, 2012 .

Dr Brian Robert Kreiser is an American biologist at the Department of Biological Sciences, University of Southern Mississippi. The University of Colorado awarded his PhD (1999). He was the friend and doctoral advisor of the senior author.

Krempf

Marbled Freshwater Whipray *Himantura krempfi* P Chabanaud, 1923
[Junior Synonym of *Fluvitrygon oxyrhyncha*]
Krempf's Flounder *Grammatobothus krempfi* P Chabanaud, 1929
Oriental Cyprinid sp. *Poropuntius krempfi* J Pellegrin & P Chevey, 1934
Sharpbelly (cyprinid) sp. *Hemiculter krempfi* J Pellegrin & P Chevey, 1938
Pangasid Catfish sp. *Pangasius krempfi* P-W Fang & J Chaux, 1949

Dr Armand-Alfred-Antoine Krempf (b.1879) was a French marine biologist, who first went to Vietnam (1903) as part of a scientific expedition to Hanoi. He was the founding Director of the Oceanographic and Fisheries Service, Nha Trang Institute of Oceanography, Indochina (Vietnam) (1922–1931). He wrote: *La pêche en Indochine* (1933) and also *Carcass on coast of Annam, 1883* (1925): the latter concerned a large, enigmatic 'armour-plated' creature reportedly washed up on the Vietnamese coast. The identity of this oddity remains a mystery. A bird is also named after him. .

Krenakore

Bristlenose Catfish sp. *Ancistrus krenakarore* RR de Oliveira, LH Rapp Py-Daniel, CH Zawadzki & J Zuanon, 2016
Tetra sp. *Hyphessobrycon krenakore* TF Teixeira, AL Netto-Ferreira, JLO Birindelli & LM Sousa, 2016

The Krenakore (also given as Kreen-Akarore, Krenakarore, and other variants) are an indigenous people who originally inhabited the Rio Tapajós basin, Pará, Brazil, where *Ancistrus krenakarore* occurs, suffered a loss of two-thirds of their original population after contact with the modern world, when the Cuiabá-Santarém road began to be built across their territory (1973). After a long struggle, they were the first Brazilian native nation to be indemnified by the government and regain the right to live in part of their original territory, having been (1960s) forced to transfer to Parque Nacional do Xingu.

Kretser

Ornate Paradisefish *Malpulutta kretseri* PEP Deraniyagala, 1937
[Alt. Spotted Gourami]

Oswald Leslie De Kretser (b.1910) was a Sri Lankan lawyer who discovered this species. He became (1940) a Crown Counsel and was active in the Sri Lankan legal system until his retirement (1972).

Kreyenberg

Chinese Cyprinid sp. *Acrossocheilus kreyenbergii* CT Regan, 1908

Dr Martin Kreyenberg (1872–1914) was a German physician and naturalist. He was aboard the German ship 'Jaguar' (1901–1905), during which he made herpetological and ichthyological collections at various coastal locations from Hong Kong to Korea and Australia. He stayed on for a while in China but returned to Germany (1906) when he was taken ill. However, he returned to China (1908) as a physician to a railway building project which was under German management there. He collected many zoological specimens which he donated to the Magdeburg Museum, including the cyprinid type. After the project was completed, he travelled and continued to collect. He wanted to take an extended collecting expedition, but had insufficient funds. At this point he was persuaded to invest in a large coconut plantation on an island near Manila, with two German merchants. He liked the idea of studying the fauna of a tropical island, but it was not to be: not only was the plantation unprofitable, it was devastated by two tropical storms. He was on passage to Manila aboard the steamer 'Sontua' when he suffered from appendicitis and died. He was buried in Manila. A dragonfly and a number of aquatic organisms are also named after him.

Krieg

Three-spot Tetra *Serrapinnus kriegi* O Schindler, 1937

Professor Dr Hans Krieg (1888–1970) was a physician as well as an ethnographer, anthropologist and zoologist who was Director of the Munich Museum. He made a number of expeditions to South America, including being a member of the German Gran Chaco expedition (1925). He made a second expedition to the same region (1932), and a third (1936) after the War of the Chaco. He led an expedition to Patagonia (1937–1938). He brought back important collections from Argentina, Bolivia, Brazil and Paraguay. He was also the first President of Deutscher Naturschutzring (German Conservation Ring). He wrote: *Zwischen Anden und Atlantik. Reisen einen Biologen in Südamerika* (1948). A reptile, an amphibian and two birds are named after him.

Kriete

Kriete's Tonguefish *Symphurus billykrietei* TA Munroe, 1998

William H 'Billy' Kriete Jr was a fisheries scientist at the Virginia Institute of Marine Science. Among his published work is the co-written: *A mark-recapture study of striped bass in the James River, Virginia : Annual Report 1987* (1987) and *Biology and management of river herring and shad in Virginia : Annual report, Anadromous Fish Project 1975* (1975).

Krishna

Indian Blind Catfish *Horaglanis krishnai* AGK Menon, 1950

Mr N Krishna Pillay collected the holotype after draining a well (1948) in Kottayam, Kerala.

Krishnamurti

Armoured Catfish sp. *Hypostomus krishnamurtii* CH Zawadzki, IS Penido & PHF Lucinda, 2020

Jiddu Krishnamurti (1895–1986) was an Indian philosopher, writer and lecturer. He travelled the world, encouraging his audiences to abandon dogmas and institutional authorities and to find their own answers. He wrote many books including *The First and Last Freedom* (1954), *The Only Revolution* (1970) and *Krishnamurti's Notebook* (1976). He was honoured in the catfish name for his life-long dedication to the improvement of mankind.

Kristin

Lipstick Dottyback *Pseudochromis kristinae* AC Gill, 2004

Kristin Marie ('Molly') Gill is the wife of the species' author, Anthony Gill, who named the fish "*…in appreciation of her love, encouragement and support.*".

Kristina

Southern Tide-water Goby *Eucyclogobius kristinae* CC Swift, B Spies, RA Ellingson & DK Jacobs, 2016

Kristina D Y Louie (1974–2004) was a young research zoologist whose untimely death from encephalitis cut short a promising career dedicated to conservation genetics. Her PhD dissertation and associated work contributed to the authors' studies of eastern Pacific phylo-geography, as well as to a novel re-interpretation of the placement of Wallace's Line across the islands of Indonesia.

Krogius

Dal'nee Char *Salvelinus krogiusae* MK Glubokovsky, SV Frolov, VV Efremov, IG Rybnikova & ON Katugin, 1993

Faina Vladimirovna Krogius (1902–1989) was a Russian ichthyologist. She worked for about 50 years on Dal'nee Lake, Kamchatka, and on lake ecology in general. She wrote *Means of Restoring and Increasing the Schools of Kamchatka Salmon* (1954). She collected the holotype of this species (1938).

Krone

Armoured Catfish genus *Kronichthys* A de Miranda Ribeiro, 1908
Silverside genus *Kronia* A de Miranda-Ribeiro, 1915 NCR
[Now in *Odontesthes*]
Krone's Blind Catfish *Pimelodella kronei* A de Miranda Ribeiro, 1907
Krone's Flat Suckermouth Catfish *Harttia kronei* A de Miranda Ribeiro, 1908
Armoured Catfish sp. *Rineloricaria kronei* A de Miranda Ribeiro, 1911

Sigismund Ernst Richard (Ricardo) Krone (1861–1917) was born in Dresden, Germany, but left there for Brazil (1884), settling in Iguape where he opened a pharmacy. He was a keen cave explorer who discovered 41 new caves and studied every aspect of the Ribeira Valley. He wrote his great work on the valley's caves, *As Grutas Calcáreas do Vale do Rio Ribeira de Iguape* (1906), which also explored the geology, ecology and palaeontology of the area. One etymology reads: "*We name this species after Ricardo Krone of Iguape, who was the premier zoologist of the first agricultural boom of the Ribeira Valley at the turn of the* [19th/20th] *century*." A bird is also named after him.

Kronos

Neotropical Cichlid genus *Kronoheros* O Říčan & L Piálek, 2016

In Greek mythology, Kronos (or Cronus) was leader of the Titans and father of Zeus. The name alludes to *Kronoheros umbrifer* being the largest Neotropical cichlid. The name is combined with *Heros*, an old name used for various Neotropical cichlids.

Krøyer

Krøyer's Deep-sea Anglerfish *Ceratias holboelli* HN Krøyer, 1845

[Alt. Longray Seadevil]

Lancet Fish *Notoscopelus kroyeri* AW Malm, 1861

Henrik Nikolai Krøyer (1799–1870) was a Danish marine biologist. In his youth he was idealistic and, like Lord Byron, went south to help the Greeks fight the Turks. He was quickly disillusioned when he found that nineteenth-century Greeks had little in common with Pericles, but the journey fired an interest in Mediterranean fauna. He started studying medicine at the University of Copenhagen (1817) and later changed to the study of history and philology. He taught school (1827–1830) at Stavanger, Norway. With his wife, he surveyed the Danish coast (1835–1836) in an open boat. He made many collecting expeditions to Norway, Spitzbergen, Madeira and North and South America. He founded the journal 'Naturhistorisk Tidsskrift'.

Krug

Brazilian Cave Catfish sp. *Rhamdiopsis krugi* FA Bockmann & RMC Castro, 2010

Luiz Krug is an extremely knowledgeable, independent tour guide based in Lençóis, Bahia, Brazil. He has been guiding in the Chapada Diamantina since 1982 specialising in geology and caving. He was honoured for calling the authors' attention to this catfish and helping to collect the type series, and for his conservation efforts.

Krumholz

Spotfin Gambusia *Gambusia krumholzi* WL Minckley, 1963

Dr Louis August Krumholz (1909–1981) was a lecturer at the Department of Biology, University of Louisville, Kentucky. He was honoured because he "*...contributed to our knowledge of many phases of aquatic biology and to the biology of the gambusias in his paper on G a affinis* (1948)".

Krupp

Sole sp. *Aseraggodes kruppi* JE Randall & SV Bogorodsky, 2013

Aden Sweeper *Pempheris kruppi* JE Randall, BC Victor & MS Aideed, 2015

Professor Dr Fareed Krupp was Curator of Fishes at the Senckenberg Research Institute, Frankfurt, and a leading authority on the fishes of the Red Sea, which he has researched for more than two decades. He is currently Director of the Qatar Museum of Nature & Science, Dohar, having previously been Project Manager there.

Krusenstern

Red-spotted Bandfish *Acanthocepola krusensternii* CJ Temminck & H Schlegel, 1845

Adam Johann von Krusenstern (1770–1846) was a Russian (born in what is now Estonia) admiral and explorer who led the first (1803–1806) Russian circumnavigation of the globe. His work won him an honorary membership in the Russian Academy of Sciences.

Kruyt

Duckbilled Buntingi *Adrianichthys kruyti* M Weber, 1913

Dr Albert Christian Kruyt (1889–1949) was a Dutch ethnographer, theologian and medical missionary in Sulawesi; known as one of the 'spiritual pioneers of central Celebes'. He grew up in Indonesia but was sent to the Netherlands (1877) to be educated as a missionary before returning (1890). He started his work running the mission (1892) and stayed until 1932 when he left for the Netherlands. He co-authored a book entitled *De Bare'e-sprekende Toradja's van Midden-Celebes* (The Bare'e speaking Toraja of Central Sulawesi), which is considered one of the best publications in the field of ethnology. He sent fish specimens to Weber, including the type.

Krynicki

Krynicki's Loach *Oxynoemacheilus merga* J Krynicki, 1840

Professor Johann Krynicki (1797–1838) was a Polish entomologist who was a Professor at Kharkov University in the Ukraine, where his collection is housed. A bird is also named after him.

Krysanov

African Rivuline sp. *Nothobranchius krysanovi* KM Shidlovskiy, BR Watters & RH Wildekamp, 2010

Eugeny Y Krysanov is a scientist at the Severtsov Institute of Ecology and Evolution, Russian Academy of Sciences. He carried out cytological studies on *Nothobranchius* species and looked at the effects of radiation on their chromosomes in the Chernobyl area. Among his published papers is: *Divergent karyotypes of the annual killifish genus Nothobranchius (Cyprinodontiformes, Nothobranchiidae)* (2016).

Kuarup

Armoured Catfish sp. *Hypostomus kuarup* CH Zawadzki, JLO Birindelli & FCT Lima, 2012

Kuarup, or Quarup, is an origin myth and a festivity shared by most of the ethnic groups living in the upper portion of the Xingu Indigenous Park, Brazil.

Kubota

Kubota's Rasbora *Microdevario kubotai* M Kottelat & KE Witte, 1999
Burmese Border Loach *Botia kubotai* M Kottelat, 2004

Katsuma Kubota is a fish collector who is Managing Director of the Siam Pet Fish Trading Co., Bankgkok. He studied at Tokai University. He was honoured in the loach's name for "*...help with various projects and for the gift of valuable material, including the first known specimens of this species with locality information.*"

Kuder

Goby sp. *Callogobius kuderi* AWCT Herre, 1943
[Sometimes regarded as a Junior Synonym of *C. maculipinnis*]

Edward M Kuder was an American colonial administrator who was Division Superintendent of Schools in the Philippines (1924–1941). During this time, he learned three local languages. He was interned (1944–1946) by the Japanese during WW2. His interest in Asian Studies resulted in a number of papers, such as *The Moros in the Philippines* (1945). His 'generous cooperation' made possible Herre's visit to a group of small islands west of Jolo, Sulu Province, the type locality.

Kuderski

Eel Cod sp. *Muraenolepis kuderskii* AV Balushkin & VP Prirodina, 2007

Leonid Aleksandrovich Kudersky (b.1927). This species was named after him to celebrate his 80th birthday and the complimentary etymology refers to him as a "*...famous ichthyologist, organizer of Russian fishery science, and an outstanding expert of the freshwater fishes of Russia.*"

Kudo

Kudo's Pygmygoby *Trimma kudoi* T Suzuki & H Senou, 2008

Takahiro Kudo is a senior researcher at the Kanagawa Prefectural Fisheries Research Institute. Among his published papers is: *A New Record of American Lobster* Homarus americanus *H. Milne Edwards, 1837 (Crustacea, Nephropidae) from Tokyo Bay* (2013); and the co-written: *A New Species of the Genus* Chromis *(Perciformes: Pomacentridae) from Taiwan and Japan* (2007). He was honoured as he helped collect the type and supplied it to the authors.

Kufferath

Lake Tanganyika Cichlid sp. *Trematocara kufferathi* M Poll, 1948

Jean Kufferath was a chemist who was a member of the Belgian Hydrobiological Mission to Lake Tanganyika (1946–1947). During the mission he collected the holotype.

Kuhl

Flagtail genus *Kuhlia* TN Gill, 1861
Offshore Rockfish *Pontinus kuhlii* S Bowdich, 1825
Shortfin Devil Ray *Mobula kuhlii* J Müller & FGJ Henle, 1841
Kuhl's Stingray *Neotrygon kuhlii* J Müller & FGJ Henle, 1841
[Alt. Blue-spotted Stingray]
Kuhl's Loach *Pangio kuhlii* A Valenciennes, 1846
[Alt. Coolie Loach]

Dr Heinrich Kuhl (1797–1821) was a German naturalist and zoologist. He became an assistant to Coenraad Temminck at the Rijksmuseum van Natuurlijke Historie, Leiden. He travelled to Java (1820) with his friend Johan Coenraad van Hasselt to study the fauna of the Dutch East Indies. After less than a year in Java, Kuhl died in Buitenzorg (Bogor) of a liver infection brought on by the tropical climate and overexertion. His collections are housed at Leiden and were studied by Müller and Henle. Seven birds, six mammals, three reptiles and an amphibian are also named after him as well as other fishes.

Kühne

Betta (fighting fish) sp. *Betta kuehnei* I Schindler & J Schmidt, 2008
Mountain Minnow sp. *Tanichthys kuehnei* J Bohlen, T Dvorák, HN Thang & V Šlechtová, 2019

Jens Kühne is a German aquarist and tour guide (Mahachai Tours, Thailand) who is also a trained chef. He was honoured in the betta's name for "*…his contributions to increase the knowledge about the fighting fishes*", and in the mountain minnow's name for his efforts to locate the species in the field (in Vietnam).

Kuhnt

Beira Killifish *Nothobranchius kuhntae* E Ahl, 1926

Mrs Bertha Kuhnt was a wholesaler of aquarium fish, based in the Konradshöhe district of Berlin. Her company imported this species.

Kuiter

Anglerfish genus *Kuiterichthys* TW Pietsch, 1984
Black Leopard Wrasse *Macropharyngodon kuiteri* JE Randall, 1978
Clingfish sp. *Kopua kuiteri* JB Hutchins, 1991
Orange-black Dragonet *Dactylopus kuiteri* R Fricke, 1992
Kuiter's Damselfish *Chrysiptera kuiteri* GR Allen & A Rajasuriya, 1995
Kuiter's Weedfish *Heteroclinus kuiteri* DF Hoese & DS Rennis, 2006

Rudolf 'Rudie' Herman Kuiter (b.1943) is an Australian underwater photographer and aquarist. He collected the damselfish type and was honoured for his "*…many valuable contributions to our knowledge of tropical reef fishes of the Australian-Indonesian region.*" The wrasse was discovered when John Randall saw a juvenile specimen in Kuiter's home aquarium.

Kujunju

Lake Victoria Cichlid sp. *Haplochromis kujunjui* MJP van Oijen, 1991

This is a toponym referring to Kujunju point, a landmark northeast of Standieri Island, Lake Victoria, Tanzania, where this species occurs.

Kükenthal

Halfbeak sp. *Hemirhamphodon kuekenthali* F Steindachner, 1901
Borneo Barb sp. *Osteochilus kuekenthali* E Ahl, 1922

Professor Dr Wilhelm 'Willy' Georg Kükenthal (1861–1922) led the Bremen Geographical Society's expedition in the yacht 'Berentine' to Kong Karls Land in the Arctic (1889). He was there again (1893–1894) but the party ran aground, and the ship was crushed by ice. Luckily the 'Cecilie Maline', a sealing vessel, saved everyone four days after they were stranded. With support from the Senckenberg Natural History Society he also participated in an expedition to the Moluccas and to Borneo (1894) where he collected the type of an eponymous dragonfly. He specialised in the study of Octocorallia, a taxonomic subclass that includes sea pens, sea fans and soft corals. He also conducted embryological and comparative anatomical investigations of whales and other marine mammals. His large collection of zoological specimens is now housed at the Senckenberg Museum in Frankfurt. He is remembered in the names of about twenty other taxa, including two reptiles.

Kuku(j)ev

Mid-Atlantic Skate *Rajella kukujevi* VN Dolganov, 1985
Ridgehead sp. *Poromitra kukuevi* AN Kotlyar, 2008
Barbeled Dragonfish sp. *Eustomias kukuevi* AM Prokofiev, 2018

Dr (Y)Efim Izrailevich Kukuev (b.1947) is a Russian ichthyologist. He enrolled at the Kaliningrad Technical Institute, Faculty of Ichthyology and Fishery (1966) and studied the collections of Atlantic fishes at the Atlantic Scientific Research Institute of Marine Fisheries & Oceanography (AtlantNIRO). His doctorate was awarded by the P. P. Shirshov Institute of Oceanology of the Russian Academy of Sciences, Moscow (1980). He is presently the head of the sector of fauna taxonomy in FSUE 'AtlantNIRO'. He has written or co-written more than 160 scientific papers, including: *Status of fanfishes of the genus* Pteraclis *from the South-Eastern Pacific Ocean (Perciformes: Bramidae)* (2009) and has described 14 new species involving five different fish families. Prokoviev's etymology says he named the fish: "*…in honor of associate, friend and occasional coauthor Efim Izrailevich Kukuev (b. 1947), who has made a 'large contribution'* (translation) *to the study of mesobathypelagic fishes of the Atlantic Ocean.*"

Kulbicki

Kulbicki's Triplefin *Springerichthys kulbickii* R Fricke & JE Randall, 1994
Kulbicki's Pipefish *Festucalex kulbickii* R Fricke, 2004
[Alt. New Caledonian Pipefish]

Dr Michel Kulbicki is a Research Director who was working at IRD Noumea, New Caledonia (2016). He collected the holotypes of both species. He and senior author Fricke co-authored: *Checklist of the shore fishes of New Caledonia* (2007).

Kulkarni

Dwarf Mahseer *Tor kulkarnii* AGK Menon, 1992

Dr C V Kulkarni was Director of Fisheries, Government of Maharashtra, Mumbai, India. He also served as Director, Central Institute of Fisheries Education, Mumbai (1962–1963). He co-wrote: *Mahseer - the mighty game fish of India* (1991), and was honoured "*...for his outstanding contributions in masheer conservation in India.*"

Kullander

Characin sp. *Astyanax kullanderi* WJEM Costa, 1995
Dwarf Cichlid sp. *Apistogramma kullanderi* HR Varella & MH Sabaj Pérez, 2014

Dr Sven Oscar Kullander (b.1952) is a Swedish biologist and ichthyologist. The University of Umeå awarded his bachelor's degree (1977) and the University of Stockholm his doctorate (1984). He was appointed (1990) Senior Curator for Ichthyology and Herpetology at the Swedish Museum of Natural History, Stockholm, having previously been there as a Research Associate (1983–1990). He has been on more than 30 collecting expeditions for freshwater fishes in South America, Myanmar, Zambia, Gambia, and China. He wrote: *Cichlid fishes of the Amazon River drainage of Peru* (1886). His wife, Fang Fang, was also an ichthyologist at the Stockholm museum. The dwarf cichlid was named after him because his forty years of "*...careful and comprehensive work have transformed the taxonomy of Neotropical Cichlidae and continue to inspire all those fascinated with its diversity.*" (Also see **Didi** & **Tiantian**).

Kullmann

Stone Loach sp. *Triplophysa kullmanni* PM Bănărescu, TT Nalbant & W Ladiges, 1975

Dr Ernst Josef Kullmann (1931–1996) was a zoologist and arachnologist who studied medicine and anatomy as well as zoology at the University of Bonn. There he gained his doctorate (1957) and became successively a lecturer (1964), Associate Professor (1968), and Full Professor (1970). He was a member of the German group researching at the University of Kabul, Afghanistan (1962–1966) and was instrumental in founding the zoo in Kabul. He was Professor of General Zoology at the University of Kiel (1972–1975). He was the director (1975–1981) of the Cologne Zoo (Germany). He led the Afghan expedition that collected the holotype (1971).

Külpmann

Neotropical Rivuline sp. *Cynodonichthys kuelpmanni* HO Berkenkamp & V Etzel, 1993

Volker Külpmann is a German aquarist who makes overseas collecting trips. He participated in field work in Panama and helped collect the type.

Kuma

Cusk-eel sp. *Monomitopus kumae* DS Jordan & CL Hubbs, 1925

Kumakichi Aoki was a highly skilled fisherman who was assistant to Kakichi Mitsukuri at the Marine Laboratory at Misaki during the late 19th and early 20th centuries. He was known as 'Kuma-San', an expert at traditional Japanese long-line fishing, and was described as "*one of the best [fish] collectors in Japan*". Alan Owston (q.v.) used to sail in the region of the Marine Laboratory and often took staff from the station on board his vessel 'Golden Hind' on dredging and fishing trips. He collected the holotype of this species.

Kumba

Rattail genus *Kumba* NB Marshall, 1973

KUMBA is an anagram of the initial letters of the **M**arine **B**iological **A**ssociation of the **U**nited **K**ingdom.

Kumpera

Four-eyed Flounder *Ancylopsetta kumperae* JC Tyler, 1959

Helga O Tyler née Kumpera was the wife of the flounder's describer, James Tyler.

Kuna

Kuna Goby *Coryphopterus kuna* BC Victor, 2007

This is not a true eponym as it is named for the Kuna indigenous people of the Kuna Yala region of Atlantic Panama, where the type was collected. They were honoured for their cooperation in marine biological research.

Kunaloa

Kuna Loa (snake-eel) *Ophichthus kunaloa* JE McCosker, 1979

Kuna Loa was the Long Eel in Hawaiian mythology. In the legends, the demi-god Maui cut Kuna Loa into pieces, and its tail evolved into the conger eel.

Kunze

Neotropical Rivuline sp. *Melanorivulus kunzei* WJEM Costa, 2012

Eduardo Kunze Bastos was the author of an unpublished dissertation (1979) comprising a pioneering study on the ecology of *Rivulus punctatus* from central Brazil. This provided a stimulus for Costa to study the taxonomy of the rivulines that he was later (2006) to place in the genus *Melanorivulus*. Kunze wrote: *Aspectos da Fauna Brasileira* (1990).

Kuo

Hagfish sp. *Myxine kuoi* HK Mok, 2002

Dr Chien-Hsien Kuo is a Taiwanese molecular biologist. He was a researcher at the Department of Biology, National Taiwan Normal University, Taipei and is now Assistant Professor, Department of Aquatic Bioscience, National Chiayi University, Taiwan. Among his many published papers is the co-written: *Phylogeny of hagfish based on the mitochondrial 16S rRNA gene* (2003). He was honoured in the species' name for his contributions to hagfish taxonomy.

Kura

Kura Shad *Alosa curensis* EK Suvorov, 1907

This is a toponym; it refers to the Kura River, Azerbaijan, where the holotype was collected.

Kural

Indian Barb sp. *Hypselobarbus kurali* AGK Menon & K Rema Devi, 1995

Not an eponym but based on the fish's local name in Kerala, India.

Kurchatov

Kurchatov's Snailfish *Notoliparis kurchatovi* AP Andriashev, 1975

Named after the Russian research vessel 'Akademik Kurchatov' rather than the man behind the ship's name: Igor Kurchatov (1903–1960) the "father of the Soviet atomic bomb."

Kurematsu

Chinese Cyprinid sp. *Ancherythroculter kurematsui* S Kimura, 1934

U Kurematsu was a member of the Japanese General Council of Chengtu (now Chingdu), capital of Sichuan Province, China, during the Japanese invasion and occupation.

Kuroda

Goby sp. *Rhinogobius kurodai* S Tanaka, 1908

Dr Nagamichi Kuroda (1889–1978) was a Japanese ornithologist and ichthyologist. He published widely, mainly on birds, including: *Birds of the Island of Java* (1933–1936) and *Parrots of the World in Life Colours* (1975). His ichthyological writings include: *Fishes of Lake Biwa, with their distribution records* (1953). Most of his extensive collection of specimens was destroyed in WW2; that which survived is now in the Yamashina Institute for Ornithology at Chiba, Japan. He discovered this goby. Eleven birds are named after him.

Kuroiwa

Goby sp. *Tridentiger kuroiwae* DS Jordan & S Tanaka, 1927

Hisashi Kuroiwa (1858–1930) of the Imperial University, Tokyo, was a Japanese botanist and botanical collector with a special interest in aquatic species. He wrote or co-wrote a number of books and papers such as: *A list of phanerogams collected in the southern part of Isl. Okinawa one of the Loochoo Chain* (1900). He presented a collection of fishes from the Ryukyu Islands, Japan, including this species. An odonata is named after him.

Kuronuma

Kuronuma's Grenadier *Spicomacrurus kuronumai* T Kamohara, 1938
Poacher sp. *Occella kuronumai* HW Freeman, 1951

Dr Katsuzô Kuronuma (1908–1992) was a Japanese ichthyologist, who retired as President Emeritus, Tokyo University of Fisheries. He graduated from the Imperial Fisheries Institute, Tokyo, and was awarded a master's degree by the University of Michigan (1935). The grenadier etymology says that Kuronuma helped Kamohara in 'various ways' whilst at the University of Michigan. He co-wrote: *Fishes of the Arabian Gulf* (1986).

Kurt

Kurt's Coralblenny *Ecsenius kurti* VG Springer, 1988

Kurt A Bruwelheide is (1987–present) IT Operations Branch Chief at the Smithsonian, having formerly (1977–1987) been a museum technician in the Division of Fishes there. He studied at the University of Maryland, which awarded his BSc (1976). He is also a keen angler. He was honoured for his work on the early part Springer's revision of the genus *Ecsenius*. He also took photos of many of the type specimens of species that Springer described.

Kuru

Tigris Chub *Petroleuciscus kurui* NG Bogutskaya, 1995
Loach sp. *Cobitis kurui* F Erk'akan, FG Atalay-Ekmekçi & TT Nalbant, 1998
Riffle Minnow sp. *Alburnoides kurui* D Turan, C Kaya, E Bayçelebi, Y Bektas & FG Ekmekçi, 2017

Dr Mustafa Kuru (b.1940) is a Turkish ichthyologist whose bachelor's degree (1964) and doctorate (1971) were awarded by the Erzurum Ataturk University and where he taught (1965–1989), eventually as Professor (1981–1989). He was Professor at Hacettepe University (1989–1992) and at Gazi University (1992–1993). He is currently Vice-Rector of Baskent University. He collected the holotype.

Kurz

Torrent Catfish sp. *Amblyceps kurzii* F Day, 1872

S Kurz collected the holotype in Burma (Myanmar). Though Day gives no further details, this probably refers to the German botanist and garden director Wilhelm Sulpiz Kurz (1834–1878). He worked in Bogor, West Java, and later in Calcutta where he was curator of the herbarium (1864–1872). He took part in botanical explorations of the Andaman Islands, Burma, Singapore and Malaysia.

Kusakari

Saddle-back Snapper *Paracaesio kusakarii* T Abe, 1960

Mr T Kusakari worked at the Hachije Branch, Tokyo Prefectural Fisheries Experiment Station. We have been unable to find further biographical details.

Kuschakewitsch

Chu Sharpray *Capoetobrama kuschakewitschi* KF Kessler, 1872
Kuschakewitsch's Loach *Iskandaria kuschakewitschi* SM Herzenstein, 1890
[Alt. Tadzhik Loach; Syn. *Nemacheilus kuschakewitschi*]

Colonel Aleksandr Aleksandrovich Kushakevich (surname also given as Kuschakewitsch) was a botanist, entomologist, traveler and explorer through Middle Asia (1887–1878), who collected the holotypes. He became Head of the Khodjent District of Tajikistan and was a noted collector of ethnographic artefacts. (Note that the loach is a double eponym in as much as *Iskandar* was the Arabian nickname of Alexander of Macedonia).

Kushavali

Indian Barb sp. *Hypselobarbus kushavali* M Arunachalam, S Chinnaraja, P Sivakumar & RL Mayden, 2016

A toponym referring to the village of Kushavali in the Western Ghats of India, where the Kali River (type locality) originates.

Kusnetzov

See **Kuznetsov, II**

Kussakin

Snailfish sp. *Liparis kussakini* VI Pinchuk, 1976

Professor Oleg Grigoryevich Kussakin (1930–2001) was a Russian marine biologist. He worked for many years in Vladivostok in the Institute of Marine Biology. His main interests were in isopods and littoral faunal communities. Several marine invertebrates are also named after him.

Kuwamura

Combtooth Blenny sp. *Cirripectes kuwamurai* R Fukao, 1984

Dr Tetsuo Kuwamura (b.1950) is a Japanese scientist. His degress, including Doctor of Science, were awarded by Kyoto University. He is Professor in the School of International Liberal Studies at Chukyo University. He started there as an Instructor (1980–1981), Associate Professor (1981–1992) and Professor (1992–2008) before his current position. He has been a member of the Ichthyological Society of Japan since 1996. Among his publications are the book: *Fish Communities in Lake Tanganyika* (1997) and such papers as the co-authored: *Aggressive mimicry of the cleaner wrasse by Aspidontus taeniatus functions mainly for small blennies* (2018). He collected the holotype.

Kuznetsov, ID

Sculpin sp. *Cottus kuznetzovi* LS Berg, 1903

Innokentiy Dmitrievich Kuznetsov (1863–1921) was a Russian ichthyologist and fisheries scientist. St-Petersburg University awarded his first degree (1885) and zoology master's (1890). He then served in the Department of Arable Farming as inspector of a fishery (1885–1917) and as consultant to a fishery in Petrograd (1917–1919). His major publication was *Sketch on Russian fishing* (1902). He took part in a number of expeditions to: Azov Sea (1886), Volga (1890), Caspian Sea (1894) and Lake Baikal (1908) and led the Pskov scientific-industrial expedition of Department of Arable Farming (1912–1913); one of the biggest expeditions on the study of fauna of lakes and fishing in brackish waters since the times of Karl Baer.

Kuznetsov, II

Stone Char *Salvelinus kuznetzovi* AY Taranetz, 1933
Snailfish sp. *Liparis kusnetzovi* AY Taranetz, 1935

Ivan Ivanovich Kuznetsov (1885–1962) was a Russian fisheries biologist and fish breeder (the founder of fish acclimatisation in the Far East), who worked in Kamchatka (where the char occurs). He was educated at the Samara Agricultural College. On vacation, he helped his father plant a forest now marked on topographic maps as 'Kuznetsovskaya'. After college he was seconded (1908–1917) to the Amur Ichthyological Expedition. After the civil war he moved to Vladivostok, where he continued work on studying the biology of salmonids and organised checkpoints on spawning rivers. He became a senior researcher at the Pacific Experimental Station (later TINRO). He visited Kamchatka (1923), where he first worked on the Bolshoy River, and (1926–1928 & 1930) conducted research in the basin of the Kamchatka River: he compiled maps of spawning rivers, monitored salmon spawning migrations, their distribution and timing of spawning. He moved (1940) to Khabarovsk, where he was appointed chief specialist in the reproduction of fish stocks in the Amurrybvod system. He became seriously ill and almost blind, but managed to complete his manuscript *Materials on the study of natural reproduction of Far Eastern salmons and the cause of fluctuations in their stocks*. After recovering his health, he wrote further works.

Kwinti

Armoured Catfish sp. *Pseudancistrus kwinti* PW Willink, JH Mol & B Chernoff, 2010

The Kwinti people live along the Coppename River, Suriname, the area where this catfish is found.

Kyburz

Kyburz's Tetra *Pseudochalceus kyburzi* LP Schultz, 1966

William Alfredo Kyburz (1900–1965) was an ornamental fish exporter in Bitaco, Colombia. He helped collect the holotype and supplied habitat information.

L

Laan

Armoured Catfish sp. *Rhadinoloricaria laani* IJH Isbrücker & H Nijssen, 1988
[Syn. *Apistoloricaria laani*]

Louis André van der Laan of the Zoölogisch Museum, Amsterdam, was honoured for providing the authors with excellent photographs of fishes for many years.

La Barca

Sharpnose Silverside *Chirostoma labarcae* SE Meek, 1902

This is a toponym after La Barca; a place on the shore of Lake Chapala, Mexico.

Labarre

African Rivuline sp. *Aphyosemion labarrei* M Poll, 1951

Clement Labarre was the original collector of the species in what is now Zaire. Nothing more was recorded of him.

Laboute

Laboute's Fairy Wrasse *Cirrhilabrus laboutei* JE Randall & R Lubbock, 1982

Pierre Laboute (b.1942) is a French marine biologist, diver and underwater photographer. He worked for the IRD (L'Institut de Recherche pour le Développement) in New Caledonia (1996–2002), and set up his own diving and marine consultancy company there. He was honoured as he had published the first images of this species in a field guide to the marine fauna of New Caledonia. He has written or co-written a number of books on the marine life of the area, such as: *Le plus beau lagon du monde: Nouvelle-caledonie* (1991) and *Poissons de Nouvelle-Calédonie* (2002). He has around fifty publications to his name. A sea-snake is also named after him.

Labrador

Freshwater Stingray sp. *Potamotrygon labradori* MN Castex, I Maciel & G Martínez Achenbach, 1963
[Junior Synonym of *Potamotrygon motoro*]

José Sánchez Labrador (1718–1798) was a Spanish naturalist and Jesuit monk in Argentina. He was expelled to Italy by the authorities, having been accused of espionage (1767). He wrote: *Historia de las regiones del Rio de la Plata.*

Lacandon

Priapella (Poeciliid) sp. *Priapella lacandonae* MK Meyer, S Schories & M Schartl, 2011

The Lacandon are a Mayan tribe living in the Mexico/Guatemala border region.

Lacépède

Broadnose Sevengill Shark *Notorynchus cepedianus* F Péron, 1807
Crested Oarfish *Lophotus lacepede* M-E Giorna, 1809
American Gizzard Shad *Dorosoma cepedianum* CA Lesueur, 1818
Eel Goby sp. *Odontamblyopus lacepedii* CJ Temminck & H Schlegel, 1845

Bernard Germaine Etienne de la Ville, Comte de Lacépède (also spelled La Cepède) (1756–1825), was a French naturalist and politician. He came to the attention of Buffon, whose work on the classification of animals he was encouraged to continue. Buffon also got him a job at the Jardin du Roi (later Jardin des Plantes) (1785). Lacépède was active in politics and during the 'Terror' lived in Normandy to avoid the guillotine. After his return to Paris he gave up scientific work for a political career and held several offices of state. He was also a good musician and composer. He wrote poetry, political treatises and even extended Buffon's work. His own zoological output included: *Histoire naturelle des quadrupèdes ovipares, serpents, poissons et cétacées* (1825) and *Histoire Naturelle des Poissons* (5 vols., 1798–1803). Two mammals and two reptiles are named after him.

Lacerda, FJM

Western Bottlenose Mormyrid *Mormyrus lacerda* F de L de Castelnau, 1861

Dr Francisco José Maria de Lacerda (1753–1798) was a Portuguese explorer born in São Paulo, Brazil. The University of Coimbra, Portugal, awarded his doctorate in mathematics and astronomy (1777). He was given the task of fixing the borders of Portuguese territory in South America in relation to all the Spanish possessions around it. He left Lisbon (1780) for Belem and explored and mapped until he arrived in São Paulo (1790). He returned to Portugal (1791) and became a Professor at the Naval Academy. He was appointed Governor of the Portuguese East African colony in what is now Mozambique (1797) and given a specific task; namely to cross Africa from Mozambique to Angola (also a Portuguese colony) in what was the first scientific expedition to attempt to traverse the continent. He set out (1797) and reached Kazembe (now in Zambia) where he died of fever (1798), but he had kept diaries and comprehensive records. His Brazilian years were eventually published in São Paulo over 40 years after his death as: *Diário de viagem pelas capitanias do Pará, Rio Negro, Matto-Grosso, Cuyabá e S. Paulo, nos anos de 1780 a 1790* (1841) and his African diaries were saved and published in Lisbon (1844). These latter diaries were translated by Sir Richard Burton and published in London as: *The Lands of Kazembe: Lacerda's journey to Cazembe in 1798* (1873).

Lacerda, JB

Trahira (wolf tetra) sp. *Hoplias lacerdae* A Miranda Ribeiro, 1908

Dr João Batista de Lacerda (1845–1915) was a physician, biomedical scientist and anthropologist. He qualified as a physician at the medical school in Rio de Janeiro but joined the newly created National Museum of Natural History of Rio de Janeiro as Associate Director - anthropology, zoology and palaeontology; eventually becoming general Director, National Museum of Rio Janeiro. He was very interested in promoting harmony and good will among different races and at the Universal Races Congress, London (1911) he delivered a paper: *The Metis, or half-breeds of Brazil* using the example of fertility to show that white and black humankind are just two 'races' and not two separate species.

Lacerda, MTC

Neotropical Rivuline sp. *Plesiolebias lacerdai* WJEM Costa, 1989
Lacerda's Cory *Corydoras lacerdai* H Hieronimus, 1995
[Syn. *Scleromystax lacerdai*]

Marco Túlio Cortes de Lacerda, who works at the Rio de Janeiro botanical gardens, collected the *Corydoras* holotype. He originally graduated from the Military Engineering Institute (1988). He has worked in zoology and ichthyology (1983–2004), but began to concentrate on cultivating unconventional tropical fruits (1999) and is co-author of: *Brazilian Fruit – Native & Exotic* (2006) and a number of papers such as *Zwei neue Cynolebias aus dem Tocantins* (1991) and *Mato Grosso – die große Weite* (2006). He also became owner of E-jardim.com (2007).

Lach de Bère

Red-light Dwarfgoby *Eviota lachdeberei* L Giltay, 1933

Lieutenant-Colonel Philip Frederik Lambertus Christiaan Lach de Bère (1859–1936) of the Royal Netherlands East Indies Army was born in Sumatra and lived in Indonesia, where this goby was collected. It seems he retired to the Netherlands. He was keenly interested in genealogy and made a collection of many documents. Giltay's original text does not identify the person honoured in the binomial, so it remains a 'best guess' that the Lieutenant-Colonel was the person intended.

Lachner

Cardinalfish genus *Lachneratus* TH Fraser & PJ Struhsaker, 1991
Greater Jumprock *Moxostoma lachneri* CR Robins & EC Raney, 1956
Whitestar Cardinalfish *Apogon lachneri* JE Böhlke, 1959
Aqaba Cardinalfish *Cheilodipterus lachneri* W Klausewitz, 1959
Ouachita Madtom *Noturus lachneri* WR Taylor, 1969
Asian Gudgeon sp. *Mesogobio lachneri* PM Bănărescu & TT Nalbant, 1973
Lachner's Tongue-Sole *Cynoglossus lachneri* AGK Menon, 1977
Lachner's Shrimp-Goby *Myersina lachneri* DF Hoese & HR Lubbock, 1982
Lachner's Dwarfgoby *Sueviota lachneri* R Winterbottom & DF Hoese, 1988
Bintuni Goby *Stenogobius lachneri* GR Allen, 1991
Tombigbee Darter *Etheostoma lachneri* RD Suttkus & RM Bailey, 1994

Dr Ernest Albert Lachner (1916–1996) was an American ichthyologist and biologist who was Curator of Fishes, Smithsonian National Museum of Natural History (1949–1983) retiring as Curator Emeritus. His doctorate was awarded by Cornell University. He served in the US Army Air Force in WW2. His published work includes the co-authored: *Two New Gobiid Fishes of the Genus* Gobiopsis *and a Redescription of* Feia nympha *Smith* (1979). In one of the etymologies he is honoured as the person "*...who has added much to the knowledge of catostomid fishes*" and: "*...for facilitating the senior author's visits to several museums in the United States.*"

Lacmi

Snailfish sp. *Careproctus lacmi* AP Andriashev & DL Stein, 1998

The binomial refers to the Natural History Museum of Los Angeles County (LACM is the approved abbreviation for referencing this museum's specimens). The museum was the original repository for all liparids (snailfish) collected during the *Eltanin* expeditions in the Southern Ocean.

Lacorte

Rainbow Tetra *Nematobrycon lacortei* SH Weitzman & WL Fink, 1971

Neotropical Rivuline sp. *Maratecoara lacortei* KJ Lazara, 1991

Rosario S LaCorte is an American aquarist who served in the US Army Corps (1947–1951) and made five collecting expeditions to South America (1977–1988). He wrote: *Enjoy your Cichlids* (1969) and *An Aquarist's Journey* (2018).

Lacrymatus

White-spotted Devil *Plectroglyphidodon lacrymatus* JRC Quoy & JP Gaimard, 1825

The authors gave no etymological explanation, but the meaning of *lacrymatus* is 'with tears' and perhaps the allusion is explained as the authors dedicated this species to the "*…memory of Mr. Vidal, a young naval surgeon, who died of yellow fever*" (translation).

Ladd

Brownboy Goby *Bathygobius laddi* HW Fowler, 1931

Ladd's Dragonet *Synchiropus laddi* LP Schultz, 1960

Dr Harry Stephen Ladd (1899–1982) was an American geologist with the US Geological Survey and a research associate of the Smithsonian Institution. He was a major contributor to our knowledge of the geology and palaeontology of the islands of the Pacific, to the understanding of coral reefs, and was a founder of the modern science of paleoecology. Washington University in St Louis awarded his first degree and the University of Iowa his MSc (1924) and PhD (1925). The goby etymology says that the species is named in honour of: "*Dr. H. L. Ladd, who collected the type and other fishes for the Bishop Museum (Honolulu)*". However, Fowler's use of the initials 'H. L.' seem to be an error and we believe this refers to Harry Stephen, as he went to Fiji (the type locality) on a Bishop Museum fellowship (1925). Over the next decade he spent almost three years there undertaking field research. He taught, worked as a state geologist and for an oil company, then joined the Geological Survey (1940–1969, apart from the war years). Whilst working for the Survey, he visited Bikini Atoll (1946 and 1947), the type locality of the dragonet.

Ladiges

Jelly Bean Tetra genus *Ladigesia* J Géry, 1968

Minnow genus *Ladigesocypris* M Karaman, 1972

Celebes Rainbowfish *Marosatherina ladigesi* E Ahl, 1936

Shellear sp. *Parakneria ladigesi* M Poll, 1967

Snowtrout (cyprinid) sp. *Schizocypris ladigesi* M Karaman, 1969

Dr Werner Ladiges (1910–1984) was a German zoologist, ichthyologist and aquarist who was Director and Professor, Zoologisches Staats Instituut und Zoologisches Museum, Hamburg. He studied at the universities of Innsbruch and Hamburg, which awarded his doctorate (1934). He wrote: *Tropische Fische (*1934). He was honoured in the snowtrout's name for offering Karaman an 'employment opportunity'. (Also see **Brüning**)

Ladislav

Gudgeon genus *Ladislavia* B Dybowski, 1869

Władysław Ladislaus Taczanowski (1819–1890). (See **Taczanowski**)

Laessoe

Squeaker Catfish sp. *Synodontis laessoei* JR Norman, 1923

Major Harold Henry Alexander de Laessoe (d.1948) was a pioneer, administrator and explorer in Rhodesia (1896–1914), and it appears was in Tanganyika (Tanzania) in the 1920s. In WW1, he served with the13th Battalion, Rifle Brigade. He clearly served with distinction, being awarded both the Distinguished Service Order (1918) and the Military Cross (1917), but in WW2 he appears to have been detained under Defence Regulation 18B (1940–1944). He was a close associate of Oswald Mosley. He took legal action against the Home Secretary for damages for false imprisonment and breach of statutory duty. He helped to collect the holotype in Angola.

Laetari

Burrow Splitfin Goby *Psilotris laetarii* JL Van Tassell & F Young, 2016

Heath Jens Laetari (1978–2006) was Vice President of Dive Operations, Partner & Acquisition Manager for Dynasty Marine (a supplier of live Caribbean marine life), who was lost at sea during a free dive in the Florida Keys. He learned to swim when three months old and gained his PADI open water diver certificate aged thirteen. He was awarded a degree in marine biology (2001), before which he had earned his PADI dive master rating. He began work at Dynasty Marine (2001).

Lagler

Mekong Barb sp. *Cyclocheilichthys lagleri* S Sontirat, 1989

Mekong Barb sp. *Hypsibarbus lagleri* WJ Rainboth, 1996

Dr Karl Frank Lagler (1912–1985) was an American zoologist and fishery biologist whose bachelor's degree was awarded by the University of Rochester, New York, his master's by Cornell University (1936), and his doctorate by the University of Michigan (1940). He taught botany at the University of Rochester (1934–1935) and joined the faculty of the University of Michigan (1939). There he became both Professor of Fisheries in the School of Natural Resources and Professor of Zoology in the College of Literature, Science, and the Arts. Additionally, he was Chairman and Founder of the Department of Fisheries (1950–1965). His outside activities included much travel in Europe, Asia, Africa and the Americas and consultancy work. He was Field Director of the Gambia River basin environmental and economic studies (1982–1984). Among his 150-plus publications is the co-written: *Ichthyology* (1962).

Lagowski

Lagowski's Minnow *Rhynchocypris lagowskii* B Dybowski, 1869
[Alt. Amur Minnow; Syn. *Phoxinus lagowskii*]

Dybowski did not identify the person honoured in the binomial. One source suggests this fish is named after a Siberian revolutionary, Mikhail Fedorovich Lagowski (1856–1903). This idea is not unattractive as Dybowski himself was in exile in Siberia (1869), but seems unlikely since the 'revolutionary' would only have been a young teenage boy at the time. Perhaps it refers to Colonel P Lagowski, a hero of the Polish revolution (1830)?

Lahille

Characin genus *Lahilliella* CH Eigenmann & CH Kennedy, 1903 NCR
[Now included in *Schizodon*]
Rosefish sp. *Helicolenus lahillei* JR Norman, 1937

Dr Fernando Lahille (1861–1940) was a French physician, naturalist and marine biologist. His PhD and MD were both awarded by the University of Paris (1891 & 1893). He was at the Faculty of Science at Toulouse (1890–1892) as a 'free' professor working closely with renowned naturalists. He went to Argentina (1893) at Moreno's invitation. He was Curator of Zoology, La Plata Museum of Natural History (1893–1899), setting up the first coastal marine laboratory in South America, which later became National Institute of Fisheries Research and Development (INIDEP). He led an expedition to Tierra del Fuego (1896) which studied its marine fauna and climate. He was the first head of the Fish and Game Department, Ministry of Agriculture, and was Professor of Zoology at Escuela Normal de Profesores Mariano Acosta (1904–1930) and at University of Buenos Aires (1910–1930). He wrote over three hundred scientific papers, including: *Los peces argentinos de cara torcida* (1939). A bird and Lahille Island in Antarctica are named after him.

Lakeside

Hagfish sp. *Rubicundus lakeside* MM Mincarone & JE McCosker, 2004
[Syn. *Eptatretus lakeside*]

This species is named after the Lakeside Foundation of California for supporting the senior author's work.

Lako

Plated Catfish sp. *Aspidoras lakoi* A Miranda Ribeiro, 1949

Carlos (Károly) Lako (1895–1960) was a Hungarian who lived in Brazil (1920–1960) and worked as an osteologist and taxidermist at the Museu Nacional, Rio de Janeiro. He collected the holotype.

Laland

Yellowtail Amberjack *Seriola lalandi* A Valenciennes, 1833
Brazilian Sharpnose Shark *Rhizoprionodon lalandii* JP Müller & FGJ Henle, 1839

Pierre Antoine Delalande (1787–1823). (See **Delalande**)

Lalanne

Seychelles Spurdog *Squalus lalannei* A Baranes, 2003

Maurice Jean Leonard Loustau-Lalanne (b.1955) has held a number of posts in the government of the Seychelles and was the Seychelles Principal Secretary for the Environment at the time of the discovery of this species. He became (2010) Ambassador and Principal Secretary, Ministry of Foreign Affairs. He was Interim Chairman, Air Seychelles (2011) and is Chairman of both the Seychelles Islands Foundation and the Botanical Gardens Foundation. He was honoured for his "*...help in organizing the expedition that collected the type, his kindness, and his friendship.*"

Laman

Loach Catfish sp. *Amphilius lamani* AJE Lönnberg & CH Rendahl, 1920
African Barb sp. *Enteromius lamani* AJE Lönnberg & CH Rendahl, 1920

Karl Edvard Laman né Ersson (1867–1944) was a Swedish missionary and ethnographer. He changed his name in honour of his great-aunt and great-uncle who paid for his education. He was ordained as a missionary (1890) and worked in the Kingdom of Kongo (1891–1919). He donated a small collection of Congo fishes, including the holotypes of these species, to the Natural History Museum in Stockholm. His most notable work was to translate the Bible into Kikongo.

Lamarck

Black-striped Angelfish *Genicanthus lamarck* BGE Lacepède, 1802

Jean-Baptiste Pierre Antoine de Monet, Chevalier de Lamarck (1744–1829), usually known simply as Lamarck, was a French naturalist and an early proponent of the idea that biological evolution occurred. However, he is today largely remembered for his belief in the inheritance of acquired characteristics – a rejected theory now often called 'Lamarckism'.

Lamarre

Bagrid Catfish sp. *Sperata lamarrii* A Valenciennes, 1840

Christophe-Augustin Lamarre-Picquot (1785–1873) (also given as Lamare-Picot, and other variants) was a French naturalist, pharmacist and explorer. He made several expeditions, including three to India (1821–1823, 1825–1826 and 1828–1829) to collect ethnographical and natural history specimens including the catfish type specimen. Later (1848) he was commissioned by the French ministry of agriculture to search for edible plants in North America that might have the potential to replace potatoes. Although he tried to introduce breadroot (*Psoralea*) to France, it could never really compete with the humble spud

Lambalot

Toadfish sp. *Triathalassothia lambaloti* NA de Menezes & JL de Figueiredo, 1998

Dr Raoul Pierre Lambalot was a volunteer in the Fish Section, Museum of Zoology, University of São Paulo, Brazil. He was honoured for depositing specimens of marine fishes he collected along the São Paulo coast.

Lambari

Characin sp. *Bryconamericus lambari* LR Malabarba & A Kindel, 1995

This is not an eponym but derives from a local word used to denote small characins in southern Brazil.

Lambert, A

African Rivuline sp. *Aphyosemion lamberti* AC Radda & JH Huber, 1977

André Lambert was a "…well known French killifish lover" (translation), who bred this species in his aquarium. (Because the fish was also named after J.G. Lambert (below), the authors should really have used the plural form in the binomial = *lambertorum*)

Lambert, JG

Lambert's Lampeye *Rhexipanchax lamberti* J Daget, 1962
Squeaker Catfish sp. *Microsynodontis lamberti* M Poll & J-P Gosse, 1963
African Tetra sp. *Nannopetersius lamberti* M Poll, 1967
African Rivuline sp. *Aphyosemion lamberti* AC Radda & J-H Huber, 1977

Jacques G Lambert (1923–2013) was a Belgian ichthyologist and botanist. He wrote a number of papers including: *Poissons Siluriformes et Cyprinodontiformes récoltés en Guinée Française, avec la description d'une nouvelle espèce de Microsynodontis* (1958) and *Les poissons d'aquarium du bassin du Congo* (1960) and described four new taxa. He was honoured in the name of the rivuline (with André Lambert, above) for his work on the systematics and distribution of killifishes and his research in Gabon (where this species is endemic).

Lambert (van Tujil)

Corydoras sp. *Corydoras lamberti* H Nijssen & IJH Isbrücker, 1986

Lambertus van Tuijl (1944–2012) was a technician in the Department of Ichthyology, Zoölogisch Museum, Amsterdam, where the authors were curators of fishes.

Lambour

Mormyrid sp. *Marcusenius lambouri* J Pellegrin, 1904

Jean-Baptiste Lambour was a 'préparateur' in the herpetological department of the Muséum National d'Histoire Naturelle (1891–1924) and Pellegrin's colleague.

Lamilla

Warrah Skate *Dipturus lamillai* FJ Concha, JN caira, DA Ebert & JHW Pompert, 2019

Professor Dr Julio Lamilla Gómez (d.2016) was a Chillean research biologist and teacher at the Faculty of Sciences and of the Institute of Marine and Limnological Sciences at the Universidad Austral de Chile (1979–2011) where he was awarded his MSc (1979) and PhD (1994). He published c.50 papers. The etymology says: "*…in memory of Julio Lamilla, a Chilean biologist who devoted his life to teaching and research focused on the* *biology and conservation of chondrichthyans, especially batoids.*" He died of a heart attack while at work.

Lamna

Mackerel Shark genus *Lamna* G Cuvier, 1816

The genus name comes from the Greek *lamia*, a large and voracious shark. However, originally this derives from Lamia in Greek mythology: a beautiful woman who revenged the murder of her own children (by the goddess Hera) by killing the children of others. Her face became a nightmarish mask, and in some versions of the myth she was also deprived of the ability to sleep.

Lamonte

Armoured Catfish genus *Lamontichthys* A Miranda Ribeiro, 1939

Francesca Raimonde La Monte (1895–1982) was an ichthyologist who worked at the AMNH (1920–1968). She was an Assistant Curator (1947–1968) and retired as Curator Emeritus. She was regularly consulted by Ernest Hemingway. She was a founder member of the International Game Fish Association (1939) and was a member of Michael Lerner's scientific expedition (1936–1941). She wrote: *North American Game Fishes* (1945).

Lamotte

Brook Lamprey sp. *Lampetra lamottei* CA Lesueur, 1827 NCR
[Originally described as *Petromyzon lamottenii*, which is regarded as ambiguous and replaced by *Lethenteron appendix* (DeKay 1842)]

Monsieur Lamotte (or La Motte) was a mineralogist who was one of the party led by the French explorer Philip Francis Renault which discovered (1720) the location of what became the Mine Lamotte; at one time an important source of lead. Lesueur acquired the holotype, as he described in his *American ichthyology, or natural history of the fishes of North America, with coloured figures from drawings executed from nature* (1827), in a cave near the mine. It may not be a true eponym but may have been named for the mine.

Lamotte, MG

Squeaker Catfish sp. *Chiloglanis lamottei* J Daget, 1948
Red-spotted Panchax *Epiplatys lamottei* J Daget, 1954
Airbreathing Catfish sp. *Clarias lamottei* J Daget & P Planquette, 1967

Dr Maxime Georges Lamotte (1920–2007) was a French biologist and geneticist. He collected in West Africa (1942 & 1946), and also trained others to collect and conserve specimens. He was Professor of Zoology at the Faculty of Sciences, Lille (1952–1955), and later at the Faculty of Sciences, Paris. He was President of the Zoological Society of France (1978–1983). Four mammals and three amphibians are named after him.

Lampe

African Loach Catfish sp. *Amphilius lampei* V Pietschmann, 1913

Eduard Lampe (1871–1919) collected for Museum Wiesbaden in the first two decades of the 20th century and worked at the museum cataloguing the collection and describing specimens. He was also interested in meteorology, producing regular yearbooks on observations made at Wiesbaden. A reptile is also named after him.

Lamprecht

Cameroon Cichlid sp. *Sarotherodon lamprechti* D Neumann, MLJ Stiassny & UK Schliewen, 2011

Jürg Lamprecht (1941–2000) was a Swiss fish behaviourist, researcher and teacher. He studied biology and anthropology at the University of Zürich, then (1967) joined the Max Planck Institute of Behavioural Physiology in Seewiesen, Germany. He worked for his PhD on haplotilapiine cichlids there. He supported the first author as a teacher and also supported the third author as a mentor, teacher and friend. He wrote several books and a number of papers. Among his books was *Biologische Forschung: Von der Planung bis zur Publikation* (1999). He died of cancer aged fifty-nine.

Lan

Chinese Cyprinid genus *Lanlabeo* M Yao, Y He & Z-G Peng, 2018
Chinese Loach sp. *Sinibotia lani* T-J Wu & J Yang, 2019
Bagrid Catfish sp. *Tachysurus lani* J-L Cheng, W-H Shao, JA López & E Zhang, 2021

Jia-Hu Lan is a Chinese ichthyologist at the Du'an Fisheries Technology Extension Department, part of the Aquatic Bureau in Du'an, China. The first part of the genus' name is the Chinese family name Lan, and honours him for his contributions to the discovery of fish diversity in southern China. Amongst his publications are the co-written: *A new cavefish species,* Sinocyclocheilus brevibarbatus *(Teleostei: Cypriniformes: Cyprinidae), from Guangxi, China* (2009) and *Cave Fishes of Guangxi, China* (2013). The etymology for the loach says: "*The new species is named in appreciation of Mr. Lan Jiahu, Du'an Fisheries Technology Extension Department, Guangxi, China, who collected the type specimens of the new species. Since 1987, his explorations in south China have revealed about 50 new species.*"

Lana

Lana's Sawshark *Pristiophorus lanae DA* Ebert & HA Wilms, 2013

Lana Ebert is a marine biologist, shark enthusiast and the niece of senior author David Ebert. David A Ebert is program Manager, Pacific Shark Research Center, Moss Landing Marine Laboratories, California. The etymology reads: "*The species name* lanae *is after shark enthusiast Lana Ebert on the occasion of her graduation from the University of San Francisco.*"

Landford

Sleeper sp. *Allomogurnda landfordi* GR Allen, 2003
[Binomial sometimes corrected to *landfordorum*]

Alan and Julia Landford, formerly of Bulolo, Papua New Guinea, were honoured for their assistance with collecting paratypes of this species and of *A. flavimarginata*. Now living in

Australia, Julia is the founder of NatureArt Lab, Canberra's first dedicated natural history art school.

Landon, HM

Characin genus *Landonia* CH Eigenmann & AW Henn, 1914
Characin sp. *Pterobrycon landoni* CH Eigenmann, 1913
Armoured Catfish sp. *Hemiancistrus landoni* CH Eigenmann, 1916

Hugh McKennan Landon (1867–1947) was an Indianapolis businessman and philanthropist who helped finance the Chocó, Colombia, expedition that collected the holotypes of both *Hemiancistrus landoni* and *Pterobrycon landoni.*

Landon, RR

Cardinalfish sp. *Apogonichthys landoni* AW Herre, 1934

Major Robert Roberts Landon (1873–1938) of Cebu, Philippines, was one of the founders of the Visayan Electric Company. He was an engineer who was part of the American forces which came to the Philippines (1900) after the islands were sold by Spain to the USA. Herre described him as his friend "*…to whose co-operation I am greatly indebted.*"

Lane

Characin sp. *Characidium lanei* HP Travassos, 1967

John Lane (1905–1963) was a Brazilian medical entomologist at the University of São Paulo. His brother, Dr Frederico Lane Jr (1901–1979) was also a Brazilian entomologist; they were the sons of immigrants from the USA. John Lane collected the type specimen.

Lanfear

Royal Highhat *Pareques lanfeari* O Barton, 1947

"Lanfear B. Norrie of New York City" was honoured in the binomial with no further information or explanation being given. However, we believe this to be the mining engineer and minerals prospector Lanfear Barbey Norrie (1896–1977) who owned mining properties in the USA and Canada. He was the son of Ambrose Lanfear Norrie, who famously discovered the iron ore of the Gogebic range of upper Michigan and founded the town of Ironwood there. The Norrie mine was at one time considered to be the greatest iron mine in the world.

Lang, HO

African Characin sp. *Distichodus langi* JT Nichols & L Griscom, 1917
Blenny sp. *Hypleurochilus langi* HW Fowler, 1923

Herbert Otto Henry Lang (1879–1957) was born in Germany and trained as a taxidermist. He later worked as such in both the Natural History Museum, Universität Zürich, and a commercial establishment in Paris. He moved to the USA (1903) and joined the AMNH as a taxidermist. He led the museum's Congo expedition (1909–1915). On returning to New York he became Assistant in Mammalogy, then Assistant Curator (1919). He returned to Portuguese West Africa (now Angola) for the museum (1925) with Rudyerd Boulton.

They covered 6,500 kilometres (4,000 miles) and collected 1,200 mammal specimens as well as other taxa. He stayed on in Africa after the Angola expedition and took a job with the Transvaal Museum, South Africa. He made a number of further expeditions, including one for the AMNH to the Kalahari Desert. Later, he took over the management of a hotel in Pretoria (1935). Two mammals, three reptiles and two birds are also named after him.

Lang, MA

Lang's Blenny *Starksia langi* CC Baldwin & CI Castillo, 2011

Dr Michael A Lang is a marine biologist, environmental physiologist, author and lecturer as well as a diver trainer. San Diego State University awarded his BSc and the Norwegian University of Science and Technology in Trondheim, his DPhil. He is Vice President of OxyHeal Health Group, UCSD Emergency Medicine Adjunct Faculty, Senior Research Fellow at The Ocean Foundation and Director of the Smithsonian Marine Science Network (MSN) and Smithsonian Science Diving Program. Before this he was Marine Collector/ Curator at San Diego State University, Director of the Smithsonian Marine Science Network, Smithsonian Scientific Diving Officer, and National Science Foundation Polar Diving Safety Officer. He has published more than 50 scientific papers and articles and over 300 seminars on marine biology and diving and is also author of *Research and Discoveries: The Revolution of Science through Scuba* (2013). He was honoured in gratitude for the support MSN provided for the authors' Caribbean fish diversity studies "*…and in recognition of the contributions Michael has made to science diving.*"

Lange, EA

Siamese Algae-eater *Crossocheilus langei* P Bleeker, 1860

E A Lange was an acting health officer and hospital inspector, Dutch East Indian Army, who was stationed at Palembang, Sumatra and collected there (1858–1859). It was he who sent the holotype to Bleeker (q.v.).

Lange, RB

Characin sp. *Deuterodon langei* HP Travassos, 1957

Armoured Catfish sp. *Rineloricaria langei* LFS Ingenito, MS Ghazzi, LF Duboc & V Abilhoa, 2008

Professor Rudolf Bruno Lange (1922–2016) was a Brazilian zoologist and botanist who was one of the first curators of the zoological collection at the Paraná Museum, where his collection of arachnids is held. Other taxa, including an amphibian, are named after him.

Langeani

Tetra sp. *Hyphessobrycon langeanii* FCT Lima & CR Moreira, 2003

Halftooth Characin sp. *Hemiodus langeanii* HD Beltrão & JAS Zuanon, 2012

Armoured Catfish sp. *Neoplecostomus langeanii* FR Roxo, C Oliveira & CH Zawadzki, 2012

Dr Francisco Langeani Neto is a Brazilian ichthyologist who specialises in Neotropical fish at Universidade Estadual Paulista, Instituto de Biociências, Letras e Ciências Exatas, where he is an adjunct professor. Universidade Federal de São Carlos awarded his bachelor's degree (1981) and Universidade de São Paulo his master's (1989) and doctorate (1996); he also has a doctorate (2009) from Universidade Estadual Paulista. He co-wrote: *Reconciling more than 150 years of taxonomic confusion: the true identity of* Moenkhausia lepidura, *with a key to the species of the* M. lepidura *group (Characiformes: Characidae)* (2016).

Langsdorff

Ghost Flathead sp. *Hoplichthys langsdorfii* G Cuvier, 1829
Japanese Silver Crucian Carp *Carassius langsdorfii* CJ Temminck & H Schlegel, 1846

Baron Georg Heinrich von Langsdorff (otherwise Grigoriy Ivanovich) (1774–1852) was a German physician, botanist, zoologist, traveller, ethnographer and diplomat. He graduated as a Doctor of Medicine at Göttingen University (1797) and was then sent to Portugal. He was elected as a corresponding member of the Academy of Science at St Petersburg (1803) and joined Krusenstern's round-theworld expedition (1803–1806) on board the 'Nadezhda', representing the Academy. He continued to travel widely, to Japan (1804–1805); northwestern America (1805–1806); and Kamchatka, Siberia and European Russia (1806–1808). He became Associate Professor of Botany at the Academy, moving to zoology (1809). He wrote: *Remarks and Observations on a Voyage Around the World from 1803-1807* (1812) and was elected as an Extraordinary Member of the Academy before being appointed as Russian Consul General in Brazil (1813–1817). He was also Chargé d'Affaires for Russia to Portugal in Rio, where the Portuguese government was in exile during the Napoleonic Wars. Whilst in Brazil he travelled to Minas Gerais with the French scientist Saint**Hilaire**. He returned to Russia (1821) to organise an expedition in Brazil, which he led (1822–1828), but he caught a tropical fever that caused a psychological breakdown. He returned to Germany (1830), retiring (1831) and living there until his death. Two birds, an amphibian and a reptile are also named after him.

Langson

Oriental Cyprinid sp. *Luciocyprinus langsoni* L Vaillant, 1904

This is a toponym; referring to Lang-Son, northern Vietnam (the type locality).

Lankester

Priapiumfish sp. *Neostethus lankesteri* CT Regan, 1916

Sir Edwin Ray Lankester (1847–1929) was a British invertebrate zoologist and evolutionary biologist. He graduated in biology from Oxford University (1870) and was appointed Jodrell Professor of Zoology and curator of (what is now) the Grant Museum of Zoology at University College London (1874–1890). He then became Linacre Professor of Comparative Anatomy at Merton College, Oxford (1891–1898), and director of the BMNH (1898–1907). He was a founder (1884) of the Marine Biological Association and served as its second President (1890–1929). Among his many publications are: *On comparative longevity in man and the lower animals* (1870) and *Extinct Animals* (1905).

Laperouse

Laperouse Snailfish *Careproctus laperousei* N Chernova, R Thiel & I Eidus, 2020

Jean-François Laperouse (Chevalier Jean-François de Galaup, comte de La Pérouse) (1741–1788) was a French Naval officer, seafarer and explorer. He had an illustrious naval career being wounded in one battle and taken prisoner in another. He was appointed (1785) to lead a circumnavigation and sailed as commander of 'La Boussole' accompanied by 'L'Astrolabe'. They visited Chile & Hawaii, Alaska & California then East Asia including Asian Russia and Japan and discovered the Bussol Strait (1787) a channel between the Kuril Islands near where this species was later taken. They sailed the South Pacific and were attacked in Samoa losing twelve men including the commander of 'L'Astrolabe'. They went on to Australia landing there and re-provisioning (December 1787). They left Port Jackson (March 1788) setting sail for New Caledonia and not heard of again. It was later confirmed (1964) that both ships were wrecked off Vanikoro Island, Solomon Isles. A bird is also named after him.

Laprade

Nile Bichir ssp. *Polypterus bichir lapradei* F Steindachner, 1869

Émile Pinet-Laprade (1822–1869) was a former Governor of Senegal (1863–1865) and regarded as the founder of the city of Dakar. He originally went to Senegal (1849) and was still living there when he caught cholera and died in the same year that the holotype, originally collected in Senegal, was described. His body was eventually returned to France and buried (1872). The description lacks an etymology so the attribution to Émile Pinet-Laprade cannot be 100% certain.

Lara

Banjo Catfish sp. *Bunocephalus larai* R Ihering, 1930

Rodolpho Lara Campos sponsored the expedition that collected the holotype. He had a boat especially built for the expedition to Rio Piracicaba (1928).

Larissa

Neotropical Rivuline sp. *Melanorivulus larissae* EV Ywamoto, DTB Nielsen & C Oliveira, 2020

Larissa da Silva Sobral is the daughter of Daniel Sobral dos Santos, who discovered this species in Brazil.

La Rivers

Big Smokey Valley Speckled Dace *Rhinichthys osculus lariversi* T Lugaski, 1972

Dr Ira John La Rivers (1915–1977) was an American zoologist and entomologist. The University of Nevada awarded his bachelor's degree (1937), North Caroline State College his master's (1938), and The University of California his doctorate (1948). He worked as a field entomologist and biologist for state government organisations in Nevada and California (1936–1948). He then returned to the University of Nevada as Assistant Professor of Biology (1948–1954), Director of the Museum of Biology (1953–1961), Associate Professor (1954–1961) and Professor (1961–1977). He wrote: *Fishes and Fisheries of Nevada* (1962).

Larnaudie

Spot Pangasius *Pangasius larnaudii* MF Bocourt, 1866
[Alt. One-spot Pangasius, Black-eared Catfish]

Father R P Larnaudie (d.1899) was a Jesuit missionary priest who spent 15 years in Siam (Thailand) and accompanied the Siamese ambassadors to Paris (1861), acting as their interpreter. He was a linguist fluent also in Japanese and at least one Chinese dialect. Bocourt honoured him for his care and hospitality during Bocourt's seven months in Thailand.

Larrec

Loach sp. *Rhyacoschistura larreci* M Kottelat, 2019

Named after LARReC (Living Aquatic Resources Research Center), Vientiane, Laos, for its 20th anniversary and appreciation to several of its staff for 20 years of collaboration in the field.

Larsen, CA

Painted Notothen *Nototheniops larseni* E Lönnberg, 1905

Carl Anton Larsen (1860–1924) was a Norwegian Antarctic explorer. He led an expedition to Antarctica, in command of the 'Jason' (1892–1894), and later captained the ship 'Antarctic' as part of the Swedish Antarctic Expedition (1901–1904). After several years' residence on South Georgia, he took British citizenship (1910).

Larsen, V

Pale Snipe-Eel *Nemichthys larseni* JG Nielsen & DG Smith, 1978

Verner Larsen is a Danish ichthyologist, who began revising *Nemichthys* whilst still a student at the University of Copenhagen, and who, according to the etymology, "*...generously handed over his material*" to the authors "*...when he was unable to finish it*". He appears to have become a schoolmaster at Viborg Cathedral School (retired 2011) and is mentioned in the local newspaper (2013) for being a 'passionate angler'. He co-wrote with the senior author: *Notes on the Bathylaconidae (Pisces, Isospondyli) with a new species from the Atlantic Ocean* (1969).

Larson

Goby genus *Larsonella* JE Randall & H Senou, 2001
Helen's Pygmy Pipehorse *Idiotropiscis larsonae* CE Dawson, 1984
Larson's Sueviota (goby) *Sueviota larsonae* R Winterbottom & DF Hoese, 1988
Western Australian Blackhead Triplefin *Enneapterygius larsonae* R Fricke, 1994
Goby sp. *Stiphodon larson* RE Watson, 1996
Swordtip Gurnard *Lepidotrigla larsoni* L del Cerro & D Lloris I Samo, 1997
Larson's Shrimpgoby *Stonogobiops larsonae* GR Allen, 1999
Larson's Cusk *Acarobythites larsonae* Y Machida, 2000
Goby sp. *Pleurosicya larsonae* DW Greenfield & JE Randall, 2004
Viviparous Brotula sp. *Beaglichthys larsonae* W Schwarzhans & PR Møller, 2007
Goby sp. *Trypauchenichthys larsonae* EO Murdy, 2008

Dr Helen K Larson was Curator of Fishes, Museum and Art Gallery of the Northern Territory, Darwin, Australia (1981–2009), retiring as Emeritus Curator, and is now a Research Associate at the Museum of Tropical Queensland, Townsville. She has been actively publishing on fish (120 papers since 1975) and among her longer works co-wrote: *Freshwater fishes of the Northern Territory* (1990). She is working on *Gobioids of the World* with Douglas Hoese. Her work focuses on the taxonomy and systematics of Indo-Pacific coral reef and estuarine fishes. She is Co-Chair of a group within the Marine Fishes Red List Authority of the International Union for the Conservation of Nature. She was acknowledged for her many contributions to the knowledge of the fishes of the Northern Territory. She collected the holotype and paratypes of the Swordtip Gurnard, but the authors used the masculine form *larsoni* in the binomial instead of the feminine *larsonae*.

Larue

Bigeye Hatchetfish *Polyipnus laruei* E Vourey, C Dupoux & AS Harold, 2017

Pierre and William Larue are yachtsmen who collected (and photographed) the holotype off the coast of New Caledonia (2014). The specimen was found dead and floating on the surface.

Lasso, CA

Characin sp. *Creagrutus lassoi* RP Vari & AS Harold, 2001
Characin sp. *Bryconamericus lassorum* C Román-Valencia, 2002

Dr Carlos Andres Lasso-Alcalá, brother to OM Lasso Alcalá (below), is a Venezuelan ichthyologist. He was at the Museo de Historia Natural La Salle (Caracas) and the Asociación Amigos de Doñana, Seville, which awarded his doctorate (1996). He was recognised for his contributions to the knowledge of Venezuelan fishes and his assistance to the authors. He is now (2017) a senior researcher at Instituto Humboldt, Biología de la Conservación, Bogota, Colombia. He co-wrote: *Aequidens superomaculatum* (Teleostei: *Cichlidae) una nueva especie del alto Orinoco y Río Negro, Venezuela* (2016). *Bryconamericus lassorum* is named after both brothers, who together wrote: *Revisión taxonómica del género* Awaous *Valenciennes 1837 (Pisces: Perciformes, Gobiidae) en Venezuela, con notas sobre su distribución y habitat* (2007).

Lasso, OM

Characin sp. *Bryconamericus lassorum* C Román-Valencia, 2002

Oscar Miguel Lasso-Alcalá (b.1972), brother to CA Lasso Alcalá (above), is a Venezuelan ichthyologist, zoologist and environmentalist who was at the Museo de Historia Natural La Salle, Fundación La Salle de Ciencias, Venezuela, where he was Assistant Curator (2001–2005). He is now (2017) a Professor at the Universidad Central de Venezuela. The fish is named after both brothers. An amphibian is named after him alone.

Last

Lagoon Goby *Tasmanogobius lasti* DF Hoese, 1991
Trawl Perchlet *Plectranthias lasti* JE Randall & DF Hoese, 1995
Rough-snout Whiptail *Coelorinchus lasti* T Iwamoto & A Williams, 1999
Rusty Snailfish *Paraliparis lasti* DL Stein, NV Chernova & AP Andriashev, 2001
Last's Numbfish *Narcinops lasti* MR de Carvalho & B Séret, 2002

Dr Peter Robert Last is an eminent Australian ichthyologist who is the senior taxonomist and Senior Principal Research Scientist at CSIRO Marine and Atmospheric Research Hobart. The University of Tasmania awarded both his bachelor's degree (1975) and his doctorate (1983). He joined the Tasmanian Fisheries Development Authority as a Research Scientist (1978) and then moved to the CSIRO Division of Fisheries (1984) as the Curator of the Australian National Fish Collection. The Museum National d'Histoire Naturelle, Paris made him an honorary professor (1997). He has written or co-written over 220 papers, which include descriptions of 153 species. He is senior author of: *Sharks and Rays of Australia* (2009) and *Sharks and Rays of Borneo* (2010).

Lastarria

Andean Pupfish sp. *Orestias lastarriae* RA Philippi, 1876

Señor Don Demetrio Lastarria of Valparaiso collected this species. We have no further biographical details.

Latham

Rough Scad *Trachurus lathami* JT Nichols, 1920

Roy A Latham (1881–1979) of Orient, Long Island, New York, USA, was a farmer and amateur self-eductaed naturalist who collected from a very early age and continued into his late nineties! Over the years he filled the two floors and multiple rooms of his farm house with specimens and displays of birds, insects, plants, mammals, reptiles and fish, which included an 'Indian Room' filled with thousands of local artifacts, as well as a library which eventually housed over 2,000 volumes, most pertaining to Natural History. When he died his collection was divided between Cornell and New York State Museum. He published widely, such as: *Migration Notes of Fishes from Orient* (1916) and *The Flora of the Town of Southold and Gardiner's Island* (1917). He collected the type.

Latimer

Coelacanth genus *Latimeria* JLB Smith, 1939

Miss Marjorie Courtenay-Latimer (1907–2004) was the first Curator of the Museum at East London, South Africa, which was based on the Latimer family collection and holds the only remaining Dodo egg. She was primarily an ichthyologist best remembered for her rediscovery (1938) of the Coelacanth, a fish that had been believed to be extinct for 70 million years. She was highly regarded in her country, and clay casts of her footprints were placed in Heroes Park alongside those of Nelson Mandela and Walter Sisulu (2003). Two birds are named after her.

Laura (Albini)

Thin-barred Goby *Psilotris laurae* JL van Tassell, L Tornabene & CC Baldwin, 2016

Laura Albini was the wife of Adriaan 'Dutch' Schrier, owner of Substation Curacao, through whose efforts new, tropical, deep-water species are being discovered; Laura generously fed and hosted numerous researchers during their visits to Curacao.

Laura (Hubbs)

Tonguetied Minnow *Exoglossum laurae* CL Hubbs, 1931
Hagfish sp. *Eptatretus laurahubbsae* CB McMillan & RL Wisner, 1984
Twoline Prickleback *Esselenichthys laurae* WI Follett & ME Anderson, 1990

Laura Cornelia Hubbs née Clark (1893–1988) was the wife of Carl Levitt Hubbs (q.v.) and a noted ichthyologist in her own right, as well as contributing to the work of her husband. Both her bachelor's degree (1915) and her master's (1916) were awarded by Stanford University. She worked part-time at the University of Michigan Museum of Zoology (1929–1944).

Laurent

Rippled Klipfish *Pavoclinus laurentii* JDF Gilchrist & WW Thompson, 1908

Master Lawrence Robinson caught three specimens at Winkle Spruit, Natal, South Africa. Use of the term 'Master' in this context suggests Lawrence was a boy at the time. Here 'Laurent' is used as a 'latinization' of the name Lawrence.

Laurent, P

Three-barbeled Catfish sp. *Pimelodella laurenti* HW Fowler, 1941

Philip Laurent (1858–1942) was an industrialist and amateur naturalist, and a well-known collector of natural history objects including butterflies and other insects. At least one moth is named after him. His collection of over 40,000 pinned insects, mostly Lepidoptera and Coleoptera from Florida and the western USA, went to the Academy of Natural Science in Philadelphia.

Lauretta

Lauretta Whitefish *Coregonus laurettae* TH Bean, 1881
[Alt. Bering Cisco]

Mrs Lauretta Bean was the wife of the author, Tarleton Bean. (Also see **Bean, TH**)

Laureys

Guinean Codling *Laemonema laureysi* M Poll, 1953

J Laureys commanded the expedition trawler from which the holotype was collected. We have been unable to find any more details about him.

Lauro

Characin sp. *Characidium lauroi* CH Travassos, 1949

Dr Lauro Pereira Travassos (1890–1970) was a prominent Brazilian helminthologist-entomologist. He graduated as a physician from the School of Medicine, Rio de Janeiro (1913). He became Professor of Parasitology, Faculty of Medicine of São Paulo (1926). He worked in Hamburg at the Tropen Institute (1929) at the invitation of Fülleborn. He was a full professor at the National Veterinary School, Federal Rural University of Rio de Janeiro (1930–1937), and used to travel to Europe by signing on as a ship's doctor on a

Lloyd Brasileiro ship going to a European port for repairs. While the ship was in dry-dock, he took the chance to attend medical centres and study before re-boarding the ship for the return journey. He collected the characin holotype and it was described by one of his three sons, all of whom became scientists.

Láurusson

Iceland Catshark *Apristurus laurussonii* B Sæmundsson, 1922

Gísli Láurusson (1865–1935) was an Icelander described as 'Hr Direktur'. He was a goldsmith, watchmaker and farmer with an interest in wildlife, particularly of that in the surrounding seas and also birds and their conservation. He was on the board of several companies, including two trawlers (Draupnir and Herjólfur). He is mentioned as supplying Sæmundsson with a fish of a different species. He was honoured "*...for his long and invaluable support of the study of Icelandic fishes, and for carefully encouraging the skilful fishermen of Vestmannaeyjar Island*" (which is close to the type locality).

Laury

Pencil Catfish sp. *Trichomycterus lauryi* LM Donin, J Ferrer & TP Carvalho, 2020

Laury João Donin was the late father of the senior author, Laura Donin.

Lauvergne

Lauvergne's Mullet *Planiliza lauvergnii* F Eydoux & LFA Souleyet, 1850

Barthélemy Lauvergne (1805–1871) was a painter and draughtsman who was on the voyage that collected the holotype. He went around the world three times: first on 'L'Astrolabe' (1826–1829) as secretary to Jules Dumont d'Urville, then on 'La Favorite' (1830–1832) and finally on 'Bonite' (1836–1837). He became attached to the Navy's Marine Scientific Committee Commissariat (1838–1839). He joined (1839) the Northern Scientific Commission at Hammerfest, Norway; travelling to Finland, Spitzberg Island and Russia. Hundreds of lithographs were published (1841) of the drawings from his voyages. He was commissioned to paint port scenes from Algeria (1842) and was also appointed (1851) to paint the future Napoleon III aboard the ship 'Napoleon'. He retired early (1863). Although the patronym is not identified, he seems the most probable honoree.

Lauzanne

African Barb sp. *Enteromius lauzannei* C Lévêque & D Paugy, 1982
Headstander Characin sp. *Rhytiodus lauzannei* J Géry, 1987
Pencil Catfish sp. *Trichomycterus lauzannii* G Miranda & L Fernández, 2020

Dr Laurent Lauzanne was a French ichthyologist and hydro-biologist who was a friend of the barb's authors. He worked for ORSTOM in Chad and in the Bolivian portion of the Amazon River system. He described a number of new fishes including several *Orestias* species from Lago Pequeño (1981), leading him to review all such species from there (1982). Among his publications he wrote: *Trophic relations of fishes in Lake Chad* (1983) and co-wrote: *Peces del Rio Mamore* (1985). The catfish etymology honours him as: "*one of the first ichthyologists to work systematically on Bolivian ichthyofauna.*"

Laval

Laval Eelpout *Lycodes lavalaei* VD Vladykov & J-L Tremblay, 1936
[Alt. Newfoundland Eelpout]

Named for the Université Laval, a French-language university in Quebec, Canada.

Lavenberg

Gulper Eel sp. *Saccopharynx lavenbergi* JG Nielsen & EOG Bertelsen, 1985
Goby sp. *Lythrypnus lavenbergi* WA Bussing, 1990
Snaketooth sp. *Pseudoscopelus lavenbergi* MRS Melo, HJ Walker & C Klepadlo, 2007
Black Swallower sp. *Chiasmodon lavenbergi* AM Prokofiev, 2008

Dr Robert 'Bob' John Lavenberg was Curator of Ichthyology at the Natural History Museum of Los Angeles County and is now Emeritus Curator. The authors of the gulper eel wrote that he was honoured "*...for his contributions to oceanic ichthyology and for making material from his museum available*." Among his publications he co-wrote: *Tidepool and Nearshore Fishes of California* (1975) and: *The Fishes of the Galápagos Islands* (1997). He also made and appeared in a documentary film: *Legend of Loch Ness* (1976). (See also **John Bob**)

Lavett Smith

Clingfish sp. *Tomicodon lavettsmithi* JT Williams & JC Tyler, 2003

Clarence Lavett 'Smitty' Smith Jr (1928–2015) was Curator, then Emeritus Curator, of Ichthyology at AMNH (1962–1997). Cornell University awarded his BSc, Tulane University his MSc and The University of Michigan his PhD. He then served in the Army Medical Corps at Walter Reed Hospital in Washington, D.C. and the Tropical Medical Research Laboratory in San Juan, Puerto Rico. He also taught at numerous universities, including Ohio State, Michigan, Oklahoma, Alabama, Hawaii, Guam, City College NY, CW Post, and others. As well as 35 years spent at the American Museum, he worked a total of 18 summers over five decades at Ohio State's F.T. Stone Lab which is on an island in western Lake Erie. Most of his summers were spent teaching at biological stations, but he also undertook research at marine laboratories in several tropical locations, such as the American Museum's Lerner Marine Laboratory in the Bahamas and the Smithsonian's Research Station on Carrie Bow Cay, Belize. He specialised in coral reef fishes, and in the freshwater fishes of New York State and the Great Lakes. He wrote more than 100 books and papers, which he illustrated himself.

Lavinia

Hitch (Californian cyprinid) genus *Lavinia* CF Girard, 1854

Girard gave no reasoning for his naming of this genus. It could be a 'private' eponym, or taken from Roman mythology (Lavinia was the last wife of Aeneas), or possibly in honour of Lavinia Bowen, an artist, lithographer and colourist who worked on natural history subjects.

Lawrence (Page)

Headwater Darter *Etheostoma lawrencei* PA Ceas & BM Burr, 2002

Professor Dr Lawrence 'Larry' Merle Page (b.1944) (see **Page**)

Laws

Lake Malawi Cichlid sp. *Gephyrochromis lawsi* G Fryer, 1957

Robert Laws (1851–1934) was a Scottish missionary and educator who headed the Livingston mission in the Nyasaland Protectorate (now Malawi) for more than fifty years (1875–1927). He was apprenticed to a cabinet maker, but after reading David Livingstone's *Travels* he resolved to become a missionary. Working by day, he attended evening classes and gained admission to the University of Aberdeen. He spent seven years there, earning degrees in Arts, Medicine and Theology. Fryer says of him: "*…to whose pioneering missionary endeavour the present peace and prosperity of the Nyasan peoples owe so much.*" On his death a journalist wrote of him: "*Nothing impressed me more about Dr Laws than his humility. He was a great man who was unconscious of his greatness.*"

Layard

Thinspine Sea Catfish *Plicofollis layardi* A Günther 1866
Sri Lankan Barb sp. *Puntius layardi* A Günther, 1868

Edgar Leopold Layard (1824–1900) was born in Florence, Italy. He spent ten years as a Civil Servant in Ceylon (Sri Lanka), where he studied the local fauna. He went to the Cape Colony, South Africa (1854), again as a civil servant on the staff of the Governor, Sir George Grey. He became Curator of the South African Museum (1855) in his spare time. Later he was Honorary British Consul in New Caledonia. From here, he and his son Edgar Leopold Calthrop Layard (known as Leopold Layard) (b.1848) made collecting trips all over the South-West Pacific. Layard wrote: *The Birds of South Africa* (1867), which was later updated by Richard Bowdler Sharpe. He presented the barb type to the BMNH. Twenty birds, two mammals and a reptile are also named after him.

Lazarev

Pelagic Eelpout sp. *Melanostigma lazarevi* MV Orlovskaya & AV Balushkin, 2020

Admiral Mikhail Petrovich Lazarev (1788–1851) was a Russian noble who became fleet commander and explorer. As part of his training he served for five years in the British Royal Navy. Aboard *Suvorov* he circumnavigated the globe (1813–1816). As commander (1819–1821) of the ship *Mirny*, he took part in the discovery of the Antarctic mainland (January 1820) and numerous islands. He again circumnavigated commanding *Kreyser* (1822–1825). Four naval ships including a battle cruiser were also named after him as were an atoll, two capes, an island, a bay, a sea port, a settlement and a minor planet.

Lazzarotto

Neotropical Rivuline sp. *Atlantirivulus lazzarotoi* WJEM Costa, 2007

Dr Henrique Lazzarotto is a biologist and ecologist who is (since 2016) a Colaborator in the Ecology Department of the Universidade Federal do Rio de Janeiro, Brazil. The Federal University of Rio de Janeiro awarded his BSc (2002), his MSc (2006) and his PhD (2014); after which he was a post-doctoral researcher at the California Academy of Sciences, Department of Ichthyology. He was the first to collect the species and collected the paratypes.

Lea

Lea's Cardinalfish *Taeniamia leai* ER Waite, 1916

Arthur Mills Lea (1868–1932) was an Australian entomologist with an enduring passion for beetles. By 1911, when he applied for the position of entomologist at the South Australian Museum, he could cite published descriptions of 1,853 new species. He made a series of collecting trips to many different parts of Australia and offshore islands. He also collected the holotype of this cardinalfish whilst on Norfolk Island (December 1915).

Leal

Neotropical Rivuline sp. *Melanorivulus leali* WJEM Costa, 2013

Fabiano Leal de Souza (b.1976) is a tropical fish hobbyist who breeds killifish. The Universidade Federal Fluminense awarded his history BA (2008) and he is presently studying at the Universidade Salgado de Oliveira for his MSc. His interest in ichthyology derives from a paternal family tradition, in which the men have a Natural History hobby, mainly related to zoology. Following this tradition, he has developed research with several Neotropical fish species and was very successful in the reproduction of aplocheiloid killifishes, providing information and material for several researchers. Among his publications is the co-written: *Egg Surface Morphology in the Neotropical seasonal killifish genus* Leptolebias *(Teleostei: Aplocheiloidei: Rivulidae)* (2008). The etymology says that the name honours: "*Fabiano Leal for his dedication in breeding aplocheiloid killifishes and for providing material of the new species.*"

Leavell

Goby sp. *Rhinogobius leavelli* AWCT Herre, 1935

Dr George Walne Leavell (1882–1957) of the Baptist Hospital, Wuchow, Kwangsi Province, China, was an American medical missionary for twenty years. He studied at the University of Mississippi, then did medical training at the University of Louisville. After completing his missionary work, he practiced medicine in Bristol, Virginia.

Le Bail

Characin sp. *Leporinus lebaili* J Géry & P Planquette, 1983
Serrasalmid Characin sp. *Tometes lebaili* M Jégu, P Keith & E Belmont-Jégu, 2002

Pierre-Yves Le Bail (b.1954) is a French ichthyologist who is the Research Director, Institut National de la Recherche Agronomique, Paris (1979–present). He studied to be an agronomist engineer in fisheries at the Agrocampus Ouest school (1978) followed by a PhD there (1981). His military service was in Guyana, during which he studied the reproductive biology of potentially farmable fish. Already an aquarist, he discovered the richness of the area's freshwater fauna and maintains an interest in it. He has written over 130 peer-reviewed papers such as: *Un nouveau Tetragonopterinae (Pisces, Characoidei, Characidae) de la Guyane: Astyanax leopoldi* sp. (1998, with Géry & Planquette), and *Etat des lieux sur les poissons d'eau douce et estuariens de Guyane française* (2015, with Vigouroux). He has discovered about fifty new species in all. This led to him compiling, with colleagues, the three-volume *Atlas des poissons dulçaquicoles de Guyane Française* (1996 >).

Lebeck

Long-nosed Stargazer *Ichthyscopus lebeck* ME Bloch & JG Schneider, 1801

Heinrich Julius Lebeck (1772–1801) was a botanist and naturalist born in Ceylon (Sri Lanka) but of Dutch descent. He was taught at the missionary school of Dr Christoph Samuel John (q.v.) and then became his assistant. John described him (1789) as "*...more of an organiser, sketcher and lover of the animal kingdom...* (who) *...puts my molluscs, snakes and collection of fishes in order and writes descriptions according to Linné, which I correct.*" He later studied at Uppsala (1794–1795) before returning to India, where John regarded him as his outstanding acolyte. He wrote a number of papers including the description of the Gangetic dolphin. He worked as a 'mint master' for the Dutch East India Company until his death in Java.

Lebedev

Spiny Loach sp. *Cobitis lebedevi* ED Vasil'eva & VP Vasil'ev, 1985
[Perhaps a junior synonym of *Cobitis choii* (Kim & Son, 1984)]

Vladimir Dmitrievich Lebedev (1915–1975) was a Russian ichthyologist and marine scientist in Odessa (now in Ukraine). He discovered the first fossil remain of *Cobitis* based on which he determined changes in the biological aspects of freshwater fish populations, and confirmed his findings through expeditions in European Russia, Siberia and Central Asia.

Lécluse

Sharp-headed Wrasse *Cymolutes lecluse* JRC Quoy & JP Gaimard, 1824

M de Lécluse was a naval surgeon. The authors named this species in his memory as he had died of yellow fever. We can find nothing further.

Lecointe

Rattail sp. *Coryphaenoides lecointei* LAMJ Dollo, 1900

Lieutenant Georges Lecointe (1869–1929) was a Belgian naval officer, explorer, scientist and astronomer. After service on secondment to the French navy – Belgium at that time had no navy of its own – he was appointed to be commander of 'Belgica' and second-in-command of the Belgian Antarctic Expedition (1897–1899), during which the holotype was collected. The great Norwegian explorer Roald Amundsen was first mate on the 'Belgica' and was, de facto, the expedition leader when Lecointe was severely stricken by scurvy when the 'Belgica' was frozen in the Weddell Sea (1898–1899) and the expedition was forced to overwinter and lived off raw fish and penguin meat. He served in the Belgian Legion in China during the Boxer War and in the Belgian artillery in WW1. He was based at the Royal Observatory until illness forced him to resign (1925). He wrote: *Au Pays des Manchots* (1904). An asteroid 3755 Lecointe is named after him.

Le Conte

Silver Loach *Yasuhikotakia lecontei* HW Fowler, 1937

Dr John Lawrence Le Conte (1825–1883) was an American biologist and the most important American entomologist of the 19th century. He undertook a number of collecting expeditions to the western USA, one with Louis Agassiz (1848) to the Rocky Mountains, and to Central America including Honduras (1857) and Panama (1867). He also visited Europe, Algeria and Egypt. He founded the Entomological Society of America and was cofounder of the National Academy of Science. He was also a physician during the American Civil War. His father, John Eaton Le Conte (1784–1860), was also a naturalist and US Army engineer, and some of his writings are addressed to his son who may also have contributed some illustrations. Two birds, two reptiles and two mammals are named after him. Fowler's etymology describes him as a 'distinguished entomologist of past generations', and a contributor to the fish collection of the Academy of Natural Sciences of Philadelphia.

Lecornet

Lanternbelly sp. *Acropoma lecorneti* P Fourmanoir, 1988

D Lecornet was the fisherman on board the *Thalassa*, who caught the holotype off New Caledonia. Nothing further was recorded.

Le Crom

Perchlet sp. *Chelidoperca lecromi* P Fourmanoir, 1982

Albert Le Crom was in charge of the deep trawling carried out by the R/V 'Vauban', which caught the holotype in the îles Chesterfield, Coral Sea (1976–1986). This 'master' sailor also repaired the gear as necessary. He was later aboard the R/V 'Alis' (1987–1993). He is also commemorated in the name of at least one crab, the etymology for which says: "*This species is dedicated to Mr Albert LE CROM, fishing master, who participated in almost all the expeditions whose materials are studied herein.*"

Leda

Speckled Deepwater Scorpionfish *Pontinus leda* WN Eschmeyer, 1969

In Greek mythology, Leda was the mother of Helen (of Troy) and of the twins Castor and Pollux. Eschmeyer's name was doubtless inspired by the fact that two other members of this genus had, a century earlier, been named as *P. castor* and *P. pollux*.

Leda

Neotropical Silverside sp. *Odontesthes ledae* LR Malabarba & BS Dyer, 2002

Dr Leda Francisca Armani Jardim is a Brazilian ichthyologist who was Professor of Ichthyology at Universidade Federal do Rio Grande do Sul, in Porto Alegre. She was also the first President (1982) of the Brazilian Ichthyological Society. The etymology reads thus: "*Named in honor of the ichthyologist Leda Francisca Armani Jardim, now retired. She was the former advisor of several young undergraduate southern Brazilian students at the Universidade Federal do Rio Grande do Sul, in projects related to ichthyology.*"

Lee, LA

Lee's Tonguefish *Symphurus leei* DS Jordan & CH Bollman, 1890

Leslie Alexander Lee* (1852–1908) was a geologist and palaeontologist, as well as a zoological and botanical collector operating (1885–1892) in Canada, the USA and South America, particularly Ecuador and Argentina. He was Instructor of Natural History (1876–1881) and Professor of Biology and Geology (1881–1908) at Bowden College and State Geologist of Maine. He was a regular member of the group of naturalists used by the US Fish Commission on the cruises of their research vessel 'Albatross'. He was aboard when they called at Cozumel Island off the Yucatan Peninsula (1885), and again in the Bahamas and the West Indies (1886) and the Galapagos Islands (1888). The death by drowning of his son (1907) led to his health deteriorating shortly after the tragedy, and in turn to his own demise.

Lee, T

Lee's Tonguefish *Symphurus leei* DS Jordan & CH Bollman, 1890

Dr Thomas Lee* was a regular member of the group of naturalists used by the US Fish Commission on the cruises of their research vessel, Albatross. He was aboard when they called at Cozumel Island off the Yucatan Peninsula (1885), and again in the Bahamas and the West Indies (1886) and the Galapagos (1888) en route from New York to San Francisco via Cape Horn. A reptile is also named after him.

*NB - the binomial should really be the plural *leeorum*, as the etymology says: "*...We have named the species for Prof. L. A. Lee and Mr. Thomas Lee, naturalists on board the Albatross when the species was discovered.*"

Leeds

Bannerfin Shiner *Cyprinella leedsi* HW Fowler, 1942

Arthur Newlin Leeds (1870–1939) was a textile manufacturer and botanist at the Academy of Natural Sciences of Philadelphia where he became Research Associate (1928) and, eventually, Research Fellow. Haverford College awarded him a master's degree in English (1890). He retired from his business (1926) and made a leisurely trip round the world in the company of Charles W Townsend. According to the etymology, he "*...had been much attracted to the charm of the Ohoopee*" (a river in Georgia, the type locality), and was present when the holotype was collected.

Leer

Tapah Catfish *Wallagonia leerii* P Bleeker, 1851
[Alt. Helicopter Catfish, Striped Wallago Catfish]
Pearl Gourami *Trichopodus leerii* P Bleeker, 1852

Lieutenant-Colonel J M van Leer was a Dutch army physician who was a colleague of Bleeker's. He was for a time the Directing Surgeon at Samarang, East Sumatra. He provided the holotypes of these species.

Lefroy

Mottled Mojarra *Ulaema lefroyi* GB Goode, 1874

Sir John Henry Lefroy (1817–1890) was a British military officer, scientist and civil servant. He was at the Royal Military Academy (1831–1834) leaving as Second Lieutenant. He was sent to Saint Helena (1839) to set up and supervise an observatory, then (1842) was sent to Toronto as Superintendent of the Magnetic and Meteorological Observatory there (1842–1853). During his time in Canada he took measurements at over 300 stations in an attempt to map the geo-magnetic activity of British North America from Montreal to the Arctic Circle. He later returned to England, serving in various army posts until retiring (1870) as Major General. He entered the Colonial Service and was appointed Governor of Bermuda (1871–1877) where, according to the etymology, he was "*...taking an active part in adding to our knowledge of* [Bermuda's] *natural history.*" He returned to England in poor health but was later appointed Administrator of Tasmania (1880–1881).

Leftwich

Oyster Goby *Arenigobius leftwichi* JD Ogilby, 1910

Richard W Leftwich, Jr. (1856–1914) was an oyster merchant of Maryborough (Queensland, Australia), to whom Ogilby was "*indebted for many kindnesses.*"

Legand

Cucumberfish sp. *Paraulopus legandi* P Fourmanoir & J Rivaton, 1979

Michel Legand is an oceanographer and ichthyologist who was the Director, ORSTOM, Noumea, New Caledonia. He studied for many years the meso-bathypelagic and deep-benthic fishes of New Caledonia. He co-wrote: *Atlas de la Nouvelle Caledonie et Dependances* (1981). He was awarded La Médaille de la Société d'Océanographie de France (1969). (Also see **Michel**)

Legendre, J

Sleeper sp. *Ratsirakia legendrei* J Pellegrin, 1919

Jean Legendre was a physician with the French Colonial Troops in Madagascar (1922–1924). He observed this sleeper and provided 'interesting details' (translation) of its biology to the author.

Legendre, M

Red Indonesian Arowana *Scleropages legendrei* L Pouyaud, Sudarto & GG Teugels, 2003 [Taxonomy disputed: may be a colour strain of *S. formosus*]

Dr Marc Legendre (b.1955) is a Senior Researcher in Physiology and Reproduction of Fishes. He is Director of Research at the Institut des Sciences de l'Evolution (ISEM - CNRS, IRD, UM2), Université Montpellier II, France. He is author or co-author of 70 scientific publications, mainly in the field of the biology and culture of tilapias and tropical catfishes.

Leggett

Leggett's Rainbowfish *Glossolepis leggetti* GR Allen & SJ Renyaan, 1998

Ray Leggett (b.1936) is an Australian aquarist and naturalist and well-known member of the Queensland Naturalists Club. He has collected over 8000 freshwater fish of 99 species, mostly in northern Queensland and the Kimberley. He has also managed to breed 189 different fish species in captivity. He co-wrote: *Australian Native Fishes for Aquariums* (1987) and is a regular contributor to papers on fish and other natural history topics. He was actively involved in the establishment of the Australian and New Guinea Fish Association and is or has been an active member of the Aquarium and Terrarium Society, Queensland Finch Society, Society for Growing Australian Plants and the Royal Geographic Society of Queensland. He also spent a decade at the Queensland Museum as an interpretive officer answering queries from the public. He also designed the Surat Public Aquarium.

Le Gras

Goldie (seaperch) *Callanthias legras* JLB Smith, 1948

M G le Gras, of Port Elizabeth, South Africa, was honoured because it was he "...*who has collected many valuable fishes*".

Leh

Priapiumfish sp, *Phallostethus lehi* LR Parenti, 1996
Betta sp. *Betta lehi* HH Tan & PKL Ng, 2005

Dr Charles Leh Moi Ung is Curator of Zoology, Sarawak Museum Department and works with the Ministry of Environment and Tourism. He helped with the first scientific collection of *Phallostethus* in Borneo. The University of Malaya awarded his BSC in zoology (1980), his MSC (1983) and PhD on Fisheries Biology and Coastal Resources (1992). He is also Scientific Editor of the Sarawak Museum Journal (since 1982) and has been an adjunct lecturer at the University of Malay (since 2001). He is also advisor (2014) to the Core Team National Biodiversity Centre. He has named ten species and has eight species and one genera of fauna named after him. He has written or co-authored a number of papers such as: *Zooarchaeology in Sarawak in the 21st Century* (2013).

Lehmann

Zeravshan Dace *Leuciscus lehmanni* JF Brandt, 1852

Alexander Lehmann (1814–1842) was a Russian explorer, biologist and collector of Baltic-German descent from Dorpat (now Tartu, Estonia). He traveled to Siberia (1839–1840) and to Turkestan as leader of an expedition (1841–1842) mounted to collect for the St Petersburg Botanical Garden. During this expedition the dace holotype was collected. Lehmann was taken sick and died on his way home. His notes and accounts were published after his death as: *Alexander Lehmann's Reise nach Buchara und Samarkand in den Jahren 1841 und 1842* (1852). A reptile is also named after him, as is a dragonfly.

Le Hoa

Vietnamese Cyprinid sp. *Hemibarbus lehoai* VH Nguyen, 2001
[Probably a junior synonym of *Hemibarbus maculatus*]

Le Hoa Doan is a Vietnamese fisheries biologist who was with the Vietnamese government Fisheries Division. He and Nguyen Van Hao undertook a survey of aquaculture (1959–1966) in northern Vietnam, taking specimens from a number of rivers, reservoirs and lakes and analysing the results (1963–1967). He was the second author (VH Nguyen was the senior author) of *Hemibarbus longianalis* (1969): a pre-occupied name for which *H. lehoai* was proposed as a replacement. They also wrote: *Some data on composition, origin, and distribution of cyprinid fish species in northern Vietnam with descriptions of New Taxa* (2008).

Leiby

String Eel *Gordiichthys leibyi* JE McCosker & JE Böhlke, 1984

Dr Mark M Leiby (b.1944) is an ichthyologist who is a retired Specimens Collections Manager at the Florida Department of Natural Resources, Marine Research Laboratory, St Petersburg. He wrote: *Leptocephalus larvae of the tribe Callenchelyini (Anguilliformes, Ophichthidae, Ophichthinae) in the western North Atlantic* (1984) and was one of the authors of part 9 of *Fishes of the Western North Atlantic* (2010). The etymology described him as a "*...friend and apodal ichthyologist.*"

Leichhardt

Spotted Bonytongue *Scleropages leichardti* AKLG Günther, 1864
[Alt. Southern Saratoga, Spotted Australian Arowana]
Leichhardt's Velvetfish *Kanekonia leichhardti* JW Johnson, 2013

Friedrich Wilhelm Ludwig Leichhardt (1813–1848) was a Prussian naturalist and explorer noted for having made a three-thousand-mile journey across Australia. He studied natural sciences and languages at the Universities of Berlin and Göttingen (1831–1836), after which he studied natural sciences at the British Museum in London and Le Jardin des Plantes in Paris. He went to Sydney (1842) and led three expeditions. The first from Moreton Bay to Port Essington lasted 14 months (1844–1845) and covered about 3,000 miles. The second (1846) was planned to cross from Darling Downs to Perth but only lasted 6 months due to malaria and famine. The final expedition (1848) started out to traverse Australia to the Swan River. He was last seen on Darling Downs shortly after the start of the journey and then vanished. His fate has never been established. A number of Australian features are named after him, such as the eponymous Sydney suburb and highway as well as a reptile.

Leighton

Dwarf Royal Twig Catfish *Sturisomatichthys leightoni* CT Regan, 1912

Lieutenant-Colonel Sir Bryan Baldwin Mawddwy Leighton, 9th Bt. (1868–1919) was a soldier who was commissioned as a 2nd Lieutenant in the Shropshire Yeomanry (1891) and served in Bechuanaland (Botswana) (1896–1897). He fought for the Americans in the Spanish-American War (1898) and was mentioned by Theodore Roosevelt in his account of the war. He served in South Africa (1899–1900) during the Boer War and worked as a war correspondent in Shanghai and Chefoo during the Russo-Japanese War (1904). He was with Turkish forces in the Balkan War (1913) and made a film of it at the time. On the outbreak of WW1, he was put in command of the Westmoreland and Cumberland Yeomanry but switched to the newly formed Royal Flying Corp (1915) in which he was gazetted as a Flying Officer and saw service in France, including testing new designs of parachute. He died of pneumonia, a victim of the Spanish flu pandemic. He presented the holotype to the BMNH.

Leis

Leis' Coralbrotula *Diancistrus leisi* W Schwarzhans, PR Møller & JG Nielsen, 2005
Cardinalfish sp. *Foa leisi* TH Fraser & JE Randall, 2011

Dr Jeffrey Martin Leis (b.1949) is a larval-fish biologist and Adjunct Professor at the Institute for Marine and Antarctic Studies, University of Tasmania. The University of Arizona awarded his Zoology BSc (1971) and the University of Hawaii his PhD in Biological Oceanography (1978). He was employed at the Australian Museum, Sydney (1979–2014) and when undertaking scientific expeditions for them collected the types of both eponymous fish species, in the northern Great Barrier Reef and Taiaro Atoll in the Tuamotu Islands respectively. His research interests include fish systematics, taxonomy and ecology of marine fish larvae, and the role of behaviour in larval-fish dispersal. His major publications include the co-authored: *The larvae of Indo-Pacific coastal fishes. An identification guide to marine fish larvae* (2000), *How Nemo finds home: the neuroecology of dispersal and of population connectivity in larvae of marine fishes* (2011) and *Taxonomy and systematics of larval Indo-Pacific fishes: a review of progress since 1981* (2015). He is a SCUBA diver (since 1969), loves the Australian outdoors, and a good red wine.

Leitao

Neotropical Rivuline sp. *Mucurilebias leitaoi* CAG da Cruz & OL Peixoto, 1992

Dr Antenor Leitão de Carvalho (1910–1985). (See **Carvalho, AL**)

Lek

African Killifish sp. *Hylopanchax leki* JA van der Zee, R Sonnenberg & UR Schliewen, 2013

The binomial is not an eponym but a word in the Lingala language for 'smaller sibling', referring to the fact that it is slightly smaller than a related species found in the same area.

Leleup

Lemon Cichlid *Neolamprologus leleupi* M Poll, 1956
African Barb sp. *Labeobarbus leleupanus* H Matthes, 1959
[Syn. *Varicorhinus leleupanus*]
Congo Cyprinid sp. *Opsaridium leleupi* H Matthes, 1965
Robber (African Tetra) sp. *Rhabdalestes leleupi* M Poll, 1967

Dr Narcisse Leleup (1912–2001) was an entomologist who collected in the Congo for the Tervuren Museum, Belgium (1940s & 1950s), and led the Belgian Zoological Expedition to the Galápagos Islands and Ecuador (1964–1965). He co-wrote: *La géographie et l'ecologie des grottes du Bas-Congo. Les habitats de Caecobarbus geertsi* (1954). He collected, or helped to collect, the holotypes of all these eponymous species. A reptile and two amphibians are also named after him.

Leloup

Lake Tanganyika Cichlid sp. *Neolamprologus leloupi* M Poll, 1948

Eugène Henri Joseph Leloup (1902–1981) was a Belgian biologist, mainly specialising in aquatic invertebrates. He was director (1947–1967) of the Marine Science Institute (ZWI) in Ostend. He was the leader of the Belgian Hydrobiological Mission to Lake Tanganyika (1946–1947), during which the cichlid holotype was collected.

Lemaire

Lake Tanganyika Cichlid sp. *Grammatotria lemairii* GA Boulenger, 1899
Lemaire's Lamprologus (cichlid) *Lamprologus lemairii* GA Boulenger,1899

Captain Charles François Alexandre Lemaire (1863–1925) was a military officer in the second Belgian Artillery, a colonial administrator and explorer who led several expeditions in the Belgian Congo (DRC). He was District Commission of Ecuador before heading the Katanga Scientific mission – as the leader of the Congo Free State Expedition (1898) which collected fishes in Lake Tanganyika, including the cichlid types. Among a number of publications, he wrote a journal of that and further expeditions later published as: *Reconnaissance menée aux sources du Yé-Yi (avril-mai 1903); journal de route de Charles Lemaire* (1953). Lemaire was a keen supporter of Esperanto, writing at least one of his books in it, and published with Esperanto and French texts on opposite pages, thus *Tra Mez-Afriko. A travers l'Afrique centrale. Parolado kun projekcioj donita al la dua Universala kongreso d'Esperanto, en Genevo, je la sabato la de septembro 1906a, conférence avec projections donnée au 2me Congrès universel d'Esperanto, à Genève, le samedi 1er septembre 1906.* His diaries were published later (1986), revealing that he was the leader in murder and maiming of Congolese natives by men under his command. Joseph Conrad's novel *Heart of Darkness*, which is based on his own experiences in the Belgian Congo, has as one of its main characters a terrible man called Kurtz, and it is thought that Lemaire may have been the inspiration for him. Two amphibians are also named after him.

Lemaitre

Lemaitre's Mosquitofish *Gambusia lemaitrei* HW Fowler, 1950

Ernesto D *Lemaitre* was *a Colombian naturalist and angler. He accompanied Fowler on his trips to the fish market in Cartegena and on his collecting trips to Totumo (the mosquitofish's type locality). He also shared his notes and sketches with Fowler, who acknowledged that Lemaitre had assisted him "in every way".*

Lemasson

Oriental Cyprinid sp. *Decorus lemassoni* J Pellegrin & P Chevey, 1936
[Syn. *Bangana lemassoni*]

Jean L Lemasson was an aquatic engineer who was Chief of the Fishing and Hunting Service in Hanoi, Indochina (Vietnam) (ca.1934–ca.1948). He wrote: *Essais de mise au point de méthodes de rizipisciculture dans le delta et la moyenne région du Tonkin* (1942). He collected the holotype.

Le May

Lace Goby *Oxyurichthys lemayi* JLB Smith, 1947
[Possibly a junior synonym of *Oxyurichthys notonema*]

Basil C Le May was a South African businessman in the diamond mining industry, philanthropist and angler. He was thanked by the author for his assistance, but no further details are given. However, he was a corporate sponsor of the Sea Fishes Trust which published *Smith's Sea Fishes* (1996) by MM Smith, which is dedicated to JLB Smith and published by the JLB. Smith Institute of Ichthyology. Moreover, he and his brother Hugh had supported the publication of JLB Smith's book: *Sea Fishes of Southern Africa* (1946) and continued to support her work thereafter.

Lemming

Iberian Arched-mouth Nase *Iberochondrostoma lemmingii* F Steindachner, 1866

There is little prospect of finding out who or what is referred to here! The original text does not help, plus there is the fact that originally Steindachner spelt the binomial as *leminingii* - presumably in error and not by design?

Lempriere

Thornback Skate *Dentiraja lemprieri* J Richardson, 1845

Deputy Assistant Commissary-General Thomas James Lempriere (1796–1852) was born in Hamburg, the son of a British banker and merchant. He emigrated to Van Diemen's Land (Tasmania) (1822) where he followed in his father's footsteps as a merchant and banker. He joined the Commissariat Department (1826) as a storekeeper at the penal settlements on Sarah Island and Maria Island, Macquarie Harbour. He became Deputy Assistant Commissary General (1837) and Assistant Commissary General (1844) and Coroner for Tasmania (1846). He was recalled to England (1849) for transfer as assistant commissary general in Hong Kong, but his health failed and he was invalided home (1851) but died on the voyage and is buried in Aden. He cut a tide gauge into a rock near Port Arthur and it shows that the sea level there has risen about 13.5 cm since the 1840s. He is honoured, according to the original description, as someone for '*…whose exertions the Ichthyology of Van Diemen's Land is much indebted*.'

Lendl

Anatolian Loach *Seminemacheilus lendlii* B Hankó, 1925

Dr Adolf Lendl (1862–1943) was a Hungarian zoologist, politician and collector in Asia Minor, including Turkey where the loach is endemic. He received his doctorate (1887) and became a professor (1888), joining the Zoology Department, National Museum of Hungary, Budapest (1890). He was elected to the Hungarian parliament (1901). He was in Turkey (1906) and Argentina (1909). He was Director, Budapest Zoo (1911–1919). A bird is also named after him.

Lengerich

Rosefish sp. *Helicolenus lengerichi* JR Norman, 1937

Dr Juan Lengerich was a marine biologist who was head of the Industrial Section of the Marine Biology station at the University of Chile, and a member of the Chilean Antarctic Mission. He collected the holotype.

Lenice

Bumblebee Catfish sp. *Microglanis leniceae* **OA** Shibatta, 2016

Dr Lenice Souza-Shibatta is a biologist and ichthyologist at the State University of Londrina (UEL). She is married to the author, Dr Oscar Akio Shibatta. She used to teach at UniFil Londrina. She studied at the Federal University of Mato Grosso do Sul (BSc., 1999), then at The George Washington University and at UEL, which awarded her MSc (2003) and PhD. She has published more than a dozen scientific papers such as: *Genetic Diversity of the Endangered Neotropical Cichlid Fish (Gymnogeophagus setequedas) in Brazil* (2018). Her foci are molecular biology, genetics and evolution. She was honoured for her dedication to the study of the biogeography and evolution of neotropical fishes. (Also see **Shibatta**)

Lennard

Blue-and-yellow Wrasse *Anampses lennardi* TD Scott, 1959

Fynes Barrett-Lennard (1915–2008) was a land-owner in Australia who collected, particularly across Western Australia (1950–1960), sending many herpetological and ichthyological specimens to the Western Australia Museum. He was thanked for his "*... valuable assistance in the collecting of many of these specimens, and for the most useful notes and colour photographs which he has placed at my disposal.*"

Lennon

Chacambero Splitfin *Ilyodon lennoni* MK Meyer & W. Förster, 1983

John Lennon (1940–1980) of the Beatles needs no biography here.

Leo Berg

Caspian Stellate Goby *Benthophilus leobergius* LS Berg, 1949
Snailfish sp. *Paraliparis leobergi* AP Andriashev, 1982
Azov Shemaya *Alburnus leobergi* J Freyhof & M Kottelat, 2007
Lanternbelly sp. *Acropoma leobergi* AM Prokofiev, 2018

Lev (Leo) Semionovitch (Semenovich) Berg (1876–1950). (See **Berg**)

NB - The lanternbelly is named after the man, although the type was caught by the vessel that also bears his name.

Leo Hoigne

Neotropical Rivuline sp. *Austrofundulus leohoignei* T Hrbek, DC Taphorn & JE Thomerson, 2005

Emil 'Leo' Hoigne (d.1996) discovered (1969) this species. The authors say: "*... We take great pleasure in naming this species in honor of the late Mr. Leo Hoigne who discovered this species, and many other annual killifishes in Venezuela. It was our privilege to know him and share his delight in discovering and keeping annual killifishes.*" (See **Hoigne**)

Leon

Neotropical Rivuline sp. *Austrofundulus leoni* T Hrbek, DC Taphorn & JE Thomerson, 2005

Bristlenose Catfish sp. *Ancistrus leoni* L De Souza, DC Taphorn & JW Armbruster, 2019

Oscar León Mata (1964–2018) was a Venezuelan ichthyologist. He collected the bristlenose catfish type series and, according to the authors their 'fallen colleague' "*...dedicated much of his too-short life to Venezuelan ichthyology*" and "*...was invaluable during many expeditions to Venezuela, which would not have succeeded without him. He is sorely missed by his family and friends*". The etymology of the rivuline says that the name derives from the Latin *leo* (lion) because of its relatively large size, but also "*...for the family León Mata who has been instrumental to conducting research in the Maracaibo basin*" (Venezuela).

Leonard

Hillstream Loach sp. *Pseudohomaloptera leonardi* SL Hora, 1941

George Russell Leonard (b.1909) was a member of the Game Department, Federated Malay States, who was the Superintendent of King George V National Park, Malaya (now Taman Negara, Malaysia). He made a study of the snakes in the park (before 1941). He helped Hora when he was collecting fishes at Kuala Taham, Pahang. He must have escaped after the fall of Singapore (1942), as he enlisted in the Australian Army in Melbourne (1942) and with the rank of Major, Special Operations Executive, commanded a group of commandos who were parachuted (1944) into Pahang (where he had worked before WW2) to contact Chinese guerrillas and to harass the Japanese occupation forces. He was demobilised (1946), was decorated with the MBE, and returned to Malaya as a planter in Selangor. A reptile is also named after him.

Leonidas

Characin sp. *Astyanax leonidas* M de las Azpelicueta, JR Casciotta & AE Almirón, 2002

King Leonidas of Sparta (d.480 BC) led the Greek forces at the Battle of Thermopylae against the invading Persian army, and died with his 300 Spartan troops. This epithet has an apparent political motive, as it is "*dedicated to all the academic teachers of Argentina that stand in defence of a free and independent education*".

Leontina

Lebanese Loach *Oxynoemacheilus leontinae* LCE Lortet, 1883

The binomial is not explained and may well not be an eponym. The intended meaning may perhaps be 'of a lion', referring to the loach's high, rounded head, which might fancifully be likened to that of a lion.

Leopold, A

Bavispe Sucker *Catostomus leopoldi* DJ Siebert & WL Minckley, 1986

Aldo Leopold (1887–1948) was an ecologist, author, environmentalist and biologist. He worked for the US Forest Service (1909–1933) and made extensive studies of North American wildlife. He was the first Professor of Game Management in the Agricultural Economics Department at the University of Wisconsin-Madison (1933–1948). He wrote the posthumously-published: *A Sand County Almanac* (1949). He died of a heart attack while battling a wild fire on a neighbour's property. A mammal is also named after him.

Leopold (King)

Leopold's Angelfish *Pterophyllum leopoldi* JP Gosse, 1963
[Alt. Teardrop Angelfish, Roman-nosed Angelfish]
White-blotched River Stingray *Potamotrygon leopoldi* MN Castex & HP Castelo, 1970
[Alt. Xingu River Ray, Polka-Dot Stingray]
Leopold's Tetra *Astyanax leopoldi* J Géry, P Planquette & PY Le Bail, 1988

Leopold Filips Karel Albert Meinrad Hubertus Maria Miguel, King Leopold III of Belgium (1901–1983) was King of the Belgians (1934–1951) before abdicating in favour of his son. After this he followed his passion for anthropology and entomology, travelling the world. Among other places, he visited Senegal and explored the Orinoco and Amazon with Heinrich Harrier. He spent time with the indigenous peoples of the area, and wrote: *La Fête indienne, souvenirs d'un voyage chez les Indiens du Haut-Xingu* (1967).

Leopold (Lake)

African Tetra sp. *Alestopetersius leopoldianus* GA Boulenger, 1899
Mormyrid sp. *Marcusenius leopoldianus* GA Boulenger, 1899

Lake Leopold (now Lake Mai-Ndombe), Democratic Republic of the Congo, is the holotype locality.

Lepechin

Char sp. *Salvelinus lepechini* JF Gmelin, 1789

Dr Ivan Ivanovich Lepechin (also spelled Lepyokhin) (1740–1802) was a Russian physician, naturalist and explorer. After studying at the Academy of Sciences of St Petersburg, he qualified as a physician at the University of Strasbourg. He explored the Volga region and Caspian Sea (1768), the Urals (1769–1774) and Siberia (1774–1775). He was Secretary of the Russian Academy (1783–1802) and in charge of the St Petersburg Botanical Garden (1774–1802). He was the first person to describe this species (1780) but it appears he never gave it a name.

Lépiney

Maghreb Barb sp. *Luciobarbus lepineyi* J Pellegrin, 1939
[Sometimes considered a junior synonym of *Luciobarbus callensis* or *L. pallaryi*]

Jacques Granjon de Lépiney (1896–1942) was an entomologist who was President of the French Entomological Society (1941). He visited French Sudan (1933–1934) and was Director of L'Institut scientifique chérifien, Rabat, Morocco (1938–1942). He wrote: *A study of the biological complex of Lymantria dispar* (1930). He collected the holotype. He is much more famous as a daring mountaineer and Alpinist who pioneered a number of climbs and first ascents in the Alps in the period between the wars. He was killed in a climbing accident in Moroccan High Atlas Mountains.

Lerikim

Dusty Snailfish *Careproctus lerikimae* JW Orr, Y Kai & T Nakabo, 2015

This is a composite name drawn from the letters or syllables in the forenames of Libby Logerwell, Erika Acuña and Kim Randall who are all scientists who worked for the Alaska Fisheries Science Center. They were honoured for collecting or coordinating the collection of the first representatives of the new species at sea (2008). The specific epithet is an amalgamation of the collectors' names.

Lerma

Olive Skiffia *Skiffia lermae* SE Meek, 1902

This is a toponym referring to the basin of the Rio Lerma, Mexico.

Leschenault

Characin sp. *Leporinus leschenaulti* A Valenciennes, 1850

Jean Baptiste Louis Claude Theodore Leschenault de la Tour (1773–1826) was a French botanist who served as naturalist to two Kings of France; Louis XVIII (1814–1824) and Charles X (1824–1830). He was botanist on the voyage of Casuarina, Géographe, and Naturaliste (1801–1803), and he collected in Australia (1801–1802). The town now called Bunbury in Western Australia was seen from the Casuarina (1803) and named Port Leschenault after him by French explorer Captain De Freycinet. He also collected in Java (1803–1806) and India (1816–1822) and visited the Cape Verde Islands, the Cape of Good Hope, Ceylon (Sri Lanka), Brazil, and British Guiana (Guyana). Interestingly, he wrote one of the first descriptions of coconuts and the extraction of their oil (1803). Five birds, four reptiles and a mammal are named after him.

Lesley

Mottled Twister *Bellapiscis lesleyae* GS Hardy, 1987

Dr Lesley Anne Bolton-Ritchie is a New Zealand marine biologist. She is a Senior Scientist on coastal water quality and ecology for the Canterbury Regional Council, NZ (2003–present). She wrote: *Factors influencing the water quality of Lyttelton Harbour/Whakaraupo* (2011). In honouring her, Hardy said she was a "*…marine biologist and companion on several marine surveys.*"

Leslie (Clift)

Leslie's Cardinalfish *Ostorhinchus leslie* J Schultz & JE Randall, 2006

Leslie Whaylen Clift (b.1970) is an aquatic scientist, a diver and an environmental planner. She joined the Flower Garden Banks National Marine Sanctuary, Galveston, Texas (2015). She discovered this species and realised it was undescribed, and persisted in ensuring specimens and photographs were obtained.

Leslie, R

African Gulper Shark *Centrophorus lesliei* WT White, DA Ebert & GJP Naylor, 2017

Dr Robin Leslie works for the Fisheries Branch, Department of Agriculture, Forestry and Fisheries in South Africa. According to the etymology he: "*…has contributed greatly to our* *knowledge of southern African chondrichthyans and provided numerous important specimens and tissue samples for various projects*"

Leslie Knapp

Knapp's Deepwater Flathead *Bembras leslieknappi* H Imamura, PN Psomadakis & H Thein, 2018

Dr Leslie William Knapp (1929–2017). The etymology states that the flathead is: "*Named in honor of the late Dr. Leslie W. Knapp, who contributed greatly to fish taxonomy, especially that of deepwater flatheads (Bembridae) and flatheads (Platycephalidae).*" (See **Knapp**)

Lessa

Lessa's Butterfly Ray *Gymnura lessae* L Yokota & MR de Carvalho, 2017

Dr Rosângela Paula Teixeira Lessa, a lecturer at the Federal Rural University of Pernambuco (UFRPE), is a Brazilian researcher working for the conservation of sharks and rays in her country. The Federal University of Rio Grande awarded her BSc (1977), and the Université de Bretagne Occidentale her MSc (1979) and Phd (1982). She was professor in the Department of Biology of the Federal University of Maranhão (1983–1985) and Research Fellow at CNPq from (1989–present). Among her publications are: *Occurence and biology of the daggernose shark* Isogomphodon oxyrhynchus *(Chondrichthyes: Carcharhinidae) off the Maranhao coast (Brazil)* (1999) and *Age and growth of the blue shark* Prionace glauca *(Linnaeus, 1758) off northeastern Brazil* (2004).

Lesson

Lesson's Thicklip *Plectorhinchus lessonii* G Cuvier, 1830

Oceania Fantail Ray *Taeniura lessoni* PR Last, WT White & G Naylor, 2016

René Primevère Lesson (1794–1849) was a French naturalist and surgeon. He served on Duperrey's round-the-world voyage of 'La Coquille' (1822–1825), during which he collected natural history specimens. On returning to Paris he spent seven years preparing the section on vertebrates for the official account of the expedition: *Voyage autour du monde entrepris par ordre du Gouvernement sur la corvette La Coquille* (1826 to 1839).

Lestrade

Lake Tanganyika Cichlid genus *Lestradea* M Poll, 1943

Arthur Lestrade (1897–1990) was a Belgian ethnographer, protestant missionary, colonial administrator and customs officer in Rwanda (1922–1931, 1933–1940, and 1945–1955). He refused to collaborate with the German forces occupying Belgium in WW1 and was imprisoned in Germany (1916–1917). He was on holiday in Belgium when the country was invaded in WW2 and spent the occupation there (1940–1944) at home. After retirement (1955) he was active in local affairs. He wrote: *A la rencontre du Rwanda* (1978). He collected an important series of fishes from Lake Tanganyika and sent specimens to the Royal Museum for Central Africa, Tervuren, Belgium; they included the type of this genus. A mammal is also named after him.

Lesueur

Pygmy-Sandgazer genus *Lesueurina* HW Fowler, 1908
Goby genus *Lesueurigobius* GP Whitley, 1950

Charles Alexandre Lesueur (Le Sueur) (1778–1846) was a French naturalist, artist and explorer. At 23 he set sail for Australia and Tasmania aboard *Le Géographe* as an assistant gunner. Baudin (q.v.) appointed him as an official expedition artist when the original artists jumped ship in Mauritius. During the next four years he and fellow naturalist François Péron (q.v.) collected more than 100,000 zoological specimens representing 2,500 new species, and Lesueur had made 1,500 drawings. From these drawings he produced a series of watercolours on vellum, which were published (1807–1816) in the expedition's official report, *Voyage de Découvertes aux Terres Australes*. Lesueur lived in the USA (1815–1837) and undertook some local travels and collecting. He was appointed Curator of the Natural History Museum in Le Havre (1845), which was created to house his drawings and paintings. Two birds, an amphibian, a mammal and three reptiles are also named after him. (Also see **Sueur**)

Leticia (Lucena)

Characin sp. *Charax leticiae* CAS de Lucena, 1987

Leticia Lucena is the daughter of the author, Carlos Alberto Santos De Lucena.

Leticia (Lucinda)

Poeciliid livebearer sp. *Phalloceros leticiae* PHF Lucinda, 2008

Leticia M Lucinda is the daughter of the author, Paulo Lucinda (q.v.).

Letourneux

African Jewelfish *Hemichromis letourneuxi* HE Sauvage, 1880
Corfu Toothcarp *Valencia letourneuxi* HE Sauvage, 1880

Aristide Horace Letourneux (1820–1890) was a botanist and entomologist in Algiers. He studied law and was awarded his doctorate, becoming an official in the public prosecutor's office. However, he moved to Nantes (1850) where his uncle was a magistrate after his family lost everything during the (1848) revolution. He began to collect molluscs with his uncle who was a keen natural historian. He then became a prosecutor in Bône, Algeria, where he studied Arabic and made archaeological, botanical and malacological collections. He became, among others, adviser of the court of Algiers, president of the society of climatology of Algiers, vice-president of the historical society of Algiers and member of the botanical society of France. He was co-author of: *Kabylie and Kabyle customs* (1868). He was sent to Egypt (1876) and retired (1880) before moving back to Algiers (1881). He was invited to participate in the Scientific Exploration of Tunisia (1883–1885) and wrote the reports on malacology, botany and entomology. He died following an accident while horse-riding.

Levanidov

Char sp. *Salvelinus levanidovi* IA Chereshnev, MB Skopets & PK Gudkov, 1989

Dr Vladimir Yakovlevich Levanidov (1913–1981) was a Russian ichthyologist, and hydrobiologist who was described as "*...a leading expert on and student of the biology of salmonid ecosystems of the Russian Far East.*" He was head of what is now known as the Laboratory of Salmonids Studies in the Kamchatka Research Institute of Fisheries and Oceanography (1962–1971). He moved to Vladivostok and founded the Laboratory of Freshwater Hydrobiology (1972), where he worked until his death.

Leveque

Squeaker Catfish sp. *Synodontis levequei* D Paugy, 1987
Claroteid Catfish sp. *Chrysichthys levequei* LM Risch, 1988
African Cyprinid sp. *Raiamas levequei* GJ Howes & G Teugels, 1989
Churchill (mormyrid) sp. *Petrocephalus levequei* R Bigorne & D Paugy, 1990

Dr Christian Lévêque is a French ichthyologist and hydrobiologist who is Professor Emeritus at IRD (formerly ORSTOM). The University of Lille awarded his bachelor's degree (1962) and the University of Paris VI his doctorate (1972). He worked at the ORSTOM centre at Port Lamy, Lake Chad (1965–1970). He was head of hydrobiology at the ORSTOM laboratory, Bouake, Ivory Coast and was Project Manager of the Interdisciplinary Research Programme on the Environment (PIREN) in charge of inland waters. He started a programme of research into the freshwater fishes of West Africa. He was Deputy Director of the 'Water Continental' department of ORSTOM (1987–1993), Permanent delegate of Environment ORSTOM (1993–1996) and Deputy Director of the Directorate of Strategy and Programming (ORSTOM) (1996–1998). He was National Biodiversity Programme Manager (1996–1999) and Deputy Scientific Director, Institute of Sciences of the Universe (1998–2003). After officially retiring (2006) he has undertaken a number of other responsibilities, including being (since 2007) Chairman of the Water Group of the Academy of Agriculture. Among his publications is the co-written: *State of health of aquatic ecosystems. Biological variables as indicators* (1997).

Lever

Lever's Goby *Redigobius leveri* HW Fowler, 1943

Robert John Aylwin Wallace Lever (1905–1969) was Government Entomologist on Fiji (1937). He studied at the University of Durham, then trained in tropical entomology at Imperial College London, following which he spent a year at Imperial College in the West Indies. He was appointed (1930) Government Entomologist to the British Solomon Islands Protectorate, during which he made entomological and anthropological collections for the BMNH. He was then posted to Fiji (1937) and Malaya. Lever collected the holotype in Fiji (January 1943). An odonata is also named after him.

Leverett

Butterfly Hillstream Loach *Beaufortia leveretti* JT Nichols & CH Pope, 1927

Rev William J Leverett was an American Presbyterian missionary at Nodoa, Hainan, China (1894–1924). Yale awarded his bachelor's degree (1891) and he then studied at Auburn Theological Seminary and was ordained as a priest (1894). A skink is also named after him in recognition of his invaluable aid to the work of the expedition of the AMNH (1923).

Levinson

Hourglass Basslet *Lipogramma levinsoni* CC Baldwin, A Nonaka & DR Robertson, 2016

Dr Frank Levinson is a fibre-optics entrepreneur with a PhD in astronomy from the University of Virginia (1980). He then worked for Bell Systems for nearly four years, following which he started his own company (1984). He moved on to work for another company before finally (1988) starting up the company (Finisar) which made his fortune, taking it public ten years later. He was honoured "*…for his generous, continuing support of research on neo-tropical biology at the Smithsonian Tropical Research Institute (Panama), where the third author works.*"

Leviton

Sand-diver sp. *Pteropsaron levitoni* T Iwamoto, 2014

Dr Alan Edward Leviton (b.1930) is Curator Emeritus of the California Academy of Sciences. He took his bachelor's (1949), master's (1953), and doctorate (1960) at Stanford. He became Curator, Department of Herpetology, California Academy of Sciences (1957). He lectured in biology at Stanford (1962–1970). He was (1969) Adjunct Professor of Biological Sciences, San Francisco State University. His specialty is the herpetofauna of Asia and the Arabian Peninsula. He was honoured in the name because he "*…strongly promoted this Hearst Expedition volume, used his considerable technical knowledge of desk-top publishing to produce this work, and who provided much advice and support with this paper.*" Ten reptiles are also named after him.

Lew

Pencil Catfish sp. *Trichomycterus lewi* CA Lasso-Alcalá & F Provenzano Rizzi, 2003

Daniel Lew is a zoologist and ecologist who was with Fundacion LaSalle (2006) and is now at the Venezuelan Institute for Scientific Research, Caracas. He took part in the expedition that collected the holotype and was honoured for "*…contributions to the knowledge and conservation of biodiversity in the Guyana Shield of Venezuela.*" He co-wrote: *Fauna de Venezuela* (1997).

Lewin

Australian Longfin Pike *Dinolestes lewini* E Griffith & CH Smith, 1834
Scalloped Hammerhead *Sphyrna lewini* E Griffith & CH Smith, 1834

John William Lewin (1770–1819) was an English naturalist and engraver. He went to Sydney, Australia (1800) and collected widely there until his death. His father (some sources say his older brother), William Lewin, was the author of a seven-volume work: *Birds of Great Britain.* Lewin accompanied James Grant on his survey expeditions to the Bass Strait and then to the Hunter River. In Sydney, he earned a meagre living as a portrait artist. Governor Macquarie appointed Lewin to the position of City Coroner (1814). He also accompanied Macquarie and made drawings during the construction of the road across the Blue Mountains. Macquarie commissioned Lewin to draw plants collected by the Surveyor-

General, Henry Oxley, when exploring the country beyond Bathurst, the Liverpool Plains and New England District. As well as natural history, Lewin also painted landscapes and portraits of Aboriginals. He wrote: *Prodromus Entomology, Natural History of Lepidopterous Insects of New South Wales* (1805) and *Birds of New Holland* (1808) the 1813 edition of which was the first illustrated book to be engraved and printed in Australia. Five birds are named after him.

NB – there is no etymology for the hammerhead in the original text, and José Castro in *The Sharks of North America* (2011) suggests the species was named after Danish military surgeon and anatomist Ludwig Lewin Jacobson (1783–1843). However, no supporting evidence is provided for this opinion.

Lewis, AD

Lewis' Round Herring *Spratelloides lewisi* T Wongratana, 1983
Slender Pinjalo *Pinjalo lewisi* JE Randall, GR Allen & WD Anderson, 1987
Lewis' Wriggler *Rotuma lewisi* VG Springer, 1988

Dr Anthony David Lewis (b.1948) is an Australian fisheries biologist and independent fisheries adviser who has joined the International Pole & Line Federation as a trustee (2014). The University of Queensland awarded his bachelor's degree (1968) and 1st class honours (1970) and the Australian National University his doctorate (1981). He has worked in most of the western Pacific Island countries including Papua New Guinea (1971–1977 & 2002–2003), Fiji (1981–1987) and New Caledonia (1977–1978 & 1988–2002), and was based in Noumea as Manager of the Pacific Community Oceanic Fisheries Programme. Since 2002 most of his work has been in Indonesia, the Philippines and Vietnam.

Lewis, M

Westslope Cutthroat Trout *Oncorhynchus clarkii lewisi* CF Girard, 1856

Captain Meriwether Lewis (1774–1809) was one half of the 'Lewis and Clark' duo, whose famous expedition collected six specimens of this trout on the Missouri River, Montana. Lewis was chosen to lead the expedition by President Thomas Jefferson; he was the latter's private secretary at the time. Jefferson wrote: 'It was impossible to find a character who, to a complete science in botany, natural history, mineralogy & astronomy, joined the firmness of constitution & character, prudence, habits adapted to the woods, & a familiarity with the Indian manners & character, requisite for this undertaking. All the latter qualifications Capt. Lewis has.' Before becoming Jefferson's secretary, he grew up in the country, managed the family plantation, and spent time in the army as an ordinary soldier and then an officer. Lewis chose Clark (q.v.), a friend he had made whilst in the army, to accompany him. The famous journey of exploration took a year and a half and covered more than 4,000 miles to the Pacific Ocean. Lewis was fascinated with the Native Americans, plants, animals, fossils, geological formations, topography and other facets of the trip, all of which he recorded in his journal entries. As a reward for his success he was appointed to the Governorship of the Louisiana Territory. Lewis began a journey to Washington (1809) to clear his name, having been publicly accused of misusing public money, but in a Tennessee inn he met his death from two gunshot wounds to the head and chest, and it is still not known whether this was murder or suicide. As he was bipolar, suicide appears more likely, although his family believed it was murder.

Lex

Armoured Catfish sp. *Hypostomus lexi* R Ihering, 1911

Belisária Fausto Lex (1878–1950) was a teacher, a geologist and a biologist. He graduated from the State Gymnasium, Sãp Paulo (1902) and taught in a local school and worked at the Natural History section, Museu Paulista, São Paulo, Brazil. He taught in a very isolated school at Barretos (1907–1915). His career as a teacher continued until he transferred to PIracicaba (1932) and retired (1937) as Director of the Normal School. Over the years he presented many specimens to the Museu Paulista including the holotype of this catfish.

Lexa

Armoured Catfish sp. *Chaetostoma lexa* NJ Salcedo-Maútua, 2013
[Syn. *Loraxichthys lexa*]

Alexandra 'Lex' Keane was described as a sustainability activist. She was a Political Science student at the College of Charleston, South Carolina, where Salcedo taught at the time. She graduated (2014) and is now Restorative Agriculture Director at the College of Charleston.

Leyland

Leyland's Skate *Dentiraja healdi* PR Last, WT White & JJ Pogonoski, 2008
[Alt. Heald's Skate]
Painted Maskray *Neotrygon leylandi* PR Last, 1987

Guy Geoffrey Leyland (b.1950) is Principal Executive Officer, Western Australia Fishing Industry Council. His BSc (Zoology) was awarded by the University of Western Australia. He supplied the majority of the Australian material used by Last in his revision of the genus. He co-wrote: *Continental shelf fishes of northern and north-western Australia: an illustrated guide* (1985) and contributed to *Trawled Fishes of Southern Indonesia and Northwestern Australia* (1984) whilst he was working for CSIRO, Division of Fisheries Research. His work with the fishing industry in Western Australia has focused on shoring up the statutory basis for rights-based fisheries management, including quota-based management and statutory compensation for fishers displaced from marine reserves. He is currently the fishing industry project leader for facilitating third party environmental certification of Western Australian fisheries through the London-based Marine Stewardship Council, which provides independent validation of the performance of fisheries.

Li

Chinese Gudgeon sp. *Gobiobotia lii* X Chen, M Wang, L Cao & E Zhang, 2022

Shi-Zhen Li, or Li Shizhen (1518-1593), was a Chinese naturalist, physician and herbalist most famous for his work: Compendium of Materia Medica. He was born in Qichun County, Hubei, where the gudgeon holotype was caught.

Liana

Pencil Catfish sp. *Copionodon lianae* CM Campanario & MCC de Pinna, 2000

Dr Liana de Figueiredo Mendes is a Brazilian zoologist and ichthyologist. The University of São Paulo awarded all three of her degrees, bachelor's (1990), master's (1996) and doctorate

(2000). She is currently an associate professor in the Laboratório do Oceano, Depto. de Botânica, Ecologia e Zoologia Universidade Federal do Rio Grande do Norte. She co-wrote: *Fish fauna of Pratagi River coastal microbasin, extreme north Atlantic Forest, Rio Grande do Norte state, northeastern Brazil. Check List* (2014). She collected the first known specimens of this species and brought them to the authors' attention.

Liban

Levantine Minnow *Pseudophoxinus libani* L Lortet, 1883

This is a toponym, with *liban* meaning Lebanon, where the species is endemic.

Lichtenstein

Silverside sp. *Atherina lichtensteinii* A Valenciennes, 1835 NCR
[Junior Synonym of *Odontesthes bonariensis*]
Lichtenstein's Seahorse *Hippocampus lichtensteinii* JJ Kaup, 1856

Martin Heinrich Carl Lichtenstein (1780–1857) was a physician, naturalist and traveler, who become head of the Museum für Naturkunde Berlin (1813) and founded the Berlin Zoo (1844). He traveled in South Africa (1802–1806), and while there became personal physician to the Dutch Governor of the Cape of Good Hope. Lichtenstein studied many species sent to the Berlin Museum by others, and "*...while he gave every species, or what he judged to be a species, a name, this was done without consulting the recent English and French literature. His only aim was to give the specimens in question a distinguishing mark for his personal needs. These names were used in Lichtenstein's registers and reappeared on the labels of the mounted specimens, but only exceptionally were they published by himself in connection with a scientific description.*" This caused much unnecessary confusion and trouble to others. He died at sea off Kiel – not of illness, as is sometimes reported, but rather when he fought a duel and came out second best. He wrote: *Reisen in Sudlichen Africa* (1810). Nine birds, five reptiles and two mammals are named after him.

Lidwill

Dwarf Tiger Goby *Pandaka lidwilli* AR McCulloch, 1917

Dr Mark Cowley Lidwill (1878–1969) was an English-born anaesthesiologist, cardiologist and co-inventor of the cardiac pacemaker, first successfully used (1928). He had emigrated to Australia (1894) and graduated as a physician at Melbourne University (1905). He was an avid rod-and-line fisherman and was the first known person to catch a Black Marlin *Tetrapterus indicus.* This particular specimen is on permanent display at the Australian Museum, Sydney. He discovered this species of goby.

Liebrechts

African Tetra sp. *Alestes liebrechtsii* GA Boulenger, 1898

Charles Liebrechts (1858–1938) was a Belgian army officer and colonial administrator, whose first mission was to try and sell Belgian-made cannons in the Congo! He eventually became Secretary-General of Home and Military Affairs of the Congo Free State (1900–1908).

Liechtenstein, JMFP

Liechtenstein's Goby *Corcyrogobius liechtensteini* C Kolombatovi, 1891

Johann Maria Franz Placidus (1840–1929), Prince Johann II of Liechtenstein (1858–1929) was a prominent patron of the arts and sciences during his long reign (70 years). The etymology says he was honoured for his support of the natural sciences.

Liem, D

Sulawesi Halfbeak sp. *Nomorhamphus liemi* D Vogt, 1978

Dr Dig Liem was, according to the description, a long-time friend of the author in Indonesia and we presume him to be an aquarist and collector. (We cannot discount the possibility that 'Dig' is a nickname for Dr K F Liem [below]).

Liem, KF

Lake Malawi Cichlid sp. *Caprichromis liemi* KR McKaye & CA MacKenzie, 1982
Barbeled Dragonfish sp. *Photostomias liemi* CP Kenaley, 2009

Dr Karel Frederik Liem (1935–2009), was Curator of Ichthyology and Henry Bryant Bigelow Professor of Ichthyology at the Museum of Comparative Zoology, Harvard University (1972–2009). The University of Indonesia awarded his bachelor's degree and his master's (1958) and the University of Illinois his doctorate (1961). Before moving to Harvard, he taught at Leiden University, the Netherlands and at the University of Illinois and also held a curator's position at the Field Museum, Chicago. He co-wrote: *Life: An Introduction to Biology* (1991). The cichlid was named after him for his pioneering studies and insight into the feeding mechanism of cichlids.

Lifalili

Blood-red Jewel Cichlid *Hemichromis lifalili* PV Loiselle, 1979

The binomial is derived from one of the fish's vernacular names near Lake Tumba, Democratic Republic of Congo; the type locality.

Light

Tonguefish sp. *Cynoglossus lighti* JR Norman, 1925
Common Frog Flounder *Pleuronichthys lighti* HW Wu, 1929
Light's Bitterling *Rhodeus lighti* HW Wu, 1931
Rockskipper Blenny sp. *Entomacrodus lighti* AWCT Herre, 1938

Dr Sol Felty Light (1886–1947) was an American zoologist who taught at the University of California, Berkeley. Park College, Missouri awarded his bachelor's degree (1908), after which he taught in schools in Japan (1908–1910) and Manila (1910–1912). He then taught at the University of the Philippines (1912–1922), where he became a Full Professor. He was Chairman of the Department of Zoology at the University of Amoy (1922–1924). He returned to the USA (1924) to do a doctorate (awarded 1926) at the University of California, where he remained until his death. Wu was his student at the University of Amoy. He wrote: *Isoptera of Guam* (1946). An amphibian is also named after him.

Lika

Lika Minnow *Phoxinus likai* PG Bianco & S De Bonis, 2015

This is a toponym, referring to the Lika-Dinaric karstic region of Croatia.

Likoma

Lake Malawi Cichlid sp. *Copadichromis likomae* TD Iles, 1960

This is a toponym referring to the type locality; Likoma Island, Lake Malawi.

Lilia

Armoured Catfish sp. *Corumbataia liliai* GSC Silva, FF Roxo, CS Souza & C Oliveira, 2018

Lilian Maria Costa e Silva is the sister of the senior author, Gabriel Silva of the Department of Morphology at Instituto de Biociências, Universidade Estadual Paulista.

Lilith

Eartheater sp. *Satanoperca lilith* SO Kullander & EJG Ferreira, 1988

Lilith was a nocturnal female demon in Babylonian and Jewish folklore. The use of her name continued a tradition of naming these cichlids after demons (e.g. *S. daemon*), in keeping with the Tupi name for these fish – 'jurupari' (also meaning a malignant forest spirit).

Lillie

Crested Bellowsfish *Notopogon lilliei* CT Regan, 1914

Dr Dennis Gascoigne Lillie (1884–1963) was a Cambridge-educated marine biologist and polar explorer. He was on Captain Scott's Terra Nova Antarctic Expedition (1910–1913). His observational studies on birds and whales, and his caricature sketches of fellow members of the expedition, were published and much sought after. He was awarded the Polar Medal along with fellow expedition survivors. A conscientious objector, he later served as a military bacteriologist (WW1). He suffered periods of mental ill-health (1918–1921 & 1921–1963) briefly interrupted by a few months when he lectured at Cambridge. He never fully recovered.

Lilljeborg

Norway Bullhead *Micrenophrys lilljeborgii* R Collett, 1875

Wilhelm Lilljeborg (1816–1908) was a Swedish zoologist who became Professor of Zoology at Uppsala University and a member of the Swedish Academy of Science (1861). He is known for his work on water fleas and whales. The original contains no proper etymology, but this man seems a very likely candidate.

Lim

Loach genus *Kottelatlimia* TT Nalbant, 1994

Glassfish sp. *Gymnochanda limi* M Kottelat, 1995

Kelvin Kok Peng Lim is collections manager for herpetology and ichthyology at Raffles Museum of Biodiversity Research, Department of Biological Sciences, National University of Singapore. He co-wrote: *A guide to the common marine fishes of Singapore* (1998). A reptile is also named after him. The loach genus *Kottelatlimia* is named in honour of both Kim and Kottelat (q.v.) as they described the 'extremely interesting' type species, *K. katik*.

Lima

Freshwater Stingray sp. *Potamotrygon limai* JP Fontenelle, JPCB da Silva & MR de Carvalho, 2014

Dr José Lima de Figueiredo (b.1943) aka 'Zé Lima' is a Brazilian ichthyologist who is a former researcher and Curator of Fishes at the Zoological Museum of the University of São Paulo, He is also a professor at the University of São Paulo, which had awarded his bachelor's degree (1969) and his doctorate (1981). He co-wrote: *The northernmost record of Bassanago albescens and comments on the occurrence of Rhynchoconger guppyi (Teleostei: Anguilliformes: Congridae) along the Brazilian coast* (2011). (See **Figueiredo**)

Limbaugh

Clipperton Angelfish *Holacanthus limbaughi* WJ Baldwin, 1963
Yellowface Pike-Blenny *Chaenopsis limbaughi* CR Robins & JE Randall, 1965
Limbaugh's Damselfish *Chromis limbaughi* DW Greenfield & LP Woods, 1980
Wide-banded Cleaner Goby *Tigrigobius limbaughi* DF Hoese & S Reader, 2001
[Syn. *Elacatinus limbaughi*]
Canyon Sculpin *Icelinus limbaughi* RH Rosenblatt & WL Smith, 2004

Conrad 'Connie' Limbaugh (1925–1960) was a zoologist, diver and underwater photographer. He was honoured for his pioneering (1961) work on the symbiotic cleaning behaviour of marine organisms (like the goby). He photographed (1954) the sculpin underwater, then collected the first two specimens, though the species was not formally described for another 50 years. Tragically, he died after losing his way while diving in the labyrinth of an underground river in France. (NB – The goby authors give his forename as 'Clyde')

Lim Boon Keng

Amoy Fanray *Platyrhina limboonkengi* DS Tang, 1933
[Junior Synonym of *Platyrhina sinensis*]

Dr Lim Boon Keng (1869–1957) was a Singaporean Chinese physician who, in addition to being very interested in natural history, promoted social and educational reforms in Singapore in the early part of the 20th century. During the Japanese occupation of Singapore in WW2 he was expected to co-operate, but developed the defence mechanism of pretending to be in a drunken stupor to avoid it. An area of Singapore and a station on the MRT system there are named Boon Keng after him.

Limmon

Sulawesi Goby sp. *Schismatogobius limmoni* P Keith & N Hubert, 2021

Gino Valentino Limmon is a marine biotechnologist. The University of Hasanuddin awarded his BSc (1990), McMaster University, Ontario, Canada awarded his MSc (1996) and

Friedrich-Schiller University, Jena, Germany, his PhD in molecular biology (2003). He was a Postdoctoral Researcher at the National Institute of Environmental Health Sciences, USA (2004–2008) after which he worked (2008–2012) as a research scientist at the Singapore-MIT Alliance for Research and Technology Centre. He then became (2013) Director of the Marine Science Centre of Excellence, Pattimura University, Maluku, Indonesia. His masters' thesis was *An Assessment of Coral Reefs in Ambon* (1996) and he has written or co-written c.36 papers such as: *Assessing species diversity of Coral Triangle artisanal fisheries: A DNA barcode reference library for the shore fishes retailed at Ambon harbor (Indonesia)* (2020).

Lin, R-D

Chinese Barb genus *Linichthys* E Zhang & F Fang, 2005

Lin Ren-Duan was an ichthyologist at the Institute of Hydrobiology, Chinese Academy of Sciences, Wuhan. He was co-describer (1986) of *Linichthys laticeps* under the name *Barbodes laticeps.*

Lin, S-Y

Hillstream Loach genus *Liniparhomaloptera* PW Fang, 1935
Barb sp. *Osteochilus lini* HW Fowler, 1935
Chinese Cyprinid sp. *Onychostoma lini* HW Wu, 1939
Chinese Cyprinid sp. *Ancherythroculter lini* Y-L Luo, 1994
Garnet Minnow *Aphyocypris lini* SH Weitzman & LL Chan, 1966

Lin Shu-Yen (1903–1974) was a Chinese ichthyologist who was Director of the Fisheries Experiment Station, Canton (Guangzhou), China in the 1930s. He moved to Taiwan after the Communist revolution and was (1964) Professor of Fish Culture, National Taiwan University, Taipei. He wrote: *Three new fresh-water fishes of Kwangtung province* (1934). In the name of the loach genus, 'Lin' is attached to *Parhomaloptera*, a related genus described in 1902.

Linam

LInam's Pimelodella (catfish) *Pimelodella linami* Schultz, 1944

Henry Edward Linam (1899–1972) went to Venezuela (1925) to work for Standard Oil Co. of Venezuela and was its manager and president (1932–1943), after which he became an independent oil operator. He was in charge of operations for the Rubber Development Corporation (1942–1943) as part of the US Department of Commerce. When Schultz went to Venezuela, Linam invited him to stay at their camps in the Maracaibo Basin.

Linda

Yellowtail Dottyback *Pseudochromis linda* JE Randall & BE Stanaland, 1989

Linda J McCarthy (b.1952) is an ichthyologist. She was honoured as she was the first to determine that this species is distinct from *P. olivaceus* of the Red Sea. She also collected many of the type series. Interestingly, she co-wrote (with Randall): *Solea stanalandi, a new sole from the Persian Gulf* (1988), naming a species after Brock E Stanaland. With Randall, Stanaland and others she wrote: *Fifty-One New Records of Fishes From The Arabian Gulf* (1994).

Linda(s)

Blackedge Pearlfish *Pyramodon lindas* DF Markle & JE Olney, 1990

The etymology says that the binomial is in honour of "*…two mature women who supported and endured this project*." It is not stated, but we presume that the two women, both named Linda, are the authors' partners.

Lindalva

Ghost Knifefish sp. *Apteronotus lindalvae* CDCM de Santana & C Cox Fernandes, 2012

Lindalva Sales da Costa Serrão has been contributing to the organisation of INPA's (q.v.) fish collection for more than 20 years.

Lindberg, GU

Notothen genus *Lindbergichthys* AV Balushkin, 1979
Lumpfish sp. *Cyclopteropsis lindbergi* VK Soldatov, 1930
Goby sp. *Rhinogobius lindbergi* LS Berg, 1933
Lindberg's Dace *Leuciscus lindbergi* AP Zanin & G Eremejev, 1934
Snailfish sp. *Genioliparis lindbergi* AP Andriashev & AV Neyelov, 1976
Commander Skate *Bathyraja lindbergi* R Ishiyama & H Ishihara, 1977
Eelpout sp. *Hadropogonichthys lindbergi* VV Fedorov, 1982

Dr Georgii Ustinovich Lindberg (1894–1976) was a Russian ichthyologist and marine biologist who was a lecturer at Chita University and who was invited (1923) to the Far-Eastern University in Vladivostok where he established a coastal biological station (1924). He started work at the Zoological Institute, Russian Academy of Science, Leningrad (St Petersburg) (1932), which awarded his doctorate (1938). Among his many publications and descriptions of new species he wrote: *Description of a new species Bothragonus occidentalis (Agonidae Pisces) from the Sea of Japan. Izvestia Akademii nauk Soiuza Sotsialisticheskikh Reespublik* (1935). He was honoured in the name of the skate for his 'great work' on western North Pacific zoogeography.

Lindberg, KC

Stone Loach sp. *Paraschistura lindbergi* PM Bănărescu & MR Mirza, 1965

Dr Knut Carl Lindberg (1892–1962) was a Swedish physician and zoologist. He qualified as a physician at the Sorbonne in Paris and became (1924–1926) medical officer with the American Presbyterian Mission in Allahabad, Uttar Pradesh, India. He then took further qualifications in London and worked as Chief Medical Officer with the Barsi Light Railways at Kurdvadi near Bombay (Mumbai) (1927–1947). After returning to Sweden (1947) he concentrated on zoology and botany and went on several expeditions (1947–1962) to Portugal, Greece, Turkey, Iran and Afghanistan. He was killed in a road accident near Lund, Sweden. He collected the holotype of the eponymous species.

Linde

Tetra sp. *Axelrodia lindeae* J Géry, 1973

Mrs Linde Geisler collected the holotype with her husband, German biologist and aquarist Rolf Geisler (q.v.).

Lindman

Lindman's Grenadier Anchovy *Coilia lindmani* P Bleeker, 1857

Colonel Lazarus Lindman (1814–1877) was a military health officer in the Dutch East Indies (Indonesia). He collected the type in Sumatra.

Lindsay

Boxfish sp. *Paracanthostracion lindsayi* WJ Phillipps, 1932
[Perhaps a Junior Synonym of *Lactoria diaphana*]

Charles John Lindsay (1902–1966) was Phillipps' colleague at the Dominion Museum, Wellington, New Zealand. He was taxidermist and Curator of Technology at the museum for more than 40 years. He was also an avid collector, and authority on, ancient firearms, with specialised interest in hand weapons, armour (and typewriters!). He helped in the preservation of many historical pioneer mechanical inventions. He photographed the type specimen.

Link, E

Yellow-lined Basslet *Gramma linki* WA Starck & PL Colin, 1978

Edwin Albert Link (1904–1981) was a pioneer in aviation (he took his first lesson in 1920), underwater archaeology and submersibles. He was the developer of the diver lock-out submarine *Deep Diver*, which collected the type. The etymology says that his *...imaginative developments in undersea technology and generous support of marine science have made untouched realms of the sea accessible.*" He is best known for having invented the flight simulator, which he took commercial (1929) and later sold (1954), among twenty-seven patents for aeronautics, navigation and oceanographic equipment. He died in his sleep while undergoing treatment for cancer.

Link, F

Link's Goby *Amblygobius linki* AW Herre, 1927

Captain Francis Link was a Police Officer and long-time resident of Jolo, in the Philippines. He accompanied Philip Taylor on a number of herpetological collecting trips. He was honoured for his "*...indefatigable labours in advancing our knowledge of the Sulu Archipelago, and its fauna, flora, and people.*" Among his achievements was creating (1920s) a dictionary of Tausug, the language used in Sulu, and later (1931) translating the New Testament into that language.

Linke

Cameroon Cichlid sp. *Chromidotilapia linkei* W Staeck, 1980
Dwarf Cichlid sp. *Apistogramma linkei* I Koslowski, 1985
Moonspot Licorice Gourami *Parosphromenus linkei* M Kottelat, 1991

Horst Linke is a German aquarist, photographer and author who has travelled widely collecting fish. He has written a number of books on the identification, care and breeding of various aquarium fish, such as: *Cichlids From West Africa* (1995) and *Labyrinthfische* (2017) as well as collaborating on scientific papers such as: *Two new species of the genus* Parosphromenus *(Teleostei: Osphronemidae) from Sumatra* (2012). He collected both cichlid types.

Linn

Elephant-nose Cichlid *Nimbochromis linni* WE Burgess & HR Axelrod, 1974

Dr DeVon Wayne Linn was Chief Fisheries Officer, Malawi (1973–1975) while serving in the US Peace Corps. He was honoured as his assistance to the junior author made Axelrod's field trip to Lake Malawi possible.

Linnell, George

Thorny Catfish sp. *Leptodoras linnelli* CH Eigenmann, 1912

George Linnell was an executive of the Essequibo Exploring Company. He provided Eigenmann with boats and crewmen for his expedition to British Guiana (Guyana).

Linnell, Gunnar

Blackfin Tilapia *Sarotherodon linnellii* E Lönnberg, 1903

Gunnar Linnell was a Swedish plantation owner in Cameroon, who was Lönnberg's friend. He sent a collection of fishes and crustaceans from that country, including the type of this cichlid.

Lins

Blind Cusk Eel sp. *Barathronus linsi* JG Nielsen, MM Mincarone & F di Dario, 2015

Dr Jorge Eduardo Lins de Oliveira is an oceanographer, marine biologist and limnologist. He is head of the biology laboratory and full professor at Universidade Federal do Rio Grande do Norte, Natal, Rio Grande do Norte. He joined the faculty (1988) and was responsible (1993) for managing the Fisheries Biology Laboratory, and (since 2004) also by the Scientific Diving Laboratory. The Federal University of Rio Grande do Norte awarded his bachelor's degree in Marine Biology (1980) and Université Pierre et Marie Curie, Paris, his doctorate (1991). He also has a SCUBA instructor qualification.

LIPI

Filamented Pygmy Sand-Goby *Grallenia lipi* K Shibukawa & A Iwata, 2007

This is an acronym of **L**embaga **I**lmu **P**engetahuan **I**ndonesia (Indonesian Institute of Science), referring to the fact that all examined specimens were captured during the JSPS (Japan Society for the Promotion of Science)-LIPI cooperative research on marine science at Ambon Island (1999).

Lippe

West African Spadefish *Chaetodipterus lippei* F Steindachner, 1895

Dr Lippe collected a specimen of this fish in Freetown (Sierra Leone) during a voyage of the S M 'Helgoland'. We know nothing further about him.

Lisa

Naked Silverside *Atherinella lisa* SE Meek, 1904

This is not an eponym but is based on a local name for the fish in Mexico.

Lissner

Jordan Bream *Acanthobrama lissneri* E Tortonese, 1952

Dr Helmut Lissner (1895–1951) was a Polish-born Israeli ichthyologist. The University of Leipzig awarded his doctorate (1922). He was the first Director of the Haifa Sea Fisheries Research Station. He studied the fishes of Lake Tiberias (or Sea of Galilee), Israel, which is the type locality. He wrote: *Sardine fishing in Israel (1949)*

Litvinov

Smalleye Lantern Shark *Etmopterus litvinovi* NV Parin & AN Kotlyar, 1990
Manefish sp. *Caristius litvinovi* EI Kukuev, NV Parin & IA Trunov, 2013
Dragonfish sp. *Photonectes litvinovi* AM Prokofiev, 2014

Dr Feodor Fedorovich Litvinov (1954–2011) was a Russian ichthyologist and fisheries expert at the Atlantic Scientific Research Institute of Marine Fisheries and Oceanography, Kaliningrad, Russia, where he was Senior Scientist. He contributed to longer works and wrote a number of papers such as: *Ecological characteristics of the spiny dogfish* Squalus mitsukurii *from Nazca and Sala-y-Gomez submarine ridges* (1990).

Litz

Neotropical Rivuline sp. *Austrolebias litzi* WJEM Costa, 2006

Thomas Otto Litz (b.1965) is a German aquarist and amateur ichthyologist with a particular interest and expertise in the freshwater fish of Argentina and Uruguay. He studied chemistry and pharmacy and obtained his PhD in pharmaceutical chemistry (1995). He works in the pharmaceutical industry. Starting as an aquarist at age six, and being interested in South American killifish in the early 1980s, he became a member of the German Killifish Assiciation (DKG); later being elected board member in different capacities. He travelled several times to Argentina, Brazil and Uruguay to study the flora and fauna of these countries and to meet the local people. Also interested in literature since his childhood, Litz has been compiling a library, both for aquarium and ichthyological literature, and specialising in killifishes and Neotropical fish.

Liu, C-C

Chinese Cyprinid sp. *Anabarilius liui* HW Chang, 1944
Hillstream Loach sp. *Beaufortia liui* HW Chang, 1944

Liu Ch'eng-Chao (1900–1976) was a Chinese herpetologist who graduated at Peking University (1929) and then taught at the University of the North-East in Shenyang. Its libraries were destroyed during the Sino-Japanese war (1931). He studied at Cornell University (1932–1934) and taught at the University of Suzhou (1935–1939) and at Chengdu in Sichuan (1939–1950), then becoming Head of the Biology Department, University of Yenching (1950–1951). He finally returned to Chengdu (1951) as Director of the Medical School. He led the expedition that collected *Beaufortia liui.* He wrote: *Amphibians of Western China* (1960). Six amphibians and two reptiles are named after him.

Liu, J-K

Asian Perch sp. *Coreoperca liui* L Cao & X-F Liang, 2013

Liu Jian-Kang (1917–>2016) was a Chinese ichthyologist and ecologist who was a pioneer of sexual determination in vertebrates. He graduated from the Biology Department of Soochow University (1938) and accepted an offer from the Institute of Zoology and Botany of Academia Sinica, where he began his innovative studies on fish biology directed by ichthyologist Dr. Xian-Wen Wu. He published a total of 19 research articles (1939–1945) then continued research at McGill University, Canada, which awarded his PhD (1947). He worked in the USA (1947–1949) before returning to Shanghai where he was offered a faculty position at the current Institute of Hydrobiology, Chinese Academy of Sciences, rising to Director and later being elected to the Chinese Academy of Science (1981).

Livingstone, C

Livingstone's Bulldog Mormyrid *Marcusenius livingstonii* GA Boulenger, 1899

Rev Charles Livingstone (1821–1873) was the younger brother of David Livingstone, (below). Charles accompanied David, as his mission's Secretary, on the second Zambezi expedition, which sailed from Liverpool on 'HMS Pearl' (March 1852) to the mouth of the river. The party, which included Dr (afterwards Sir) John Kirk, ascended the river in a steam launch, the *MaRobert*, reaching Tete in September. Charles Livingstone and Kirk both became ill and returned to England (1863). The Livingstones were co-authors of: *Narrative of an Expedition to the Zambesi and Its Tributaries* (1866). Two birds are also named after him. He collected the mormyrid holotype.

Livingstone, D

Livingstone's Cichlid *Nimbochromis livingstonii* A Günther, 1894
Lake Malawi Cichlid sp. *Pseudotropheus livingstonii* GA Boulenger, 1899
[Syn. *Maylandia livingstonii*]

David Livingstone (1813–1873) was a Scottish doctor and missionary, and undoubtedly the most famous African explorer of all time. Livingstone is remembered as the first European to have gone into the heart of Africa, and as someone who came to be regarded as a saint in his own lifetime. He worked in a cotton mill from the age of ten, earning extra income by selling tea from farm to farm. He studied Latin and Greek on his own, and elected to become a missionary when he was persuaded that science and theology were not in opposition. He trained at the London Missionary Society and, in medicine, in Glasgow. Livingstone left for South Africa in 1840. His many expeditions brought him fame as a surgeon and scientist over the next few years, but his missionary efforts were less successful. He sympathised with the lot of the indigenous peoples and so made enemies among white settlers. It annoyed some of them that he learned the languages and tribal customs of the peoples he tried to convert. Nevertheless, his indictment of the slave trade did much to make anti-slavery laws enforced. In 1853 his expedition into the interior of the continent lasted three years, during which he discovered the Victoria Falls; a find which sealed his fame on his return to Britain (1856). His last expedition (1866) was to search for the source of the Nile. False reports of his death, and the public's desire to know where the lost explorer was, led to Stanley's equally famous mission to find him. He collected the *Pseudotropheus* cichlid during his Zambezi expedition. Five mammals are also named after him.

Llanero

Armoured Catfish sp. *Lamontichthys llanero* DC Taphorn Baechle & CG Lilyestrom, 1984

The Llanero are the people who occupy the plains in the region of the Apure River drainage in Venezuela, where this catfish is found.

Lloris

Grotto Goby sp. *Speleogobius llorisi* M Kovačić, F Ordines & UK Schliewen, 2016

Dr Domènec Lloris I Samo is a Spanish ichthyologist whose bachelor's degree (1977) and doctorate in biology were awarded by the University of Barcelona (1984). He is retired from the Institute of Marine Sciences, Barcelona, where he worked (1973–2012) and where he founded the Biological Collections of the Institut de Ciències del Mar. He co-wrote: *Encyclopaedia of living marine resources of the Mediterranean* (1999). He was honoured for his 'outstanding' contribution to the knowledge of fish species.

Lloyd

Lloyd's Electric Ray *Heteronarce mollis* RE Lloyd, 1907
Lloyd's Slickhead *Narcetes lloydi* HW Fowler, 1934

Captain Dr Richard E Lloyd MB DSc (b.1875) was a physician in the Indian Medical Service and was Surgeon Naturalist of the Marine Survey of India. He was also an accomplished artist. He took part in a cruise of the research vessel 'Investigator' (1906) and wrote up the results of this and previous voyages (1909). He was (1908) Acting Professor of Biology at the Medical College of Bengal and wrote a biology textbook for students there: *An Introduction to Biology for Students in India*. He was promoted to Major (1913). He also wrote: *A description of the deep-sea fish caught by the R.I.M.S. ship 'Investigator' since the year 1900, with supposed evidence of mutation in* Malthopsis (1909).

Loat

Nile Killifish *Micropanchax loati* GA Boulenger, 1901
Nile Cichlid sp. *Haplochromis loati* PH Greenwood, 1971

William Leonard Stevenson Loat (1871–1932) was a British archaeologist and naturalist. He first visited Egypt (1899) as assistant to Boulenger in the ichthyological survey of the Nile, during which he collected the type material of the killifish. He was Superintendent of the survey party that first collected the cichlid species at Gondokoro, South Sudan. He excavated at Gurob (1903) and at Abydos (1908–1909 & 1912–1913), publishing an account as: *The Cemeteries of Abydos* (1913). WW1 put a stop to his Egyptological work, and he afterwards married and settled at Cornwall, taking up horticulture. He presented some Egyptian antiques to the Penzance Museum and bequeathed an extensive collection of objects from China, Egypt, Peru, Africa, Australia, Assam, Polynesia and North America to the British Museum. He later visited the Andes (1927).

Lobel

Lobel's Lizardfish *Synodus lobeli* RS Waples Jr & JE Randall, 1989
Caribbean Neon Goby *Elacatinus lobeli* JE Randall & PL Colin, 2009

Dr Phillip S 'Phil' Lobel is a biologist and ichthyologist who is Professor of Biology at Boston University. He teaches ichthyology and scientific diving. His BSc in zoology was awarded by the University of Hawaii (1975) and his doctorate by Harvard (1979). He is interested in fundamental concepts of fish biology and in applying this knowledge to scientific issues and to societal concerns of fisheries management and conservation. His main study site (1983–2003) was Johnston Atoll, Central Pacific Ocean, conducting research as part of the US Army marine ecological monitoring program evaluating operation of the Johnston Atoll Chemical Weapons Disposal System, since when he has worked mainly in Belize on Central American fish bioacoustics and discovering new species. Among other papers he wrote: *Transport of reef lizardfish larvae by an ocean eddy in Hawaiian waters* (2011) and *A review of the hamlets (Serraanidae, Hypoplectrus) with description of two new species* (2011).

Lo Bianco

Lanternfish genus *Lobianchia* MA Gatti, 1904

Salvatore Lo Bianco (1860–1910) was an Italian marine biologist. He worked as a preparator, at Stazione Zoologica Anton Dohrn, Naples (1874), later becoming in charge of the station's conservation department (1881). The University of Naples awarded him an honorary degree in natural sciences (1895). He wrote: *Biological news concerning the period of sexual maturation of animals in the Gulf of Naples* (1888). He died very suddenly and young of a cerebral apoplexy.

Lockington

Medusafish sp. *Icichthys lockingtoni* DS Jordan & CH Gilbert, 1880

William Neale Lockington (1840–1902) was a British zoologist and ichthyologist, as well as an architect. He spent much of his career in the USA where he was at the California Academy of Sciences as Curator of the Museum (1875–1881). Among his published papers are: *Description of a new species of Agonidae (Brachyopsis verrucosus) from the coast of California* (1880) and *List of the fishes collected by M. W. J. Fisher upon the coasts of Lower California, 1876–1877, with descriptions of new species* (1881). He was honoured "*...in recognition of his important work in Californian ichthyology.*"

Loftin

Characin sp. *Roeboides loftini* CAS de Lucena, 2011

Dr Horace Greeley Loftin (b.1927) was an American ornithologist. He was an Assistant Professor in the Biology Department, Florida State University (1966–1970) becoming full Professor at the Canal Zone branch (1993–1995). The University has a research station in Panama where he ringed birds and contributed to the knowledge of the distribution of the freshwater fishes of Panama. He wrote: *Notes on autumn bird migrants in Panama* (1963).

Lohberger

Leka Keppe (Cameroon cichlid) *Sarotherodon lohbergeri* M Holly, 1930

Karl Lohberger was (probably) Holly's colleague at Naturhistorisches Staatsmuseum (Vienna). He published six papers on fishes (1929–1930) and then seems to have dropped from the ichthyological record. He named a barbel after Holly.

Loise

Characin sp. *Hemibrycon loisae* J Géry, 1964
[Syn. *Bryconamericus loisae*]

Loise Socolof (1925–2000) was the wife of Ross Socolof (1925–2009), an aquarium-fish exporter, breeder and wholesaler and who collected the holotype. She wrote: *Gerbils as pets* (1966).

Loiselle

Central American Cichlid sp. *Parachromis loisellei* WA Bussing, 1989
Madagascan Cichlid sp. *Ptychochromis loisellei* MLJ Stiassny & JS Sparks, 2006
Madagascan Cichlid sp. *Paretroplus loisellei* JS Sparks & RC Schelly, 2011

Dr Paul V Loiselle is Emeritus Curator of Freshwater Fishes at the New York Aquarium and a very enthusiastic aquarist. He has a Master's degree from Occidental College in Los Angeles, and a doctorate from the University of California at Berkeley. His professional background includes five years as a Peace Corps fisheries biologist in West Africa, where he carried out faunal and environmental impact surveys in Togo and Ghana. During the course of his career, he has observed the behaviour of cichlids in Lakes Victoria and Tanganyika, in México and in Central America. He is the inspiration for the creation of CARES (Conservation, Awareness, Recognition and Responsibility, Encouragement and Education, and Support and Sharing) an organisation of aquarists promoting conservation of rare freshwater fish. He has served as CARES Technical Editor of Freshwater Fishes, Madagascar Regional Coordinator, and CARES Speaker and Consultant since its inception (2004). He was also a founding member of the American Cichlid Association. He has numerous publications. He collected the *Ptychochromis* type and directed the authors' attention to the *Paretroplus* being a new taxon. He was honoured for his many contributions to the understanding and conservation of Madagascar's freshwater fishes and for his efforts to document, preserve, and educate the public regarding Madagascar's unique and severely threatened freshwater ichthyofauna.

Loki

Loki Whip-Goby *Bryaninops loki* HK Larson, 1985

Loki was a legendary Old Norse God who was a rather enigmatic trickster, presumably referring to the goby's cryptic coloration, matching the gorgonian sea fans and whips on which it is most often found.

Lombardero

Catfish sp. *Xyliphius lombarderoi* FJJ Risso & ENP de Risso, 1964

[Perhaps a junior synonym of *X. barbatus*]

Dr Oscar J Lombardero (1921–2001) was a zoologist, parasitologist and veterinary surgeon who qualified at the University of Buenos Aires (1947). He was Professor of Parasitology and Parasitic Diseases, Faculty of Veterinary Sciences and Dean of Faculty of Natural Sciences, Universidad del Nordeste (Argentina). He wrote: *Lessons in Parasitology* (1990).

Lombardi

Clingfish sp. *Derilissus lombardii* JS Sparks & DF Gruber, 2012

Michael Robert Lombardi (b.1979), was part of the diving team in the Bahamas that collected the type while exploring and documenting the natural history of deep Mesophotic coral reefs. His BSc (Marine Biology) was awarded by the University of New Hampshire (2000). He is now a diving contractor and technologist by trade. He has more than 20 years experience of 'mud diving' inshore and his undersea work spans both academic and industry sectors. He created 'Ocean Opportunity Inc.' as a not for profit organisation to apply his innovative, grassroots model for business and exploration to many projects, including the 'Diving a Dream Project' (2004–2007) which was aired on NBC TV. Other roles have been as Diving Safety Officer for NOAA's Caribbean Marine Research Center (1999–2006), interim Diving Safety Officer (2011–2012) for the University of Rhode Island, and a contract position (2009–2014) as Diving Safety Officer for the American Museum of Natural History. He is a lifetime member of the American Academy of Underwater Sciences (AAUS), the Marine Technology Society (MTS), and the National Marine Educators Association (NMEA). He runs (since 2012) the 'Lombardi Undersea Resource Center' in Rhode Island. His publications include: *A Visual Profile of the Vertical Mesophotic Coral Ecosystem of the Tongue of the Ocean* (TOTO), *Andros, Bahamas to 100 meters* (2011).

Lombardo

Lake Malawi Cichlid sp. *Maylandia lombardoi* WE Burgess, 1977

John Lombardo of 'African Fish Imports' first brought this species to Burgess' attention and provided the holotype. He also coined the name 'Peacocks' for the *Aulonocara* genus. (Also see **Lori**)

Lombarte

Goby sp. *Buenia lombartei* M Kovačić, F Ordines & UK Schliewen, 2018

Dr Antoni Lombarte is an evolutionary marine biologist and ecologist. He is currently a Researcher at the Institut de Ciències del Mar-CSIC, Barcelona. He found the first specimen of the goby species during his fieldwork; the paratype was among trawl debris and he spotted it despite its small size. He has written or co-written many scientific papers such as: *Identifying sagittal otoliths of Mediterranean Sea gobies: Variability among phylogenetic lineages* (2018).

Long

Aleutian Dotted Skate *Rhinoraja longi* W Raschi & JD McEachran, 1991
[Junior Synonym of *Bathyraja taranetzi*]

James John Long (b.1952) is a Fisheries Biologist (1976–1993), living in Shoreline, Washington State. He sent the authors the type series of this skate along with many other fishes from the Aleutian Islands. His extensive collections of Alaskan marine fishes are deposited at Oregon State University in Corvallis, Oregon State.

Longhurst

Viviparous Brotula sp. *Grammonus longhursti* DM Cohen, 1964

Dr Alan Reece Longhurst (b.1925) is a British-born Canadian oceanographer, zoologist and marine biologist. Bedford College, University of London awarded his doctorate (1952). He worked for the West African Fisheries Research Institute, Freetown, Sierra Leone and the Federal Fisheries Service, Lagos, Nigeria (1954–1963). He was the first Director of the Southwest Fisheries Science Center, US National Marine Fisheries Service, La Jolla, California (1967–1971). He returned to England (1971) as Deputy Director, Institute for Marine Environmental Research, Plymouth. He was at the Marine Ecology Laboratory at the Bedford Institute of Oceanography, Nova Scotia, as Director (1977–1979) and as Director-General (1980–1986). He wrote: *Ecological Geography of the Sea* (1998). He retired to Cajarc, France (1995), where he and his wife run a gallery of contemporary art. He collected the holotype of this species.

Longley

Rattail sp. *Hymenocephalus billsam** NB Marshall & T Iwamoto, 1973

William 'Bill' Harding Longley (1881–1937) was a Canadian-born botanist and ichthyologist who studied colours and patterns of tropical reef fish. He was Professor of Botany, Goucher College, Baltimore (1911–1937) and was Director of the Dry Tortugas Laboratory, Carnegie Institution, Washington D.C. (1922–1937). He co-wrote: *Systematic catalogue of the fishes of Tortugas, Florida: with observations on color, habits, and local distribution* (1941).

*The binomial honours two ichthyologists: William '**Bill**' Longley and Samuel '**Sam**' Hildebrand. (See also **Billsam** & **Hildebrand**)

Lönnberg

Spiny Eel sp. *Mastacembelus loennbergii* GA Boulenger, 1898
African Rivuline sp. *Aphyosemion loennbergii* GA Boulenger, 1903
African Shovelnose Catfish *Parauchenoglanis loennbergi* L Keilback, 1910
[Junior Synonym of *Parauchenoglanis monkei*]
Scaly Rockcod *Trematomus loennbergii* CT Regan, 1913

Professor Axel Johan Einar Lönnberg (1865–1942) was a Swedish zoologist who mainly worked at the Vertebrate Department of the Swedish Museum of Natural History in Stockholm. He travelled in the Caspian Sea region (1899) and British East Africa (1910–1911) and was the last prefect of the Kristineberg Marine Zoological Station (1925–1942). An early conservationist, he worked for laws protecting reindeer and waterbirds. He founded the biological journal *Fauna och Flora*. Three amphibians and eight birds are named after him.

Lonsdale

Haweswater Char *Salvelinus lonsdalii* CT Regan, 1909

Hugh Cecil Lowther, 5th Earl of Lonsdale (1857–1944), was an English nobleman and sportsman. He was enormously wealthy but, despite that, overspent to the extent that he impoverished the estate and his successor. He was known as the Yellow Earl as he loved the colour; he was founder and first president of the Automobile Association, which adopted his favourite colour as their own. He is probably best known for the 'Lonsdale Belt' awarded to outstanding champion British boxers. He explored the Arctic regions of Canada (1888) reaching Kodiak, Alaska (1889). He made a collection of Inuit artefacts that is now in the BM. He also presented the holotype of this species.

Loock

Lake Tanganyika Cichlid sp. *Interochromis loocki* MFL Poll, 1949

E Van Loock was Director of the Great Lakes Railways Company. He was honoured for his support of the Belgian Hydrobiological Mission to Lake Tanganyika (1946–1947), during which the endemic species was collected.

Lootens

African Characin sp. *Monostichodus lootensi* MFL Poll & J Daget, 1968

Révérend Père Germain Lootens (1910–1976) was a Catholic missionary and naturalist in the Congo (1936–1976). He collected the holotype.

Lopes, B

Characin sp. *Moenkhausia lopesi* HA Britski & KZ de Silimon, 2001

Balzac Santana Lopes was the head of the station at the Fisheries Station of the Mato Grosso Company of Research, Assistance and Rural Extension. He was the authors' collecting companion in the Pantanal. He co-wrote: *Pantanal Fish* (2007).

Lopes, E

Neotropical Rivuline sp. *Hypsolebias lopesi* DTB Nielsen, OA Shibatta, R dos R Suzart & AF Martin, 2010

Edson Lopes is an aquarist who is proprietor of Aqua Mundo Aquarios in Sao Paulo, Brazil. He was honoured "*...because of his contribution to our knowledge on raising and breeding annual fishes in captivity.*"

Lopez

Sand Smelt sp. *Atherina lopeziana* M Rossignol & J Blache, 1961

This is a toponym; the species was originally collected near Cape Lopez, Gabon

Lopez, G

Barb sp. *Hampala lopezi* AW Herre, 1924

Elongate Unicornfish *Naso lopezi* AW Herre, 1927

G A Lopez was a collector for the Philippines Bureau of Science. He was accustomed to act as a field assistant to scientists and, as an example, collected herpetofauna on the island of Palawan for Edward Harrison Taylor (1889–1978). He collected or obtained the holotype of the barb species.

Lopez, MT (Boegeholz)

Three-barbeled Catfish sp. *Chasmocranus lopezi* A Miranda Ribeiro, 1968
[Binomial sometimes corrected to feminine *lopezae*]

Maria Theresa Lopez Boegeholz (1927–2006) was a Chilean marine scientist who originally trained as a teacher and worked at the University of Chile as a teaching assistant (1952). She became (1966) Professor of Zoology, Centro de Investigaciones Zoologicas de Universidade de Chile, Santiago de Chile. She was also a visiting professor at the Universidad Austral and a full professor at the University of Concepción (1966–1998). She co-wrote: *Education, foundation for environmental education and sustainable development* (1988) and was the person who collected the holotype.

Loppé

Loppé's Tadpole Fish *Ijimaia loppei* L Roule, 1922
Cadenat's Rockfish *Scorpaena loppei* J Cadenat, 1943

Dr Etienne Loppé (1883–1954) was a physician who qualified in Paris. He was appointed Director and Chief Curator, Lafaille Museum of Natural History, La Rochelle, France (1915–1954). Roule honoured him as a thank you for supplyimg two specimens of the tadpole-fish for Roule to study.

Lord, C

Lord's Sicyopus (goby) *Sicyopus lord* P Keith, G Marquet & L Taillebois, 2011

Clara Lord is an assistant professor at L'Unité Mixte de Recherche (UMR) BOREA – 'Biologie des Organismes et Ecosystèmes Aquatiques'. She has a particular interest in the Sicydiinae subfamily of gobies. It might have been more correct for the authors to have styled the binomial as *lordae*.

Lord, CE

Lord's Goby *Tasmanogobius lordi* EOG Scott, 1935
[Alt. Lord's Tasman(ian) Goby]

Clive Errol Lord (1889–1933) was a Tasmanian naturalist and museum director. He trained and practised as an architect, but had an interest in natural history which led to him becoming a Foundation Member of the Tasmanian Field Naturalists' Club (1904), later serving as its Honorary Secretary and President. He worked for the Tasmanian Museum as Assistant Curator (1917–1920), Curator (1921–1922) and Director (1923–1933) of both the Tasmanian Museum and Art Gallery. He co-wrote: *A Synopsis of the Vertebrate Animals of Tasmania* (1924).

Lord, JK

Southern Dolly Varden Trout *Salvelinus malma lordii* A Günther, 1866

John Keast Lord (1819–1872) was an English veterinarian, traveller, naturalist, journalist and author. He qualified as a veterinary surgeon (1844) at the Royal Veterinary College, London. He practised as a vet in Tavistock, Devon, but started to drink heavily and then, suddenly, disappeared. Rumour has it that he made a whaling voyage, was shipwrecked and then worked as a trapper in Minnesota and the Hudson Bay area. It is known that he officially re-appeared when he was appointed (1855) to the British Army in the East as a veterinary surgeon, attached to Turkish forces with whom he served in the Crimean War (1855–1856). He was appointed naturalist to the British North American Boundary Commission (1858), which delineated the 49th parallel of latitude as the boundary between the newly-formed colony of British Columbia and the USA. He lived on Vancouver Island and amassed a varied collection which is now in the BMNH. He was employed by the Viceroy in Egypt for archaeological and scientific research. He was appointed the first manager of the Brighton Aquarium (1872) but died after only four months in the job. He wrote: *The Naturalist in Vancouver's Island* (1866, 2 volumes). He presented the holotype of this subspecies to the BMNH.

Lorentz, HA

Lorentz's Rainbowfish *Chilatherina lorentzii* M Weber, 1907
Lorentz's Grunter *Pingalla lorentzi* M Weber, 1910
Primitive Archerfish *Toxotes lorentzi* M Weber, 1910
Borneo Loach sp. *Lepidocephalichthys lorentzi* M Weber & LF de Beaufort, 1916

Hendrikus Albertus Lorentz (1871–1944) was a Dutch explorer and (later) diplomat. He studied law and biology at Utrecht University. He participated in the expedition (1901) of Professor Wichmann to northern (Dutch) New Guinea and he himself led expeditions (1905–1906 & 1909–1910) in southern New Guinea, leading to the discovery of Wilhelmina Peak (named after the Dutch Queen Wilhelmina) in the Snow Mountains. Upon his return, he entered the Dutch consular services, becoming Ambassador in Pretoria, South Africa (1929). The Lorentz River in New Guinea is named after him, and the Lorentz River Sole (*Leptachirus lorentz*) is named after that river.

Lorentz, PG

American Sole sp. *Catathyridium lorentzii* H Weyenbergh, 1877

Dr Paul (Pablo) Günther Lorentz (1835–1881) was a German botanist and mycologist. He emigrated to Argentina (1870) and became a naturalised Argentine citizen. He was Professor of Botany at the National Academy of Sciences (Cordoba), and a colleague of the fish's author, Hendrik Weyenbergh. He died in Concepción del Uruguay at the age of 46, a victim of liver disease.

Lorenz

Black-tail Sergeant (damselfish) *Abudefduf lorenzi* DA Hensley & GR Allen, 1977

Dr Konrad Lorenz (1903–1989) was an Austrian zoologist, ethnologist and ornithologist. He shared the Nobel Prize in Physiology or Medicine with Nikolaas Tinbergen and Karl von Frisch (1973). He is most famed for having studied instinctive behaviour in animals, especially in greylag geese and jackdaws. Working with geese, he investigated the principle of imprinting, the process by which some nidifugous birds bond instinctively with the first moving object that they see within the first hours of hatching. Although Lorenz did not discover the topic, he became widely known for his descriptions of imprinting as an instinctive bond. He is often regarded as one of the founders of modern ethology, the study of animal behaviour, and was honoured in the binomial for his contributions to the science of ethology. His work was interrupted by the onset of WW2, as he was recruited (1941) into the German army as a medic. He was sent to the Eastern Front (1944) where he was captured and spent four years as a Soviet POW. He later regretted his membership in the Nazi party. He wrote numerous books, some of which, such as *King Solomon's Ring*, *On Aggression*, and *Man Meets Dog*, achieved popularity.

Loreto

Royal Gramma *Gramma loreto* F Poey, 1868

Señorita Loreto Martínez caught the type specimen while fishing in the bay at Matanzas, Cuba. According to Poey, she "...*took advantage of the town where she lived* (Playa de Judíos) *to enrich museums and educate lovers of natural history*" (translation). Nothing more seems to be recorded of her.

Lori (Colin)

Linesnout Goby *Elacatinus lori* PL Colin, 2002

Lori Jane Bell Colin is Co-Director and Manager and Research Scientist, Coral Reef Research Foundation (Koror, Palau). In private life she is also the wife of the author, Patrick Colin. She was honoured for her numerous contributions to the biology of coral reef fishes.

Lori (Lombardo)

Lake Malawi Cichlid sp. *Melanochromis loriae* DS Johnson, 1975

Lori Lombardo is the daughter of John Lombardo (see **Lombardo**).

Lori (Randall)

Lori's Anthias *Pseudanthias lori* HR Lubbock & JE Randall, 1976

Lori Randall is the daughter of the second author, John E Randall.

Lorna

African Barb sp. *Enteromius lornae* CK Ricardo-Bertram, 1943

Lady Lorna Gore-Browne née Goldman (1908–2002) was the wife (1927–1950) of Lieutenant-Colonel Sir Stewart Gore-Brown (1883–1967). After their divorce (1950) she moved to London and called herself 'Mrs Browne'. Gore-Browne was an early settler in Northern Rhodesia (Zambia) and lived on his estate, Shiwa Ngandu (1920–1967), where the author stayed during her expedition to the area.

Lortet

Lortet's Barbel *Barbus lorteti* HE Sauvage, 1882

Redeye Puffer *Carinotetraodon lorteti* G Tirant, 1885

Dr Louis Charles Émile Lortet (1836–1909) was a French botanist, zoologist, palaeontologist, Egyptologist, anthropologist and physician who qualified in medicine (1861) and in natural sciences (1867). He was Director of the Natural History Museum, Lyon (1868–1909). He travelled widely in the Middle East, studying the mummified animals found in ancient Egyptian tombs, and helped excavate a Phoenician necropolis (1880). He wrote: *La Syrie d'aujourd'hui. Voyages dans La Phénicie, Le Liban et La Judée 1875–1880* (1881). The barbel is assumed to be named after him (the etymology does not specify, but it is difficult to see who else it could be).

Losse

Gulf Herring *Herklotsichthys lossei* T Wongratana, 1983

Georg F Losse was a German biologist, ichthyologist, collector and Fishery Officer. He was in East Africa, based at East African Marine Fisheries Research Organization, Zanzibar in the 1960s. He was involved in work with and on behalf of the FAO (Department of Fisheries) in Rome, and also in the FAO Fishery Development Project in Aden, South Yemen (1971). He wrote: *The elopoid and clupeoid fishes of East African coastal waters* (1968) and *The small-scale fishery on Lake Kariba in Zambia* (1998). He was honoured, to quote from the etymology, as the person "*...who collected most of the type material, and in recognition of his most useful studies of East African clupeoids.*"

Lotte

Goby genus *Lotilia* W Klausewitz, 1960

Charlotte 'Lotte' Hildregard Hass née Baierl (1928–2015) was an Austrian diver, model and actress. She was the (second) wife and erstwhile secretary of biologist, underwater cinematographer, TV personality and scuba-diving pioneer Hans Hass (q.v.), who led the expedition that collected the holotype. She trained as a diver because she wanted to appear in his films. He was initially reluctant to have women involved, but relented and her first appearance was in the Oscar-winning *Under the Red Sea* (1950). To the delight of the audience she wore a low-cut swimming costume. She wrote of her experiences during the expedition to the Red Sea in her book: *A Girl on the Ocean Floor* (1970). The etymology gives no explanation, but Wolfgang Klausewitz revealed that Lotte could not join her husband in an expedition because she was pregnant or taking care of a new-born, so Hass asked Klausewitz (they were friends) to name this goby after her. Klausewitz named it as *Lotilia graciliosa*.

Louise (Chaplin)

Spotlight Goby *Elacatinus louisae* JE Böhlke & CR Robins, 1968

Louise Davis Chaplin née Catherwood (1906–1983) came from an upper-class Philadelphia family and was described in a newspaper report (1955) with the headline '*Philadelphia socialite finds giving money away hard work*'. She co-founded the Catherwood Foundation, which endowed scholarships and financed biological research at the Philadelphia Academy of Sciences and, with her husband, ichthyologist Charles C G Chaplin (1906–1991), made possible the Academy of Natural Sciences of Philadelphia's shorefish programme in the Bahamas, during which the type was collected. A bird is also named after her.

Louise (Wrobel)

Goatfish sp. *Parupeneus louise* JE Randall, 2004

Louise Wrobel is an ichthyologist at 'Service des Ressources Marines' in Papeete, Tahiti, French Polynesia. She and Randall co-wrote: *A new species of soldierfish of the genus Ostichthys and records of O. archiepiscopus and O. sandix from Tahiti* (1988), in which it is stated that she had been "*...monitoring the deeper water catches of fishermen in Tahiti and Moorea...in recent years*".

Louka

West African Cichlid sp. *Coptodon louka* DFE Thys van den Audenaerde, 1969

Louka Thys van den Audenaerde is the wife of the author, Belgian ichthyologist Dirk Frans Elisabeth Thys van den Audenaerde (see **Thys**). His etymology says: "*...dedicated to my wife Lou, who since many years has been assisting me in my Tilapia-research.*" (*Tilapia* was the genus in which this species was originally placed).

Lourens

African Rivuline sp. *Nothobranchius lourensi* RH Wildekamp, 1977

Jan Lourens worked as a biologist for the United Nations Development Programme (Dar es Salaam, Tanzania), during which time he discovered this species and collected the type (1976).

Lout

Yellow-edged Lyretail Grouper *Variola louti* P Forsskål, 1775

This is not an eponym but, according to the author, *Louti* was an Arabic name for the fish.

Loveridge

African Barb sp. *Enteromius loveridgii* GA Boulenger, 1916
African Cyprinid sp. *Opsaridium loveridgii* JR Norman, 1922
African Cyprinid sp. *Xenobarbus loveridgei* JR Norman, 1923

Arthur Loveridge (1891–1980) was a British herpetologist and ornithologist who became Curator of Nairobi Museum (1914). He worked for the Museum of Comparative Zoology, Harvard (1924–1957) and made several field trips to East Africa. His writings include: *Many Happy Days I've Squandered* (1949) and *I Drank the Zambesi* (1954). He retired to the island of St Helena. He collected the cyprinid types and presented them to the BMNH. Seven amphibians, thirteen reptiles and four birds are also named after him.

Lovett

Tasmanian Whitebait genus *Lovettia* AR McCulloch, 1915

Edward Frederick Lovett (1857–1943) of Hobart, Tasmania, supplied the author with 'beautifully preserved specimens' of *L. sealii*, as well as many other Tasmanian fishes.

Lowe, G

Lowe's Leatherjacket *Paramonacanthus lowei* JB Hutchins, 1997

Graham Lowe of Bundaberg, Queensland, collected the holotype. Hutchins wrote: "...*This species is named in honour of Mr Graham Lowe who not only collected the holotype, but also provided additional monacanthid material and information for this study*." He seems to have been a fisherman and amateur collector.

Lowe, RH

African Airbreathing Catfish sp. *Dinotopterus loweae* PBN Jackson, 1959
Characin sp. *Moenkhausia loweae* J Géry, 1992
Lowe's Tetra *Hyphessobrycon loweae* JEM Costa & J Géry, 1994

Dr Rosemary Helen Lowe-McConnell (1921–2014) was a British biologist and a pioneer in the study of tropical fish ecology. The University of Liverpool awarded her bachelor's and master's degrees and her doctorate. She wrote: *Recent research in the African great lakes: Fisheries, biodiversity and cichild evolution* (2003). She collected the types of the *Moenkhausia* and *Hyphessobrycon* species. Her survey of the *Tilapia* of Lake Malawi "drew attention to the large number of clariid [catfish] species existing in the lake".

Lowe, RT

Lanternfish genus *Loweina* HW Fowler, 1925
Beardfish *Polymixia lowei* A Günther, 1859
Needlefish sp. *Platybelone lovii* A Günther, 1866
Hammerjaw *Omosudis lowii* A Günther, 1887

Rev. Richard Thomas Lowe (1802–1874) was a British botanist, ichthyologist and malacologist as well as being a clergyman. He graduated from Christ's College, Cambridge (1825) and then took holy orders. He became a pastor in Madeira (1832), where he also became an expert on the local flora and fauna and wrote a book on its plant life. He sent the needlefish holotype to Günther at the BMNH. He died off the Scilly Isles in a shipwreck. Fowler does not identify who the genus is named after, so in this case we can only provide a 'best guess'. Günther, meanwhile, appears to have tried as many different spellings for a binomial honouring one man as he could think of.

Lowe, WP

Lowe's Bichir *Polypterus lowei* GA Boulenger, 1911
[Junior Synonym of *Polypterus palmas*]

Willoughby Prescott Lowe (1872–1949) was a British naturalist, explorer and collector for the BMNH. He was in the USA in the late 19th century, and in many parts of Africa and the Far East (1907–1935). He is reputed to have sent over 10,000 specimens to the Bird Room at the BMNH. He is also notorious for having shot eight specimens of Miss Waldron's Red Colobus in Ghana (1933); the monkey was already rare, and is now thought extinct. Five mammals and three birds are also named after him.

Lozano

Lozano's Goby *Pomatoschistus lozanoi* F de Buen, 1923
Cusk-eel sp. *Ophidion lozanoi* J Matallanas García, 1990
Iberian Gudgeon sp. *Gobio lozanoi* I Doadrio Villarejo & MJ Madeira, 2004

Dr Luis Lozano Rey (1878–1958) was a Spanish naturalist. He was Professor of Vertebrate Zoology, University of Madrid, and in charge of vertebrates at the National Museum of Natural Science. He was also technical adviser to the General Directorate of Fisheries and an Assistant Professor at the Spanish Oceanography Institute. He was honoured in recognition of his contribution to the knowledge of Iberian freshwater fishes.

Lozi

Banded Neolebias *Neolebias lozii* KO Winemiller & LC Kelso-Winemiller, 1993

The Lozi tribe are described by the authors as being "*...the traditional caretakers of the Barotse floodplain and its fishery resources.*" The floodplain is in the Western Province of Zambia, where the species occurs. The Lozi are also known by several other names, such as the Barotsi, Barotose, Malozi, Silozi and Kololo.

Lualaba

African Tetra sp. *Micralestes lualabae* M Poll, 1967

This is a toponym; the Lualaba River, Democratic Republic of Congo, is where the holotype was collected.

Lubbock

Dottyback genus *Lubbockichthys* TN Gill & AJ Edwards, 1999
Lubbock's Fairy Wrasse *Cirrhilabrus lubbocki* JE Randall & KE Carpenter, 1980
Dottyback sp. *Pectinochromis lubbocki* AJ Edwards & JE Randall, 1983
Lubbock's Chromis *Chromis lubbocki* AJ Edwards, 1986
Goby sp. *Corcyrogobius lubbocki* PJ Miller, 1988
Lubbock's Coral-Blenny *Ecsenius lubbocki* VG Springer, 1988
Tonguefish sp. *Symphurus lubbocki* TA Munroe, 1990
Ascension Yellowtail Damselfish *Stegastes lubbocki* GR Allen & KN Smith, 1992

Dr Hugh Roger Lubbock (1951–1981) was an English marine biologist at Cambridge University. He led (1980) the Cambridge Expedition, with A J Edwards, to Saint Paul's Rocks (Cape Verde Archipelago) where he collected the *Chromis*, recognising it as a new species. Although his professional life spanned relatively few years, Dr Lubbock made numerous valuable contributions to the knowledge of reef fish taxonomy and the systematics of pseudochromid fishes. With J E Randall, he co-wrote: *Labrid fishes of the genus* Paracheilinus, *with descriptions of three new species from the Philippines* (1981). He died in a car crash in Rio de Janeiro just before his 30th birthday.

Lucanus

West African Cichlid sp. *Enigmatochromis lucanusi* A Lamboj, 2009

Oliver Lucanus (b.1970) is a German born Canadian collector, aquarist, videographer, explorer and tropical fish importer. He has travelled extensively in Africa and South America over 25 years, photographing and collecting fish. He has lectured and written extensively about fish and their habitats, with a focus on gathering and relaying information for aquarists. He has co-published many articles and papers, and his books include: *The Amazon Below Water* (2016), which showcases more than 150 species of fish in their natural habitats. Despite lacking formal qualifications, he is also one of the world's leading experts on the Xingu and its ecology and endemic fish. He co-founded the Fish & Forest project at McGill University that is based in the remote sensing lab. He was also a pilot for CABO (Canadian Airborne Biodiversity Observatory). His research interests include biodiversity mapping, hyperspectral imaging, fish biology and photography. He is a friend of the author Anton Lamboj, and provided information about the type locality in the Foto River, Guinea, giving a detailed description of the habitat where he first collected it (2006) and its distribution.

Lucas

Leister Sculpin *Enophrys lucasi* DS Jordan & CH Gilbert, 1898
[Syn. *Ceratocottus lucasi*]

Frederick Augustus Lucas (1852–1929) was an American naturalist and curator. He entered (1871) Ward's Natural Science Establishment in Rochester, New York, where he remained for eleven years working at taxidermy, osteology, and museum technique. He joined the Smithsonian Institution in 1882 as an 'osteological preparatory' and became Assistant Curator (1887) and Curator (1893–1904). He was then Curator-in-Chief of the Museum of the Brooklyn Institute of Arts and Sciences (1904–1911), going on to become the Director of the American Museum of Natural History (1911–1923), then Director Emeritus until his death. Late in life (1909) he was awarded an honorary D.Sc. degree from the University of Pittsburgh. A mammal is also named after him.

Lucas Bah

Loach sp. *Syncrossus lucasbahi* HW Fowler, 1937

Lucas Bah collected Thai specimens for the Academy of Natural Sciences of Philadelphia and was also employed as a collector by the ornithologist, Meyer de Schauensee in the 1930s.

Lucena, CAS de

Driftwood Catfish sp. *Trachelyopterus lucenai* JJ Bertoletti, JFP da Silva & EHL Pereira, 1995
Poeciliid livebearer sp. *Phalloceros lucenorum* PHF Lucinda, 2008
Pike-Cichlid sp. *Crenicichla lucenai* JL Mattos, I Schindler, FP Ottoni & MM Cheffe, 2014
Tetra sp. *Hyphessobrycon lucenorum* WM Ohara & FCT Lima, 2015

Dr Carlos Alberto Santos de Lucena is a zoologist and ichthyologist, as is his wife Dr Zilda Margarete Seixas de Lucena (below). He has a bachelor's degree (1976) and master's (1985), both awarded by Pontifícia Universidade Católica do Rio Grande do Sul, Porto Alegre,

Brazil and a doctorate (1993) awarded by the University of São Paulo. He is currently (2017) Curator II of the Museum of Science and Technology of the Pontifical Catholic University of Rio Grande do Sul. He co-wrote: *Pimelodus quadratus, a new long-whiskered catfish from the Rio Tocantins drainage, Brazil (Siluriformes: Pimelodidae)* (2016). The livebearer and tetra are named after both husband and wife.

Lucena, ZMS de

Poeciliid livebearer sp. *Phalloceros lucenorum* PHF Lucinda, 2008
Tetra sp. *Hyphessobrycon lucenorum* WM Ohara & FCT Lima, 2015

Dr Zilda Margarete Seixas de Lucena is a zoologist and ichthyologist, as is her husband Dr Carlos Alberto Santos de Lucena (above). She has a bachelor's degree from Santa Úrsula University (1978), a Master's degree from the Pontifical Catholic University of Rio Grande do Sul (1986) and a Doctorate awarded by the Federal University of Rio Grande do Sul (2003). She is currently (2017) a curator of the fish collection and was (2007–2015), also co-ordinator of collections at the Museum of Science and Technology of the Pontifical Catholic University of Rio Grande do Sul. She co-wrote: *Moenkhausia tergimacula, A new species from upper Rio Tocantins, Brazil (Osteichthyes: Characidae) (*1999). These species are named after both husband and wife. (See also **Margarete**).

Lucero

Shovelnose Catfish sp. *Platystoma luceri* H Weyenbergh, 1877 NCR
[Junior Synonym of *Sorubim lima*]

Dr Manuel Lucio Lucero (1814–1878) was an Argentine lawyer, politician and naturalist. He qualified as a lawyer (1838) at the University of Córdoba where he taught (1838–1840). He was on the losing side in the 1840 revolution and fled into exile in Chile (1840–1847). He was a deputy and senator in the Argentine parliament (1854–1862) after which he worked as a private lawyer. He returned to the University of Córdoba as Director and Professor (1874–1878), where he founded the faculties of medicine and of exact sciences.

Lucetius

Panama Lightfish *Vinciguerria lucetia* S Garman, 1899

In Roman mythology, Lucetius ('light bringer') was an epithet given to the god Jupiter.

Lucia

Saint Lucia Mullet *Chelon luciae* ML Penrith & MJ Penrith, 1967

This is a toponym; the fish is named after Saint Lucia, South Africa.

Lucia

Long-whiskered Catfish sp. *Pimelodus luciae* MS Rocha & FRV Ribeiro, 2010
Glass Knifefish sp. *Archolaemus luciae* RP Vari, CDCM de Santana & WB Wosiacki, 2012
Armoured Catfish sp. *Loricaria luciae* MR Thomas, MS Rodriguez, MR Cavallaro, O Froehlich & RM Corrêa e Castro, 2013

Dr Lúcia Helena Rapp Py-Daniel is a Brazilian ichthyologist. (See **Py-Daniel**)

Lucia (Lucy)

Spotfin Killifish *Fundulus luciae* SF Baird, 1855

Lucy Hunter Baird (1848–1913) was the only child of the author, Spencer Baird.

Luciano

Armoured Catfish sp. *Corumbataia lucianoi* GSC Silva, FF Roxo, CS Souza & C Oliveira, 2018

Luciano de Souza da Costa e Silva is the brother of the senior author, Gabriel Silva of the Department of Morphology at Instituto de Biociências, Universidade Estadual Paulista.

Luciene

Serrasalmid Characin sp. *Myloplus lucienae* MC Andrade, RP Ota, DA Bastos & M Jégu, 2016

Dr Luciene Maria Kassar Borges is a zoologist and ichthyologist who has two bachelor's degrees from the Federal University of Mato Grosso (1973 & 1976) and a master's from the National Institute of Amazonian Research (1983). She has been employed by the University of the State of Mato Grosso since 1976 and is now a 'Profesora interina'. She wrote: *Contribution to the knowledge of the genus Myleus Müller & Troschel, 1844 (Pisces – Characidae) of the Negro River, AM* (1986) which is described in the etymology as 'her pioneer attempt' (a 1986 PhD dissertation) to organise what is known about herbivorous *Serrasalmidae* from the Rio Negro basin.

Lucifer

Deep-sea Angler sp. *Linophryne lucifer* R Collett, 1886

Blackbelly Lantern Shark *Etmopterus lucifer* DS Jordan & JO Snyder, 1902

Dragonfish sp. *Astronesthes lucifer* CH Gilbert, 1905

Rattail sp. *Lucigadus lucifer* HM Smith & L Radcliffe, 1912

The fallen angel Lucifer, of the Judeo-Christian tradition, is referred to in the fourteenth chapter of the Biblical book of Isaiah: '*How art thou fallen from heaven, O Lucifer, son of the morning! How art thou cut down to the ground, which didst weaken the nations!*' The meaning of the name Lucifer is 'Light-bearer', and the name was given to the Morning Star (Venus). It seems the fish were generally thus named because of their possession of luminescent organs, without reference to any 'demonic' quality.

Lucifer (Mountains)

Armoured Catfish sp. *Harttiella lucifer* R Covain & S Fisch-Muller, 2012

This is a toponym; referring to the Lucifer Mountains, French Guiana, where the holotype was collected.

Lucille

Panama Triplefin *Axoclinus lucillae* HW Fowler, 1944

Louise 'Lulu' Miriam Vanderbilt née Parsons (1912–2013) was the first wife of the yachtsman and explorer George Washington Vanderbilt III who financed and led the expedition that collected the holotype. She was with him in Sumatra (1937 & 1939). She was well known as an expert golfer, horsewoman and rifle shot. They divorced (1946) and she married Ronald Balcom (1949). Oddly, her name *wasn't* Lucille (it might be expected that the binomial would be *louisae*), though the etymology states that the fish is named after "Mrs. George Vanderbilt". However, her daughter was Lucille Margaret (b.1938), suggesting that Fowler may have intended the Vanderbilts' young child.

Lucinda

Characin sp. *Serrapinnus lucindai* LR Malabarba & FC Jerep, 2014

Dr Paulo Henrique Franco Lucinda is a Brazilian zoologist, bio-geographer and ichthyologist at Universidade Federal do Tocantins, Porto Nacional, Brazil, where he is Associate Professor I. The Federal University of Viçosa awarded his bachelor's degree (1991), the Federal University of Paraná his master's (1994) and the Pontifical Catholic University of Rio Grande do Sul his doctorate (2003). He did post-doctoral work at the University of Michigan Museum of Zoology, Ann Arbor, and has been a visiting scholar at Museum of Comparative Zoology, Harvard, the Field Museum of Natural History, Chicago and at the Smithsonian. He co-wrote: *Pimelodus quadratus, a new long-whiskered catfish from the Rio Tocantins drainage, Brazil (Siluriformes: Pimelodidae)* (2016).

Lucretia

Lucretia's Goby *Parrella lucretiae* CH Eigenmann & RS Eigenmann, 1888

Lucretia M Smith née Gray (1817–1911) was the mother of the second author Rosa Eigenmann née Smith (1858–1947) and mother-in-law of her husband (1887) Carl Henry Eigenmann (1863–1927).

Lucula

Angolan Cichlid sp. *Haplochromis lucullae* GA Boulenger, 1913
[Syn. *Thoracochromis lucullae*]

This is a toponym referring to the Lucula River, Angola; the type locality.

Lucullus

Morid Cod sp. *Svetovidovia lucullus* AS Jensen, 1953

Lucius Licinius Lucullus (118–57/56 BC) was a Roman consul and general famous for his wealth, luxury and banquets. Why the name was chosen is not explained.

Luebbert

Lübbert's Guitarfish *Rhynchobatus luebberti* EME Ehrenbaum, 1915
[Alt. African Wedgefish]

Hans Julius Lübbert (1870–1951) studied politics and law at universities in Kiel and Rome, but caught severe malaria and had to change course. He was in the civil service (1904–1919) working as a fisheries inspector at Cuxhaven, Lower Saxony, Germany and became Director of the Hamburg State Fisheries Service (1922). In retirement, he researched and taught, but as he was of Jewish descent he was expelled from all societies and public bodies by the Nazi regime. In the aftermath of WW2, he was re-instated as Fisheries Director (1945) to rebuild the local fishing industry. With Ernst Ehrenbaum he co-wrote: *Handbuch der Seefischerei Nordeuropas* (1930).

Lueke

African Rivuline sp. *Nothobranchius luekei* L Seegers, 1984

Karl-Heinz Lüke is a German aquarist who was first to breed this species.

Lueling

Characin sp. *Cheirodon luelingi* J Géry, 1964
Tetra sp. *Hemigrammus luelingi* J Géry, 1964
Dwarf Cichlid sp. *Apistogramma luelingi* SO Kullander, 1976
Neotropical Rivuline sp. *Atlantirivulus luelingi* L Seegers, 1984

Dr Karl-Heinz Lüling (1913–1984) was a German ichthyologist at Forschungsmuseum Alexander Koenig (Bonn) where he was the first Curator and Chairman of the Ichthyology Department until retiring (1978). He was a specialist on lungfishes. He wrote: *South American fish and their habitat* (1978). He collected the cichlid (1965) and sent it to Kullander.

Luetken

Dr Christian Frederik Lütken (1827–1901). (See under **Lütken**)

Lufira

Kneria sp. *Parakneria lufirae* M Poll, 1965
Squeaker Catfish sp. *Synodontis lufirae* M Poll, 1971
Squeaker Catfish sp. *Chiloglanis lufirae* M Poll, 1976

This is a toponym; it refers to the Lufira River, Democratic Republic of Congo, where these fish are found.

Lugo

Tufa Darter *Etheostoma lugoi* SM Norris & WL Minckley, 1997

José 'Pepe' Lugo Guajardo has acted as a guide for many scientists studying the fauna of Mexico's Cuatro Cienegas Basin. He also has a reptile named after him.

Luita Lima

Neotropical Rivuline sp. *Anablepsoides luitalimae* DTB Nielsen, 2016

Luita Lima is the aunt of the discoverer, Gilson Pontes Lima.

Luitpold

Green Goodea *Goodea luitpoldii* F Steindachner, 1894

Prince Regent Luitpold of Bavaria (1821–1912) seems to have been honoured in the binomial because his daughter, Princess Therese of Bavaria (1850–1925), collected the type specimen (though one wonders why Steindachner didn't name it *G. theresae*). She was an ethnologist, naturalist and travel writer. She visited Mexico, where the species occurs (1893), and later made a tour of South America (1898).

Luja

Congo Cyprinid sp. *Leptocypris lujae* GA Boulenger, 1909
African Rivuline sp. *Aphyosemion lujae* GA Boulenger, 1911
African Barb sp. *Enteromius lujae* GA Boulenger, 1913

Édouard-Pierre Luja (1875–1953) was a horticulturalist, botanist and entomologist from Luxembourg. He studied in Belgium, France and England but began to turn to biology and exploration. He went to the Congo and Mozambique and to South America, visiting Brazil in particular. There he collected for a number of individuals and institutions specialising in insects (and especially ants), but all manner of animals and plants. He discovered at least 80 new plants and 120 animals including fish. He collected all three holotypes at Kasai in the Congo. A dragonfly is also named after him.

Lujan

Loach Catfish sp. *Amphilius lujani* AW Thomson & LM Page, 2015
Armoured Catfish sp. *Peckoltia lujani* JW Armbruster, DC Werneke & M Tan, 2015

Dr Nathan Keller Lujan (b.1976) is an ichthyologist, taxonomist, ecologist and evolutionary biologist who is a postdoctoral researcher at the Canadian Department of Fisheries and Oceans, Toronto. He is also an Associate Editor of the *Journal of Fish Biology* as well as a Research Associate at the Noonan Lab, University of Mississippi and an International Research Fellow of the Royal Ontario Museum. Calvin College awarded his BSc (2000) and Auburn University his PhD in Evolutionary Biology (2009). He was a field entomologist (2000–2002) at the Tennessee Valley Authority and undertook some teaching while furthering his studies. He was a postdoctoral fellow at Texas A&M University (2009–2011), then at the Royal Ontario Museum (2011–2014), Philadelphia Academy of Natural Science (2014–2015) and is now Canada DFO postdoctoral fellow at the University of Toronto. He has published around fifty papers, articles and book chapters, and has conducted aquatic biodiversity inventories throughout eastern North America, in seven South American countries, and in Uganda. According to Jonathan Armbruster who supervised his PhD, Lujan has made expeditions to 'some of the most remote regions of South America', giving Armbruster 'more taxonomic work in the last decade than he had thought possible'. This included collecting the best specimens known of *Peckoltia lujani*. In naming the species after him the authors said it was Lujan "*…who collected the holotype and most paratypes of this species, and who has made excellent contributions to our knowledge of freshwater fishes.*"

Lukoma

Lake Malawi Cichlid sp. *Placidochromis lukomae* M Hanssens, 2004

This is a toponym referring to Lukoma Bay, Lake Malawi, Tanzania; the type locality.

Lukula

Red-spot Mudsucker *Labeo lukulae* GA Boulenger, 1902

A toponym referring to the Lukula (Lakula) River, Democratic Republic of Congo; the type locality.

Lulua

Congo Cichlid sp. *Ctenochromis luluae* HW Fowler, 1930
African Barb sp. *Enteromius luluae* HW Fowler, 1930
Labeo sp. *Labeo luluae* HW Fowler, 1930

These are toponyms referring to the Lulua River, Democratic Republic of Congo, where the fish are found.

Lumbantobing

Freshwater Goby sp. *Pseudogobiopsis lumbantobing* HK Larson, RK Hadiaty & N Hubert, 2017

Dr Daniel Natanael Lumbantobing is an Indonesian ichthyologist whose doctorate was awarded by the George Washington University, Washington DC, and is now at the Florida Museum of Natural History, Gainsville, where he is a Post-Doctoral Research Fellow. He wrote: *Four new species of Rasbora of the Sumatrana group (Teleostei: Cyprinidae) from northern Sumatra, Indonesia (2014).* He collected the first specimens of this species (2012).

Lumbee

Sandhills Chub *Semotilus lumbee* FF Snelson Jr & RD Suttkus, 1978

Not a true eponym as it refers to a people; the Lumbee Indians who inhabited the Lumber River system in North Carolina, USA, the type locality.

Lumholtz

Borneo Catfish sp. *Kryptopterus lumholtzi* CH Rendahl, 1922

Dr Carl Sophus Lumholtz (1851–1922) was a Norwegian naturalist, ethnologist, humanist and explorer. Having just graduated with a natural science degree (1880), he set off for northeastern Australia, where he spent time (1880–1884) living with Aboriginal people. He organised a number of expeditions, including one to explore the Sierra Madre, Mexico (1890), for the American Museum of Natural History. He visited Borneo (1914), but a planned a trip to New Guinea was prevented by the outbreak of WW1. Lumholtz National Park in Queensland is named after him, as are also a tree kangaroo, a reptile and two birds.

Lumnitzer

Sydney's Pygmy Pipehorse *Idiotropiscis lumnitzeri* RH Kuiter, 2004

Ákos Gábor Lumnitzer (b.1972) was born in Budapest, Hungary, and emigrated to Australia (1985). He has a university business qualification and works in the logistics/transport sector. In 'private life' he is a diver and underwater photographer (since 1996). He discovered the pipehorse by photographing it and sending the photos to Mark McGrouther at the Australian Museum, who confirmed it was an undescribed form. He then collected the type specimens (1997) and took them to the Museum. He has made over 2000 dives, mostly in New South Wales. Since this award-winning nature photographer moved inland (2001) he no longer dives, and has turned his passion instead to bird photography.

Luna

Armoured Catfish sp. *Hypancistrus lunaorum* JW Armbruster, NK Lujan & DC Taphorn Baechle, 2007

The Luna family founded the village of Macurucu on the Río Orinoco near the mouth of the Ventauri (Amazonas, Venezuela). They were thanked as their "...*progressive interest in the development of Macurucu via promotion of scientific research in the nearby region has been indispensable to the completion of recent fieldwork*."

Lund

Characin sp. *Cyphocharax lundi* GM Dutra, I de S Penido, GCG Mello & TC Pessali, 2016

[Junior Synonym of *Cyphocharax albula*]

Dr Peter Wilhelm Lund (1801–1880) was a Danish physician, botanist, zoologist and palaeontologist who lived and worked in Lagoa Santa, Minas Gerais, Brazil. He first travelled to Brazil (1833) and settled there for health reasons. His interest in fossils led him to explore many of the caves of the area. He assembled one of the most important mammal collections from a single locality in the Neotropics, and made outstanding contributions towards describing the Pleistocene and recent mammal faunas of Brazil. He regularly corresponded with Charles Darwin. *Two mammals, an amphibian and a reptile are named after him.*

Lundberg

Phantom Blindcat *Prietella lundbergi* SJ Walsh & CR Gilbert, 1995

Pencil Catfish sp. *Typhlobelus lundbergi* SA Schaefer, F Provenzano, MCC de Pinna & JN Baskin, 2005

Glass Knifefish sp. *Rhabdolichops lundbergi* SB Correa, WGR Crampton & JS Albert, 2006

Armoured Catfish sp. *Loricaria lundbergi* MR Thomas & LH Rapp Py-Daniel, 2008

Banjo Catfish sp. *Pseudobunocephalus lundbergi* JP Friel, 2008

Bumblebee Catfish sp. *Microglanis lundbergi* LR Jarduli & OA Shibatta, 2013

Dr John Graham Lundberg (b.1942) is an American ichthyologist. His bachelor's degree was awarded by Fairleigh Dickinson University (1964) and his doctorate by the University of Michigan (1970). He was Full Professor at Duke University (1970–1992) and at the University of Arizona (1992–2000). He moved to the Academy of Natural Sciences, Philadelphia (2000) as Chairman and Curator of Ichthyology. He is now Emeritus Professor, Department of Biodiversity, Earth & Environmental Science and Emeritus Curator, Academy of Natural Sciences of Drexel University. His focus is in tropical fish diversification and evolution. Among his published papers and contributions, he wrote: *The Temporal Context for Diversification of Neotropical Fishes* (1998) and co-wrote: *Tributaries Enhance the Diversity of Electric Fishes in Amazon River Channels* (2004). (Also see **Joannis**)

Lung

Neotropical Rivuline sp. *Anablepsoides lungi* HO Berkenkamp, 1984

Karl Lung was a German aquarist who collected this species in French Guiana. We have been unable to unearth more.

Lusher

Streaky Clingfish *Lissonanchus lusheri* JLB Smith, 1966
[Binomial sometimes corrected to the feminine *lusherae*]

Mrs D N Lusher of Brakpan, South Africa, was honoured in the name as she found the type specimen and presented it to the author, having collected it at Ponte Zavora in southern Mozambique.

Luther, AF

Luther's Spiny Loach *Cobitis lutheri* CH Rendahl, 1935

Alexander Ferdinand Luther (1877–1970) was a Finnish zoologist and developmental biologist who was an expert on Turbellarians (flatworms). He worked at Spemann's laboratory in Rostock, Germany (1912–1914). He became Professor Extraordinary of Zoology at the University of Helsinki (1918) and head of the Tvärminne Zoological Station (1919). He collected the holotype of this species and sent it to his friend, Carl Rendahl.

Luther, W

Luther's Prawn-Goby *Cryptocentrus lutheri* W Klausewitz, 1960

Wolfgang Luther worked at the Zoologischen Institut der Technischen Hochschule, Darmstadt, Germany. The fish was named in recognition of his early studies of the symbiotic relationship between some gobies and crustaceans. He co-wrote: *Field Guide to the Mediterranean Sea Shore* (1976).

Luthuli

Snake-Eel genus *Luthulenchelys* JE McCosker, 2007

Inkosi (Chief) Albert John 'Mvumbi' Luthuli (c.1898–1967) of KwaZulu-Natal was a politician, anti-apartheid activist, lay preacher and a teacher who was Africa's first winner of the Nobel Peace Prize (1960) and former President of the African National Congress (1952–1967). He was born in Southern Rhodesia (Zimbabwe), but was taken to South Africa as a child after his father died. His autobiography was called: *Let My People Go* (1962).

Lutke

Lutke's Halfbeak *Hemiramphus lutkei* A Valenciennes, 1847

Count Frédéric Benjamin Lütke (1797–1882) was a Russian (of German descent) explorer and navigator. His name is 'Russified' as Fyodor Petrovich Litke. He began his career in the Imperial Russian Navy (1813) and took part in a world cruise on board the *Kamchatka* (1817–1819). He led an expedition (1821–1824) to explore the coastline of Novaya Zemlya, the White Sea and parts of the Barents Sea. He then led another world cruise (1826–1829) on the *Senyavin*. Though the original description does not specify who the fish was named after, there can be no real doubt that Count Frédéric was intended.

Lütken

Gilded Catfish *Zungaro luetkeni* F Steindachner, 1876
[Junior Synonym of *Zungaro zungaro*]
Armoured Catfish sp. *Hypostomus luetkeni* F Steindachner, 1877
Lütken's Eelpout *Lycodes luetkenii* R Collett, 1880
Pricklefish *Acanthochaenus luetkenii* TN Gill, 1884
Tetra sp. *Hyphessobrycon luetkenii* GA Boulenger, 1887
Lütken's Lanternfish *Diaphus luetkeni* AB Brauer, 1904
Dreamer sp. *Oneirodes luetkeni* CT Regan, 1925
Snaggletooth sp. *Astronesthes luetkeni* CT Regan & E Trewavas, 1929
Naked Barracuda *Lestrolepis luetkeni* V Ege, 1933
Pencil Catfish sp. *Trichomycterus luetkeni* AM Katz & WJEM Costa, 2021

Professor Christian Frederik Lütken (1827–1901) was a Danish naturalist, specialising in marine zoology. As a young man he was a professional soldier in the Danish army, but resigned his commission (1852) to concentrate on natural history and taking his master's degree (1853). He became assistant to Japetus Steenstrup at the University of Copenhagen Zoological Museum (1853–1885), replacing him as Professor of Zoology and Director there (1885–1899). He also taught at the Polytechnic School (1877–1881) before returning to the University. He suffered a stroke that left him paralysed (1899) and forced him to retire. He co-wrote: *Bidrag til Kundskab om Brasiliens Padder og Krybdyr* (1861) and some very successful books popularising natural history. Two amphibians are named after him, as are a number of marine organisms.

Lutz

Oaxaca Livebearer *Poeciliopsis lutzi* SE Meek, 1902

Frank Eugene Lutz (1879–1943) was an America entomologist. Haverford College awarded his BA (1900) and the University of Chicago his MA (1902). He then accompanied Meek (who was working out of the Field Museum, Chicago) to Mexico as his volunteer assistant. He studied at University College London, and then was resident investigator at the Carnegie's Station for Experimental Evolution at Long Island (1904–1909). The University of Chicago awarded his PhD (1907). He became Assistant Curator of Invertebrate Zoology at AMNH, New York, and later (1916) Associate Curator. He wrote numerous articles and papers, and several books including: *Field Book of Insects* (1917).

Luzardo

Neotropical Rivuline sp. *Austrolebias luzardoi* E Perujo, PA Calvino, H Salvia & F Prieto, 2005
[Junior Synonym of *Austrolebias periodicus*]

Hector Luzardo (1937–2004) was a Uruguayan-born naturalist, aquarist, fish breeder and killifish enthusiast who spent most of his life in Argentina.

Lyons, E

Lyon's Cichlid *Amphilophus lyonsi* JP Gosse, 1966

E Lyons was the Belgian ambassador to Panama. The type was collected in Costa Rica during the collecting trip of King Leopold of Belgium (1965).

Lyons, JD

Tamazula Redtail Splitfin *Xenotoca lyonsi* O Domínguez-Domínguez, DM Bernal-Zuñiga & KR Piller, 2016

Dr John D Lyons is an American fish biologist and ichthyologist who is Curator of Fishes at the University of Wisconsin Zoological Museum and was Fisheries Research Supervisor for the Wisconsin Department of Natural Resources (1986–2017). Union College, NY awarded his BSc and the University of Wisconsin-Madison awarded his MSc and PhD. As curator he has conducted conservation-oriented work on freshwater fish throughout Mexico. His long-term focus has been the goodeid fishes in central Mexico. He is Chair of the North American branch of the Goodeid Working Group (NAGWG) and an honorary member of the Mexican Ichthyological Society (SIMAC). The etymology says the fish was named after "*...the prominent North American ichthyologist, Dr. John Lyons, who has made substantial contributions to our understanding of the distribution, ecology, diversity, and conservation status of fishes in Mexico, and to goodeids in particular.*" Among his c.180 publications is the co-authored: *Notropis calabazas (Teleostei; Cyprinidae): New Species from the Río Pánuco Basin of Central México* (2004).

M

Maack

Armoured Catfish sp. *Rineloricaria maacki* LFS Ingenito, MS Ghazzi, LF Duboc & V Abilhoa, 2008

Reinhard Maack (1892–1969) was a German explorer, geographer and geologist. He arrived in South West Africa (Namibia) (1911) and fought in WW1 in the German Colonial Army. He was headmaster of Windhoek School and discovered the famous 'White Lady' rock painting (1918). He left Africa for Brazil (1923) and worked as a mining engineer for a number of companies, including prospecting for diamonds. When Brazil entered WW2 on the side of the allies he was interned as an enemy alien and only released (1944). He was a landowner and became a Brazilian citizen (1949). He was honoured for having made "… *some of the most important contributions to the knowledge of the geology and physiography of the rio Iguaçu basin and Paraná State.*"

Maafu

Damsel sp. *Pomacentrus maafu* GR Allen & JA Drew, 2012

Enele Ma'afu'out'itonga, commonly known as Ma'afu (1826–1881), was a Tongan prince and warrior who, at age 22, migrated to the Lau Group of Fiji and eventually became leader of the resident Tongan community and Fijian Chief; he was a "*…man of two kingdoms and as such his influence reflects the geographic distribution*" of this species, which occurs in both Fiji and Tonga.

Maass

Malay Cyprinid sp. *Malayochela maassi* M Weber & LF de Beaufort, 1912
Forktail Loach *Vaillantella maassi* M Weber & LF de Beaufort, 1912
[Alt. Spiny-eel Loach]

Alfred Maass (1863–1936) was a German traveller, naturalist and anthropologist. He wrote the two-volume: *Durch zentral Sumatra* (1910 & 1912). He led an expedition to Sumatra (1907), during which the holotypes of both species were collected.

Mabahiss

Lightfish sp. *Vinciguerria mabahiss* RK Johnson & RM Feltes, 1984

'Mabahiss' is an Egyptian research ship. The name is also apparently "*...for her captain and crew, for the scientists aboard, for the organizing committee and supporters, and for scientists serving as authors of the eleven volumes (1935–1967) issued as Scientific Reports of the John Murray Expedition (1933–1934) to the Red Sea, where this species occurs.*"

MacCain

McCain's Skate *Bathyraja maccaini* S Springer, 1971

Dr John Charles McCain (b.1939). (See under **McCain**)

MacClure

Redlip Blenny *Ophioblennius macclurei* CF Silvester, 1915

Dr Charles Freeman Williams McClure (1865–1955). (See under **McClure**)

MacConnel

Lake Turkana Cichlid sp. *Haplochromis macconneli* PH Greenwood, 1974

Frederick Vavasour McConnel (1868–1914). (See under **McConnel**)

MacCoy

Southern Bluefin Tuna *Thunnus maccoyii* FL Castelnau, 1872

Sir Frederick McCoy (1817–1899). (See under **McCoy**)

MacCulloch

Australian Freshwater 'Cod' genus *Maccullochella* GP Whitley, 1929
MacCulloch's Rainbowfish *Melanotaenia maccullochi* JD Ogilby, 1915
McCulloch's Hardyhead *Atherion maccullochi* DS Jordan & CL Hubbs, 1919
McCulloch's Tongue Sole *Cynoglossus maccullochi* JR Norman, 1926
Half-banded Seaperch *Hypoplectrodes maccullochi* GP Whitley, 1929
McCulloch's Bandfish *Owstonia maccullochi* GP Whitley, 1934
Goby sp. *Nesogobius maccullochi* DF Hoese & HK Larson, 2006

Allan Riverstone McCulloch (1885–1925). (See under **McCulloch**)

MacCune

Sleeper sp. *Mogurnda maccuneae* AP Jenkins, PM Buston & GR Allen, 2000

Amy McCune (See under **McCune**)

MacDonald, D

Australian Darter Dragonet *Callionymus macdonaldi* JD Ogilby, 1911

Captain Donald MacDonald is described in the etymology as: "*...late first officer of the FIS 'Endeavour', and now chief pilot at Keppel Bay, to whom I am indebted for many favours.*"

MacDonald, M

Totoaba *Totoaba macdonaldi* CH Gilbert, 1890
Mexican Rockfish *Sebastes macdonaldi* CH Eigenmann & CH Beeson, 1893
Rakery Beaconlamp *Lampanyctus macdonaldi* GB Goode & TH Bean, 1896
Yellowfin Cutthroat Trout *Oncorhynchus clarkii macdonaldi* DS Jordan & CH Evermann, 1890

Colonel Marshall 'Marsh' McDonald (1836–1895). (See under **McDonald**)

MacEachran

Madagascar Pygmy Skate *Fenestraja maceachrani* B Séret, 1989

Dr John D McEachran (b.1941). (See under **McEachran**)

Machado, A

Three-barbeled Catfish sp. *Gladioglanis machadoi* CJ Ferraris & F Mago-Leccia, 1989

Characin sp. *Creagrutus machadoi* RP Vari & AS Harold, 2001

Armoured Catfish sp. *Hypoptopoma machadoi* AE Aquino & SA Schaefer, 2010

Dr Antonio Machado-Allison (b.1945) is a Venezuelan zoologist and biologist. Universidad Central de Venezuela awarded his bachelor's degree (1971) and the George Washington University his doctorate (1982). He joined the faculty at the Universidad Central de Venezuela (1971), becoming Professor (1991) and is now retired as Professor Emeritus. He wrote: *The fish of the plains of Venezuela* (2005). The etymology for the *Creagrutus* thanks him for his laboratory and field assistance to the authors, and his many contributions to our knowledge of Neotropical fishes.

Machado, A de Barros

African Barb sp. *Enteromius machadoi* M Poll, 1967

Cunene Dwarf Happy (cichlid) *Orthochromis machadoi* M Poll, 1967

Airbreathing Catfish sp. *Platyclarias machadoi* M Poll, 1977

Dr António de Barros Machado (1912–2002) was a Portuguese zoologist and arachnologist. The University of Porto awarded his BSc (1929) and he was appointed as an Assistant in Zoology there (1934). He then worked at the MNCN in Madrid and studied for his PhD, but this was interrupted by the civil war and he returned to Porto (1936). He then worked at the NMHN, Paris, before becoming a free researcher (1936–1947) and taught biology and geography in secondary schools (1937–1947). He went to Angola as he was unable to get work in a Portuguese University because of his politics, which were radical and disliked by the Salazar régime to the extent that he was prevented from taking up a post as an assistant in zoology at the University of Coimbra (1945). He was Director of the Dundo Laboratory and Museum of Zoology and Anthropology in Angola (1947–1974). He became the President of the Portuguese Society of Ethnology (1978) and finally received a doctorate from the Abel Salazar Biomedical Institute of the University of Porto (1990), becoming Professor Emeritus there. Two odonata are also named after him.

Machado, F

Tetra sp. *Hemigrammus machadoi* RP Ota, FCT Lima & CS Pavanelli, 2014

Characin sp. *Gephyrocharax machadoi* KM Ferreira, E de Faria, AC Ribeiro, JC Santana, I Quagio-Grassioto & NA Menezes, 2018

Dr Francisco de Arruda Machado is a Brazilian ichthyologist and ecologist whose master's degree (1983) and doctorate (2003) were both awarded by the State University of Campinas. He worked at Universidade Federal de Mato Grosso, where he was Associate Professor III but is now retired. He is presently Special Adviser in the Public Ministry of the State of

Mato Grosso to advise on environmental issues related to fish in particular and animals in general. He wrote: *Natural history of Pantanal fish: with emphasis on food habits and defence against predators* (2003). The etymology mentions his 'tireless dedication' in surveying the fishes of Mato Grosso, and his 'struggle for their conservation'.

Machado (da Silva)

Pencil Catfish sp. *Plectrochilus machadoi* A de Miranda Ribeiro, 1917
Sea Catfish sp. *Genidens machadoi* A de Miranda Ribeiro, 1918
Crestfish sp. *Lophotus machadoi* A de Miranda Ribeiro, 1927

Rev Francisco Machado da Silva collected for and/or donated specimens to the Museu Urbis of Rio de Janeiro.

Machida

Machida's Coralbrotula *Diancistrus machidai* W Schwarzhans, PR Møller & JG Nielsen, 2005
Cusk-eel sp. *Neobythites machidai* S Ohashi, JG Nielsen & M Yabe, 2012
Machida's Snake-eel *Ophichthus machidai* JE McCosker, S Ide & H Endo, 2012

Dr Yoshihiko Machida is a Japanese ichthyologist, whose doctorate was awarded by Kyushu University (1974). He became Professor at the Department of Biology, Kochi University, Japan (1989), and is now retired as Professor Emeritus and a researcher. He has published many papers including descriptions of new species, such as: *Parabrosmolus novaeguineae, a new genus and species of the subfamily Brosmophycinae from Papua New Guinea (Bythitidae, Ophidiiformes)* (1996). He was honoured in the snake-eel's name as the person who guided second author (Sachiko Ide) through her thesis at Kochi University.

Machlan

Weasel Shark sp. *Hemigaleus machlani* AW Herre, 1929
[Probably a Junior Synonym of *H. microstoma*]

Perry Lester Machlan (1880–1941) was an American who moved to the Philippines (c.1913) and was Acting Collector of Customs, Sitankai, when Herre described the species. Herre honoured his: "*esteemed friend... ...for assisting in his study of the fishes of the Sulu Archipelago.*"

Machris

Scrapetooth Characin sp. *Apareiodon machrisi* H Travassos, 1957

Maurice Alfred Machris (1905–1980) was an American millionaire businessman in the oil industry, a collector, trophy hunter and sponsor. He sponsored and went on many expeditions, including the Los Angeles County Museum expedition to Brazil (1956) which was co-sponsored by his wife and which he and his wife co-led with Jean Delacour. He was President of the Shikari Safari Club. A bird is also named after him, as are some insects collected on the Brazil expedition.

Mack

Whiskery Shark *Furgaleus macki* GP Whitley, 1943

George Mack (1899–1963) was a British-born ichthyologist and ornithologist who emigrated from the UK to Western Australia (1919) and worked at the National Museum of Victoria, Melbourne (1923–1945) and at the Queensland Museum (1946–1963). He became Director just before his death.

Mackay, CL

Pighead Prickleback *Acantholumpenus mackayi* CH Gilbert, 1896

Charles Leslie McKay (1855–1883) was an amateur naturalist in the US Army Signal Corps, who collected specimens in the Nushagak area. He disappeared (April 1883) when out on a collecting trip in a kayak. Although no etymology is given, the holotype was taken by the Steamer 'Albatross' during an expedition (1890–1891) to Alaskan waters, near the mouth of the Nushagak River, so we believe it was named in memory of this naturalist.

Mackay, DJ

Mackay's Torpedo *Torpedo mackayana* J Metzelaar, 1919
[Alt. Ringed Torpedo, West African Torpedo]

Donald Jacob Baron Mackay (1839–1921), who was born a Dutch national in the Hague, became 11th Lord Reay on the death of his father (1876) and changed his nationality to British (1877). He had a distinguished career in politics and as a colonial administrator and governor, including being Governor of Bombay (1885–1890) and serving as Under-Secretary of State for India (1894–1895). He maintained very close contact with the Dutch community and the Netherlands and was one of the promoters of an expedition that collected fish in the Dutch West Indies (1904–1905).

Macklot

Hardnose Shark *Carcharhinus macloti* JP Müller & FGJ Henle, 1839

Heinrich Christian Macklot (1799–1832) was a German naturalist and taxidermist. After studying medicine at the University of Heidelberg, he was appointed to assist members of the Dutch Natural Science Commission in collecting specimens for the Leiden Museum. He took part in an expedition to New Guinea and Timor (1828–1830). Three birds, a reptile and a mammal are named after him.

MacLaren

Airbreathing Catfish sp. *Clarias maclareni* E Trewavas, 1962
Barombi Mbo cichlid sp. *Pungu maclareni* E Trewavas, 1962

Major Peter I R MacLaren (d.1956) was an ornithologist and fisheries officer. He was a Major in the Royal Indian Army Service Corps, stationed in Assam (1945), and published extensively on birds of the Middle East and India, including: *Spring Passage of Phalaropes in Iraq* (1946). In the late 1940s he was Fisheries Development Officer in NIgeria and in the 1950s he filled a similar position in Northern Rhodesia (Zambia). He was attacked and killed by a crocodile.

Macleay

Macleay's Crested Pipefish *Histiogamphelus cristatus* WJ Macleay, 1881

Macleay's Glass Perchlet *Ambassis macleayi* FL Castelnau, 1878
Australian Sandpaper Fish *Paratrachichthys macleayi* RM Johnston, 1881
Narrow-banded Sole *Synclidopus macleayanus* EP Ramsay, 1881
Australian Marbled Catshark *Atelomycterus macleayi* GP Whitley, 1939

Sir William John Macleay (1820–1891) was a Scottish medical student who followed his uncle Alexander to Sidney (1838), where he became an all-round naturalist. He wrote widely on entomology, ichthyology and zoology and took part in several collecting expeditions. He published a two-volume *Descriptive Catalogue of Australian Fishes* (1881). The whole of the Macleay family were avid naturalists and collectors, so prolifically that the Macleay Museum University of Sydney was built (1887) to house their vast collection. Alexander Macleay's insect collection was added to by his son, William Sharp Macleay (1792–1865), and expanded to include all aspects of natural history by William's cousin, William John Macleay. There is no etymology in Whitley's description of the Catshark, so which Macleay it was named after remains a mystery: perhaps the family in general was intended. A mammal, a reptile and two birds are named after him.

Macloviana

Patagonian Skate *Bathyraja macloviana* JR Norman, 1937

The original description has no etymology. However, the skate is not named for a person called MacLove or some such, but is derived from the Latin name for the town St Malo in Brittany, France. People from this town are called 'Malouines'. The Spanish name 'Malvinas' (for the Falkland Islands) is a derivative of the French name les Malouines. Louis Antoine de Bougainville christened the islands (1764) in reference to (Brittany) Saint Malo, from which his expedition departed. Other taxa of fauna and flora from the Falklands are also called *macloviana*/*maclovianus* after these islands.

MacMahon

Pakistani Carp sp. *Tariqilabeo macmahoni* E Zugmayer, 1912
[Syn. *Labeo macmahoni*]

Sir Arthur Henry McMahon (1862–1949). (See under **McMahon**)

Macmaster

Macmaster's Dwarf Cichlid *Apistogramma macmasteri* SO Kullander, 1979

Mark McMaster (see under **McMaster**).

MacMillan

McMillan's Catshark *Parmaturus macmillani* GS Hardy, 1985

Peter John McMillan (b.1955) (See **McMillan, PJ**)

MacNeill

Shorttail Torpedo *Tetronarce macneilli* GP Whitley, 1932
Blue Devilfish *Assessor macneilli* GP Whitley, 1935

Francis 'Frank' Alexander McNeill (1896–1969). (See **McNeill**)

Macoun

Pacific Viperfish *Chauliodus macouni* TH Bean, 1890

John C Macoun (1831–1920) was a self-taught botanist who emigrated from Northern ireland to Canada (1850). He farmed for some years and became a schoolteacher (1856), rising to become Professor of Botany and Geology, Albert College, Belleville, Ontario (1868). He took part in five surveying expeditions (1872–1881) concerned with proposed routes of the Canadian Pacific Railway. He moved to Ottawa to work for the Geological Survey of Canada (1881–1912), becoming Assistant Director (1887). Not everyone was a fan of this temperamental and outspoken character: in 1883 his superior, Lindsay Alexander Russell, deputy minister of the interior, called him "a good specialist and honest fool outside of that."

MacRae

Pencil Catfish sp. *Hatcheria macraei* CF Girard, 1855

Lieutenant Archibald MacRae (1820–1855) was an American naval officer. He was a member of the US Naval Astronomical Expedition to the Southern Hemisphere (1849–1852). He wrote: *Report of a journey across the Andes and pampas of the Argentine provinces* (1855). He collected the holotype of this species. He committed suicide whilst on board the U.S. Surveying schooner *Ewing*, for unknown reasons.

Macturk

Three-barbeled Catfish sp. *Pimelodella macturki* CH Eigenmann, 1912

Michael McTurk (1843–1915) (See under **McTurk**)

Macushi

Armoured Catfish sp. *Hypostomus macushi* JW Armbruster & LS de Souza, 2005

The Macushi people of the northern Rupununi, southern Guyana, gave the authors a great deal of help and hospitality. They collected most of the best specimens in the type series.

Madhava

Sri Lankan Loach sp. *Schistura madhavai* H Sudasinghe, 2017

Dr Madhava Meegaskumbura is a Sri Lankan evolutionary biologist who is Professor at Guanxi University College of Forestry (2018–2019). He was Professor of Molecular Biology at University of Peradeniya (2016), having worked there as a lecturer (2009–2016). He is also a Research Associate of Harvard University, Museum of Comparative Zoology (2007–present) and of the University of Boston Department of Biology (2001–present). He was a post-doctoral student at Harvard (2007–2009) after being a Research Assistant at Boston (2000–2007), where he completed his PhD (2007), and Research Biologist at the Wildlife Heritage Trust of Sri Lanka (1998–2000). He has discovered nearly 100 previously undescribed frogs! His publications include: *Description of eight new species of shrub frogs (Ranidae: Rhacophorinae:* Philautus*) from Sri Lanka* (2005) and *Life among crevices: osteology of Nannophrys marmorata (Anura: Dicroglossidae)* (2015).

Madhusoodana

Spotted Barb sp. *Puntius madhusoodani* K Krishnakumar, FG Benno Pereira & KV Radhakrishnan, 2012

Professor Dr B Madhusoodana Kurup is a fishery scientist who was founding Vice-Chancellor, Kerala University of Fisheries and Ocean Sciences (2011). He has three decades of research and teaching experience at Cochin University of Science & Technology and Kerala Agricultural University. He was Associate and Assistant Professor in Kerala Agricultural University, Professor in CUSAT for more than 12 years and Director of School of Industrial Fisheries (2008–2011). He also served as technical advisor to the Minister for Fisheries, Government of Kerala (2006–2011), with the rank of Government Secretary. Cochin University awarded his master's in Marine Biology and his PhD in Fisheries Science. He has written over 300 papers, co-authored two books and edited eight others.

Maekawa

Maekawa's Perchlet *Plectranthias maekawa* H Wada, H Seou & H Motomura, 2018

Takanori Maekawa is an amateur ichthyologist working in the industry who has often presented fish gleaned as by-catch from catches and fish markets that he recognises as being possibly new to science. The etymology says that the species is named in honour of "*…Mr. Takanori Maekawa (and the Maekawa Fisheries Co., Ltd.), who has kindly supported our ichthyofaunal research in the Ryukyu Islands.* "

Maeotis

Black Sea Brill *Scophthalmus maeoticus* PS Pallas, 1814
Azov Percarina *Percarina maeotica* ID Kuznetsov, 1888
Black Sea Shad *Alosa maeotica* O von Grimm, 1901

This is a toponym referring to Maeotis, the ancient name for the Sea of Azov, where these species occur.

Maes

African Loach Catfish sp. *Amphilius maesii* GA Boulenger, 1919

Dr Joseph Yvon Maes (1882–1960) was an anthropologist and ethnographer at the Belgium Royal Museum of Central Africa for which he wrote a guide: *Le Museedu Congo Belgea Tervuren: Guide illustree du visiteu* (1925). He led an expedition to Bandundu, Congo (1913–1914) where he collected odonata. He wrote many books, mostly on anthropology and native artefacts.

Maesen

Maesen's Killifish *Epiplatys maeseni* M Poll, 1941

'A Maesen' was one of the collectors of this species, probably referring to Belgian ethnologist Albert Maesen (1915–1992), who explored the Ivory Coast (the type locality) with his professor Frans Olbrechts (q.v.) in 1938–39. He worked at the Museum of Central Africa in Tervuren, Belgium, as a researcher (1941), Assistant Curator (1949), Head of Ethnology (1954) and finally Director (1980), so would have been a contemporary of Max Poll.

Magdalena (Bay)

Magdalena Blenny *Paraclinus magdalenae* RH Rosenblatt & TD Parr, 1969

A toponym referring to the type locality; Magdalena Bay, Baja California, Mexico.

Magdalena (River)

Magdalena River Stingray *Potamotrygon magdalenae* AHA Duméril, 1865
Toothless Characin sp. *Cyphocharax magdalenae* F Steindachner, 1878
Narrow Hatchetfish sp. *Triportheus magdalenae* F Steindachner, 1878
Characin sp. *Cynopotamus magdalenae* F Steindachner, 1879
Flannel-mouth Characin sp. *Prochilodus magdalenae* F Steindachner, 1879
Armoured Catfish sp. *Rineloricaria magdalenae* F Steindachner, 1879
Banjo Catfish sp. *Xyliphius magdalenae* CH Eigenmann, 1912
Characin sp. *Creagrutus magdalenae* CH Eigenmann, 1913
Striped Hoplo Catfish *Hoplosternum magdalenae* CH Eigenmann, 1913
Characin sp. *Astyanax magdalenae* CH Eigenmann & AW Henn, 1916
Magdalena Rivulus *Cynodonichthys magdalenae* CH Eigenmann & AW Henn, 1916

This is a toponym referring to the Río Magdalena, western Colombia, where these species are found.

Magdalene

Bunjako Barb *Barbus magdalenae* GA Boulenger, 1906

Mrs Magdalene Minchin was the wife of British zoologist Edward Alfred Minchin. The holotype was collected by Edward Degen (q.v.) when in Uganda as Minchin's assistant.

Maggie Walker

Queensland Frogfish *Histiophryne maggiewalker* RJ Arnold & Pietsch, 2011

Margaret 'Maggie' Walker (b.1953) was a graduate of, and benefactor to, the University of Washington and The Burke Museum of Natural History and Culture. The fish was discovered by Rachel Arnold (also a graduate there), and described by her and Ted Pietsch. Maggie has, over twenty years, helped raise over $280m dollars for the colleges of arts and science and served on many of the university's boards and community organisations, as did her late husband Doug (Douglas W Walker 1950–2016) who used the fortune he made as a software entrepreneur to further conservation and outdoor pursuits. He died in an avalanche whilst hiking on Granite Mountain (2016). She is also a member of the board of National Audubon.

Magister

Azov Tadpole Goby *Benthophilus magistri* BS Iljin, 1927

The binomial means 'teacher'; the allusion is not explained by Iljin but may well refer to Nikolai Mikhailovich Knipovich (1862–1938), a Caspian Sea biologist and Iljin's teacher whom Iljin had honoured with the new genus *Knipowitschia* in the same paper. (See **Knipovich**)

Magnus

Blenny sp. *Alticus magnusi* W Klausewitz, 1964

Goby sp. *Cryptocentroides magnusi* W Klausewitz, 1968
[Syn. *Amblygobius magnusi*]
Magnus' Prawn-Goby *Amblyeleotris sungami* W Klausewitz, 1969

Dr Dietrich B E Magnus is an ethologist who was professor at the Zoologisches Institut der Technischen Hochschule, Darmstadt, Germany. He collected the *Amblyeleotris* type (Sungam is Magnus spelled backwards; why Klausewitz reversed the name is not explained). He is best remembered for the paper: *Experimental analysis of some "overoptimal" sign-stimuli in the mating-behaviour of the fritillary butterfly Argynnis paphia L. (Lepidoptera: Nymphalidae)* (1958). Another etymology records that he had "*...collected the type while studying the symbiosis between different gobies and the shrimp Alpheus djiboutensis*." (Also see **Sungam**)

Mago

Three-barbeled Catfish sp. *Brachyglanis magoi* A Fernández-Yépez, 1967
Stardrum sp. *Stellifer magoi* O Aguilera, 1983
Scale-eating Characin sp. *Serrabrycon magoi* RP Vari, 1986
Driftwood Catfish sp. *Ageneiosus magoi* GO Castillo & O Brull G, 1989
Characin sp. *Creagrutus magoi* RP Vari & AS Harold, 2001
Characin sp. *Bryconops magoi* B Chernoff & A Machado-Allison, 2005
Pencil Catfish sp. *Pygidianops magoi* SA Schaefer, F Provenzano, M de Pinna & JN Baskin, 2005
Ghost Knifefish sp. *Apteronotus magoi* CD de Santana, O Castillo & D Taphorn, 2006
Glass Knifefish sp. *Eigenmannia magoi* EE Herrera-Collazos, AM Galindo-Cuervo, JA Maldonado-Ocampo & M Rincón-Sandoval, 2020

Dr Francisco Mago Leccia (1931–2004) was a distinguished Venezuelan ichthyologist who specialised in Neotropical freshwater electric fishes. He attained the degrees of Docent in Biology and Chemistry, graduating from the Instituto Pedagógico de Caracas, MSc. (Marine Biology) from the University of Miami, Florida, USA, and Doctor in Sciences from the University of Central Venezuela. He was a founding member of the Instituto Oceanográfico de la Universidad de Oriente in Cumaná Sucre state and of the Instituto de Zoologia Tropical (IZT), University of Central Venezuela, Caracas. He was also Professor of Animal Biology, Vertebrate Biology and Systematic Ichthyology at the Biology School of Sciences Faculty there. He also became Director of its museum. He wrote five books, chapters in others' works, and numerous articles. His best-known book is: *Electric Fishes of the continental waters of América* (1994). He described at least 23 species and is honoured in the name of a damselfly and at least three other insects.

Mahabali

Snakehead sp. *Aenigmachanna mahabali* RG Kumar, VS Basheer & C Ravi, 2019

Mahabali was a king in the mythology of Kerala. He was banished to the netherworld by Vamana, an avatar of Vishnu, but is allowed to return to the mortal realm once a year, which is celebrated in the festival of Onam.

Mahidol

Goby genus *Mahidolia* HM Smith, 1932

Mahidol Adulyadej (1892–1929), Prince of Songkla, was honoured for his interest in the fishes and fisheries of Siam (Thailand).

Mahmudbekov

Small-spine Tadpole Goby *Benthophilus mahmudbejovi* DB Ragimov, 1976

A A Mahmudbekov was an Azerbaijanian ichthyologist, who devoted his life to the study of the Caspian Sea (where this goby occurs). Among his publications was: *On the standard weight of sturgeon fingerlings produced by Kura fish growing plants* (1966).

Mahnert

Firehead Tetra *Hemigrammus mahnerti* A Uj & J Géry, 1989
Burmese-border Sand Loach *Schistura mahnerti* M Kottelat, 1990
[Alt. Red-tailed Zebra Loach]

Dr Volker Mahnert (1943–2018) was an Austrian ichthyologist, arachnologist and parasitologist. His doctorate was awarded by the University of Innsbruck (1971) and that year he joined the staff of the Geneva Museum of Natural Sciences, as Curator of the newly-created Department of Herpetology & Ichthyology. He later became the museum Director (1989) until officially retiring (2005), although he continued his research. He became (1990) Associate Professor of Zoology and Animal Biology, University of Geneva, and (1991) a member of the International Commission on Zoological Nomenclature. He undertook field work in Greece (1971, 1972 & 1973), Kenya (1975 & 1977), Paraguay (1979 & 1990) and Ivory Coast (1980). He published more than 200 papers and described no fewer than 351 pseudoscorpions. He often collaborated with Jaques Géry (q.v.). An amphibian is also named after him, among at least fifty species and genera including pseudoscorpions, beetles, parasites, and even a mineral.

Mai

Vietnamese Gudgeon sp. *Parasqualidus maii* A Doi, 2000

Professor Mai Dinh Yên (b.1933) is a freshwater fish taxonomist who is the retired head of the ichthyology section of the Institute of Zoology, Hanoi Science University, Vietnam, where he was originally a lecturer (1956–1986) and in charge of vertebrate animal studies at the Biology Department. He was a visiting professor at Tlemcen University, Algeria (1986–1989). He wrote: *Identification of the Fresh-Water Fishes of North Viet Nam* (1978).

Maindron

Goby sp. *Nematogobius maindroni* H Sauvage, 1880

Maurice Maindron (1857–1911) was a French naturalist, collector and entomologist. He joined the staff of Muséum National d'Histoire Naturelle, Paris (1875), and started on 25 years of almost continual travel. He was in New Guinea (1876–1877), Senegal (1879 & 1904), India (1880–1881, 1896 & 1900–1901), Indonesia (1884–1885), Djibouti and Somalia (1893), and Arabia (1896). He wrote many books of fiction, including: *The Tree of Science* (1906), which is, apparently, partly autobiographical. He collected the goby type in Senegal. A reptile is also named after him.

Mainland

Mainland's Goby *Psilogobius mainlandi* WJ Baldwin, 1972

Dr Gordon B Mainland (d.1962) was an American zoologist. He was honoured for his studies on Hawaiian fishes while a master's student at the University of Hawaii (he had graduated there 1938). He described this goby in his unpublished master's thesis (1939). He went on to be involved in the use of pest parasites in fruit production in Hawaii (1950). He co-wrote *The Drosophilidae of Mexico* (1944) when he was an Instructor at the University of Texas which awarded his PhD. He then (1944) became Assistant Professor of Biology at Illinois Institute of Technology, and later was Associate Professor of Biology at Jackson State University (1957–1962).

Mairasi

Lake Furnusu Rainbowfish *Melanotaenia mairasi* GR Allen & RK Hadiaty, 2011

Mairasi is the tribal name of the local landowners of the type locality, in Indonesian New Guinea.

Maisome

Lake Victoria Cichlid sp. *Haplochromis maisomei* MJP van Oijen, 1991

This is a toponym referring to Maisome Island, Lake Victoria, Tanzania; the type locality.

Majima, Toyohiko

Striped Siphonfish *Siphamia majimai* K Matsubara & T Iwai, 1958

Toyohiko Majima was a Japanese shell collector. He helped the authors with their icthyological collections made at Amami Oshima Island, Japan.

Majima, Toyoji

Oriental Gudgeon ssp. *Squalidus gracilis majimae* DS Jordan & CL Hubbs, 1925

Toyoji Majima was at the Imperial University of Hokkaido, Japan. He often collected specimens in the fish market at Sapporo.

Makax

Dwarf Gulf Pipefish *Syngnathus makaxi* ES Herald & CE Dawson, 1972

A toponym referring to the type locality; Laguna Makax, Quintana Roo, Mexico.

Makonde

African Rivuline sp. *Nothobranchius makondorum* RH Wildekamp, KM Shidlovskiy & BR Watters, 2009

The Makonde are a people who live in southeast Tanzania and northeast Mozambique; the area where this fish occurs.

Makushok

Eelpout sp. *Lycenchelys makushok* VV Fedorov & AP Andriashev, 1993

Viktor Markelovich Makushok (1924–1993) was a Russian ichthyologist who worked at the Institute of Oceanography, Russian Academy of Sciences (1959–1984). Immediately after graduating from secondary school (1941) he volunteered for the army and, after completing a translator's course at the Military Institute of Foreign Languages in Saratov (1943), he was sent to the 1st Special Regiment of Regions, with whom he served from Belorussia to Berlin. He was decorated, including the Order of the Red Star. After demobilization (1947), he enrolled in the Biology and Soil Faculty of Moscow State University, graduating in the Department of Zoology of Vertebrates (1952), after which he studied at the Zoological Institute of the Academy of Sciences in Leningrad, where he began to investigate the morphology and taxonomy of *Stichaeiodae*. This was the period of his greatest creativity. After graduating and preparing a large summary work for publication, he moved (1957) to the Comprehensive Antarctic Expedition and became the first Russian biologist to winter on the Antarctic continent. He returned to Leningrad (1959) and defended his PhD thesis, then joined the Institute of Oceanology until his retirement (1984). He took part in three large expeditions – the 33rd and 39th cruises of the research vessel *Vityaz* to the Indian Ocean (1960s) and the Kurile-Kamchatka Trench area (1966) and on a joint flight of IOAN-TINRO to the BMRT 'Equator' to the Alaska Gulf (1969). He published several scientific and popular science articles, including essays on the most diverse groups of fish for the book *Animal Life* (1971, 1983) and for the Great Soviet Encyclopedia.

Malabarba

Bumblebee Catfish sp. *Microglanis malabarbai* VA Bertaco & AR Cardoso, 2005
Neotropical Livebearer sp. *Phalloceros malabarbai* PHF Lucinda, 2008
Armoured Catfish sp. *Rineloricaria malabarbai* MS Rodriguez & RE Reis, 2008
Characin sp. *Serrapinnus malabarbai* FC Jerep, FCP Dagosta & WM Ohara, 2018

Dr Luiz Roberto Malabarba is a Brazilian ichthyologist and biologist. The Federal University of Rio Grande do Sul awarded his bachelor's degree (1983), the Pontifical Catholic University of Rio Grande do Sul his master's (1988) and the University of São Paulo his doctorate (1994). He did post-doctoral study at the Smithsonian (1999). He is currently a full professor; Federal University of Rio Grande do Sul, where he has had two stints as Coordinator of the Graduate Program in Animal Biology (2003–2007 & since 2015). He has been President of the Brazilian Society of Ichthyology since 2015. He co-wrote: *New species of Scleromystax Günther, 1864 (Siluriformes: Callichthyidae) – extending the southern distribution of genera endemic to the Atlantic Forest* (2016).

Malaise

Sisorid Catfish sp. *Glyptosternon malaisei* H Rendahl & G Vestergren, 1941
Stone Loach sp. *Schistura malaisei* M Kottelat, 1990

René Edmond Malaise (1892–1978) was a Swedish entomologist, art collector and explorer who is remembered for inventing the Malaise trap. He was a member of the Swedish expedition to Kamchatka (1920–1922) and returned to Sweden (1923) via Japan where he witnessed the great earthquake of that year. He returned to Kamchatka (1924) and stayed in the Soviet Union until 1930. He went on an expedition to northern Burma (Myanmar) (1933–1935) during which he collected about 100,000 insects, many previously unknown to science including at least three damselflies named after him. He supervised the entomological department of the Naturhistoriska Museet, Stockholm (1952–1958). He wrote: *Atlantis, en geologisk verklighet* (1954). He collected the holotypes of the two eponymous species and made a notable collection of fishes in Burma.

Malaisse

Shellear sp. *Parakneria malaissei* M Poll, 1969
African Rivuline sp. *Nothobranchius malaissei* RH Wildekamp, 1978

Professor Dr François Malaisse is a Belgian ecologist at Agro-Bio Tech, Université de Liège (1987–present). He formerly worked at the Université de Mons, Belgium (1994–1999) and the University of Lubumbashi, Democratic Republic of the Congo (Université National du Zaire) (1965–1986) where he collected the holotype of the shellear. He studied Ecology, Botany and Ethnobiology at the Faculté Universitaire des Sciences Agronomiques de Gembloux, Namur, Belgium (1951), and Botany at the Université Libre de Bruxelles. During the course of his research he has visited more then twenty African countries. He has published (1963–2010) an incredible 371 scientific papers, reports, books and articles such as: *Human consumption of Lepidoptera, termites, Orthoptera, and ants in Africa* (2005) and *How to Live and Survive in Zambezian Open Forest (Miombo Ecoregion)* (2010).

Malcolm (Francis)

Mottled Triplefin *Forsterygion malcolmi* GS Hardy, 1987

Dr Malcolm Philip Francis (b.1954). (See **Francis, MP**)

Malcolm (Smith)

Goldfin Tinfoil Barb *Hypsibarbus malcolmi* HM Smith, 1931

Dr Malcolm Arthur Smith (1875–1958) was an English physician and herpetologist. He practised medicine in Bangkok (1902–1924), including five years as Court Physician, publishing his memoirs under the title: *A Physician at the Court of Siam* (1947). He visited French Indo-China (1918). He was President of the British Herpetological Society (1949–1954). Seven reptiles, three amphibians and a mammal are named after him.

Maldonado, E

Pencil Catfish sp. *Bullockia maldonadoi* CH Eigenmann, 1920

Ernesto Maldonado was a Chilean agronomist and agricultural engineer who was the Director, Forests, Fishing and Hunting, Santiago de Chile. He wrote: *Contribución al estudio de la industria maderera y bosques chilenos* (1925)

Maldonado, J

Pencil Catfish sp. *Trichomycterus maldonadoi* CA Ardila Rodríguez, 2011
Armoured Catfish sp. *Ancistrus maldonadoi* AG Bifi & H Ortega, 2020

Dr Javier Alejandro Maldonado Ocampo (1977–2019) was a prodigious and prestigious Colombian biologist, biogeographer and ichthyologist. The Federal University of Rio de Janeiro awarded his doctorate. He co-wrote: *Checklist of the freshwater fishes of Colombia*

(2008). This species was named after him to recognise his dedication to Colombian ichthyology. Under his management (2015), eight countries launched a large-scale project aimed at creating the largest and most complete database of freshwater fishes in the Amazon; the most diverse assemblage of freshwater fishes in the world. He and his collaborators had documented over 2,300 species of fish. Many more await discovery. He described or co-described 10 species of fish from the Amazon, all but three of them knifefishes (Gymnotiformes). His greatest passion was discovering unknown fishes from the Amazon rainforests of Colombia, Brazil and Peru. He was a full professor, Pontificia Universidad Javeriana, Bogotá. According to his students there he loved 'Going places where nobody has gone before to collect fish'. When attempting to cross the río Vaupés, Colombia, in a small boat with two other researchers their boat overturned. The others made it to safety but he was seen swimming downriver as he was swept away. Three days later a Brazilian soldier who was part of the search effort found his body 75km from the accident site. More fishes will likely be named for him. (Also see **Xaveriellus**)

Malekula

Striped Barbel-Goby *Gobiopsis malekulae* AW Herre, 1935

This is a toponym referring to Malekula Island, Vanuatu; the type locality.

Malinche

HIghland Swordtail *Xiphophorus malinche* M Rauchenberger, KD Kallman & DC Morizot, 1990

La Malinche (c.1496–c.1529), also known as Malintzin or Doña Marina, was a Nahuan Indian slave who was the interpreter, secretary, and mistress of Hernando Cortés. In Mexico, her reputation has changed over the years along with social and political perspectives, especially after the Mexican Revolution, when she was portrayed in dramas, novels, and paintings as a scheming temptress and a traitor to the native population.

Mallet

Borneo Noodlefish sp. *Sundasalanx malleti* DJ Siebert & O Crimmen, 1997

John Valentine Granville Mallet (b.1930) is a ceramics historian who was Keeper of Ceramics and Glass at the Victoria & Albert Museum, London, until that institution was, for financial reasons, restructured (1989) and he was forced to retire. Paxos Festival Trust Ltd then employed him as Museum Director and Curator (1990–1997) and as Ceramic Historian (1997–2004). He was honoured as the former Prime Warden of the Worshipful Company of Fishmongers (a London livery company) "*...whose enthusiasm, encouragement, and support have made possible the continuation of a research programme on freshwater fishes of Southeast Asia*"

Malloch

Malloch's Char *Salvelinus mallochi* CT Regan, 1909

Peter Duncan Malloch (1853–1921) was well-known in Perth, Scotland, as a skilled fishing tackle maker and amateur naturalist. He started a taxidermy and fishing tackle business and shop (1871) which is still in business (2017). He was the first person to study scale markings on salmon and proved that every period passed, whether at sea or in a river, could be explained by the markings on the scales themselves. He wrote: *Life-History and Habits of the Salmon, Sea-Trout, Trout, and Other Freshwater Fish* (1910), which is still regarded as the standard work on the subject.

Malsmith

Sleeper sp. *Mogurnda malsmithi* GR Allen & M Jebb, 1993

Malcolm Smith is the owner of Pacific Helicopters, Ltd. He was honoured for generously donating the use of one of his aircraft, which helped the authors collect this species in Papua New Guinea.

Maltzan

Goby sp. *Wheelerigobius maltzani* F Steindachner, 1881

Hermann Friedrich Freiherr von Maltzan (1843–1891) was a German malacologist and Steindachner's 'dear' (translation) friend, who provided the type. He undertook (1864–1865) an extended scientific journey to France, Spain, Italy and Egypt, during which he collected numerous zoological specimens. In 1879, he traveled to Portugal and conducted extensive zoological research in the Algarve.

Malumbres

African Rivuline sp. *Aphyosemion malumbresi* O Legros & F Zentz, 2006

Dr Francisco J Malumbres is an aquarist from Madrid, who has initiated and led ichthyological expeditions to Equatorial Guinea. Among his publications is the co-written: *Identification of Micropanchax scheli (Cyprinodontiformes: Poeciliidae: Aplocheilichthyinae) with the description of a new species of the genus Poropanchax* (2004). He was co-discoverer of the species.

Malvolio

Snake-eel genus *Malvoliophis* GP Whitley, 1934

Malvolio was Lady Olivia's pompous steward in Shakespeare's *Twelfth Night*. In this case, the first part of the genus' name refers to the banded coloration of *M. pinguis*, suggestive of cross-gartered legs and yellow socks worn by the character; combined with *ophis* meaning snake, conventionally used for generic names of snake-eels due to their snake-like appearance.

Mamonekene

Central African Cichlid sp. *Chromidotilapia mamonekenei* A Lamboj, 1999
African Loach Catfish sp. *Amphilius mamonekenensis* PH Skelton, 2007

Dr Victor Mamonekene (b.1959) is a Congolese ichthyologist who is a professor at University of Brazzaville and is a field associate of the American Museum of Natural History. He was the person mainly responsible for the maintenance of the Biosphere Reserve in Dimonika, Republic of the Congo, the cichlid's type locality. Without him Lamboj's collections would not have been possible. He studied Natural Sciences and Aquatic Ecology at the University of Provence Aix-Marseille. His 30+ publications include the co-written: *Distichodus teugelsi a new distichodontid from the middle Congo River basin, Africa (Characiformes: Distichodontidae)* (2008) and *Diversity and distribution of fish species along the Loua River, lower Congo river basin (Republic of the Congo, Central Africa)* (2019). Skelton's binomial is unusual in using the *–ensis* ending, normally associated with toponyms rather than eponyms.

Mana

Papuan Guitarfish *Rhinobatos manai* WT White, PR Last & GJP Naylor, 2016

Dr Ralph Reeves Mana is Associate Professor in the Department of Biological Sciences at the University of Papua New Guinea. Among his published papers are the co-authored: *Description of a new species of deepwater catshark* Apristurus yangi *n.sp (Carcharhiniformes: Pentanchidae) from Papua New Guinea* (2017), which he wrote with William White and Gavin Naylor, and *A new species of velvet skate,* Notoraja sereti *n.sp. (Rajiformes: Arhynchobatidae) from Papua New Guinea* (2017) that he wrote with William White and Peter Last.

Manabe

Northern Tubelip Wrasse *Labropsis manabei* PJ Schmidt, 1931

Mr Manabe was Director of the Electrical Station of Naze. Schmidt said he was very much indebted to him for the success of his collecting work on Japan's Anami-Oshima Island.

Manabe, Y

Eel sp. *Anguilla manabei,* DS Jordan, 1913
[Junior Synonym of *Anguilla japonica*]

Yoshiro Manabe was a student of David Starr Jordan who became a science teacher at the Kwansei Gakuin University at Kobe. He made a collection of fishes for the college among which Jordan identified the apparently new eel.

Mánamo

Manamo Anchovy *Anchoviella manamensis* F Cervigón, 1982

This is a toponym referring to Caño Mánamo, near Tucupita, Orinoco Delta (Venezuela); the type locality.

Manco

Suckermouth Catfish sp. *Astroblepus mancoi* CH Eigenmann, 1928

Ayar Manco has been called the 'Moses of the Peruvians', because he led the exodus from Tampu-tocco to Cuzco about 1100 A.D. Whether he actually existed is unclear, but he mainly features in two legends concerning the origin of the Inca empire.

Mandela

African Barb sp. *Enteromius mandelai* MJ Kambikambi, WT Kadye & A Chakona, 2021

Nelson Rolihlahla Mandela (1918–2013) was a revolutionary, political prisoner, statesman and philanthropist who became South Africa's first democratically elected head of state

(1994–1999). Despite being imprisoned for twenty-seven years for his beliefs by a racist regime, when elected he spent his time dismantling the racist state apparatus and initiating racial reconciliation. After leaving office he continued to champion democracy and social justice and received over 250 awards, including the Nobel Peace Prize. Many regard him as the greatest man of the twentieth century. He was from the Eastern Cape Province, where this fish species is endemic.

Mandelburger

Pike-Cichlid sp. *Crenicichla mandelburgeri* SO Kullander, 2009

Juan Darío Mandelburger is a Paraguayan ichthyologist who was Director General of Biodiversity at the Ministry of Environment and Sustainable Development (2006–2008). Colegio Nacional de la Capital awarded his BSc (1980) and Universidad Nacional de Asunción his master's. He was co-coordinator of the Proyecto Vertebrados del Paraguay (1992–1999).

Mandeville

Squeaker Catfish sp. *Atopochilus mandevillei* M Poll, 1959
African Glass Catfish sp. *Pareutropius mandevillei* M Poll, 1959
Loach Catfish sp. *Zaireichthys mandevillei* M Poll, 1959

J Th Mandeville was the fisheries agent for the government of Leopoldville, Congo (Kinshasa, DRC). He collected some of the type and paratype specimens.

Manibui

Manibui Rainbowfish *Melanotaenia manibuii* Kadarusman, J Slembrouck & L Pouyard, 2015

Alfons Manibui is the Bupati (local leader) from Bintuni, a town near the type locality in western New Guinea.

Manikfan

Damselfish sp. *Abudefduf manikfani* S Jones & M Kumaran, 1970

Ali Manikfan (b.1938) is an Indian ecologist, marine researcher and Muslim scholar. He believes that formal education is artificial and pointless, and that the best way to acquire wisdom is by observing our environment. He collected the damselfish holotype and was also honoured for the extensive collections of fishes he made from the Laccadive archipelago.

Mann, M

Feather-barbelled Squeaker (catfish) *Synodontis manni* L De Vos, 2001

Michael 'Mike' J Mann was the FAO Fisheries Officer in Kenya. He was the first to report the presence of this species in the Tana River, Kenya (1968). He wrote: *A Preliminary Report on a Survey of the Fisheries of the Tana River, Kenya* (1967).

Mann, W

Bahama Gambusia *Gambusia manni* CL Hubbs, 1927

William Montana Mann (1886–1960) was an American entomologist. He attended Staunton Military Academy, Virginia (1902–1905), then worked as a rancher in Texas and New Mexico, all the while collecting entomological specimens. Washington State College (1909–1911) awarded his BA and the Bussey Institution, Harvard, his DSc.in Entomology (1915). He collected overseas, including on the Stanford Expedition to Brazil (1911), Haiti (1912), Cuba and Mexico (1913), on the Philip Expedition to the Middle East (1914), Fiji & Solomon Islands (1915–1916). He was an entomologist for the Bureau of Entomology, US Department of Agriculture (1916–1925) and made further collecting trips to Spain, Mexico and Cuba. He was Superintendent of NZP (1925) becoming Director (1927) until retirement (1956), then Director Emeritus. He was an Honorary Curator at the USNM to which he left his collection. He wrote or co-wrote at least two dozen articles and an autobiography: *Ant Hill Odyssey* (1948). Four reptiles and a dragonfly are also named after him.

Manocherian

Manocherian's Catshark *Apristurus manocheriani* JA Cordova & DA Ebert, 2021

Greg Manocherian is an American businessman, the founder (1994) of RoseCo Holdings LLC, a real estate investment and development company. He also has a strong commitment to shark conservation and research and is a board member of ConservAmerica, as well as supporting biomedical research as Vice-Chairman of ACT for NIH: Advancing Cures Today.

Manuel

Manuel's Piranha *Serrasalmus manueli* A Fernández-Yépez & MV Ramírez, 1967

Manuel Ramirez was the second author's son, who died young, and who since early childhood had accompanied the authors on their fishing expeditions.

Maraldi

Gadella *Gadella maraldi* A Risso, 1810

Giacomo Filippo Maraldi, also known as Jacques Maraldi (1665–1729), was a French-Italian astronomer and mathematician who worked for much of his life at the Paris Observatory (1687–1718). Craters on both the Moon and on Mars are named after him.

Marasri

Marasri's Thryssa *Thryssa marasriae* T Wongratana, 1987

Marasri Ladpli was the wife of the author. He dedicated the species to her for her "*...patient sharing of my study of fishes, her encouragement, and her tolerance of my trips away from home.*"

Marcella

Marcella Butterflyfish *Prognathodes marcellae* M Poll, 1950

Marcelle Aen Poll née den Boom is the wife of the author, Max Poll (q.v.).

Marcellino

Molly sp. *Poecilia marcellinoi* FN Poeser, 1995

Marcellino JC Rozemeyer (Rozemeijer) is a marine ecologist at the Marine Research unit at Wagenungen Universirty. He was formerly at the University of Amsterdam where he was a colleague of Poeser's. The etymology says that he "*...has helped me throughout my studies.*" His publications include: *Marine and coastal ecological potential for the economic development of Colombia* (2013). In another paper Poeser says: "*The whole project, comprising about 16 years of study, was pushed and monitored continuously by my paranymphs, Marcellino Rozemeijer and Michel Koper, involving discussions and other friendly talks. May they always be responsible for drinks and laughter.*"

Marcgrave

Leporinus (characin) sp. *Leporinus marcgravii* CF Lütken, 1875

Georg Marcgrave (also Marcgraf) (1610–1644) was a German naturalist and astronomer. He studied botany, astronomy, mathematics and medicine in Germany and Switzerland until 1633. He then practiced medicine in the Netherlands (1633–1637). The Dutch West India Company offered him the position of Personal Physician to Count Johan Maurits van NassauSiegen (1637), who was at that time in South America in the capacity of Governor of the colony of Dutch Brazil. When Marcgrave arrived (1638) he undertook the first zoological, botanical and astronomical expedition in Brazil, exploring various parts of the colony to study natural history and geography. Cuvier praised him as one of the best and most diligent observers and recorders of the era. He was sent to Angola by the West India Company, but died shortly after his arrival. He was co-author, with Willem Piso, of: *Historia Naturalis Brasiliae,* an 8–volume work on the botany and zoology of Brazil published posthumously (1648). A mammal and a bird are named after him.

Marche

Mormyrid sp. *Ivindomyrus marchei* HE Sauvage, 1879

Spiny Eel sp. *Mastacembelus marchei* HE Sauvage, 1879

AntoineAlfred Marche (1844–1898) was a French explorer and naturalist. He wrote *Trois Voyages dans l'Afrique Occidentale* (1879) during one of which voyages he collected specimens from the Ogooué (Ogowe) River in Gabon under the command of French explorer Pierre Savorgnan de Brazza. After his visit to the Philippines (1879–1883) he wrote: *Luçon et Palaouan, six années de voyages aux Philippines* (1887). He also conducted archaeological explorations and removed a vast number of artifacts, sending them to the Musée de l'Homme in Paris. A mammal and five birds are also named after him.

Marchena

Lizardfish sp. *Synodus marchenae* SF Hildebrand, 1946

A toponym referring to the type locality; Marchena Island in the Galápagos.

Marcia

Marcia's Anthias *Pseudanthias marcia* JE Randall & JP Hoover, 1993

Marcia Hoover is the wife of the second author John Hoover.

Marcos

Marcos' Tilefish *Hoplolatilus marcosi* WE Burgess, 1978
[Alt. Redback Sand Tilefish, Skunk Tilefish]

His Excellency Ferdinand Emmanuel Edralin Marcos Sr (1917–1989) was the tenth President of the Philippines (1965–1986). For part of which, he ruled as a dictator under martial law (1972–1981). His regime was infamous for its brutality, corruption and extravagance and the family stole between five and ten billion dollars! He was an attorney before he went into politics. He was honoured at the request of Earl and Gloria Kennedy who discoverered the species. The Kennedys sent specimens to Burgess. Earl Kennedy has been described as the 'founding father of the Philippine marine aquarium fish trade'.

Marcusen

Mormyrid genus *Marcusenius* TN Gill, 1862

Dr Johann Andreas Marcusen (1817–1894) was a Latvian of German descent who was a physician, ichthyologist and zoologist. He studied at the St Petersburg Medical Academy (1833–1841) and worked in the Workers' Hospital, St Peterburg (1845–1847). He moved to Dorpat, Estonia, where he worked as a junior doctor (1847–1849) and qualified as a physician (1848). He worked in St Petersburg at a military hospital (1849–1851). He travelled to the Mediterranean and visited both Syria and Egypt (1851–1854). He was Assistant Professor of Medicine at the St Petersburg Medical Academy (1854–1858), Professor of Zoology at the Richelieu Lyceum at Odessa (1858–1865) and Professor of Zoology and Comparative Anatomy at the University of Odessa (1865–1873). He retired (1873) to Switzerland and died in Bern. He was a specialist in the Mormyridae and wrote a treatise on the anatomy and other aspects of them: *Die Familie der Mormyren. Eine anatomisch-zoologische Abhandlung* (1864) and *Zur Fauna des schwarzen Meeres* (1867).

Marée

Spotjaw Moray *Gymnothorax mareei* M Poll, 1953

Major I Marée was administrator of Banana, a seaport in Bas-Congo (now Kongo Central), Democratic Republic of the Congo. As a Lieutenant, he collected a great number of crustaceans (1948–1950). He was a long-term resident (<1937–>1953) and was very helpful to the curator of the museum in Leopoldville in collecting native artefacts and artwork.

Mareike

Mareike's Tube-snouted Ghost Knifefish *Sternarchorhynchus mareikeae* CD de Santana & RP Vari, 2010

Dr Mareike Roeder is a German biologist who specialised in Tropical Silviculture and Forest Ecology. The University of Marburg awarded her master's (2003) and the University of Göttingen her doctorate. She is currently doing post-doctoral studies at Xishuangbanna Tropical Botanical Garden, Mengla, China. She co-wrote: *Seed and Germination Characteristics of 20 Amazonian Liana Species* (2013). The etymology says that she has "…

greatly added to the senior author's life." Perhaps a hint that she is Mrs de Santana in her private life?

Margaret (Bradbury)

Batfish sp. *Halieutopsis margaretae* H-C Ho & K-T Shao, 2007

Dr Margaret G Bradbury (1927–2010) was honoured: "*in recognition of her enormous contribution to our knowledge of the batfish family, Ogcocephalidae.*" (See **Bradbury**)

Margaret (Smith)

Lake Tanganyika Cichlid sp. *Pseudosimochromis margaretae* GS Axelrod & JA Harrison, 1978
Half-scaled Jawfish *Opistognathus margaretae* WF Smith-Vaniz, 1983
Smooth-scale Goby *Heterelotris margaretae* DF Hoese, 1986
Smith's Dogfish Shark *Squalus margaretsmithae* S de FL Viana, MW Lisher & MR de Carvalho, 2017

Professor Margaret Mary Smith née Macdonald (1916–1987) (See **Smith, MM**)

Margaret (Whitby Smith)

Margaret's Dragonet *Callionymus margaretae* CT Regan, 1905

Margaret Whitby Smith. Regan says: "*Mr. Townsend writes that Mr. and Mrs. Whitby Smith have taken great interest in his collecting*", and because of this the new dragonet is being named after Mrs. Smith. She also had a clam named after her, also at the request of F W Townsend (q.v.) who was a notable conchologist and collector around the Persian Gulf (1893–1905).

Margarete

Characin sp. *Roeboides margareteae* CAS Lucena, 2003

Zilda Margarete S Lucena (see **Lucenorum**).

Margaretha

Margaretha's Goatfish *Upeneus margarethae* F Uiblein & PC Heemstra, 2010

Margaretha Uiblein née Feichtinger was the mother of the senior author, Franz Uiblein.

Margarita

Allegheny Pearl Dace *Margariscus margarita* ED Cope, 1867
Daisy Stingray *Fontitrygon margarita* A Günther, 1870
Barbeled Dragonfish sp. *Photonectes margarita* GB Goode & TH Bean, 1896

These are not eponyms but relate to the meanings of 'margarita'; pearl and daisy. The dragonfish was so-named due to having a "pearl-colored spot above the maxilla".

Margarita (Island)

Mullet sp. *Mugil margaritae* NA Menezes, M Nirchio, C de Oliveira & R Siccha-Ramirez, 2015

This is a toponym referring to the type locality; Isla Margarita, Venezuela.

Margarita

Margarita Blenny *Malacoctenus margaritae* HW Fowler, 1944

This is a toponym referring to the Pearl Islands (Archipiélago de las Perlas) in the Gulf of Panama; the type locality.

Margaritella

Pearl Stingray *Fontitrygon margaritella* LJV Compagno & TR Roberts, 1984

Margaritella is the diminutive of Margarita and is used to show this species is smaller than its relative (see **Margarita** above). The vernacular name is not only used to convey 'Little Pearl' but to reflect the fact that this species has a pearl spine.

Margary

Glass Barb *Poropuntius margarianus* J Anderson, 1879

Augustus Raymond Margary (1846–1875) was a British diplomat and explorer, who served as an interpreter in the British consular service in China, variously (1867–1974) in Beijing, Taiwan, Shanghai and Yantai. He travelled overland from Shanghai to Bhamo (Upper Burma), a journey of 1,800 miles (about 2,900 kms) in six months to prospect possible trade routes, but all his companions were murdered at Tengchong, Yunnan, on his way back to Shanghai.

Margit

Characin sp. *Moenkhausia margitae* A Zarske & J Géry, 2001

Bleeding Blue Tetra *Hyphessobrycon margitae* A Zarske, 2016

Margit Zarske is the senior author's wife. She is a graduate teacher who, according to her husband, "*...for many years has endured and supported my passion for ichthyology*".

Margrethe

Portholefish genus *Margrethia* P Jespersen & AV Tåning, 1919

The *Margrethe* was a Danish research ship

Maria

Kinshasa Mormyrid *Mormyrops mariae* L Schilthuis, 1891

Schilthuis does not identify the person he is naming this fish after: perhaps a female relative.

Maria

African Barb sp. *Labeobarbus mariae* M Holly, 1926

[Syn. *Varicorhinus mariae*]

Barombi-Mbo Cichlid sp. *Stomatepia mariae* M Holly, 1930

It is not known after whom Maximilian Holly named these species. One possibility is that their binomials refer to Holly's mother-in-law, Maria Adolfine Sperat, whom he also honoured in the bagrid catfish genus *Sperata* (1939), noting how she had supported his studies with "great understanding" (translation). (Also see **Holly** & **Sperat**)

Maria, Brother A

Ghost Knifefish sp. *Apteronotus mariae* CH Eigenmann & HG Fisher, 1914
Suckermouth Catfish sp. *Astroblepus mariae* HW Fowler, 1919
Orinoco Cichlid sp. *Bujurquina mariae* CH Eigenmann, 1922
Flannel-mouth Characin sp. *Prochilodus mariae* CH Eigenmann, 1922

Brother Apolinar María (1867–1949) (See **Apolinar**)

Maria, Brother N

Three-barbeled Catfish sp. *Nemuroglanis mariai* LP Schultz, 1944

Brother Niceforo Maria (1888–1980), né Antoine Rouhaire Siauzade (see **Niceforo**).

Maria (Berg)

Ukranian Brook Lamprey *Eudontomyzon mariae* LS Berg, 1931

Maria Mikhailovna Berg née Ivanova was (1922) the second wife of the author. He honoured her not just for the family connection but because it was she "*...who examined many thousands of river lampreys from the mouth of the Neva and other streams, falling into the Finnish Gulf.*"

Maria (Darlington)

Pinewoods Darter *Etheostoma mariae* HW Fowler, 1947

Mary Darlington (d. 1951) was described by Fowler as "*...my generous sponsor*". She was the wife of entomologist Emlen P Darlington.

Maria (George)

Secretary Blenny *Acanthemblemaria maria* JE Böhlke, 1961

Mrs Mary George, whom Böhlke extolled as "*...my secretary for the past five years and now parent and housewife, in appreciation of her assistance in all the activities of the department during that period.*"

Maria (Horn)

Bigeye Lates *Lates mariae* F Steindachner, 1909

Marie Horn, with her husband Adolf, collected fishes at Lake Tanganyika (1908). [There is also a mention of the brothers Adolf and Albin Horn visiting Tanganyika in 1913, but almost nothing else is recorded about them]

Maria (Howes)

Maria's Oto (catfish) *Otocinclus mariae* HW Fowler, 1940

Mrs Maria Howes was the wife of Arthur Howes. Fowler wrote that he was indebted to Howes 'for many American fishes.' Fowler had a ring of friends and acquaintances who supplied him with specimens. We assume that Arthur Howes was one of them. (Also see **Howes, A**)

Maria (Kingsley)

Spotted Tilapia *Pelmatolapia mariae* GA Boulenger, 1899

Mary Henrietta Kingsley (1862–1900). (See **Kingsley**)

Maria (van Nagell)

Maria's Tandan *Oloplotosus mariae* M Weber, 1913

Marie Louise Clémence, Baroness van Nagell (1885–1981), was the wife of the Dutch explorer Hendrikus Albertus Lorentz (1871–1944). He collected the holotype in New Guinea.

Maria Elena

Twig Catfish sp. *Farlowella mariaelenae* FJ Martín Salazar, 1964

Mrs María Elena Salazar was the author's wife and 'great companion'.

Maria Joris

Maria Joris's Sand-Dab *Citharichthys mariajorisae* AM van der Heiden & S Mussot-Pérez, 1995

[Alt. Five-rayed Sand-Dab]

Maria Joris (1917–1995) was the maiden name of the senior author's mother, who was honoured for her "*teaching and perpetual support*". Unable to continue her own education during and after World War II, she nevertheless encouraged her six children to study foreign languages, learn music, play an instrument, earn a university degree, and follow their dreams. (Albert van der Heiden, personal comment)

Mariana (Gadig)

Brazilean Large-eyed Stingray *Hypanus marianae* UL Gomes, RS Rosa & OBF Gadig, 2000

Mariana Ramos de Oliveira Gadig Gonçalves is the daughter of Otto Gadig, the third author. She works at the Epidemiology department, Instituto Adolpho Lutz, Santos, Brazil.

Mariana (Wosiacki)

Plated Catfish sp. *Aspidoras marianae* MDV Leão, MR Britto & WB Wosiacki, 2015

Mariana P Wosiacki is the third author's daughter.

Marianne

Lisikili Mormyrid *Pollimyrus marianne* B Kramer, H van der Bank, N Flint, H Sauer-Gürth & M Wink, 2003

Marianne Elfriede Kramer (1914–2000) was the mother of senior author Bernd Kramer.

Marijean

Lonely Clingfish *Gobiesox marijeanae* JC Briggs, 1960

Named after the yacht 'Marijean', in recognition of the many valuable fish collections which have been made during the cruises of this vessel to the tropical eastern Pacific.

Marilyn (Gilmore)

Orange-bellied Goby *Varicus marilynae* RG Gilmore, 1979

Marilyn Gilmore, wife of the author Dr R Grant Gilmore (q.v.), whose etymology stated she "...*has aided me considerably in my ichthyological studies.*"

Marilyn (Hardy)

Pufferfish genus *Marilyna* GS Hardy, 1982
Stout Rockfish *Acanthoclinus marilynae* GS Hardy, 1985

Marilyn Hardy was the wife of the author. According to the etymology she: "...*assisted uncomplainingly at poison stations in the hot, muddy, and potentially dangerous mangrove swamps of North Queensland.*"

Marilyn (Weitzman)

Greenstripe Pencilfish *Nannostomus marilynae* SH Weitzman & JS Cobb, 1975
Characin sp. *Lebiasina marilynae* AL Netto-Ferreira, 2012
Pyrrhulina (characin) sp. *Pyrrhulina marilynae* AL Netto-Ferreira & MMF Marinho, 2013

Mrs Marilyn Jean Weitzman née Sohner (b.1926) is a Research Associate at the Smithsonian, where she has devoted her career to the study of fishes of the *Lebiasinidae* and *Characidae* families. She co-wrote, with Smithsonian Curator Emeritus Stanley H Weitzman (1927–2017): *Bio-geography and evolutionary diversification in neo-tropical freshwater fishes, with comments on the Refuge Theory* (1982). (See **Weitzman, MJ** & **Weitzman, SH**)

Marina (Deynat)

Gold-dust River Stingray *Potamotrygon marinae* P Deynat, 2006

Marina Deynat is the daughter of the author, Pascal Deynat.

Marina (Lloris)

Eelpout sp. *Pogonolycus marinae* D Lloris, 1988

Marina Lloris is the daughter of the eelpout's describer.

Marina (Winterbottom)

Princess Pygmy Goby *Trimma marinae* R Winterbottom, 2005

Marina is one of many names attributed to the goddess Aphrodite, whose legendary beauty is reminiscent of this "*gorgeous little species*"; it is also, coincidentally, the name of Winterbottom's daughter, "*whose cheerful assistance in collecting and documenting coral reef fishes for my research program is much appreciated.*"

Marina (Wong)

Rasbora sp. *Rasbora marinae* TH Hui & M Kottelat, 2020

Dr Marina Wong worked (now retired) as a Curator at the Brunei National Museum where she supervised the development of a national park. The University of Michigan awarded her PhD for research on Malaysian rainforest birds. For over two decades she was a tropical biologist in Southeast Asia and South America. Returning to the USA she taught high school Spanish and is involved in local conservation in Rhode Island. She was honoured "*...in appreciation of her contributions to the knowledge of the natural history of Southeast Asia and her generous help in organising fieldwork in Brunei for the first author and team.*" She wrote: *Common Seashore Life of Brunei* (1996) & *Birds of Pelong Rocks* (1996).

Marini

Marini's Grenadier *Coelorinchus marinii* CL Hubbs, 1934
Marini's Anchovy *Anchoa marinii* SF Hildebrand, 1943

Dr Tomás Leandro Marini (1902–1984) was an Argentine ichthyologist and fisheries biologist. The University of Buenos Aires awarded his doctorate (1927). He was Chief of Experimental Work in Agricultural Zoology, University of Buenos Aires (1927), then Chief of the Division of Fisheries and Pisciculture at the Argentine Department of Agriculture (1930). He became Director General of Fisheries, Buenos Aires (1967). He provided Hubbs with the type specimen of the grenadier.

Marion (Ellis)

Characin sp. *Astyanax marionae* CH Eigenmann, 1911

Marion Lee Ellis née Durbin (1887–1972) (See **Durbin** & **Ellis, ML**)

Marion (Grey)

Marion's Spiderfish *Bathytyphlops marionae* GW Mead, 1958

Marion Grey (1911–1964) (see **Grey, M**).

Maritz

Sandperch sp. *Parapercis maritzi* ME Anderson, 1992

Willie Maritz was Curator of the East London Aquarium, South Africa. He is currently self-employed in the fields of landscaping and environmental management. After a 3-year stint in the army, where he was a platoon seargent in the infantry (31 bn in the Caprivi), he studied first at The University of Pretoria for his BSc (1986) and then at UPE for his Honours (1988). At UPE he focused on the field of Marine Biology for the next 15 years. He was then employed as Curator of the East London Aquarium and took part in Scuba diving, boating, fishing, field trips to Transkei and internationally diving in the Mediterannean, Florida Keys, Malta, the Great Barrier Reef in Australia and the entire South African coastline. He was appointed as General Manager of the Katberg Eco Golf Estate in the Eastern Cape (2008–2012), then re-located to the Western Cape as an estates manager, before (2018) becoming self-employed.

Marjorie (Awai)

Marjorie's Fairy Wrasse *Cirrhilabrus marjorie* GR Allen, JE Randall & BA Carlson, 2003

Marjorie Awai was Curator of the Florida Aquarium, and former Curatorial Assistant in the Ichthyology Department at the Bishop Museum. She and her husband, ichthyologist and co-author of the wrasse Dr Druce Carlson (q.v.), first observed the species when diving off the north coast of Viti Levu, Fiji (2000). Among other papers and articles, she co-wrote (with Carson et al.): *Hatching and early growth of Nautilus belauensis and implications on the distribution of Nautilus* (1992). She is also a fine underwater photographer and has illustrated many articles etc., including *A Guide to Hawaiian Marine Life* (1989). (Also see **Carlson**)

Marjorie (Frewer)

Marjorie's Hardyhead *Craterocephalus marjoriae* GP Whitley, 1948

Whitley provided no etymology, but the most likely candidate is his sister Marjorie Clare Frewer née Whitley.

Marjorie (McPhail)

Pearly Prickleback *Bryozoichthys marjorius* JD McPhail, 1970

Mrs D Marjorie McPhail is the wife of the author, Canadian ichthyologist John D McPhail – co-author of *Freshwater Fishes of Northwestern Canada and Alaska* (1970). She is mother to their daughter, appropriately called Pearl.

Mark, EL

Characin genus *Markiana* CH Eigenmann, 1903

Dr Edward Laurens Mark (1847–1946) was an anatomist and histologist who was head of Harvard University's zoology department, where Eigenmann studied under him. He graduated from the University of Michigan (1871) and his doctorate was awarded by the University of Leipzig (1876). He was an instructor at Michigan and on the US Northwest Boundary Survey (1871–1873). He joined the faculty at Harvard (1877) and was Hersey Professor of Anatomy (1885–1921). He was also Director, Bermuda Biological Station for Research (1902–1931). He is also remembered for his frequent power struggles with Alexander Agassiz.

Mark (Pote)

Rippled Blaasop *Pelagocephalus marki* PC Heemstra & MM Smith, 1981

Mark Pote was a schoolboy when (as the etymology states): "*On the morning of 17 November 1979, Mark Pote found an unusual littIe puffer fish alive in a tide pool at Port Alfred on the southeast coast of South Africa. Despite his efforts to keep it alive, the fish died the next day and was then donated to the JLB Smith Institute of Ichthyology.*"

Mark (Sabaj)

Thorny Catfish sp. *Leptodoras marki* JLO Birindelli & LM de Sousa, 2010

Dr Mark Henry Sabaj Pérez (b.1969). (See **Sabaj**)

Markert

Congo Cichlid sp. *Lamprologus markerti* S Tougas & MLJ Stiassny, 2014

Dr Jeffrey A Markert is an evolutionary geneticist at the Department of Biology, Providence College, Rhode Island, USA. His initial analyses of cichlid population structure in the region of the large Inga Rapids (Democratic Republic of Congo) stimulated the authors' morphological study. Hampshire College awarded his BA (1989), the University of Vermont his MSc (1994) and the University of New Hampshire his PhD (1998). He was a teaching assistamnt at Vermont and New Hampshire before becoming an instructor in biology at the University of Rhode Island (2008–2009). He became an adjunct professor at Providence college (2010) and is now a laboratory co-ordinator and instructor there (since 2011). Among his publications is the co-written: *Phylogeny of a rapidly evolving clade: The cichlid fishes of Lake Malawi, East Africa* (1999) and *Fine-scale genetic isolation and morphological divergence mediated by large high-energy rapids in two cichlid genera from the Lower Congo River* (2010).

Markevich

Prickleback sp. *Alectrias markevichi* BA Sheiko, 2012

Aleksandr Igorevich Markevich is a worker at the Far Eastern Marine Biosphere Reserve, Russian Academy of Sciences (Vladivostok). The author gives no further biography.

Markle

Nova Scotia Skate *Breviraja marklei* JD McEachran & T Miyake, 1987

Dr Douglas Frank Markle (b.1947) is an ichthyologist who is Professor Emeritus of Fisheries, Oregon State University, which he joined (1985). Cornell University awarded his bachelor's degree (1969) and the College of William and Mary, Virginia Institute of Marine Science, Gloucester Point, Virginia, his master's (1972) and doctorate (1976). He sent the describers the type series of the species. His major collaboration has been with John Edward Olney on larval taxonomy and systematics of *Carapidae* and he has carried out other systematics work was on *Gadiformes, Alepocephalidae*, and *Catostomidae* and early life history ecology of *Pleuronectidae* and *Catostomidae.* His major publications include *Audubon's hoax: Ohio River fish described by Rafinesque* (1997), and as co-author *Systematics of pearlfishes* (*Pisces: Carapidae*) (1990).

Mark Smith

Lake Tanganyika Cichlid sp. *Eretmodus marksmithi* WE Burgess, 2012
Lake Tanganyika Cichlid sp. *Julidochromis marksmithi* WE Burgess, 2014

Mark Smith is an American aquarist who specialises in Rift Valley cichlids. He is also a photographer of aquatic wildlife, an explorer and a writer. Among his books is: *Lake Malawi Cichlids* (2001).

Marley

Flounder genus *Marleyella* HW Fowler, 1925
Sand Dragonet *Callionymus marleyi* CT Regan, 1919
Double-sash Butterflyfish *Chaetodon marleyi* CT Regan, 1921
Threeline Tongue-Sole *Cynoglossus marleyi* CT Regan, 1921
African Blackmouth Croaker *Atrobucca marleyi* JR Norman, 1922

Black Snoek *Thyrsitoides marleyi* HW Fowler, 1929
Pufferfish sp. *Torquigener marleyi* HW Fowler, 1929
Longnose Pygmy Shark *Heteroscymnoides marleyi* HW Fowler, 1934
Sand-diver sp. *Trichonotus marleyi* JLB Smith, 1936

Harold Walter Bell Marley (1872–1945) was Principal Fisheries Officer (1918–1937) at Durban, South Africa, and a naturalist with a particular interest in entomology. He was born in England and went to South Africa at the time of the Boer War, then decided to stay on. He collected continuously for c.50 years in nearly every area south of the Zambesi River. Museums in many parts of the world have specimens that he sent. He also collected birds' eggs, and his collection, now in the Pretoria Museum, is regarded as one of the most complete ever assembled. He contracted Blackwater Fever (1944) while collecting in northern Zululand and died soon after his return to Durban. A mammal and three birds are also named after him. He was honoured because he had "*...collected many interesting South African fishes*" for Fowler.

Marlier

Lake Tanganyika Cichlid sp. *Spathodus marlieri* M Poll, 1950
Squeaker Catfish sp. *Chiloglanis marlieri* M Poll, 1952
Lake Tanganyika Cichlid sp. *Julidochromis marlieri* M Poll, 1956

Dr Georges Marlier was a Belgian entomologist and zoologist whose doctorate was awarded by the University of Brussels. He was Head of the Tanganika Scientific Research Centre, Uvira, Belgian Congo (1954). He became a Professor at Institut Royal des Sciences Naturelles, Brussels. He wrote: *A new Pelagic Trichoptera from Lake Tanganika* (1955). He collected the holotypes of these species and Poll said he named the fish "*...in honor of and gratitude for his valuable collaboration.*" (translation)

Maroni

Keyhole Cichlid *Cleithracara maronii* F Steindachner, 1881

A toponym referring to the Maroni (or Marowijne) River, which forms the boundary between French Guiana and Suriname.

Marqués

Driftwood Catfish sp. *Ageneiosus marquesi* FJ Risso & ENP de Risso, 1964 NCR
[Junior Synonym of *Ageneiosus militaris*]

Argentino Marqués was an Argentinean who donated the holotype of this species. As the authors explicitly dedicated the new species not only to this person, but also to his wife, the correct form should have been the plural *marquesorum*.

Marquet, G

Goby sp. *Stenogobius marqueti* RE Watson, 1991

Gérard Marquet (b.1948) is an aquatic biologist formerly at MNHN, Paris. He was honoured for his extensive collection efforts in freshwaters throughout French Polynesia and the discovery of four new species of *Stenogobius*. Among his published books and papers is the co-written: *Stenogobius (Insularigobius) keletaona, a new species of freshwater goby from Futuna Island (Teleostei: Gobiidae)* (2006).

Marquet, JP

African Minnow sp. *Raiamas marqueti* BK Manda, J Snoeks, AC Manda & E Vreven, 2018

Jean-Pierre Marquet was formerly technical assistant of the BTC (Belgian Technical Cooperation) project 'PRODEPAAK' (Projet de Développement de la Pêche Artisanale et de l'Aquaculture au Katanga, 2008–2013). He was honoured in recognition of his 'remarkable efforts' in fish collecting, and his provision of logistic support for the Katanga Expedition (2012), which resulted in the discovery of this new species.

Marr, JC

Marr's Fusilier *Pterocaesio marri* LP Schultz, 1953

Professor John C Marr was head of the South Pacific Fishery Investigations of the US Fish and Wildlife Service. Among his published works are the report of the *Pacific Tuna Biology Conference 1961, Fishery and Resources Management in South East Asia* and *A Plan for Fishery Development in the Indian Ocean* (1976).

Marr, JWS

Deep-water Dragonfish sp. *Bathydraco marri* JR Norman, 1938

James William Slesser Marr (1902–1965) was a Scottish marine biologist and polar explorer. He participated (1925) in the British Arctic Expedition to Franz Josef Land, and joined the Discovery Investigations (1928–1929). On the outbreak of WWII, he conducted research in the Antarctic into the feasibility of whale meat for human consumption, and on his return (1940) was commissioned in the Royal Naval Volunteer Reserve, serving in Iceland, the Far East and South Africa.

Marrero

Marrero's Tube-snouted Ghost Knifefish *Sternarchorhynchus marreroi* CD de Santana & RP Vari, 2010

Dr Críspulo Marrero (b.1954) is an aquatic biologist who is Professor Emeritus, Universidad Nacional Experimental de los Llanos Occidentales, Venezuela, where he was a member of the faculty and Associate Researcher (1988–2017). University Central Venezuela, Caracas, awarded his bachelor's degree (1980) and his doctorate (1986) and he was an Associate Researcher there (1986–1988). Among his publications, he wrote: *Preliminary notes on natural history of low land fishes. 1 Food habits comparison of three Gymnotiform fish especies in the Apure river (Edo. Apure, Venezuela)* (1987) and *Tube-snouted gymnotiform and mormyriform fishes: convergence of specialized foraging mode in teleosts* (1993). His current field of work is in wetlands, aquatic biota and environmental impact on aquatic ecosystems. An aquatic insect is also named after him.

Marsh, EG

Marsh's Gambusia *Gambusia marshi* WL Minckley & JE Craddock, 1962

Ernest G Marsh Jr (1914–1983) was an American naturalist who worked at the Gus Engling Wildlife Refuge, Texas, and collected in Coahuila, Mexico. He was the first to discover the Cuatro Ciénegas area for zoology, and "... *contributed many specimens of fishes, and other vertebrates, from northern Mexico.*" He began (1936) collecting natural history specimens in this area of Coahuila State, Mexico, continuing into the late 1940s. Although the fish which were sent to the University of Michigan Museum of Zoology were mostly identified by Hubbs, much material was not properly examined until the 1960s. Among his articles were: *Bird Records from Northern Coahuila* (1938) and *Bobwhites At Large* (1946).

Marsh, RO

Characin sp. *Characidium marshi* CM Breder, 1925

Richard Oglesby Marsh (1883–1953) was an engineer, American diplomat – variously in Panama (1910) and St Petersburg (1912) – and amateur ethnologist. He worked in Panama under George Goethals (1923) and took part in a number of expeditions to Panama including leading and providing financial support for the Marsh-Darien Expedition (1924–1925). He wrote: *White Indians of Darien* (1934).

Marsha

Marsha's Lanternshark *Etmopterus marshae* DA Ebert & KE van Hees, 2018

Marsha Englebrecht (b.1963) is an American shark biologist, who is Facilities Curator for the Aquarium of the Bay (San Francisco, California). Previously she spent a decade working at Marine World/Africa USA in Vallejo. She grew up on her father's ranch and has said "*I had fish, horses, dogs, cats, quite a menagerie over the years. I grew up with a lot of space around me and the opportunity to see all the wildlife.*" She was honoured for her innovative contributions in the field of elasmobranch husbandry.

Marshall, B

Loach Catfish sp. *Amphilius marshalli* DN Mazungula & A Chakona, 2021

Brian E Marshall works as a lecturer at the Department of Biological Sciences, University of Zimbabwe, Harare. He was honoured for his contributions to the fields of aquatic biodiversity and freshwater ecology in Southern Africa, "...in particular the building of ichthyological capacity through training of several currently practicing researchers in the region, including the second author." Among his c.70 publications are: *A Review of Zooplankton Ecology in Lake Kariba* (1997) and *Crayfish, Catfish and Snails: the perils of uncontrolled biological control* (2019).

Marshall, N

Rattail sp. *Coryphaenoides marshalli* T Iwamoto, 1970

Norman Bertram Marshall (1915–1996) was an ichthyologist and marine biologist at the BMNH (1947–1972) and became Professor of Zoology, Queen Mary College, University of London (1972–1977), retiring as Professor Emeritus. Cambridge awarded his bachelor's degree (1937). After graduating he worked on plankton research at the Department of Oceanography, University of Hull (1937–1941), after which he was in army operations research. He was based in Antarctica (1944–1946) as part of Operation Tabarin, to establish a permanent base at Hope Bay. He wrote: *Aspects of Deep Sea Biology* (1954). Marshall Peak in Antarctica is named after him.

Marsigli

Siberian Sterlet *Acipenser ruthenus marsiglii* JF Brandt, 1833

Luigi Ferdinando Marsigli (sometimes Marsili) (1658–1730) was an Italian solider and naturalist. As he came from a well-to-do family he was privately educated, but also studied mathematics, anatomy and natural history in Bologna. He continued his scientific studies while collecting data on the Turkish military organisation during his travels through the Ottoman Empire (c.1680–c.1682). He then entered the service of Emperor Leopold and fought against the Turks and was wounded and taken prisoner. Sold to a Pasha, whom he accompanied to the Battle of Vienna, his release was purchased (1684) and he returned to serve as a military engineer. He was commissioned to lead the Habsburg border demarcation commission and mapped the 850km border in the Hungarian Empire in what is today Croatia, Serbia and Romania. During the twenty years he spent in Hungary he collected scientific information, specimens, antiques, took measurements and observations for his work on the Danube. He was (unfairly) disgraced during the War of the Spanish Succession (1703) and had to leave the Hapsburg army. He journeyed to Switzerland and then France where he mapped, observed the heavens, studied the mines and rivers and collected fossils, birds and fish. When he returned to Bologna he donated his collection to its Senate (1712). He then opened his Institute of Sciences and Arts (1715). He published, among others, a major work on the Danube *Danubius Pannonico-Mysicus* (1726), and atlas (1744) and a treatise on the oceans: *Histoire physique de la mer* (1725) and is considered the father of oceanography. He was honoured as the man who previously identified this sturgeon using a pre-Linnaean name (*Antaceus glaber*) in 1726.

Marta

Characin sp. *Gephyrocharax martae* G Dahl, 1943

Marta Petronella Dahl née Althén (b.1905), who collected the holotype, was the wife of the author (Professor George Dahl). (See also **Dahl**)

Martelli

Goby sp. *Luciogobius martellii* L Di Caporiacco, 1948

A Martelli was a technician at La Specolo, Museum of Zoology and Natural History in Florence, Italy.

Martens

Martens' Pipefish *Doryichthys martensii* W Peters, 1868

Barbeled Dragonfish sp. *Astronesthes martensii* CB Klunzinger, 1871

Dr Eduard von Martens (1831–1904) was a German zoologist who spent most of his career at the Museum für Naturkunde, Berlin. He embarked (1860) on the *Thetis* expedition to East

Asia and published the results in two volumes. His main interest was in molluscs. (Klunzinger gives no etymology, but it seems very likely that Eduard von Martens is the person intended)

Martenstyn

Martenstyn's Barb *Systomus martenstyni* M Kottelat & R Pethiyagoda, 1991

Goby sp. *Stiphodon martenstyni* RE Watson, 1998

Cedric Douglas Martenstyn (1946–1996) was a naturalist and diver, a Lieutenant Commander in the Sri Lankan Navy and Commander of the Navy's Special Boat Service during the civil war against the Tamil insurgents. He had previously been Director of the National Marine Mammal Programme. The helicopter in which he was travelling disappeared over the Indian Ocean (1996) and he was listed as Missing in Action and presumed dead. The etymology for the barb states that Martenstyn had contributed to the book in which the description appeared by "*collecting many of the more hard-to-get species…almost all of them from remote locations.*"

Marteyne

Red Star-eye Dwarf-goby *Eviota marteynae* DW Greenfield & MV Erdmann, 2020

Marteyne van Well is the General Manager (2011–present) of the Six Senses Resort on Laamu Atoll, Maldive Islands. Her bachelor's degree in Hospitality Management is from the Hotelschool The Hague and she has further diplomas from ESSEC Business School (1988) and Cornell (2011). She was honoured for "*her deep commitment to sustainability in the hospitality industry and her strong support for marine conservation initiatives that protect the reefs where this species is found.*"

Martha (Joynt)

Driftwood Catfish sp. *Tatia marthae* RP Vari & CJ Ferraris, 2013

Martha Joynt is a Management Support Specialist, Smithsonian, and was honoured for "*…significant assistance to both authors over the years, particularly the senior author.*" She has a bachelor's degree in liberal arts, awarded by Simon's Rock Early College (1976).

Martha (Myers)

Blackwing Hatchetfish *Carnegiella marthae* GS Myers, 1927

Mrs Martha Ruth Myers née Frisinger was the first wife (1926) of the author, Dr George Sprague Myers (1905–1985). (Also see **Myers**)

Martha (Sands)

Three-barbeled Catfish sp. *Brachyrhamdia marthae* DD Sands & BK Black, 1985

Mrs Martha Elizabeth Sands is the senior author's wife.

Martin, B

Armoured Catfish sp. *Ancistrus martini* LP Schultz, 1944

Bethea Martin was a geologist employed by Lago Petroleum Corporation. He helped Schultz in the collection of fishes in Venezuela (1942).

Martin, J

Lake Victoria Cichlid sp. *Haplochromis martini* GA Boulenger, 1906

James Martin, Transport Officer in the Uganda Protectorate, was honoured in the cichlid's name for "*...much assistance rendered to Mr. Degen*" (Edward Degen was the Swiss ornithologist who collected the type).

Martin (Mortenthaler)

Dwarf Cichlid sp. *Apistogramma martini* U Römer, I Hahn, E Römer, DP Soares & M Wöhler, 2003

Martin Mortenthaler (1961–2018) was the manager and owner of Rio Momon EIRL (formerly Aquarium Rio Momon), being an aquarium-fish exporter based in Iquitos, Peru. He was born in Vienna, Austria, and worked in a pet shop there (1975–1978) and took over the business, as Aquarium Wien (1992). He moved to live in Peru and settled in Iquitos in 1993, where he founded his fish export company which is now managed by his son. He collected and discovered various undescribed species of fish. (See also **Mortenthaler**)

Martin (Salazar)

Twig Catfish sp. *Farlowella martini* A Fernández-Yépez, 1972

Armoured Catfish sp. *Acestridium martini* ME Retzer, LG Nico & F Provenzano, 1999

Dr Felipe José Martín Salazar (b.1930) was a Venezuelan ichthyologist. He was head of the Fisheries and Wildlife Division, Venezuelan Ministry of Agriculture (1960). He wrote: *Las especies del género* Farlowella *de Venezuela (Piscis-Nematognalhi-Loricariidae) con descripción de 5 especies y 1 subespecie nuevas* (1964).

Martina

Martina's Dragonet *Callionymus martinae* R Fricke, 1981

Miss Martina Wolf of Braunschweig was honoured by Ronald Fricke "*...for her continued interest in my studies.*" We have failed to find further biographical details.

Martine

Sole sp. *Aseraggodes martine* JE Randall & SV Bogorodsky, 2013

Martine Desoutter-Méniger (see **Desouter**).

Martinez, A

Three-barbeled Catfish sp. *Pimelodella martinezi* A Fernández-Yépez, 1970

Dr Alfonzo Martinez M was a physician in practice at San Fernando de Apure, Venezuela. He was a student of the gamefish of Colombia.

Martinez, AJ

Suckermouth Catfish sp. *Astroblepus martinezi* CA Ardila Rodríguez , 2013

Antonio José Martinez Negrete is the Administrator, Parque Nacional Natural Paramillo, Cordoba, Colombia. He was honoured for scientific research and for his conservation work.

Martini

Hagfish sp. *Myxine martinii* MM Mincarone, D Plachetzki, CL McCord, TM Winegard, B Fernholm, CJ Gonzalez & DS Fudge, 2021

Frederic H 'Ric' Martini (b.1947) is an American anatomist and physiologist who taught for 23 years at the Shoals Marine Laboratory, Maine. He received his PhD in Comparative and Functional Anatomy from Cornell University (1974). He has studied the ecology and physiology of hagfish, and published reports on hagfish populations in the North Atlantic and South-west Pacific. He has now retired from teaching but maintains an affiliation with the Oceanography Department at the University of Hawaii at Manoa.

Martins

Neotropical Rivuline sp. *Hypsolebias martinsi* R Britzke, DTB Nielsen & C Oliveira, 2016

Dr Itamar Alves Martins is Assistant Professor and Coordinator of the Zoology Laboratory at the University of Taubaté, Brazil, where he was a post-graduate research fellow (2006–2012). His BSC (1992), master's (1996) and PhD (2001) were all awarded by the Paulista State University Júlio de Mesquita Filho. He has published widely, such as the co-authored paper: *Ichthyofauna of the Una river in the Paraíba do Sul Paulista River Valley, Southeastern Brazil* (2018). He was honoured for his contributions to South American ichthyology and herpetology.

Martorell

African Barb sp. *Enteromius martorelli* B Román, 1971

Fernando Martorell was a teacher at the Escuela de Artes y Oficios La Salle, Bata, Equatorial Guinea. He is recorded as having collected dragonfly specimens, and perhaps also helped Román collect the barb, though the original text only remarks that the fish is being named after Martorell as a sign of Román's gratitude.

Marvel

Eyipantla Silverside *Atherinella marvelae* B Chernoff & RR Miller, 1982

Marvel B Parrington (1910–1986) was a museum cataloguer at the University of Michigan (1965–1982). The etymology says her: "*...dedication and hard work have contributed significantly to ichthyological efforts at the University of Michigan [Museum of Zoology] for the past 17 years.*"

Mary (Chin)

Loach sp. *Pangio mariarum* RF Inger & PK Chin, 1962

Mrs Mary Chin is the wife of the junior author. The binomial honours both Phui-Kong Chin's wife and Robert Frederick Inger's wife – they were both called Mary.

Mary (Inger)

Loach sp. *Pangio mariarum* RF Inger & PK Chin, 1962

Mrs Mary Lee Inger née Ballew (1918–1985) was the first wife of the senior author. The binomial honours both Robert Frederick Inger's wife and Phui-Kong Chin's wife - they were both called Mary.

Maryann

Oblique-swimming Triplefin *Forsterygion maryannae* GS Hardy, 1987

Maryann W Williams was, according to the etymology, a "*…gentle lady and fine underwater photographer… …who provided photographs of this fish in the wild.*"

Maslenikov

Blushing Snailfish *Careproctus maslenikovae* JW Orr, 2021

Katherine Pearson Maslenikov is (2001–present) Ichthyology Collections Manager of the Burke Museum's fish collection at the School of Aquatic and Fishery Sciences, University of Washington, Seattle. She was previously a Fisheries Biologist of the National Marine Fisheries Service's Alaska Fisheries Science Centre, Seattle (summers of 1999–2011). She studied marine biology at The Evergreen State College (1991–1995) and Ecology and Evolutionary Biology at the University of Kansas (1996–1998). Her publications include: *Specimens by the Millions: Managing Large, Specialized Collections at the University of Washington Burke Museum Fish Collection* (2021). The author describes her as a "… diligent collector of many snailfish types."

Maslovsky

Morid Cod sp. *Physiculus maslowskii* IA Trunov, 1991

Alexandr Davidovich Maslovskiy (1897–1969) was Head of the Museum of Nature of Kharkov University (Ukraine). He served in WW2, and afterwards became an Associate Professor in the Department of Hydrobiology, Kharkov State University. He participated in works connected with forming the Krasnyi Oskol water reservoir in Kharkov Region (1963). Trunov named this species after him to honour his first teacher.

Mason

Goby sp. *Acentrogobius masoni* F Day, 1873

James Wood-Mason (1846–1893) was a British zoologist in India. (See **Wood-Mason**)

Massutí

Goby sp. *Buenia massutii* M Kovacic, F Ordines & UK Schliewen, 2017

Enric Massutí is a marine biologist who is Director (2009) of Instituto Español de Oceanografía in the Centre Oceanogràfic de les Balears. The University of Barcelona awarded his BSc (1987) and the University of the Balearic Islands his PhD (1997). Among his over 200 published articles, papers and books are the co-written: *Mediterranean and Atlantic deep-sea fish assemblages: Differences in biomass composition and size-related structure* (2004) and *Large-scale distribution of a deep-sea megafauna community along Mediterranean trawlable grounds* (2019). He was honoured for his dedication to the study of benthic habitats in the circalittoral bottoms of the Balearic Islands.

Masters

Masters' Catfish *Hexanematichthys mastersi* JD Ogilby, 1898

George Masters (1837–1912) was an entomologist who emigrated from England to Australia. He was Assistant Curator of the Australian Museum (1864), apparently on condition that he sold his private collection and made no new one; an agreement he ignored. He became Curator of the Macleay collection in Sydney (1874–1912) until his death in a carriage accident. He certainly collected widely, as there are skeletons of Thylacines ('Tasmanian Tigers') which he collected (1870s). Masters collected with Macleay (q.v.) when he was the latter's personal curator, including accompanying him to New Guinea (1875). He helped to plan and provision the expedition, which Macleay paid for. In his diary for that year Macleay notes: 'Masters continues still getting things ready for the expedition. It ought to be well found in everything from the bills coming in'. Masters' great knowledge of Australian fauna went largely unrecorded, as he disliked writing. A reptile and six birds are named after him.

Masuda

Velvetfish sp. *Cocotropus masudai* K Matsubara, 1943
Masuda's Hogfish *Bodianus masudai* C Araga & T Yoshino, 1975
[Alt. Candystripe Hogfish]
Masuda's Dragonet *Foetorepus masudai* T Nakabo, 1987
Masuda's Dwarfgoby *Eviota masudai* K Matsuura & H Senou, 2006

Hajime Masuda (1921–2005) was a Japanese ichthyologist. Among his publications were the co-authored papers: *Coastal fishes of southern Japan* (1975) and *Two new anthiine fishes from Sagami Bay, Japan* (1980), and the book *Sea fishes of the world (Indo-Pacific region)* (1987). He was honoured in the goby's binomial because he "...*contributed greatly to Japanese and Indo-west Pacific ichthyology through his many books containing excellent photographs*."

Masui, M

Masui's Shrimp-Goby *Amblyeleotris masuii* Y Aonuma & T Yoshino, 1996

M Masui is a collector for Umikawa Coral Fish Shop (Okinawa-jima Island, Japan). He was honoured because he "...*kindly offered us important specimens and collecting data*" and collected the type specimen.

Masui, T

Mormyrid sp. *Mormyrops masuianus* GA Boulenger, 1898

Jean-Baptiste Théodore Masui (b.1863) was a Belgian army officer and museum administrator. He served as a lieutenant in an artillery regiment, and rose to become Director of the Musée royal d'Afrique centrale (then Musée du Congo) in Tervuren. The museum mounted the Colonial Exhibition (1897) about the newly formed Congo FreeState.

Masya

Arrow Loach *Nemacheilus masyae* HM Smith, 1933
Spotted Barb sp. *Puntius masyai* HM Smith, 1945

Luang Masya Chitrakarn (1896–1965), also known as Prasop Teeranunt, was a Thai ichthyologist and illustrator of fishes, renowned for his skill with both brush and pen who worked for the Siamese Department of Fisheries. He helped collect the holotype of the loach.

Matahari

Sunburst Dottyback *Pseudochromis matahari* AC Gill, MV Erdmann & GR Allen, 2009

Perhaps disappointingly, this name has nothing to do with the Dutch spy and exotic dancer Margreet MacLeod – better known by her stage name Mata Hari – but derives from an Indonesian term for 'sun', with reference to the fish's coloration.

Matallanas

Striped Rabbitfish *Hydrolagus matallanasi* JMR Soto & CM Vooren, 2004

Jesús Matallanas Garcia is a Spanish ichthyologist and Professor in the Faculty of Biosciences at the University of Barcelona. Many of his published papers have described new species, such as: *Description of Gosztonyia antarctica, a new genus and species of Zoarcidae (Teleostei: Perciformes) from the Antarctic Ocean* (2009). He was honoured for his "*…extensive work and tireless dedication to ichthyology.*"

Mate

Tetra sp. *Hemigrammus matei* CH Eigenmann, 1918

The original text does not provide an etymology, but probably named after Paul Matte (1854–1922). (See **Matte**)

Mathotho

Lake Malawi Cichlid sp. *Labidochromis mathotho* WE Burgess & HR Axelrod, 1976

A J Mathotho was Chief Fisheries Officer, Malawi, and went on to become Permanent Secretary at the Malawi Ministry of Agriculture. He co-wrote: *Africa's great lakes and their fisheries potential* (1973). The authors say that: "*…without whose help the discovery* [of this and other new species] *would not have been possible.*"

Matinta Pereira

Glass Knifefish sp. *Eigenmannia matintapereira* LA Peixoto, GM Dutra & WB Wosiacki, 2015

Matinta Pereira is a figure in Brazilian mythology, a witch who turns into a bird at night. Her blackened appearance is said to resemble the colour pattern of this fish species.

Matsubara, K

Duckbill genus *Matsubaraea* I Taki, 1953
Velvetfish genus *Matsubarichthys* SG Poss & GD Johnson, 1991
Loach sp. *Cobitis matsubarae* Y Okada & H Ikeda, 1939
Pitted Stingray *Bathytoshia matsubarai* Y Miyosi, 1939
Sculpin sp. *Astrocottus matsubarae* M Katayama, 1942
Dusky-purple Skate *Bathyraja matsubarai* R Ishiyama, 1952

Japanese Snake-Blenny *Xiphasia matsubarai* Y Okada & K Suzuki, 1952
Sculpin sp. *Myoxocephalus matsubarai* M Watanabe, 1958
Marine Hatchetfish sp. *Polyipnus matsubarai* LP Schultz, 1961
Japanese Blue-spotted Seabream *Amamiichthys matsubarai* M Akazaki, 1962
Sole sp. *Heteromycteris matsubarai* A Ochiai, 1963
Shokotsu Lamprey *Lethenteron matsubarai* VD Vladykov & E Kott, 1978
Rattail sp. *Coelorinchus matsubarai* O Okamura, 1982
Eelpout sp. *Lycodes matsubarai* M Toyoshima, 1985

Dr Kiyomatsu Matsubara (1907–1968) was originally called Kiyomatsu Sakamoto but on marrying he took his wife's name as the family surname. He was a Japanese herpetologist and ichthyologist. The Imperial Fisheries Institute (now the Tokyo University of Fisheries) awarded his bachelor's degree (1929). He was Professor of Fisheries at the Imperial Fisheries Institute (1943–1947) and was Professor in the Department of Fisheries, Kyoto University (1947–1968). He wrote: *Fish Morphology and Hierarchy* (1955). He was a student of Carl Hubbs, who wrote his obituary. Miyosi honoured him in the stingray name as the person who collected one of the paratypes at a fish market, and "*…to whom the author is much indebted for many favours*".

Matsubara, S

Rockfish sp. *Sebastes matsubarai* FM Hilgendorf, 1880

Shin'nosuke Matsubara (b.1858) was Director of the Imperial Fisheries Institute, Tokyo. He was (1875) Associate Professor at the College of Medicine, Tokyo, his alma mater. He was also a general affairs official at the Ministry of Agricultcure. He wrote a book on botany (1882) and *Goldfish and Their Culture in Japan* (1908).

Matsui, T

Tubeshoulder genus *Matsuichthys* YI Sazonov, 1992

Tetsuo Matsui worked at the Scripps Institution of Oceanography. He co-wrote: *Review of the Deep-Sea Fish Family Platyroctidae* (1987) and co-described the only known member of the genus – *M. aequipinnis* (under the name *Barbantus aequipinnis*). In 2011 he was recorded, aged 80, as helping the homeless in San Diego and described as a "retired ichthyologist".

Matsui, Y

Poacher sp. *Percis matsuii* K Matsubara, 1936

Dr Yoshiichi Matsui (1891–1976) was a fish geneticist and fisheries scientist; a graduate of the Fisheries School of the Ministry of Agriculture and Forestry. He was an engineer at the Aichi Fisheries Research Institute and the Fisheries School, also serving as technical advisor to fisheries in Mexico. He was a Lecturer at the Department of Physics of Kyoto University (1958) and later appointed Professor at the Faculty of Agriculture of Kinki University and Director of the Shirahama Marine Laboratory of the University. His research interests spanned freshwater fish and pearls. His *A Genetic Research of Goldfish Produced in Japan* (1934) is known internationally. His other books include *The Encyclopedia of Goldfish* (1968) and *The Encyclopedia of Pearl* (1965). In Kiyomatsu Matsubara's introduction to the paper describing the species he says: "*Through the courtesy of Dr. Yoshiichi Matsui, chief-*

expert of the Toyohashi Branch of the Fisheries Experimental Station, a number of specimens of fish obtained by motor-trawlers near the sea-coast of Owase, Mie-ken, at depths ranging from 100 to 200 fathoms, from October 1932 to March 1933, have of late come under my examination. They comprise many interesting forms, including two new species which are described in the present paper. It is my pleasant duty to return herewith my gratitude to Dr. Y. Matsui for his favour of valuable material..."

Matsuno

Matsuno's Pygmygoby *Trimma matsunoi* T Suzuki, J Sakaue & H Senou, 2012

Kazushi Matsuno is a professional diver at AQUAS Kashiwajima Diving Service (Kochi, Japan). He discovered this goby.

Matsushima

Snailfish sp. *Crystallias matsushimae* DS Jordan & JO Snyder, 1902

This is a toponym, referring to Matsushima Bay, Japan.

Matsuura

Filefish sp. *Paramonacanthus matsuurai* JB Hutchins, 1997

Lombok Sole *Aseraggodes matsuurai* JE Randall & M Desoutter-Meniger, 2007

Dr Keiichi Matsuura works in the Department of Zoology at the National Museum of Nature and Science, Tokyo. He has published more than 80 works, such as the paper: *Underwater observations of the rare deep-sea fish Triodon macropterus (Actinopterygii, Tetraodontiformes, Triodontidae), with comments on the fine structure of the scales* (2016) and the book chapter *Fishes of Ha Long Bay in Northern, Vietnam fishes-small* (2018). He collected the sole holotype (1994). Hutchins says that Matsuura "*...helped the author on many occasions with information and specimens of monacanthids.*"

Matt

Splendid Rockfish *Acanthoclinus matti* GS Hardy, 1985

Matthew 'Matt' Hardy is the son of the author Graham Hardy, who says his offspring "*... even from an early age, has shown considerable awareness and appreciation of his natural surroundings.*"

Matte

Tetra sp. *Hemigrammus matei* CH Eigenmann, 1918

Paul Matte (1854–1922) is the most likely candidate as the person after whom this tetra was named, despite Eigenmann providing no etymology and spelling the binomial with a single *t*. The label on the type specimens identify that Matte was the source of the material. He was a German importer and fish breeder who was an expert on tropical fish. Among his achievements was to import the first Zebra Fish (*Danio rerio*) into Europe (1905).

Matthes

African Barb sp. *Clypeobarbus matthesi* M Poll & J-P Gosse, 1963

Squeaker Catfish sp. *Synodontis matthesi* M Poll, 1971
Matthes' Lampeye *Lacustricola matthesi* L Seegers, 1996

Dr Hubert Matthes is a Dutch ichthyologist who, early in life, moved with his parents to the then Belgian Congo for several years. From this experience he found the inspiration for his dissertation: *Les poissons du Lac Tumba et de la region d'Ikela. Etude systématique, écologique et zoogéographique. Résultats scientifiques d'une mission I.R.S.A.C. dans la Cuvette Centrale congolaise* (1964). He worked (1957–1964) as a Research Associate in Limnology at the Institut pour Ia Recherche Scientihque en Afrique Centrale (IRSAC) in DRC, which was part of MRAC in Tervuren. At the same time he also studied biology under Professor Hendrik Engel, Director of the Zoölogisch Museum, Universiteit van Amsterdam and defended his dissertation there. When promoted, he had no official connections with ZMA anymore, but irregularly visited the department of fishes and other museums attempting to identify his fish collection from Central Africa, which were divided between Tervuren and Amsterdam. His longer works include: *A bibliography of African freshwanter fish* (1973).

Mattos

Brazilian Cichlid sp. *Australoheros mattosi* FP Ottoni, 2012

José Leonardo de Oliveira Mattos is a Brazilian ichthyologist who is head of Laboratory of Systematics and Evolutionary Biology in the Biology Institute of the Zoologoy Department, Universidade Federal do Rio de Janeiro. He is author of more than 60 articles and papers, such as: *Microcambeva draco, a new species from northeastern Brazil (Siluriformes: Trichomycteridae)* (2010) and the co-written: *Phylogeny and species delimitation based on molecular approaches on the species of the* Australoheros autrani *group (Teleostei, Cichlidae), with biogeographic comments* (2019).

Mattozo

Papermouth (barb) *Enteromius mattozi* ARP Guimarães, 1884

Dr Fernando dos Santos Mattozo (1849–1921) was a Portuguese academic, politician and diplomat. He graduated in medicine and philosophy from the University of Coimbra (1874) and became Professor of Zoology and Comparative Anatomy at Escola Politécnica de Lisboa (1880). He served as a minister in several capacities including being Minister for Foreign Affairs (1901–1903) and was Envoy and Minister Plenipotentiary to Brazil (1901).

Maude

Maude's Shrimp-Goby *Cryptocentrus maudae* HW Fowler, 1937

Maude de Schauensee was the daughter of ornithologist Rodolphe Meyer de Schauensee (q.v.) and his wife Williamina, who presented the Academy of Natural Sciences (Philadelphia) with an "elaborate gift" of a collection of fishes from Thailand, including the holotype of the goby.

Mauge

Maugé's Dragonet *Draculo maugei* JLB Smith, 1966
Perchlet sp. *Plectranthias maugei* JE Randall, 1980

André L Maugé (1922–2008) of the MNHN (Paris) was a somewhat atypical ichthyologist, beginning his working life as an accountant. He went to Madagascar (1955), became the General Secretary of the Institut Scientifique de Madagascar (1958), and then of the Marine Station at Tulear (1964). His developing interest in marine life was largely self-taught. He joined the Laboratoire d'ichtyologie générale et appliquée du Muséum de Paris (1974).

Maul

Searsid genus *Maulisia* AE Parr, 1960
Goby genus *Mauligobius* Miller 1984
Stareye Lightfish *Pollichthys mauli* M Poll, 1953
Maul's Searsid *Maulisia mauli* AE Parr, 1960
Maul's Waryfish *Scopelosaurus mauli* E Bertelsen, G Krefft & NB Marshall, 1976
Footballfish sp. *Himantolophus mauli* E Bertelsen & G Krefft, 1988
Manefish sp. *Platyberyx mauli* EI Kukuev, NV Parin & IA Trunov, 2012
Three-bearded Rockling sp. *Gaidropsarus mauli* M Biscoito & L Saldanha, 2018

Günther Edmund Maul (1909–1997) was a German ichthyologist and taxidermist who lived and worked most of his life in Madeira. He started work as a taxidermist at the Museu Municipal do Funchal (1930) and went on to become Director (1940–1979) for the rest of his working life, although he continued his research after he retired. He opened the Museum's aquarium to the public (1959). He took part in several expeditions to the Salvage Islands (1963), notably with the French bathyscaphe 'Archimède' (1966). Biscoito & Saldanha said of him that the fish was named: "*...in recognition of his outstanding contribution to the knowledge of the Atlantic ichthyofauna.*"

Maule

Neotropical Silverside sp. *Odontesthes mauleanum* F Steindachner, 1896

This is a toponym referring to the river Maule in Chile.

Maurice

Longsnout Scraper (barb) *Capoeta mauricii* F Küçük, D Turan, C Şahin & I Gülle, 2009

Dr Maurice Kottelat (See Kottelat)

Maurine

Maurine's Damselfish *Chrysiptera maurineae* GR Allen, MV Erdmann & NK Dita Cahyani, 2015

Maurine Shimlock is an American journalist, underwater photographer and marine tourism consultant. In 2008 Conservation International hired her and her partner Burt Jones to consult on sustainable marine tourism initiatives in the most bio-diverse tropical reef environment on earth: Indonesia's Raja Ampat and the surrounding Bird's Head Seascape. She co-wrote: *Diving Indonesia's Raja Ampat* (2009) and *Diving Indonesia's Bird's Head Seascape* (2011). She was described by the authors as "*...our dear friend of many years, who has zealously promoted marine conservation of Cenderawasih Bay* (the type locality) *and the surrounding Bird's Head region by means of her excellent journalism and photography.*"

Maurolico

Lightfish genus *Maurolicus* A Cocco,1838

Francesco Maurolico (1494–1575) was an Italian mathematician and astronomer from Messina, Sicily. He was ordained as a priest (1521), and became a Benedictine monk (1550). The etymology is not explained by Cocco but it seems likely that this genus is named after Maurolico: Cocco named a species (*M. amethystinopunctatus*) where the holotype originated from Messina, and presumably coined the genus name after a famous son of that city.

Mawson

Mawson's Dragonfish *Cygnodraco mawsoni* ER Waite, 1916
Antarctic Toothfish *Dissostichus mawsoni* JR Norman, 1937
Snailfish sp. *Paraliparis mawsoni* AP Andriashev, 1986

Sir Douglas Mawson (1882–1958) was an Australian geologist and Antarctic explorer. He joined Ernest Shackleton's 'Nimrod' expedition to Antarctica (1907–1909), while still a doctoral student, during which they completed the longest ever Antarctic man-hauling sledge journey (122 days). He then took part in the Australian Antarctic Expedition (1911–1914). He was the lone survivor of a three-man sledge team, managing to return to base after 30 days by himself, missing the pick-up boat by 1 day! He had to overwinter at the station with the volunteers who had stayed. He took part in a further Antarctic expedition (1929–1931). The Mawson Research Station in Antarctica was named (1954) after him, and the snailfish may be named after that station.

Max Weber

Pygmy Snapper *Lutjanus maxweberi* CML Popta, 1921
Max Weber's Pipefish *Cosmocampus maxweberi* GP Whitley, 1933

Max Wilhelm Carl Weber van Bosse (1852–1937). (See **Weber, M**)

Maya

Cichlid genus *Mayaheros* O Říčan & L Piálek, 2016
Maya Pupfish *Cyprinodon maya* JM Humphries & RR Miller, 1981
Mayan Swordtail *Xiphophorus mayae* MK Meyer & M Schartl, 2002
Maya Hamlet *Hypoplectrus maya* PS Lobel, 2011

Named after the Mayan people who inhabit the areas of Central America where these fish occur. In the name of the cichlid genus, *Maya* is combined with *Heros*, an older name for some South American cichlids. The coined word can be taken to mean 'hero of the Mayans'. The hamlet was named for both the Mayan people and the author's daughter (see next entry).

Maya (Lobel)

Maya Hamlet *Hypoplectrus maya* PS Lobel, 2011

Maya Rose Lobel is the daughter of the author, Phillip Lobel (q.v.). He wrote that the binomial had a "dual purpose" in honouring both his daughter and "the Maya people of Belize."

Maydell

Cunene Kneria *Kneria maydelli* W Ladiges & J Voelker, 1961
Bagrid Catfish sp. *Hemibagrus maydelli* F Rössel, 1964
Goalpara Loach *Neoeucirrhichthys maydelli* PM Bănărescu & TT Nalbant, 1968

Gustav Adolf Baron von Maydell (1919–1959) was a German ecologist, bio-geographer and collector. He led the German India-Expedition (1955–1958), a Zoological Expedition of the University of Hamburg Zoological Institute and Museum. He also collected for a wide range of other institutions, including the Zoological Survey of India.

Mayden

Black River Madtom *Noturus maydeni* JJD Egge, 2006
Redlips Darter *Etheostoma maydeni* SL Powers & BR Kuhajda, 2012

Dr Richard Lee Mayden (b.1955) is an American ichthyologist and systematic biologist. Southwestern Illinois College awarded his bachelor's degree and the University of Illinois his master's. The University of Kansas awarded his doctorate (1987), after which he was a Research Fellow at the University of California, Los Angeles (1988–1990). He then became an Assistant Professor and Curator of Fishes at the Ichthyological Collection, University of Alabama (1987–2001) before taking up (2001) his present position as Professor of Natural Sciences, Department of Biology, Saint Louis University, Missouri. He co-wrote: *Fishes of Alabama* (2004).

Mayer

Angolan Deepwater Cardinalfish *Epigonus mayeri* M Okamoto, 2011

Garry Franklin Mayer works at the National Oceanic and Atmospheric Administration (NOAA), National Marine Fisheries Service. He has written various texts on cardinalfish, including: *A Revision of the cardinalfish genus* Epigonus *(Perciformes, Apogonidae) with descriptions of two new species* (1974).

Mayi

False Silver Whiptail *Coelorinchus mayiae* T Iwamoto & A Williams, 1999

Not an eponym, but from the Australian Yindjibarndi language, *mayi*, meaning younger sister. This refers to the putative sister-species relationship with *C. argentatus*.

Mayland

Lake Malawi Cichlid genus *Maylandia* MK Meyer & W Förster, 1984
Mayland's Rainbowfish *Melanotaenia maylandi* GR Allen, 1982
Balsas Molly *Poecilia maylandi* MK Meyer, 1983
Sulphurhead Peacock Cichlid *Aulonocara maylandi* E Trewavas, 1984

Hans-Joachim Mayland (1928–2004) was a German ichthyologist, aquarist, photographer and writer. A businessman by training, he decided early on to follow a career in journalism and initially worked as a sports journalist, writing on football. His initial passion for the terrarium hobby later turned to a preference for fishkeeping, and this soon led to his energies being directed entirely to writing about fish (mid-1960s). He was highly successful as an author of aquarium books and articles and made a huge contribution to the advancement of the aquarium hobby in Europe. Well over 70 books were published, some being translated into other languages including Chinese, Polish and Dutch. Among his publications are the co-written: *South American Dwarf Cichlids* (1996) and *South American Cichlids IV* (1998). He made more than 40 overseas field trips focusing on cichlids, particulary those from the African Rift Valley and South America. He contacted and corresponded with many top ichthyologists, and brought the *Aulonocara* species to Trewavas' attention.

Mayrink

Pellona sp. *Pellona mayrinki* SY Pinto, 1972 NCR
[Junior Synonym of *Pellona harroweri*]

Wilson Mayrink (1925-2017) was a Brazilian physician and parasitologist. The Federal University of Minas Gerais awarded his first degree (1951) and doctorate (1960). He became Professor in the biology department there (1967) until retirement, when he became Professor Emeritus. He spent almost 50 years searching for a cure or prevention for leishmaniasis, which resulted in an effective vaccine. He published around 150 scientific papers and books.

Maytag

Maytag Bass *Serranus maytagi* CR Robins & WA Starck, 1961

Robert Elmer Maytag (1923–1962) was an American naturalist, conservationist and philanthropist who was the founder of the Phoenix Zoo and Arizona Zoological Socity. His father (Elmer Henry Maytag, 1883–1940) ran the Maytag Corporation and founded Maytag Dairy Farms. The authors stated: *It is a pleasure to name this seabass for Mr Robert E Maytag who has so generously supported ichthyological research at the Marine Laboratory.* He died of Pneumonia at just 38.

Mazaruni

Guianan Cichlid genus *Mazarunia* SO Kullander, 1990
Neotropical Rivuline sp. *Anablepsoides mazaruni* GS Myers, 1924
Guianan Cichlid sp. *Mazarunia mazarunii* SO Kullander, 1990

This is a toponym referring to the Mazaruni river system in Guyana.

Mbeni

Lake Malawi Cichlid sp. *Labidochromis mbenjii* DSC Lewis, 1982
Lake Malawi Cichlid sp. *Copadichromis mbenjii* AF Konings, 1990
Lake Malawi Cichlid sp. *Maylandia mbenjii* J Stauffer Jr, NJ Bowers, KA Kellogg & KR McKaye, 1997

This is a toponym referring to Mbenji Island, Lake Malawi, the type locality.

Mbigua

Neotropical Wolf-Fish sp. *Hoplias mbigua* MM Azpelicueta, MF Benítez, DR Aichino & CMD Mendez, 2015

Isabelino Rodríguez has the nickname of *Mbigua*. He is a Paraguayan citizen who worked during many years in the Proyecto Biología Pesquera Regional in northern Argentina. *Mbigua* is a guaraní word meaning 'cormorant', and was given due to Isabelino's skill at catching fish. He was honoured in the name as he taught some fishing skills to the authors.

M'Boycy

Pencil Catfish sp. *Cambeva mboycy* BW Wosiacki & JC Garavello, 2004

M'Boy cy is a character in Tupí-Guaraní mythology. He is part of the legend of the origin of the Iguaçu waterfalls, Paraná, Brazil, which is close to where this catfish can be found.

Mbudya

Goby sp. *Acentrogobius mbudyae* TT Nalbant & R Mayer, 1975

This is a toponym referring to Mbudya Island, Tanzania, the only area where the species is known to occur.

McAdams

McAdam's Scorpionfish *Parascorpaena mcadamsi* HW Fowler, 1938

Fred McAdams of Cape May, New Jersey, of whom Fowler wrote: "*...to whom I am indebted for many interesting off-shore American fishes*." We know nothing further.

McAllister

Eelpout sp. *Oidiphorus mcallisteri* ME Anderson, 1988
McAllister's Eelpout *Lycodes mcallisteri* PR Møller, 2001
Stardrum sp. *Stellifer macallisteri* NL Chao, A Carvalho-Filho & J de Andrade Santos, 2021

Dr Donald Evan McAllister (1934–2001) was a Canadian ichthyologist. He was appointed (1958) as Curator of Fishes, National Museum of Canada, and massively increased the museum's ichthyological collection.

McAlpin

Arctic Lumpsucker *Cyclopteropsis mcalpini* HW Fowler, 1914

Charles Williston McAlpin (1865–1942) was heir to one of the largest manufacturers of tobacco in the USA. He graduated from Princeton (1888) and was elected its first Secretary (1900–1917). After retirement, he devoted most of his time to philanthropy and collecting engraved portraits of Washington. Fowler wrote of him: "*...to whom the University is indebted for assistance in securing the present collection.*"

McCain

McCain's Skate *Bathyraja maccaini* S Springer, 1971

Dr John Charles McCain (b.1939) was formerly a Professor and Senior Research Scientist at the University of Petroleum and Minerals, Dhahran, Saudi Arabia, where he was the principal investigator of the Northern Area Marine Environmental Baseline Study and investigated the effects of oil spills in Persian Gulf. Texas Christian University awarded his bachelor's degree (1962), the College of William and Mary, Virginia, his master's (1964) and George Washington University his doctorate (1967). He collected the holotype (1967) when aboard the marine vessel 'Hero'. He co-wrote: *The Gulf War Aftermath: An Environmental Tragedy* (1993).

McClelland

McClelland's Unicorn-Cod *Bregmaceros mcclellandi* W Thompson, 1840
Cauvery Garra *Garra mcclellandi* TC Jerdon, 1849

John McClelland (1805–1875) was a British doctor with interests in geology and natural history. He was sent on a mission (1835) to ascertain if tea could be grown in northeastern India. He was appointed (1836) Secretary of the 'Coal Committee', the forerunner of the Geological Survey of India, to advise the government how best to exploit Indian coal reserves. He was editor of the *Calcutta Journal of Natural History* (1841–1847). Three birds are named after him.

McClure

Redlip Blenny *Ophioblennius macclurei* CF Silvester, 1915

Dr Charles Freeman Williams McClure (1865–1955) was an American anatomist and embryologist. Princeton awarded his BA (1888) and his MA (1892); Columbia University gave him an honourary DSc (1908). He taught at Princeton whilst also taking several study trips to Europe. He established a collection to illustrate vertebrate morphology and Charles Silvester became an assistant to him there. Together over two decades they expanded the collection, particularly with material from the Arctic. He took part in the Perry Relief Expedition to Greenland, as did Silvester (1899), and took many trips to the West Indies. He wrote c.50 papers (1889–1929). He was honoured in the blenny's name for his studies on the development of the lymphatic system in fishes. Silvester wrote his obituary.

McConnaughey

Shorthead Hagfish *Eptatretus mcconnaugheyi* RL Wisner & CB McMillan, 1990

Ronald R McConnaughey is a marine technician and diver. He was at Scripps Institution of Oceanography, rising to Manager of Scientific Collections and Experimental Aquariums (retired 2000). Among his publications is: *A Tropical Eastern Pacific Barnacle, Megabalanus coccopoma (Darwin), in Southern California, following El Nino 1982–83* (1987). He was honoured because he had helped developed the gear (a three-chambered trap) used by the authors to capture the holotype.

McConnell

Lake Turkana Cichlid sp. *Haplochromis macconneli* PH Greenwood, 1974

Richard B McConnell was Officer in Charge of the Fisheries Department at Lake Rudolf (now Turkana). He was honoured for the assistance he gave to the Lake Rudolf Research Project team. He married (1953) Rosemary Lowe (q.v.)

McCosker

Tufted Blenny genus *Mccoskerichthys* RH Rosenblatt & JS Stephens, 1978
McCosker's Flasher Wrasse *Paracheilinus mccoskeri* JE Randall & ML Harmelin-Vivien, 1977
Hagfish sp. *Myxine mccoskeri* RL Wisner & CB McMillan, 1995
Manyband Moray *Gymnothorax mccoskeri* DG Smith & EB Böhlke, 1997
Hagfish sp. *Eptatretus mccoskeri* CB McMillan, 1999
McCosker's Coralbrotula *Ogilbia mccoskeri* PR Møller, W Schwarzhans & JG Nielsen, 2005
Galápagos Ghostshark *Hydrolagus mccoskeri* LAK Barnett, DA Didier, DJ Long & DA Ebert, 2006
McCosker's Grenadier *Spicomacrurus mccoskeri* T Iwamoto, K-T Shao & H-C Ho, 2011
McCosker's Worm-Eel *Neenchelys mccoskeri* Y Hibino, H-C Ho & S Kimura, 2012
Taiwanese Flashlightfish *Protoblepharon mccoskeri* H-C Ho & GD Johnson, 2012
Scorpionfish sp. *Phenacoscorpius mccoskeri* K Wibowo & H Motomura, 2017
Snake-Eel sp. *Ophichthus johnmccoskeri* A Mohapatra, D Ray, S Mohanty & SS Mishra, 2018
Snake-Eel sp. *Ophichthus mccoskeri* KS Sumod, Y Hibino, H Manjabrayakath & VN Sanjeevan, 2019

Dr John Edward McCosker (b.1945) is an ichthyologist and evolutionary biologist who is Senior Scientist and first Professor of Aquatic Research at California Academy of Sciences, San Francisco. The Scripps Institution of Oceanography awarded his doctorate (1973). He was Director of the Steinhart Aquarium (1973–1994) and is an Adjunct Professor in Marine Biology at San Francisco State University. His research has included the marine life of the Galápagos, the biology of the coelacanth, the biology and systematics of snake-eels and moray eels, and white shark attack behaviour. He has written more than 270 articles and books including *The History of Steinhart Aquarium: A Very Fishy Tale* (1999) and, as co-author, *Great White Shark* (1991). He has appeared in a number of films, including: *Galapagos: Beyond Darwin* (1996) and *Naked Science* (2004). In the etymology for the worm-eel, we read that McCosker was a "*...friend, colleague and fellow eel enthusiast... ...who visited several Australian museums and informed the authors of the presence of this species in those collections.*" (Also see **John McCosker**)

McCoy

Southern Bluefin Tuna *Thunnus maccoyii* FL Castelnau, 1872

Sir Frederick McCoy (1817–1899) was an Irish paleontologist and naturalist. He was educated at Cambridge and worked at the Woodwardian Museum on its fossil collection (1846–1850). He became Professor of Geology at Queen's College, Belfast (1850). When the Chair of Natural Science at the University of Melbourne was created (1854), McCoy was its first occupant. He taught many different subjects for about 30 years. He established, and was the first Director of, the National Museum of Natural History and Geology, Melbourne. Although Castelnau gives no explicit etymology, there is little doubt that Frederick McCoy is the person intended.

McCulloch

Australian Freshwater 'Cod' genus *Maccullochella* GP Whitley, 1929
McCulloch's Rainbowfish *Melanotaenia maccullochi* JD Ogilby, 1915
McCulloch's Hardyhead *Atherion maccullochi* DS Jordan & CL Hubbs, 1919
McCulloch's Tongue Sole *Cynoglossus maccullochi* JR Norman, 1926
Whitesnout Anemonefish *Amphiprion mccullochi* GP Whitley, 1929
Half-banded Seaperch *Hypoplectrodes maccullochi* GP Whitley, 1929
McCulloch's Scalyfin *Parma mccullochi* GP Whitley, 1929
McCulloch's Bandfish *Owstonia maccullochi* GP Whitley, 1934
Goby sp. *Nesogobius maccullochi* DF Hoese & HK Larson, 2006

Allan Riverstone McCulloch (1885–1925) was a noted Australian ichthyologist. He started his career at 13 as an unpaid assistant to Edgar Waite (q.v.) at the Australian Museum, Sydney. Encouraged by Waite to study, he became 'Mechanical Assistant' (1901) and later (1906–1925) Curator of Fishes there. He was a prolific collector, taking over 40,000 specimens in Queensland, Lord Howe Island, New Guinea, the Great Barrier Reef and a number of Pacific Islands, and wrote more than 100 papers, many of which he illustrated himself. He wrote: *Check List of Fishes and Fish-like Animals of New South Wales* (1922). His health was poor and despite taking a year off, he died while in Hawaii. Two odonata are named after him.

McCune

Papuan Sleeper-Gudgeon sp. *Mogurnda maccuneae* AP Jenkins, PM Buston & GR Allen, 2000

Professor Dr Amy R McCune is an evolutionary biologist at Cornell University who is Senior Associate Dean of College of Agriculture and Life Sciences; Faculty Curator of Ichthyology. Her BA in Biology was awarded by Brown University (1976) and her PhD by Yale (1982) after which she spent a year as Miller Postdoctoral Fellow at UC, Berkeley. She has studied a variety of fishes, from ancient forms like lungfishes, bichir, sturgeon and their fossil relatives, to a variety of living teleost fishes. She has published numerous papers such as: *Morphological anomalies in the Semionotus complex: Relaxed selection during colonization of an expanding lake* (1990) and *Using Genetic Networks and Homology to Understand the Evolution of Phenotypic Traits* (2012). She was honoured as a person "*…whose teaching and research in the field of ichthyology are truly inspirational*".

McDade

Squarenose Unicornfish *Naso mcdadei* JW Johnson, 2002

Michael McDade is a spearfisherman and records officer for the Australian Underwater Federation. He has collected and donated specimens to the Queensland Museum, including much of the type material of this species.

McDonald

Totoaba *Totoaba macdonaldi* CH Gilbert, 1890
Mexican Rockfish *Sebastes macdonaldi* CH Eigenmann & CH Beeson, 1893

Rakery Beaconlamp *Lampanyctus macdonaldi* GB Goode & TH Bean, 1896
Yellowfin Cutthroat Trout *Oncorhynchus clarkii macdonaldi* DS Jordan & CH Evermann, 1890

Colonel Marshall 'Marsh' McDonald (1836–1895) was an American naturalist and ichthyologist. He served in the Confederacy Army in the American Civil War, including being Captain of Artillery at the siege of Vicksburg (1863), and was a professor at the Virginia Military Institute. He became Commissioner of Fish and Fisheries for the State of Virginia (1875). He devised a system of automatic fishhatching jars (1881). He was appointed head of the US Fish Commission (1887) upon the death of Spencer Fullerton Baird. A bird is also named after him.

McDowall

McDowall's Galaxias *Galaxias mcdowalli* TA Raadik, 2014

Dr Robert 'Bob' Montgomery McDowall (1939–2011) was a distinguished New Zealand freshwater ichthyologist. Victoria University of Wellington awarded his bachelor's degree (1958) and he went on to study at Harvard University. He moved to the Christchurch freshwater fisheries laboratory (1978) and became Assistant Director (1983). He was elected a Fellow of the Royal Society of New Zealand (1984). His last book was: *Ikawai: freshwater fishes in Māori culture and economy* published posthumously (2011).

McDowell

Snake-eel sp. *Cirricaecula macdowelli* JE McCosker & JE Randall, 1993

Michael McDowell is an Australian Tour Operator (President of Deep Ocean Expeditions), diver, climber and explorer. He founded a number of adventure travel companies including Deep Ocean Quest and Quark Expeditions. He is described in the etymology as: "*…tour operator and bon vivant, who has taken us to remote outposts in search of rare specimens… …diver, explorer and friend.*"

McEachran

Madagascar Pygmy Skate *Fenestraja maceachrani* B Séret, 1989

Dr John D McEachran (b.1941) is Professor of Ichthyology, Texas A&M University, and Curator of Fishes at the Texas Cooperative Wildlife Collection. Michigan State University awarded his bachelor's degree (1965) and the William and Mary College, Virginia, his master's (1968) and his doctorate (1973). During his 34-year career at Texas A&M he taught ichthyology, herpetology, conservation of marine resources, museums and their functions, systematics of vertebrates and evolutionary mechanisms of vertebrates, at both undergraduate and graduate level. His research was mainly focused on systematics and evolutionary relationships of skates and rays (*Batoidea*) and biogeography of fishes of the Gulf of Mexico. Among his many publications he co-wrote the two-volume: *Fishes of the Gulf of Mexico* (1998 & 2006) and wrote: *Revision of the South American skate genus* Sympterygia *(Chondrichthyes, Rajiformes)* (1982). He was honoured in the skate's name for his major contributions to skate and ray systematics.

McGinnis

Lanternfish sp. *Protomyctophum mcginnisi* AM Prokofiev, 2004

Richard Frank McGinnis was a researcher associated with the University of Southern California. He was honoured for his contribution to the study of lanternfishes of the Southern Hemisphere, referring in particular to his monograph: *Biogeography of lanternfishes (Myctophidae) south of 30° S* (1974).

McGinty, P

Tricorn Batfish *Zalieutes mcgintyi* HW Fowler, 1952

Paul Ladue McGinty (1906–1985), operated the yacht 'Triton' with his brother JTL McGinty (below) from which the holotype was caught.

McGinty, T

Armoured Gurnard sp. *Peristedion mcgintyi* HW Fowler, 1952
[Junior synonym of *Peristedion miniatum*]

John Thomas 'Tom' Ladue McGinty (1907–1986) became Police Chief in Ocean Ridge (1949). He was an amateur malacologist and naturalist. He amassed a collection of over three million shells! He published more than 40 papers on marine molluscs of Florida and the West Indies. He also wrote a pioneering work (1941) on diving for molluscs. He was a Research Associate of ANSP and PRI. He operated the yacht 'Triton' both dredging and diving for molluscs in waters off Florida, the West Indies and Mexico, with his brother PL McGinty (above) from which the holotype was caught.

McGowan

Pearleye sp. *Scopelarchoides climax* RK Johnson, 1974

Dr John A McGowan was an associate professor of oceanography at the University of California, San Diego, and is now Professor Emeritus of Oceanography, Scripps institute of Oceanography, University of California, San Diego. In WW2, he served in the US Navy in the Pacific. Oregon State University awarded his bachelor's and master's degrees and the Scripps Institute his doctorate. He led the 'Climax' Expeditions to the central Pacific Ocean after which this species is named, during which the holotype was collected. The etymology makes it clear that the dedication is to include the scientists and crews of the expedition, and in particular the expedition's leader. (Also see **Climax**)

McGrouther

McGrouther's Coralbrotula *Diancistrus mcgroutheri* W Schwarzhans, PR Møller & JG Nielsen, 2005
McGrouther's Flathead. *Rogadius mcgroutheri* H Imamura, 2007
McGrouther's Ghost Flathead *Hoplichthys mcgroutheri* Y Nagano, H Imamura & M Yabe, 2014
McGrouther's Triangular Batfish *Malthopsis mcgroutheri* H-C Ho & P Last, 2021

Mark Andrew McGrouther (b.1958), whose honours degree was awarded by Sydney University (1980), worked at the Australian Museum, Sydney (1981–2018) and was

Ichthyology Collection Manager until retirement (1987–2018). He continued at the Australian Museum as a Senior Fellow. During his career, he has organised or participated in more than 30 collecting trips and expeditions within Australia and throughout the Pacific region. He co-wrote: *Sydney Harbour: its diverse biodiversity* (2013).

McMahon

Pakistani Carp sp. *Tariqilabeo macmahoni* E Zugmayer, 1912
[Syn. *Labeo macmahoni*]

Sir Arthur Henry McMahon (1862–1949) was a soldier and colonial administrator. He was commissioned from Sandhurst (1882) and was posted to the India Staff Corps, entering the Punjab frontier force (1885). He transferred to the political department (1890) and acted as political agent for a number of small states. He was Commissioner for Baluchistan (1901–1903) and for Seistan (1903–1905), and returned to Baluchistan (1905) to act as arbitrator in the boundary dispute between Iran and Afghanistan (1906). He became Foreign Secretary to the Government of India (1911). He left India (1914) and became the first British High Commissioner for Egypt under the British Protectorate (1914–1916). He made a collection of reptiles in Baluchistan (1896) when he was arbitrating a boundary dispute. He asked Zugmayer to establish a fish collection for a national museum for Baluchistan at Quetta. Two reptiles are named after him. The original text does not identify him, but he is very likely the man intended.

McMaster

McMaster's Dwarf Cichlid *Apistogramma macmasteri* SO Kullander, 1979

Mark McMaster is an aquarist and a well-known dwarf cichlid enthusiast. He was honoured as he brought this cichlid to Kullander's attention (1973), as rhe species had been popular with aquarists from at least the early 1960s. Among his articles are: *A Checklist of the Named Species of Apistoqramma with Notes* (1974) and *On the Hobby Species of Apistogramma* (1977).

McMillan, CB

Hagfish sp. *Myxine mcmillanae* DA Hensley, 1991

Dr Charmion B McMillan (b.1925) was a marine biologist at the Marine Biology Research Division, Scripps Institution of Oceanography, University of California San Diego. She published widely, including papers describing many new species such as: *Three new species of seven-gilled hagfishes (Myxinidae, eptatretus) from the Pacific Ocean* (1984). She was honoured in the name of the hagfish for her "*...fine contributions to hagfish science.*"

McMillan, DA

Hagfish sp. *Eptatretus grouseri* CB McMillan, 1999

David 'Grouser' Allen McMillan (b.1957) is a Chief Engineer in the US Merchant Marine and the author's son. She honoured him in the name for his: "*...continued encouragement of Mom's hagfish studies and for his knowledge and love of ships and the sea.*"

McMillan, PJ

McMillan's Catshark *Parmaturus macmillani* GS Hardy, 1985

McMillan's Whiptail *Coryphaenoides mcmillani* T Iwamoto & YN Shcherbachev, 1991

Peter John McMillan (b.1955) is a Fisheries biologist who has a master's degree in zoology. He is a specialist in deep-sea species, and collected the catshark holotype and "*many examples of undescribed or poorly known marine fish and invertebrate species*" from New Zealand waters. He has worked for the New Zealand National Institute of Water and Atmospheric Research since 1980. He is an Honorary Research Associate of Museum of New Zealand, Te Papa Tongarewa., Wellington. He co-wrote *Two new species of* Coelorinchus (*Teleostei, Gadiformes, Macrouridae*) *from the Tasman Sea* (2009).

McNeill

Shorttail Torpedo *Tetronarce macneilli* GP Whitley, 1932
[Perhaps a Junior Synonym of *T. nobiliana*]
Blue Devilfish *Assessor macneilli* GP Whitley, 1935

Francis Alexander 'Frank' McNeill (1896–1969) was an Australian ichthyologist and carcinologist at the Australian Museum in Sydney. He served (1914–1918) with the Australian Light Horse in WW1 including fighting at Gallipoli (1915). He became Zoologist in charge of Lower Invertebrates (1921) and was the Australian Museum's representative on the Great Barrier Reef Committee (1945). He collected the holotypes of the species named after him. Whitley wrote his obituary.

McTurk

Three-barbeled Catfish sp. *Pimelodella macturki* CH Eigenmann, 1912

Michael McTurk (1843–1915) was a member of an English family that settled in British Guiana (Guyana) in the early 19th century. He was born in LIverpool and ran away to sea (1860) to get to British Guiana to see his family there, and there he stayed. He was the Surveyor for the County of Essequibo, British Guiana (1872). He was a guide for the expedition that made an unsuccessful attempt to climb Mount Roraima (1878). He was appointed (1896) Commissioner for the Essequebo and Pomeron Rivers District and Protector of Aboriginal Indians. Eigenmann named this species after him as he had delivered a parcel of letters to Eigenmann during his 1908 expedition to British Guiana.

Mead

Cutthroat Eel genus *Meadia* JE Böhlke, 1951
Short-tail Eel sp. *Coloconger meadi* RH Kanazawa, 1957
Spotted Spikefish *Hollardia meadi* JC Tyler, 1966
Tripodfish sp. *Ipnops meadi* JG Nielsen, 1966
Blotched Catshark *Scyliorhinus meadi* S Springer, 1966
Blackring Waryfish *Scopelosaurus meadi* E Bertelsen, G Krefft & NB Marshall, 1976
Mead's Lanternfish *Diaphus meadi* BG Nafpaktitis, 1978
Whipnose Angler sp. *Gigantactis meadi* E Bertelsen, TW Pietsch & RJ Lavenberg, 1981

Dr Giles Willis Mead Jr (1928–2003) was an American zoologist and curator. His BA (1949), MA (1952) and doctorate in ichthyology (1953) were all awarded by Stanford University. He initially worked at the Smithsonian for the US Fish and Wildlife Service as Laboratory Director in charge of fish taxonomy. He was Curator of Fishes at the Museum of

Comparative Zoology (1960–1970) and Professor of Biology at Harvard. He was Director of the Los Angeles County Natural History Museum (1970–1978). His family was wealthy – his father was a co-founder of the chemical giant Union Carbide – and he was able to retire to the Mead family ranch and vineyard in the Napa Valley, California (1978) managing the vineyards and supporting conservation in the valley. He continued to travel, collect and study throughout his life. Although the patronym is not identified in the etymology of the eel genus, it is almost certainly in honour of Giles Mead, as he was Böhlke's frequent collaborator and Stanford University colleague.

Meany

Puget Sound Sculpin *Ruscarius meanyi* DS Jordan & EC Starks, 1895

Edmond Stephen Meany (1862–1935) was Professor of Botany and History and became Secretary of the University of Washington, from which he graduated (1885) and where the theatre is named after him. He was awarded Master of Letters by the University of Wisconsin (1901) and Honorary Doctor of Laws by the College of Puget Sound (1926). He was a Washington state legislator for the 1891 & 1893 sessions. He was honoured in the name: "*…in recognition of his work in the Young Naturalists' Society*." He died of a stroke on campus, minutes before he was to give a lecture.

Mechthilda

Molly sp. *Poecilia mechthildae* MK Meyer, V Etzel & D Bork, 2002

Mechthild Etzel is the daughter of the second author.

Medem

Characin sp. *Brycon medemi* G Dahl, 1960

Professor Dr Federico Medem (1912–1984) was born in Riga as Friedrich Johann Comte von Medem. He was of German origin but thought of himself as a Latvian. His family left Latvia after the Russian Revolution (1917) and moved to Germany. Medem studied at Humboldt-Universität Berlin and at Eberhard Karls Universität Tübingen. He worked for his doctorate at the marine biology station in Naples run by Gustav Kramer. He served in the Wehrmacht during WW2 and fought on the Russian front. After the war, he worked in Germany and Switzerland. He moved to Colombia (1950), changed his name, and became a herpetologist and ardent conservationist at the research station at Villavicencio and at Universidad Nacional de Colombia, Bogotá. There is a herbarium named after him at Instituto Alexander von Humboldt. He wrote numerous scientific papers (1950–1984), mostly on Colombian reptiles. He published two volumes that make up his: *Los Crocodylia de Sur America* (1981–1983). Three amphibians, five reptiles and a mammal are named after him.

Meder

Snailfish sp. *Careproctus mederi* PY Schmidt, 1916

Dr Gerhard Richardovich Meder (b.1865) was a Russian naval physician aboard the hydrographical vessel 'Okhotsk' (1914–1916). The collecting of marine specimens was a duty of the medical staff which is how his name became associated with the fish.

Medusa

Blenny genus *Medusablennius* VG Springer, 1966
Medusa Blenny *Acanthemblemaria medusa* WF Smith-Vaniz & FJ Palacio, 1974
Barbeled Dragonfish sp. *Eustomias medusa* RH Gibbs, TA Clarke & JR Gomon, 1983

In Greek mythology, Medusa was a Gorgon whose hair was made up of living snakes. The name is used fancifully for taxa with numerous filaments or barbels.

Meek

Sicklefin Chub *Macrhybopsis meeki* DS Jordan & BW Evermann, 1896
Goby sp. *Microgobius meeki* BW Evermann & MC Marsh, 1899
Conger Eel sp. *Ariosoma meeki* DS Jordan & JO Snyder, 1900
Hawaiian Bigeye *Priacanthus meeki* OP Jenkins, 1903
Silverside sp. *Atherinella meeki* N Miller, 1907
Three-barbeled Catfish sp. *Pimelodella meeki* CH Eigenmann, 1910
Characin sp. *Brycon meeki* CH Eigenmann & SF Hildebrand, 1918
Firemouth Cichlid *Thorichthys meeki* WL Brind, 1918
Zirahuen Allotoca *Allotoca meeki* J Álvarez, 1959
Mezquital Pupfish *Cyprinodon meeki* RR Miller, 1976
American Halfbeak *Hyporhamphus meeki* HM Banford & BB Collette, 1993
Glass Knifefish sp. *Eigenmannia meeki* GM Dutra, CD de Santana & WB Wosiacki, 2017

Seth Eugene Meek (1859–1914) was an ichthyologist at the Field Museum of Natural History in Chicago. He was the first to compile a book on Mexican fishes and also wrote one on the fish of Panama. He was a long-time collaborator of C H Gilbert (q.v.) and collected with him and D S Jordan (q.v.). He became Professor of Biology and Geology at Arkansas Industrial University (1884). He was the first to recognise the distinctiveness of the conger eel species.

Meel

Lake Tanganyika Cichlid sp. *Lepidiolamprologus meeli* M Poll, 1948

Ludo Isabelle Joseph Hyacinthe van Meel (1908–1990) was a botanist who was a member of the Belgian Hydrobiological Mission to Lake Tanganyika (1946–1947), during which the type was collected. He also collected in Burundi, Tanzania and Zambia.

Meerdervoort

Bigeye Skate *Okamejei meerdervoortii* P Bleeker, 1860
Flathead sp. *Suggrundus meerdervoortii* P Bleeker, 1860

Dr. Johannes Lijdius Catharinus Pompe van Meerdervoort (1829–1908) was a Dutch physician who qualified at Utrecht and became a naval surgeon (1849). He lived at Dejima, a Dutch enclave in Nagasaki Harbour (1857–1863) and was invited by the Japanese authorities to teach western medicine, chemistry and photography in the Kaigun Denshujo Naval Academy. He established the first western-style hospital and medical school in Japan (1861). He wrote: *Five Years in Japan* (1868) after returning to the Netherlands. He collected for Bleeker, who wrote an article entitled: *Fish Species of Japan, gathered in Dejima by Jhr. J. L. C. Pompe van Meerdervoort* (1859).

Mees

Three-barbeled Catfish sp. *Brachyrhamdia meesi* DD Sands & BK Black, 1985

Driftwood Catfish sp. *Tatia meesi* LM Sarmento-Soares & RF Martins-Pinheiro, 2008

Dr Gerlof Fokko Mees (1926–2013) was born in the Netherlands, lived for some years in the East Indies, and retired to Australia. He studied biology at the University of Leiden. As a student, he collected birds on Trinidad and Tobago (1953–1954) and during his first year made a detailed study of whiteeyes (*Zosterops*) which had attracted his attention during his service as a volunteer in the Dutch army during the Indonesian independence war (1946–1949). This study ultimately resulted in a 762-page monograph: *A Systematic Review of the Indo-Australian Zosteropidae* (1957–1969). Volume I was his PhD thesis. He became Curator of Vertebrate Animals at the Western Australian Museum in Perth (1958), dividing his attention between ichthyology and ornithology. He returned to Holland (1963) to become Curator of Birds at the University of Leiden. He made expeditions to Suriname (1965–1990), then returned to Australia (1991). Four birds are named after him.

Meggitt

Meggitt's Goby *Bathygobius meggitti* SL Hora & DD Mukerji, 1936

F J Meggitt was a parasitologist at the University College, Rangoon (now in Myanmar), who was a Research Associate of the University of Birmingham. His published papers include: *Report on a Collection of Cestoda, mainly from Egypt* (1930). He provided a collection of fishes from Tavoy District, Myanmar, including the holotype of this goby.

Mehmet

Turkish Minnow sp. *Pseudophoxinus mehmeti* FG Ekmekçi, MA Atalay, B Yoğurtçuoğlu, D Turan & F Küçük, 2016

Mehmet Ekmekçi is a hydrogeologist, and the husband of the first author.

Meidinger

Pearlfish (roach sp.) *Rutilus meidingeri* JJ Heckel, 1851

[Syn. *Rutilus frisii meidingeri*]

Baron Carl von Meidinger (1750–1820) was an Austrian nobleman. He wrote the five-part: *Icones piscium Austriae indigenorum* (1785–1794) in which he illustrated, but misidentified, this species. His engravings are much reprinted and still make popular posters.

Meingangbi

Indian Barb sp. *Pethia meingangbii* L Arunkumar & H Tombi Singh, 2003

This is not an eponym but taken from the Manipuri word *meingangbi*, meaning 'red-coloured tail'.

Meinken

Rasbora sp. *Rasbora meinkeni* LF de Beaufort, 1931
Dwarf Cichlid sp. *Apistogramma meinkeni* SO Kullander, 1980
Spotted Splashing Tetra *Copella meinkeni* A Zarske & J Géry, 2006 NCR
[Junior Synonym of *Copella nattereri*]

Dr Hermann Meinken (1896–1976) was a German amateur ichthyologist and a pioneer in the keeping of tropical fish in aquaria. He gave de Beaufort a breeding pair of *Rasbora meinkeni* on which de Beaufort based his description. He co-wrote: *Iriatherina werneri, a new atherinid fish from New Guinea* (1975).

Meity

Meity's Pygmygoby *Trimma meityae* R Winterbottom & MV Erdmann, 2018

Meity Mongdong is an Indonesian marine conservationist who is a senior manager of Conservation International Indonesia; she is the capacity-building manager of Bird's Head Seascape, including the Raja Ampat Islands, the epicenter of marine biodiversity in northwest Papua Barat province, Indonesia. The authors describe her as: "*...one of Indonesia's foremost marine conservationists, who has dedicated the past several decades of her career towards expanding and improving the management of marine protected areas in West Papua, including the Cendrawasih Bay National Park,*" (where this goby is found).

Mejia

Neotropical Rivuline sp. *Anablepsoides mejiai* FBM Vermeulen, 2020
[Syn. *Rivulus mejiai*]

Daniel Mejia-Vargas is a Colombian biologist at the Universidad de los Andes. He discovered the rivuline species. Among his published papers is the co-written: *A new species of Andinobates (Anura: Dendrobatidae) from the Urabá region of Colombia* (2017).

Melanie

Gabonese Cichlid sp. *Chromidotilapia melaniae* A Lamboj, 2003
Xingu Peacock Bass (cichlid) *Cichla melaniae* SO Kullander & EJG Ferreira, 2006

Dr Melanie Lisa Jane Stiassny (b.1953) is Curator of Fishes at the American Museum of Natural History. (See **Stiaassny**)

Melin

Bandit Corydoras *Corydoras melini* E Lönnberg & H Rendahl, 1930
Armoured Catfish sp. *Rineloricaria melini* O Schindler, 1959

Douglas Melin (1895–1946) was a Swedish zoologist and herpetologist who was attached to the Department of Zoology, University of Uppsala. He led the Swedish Amazon Expedition (1923–1925). He was also an athlete: he competed in the standing long jump for Sweden at the 1912 Olympic Games. He collected the holotypes of both species.

Melissa

Torrent Minnow sp. *Psilorhynchus melissa* KW Conway & M Kottelat, 2010

Melissa is the classical Greek name of the honey bee, and is here used in allusion to the species' black and yellow colour pattern.

Melland

Snaileater Cichlid *Sargochromis mellandi* GA Boulenger, 1905

Frank Hulme Melland (1879–1939) was an explorer and big-game hunter who collected a series of fishes from Lake Bangweulu, Zambia, including the type of this cichlid. In his book *In Witch-Bound Africa* (1923), Melland tells how he showed a picture of a pterodactyl to locals and they identified it as a dangerous (living) creature they called *kongamato*. No one has yet found a specimen of this featherless flying fiend!

Melliss

Melliss's Scorpionfish *Scorpaena mellissii* A Günther, 1868
Silver Conger *Ariosoma mellissii* A Günther, 1870
St. Helena Flounder *Bothus mellissi* JR Norman, 1931

John Charles Melliss (1835–1911) was an amateur naturalist. After training as an engineer and serving as an officer in the Royal Engineers, he was appointed as government surveyor on St. Helena island (1860–1871); the island where he was born. However, he was made redundant during government cutbacks and returned to London where he set up a company. He wrote: *St. Helena: A Physical, Historical and Topographical Description of the Island, including the Geology, Fauna, Flora and Meteorology* (1875). (To commemorate the book's centenary in 1975, the St. Helena Post Office published a series of stamps.)

Melrose

Oriental Cyprinid sp. *Sinibrama melrosei* JT Nichols & CH Pope, 1927

Mrs J C Melrose, of the American Presbyterian Mission of Hainan, was honoured in the binomial "*...in appreciation of her interest in the work*" [of the 1922/1923 Hainan collection made by Clifford Pope]. She may have been even more honoured if the authors had used the correct form for a female eponym, which would have been *melrosae*.

Menchaca

Freshwater Stingray sp. *Potamotrygon menchacai* G Martinez Achenbach, 1967 NCR
[Junior Synonym of *Potamotrygon falkneri*]

Dr Manuel J Menchaca (1876–1969) was an Argentine politician and physician. During his tenure as Governor of Santa Fé Province he started the Provincial Museum of Natural Sciences 'Florentino Ameghino', which became the centre for investigating the genus *Potamotrygon*.

Mendeleev

Mendeleev's Clingfish *Lepadicyathus mendeleevi* AM Prokofiev, 2005

This is not a direct eponym, as the fish is named after the Russian research vessel 'Dimitri Mendeleev'; itself named after the Russian chemist and inventor of that name (1834–1907).

Mendelssohn

Mendelssohn's Pygmygoby *Trimma mendelssohni* M Goren, 1978

Heinrich Mendelssohn (1910–2002) of Tel Aviv University studied zoology and medicine at the University of Berlin (1928–1933). He had to terminate his studies with the rise of the Nazis, and with family funding he emigrated (1933) to Palestine where he was joined by his parents and sister (1935). There he earned his master's (1935) and PhD (1940) with a dissertation on the population density of wild birds. He worked (1935–1953) in teacher training at the Institute of Biological Education, then Tel Aviv University (1953–1956) where he was Dean and Institute Director of the Zoological Department, becoming (1961) Dean of the Faculty of Natural Sciences. He was appointed professor (1970), later taking the chair of the Department of Wildlife Reserves (1977). He was co-author of *Plants and animals of the land of Israel: an illustrated encyclopedia. Vol.7: Mammals* (1987). He was honoured in the goby's name for his 'invaluable' contributions to zoological research and nature conservation in Israel.

Mendes

Mendes' Tube-snouted Ghost Knifefish *Sternarchorhynchus mendesi* CD de Santana & RP Vari, 2010

Dr George Nilson Mendes is a fisheries scientist and oceanographer who is an Associate Professor at Universidade Federal de Pernambuco, Brazil. The Federal Rural University of Pernambuco awarded his bachelor's degree (1979) and his master's degree (1988). The Federal University of São Carlos awarded his doctorate (1996). He wrote: *View Profile Ornamental fish farming, a lucrative alternative* (2006).

Mendez, E

Panama Hillstream Catfish sp. *Astroblepus mendezi* CA Ardila Rodríguez, 2014

Dr Eustorgio Mendez Cedeño (1927-2016) was a Panamanian parasitologist and medical entomologist. He looked after the zoological collection at the Gorgas Memorial Institute for Health Studies, Panama for 40 years. The zoological collection is now named after him. He wrote: *Los Principales `Mamiferos Silvestres de Panama* (1970).

Mendez, F

Dwarf Cichlid sp. *Apistogramma mendezi* U Römer, 1994

Francisco Alves 'Chico' Mendes (1944–1988) was a Brazilian rubber tapper, union leader and environmentalist. He fought to preserve the Amazonian rainforest, and for the rights of indigenous peoples. He was murdered by a rancher whom Mendes had prevented from logging a protected area.

Mendoza

Mendoza's Hagfish *Eptatretus mendozai* DA Hensley, 1985

Luis H 'Uchy' Mendoza was captain of the research vessel 'Crawford' which belonged to the Department of Marine Sciences, University of Puerto Rico. The holotype was taken during a cruise of this vessel. Hensley's etymology described Mendoza as "*...a man with a beautiful combination of experiential knowledge and academic curiosity of the sea*".

Menezes, N

Anthias sp. *Anthias menezesi* WD Anderson & PC Heemstra, 1980
Snake-eel sp. *Ophichthus menezesi* JE McCosker & EB Böhlke, 1984
Armoured Catfish sp. *Pyxiloricaria menezesi* IJH Isbrücker & H Nijssen, 1984
Pike-Cichlid sp. *Crenicichla menezesi* A Ploeg, 1991
Characin sp. *Oligosarcus menezesi* AM Miquelarena & LC Protogino, 1996
Hagfish sp. *Eptatretus menezesi* MM Mincarone, 2000
Characin sp. *Creagrutus menezesi* RP Vari & AS Harold, 2001
American Sole sp. *Apionichthys menezesi* RTC Ramos, 2003
Driftwood Catfish sp. *Trachycorystes menezesi* HA Britski & A Akama, 2011
Bluntnose Knifefish sp. *Brachyhypopomus menezesi* WGR Crampton, CD de Santana, JC Waddell & NR Lovejoy, 2017
Stardrum sp. *Stellifer menezesi* NL Chao, A Carvalho-Filho & J de Andrade Santos, 2021

Professor Dr Naércio Aquino Menezes (b.1937) is a Brazilian zoologist who is Professor, Department of Zoology, Institute of Biosciences, University of São Paulo and at Museu de Zoologia da Universidade de São Paulo. The University of São Paulo awarded his bachelor's degree (1962) and Harvard his PhD (1968). His specialist areas are systematics, biogeography and evolution of fishes (mainly of marine fish around the coasts of Brazil) and Neotropical freshwater fish. He was honoured in the name of the hagfish for: "*...his extensive contribution to Brazilian ichthyology*", and in that of the snake-eel for his "*generous contributions*" of specimens to the authors' eel research.

Menezes, R

Menezes' Catfish *Aspidoras menezesi* H Nijssen & IJH Isbrücker, 1976
Driftwood Catfish sp. *Auchenipterus menezesi* CJ Ferraris & RP Vari, 1999

Rui Simões de Menezes (1917–2001) was a Brazilian zoologist and ichthyologist at the Universidade Federal do Ceara, Brazil where he was instrumental in founding the Institute of Marine Sciences (1961). He collected the holotype of the *Aspidoras* species.

Meng

Skate sp. *Okamejei mengae* C-H Jeong, T Nakabo & HL Wu, 2007

Professor Dr Qing-Wen Meng was President of Shanghai Fisheries College (now Shanghai Ocean University). She wrote: *Studies on The Blood Vessels of Eye, Gas Bladder and Kidney of Snakehead* (1990). She was honoured for her 'great contributions to elasmobranch studies in China'. A fossil lamprey from the Early Cretaceous epoch is also named after her.

Mengila

African Rivuline sp. *Aphyosemion mengilai* S Valdesalici & W Eberl, 2014

François Mengila, a Gabonese man born in Lambaréné and living in Libreville, was honoured in appreciation of his continued help as driver and guide during the collecting trips in Gabon (since 2002) of the junior author, Wolfgang Eberl.

Menne

Menne's Coralbrotula *Diancistrus mennei* W Schwarzhans, PR Møller & JG Nielsen, 2005

Tammes Menne (b.1947) was Assistant Curator and fish-collection manager at the Natural History Museum of Denmark. He has had a life long interest in biology. He qualified as a school teacher but never taught school and did a variety of jobs including being a bricklayer, layout in a print shop, and for seven years was a rigger at a shipyard. He took care of the fish collection (1984 until retirement) and still does in one day a week as a volunteer. He also plays Danish folk music on a nyckelharpa.

Menni

South Brazilian Skate *Dipturus mennii* UL Gomes & C Paragó, 2001

Characin sp. *Bryconamericus mennii* AM Miquelarena, LC Protogino, R Filiberto & HL López, 2002

Dr Roberto Carlos Menni is an ichthyologist and professor at Museo de La Plata, Universidad Nacional de La Plata, Argentina. Among his numerous publications, both as sole and as joint author, is: *Peces y ambientes en la Argentina continental* (2003). He was honoured in the name of the skate for his contributions to the study of South American rays.

Menon

Garra sp. *Garra menoni* K Rema Devi & TJ Indra, 1984

Loach sp. *Ghatsa menoni* CP Shaji & PS Easa, 1995

Periyar Blotched Loach *Mesonoemacheilus menoni* VJ Zacharias & KC Minimol, 1999

Goby sp. *Awaouichthys menoni* TK Chatterjee & SS Mishra, 2013

Idukki Catfish *Mystus menoni* M Plamoottil & NP Abraham, 2013

Indian Barb sp. *Hypselobarbus menoni* M Arunachalam, S Chinnaraja, A Chandran & RL Mayden, 2014

Dr Ambat Gopalan Kutty Menon (1921–2002) was the foremost Indian ichthyologist and zoo-geographer of the 20th century. His first degree was awarded by Madura College, his master's from the Presidency College, Chennai, and his PhD from Madras University (1952). He published more than 100 papers, which included descriptions of 43 new species. He spent many years in research, rising to Deputy Director of the Southern Regional Station of the Zoological Survey of India (ZSI) when he retired (1978). He was Associate Professor of Marine Biology at the University of Dare es Salaam, Tanzania (1978–1981) then returned to India as Professor Emeritus at ZSI during which time he completed two books on *Homalopteridae* (1987) and *Cobitidae* (1992). He continued his work well into his late seventies, publishing: *Threatened Fishes of India* (2004) and *Checklist of the Freshwater Fishes of India* (1999). He was also Founder President of the Indian Society of Ichthyologists. He left all his collection and library to ZSI. He helped to collect the holotype of *Hypselobarbus menoni* (1990).

Menzies

Snailfish genus *Menziesichthys* TT Nalbant & RF Mayer, 1971

Dr Robert James Menzies (1923–1976) was a carcinologist and oceanographer who began his career as a member of an expedition to Baja California, Mexico, under the auspices of the San Diego Society of Natural History (1941). He enrolled in the United States Navy Officer Training Program at Arizona State College and then transferred to College of the Pacific where he received his BA (1945). He stayed on the program as a student in the Medical College of the University of Nebraska (Omaha). He then studied Crustacea with the Allen Hancock Foundation (1946) and the Pacific Marine Station of the College of the Pacific (1947–1949), earning his MA (1948). He returned to the Hancock Foundation which awarded his PhD (1951). He was associated with the University of California, Davis (1951–1952), the Scripps Institute of Oceanography (1953–1955), and the Biology Program at Lamont Geological Observatory of Columbia University (1955–1960). He became (1961) the Director of the Estacion Cientifico de Investigaciones Marinas de Margarita of the Lasalle Foundation in Caracas, Venezuela. He returned to USC (1961–1962) and joined the faculty of Duke University (1962) as a professor of zoology and Director of the Oceanography Program. He became (1968) Professor of Oceanography at Florida State University, where he remained until his death. Menzies was an active oceanographer who participated or led numerous oceanographic cruises all over the world, and is considered to be an expert on Isopoda. He was the head scientist of the eleventh cruise of the R/V *Anton Bruun* during which the type was taken.

Mephisto

Spikefish genus *Mephisto* JC Tyler, 1966

Mephisto (more commonly known in English as Mephistopheles) is the demon to whom Dr Faustus sells his soul. Goethe's *Faust* is perhaps the most famous literature featuring him. Here it refers to the type species' (*fraserbrunneri*) reddish exterior, black interior and horn-like spines.

Mercer

Mercer's Tusked Silverside *Dentatherina merceri* JM Patten & W Ivantsoff, 1983

Professor Dr Frank V Mercer was a founding Professor of the School of Biological Science at Macquarie University, Sydney (1964), having been formerly a Professor of Botany at Sydney University. Macquarie's biological gardens (and a stuffed Kodiak Bear) are named after him: welcoming visitors to the Department of Biological Sciences is 'Frank the Bear,' a three metre Kodiak who began his life in Seattle Zoo before moving to Taronga Zoo, where he died. The etymology thanks Mercer "*…whose help and encouragement in the study of the family Atherinidae will not be forgotten.*"

Mercury

Mercury Anchovy *Stolephorus mercurius* H Hata, S Lavoué & H Motomura, 2021

Mercury was the Roman god of travel, commerce and thieves. The authors describe him as 'the guardian deity of merchants,' and chose the name "*…in reference to the commercial importance of the new species in Southeast Asian fisheries.*"

Mercy

Mercy's Garden Eel *Heteroconger mercyae* GR Allen & MV Erdmann, 2009

Mercy Paine (b.1998) discovered the eel colony and helped the authors to collect the type specimens during a diving cruise with her family. Dexter and Susan Paine and their three children, Mercy, Sam and Honor, invited the authors to participate on the live-aboard dive vessel 'Silolona' on its cruise to Banda and West Papua. Mercy is considering a career in marine biology. Her account of the discovery, *Harvesting Garden Eels: Swimming with the Fishes*, won first place in a Science Writing Contest (2012).

Meredith

Queensland Yellowtail Angelfish *Chaetodontoplus meredithi* RH Kuiter, 1990

John G Meredith was a keen marine aquarist living in Sydney, New South Wales where he was a PADI dive instructor too. He collected the holotype with a handnet at Sydney Harbour (1985) and kept a number of juveniles that grew to a good size. He has since moved to Bendigo, Victoria where he is a consultant.

Merlin

Orangetail Shiner *Pteronotropis merlini* RD Suttkus & MF Mettee, 2001

Lieutenant-Colonel Merlin G Suttkus (1919–1986) was an officer in the United States Air Force and the senior author's brother. He helped him collect fish in the late 1940s.

Merona

Sierra Leone Mormyrid *Marcusenius meronai* R Bigorne & D Paugy, 1990

Dr Bernard de Mérona is a biologist with a particular interest in tropical river systems. He has worked for ORSTOM (now known as IRD), a French government organisation in Ivory Coast, Brazil and French Guiana. He is presently retired as a Research Director. He has written more than 40 scientific publications and 10 books and/or chapters mostly as a co-author, including: *La croissance des poissons d'eau douce africains* (1988), *Fish communities in rivers* (1988), *La pêche en Amazonie* (1991), *Les poissons d'eau douce de Guyane* (2001) and *Alteration to fish diversity downstream from Petit-Saut Dam in French Guiana – Implication of ecological strategies of fish species* (2003).

Merrett

Spiny Eel sp. *Polyacanthonotus merretti* KJ Sulak, RE Crabtree & J-C Hureau, 1984
Merrett's Snailfish *Careproctus merretti* AP Andriashev & NV Chernova, 1988
Merrett's Whiptail *Nezumia merretti* T Iwamoto & A Williams, 1999
Blind Cusk Eel sp. *Paraphyonus merretti* JG Nielsen, 2015

Dr Nigel Robert Merrett (b.1940) is a British zoologist and ichthyologist. Bristol University awarded his bachelor's degree (1963) and doctorate (1992) and the University of East Africa, Dar-es-Salaam, his master's (1968). After a temporary job attached to the Southern Harvester Whaling Expedition in Antarctica (1962–1963), he was scientist at the East African Marine Fisheries Research Organization (1963–1968). He was ichthyologist at what is now the National Oceanographic Centre, Southampton University (1968–1989) and head of the fish group in the Zoology Department, BMNH (1989–1999). In his career he accumulated about six years of sea time on more than 50 scientific cruises, including the 'Vitjaz' cruise (NV Parin deep-sea expedition) (1998). Among some 100 scientific papers and longer works he co-wrote: *"Southern Harvester" Whaling Expedition, 1962–1963: bird observations* and *Deep-Sea Demersal Fish and Fisheries* (1997). He was honoured in the spiny eel's binomial for his considerable contributions to the systematics and ecology of deep-sea bottom fishes. (See also **Nigel)**

Merriam

Ash Meadows Killifish *Empetrichthys merriami* H Gilbert, 1893 (Extinct)

Dr Clinton Hart Merriam (1855–1942) was an American naturalist and physician. His father was a Congressman, through whom Merriam met Baird of the USNM (1871). This led to an invitation to work as a naturalist in Yellowstone, Wyoming, as a member of the Hayden Geological Survey. The experience sparked his interest and guided his further education; he studied biology and anatomy at Yale, finally graduating as a physician (1879). His natural history passtime continued while he practised medicine, until he opted for scientific work full-time (1883). When he became Chairman of the Bird Migration Committee of the AOU they successfully applied to Congress for funds to study birds on the grounds that such work would benefit farmers. Merriam became the first chief of the United States Biological Survey's (USBS) Division of Economic Ornithology and Mammalogy. He is most famed for his 'life zone' theory, which hypothesised that 'temperature extremes were the principal desiderata in determining the geographic distribution of organisms'. Later in life, Merriam's focus shifted to studying the Native American tribes of the western USA. Ten mammals, two birds and a reptile are also named after him. He was one of the collectors of the type.

Mertens, KH

Atoll Butterflyfish *Chaetodon mertensii* G Cuvier, 1831
Kamchatka Grayling *Thymallus mertensii* A Valenciennes, 1848

Karl Heinrich Mertens (1796–1830) was a German botanist, naturalist and explorer. He was on the expeditions of the Russian ship 'Senyavin', commanded by Captain Fedor Petrovich Litke, to the coasts of Russian America (now Alaska) and Asia and to Iceland (1830). Various species are named after him, as is Cape Mertens east of the Chukot Peninsula. Valenciennes proposed the grayling species based on a plate that Mertens had illustrated.

Mertens, P

Mertens' Moonflounder *Monolene mertensi* M Poll, 1959

Mrs P Mertens was an artist-designer, and author of very accurate illustrations of all the ichthyological publications relating to the fish taken on the Belgian oceanographic expedition of the South Atlantic. Poll says that "*...no systematic work can do without a faithful and careful iconography. Once more, I will say what a valuable supplement it was for me that the accurate but artistic realization of the figures representing all the species that were executed by Mrs. P. Mertens*" (translation). He should also have used the feminine form in his binomial; *mertensae*.

Mertens, R

Mertens' Prawn-Goby *Vanderhorstia mertensi* W Klausewitz, 1974

Dr Robert Friedrich Wilhelm Mertens (1894–1975) was a Russia-born German herpetologist who moved to Germany (1912) and earned his doctorate at the University of Lepzig (1915). He started as Assistant (1919) and eventually retired (1960) as Director Emeritus, Naturmuseum Senckenberg (Frankfurt). He also lectured at Goethe University (1932) becoming professor there (1939). He collected specimens in thirty countries. During WW2 he scattered the museums' collections in order to preserve them, and also had German soldiers bring specimens back during their service. He wrote many articles and several books, such as *The Life of Amphibians and Reptiles* (1959). He died after being bitten by his pet snake, suffering over 18 days and all the while keeping a journal noting that this was a singularly appropriate end to a herpetologist. Thirteen amphibians, seven reptiles, a bird and a mammal are also named after him.

Merton

Chequered Mangrove Goby *Mugilogobius mertoni* M Weber, 1911
Merton's Sleeper-Goby *Oxyeleotris mertoni* M Weber, 1911

Dr Hugo Philip Ralph Merton (1879–1940) was a German zoologist and explorer. He studied at Bonn, Berlin and Heidelberg (1898–1905) culminating in the latter awarding his PhD (1905). He undertook research in Naples (1905–1906), then went on an expedition (1907–1908) for the Senckenberg Nature Research Society to the Moluccas, visiting the Kai and Aru Islands with Roux (1907–1908). He became Deputy Director of Seckenberg Museum, Frankfurt (1909–1913). He was in the military during WW1 (1914–1918) later (1920–1935) becoming a Professor at the University of Heidelberg but was not allowed to continue when the Nazis took power due to his Jewish ancestry. He was visiting Professor at the Institute of Animal Genetics, Edinburgh (1937). He returned to Germany (1938) was interned and sent to Dachau concentration camp where he was seriously ill. He was allowed to emigrate, taking up the post of Assistant at the Institute of Animal Genetics back in Edinburgh (1939–1940) where he died, probably as a result of his time at Dachau. An amphibian, a reptile, an odonata and two birds are also named after him.

Merzbacher

Zhungarian Ide *Leuciscus merzbacheri* E Zugmayer, 1912

Dr Gottfried Merzbacher (1843–1926) was a furrier in Munich (1868–1888) who sold his business to become a mountaineer, naturalist and explorer in the Caucasus (1890–1892) and Tien Shan (1902–1903 & 1907–1908). The University of Munich awarded his honorary doctorate (1902) and made him an honorary professor (1907). Two birds and Merzbacher Glacier Lake are named after him. He collected the holotype of this species.

Meseda

Meseda Waspfish *Neocentropogon mesedai* W Klausewitz, 1985

This is not an eponym but an acronym that relates to a number of Red Sea explorations, known as: **Me**talliferous **Sed**iments **A**tlantis-II-Deep, or MESEDA, which consisted of three cruises (1977–1981). The type was taken during one of these.

Mesmaekers

African Characin sp. *Monostichodus mesmaekersi* M Poll, 1959
Marbled Lungfish ssp. *Protopterus aethiopicus mesmaekersi* M Poll, 1961

I Mesmaekers was the commander of the port of Boma, Belgian Congo (Democratic Republic of the Congo). He is recorded (1948–1955) as having collected and supplied in that period five skulls of the West African Manatee (*Trichechus senegalensis*) to the Royal Museum of Central Africa, Belgium, and the AMNH. He was honoured for facilitating the shipment of lungfish specimens and their mucus cocoons.

Messier

Hair-lip Brotula *Cataetyx messieri* A Günther, 1878

This is a toponym referring to the Messier Strait, Patagonia, Chile, where the holoype was collected.

Mesude

Isikli Loach *Oxynoemacheilus mesudae* F Erk'akan, 2012
[Junior synonym of O. germeniccus]

Mesude Kaynak is the author's mother.

Meta

Barje Sculpin *Cottus metae* J Freyhoff, M Kottelat & A Nolte, 2005

Meta Povs was thanked for her "*...continuous help with several projects*." We can find no biography.

Meta

Brown-banded Copella *Copella metae* CH Eigenmann, 1914
Masked Corydoras *Corydoras metae* CH Eigenmann, 1914
Purple Tetra *Hyphessobrycon metae* CH Eigenmann & AW Henn, 1914
Pencil Catfish sp. *Ituglanis metae* CH Eigenmann, 1917
Three-barbeled Catfish sp. *Pimelodella metae* CH Eigenmann, 1917
Yellow Acara *Aequidens metae* CH Eigenmann, 1922
Characin sp. *Charax metae* CH Eigenmann, 1922
Characin sp. *Moenkhausia metae* CH Eigenmann, 1922
Characin sp. *Hemibrycon metae* GS Myers, 1930
Characin sp. *Tyttocharax metae* C Román-Valencia, CA García-Alzate, RI Ruiz-C & DCB Taphorn, 2012

This is a toponym; generally referring to the Río Meta in the Colombian and Venezuelan Orinoco River system. In the case of the *Tyttocharax* it refers to Meta State, eastern Colombia.

Meteor

Blind Cusk-eel genus *Meteoria* JG Nielsen, 1969
Meteor Goby *Palutrus meteori* W Klausewitz & CD Zander, 1967
Hatchetfish sp. *Polyipnus meteori* A Kotthaus, 1967

Cusk-eel sp. *Neobythites meteori* JG Nielsen, 1995
Duckbill Eel sp. *Saurenchelys meteori* W Klausewitz & U Zajonz, 2000

These fish are named after the German research vessel 'Meteor', from which the holotypes were collected.

Methuen

Striped Silver Biddy *Gerres methueni* CT Regan, 1920

Major Paul Ayshford Methuen, 4th Baron Methuen (1886–1974), was an English painter, zoologist and landowner. He worked (1910–1914) in the Transvaal Museum, Pretoria, where he published several scientific papers with the South African herpetologist John Hewitt, with whom he collected and described a number of southern African and Madagascan taxa. He made a collection of fishes in Madagascar (1911) which included the *Gerres* holotype.

Metz

Oriental Cyprinid genus *Metzia* DS Jordan & WF Thompson, 1914
Striped Kelpfish *Gibbonsia metzi* CL Hubbs, 1927

Dr Charles William Metz (1889–1975) was an American zoologist, geneticist and taxonomist whose doctorate was awarded by Stanford University. He co-wrote: *A catalog of the fishes known from the waters of Korea* (1913).

Metzelaar

Sponge Goby *Evermannichthys metzelaari* CL Hubbs, 1923

Dr Jan Metzelaar (1891–1929) was a Dutch-born American ichthyologist who studied at the University of Amsterdam, being awarded his doctorate (1919) with a thesis on the systematics of tropical Atlantic fish. He emigrated to the USA (1923) and was almost immediately employed as a fisheries expert at the Michigan Department of Conservation. He became an American citizen on October 2, 1929, but two days later he drowned in Grand Lake. He translated the *Philosophie zoologique* of Jean-Baptiste de Lamarck (1744–1829) into Dutch. He described this goby (1919) but used a preoccupied name.

Meunier

Characin sp. *Jupiaba meunieri* J Géry, P Planquette & P-Y Le Bail, 1996

Dr François Jean Meunier (b.1942) is a French fish osteologist at MNHN, Paris. His doctorate was awarded in Paris (1982). He co-wrote: *From natural history to earth sciences* (2014). He helped collect the holotype.

Meyburg

Characin sp. *Microschemobrycon meyburgi* H Meinken, 1975

Dr Otto August Gert Meyburg (1910–1996) was a physician in Bremen, Germany. He and his wife, Erika, made a collection near Lake Victoria, East Africa (1971). He presumably also collected in Brazil, where the characin originates.

Meyen

Blotched Fantail Ray *Taeniurops meyeni* J Müller & FGJ Henle, 1841
[Alt. Round Ribbontail Ray]

Dr Franz Julius Ferdinand Meyen (1804–1840) was a German surgeon, a botanist and collector who was a Professor in Berlin. He suggested that new cells were created through cell division rather than by the creation of new free cells. He published this theory in his work on plant anatomy, *Phytotomie* (1830). He took part in the circumnavigation (1830–1832) by 'Prinzess Louise', an expedition that included considerable time in South America (1830–1832). A mammal and two birds are named after him.

Meyer, AB

Bigeye sp. *Pristigenys meyeri* A Günther, 1872
Toothless Characin sp. *Curimatella meyeri* F Steindachner, 1882

Dr Adolf Bernhard (né Aron Baruch) Meyer (1840–1911) was an anthropologist, entomologist and ornithologist who studied at the Universities of Göttingen, Vienna, Zürich and Berlin. He collected in the East Indies, New Guinea and the Philippines. He was Professor at the Ethnographische Museum, Dresden, becoming Director, Staatlisches Museum für Tierkunde, Dresden (1874–1905). He wrote: *The Birds of the Celebes and Neighbouring Islands* (1898). He was very interested in the evolution debate and corresponded with Wallace. One amphibian, seventeen birds, two reptiles and two mammals are named after him, as is a dragonfly.

Meyer, JF

Meyer's Butterflyfish *Chaetodon meyeri* ME Bloch & JG Schneider, 1801
[Alt. Maypole Butterflyfish, Scrawled Butterflyfish]

The authors indicated that the type specimen was held in *Museo Meyeri Lugduni Batavorum* (Meyer's Museum in Leiden). Unfortunately, the type has been lost and the identity of Meyer remains uncertain. A likely candidate is Johan Frederik Meijer, who was 'bookkeeper' at Amboina (Ambon Island, Indonesia; the type locality) and who maintained a small natural history cabinet in the Netherlands (1700s).

Meyer, MK

Marbled Swordtail *Xiphophorus meyeri* M Schartl & JH Schröder, 1988

Manfred K Meyer is a German ichthyologist. Among his published papers is: *Notes on the subgenus* Mollienesia *LESUEUR, 1821, with a description of a new species of* Poecilia *BLOCH & SCHNEIDER, 1801 (Cyprinodontiformes: Poeciliidae) from Venezuela* (2000). He has described at least five *Poecilia* species including one with Manfred Schartl. The etymology reads: "*Named in honour of M.K. Meyer, Bad Nauheim, FRG, who has contributed much to the taxonomy of poeciliid fish and who has brought the new fish to our attention.*"

Meyer-Waarden

Austral Lightfish *Woodsia meyerwaardeni* G Krefft, 1973

Dr Paul-Friedrich Meyer-Waarden (1902–1975) was Executive Director, Federal Research Centre for Fisheries. He was honoured to celebrate his 70th birthday and to acknowledge his contribution to the publication of Krefft's series of papers on fishes collected during research cruises of the 'Walther Herwig' in South America. He studied in Berlin and at Rostock where the University of Rostock awarded his doctorate (1929) and he held another doctorate from the University of Berlin (1943). He taught at the University of Greifswald (1943).

Miang

Characin sp. *Moenkhausia miangi* F Steindachner, 1915

A toponym referring to the type locality; the Miang River on the Venezuela/Brazil border.

Mibang

Sisorid Catfish sp. *Glyptothorax mibangi* A Darshan, R Dutta, A Kachari, B Gogoi & DN Das, 2015

Tamo Mibang is Vice-Chancellor of, and Professor at, Rajiv Gandhi University, Doimukh, India. The etymology says that his "*...patronage has continually been extended to freshwater-fish research and conservation in the Eastern Himalyan region of India.*" He has written about traditional societies in Arunachal Pradesh, including the co-written: *Tribal Villages in Arunachal Pradesh* (2004).

Michael (Goulding)

Characin sp. *Charax michaeli* CAS de Lucena, 1989
Brazilian Cichlid sp. *Aequidens michaeli* SO Kullander, 1995
Dr Michael Goulding (b.1950). (See **Goulding**)

Michael (Graham)

Lake Victoria Cichlid sp. *Haplochromis michaeli* E Trewavas, 1928
Michael Graham (1898–1972) was a British fisheries scientist. (See **Graham, M**)

Michael, SW

Michael's Epaulette Shark *Hemiscyllium michaeli* GR Allen & CL Dudgeon, 2010
[Alt. Leopard Epaulette Shark]

Scott W Michael is an author and a photographer who specializes in elasmobranchs and coral reef fishes. He wrote: *Reef Sharks and Rays of the World* (1994). He was the first person to recognise this species' unique colour pattern and to bring the difference between this species and *H. freycineti* to the authors' attention. They also honoured him for contributing information and photographs to the first author's research on Indo-Pacific fishes.

Michael Sars

Michael Sars Smooth-head *Bathytroctes michaelsarsi* E Koefoed, 1927
Bigfin Pearleye *Scopelarchus michaelsarsi* E Koefoed, 1955

'Michael Sars' was the name of the Norwegian research vessel from which the holotypes were collected. The vessel was named after the Norwegian theologian and biologist Michael Sars (1805–1869).

Michaux

Central Stoneroller ssp. *Campostoma anomalum michauxi* HW Fowler, 1945

André Michaux (1746–1802) was a French explorer and botanist. He travelled and collected in England (1779), Spain (1780) and Persia (Iran) (1782–1784). He was appointed Royal Botanist by King Louis XVI and sent to the USA, where he worked and collected (1785–1791). The French Revolution meant he no longer had support from the King, but Thomas Jefferson asked him to explore to the west (10 years before the more famous Lewis and Clark Expedition). He travelled through parts of the USA and Canada. He was shipwrecked when returning to France (1796). He was a member of Baudin's expedition to Australia (1800), but disembarked at Mauritius from where he went to Madagascar, dying there from a tropical fever.

Michel

Michel's Ghost Goby *Pleurosicya micheli* P Fourmanoir, 1971

The patronym is not identified, but is possibly in honor of Fourmanoir's colleague and occasional collaborator, marine biologist Michel Legand, French Institute of Oceania, Nouméa, New Caledonia. Fourmanoir certainly later honoured him in the name of a cucumberfish using his surname. (See also **Legand**)

Midas

Midas Blenny *Ecsenius midas* WA Starck, 1969
[Alt. Persian Blenny]
Betta sp. *Betta midas* HH Tan, 2009

In Greek mythology, Midas was a king of Phrygia whose touch could turn anything to gold (a gift from the gods which had to be renounced once the impracticalities sunk in). The fish are named after him because of their golden coloration.

Middendorff

Eelpout sp. *Hadropareia middendorffii* PJ Schmidt, 1904

Alexander Theodor (Aleksander Fedorovich) von Middendorff (1815–1894) was an Estonian of German extraction who was a traveller and naturalist. He qualified as a physician (1837) and became (1839) Assistant Professor of Zoology at Kiev University. After journeying throughout Siberia and the surrounding regions on an Imperial Academy expedition (1842–1845) he was made a member of the Imperial Academy of Sciences at St Petersburg (1845). His accounts of the Amur River and other remote regions were the fullest made by a naturalist and anthropologist of these little-explored areas. As a naturalist, he wrote on the spread of permafrost and how this affected the distribution of plants and animals. Four birds and two mammals are also named after him.

Midgley

Midgley's Grunter *Pingalla midgleyi* GR Allen & JR Merrick, 1984
Silver Cobbler (catfish) *Neoarius midgleyi* PJ Kailola & BE Pierce, 1988
[Binomial sometimes corrected to *midgleyorum*]

Dr Stephen Hamar Midgley (1918–2014) was an amateur ichthyologist and limnologist whose honorary doctorate was awarded by the University of Queensland (1994). In WW2, he served in the Australian Imperial Force as a Captain (1939–1945). He worked for Australian Fisheries (1970s & 1980s). The catfish's binomial equally celebrates his wife, Mary, who was also a limnologist and ichthyologist and who was awarded an honorary doctorate by the University of Queensland at the same time as her husband. He co-wrote: *Field guide to the freshwater fishes of Australia* (1968).

Mieg

Ebro Nase *Parachondrostoma miegii* F Steindachner, 1866

Juan Mieg (1779–1859) was a Swiss entomologist, ornithologist, general naturalist and artist. He went to Spain (1814), accompanying Fernando VII on his return from exile after the fall of Napoleon Bonaparte. He became Professor of Physics and Chemistry and Director of the National Museum of Natural Sciences. Steindachner does not identify him, but it seems likely that this species is named after him.

Mikado

Sakhalin Sturgeon *Acipenser mikadoi* FM Hilgendorf, 1892

'Mikado' is a (now obsolete) term for the Emperor of Japan. Hilgendorf lectured at Tokyo College of Medicine (1873–1876) and described this sturgeon from one he saw at a fish market. At the time Hilgendorf was in Japan, the emperor was Meiji (reigned 1867–1912).

Mike

Wachters' Killi ssp. *Aphyosemion wachtersi mikeae* AC Radda, 1980

Mike Wachters is the wife of Walter Wachters. (See **Wachters**)

Mikhailin

Mikhailin's Scabbardfish *Aphanopus mikhailini* NV Parin, 1983

Slimehead sp. *Hoplostethus mikhailini* AN Kotlyar, 1986

Barbeled Dragonfish sp. *Eustomias mikhailini* AM Prokofiev, 2020

Sergey Vladimirovich Mikhailin (1943–1981) was a Russian ichthyologist at the All-Union Research Institute of Marine Fisheries and Oceanography, Moscow. He made a great contribution to the study of trichiuroid fish and co-wrote a number of articles with Parin, such as: *Lepidus calcar, a New Trichiurid Fish from the Hawaiian Underwater Ridge* (1982) and worked with him in the 1960s and 1970s. He is honoured in the slimehead's name for his great contribution to the study of fishes of southern Africa. In the case of the scabbardfish it was named after him because he had supplied material to Parin on the *Aphanopus* genus and because he had tragically died at the height of his career, when he was saving people from a burning train.

Miki

Miki's Dwarfgoby *Eviota mikiae* GR Allen, 2001

Miki Tonozuka of Bali, Indonesia, helped to collect the type. She ran 'Dive & Dive' with her late husband Takamasa Tonozuka (q.v.). She was honoured for her assistance in the field during the Weh Island survey. (Also see **Tonozuka**)

Mikolji

Oscar cichlid sp. *Astronotus mikoljii* AP Lozano, OM Lasso-Alcalá, PS Bittencourt, DC Taphorn, N Perez & IP Farias, 2022

Ivan Mikolji (b.1972) is a Venezuelan explorer, artist, author, underwater photographer and audiovisual producer. He was honoured "in recognition for being a tireless and enthusiastic diffuser of the biodiversity and natural history of freshwater fishes, conservation of aquatic ecosystems of Venezuela and Colombia." Since 2020, Mikolji has been recognized as Associate Researcher of the Museo de Historia Natural La Salle, from the Fundación La Salle de Ciencias Naturales, Caracas, Venezuela.

Mildred

Argentine sp. *Glossanodon mildredae* DM Cohen & SP Atsaides, 1969

Mrs Mildred H Carrington (1908–1988) was a scientific illustrator with the US Fish and Wildlife Service (1951–1971) "...*whose tasteful and accurate drawings have contributed greatly to the progress of ichthyology*." She was a graduate of the University of Maryland. She was a co-compiler of: *A Bibliography of American Gobies* (1965).

Miles

Southern Velvetfish *Aploactisoma milesii* J Richardson, 1850

No etymology is given by Richardson. However, in the same year Valenciennes had named another fish species *Aulopus milesii* (a junior synonym of *Latropiscis purpurissatus*), saying that specimens had been sent to the national collections by Mr. Miles, an English naturalist "who has been very successfully engaged with the ichthyology of the coasts of New Holland" (translation). The velvetfish is also from Australian coasts, so this is probably the same 'Mr. Miles' – but his full identity is still a mystery.

Miles, CW

Miles' Rubbernose Pleco *Chaetostoma milesi* HW Fowler, 1941

Ghost Knifefish sp. *Apteronotus milesi* CD de Santana & JA Maldonado-Ocampo, 2005

Cecil W Miles combined being Secretary of the Dorada Railway, Mariqueta, Colombia with being a collector and ichthyologist who sent Fowler a considerable number of specimens of Colombian freshwater fishes in the 1940s. He wrote: *Estudio economico y ecologico de los peces de agua dulce de Valle del Cauca* (1943).

Miles, SB

Bighead Lotak (barb) *Cyprinion milesi* F Day, 1880

Colonel Samuel Barrett Miles (1838–1914) was a scholar, Arabist antiquarian and traveller in the Middle East. He was amongst the first Europeans to travel widely in Oman, where he was British Political Agent five times (1872–1887). He wrote: *Countries and Tribes of the*

Persian Gulf, published posthumously and incomplete, only covering Oman (1919). The etymology just names a 'Colonel Miles', but it is probably intended to honour this person. A bird is also named after him.

Milford

Finless Flounder *Neoachiropsetta milfordi* MJ Penrith, 1965

C S Milford was the Managing Director of a South African commercial fishing company which at its height had fifty-four deep-sea trawlers. The etymology says: "*The species is named for Mr. C. S. Milford, managing director of the trawling firm, Messrs. Irvin and Johnson (Pty.) Ltd., in recognition of generous support for marine biological research.*"

Milius

Australian Ghost Shark *Callorhinchus milii* JBGM Bory de Saint-Vincent, 1823
Green-striped Coral Bream *Scaevius milii* JBGM Bory de Saint-Vincent, 1823

Baron Pierre Bernard Milius (1773–1829) was a French naval officer, naturalist and civil servant who took part in an exploratory voyage (1804) of the Mascarene Islands, Indian Ocean, under Nicolas Baudin, during which he became friends with Bory. He was Governor of Bourbon (now Réunion) (1818–1821), where he established a port and undertook agricultural projects. He was also despotic and despised by the locals. He was later appointed as chief administrator of French Guiana (1823–1825). He was present at the Battle of Navarino (1827) during the War of Greek Independence.

Miller, GC

Guinean Flyingfish *Cheilopogon milleri* RH Gibbs & JC Staiger, 1970
Miller's Grenadier *Nezumia milleri* T Iwamoto, 1973
Armoured Gurnard sp. *Satyrichthys milleri* T Kawai, 2013

Dr George C Miller is a zoologist, who was at the National Marine Fisheries Service, Southeast Fisheries Center, Miami Laboratory, Florida, and is now retired. For a time in the 1950s he was based in Liberia, covering much of West Africa and his collecting efforts off Angola provided specimens for Iwamoto, including the holotype of the grenadier, and where he also collected the flyingfish. He wrote: *Commercial Fishery and Biology of the Fresh-water Shrimp, Macrobrachium, in the Lower St. Paul River, Liberia, 1952–53* (1971).

Miller, N

Tetra sp. *Hyphessobrycon milleri* ML Durbin, 1908

Newton Miller was an American zoologist who was Durbin's colleague at Indiana University, where he was awarded his bachelor's degree (1904), and was later at Clark University. He wrote: *The fishes of the Motagua River, Guatemala* (1907) and articles such as: *The American Toad (Bufo Lentiginosus Americanus, Le Conte). II A Study in Dynamic Biology* (1909). He collected the tetra holotype.

Miller, PJ

Goby genus *Millerigobius* H Bath, 1973

Acheron Spring Goby *Knipowitschia milleri* H Ahnelt & PG Bianco, 1990
Freshwater Goby sp. *Rhinogobius milleri* I-S Chen & M Kottelat, 2001

Dr Peter J Miller is an ichthyologist and taxonomist specialising in gobies. He works at the School of Biological Sciences, University of Bristol, where he is Emeritus Professor and Senior Research Fellow. He wrote: *Fries' goby, a European oddity* (2015). The etymology for the *Rhinogobius* mentions Miller's "*very kind support to the studies and researches*" of the first author.

Miller, RJ

Ctenopoma sp. *Microctenopoma milleri* SM Norris & ME Douglas, 1991

Rudolph J Miller (1934–2017) was Professor Emeritus of Zoology at Oklahoma State University, as well as an artist and expert in fish behaviour. He was born in Czechoslovakia, but went to the USA with his mother when he was a young child. Oklahoma State University offered him a temporary position (1962) without an interview. He took the job and never left, staying on to teach ethology, vertebrate natural history and ichthyology, until he retired (1990).

Miller, RR

Mexican Rivuline genus *Millerichthys* WJEM Costa, 1995
Miller's Clingfish *Gobiesox milleri* JC Briggs, 1955
Catemaco Platyfish *Xiphophorus milleri* DE Rosen, 1960
Miller's Damselfish *Pomacentrus milleri* WR Taylor, 1964
Miller's Silverside *Atherinella milleri* WA Bussing, 1979
Grijalva Gambusia *Heterophallus milleri* AC Radda, 1987
Mexican Redhorse ssp. *Moxostoma austrinum milleri* CR Robins & EC Raney, 1957
Cottonball Marsh Pupfish *Cyprinodon salinus milleri* JF LaBounty & JE Deacon, 1972

Robert Rush Miller (1916–2003) was an American ichthyologist and conservationist. The University of California, Berkeley awarded his bachelor's degree (1938), the University of Michigan his master's (1943) and his PhD (1944): the latter being the year he took the position of Associate Curator at the US National Museum. During his time there he joined the Arnhem Land expedition in Australia. He taught (1948) at the University of Michigan, getting tenure (1954). He was ichthyological editor of 'Copeia' (1950–1955). Among other scientific papers he wrote: *Xiphophorus gordoni, A New Species of Platyfish from Coahuila, Mexico* (1963). He married a fellow ichthyologist, Frances Hubbs (1919–1987), the daughter of his collaborator C L Hubbs (q.v.).

Miller, SW

Stout Blacksmelt *Pseudobathylagus milleri* DS Jordan & CH Gilbert, 1898

Dr Samuel Walter Miller (1864–1949) was a philologist, classics scholar and archaeologist. He gained a master's degree at the University of Michigan (1884) and undertook doctoral studies at the University of Leipzig (1884–1885). He joined the American School of Classical Studies, Athens (1886). Having survived being robbed, beaten and left for dead, he was commissioned in the Greek army to find and arrest his attackers, which he did - they each got sentenced to 10 years imprisonment. He returned to the University of Michigan to teach classics as an acting associate professor (1886–1889). He and his family were in Leipzig (1890–1891) where he taught archaeology. He was an associate professor of Greek, University of Missouri, Columbia (1891–1892). He was one of the founders of the Classics Department at Stanford University where he was Professor of Latin and Archaeology (1892–1902). He became Professor of Greek and Philology, Tulane University, New Orleans (1902–1910). He was again at the University of Missouri as Professor of Latin and later of Classical Languages and Archaeology (1911–1936), retiring as Professor Emeritus. During WW1, he served in France and Italy with the YMCA. In retirement, he went on teaching at Southwestern University, Memphis and Washington University, St Louis. He was honoured for his 'intelligent interest' in zoological nomenclature; Miller reviewed and verified the name etymologies in Jordan and Evermann's: *Fishes of North and Middle America* (1896–1900). The fascinating thing about his career of 50 years as an academic and university professor is that he never completed his doctorate – the doctorate he had was honorary, awarded by the University of Michigan (1932).

Milly

Miller Drum *Umbrina milliae* RV Miller, 1971

Mildred 'Milly' Miller was the wife of the author, American ichthyologist Robert Victor Miller.

Milorad

Konavle Dace *Telestes miloradi* NG Bogutskaya, P Zupančič, I Bogut & AM Naseka, 2012

Dr MIlorad Mrakovčić (b.1949) is a Croatian zoologist who is a Professor in the Zoology Department, College of Natural Sciences and Mathematics, University of Zagreb (since 2003). He co-wrote: *Sexual diversity of five* Cobitis *species (Cypriniformes, Actinopterygii) in the Adriatic watershed* (2015).

Milvertz

African Rivuline sp. *Nothobranchius milvertzi* B Nagy, 2014

Finn Christian Milvertz of Solrød Strand, Denmark, is an aquarist; a 'renowned' killifish breeder and collector. He co-collected the type with Béla Nagy at ephemeral pools of the Lushiba Marsh, Zambia.

Minamori

Japanese Loach sp. *Cobitis minamorii* J Nakajima, 2012

Dr Sumio Minamori (b.1917) was a Japanese biologist and geneticist at the Zoological Laboratory, Faculty of Science, Hiroshima University, Japan. He was honoured as a 'pioneer' in the study of Japanese loach speciation. He co-wrote: *Relation of Infection to Population Structure in Drosophila melanogaster* (1968).

Mincarone

Southern Sawtail Catshark *Galeus mincaronei* JMR Soto, 2001

Dr Michael Maia Mincarone (b.1971) is a Brazilian ichthyologist who is a Professor at Universidade Federal do Rio de Janeiro. The Universidade do Vale do Itajaí awarded his bachelor's degree in oceanography (1997) and Pontifícia Universidade Católica do Rio Grande do Sul his doctorate in zoology (2007). His research is focused on the systematics and conservation of deep-sea fishes. Soto honoured him for his "*...extensive work and tireless dedication*" to the Museu Oceanográfico Univali, Brazil.

Minckley

Minckley's Cichlid *Herichthys minckleyi* I Kornfield & JN Taylor, 1983

Wendell Lee Minckley (1935–2001) was a college professor who spent most of his career at Arizona State University. He studied the fauna of Cuatro Ciénegas, Coahuila, México, where this cichlid is endemic, for many years. With Robert Rush Miller (q.v.) he discovered and named the Northern Platyfish *Xiphophorus gordoni*. He died from complications associated with treatment for cancer.

Mindi

Freshwater Pipefish sp. *Pseudophallus mindii* SE Meek & SF Hildebrand, 1923

This is a toponym referring to the type locality; a creek near *Mindi*, Panama Canal Zone.

Miners

Miners' Dwarfgoby *Sueviota minersorum* DW Greenfield, MV Erdmann & IV Utama, 2019

Andrew and Marit Miners founded (2008) the Misool EcoResort in Raja Ampat Islands, Indonesia, and its charitable arm, the Misool Foundation; the joint mission of which "*...is to safeguard the most biodiverse reefs on Earth through the empowerment of local communities, providing a structure by which they are able to reclaim their traditional tenureship of reefs*". Andrew was born in the UK and grew up in Cornwall spending his youth swimming, surfing and taking part in lifeguard competitions. His psychology degree was awarded (1993) by Lancaster University and he became a certified diving instructor and dive master (1994). He started a business in Indonesia (1999) with a 60-foot boat 'Felidae'. Marit was born in Sweden but moved to the USA as a toddler. She earned an Anthropology degree from Columbia University. She discovered scuba diving on a trip to Thailand and met Andrew (2005) in Bangkok. The etymology says their: "*...superlative efforts in conservation and sustainable economic development in the region have dramatically improved the health and biomass of the thriving reefs where this new species was discovered. It is a pleasure to name this unique and apparently rare goby in their honor.*"

Ming

Singapore Banded Goby *Hemigobius mingi* AWCT Herre, 1936

Mr. Ming is described in the etymology as a: "*...chemist in the Department of Fisheries, S.S. and F.M.S., who was of much assistance to me during my stay in Singapore.*" Herre did not record Ming's full name.

Minotaur

Bullneck Seahorse *Hippocampus minotaur* MF Gomon, 1997

Obviously an allusion to the bull-headed Minotaur of Greek mythology.

Mirian

Pike-Cichlid sp. *Cichla mirianae* SO Kullander & EJG Ferreira, 2006

Mirian Leal-Carvalho is a zoologist workung at the Ichthyology department of Museu Paraense Emīlio Goeldi, Belèm, Brazil. She is co-author of: *Rio Negro, Rich Life in Poor Water. Amazonian Diversity and Foodchain Ecology as Seen Through Fish Communities* (1988) and participated in the collection of part of the type series.

Mirini

Three-barbeled Catfish sp. *Imparfinis mirini* JD Haseman, 1911

This is a toponym referring to the Rio Piracicaba-mirini, Brazil, the type locality.

Miry

Miry's Demoiselle *Neopomacentrus miryae* M Dor & GR Allen, 1977

Miry Dor was the wife of senior author Menachem Dor (1901–1998) (q.v.), and was honoured in memorium.

Mishky

Three-barbeled Catfish sp. *Imparfinis mishky* A Almirón, J Casciotta, J Bechara, F Ruíz Díaz, C Bruno, S D'Ambrosio, P Solimano & P Soneira, 2007

Mishky is the Quichua word for sweet. The authors say: "*The epithet was dedicated to Patricia Garcia Tartalo, our friend and student who died tragically in February, 2006.*" We have not been able to find anything more about her.

Mishra

Moray Eel sp. *Gymnothorax mishrai* D Ray, A Mohapatra & DG Smith, 2015

Subhrendu Sekhar Mishra was Officer in Charge of the Fish Section, Zoological Survey of India (2002–2015) and is now Principle Investigator there. His published papers include: *First record of Lagocephalus guentheri Miranda Ribeiro 1915 (Tetraodontiformes: Tetraodontidae) from the West Coast of India* (2018) and *Ophichthus chilkensis Chaudhuri, 1916 (Anguilliformes: Ophichthidae) – resurrection as a valid species from India, with re-description* (2019) written with Ray and Mohapatra among others.

Mitchell

Long-spined Anglerfish *Echinophryne mitchellii* A Morton, 1897
[Alt. Spinycoat Anglerfish]

Alexander Morton gave no etymology or any clue to Mitchell's identity.

Mitchell, D

Mitchell's Emperor *Lethrinus mitchelli* GR Allen, BC Victor & MV Erdmann, 2021

David Kym Mitchell became (2016) Director at Eco Custodian Advocates, Alotau Milne Bay Province, Papua New Guinea. He is a conservationist who has dedicated his career to the nature and local indigenous communities of Milne Bay, Papua New Guinea, where the fish species occurs. He was Director of Conservation International at Alotau (2010–2016) where he had been a conservation strategy specialist (2000–2010). He took his degree in agricultural science at the University of Adelaide (1977–1981) and undertook a number of voluntary positions and studied further at Curtin University (Western Australia) (1983–1988). He also took a course to teach English as a foreign language at Massey University (1986–1987), and later a diploma in wildlife management at the University of Otago, New Zealand (2003–2004).

Mitchell, J

Indian Catfish sp. *Pseudeutropius mitchelli* A Günther, 1864

Captain Jesse Mitchell (d.1872) was a British army officer who was adjutant of the 1st Native Veteran Battalion, Indian Army, and Commandant of the Madras Mounted Police. He also undertook the duties of being the Superintendent of the Government Museum, Madras (Chennai) (1859–1872) during which period he added over 72,000 specimens to the collection. He was a pioneer of photomicrography in India, exhibiting his work as early as 1857. He was the instigator of a library at the Museum (1860); this was largely stocked with surplus volumes which he sourced from Haileybury College in England. This library expanded and became known as the Connemara Public Library and is now one of India's four depository libraries. He originally landed in India (1829) as a member of the Honourable East India Company's Madras Army and was posted to the Madras Horse Artillery, based in Bangalore. He presented the holotype of this catfish species to the BMNH.

Mitchill

Bay Anchovy *Anchoa mitchilli* A Valenciennes, 1848

Dr Samuel Latham Mitchill (1764–1831) was an American physician, naturalist and politician. His wealthy uncle paid for him to attend Edinburgh University where he graduated (1786). He taught chemistry, botany, zoology, mineralogy and natural history at Columbia College (1792–1801) during which time he collected plants and animals with a particular emphasis on aquatic species. He went on to teach at the New York College of Physicians and Surgeons (1807–1826) and then helped to organise the Rutgers Medical College, serving as its Vice President (1827–1830). Concurrently he entered politics, serving in the New York State Assembly (1791 & 1798), the US House of Representatives (1801–1804) and the US Senate (1804–1809 & 1810–1913). As a naturalist, he studied the fishes of New York Harbour. Because of his breadth of knowledge, he was known as a 'living encyclopaedia,' and a 'stalking library'.

Mitre

Pacu sp. *Myletes mitrei* C Berg, 1895 NCR

[Junior Synonym of *Piaractus mesopotamicus*]

Bartolomé Mitre Martinez (1821–1906) was an Argentine politician, soldier and author. He was the 6th President of Argentina (1862–1868).

Mitrofanov

Stone Loach ssp. *Triplophysa coniptera mitrofanovi* AM Prokofiev, 2017

Valery Petrovich Mitrofanov (1932–2001) was a Russian ichthyologist. He wrote an important report on the loaches of Kazakhstan (1989) and co-authored: *The impact on fish stocks of river regulation in Central Asia and Kazakhstan* (1998).

Mitsui

Pitgum Lanternfish *Opostomias mitsuii* S Imai, 1941

Takanaga Mitsui (b.1892) was a member of the powerful MItsui clan which owned the House of Mitsui (Tokugawa shogunate period (1603–1867)) which evolved into the Mitsui group, one of the biggest trading groups in the world. He founded the Mitsui Institute of Marine Biology (1937).

Mitsukuri

Goblin Shark genus *Mitsukurina* DS Jordan, 1898
Honnibe Croaker *Nibea mitsukurii* DS Jordan & JO Snyder, 1900
Lamprey sp. *Lethenteron mitsukurii* S Hatta, 1901
Sand-lance sp. *Bleekeria mitsukurii* DS Jordan & BW Evermann, 1902
Shortspine Spurdog *Squalus mitsukurii* DS Jordan & JO Snyder, 1903
Spookfish *Hydrolagus mitsukurii* DS Jordan & JO Snyder, 1904

Dr Kakichi Mitsukuri (1857–1909) was a Japanese zoologist who first went to the USA (1873) and achieved doctorates from Yale (1879) and from Johns Hopkins University (1883). He worked for a time at the Naples Zoological station and founded the Misaki Marine Laboratory (1887). He was a Professor at the College of Science, Imperial University of Tokyo (1882–1909), and was Dean of the College of Science there (1901–1909). Among his publications is a Japanese-English dictionary. He was with Jordan and Snyder at Misaki, Japan, when some of the holotypes was taken.

Mivart

Toothless Characin sp. *Curimata mivartii* F Steindachner, 1878

Dr St. George Jackson Mivart (1827–1900) was best known as a naturalist, but he was also known for defending his Catholic faith from 'scientific attacks'. He was a physician and lawyer who practiced at the bar for a short while (1851) before following his natural scientific bent for research. He lectured in zoology at St. Mary's Hospital, London (1862). He was Professor of Biology at University College, Kensington (a short-lived Catholic University) (1874–1877). His publications included: *Genesis of Species* (1871). By maintaining the creationist theory of the origin of the human soul, he attempted to reconcile his scientific evolutionism with the Catholic faith. However, Catholic authorities decided his orthodoxy to be questionable. In January 1900, after admonition and three formal notifications requiring him to sign a profession of faith, he was banned from receiving the sacraments by Cardinal Vaughan. He died of diabetes that same year, and a struggle ensued between his friends and the Roman Catholic Church as to who should bury him. Eventually he was buried in Kensal Green Catholic Cemetery (1904). His father was a wealthy man, owning the London hotel now known as Claridges. A reptile is also named after him.

Mixe

Mixe Swordtail *Xiphophorus mixei* KD Kallman, RB Walter, DC Morizot & S Kazianis, 2004

The Mixe are an indigenous people who inhabit the eastern highlands of the state of Oaxaca, southern Mexico.

Miyabe

Miyabe Char *Salvelinus malma miyabei* M Oshima, 1938

Dr Kingo Miyabe (1860–1951) was a Japanese botanist, taxonomist and mycologist who graduated at the Sapporo Agricultural College (1881). After he did post-graduate work in botany at Harvard (1886–1889), where his doctorate was awarded (1889), he returned to Sapporo. He was a professor (1907) at Tohoku Imperial University, Sapporo (later re-named Hokkaido Imperial University). He was made Professor Emeritus (1927). He became President, Botanical Society of Japan (1936). He wrote: *Flora of Sakhalin* (1915). (This identification is slightly speculative, as the original text is silent on etymology, but no one else seems to fit the bill).

Mizenko

Mizenko's Snapper *Lutjanus mizenkoi* GR Allen & FH Talbot, 1985
[Alt. Samoan Snapper]

David Mizenko worked for the National Marine Fisheries Service. While at the School of Oceanography, University of Rhode Island, he wrote his master's thesis on Samoan snapper populations: *The biology of Western Samoan reef-slope snapper (Pisces: Lutjanidae) populations of: Lutjanus kasmira, Lutjanus rufolineatus, and Pristipomoides multidens* (1984).

Mizuno

Freshwater Goby sp. *Rhinogobius mizunoi* T Suzuki, K Shibukawa & M Aizawa, 2017

Nobuhiko Mizuno was a Professor, Department of Biology, Ehime University, Japan. He is an expert on the freshwater fishes of Japan. He co-wrote: *Mechanisms of embryonic drift in the amphidromous goby, Rhinogobius brunneus (1991).*

Mobbs

Cave Loach sp. *Schistura mobbsi* M Kottelat & C Leisher, 2012

Jerry Mobbs has been a cave diver since 1995. He was the discoverer (2002) and first explorer of the Phuong Hoang cave system, Vietnam, where this species is endemic. His day job is in the telecommunications industry and he has worked in Indonesia, Iran, Pakistan and Vietnam.

Mochigarei

Dusky Sole *Lepidopsetta mochigarei* JO Snyder, 1911

This is not an eponym but a Japanese name meaning 'rice-cake flounder' (*mochi* = Japanese rice cake traditionally eaten at New Year).

Mochizuki

Mochizuki's Seabass *Malakichthys mochizuki* Y Yamanoue & K Matsuura, 2002
Seabass sp. *Parascombrops mochizukii* WW Schwarzhans & AM Prokofiev, 2017

Kenji Mochizuki researched (1980s) at the Department of Fisheries, University Museum, University of Tokyo. Among his published papers are a catalogue of the fishes in the museum and the co-writen: *A new lophiid anglerfish,* Lophiodes fimbriatus from the coastal waters of Japan (1985) and *Trigonognathus kabeyai, a new genus and species of the squalid sharks from Japan* (1990).

Mocquard

Congo Cichlid sp. *Lamprologus mocquardi* J Pellegrin, 1903

François Mocquard (1834–1917) was a French herpetologist. He originally studied medicine and mathematics, but made a career change to the natural sciences when middle-aged. He did much research on reptiles in Mexico and Central America (1870–1909). With Bocourt and Duméril he published: Études sur les reptiles. Mission scientifique au Mexique et dans l'Amerique Centrale. Recherches zoologiques pour servir a l'histoire de la fauna de l'Amerique Centrale et du Mexique (1883). Four amphibians and at least 18 reptiles are also named after him.

Modigliani

Loach sp. *Homalopterula modiglianii* A Perugia, 1893

Emilio Modigliani (1860–1932) was an Italian explorer, zoologist and anthropologist who collected in Sumatra (1886–1894). Four reptiles, three birds and two dragonflies are named after him, as is an amphibian that was only described after the museum specimens were close to 100 years old!

Moebius

African Schilbid Catfish sp. *Schilbe moebiusii* GJ Pfeffer, 1896

Professor Dr Karl August Möbius (1825–1908) was a German zoologist and pioneer in the field of ecology. He qualified as a teacher at the age of 19 but later (1849) studied natural science and philosophy at Humboldt University, Berlin. After graduating, he taught sciences in high school. He opened Germany's first seawater aquarium (1863). After obtaining his doctorate from the University of Halle (1868) he was appointed Professor of Zoology at the University of Kiel and Director of its Zoological Museum. His: *Die Auster und die Austernwirtschaft* (1870) broke new ground in describing the ecosystem interactions between organisms in oyster beds. He became Director of the Zoology Museum in Berlin (1888) and Professor of Systematics and Geographical Zoology at the Kaiser Wilhelm University, teaching there until (1905) retiring aged 80!

Moeller

Eelpout sp. *Pyrolycus moelleri* ME Anderson, 2006

Dr Peter D Rask Møller is Associate Professor at the Natural History Museum of Denmark, and worked at the Zoological Museum, University of Copenhagen. He has written more than 175 papers such as: *Distribution and abundance of eelpouts* (*Pisces, Zoarcidae*) *in West Greenland waters* (1999). He was honoured for his contributions to eelpout systematics.

Moenkhaus

Characin genus *Moenkhausia* CH Eigenmann, 1903

Characin sp. *Knodus moenkhausii* CH Eigenmann & CH Kennedy, 1903

Dr William J Moenkhaus (1871–1947) was an American geneticist and ichthyologist who became Professor of Physiology at Indiana University Medical School (1904–1941), where he was Eigenmann's colleague. He was awarded both his bachelor's degree (1894) and master's (1895) by Indiana University. He was at Harvard (1896–1898) and worked as Assistant Director at the Museu Paulista, São Paulo, Brazil (1899). He transferred to the University of Chicago (1900) where he was awarded his doctorate (1903). He wrote: *Description of a New Species of Darter from Tippecanoe Lake* (1903).

Moesch

Bagrid Catfish sp. *Pseudomystus moeschii* GA Boulenger, 1890

Casimir Moesch (aka Mösch) (1827–1899) was a Swiss geologist and palaeontologist. Originally intending to be a pharmacist, he studied in Munich and switched to natural sciences. He moved to Zurich (1864), was appointed Director of the zoological collection of higher animals at the Polytechnic (1866), and lectured at the University of Zurich (1868–1874) and at Jena. He travelled to Sumatra (1889) and wrote a report: *To and from the pepperland, pictures and naturalistic sketches of Sumatra* (1897). He collected the holotype of this species.

Moffitt

Goatfish sp. *Parupeneus moffitti* JE Randall & RF Myers, 1993

Robert B Moffitt worked at the Honolulu Laboratory, Southwest Fisheries Science Center, National Marine Fisheries Service. Among his publications are: *Deepwater Demersal Fish* (1993) and *Habitat and Life History of Juvenile Hawaiian Pink Snapper, Pristipomoides filamentosus* (1996). He was honoured in the name because he collected and photographed the first specimen.

Mogk

Lefteye Flounder sp. *Engyprosopon mogkii* P Bleeker, 1854

Dr C W F Mogk was a Surgeon Lieutenant Colonel in the Dutch East Indies, stationed in Samarang, North Celebes (1853). He discovered this species.

Mohnike

Japanese Seahorse *Hippocampus mohnikei* P Bleeker, 1853

Otto Gottlieb Johann Mohnike (1814–1887) was a German physician and naturalist. He was educated in medicine at Greifswald and Bonn Universities, then was a Dutch military physician (1844–1869) afterward practicing in Bonn. He is particularly remembered for implementing the first successful nationwide smallpox vaccination in Japan. Among his publications is: *Die Cetoniden der Philippinischen Inseln* (1873).

Moïse

Characin sp. *Moenkhausia moisae* J Géry, P Planquette & P-Y Le Bail, 1995

Moïse Berniac-Bereau is a French microbiologist. She was a senior researcher with INRA (Institut national de la Recherche agronomique, Guyana) from its inception (1975).

Mojica

Suckermouth Catfish sp. *Astroblepus mojicai* CA Ardila Rodríguez, 2015

Dr José Iván Mojica Corzo is a Colombian zoologist, ecologist and marine biologist who is Professor and Director, Museo de Ictiología del Instituto de Ciencias Naturales de la Universidad Nacional de Colombia, Bogotá. Pontificia Universidad Javeriana, Bogotá, awarded his bachelor's degree (1986) and the University of Madrid his doctorate (2009). He wrote: *Decreto da vía libre a cría de especies invasoras* (2016)

Mok

Hagfish sp. *Eptatretus moki* CB McMillan & RL Wisner, 2004

Dr Michael Hin-Kiu Mok (b.1947) is a Taiwanese marine biologist and ichthyologist who is a Research Associate, Vertebrate Zoology, Ichthyology, AMNH and after his retirement has become an Adjunct Professor, Department of Oceanography, National Sun Yat-sen University. The National Taiwan University awarded his bachelor's degree (1967) and the City University of New York his master's (1975) and his PhD (1978). After being a Post-Doctoral Research Fellow at the Harbor Branch Oceanographic Institution, Florida (1978–1980), he taught at Ocean College Taiwan (1980–1981). He joined National Sun Yat-Sen University, Taiwan (1981) and was Professor, Institute of Marine Biology (1983–2014), Dean of the College of Marine Sciences (1992–1998), Director of the Institute of Marine Biology (1984–1989) and of the Institute of Undersea Technology (1966–1999). Among his publications is: *Osteology and phylogeny of squamipinnes* (1983) and *Study on the Accumulation of Heavy Metals in Shallow-water and Deep-sea Hagfishes* (2010), as well as a number of papers describing new hagfish species. He was honoured in the binomial for his many outstanding contributions to our knowledge of hagfishes.

Mol

Long-whiskered Catfish sp. *Pimelabditus moli* BM Parisi & JS Lundberg, 2009

Dr Jan H A Mol. (See **Jan Mol**)

Molavi

Stone Loach sp. *Paracobitis molavii* J Freyhof, HR Esmaeili, G Sayyadzadeh & M Geiger, 2014

Jalal ad-Din Muhammad Balkhi (1207–1273), also known as Molavi, was a Persian poet, theologian and Sufi mystic.

Molesworth

Blunt-nosed Snowtrout *Schizothorax molesworthi* BL Chaudhuri, 1913
[Syn. *Oreinus molesworthi*]

Brigadier Alec Lindsay Mortimer Molesworth (1881–1939) was an officer in the 8th Gurkha Rifles in the Indian Army (1904–1937), a naturalist and collector. A bird is also named after him.

Molina, AJ

Snailfish sp. *Paraliparis molinai* DL Stein, R Meléndez C & I Kong U., 1991

Abate Juan Ignacio Molina (1740–1829) is considered "*...the first Chilean naturalist and ichthyologist*". He was also a Jesuit priest, and was forced to leave Chile (1768) when the Jesuits were expelled from the Spanish Empire. He moved to Bologna, Italy, and became a professor of natural sciences there.

Molina, C

Three-barbeled Catfish sp. *Cetopsorhamdia molinae* CW Miles, 1943

Dr Ciro Molina Garcés (1891–1953) was Secretary of Agriculture and Development, Valle del Cauca, Colombia. HIs doctorate in philosophy and letters was awarded by the College of Our Lady of the Rosary (1914). He was in advance of his time in understanding the value of systematic research in all branches of science to the state and national economies of Colombia.

Moller

Moller's Lantern Shark *Etmopterus molleri* GP Whitley, 1939
[Alt. Slendertail Lanternshark]

Captain Knud Moller was in command of the trawler that brought the holotype to the surface (1933). He commanded a number of trawlers in his career, including 'Durraween' (1929), and 'Ben Bow' (1947) and is said to have been assiduous in remarking unusual specimens and sending them to the Australian Museum in Sydney.

Mollison

Mollison's Pipefish *Mitotichthys mollisoni* EOG Scott, 1955

Bruce Mollison found the holotype (1953) and deposited it with the Queen Victoria Museum, Launceston, Tasmania. He may well have been a professional fisherman, as the pipefish 'came up fastened to handline' on Mollison's boat. His other catch at the time included '70 lbs. of flathead, box of perch' and other fish.

Moltrecht

Moltrecht's Minnow *Aphyocypris moltrechti* CT Regan, 1908

Dr Arnold Christian-Alexander Moltrecht (1873–1952) was a Latvian entomologist and lepidopterist who qualified at Tartu University, Estonia, as a physician and ophthalmologist (1899). He worked in St Petersburg before being put in charge (1906) of mobile eye clinics in Russia's Far East. Based in Vladivostock, he travelled in the Amur region, parts of China, and visited Formosa (Taiwan) to collect birds, butterflies and fishes (1907). An amphibian is also named after him.

Mona

Pricklefish sp. *Stephanoberyx monae* TN Gill, 1883

No etymology is given in the original description, but apparently named after the author's niece, Mona.

Monard

African Characin sp. *Nannocharax monardi* J Pellegrin, 1936

Professor Dr Albert Monard (1886–1952) was a Swiss naturalist and explorer who made six expeditions to Africa (1928–1947), mostly to Angola. He taught at a high school in La Chaux-de-Fonds and was Curator of its Natural History Museum (1920–1952). His best-known work was: *The Little Swiss Botanist* (1919), still used as a school textbook in the French-speaking cantons of Switzerland. Two mammals and a reptile are also named after him as well as five Odonata.

Mondolfi

Pencil Catfish sp. *Trichomycterus mondolfi* LP Schultz, 1945

Professor Dr Edgardo Mondolfi (1918–1999) was a Venezuelan mammologist, ecologist and biologist. He was Professor, Metropolitan University, Caracas. He had a lifelong interest in manatees and made studies of their distribution. He was on the board of several conservation bodies and, in later life, was Venezuela's Ambassador to Kenya. He co-wrote: *Jaguar* (1993). He died from dengue fever. A mammal and two amphibians are named after him.

Monica

Pike-Cichlid sp. *Crenicichla monicae* SO Kullander & HR Varella, 2015

Characin sp. *Moenkhausia monicae* MMF Marinho, FCP Dagosta, P Camelier & FCT Lima, 2016

Dr Mônica Toledo-Piza Ragazzo is a Brazilian zoologist and ichthyologist at Universidade de São Paulo where she is a Professor in the Department of Zoology. The Universidade de São Paulo awarded her bachelor's degree (1988) and her master's (1992), and the City University of New York her doctorate (1997) which she followed with post-doctoral studies at Universidade de São Paulo (1997–1999). She co-wrote: *Skeletal development and ossification sequence of the characiform Salminus brasiliensis (Ostariophysi: Characidae)* (2014).

Monika (Bacardi)

Cenderawasih Fusilier sp. *Pterocaesio monikae* GR Allen & MV Erdmann, 2008

Lady Monika Gomez Del Campo Bacardi (b.1968) is an Italian film producer, philanthropist and avid marine conservationist. She is co-founder of the AMBI Group of companies and is married to Lord Luis Bacardi descendant of the founder of the Bacardi company. She was honoured as an "*avid marine conservationist who successfully bid to support the conservation of this species at the Blue Auction in Monaco on 20 September 2007 and has given generously to support Conservation International's Bird's Head Seascape initiative.*"

Monika (Etzel)

Lampeye Killifish sp. *Aplocheilichthys monikae* HO Berkenkamp & V Etzel, 1976 NCR [Junior Synonym of *Rhexipanchax schioetzi*]

Neotropical Rivuline sp. *Cynodonichthys monikae* HO Berkenkamp & V Etzel, 1995

Monika Etzel is the wife of the junior author, Vollrad Etzel (q.v.) (Also see **Etzel**).

Monke

Dotted Catfish *Parauchenoglanis monkei* L Keilhack, 1910

Dr H Monke collected the holotype in Cameroon. No further details are known.

Monod

Grass-eater (characin) *Ichthyborus monodi* J Pellegrin, 1927

Guinean Tongue-Sole *Cynoglossus monodi* P Chabanaud, 1949

Atlantic Rubyfish *Erythrocles monodi* M Poll & J Cadenat, 1954

Snake-eel sp. *Apterichtus monodi* C Roux, 1966

Goosefish sp. *Lophiodes monodi* Y Le Danois, 1971

Théodore André Monod (1902–2000) was a French naturalist, humanist and philosopher who was most well-known for his treks through the Sahara Desert. After studying for his bachelor's and master's degrees he became an Assistant (1921) in the ichthyology department at the Museum of Natural History (MNHN) in Paris. Here he was often assigned work in the Sahara, usually in Mauritania. He rose to become a professor at the MNHN, and the founder of the Institut Français d'Afrique Noire, Dakar, Senegal where he was Director (1938–1965). Monod was an activist in pacifist and anti-nuclear campaigns. He was a protestant who was keen to bring all faiths together. He wrote many papers and two books, including: *Révérence à la vie* (1999). A bird is also named after him.

Montag

Tetra sp. *Hyphessobrycon montagi* FCT Lima, DP Coutinho & WB Wosiacki, 2014

Dr Luciano Fogaça de Assis Montag (b.1975) is a Brazilian zoologist who is a Professor at the Federal University of Pará. He is also a researcher at the Emílio Goeldi Museum. The Pontifical Catholic University of São Paulo awarded his bachelor's degree (1997), the Federal University of Pará his master's (2001), and the Federal University of Pará his doctorate (2006). His post-doctoral degree was awarded at Texas A&M University (2018). His research focus is on the ecology, natural history and conservation of fish from Amazon streams. Among his publications are: *Contrasting associations between habitat conditions and stream aquatic biodiversity in a forest reserve and its surrounding area in the Eastern Amazon* (2019) and *Land cover, riparian zones and instream habitat influence stream fish assemblages in the eastern Amazon* (2019). He collected part of the type series.

Montagu

Montagu's Blenny *Coryphoblennius galerita* Linnaeus, 1758
Montagu's Seasnail *Liparis montagui* E Donovan, 1804
Spotted Ray *Raja montagui* HW Fowler, 1910

Colonel George Montagu (1751–1815) was a soldier and natural history writer. He attained the rank of Captain in the British army whilst still young, and later served in the American Revolution as a Lieutenant-colonel in the English militia. His military career was curtailed when he was court-martialled and cashiered for causing trouble among his brother officers by what was described as 'provocative marital skirmishing'. Montagu then devoted himself to science, particularly biology. In his own words; '*I have delighted in being an ornithologist from infancy, and, was I not bound by conjugal attachment, should like to ride my hobby to distant parts*.' Montagu was among the first members of the Linnean Society. He was also an expert on shells and wrote (1803) *Testacea Britannica, a History of British Marine, Land and Freshwater Shells*. He wrote many papers on the birds of southern England, but his greatest work was the *Ornithological Dictionary or Alphabetical Synopsis of British Birds* (1802). He was renowned for his meticulous work and observations bordering on the clinical. He died of lockjaw (tetanus) after stepping on a rusty nail. Fowler named the Spotted Ray after Montagu as the latter had described the species but had used a pre-occupied name (*Raja maculata*). His name became attached to the blenny when John Fleming named a species (1828) as *Blennius montagui*, but this was later shown to be a junior synonym of *Blennius* (now *Coryphoblennius*) *galerita*.

Montalban

Indonesian Greeneye Spurdog *Squalus montalbani* GP Whitley, 1931
[Alt. Philippine Spurdog]

Whitley's original description gives no etymology. However, as the binomial is a replacement name for *S. philippinus* (as that was pre-occupied), it may well be intended to honour Dr. Heraclio R Montalban. He was Director of Fisheries in the Philippines and wrote, among other works: *Pomacentridae of the Philippine Islands* (1927).

Montano

Montano's Rockskipper *Alticus montanoi* H-E Sauvage, 1880
Philippine Barb sp. *Barbodes montanoi* H-E Sauvage, 1881

Dr Joseph Montano (1844–c.1886) was a French anthropologist who spent some time undertaking a scientific survey in the Philippines (1879–1881). He accompanied Don Joaquin Rajal, the Spanish Governor of Davao, on an expedition to Mount Apo on Mindanao (1880), the first ascent of the peak. Montano wrote: *Voyages aux Philippines et en Malaisie* (1886). A bird is also named after him.

Monte

Croaker sp. *Plagioscion montei* L Soares & L Casatti, 2000

Professor Sebasti*ão* Monte was a former head of the Department of Oceanography and Limnology at the Universidade Federal do Rio Grande do Norte, Brazil.

Monteiro

Monteiro's Bulldog Mormyrid *Marcusenius monteiri* A Günther, 1873

Joachim John (João José) Monteiro (1833–1878) was a Portuguese mining engineer and entomologist who had English ancestors – hence 'John' and alternatives to it. He collected natural history specimens in Angola, including the mormyrid holotype (1860–1875). He wrote: *Angola and the River Congo* (1875). Five birds are named after him.

Montes

Pencil Catfish sp. *Trichomycterus montesi* CA Ardila Rodríguez, 2016

Andrés Camilo Montes Correa is a Colombian research biologist and herpetologist in the Faculty of Science at the Universidad del Magdalena where he began studying in 2010. He has already published more than thirty papers, mostly on herpetology, such as: *Herpetofauna del campus de la Universidad del Magdalena, Santa Marta – Colombia* (2015). He collected the catfish holotype with the author.

Montezuma

Montezuma Swordtail *Xiphophorus montezumae* DS Jordan & JO Snyder, 1899

Montezuma II (1480–1520), also known as Moctezuma, was Emperor of the Aztecs at the time of the Spanish conquest of Mexico. He also has two birds, an amphibian and a reptile named after him.

Mooi

Mooi's Dottyback *Pseudochromis mooii* AC Gill, 2004

Dr Randall D Mooi is (2004–present) Curator of Zoology, Manitoba Museum (Winnipeg), Canada. The University of Totonto awarded his PhD, after which he was a post-doctoral fellow at the Division of Fishes at the Smithsonian Institution investigating the biology of deep-sea fish families and examining relationships of perch-like fishes. He was then Curator of Fishes and Section Head of Vertebrate Zoology at the Milwaukee Public Museum. He undertook several expeditions to Indo-Pacific coral reefs. Gill described him as a 'good friend and colleague' and said in his etymology that he has been "*...a constant source of encouragement throughout this study, and who has contributed significantly to our understanding of the systematics and biogeography of perciform fishes*".

Moolenburgh (Mrs)

Glass Catfish sp. *Pseudeutropius moolenburghae* M Weber & LF de Beaufort, 1913

Mrs. Moolenburgh, who supplied with her husband, Pieter Moolenburgh (below), a large collection of fishes from Sumatra to the authors, including eponymous type.

Moolenburgh, PE

Sheatfish (catfish) sp. *Hemisilurus moolenburghi* M Weber & LF de Beaufort, 1913

Pieter Eliza Moolenburgh (1872–1944) was an ethnographer and government official in the Dutch East Indies. He is recorded as being in Dutch New Guinea (West Papua) (1901) and having discovered some rock engravings there (1903). He gave a large collection of fishes from Sumatra to the authors, presumably including the holotype of this one. (See also **Moolenburgh (Mrs)**)

Moon

Titicaca Pupfish sp. *Orestias mooni* VV Tchernavin, 1944

Harold Philip Moon (1919–1982) was an English biologist. He made a detailed study of Lake District waters while a student at Kings College, Cambridge (1929–1932). He spent time at the Freshwater Biological Association (1930s) where he carried our pioneering work on littoral invertebrates and was on the FBA council off and on for some years (1946–1963). He also surveyed the River Avon and southern England watermeadows whilst at Southampton University. He later became Professor of Zoology at Leicester University. Among his published works was a biography of the explorer and scientist *Henry Walter Bates FRS, 1825–1892* (1976). He was a member of the Percy Sladen Trust Titicaca Expedition (1937) during which the type was collected.

Moore, CB

Speckled Stargazer *Dactyloscopus moorei* HW Fowler, 1906

Clarence Bloomfield Moore (1852–1936) was an American archaeologist, particularly noted for his studies of Native American sites. He accessed some of these sites by water, in his own steamboat – the 'Gopher'. He presumably helped finance Fowler's trip to the Florida Keys during which the stargazer was discovered, as the etymology notes of Moore that "… *through whose interest in Zoology the expedition to the Florida Keys was realized.*"

Moore, GA

Yellowcheek Darter *Etheostoma moorei* EC Raney & RD Suttkus, 1964

Dr George Azro Moore (1899–1998) was an American zoologist. His BSc was awarded by the Oklahoma Agriculture & Mechanical College and his MSc (1942) by the University of Oklahoma with his dissertation being *Studies on the Adaptations of Fishes of Silty Waters of the Great Plains*. He became Professor of Zoology at Oklahoma State University and President of the American Society of Ichthyologists and Herpetologists.

Moore, GE

Snake-eel sp. *Yirrkala moorei* JE McCosker, 2006

Dr Gordon Earle Moore (b.1929) is an American businessman, billionaire and philanthropist. Along with his wife, Betty, he has made significant contributions to conservation. They also made a generous donation to the University of Cambridge for the construction of a new library for the physical sciences, technology and mathematics. Moore was Director of Development at Fairchild Semiconductor (1960s) and made very accurate predictions concerning the growth of computing power, which have become known as Moore's Law. This, combined with Intel Corporation's microprocessor, is the basis for today's microcomputer revolution. Moore is Chairman Emeritus of Intel Corporation, which he cofounded (1968). He and his wife set up (2000) the Gordon E and Betty I Moore Foundation to fund scientific, educational and environmental ventures. An amphibian, a bird and a mammal are also named after them. McCosker honoured him in the name for "*...his interest in fishes, his love of fishing, and his support of biodiversity research and conservation.*"

Moore, JES

Mottled Spiny Eel *Mastacembelus moorii* GA Boulenger, 1898
Blunthead Cichlid *Tropheus moorii* GA Boulenger, 1898
Lake Tanganyika Cichlid sp. *Variabilichromis moorii* GA Boulenger, 1898
Lake Rukwa Minnow *Raiamas moorii* GA Boulenger, 1900
Lake Malawi Cichlid sp. *Gephyrochromis moorii* GA Boulenger, 1901
Hump-head Cichlid sp. *Cyrtocara moorii* GA Boulenger, 1902

John Edmund Sharrock Moore (1870–1947) was a British biologist, zoologist and cytologist who studied at the Royal College of Science, South Kensington (now part of Imperial College, London). He worked at the marine research station in Naples (1893–1894). He was leader of the first two Tanganyika (Tanzania) Expeditions (1894–1897 & 1899–1900). He was the first person to reach the snowline of the Ruwenzori Mountains and proved the existence of permanent glaciers there. He was Professor of Experimental and Pathological Cytology, University of Liverpool (1906–1909). He wrote: *The freshwater fauna of Lake Tanganyika* (1897). He is also remembered for creating the biological term 'synapsis' (1892) and being co-inventor of the term 'meiosis' (1905). He collected the cichlid types at Lake Malawi.

Moore, JP

Jaguar Dottyback *Pseudochromis moorei* HW Fowler, 1931

John Percy Moore (1869–1965) was an American zoologist who specialised in leeches. He became (1902) Assistant Curator at the Academy of Natural Sciences, Philadelphia (where Fowler also worked). He was also an instructor in zoology at the Marine Biological Laboratory at Woods Hole, Massachusetts. An amphibian is also named after him.

Moore, RA

Mekong Catfish sp. *Micronema moorei* HM Smith, 1945
[Possibly a junior synonym of *Micronema* (or *Kryptopterus*) *cheveyi*]

R Adey Moore worked at the *Bangkok Times*, and served for many years as the honorary secretary of the Siam Society. Smith named the catfish after him "*...in slight recognition of his sustained interest in the promotion of zoological science in Thailand.*"

Moore, SLM

Characin sp. *Astyanacinus moorii* GA Boulenger, 1892

Spencer Le Marchant Moore (1850–1931) was an English botanist. He worked at Kew (1870–1879) and as an unofficial volunteer at the BNHH (1896–1931). He was a member of an expedition to the remoter parts of Western Australia (1894–1895). He helped collected the characin holotype on an expedition to the Matto Grosso, Brazil.

Moore, TJ

Cameroon Croaker *Pseudotolithus moorii* A Günther, 1865
Talagouga Mormyrid *Marcusenius moorii* A Günther, 1867

Thomas John Moore (1824–1892) was a British museum curator. He was an assistant at Lord Derby's menagerie and aviary (1843). When Lord Derby died (1851), the collection of stuffed animals was transferred to what became the new Free Public Museum of Liverpool with Moore appointed as the fist curator (1851–1892). He wrote a few reports and papers, the most important being: *Report on the Seals and Whales of the Liverpool District* (1889). He loaned Günther specimens collected by R B N Walker in Gabon, and the holotype of the croaker collected by J Lewis Ingram in Gambia.

Moore

Dorada (characin) *Brycon moorei* F Steindachner, 1878

We have been unable to identify this person and have little hope of doing so, as there is nothing in the original text or among Steindachner's known associates and collectors to even make a reasonable guess. It could be named after the same person Steindachner honoured in the thorny catfish name *Oxydoras* [= *Hemidoras*] *morei* (see **More**), but that person is also unknown.

Morales C

Characin sp. *Leporinus moralesi* HW Fowler, 1942

Dr Carlos Morales Macedo (1888–1952) was a biologist who was Professor and Director of the Museo de Historia Natural, Lima, Peru (1938–1947).

Morales S

Papaloapan Chub *Graodus moralesi* F de Buen, 1955
[Syn. *Notropis moralesi*]

Salvador Morales was a Mexican water resources engineer who helped collect the holotype.

More

Thorny Catfish sp. *Hemidoras morei* F Steindachner, 1881
[Syn. *Opsodoras morei*]

This binomial cannot be identified. It may be that *morei* was a mistake for *moorei*, as Steindachner named a characin *Brycon moorei* in 1878 (see above). However, who *moorei* represents is also unknown, so which spelling (if either) is the mistake?

Moreira, C

Three-barbeled Catfish sp. *Rhamdiopsis moreirai* JD Haseman, 1911

Professor Carlos Moreira (1869–1946) was a Brazilian zoologist who was a specialist in crustaceans. He became Director, Museu Nacional, Rio de Janeiro (1916). A bird and two amphibians are named after him.

Moreira, O

Scrapetooth Characin sp. *Parodon moreirai* LFS Ingenito & PA Buckup, 2005

Dr Orlando Moreira-Filho is a Brazilian zoologist, ecologist and ichthyologist. The Federal University of São Carlos, where he is currently Associate Professor 3, awarded his bachelor's degree (1976), master's (1963) and doctorate (1989). He co-wrote: *Chromosomal evidence of downstream dispersal of Astyanax fasciatus (Characiformes, Characidae) associated with river shed interconnection* (2009). He collected the type series, recognising them as a new species.

Moreland

Urchin Clingfish *Dellichthys morelandi* JC Briggs, 1955
New Zealand Flathead *Bembrops morelandi* JS Nelson, 1978

John 'Jock' Munne Moreland (1921–2012) was Curator of Fishes, Dominion Museum (now Te Papa), Wellington. He joined the museum after serving in the army in the Pacific and Egypt. He was awarded his BSc in zoology at Victoria University of Wellington (1958). He was appointed as Assistant Zoologist at the museum where he developed the bird collection, publishing: *A guide to the larger oceanic birds (albatrosses and giant petrel) of New Zealand waters* (1957). He then began studying fish and went on to become the first Curator of Fishes at the museum. He published a number of papers and several books, including (being a keen angler) the *New Zealand Sea Anglers' Guide* (1960). He named several new taxa and took part in a number of collecting trips and voyages. He was honoured in the name of the clingfish "*...for the loan of fishes he helped collect, and because of his interest in the shore fishes of New Zealand*" and in the flathead for "*...bringing the two specimens of the new Bembrops to my attention, for the loan of the material, and for reading the manuscript.*"

Moreno

Whale Catfish sp. *Cetopsidium morenoi* A Fernández-Yépez, 1972

José Moreno, according to the etymology "*...has collected fishes for science for over 20 years.*" Beyond assuming that he did his collecting in and around the Orinoco River and basin, we have been unable to progress further.

Moresby

Dark Blind Ray *Benthobatis moresbyi* AW Alcock, 1898

Captain Robert Moresby (1794–1854) of the Indian Navy was a hydrographer, marine surveyor and draughtsman. He surveyed dangerous waters and reefs in the 1820s and 1830s, including the most comprehensive survey of the Red Sea. After the completion of the Red Sea Survey, he was sent to chart various coral island groups lying across the track of India-to-South Africa trade routes. His health gave way and he had to give up surveying. He transferred to the Merchant Navy and worked for P&O, commanding their steamer 'Hindostan' (1842) and later commanded 'Ripon' which was on the England to Alexandria run (1852). Moresby came from a distinguished naval family: his elder brother was Admiral of the Fleet Sir Fairfax Moresby, after whom Port Moresby in Papua New Guinea was named.

Morgans

Flagfin Perchlet *Plectranthias morgansi* JLB Smith, 1961

Dr John Frederick Croil Morgans was a research assistant at the Zoology Department of the University of Cape Town. His BA was awarded by St Catherine's College, Cambridge (1949). At some point he moved to Zanzibar, as Smith records that Morgans sent him fish specimens from there. He took part in an underwater archeological expedition to the Greek island of Chios, for which he was diver/photographer and marine biologist. (See also **Croil**).

Mori, L & PP de Mori

Amazonian Cichlid sp. *Bujurquina moriorum* SO Kullander, 1986

Luis Mori Pinedo and Palmira Padilla de Mori of the Proyecto de Asentamiento Rural Integral (PARI) were jointly honoured for their assistance to ichthyologists collecting in the Jenaro Herrera region of Peru. (As the plural form *–orum* suggests, more than one 'Mori' is intended.)

Mori, T

Korean Lamprey *Eudontomyzon morii* LS Berg, 1931

Rainbow Gudgeon ssp. *Sarcocheilichthys nigripinnis morii* DS Jordan & CL Hubbs, 1925

Dr Tamezo Mori (1884–1962) was a Japanese ichthyologist, entomologist and ornithologist. He taught at Seoul Higher Common School for Keiji Imperial University, Seoul (1909–1945). He was expelled from Korea (1945) by the American authorities and became Director of Zoology at the Agricultural University Hyogo, retiring as Professor Emeritus (1961). Two birds are named after him.

Moricand

Brown-striped Grunt *Anisotremus moricandi* C Ranzani, 1842

Moïse Etienne (Stéfano) Moricand (1779–1854) was an amateur naturalist from Geneva, particularly interested in botany and malacology, who was also a museum administrator. He started out as a commercial traveller for Swiss watches, at twelve years of age. At some stage he began collecting plants and minerals in Tuscany for Italian scholars, continuing when he returned to Geneva (1814) where he met Augustin Pyramus de Candolle (q.v.) who encouraged him to study botany. He studied the plants of South America, publishing *Plantes nouvelles d'Amérique* (1833–1846) on the collections of others. Much of the material went to Geneva Natural History Museum where he became treasurer and secretary. Part of the collection was zoological, including many fish from Brazil, but his herbarium numbered 54,000 species at his death. Throughout this time he continued to collect widely.

Morichen

Cutlassfish sp. *Evoxymetopon moricheni* R Fricke, D Golani & B Appelbaum-Golani, 2014

Mordechai 'Mori' Chen discovered this species while snorkeling in the Gulf of Aqaba. He found the (dead) holotype, at first imagining the long, shiny object to be a strip of metal.

Morishita

Shrimp-Goby sp. *Amblyeleotris morishitai* H Senou & Y Aonuma, 2007

Osamu Morishita is a diver and underwater photographer who discovered this goby.

Morlet

Hora's Loach *Yasuhikotakia morleti* G Tirant, 1885
[Alt. Skunk Botia]

Laureut-Joseph Morlet (1823–1893) was a French biologist and conchologist who worked at MNHN Paris. He wrote: *Diagnoses de mollusques terrestres et fluviatiles du Tonkin* (1886).

Morrigu

Seapen Goby *Lobulogobius morrigu* HK Larson, 1983

Morrigu, or The Morrigan (thought to mean 'Phantom Queen'), is a war goddess in old Irish mythology. The name was given "in reference to the jaws and dentition" of this goby, which has several rows of sharp, pointed teeth.

Morris, P

Thorny Catfish sp. *Hemidoras morrisi* CH Eigenmann, 1925

Percival Morris was Eigenmann's assistant and interpreter in the Iquitos area of Peru.

Morris, T

Oaxaca Cave Sleeper *Caecieleotris morrisi* SJ Walsh & P Chakrabarty, 2016

Thomas L Morris is a "*...renowned cave diver and speleobiologist, intrepid explorer, and respected conservationist devoted to the protection of karst habitats and their associated biotas*". He is also a 'good friend and colleague' of the authors. He discovered this species and collected the type.

Morrison, H

Morrison's (Fairy-)Wrasse *Cirrhilabrus morrisoni* GR Allen, 1999

Hugh Morrison (b.1951) is the owner of Perth Diving Academy and has led diving expeditions to many parts of the world. He has also been collecting seashells since childhood.

Morrison, JPE

Morrison's Dragonet *Synchiropus morrisoni* LP Schultz, 1960

Dr Joseph Paul Eldred Morrison (1906–1983) was an American malacologist and collector. The University of Chicago awarded his bachelor's degree (1926), the University of Wisconsin his master's (1929) and doctorate (1931). He worked at the USNM (1934–1975), retiring as Associate Curator. A subspecies of bird is also named after him.

Morrison, S

Morrison's Mudbrotula *Dermatopsoides morrisonae* PR Møller & W Schwarzhans, 2006
Western Yellowfin Seabream *Acanthopagrus morrisoni* Y Iwatsuki, 2013

Susan 'Sue' M Morrison worked at the Fish Section, Aquatic Zoology, Western Australian Museum, Perth. As a Research Officer, she was seconded (2005) to the Western Australia Department of Fisheries. She co-wrote: *Echinoderms of the Dampier Archipelago, Western Australia* (2004).

Morrow, JE

Persian-Carpet Sole *Pardachirus morrowi* P Chabanaud, 1954
[Syn. *Aseraggodes morrowi*]

Dr James E Morrow was an American ichthyologist with a particular interest in the anatomy and life history of the billfishes. He took part in a number of expeditions by Yale's Bingham Oceanographic Laboratory (late 1940s to late 1950s). He also acted as unofficial 'curator of fishes' (1949–1960) until he left Yale. Among his published papers is: *Fishes from East Africa, with new records and descriptions of two new species* (1954)

Morrow, WC

False Jaguar Catfish *Liosomadoras morrowi* HW Fowler, 1940
Armoured Catfish sp. *Rineloricaria morrowi* HW Fowler, 1940

William C Morrow was Curator of Fishes at Philadelphia Academy of Natural Sciences and led their expedition to the Ucayali River basin, Peru (1937) during which the holotypes of these species were collected.

Mortensen

Bombay-Duck Lizardfish sp. *Harpadon mortenseni* JDF Hardenberg, 1933
Mortensen's Darter Dragonet *Callionymus mortenseni* Suwardji, 1965

Dr Ole Theodor Jensen Mortensen (1868–1952) was a Danish echinoderm biologist who was Professor of Zoology, University of Copenhagen. He graduated in theology (1890) and then took a doctorate (1898). He made zoological expeditions to the Faeroe Islands and to Siam (Thailand) (1899–1900). He was in New Zealand waters (1914–1915) on government vessels. He collected the holotype of the lizardfish. An amphibian is named after him.

Mortenthaler

Coral Red Pencilfish *Nannostomus mortenthaleri* H-J Paepke & K Arendt, 2001

Martin Mortenthaler (1961–2018) was an Austrian fish exporter and the owner of Aquarium Rio Momon, Iquitos, Peru. He collected the holotype. (See also **Martin (Mortenthaler)**)

Mortiaux

Lake Tanganyika Catfish sp. *Tanganikallabes mortiauxi* M Poll, 1943

T Mortiaux, a health worker in Albertville, Belgian Congo (Kalemie, Democratic Republic of the Congo), on the western shore of Lake Tanganyika, collected the holotype.

Mortimer

Kariba Tilapia *Oreochromis mortimeri* E Trewavas, 1966
Mortimer's Hap (cichlid) *Sargochromis mortimeri* G Bell-Cross, 1975

M A E Mortimer was a Research and Administrative Officer of the Zambian Department of Game and Fisheries. Trewavas honoured him for his work on the Tilapia (original genus) of Zambia; in addition, he arranged a 'memorable trip' to the Luangwa Valley for Trewavas so she could study this species in its environment. Among his publications from around that time are: *Fishing in Lundazi Dam* (1959), *Lake Lusiwasi, Northern Rhodesia* (1960) and *Fish production from a stream in Northern Rhodesia* (1965) as well as the National Resources Handbook: *The Fish and Fisheries of Zambia* (1965). He became Chief Fisheries Officer of the Zambian Department of Wildlife, Fisheries and National Parks, and has 'contributed much' to our knowledge of Zambian fishes.

Mosavi

Dwarf Cardinalfish *Apogon mosavi* G Dale, 1977

'Mosavi' is an acronym for Mount Saint Vincent and, in the author's words, "recognizes the College of Mt. St. Vincent's annual summer course in marine biology, conducted in the Bahamas." The author, George Dale, taught on such a course (1975) and studied the local cardinalfish. The college is located in New York.

Moseley

Beaked Sandfish sp. *Gonorynchus moseleyi* DS Jordan & JO Snyder, 1923

Dr Edwin Lincoln Moseley (1865–1948) was an American naturalist, botanist, teacher, meteorologist and collector in the Philippines, China and Japan. He taught at Bowling Green State University, Ohio (1914–1936), as Professor of Science until becoming Professor Emeritus (1936–1948). He collected the holotype of the sandfish.

Moser, HG

White-speckled Rockfish *Sebastes moseri* B Eitner, 1999

H Geoffrey Moser is an American fisheries biologist. He has published more than fifty papers such as: *Early life history of sablefish,* Anoplopoma fimbria, *off Washington, Oregon, and California, with application to biomass estimation* (1994) and *Larvae and Pelagic Juveniles of Blackgill Rockfish,* Sebastes melanostomus, *Taken in Midwater Trawls off Southern California and Baja California* (2011). He was honoured "*…in recognition of his extensive contributions to marine fish biology and his many years of service as a U.S. National Marine Fisheries Service biologist. He also is a nice guy who loves rockfishes.*"

Moser, JF

Barfin Flounder *Verasper moseri* DS Jordan & CH Gilbert, 1898
Snake-eel sp. *Apterichtus moseri* DS Jordan & JO Snyder, 1901

Jefferson Franklin Moser (1848–1934) was a US Navy officer who was commander of the Fish Commission steamer 'Albatross'. He entered (1864) the U.S. Naval Academy and graduated as Midshipman (1868). He was promoted to Ensign (1869); Master (1870);

Lieutenant (1872); and Lieutenant Commander (1893) and later Rear Admiral until retiring (1904). He was employed on special duty under the Government on expeditions looking for inter-oceanic canal routes on the Isthmus of Darien and in Nicaragua, and also on the US Coast and Geodetic Survey and the Fish Commission. He commanded the US Bureau of Fisheries steamer Albatross during cruises in Alaskan waters (1897 & 1898); a report of which was published (1899). He was on the Albatross, (1900–1901) when it visited most of the canneries and salmon streams on the Alaskan coast. He made sketch maps of the streams with their tributary lakes, added to the chart of Bristol Bay, and made reconnaissance charts of Alitak Bay, the southwestern coast of Kodiak Island, and Afognak Bay. Moser is credited with having first reported and named many places and features in Alaska, and a number were later named after him. On retirement he became the General Superintendent of the Alaska Packers Association of San Francisco, during which he had charge of a large fleet of fishing steamers and the largest fleet of deep-sea sailing vessels in the world. He wrote: *The Salmon and Salmon Fisheries of Alaska* (1897) and *Alaska Salmon* (1900–1901). He was honoured in the snake-eel name for his valued services to ichthyology.

Moskalev

Moskalev's Tadpole (snailfish) *Careproctus moskalevi* AP Andriashev & NV Chernova, 2011

Dr Lev Ivanovich Moskalev (b.1935) is a Russian hydrobiologist at the Shirshov Institute of Oceanology. With interests in malacology and ichthyology, he has published widely, such as *On the Generic Classification in the Family Acmaeidae (Gastropoda, Prosobranchia) Based on the Radula* (1966) and *Neopilina starobogatovi, a new monoplacophoran species from the Bering Sea, with notes on the taxonomy of the family Neopilinidae (Mollusca: Monoplacophora)* (2007). Several other marine organisms are named after him and he has described 17 new species.

Moszkowski

Labeo sp. *Labeo moszkowskii* E Ahl, 1922

Dr Max Moszkowski (1873–1939) was a German physician, botanist and ethnologist. He travelled to Ceylon (Sri Lanka) and Sumatra (1907), then undertook an expedition to New Guinea (1910–1913) in the area of the Van Rees Mountains and the Mamberano River. He tried to reach the Snow Mountains in Central New Guinea but failed due to food shortages. He wrote the articles: *Expedition zur Erforschung des Mamberamo in Hollandish NeuGuinea* and *Wirtschaftsleben der primitiven Völker* (1911). Being interested in tribal customs and languages, he wrote *Wörterverzeichnisse der Sprachen vom Zentralgebirge, vom Südfluß, des Tori, des Sidjuai, des Borumesu, des Pauwoi* (1913). He wrote a book: *Inst Unerforschte Neuguinea, Erlebnisse mit Kopfjägern und Kannibalen* (1928). Two birds are named after him. He collected the holotype of this species.

Mouchez

Rosefish *Helicolenus mouchezi* H-E Sauvage, 1875

Ernest Amédée Barthélemy Mouchez (1821–1892) was born in Spain, but made his career in the French Navy. He worked on hydrographic studies along the coasts of China and South America. He was appointed (1874) to lead a mission to observe the transit of Venus from St Paul Island in the Indian Ocean (where the rosefish occurs).

Mould

Blacknose Skate *Breviraja mouldi* JD McEachran & RE Matheson, 1995

B Mould noticed that McEachran & Matheson's name *Breviraja schroederi* (1985) was pre-occupied by *Breviraja schroederi* (Krefft 1968) and brought this to the describers' attention so they re-named the skate after him. The etymology only mentions his family name and initial. We think that he is Brian Mould, a biodiversity researcher and taxonomist at the University of Nottingham, who wrote *Classification of the recent Elasmobranchii* (1997). We have been completely unable to trace him and the University of Nottingham advised us that no one there remembers him and they have no written record that he was ever there at all.

Moulton, JM

Scorpionfish sp. *Parascorpaena moultoni* GP Whitley, 1961

Professor James M Moulton (b.1921) was Professor of Biology at Bowdoin College, Brunswick, Maine, USA, and specialised in acoustical biology. When he was Associate Professor of Zoology, he visited Queensland (1960–1961) "*...to study underwater noises made by animals.*" He collected the holotype during that visit.

Moulton, P

Moulton's Handfish *Sympterichthys moultoni* PR Last & DC Gledhill, 2009

Peter Moulton was an Australian fisheries biologist with the former Victorian Institute of Marine Science (now the Marine and Freshwater Resources Institute). According to the authors, his: "*...interest in temperate Australian fishes led to the collection of the first specimen of this species, as well as many specimens of hitherto poorly known handfishes.*"

Mourlan

Star Pearlfish *Carapus mourlani* G Petit, 1934

Roger Mourlan (1912–1987) made two documentary films on expeditions of which he was a member - Madagascar (1932) with Georges Petit, during which expedition the pearfish holotype was collected, and Marcel Griaule's Sahara-Sudan expedition (1935). He went on no more scientific expeditions, but continued to make documentaries, based with his father at their studio in Montfermeil (Paris). He wrote: *Mon premier film* (1933).

Mou-Tham

Moutham's Goatfish *Upeneus mouthami* JE Randall & M Kulbicki, 2006

Gerard Mou-Tham is a colleague of the second author. As a former professional fisherman, Gerard Mou-Tham joined ORSTOM, now IRD, as a scientific diver (1982) and worked with Michel Kulbicki on a variety of projects, collecting fish data throughout the Pacific. He: "...

assisted in the collection of many of the goatfishes in this report and provided two of the color photographs for our illustrations."

Mowbray

Bermuda Goby *Lythrypnus mowbrayi* TH Bean, 1906
Cave Bass *Liopropoma mowbrayi* LP Woods & RH Kanazawa, 1951

Louis Leon Mowbray (1877–1952) was a Bermudan naturalist who collected the types of both these species. He became (1914) superintendent of the New York Aquarium. He moved to Miami (1919–1923) where he ran a new aquarium, rejoined the New York Aquarium (1923–1926), then returned to Bermuda. He became (1928–1944) Director of the Bermuda Aquarium, Museum and Zoo. He was honoured by Bean for his 'intellegent and effective work' in collecting Bermudian fishes.

Moyer

Leopard Wrasse sp. *Macropharyngodon moyeri* JW Shepard & KA Meyer, 1978
Moyer's Dragonet *Synchiropus moyeri* MJ Zaiser & R Fricke, 1985

Dr Jack Thomson Moyer (1929–2004) was an American marine biologist who lived for most of his life on Miyake-jima, Japan, where he became Director of the Tatsuo Tanaka Memorial Biological Station. Later he became a teacher at the American School in Japan (ASIJ). He committed suicide. After his death, multiple allegations surfaced that Moyer had sexually molested students during his tenure at the ASIJ. ASIJ Board of Directors released a letter (June 2015) admitting that an independent investigation found that Moyer's abuse of students was extensive and had been covered up by faculty members for years.

Mozino

Kelp Poacher *Agonomalus mozinoi* NJ Wilimovsky & DE Wilson, 1979

Dr José Mariano Moziño aka Mociño (1757–1820) was a Mexican physician and naturalist. He joined (1791) the Royal Botanical Expedition, which explored the Pacific Northwest. During the expedition he created an important natural history collection. One of the world's most beautiful birds, the Resplendent Quetzal (*Pharomachrus mocinno*), derives its binomial from a version of Moziño's surname.

MRAC

Gabon Cichlid sp. *Chromidotilapia mrac* A Lamboj, 2002

This is an abbreviation for Musée Royal de l'Afrique Centrale (Tervuren, Belgium); which, according to the author, for "*...more than a hundred years this museum has been one of the most important institutions working on African fishes.*"

Mrakovi

Freshwater Goby sp. *Knipowitschia mrakovcici* PJ Miller, 2009

Milorad Mrakovic is a biologist at the University of Zagreb, Croatia. He provided the original type material.

Mrs Schwartz

Mrs Schwartz's Cory *Corydoras robineae* WE Burgess, 1983
[Alt. Bannertail Cory]

Robine Schwartz (See **Robine**).

Msaka

Lake Malawi Cichlid sp. *Placidochromis msakae* M Hanssens, 2004

This is a toponym referring to Msaka, Lake Malawi, Malawi, the type locality.

Muhlis

Marmara Sprat *Clupeonella muhlisi* W Neu, 1934

Professor Dr Rashad Muhlis Erkmen (1891–1985) was a Turkish Agricultural School Professor and politician (MP), who became the Minister of Agriculture of Turkey when the sprat (endemic to Apolyont Lake in Turkey) was discovered.

Muirapinima

Glass Knifefish sp. *Eigenmannia muirapinima* LAW Peixoto, GM Dutra & WB Wosiacki, 2015

The Muirapinima are an indigenous people who live in Pará, Brazil, near to where the knifefish holotype was collected.

Mulhall

Roundsnout Gurnard *Lepidotrigla mulhalli* W Macleay, 1884

Thomas Mulhall (d.1897) was Inspector in the Department of Fisheries in New South Wales. He supplied a number of fish to Macleay (often seeing them in Sydney fish market and recognising them as unusual or unknown) who wrote of the gurnard: "*I name it after Mr. Sub-Inspector Mulhall, to whom I am indebted for much of my knowledge of the Fishes of this country.*"

Mullan

Arabian Blackspot Threadfin *Polydactylus mullani* SL Hora, 1926

Jai Phirozshah Mullan (d.1957) was an entomologist at St. Xavier's College, Bombay, India. He discovered this species and sent specimens to Hora for study.

Müller FJH

Müller's Flounder *Arnoglossus muelleri* CB Klunzinger, 1872
Spot-fin Beachsalmon *Leptobrama muelleri* F Steindachner, 1878
Blackfin Coralfish *Chelmon muelleri* CB Klunzinger, 1879
Tufted Sole *Dexillus muelleri* F Steindachner, 1879
[Syn. *Brachirus muelleri*]
Müller's Rockskipper *Istiblennius muelleri* CB Klunzinger, 1879
Kimberley Catfish *Paraplotosus muelleri* CB Klunzinger, 1879
New Zealand Sprat *Sprattus muelleri* CB Klunzinger, 1879

Baron Dr Sir Ferdinand Jacob Heinrich von Müller (1825–1896) was a German-born Australian botanist, geographer, explorer, physician, and naturalist. He was born at Rostock and, after education in Schleswig, was apprenticed to a chemist (1840). He studied botany at Christian-Albrechts-Universität zu Kiel, receiving his doctorate (1847). He had intended to practice medicine but was advised to go to a warmer climate for his health and left for Australia (1847). He first found employment in Adelaide as a chemist and contributed a few papers on botanical subjects to German periodicals. He moved to Melbourne (1851) and travelled within Victoria (1848–1852), describing a large number of plants. After he sent a paper to the Linnean Society at London on "The Flora of South Australia" (1852), he was appointed Government Botanist (1853). He was expedition naturalist for the exploration of the Victoria River and other parts of North Australia, and he was one of the four who reached Termination Lake (1856), continuing with Gregory's expedition overland to Moreton Bay. He was a member of the Victorian Institute for the Advancement of Science, later renamed the Royal Society of Victoria (1854–1872). He was a member of the society's "Exploration Committee," which established the Burke and Wills expedition (1860). He was Director of the Melbourne Botanic Gardens (1857–1873) and the benefactor of explorer Ernest Giles, the discoverer of Lake Amadeus and Kata Tjuta. Giles had originally wanted to name both after Müller, who found that embarrassing and prevailed upon Giles to change his mind. He wrote the 11-volume *Fragmenta phytographica Australiae* (1862–1881). A bird and a reptile are named after him.

Müller, JP

Titicaca Pupfish sp. *Orestias mulleri* A Valenciennes, 1846
Characin sp. *Creagrutus muelleri* A Günther, 1859
Three-barbeled Catfish sp. *Rhamdia muelleri* A Günther, 1864
Ghost Knifefish sp. *Sternarchorhamphus muelleri* F Steindachner, 1881

Professor Johannes Peter Müller (1801–1858) was a German physiologist, ichthyologist and comparative anatomist. He entered the University of Bonn (1819), graduated and started teaching (1824), becoming professor (1830). He was Professor of Anatomy and Physiology at Humboldt University, Berlin (1833–1858) and a member of the Royal Prussian Academy of Science, Berlin. He wrote: *Handbuch der Physiologie des Menschen* (1833–1840). He also co-wrote a study on characins (1844), which is cited by Günther. Among his discoveries was (1835) that caecilians are amphibians and not snakes. A reptile and an amphibian are also named after him.

NB - Valenciennes gives no first names of the 'Muller' he is naming the pupfish after, and it is possible that Otto Friedrich Müller (below) was intended. However, Johannes Müller – along with Troschel (q.v.) – had written *Synopsis generum et specierum familiae Characinorum* (1844), and perhaps this prompted Valenciennes to name a South American fish after him.

Müller, OF

Mueller's Pearlside *Maurolicus muelleri* JF Gmelin, 1789
[Alt. Silvery Lightfish]

Otto Friedrich Müller (1730–1784) was a Danish naturalist who was employed by King Frederick V to continue compiling the *Flora Danica*, and Müller added two volumes to it.

He gave a brief description of this species in his *Zoologiae Danicae Prodromus* (1766), the first survey of the fauna of the combined kingdoms of Denmark and Norway.

Muller, S

Armoured Catfish sp. *Ancistrus mullerae* AG Bifi, CS Pavanelli & CH Zawadzki, 2009

Dr Sonia Fisch-Muller. (See **Sonia**)

Munduruku

Characin sp. *Bryconops munduruku* C Silva-Oliveira, ALC Canto & FR Ribeiro, 2015

The Munduruku Indians settled on the right margin of the Tapajós River, giving rise to what is now the city of Aveiro, Brazil, the type locality.

Mundy

Lizardfish sp. *Synodus mundyi* JE Randall, 2009
Bandfish sp. *Owstonia mundyi* WF Smith-Vaniz & GD Johnson, 2016

Bruce C Mundy is a fisheries biologist who works at the National Marine Fishery Service, Pacific Islands Fisheries Science Center, Honolulu. He wrote: *Checklist of the fishes of the Hawaiian Archipelago* (2005). He arranged for the bandfish's describers to receive the type specimens.

Munk

Munk's Devil Ray *Mobula munkiana* G Notarbartolo di Sciara, 1987
[Alt. Pygmy Devil Ray]

Walter Heinrich Munk (1917–2019) was an Austrian-born American oceanographer. His parents divorced when he was a child, and his family sent him to a boys' school in upper New York State (1932). He worked in a bank, which he hated, and studied at Columbia University for three years before resigning to go to the California Institute of Technology, which awarded his bachelor's degree in physics (1939). He went to Scripps Institute of Oceanography (1939) and completed a master's degree in geophysics (1940). However, he did not obtain his doctorate from the University of California, Los Angeles until 1947 as WW2 interrupted his progress. He became an American citizen (1939) and enlisted in the US Army as a private in the ski troops, but was later released to do research at Scripps in relation to amphibious warfare. He was involved in many important research projects, including the Mohole Project and analysis of the currents, diffusion, and water exchanges at Bikini Atoll in the South Pacific, where the USA was testing nuclear weapons. He developed plans for a La Jolla branch of the Institute of Geophysics (1956) and oversaw its construction (1959–1963). It is now the centre of the campus of the Scripps Institute of Oceanography, La Jolla, California. He became an Assistant Professor at Scripps (1947) and a full Professor (1954). He died of pneumonia at the age of 101.

Munro

Munro's Sole *Zebrias munroi* GP Whitley, 1966
Australian Spotted Mackerel *Scomberomorus munroi* BB Collette & JL Russo, 1980

Robust Pygmy-Stargazer *Crapatalus munroi* PR Last & GJ Edgar, 1987
Munro's Hardyhead *Craterocephalus munroi* LELM Crowley & W Ivantsoff, 1988
Munro's Goby *Glossogobius munroi* DF Hoese & GR Allen, 2012

Ian Stafford Ross Munro (1919–1994) was an Australian ichthyologist, aquarist and marine biologist. A large but shy man, he was often referred to as a 'gentle giant'. He graduated with a BSc from the University of Queensland (1941). Following military service and completing his MSc, he worked for CSIRO (1943–1984) retiring as Principal Scientist. He undertook extended cruises on the RV *Fairwind* studying marine fauna off New Guinea, and also worked for a while in Sri Lanka. For two decades he led the Gulf of Carpentaria Prawn Survey (1963–1984). He continued to research and write in retirement. He wrote around 100 papers and several books, including: *The Fishes of New Guinea* (1967) and *Handbook of Australian Fishes* (1956–1961) much of which he skilfully illustrated.

Mura

Armoured Catfish sp. *Otocinclus mura* SA Schaefer, 1997

Mura is the Portuguese name for an indigenous Brazilian people who inhabited part of the reaches of the Rio Solimões, Amazonas. They were a constant threat to river travel and fiercely resisted colonial assimilation. Frontier colonists consistently attacked them towards the end of the 18th century, and introduced diseases such as measles also took a heavy toll.

Murdy

Murdy's Sandgoby *Istigobius murdyi* DF Hoese & MV Erdmann, 2018

Dr Edward O Murdy is an American ichthyologist who is Adjunct Professor of Marine Biology at the Department of Biological sciences, George Washington University, and Senior International Analyst at the National Science Foundation. He has been (1996–1999) Director of the National Science Foundation, Tokyo. Among his publications is: *Field Guide to the Fishes of the Chesapeake Bay* (2013). He was honoured because he "pioneered research on the genus". (Also see **Rebecca**)

Muria

Brazilian Cichlid sp. *Australoheros muriae* FP Ottoni & WJEM Costa, 2008

This is a toponym referring to the Rio Muriaé basin in southeastern Brazil where the cichlid was discovered.

Murie, J

Ocellated Ctenopoma *Ctenopoma muriei* GA Boulenger, 1906

Dr James Murie (1832–1925) was a Scottish physician and naturalist. He accompanied John Petherick to Gondokoro, on the White Nile, as part of an expedition (1861–1862) to support Speke and Grant. Murie made collections of fishes and flora during this expedition. He later settled in London, becoming a lecturer on comparative anatomy at the Middlesex Hospital. He also wrote: *Report on the Sea Fisheries and fishing industries of the Thames Estuary* (1903).

Murie, OJ

Snake River Sucker *Chasmistes muriei* RR Miller & GR Smith, 1981

Dr Olaus Johan Murie (1889–1963) was an American zoologist, conservationist and ecologist. Pacific University, Oregon, awarded his bachelor's degree (1912). He was an Oregon State conservation official (1912–1920) and took part in expeditions to Hudson Bay and Labrador (1914–1917). He worked as a wildlife biologist for the US Bureau of Biological Survey (1920–1945) and was then Director of the Wilderness Society (1945–1963). He and his wife began the campaign (1956) to protect what is now the Arctic National Wildlife Refuge. He wrote: *The Elk of North America* (1951).

Muriel (Matallanas)

Snailfish sp. *Paraliparis murieli* J Matallanas, 1984

Muriel Matallanas is the daughter of the author, Jesús Matallanas. (It would have been more correct to use the feminine form in the binomial = *murielae*)

Muriel (Vanderbilt)

Searobin sp. *Prionotus murielae* L Mowbray, 1928

Muriel Vanderbilt (1900–1972) was the daughter of William Kissam Vanderbilt II (1878–1944). He collected the type and dedicated it to his daughter. She continued her father's interest in horse racing and was a noted owner and breeder. She married three times and divorced twice.

Murphy

Chilean Jack Mackerel *Trachurus murphyi* JT Nichols, 1920
[Alt. Inca Scad]

Robert Cushman Murphy (1887–1973) was an American naturalist who became Curator of Birds, AMNH, and was a world authority on marine birds. He is also famed for persuading Rachel Carson to write: *Silent Spring*, after he was unable to persuade the US government to stop spraying DDT. His: *Oceanic Birds of South America* (1936) was awarded the John Burroughs Medal for excellence in natural history and the Brewster Medal of the AOU. Murphy also spent time excavating the remains of moas in New Zealand. On Bermuda (1951) he 'slipped a noose onto a pole, slid it down a tunnel between some ocean side rocks, and pulled out a sea bird called a cahow' – the first living member of that species (Bermuda Petrel *Pterodroma cahow*) seen alive since the early 17th century. "*As a scientist*," he once said, "*I'd as soon have a louse named for me as a mountain.*" According to his obituary in the *New York Times*, there are also two mountains, a spider, a lizard and a louse named in his honour as well as a country park and a school in New York State. He collected the holotype of this jack mackerel.

Murray

Abyssal Rattail *Coryphaenoides murrayi* A Günther, 1878
Murray's Grideye Fish *Ipnops murrayi* A Günther, 1878
Murray's Skate *Bathyraja murrayi* A Günther, 1880

Murray's Armoured Gurnard *Paraheminodus murrayi* A Günther, 1880
Murray's Abyssal Anglerfish *Melanocetus murrayi* A Günther, 1887
Roughnose Grenadier *Trachyrincus murrayi* A Günther, 1887
Moustache Sculpin *Triglops murrayi* A Günther, 1888
Murray's Smooth-head *Conocara murrayi* E Koefoed, 1927
Roundhead Grenadier *Odontomacrurus murrayi* JR Norman, 1939
Murray's Deepsea Batfish *Halieutopsis murrayi* H-C Ho, 2021

Sir John Murray (1841–1914) was a Canadian marine naturalist often described as the founder of modern oceanography. He explored the Faroe Channel (1880–1882) and took part in and financed expeditions to Christmas Island. He was in charge of collections on the HMS 'Challenger' expedition (1874–1876) and edited: *Report on the Scientific Results of the Voyage of "HMS Challenger"* (1880–1895). A motorcar killed him as he was crossing a street. A bird and a reptile are named after him.

Murray Stuart

Torrent Catfish sp. *Amblyceps murraystuarti* BL Chaudhuri, 1919

Dr Murray Stuart worked for the Geological Survey of India. He collected the holotype and made a small collection of fishes in northern Burma (Myanmar). He was also interested in fossil fuels and wrote: *The Geology of Oil, Oil-Shale and Coal* (1926). An amphibian is named after him. (See also **Stuart**)

Murty

Sand-lance sp. *Bleekeria murtii* KK Joshi, PU Zacharia & P Kanthan, 2012

Dr V Sriramachandra Murty is a well-known fish taxonomist and marine fisheries scientist. The Government Arts College Rajahmundry awarded his BSc (1962), Vikram University his MSc (1965) and Andhra University his PhD (1973). He was (1965–2003) Head of Demersal Fisheries Division, Central Marine Fisheries Research Institute, Kochi, India.

Musafiri

African Rivuline sp. *Aphyosemion musafirii* R van der Zee & R Sonneberg, 2011

Dr Jean Musafiri is coordinator for the national tuberculosis and leprosy control programme in the 'Province Orientale Occidentale', a large forested area around Kisangani (DRC). He made it possible for A Van Deun, Institute of Tropical Medicine, Antwerp, to collect the type material of this rivuline.

Musick

Stardrum sp. *Stellifer musicki* NL Chao, A Carvalho-Filho & J de Andrade Santos, 2021

John 'Jack' Andrew Musick (1941–2021) was an American ichthyologist. Rutgers University awarded his first degree and Harvard his PhD. He undertook post-doctoral work at Woods Hole Oceanographic Institution, then began work as an instructor at what became the Virginia Institute of Marine Science (VIMS) at College of William and Mary, and where he worked for the next 42 years (1966–2008). There he established what is now the Nunnally Ichthyology Collection, which today contains approximately 350,000 specimens used for study and research by scientists around the world. He rose through every academic post until becoming Professor of Marine Science (1999) and head of Vertebrate Ecology and Systematics. He was the major professor of the senior author and many other students, including several Brazilian ichthyologists (this was what led to him being honoured in the stardrum's binomial). He wrote around 170 scientific papers, wrote several books and edited others, writing eight trade books, four of which were co-written by his wife Beverly McMillan, as science writer. They co-wrote: *The Shark Chronicles* (2003) about his life and times.

Musorstom

Searobin sp. *Lepidotrigla musorstom* L del Cerro & D Lloris, 1997
Grenadier sp. *Kumba musorstom* NR Merrett & T Iwamoto, 2000
Cusk-eel sp. *Neobythites musorstomi* JG Nielsen, 2002
Lanternbelly sp. *Acropoma musorstom* M Okamoto, JE Randall & H Motomura, 2021

This binomial commemorates the exploratory cruises to the Indo-West Pacific region jointly sponsored by ORSTOM and MNHN.

Mutambue

African Tetra sp. *Nannopetersius mutambuei* S Wamuini Lunkayilakio & E Vreven, 2008

Dr Mutambue Shango is a biologist and aquarist who is the General Academic Secretary and Professor, École Régionale Post-universitaire d'Aménagement et de Gestion intégrés des Forêts et Territoires Tropicaux, Kinshasa, Democratic Republic of Congo. He has doctorates (1984 & 1992) from the University of Toulouse, France. He wrote: *L' élevage intensif des poissons: Orechromis niloticus, Clarias gariepinus, Parachanna insignis* (1985). He collected many fishes from the Inkisi River basin (1985–1986).

Mutis

Pencil Catfish sp. *Eremophilus mutisii* FHA von Humboldt, 1805

Rev Dr José Celestino Bruno Mutis y Bosio (1732–1808) was a Spanish priest, mathematician and botanist. He graduated as a physician at the University of Seville (1755) and received his doctorate (1757). He was Professor of Anatomy, University of Madrid (1757–1760). He went to New Granada (now part of Colombia) (1761) as physician to the governor. He was leader of the Royal Botanical Expedition (1783–1808) during which period a large area was explored and assessed. Humboldt stayed with Mutis for two months (1801), before setting off overland to Ecuador.

Muysca

Headstander Characin sp. *Megaleporinus muyscorum* F Steindachner, 1900

The Muisca (also spelled Muysca) are an indigenous people of Colombia's Eastern Range, an area that includes the holotype locality.

Muzuspi

Armoured Catfish sp. *Hypoptopoma muzuspi* AE Aquino & SA Schaefer, 2010

The binomial honours a museum: the Museu de Zoologia, Universidade de São Paulo, Brazil (MZUSP).

Mvogo

Cameroon Tilapia sp. *Sarotherodon mvogoi* DFE Thys van den Audenaerde, 1965

Léon Mvogo is a hydrobiologist at Station de Pisciculture de Melen (Yaounde, Cameroon). He collected the type.

Mya

Mya's Klipfish *Pavoclinus myae* MS Christensen, 1978

Mya van Harten was honoured for her 'continuous help and support' during the course of Christensen's studies; she was his fiancée at the time.

Myers, GS

Goby genus *Myersina* AW Herre, 1934
Asian catfish genus *Myersglanis* SL Hora & EG Silas, 1952
Gambusia sp. *Gambusia myersi* E Ahl, 1925 NCR
[Junior Synonym of *Gambusia holbrooki*]
Snake-eel sp. *Bascanichthys myersi* AW Herre, 1932
Myers' Hillstream Loach *Pseudogastromyzon myersi* AW Herre, 1932
Flabby Whalefish sp. *Gyrinomimus myersi* AE Parr, 1934
Mexican Rivuline sp. *Cynodonichthys myersi* CL Hubbs, 1936
Paleback Goby *Gobulus myersi* I Ginsburg, 1939
Amazonian Cichlid sp. *Caquetaia myersi* LP Schultz, 1944
Armoured Catfish sp. *Hypostomus myersi* WA Gosline, 1947
Myers' Loach *Pangio myersi* RR Harry, 1949
Characin sp. *Astyanax myersi* A Fernández-Yépez, 1950
Pygmy Hatchetfish *Carnegiella myersi* A Fernández-Yépez, 1950
Hummingbird Lampeye *Poropanchax myersi* M Poll, 1952
Characin sp. *Planaltina myersi* JE Böhlke, 1954
Silver Rasbora *Rasbora myersi* MR Brittan, 1954
Myers' Clingfish *Tomicodon myersi* JC Briggs, 1955
Characin sp. *Lignobrycon myersi* A Miranda Ribeiro, 1956
Neotropical Rivuline sp. *Austrofundulus myersi* G Dahl, 1958
Myers' Icefish *Chionodraco myersi* HH DeWitt & JC Tyler, 1960
Gargoyle Cusk *Xyelacyba myersi* DM Cohen, 1961
Reef-sand Blenny *Ekemblemaria myersi* JS Stephens, 1963
Characin sp. *Rhinopetitia myersi* J Géry, 1964
Characin sp. *Brittanichthys myersi* J Géry, 1965
Characin sp. *Chrysobrycon myersi* SH Weitzman & JE Thomerson, 1970
Thorny Catfish sp. *Leptodoras myersi* JE Böhlke, 1970
Neotropical Rivuline sp. *Xenurolebias myersi* AL de Carvalho, 1971
Myers' Pomfret *Brama myersi* GW Mead, 1972
Myers' Pipefish *Minyichthys myersi* ES Herald & JE Randall, 1972

Toothless Characin sp. *Curimatopsis myersi* RP Vari, 1982
Myers' Grenadier *Coryphaenoides myersi* T Iwamoto & YI Sazonov, 1988
Bagrid Catfish sp. *Pseudomystus myersi* TR Roberts, 1989

Dr George Sprague Myers (1905–1985), an American ichthyologist, bio-geographer and herpetologist, was Professor Emeritus of Biological Sciences at Stanford, California. He was a keen natural historian with a life-long interest in fish and amphibians. He published his first paper on ichthyology aged 15, eventually writing over six hundred scientific papers and articles. He was a volunteer assistant at the AMNH, New York (1922–1924). He enrolled (1924) at Indiana University part-time, but when his sponsor fell ill he transferred to Stanford and graduated from there (1930), eventually completing his master's and doctorate (1933). He was Assistant Curator at the Smithsonian but was invited (1936) to return to Stanford as Assistant Professor in Biological Sciences and Curator of Zoological Collections. He developed courses in systematics for ichthyology and vertebrate palaeontology and was appointed full Professor (1938). During the Second World War he spent over two years in Brazil on US State Department funds to aid the Museo Nacional and Divisão de Caça e Pesca – a programme to maintain good relations with Latin America. He amassed an extensive library on ichthyology, herpetology, biogeography, the history of biology and exploration, and, as a sidelight, the American Civil War. After retirement (1970) he became Visiting Professor of Ichthyology at the Museum of Comparative Zoology, Harvard. Herre honoured him because he had been so helpful in Herre's study of Chinese fishes. A genus of reptile and two amphibians are also named after him.

Myers, PVN

Characin sp. *Roeboides myersii* TN Gill, 1870

Philip van Ness Myers (1846–1937) was an American academic and historian who became Professor of History and Political Economy, University of Cincinatti and was Dean there (1895). He was a member of James Orton's first Andean expedition (1867) during which the holotype was collected. Orton specifically requested this species be name after Myers, who co-wrote: *Life and Nature Under the Tropics* (1871). He also wrote a number of history books, such as: *Remains of Lost Empires* (1875) and *Outlines of Mediaeval History* (1886).

Myers, RF

Fairy Basslet sp. *Selenanthias myersi* JE Randall, 1995
Dottyback sp. *Lubbockichthys myersi* AC Gill & AJ Edwards, 2006

Robert F 'Rob' Myers (b.1953) is a coral-reef marine biologist and underwater photographer, who collected the dottyback holotype. He was honoured in its name for his important contributions to our understanding of Micronesian fish fauna. He started photographing marine life (1970) and has written or illustrated with his photographs a number of books such as: *Micronesian Reef Fishes: A Practical Guide to the Identification on the Coral Reef Fishes of the Tropical Central and Western Pacific* (1991), *Micronesian Reef Fishes: A Comprehensive Guide to the Coral Reef Fishes of Micronesia* (1999) and *Coral Reef Guide Red Sea* (2004). He currently serves with the IUCN (International Union for the Conservation of Nature) as a member of the Global Marine Species Assessment group. Current projects include a Caribbean field guide as well as fish-id applications.

Myrna

Semaphore Tetra *Pterobrycon myrnae* WA Bussing, 1974
Topaz Cichlid *Amatitlania myrnae* PV Loiselle, 1997

Myrna Isabel Bussing née López Sanchez (b.1937) is the widow of William Busing (q.v.) who named the tetra after her. She is an ichthyologist and DIrector, Zoological Museum, San Jose, Costa Rica. She co-wrote: *A new species of cichlid fish,* Cichlasoma rhytisma from the Río Sixaola drainage, Costa Rica (1983). Loiselle calls her a "…dedicated and enthusiastic student of Central American fishes whose invaluable assistance greatly contributed to the success of the Atlantic Coast Cichlid's collecting trip to Costa Rica."

Mysi

Snake-eel sp. *Apterichtus mysi* JE McCosker & Y Hibino, 2015

Mysi Dang Hoang is a Curatorial and Administrative Assistant at the California Academy of Sciences. She was a co-author of *Catalog of Fishes* (1998).

N

Naccari

Adriatic Sturgeon *Acipenser naccarii* CL Bonaparte, 1836

Fortunato Luigi Naccari (1793–1860) was an Italian philosopher, librarian and professor of natural history. Among other topics he wrote about algae: *Algologia adriatica* (1828). He also wrote: *Flora Veneta: Description of Plants found In Venice province, Arranged According to the Linnaean System* (1826–1828).

Nachtigall

Neotropical Rivuline sp. *Austrolebias nachtigalli* WJEM Costa & MM Cheffe, 2006

Dr Giovani Nachtigall Mauricio is a Brazilian ornithologist who is an associate professor at the Universidade Federal de Pelotas, Brazil. The Catholic University of Pelotas awarded his BSc (1997) and the Pontifical Catholic University of Rio Grande do Sul his MSc (2003) and PhD (2010). He then worked at the Laboratorio de Aves Aquaricas e Tartarugas Marinhas (2011–2013) as a post-doctoral researcher. Among his publications is: *Review of the breeding status of birds in Rio Grande do Sul, Brazil* (2013).

Nachtrieb

Northern Pearl Dace *Margariscus nachtriebi* UO Cox, 1896

Henry Francis Nachtrieb (1859–1942) was state zoologist of Minnesota, USA. The University of Minnesota awarded his bachelor's degree (1882). He went to Johns Hopkins University graduate school (1883–1885), then returned to the University of Minnesota as an instructor in biology, becoming an assistant professor (1886) and full professor and head of the animal biology department until his retirement (1887–1925). Alongside this he served as the State Zoologist for the Geological and Natural History Survey of Minnesota. Thereafter he was Professor Emeritus of Animal Biology. His annual reports were published by the state such as: *Report of the state zoologist* (1892–1895).

Nadeshny

Scale-eye Plaice *Acanthopsetta nadeshnyi* PJ Schmidt, 1904

'Nadeshny' was a Russian icebreaker based in Vladivostok.

Nagareda

Viviparous Brotula sp. *Grammonus nagaredai* JE Randall & MJ Hughes, 2008

Bronson Hiroki Nagareda (b.1976) is a biologist and aquarist. He collected the holotype (2007) whilst collecting aquarium fishes in Hawaii. He also provided a series of photographs taken in his aquarium. He wrote: *Feeding biology and age structure of Atlantic batfishes (Lophiiformes: Ogcocephalidae)* (2005).

Nägeli

Toothless Characin sp. *Cyphocharax naegelii* F Steindachner, 1881

Dr Carl Wilhelm von Nägeli (1817–1891) was a physician (graduated at the University of Zurich) and a botanist. He was a professor at the University of Zurich and then at the University of Freiburg, with his career ending as Professor of Botany, University of Munich (1857–1891). He wrote: *Untersuchungen über niedere Pilze aus dem Pflanzen-physiologischen Institut in München* (1882). Although there is no etymology in the original text, we believe this person to be the most likely candidate.

Naggs

Rasbora sp. *Rasbora naggsi* A Silva, K Maduwage & RD Pethiyagoda, 2010

Professor Fred Naggs is a malacologist who is Biodiversity & Conservation Officer in the Mollusca Research Group, Invertebrates section of the Life Sciences Department of the Natural History Museum, London. His main focus is the mechanisms effecting snail species distribution, interactions and their systematics. The University of Westminster awarded his master's degree (1982). He has been Visiting Professor at Chulalongkorn University, Bangkok, Thailand (since 2010). He was Acting Curator of Darwin's House, Down House, Down, Kent (1989), which is now a museum. He has been involved with projects in Sri Lanka for many years (1996–2015) as well as India, Nepal, Thailand, Malaysia and Vietnam where he made field trips (2007 & 2008). He has written or collaborated in many publications on snails such as: *A coloured guide to the land and freshwater Mollusca of Sri Lanka* (1996) and *Evolutionary relationships among the Pulmonata land snails and slugs (Pulmonata, Stylommatophora)* (2006). He was honoured for his support of biodiversity exploration and research in Sri Lanka.

Nagy

Nagy's Licorice Gourami *Parosphromenus nagyi* D Schaller, 1985

Peter Nagy of Felsö Gör, Salzburg, is an Austrian aquarist who first brought this fish to Europe (1979), having collected it in peninsular Malaysia. (Also see **Tussy**)

Nahacky

Nahacky's Dwarf Angelfish *Centropyge nahackyi* RK Kosaki, 1989

Nahacky's Fairy Wrasse *Cirrhilabrus nahackyi* F Walsh & H Tanaka, 2012

Anthony Nahacky is a collector of tropical marine fish for the aquarium trade. As a teenager, he worked evenings in his parents' aquarium store in Florida, and later (1967) left Florida to attend the University of Hawaii. He then started a fish-export business in Hawaii (1971).

Naia

Naia Pipefish *Dunckerocampus naia* GR Allen & RH Kuiter, 2004

Nai'a was the name of the Fiji-based dive vessel from which specimens were collected.

Naipi

Pencil Catfish sp. *Cambeva naipi* WB Wosiacki & JC Garavello, 2004

Naipi is a character in Tupí-Guaraní mythology. She was a beautiful young woman who was transformed into a rock, as part of the legend telling of the origin of the Iguaçu waterfalls, Paraná, Brazil.

Naka

Tongan Spiny Basslet *Acanthoplesiops naka* R Mooi & AC Gill, 2004

The binomial is a construct from the first letters of the names of the two authors' (Randall Mooi and Anthony Gill) four children: Aaron and Adam (Mooi) and Nat and Kelly (Gill).

Nakabo

Nakabo's Slope Dragonet *Centrodraco nakaboi* R Fricke, 1992

Dr Tetsuji Nakabo (b.1949) is a Japanese ichthyologist who is a professor at Kyoto University which awarded his PhD (1981). He was also Director of the Museum there. His more than 120 publications over many years include: *Revision of the family Draconettidae* (1982) and the co-written *Taxonomic review of the* Sebastes vulpes *complex (Scorpaenoidei: Sebastidae)* (2018), as well as a number of books such as *The Fishes of the Japanese Archipelago* (1984).

Nakahara

Longfin sp. *Plesiops nakaharae* S Tanaka, 1917

Mr Kosaku Nakahara (no other information is given in the etymology), who was "*…fortunate enough to obtain this rare species*" from a fishmonger at Tomita, Ise Province, Japan.

Nakamura, H

Big-eyed Sixgill Shark *Hexanchus nakamurai* H-T Teng, 1962

Dr Hiroshi Nakamura was a Japanese ichthyologist. He was a technician (1936–1947) at the Fisheries Experiment Station of the Taiwan Government-General during the Japanese occupation of Taiwan. He wrote many papers, from the early 1930s onwards, and several longer works such as: *Report of an Investigation of the Spearfishes of Formosan Waters* (1937) and *The Tunas and their Fisheries* (1949).

Nakamura, I

Nakamura's Escolar *Rexea nakamurai* NV Parin, 1989

Dr Izumi Nakamura (1939–2021) was a Japanese ichthyologist at the Kyoto University Aquatic Natural History Museum's Maizuru Fisheries Research Station and an expert on billfishes. His doctoral thesis was: *Systematics of the Billfishes (Xiphiidae and Istiophoridae)* (1980). He was a student of Kiyomatsu Matsubara (q.v.). He co-wrote articles with Parin, such as: *Snake mackerels and cutlassfishes of the world* (1993). He was a council member of the Ichthyological Society of Japan.

Nakamura, M

Redfin (cyprind) sp. *Tribolodon nakamurai* A Doi & H Shinzawa, 2000

Blotched Tabira Bitterling *Acheilognathus tabira nakamurae* R Arai, H Fujikawa & Y Nagata, 2007

Dr Morizumi Nakamura (1914–1998) was a Japanese ichthyologist at the National Science Museum, Tokyo. Perhaps his most well-known publication was the colour illustrated: *Keys to the Freshwater Fishes of Japan* (1963). He was honoured for his contributions to the systematics of Japanese bitterlings and for his contribution to the authors' knowledge of the classification of Japanese cyprinid fishes.

Nakamura (?)

Nakamura's Eelpout *Lycodes nakamurae* S Tanaka, 1914

Figures and descriptions of the fishes of Japan including Riukiu Islands, Bonin Islands, Formosa, Kurile Islands, Korea and southern Sakhalin. v. 18: 295–318

Nakasathian

Sisorid Catfish sp. *Oreoglanis nakasathieni* C Vidthayanon, P Saenjundaeng & HH Ng, 2009

Seub Nakhasathien (1949–1990) was a Thai wildlife biologist and conservationist. He obtained a master's degrees from Keletsart University, Thailand, and from the University of London. He became Superintendent (1989) of the Huai Kha Khaeng Wildlife Sanctuary. However, following disagreements with higher-ups and the death of some Sanctuary employees at the hands of encroachers, he became despondent and committed suicide. His death jolted Thailand into action, and ten days after his cremation, the Seub Nakhasathien Foundation was established with a goal to protect natural sanctuaries and their flora and fauna.

Nakaya

Milk-eye Catshark *Apristurus nakayai* SP Iglésias, 2013

Dr Kazuhiro Nakaya (b.1945) is a Japanese ichthyologist and marine scientist who specialises in the taxonomy and evolution of sharks and rays, and also the fish of Lake Tanganyika. Hokkaido University awarded his BA (1968) and PhD (1974) and he is Professor of Marine Environment and Resources at the Marine Laboratory for Biodiversity there. He has described twelve new species of catsharks and published widely, such as: *Taxonomy, comparative anatomy and phylogeny of Japanese catsharks, Scyliorhinidae* (1975) and *Description of familial allocation of the African fluvial genus Teleogramma to the Cichlidae* (2002).

Nakayama

Seabass sp. *Parascombrops nakayamai* WW Schwarzhans & AM Prokofiev, 2017

Dr Naohide Nakayama is a Japanese ichthyologist who is (2018) Assistant Professor in the Department of Marine Biology at the School of Marine Science and Technology, Tokai University. He was Assistant Professor (2017–2018) at Kyoto University and a researcher there (2016–2017). Among his publications are: *First record of the grenadier Coelorinchus sheni (Actinopterygii: Gadiformes: Macrouridae) from Japan* (2018) and *A new grenadier of the genus* Nezumia *(Pisces: Gadiformes: Macrouridae) from southern Japan* (2012). He was the first to recognise this fish as a distinct species.

Nalani

Bandfish sp. *Owstonia nalani* WF Smith-Vaniz & GD Johnson, 2016

Nalani Schell is curator of fishes, MNHN. She was honoured "*...in appreciation for her outstanding assistance facilitating loans and examining type specimens of* Owstonia *in Paris.*"

Nalbant

Eelpout genus *Nalbantichthys* LP Schultz, 1967

Loach sp. *Schistura nalbanti* PM Bănărescu & MR Mirza, 1972

Spined Loach sp. *Cobitis nalbanti* ED Vasil'eva, D Kim, VP Vasil'ev, M-H Ko & Y-J Won, 2016

Dr Teodor T Nalbant (1933–2011) was a Romanian ichthyologist who worked at the Grigore Antipa National Museum of Natural History, Bucharest, which houses his collection of more than 10,000 specimens – the majority of which are of the families Cyprinidae, Cobitidae and Gobiidae. He described several new fish species and wrote more than 25 papers, including: *The loaches of Iran and adjacent regions with description of six new species (Cobitoidea)* (1998) and co-wrote the report on species collected by the 3rd Danish Expedition to Central Asia (1996).

Nalsen

Piranha sp. *Serrasalmus nalseni* A Fernández-Yépez, 1969

Bo Jaime Nalsen (b.1939) worked at the fish research station, Oficina Nacional de Pesca, Venezuela. He helped collect the holotype.

Nalu

Sodwana Pygmy Seahorse *Hippocampus nalu* G Short, L Claassens, R Smith, M De Brauwer, H Hamilton, M Stat & D Harasti, 2020

Savannah Nalu Olivier is an Instructor and Dive Master at Pisces Diving, Sodwana Bay, South Africa. She discovered the new species there. The etymology goes on to say more about the scientific name: "*In the South African languages, Xhosa and Zulu, nalu refers to the expression 'here it is' and therefore we extend its meaning in this case to the simple fact that H. nalu was there all along until its discovery. Additionally, the species name nalu is also the Hawaiian word that refers to the waves or surf of the moana (ocean), for that reason we find the name relevant as H. nalu was observed moving about in strong surge to different locations in the sandy habitat.*"

Namak

Riffle Minnow sp. *Alburnoides namaki* NG Bogutskaya & BW Coad, 2009

This is a toponym, referring to the Namak Lake basin in Iran where the species is endemic.

Namiye

Black Combtooth Blenny *Ecsenius namiyei* DS Jordan & BW Evermann, 1902

Motoyoshi Namiye (1854–1918) was a Japanese naturalist and herpetologist. He was a member of the Faculty of Zoology at the Tokyo Educational Museum, and a Corresponding Fellow of the AOU. Stejneger wrote a report on a collection of birds Namiye had made in the Riu Kiu Islands (1886). He wrote: *Oviposition of a blind snake from Okinawa* (1912). He is also remembered in the names of six birds and an amphibian. The authors give the name 'Motokiche Namiye' in their etymology, stating that he was a curator in the University of Tokyo's museum, and the author of the earliest systematic account of Japanese vertebrates by a Japanese national. We believe this to be either an error or perhaps a variant of his forename.

Namnas

Goby sp. *Gobiopsis namnas* K Shibukawa, 2010

This is an abbreviation of National Museum of Nature and Science (Tokyo, Japan), which conducted the deep-water biological survey that collected the type series.

Naná

Characin sp. *Characidium nana* MB Mendonça & AL Netto-Ferreira, 2015

Mariana Barreira Mendonça is the senior author's sister. She is known to her family and friends as 'Naná'.*

*Note that several other species have the binomial *nana*, such as the Dwarf Stingray, *Urotrygon nana*. In these cases, the word is derived from the Latin for 'dwarf'.

Nancy

African Dwarf Sawshark *Pristiophorus nancyae* DA Ebert & G Cailliet, 2011

Mrs Nancy Ann Burnett née Packard (b.1943) of the Packard family of Hewlett-Packard fame, is a marine biologist who has donated generously to organisations researching the oceans, including being one of the founders of the Monterey Bay Aquarium. She has a bachelor's degree in biology from Stanford University and a master's from San Francisco State University. She has served as chairperson of the Sea Studios Foundation and was the Executive Director for 'The Shape of Life', a revolutionary eight-part television series detailing the dramatic rise of the animal kingdom through the breakthroughs of scientific discovery. She was honoured for her "*...gracious support of chondrichthyan research*" at the Pacific Shark Research Center at Moss Landing Marine Laboratories.

Nanda

Armoured Catfish sp. *Parotocinclus nandae* P Lehmann, P Camelier & A Zanata, 2020

'Nanda' is the affectionate nickname of Maria Fernanda Boaz Lehmann (b.2010), daughter of the senior author, Brazilian ichthyologist Pablo Lehmann.

Nani

Air-sac Catfish sp. *Heteropneustes nani* MS Hossain, S Sarker, SM Sharifuzzaman & SR Chowdhury, 2013

Dr Nani Gopal Das (b.1944) is a fisheries and aquaculture scientist. He was an Assistant Professor, Institute of Marine Sciences, Chittagong University, Bangladesh (1986) and today is a Full Professor. He co-wrote: *Growth Performance and Survival Rate of Macrobrachium rosenbergii (De Man, 1979) Larvae Using Different Doses of Probiotics* (2014). The etymology acknowledges his 'continuous contribution' to the authors' research.

Nani (Caputo)

Hagfish sp. *Eptatretus nanii* RL Wisner & CB McMillan, 1988

Alberto Nani Caputo (1913–1989) was an Argentinian zoologist and ichthyologist. He worked in the ichthyology department at the Museo Argentino de Ciencias Naturales, Buenos Aires; first as a student assistant (1931) then on the staff (1937). He was Associate Professor at the Institute of Biology at the University of Buenos Aires (1959–1966) and thereafter worked at the National Fisheries Directorate. He took part in six research trips to Argentinian Antarctica (1942–1947). Among his published works are descriptions of new species such as: *Hypophthalmus oremaculatus una nueva especie del orden Nematognathi (Pisces, Hypophthal)* (1947).

Nanna

Moon-spotted Shrimp-Goby *Vanderhorstia nannai* R Winterbottom, A Iwata & T Kozawa, 2005

This refers to the Sumerian moon god Sin, who had several different names corresponding to phases of the moon; the name 'Nanna' represented the full moon, referring to moon-like lateral spots on this goby.

Nannings

African Barb sp. *Labeobarbus nanningsi* LF de Beaufort, 1933

Mr. and Mrs. Nannings collected the type specimen. Unfortunately, de Beaufort did not give their forenames in his etymology. We do know that a P A Nannings was collecting for malacologists in the same area of Angola (Lunda) a few years earlier (1927), and believe this was probably Petrus Albertus Nannings (1883–1960) who is known to have been at Silva Porto (now Kuito), Angola in 1944.

Nansen

Pencil Smelt genus *Nansenia* DS Jordan & BW Evermann, 1896

Nansen's Snake-Eel *Ophichthus nansen* JE McCosker & PN Psomadakis, 2018

Fridtjof Nansen (1861–1930) was a Norwegian explorer, scientist, diplomat and later recipient of the Nobel Peace Prize. He is regarded as probably the most famous Norwegian of all time, so there is no need for an extensive entry here. The etymology for the snake-eel states it is named both for the man and a programme named after him: "*Named after the EAF-Nansen Programme and in honor of Dr. Fridtjof Nansen, the famous Norwegian explorer and scientist for whom the programme and research vessel were named. Since 1975, the EAF-Nansen Programme has contributed to increasing the knowledge of global marine biodiversity while supporting developing countries in fisheries research and sustainable management of their resources throughout surveys at sea and capacity building.*"

Nansen (ship)

Nansen Goatfish *Parupeneus nansen* JE Randall & E Heemstra, 2009

The species is named after RV 'Dr Fridtjof Nansen', the research vessel from which the holotype was taken, though the binomial is also intended "*...to honour the famous Norwegian explorer and scientist for whom the vessel was named.*" (See entry above)

Naoko

Naoko's Fairy Wrasse *Cirrhilabrus naokoae* JE Randall & H Tanaka, 2009

Naoko Tanaka is the wife of the second author.

Narayan

Narayan Barb *Pethia narayani* SL Hora, 1937
[Syn. *Puntius narayani*]

Professor C R Narayan Rao (1882–1960) was an Indian herpetologist and zoologist. He was educated at Madras Christian College and, after gaining his diploma, became a teacher in Coimbatore and Ernakulam. He moved to Central College, Bangalore, University of Mysore (1909). Here he was Head of the Department of Zoology, which he created, until his retirement (1937). He co-founded (1932) and was first Editor of *Current Science*. He was influential in the decision that led to the foundation of the Indian Academy of Sciences. His speciality was frogs and their taxonomy. He was honoured as the person who provided a 'valuable' collection of fishes from Cauvery (=Kaveri) River, India.

Narcissus

Long-nosed Skunk Cory *Corydoras narcissus* H Nijssen & IJH Isbrücker, 1980

Narcissus, in Greek mythology, fell in love with his own reflection in a pool of water. Here, it is believed to be a swipe at Herbert Axelrod (q.v.), a man not noted for his humility. The authors wrote that the binomial honoured "*...those who recently collected undescribed Corydoras species and kindly suggested new names for them*" – Axelrod being one of those collectors.

Nares

Goosefish sp. *Lophiodes naresi* A Günther, 1880
Southern Flounder sp. *Thysanopsetta naresi* A Günther, 1880
Pharao Flyingfish *Cypselurus naresii* A Günther, 1889

Vice-Admiral Sir George Strong Nares (1831–1915) was a Royal Navy officer and Arctic explorer. He commanded the first ship to pass through the Suez Canal the day before it officially opened (1869); led the Challenger Expedition (1872–1876) and also led the British Arctic Expedition (1875–1876). In later life he worked for the Board of Trade and as a conservator of the River Mersey. Educated at the Royal Naval School, he joined the Royal Navy (1845) serving aboard 'Canopus' and 'Havannah' as midshipman and mate. He passed his lieutenant's exam (1852) and became second mate on the 'Resolute' during the search for Sir John Franklin in the Arctic (1852–1854). He served on a number of ships as gunnery officer in the Mediterranean and in the Black Sea. He was an instructor on the 'Illustrious'

(1859) and wrote a best-selling book: *The Naval Cadet's Guide*. Promoted to Commander (1862) he first took command of the training ship 'Boscawen'. He was promoted Captain (1869), though his later promotions to Rear-Admiral and Vice-Admiral were post retirement (1886).

Naresh

Assam River Catfish sp. *Erethistes nareshi* BK Mahapatra & S Kar, 2015

Dr Naresh Chandra Datta (1934–2018) was an Indian ichthyologist who was Professor and Head of the Department of Zoology, University of Calcutta. Bangabasi College awarded his BSc (1953) and the University of Calcutta awarded both his MSc (1955) and PhD (1972). Revered as the 'Professor of Professors' he was very influential in education, research, zoology, ichthyology and fisheries science. He spent his entire academic career at the University of Calcutta and founded its Fishery and Ecology Research Unit in the zoology department. Among his publications was the co-written book: *Aspects of History of Science* (2006).

Narinari

Spotted Eagle Ray *Aetobatus narinari* BA Euphrasen, 1790

Narinari was a word used in Brazil in the 17th and 18th centuries and means 'stingray'. The Swedish naturalist Bengt Euphrasén referred in his description to Francis Willughby's book: *De Historia Piscium* (1686) as his source, but in fact Willughby copied both the description and the name from Georg Marcgrave (1610–1644) who was co-author with Willem Piso (1611–1678) of *Historia Naturalis Brasiliae* (1648).

Narutobie

Naru Eagle Ray *Aetobatus narutobiei* WT White, K Furumitsu & A Yamaguchi, 2013

Though it looks as if it could be an eponym, the binomial combines 'Naru' (Naru Island, Japan) with 'tobi-ei' – the Japanese name for eagle rays.

Naseer

Stone Loach sp. *Paraschistura naseeri* N ud-Ahmad & MR Mirza, 1963

Punjabi Baril (cyprinid) *Barilius naseeri* MR Mirza, M Rafiq & FA Awan, 1986

Dr Khan Naseerud-Din Ahmad was (1963) Professor and Head of the Department of Zoology, Government College Lahore, Pakistan, where he taught the junior author of the loach species. Because the junior author (Mirza) did not specify solo authorship, the odd situation arises whereby the 'official' senior author is the same person as the man honoured in the binomial. The etymology of the cyprinid reads: "*...in memory of the authors' late professor, Khan Naseerud-Din Ahmad, former head of the Department of Zoology, Government College, Lahore, Pakistan.*"

Nash

Nash's Barb *Osteochilus nashii* F Day, 1869

Dr John Pearson Nash (1828–1885) was a Surgeon in Her Majesty's Madras Army. He was commissioned (1854) and was quite prominent in the profession, publishing a number of case studies in the Lancet. He was ill and on home leave (1873) and retired (1878). He was also an early amateur photographer: at the Madras Photographic Society meeting of 25 September 1856, Nash "*...exhibited four positives on glass, views of the banqueting Hall, the Nabob of the Carnatic's Palace, Government House and St George's Cathedral, Madras, taken by the collodion process in Newton's Camera in 3 seconds, the view of the Banqueting Hall and the Nabob's Palace were remarkably clear, the Cathedral was also well focused and picturesque.*"

Nasreddin

Central Anatolian Bleak *Alburnus nasreddini* F Battalgil, 1943
[Alt. Eber Bleak; now thought to be a Junior Synonym of *A. escherichii*]
Turkish loach sp. *Oxynoemacheilus nasreddini* B Yoğurtçuoğlu, C Kaya & J Freyhof, 2021

Nasrettin Hoca (aka Nasreddin Hodja) was a 13th-century populist philosopher and humourist. He appears in thousands of stories, sometimes witty, sometimes wise, but often as the butt of a joke. An example: Nasreddin had lost his ring in the living room. He searched for it for a while, could not find it, and went out into the yard and began to look there. His wife asked: "Husband, you lost your ring in the room, why are you looking for it in the yard?" He replied: "The room is too dark to see well. I came out here to look because there is more light." The town of Akşehir, Turkey, where he died, is the type locality of the bleak. The loach is also endemic to this general area.

Natagaima

Tetra sp. *Hyphessobrycon natagaima* CA García-Alzate, DC Taohorn, C Roman-Valencia & FA Villa-Nabarro, 2015

The Natagaima people live in a county of the same name in Colombia which is the type locality. The legend is that a chief called Nataga and a princess named Aima were married and so originated the tribe.

Natalia

Darkgill Snailfish *Psednos nataliae* DL Stein & AP Andriashev, 2001

Dr Natalia V Chernova is a marine biologist and ichthyologist who is (1991–present) a Senior Researcher at the Zoological Institute, Russian Academy of Sciences, St Petersburg. Among her many published works are: *A redescription of the Shantar snailfish, Liparis schantarensis (Scorpaeniformes: Liparidae), with new records from the southeastern Kamchatka Peninsula* (2001) and *New species of the genus* Careproctus *(Liparidae) from the Kara Sea and identification key for congeners of the North Atlantic and Arctic* (2014). She has authored papers with the senior author. The etymology says: "*...The new species is named after Dr Natalia V Chernova, Zoological Institute, Russian Academy of Sciences, St. Petersburg, in honour of her contributions to knowledge of Arctic liparids and other fishes.*"

NatGeo

Pencil Catfish sp. *Ammoglanis natgeorum* E Henschel, NK Lujan & JN Baskin, 2020

NatGeo is an abbreviation for the National Geographic Society. The binomial honours the employees of that society, without whose support the authors' research would not have been possible.

Natterer

Red-bellied Piranha *Pygocentrus nattereri* R Kner, 1858
[Alt. Red Piranha; Syn. *Rooseveltiella nattereri*]
Characin sp. *Brycon nattereri* A Günther, 1864
Ghost Knifefish sp. *Sternarchogiton nattereri* F Steindachner, 1868
American Sole sp. *Apionichthys nattereri* F Steindachner, 1876
Spotted Tetra *Copella nattereri* F Steindachner, 1876
Leporinus sp. *Leporinus nattereri* F Steindachner, 1876
Copper-Joe Toadfish *Thalassophryne nattereri* F Steindachner, 1876
Natterer's Cory *Corydoras nattereri* F Steindachner, 1876
Natterer's Anchovy *Anchoviella nattereri* F Steindachner, 1879
Thorny Catfish sp. *Trachydoras nattereri* F Steindachner, 1881
Dawn Tetra *Aphyocharax nattereri* F Steindachner, 1882
Twig Catfish sp. *Farlowella nattereri* F Steindachner, 1910

Dr Johann Natterer (1787–1843) was an Austrian naturalist and collector. He studied botany, zoology, mineralogy, chemistry and anatomy and was appointed as a taxidermist to what became the Natural History Museum in Vienna. As a zoologist, he took part with Spix (q.v.) and others in the expedition to Brazil (1817), which started on the occasion of Archduchess Leopoldina's wedding to Dom Pedro, the Brazilian Crown Prince. The entire suite travelled in two Austrian frigates, 'Austria' and 'Principesse Augusta'. Natterer explored a potential river route to Paraguay (1818–1819), and subsequently undertook (1821–1835) five expeditions, exploring the Mato Grosso and the Amazon Basin before returning to Vienna. He accumulated a huge collection numbering 12,293 birds and c.24,000 insects, which can still be seen today in the Vienna Museum. He lost the majority of his possessions in the Civil War then being waged in Brazil, so his total collection must have been staggering. He ended his career at the Austrian Imperial Museum of Natural History, dying of a lung ailment. He did not publish an account of his travels and, unfortunately for posterity, his notebooks and diary were destroyed by fire (1848). Fortunately, August von Pelzeln was able to reconstruct some of Natterer's itinerary and information and wrote up the collection (1868–1871). Natterer's specimens are beautifully prepared and his tiny labels are always meticulous and clear (a rarity for the period). He never received the credit in Austria that he should have, but abroad he was held in high esteem; for instance, being made an honorary Doctor of Philosophy at Heidelberg University. Twenty-three birds, three reptiles, two amphibians and two mammals are also named after him.

Naudé

Naudé's Rubble Goby *Trimma naudei* JLB Smith, 1957
[Alt. Naudé's Pygmygoby]

Dr Stefan Meiring Naudé (1904–1985) was a South African physicist who was Professor of Physics at Stellenbosch University (1934–1946). Stellenbosch University awarded his MSc (1925) and Berlin University – where he studied under Albert Einstein and Max Planck – his

PhD (1928). He was a post-doctoral student at Chicago University. He discovered isotype N15 (1932) and then returned to South Africa lecturing at Cape Town and Stellenbosch before taking the Physics chair at the latter. He was President of the South African Council for Scientific and Industrial Research and became Scientific Advisor to the South African Prime Minister (1971). He wrote c.20 books, mostly on physics. A research vessel was also named after him.

Naumann

Loach sp. *Paraschistura naumanni* J Freyhof, G Sayyadzadeh, HR Esmaeili & M Geiger, 2015

Dr Clas Michael Naumann zu Königsbrück (1939–2004) was a German zoologist and the senior author's supervisor. His two doctorates were awarded by the University of Bonn (1970) and the University of Munich (1977). He became Professor at the University of Bielefeld (1977–1989) and then at the University of Bonn for the rest of his life (1989–2004). Among other works he wrote: *Die Kirghisen des Afghanischen Pamir* (1977). He is described as someone who loved Iran, its people, nature and culture. He also spent time in Afghanistan, teaching in the zoology department of Kabul University (1970–1972).

Navarra

Blackish Stingray *Hemitrygon navarrae* F Steindachner, 1892

Bruno R Navarra (1850–1911) was a scholar, variously described as a French traveller and German soldier, who was an expert on Chinese history and culture including music. He supplied the Imperial Court Museum of Natural History (Vienna) with fish specimens collected in Shanghai. He was editor (1896–1899) of *Der Ostasiatische Lloyd*, a German language daily newspaper published in Shanghai. He wrote more than ninety publications, with his most influential book being: *China und die Chinesen* (1901) and translated Sun Tzu's: *Art of War* (1910).

Navarro

Long-whiskered Catfish sp. *Pimelodus navarroi* LP Schultz, 1944

Rafael Navarro was a Venezuelan who acted as Schultz's assistant in collecting fishes for the Smithsonian from the Maracaibo Basin, Venezuela (1942). Schultz reported that Navarro was unfortunate in that he stepped on a stingray and suffered greatly from its stinger.

Navjot Sodhi

Banded Barb sp. *Puntigrus navjotsodhii* HH Tan, 2012

Dr Navjot Singh Sodhi (1962–2011) was an Indian-born tropical biologist, ecologist and conservationist with interests in ornithology, ichthyology and malacology. His bachelor's and master's degrees were awarded by Panjab University and his doctorate by the University of Saskatchewan (1991). He was then a post-doctoral researcher at the University of Alberta, and then the National Institute of Environmental Studies, Japan (1991–1995). He was appointed Assistant Professor at the Department of Biological Sciences, National University of Singapore (1995), Associate Professor (2001) and full Professor of Conservation Ecology

(2007). He was at Harvard as Charles Bullard Fellow (2002), and as Sarah and Daniel Hardy Fellow in Conservation Biology (2008–2009). Among his written work is the book: *Conservation Biology for All* (2012). He was a council member of the 'Association for Tropical Biology and Conservation'. He was honoured for his 'inputs' to conservation and ecological research in Southeast Asia, and for his 'considerable contributions and services' to the editorship of the Raffles Bulletin of Zoology. He died of an aggressive lymphoma cancer. A bird, a crab and a snail are also named after him.

Naylor

Dusky-snout Catshark *Bythaelurus naylori* DA Ebert & PJ Clerkin, 2015

Dr Gavin Naylor is Program Director of the Florida Program for Shark Research at the Florida Museum of Natural History, where he is also Curator and Professor. He is also associated with the Biology Department of the College of Charleston. He has published around 150 papers. He was honoured "*…for his contributions and innovative molecular research into the higher classification of chondrichthyans.*"

Nazir

Mahseer genus *Naziritor* MR Mirza & MN Javed, 1985
Loach sp. *Triplophysa naziri* N Ahmad & MR Mirza, 1963
Sisorid Catfish sp. *Glyptothorax naziri* MR Mirza & IU Naik, 1969
Indus Garua (catfish) *Clupisoma naziri* MR Mirza & MI Awan, 1973

Dr Nazir Ahmad (1910–1985) was a Pakistani zoologist. He became Director of Fisheries, East Pakistan (1955–1960) and West Pakistan (1960–1969) and previously Assistant Director of Fisheries East Bengal (1940s). He wrote, among other books: *Fish Wealth of Pakistan* (1955) and *Fishing Gear of East Pakistan* (1961). He was Head of the Department of Zoology (1956) and became Principal of the Government College of Lahore (1959–1965). He was a Founding Member of the Pakistan Academy of Science. In the genus, Nazir's name is combined with *Tor* (another Mahseer genus, based on a local name for these fish - *tora*).

NEAMB

Brazilian Cichlid sp. *Geophagus neambi* PHF Lucinda, CAS Lucena & NC Assis, 2010

This honours an institution: **N**úcleo de **E**studos **Amb**ientais (Neamb), Universidade Federal do Tocantins (Brazil), for its effort in studying the Rio Tocantins ichthyofauna.

Nebeshwar

Stone Loach sp. *Schistura nebeshwari* Y Lokeshwor & W Vishwanath Singh, 2013

Dr Kongrailakpam Nebeshwar Sharma is an Indian biologist and ichthyologist. He was at the Department of Life Sciences, Manipur University, India and the Centre of Biodiversity, Rajiv Gandhi University, Itanagar, India. Among his publications are: *A New Nemacheiline Fish of the Genus* Schistura *McClelland (Cypriniformes: balitoridae) from Manipur, India* (2005) and *Ichthyological survey and review of the checklist of fish fauna of Arunachal Pradesh, India* (2009). He was honoured in the name of the loach for his assistance to the authors during field work in Mizoram, India.

Neblina

Marble Knifefish *Hypopygus neblinae* F Mago-Leccia, 1994

This is a toponym referring to La Neblina National Park, Amazonas, Venezuela, which is where Mago-Leccia first noticed the differences between this species and *H. lepturus.*

Neef

Sidespot Barb *Enteromius neefi* PH Greenwood, 1962

Graham Bell-Cross (1927–1998). He collected the barb type. He often called Greenwood 'Oom', the Afrikaans word for uncle and 'neef' is the Afrikaans word for nephew, so the eponym is a humorous acknowledgment of their working relationship. (see **Bell-Cross**).

Neelesh

Neelesh's Hillstream Loach *Indoreonectes neeleshi* P Kumkar, M Pise, PA Gorule, CR Verma & L Kalous, 2021

Dr Neelesh Dahanukar is a researcher (2010) and fellow (2012>present) from the Indian Institute of Science Education and Research (IISER), Pune, India. He was educated at Savitribai Phule Pune University with an initial degree in microbiology (2001), Master's (2003) and PhD (2010). He was honoured for his "*…remarkable contributions to the understanding of the systematics and evolution of Indian freshwater fishes.*" His (220) published works include: *The status and distribution of freshwater fishes of the Western Ghats* (2011).

Neelov

Snailfish sp. *Paraliparis neelovi* AP Andriashev, 1982

Alexei Vladimirovich Neyelov is a Russian marine biologist at the Laboratory of Ichthyology, Russian Academy of Sciences in St. Petersburg. His main interests are connected with research into the morphology of fishes, biota of Antarctica and problems of studying polar fauna. Among his publications is: *Parallelism in the structure of the seismosensory system in two ecologically different representatives of cottoid fish (Cottoidei)* (1979). He has described five other snailfishes with Anatoly Andriyashev, his colleague at the Academy.

Negodagua

Tetra sp. Characin sp. *Hyphessobrycon negodagua* FCT Lima & P Gerhard, 2001

Nego D'água is a creature of Brazilian legend. He is said to be man-like, with webbed hands and feet. He sometimes attacks inattentive fishermen at night, or overturns canoes if the fishermen don't give him a fish.

Neil

Neil's Grunter *Scortum neili* GR Allen, HK Larson & SH Midgley, 1993
[Alt. Angalarri Grunter]

Arthur Neil is a Queensland resident who assisted the second author and "*…was instrumental* *in collecting most of the type specimens of S. neili"*. We have found no biographical information for him.

Neill

Bay of Bengal Hogfish *Bodianus neilli* F Day, 1867
Snow Trout sp. *Osteobrama neilli* F Day, 1873

Dr Andrew Charles Brisbane Neill (1814–1891) was a Scottish physician who qualified at the University of Glasgow (1837) and served in the Madras Medical Service (1838–1858). He was also a keen early photographer and photographed at Lucknow during the Mutiny (1857). He was honoured as the 'esteemed friend' of the author and for helping Day bring his: *Fishes of Malabar* (1865) to press.

Neiva

Driftwood Catfish sp. *Tatia neivai* R Ihering, 1930

Dr Arthur Neiva (1880–1943) was an epidemiologist and biologist. He qualified in medicine in Rio de Janeiro (1903), and did entomological research at the Institute of Manguinhos. He organised the Medical Section of Zoology and Parasitology, Instituto Bacteriológico, Buenos Aires, for the Argentine government (1915). He returned to Brazil (1916), becoming Director of Public Health for São Paulo State, a member of the staff at Instituto Butantan, São Paulo, and then Director of the National Museum, Rio de Janeiro (1923). He was first Director, Institute Bacteriológico (1928–1932). After the revolution (1930) he held a number of appointments, including being DirectorGeneral of Research, Ministry of Agriculture. He entered politics (1933–1937) and then gave it up to resume his original research at the Institute of Manguinhos. A reptile and a bird are also named after him.

Nelson (Abraham)

Travancore Yellow Barb *Puntius nelsoni* M Plamootil, 2015

Dr Nelson P Abraham is an Indian zoologist and ichthyologist who is an Associate Professor at St. Thomas College, Kozhencherry, Kerala (since 1993). His doctorate was awarded by Kerala University. He co-wrote: Macrognathus fasciatus *(*synbranchiformes; mastacembelidae*) – A new fish species from Kerala, India* (2014).

Nelson, D

Thorny Catfish sp. *Leptodoras nelsoni* MH Sabaj Pérez, 2005

Douglas W Nelson is a researcher and technician at the University of Michigan Museum of Zoology where he has worked for over 30 years, including as Collection Manager of Fishes since 1993. He wrote: *Two New Species of the Cottid Genus* Artediellus *from the Western North Pacific Ocean and the Sea of Japan* (1986).

Nelson, EW

Alaska Whitefish *Coregonus nelsonii* TH Bean, 1884
Lacandon Sea Catfish *Potamarius nelsoni* BW Evermann & EL Goldsborough, 1902

Baja California Rainbow Trout *Oncorhynchus mykiss nelsoni* BW Evermann, 1908

Edward William Nelson (1855–1934) was an outstanding American naturalist. As a young man, he was sent as a weather observer to Alaska (1877). Although his major objective was to make meteorological observations, he was also tasked to '... *obtain all the information possible on the geography, ethnology, and zoology of the surrounding region*'. With the help of Inuits, dog sleds, and kayaks, he explored (1877–1881) areas where no Europeans had been before. He was the naturalist on board the 'Corwin' during its search for the missing Arctic exploration vessel 'Jeanette' (1881). This expedition was the first to reach and explore Wrangel Island. He was a member of the Death Valley Expedition (1890) and then conducted a field survey of Mexico (1891–1905) before returning to the Bureau of Biological Survey (1906–1929), including being Chief of the Bureau (1916–1927), and President of both the American Ornithologists' Union and the American Society of Mammologists. His greatest lasting contribution was the Migratory Bird Treaty, which is still in force today. Fifteen mammals, twenty-four birds, an amphibian and five reptiles are named after him.

Nelson, GJ

Nelson's Anchovy *Stolephorus nelsoni* T Wongratana, 1987
Hagfish sp. *Eptatretus nelsoni* CH Kuo, KF Huang & HK Mok, 1994
Eastern Numbfish *Narcinops nelsoni* MR de Carvalho, 2008

Dr Gareth (Gary) Jon Nelson (b.1937) is Emeritus Curator Vertebrate Zoology, Ichthyology, at the AMNH where he worked for over three decades (1967–1998), having been chairman of ichthyology (1982–1987) and of ichthyology and herpetology (1987–1993). In retirement, he moved to Australia and is Professorial Fellow, School of Botany, University of Melbourne and an honorary associate in ichthyology at Museum Victoria, Melbourne. After serving in the US Army (1955–1958) he gained a bachelor's degree at Roosevelt University, Chicago (1962). The University of Hawaii awarded his doctorate (1966), after which he did post-doctoral research at the BMNH and at the Swedish Museum of Natural History, Stockholm. He was an adjunct associate professor at Long Island University (1973–1974) and at New York University (1973–1978). Among his more than 270 publications are: *Outline of a theory of comparative biology* (1974), *Identity of the anchovy* Engraulis clarki *with notes on the species groups of* Anchoa (1986) and *Cladistics and evolutionary models* (1989). He was honoured in the anchovy's binomial for his: "...*knowledge and classic works on the comparative anatomy and cladistic relationships among clupeoid fishes*," and for providing specimens of this species and suggesting that it might be an undescribed form.

Nelson, JS

Indian Cyprinid sp. *Barilius nelsoni* RP Barman, 1988
Duckbill sp. *Bembrops nelsoni* BA Thompson & RD Suttkus, 2002
Sandburrower sp. *Myopsaron nelsoni* K Shibukawa, 2010

Dr Joseph Schiesser 'Tiger Joe' Nelson (1937–2011) was an American ichthyologist, aquarist and keen amateur astronomer. He studied ichthyology at the University of British Columbia, which awarded his bachelor's degree (1960), then at the University of Alberta for his MSc (1962), but returned to UBC for his PhD (1965). He returned to the United States as a Research Associate at Indiana University, where he became Assistant Director of the

Indiana University Biological Stations. He went back to the University of Alberta (1968) as Assistant Professor, progressing through Associate Professor and then full Professor, becoming Dean among other administrative offices and Curator of Ichthyology for the University of Alberta Museum of Zoology (UAMZ). He published more than 120 papers but is best known for his: *Fishes of Alberta* (1970) and *Fishes of the World* (1976). He was honoured in the *Barilius* binomial "*...in recognition of his valuable contribution to the study of the fishes of the world.*"

Nelson (Lee)

Viviparous Brotula sp. *Thalassobathia nelsoni* RS Lee, 1974

Nelson B Lee was the father of the author, Richard Lee.

Nelson, PR

Sisorid Catfish sp. *Glyptothorax nelsoni* DN Ganguly, NC Datta & S Sen, 1972

Philip R Nelson (1918–2008) was a marine biologist and fisheries scientist. He graduated with a bachelor's degree from the University of Washington, Seattle. He was in the US army in WW2 and served in the South Pacific. He worked for the US Fish and Wildlife Service and later for the National Marine Fisheries Service of which he was (1972) Chief, Branch of Inland Fisheries, Washington DC. He co-wrote: *Life history of the threespine stickleback* Gasterosteus aculeatus *Linnaeus in Karluk Lake and Bare Lake, Kodiak Island, Alaska* (1959).

Nemnez

Snailfish sp. *Psednos nemnezi* DL Stein, 2012

An invented word, created from the initials NMNZ, the National Museum of New Zealand.

Nemoto

Pearleye sp. *Rosenblattichthys nemotoi* M Okiyama & RK Johnson, 1986

Dr Takahisa Nemoto (1930–1990), was a Japanese biological oceanographer. He graduated from the Department of Fishery, Faculty of Agriculture, University of Tokyo (1953) which also awarded his PhD (1963). He worked at the Whales' Research Institute (1953–1966). He was an Associate Professor, Ocean Research Institute, University of Tokyo (1967), was promoted to Professor (1982) and was then the Director of the Ocean Research Institute (1986–1990). This species was named after him in recognition of his great contributions to Antarctic biology, including directing the cruise during which the holotype was collected.

Neptune

Tetra sp. *Hemigrammus neptunus* A Zarske & J Géry, 2002

Neptune was the Roman god of the sea. He is often depicted holding a trident, and the name was inspired by the trident-like mark on the tetra's caudal fin.

Nesaie

Rattail sp. *Hymenocephalus nesaeae* NR Merrett & T Iwamoto, 2000

In Greek mythology, Nesaie was one of the Nereids, a sea nymph. The junior author later commented there is no particular allusion or meaning, simply: "*...a nice name that might have some bearing on the creature.*"

Netto Ferreira

Suckermouth Catfish sp. *Astroblepus nettoferreirai* CA Ardila Rodríguez, 2015

Dr André Luiz Netto Ferreira (b.1982) is a Brazilian ichthyologist. The Federal University of Rio de Janeiro awarded his bachelor's degree (2004) and his master's (2006). The Universidad de São Paulo awarded his doctorate (2010) and hosted his post-doctoral work (2011–2013). He was a visiting professor, Federal University of Pará (2013–2015) and a post-doctoral fellow at the National Museum of Natural History/Smithsonian Institution (2015) and Museu Paraense Emilio Goeldi (2016–2017). Currently, he is a professor at the Zoology Department, Federal University of Rio Grande do Sul (2018). His career has been mostly dedicated to the study of systematics of Characifirmes, and to date he has described two genera and 28 species. Among those, he co-wrote: *Phallobrycon synarmacanthus, a new species of Stevardiinae from the Xingu basin, Brazil (Teleostei: Characidae)* (2016).

Neumann, M

Shrimp-Goby sp. *Amblyeleotris neumanni* JE Randall & JL Earle, 2006

Mike Neumann is a diver and underwater photographer. He is, according to the etymology a "*...fellow diver, underwater photographer and good friend*" of the principle author, Jack Randall. He helped collect and photograph this goby. We believe this to be the same person who co-owns *Beqa Adventure Divers in Fiji. Known as 'Big Mike', he was a Swiss executive for Deutsche Bank who on retirement has enjoyed a* second career devoted to ocean-related environmental projects, with a particularly avid interest in sharks. As a life-long diver and trained biologist, he travelled the world, seeing conservation programs that had succeeded and those that were disappointing failures: so he met with the chiefs of the villages that had ownership rights to the reefs and, by way of consensus with local leaders and with the villagers' blessing, ensured the reefs' conservation.

Neumann, OR

African Rivuline sp. *Nothobranchius neumanni* FM Hilgendorf, 1905
Neumann's Suckermouth *Chiloglanis neumanni* GA Boulenger, 1911

Professor Oskar Rudolph Neumann (1867–1946) was a German ornithologist who collected widely in east and northeast Africa (1892–1894). He travelled through Somaliland and southern Ethiopia (1900–1901). After becoming bankrupt (1908), he worked for a few months that year at the Rothschild Museum in Tring, England, but due to Walter Rothschild's own financial difficulties he had to leave, after which he became a stockbroker in Berlin. Although he received the Iron Cross (WW1) when an officer, he had to flee from the Nazi regime (1941). Via Switzerland and Cuba, he reached the USA, where he worked the last few years of his life for the Field Museum of Natural History in Chicago. The results of his expeditions to Africa were published in the *Journal für Ornithologie*. Thirty-five birds, three mammals, two reptiles and an amphibian are named after him.

Neumann, W

African Rivuline sp. *Epiplatys neumanni* HO Berkenkamp, 1993

Dr Werner Neumann of Zwickau, Germany, was an aquarist with an interest in killifish. He wrote *Die Hechtlinge* (1983), which deals with the aquarium care of pike-like killifishes (*Epiplatys, Aplocheilus, Pachypanchax, Episemion*).

Neumayer

Neumayer's Barb *Enteromius neumayeri* JG Fischer, 1884

Dr George Balthasar von Neumayer (1826–1909) was a scientist, magnetician, hydrographer, oceanographer and meteorologist. He studied at Munich University culminating in a PhD (1849). Wanting practical experience of navigation, he sailed to South America and in doing so acquired a Mate's certificate. On his return to Europe he was offered the chair in physics at Hamburg. Still fascinated with magnetism he sailed again, this time for southern waters, reaching Sydney, Australia (1852). He worked there as a miner and sailed coastal waters as well as researching at the Hobart magnetic observatory. He then returned to Germany (1854), but was back in Australia (1857) and set up an astronomical observatory there. Thereafter (1858–1864) he conducted a complete magnetic survey of Victoria State. He wanted to mount an expedition to the Australian interior, but when this failed to materialise, he once again returned (1872) to Germany. Eventually he became Director of the Hamburg oceanic Observatory (1876–1903) and was vice-chairman of the Geographical Society in Hamburg.

Nevelsky

Prickleback sp. *Stichaeopsis nevelskoi* PJ Schmidt, 1904

Admiral Gennady Ivanovich Nevelsky (1813–1876) was a Russian navigator who led an expedition to the Russian Far East (1849–1855), exploring Sakhalin and the Amur River delta and establishing a settlement there. A number of geographical features such as the Nevelskoy Strait, a city, streets, monuments, vessels and a school and a university are also named after him. Schmidt wrote the report on the fish collected on the Imperial Russian Geographical Society expedition (1900–1901) which collected fish from waters named after the admiral. He wrote: "*…the new species described by me, which I name in honour of Admiral Nevelsky, who discovered the strait of his name…*" (translation).

Newberry

Northern Tide-water Goby *Eucyclogobius newberryi* CF Girard, 1856

Great Basin Redband Trout *Oncorhynchus mykiss newberrii* CF Girard, 1859

Dr John Strong Newberry (1822–1892) was an American physician, geologist, author and explorer. He graduated from Western Reserve University (1846) and qualified as a physician at Cleveland Medical School, Ohio (1848). He studied palaeontology and medicine in Paris (1849–1850) and started in private practice at Cleveland (1851). He joined Lieutenant Williamson's expedition to explore between San Francisco and the Columbia River (1855) and was the geologist on the expedition led by Lieutenant Ives to explore the Colorado River (1857–1858). He was the naturalist on the expedition led by Captain Macomb, which explored (1859) south-western Colorado and adjacent parts of Utah, Arizona, and New

Mexico and during which he was the first geologist to visit the Grand Canyon. He was made a professor at Colombian University (now George Washington University, Washington D.C.) (1857). He was elected (1861) to the United States Sanitary Commission to utilise his medical knowledge and experience in the army. He resigned his post with the army (1861) and became secretary of the Western Department of the Sanitary Commission, having supervision of all the work of the commission in the Mississippi Valley, and had control and management of the needs of more than 1,000,000 soldiers (1861–1866). He became Professor of Geology and Palaeontology, School of Mines, Columbia College (now Columbia University, New York) (1866–1890). He wrote: *Fossil Fishes and Fossil Plants of the Triassic Rocks of New Jersey and the Connecticut Valley* (1888). Newberry Crater in Oregon was named after him (1903). He collected the holotype of the trout subspecies.

Newbold

Tetra sp. *Hemigrammus newboldi* A Fernández-Yépez, 1949

Philip Newbold was a friend of the author. He died while working on experiments on Lake Maracaibo. We have been unable to find anything more about him, or the nature of his experiments.

Newman

Snailfish sp. *Liparis newmani* DM Cohen, 1960

Merrill Edward Newman (1928-2022) graduated in zoology at the University of California (1950) and was awarded his master's by Stanford. He was in the US Army serving in Korea as a member of the White Tigers, where he advised a UNPIK guerrilla unit (c.1950–c.1953). After serving in the army, Newman was a high school teacher and, later, worked in technology and financial consulting and as an executive for technology companies in Silicon Valley. He is most well-known for being arrested whilst on a trip to North Korea (2013). The Korean Central News Agency released a video showing Newman signing a letter of apology and confession for war crimes committed during the Korean War. He was released a month later. Newman had collected a number of fish species from the Yellow Sea off the Korean west coast (1953) which Daniel Cohen was able to study.

Newnes

Dusky Rockcod *Trematomus newnesi* GA Boulenger, 1902

Sir George Newnes (1851–1910) was a British newspaper publisher and Liberal politician. He sponsored the Southern Cross Expedition (1898–1900) to Antarctica.

Newton, F

Newton's Wrasse *Thalassoma newtoni* B Osório, 1891

Colonel Francisco Xavier Aguilar O'Kelly Azeredo Newton (1864–1909) was a Portuguese explorer and naturalist who collected on São Tomé and other islands in the Gulf of Guinea (1885–1895), and in Timor (1896). He wrote accounts of his travels and findings, and was ahead of his time in recording especially meticulous detailed information on the localities and ecology of the specimens he collected. Three birds, two amphibians, two reptiles and a mammal are also named after him.

Newton (dos Santos)

Characin sp. *Moenkhausia newtoni* HP Travassos, 1964

Dr Newton Dias dos Santos (1916–1989) was a Brazilian entomologist at Museu Nacional, Rio de Janeiro, which he joined (1939) and where he was Professor and the Director (1961–1963). He graduated at Escola de Ciências da Universidade do Distrito Federal, Rio de Janeiro (1938), took a medical qualification at the Faculdade Nacional de Medicina (1940) and was awarded a doctorate by Faculdade Nacional de Filosofia (1950). He was honoured for his contributions to the museum's fish collection. He wrote 124 scientific papers on odonata and a book, *Science Practice* (1955). Seven odonata are named after him.

Neyelov

Sculpin sp. *Artediellus neyelovi* F Muto, M Yabe & K Amaoka, 1994

Hopbeard Plunderfish *Pogonophryne neyelovi* GA Shandikov & RR Eakin, 2013

Dr Alexei V Neyelov of the Zoological Institute, Russian Academy of Sciences, St Petersburg is a Russian ichthyologist. Among his publications is a description of another Antarctic plunderfish: *Description of Harpagifer permitini sp. nova (Harpagiferidae) from the sublittoral zone of South Georgia and redescription of the littoral H. georgianus Nybelin* (2006). According to the etymology he "*...contributed significantly to the knowledge of Antarctic fishes*".

Nezahualcoyot

Mountain Swordtail *Xiphophorus nezahualcoyotl* M Rauchengerger, KD Kallman & DC Morizot, 1990

Nezahualcoyotl (1402–1472) was a philosopher king of the Texcocan people in Mexico. Under his rule, Texcoco became "the Athens of the Western World", and the remains of hilltop gardens, sculptures and an aqueduct system testify to the impressive engineering skills and aesthetic architecture of his reign.

Nguti

Cameroon Cichlid sp. *Etia nguti* UK Schliewen & MLJ Stiassny, 2003

This is a toponym, being the name of a village in southwestern Cameroon where most of the type series was collected.

Nguyễn, HD

Vietnamese Cyprinid sp. *Opsariichthys duchuunguyeni* T-Q Huynh & I-S Chen, 2014

Dr Nguyễn Huu Duc is a Vietnamese ichthyologist who is Associate Professor at the Hanoi National University of Education. He co-wrote: *Ecological attributes of a tropical river basin vulnerable to the impacts of clustered hydropower developments* (2008). He was honoured for his contributions to Vietnamese freshwater fish research. (Also see **Duchuu Nguyen**)

Nguyễn, VH

Bitterling sp. *Acheilognathus nguyenvanhaoi* HD Nguyen, DH Tran & TT Tat, 2013

Nguyễn Văn Hảo is a Vietnamese ichthyologist at the Hanoi University of Education, Vietnam. He co-wrote: *Two new species of* Oreias *Sauvage, 1874 discovered in Son La City, Vietnam* (2010).

Niceforo

Suckermouth Catfish sp. *Astroblepus nicefori* GS Myers, 1932
Armoured Catfish sp. *Hypostomus niceforoi* HW Fowler, 1943
Leporinus sp. *Leporinus niceforoi* HW Fowler, 1943

Brother Nicéforo Maria (1888–1980) was a Frenchman originally named Antoine Rouhaire, who became a missionary in Colombia under his monastic name. He went to Medellin (1908) and was given the task of forming a natural history museum (1913). He was primarily a herpetologist and an excellent taxidermist. He is also remembered in the names of a mammal, four birds, ten amphibians, and eight reptiles.

Nichof

Banded Eagle Ray *Aetomylaeus nichofii* ME Bloch & JG Schneider, 1801

This is apparently an error for (Johan) Nieuhof (1618–1672), on whose illustration of this ray (copied by Francis Willughby, 1686) Bloch and Schneider based their description. (See **Nieuhof**).

Nicholls

NIcholls' Pupfish *Cyprinodon nichollsi* ML Smith, 1989
[Alt. Jaragua Pupfish]

Kenneth W Nicholls helped support Smith's fieldwork. We can find no further information about him.

Nichols, HE

Spinster Wrasse *Halichoeres nicholsi* DS Jordan & CH Gilbert, 1882
Blackeye Goby *Rhinogobiops nicholsii* TH Bean, 1882

Captain Henry Ezra Nichols (d.1899) was a US Navy officer at the Department of Alaska. He graduated from the US Naval Academy (1865) and spent the rest of his life in the navy, rising to Captain (1899) a few months before his death. He was commander of the United States Coast and Geodetic Survey steamer 'Hassler', which (1880) made a voyage along the Pacific coast of Mexico, during which the wrasse holotype was taken. He also served on 'USS Pinta' (1884). He obtained the goby type and preserved 31 species in total, all 'in excellent condition', leading Bean to write that: "*…no better-preserved lot of fishes has been received from any other collector.*"

Nichols, JT

Parrotfish genus *Nicholsina* HW Fowler, 1915
Nichols' Lanternfish *Gymnoscopelus nicholsi* CH Gilbert, 1911
Indian Cyprinid sp. *Chagunius nicholsi* GS Myers, 1924
Congo Cichlid sp. *Pseudocrenilabrus nicholsi* J Pellegrin, 1928

Pearleye sp. *Scopelarchoides nicholsi* AE Parr, 1929
Limia (livebearer) sp. *Poecilia nicholsi* GS Myers, 1931
Yellowfin Bass *Anthias nicholsi* FE Firth, 1933
Stone Loach sp. *Schistura nicholsi* HM SmIth, 1933
Chinese Gudgeon sp. *Gnathopogon nicholsi* PW Fang, 1943
Highfin Blenny *Lupinoblennius nicholsi* WN Tavolga, 1954
Chinese Gudgeon sp. *Gobiobotia nicholsi* PM Bănărescu & TT Nalbant, 1966

John Treadwell Nichols (1883–1958) was an American ichthyologist, mammalogist and ornithologist. He studied vertebrate zoology at Harvard, graduating AB (1906). He then joined the AMNH (1907) as an assistant in the mammalogy department. He founded the journal of the American Society of Ichthyologists and Herpetologists, '*Copeia*' (1913). He became first assistant curator (1913–1952) then associate curator and finally Curator of Ichthyology (1952–1958). He wrote over 1,000 articles and scientific papers and a number of books, including: *Fishes and Shells of the Pacific World* (1945). He was honoured for his extensive contributions to the ichthyology of China in general and Chinese loaches in particular. Two reptiles are also named after him. (see also **John Treadwell**)

Nichols, M

Nichols' Worm Eel *Scolecenchelys nicholsae* ER Waite, 1904

Mrs. Thomas (Mary) Nichols née Andrews (1846–1923) was the wife of a Whaler Captain and collected for the Australian Museum. She provided many local fishes for Waite from Lord Howe Island, where she lived. She also collected seed for botanical gardens. She sometimes went to sea with her husband. She was also said to have contributed much to our knowledge of the island fauna. She went on at least one conch collecting expedition, spending three days on HMS Tambo (1908) with a number of conchologists. There followed evenings "…spent visiting the homes of Mrs Nichols' numerous offspring, who all produced some form of musical entertainment, cocoa and cake." She also started a guesthouse on the island, which employed 25 people and catered for 85 guests! Her daughters (she had eleven children) are included in the etymology, that says the fish is named for the many kindnesses Waite received from them all. Thus the binomial probably should be *nicholsarum* since it is honouring more than one woman.

Nico

Neotropical Rivuline sp. *Laimosemion nicoi* JE Thomerson & DC Taphorn Baechle, 1992 [Syn. *Rivulus nicoi*]
Banjo Catfish sp. *Acanthobunocephalus nicoi* J Friel, 1995
Armoured Catfish sp. *Pseudolithoxus nicoi* JW Armbruster & F Provenzano, 2000

Dr Leo G Nico is an American ichthyologist. Southern Illinois University awarded his BA (1979) and his MSc (1982). He then spent the next two years in Venezuela where he was affiliated with the University of the Llanos (UNELLEZ) and collaborated with professor Don Taphorn, conducting research and field work on freshwater fishes. He continued to conduct field work in South America and intermittently continued to travel in the Neotropics, spending over thirty years in that region of the world (1980–2010). The University of Florida awarded his PhD (1991). He is a Research Fishery Biologist with the

US Department of Interior and US Geological Survey (since 1993). He found the rivuline species in a temporary pool close to the Rio Venturari in southwest Venezuela. His published work includes: *Parasites of Imported and Non-Native Wild Asian Swamp Eels* (2016).

Nicolai

Spotted Notothen *Trematomus nicolai* GA Boulenger, 1902

Nicolai Hanson (1870–1899) was a Norwegian zoologist and Antarctic explorer. (See **Hanson**).

Nicolas

Tetra sp. *Hyphessobrycon nicolasi* AM Miquelarena & HL López, 2010
2010

Nicolás Bonelli is the grandson of senior author Amalia Miquelarena. The etymology says: "*This species is dedicated to Nicolás Bonelli, whose affection and company we have enjoyed for the last few years.*"

Nicolaus

Riffle Minnow sp. *Alburnoides nicolausi* NG Bogutskaya & BW Coad, 2009

Nicolaus refers to sons of both authors: Nikolay is Nina Bogutskaya's eldest son and Nicholas is Brian Coad's son.

Niebuhr

Niebuhr's Goatfish *Upeneus niebuhri* P Guézé, 1976

Carsten Niebuhr (1733–1815) was a German traveller, orientalist, cartographer and mathematician. Originally Niebuhr had intended to become a surveyor, but (1757) studied mathematics, cartography and navigational astronomy at the University of Göttingen. Niebuhr, Forsskål (a student of Linnaeus) and others were appointed (1760) by Frederick V of Denmark to explore in Egypt and the Arabian Peninsula (1761–1767). They first went to Egypt where they stayed for about a year, before moving on to South Arabia (Arabia Felix, present-day Yemen) (1762). Peter Forsskål collected botanical and zoological specimens, but caught malaria and died aged just 31 (1763). He was buried in Yemen. The only survivor of the expedition was Carsten Niebuhr who was charged with editing all the manuscripts of the journey, which were published as *Descriptiones Animalium – Avium, amphiborum, insectorum, vermium quæ in itinere orientali observavit Petrus Forskål* (1775). This contained an account of a species of goatfish, the first in that family to be described. Niebuhr travelled on from Yemen to Bombay where he stayed for over a year, eventually returning home by way of Muscat, Persia, Cyprus, Syria, and Turkey creating maps all the while. After his return to Denmark he held a military post and was then a civil servant.

Niederlein

Pike-Cichlid sp. *Crenicichla niederleinii* EL Holmberg, 1891

Gustav Niederlein (1858–1924) was a German botanist who emigrated to Argentina. He participated in the 'Conquest of the Desert' campaign (1870–1884) that destroyed indigenous cultures in Patagonia to make their land available for white settlers, and advertised Argentina's suitability for settlement as a curator of the Argentinean expositions in the World Fairs (1889 & 1893). By virtue of this, he was invited by the Argentinean Ministry of Agriculture to organise a herbarium. He was hired as an advisor and contributor, among others by William P Wilson's Commercial Museum in Philadelphia and the French colonial museum in Paris. He was in contact (1894) with Count von Linden, the founder of the Linden-Museum, who was interested in acquiring objects from Niederlein. After the American conquest of the Philippines, the US government commissioned Niederlein and Wilson to create a Philippine exposition for the (1904) World Fair that would provide a justification for the American 'civilizing mission' there. Niederlein did this by displaying members of indigenous groups in a human 'zoo'. Among them were Cordillerans who had to perform 'headhunting dances and savage rituals' for the visitors. Later, Niederlein sent some of their objects to Count von Linden, hoping that von Linden would recommend him for an order of merit from the King of Württemberg.

Nielsen, DTB

Neotropical Rivuline sp. *Hypsolebias nielseni* WJEM Costa, 2005
[Syn. *Simpsonichthys nielseni*]

Dr Dalton Tavares Bressane Nielsen is a Brazilian ichthyologist at the Laboratório de Zoologia, Departamento de Biologia, Universidade de Taubaté, Brazil. The University of Taubaté awarded his BSc (2002). He has written a number of papers with Wilson Costa, such as: *Cynolebias paraguasuensis n. sp. (Teleostei: Cyprinodontiformes: Rivulidae), a new seasonal killifish from the Brazilian Caatinga. Paraguaçu River basin* (2007) and described at least ten species in this genus, most with Costa.

Nielsen, JG

Viviparous Brotula genus *Nielsenichthys* W Schwarzhans & PR Møller, 2011
Scorpionfish sp. *Neoscorpaena nielseni* JLB Smith, 1964
Cusk-eel sp. *Brotulotaenia nielseni* DM Cohen, 1974
Lanternfish sp. *Diaphus nielseni* BG Nafpaktitis, 1978
Slickhead sp. *Bathylaco nielseni* YI Sazonov & AN Ivanov, 1980
Nielsen's Righteye Flounder *Samariscus nielseni* J-C Quéro, DA Hensley & AL Maugé, 1989
Conger Eel sp. *Bassanago nielseni* ES Karmovskaya, 1990
Lefteye Flounder sp. *Tosarhombus nielseni* K Amaoka & J Rivaton, 1991
Morid Cod sp. *Physiculus nielseni* YN Shcherbachev, 1993
Stone Loach sp. *Paraschistura nielseni* TT Nalbant & BG Bianco, 1998
Nielsen's Brotula *Cataetyx nielseni* AV Balushkin & AM Prokofiev, 2005
Cusk-eel sp. *Bassozetus nielseni* S Tomiyama, M Takami & A Fukui, 2018

Jørgen G Nielsen (b.1932) is a Danish zoologist. (The middle initial does not stand for a name, he chose the letter 'G' himself when another Jørgen Nielsen was employed at the museum, in order that everyone could tell who was who – he based this on a family name all his brothers had, 'Gissel'). The University of Copenhagen awarded his natural history degree (1958) after six years of study covering zoology, botany, geology, geography, physiology etc. His thesis was about flatfish distribution and he took employment at University Zoological Museum (ZMUC) sorting the latest fish collection. It was in that year that he wrote his first scientific paper. There followed compulsory military service in the Danish navy. He was immediately hired (1959) as ichthyologist at the museum, where he stayed for the rest of his career until retiring (2002). The University conferred on him a DSc (1969). He has participated in many scientific cruises (1960–2007) and published over 160 papers.

Nierstrasz

Armoured Gurnard sp. *Peristedion nierstraszi* M Weber, 1913
Rough Sillago *Sillago nierstraszi* JDF Hardenberg, 1941

Hugo Frederik Nierstrasz (1872–1937) was a Dutch marine biologist. He studied (1892) medicine at Utrecht University but switched to biology, researching marine animals (1898), and is known most for malacology and carcinology. He took part in the Siboga Expedition (1899–1900) to the Dutch East Indies (Indonesia). When he returned, he taught biology classes and later (1904) became a lecturer in zoology at Utrecht University, rising to become Professor of Zoology (1910). Among his published works were reports on various taxa collected on the Siboga expedition.

Nieto

Pencil Catfish sp. *Trichomycterus nietoi* CA Ardila Rodríguez, 2014

Luis Eduardo Nieto Alvarado is a Colombian ichthyologist. Universidad del Magdalena, where he is a member of the faculty and was awarded his bachelor's degree (1988) and his master's (1993). He co-wrote: *Geographical extension of the sharpnose sevengill shark* Heptranchias perlo *Bonaterre* (Hexanchiformes: Hexanchidae) *for the continental Colombian Caribbean* (2016).

Nieuhof

Nieuhof's Eagle Ray *Aetomylaeus nichofii** ME Bloch & JG Schneider, 1801
[Alt. Banded Eagle Ray]
Slender Walking Catfish *Clarias nieuhofii* A Valenciennes, 1840
Silver Tripodfish *Triacanthus nieuhofii* P Bleeker, 1852

Johan Nieuhoff (1618–1672) was a Dutch traveller who wrote about his journeys to India, China and Brazil. His first trip was to Brazil (1640–1649), after which he joined the Dutch East India Company living in Batavia for several years before being put in charge of an expedition to China (1654–1657). He was in India (1663) and then in Ceylon (Sri Lanka). He returned to the Netherlands (1672) before going to Madagascar where he disappeared. His *Voyages & Travels to the East Indies 1653–1670* was not published until 1988. There is no formal etymology in Bloch & Schneider but they do refer to: *De Historia Piscium libri quator* Willughby, Raius & Nieuhof (1686), which is probably a reference to the first informal description of the ray.

** binomial sometimes 'corrected' to nieuhofii but original spelling should stand*

Nieuwenhuis

Sisorid Catfish sp. *Glyptothorax nieuwenhuisi* LL Vaillant, 1902
Shark Catfish sp. *Pangasius nieuwenhuisii* CML Popta, 1904
Borneo River Loach sp. *Neogastromyzon nieuwenhuisii* CML Popta, 1905

Dr Anton Willem Nieuwenhuis (1864–1953) was an explorer and ethnographer who was a physician in the Dutch East Indian Army (1889–1901) and in Borneo (1893–1900). He led expeditions to central Borneo (1894 & 1896–1897 & 1898–1900) during the last of which he collected the loach type. He was also an ethnologist and specialised in collecting tribal items and artefacts. He became Professor of Geography and Ethnology at Leiden University (1904–1934). He wrote: *In Central Borneo* (1900). A reptile and two birds are also named after him.

Nigel

Onejaw sp. *Monognathus nigeli* E Bertelsen & JG Nielsen, 1987
Cutthroat Eel sp. *Ilyophis nigeli* YN Shcherbachev & KY Sulak, 1997

Nigel Robert Merrett (b.1940) is a British zoologist and ichthyologist. Shcherbachev & Sulak honoured him for his substantial contributions to the knowledge of *Ilyophis* and other synaphobranchid eels, and Bertelsen & Nielsen described him as very helpful during the authors' revision of the Family Monognathidae. (Also see **Merrett**)

Nigri

Niger Hind *Cephalopholis nigri* ACLG Günther, 1859
Guinean Striped Mojarra *Gerres nigri* ACLG Günther, 1859
Goby sp. *Mauligobius nigri* ACLG Günther, 1861
African Knifefish *Xenomystus nigri* ACLG Günther, 1868

This is a toponym referring to the River Niger; the river (or its mouth, in the case of marine species) being the type locality.

Nijssen

Armoured Catfish sp. *Metaloricaria nijsseni* M Boeseman, 1976
Goby sp. *Oligolepis nijsseni* AGK Menon & N Govindan, 1977
Panda Dwarf Cichlid *Apistogramma nijsseni* SO Kullander, 1979
Nijssen's Cory *Corydoras nijsseni* DD Sands, 1989
Leporinus sp. *Leporinus nijsseni* JC Garavello, 1990
Nijssen's Hypopygus (knifefish) *Hypopygus nijsseni* CDCM de Santana & WGR Crampton, 2011

Dr Han Nijssen (1935–2013) was a Dutch ichthyologist who was an expert on the genus *Corydoras*. The University of Amsterdam awarded his doctorate (1970) and he became a curator at the Amsterdam Zoological Museum. He co-wrote: *The Cichlids of Surinam: Teleostei: Labroidei* (1989). Sands honoured him as he had encouraged Sands in his 'early and later ignorance' and wrote the introduction to Sands' first book. Kullander described him as the "*...author of many papers on South American fishes*" (he also brought the dwarf cichlid species to Kullander's attention).

Nik

Brown Brotula *Cataetyx niki* DM Cohen, 1981

Professor Dr Nikolai Vasilyevich Parin (1932–2012). (See **Parin**, also see **Nikolay (Parin)** & **Nik Parin**)

Niki

Sand-diver sp. *Trichonotus nikii* E Clark & K von Schmidt, 1966

Niki Konstantinou is the son of the late senior author, Eugenie Clark (see **Clark, E**).

Nikiforov

African killifish sp. *Nothobranchius nikiforovi* B Nagy, BR Watters & AA Raspopova, 2021

Andrei Nikiforov is a Russian killifish enthusiast who participated in several field surveys aimed at research on *Nothobranchius* species in Tanzania, and co-discovered and co-collected the type specimens of this one.

Nikki

Sabre Goby *Antilligobius nikkiae* JL Van Tassell, L Tornabene & PL Colin, 2012

Nicole Laura Schrier is the daughter of Adriaan 'Dutch' Schrier, owner and Director of the Curaçao Sea Aquarium (opened 1984), and supporter of SECORE (coral reef conservation) who collected many of the type specimens.

Nikolaj and Olivia

Viviparous Brotula sp. *Saccogaster nikoliviae* JG Nielsen, W Schwarzhans & DM Cohen, 2012

Nikolaj and Olivia are two of the grandchildren of the senior author, Jørgen Nielsen (Also see **Nielsen, JG**).

Nikolay (Kotlyar)

Bigscale sp. *Melamphaes nikolayi* AN Kotlyar, 2012

Nikolay Efimovich Kotlyar (1908–1994) was the father of the author, Aleksandr Kotlyar.

Nikolay (Parin)

Nikolay's Lanternfish *Bolinichthys nikolayi* VE Becker, 1978

Professor Dr Nikolai Vasilyevich Parin (1932–2012) (See **Parin**, also see **Nikolay (Parin)** & **Nik Parin**)

Nikolsky, A

Upper Ob Grayling *Thymallus nikolskyi* NF Kaschenko, 1899

Fat Sculpin *Batrachocottus nikolskii* LS Berg, 1900

Dr Alexander Mikhailovich Nikolsky (1858–1942) was a Russian herpetologist. He studied at St Petersburg University (1877–1881), earning his doctorate later (1887). He became

Assistant Professor and a curator of the zoological collection there. He was Director of the Department of Herpetology, Natural History Museum, Russian Academy of Sciences (1895–1903), then Professor, Kharkov University, Ukraine (1903). He made a number of expeditions to the Caucasus Mountains, Iran, Siberia and Japan (1881–1891). Today in Russia the A M Nikolsky Herpetological Society commemorates him. Eight reptiles and three birds are also named after him.

Nikolsky, G

Weatherfish (loach) sp. *Misgurnus nikolskyi* ED Vasil'eva, 2001

Georgy Vasil'evich Nikolskyi (1910–1977) was a Russian ichthyologist who became Head of the Ichthyology Department at Moscow University and the Laboratory of Ichthyology Institute of Animal Morphology of the USSR. He is always associated with the fishing industry and showed great interest in the formation of fish fauna reservoirs and finding ways to increase their fish production. He led a number of expeditions, to the Far East (1945–1949), Amur (1950–1955) and China (1957–1958). Later (1960s) he oversaw expeditions to the Barents Sea and White Sea. His most significant work was the taxonomy textbook: *Private Ichthyology* and he also published over 400 papers and longer works including: *Fishes of Tajikistan* (1938) and *Fishes of the Amur Basin* (1956). He was the author Vasil'eva's teacher.

Nik Parin

North Pacific Daggertooth *Anotopterus nikparini* YI Kukuev, 1998

Professor Dr Nikolai Vasilyevich Parin (1932–2012). (see **Parin**, also see **Nik**)

Nilsson

Whitefish sp. *Coregonus nilssoni* A Valenciennes, 1848

Dr Sven Nilsson (1787–1883) was a Swedish zoologist and archaeologist. He began studying for the priesthood (1806) at Lunds Universitet, but was persuaded by Anders Jahan Retzius, Professor of Zoology, to switch to natural history. He was Director, Naturhistoriska Riksmuseet, Stockholm, and endeavoured to assemble a complete collection of the vertebrates of Sweden (1822–1831). He was Professor of Zoology and Curator of the Museum, Lunds Universitet (1831–1856), where he had taken his doctorate. He wrote: *Illuminerade figurer till Skandinaviens fauna* (1832–1840). A mammal and a reptile are also named after him.

Nimasow

Flying Barb sp. *Esomus nimasowi* S Abujam, B Gogoi, AN Das, DN Das & SP Biswas, 2021

Dr Gibji Nimasow (b.1976) is Senior Assistant Professor (2009–present) in the Department of Geography at Rajiv Gandhi University, Arunachal Pradesh, India. It was here he took his BA, MA, M Phil and PhD (2001–20018). He was honoured for his "*...constant encouragement and interest in fishery related works.*" Among his publications is the co-written: *Ethnomedicinal knowledge among the Adi tribes of lower Dibang valley district of Arunachal Pradesh, India* (2012).

Nina

Onaç Spring Minnow *Pseudophoxinus ninae* JA Freyhof & M Özuluğ, 2006
Nina's Lamprey *Lethenteron ninae* AM Naseka, SB Tuniyev & CB Renaud, 2009
[Alt. Western Transcaucasian Lamprey]
Nina's Chub Petroleuciscus ninae D Turan, G Kalayci, C Kaya, Y Bektaş & F Küçük, 2018

Dr Nina Gidalevna Bogutskaya (b.1958) is a Russian ichthyologist who has chiefly focused on cypriniform fish. Leningrad University awarded her MSc (1981) and PhD, under Svetovidov (q.v.) (1988), in combination with the Zoological Institute of the Russian Academy of Sciences. She then worked there as a researcher in the Laboratory of Ichthyology (1988–1997) until she got the position of the Scientific Secretary of the Zoological Institute (1997). Over the past 6 years (2012) she has lectured in Systematic Ichthyology at St. Petersburg State University. She has also been the principal investigator of a variety of research projects and the head of many expeditions supported by the Russian Foundation for Basic Research. She was honoured in the name of the chub for her contribution to the knowledge of the fishes of Europe and Asia. Among other papers she co-wrote: *Resolving taxonomic uncertainties using molecular systematics:* Salmo dentex *and the Balkan trout community* (2010).

Niobe

Armoured Catfish genus *Niobichthys* SA Schaefer & F Provenzano Rizzi, 1998

In Greek mythology, Niobe was the daughter of Tantalus and wife of Amphion, King of Thebes, by whom she had 14 children. She made the mistake of teasing the goddess Latona for only having two. Latona's children were Artemis and Apollo, gods who did not take kindly to anyone insulting their mother. They killed all of Niobe's sons and daughters; it is unwise to mock the gods! Niobe, who is the personification of maternal grief, wept continuously until she died and was transformed into a stone, from which water ran. This is a fanciful reference to the permanent cloud-mist surrounding Cerra La Neblina, Venezuela, the type locality. Three mammals and a bird are named after her.

Nion

Neotropical Rivuline sp. *Austrolebias nioni* HO Berkenkamp, JJ Reichert & F Prieto, 1997

Dr Heber Nion is a Uruguayan fisheries scientist who was Director of the National Fisheries Institute, Montevideo, Uruguay. He is also a lifelong aquarist. His best-known work is: *Peces del Uruguay : lista sistemática y nombres communes* (2002).

NIRCH

Caspian Marine Shad ssp. *Alosa braschnikowi nirchi* AV Morozov, 1928

Morozov's employer was the Scientific Institute of Fisheries (Moscow), whose initials, when transliterated into English, spell the acronym NIRCH (N=Scientific, I=Institute, R=Fish, Ch=Industry).

Nishimura

Eelpout sp. *Lycodes nishimurai* G Shinohara & SM Shirai, 2005

Dr Saburo Nishimura (1930–2001) was a Japanese biologist and marine historian. He was a Researcher at Japan's National Fisheries Research Institute and Assistant Professor at Seto Marine Biological Laboratory, Kyoto University. He was then Associate Professor at the Department of Liberal Arts and Sciences, Kyoto University (1980) becoming Professor, Faculty of Integrated Human Studies there (1992–1994). From retirement until his death he was Emeritus Professor there. His best-known works are the first and second parts of *The zoogeographical aspects of the Japan Sea* (1965). He was honoured for his zoogeographic studies in the Sea of Japan.

Niwa

Loach genus *Niwaella* TT Nalbant, 1963

Hisashi Niwa was a Japanese ichthyologist. He published a Hokkaido fishing calendar (1949). Nalbant honoured him in the genus name simply because Niwa was the person to first describe a species within that genus; *N. delicata* (named by Niwa as *Cobitis delicata*).

Niwa

Threadtail Anthias *Tosana niwae* HM Smith &TEB Pope, 1906

H Niwa was the Director of the Fishery Experiment Station of Kōchi Prefecture at Susaki, Shikoku, Japan. A number of the fish that the authors described were collected from the experimental station thanks to its Director.

Nkata/Nkhata

Lake Malawi Cichlid sp. *Copadichromis nkatae* TD Iles, 1960

Lake Malawi Cichlid sp. *Placidochromis nkhatae* M Hanssens, 2004

This is a toponym referring to Nkhata Bay, Lake Malawi, the type locality.

Nkhotakota

Lake Malawi Cichlid sp. *Placidochromis nkhotakotae* M Hanssens, 2004

This is a toponym referring to Nkhotakota, a port on Lake Malawi, Malawi; the only known area of the cichlid's occurrence.

Nobili

Atlantic Torpedo *Tetronarce nobiliana* CL Bonaparte, 1835

[Alt. Atlantic Electric Ray]

Leopoldo Nobili (1784–1835) was an Italian physicist who studied animal electricity and invented a number of instruments used in thermodynamics and electrochemistry. He was educated at the Military Academy of Moderna and then served in Napoleon's campaign in Russia as an artillery officer, being awarded the Légion D'Honneur. He was later appointed as Professor of Physics at the Regal Museum of Physics and Natural History in Florence.

Noboli

African Characin sp. *Distichodus noboli* GA Boulenger, 1899

Noboli is the local vernacular name for this species in the Upper Congo.

Noel

Rosy Jewelfish *Anthias noeli* WD Anderson & CC Baldwin, 2000

Noel Archambault (d.1998) was a Canadian IMAX cameraman who died in an aircraft crash in the Galapagos islands, where the jewelfish is found. Among his film credits is: *Into The Deep* (1994).

Noel Kempff

Corydoras sp. *Corydoras noelkempffi* J Knaack, 2004

Professor Noel Kempff Mercado (1924–1986) was a Bolivian biologist, ornithologist and conservationist. The University of Santa Cruz awarded his bachelor's degree (1946). He and some colleagues were murdered by a gang of cocaine producers and smugglers, whom they had taken by surprise in a national park - that park in Bolivia is now named after him, as is an amphibian and a bird.

Nolf

Cusk-eel sp. *Luciobrotula nolfi* DM Cohen, 1981

Dirk Nolf is a Belgian ichthyo-palaeontologist who works at Département Paléontologie, Koninklijk Belgisch Instituut voor Natuurwetenschappen, Brussels. He wrote: *The Diversity of Fish Otoliths, Past and Present* (2013). As the etymology states, he called Cohen's attention to the fact that *Luciobrotula* from the eastern and western Atlantic are different and provided the information on otoliths included in Cohen's paper.

Nom

Stone Loach sp. *Schistura nomi* M Kottelat, 2000

Mr. Nom (forename not given) was honoured for his help (driving) during Kottelat's (1999) field work in Laos.

Nomura

Nomura's Dwarfgoby *Trimma nomurai* T Suzuki & H Senou, 2007

Tomoyuki Nomura is a volunteer in the Fish Division of the Kanagawa Prefectural Museum of Natural History, Japan. He was the third author (Senou being the senior author) of: *Coastal Fishes of Ie-jima, the Ryukyu Islands, Okinawa, Japan* (2006). He collected the type and provided it to the authors.

Nonato

Goby sp. *Cristatogobius nonatoae* GL Ablan, 1940

Susana G Ablan née Nonato was married (1945) to the author Guillermo L Alban (b.1904); she collected the type specimen. Ablan was the long-time Region One Director of Philippine Fisheries Commission based in Dagutan City.

Nonsuch

Bigeye Lightfish *Woodsia nonsuchae* W Beebe, 1932

This is a toponym referring to Nonsuch Island, Bermuda, close to where the holotype was collected.

Nora

Longsnout Pipefish *Leptonotus norae* ER Waite, 1910

The brief description gives no etymology. However, Waite wrote on the scientific results of the New Zealand Government Trawling Expedition (1907) undertaken by the ship 'Nora Niven', so we believe this to be a reference to that vessel.

Norbert

Dward Cichlid sp. *Apistogramma norberti* W Staeck, 1991

Norbert Wiesheu is an aquarist who discovered this species. He was the first to keep it in an aquarium, and, according to Staeck "*...did not shy away from the hardships and costs of subsequently determining its exact location*" (translation).

Nordenskjöld

Nordenskjold's Bigscale *Sio nordenskjoldii* AJE Lönnberg, 1905

Dr Nils Otto Gustaf Nordenskjöld (1869–1928) was a Finnish-Swedish geologist, geographer and polar explorer. He taught as a lecturer and associate professor at Uppsala University where he had been awarded his doctorate in geology (1894). He led mineralogy expeditions to Patagonia (1890s) and to Alaska and the Klondike (1898). He was the leader of the Swedish Antarctic Expedition (1901–1904). He explored Greenland (1909) and Chile and Peru (1920s). He became Professor of Geography at the University of Gothenburg (1905). He was hit by a bus and killed in Gothenburg.

Nordmann

Danube Shad *Alosa caspia nordmanni* G Antipa, 1904

Alexander von Nordmann (1803–1866) was a Finnish-born biologist who was interested in everything from palaeontology to botany and birds to molluscs. He collected extensively in southern Russia. He went to Berlin (1827) to study with the famous parasitologist and anatomist Karl Rudolphi. He became a Professor at Odessa (1832), finally becoming Professor of Zoology at Helsinki University (1849). Antipa frequently cited Nordmann's work (1840) on Black Sea fishes. Two birds and the Nordmann Fir *Abies nordmanniana* are also named after him.

Norma

Viviparous Brotula sp. *Parasaccogaster normae* DM Cohen & JG Nielsen, 1972

Dr Norma Victoria Chirichigno Fonseca (b.1929) is a Peruvian ichthyologist who worked at the Instituto del Mar del Perú, Lima. She is now a professor emeritus at the Faculty of Oceanography at the Universidad Nacional Federico Villarreal, Lima. She wrote: *Nuevos tiburones para la fauna del Perú* (1963). She identified this fish as undescribed and 'graciously' placed her specimens at the authors' disposal.

Norman

Mote Sculpin genus *Normanichthys* HW Clark, 1937
Searsid genus *Normichthys* AE Parr, 1951
Norman's Lampeye *Poropanchax normani* E Ahl, 1928
[Syn. *Aplocheilichthys normani*]
Sole sp. *Aseraggodes normani* P Chabanaud, 1930
Loach sp. *Annamia normani* SL Hora, 1931
Barb sp. *Poropuntius normani* HM Smith, 1931
Paperbones *Luciosudis normani* AF Fraser-Brunner, 1931
Norman's Lanternfish *Protomyctophum normani* AV Tåning, 1932
Squeaker Catfish sp. *Chiloglanis normani* J Pellegrin, 1933
Short-jaw Lizardfish *Saurida normani* WH Longley, 1935
Norman's Rockfish *Scorpaena normani* J Cadenat, 1943
Blenny sp. *Blennius normani* M Poll, 1949
Norman's Tonguefish *Symphurus normani* P Chabanaud, 1950
Sheep Pacu *Acnodon normani* WA Gosline III, 1951
Norman's Smooth-head *Micrognathus normani* AE Parr, 1951
Shortfin Sand Skate *Psammobatis normani* JD McEachran, 1983
Morid Cod sp. *Physiculus normani* R Brüss, 1986
Righteye Flounder sp. *Poecilopsetta normani* VP Foroshchuk & VV Fedorov, 1992

John Roxborough Norman (1898–1944) was an English ichthyologist who started his working life as a bank clerk. He had rheumatic fever during his military service (WW1), which affected him ever after. He started work at the BMNH (1921) under Charles Tate Regan. He was Curator of Zoology (1939–1944) at Tring. He wrote, among other books: *A History of Fishes* (1931) and *A Draft Synopsis of the Orders, Families and Genera of Recent Fishes* (1957). He often lent specimens to ichthyologists to study and was honoured by Hora for loaning him the loach type.

Noronha

Noronha's Scorpionfish *Pteroidichthys noronhai* HW Fowler, 1938
Bigeye Sand Tiger Shark *Odontaspis noronhai* GE Maul, 1955

Adolfo César de Noronha (1873–1963) was a naturalist and librarian who was Director of the Funchal Museum in Madeira (where the type specimen of the shark is housed). He particularly studied the ichthyology, ornithology and malacology of Madeira and worked at the Funchal Municipal Libraray where he was appointed Librarian (1914) and Director (1928). He wrote: *A new species of deep-water shark (Squaliobis Sarmienti) from Madeira* (1926).

Norris

Narrowfin Smooth-Hound *Mustelus norrisi* S Springer, 1939

Dr Harry Waldo Norris (1862–1946) was a zoologist and anatomist who was Professor of Biology and Zoology and Curator of the Museum at Grinnell College, Iowa (1891–1931) and was Research Professor of Zoology (1931–1941), retiring as Emeritus Professor. Iowa College (now Grinnell College) awarded his bachelor's degree (1880), his master's (1886) and his Doctor of Science degree (1924). His graduate work included stints at Cornell University, the University of Nebraska and at Freiburg, Germany. He wrote: *On the cranial nerves in elasmo-branch fishes (*1926).

Norton

Hake Weakfish *Cynoscion nortoni* P Béarez, 2001

Presley Norton Yoder (1932–1993) was an independently wealthy US & Ecuadorian archaeologist and founder of the Research Centre of Salango, Ecuador. Having dual citizenship, he joined the US army and was stationed in Hawaii (1950–1952). He also studied literature and history at Brown University (1950–1953), then was at the Sorbonne (1953–1955). He started a TV channel in Ecuador (1962) and published quite widely, mostly on archaeological matters but also more generally. Béarez wrote that Norton had "*...permitted and encouraged my investigations in Ecuador*." He died of a heart attack.

Nosferatu

Central American Cichlid genus *Nosferatu* M De la Maza-Benignos, CP Ornelas-Garcia, M de L Lozano-Vilano, ME Garcia-Ramírez & I Doadrio, 2015

Nosferatu was the title of a silent German horror film made (1922) by director Friedrich Murnau. The authors write that "*...the pair of well-developed recurved fangs in the upper jaw...*" are "*...reminiscent of those in Murnau's vampire Nosferatu*." In fact, the vampire's name was Orlok – *Nosferatu* is just another word for a vampire.

Nott

Bayou Topminnow *Fundulus nottii* L Agassiz, 1854

Dr Josiah Clark Nott (1804–1873) was an American physician particularly noted for his studies into the aetiology of yellow fever (having lost four children to the disease). His medical degree was awarded by the University of Pennsylvania (1827) which was followed by post-graduate study in Paris. He began medical practice in Mobile, Alabama (1833). He was founder of the Medical College of Alabama (1858) and served as professor of surgery. He was also well known for being an author and proponent of racist theories. A slave owner himself, he claimed that "*...the negro achieves his greatest perfection, physical and moral, and also greatest longevity, in a state of slavery*". He was a friend of Louis Agassiz, and discovered the eponymous species.

Nouhuys

Mountain Hardyhead *Craterocephalus nouhuysi* M Weber, 1910

Captain Jan Willem van Nouhuys (1869–1963) was a Dutch naval officer commanding the ships on expeditions in New Guinea (1903, 1907 & 1909). He carried out independent biological and geological research on the islands of Sula, Indonesia, and was one of the first to reach the permanent snows in the mountains of tropical New Guinea. He returned to the Netherlands and was Director of the Museum of Asian and Caribbean Studies and the Maritime Museum Prins Hendrik (1915–1934). A mammal and two birds are also named after him.

Nourissat

Blue-mouth Cichlid *Wajpamheros nourissati* R Allgayer, 1989

Lamena (Madagascan cichlid) *Paretroplus nourissati* R Allgayer, 1998

Jean Claude Nourissat (1942–2003) was a French cichlid aquarist and collector who risked his life looking for new species in Madagascar, including the *Paretroplus* (he died from malaria three days after returning from his last trip there). He was founder and long-time president of the French Cichlid Association (Association France Cichlid).

Novaes

Armoured Catfish sp. *Furcodontichthys novaesi* LH Rapp Py-Daniel, 1981

Dr Fernando da Costa Novaes (1927–2004) was a Brazilian ornithologist based at the Museu Paraense Emílio Goeldi, Belém, where he assembled a huge collection of bird skins and skeletons. Novaes wrote several books and many papers on the avifauna of various areas of Brazil. Three birds and a mammal are also named after him.

Novick

Barbeled Dragonfish sp. *Bathophilus novicki* MA Barnett & RH Gibbs Jr, 1968

Dr Alvin Novick (1925–2005) was an American physician and biologist who was Professor of Biology at Yale. During WW2 he served in the US army and was a prisoner of war in Germany. After the war he graduated from Harvard Medical School. He was a specialist in the echo-location system used by bats and wrote: *The World of Bats* (1969). In 1982, Novick shifted the focus of his work to the ethical and public policy aspects of the AIDS crisis, at a time when there was great fear and stigmatization over the disease. The etymology states that it was he: "*...who taught the senior author how to see in the dark.*" He died from prostate cancer.

Novikov

Jawfish sp. *Stalix novikovi* AM Prokofiev, 2015

Dr Georgii Gennadievich Novikov (1942–2007) was a Russian ichthyologist who was professor in the Department of Biology at Moscow State University and head of the Laboratory of Fish Ontogenesis of the Department of Ichthyology there. His spent his entire forty-year (1967–2007) scientific career there, having graduated there (1965) and going on to study for his doctorate (1970) at the ichthyology lab of the Russian Academy of Science. He defended a second doctoral thesis at Moscow University (1992). Novikov undertook many expeditions and trips including (1975) the joint Russo-American Expedition at R/V 'Akademik Petrovskii' to the central Atlantic and (1994) the Russian-Swedish Expedition Tundra-Ecology-94 to the Arctic Regions on the R/V 'Akademik Fedorov'. He also took lecture tours, including to Cuba (1982–1983), Japan (1989–1990) and China. His more than 250 works ranged from: *Module-Type Fish-Cultural Farms* (1971) to *The Selective Effect of the Duration of the Period of Critical Temperatures on Some Allozyme Loci of Atlantic Salmon* Salmo salar *L. (Salmonidae)* (2006). The author says of him "*... with whom the first scheduled trawlings in Nha Trang Bay in May 2005 were organized.*"

Novoa

Orinoco Lined Sole *Achirus novoae* F Cervigón, 1982

Daniel Francisco Novoa Raffalli (1946–2006) was President of the Venezuelan Ministry of Agriculture's National Institute of Fish and Fisheries. He had a particular interest in, and knowledge of, the Orinoco Delta region. He was the editor of the work in which the species' description was published.

Nozawa

Prickleback sp. *Stichaeus nozawae* DS Jordan &JO Snyder, 1902
Swallow-tail Pomfret *Pampus nozawae* C Ishikawa, 1904
Freshwater Sculpin sp. *Cottus nozawae* JO Snyder, 1911

Professor Sunziro Nozawa was a Japanese zoologist, curator at the museum in Sapporo, Hokkaido, and also Director of the fisheries bureau there. He loaned the holotype of the prickleback species to the authors.

Numberi

Papuan Whiptail *Pentapodus numberii* GR Allen & MV Erdmann, 2009

Freddy Numberi (b.1947) is a Papuan Indonesian retired Vice-Admiral in the Indonesian Navy who became an ambassador to Italy, Albania and Malta and then Governor of Papua (1998–2000). He was afterwards Minister of Marine Affairs and Fisheries (2004–2009) and Minister of Transport (2009–2001). According to the authors, he "*...has played a critical role in championing marine conservation initiatives in his native Papua.*"

Nupelia

Characin sp. *Characidium nupelia* WJ da Graça, CS Pavanelli & PA Buckup, 2008

Nupélia (Núcleo de Pesquisas em Limnologia, Ictiologia e Aqüicultura) is an organisation that was involved in the survey and ecological research of fishes from the Manso Reservoir region, Mato Grosso, Brazil, during which this species was discovered.

Nutting

Nutting's Moray *Gymnothorax nuttingi* JO Snyder, 1904
Nutting's Hatchetfish *Polyipnus nuttingi* CH Gilbert, 1905

Charles Cleveland Nutting (1858–1927) was an American naturalist and collector. He became Curator of the Museum of Natural History of the University of Iowa (1886), and worked to build up the museum's collection further when he became Professor of Systematic Zoology (1888). He generated public and private support to finance several expeditions. His journeys to the Bay of Fundy, the Bahamas, Nicaragua and Costa Rica added many more specimens, including seabirds, to the collection. On these trips, he was said to be an "*energetic, forceful character, his organizing abilities and his enthusiasm for collecting helped to ensure the success of his trips.*" As well as birds, three of which are named after him, he was particularly interested in hydrozoans (relatives of jellyfish and corals). He collected the type of the moray and was the naturalist of the 'Albatross' Hawaiian expedition (1902) which collected the holotype of the hatchetfish.

Nyanza

Nyanza Barb *Enteromius nyanzae* PJP Whitehead, 1960
Lake Victoria Cichlid sp. *Haplochromis nyanzae* PH Greenwood, 1962

This is a toponym referring to Nyanza, a local name for Lake Victoria, where these species are found.

Nybelin

Blind Cusk-Eel genus *Nybelinella* JG Nielsen, 1972
Nybelin's Sculpin *Triglops nybelini* AS Jensen, 1944
Cod Icefish sp. *Nototheniops nybelini* AV Balushkin, 1976
Spiny Plunderfish sp. *Harpagifer nybelini* VP Prirodina, 2002

Orvar Nybelin (1892–1982) was a Swedish zoologist, marine biologist and ichthyologist. He took part in the expeditions of both 'Skagerak' and 'Albatross'. He was head of the Museum of Natural History in Gothenburg (1937–1958). He wrote: *Våra fiskar och hur man känner igen dem. Illustrerad fickbok med beskrivningar av deras utseende, levnadssätt och förekomst* (1933 & 1937). He 'kindly placed' his specimens at Nielsen's disposal.

Nyerere

Lake Victoria Cichlid sp. *Pundamilia nyererei* ELM Witte-Maas & F Witte, 1985
[Syn. *Haplochromis nyererei*]

Mwalimu Julius Nyerere (1922–1999) was President (1963–1985) of Tanzania, and was regarded by the people as a 'great teacher' as he transformed his independent nation with reforms to its education, health and economy (Mwalimu means 'teacher').

Nystrom

Conger Eel sp. *Gnathophis nystromi* DS Jordan & JO Snyder, 1901
[Junior Synonym of *Gnathophis heterognathus*]

Edvard Thorbjörn Nyström (1863–1950) was a Swedish veterinarian. Uppsala University awarded his bachelor's degree (1892). That year he lectured in domestic animals' anatomy, physiology and medicine at the Ultuna lantbruksinstitut. Later (1898) he became the veterinarian for a stud farm, but a year after this became Professor of Animal Husbandry and Veterinary Law (1899) at the Veterinary Institute in Stockholm. He wrote a number of scientific papers including one on the history of the Japanese fish collection at Uppsala University Zoological Museum, on domestic animal diseases, breeding and care etc. He also wrote a book about his 30 years as secretary of the Teachers College at Veterinary College.

Nyx

Eelpout sp. *Bothrocara nyx* DE Stevenson & ME Anderson, 2005

Nyx was the goddess of night and darkness in Greek mythology. The choice of name is in allusion "*...to the dark conditions prevalent in the deep waters and northern latitudes inhabited by this species, as well as the heavily pigmented lining of the mouth and visceral cavity.*"

O

Oatea

Goby sp. *Stiphodon oatea* P Keith, E Feunteun & E Vigneux, 2010

Oatea is a mythical religious figure in central Polynesia. He created the Marquesas Islands, where this goby appears to be endemic.

Oates

Burmese Barb sp. *Hypsibarbus oatesii* GA Boulenger, 1893
Oates' Spiny-eel *Mastacembelus oatesii* GA Boulenger, 1893

Eugene William Oates (1845–1911) was a civil servant in the public works department in India and Burma (1867–1899), and also an amateur naturalist and ornithologist. He wrote the two-volume: *A Manual of the Game Birds of India* (1883–1899) among other works on birds of the Indian sub-continent. When he retired to England he compiled a catalogue of the birds eggs in BMNH. He was also secretary of the BOU (1898–1901). He collected the holotypes of these fish. Seven birds, an amphibian and a reptile are also named after him.

Obama

Spangled Darter *Etheostoma obama* RL Mayden & SR Layman, 2012
Congo Cichlid sp. *Teleogramma obamaorum* ML Stiassny & SE Alter, 2015
Hawaiian Basslet sp. *Tosanoides obama* RL Pyle, BD Green & RK Kosaki, 2016

Barack Hussein Obama II (b.1961) was 44th US President (2009–2017) and his wife Michelle LaVaughn Robinson Obama (b.1964) was First Lady. They were both honoured in the cichlid's binomial for their "*...commitment to science education, development, gender equality, and self-reliance for all peoples of African nations, and their dedication to environmental conservation in Africa and beyond.*" A bird is also named after him.

Obbes

Sulu Velvetfish *Paraploactis obbesi* MCW Weber, 1913
Obbes' Catfish *Porochilus obbesi* MWC Weber, 1913

Joan François Obbes (1869–1963) was a Dutch artist. He did the illustrations in Weber's monograph in which the original description of these species was published.

Obein

Stone Loach sp. *Schistura obeini* M Kottelat, 1998

François Obein is a French Environmental Specialist whose forte is sustainable hydropower development. The University of Montpellier awarded his bachelor's degree in Agronomy (1988). His master's in tropical forest biology was awarded by Paris VI University (1990).

He has worked in many countries for a variety of organisations including the UN. He is particularly associated with his work in Laos for the Nam Theun 2 Electricity Consortium (Vientiane, Laos). He wrote *Hydropower opportunities in Lao P.D.R* (1997). He was honoured for the assistance he gave the author in the field.

Obermuller

Characin sp. *Pyrrhulina obermulleri* GS Myers, 1926

August Obermüller was credited by Myers as being the person who introduced him to South American characins, at Obermüller's aquarium business in Jersey City, New Jersey.

Occlo

Armoured Catfish sp. *Ancistrus occloi* CH Eigenmann, 1928

In Inca mythology Mama Occlo was a mother and fertility goddess.

Ochiai

Japanese Lowfin Deepwater Dragonet *Callionymus ochiaii* R Fricke, 1981
Tongue-sole sp. *Cynoglossus ochiaii* K Yokogawa, H Endo & H Sakaji, 2008
Flathead sp. *Inegocia ochiaii* H Imamura, 2010

Dr Akira Ochiai (1923–2017) was a fish biologist and systematist who was Emeritus Professor of Kochi University and an Honorary Member of the Japanese Society of Fisheries Science. He graduated from Kyoto University (1950), becoming an assistant professor that year. He then held several positions in the Department of Fisheries, Faculty of Agriculture, Kyoto University. He was appointed (1965) Professor at the Laboratory of Aquaculture Science at Kochi University. He wrote 21 books, 73 papers and 44 reviews as well as ten reports. He was honoured in the name of the tongue-sole "*...for his great contributions to the taxonomy of Japanese cynoglossids.*"

Ochoterena

Chapala Catfish *Ictalurus ochoterenai* F de Buen y Lozano, 1946

Dr Isaac Ochoterena Mendieta (1885–1950) was Professor of Histology and Embryology at, and Director of, Instituto de Biologia, Universidad Nacional de Mexico, Mexico City. He was a Lieutenant Colonel in the Mexican army, having been Professor of Histology at the Mexican Army Medical School. Three reptiles and a bird are named after him.

Ochriamkin

Prickleback sp. *Stichaeus ochriamkini* A Taranetz, 1935

Dmitriy Ivanovich Ochryamkin (aka Okhryamkin) was a Russian fisheries scientist who was very active in the Russian Far East (1930s). Along with Taranetz and Moiseev, he co-authored: *The Commercial Caught Flatfish of Primorie* (1937) and wrote: *Sea fishes of the Far East* (1931), as well as many articles. There is a reference to a 'D M Okhryamkin' being killed during WWII: it is not clear if the initial 'M' is an error, or whether this refers to a different person, but we find nothing more about him after the 1930s.

O'Connor

Snowtrout sp. *Schizothorax oconnori* RE Lloyd, 1908

Captain William Frederick Travers O'Connor (1870–1953) was a soldier, interpreter, commercial attaché and writer. He trained at the Royal Military Academy, Woolwich. He served in the Swat Valley (1897–1898), in Gilgit (1899–1903), and was then appointed as Interpreter and Secretary to the Younghusband Mission to Lhasa (1903–1904). After their withdrawal, he remained in Lhasa as a commercial agent (1904–1908). He was a friend of the Panchen Lama and drove one of the first cars to be brought there: when he left, he gifted the car to the Lama. He then became British Consul in Shiraz, Persia (now Iran) (1912–1915) and later (1918–1920) British Resident in Nepal. He wrote several books including: *Report on Tibet* (1903). Although he is not identified in the etymology, he was part of the (1904) British expedition to Tibet during which the snowtrout type was collected.

Oda

Oda's Skate *Rhinoraja odai* R Ishiyama, 1958

Mikiji Oda discovered the type specimen at the Miya fish market; it had been netted near Mikomoto-jima, off Izu Peninsula.

Odd

Deep-sea Tripodfish sp. *Bathypterois oddi* KL Sulak, 1977

The author decided there were enough taxa named after Greek and Roman gods and classical heroes and decided it was time to introduce Odd, the legendary Icelandic hero of Bandamanna Saga. He is symbolic of good fortune and of the Scandinavian seafaring spirit. Sulak has also said that he liked the play on words because tripodfishes are indeed rather 'odd'.

Odón de Buen

Goby genus *Odondebuenia* F de Buen y Lozano, 1930

Odón de Buen y del Cos (1863–1945) was the author's father. He was a naturalist, politician and founder of the Spanish Institute of Oceanography. When he was a naturalist at the Madrid Natural History Museum, he was invited to undertake a scientific expedition on the frigate 'Blanca' and discovered a love for oceanography and marine science. In the 1890s, he promoted Darwin's theories in his lectures at the University of Barcelona and also published textbooks supportive of evolution. In 1895, the Bishop of Barcelona obtained a ban from the government on the use of his textbooks at the university and a suspension of his classes. However, the majority of students supported de Buen, and the issue caused enough controversy to make the government retreat. Like many other scientists, artists and intellectuals, he left Spain after the civil war and settled in Mexico (1939) for the rest of his life. "*Mis innovaciones científicas me produjeron graves disgustos, desataron contra mí todo género de asechanzas*", ("*My scientific innovations caused me serious annoyances, unleashed all kinds of snares against me*,") he said at the end of his career. (Also see **De Buen, F**)

Odysseus

Batfish sp. *Halicmetus odysseus* AM Prokofiev, 2020

The *Odissey* (latinized as *Odyssey*) was the Soviet research vessel from which the holotype was collected (1984).

Oedipus

Cave Loach sp. *Schistura oedipus* M Kottelat, 1988

This is not a true eponym as Oedipus was a mythical Theban king who tore out his eyes (after discovering that the woman he married was his own mother). The binomial is here used in fanciful reference to the loach's degenerate eyes.

Oeser

African Rivuline sp. *Fundulopanchax oeseri* H Schmidt, 1928

Dr Richard Öser was a German physician who collected plants and animals in Africa. He was particularly interested in bromeliads; a passion that grew from an earlier interest in tropical frogs, when he was looking for plants that would provide the amphibians with a proper environment. He found bromeliads ideal for that purpose. In 1969 he was reported to be in retirement and living in the Black Forest. He was the original collector of the fish on the island of Bioko.

Oestergaard

African Rivuline sp. *Nothobranchius oestergaardi* S Valdesalici & G Amato, 2011

Kaj Østergaard is a Danish friend of the authors who collected the type during a species survey in northern Zambia. He was honoured "*…for his contributions over a long period of time on field investigations that have led to the discovery of many new populations of* Nothobranchius *species in numerous countries.*"

Ogilby, JD

Brotula genus *Ogilbia* DS Jordan & SW Evermann, 1898
Dottyback Genus *Ogilbyina* HW Fowler, 1931
Coral-Brotula genus *Ogilbichthys* PR Møller, W Schwarzhans & JG Nielsen, 2004
Ogilby's Hardyhead *Atherinomorus vaigiensis* JRC Quoy & JP Gaimard, 1825
[*Pranesus obilbyi* is a junior synonym]
Ogilby's Ghostshark *Chimaera ogilbyi* ER Waite, 1898
[Syn. *Hydrolagus ogilbyi*]
Ogilby's Flyingfish *Cypsilurus ogilbyi* DS Jordan & JO Snyder, 1908 NCR
[Junior Synonym of *Cheilopogon unicolor*]
Ogilby's Rainbowfish *Melanotaenia ogilbyi* MCW Weber, 1910
Ogilby's Ghost Flathead *Hoplichthys ogilbyi* AR McCulloch, 1914
Tongue-sole sp. *Cynoglossus ogilbyi* JR Norman, 1926
Gulf Grunter *Scortum ogilbyi* Whitley, 1951
Ogilby's Stinkfish *Callionymus ogilbyi* R Fricke, 2002
[Alt. East Australian Longtail Dragonet; Syn. *Calliurichthys ogilbyi*]

James Douglas Ogilby (1853–1925) was an Irish-born Australian ichthyologist and taxonomist, son of the famous zoologist William Ogilby (below). He worked at BMNH before emigrating to Australia (1884) and joining the Australian Museum, Sydney (1885). He was sacked (1890) for being drunk on the job. The contemporary report criticized his "*extreme and undiscriminating affinity for alcohol*." Though sacked as a permanent employee, he went on working on a contract basis. He worked for the Queensland Museum (1901–1904 & 1913–1920). The Gulf Grunter was named after him as he had prepared the original specimen, although he had not separated it from *S. hilli*. A reptile is also named after him.

Ogilby, W

Vatani Rohtee *Rohtee ogilbii* WH Sykes, 1839

William Ogilby (1808–1873, or possibly 1804–1873) was a Cambridge-educated Irish barrister and zoologist. He practiced at the bar in London (1832–1846), then returned to Ireland. He wrote such papers as: *Descriptions of Mammalia and Birds from the Gambia* (1835), *Exhibition of the Skins of Two Species of the Genus Kemas* (1838), and *Observations on the History and Classification of the Marsupial Quadrupeds of New Holland* (1839), all published in the *Proceedings of the Zoological Society of London*. He was Honorary Secretary of the Zoological Society (1839–1846) and crossed swords a number of times with Gray of the British Museum. He started building a castle (Altnachree Castle) in Ireland in the 1840s, just as the Great Famine struck. The castle was completed in the 1860s. Unlike many landlords, he kept on all his workers and fed them by importing grain. A mammal is also named after him. His son, James Douglas Ogilby (1853–1925) (above), was a notable ichthyologist who described many Australian fishes.

Ogilvie, CS

Lizard Loach sp. *Homaloptera ogilviei* ER Alfred, 1967

Charles Symon Ogilvie (b.1896) of the Malayan Game Department was Superintendent of King George V National Park, Malaya (1948–1954), and assisted Alfred's collecting expedition there. He was a clerk in the Union Bank of Scotland when WW1 broke out. He was gazetted as a 2nd Lieutenant (1914) in the Scottish Horse, and eventually rose to be a Major serving in the Machine Gun Corps (1918). He was a keen amateur ornithologist and ichthyologist and, according to Alfred, an unfailing source of information, inspiration, and assistance. He was also something of an ethnographer and wrote a number of articles such as: *The 'Che Wong': a little known primitive people* (1948) and *Some of our rivers and their game fishes* (1953)

Ogilvie, JW

Pike-Characin sp. *Roestes ogilviei* HW Fowler, 1914

John W Ogilvie was a Scottish adventurer who went to South America in search of gold and rubber. He is believed to have first arrived in British Guiana (Guyana) c.1900 but there are no reliable records before his meeting with the US anthropologist William Curtis Farabee (1913) when Ogilvie was living near the Rupununi river. Unlike many other settlers, Ogilvie was on excellent terms with the local native peoples of the region and was fluent in several of their languages. He is believed to have married and settled down as a cattle rancher among the Wapishana people. What is known is that he agreed to help Farabee and led an expedition of extraordinary length and hardship, involving (1913–1914) traversing previous unexplored jungle and walking from Manaus (Brazil) to Georgetown (British Guiana). It is known that he eventually retired, returned to Scotland, and lived in Edinburgh in a house called 'Rupununi'. His widow gave his papers to the Edinburgh University Museum (1975). He collected the holotype of this species (1911).

Ohe

Lanternbelly sp. *Parascombrops ohei* W Schwarzhans & AM Prokofiev, 2017

Fumio Ohe was a Japanese palaeontologist, ichthyologist and evolutionary marine biologist at the Department of Earth and Planetary Science, Nagoya University, Japan (2005) but has since retired. He was also a member of the Tokai Fossil Society. He co-authored: *Trigonognathus kabeyai, a new genus and species of the squalid sharks from Japan* (1990) among other papers, and also wrote a number of papers on fossil fish.

Oiticica

Characin sp. *Characidium oiticicai* HP Travassos, 1967

José Oiticica Filho (1906–1964) was an entomologist and photographer at the Museu Nacional, Universidade Federal do Rio de Janeiro, Brazil (1943–1964) where Travassos was his colleague. He graduated from the National School of Engineering, Rio de Janeiro (1930).

Okamoto

Okamoto's Deepwater Cardinalfish *Epigonus okamotoi* R Fricke, 2017

Okamoto's Brotula *Bidenichthys okamotoi* PR Møller, W Schwarzhans, H Lauridsen & JG Nielsen, 2021

Makoto Okamoto is a Japanese ichthyologist at the Seikai National Fisheries Research Institute. He has described, or co-described, several new species of *Epigonus* cardinalfishes. He was honoured in the cardinalfish's binomial "*...in appreciation of his valuable and thorough research on the family Epigonidae*."

Okamura

Sandperch sp. *Parapercis okamurai* T Kamohara, 1960

Rattail sp. *Malacocephalus okamurai* T Iwamoto & T Arai, 1987

Okinawa Chromis *Chromis okamurai* T Yamakawa & JE Randall, 1989

Piedtip Cucumberfish *Paraulopus okamurai* T Sato & T Nakabo, 2002

Okamura's Grenadier *Coelorinchus okamurai* **N** Nakayama & H Endo, 2017

Okamura's Deepsea Batfish *Halieutopsis okamurai* H-C Ho, 2021

Dr Osamu Okamura was Professor Emeritus at the Laboratory of Natural Science, Kochi University, where he was Professor of Zoology (1989). He described *Paraulopus filamentosus* (1982) and was the first to recognise *P. okamurai* as a new species. He co-wrote: *Erythrocles microceps, a new emmelichthyid fish from Kochi, Japan* (1998). By the time of the grenadier's description, he was 'the late' Dr Okamura.

Okazaki

Clinid Blenny sp. *Neoclinus okazakii* R Fukao, 1987

Dr Toshio Okazaki was a Japanese geneticist at the Yamazaki University of Animal Health Technology. He was credited by Fukao with the research leading to the identification of *Neoclinus okazakii* as a distinct species. Among his published works is: *Genetic differentiation between two types of dark chub, Zacco temmincki, in Japan* (1991). (See also **Chihiroe**)

Okumus

Okumus' Trout *Salmo okumusi* D Turan, M Kottelat & S Engin, 2014

Dr Ibrahim Okumus (1960–2008) was a fisheries scientist who was Professor of Aquaculture and Dean of the Faculty of Fisheries, Rize University, Turkey (2007–2008) before which he was at the Fisheries Department, Karadeniz Technical University (1994–2007). He graduated at Cukurova University with a master's degree (1986) and studied at the Humberside College of Higher Education, Grimsby, England (1989–1990), and at the University of Stirling, Scotland (1990–1993) where he was awarded his doctorate.

Olalla

Long-whiskered Catfish sp. *Platysilurus olallae* G Orcés V, 1977

R Olalla (either Ramón or his brother Rosalino) is honoured. The whole Olalla family seems to have been professional animal collectors. In addition to Ramón and Rosalino, there were two other brothers, Alfonso and Manuel, plus their father Carlos. The next generation were also collectors, but we have no names for any of them except Jorge and we do not know whose son he was. Various members of the Olalla family have five birds, two mammals and an amphibian named after them.

Olaso

Eelpout sp. *Bellingshausenia olasoi* J Matallanas, 2009

Dr Ignacio Olaso is a Researcher who was (2002–2003) Principle Investigator at the Department of Fishery of the Spanish Institute of Oceanography (IEO). He was described by the author as an "*…expert in feeding ecology of fish*".

Olbrechts

African Rivuline sp. *Epiplatys olbrechtsi* M Poll, 1941

African Characin sp. *Congocharax olbrechtsi* M Poll, 1954

Frans Maria S Olbrechts (1899–1958) was a Belgian philologist and ethnologist who studied at the University of Louvain, Belgium, and at Columbia University, New York (1925–1926 & 1928–1929). He organised the Department of Ethnology, Musées Royaux d'Art et Histoire, Brussels (1929) and taught ethnology, becoming a Professor at the University of Ghent (1932). He was later Director, Musée Royal du Congo Belge (1947–1958).

Olfers

Marine Hatchetfish sp. *Argyropelecus olfersii* G Cuvier, 1829

Ignaz Franz Werner Maria von Olfers (1793–1871) was a German naturalist, historian and diplomat; in which latter capacity he went to Brazil (1816). While in Brazil he described a number of new mammal species. Back in Germany he was made director of the royal art collections (1839), was influential in the establishment of the Neues Museum, Berlin, and became Director General of the Königlichen *Museen*. A reptile and an amphibian are named after him.

Olga

Flyingfish sp. *Cheilopogon olgae* NV Parin, 2009

Olga Vladimirovna Parina was the wife (1966) of the author Professor Dr Nikolai 'Nik' Vasilyevich Parin (1932–2012). They met when she was working at the Laboratory of Microbiology, part of the P. P. Shirshov Institute of Oceanology in Moscow. (Also see **Olpar**)

Oliva

Velvet Catfish genus *Olivaichthys* MGE Arratia Fuentes, 1987

Dr Rubén Oliva (1950–2010) was a naturalist and agronomist, whose doctorate was awarded by the University of Oregon. He and his wife, Professor Maria Beatriz Oliva née Peñafort, founded (2005) and ran Vivero Silvestra, a nursery specialised in xerophilous plants, which she has continued to run after his death. The etymology which is clear in that both husband and wife are included, is otherwise somewhat cryptic: they are said to have "*…expended much effort, patience and money*" seeking diplomystid catfishes in Argentina, but in what capacity is not stated.

Oliveira, C

Tetra sp. *Ctenobrycon oliverai* RC Benine, GAM Lopes & E Ron, 2010

Bumblebee Catfish sp. *Microglanis oliveirai* WBG Ruiz & OA Shibatta, 2011

Armoured Catfish sp. *Curculionichthys oliveirai* FF Roxo, CH Zawadzki & WP Troy, 2014

Dr Claudio de Oliveira is an ichthyologist at the laboratory of Biology and fish genetics, Universidade Estadual Paulista, São Paulo, Brazil, and Professor (1989) at Paulista Júlio de Mesquita Filho State University. The University of São Paulo awarded his doctorate (1991). He is also a Research Associate at the Smithsonian. He co-wrote: *Description of two new species of annual fishes of the Hypsolebias antenori* species group (*Cyprinodontiformes: Rivulidae*), *from Northeast Brazil* (2016). He collected the holotype of *Ctenobrycon oliverai*, but the binomial mis-spells his surname. He was honoured for "*…his dedication and contributions to the study of neotropical freshwater fishes*."

Oliveira, J

Whale Catfish sp. *Cetopsis oliveirai* JG Lundberg & LH Rapp Py-Daniel, 1994

Dr José Carlos de Oliveira is a Brazilian zoologist and ichthyologist. The State University Paulista Júlio de Mesquita Filho awarded his bachelor›s degree (1976) and the University of São Paulo awarded both his master›s (1981) and doctorate (1988). He is at present an associate professor at the Federal University of Juiz de Fora, Brazil. He co-wrote: *A new species of 'Whale Catfish' (Siluriformes: Cetopsidae) from the western portions of the Amazon basin* (2001).

Oliver (Jenkins)

Snake-eel sp. *Cirrhimuraena oliveri* A Seale, 1910

Dr Oliver Peebles Jenkins (1850–1935) worked with Alvin Seale at Stanford University. The patronym is not identified in Seale's text, but almost certainly honours this man. (See **Jenkins, OP**)

Oliver, WRB

Hawknose Grenadier *Coelorinchus oliverianus* WJ Phillipps, 1927

Dr Walter Reginald Brook Oliver (1883–1957) was a New Zealand ornithologist and avian palaeontologist. He wrote the definitive: *New Zealand Birds* (1930) and revised it several times. He also wrote: *The Moas of New Zealand and Australia* (1949). His research notes for: *New Zealand Birds* (1920–1960) and his collected works (1910–1957) are in the archives of the Te Papa Tongarewa, the Museum of New Zealand at Wellington, where he was Director (1928) until his retirement (1947). He was described as: "*…our last true biologist equally authoritative about animals or plants*" by Professor G T S Baylis. A reptile and four birds are also named after him.

Olivier

Steenbras (seabream) *Lithognathus olivieri* MJ Penrith & M-L Penrith, 1969

Dr P G Olivier of the National Museum of Namibia caught the first known specimens. Among his publications are: *New species of the genus* Pimeliaphilus *(Acari: Pterygosomidae) from South West Africa* (1977) and *A Bibliography of Pans and Related Deposits* (1990). Early in his career he was associated with Rand Afrikaans University and its library and also collected for the Namibia Museum. His focus appears to have been on invertebrates such as millipedes, mites and scorpions.

Ollie

Torrent Minnow sp. *Psilorhynchus olliei* KW Conway & R Britz, 2015

Oliver 'Ollie' Crimmen is a friend and colleague of the authors (see **Crimmen**).

Olmsted

Tessellated Darter *Etheostoma olmstedi* DH Storer, 1842

Charles Hyde Olmsted (1798–1878) was once President of the Hartford Natural History Society in Connecticut. He discovered this species. He is famed for establishing the Ox-Cart Library, shipping books from his father's personal collection in Connecticut to Lenox in Ohio (in return for Lenox renaming itself as Olmsted). His cousin, Frederick Law Olmsted (1822–1903) was the designer of Central Park, New York City.

Olney

Yellow-spotted Golden Bass *Liopropoma olneyi* C Baldwin & GD Johnson, 2014

John Edward Olney Sr (1947–2010) was an American ichthyologist, fisheries biologist and naturalist who worked at the Virginia Institute of Marine Science. He was regarded as an expert on the identification of early/larval stages of fishes. His interest in fish started when aged eight he was given an aquarium. The College of William & Mary awarded his BSc (1971) following which he taught high school biology. He was then an Assistant Marine Scientist at VIMS (1972) where he undertook his MA (1974–1978). He became an Instructor there (1979) and Assistant Professor (1982). He took leave of absence to do a PhD at the University of Maryland, his thesis being *Community Structure, small-scale patchiness, transport and feeding of larval fishes in an estuarine plane* (1996). He spent the rest of his career at VIMS. Much of his career and publications focused on larval fish such as: *Preliminary guide to the identification of early life history stages of carapid fishes of the western central North Atlantic* (2003). He died after a year-long battle with cancer.

Olokun

False Moray sp. *Chlopsis olokun* CR Robins & CH Robins, 1966

In the culture of the Yoruba-speaking peoples 'Olokun' is a deity of the sea. The type was collected in the waters off the Ivory Coast, where Yoruba is spoken.

Olpar

Darkbar Flyingfish *Cypselurus olpar* IB Shakhovskoy & NV Parin, 2019

Olga Vladimirovna Parina was the wife of junior author Nikolai Parin. She "*...provided both authors with great help and moral support during the work on this review.*" The name 'olpar' is formed by a combination of the first two letters of her first name and the first three letters of her surname. (See **Olga**)

Olrik

Arctic Alligatorfish *Aspidophoroides olrikii* CF Lütken, 1877

Christian Søren Marcus Olrik (1815–1870) was a Danish Greenlander who was a zoologist, botanist, teacher and Inspector of North Greenland (where the species occurs). He began his teaching career in Copenhagen. He returned to Greenland and was appointed Inspector of the North (1846). During his two-decade tenure as inspector he encouraged the self-sufficiency of the Greenlandic economy and was a member of the Greenland Trade Commission before returning again to Copenhagen. He was a popular contact for scientific expeditions to Greenland, as he was a trained and experienced botanist. A marine worm and a leech are also named after him.

Olsen

Spreadfin Skate *Dipturus olseni* HB Bigelow & WC Schroeder, 1951

Dr Yngve H Olsen (1905–2000) was a Danish American ichthyologist who worked (until at least 1963) at the Bingham Oceanic Laboratory, Yale University. He took part in the Vermillion Sea Expedition (1959). Gordon Arthur Riley, in his reminiscences, recalled that Olsen was (1937) assistant to Albert E Parr and preferred to work in the laboratory as he was prone to seasickness. When Parr became the Director of the Laboratory and Museum and had no time for cruising, Olsen turned to editorial work and was honoured for his work as editor of the monograph series: *Fishes of the Western North Atlantic*.

Olson

Sand Stargazer sp. *Storrsia olsoni* CE Dawson, 1982

Storrs Lovejoy Olson (b.1944) is an American biologist and avian palaeontologist who was Curator of the Division of Birds in the USNM (1975–2009). Olson's main specialisation is fossil birds and he was also *de facto* curator of the fossil bird collection in the Department of Paleobiology, which is the largest in the world by far. He graduated (1966) from the Florida State University and went on to achieve his doctorate (1972) at Johns Hopkins University. He is the author of well over 300 publications in a variety of scientific journals. His interests are primarily in avian palaeontology and systematics, avifaunas of oceanic islands prior to human-caused extinction events, and biogeography and systematics of Neotropical birds, especially those of the Panamanian isthmus. Olson has received many honours from ornithological institutions. He collected the holotype in Brazil (1973). Five birds are also named after him.

Omar

Omar Electric Fish *Gymnotus omarorum* MM Richer-de-Forges, WGR Crampton & JS Albert, 2009

The two people after whom this species is named are Omar Macadar and Omar Trujillo-Cenoz, both of whom are Uruguayan scientists and pioneers in the study of the anatomy and physiology of electrogenesis in knifefishes and who have worked closely together. Both of them are, officially but not really, retired.

Dr Omar Macadar is physician and neurologist at Department of Neuroscience, Instituto de Investigaciones Biologicas Clemente Estable, Montevideo. He qualified as a physician at Universidad de la Republica, Uruguay (1969), which has also honoured him with an honorary doctorate. He undertook postgraduate research at the University of California. He co-wrote: *Vasotocin actions on electric behavior: interspecific, seasonal, and social context-dependent differences* (2010).

Dr Omar Trujillo-Cenoz (b.1933) was at the Department of Comparative Neuroanatomy, Instituto de Investigaciones Biologicas Clemente Estable, Montevideo. He qualified as a Doctor of Medical Science at Universidad de la Republica, Uruguay (1968), He co-wrote: *Electroreception in* Gymnotus carapo*: pre-receptor processing and the distribution of electroreceptor types* (2000).

Omont

Mexican Cichlid sp. *Paraneetroplus omonti* R Allgayer, 1988
[Possibly a Junior Synonym of *Paraneetroplus gibbiceps*]

Jean-Marie Omont is a French aquarist whom Allgayer described as a "*...traveller devoted to the cause of the cichlid family*" (translation).

Ona

Nyanga Elephantfish *Cryptomyrus ona* JP Sullivan, S Lavoué & CD Hopkins, 2016

Marc Ona Essangui is a Gabonese environmental and civic activist, founder and Executive

Director of the NGO 'Brainforest' and recipient of the 2009 Goldman Environmental Prize. He was honoured for "...*his efforts to protect Gabon's equatorial forests and wetlands.*"

Onuma

Damselfish sp. *Chromis onumai* H Senou & T Kudo, 2007

Hisashi Onuma "*who was the first to discover the species*" was also thanked for his assistance in collecting the new damselfish at Izu-oshima Island, Japan, "*and for his excellent underwater photographs.*"

Oort

Java Rockskipper *Praealticus oortii* P Bleeker, 1851

Pieter van Oort (1804–1834) was a Dutch naturalist and collector in the East Indies (Indonesia). He travelled to the west of Java (1826–1828), to the islands of Ambon, New Guinea and Timor (1828–1829), and finally to the west coast of Sumatra (1833–1834), where he died of malaria near Padang. A bird is also named after him.

Opdenbosch

Ivindo Mormyrid *Ivindomyrus opdenboschi* L Taverne & J Géry, 1975

Armand Opdenbosch was a Belgian who was the Chief Technician (Taxidermist) of the MRAC, Tervuren (1929–1977). He was a close friend of Max Poll, then Curator of Vertebrates, whom he accompanied on an expedition to the Belgian Congo (1956) to collect good-quality mammal specimens to be exhibited in the Belgian Congo's pavilion at the Brussels Exposition (1958). He was honoured for his invaluable technical aid in the study of mormyrid systematics at MRAC. He is also commemorated in the name of a monkey species.

Ophuysen

Spotstripe Snapper *Lutjanus ophuysenii* P Bleeker, 1860

Johannes Adrianus Wilhelmus Van Ophuysen (1820–1890) was a Dutch colonial administrator and amateur naturalist. He was Assistant Resident of Benkoelen (now Benkulu, a province of Sumatra), Indonesia. He collected the type, which Bleeker described in the paper: *Fish species from the fresh waters of Benkoelen, collected by J A W van Ophuysen.*

Opic

Snake-eel sp. *Hemerorhinus opici* J Blache & M-L Bauchot, 1972

Pierre Opic is a French artist and illustrator who was formerly a technician at IRD Laboratory (1965–1996). He provided the illustrations for J Blache's African anguilliform monographs. He also illustrated Seret's: *Poissons de mer de l'ouest Africain tropical* (2007).

Opud

Sulawesi Rainbowfish sp. *Telmatherina opudi* M Kottelat, 1991

The name is based on a local name rather than an eponym. One reference says: "*Telmatherina* is known by the community as the opudi fish".

Orbigny

Argentine Conger *Conger orbignianus* A Valenciennes, 1837
Large-tooth Flounder sp. *Paralichthys orbignyanus* A Valenciennes, 1839
River Pellona sp. *Pellona orbignyana* A Valenciennes, 1847
[Junior Synonym of *Pellona flavipinnis*]
Characin sp. *Astyanax orbignyanus* A Valenciennes, 1850
Characin sp. *Brycon orbignyanus* A Valenciennes, 1850
Smooth-back River Stingray *Potamotrygon orbignyi* FL de Castelnau, 1855

Alcide Charles Victor Dessalines d'Orbigny (1802–1857) was a traveller, collector, illustrator and naturalist. His father, Charles-Marie Dessalines d'Orbigny (1770–1856) was a ship's surgeon. Alcide went to the Academy of Science in Paris to pursue his methodical paintings and classification of natural history specimens. The Muséum National d'Histoire Naturelle (Paris) sent him to South America (1826) where the Spanish briefly imprisoned him, mistaking his compass and barometer, which had been supplied by Humboldt (q.v.), for 'instruments of espionage'. After he left prison, he lived for a year with the Guarani Indians, learning their language. He spent five years in Argentina and then travelled north along the Chilean and Peruvian coasts, before moving into Bolivia and returning to France (1834). Once home he donated thousands of animal specimens, as well as plants, samples of rocks, fossils, land surveys, pre-Colombian pottery, etc. to the MNHN. His fossil collection led him to determine that there were many geological layers, revealing that they must have been laid down over millions of years. This was the first time such an idea was put forward. He was the author of *Dictionnaire Universel d'Histoire Naturelle* (1861). He is also remembered in the names of ten birds, two amphibians, two mammals and five reptiles, as well as other taxa. (Also see **D'Orbigny**)

Orces

Characin sp. *Boehlkea orcesi* JE Böhlke, 1958
[Syn. *Hemibrycon orcesi*]

Professor Gustavo Edmundo Orcés-Villagomez (1902–1999) was an Ecuadorian zoologist and herpetologist who taught at Meji College and then at the National Polytechnic, Quito. At first, he was a pioneer herpetologist describing four new species and seven new fish species. He later went on to study birds, amassing and classifying the most valuable ornithological series for Ecuador. The Fundación Herpetológica Gustavo Orces at the Museum of Natural History was founded (1989) to hold his collection. A bird, five reptiles, two amphibians and two mammals are also named after him.

Orcutt

Arroyo Chub *Gila orcuttii* CH Eigenmann & RS Eigenmann, 1890

Charles Russell Orcutt (1864–1929) was primarily a botanist and malacologist. He combined collecting with publishing scientific journals. He had no formal schooling, being taught on the farm by his parents. The family moved (1879) from Vermont to San Diego. He started (1884) to publish 'The West American Scientist' to get his own work and notes before the public. It continued to appear sporadically (1884–1919). He accumulated a large, if eclectic, collection, which ended up with the San Diego Society of Natural History. He collected for the Smithsonian (1927–1929) in Baja California, Mexico, Central America and the Caribbean. He collected the type of the chub using a blanket as a seine net. A reptile and an amphibian are also named after him.

Ordway

Ordway's Brotula *Brotula ordwayi* SF Hildebrand & FO Barton, 1949
[Alt. Fore-spotted Brotula, Ordway's Bearded Cusk-eel]

Samuel Hanson Ordway, Jr. (1900–1971) was a lawyer, conservationist and trustee of the New York Zoological Society. He graduated from Harvard (1924) and practised as a lawyer in New York (1925–1958). He wrote: *A Conservation Handbook* (1949). The 7,800-acre Ordway Prairie in South Dakota was bought by the Nature Conservancy, of which he was a founder, as a memorial to him.

Oregon

Hooktail Skate *Dipturus oregoni* HB Bigelow & WC Schroeder, 1958
Cutthroat Eel sp. *Synaphobranchus oregoni* PHJ Castle, 1960
Flapnose Conger *Parabathymyrus oregoni* DG Smith & RH Kanazawa, 1977

These fish are named after the US National Marine Fisheries Service research vessel 'Oregon', from which the holotypes were collected.

Oren

Tilefish sp. *Hoplolatilus oreni* E Clark & A Ben-Tuvia, 1973

Otto Haim Oren (1921–1983) was a Croatian-born Israeli chemist and oceanographer at the Haifa Sea Fishery Research Station (which published the authors' paper). He was Director, Foreign Relations Department, Israel Oceanographic and Limnological Research Ltd., Haifa (1973). Among his published works are: *Some hydrographical features observed by the coast of Israel* (1952) and *Changes in temperature of the Mediterranean Sea in relation to the catch of the Israel trawl fishery during the years 1954–55 and 1955–56* (1957).

Orestes

Thorny Catfish sp. *Hassar orestis* F Steindachner, 1875

Orestes Hawley St John (1841–1921) was a palaeontologist who was a member of the Thayer Expedition to Brazil. The expedition was led by Louis Agassiz, who decided the fish should be named after St John as he had collected the holotype and, presumably, Steindachner obliged. St John was also Agassiz' pupil and became (1871) an assistant in palaeontology at the Museum of Comparative Anatomy, Harvard, after having been a member of the Iowa Geological Survey (1866–1871). He was regarded as one of his generation's leading experts on fossil fishes.

Orestias

Andean Pupfish genus *Orestias* A Valenciennes, 1838

Orestis or Orestes is a Greek name meaning "he who stands on the mountain". Presumably the fish genus is not named after the famous Orestes of Greek mythology, the son of Clytemnestra and Agamemnon, but was inspired by the mountain habitat of these pupfish.

ORI

Black Leg-Skate *Indobatis ori* JH Wallace, 1967
[Syn. *Anacanthobatis ori*]
Bronze Whiptail *Lucigadus ori* JLB Smith, 1968
Natal Snakelet *Natalichthys ori* R Winterbottom, 1980
Snake-eel sp. *Yirrkala ori* JE McCosker, 2011

This is an acronym for the Oceanographic Research Institute (ORI) of South Africa, honoured for its contributions to the ichthyology of South African waters. Its research vessel *David Davies* collected the type of the whiptail.

Oriana

Barbeled Plunderfish sp. *Artedidraco orianae* CT Regan, 1914

Oriana Fanny Wilson, née Souper (c.1874–1945) was an English naturalist and humanitarian; the wife of English polar explorer Edward Adrian Wilson (q.v.). He took part in the Terra Nova Expedition (1910–1912) during which the fish holotype was collected. The original brief description contains no etymology, but it seems highly likely that this fish is named after her. At some point she destroyed much of her personal correspondence, so many details of her later life are now unknown. However, after being widowed (1912) she appears to have travelled extensively through East Africa (based on surviving correspondence), and also travelled to an area of Australia's Northern Territory that had not previously been visited by a Western woman. It was here that she collected some small mammals which she sent to the British Museum. Among them were three specimens of a bat which Oldfield Thomas (q.v.) named after her.

Origuela

Neotropical Rivuline sp. *Anablepsoides origuelai* DTB Neilsen & RP Veiga, 2021

Fábio Origuela de Lira is a Brazilian environmentalist, anthropologist and archaeologist. His degree in anthropology was awarded by State University of Rio de Janeiro (2003). After specialising in Brazilian archaeology at the Institute of Brazilian Archaeology he took a masters in landscape history at the Federal University of Viçosa. He was Coordinator of the National Inventory of Cultural References at IPHAN/Amazonas (2005–2008) and Head of the Cultural Heritage of Intangible Nature Nucleus – Manaus City Hall (2006–2007), since when he has been the Coordinator responsible for archaeological and cultural heritage licensing projects in the north, northeast, southeast and south regions of Brazil. He has been a Partner-Director of Meandros Socioambiental (since 2015). He is also a specialist in rivuline fish and is the Technical Coordinator of the Killifish Brazil Institute for the preservation of annual fish species. He co-wrote: Hypsolebias trifasciatus, *a new species of annual fish (Cyprinodontiformes: Rivulidae) from the rio Preto, rio São Francisco basin, northeastern Brazil* (2014).

Orontes

Garra sp. *Garra orontesi* E Bayçelebi, C Kaya, D Turan & J Freyhof, 2021

Orontes, according to the ancient Greek epic poem *Dionysiaca* (circa 400 CE), was an Indian military leader who killed himself and fell into a river after losing to the god Dionysus in single combat. The river was thereafter known as the Orontes (in modern Turkey and Syria) and is where this species occurs.

Orpheus

Orpheus Dace *Squalius orpheus* M Kottelat & PS Economidis, 2006

This is not a true eponym as it refers to Orpheus, the legendary Thracian musician and poet, son of King Oeagrus (or, in some versions, the god Apollo) and the Muse Calliope. The name reflects the dace's occurrence in Thrace, Greece.

Orr

Mangrove Molly *Poecilia orri* HW Fowler, 1943

G A Bisler Orr (1916–1956) was an American pilot and collector of fish. Fowler reported on a collection of fishes made by him in the Bay Islands, Honduras (1942). He also examined other specimens collected by Orr along the North Carolina coast (1941). Described as a pilot and 'chief African representative of the Aero Service Corporation' he was killed in an air crash in Northern Rhodesia (now Zambia) (1956).

Orso

Cameroon Cichlid sp. *Parananochromis orsorum* A Lamboj, 2014

Rose and Anthony 'Tony' Orso of Vernon, New Jersey, USA are aquarium-fish importers and breeders, and lovers of tropical fish. He was a chemistry and physics teacher but has been running a wholesale tropical fish business for many years (from 1981). He was joined by his wife Rose when they started dating (2001). They import from South America, Africa, Europe and Asia. Tony got his first aquarium when seven years old. They helped Lamboj import a number of new cichlid species over the years and donated specimens for scientific research; this provided him with the initial stimulus to check collections for additional species of *Parananochromis*.

Ørsted

Mexican Moonfish *Paraselene orstedii* CF Lütken, 1880

Professor Anders Sandoe Ørsted (aka Oersted) (1816–1872) was a Danish naturalist, botanist, mycologist and marine biologist who travelled (1845–1848) in the Caribbean and Central America. He became Professor of Botany at the University of Copenhagen (1851–1862). The orchid genus *Oerstedella* is named after him.

Ortega, A

Pencilfish sp. *Lebiasina ortegai* CA Ardila Rodríguez, 2008

Dr Armando Ortega Lara is a Colombian biologist and ichthyologist who is Managing Director, Foundation for Research and Sustainable Development. The Universidad del Valle Cali awarded his bachelor's degree (1996), his master's (2005) and his doctorate (2014). He co-wrote: *Fishes of the Andes of Colombia* (2005).

Ortega, H

Tetra sp. *Cheirodon ortegai* RP Vari & J Géry, 1980
Amazonian Cichlid sp. *Bujurquina ortegai* SO Kullander, 1986
Characin sp. *Creagrutus ortegai* RP Vari & AS Harold, 2001
Neotropical Rivuline sp. *Moema ortegai* WJEM Costa, 2003
[Described the same year as *M. quiii*, which name probably has precedence]
Loreto Panda Cory *Corydoras ortegai* MR Britto, FCT Lima & MH Hidalgo, 2007
Ortega's Hypopygus (knifefish) *Hypopygus ortegai* CDCM de Santana & WGR Crampton, 2011
Suckermouth Catfish sp. *Astroblepus ortegai* CA Ardila Rodríguez, 2012
Dwarf Cichlid sp. *Apistogramma ortegai* R Britzke, C de Oliveira & SO Kullander, 2014
Sea Catfish sp. *Chinchaysuyoa ortegai* AP Marceniuk, J Marchena, C Oliveira & R Betancur-R, 2019

Dr Hernán Ortega Torres is a Peruvian ichthyologist at the National University of San Marcos, Lima, where he is Professor and Curator of the fish collection in the university's Museum of Natural History. He was co-leader of the Peruvian leg of the South American Transcontinental Catfish Expedition (1904). He co-wrote: *Hemibrycon tridens. The IUCN Red List of Threatened Species* (2016). He was honoured in the dwarf cichlid's binomial for his "*...life-long dedication and contribution to the study of the fishes of Peru.*"

Ortenburger

Kiamichi Shiner *Notropis ortenburgeri* CL Hubbs, 1927

Professor Dr Arthur Irving Ortenburger (1898–1961) was an American herpetologist and naturalist. The University of Michigan, where he had done his undergraduate work, awarded his PhD (1925). He also initially worked at the AMNH. He was a member of the Faculty of Zoology at the University of Oklahoma (1924–1958), rising to full Professor and Curator of the University's Museum of Zoology. In his early days at the University he directed the Oklahoma Biological Survey. Among his over 150 published works are: *A Key to the Snakes of Oklahoma (*1927) and *The Ecology of the Western Oklahoma Salt Plains* (1933). In his retirement to an Island in British Columbia he studied the whaling industry and whale embryos. In honouring him in the name of the shiner, Hubb's etymology says that Ortenburger "*...is initiating an Oklahoma Fish Survey.*"

Orti

Characin sp. *Acrobrycon ortii* DK Arcila-Mesa, RP Vari & NA Menezes, 2014

Dr Guillermo Ortí is an Argentine evolutionary biologist, whose bachelor's degree was awarded by the University of Buenos Aires (1981). Stony Brook University, New York, awarded his doctorate (1995). His post-doctoral studies (1995–1997) were at the Department of Genetics, University of Georgia, Athens, USA. He was an Associate

Professor, University of Nebraska, Lincoln (2003–2009), since when he has been at George Washington University where he is now Weintraub Professor of Biology. He co-wrote: *The populations of Odontesthes (Teleostei: Atheriniformes) in the Andean Region: body shape and hybrid individuals* (2013).

Ortmann

Long-whiskered Catfish sp. *Pimelodus ortmanni* JD Haseman, 1911
Dwarf Cichlid sp. *Apistogramma ortmanni* CH Eigenmann, 1912

Dr Arnold Edward Ortmann (1863–1927) was a Prussian-born US naturalist, zoologist and specifically a malacologist who became (1903) Curator of Invertebrate Zoology at the Carnegie Museum. He studied at the University of Kiel and the University of Strasbourg before being awarded his PhD by the University of Jena (1885). He then (1886) became an instructor at the University of Strasbourg. He took part in an expedition to Zanzibar (1890–1891). He emigrated to the USA (1894) where he became Curator of Invertebrate Palaeontology at Princeton University (1894–1903). He took part in the Peary Relief Expedition (1899) and (1900) became a naturalized US citizen. Sometime after joining the Carnegie Museum he also became (1910) Professor of Physical Geography at the University of Pittsburgh, where he was also awarded a ScD (1911) and went on to take the Chair of Zoology there (1925). He formulated 'Ortmann's Law of Stream Position', which states that a species of mussel can have a different appearance depending on where in a river system the individuals live. Among his publications was: *Grundzüge der Marinen Tiergeographie* (Foundations of marine animal geography) (1886).

Osbeck

Osbeck's Grenadier Anchovy *Coilia mystus* C Linnaeus, 1758

Pehr (Peter) Osbeck (1723–1805) was a Swedish pupil of Linnaeus. He became Chaplain on a Swedish East Indiaman, 'The Prince Charles', which sailed to Canton, China (1750) also visiting Java before returning to Gothenburg (1752). A very keen naturalist, Osbeck made an extensive collection during that time which he then passed on to Linnaeus, who described all the plants the following year. Osbeck wrote an account of his journey: *Voyage to China and the East Indies* (1757).

Osborn

Mormyrid ssp. *Pollimyrus isidori osborni* JT Nichols & L Griscom, 1917

Dr Henry Fairfield Osborn (1857–1935) was an American zoologist, palaeontologist, humanist and evolutionist. He was Professor of Natural Sciences at Princeton (1881–1891), where he had graduated (1880). He was Professor of Biology and Zoology at Columbia University (1891–1907), and worked at the American Museum of Natural History (1908–1933) where he was to become President of the Board of Trustees. Osborn was a leading proponent of Darwin's theory of evolution and was called as an expert witness at the famous 'Scopes monkey trial' (1925). However, his beliefs clearly led him down some dark avenues, as he was a confirmed racist, once saying, "*The Negroid stock is even more ancient than the Caucasian and Mongolians as may be proved by an examination not only of the brain, of the hair, of the bodily characteristics, but of the instincts, the intelligence. The standard intelligence of the average adult Negro is similar to that of the eleven-year-old youth of the species Homo sapiens.*" He was honoured in the name of the mormyrid for "*...continuing and inspiring support*" that led to the "*...great success*" of the authors' Congo Expedition. Four mammals and a reptile are also named after him.

Osburn

Candy Darter *Etheostoma osburni* CL Hubbs & MB Trautman, 1932

Dr Raymond Carroll Osburn (1872–1955) was an American zoologist who taught at the Ohio State University. He was home schooled because of a physical disability and entered Ohio State University, graduating BSc (1898). He taught at Starling Medical College while gaining his MSc from OSU and went on to Columbia University which awarded his PhD (1906). He took various posts as Assistant Professor at Barnard College, the chair of Biology at Connecticut College for Women, Scientific Investigator for the US Fisheries Department and Associate Director of the New York Aquarium. He then became Professor of Zoology and Entomology at OSU for 25 years. He wrote many papers and at least three books. He was honoured in the binomial "*...in recognition of the contributions he has made, by study and encouragement, to the advancement of our knowledge of the freshwater fishes of interior North America.*"

Oscar

Neotropical Rivuline sp. *Renova oscari* JE Thomerson & DC Taphorn, 1995

Oscar Leon Mata is a Venezuelan scientist and engineer at the Museo de Ciencias Naturales in Guanare where Donald Taphorn also works. He was a co-discoverer of the species (1991) at Isla Raton in the Rio Orinoco on the Colombian border and collected another series later (1992). He illustrated Taphorn's PhD dissertation (1990). They seem to have been friends for many years and he has often helped Taphorn collect fish.

Oschanin

Sisorid Catfish sp. *Glyptosternon oschanini* SM Herzenstein, 1889

Vasili Fedorovich Oschanin (1844–1917) was a Russian zoologist and entomologist who was also President of the Russian Geographic Society. He led an expedition to Buchara, Karategin and northwest Pamir (1878). He wrote a number of papers and the book: *List of the Palaearctic hemipteras with particular reference to their distribution in Russian empires* (1904). He collected part of the type series.

Osgood

Tetra sp. *Copeina osgoodi* CH Eigenmann, 1922

Characin sp. *Bryconamericus osgoodi* CH Eigenmann & WR Allen, 1942

Dr Wilfred Hudson Osgood (1875–1947) was an American ornithologist and mammalogist. He studied in California at Santa Clara and San Jose, then taught at a school in Arizona for a year before moving to the newly formed Stanford University. He then began working as a biologist in the US Department of Agriculture (1897–1909) in charge of the US biological investigation in Canada. He worked at the Field Museum, Chicago, as Assistant Curator,

Mammalogy and Ornithology (1909–1921) and Curator of Zoology (1921–1940). He conducted biological explorations and surveys of many areas of North and South America, Ethiopia and Indo-China. He spent several years (1906, 1910 & 1930) studying in European museums. He led the Field Museum Abyssinian Expedition (1926–1927) and their Magellanic Expedition (1939–1940). He co-wrote: *Artist and Naturalist in Ethiopia* (1936), and wrote *Mammals of Chile* (1943). Ten mammals, five birds and an amphibian are also named after him.

Osher

Galapagos Thornyhead *Trachyscorpia osheri* JE McCosker, 2008

Bernard Osher (b.1927) is an American businessman and philanthropist. He was described by John McCosker as a fisherman, amateur ichthyologist and supporter of research and education. His BA was awarded by Bowdoin College and he started work running a large hardware store, going on to establish World Savings and other businesses which he has sold. He has been called 'the quiet philanthropist' and was listed in Forbes (2005) as the 584th richest man in the world and the eleventh most philanthropic. He intends to give his whole fortune away through the foundation he established (1977), which supports over 120 US Universities and many research fellowships. His other passions are art, opera and fly-fishing.

Osima

Silk Sculpin *Furcina osimae* DS Jordan & EC Starks, 1904

This is a toponym; Oshima is a subprefecture in southern Hokkaido, Japan.

Osse

Lake Tana Barb sp. *Labeobarbus osseensis* LAJ Nagelkerke & FA Sibbing, 2000

Dr Jan W M Osse (b.1935) is a Dutch zoologist who studied biology at Leiden (1961) and was a member of the scientific staff there (until 1971). He was appointed Professor of Zoology at Wageningen University & Research Centre (1972) and retired (2000) as Emeritus Professor but remained active in research. Among his more than 120 published papers is: *Allometric Growth in Fish Larvae: Timing and Function* (2004). Despite the construction of the scientific name seeming to denote a place, the name is a patronym honouring Osse, who helped initiate the authors' research on the cyprinids of Lake Tana (Ethiopia), and for his "*…knowledge on many aspects of biology, his stimulating criticism of the work, and his original ideas about approaching practical and scientific challenges in the field.*"

Ostenfeld

Ostenfeld's Lanternfish *Diaphus ostenfeldi* AV Tåning, 1932

Carl Hansen Ostenfeld (1873–1931) was a Danish botanist. He was Professor of Botany, Royal Veterinary and Agricultural University and Keeper of the Botanical Museum (1900–1918). He was Professor of Botany, University of Copenhagen (his alma mater) and Director, Copenhagen Botanical Garden (1923–1931). He was also a director of the Carlsberg Foundation (q.v.) and chairman of the committee that edited the oceanographic reports of the 'Dana' (q.v.) expeditions. He co-wrote: *Plankton fra det Røde Hav og Adenbugten* (1901). Around a dozen plants are also named after him.

Osvaldo

Armoured Catfish sp. *Rineloricaria osvaldoi* I Fichberg & CC Chamon, 2008

Dr Osvaldo Takeshi Oyakawa. (See **Oyakawa**)

Otaki

Fat Greenling *Hexagrammos otakii* DS Jordan & EC Starks, 1895
Scalyhead Goby *Hazeus otakii* DS Jordan & JO Snyder, 1901

Keinosuke Otaki (d.1911) was Professor of English at the Imperial Military Academy of Tokyo for a decade, and a former zoology graduate (1894) student at Stanford University (where Jordan was president). He was, in fact, the first Japanese student at Stanford. He worked briefly for the US Fish Commission and then returned to Japan and made collections of fish there (1885 & 1896) when working as an assistant at the Imperial Fisheries Bureau of Japan. He also accompanied (1900) Jordan and Snyder in their travels through northern Japan, when he was teaching at the military academy, and to whom they were 'indebted for many favours' as he acted as their interpreter. He later became Professor of Ichthyology and fishery matters at the Agricultural College at Sapporo. He collected many fish later described by Jordan. With others he wrote a series of popular essays entitled *Fishes of Japan*. He died as a result of injuries sustained in a tramway accident that had happened three years before. He has been described as one of the most active and efficient of Japanese naturalists.

Otto Gartner

African Rivuline sp. *Aphyosemion ottogartneri* AC Radda, 1980
[Syn. *Aphyosemion ogoense ottogartneri*]

'Professor' Otto Gartner (1925–2018) was an Austrian policeman (retired 1980) and aquarist who collected in Africa with Alfred Radda. His first article on fish was *Kleine Fische aus Afrika* (1971). He visited West and Central Africa a number of times in the 1970s and 1980s, collecting fish and plants.

Ottoni

Neotropical Rivuline sp. *Anablepsoides ottonii* WJEM Costa, PHN Bragança & PF Amorim, 2013

Dr Felipe Polivanov Ottoni is a Brazilian ichthyologist who is Professor of Zoology and head of the laboratory at Universidade Federal do Maranhão. The etymology says: "*Named after the ichthyologist Felipe Ottoni for his constant enthusiasm and friendship, and important participation during expeditions to northern Brazil.*" His main current projects are 'Total evidence phylogeny and species delimitation of the clade Laetacara + Rondonacara (Teleostei: Cichlidae: Cichlasomatini)'; "Diversity and distribution of species of *Australoheros* from the Atlantic forest, eastern Brazil (Teleostei: Cichlidae: Heroini)"; "Biodiversity estimates, taxonomy, systematics, biogeography, genetic, ecology and conservation of brackish and freshwater fishes from the Ceará, Maranhão, Pará and Piauí states"; "Phylogeny and taxonomy of Neotropical cichlids (Teleostei: Ovalentaria)". Among his published papers is: *Taxonomic revision of the genus* Australoheros *Rícan & Kullander, 2006 (Teleostei: Cichlidae) with descriptions of nine new species from southeastern Brazil* (2008), co-written with Costa.

Otto Schmidt

African Rivuline sp. *Nothobranchius ottoschmidti* BR Watters, B Nagy & DU Bellstedt, 2019

Otto Schmidt is a "*keen birder and fish enthusiast*", who was honoured for "*...his long-time and significant contributions to the study of fishes of the genus Nothobranchius.*" He was co-collector of the type with EB Watters and B Cooper.

Oudot

Bamako Mormyrid *Mormyrops oudoti* J Daget, 1954

'Monsieur Oudot' was honoured because he obtained the type and several other 'rare and interesting species' while shopping in Bamako, Mali, on the Niger River. We have found no further biographical details.

Oumat

Saffron Anthias *Pseudanthias oumati* JT Williams, E Delrieu-Trottin & S Planes, 2013

Not an eponym, but based on the word for 'sun' in the language of the Marquesas Islands and referring to the fish's colour.

Ouwens

Ouwens' Goby *Sicyopterus ouwensi* M Weber, 1913

Pieter Antonie Ouwens (1849–1922) was a Dutch naturalist and Director of the Java Zoological Museum and Botanical Gardens. He is best known for writing the first formal description of the Komodo Dragon, the world's largest lizard (1912).

Ovcharov

Lanternfish sp. *Myctophum ovcharovi* SA Tsarin, 1993

Oleg Petrovich Ovcharov of the Ukrainian Academy of Sciences made particular studies of how fish move through water and their growth patterns. He also first noted that a related species of lanternfish (*M. asperum*) in the Indian Ocean did not migrate to the surface in its diurnal vertical migrations. He wrote a number of papers, such as: *Features of the flow around the body of the sevryuga Acipenser stellatus* (1976) and *Age and growth of the lantern fish Myctophum nitidulum (Myctophidae) from the tropical Atlantic* (1992).

Oven

Halosaur sp. *Halosaurus ovenii* JY Johnson, 1864

Professor Sir Richard Owen (1804–1892). Apparently Johnson was a Latin-purist, who could not bring himself to use a *w*, a letter that does not exist in Latin, in the binomial. (See **Owen, R**)

Owashi

Eelpout sp. *Eulophias owashii* Y Okada & K Suzuki, 1954

A toponym: the type was caught off Owashi (Owase) on the Pacific coast of southern Honshu, Japan.

Owen, R

Pupfish sp. *Orestias owenii* A Valenciennes, 1846
[Junior Synonym of *Orestias jussiei*]
Halosaur sp. *Halosaurus ovenii* JY Johnson, 1864

Professor Dr Sir Richard Owen (1804–1892) was a British anatomist and biologist who studied medicine at the University of Edinburgh. He became Assistant Curator of the Hunterian Collection at the Royal College of Surgeons and Superintendent of the Natural History Departments of the British Museum. His fame as a scientist led to his appointment to teach natural history to Queen Victoria's children. He was largely responsible for the creation of the Natural History Museum in London, having successfully separated it from the British Museum. He was the first person to conceptualise what a moa looked like (especially its size), based on his study of a few bones sent from New Zealand. He also named the *Dinosauria* ('terrible lizards'), so creating the present-day dinosaur industry. Despite his fame, Owen had many critics. The palaeontologist Gideon Mantell (father of Walter Mantell) said of him that it was 'a pity a man so talented should be so dastardly and envious'. When Mantell suffered an accident that left him permanently crippled, Owen exploited the opportunity by renaming several dinosaurs, which had already been named by Mantell, even claiming credit for their discovery himself. Owen's unwillingness to come off the fence concerning the debate over evolutionary theory also became increasingly damaging to his reputation. He was honoured as the person "*...whose investigations in regard to the skeletons of fishes are not the least valuable part of his many contributions to zoological science.*" Two mammals, a bird and a reptile are named after him.

Owen, RJ

African Barb sp. *Enteromius owenae* CK Ricardo-Bertram, 1943

Rachel Janet Trant née Owen (1912–2005) was born in the USA. She studied fishes with the author in Lake Rukwa (Tanzania) and in the Bangweulu Region of Zambia (1936–1937). She was a member of the Women's Land Army in WW2, married (1945) Ion Fitzgibbon Trant, High Sheriff of Montgomeryshire (1973) and became a Justice of the Peace. J P Harding wrote: *Cladocera and Copepoda collected from East African lakes by Miss CK Ricardo and Miss RJ Owen* (1942).

Owroewefi

Guianan Cichlid sp. *Guianacara owroewefi* SO Kullander & H Nijssen, 1989

Not an eponym but derived from a local name for cichlids of this type; *owroe wefi* meaning 'old wife'.

Owston

Bandfish genus *Owstonia* S Tanaka, 1908
Goblin Shark *Mitsukurina owstoni* DS Jordan, 1898
Snailfish sp. *Liparis owstoni* DS Jordan & JO Snyder, 1904
Owston's Chimaera *Chimaera owstoni* S Tanaka, 1905
Roughskin Dogfish *Centroscymnus owstonii* S Garman, 1906

Owston's Slickhead *Alepocephalus owstoni* S Tanaka, 1908
Rockfish sp. *Sebastes owstoni* DS Jordan & WF Thompson, 1914

Alan Owston (1853–1915) was an English businessman who was a collector of Asian wildlife, as well as a yachtsman. He left for the Orient when still quite young and worked as a merchant in Yokohama, Japan, all the while collecting local fauna. He had no formal training, but collected both for pleasure and to sell, especially to wealthy collectors like Lord Rothschild and museums and universities in the USA and Britain eager to acquire specimens from Japan. Owston's most active collecting period was in the early years of the 20th century. The Carnegie Museum of Natural History, Pittsburgh, acquired a collection of 1,364 of his Asian fishes. He died of lung cancer in Yokohama. Eleven birds, an amphibian and a mammal are named after him, as well as other marine organisms.

Oyakawa

Thorny Catfish sp. *Leptodoras oyakawai* JLO Birindelli, LM de Sousa & MH Sabaj Pérez, 2008
Characin sp. *Probolodus oyakawai* O Santos & RMC Castro, 2014

Dr Osvaldo Takeshi Oyakawa is a marine biologist and limnologist at the Museu de Zoologia da Universidade de São Paulo where his doctorate was awarded (1998). Among his c.40 papers he co-wrote: *A new species of* Characidium *Reinhardt, 1867 (Characiformes: Crenuchidae) endemic to the Atlantic Forest in Paraná State, southern Brazil* (2016). He collected much of the type series of *Probolodus oyakawai*.

Oyama

Sculpin sp. *Astrocottus oyamai* M Watanabe, 1958
[Status unclear: holotype (and only specimen) apparently lost]

Dr Katsura Oyama (1917–1995) was a Japanese palaeontologist, zoologist and malacologist who supplied Watanabe with the type. The Imperial University Tokyo awarded his BSc (1941). After the award of his PhD (1955) he was a post-doctoral researcher at Stanford University (1955–1957). He became a research assistant at the Natural Resources Institute at the Interior Ministry, then the Imperial Navy's Macassar Research Institute (1944–1946) and then returned to the Institute (1946–1947). He spent the rest of his career with the Geological Survey of Japan (1947–1979). After retiring he founded a research laboratory at Toba Aquarium, Mie, which also housed his vast library of paleontological and malacological literature. The etymology reads thus: "*The specimen, which was the gift of Dr. Katsura Oyama, one of the authorities in conchology, was named for the consignor.*"

Oyens

Oyens' Goby *Redigobius oyensi* LF de Beaufort, 1913

Ferdinand August Hendrik Weckherlin de Marez Oyens (1883–1941) was a geologist and palaeontologist who became the Curator of the Amsterdam Geological Institute. He wrote: *Geological Expedition of the University of Amsterdam to the Lesser Sunda Islands in the South Eastern Part of the Netherlands East Indies 1937* (1940). He collected one of the three specimens that de Beaufort examined.

Ozawa, K

Yellow-spotted Duckbill *Acanthaphritis ozawai* RJ McKay, 1971

Keijiro Ozawa was Captain of the *Umitaka Maru*, the vessel from which the holotype was taken (1969). Ozawa donated the type, along with other fishes, to the Western Australian Museum.

Ozawa, T

Onejaw sp. *Monognathus ozawai* E Bertelsen & JG Nielsen, 1987

Dr Takakazu Ozawa is a Japanese ichthyologist at the Faculty of Fisheries, Kagoshima University. He wrote the collection of papers: *Studies on the Oceanic Ichthyoplankton in the Western North Pacific* (1986) and other articles in the Japanese Journal of Ichthyology such as: *Occurrence and Abundance of Bregmacerotid Larvae in Kagoshima Bay, Southern Japan, with Descriptions of Ontogenetic Larval Characters* (1992). He was honoured in the binomial as he kindly let the authors describe the only Japanese specimen of this genus.

P

Paamiut

Paamiut Eelpout *Lycodes paamiuti* PR Møller, 2001

The R/V Paamiut is a research vessel of the Greenland Institute of Natural Resources.

Paccagnella

Royal Dottyback *Pictichromis paccagnellae* HR Axelrod, 1973

The Paccagnella family of Bologna, Italy, were aquarium fish wholesalers who provided the type to Axelrod.

[NB. Since the name honours more than one person, the binomial should be *paccagnellorum*]

Pachacuti

Characin sp. *Bryconamericus pachacuti* CH Eigenmann, 1927

Pachacuti Inca Yupanqui (1438–1471/72) was the ninth ruler of the Kingdom of Cuzco, which he transformed into the Inca Empire.

Paepke

Rattail sp. *Kuronezumia paepkei* YN Shcherbachev, YI Sazonov & T Iwamoto, 1992
Tetra sp. *Hyphessobrycon paepkei* A Zarske, 2014

Dr Hans-Joachim Paepke (b.1934) is a German zoologist, ichthyologist, fish collector and writer. He was Curator of the ichthyology collection at the Museum für Naturkunde der Humboldt-Universität, Berlin, where he worked (1953–1999). Among his full-length written works are: *Die Stichlinge: Gasterosteidae* (1996), *Die Segelflosser* (2003) and *Die Paradiesfische: Gattung Macropodus* (2005). He has described at least one new fish species.

Page

Armoured Catfish sp. *Hypostomus pagei* JW Armbruster, 2003
Loach Catfish sp. *Amphilius pagei* AW Thomson & ER Swartz, 2018

Professor Dr Lawrence 'Larry' Merle Page (b.1944) is an American zoologist who is (2002–present) Curator of Fishes at the Florida Museum of Natural History and Director (2011–present), iDigBio (Coordinating Center for Advancing Digitization of Biodiversity Collections). The University of Illinois awarded his PhD (1972). He was a Professional Scientist for the Illinois Natural History Survey (1972–2000). He was a Graduate Faculty Member University of Illinois (1998–2003) and also an Affiliate Professor, Department of Animal Biology there during the same period. At the same time (1989–1996) he was Director, Center for Biodiversity, Illinois Natural History Survey and Professor, Department of Natural Resources and Environmental Sciences, University of Illinois (1994–2003). Since 2001 he has been Principal Scientist Emeritus, Illinois Natural History. His interests are primarily systematics, evolution, and ecology of freshwater fishes; and the protection of aquatic natural areas. He has a long list of publications from the late 1960s through to the present day, including the *Peterson Field Guide to Freshwater Fishes of North America and Mexico* (1991, 2011) and *Revision of the horseface loaches (Cobitidae, Acantopsis), with descriptions of three new species from Southeast Asia* (2017). Thomson and Swartz honoured him in the catfish's name "*...for his excellent contributions to the study of freshwater fishes.*" (See also **Lawrence (Page)**)

Pagenstecher

African Barb sp. *Labeobarbus pagenstecheri* JG Fischer, 1884

Dr Arnold Andreas Friedrich Pagenstecher (1837–1913) was a German entomologist, particularly interested in Lepidoptera. He studied medicine at Wurzburg, Berlin and Utrecht then worked as an Assistant to his cousin Alexander Pagenstecher (1828–1879) at his ophthalmology clinic in Wiesbaden. He established a general practice there (1863). He became (1876) a *Sanitätsrat* (medical officer), followed by an appointment as *Geheimen Sanitätsrat* (privy medical counsellor) (1896). He was also Director of the Naturhistorischen Museum in Hamburg. He is best known for his extensive studies of Lepidoptera species of the Malay Peninsula. He wrote: *Die geographische Verbreitung der Schmetterlinge* (1909), looking at the causes of geographical distribution of lepidoptera. A dragonfly is also named after him.

Paiva

Paiva's Blenny *Lupinoblennius paivai* SY Pinto, 1958

Dr João de Paiva Carvalho (1903–1961) was a Brazilian marine biologist. He started out (1925) as an ornithologist but was appointed (1932) to the Department of Marine Fisheries and started research in that field, which continued for the rest of his career. He worked in the Zoology Department at the University of Sao Paulo (1940–1945). What is now the Oceanographic Institute there then started up and he led it until his death. He also edited their bulletin (1950–1961) and wrote over 150 papers. His focus was on parasites and shellfish in particular. Around ten marine species are named in his honour.

Pakhorukov

Pakhorukov's Rockling *Gaidropsarus pakhorukovi* YN Shcherbachev, 1995

N P Pakhorukov is a Ukrainian ichthyologist at the Institute of the Biology of Southern Seas, National Academy of Sciences of the Ukraine, Sevastopol, Crimea. He wrote: *Observations of deep-sea fishes from submersibles in the Sierra Leone Rise area (Atlantic Ocean)* (1999). He collected the holotype of this species.

Palacios

Bogardilla (Spanish cyprinid) *Iberocypris palaciosi* I Doadrio, 1980

Dr Fernando Palacios Arribas is an evolutionary biologist and conservationist who is a Research Scientist at the Museo Nacional de Ciencias Naturales (Madrid, Spain). He is Director of Consejo Superior de Investigaciones Científicas (CSIC). The University of Madrid awarded his PhD. He is co-founder of the Foundation for Development and Nature. He was honoured for his 'tireless' (translation) research of Spanish vertebrates.

Palacký

Goby sp. *Drombus palackyi* DS Jordan & A Seale, 1905

Jan Palacký (1830–1908) was a Czech geographer, biogeographer and politician. He was the author of *Die Verbreitung der Fische* (1895), a work containing a list of the fishes of the Philippines, where the goby holotype was collected.

Palaeostomi

Caspian Shad ssp. *Alosa caspia palaeostomi* A Sadowsky, 1934
[Syn. *Alosa tanaica palaeostomi*]

This is a toponym referring to Palaeostomi Lake, west Georgia (Eurasia), the type locality.

Paleracio

Peri's Snake-eel *Myrichthys paleracio* JE McCosker & GR Allen, 2012
[Alt. Whitenose Snake-eel]

Christopher 'Peri' Paleracio. (See **Peri**)

Pallaoro

Mediterranean Goby sp. *Zebrus pallaoroi* M Kovačić, R Šanda & J Vukić, 2021

Armin Pallaoro (1955–2020) was a Croatian ichthyologist who was a scientific adviser to the Institute of Oceanography and Fisheries in Split. The University of Zagreb awarded his biology BSc (1983), MSc in ecology (1988) and PhD (1996). He was a scientific assistant at the University of Split (1985–1986) and thereafter spent his career at the Institute of Oceanography and Fisheries. He published 104 scientific papers and 50 studies and reports, and was co-author of *Red Book of Sea Fishes of Croatia* (2008). He was described by his colleagues as "…devoted to ichthyology, fisheries ecology and biology."

Pallary

Zousfana Barbel *Luciobarbus pallaryi* J Pellegrin, 1919
Sidi Ali Trout *Salmo pallaryi* J Pellegrin, 1924 (Extinct)

Paul Maurice Pallary (1869–1942) was a French-Algerian malacologist who collected the type specimen of the barbel. His pioneering work was on the molluscs of the western Mediterranean Sea and the Middle East. He was also interested in all aspects of zoology and geology and the pre-history of Northern Africa, and became known as the 'Dean of North African Pre-history'. He was the co-discoverer (1892) of the Neolithic caves at Cuartel and Kouchet El Djit. Among his published papers are: *Les cyclostomes du nord-ouest de l'Afrique* (1898) and *Deuxième addition à la faune malacologique de la Syrie* (1939). He named more than 100 molluscs and ten species are named after him.

Pallas

Pacific Herring *Clupea pallasii* A Valenciennes, 1847
Aspsik (whitefish sp.) *Coregonus pallasii* A Valenciennes, 1848
East Siberian Grayling *Thymallus pallasii* A Valenciennes, 1848
Caspian Sand Goby *Neogobius pallasi* LS Berg, 1916

Peter Simon Pallas (1741–1811) was a German-born Russian (1767) explorer, zoologist and one of greatest 18th century naturalists. He earned his doctorate from the University of Leiden at the age of 19! He went to London (1761) to study the English hospital system and was enchanted by the Sussex coast and the countryside in Oxfordshire. The Empress Catherine II summoned him to Russia (1767) to become the Professor of Natural History at the St Petersburg Academy of Sciences and to investigate Russia's natural environment. He was also a geographer and traveller and explored widely in lesser-known areas of Russia. He headed an Academy of Sciences expedition (1768–1774) which studied many regions of Russia, including southern Siberia (Altai, Lake Baikal and the region to the east of Baikal). He described many new species of mammals, birds, fish, insects and fossils. His works include *A Journey through Various Provinces of the Russian State* (1771), *Flora of Russia* (1774), and *A History of the Mongolian People and Asian-Russian Fauna* (1811), as well as works on zoology, palaeontology, botany, ethnography, etc. He found a mass of iron weighing 700 kilograms (1772) that was a new class of meteorite – which are now named after him, as 'pallasites'. Seven mammals and three reptiles are also named after him, as is a volcano in the Kuril Islands. He was honoured in the herring's name as he had identified this herring as a form of *Clupea harengus* in his *Zoographia Rosso-Asiatica* (1811).

Pallon

Scale-rayed Wrasse *Acantholabrus palloni* A Risso, 1810

Probably a toponym. Risso does not provide an etymology, but described this species in his work *Ichthyologie de Nice, ou histoire naturelle des poissons du Département des Alpes Maritimes*. The Paillon (earlier spelt Pallon) is a river which enters the Mediterranean at Nice, though its course through the city is now covered.

Palmer

Emperor Tetra *Nematobrycon palmeri* CH Eigenmann, 1911
Armoured Catfish sp. *Chaetostoma palmeri* CT Regan, 1912

Mervyn George Palmer (1882–1954) was an English naturalist, traveller and collector in Central and South America. After graduating, he became an analytical chemist before deciding upon a career as a freelance collector and naturalist. He collected for the BMNH, London (1904–1910) in Colombia, Ecuador and Nicaragua. During this time, he discovered over 60 species new to science, learnt to speak Spanish and two South American Indian languages, married a South American woman, undertook archaeological digs, and explored and mapped the Río Segovia between Nicaragua and Honduras. He worked for commercial concerns in Ecuador (1910–1918) before moving to London with the same company. After suffering malaria and yellow fever, he was declared unfit for overseas army service in WW1. He was in Venezuela (1919–1921), later being based in London, but with frequent visits to

South America. He lived in Ilfracombe, Devon, England (1932–1954), where he founded and was curator of a museum and ran a library and field club, as he 'wanted something to do'. He was also at one time the Editor of the Natural Science Gazette. He wrote: *Through Unknown Nicaragua – The Adventures of a Naturalist on a Wild-Goose Chase* (1945). A reptile, two amphibians and a bird are named after him.

Palmqvist

African Rivuline sp. *Nothobranchius palmqvisti* AJE Lönnberg, 1907

Gustaf Palmqvist was the patron who funded the Kilimanjaro-Meru expedition (1905–1906) led by Yngve Sjöstedt (q.v.) which collected the type (and 27 other species in the same genus).

Pam (McCosker)

Damselfish sp. *Chromis pamae* JE Randall & JE McCosker, 1992

Pamela J McCosker is the wife of the second author. She was compared with the fish, which was described as a "*...slender specimen of comparable beauty*."

Pam (Rorke)

Pam's Dwarfgoby *Eviota pamae* GR Allen, WM Brooks & MV Erdmann, 2013

Pamela Scott Rorke is the second author's wife and a diving member of the expedition that discovered this goby.

Pame

Mexican Cichlid sp. *Nosferatu pame* M De la Maza-Benignos & M de L Lozano-Vilano, 2013

[Syn. *Herichthys pame*]

This is named for the Pame people of México, whose territory includes five municipalities in the state of San Luis Potosí, where this cichlid is found.

Pan

Pan Sole *Brachirus pan* F Hamilton, 1822

Driftwood Catfish sp. *Gelanoglanis pan* BB Calegari, RE Reis & RP Vari, 2014

Pan was the Greek God of fertility and male sexuality. The catfish was so named in reference to the large gonopodium of the males of this species. The reason behind the sole's name is unclear, and may not refer to the god. One suggestion is that Hamilton found it to be a local name for the fish in India.

Pancho Villa

Mexican Cichlid sp. *Thorichthys panchovillai* LF Del Moral-Flores, E López-Segovia & T Hernández-Arellano, 2017

José Doroteo Arango Arámbula (1878–1923) is better known as 'Francisco Villa' and 'Pancho Villa', whom the etymology describes as the "*...historical, chief and fundamental pillar of the Mexican Revolution.*" (translation)

Pandora

Rio Grande Chub *Gila pandora* ED Cope, 1872

Although the etymology is not explained; Cope was unsure of the 'truer affinities' of the species and mentions several genera to which it might belong; so perhaps its taxonomic ambiguity was a Pandora's box: i.e., a source of troubles for Cope (Mark Sabaj Pérez, pers. comm.)

Panduro

Panduro's Dwarf Cichlid *Apistogramma panduro* U Römer, 1997
[Alt. Blue Panda Apisto]

Jesus Victoriano Panduro Pinedo and Noronha Jorge Luis Panduro Pinedo are Peruvian ornamental-fish exporters. They were the first to recognise this cichlid as a new species, and they collected and shipped the type specimens to Germany.

Panizza

Adriatic Dwarf Goby *Knipowitschia panizzae* D Verga, 1841

Dr Bartolomeo Panizza (1785–1867) was an Italian anatomist and physician who became (1809) a professor at the University of Pavia. Among other things, he studied post-reproductive mortality in male Sea Lampreys. His most famous work was *Osservazioni sul nervo ottico* (1855), a treatise that established the posterior cortex as the part of the brain which controls vision. The patronym is not identified by Verga, but is almost certainly in honour of Bartolomeo Panizza.

Panos

Acheloos Roach *Leucos panosi* NG Bogutskaya & K Iliadou, 2006

Professor Dr Panos Stavros Economidis is a Greek ichthyologist. He was formerly Director of the Ichthyology and Zoology Laboratories in the School of Biology, Aristotle University of Thessaloniki, and was the main contributor to its fish collection. He is now Emeritus Professor. He is an expert in fish taxonomy and conservation, and in fisheries management with more than four decades of experience. He has written or co-written many papers, including: *Fish Fauna of Greece in a Protected Greek Lake: Fish biodiversity, impact of introduced fish species on the ecosystem* (2008).

Pantalone

Pantalone Dwarf Cichlid *Apistogramma pantalone* U Römer, E Römer, DP Soares & IJ Hahn, 2006

Pantalone (or Pantaloon) is a principal character in the Comedia dell'Arte, an Italian early form of professional theatre. He is an elderly and clumsy-looking gentleman who constantly and hotly pursues young girls with whom he was infatuated, usually without success. The name is in reference to this cichlid's unusual courtship behaviour, apparently unique in the genus, in which males swim around females (regardless of their readiness to spawn) in a "*…sometimes rather violent and clumsy-looking zig-zag dance*" as he "*…tries to impress her with his passionate courtship*."

Pantulu

Coromandel Hairtail *Lepturacanthus pantului* MV Gupta, 1966

Dr V R Pantulu was a biologist at the Central Inland Fisheries Research Station, Calcutta and had been a fisheries scientist in West Bengal (type locality). He went on to become Chief of the UN Environmental Unit, part of the Mekong Secretariat, Thailand. Among his published works are his PhD thesis at the University of Calcutta: *Contribution to the study of the biology and fishery of some estuarine Fishes* (1966) anf *Fish of the lower Mekong basin* (1986).

Panzer

Pencil Catfish sp. *Stegophilus panzeri* E Ahl, 1931

Werner Panzer (1901–1976) was a zoologist and entomologist. He was at Freiburg until 1937 when he moved to Danzig (Gdansk). He was a graduate student and traveling companion of Hans Böker (1886–1939), the zoologist who collected the holotype.

Paoi

Chinese Cyprinid sp. *Hongshuia paoli* E Zhang, X Qiang & J-H Lan, 2008

This is a toponym, referring to 'Pao Li' in Guangxi Province, China, which is the type locality.

Papachibe

Characin sp. *Characidium papachibe* LAW Peixoto & WB Wosiacki, 2013

Papa-chibé is a name traditionally associated with people from the Brazilian state of Pará, which is where this species occurs.

Pappenheim

Pappenheim's Stonebasher (mormyrid) *Hippopotamyrus pappenheimi* GA Boulenger, 1910
Eightbarbel Gudgeon *Gobiobotia pappenheimi* M Kreyenberg, 1911
Crocodile Icefish sp. *Cryodraco pappenheimi* CT Regan, 1913
Albertine Rift Cichlid sp. *Haplochromis pappenheimi* GA Boulenger, 1914
Characin sp. *Aphyocharax pappenheimi* E Ahl, 1923
[Junior Synonym of *Aphyocharax dentatus*]
Three-barbeled Catfish sp. *Pimelodella pappenheimi* E Ahl, 1925
Bangkok Halfbeak *Zenarchopterus pappenheimi* EW Mohr, 1926
Slickhead sp. *Bathytroctes pappenheimi* HW Fowler, 1934
Stone Loach sp. *Triplophysa pappenheimi* PW Fang, 1935

Professor Dr Eugen Julius Adolph Paul Pappenheim (1878–1945) was a German zoologist who was Curator of Fishes, Königliche Zoologische Museum, Berlin. Among his many scientific papers is: *Neue und ungenügend bekannte elektrische Fische (Fam. Mormyridae) aus den deutsc afrikanischen Schutzgebieten* (1906). Boulenger honoured Pappenheim for his contributions to the knowledge of mormyrids. Fowler honoured him in the slickhead's binomial as the "*…investigator of the deep-sea fishes obtained by the German South Polar Expedition, 1914*".

Para

Para Molly *Poecilia parae* CH Eigenmann, 1894
Leporinus sp. *Leporinus parae* CH Eigenmann, 1907

This is a toponym, referring to Pará State in Brazil.

Parakana

Armoured Catfish sp. *Lamontichthys parakana* A de Carvalho Paixão & M Toledo-Piza, 2009

The Parakanã are an indigenous people who used to live in the area of the lower Rio Tocantins, Pará, Brazil, where the holotype was collected.

Paramarshall

Rattail sp. *Coryphaenoides paramarshalli* NR Merrett, 1983

Named because of its similarity to *Coryphaenoides marshalli* (see **Marshall**).

Parapietsch

Dreamer (anglerfish) sp. *Oneirodes parapietschi* AM Prokofiev, 2014

Named because of its similarity to *Oneirodes pietschi* (see **Pietsch**).

Parbevs

Glasshead Barrel-eye sp. *Rhynchohyalus parbevs* AM Prokofiev & EI Kukuev, 2020

Professor Dr Nikolai Vasilyevich Parin (1932–2012) (q.v.), Dr Tat'yana Nikolaevna Belyanina (q.v.) and Dr Sergei Afanasievich Evseenko (b.1949) (q.v.).The binomial is a combination of the first letters (**Par** + **b** + **evs**) of the last names of these three Russian ichthyologists who collaborated on a revision of barrel-eyes: *Materials to the revision of the genus Dolichopteryx and closely related taxa (Ioichthys, Bathylychnops) with the separation of a new genus Dolichopteroides and description of three new species* (2009).

Pardi

Somalian Giant Catfish genus *Pardiglanis* M Poll, B Lanza & A Romoli Sassi, 1972

Dr Leo Pardi (1915–1990) was an Italian ethologist and zoologist. The University of Pisa awarded his bachelor's degree (1938) and doctorate (1943), and was where he worked until becoming Professor of Zoology, University of Turin (1953–1962). He was Professor of Zoology, University of Florence (1962–1980) and Professor of Ethology (1982–1987). He was Director, Zoological Museum, University of Florence (1963–1972) and Director, Centre for the Study of Tropical Wildlife and Ecology, University of Florence (1971–1985). He retired as Professor Emeritus (1988). The University of Florence sponsored the expedition to Somalia that collected the holotype.

Paresi

Neotropical Rivuline sp. *Melanorivulus paresi* WJEM Costa, 2008
Armoured Catfish sp. *Curculionichthys paresi* F Roxo, C Zawadzki & W Troy, 2014

These binomials commemorate the indigenous Paresi people who used to inhabit much of the Mato Grosso State, Brazil.

Parfait

Blind Cusk Eel sp. *Barathronus parfaiti* LL Vaillant, 1888

Captain Jacques Théophile Parfait (1839–1915) was master of the 'Talisman', the French research vessel from which the holotype was collected.

Parham

Parham's Riffle Minnow *Alburnoides parhami* H Mousavi-Sabet, S Vatandoust & I Doadrio, 2015

Saeid Parham (1980–2009) was an Iranian conservation officer who was killed in a battle with illegal hunters near the border with Turkmenistan. The authors' intention is that this species' name shall also honour all Iranian conservation officers, who have sacrificed their lives for the sake of preserving the environment.

Parin

Slimehead genus *Parinoberyx* AN Kotlyar, 1984
Oceanic Basslet sp. *Howella parini* BI Fedoryako, 1976
Tubeshoulder sp. *Pectinantus parini* YI Sazonov, 1976
Parin's Grenadier *Nezumia parini* CL Hubbs & T Iwamoto, 1977
Hatchetfish sp. *Polyipnus parini* OD Borodulina, 1979
Lightfish sp. *Ichthyococcus parini* VA Mukhacheva, 1980
Leftvent Anglerfish sp. *Linophryne parini* E Bertelsen, 1980
Parin's Spinyfin *Diretmichthys parini* A Post & J-C Quéro, 1981
Splendid Perch sp. *Callanthias parini* WD Anderson & GD Johnson, 1984
Pocket Shark *Mollisquama parini* VN Dolganov, 1984
Parin's Rockling *Gaidropsarus parini* AN Svetovidov, 1986
Parin's Ariomma *Ariomma parini* AS Piotrovsky, 1987
Deepwater Cardinalfish sp. *Epigonus parini* AA Abramov, 1987
Deep-sea Tripodfish sp. *Bathypterois parini* YN Shcherbachev & KD Sulak, 1988
Flabby Whalefish sp. *Cetichthys parini* JR Paxton, 1989
Deep-sea Smelt sp. *Bathylagichthys parini* SG Kobyliansky, 1990
Conger Eel sp. *Gnathophis parini* ES Karmovskaya, 1990
Pearlfish sp. *Pyramodon parini* DF Markle & JE Olney, 1990
Snailfish sp. *Volodichthys parini* AP Andriashev & VP Prirodina, 1990
Moray sp. *Gymnothorax parini* BB Collette, DG Smith & EB Böhlke, 1991
Morid Cod sp. *Physiculus parini* KD Paulin, 1991
Parin's Anthias *Plectranthias parini* WD Anderson & JE Randall, 1991
Lanternfish sp. *Diaphus parini* VE Becker, 1992
Parin's Wolf Eelpout *Lycenchelys parini* VV Fedorov, 1995
Bigscale sp. *Melamphaes parini* AN Kotlyar, 1999
Winged Spookfish *Dolichopteryx parini* SG Kobyliansky & VV Fedorov, 2001
Parin's Dragonfish *Eustomias parini* TA Clarke, 2001

Sala y Gómez Slopefish *Symphysanodon parini* WD Anderson & VG Springer, 2005
Snaketooth sp. *Pseudoscopelus parini* AM Prokofiev & EI Kukuev, 2006
Dwarf False Catshark *Planonasus parini* S Weigmann, MFW Stehmann & R Thiel, 2013
Lanternshark sp. *Etmopterus parini* VN Dolganov & AA Balanov, 2018

Professor Dr Nikolai Vasilyevich Parin (1932–2012) of the P.P. Shirov Institute of Oceanology, Russian Academy of Sciences, was an oceanographer and ichthyologist who was an expert on flying fishes. He graduated from the Institute of Fisheries (1955) and later was awarded a PhD (1961) and a DSc (1967). He became Chief of the Laboratory of Oceanic Ichthyofauna (1973) and studied pelagic fishes for the rest of his life, taking part in 20 cruises and notching up eight years of sea time. During his career he published (1958–2012) numerous articles and papers, including: *Ichthyofauna of the epipelagic zone* (1970) and co-wrote: *Biology of the Nazca and Sala-y-Gómez submarine ridges, an outpost of the Indo-West Pacific fauna in the Eastern Pacific Ocean: Composition and distribution of the fauna, its communities and history* (1997). He was chief scientist on Cruise 17 of the research vessel 'Vityaz' (1988–1989), during which the Dwarf False Catshark holotype was collected. (Also see **Parbevs**)

Paris

Characin sp. *Andromakhe paris* M de las M Azpelicueta, A Almirón & JR Casciotta, 2002
[Syn. *Astyanax paris*]

Paris is a mythological character in the story of the Trojan War. He was a son of King Priam of Troy who ran off with Helen, wife of Greek king Menelaus (see Homer's *Iliad* for the full story). Astyanax, son of Hector, was his nephew.

Park

Guinean Sea Catfish *Carlarius parkii* A Günther, 1864

Nobody knows who Park – or perhaps, Parke – was. The holotype was collected in Lagos, Nigeria, so Park(e) may have been a trader or an administrator. However, we would like to float the theory that the fish is named after the renowned Scottish explorer Mungo Park (1771–1806). He was the first European to explore the central part of the Niger River. His correspondence with Sir Joseph Banks on the fish he sent to the BM from his expeditions is preserved in the library of the BMNH.

Parker, CS

Gillbacker Sea Catfish *Sciades parkeri* TS Traill, 1832

Charles Stewart Parker (1800–1868) was a Liverpool-based merchant and a friend of the author. His major trading area was Demerara, British Guiana (Guyana) and he wrote to Traill from there (1825). When the British government emancipated the slaves in British Guiana in the 1830s, Parker was compensated for over 400 slaves he shared ownership of on 16 estates. Traill's etymology states that Parker "*...favoured the author with a drawing of the catfish and its skin.*"

Parker, TJ

Streamer Fish *Agrostichthys parkeri* WB Benham, 1904

Thomas Jeffery Parker (1850–1897) was a British zoologist who graduated from the University of London (1868). He worked with Thomas Huxley (1872). He emigrated to New Zealand (1880) to become Professor of Zoology, University of Otago and Curator, Otago Museum. He was succeeded at the University of Otago by the author. He suffered from diabetes in his later years. He co-wrote: *A Text-book of Zoölogy* (1897) that continued to be in use into the 1960s.

Parkinson

Parkinson's Rainbowfish *Melanotaenia parkinsoni* GR Allen, 1980

Brian J Parkinson (b.1944) is a New Zealander who is a naturalist, dealer in the shells of New Guinea, collector and author. He has been a reporter, a teacher of prostitutes in Bangkok, a uranium prospector, and started work inseminating cows. Although he has no formal qualifications he is an expert on New Zealand flora and fauna. His twenty books cover shells, butterflies and seabirds including: *Common Seashells of New Zealand* (1999). He was a regular companion on the author's many collecting trips to Papua New Guinea. When asked how many species he has discovered, Parkinson waves a hand and mutters "about 30 or 40". Seven of those are named after him, including *Oliva parkinsoni*, a carnivorous seashell of the Indian Ocean; *Megalacron parkinsoni*, a tree-dwelling snail of New Guinea; and this fish which he discovered in a creek behind a New Guinea airport while waiting for a flight.

Parko

Pencil Catfish sp. *Ituglanis parkoi* P de Miranda Ribeiro, 1944

Alexandre Parko was a Polish amateur naturalist. He collected specimens for Museu Nacional, Rio de Janeiro, including the holotype of this species.

Parlette

Neotropical Rivuline sp. *Anablepsoides parlettei* S Valdesalici & I Schindler, 2011

Casey Parlette (b.1979) is an American sculptor and jewellery maker and, at heart, a naturalist. UCLA awarded his anthropology degree (2003), after which he was recruited to be a commercial diver at a naval base. He later pursued a career in ocean lifeguarding. During breaks from his work schedule he operated underwater cameras in exotic locations on behalf of *Inside Sportfishing* on FOX Sports. During one of his travels, Casey ended up in a remote region of the Peruvian Amazon. The diverse wildlife and a passion for exploring inspired him to stay for eight months, exploring remote tributaries of the Amazon. During that time, he co-collected the rivuline holotype and provided the authors with type locality data, photographs and specimens for *Anablepsoides palettei* and *A. lineasoppilatae*, both of which were new species. Upon returning to the USA, Casey became a career lifeguard for Laguna Beach. Throughout his life he had made wildlife inspired sculptures, which eventually grew into a full-time profession. His wood, metal and stone sculptures reflect his love of nature. Along with the sculpture work he has launched a line of nature-inspired jewelry named 'parlettei jewelry' after this 'colorful little fish'.

Parnaiba

Characin sp. *Brachychalcinus parnaibae* RE Reis, 1989
Armoured Catfish sp. *Pterygoplichthys parnaibae* C Weber, 1991
Brazilian Cichlid sp. *Geophagus parnaibae* W Staeck & I Schindler, 2006

These species are named after the Parnaíba River basin, Brazil.

Parr

Barbeled Dragonfish sp. *Eustomias parri* CT Regan & E Trewavas, 1930
Parr's Lanternfish *Diaphus parri* AV Tåning, 1932
Flabby Whalefish sp. *Cetomimoides parri* E Koefoed, 1955
Cusk-eel sp. *Barathrites parri* O Nybelin, 1957
Parr's Combtooth Whalefish *Gyrinomimus parri* HB Bigelow, 1961
Snaketooth sp. *Kali parri* RK Johnson & DM Cohen, 1974

Albert Eide Parr (1900–1991) was a Norwegian-born zoologist, oceanographer and innovative museum director. He graduated from the Royal University of Oslo and served in the Norwegian Merchant Marine before working in fisheries research at the Museum of Bergen. He migrated to the USA (1926), taking a post at the New York Aquarium and was recruited (1927) by Bingham to curate his collection. Bingham donated his collection to Yale and Parr went along, becoming Director of Yale's Bingham Oceanographic Laboratory (1931). He became Director of Marine Research (1937–1942) and Professor of Oceanography at Yale (1938) and that year was made Director of the Yale Peabody Museum – there he created the famed 110–foot long mural of the Age of Reptiles. He pioneered the interpretive function of museums as opposed to what he called the 'dead circus' approach. He was then offered the post of Director of AMNH (1942–1959). After retirement he was Senior Scientist there, and was named Director Emeritus of the museum in 1968. His particular specialism was the classification of *Alepocephalidae* (deep-water marine smelts known as 'slickheads' or 'nakedheads').

Parry

Tonala Catfish *Rhamdia parryi* CH Eigenmann & RS Eigenmann, 1888

Dr Charles Christopher Parry (1823–1890) was an English botanist, mountaineer and geologist who emigrated with his parents to the USA (1832). He qualified as a physician at Columbia University, New York, and practised as a doctor at Davenport, Iowa (1846–1848). He was a member of the US-Mexican Boundary Commission (1848–1955). He made important botanical collections along the Mexico-California border and later in Utah, Colorado and other western states. He was the first person to make barometric measurements of the heights of many of the mountains in Colorado, where Parry Peak is named after him.

Parvé

Goby sp. *Sicyopterus parvei* P Bleeker, 1853

H A Steijn Parvé was a Dutch civil servant in the colonial government of Western Java, Indonesia. He collected several species (and discovered this goby) for Bleeker. He is recorded as being the Resident at Tapanoeli (1862–1863) and living in Djokja, Central Java, from where he sent (1874) plants to Buitenzorg (Bogor).

Paryag

Paryag's Guianan Rivuline *Laimosemion paryagi* FBM Vermeulen, WH Suijker & GE Collier, 2012
[Alt. Paryag's Killifish]

Subhas Chand Paryag, from Georgetown, Guyana, was co-collector of the new species. He was a local helper during most of the expeditions in Guyana made by the first and second authors.

Paschen

African Rivuline sp. *Aphyosemion pascheni* CGE Ahl, 1928

Mr Paschen was the collector of the holotype, but the etymology gives no forename(s). It may refer to Hans Paschen, a German settler and trader in Yaounde, Cameroon, who has mainly entered history for his shooting of a large gorilla – then the largest gorilla known – in 1900. It was purchased by Rothschild for his museum in Tring (United Kingdom).

Pascual

Pencil Catfish sp. *Cambeva pascuali* LE Ochoa, GS Silva, GJ Costa e Silva, C Oliveira & A Datovo, 2017

José Pascual Ochoa is the senior author's father.

Paska

Paska's Blue-eye *Pseudomugil paskai* GR Allen & W Ivantsoff, 1986

John Paska was a fisheries technician for the Papua New Guinea Ministry of Fisheries. He was honoured for the assistance he gave Gery Allen during his visits there, and was a co-collector of the type specimen.

Passarelli

Pencil Catfish sp. *Homodiaetus passarellii* P de Miranda Ribeiro, 1944

António Passarelli Filho was a commercial collector who provided material for the Museu Nacional, Rio de Janeiro, Brazil (1940s). An amphibian is also named after him.

Passaro

African Rivuline sp. *Aphyosemion passaroi* JH Huber, 1994

Guido Passaro of Ludwigsburg is a German aquarist. He, along with Wolfgang Eberl, first collected this species (1993) in a shallow stream by a roadside in Cameroon. They took specimens back to Germany, where they were bred for the aquarium trade.

Pastinha

Armoured Catfish sp. *Hypostomus pastinhai* CH Zawadzki & I de S Penido, 2021

Vicente Ferreira Pastinha, commonly called Mestre Pastinha (1889–1981), was a master practitioner of the Brazilian martial art of capoeira, and 'symbolic patron' of the Angola style of this art.

Pataxó

Bumblebee Catfish sp. *Microglanis pataxo* LM Sarmento-Soares, RF Martins-Pinheiro, AT Aranda & CC Chamon, 2006
Neotropical Rivuline sp. *Xenurolebias pataxo* WJEM Costa, 2014

The Pataxó are an indigenous people of northeastern Brazil. They live in areas where these species occur.

Pathirana, A

Barred Danio *Devario pathirana* M Kottelat & R Pethiyagoda, 1990

Ananda 'Andy' Pathirana has been a Sri Lankan aquarium-fish breeder, trader and exporter for over forty years. He is Chairman and CEO of Aquamarine International (Pvt) Ltd. In addition to his business interests he has donated numerous artefacts of archaeological significance that he has discovered in the seas around Sri Lanka in 30 years of diving. He has bred captive endangered species and released them back into the wild. He discovered this species and was honoured for calling the authors' attention to it.

Pathirana, Y

Trincomalee Sweeper *Pempheris pathirana* JE Randall & BC Victor, 2015

Yohan Pathirana of Aquamarines International, Sri Lanka, provided the authors with specimens and photographs of *Pempheris* from that country. He is the son of Ananda Pathirana (above).

Patia

Toothless Characin sp. *Pseudocurimata patiae* CH Eigenmann, 1914
Armoured Catfish sp. *Chaetostoma patiae* HW Fowler, 1945

This is a toponym, referring to the Patia River basin, Colombia.

Patimar

Goby sp. *Ponticola patimari* S Eagderi, N Nikmehr & H Poorbagher, 2020

Dr Rahman Patimar is an Iranian ichthyologist who works at the Faculty of Agriculture and Natural Resources, Gonbad Kavous University (2004) currently as Associate Professor. Gonbad Kavous University awarded his bacheor's degree and doctorate (2008) whereas Tarbiat Modares University of Tehran awarded his Master's. He is also editor-in-chief of the Iranian Journal of Fish Ecology and has written or co-written around 150 published papers such as: *Ecological Assessment of Organic Pollution in the Gorgan Bay, Using Palmer Algal*

Index (2020). He was honoured for "*…his long and outstanding contributions in biological studies of Iranian fishes.*"

Patot

Patoti's Rainbowfish *Melanotaenia patoti* M Weber, 1907
Tiger Betta *Betta patoti* M Weber & LF de Beaufort, 1922

W J Tissot van Patot was a collector in the Dutch East Indies (Indonesia) for the Zoölogisch Museum (Amsterdam) and other European museums. He collected the rainbowfish in the Aru Islands and the betta in Borneo.

Patricia (Eggleston)

Goby sp. *Egglestonichthys patriciae* PJ Miller & P Wongrat, 1979

Mrs Patricia Eggleston is, we believe, the wife of David Eggleston who collected the type and after whom the genus is named. (See **Eggleston**)

Patricia (Evans)

Ghost Knifefish sp. *Sternarchella patriciae* KM Evans, WGR Crampton & JS Albert, 2017

Patricia Evans is a civil-rights activist and community leader in Philadelphia, Pennsylvania, USA. She is also the senior author's mother.

Patricia (Fromm)

Neotropical Rivuline sp. *Austrolebias patriciae* JH Huber, 1995

Patricia Fromm is an American aquarist from New Jersey. Together with her husband Daniel she collected in Costa Rica and Paraguay (1993). They collected the type specimens of this species, and Huber followed Daniel's request to name it after Patricia. (See also **Fromm**)

Patricia (Gonçalves)

Characin sp. *Bryconamericus patriciae* JFP da Silva, 2004

Patrícia L da Silva née Gonçalves is the author's wife.

Patricia (Kailola)

Black-banded Flathead *Rogadius patriciae* LW Knapp, 1987
Whipfin Sea Catfish *Netuma patriciae* Y Takahashi, S Kimura & H Motomura, 2019

Dr Patricia 'Tricia' J Kailola. She was honoured in the flathead name "*…in recognition of her many contributions to the knowledge of the fishes of north-western Australia and southern Indonesia.*" The etymology for the catfish says that "*…the specific name patriciae is in honor of Patricia J. Kailola, the University of the South Pacific and Pacific Dialogue Ltd, in recognition of her research on the catfish family Ariidae.*" (See **Kailola** & also see **Tricia**)

Patricia (Yazgi)

Madagascan Killifish sp. *Pachypanchax patriciae* PV Loiselle, 2006

Patricia Griffith Yazgi (1946–2006) was honoured in recognition of her support in the documentation and conservation of Malagasy freshwater fish. She ran the conservation charity, 'Friends of Fishes'.

Patrick

Peruvian Cichlid sp. *Aequidens patricki* SO Kullander, 1984

Dr Patrick de Rham (b.1936) helped collect the type. (See **De Rham** & **Rham**)

Patrick Yap

Rasbora sp. *Rasbora patrickyapi* HH Tan, 2009

Patrick Yap Boon Hiang of Aquaculture Technologies of Singapore is a freshwater fish enthusiast and exporter, and a long-time supporter of the Raffles Museum of Biodiversity Research. He generously donated much fish material to help Tan's research.

Patrizi

Patrizi's Notho (African killifish) *Nothobranchius patrizii* D Vinciguerra, 1927
Buta Mormyrid *Stomatorhinus patrizii* D Vinciguerra, 1928

Marquis Don Saverio Patrizi Naro Montoro (1902–1957) was an Italian explorer, zoologist, entomologist, collector and speleologist. He was the Italian signatory of a 1930s League of Nations treaty on African mammal preservation. During the Italian occupation of Ethiopia, Patrizi collected and prepared almost all the initial mammal and bird specimens currently displayed at Ethiopia's Zoological Natural History Museum. They were donated to the University College of Addis Ababa following their discovery in storage at Akaki (1955). Patrizi collected in central Africa (1920s) as well as in Ethiopia (1930s and 1940s). He collected in Kenya (1946–1947), having been interned there (WW2). A mammal is also named after him. He was honoured in the mormyrid's name as he collected the type, and for donating specimens from his Congo expedition to the Museum Civico di Storia Naturale de Genova.

Patterson

Toothless Blindcat *Trogloglanis pattersoni* CH Eigenmann, 1919

Dr John Thomas 'Pat' Patterson (1878–1960) was an American embryologist and geneticist at the University of Texas, Austin, where he joined the faculty (1908), having been awarded his doctorate by the University of Chicago (1908). He officially retired as Professor Emeritus of Zoology (1950) but continued regular attendance at his office until 1958. Among his numerous publications is: *The Drosophilidae of the Southwest* (1943). He sent the holotype of the blindcat (a Texan troglodytic catfish) to Eigenmann.

Patzner

Patzner's Blenny *Salarias patzneri* H Bath, 1992
Goby sp. *Lebetus patzneri* UK Schliewen, M Kovačić & F Ordines, 2019

Dr Robert Arthur Patzner (b.1945) is an Austrian marine biologist, ecologist and conservationist who is (2019) professor emeritus at the University of Salzburg's Department of Ecology & Evolution. He was Associate Professor there (1982–2009). His doctorate was conferred by the University of Oslo (1972) and he was a post-doctoral researcher at the University of Tokyo (1975–1976). He has co-authored more than 400 papers and books, including: *The Biology of Blennies* (2009) and *The Biology of Gobies* (2011).

Paucke

Freshwater Stingray sp. *Potamotrygon pauckei* MN Castex, 1963
[Junior Synonym of *Potamotrygon motoro*]

Florian Paucke (1719–1780) was a Silesian Jesuit missionary and ethologist who was also an accomplished artist. He asked his superiors to be sent to the Americas, and left Europe (1748) for Buenos Aires. He lived for some time among the Mocovi indigenous tribe, making hundreds of drawings of fauna, flora and people. He was accused by the Spanish authorities of being a British spy, was expelled from (what is now) Argentina (1767), and returned to Bohemia. His original manuscripts are in the Library of the Zwettl Abbey, Austria.

Paugy

Kolente Stonebasher (mormyrid) *Hippopotamyrus paugyi* C Lévêque & R Bigorne, 1985

Dr Didier Paugy is a French hydrobiologist who is Senior Scientist at the IRD (Institut de Recherche pour le Développement) and whose specialist area is African freshwater ichthyology, mainly ecology and biogeography. He carried out research in West Africa (1975–1988). He has written many scientific papers and parts of longer works, such as the chapter: *Faune ichtyologique des eaux douces d' Afrique de l'Ouest* of the report: *Diversite biologique des poissons des eaux douces et saumâtres d' Afrique* (1994) and with Lévêque the book: *Guide des poissons d'eau douce de la zone du programme de lutte contre l'onchocercose en Afrique du l'Ouest* (1984). He was a friend of the authors, and returned the favour to Lévêque by naming a catfish *Synodontis levequei* after him (1987).

Paul

Yellow spotted Gurnard *Pterygotrigla pauli* GS Hardy, 1982

Dr L J 'Larry' Paul was a scientist at the Fisheries Research Division, Ministry of Agriculture and Fisheries, and honorary Research Associate of the National Museum of New Zealand. He was honoured in recognition of his contribution to New Zealand ichthyology and his continuing interest in the development of the museum's collections. He wrote: *A Bibliography of Literature about New Zealand's Marine and Freshwater Commercial Fisheries 1840–1975* (1979).

Paul Müller

Dwarf Cichlid sp. *Apistogramma paulmuelleri* U Römer, J Beninde, F Duponchelle, CRG Dávila, AV Díaz & J-F Renno, 2013

Dr Paul Müller (1940–2010) was a German herpetologist and zoographer who became Professor of Biology at the University of Trier. His specialism was the biogeography of the Neotropics. His PhD (1966) at the University of Saarbrücken was on the birds and other vertebrates of the Ilha de São Sebastião (Brazil). Subsequently, his studies focussed on herpetofauna, tropical ecology and sustainability of hunting. He was also awarded honorary doctorates by the Yokohama University and Chiang Mai University.

Paula (Keener)

Dwarf Spinyhead Blenny *Acanthemblemaria paula* GD Johnson & EB Brothers, 1989

Paula is Latin for 'little', and is in reference to the species' diminutive size. However, the etymology also says that: "*The name was chosen to honor Paula Keener, who participated in the collection that resulted in recognition of this species.*" She is a marine biologist who was a research team member in the Smithsonian Institution's Western Atlantic Mangrove Program off Belize, Central America. She later became Director of the National Oceanic and Atmospheric Administration's (NOAA's) Ocean Exploration's Education Program.

Paula (Pezold)

Jester Goby *Oxyurichthys paulae* F Pezold, 1998

Paula Arledge Pezold was the author's 'academic companion and spouse'.

Paulay

Zebra Sweetlips *Plectorhinchus paulayi* F Steindachner, 1895

Dr Stefan Paulay (1839–1913) was a ship's surgeon and botanist who took part in expeditions of the Vienna Academy of Sciences (1896–1900) and collected, often with Oskar Simony, for the NHMW in the Azores, Madeira, Cape Verde Islands, Socotra, Yemen, and other areas. While the original text identifies the provider of the holotype only as 'Dr Paulay', this most probably refers to Stefan.

Paulian

Madagascan Cave Sleeper sp. *Typhleotris pauliani* J Arnoult, 1959

Dr Renaud Maurice Adrien Paulian (1913–2003) was a French zoologist, considered one of the greatest entomologists of the 20th century and the leading European expert on scarab beetles. He was Deputy Director of the Institut de Recherche Scientifique de Madagascar (1947–1961), then became Director of the Institut Scientifique de Congo-Brazzaville (1961–1966) and head of the local university. He was Head of the Université d'Abidjan in the Ivory Coast (1966–1969) before returning to France, where he became Rector of the Academy of Amiens and then of the Academy of Bordeaux. He was elected correspondent of the Academy of Sciences (1975). He initiated (1956) the important series: *Faune de Madagascar*, over 90 volumes of which have been published. He wrote over 350 papers and a number of books, including: *Madagascar, un sanctuaire de la Nature* (1981). Among other taxa, three odonata, two reptiles, two amphibians, a mammal and a bird are also named after him.

Paulus

Pygmy Sculpin *Cottus paulus* JD Williams, 2000

Not an eponym, but from the Latin *paulus* meaning 'little'.

Pautzke

Slimy Cusk-eel *Brosmophyciops pautzkei* LP Schultz, 1960

Clarence F Pautzke (1907–1971) was the chief biologist, Game Department, State of Washington. The University of Washington awarded his bachelor's degree in aquatic biology (1931), after which he joined the Washington Department of Game, becoming head of its fisheries programme. He was Assistant Director of the state's Department of Fisheries (19157–1960) and, after it achieved statehood, he became Alaska's Assistant Commissioner of Fish and Game. He was appointed by President Kennedy to be the US Fish and Wildlife Service Commissioner (1961) and later he was briefly Assistant Secretary of the Department of the Interior towards the end of President Lyndon Johnson's administration. He is noted in the etymology as having been on Bikini Atoll (site of atomic explosions) (1946 & 1947) when the holotype was collected. He died unexpectedly while undergoing surgery.

Pavanelli

Corydoras sp. *Corydoras pavanelliae* LFC Tencatt & WM Ohara, 2016

Dr Carla Simone Pavanelli (b.1967) is a Brazilian biologist. The State University of Maringá Brazil awarded her bachelor's degree (1990), the Federal University of Rio de Janeiro her master's (1994) and the Federal University of São Carlos her doctorate (1999). She undertook post-doctoral work at the Smithsonian (2006). She works at the State University of Maringá and is Curator of the Ichthyology Collection of the Centre for Research in Limnology, Ichthyology and Aquaculture. She co-wrote: *First record of Megalechis picta (Müller and Troschel, 1849) (Siluriformes: Callichthyidae) in the upper Rio Paraná basin, Brazil* (2013). The etymology states that she is an "*…advisor of the first author and dear friend, for her extensive contributions to the knowledge of the ecology and taxonomy of the Neotropical fishes.*"

Pavie

Sidestripe Rasbora *Rasbora paviana* G Tirant, 1885
Sisorid Catfish sp. *Pseudecheneis paviei* LL Vaillant, 1892

Auguste Jean-Marie Pavie (1847–1925) was a French colonial civil servant, explorer and diplomat who was instrumental in establishing French colonial control over Laos. He was a civil servant in Cambodia and Vietnam before becoming the first Governor-General for the colony of Laos (1885). He joined the army at 17 (1864) and was posted to Indo-China in the infantry (1869), but was recalled to fight in the Franco-Prussian war during which he reached the position of Sergeant Major. He then (1871) returned to Cambodia in charge of a small telegraph office. This posting allowed him to develop a deep understanding of the local culture and language. He 'went native' and wandered, dressed in local garb, recording everything he saw. Brought to the attention of the Governor (1879) he was entrusted with a five-year mission to explore the Gulf of Siam. These 'Missions Pavie' extended (1879–1895), being so successful that he was appointed as Vice-Consul in Luang Prabang, then Consul (1889) and Consul-General (1891). The missions gathered vast amounts of scientific data and specimens from Archaeology to Zoology. On his return to France he wrote a multi-volume work: *La mission Pavie, A la conquête des coeurs* and *Contes du Cambodge, du Laos et du Siam* (1898–1921). Moreover, he was Tirant's friend and, according to the etymology, a "*...tireless explorer of the southern and western provinces of Cambodia.*" (translation) Other taxa, including a dragonfly, other insects and gastropods are also named after him.

Pawnee

Pawnee Dragonfish *Bathophilus pawneei* AE Parr, 1927

'Pawnee II' was a yacht owned by Harry Payne Bingham (q.v.). She was specially designed for deep-sea trawling and research, and the holotype of this species was collected from her.

Paxman

Paxman's Leatherjacket *Colurodontis paxmani* JB Hutchins, 1977

Barry Paxman, with his brother Frank, was a SCUBA diver. The filefish was "*...named in grateful appreciation to Mr B. Paxman who was instrumental in obtaining many monacanthid specimens for the collections of the Western Australian Museum.*"

Paxton

Cardinalfish subfamily *Paxtoninae* TH Fraser & K Mabuchi, 2014
Cardinalfish genus *Paxton* CC Baldwin & GD Johnson, 1999
Paxton's Cardinalfish *Paxton concilians* CC Baldwin & GD Johnson, 1999
Paxton's Pipefish *Corythoichthys paxtoni* CE Dawson, 1977
Paxton's Whipnose *Gigantactis paxtoni* E Bertelsen, TW Pietsch & RJ Lavenberg, 1981
Orange-lined Wirrah *Acanthistius paxtoni* JB Hutchins & RH Kuiter, 1982
Paxton's Toadfish *Torquigener paxtoni* GS Hardy, 1983
Baldhead Cusk *Fiordichthys paxtoni* JG Nielsen & DM Cohen, 1986
[Syn. *Bidenichthys paxtoni*]
Paxton's Escolar *Rexichthys johnpaxtoni* NV Parin & DA Astakhov, 1987
Humpback Hairfin Anchovy *Setipinna paxtoni* T Wongratana, 1987
Paxton's Tilefish *Branchiostegus paxtoni* JK Dooley & PJ Kailola, 1988
Paxton's Hatchetfish *Polyipnus paxtoni* AS Harold, 1989
Spinycheek Seabass *Ostracoberyx paxtoni* J-C Quéro & C Ozouf-Costaz, 1991
Duckbill Eel sp. *Nettenchelys paxtoni* ES Karmovskaya, 1999
Thin-barbel Whiptail *Ventrifossa paxtoni* T Iwamoto & A Williams, 1999
Paxton's Dragonet *Foetorepus paxtoni* R Fricke, 2000
Blunt-tooth Snailfish *Careproctus paxtoni* DL Stein, NV Chernova & AP Andriashev, 2001
Barbeled Dragonfish sp. *Eustomias paxtoni* TA Clarke, 2001
Slickhead sp. *Conocara paxtoni* YI Sazonov, A Williams & SG Kobyliansky, 2009
Snaketooth sp. *Pseudoscopelus paxtoni* MRS Melo, 2010
Dragonfish sp. *Photonectes paxtoni* AJ Flynn & C Klepadlo, 2012

Dr John Richard Paxton (b.1938) is an American-born Australian ichthyologist. He is Founder and President of the Australian Society for Fish Biology and was Curator of Fishes (1968) at the Australian Museum, Sydney, where he has spent his entire post-student career and was Principal Research Scientist when he retired (1998). He is currently a Senior Fellow there, where he continues his research on deep-sea fishes. The University of

Southern California awarded his BA in Zoology (1960), his MSc in Biology (1965) and his PhD in Biological Sciences (1968) before migrating to Sydney. In his time at the museum he has expanded the collection of about 80,000 specimens to over one million. In addition to writing more than 100 scientific papers he was lead editor for: *Encyclopaedia of Fishes* (1994) and is about to complete a massive publication on the myctophids of Australia with Alan Williams. Wongratana honoured him as the person who encouraged him to broaden his knowledge of Australian clupeoid fishes. (Also see **John Bob** & **John Paxton**)

Payne, GH

Payne's Tuskfish *Choerodon paynei* GP Whitley, 1945

Flight-Lieutenant George Herbert Payne was an officer in the Royal Australian Air Force in Melbourne with whom the author "*...was associated in some experiments on sharks in Western Australia in 1944.*" They were assigned the testing of chemical shark repellents from the USA; Whitley found and identified the sharks while Payne was in charge of the chemicals. Together they wrote: *Testing a Shark Repellant* (1945).

Payne, IA

West African Cichlid sp. *Hemichromis paynei* PV Loiselle, 1979
Payne's Catfish *Mochokiella paynei* GJ Howes, 1980

Ian A Payne, who collected the catfish holotype, was a fisheries biologist at the Zoology Department, Fourah Bay College University of Sierra Leone, Freetown, Sierra Leone (1975) and at Department of Biological Sciences, Faculty of Applied Science, Coventry (Lanchester) Polytechnic (now Coventry University) (1976). He co-wrote: *Deforestation, the Decline of the Horse, and the spread of the Tsetse Fly and Trypanosomiasis (nagana) in Nineteenth Century Sierra Leone* (1975). He was honoured "*... for his interest in the systematics and ecology of hemichromid cichlids of Sierra Leone.*"

Payton

Moroccan Barb sp. '*Barbus*' *paytonii* GA Boulenger, 1911
[Probably a junior synonym of *Carasobarbus fritschii*]

Sir Charles Alfred Payton (1843–1926) was a British adventurer, writer, fisherman and diplomat who was British Consul to Morocco, where this barb is found. He matriculated from New College London (1860) and afterwards became an insurance clerk. He prospected for gold in California (1864), without success. He had many jobs including making explosives, owning a Cornish clay mine, digging for diamonds at Kimberley (South Africa), coal merchant salesman in Europe and a merchant in Mogador, Morocco where he was appointed Consul (1880). This was later expanded (1890) to include all of Southern Morocco. He transferred as Consul at Genoa, Italy (19893). He was transferred to Calais (1897), covering several départments and serving out the rest of his career there, appointed MVO (1906) and promoted to Consul-General (1911) before retiring (1913). The following year he was knighted. Among his published works are: *The diamond diggings of South Africa: a personal and practical account* (1872), *The Rod on the Rivieras*, in *Sport on the Rivieras, with chapters on river and sea fishing in the south of Europe* (1911), and *Days of a knight: an octogenarian's medley of memories (life, travel, sport, adventure)* (1924). Although he is not identified in the etymology, his interest in fishing and his presence at the type locality make it almost certain that he is the person honoured.

Peaolopes

Mozambique Knifejaw *Oplegnathus peaolopesi* JLB Smith, 1947

Peão Lopes of the Lourenco Marques Museum was described as "*...a most able collector and technician*". He collected eight of the records of fishes in or near Delagoa Bay that Smith discussed in his paper, presumably including the knifejaw holotype.

Pearcy

Eelpout sp. *Lycenchelys pearcyi* ME Anderson, 1995
Snailfish sp. *Paraliparis pearcyi* DL Stein, 2012

Professor Dr William G 'Bill' Pearcy was an American oceanographer who spent much of his career as Professor of Oceanography at Oregon State University, where he stayed until his retirement (1990). Iowa State University awarded his BSc and MSc. His first study of oceanography was as a post-graduate in Hawaii. He was then in the Naval Air Force, stationed at Virginia Beach as well as serving periods on aircraft carriers. After service he worked on his doctoral thesis at Yale. He took part in a number of research voyages. As a biological oceanographer he studied the ocean ecology of Pacific salmon, culminating in the book: *Ocean Ecology of North Pacific Salmonids (Books in Recruitment Fishery Oceanography)* (1992) and wrote more than 130 papers such as *Effects of the 1983 El Nino on coastal nekton off Oregon and Washington* (1985) and *Ocean distribution of the American shad (Alosa sapidissima) along the Pacific coast of North America* (2011).

Pearse

Armoured Catfish sp. *Chaetostoma pearsei* CH Eigenmann, 1920
Pantano Cichlid *Cincelichthys pearsei* CL Hubbs, 1936
Mexican Blind Brotula *Typhliasina pearsei* CL Hubbs, 1938

Dr Arthur Sperry Pearse (1877–1956) was an American zoologist. The University of Nebraska awarded his bachelor's degree (1900), while his master's (1904) and doctorate (1908) were from Harvard. He taught (1900–1904) in Omaha High School, as an assistant at Harvard (1904–1907), taught zoology at Lake View High School, Chicago (1907), at Harvard (1908), the University of Michigan (1909–1910), in the Philippines (1911), and the University of Wisconsin (1912–1927), becoming Professor (1919). He was at Duke University (1929–1945), where he was Director, Marine Biology Laboratory (1938–1945). He directed the journal *Ecology Monographs* (1931–1951) and wrote: *The Migration of Animals from Sea to Land* (1936). Pearse was the leader of the Yucatán expedition during which the cichlid holotype was collected. His health deteriorated in his later years and he was invalid from 1953, suffering from coronary heart disease.

Pearson

Characin sp. *Monotocheirodon pearsoni* CH Eigenmann, 1924
Pencil Catfish sp. *Tridentopsis pearsoni* GS Myers, 1925
Characin sp. *Creagrutus pearsoni* V Mahnert & J Géry, 1988

Characin sp. *Leporinus pearsoni* HW Fowler, 1940
Whale Catfish sp. *Cetopsis pearsoni* RP Vari, CJ Ferraris & MCC de Pinna, 2005

Dr Nathan Everett Pearson (1895–1982) was an ichthyologist at Indiana University, which also awarded his doctorate. He was on the Mulford Expedition (1921–1922) to Bolivia. He wrote: *Fishes of the Rio Beni Basin (*1924). An amphibian is also named after him.

Peckolt

Armoured Catfish genus *Peckoltia* A Miranda-Ribeiro, 1912
Armoured Catfish genus *Peckoltichthys* A Miranda-Ribeiro, 1917

Gustavo Peckolt (1861–1923) was a Brazilian-born German botanist and pharmacist. He became a member of the Natural History Commission of Rondon, and published important books about Brazilian plants with his father Theodor (1822–1912), including: *History of medicinal and useful plants of Brazil* (1888).

Pedaschenko

Stone Loach sp. *Triplophysa pedaschenkoi* LS Berg, 1931

Dr Dmitry D Pedaschenko (1868–1927) was a Russian zoologist who was a senior assistant in the zoological department, St Petersburg Imperial University, who collected some of the fishes that Berg studied from Issyk Kul, Kyrgyzstan. He also collected in Turkestan (1904–1906).

(NB. Technically, the binomial is not available, as it was first proposed as an infra-specific taxon: *Diplophysa strauchi ulacholica pedaschenkoi*. However, the name is widely accepted as available in current literature)

Pedro

Characin sp. *Deuterodon pedri* CH Eigenmann, 1908
[Syn. *Astyanax pedri*]

Dom Pedro II (1825–1891), nicknamed 'the Magnanimous', was the last Emperor of Brazil. He was also a patron of the arts and sciences, with wide-ranging interests. He collected several specimens of this fish, but as they were in a poor condition they were not included in the type series.

Peel

Murray Cod *Maccullochella peelii* TL Mitchell, 1838

Despite appearing to be an eponym, this is actually a toponym referring to the Peel River, New South Wales, where the species was first caught (by Europeans). The river was named after Sir Robert Peel (1788–1850).

Pekkola

Ctenopoma sp. *Microctenopoma pekkolai* H Rendahl, 1935

Wäinö Pekkola was a preparator at the Zoological Institute, University of Turku, Finland. He collected fishes from the White Nile in Sudan (1914), including the type of this species.

Pel

Pebbletooth Moray *Echidna peli* JJ Kaup, 1856
Boe Drum *Pteroscion peli* P Bleeker, 1863
West African Cownose Ray *Rhinoptera peli* P Bleeker, 1863

Hendrik Severinus Pel (1818–1876) was the Dutch Governor of the Gold Coast (Ghana) (c.1840–1850). He was also an amateur naturalist and trained taxidermist, and acted as such for the Leiden Museum, to which he spent shipments of animal specimens. He was honoured as the person whose 'enlightened zeal' led to the deposition of natural history specimens at the Leiden Museum, including the types of these fishes. He is remembered in the names of other taxa, including two mammals and two birds.

Pele

Scorpionfish sp. *Scorpaena pele* WN Eschmeyer & JE Randall, 1975

In Hawaiian myths, Pele is the goddess of fire and volcanoes. The name was inspired by the fish's red coloration.

Pelicice

Armoured Catfish sp. *Pareiorhina pelicicei* VM Azevedo-Santos & FF Roxo, 2015

Fernando Mayer Pelicice of the Universidade Federal do Tocantins, Brazil, was honoured in the binomial for his scientific contributions to fish ecology and the impacts of dams on Neotropical fishes. Among his papers on these subjects is the co-written: *Large reservoirs as ecological barriers to downstream movements of Neotropical migratory fish* (2015).

Pelicier

Pelicier's Wrasse *Halichoeres pelicieri* JE Randall & MM Smith, 1982
Mauritian Gregory *Stegastes pelicieri* GR Allen & AR Emery, 1985
Pelicier's Perchlet *Plectranthias pelicieri* JE Randall & T Shimizu, 1994

Daniel Pelicier (1946–2018) was an aquarium fish collector and exporter in Flic en Flac (a village in Mauritius). He collected the type specimens of the perchlet, and provided a boat, diving equipment and collecting assistance during Gerald Ray Allen's visit to Mauritius (1979). He died during a dive.

Pellegrin

Ctenopoma sp. *Ctenopoma pellegrini* GA Boulenger, 1902
Sharkminnow sp. *Luciosoma pellegrinii* CML Popta, 1905
Characin sp. *Astyanax pellegrini* CH Eigenmann, 1907
Characin sp. *Leporinus pellegrinii* F Steindachner, 1910
Yellowhump Eartheater *Geophagus pellegrini* CT Regan, 1912
Lake Victoria Cichlid sp. *Haplochromis pellegrini* CT Regan, 1922
Chubby Clingfish *Apletodon pellegrini* P Chabanaud, 1925
Chinese Bream sp. *Megalobrama pellegrini* TL Tchang, 1930
Madagascar Rainbowfish sp. *Rheocles pellegrini* JT Nichols & FR La Monte, 1931
Tensift Trout *Salmo pellegrini* F Werner, 1931

Barbless Carp *Cyprinus pellegrini* T-L Tchang, 1933
Loach sp. *Leptobotia pellegrini* PW Fang, 1936
Cape Verde Weever *Trachinus pellegrini* J Cadenat, 1937
Labeo sp. *Labeo pellegrini* G Zolezzi, 1939
Pellegrin's Barb *Enteromius pellegrini* M Poll, 1939
Nuon Mormyrid *Petrocephalus pellegrini* M Poll, 1941
African Cyprinid sp. *Labeobarbus pellegrini* L Bertin & R Estève, 1948
Madagascan Sleeper sp. *Eleotris pellegrini* AL Maugé, 1984
Pike-Cichlid sp. *Crenicichla pellegrini* A Ploeg, 1991
Mimic (Speckled) Garden Eel *Heteroconger pellegrini* PHJ Castle, 1999

Dr Jacques J Pellegrin (1873–1944) was a French zoologist. He was an ichthyologist at the Muséum National d'Histoire Naturelle (MNHN), Paris, where he became Assistant Curator of Zoology (1894) when Vaillant relinquished the post. During this time he continued to study and was awarded his MD (1899) and his doctorate (1904), becoming Assistant Professor (1908). He undertook a number of overseas trips collecting for the Museum, and was appointed as Deputy Director (1937) and Curator of Herpetology and Ichthyology. He published more than 600 books and scientific papers including: *Poissons des eaux douces de l'Afrique occidentale* (1923) and discovered over 350 species new to science. He was honoured in the binomial of the garden eel as the person who described the first garden eel collected in the Gulf of California (*H. digueti*). The pike-cichlid was named for his contribution to the knowledge of the genus *Crenicichla* and of cichlids in general. He was killed when fighting for the French Partisans during World War II.

Pelzam

Transcaspian Marinka *Schizothorax pelzami* KF Kessler, 1870

Dr Emmanuel Danilovich (or possibly Evgeniy Dmitrievich) Pelzam (1837–fl.1886) worked as an assistant at Kazan University Museum, Russia. He collected along the eastern coast of the Caspian Sea for six months (1867) and again the following year (1868), and again two years later (1870). He also collected by dredging along the Petchora River (1874). During his time there he collected the marinka (cyprinid) type. Sometime later he collected in northern Persia. He wrote: *Account of artificial fertilization of Ganoids and rearing of the fry of the Sterlet* (1875). There is also a work about his botanical work: *Catalogue of Plants, collected in 1874 in the vicinity of the Petchora River and the Timan Mountains by E. D. Pel'tzam* (1878). A bird is named after him.

Pemon

Whale Catfish sp. *Cetopsidium pemon* RP Vari, CJ Ferraris & MCC de Pinna, 2005
Ghost Knifefish sp. *Apteronotus pemon* CD de Santana & RP Vari, 2013

The Pemon are an eastern Venezuelan Amerindian tribe, whose traditional territories included the localities of the holotypes of these species.

Pence

Pence's Squirrelfish *Neoniphon pencei* JM Copus, RL Pyle & JL Earle, 2015

David Franklin Pence (b.1956) is an American marine scientist and diver. He received his B.A. (1979) from Miami University (Ohio) and his M.S. (1989) from North Carolina State University. He has served as the Diving Safety Officer for the University of Hawaii from 1995 to present, after previous work in marine science education and biological oceanography. Academically trained in ichthyology, behavioural ecology and molecular biology, and active in the training of scientific research divers, Pence facilitated the early adoption of advanced diving methods by university research scientists to explore novel habitats, including deep coral reef environments. While leading a 2014 expedition to Rarotonga, Cook Islands, Mr. Pence collected the first specimens of this squirrelfish from a depth of approximately 110m., using a closed-circuit mixed gas rebreather.

Peng

Oriental River Loach sp. *Hemimyzon pengi* S Huang, 1982

This patronym – if indeed it is an eponym – is not identified in the original text.

Pengelley

Jamaican Killifish *Cubanichthys pengelleyi* HW Fowler, 1939

Dr Charles Edward Pengelley (1888–1966) of Mandeville, Jamaica, was a physician and an aquarist. He discovered this species and sent it, along with another new fish, to fellow aquarist and author Dr William Thornton Innes (q.v.) and they were gifted to the Academy of Natural Sciences of Phladelphia. Pengelley added notes on observations made of them as aquarium fish.

Penggal

Indonesian Shovelnose Ray *Rhinobatos penggali* PR Last, WT White & Fahmi, 2006

Penggali is the Indonesian for 'shovel' and refers to the shape of this species' head.

Penn

Penn's Thrush Eel *Moringua penni* LP Schultz, 1953

Dr George Henry Penn (1918–1963) was an American invertebrate zoologist and naturalist. He joined (1947) the staff of the Zoology Department at Tulane University, Texas, as Associate Professor of Zoology and built a large collection of crustaceans and aquatic insects. He was particularly interested in crayfish and wrote a number of papers about them, such as: *A new burrowing crawfish of the genus* Procambarus *from Louisiana and Mississippi (Decapoda, Astacidae)* (1953). According to the etymology he collected the type specimen as a Lieutenant in the US Navy.

Pennant

Gwyniad *Coregonus pennantii* A Valenciennes, 1848

Dr Thomas Pennant (1726–1798) was a highly regarded Welsh naturalist, antiquary and traveller around the British Isles. His early work, *Tour in Scotland* (1771), was instrumental in encouraging tourism in the Highlands. Gilbert White published his 'Natural History of Selbourne' in the form of letters to Thomas Pennant and Daines Barrington. Pennant published on the Arctic, Britain and India, and wrote about quadrupeds as well as birds. Among his other publications were: *Genera of Birds* (1773), and *Arctic Zoology* (1784, 1785 & 1787). He was said to make 'dry and technical material interesting'. He wrote about this species (1769) but referred to it as *C. lavaretus*. Three mammals and two birds are named after him.

Pennell

Sharp-spined Notothen *Trematomus pennellii* CT Regan, 1914

Commander Harry Lewin Lee Pennell, R.N. (1882–1916) served (as Navigator, at the time he was a Lieutenant) on the 'Terra Nova', British Antarctic Expedition (1910–1913). He died during the Battle of Jutland.

Pennock

Spotted Algae-Eater *Gyrinocheilus pennocki* HW Fowler, 1937

Charles John Pennock (1857–1935) was an American ornithologist to whom Fowler was indebted for various North American fishes. When the family moved briefly to New York he attended Cornell University, and was employed thereafter for a short time at Princeton University Museum before moving back to Pennsylvania. He started work in the family business making agricultural machinery, but soon moved on to various occupations including raising carnations, lumber, coal, fibre manufacture and, eventually, real estate and insurance. He was a JP and involved in local politics. He wrote the list of birds for Chester County, Pennsylvania. The one constant in his life was the love for and study of birds. Curiously he took the wrong train one day, suffering amnesia and, without his usual beard, he lived for six years in Florida under the name of John Williams keeping books for a local fishing company...but he still continued his ornithological pursuits. When his handwriting was recognised on a paper he submitted to the journal 'Auk', members of his family visited him in Florida and he returned with them to the family home.

Penrose

Brazilian Wrasse *Halichoeres penrosei* EC Starks, 1913

Richard Alexander Fullerton Penrose Jr. (1863–1931) was an American mining geologist and sometime lecturer in geology at Stanford University. After his father's death (1908), he made a career change, becoming a mining investor and entrepreneur. He was honoured in the wrasse's binomial "*...in recognition of his interest in the Stanford Expedition to Brazil.*"

Pentland

Boga (Andean pupfish) *Orestias pentlandii* A Valenciennes, 1846

Joseph Barclay Pentland (1796–1873) was an Irish explorer and diplomat (1836–1839) in Bolivia. He worked for a time in Paris, studying with Cuvier (q.v.). He was particularly interested in geology, and the mineral pentlandite (a nickel iron sulphide) is named after him, as are two birds.

Pepper

Blacktongue Rattail *Gadomus pepperi* T Iwamoto & A Williams, 1999

Roger Pepper is an English-born fishing master in Australia, where he has been Master of the fishing research vessels 'Southern Surveyor' and 'Soela'. He has made contributions to many scientific fishing expeditions.

Pequeño

Eelpout sp. *Lycenchelys pequenoi* ME Anderson, 1995
Hagfish sp. *Myxine pequenoi* RL Wisner & CB McMillan, 1995
Ridgehead sp. *Scopeloberyx pequenoi* AN Kotlyar, 2004

Dr German Enrique Pequeño Reyes (b.1941) was Professor of Zoology at and Director of the Instituto de Zoología, Ernst F Kilian, Universidad Austral de Chile (1973–2007). The University of Chile awarded his bachelor's degree (1967) and Oregon State University awarded his PhD (1984). He was an Assistant in the ichthyology department of the National Museum of Natural History (1965–1970). He undertook post-doctoral studies in Barcelona (1986 & 1988) and in London (1988). He received his Professional Diploma as Profesor de Estado en Biología y Ciencias in the Universidad de Chile, Santiago (1968–1972). Interestingly he is a descendant of the Chilean hero Bernardo O'Higgins. He has described a number of fish species. He was honoured in the name of the hagfish for his work on Chilean fishes and for providing the authors with the holotype. Among his many publications are: *Sinopsis de Macrouriformes de Chile (Pisces: Teleostomi)* (1971) and *Las colecciones de animales chilenos y el problema de su ordenación* (1979).

Percival

Percival's Goby *Awaous percivali* GA Boulenger, 1901
Ewaso Nyiro Labeo *Labeo percivali* GA Boulenger, 1912
Buffalo Springs Tilapia *Oreochromis spilurus percivali* GA Boulenger, 1912

Arthur Blayney Percival (1875–1941) was a British game warden in East Africa (1901–1928) who retired there, dying in Kenya. With the taxidermist W Dodson, he took part in a Royal Society expedition to Arabia (1899). He was appointed (1900) Assistant Collector, which was purely an administrative job, in Kenya and one year later he was made Ranger for Game Preservation. He was instrumental in the establishment of two big game reserves and largely the author of the codified game laws in the East African Game Ordinance of 1906. He was one of the founders of the East Africa and Uganda Natural History Society (1909) and was known as one of the most knowledgeable wildlife experts and hunters in East Africa. He wrote: *A Game Ranger's Notebook* (1924) and *A Game Ranger on Safari* (1928). Five birds, four mammals and a reptile are also named after him.

PERD

Brazilian Cichlid sp. *Australoheros perdi* FP Ottoni, AQ Lezama, ML Triques, EN Fragoso-Moura, CCT Lucas & FAR Barbosa, 2011

PERD is an acronym formed from **P**arque **E**stadual do **R**io **D**oce, in Minas Gerais, Brazil; the type locality.

Perez, L

Anchovy sp. *Anchoviella perezi* F Cervigón, 1987

Luis Pérez is a fisheries lecturer in Venezuela, having completed a five-year course of academic study. He has collected around the Lower Orinoco and written a number of papers about its ecology, such as: *Ecología y factibilidad de cultivo de los Engraulidae dulceacuícolas del Río Orinoco* (1984). He was honoured in the binomial as the person who provided the type material.

Pérez Ponce de León

Grunt sp. *Anisotremus perezponcedeleoni* EA Acevedo-Álvarez, G Ruiz-Campos & O Domínguez-Domínguez, 2021

Dr Gerardo Pérez Ponce de León is a Mexican parasitologist who is Senior Researcher at the Biology Institute of the Universidad Nacional Autónoma de Mexico. His degrees were all awarded by that university: BSc (1986), MSc (1989) and PhD (1992). He undertook post doctoral work at the University of Toronto (1995). He has contributed to around 350 scientific papers since the 1980s, such as: *Update on the distribution of the co-invasive* Schyzocotyle acheilognathi (= Bothriocephalus acheilognathi), *the Asian fish tapeworm, in freshwater fishes of Mexico* (2018). The authors say he has: "…strongly contributed to the study and knowledge of the systematics and phylogeny of helminth parasites of fishes in Latin America."

Perez (Arcas)

Caribbean Reef Shark *Carcharhinus perezii* F Poey, 1876

Professor Laureano Perez Arcas (1824–1894) was a Spanish entomologist and malacologist who became Professor of Zoology at the University of Madrid and Poey's friend and companion. He was co-founder of the Spanish Society of Natural History. Poey used Arcas' book: *Elementos de Zoología* when at the University of Havana.

Peri

Peri's Snake-eel *Myrichthys paleracio* JE McCosker & GR Allen, 2012
[Alt. Whitenose Snake-eel]

Christopher 'Peri' Paleracio is a dive guide and underwater photographer in Anilao, Batangas, Philippines. He studied economics and engineering but took up photography and diving when working with a local marine conservation group that focussed on coastal resource management. He collected the snake-eel holotype and submitted it for examination at the California Academy of Sciences (2009). He had been seeing that particular kind of eel for years, but thought it was just a colour variation of another snake-eel that was already identified. Every year he also contributes an average of five unidentified species of nudibranch.

Peringuey

Southern Barred Minnow *Opsaridium peringueyi* JDF Gilchrist & WW Thompson, 1913
African Tetra sp. *Alestes peringueyi* GA Boulenger, 1923

[Syn. *Brycinus peringueyi*]

Dr Louis Albert Péringuey (1855–1924) was a French entomologist and naturalist. He left France (1879) for South Africa, where he became a Scientific Assistant, South African Museum (1884), was in charge of the Invertebrates Collection (1885), and became the museum's Director (1906–1924). He dropped dead while walking home from the museum. Two reptiles, two birds and a dragonfly are also named after him. Oddly, the patronym is not identified in either of the original descriptions, though there can be no real doubt as to the person honoured in these fishes' names.

Peris

Welsh Char sp. *Salvelinus perisii* A Günther, 1865

This is a toponym; it refers to Llyn Peris (aka Llanberris), a lake in Snowdonia, Wales, where the holotype was obtained.

Perminov

Scaly-belly Sculpin *Icelus perminovi* AY Taranetz, 1936

G I Perminov left an ichthyological career as a young man and joined the Russian military, retiring with the rank of colonel. He spent time at TIRH (Pacific Institute of Fishing Industry) and said of former friends and colleagues there: "...*The memory of the ichthyologists of Tirkhiv at that time should be preserved, especially of such remarkable researchers as A. Ya. Taranetz and G. U. Lindberg. I would very much like to see the names of these outstanding scientists on the boards of our scientific search vessels.*"

Permitin

Fine-spotted Plunderfish *Pogonophryne permitini* AP Andriashev, 1967
Permitin's Spiny Plunderfish *Harpagifer permitini* AV Neyelov & VP Prirodina, 2006

Dr Yuri Efimovich Permitin (b.1925) was a Russian ichthyologist and marine fisheries scientist. He was drafted into the army (1943) but moved to Moscow (1944) enrolling at the Moscow Technical Institute of Fish Industry, where he graduated with a degree in ichthyology and fish farming. He worked at the Institute of Oceanology of the Academy of Sciences of the USSR (1952) and made several voyages on the *Vityaz* expedition ship to the Pacific Ocean. He was then on an Antarctic exploratory voyage (1956) studying the ichthyofauna. He moved (1964) to the All-Union Scientific Research Institute of Marine Fisheries (VNIRO) as a senior researcher. On the vessel *Akademik Knipovich* he participated in several voyages to the Western Antarctic, in the Scotia Sea, to study the composition and biology of the fish of that region, where he discovered five new fish including the two named after him. He delivered scientific reports on Antarctic ichthyofauna at symposiums in Cambridge (England) and at Harvard University (USA) at a congress of American ichthyologists. His PhD thesis on the Antarctic ichthyofauna was a first for a Russian. He wrote c.50 scientific articles on this topic alone. During sea expeditions and work in Africa, he was an avid hunter. After retirement (1985), he became actively involved in his father's literary heritage.

Péron

Notched-fin Threadfin Bream *Nemipterus peronii* A Valenciennes, 1830
Pot-bellied Leatherjacket *Pseudomonacanthus peroni* HLGM Hollard, 1854

François Péron (1775–1810) was a French voyager and naturalist. Originally intending to become a priest, he was a reluctant army volunteer (1792) to fight Prussia. He was wounded and taken prisoner (1793), only being repatriated after over a year (1794) and invalided out as he had lost an eye. He became a Town Clerk then gained a scholarship to study medicine in Paris. After an unhappy love affair, he became an anthropological observer on Baudin's scientific expedition with the ships 'Geographe' and 'Naturaliste' (1800–1804), which visited New Holland, Maria Island, Van Diemen's Land (Tasmania) and Timor, Indonesia. He was always clashing with Baudin but was soon the only zoologist left on the expedition, and with Lesueur (q.v.) collected more than 100,000 zoological specimens. As well as collecting, Péron also conducted pioneering experiments on seawater temperatures at depth. He died at the age of 35 of tuberculosis. The Peron Peninsula, Western Australia is named after him as are two mammals, six birds, three reptiles and two amphibians.

Perotae

Parrot Grunt *Pomadasys perotaei* G Cuvier, 1830

Cuvier called this species the 'Pristipome de Pérotet', with the binomial being a somewhat curious variation on the name Pérotet (see **Perrottet** below)

Pérouse

Longfinned Mullet *Osteomugil perusii* A Valenciennes, 1836

Jean-François de Galaup La Pérouse (1741–1788) was a French naval officer and explorer. It was, according to Valencienne: "*a name that will remind all scientists of the coasts where this species can be found*" (translation), referring to Vanikoro Island in the South Pacific, the type locality, where Pérouse was stranded after both his ships struck reefs. He and his crew were never seen again.

Perrier

Lake Victoria Cichlid sp. *Haplochromis perrieri* J Pellegrin, 1909

Jean Octave Edmond Perrier (1844–1921) was a French invertebrate zoologist best known for his studies of annelids and echinoderms. He started professional life as a schoolteacher for three years in Agen college, following which his PhD was awarded (1869) and he taught zoology at the École normale supérieure (1872). He took the chair of Natural History at the Muséum National d'Histoire Naturelle, Paris (1876) and participated in a number of sea expeditions. He became (1900–1919) Director of the MNHN. Among his best-known writings are: *La Philosophie zoologique avant Darwin* (1884) and *La Terre avant l'Histoire. Les Origines de la Vie et de l'Homme* (1920).

Perrottet

Large-tooth Sawfish *Pristis perotteti* JP Müller & FGJ Henle, 1841
[Junior Synonym of *Pristis pristis*]

George Samuel Perrottet (1790–1870), also known as Georges Guerrard-Samuel Perrottet (and whose surname is sometimes given as Pérotet or Perrotet), was a Swiss-born French botanist and horticulturalist. He was a gardener at the Jardin des Plantes. He became the naturalist on *Rhône* during an expedition (1819–1821) to Reunion, Java, and the Philippines and was sent to Cayenne (French Guiana) to introduce plants that were thought would be useful there. Perrotet made large mineralogical and botanical collections in Cayenne before returning to France. He made a number of voyages to Africa and South America (1822–1832), including one circumnavigation, and wrote: *Souvenirs d'un voyage autour du monde* (1831). He explored in Senegambia (1824–1829) and was an administrator at a government trading post. He also explored Cape Verde (1929). He co-wrote a work on the plant life of that part of Africa as *Florae Senegambiae Tentamen* (1830–1833). He became correspondent of the MNHN, Paris (1832) and was then assigned to a botanical garden in Pondicherry (1834–1839) after which he returned to France and cultivated silk worms. He returned to Pondicherry and lived there for the rest of his life (1843–1870). Four reptiles are named after him. (Also see **Perotae** above)

Perry (Gilbert)

Dwarf Lantern Shark *Etmopterus perryi* S Springer & GH Burgess, 1985

Perry Webster Gilbert (1912–2000) was an American biologist with a particular interest in sharks. He studied for his first degree at Dartmouth College, graduating in Zoology (1934) and was given an instructorship there. He entered the programme at Cornell University (1936), which awarded his PhD (1940) in comparative vertebrate anatomy. He was an instructor there until becoming Assistant Professor (1943–1946) gaining tenure (1946–1952) then full Professor (1952–1978) until retirement, after which he continued to research as Professor Emeritus. His many dissections of sharks continued to fuel an interest that he pursued whenever time (and sabbaticals) allowed, including undertaking research trips to the Bahamas, Tahiti, Australia, Belize, South Africa and Japan as well as within the USA. He published on virtually every aspect of shark biology, writing over 150 papers and editing two books including: *Sharks, Skates and Rays* (1967). He was honoured for his contributions to the knowledge of elasmobranch reproduction and other aspects of shark biology.

Perry, G

Pigfish genus *Perryena* GP Whitley, 1940

George Perry (1771-1823) was an eminent English architect and stonemason, as well as a naturalist and malacologist. He wrote: *Arcana; or the museum of natural history* published in 22 monthly parts (1810–1811) and *Conchology, or the natural history of shells* (1811). Whitley coined the name because Perry had named the related genus *Congiopodus*.

Perry, M

Sakhalin Taimen *Parahucho perryi* JC Brevoort, 1856
[Alt. Japanese Huchen]

Commodore Matthew Calbraith Perry (1794–1858) of the US Navy, is very famous for having commanded the United States Japan Expedition (1852–1854) that opened Japan to the outside world after more than 200 years of seclusion. The etymology says that it was "… *to* [Parry's] *efforts alone we owe the scanty yet interesting zoological collections and drawings, made under disadvantageous circumstances, while the squadron was in those distant seas*."

Persephone

Fighting-fish sp. *Betta persephone* D Schaller, 1986

In Greek mythology, Persephone was the queen of the Underworld - the equivalent of the Roman goddess Proserpine (q.v.). The name was given in fanciful allusion to the fish's dark and sombre colouration.

Perugia

Flat-whiskered Catfish genus *Perugia* CH Eigenmann & AA Norris, 1900
[Junior Synonym of *Pinirampus*]
Honeycomb Driftwood Catfish *Duringlanis perugiae* F Steindachner, 1882
[Syn. *Centromochlus perugiae*]
Perugia's Limia *Limia perugiae* BW Evermann & HW Clark, 1906
Neotropical Silverside sp. *Odontesthes perugiae* BW Evermann & WC Kendall, 1906

Dr Alberto Perugia (1847–1897) was an Italian ichthyologist at the Civic Museum of Natural History in Genoa. He wrote: *Descrizione di due nuove specie di Pesci raccolti in Sarawak dai Sig. G. Doria ed O. Beccari* (1892).

Pes

Pale Dottyback *Pseudochromis pesi* R Lubbock, 1975

P.E.S. are the chief initials of **P**eter F **E**therington-**S**mith, who helped Lubbock collect the holotype and other pseodochromids. We have been unable to add any biography.

Pesta

Eğirdir Longsnout Scraper *Capoeta pestai* V Pietschmann, 1933
[Alt. Eğirdir Barb]

Dr Otto Pesta (1885–1974) was an Austrian zoologist, limnologist, hydrobiologist and specialist in crustaceans. After completing his PhD (1907) he was appointed Curator at the Natural History Museum of Vienna (briefly Acting Director, 1938–1939) and a Professor at the Universität für Bodenkultur, Vienna (1922) (now University of Natural Resources and Life Sciences) and then at the University of Vienna (1927). Among his written works are: *Die Decapodenfauna der Adria : Versuch einer Monographie* (1918) and *Der Hochgebirgssee der Alpen* (1929). The patronym is not identified in the etymology, but the probability is that it honours Pietschmann's colleague at the Museum.

Petard

Pinkeye Mullet *Trachystoma petardi* FL de Castelnau, 1875

Mr. Petard (forename not known) sent fishes to Castelnau that he had collected from the Richmond River, New South Wales, including the type of this one.

Petchkovsky

African Barb sp. *Enteromius petchkovskyi* M Poll, 1967

Monsieur de Petchkovsky was thanked for his help in collecting fishes, but the author declined to give any first name(s) or other details.

Peten

Central American Cichlid genus *Petenia* A Günther, 1862
Threadfin Shad *Dorosoma petenense* A Günther, 1867

This is a toponym referring to Lake Petén, Guatemala, the type locality.

Petényi

Romanian Barbel *Barbus petenyi* JJ Heckel, 1852

Salamon János Petényi (1799–1855) was a Hungarian Lutheran pastor, zoologist, palaeontologist and ornithologist. He followed his father into religious orders, passed his exams as a pastor (1826) and was given a parish (1833) at Cinkota, a Budapest suburb. However, he preferred a life of science and he worked at the Hungarian National Museum, becoming (1834) Curator of its zoological collection. Many consider him to be the founder of Hungarian ornithology. (The barb is from nearby Romania and Bulgaria).

Peter Davies

Lake Malawi Cichlid sp. *Alticorpus peterdaviesi* WE Burgess & HR Axelrod, 1973

Peter Davies was a fish exporter of Lake Malawi. He was honoured for his help in securing many fishes of the lake for photography and study. (Also see **Henny Davies**)

Peters, JA

Needlefish sp. *Potamorrhaphis petersi* BB Collette, 1974

Dr James Arthur Peters (1922–1972) was a zoologist who specialised in Ecuadorean herpetofauna. He attended the University of Michigan and was awarded a bachelor's degree (1948), a master's (1950), and a doctorate (1952). He taught at Brown University as an Associate Professor (1952–1958), leaving to become a Fulbright Lecturer at Universidad Centrale de Ecuador (1958–1959). He was a Professor at San Fernando Valley State College (1959–1964) and was at the Smithsonian as Assistant Curator, Reptiles and Amphibians (1965–1972), and as Curator for the last few years of his life. He co-wrote: *Catalogue of the Neotropical Squamata* (1970). Five amphibians and nine reptiles are named after him.

Peters, WHK

African Characin genus *Petersius* FM Hilgendorf, 1894
Peters' Toby *Canthigaster petersii* GG Bianconi, 1854
Peters' Elephantnose Fish *Gnathonemus petersii* A Günther, 1862
Peters' Banded Croaker *Paralonchurus petersii* F Bocourt, 1869
Peters' Goby *Oxyurichthys petersii* CB Klunzinger, 1871
Lampfish *Dinoperca petersi* F Day, 1875
Peters' Clingfish *Tomicodon petersii* S Garman, 1875

Blenny sp. *Enchelyurus petersi* R Kossmann & H Räuber, 1877
Prickly Fanfish *Pterycombus petersii* FM Hilgendorf, 1878
Ice Goby *Leucopsarion petersii* FM Hilgendorf, 1880
African Rivuline sp. *Nimbapanchax petersi* H-E Sauvage, 1882
African Characin sp. *Distichodus petersii* GJ Pfeffer, 1896
Zambezi Mormyrid *Petrocephalus petersi* B Kramer, R Bills, P Skelton & M Wink, 2012
Peters' Dragonet *Callionymus petersi* R Fricke, 2016

Wilhelm Karl Hartwig Peters (1815–1883) was a German zoologist and traveller who made some very important collections in Mozambique. Most of his published works are on herpetology. He conducted the first major fish survey of the lower Zambezi region (1842–1848) and discovered many new species. For many years he was the Head of the Berlin Zoological Museum. He was elected a corresponding member of the Russian Academy of Sciences (1876). As many as 23 mammals, 18 amphibians and 39 reptiles are named after him. The dragonet was named after Peters as he "*...described the fishes collected by S.M.S. Gazelle in New Ireland in July, 1875, and was the first to observe this new species (although he misidentified it as* Callionymus calauropomus [non *Richardson, 1844*])."

Petersen

Starry Flying Gurnard *Dactyloptena peterseni* E Nyström, 1887
Goby sp. *Oxyurichthys petersenii* F Steindachner, 1893
[Junior Synonym of *Oxyurichthys auchenolepis* Bleeker, 1876]

Julius W Petersen was a businessman who made a collection of fish in Asia while conducting business there. Most of the fish Edvard Nyström studied were obtained (1883) from the Director of the telegraph company; this being Petersen who had at that time settled in Nagasaki, Japan. Steindachner says little in the description of the goby, only that 'Director Petersen' supplied the holotype from Swatow (now Shantou), China. We believe this is most likely the same person. However, we cannot dismiss the possibility that he had in mind a Mr Petersen who was Danish and was the keeper of Lamoks Lighthouse in Swatow. He was aboard the 'Namoa' when it was attacked and hijacked by Chinese 'pirayes' (1890) and he was killed.

Peterson, CB

Coastal Shiner *Notropis petersoni* HW Fowler, 1942

C Bernard Peterson (1906–1963) was Fowler's editor at the Academy of Natural Sciences (Philadelphia). He co-wrote: *The Earliest Account of the Association of Human Artifacts With Fossil Mammals in North America* (1944). He collected fish specimens, including the shiner holotype at Crane Creek, North Carolina (August 1940).

Peterson, E

Peterson's Grenadier *Ventrifossa petersonii* AW Alcock, 1891

E Peterson was the gunner of the 'Investigator', the ship from which the holotype was collected. To judge by the etymology, he appears to have been lucky to have survived since his "*...unabating zeal on behalf of our zoological collections led on one occasion to his getting his fingers almost amputated by the dredging-wire, and on another occasion to his falling overboard almost into the mouth of a shark.*"

Petit, A

Flagtail sp. *Kuhlia petiti* LP Schultz, 1943

Arthur Petit was a Pharmacist's Mate (a rank in the US Navy), who helped Leonard Schultz collect fishes on Canton and Enderbury Islands (now in Kiribati).

Petit, G

Tetra genus *Petitella* J Géry & H Boutière, 1964
Madagascar Guitarfish *Glaucostegus petiti* P Chabanaud, 1929
[Syn. *Rhinobatos petiti*]
Madagascar Cichlid sp. *Paretroplus petiti* J Pellegrin, 1929
Lez Sculpin *Cottus petiti* MC Bacescu & L Bcescu-Mester, 1964
Sand Knifefish sp. *Gymnorhamphichthys petiti* J Géry & T-T Vu, 1964

Professor Georges Jean-Jacques Petit (1892–1973) was a French marine biologist at the Muséum National d'Histoire Naturelle. He became the Director of the marine research stations at Banyuls-sur-mer and Villefranche-sur-mer. He collected the types of the Madagascan species.

Petitjean

African Barb sp. *Labeobarbus petitjeani* J Daget, 1962

M. (Monsieur) Petitjean (forename not given) was honoured for his role in facilitating Daget's missions in Guinea, but without further information from the original text we have been unable to unearth more details.

Petracini

Corydoras sp. *Corydoras petracinii* PA Calviño & F Alonso, 2010

Roberto Petracini (1941–2016) was an Argentine aquarist. He was honoured for his contribution to the development, knowledge and spread of aquarism in Argentina and throughout South and Central America.

Petru Banarescu

Stone Loach sp. *Mesonoemacheilus petrubanarescui* AGK Menon, 1984
Riffle Minnow sp. *Alburnoides petrubanarescui* NG Bogutskaya & BW Coad, 2009

Dr Petru Mihai Bănărescu (1921–2009) (See **Bănărescu**)

Peugeot

Tetra sp. *Hyphessobrycon peugeoti* LFS Ingenito, FCT Lima & PA Buckup, 2013

The fish is named for a family rather than an individual: the Peugeot family (best known now for their cars) invented the Peugeot pepper mill mechanism (1842) and their manufacturing business led to the establishment of a carbon sink reforestation project in central Brazil, and eventually to the discovery of this tetra species.

Pezold

Squeaker Catfish sp. *Chiloglanis pezoldi* RC Schmidt & HL Bart, 2017

Dr Frank L Pezold, Dean and Professor of Biology (2006–present) at Texas A&M University, led the expedition to Guinea (2003) that collected this species. The University of New Orleans awarded his BA (1974) and MSc (1979) and the University of Texas, Austen his PhD (1984). He was professor at the University of Louisiana (1996–2005) and Director of its natural history museum (2002–2005). His major contributions have been in the systematics and classification of gobies. He has written a book and around 150 papers and articles. He has conducted field work in freshwater and coastal waters of the southeastern USA, Mexico, Venezuela, West Africa and Micronesia.

Pfaff

Pfaff's Lampeye *Micropanchax pfaffi* J Daget, 1954

Dr Johannes Rasch Pfaff (d.1959) was a Swedish ichthyologist and collector who was Curator of Vertebrates at the University of Copenhagen Zoological Museum. He wrote: *Report on the fishes collected by Mr. Harry Madsen during Professor O. Olufsen's Expedition to French Sudan in the years 1927–28* (1933).

Pfeffer

Lake Tanganyika Cichlid sp. *Gnathochromis pfefferi* GA Boulenger, 1898

Dr Georg Johann Pfeffer (1854–1931) was a zoologist and malacologist who became Curator, Naturhistorisches Museum zu Hamburg (1887). He wrote: *Die Cephalopoden der Plankton-Expedition. Zugleich eine Monographische Übersicht der Oegopsiden Cephalopoden* (1912). He was, according to the author, a 'distinguished' German zoologist "*...who has much contributed to our knowledge of East African ichthyology*." Three reptiles are also named after him.

Pfeiffer

Stone Loach sp. *Nemacheilus pfeifferae* P Bleeker, 1853

Ida Laura Pfeiffer née Reyer (1797–1858) was an Austrian traveller and travel writer. She was a 'tomboy' and was given an education usually reserved for boys. She was taken on her first long trip to the Middle East when only five. After her (much older) husband died and his children left home, she was able to travel. Among the first female explorers, her books were very popular and were translated into seven languages. In one of her books: *Visit to Iceland* she wrote: "*When I was but a little child, I had already a strong desire to see the world. Whenever I met a travelling-carriage, I would stop involuntarily, and gaze after it until it had disappeared; I used even to envy the postilion, for I thought he also must have accomplished the whole long journey*." Her travels included along the Danube to the Black Sea and Istanbul, Palestine and Egypt returning via Italy (1842), Scandinavia and Iceland (1845), around the world via Brazil, Chile and other parts of South America, Tahiti, China, India, Persia, Turkey & Greece (1846–1848), a second round the world trip (1851–1854) and to Madagascar (1857). Each trip was financed by the sales of her book about the previous trip. During her travels she collected plants, insects, molluscs, marine life and mineral specimens. The carefully documented specimens were sold to the Vienna Natural History Museum. She collected the loach type.

Pflaum

Striped Sand-Goby *Acentrogobius pflaumii* P Bleeker, 1853

Dr A K J L W Pflaum was a Surgeon Major in the Royal Dutch East Indies Army who had been stationed at Samarang, Celebes (Sulawesi). He provided Pieter Bleeker with the type specimen of the goby.

Pflueger

Orange Goatfish *Mulloidichthys pfluegeri* F Steindachner, 1900

Unfortunately, Steindachner says nothing in his original brief text about the person he was honouring in the binomial. The holotype came from Honolulu, and we know that members of the Pflüger family were resident there - e.g. Wilhelm Jolani Pflüger (b.1871 in Honolulu) – so perhaps some member of that family provided the holotype. Another, less likely, possibility is that Steindachner intended Dr Eduard Friedrich Wilhem Pflüger (1829–1910), a German physiologist at the University of Bonn.

Pflueger, A

Longbill Spearfish *Tetrapturus pfluegeri* CR Robins & DP de Sylva, 1963

Albert Pflueger Sr. (d.1962) was a taxidermist based in Miami who invented a new method of mounting fish specimens, especially large 'sport' specimens like tarpon and sailfish.

Phaeton

Phaeton Dragonet *Synchiropus phaeton* A Günther, 1861

In Greek mythology, Phaëton was the son of the sun god Helios. The name means 'shining one'. The reason for applying this name to the dragonet was not explained.

Phamhring

Stone Loach sp. *Schistura phamhringi* B Shangningham, Y Lokeshwor & W Vishwanath, 2014

B D Phamhring Anal (d.2014) was an Indian collector who obtained a number of new fishes in the region of Manipur, including the holotype of this species (2013) which he submitted to Manipur University. He died following a heart attack.

Phelps

Suckermouth Catfish sp. *Astroblepus phelpsi* LP Schultz, 1944
Armoured Catfish sp. *Spatuloricaria phelpsi* LP Schultz, 1944

The fish are named after W H Phelps Sr. or after his son W H Phelps Jr. – well-known leaders in furthering the development of the biological sciences in Venezuela.

William Henry Phelps Sr. (1875–1965) was an Americanborn Venezuelan ornithologist. He first visited Venezuela as a Harvard student (1896). He wrote: *Lista de las Aves de*

Venezuela (1958). His son William 'Billy' Henry Phelps Jr (1902–1988) co-wrote: *A Guide to the Birds of Venezuela* (1978). The 'Colección Ornitológica Phelps' in Caracas consists of over 75,000 skins, mostly of Venezuelan origin. William, Billy and Billy's wife Kathleen built up the collection. The Phelps family had their own specially equipped yacht, 'Ornis', in which they made 49 trips to the Caribbean islands and to the hinterland of Venezuela. Phelps Sr. had married the daughter of wealthy British settlers called Tucker in San Antonio de Maturin where he started selling coffee – the first of many successful business ventures. Eight birds and a reptile are named after various members of the family.

Philip

Aden Ringed Skate *Orbiraja philipi* RE Lloyd, 1906
[Syn. *Okamejei philipi*]

The description has no etymology and we have been completely unsuccessful at identifying 'Philip'. If any reader knows, we would love to hear from them.

Philippe

African Characin sp. *Neolebias philippei* M Poll & J-P Gosse, 1963

R Philippe collected the type in the Congo (DRC), but no more information is given.

Philippi

Chalapo Clinid *Labrisomus philippii* F Steindachner, 1866
Guitarfish sp. *Tarsistes philippii* DS Jordan, 1919
[Status uncertain: known from a single dried head]

Professor Dr Rodulfo Amando (sometimes Rudolph Amandus) Philippi (1808–1904) {Krumweide}* was a German-born Chilean naturalist, principally interested in zoology and palaeontology. He was educated in Berlin, qualifying as a physician with further study in zoology. He was then Professor of Natural History and Geography at the Polytechnic of Kassel. Believing he was mortally ill, he moved to southern Italy – where he recovered and began working. His brother worked for the Chilean government and invited him to join him there (1851). Rodulfo became a professor of zoology and botany and the Director of the Museo Nacional de Historia Natural, Santiago (1853–1883). He organised over sixty expeditions within Chile, published 456 scientific papers, and described more than 6,000 plants before retiring (1896). He is commemorated in the names of numerous plants, an amphibian, a mammal and three reptiles.

*{Krumweide}, derived from his maternal line, is used to distinguish him in scientific descriptions from his zoologist grandson who had exactly the same name.

Phillip

Medium-snouted Pipefish *Vanacampus phillipi* AHS Lucas, 1891
[Alt. Port Phillip Pipefish]

This is a toponym referring to Port Phillip, Victoria, Australia.

Phillips, JB

Chameleon Rockfish *Sebastes phillipsi* JE Fitch, 1964

Julius B Phillips (b.1905–retired 1968) worked at the California Division of Fish and Game for about 40 years. He was an expert on rockfish and the author of: *A review of the rockfishes of California* (1957).

Phillips, RJ

Phillips' Lanternfish *Diaphus phillipsi* HW Fowler, 1934

Dr Richard J Phillips of Philadelphia collected many local fishes for Fowler (q.v): clearly another of the wide circle of friends and contacts whom Fowler cultivated. He co-wrote with Fowler: *A new fish of the genus* Paralepis *from New Jersey* (1910). By the time of Fowler's description of the lanternfish, he was 'the late' Dr Phillips.

Phillips, WWA

Phillips' Garra *Garra phillipsi* PEP Deraniyagala, 1933

Major William 'Bill' Watt Addison Phillips (1892–1981) was, from the age of nineteen, a tea and rubber planter in Ceylon (Sri Lanka). He saw action in Mesopotamia during WW1 with the 24th Punjab Regiment, but after a four-month siege he was taken prisoner along with all other survivors. Whilst a prisoner in Turkey he developed a serious interest in zoology, having always been keen on natural history and keeping diaries of the birds, bats and snakes he saw as a planter (He would pay the locals 5c for live snakes that they brought to him). He was Secretary and later Chairman of the Ceylon Bird Club, and wrote: *Check List of the Birds of Ceylon* (1952). He returned to England in 1956. He collected the garra type. His daughter wrote his biography: *W.W.A. Philips - A Naturalist's Life* (2002). A reptile and three birds are also named after him.

Phil Pister

Mexican Killifish sp. *Fundulus philpisteri* ME García-Ramírez, S Contreras-Balderas & M de Lozano-Vilano, 2007

Edwin Philip 'Phil' Pister (b.1929) is a fisheries biologist, ichthyologist and ecologist, and an expert on desert fishes and their conservation. He graduated from Berkeley having switched from pre-med to Life Sciences. He was then an Officer in the US Army during the Korean War, although he did not see action. Shortly after re-entering civilian life he worked (until 1990) for the California Department of Fish and Game. He is credited with saving the Owens Pupfish *Cyprinodon radiosus* by transferring the entire remaining population into several buckets and transporting them to a safe location (1969)! He has written scientific papers and popular articles, such as: *The Rights of Species and Ecosystems* (1995). He is still active in the Desert Fish Council that he helped to found (1969). (Also see **Pister**).

Phoebe

Tattler (sea bass) *Serranus phoebe* F Poey, 1851

Poey calls this species the 'Serrano Diana', with Diana being a local name given to the fish in Cuba. Diana was also the Roman goddess of the moon, her Greek equivalent being Artemis. 'Phoebe' was one of the epithets of Artemis. Clearly, Poey wished to display his knowledge of classical mythology.

Phuong

Freshwater Goby sp. *Rhinogobius phuongae* M Endruweit, 2018

Thi Dieu Phuong Nguyen (b.1975) was at the Research Institute for Aquaculture No. 1 (Bac Ninh, Viêt Nam) as an MSc student (1998). Among her publications is the co-written: *Typical livelihood of aquatic producer households in peri-urban Hanoi* (2006) and *Tilapia seed production in rice field* (2007). She has written at least one paper with the goby's describer Marco Endruweit: *Sewellia hypsicrateae, a new species of loach from central Vietnam (Teleostei: Balitoridae)* (2016). She was honoured for her enthusiastic interest in Vietnamese fishes.

Phyllis

Lanternfish sp. *Nannobrachium phyllisae* BJ Zahuranec, 2000

Phyllis Elaine Fabian (b.1939) is a registered nurse who is the author's former wife. They may have broken up but he clearly had a soft spot for her, saying in the etymology "*...as a token of recognition for her many years of support, which culminated in this study.*"

Picart

African Halfbeak *Hyporhamphus picarti* A Valenciennes, 1847

Monsieur Picart is said to have contributed natural history specimens from Cádiz, Spain, to the 'King's Cabinet', but no further elaboration is made.

Picasso

Picasso Triggerfish *Rhinecanthus* spp.

Pablo Ruiz Picasso (1881–1973) was a Spanish painter and sculptor. The vivid patterns and colours of *Rhinecanthus* triggerfish have led to them being popularly known as picassofish.

Pichardo

Bobo Mullet *Joturus pichardi* F Poey, 1860

Esteban Pichardo (1799–1879) was a Cuban lexicographer-geographer who was, according to Poey, the 'estimable auteur' of *Diccionario Provincial de voces Cubanos* and *Geografia de la isla de Cuba* (he also recommended the vernacular name *Joturo*).

Pickle, CB

Three-barbeled Catfish sp. *Cetopsorhamdia picklei* LP Schultz, 1944

Chesley B Pickle was an executive at Lago Petroleum Corporation, Lagunillas. He helped Schultz collect fishes at the southern end of Lago Maracaibo, Venezuela.

Pickle

Flabby Whalefish sp. *Cetomimus picklei* JDF Gilchrist, 1922

This is named after a boat; the 'Pickle', from which the holotype was collected; a South African marine survey vessel.

Piedrabuena

Eelpout genus *Piedrabuenia* AE Gosztonyi, 1977

Luis Piedrabuena (1833–1883) was an Argentine sailor regarded as a hero of Patagonia. He saved the lives of many shipwrecked sailors, and the National Government gave him the title (1864) of 'unpaid honorary captain' to defend Argentine sovereignty in Patagonia.

Pienaar

African Rivuline sp. *Nothobranchius pienaari* KM Shidlovskiy, BR Watters & RH Wildekamp, 2010

Dr Uys de Villiers 'Tol' Pienaar (1930–2011) was a South African biologist and game warden, whose doctorate was awarded by Wits Medical School (1953). He was appointed as a junior ranger at the Kruger National Park (1955) and became Nature Conservator (1970) and Chief Warden (1978). He was also Chief Director of South African National Parks (1987–1991). He wrote *The Freshwater Fishes of the Kruger National Park* (1968). A reptile is also named after him.

Pierre

Labeo sp. *Labeo pierrei* HE Sauvage, 1880

Yellow-eyed Silver Barb *Hypsibarbus pierrei* HE Sauvage, 1880

Jean Baptiste Louis Pierre (1833–1905) was a French botanist and the first Director (1864–1877) of Saigon Botanic Garden, who made many collections in tropical Asia. He initially worked in the botanical gardens at Calcutta, India, but then (1864) founded the Saigon Zoo and Botanical Gardens. His publications include: *Flore forestière de la Cochinchine* (1880–1907). Several plants are also named after him. Sauvage writes that 'Pierre' (no other information given) collected the types of these fish, but given the location and his standard abbreviation as a collector, we believe the above is intended.

Pierson

Blenny sp. *Parahypsos piersoni* CH Gilbert & EC Starks, 1904

C J Pierson was a member of the Panama Expedition (1896) of the Leland Stanford Junior University during which the type was collected. We have been unable to uncover any further biography.

Pierucci

Pierucci's Rainbowfish *Melanotaenia pierucciae* GR Allen & SJ Renyaan, 1996

Silverside sp. *Bleheratherina pierucciae* Aarn & W Ivantsoff, 2009

Paola Perucci was with Heiko Bleher (q.v.) when he collected the first specimen of the rainbowfish (1995). She is described by Bleher as being his 'Italian friend' and seems to accompany him to various remote parts of the world, collecting fish. No more is recorded.

Pietsch

Dreamer Anglerfish genus *Pietschichthys* VE Kharin, 1989
[Junior Synonym of *Dermatias*]
Sculpin sp. *Icelinus pietschi* M Yabe, A Soma & K Amaoka, 2001
Dreamer sp. *Oneirodes pietschi* HC Ho & KT Shao, 2004
Pietsch's Frogfish *Kuiterichthys pietschi* RJ Arnold, 2013
Manefish sp. *Platyberyx pietschi* DE Stevenson & CP Kenaley, 2013
Thin Eel sp. *Myroconger pietschi* VC Espindola, RA Caires, KA Tighe, MCC de Pinna & MRS de Melo, 2021

Dr Theodore 'Ted' Wells Pietsch III (b.1945) was a professor of ichthyology at the University of Washington, Seattle (1978 onwards) and is an American systematist and evolutionary biologist especially known for his studies of anglerfishes. His BA in zoology was awarded by the University of Michigan and his MSc and PhD by the University of Southern California. He was a post-doctoral fellow at Harvard (1973–1975) He has described 65 species and 12 genera of fish and published over 200 papers and a dozen books, including a novel: *The Curious Death of Peter Artedi: A Mystery in the History of Science*. He was principal investigator for the International Kuril Island Project (1994–1999) during which the sculpin species was discovered.

Pietschmann

Labeo sp. *Labeo pietschmanni* B Machan, 1930
Infantfish sp. *Schindleria pietschmanni* K Schindler, 1931
African Airbreathing Catfish sp. *Clariallabes pietschmanni* H Güntert, 1938

Dr Viktor Pietschmann (1881–1956) was an Austrian ichthyologist who worked as Curator of the Fish Collection at the Vienna Museum of Natural History (1919–1946) where Machan also worked. He graduated from the University of Vienna (1904). He made several collecting trips, notably to Greenland, Armenia, Hawaii, Romania, Poland and the Middle East. He was a member of the Nazi party (from 1932) so was forced to resign (1946). He collected the type of the Infantfish.

Pijpers

Three-barbeled Catfish sp. *Imparfinis pijpersi* JJ Hoedeman, 1961

H P Pijpers was a corporal in the army of Suriname and is recorded as having collected fish taxa there during the Sipaliwini & Wilhelmina Mountains Expeditions (1960–1962), including with Dr. Geijskes, Director of the Surinam Museum. He is mentioned in several papers as part of collecting expeditions, but his status is not spelt out.

Pike

Pike's Moray *Gymnothorax pikei* R Bliss Jr, 1883
Blacklip Damsel *Pomacentrus pikei* R Bliss Jr, 1883

Colonel Nicholas Pike (1817–1905) was a soldier, naturalist and author who was United States Consul in Mauritius. He was a competent artist and donated many paintings of the fish of the area to Cambridge University, which were later (1929) published by AMNH. He collected birds, fishes, shells and algae there. He wrote about his time as Consul as: *Sub-tropical rambles in the land of the Aphanapteryx; personal experiences, adventures, and wanderings in and around the island of Mauritius* (1873). He had written an occasional scientific paper such as: *Notes on the hermit spadefoot (Scaphiopus holbrookii Harlan; S. solitarius Holbr.* (1886). One claim to fame (or infamy) is that when Director of the Brooklyn Institute (1853) he introduced House Sparrows to New York, which he had collected in England (1851). He was known as an authority on venomous reptiles and left a collection of spiders and birds that he had mounted himself. He died in New York of 'paralysis' (presumable a stroke). He was honoured in the moray's name as he had supplied Bliss with the type and for the damselfish because he had provided Harvard's Museum of Comparative Zoology with a 'large and valuable' collection of fishes from Mauritius, including the damsel type.

Pillai, NGK

Basslet sp. *Pseudanthias pillai* PC Heemstra & KV Akhilesh, 2012

Dr N Gopala Krishna Pillai is Scientist Emeritus, Indian Council of Agricultural Research (ICAR) where he worked (1978–2015). He worked (1996–2010) for Central Marine Fisheries Research Institute (CMFRI), India, where he was Principal Scientist and Head of the Division of Pelagic Fisheries,and later its Director (2007). He has published more than 150 papers. He was honoured in the name for his "*...valuable contributions to the better understanding of marine fishes and the fisheries of India.*"

Pillai, RS

Spineless Eel genus *Pillaia* GM Yazdani, 1972

Indian Loach sp. *Ghatsa pillaii* TJ Indra & K Rema Devi, 1981

Dr Raghavan Sridharan Pillai is a herpetologist who worked for the Zoological Survey of India, which he joined *(1963).* He was Deputy Director and Officer-in-Charge, Southern Regional Station, Zoological Survey of India. *He was a member of the* Department of Zoology, University College, Trivandrum, India (1960). He co-wrote: *Gymnophiona (Amphibia) of India: A Taxonomic Study* (1999). He collected the type of the loach species. Two amphibians are also named after him.

Pillet

Neotropical Rivuline sp. *Spectrolebias pilleti* DTB Nielsen & R Brousseau, 2013

Didier Pillet is a French aquarist and environmentalist. He was one of the collectors of the species along with Michel Beuchey (q.v.), Jean Marc Beltramon and Christine Lambert, with whom he often collects in the Amazon.

Pilsbry

Duskystripe Shiner *Luxilus pilsbryi* HW Fowler, 1904

Dr Henry Augustus Pilsbry (1862–1957) was a conchologist and malacologist, an expert on barnacles, and a leading light in the Philadelphia Academy of Sciences. He was awarded his bachelor's degree (1882) and an honorary doctorate (1899) at the University of Iowa. He worked as a newspaper reporter in Iowa (1883–1887). He was a conservator of conchology at Philadelphia (1888–1895). He edited: *Manual of Conchology* (1889–1932) and founded *Nautilus* (1889), a journal he edited until his death. He collected in much of the Americas and in Australia, Japan, and the Pacific islands. He collected the shiner type. Two reptiles are also named after him.

Pinchuk

Tadpole Goby sp. *Benthophilus pinchuki* DB Ragimov, 1982

Vitaly Iustinovich Pinchuk (1931–1992) was a Ukrainian-Russian ichthyologist at the USSR Academy of Sciences, who collaborated with Ragimov on the description of *B. svetovidovi* (1979).

Pinda

Pinda Moray *Gymnothorax pindae* JLB Smith, 1962

This is a toponym referring to Pinda in Mozambique, the type locality.

Pinetorum

Righteye Flounder sp. *Cleisthenes pinetorum* DS Jordan & EC Starks, 1904
Sculpin sp. *Ricuzenius pinetorum* DS Jordan & EC Starks, 1904

This is not an eponym, but a form of toponym. *Pinetorum* means 'of the pine trees'. The holotypes of these species were caught in Matsushima Bay, Japan. *Matsushima* = 'Pine Island'.

Ping

Loach sp. *Beaufortia pingi* P-W Fang, 1930
Chinese Cyprinid sp. *Percocypris pingi* T-L Tchang, 1930

Dr Chih Ping (1886–1965), whose doctorate was awarded (1918) by Cornell, became a zoology investigator at Wistar Institute, Philadelphia, then Professor of Zoology, College of Agriculture, National Southeastern University, Nanking (1921). He was Director of the Biological Laboratory at Nanking (Nanjing) (1935). He was honoured in the case of the loach for his zeal in encouraging the development of zoology in China. Two amphibians and a bird are also named after him. (See also **Chiping**)

Pingel

Ribbed Sculpin *Triglops pingelii* JCH Reinhardt, 1837

Dr Peter Christian Pingel (1793–1852) was a Danish geologist and Arctic scientist. He studied at the University of Copenhagen (1810–1813) then in Germany, continuing his studies until being awarded his PhD (1817) at the University of Jena. He travelled in Germany (1818–1820) studying its geology, then similarly in Denmark, Norway and Sweden. He began researching in Greenland (1828–1829) then was a mineralogical assistant at the Natural History Museum (1829–1842). He was promoted to Inspector (1842–1847) then Superintendant.

Pinheiro

Pinheiro's Cory *Corydoras pinheiroi* J Dinkelmeyer, 1995

Mario Pinheiro, who provided the holotype, was manager of Trop Rio, an aquarium-fish exporter in Rio de Janeiro, Brazil.

Pinocchio

Pinocchio Catshark *Apristurus australis* K Sato, K Nakaya & M Yorozu, 2008
Pinocchio Dwarf-Goby *Eviota pinocchioi* DW Greenfield & R Winterbottom, 2012

Pinocchio is a fictional character created by Carlo Collodi in his childrens' novel: *The Adventures of Pinocchio* (1883). One of his main features was his nose that grew longer and longer when he was under stress and particularly when he was telling lies. The catshark has a long snout, the goby has exceptionally long anterior tubular nares.

Pinto, AA

Snakehead Goby *Gobiopsis pinto* JLB Smith, 1947

Adolfo Abranches Pinto (1895–1981) was Military Commander of Mozambique, where this goby is endemic. Smith proposed a new genus for this species, *Abranches*, so that its original name, *Abranches pinto*, would match that of the honoree. However, the genus is no longer considered valid.

Pinto, OM de

Characin sp. *Oligosarcus pintoi* A Amaral Campos, 1945

Olivério Mário de Oliveira de Pinto (1896–1981) was a Brazilian ornithologist. He studied medicine at the Faculdade de Medicina da Bahia, and after graduating (1921) settled to practice in Araraquara, founding a clinic there. He also taught Natural Sciences at the recently founded School of Odontology and Pharmacy in the city. He started making scientific drawings for the museum director and was appointed as a researcher in zoology there. He was Director of the Zoological Museum, University of São Paulo (1939–1956) and led expeditions under its auspices. He wrote: *Catalogo das Aves do Brasil* (1938). Five birds and an amphibian are also named after him.

Piorski

Tetra sp. *Hyphessobrycon piorskii* EC Guimarães, PS De Brito, LM Feitosa, LF Carvalho-Costa & FP Ottoni, 2018

Nivaldo Magalhães Piorski is a Brazilian ichthyologist who is a consultant and professor at the Federal University of Maranhão, where he graduated BSc (1991). His MSc was awarded by the Paulista State University Júlio de Mesquita Filho (1997) and his PhD (2010) was awarded by the Federal University of São Carlos. His focus is freshwater fish ecology. He was honoured for his contributions to the ichthyological knowledge of Maranhão State, Brazil, where the tetra occurs.

Piquit

Peacock Bass sp. *Cichla piquiti* SO Kullander & EJG Ferreira, 2006

This is not an eponym: *Piquiti* is from the Tupi-Guarani word meaning 'striped'.

Pira

Neotropical Killifish sp. *Cnesterodon pirai* G Aguilera, JM Mirande & M Azpelicueta, 2009

This is not an eponym: *Pirai* is derived from the Guarani words that translate as 'little fish'.

Pires

Driftwood Catfish sp. *Tocantinsia piresi* A de Miranda Ribeiro, 1920

Antenor Pires was a taxidermist and an assistant at the Rondon Commission. He was a member of the Paranatinga River expedition (1914–1915) during which the holotype was collected. He was a friend of the author, Miranda Ribeiro.

Piri

Loach sp. *Cobitis pirii* J Freyhof, E Bayçelebi & M Geiger, 2018

Ahmed Muhiddin Piri, better known as Piri Reis (1465–1553), was an Ottoman admiral, navigator, geographer and cartographer. He gained belated fame as a cartographer when a part of his first world map (prepared in 1513) was discovered in 1929 at the Topkapi Palace, Istanbul. His map is the oldest known Turkish atlas showing the New World, and one of the oldest maps of America still in existence. He also wrote the *Kitāb-ı Ba⊠rīye*, or 'Book of the Sea', which gives seafarers information on the Mediterranean coasts, islands, crossings, straits, etc. and where to take refuge in the event of a storm.

Pirie

Dogtooth Grenadier *Cynomacrurus piriei* L Dollo, 1909

Dr James Hunter Harvey Pirie (1878–1965) was a Scottish physician, bacteriologist and philatelist. He qualified as a physician at the University of Edinburgh. He was the physician and geologist on the Scottish National Antarctic Expedition (1902–1904) on board 'Scotia', during which the holotype was collected. After working in private practice in Scotland, he joined (1913) the Colonial Medical Service in Kenya as a bacteriologist and became Deputy Director, South African Institute for Medical Research, Johannesburg (1926–1941). Today he is more remembered for having been a noted philatelist who wrote important works on the stamps of Swaziland, as well as collecting the stamps of the polar regions. He co-wrote: *The voyage of the Scotia, being a record of a voyage of exploration in the Antarctic Seas* (1906).

Piso

Spinycheek Sleeper *Eleotris pisonis* JF Gmelin, 1789

Willem Piso (1611–1678) was a Dutch physician and naturalist who practiced medicine in Amsterdam. He participated in an expedition to 'Dutch' Brazil (1637–1644). With Georg

Marcgrave, he wrote about this fish and provided a pre-Linnaean name (1648). He became one of the founders of tropical medicine. He wrote *Historia Naturalis Brasiliae* (1648). A minor planet and a plant genus are also named after him.

Pister

Palomas Pupfish *Cyprinodon pisteri* RR Miller & WL Minckley, 2002

Edwin Philip 'Phil' Pister. (See **Phili Pister**)

Pitman

Lake Victoria Cichlid sp. *Haplochromis pitmani* HW Fowler, 1936

Colonel Charles Robert Senhouse Pitman (1890–1975) was a British colonial administrator and naturalist who, when he retired as an officer of the Indian Army, settled in Uganda (1925–1951) and became a game warden and wrote about his work in *A game warden among his charges* (1931). He also wrote: *A Guide to the Snakes of Uganda* (1974) as well as scientific papers such as: *The Breeding of the Standard-wing Nightjar* (1929) and *The gorillas of the Kayonsa region* (1935). He was an all-round naturalist with particular interests in malacology, conchology and ornithology. He also laid the foundations of the Kafue National Park in Northern Rhodesia (now Zambia) and was seconded (1931–1932) to Northern Rhodesia to undertake a game survey. Fowler says he "*...cordially assisted the George Vanderbilt African Expedition*" (1934) during its work at Kisubi Mission, Kitala, Lake Victoria. A mammal is also named after him.

Pittier

Diamond Tetra *Moenkhausia pittieri* CH Eigenmann, 1920

Dr Henri François Pittier (1857–1950) was a Swiss-born geographer, linguist, ethnographer and botanist who moved to Costa Rica (1887). He held several degrees from Swiss polytechnics and universities, including a PhD from Jena. He was a member of the Swiss Topographical Survey for two years and taught in Swiss universities for five years. He was then invited by the Costa Rican government to found and direct the country's first geographic institute, the Instituto Físico-Geográfico, and spent eighteen years there. He was offered a position (1904) as a botanist in the United States Department of Agriculture (USDA). Over the next fourteen years, he explored and collected for the USDA in all the countries of Central America (1910–1914) leading the botanical survey of Panama, serving (1915–1916) as director of the Agricultural Station at Matías Hernandez in Panama City. He left the USDA (1919) to become Director of the Museu Commercial in Caracas, Venezuela, and was later (1933) Director of the National Observatory and (1936) the botanical services of the Ministry of Agriculture. He collected (1887–1918) plant specimens in Central America and Venezuela. He published over 300 papers. The Henri Pittier National Park in Venezuela is named after him, as are a bird, an amphibian and a mammal.

Pixi

Barred Fingerfin *Cheilodactylus pixi* MM Smith, 1980

Pixie John was a South African of whom we know little beyond what Smith's etymology says: "*The specific name* pixi *is a contraction of the name Pixie in honour of Mr Pixie John, formerly of Port Alfred, who sent me the first specimen and always evinced a lively interest in fish and their habits.*" We have also found a reference to a Port Alfred fisherman named Ronnie Samuel (d. 1997) who donated ichthyological specimens to Rhodes University and who had a relative named Pixie John; "a rough man renowned as a beachcomber". It seems likely this was the finder of the fingerfin holotype.

Planer

European Brook Lamprey *Lampetra planeri* ME Bloch, 1784

Professor Dr Johann Jacob Planer (1743–1789) was a German physician and botanist. He gave Bloch the holotype. A genus of American tree, *Planera*, is named after him.

Planes

Damselfish sp. *Chromis planesi* D Lecchini & JT Williams, 2004

Dr Serge Planes (b.1966) is a coral-reef fish ecologist at the Universite de Perpignan (France), who is Research Director of Centre de Recherches Insulaires et Observatoire de l'Environnement (CNRS). He was appointed Research Scientist (1993), Senior (1997) and Principal (2004). He has published around 65 papers since gaining his PhD (1989) such as: *Larval dispersal connects fish populations in a network of marine protected areas* (2009). He discovered and captured the first specimens.

Planquette

Armoured Catfish sp. *Paralithoxus planquettei* M Boeseman, 1982
Characin sp. *Creagrutus planquettei* J Géry & J-F Renno, 1989
Serrasalmid Characin sp. *Myloplus planquettei* M Jégu, P Keith & P-Y Le Bail, 2003

Paul Planquette (1940–1996) was a native of French Guiana and worked at the Institut National de la Recherche Agronomique, Kourou. His first posting with INRA was to Lamto, Ivory Coast (1963), followed by various posting to other places including Guadeloupe (1976–1979) before returning to French Guiana (1980). He co-wrote: *Atlas des poissons d'eau douce de Guyane*, (Tome 1) (1996). He suffered a fatal heart attack whilst working in his laboratory. He collected the catfish holotype.

Plate

Splendid Perch sp. *Callanthias platei* F Steindachner, 1898
Galaxias sp. *Galaxias platei* F Steindachner, 1898

Dr Ludwig Hermann Plate (1862–1937) was a German zoologist and geneticist. He studied under Haeckel at Friedrich Schiller University, Jena, which awarded his doctorate (1886). He qualified in zoology at Philipps University, Marburg (1888), then taught at the Veterinary High School in Berlin (1898–1905) and at the Agricultural College (1905–1908) as Professor of Zoology. He collected along the coast of Chile and in the Juan Fernández Archipelago (1893–1895) and also collected in the West Indies and the Red Sea. He was Haeckel's successor as Professor at Jena (1909–1935) but became embroiled in an unpleasant case when he accused Haeckel and his circle of slandering him. He was a convinced Darwinist but was also a virulent antiSemite. He led the expedition that collected the galaxias holotype. A reptile is also named after him.

Playfair

Golden Panchax *Pachypanchax playfairii* A Günther, 1866
Fringelip Snake-eel *Cirrhimuraena playfairii* A Günther, 1870
White-barred Rubberlip *Plectorhinchus playfairi* J Pellegrin, 1914

Lieutenant-Colonel Sir Robert Lambert Playfair (1828–1899) was an engineer, soldier, diplomat, linguist and author. He joined the Madras Artillery (1846) and rose to captain (1858). He transferred to the Madras staff corps (186i) and given the local rank of Lieutenant-Colonel at Zanzibar. He was then (1866) promoted to Major, and left the corps (1867) with the rank of Lieutenant-Colonel. During his service he served in various administrative roles, for example as Assistant Resident in Aden. Wherever he was posted he explored the area, collected and studied the local history and language and was consequently elected a Fellow of the Royal Geographical Society. His first publication was: *Selections from the Records of the Bombay Government* (1859). Whilst in Zanzibar he was nominated consul and was then posted (1885) to Algeria as Consul General at Algia & Tunis, where he remained for the rest of his career until retirement (1896). He later (1899) received the honorary degree of LL.D. from the university of St. Andrews. He acquired an extensive knowledge of Algeria and the Mediterranean generally, visiting among other places the Balearic Islands and Tunis, where he explored the previously almost unknown Khomair country (1876). He wrote several volumes for the: *A Handbook for Travellers* series. There followed many other books such as: *The Bibliography of Tunisia* (1889). Günther honoured him because he presented the type specimens of the panchax and snake-eel to the BMNH.

Pleske

Stone Loach sp. *Lefua pleskei* SM Herzenstein, 1888

Fedor Dimitrievich Pleske (1858–1932) was a Russian zoologist, ornithologist, geographer and ethnographer. From childhood, he gathered a collection of birds and insects in the different provinces of European Russia. He also analysed ornithological collections of other travellers in central Asia, including Przewalski (q.v.), describing several new species. He became Scientific Secretary of the St Petersburg Natural History Society (1881) and joined their expedition to the Kola Peninsula. He graduated from St Petersburg University (1882) and became a Fellow of the Russian Imperial Academy of Science, St Petersburg (1886), then Scientific Keeper (Curator) at their Zoological Museum, and finally Director (1892–1896) before retiring through ill health. He was an active member of the USSR Zoological Museum (1918). He wrote: *Ornithological Fauna of Imperial Russia* (1891) and *Ornithographia Rossica* (1898), as well as works on the systematics of Arctic birds, and other taxa such as Diptera (gadflies, horseflies etc.). His final monograph: *The Birds of the Eurasian Tundra* (1928) was written in English and edited in the USA. It was devoted largely to the heroic role of Admiral Aleksandr Vasiliyevich Kolchak (1874–1920): since Kolchak later (1918) commanded the White Russian forces trying to overthrow the communist regime in Russia, Pleske would have risked a long spell in the gulag, or even a bullet, for mentioning

Kolchak in any positive light. Five birds, a reptile and an amphibian are also named after him.

Plessis

Plessis' Morwong *Cheilodactylus plessisi* JE Randall, 1983

Yves Plessis (1921–1989) was a French naturalist who collected a specimen of the morwong on Rapa, French Polynesia. and realised it was an undescribed taxon. Plessis visited the island as part of a multi-disciplinary scientific mission.

Ploeg

Glassfish sp. *Gymnochanda ploegi* HH Tan & KKP Lim, 2014

Pike-Cichlid sp. *Crenicichla ploegi* H Varella, M Loeb, FCT Lima & SO Kullander, 2018

Dr Alex A Ploeg (d.2014) was a Dutch ichthyologist who was Secretary General of Ornamental Fish International (2004–2014) and Secretary General of the European Pet Organization (2006–2014). He had kept ornamental fish since he was six years old. He obtained a doctorate (1991) from the University of Amsterdam, with his PhD thesis being on the taxonomic revision, biogeography and phylogeny of *Crenicichla*. That same year he entered the ornamental aquatic industry as a fish breeder on Bonaire (Netherlands Antilles). He was also a publisher of books (AQUALOG/Germany) on ornamental fish (2000–2004). After that he worked as interlocutor between the ornamental fish industry and other institutions worldwide. Unfortunately, Dr Ploeg perished with his wife Edith and their son Robert in the Malaysia Airlines 17 (MH17) airplane which was shot down in the Ukraine on 17 July 2014. He was honoured by Tan and Lim as a "*...good friend, fellow taxonomist, advisor and fellow conservationist against alien aquatic species; for his services to the ornamental fish trade in this region and abroad in his role as the Secretary General of the Ornamental Fish International.*"

Plumier

Sand Tilefish *Malacanthus plumieri* ME Bloch, 1786

Sirajo Goby *Sicydium plumieri* ME Bloch, 1786

Spotted Scorpionfish *Scorpaena plumieri* ME Bloch, 1789

White Grunt *Haemulon plumierii* BGE Lacepède, 1801

Striped Mojarra *Eugerres plumieri* G Cuvier, 1830

Charles Plumier (1646–1704) was a Franciscan monk, botanist and naturalist. He made three trips to the French West Indies (1689–1691, 1693 & 1695), during one he discovered this goby in Martinique. He died of pleurisy when about to embark on a fourth journey to the Indies. He wrote: *Description des Plantes d'Amérique* (1693) and in the same year was appointed Royal botanist to King Louis XIV. Bloch, Lacepède and Cuvier based their descriptions on Plumier's drawings and manuscript.

Plunket

Lord Plunket's Shark *Scymnodon plunketi* ER Waite, 1910

[Alt. Plunket's Dogfish; Syn. *Proscymnodon plunketi*]

Sir William Lee Plunket, 5th Baron Plunket of Newton, County Cork (1864–1920) was an Anglo-Irishman who was the 16th Governor of New Zealand (1904–1910). He served as a Diplomat in Rome and Constantinople (Istanbul) (1889–1894). The Plunket Society in New Zealand is named after his wife, Victoria. He was honoured for his interest in the Canterbury Museum and: "*...gratefully remembering His Excellency's kindness when, as his guest, [Waite] accompanied him on his cruise to the southern islands of New Zealand in 1907.*"

Plutarco

Characin sp. *Hemibrycon plutarcoi* C Román-Valencia, 2001

Dr Plutarco Cala Cala (b.1938). (See **Cala**)

Pluto

Pluto Skate *Fenestraja plutonia* S Garman, 1881
[Alt. Underworld Windowskate]
Narcissus Pencil Wrasse *Pseudojuloides pluto* Y-K Tea, BD Greene, JL Earle & AC Gill, 2020

Pluto was the ruler of the underworld in classical mythology. In zoology, his name is sometimes attached to animals of very dark coloration, which is true of the male pencil wrasse but not particularly so with the skate. The name was probably given fancifully.

Pobeguin

African Barb sp. *Enteromius pobeguini* J Pellegrin, 1911

Charles Henri Oliver Pobeguin (1856–1951) was a French botanist and colonial administrator in Africa. He joined the army (1874) and was posted to Algeria (1876–1878) leaving with the rank of Sergeant. He had a variety of jobs before joining the colonial service in the Congo (1886–1891) making mapping surveys. He was the first civilian administrator in Grand Lahou, Côte d'Ivoire (1892), then of Tiassale (1893) when he mapped the west coast. He returned to France for two years and was then appointed Director in Guinea (1899–1911). Among other things he wrote: *Essay on the flora of French Guinea, forestry products, agricultural and industrial* (1906) and *The Medicinal Plants of Guinea* (1912). He sent the type specimen to Muséum d'Histoire Naturelle, Paris (MNHN) (1904).

Pöch

Dashtail Barb *Enteromius poechii* F Steindachner, 1911

Dr Rudolf Pöch (1870–1921) was an Austrian physician, anthropologist, ethnologist, explorer, natural historian and photographer. He was in India (1897) to study the plague, which sparked his interest in ethnology so he trained (1900–1901) at the Royal Museum of Ethnology in Berlin. He was in Africa to treat malaria, but this further hardened his interest in anthropology and he planned collecting and recording trips. He spent two years in Melanesia (1904–1906), returning with thousands of artefacts, photographs and film recordings particularly from New Guinea. The Welt Museum in Vienna holds some 5000 objects from Melanesia and South Africa collected by him. He became Professor of

Anthropology and Ethnology (1913) and founded (1919) the Institute of Anthropology and Ethnology at the University of Vienna. He collected the barb type.

Poco

Saddle Squirrelfish *Sargocentron poco* LP Woods, 1965

Mary Ann 'Poco' Holloway, according to Woods, "…has prepared illustrations of many species of berycids." We have been unable to find biographical details.

Poetzschke

Characin sp. *Astyanax poetzschkei* E Ahl, 1932

Paul Pötzschke (1881–1957) was co-owner of Scholze & Pötzschke, an aquarium supply and tropical fish importation firm in Berlin. He gave a large number of 'valuable objects' to the Zoological Museum, Berlin, including the holotype of this species.

Poey

Poey's Anchovy *Anchoviella perfasciata* F Poey, 1860
Shortchin Stargazer *Dactyloscopus poeyi* TN Gill, 1861
Pacific Sabretooth Anchovy *Lycengraulis poeyi* R Kner, 1863
Blackear Wrasse *Halichoeres poeyi* F Steindachner, 1867
Curved Sweeper *Pempheris poeyi* TH Bean, 1885
Poey's Scabbardfish *Evoxymetopon poeyi* A Günther, 1887
Offshore Lizardfish *Synodus poeyi* DS Jordan, 1887
Three-barbeled Catfish sp. *Rhamdia poeyi* CH Eigenmann & RS Eigenmann, 1888
Cuban Legskate *Cruriraja poeyi* HB Bigelow & WC Schroeder, 1948
Mexican Snook *Centropomus poeyi* H Chávez, 1961

Professor Felipe Poey y Aloy (1799–1891) was a Cuban zoologist, naturalist and artist. He was brought up in France (1804–1807) and later Spain. He qualified as a lawyer in Madrid, but his ideas were too liberal for the age, and he was forced to return to Cuba (1823). He returned to France (1825) and was one of the founders of the Société Entomologique de France (1832), finally returning to Cuba (1833). He concentrated on natural history, describing 85 species of Cuban fish. His *Memorias sobre la historia natural de la isla de Cuba, acompañadas de sumarios latinos y extractos en francés,* (1858), depicts mainly fishes and snails but also some mammals, hymenoptera, and lepidoptera; the drawings are all done by Poey. He founded the Museum of Natural History, Havana (1839) and became its first director. He also was appointed the first Professor of Zoology and Comparative Anatomy at the University of Havana (1842). The Museum merged (1849) with the University of Havana and was later named in his honour. Poey, who also supplied Cuvier and Valenciennes in Paris with Cuban fish specimens, prepared many of the Museum's exhibits, especially of fishes.

Pogonoski

Pogonoski's Sandperch *Parapercis pogonoskii* JW Johnson & J Worthington Wilmer, 2018

John James Pogonoski is an Australian ichthyologist with CSIRO Marine Research, Hobart, Tasmania, Australia. He previously (2004) worked at the Australian Museum, Sydney. Among his c.35 papers he co-wrote: *Revision of the genus* Parequula *(Pisces: Gerreidae) with a new species from southwestern Australia* (2012) and has described a dozen species of fish. His BSc is in environmental science. He was honoured for his contributions to Australian ichthyology and his helpful co-operation with taxonomic research on CSIRO fish collections.

Pohl

Long-whiskered Catfish sp. *Pimelodus pohli* FRV Ribeiro & CAS de Lucena, 2006

Johann Baptist Emanuel Pohl (1782–1834) was a physician, entomologist, botanist and geologist, born in what is now the Czech Republic. He graduated as a physician at Prague University (1808). He was a member of the Austrian Mission to Brazil (1817–1821) and made large collections, including some 4000 specimens of plants. These were housed with the rest of the expedition's collections in the Brazil Museum of Vienna. After returning to Europe, Pohl was Curator of the Vienna Natural History Museum and of the Brazil Museum (1821–1834). He wrote: *Reisen im Innern von Brasilien* (1837)

Pohle

Pohle's Tilefish *Hoplolatilus pohle* JL Earle & R Pyle, 1997

John Frederick Pohle (1935–2016) was a US Airforce officer and research scuba diver. He was "*...the first to discover the mounds on the reef slope where the type specimens were collected*" whilst scuba diving in Papua New Guinea (1995). He served in the US Air Force before beginning a second career as a research scuba diver.

Poilane

Garra sp. *Garra poilanei* G Petit & T-L Tchang, 1933

Sisorid Catfish sp. *Pareuchiloglanis poilanei* J Pellegrin, 1936

Eugène Poilane (1887–1964) was a French botanist at the Paris herbarium who was in Cochinchina (Vietnam) and neighbouring areas of French Indochina (1909–1964). He started a coffee plantation (1918). He fathered 10 children, five of them after the age of 60. He was assassinated by Viet Cong troops. He collected the type specimens of these fish. Two reptiles are also named after him.

Pointon

Mekong Cyprinid sp. *Oxygaster pointoni* HW Fowler, 1934

Arnold Cecil 'Peter' Pointon (1898–1982) was Forestry Manager of the Chiang Mai branch of the Bombay-Burma Company, Ltd. He served in the British Army in WW1 as a 2nd Lieutenant, Worcestershire Regiment, and was awarded the MC. In WW2 he was a Lieutenant-Colonel in command of the Special Operations Executive Force 136's operatives in Thailand as guerrilla fighters against the Japanese, for which he was awarded the OBE. He wrote: *The Bombay-Burmah Trading Corporation 1863–1963* (1964). Fowler gives no particular reason for honouring Pointon in the binomial, but presumably he gave assistance to Rodolphe Meyer de Schauensee (q.v.) on the latter's collecting trip to Thailand during which the holotype was obtained.

Poit

Blenny sp. *Scartella poiti* CA Rangel, JL Gasparini & RZP Guimarães, 2004

Named after the **P**osto **O**ceanográfico da **I**lha da **T**rindade (Oceanographic Post of Trindade Island) of the Brazilian Navy, "in recognition of their extensive help during all trips by the authors."

Pojer

Squeaker Catfish sp. *Chiloglanis pojeri* M Poll, 1944
Yellowfish sp. *Labeobarbus pojeri* M Poll, 1944

Dr G Pojer was a Belgian scientist who made a collection of fish in the area of Albertville (now Kalemie) on the western side of LakeTanganyika. We have found no more biographical information.

Polarstern

Snailfish sp. *Careproctus polarsterni* G Duhamel, 1992

Named after the German research vessel 'Polarstern'.

Poliak

African Rivuline sp. *Aphyosemion poliaki* J-L Amiet, 1991

Daniel Poliak was a French aquarist and killifish hobbyist who was honoured for "*... contribtutions to the knowledge of the subgenus Chromaphyosemion, both in the field and in the aquarium.*"

Poljakow

Balkhash Minnow *Rhynchocypris poljakowii* KF Kessler, 1879
[Syn. *Phoxinus poljakowii*]

Ivan Semonovich Poljakow (1845–1887) was a Russian zoologist, anthropologist, explorer and ethnographer. His first expedition was to the Yablonoi Mountains, Siberia (1866). He graduated at the University of St Petersburg (1870). He explored Lake Onega (1871–1873) and the lakes that feed the Volga (1874). He was appointed conservator at the Zoological Museum of the University of St Petersburg (1875) and explored the River Ob (1876–1877). He was in central Russia (1878) and the eastern Caucasus (1879). His final expedition was to Sakhalin, Japan and southern China (1881–1884). He collected the minnow type.

Poll

Lightfish genus *Pollichthys* MG Grey, 1959
Mormyrid genus *Pollimyrus* L Taverne, 1971
Western Shellear *Kneria polli* E Trewavas, 1936
Sole sp. *Bathysolea polli* P Chabanaud, 1950

Benguela Hake *Merluccius polli* J Cadenat, 1950
White-spotted Stargazer *Uranoscopus polli* J Cadenat, 1951
African Lantern Shark *Etmopterus polli* HB Bigelow, WC Schroeder & S Springer, 1953
Argentine sp. *Glossanodon polli* DM Cohen, 1958
Squeaker Catfish sp. *Microsynodontis polli* JG Lambert, 1958
African Sawtail Catshark *Galeus polli* J Cadenat, 1959
African Characin sp. *Phenacogrammus polli* JG Lambert, 1961
Round Hatchetfish *Polyipnus polli* LP Schultz, 1961
Congo Cichlid sp. *Haplochromis polli* DFE Thys van den Audenaerde, 1964
[Syn. *Ctenochromis polli*]
Lightfish sp. *Ichthyococcus polli* J Blache, 1964
Ikela Mormyrid *Stomatorhinus polli* H Matthes, 1964
Songatana Cichlid *Oxylapia polli* A Kiener & AL Maugé, 1966
Rattail sp. *Coelorinchus polli* NB Marshall & T Iwamoto, 1973
Lake Malawi Cichlid sp. *Placidochromis polli* WE Burgess & HR Axelrod, 1973
Lake Tanganyika Cichlid sp. *Tropheus polli* GS Axelrod, 1977
African Rivuline sp. *Nothobranchius polli* RH Wildekamp, 1978
Poll's Upside-down Catfish *Synodontis polli* J-P Gosse, 1982
African Rivuline sp. *Aphyosemion polli* AC Radda & E Pürzl, 1987
African Catfish sp. *Chrysichthys polli* L Risch, 1987
Poll's Bichir *Polypterus polli* J-P Gosse, 1988
[Syn. *Polypterus palmas polli*]
Labeo sp. *Labeo polli* SM Tshibwabwa, 1997
Spiny Eel sp. *Mastacembelus polli* EJ Vreven, 2005
African Characin sp. *Distichodus polli* E Abwe, J Snoeks, AC Manda & E Vreven, 2019

Dr Max Fernand Leon Poll (1908–1991) was a Belgian ichthyologist, and 'connoisseur of the fish fauna'. He worked in the Congo and in the Musée Royal du Congo Belge, Tervuren, and was professor at the Université Libre de Bruxelles. He led expeditions to Lake Tanganyika (1946–1947) and a Congo River survey (1953). Taverne, who named the mormyrid genus after Poll, was his student. An amphibian and a mammal are also named after him.

Pollen

Harlequin Hind *Cephalopholis polleni* P Bleeker, 1868
Madagascar Cichlid sp. *Paratilapia polleni* P Bleeker, 1868

François Paul Louis Pollen (1842–1886) was a Dutch naturalist and merchant, who made major contributions to the study of Malagasy fauna. He went to study medicine at Leiden (1862) but was encouraged by Hermann Schlegel to study zoology instead. Together with the Dutch explorer and naturalist Douwe Casparus van Dam (1827–1898), he made an expedition to Madagascar (1863–1866). They collected specimens of local fauna for the museum in Leiden, financing the trip and the fieldwork of others from his trading and inherited wealth. He also visited the Comoro Islands and Reunion. He wrote: *Récherches sur la Faune de Madagascar et de ses Dépendances – d'après les déscouvertes de F.P.L. Pollen et D.C. van Dam* (1868). Two birds, two reptiles and a dragonfly are among the other taxa named after him.

Pollux

Japanese Fluvial Sculpin *Cottus pollux* A Günther, 1873

One of the 'heavenly twins' in Greek mythology (see also **Castor**). The etymology is not explicit, but apparently so named because the author viewed this species as a 'twin' to *Cottus gobio*.

Polonorum

Lake Junin Pupfish sp. *Orestias polonorum* VV Tchernavin, 1944

Polonorum is Latin for 'of the Poles' and commemorates the expedition funded by Branicki to South America (1866–1867) during which the first example of this species was collected by the Polish naturalist Constantin Jelski (q.v.).

Polyakov

Stickleback sp. *Pungitius polyakovi* SV Shedko, MB Shedko & TW Pietsch, 2005

Ivan Semenovich Polyakov (1847–1887) was a Russian zoologist, anthropologist and ethnographer. He trained at Irkutsk Military School (1859) and later took part in the Vitimo-Olekminsk expedition (1866). He then moved (1867) to St Petersburg where he graduated (1874). He became the Curator of the Zoological Museum at the Imperial Academy of Sciences. He made a number of expeditions (1866–1880), all the while publishing papers and reports such as: *Report on the Olekminsko-Vitimskoy expedition of 1866* (1873).

Pomo

Russian River Tule Perch *Hysterocarpus traskii pomo* JD Hopkirk, 1974

Not an eponym but a name for the Native American people who, prior to the arrival of the Spanish, occupied the Russian River drainage and adjacent regions (California, USA), where this subspecies occurs.

Pontoh

Pontoh's Pygmy Seahorse *Hippocampus pontohi* SA Lourie & RH Kuiter, 2008

[Alt. Weedy Pygmy Seahorse]

Hence Pontoh is an Indonesian dive guide who first brought the authors' attention to this species. We can find no further biography.

Pope

Lemon Damsel *Pomacentrus popei* BW Evermann & A Seale, 1907

(Jr Syn *Pomacentrus moluccensis*)

Pope's Ponyfish *Equulites popei* GP Whitley, 1932

Thomas Edmund Burt Pope (1879–1958) was a scientific assistant in the US Bureau of Fisheries. Brown University awarded his degree in zoology (1903). Among his publications are *Devils Lake, North Dakota: A Study of Physical and Biological Conditions, with a View to the Acclimatization of Fish* (1908) and the co-written *Oyster Culture Experiments and Investigations in Louisiana* (1910) as well as *The Amphibians & Reptiles of Wisconsin* (1928).

The ponyfish is a replacement name for *Leiognathus elongatus*, which was co-described by Pope (1906), but later found to be preoccupied in *Leiognathus* by *Equula elongata* (Günther 1874).

Popelin

Lake Tanganyika Cichlid sp. *Chalinochromis popelini* P Brichard, 1989

Captain Émile Gustave Alexandre Popelin (1847–1881) was a Belgian officer who died of a liver abscess on the shores of Lake Tanganyika (where this cichlid is endemic). The patronym is not identified by Brichard, but the binomial most probably honours this person.

Popov

Lumpfish sp. *Cyclopteropsis popovi* VK Soldatov, 1929
Aleutian Pout *Gymnelus popovi* AY Taranetz & AP Andriashev, 1935

Alexander Mikhailovich Popov (d.1942) was a Russian ichthyologist at the Zoological Institute of the Academy of Sciences. During a stay in the Russian Far East (1930–1932) he studied the commercial resources of marine fish and invertebrates of Kamchatka and wrote a number of articles on marine fish, including one of the first works on the fauna of the Sea of Okhotsk. However, soon after returning to Leningrad (1932), Popov left his post of Head of the Ichthyology Department at the Zoological Institute due to an incurable disease. There is no reliable information about the further fate of Popov in the pre-war period, but it is recorded that during WWII (1942) he was shot by German soldiers in the city of Orel.

Popta

Characin genus *Poptella* CH Eigenmann, 1908
African Characin sp. *Brycinus poptae* J Pellegrin, 1906
Popta's Buntingi *Adrianichthys poptae* M Weber & LF de Beaufort, 1922
Marianas Rockskipper *Praealticus poptae* HW Fowler, 1925
Gourami sp. *Trichopodus poptae* BW Low, HH Tan & R Britz, 2014

Dr Canna Maria Louise Popta (1860–1929) was one of the first Dutch women to take up ichthyology and one of the first women to study at Leiden, which awarded her degree in Geology, Zoology and Botany. She went on to the University of Bern, which awarded her doctorate (1898). She became lab assistant to the Curator of Reptiles, Amphibians and Fish at the Rijksmuseum van Natuurlijke Historie, Leiden (1889). Here she concentrated on the study of fish and was appointed Curator of Fishes (1891), a position she held until retiring at the age of 68 (1928). She published more than 40 papers. The gourami was named "*…in honour for her pioneer taxonomic work on the freshwater fish fauna of Borneo*".

Por

Por's Goatfish *Upeneus pori* A Ben-Tuvia & D Golani, 1989

Dr Francis Dov Por né Francisc Bernard Pór (1927–2014) was a Romanian hydrobiologist and biogeographer. In his early years, he was imprisoned in Romania (1948) as a result of his participation in the Zionist Socialist Youth Movement. During this time, he managed to graduate from the University of Bucharest. After his release he worked as Curator of

Invertebrates at the Grigore Antipa Museum of Natural Science in Bucharest. He emigrated to Israel (1960), finished his PhD (1963), and later became Lecturer (1964), Associate Professor (1969) and Professor of Zoology (1976–1996) and thereafter professor emeritus at the Hebrew University of Jerusalem. He travelled to Brazil off and on for many years (1977–1997), sometimes teaching as guest professor at the University of Sao Paulo as well as other overseas universities. He was honoured for his contribution to the field of Lessepsian migration of organisms.

Port Jackson

Port Jackson Shark *Heterodontus portusjacksoni* FAA Meyer, 1793

As the name implies, this shark is named after a place: Port Jackson, Sydney, New South Wales, near Botany Bay, where the type specimen was collected. The inlet was named by James Cook after Sir George Jackson (1725–1822), a Lord Commissioner of the British Admiralty.

Portenoy

Blenny sp. *Ecsenius portenoyi* VG Springer, 1988

Norman S Portenoy of Bethesda, Maryland, was honoured for his many years of support of ichthyological exploration by the staff of the Smithsonian Institution.

Porthos

Stone Loach sp. *Schistura porthos* M Kottelat, 2000

Not a true eponym but one of the characters in Alexander Dumas': *The Three Musketeers*, joining the two other musketeeer *Schistura* species (*aramis*, *athos*) all found in the Nam Ou basin, northern Laos.

Portillo

Ulúan Killifish *Tlaloc portillorum* WA Matamoros & JF Schaefer, 2010
[Syn. *Profundulus portillorum*]

Danilo and Héctor Portillo guided the senior author to the type locality in Honduras. They are brothers who have been life-long fieldwork collaborators. Héctor Orlando Portillo Reyes (b.1963) is an ecologist and environmental consultant who works on conservation projects in the Moskitia area of eastern Honduras. He studied for his first degree in biology at the Universidad Nacional Autónoma de Honduras (1990) and has a master's in the management of protected areas (2011). He has written around 30 papers, including with the senior author, such as: *Distribution and conservation status of the Giant Anteater (Myrmecophaga tridactyla) in Honduras* (2010). Danilo Portillo Reyes (b.1965) is also a passionate conservationist who also went to the site despite being a wheelchair user.

Portugal

Neotropical Rivuline sp. *Moema portugali* WJEM Costa, 1989

Luiz Paulo Stockler Portugal is a Brazilian biologist who discovered this species (1987). He studied for his tropical biology degree at the University of Brazil (UFRJ) (1978–1981) and at the Instituto Nacional de Pesquisas da Amazônia (INPA) (1981–1982). He was awarded his MSc in biology by the Universidade de São Paulo (USP) (1988) then went on to study the History of Science at King's College, London (1993–1995). He has also worked in food research for FMB Alimentos e Bebidas Ltda (2010–2012).

Posada

Characin sp. *Brycon posadae* HW Fowler, 1945

Dr Andrés Posada Arango (1839–1923) was a Colombian physician, zoologist, botanist, educationalist and promoter of natural sciences. He graduated (1859) in Bogotá and served as an army surgeon (1860–1862) in the army of the Confederation of Granada (long gone, and now part of a number of South American countries, including Colombia). He went to Europe and stayed in Paris (1868–1872). He became the first Professor of Physical and Natural Sciences, Universidad de *Antioquia, Colombia. Later in life he held many important appointments, but a fire (1921) destroyed his home and most of the papers recording his life's work.*

Poseidon

Clingfish genus *Posidonichthys* JC Briggs, 1993

Not an eponym, but named after the seagrass genus *Posidonia*. The clingfish is usually found living in this seagrass. The grass itself probably takes its name from the Greek god of the sea, Poseidon.

Pospisil

Armoured Catfish sp. *Hypostomus pospisili* LP Schultz, 1944
[Junior Synonym of *Hypostomus hondae*]

Frank J Pospisil was a geologist with the Lago Petroleum Corporation, Venezuela. He was honoured in the binomial as Pospisil "*...made it possible for [the author] to collect fishes in the Rio Machango and also in the Andes of Venezuela.*"

Poss

Waspfish sp. *Ocosia possi* SA Mandritsa & SI Usachev, 1990
Poss' Scorpionfish *Scorpaenopsis possi* JE Randall & WN Eschmeyer, 2001
Velvetfish sp. *Cocotropus possi* H Imamura & G Shinohara, 2008

Dr Stuart G Poss is an American ichthyologist who is a Research Associate at the California Academy of Sciences. He is an expert on taxonomy of Scorpaenoid fishes. The University of California at Los Angeles awarded his BA (1972) and San Francisco State University his MSc in Marine Biology (1976). The University of Michigan conferred his PhD in Evolutionary Biology (1981) and he was a post-doctoral fellow of the Academy of Natural Sciences, Philadelphia (1981–1982). He was then Collections Manager, California Academy of Sciences (1983–1985), before becoming Curator and Senior Scientist at the Gulf Coast Research Laboratory, University of Southern Mississippi, and both Associate Professor of Coastal Science, and Adjunct Associate Professor of Marine Sciences there until retirement (1985–2000). Among his many papers are several describing new species, including the co-written: *Matsubarichthys inusitatus, a new genus and species of velvetfish (Scorpaeniformes: Aploactinidae) from the Great Barrier Reef* (1991).

Post

Dreamer Anglerfish sp. *Oneirodes posti* E Bertelsen & DB Grobecker, 1980
Barbeled Dragonfish sp. *Eustomias posti* RH Gibbs, TA Clarke & JR Gomon, 1983

Alfred Post (b.1935) is a German zoologist who worked at Hamburg's Institut für Seefischerei. He was honoured for his contributions to our knowledge of deep-sea fishes and his continuing services to the ichthyological community.

Postel

Striped-fin Grouper *Epinephelus posteli* P Fourmanoir & A Crosnier, 1964
Goatfish sp. *Parupeneus posteli* P Fourmanoir & P Guézé, 1967

Emile Postel was the co-author with Fourmanoir and Guézé of: *Serranidés de la Réunion* (1963). The original texts have no etymology, but there is no reason to doubt he was the person honoured. He was an ichthyologist associated with the MNHN, Paris, who seems to have been active (c.1950–1965), but we cannot find full details.

Potanin

Altai Osman *Oreoleuciscus potanini* KF Kessler, 1879
Chinese Cyprinid sp. *Gymnocypris potanini* SM Herzenstein, 1891
Stone Loach sp. *Homatula potanini* A Günther, 1896

Grigory Nikolayaevich Potanin (1835–1920) was a Russian botanist, ethnologist and explorer in Central Asia (1861–1892). He graduated from Tomsk military academy (1952) and joined the Russian army (1953–1958), then studied at the University of St Petersburg (1859–1862). He became involved in the Society for Siberian Independence and in student riots (1861) leading to his imprisonment (1865–1874), and he was later exiled and sentenced to hard labour. He went on a number of expeditions, to Zaisan and the Tarbagatai Range (1863–1864), Mongolia and Tuva (1876–1877 & 1879–1880), China, Tibet and Mongolia (1884–1886 & 1892–1893) and the Greater Khingan Mountains (1899). Two reptiles and two amphibians are also named after him. (See also **Alexandra (Potanina)**)

Potsch

Pencil Catfish sp. *Trichomycterus potschi* MA Barbosa & WJEM Costa, 2003

Dr Sergio Potsch de Carvalho e Silva is a Brazilian herpetologist in the Department of Zoology, Institute of Biology and Laboratory Head, Federal University of Rio de Janeiro. Universidade Santa Ursula awarded his bachelor's degree (1981), the Federal University of Rio de Janeiro his master's (1986) and the University of São Paulo his doctorate (1994). He co-wrote: *Anuran fauna of the high-elevation areas of the parque nacional de serra dos* órgãos *(PARNASO), South-eastern Brazil* (2016). A reptile is also named after him.

Pott

Velvetfish sp. *Erisphex pottii* F Steindachner, 1896

Tonguefish sp. *Cynoglossus pottii* F Steindachner, 1902

Captain Constantin Edler von Pott of the ship 'Pola' was honoured for zoological collections made by that vessel in the Red Sea. This refers to the Austrian Academy of Science scientific voyage under SM Schiff (1890–1898) covering the Adriatic, eastern Mediterranean and Red Seas (1895–1896). There seems to be no etymology for the tonguefish, but the description of the velvetfish thanks the captain. Interestingly, one of the captain's descendants, Paul Edler von Pott, wrote a history of the voyage: *Expedition S. M. Schiff 'Pola' in das Rote Meer* (2013).

Potter, FA

Potter's Angelfish *Centropyge potteri* DS Jordan & CW Metz, 1912
[Alt. Russet Angelfish]

Frederic A Potter was the first Director (1904–1940) of the Honolulu Aquarium (now the Waikiki Aquarium). It was established as a commercial venture by the Honolulu Rapid Transit and Land Company, who wished to "show the world the riches of Hawaii's reefs." Potter was a clerk working for the company and was transferred to manage the aquarium.

Potter, GE

Chub Shiner *Notropis potteri* CL Hubbs & K Bonham, 1951

Dr George Edwin Potter (1898–1962) was Professor of Zoology at the Agricultural and Mechanical College of Texas (1939–1962) and before that at Baylor University (1927–1939). His doctorate was awarded by Ottawa University, Ottawa, Kansas, which also awarded him an honorary DSc degree (1946). He wrote, among other works: *Laboratory Outline for General Zoology* (1929) and *Textbook of Zoology* (1938). He collected the type and sent it to Hubbs for study.

Potts

Chihuahua Darter *Etheostoma pottsii* CF Girard, 1859

John Potts, called by the author "*...our esteemed friend from Chihuahua*", was an amateur naturalist and manager of a mine. Unfortunately, nothing more was recorded.

Poulson

Alabama Cavefish *Speoplatyrhinus poulsoni* JE Cooper & RA Kuehne, 1974

Dr Thomas Layman 'Tom' Poulson (b.1934) is a speleologist and biologist who is now Professor Emeritus, Department of Biological Sciences, University of Illinois, Chicago (1973–2000). He also taught at Yale and Notre Dame. The University of MIchigan awarded his doctorate (1961). He taught at Yale and was Professor of Biology (1973). His foci are cave biology (aquatic & terrestrial), old-growth forests, plant functional morphology, dunes succession, biological integrity of streams and wetlands including the Greater Everglades Ecosystem. He co-wrote: *The Life of the Cave* (1966).

Pourtales

Blackspot Porgy *Archosargus pourtalesii* F Steindachner, 1881

Louis François de Pourtalès (1824–1880) was a marine biologist born in Switzerland who spent most of his career in the USA. He presented his extensive collections to the Harvard Museum. No etymology is given, but we presume this to be the person intended.

Powell, CB

African Characin sp. *Neolebias powelli* GG Teugels & TR Roberts, 1990
African Rivuline sp. *Fundulopanchax powelli* JR van der Zee & RH Wildekamp, 1994

Charles Bruce Powell (1943–1998) was a Canadian biologist who was a lecturer at the University of Port Harcourt (Nigeria). His BSc was awarded by the Acadia University, Nova Scotia (1965) and his MSc by the University of Calgary (1967) where he was a post-graduate Research Assistant (1965–1966) and a Demonstrator (1966–1967). He took a post as Assistant Lecturer, later Lecturer in Zoology, at the University of Ibadan, Nigeria (1967) and began a PhD study of the toad *Bufo regularis*. He never completed it because he became fascinated instead by shrimps and crabs. He transferred to the University of Benin (1973) and then to the University of Port Harcourt (1976) and finally the River State University of Science & Technology (1982) in the same city. Among his published papers is: *Fresh and brackish water shrimps of economic importance in the Niger Delta* (1985). He went on to research mammals in the Niger Delta, discovering species previously not recorded there and pressing for better conservation. When he was diagnosed with lung cancer, he was sent to a London hospital where he continued to work until he succumbed. He discovered both of these fish species.

Powell, FT

Indian Ringed Skate *Orbiraja powelli* AW Alcock, 1898
[Syn. *Okamejei powelli*]

Lieutenant Frederick Thomas Powell (1806–1859) of the Indian Navy was one of Robert Moresby's officers on 'HMS Palinurus' (1829–1833) and on 'HMS Benares' surveying the Chagos Archipelago (1837–1838). While Moresby was on sick leave, Powell commanded 'HMS Benares' in undertaking other Indian coastal surveys. He served in the Indian Navy for more than 30 years, and was commanding 'Oriental' on the Suez/India leg of the same journey as 'Ripon' started from England (See **Moresby**).

Powell, R

False Moray genus *Powellichthys* JLB Smith, 1966
Golden Grouper *Saloptia powelli* JLB Smith, 1964

Ronald Powell was a Fisheries Officer, Cook Islands, and later of the South Pacific Commission. Among other things he was responsible for importing trochus there (1956) to establish a shell industry. He wrote: *"Akule" night fishing gear* and *Hawaiian "Opelu" Hoop Net Fishing Gear* (both 1968). He provided the type specimen of *Powellichthys ventriosus*.

Power, J

Power's Deep-water Bristlemouth *Vinciguerria poweriae* A Cocco, 1838

Jeanne Villepreux-Power, also known as Jeanette Power (1794–1871), was a French marine biologist and pioneer of research into aquatic organisms, who started out by walking more than 400 kilometres to Paris where she became a dressmaker. After marrying (1818), she moved to Sicily. She is credited with inventing the aquarium, and used it as a tool to study marine life systematically. She left Sicily (1843) but many of her records and drawings were lost in a shipwreck. Her most famous study was of the pelagic octopus *Argonauta argo* and she demonstrated that this cephalopod produces its own shell, rather than acquiring it from a different organism in the way that a hermit crab does. She wrote: *Observations et expériences physiques sur plusieurs animaux marins et terrestres* (1839).

Power, P

Northern Blue Devil *Paraplesiops poweri* JD Ogilby, 1908

Percy Power caught the type specimen and presented it to the Amateur Fishermen's Association of Queensland. Unfortunately, nothing further was recorded of him.

Prabhu

Quilon Electric Ray *Heteronarce prabhui* PK Talwar, 1981

Madhav Sudhakar Prabhu (b.1922) was Director of the Central Marine Fisheries Research Institute (1970) Mandapam Camp, India. The holotype was taken nearby. He wrote: *Mackerel and oil sardine tagging programme, 1966–67 to 1968–69* (1970).

Prabin

Bagrid Catfish sp. *Mystus prabini* A Darshan, S Abujam, R Kumar, J Parhi, YS Singh, W Vishwanath, DN Das & PK Pandey, 2019

Prabin Kumar Mahanta was associated with the Directorate of Cold Water Fisheries Research, in the Department of Zoology, Kumaun University, Nainital, Uttarakhand, India. The etymology says that: "*The species is named for the late Prabin Kumar Mahanta, for his substantial contribution to the development of the cold-water fisheries sector in the Himalayan regions of India.*"

Prada

Suckermouth Catfish sp. *Astroblepus pradai* CA Ardila Rodríguez, 2015

Dr Saúl Prada-Pedreros is an ichthyologist who is a researcher at the Department of BIology, Pontifical Xavierian University, Bogotá, Colombia. The National University of Colombia awarded his bachelor's degree (1986), the National Institute of Amazonian Research his master's (1992) and Universidade Estadual Paulista Júlio de Mesquita Filho his doctorate (2003). He is presently president of the Colombian Association of Ichthyologists. He co-wrote: *A new catfish species of the genus* Trichomycterus *(Siluriformes: Trichomycteridae) from the Río Orinoco versant of Páramo de Cruz Verde, Eastern Cordillera of Colombia* (2014).

Pradhan

Goan Melon Barb *Haludaria pradhani* R Tilak, 1973

[Syn. Haludaria fasciata pradhani]

Dr K S Pradhan (b.1918) was Superintending Zoologist of the Zoological Survey of India. Among other reports and papers, he wrote: *A guide to the zoological galleries [of the Indian Museum]* (1965). He collected the barb type.

Praeli

Suckermouth Catfish sp. *Astroblepus praeliorum* WR Allen, 1942

The Hermanos Praeli were, according to the etymology, "*...merchants of Tarma and La Merced [Peru], who were instrumental in procuring facilities and who aided the collecting in person*". The Hermanos (Brothers) were Juan and Ecuador Praeli, but we have not been able to find out more about them.

Prahl

Gorgona Guitarfish *Pseudobatos prahli* A Acero P & R Franke, 1995

Henry von Prahl (1948–1989) was a pioneering Colombian marine biologist who studied Gorgona Island (the type locality) and whose family emigrated from Germany (1953). He was killed by a bomb, which exploded on Avianca Airlines Flight 203 over Bogotá.

Prashad

Stone Loach sp. *Mustura prashadi* SL Hora, 1921
[Syn. *Schistura prashadi*]
Sisorid Catfish sp. *Glyptothorax prashadi* DD Mukerji, 1932
Indawgyi Stream Catfish *Akysis prashadi* SL Hora, 1936

Dr. Baini Prashad (1894–1969) was an Indian zoologist and malacologist who was Director of the Zoological Survey of India, Indian Museum, Calcutta. He was honoured by Hora as he had given him 'every possible encouragement'. He also has a gecko named after him.

Prashar

Stone Loach sp. *Paraschistura prashari* Hora, 1933

Prashar Bhatia was a friend of Hora. He collected the holotype (1919); it was one of a small collection he made at Hora's request near Kohat in the North-West Frontier Province.

Prater

Asian Schilbeid Catfish sp. *Clupisoma prateri* SL Hora, 1937
Deolali Minnow *Parapsilorhynchus prateri* SL Hora & KS Misra, 1938

Stanley Henry Prater (1890–1960) was an Indian-born British naturalist. He joined the Bombay Natural History Society (BNHS) in 1907, and developed his extensive knowledge of the mammals of the subcontinent during the Society's Mammal Survey (1911–1923) - during which he was also badly wounded when he was accidentally shot in the thigh. He became Curator of the Bombay Natural History Society and of the Prince of Wales Museum of Western India (1923–1947). He trained in taxidermy in England (1923). He also visited several American museums (1927) to learn more about museum exhibition techniques. He wrote: *The Book of Indian Animals* (1948), since updated and republished several times. He was also a parliamentarian – he was President of Anglo-Indian and Domiciled European Association, and their representative in the Bombay legislative assembly (1930–1947). After independence, he represented that community in the Indian Constituent Assembly. He emigrated with his family to England (c.1950), but died there after a very long and debilitating illness a decade later.

Pratt

Bagrid Catfish sp. *Tachysurus pratti* A Günther, 1892
[Syn. *Pseudobagrus pratti*]

Antwerp Edgar Pratt (1852–1924), after whom the catfish is named, was one of a family of great travellers and explorers. Antwerp Edgar and his sons, Felix and Charles, and a cousin, Joseph, all explored - sometimes alone and sometimes together. Antwerp and Felix collected Lepidoptera in the Colombian Andes in the 1890s and later spent two years in New Guinea. While travelling (1891) to Tibet, Antwerp visited Tatsienlu, China, and there met two famous naturalists, both of whom were Lazarite missionaries – Bishop Felix Biet and Père Jean André Soulie. He wrote: *To the Snows of Tibet through China* (1892). Three mammals, two amphibians and two reptiles are named after him.

Pravdin

Whitefish sp. *Coregonus pravdinellus* GD Dulkeit, 1949
Snailfish sp. *Liparis pravdini* PY Schmidt, 1951

Professor Ivan Fedorovich Pravdin (1880–1963) was a Soviet ichthyologist and biologist from Karelian-Finnish SSR, noted for his studies of Russian lakes and rivers. He studied (1895–1901) at the Kostroma Theological Seminary and moved to St Petersburg (1910) where he became interested in ichthyology, and later (1921) passed exams in science (chemistry, biology, physiology and agronomy) of the Physics and Mathematics Faculty of Petrograd University. He took part (1925) in the organisation of the Pacific Research Station in Vladivostok. He was appointed (1931) the first Director of the Karelian Fisheries Research Station in Petrozavodsk, and continued to act as scientific supervisor there (1933–1941) whilst also engaged in work at Leningrad University. After several months spent in besieged Leningrad (1942), he and his family were evacuated to Saratov. His scientific work focused on the study of the Caspian, Aral, White, Japan and Okhotsk seas, the Ladoga and Onega lakes, lakes of the Leningrad region, Karelia and the Kola Peninsula, the Volga, Volkhov, Svir, Amur, Syr-Darya rivers, and on intraspecific taxonomy of fish. He was the author of over 300 papers and took part in the compilation of the *Atlas of commercial fish of the USSR* (1949).

The whitefish binomial, *pravdinellus*, is a very unusual form of eponym, being a diminutive (i.e. literally "little Pravdin"). Probably intended as a form of endearment.

Prenant

Snowtrout sp. *Schizothorax prenanti* T-L Tchang, 1930

Auguste Prenant (1861–1927) was a French physician and embryologist. The patronym is not identified in Tchang's text, so could refer either to Auguste or to his son, parasitologist Marcel Prenant (1893–1983).

Pretorius

Pondoland Sailfin Goby *Myersina pretoriusi* JLB Smith, 1958

P J G Pretorius of Bizana, Eastern Cape, South Africa found and preserved the type specimen. We can find no further biographical data.

Priapus

Neotropical livebearer genus *Priapella* CT Regan, 1913
Neotropical livebearer genus *Priapichthys* CT Regan, 1913
Phallic Catshark *Galeus priapus* B Séret & PR Last, 2008
Micro-Cyprinid sp. *Danionella priapus* R Britz, 2009

This is not a true eponym as it is after Priapus, the Greek god of fertility. In the *Danionella* the name refers to a conical projection of genital papilla in males, which superficially resembles the penis of mammals. In the catshark, the name was chosen due to the males' long claspers.

Price, DS

Price's Damselfish *Chrysiptera pricei* GR Allen & M Adrim 1992
Price's Rainbowfish *Chilatherina pricei* GR Allen & SJ Renyaan, 1996

David Samuel Price (b.1959) is a New Zealander missionary, aquarist, naturalist, ecologist, linguist, translator and community development consultant who is (2012–current) Senior Environmental Consultant for LEAD Asia, Bangkok, Thailand. Massey University, New Zealand awarded his BSc (2005) and his Post Graduate Diploma in Science (2012). He was Linguist/Translator/Community Development Consultant for Ambai Language Development Program with SIL, Papua, Indonesia (1986–2010). He also owns Laughing Frog Photography (2011–present). After living in Asia for many years, and latterly in California, he returned to New Zealand (2019). He provided accommodation and logistical assistance for Gerald R. Allen during the latter's visit to Irian Jaya.

Price, GC

Seafan Blenny *Emblemariopsis pricei* DW Greenfield, 1975

George Cadle Price (1919–2011) was the premier of Belize (1961–1984 & 1989–1993). He served as First Minister and Premier under British rule until independence (1981) and was the nation's first prime minister after independence. The etymology says: "*I am pleased to name this species for the Honorable George C. Price, Premier of the emerging Central American nation of Belize.*"

Price, WW

Yaqui Catfish *Ictalurus pricei* CL Rutter, 1896

William Wightman Price (1871–1922) appears to have been a wild child, as he ran away from home at age 8 and lived for a few days with a band of Native Americans in Wisconsin. The result was that his father moved him to California (1880). When his father died (1885), William departed for Arizona and spent 18 months exploring and living rough. He returned to California (1887) and entered Oakland High School, paying his fees by selling bird and mammal skins. He collected in California, Nevada and Arizona for Stanford (1892–1895).

He founded a camp for boys (1897) near Lake Tahoe and named it after Agassiz. He took a bachelor's degree in economics at Stanford (1898) and a master's (1899). He worked for the Red Cross as Assistant Field Director in charge of the Palo Alto Base Hospital (1917–1919). A reptile is also named after him.

Pridi

Dr Pridi's Loach *Schistura pridii* C Vidthayanon, 2003

Pridi Banomyong (also known as Pridi Phanomyong) (1900–1983) was a Thai politician and statesman. He was Prime Minister of Thailand (1946–1947). In November 1947, a coup by army troops forced him to flee the country but he secretly returned (1949) to stage a prodemocracy coup. When it failed, he left for China, never to return. From China he travelled to France, where he spent the remainder of his life. He was honoured in the binomial because he had founded Thammasart University, which plays an important role in the development of social sciences in Thailand.

PRIDIS

Trindade Cleaner Goby *Elacatinus pridisi* RZP Guimarães, JL Gasparini & LA Rocha, 2004

Not an eponym but in honour of the Brazilian Navy First District (Primeiro Distrito Naval, Marinha do Brasil, or 'PRIDIS'), for the "*...impeccable logistic support*" provided during the authors' field trips to Trindade Island, off Brazil, the type locality.

Priede

Eelpout sp. *Pachycara priedei* PR Møller & N King, 2007

Dr Imants George 'Monty' Priede (b.1948) is Professor Emeritus at Oceanlab, a field research station of the University of Aberdeen. His BSc was awarded by University College of North Wales, Bangor (1970), and his PhD by the Department of Biology, University of Stirling (1973). The University of Aberdeen conferred his DSc (1996) in Fish Biology, Energetics and Telemetry Studies on Free-Living Animals. He has published over 150 papers such as: *Colonisation of the deep-sea by fishes* (2013) and the book: *Deep-Sea Fishes: Biology, Diversity, Ecology and Fisheries* (2017). He pioneered satellite tracking of sharks and the use of robotic lander vehicles to study deep-sea life. His team at Oceanlab discovered life at all depths down to the bottom of the deepest trenches. The Fisheries Society of the British Isles awarded him the Beverton Medal (2011).

Pringle

Black Butterfish *Hyperoglyphe pringlei* JLB Smith, 1949

John Adams Pringle (1910–2002) was a South African zoologist. His first appointment was as Assistant at the Port Elizabeth Museum, later becoming its Director. After many years at Port Elizabeth, he was appointed Director of the Natal Museum (1953). He helped to source "rare and valuable fishes" for J.L.B. Smith.

Procne

Swallowtail Shiner *Notropis procne* ED Cope, 1865
Swallowtail Fang-Blenny *Meiacanthus procne* WF Smith-Vaniz, 1976

Not a true eponym as this is after a character in Greek mythology; Procne (sometimes Progne – see below), a daughter of King Pandion of Athens. She was metamorphosed into a swallow. The name is used for species which have some 'swallow-like' attribute.

Progne

Flyingfish genus *Prognichthys* CM Breder, 1928

Not a true eponym, Progne (sometimes Procne – see above) was metamorphosed into a swallow. Pliny likened flyingfish to the swallow, and the genus name can be translated as 'swallow-fish'.

Prometheus

Roudi Escolar *Promethichthys prometheus* G Cuvier, 1832

In Greek mythology, Prometheus stole fire from heaven to give to mankind: for which he was bound to a rock where an eagle was sent to feed upon his liver (which then grew back, to be devoured anew the next day). In his description of the escolar holotype, Cuvier says he cannot give details of the fish's visceral organs due to their badly-preserved state. Presumably this made him think of the state of Prometheus's liver!

Proserpine

Proserpine Shiner *Cyprinella proserpina* CF Girard, 1856
Pomegranate Pencil Wrasse *Pseudojuloides proserpina* Y-K Tea, BD Greene, JL Earle & AC Gill, 2020

Proserpine was a Roman goddess, equivalent to the Greek Persephone. She was the queen of the Underworld, and her name seems to have been used by Girard in fanciful reference to the type locality: Devils River, Texas. Thus, one could say this is a pseudo-eponym and a pseudo-toponym. The wrasse was named "*…in reference to both its haunting coloration and close relationship to P. pluto.*"

Proteus

Amazonian Cichlid sp. *Crenicichla proteus* ED Cope, 1872
Tetra sp. *Hyphessobrycon proteus* CH Eigenmann, 1913
Socorro Blenny *Hypsoblennius proteus* RJ Krejsa, 1960
Colombo Damsel *Pomacentrus proteus* GR Allen, 1991

Proteus was a sea-god in Greek mythology who could change his form, referring to these fish's colour changes from juvenile to adult and their colour morphs

Proudlove

Cave Loach sp. *Eidinemacheilus proudlovei* J Freyhof, YSAbdulllah, K Ararat, H Ibrahim, & MF Geiger, 2016

Graham S Proudlove is an independent writer and editor, spelaeologist, entomologist and biologist formerly at the Department of Zoology at The Manchester Museum, where he is also Honorary Curator of Myriapoda. Manchester University awarded his bachelor's degree in zoology. He wrote: *The Conservation Status of Hypogean Fishes* (2001), *Essential Sources in Cave Science* (2006) and *Subterranean fishes of the world. An account of the subterranean (hypogean) fishes described to 2003 with a bibliography 1541–2004* (2007).

Provenzano

Pencilfish sp. *Lebiasina provenzanoi* CA Ardila Rodríguez, 1999

Characin sp. *Creagrutus provenzanoi* RP Vari & AS Harold, 2001

Three-barbeled Catfish sp. *Phenacorhamdia provenzanoi* C DoNascimiento & N Milani, 2008

Bluntnose Knifefish sp. *Brachyhypopomus provenzanoi* WGR Crampton, CD de Santana, JC Waddell & NR Lovejoy, 2017

Dr Francisco Provenzano Rizzi is an ichthyologist at Museum of Biology, Instituto de Zoología Tropical de la Universidad Central de Venezuela, where both his bachelor's degree (1980) and his doctorate (2012) were awarded. He co-wrote: *Parodon orinocensis (Bonilla et al., 1999) (Characiformes: Parodontidae): Emendations and generic reallocation* (2012).

Prugh

Damselfish sp. *Chrysiptera prughi* HW Fowler, 1946

[Junior Synonym of *Plectroglyphidodon imparipennis*]

Captain Byron J Prugh was a good friend of Captain Tinkham (q.v.) who collected this species. In that collection he was aided by an especially designed net with floats, made for him by Prugh, so he could collect from pools in the reef. Prugh also accompanied Tinkham a number of times, and "*...greatly contributed in forming the Riu Kiu collection of fishes.*"

Pruvost

Spikefish sp. *Bathyphylax pruvosti* F Santini, 2006

Viviparous Brotula sp. *Didymothallus pruvosti* W Schwarzhans & PR Møller, 2007

Patrice Pruvost (b.1966) is a biologist and ecologist who works at MNHN (1995 – present), where he is currently one of the curators of the fish collection. His main interest is in the ichthyofauna of the Antarctic and sub-Antarctic. He has written more than 100 papers such as: *Parabothus rotundifrons (Pleuronectiformes: Bothidae), a new bothid flatfish from Saya de Malha Bank (Indian Ocean)* (2016) and co-wrote: *A short history of the fisheries of the Crozet Islands* (2015). He was honoured in the brotula's binomial as a thank you for his support of the authors' work.

Pruvot

Goby sp. *Vanneaugobius pruvoti* L Fage, 1907

Dr Georges Florentin Pruvôt (1852–1924) was a French physician and zoologist, and also Fage's teacher. He qualified as a physician (1882) and then earned a doctorate in natural sciences (1885). He became Professor of Zoology, Anatomy and Physiology at the Paris

Faculty of Science. Later (1900) he was named director of the laboratory at Banyuls-sur-Mer, southern France. He wrote: *Sondages exécutés d'août à octobre 1893 à bord du "Roland", navire du laboratoire Arago, par M. G. Pruvot... sous la direction de M. H. de Lacaze* (1894). He helped collect this goby and urged Fage to study it.

Przewalski

Scaleless Carp *Gymnocypris przewalskii* KF Kessler, 1876

General Nikolai Mikhailovitch Przewalski (1839–1888) was a Russian Cossack naturalist who explored Central Asia. He was undoubtedly one of the greatest explorers the world has ever seen, making five major expeditions; one to the Russian Far East and the others to Mongolia where he collected the carp type. He wrote: *Mongolia, and the Tangut Country* (1875) (in which Kessler's description appears) and *From Kulja, across the Tian Shan to LobNor* (1879). He is best known for having the wild horse *Equus przewalskii*, which he discovered, named after him. The Russian Academy of Sciences instituted the Przewalski Gold Medal (1946). There are at leasthalfa-dozen different spellings of his name including Przewalski and Prjevalsky (pronounced 'Shev-al-ski'). He died of typhus aged 49 whilst preparing for another expedition. Tsar Alexander II decreed that the town where he died, Karakol, should immediately have its name changed to Przhevalsk. Sixteen birds, five mammals and five reptiles are also named after him.

Puck

Dreamer (Anglerfish) genus *Puck* TW Pietsch, 1978
Pallid Cusk-eel *Ophidion puck* Lea & Robins, 2003

Puck is a fairy in the service of Oberon (see Shakespeare's: *A Midsummers Night's Dream*), one of three Shakespearean names coined in the same paper (see *O. antipholus* and *O. dromio*), indicating that these species are "*...part of a larger story*". The name was already old when Shakespeare used it: it probably derives from the Old Norse *pook* or *puki* for a nature spirit.

Puerzl

African Rivuline sp. *Fundulopanchax puerzli* AC Radda & JJ Scheel, 1974

Eduard Pürzl is an Austrian aquarist, collector and fish photographer. He collected the type in Cameroon, and kept the species in his aquarium.

Puetz

Panama Livebearer sp. *Priapichthys puetzi* MK Meyer & V Etzel, 1996
African Rivuline ssp. *Epiplatys olbrechtsi puetzi* HO Berkenkamp & V Etzel, 1985

Wilfried Pütz (b.1940) of Würselen, Germany, is an aquarist. He collected in the field with the junior author, Vollrad Etzel.

Pun

Titicaca pupfish sp. *Orestias puni* VV Tchernavin, 1944

This is a toponym referring to Bahia de Puno, Lake Titicaca.

Punu

Squeaker Catfish sp. *Synodontis punu* E Vreven & L Milondo, 2009

The Punu people, who live in part of Gabon and the Republic of the Congo where this catfish occurs, were notable for their help with collecting and their 'spontaneous hospitality'.

Purus

Characin sp. *Iguanodectes purusii* F Steindachner, 1908

This is a toponym, referring to the Rio Purus, Brazil.

Putnam, FW

American Smooth Flounder *Pleuronectes putnami* TN Gill, 1864

Professor Frederic Ward Putnam (1839–1915) was an anthropologist and naturalist. He went to Harvard (1856) and studied under Louis Agassiz. He worked as Agassiz's assistant (1857–1864) and became immersed in ichthyology and herpetology. He worked in a variety of museums and institutions (1864–1875). He was Curator, Peabody Museum of American Archaeology and Ethnology, Harvard (1875–1909). He was made Peabody Professor (1888), then Emeritus Curator and Emeritus Professor (1909–1915).

Putnam, ML

Sawtooth Barracuda *Sphyraena putnamae* DS Jordan & A Seale, 1905

Mary Louisa Duncan Putnam (1832–1903) was the first woman elected to full membership of the Davenport Academy of Sciences (1869). She was elected as president (1879), raising funds and pushing the academy to expand its role in sponsoring cultural events.

Putra

Putra's Pygmygoby *Trimma putrai* R Winterbottom, MV Erdmann & R Mambrasar, 2019

Ketut Sarjana Putra (b.1963) is Vice President of Conservation International Indonesia, having joined in 2004, and is one of Indonesia's foremost marine conservationists. His master's degree in integrated coastal management was awarded (1992) by the University of Newcastle-upon-Tyne, UK. His work over the past three decades has ranged from sea-turtle conservation to pioneering Sustainable Landscapes Partnerships, Bird's Head Seascape (BHS) in West Papua, which became a global model for community-based marine conservation; various sustainable conservation financing mechanisms and has focused on both the Lesser Sunda Islands and West Papua. He has recently initiated the Blue Carbon Program with a pilot project in Kaimana, West Papua, and driving for conservation province policy in West Papua Province.

Puzanov

Eelpout genus *Puzanovia* VV Fedorov, 1975

Ivan Ivanovich Puzanov (1885–1971) was a Russian zoologist and zoogeographer. He was awarded his doctorate (1938). He was a professor at the University of the Crimea, and later

(1934–1947) at the University of Gorky and Professor of Vertebrate Zoology at the State University of Odessa. He made investigations of the Black Sea coast, the Red Sea, eastern Sudan and Ceylon. Among his major works were: *The Great Canyon of Crimea* (1954) and *Over Untrod Crimea* (1960).

Py-Daniel

Pike-Cichlid sp. *Crenicichla pydanielae* A Ploeg, 1991

Dr Lúcia Helena Rapp Py-Daniel is a Brazilian ichthyologist who is Curator of Fishes, Instituto Nacional de Pesquisas da Amazônia, Manaus. Among her published papers is: *Ictiofauna do rio Jufari e arquipélago de Mariuá, Médio Rio Negro, Bacia Amazônica, Brasil* (2017), and she co-wrote: *A new Silver Dollar species of Metynnis Cope, 1878 (Characiformes: Serrasalmidae) from Northwestern Brazil and Southern Venezuela* (2016). She was honoured by Ploeg for her hospitality when he visited Manaus (1987 & 1989). (Also see **Lucia**)

Pyle, RL

Pyle's Sand-lance *Ammodytoides pylei* JE Randall, H Ida & JL Earle, 1994
Pyle's Wrasse *Cirrhilabrus pylei* GR Allen & JE Randall, 1996
Orange-spotted Soapfish *Belonoperca pylei* CC Baldwin & WL Smith, 1998
Twilight Fang-Blenny *Petroscirtes pylei* W Smith-Vaniz, 2005
Lizardfish sp. *Synodus pylei* JE Randall, 2009

Dr Richard Lawrence Pyle (b.1967) is an ichthyologist and marine biologist at Bishop Museum, Honolulu, which he joined as an Ichthyology Collections Technician (1986). He attended the University of Hawai'i at Mānoa, Honolulu, which awarded both his bachelor's degree and his doctorate. He has written nearly 200 scientific, technical and popular publications on topics including ichthyology, technical diving, scientific nomenclature, and biodiversity database systems. He is a Commissioner on the International Commission on Zoological Nomenclature (ICZN) (2006–present) and his research focus is on using advanced mixed-gas closed-circuit rebreathers to explore deep coral reef environments (known as Mesophotic Coral Ecosystems). He has discovered more than 100 new species of fishes on deep coral reefs, over 30 of which he has written or co-written the original descriptions.

Pyle, RM

Pyle's Dottyback *Pseudochromis pylei* JE Randall & JE McCosker, 1989

Robert M Pyle, a travel agent by profession, was honoured for helping Randall collect and photograph fishes, including an underwater photograph of this species. (He is no relation to ichthyologist Richard L. Pyle, also mentioned in the description)

Pylzov

Tibetan Cyprinid sp. *Schizopygopsis pylzovi* KF Kessler, 1876

Lieutenant Mikhail Alexandrovich Pylzow (b.1850) was a Russian explorer in Central Asia. He took part in Przewalski's expedition to Mongolia (1870–1873) during which the holotype was collected. A reptile and a bird are also named after him.

Pynand

Characin sp. *Astyanax pynandi* JR Casciotta, AE Almiron, JA Bechara, JP Roux & F Ruiz Diaz, 2003

Not an eponym, but a word from the Guarani language meaning 'people without shoes'. The binomial was bestowed in honour of the 'descalzos' [of Argentina] "*…that every day struggle to recover their dignity in an unjust world*".

Pyragy

Scraper Barb sp. *Capoeta pyragyi* A Jouladeh-Roudbar, S Eagderi, L Murillo-Ramos, HR Ghanavi & I Doadrio, 2017

Magtymguly Pyragy (1724–c.1807) was a Turkmen spiritual leader and philosophical poet. Though he promoted Turkmen unity, he had little success in his own time due to local rivalries and tribal loyalties.

Pyrineus

Armoured Catfish sp. *Hypostomus pyrineusi* A de Miranda Ribeiro, 1920

Eartheater (cichlid) sp. Geophagus pyrineusi GC Deprá, WM Ohara & HP Silva, 2022

Lieutenant Antonio Pyrineus de Souza (d.1936) was a naturalist and geographer with the Rondon Commission, as was his friend the author, Miranda Ribeiro. They were both members of Rondon's expedition to what is now Rondonia (1909) to install telegraph poles from Mato Grosso to Amazonas. De Souza led the 1915–1916 expedition on which he reported in: *Exploration of the Paranatinga River and its topographic survey as well as the S. Manoel and Telles Pires rivers* (1916).

Q

Qantas

Slender Dwarf Monocle Bream *Parascolopsis qantasi* BC Russell & T Gloerfelt-Tarp, 1984

Qantas is an Australian airline. The junior author explains how the name came about: We were working (1979–1984) on a fisheries bottom trawl survey stretching across Southern Indonesia. This was conducted by the Food & Agriculture Organisation of the United Nations and the German Technical Assistance, plus CSIRO and covered the North West Shelf of Australia. Every day we discovered many undescribed fish. With limited interest from the Indonesian museums, Australia put their hands up to take these fish to complement their own museum collections. We got these fish to Australia by preserving them in formaldehyde, then wrapping them up and putting them in drums and asking Qantas – flying from Bali to various destinations in Australia – if they would freight them on board. Qantas gladly assisted and they did this for free. They would pick them up and fly them to all the museums around the country whose staff gratefully accepted them. 84 new species were found and we thought why not say thank you to Qantas and name a fish after them. It was named with little fanfare. There were some grumblings in the scientific fraternity – "you can't name a fish after Qantas" – but I said, "Well, give it back then! But they did not want to relinquish a new species."

Quasimodo

Toothless Characin sp. *Steindachnerina quasimodoi* RP Vari & AM Williams Vari, 1989
Stone Loach sp. *Schistura quasimodo* M Kottelat, 2000
Humpback Western Dogfish *Squalus quasimodo* ST de F Viana, MR de Carvalho & UL Gomes, 2016
Lake Edward cichlid sp. *Haplochromis quasimodo* N Vranken, M Van Steenberge, A Heylen, E Decru & J Snoeks, 2022

Not a true eponym but the name of the title character in Victor Hugo's: *The Hunchback of Notre Dame*, referring to the prominent dorsal profile of these species.

Quechua

Pencil Catfish sp. *Trichomycterus quechuorum* F Steindachner, 1900
[Perhaps a junior synonym of *Trichomycterus rivulatus*]

Though the etymology is not explained by Steindachner, the binomial is presumably intended to mean 'of the Quechua'. The Quechua are an indigenous people of South America, particularly of Peru.

Quekett

Flapnose Houndshark *Scylliogaleus quecketti* GA Boulenger, 1902
Spot-fin Cardinalfish *Jaydia queketti* JDF Gilchrist, 1903
Lesser Gurnard *Chelidonichthys queketti* CT Regan, 1904

John Frederick Whitlie Quekett (1849–1913) was born in London, but emigrated to Natal (1871) and was appointed to organise the collection of the Natal Society (1886). He became Curator of the Durban Museum (1895–1909). His main interest was in conchology. He sent the houndshark holotype to Boulenger who named the species after him, but misspelt his surname, as he did with a frog species also named after him.

Quélen

Silver Catfish *Rhamdia quelen* JRC Quoy & JP Gaimard, 1824

Abbé Florentin-Louis de Quélen de la Villeglée (b.1762) was the chaplain on board the 'Uranie' circumnavigation expedition (1817–1820) commanded by Freycinet (q.v.). Quoy and Gaimard were the expedition's naturalists, and the holotype was collected during the expedition.

Quignard

Languedoc Loach *Barbatula quignardi* L Bacescu-Meşter, 1967

Professor Dr Jean-Pierre Quignard (1938–2013) was a French ichthyologist who was Professor of Ecology, Genetics and Zoology at Montpelier University. He had nearly two hundred publications to his name, including a number of popular books such as: *Pas si bêtes les poissons : Scènes de leur vie intime* (2006) and *Les poissons font-ils l'amour ? : Et autres questions insolites sur les poisons* (2009). He was, at the time of the loach's description, an ichthyologist at the Marine Station, Sète, France, by whose courtesy the author obtained the types.

Quii

Neotropical Rivuline sp. *Moema quiii* JH Huber, 2003

'Qui-i-i' is the name given by a shaman friend to Belinda Peck, whose husband, Lance, was one of the collectors of the type. Lance Peck is a commercial fish-exporter based in Peru.

Quiroga

Neotropical Rivuline sp. *Austrolebias quirogai* M Loureiro, A Duarte & M Zarucki, 2011

Horacio Quiroga (1878–1937) was a Uruguayan writer whose "*...tales and fables based on his life in the Misiones rainforest inspired the authors to explore nature and its mysteries.*" He also collected snakes in the early 1900s. After a sojourn in Paris, he returned to teach in Argentine schools and tour the wilds of Argentina as a photographer. He settled (1904) in Chaco Province, tried and failed to grow cotton, so returned to teaching in Buenos Aires. He was registrar in the San Ignacio district of Misiones (1909–1915). He returned to Buenos Aires and worked (1919–1925) in the Uruguayan consulate, then went again to San Ignacio, where he became Honorary Uruguayan Consul (1935). Now regarded as one of the greatest Uruguayan writers, he wrote works dealing with anthropomorphic, intelligent animals, a jungle that seems to be alive, fate, and bizarre coincidences set against a backdrop of despair—which is perhaps understandable, given his famously unhappy life. His father was killed in an accidental shooting when Quiroga was young. His stepfather committed suicide (1900). His first wife poisoned herself (1915), and his second marriage failed. He found he had cancer and swallowed cyanide. Later both his children also committed suicide.

Quispe

Suckermouth Catfish sp. *Astroblepus quispei* CA Ardila Rodríguez, 2012

Roberto Quispe is an ichthyologist at Museo de Historia Natural de la Universidad Nacional Mayor de San Marcos, Peru, where he was awarded a bachelor's degree (1991) and a master's (2009). He co-wrote: *Aquatic Biodiversity in the Amazon: Habitat Specialization and Geographic Isolation Promote Species Richness* (2011).

Quoy

Longfin Grouper *Epinephelus quoyanus* A Valenciennes, 1830
Galapagos Bullhead Shark *Heterodontus quoyi* CPP Fréminville, 1840
Quoy's Parrotfish *Scarus quoyi* A Valenciennes, 1840
Quoy's Garfish *Hyporhamphus quoyi* A Valenciennes, 1847

Jean René Constant Quoy (1790–1869) was a French naval surgeon and zoologist who named many species, often with Joseph Paul Gaimard (q.v.). He took part in a number of voyages of discovery, including two circumnavigations aboard *L'Astrolabe* (1817–1820 & 1826–1829) with Jules Dumont d'Urville. He was on board the 'Uranie' when she was wrecked but continued his journey on the rescuing vessel, which the expedition bought and re-named 'Physicienne'. He became chief medical officer of the naval hospitals at Toulon (1835–1837) and Brest (1838–1848). Fréminville named the Bullhead Shark "*...for his friendship, untiring zeal and wide knowledge of zoology*". Quoy is commemorated in the names of other taxa, including five birds, two amphibians and a reptile.

R

Rabaud

Rock Carp *Procypris rabaudi* TL Tchang, 1930

Dr Étienne Antoine Prosper Jules Rabaud (1868–1956) was a zoologist at l'Université de Paris. He qualified as a physician (1898) and was Head of Research at the Paris Faculty of Medicine (1896–1905). However, he was more interested in zoology and lectured at Paris University (1907–1923) where he became Professor until retiring (1938). During WW1 he returned to medicine to treat wounded soldiers. He was particularly keen to study actual animal behaviour and counteract the anthropomorphism of the day. He published more than 400 books, papers and articles including: *Elementary Anatomy of the pharynx, larynx, ear and nose* (1901) and *Randomness and the Life of Species* (1953).

Rabaut

Rust Corydoras *Corydoras rabauti* F La Monte, 1941

Auguste Rabaut (1894–1985) was an explorer and natural history collector who lived and travelled in South America (1930s and 1940s) for the Smithsonian and for the AMNH. He was born in France, educated in Spain, and trained to be a cutter of fine gem stones which first induced him to search for emeralds and diamonds in South America. *The Travels of Auguste Rabout* (2016) by Kathleen Demarre is a series of short stories he told to his daughter and grandchildren. He collected the holotype of this species but today is more remembered for discovering the Neon Tetra, *Paracheirodon innesi*.

Racenis

Bluntnose Knifefish genus *Racenisia* F Mago-Leccia, 1994

Dr Janis Rácenis (1915–1980) was a Latvian-born Venezuelan entomologist and ornithologist. The University of Latvia awarded his first degree (1943) and he began to study ornithology. He moved to Germany and finished his education by being awarded his PhD by the University of Erlangen (1947). He and his wife Gaida (q.v.) left soon after (1948) to work in the Department of Science at the Central University of Venezuela, Caracas, where he taught. He stayed there until retirement (1976). During this time, he served in various management positions including in the School of Biology and the Biology Museum of the Central University of Venezuela (MBUCV) becoming its Founder Director (1949). He was also founder Editor and Director of the journal 'Acta Biologica Venezuelica' and founding Director (1965) of the Institute of Tropical Zoology. He described more than 20 new species. He wrote many papers such as: *Los odonatos de la región del Auyantepui y de la Sierra de Lema, en la Guayana Venezolana. 2 Las familias Gomphidae, Aeshnidae y Corduliidae* (1970). He was also honoured in the names of a reptile, a spider, a crustacean, a scorpion and several insects.

Rachow

Characin genus *Rachoviscus* GS Myers, 1926
Neotropical Killifish genus *Rachovia* GS Myers, 1927
Characin sp. *Iguanodectes rachovii* CT Regan, 1912
Rachow's Darter Tetra *Characidium rachovii* CT Regan, 1913
Coatzacoalcos Gambusia *Heterophallus rachovii* CT Regan, 1914
Blue-fin Notho *Nothobranchius rachovii* CGE Ahl, 1926
Fanning Pyrrhulina *Pyrrhulina rachoviana* GS Myers, 1926
[Probably a Junior Synonym of *Pyrrhulina australis*]

Arthur Rachow (1884–1960) was a self-taught pioneer German aquarist. From a poor family, he began working at just thirteen years old. By the time of WW1, he had established himself as a merchant. He often wrote articles in aquarist publications. Wanting to be accurate in offering aquaria fish for sale he consulted leading ichthyologists. He even described one new species himself: *Aphyosemion australe* (1921). His published works include: *Manual of Ornamental Fish* (1928). Together with Holly & Meinken he edited: *Die Aquarienfische in Wort und Bild* (1934–1967). He kept an extensive library which was lost to bombing during WW2.

Racovitza

Antarctic Dragonfish genus *Racovitzia* L Dollo, 1900
Petzea Rudd *Scardinius racovitzai* GJ Müller, 1958

Professor Dr Emil G Racoviță (sometimes Racovitza) (1868–1947) was a Romanian zoologist and especially a pioneering bio-speleologist. He also explored in Antarctica. He studied law at Paris University, which awarded his degree (1889), but he then turned to natural science, gaining a BSc (1891) and his PhD (1896) from the Sorbonne. He was selected to be part of the Antarctic expedition aboard 'Belgica' (1897–1899) and he published his diary of the trip that year and then his scientific work as: *La vie des animaux et des plantes dans l'Antarctique* (1900). He spent a number of years exploring caves across Europe (France, Spain, Italy & Slovenia) and Algeria and is recognised as one of the founders of speleology. He became head of the Biology Department at the Upper Dacia University (1919) and founded the world's first Speleological Institute there (1920). He wrote: *Essai sur les problèmes biospéologique* (1907) and later: *Speleology: A new science of the old underworld mysteries* (1927). He was honoured in the rudd's name on the tenth anniversary of his death.

Radcliffe

Pygmy Ribbontail Catshark *Eridacnis radcliffei* HM Smith, 1913
Cardinalfish sp. *Ostorhinchus radcliffei* HW Fowler, 1918
Rattail sp. *Coelorinchus radcliffei* CH Gilbert & CL Hubbs, 1920

Lewis Radcliffe (1880–1950) was an American naturalist, particularly interested in malacology and ichthyology. He was Assistant Naturalist to Hugh McCormick Smith (q.v.) on the Philippines Expedition (1907–1910). He later became Scientific Assistant, then Deputy Commissioner, of the US Bureau of Fisheries (to 1932). He was Director of the Oyster Institute of North America when he died. He was a member of the Bureau of Fisheries team that collected the catshark holotype from the steamer 'Albatross'.

Radda

African Rivuline sp. *Aphyosemion raddai* JJ Scheel, 1975

Neotropical Killifish sp. *Cnesterodon raddai* MK Meyer & VMF Etzel, 2001

Dr Alfred C Radda (1936–2022) was an Austrian zoologist and Professor of Virology. He was also an ichthyologist who studied Cyprinodontiform fish, and an expert on killifish much admired by aquarists. His doctorate in zoology and botany was awarded by the University of Vienna. He has made many collecting expeditions in Africa (including eight visits to Cameroon) as well as Latin America, the Caribbean and Sri Lanka. He co-wrote: *Color Atlas of Cyprinodonts of the Rain Forests of Tropical Africa* (1987). Later in his professional career he developed an interest in ethnology and wrote a series of books on his experiences travelling and meeting some disappearing indigenous peoples.

Radović

Norin Goby *Knipowitschia radovici* M Kovaćîc, 2005

Dragan Radović (1959–2017) was Kovacîć's friend; a Croatian ornithologist. He graduated from Zagreb University (1987) where he studied ichthyology and ornithology and always retained his interest in fish. He went on to further study ornithology at the Croatian Academy of Science, working there and becoming Director (1996–2009). His particular interest was the protection of wetlands in Croatia. He was also the long-time manager of the ringing station in Zagreb. He was a founder member and President of the Croatian Ornithology Society. He encouraged and helped Kovacîć to collect samples along rivers and lakes of the Adriatic coast.

Raffles

Latticed Butterflyfish *Chaetodon rafflesii* Anonymous [ET Bennett], 1830

Sir Thomas Stamford Bingley Raffles (1781–1826) was a colonial officer, Lieutenant-Governor of Java (1811–1815) and founder of the city-state of Singapore (1819). He was noted for his liberal attitude toward peoples under colonial rule, his rigorous suppression of the slave trade, and his zeal in collecting historical and scientific information. He was also the first President of the Zoological Society of London, and wrote a *History of Java* (1817). He employed zoologists and botanists to collect specimens, paying them out of his own pocket. On his return journey to England (1824) on 'HMS Fame', he lost a huge collection of specimens, notes and drawings to a fire. As the local vicar of the parish where Raffles died was a man whose family had made money out of the Jamaica slave trade, and disliked abolitionists, he refused to allow Raffles to be buried in the local church. Two mammals are named after him, as is the plant genus *Rafflesia* and many other taxa.

Rafinesque

White-spotted Lanternfish *Diaphus rafinesquii* A Cocco, 1838

Kentucky Darter *Etheostoma rafinesquei* LM Page & BM Burr, 1982

Yazoo Shiner *Notropis rafinesquei* RD Suttkus, 1990

Professor Constantine Samuel Rafinesque-Schmaltz (1783–1840) was born in Constantinople to a French father and German mother. He was sent to live in Tuscany to escape the turmoil of the French Revolution. His father was a merchant who died in Philadelphia (1793), leaving the family very badly off. Despite being unable to attend a university, Rafinesque was a highly gifted individual, accomplished as a botanist, geologist, historian, poet, philosopher, philologist, economist, merchant, manufacturer, professor, architect, author and editor. He was apprenticed (1802) to a merchant house in Philadelphia, and for the next two years he roamed the fields and woods and made collections of plants and animals. He was in Sicily (1805–1815) as Secretary to the US Consul and carried on a lucrative trade in commodities. He scoured the island for plants and collected previously unrecorded fishes from the stalls of the Palermo market. He sailed for New York (1815) but was shipwrecked in Long Island Sound, losing all his unpublished manuscripts and collections. He sailed down the Ohio River (1818) and conducted a comprehensive survey of the fish species there, published as: *Ichthyologia Ohiensis* (1820). He visited Henderson, Kentucky, and stayed for eight days with John James Audubon. He was Professor of Botany and Natural Science, University of Transylvania (Lexington, Kentucky) (1819–1826). He returned to Philadelphia (1826) with 40 crates of specimens. He had a remarkable gift for inventing scientific names, some 6,700 in botany alone. He died in poverty but was later re-interred in Lexington. He was honoured in the shiner name as "*...one of our early American naturalists.*" A mammal and an amphibian are also named after him.

Ragimov

Tadpole Goby sp. *Benthophilus ragimovi* VS Boldyrev & NG Bogutskaya, 2004

Dr Dadaş Bəhmən oğlu Rahimov (1932–2004) was an Azerbaijani zoologist. He graduated (1949) from the Lankaran Agricultural Engineering College of Agronomy. Two years later he entered the biology faculty of Azerbaijan State University. He became a postgraduate of the Institute of Zoology at the National Academy of Sciences of the Azerbaijan Republic. He studied the fish of Azerbaijan lakes (1976–1980) and was later (1982) Head of the Ichthyology Laboratory, where he led scientific research on the study of fish fauna in Azerbaijan for the rest of his life. Late in life he was awarded his PhD (1991). He was the author of more than 100 scientific articles and described two new species and seven sub-species of fish of the Caspian Sea. In the etymology of the goby he was described as: "*...a well-known ichthyologist and expert in Caspian gobiid fishes.*"

Rahman

Mountain Carp sp. *Psilorhynchus rahmani* KW Conway & RL Mayden, 2008

A K Ataur Rahman (b.1937) is a Bangladeshi biologist and ichthyologist at the Department of Zoology, University of Dhaka. His master's degree was awarded by the University of Michigan and he also studied in China, Denmark, India and the Philippines. He was Director General, Department of Fisheries, Dhaka, and is now retired. He wrote: *Freshwater fishes of Bangladesh* (1989). He was honoured for his contribution to the knowledge of the fishes of Bangladesh.

Raimbault

African Barb sp. *Enteromius raimbaulti* J Daget, 1962

R Raimbault was Inspector, Eaux et Forêts (Waters and Forests), in French West Africa. He was honoured for his role in assisting Daget's missions in Guinea, but we can shed no further light on him.

Raimund

Plated Catfish sp. *Aspidoras raimundi* F Steindachner, 1907

Steindachner gave no help to future researchers in his text! It has been suggested that the binomial refers to Raimund Banowsky (d.1885) who was contemporary with Steindachner at the Zoological and Botanical Society of Vienna. However, it might equally refer to some unknown Brazilian associate of the author.

Rainford

Flathead Perch genus *Rainfordia* AR McCulloch, 1923
Rainford's Dartfish *Parioglossus rainfordi* AR McCulloch, 1921
Rainford's Butterflyfish *Chaetodon rainfordi* AR McCulloch, 1923
Rainford's Goby *Koumansetta rainfordi* GP Whitley, 1940

Edward Henry Rainford (1853–1938) was an English-born Australian amateur naturalist who was viticulturist at the Queensland Agricultural Department, based at Bowen in north Queensland. He collected specimens of fauna and flora for the Australian Museum.

Rajagopalan

Razorfish (wrasse) sp. *Xyrichtys rajagopalani* K Venkataramanujam, VK Venkataramani & N Ramanathan, 1987

Dr V Rajagopalan (d.2017) was an Indian agricultural economist. He served as Vice-Chancellor of the Tamil Nadu Agricultural University, Tuticorin, India (1983–1988). The authors were based at that university's Fisheries College, and gave credit to Dr. Rajagopalan for "*...dedicated and devoted service to the Institute in building [it] up as a centre of national importance.*"

Rajeev

Rajeev's Hillstream Loach *Indoreonectes rajeevi* P Kumkar, M Pise, PA Gorule, CR Verma & L Kalous, 2021

Dr Rajeev Raghavan is an aquatic conservation biologist, particularly of freshwater fishes of south Asia. He works as an assistant professor at the Kerala University of Fisheries and Ocean Studies (KUFOS), Kochi, India. He previously lectured at St Albert's College, Kochi. His initial degree and Master's were in aquaculture and his doctorate was in Fish Ecology. He was honoured for his "...remarkable contributions to the understanding of the systematics and evolution of Indian freshwater fishes." Rajeev's expertise has been used by various international organisations, including the World Bank and the International Union for Conservation of Nature (IUCN). Among his (185) published works is: *The status and distribution of freshwater fishes of the Western Ghats* (2011).

Raju

Onejaw sp. *Monognathus rajui* E Bertelsen & JG Nielsen, 1987

Solomon N Raju is an ichthyologist who worked at the Scripps Institute of Oceanography, Marine Life Research Group (c.1971–c.1985). He was honoured in the name of the onejaw species for his contributions to the systematics of this 'rare family' of fishes. Among his publications is: *Congrid Eels of the Eastern Pacific and Key to Their Leptocephali* (1985) and the co-written: *Some rare leptocephali from the Atlantic and Indo-Pacific Oceans* (1975). He named *Monognathus jesse* after his wife. (See **Jesse**)

Raleigh

Pacific Longnose Chimaera *Harriotta raleighana* GB Goode & TH Bean, 1895
[Alt. Narrownose Chimaera]

Sir Walter Raleigh (c.1554–1618), the English explorer, writer, poet, soldier and spy, needs no long biography here. The authors say of him: "*…by whom the first English scientific expedition was sent to the New World.*"

Ramachandra

Danionine sp. *Betadevario ramachandrani* PK Pramod, F Fang, K Rema Devi, T-Y Liao, TJ Indra, KS Jameela Beevi & SO Kullander, 2010

Professor Dr Alappat Ramachandran (b.1957) is Professor of Fisheries Management (since 1999) and Director (since 2011), School of Industrial Fisheries, Cochin University of Science and Technology, Kochi, India. Cochin University awarded his MSc (1982) and PhD (1984). He also undertook an advanced postgraduate course in Marine Food Processing Technology (1990–1991). He has had more than fifty papers published, including: *Deep Sea Fishing Policies In India from 1981 to 2014 – An Analysis* (2015) and *New Species of Fish from Western Ghats* (2010). He was a researcher in various capacities (1994–2006). He was honoured for his contributions to fisheries and seafood production management and studies on indigenous ornamental fishes.

Rama Rao

Waspfish sp. *Ocosia ramaraoi* SG Poss & WN Eschmeyer, 1975
Rama Rao's Scorpionfish *Scorpaenopsis ramaraoi* JE Randall & WN Eschmeyer, 2001

Dr Kaza V Rama Rao (1940–<2001) was an Indian ichthyologist who was a project coordinator at the Zoological Survey of India. Among his published papers are: *A systematic review of the family Scorpaenidae (Pisces) of the Indian seas and adjacent waters* (1970), which was his PhD dissertation at the University of Gorakhpur, and *Fauna of Chilka Lake* (1995). In a paper of 2001 he was described as 'the late Rama Rao'.

Rambai

Queen of Siam Goby *Mugilogobius rambaiae* HM Smith, 1945

Her Majesty Rambai Barni (1904–1984) was a former Queen of Siam (now Thailand).

Rambarran

Three-barbeled Catfish sp. *Brachyrhamdia rambarrani* HR Axelrod & WE Burgess, 1987

Harry Rambarran is an aquarium-fish exporter. He was co-manager of International Fisheries, Inc., Hialeah, Florida when the company supplied the holotype of this species. He is now President, Ornamental Fish Distributors, Miami, which appears to be the same company as in 1987, but re-branded (1995).

Ramirez, H

Glass Blenny sp. *Emblemariopsis ramirezi* F Cervigón, 1999

Humberto Ramirez discovered this species on the Venezuelan coast, and "*...also had the courtesy of putting the species at* [the author's] *disposal for study*" (translation). No further biography is given.

Ramirez, MV

Ramirez's Dwarf Cichlid *Mikrogeophagus ramirezi* GS Myers & RRR Harry, 1948
[Alt. Ram Cichlid, (Dwarf) Butterfly Cichlid]

Manuel Vicente Ramirez was a Venezuelan collector and fish explorer. He collected the cichlid type with Herman Blass (Franjo Fisheries, Miami, Florida, USA) in Venezuela. It was Blass who popularized the fish in the aquarium trade, but it seems very little has been recorded of Señor Ramirez himself.

Ramiro

Pencil Catfish sp. *Ituglanis ramiroi* ME Bichuette & E Trajano, 2004

Ramiro Hilário dos Santos is a local guide and dedicated protector of the caves in Terra Ronca State Park, Goiás, Brazil. He has discovered and explored many caves as well as discovering the holotype of this subterranean species.

Ramsay, AG

Longtail Southern Cod *Patagonotothen ramsayi* CT Regan, 1913

Allan George Ramsay (1878–1903) was chief engineer of the *Scotia*, during the Scottish National Antarctic Expedition (1902–1904), and became the only person to die during that expedition. He fell ill when the vessel was in the Falkland Islands, but kept quiet in the knowledge that it would be almost impossible to find a replacement engineer.

Ramsay, EP

Ramsay's Mullet *Liza ramsayi* W Macleay, 1883
Ramsay's Sunfish *Mola ramsayi* EH Giglioli, 1883
[Alt. Bumphead Sunfish; Junior Synonym of *Mola alexandrini*]
Spotted Grubfish *Parapercis ramsayi* F Steindachner, 1883
Sole sp. *Aseraggodes ramsaii* JD Ogilby, 1889

Edward Pearson Ramsay (1842–1916) was an Australian naturalist, oologist, ornithologist and (particularly) a marine zoologist. He corresponded with and sent specimens to John Gould (q.v.), who persuaded him not to continue publishing on his oological studies, which he had done (1882 and 1883), as Gould anticipated producing a work on this subject himself – a project that never materialised. Ramsay became Curator of the Australian Museum

(1874). Two reptiles and eleven birds are also named after him. Ogilby gives no etymology for the sole, but it seems near-certain that Edward Ramsay was intended.

Ramsden

Joturo (Cuban Cichlid) *Nandopsis ramsdeni* HW Fowler, 1938
Toothy Topminnow ssp. *Girardinus denticulatus ramsdeni* LR Rivas, 1944

Dr Charles Theodore Ramsden (1876–1951) was a Cuban entomologist, herpetologist and general naturalist who received his doctorate from the University of La Habana (1917). He collected mainly in eastern Cuba and co-wrote: *The Herpetology of Cuba* (1919). The museum at Oriente University, Santiago de Cuba, is named after him. He collected the cichlid type. Two birds and a reptile are also named after him.

Ramsey

Alabama Darter *Etheostoma ramseyi* RD Suttkus & RM Bailey, 1994

Dr John S Ramsey was Associate Professor of Fisheries at Auburn University, Alabama, and contributed extensively to knowledge of Alabama's fishes. It was under his leadership that the Cooperative Fisheries Unit became operative (1967). Among his published papers is: *Habitat use and movements of shovelnose sturgeon in Pool 13 of the upper Mississippi River during extreme low flow conditions* (1997). He went on to be a District Fisheries Biologist with the US Fish and Wildlife Service.

Ramzan

Snow Carp sp. *Racoma ramzani* MN Javed, Azizullah & K Pervaiz, 2012
[Syn. *Schizothorax ramzani*]

Dr Muhammad Ramzan Mirza (b.1936) is a Pakistani zoologist, ichthyologist and limnologist. He is an independent researcher at the Department of Zoology, Government College University, Lahore. He co-wrote: *Fishes of Pakistan and fish culture* (1993).

Rancurel

Devil Ray sp. *Mobula rancureli* J Cadenat, 1959
[Junior Synonym of *Mobula mobular*]
Rancurel's Lampeye *Poropanchax rancureli* J Daget, 1965
Cyrano Spurdog *Squalus rancureli* P Fourmanoir, 1979
[Junior Synonym of *Squalus melanurus*]

Dr Paul G Rancurel was a zoologist who worked for ORSTOM. His particular interest was in aquatic invertebrates. He worked in West Africa (1950s & 1960s) and was at the Centre de recherches océanographiques, Abidjan, Ivory Coast (1967) and later in Noumea (New Caledonia) and Marseille. Among his written papers is: *Topographie générale du plateau continental de la Côte d'Ivoire et du Libéria* (1968). He collected the spurdog holotype, and Fourmanoir wrote in his etymology: '*Nous dédions ce nouveau Squale à P. RANCUREL océanographe O.R.S.T.O.M., qui est le premier à l'avoir capturé et à avoir signalé ses caractères principaux*'.

Rand

Kubuna Hardyhead *Craterocephalus randi* JT Nichols & HC Raven, 1934

Austin Loomer Rand (1905–1982) was a Canadian zoologist and ornithologist who collected in several countries. His first degree was awarded by Acadia University, which also awarded him an honorary DSc (1961). He was on a bird-collecting expedition to Madagascar (1929) which Richard Archbold wrote up as: *The Distribution and Habits of Madagascar Birds* (1936). Archbold subsequently financed a series of biological expeditions to New Guinea (1930s), which Rand co-led and which ultimately resulted in his: *Handbook of New Guinea Birds* (1967). He named a number of birds after Archbold, and became Assistant Zoologist at the National Museum of Canada (1942–1947), then Curator of Birds at the Field Museum in Chicago (1947–1955), and finally Chief Curator of Zoology (1955–1970). He was President of the American Ornithologists' Union (1962–1964) and 14 birds are also named after him.

Randall

Snapper genus *Randallichthys* WD Anderson, HT Kami & GD Johnson, 1977
Ocellated Pipefish *Bryx randalli* ES Herald, 1965
Hornless Blenny *Emblemariopsis randalli* F Cervigón, 1965
Blenny sp. *Entomacrodus randalli* VG Springer, 1967
Brown Soapfish *Rypticus randalli* WR Courtenay, 1967
Yellownose Goby *Elacatinus randalli* JE Böhlke & CR Robins, 1968
Randall's Frogfish *Antennarius randalli* GR Allen, 1970
Randall's Chromis *Chromis randalli* DW Greenfield & DA Hensley, 1970
Soldierfish sp. *Myripristis randalli* DW Greenfield, 1974
Randall's Assessor *Assessor randalli* GR Allen & RH Kuiter, 1976
Hawkfish sp. *Cirrhitichthys randalli* A Kotthaus, 1976
Randall's Viviparous Brotula *Ematops randalli* DM Cohen & JP Wourms, 1976
Dottyback sp. *Chlidichthys randalli* R Lubbock, 1977
Randall's Prawn-Goby *Amblyeleotris randalli* DF Hoese & RC Steene, 1978
Randall's Fairy Basslet *Pseudanthias randalli* R Lubbock & GR Allen, 1978
Perchlet sp. *Plectranthias randalli* P Fourmanoir & J Rivaton, 1980
Lizardfish sp. *Synodus randalli* RF Cressey, 1981
Randall's Fold Dragonet *Diplogrammus randalli* R Fricke, 1983
Goby sp. *Psilogobius randalli* M Goren & I Karplus, 1983
Randall's Puffer *Torquigener randalli* GS Hardy, 1983
Snake-Eel sp. *Gordiichthys randalli* JE McCosker & JE Böhlke, 1984
Dragonet sp. *Synchiropus randalli* GT Clark & R Fricke, 1985
Randall's Threadfin Bream *Nemipterus randalli* BC Russell, 1986
Filefish sp. *Pervagor randalli* JB Hutchins, 1986
Randall's Fusilier *Pterocaesio randalli* KE Carpenter, 1987
Blenny sp. *Cirripectes randalli* JT Williams, 1988
Variegated Spinefoot *Siganus randalli* DJ Woodland, 1990
Blenny sp. *Ecsenius randalli* VG Springer, 1991
Mauritian Damsel *Plectroglyphidodon randalli* GR Allen, 1991

Coastal Stream Goby sp. *Stenogobius randalli* RE Watson, 1991
Randall's Goby *Priolepis randalli* R Winterbottom & ME Burridge, 1992
Greenband Goby *Valenciennea randalli* DF Hoese & HK Larson, 1994
Randall's Fairy Wrasse *Cirrhilabrus randalli* GR Allen, 1995
Slender Weasel Shark *Paragaleus randalli* LJV Compagno, F Krupp & KE Carpenter, 1996
Rapa Triplefin *Enneapterygius randalli* R Fricke, 1997
Randall's Moray *Gymnothorax randalli* DG Smith & EB Böhlke, 1997
Snake-Eel sp. *Callechelys randalli* JE McCosker, 1998
Brazilian Dartfish *Ptereleotris randalli* JLR Gasparini, LA Rocha & SR Floeter, 2001
Randall's Triplefin *Helcogramma randalli* JT Williams & JC Howe, 2003
Randall's Pygmy-Goby *Trimma randalli* R Winterbottom & M Zur, 2007
Middle East Black Seabream *Acanthopagrus randalli* Y Iwatsuki & KE Carpenter, 2009
Randall's Dwarf-Goby *Eviota randalli* DW Greenfield, 2009
Gold-specs Jawfish *Opistognathus randalli* WF Smith-Vaniz, 2009
[Alt. Black Cap Jawfish]
Randall's Tilefish *Hoplolatilus randalli* GR Allen, MV Erdmann & AM Hamilton, 2010
Sandperch sp. *Parapercis randalli* H-C Ho & K-T Shao, 2010
Tan Hamlet *Hypoplectrus randallorum* PS Lobel, 2011
Lombok Swallowtail *Odontanthias randalli* WT White, 2011
Goatfish sp. *Upeneus randalli* F Uiblein & PC Heemstra, 2011
Arabian Whipray *Maculabatis randalli* PR Last, BM Manjaji-Matsumoto & ABM Moore, 2012
Basslet sp. *Liopropoma randalli* KV Akhilesh, KK Bineesh & WT White, 2012
Cardinalfish sp. *Siphamia randalli* O Gon & GR Allen, 2012
Snub-snouted Flathead *Thysanophrys randalli* LW Knapp, 2013
Conger sp. *Rhynchoconger randalli* S Acharya, SR Mohanty, D Ray, SS Mishra & A Mohapatra, 2022

Dr John 'Jack' Ernest Randall Jr. (1924–2020) was one of the most respected modern ichthyologists and a world authority on coral reef fishes. He described over 830 species (97% of which are still valid) and wrote 17 books, such as *Sharks of Arabia* (1986), and over 900 scientific and popular articles (more than any other ichthyologist, ever). He participated in hundreds of scientific expeditions and has made countless SCUBA dives to collect specimens. He developed new photographic methods, both underwater and of newly collected specimens, and his photographs have been published widely. The University of California, Los Angeles awarded his bachelor's degree (1950) and the University of Hawai'i-Manoa awarded his doctorate (1950). He was a Research fellow at the Bishop Museum (1955–1957). He worked at the Marine Laboratory, University of Miami (1957–1961), and then as Professor of Zoology (1961–1965) and Director of the Institute of Marine Biology, University of Puerto Rico (1965). He returned to Hawaii as Director of the Oceanic Institute (1965) and became Senior Ichthyologist at the Bishop Museum in Honolulu, Hawaii (1967). He was among the first to photograph the Arabian Whipray and he was honoured in its name for this and for his 'legendary' work on the taxonomy of Indo-Pacific fish. Others have called him the 'guru of the alpha-taxonomy of Indo-

Pacific marine fishes', their 'frequent accomplice in ichthyological pursuits', outstanding contributor and more. His family informed the scientific community of his recent death saying that he: "…left for his final diving adventure on Sunday 26th April 2020." The Tan Hamlet is named after both John and Helen Randall. (See also **Jack Randall**, **John Randall** and also **Helen (Randal)**)

Raney

Fluvial Shiner *Notropis edwardraneyi* RD Suttkus & GH Clemmer, 1968
Bull Chub *Nocomis raneyi* EA Lachner & RE Jenkins, 1971
Bahama Duckbill *Bembrops raneyi* BA Thompson & RD Suttkus, 1998

Dr Edward Cowden Raney (1909–1984) was an American ichthyologist who was Professor of Zoology at Cornell University (1952–1971). Slippery Rock College, Pennsylvania, awarded his BSc in education (1931). He taught science in schools (1931–1935) while studying for his MSc and went on to do his PhD at Cornell (1938). He then (1936–1952) became an Instructor, Assistant Professor and Associate Professor until taking the zoology chair (1952). During WW2, he enlisted in the United States Navy (1942) and served as an executive officer on a destroyer escort until his return to Cornell as an instructor (1945). He wrote over 140 papers dealing with the systematics, behaviour, and ecology of fishes and other vertebrates. He was a member of over 30 professional societies, and he served as secretary (1948–1951) and president (1955–1956) of the American Society of Ichthyologists and Herpetologists. He collected on many field trips in the eastern USA. He was honoured for his contributions to North American ichthyology and "*…his guidance and imparted enthusiasm toward a multitude of students*" and it was said his: "*…enthusiasm and guidance placed many American students on the professional pathway to ichthyology.*" (Also see **Edward Rainey**)

Ransonnet

Pygmy Sweeper *Parapriacanthus ransonneti* F Steindachner, 1870
Karimunjawa Dottyback *Pseudochromis ransonneti* F Steindachner, 1870
Dwarf Flathead *Elates ransonnettii* F Steindachner, 1876
Surfperch sp. *Neoditrema ransonnetii* F Steindachner, 1883

Baron Eugen von Ransonnet-Villez (1838–1926) was an Austrian diplomat, painter, lithographer, biologist and explorer, who was also a pioneer of underwater oil painting. He was educated at the Academy of Fine Arts, Vienna (1849–1855) then studied law (1855–1858), becoming an official in the Ministry of Foreign Affairs. He was (1860) in Egypt, Palestine, India and Japan where he began to dive. One of the very first underwater paintings was completed while he sat in a diving bell in Ceylon (Sri Lanka) (1864–1865). During his travels he painted, sketched and took photographs but also collected natural history specimens and continued to study the natural sciences. Among other works. he published: *Sketches of the inhabitants, animal life and vegetation in the lowlands and high mountains of Ceylon, as well as the submarine scenery near the coast, taken in a diving bell* (1867). He donated (1892) the only underwater oil painting to the Natural History Museum in Vienna, together with over 5000 zoological artefacts that he had collected during his explorations.

Ranta

Ranta Goby *Feia ranta* R Winterbottom, 2003

This is an arbitrary combination of letters reflecting the first three letters of the forenames of Randall D Mooi and Anthony C Gill, two specialists in Indo-Pacific fish systematics who have worked on the genus *Feia*. Mooi is additionally recognised for his help and 'cheerful companionship' on collecting trips to the Philippines, Thailand and French Polynesia.

Ranzani

Slender Sunfish genus *Ranzania* GD Nardo, 1840

Camillo Ranzani (1775–1841) was an Italian priest and naturalist who was director of the Museum of Natural History of Bologna (1803–1841).

Rao, AA

Chameleonfish sp. *Badis andrewraoi* S Valdesalici & S Van der Voort, 2015

Andrew Arunava Rao is an ornamental-fish collector and breeder who owns and runs Malabar Tropicals. (See **Andrew (Rao)**)

Rao, HS

Stone Loach sp. *Physoschistura raoi* SL Hora, 1929

Rao's Hover-Goby *Parioglossus raoi* AW Herre 1939

Professor Dr H Srinivasa Rao (1894–1971) was an Indian zoologist who specialised in faunistic morphology. The University of Madras awarded his PhD (1925), and he shortly afterwards became Assistant Superintendent of the Zoological Survey of India. He was in charge of the invertebrate collections at the Indian Museum. He was elected Fellow of the Indian Academy of Sciences, Bangalore. Among his published papers are: *On certain Succineid Molluscs from the Western Ghats* in *Records of the Indian Museum* (1925) and: *On the habitat and habits of Trochus niloticus Linn, in the Andaman Seas* (1937). He helped collect the loach type.

Rapson

Rapson's Ponyfish *Eubleekeria rapsoni* ISR Munro, 1964

Dr Alan Morris Rapson (1912–2001) was (1952–1971) Chief of the Division of Fisheries, Department of Agriculture, Stock & Fisheries, PNG. He wrote: *A brief history of fisheries of Papua and New Guinea* (1968). He was educated in New Zealand then was a Research Scientist on the vessel *Discoverer 2*, studying ocean currents to enable submarine operations in WWII. After the war he moved to Sydney, where he worked for the CSIRO Oceanographic Research Unit. He was appointed (1952) head of the Department of Fisheries at Konedobu (a suburb of Port Moresby) during the course of which he made visits to New Guinea Highlands' lakes, to Bikini Atoll to study the effects of the atom bomb tests, to the Fly River to check for deer possibly suffering from foot and mouth disease, to the Solomon Islands, and to Japan during which he became an accomplished writer, completing a diary of every trip, writing official reports and presenting papers to scientific meetings. He retired aged 59 but spent the next 30 years continuing to write, travel and photograph and campaign for conservation. The original paper describing the species does not contain an etymology, but we are confident that the fish is named after this person, as Rapson's paper *Small fish trawling in Papua* (1955) is referenced and his collection of the same date is examined in detail.

Raquel

Characin sp. *Hemibrycon raqueliae* C Román-Valencia & DK Arcila-Mesa, 2010

Raquel Ivveth Ruiz Calderón is a Colombian biologist. The Univeridad Del Quindio, Uniquindio awarded her bachelor's degree (2004). Whilst she was teaching at the University of Gran Colombia (2007) and the University of Quindio (2008–2010), she was enrolled (2005) at the Central University of Venezuela to do a doctorate. Her thesis and defence of it were rejected by the jury (2010). After this, she does not appear in Academe but was incensed at her treatment and the favouritism shown to other candidates, and started a legal action (2011) against the University to get the courts to over-turn the jury and confirm her doctorate. The courts (2012) would not make a judgement and stated that they did not have the knowledge and experience necessary to adjudicate on scientific matters of this kind. This species was named after her for her: "*...generous contribution of works for the preservation and study of Neotropical fishes*."

Rara

Snakehead sp. *Channa rara* R Britz, N Dahanukar, VK Anoop & A Ali, 2019

Dr Rajeev Raghavan (b.1979) is an aquatic biologist and conservationist, particularly known for his work on the freshwater fish of the Western Ghats. He is Assistant Professor at the Department of Fisheries Resource Management, Kerala University of Fisheries and Ocean Studies, as well as being the India and South Asia Coordinator of the IUCN's Freshwater Fish Specialist Group. Previously he worked as a Lecturer at St. Albert's College, Kochi, India (where he set up the Conservation Research Group, a multidisciplinary network of conservation biologist working on various aspects of the biodiversity of the Western Ghats) and also held affiliate positions at the Zoo Outreach Organisation (ZOO), Coimbatore, India. The etymology says that species name "*...honours the Indian ichthyologist Rajeev Raghavan for his contributions to the elucidation of taxonomically difficult Indian freshwater fishes and their conservation; it is formed from the first two letters of his first name and surname*."

Raredon

Raredon's Sea Catfish *Cathorops raredonae* AP Marceniuk, R Betancur-Rodriguez & P Acero, 2009

[Alt. Curator Sea Catfish]

Deepwater Bandfish sp. *Owstonia raredonae* W Smith-Vaniz & GD Johnson, 2016

Sandra Jose Raredon (b.1954) is a French-born museum specialist who emigrated to the USA (1964) and has been associated with the Smithsonian, Washington DC, since high school when she participated on a summer dig in Israel with Smithsonian archaeologists. She continued her studies in Archaeology, History and Spanish at Madison College in Harrisonburg, Virginia, and then entered the archaeology graduate program at the University of Tennessee, Knoxville. She was hired (1983) at the Smithsonian Institution's National

Museum of Natural History, and five years later became a permanent staff of the Division of Fishes of the Department of Vertebrate Zoology and is currently a museum specialist, curating the National collection of fishes, assisting visiting researchers and specialising in digital radiography and digital photography. (Also see **Sandra**)

Rasor

Short River Garfish *Zenarchopterus rasori* CML Popta, 1912

August Rasor was treasurer of the Frankfurt Geographical and Statistical Association that mounted an expedition (1911) to the Lesser Sunda Islands (Indonesia).

Rass

Eelpout sp. *Lycenchelys rassi* AP Andriashev, 1955
Blind Cusk Eel sp. *Paraphyonus rassi* JG Nielsen, 1975
[Syn. *Aphyonus rassi*]
Rass' Snailfish *Pseudonotoliparis rassi* DL Pitruk, 1991

Dr Teodor Saulovich Rass (1904–2001) was a Russian zoologist and ichthyologist. Moscow State University awarded his bachelor's degree (1925), his master's (1929) and his doctorate with title of Professor (1940). He was at the Murmansk Biological station as a researcher (1925–1929) and as a senior research scientist at the State Institute of Oceanography (later called the All-Union Institute of Marine Fisheries and Oceanography (VNIRO)) (1929–1948). He worked at the P P Shirshov Institute of Oceanology, USSR Academy of Sciences (1948–1987), as head of the Nekton Laboratory and (1987–2001) as a leading research scientist. He participated in 12 scientific cruises to the Sea of Okhotsk, the Bering Sea and Kuril-Kamchatka waters, as well as the central part of the Pacific Ocean (1957–1958), the Indian Ocean (1959–1960), the southeast Pacific (1968), Caribbean Sea and Gulf of Mexico (1973), and Mediterranean Sea (1979). He wrote: *Fishes of the Pacific and Indian Oceans; biology and distribution* (1966). He was honoured as he lent Nielsen specimens from many of the cruises on which he went.

Rathbun

Ronquil genus *Rathbunella* DS Jordan & BW Evermann, 1896
Speckled Killifish *Fundulus rathbuni* DS Jordan & SE Meek, 1889
Bearded Banded Croaker *Paralonchurus rathbuni* DS Jordan & CH Bollman, 1890
Highfin Scorpionfish *Pontinus rathbuni* GB Goode & TH Bean, 1896
Redflank Bloodfin *Aphyocharax rathbuni* CH Eigenmann, 1907

Dr Richard Rathbun (1852–1918) was an American zoologist, ichthyologist, geologist and museum administrator. He received a master's degree in science from Indiana University (1883) and an honorary Doctorate of Science from Bowdoin University (1894). He worked (1873–1875) as a voluntary assistant to Baird at the US Fish Commission, whilst being Assistant, Zoology, Boston Society of Natural History (1874–1875). He was Assistant Geologist, Geological Commission of Brazil (1875). He was Scientific Assistant, US Fish Commission (1878) and was Assistant, Zoology at Yale (1879–1880). He was later Curator, Marine Invertebrates, AMNH (1880–1914). He was a member of the joint US/

British Fisheries Commission (1892–1896). He became Assistant Secretary, Smithsonian Institution and was in charge (1899–1918) of the US National Museum.

Rathke

Sixbar Panchax ssp. *Epiplatys sexfasciatus rathkei* AC Radda, 1970

Karl Heinz Rathke (1932–2017) was a missionary in Kumba, Cameroon, the type locality. He was Radda's 'gracious host' there.

Ratmanov

Eelpout sp. *Lycenchelys ratmanovi* AP Andriashev, 1955

Georgy Efimovich Ratmanov (1900–1940) was a Russian oceanographer. He graduated from the University of Leningrad (1926). He participated in the expedition under the guidance of Professor K M Deriugina to study the seas of the North Pacific (1932–1933). He also conducted the first detailed studies of the currents of the Bering Sea and water exchange and, through it, the hydrological regime of the Bering and Chukchi seas. One of the islands of Diomede (Big Diomede) is also named after him.

Ratsiraka

Malagasy Sleeper genus *Ratsirakia* AL Maugé, 1984

Didier Ignace Ratsiraka (1936–2021) was President of Madagascar (1975–1993 & 1997–2002).

Raymond

Raymond's Grunter *Hephaestus raymondi* GF Mees & KJ Kailola, 1977
[Syn. *Therapon raymondi*]

Raymond Moore was an Australian fisheries biologist. The etymology says he was: "*a fisheries biologist who, in the course of an extensive study of the Barramundi,* Lates calcarifer *(Bloch), carried out and directed numerous collecting surveys of the rivers, creeks and swamps of western Papua. It was on one such survey that the specimens of this new species were obtained.*"

Razi

Scraper Barb sp. *Capoeta razii* A Jouladeh-Roudbar, S Eagderi, HR Ghanavi & I Doadrio, 2017

Abu Bakr Muhammad ibn Zakariyya al-Razi, also known by the Latinized version of his name, Rhazes (c.865–925), was a Persian polymath, physician, alchemist and philosopher. He made important contributions to medicine, and has been described as "*…probably the greatest and most original of all the Muslim physicians*".

Reader

Reader's Dwarf-Goby *Eviota readerae* AC Gill & SL Jewett, 2004
[Alt. Sally's Eviota]
Goby sp. *Hetereleotris readerae* DF Hoese & HK Larson, 2005

[Syn. *Pascua readerae*]
Red-spotted Pygmygoby *Trimma readerae* R Winterbottom & DF Hoese, 2015

Sally E Reader is an ichthyologist at the Australian Museum Research Institute. She helped collect the type series of the *Eviota* and assisted Hoese with the collection of most of the type specimens of the *Pascua*, and 'kindly arranged the loan of specimens' for his study. She was honoured in the name of the pygmygoby for the many years she has spent gathering, analysing and documenting data on *Trimma* for Hoese's research program. Among her published papers, several were co-written with Douglas Hoese such as: *Revision of the eastern Pacific species of* Gobulus *(Perciformes: Gobiidae), with description of a new species* (2001) and *A preliminary review of the eastern Pacific species of* Elacatinus *(Perciformes: Gobiidae)* (2001). (See also **Sally**)

Rebains

Rebains' Portholefish *Diplophos rebainsi* G Krefft & NV Parin, 1972

Eduard Rebains was captain of the Russian research vessel *Akademic Kurchatov* from which the holotype of this species was collected.

Rebecca

Goby sp. *Odontamblyopus rebecca* EO Murdy & K Shibukawa, 2003

Rebecca Rootes is the first author's 'life partner and spouse'. (Also see **Murdy**)

Rebel

Western Balkan Barbel *Barbus rebeli* O Koller, 1926
Squeaker Catfish sp. *Synodontis rebeli* M Holly, 1926

Hans Rebel (1861–1940) was an Austrian lawyer, entomologist and lepidopterist. He gave up law to pursue his interest in butterflies, and became Keeper of the Lepidoptera Collection (1897–1932) at the Naturhistorisches Museum in Vienna. He enriched the collections by making many collecting trips in Austro-Hungary and five trips in the Balkans. He directed the Department of Zoology (1923) and became the museum's Director General (1925). He published more than 300 papers. Holly's description of the catfish does not identify the patronym, but it is most probably in honour of this man. Vladimir Nabokov (1899–1977), famous as the author of: *Lolita*, included Rebel as a character in his short story: *The Aurelian*.

Rebentisch

Many-fingered Grenadier Anchovy *Coilia rebentischii* P Bleeker, 1858

Johann Heinrich Andreaus Bernhard Sonnemann Rebentisch was a Dutch military health officer (he was Surgeon Colonel in Sinkawang, a city in Indonesian Borneo) who collected (1858–1859) specimens of fish for Bleeker, including the holotype of this anchovy. He wrote: *Reptiliën van Borneo* (1859).

Recep

Recep's Chub *Alburnoides recepi* D Turan, C Kaya, FG Ekmekçi & E Doğan, 2014

Recep Buyurucu was honoured as the person: "*...who has contributed greatly to the authors' sampling for many years*" (he also helped collect the type). He has collected with Turan on a number of occasions and is his brother-in-law.

Reddell

Blind Whiskered Catfish *Rhamdia reddelli* RR Miller, 1984

James R Reddell is an entomologist, arachnologist and herpetologist and a pioneer in exploring Latin American caves. He works at University of Texas, Austin, where he is Curator of Arthropods at the Texas Memorial Museum. He co-wrote: *Up high and down low: Molecular systematics and insight into the diversification of the ground beetle genus* Rhadine *LeConte* (2016). He collected the holotype of this species.

Redding

Bronze-striped Grunt *Orthopristis reddingi* DS Jordan & JA Richardson, 1895

Benjamin Barnard Redding (1824–1882) was a Canadian-born politician. After joining the California gold-rush as a young man, he later served as mayor of Sacramento, Secretary of State for California, and Fish Commissioner of California. The etymology calls him "*...a man deeply interested in scientific research*".

Reed, EC

Croaker sp. *Umbrina reedi* A Günther, 1880

Edwyn Charles Reed (1841–1910) was an English naturalist whose first post was as Secretary and Naturalist for the Bristol Museum. He took part in an expedition to Brazil but returned (1869) in poor health. Advised to move to a dry climate, and acting on advice from Charles Darwin, he made a journey of exploration to Chile and spent the rest of his life there. He was naturalist at the Museo Nacional for the next seven years. He was given a commission to study collections in Europe (1873) but he moved permanently to Valparaíso (1876), being appointed Curator of Museo Historia Natural de Valparaíso but leaving after one year. For the next seven years, he was Professor of Natural History and Physical Geography at the local naval academy (1877–1884). During that time Reed founded an observatory in Valparaíso and a museum at the *San Rafael Arcángel del Valparaíso* seminary. Again, because of health reasons he resigned and moved to the mountain towns of Los Andes and Baños de Cauquenes, where he founded a regional natural history museum (1895). In 1902, he was appointed the director of the newly-founded *Museo de Concepción*, a post which he held until his death. He presented the skin of a specimen of this fish to the British Museum 'some years ago' (from Günther's perspective).

Reed, EP

Pinecone Fish sp. *Monocentris reedi* LP Schultz, 1956

Dr Edwyn Pastor Reed Rosa (1880–1966) was an Anglo-Chilean physician in Valparaiso. He was Chief of the biological department, Dirección General de Pesca y Gaza, Valparaiso, Chile. He secured the holotype and sent it to Schultz for identification. He wrote:

Entomological Notes (1932). His nickname was 'The General', which he acquired one very hot day after greeting his friends with the words: "*It's as hot as Jesus Christ and General Jackson*" – a very mixed metaphor!

Reed, GB

Yellowmouth Rockfish *Sebastes reedi* SJ Westrheim & H Tsuyuki, 1967

The 'G.B. Reed' was a research vessel of the Canadian Department of Fisheries and Oceans. The vessel was named after Dr Guilford Bevil Reed (1887–1955), a medical researcher who became joint Chairman (1947–1953) of the Fisheries Research Board of Canada.

Reeves, C

Chinese Loach sp. *Sinibotia reevesae* HW Chang, 1944

Dr Cora Daisy Reeves (1873–1953) was (1917–1941) Professor of Zoology and Head of Department of Biology, Ginling College (Nanjing, China). The University of Michigan awarded her bachelor's degree (1906) and PhD (1917). She wrote: *Discrimination of light of different wave-lengths by fish, (Behavior Monographs)* (1919) and: *Manual of the vertebrate animals of northeastern and central China, exclusive of birds* (1933). She was honoured for her 'valuable suggestion' and her 'kindness' in allowing Chang to study the fish collection of her department.

Reeves, J

Noodlefish sp. *Salanx reevesii* JE Gray, 1831
[Syn. *Leucosoma reevesii*]
Reeves' Moray *Gymnothorax reevesii* J Richardson, 1845
Reeves' Shad *Tenualosa reevesii* J Richardson, 1846

John Reeves (1774–1856) was an English amateur naturalist and collector who served in China, chiefly Canton and Macao, as a civil servant (1812–1831). The East India Company employed him as an 'Inspector of Tea'. He sent the first specimens of the Reeves' Muntjac *Muntiacus reevesi* back to England, where escapees from collections have established it as a feral species. Reeves illustrated many specimens collected by the HMS 'Sulphur' in the China Sea, including the moray, and also commissioned local artists to paint accurate pictures of Chinese flora and fauna; this collection of over 2,000 paintings is now in the BMNH as are many specimens he collected. He is also commemorated in the names of a bird and three reptiles.

Regan

Armoured Catfish genus *Reganella* CH Eigenmann, 1905
Lake Tanganyika Cichlid genus *Reganochromis* GP Whitley, 1929
Regan's Anchovy *Anchoa argentivittata* CT Regan, 1904
Regan's Fangtooth Pellonuline *Odaxothrissa vittata* CT Regan, 1917
Giant Whitespot Pleco *Hypostomus regani* R Ihering, 1905
Central African Cichlid sp. *Chromidotilapia regani* J Pellegrin, 1906
South African Zebra Sole *Zebrias regani* JDF Gilchrist, 1906
Thorny Catfish sp. *Anadoras regani* F Steindachner, 1908

Regan's Ghost Flathead *Hoplichthys regani* DS Jordan, 1908
Suckermouth Catfish sp. *Astroblepus regani* J Pellegrin, 1909
Pink Flabby Whalefish *Cetostoma regani* EJG Zugmayer, 1914
Pencil Catfish sp. *Trichomycterus regani* CH Eigenmann, 1917
Izak Catshark *Holohalaelurus regani* JDF Gilchrist, 1922
Forlon Gambusia *Gambusia regani* CL Hubbs, 1926
Tonguefish sp. *Symphurus regani* MCW Weber & LF de Beaufort, 1929
Regan's Lanternfish *Diaphus regani* AV Tåning, 1932
Eelpout sp. *Lycozoarces regani* AM Popov, 1933
Chinese Cyprinid sp. *Percocypris regani* TL Tchang, 1935
Bristlenose Pleco *Pareiorhaphis regani* LP Giltay, 1936
Ariake Dwarf Icefish *Neosalanx reganius* Y Wakiya & N Takahasi, 1937
Convict Julie *Julidochromis regani* M Poll, 1942
[Alt. Lake Tanganyika Convict Cichlid]
Cardinalfish sp. *Apogonichthyoides regani* GP Whitley, 1951
Regan's Cichlid *Maskaheros regani* RR Miller, *1974*
[Alt. Almoloya cichlid]
Regan's Deepwater Dragonet *Callionymus regani* T Nakabo, 1979
Dwarf Cichlid sp. *Apistogramma regani* SO Kullander, 1980
Dwarf Pike-Cichlid *Crenicichla regani* A Ploeg, 1989
Congo Cichlid sp. *Tylochromis regani* MLJ Stiassny, 1989
Regan's Flatfish *Engyprosopon regani* DA Hensley & AY Suzumoto, 1990
Bluntnose Knifefish sp. *Brachyhypopomus regani* WGR Crampton, CDCM de Santana, JC Waddell & NR Lovejoy, 2017

Charles Tate Regan (1878–1943) was a British ichthyologist. He was educated (1897–1900) at Queen's College, Cambridge, where he graduated with his BA (1900) proceeding to MA (1907) before joining (1901) the staff of the BMNH as an Assistant. He became Keeper of Zoology (1921) and later Director of the museum (1927–1938). He published his first paper in 1902 and over the years published at least 250 papers on fish, many of which he illustrated. Among his longer works is: *The Freshwater Fishes of the British Isles* (1911). (See also **Tateregan**)

Regina

Fowler's Danio *Devario regina* HW Fowler, 1934

Regina, meaning a queen, refers in this case to Her Majesty Rambai Barni (1904–1984), Queen of Siam. (See also **Rambai**)

Reiche

Indo-Pacific Tropical Sand Goby *Favonigobius reichei* P Bleeker, 1854

Reiche (no forename given) collected the type. This possibly refers to a 'M. Th. Reiche' who was in the Civil Medical Service, Dutch East Indies, but nothing can be established with certainty.

Reichelt

Pufferfish genus *Reicheltia* GS Hardy, 1982

John and Bonnie Reichelt were friends of the author, who "*...assisted in Seine netting along the southern New South Wales coast, whereby new locality records for* R. halsteadi *were obtained.*"

Reichert, KB

Amur Pike *Esox reichertii* BT Dybowski, 1869

Dr Karl Bogislaus Reichert (1811–1883) was the author's professor of anatomy, who was also an embryologist and histologist. He was a Baltic German from East Prussia (now in Kaliningrad, Russia). He studied at the University of Konigsberg (1831) and the University of Berlin, which awarded his doctorate (1836) and where he worked as an assistant (1839–1843). He was Professor of Anatomy at the University of Dorpat (Tartu) (1843–1853), Professor of Physiology, University of Breslau (Wroclaw, Poland) (1853–1858), and became Professor of Anatomy, University of Berlin (1858). He wrote: *Das Entwicklungsleben im Wirbelthierreiche* (1840). Dybowski's text has no etymology, but it seems highly likely that Karl Bogislaus is intended.

Reichert (Lang)

Neotropical Rivuline sp. *Austrolebias reicherti* M Loureiro Barrella & GB García, 2004

Juan Jorge Reichert Lang (1929–2000) was a fish hobbyist and expert on the freshwater fish of Uruguay who described a number of fish species. He wrote (and illustrated) the: *Illustrated Atlas of Freshwater Fish of Uruguay* (2001). Others said of him that he "*...always and unconditionally shared his knowledge on the freshwater fishes of Uruguay.*" (Also see **Juan Lang**)

Reid, ED

Reid's Damselfish *Pomacentrus reidi* HW Fowler & TH Bean, 1928
Reid's Seahorse *Hippocampus reidi* I Ginsburg, 1933
[Alt. Slender Seahorse, Longsnout Seahorse]
Stone Loach sp. *Schistura reidi* HM Smith, 1945

Earl Desmond Reid (1885–1960) was Assistant Curator, Division of Fishes at the Smithsonian (United States National Museum, USNM) where Isaac Ginsburg also worked. Among his published papers are: *Two new congrid Eels and a new flatfish* (1934) and *A new genus and species of eel from the Puerto Rican Deep* (1940).

Reid, GM

Congo Labeo sp. *Labeo reidi* SM Tshibwabwa, 1997
African Characin sp. *Nannocharax reidi* RP Vari & CJ Ferraris, 2004
Loach Catfish sp. *Doumea reidi* CJ Ferraris, PH Skelton & RP Vari, 2010

Professor Dr Gordon McGregor Reid (b.1948) is a British ichthyologist, zoologist and zoo director. He was Director General and Chief Executive of the North of England Zoological Society (Chester Zoo) (1995–2010), and remains Director Emeritus. He started professional life as a 16-year-old technician in the Zoology Department of Glasgow University. He completed a PhD at Kings College London with a thesis on tropical fish, after which he did research in Nigeria and then moved to Liverpool where he became Keeper of Collections for the Liverpool Museum. He was later Keeper of Natural History Collections at the Horniman Museum in London. Whilst in London, he became a government-appointed Inspector of Zoos and was then appointed as Chief Curator of Chester Zoo. Manchester Metropolitan University made him an honorary doctor of science (2008). He wrote: *A systematic study of labeine cyprinid fishes with particular reference to the comparative morphology, functional morphology and morphometrics of African* Labeo *species* (1978) and *Introduction to Freshwater Fishes and Their Conservation* (2013).

Reighard

Shortnose Cisco *Coregonus reighardi* WM Koelz, 1924

Professor Jacob Ellsworth Reighard (1861–1942) was an American zoologist. He studied biological science at the University of Michigan (1878–1882), where he then worked as instructor in zoology (1886) and acting assistant professor (1887–1888). He became assistant professor (1889–1891), full Professor of Animal Morphology (1892), Professor of Zoology (1895) and Director of the Biological Station and Professor Emeritus (1927–1942). He was a co-founder of the University of Michigan Biological Station on Douglas Lake (1909) and its first Director. He also directed the scientific work of the Michigan Fish Commission (1890–1895). He wrote: *The photography of aquatic animals in their natural environment* (1908). The author, Koelz, studied under Reighard at the University of Michigan.

Reiko

Bumblebee Catfish sp. *Microglanis reikoae* WBG Ruiz, 2016

Reiko Sugizaki Matsushima was the grandmother of the author, William Ruiz.

Rein

Giant Atlas Barbel *Labeobarbus reinii* A Günther, 1874
Japanese Catfish sp. *Liobagrus reinii* FM Hilgendorf, 1878
Sculpin sp. *Cottus reinii* FM Hilgendorf, 1879

Dr Johannes Justus Rein (1835–1918) was a German geographer, traveller, natural history collector and Japanologist. He was on an expedition to Morocco (1872–1873) during which he helped collect the barbel type. He also travelled in the Americas, Great Britain and Spain, and was in Japan (1873–1875). He was appointed Professor of Geography at the University of Marburg (1876) and to the same position at the University of Bonn (1883).

Reina

Armoured Catfish sp. *Sturisomatichthys reinae* A Londoño-Burbano & RE Reis, 2019

Ruth Gisela Reina (1977–2016) was formerly the curator of fishes, Smithsonian Tropical Research Institute (Panama). She was honoured for her contributions to the knowledge of fishes and invaluable help and assistance for several ichthyologists around the world and was described by her contemporaries as a hard working and dedicated scientist. She co-wrote:

Molecular Phylogeny and Biogeography of the Amphidromous Fish Genus Dormitator Gill 1861 (Teleostei: Eleotridae) (2016). She died rescuing her son on a beach in Colón.

Reinhardt, JCH

Greenland Halibut genus *Reinhardtius* TN Gill, 1861
Sea Tadpole *Careproctus reinhardti* HN Krøyer, 1862

Dr Johannes Christopher Hagemann Reinhardt (1778–1845) was a Norwegian zoologist who became Professor of Zoology (1814) at the University of Copenhagen. He studied theology, zoology, botany, mineralogy and anatomy in Copenhagen, Freiberg, Göttingen and Paris. His doctorate was an honorary PhD, awarded by the University of Copenhagen (1836). When the Danish Royal Museum of Natural History was founded (1805), he was invited to be its 'inspector' of newly purchased collections of the Society for Natural History. He accepted but only returned to Copenhagen (late 1806) after studying the museums in Paris and attending Cuvier's lectures. He started lecturing in the Museum (1809) and was eventually employed by the University (1813), being appointed a 'professor extraordinarius' (1814). He became a member of the Royal Danish Academy of Sciences and Letters (1821), a full professor of the University (1830) and a titular councillor of state (1839). He described a number of taxa including the Black Dogfish *Centroscyllium fabricii* (1825). A bird is also named after him. His son was Dr Johannes Theodor Reinhardt (1816–1882) (below).

Reinhardt, JT

Speckled Longfin Eel *Anguilla reinhardtii* F Steindachner, 1867
Long-whiskered Catfish sp. *Bagropsis reinhardti* CF Lütken, 1874
Headstander Characin sp. *Megaleporinus reinhardti* CF Lütken, 1875
Pencil Catfish sp. *Ochmacanthus reinhardtii* F Steindachner, 1882
Reinhardt's Lanternfish *Hygophum reinhardtii* CF Lütken, 1892
Pencil Catfish sp. *Trichomycterus reinhardti* CH Eigenmann, 1917

Dr Johannes Theodor Reinhardt (1816–1882) was a Danish zoologist who was Director of the National Natural History Museum, Copenhagen. He wrote a 'List of birds hitherto observed in Greenland' in *Ibis* (1861), which included the now extinct Eskimo Curlew *Numenius borealis* and Great Auk *Pinguinus impennis*. His father was the zoologist J.C.H. Reinhardt (above). An amphibian, a bird and five reptiles are also named after him. The patronym for *Anguilla reinhardtii* is not explained, and could possibly refer to J.C.H. Reinhardt rather than J.T. Reinhardt.

Reinhardt, T

Lanternfish sp. *Lampanyctus reinhardti* DS Jordan, 1921

Tom Reinhardt was a boatman in Hawaii. After an eruption and lava flow by Mauna Loa (1919), he found a number of fish floating in the sea, which had been 'boiled' by lava entering the ocean. Among them were three examples of this species.

Reis

Armoured Catfish sp. *Ancistrus reisi* S Fisch-Muller, AR Cardoso, JFP da Silva & VA Bertaco, 2005

Neotropical Livebearer sp. *Phalloceros reisi* PHF Lucinda, 2008
Armoured Catfish sp. *Rineloricaria reisi* M S-A Ghazzi, 2008
Plated Catfish sp. *Scleromystax reisi* MR Britto, CK Fukakusa & LR Malabarba, 2016

Dr Roberto Esser dos Reis is a Brazilian ichthyologist. The Federal University of Rio Grande do Sul awarded his bachelor's degree (1983), the Pontifical Catholic University of Rio Grande do Sul his master's (1988) and the University of São Paulo his doctorate (1993). He undertook post-doctoral study at the University of Michigan, Ann Arbor (1995–1996) and the University of Central Florida, Orlando (2013–2014). He is currently Professor of Zoology and Curator of Fishes at Pontificia Universidade Católica do Rio Grande do Sul. He wrote: *Anatomy and phylogenetic analysis of the neotropical callichthyid catfishes (Ostariophysi, Siluriformes)* (1998). One etymology remarks he is honoured: "*...for his many contributions to neotropical ichthyology, including studies of callichthyid fishes.*"

Reissner

Far Eastern Brook Lamprey *Lethenteron reissneri* B Dybowski, 1869

Ernst Reissner (1824–1878) was an Estonian (Baltic German) anatomist. The University of Dorpat awarded his medical degree (1851) and he became Professor of Anatomy there (1855) until retiring early in poor health (1875). He is also commemorated in the name of an anatomical detail: 'Reissner's membrane', part of the cochlea in the inner ear. We believe this to be the man honoured, although the original text does not make the etymology clear.

Reitz

Clingfish sp. *Tomicodon reitzae* JC Briggs, 2001

Dr Elizabeth Jean Reitz (b.1946) is an anthropologist and zoo-archaeologist and Director of the Georgia Museum of Natural History, Professor Emerita of the University of Georgia and a Research Associate of the AMNH. Her PhD was awarded by the University of Florida (1979). Among other books and many papers, her most well-known work is the book: *Zooarchaeology* (1999).

Rema Devi

Loach sp. *Mesonoemacheilus remadevii* CP Shaji, 2002
Mahseer sp. *Tor remadevii* BM Kurup & KV Radhakrishnan, 2011
[*NB. The binomials of these species are sometimes amended to *remadevae* / *remadeviae* to correctly reflect female gender]

Dr Karunakaran Rema Devi is a freshwater fish taxonomist who is Officer in Charge, at the Zoological Survey of India, Southern Regional Centre, Chennai, Tamil Nadu, India. An example of her written work is the co-written: *Records of the Zoological Survey of India : Fishes of River Pennar and its Branches: Occasional Paper No. 329* (2011). (See also **Devi**)

Rendahl

Sole genus *Rendahlia* P Chabanaud, 1930
Three-barbeled Catfish sp. *Pimelodella rendahli* CGE Ahl, 1925
Rendahl's Catfish *Porochilus rendahli* GP Whitley, 1928

Chinese Carp sp. *Acrossocheilus rendahli* SY Lin, 1931
Chinese Cyprinid sp. *Decorus rendahli* S Kimura, 1934
Bagrid Catfish sp. *Pseudobagrus rendahli* J Pellegrin & P-W Fang, 1940
Rendahl's Messmate *Echiodon rendahli* GP Whitley, 1941
Sand-burrower sp. *Limnichthys rendahli* AW Parrott, 1958
Loach sp. *Schistura rendahli* PM Bănărescu & TT Nalbant, 1968

Dr Carl Hialmar Rendahl (1891–1969) was a Swedish herpetologist, ichthyologist and artist at Naturhistoriska Riksmuseet, Stockholm (1912–1958), where he also was Professor of Natural History (1933–1958), retiring as Professor Emeritus. Stockholm University awarded his bachelor's degree (1916) and his doctorate in zoology (1924). He described many Chinese fish species and was honoured for his 'great contribution' to the systematics of Central Asian fish. Two reptiles are also named after him. (See also **Hialmar**)

Rendall

Redbreast Tilapia *Coptodon rendalli* GA Boulenger, 1897

Dr Percy Rendall (1861–1948) was a medical practitioner and itinerant naturalist who collected over much of Africa and in Trinidad and other Caribbean locations in the late 19th century. He made a collection of new fish species from the Upper Shiré River, British Central Africa, which was presented to the British Museum by Sir Harry Johnston (q.v.). He published *Notes on the ornithology of the Gambia* (1892) and *Natural history notes from the West Indies* (1897). Two birds and a mammal are also named after him. He collected the tilapia holotype.

Renesto

Armoured Catfish sp. *Hypostomus renestoi* CH Zawadzki, HP da Silva & WP Troy, 2018

Dr Erasmo Renesto is a Brazilian evolutionary biologist and ichthyologist who is an Associate Professor in the Department of Biotechnology at Universidade Estadual de Maringá. Universidade Estadual Paulista Júlio de Mesquita Filho awarded his first degree in natural history (1970) and Universidade Estadual Paulista Júlio de Mesquita Filho his master's (1973) and PhD (1979). He was honoured for his contributions to the genetics of Neotropical fishes.

Rengifo

Suckermouth Catfish sp. *Astroblepus rengifoi* G Dahl, 1960

Dr Santiago Rengifo Salcedo (1913–1965) was a Colombian medical entomologist and parasitologist. He graduated from Universidad Nacional de Colombia as a doctor (1944) and was awarded a master's degree in Public Health by Johns Hopkins University, Baltimore. He worked as Director de la División de Enfermedades Comunicables del Ministerio de Higiene y Jefe de Investigaciones de Malaria (1947–1952). He was responsible for founding the medical school at the Universidad del Valle (1952) and its Department of Preventative Medicine (1956). He wrote: *Notas entomoiógicas regionales. Trabajo presentado para optar al título en medicina y cirugía* (1944). Dahl named this species after him for his: "…*ceaseless [sic] work for the advancement of biological science in Colombia."* He was killed in a street accident.

Rennie

Scaly Sand-lance *Ammodytoides renniei* JLB Smith, 1957

Master John Rennie of Grahamstown, South Africa, discovered three specimens of this fish "found on the beach after an onset of cold water". We assume the use of the term 'Master' implies that John was a boy when he made this discovery.

Renny

Renny's Ricefish *Oryzias hadiatyae* F Herder & S Chapuis, 2010

Flasher Wrasse sp. *Paracheilinus rennyae* GR Allen, MV Erdmann & NLA Yusmalinda, 2013

Renny's Catfish *Clarias rennyae* BW Low, HH Ng & HH Tan, 2022

Dr Renny Kumia Hadiaty (1960–2019) was an Indonesian ichthyologist (see **Hadiaty**).

Retout

Red-tipped Grouper *Epinephelus retouti* P Bleeker, 1868

Mr. Retout (no other names given) of Mauritius contributed to the natural history collections made by François Pollen (q.v.) for the Leiden Museum.

Retz

Retz's Pipefish *Microphis retzii* P Bleeker, 1856

[Alt. Ragged-tailed Pipefish]

Anders Adolph Retzius (1796–1860) was Professor in Anatomy and Physiology at the Karolinska Institute, Stockholm. Bleeker gave no first name, but he did indicate that he honoured Retzius for discovering that male syngnathids incubate the eggs, not the females (although Retzius never claimed it was his discovery). He did, however, write about the gills of different species of the Syngnathidae (1835) so definitely studied them. We have not been able to absolutely confirm which Retzius Bleeker intended, so there is a possibility that it was named after his father, Anders Jåhan Retzius (1742–1821), a professor of natural history who also sometimes published on fishes, although he is best known for his work on flora.

Retzer

Retzer's Tube-snouted Ghost Knifefish *Sternarchorhynchus retzeri* CD de Santana & RP Vari, 2010

Dr Michael 'Mike' E Retzer is an American ichthyologist who is Curator of Fishes, Illinois Natural History Survey, Champaign. He is also (2017) on the staff of Parkland College, Illinois. He co-wrote: *Two new species of* Acestridium *(Siluriformes; Loricariidae) from southern Venezuela, with observations on camouflage and color change* (1999).

Revelle

Bearded Dottyback *Pseudoplesiops revellei* LP Schultz, 1953

Commander Roger Randall Dougan Revelle (1909–1991) of the US Naval Reserve was in charge of oceanographic studies during Operation Crossroads (the post-war atomic tests) and the Bikini Scientific Re-survey. He graduated in geology from Pomona College (1929) and went on to complete a PhD on oceanography awarded by the University of California, Berkeley (1936). He worked at the Scripps Institute of Oceanography with a break serving as an oceanographer in the navy in WW2, going on to become Director of the Institution (1950–1964). (The Institution's latest research vessel was named after him.) He was one of the first scientists (1956) to study and raise concerns about global warming. He was a Science Advisor during the Kennedy Administration and was President of the American Association for the Advancement of Science (1974).

Rex

Sulawesi Goby sp. *Mugilogobius rexi* HK Larson, 2001

Rex Williams is a technical officer at the Museum and Art Gallery of the Northern Territory in Darwin, Australia. He was honoured for his: "*...careful work and commitment*" to the museum's fish collection.

Rey

Suckerlip Blenny *Andamia reyi* H-E Sauvage, 1880

Dr Paul Rey (b.1849) was a French physician who qualified in Paris (1879). He wrote a number of papers such as (translation) *Historical and Critical Study on the Toxicity of Copper* (1879). He collected the blenny type in Luzon (Philippines) with Dr Joseph Montano. A paper was written about that voyage in the Bulletin de la *Société académique indochinoise*: *Voyage du Docteur Montano et du Docteur Paul Rey a Lucon, a Jolo, a la cote N. de Borneo et a Mindanao* (1882).

Reyes

Three-barbeled Catfish sp. *Pimelodella reyesi* G Dahl, 1964

Dr Hernando Reyes Duarte (1926–2012) trained as a lawyer, but became better known as an author. He was Executive Director, Corporación Autónoma Regional para los Valles del Magdalena, Sinú y San Jorge (Colombia), the regional environmental authority that sponsored Dahl's research and published his report.

Reygel

Spiny Eel sp. *Mastacembelus reygeli* EJ Vreven & J Snoeks, 2009

Alain Marie Robert Godelieve Reygel (b.1956) is a Belgian artist and scientific illustrator in the Department of Biology at Tervuren (MRAC) (1979–present). He studied Publicity & Design at the Art Academy of Leuven. He is also Collection Manager of Ornithology there (2012–present) following the retirement of Dr Michel Louette, being a passionate birdwatcher and amateur ornithologist.

Reynald

Reynald's Grenadier Anchovy *Coilia reynaldi* A Valenciennes, 1848

Valenciennes did not explain who Reynald was. Despite the spelling, the binomial may refer to Auguste Adolphe Marc Reynaud (1804–1872), a French naval surgeon, naturalist and explorer on board 'La Chevrette' (1827–1828). He was later Chief Inspector of Medical Services (1858–1872). He collected in Burma (Myanmar) and Madagascar. Two birds are named after him. The holotype was collected in Rangoon while the vessel was there.

Reynolds, JN

Reynolds' Cory *Corydoras reynoldsi* GS Myers & SH Weitzman, 1960

Colonel John N Reynolds (1912–1987) of the U.S. Air Force helped to collect the holotype. He was described in a Florida law report (1958) as the Base Commander, Elgin Air Force Base, Florida. The etymology describes him as: "*...an ardent aquarist and a fine fish collector.*"

Reynolds, LC

Reynolds' Anglerfish *Echinophryne reynoldsi* TW Pietsch III & RH Kuiter, 1984
[Alt. Sponge Anglerfish]

Lewis C 'Lou' Reynolds of Melbourne collected the holotype (1983) and a paratype in southern Australia. We have no further details of him.

Rham

Armoured Catfish sp. *Rhadinoloricaria rhami* IJH Isbrücker & H Nijssen, 1983

Dr Patrick Henri de Rham (b.1936) is a Swiss zoologist, botanist, ichthyologist and aquarist in Lausanne, and was Expert Ecologist to the Coopération Technique Suisse. He is a Director of the Aquatic Conservation Network. (See also **Patrick** and **De Rham**)

Rharhabe

Triplefin sp. *Helcogramma rharhabe* W Holleman, 2007

Rharhabe (c.1715–c.1782) was the eldest son of Phalo, paramount chief of amaXhosa, a Bantu ethnic group in the Eastern Cape of South Africa (where this blenny occurs). ca.1750, Rharhabe and his father quelled an uprising by Rharhabe's half-brother Gcaleka, and subsequently led a break-away group which Rharhabe ruled as paramount chief (1775–1787).

Rheinhart

Bitterling sp. *Rhodeus rheinardtii* G Tirant, 1883

Lieutenant-Colonel Pierre-Paul Rheinart (1840–1902) was an officer in the French army and an administrator in Vietnam. He was Chargé d'Affaires in Annam (1879–1889) and Résident-Générale in Annam-Tonkin (1889). He also explored in Laos (1859) and sent the first specimen of an eponymous pheasant, *Rheinardia*, to Paris. Both the pheasant's genus and the bitterling's binomial manage to misspell Rheinart's name with incorrect use of a *d*. The patronym is not explained in Tirant's text, but, as the type locality for the bitterling is where Rheinart was in charge at the right period, and he had a history of sending zoological specimens to Paris, we are sure this is who it is named after.

Rhen

Mormyrid sp. *Marcusenius rheni* HW Fowler, 1936

[Perhaps a junior synonym of *M. victoriae*]

James Abram Garfield Rehn (1881–1965) was an American entomologist who served as zoologist on the George Vanderbilt African Expedition (1934). Fowler mis-spelt his surname in the binomial – quite an achievement for a 4-letter name!

Rhoades

Lake Malawi Cichlid sp. *Buccochromis rhoadesii* GA Boulenger, 1908

Lake Malawi Cichlid sp. *Chilotilapia rhoadesii* GA Boulenger, 1908

Captain Edmund L Rhoades was commander of the British anti-slavery gunboat 'Gwendolen' (aka 'Gwen') on Lake Nyasa (Malawi). When news arrived that war was declared, the gunboat was commandeered and ordered to sink the only German boat on the lake, 'Hermann von Wissman'. It was still hauled up but Rhoades managed to destroy it. Its captain rowed out to them, furious and thinking Rhoades had gone mad, as news of the war had not yet reached him. Rhoades presented a large collection of well-preserved fishes from Lake Malawi, including types of the cichlids, to the British Museum, with sketches of their coloration in life made by Rhoades himself.

Rhoads

Characin genus *Rhoadsia* HW Fowler, 1911

Samuel Nicholson Rhoads (1862–1952) was an American vertebrate taxonomist. He was interested in all areas of zoology including mammalogy, ornithology, malacology, etc. He collected widely in the USA (it was he who identified the Wood Bison as a subspecies of the American Bison) and also in Ecuador, donating a substantial part of his collections to the Academy of Natural Sciences of Philadelphia. He wrote a number of scientific papers such as: *Geographic Variation in Bassariscus astutus, with Description of a New Subspecies* (1893) as well as monographs such as *The Mammals of Pennsylvania and New Jersey* (1903). He suffered a mental breakdown (1926) and spent the rest of his life in sanitaria.

Rhodion

Riverine Goby *Ponticola rhodioni* ED Vasil'eva & VP Vasil'ev, 1994

Officer Rhodion Denisovich Medvedev was the senior author's brother. He died in Abkhazia (a partially recognised republic in the Caucasus, claiming independence from Georgia) on 23 April 1994.

Ribeira

Characin sp. *Astyanax ribeirae* CH Eigenmann, 1911

Pencil Catfish sp. *Microcambeva ribeirae* WJEM Costa, SMQ Lima & CRSF Bizerril, 2004

Brazilian Cichlid sp. *Australoheros ribeirae* FP Ottoni, OT Oyakawa & WJEM Costa, 2008

In the case of the catfish and cichlid, this is a toponym referring to the Rio Ribeira de Iguape, a river basin in Brazil, which is the type locality. The same is probably true for the *Astyanax*, but as the etymology is not explained it could possibly be intended to honour ichthyologist-herpetologist Alípio de Miranda Ribeiro (below), whose work was often cited in Eigenmann's monograph.

Ribeiro, A

Characin sp. *Mixobrycon ribeiroi* CH Eigenmann, 1907
Driftwood Catfish sp. *Glanidium ribeiroi* JD Haseman, 1911
Ribeiro's Searobin *Bellator ribeiroi* GC Miller, 1965

Dr Alipio de Miranda-Ribeiro (1874–1939) was a Brazilian ichthyologist and herpetologist. Miller's etymology says that Miranda-Ribeiro "*...contributed greatly to the knowledge of the marine fauna of South America.*" (See **Alipio**)

Ribeiro, P

Clinid genus *Ribeiroclinus* SY Pinto, 1965

Paulo de Miranda-Ribeiro (1901–1965) was a Brazilian zoologist and the son of Alipio (above). Like his father, he worked at the Museu Nacional in Rio de Janeiro.

Ribeiro, SH

Painted Catfish *Pseudolaguvia ribeiroi* SL Hora, 1921

Sydney H Ribeiro was an Entomological Assistant, Zoological Survey of India, based at the India Museum, Calcutta (Kolkata). He co-wrote: *On a collection of ants (Formicidae) from the Andaman Islands* (1925). He collected the holotype of this species.

Ribera

Suckermouth Catfish sp. *Astroblepus riberae* L Cardona Pascual & G Guerao Serra, 1994

Carles Ribera Almerje is an arachnologist who specialises in cavernicolous spiders. He works at the Department of Biology, University of Barcelona, where he is an Associate Professor. He co-wrote: *Description of three new troglobiontic species of Cybaeodes (Araneae, Liocranidae) endemic to the Iberian Peninsula* (2015). He collected the holotype of this species from Ninabamba caves in Peru.

Ricardo

Ricardo's Squeaker *Synodontis ricardoae* L Seegers, 1996

Cicely Kate Bertram née Ricardo (1912–1999) was a British ichthyologist who, together with Ms R J Owen, collected in the Lake Rukwa drainage. She was part of the 'Cambridge School' of biologists and contributed two important reports on East African freshwater fish: *Report on the Fish and Fisheries of Lake Rukwa in Tanganyika Territory and the Bangweulu Region in Northern Rhodesia* (1939) and *Report on the Fish and Fisheries of Lake Nyasa* (1942). She became only the second ever President of Lucy Cavendish College, University of Cambridge (1970–1979). She was also a JP for twenty years.

Rice

Spoonhead Sculpin *Cottus ricei* EW Nelson, 1876

F L Rice of Evanston, Illinois, found the holotype on the shores of Lake Michigan and passed it to the author for identification. No further details are given about him.

Richard, J

Richard's Whiptail *Sphagemacrurus richardi* MCW Weber, 1913

Dr Jules Richard (1863–1945) was Director, Monaco Oceanographic Museum (1900–1945). He developed a variety of instruments for taking water samples and published a series of papers on them. The author cites these in his report (1902) on the 'Siboga' expedition. This is the main reason for thinking this species is named after Jules Richard as the original text does not help. He wrote: *L'océanographie* (1907).

Richard (Winterbottom)

Rick's Dwarf-goby *Eviota richardi* DW Greenfield & JE Randall, 2016

Dr Richard Winterbottom (See **Winterbottom**)

Richards

Armoured Searobin sp. *Peristedion richardsi* T Kawai, 2016

Dr William J 'Bill' Richards is a zoologist who was the Director (1977) of the Miami Laboratory of the National Marine Fisheries Service, Southeast Fisheries Science Center (NOAA), Florida. In the paper on the searobin, Toshio Kawai says: "*I sincerely thank William J. Richards for his critical reading of a draft manuscript and valuable suggestions*" and honoured him "*…for his numerous contributions to ichthyology.*" He wrote on fisheries, fish taxonomy and systematics, particularly sea robins, including: *Stemonosudis rothschildi, a new paralepidid fish from the central Pacific* (1967) and *Preliminary guide to the identification of the early life history stages of scombroid fishes of the western central Atlantic* (1989).

Richardson, J

Shiner genus *Richardsonius* CF Girard, 1856
Waspfish genus *Richardsonichthys* JLB Smith, 1958
Common Snowtrout *Schizothorax richardsonii* JE Gray, 1832
Pelagic Armourhead *Pentaceros richardsoni* A Smith, 1844
South African Mullet *Chelon richardsonii* A Smith, 1846
Richardson's Snaggletooth *Astronesthes richardsoni* F Poey, 1852
Richardson's Moray *Gymnothorax richardsonii* P Bleeker, 1852
Tiger Flathead *Platycephalus richardsoni* FL Castelnau, 1872
Grunt Sculpin *Rhamphocottus richardsonii* A Günther, 1874
Dead Sea Toothcarp *Aphanius richardsoni* GA Boulenger, 1907
[Syn. *Aphanius dispar richardsoni*]
Lanternfish sp. *Diaphus richardsoni* AV Tåning, 1932
Western Brook Lamprey *Lampetra richardsoni* VD Vladykov & WI Follett, 1965
Richardson's Sardinella *Sardinella richardsoni* T Wongratana, 1983

Robust Icefish *Channichthys richardsoni* GA Shandikov, 2011

Sir John Richardson (1787–1865) was a Scottish naval surgeon, naturalist and Arctic explorer. He was a friend of Sir John Franklin, to whom he was also related by marriage, and took part in Franklin's expeditions (1819–1822 and 1825–1827). He also participated (1847) in the vain search for Franklin and his colleagues. The Richardson Mountains in Canada are also named after him, as are eight amphibians, five mammals, four reptiles and a dragonfly. He wrote accounts dealing with the natural history, and especially the ichthyology, of several Arctic voyages, and was the author of *Icones Piscium* (1843), *Catalogue of Apodal Fish in the British Museum* (1856), and the second edition of Yarrell's *History of British Fishes* (1860). There is no etymology given for the armourhead or the mullet species, so it is possible that Smith had another Richardson in mind.

Richardson, LR

Richardson's Skate *Bathyraja richardsoni* JAF Garrick, 1961

Professor Laurence 'Larry' Robert Richardson (1911–1988) was a British zoologist. His master's degree and doctorate were awarded by McGill University, Montreal, Canada. He was working in New Zealand after WWII and became a professor in the Department of Zoology, Victoria University of Wellington. He fell out with his colleague and was reprimanded. In a fit of anger, he resigned (1964) and when he tried to withdraw his resignation, found that he could not. He left New Zealand and settled in Australia. He became a Fellow of the Royal Society of New Zealand (1959). He was honoured in the skate's name: "*...for his extensive contribution to deep water research in New Zealand, and especially in Cook Strait where the type specimen was taken.*" He is also honoured in the cephalopod name *Megalocranchia richardsoni*.

Richardson, RE

Richardson's Reef-Damsel *Pomachromis richardsoni* JO Snyder, 1909

Robert Earl Richardson (1877–1935) was an American ichthyologist who was part of the Illinois Natural History Survey. He co-wrote, with the head of the survey, the entomologist Stephen Alfred Forbes (q.v.): *On a new Shovelnose Sturgeon from the Mississippi* River (1905) and *Fishes of Illinois* (1908). With David Starr Jordan he co-wrote: *Check-list of the Species of Fishes known from the Philippine Archipelago* (1910).

Riche

Prickly Toadfish *Contusus richei* CPP Fréminville, 1813

Dr Claude Antoine Gaspard Riche (1762–1797) was one of the naturalists attached to the expedition of General Bruni d'Entrecasteaux which went in search of the lost ships of. Jean-François de Galaup, Comte de La Pérouse. He studied medicine in Montpelier and received his doctorate (1787) on the geology and botany of the mountains of Languedoc. He moved to Paris (1788) and wrote on anatomy. He was elected to the Société d'Histoire Naturelle (1791) and the same year joined the expedition (probably because he thought it would help his TB), which went to Tenerife, the Cape of Good Hope, Tasmania, New Ireland, the Admiralty Islands and Western Australia. The expedition split along royalist or republican lines, having heard of the execution of Louis XVI. Riche's main responsibilities were the study of minerals, birds and invertebrates, but he also assumed responsibility for meteorological observations and chemical studies. However, he lost his collections when the expedition disintegrated in the Dutch East Indies. He returned to France (1796) but died less than a year later. Cape Riche in Australia is also named after him.

Richer

Richer's Dragonet *Synchiropus richeri* R Fricke, 2000
Velvetfish sp. *Cocotropus richeri* R Fricke, 2004

Dr Bertrand Richer de Forges (b.1948) is a French marine biologist. He is Professor Emeritus at the National University of Singapore where he was an associate professor (2012–2013). He was formerly Research Director at the Institute of Research for Development (ORSTOM) (2002–2008). He holds two doctorates from the Paris VI University (1976) and the MNHN Paris (1998). He is known as a taxonomist of deep-sea crabs, and author of several books on coral reef fauna. He has been involved in research for 40 years, five of which were spent at sea, sampling the deep-sea benthos of the Indo-Pacific.

Richerson

Titicaca Pupfish sp. *Orestias richersoni* LR Parenti, 1984

Dr Peter James 'Pete' Richerson (b.1943) started his professional career as a limnologist but went on to become an authority on how culture affects human evolution and to co-write the seminal books: *Culture and the Evolutionary Process* (1985) and *Not By Genes Alone: How Culture Transformed Human Evolution* (2005). His BSc in entomology was awarded by UC Davis (1965), as was his PhD in zoology (1969). He was Professor of Environmental Science (1971–2006) at the University of California, Davis, Department of Environmental Science and Policy and remains professor emeritus there. He is also a Visiting Professor at the Institute of Archaeology at University College London. He is currently President Elect of the Evolutionary Anthropology Society and past president of the Human Behavior and Evolution Society and the Society for Human Ecology. He was honoured in the fish's binomial because he had made a limnological study of Lake Titicaca while researching in collaboration with the Instituto del Mar del Perú (1981).

Richmond

Richmond's Wrasse *Halichoeres richmondi* HW Fowler & BA Bean, 1928

Charles Wallace Richmond (1868–1932) was an American ornithologist who specialised in being a nomenclaturist and bibliographer. He was a colleague of Ridgway (see below) at the USNM. He is best known for *The Richmond Index to the Genera and Species of Birds*; his life's work lasting 40 years (started 1889) and culminating in 70,000 file cards. A mammal, an amphibian and thirteen birds are named after him.

Richmond (River)

Australian Freshwater Herring *Potamalosa richmondia* WJ Macleay, 1879

This is a toponym referring to Richmond River, New South Wales, Australia, the type locality.

Ridgway

Halosaur sp. *Halosaurus ridgwayi* HW Fowler, 1934

Robert Ridgway (1850–1929) was an American ornithologist. At just seventeen he was appointed zoologist on a geological survey of the 40th parallel. He was Curator of Birds at the United States National Museum (USNM) (1880–1929) and Founder President of the American Ornithologists' Union. He was also a fine illustrator, famed for having sketched and collected birds around his home in Richland County, Illinois. Because he encountered an almost infinite number of colours, and needed accuracy for a scientific description, he realised that this would only be possible through some form of standardisation. He therefore proposed a colour system, which was published (1912) under the title *Colour Standards and Nomenclature.* He developed an 18–acre area near his home as a bird sanctuary, which he called Bird Haven. Although he lacked a formal post-secondary education, Ridgway received an honorary master's degree in science from Indiana University (1884). This was partly in gratitude for his supplying them with bird specimens after their museum burned down. Fowler honoured him in the halosaur's name "*…with pleasant memories of by gone days in his department of ornithology in the Smithsonian Institution.*" Thirty-nine birds are named after him.

Rieffel

Spotted Armoured-Gurnard *Satyrichthys rieffeli* JJ von Kaup, 1859

Franz Joseph Friedrich Thaddäus Freiherr von Rieffel (1815–1858) was a government official and politician in Darmstadt, in the Grand Duchy of Hesse, responsible for public finances. Johann Kaup says in his paper "I have named this very interesting species in honour of the memory of my true and excellent friend Dr Rieffel, who has done so much for our Museum and University." Despite Kaup's effusive comments, it has proved difficult to identify Rieffel, but we believe him to be the above: he had died shortly before Kaup's description, and Kaup lived and worked in Darmstadt. The university mentioned by Kaup is probably the Justus-Liebig-Universität Gießen, and the museum the Natural History Cabinet of Ludwig III, now Hessisches Landesmuseum Darmstadt.

Riekert

Toadfish genus *Riekertia* JLB Smith, 1952

Dr C Riekert of Bizana, Eastern Cape, South Africa, sent 'valuable specimens' to Smith. We have not been able to find more about him.

Riera

Slender Electric Ray *Narcine rierai* D Lloris & JA Rucabado, 1991

Ignacio Riera Julia was Chief of the Spanish Fisheries Office and adviser to the General Direction of International Fisheries Relations, Ministry of Agriculture and Fisheries, Mahé, Seychelles (1989). He was a good friend of the senior author Lloris.

Riese

Ruby Tetra *Axelrodia riesei* J Géry, 1966

William Riese was a tropical fish exporter who helped collect the type (1964) with Axelrod (q.v.) after whom the genus is named. Thus the scientific name may uniquely honour two collectors in its two parts.

Rift

Roughy sp. *Hoplostethus rifti* AN Kotlyar, 1986

''Rift' was the USSR Academy of Sciences Institute of Oceanology scientific research vessel from which the holotype was collected.

Riggenbach

African Rivuline sp. *Aphyosemion riggenbachi* E Ahl, 1924
Ctenopoma sp. *Ctenopoma riggenbachi* E Ahl, 1927
[Junior Synonym of *Ctenopoma petherici*]

Fritz Wilhelm Riggenbach (1864–1944) was a Swiss zoologist and collector. He travelled to Morocco a number of times (1888–1893 & 1894–1909), to New Guinea (1910) and to Senegal and Cameroon while employed on several of the German Central African Expeditions (1902–1911), which were led by Adolf Friedrich Duke of Mecklenburg. He then worked for a US insurance company (1911–1933) and became an art dealer. A mammal, a reptile, an amphibian and eight birds are also named after him.

Riise

Swordtail Characin *Corynopoma riisei* TN Gill, 1858

Albert Heinrich Riise (1810–1882) was a Danish pharmacist, botanist, and collector who supplied material to the Zoologisk Museum, Copenhagen. He was an apprentice pharmacist (1824–1830), then moved to Copenhagen where he graduated (1832) and worked (1832–1838). The Danish King appointed him to be a pharmacist in the Danish West Indies (now U.S. Virgin Islands) with an exclusive license to open a retail shop, 'A H Riise' in St. Thomas, which is still in business. Riise's interest in botany led him to utilize Caribbean exotic plants for the preparation of pharmaceutical, cosmetic and alcoholic products – especially rum. There were epidemics of cholera, smallpox, and yellow fever in St. Thomas (1868), so Riise returned to Denmark with his family. A bird and a reptile are also named after him.

Ringuelet

Eelpout sp. *Piedrabuenia ringueleti* AE Gosztonyi, 1977
Armoured Catfish sp. *Hisonotus ringueleti* AE Aquino, SA Schaefer & AM Miquelarena, 2001

Dr Raúl Adolfo Ringuelet (1914–1982) was an Argentine zoologist, ichthyologist and arachnologist. He graduated with a doctorate (1939) at the Museum of Natural Sciences of La Plata, Buenos Aires, where he was a Professor (in many categories of zoology) (1944–1978), retiring as Extraordinary Professor Emeritus (1980). He was also Professor of Systematic Zoology, Faculty of Natural Sciences, University of Buenos Aires (1956–1964). He wrote: *Los Peces de Agua Dulce de la República Argentina* (1967) which, according to the catfish etymology: "*...set the standard for systematics research conducted during the last decades of the 20th century in the Austral region of the Neotropics.*"

Rioja

Toluca Silverside *Chirostoma riojai* A Solórzano Preciado & Y López, 1966

Dr Enrique Rioja Lo Blanco (1895–1963) was a Spanish biologist who left Spain to exile in Mexico following the civil war (1939). The University of Madrid awarded his first degree (1915) and his PhD (1916). He took up research at Madrid's Museum of natural Science but supported himself teaching in middle schools. He was a professor of natural science at the School of Industrial Engineers (1937–1939). An ardent republican, he left for France as soon as it was clear that Franco would win, and soon after left Europe for Mexico. From his arrival until his death, he was linked to the Mexican Institute of Biology, although only later (1956) became full-time professor. He was honoured for his contributions to Mexican hydrobiology.

Ripley

Hill-stream Loach sp. *Homalopterula ripleyi* HW Fowler, 1940

Dr Sidney Dillon Ripley II (1913–2001) was an eminent American ornithologist who was awarded the Presidential Medal of Freedom, the highest civilian honour of the United States. He was privately educated and travelled widely in both Europe and India. He graduated in history at Yale, but then studied zoology at Columbia University where he specialised in ornithology. He joined an expedition to New Guinea (1936), where he spent 18 months collecting bird specimens. Always a bird lover, at 17 he built a pond to attract waterfowl. His service during WW2 was with the Office of Strategic Services, coordinating United States and British intelligence efforts in Southeast Asia. After the war he taught at Yale, and was then Director of the Peabody Museum of Natural History, until becoming Secretary the USNM (1964–1984). On retirement, he was reported to have said: '*I shall enjoy my freedom from the tyranny of the In and Out boxes*'. He travelled 'around the world from Patagonia to Pakistan'. Ripley was instrumental in the founding of the Charles Darwin Foundation for the Galápagos Islands (1959). He is recognised as one of the giants of Indian ornithology. Among his many publications, he co-wrote, with Salim Ali the 10-volume: *Handbook of the Birds of India* (1968–1975) and (as sole author): *Rails of the World.* He acted as a field representative for the Academy of Natural Sciences of Philadelphia during the Sumatran expedition that collected the loach type.

Ripon

Lake Victoria Cichlid sp. *Haplochromis riponianus* GA Boulenger, 1911

A toponym, referring to the type locality: Ripon Falls at the northern end of Lake Victoria. The falls are named after George Robinson, 1st Marquess of Ripon (1827–1909).

Risdawati

Goby sp. *Schismatogobius risdawatiae* P Keith, H Darhuddin, T Sukmono & N Hubert, 2017

Dr Renny Risdawati (b.1967) is a biologist at Padang University, Indonesia, where she is a lecturer. She helped the authors collect freshwater fishes in Padang.

Risso

Risso's Dragonet *Callionymus risso* CA Lesueur, 1814
Risso's Lanternfish *Electrona risso* A Cocco, 1829
Spotted Barracudina *Arctozenus risso* CLJL Bonaparte, 1840
Smallmouth Spiny Eel *Polyacanthonotus rissoanus* F De Filippi & JB Vérany, 1857

Professor Giovanni Antonio Risso (1777–1845) (aka Joseph Antoine Risso) was an Italian-French naturalist and the first to describe deep-sea fishes as distinctive fauna and describe depth zonation with different species living in discrete depth strata. He was (until 1826) an apothecary in his native town of Nice (then in Italy). He later became Professor of Botany and Chemistry at the University of Nizza (Nice). He had an outstanding knowledge of ichthyology and published *Ichthyologie de Nice* (1810). Other works included: *Histoire naturelle de l'Europe méridionale* (1826) and *Histoire Naturelle des Orangers* (1818–1822). Risso's Dolphin is also named after him, as is a genus of marine gastropods, *Rissoa*.

Rita

Darter Characin sp. *Klausewitzia ritae* J Géry, 1965

Rita Klausewitz née Willmann (d.1995) was the wife of ichthyologist Dr Wolfgang Klausewitz (q.v.).

Ritter

Whitefin Dogfish *Centroscyllium ritteri* DS Jordan & HW Fowler, 1903
Ritter's Turbot *Pleuronichthys ritteri* EC Starks & EL Morris, 1907
[Alt. Spotted Turbot]
Broadfin Lampfish *Nannobrachium ritteri* CH Gilbert, 1915
Stripefin Poacher *Xeneretmus ritteri* CH Gilbert, 1915

Dr William Emerson Ritter (1856–1944) was an American marine biologist and tunicatologist. He financed his own education by taking a series of teaching jobs and eventually obtained a bachelor's degree from the University of California, Berkeley (1888). He was then able to move to Harvard for his master's degree (1891) and doctorate (1893). He worked at the University of California, Berkeley, (1891–1924) being a full Professor (1902) and retiring as Emeritus Professor. He was a member of the Harriman Alaska Expedition (1899). He met the newspaper proprietor, E W Scripps, (1903) and their friendship and partnership led to what is now the Scripps Institute of Oceanography, of which Ritter became the first Director until retiring (1922). He wrote: *Organization in scientific research* (1905). He was honoured for his work on the tunicates and enteropneusta (acorn worms) of the Pacific Ocean.

Ritva

Spineless Eel sp. *Chaudhuria ritvae* R Britz, 2010

Ritva Roesler (b.1970) is a Finnish biologist, artist and scientific illustrator who is based in London. She has a master's degree in biology awarded by Tübingen University (1998.)

Ralph Britz wrote in his etymology: "*Named after my wife Ritva Roesler, who helped collect the species, honouring her continuing support of my work on Myanmar freshwater fishes.*"

Rivas, LA

Characin genus *Rivasella* A Fernández-Yépez, 1948
[Now in *Steindachnerina*]
Thorny Catfish sp. *Pterodoras rivasi* A Fernández-Yépez, 1950

Luis A Rivas L collected fishes with the author in the Orinoco basin, but we have no further details.

Rivas, LR

Spotjaw Blenny *Acanthemblemaria rivasi* JS Stephens, 1970
Rivas' Limia *Limia rivasi* LR Franz & GH Burgess, 1983

Dr Luis René Rivas y Díaz (1916–1986) was an ichthyologist and marine biologist. He was an Associate Professor, and Curator of Fishes, Department of Zoology, University of Miami. He was a specialist in the marine fishes of Cuba, the USA and Canada. As a friend and associate of Ernest Hemingway, he served as technical advisor for the motion picture production of: *The Old Man and the Sea*. He wrote: *Check list of the Florida game and commercial marine fishes including those of the Gulf of Mexico and the West Indies, with approved common names* (1949). In the limia etymology, he was honoured: "*…in recognition of his long standing interest in the systematics of poeciliid fishes of the Greater Antilles.*"

Rivas, TS

Suckermouth Catfish sp. *Astroblepus rivasae* CA Ardila Rodriguez, 2018

Tulia Sofia Rivas Lara is a Colombian biologist at the Universidad Tecnológica del Chocó 'Diego Luis Cordoba'. She was honoured for having "contributed greatly" (translation) to the knowledge of the fishes of Chocó, Colombia (where this catfish occurs).

Rivaton

Lanternfish sp. *Diaphus rivatoni* P Bourret, 1985
New Caledonian Longtail Dragonet *Callionymus rivatoni* R Fricke, 1993

Jacques Rivaton (1921–2009) was a French zoologist who worked at the marine biology laboratory in Noumea, New Caledonia, for many years; first for ORSTOM and then after it had been metamorphosed, for IRD. He co-wrote: *Pisces, Pleuronectiformes: flatfishes from the waters around New Caledonia – A revision of the genus Engyprosopon* (1993).

River

Rivero's Catshark *Apristurus riveri* HB Bigelow & WC Schroeder, 1944
[Alt Broadgill Catshark]

Dr Luis Hugo Howell-Rivero (1899–1986) (See **Rivero**)

Riverend

Caribbean Pupfish sp. *Cyprinodon riverendi* F Poey, 1860

Luis Le Riverend is credited with finding the fish, and the author comments that Riverend has extended knowledge of Cuban natural history by his collections "*...and by the use of his aquariums*" (translation). It is tempting to speculate that he was a relative of Dr Julio Jacinto Le Riverend (1794–1864), who taught hygiene at the Royal Academy of Sciences, Havana, where the author, Poey, taught zoology.

Rivero

Oriental Cyprinid sp. *Paralaubuca riveroi* HW Fowler, 1935

Dr Luis Hugo Howell-Rivero (1899–1986) was a Cuban biologist and anthropologist at the University of Havana. The Instituto de Segunda Enseñanza, Havana, awarded his bachelor's degree (1925) and the University of Havana his doctorate (1930). He researched West Indian fishes at Harvard's Museum of Comparative Zoology (1934–1935). He wrote: *Some new, rare, and little known fishes from Cuba* (1936). Fowler mentions that he was indebted for collections of Cuban fishes made by Rivero (although this Asian fish has nothing to do with Cuba!). (Also see **River** above)

Rivers Anderson

Armoured Searobin sp. *Peristedion riversandersoni* AW Alcock, 1894

Viviparous Brotula sp. *Diplacanthopoma riversandersoni* AW Alcock, 1895

Lieutenant-Colonel Adam Rivers Steele Anderson (1863–1924) was a naturalist and physician whose bachelor's degree was awarded by Cambridge (1882). He acquired his medical qualifications at St Mary's Hospital, London (1889) and he worked there as a resident medical officer (1886–1889). He served in the Indian Medical Service (1889–1918), and was seconded to the Indian Marine Survey (1893–1900) as surgeon-naturalist on board the survey ship 'Investigator'. He was senior medical officer at Port Blair penal settlement, Andaman Islands (1900–1906). He transferred to civilian service in Bengal (1906) and became civil surgeon at Dakka (1912–1916) returning to military duties (1916–1918).

Rivoli

Longfin Yellowtail *Seriola rivoliana* A Valenciennes, 1833

François Victor Masséna Prince d'Essling, 2nd Duc de Rivoli (1799–1863), was the son of one of Napoleon's marshals and an amateur ornithologist. He amassed a huge collection of hummingbirds. According to Valenciennes' text, the Duke of Rivoli gave a specimen of this fish to the 'Cabinet du Roi' (The King's Cabinet being a collection of works of art and 'curiosities' of natural history).

Ro

Whale Catfish sp. *Cetopsidium roae* RP Vari, CJ Ferraris Jr & MCC de Pinna, 2005

Dr Rosemary Helen Lowe-McConnell (1921–2014) was a British biologist and ichthyologist. The University of Liverpool awarded her bachelor's and master's degrees and her doctorate.

She wrote: *Recent research in the African great lakes: Fisheries, biodiversity and cichild evolution* (2003). 'Ro' was her nickname.

Robb

Squeaker Catfish sp. *Synodontis robbianus* JA Smith, 1875

Rev. Alexander Robb (1824–1901) was a Scottish Presbyterian missionary. He was sent from the mission in Jamaica to Calabar, Nigeria (1856), where he served until 1875 when his health deteriorated and he returned to Jamaica and finally moved to Philip Island, Victoria, Australia (1888). He wrote: *The Heathen World and the Duty of the Church* (1863). He provided specimens from the "*Old Calavar district of tropical Africa*", including the type of this catfish.

Robbers

Silverside sp. *Atherinella robbersi* HW Fowler, 1950

Raymond J Robbers collected the original specimen of the silverside species. He was Project Manager for a construction company (Compañia Constructora de Carreteras) at Totumo, Bolivar State, Colombia. He invited Henry Fowler to collect at the large shallow lake there and showed him around the area.

Robby

Striped Cardinalfish *Apogon robbyi* CR Gilbert & JC Tyler, 1997

Dr Charles Richard 'Dick' Robins (1928–2020) (See **Robins, CH & CR**)

Robert

Slender Halfbeak *Hyporhamphus roberti* A Valenciennes, 1846

Monsieur Robert is said in the original text to have been one of the collectors of the fish along with Monsieur Poiteau – probably the French botanist Pierre-Antoine Poiteau (1766–1854) – in Cayenne, French Guiana. It has been impossible to further identify Robert.

Robert

Glassy Perchlet sp. *Parambassis roberti* AK Datta & BL Chaudhuri, 1993

Dr Tyson Royal Roberts (q.v.). Despite the binomial being misspelled *roberti* in the original, it is sometimes corrected to *robertsi*. (See **Roberts, TR**)

Roberta

Roberta's Toothcarp *Valencia robertae* J Freyhof, H Kärst & M Geiger, 2014

Roberta Barbieri is an ichthyologist and fish biologist who is Research Associate at the Hellenic Centre for Marine Research, Athens. The etymology reads thus: "*The species is named for Roberta Barbieri (Athens), who studied the Greek* Valencia *species for many years and is engaged in the conservation of the two species.*" Among her publications are the co-written: *Freshwater larval fish from lake Trichonis (Greece)* (1994) and *Freshwater fishes and lampreys of Greece: An annotated checklist* (2015).

Roberts, CD

Filamentous Perchlet *Plectranthias robertsi* JE Randall & DF Hoese, 1995
Roberts' Gurnard *Pterygotrigla robertsi* L del Cerro & D Lloris, 1997
Eelpout sp. *Seleniolycus robertsi* PR Møller & AL Stewart, 2006
Roberts' Eucla Cod *Euclichthys robertsti* PR Last & JJ Pogonoski, 2020

Dr Clive Douglas Roberts (b. 1952) was born in England and received a BSc (Zoology) at Queens University of Belfast. He moved to Wellington, New Zealand (1980) to obtain his PhD at Victoria University, researching the biology and taxonomy of wreckfish. He became Curator of Fishes at the Museum of New Zealand Te Papa Tongarewa, holding that post until retirement (2018). He is the senior editor of: *The Fishes of New Zealand* (2015). (Also see **Clive R**)

Roberts, J

Characin sp. *Moenkhausia robertsi* J Géry, 1964

Jack Roberts was a dealer in, and breeder of, tropical fish. He owned Roberts' Fish Farm in Miami, Florida, in partnership with his brother, Frank. Jack collected the holotype. (See also **Jack Roberts**).

Roberts, JD

Bigmouth Skate *Amblyraja robertsi* PA Hulley, 1970

Dr J Douglas 'JD' Roberts (d.1982) was a British-born South African businessman who was Chairman of Murray & Roberts, South Africa's leading engineering, contracting and construction services company. In honouring him Hulley wrote that "*...by his kind generosity, made the study of the 'Walther Herwig' material in Hamburg possible.*" Roberts was generous in sponsoring research and providing funds so that important collections stayed in South Africa, such as the Harald Pager collection of San rock paintings.

Roberts, TR

West African Cichlid sp. *Limbochromis robertsi* DFE Thys van den Audenaerde & PV Loiselle, 1971
Squeaker Catfish sp. *Synodontis robertsi* M Poll, 1974
Roberts' Blenny *Omobranchus robertsi* VG Springer, 1981
Roberts' River Garfish *Zenarchopterus robertsi* BB Collette, 1982
Yellowfish sp. *Labeobarbus robertsi* KE Banister, 1984
Priapiumfish sp. *Neostethus robertsi* LR Parenti, 1989
Congo Cichlid sp. *Tylochromis robertsi* MLJ Stiassny, 1989
Stone Loach sp. *Schistura robertsi* M Kottelat, 1990
Roberts' Loach *Aperioptus robertsi* DJ Siebert, 1991
Glassy Perchlet sp. *Parambassis roberti* AK Datta & BL Chaudhuri, 1993
[Binomial sometimes corrected to *robertsi*]
Spiny Eel sp. *Mastacembelus robertsi* EJ Vreven & G Teugels, 1996
Halftooth Characin sp. *Argonectes robertsi* F Langeani-Neto, 1999
Needlefish sp. *Dermogenys robertsi* AD Meisner, 2001

Roberts' Sole *Leptachirus robertsi* JE Randall, 2007
Goby sp. *Glossogobius robertsi* DF Hoese & GR Allen, 2009
Garra sp. *Garra robertsi* RJ Thoni & RL Mayden, 2015

Dr Tyson Royal Roberts (b.1940) is an American naturalist and evolutionary biologist who has devoted most of his life to studying biodiversity, particularly that of fishes, throughout the tropics. Stanford University awarded his bachelor's degree (1961) and his doctorate (1968). Although for many years associated with the Department of Ichthyology, California Academy of Sciences, San Francisco he never held a position or an honorary appointment there. He was at the Museum of Comparative Zoology, Harvard University (1969–1975) as Associate Professor of Biology and Associate Curator of Fishes. While there he submitted a five-year plan for total renovation of the MCZ fish collection to the National Science Foundation that was implemented by others after he left. He has been actively involved in building research collections of tropical marine and freshwater fishes at several institutions. He is currently Research Associate of the Smithsonian Tropical Research Institute, Panama, and Advisor and Research Associate of the Institute for Molecular Biosciences, Mahidol University, Thailand. His publications include three faunal monographs: *Ichthyological Survey of the Fly River of Papua New Guinea* (1975); *Fishes of the Lower Rapids of the Congo River* (1976); and *Freshwater Fishes of Western Borneo (Kalimantan Barat, Indonesia)* (1989). He has described more than 30 genera and nearly 200 species of fish. He perhaps is best known for monographs on distribution of the freshwater fishes of Africa (1975) and on the oceanic oarfishes (*Regalecidae*) (2012), and for assessments of negative environmental impacts of hydropower dams in the Mekong basin. (See also **Tyson**)

Robertson, DR

Clipperton Rainbow Wrasse *Thalassoma robertsoni* GR Allen, 1995
Robertson's Coralbrotula *Ogilbia robertsoni* PR Møller, W Schwarzhans & JG Nielsen, 2005
Robertson's Blenny *Starksia robertsoni* CC Baldwin, BC Victor & CI Castillo, 2011

Dr David Ross Robertson (b.1946) is an ichthyologist whose bachelor's degree (1966) and doctorate (1974) were awarded by The University of Queensland, Australia. He worked as a staff scientist at the Smithsonian Tropical Research Institute, Panama. He co-wrote: *Defining and Dividing the Greater Caribbean: Insights from the Biogeography of Shorefishes* (2014). He collected the coral brotula holotype.

Robertson (F de Azevedo)

Armoured Catfish sp. *Hypostomus robertsoni* AC Dias & CH Zawadzki, 2021

Robertson Fonseca de Azevedo is a Brazilian environmental attorney, honoured for his efforts to preserve natural landscapes in Paraná State, Brazil. Robertson fought to prevent unnecessary small hydroelectric power plants in the high-gradient stretches along two main Upper Paraná River left tributaries, the Ivaí and Piquiri Rivers, where this catfish occurs.

Robertson, J

False Firemouth Cichlid *Cribroheros robertsoni* CT Regan, 1905
[Syn. *Cichlosoma robertsoni*]

Rev. John Robertson was a British priest and naturalist. His parish was Stann Creek in British Honduras (now Belize) where he collected botanical specimens (1883–1884) which he sent to the BMNH. He also collected amphibian and fish specimens (1890–1893) that he sent to both the BMNH and the Field Museum.

Robertson, JT

African Rivuline sp. *Fundulopanchax robertsoni* AC Radda & JJ Scheel, 1974

John T Robertson was an aquarist and manager of the CDC rubber plantation in Ekona, Cameroon. The description says: *Mr. J. T. ROBERTSON, manager of planting Cameroon Development Corporation in Ekona, Mr. ROBERTSON himself is an aquarist and a good connoisseur of the watercourses of western Cameroon, and he drew our attention to the occurrence of this species in the area mentioned* (translation).

Robin

Armoured Catfish sp. *Hypostomus robinii* A Valenciennes, 1840

Charles-Cesár Robin was a French writer, naturalist and explorer. He is known to have been in Louisiana at the time that it was sold by France to the USA (1803), but beyond what he published in his: *Voyages dans l'intérieur de la Louisiane, de la Floride Occidentale, and the Saint-Domingue de la Martinique et de Saint-Domingue. 1802–1806* (1807) little is known about him. He sent the holotype (a dried example) of this species from Trinidad. 'Monsieur Robin' is not fully identified in the original text, but the majority opinion is that Valenciennes was honouring the above.

Robine

Banner-tail Corydoras *Corydoras robineae* WE Burgess, 1983

Robine Schwartz was the mother of aquarium-fish collector and exporter Adolfo Schwartz, Turkys Aquarium, Manaus, Brazil. Adolfo supplied the holotype and specifically asked that it should be named after his mother. (Also see **Mrs Schwartz**)

Robins, CH & CR

False Moray genus *Robinsia* JE Böhlke & Smith DG, 1967
Goby genus *Robinsichthys* RS Birdsong, 1988
Roughlip Cardinalfish *Apogon robinsi* JE Böhlke & JE Randall, 1968
Twospot Flounder *Bothus robinsi* RW Topp & FH Hoff, 1972
Teacher Brotula *Calamopteryx robinsorum* DM Cohen, 1973
Goby sp. *Priolepis robinsi* J Garzón-Ferreira & P Acero 1991
Colonial Cusk-eel *Ophidion robinsi* MP Fahay, 1992
Rattail sp. *Trachonurus robinsi* T Iwamoto, 1997
Hagfish sp. *Myxine robinsorum* RL Wisner & CB McMillan, 1995
West Indian Lantern Shark *Etmopterus robinsi* PJ Schofield & GH Burgess, 1997
Moray sp. *Gymnothorax robinsi* EB Böhlke, 1997
Searobin sp. *Lepidotrigla robinsi* WJ Richards, 1997
Yellowbar Basslet *Lipogramma robinsi* RG Gilmore Jr, 1997
Cutthroat Eel sp. *Ilyophis robinsae* KJ Sulak & YN Shcherbachev, 1997

Spotfin Jawfish *Opistognathus robinsi* WF Smith-Vaniz 1997
Cutthroat Eel sp. *Atractodenchelys robinsorum* ES Karmovskaya, 2003
Cutthroat Eel sp. *Dysomma robinsorum* HC Ho & KA Tighe, 2018

Dr Charles Richard 'Dick' Robins (1928–2020) was an American systematic ichthyologist whom one etymology honoured as an 'eelologist'; as well as a colleague, advisor and friend, for his contributions to the knowledge of fishes (including anguilliforms), and for help and encouragement, both scientific and personal, over the past 37 years. He studied ornithology at Cornell University, gaining a bachelor's degree (1949) and it was at Cornell he became fascinated by ichthyology. He did not do a master's degree but instead went straight for a doctorate at Cornell (awarded (1954). He served in the US Army (1954–1956) and was assigned to Fort Detrick biological warfare laboratory. On being discharged he joined the Marine Laboratory of the University of Miami (1956) and did much to develop its ichthyology collection, becoming Professor, Marine Biology and Fisheries Division. He became Faculty Curator Emeritus, Division of Ichthyology, at the University of Kansas (1994). During his time at the university he undertook a number of overseas expeditions to Africa and the Caribbean. He wrote more than 200 papers and longer works including the co-written: *A Field Guide To Atlantic Coast Fishes Of North America* (1986).

The cutthroat eel species *Ilyophis robinsae* is named for Mrs Catherine Robins née Hale, wife (1965) of the above. The University of Tulsa awarded her zoology degree and the University of Florida her master's and then doctorate (1964). She has now become an award-winning sculptor. "*After 20 years researching eels I felt boxed in and isolated from far-flung colleagues,*" she says. "*I took a bronze-casting workshop as a respite and was instantly hooked.*" Not surprisingly, many of her sculptures are of animals.

Species with the binomial *robinsorum* are named after both Catherine and Charles.

Robinson, GS

Three-barbeled Catfish sp. *Pimelodella robinsoni* HW Fowler, 1941

Dr George S Robinson of Philadelphia appears to have been one of Fowler's wide circle of friends and acquaintances for whom he would make a comment such as 'indebted for many local fishes' – and leave the future researcher guessing! By the time of the catfish's description, he was 'the late' Dr Robinson.

Robinson, JBR

Natal Knifejaw *Oplegnathus robinsoni* CT Regan, 1916
African Deep-water Flathead *Parabembras robinsoni* CT Regan, 1921
Smallscale Grubfish *Parapercis robinsoni* HW Fowler, 1929

John Benjamin Romer Robinson (1869–1949) was a South African attorney and businessman. He was also a keen angler with an interest in marine fishes. He presented many specimens to the South African Museum and Durban Museum. After retiring from his law practice (1932) he became manager of *The Natal Mercury*.

Robson

Glass Knifefish sp. Eigenmannia robsoni GM Dutra, TPA Ramos & NA Menezes, 2022

Robson Tamar da Costa Ramos is a Brazilian ichthyologist, working at the Laboratório de Sistemática e Morfologia de Peixes, Universidade Federal da Paraíba, João Pessoa. He was honoured for his contributions to studies of the Caatinga ecoregion.

Rocadas

Yellowfish sp. *Labeobarbus rocadasi* GA Boulenger, 1910

Lieutenant-Colonel José Augusto Alves Roçadas (1865–1926) was Governor of Macau (1908–1909) and Governor-General of Angola (1909–1910), having served previously as a district governor (1905–1908). He commanded Portuguese forces in southern Angola in WW1 when German forces attacked (1914). He was honoured for the 'kind assistance' he offered explorer William John Ansorge (1850–1913), who collected the yellowfish type.

Rocca

Three-barbeled Catfish sp. *Pimelodella roccae* CH Eigenmann, 1917

Rocca, or Inca Rocca (reigned c.1134–1197), was, according to Eigenmann: "*...the first of the great Incas, proclaimed sovereign by the people, under direction of his mother, Siuyacu, the ladies of the court having an active part in shaping history, then as always.*"

Rocha, F

Plated Catfish sp. *Aspidoras rochai* R Ihering, 1907

Dr Francisco Dias da Rocha (1869–1960) was a Brazilian naturalist and pharmacist. He travelled to Europe with his father (1886). Francisco wanted to stay in Portugal to train as a physician but instead entered his father's business, selling imported goods. However, in his spare time he studied science and collected natural history specimens. The collections were housed in the Museu Roch that he founded (1887) at his home in Fortaleza and of which he was owner and Director thereafter. He gave up business (1898) to concentrate on collecting and studying the local flora and fauna. The museum had sections for Archaeology, Botany, Minerology and Zoology and the gardens were for native plants including cacti. He published the Boletim do Museu Rocha (1908–1950). The journal included his own articles, mostly concerning insects which he collected in his home state, such as: *Subsídio para o estudo da fauna cearense (catálogo das espécies animais por mim coligidas e anotadas)* (1950). The museum was later nationalised (1947). He also founded the School of Pharmacy and the School of Agriculture that grew into the Federal University of Ceará. He offered the holotype of this species to Ihering.

Rocha, L

Trindade Parrotfish *Sparisoma rocha* HT Pinheiro, JL Gasparini & I Sazima, 2010

Rocha's Sweeper *Pempheris rochai* JE Randall & BC Victor, 2015

Dr Luiz Alvez Rocha is a Brazilian ichthyologist and Associate Curator of Ichthyology and Follett Chair of Ichthyology at the California Academy of Sciences. The Universidade

Federal da Paraíba, João Pessoa, Paraíba, Brazil, awarded his BSc (1996) and MSc (1999), whereas his PhD in Fisheries and Aquatic Sciences was awarded by the University of Florida (2003). He undertook post-doctoral research at the Hawaii Institute of Marine Biology, University of Hawaii, with his research focusing on the evolution, phylo-geography, biogeography, systematics and behavioural ecology of coral reef fishes. He is also an excellent fish photographer. The parrotfish's authors say he "*...pioneered molecular genetics and phylogeography of Brazilian reef fishes.*"

Roche

Bullet Tuna *Auxis rochei* A Risso, 1810
Roche's Snake Blenny *Ophidion rochei* JP Müller, 1845

Dr François-Etienne de la Roche (or Delaroche) (1780–1813) was a Swiss physician, chemist, botanist and ichthyologist, who qualified as a physician in Paris (1808) and worked at the Hospice Necker. He collected and studied fish on an expedition to the Balearic Islands (1807/1808), and also made a study of the swim bladders of fish. His death at a young age was due to typhus.

Rochebrune

Lesser Guinean Devil Ray *Mobula rochebrunei* L Vaillant, 1879
[Junior Synonym of *Mobula hypostoma*]

Dr Alphonse Trémeau de Rochebrune (1836–1912) was a French military doctor who took a zoological doctorate (1874) and became a botanist, malacologist and ichthyologist. He was in Senegal (1875–1883) and wrote: *Fauna de Senegal* (1883). He worked at the Muséum National d'Histoire Naturelle, Paris (1884–1911).

Rocio

Central American Cichlid genus *Rocio* JJ Schmitter-Soto, 2007

Rocío is a Spanish word for 'morning dew', an "image evoked by the resplendent spots on cheek and sides of some species"; Rocío is also the name of Juan Schmitter-Soto's wife.

Rodrigues

Pencil Catfish sp. *Glaphyropoma rodriguesi* MCC de Pinna, 1992

Dr Miguel Trefaut Urbano Rodrigues (b.1953) is a Brazilian herpetologist. He took a degree in quantitative biology, Université Paris VII (1978), and obtained his doctorate from Universidad de São Paulo (1984). Since 1996 he has been Professor of Biological Sciences, Universidad de São Paulo, and was Director of the university's Zoological Museum (1997–2001). He has described many new species of reptile and has six reptiles and an amphibian named after him. (See also **Trefaut**)

Rodriguez

Armoured Catfish sp. *Rineloricaria rodriquezae* GJ da Costa-Silva, C Oliveira & G da Costa e Silva, 2021

Mónica Sonia Rodriguez is a Brazilian ichthyologist at the Universidade Federal de Viçosa, Minas Gerais. She was honoured for her contribution to knowledge to the genus Rineloricaria.

Rodway

Gold Tetra *Hemigrammus rodwayi* ML Durbin, 1909

James Rodway (1848–1926) was an English botanist, historian and travel writer. Originally working as a pharmacy assistant in Hertfordshire, he responded to a newspaper advertisement calling for clerks to work in British Guiana (Guyana) and moved to that colony (1870). He built up a reputation as a botanist and a chronicler of the colony's history, and eventually became Curator of the British Guiana Museum. He wrote: *Guiana: British, Dutch, and French* (1912), as well as a novel: *In Guiana Wilds: A Study of Two Women*' (1898).

Roe

Antilles Lanternfish *Diaphus roei* BG Nafpaktitis, 1974

Richard B Roe (1936–2016) was an American marine biologist who worked in fisheries for about 40 years. He was also a lifelong fisherman, hunter and amateur ornithologist. Rutgers University awarded his bachelor's degree (1958) and Utah State University his master's (1962). He was a researcher with exploratory teams of marine biologists in the Gulf of Mexico, the Caribbean and the tropical Atlantic. He directed the National Marine Fisheries Service's Office of Marine Mammals and Endangered Species. He oversaw the modernisation of the Fisheries Services' research fleet before his retirement (1995). Whilst he was based at the National Marine Fisheries Service, Southeast Fisheries Center, Pascagoula, Mississippi he provided specimens collected from the research vessel 'Oregon'. He died from pancreatic cancer.

Roelandt

Cutlassfish sp. *Lepturacanthus roelandti* P Bleeker, 1860

J W Roelandt is listed as being an 'Apothecary 3rd Class' in Sinkawang, Borneo. He was one of many medical personnel in the Dutch East Indies who supplied fish specimens to Bleeker.

Roemer

(See under **Römer**)

Rofen

Rofen's Rockskipper *Entomacrodus rofeni* VG Springer, 1967
Goby sp. *Thorogobius rofeni* PJ Miller 1988
Rofen's Barracudina *Lestidium rofeni* H-C Ho, K Graham & B Russell, 2020

Dr Robert Rees (Harry-)Rofen (1925–2015) was an American ichthyologist. He was born Robert Rees Harry Jr., but changed his name (1958) to Robert Rees Rofen and is known by that name or as Harry-Rofen. He graduated MA (1949) and conducted research at Stanford which also awarded his PhD (1952). He was Director of Stanford's Vanderbilt Foundation which studied the distributions of Pacific fish (1951–1962), during which period he went on twelve expeditions in the Indo-Pacific region. After the death of George Vanderbilt (1961) the programme ended and he set up a private laboratory, Aquatic Research Institute, in Stockton, California (1962) and was still its Director (2013). In private life, Rofen ran a pet supply company focusing primarily on water conditioners, treatments and medications for fish. He was also involved with the origins of brine shrimp harvesting in San Francisco Bay. He was honoured in the goby's binomial for his work on the systematics of both gobioid and deep-sea fishes, including earlier examination of the type material. He has published widely as RR Harry-Rofen, RR Harry and RR Rofen, such as: *Antarctic Fishes* (1959) and *Handbook of the food fishes of the Gulf of Thailand* (1963) (Robert R Harry), *The whale-fishes: families Cetomimidae, Barbourisiidae and Rondeletiidae. (Order Cetunculi)* (1959) (Robert R Rofen) and *Project Coral Fish Looks at Palau* (1957) (Robert R Harry-Rofen) as well as being one of the authors of *Fishes of the Western Atlantic (Part 5)* 2018. (See also **Harry, RR**)

Roger

Pigsnout Grunt *Pomadasys rogerii* G Cuvier, 1830

Baron Jacques-François Roger (1787–1849) was a jurist, agronomist and colonial administrator who became Governor of Senegal (1821–1827). He was a liberal politician afterwards in France and an abolitionist. He supplied the holotype to Cuvier. Among other shorter works he wrote *Fables Sénégalaises* (1828).

Rogers, GW

Roger's Round Ray *Urotrygon rogersi* DS Jordan & EC Starks, 1895
[Alt. Thorny Stingray]

Dr George Warren Rogers was a physician from Vermont who lived in Mazatlan, Sinaloa, Mexico. He was described in the etymology as 'a scholarly physician' who assisted the authors.

Rogers, MA

Thorny Catfish sp. *Leptodoras rogersae* MH Sabaj Pérez, 2005

Mary Anne Rogers is a biologist who was Collection Manager of Fishes, Field Museum of Natural History, Chicago (1989–2010). She wrote: *Evolutionary differentiation within the northern Great Basin pocket gopher,* Thomomys townsendii. *I. Morphological variation* (1991).

Rohan, J

African Barb sp. *Enteromius rohani* J Pellegrin, 1921

Jacques Fernand de Rohan-Chabot, Comte de Jarnac (1889–1958), was a French explorer who led the Rohan-Chabot expedition to Angola and Rhodesia (Zimbabwe/Zambia) (1912–1914). He served in the French army with distinction in WW1. He became Director-General of the French Red Cross and was arrested by the Germans as a spy and resistance leader (1944), but was saved by the intervention of Pierre Laval. He collected the barb type. A bird and a reptile are also named after him.

Rohan (Pethiyagoda)

Rohan's Barb *Dawkinsia rohani* K Rema Devi, TJ Indra & JDM Knight, 2010
Fire Rasbora sp. *Rasboroides rohani* S Batuwita, M de Silva & U Edirisinghe, 2013
[Junior Synonym of *Rasboroides pallidus*; based on a (human-)translocated population of the latter]

Tilak Rohan David Pethiyagoda (b.1955) (abbreviated to Rohan Pett by deed poll in 2010) is a Sri Lankan biologist. Kings College, London awarded his BSc in electronics (1977) and Sussex his MPhil in biomedical engineering (1980). He worked as an engineer in the Division of Biomedical Engineering of the Ministry of Health in Sri Lanka (1981–1987), becoming Director (1982). He then resigned government service to research freshwater fish, leading to his first book: *Freshwater fishes of Sri Lanka* (1990). With colleagues from the Wildlife Heritage Trust that he founded he has discovered over 100 vertebrates in Sri Lanka, including fish, amphibians and reptiles. Concerned by the rapid loss of montane forest in Sri Lanka, he began (1998) an on-going project to restore abandoned tea plantations to natural forest. He has received many honours and sits on many national and international committees. A reptile and a dragonfly are named after him.

Roig

Pencil Catfish sp. *Trichomycterus roigi* AGE Arratia Fuentes & SA Menu-Marque, 1984

Dr Arturo Hernán Roig-Alsina is an Argentine biologist, zoologist, agronomist and entomologist at the Argentine Museum of Natural Sciences 'Bernardino Rivadavia'. He is also a research associate at AMNH. A reptile is named after him. He collected the holotype of this species.

Roissal

Five-spotted Wrasse *Symphodus roissali* A Risso, 1810

Clément Honoré Claude Roissal (also spelt Roassal) (1781–1850) was a close friend of the author, Risso, and an amateur artist. Both Risso and Roissal came from Nice (now in France, but then in the Duchy of Savoy).

Rolland

Rolland's Demoiselle *Chrysiptera rollandi* GP Whitley, 1961

Jean Rolland (1933–2018) was a highly skilled welder by trade. He was founder (1975) of the Association Témoignage d'Un Passé, which seeks to preserve the history and heritage of New Caledonia. He was a volunteer helping to install the local aquarium in Nouméa, New Caledonia, and the damselfish's name was dedicated at the request of Rene Catala of the Aquarium, who sent the type to Whitley.

Roloff

African Rivuline genus *Roloffia* HS Clausen, 1966
[Junior Synonym of *Callopanchax*]
African Rivuline sp. *Scriptaphyosemion roloffi* E Roloff, 1936
Hispaniolan Rivulus *Rivulus roloffi* E Roloff, 1938

Jelly Bean Tetra *Ladigesia roloffi* J Géry, 1968
African Dwarf Cichlid sp. *Pelvicachromis roloffi* DFE Thys van den Audenaerde, 1968
Characin sp. *Odontostilbe roloffi* J Géry, 1972
African Rivuline sp. *Epiplatys roloffi* R Romand, 1978

Erhard Roloff (1903–1980) was a German aquarist and amateur ichthyologist. He travelled and collected in Sierra Leone (six visits), Liberia, Mozambique, Ecuador, Dominican Republic and Sri Lanka. In two of the above cases, it looks as though Roloff named species after himself, but this is not the case. The species were dedicated to Roloff by others, but on account of delays in their formal publication, Roloff's use of the names in print preceded what should have been the 'official' descriptions.

Roman

Driftwood Catfish sp. *Duringlanis romani* GF Mees, 1988
[Syn. *Centromochlus romani*]

Brother Dr Benigno Roman Gonzalez (1913–1993) was a Venezuelan herpetologist, ichthyologist, and Jesuit monk whose doctorate was awarded by Universidad de Barcelona (1969). He wrote: *Deux sous-espèces de la vipère Echis carinatus (Schneider) dans les territoires de Haute-Volta et du Niger* (1972). The Oceanologic Museum, Fundación La Salle, Venezuela is named after him, as are two reptiles. He collected the holotype of this species.

Romand

Lake Malawi Cichlid sp. *Tropheops romandi* J Colombé, 1979

Dr Raymond Romand is an evolutionary neurobiologist and geneticist at the Department of Biology and Stem Cell Development of the Institut de Génétique et Biologie Moléculaire et Cellulaire (llkirch-Graffenstaden, France). He is particularly interested in the auditory mechanisms in vertebrates. He is also an ichthyologist and aquarist. He was honoured for his work on African fishes. Among those works are: *Aphyosemion cauveti, a new species of killifish (Cyprinodontidae) from Guinea, West Africa* (1971), *Feeding biology of a small Cyprinodontidae from West Africa, Aplocheilichthys normani Ahl, 1928* (1985) and *Collecting a Colorful Killifish in West Africa: Aphyosemion geryi* (1998). He has collected fish in West Africa.

Romer, LS

Römer's Grunter *Hephaestus roemeri* M Weber, 1910

Lucien Sophie Albert Marie von Römer (1873–1965) was a Dutch physician and botanist. He joined an expedition to New Guinea and later settled in the Dutch East Indies where he worked as a public health specialist.

Römer, U

Neotropical Rivuline sp. *Laimosemion romeri* WJEM Costa, 2003

Dr Uwe Römer (b.1959) is a German aquarist and ichthyologist. In private life he is a teacher who was (2006) a lecturer in the Department of Biogeography at the University of Trier. Among his publications is the *Cichlid Atlas* (2000) and *Cichlid Atlas 2* (2007). He collected

(1995) the type specimen of the rivuline. He has a special interest in the dwarf cichlids of the genus *Apistogramma*, and has described or co-described several new species, including *Apistogramma atahualpa* and *A. panduro*.

Romero

Pencil Catfish sp. *Trichomycterus romeroi* HW Fowler, 1941

Augusto Romero Padilla was a Colombian fish culturist, about whom we could find no further biographical details.

Rondelet

Redmouth Whalefish genus *Rondeletia* GB Goode & TH Bean, 1895
Black-wing Flyingfish *Hirundichthys rondeletii* A Valenciennes, 1847
Rondelet's Ray *Raja rondeleti* P Bougis, 1959
Hérault Bullhead *Cottus rondeleti* J Freyhof, M Kottelat & A Nolte, 2005

Dr Guillaume Rondelet (1507–1566) was a French physician, botanist and zoologist. He qualified at the University of Montpellier as a physician (1537) but was not a success as a doctor and was always short of money. Cardinal François de Tournon employed him as his personal physician (1538) and with such a powerful patron his fortunes revived. He became Regius Professor of Medicine at the University of Montpellier (1545) and Chancellor of the University (1556–1566). He is satirised as 'Rondibilis' in: *La vie de Gargantua et Pantygruel* written by his great friend François Rabelais. He is also noteworthy for having expelled Nostradamus from the University for slandering doctors and being an apothecary. He wrote: *Libri de piscibus marinis in quibus verae piscium effigies expressae sunt* (1554). He recognised the distinctiveness of this ray (1554), but assumed it was a juvenile *Raja fullonica*.

Rondon

Amazonian Cichlid genus *Rondonacara* FP Ottoni & JL de O Mattos, 2015
Armoured Catfish sp. *Hypostomus rondoni* A Miranda Ribeiro, 1912
Driftwood Catfish sp. *Tympanopleura rondoni* A Miranda Ribeiro, 1914
Acara Cichlid *Aequidens rondoni* A Miranda Ribeiro, 1918
Dwarf Cichlid sp. *Heterogramma rondoni* A Miranda Ribeiro, 1918
Mousetail Knifefish *Gymnorhamphichthys rondoni* A Miranda Ribeiro, 1920
[Alt. Longbilled/Elephantnose/Thermometer Knifefish]
Armoured Catfish sp. *Harttia rondoni* OT Oyakawa, I Fichberg & L Rapp Py-Daniel, 2018
Tetra sp. *Moenkhausia rondoni* K Mathubara & M Toledo-Piza, 2020

Marshall Cândido Mariano da Silva Rondon (1865–1958) was a Brazilian army officer (1881–1955) engineer and explorer, whose Rondon Commission was responsible for installing telegraph poles from Mato Grosso to Amazonas. He led a number of expeditions to Mato Grosso and the Amazon, including one with the former US President Theodore Roosevelt (1914). He was the first Director of the Brazilian Indian Protection Bureau (1910–1915) and promoted the creation of the Xingu National Park (1961). The Brazilian state of Rondônia is also named after him. The etymology for the armoured catfish states that the name: "*...honors Cândido Mariano da Silva Rondon (Marechal Rondon), a military and frontiersmen who was responsible for the creation of Indian Protection Service (SPI), which was subsequently replaced by the National Indian Foundation (FUNAI). Rondon and Noel Nutels participated, together with the Villas Boas brothers, in the implementation of the Parque Indígena do Xingu.*" Authors of the tetra said they named it after him as he was "*... famous for having explored the Western portion of the Amazon basin. As an indigenist he gave support to the native populations of Brazil all his life. Marshal Rondon gives name to the state of Rondônia, Brazil, where Moenkhausia rondoni is widely distributed.*"

Ronquillo

Ronquillo's Anchovy *Stolephorus ronquilloi* T Wongratana, 1983

Professor Inocencio Aricayos Ronquillo (b.1918) was a Filipino marine biologist. He was a member (1972) and chair (1982–1987) of UNESCO's Intergovernmental Oceanographic Commission. He was honoured as the person who collected the holotype and whose studies of *Stolephorus* 'broke the ground' for Wongratana. Among his many scientific papers (1949–1987) is the co-authored: *On the biology of anchovies (Stolephorus Lacépède) in Philippine waters* (1970).

Roos

Asian Catfish sp. *Clupisoma roosae* CJ Ferraris Jr, 2004

Dr Anna Roos (b.1961) is an ecotoxicologist, researcher and curator at Department of Environmental Research and Monitoring, Swedish Museum of Natural History (1988–present), and at the Climate Centre, Greenland Institute of Natural Resources (2016–present) splitting her time between Stockholm and Nuuk, Greenland. She is an expert on otters, seals, whales and dolphins. Her PhD was awarded by Uppsala University (2013). She wrote: *Increasing Concentrations of Perfluoroalkyl Acids in Scandinavian Otters (Lutra lutra) between 1972 and 2011: A New Threat to the Otter Population?* (2013) and *Population structure and recent temporal changes in genetic variation in Eurasian otters from Sweden* (2015). She collected the holotype of this species in Myanmar together with Dr Fang Fang Kullander of the Swedish Museum of Natural History.

Roosevelt

Roosevelt's Goby *Chriolepis roosevelti* I Ginsburg, 1939

Franklin D Roosevelt (1882–1945) was President of the United States (1933–1945).

Rosa, AL

African Barb sp. *Labeobarbus rosae* GA Boulenger, 1910

Padre Anastacio Luis Rosa was a Portuguese parish priest in Angola. He was honoured for his 'helpful courtesy during a seven years' friendship' with explorer William John Ansorge (q.v.), who collected the type.

Rosa (Holub)

Rednose Labeo *Labeo rosae* F Steindachner, 1894

Rosa Holub née Hoff (1865–1958) was the Viennese wife of Czech physician, cartographer and ethnographer Emil Holub (1847–1902), who collected the type. She participated in his later African explorations (1883–1886).

Rosa, JG

Armoured Catfish sp. *Pareiorhina rosai GSC* Silva, FF Roxo & OT Oyakawa, 2016

Dr João Guimarães Rosa (1908–1967) was a Brazilian diplomat and writer who documented the history of people living near the Rio das Velhas and Rio Paraopeba in the Brazilian Savanna, Minas Gerais, where this catfish occurs. He originally qualified as a physician (1930) and graduated (1930). He practised (1930–1932) until becoming involved with the Constitutional Revolution (1932), which led him to becoming a civil servant and a diplomat (1934). He served as Assistant-Consul in Hamburg, Germany, where, with his wife, he was active in helping Jews escape the Third Reich. He wrote: *The Devil to Pay in the Backlands* (1956).

Rosa, R de S

American Sole sp. *Apionichthys rosai* RTC Ramos, 2003

Rosa's Round Ray *Heliotrygon rosai* MR de Carvalho & NR Lovejoy, 2011

Dr Ricardo de Souza Rosa (b.1954) is a Brazilian ichthyologist who is a Professor at the Department of Systematics and Ecology, Federal University of Paraíba, Brazil. He was a volunteer (1972–1976) at the ichthyology department of the Zoology Museum of the University of São Paulo, which awarded his bachelor's degree in biology (1976). Virginia Institute of Marine Science, College of William and Mary, Virginia awarded his doctorate (1985), after which he carried out post-doctoral research at the University of Alberta, Canada (1990). He wrote: *A Systematic Revision of the South American Freshwater Stingrays (Chondrichthyes: Potamotrygonidae)* (1985) for which he received the Rodolpho von Ihering Award presented by the Brazilian Zoology Society. He was also honoured in the name of the ray for this revision, which "*...represents a landmark in our understanding of the taxonomy and diversity of this family.*" In addition to his contribution to the knowledge of freshwater stingrays, he also worked on the taxonomy of other marine and freshwater Neotropical fish groups, as well as on the biology of tropical sharks. His major findings on South American stingrays are included in a book: *Rayas de agua dulce* (*Potamotrygonidae*) *de Suramérica. Parte I* edited by Carlos A. Lasso *et al.* (2013).

Rosa (Smith)

California Halfbeak *Hyporhamphus rosae* DS Jordan & CH Gilbert, 1880

Ozark Cavefish *Amblyopsis rosae* CH Eigenmann, 1898

Characin sp. *Deuterodon rosae* F Steindachner, 1908

Three-barbeled Catfish sp. *Chasmocranus rosae* CH Eigenmann, 1922

Rosa Smith Eigenmann (1858–1947) was the wife of ichthyologist Carl H Eigenmann and an ichthyologist in her own right. The etymologies of the characin and catfish are not explained in the original texts, so though identification with Rosa Smith is likely, there remains an element of uncertainty. (See **Eigenmann, R**).

Rosa Maria

Sand Knifefish sp. *Gymnorhamphichthys rosamariae* HO Schwassmann, 1989

Though this looks like an eponym, it is in fact a toponym; a place on the upper Rio Negro, Brazil, where the holotype of this species was collected.

Rosa Pinto

Basketfish *Kentrocapros rosapinto* JLB Smith, 1949

Dr Antonio Augusto da Rosa Pinto (1904-1986) was a Portuguese ornithologist. He wrote on several national parks and their avi¬faunas, and a standard work on the ornithology of Angola, Ornitologia de Angola. He was Director of the Museum in Lourenço Marques (Maputo), Mozambique (1953-1960), the type locality. Four birds are also named after him.

Rosamma

Stone Loach sp. *Schistura rosammai* N Sen, 2009

[Syn. *Aborichthys rosammai*. Binomial sometimes amended to the feminine *rosammae*]

Dr Rosamma Mathew (b.1952) is an entomologist and zoologist with a particular focus on systematics who worked for the Zoological Survey of India until retiring. She was in charge of the Wildlife Sub-regional Office, Cochin (Kochi) (2005). She collected the loach holotype. She co-wrote: *Studies on Lower Vertebrates of Nagaland* (2008) and *Rediscovery of Rhacophorus nuso Annandale, 1912 (Amphibia: Anura: Rhacophoridae) from Mizoram, North East India* (2008).

Rosa Pinto

Basketfish *Kentrocapros rosapinto* JLB Smith, 1949

Dr Antonio Augusto da Rosa Pinto (1904–1986) was a Portuguese ornithologist. He wrote on several national parks and their avifaunas, and a standard work on the ornithology of Angola, *Ornitologia de Angola*. He was Director of the Museum in Lourenço Marques (Maputo), Mozambique (1953–1960), the type locality. Four birds are also named after him.

Rose

Rose Island Basslet *Pseudoplesiops rosae* LP Schultz, 1943

[Alt. Large-scaled Dottyback, Rose Dottyback]

As the common name shows, this is a toponym after the type locality in American Samoa.

Rose, JN

Suckermouth Catfish sp. *Astroblepus rosei* CH Eigenmann, 1922

Dr Joseph Nelson Rose (1862–1928) was a botanist who worked for the US Department of Agriculture and became an assistant curator at the Smithsonian (1896). Wabash College, Crawfordsville, Indiana awarded his bachelor's degree, his master's and his doctorate (1889). He co-wrote: *The Cactaceae* (4 volumes, 1919–1923).

Rose (Mousavi-Sabet)

Iranian Garra sp. *Garra roseae* H Mousavi-Sabet, M ASaemi-Komsari, I Doadrio & J Freyhof, 2019

Rose Mousavi-Sabet is the daughter of the first author, Hamed Mousavi-Sabet.

Rosemary

Pike-Cichlid sp. *Crenicichla rosemariae* SO Kullander, 1997

Rosemary Helen 'Ro' Lowe-McConnell (née Lowe) (1921–2014) was a pioneering tropical fish biologist, ichthyologist, limnologist and ecologist. The University of Liverpool awarded her BSc, MSc and DSc. She was an early adopter of SCUBA to study fish *in situ*. She conducted research of tilapia fisheries in Lake Nyasa (1945), relying on local fisherman to help. After marrying Richard McConnell (1953) she left her position at the East African Fisheries Research Organisation (which she helped found and of which she was acting director), as it was funded by the British Colonial Service which did not allow married women to hold permanent posts. Undaunted, Ro continued her research. She worked for the UK Freshwater Biological Association and was later a Research Associate at the BMNH. Among her many scientific contributions was her book: *Fish Communities in Tropical Fresh Waters* (1987). She was honoured as she had collected the type and was a "*…persistent inspiratrix to students of tropical fish ecology*." (Also see **McConnel, RB**)

Rosen

Rosen's Platy *Xiphophorus roseni* MK Meyer & L Wischnath, 1981 (NCR)
[Believed to be of hybrid origin]
Neotropical Livebearer sp. *Brachyrhaphis roseni* WA Bussing, 1988
Toothless Characin sp. *Curimata roseni* RP Vari, 1989
Cutthroat Eel sp. *Meadia roseni* MH-K Mok, C-Y Lee & H-L Chan, 1991
Rosen's Tube-snouted Ghost Knifefish *Sternarchorhynchus roseni* F Mago-Leccia, 1994
Swamp Eel sp. *Rakthamichthys roseni* RM Bailey & GC Gans, 1998
Rosen's Buntingi *Adrianichthys roseni* LR Parenti & B Soeroto, 2004

Dr Donn Eric Rosen (1929–1986) was an American ichthyologist. New York University awarded his BS (1955), MS (1957) and PhD (1959). He was Chairman of the Ichthyology Department, American Museum of Natural History, which he headed for ten years (1965–1975) having earlier (1961) joined the staff there. During that time, he re-organised and tripled the collection. Perhaps his most well-known publications were: *Phyletic Studies of Teleostean Fishes, with a Provisional Classification of Living Forms* (1966) and: *Modes of Reproduction in Fishes* (1966). One etymology honours him for his 'tremendous contribution' to fish systematics. Baily and Gans say: "*This species* [Monopterus roseni] *is named for the late Dr. Donn E. Rosen, accomplished ichthyologist, discerning student of the Synbranchidae and personal friend and associate of both of us. He was a field companion with one of us (RMB) on five expeditions to Guatemala where the many memorable months of ichthyological research included field investigation of two species of synbranchids.*"

Rosenberg, H

Rosenberg's Jawfish *Opistognathus rosenbergii* P Bleeker, 1856

Baron Carl (Karl) Benjamin Hermann von Rosenberg (1817–1888) was a German naturalist and cartographer. He went to the Dutch East Indies as a military cartographer, stationed first on Sumatra (1840–1856) and later in the Moluccas and New Guinea. He returned to Europe (1871) and later published a detailed account of the natural history and peoples of the regions he had visited: *Der malayische Archipel* (1878). Seven birds and two mammals are named after him, among other taxa.

Rosenberg, WFH

Goby sp. *Sicydium rosenbergii* GA Boulenger, 1899

William Frederick Henry Rosenberg (1868–1957), who described himself as 'a traveller-naturalist', was an English natural history dealer who had a shop in London (1930–1948). He collected in South America, including Ecuador (1896–1898). Eight birds and an amphibian are named after him.

Rosenblatt

Deepwater Cardinalfish genus *Rosenblattia* GW Mead, 1965
Pearleye genus *Rosenblattichthys* RK Johnson, 1974
Spikefin Blenny *Coralliozetus rosenblatti* JS Stephens, 1963
Green-blotched Rockfish *Sebastes rosenblatti* LC Chen, 1971
Sailfin Snake-eel *Letharchus rosenblatti* JE McCosker, 1974
Dreamer (anglerfish) sp. *Oneirodes rosenblatti* TW Pietsch III, 1974
Onejaw sp. *Monognathus rosenblatti* E Bertelsen & JG Nielsen, 1987
Blue-spotted Jawfish *Opistognathus rosenblatti* GR Allen & DR Robertson, 1991
Oval Puffer *Sphoeroides rosenblatti* WA Bussing, 1996
Cook Islands Flashlightfish *Protoblepharon rosenblatti* CC Baldwin, GD Johnson & JR Paxton, 1997
Spotted Rainbow Wrasse *Suezichthys rosenblatti* BC Russell & MW Westneat, 2013

Dr Richard Heinrich 'Dick' Rosenblatt (1930–2014) was an American ichthyologist and oceanographer. UCLA awarded his degrees, all in zoology: bachelor's (1953), master's (1954) and PhD (1959). He worked at the Scripps Institution of Oceanography as Assistant Researcher (1958–1965), Curator of Marine Vertebrates and Director of the aquarium and museum there (1961–1965). He was appointed as Full Professor of Marine Biology (1972) and retired as Professor Emeritus of the University of California, San Diego, and Curator Emeritus of Scripps' Marine Vertebrate Collection. The etymology for the snake-eel honours him: "*...in recognition of his contributions to the study of apodal fishes and to the education of ichthyologists.*"

Rosenstock

African Rivuline sp. *Nothobranchius rosenstocki* S Valdesalici & RH Wildekamp, 2005

John Rosenstock (b.1945) is a Danish psychologist and killifish aquarist. He was a Research Assistant, then Department Head, at the Danish Mental Health Research

Institute (1968–1971); then Head of Section at Mellemfolklige Samvirke (1971–2004) whilst also being a consultant in Danida (1980–2000). He has written a book on the psychology of Fundamentalism (2017), as well as publishing a book (2011) on his family's genealogy after 15 years of research. A lifelong aquarist, he discovered and collected this species in Zambia (1997). The etymology reads: "*Described in honour of John Rosenstock of Hellerup, Denmark, who collected this new species during one of his many journeys in Africa for the Danish Development Assistance Organisation.*" He is also known as an excellent storyteller.

Rosita

Molly sp. *Poecilia rositae* MK Meyer, K Schneider, AC Radda, B Wilde & M Schartl, 2004
Dwarf Cichlid sp. *Apistogramma rositae* U Römer, U Römer & IJ Hahn, 2006

Rosita Bonhaus is the "*long-term partner*" of aquarium-book publisher Hans-Albrecht Baensch (q.v.), who helped in the preparation of the book in which the description of the cichlid appeared; in addition, the name also "*…highlights the similarity and apparent close systematic links*" between *A. rositae* and *A. baenschi*. (Also see **Baensch, HA**)

Ross

Black Seadevil sp. *Melanocetus rossi* AV Balushkin & VV Fedorov, 1981

This is a toponym referring to the Ross Sea where the holotype was discovered.

Ross, CA

Halfbeak sp. *Nomorhamphus rossi* AD Meisner, 2001

Charles Andrew 'Andy' Ross (1953–2011) was a herpetologist and palaeontologist at the Department of Vertebrate Zoology, the Smithsonian, where he was a specialist on crocodilians. He collected the type specimens of this fish in the Philippines. He co-wrote: *Four New Species of* Lycodon *(Serpentes: Colubridae) from the Northern Philippines* (1994).

Ross, JC

Marbled Rockcod *Notothenia rossii* J Richardson, 1844
Threespot Eelpout *Lycodes rossi* AJ Malmgren, 1865
Snailfish sp. *Paraliparis rossi* NV Chernova & JT Eastman, 2001

Rear-Admiral Sir James Clark Ross (1800–1862) discovered the Ross Sea and the Ross Ice Shelf and an island (1841) which Scott, on his first expedition, named Ross Island in his honour. Ross joined the Royal Navy aged 12! He was also a member of several important expeditions to the Arctic. He commanded 'Erebus' and 'Terror' during the Antarctic expedition (1839–1843). It was while close to the Magnetic South Pole that he broke through a wide expanse of pack ice and into a large and clear sea that later bore his name. A mammal and two birds are named after him.

Ross, SW

Snailfish sp. *Psednos rossi* NV Chernova & DL Stein, 2004

Dr Steve W Ross is a professor at the University of North Carolina Center for Marine Science. He has spent most of his career involved in marine science of the southeast region where he is Principal Investigator and cruise Chief Scientist of the USA. He earned a bachelor's degree in zoology from Duke University, a master's from UNC-Chapel Hill, and a Ph.D. from North Carolina State University. He was the Research Coordinator for the NC Coastal Reserve Program for 13 years. He has also led offshore studies for the US Geological Survey. His area of specialization is in the ecology and life history studies (age, growth, feeding, reproduction) of fish. He has conducted numerous projects in estuaries and offshore waters and has served as chief scientist on many cruises, including those using submersibles and ROVs. He was honoured by the authors as it was Ross "*...who initially notified us of the captures and furnished the specimens to us for examination.*"

Ross (Robertson)

Blackfin Spectre Goby *Akko rossi* JL Van Tassell & CC Baldwin, 2004

Dr D Ross Robertson is an ichthyologist who works at the Smithsonian Tropical Research Institute, Panama (since 1975), where he conducts field research on a broad range of topics relating to the reproductive biology, demography, population biology, behaviour, community ecology, evolution and biogeography of tropical reef fishes, with an emphasis on those species that live in the Neotropical seas on both sides of the Americas. The University of Queensland awarded his BSc (1966) and PhD (1974). His many publications include such papers as: *The roles of female mate choice and predation in the mating systems of some tropical Labroid fishes* (1977) and *An Indo-Pacific damselfish (Neopomacentrus cyanomos) in the Gulf of Mexico: origin and mode of introduction* (2018). His books include: *Fishes: Greater Caribbean* (2015). This friend and colleague of the authors was honoured "*...for substantial contributions to our understanding of the diversity of tropical eastern Pacific shorefishes.*"

Rössel

Characin sp. *Characidium roesseli* J Géry, 1965

Dr Fritz Rössel was a German catfish specialist who worked at the Senckenberg Museum, Frankfurt (1964). He wrote: *Welse (Siluroidea) collected from the German expedition in India in 1955/58* (1964).

Rossignol

Rossignol's Toadfish *Perulibatrachus rossignoli* C Roux, 1957

Martial Rossignol was an oceanographer and biologist who worked for ORSTOM in the Gulf of Guinea and the Caribbean. He collaborated with the author in a marine faunal survey of Pointe-Noire, Republic of the Congo, where this toadfish occurs. He co-wrote: *Coastal marine hydrology of the peninsula of Cape Verde* (1965).

Rosso

Neotropical Rivuline sp. *Melanorivulus rossoi* WJEM Costa, 2005

Aldovan Rosso is an aquarist and killifish hobbyist who lived in the region of Brazil where this species occurs and collected the holotype.

Rossoperegrin

Loach sp. *Triplophysa rossoperegrinatorum* AM Prokofiev, 2001

Not an individual eponym but "in honour of famous Russian explorers of Central Asia" [*rosso peregrinatorum* = 'of Russian travellers'].

Roswitha

Neotropical Livebearer sp. *Brachyrhaphis roswithae* MK Meyer & VMF Etzel, 1998

Roswitha Erasmy née Etzel of Cuxhaven is, we believe, the daughter of the junior author; the late Dr *Vollrad Max Friedrich Etzel* (1944–2012) (q.v.).

Rothrock

White-barred Prickleback *Poroclinus rothrocki* TH Bean, 1890

Joseph Trimble Rothrock (1839–1922) was an American physician, botanist and an early environmentalist. He served as the first President of the Pennsylvania Forestry Association, which supported the creation of state parks and state forests.

Rothschild

Rothchild's Barracudina *Stemonosudis rothschildi* WJ Richards, 1967

Dr Brian J Rothschild (b.1934) was at the Bureau of Commercial Fisheries, Honolulu, Hawaii. He supplied the holotype (which was taken from the stomach of a lancetfish). Rutgers University awarded his bachelor's degree, the University of Maine his master's and Cornell University his doctorate (1962). He is now Montgomery Charter Professor of Marine Science and former Dean of the School for Marine Science and Technology at the University of Massachusetts Dartmouth. Previously he was a Professor at the University of Maryland and the University of Washington, in addition to many other affiliations with organisations such as the Scripps Institution of Oceanography and Woods Hole Oceanographic Institution. He wrote: *Dynamics of Marine Fish Populations* (1987).

Rouanet

Labeo sp. *Labeo rouaneti* J Daget, 1962

Raymond Rouanet was Curator, Eaux et Forêts (Waters and Forests) in Guinea. He wrote: *Justification des projets de budget local du service des eaux et forêts, Conakry* (1957). He helped in facilitating Daget's expeditions in Guinea.

Roule

Slickhead genus *Rouleina* DS Jordan, 1923
Roule's Goby *Gobius roulei* F de Buen, 1928
Slender Mandarinfish *Siniperca roulei* HW Wu, 1930
Chinese Cyprinid sp. *Atrilinea roulei* HW Wu, 1931
Tonguefish sp. *Cynoglossus roulei* HW Wu, 1932
Blind Cusk-Eel sp. *Barathronus roulei* JG Nielsen, 2019

Professor Dr Louis Roule (1861–1942) was a French ichthyologist and herpetologist. He was at the Marine Biology Station, Faculty of Sciences, Université de Toulouse (1885–1910).

He became Professor of Zoology, MNHN, Paris (1910), which was where Wu conducted his ichthyological studies. Roule also had a strong interest in the work of earlier French naturalists such as Buffon and Cuvier, and published some works on them, such as: *Buffon et la description de la nature* (1924). Two reptiles are also named after him.

Roulin

Freshwater Stingray sp. *Potamotrygon roulini* TR Roberts, 2020

François Désiré Roulin (1796–1874) was a French physician, naturalist and explorer. After studying medicine at the University of Paris he visited Colombia (1822–1828), studying the natural history of that country. He reported on the presence of stingrays in the Orinoco Basin.

Rousseau

Dorado Catfish *Brachyplatystoma rousseauxii* FL de Castelnau, 1855

Louis Rousseau (1811–1874) was a French malacologist, traveller, collector and a pioneer of zoological photography. He worked as an assistant naturalist in the Department of Malacology, MNHN, Paris. He took up photography (early 1850s) and was a founding member of the Société française de photographie (1854).

Rousselle

African Barb sp. *Enteromius roussellei* W Ladiges & J Voelker, 1961

Ardo Rousselle was a plantation owner in Angola, and it was on his plantation that the barb type was collected.

Roux, C

Yellowtail Sardinella *Sardinella rouxi* M Poll, 1953
African Barb sp. *Enteromius rouxi* J Daget, 1961
Skate sp. *Raja rouxi* C Capapé, 1977

Professor Dr Charles Roux (b.1920) was a French marine zoologist. He was Deputy Director of the Overseas Fisheries Laboratory until moving (1970) to the Reptiles & Fish Zoology Lab until the Fisheries Lab merged with it (1974) and he moved back before retiring (1985). He became Professor and Deputy Director, MNHN, Paris (1987) and Deputy Director, then Director, of the Centre Oceanographique de Pointe-Noire. His doctorate was awarded in 1973. He wrote a number of papers and books including co-writing: *Ocean Dwellers* (1982). He collected the barb type.

Roux, JB

Roux's Pygmy-goby *Pandaka rouxi* MWC Weber, 1911
Congo Cichlid sp. *Steatocranus rouxi* J Pellegrin, 1928

Dr Jean B Roux (1876–1939) was a Swiss zoologist and herpetologist who gained his doctorate at Geneva University (1899) and initially focused his studies on protozoa. He turned towards herpetology on becoming Curator, Natural History Museum, Basel (1902–1930). He was in New Guinea and Australia (1907–1908), and in New Caledonia and the Loyalty Islands (1911–1912) with Fritz Sarasin. Six reptiles, an amphibian and a bird are named after him. He was a member of the expedition during which the pygmy-goby holotype was collected.

Roux, JLFP

Long-striped Blenny *Parablennius rouxi* A Cocco, 1833

Jean Louis Florent Polydore Roux (1792–1833) was a French artist and naturalist. He was appointed (1819) as curator of the Muséum d'histoire naturel de Marseille. He joined Charles von Hügel, who was travelling for the Austrian government, on an excursion to Egypt (1831). From there he moved on to Bombay, where he later died of plague. Although Cocco did not explain his etymology, it seems likely this is the person intended. A mammal and a reptile are also named after him.

Rowley

Mountain Carp sp. *Psilorhynchus rowleyi* SL Hora & KS MIsra, 1941

Major Guy Shafto Rowley (1889–1976) of the 8th King's Royal Irish Hussars was commissioned in 1909 and retired from the army (1929) having served in France in WW1 (1914–1918) and subsequently in Germany and Uganda. He became a *shikari* (big game hunter) and explorer. He was a member of the Vernay-Hopwood Chindwin expedition that left Rangoon to explore the upper reaches of the Chindwin River on behalf of the American Museum of Natural History. Arthur Vernay, an established New York City-based dealer in English antiques, an intrepid field associate in the AMNH's Department of Mammalogy, and a museum trustee, financed the expedition. Henry C Raven, a comparative anatomist at the museum, joined as the lead scientist, principal filmmaker and photographer. For three months, the party traversed northern Burma, gathering biological and anthropological specimens. The expedition collected the mountain carp type.

Roxas

Goby sp. *Schismatogobius roxasi* AW Herre, 1936

Dr Hilario Atanacio Roxas (b.1896) was a zoologist who was chief of the Philippine Fish and Game Association and the first Filipino to be first head of the University of the Philippines' Department of Biology. He co-wrote: *A check list of Philippine fishes* (1937).

Roxo

Three-barbeled Catfish sp. *Phenacorhamdia roxoi* GSC Silva, 2020

Dr Fábio Fernandes Roxo is a post-doctoral researcher at the Laboratório de Ictiologia de Botucatu, Setor Zoologia, Universidade Estadual Paulista, Brazil. Among his publications is: *Molecular Phylogeny and Biogeographic History of the Armored Neotropical Catfish Subfamilies Hypoptopomatinae, Neoplecostominae and Otothyrinae (Siluriformes: Loricariidae)* (2014). He was honoured in recognition of his contributions to our knowledge of Neotropical ichthyology.

Roy

Snakehead sp. *Channa royi* J Praveenraj, JDM Knight, R Kiruba-Sankar, B Halalludin,

JJA Raymond & VR Thakur, 2018
[Junior Synonym of *Channa harcourtbutleri*]

Dr Sibnarayan Dam Roy (b.1959) is Director (since 2012) of ICAR-Central Island Agricultural Research Institute. He studied for his BFSc and MFSc at the Mangalore University of Agricultural Science (1977–1984). His PhD was awarded by Ambedkar University, Agra. He began his career as a Fisheries Officer for the Department of Fisheries Arunachal Pradesh (1984–1989), becoming Scientist (1989–1999) then Director of Fisheries A&N Islands (1999–2000). He then joined ICAR as Senior Scientist (2000–2005), Head of Fisheries Science (2005–2010), then Head of the Aquaculture Division (2010–2012). He has written seven books and 39 research papers as well as seminars, book chapters and articles. The etymology says that: "*The species is named after Dr. S. Dam Roy, in appreciation for his immense encouragement and support for the exploration of the freshwater fishes of Andaman and Nicobar Islands.*" Unfortunately, not long after its description, other researchers concluded that the 'new' species was a synonym of one named in 1918.

Royal

Swamp Eel sp. *Synbranchus royal* MH Sabaj, M Arce H. & LM de Sousa, 2022

Tyson Royal Roberts (b. 1940) is an American ichthyologist. (See Roberts, TR).

Royaux

Squeaker Catfish sp. *Euchilichthys royauxi* GA Boulenger, 1902
Royal Sprat *Microthrissa royauxi* GA Boulenger, 1902

Colonel Louis-Joseph Royaux (1866–1936) was a Belgian colonial policeman. He enlisted (1888) and was sent overseas (1892) to the then Belgian Congo as a Sergeant. A skilled administrator, he soon became a station chief. On his second station, he was seriously wounded in action against local indigenous people. He returned to Europe (1895) but went back again (1896–1902) as Lieutenant. then captain. He was appointed as an area manager (1897) and again returned to Europe (1902) but went back to the Congo (1907). Although not required to by virtue of age, he volunteered to fight back in Europe when WW1 broke out (1914). He was again wounded and became an instructor promoted to Major until the end of hostilities. He retired (1922) but worked in trade and commerce. He was also good at learning local languages wherever he was stationed. When a Captain he led the expedition that collected the types of the two fish named after him.

Royero

Whale Catfish sp. *Denticetopsis royeroi* CJ Ferraris Jr, 1996

Dr Ramiro Royero-Leon (b.1958) is an ichthyologist and parasitologist at the School of Biology, Universidad Central de Venezuela. He was at Bristol University, England (1996) and is associated with the Department of Parasitology, The Czech Academy of Sciences, Prague. He co-wrote: *Some nematodes of freshwater fishes in Venezuela* (1997). He accompanied Ferraris on the latter's field work in Venezuela.

Royle

African Barb sp. *Labeobarbus roylii* GA Boulenger, 1912

Harry Royle was an agent with the Liverpool firm Hutton and Cookson, whose 'warm welcome' (translation) and other services (unspecified) greatly assisted the explorer William John Ansorge (1850–1913), who collected the barb holotype.

Ruaha

Kneria sp. *Kneria ruaha* L Seegers, 1995

This is a toponym referring to the Ruaha River drainage in Tanzania, the type locality.

Ruasa

African Barb sp. *Labeobarbus ruasae* P Pappenheim, 1914

This species is named after the locality where the holotype was acquired, Ruasa in northwest Rwanda.

Rubbioli

Pencil Catfish sp. *Trichomycterus rubbioli* ME Bichuette & PP Rizzato, 2012

Ezio Luiz Rubbioli (b.1964) is a Brazilian speleologist who was the first person to explore the Serra do Ramalho caves. He co-wrote: *Mapping of Caverns* (2005). He brought this species to the authors' attention.

Rubenstein

African Characin sp. *Nannocharax rubensteini* FC Jerep & RP Vari, 2013

David Mark Rubenstein (b.1949) is an American philanthropist who founded the Rubenstein Fellowships of the Encyclopedia of Life at the Smithsonian Institution. He is also famous for having bought the last privately-owned copy of the Magna Carta (for $21.3 million).

Rubinoff

Rubinoff's Triplefin *Axoclinus rubinoffi* GR Allen & DR Robertson, 1992

Dr Ira Rubinoff (b.1938) is an American marine biologist who was Director of the Smithsonian Tropical Research Institute in Panama (1973–2007) (now Director Emeritus). He is married to Anabella Guardia de Rubinoff, the former Panamanian Ambassador to Austria. Queen's College, New York awarded his BSc (1959) and Harvard his MA (1961) and PhD (1963), after which he began work at the Smithsonian TRI (1964). He pioneered research on the effect of the Panama Canal on the ocean's ecology and the evolution of animals within it. He has published many papers such as: *A strategy for preserving tropical forests* (1983) and *A Century of the Smithsonian Institution on the Isthmus of Panama* (2013). He was decorated with the Order of Vasco Nunez de Balboa by the Republic of Panama and received the Joseph Henry Medal from the Board of Regents of the Smithsonian Institution. He has said: "*Understanding the biodiversity of Planet Earth is the most interesting of human endeavours. Where better to pursue this interest than in the tropics?*"

Rudd

Silver Labeo *Labeo ruddi* GA Boulenger, 1907

Charles Dunell Rudd (1844–1916) was an associate of Cecil Rhodes and attended to their mining business while Rhodes got himself into politics. Rudd obtained the concession (1883) for Rhodes to go into Mashonaland to establish mining. He co-founded (1888) the De Beers Mining Company with Rhodes. Rudd financed Captain C H B Grant to collect zoological specimens in southern Africa. Publications of the Zoological Society of London in the first decade of the 20th century carry many references to 'the Rudd Exploration of South Africa' and descriptions of new species unearthed by Captain Grant. He financed the expedition that collected the silver labeo type. He is also commemorated in the names of two birds and two reptiles.

Rudjakov

Pearlside sp. *Maurolicus rudjakovi* NV Parin & SG Kobyliansky 1993

Yuri Alexandrovich Rudjakov (b.1938) is a researcher of suprabenthic plankton. He took part in cruises to Nazca and Sala y Gomez ridges, Eastern South Pacific, where this species can be found. (Also see **Yuri (Rudjakov)**)

Rudolph

Armoured Catfish sp. *Pareiorhina rudolphi* A Miranda Ribeiro, 1911

Three-barbeled Catfish sp. *Pimelodella rudolphi* A Miranda Ribeiro, 1918

Rodolpho Teodoro Gaspar Wilhelm von Ihering (1883–1939). **(See Ihering, RTGW)**

Rueppell

See under **Rüppell**

Ruetzler

Ruetzler's Blenny *Emblemariopsis ruetzleri* DM Tyler & JC Tyler, 1997

Klaus Ruetzler (b.1936) is an Austrian-born zoologist who became an American citizen (1973) and Curator of Invertebrate Zoology at the US National Museum. The etymology says he "*...has so effectively directed the Smithsonian's Caribbean Coral Reef Ecosystem (CCRE) Program and its marine laboratory at Carrie Bow Cay, Belize, since its inception in 1972*".

Ruffo

Chub sp. *Squalius ruffoi* PG Bianco & F Recchia, 1983

Sandro Ruffo (1915–2010) was an Italian naturalist who became Professor of Zoology at Modena and Director of the Museo Civico di Storia Naturale di Verona (1964). He was drafted during WW2 (1939–1945) and helped with the post-war restoration of the museum in Verona, being appointed conservator zoologist. He had a keen interest in bio-speleology.

Rugger

Rugby Hatchetfish *Polyipnus ruggeri* RC Baird, 1971

This is one of the strangest binomials we have ever encountered. 'Rugger' is slang for rugby football, and this fish is specifically named in honour of New Zealand's national sport (some would say 'religion').

Rukhopf

African Rivuline sp. *Epiplatys ruhkopfi* HO Berkenkamp & VMF Etzel, 1980

Willi Ruhkopf is a German aquarist. He was at the time an active member of the Deutsche Killifisch Gemeinschaf.

Rumengan

Lembeh Sea Dragon *Kyonemichthys rumengani* MF Gomon, 2007
[Alt. Thread Pipefish]

Noldy Rumengan is an Indonesian diver and dive guide in Sulawesi. He discovered the new pipefish (2006) on a coral ridge in Lembeh Strait, and took his friend, photographer William Tan, to see and photograph it. He advised Rudie Kuiter in Australia who asked them to send a specimen to Martin Gomon, Senior Curator (ichthyology) at Museum Victoria, who wrote the paper naming it.

Rumsfeld

Deep-blackfin Butterflyfish *Roa rumsfeldi* LA Rocha, HT Pinheiro, M Wandell, CR Rocha & B Shepherd, 2017

Donald Henry Rumsfeld (1932-2021) was an American politician who immortalized the quote (2002): "*…there are known knowns; there are things we know we know. We also know there are known unknowns; that is to say we know there are some things we do not know. But there are also unknown unknowns – the ones we don't know we don't know.*" The authors think the quote applies well to the taxonomy of MCE (mesophotic coral ecosystems) species such as this butterflyfish.

Rungthip

Piebald Horseface Loach *Acantopsis rungthipae* DA Boyd, P Nithirojpakdee & LM Page, 2017

Rungthip 'Kae' Plongsesthee (1978–2014), was a: "*…dear friend, close colleague, a Ph.D. student of Dr F W H Beamish at Burapha University, Bangsaen, Thailand, and an extremely enthusiastic ichthyologist who is greatly missed by her many friends.*" She died tragically young from breast cancer.

Rupir

Sisorid catfish sp. *Glyptothorax rupiri* L Kosygin, P Singh & S Rath, 2021

Rupir Boli is a forest officer working in the Mouling National Park for the Forest Department, Government of Arunachal Pradesh, India. He was honoured for his help in collecting specimens during the senior author's survey of Arunachal Pradesh.

Rüppell

Mongoose Loach *Nemachilichthys rueppelli* WH Sykes, 1839
[Binomial originally given as *rupelli*, but usually corrected]
Rüppell's Scaldback *Arnoglossus rueppelii* A Cocco, 1844
Banded Moray *Gymnothorax rueppelliae* J McClelland, 1844
Western Gobbleguts (cardinalfish) *Ostorhinchus rueppellii* A Günther, 1859
Klunzinger's Wrasse *Thalassoma rueppellii* CB Klunzinger, 1871
Nile Catfish sp. *Chrysichthys rueppelli* GA Boulenger, 1907

Wilhelm Peter Eduard Simon Rüppell (1794–1884) was a German collector. He went to Egypt and ascended the Nile as far as Aswan (1817), and later made two extended expeditions to northern and eastern Africa, Sudan (1821–1827) and Ethiopia (1830–1834). Although he brought back large zoological and ethnographical collections, his expeditions impoverished him. He wrote: *Reisen in Nubien, Kordofan und dem Petraischen Arabien* (1829), *Reise in Abyssinien* (1838–1840) and *Systematische Uebersicht der Vögel NordOstAfrikas* (1845). His work on Red Sea fishes (1828–1830) contains many species referenced by McClelland, who strangely used a spelling of the moray's binomial that would normally reflect a feminine gender. Rüppell was a collector in the broadest sense, and presented his collection of coins and rare manuscripts to the Historical Museum in Frankfurt (his home town). Eleven birds, five mammals, two reptiles, an amphibian and a dragonfly are named after him.

Rusby

Three-barbeled Catfish sp. *Rhamdella rusbyi* NE Pearson, 1924

Dr Henry Hurd Rusby (1855–1940) was an American pharmacist, explorer and botanist. He graduated as a physician (1884) but abandoned medicine for botany. He was Professor of Botany and Materia Medica, School of Pharmacy, University of Columbia, New York (1889–1930), being Dean of the Faculty (1904–1930) and Dean Emeritus (1930–1940). He took part in expeditions (1880–1921), culminating as leader of the 1921 Mulford Biological Exploration of the Amazon basin during which the holotype was collected.

Ruschi

Armoured Catfish sp. *Pareiorhaphis ruschii* EHL Pereira, A Lehmann & RE Dos Reis, 2012

Augusto 'Gutti' Ruschi (1915–1986) was a Brazilian naturalist who dedicated his life to conservation. He was considered to be a world authority on hummingbirds but is probably even more famous for his second love: orchids. He studied law and agronomy, so was largely self-taught in natural history. He learned English, French, Latin and German in order to read standard botanical texts. He lived on Santa Lucia Reserve, Santa Teresa, in a nineteenth-century house built by his grandfather. He transformed his house into the 'Professor Mello Leitão Biology Museum' and, before he died, he asked the Pro-Memória Foundation to bury him in the Reserve. It was thought that touching poisonous frogs during years of research caused the liver disease from which he died. However, the doctor who treated him said his problems were caused by an overdose of anti-malarial drugs. He wrote over 400 scientific articles and books. A mammal and an amphibian are named after him.

RUSI

Dottyback genus *Rusichthys* R Winterbottom, 1979

Rusi Blenny *Mimoblennius rusi* VG Springer & AE Spreitzer, 1978

RUSI is an acronym standing for Rhodes University, Smith Institute (of Ichthyology). Winterbottom honoured the Institute for the 'tremendous' contributions to ichthyology made by the late J L B Smith, his wife Margaret Mary Smith, and its staff and students.

Russel

Russel's Smooth-back Herring *Raconda russeliana* JE Gray, 1831

[Alt. Raconda]

Gray's original text does not record who is honoured in the binomial. It is likely to refer to surgeon-herpetologist Patrick Russell (q.v.), who provided many specimens from India to the British Museum (Natural History), or possibly to his half-brother Alexander Russell (1715–1768) who also collected in India. Gray described this species in: *Description of Twelve New Genera of Fish, discovered by General Hardwicke in India* (1831). It could honour John Russell (sometimes Russel) Reeves (1774–1856) who was one of the artists sometimes used in India by Hardwicke and the former's 1829 notebooks record him making a collection of fish paintings for Hardwicke around that time. However, given that in the same paper Gray named a species as *reevesii* after him this looks less likely.

Russell, BC

Smooth Gurnard *Lepidotrigla russelli* L del Cerro & D Lloris, 1995

Dr Barry C Russell is an Australian ichthyologist who was Director of Research and Collections (now Curator Emeritus) at the Museum of the Northern Territory, Darwin (1982–present). He is Research Associate Charles Darwin University (2008–present). His publications range from *A preliminary annotated checklist of fishes of the Poor Knights Islands* (1971) to *Scolopsis igcarensis Mishra, Biswas, Russell, Satpathy & Selvanayagam, 2013, a junior synonym of S. vosmeri (Bloch, 1792) (Perciformes: Nemipteridae)* (2013). He loaned gurnard specimens to the authors.

Russell, P

Oarfish sp. *Regalecus russelii* G Cuvier, 1816

Goatee Croaker *Dendrophysa russelii* G Cuvier, 1829

Indian Scad *Decapterus russelli* WPES Rüppell, 1830

Russell's Lionfish *Pterois russelii* ET Bennett, 1831

Large-tooth Flounder sp. *Pseudorhombus russellii* JE Gray, 1834

Eclipse Parrotfish *Scarus russelii* A Valenciennes, 1840

Russell's Snapper *Lutjanus russellii* P Bleeker, 1849

Dr Patrick Russell (1727–1805) was a British surgeon and naturalist. He first went to India in 1781 to look after his brother, who was employed by the Honourable East India Company in Vizagapatnam. He became fascinated by the plants in the region and was appointed to be the company's Botanist and Naturalist, Madras Presidency (1785). He spent six years in Madras (Chennai) and sent a large collection of snakes to the British Museum (1791). One of his major concerns was snakebite, and he tried to find a way for people to identify poisonous snakes without first getting bitten and seeing what happened. His chief ichthyological work was: *Descriptions and Figures of Two Hundred Fishes Collected at Visagapatam on the Coast of Coromandel* (1803). Four reptiles are named after him.

Rutten

Blue Cardinal Rasbora *Rasbora rutteni* MCW Weber & LF de Beaufort, 1916

Louis Martin Robert Rutten (1884–1946) was a Dutch geologist and palaeontologist. He mapped large parts of the Dutch East Indies (Indonesia), Cuba, and the Netherlands Antilles. He graduated from Utrecht University (1909), joined a company later known as Royal Dutch Shell, and was sent prospecting for oil in Borneo. He travelled on business to Argentina, Cuba, Mexico, and Peru (1919–1921). He became Professor of Crystallography, Geology, and Palaeontology, Utrecht University (1921). He led expeditions by his students to the Netherlands Antilles (1930) and to Cuba (1933 and 1938). He collected the rasbora type.

Rutter

Ringtail Snailfish *Liparis rutteri* CH Gilbert & JO Snyder, 1898

Cloudsley Louis Rutter (1867–1903) was an American ichthyologist and a pioneer in studies of pacific salmon. He worked as an Assistant at the US Fish Commission (1893–1894 & 1897–1902) and was also Curator of Fishes at the California Academy of Science (1900–1902). He often researched aboard the fisheries steamer 'Albatross' on which he was Resident Naturalist (1900–1902). He died aged 36 of erysipelas apparently contracted while travelling on a train from California to Indiana.

Ruud Wildekamp

African Rivuline sp. *Nothobranchius ruudwildekampi* WJEM Costa, 2009

Rudolf 'Ruud' Hans Wildekamp (1945–2019) was honoured for his taxonomic work on the genus *Nothobranchius.* (See **Wildekamp**)

Ruwet

Okavango Tilapia *Tilapia ruweti* M Poll & DFE Thys van den Audenaerde, 1965

Professor Emeritus Dr Jean-Claude Ruwet (d.2007) was a Belgian ethologist and animal behaviourist at the University of Liège. He published extensively, mostly in the *Belgian Journal of Zoology*. He wrote: *Introduction to Ethology; The Biology of Behaviour* (1972). He obtained the type specimen of the tilapia. A bird is also named after him.

S

Saadi

Stone Loach sp. *Turcinoemacheilus saadii* HR Esmaeili, G Sayyadzadeh, M Özuluğ, M Geiger & J Freyhof, 2014

Abū-Muhammad Muslih al-Dīn bin Abdallāh Shīrāzī, Saadi Shirazi (better known by his pen-name Saʿdī, or Saadi) (c.1210–ca 1291) was one of the major Persian poets. He travelled and wandered widely for about 30 years in Central Asia and North Africa, from Sindh (now in Pakistan) and India (Delhi and Gujerat) to Egypt. He fought against the Crusaders, was captured at Acre and spent seven years there as a slave labourer.

Sabaj

Para Pleco *Peckoltia sabaji Hemiancistrus sabaji* JW Armbruster, 2003
[Syn. *Hemiancistrus sabaji*]
Sabaj's Coralbrotula *Ogilbia sabaji* PR Møller, W Schwarzhans & JG Nielsen, 2005
Armoured Catfish sp. *Curculionichthys sabaji* FF Roxo, GSC Silva, LE Ochoa & C Oliveira, 2015

Dr Mark Henry Sabaj Pérez (b.1969) is an ichthyologist and collection manager of fishes at the Academy of Natural Sciences of Philadelphia (2000–present). The University of Richmond, Virginia, awarded his bachelor's degree (1990) and master's (1992). The University of Illinois, Urbana-Champaign awarded his doctorate (2002), with a thesis on the taxonomy of the neotropical thorny catfishes (*Siluriformes: Doradidae*) and revision of the genus *Leptodoras*. He has co-authored 43 peer-reviewed publications, including the descriptions of 22 new species and one new subspecies of freshwater fish from Asia and South America, one new genus of Doradidae, and two new species of North American crayfish. He has joined or led 30 expeditions to 12 countries in Asia, Europe and North and South America. (Also see **Mark**)

Sabanejew

Spined Loach genus *Sabanejewia* V Vladykov, 1929

Leonid Pavlovich Sabanejew, aka Sabaneev (1844–1898), was a Russian zoologist and an expert on freshwater fish biology. He was educated at Moscow University. He was on friendly terms with Grand Duke Alexander Alexandrovich and held the post of 'Stallmeister' (stable master) at his court. He also made a study of hunting in Russia and set up the Hunter's Gazette (1888) and was the author of the enormously popular Hunter's Calendar. He wrote: *Les poissons de la Russie* (1875). His son of the same name (1881–1968) was a noted musician.

Sabina

Congo Cichlid sp. *Congochromis sabinae* A Lamboj, 2005

Sabina is the daughter of the author Anton Lamboj.

Sabine (River)

Sabine Shiner *Notropis sabinae* DS Jordan & CH Gilbert, 1886

A toponym referring to the type locality; the Sabine River, Texas.

Saccharae

Mandarin Sleeper Goby *Sineleotris saccharae* AW Herre, 1940

B E Sugars was Secretary of the Hong Kong Aquarium Society, and collected the type. The binomial is a play on words, being Latin for 'sugar'.

Sacchi

Giraffe Catfish sp. *Auchenoglanis sacchii* D Vinciguerra, 1898

Maurizio Sacchi (1864–1897) was an Italian explorer and geographer. He graduated in physics at the University of Pavia and became an Assistant at the Institute of Florence before moving to Rome as Assistant at the Central Meteorology and Geoinformation office. He led the expedition during which the catfish holotype was collected (1895–1897) but was killed by Abyssinians on the shores of Lake Regina Margherita, Ethiopia.

Sach

Sailfin Poacher *Podothecus sachi* DS Jordan & JO Snyder, 1901

Not an eponym but based on the local name of the fish in the Aomori area of northern Japan.

Sachs, C

Ghost Knifefish sp. *Adontosternarchus sachsi* W Peters, 1877

Dr Carl Sachs (1853–1878) was a physician and researcher into electric fishes. He graduated as a physician in Berlin (1876) and travelled to Venezuela to investigate electric eels, returning to Europe (1877) with live specimens that died on the last leg of the journey – a railway train from Bremen to Berlin. He was killed in a mountaineering accident in the Austrian Tyrol. He wrote the posthumously published: *Aus den Llanos: Schilderungen einer naturwissenschaftlichen Reise nach Venezuela* (1879). He collected the holotype of this species.

Sachs, W

Gold-finned Barb *Puntius sachsii* E Ahl, 1923

[Syn. *Barbodes semifasciolatus sachsii*]

Walter Bernhard Sachs was a German aquarist and friend of the author. He was co-author of: *Die Wunder des Meeres* (1926).

Sacrimontis

Nigerian cichlid sp. *Pelvicachromis sacrimontis* J Paulo, 1977

Walter Heiligenberg (1938–1994) was a German American neuroethologist who specialised in the motivational behaviours of cichlids and crickets. He moved from Germany to California (1972) and worked at the Scripps Institution of Oceanography. He was killed in the crash of US Air Flight 427 on his way to deliver a lecture at the University of Pittsburgh. This is a clever play on words that taxonomists are fond of; *sacrum* means holy and *montis* means mountain: a Latin transliteration of the German surname Heiligenberg.

Sadler

Sadler's Robber (African tetra) *Brycinus sadleri* GA Boulenger, 1906

Colonel Sir James Hayes Sadler (1827–1910) was a British Army officer who served in a number of artillery regiments (1854–1882). After resigning his commission, he became a diplomat. He was Consul for most of the Midwestern states of the USA (1887) based in Chicago; and later Chief Political Resident of the Persian Gulf (1893 & 1893–1894). He was Resident and Consul-General in British Somaliland (now part of Somalia) (1897–1902) after which he was Commissioner, Commander-in-Chief and Consul General for the Uganda Protectorate. He wrote: *Present-day administration in Uganda* (1905). During his tenure of office, the holotype was collected.

Sadowsky

Brazillian Skate *Rajella sadowskii* G Krefft & MFW Stehmann, 1974

Victor Sadowsky (1909–1990) was a Latvian-born Brazilian ichthyologist at Instituto Oceanográfico da Universidade de São Paulo. He first arrived in Brazil (1949) and settled (1951) at Cananeia (São Paulo state), where the local museum is named in his honour. He wrote: *First record of broad-snouted seven-gilled shark from Cananéia, coast of Brazil* (1969) and 'considerably added to the knowledge of southwestern Atlantic elasmobranchs'.

Sainsbury

Goldeneye Shovelnose Ray *Rhinobatos sainsburyi* PR Last, 2004

Sainsbury's Flathead *Sunagocia sainsburyi* LW Knapp & H Imamura, 2004

Dr Keith John Sainsbury (b.1952) is a marine ecologist and mathematical modeler with a research focus on the assessment, ecology, economics, exploitation and conservation of marine resources and ecosystems. The University of Canterbury, New Zealand, awarded both his bachelor's degree (1972) and doctorate (1977). He led research programs through CSIRO, Australia, (1978–2006). This included (late 1970s–late 1980s) several ecological surveys of the offshore marine ecosystems off northern Australia and southern Indonesia that resulted in the discovery of many new species including the shovelnose ray. He was vice-chair of the Board of Trustees for the Marine Stewardship Council (2005–2014) and is the Director of SainSolutions Pty Ltd, which provides scientific advice for marine resource management. He is also Professor of Marine System Science, Institute of Marine and Antarctic Science, University of Tasmania and (since 1995) has been closely associated with the work of Food and Agriculture Organisation of the United Nations (FAO). His work on sustainable management of fisheries has been recognised by the awards of the Japan Prize (2004) and the Swedish Seafood award (2012). He wrote: *Best practice reference points for Australian Fisheries* (2008) and co-wrote: *Continental shelf fishes of northern and north-western Australia* (1985). He is married to fellow ichthyologist Glenys Jones (b.1952). (Also see **Glenys (Jones)**)

Saint-Hilaire

Pencil Catfish sp. *Trichomycterus sainthilairei* AM Katz & WJEM Costa, 2021

Augustin François César Prouvençal de Saint-Hilaire (1779–1853) was a French botanist who travelled in Brazil (1816–1822 and 1830). He not only collected a huge number of plant specimens, but also insects, birds, reptiles and fish. He devoted himself to the study, classification, description and publication of this huge material, but was considerably impaired by his ill health, due to diseases contracted during his travels.

Saint Paul

Saint Paul's Gregory *Stegastes sanctipauli* R Lubbock & AJ Edwards, 1981

This is a toponym rather than an eponym: this species is found at Saint Paul's Rocks in the equatorial Atlantic (under Brazilian sovereignty).

Sainthouse

African Rivuline sp. *Nothobranchius sainthousei* B Nagy, FPD Cotterill & DU Bellstedt, 2016

Ian Sainthouse is a British aquarist and killifish enthusiast. He was honoured as a: "... *renowned breeder and collector of killifish*" and for "...*his special longstanding dedication to researches on the genus Nothobranchius*".

Saisi

Sisorid Catfish sp. *Glyptothorax saisii* JT Jenkins, 1910

We have been completely unable to trace the source of this eponym, if indeed it is an eponym at all. The holotype was acquired in a stream in the Paresnath Hills, Jharkhand, India - should that stir any memory in any reader!

Saito, J

Tilefish sp. *Branchiostegus saitoi* JK Dooley & Y Iwatsuki, 2012

Jiro Saito is a Japanese amateur angler who caught specimens of this fish in the Philippines (2009 & 2011) and sent frozen examples to the junior author, Yukio Iwatsuki. The etymology says that without Saito's "*considerable efforts and interest, this species would have remained unknown.*"

Saito, S

Warbonnet sp. *Chirolophis saitone* DS Jordan & JO Snyder, 1902
[Binomial sometimes amended to *saitonis*]

Sotaro Saito was director of the Museum of Aomori, Japan. He presented the type specimen to the authors, but we have been unable to find further details about him.

Saiz

Tetra sp. *Hyphessobrycon saizi* J Géry, 1964

Emilio Saiz collected the type in the Upper River Meta basin, Colombia. We know nothing more about him.

Sajica

T-Bar Cichlid *Cryptoheros sajica* WA Bussing, 1974
[Syn. *Amatitlania sajica*]

Salvador **Ji**ménez **Ca**nossa (1922–1986) was Director of the Library of Congress of Costa Rica. He was also a "*...friend and experienced field collector*" who accompanied Bussing on most of his early collecting trips in Costa Rica, and "*...through his enthusiasm and curiosity of nature, contributed greatly to their success.*" The name is in the form of an acronym utilizing the initial two letters of his given name, and his paternal and maternal surnames.

Sajor

Japanese Halfbeak *Hyporhamphus sajori* CJ Temminck & H Schlegel, 1846

This is not an eponym; rather *sajori* is the Japanese name for this fish.

Sakaizumi

Northern Medaka *Oryzias sakaizumii* T Asai, H Senou & K Hosoya, 2012

Dr Mitsuru Sakaizumi is a zoologist and molecular geneticist who heads a lab in the Faculty of Science, Niigata University, Japan. He was a pioneer of molecular phylogenetic studies of Japanese ricefish, contributing much to that field. Among his published papers is: *Diversity of sex-determining mechanism – Lessons from medaka fishes* (2009) and the co-written: *Turnover of Sex Chromosomes in* Celebensis *Group Medaka Fishes* (2015).

Sakaramy

Madagascar Panchax *Pachypanchax sakaramyi* M Holly, 1928

This is a toponym, named after the Sakaramy River in Madagascar.

Saksono

Anton's Damselfish *Pomacentrus saksonoi* GR Allen, 1995

Anton Saksono is the owner of the Pulau Putri Resort (Seribu Islands, Indonesia), who graciously provided accommodation, boat transport, and logistics assistance, without which the type would not have been collected.

Sala

African Airbreathing Catfish sp. *Clarias salae* AAW Hubrecht, 1881

Carolus Franciscus Sala (1839–1881) was a Dutch soldier and collector. He served in the Royal Netherland East India Army (Indonesia), and collected zoological specimens in Angola. He accompanied Büttikofer (q.v.) on his first expedition in what is today part of LIberia but died from a tropical disease during the expedition.

Salda

Lake Salda Killifish *Aphanius saldae* F Akşiray, 1955

This is a toponym, referring to the type locality: Lake Salda in southwestern Turkey.

Saldanha

Cusk-eel sp. *Ophidion saldanhai* J Matallanas & A Brito, 1999
Cutthroat Eel sp. *Ilyophis saldanhai* ES Karmovskaya & NV Parin, 1999
Short-tail Eel sp. *Coloconger saldanhai* JC Quéro, 2001
Eelpout sp. *Pachycara saldanhai* M Biscoito & AJ Almeida, 2004

Professor Dr Luiz Viera Calds Saldanha (1937–1997) was a prominent Portuguese ichthyologist. Lisbon University awarded his bachelor's degree in biology (1961), which he studied for while undertaking his compulsory seven-year military service. A year after graduating he was called up again and was part of a combat unit in Angola (1962–1965). During this time, he collected many specimens that he later deposited in the Museau Bocage where he was appointed Naturalist (1965). He then worked as a Researcher in the Anthropological and Zoological Laboratory there (1970–1974), during which time he also undertook foreign internships in France, UK and USA and the University of Lisbon awarded his PhD (1974). He was Assistant Professor (1975–1978) and then full Professor (1979–1997). His academic career focused on teaching academic disciplines in Biological Oceanography and Marine Biology, supervising PhD theses and graduate students and participating in selection juries of universities and research institutions in Portugal and abroad. His scientific interests focused on Marine Ecology and Biological Oceanography, in regard to deep-sea fauna and processes, as well as coastal, tropical and polar ecosystems, which earned him the nickname of 'Man of the Seven Seas'. He published more than 130 papers. Karmovskaya & Parin described him as a friend who made substantial contributions to anguilliform studies.

Salesse

African Barb sp. *Enteromius salessei* J Pellegrin, 1908

Captain Eugène Pierre Mathieu Salesse (1858–1932) was a French military engineer. He was put in charge of work on the French Guinea railway (1886). He had surveyed and mapped the route on two 'missions' (1895 & 1897–1898) from the capital Conakry to Niger. He resigned from the army (1902) and joined the colonial administration as 'Secretary General of the Colonies, second class'. He was appointed the Director of Railways in Conakry, Guinea (1904–1910). He 'greatly facilitated' (translation) the ichthyological research of Dr Robert Wurtz (q.v.) during the latter's travels through Senegal and French Guinea.

Saliha

Stone Loach sp. *Paracobitis salihae* C Kaya, D Turann, G Kalayci, E Bayçelebi & J Freyhof, 2020

Saliha Kaya (1939–2015) was the mother of the first author, Cüneyt Kaya.

Salle

Aztec Shiner *Aztecula sallaei* A Günther, 1868

Large-eye Silverside *Atherinella sallei* CT Regan, 1903

Auguste Sallé (1820–1896) was a French traveller, taxonomist and entomologist who collected in the southern USA, West Indies and Central America including Mexico (1846–1856) where he collected many fish which he sent to the BMNH. His mother accompanied him on his expeditions. He became a very successful natural history dealer in Paris, specialising in insects. Many other taxa, including a dragonfly, two reptiles and seven birds are also named after him. Günther's description of the shiner does not contain an etymology, but there is little doubt whom he intended to honour.

Sally

Sally's Eviota *Eviota readerae* AC Gill & SL Jewett, 2004
[Alt. Reader's Dwarf-Goby]

Sally E Reader (See **Reader**)

Salmacis

Corydoras catfish sp. *Scleromystax salmacis* MR Britto & RE Reis, 2005

Salmacis is a fictional character in Ovid's: *The Fountain of Salmacis*. Those who entered the fountain belonging to the naiad Salmacis were believed to become hermaphrodites. The name was chosen due to the 'very subtle' differences between the sexes of this species.

Salsbury

Oriental Barb sp. *Osteochilus salsburyi* JT Nichols & CH Pope, 1927

Dr Clarence Grant Salsbury (1885–1980) was a Canadian-born American missionary and physician. He was educated at the Brooklyn Union Missionary Training Institute where he was known as 'Mr Canada' because of his imposing height. There he met and married (1909) his very petite wife. He went on to train as a physician at the Boston College of Physicians and Surgeons where he qualified (1913). The Presbyterian Board of Missions assigned the couple to their mission in Hainan Island, China, where he was resident physician (1914–1926). They returned for a year's leave intending to return but were instead assigned to the Ganado Mission on the Navajo Reservation in Arizona where he was Superintendent and Medical Director for twenty-three years (1927–1950). He founded (1930) the Sage Memorial Hospital School of Nursing there, which was the first school to focus on the education of Native American women in the nursing profession. He was also Head of Arizona State Health department for eleven years. He was honoured for his interest and aid in Nichols' work. He wrote a memoir: *Salsbury Story: A Medical Missionary's Lifetime of Public Service* (1969) and his wife Cora wrote a history of the Ganado Mission: *Forty Years in The Desert* (1948).

Salvador

Bocochi Pupfish *Cyprinodon salvadori* M de Lozano-Vilano, 2002
Dr Salvador Contreras Balderas (1936–2009). (See **Contreras**)

Salvat

Dottyback sp. *Ogilbyina salvati* Y Plessis & P Fourmanoir, 1966

Bernard Salvat is an honorary professor at École Pratique des Hautes Études (EPHE), Paris. He is a coral-reef biologist and a friend of the authors. He is also a researcher at the Centre de Recherches Insulaires et Observatoire de l'Environnement (CRIOBE). He is an author of several books, such as: *Coquillages de Polynésie (Nature tropicale)* (1975) and *Coquillages De Nouvelle Caledonie* (1988), with co-authors Claude Rives and Pierre Reverce. He was about to take over the authors' work in New Caledonia (the type locality) when this dottyback was first collected.

Salvin

Yellow-belly Cichlid *Trichromis salvini* A Günther, 1862
Goby sp. *Sicydium salvini* WR Ogilvie-Grant, 1884

Osbert Salvin (1835–1898) was an English naturalist particularly interested in herpetology and ornithology. He made the first of several visits to Guatemala (1857). He was the first European to record observing a Resplendent Quetzal, pronouncing it: "*unequalled for splendour among the birds of the New World*" ...and promptly shot it. Salvin co-authored the 40-volume: *Biologia Centrali Americana* (1879), a near-complete catalogue of Middle American species. He collected the cichlid type with Frederick DuCane Godman. Thirty-eight birds, four reptiles and two mammals are named after him.

Salween

Salween's Barb *Hypsibarbus salweenensis* WJ Rainboth, 1996

The apparent eponym is misleading as the binomial refers to the Salween River in Myanmar.

SAM

Nail Snakelet *Natalichthys sam* R Winterbottom, 1980

SAM is an acronym for the South African Museum (Cape Town), whose ichthyologists (P. A. Hulley and E. Louw) have, according to the author, always been "*most co-operative and good company.*"

Sam Price

Sam Price's Gudgeon *Allomogurnda sampricei* GR Allen, 2003

Samuel Price was a 'keen teenage naturalist' from Jayapura, Indonesia, when he collected the holotype.

Samadi

Papuan Lanternshark *Etmopterus samadiae* WT White, DA Ebert, RR Mana & S. Corrigan, 2017

Dr Sarah Samadi is an evolutionary biologist who is a professor at the L'Institut de Systématique, Évolution, Biodiversité, Muséum National d'Histoire Naturelle. Her MSc was awarded by Ecole Normale Supérieure, Paris (1993) and her PhD by the Université Montpellier II (1997). She was Assistant Professor at University Paris 7 (1998–1999) and at EPHE (1999–2000), then took a research post at MNHN (2000–2002). She was Permanent Researcher at IRD in the research unit UMR7138 (2002–2012) and Co-Coordinator with

Jean-Noël Labat and Line Le Gall of the MNHN project 'Taxonomie moléculaire : DNA Barcode et gestion durable des collections' (2009–2012) before assuming her current post and duties among which are curating the DNA collection of aquatic animals. She has led or taken part in at least eleven investigative cruises (2003–2014). Among her publications are: *Several deep-sea mussels and their associated symbionts are able to live both on wood and on whale fall* (2009) and the co-written *Incorporation of deep-sea and small-sized species provides new insights into gastropods phylogeny* (2019).

Samborski

African Rivuline ssp. *Epiplatys chaperi samborskii* W Neumann, 2003

Christoph Samborski is a German aquarist who has collected widely, including in West Africa, French Guiana and Sri Lanka. He is currently chairman of the Rhein-Main regional group of the German Killifish Society. He collected the holotype in Ghana.

Samii

Samii's Riffle Minnow *Alburnoides samiii* H Mousavi-Sabet, S Vatandoust & I Doadrio, 2015

Dr Maajid Samii (b.1937) is an Iranian neurosurgeon and scientist who is President of the International Neuroscience Institute. After completing his high school education in Iran, he moved to Germany where he started his medical studies at the University of Mainz. He was subsequently a professor at a number of institutions in the Netherlands and Germany. He is now retired.

Samuel

Ghost Knifefish sp. *Compsaraia samueli* JS Albert & WGR Crampton, 2009

Dr Samuel Albert is the senior author's father. He accompanied his son on an electric-fish collecting trip to Peru, and found and bought the type specimens from a fish market near Iquitos, after he realised that they were different from all the other knifefish they had been collecting.

Samurai

Characin sp. *Characidium samurai* AM Zanata & P Camelier, 2014

This species is named to honour the Japanese warrior caste of the 11th-19th-centuries, specifically for their expertise in martial arts, attested today by the term 'black belt', referring to the conspicuous midlateral black band that this species sports.

Sanagal

Spiny Eel sp. *Mastacembelus sanagali* DFE Thys van den Audenaerde, 1972

This is a toponym referring to the Sanaga River, Cameroon, where the species is endemic.

Sanches

Corydoras sp. *Corydoras sanchesi* H Nijssen & IJH Isbrücker, 1967

Gijsbert Harry Sanches (b.1919) was Commissioner of Brokopondo District, Suriname. He was honoured for his 'valuable assistance' during Martin Boeseman's survey of fishes in Suriname.

Sanchez

Sanchez's Piranha *Serrasalmus sanchezi* J Géry, 1964

Dr Jorge Sánchez Romero was a Peruvian fisheries biologist and an expert on Amazon fishes. He was a founder member of the Facultade Ciencias Biologicas, Universidad Nacional Mayor de San Marcos (1948) when he was employed by the Ministry of Agriculture, Fishing and Hunting and a member of Instituto del Mar del Peru (IMARPE) of which he was the Scientific Director for many years. The main auditorium at IMARPE is named in his honour.

Sanda

Tufted Blenny *Mccoskerichthys sandae* RH Rosenblatt & JS Stephens, 1978

Despite the spelling *sandae*, this species is named after Sandra McCosker (see **Sandra (McCosker)**).

Sandager

Sandager's Wrasse *Coris sandeyeri* J Hector, 1884

Andreas Fleming Stewart Sandager was a New Zealand lighthouse keeper and amateur naturalist who collected the holotype. At that time, 'Sandager' was sometimes spelt as 'Sandeyer', thus explaining the binomial.

Sande

Sande's Mouth Almighty *Glossamia sandei* M Weber, 1907

Dr Gijsbert van der Sande (1863–1910) was a Dutch Navy physician and an anthropologist and ethnographer. He qualified as a physician in Amsterdam and joined the Navy (1890). He was posted to the East Indies (Indonesia) (1892–1901 and 1903–1909) where he collected for the Rijksmuseum. He returned to the East Indies and took part in the Dutch North New Guinea Expedition (1903), during which the holotype of this species was collected. He was part of a hydrographic expedition to Sunda, Flores and Timor (1907). He practiced as a civilian doctor at Surabaya, East Java (1909–1910). He wrote: *The Papuans of Dutch New Guinea* (1905).

Sandelie

Kurper (African anabantid) genus *Sandelia* FL Castelnau, 1861

Mgolombane Sandile (1820–1878) was a Chief of the Ngqika tribe in South Africa and led his people in the Frontier Wars with European settlers. He was captured during the War of the Axe in 1847, but on his release he was granted land in 'British Kaffraria' for his people. Castelnau used an alternative spelling of his name – 'Sandelie'.

Sanders

African Barb sp. *Labeobarbus sandersi* GA Boulenger, 1912

[Syn. *Varicorhinus sandersi*]

M C Sanders assisted the explorer William John Ansorge (1850–1913) in the Congo expedition that collected many fishes, including the type of this one (1910). Unfortunately we do not have further details about him.

Sandeyer

Sandager's Wrasse *Coris sandeyeri* J Hector, 1884

(See **Sandager**)

Sandoval

Pencil Catfish sp. *Trichomycterus sandovali* CA Ardila Rodríguez, 2006

Juan Sandoval Tarazona is a poet from Floridablanca, Colombia, and also the namesake of the cave (Don Juan Cave) where this species can be found.

Sandra (McCosker)

Tufted Blenny *Mccoskerichthys sandae* RH Rosenblatt & JS Stephens, 1978

Sandra McCosker is an anthropologist and was the wife of ichthyologist John Edward McCosker (q.v.). She participated in the first collection of this species. (It might have been expected that the binomial would be *sandrae*, but the *r* is omitted)

Sandra (Raredon)

Whale Catfish sp. *Cetopsis sandrae* RP Vari, CJ Ferraris & MCC de Pinna, 2005
Eeltail Catfish sp. *Euristhmus sandrae* EO Murdy & CJ Ferraris, 2006

Sandra Jose Raredon (b.1954). (See **Raredon**)

Sangrey

Sangrey's Blenny *Starksia sangreyae* CI Castillo & CC Baldwin, 2011

Mary Sangrey is Head of the Office of Academic Services at the Smithsonian Institution, NMNH (1982–present) and was Director of Research Program (1983–2008). She has a degree in Wildlife Management and Biology from the University of Wisconsin (1983). She has experience in Botany and curated the US National Herbarium Poaceae (grasses) collection. She designed, developed and administered a variety of Smithsonian's academic programs, including the Research Training Program, Louis Stokes Alliance for Minority Participation, Minorities in Science and Technology; PI for NSF Research Experiences for Undergraduates site awards (1988–2004). Her current focus is administration of academic programs and appointments, especially internships and fellowships and fostering research experiences for undergraduates. She was honoured for her many years coordinating the intern program at the Smithsonian's National Museum of Natural History.

Sanitwongse

Giant Pangasius *Pangasius sanitwongsei* HM Smith, 1931
[Alt. Chao Phraya Giant Catfish]

Dr Yai Suapan Sanitwongse qualified as a physician at Edinburgh University (1885). He was a medical officer in Siamese (Thai) government service, Bangkok (1885–1889). He was secretary of a special mission sent by the government to Europe (1890–1900). He resigned (1900) and abandoned medicine to become a director of the Siam Canals, Land, and Irrigation Company and devoted all his time and energies to irrigation projects. He was a keen amateur ichthyologist and brought this species to the author's attention.

Sanjay Molur

Sanjay's Black-tip Barb *Pethia sanjaymoluri* U Katwate, S Jadhav, P Kumkar, R Raghavan & N Dahanukar, 2016

Sanjay Molur is Director of the Zoo Outreach Organisation, India. He was honoured for his contribution to the conservation of endangered fauna in South Asia.

Sankey

Striped Dottyback *Pseudochromis sankeyi* HR Lubbock, 1975
[Alt. Schooling Dottyback]

Richard D Sankey was an English collector and wholesaler of marine fish for the aquarium hobby, and the founder of the Tropical Marine Centre (1970). He provided specimens for Lubbock's study and 'much useful advice' on aquarium maintenance.

Sant

Spot-tail Golden Bass *Liopropoma santi* C Baldwin & DR Robertson, 2014

Roger West Sant (b.1931) is an American businessman and philanthropist who is Co-Founder & Chairman Emeritus of the AES Corporation and Chairman of the Summit Foundation, which he and his late wife (d.2018) Victoria founded (1991) when AES went public. One of the three aims of the Foundation is ocean conservation. They provided 'generous funding' to the Smithsonian Institution's National Museum of Natural History of which he served as a Regent (2001–2013). Before starting AES (1981), he was Assistant Administrator for Energy Conservation and the Environment at the Federal Energy Administration. He was also the Director of the Energy Productivity Center and a lecturer in Finance at the Stanford Graduate School of Business. He is Vice-Chair of the board of the World Wildlife Fund US, which he previously chaired (1994–2000). He was awarded his BSc by Brigham Young University and his MBA (1960) by the Harvard Business School and is a Fellow of the American Academy of Arts and Sciences. He is co-author of *Creating Abundance – America's Least-Cost Energy Strategy* (1982).

Santana, C

Santana's Dwarfgoby *Eviota santanai* DW Greenfield & MV Erdmann, 2013

Connisso Antonino 'Nino Konis' Santana (1957–1988), a national hero in Timor-Leste's struggle for independence, was renowned for his environmental awareness. The type locality is in Tutuala, just offshore of Santana's birthplace, and is located within the Nino Konis Santana National Park.

Santana, E

Characin sp. *Hypomasticus santanai* JLO Birindelli, BF Melo, LR Ribeiro-Silva, D Diniz & C Oliveira, 2020

Edson Santana has been a technician at the Museu de Zoologia da Universidade Estadual de Londrina (MZUEL), Brazil, since 1993. He was a member of the expedition that resulted in the discovery of this species, and was honoured "*...for his help in collecting fishes, preparing vertebrate specimens and maintaining the collections of the MZUEL*".

Santana (River)

Neotropical Rivuline sp. *Simpsonichthys santanae* OA Shibatta & JC Garavello, 1992

This is a toponym referring to the Santana River, in the basin of the Paraná, Brazil.

Santos, ACA

Eartheater Cichlid sp. *Geophagus santosi* JLO Mattos & WJEM Costa, 2018

Dr Alexandre Clistenes Alcântara Santos is a Brazilian ichthyologist who is Professor at the Feira de Santana State University (UEFS) and a master's and doctoral advisor in the Postgraduate Programs in Zoology at UEFS and in Ecology and Bio-monitoring at the Federal University of Bahia. The Federal Rural University of Rio de Janeiro awarded his biology BSc (1986), MSc (1996) and his PhD was achieved at the National Museum (2003). Among his publications are the co-written: *Composition and Seasonal Variation of the Ichthyofauna from Upper Rio Paraguaçu (Chapada Diamantina, Bahia, Brazil)* (2007) and *Efeitos da regularização dos reservatórios na ictiofauna do baixo curso do rio São Francisco* (2016). He is also a friend of the authors, "*who is dedicated to the study of aquatic ecosystems of northeast Brazil.*"

Santos, GM

Pike-Cichlid sp. *Crenicichla santosi* A Ploeg, 1991

Glass Knifefish sp. *Archolaemus santosi* RP Vari, CD de Santana & WB Wosiacki, 2012

Characin sp. *Leporinus santosi* HA Britski & JLO Birindelli, 2013

Dr Geraldo Mendes dos Santos is a Brazilian aquatic biologist and philosopher. The Federal University of Mato Grosso awarded his bachelor's degree (1975) and Instituto Nacional de Pesquisas da Amazônia, Manaus his master's (1979) and doctorate (1991). He is Senior Researcher III and Professor at the Graduate School there now. He also graduated in philosophy at the Federal University of Amazonas (2007). He is Executive Secretary of the Amazon Strategic Studies Group. Among his written work is: *Effects of pH, Calcium and temperature on the ionic homeostasis of tambaqui fingerlings, Colossoma macropomum (Characiformes, Serrasalmidae)* (1998). He is recognised for his knowledge of "*...the fish fauna of the lower Rio Tocantins through his studies on anostomids and on the impacts of the Tucuruí dam*".

Santos, MD

Pogi Perchlet *Chelidoperca santosi* JT Williams & KE Carpenter, 2015

Dr Mudjekeewis Dalisay Santos is a marine evolutionary biologist who is Scientist II at the National Fisheries Research and Development Institute, Manila, Philippines (1996–

present). His PhD in Applied Marine Biosciences was awarded by Tokyo University of Marine Science and Technology (2008). He was a lecturer at the Department of Biology of the Ateneo de Manila University (2010–2013) and, part-time, is also a faculty member of the University of Santo Tomas, Graduate School (2015–present) and Ateneo de Manila University (2012–present). Among his publications are the co-written: *Milkfish and Tilapia as Biofactories: Potentials and Opportunities* (2014) and *Identification of Fish Prey of an Irrawaddy Dolphin (Orcaella brevirostris) using Mitochondrial Cytochrome c Oxidase 1 Sequence Analysis* (2019).

Sanzo

Sanzo's Goby *Lesueurigobius sanzi F de Buen,* 1918
Conger Eel sp. *Ariosoma sanzoi* U D'Ancona, 1928
Deepwater Cardinalfish sp. *Microichthys sanzoi* A Sparta, 1950

Professor Luigi Sanzo (1874–1940) was an Italian marine biologist. He started studying pharmacology at Messina University, going on to complete a medical degree (1899) followed by a degree in natural science (1901). He moved from medicine to zoology (1907) when he became Lecturer & Assistant of Comparative Anatomy & Physiology at the Institute of Zoology, Palermo University, working in marine biology. He became the Director of the Central Institute of Biology in Messina (1916–1940). He published 122 papers and books (1900–1940). He collected the eel holotype and the other Red Sea *leptocephali* featured in D'Ancona's monograph. He was Fernando de Buen's professor, which led to him being honoured in the goby's name.

Sapozhnikov

Saposchnikov's Shad *Alosa saposchnikowii* O von Grimm, 1887

Alexandre Alexandrovich Sapozhnikov (1827–1887) ran Sapozhnikov Brothers, the oldest fishery company (1796) in Astrakhan The family was renowned for its benevolence including funding, opening and running an orphanage at the time of a cholera epidemic (1830), founding hospitals and writing off enormous debts. They appear to have been responsible for the word 'pilchard', as they opened a factory (1884) to produce tins of small herrings in oil and tomato which were marketed under the name 'Pilcher'. He helped Grimm with his Caspian Sea research, and the company went on to sponsor fisheries research there. Russian artist Vasily Tropinin painted his portrait (1852).

Sara

Caspian Marine Shad ssp. *Alosa braschnikowi sarensis* AA Mikhailovsky, 1941

This is a toponym referring to Sara Island, Caspian Sea, Azerbaijan; the type locality.

Sarah

Sarah's Fairygoby *Tryssogobius sarah* GR Allen & MV Erdmann, 2012

Sarah Crow accompanied Mark Erdmann on his surveys. The etymology says the goby is named: "*...in honour of Ms. Sarah Crow, an aspiring young marine biologist who has accompanied the second author on his surveys of cryptic fish biodiversity. It is a pleasure to* *name this species after this precocious and inquisitive young ichthyologist, with the hope that this will further inspire her future investigations of the marine realm."*

Sarasin

Sarasins' Mangrove Goby *Mugilogobius sarasinorum* GA Boulenger, 1897
Sarasins' Minnow *Oryzias sarasinorum* CML Popta, 1905
Goby sp. *Sicyopterus sarasini* M Weber & LF de Beaufort, 1915
Sailfin Silversides sp. *Telmatherina sarasinorum* M Kottelat, 1991

Paul Benedikt Sarasin (1856–1929) and Karl Friederich 'Fritz' Sarasin (1859–1942) were cousins. They were Swiss zoologists, explorers and collectors. who wrote: *Reisen in Celebes* (1905). Between them, they have five birds, five reptiles and a mammal named after them, and a butterfly is named after Karl. The cousins (and Jean Roux) co-edited: *Nova Caledonia. Forschungen in Neu-Caledonien und auf den Loyalty-Inseln. Recherches scientifiques en Nouvelle-Calédonie et aux iles Loyalty* (1913–1918). *Sicyopterus sarasini* is named in honour of Fritz alone.

Sardinha

Squeaker Catfish sp. *Chiloglanis sardinhai* W Ladiges & J Voelker, 1961

Augusto Manuel Sardinha was a forestry expert and engineer in Angola. He sponsored the authors and was their companion during their collecting expedition. He wrote: *Elementary notions on fish farming* (1965).

Sarmaticus

Ukrainian Gudgeon *Gobio sarmaticus* LS Berg, 1949
Bleak sp. *Alburnus sarmaticus* J Freyhof & M Kottelat, 2007

These are not true eponyms but refer to the Sarmatians, a group of tribes that inhabited southern Russia, Ukraine and the eastern Balkans from 5th century BC to 4th century AD. The Sarmatians occupied the area of distribution of these fish.

Sarmiento, J

Pencil Catfish sp. *Stenolicmus sarmientoi* MCC de Pinna & WC Starnes, 1990

Jaime Sarmiento Tavel (b.1955) is a zoologist and ichthyologist who is a member of the faculty at Museo Nacional de Historia Natural, La Paz, Bolivia where he is a researcher and curator who has spent much of his career (1977–present) researching Bolivian fish. He was educated at Universidad Mayor de San Andrés. He co-wrote: *Fishes of the Rios Tahuamani, Manuripi and Nareuda, Depto. Pando, Bolivia: diversity, distribution, critical habitats and economic value* (1999) and *Unexpected fish diversity gradients in the Amazon basin* (2018) among other papers.

Sarmiento, M

Bandfish sp. *Owstonia sarmiento* Y-C Liao, RB Reyes & K-T Shao, 2009

Malcolm Sarmiento Jr was (until 2018) President & Executive Director of the Bureau of Fisheries and Aquatic Resources (Philippines) for three decades. He was instrumental in making possible the series of deep-sea expeditions (2007) during which the holotype was collected.

Sarraf

Neotropical Livebearer sp. *Poecilia sarrafae* PHN Bragança & WJEM Costa, 2011

Alessandra Sarraf Pereira was a Brazilian ichthyologist at the Biological Institute of the Federal University of Rio de Janeiro. She has published papers with Wilson Costa such as: *Poecilia (Lebistes) minima, a new species of neotropical poeciliid fish from the Brazilian Amazon* (1998). It was she "*...who first studied the new species in her unpublished revision of Micropoecilia*" (1998) – we believe this to be her Masters' dissertation.

Sars

Sars' Wolf-Eel *Lycenchelys sarsii* R Collett, 1871

Georg Ossian Sars (1837–1927) was a Norwegian biologist. He was commissioned by the Norwegian government (1864) to investigate fisheries around the country's coast. His main specialism was in crustaceans, of which he named many new species.

Sasaki

Croaker sp. *Johnius sasakii* N Hanafi, M-H Chen, YG Seah, C-W Chang, SYV Liu & NL Chao, 2022

Dr Kunio Sasaki is a Japanese ichthyologist at Kochi University. He was Editor-in-Chief of the publication 'Ichthyological Research' (2010-2014). Among his more than sixty scientific papers is the co-written: The Lateral Line System in the Nurseryfish Kurtus gulliveri (Percomorpha: Kurtidae): A Distribution and Innervation of Superficial Neuromasts Unique within Percomorphs (2021). He was honoured for his 'significant contributions' to the taxonomy of Johnius and related genera.

Sasal

Goby sp. *Smilosicyopus sasali* P Keith & G Marquet, 2005

Dr Pierre Sasal is a French ecologist, Associate Researcher and scientific diver at the French National Centre for Scientific Research, which he joined (2000). The University of Marseille awarded his doctorate (1997), after which he did post-doctoral research at the University of Windsor, Ontario, Canada (1998–1999) and at the University of Perpignan (1999–2000), combining it with being an Associate Professor there (1997–1998)

Sasha

Freshwater Hardyhead genus *Sashatherina* W Ivantsoff & GR Allen, 2011

Sasha Ivantsoff is the senior author's son, "*...who has assisted his father frequently on collecting trips*". His name is coupled with *Atherina*, a related genus within the silverside family (Atherinidae).

Satan

Neotropical Cichlid genus *Satanoperca* A Günther, 1862

Blindcat (catfish) genus *Satan* CL Hubbs & RM Bailey, 1947

Named after the devil, and referring in the case of the blindcat to the species' underground habitat. Günther's use of a name meaning 'Satan perch' for the cichlids was probably inspired by Heckel (q.v.) having named a species as '*Geophagus daemon*'.

Satapoomin

Andaman Sole *Aseraggodes satapoomini* JE Randall & M Desoutter-Meniger, 2007

Ukrit Satapoomin is a Thai marine biologist who was head of the Phuket Marine Biological Center's survey team and research group. He was one of the collectors of the holotype.

Sato, T

Blacknape Large-eye Bream *Gymnocranius satoi* P Borsa, P Béarez, S Paijo & W-J Chen, 2013

Professor Torao Sato (b.1945) is a Japanese ichthyologist with expertise in the Lethrinidae (the family of fishes to which this species belongs). His written work includes: *A Synopsis of the Sparoid Fish Genus* Lethrinus *with the Description of a New Species* (1978).

Sato, Y

Characin sp. *Characidium satoi* MRS Melo & OT Oyakawa, 2015

Dr Yoshimi Sato is a biologist at Companhia de Desenvolvimento dos Vales do São Francisco e do Parnaíba (CODEVASF) who led research at Três Marias Integrated Fisheries and Aquaculture Center. He was honoured for his contributions to the knowledge and conservation of the ichthyofauna of the Rio São Francisco drainage, Minas Gerais, Brazil. His written work includes: *Reproductive biology and induced reproduction of two species of Characidae (Osteichthyes, Characiformes) from the São Francisco basin, Minas Gerais, Brazil* (2006).

Satomi

Satomi's Pygmy Seahorse *Hippocampus satomiae* SA Lourie & RH Kuiter, 2008

Satomi Onishi is a dive guide and underwater photographer in West Bali, Indonesia. She collected the holotype off Derewan Island in Indonesian Borneo.

Satterlee

Twinray Dragonfish *Eustomias satterleei* W Beebe, 1933

Herbert Livingston Satterlee (1863–1947) was an American lawyer, businessman and writer. He graduated from Columbia University, New York, and was US Assistant Secretary of the Navy (1908–1909). He was a patron of the New York Zoological Society, where the author worked. He wrote: a biography of his father-in-law: *J Pierpoint Morgan: An Intimate Portrait* (1939). He committed suicide, presumably to avoid the drawbacks of failing health.

Satunin

White Kura Bleak *Leucalburnus satunini* LS Berg, 1910

Colchic Spined Loach *Cobitis satunini* NA Gladkov, 1935

Konstantin Alexeevitsch Satunin (1863–1915) was an eminent Russian zoologist who studied the fauna of the Caucasus region. Like his predecessor Radde, he was initially most interested in birds, writing: *A Systematic Catalogue of the Birds of the Caucasian Region* (1912). However, he was an allrounder, collecting fish, insects and mammals and writing on an extensive range of topics. He wrote: *New mammals from Transcaucasia* (1914) and *Mammals of the Caucasian Region* (1915). A mammal and two birds are also named after him. While the patronym is not identified in the description of the loach, it seems almost certain that he was the man intended.

Saul

Whale Catfish sp. *Denticetopsis sauli* CJ Ferraris, 1996

American Sole sp. *Apionichthys sauli* RTC Ramos, 2003

William 'Bill' G Saul (b.1944) was collection manager of the Ichthyology Department of the Academy of Drexel University; formerly known as the Academy of Natural Sciences of Philadelphia (1972–1999). He graduated from the University of Kansas with a bachelor's degree (1966) worked as a Curatorial Assistant, Fish Division, Kansas University Museum of Natural History (1969–1970), served in the US Army (1970–1972) and was awarded a master's degree (1970) by the University of Kansas. He wrote: *An Ecological Study of Fishes at a site in the Upper Amazonian Ecuador* (1975). He helped collect the type series of the catfish species and brought it to the author's attention.

Saulos

Greenface Aulonocara *Aulonocara saulosi* MK Meyer, R Riehl & H Zetzsche, 1987

Lake Malawi Cichlid sp. *Chindongo saulosi* A Konings, 1990

[Syn. *Pseudotropheus saulosi*]

Saulos Mwale of Salima, Malawi, is a diver and mechanic working for cichlid exporter Stuart M Grant. He is also the discoverer of many Malawian cichlids. According to Konings he caught more than 3,000 specimens single-handed during a ten-week collecting expedition in Lake Malawi.

Saumarez

Saumarez Gurnard *Pterygotrigla saumarez* PR Last & WJ Richards, 2012

This is a toponym referring to the Saumarez Reef in the Coral Sea, the general collecting area for most of the type series.

Sauri

Goby sp. *Schismatogobius saurii* P Keith, C Lord, H Darhuddin, G Limmon, T Sukmono, R Hadiaty & N Hubert, 2017

Sopian Sauri is an Indonesian zoologist and ichthyologist who works at LIPI (Indonesian Institute of Sciences), Zoology Division, Museum Zoologicum Bogoriense, Djakarta. He co-wrote: *A new* Stiphodon *(Gobiidae) from Indonesia* (2015). He was honoured as thanks for all the help he gave the authors in collecting freshwater fishes all around Indonesia.

Sauron

Driftwood Catfish subgenus *Sauronglanis* S Grant, 2015

Sauron is a fictional character, the Dark Lord in Tolkien's: *Lord of the Rings* trilogy.

Sauter

Blacktip Sawtail Catshark *Galeus sauteri* DS Jordan & RE Richardson, 1909

Dr Hans Sauter (1871–1943) was a German entomologist who also became interested in herpetology. He studied biology at the Universities of Munich and Tübingen. He was in Formosa (Taiwan) under Japanese occupation, collecting insects (1902–1904). He was in Tokyo (1905) before returning to Taiwan for the rest of his life. He worked for a British trading company but entomology was his passion. Although Japan and Germany were enemies in WW1, he kept his job and continued collecting, albeit kept under observation. He was said to be the first person to offer private piano lessons in Taiwan, and gave German and English lessons. He collected the holotype from a fish market in Taiwan. Three reptiles and an amphibian are also named after him.

Sauvage

Round Herring genus *Sauvagella* L Bertin, 1940

Chinese Cyprinid sp. *Hemiculterella sauvagei* NA Warpachowski, 1887

Sauvage's Mormyrid *Petrocephalus sauvagii* GA Boulenger, 1887

Lake Victoria Cichlid sp. *Haplochromis sauvagei* GJ Pfeffer, 1896

Flagtail sp. *Kuhlia sauvagii* CT Regan, 1913

Dr Louis César Henri Émile Sauvage (1842–1917) was a French palaeontologist, herpetologist and ichthyologist. He was an expert on Mesozoic fish and reptiles, and wrote: *Vertébrés fossiles du Portugal: Contributions à l'étude des poissons et des reptiles du jurassique et du crétacique* (1897). Two reptiles and a frog are also named after him. Boulenger honoured him as Sauvage: *"...has added much to our knowledge of the fishes of tropical Africa."* He collected the 15 specimens that Bertin used when proposing the eponymous genus.

Savage

Savage Tetra *Hyphessobrycon savagei* WA Bussing, 1967

Whipnose Angler sp. *Gigantactis savagei* E Bertelsen, TW Pietsch & RJ Lavenberg, 1981

Savage's Bird-snouted Whalefish *Rhamphocetichthys savagei* JR Paxton, 1989

Dr Jay Mathers Savage (b.1928) is an American herpetologist and educator who is known particularly for his work on the herpetofauna of Central America. He took all his degrees at Stanford; AB (1950), MA (1954) and PhD (1955). He was an assistant professor at Pomona College (1954–1956) before joining the faculty at the University of Southern California, Los Angeles as Assistant Professor (1957–1959), Associate Professor in Biology (1959–1964), Professor of Biology (1965–1982). He then took the chair of biology at the University of Miami, Florida (1982–1999). He is also Adjunct Professor at San Diego State University. He was President of the American Society of Ichthyologists and Herpetologists (1982), and after being on the board of directors of the Organisation of Tropical Studies for ten years he was President (1973–1980) and has held many other high offices. Among his c.200

publications is: *The Amphibians and Reptiles of Costa Rica: a Herpetofauna Between Two Continents, Between Two Seas* (2002). Eight amphibians and six reptiles are also named after him, as well as other taxa.

Savigny

Vairone (Italian cyprinid) sp. *Telestes savigny* CL Bonaparte, 1840

Marie Jules César Lelorgne de Savigny (1777–1851) was a French zoologist and artist. He studied medicine, but under Geoffroy SaintHilaire's influence turned to zoology. He was in Egypt (1798–1800) and undertook several expeditions, including one studying the avifauna of Lake Manzala, sending specimens to SaintHilaire. He wrote: *Description d'Egypte; ou Recueil des Observations et des Recherches qui ont été Faites en Egypte Pendant l'Expédition de l'Armée Française* (1798–1801), *Histoire Naturelle et Mythologique de l'Ibis* (1805) and *Système des Oiseaux de l'Égypte et de la Syrie* (1810), derived from research conducted during Napoleon's occupation of Egypt (1790s). Savigny fell out with SaintHilaire, who blocked him from becoming Professor at the Natural History Museum. He became ill (1823) and spent all his latter years in poor health. Three birds, two reptiles and an amphibian are also named after him.

Savinov

Lenok (Asiatic trout) sp. *Brachymystax savinovi* VP Mitrofanov, 1959

E F Savinov was a mammalogist and one of Mitrofanov's colleagues at the Academy of Sciences Kazakhstan USSR, Institute of Zoology. He wrote: *Siberian mountain goat and mountain sheep in Kazakhstan* (1975). The patronym is not identified in Mitrofanov's text, but the above seems a likely candidate for this naming.

Savorgnan

Squeaker Catfish sp. *Atopochilus savorgnani* H-E Sauvage, 1879

Count Pierre Paul François Camille Savorgnan de Brazza (1852–1905) was the elder brother of Jacques (or Giacomo) C Savorgnan de Brazza. The brothers were born in Rio de Janeiro of Italian descent. Count Pierre was a distinguished explorer who entered the French Navy (1870) and served in Gabon. The French government gave him 100,000 francs for exploring the country north of the Congo (1878), where he secured vast grants of land for France and founded stations, including that of Brazzaville on the north shore of Stanley Pool. He returned (1883) largely unsubsidised by the French government, established further stations and continued to explore (1886–1897). He was Commissioner-General of the French Congo (1896–1898). A bird is named after him.

Savory

Lake Tanganyika Cichlid sp. *Neolamprologus savoryi* M Poll, 1949

Bryan Wyman Savory (1904–1988) was District Commissioner of Kigoma (Tanganyika Territory). He was honoured for the: 'excellent reception' (translation) he gave the Belgian Hydrobiological Mission to Lake Tanganyika (1946–1947), during which the type was collected.

Sawada

Loach sp. *Barbatula sawadai* AM Prokofiev, 2007

Dr Yukio Sawada was a Japanese biologist and ichthyologist at the Laboratory of Marine Zoology, Faculty of Fisheries, Hokkaido University, which awarded his doctorate (1981). He was at the Seibu College of Medical Tchnology, Tokorozawa, Japan (1983). He was honoured because of his study of the osteology of loaches. Among his papers are: *Transfer of Cobitis multifasciata to the Genus Niwaella (Cobitidae)* (1977) and *Phylogeny and zoogeography of the superfamily Cobitoida (Cyprinoidei, Cypriniformes)* (1982).

Say

Bluntnose Stingray *Hypanus say* CA Lesueur, 1817
Pirate Perch *Aphredoderus sayanus* J Gilliams, 1824

Thomas Say (1787–1834) was a self-taught American naturalist whose primary interest was entomology. He described over 1,000 new species of beetles and over 400 new insects of other orders. He became a charter member and founder of the Academy of Natural Sciences of Philadelphia (where he became friends with Lesueur) (1812) and was appointed chief zoologist with Major Stephen H Long's expeditions, which explored the Rocky Mountains (1819–1820). He lived at the utopian village of 'New Harmony' in Indiana (1826–1834). Say wrote: *American Entomology, or Descriptions of the Insects of North America* (1824–1828) and *American Conchology* (1830–1834). Say's death was one of the things that prompted Lesueur to return to France. A mammal, a bird genus and a bird species and a reptile are named after him.

Saylor

Laurel Dace *Chrosomus saylori* CE Skelton, 2001

Charles F 'Charlie' Saylor (b.1948) is an ichthyologist and aqua-biologist who worked for most of his career (1972–2013) at Tennessee Valley Authority. He was part of the crew that first collected the Laurel Dace. Among his scientific papers are: *An investigation of sauger spawning in the vicinity of the Clinch River Breeder Reactor Plant* (1983) and *Fish Hosts for Four Species of Freshwater Mussels (Pelecypoda: Unionidae) in the Upper Tennessee River Drainage* (1995). He is a life-long angler. Skelton honoured him for his contributions to the knowledge of southeastern USA fishes.

Sayona

Glass Knifefish sp. *Eigenmannia sayona* LA Peixoto & BT Waltz, 2017

In Venezuelan folklore, La Sayona is a spirit of philanderous vengeance. Apparently, the name is a homage to the Venezuelan people with no meaning or significance to the fish itself.

Sazima

Characin sp. *Roeboides sazimai* CAS de Lucena, 2007
Wrasse sp. *Halichoeres sazimai* OJ Luiz, CEL Ferreira & LA Rocha, 2009
Characin sp. *Probolodus sazimai* O Santos & RMC Castro, 2014

Professor Dr Ivan Sazima is a Brazilian ichthyologist, herpetologist, ornithologist and marine biologist. He is now retired from the Department of Zoology and Museum of Natural History, University of Campinas, Brazil where he worked (1973–2008). His three degrees, bachelor's in biological sciences (1971), master's (1975), and doctorate (1980) in zoology were all awarded by the University of São Paulo. Among his more than 340 works, many describing new species, he co-wrote: *Corydoras desana, a new plated catfish from the upper rio Negro, Brazil, with comments on mimicry within the Corydoradinae (Ostariophysi: Siluriformes: Callichthyidae)* (2017). His current interests include natural history and feeding behaviour of birds from urbanized areas and the Atlantic Forest, natural history and behaviour of reef and freshwater fish, natural history and behaviour of reptiles and amphibians of the Atlantic Forest, pollination by vertebrates, cleaning symbiosis between vertebrates, associations between birds and mammals, and associations between birds and insects. Four amphibians and two reptiles are named after him.

Sazonov

Morid Cod sp. *Physiculus sazonovi* KD Paulin, 1991
Beardfish sp. *Polymixia sazonovi* AN Kotlyar, 1992
Dark Smiling Whiptail *Ventrifossa sazonovi* T Iwamoto & A Williams, 1999
Conger Eel sp. *Ariosoma sazonovi* ES Karmovskaya, 2004
Rattail sp. *Hymenocephalus sazonovi* W Schwarzhans, 2014

Dr Yuriya (also spelled Yurii) Igorevich Sazonov (1950–2002) was a Russian ichthyologist, marine biologist and botanist at the Marine Biological Laboratory, Department of Ichthyology, Zoological Museum, who graduated from Moscow State University (1972). His doctorate was awarded (1989). He was Senior Research Scientist and Curator of the marine fish collection, Zoological Museum, Moscow State University (1990). He was a participant in four ocean expeditions: Equatorial Pacific Ocean (1973), Indian Ocean (1974), Australia and New Zealand (1976) and south-eastern Pacific Ocean (1987). Among his co-authored papers is: *Indian Ocean grenadiers of the subgenus Con'phaenoides, Genus Conphaenoides (Macrouridae, Gadiformes, Pisces)* (1995) and he wrote: *A New Species of the Genus* Conocura *(Alepocephidae) from the Arabian Sea* (2002). He died of severe cold contracted when he continued to work at his laboratory at the Zoological Museum of the University of Moscow during roof repairs being performed during the winter.

Scarlato

Goby sp. *Reptiliceps scarlatoi* AM Prokofiev, 2007

Dr Orest Alexandrovich Scarlato (1920–1994) was a Russian zoologist and malacologist. The University of Leningrad, USSR (St Petersburg, Russia) awarded his bachelor's degree (1950) and his doctorate (1974). HIs career was spent at the Russian Academy of Sciences as a researcher and administrator (1953–1974), culminating as DIrector (1974–1994). He wrote: *Bivalve Molluscs of Far Eastern Seas of the Union of the Soviet Socialist Republics* (1960). He collected the holotype of this species (1957).

Scarll

Michoacan Livebearer *Poeciliopsis scarlli* MK Meyer, R Riehl, JA Dawes & I Dibble, 1985

John Richard Scarll (b.1945) from Doncaster, England, was one of the collectors of the holotype in Mexico (1982). He and his wife Joan bought Belton Fish Farm (1976), a wholesale ornamental fish business, having previously run a small aquatic shop. He is still (2019) Managing Director there. They import most of their fish through Singapore.

Schaap

Short-snout Sand-Dragonet *Callionymus schaapii* P Bleeker, 1852

Dirk François Schaap (1816–1864) was a Dutch colonial administrator in Indonesia who became Resident at Bangka, Sumatra, Dutch East Indies; the type locality of this species.

Schaefer

Wood-eating Catfish sp. *Panaque schaeferi* NK Lujan, M Hidalgo & DJ Stewart, 2010
Armoured Catfish sp. *Rhinolekos schaeferi* F de O Martins & F Langeani, 2011

Dr Scott A Schaefer joined the staff at AMNH (1996) and was Curator, Department of Ichthyology, Division of Vertebrate Zoology (2001–2008). He is now Dean of Science for Collections, Exhibitions, and the Public Understanding of Science (since 2015) and (since 2008) Professor, Richard Gilder Graduate School. Ohio State University awarded his bachelor's degree (1980), the University of South Carolina his master's (1982) and the University of Chicago his doctorate (1986). He did post-doctoral study at the Smithsonian (1987–1988). HIs earlier career included working at the Academy of Natural Sciences of Philadelphia (1988–1996). He co-wrote: *Systematics of the genus* Hypoptopoma *Günther, 1868 (Siluriformes, Loricariidae)* (2010).

Schaefers

Goosefish sp. *Sladenia schaefersi* JH Caruso & HR Bullis, 1976

Dr Edward A Schaefers was an American fisheries scientist, described in the etymology as "*…former chief of the Exploratory Fishing and Gear Research Branch of the former Bureau of Commercial Fisheries, in recognition of his outstanding contributions to marine science.*"

Schal

Wahrindi (catfish) *Synodontis schall* ME Bloch & JG Schneider, 1801

Schal is an Arabian vernacular for squeaker catfishes in Egypt.

Schaller

Schaller's Croaking Gourami *Trichopsis schalleri* W Ladiges, 1962
[Alt. Threestripe Gourami]
Betta sp. *Betta schalleri* M Kottelat & PKL Ng, 1994

Dietrich Schaller is a German aquarist and importer of tropical fish. He has written a number of articles such as, with Maurice Kottelat: *Betta strohi sp. n., ein neuer Kampffisch aus Südborneo (Osteichthyes: Belontiidae)* (1989). He was honoured in the gourami's name as the first to import the fish to Europe.

Scharff

Scharff's Char *Salvelinus scharffi* CT Regan, 1908

Dr Robert Francis Scharff (1858–1934) was a British zoologist. University College, London awarded his bachelor's degree, Edinburgh University his master's (1885) and Heidelberg his doctorate; after which he studied at the St Andrews Biological Station, New Brunswick, Canada and at Stazione Zoologica, Naples. He was an Assistant in the Natural History Division, National Museum of Ireland (1887–1890), becoming Keeper (1890–1916) and Acting Director (1916–1922). He wrote: *The History of the European Fauna* (1899). Regan wrote in the etymology that he was "*...indebted for the opportunity of describing it, in recognition of the favours I have received from him during my work on Irish fishes.*"

Schauensee

Tetra sp. *Hyphessobrycon schauenseei* HW Fowler, 1926

Dr Rodolphe Meyer de Schauensee (1901–1984) was an Italian-born American ornithologist. He worked at the Academy of Natural Sciences, Philadelphia, for nearly fifty years. He went on many expeditions, including to Burma, Southern Africa, Brazil, Indonesia, Guatemala and at least three to Siam (Thailand), where he collected anything and everything. He wrote, among a considerable output: *The Birds of Colombia* (1964), *The Species of Birds of South America with their Distribution* (1966), *A Guide to the Birds of America* (1971) and *The Birds of China* (1984). Two reptiles and seven birds are named after him.

Schauinsland

Red-spotted Sandperch *Parapercis schauinslandii* F Steindachner, 1900

Dr Hugo Hermann Schauinsland (1857–1937) was a German zoologist and explorer. He studied natural sciences at Geneva (1879) and zoology (1880–1883) at Königsberg (now Kaliningrad in Russia), where he was awarded his doctorate. He was involved (1883–1885) in research in Naples and Munich, and in the latter year became a Professor at the University of Munich. He became (1887) the Superintendent of the Municipal Collections of Natural History and Ethnology in Bremen. He travelled in East Asia and the Pacific (1896–1897), returning home via New Zealand, Australia, Ceylon (Sri Lanka) and Egypt. A mammal is also named after him.

Scheel

African Rivuline sp. *Fundulopanchax scheeli* AC Radda, 1970
Scheel's Lampeye *Poropanchax scheeli* B Roman, 1971
[Syn. *Micropanchax scheeli*]

Jørgen Jacob Scheel (1916–1989) was a Danish ichthyologist who specialised in the Aplocheiloidei. Among his published papers are: *Another new species of* Aphyosemion *from the Cameroons* (1975) and *Revue systématique de la superespèce Aphyosemion elegans* (1981) co-written with JH Huber. (Also see **Jørgen Scheel**)

Scheele

Tropical Conger *Ariosoma scheelei* PH Strömman, 1896

Captain George von Schéele was a seaman and amateur naturalist. He was Chief Officer of the Swedish ship 'Monarch' which undertook a three-year circumnavigation. Pehr Hugo Strömman, then Curator at Upsala University Museum, approached him and asked that he skim the surface of the sea throughout the time and collect anything he netted. This was done using small muslin nets that just collected small specimens which were pickled in alcohol and shipped back periodically to Upsala in several airtight cast-iron casks that were supplied by the museum. He collected the conger holotype.

Scheffers

Scheffers' Kivu (cichlid) *Haplochromis scheffersi* J Snoeks, LDG de Vos & DFE Thys van den Audenaerde, 1987

W J Scheffers was Director of the FAO Project for Fisheries Development at Lake Kivu, Rwanda, whose benevolence and kindness greatly facilitated the authors' work in the field. He co-wrote: *Répartition de la production de L. miodon et du chiffre d'affaires pour les différents produits* (1984).

Schelkovnikov

Common Roach ssp. *Rutilus rutilus schelkovnikovi* AN Derjavin, 1926

Alexander Bebutovich Schelkovnikov (aka Shelkovnikov) (1870–1933) was a Russian zoologist who founded (1922) the Herbarium of Armenia, which later became part of the Institute of Botany of the Armenian National Academy of Sciences. He worked for a time as an official in the Armenian agricultural administration. His collections, mainly of zoological and botanical specimens from Georgia, Armenia, and Azerbaijan, are held across parts of the old Soviet Union, particularly in Yerevan, St. Petersburg, and Moscow. He was a graduate of a military school in St. Petersburg, so had no formal zoological training, but was a gifted amateur who contributed much to the knowledge of the natural history of the Caucasus and Persia (now Iran). He spent part of his career as the unofficial Curator of Scientific Collections at the Moscow State University. More than 30 plant species are named after him, as are some 20 invertebrates and two mammals. He helped Derjavin collect the roach type.

Schelly

African Tetra sp. *Micralestes schelly* MLJ Stiassny & V Mamonekene, 2007

Dr Robert C Schelly is an American ichthyologist researching at the AMNH. Among his publications are: *Revision of the Congo River Lamprologus Schilthuis, 1891 (Teleostei: Cichlidae), with Descriptions of Two New Species* (2004) and *Nanochromis teugelsi, a new cichlid species (Teleostei: Cichlidae) from the Kasai Region and central Congo basin* (2006). He collected the type series (2002) in the Congo River.

Scherer

Dwarf Hatchetfish *Carnegiella schereri* A Fernández-Yépez, 1950

William G Scherer of the Evangelical Mission at Pebas, Peru, collected the holotype and according to the etymology provided the authors with "…*magnificent help with his collection of fishes from the Peruvian Amazon*". He had been sending large collections since 1933 to the Natural History Museum of Stanford University.

Scherzer

Leopard Mandarin Fish *Siniperca scherzeri* F Steindachner, 1892

A Scherzer collected natural history material with Bruno Navarra around Shanghai sending them to, among other places, the Vienna Natural History Museum where Franz Steindachner was curator of the fish collection (1860) and director of the whole zoology department (1887–1898). The original text mentions the source of the specimen as "*Herrn A. Scherzer und B.R. Navarra in Shanghai dem kaiserlichen Hofmuseum durch die gutige Vermittlung des Herrn Generalconsuls J Haas…*" Steindachner named another fish after Navarra (q.v.).

Schiff

Barbeled Dragonfish sp. *Eustomias schiffi* W Beebe, 1932

Mortimer Loeb Schiff (1877–1931) was an American banker, philanthropist and collector of books and fine art. He was greatly interested in the boy scout movement and at the time of his death had been appointed President of the Boy Scouts of America. The etymology says that his "*...interest in the work of this expedition* [to Bermuda] *was very deep and sincere*."

Schiffermüller

Austrian Lakes Trout *Salmo schiefermuelleri* ME Bloch, 1784

Johann Ignaz Schiffermüller (1727–1806) was an Austrian theologian and naturalist whose main interest was in Lepidoptera. He taught at the Theresianum College, Vienna. He wrote: *Versuch eines Farbensystems* (1772). He provided the holotype of the trout, and was rewarded by Bloch misspelling his name!

Schilthuis

Kutu Mormyrid *Marcusenius schilthuisiae* GA Boulenger, 1899

Lubbina Schilthuis (1863–1951), sometimes known as Louise Schilthuis, was a Dutch ichthyologist and herpetologist who became Curator at the Museum of Zoology, University of Utrecht. She wrote: *On a Collection of Fishes from the Congo; with Descriptions of Some New Species* (1890). She originally identified the eponymous species (1891) under the name *Mormyrus grandisquamis* (Peters 1876).

Schindler, I

Neotropical Rivuline sp. *Pituna schindleri* WJEM Costa, 2007

Ingo Schindler is a German ichthyologist, evolutionary biologist and zoologist. He was the first collector of the species. Among his publications are the co-authored: *Rivulus monticola, a new killifish (Cyprinodontiformes: Rivulidae) from the eastern slopes of the Cordillera de Allcuquiro, Ecuador* (1997) and *Description of Andinoacara stalsbergi sp. n.*

(Teleostei: Cichlidae: Cichlasomatini) from Pacific coastal rivers in Peru, and annotations on the phylogeny of the genus (2009).

Schindler, O

Infantfish genus *Schindleria* LP Giltay, 1934
Schindler's fish *Schindleria praematura* O Schindler, 1930
Characin sp. *Oligosarcus schindleri* NA de Menezes & J Géry, 1983
Characin sp. *Characidium schindleri* A Zarske & J Géry, 2001

Otto Schindler (1906–1959) was an ichthyologist at Zoologische Staatssammlung München, Germany, where he was the first post-war Curator of Ichthyology (1949). He sought to reconstruct the museum's collection, as it had been completely destroyed by Allied bombing (1944). It helped that he had worked on the collection as an Assistant (1931–1939). He died of a heart attack whilst on a collecting trip in France. He described the first known species in the eponymous genus under the name *Hemiramphus praematurus.*

Schioetz

Schitz' Lampeye *Rhexipanchax schioetzi* JJ Scheel, 1968
African Rivuline sp. *Aphyosemion schioetzi* JH Huber & JJ Scheel, 1981

Dr Arne Schiøtz (1932–2019) was a Danish herpetologist who was an authority on African tree frogs. He received a master's degree in zoology (1959) and a doctorate (1967). He was Curator of Herpetology at Copenhagen Zoo (1959–1964) and Director, Danish Aquarium (1964–1996). He was 'on leave' from the Aquarium (1980–1983) to be Director of Conservation, WWF-International, and (1990–1993) as Senior Technical Advisor to the Government of Bhutan. He contributed much of the comprehensive collection of treefrogs in the Zoological Museum, Copenhagen. He wrote: *Treefrogs of Africa* (1999). He collected in Ghana and Zaire for JJ Scheel (q.v.) Four amphibians are also named after him.

Schischkov

Black Sea Bleak *Alburnus schischkovi* P Drensky, 1943

Dr Georgi Chichkoff (also spelled Schischkov) (1865–1943) was a Bulgarian biologist who headed the Zoological Institute at the Higher School in Sofia (now St. Kliment Ohridski University of Sofia). He wrote about the local ichthyofauna, including: *The fishes in our rivers of the Aegean watershed* (1939 – in Bulgarian).

Schlegel

Brown Guitarfish *Rhinobatos schlegelii* JP Müller & FGJ Henle, 1841
Japanese Rubyfish *Erythrocles schlegelii* J Richardson, 1846
Giant Skarkminnow *Osteochilus schlegelii* P Bleeker, 1851
Blackhead Seabream *Acanthopagrus schlegelii* P Bleeker, 1854
Bandfish sp. *Cepola schlegelii* P Bleeker, 1854
Schlegel's Cardinalfish *Ostorhinchus schlegeli* P Bleeker, 1855
Seaweed Pipefish *Syngnathus schlegeli* JJ Kaup, 1856
Sunrise Perch *Caprodon schlegelii* A Günther, 1859
Goby sp. *Porogobius schlegelii* A Günther, 1861

Yellowband Parrotfish *Scarus schlegeli* P Bleeker, 1861
Korean Rockfish *Sebastes schlegelii* FM Hilgendorf, 1880

Hermann Schlegel (1804–1884) was a German-born zoologist who spent much of his life in the Netherlands. He was the first person (1844) to use trinomials to describe separate races of species. Schlegel made a trip on foot through large parts of Germany and Austria (1824–1825). While he was in Vienna (1825) he received a letter from Jacob Coenraad Temminck (q.v.) who was looking for a researcher to explore parts of Indonesia. Schlegel went to Leiden and worked so hard and well for Temminck that the latter decided that Schlegel had to stay at Leiden, as he was too valuable to risk on an overseas assignment. He succeeded Temminck as Director at the Museum (1858). He was an opponent of Darwin's theories, believing that species were fixed and immutable. Sixteen birds, two amphibians, ten reptiles and four mammals are also named after him.

Schleser

Neotropical Rivuline sp. *Moema schleseri* WJEM Costa, 2003
[Syn. *Aphyolebias schleseri*]

David M Schleser trained as a dentist before switching careers to aquatic biology. He is also a photographer and author. He worked at the Dallas Aquarium, Texas, before helping to establish 'Nature's Images, Inc.' – a natural history photography and writing company. He wrote: *Piranhas (Complete Pet Owners Manual)* (1997) and *North American Native Fishes for the home aquarium* (1998).

Schliewen

Andromeda Goby *Didogobius schlieweni* PJ Miller, 1993

Dr Ulrich Schliewen is a German ichthyologist and aquarist who is Head of the Ichthyology Section of the Bavarian State Collection of Zoology and who was at the Zoolog Staatssammlung München. He is particularly interested in the freshwater fish of west and central Africa and Sulawesi, Indonesia. Among his publications are: *Cichlid species flocks in small Cameroonian lakes* (2005) and *Diversity and distribution of marine, euryhaline and amphidromous gobies from Western, Central and Southern Africa* (2011). He has also written a number of books on keeping tropical fish, such as: *Aquarium Fish* (2000), *Tropical Freshwater Aquarium Fish* (2005) and *My Aquarium* (2008). He collected the type and suggested the common name 'Andromeda Goby' because of its nebula-like pattern of light and dark markings.

Schlosser

Giant Mudskipper *Periophthalmodon schlosseri* PS Pallas, 1770

Dr Johann Albert Schlosser (1733–1769) was a Dutch physician and naturalist who was the author's "very close friend". The University of Leiden awarded his doctorate (1753). He travelled in France and England (1755–1756), including spending a year (1755–1756) in England during which he was elected (1756) a Fellow of the Royal Society. He returned to Amsterdam (1756) and set up in practice as a physician. It is interesting to note that this is the first instance in ichthyological literature to use the patronymic "*i*"

Schlupp

African Rivuline sp. *Aphyosemion schluppi* AC Radda & JH Huber, 1978

Father Gerard Schlupp was a priest at the Zanaga mission post. He provided hospitality to Belgian aquarists John Buytaert and Walter Wachters, who collected the type.

Schmarda

Tetra sp. *Hemigrammus schmardae* F Steindachner, 1882
Chupare Whipray *Styracura schmardae* F Werner, 1904

Dr Ludwig Karl Schmarda (1819–1908) was an Austrian physician, naturalist and traveller. After qualifying as a physician in Vienna (1843) he became an army surgeon. He taught at the Joanneum, Graz (1848–1850). He was Professor of Natural History, Karl-Franzens-Universität, Graz (1850–1852) and Professor of Zoology and Director of the Zoological Museum, Univerzita Karlova v Praze (Charles University, Prague) (1852–1862). He undertook a private circumnavigation of the world (1853–1857) by 'hitch-hiking' on various sailing ships; it was a rough adventure as he suffered scurvy on the voyage from South Africa and Australia, lost his collections in a fire in Chile and was robbed in Panama. He was Professor of Zoology, Universität Wien (1862–1883). After retirement, he travelled in Spain and North Africa (1884–1887). He wrote: *Die Geographische Verbreitung der Thiere (1853). He collected the holotype of the whipray. An amphibian is also named after him.*

Schmidt

Duckbill Eel sp. *Hoplunnis schmidti* JJ Kaup, 1860

Dr Schmidt (forename not given) was one of the directors of the Hamburg Museum, who provided specimens to Kaup.

Schmidt, C

Sisorid Catfish sp. *Glyptothorax schmidti* W Volz, 1904

Dr Carl Schmidt (1862–1923) was Professor of Geology, Basel University, Switzerland. He did work for the oil industry, including being the first person (1899) to map the Jerudong structure in Brunei where Shell unsuccessfully drilled for oil (1915). The etymology contains the cryptic phrase: "*…to whom Volz owed his 'trip around the world.*"

Schmidt, EJ

Schmidt's Pipefish *Syngnathus schmidti* AM Popov, 1927
Schmidt's Dragonfish *Eustomias schmidti* CT Regan & E Trewavas, 1930
Slickhead sp. *Microphotolepis schmidti* F Angel & ML Verrier, 1931
Lanternfish sp. *Diaphus schmidti* AV Tåning, 1932
Dreamer sp. *Oneirodes schmidti* CT Regan & E Trewavas, 1932
Whiptail Gulper *Saccopharynx schmidti* L Bertin, 1934
Viperfish sp. *Chauliodus schmidti* V Ege, 1948
Sawtooth Eel sp. *Serrivomer schmidti* M-L Bauchot-Boutin, 1953
Short-finned Eel ssp. *Anguilla australis schmidtii* WJ Phillipps, 1925

Dr Ernst Johannes Schmidt (1877–1933) was a Danish biologist who became the Director of the Laboratoire Carlsberg (1909–1933). The University of Copenhagen awarded his MS (1898). He was given a grant by the Carlsberg Foundation to study the flora of the coastal areas of Ko Chang Island and then mainland Siam (Thailand), after which he was awarded his PhD (1903) the year he married Ingeborg Kühle who was the daughter of the chief director of the Carlsberg Brewery in Copenhagen. By this time he had already switched his interest to marine zoology, and worked part-time at the university's Botanical Institute and part-time for the 'Danish Commission for Investigation of the Sea' (1902–1909). Although he worked on oceanography and ichthyology, he continued to study plant physiology and genetics. One etymology cites his 'great work' in investigating the life history of the common eel *A. anguilla* as the reason he was honoured; another because he led the Danish Deep-Sea Expedition (1928–1930) aboard the 'Dana', from which the lanternfish type was collected. He is credited as being the person who discovered that eels in European rivers migrate to spawn in the Sargasso Sea.

Schmidt, KP

Dusky Cusk-eel *Parophidion schmidti* LP Woods & RH Kanazawa, 1951
Striped Jambeau *Parahollardia schmidti* LP Woods, 1959

Karl Patterson Schmidt (1890–1957) was a herpetologist. He graduated from Cornell (1916) and worked as a Scientific Assistant in herpetology at the American Museum of Natural History until 1922. He was Assistant Curator of the newly founded Department of Amphibians and Reptiles, Field Museum (1922–1940), then Curator of Zoology (1941–1955), becoming Emeritus Curator in the latter year. He undertook many expeditions, to destinations including Santo Domingo (1916), Puerto Rico (1919), Central America (1923), Brazil (1926), and Guatemala (1933). He edited the journal *Copeia* (1937–1949). He co-wrote: *Field Book of Snakes of the US and Canada* (1941). He was bitten by a Boomslang. Believing that the juvenile snake could not inject a fatal dose of venom, he went home to his wife and received no medical treatment. He kept a careful note of the development of the symptoms he experienced…until he died. Ten amphibians and a mammal are also named after him.

Schmidt, M

Pencil Catfish sp. *Pygidium schmidti* C Berg, 1897
[Junior Synonym of *Trichomycterus borellii*]

Dr Max Schmidt was a German-born physician. He was the first medical practitioner to be in Andalgalá, Catamarca, Argentina, where he is first mentioned (1891). He also acted as observer at the Andalgala meteorological and seismological station. He collected the holotype of this species, though it is no longer viewed as being valid.

Schmidt, PY

Schmidt's Dace *Leuciscus schmidti* SM Herzenstein, 1896
Sculpin sp. *Cottiusculus schmidti* DS Jordan & EC Starks, 1904
Eelpout sp. *Lycodes schmidti* VI Gratzianov, 1907
Schmidt's Cod *Lepidion schmidti* AN Svetovidov, 1936

Lumpfish sp. *Eumicrotremus schmidti* GU Lindberg & MI Legeza, 1955
Northern Smoothtongue *Leuroglossus schmidti* TS Rass, 1955
Browneye Skate *Okamejei schmidti* R Ishiyama, 1958
Char sp. *Salvelinus schmidti* RM Viktorovsky, 1978
Snailfish sp. *Liparis schmidti* GU Lindberg & ZV Krasyukova, 1987
Snailfish sp. *Careproctus schmidti* NV Chernova, EV Vedischeva & AV Datskii, 2021

Dr Petr Yulievich (Julievich) Schmidt (1872–1949) was a Soviet Russian ichthyologist who described a great many fish taxa. He wrote many papers over at least four decades such as: *On a new flat-fish of the genus Arnoglossus from the Black Sea* (1915) and *On the genera Davidojordania Popov and Bilabria n. (Pisces, Zoarcidae)* (1936). He also wrote several books including: *Fishes of the eastern seas of the Russian Empire. Scientific results of the Korea-Sakhalin Expedition of the Emperor Russian Geographical Society 1900–1901* (1904). The patronym for the dace (1896) is identified only as 'P. Schmidt', whom we assume is the same person as the above.

Schmidt, U

Eelpout sp. *Notolycodes schmidti* AE Gosztonyi, 1977

Dr Ulrich Schmidt was Director (1963–1974) of The Sea Fisheries Institute in Hamburg. The etymology says that it was named for "*Prof. Dr. U. SCHMIDT, formerly Director of the Institut fur Seefischerei, Hamburg, who made the author's trips to Europe possible.*" He undertook a number of research cruises such as to the North Sea around Iceland (1954) aboard the trawler 'Hans Böckler' and again aboard the trawler 'Bremen'.

Schmitt, G

African Rivuline sp. *Scriptaphyosemion schmitti* R Romand, 1979

Gerald Schmitt is a French killifish keeper and collector who helped Romand collect the type.

Schmitt, WL

Pencil Smelt sp. *Nansenia schmitti* HW Fowler, 1934
Narrownose Smooth-Hound *Mustelus schmitti* S Springer, 1939
Yellowmouth Pike-Blenny *Chaenopsis schmitti* JE Böhlke, 1957
Schmitt's Toadfish *Daector schmitti* BB Collette, 1968

Dr Waldo LaSalle Schmitt (1887–1977) was an expert on crustaceans who worked at the Smithsonian (1915–1957). He retired as Head Curator of Zoology but continued his association with the Smithsonian as Honorary Research Associate (1957–1977). His doctorate was awarded by George Washington University (1922) and the University of Southern California awarded him an honorary doctorate (1948). He had been on the staff of the US Bureau of Fisheries and was naturalist on board the 'Albatross' (1911–1914), continuing to take part in expeditions until 1963 to Marguerite Bay and the Weddell Sea. He collected the smooth-hound holotype.

Schnakenbeck

Schnakenbeck's Searsid *Sagamichthys schnakenbecki* G Krefft, 1953

Professor Dr Werner Schnakenbeck (1887–1971) was a German zoologist and ichthyologist. He studied at the University of Freiburg and then worked at the zoological gardens in Halle (1919) whilst studying for his doctorate (1920). He was a research assistant at the German Scientific Commission for Marine Research, Biology Institute, Heligoland (1921) and at the Department of Fisheries Biology, Zoological Museum, Hamburg (1923) and in charge of it (1931–1957). He also taught on the subject of fisheries at the University of Hamburg (1936–1945). He wrote: *Deutsche Fischerei in Nordsee und Nordmeer* (1947).

Schneider

Sheatfish (catfish) sp. *Silurichthys schneideri* W Volz, 1904

Gustav Schneider (1867–1948) was a Swiss zoologist, botanist, collector and trader. He trained at, and became the Custodian of, the Zoological Institute, Basel where he lived. He travelled quite widely in Sumatra (1897–1899) and later to the USA, Singapore, Malaysia and Bermuda, to make zoological collections, as well in other parts of Indonesia. He visited Sumatra for the first time (1888) in the company of the geologist and Director of the Zoological Museum Zürich, Professor C Mösch. He was later custodian of the museums at Kolmar and Mülhausen. He also traded in natural curiosities. He wrote among other works: *Conférence sur Sumatra* (1900) and *Die Orang Mamma auf Sumatra* (1958). He collected the holotype of this species.

Schoenlein

Blackspot Tuskfish *Choerodon schoenleinii* A Valenciennes, 1839
Striped Panray *Zanobatus schoenleinii* JP Müller & FGJ Henle, 1841

Dr Johann Lukas Schönlein (1793–1864) was a German naturalist and professor of medicine. He studied at Landshut, Jena, Göttingen and Würzburg and taught at Würzburg and Zurich until he was appointed to Berlin (1839). Here, among his duties, he was physician to King Frederick William IV of Prussia. He was one of the first German medical professors to lecture in his native tongue instead of Latin. The ray specimen, which was in the Berlin Anatomical Museum, was supplied to the authors by Schönlein who was a friend and colleague of the junior author.

Schoepf

Orange Filefish *Aluterus schoepfii* JJ Walbaum, 1792
Striped Burrfish *Chilomycterus schoepfii* JJ Walbaum, 1792

Dr Johann David Schoepf (1752–1800) was a German botanist, zoologist and physician. He wrote (1788) a two-volume book about his travels in America (1783–1784) while he was serving as a surgeon to Hessiam troops attached to the British army in New York during the American Revolutionary War. That year he also wrote a paper: *Fishes of New York* (1784).

Scholes

Indo-Pacific Short-tail Conger *Coloconger scholesi* WL Chan, 1967

Patrick Scholes (c.1946–2011) was Chief Scientist at the Ministry of Agriculture, Fisheries and Food Lowestoft Fisheries Laboratory, England. He went to Cambridge University following national service in the Royal Navy, during which he served on an aircraft carrier involved in the Blockade of Israel. The naval experience put him in good stead for his years working on small MAFF vessels operating out of Lowestoft and Grimsby. He first joined the plankton laboratory but did not enjoy it, so took a four-year secondment to Hong Kong where he undertook an extensive trawl survey of the South China Sea during which the eponymous eel was discovered. When he returned to the Centre for Environment, Fisheries and Aquaculture Science (CEFAS) he devised a system for supplying it with permanent clean seawater. He wrote a number of papers such as the co-written: *The effect of low temperature on cod, Gadus morhua (1974). When* he retired (1990s) he moved to Portugal for the rest of his life. Sadly, he suffered from dementia in his last years.

Scholze

Blackline Tetra *Hyphessobrycon scholzei* E Ahl, 1937

Arthur Scholze (1881–1956) was co-owner of Scholze & Pötzschke, a Berlin aquarium supply and tropical fish importation company, in which capacity he gave specimens to the Berlin Zoological Museum, including the holotype of this species.

Schomburgk, RH

Black-barred Myleus *Myloplus schomburgkii* W Jardine, 1841
[Alt. Disc Tetra; Syn. *Myleus schomburgkii*]
Glassy Sweeper *Pempheris schomburgkii* JP Müller & FH Troschel, 1848
Guyana Leaffish *Polycentrus schomburgkii* JP Müller & FH Troschel, 1849
Characin sp. *Mylesinus schomburgkii* A Valenciennes, 1850
Three-barbeled Catfish sp. *Rhamdia schomburgkii* P Bleeker, 1858
Amazon Croaker *Pachyurus schomburgkii* A Günther, 1860
Armoured Catfish sp. *Lasiancistrus schomburgkii* A Günther, 1864

Sir Robert Hermann Schomburgk (1804–1865) was a German businessman who was asked to supervise transporting Saxon sheep to Virginia, USA (1828), where he stayed. He lost his fortune partly because he failed as a tobacco grower, partly because he lost all his belongings in a fire on the Caribbean island of St Thomas. Afterwards he went to the British Virgin Islands to map, at his own expense, the coast of the island of Anega, notorious for its shipwrecks. The results published by him drew the attention of the Royal Geographical Society, who asked him to explore British Guiana (1835–1839), during which he discovered the famous giant water lily *Victoria regina*. The British Government asked him to undertake a second expedition (1841–1844) to Guiana to investigate its southern borders with Venezuela and the Dutch colony of Suriname. He also urged the British government to establish the borders with Brazil, because of the repeated enslavement of local Indian tribes by Portuguese Brazilian slave drivers. Upon his return, Queen Victoria knighted him and he became Consul at Barbados (1846), Consul at the Dominican Republic (1848–1857), and finally ConsulGeneral in Bangkok (Thailand) (1857–1864). Hampered by health problems, he retired to Germany but died the next year. His brother Richard (see below)-accompanied him during his second expedition to Guiana and Venezuela. A mammal and three birds are also named him.

Schomburgk, RM

Yellowfin Whiting *Sillago schomburgkii* W Peters, 1864

Richard Moritz Schomburgk (1811–1891) was a gardener and botanist, who accompanied his brother Robert (see above) during his second expedition to Guiana and Venezuela. He was also a collector, who emigrated, with a third brother, Otto (1849), to southern Australia to escape the political turmoil in Europe (1848). He became Director of the Botanical Gardens, Adelaide (1866–1891), until his death from a heart attack. Otto (1810–1857) edited Richard's travelogue: *R H Schomburgk's Reisen in Guiana und am Orinoco während 1835–1839* (1841). A reptile is also named after him.

Schonfold

European Smelt ssp. *Osmerus eperlanus schonfoldi* DE McAllister, 1984

The name *Eperlanus Schonfoldii* was first used, but not formally described, by John Rutty (1772). He did not clarify Schonfold's identity. It is perhaps in honour of the Dutch naturalist Stephan Schonevelde (also spelled Schonefeld, Schone Velde, Schoenfeld, and presumably Schonfold) (d.1632), whose book (1624) on marine animals is cited by Rutty.

Schoppe

Palawan Barb sp. *Cyclocheilichthys schoppeae* M Cervancia & M Kottelat, 2007

Dr Sabine Schoppe is a German zoologist, ecologist and marine biologist. She did her first degree in Biology in Germany (1990) but studied for her master's and PhD in Colombia, although Justus Liebig University in Giessen, Hesse, Germany awarded her PhD (1993). She has been working on wildlife conservation in Southeast Asia since 1994. She undertook ornithological surveys in Panay & Negros, Philippines (1994), and worked in community-based coastal resource management in Leyte, Philippines (1995–1999). She was guest professor in aquatic and marine biology at the Western Philippines University, Palawan, Philippines (1999–2005). She was then the freshwater turtle trade consultant of Traffic Southeast Asia in Malaysia and Indonesia (2006), and since February 2007 Co-manager of the Philippine Cockatoo Conservation Program. She was a founding member of the Katala Foundation, Palawan, Philippines (2007–2015) (Secretary since 2008) and is the Director of the Philippine Freshwater Turtle Conservation Program. Among her published papers is: *Siebenrockiella leytensis (Taylor 1920) – Palawan Forest Turtle, Philippine Forest Turtle* (2012). Her books include: *Status, Trade Dynamics and Management of the Southeast Asian Box Turtle in Indonesia* (A Traffic Southeast Asia Report) (2009). She was honoured in the barb's binomial for her "*...lasting help and support to the studies and research of the first author.*"

Schott

Ghost Knifefish sp. *Sternarchella schotti* F Steindachner, 1868

Arthur Carl Victor Schott (1814–1875) was born in Stuttgart, Germany, where he was apprenticed at the Royal Gardens. He studied at the Institute of Agriculture, Hohenheim. He spent 10 years in Hungary, managing a mining property and studying geology, botany and zoology. He travelled in Europe and the Near East, then went to the USA (1850) where the Corps of Topographical Engineers in Washington employed him. He was a member of the U.S.-Mexican border survey (1853–1855) and collected animals, fossils and minerals in the Rio Grande valley. He was naturalist and geologist on Lieutenant Michler's survey of the Isthmus of Darien (1857), and surveyed in the Yucatan Peninsula (1864–1866). Two reptiles and a bird are named after him.

Schouteden

African Characin sp. *Alestes schoutedeni* GA Boulenger, 1912
[Syn. *Brycinus schoutedeni*]
African Rivuline sp. *Aphyosemion schoutedeni* GA Boulenger, 1920
African Tetra sp. *Clupeocharax schoutedeni* J Pellegrin, 1926
Spotted Congo Puffer *Tetraodon schoutedeni* J Pellegrin, 1926
Vermiculated Synodontis *Synodontis schoutedeni* LR David, 1936
African Characin sp. *Nannocharax schoutedeni* M Poll, 1939
Lake Tanganyika Cichlid sp. *Cardiopharynx schoutedeni* M Poll, 1942
Yangambi Mormyrid *Petrocephalus schoutedeni* M Poll, 1954
Schouteden's Elephantfish *Stomatorhinus schoutedeni* M Poll, 1945
African Barb sp. *Clypeobarbus schoutedeni* M Poll & JG Lambert, 1961
Congo Cichlid sp. *Chromidotilapia schoutedeni* M Poll & DFE Thys van den Audenaerde, 1967

Henri Eugene Alphonse Hubert Schouteden (1881–1971) was a Belgian zoologist who undertook many expeditions to the Congo. He published on both ornithology and entomology, including: *De Vogels van BelgischCongo en van RuandaUrundi* (1948). He was made an Honorary Director of the Musée Royal de l'Afrique Centrale. He founded (1911) the journal *Revue de Zoologie africaine*, later the *Revue de Zoologie et de Botanique africaine*, which made important contributions to our knowledge of African zoology. Poll honoured him in the name of the barb on the occasion of his 80th birthday. Fourteen birds, three reptiles, two amphibians and two mammals are named after him.

Schrank

Grunt sp. *Haemulon schrankii* L Agassiz, 1831

Dr Franz von Paula Schrank (1747–1835) was a German Jesuit priest who became interested in studying nature whilst at college in Oedenburg, Hungary, and became a botanist and entomologist. He was (1784–1809) Professor of Botany and Zoology at the University of Ingolstadt, and was founder and first Director (1809–1832) of the Botanical Gardens in Munich. Among his entomological works is: *Enumeratio insectorum Austriæ indigenorum* (1781).

Schreiber

African Rivuline ssp. *Epiplatys chaperi schreiberi* HO Berkenkamp, 1975

Gerhard Schreiber was a German killifish hobbyist who collected the holotype. We have been unable to unearth further biographical details.

Schreiner

Sweeper sp. *Pempheris schreineri* A de Miranda Ribeiro, 1915

Carlos Shreiner (1849–1896) was a German-born Brazilian ornithologist and collector. He was an Assistant Naturalist, a colleague of the author who also worked at the Museu Nacional do Rio de Janeiro (1872), a travelling naturalist (1889) and sub-Director of the Zoological Section (1895). Miranda Ribeiro (q.v.) collected with him and wrote his obituary. A bird is also named asfter him.

Schreitmüller

Whiptail Catfish sp. *Farlowella schreitmuelleri* E Ahl, 1937

Wilhelm Schreitmüller (1870–1945) was a German aquarist, who was also a talented illustrator. He wrote: *Zierfische, Seetiere und ihre Pflege und zucht* ('Ornamental Fish, Marine Animals & their care and breeding') (1934). He provided the holotype of this species.

Schrenck

Japanese Sturgeon *Acipenser schrenckii* JF von Brandt, 1869
Balkhash Perch *Perca schrenkii* KF Kessler, 1874
Cresthead Flounder *Pseudopleuronectes schrenki* PJ Schmidt, 1904

Peter Leopold Ivanovich von Schrenck (aka Schenk or Shrenk) (1826–1894) was a Russo-German zoologist, geographer and ethnographer who was Director (1879) of the Imperial Academy of Sciences, St Petersburg. He explored the Amur River and Sakhalin Island (1854–1856), the results of which he published in: *Reisen und Forschungen im Amur-Lande in den Jahren 1854–1856* (four vols. 1860–1900). He believed that mammoths he found preserved in the permafrost must have died recently, and thought they were subterranean animals that ate earth! Among other taxa, two birds, an amphibian and a reptile are named after him.

Schreyen

Schreyen's Mormyrid *Pollimyrus schreyeni* M Poll, 1972
Lake Tanganyika Cichlid sp. *Neolamprologus schreyeni* M Poll, 1974

André Schreyen was a Belgian collaborator (and nephew) of Congo-based, then Burundi-based, aquarium fish exporter Pierre Brichard (1921–1990) and later of Brichard's daughter, Mireille, who is married to Jacky Schreyen who, we think, is his son. André Schreyen and his wife, Janine, owned a soap factory in Burundi, which was attacked by rebels (1995). He collected the holotypes of these fish.

Schrier

Curaçao Jawfish *Opistognathus schrieri* WF Smith-Vaniz, 2017
Maori Basslet *Lipogramma schrieri* CC Baldwin, A Nonaka & DR Robertson, 2018

Adriaan 'Dutch' Schrier is owner of the Substation Curaçao in Willemstad, whose *Curasub* submersible collected the types. Although it was not built originally for scientific research, the authors of the basslet description say: "*Dutch's enthusiastic support of research use of his sub has exponentially expanded our understanding of fish and invertebrate faunas of Caribbean mesophotic and deeper reefs*"

Schroeder

Catshark genus *Schroederichthys* S Springer, 1966
Oriental Barb sp. *Discherodontus schroederi* HM Smith, 1945
Rosette River Stingray *Potamotrygon schroederi* A Fernández-Yépez, 1958
[Alt. Flower Ray]
Bahamas Sawshark *Pristiophorus schroederi* S Springer & HR Bullis, 1960
Deep-sea Scalyfin sp. *Bathyclupea schroederi* MM Dick, 1962
Whitemouth Skate *Bathyraja schroederi* G Krefft, 1968
Schroeder's Coral-Blenny *Ecsenius schroederi* JF McKinney & VG Springer, 1976

William Charles Schroeder (1894–1977) was an American oceanographer and ichthyologist. He joined the Woods Hole Oceanographic Institution (1932), initially as a business manager in connection with a ship they had acquired: the 'Atlantis'. He was also an associate curator of ichthyology at the Harvard Museum of Comparative Zoology (1937) and went with the 'Atlantis' on a collecting trip to waters off Central and South America. He had a life-long collaboration with H.B. Bigelow and together they greatly enhanced and expanded knowledge of the fishes of the North Atlantic. Together they described 42 new fish species. They wrote: *Fishes of the Western North Atlantic* (1953) and *Fishes of the Gulf of Maine* (1953). Smith honoured him in the name of the barb for making a collection of Thai fishes available to him. Dick mentions that Schroeder "collected the holotype and several of the paratypes" of the scalyfin.

Schubart

Characin sp. *Characidium schubarti* H Travassos, 1955
Three-barbeled Catfish sp. *Imparfinis schubarti* AL Gomes, 1956
Characin sp. *Astyanax schubarti* HA Britski, 1964

Dr Otto Rudolf Julius Schubart (1900–1962) was a German biologist who emigrated to Brazil (1934). He was regarded as the grand old man of Brazilian diplopodology (millipedes and centipedes). He published widely on myriapods (1923–1962) and some are named after him such as *Ommatoiulus schubarti*. He worked at the Museu Paulista in São Paulo, where a street is named after him as is a reptile and two amphibians.

Schubotz

East African Cichlid genus *Schubotzia* GA Boulenger, 1914
[Often included in *Haplochromis*]
East African Cichlid sp. *Haplochromis schubotzi* GA Boulenger, 1914
East African Cichlid sp. *Haplochromis schubotziellus* PH Greenwood, 1973

Johann G Hermann Schubotz (1881–1955) was a German naturalist who was an Assistant in Zoology at the Humboldt University, Berlin, and participated in the two expeditions that Duke Adolf Friedrich zu Mecklenburg undertook to Equatorial Africa (1907–1908 & 1910–1911). On both expeditions he collected many fish including the types of those described by Boulenger. (He also edited the publication in which the description appeared.) He was appointed Professor of Zoology (1916), then Cultural Attaché at the German Embassy, Stockholm (1919–1926). He emigrated from Germany to South West Africa (Namibia) (1935) where he bred sheep. He returned to Germany (1952) and worked at the Broadcoasting Corporation ('Rundfunk'). Ten birds, a reptile and two amphibians are also named after him.

Schuhmacher

Schuhmacher's Stingray *Potamotrygon schuhmacheri* MN Castex, 1964

Roberto Schuhmacher (1947–1964) was a high school student of Castex who died tragically young in an accident.

Schultz, H

Driftwood Catfish sp. *Centromochlus schultzi* F Rössel, 1962
Three-barbeled Catfish sp. *Leptorhamdia schultzi* P de Miranda Ribeiro, 1964
Harald Schultz (1909–1966). (See **Harald** & **Harald Schultz**)

Schultz, LP

Snake-eel genus *Schultzidia* WA Gosline III, 1951
School Bass genus *Schultzea* LP Woods, 1958
Pencil Catfish genus *Schultzichthys* G Dahl, 1960
Characin genus *Schultzites* J Géry, 1964
Red Gunnel *Pholis schultzi* LP Schultz, 1931
Goby sp. *Gobiosoma schultzi* I Ginsburg, 1944
Stone Loach sp. *Schistura schultzi* HM Smith, 1945
Characin sp. *Moenkhausia schultzi* A Fernández-Yépez, 1950
Smooth-lip Clingfish *Gobiesox schultzi* JC Briggs, 1951
Chimalapa Silverside *Atherinella schultzi* J Álvarez del Villar & JC Carranza, 1952
Schultz's Pipefish *Corythoichthys schultzi* ES Herald, 1953
Fringefin Lantern Shark *Etmopterus schultzi* HB Bigelow, WC Schroeder, & S Springer, 1953
Bumblebee Catfish sp. *Pseudopimelodus schultzi* G Dahl, 1955
Tonguefish sp. *Symphurus schultzi* P Chabanaud, 1955
Dwarf Sawtail Catshark *Galeus schultzi* S Springer, 1979
Moray Eel sp. *Anarchias schultzi* JS Reece, DG Smith & E Holm, 2010

Dr Leonard Peter Schultz (1901–1986) was an ichthyologist and field naturalist who was Assistant Curator of fishes at the Smithsonian (1936–1938), then Curator (1938–1968), during which time he was Ginsburg's boss, and Senior Zoologist (1965–1968). Albion College awarded his bachelor's degree (1924), the University of Michigan, where he taught (1925–1927), his master's (1926) and the University of Washington, where he also taught (1928–1936), his doctorate (1932). After retiring, he retained his connection with the Smithsonian as Zoologist Emeritus (1968–1986). He was honoured in the moray eel's binomial because he had described the related species *A. cantonensis*. He was lead scientist studying the effect on marine organisms of the Bikini Atoll US Navy nuclear tests.

Schultze, JF

Maimed Snake-Eel *Muraenichthys schultzei* P Bleeker, 1857

Jan Francois Schultze (b.1817) was Assistant-Resident of Ambal, Java (1856). He provided Bleeker with an important collection of fishes from the south coast of Java.

Schultze, W

Goby sp. *Rhinogobius schultzei* AWCT Herre, 1927

Wilhelm 'Willy' Schultze was an entomologist at the Philippine Bureau of Science. Among his many published papers he wrote: *New and little-known Lepidoptera of the Philippine Islands* (1908), *A Catalogue of Philippine Coleoptera* (1915) and *New and rare Philippine Lepidoptera* (1925). He was also co-editor, with Herre and others of *Distribution of Life in the Philippines* (1928). He collected the holotype of this species (1923). He moved back to Bavaria but died during the Second World War either in action or during an air raid.

Schunck

Neotropical Rivuline sp. *Melanorivulus schuncki* WJEM Costa & AC de Luca, 2011

Fábio Schunck helped collect the type in Brazil. He was honoured for his "*...great enthusiasm and efforts directed to collecting trips.*" He graduated (2001) in biology, specialising in ornithology, from the University of Santo Amaro-UNISA. He took a course (2002) on eco-tourism and later (2003–2004) worked for the Instituto de Biologia da Concevação (IBC) on a conservation project for the Red-tailed Parrot. He then commenced (2004) his collaboration as an ornithologist with the Museum of Zoology at the University of Sâo Paulo, where he also worked (2014–2019) on his doctorate.

Schwanefeld

Schwanefeld's Barb *Barbonymus schwanenfeldii* P Bleeker, 1853
[Alt. Tinfoil Barb]
Indonesian Cyprinid sp. *Lobocheilos schwanenfeldii* P Bleeker, 1854

Lieutenant Colonel H W Schwanefeld was a Dutch military surgeon. He collected both of the types in West Sumatra. Bleeker misspelled his name in the binomials; a mistake he subsequently corrected but the original spelling must, by convention, be retained.

Schwartz

Schwartz's Cory *Corydoras schwartzi* F Rössel, 1963

Hans-Willi Schwartz (1909–1981) was an aquarium-fish exporter in Manaus, Brazil. He helped collect the holotype of this catfish.

Schwarzhans

Viviparous Brotula sp. *Tuamotuichthys schwarzhansi* JG Nielsen & PR Møller, 2008

Dr Werner Wilhelm Schwarzhans is a retired German oil geologist. He is now a research associate emeritus in ichthyology and palaeontology at the University of Copenhagen, Natural History Museum. His current project is 'Evolution and Diversity Patterns of Southern Ocean Fishes.' He co-wrote: *Description of* Karaganops n. gen. perratus *(Daniltshenko 1970) with otoliths in situ, an endemic Karaganian (Middle Miocene) herring* (Clupeidae) *in the Eastern Paratethys* (2017). (See also **Werner**)

Schwassmann

Ghost Knifefish sp. *Sternarchorhynchus schwassmanni* CD de Santana & RP Vari, 2010

Dr Horst Otto Schwassmann (b.1922) is a retired American zoologist and ichthyologist at the University of Florida, Gainsville, where he is Professor Emeritus. He wrote: *Seasonality of Reproduction in Amazonian Fishes* (1992).

Schwebisch

Schwebisch's Tilapia *Oreochromis schwebischi* HE Sauvage, 1884

Dr Paul-Victor Schwebisch was a physician who was assistant medical officer on the expedition that collected the type in West Africa.

Schwenk

Black-stripe Sweeper *Pempheris schwenkii* P Bleeker, 1855

H Schwenk was a Major of Infantry in the Dutch East Indies who, we believe, collected the type and sent it to Pieter Bleeker.

Schwetz

Congo Cichlid genus *Schwetzochromis* M Poll, 1948
Congo Cichlid sp. *Cyclopharynx schwetzi* M Poll, 1948
Congo Cichlid sp. *Haplochromis schwetzi* M Poll, 1967
[Syn. *Thoracochromis schwetzi*]

Yakov (later Jacques) Schwetz (1847–1957) was a Russian-born Belgian physician-entomologist. He collected the holotypes during a medical survey of the Fwa River, in the Democratic Republic of Congo (1936). (While studying schistosomiasis and other diseases, he also collected fish)

Schyri

Grunt sp. *Pomadasys schyrii* F Steindachner, 1900

Schyri is a title, meaning lord or leader, amongst some indigenous Ecuadorian peoples. It originated with the Cara culture, which flourished in coastal Ecuador (where the grunt species occurs) in pre-Colombian times. Steindachner does not supply an etymology, but this seems the most likely origin of the binomial.

Sclater, PL

Andalusian Barbel *Luciobarbus sclateri* A Günther, 1868
Goby sp. *Callogobius sclateri* F Steindachner, 1879

Dr Philip Lutley Sclater (1829–1913) was a British lawyer, zoologist and, in particular, ornithologist. He was a graduate of Oxford University and practiced law for many years. He was the founder and first Editor of *Ibis* (1858–1865) and again (1877–1912), the journal of the BOU. He was also Secretary of the Zoological Society of London (1860–1902). Sclater was a pioneer in zoogeography; his study of bird distribution resulted in the classification of the bio-geographical regions of the world into six major categories. He later adapted his scheme for mammals, and it is still the basis for work in biogeography. He wrote: *A Monograph of the Birds Forming the Tanagrine Genus Calliste* (1857), *Exotic Ornithology* (1866), *The Curassows* (1875), *A Monograph of the Jacamars and Puffbirds* (1879), *Birds of the Challenger Expedition* (1881) and *Argentine Ornithology* (1888). Fifty-nine birds, six mammals and a reptile are also named after him. He presented the type of the barbel to the BMNH.

Sclater, WL

Common Rock Catfish *Austroglanis sclateri* GA Boulenger, 1901

William Lutley Sclater (1863–1944) was the son of Philip Lutley Sclater (above). Like his father, he was educated at Oxford, obtaining a first-class degree in Natural Science (1885). He was President of the BOU (1928–1933). For a few years he was Deputy Superintendent of the Indian Museum in Calcutta, the first Director of the South African Museum, Cape Town (1896) and President of the South African Ornithologists' Union. Sclater resigned from the South African Museum (1906) and worked at the BMNH (1906–1936). He succeeded his father as Editor of *Ibis* (1913–1930). He was killed by a V1 flying bomb (July 1944) in London. He wrote: *Systema Avium Aethiopicarum* (1924). He supplied the holotype of this species. Two mammals are named after him. (Also see **Carlotta**)

Scofield

Scofield's Anchovy *Anchoa scofieldi* DS Jordan & GB Culver, 1895

Norman Bishop 'N B' Scofield (1869–1958) was an American fisheries biologist. He was a member of the Hopkins Expedition that collected the type in Sinaloa and had previously been Jordan's student. He was awarded his BS in biology 1890 when he left the Mid-West for California where he registered at Stanford, which awarded his MSc (1895). He was employed at the Department of Fish & Game (1897–1899). He pursued business in the east of the USA before returning to California (1908) and was again employed at the Department of Fish & Game (1908–1939) until retirement, becoming Director (c.1926). He was the first administrator of the Californian Department of Commercial Fisheries.

Scortecci

Goby sp. *Gobius scorteccii* M Poll, 1961

Professor Dr Giuseppe Scortecci (1898–1973) was an Italian zoologist and herpetologist. After taking his doctorate at Università degli Studi di Firenze (1921), he joined the staff of its Institute of Comparative Anatomy. He became Professor of Zoology, Università degli Studi di Genova (1942). Before WW2 he explored the Sahara, Italian Somaliland (Somalia), and Ethiopia. He produced around 50 publications on herpetology, particularly that of desert regions. Twelve reptiles, an amphibian and a bird are named after him.

Scott, DC

Cherokee Darter *Etheostoma scotti* BH Bauer, DA Etnier & NM Burkhead, 1995

Donald Charles Scott (1920–1997) was an American biologist. The University of Michigan awarded his BSc in zoology (1942); Indiana University his doctorate (1947). He joined the faculty of the University of Georgia and spent the rest of his career there (1947–1981). He co-authored: *The Freshwater Fishes of Georgia* (1971).

Scott, DR

Peach-skin Snailfish *Careproctus scottae* WM Chapman & AC DeLacy, 1934

Miss Dorothy Ruth Rustad née Scott (1909–1998) collected a number of fish, including the snailfish type material, near Petersburg, southeast Alaska and sent them to the authors. At their request she obtained a further five examples of the species. She graduated from the College of Puget Sound (1929) before becoming a teacher. She taught in Petersburg and married there (1934).

Scott, EOG

Clingfish sp. *Alabes scotti* JB Hutchins & S Morrison, 2004

Eric Oswald Gale Scott (1899–1986) was an Australian ichthyologist who was a teacher and museum director at the Queen Victoria Museum and Art Gallery, Launceston, Tasmania. After his health failed after two years at the University of Tasmania, he taught school (1918–1923) then took time out (1924) to visit museums in England and Europe before returning to teaching (1925–1929) and resuming his degree course. He then became assistant curator (1930–1938) at QVMAG where his father was Director. A grant allowed him to again study overseas museums, but while he was away his father died and on Eric's return he was appointed Director (1938–1942). He was a pacifist and was gaoled for a month for refusing to take an army medical. Denied being allowed to register as a conscientious objector, he was gaoled for two further terms in the war years. He said at the time '*I believe in the Lord Jesus Christ and I am doing what I am doing because I believe in His teachings*'. He then taught language and science in private schools in Launceston (1949–1964). He once again became an honorary associate of the museum (1965) until his death. He wrote: *Fishes of Tasmania* (1983). He died after being hit by a car.

Scott, P & P

Scotts' Fairy Wrasse *Cirrhilabrus scottorum* JE Randall & RM Pyle, 1989

Sir Peter Markham Scott (1909–1989) and his wife Philippa (1918–2010) were honoured "*…in recognition of their great contribution in nature conservation.*" Sir Peter is particularly noted as an ornithologist and founder of the Wildfowl and Wetlands Trust (WWT). He was also one of the founders of the World Wide Fund for Nature (WWF). His father was the explorer Robert F. Scott (see next entry).

Scott, RF

Crowned Rockcod *Trematomus scotti* GA Boulenger, 1907

Saddleback Plunderfish *Pogonophryne scotti* CT Regan, 1914

Robert Falcon Scott (1868–1912) was a British naval officer and Antarctic explorer. His biographies are extensive, so need not be repeated here.

Scott, W Ber

Pike-Cichlid sp. *Crenicichla scottii* CH Eigenmann, 1907

Dr William Berryman Scott (1858–1947) was a vertebrate paleontologist who became Professor of Geology and Paleontology at Princeton University. He graduated from

Princeton (1877) and was awarded his PhD by the University of Heidelberg (1880). He was once (1925) President of the Geological Society of America. He was widely published, including *A history of Land Mammals in the Western Hemisphere* (1913) and *The Theory of Evolution, With Special Reference to the Evidence Upon Which it is Founded* (1920). He collected the cichlid type.

Scott, W Bev

Purple Chromis *Chromis scotti* AR Emery, 1968
[Alt. Purple Reeffish]

Dr William Beverly 'Bev' Scott (1917–2014) was Curator of Ichthyology and Herpetology, Royal Ontario Museum (Toronto, Canada) (1950–1974) and then Scientist Emeritus. As a boy he was both an angler and a keeper of aquaria. He enrolled at the University of Toronto (1938), and after graduating with a zoology degree he started work for the department of Agriculture as a seed analyst – but soon returned to the University where he earned his master's and doctorate (1950). On retirement he became Director of the Huntsman Marine Science Centre (1976–1982). He was a long-time supporter of the Ontario Chapter of the American Fisheries Society. He co-wrote *The Freshwater Fishes of Canada* (1973). In the etymology, author Alan Emery writes that Scott "*...first introduced me to the study of fishes.*"

Scovell

Gulf Pipefish *Syngnathus scovelli* BW Evermann & WC Kendall, 1896

Dr Josiah Thomas Scovell (1841–1915) was an American physician. He graduated MD from Rush Medical College, Chicago (1869). He was Professor of natural Science at State Normal School of Indiana (1871–1881). He wrote a number of textbooks such as: *Practical Lessons in Science* (1895). He helped collect the original specimens.

Scratchley

Freshwater Anchovy *Thryssa scratchleyi* EP Ramsay & JD Ogilby, 1886

Major-General Sir Peter Henry Scratchley (1835–1885) was the first High Commissioner of New Guinea. He was in the British Army (1854–1882), being commissioned into the Royal Engineers (1854) and serving in both the Crimean War and the Indian Mutiny. He was in Australia (1860–1863) in charge of building defences for Melbourne and Geelong. He served in England (1864–1877). He was commissioner of defences for all six Australian colonies and for New Zealand (1877–1882). After retiring he was special commissioner in New Guinea (1884–1885). According to the etymology, the anchovy was named after him because his: "*...death [from malaria] at this critical period in the affairs of the young colony is greatly to be deplored.*" A bird is also named after him.

Scripps

Basketweave Cusk-eel *Ophidion scrippsae* CL Hubbs, 1916

Ellen Browning Scripps (1836–1932) was an Englishborn philanthropist, newspaper owner and columnist who emigrated to the USA with her father (1844). Among her many foundations is the Scripps Institute of Oceanography, California (1903).

Scruggs

Banjo Catfish sp. *Acanthobunocephalus scruggsi* TP Carvalho & RE Reis, 2020

Earl Eugene Scruggs (1924–2012) was an American banjo player, noted for popularizing a three-finger banjo picking style, now called 'Scruggs style', which is a defining characteristic of bluegrass music. The name is an allusion to the common name used for these catfish, which are shaped somewhat like the musical instrument.

Scudder

Grey Grunt *Haemulon scudderii* TN Gill, 1862

Samuel Hubbard Scudder (1837–1911) was an American entomologist and palaeontologist. He studied at Harvard under Louis Agassiz. He served the Boston Society of Natural History in various roles, including as President (1880–1887). Besides his many works on insects, he also published (1882–1884) *Nomenclator Zoologicus* – a comprehensive list of all generic names in zoology.

Seal

Tasmanian Whitebait *Lovettia sealii* RM Johnston, 1883

Matthew Seal (1834–1897) was President of the Tasmanian Fisheries Commission and Chairman of the Fisheries Board. This species is probably named after him, though this is not made clear in the original description.

Sealark

Tonguefish sp. *Cynoglossus sealarki* CT Regan, 1908

'HMS Sealark' was a naval vessel that was part of the Percy Sladen Trust Expedition (1905).

Seale

Seale's Moray Eel *Scuticaria okinawae* DS Jordan & JO Snyder, 1901
[Alt. Short-tailed Snake Moray: *Uropterygius sealei* is a junior synonym]
Seale's Rockskipper *Entomacrodus sealei* WA Bryan & AWCT Herre, 1903
Eight-finger Threadfin *Filimanus sealei* DS Jordan & RE Richardson, 1910
Blackspot Shark *Carcharhinus sealei* V Pietschmann, 1913
Meteor Cardinalfish *Ostorhinchus sealei* HW Fowler, 1918
Bornean Spotted Barb *Barbodes sealei* AWCT Herre, 1933
[Syn. *Puntius sealei*]
Sailor Flyingfish *Prognichthys sealei* T Abe, 1955

Alvin Seale (1871–1958) was an ichthyologist and designer of aquaria. He rode his bicycle from Indiana to California (1892) to study under Jordan at Stanford, from which he only graduated (1905) as his studies were interrupted by expeditions to Alaska (1896–1898) and Hawaii (1900–1902) as a field naturalist for the Bishop Museum, where he then worked as Curator of Fishes (1902–1904). He was again in Alaska (1906) and the Philippines (1907–1917) as chief of the Division of Fishes, Philippine Bureau of Science. He was ichthyologist at the Harvard Museum of Comparative Anatomy (1917–1920) and then retired to a ranch in California. He was persuaded to help the California Academy of Sciences with the planning and building of an aquarium in San Francisco, and came out of retirement (1921). He was the first superintendent of the Steinhart Aquarium (1923–1941).

Searcher

Searcher Stargazer *Gillellus searcheri* CE Dawson, 1977

The R/V Searcher was a vessel owned by the Janss Foundation, destroyed by fire in May 1972. The author commented that collections made from this vessel "have contributed much to our knowledge of Pacific dactyloscopids."

Sears

Searsid (Tubeshoulder) genus *Searsia* AE Parr, 1937

Henry Sears (1913–1982) was an American naval commander during WWII. Earlier in his life he had worked on the ship 'Atlantis' and recorded and preserved specimens of previously-unknown fish. His wealthy uncle, David Sears IV, left him a significant inheritance; part of which he used to create the Sears Foundation, which published the *Journal of Marine Research*. Although Parr did not identify the patronym, there can be little doubt that Henry Sears is intended.

Seba

Emperor Red Snapper *Lutjanus sebae* G Cuvier, 1816
African Moony *Monodactylus sebae* G Cuvier, 1829
Seba's Anemonefish *Amphiprion sebae* P Bleeker, 1853

Albertus Seba (1665–1736) was a very wealthy Dutch apothecary, zoologist and natural history collector, who published a lavish series of illustrations depicting, in part, marine life of the Indo-Pacific. The anemonefish species was featured, which, according to Bleeker's etymology "*...however imperfect, makes* [it] *perfectly well recognizable*" (translation). Seba also founded the museum in Amsterdam. At one stage (1717) he sold his entire collection to the Russian Tsar, Peter the Great, and then began collecting all over again. Linnaeus (q.v.) visited him (1735) and Seba's broad collection-cataloguing systems influenced him in the shaping of his own system, and many of Seba's animals became Linnaeus's holotypes. A bird, a mammal and three reptiles are also named after him.

Sebastian

Tetra sp. *Hyphessobrycon sebastiani* CA García-Alzate, C Román-Valencia & DC Taphorn, 2010

Sebastian García-Alzate is the senior author's younger brother.

Sebree

Sebree's Dwarfgoby *Eviota sebreei* DS Jordan & A Seale, 1906

Captain Uriel Sebree (1848–1922) was a US Navy commandant at the American Tutuila Naval Station (American Samoa). Through him the gunboat 'Wheeling' and its equipment were placed at the authors' disposal.

Sechura

Peruvian Puffer *Sphoeroides sechurae* SF Hildebrand, 1946
Sechura Lizardfish *Synodus sechurae* SF Hildebrand, 1946

This is a toponym referring to Sechura Bay, Peru.

Sedor

Sedor's Coralbrotula *Ogilbia sedorae* PR Møller, W Schwarzhans & JG Nielsen, 2005
[Alt. Sedora's Brotula (in error)]

Dr Allegra Noelle Sedor is a biologist whose master's degree (1985) and doctorate (1988) were awarded by the University of Southern California.

Seeber

Clanwilliam Sandfish *Labeo seeberi* JDF Gilchrist & WW Thompson, 1911
Clanwilliam Yellowfish *Labeobarbus seeberi* JDF Gilchrist & WW Thompson, 1913

C R Seeber (fl.1900–fl.1940) was a South African policeman and angler. He was Chief Constable at Clanwilliam, Western Cape, South Africa. He sent fishes to Gilchrist (1906), including the types of these species.

Seegers

African Rivuline sp. *Aphyosemion seegersi* J-H Huber, 1980
African Rivuline sp. *Nothobranchius seegersi* S Valdesalici & K Kardashev, 2011

Dr Lothar Seegers (1947–2018) was a German biologist and enthusiastic aquarist. Among his many published papers is: *The Oreochromis alcalicus flock (Teleostei: Cichlidae) from lakes Natron and Magadi, Tanzania and Kenya, with descriptions of two new species* (1999), and among his longer works are: *The Fishes of the Lake Rukwa Drainage* (1996) and the comprehensive: *The Catfishes of Africa* (2008). Huber honoured him in part for Seegers' research into the egg-surface structures of killifish.

Seemann

Tete Sea Catfish *Sciades seemanni* A Günther, 1864

Dr Berthold Carl Seemann (1825–1871) was a German botanist, traveller and collector. His doctorate was awarded by the University of Göttingen (1853). He studied at Kew (1844) and was, on Sir William Hooker's recommendation, appointed as naturalist on the HMS Herald Expedition (1847–1851) to the US West Coast and the Pacific, returning via the Cape of Good Hope. He travelled to Fiji (1859) and travelled in Latin America, variously Venezuela (1864) and Nicaragua (1866–1867). Finally, he managed a sugar plantation in Panama and a gold mine in NIcaragua where he died from a fever. He founded and edited: *Bonplandia* (1853–1862) and the *Journal of Botany, British and Foreign* (1863–1871). He collected the holotype of this species.

Sefton

Hidden Blenny *Cryptotrema seftoni* CL Hubbs, 1954

Joseph Weller Sefton Jr. (1881–1966) was an American amateur ornithologist, philanthropist and banker who was President of the San Diego Trust & Savings Bank, California, which his father had founded. Stanford awarded his bachelor's degree (1904). He was president of the local natural history society and formed a foundation that bought and equipped a vessel for oceanic research (1948). Hubbs' etymology says that Sefton "*... has done much to promote the investigation of the marine fauna and flora of the coasts of California and Baja California.*"

Seheli

Bluespot Mullet *Crenimugil seheli* PS Forsskål, 1775

Not an eponym, but recorded by Forsskål as a local Arabic name for this species.

Seigel

Six-spot Prickleback *Kasatkia seigeli* M Posner & RJ Lavenberg, 1999

Jeffrey Alan Seigel was collections manager of the ichthyology section at the Los Angeles County Museum of Natural History. Among his published papers is: *Revision of the Dalatiid Shark Genus Squaliolus: Anatomy, Systematics, Ecology* (1978).

Seiter

Spiny Eel sp. *Mastacembelus seiteri* DFE Thys van den Audenaerde, 1972

A Seiter was the the operator of the ferry across the Sanaga River at Nachtigal, Cameroon. He was a long-term resident there who was very familiar with its ichthyofauna as he was a keen angler. The author noted: "*En 1964, lors d'une premiere expédition sur Ie terrain darns Ie but d'inventorier les poissons du Sud-Cameroun, nous avons fait la connaissance de Mr A. SEITER, exploitant-responsable du bac sur la Sanaga à Nachtigal. Résidant depuis longtemps à cet endroit, Mr SEITER connaissait admirablement bien les biotopes du fleuve et, en pecheur convaincu, s'intéressa tout de suite à nos recherches.*"

Seki

Sculpin sp. *Icelus sekii* O Tsuruoka, H Munehara & M Yabe, 2006

Katsunori Seki (b.1954) is a diving photographer at Shiretoko Diving Kikaku. He showed the authors a photograph (1999) of an unidentified sculpin and later helped them obtain specimens.

Selena

Armoured Catfish sp. *Neoplecostomus selenae* CH Zawadzki, CS Pavanelli & F Langeani, 2008

Selena Canhoto Zawadzki is the senior author's daughter.

Selheim

Selheim's Sole *Brachirus selheimi* WJ Macleay, 1882
[Alt. Australian Freshwater Sole]
Giant Gudgeon *Oxyeleotris selheimi* WJ Macleay, 1884

[Alt. Black-banded Gudgeon]

Philip Frederic Sellheim (1832–1899) was a German-born pastoralist and mining official who lived in the Palmer District of Queensland, Australia. He was in charge of the Palmer River goldfields. Because Macleay refers only to 'Mr. Selheim' (with one *l*), the identification with Sellheim must remain provisional.

Sembra

Sembra Rainbowfish *Melanotaenia sembrae* Kadarusman, O Carman & L Pouyard, 2015

This is a toponym referring to the Sembra River, West Papua.

Semenov

Semenov's Eelpout *Lycodes semenovi* AM Popov, 1931

Pyotr Petrovich Semenov-Tian-Shansky (1827–1914) was a Russian statistician, geographer and entomologist who explored in Central Asia (1856–1857). The original description appears to have no etymology and, since Semenov is quite a common Russian surname, we cannot ascertain his identity with certainty.

Semon

Australian Smelt *Retropinna semoni* M Weber, 1895
Goby sp. *Stiphodon semoni* M Weber, 1895

Dr Richard Wolfgang Semon (1859–1918) was a German embryologist, evolutionary biologist and physiologist. He was particularly interested in memory and whether it could be hereditary. He took doctorates at Friedrich-Schiller-Universität Jena in zoology (1883) and medicine (1896). He was an Associate Professor at Friedrich-Schiller-Universität Jena (1892–1897) and led an expedition to Australia (1892–1893) to investigate monotreme reproduction (science had been shaken by the revelation that some mammals were egg-layers). His party discovered 207 new species and 24 new genera. He used native Queenslanders as trappers and collectors to help him study the platypus and the Australian lungfish. He was forced to resign (1897) because he had an affair with the Professor of Pathology's wife, whom he married after moving to Munich, where he worked as a private scholar. He collected the smelt holotype. He committed suicide, depressed by the death of his wife and by Germany's role and defeat in WW1. A reptile and a mammal are named after him.

Sena

Pacu sp. *Acnodon senai* M Jégu & GM dos Santos, 1990

Anazildo Mateus de Sena was a Brazilian fisheries worker at Instituto Nacional de Pesquisas da Amazônia, Manaus. During one of his last collecting trips, he collected the holotype of this species (1987).

Senckenberg

Seckenberg Rainbowfish *Melanotaenia senckenbergianus* M Weber, 1911

Probably named for the Senckenberg Museum, Frankfurt, Germany (the original description was published in the museum journal: *Abhandlungen der Senckenbergischen Naturforschenden Gesellschaft*) rather than directly after Johann Christian Senckenberg (1707–1772) who endowed the institution.

Senna Braga

Serrasalmid Characin sp. *Utiaritichthys sennaebragai* A de Miranda Ribeiro, 1937

Colonel Senna Braga was the Commander of the 5º Militar Grupamento de Engenharia (i.e. an engineering corps). He helped the author to obtain specimens when Miranda Ribeiro visited the type locality; Salto de Utiarity, Brazil (1909).

Senou

Triplefin sp. *Enneapterygius senoui* H Motomura, S Harazaki & GS Hardy, 2005
Mabul Sole *Aseraggodes senoui* JE Randall & M Desoutter-Meniger, 2007
Senou's Goby *Asterropteryx senoui* K Shibukawa & T Suzuki, 2007
Dartfish sp. *Parioglossus senoui* T Suzuki, T Yonezawa & J Sakaue, 2010
Cardinalfish sp. *Siphamia senoui* O Gon & GR Allen, 2012

Dr Hiroshi Senou of the Division of Fishes, Kanagawa Prefectural Museum of Natural History, has been a leader in the drive to find more about the tropical reef fishes occurring in southern Japan and the adjacent Ryukyu Archipelago. He co-wrote: *Two new soles of the genus* Aseraggodes *(Pleuronectiformes: Soleidae) from Taiwan and Japan* (2007). He collected and photographed the holotype of the cardinalfish and some of the paratypes of the goby. He was honoured for his great contribution to our knowledge of systematics and distribution of Japanese fishes.

Seraphima

Snailfish sp. *Careproctus seraphimae* PY Schmidt, 1950

Serafima Somova-Generozova (1907–c.1960) was a Russian ichthyologist and the wife of another ichthyologist, Mikhail Mikhailovitch Somov (q.v.). They met in Vladivostok when Serafima arrived there from Astrakhan to work at the Pacific Fishery Institute.

Seret

Short-tooth Whiptail *Coelorinchus sereti* T Iwamoto & NR Merrett, 1997
Searobin sp. *Lepidotrigla sereti* L del Cerro & D Lloris, 1997
Séret's Dragonet *Callionymus sereti* R Fricke, 1998
Cusk-eel sp. *Neobythites sereti* JG Nielsen, 2002
Conger Eel sp. *Ariosoma sereti* ES Karmovskaya, 2004
Butterfly Ray sp. *Gymnura sereti* L Yokota & MR de Carvalho, 2017
Papuan Velvet Skate *Notoraja sereti* WT White, PR Last & RR Mana, 2017

Dr Bernard Séret (b.1949) is an ichthyologist and marine biologist who is a senior scientist at IRD (Institut de Recherche pour le Développement) and is currently hosted by the Department of Systematics and Evolution, Muséum National d'Histoire Naturelle, Paris. His current researches concern the taxonomy, eco-biology, fisheries and conservation of chondrichthyan fishes (sharks, rays and chimaeras). So far, he has described more than 40 new species and published more than 120 scientific papers. He took part in the elaboration of the FAO international plan of action for the conservation and management of shark populations and elaborated (with F. Serena) the conservation plan for Mediterranean Sea cartilaginous fishes. He is the French representative to the ICES working group on elasmobranch fishes and the expert for the French ministry of ecology for the international conventions (e.g. CITES, CMS, OSPAR). He is a member of the IUCN Shark Specialist Group and the scientific chair of the European Elasmobranch Association. He is involved in the European programmes MADE (Mitigating adverse ecological impacts of open ocean fisheries) and CPOA-Shark (Provision of scientific advice for the purpose of the implementation of the CPOA Sharks). He is also leader of a programme concerning the sharks of the Scattered Islands (SW Indian Ocean). He wrote: *Guide d'identification des principales espèces de requins et de raies de l'Atlantique tropical oriental, à l'usage des enquêteurs des pêches* (2006) and *Guide des requins, des raies, et des chimères des pêches françaises* (2010). He was honoured for his "*...significant contributions to the study and collection of deepwater fishes, and for giving Karmovskaya the opportunity to study his material on eels.*"

Sergey

Yellow Sweeper *Pempheris sergey* JE Randall & BC Victor, 2015

Sergey V Bogorodsky is a Russian ichthyologist. He has helped describe many new fish species from the Red Sea in papers such as: *Survey of demersal fishes from southern Saudi Arabia, with five new records for the Red Sea* (2014) and *Two new species of Hetereleotris (Perciformes: Gobiidae) from the Red Sea* (2019). He collected the holotype (2013).

Seripierri

American Sole sp. *Apionichthys seripierriae* RTC Ramos, 2003

Dione Seripierri is head librarian at Museu de Zoologia da Universidade de São Paulo. She co-wrote: *The Entomological Collection of Ricardo von Diringshofen (1900–1986) and its incorporation to the Museu de Zoologia da Universidade de São Paulo* (2016).

Seshaiya

Goby sp. *Callogobius seshaiyai* J Jacob & K Rangarajan, 1960

Professor Rebala Venkata Seshaiya (1898–1973) was an Indian marine biologist and zoologist. The University of Madras awarded his BSc (1919). He was Professor of Zoology and Head, Department of Marine Biology, Advanced Study in Zoology and Marine Biology, Annamalai University, Porto Novo, India and founder (1957) Director of the Centre of Advanced Study in Marine Biology, there, which he modelled on Woods Hole, USA. He made contributions to the study of gastropods, embryology, marine biology and the development and composition of teleost scales, leaving a deep imprint on Indian zoology. He was honoured in the name "*...in appreciation of his devotion to zoology.*"

Setna

Malabar Ricefish *Oryzias setnai* CV Kulkarni, 1940

Indigo Barb *Pethia setnai* BF Chhapgar & SR Sane, 1992

Dr Sam Bomansha Setna FMRS (1895–1969) was a marine biologist at the Royal Institute of Science and Department of Industries, Bombay. Madras University awarded his MSc and Cambridge his PhD (1928). He was known as the 'father of fishing mechanisation' in India. He stayed for five months on the Andaman Islands (1930–1931). He was appointed as the first Director of Fisheries (1945–1953) for the then state of Bombay. He set up the first India fishing company (1955) and was Managing Director for the rest of his life. He was honoured in the etymology, which says that he: "*...was the first Director of Fisheries of the erstwhile Bombay State, whose dynamism led to the establishment of a separate Department of Fisheries, which was prior to 1945, only a Section of the Industries Department.*" He established the Taraporewala Aquarium and Marine Biology Research Station and served for a time as editor of the *Journal of the Bombay Natural History Society*. He wrote a number of papers such as his: *Fishing for Bombay Duck (Harpodon nehereus). Destructive netting methods* (1932) and the co-written: *Breeding Habits of Bombay Elasmobranchs* (1949).

Seuss

Seussi Cory *Corydoras seussi* J Dinkelmeyer, 1996

Dr Werner Seuss is a German ichthyologist and aquarist, who wrote: *Corydoras: The Most Popular Armoured Catfishes of South America* (1995). He should not be confused with Dr Seuss, the well-known author of childrens' books, such as: *The Cat in the Hat*!

Severi

Severi's Tube-snouted Ghost Knifefish *Sternarchorhynchus severii* CD de Santana & A Nogueira, 2006

Dr William Severi is a Brazilian biologist at the Department of Fisheries, Federal Rural University of Pernambuco, where his bachelor's degree was awarded (1980) and where he is currently Associate Professor IV. The Federal University of São Carlos awarded his master's (1991) and his doctorate (1997). He co-wrote: *Morphological development of larvae and juveniles of* Prochilodus argenteus (2017),

Severns

Severns' Wrasse *Pseudojuloides severnsi* DR Bellwood & JE Randall, 2000
Severns' Pygmy Seahorse *Hippocampus severnsi* SA Lourie & RH Kuiter, 2008

R Michael 'Mike' Severns is the dive guide who collected the first specimens of both fish. He operates *Mike Severns Diving* in Maui, Hawaii. He is also very knowledgeable about the location of fossil deposits and birds. A bird is also named after him.

Severtzov

Osman (Asian cyprinid) sp. *Diptychus sewerzowi* KF Kessler, 1872
Severtsov's Loach *Triplophysa sewerzowi* GV Nikolskii, 1938

Nikolai Alekseevich Severtsov (1827–1885) [sometimes Sewertsov, Severtzow, or Severzow] was a Russian zoologist and zoographer who explored in Central Asia. He is considered to be one of the Russian pioneers of ecology and evolutionary science. He wrote works on the zoogeographical division of the regions of the Palaearctic, and on the birds of Russia and Turkestan, including mapping migration routes. After becoming acquainted with Darwin's theory of natural selection he tested it against his own observations and became an eager supporter and propagandist for Darwinism. He made extensive collections on his travels, including 12,000 bird skins. He wrote several fulllength works including: *Ornithology and Ornithological Geography of European and Asian Russia* (1867). A mountain peak in PamiroAlai and several glaciers in Pamir and Zailijaskoe are named after him, as are six birds and four mammals.

Sewall

Blenny sp. *Omobranchus sewalli* HW Fowler, 1931

Arthur Wollaston Sewall (1860–1939), of the Barber Asphalt Company, secured a collection of fishes from Trinidad and Venezuela for the Academy of Natural Sciences of Philadelphia – including the holotype of this one.

Seward

Goby sp. *Amblygobius sewardii* RL Playfair, 1867

Brigadier-Surgeon George Edwin Seward (1826–1917) of the Bombay Medical Services was Surgeon to the Zanzibar Political Agency and acting consul (1856–1867). His MD was awarded by Edinburgh University (1855). He became Surgeon (1867) and Surgeon-Major (1873), retiring (1884). He served in the navy during the Indian Mutiny. While serving in Zanzibar he saw Dr Livingstone start on his last journey into the African interior.

Sewell

Hillstream Loach genus *Sewellia* SL Hora, 1932
Barb sp. *Systomus sewelli* B Prashad & DD Mukerji, 1929
Deep-sea Tripodfish *Bathytyphlops sewelli* JR Norman, 1939

Lieutenant-Colonel Robert Beresford Seymour Sewell CIE., FRS., FLS (1880–1964) was a British physician, zoologist and naturalist. He studied at Cambridge (Christ's College) and St Bartholomew's Hospital, London, receiving his BA from Cambridge (1902) and qualified M.C.R.S & L.R.C.P. (1907). He was commissioned (1908) and served with the Indian medical Service (1908–1935) rising to Captain (1911), Major (1919) and Lieutenant-Colonel (1927). During WW1, he was 'mentioned in despatches'. He became Director of the Zoological Survey of India (1925–1935) until retirement. He was the editor of: *The Fauna of British India* (1933–1963).

Seybold

African Darter Tetra sp. *Nannocharax seyboldi* LP Schultz, 1942

George Seybold, of the Firestone Plantation, Liberia, "extended much help" to entomologist William M Mann, who collected the holotype while collecting animals for the National Zoo (Washington, D.C).

Seymour, AG

African Barb sp. *Enteromius seymouri* D Tweddle & PH Skelton, 2008

Dr Anthony G 'Tony' Seymour (1948–2006) was a British marine zoologist and an independent fisheries and environmental consultant. He was for many years Senior Fishery Officer, Malawi Government Fisheries Department. Bangor University awarded his BSc in Zoology (1969) and his PhD (1976). After more than 10 years in Malawi he returned to North Wales (1988) and had a fishing vessel built to his own design. He then successfully took to commercial fishing, which enabled him to excel as a consultant on fishing issues. Among his papers is the unpublished: *Commercial gillnetting in northern Lake Malawi: a simple case study and projected business model* (1984) and he wrote an important report (2005) on Decentralisation and Fisheries. Tweddle described him as a: "*...close friend and colleague*" and honoured him for "*...his many years of environmental management and conservation service, and for his long-term commitment to supporting Lake Malawi's fishermen.*" He died when he suffered an embolism.

Seymour, EJ

African Rivuline sp. *Pronothobranchius seymouri* PV Loiselle & D Blair, 1972

Edward J 'Ted' Seymour (d.1969) was Technical Editor of the British Killifish Association. He died in September 1969 "*after a long career of service to the killifish hobby.*"

Shackleton

Plunderfish sp. *Artedidraco shackletoni* ER Waite, 1911

Ernest Henry Shackleton (1874–1922) was a British polar explorer who led three expeditions to the Antarctic. His biography is widely accessible.

Shajarian

Scraper Barb sp. *Capoeta shajariani* A Jouladeh-Roudbar, S Eagderi, L Murillo-Ramos, HR Ghanavi & I Doadrio, 2017

Mohammad-Reza Shajarian (b.1940) is an acclaimed Iranian classical singer, composer and master of Persian traditional music.

Shaji

Indian Subterranean Catfish sp. *Kryptoglanis shajii* M Vincent & J Thomas, 2011

Dr C P Shaji is an Indian ichthyologist and taxonomist from the state of Kerala who is Principal Scientific Officer at the Kerala State Biodiversity Board. The Kerala Forest Research Institute awarded his doctorate (2001). He was one of the authors of: *The status and distribution of freshwater fishes of the Western Ghats* (2011). He was described, in the etymology of this catfish as a: "*...distinguished fish taxonomist who significantly contributed to the documentation of freshwater fish diversity of the Kerala region of Western Ghats by the Kerala Forest Research Institute.*"

Shaka

Zulu Snakelet *Halimuraena shakai* R Winterbottom, 1978

King Shaka (c.1787–1828) of the Zulus "*...raised his people from a small tribe to a powerful nation*". Apparently, the hastate body of this species is a "*...perhaps fanciful reminder of the short stabbing spear or 'iKlwa' which Shaka developed and used with such devastating effect.*"

Shao

Doublespine Seadevil sp. *Bufoceratias shaoi* TW Pietsch III, HC Ho & HM Chen, 2004
Shao's Moray *Gymnothorax shaoi* H-M Chen & K-H Loh, 2007
Triplefin sp. *Enneapterygius shaoi* M-C Chiang & L-S Chen, 2008
Sandperch sp. *Parapercis shaoi* JE Randall, 2008
Long-bodied Snake-eel *Ophichthus shaoi* JE McCosker & H-C Ho, 2015

Dr Kwang-Tsao Shao (b.1951) is a Taiwanese ichthyologist. He is Adjunct Professor (since 1994) at the Institute of Marine Biology, National Taiwan Ocean University where he was Director (1991–1994). He was Vice Director (1995–1996) and Director (1996–2002) of the Institute of Zoology Academia Sinica and he is now Research Fellow there (since 2003) & Executive Officer for Systematics and Biodiversity Information (since 2008) at the Biodiversity Research Center, Academia Sinica, Taiwan and was Acting Director there (2003–2007). The National Taiwan University awarded his BSc in Fishery Biology 1972) and his MSc in Marine Biology & Fisheries (1976), and the State University of New York, Stoney Brook awarded his PhD (1983). He has published extensively including 117 papers (2006–2014). Chen honoured him because he contributed greatly to the establishing of a Taiwanese fish database and for supervising and supporting the authors' muraenid studies.

Shapur

Sisorid catfish sp. *Glyptothorax shapuri* H Mousavi-Sabet, S Eagderi, S Vatandoust & J Freyhof, 2021

Shapur I was the second Sasanian King of Iran (where this catfish is endemic). The dating of his reign is disputed, but it is generally agreed that he ruled from 240 to 270, with his father as co-regent until the death of the latter in 242. Shapur consolidated and expanded his empire, including waging war against the Roman Empire.

Sharma

Chennai Sawfin Barb *Pethia sharmai* AGK Menon & K Rema Devi, 1993

Dr Vinod Prakash Sharma (1938–2015) was an Indian medical entomologist. Allahabad University awarded his MSc (1960), DPhil (1964) and DSc (1979). He was Postdoctoral Research Associate at the University of Notre Dame and Purdue University, USA (1965–1968). When he returned to India, he worked as Pool Officer at Forest Research Institute, Dehradun (1969). Thereafter, he was Senior Scientist at the WHO/ICMR Research Unit on Genetic Control of Culicine Mosquitoes, New Delhi (1970–1975). On closure of this facility, he worked as Deputy Director, Vector Control Research Centre, and Malaria Research Unit (1976–1978); Deputy Director, Malaria Research Centre (National Institute of Malaria Research), New Delhi (1978–1982). He retired as the Additional Director-General, ICMR (1998) and became Meghnad Saha Distinguished Fellow of the National Academy of Sciences (India) at the Centre for Rural Development and Technology, IIT Delhi. He also led the *Safe Water* campaign initiated by National Academy of Sciences. He was honoured for his "*…keen interest in the study of indigenous larvivorous fishes of India*."

NB. *Puntius sharmai* was described from Mogappair, West Annanagar, Madras, Tamil Nadu, India. Jayaram (1999) considered *Puntius sharmai* as a juvenile of *Puntius fraseri* and suggested that it may have been accidentally introduced for larvicidal work.

Sharpe

Shovelnose Catfish *Chrysichthys sharpii* GA Boulenger, 1901

Sir Alfred Sharpe (1853–1935) was the Crown's Commissioner and Consul-General for the British Central Africa Protectorate (1896–1910) and Nyasaland (Malawi) (1907). Sharpe started his career in Fiji but transferred to Africa as a professional hunter, and worked for Cecil Rhodes (1890–1896). He was an amateur naturalist, and two mammals and two birds are also named after him.

Sharpey

Binni (Middle Eastern Barbel sp.) *Mesopotamichthys sharpeyi* A Günther, 1874

Dr William Sharpey (1802–1880) was a British anatomist and physiologist, known as the 'father of modern physiology'. He studied humanities at Edinburgh (1817) but switched to medicine (1818), being admitted to the Edinburgh College of Surgeons (1821). He continued to study medicine in London and Paris. He graduated MD from Edinburgh (1823) before returning to Paris for another year and then settling to practice medicine in Arbroath. However, by 1826 he devoted himself full time to science. Over the next five years he travelled in Europe, continuing to practice and learn anatomy. He obtained a fellowship at the Edinburgh College of Surgeons, which enabled him to teach (1830). For six years he taught anatomy and continued to research and write. He became an examiner at London University (1840). Various posts followed, including being a University College Professor until retiring (1872) because of failing eyesight. He lived on a government pension until dying of bronchitis. He presented the type, which was collected by his nephew in Baghdad, to the BMNH.

Shaw

Indian River Catfish sp. *Pseudolaguvia shawi* SL Hora, 1921

G E Shaw was a naturalist who was the superintendent of the Cinchona Plantation, Darjeeling, India and a quinologist (grower of cinchona trees for quinine, an early anti-malarial medicine). He co-wrote: *The Fishes of Northern Bengal* (1937). He collected the holotype of this species.

Shcherbachev

Eelpout sp. *Pachycara shcherbachevi* ME Anderson, 1989
Scaleline Cusk *Lamprogrammus shcherbachevi* DM Cohen & BA Rohr, 1993
False Duckbill Whiptail *Coelorinchus shcherbachevi* T Iwamoto & NR Merrett, 1997
Bigscale sp. *Melamphaes shcherbachevi* AN Kotlyar, 2015

Yuri Nikolaevich Shcherbachev works at the P P Shirov Institute of Oceanography, Russian Academy of Sciences. Among his published papers is: *Etmopterus molleri (Squalidae) in the Southwest Indian Ocean* (1989). He was honoured for his pioneering contributions to knowledge of the deep-sea fishes of the Indian Ocean. (See also **Yuri**)

Sheard

Weed-Whiting genus *Sheardichthys* GP Whitley, 1947
[Sometimes included in *Siphonognathus*]

Dr Keith Sheard (1903–1965) was an Australian marine zoologist. Hi initially trained as a teacher but later in life studied for his Master of Science degree (1951) and doctorate (1954). He was assistant in zoology at the South Australian Museum (1933–1942) and later employed by CSIRO Division of Fisheries and Oceanography (1942–1961), where he specialised in crustaceans.

Shebbeare

Stone Loach sp. *Schistura shebbearei* SL Hora, 1935

Edward Oswald Shebbeare (1884–1964) was an English colonial civil servant who joined the Indian Forrest Service (1906–1938) spending almost all his time in Bengal, retiring as Senior Conservator of Forests. He was the transport officer on the Everest expeditions (1924 & 1933) and the German expeditions to Kangchenjunga (1929 & 1931) and was a founder member of the Himalayan Club. He was Assistant Editor of their Journal (1930–1933), Vice-President (1933–1934) and a member of its Committee (1936–1938). He went on to Malaya as a Chief Game Warden. He was a POW in Singapore (1942–1945) then returned to being a Game Warden until retiring (1947) and returning to England. He wrote: *Trees of the Duars and Terai* (1957). He was a friend of the author and sent a collection of fishes to the Indian Museum, including the type of this loach.

Shehab

Stone Loach sp. *Oxynoemacheilus shehabi* J Freyhof & MF Geiger, 2021

Adwan Shehab (1967–2015) was described as "one of Syria's most active and renowned zoologists". He hosted and assisted the authors' team during field-work in Syria in 2008. He received his doctorate (1999) from the University of Damascus, and after graduating joined the General Commission for Scientific Agricultural Research. He was killed in the streets of Dara'a, southwest Syria, during the ongoing conflict in that country.

Sheiko

Rasptooth Dogfish *Miroscyllium sheikoi* VN Dolganov, 1986

Boris Anatolievich Sheiko (b.1957) is a Russian ichthyologist who graduated from Rostov State University. He was a junior research fellow, Pacific Research Institute of Fisheries and Oceanography, Vladivostok (1980–1982), a research fellow, Azov Research Institute of Fisheries (Rostov on Don) (1983–1987), a post-graduate student, Zoological Institute, St Petersburg (1987–1990), research fellow, Kamchatka Institute of Ecology, Petropavlovsk-Kamchatsky (1991–2000) and became a junior research fellow, Zoological Institute, the Russian Academy of Sciences, St Petersburg (2000) and, since 2012, a research fellow. He is also a veteran of many oceanographic expeditions to the Northwestern Pacific Ocean and Northern Bering Sea and his major scientific interests include species composition, vertical distribution and dispersion of marine fishes in the North Pacific. He is also interested in taxonomy of fishes of families *Cottidae, Agonidae, Cyclopteridae, Zoarcidae, Stichaeidae* and the nomenclature of Latin names of fishes. He wrote: *Ichthyofauna of Peter the Great Bay (Sea of Japan): species composition, ichthyocenes, zoogeography* (1983), *A Catalog of Fishes of the Family* Agonidae *(Scorpaeniformes: Cottoidei)* (1993) and *Alectrias markevichi sp. nov. - a new species of cockscombs (Perciformes: Stichaeidae: Alectriinae) from the sublittoral of the Sea of Japan and adjacent waters* (2012).

Sheila

Sheila's Damselfish *Chrysiptera sheila* JE Randall, 1994

Sheila McLeish, along with her husband Ian McLeish (Office of the Advisor for Conservation and the Environment of the Sultanate of Oman), helped collect some of the type series and provided logistical support.

Shelford

Borneo Loach *Pangio shelfordii* CML Popta, 1903

Robert Walter Campbell Shelford (1872–1912) was a Singapore-born, English-educated naturalist, entomologist and Curator of the Sarawak Museum (1897–1905). He became an Assistant Curator at the Oxford University Museum. His special interest was entomology, particularly insect mimicry. A number of plants and two reptiles are also named after him. He presented the type of the loach to the Leiden Museum (Netherlands).

Sheljuzhko

Fourspine Leaffish *Afronandus sheljuzhkoi* H Meinken, 1954
African Rivuline ssp. *Epiplatys chaperi sheljuzhkoi* M Poll, 1953

Leo Sheljuzhko (1890–1969) was a Ukrainian-German entomologist who collected fish for the German importers, Werner Aquarium. He was one of the first aquarium amateurs and tropical-fish breeders in the Russian Empire. He collected the leaffish type and sent specimens to Meinken.

Shelkovnikov

(See under **Schelkovnikov**)

Shen, D

Shen's Basslet *Pseudanthias sheni* JE Randall & GR Allen, 1989

David Shen. He showed his underwater photograph of the male of this species to the senior author and expressed his belief that it represented an undescribed species allied to *P. pleurotaenia*. (See **David Shen**)

Shen, S-C

Jawfish sp. *Stalix sheni* WF Smith-Vaniz, 1989

Shen's Soldierfish *Ostichthys sheni* JP Chen, KT Shao & HK Mok, 1990
Hagfish sp. *Eptatretus sheni* CH Kuo, KF Huang & HK Mok, 1994
Perchlet sp. *Plectranthias sheni* JP Chen & KT Shao, 2002
Grenadier sp. *Coelorinchus sheni* ML Chiou, KT Shao & T Iwamoto, 2004
Goby sp. *Callogobius sheni* IS Chen, JP Chen & LS Fang, 2006
Triplefin sp. *Enneapterygius sheni* MC Chiang & IS Chen, 2008
Shen's River Loach *Hemimyzon sheni* IS Chen & LS Fang, 2009

Dr Shieh-Chieh Shen is an ichthyologist who was a professor in the Department of Zoology, College of Science, National Taiwan University (retired 1999). His doctorate was awarded by Tokyo University. Among his written works are: *Coastal fishes of Taiwan* (1984) (among which co-authors are HK Mok & KT Shao) and *Catalogue of the fish specimens deposited in the Museum of the Department of Zoology, National Taiwan University (Acta zoologica Taiwanica)* (1988) and the co-written description of the bitterling species: *Rhodeus haradai, a New Bitterling from Hainan Island, China, with Notes on the Synonymy of Rhodeus spinalis (Pisces, Cyprinidae)* (1990). He was honoured for his great contribution to ichthyology in Taiwan. Shen collected the perchlet in a Taiwanese fish market. (Also see **Hsiojen Lin Shen**)

Shepard

Shepard's Pygmy Angelfish *Centropyge shepardi* JE Randall & F Yasuda, 1979
[Alt. Mango Angelfish]

John W Shepard worked at the Marine Laboratory, UOG Station, Mangilao, Guam (part of the University of Guam). Among his published works are the co-written papers: *A New Species of the Labrid Fish Genus* Macropharyngodon *from Southern Japan* (1977) and *New Records of Fishes from Guam, with Notes on the lchthyofauna of the Southern Marianas* (1980). He was the first collector of the species.

Sheppard

Sheppard's Dwarfgoby *Trimma sheppardi* R Winterbottom, 1984

Professor Dr Charles Reginald Connon Sheppard (b.1949) is an ecologist who was scientific leader on the Chagos Expedition (1978) during which the type was collected. The University of Durham awarded his PhD (1976). He is Professor Emeritus in the Life Sciences at Warwick University and was chairman of the Chagos Conservation Trust. With his wife Anne he has spent over four decades working for Chagos conservation, as well as that of Middle Eastern and Caribbean countries. He has worked with several aid and development agencies as well as having been a researcher on tropical marine systems. He is author of over 200 articles and is author or editor of 16 books, including *Coral Reefs: a very short introduction* (2014), and has received several scientific accolades. He was honoured in the goby's name for his many efforts to ensure the participating scientists had everything they needed on the expedition where this species was discovered. He also has a coral species named after him.

Sherborn

Sherborn's Pelagic Bass *Howella sherborni* JR Norman, 1930

Charles Davies Sherborn (1861–1942) was an English geologist, palaeontologist and bibliographer. He wrote the 11-volume work *Index Animalium*, which catalogued the 444,000 names of every living and extinct animal discovered (1758–1850). The author honoured Sherborn's "*…extensive and unrivalled knowledge of matters of nomenclature.*"

Sherman

Three-barbeled Catfish sp. *Cetopsorhamdia shermani* LP Schultz, 1944

Roger Hiram Sherman (1904–1954) worked for the Standard Oil Company of Venezuela and greatly helped the author during his visit to Venezuela. Sherman was co-ordinator of all oil-production activities of Standard Oil Co, New Jersey (now Exxon) when he died.

Sherwood

Sherwood's Dogfish *Scymnodalatias sherwoodi* G Archey, 1921

C W Sherwood discovered the holotype lying washed up on the beach at New Brighton, New Zealand (1920) and presented it to the Canterbury Museum. Nothing further was recorded.

Shewell Keim

Thorny Catfish sp. *Hassar shewellkeimi* MH Sabaj Pérez & JLO Birindelli, 2013

Shewell 'Bud' DeBenneville Keim (1918–2014), after serving in the US Army in WW2, graduated as an electrical engineer (1949) and spent his working life as an electronics engineer in the US Defense Department. He donated money to the Academy of Natural Sciences, Philadelphia, to fund the preservation of his uncle's (Henry Weed Fowler) (q.v.) fish collection.

Shibatta

Neotropical Rivuline sp. *Hypsolebias shibattai* DTB Nielsen, M Martins, LM de Araujo & R Suzart, 2014

Characin sp. *Tyttobrycon shibattai* VP Abrahão, M Pastana & M Marinho, 2019

Dr Oscar Akio Shibatta is a professor at the Departamento de Biologia Animal e Vegetal, Centro de Ciências Biológicas, at the Universidade Estadual de Londrina, Brazil. He studied at the Universidade Federal de São Carlos (1989–1993). He has more than 100 publications including: *Peixes de água doce* (2009) and *A new species of bumblebee catfish of the genus Microglanis (Siluriformes: Pseudopimelodidae) from the upper rio Paraguay basin, Brazil* (2016). The senior author of the characin description writes that "*Shibatta was responsible for encouraging his studies with nervous system of Neotropical fishes.*" (Also see **Lenice**)

Shibukawa

Shibukawa's Dwarf-goby *Eviota shibukawai* T Suzuki & DW Greenfield, 2014

Dr Koichi Shibukawa is a Japanese ichthyologist who is a research scientist at the Nagao Natural Environment Foundation, Tokyo. He collected and photographed the type.

Among his publications are: *A new gobiid fish, Acentrogobius insularis, from the Ryukyu Islands* (1996) and *Description of a new species of Tomiyamichthys from Australia with a discussion of the generic name* (2016). One of his claims to fame is discovering a fish species with its genitals on its head – *P. cuulong* is a species of priapiumfish, a little-known group of Asian fish all similarly endowed. Shibukawa saw one (2009) swimming alone in a canal near the Mekong river in Vietnam, and managed to catch it in a net. He was honoured in the name of the goby for his "*...great contribution to our knowledge of the systematics of the Gobioidei.*"

Shideler

Characin sp. *Moenkhausia shideleri* CH Eigenmann, 1909

S E Shideler was a volunteer assistant during Eigenmann's Guyana expedition (1908) during which he collected the holotype. Initially the expedition was funded by Eigenmann but he was later promised the support of the Carnegie Institute. They collected 25,000 specimens which included 28 new genera and 128 new species.

Shigemi

Snailfish sp. *Careproctus shigemii* K Matsuzaki, T Mori, M Kamiunten, T Yanagimoto & Y Kai, 2020

Shigemi Fujimoto is described in the etymology as 'the late' fisherman of Rausu, Hokkaido, Japan, "*...who assisted our team in collecting various marine organisms, including the present new species, and contributed significantly to our efforts to understand the marine biodiversity of Rausu and Shiretoko Peninsula, a World Heritage Area.*"

Shiho

Shiho's Seahorse *Hippocampus sindonis* DS Jordan & JO Snyder, 1901

'Shiho' is a common Japanese forename. The binomial is derived from the name of Michitaro Sindo (q.v.), but it is unclear why the name Shiho is attached to this species.

Shimada

Shimada's Dwarf-goby *Eviota shimadai* DW Greenfield & JE Randall, 2010

Kazuhiko Shimada works at the Okinawa Prefectural Fisheries and Ocean Research Center. He previously (1993) recognised this species in Japan as being undescribed. Among his written work is: *A new creediid fish* Creedia bilineatus *from the Yaeyama Islands, Japan* (1987).

Shimazu

Bitterling sp. *Tanakia shimazui* S Tanaka, 1908

Mr. Shimazu (no other name was given in the etymology), described as being a naturalist in Tokyo, collected the bitterling type.

Shimizu

Shimizu's Squirrelfish *Sargocentron shimizui* JE Randall, 1998

Takeshi Shimizu is an ichthyologist at the Laboratory of Marine Zoology, Hokkaido University, Japan. He co-wrote: *A Revision of the Holocentrid Fish Genus* Ostichthys *with Descriptions of Four New Species and a Related New Genus* (2010). He was honoured for his systematic research on the Holocentridae.

Shimon

Shimoni Sweeper *Pempheris shimoni* JE Randall & BC Victor, 2015

This is a toponym referring to the type locality; the port town of Shimoni near the Kenya/Tanzania border.

Shinohara

Viviparous Brotula sp *Alionematichthys shinoharai* PR Møller & W Schwarzhans, 2008
Shinohara's Grenadier *Nezumia shinoharai* N Nakayama & H Endo, 2012

Dr Gento Shinohara is a Japanese research ichthyologist at the Fish Section, Department of Zoology, National Museum of Nature and Science, Ibaraki, Japan. He was (until 2010) editor-in-chief at Ichthyological Research. Among his c.50 (1992–2008) published papers are: *A psychrolutid, Malacocottus gibber, collected from the mesopelagic zone of the Sea of Japan, with comments on its intraspecific variation* (1992) and *Careproctus rotundifrons, a new snailfish (Scorpaeniformes: Liparidae) from Japan* (2008).

Shingon

Snakehead sp. *Channa shingon* M Endruweit, 2017

Shingon, also referred to as 'Lady Humpback', is one of the 37 officially recognised spirits (nats) in Myanmar. The name is an allusion to the species' humpbacked appearance.

Shirai, K

Japanese Cyprinid sp. *Pungtungia shiraii* M Oshima, 1957

Kunihko Shirai (DNF) was a Japanese ornithologist and collector. He was with the Bureau of Game and Hunting of the Ministry of Agriculture and Forestry when he made a collection of fish downstream of the Tame River including the type of this cyprinid, which he "*... kindly forwarded to the author for identification.*" He was at one time Curator of the Nagasaki Aquarium (1967). A bird is named after him.

Shirai, S

Shirai's Spurdog *Squalus shiraii* STFL Viana & MR de Carvalho, 2020

Dr Shigeru M Shirai is a Japanese ichthyologist. Tokyo University awarded his PhD and he currently works at the department of Aqua Bioscience and Industry at Tokyo University of Agriculture. Among other works he wrote: *Squalean Phylogeny* (1992), which was originally his doctoral thesis. He is editor of the Japanese Journal of Ichthyological Research. He was honoured for his "*...valuable contributions to Systematics of Squaliformes.*"

Shirleen

Hurghada Sweeper *Pempheris shirleen* JE Randall & BC Victor, 2015

Shirleen Smith is a Museum Specialist at the Division of Fishes, United States National Museum of Natural History. In the etymology she was thanked "*...in appreciation of the many loans of Pempheris and other fishes that she has prepared for the senior author.*"

Shirley

Shirley's Coral-Blenny *Ecsenius shirleyae* VG Springer & GR Allen, 2004

Shirley Springer was the wife of the senior author.

Shojima

Butterfish sp. *Psenopsis shojimai* A Ochiai & K Mori, 1965

Dr Yoichi Shojima worked at the Seikai Regional Fisheries Research Laboratory. It was named after him as it was he "*...who originally collected and is the first to describe the species from our areas.*"

Shropshire

Sleeper sp. *Erotelis shropshirei* SF Hildebrand, 1938

James B Shropshire was an entomologist and expert on malaria and mosquitoes who was in the Panama Canal Zone (1915). He became Chief Sanitary Inspector in the US Army in the Panama Canal Zone (1924). He wrote: *Mosquito Fighters in Panama* (1931). He collected the holotype of a reptile named after him as well as the holotype of this fish (1937), whilst assisting the author.

Shubnikov

Metavay Sawbelly *Hoplostethus shubnikovi* AN Kotlyar, 1980

Dar Alexeevich Shubnikov worked at the All-Russian Research Institute of Fisheries and Oceanography (VNIRO). He co-wrote: *Materials of fish resources on the shelf of the Indian Peninsular* (1973) He was honoured for his help in Kotlyar's study of trachichthyids.

Shufeldt

American Freshwater Goby *Ctenogobius shufeldti* DS Jordan & CH Eigenmann, 1887

Professor Dr Robert Wilson Shufeldt (1850–1934) was an American surgeon and zoologist who was an expert in extant and fossil bird bones (1885–1925). He made a number of expeditions to Africa, but was profoundly racist and published: *America's Greatest Problem: The Negro* (1915). Shufeldt served as a major in the medical corps in WW1. He collected the holotype of this goby. A bird is also named after him.

Shumard

Silverband Shiner *Notropis shumardi* CF Girard, 1856
River Darter *Percina shumardi* CF Girard, 1859

Dr George Getz Shumard (1823–1867) was an American surgeon, naturalist and geologist. He graduated from the medical school in Louisville, Kentucky. He practiced as a surgeon in Arkansas, but at the same time was assistant geologist to his brother in Texas. His brother, Dr Benjamin Franklin Shumard (1820–1869), was the first state geologist of Texas. G G

Shumard was part of the US Pacific Railroad Survey. He also explored the Red River of the South (1852). His diary of the geology and palaeontology in the region was presented to President Franklin Pierce. He also explored the Wichita River and Brazos River (1854) and the Delaware Creek and Devil's River (1855). He accompanied Girard on the boundary and railroad surveys (1855–1856). He was Assistant State Geologist for Texas (1858–1861) when his brother headed the Texas geological survey (until 1859, after which he collected insects (1860) for LeConte). His pamphlet: *Notice of Fossils from the Permian Strata of Texas to New Mexico* was given in an address to the Portland Natural History Society (1858) by Dr BF Shumard announcing the Permian age fossils collected by Dr G G Shumard in the white limestone of New Mexico. GG Shumard moved (1861) to Cincinnati, Ohio where he served as Ohio State Surgeon until his death of 'general paralysis' (stroke). He collected the types of both the eponymous shiner and the darter.

Shunkan

Goby sp. *Callogobius shunkan* K Takagi, 1957

Shunkan Sôzu (1142–1179) was a Buddhist sub-bishop and a tragic hero in Japanese historical literature.

Shuntov

Righteye Flounder sp. *Microstomus shuntovi* LA Borets, 1983
Longnose Deep-sea Skate *Bathyraja shuntovi* VN Dolganov, 1985

Dr Vyacheslav Petrovich Shuntov (b.1937) is an ichthyologist with an interest in marine ornithology. Kazan State University awarded his bachelor's degree (1959) and his doctorate was awarded (1973). He worked at the Pacific Scientific Research Fisheries Centre, Vladivostok (1959–1999). He became a professor (1983) and an academician of the Russian Academy of Natural Sciences (1995). He was Deputy Editor-in-Chief of the Russian Journal of Marine Biology (2000).

Siah

Oriental Cyprinid sp. *Longiculter siahi* HW Fowler, 1937

Mr. Y Siah was a bird collector and taxidermist in Bangkok, Thailand. He collected birds for de Schauensee on Doi Suthem (1933). He assisted Fowler in forming his collection of Thai fishes and may have collected the type. He was a mixture of taxidermist, naturalist, collector and maker of crocodile handbags and wallets and kept a shop. William and Lucile Mann in the archives of the Smithsonian (1937) note: "*…we called at the shop of P. Siah, a taxidermist, who has a few animals in his back yard, one monkey, a few birds, and quite a number of snakes, including two huge pythons, a king cobra, and an albino cobra.*" He also presented at least two papers to the Siam Society: *Observations on the Male Coloration of Banteng in Thailand* (1954) and *Siamese Crocodile* (1957).

Siammakuti

Sole sp. *Soleichthys siammakuti* T Wongratana, 1975

This refers to the Royal Title of His Royal Highness Prince Vajiralongkorn (now Maha Vajiralongkorn Bodindradebayavarangkun) (b.1952) in commemoration of his investiture ceremony (1972) as the Crown Prince of Thailand (when the work on the species was taking place). His full title is Siammakutrajakumarn, abbreviated to the form of 'Siammakuti'. He became King of Thailand (2016).

Sibayi

Barebreast Goby *Silhouettea sibayi* FL Farquharson, 1970

This is a toponym; Lake Sibayi (also spelled Sibhayi), on the east coast of South Africa, is the type locality.

Siboga

Pale Catshark *Apristurus sibogae* MCW Weber, 1913
Siboga Snake-Eel *Bascanichthys sibogae* MCW Weber, 1913
Tonguefish sp. *Cynoglossus sibogae* MCW Weber, 1913
Siboga Skate *Fenestraja sibogae* MCW Weber, 1913
Lanternfish sp. *Lampanyctus sibogae* MCW Weber, 1913
Bandfish sp. *Sphenanthias sibogae* MCW Weber, 1913
Siboga Worm Eel *Muraenichthys sibogae* MCW Weber & LF de Beaufort, 1916
Rattail sp. *Coryphaenoides sibogae* MCW Weber & LF de Beaufort, 1929
Siboga Blenny *Salarias sibogai* H Bath, 1992
Flathead sp. *Onigocia sibogae* H Imamura, 2011

These species are named after the ship 'Siboga'; some after the Indonesian expedition (1898–1899) of the same name, during which the holotypes were collected. The naming of the flathead (2011) is explained by this being a replacement name for the species *Platycephalus grandisquamis* (Weber, 1913), which was found to be preoccupied.

Sidibe

African Rivuline sp. *Callopanchax sidibeorum* R Sonnenberg & E Busch, 2010

Samba Sidibe and the Sidibe family, first collectors of this species (Guinea). No more is recorded.

Sidlauskas

Leporinus (characin) sp. *Leporinus sidlauskasi* HA Britski & JL Birindelli, 2019

Dr Brian Lee Sidlauskas (b.1976) is an Associate Professor at Oregon State University & Curator of Fishes at Oregon State Ichthyology Collection. He is probably best known for popularising the concept of 'phylomorphospaces', which are multidimensional inferences and visualizations of morphological diversification. He is also well-known for phylogenetic, systematic and evolutionary studies on characiform fishes, particularly Anostomidae and its related families Curimatidae, Prochilodontidae and Chilodontidae, and was recently awarded a Fulbright fellowship to continue that work in Brazil (2020). His BA Biological Sciences was awarded by Cornell University (1998) and his MSc (2003) and PhD (2006) in Evolutionary Biology by the University of Chicago. He was hired as an

Assistant Professor at OSU (2009) after having been a Postdoctoral Fellow at the National Evolutionary Synthesis Center (2006–2009). His fieldwork has included collecting fishes in Peru, Venezuela, Guyana and Gabon. His most influential publications as lead or senior author include: *Continuous and arrested morphological diversification in sister clades of characiform fishes: a phylomorphospace approach* (2008), *Phylogenetic relationships within the South American fish family Anostomidae (Teleostei, Ostariophysi, Characiformes)*(2008) and *Ancient and contingent body shape diversification in a hyperdiverse continental fish radiation* (2019). In his spare time, he enjoys hiking, boardgaming, and playing music on ancient instruments (harp, lyre, woodwinds). The etymology says he was honoured in the name "*...for the important contributions to our knowledge of the systematics of Anostomidae.*"

Sidthimunk

Aree's Clown Loach *Ambastaia sidthimunki* W Klausewitz, 1959
[Alt. Dwarf Chain Loach, Ladderback Loach]

Aree Sidthimunk (d.2012) was a researcher at Department of Fisheries, Ministry of Agriculture, Thailand. He was Director General, Department of Fisheries of Thailand (1978–1983). He wrote: *Age and growth of the climbing perch, Anabas testudineus* (1975). The patronym is not identified in the etymology. However, given the use of the common name as well as the binomial, it seems almost certain to be named in his honour.

Siebold, K

Central American Cichlid sp. *Talamancaheros sieboldii* R Kner, 1863
Colchic Khramulya *Capoeta sieboldii* F Steindachner, 1864

Dr Karl Theodor Ernst von Siebold (1804–1885) studied at Universität Berlin and Georg-August-Universität Göttingen, becoming a physician who practised in East Prussia (1831–1834) at Heilsberg (Lidzbark Warmiński, Poland) and Königsberg (Kaliningrad, Russia). He was headmaster in Danzig (Gdansk, Poland) (1834–1840) and subsequently Professor of Zoology, Comparative Anatomy, and Veterinary Science, Friedrich-Alexander-Universität Erlangen (1840–1845); Zoology and Physiology, Albert-Ludwigs-Universität Freiburg (1845–1850); Physiology, Universität Breslau (Wroclaw, Poland) (1850–1853); and Zoology and Comparative Anatomy, Maximilians-Universität, Munich (1853). He co-wrote: *Lehrbuch der Vergleichenden Anatomie* (1845–1848). He was a cousin of P F B Siebold (q.v.). The khramulya etymology does not identify this patronym, but everything points to it being in honour of him, especially as he expanded the fish collection at the Bavarian State Collection of Zoology (München, Germany) when Steindachner was curating in Vienna. The cichlid is more straightforward: he invited Kner to study a collection of Central American fishes made by German explorer and geographer Moritz Wagner (1813–1887), including the type of this one.

Siebold, P

Japanese Cyprinid sp. *Nipponocypris sieboldii* CJ Temminck & H Schlegel, 1846
Lavender Jobfish *Pristipomoides sieboldii* P Bleeker, 1855
Wrasse sp. *Pseudolabrus sieboldi* K Mabuchi & T Nakabo, 1997

Dr Philipp Franz Balthasar von Siebold (1796–1866) was a German physician, biologist, traveller and medical officer to the Dutch East Indian Army in Batavia (Java) and at the Dutch Trading Post, Dejima Island, Nagasaki, Japan. He taught Western medicine and treated Japanese patients, accepting ethnographic and art objects as payment, and established a boarding school in the outskirts of Nagasaki. Using local Japanese agents, he collected in the interior (1823–1829). With the connivance of the Imperial librarian and astronomer, he copied a map of the northern regions of Japan, so upsetting the government that all his known Japanese contacts were imprisoned, his house was searched and many possessions confiscated. He packed all of his manuscripts, maps and books in a large leadlined chest, which was then hidden. He was banished from Japan (1829) and was forced to leave behind his young Japanese mistress and a two-year-old daughter (shades *of Madame Butterfly*). He returned to Holland, prepared his Japanese materials for publication, and was appointed by the King to advise on Japanese affairs (1831). The Japanese ban was lifted (1859) and he became (1861) chief negotiator for all European nations who were trying to establish trade links. However, his mission was a failure and he was pensioned off (1863). His eldest son served the Japanese government as an interpreter and adviser for external and financial affairs (1870–1911). He wrote much on Japan, including: *Fauna Japonica – Aves* (1844). He supervised the whole of the series: *Fauna Japonica* (1833–1850) and collected many of the fishes described in the fish section. In Nagasaki, the Siebold Memorial Museum was founded to honour his contributions to the modernisation of Japan. Thee birds, two reptiles and an amphibian are named after him.

Siegfried

Neotropical Rivuline sp. *Cynodonichthys siegfriedi* WA Bussing, 1980

Peter Siegfried was co-discoverer of the fish in Costa Rica (1974) with the author, the late Bill Busing. According to the etymology, he "*...dedicated considerable effort*" and "*kindly volunteered his time on numerous occasions*" in collecting several series of this and other Costa Rican *Rivulus* (original genus) for Bussing's study.

Sifontes

Thorny Catfish sp. *Oxydoras sifontesi* A Fernández-Yépez, 1968

Ernesto Sifontes (1881–1959) was a Venezuelan hydrologist and meteorologist. He was a member of the (1912) expedition to Auyantepuy and the Angel Falls. He was director of the Ciudad Bolivar meteorological station (until 1940). His great interest and study was the River Orinoco, in which this species is found, and he decided to navigate it from Ciudad Bolivar (1950) travelling in a motor boat and making the most exact measurements of the behaviour of the river and its environment. He wrote: *From Orinoco to Avila* (1958).

Sikora

Madagascar Rainbowfish sp. *Rheocles sikorae* H-E Sauvage, 1891

Franz Sikora (1863–1902) was an Austrian explorer and collector who was based in Réunion and collected in Madagascar for seven years (1890s). He discovered fossil remains of giant lemurs and early human settlers at Andrahomana Cave, Madagascar (1899).

Silas

Indian Swellshark *Cephaloscyllium silasi* PK Talwar, 1974
Loach sp. *Ghatsa silasi* BM Kurup & KV Radhakrishnan, 2011
[Syn. *Homaloptera silasi*]
Shark Catfish sp. *Pangasius silasi* AK Dwivedi, BK Gupta, RK Singh, V Mohindra, S Chandra, S Easawarn, J Jena & KA Lal, 2017

Dr Eric Godwin Silas (1928–2018) was a Sri Lankan-born Indian ichthyologist and fisheries scientist. He held six degrees, including a MSc (1951), PhD (1954) and a DSc (1972), all awarded by Madras University (Chennai). He worked for the Zoological Survey of India, Calcutta (1949–1955) and at the Scripps Institution of Oceanography, La Jolla, California (1955–1956). He concentrated on Indian freshwater fishes (1956–1958) and later on sea fishes, mainly Indian Ocean pelagic fishes. He was based at the Central Marine Fisheries Research Institute, Cochin (1959–1993), being its Director (from 1975). He was a former Vice-Chancellor, Kerala Agricultural University, Cochin. He was Chairman of the Scientific Consultative Committee of the Rajiv Ganghi Centre for Aquaculture. Among his many publications is: *Fishes from the high range of Travancore* (1951). The etymology for the swellshark says that Silas' "*…excellent publications on the ichthyofauna of the continental shelf of the south-west coast of India have added much to our knowledge of the fauna of this region.*" He was honoured in the loach's name for his outstanding contributions to the taxonomy of freshwater fishes of the Western Ghats.

Silenus

Prowfish *Zaprora silenus* DS Jordan, 1896

Silenus, in Greek mythology, was a companion and tutor to the wine-god Dionysus. He was usually drunk, and the binomial may have been inspired by the fact that the holotype was preserved in alcohol.

Silimon

Characin sp. *Hemigrammus silimoni* HA Britski & FCT Lima, 2008

Keve Zobogany de Szonyi de Silimon is a Brazilian biologist. He collected the type series and is honoured for his 'long and continuous' efforts in documenting fishes from Mato Grosso, Brazil. He has spent more than four decades researching fish in Mato Grosso. He is a dedicated environmentalist and a freelance consultant and designer of environmental projects.

Sillner

Garden Eel sp. *Gorgasia sillneri* W Klausewitz, 1962

Ludwig Sillner (1914–1973) was a German pioneering underwater photographer. The Underwater Society of America named him Photographer of the Year (1968). He was formerly an underwater cameraman for Jacques Cousteau. He collected the type of this eel and made important field observations on ecology. He wrote: *Ein kleiner Sprung ins große Meer. Die Abenteuer eines Sporttauchers an und in den Meeren des Nahen Ostens* (1968). He died in a traffic accident.

Silolona

Silolona Damselfish *Amblyglyphidodon silolona* GR Allen, M van Nydeck Erdmann & JA Drew, 2012

Named for the luxury charter vessel *Silolona*, in recognition of owner Patti Seery's generosity in providing opportunities for the authors to do field research in the East Indian region.

Silustan

Lake Titicaca Pupfish sp. *Orestias silustani* WR Allen, 1942

This is a toponym (wrongly spelt) referring to Sillustani, a port in the Peruvian area around Lake Titicaca and the site of a pre-Incan civilisation.

Silveira Martins

Goby sp. *Gobius silveiraemartinsi* HFA Ihering, 1893

This is a toponym; Silveira Martins is a municipality in the state of Rio Grande do Sul, Brazil, where this species (of uncertain validity) appears to be endemic.

Silvestri

Leporinus (characin) sp. *Leporinus silvestrii* GA Boulenger, 1902

Professor Dr Filippo Silvestri (1873–1949) was an Italian entomologist. He attended Università degli Studi di Roma 'La Sapienza' (1892), later moving to Università degli Studi di Palermo, where he graduated (1896). He worked at the Institute of Comparative Anatomy, Rome (1896–1902), then went to Laboratorio di Zoologia Generale e Agraria della R. Scuola Superiore d'Agricoltura, Portico, becoming Director (1904–1949). He visited South America (c.1900). A reptile and two birds are named after him.

Silvia

African Dwarf Cichlid sp. *Pelvicachromis silviae* A Lamboj, 2013

Silvia Lamboj, wife of the author, was honoured "*...as a 'thank you' in accepting long absences of mine in mind and body, and in endurance of my usage of many resources for my work. Without her understanding and support, all my works would never have been possible.*"

Silvie

Sisorid Catfish sp. *Glyptothorax silviae* BW Coad, 1981

Mrs Sylvie Coad is the wife of the author. He honoured her "*...for her assistance with field work in Iran under trying conditions.*"

Silvina

Pencil Catfish genus *Silvinichthys* G Arratia, 1998

Pencil Catfish sp. *Trichomycterus pseudosilvinichthys* L Fernández & RP Vari, 2004

Dr Silvina Adela Menu-Marque is an Argentinean zoologist at the Department of Biological Sciences, University of Buenos Aires, where her doctorate was awarded (2002). She is a specialist in the study of copepods. She co-wrote: *Morphological comparison of Mesocyclops araucanus Campos et al., 1974, and M. longisetus Thiébaud, 1912 and first description of their males* (2002). The *Trichomycterus* species' binomial is not a true eponym as it reflects the resemblance of this taxon to those in the genus *Silvinichthys.*

Sim

Betta (Fighting Fish) sp. *Betta simorum* HH Tan & PKL Ng, 1996

Thomas G K Sim & Farah Sim were the owners and operators of Sindo Aquarium in Indonesia, a tropical fish shop. The etymology says: "*The fish is named after Thomas G. K. Sim and his wife, Farah, proprietors of Sindo Aquarium Pte. Ltd., for being such excellent hosts during our stays in Jambi.*"

Simeons

African Airbreathing Catfish sp. *Clariallabes simeonsi* M Poll, 1941

H M Simeons collected the holotype or provided the collection that contained it to the Musée Royal d'Histoire Naturelle de Bruxelles. Poll's etymology does not help and we have been unable to find anything about him.

Simon

Chinese Cyprinid sp. *Pseudobrama simoni* P Bleeker, 1864

Gabriel Eugène Simon (1829–1896) was an agronomist and a consular official. He was first in China for the Ministry of Agriculture (1861–1864). He was French consul in Ningpo, China (1864–1868) before transferring to Fuzhou, China (1868–1869 & 1870–1872). He sent the first shipment of Chinese tropical fish back to Europe. He wrote a Pioneering Sociological Treatise on China: *La cité chinoise... Le village abandonné* (1885). He 'reported' (translation) a collection of Chinese fishes, including the type of this one. He also said that, despite claims, there were more infanticides in Europe in general – and France in particular – than in China. He became French Consul (1872) in Sydney, New South Wales. He sent the first apricot trees to be seen in Europe to MNHN, Paris.

Simon Birch

Rasbora sp. *Rasbora simonbirchi* R Britz & HH Tan, 2018

Simon Birch (1921–1995) was Prime Warden of the Fishmongers' Company London (1970–1971) and was also a High Sheriff of the City of London. He read architecture at Trinity College Cambridge but joined the Coldstream Guards (1941), finishing the war as a Captain. After the war he continued his architectural training at the Regent Street Polytechnic and sold his paintings, mainly watercolour still-lifes, through the Roland Ward gallery in Bond Street. However, with the advent of a family, his father persuaded him to go into the City, where he became a successful stockbroker with Rowe Swann, later Sheppards & Chase. When he retired in 1983, he was able to indulge and make use of his knowledge of art. Through his friend Peter Chance, the chairman of Christie's, he set up a branch of the firm in the City. He was an "*...enthusiastic supporter of ichthyological explorations and taxonomic research, who was instrumental in securing funding for the collecting trip during which this species was discovered.*"

Simone

Lake Malawi Cichlid sp. *Labeotropheus simoneae* MJ Pauers, 2016

Simone Josephine Pauers is the daughter of the author American zoologist Michael J Pauers. As he wrote, she is his: "*...beautiful daughter, Simone Josephine Pauers, whose rosy cheeks remind me of the bright orange opercula of the males of this species.*"

Simons

Suckermouth Catfish sp. *Astroblepus simonsii* CT Regan, 1904

Perry Oveitt Simons (1869–1901) was an American citizen who collected in South America, taking reptiles and amphibians in Peru (c.1900) and birds in Bolivia (1901). When crossing the Andes his lone guide murdered him. Chubb studied Simons' ornithological collections extensively in the second decade of the 20th century. There are more than a dozen holotypes in the BMNH which he collected. Four reptiles, seven birds, two amphibians and a mammal are named after him.

Simony

Simony's Frostfish *Benthodesmus simonyi* F Steindachner, 1891
Simony's Blenny *Antennablennius simonyi* F Steindachner, 1902

Dr Oskar Simony (1852–1915) was an Austrian mathematician, physicist and naturalist. The University of Vienna awarded his doctorate (1874). He was Professor of Mathematics, Physics, and Mechanics at an academy in Vienna (1875–1912). He went to the Canary Islands (1888) and made a botanical collection on Socotra Island (Gulf of Aden) (1899). After being semi-paralysed by a stroke he committed suicide by throwing himself out of a window. Three reptiles are named after him.

Simpson, CD

Labeo sp. *Labeo simpsoni* CK Ricardo-Bertram, 1943

Charles 'Chambeshi' 'Charlie' or 'Butty' D Simpson (d.1937) was Superintendent of the African Lakes Corporation, Nyasaland. He was nicknamed for the river that is the type locality of this species. He is believed to have arrived in Africa in 1900 from Scotland in connection with the Afican Laakes Corporation steamer 'Scotia' which had been sent from Scotland in pieces, carried overland, and reconstructed for service on Lake Mweru. In 1918, he ran a rubber factory at Kasama, Northern Rhodesia (Zambia) on the banks of the Chanbeshi river. He buried all the cash in the town, £10,000 (a lot of money in 1918), under a goat pen so that the German forces, which were attacking, would not find it. The Germans captured the town on 12th November. This was the last military action in WW1 – as news of the Armistice in France (11th November 1918) did not reach the warring forces in East Africa until 14th November 1918.

Simpson, CJ

Neotropical Rivuline genus *Simpsonichthys* AL de Carvalho, 1959

Charles J Simpson, California, was a friend of the author. We have been unable to find further details.

Simpson, MJA

Lake Nabugabo Cichlid sp. *Haplochromis simpsoni* PH Greenwood, 1965

Michael J A Simpson was a biologist at the Sub-Department of Animal Behaviour at Cambridge University. His publications (1960s-1990s) include: *Ethological Isolating Mechanisms in Four Sympatric Species of Poeciliid Fish* (1966) and *The Interpretation of Individual Differences in Rhesus Monkey Infants* (1980). He was the leader of the Cambridge Nabugabo Biological Survey in Uganda (1962), during which the type was collected.

Simushira

Snailfish sp. *Polypera simushirae* CH Gilbert & CV Burke, 1912

This is a toponym referring to the type locality; Simushir Island, Kuril Islands (formerly belonging to Japan, but ceded to Russia after WWII).

Sinclair

Sinclair's Damselfish *Chrysiptera sinclairi* GR Allen, 1987
Western Blue Devil *Paraplesiops sinclairi* JB Hutchins, 1987

Nick Sinclair (d.c.1987) was an Australian museum technical officer. While a member of the Western Australian Museum's Department of Ichthyology, he was involved in the collection of the holotype and two paratypes of the Western Blue Devil. He was described by Hutchins as: "*a friend and loyal workmate who is greatly missed by all who knew him.*" He often collected with Hutchins, such as at Rottnest Island (1980).

Sindo

Shiho's Seahorse *Hippocampus sindonis* DS Jordan & JO Snyder, 1901
Rock Damselfish *Plectroglyphidodon sindonis* DS Jordan & BW Evermann, 1903

Michitaro Sindo, originally from Yamaguchi, Japan, was one of Jordan's students at Stanford University, and became Assistant Curator of fishes there. He collected with David Jordan in Hawaii (1901) and also did field work in Samoa with Jordan and Vernon Kellog (1902). He co-wrote some papers with Jordan, such as: *A review of the Japanese species of surf-fishes or Embiotocidæ* (1902) and *A review of the pediculate fishes or anglers of Japan* (1902). He was aboard the Pacific research cruise of the Albatross (1906) which also visited Alaska and Japan. He returned to Japan to find that people who had been largely educated overseas did not find appropriate employment easily, as he wrote to Jordan (1908).

Singh

Stone Loach sp. *Schistura singhi* AGK Menon, 1987
[Syn. *Nemacheilus singhi*]

Dr Kalika Prasad Singh is an ecologist who is Professor Emeritus, Department of Botany, at Banaras Hindu University, Varanasi, India. Agra University awarded his bachelor's degree (1964) and both his MSc (1966) and PhD (1970) were awarded by Banaras Hindu University. His areas of interest are ecology and environmental science; tropical forests, savannahs and agro-ecosystems; ecosystem ecology and soil ecology. He collected the loach holotype. He co-wrote: *A new technique for single spore isolation of tweo predacious fungi forming contriction ring* (2004),

Sini

Earthworm Eel sp. *Pillaiabrachia siniae* R Britz, 2016

Sini Britz is the author's daughter. She is, according to the etymology, "*a remarkable little girl... for supporting the author's fieldwork through patience and understanding*". Sounds like a devoted father writing about his daughter.

Sinkler

Chao Phraya Cyprinid sp. *Altigena sinkleri* HW Fowler, 1934

[Syn. *Bangana sinkleri*]

James Mauran Rhodes Sinkler (1905–1981) of Philadelphia, USA, helped Rodolphe Meyer de Schauensee collect fishes in Thailand on the third De Schauensee Siamese Expedition (1932). They not only collected from freshwater and the sea, but also in fish markets. Not only were Sinkler and Meyer de Schauensee friends but also their wives appear to have been great friends.

SIO

Bigscale genus *Sio* Moss, 1962
Twospine Driftfish *Psenes sio* RL Haedrich, 1970
Fathead sp. *Psychrolutes sio* JS Nelson, 1980

SIO is the official abbreviation of the Scripps Institution of Oceanography, California.

Sipahiler

Loach sp. *Cobitis sipahilerae* F Erkakan, F Özdemir & SC Özeren, 2017

Dr Füsun Sipahiler is a Turkish entomologist and, in particular, a trichopterologist who is a professor emeritus at Hacettepe University, Ankara, Turkey. She is also a friend of the senior author, Professor Füsun Erkakan. Among her publications is: *Revision of the Rhyacophila stigmatica Species Group in Turkey with descriptions of three new species (Trichoptera, Rhyacophilidae)* (2013). A caddis-fly genus is also named after her.

Sipaliwini

Cory (catfish) sp. *Corydoras sipaliwini* JJ Hoedeman, 1965
Armoured Catfish sp. *Hypostomus sipaliwinii* M Boeseman, 1968
Pike-Cichlid sp. *Crenicichla sipaliwini* A Ploeg, 1987

This is a toponym referring to the Sipaliwini River, Corantijn River system in Suriname, where these species were first collected.

Sirhan

Azraq Toothcarp *Aphanius sirhani* W Villwock, A Scholl & F Krupp, 1983

This is a toponym referring to the Wadi Sirhan depression, Jordan, where the fish is found in only one oasis.

Sirindhorn

Crown Scaly Stream Loach *Schistura sirindhornae* A Suvarnaraksha, 2015

Princess Maha Chakri Sirindhorn (b.1955) of Thailand was honoured in the loach's binomial for her 60th birthday anniversary, as well as for her biodiversity conservation projects, and many projects located in Nan Province, the type locality of this species. A bird and a dragonfly are also named after her.

Sirisha

Lobejaw Ilisha *Ilisha sirishai* BV Seshagiri Rao, 1975

V S Sirisha is the daughter of Seshagiri Rao's cousin. According to the etymology she "*...is keen in learning about fishes.*"

Sistan

Sistan's Loach *Paracobitis rhadinaeus* CT Regan, 1906

The common name is a topographical reference; Sistan is an area of Iran where the species occurs.

Sjölanders

Shellear sp. *Kneria sjolandersi* M Poll, 1967

David Sjölanders (1886–1954), aka Sjölander, was a Swedish adventurer, conservationist and nature photographer, who led the Angola Expedition (1948–1949) that collected the holotype. He was a 'conservator' at the Gothenburg Museum of Natural History (1925–1952). He was nationally famous for his travels to China and Angola, and for his faithfully assembled mammal specimens. He is considered as one of the world's foremost curators, especially famous for his 'stuffed' elephant.

Sjostedt

Blue Gularis *Fundulopanchax sjostedti* E Lönnberg, 1895

Bror Yngve Sjöstedt (1866–1948) was a Swedish entomologist and ornithologist. He was in Cameroon (1890–1891) collecting for the Uppsala University Zoological Department and for the State Natural History Museum. He joined the entomology section of the National Natural History Museum (1897) and went on an expedition to the USA and Canada to visit entomological stations and to study their methods (1898). He was part of the Swedish Zoological Expedition to Mount Kilimanjaro (1905–1906). He published extensively, including: *Zur ornithologie Kameruns* (1895), and he edited the *Wissenschaftliche Ergenbisse der Schwedischen Expedition nach dem Kilimanjaro* (1905–1906). An amphibian, a reptile and four birds are also named after him.

Skelton

Giant Redfin *Pseudobarbus skeltoni* A Chakona & ER Swartz, 2013

African Loach Catfish sp. *Doumea skeltoni* CJ Ferraris & RP Vari, 2014

African Rivuline sp. Nothobranchius skeltoni BR Watters, B Nagy & DU Bellstedt, 2019

Professor Dr Paul Harvey Skelton (b.1948) is a South African ichthyologist. Rhodes University, Grahamstown awarded his BSc (1969) and honours (1970), and his PhD (1980). He is a Research Associate at the South African Institute for Aquatic Biodiversity now devoting his time to systematic ichthyology, having been working at the institution (1984–2011), as Curator of Freshwater Fishes (1984–1985) and Director (1995–1998) of the JLB Smith Institute Grahamstown, South Africa – which metamorphosed into the South African Institute for Aquatic Biodiversity with him as Managing Director (1999–2011). He was previously Curator of Fishes at Albany Museum (1972–1983). He wrote: *A Complete Guide to Freshwater Fishes of Southern Africa* (2001). He was honoured for his "...*lifelong service to taxonomic and systematic research on freshwater fishes in southern Africa, his contribution to the taxonomic revision and systematics of* Pseudobarbus *and specifically for his mentoring of students on this group of fishes*."

Sket

Pencil Catfish sp. *Trichomycterus sketi* CA Castellanos-Morales, 2011

Dr Boris Sket (b.1936) is a Slovenian zoologist and speleo-biologist. The University of Ljubljana awarded his bachelor's degree (1958) and his PhD (1961). He was a Research Assistant at the former Natural Sciences Faculty there (1959–1969). He went on to become Professor of Invertebrate Zoology and Speleo-biology at the Biotechnical Faculty (1969–2006), University of Ljubljana. In that time, he held other positions including Rector (1989–1991). He is currently a scientific councillor and still lectures in speleo-biology to graduate and post-graduate students. He is an Associate Member of the Slovenian Academy of Science. He co-wrote: *A new blind cave loach of Paracobitis with comments on its evolution characters* (1998).

Skolkov

Amur Bream sp. *Megalobrama skolkovii* BN Dybowski, 1872

Adjutant General I G Skolkov (1814–1873) led an 1869 expedition to the Amur drainage basin where this species is found. The patronym is not identified in the original text, but Dybowski was the physician and naturalist on the General's inspection team when he visited the Amur and Ussuri regions.

Skopets

Eelpout sp. *Magadanichthys skopetsi* G Shinohara, MV Nazarkin & IA Chereshnev, 2004

Dr Mikhail B Skopets (b.1954) is a Russian fisheries biologist who is a fly-fishing guide and free-lance jopurnalist. He has worked at the Russian Academy of Sciences as Russian Far East Program Coordinator, Magadan. He also worked (1994–2005) at the Wild Salmon Centre, Oregon, USA. He studied at the Ural State University, which awarded his biology degree (1977), and the Institute of Marine Biology in Vladivostok awarded his PhD. He wrote *Fly Fishing Russia – The Far East* (2018) as well as around 140 articles and scientific papers. He collected the holotype (1997) during a Japan-Russia Joint Research Expedition to the Russian Far East.

Skottsberg

Plunderfish sp. *Artedidraco skottsbergi* E Lönnberg, 1905

D Carl Johan Fredrik Skottsberg (1880–1963) was a Swedish botanist and Antarctic explorer. His doctorate was awarded by Uppsala University (1907). He served as official botanist to the Swedish Antarctic Expedition (1901–1903), and later led the Swedish Magellanic Expedition to Patagonia (1907–1909). He was a curator at Uppsala University Botanical Museum (1909–1914) and was in charge of the development of the botanical gardens in Gothenburg (1915), then Director and Professor (1919). A bird is also named after him.

Skywalker

Emerald Gudgeon *Romanogobio skywalkeri* T Friedrich, C Wiesner, L Zangl, D Daill, J Freyhof & S Koblmüller, 2018

Luke Skywalker was the hero of the movie "Star Wars: Episode IV-A New Hope" (Lucasfilms, Twentieth Century Fox, 1977) as well as other 'Star Wars' books and films. No explanation is given for the choice of binomial, though a museum press release at the time of the description claimed the gudgeon was named because the colour of its scales resembled Luke's green lightsaber.

Sladen, EB

Razorbelly Minnow sp. *Salmostoma sladoni* F Day, 1870

Lieutenant Colonel Sir Edward Bosc Sladen (1827–1890) was British Chief Commissioner at the Court of Mandalay (1856–1886). He led a political mission (1868) sent to the Chinese frontier to inquire into the cause of the cessation of overland trade between Burma (now Myanmar) and China. He was Commissioner of the Arakan division (1876–1885). The patronym is not identified in the etymology, but, although the binomial has a different spelling, the type was collected in Myanmar and he was in the right place at the right time so we believe he was the intended honouree. A mammal is also named after him.

Sladen, WP

Goosefish genus *Sladenia* CT Regan, 1908

Sladen's Hatchetfish *Argyropelecus sladeni* CT Regan, 1908

Walter Percy Sladen (1849–1900) was a self-taught British academic biologist whose main interest was ichthyology. He is best known for the work he did on the specimens brought back by the 'Challenger' expedition, which took years and broke his health. His travelling was restricted to visiting collections of material in European museums and working in Naples with Dr Anton Dohrn, the founder of the Stazione Zoologica di Messina 'Anton Dohrn'. He gained a reputation as a taxonomic 'splitter', as he declared many specimens to belong to new genera or species. Exeter Museum holds his huge collection of echinoderms. He resigned as Secretary of the Linnean Society after 10 years (1895) because of illness and in order to manage his uncle's country estate. He collapsed and died in Florence when walking back to his hotel at the end of a six-week holiday in Italy. Though the etymology is silent, the Percy Sladen Memorial Trust, which funded the expedition during which the type specimens were caught, is perhaps also intended in these names.

Slartibartfast

Viviparous Brotula sp. *Bidenichthys slartibartfasti* CD Paulin, 1995
[Syn. *Fiordichthys slartibartfasti*]

Slartibartfast is a designer of fjords in the first and third books of Douglas Adams' *Hitchhiker's Guide to the Galaxy* series. He was keen to advise that he had once won a prize for designing Norway! The allusion here is to the brotula's distribution in the Fiordland region of New Zealand.

Sloane

Sloane's Viperfish *Chauliodus sloani* ME Bloch & JG Schneider, 1801

Sir Hans Sloane (1660–1753) was a physician and avid collector. His collections, with the contents of George II's royal library, became the basis of today's British Museum, British Library, and Natural History Museum. He graduated in medicine at the University of Orange (1683). He traveled to Jamaica (1687–1688), where he made a natural history collection and invented the practice of drinking chocolate by mixing it with milk instead of water. His investments included buying Chelsea, London, where Sloane Square is named after him. He wrote: *Natural History of Jamaica* (1707). His *Voyage to Jamaica* (1725) is cited several times by the authors. A reptile is also named after him.

Sluiter

Sluiter's Pearlfish *Carapus sluiteri* M Weber, 1905
Chessboard Blenny *Starksia sluiteri* J Metzelaar, 1919

Dr Carel Philip Sluiter (1854–1933) was a Dutch zoologist, biologist and anatomist who studied at the Universities of Leiden and Amsterdam. He was a member of the expedition to Spitsbergen on the 'Willem Barents' (1878). He taught at Batavia (Jakarta) (1889–1891), then returned to Holland as an assistant teacher at the University of Amsterdam (1891–1898) where he became Professor of Zoology (1898). He wrote: *Die holothurien der Siboga-expedition* (1901). He found the only known specimen of the pearlfish inside the body of *Polycarpa aurata* (variously known as the Ox Heart Ascidian, the Gold-mouth Sea Squirt or the Ink-spot Sea Squirt).

Slusser

False Moray sp. *Chlopsis slusserorum* KA Tighe & JE McCosker, 2003

Marion Bridgman Slusser (1914–2013) and Willis 'Bill' Stanfield Slusser (1915–2010) were enthusiastic amateur naturalists, keen birdwatchers and botanists. She graduated from Bryn Mawr College, Pennsylvania. She was a docent at Audubon Canyon Ranch on the Bolinas Lagoon, California, and travelled extensively with her second husband Bill in the course of his work at Bechtel Corporation. Despite blindness for the last twenty years of her life, Marion volunteered at Marin General hospital for 19 years, retiring at the age of 94. They were both philanthopists, and were honoured in the eel's name for their: "...*keen interest in natural history and generous support of research and education.*"

Smale

Deep-reef Klipfish *Pavoclinus smalei* PC Heemstra & JE Wright, 1986

Dr Malcolm John Smale is an ichthyologist and marine biologist who is Senior Researcher at Bayworld Centre for Research and Education, Cape Town, who became Curator Emeritus (2015) at the Port Elizabeth Museum, South Africa. Rhode Iniversity awarded his PhD. He co-authored *Guide to the Sharks and Rays of Southern Africa* (1989) and *Otolith atlas of southern African marine fishes (Ichthyological monographs)* (1995).

Smart, D

Phra Sai Ngam Blind Cave Loach *Schistura deansmarti* C Vidthayanon & M Kottelat, 2003

Dean Smart is a British speleologist who worked for the Thai Forestry Department. Among his written works are: *Bats, Slabs, & Dinosaurs. The Story of Tham Jowlarm* (1996) and *The Caving Scene: Thailand* (1997). He was also co-author of the book: *Caves of Northern Thailand* (2005). He collected most of the loach type specimens and is a 'strong voice' for cave conservation in Thailand. (Also see **Deansmart**)

Smart, E

Garra sp. *Garra smarti* F Krupp & K Budd, 2009
[Binomial sometimes corrected to the feminine *smartae*]

Emma Smart (b.1978) is a British marine biologist who was (2003–2012) researching and working as Conservation Officer at Emirates Wildlife Society, Dubai. She was honoured for her studies of Arabian wadi fish and her contributions to the conservation of freshwater habitats in Arabia; she also collected the type specimens and provided detailed information about the type locality. She and her partner set off (2012) to circumnavigate the world in a 21-year-old 4x4 in 800 days! Family circumstances prompted a return home and they set off again in late summer 2014, returning to England (August 2016) with plans for further travel.

Smedley

Barb sp. *Poropuntius smedleyi* LF de Beaufort, 1933
[Junior Synonym of *Poropuntius normani*]

Norman Smedley (1900–1980) was a British zoologist. He joined the Durham light Infantry when just sixteen, seeing active service almost immediately in France during WW1 (1916–1918). He was Assistant Curator at the Raffles Museum, Singapore (1920s-1930s) and took part in at least one collecting expedition in the Malay States (1927) as well as representing the Straits Settlements at the Fourth Pacific Congress in Java (1928). Ill health brought an end to his colonial service and he became Curator of Doncaster Museum. During WW2, he worked for the Ministry of Information and organised the collecting of medicinal herbs in northeast England. He moved to Suffolk (1952) and became Curator of Ipswich Museum (1953–1964) where his interest turned away from the natural world and more to archaeology. For many years he had deplored the destruction of the relics of our rural past; he had accordingly made a collection of farm and craft tools which he stored at Beccles, hoping that one day he would be able to found a rural life museum. Someone donated a barn and some land and the Museum of East Anglian Life was opened at Stowmarket (1967), with Smedley as Director until he retired again (1974). He wrote a number of zoological papers, such as: *Amphibians and reptiles from the Cameron Highlands, Malay Peninsula* (1931) and *An Ocean Sunfish in Malaysian Waters* (1932). He also wrote books and articles about East Anglian life, such as: *Life and Tradition in Suffolk and Northeast Essex* (1976) and *East Anglian Crafts* (1977) and was President of the Suffolk Institute of Archaeology (1968). A damselfly is also named after him.

Smirnov

Golden Skate *Bathyraja smirnovi* VK Soldatov & MN Pavlenko, 1915

Mr. Smirnov was an Inspector of Fisheries who collected fishes from the Okhtosk Sea, whence came the holotype. No forenames are given for Smirnov by the authors and we have been unable to find anything about him.

Smirnov, AV

Stonefish sp. *Inimicus smirnovi* SA Mandrytsa, 1990
Snailfish sp. *Volodichthys smirnovi* AP Andriashev, 1991

Alexey Vladimirovich Smirnov (b.1955) is a Senior Researcher at the Zoological Institute of the Russian Academy of Sciences. He has a particular interest in echinoderms. He took part in an expedition (1991) to the Kuril, Commander and Aleutian Islands.

Smit

Squeaker Catfish sp. *Synodontis smiti* GA Boulenger, 1902

Rev Pierre Jacques Smit (1863–1960) was born in the Netherlands but moved to England (1865) when his father became an illustrator at the BMNH. He followed his father and became an illustrator, and illustrated Boulenger's work. He emigrated to South Africa (1903), worked for an insurance company in Blomfontein and was ordained as a minister in the Methodist church (1906), later serving as a missionary in the Cape (1906–1932). In retirement he returned to painting, including producing 24 plates and 230 figures for Austin Robert's: *Mammals of South Africa (*1951) and was still working when he died aged 96.

Smith, Andrew

Barbeled Houndshark *Leptocharias smithii* JP Müller & FGJ Henle, 1839
African Softnose Skate *Bathyraja smithii* JP Müller & FGJ Henle, 1841
Lake Ngami Cichlid sp. *Haplochromis smithii* FL Castelnau, 1861
Smith's Cusk-eel *Ophidion smithi* HW Fowler, 1934

Dr Sir Andrew Smith (1797–1872) was a Scotsman who joined the Army Medical Service (1819) after graduating from Edinburgh University. He was a zoologist and herpetologist famous for his scrupulous accuracy. He was in the Cape Colony, South Africa (1820–1837), and became the first superintendent of the South African Museum of Natural History, Cape Town (1825). He led the first scientific expedition into the South African interior (1834–1836) and wrote: *Report of the expedition for exploring Central Africa* (1836) the same year in which he met Darwin who later sponsored him for his Fellowship of the Royal Society). However, Smith stopped his natural history collecting and study after returning to Britain, as Principal Medical Officer at Fort Pitt, Chatham (1841) and Director General of Army Medical Services (1853). The Times newspaper accused him of incompetency in organizing medical services during the Crimean War but he was cleared by an enquiry. He retired on the grounds of ill health (1858). Much of his private collection was given to Edinburgh University and is now in the Royal Museum of Scotland. He wrote the 28-part: *Illustrations of the Zoology of South Africa* (1838–1850). He collected many South African sharks and the skate holotype and coined many of the shark names later formally described by Müller and Henle. Twelve birds, nine reptiles, two amphibians and four mammals are named after him. (NB. There is no etymology for the cichlid, but it seems likely that Andrew Smith is intended)

Smith, Anthony

Blind Loach *Paracobitis smithi* PH Greenwood, 1976
[Syn. *Eidinemacheilus smithi*]

Anthony Smith (1926–2014) was a writer, explorer, sailor, balloonist and television personality (he presented the BBC TV programme 'Tomorrow's World'). He was perhaps best known for his bestselling work *The Body* (originally published in 1968 and later renamed *The Human Body*), which has sold over 800,000 copies worldwide and tied in with a BBC TV programme. He read zoology at Oxford, was a pilot in the RAF and was science correspondent for the *Daily Telegraph*. His first overseas expedition was to Persia, exploring the Qanat underground irrigation tunnels and it was here he discovered the eponymous loach. He wrote about the expedition in *Blind White Fish in Persia* (1953). He was honoured because he: "*...took great pains (some of them physical)*" in collecting ("*not without considerable difficulty*") the type. He is also famed for having crossed the Atlantic by the raft An-Tiki (2011–2012).

Smith, CL

Smith's Butterflyfish *Chaetodon smithi* JE Randall, 1975
Sand Stargazer sp. *Platygillellus smithi* CE Dawson, 1982

Clarence Lavett Smith, Jr. (1927–2015) was an American ichthyologist who spent most of his career at the AMNH. Among his many publications was: *Fish watching – an outdoor guide to freshwater fishes* (1994).

Smith, DG

Conger Eel sp. *Gnathophis smithi* ES Karmovskaya, 1990
Malacho (ladyfish) *Elops smithi* RS McBride, CR Rocha, R Ruiz-Carus & BW Bowen, 2010
Indian White-spotted Moray *Gymnothorax smithi* FS Sumod, A Mohapatra, VN Sanjeevan, TG Kishor & KK Bineesh, 2019
Conger Eel sp. *Rhynchoconger smithi* A Mohapatra, H-C Ho, S Acharya, D Ray & SS Mishra, 2022

Dr David G Smith is an American ichthyologist who is a Museum Specialist at the Smithsonian Institution. He is also historian of ichthyology of the American Society of

Ichthyologists and Herpetologists. Cornell awarded his BS (1964), the University of Miami his MS (1967) and PhD (1971). Among his many publications is: *A checklist of the moray eels of the world* (2012). His research interests continue to be: Systematics, life history, and distribution of fishes, especially eels and eel larvae, and larvae of other oceanic fishes. One etymology describes him as: "*...the well-known specialist on eels*" and another as "*...an eminent eel expert*." (Also see **David Smith**)

Smith, EA

Whiptail Catfish sp. *Farlowella smithi* HW Fowler, 1913
Characin sp. *Knodus smithi* HW Fowler, 1913

Edgar A Smith (d.1953) collected the holotypes of these fish. He was a member of the Madeira-Mamoré expedition (1907–1912), which was commissioned by the Brazilian Government to build a railway along the banks of the Rio Madeira.

Smith, HI

Banff Longnose Dace *Rhinichthys cataractae smithi* JT Nichols, 1916

Harlan Ingersoll Smith (1872–1940) was an archaeologist. He wrote quite extensively mostly on archeology but also on herpetology, ethnology and natural history such as: *The Crawfishes of Michigan* (1910) and *The Herpetology of Michigan* (1912). He began by collecting Native American artefacts and then excavating their sites. For a time (1891–1893) he was employed as 'Assistant in Ohio and Michigan Archeology' at the AMNH and then Assistant Curator of North American Archeology there (1895–1911) when he took a post as head of the newly created Archaelogy section of the Geological Survey of Canada's Victoria Memorial Museum (1911–1937) until retirement. He collected the type of this subspecies.

Smith, HM

False Graceful Whiptail *Coelorinchus smithi* CH Gilbert & CL Hubbs, 1920
Lizardfish sp. *Scopelosaurus smithii* BA Bean, 1925
Smith's Priapium Fish *Phenacostethus smithi* GS Myers, 1928
Smith's Damsel *Pomacentrus smithi* HW Fowler & BA Bean, 1928
Green Lizard Loach *Homalopteroides smithi* SL Hora, 1932
Ghost Flathead sp. *Hoplichthys smithi* HW Fowler, 1938
Arched Sculpin *Sigmistes smithi* LP Schultz, 1938
Yellowfin Menhaden *Brevoortia smithi* SF Hildebrand, 1941
Pacific Gizzard Shad *Dorosoma smithi* CL Hubbs & RR Miller, 1941

Dr Hugh McCormick Smith (1865–1941) was an American physician (1888) who became an ichthyologist. He led an expedition that explored in the Philippines (1907–1910). He worked for the US Bureau of Fisheries (1886–1922), was Fisheries Advisor in Siam (Thailand) (1923–1934), and Curator of Zoology at the USNM (1935-1941). Hubbs and Miller described him as a: "*...worthy colleague of such masters as Jordan and Gilbert and Evermann*," all members of America's "*greatest school of ichthyologists.*" He was also Hildebrand's former chief at the U.S. Bureau of Fisheries, who honoured him for his "*... outstanding accomplishments in fishery research*" and his '*useful*' book: *The Fishes of North*

Carolina (1907). Fowler said in the damsel etymology that the honour was in: "*...slight appreciation of his interest in Philippine ichthyology.*" Three birds are also named after him.

Smith, JLB

Leafy Klipfish genus *Smithichthys* C Hubbs, 1952
Mini-Clingfish *Pherallodus smithi* JC Briggs, 1955
Disco Blenny *Meiacanthus smithi* W Klausewitz, 1962
Lefteye Flounder sp. *Tosarhombus smithi* JG Nielsen, 1964
Maputo Conger *Bathymyrus smithi* PHJ Castle, 1968
Smith's Cardinalfish *Jaydia smithi* A Kotthaus, 1970
Little Scorpionfish *Scorpaenodes smithi* WN Eschmeyer & KV Rama Rao, 1972
Goby sp. *Gobiopterus smithi* AGK Menon & PK Talwar, 1973
Blenny sp. *Omobranchus smithi* VV Rao, 1974
Smith's Short-bodied Pipefish *Choeroichthys smithi* CE Dawson, 1976
Goby sp. *Bathygobius smithi* R Fricke, 1999
Smith's Shrimpgoby *Tomiyamichthys smithi* I-S Chen & L-S Fang, 2003
Sweeper sp. *Pempheris smithorum* JE Randall & BC Victor, 2015 *

Dr James Leonard Brierley Smith (1897–1968) was a South African chemist and ichthyologist most famous for identifying a stuffed fish as a coelacanth, then thought to have been extinct for millions of years. The University of the Cape of Good Hope awarded his bachelor's degree (1916) and Stellenbosch University his master's (1918). Cambridge awarded his doctorate (1922), after which he worked at Rhodes University, Grahamstown, South Africa as senior lecturer and head of department and Professor of Ichthyology (1947). With his wife, he wrote: *Sea Fishes of Southern Africa* (1949). Castle honoured him for his 'monumental works on the fishes of Mozambique area and 'valuable study' of the genus *Bathymyrus*, Chen & Fang honoured him for his "*great ichthyological achievements for the Southern Hemisphere.*" He described *Bathygobius smithi* (1960) but used the name *Pyosicus niger* now secondarily preoccupied in *Bathygobius* by *Gobius nigri* Günther, 1861. After a long illness he committed suicide by taking cyanide.

(* The sweeper is named after both JLB Smith and MM Smith) (Also see **Smith, MM** & **Margaret (Smith)**)

Smith, KL

Onejaw sp. *Monognathus smithi* E Bertelsen & JG Nielsen, 1987

Dr Kenneth L Smith Jr. is an open ocean ecologist, Senior Scientist at the Monterey Bay Aquarium Research Institute (2006–present) and is also Adjunct Professor, Department of Ocean Sciences, University of California, Santa Cruz. Southern Illinois University awarded his BA and the University of Georgia his PhD. He was Assistant Scientist, Woods Hole Oceanographic Institution (1971–1975) then Assistant Research Biologist (1976–1979) and Associate Research Biologist (1979–1984), Research Biologist (1984–2006) and ever since Research Biologist Emeritus at Scripps Institution of Oceanography. He has over 40 years experience going to sea and studying extreme ecosystems ranging from the deep ocean to Antarctic icebergs. The main thrust of his research is to understand the impact of a changing climate on deep-sea and polar ecosystems. He has published c.130 papers (1967–

2014). He was honoured because he was the chief scientist on board the cruise from which the holotype was caught in the Central North Pacific.

Smith, MM

Smith's Pufferfish *Canthigaster smithae* GR Allen & JE Randall, 1977
Dottyback sp. *Chlidichthys smithae* R Lubbock, 1977
Chumbe Sweeper *Pempheris smithorum* JE Randall & BC Victor, 2015 *
Smith's Dogfish Shark *Squalus margaretsmithae* ST de FL Viana, MW Lisher & MR de Carvalho, 2017

Professor Margaret Mary Smith née Macdonald (1916–1987) was a South African ichthyologist. She was the second wife (1938) of James Leonard Brierley Smith (q.v.). She had a bachelor's degree from Rhodes University, Grahamstown, South Africa and became her husband's research assistant and illustrator. After his death (1968) she persuaded Rhodes University to establish the JLB Smith Institute of Ichthyology and became its first Director. She was appointed Full Professor (1981) and retired (1982). The institute was re-named the South African Institute for Aquatic Biodiversity (2000) where the library was named after her (2010). She helped collect the pufferfish type and aided Lubbock with his studies of western Indian Ocean pseudochromids. (Also see **Smith, JLB** & **Margaret (Smith)**)

(* The sweeper is named after both MM Smith and JLB Smith)

Smith, PW

Slabrock Darter *Etheostoma smithi* LM Page & ME Braasch, 1976

Philip Wayne Smith (1921–1986) was a herpetologist and ichthyologist at the Illinois Natural History Survey (1953–1979). He is best known for his two monographs: *The Amphibians and Reptiles of Illinois* (1961) and *The Fishes of Illinois* (1979).

Smith, RG

Bitterling sp. *Rhodeus smithii* CT Regan, 1908

Richard Gordon Smith (1858–1919) was an English traveller, sportsman (animal hunter) and naturalist who spent time in France, Canada, Norway, Sri Lanka, Burma, New Guinea, Fiji, China, Singapore, and Japan, largely as the result of falling out with his wife and not being able to contemplate the shame of divorce. Throughout his travels he kept diaries and embellished them with drawings and keepsakes. He wrote: *Ancient Tales and Folklore of Japan* (1918). From 1915 onward he made no further diary entries, being ravaged by beriberi and malaria. He presented to the BMNH a collection of small mammals obtained by him in Japan, including the type specimen of a vole also named after him. The collection also included some other taxa including this eponymous bitterling.

Smith, RW

Triplefin sp. *Enneanectes smithi* HR Lubbock & AJ Edwards, 1981

Roger Wellesley Smith was honoured for his considerable help to the Cambridge Expedition to St Paul's Rocks (a remote group of barren islets lying just north of the Equator on the mid-Atlantic ridge, approximately 960km from the northeast coast of Brazil), during which the type was collected.

Smith-Vaniz

Princess Anthias *Pseudanthias smithvanizi* JE Randall & R Lubbock, 1981
Cardinalfish sp. *Apogon smithvanizi* GR Allen & JE Randall, 1994
Jawfish sp. *Opistognathus smithvanizi* WA Bussing & RJ Lavenberg, 2003
Brokenbar Blenny *Starksia smithvanizi* JT Williams & JH Mounts, 2003
Muscadine Darter *Percina smithvanizi* JD Williams & SJ Walsh, 2007
Scad sp. *Decapterus smithvanizi* S Kimura, K Katahira & K Kuriiwa, 2013

Dr William Farr 'Bill' Smith-Vaniz (b.1941) is a Research Associate of the Florida Museum of Natural History, researching into the diversity and evolutionary relationships of fishes. Delta State College awarded his BSc (1963) and Auburn University awarded his MSc in fisheries biology (1966), following which he published the book *Freshwater Fishes of Alabama*. As a graduate student he made several trips to the Gulf of Aqaba studying reef blennies…including once allowing one to bite his midriff to test its venom! His PhD was awarded by the University of Miami (1975), with his dissertation being a 196–page monograph on about 50 species of Indo-Pacific sabretooth blennies. He worked at the Academy of Natural Sciences of Philadelphia for 20 years before moving to Gainesville (1992) where he worked for the US Geological Survey in the Biological Resources Division until retirement (2007). He undertook fieldwork in a number of countries including Israel, Sri Lanka and the Virgin Islands, and in a number of US states. Since retiring he has continued his fish research and is currently a Research Associate at the Florida Museum of Natural History. He was honoured for his "*…wide variety of studies, especially dealing with carangids and for setting the standards for the systematic treatment of opistognathids*." Among his published 100+ works are a mumber of books such as: *Freshwater Fishes of Alabama* (1968) and papers such as: *A new species of the fangblenny* Adelotremus *from Indonesia, with supplemental description of A. leptus (Teleostei: Blenniidae: Nemophini)* (2017). (Also see **Bill**)

Smitt

Lake Teletskoye Whitefish *Coregonus smitti* NA Warpachowski, 1900
Neotropical Silverside sp. *Odontesthes smitti* F Lahille, 1929

Professor Dr Fredrik Adam Smitt (1839–1904) was a Swedish ichthyologist. He studied at Lund and Uppsala, with the latter awarding his doctorate (1863). He took part in a number of expeditions to Svalbard (1861 & 1868). He became (1871) Professor in charge of vertebrates at the Swedish Museum of Natural History. He also taught zoology (from 1879) at Stockholm University. He wrote many scientific papers and longer works, such as *A history of Scandinavian fishes* (1892). He also promoted modernizing techniques in herring fisheries. He was honoured because he had reported this silverside as a variety of *O. regia* (1898); in addition, Smitt's account of silversides in his (1898) monograph on the fishes of Tierra del Fuego is cited many times by Lahille.

Smykala

African Characin sp. *Alestopetersius smykalai* M Poll, 1967

E R Smykala collected the holotype. We know nothing more about him except that he collected regularly in the area round d'Aba on the Lower Niger River.

Sneider

Kumawa Rainbowfish *Melanotaenia sneideri* GR Allen & RK Hadiaty, 2013

Dr Richard Sneider (b.1960) is a fish hobbyist and aquarist who is Global Chair (2013) of the IUCN Freshwater Fish Specialist Group. His first degree in Philosophy and Psychoanalysis was awarded by New College in Sarasota Florida (1981), after which he attended graduate courses in Harvard Medical School of Psychiatry, and Critical Thinking and Consciousness from University of Illinois, and obtained his PhD at Claremont Graduate School in Intellectual History (1991). He is fascinated by nature, human societies, and their interaction. This has taken him to some of the most interesting and pristine places in the Americas, Africa, and Asia, being part of biodiversity monitoring programs, often in areas not previously surveyed or collected. He has been deeply involved in the world of Aquaria for 35 years and still designs, builds, and sustains freshwater eco-systems, as well as having participated in freshwater fish field surveys in the Amazon, Mexico and Western Papua. In private life he is the CEO of One World Apparel LLC and Unger Fabrik LLC manufacturing fashion clothing in 14 countries. He discovered this species, photographed and filmed it and collected the types.

Sneidern

Armoured Catfish sp. *Rineloricaria sneiderni* HW Fowler, 1944

Kjell Eriksson von Sneidern Johansson (1910–2000) was a Swedish naturalist and taxidermist who settled in South America and worked as a collector in Colombia for the Philadelphia Academy of Natural Sciences. He became (1946) Deputy Director of the Natural History Museum at Universidad del Cauca. His son, Erik, runs shooting lodges in Colombia and Paraguay. A bird and a reptile are named after him.

Snellius

Goby sp. *Callogobius snelliusi* FP Koumans, 1953

Not a true eponym but based on the name of the Dutch hydrographic research vessel 'Snellius', which collected the type.

Snelson

Ouachita Shiner *Lythrurus snelsoni* HW Robison, 1985

Professor Dr Franklin Fielder Snelson Jr (b.1943), also known as Uncle Buck, is an ichthyologist at the Florida Museum of Natural History and professor at the University of Central Florida. North Carolina State University awarded his BSc (1965) and Cornell his PhD in ichthyology and evolutionary biology (1970). While at NCSU he worked part-time at the museum and had to survey the fish species in a stream and catalogue the species caught. Afterwards, in his words, he was 'hooked' on ichthyology. After this he was Assistant Professor UCF (1970–1974), Associate Professor there (1974–1981) and full professor ever since. He has also been Chairman, Department of Biological Sciences (1981–1988). He has published many articles, papers and reports and a book: *Ecology and Evolution of Livebearing Fishes (Poeciliidae)* (1997). Robison honoured him for 'outstanding contributions' to the knowledge of *Lythrurus*. In his spare time, he is a keen angler.

Snethlage

Ghost Pleco *Ancistomus snethlageae* F Steindachner, 1911
[Syn. *Hemiancistrus snethlageae*]

Dr Henriette Mathilde Maria Elisabeth Emilie Snethlage (1868–1929) was a German ornithologist (former assistant in zoology at the Berlin Museum specialising in ornithology) who collected in the Amazon rainforests (1905–1929), having been recommended to the Goeldi Museum. She succeeded Goeldi (q.v.) as head of the zoological section (1914) but was suspended (1917) when Brazil entered WW1 against Germany, being reinstated after the Armistice (1918). She wrote: *Catalogo das Aves Amazonicas* (1914). She also wrote on local languages. She was the first woman scientist to direct a museum in Brazil and to work in Amazonia. Eight birds, six mammals, two reptiles and an amphibian are named after her.

Snoeks

Congo Cichlid sp. *Haplochromis snoeksi* S Wamuini Lunkayilakio & E Vreven, 2010

Joseph 'Jos' Snoeks is a specialist in African cichlids; a part-time professor in the Laboratory of Biodiversity and Evolutionary Genomics, Katholieke Universiteit Leuven (Belgium), and Curator of Fishes, Musée royal de l'Afrique centrale. He was honoured in the name for support and supervision of the senior author's doctoral dissertation.

Snow

Snow's Rockskipper *Rhabdoblennius snowi* HW Fowler, 1928

Rev Benjamin Galen Snow (1817–1880) was a missionary in the Caroline Islands, who also made collections of natural history specimens and presented them to the Museum of Comparative Zoology. He graduated from Bowdoin College (1846) and later Bangor Theological Seminary (1849). He was ordained and worked as a missionary for the American Baptist Council of Foreign Missions, firstly at Kusaie, Micronesia (1852), and then on Ebon Atoll, Marshall Islands (1860). He remained in the South Pacific until retiring due to ill health (1877), with the exception of a three-year hiatus. He kept a daily diary of his time on Ebon Atoll, Marshall Islands (1866).

Snyder, AM

Cameroon Cichlid sp. *Coptodon snyderae* MLJ Stiassny, UK Schliewen & WJ Dominey, 1992

Alexandra Mary 'Lex' Snyder (b.1953) is a museum collections manager. The University of Massachusetts awarded her anthropology BA (1975). She was Senior Collections Manager, Fishes (1992–2018) at the Museum of Southwestern Biology, University of New Mexico,

Collections Manager, Fishes (1988–1992), Burke Museum, University of Washington, Curatorial Assistant, Fishes (1986–1988) at the California Academy of Sciences, San Francisco, and Collections Manager, Fishes (1980–1984) at Museum of Zoology, University of Michigan. She served as a field assistant, collecting didelphid mammals and orthopteran insects (1979–1980) on the eastern slopes of the Andes; endemic crater lake cichlids of Cameroon (1984–1985); monitoring endemic fruit fly, *Drosophila silvestris*, Upper Ola'a Forest, Hawaii (1985–1987), and collector of deepsea fishes (1989/1990) NOAA surveys of North Pacific cetaceans, and larval fishes (1993–2000) of San Juan River, Utah. Her interest in the ecology of desert fishes led to employment by the Museum of Southwestern Biology and the opportunity to collect and manage collections of native fishes from the rivers, streams and sinkholes of the southwestern deserts of the USA. Her primary interests and career foci have been preservation and archive techniques for larval fishes, museum facility design and renovations for fluid preserved collections, history of natural history and museum expeditions. She was honoured for her contribution (field assistance) to the success of Wallace J Dominey's (1985) field trip to Lake Bermin, Cameroon, where this cichlid is endemic.

Snyder, JO

Waspfish genus *Snyderina* DS Jordan & EC Starks, 1901
Pearlfish genus *Snyderidia* CH Gilbert, 1905
Snyder's Moray *Anarchias leucurus* JO Snyder, 1904
Klamath Largescale Sucker *Catostomus snyderi* CH Gilbert, 1898
Fluffy Sculpin *Oligocottus snyderi* AW Greeley, 1898
Salema Butterfish *Peprilus snyderi* CH Gilbert & EC Starks, 1904
U-mark Sandperch *Parapercis snyderi* DS Jordan & EC Starks, 1905
Crocodile Toothfish sp. *Champsodon snyderi* V Franz, 1910
Snyder's Barb *Barbodes snyderi* M Oshima, 1919
[Syn. *Puntius snyderi*]
Skipper Halfbeak *Hyporhamphus snyderi* SE Meek & SF Hildebrand, 1923
Worm-goby sp. *Taenioides snyderi* DS Jordan & CL Hubbs, 1925
Bearded Warbonnet *Chirolophis snyderi* AY Taranetz, 1938
Goby sp. *Callogobius snyderi* HW Fowler, 1946
Pufferfish sp. *Takifugu snyderi* T Abe, 1988
Owens Tui Chub *Siphateles bicolor snyderi* RR Miller, 1973

John Otterbein Snyder (1867–1943) was an American zoologist and pioneer ichthyologist of the American west. Stanford University awarded both his bachelor's degree (1897) and his master's (1899). He was an instructor and professor at Stanford University (1899–1943). He took part in the 'USS Albatross' expeditions (1900s) and organised the fish collection at the US National Museum (1925). When he was Assistant Professor of Zoology at Stanford, he published: *Notes on the fishes of the streams flowing into San Francisco Bay* in *Report of the Commissioner of Fisheries to the Secretary of Commerce and Labor for the fiscal year ending June 30, 1904* (1905). He frequenty collaborated with Gilbert and was the first to notice that the sucker species was unnamed. Fowler houred him in the goby's name for Snyder's papers on the fishes of the Riu Kiu islands (Ryukyu, Japan).

Soares

Sleeper sp. *Eleotris soaresi* RL Playfair, 1867

João da Costa Soares of Mozambique. Unfortunately, no further details are given by the author, nor any reason behind his choice of the name.

(Ferraris) Soares

Driftwood Catfish subgenus *Ferrarissoaresia* S Grant, 2015

Dr Luisa Maria Sarmento-Soares Porto is an ichthyologist at the Museu de Bologia. Her bachelor's degree (1987) and master's (1991) were awarded by the State University of Rio de Janeiro, whilst the University of São Paulo awarded her doctorate (1997). Her post-doctoral studies (2006–2008) were at the State University of Rio de Janeiro. She is an ichthyologist at the Museu de Bologia Prof. Mello Leitão, Santa Teresa, Espiritu Santo, Brazil. She co-wrote: *Trichomycterus payaya, new catfish (Siluriformes: Trichomycteridae) from headwaters of rio Itapicuru, Bahia, Brazil* (2011). This name is a combination of Carl Ferraris, Jr. (q.v.) and Dr Luisa Maria Sarmento-Soares Porto. (Also see **Ferraris**)

Socolof

Pindani *Chindongo socolofi* DS Johnson, 1974
[Alt. Powder Blue Cichlid; Syn. *Pseudotropheus socolofi*]
Lesser Bleeding-heart Tetra *Hyphessobrycon socolofi* SH Weitzman, 1977
Chiapas Cichlid *Thorichthys socolofi* RR Miller & JN Taylor, 1984

Ross Benjamin Socolof (1925–2009) was a naturalist, explorer and naturalist who was also an aquarium fish exporter, breeder and wholesaler. In WW2, he served in the US Army and was wounded in action. He was obviously popular, as the tetra etymology says that he: "*...in a variety of ways has come to the aid of various ichthyologists and fisheries biologists,*" including securing the tetra type through his contacts in Brazil. He wrote: *Confessions of a Tropical Fish Addict* (1996). According to Miller and Taylor he "*...was instrumental in making the first collections of the new species and who, over the years, has enthusiastically collected cichlids and other fishes from Middle America and generously made them available for study.*"

Soela

Soela Wrasse *Suezichthys soelae* BC Russell, 1985
Conger Eel sp. *Macrocephenchelys soela* PHJ Castle, 1990
Soela Hatchetfish *Polyipnus soelae* AS Harold, 1994
Soela Whiptail *Nezumia soela* T Iwamoto & A Williams, 1999
Soela Cusk *Neobythites soelae* JG Nielsen, 2002
Soela Gurnard *Pterygotrigla soela* WJ Richards, T Yato & PR Last, 2003

Named after the CSIRO (Australian Commonwealth Scientific and Industrial Research Organisation) fisheries research vessel 'Soela', from which some of the holotypes were collected and which has contributed substantially to collections of fishes around Australia.

Soeroto

Ricefish sp. *Oryzias soerotoi* DF Mokodongan, R Tanaka & K Yamahira, 2014

H R Bambang Soeroto (1929 –2021) was a systematic ichthyologist who was also professor (1986) of geomorphology in the Department of Biology at Sam Ratulangi University, Indonesia. After service in WW2 he studied geography, graduating from Gadjah Mada University, Yogyakarta (1955). After Independence he helped establish a veteran's college, becoming chancellor there (Universitas Pembangunan Nasional Veteran) (1958–1993). Among his published papers is: *Adrianichthys roseni and Oryzias nebulosus, two new ricefishes (Atherinomorpha: Beloniformes: Adrianichthyidae) from Lake Poso, Sulawesi, Indonesia* (2003).

Sofia

Eyeless Catfish sp. *Xyliphius sofiae* MH Sabaj Pérez, TP Carvalho & RE Reis, 2017

Sofia Sabaj is the daughter of the first author, Mark Sabaj Pérez.

Sokolov

Stone Loach sp. *Schistura sokolovi* J Freyhof & DV Serov, 2001

Vladimir Evgenevich Sokolov (1928–1998) was a mammalogist and member of the Russian Academy of Sciences. He co-wrote: *Guide to the Mammals of Mongolia* (1980) as well as at least a dozen other books on mammals. Freyhof honoured him for his 'great efforts' in the zoological exploration of central Viêt Nam. Two mammals and a reptile are also named after him.

Solander

Wahoo *Acanthocybium solandri* G Cuvier, 1832
Silver Gemfish *Rexea solandri* G Cuvier, 1832
Spotted Sharpnose Puffer *Canthigaster solandri* J Richardson, 1845

Daniel Carl [Karl] Solander (1733–1782) was a Swedish naturalist and explorer who was one of Linnaeus's pupils at Uppsala. He undertook an expedition to the extreme north of the Scandinavian Peninsula (1756). On Linnaeus's recommendation, he went to England to continue his natural history studies (1760). In London he met Sir Joseph Banks, whose influence resulted in Solander sailing on Cook's first expedition on HMS Endeavour to the Southern Ocean with Banks (1768–1771). The botanical observations of Banks and Solander were published under the aegis of the British Museum (1900–1905) as *Illustrations of the Botany of Captain Cook's Voyage Round the World*. Solander also accompanied Banks on an expedition to Iceland (1772). He is credited with an unpublished manuscript: *Descriptions of plants from various parts of the world* (1767). He was the official Curator of the Duchess of Portland's considerable collection at Bulstrode in Buckinghamshire (1779). A monument was erected at Botany Bay, New South Wales (1914), to mark the spot where Cook, Banks and Solander landed in Australia (1770). He died in London and was buried in the Swedish Church there, although his remains were removed (1913) to the Swedish Cemetery in Woking. Two birds are also named after him.

Solari

Headstander characin sp. *Leporinus solarii* EL Holmberg, 1891
[Junior Synonym of *Abramites hypselonotus*]

Constantino Solari was employed by Holmberg as a collector during his expeditions, including his expedition to the Chaco (1885).

Soldatov

Prickleback genus *Soldatovia* AY Taranetz, 1937
Soldatov's Gudgeon *Gobio soldatovi* LS Berg, 1914
Soldatov's Thicklip Gudgeon *Sarcocheilichthys soldatovi* LS Berg, 1914
Lumpfish sp. *Proeumicrotremus soldatovi* AM Popov, 1930
Eelpout sp. *Lycodes soldatovi* AY Taranetz & AP Andriashev, 1935
Soldatov's Catfish *Silurus soldatovi* GV Nikolskii & SG Soin, 1948
Eelpout sp. *Bothrocara soldatovi* PY Schmidt, 1950
Eelpout sp. *Gymnelus soldatovi* NV Chernova, 2000

Vladimir Konstantinovich Soldatov (1875–1941) was a Russian ichthyologist who was professor at the Moscow Technical Institute of Fishing Industry and Fish Farming (1919–1941). He wrote: *Fishes and Commercial Fishing* (1928). A seamount in the Pacific is named Soldatov after him. He collected the holotypes of both gudgeon species.

Solemdal

Flounder sp. *Platichthys solemdali* P Momigliano, GPJ Denys, H Jokinen & J Merila, 2018

Per Solemdal (1941–2016) was a marine scientist who spent his entire career at the Insitute of Marine Research, Bergen, Norway (1967–2011). According to the etymology: "*He was the first researcher to study the Baltic Sea flounder's eggs and sperm in connection to salinity and discovered that the specific gravity of the eggs is a fixed population characteristic which is almost unchangeable" (Solemdal, 1973) laying the foundations on which many subsequent studies on local adaptation and speciation of Baltic Sea marine fishes were built.*"

Soljan

Šoljan's Brook Lamprey *Lampetra soljani* P Tutman, J Freyhof, J Dulčić, B Glamuzina & MF Geiger, 2017

Tonko Šoljan (1907–1980) was a Croatian ichthyologist, particularly noted for his monograph *Ribe Jadrana* (Fishes of the Adriatic). He studied at the universities of Zagreb, Vienna and Graz graduating (1929) and almost immediately being awarded his PhD (1930). He was then appointed as Curator of the newly established Biology and Oceanography Instiitute in Split. At the same time, he was permanent consultant (1930–1939) to the Zagreb Chamber of Commerce on sea fishing. He became a marine biologist at the Institute of Marine Biology at Helgoland (1932), then becoming professor at the Real Gymnasium in Split (1935–1941) and director of the City Museum of Natural History, Zoovrta and the Sea Aquarium in Split (1939–1941). During the Italian occupation he had to move to Zagreb, where he took over the position of Chief of the Department of Zoology at the

Faculty of Philosophy (1943–1945) and Director of the Zoological Museum (1943–1944). He took various research and teaching posts, culminating in being Director of the Biological Institute of the University of Sarajevo (1959–1970). He and his wife were killed in a road traffic accident in Sarajevo.

Solon

Scissortail Synodontis *Synodontis soloni* GA Boulenger, 1899

Alexandre Solon. Little is known about him; only that he was a young traveller who died in the Congo shortly after helping a man named Capt. Capra to collect fish specimens.

Solovjeva

Snailfish sp. *Volodichthys solovjevae* AV Balushkin, 2012

Natalia Stepanovna Solovjeva (1911–2005) was the author's first ichthyology teacher. She was an associate professor who became head of the vertebrate department at Perm State University for over twenty yrears. Balushkin's etymology further says: "*She was a talented teacher and cultivated active interest in fish in her students. Having named a species in honor of her, I expressed my deep gratitude to my teacher, whom I will never forget. June 13, 2011, will be N.S. Solovjeva's 100th anniversary, and a lot of her students and colleagues will remember with thanks this charming woman, great teacher, and very modest and polite person.*"

Somboon

Stone Loach sp. *Schistura sombooni* M Kottelat, 1998

Somboon Phetphommasouk is a consulting engineer with the OCORP Consulting Company, Vientiane, Laos and is the liaison engineer, Nam Theun 2 Electricity Consortium (Vientiane, Laos). He was honoured for his assistance and help in the field.

Someren, VD

Someren's Suckermouth (catfish) *Chiloglanis somereni* PJP Whitehead, 1958

Dr Vernon Donald van Someren, MBE (1915–1962) was a zoologist and naturalist. He graduated BSc in Zoology at Edinburgh University and while there took part in establishing the Isle of May Bird Observatory. He won a scholarship and took his PhD at London University (1938) before war service as a Captain in the East African Army Medical Corps (1945). Afterwards he became a Salmon Research Officer with the Scotland Hydro-electricity Board. He went on to become a Senior Research Officer, Ministry of Forest Development, Game and Fisheries, Nairobi, Kenya. He wrote a number of papers such as: *Territory and Distributional Variation in Woodland Birds (1936)* and two books: *The Biology of Trout in Kenya Colony* (1952) and *A Birdwatcher in Kenya* (1958). He died suddenly in Uganda shortly before he was due to retire and return to Scotland. His uncle was Dr VGL van Someren (below).

Someren, VGL

African Barb sp. *Labeobarbus somereni* GA Boulenger, 1911

Dr Victor Gurney Logan van Someren (1886–1976) qualified at Edinburgh University in both medicine and dentistry. He was appointed medical officer in British East Africa (Kenya) and spent 40 years practising there, during which time he studied local natural history. He and his brother Robert started a survey of the birds of Kenya and Uganda (1906) and their collection ultimately exceeded 25,000 specimens. He was honorary Curator of the Natural History Museum in Nairobi (1914–1938), and a Fellow of both the Linnean Society of London and the Royal Entomological Society. His collection of butterflies and other insects is in the BMNH, and his collection of African birds is in the Field Museum, Chicago. He wrote, among other works: *Days with Birds – Studies of Habits of some East African Species* (1956), continuing to write into the 1960s. He 'obtained' the type in a 'snow-water' stream at 6000 feet on Mt. Ruwenzori, Uganda. Eight birds are named after him. His nephew was VD van Someren (above).

Somov

Snailfish sp. *Paraliparis somovi* AP Andriashev & AV Neyelov, 1979

Dr Mikhail Mikhailovitch Somov (1908–1973) was a Soviet oceonologist, geographer and polar explorer. He graduated from the Moscow Hydrometeorological Institute (1937) and became (1939) senior researcher at the Arctic and Antarctic Institure. He was head of the drift-ice station at the North Pole (1950–1951) and led the first soviet Antarctic expedition (1955–1957). Part of the sea north of Victorai Land is named the Somov Sea after him. He is also honoured in the name of an asteroid and an icebreaker.

Somphongs

Somphongs' Rasbora *Trigonostigma somphongsi* H Meinken, 1958

Barb sp. *Discherodontus somphongsi* G Benl & W Klausewitz, 1962

Somphong Lekaree (d.1971) was a Thai aquarist and fish exporter who set up the Somphongs Aquarium Company. He collected the first specimens of *Discherodontus somphongsi* and provided the authors with living and preserved material. In honouring him in his etymology, Meinken said the company "*...which has supplied the German aquarium fish hobby with many beautiful fish novelties and will, it is hoped, provide even more*" (translation). The rasbora is named after just the company and not the man.

Sondhi

Sondhi's Danio *Devario sondhii* SL Hora & DD Mukerji, 1934

Burmese Barb sp. *Osteochilus sondhii* SL Hora & DD Mukerji, 1934

Dr Ved Pall Sondhi (1903–1989) was an Indian geologist who worked at the Geological Survey of India. Punjab University, Lahore (now in Pakistan) awarded his master's degree (1925). He reported on the petroleum geology of Burma (1935) and was technical adviser to the Sino-Burma Boundary Commission (1936–1937) and in charge of survey parties covering a large part of Northen India (1942–1947). After the partition of India and Pakhistan (1947) he was Director, Geological Srvey of India (1955–1945) and later Chief Geologist of Director of Drilling, National Coal Development Corporation (1958–1965). He wrote a number of reports, scientific papers etc., such as: *Preliminary note on Antimony in Chitral* (1942) and *Landslides and Hillside Stability* (1966). He collected the types of these fish.

Sonia

Blue-eyed Redfin Pleco *Hypostomus soniae* P Hollanda Carvalho & C Weber, 2005
Whale Catfish sp. *Cetopsidium soniae* RP Vari & CJ Ferraris, 2009

Dr Sonia Fisch-Muller is a professor at the Department of Herpetology and Ichthyology, Muséum d'histoire naturelle, Geneva. She co-wrote: *Three new species of Ancistrus* (2005). She brought *Cetopsidium soniae* to the authors' attention. (See also **Muller, S**)

Sonja

African Rivuline sp. Nothobranchius sonjae BR Watters, B Nagy & DU Bellstedt, 2019

Sonja Hengstler, who helped collect the holotype, is the wife of "renowned killifish enthusiast" Holger Hengstler. (See also **Hengstler**)

Sonnerat

Tomato Hind *Cephalopholis sonnerati* A Valenciennes, 1828

Pierre Sonnerat (1748–1814) was a French explorer, naturalist and collector. He wrote *Voyage à la Nouvelle Guinée* (1776), although he never set foot on the island (only landing from a ship called *Isle de France* on nearby islands). He also wrote *Voyage aux Indes Orientales et à la Chine* (1782), both books being illustrated with engravings taken from his own drawings. He also recognised that India and China were the seats of ancient civilisations. Four birds are also named after him.

Sonoda

Marine Hatchetfish genus *Sonoda* M Grey, 1959
Silver Roughy ssp. *Hoplostethus mediterraneus sonodae* AN Kotlyar, 1986

Pearl Mitsu Sonoda (1918–2015) was a largely self-taught naturalist. Her academic career was ruined by WW2, so that she had no formal qualifications, because, as she was a California-born Japanese-American, she was interned (1942–1943). She joined the staff of the Field Museum, Chicago as secretary and assistant in Division of Mammals (1943–1955) and Division of Fishes (1955–1967), where Grey also worked when he described the genus. She returned to California as Senior Curatorial Assistant in Ichthyology, California Academy of Sciences (1967–1997). She co-wrote: *Marion Grey* (1964) (q.v.).

Sophia

Soffia Toothcarp *Esmaeilius sophiae* JJ Heckel, 1847

No explanation is given in the description, but possibly after Heckel's mother Sophia. Others suggest it honours Princess Sophie of Bavaria (Sophie Friederike Dorothea Wilhelmine) (1805–1872).

Sostra

Sostra Pygmy-goby *Trimma sostra* R Winterbottom, 2004

This is a classical reference to Sostratus, the Cnidian designer and builder of the Lighthouse of Alexandria (ca. 350 BC, one of the seven wonders of the ancient world), referring to its red and white bars, which are reminiscent of the colours of many 20th-century lighthouses.

Soto

Hagfish sp. *Myxine sotoi* MM Mincarone, 2001

Dr Jules Marcelo Rosa Soto (b.1970) is a Brazilian ichthyologist and geographer who is Curator at Museu Oceanográfico do Vale do Itajaí, Universidade do Vale do Itajaí. He wrote: *Fauna microbiana ocorrente na cavidade bucal da piranha* Serrasalmus spilopleura *(Characidae) no município de Uruguaiana, Rio Grande do Sul, Brasil* (2001). He was honoured in the name of the hagfish for his work on Brazilian marine fauna and for encouraging Mincarone to study hagfishes.

Sousa

Meteor Dragonet *Protogrammus sousai* GE Maul, 1972

Commander Manuel António Pereira Christiano de Sousa was Captain of the Port of Funchal, Madeira. Günther Maul (q.v.) said in his etymology that he and Funchal Museum was greatly indebted to de Sousa for his 'untiring interest and help' during the period he was in post.

Southwell

Southwell's Pipefish *Siokunichthys southwelli* G Duncker, 1910

Dr Thomas Southwell (1879–1962) was a British zoologist and parasitologist. He took work (1906) with the Pearl Fisheries Department in Ceylon (Sri Lanka), then moved (1911) to the Fisheries Department in India. He returned to Britain (1919) to lecture in Parasitology and Helminthology at the School of Tropical Medicine in Liverpool.

Sovvityaz

Goby genus *Sovvityazius* AM Prokofiev, 2015

'Vityaz' was a famous Soviet research vessel (see **Vityaz/Vitiaz**). The genus' name can be taken to mean 'of the Soviet (vessel) Vityaz', and the goby type specimens were collected from this ship.

Sowerby

Goby sp. *Rhinogobius sowerbyi* I Ginsburg, 1917

Arthur de Carle Sowerby (1885–1954) was a zoologist, naturalist, explorer and artist who was born in China, where his father was a Baptist Missionary. He went to Bristol University but only stayed a short time before returning to China, where he began to collect specimens for the Natural History Museum. He collected mammals (1907) for the British Museum (Natural History) during an expedition to the Ordos Desert in Mongolia and (1908) was part of the Clark Expedition to Shansi and Kansu provinces. Jointly with Robert Sterling Clark, he wrote an account: *Through Shên Kan, the Account of the Clark Expedition in North China 1908–09* (1912). Clark was a very wealthy man and he financed a number of collecting trips for Sowerby. There was a revolution in China (1911) and Sowerby led an expedition to evacuate foreign missionaries from Shensi and Sianfu provinces. During WW1, Sowerby was a technical officer in the Chinese Labour Corps and saw service in France. After the war, he settled in Shanghai and established *The China Journal of Science and Arts*, which he edited until the Japanese occupied Shanghai during WW2. The Japanese Army in Shanghai interned him for the duration but, despite that, he appears to have been able to go on writing and publishing as evidenced by: *Birds recorded from or known to inhabit the Shanghai* area (1943). He moved to the USA (1949) and lived the rest of his life in Washington D.C., spending his time in genealogical research which resulted in a family history: *The Sowerby Saga*. He collected the holotype of this goby. Three birds, an odonata, an amphibian and a reptile are also named after him.

Soyo-maru

Black Grenadier *Coryphaenoides soyoae* N Nakayama & H Endo, 2016

The 'Soyo-maru', from which the holotype was collected, is a research vessel of the National Research Institute of Fisheries Science, Japan.

Sparks

Madagascar Killifish sp. *Pachypanchax sparksorum* PV Loiselle, 2006

John Stephen Sparks (b.1963) is an American ichthyologist who is senior curator (2012–present) at the Department of Ophthalmology, American Museum of Natural History and Adjunct Professor at the CERC and the EEEB. He was awarded his BA in economics (1987) by the University of Michigan. He studied biology there (1995–1997) and obtained his MSc and PhD (2001). He was was a curatorial researcher at the Department of Fish at the University of Michigan's Museum of Zoology (1997 & 1999–2001), then Assisitant Lecturer at the Consortium for Environmental Research and Conservation (CERC) and the Department of Ecology, Evolution, and Environmental Biology (EEEB) at the University of Columbia and assistant curator AMNH (2002–2007). His research interests are very diverse and range from the cichlids of Madagascar and South Asia to the bioluminescence of marine fish and he has published more than 80 papers. He his wife Karen Riseng Sparks collected much of the type material of this species and are jointly honoured.

Sparrman

Banded Tilapia *Tilapia sparrmanii* A Smith, 1840

Dr Anders Erikson Sparrman (1748–1820) was a Swedish explorer, collector and naturalist who explored the Cape area of South Africa (1772–1776). He enrolled as a student at Uppsala University at the age of nine and studied medicine (at 14) under Linnaeus (q.v.). He went as ship's doctor to China (1765–1767) and later to Cape Town (1772) to become a tutor, but joined Captain Cook's (q.v.) second voyage (1773–1775) as an assistant naturalist to Johann and Georg Forster (q.v.). He returned to Cape Town (1775) to practise medicine and explore the interior. He returned to Sweden (1776) and was appointed Keeper of the natural history collections of the Academy of Sciences (1780) and Professor of Natural History and Pharmacology (1781). He took part in an expedition to West Africa (1787). He wrote (in English): *A Voyage to the Cape of Good Hope, Towards the Antarctic Polar*

Circle, and Round the World: But chiefly into the Country of the Hottentots and Caffres, from the Year 1772 to 1776 (1789). The patronym is not identified in the description, but is most probably in honour of the Swedish naturalist. A bird, and the Asteroid '16646 Sparrman' are also named after him.

SPE

Congo Cichlid sp. *Serranochromis spei* E Trewavas, 1964

In honour of Service Piscicole d'Elizabethville (SPE) (Elizabethville is now Lubumbashi, Democratic Republic of Congo). The word is also Latin for 'hope', referring to "hope for the future of African fishery research, management and conservation".

Specht

Black Paradise Fish *Macropodus spechti* W Schreitmüller, 1936

P Specht was a German aquarist based in the French port of Le Havre, who received the first known specimens and donated examples to the author and to the BMNH.

Spegazzini

Long-whiskered Catfish sp. *Pimelodus spegazzinii* A Perugia, 1891
[Junior synonym of *Parapimelodus valenciennis*]
Pencil Catfish sp. *Trichomycterus spegazzinii* C Berg, 1897

Carlo Luigi (Carlos Luis) Spegazzini (1858–1926) was an Italian-born Argentinean mycologist and naturalist. He was trained in oenology but from the outset his main interest was fungi. He travelled from Italy to Brazil (1879) but quickly moved from there to Argentina to escape an epidemic of yellow fever. He was a member (1881) of the Italo-Argentine expedition to Patagonia and Tierra del Fuego. The expedition was shipwrecked and Spegazzini had to swim for it, bearing all his notes on his shoulder to keep them from the sea. He later took up permanent residence in Argentina (1884). He became a Professor at the University of La Plata (1887–1912). In the same period, he was also Curator of the National Department of Agriculture Herbarium, first head of the Herbarium of the Museo de La Plata, and founder of an arboretum and an Institute of Mycology in La Plata city. He is most remembered for his study of mycological and vascular plants, but he travelled widely and collected natural history specimens wherever he went. He published about 100 papers on vascular plants, mostly in Argentinean journals, and described around 1,000 new taxa. A mammal and a reptile are also named after him.

Speigler

Speigler's Mullet *Crenimugil speigleri* P Bleeker, 1858
[Syn. *Valamugil speigleri*]

Ludwig Speigler (1824–1893) was a dutchman living in Bronbeek. He intended to join the army in the East Indies (1853). There he was recruited by Bleeker to draw for his atlas and he made thousands of drawings and watercolours, providing many of the illustrations in Bleeker's *Atlas Ichthyologique des Orientales Neerlandaises.* Bleeker named the fish to show his appreciation.

Speke

Shellear sp. *Parakneria spekii* A Günther, 1868

Lake Victoria Cichlid sp. *Haplochromis spekii* GA Boulenger, 1906

Captain John Hanning Speke (1827–1864) was a British explorer. He was the first European to see Lake Victoria (Lake Nyanza) and it was he who proved it to be the source of the Nile. Speke joined Richard Burton's expedition to discover the source of the Nile, not because he was particularly interested in finding it, but more because he wanted the chance to hunt big game. By the time he parted from Burton, who went on to Lake Tanganyika, Speke too had caught the source-location obsession. Speke hunted to supply the expedition, but he also observed the behaviour and ecology of birds and collected specimens. His own shotgun killed him when he stumbled over a stile whilst out shooting in England, although some believe it was suicide. Three mammals, two reptiles and a bird are also named after him. He presented the type of the kneria to the British Museum (Natural History).

Spekul

Stone Loach sp. *Schistura spekuli* M Kottelat, 2004

This is not a true eponym but named for SPEKUL, the Caving Club of the University of Leuven, Belgium.

Spence

Pearl Goby *Istigobius spence* JLB Smith, 1947

Charles Francis Spence (1907–1982) was a businessman in Mozambique, the type locality, whom Smith thanked for providing personal assistance. Spence later published: *The Portuguese Colony of Moçambique: an Economic Survey (1951).*

Spengler

Bandtail Puffer *Sphoeroides spengleri* ME Bloch, 1785

Lorentz Spengler (1720–1807) was a zoologist employed by Det Kongelige Danske Kunstkammer, Copenhagen, as Assistant to the Keeper (1765). He became Keeper (1777), being then 'Master Turner and Conchologist' to the King. He wrote a series of scientific papers on bivalves and other molluscs. In the 1780s his son Johan Conrad (1767–1839) became his assistant, taking over as Keeper (1807). A reptile is also named after him.

Sperat

Bagrid Catfish genus *Sperata* M Holly, 1939

Maria Adolfine Sperat was the late mother-in-law of author Maximilian Holly (q.v.) who had supported Holly's studies with 'great understanding'. (Also see **Maria**)

Spies

Stone Loach sp. *Schistura spiesi* C Vidthayanon & M Kottelat, 2003

John Spies (b.1956) is an Australian photographer, speleologist and pioneer in ecological and archaeological cave studies and conservation in Thailand, where he has lived for more than thirty years and where this species occurs. He has been a cave explorer and guide for over thirty years. His explorations have included trips to Tham Xe Bang Fai cave in Laos, and Vietnam's Hang Son Doong - the largest known cave in the world.

Spio

Red Sea Nymph Cardinalfish *Cercamia spio* TH Fraser, SV Bogorodsky, AO Mal & TJ Alpermann, 2021

Spio was a Nereid (sea nymph) in Greek mythology, the daughter of Nereus and Doris. Her name means 'the dweller in caves', and this species is usually seen in the vicinity of underwater caves.

Spira

Searobin sp. *Pterygotrigla spirai* D Golani & A Baranes, 1997

Dr Micha E Spira is a neurobiologist who was the founding Scientific Director of the Inter-university Institute for Marine Sciences, Elat, Israel (1967). It is now (1976) the Alexander Silberman Institute of Life Science where Spira is Professor Emeritus. He has published around 130 papers one of the latest being *Multisite Intracellular Recordings by MEA* (2019).

Spix

Madmango Sea Catfish *Cathorops spixii* L Agassiz, 1829

Neotropical Rivuline sp. *Melanorivulus spixi* WJEM Costa, 2016

Johann Baptist Ritter von Spix (1781–1826) was a German naturalist working in Brazil (1817–1820). He gained his PhD aged 19! He studied theology for three years in Würzburg, then medicine and the natural sciences, qualifying as a medical doctor (1806). He was awarded a scholarship by the King of Bavaria (1808) and went to Paris to study zoology. At that time Paris was *the* centre for the natural sciences, with renowned scientists such as Cuvier (q.v.), Buffon (q.v.), Lamarck and Etienne Geoffroy de SaintHilaire at the height of their reputations. The King appointed him Assistant to the Bavarian Royal Academy of Sciences (1810) with special responsibility for the natural history exhibits. A group of academicians was invited to travel to Brazil (1816) and King Maximilian I agreed that two members of the Bavarian Academy of Sciences should accompany them. When Spix went to South America (1817), Natterer (q.v.) was also on board. Spix returned (1820) with specimens of 85 species of mammal, 350 birds, 130 amphibians, 116 fish and 2,700 insects as well as 6,500 botanical items. His party also brought back 57 species of living animals, mainly monkeys, parrots and curassows. This was to form the basis for the Natural History Museum in Munich. The King awarded him a knighthood and a pension for life. When he returned, he catalogued and published his findings despite extremely poor health caused by his stay in Brazil. The 3-volume report on the expedition was later published (1823, 1828 & 1831). He wrote: *Avium Brasiliensium Species Novae* (1824). Ten mammals, eight birds, four amphibians and five reptiles are also named after him.

Splechtna

Goby sp. *Didogobius splechtnai* H Ahnelt & RA Patzner, 1995

Dr Heinz Splechtna (1933–1996) was a marine biologist who was Professor of Anatomy and Morphology at the University of Vienna (1987–1996). Amongst other publications he wrote the report: *Trends in vertebrate morphology : proceedings of the 2nd International Symposium on Vertebrate Morphology, Vienna, 1986* (1989) and co-wrote the posthumousl;y published: *The Threespine Stickleback in Austria (Gasterosteus aculeatus L., Pisces: Gasterosteidae) – Morphological Variations* (1998). He was honoured for "*...introducing generations of students to the diversity of marine life.*"

Spoorenberg

African Rivuline sp. *Fundulopanchax spoorenbergi* HO Berkenkamp, 1976

Frank Spoorenberg was a Dutch aquarist who discovered the species in an aquarium store in Amsterdam.

Sprenger

Lavender Mbuna (cichlid) *Iodotropheus sprengerae* MK Oliver & PV Loiselle, 1972

Kappy Sprenger is an amateur aquarist and wildlife rehabilitator. She began working in wildlife rehabilitation (1985) in California, but has continued since moving to Maine (2002). She specialises in the care of fish-eating birds, especially loons. According to the authors, she is an "*outstanding aquarist, aquarium writer and artist of Los Gatos, California*", who took a special interest in this species; "*...Her persistent efforts to have this fish correctly identified led to the recognition that it was undescribed.*"

Springer, S

Bull Pipefish *Syngnathus springeri* ES Herald, 1942
Ridgefin Eel *Callechelys springeri* I Ginsburg, 1951
Gulf Hagfish *Eptatretus springeri* HB Bigelow & WC Schroeder, 1952
Roughbelly Skate *Dipturus springeri* JH Wallace, 1967
Springer's Sawtail Catshark *Galeus springeri* H Konstantinou & JR Cozzi, 1998
Broadnose Wedgefish *Rhynchobatus springeri* LJV Compagno & PR Last, 2010

Stewart 'Stew' Springer (1906–1991) was a field naturalist who dropped out of Butler College (1929) but was awarded a baccalaureate by George Washington University (1964), by which time he was world-renowned as an expert on both the taxonomy and behaviour of sharks. He was originally interested in herpetology and collected, identified and described the Plateau Striped Whiptail *Cnemidophorus velox* (1928). During and after the Second World War he worked variously for the Office of Strategic Services (now CIA) and for the US Navy on shark repellents and on survival manuals. After a short period in the business of commercial shark fishing, he worked for the US Fish and Wildlife Service (1950–1971). After retirement from Government service (1971) he worked for the Mote Marine Laboratory, Sarasota, Florida (1971–1979). He wrote more than 80 papers on sharks and rays, including the extensive: *A Revision of the Catsharks, Family Scyliorhinidae* (1979). He was honoured in the eel name because he obtained the type specimen by retrieving it from the stomach of a shark! (Also see **Stewart (Springer)**)

Springer, VG

Triplefin genus *Springerichthys* S-C Shen, 1994
Tube-Blenny sp. *Coralliozetus springeri* JS Stephens & RK Johnson, 1966
Orange-spotted Blenny *Hypleurochilus springeri* JE Randall, 1966
Blenny sp. *Scartella springeri* M-L Bauchot, 1967
Wormfish sp. *Paragunnellichthys springeri* CE Dawson, 1970
Korean Cyprinid sp. *Biwia springeri* PM Bănărescu & TT Nalbant, 1973
Blue-striped Dottyback *Pseudochromis springeri* HR Lubbock, 1975
Springer's Demoiselle *Chrysiptera springeri* GR Allen & HR Lubbock, 1976
Blenny sp. *Petroscirtes springeri* WF Smith-Vaniz, 1976
Springer's Barbelgoby *Gobiopsis springeri* EA Lachner & JF McKinney, 1979
Springer's Triplefin *Helcogramma springeri* PEH Hansen, 1986
Springer's Blenny *Cirripectes springeri* JT Williams, 1988
Springer's Wriggler *Paraxenisthmus springeri* TN Gill & DF Hoese, 1993
Springer's Clingfish *Lepadichthys springeri* JC Briggs, 2001
Springer's Coralbrotula *Diancistrus springeri* W Schwarzhans, PR Møller & JG Nielsen, 2005
Clingfish sp. *Alabes springeri* JB Hutchins, 2006
Labrisomid Blenny sp. *Starksia springeri* CI Castillo & CC Baldwin, 2011
Springer's Dwarfgoby *Eviota springeri* DW Greenfield & SL Jewett, 2012
Cardinalfish sp. *Pseudamiops springeri* O Gon, SV Bogorodsky & AO Mal, 2013
Red Sea Flathead *Thysanophrys springeri* LW Knapp, 2013
Springer's Coral Blenny *Ecsenius springeri* GR Allen, MV Erdmann & S-Y Vanson Liu, 2019

Dr Victor Gruschka Springer (b.1928) is an American ichthyologist. He is Senior Scientist Emeritus, Division of Fishes at the Smithsonian, which he originally joined (1961). He specialised in anatomy, classification and fish distribution. Emory University awarded his bachelor's degree (1948). He originally intended to be a physician but hated the sight of blood and switched to marine biology (1948) at the University of Miami, which awarded his master's (1954). His academic career was interrupted by service in the US Army (1950–1952), including in Korea during the Korean War, but he was able to spend a lot of time collecting. The University of Texas awarded his doctorate (1957). Like many naturalists he is a keen philatelist, collecting and publishing on stamps with a fish or fishing theme. He wrote: *Sharks in Question: The Smithsonian Answer Book* (1989). He collected several of the types of the above fish, as well as many other species and was honoured for "*for his many contributions to modern ichthyology*" and because he "*kindly provided many specimens for Lubbock's study and helped in the preparation of the manuscript.*"

Spurrell

Three-barbeled Catfish sp. *Imparfinis spurrellii* CT Regan, 1913
Ghost Knifefish sp. *Apteronotus spurrellii* CT Regan, 1914

Professor Dr Herbert George Flaxman Spurrell (1877–1918) was a British physician and zoologist, a Fellow of the Zoological Society who collected in both Ghana and Colombia. He wrote: *Modern Man and His Forerunners: A Short Study of the Human Species Living and Extinct* (1917). He served in the Royal Army Medical Corps (WW1) as a Captain. He died in Egypt of pneumonia. Three mammals, two amphibians, a bird and three reptiles are named after him.

Squire, C

Squire's Sailfin Anthias *Rabaulichthys squirei* JE Randall & F Walsh, 2010

Cadel Squire collected most of the type series. With his brother Lyle (below), he runs 'Cairns Marine', an Australian company exporting marine fauna.

Squire, L

Squire's Fairy Wrasse *Cirrhilabrus squirei* FM Walsh, 2014

Lyle Squire is the brother of Cadel (above) with whom he runs 'Cairns Marine', an Australian company exporting marine fauna.

Staeck

Dwarf Cichlid sp. *Apistogramma staecki* I Koslowski, 1985
Neotropical Rivuline sp. *Moema staecki* L Seegers, 1987
Rivuline sp. *Laimosemion staecki* I Schindler & S Valdesalici, 2011

Dr Wolfgang Staeck (b.1939) of Berlin, Germany, is a biologist and cichlid aquarist and former president of the Deutsche Cichliden-Gesellschaft. He studied biology at the free University of Berlin and after graduation was, for ten years, a research assistant and lecturer at the Technical University of Berlin which awarded his PhD. After that he taught high school biology and later was a trainer of student teachers. He has published a dozen books on keeping cichlids and numerous articles. He has collected around the world (since 1970) in Africa, Madagascar, Central and South America. He has visited both Lake Tanganyika and Lake Malawi many times. Discovering many species, he has written at least 25 scientific descriptions. Among his books are: *African Cyclids 1: Cichlids of West Africa: A Handbook For Their Indentification, Maintenance And Breeding* (1987) and *Kleine Buntbarsche: Amerikanische Cichliden I* (2017). He collected the cichlid type with Horst Linke (q.v.).

Stahl

Eelgrass Blenny *Stathmonotus stahli* BW Evermann & MC Marsh, 1899

Augustin Stahl (1842–1917) was a physician and naturalist in Puerto Rico, who made collections of local natural history specimens. He graduated MD from the Charles University of Prague (1864) and established a medical practice in the city of Bayamón. He continued to study ethnology, botany and zoology. He had a position at the Civil Institute of Natural Sciences, but his support for Puerto Rican independence from Spain led to him being deported from Spain (1898). He wrote a number of books including the first flora for Puerto Rico, but his 720 illustrations were thought lost until being found (1922). He collected 1330 diferent plants (1882–1889) on which he based his descriptions. Several plants are also named after him.

Staiger, JC

Viviparous Brotula sp. *Saccogaster staigeri* DM Cohen & JG Nielsen, 1972

Dr Jon C Staiger is a marine biologist and ecologist who is Senior Scientist at Coastal Engineering Consultants, Baton Rouge, Louisiana, where he designs and constructs barrier island restorations in the Mississippi Delta (2009–present). He was previously a Marine Consultant (2005–2008) and a Natural Resources Manager (1986–2005). His doctorate in Marine Sciences was awarded by the University of Miami (1970). He wrote: *Atlantic Flyingfishes of the genus* Cypselurus, *with descriptions of the juveniles* (1965). He was the first person to bring this species to the authors' attention.

Staiger, KT

Northern Rock Flathead *Cymbacephalus staigeri* FL Castelnau, 1875

Karl Theodor Staiger (d.1888) was a German chemical analyst, naturalist and museum curator. He worked as a chemist for the Queensland Government (1873–1880) and was Secretary to the Queensland Museum (1876–1879). A eucalyptus tree is also named after him. Famously, a 'fish' consisting of an eel's tail, mullet's body and platypus bill was cooked and served to Staiger by some hoaxers (1882). Apparently taken in, he sent a picture of the odd specimen to Castleneau, who described it under the name *Ompax spatuloides*.

Stalsberg

Peruvian Cichlid sp. *Andinoacara stalsbergi* Z Musilová, I Schindler & W Staeck, 2009

Alf Stalsberg of Tjodalyng, Norway, is a cichlid aquarist who collected the type. He was honoured for his "*...longstanding commitment to increase the knowledge about cichlid fishes.*"

Stampfli

Smooth Flounder *Citharichthys stampflii* F Steindachner, 1894

Franz Xaver Stampfli (1847–1903) was a German (possibly Swiss) naturalist who was working in Liberia (1879–1887). Büttikofer (q.v.), who was Stampfli's companion on at least one expedition, wrote a paper (1886) entitled *Zoological researches in Liberia: a list of birds, collected by Mr. F. X. Stampfli near Monrovia, on the Messurado River, and on the Junk River with Its tributaries*. A mammal and a bird are also named after him.

Stanaland

Stanaland's Sole *Solea stanalandi* JE Randall & LJ McCarthy, 1989

Brock Edward Stanaland (b.1951) is a marine biologist and diver who, with the authors, observed this sole on the seabed at Half-moon Bay near Dhahran, Saudi Arabia (1985). The junior author was able to catch it by hand. Stanaland later caught another specimen (1986). He was born at Dhahran and returned there as an adult as a biologist. He wrote, and illustrated with his photos: *In Harm's Way* (1991) about the coral islands in the Arabian Gulf. He was a commercial fisherman in Jupiter, Florida, before moving to Okeechobee where he still fishes.

Staner

Lake Tanganyika Cichlid sp. *Greenwoodochromis staneri* M Poll, 1949
[Syn. *Limnochromis staneri*]

Dr Pierre-Joseph Staner (1901–1984) was Directeur d'administration au Ministère des Colonies (1946–1962). He held a 'doctor of science' degree. Previously he had been Secretary of the Belgian Royal Academy of Science Overseas (1926–1931) and at various colonial missions (1931–1946). Later he was President of the Agricultural Society of Ardenne & Gaume (1967–1973). He was honoured for his 'countless' (translation) services rendered on behalf of Poll's expedition (1946–1947) to Lake Tanganyika. He also collected natural history speciments, especially plants, during his tours of duty.

Stanko Karaman

Drin Brook Lamprey *Eudontomyzon stankokaramani* MS Karaman, 1974

Stanko Luka Karaman (1889–1959) was the author's father. (See **Karaman, SL**)

Stanley

Boyoma Mormyrid *Marcusenius stanleyanus* GA Boulenger, 1897

This is a toponym and refers to Stanley (Boyoma) Falls, DRC, which is the type locality.

Stanley, HM

African Barb sp. *Enteromius stanleyi* M Poll & J-P Gosse, 1974

Sir Henry Morton Stanley (1841–1904) was a Welsh (later American) journalist and explorer most famous for his expedition in search of David Livingstone and for uttering the immortal phrase, on tracking him down 'Dr Livingstone, I presume'. In the etymology, he is described as: '*du grand explorateur*' of the Congo Basin. His adventures are too well documented to need further details here.

Stanley (Weitzman)

Tetra sp. *Pseudocorynopoma stanleyi* LR Malabarba, J Chuctaya, A Hirschmann, EB de Oliveira & AT Thomaz, 2020

Professor Dr Stanley Howard Weitzman (1927–2017) was honoured for his great contribution to our knowledge of Neotropical freshwater fish (see **Weitzman, SH**).

Stannius

Armoured Catfish sp. *Chaetostoma stannii* CF Lütken, 1874

Dr Hermann Friedrich Stannius (1808–1883) was a German biologist, anatomist, physiologist and entomologist. It is thought that he acquired the holotype of this species from the German botanist Gustav Karl Wilhelm Hermann Karsten (1817–1908) who, it is believed, may have collected it in Venezuela. He wrote: *About some malformations on insects* (1835).

Stappers

Sleek Lates *Lates stappersii* GA Boulenger, 1914
Blotched Catfish *Clarias stappersii* GA Boulenger, 1915
Shellear sp. *Kneria stappersii* GA Boulenger, 1915
African Barb sp. *Labeobarbus stappersii* GA Boulenger, 1915
Stappers' Mormyrid *Pollimyrus stappersii* GA Boulenger, 1915
Lake Tanganyika Catfish sp. *Chrysichthys stappersii* GA Boulenger, 1917
[Syn. *Bathybagrus stappersii*]
Lake Tanganyika Cichlid sp. *Lamprologus stappersi* J Pellegrin, 1927
Lake Tanganyika Cichlid sp. *Astatotilapia stappersii* M Poll, 1943
Stappers' Sardine *Limnothrissa stappersii* M Poll, 1948
Congo Cichlid sp. *Serranochromis stappersi* E Trewavas, 1964

Dr Jean Hubert Louis Stappers (1883–1916) was a Belgian zoologist. He attended the Catholic University of Louvain where he received his doctorate in science and medicine. He went on to study herpetology and marine science. He was on an expedition exploring the marine life of the Antarctic Ocean (1907). He led the first Belgian government expedition to Lakes Tanganyika and Moero (1911–1913) and collected the types of the eponymous barb and cichlid. After his returned he studied crustaceans and was appointed as an Assistant Curator at the Royal Belgium Natural History Museum. At the start of WW1, he became an Army Medical Officer and died of a 'mortal illness' in Calais, December 1916.

Starck

Key Blenny *Starksia starcki* CR Gilbert, 1971
Starck's Damselfish *Chrysiptera starcki* GR Allen, 1973
Starck's Tilefish *Hoplolatilus starcki* JE Randall & JK Dooley, 1974

Dr Walter Albert Starck II (b.1939) is a marine biologist, ichthyologist and pioneer of coral reef research, who first pointed out the damselfish species to Allen while diving at Osprey Reef, Coral Sea. While still an undergraduate student (1958), Starck began what was to become a 10–year investigation of the fish fauna of Alligator Reef in the Florida Keys. The University of Miami awarded his PhD (1964). He developed (1964) the optical dome port now used universally for wide-angle underwater photography and later (1968) developed the Electrolung, the first electronically regulated, closed circuit, mixed gas scuba. He has participated in numerous other marine biological expeditions around the world, including the Bahamas, the Caribbean, the Mediterranean, the Indian Ocean and the Eastern and Western tropical Pacific. Since 1978 his home has been in the far north of Queensland, Australia. From here he carried out ten years of work on the Great Barrier Reef. He is also the editor of *The Golden Dolphin* website, where many of his scientific papers, editorials, blogs and other writings are offered. He has published widely on coral reefs, fisheries, fish management, marine protected areas, climate change, environmentalism, and more. He published a *List of Fishes of Alligator Reef Florida* (1968) in which 517 species were recorded - 45 of these were previously unrecorded from Florida and an additional eight were undescribed. He is the co-author of: *The Art of Underwater Photography* (1972). The tilefish etymology says that his "*...collecting efforts, photos, and observations added much to our knowledge of the genus Hoplolatilus.*"

Starks

Yellowbelly Pipefish *Pseudophallus starksii* DS Jordan & GB Culver, 1895
Blenny genus *Starksia* DS Jordan & BW Evermann 1896
Starks' Anchovy *Anchoa starksi* CH Gilbert & CJ Pierson, 1898
Night Smelt *Spirinchus starksi* M Fisk, 1913
Star Silverside *Atherinella starksi* SE Meek & SF Hildebrand, 1923
Flyingfish sp. *Cypselurus starksi* T Abe, 1953
Starks' Tube-snouted Ghost Knifefish *Sternarchorhynchus starksi* CD de Santana & RP Vari, 2010

Edwin Chapin Starks (1867–1932) was an American ichthyologist who was an authority on the osteology of fish. He was educated at Stanford University, where he started a course in zoology (1893). He went on many expeditions including with Jordan to Mazatlán (1894), to Panama (1895), on the Harriman Expedition to Alaska (1899) and the Stanford Expedition to Brazil (1911). He was a fieldwork assistant for the US Bureau of Biological Survey (1897–1899) and Curator of the Museum and assistant professor of zoology, University of Washington (1899–1901). Stanford appointed him as Curator of Zoology (1901) and assistant professor (1927). He retired as Professor Emeritus (1932). Gilbert honoured him both because Starks had been one of his students, but more importantly because he had been a member of the Hopkins Expedition that collected the type of the anchovy in Panama (1885).

Starnes

Whale Catfish sp. *Cetopsis starnesi* RP Vari, CJ Ferraris & MCC de Pinna, 2005
Caney Fork Darter *Nothonotus starnesi* BP Keck & TJ Near, 2013
Characin sp. *Acrobrycon starnesi* D Arcila, RP Vari & NA Menezes, 2014

Dr Wayne C Starnes is a zoologist interested in evolutionary biology, systematics and genetics. He worked at the Department of Zoology, University of Tennessee, which also awarded his PhD. He was at one time married to Lynn B Starnes. Together they wrote, among other papers: *Biology of the Blackside Dace Phoxinus cumberlandensis* (1981) and *Ecology and Life History of the Mountain Madtom, Noturus eleutherus (Pisces: Ictaluridae)* (1985). He was Research Curator of Fishes for the North Carolina Museum of Natural Sciences until retiring (2014).

Starostin

Starostin's Loach *Troglocobitis starostini* NV Parin, 1983

I V Starostin was a Russian hydrobiologist who studied the inland waters of Turkmenistan where the genus is endemic. He wrote: *Fauna vnutrennikh vodoemov Turkmenistana* (Fauna of the inner waters of Turkmenistan) (1992).

Stauch

African Barb sp. *Enteromius stauchi* J Daget, 1967

Alfred Stauch (1921–1993) was a French oceanographer and ichthyologist who was a technician at the Office de la Recherche Scientifique et Technique d'Outre-Mer. He co-wrote a number of articles such as: *Les poissons du bassins du Tchad ed du bassin adjacent du Mayo-Kabbi. Études systematiques et biologiques* (1964), as well as describing a number of marine organisms. He collected the barb type.

Stawiarsky

Pencil Catfish sp. *Cambeva stawiarski* P de Miranda Ribeiro, 1968

Professor Vitor (Victor) Stawiarsky (1903–1979) taught biology at a Buenos Aires College (1929) and was at Museu Nacional, Rio de Janeiro, Brazil (1930–1945). He collected the holotype of this species.

Stearley

Armoured Catfish sp. *Soromonichthys stearleyi* NK Lujan & JW Armbruster, 2011

Dr Ralph F Stearley is (since 1992) on the faculty of Calvin College, Grand Rapids, Michigan, where he is now Professor of Geology and Palaeontology. The University of Missouri awarded his bachelor's degree (1975), the University of Utah his master's and the University of Michigan his doctorate (1990). He was a post-doctoral researcher at Illinois State Museum (1991–1992) and taught at Wheaton College's Black Hills of South Dakota field station (1994–1996). He co-wrote: *The Bible, Rocks and Time* (2008).

Stearns, R

Spotfin Croaker *Roncador stearnsii* F Steindachner, 1876

Robert Edwards Carter Stearns (1827–1909) was an American conchologist who served as director of the museum of the California Academy of Sciences. Steindachner's original text refers to "Herrn C. R. Stearns", but we believe this to be merely an accidental reversal of his initials as other wording points to the current candidate as the intended honoree.

Stearns, S

Shortwing Searobin *Prionotus stearnsi* DS Jordan & J Swain, 1885

Silas Stearns was an amateur ichthyologist who worked at the Warren Fish Company in Pensacola, Florida. He sent collections of fishes from there to the United States National Museum. (NB. One source says: "he didn't make it past thirty", but we don't know his dates or the cause of death.)

Stedman

Stedman Barb *Systomus clavatus* J McClelland, 1845

We cannot trace the origin of the common name, which we believe was adopted long after 1845 (no mention of 'Stedman' is made in McClelland's original text). It may not even refer to a person, but we hope a reader may be able to shed light on this!

Steele

Shrimp-Goby sp. *Vanderhorstia steelei* JE Randall & PL Munday, 2008

Dr Mark Adams Steele (b.1967) is a biologist and ichthyologist who is a Professor in the Department of Biology, California State University, Northridge. The University of California,

San Dego, where he began his academic career as a Biology Department teaching assistant (1985–1989) awarded his Ecology BA (1989), and the University of California, Santa Barbara awarded his PhD in Biology (1995). He was a Postdoctoral Associate, Department of Organismic Biology, Ecology and Evolution, UCLA (1995–2000) then a lecturer there. He became Research Assistant Professor at the University of Rhode Island (1999–2001), then was Assistant Research Biologist (2002–2006) and Associate Research Biologist (2006–2008) at the Marine Science Institute of University of California, Santa Barbara. He became Assistant Professor (2007–2010), Associate Professor (2010–2015), and Professor (2015–present) at CSU. He has published c.50 scientific papers including his PhD dissertation: *The contributions of predation, competition, and recruitment to population regulation of two temperate reef fishes* (1995). His research focuses on the ecology of marine and estuarine fishes. He is an avid spearfisherman and serves as the Ichthyologist for the International Underwater Spearfishing Association. He photographed the goby in its habitat and alerted the authors of its existence.

Steenackers

Lakeweed Chub *Ischikauia steenackeri* H-E Sauvage, 1883

Francisque (aka Francis) Steenackers (note the dropped '*s*' in the binomial) (1858–1917) was French Consul in Japan; in Kobe (1885–1888), Vice-Consul Nagasaki (1891) and consul in Yokohama (1900–1906). He mounted a scientific expedition to central Japan and Lake Biwa, where he made a collection of fauna including the type of this chub, all of which he sent to the Museum of Natural History in Paris. During his time in Japan he also bought Kogo (ceramic perfume boxes) and prints for his friend Georges Clémenceau (1841–1929), the renowned politician who was Prime Minister of France and President of the Versailles Peace Conference. He was French Consul General at Zurich, Switzerland (1915). He published a book of: *One Hundred Japanese Proverbs* (1885).

Steene

Prettyfin genus *Steeneichthys* GR Allen & JE Randall, 1985
Steene's Scorpionfish *Scorpaenodes steenei* GR Allen, 1977
Steene's Prettyfin *Steeneichthys plesiopsus* GR Allen & JE Randall, 1985
Lyretail Dottyback *Pseudochromis steenei* AC Gill & JE Randall, 1992

Roger C Steene (b.1942) is an Australian naturalist and underwater photographer who has lived near the Great Barrier Reef his whole life. He greatly assisted the authors on numerous expeditions in the Indo-Pacific region, usually at his own expense. He is the sole author of a number of books such as: *Butterfly and Angelfishes of the World* (1978) and *Coral Seas* (1998); has co-authored others with Gerry Allen such as *Oceanic Wilderness* (2007) and with Gerry Allen and John Randall, such as *Fishes of the Great barrier Reef and Coral Sea* (1990).

Stefanov

Cusk-eel sp. *Neobythites stefanovi* JG Nielsen & F Uiblein, 1993

'Dmitry Stefanov' is a Russian research vessel from which a major part of the type material was caught.

Stegemann

Savanna Tetra *Hyphessobrycon stegemanni* J Géry, 1961

Carlos Stegemann was a German baker and aquarist in São Paulo, Brazil. The holotype was collected by the fish collector and ethnographer, Harald Schiltz (1909–1966), who was a close friend of Stegemann. We posit that Schultz asked Géry to commemorate his friend?

Stehmann

African Pygmy Skate *Neoraja stehmanni* PA Hulley, 1972
Eelpout sp. *Plesienchelys stehmanni* AE Gosztonyi, 1977
Snailfish sp. *Paraliparis stehmanni* AP Andriashev, 1986
Pearlside sp. *Maurolicus stehmanni* NV Parin & SG Kobyliansky, 1993
Socotra Blue-spotted Guitarfish *Acroteriobatus stehmanni* S Weigmann, DA Ebert & B Séret, 2021

Dr Matthias F W Stehmann (b.1943) is a German ichthyologist whose university career at Kiel University (1962–1969) resulted in degrees covering marine sciences, zoology and limnology and culminated with a doctorate in marine sciences. He was a research scientist at the Federal Institute for Fisheries, Hamburg (1969–2002), during which time he went on many research expeditions from the Arctic to the Antarctic, as well as the whole Atlantic basin. Having retired (2002) he set up and runs his own ICHTHYS research laboratory in Hamburg. He was a contributor on batoid fishes to a number of fundamental faunal checklists and handbooks, notably: *Check-list of the fishes of the NE Atlantic and Mediterranean* (1973), *Fishes of the NE Atlantic and Mediterranean* (1984) and *Check-list of the fishes of the eastern tropical Atlantic* (1990), all published by UNESCO. He has produced a very large number of publications on chondrichthyan fishes, too many to mention here, both as a sole and as a joint author including: *Batoid Fishes – technical terms and principal measurements, general remarks, key with picture guide to families, list of species* (1978) and *Bathyraja meridionalis sp. n. (Pisces, Elasmobranchii, Rajidae), a new deep-water skate from the eastern slope of subantarctic South Georgia Island.*(1987).

Stein

Snailfish sp. *Careproctus steini* AP Andriashev & VP Prirodina, 1990
Stein's Dwarf Snailfish *Psednos steini* N Chernova, 2001

Dr David L Stein is an American ichthyologist and oceanographer who is a world expert on snailfishes. He began his professional life as an Assistant in Oceanography at Oregon State University (1971). Among his published works are: *Paraliparis hawaiiensis, a new species of snailfish (Scorpaeniformes: Liparidae) and the first described from the Hawaiian Archipelago* (2014) and *Description of a New Hadal Notoliparis from the Kermadec Trench, New Zealand, and Redescription of Notoliparis kermadecensis (Nielsen) (Liparidae, Scorpaeniformes)* (2016).

Steinbach, G

Kululu (Cameroon cichlid) *Sarotherodon steinbachi* E Trewavas, 1962

Gerhard Steinbach (1923–2016) was an entomologist at the Humboldt University of Berlin. He took part in the expedition, led by zoologist Martin Eisentraut (see *Konia eisentrauti*), during which the type was collected.

Steinbach, J

Armoured Catfish sp. *Rineloricaria steinbachi* CT Regan, 1906
[Syn. *Ixinandria steinbachi*]

Dr José (Joseph) Steinbach Kemmerich (1875–1930) was born in Germany but went to Bolivia (1904) and stayed there for the rest of his life. He shortened his name by dropping the 'Kemmerich', became a Bolivian citizen and married. His descendants are still a prominent family there. He was a collector in Argentina and Bolivia for the Field Museum of Natural History, Chicago, but also sold various natural history collections to museums and universities all over the world. The Carnegie Museum of Natural History holds some of his collection, and many of the plants he collected are in the Darwin Institute at San Isidro in Argentina. His grandson Roy F Steinbach was a professional collector in the 1960s. Two mammals, three birds, an amphibian and a reptile are named after him.

Steinbeck

Longfin Lampfish *Lampanyctus steinbecki* RL Bolin, 1939

John Steinbeck (1902–1968) was a famous American writer who was awarded the Nobel Prize for Literature (1962) Many of his novels were set around the coast of California, such as *Cannery Row* (1945) and feature 'Doc' as a central character, a marine biologist. He was a friend of the author. He needs no lengthy biography here.

Steindachner

Luminous Hake genus *Steindachneria* GB Goode & TH Bean, 1888
Characin genus *Steindachnerina* HW Fowler, 1906
Long-whiskered Catfish genus *Steindachneridion* CH Eigenmann & RS Eigenmann, 1919
Characin sp. *Characidium steindachneri* ED Cope, 1878
Rockfish sp. *Sebastes steindachneri* FM Hilgendorf, 1880
Chere-chere Grunt *Haemulon steindachneri* DS Jordan & CH Gilbert, 1882
Minnow sp. *Phoxinus steindachneri* HE Sauvage, 1883
Sickle Pomfret *Taractichthys steindachneri* L Döderlein, 1883
Characin sp. *Leptagoniates steindachneri* GA Boulenger, 1887
Smalltooth Weakfish *Cynoscion steindachneri* DS Jordan, 1889
Pacific Cownose Ray *Rhinoptera steindachneri* BW Evermann & OP Jenkins, 1891
[Alt. Golden Cownose Ray]
Characin sp. *Bario steindachneri* CH Eigenmann, 1893
Blenny sp. *Istiblennius steindachneri* GJ Pfeffer, 1893
Armoured Catfish sp. *Hypoptopoma steindachneri* GA Boulenger, 1895
Thorny Catfish sp. *Trachydoras steindachneri* A Perugia, 1897
Large-banded Blenny *Ophioblennius steindachneri* DS Jordan & BW Evermann, 1898
Steindachner's Cichlid *Nosferatu steindachneri* DS Jordan & JO Snyder, 1899

Eelpout sp. *Neozoarces steindachneri* DS Jordan & JO Snyder, 1902
Steindachner's Moray *Gymnothorax steindachneri* DS Jordan & BW Evermann, 1903
Steindachner's Sea Catfish *Cathorops steindachneri* CH Gilbert & EC Starks, 1904
Armoured Catfish sp. *Rineloricaria steindachneri* CT Regan, 1904
Borneo Cyprinid sp. *Nematabramis steindachnerii* CML Popta, 1905
Characin sp. *Leporinus steindachneri* CH Eigenmann, 1907
Dwarf Cichlid sp. *Apistogramma steindachneri* CT Regan, 1908
African Cyprinid sp. *Raiamas steindachneri* J Pellegrin, 1908
African Barb sp. *Labeobarbus steindachneri* GA Boulenger, 1910
Steindachner's Catfish *Glyptothorax steindachneri* V Pietschmann, 1913
Arowana Tetra *Gnathocharax steindachneri* HW Fowler, 1913
Squeaker Catfish sp. *Synodontis steindachneri* GA Boulenger, 1913
Three-barbeled Catfish sp. *Pimelodella steindachneri* CH Eigenmann, 1917
Armoured Catfish sp. *Pareiorhaphis steindachneri* A Miranda Ribeiro, 1918
Redhump Eartheater Cichlid *Geophagus steindachneri* CH Eigenmann & SF Hildebrand, 1922
Msola Mormyrid *Petrocephalus steindachneri* HW Fowler, 1958
Steindachner's Barbel *Luciobarbus steindachneri* CA Almaça, 1967
Paraná Corydoras *Corydoras steindachneri* IJH Isbrücker & H Nijssen, 1973
Pencil Catfish sp. *Trichomycterus steindachneri* CL DoNascimiento Montoya, S Prada-Pedreros &J Guerrero-Kommritz, 2014

Franz Steindachner (1834–1919) was an Austrian zoologist who specialised in herpetology and ichthyology. He originally planned to become a lawyer, but became interested in fossil fish and (1860) joined the Naturhistorisches Museum in Vienna, becoming a curator (1861) and Head of the Zoology Department (1874). He went on to become Director of the Vienna Museum (1898–1919). Unlike many museum curators he also travelled and actively actively and collected in the Americas including the Galapagos Islands, Africa and the Red Sea. One of his major works was to write up the amphibian and reptile sections of the published results of the circumnavigation of the globe by the Austrian frigate 'Novara'. He was honoured by Pellegrin who had spent a 'charming holiday' in Vienna and because "*...science is indebted* (to Steindachner) *for the knowledge of so many interesting kinds of fishes, particularly from Senegal*". Among his ichthyological works is: *Beiträge zur Kenntniss der Flussfische Sudamerikas* (1879). Seven amphibians, ten reptiles and two birds are also named after him.

Steiner

Chinese Rasbora *Rasbora steineri* JT Nichols & CH Pope, 1927

Rev Dr John Franklin Steiner (1884–1957) and his wife Madeline Emma Steiner née Huscher were missionaries at the American Presbyterian Mission in Hainan (1913–1942). Nichols honoured him for "*...his interest in the authors' work.*"

Steinfort

African Rivuline sp. *Nothobranchius steinforti* RH Wildekamp, 1977

Theo Steinfort (d.2008) was a Dutch aquarist and breeder of killifish. His breeding of this species helped make it available to other killifish enthusiasts. He was one of the collectors of the type with Wildekamp and others.

Steinhardt

Steinhardt's Shrimp-Goby *Cryptocentrus steinhardti* M Goren & N Stern, 2021

Michael H Steinhardt (b.1940) is an American investor and philanthropist. He has published an autobiography: *No Bull: My Life in and out of Markets* (2001). The goby was named after him "in recognition of his immensely important contribution to the establishment and construction of the Steinhardt Museum of Natural History at Tel Aviv University, Israel."

Steinitz

Scorpionfish sp. *Scorpaenodes steinitzi* W Klausewitz & Ø Frøiland, 1970
Sole sp. *Aseraggodes steinitzi* A Joglekar, 1971
Steinitz's Goby *Gammogobius steinitzi* H Bath, 1971
Flashlight Fish *Photoblepharon steinitzi* T Abe & Y Haneda, 1973
Steinitz' Prawn-Goby *Amblyeleotris steinitzi* W Klausewitz, 1974
Blenny sp. *Omobranchus steinitzi* VG Springer & MF Gomon, 1975
Steinitz' Velvetfish *Cocotropus steinitzi* WN Eschmeyer & M Dor, 1978
Red Triplefin *Helcogramma steinitzi* E Clark, 1980

Dr Heinz Steinitz (1909–1971) was an Israeli marine biologist, ichthyologist and physician. His doctorate (1938) was the first in zoology awarded by the Hebrew University, Jerusalem, where he later became Associate Professor (1957) and Professor (1968). He was the founder of a marine laboratory in Eilat, Israel, that now bears his name. He took part in a number of expeditions, including the Lake Huleh Expeditions (1938–1940). He sent specimens of the flashlight fish to Abe and suggested he describe it. Bath wrote that his "*…unexpected death represents a great loss to ichthyology*." An amphibian is also named after him.

Steinmann

Steinmann's Balchen *Coregonus steinmanni* OM Selz, CJ Dönz, P Vonlanthen & O Seehausen, 2020

Paul Steinmann (1885–1953) was a Swiss zoologist with interests in hydrology and fisheries. He was editor of the *Schweizerische Fischerei Zeitung* and author of the book *Die Fische der Schweiz* (1936). He also wrote a monograph on the Swiss species of the genus *Coregonus* (1951).

Stejneger

Sculpin sp. *Stelgistrum stejnegeri* DS Jordan & CH Gilbert, 1898
Oriental Cyprinid sp. *Scaphognathops stejnegeri* HM Smith, 1931
Oriental Cyprinid sp. *Sikukia stejnegeri* HM Smith, 1931

Dr Leonhard Hess Stejneger (1851–1943) was a Norwegian-born ornithologist and herpetologist who settled in the USA, where he became the USNM's vertebrate expert. He was the first fulltime Curator of the Herpetology Division and held the position of Curator of the Department of Reptiles and Batrachians (1889–1943). He wrote the: *Aves* volume in the series: *Standard Natural History* (1885) and *Birds of the Commander Islands and Kamtschatka*. He had a lifelong fascination with Steller (see below), writing a biography of him (1936) and retracing many of his journeys, discovering a petrel during one of them. Two mammals, sixteen reptiles, eleven birds and ten amphibians are also named after him.

Stella

Lake Turkana Minnow *Neobola stellae* EB Worthington, 1932

Stella Worthington née Johnson (1905–1978) was the wife of the author, Dr Edgar Barton Worthington (1905–2001) (q.v.), and a member of the expedition that collected the type. She read geography at Newnham College, Cambridge; it was suggested that she should be awarded an honorary degree for sacrificing her own career in favour of her husband's. She co-wrote: *Inland waters of Africa: the result of two expeditions to the great lakes of Kenya and Uganda, with accounts of their biology, native tribes and Development* (1933). In his etymology, Worthington says that she was honoured for 'greatly assisting' with her husband's fish research. She was also a talented pianist. (Also see **Worthington**)

Steller

Whitespotted Greenling *Hexagrammos stelleri* WG von Tilesius, 1810
Steller's Sculpin *Myoxocephalus stelleri* WG von Tilesius, 1811
Blackfin Flounder *Glyptocephalus stelleri* PJ Schmidt, 1904

Georg Wilhelm Steller (originally Stöhler) (1709–1746) was a German naturalist and explorer in the Russian service. He studied medicine at Halle and went to Russia (1731–1734) as a physician in the Russian Army. He became an Assistant at the Academy of Sciences in St Petersburg (1734) and left for Kamchatka (1737) accompanying Vitus Bering (q.v.) on his second expedition (1738–1742) to Alaska on board 'St Peter', which was accompanied by the St Paul. This expedition ended when the St Peter was wrecked on a desolate island, now called Bering Island, where Bering died and the surviving crew had to spend the winter in crude huts. Steller and the Danish first lieutenant Waxell proved effective in ensuring their survival. After nine months, a boat was constructed from the wreckage of the 'St Peter', enabling the survivors to leave the island for Kamchatka (1742). Steller worked in Petropavlovsk (1742–1744) but died on his return journey from there to St Petersburg. He published *Journal of a Voyage with Bering 1741–1742* (1743) in which he informally described the large marine mammal now known as Steller's Sea-Cow. Soon afterwards the animal was hunted to extinction, so Steller's expedition members were the only scientists to see it alive. Two mammals and seven birds are also named after him.

Stensen

Panama Ghost Catshark *Apristurus stenseni* S Springer, 1979

Dr Neils Stensen (variously rendered Nicolas Steno, Nicolaus Stenonis or Nicolaus Stenonius) (1638–1686) was a Danish geologist, anatomist and author. He was born Niels Stensen but, as Linnaeus did later, Latinized his name. He went to Leiden (1660) to study medicine (1660). After a short period in Paris and Montpelier, Stensen went to Florence (1665), where he studied anatomy. He was the first person to realise that what looked like sharks' teeth embedded in rocks were in fact fossilized sharks' teeth. From that discovery, he was led to formulate his most important contribution to geology, Steno's Law of Superposition. He was the Royal Anatomist in Copenhagen (1672–1674). He converted to Roman Catholicism and abandoned science (1667) and was later ordained as a priest (1675). He became a bishop (1677) and spent the rest of his life ministering to the minority Catholic populations in Denmark, Norway, and northern Germany. He was beatified (1987), the first step on the road to sainthood. The wording in the etymology praises Stensen's "*...scientifically accurate work on elasmobranch anatomy were highly influential in the beginnings of elasmobranch systematics.*"

Stephanica

Spotted-fin Rockfish *Scorpaena stephanica* J Cadenat, 1943

The original text carries only a short footnote: *Stephanica*, de 'Port Etienne'. Port Etienne in Mauritania is now called Nouadhibou and is the final resting place of over 300 ships, the world's largest ship's graveyard. We assume that the 'Stephanica' was a vessel.

Stephanidis

Velestino Spined Loach *Cobitis stephanidisi* PS Economidis, 1992

Dr Alexander I Stephanidis (1911–1990) was a Greek ichthyologist, who appears to have been independent (his private address appearing as point of contact in a published paper of 1981). The University of Athens awarded his doctorate (1939). He wrote: *Poissons d'eau douce du Péloponnèse* (1971). He was honoured as "*...the first modern explorer of the Greek freshwater fauna.*"

Stephens, JS

Pike-Blenny sp. *Chaenopsis stephensi* CR Robins & JE Randall, 1965
Professor Blenny *Paraclinus stephensi* RH Rosenblatt & TD Parr, 1969
Pearleye Lizardfish sp. *Scopelarchus stephensi* RK Johnson, 1974
Malpelo Barnacle Blenny *Acanthemblemaria stephensi* RH Rosenblatt & JE McCosker, 1988

Dr John Stewart Stephens Jr (b.1932) is Professor Emeritus (1993) of Environmental Biology, Occidental College, Los Angeles where he was on the faculty (1959–2003), and Executive Director (1995) of the Vantuna Research Group there, which he founded (1969–1995). Stanford awarded his bachelor's degree (1954) and the University of California, Los Angeles his master's (1957) and doctorate (1960). He taught at UC, Santa Barbara (1958–1959) and Occidental College (1959–1995) being Assistant Professor (1960–1965), Associate Professor (1966–1972), Professor (1972–1974) and James Irvine Professor of Environmental Biology (1974–1993). His abiding interest is the ecology and systematics of fish. He has written or co-written numerous papers including: *A revised classification of the blennioid fishes of the American family Chaenopsidae* (1963), *A method for estimating marine habitat values based on fish guilds, with comparisons between sites in the southern California bight* (1999) and *Biogeography of the trawls caught Fishes of California and an examination of the Point*

Conceptioin faunal break (2016). He was honoured for his numerous contributions to the biology of fishes and to the education of scientists.

Stephens, KB

Yellowfin Fringehead *Neoclinus stephensae* C Hubbs, 1953

Kate Stephens née Brown (c.1853–1954) was an American naturalist. She was born in England (c.1853) and may have worked for a short time at BMNH, but moved (c.1888–1890) to the USA. There she was a dressmaker and probably taught school before she married (1898) zoologist Frank Stephens, working and collecting with him, including a number of expeditions such as in the Colorado desert (1902) and Alaska (1907). She became Curator of Collections (1910–1936), eventually specialising in Mollusks and Marine Invertebrates at the San Diego Natural History Museum. She continued her interest in conchology well into her nineties until losing her sight. Hubbs commented, at the time of his description of the fringehead, that Stephens was "*…now over 100 years old*." Five marine invertebrates are also named after her.

Stephenson

Hillstream Loach sp. *Homalopteroides stephensoni* SL Hora, 1932

Lieutenant-Colonel John Stephenson (1871–1933) was a civil surgeon, Indian Medical Service Officer and biology professor at the Government College, Lahore. He was at Manchester University and London University (1887–1890) where he studied for his zoology BSc. He took his MB in both Manchester (1893) and London (1894), all with first class honours. He was House Physician at the Manchester Royal Infirmary (1893–1894) and at the Royal Hospital for Diseases of the Chest, London. He then (1895) obtained his commission in the Indian Medical Service. He saw active service in India (1897–1898) on plague duty and was then (1898) appointed as Medical Officer in the Punjab Cavalry. He had civil surgeon posts in Peshawar, Ambala and other parts of northwest India (1900–1906). He was asked to be temporary chair of Biology in the Government College at Lahore for six months but remained there, building up the zoology department and becoming Principal (1912), until retiring from India (1919) having also been made (1918) Vice-chancellor of Punjab University. Despite his wishes he could not be spared for War Service. He took a post as a lecturer at Edinburgh University (1920–1929). Thereafter he moved to London and worked unofficially at BMNH.

Sterba

Sterba's Corydoras *Corydoras sterbai* J Knaack, 1962
Darter Characin sp. *Geryichthys sterbai* A Zarske, 1997
Seven-Rays MInt Tetra *Serrapinnus sterbai* A Zarske, 2012

Dr Günther Hans Wenzel Sterba (1922–2021) was a zoologist, ichthyologist and aquarist in the former East Germany. He fought in the Wehrmacht (1943–1945), was wounded (1944) and while still in hospital after the war enrolled (1945) at the University of Jena to study medicine. He extended his studies (1947) to include zoology and was awarded his doctorate (1949). He was a research assistant and taught at the University of Jena (1949–1958), being appointed professor (1958). He became a professor at the University of Leipzig and Director of the Zoological Institute (1959) and Professor of Zoology (1961). He retired as professor emeritus (1987). *Geryichthys sterbai* was named after him to commemorate his 75th birthday. He co-wrote: *Freshwater Fishes of the World* (1959) which has had many editions since its first publication.

Stergios

Acara (cichlid) sp. *Guianacara stergiosi* H López-Fernández, DC Taphorn Baechle & SO Kullander, 2006

Basil Stergios is a botanist at the Universidad Nacional Experimental de los Llanos Occidentales, whose numerous expeditions into remote regions of southern Venezuela have uncovered undescribed fishes, as well as plants. He has co-aurthored a number of papers such as: *Four new species of Andean Pilea (Urticaceae), with additional notes on the genus in Venezuela* (2014).

Stern

Halftooth Characin sp. *Hemiodus sterni* J Géry, 1964

Max Stern (1898–1982) and his brother, Gustav, emigrated from Germany to the USA (1926), taking with them 5,000 singing canaries (Harz Roller or Miner's Canary). They started manufacturing bird food (11932) under the Hartz Mountain brand and diversified into dealing with all kinds of pets and equipment. Max is honoured in the etymology as founder of Hartz Mountain Bird Company, which "*…has done so much for the aquarium industry*".

Stevčić

Stevcic's Goby *Gorogobius stevcici* M Kovačíc & UK Schliewen, 2008

Zdravko Števčić (1931–2018) was a Croatian biologist who graduated (1956) and earned his PhD (1965) at the Faculty of Natural Sciences and Mathematics in Zagreb. He was at the Institute for Marine Biology JAZU (1960–1999) (today called the Center for Marine Research of the Ruđer Bošković Institute in Rovinj), where he became a Scientific Adviser (1983). His main areas of interest were in carcinology and theoretical biology. He was the author of a monograph: *Revision of Brachyuran Crabs* (2005) as well as around 100 papers. He founded the Association of Mediterranean Decapodologists, and discovered numerous new taxa. He was, according to the etymology, "*…a carcinologist who encouraged and helped the first author in the beginning of his work on gobies.*"

Steve Boyes

African Climbing Perch sp. *Microctenopoma steveboyesi* PH Skelton, JR Stauffer, A Chakona & JM Wisor, 2021

Dr Rutledge Steven 'Steve' Boyes is a South African ornithologist, conservationist, and a National Geographic explorer. Stellenbosch University awarded his BSc, his MSc at the University of Natal (2002) and PhD (2009). He was at the Percy FitzPatrick Institute of African Ornithology with a Centre of Excellence Postdoctoral Fellowship. He is a founder of the Wild Bird Trust and the leader of the National Geographic Okavango Wilderness

Project, on which expeditions in Angola this fish species was discovered. Among his publications is the co-written: *Patterns of daily activity of Meyer's Parrot* (Poicephalus meyeri) *in the Okavango Delta, Botswana* (2010).

Steve Norris

African Climbing Perch sp. *Microctenopoma stevenorrisi* PH Skelton, JR Stauffer, A Chakona & JM Wisor, 2021

Dr Steven Mark 'Steve' Norris is an American ichthyologist at the Biology Department, California State University Channel Islands. He is a leading researcher of African anabantid fish. He proposed the genus *Microctenopoma* (1995). He examined specimens of this species and recognised that they were a new form, but did not describe them at the time. Among his publications is the co-written: *A New Species of* Ctenopoma *(Teleostei: Anabantidae) from Southeastern Nigeria* (1990) and the book: *Freshwater Fishes of Mexico* (2006).

Steven

Lake Malawi Cichlid sp. *Aulonocara steveni* MK Meyer, R Riehl & H Zetzsche, 1987
[Possibly a synonym of *Aulonocara stuartgranti*]

Steven Longwe is a fisherman at Salima, Malawi, who cooperated with the author and supplied him with the species. He is honoured in the name but the original etymology tells us nothing more about him.

Stevens

White-spotted Gummy Shark *Mustelus stevensi* WT White & PR Last, 2008
Stevens' Swellshark *Cephaloscyllium stevensi* E Clark & JE Randall, 2011

Dr John Donald Stevens (b.1947) is a biologist and ichthyologist who was a Senior Principal Research Scientist with CSIRO Marine and Atmospheric Research, originally in Sydney (1979–1984) and subsequently in Hobart, Tasmania, Australia (1984–2011). Both his bachelor's degree (1970) and his doctorate (1976) were awarded by London University. He spent a year on Aldabra Atoll in the Indian Ocean, carrying out his post-doctoral research on reef sharks. He has published more than 100 scientific papers and reports on sharks, contributed to and edited several books on sharks as well as co-authoring *Sharks and Rays of Australia* (2009). Clark, in honouring him in the name of the swellshark, described this work as the "*...foundation for research that led to the descriptions of 37 new chondrichthyan fishes, including 11 species of Cephaloscyllium.*" White and Last honoured him because he had "*...dedicated a lifetime to researching sharks around the world, and who has contributed greatly to our knowledge of sharks and rays in Australia.*"

Stevenson

Mahseer sp. *Neolissochilus stevensonii* F Day, 1870

Colonel Stevenson (first name not given) collected fishes for Day in Myanmar, including the type of this one. No further clue is given to further identify the man.

Stewart, AL

Scaly-headed Triplefin *Karalepis stewarti* GS Hardy, 1984
Whipbeard Plunderfish *Pogonophryne stewarti* RR Eakin, JT Eastman & TJ Near, 2009
Footballfish sp. *Himantolophus stewarti* TW Pietsch & CP Kenaley, 2011
Seamount Rudderfish *Tubbia stewarti* PR Last, RK Daley & G Duhamel, 2013
Snailfish sp. *Notoliparis stewarti* DL Stein, 2016

Andrew Louis Stewart (b.1958) is Collection Manager: Science (Fishes) at the Museum of New Zealand Te Papa Tongerewa. He worked as a Geology Technician at the New Zealand Oceanographic Institute (1978–1980) before moving to the National Museum as a Technician working on cephalopods (1980–1981). He worked at the Environment and Conservation Organisations of New Zealand Inc. (1981–1982), then as Technician (Fishes) at the National Museum (later Museum of New Zealand) and was later appointed as Collection Manager Vertebrates (1991). He specialises in deep-sea fishes, Southern Ocean fishes and New Zealand freshwater fishes. He is co-editor and author of the four-volume set: *The Fishes of New Zealand* (2015).

Stewart, D

Armoured Catfish sp. *Rineloricaria stewarti* CH Eigenmann, 1909
Three-barbeled Catfish sp. *Heptapterus stewarti* JD Haseman, 1911

Dr Douglas Stewart (1873–1926) held a bachelor's degree from Yale (1896) and a doctorate from the University of Pittsburgh (1924). He joined the staff at the Carnegie Museum (1898) and was Custodian of the mineral collection (1901–1926), becoming Director of the museum (1923–1926). He was attached to the American Red Cross in Washington DC (1918–1919).

Stewart, DJ

Mai-ndombe Dwarf Sprat *Nannothrissa stewarti* M Poll & TR Roberts, 1976
Glass Knifefish sp. *Rhabdolichops stewarti* JG Lundberg & F Mago-Leccia, 1986
Croaker sp. *Pachyurus stewarti* L Casatti & NL Chao, 2002
Long-whiskered Catfish sp. *Pimelodus stewarti* FRV Ribeiro, CAS Lucena & PHF Lucinda, 2008
Stewart's Tube-snouted Ghost Knifefish *Sternarchorhynchus stewarti* CD de Santana & RP Vari, 2010
African Airbreathing Catfish sp. *Tanganikallabes stewarti* JJ Wright & RM Bailey, 2012
Armoured Catfish sp. *Rhadinoloricaria stewarti* F Provenzano-Rizzi & R Barriga-Salazar, 2020

Dr Donald James Stewart (b.1946) is Professor of Environment & Forest Biology at the College of Environmental Science and Forestry, State University of New York, Syracuse, where he presently teaches Ichthyology & Tropical Ecology. He has previously worked at the Zoology Department of the Field Museum, Chicago; the Center for Limnology, University of Wisconsin, Madison. His research interests include the ecology, conservation, and management of freshwater fishes and aquatic systems. Among his very many publications (1971–2014) are descriptions of new fish taxa from Africa, South America and North

America, such as the co-authored: *Bagrid Catfishes from Lake Tanganyika, with a Key and Descriptions of New Taxa* (1984). He helped collect the type of the sprat.

Stewart, FH

Tibetan Cyprinid sp. *Oxygymnocypris stewartii* RE Lloyd, 1908
Stone Loach sp. *Triplophysa stewarti* SL Hora, 1922

Major Francis Hugh Stewart IMS. MA, MB., DSc. (1879–1951) studied at the universities of St Andrews, Edinburgh and Cambridge. He served in the Indian Medical Service (1904–1921). He was Surgeon-Naturalist to the Indian Marine Survey. There is a report in the annals of the Indian Museum of the fish collected by him (1904) in Tibet written by Lloyd, and another about a later (1907) trip written by Lieutenant-Colonel John Stephenson (q.v.). He made at least one other collecting trip to high altitudes in Tibet (1909) about which he wrote the scientific report.

Stewart, NH

Barbeled Dragonfish sp. *Melanostomias stewarti* HW Fowler, 1934

Dr Norman H Stewart was a zoologist who spent 10 years as a guide and teacher in Canadian forests and was appointed Instructor in Vertebrate Zoology, University of Michigan, Ann Arbor (1910). He became Assistant Professor of Biology, Bucknell University, Pennsylvania (1912), later becoming their Professor of Zoology. He wrote: *The Amphibia of Pennsylvania* (1926). Fowler wrote: that Stewart: "*…furnished* (him) *with ichthyological material*."

Stewart, R

Seamount Catshark *Bythaelurus stewarti* S Weigmann, CJ Kaschner & R Thiel, 2018

Rob Stewart (1979–2017) was a Canadian filmmaker, freelance journalist, underwater photographer and conservationist. His biology degree was awarded by the University of Western Ontario and he also studied marine zoology in Kenya and Jamaica. His best-known works were *Sharkwater* and *Revolution*. He wrote *Save the Humans* (2012). He died of hypoxia in a scuba diving incident while filming *Sharkwater Extinction* in Florida. The etymology says: "*The new species is named after the late filmmaker and shark conservationist Rob Stewart, who inspired the second author and stimulated her interest in sharks*."

Stewart, R

Assamese Snakehead *Channa stewartii* RL Playfair, 1867

Captain Robert Stewart was a colonial civil servant who became Superintendent at Cachar, Assam, where the holotype was collected. He was co-founder (1859) of the world's first polo club!

Stewart (Springer)

Argentine sp. *Argentina stewarti* DM Cohen & SP Atsaides, 1969

Stewart 'Stew' Springer (1906–1991). This fish was named after him in recognition of numerous contributions to the ichthyology of the tropical western Atlantic. (See **Springer, S**)

Steyermark

Characin sp. *Leporinus steyermarki* RF Inger, 1956

Dr Julian Alfred Steyermark (1909–1988) was an American botanist, explorer, taxonomist and plant collector who was Curator of the Missouri Botanical Gardens. His major works covered the flora of Missouri, Guatemala, and Venezuela. His bachelor's degree (1929), his master's (1930) and his doctorate (1933) were all awarded by Washington University, St Louis. He worked for the US Forest Service (1935–1937). He also worked for the Field Museum of Chicago, the Instituto Botánico in Caracas, and the Missouri Botanical Garden in St. Louis (1984–1988) right up until his death. During his life, he collected more than 138,000 plants in 26 countries, thus earning him an entry in the Guinness Book of Records! He wrote many papers and longer works including: *Flora of Missouri* (1963). A reptile and two amphibians are also named after him.

Stiassny

Lake Afdira Toothcarp *Aphanius stiassnyae* A Getahun & KJ Lazara, 2001
Electric Catfish sp. *Malapterurus stiassnyae* SM Norris, 2002
African Loach Catfish sp. *Phractura stiassny* PH Skelton, 2007
African Killifish sp. *Hypsopanchax stiassnyae* JR van der Zee, R Sonnenberg & JJ Mbimbi, 2015

Dr Melanie Lisa Jane Stiassny (b.1953) is a British-born naturalized American citizen who is Herbert R. and Evelyn Axelrod Research Curator, Department of Ichthyology, Division of Vertebrate Zoology and Professor at the Richard Gilder Graduate School at the American Museum of Natural History, and is also an Adjunct Professor at Columbia University and the City University of New York. She was educated at the University of London, which awarded her BSc (1976) and PhD (1980). After a postdoctoral fellowship at the University of Leiden (1980–1983) she moved to the United States to take a position as Assistant Professor at Harvard University before joining the faculty of the American Museum in New York City (1987). She has discovered and described over fifty fish species, in papers such as: *Revision of Sauvagella Bertin (Clupeidae: Pellonulinae: Ehiravini) with a description of a new species from the freshwaters of Madagascar and diagnosis of Ehiravini* (2002). Other publications include: *Interrelationships of Fishes* (1996), *The Fresh and Brackish Water Fishes of Lower Guinea, West-Central Africa* (2007), and *Natural Histories, Opulent Oceans* (2014). She is currently (2015) writing: *An Illustrated Guide to the Anatomy of Fishes.* (Also see **Melanie**)

Stimpson

Stimpson's Goby *Sicyopterus stimpsoni* TN Gill, 1860

William Stimpson (1832–1872) was an American engineer, conchologist, marine biologist and zoologist. He was a member of the North Pacific Exploring Expedition (1853–1856). He was Curator and Director, Chicago Academy of Sciences (1864–1871). He collected the holotype of this species. He died of tuberculosis. A bird is also named after him.

Stobbs

Stobbs' Pygmy-goby *Trimma stobbsi* R Winterbottom, 2001

Robin E Stobbs was Senior Technician at the JLB Smith Institute of ichthyology. After school in Kenya (1949–1952) he worked in commerce until being called up for national service (1953). Shortly after that he worked at a vetinerary laboratory. After marrying he trained as a medical technician in bacteriology in England (1953–1955). Some years later he moved to South Africa to work in a firm of medical diagnosticians for nine years. He then moved to Rhodes University to join the institute until retirement (1994). Winterbottom described him as: "*...friend, guru, and colleague, whose expertise in so many things was instrumental in launching* [Winterbottom's] *career (especially the fieldwork aspects) all those years ago at the JLB Smith Institute of Ichthyology in Grahamstown, South Africa.*"

Stock

Sisorid Catfish sp. *Glyptothorax stocki* MR Mirza & H Nijssen, 1978
Armoured Catfish sp. *Lithoxus stocki* H Nijssen & IJH Isbrücker, 1990
[Syn. *Paralithoxus stocki*]
Amazonian Cichlid sp. *Crenicichla stocki* A Ploeg, 1991

Dr Jan Hendrik Stock (1931–1997) was a carcinologist at Zoölogisch Museum, Amsterdam. He wrote: *The Pycnogonid family Austrodecidae* (1957). *Lithoxus stocki* was named with the dedication "*...on occasion of his retirement, with remembrance of and gratitude for his energetic and enthusiastic activities as a teacher and colleague.*" Ploeg also commemorated his PhD supervisor's birthday.

Stokell

Smelt genus *Stokellia* GP Whitley, 1955
Stokell's Smelt *Stokellia anisodon* G Stokell, 1941

Gerald Stokell (1890–1972) was a New Zealand amateur ichthyologist. For more than 40 years he collected and studied New Zealand's freshwater fishes and wrote: *Freshwater fishes of New Zealand* (1955). He was a member of the staff of the Canterbury Museum and Secretary (1938) and President (1941) of the Canterbury Branch of the Royal Society of New Zealand.

Stol

Characin sp. *Pyrrhulina stoli* M Boeseman, 1953

E C Stol (d.1975) was a Dutch naturalist who presented specimens of fish he had collected in Suriname (1951) to the Leiden Museum. He founded a pet store 'Vivarium Stol' (1932) in Leiden, which his son Eduard took over on his father's death. He collected the holotype of this species.

Stoliczka

False Osman *Schizopygopsis stoliczkai* F Steindachner, 1866
Tibetan Stone Loach *Triplophysa stoliczkai* F Steindachner, 1866
Sisorid Catfish sp. *Glyptothorax stolickai* F Steindachner, 1867
Labeo sp. *Labeo stolizkae* F Steindachner, 1870

Stoliczka's Barb *Pethia stoliczkana* F Day, 1871
Epaulet Grouper *Epinephelus stoliczkae* F Day, 1875

Dr Ferdinand Stoliczka (1838–1874) was an Austrian palaeontologist and zoologist who was born in Moravia (Czech Republic). He was educated at Prague and at the University of Vienna, where he obtained his doctorate. He collected during travels throughout India (1864–1874) as an Assistant Superintendent of the Geological Survey there. He wrote some scientific papers, such as: *Contribution towards the knowledge of Indian Arachnoidea*, in the 'Journal of the Asiatic Society'. During the last ten years of his life he published geological memoirs on the western Himalayas and Tibet, and many papers on Indian zoology, from mammals to insects and corals. He took part in the Second Yarkand Mission (1873–1874) but collapsed and died of spinal meningitis when: "*...returning loaded with the spoils and notes of nearly a year's research in one of the leastknown parts of Central Asia*", according to his obituary in *Nature*. He is commemorated in species from many phyla including nine reptiles, eight birds, three mammals and an amphibian.

Stolzmann

Characin sp. *Brycon stolzmanni* F Steindachner, 1879
Stolzmann's Weakfish *Cynoscion stolzmanni* F Steindachner, 1879

Jean Stanislas Stolzmann (or Sztolcman) (1854–1928) was a Polish zoologist and pioneer conservationist who worked as a collector in Peru (1875–1881). Oldfield Thomas described him as "*...one of the best known and most successful of Peruvian collectors.*" After returning to Poland he became Director of the Branicki Museum. Among his publications is: *Ornitologia łowiecka* (1905). Nine birds, a reptile, an amphibian and a mammal are named after him.

Stone, L

McCloud River Redband Trout *Oncorhynchus mykiss stonei* DS Jordan, 1894

Livingston Stone (1836–1913) graduated from Harvard (1857) and then studied at theological school and became a pastor, but his health issues were such that his doctors instructed him to spend as much time as he could out of doors. In the late 1860s he pursued an interest in fish and helped found (11871) the American Fish Culturists Association (now the American Fisheries Society). He became (1872) superintendent of the first U.S. fish hatchery, the Baird Hatchery, Shasta County, California. He is renowned for introducing American shad to waters off the West coast of the USA. He was offered (and declined) the position of United States Commissioner of Fisheries but instead became US Deputy Fish Commissioner for the Pacific Coast (1872–1897) and Superintendent of the Cape Vincent Hatchery in New York State (1897–1906). Ill health finally forced his retirement (1906). He wrote: *Domesticated Trout: How to Breed and Grow Them* (1872). Jordan (1896) said the name honoured Stone for his "valuable services in the propagation of salmon and trout." He collected the holotype of this subspecies.

Stone, R

Cocoa Snapper *Paracaesio stonei* U Raj & J Seeto, 1983

Robert Stone caught two specimens of this new species in Fiji (1981). Nothing more is said in the original text, but presumably this is the same Robert Stone as the Pacific Islands Forum Fisheries Agency (FFA) Development Advisor of that name.

Stone, W

Lowland Shiner *Pteronotropis stonei* HW Fowler, 1921

Dr Witmer Stone (1866–1939) was an American ornithologist who worked for over 50 years in the ornithology department of the Academy of Natural Sciences, Philadelphia (1888–1939). He edited *The Auk* (1912–1936). His important works include: *Birds of Eastern Pennsylvania and New Jersey* (1894) and *Bird Studies at Old Cape May* (1937). He collected the type of the shiner. Four birds are also named after him.

Stoneman

Lake Malawi Cichlid sp. *Aulonocara stonemani* WE Burgess & HR Axelrod, 1973

John Stoneman was a fisheries officer in eastern Africa during the 1950s and 1960s who became Chief Fisheries Officer, Malawi (1971). He was honoured for his help in making the authors' expedition a success.

Storer

Silver Chub *Macrhybopsis storeriana* JP Kirtland, 1845
Flabby Whalefish sp. *Ditropichthys storeri* GB Goode & TH Bean, 1895

Dr David Humphreys Storer (1804–1891) qualified in obstetrics at Harvard Medical School (1825) and founded the Tremont Street Medical School (1837). He was a physician at the Massachusetts General Hospital (1849–1858) and Professor (later Dean), Harvard Medical School (1854–1868). He was President of the American Medical Society (1866). The Massachusetts legislature wanted a new look at the state's natural resources, and Storer was put in charge of the Department of Zoology and Herpetology. He also collected and described molluscs. He wrote *Ichthyology and Herpetology of Massachusetts* (1839) and *A synopsis of the fishes of North America* (1846). A genus of snakes is also named after him.

Storey

Triplefin sp. *Axoclinus storeyae* VE Brock, 1940

Margaret Hamilton Storey (1900–1960) was an American museum curator, herpetologist and ichthyologist. Her father, Thomas Storey, was the founder of the Stanford University School of Health. After receiving her master's degree at Stanford (1936) she started working at the Stanford Natural History Museum, initially as a volunteer but later working as a curator and librarian. She also edited the *Stanford Ichthyological Bulletin* and *Occasional Papers*.

Storm

Armoured Sea Catfish *Hemiarius stormii* P Bleeker, 1858

Frans Jonathan Pieter Storm van 's Gravesande (1812–1875) was a Dutch colonial administrator in the Dutch East Indies (Indonesia). He was Assistant Resident and

Magistrate at Palembang (1852). He was described by Bleeker as a general tax collector in Batavia (Jakarta), who had sent him a collection of fishes from East Sumatra and was the Dutch government commissioner of Djambi, Sumatra, when he provided the holotype.

Storms

True Red Congo Tetra *Micralestes stormsi* GA Boulenger, 1902
Congo Cichlid sp. *Orthochromis stormsi* GA Boulenger, 1902

Lieutenant Maurice Joseph Auguste Marie Raphael Storms (1875–1941) is described in Boulenger's cryptic etymology as "*...a cousin of the late Raymond Storms, so well-known for his important contributions to paleoichthyology*". He enlisted in the Belgian Army Guides Regiment (1895–1896), transferred to the regular army and was posted to the Congo where he served (1897–1901). During his time in the Congo he collected on the Lindi River. On his return, he presented the holotype to the Brussels Museum.

Stott

Stott's Goatfish *Upeneichthys stotti* JB Hutchins, 1990

Chris Stott was described as an 'honorary field assistant' with the Western Australian Museum. He was involved in the collection of two paratypes of this species.

Stoumboudi

Lake Volvi Roach *Rutilus stoumboudae* PG Bianco & V Ketmaier, 2014

Dr Maria Th. Stoumboudi is a Greek ichthyologist who is (2008) Research Director of the Hellenic Centre for Marine Research, Institute of Inland Waters, Anavyssos, Greece, and was previously Principal Researcher there (2003–2008). She completed (1991) her doctorate at Aristotle University of Thessaloniki where she had been granted her BSc (1986) on fish biology, then undertook post-doctoral research at the Department of Zoology, Hebrew University, Jerusalem, Israel (1991–1993) and in the same period was post-doctoral fellow at the Hadassah School of Dental Medicine. She has published widely, such as: *The spawning behaviour of the endangered freshwater fish Ladigesocypris ghigii (Gianferrari, 1927)* (2005). She is also President of the European Ichthyological Society. In their etymology the authors honoured her as a "*...colleague and friend* [and...] *for her research on the ecology and conservation of the freshwater fishes of Greece*."

Stout

Pacific Hagfish *Eptatretus stoutii* WN Lockington, 1878

Dr Arthur Breese Stout (1814–1898) was a prominent surgeon in San Francisco. He was corresponding secretary of the California Academy of Sciences, where he had been a member (1853–1898) having been Curator of Ethnology and Osteology (1881). Among his medical writings is: *Hygiene, as regards the sewerage of San Francisco* (1868).

Strachey

Mahseer sp. *Neolissochilus stracheyi* F Day, 1871

LieutenantGeneral Sir Richard Strachey (1817–1908) went to India (1836), returned to England (1850) and went back to India (1855). He served in the Bengal Engineers, rising from Lieutenant (1841) to Lieutenant-General (1875). Frequent attacks of fever compelled him to go to Nani Tal in the Kumaon Himalayas for his health (1847). There he met Major E. Madden, under whose guidance he studied botany and geology, making expeditions into the western Himalayas for scientific purposes. He served as an administrator in several capacities (1862–1871) and was a member of the Council of India (1875–1889). He was President of the Royal Geographical Society (1887–1889). Jointly with his brother, Sir John Strachey, who was also a colonial administrator, he wrote *The Finances and Public Works of India* (1882). He wrote many monographs and papers, such as 'On the physical geography of the provinces of Kumaon and Garhwal, in the Himalaya Mountains, and of the adjoining parts of Tibet' (1851). A mammal and a bird are also named after him. Day honoured him for helping him "*…to prosecute* (his) *enquiries into the fish and fisheries of India.*"

Straelen

Bluelip Haplo (cichlid) *Astatoreochromis straeleni* M Poll, 1944
Lake Tanganyika Cichlid sp. *Perissodus straeleni* M Poll, 1948
[Syn. *Plecodus straeleni*]
Spotted Skate *Raja straeleni* M Poll, 1951

Professor Victor Émile van Straelen (1889–1964) was a Belgian geologist, palaeontologist, carcinologist and naturalist. He was Director, Belgian Royal Institute of Natural Science, Brussels (1925–1954). He travelled widely, especially in Indonesia and Belgian Congo (DRC) and became (1933) President, Institute of National Parks of the Congo. He was the first President (1959–1964) of the Darwin Foundation. He was also President of the non-profit organisation (Mbizi) that sponsored the expedition which collected the skate holotype. He also sponsored the Belgian Hydrobiological Mission to Lake Tanganyika (1946–1947), during which the cichlid type was collected. An amphibian is also named after him.

Strahan

Hooded Carpetshark *Hemiscyllium strahani* GP Whitley, 1967
Hagfish sp. *Eptatretus strahani* CB McMillan & RL Wisner, 1984

Dr Ronald Strahan (1922–2010) was an Australian zoologist, ichthyologist and research scientist who was Director of Taronga Park Zoo, Sydney (1967–1974) when the carpetshark was described; the holotype had been swimming in the zoo's aquarium since 1960. The University of New South Wales awarded his honorary doctorate (1999). He served in the Australian army in WW2 in a Mobile Entomological Research Unit, which had to identify insect vectors for tropical diseases, particularly malaria and dengue fever, which caused more casualties than combat with the Japanese. After being demobilized he eventually completed his degree in zoology at the University of Western Australia (1947). He went on to study in Europe at Oxford and then to lecture on zoology at the University of Hong Kong, only returning to Australia (1961) as Senior Lecturer, University of New South Wales. He became Research Fellow at the Australian Museum, Sydney (1974) and later head of its

National Photographic Index of Australian Wildlife. He wrote: *Taronga Zoo and Aquarium* (1974).

Strasburg

Strasburg's Damselfish *Dascyllus strasburgi* W Klausewitz, 1960
Strasburg's Blenny *Entomacrodus strasburgi* VG Springer, 1967

Donald W Strasburg (1925–2008) was a fish ecologist at the University of Hawaii and at the Bureau of Commercial Fisheries Biological Laboratory, Honolulu and an environmentalist at the Naval Research Laboratory, Washington, DC. He collected the damselfish type. He wrote some influential reports such as *Fishes of the Southern Marshall Islands* (1953) and among his published papers is: *Further Notes on the Identification and Biology of Echeneid Fishes* (1964) and the co-written: *Ecological Relationships of the Fish Fauna on Coral Reefs of the Marshall Islands* (1990).

Strauch

Spotted Thicklip Loach *Triplophysa strauchii* KF Kessler, 1874

Professor Dr Alexander Alexandrovich Strauch (1832–1893) was a RussianGerman zoologist. He finished his training as a physician in Estonia (1859), but he was also a naturalist, mainly interested in herpetology, and his doctoral dissertation was on zoology. He was sent to Algeria (1859–1860). He became Director of the Zoological Museum in St Petersburg (1879) and (1890) Permanent Secretary of the library of the Academy of Science. He wrote: *Essai d'une Erpétologie de L'Algérie* (1862). He mainly wrote on herpetological zoogeography. He died in St Petersburg and is buried in the Lutheran cemetery there. Seven reptiles, an amphibian and a bird are also named after him.

Strauss

Dottyback sp. *Anisochromis straussi* VG Springer, CL Smith & TH Fraser, 1977

Lewis H Strauss conceived, organised, produced, funded and participated in the (1976) Smithsonian expedition to St Brandon's Shoals (Indian Ocean) which included US National Museum scientists. According to the etymology, it "*...netted a scientifically, highly valuable collection of fishes and other marine organisms*," including the type of this species.

Streltsov

African Rivuline sp. *Nothobranchius streltsovi* S Valdesalici, 2016

Dr Sergey Streltsov (Russia), was a killifish aquarist.

Strickrott

Strickrott's Hagfish *Eptatretus strickrotti* PR Møller & WJ Jones, 2007

W Bruce Strickrott (b.1964) was the pilot of the Alvin submarine that was used to take the holotype. Author WJ Jones told the Woods Hole Oceanographic Institute that without Alvin pilots oceanographers would not get their jobs done: "*We saw this little thing swimming like a worm and I told Bruce, 'There is no way you are going to catch it',*" said Jones. However, Strickrott managed to get the submarine behind the hagfish and vacuum the fish up using a device known as a slurp gun: "*…I was like, 'Man, this guy has skills and deserves recognition. The naming was a way to express our gratitude.*" Strickrott said: "*…It's a feather in my cap. It's recognition from researchers for my contributions to the advancement of science.*"

Stride

Lake Malawi Cichlid sp. *Lethrinops stridei* DH Eccles & DSC Lewis, 1977

Kenneth E Stride was known as the person who brought the first successful commercial trawling to Lake Malawi with the motor launch 'Ethelwynn Trewavas'. He oversaw hundreds of productive trawl hauls in the Experimental Trawling Program (late 1960s-early 1980s).

Stroh

Father Stroh's Betta *Betta strohi* D Schaller & M Kottelat, 1989
[Alt. Father Stroh's Fighting Fish]

Father H Stroh was a 'little-known' missionary priest and amateur naturalist who discovered this species in 'the wilds of Borneo'. He also collected a number of plants that he took with him when he returned to Germany (1978).

Stroud

Driftwood Catfish sp. *Gelanoglanis stroudi* JE Böhlke, 1980

William Boulton Dixon Stroud (1917–2005) was a member of a family described as 'landed gentry in Pennsylvania's horse country'. He served in the US Navy in WW2 as a lieutenant, twice being a survivor of a ship sunk by enemy action. After WW2, he worked in textile manufacture in South Carolina, retiring in the late 1960s to devote time to his farming and environmental interests. He is remembered as a philanthropist who supported the author's field studies and collecting in the Colombian llanos, where he had a cattle-raising operation. He also co-founded the Stroud Water Research Center, Avondale (1967) on farmland that he donated, and supported the planting of over 40,000 trees on it.

Struhsaker

Struhsaker's Deep-sea Smelt *Glossanodon struhsakeri* DM Cohen, 1970
Struhsaker's Chromis *Chromis struhsakeri* JE Randall & SN Swerdloff, 1973
Golden Redbait *Emmelichthys struhsakeri* PC Heemstra & JE Randall, 1977

Dr Paul James Struhsaker (1935–2018) was a fishery biologist and ichthyologist. The Michigan State University awarded his BA (1958) and the University of Hawaii, Honolulu awarded his master's (1967) and doctorate in zoology (1973). As a marine fisheries biologist he participated in fisheries investigations in the western North Atlantic, Gulf of Mexico, western Caribbean, Alaska, US west coast and the Hawaiian Islands. He worked at the National Marine Fisheries Service, Southeast Fisheries Center (1957, 1959–1965, 1981–82) and at the Honolulu Laboratory (1969–1977). He was associated with the Seattle, Juneau and Woods Hole Laboratories for brief periods. He was a research associate at the Bishop Museum, Hawaii. Among his numerous publications he co-wrote: *Observations on the biology and distribution of the thorny stingray, Dasyatis centroura (Pisces: Dasyatidae)* (1969) and *Megamouth: A new species, genus, and family of lamnoid shark (Megachasma pelagios, family Megachasmidae) from the Hawaiian Islands.* (1983). He has contributed to such research as initiating modern bottom trawl surveys in the Hawaiian Islands and the first studies of daily growth rings in otoliths of tropical marine fishes. He collected the smelt holotype during his investigation into the biology of Hawaiian demersal fish and shrimp populations and the damselfish type during a survey for shrimps in the Hawaiian area and suspected it was an undescribed species.

Struthers

Snailfish sp. *Psednos struthersi* DL Stein, 2012

Carl D Struthers is a New Zealand ichthyological researcher who is (2005–present) Research & Technical Officer: Fishes, at the National Museum of New Zealand Te Papa Tongerewa. He is also the museums Dive Officer. He is an experienced field worker, participating in many fish collecting expedition using scuba in coastal waters and blue-water expeditions to remote locations, both within and outside New Zealand waters. He has written a number of papers and is one of the authors of *The Fishes of New Zealand* (2015).

Stuart, LC

Barred Livebearer *Carlhubbsia stuarti* DE Rosen & RM Bailey, 1959

Dr Laurence Cooper Stuart (1907–1983) of the University of Michigan's Museum of Zoology was an expert on the herpetofauna of Guatemala and described a number of new species. He collected the livebearer type. Eight reptiles, four amphibians and a bird are also named after him.

Stuart, M

Sisorid Catfish sp. *Exostoma stuarti* SL Hora, 1923

Dr Murray Stuart. (See **Murray Stuart**)

Stuart (Brooks)

Nuakata Shrimpgoby *Tomiyamichthys stuarti* GR Allen, MV Erdmann & WM Brooks, 2018

Stuart Mathews Brooks is the son of the third author William M Brooks. (Also see **Berry (Levy)**, **Bill (Brooks)** & **Jack Brooks**)

Stuart Grant

Flavescent Peacock (cichlid) *Aulonocara stuartgranti* MK Meyer & R Riehl, 1985

Lake Malawi Cichlid sp. *Iodotropheus stuartgranti* AF Konings, 1990

Stuart M Grant (1937–2007) was an English exporter of cichlids from Lake Malawi, where these eponymous species are endemic. He served in the Royal Air Force (1955) being stationed around the Mediterranean. Later he had a government post in Nyasaland (now Malawi). He was asked (1972) to establish an aquarium fish exporting business, and did so operarting out of the Eagle Inn 'rest house' on the shores of Lake Malawi. He piloted a small plane to transport fishes and supplies locally. Eventually, he employed sixty workers and exported many cichlids new to science. He purchased a large boat (1986), the 'Lady Dianan' and took hobbyists on trips around the lake. He died of heart failure and is buried on the land he owned at Kambiri Point, still owned by his Malawian widow and their children.

Stübel

Blackfish Drummer *Girella stuebeli* FH Troschel, 1866
Thorny Catfish sp. *Hemidoras stuebelii* F Steindachner, 1882
[Syn. *Opsodorus stuebelii*]
Armoured Catfish sp. *Loricariichthys stuebelii* F Steindachner, 1882
Suckermouth Catfish sp. *Astroblepus stuebeli* B Wandolleck, 1916

Dr Moritz Alphons Stübel (1835–1904) was a German geologist, vulcanologist, archaeologist, explorer, ethnologist and collector who was educated at the University of Leipzig. He visited Santorini, Greece (1866) and was in Colombia and Ecuador (1868–1874), and thereafter in Peru, Brazil, Argentina, Uruguay, Chile and Bolivia (1874–1877). A bird is named after him.

Studer

Blue-spotted Spinefoot *Siganus studeri* WKH Peters, 1877
(Junion Synonym *Siganus corallinus*)

Théophile Rudolphe Studer (1845–1922) was a Swiss zoologist, particulary an ornithologist and marine biologist. He was Curator of Zoology at the Swiss Natural History Museum, Berne (1871–1922). He was a member of the German expedition (1874–1876) aboard the frigate 'SMS Gazelle', during which type was collected. He was appointed Professor of Zoology and Anatomy at the School of Vetinary Medicine in Berne (1878). Among his publications was *Fauna Helvetica, Molluscs* (1896).

Stuhlmann

African Characin sp. *Alestes stuhlmannii* GJ Pfeffer, 1896
Kingani Mormyrid *Petrocephalus stuhlmanni* GA Boulenger, 1909
Eastcoast Lampeye *Pantanodon stuhlmanni* CGE Ahl, 1924

Professor Dr Franz Ludwig Stuhlmann (1863–1928) was a German zoologist and explorer who collected in East Africa (1888–1900). He made his career in the German Colonial Forces and Civil Service. He did not confine himself to zoological specimens, as he also collected plants and local artefacts. He was Secretary (1908–1910) of the Colonial Institute in Hamburg. The German Government published a monograph by Stuhlmann: *Dr Franz Stuhlmann: Mit Emin Pasha ins Herz von Africa* (1894). Five birds, two mammals, a reptile and an amphibian are named after him as well as insects and plants.

Sturany

Stone Loach sp. *Barbatula sturanyi* F Steindachner, 1892

Dr Rudolf Sturany (1867–1935) was an Austrian malacologist. His undertook his first degree in Zoology at the University of Leipzig (1886) and went on to receive his PhD from Vienna University (1891) while working as a volunteer Curator of Molluscs at the Natural History Museum there. He was Assistant Curator (1897) and Adjunct Curator (1901) rising to full Curator (1915). During his time there, he undertook collecting trips to Dalmatia,

Bosnia, Montengro, Croatia, Albania and Crete, collecting molluscs and insects. He had eye problems that eventually forced him to retire (1924). He accompanied Steindachner when the loach type was collected in Macedonia, and Steindachner honoured him: "*…as a token of my sincerest affection*" (translation). A snail genus is also named after him.

Styan

Torrent Catfish sp. *Liobagrus styani* CT Regan, 1908

Frederick William Styan (1838–1934) was a tea trader and collector in China for 27 years who corresponded from Kiukiang. He was a Fellow of the Zoological Society of London and was elected as a Member of the BOU (1887). A mammal, eleven birds and three reptiles are named after him.

Suarez

Suarez' Coralbrotula *Ogilbia suarezae* PR Møller, W Schwarzhans & JG Nielsen, 2005

Dr Susan Stevens Suarez (b.1949) is Professor Emeritus (2017) of Biomedical Sciences, College of Veterinary Medicine, Cornell University. Cornell awarded her bachelor's degree (1971) the University of Miami her master's (1974) and the University of Virginia her doctorate (1981). She was at the University of California, Davis as a post-doctoral researcher and as a teacher (1981–1985), and at the University of Florida (1988–1999) as Assistant and Associate Professor. She served as Associate Professor and Full Professor at Cornell University (1994–2017). She has published more than 100 peer-reviewed papers and book chapters, mostly on the subject of mammalian reproduction. Her most recent notable contribution, as co-senior author, was *Microgrooves and fluid flows provide preferential passageways for sperm over pathogen Tritrichomonas foetus* (2015).

Suckley

Spotted Spiny Dogfish *Squalus suckleyi* CF Girard, 1855

Dr George Suckley (1830–1869) was an American Army surgeon and naturalist. He was appointed as assistant surgeon and naturalist of the Pacific Railway Survey between Minnesota and the Puget Sound (1853). Later, he explored the Oregon and Washington territories, which had not yet been admitted as States of the Union. He resigned from the army (1856) to concentrate on natural history, then re-joined the Union Army and served as a surgeon throughout the Civil War. He co-wrote: *Natural History of Washington Territory* (1859). He collected the dogfish holotype. Three birds are named after him.

Sudara

Sisorid Catfish sp. *Oreoglanis sudarai* C Vidthayanon, P Saenjundaeng & HH Ng, 2009

Dr Surapol Sudara (1939–2003) was a Thai marine biologist at the Department of Marine Science, Chulalongkorn University, where he was a Professor until retiring (1999). After this, he became President, Sueb Nakhasathien Foundation where his work helped to alert the public to the threats to marine and coastal ecological systems. He died from liver cancer.

Sue

Dwarf-goby genus *Sueviota* R Winterbottom & DF Hoese, 1988

Susan Lee Jewett (formerly Susan J Karnella) (b.1945). (Also see **Jewett** & **Susan (Jewett)**). In the genus' name, 'Sue' is attached to *Eviota*, a related genus.

Suenson

Seagrass Eel *Chilorhinus suensoni* CF Lütken, 1852

Vice-Admiral Edouard Suenson (1805–1887) was a Danish naval officer who collected specimens during his voyages, including some of the type material for this species. He became a cadet (1817) and rose to Second Lieutenant (1823), Lieutenant (1831), Lieutenant Commander (1850), Captain (1855) and Vice-Admiral (1880). He served on many vessels such as the frigate Nymphen in the Mediterranean (1825), the corvette Diana in the Caribbean (1826), time in the French service (1830–1832), and on the frigate Bellona in South American waters (1844–1845). He saw active service at the Battle of Heliogoland, chased privateers, commanded steamships, was in gun-boat battles and commanded a number of vessels crewed by cadets.

Suer

Lesueur's Goby *Lesueurigobius suerii* A Risso, 1810

Charles Alexandre Lesueur (Le Sueur) (1778–1846). Risso amended the spelling to '*lesueurii*' in 1827, but his original (incorrect) spelling must be retained. (See **Lesueur**)

Suess

Red Sea Torpedo *Torpedo suessii* F Steindachner, 1898

Eduard Suess (1831–1914) was an Austrian palaeontologist, geologist and malacologist. He is famous for his hypothesis of a prehistoric mega-continent 'Gondwana' (1861) and the Tethys Ocean (1893). He wrote the three-volume: *Das Antlitz der Erde* (1885–1901) in which he was the first to posit the idea of the biosphere. Craters on both Mars and the Moon are named after him. The original description contains no etymology, but he is the most probable candidate.

Sufi

Sisorid Catfish sp. *Glyptothorax sufii* K Asghar Bashir & MR Mirza, 1975

S M K Sufi was described as: "*...one of the pioneer ichthyologists of Pakistan*." He wrote: *Revision of the Oriental Fishes of the Family Mastacembelidae* (1956).

Suidter

Swiss Whitefish (salmonid) sp. *Coregonus suidteri* V Fatio, 1885

Otto Suidter (1833–1901) was a Swiss pharmacist and naturalist in Lucerne where he inherited the pharmacy (1855) upon the death of his father. He was a student of coregonine fishes.

Sullivan

Sullivan's Churchill (mormyrid) *Petrocephalus sullivani* S Lavoué, CD Hopkins & A Kamdem Toham, 2004

Bluntnose Knifefish sp. *Brachyhypopomus sullivani* WGR Crampton, CD de Santana, JC Waddell & NR Lovejoy, 2017

Dr John P Sullivan (b.1965) is an ichthyologist who is (2013–present) a Curatorial Affiliate at Cornell University, Museum of Vertebrates Ithaca, NY, USA where he had previously been a visiting fellow (2009–2013). He is also a Research Associate at the Academy of Natural Sciences, Philadelphia. He was a U.S. State Department Fulbright Fellow at the University of Kisangani (2010) in the Democratic Republic of the Congo and an EOL Rubenstein Fellow. He is particularly interested in phylogenetic systematics and taxonomy of African electric fishes (*Mormyridae*) and catfishes (*Siluriformes*). St John's College, Annapolis awarded his BA (1987) and Duke University his PhD (1997). He co-wrote: *Brachyhypopomus bullocki, a new species of electric knifefish (Gymnotiformes: Hypopomidae) from northern South America* (2009) and *Petrocephalus boboto and Petrocephalus arnegardi, two new species of African electric fish (Osteoglossomorpha, Mormyridae) from the Congo River basin* (2014).

Sultan Sazlığı

Sultan Sazlığı Toothcarp *Aphanius danfordii* GA Boulenger, 1890

[Alt. Danford's Killifish]

This is a toponym referring to the Sultan Reedy National Park (*Sultan Sazlığı Milli Parkı*) in Turkey.

Sungam

Magnus' Prawn-Goby *Amblyeleotris sungami* W Klausewitz, 1969

'Sungam' is Magnus spelled backwards, and refers to Dr Dietrich B E Magnus. (See **Magnus**)

Sunier

Borneo Barb sp. *Barbonymus sunieri* M Weber & LF de Beaufort, 1916

[Syn. *Barbodes sunieri*]

Sunier's Righteye Flounder *Samariscus sunieri* M Weber & LF de Beaufort, 1929

Armand Louis Jean Sunier (1886–1974) was a Dutch biologist who became the Director of Artis Zoo, Amsterdam. Leiden awarded his PhD (1911) with his thesis being on fish anatomy. He almost immediately left for the Dutch East Indies, becoming a zoologist at the Department of Agriculture (1911–1921). He became (1921) head of the marine Investigations laboratory in Batavia (Jakarta). After several years he returned to Leiden where he was in charge of molluscs and crustaceans. He left (1927) to become Director of Artis (1927–1953).

Suraswadi

Sisorid Catfish sp. *Oreoglanis suraswadii* C Vidthayanon, P Saenjundaeng & HH Ng, 2009

Plodprasop Suraswadi (b.1945) is a Thai politician who has a bachelor's degree in fisheries from Kasetsart University, Bangkok (1968), and a master's degree in fisheries from Oregon

State University (1970). He was Director General of the Department of Fisheries, Thailand (1989–1997), Minister of Science and Technology (2011–2012) and since 2012 Deputy Prime MInister. He started the fisheries development and conservation programme in the area where the holotype of this species was located.

Surendranathan

Periyar Garra *Garra surendranathanii* CP Shaji, LK Arun & PS Easa, 1995

Shri. Punnavilakathu Kochukrishnan Surendranathan Asari was, until retirement (2004), Principal Chief Conservator of Forests, Kerala Forest Department. He is now (2005–2015) Managing Director of a garden resort. In honouring him the etymology says he was a "… *constant source of encouragement*" for wildlife research in Kerala, India.

Surinbinnan

Garra sp. *Garra surinbinnani* LM Page, BC Ray, S Tongnunui, DA Boyd & ZS Randall, 2019

Amphol Tapanapunnitikul (d.2019), who went by the name Surin Binnan, was a Thai conservationist. He was formerly the owner of a computer repair business and a property developer, but devoted the latter part of his life to preserving Thailand's biodiversity. He established a non-profit organisation, the Foundation of Western Forest Complex Conservation, to protect the area where the garra occurs. He died of liver cancer while the description of this fish was still in review.

Susan (Banister)

Chinese Barb sp. *Poropuntius susanae* KE Banister, 1973

Susan Banister is the widow of the author.

Susan (Jewett)

Susan's Dwarf-goby *Eviota susanae* DW Greenfield & JE Randall, 1999

Susan Lee Jewett (formerly Susan J Karnella) (b.1945). (Also see **Jewett** & **Sue**)

Susan (Jordan)

Cumberland Darter *Etheostoma susanae* DS Jordan & J Swain, 1883

Susan Bowen Jordan (1845–1885) was the first wife of the senior author.

Susan (Mondon)

Cardinalfish sp. *Apogon susanae* DW Greenfield, 2001

Susan G Mondon is a scientific illustrator at the Department of Zoology, University of Hawaii. She prepared the illustrations in the paper in which this species was described,

Susanne

Stone Loach sp. *Schistura susannae* J Freyhof & DV Serov, 2001

Susanne Klähr is a biologist and zoologist. She co-wrote" *Breeding biology and habitat of corn bunting Miliaria calandra in the intensively used agricultural landscape of the Hellwegboerde, North Rhine-Westphalia.* She was honoured for her help with the author's field work.

Susi

Susi Creek Rainbowfish *Melanotaenia susii* Kadarusman, N Hubert & L Pouyaud, 2015

This is a toponym, referring to Susi Creek in West Papua.

Susian

Loach sp. *Paraschistura susiani* J Freyhof, G Sayyadzadeh, HR Esmaeili & M Geiger, 2015

Named after the Susian people. Susa was an ancient city of the Elamite, Persian and Parthian empires, located in the lower Zagros Mountains of Khuzestan Province, Iran - the type locality.

Suttkus

Spotside Wormfish *Microdesmus suttkusi* CR Gilbert, 1966
Sea Toad sp. *Chaunax suttkusi* JH Caruso, 1989
Alabama Sturgeon *Scaphirhynchus suttkusi* JD Williams & GH Clemmer, 1991
Rocky Shiner *Notropis suttkusi* JM Humphries & RC Cashner, 1994
Gulf Logperch *Percina suttkusi* BA Thompson, 1997

Royal Dallas Suttkus aka 'Sut' (1920–2009) was a noted authority on fishes of the southeastern USA and mentor to many ichthyologists. He left Fremont Ross High School (1937) and worked in a celery garden for two years at a salary of $0.25 cents per hour to earn money for college! He went to Michigan State University (1939) and after graduating enlisted in officer school. He was posted to Wales and from there to fight in France (1944). When he left the army (1946) he entered the graduate programme at Cornell's Agriculture School. Following this he took a faculty position in Zoology at Tulane University (1950). He devoted his career at Tulane to collection building and studies of the taxonomy and natural history of specimens he collected. He was Principal Investigator of the Environmental Biology Training Program (1963–1968). He became the Tulane University Museum of Natural History's first Director (1976). He also collected marine organisms during oceanic cruises in the Gulf of Mexico, Indian Ocean, off the coasts of Peru and Venezuela, and around the Galapagos Islands. All of the specimens collected were ultimately catalogued into Tulane's natural history collections. He took the Tulane fish collection from just two mounted fish specimens to over two million (1968). He published c.125 scientific papers. His great skill and his insatiable appetite for field collecting are legendary.

Sutton, FA

Blue-eye Panaque *Panaque suttonorum* LP Schultz, 1944

Dr Fredrick Albert Sutton (1894–1950) and his wife were 'very kind' to Schultz when he stayed at the camp of the Lago Petroleum Corporation in Maracaibo, Venezuela. Dr Sutton was a geologist whose bachelor's degree was awarded by the University of Utah (1917). In addition to Venezuela he also worked in China and Tibet.

Sutton, GED

Soapfish sp. *Suttonia suttoni* JLB Smith, 1953

Guy E Drummond Sutton greatly assisted Smith's scientific work. Presumably he was a relative of **Sutton, RND** (below), with the genus being named after the elder Drummond Sutton and the species after the younger.

Sutton, HV

Sutton's Flyingfish *Cheilopogon suttoni* GP Whitley & AN Colefax, 1938

Harvey Vincent Sutton (1882–1963) was an Australian physician and athlete. He competed in the 1908 Summer Olympics. He served as a doctor during WWI and was awarded (1919) the Order of the British Empire. He became (1930) the first Director of the School of Public Health and Tropical Medicine, University of Sydney.

Sutton, RND

Soapfish genus *Suttonia* JLB Smith, 1953

R N Drummond Sutton may have been the author's teacher, as he states that this is the person "who guided my early steps in science." (By the time of Smith's description, he was the 'late' R N Drummond Sutton)

Suvatti

Danio sp. *Devario suvatti* HW Fowler, 1939
Arrowhead Puffer *Pao suvattii* S Sontirat, 1989
Suvatti's Barb *Hypsibarbus suvattii* WJ Rainboth, 1996

Dr Chote Suvatti (b.1904) was a Thai ichthyologist, biologist, botanist, malacologist and illustrator. His bachelor's degree (1933) was awarded by Cornell University and (1937) he was working in the Department of Agriculture and Fisheries, Ministery of Agriculture, Bangkok. He was Dean of the Faculty of Fisheries, Kasetart University, Bangkok and, while there he established (1964) the Kasetart University Museum of Fisheries. Among other works he wrote: *Index to Fishes of Siam* (1936), *Fauna of Thailand* (1967) and *Fishes of Thailand* (1981).

Suvorov

Shad sp. *Alosa suworowi* LS Berg, 1913

Evgenii Konstantinovich Suvorov (1880–1953) was a Russian zoologist and ichthyologist. He graduated from the University of St Petersburg (1903). He took part in and, later, led expeditions to study fish biology and fishing methods and to discover new fishing regions in many places from the Baltic to the Pacific and from the White Sea to the Caspian (1904–1920). He was Director of the Polytechnic Institute of Fish Culture, Leningrad (St Petersburg) (1921–1931). He was a professor at Leningrad University (1931–1952) and head of a sub-department (1949). Among his achievements was to induce the artificial breeding of Atlantic Salmon (1920). He was the first scientist to recognise the eponymous shad as a distinct taxon.

Suyen

Suyen's Tongue Sole *Cynoglossus suyeni* HW Fowler, 1934

Lin Su-Yen was head of the Fisheries Experiment Station, Canton. He was the author of: *Carps and Carp-like Fishes of Kwangtung and Adjacent Islands* (1931). His name is also given as Lin Shu-Yen.

Suzart

Neotropical Rivuline sp. *Simpsonichthys suzarti* WJEM Costa, 2004
[Syn. *Ophthalmolebias suzarti]*

Rogério Dos Reis Suzart is an aquarist who sent Wilson Costa the type material of this species. He has written papers with Costa such as: *Cynolebias paraguassuensis n. sp. (Teleostei: Cyprinodontiformes: Rivulidae), a new seasonal killifish from the Brazilian Caatinga, Paraguaçu River basin* (2007).

Suzuki

Anthias sp. *Rabaulichthys suzukii* H Masuda & JE Randall, 2001

Keiu Suzuki served as the senior author's assistant and helped collect the holotype.

Suzumoto

Ambon Sole *Aseraggodes suzumotoi* JE Randall & M Desoutter-Meniger, 2007

Arnold Y Suzumoto (b.1951) is Ichthyology Collections Manager at the Bishop Museum, Hawaii. The University of Hawaii awarded his Zoology BA (1974). His particular interests are the general ichthyology of the Hawaiian Islands and conservation, curation and preservation of natural history collections. Among his published papers is: *Samaretta perexilis, a New Genus and New Species of Samarid Flatfish (Pleuronectiformes: Samaridae) from the South Pacific* (2017). He was honoured by the authors for his contribution to their discovery of nine new species of *Aseraggodes.*

Sven

Pike-Cichlid sp. *Crenicichla sveni* A Ploeg, 1991
Brazilian Cichlid sp. *Geophagus sveni* PHF Lucinda, CAS de Lucena & NC Assis, 2010
Dr Sven Oscar Kullander (b.1952) (See **Kullander**)

Svetovidov

Morid Cod genus *Svetovidovia* DM Cohen, 1973
Short-beaked Garfish *Belone svetovidovi* BB Collette & NV Parin, 1970
Tadpole Goby sp. *Benthophilus svetovidovi* VI Pinchuk & DB Ragimov, 1979
Brown Grenadier Cod *Tripterophycis svetovidovi* YI Sazonov & YN Shcherbachev, 1986
Naked-head Toothfish *Gvozdarus svetovidovi* AV Balushkin, 1989
Long-finned Char *Salvethymus svetovidovi* IA Chereshnev & MB Skopets, 1990

Morid Cod sp. *Gadella svetovidovi* IA Trunov, 1992

Upper Yenisei Grayling *Thymallus svetovidovi* IB Knizhin & SJ Weiss, 2009

Anatolii Nikolaveich Svetovidov (1903–1985) was a Russian ichthyologist. He graduated from the K A Timiriazev Moscow Agricultural Academy (1925). He worked at the Zoological Institute of the Academy of Sciences of the USSR (1932). His principal works dealt with intraspecific variation, as well as with fish taxonomy and phylogeny based on comparative and functional morphology. He also studied the geographic distribution, origin, and population dynamics of fish, especially cod and herring. He was made a Corresponding Member of the Academy of Sciences of the USSR (1953). The etymology for *Thymallus svetovidovi* describes him as a: "...*famous researcher of graylings of Eurasia*" and the goby etymology says he was honoured "...*for his significant contributions to the study of Caspian fishes*."

Swain, JS

Gulf Darter *Etheostoma swaini* DS Jordan, 1884

Joseph Swain (1857–1927) was Professor of Mathematics and Biology at Indiana University (1883–1891) and a student of the author, David Starr Jordan, whom he followed (1891) to Stanford University and taught there as a professor of mathematics.

Swain, SK

Mahanadhi Minnow *Parapsilorhynchus swaini* B Baliarsingh & L Kosygin, 2017

Dr Saroj Kanta Swain (b.1963) is an ichthyologist who is Principal Scientist of the Aquaculture Production and Environment Division at the Indian Council of Agricultural Research – Central Institute of Freshwater Aquaculture (ICAR-CIFA), Bhubaneshwar, Orissa. His MAS Diploma was awarded by the University of Chile and his MSc (1986) by Ravenshaw College, Utkal University, Orissa. His PGDFSc. was awarded (1989) by CIFE, Mumbai and his PhD (1997) by Orissa University of Agriculture and Technology. He spent seven years researching fish nutrition and two years researching carp and catfish culture. More recently he has been involved in developing captive breeding technology of more than fifty species of ornamental fish. He has written well over 100 scientific papers, reports and book chapters as well as several books, such as the co-written: *Ornamental Fish Farming* (2010). He was honoured for his 'encouragement and support' to the authors, who are also at the institute.

Swales

Swales' Basslet *Liopropoma swalesi* HW Fowler & BA Bean, 1930

Bradshaw Hall Swales (1875–1928) was an American ornithologist. He qualified and practised as a lawyer in Detroit and appears to have lived most of his life in Michigan. With Alexander Wetmore he wrote: *Birds of Haiti and the Dominican Republic* (1931). He was one of the founders of the Baird Ornithological Club, and was President at the time of his death. The authors commented on "...*his general interest in natural history*", and, in addition to ornithology, they particularly noted his interest in the anthropology of Native American peoples. A bird is also named after him.

Swan

Rockhead *Bothragonus swanii* F Steindachner, 1876

James Gilchrist Swan (1818–1900) of Port Townsend, Washington, was an anthropologist, judge, political advisor, artist and schoolteacher, as well as being an Indian Agent and ethnologist and one of the most colorful personalities of Washington State's territorial period (1853–1889). He took part in a gold rush (1852) then learned Chinook lamguage, allowing him to become a translator for treaty negotiations (1854–1855). He then returned east and wrote a book: *The Northwest Coast; or, Three Years' Residence in Washington Territory* and later worked as the Governor's Secretary (1857–1858). He spent part of the next three years at Makah Indian Reservation at Neah Bay, supporting himself writing for newspapers. He was the first schoolteacher at the Makah Reservation. Under criticism for failing to teach Christianity to the Makah, he resigned (1866) and moved to Port Townsend. He was admitted to the bar (1867) and later (1882) became district court judge. The Smithsonian Institution hired Swan to collect Indian artifacts for the world's fair in Philadelphia (1876), the fair in London (1884), and the exposition in Chicago (1893). The US Fish Commission asked Swan to write a series of reports on the fish and fisheries of the northern Pacific, permitting him to visit Neah Bay intermittently (1882–1891).

Swartz

Gamtoos Redfin *Pseudobarbus swartzi* A Chakona & PH Skelton, 2017
Angolan cichlid sp. *Serranochromis swartzi* JR Stauffer, R Bills & PH Skelton, 2021

Dr Ernst Roelof Swartz was an Aquatic Biologist (2005–2010) and is now Senior Aquatic Biologist (2010–present) at the South African Institute for Aquatic Biodiversity, Grahamstown. He is also a lecturer at Rhodes University (2006–present). The University of Stellenbosch awarded his BSc (1996) honours (1997) and MSC (2000) and his PhD was awarded by the University of Pretoria (2005). He has fish survey experience in South Africa, Swaziland, Mozambique, Lesotho, Namibia and Angola. He has written or co-written articles, reports and a number of papers such as: *Invasion status of Florida bass Micropterus floridanus (Lesueur, 1822) in South Africa* (2017) and *A new species catfish, Amphilius pagei (Siluriformes: Amphiliidae) from Angola* (2018). He was honoured for his contributions to the biogeography and systematics of *Pseudobarbus*.

Swegles

Red Phantom Tetra *Hyphessobrycon sweglesi* J Géry, 1961

Kyle Swegles (b.c.1925) was a distributor of tropical fish and owner of the Rainbow Aquarium, Chicago. He collected the type series in the upper Rio Orinoco basin, Colombia.

Swierstra

Lowveld Suckermouth *Chiloglanis swierstrai* CJ van der Horst, 1931

Cornelis Jacobus Swierstra (1874–1952) was a Dutch-South African entomologist who was born in Groningen and died in Pretoria. He started his career in the Transvaal Museum (1897) as a General Assistant, becoming Deputy Director under Jan Willem Gunning and eventually succeeding him as Director (1922–1946). He was an accomplished collector of butterflies, who described many holotypes. He also managed the ethnology collection, having conducted some field studies among the Hananwa people (1912). A bird is also named after him.

Swift

Swift's Goby *Callogobius swifti* GR Allen, MV Erdmann & W Brooks, 2020

John F Swift is a conservationist and the founder of the Swift Foundation, which supports conservation around the world. He is a Board Member of Conservation International having been its Director (2008). He studied Conservation of Natural Resources and International Agriculture at the University of California, Berkeley. He has worked in Guatemala, Papua New Guinea and Ecuador and raised his family on an organic ranch. He is an 'unflagging patron' of the local environmental organisation Eco Custodian Advocates, which promotes marine management across the Milne Bay (New Guinea) communities where this new species was discovered. An amphibian is also named after him and his family.

Swinhoe

Sole sp. *Brachirus swinhonis* F Steindachner, 1867
Freshwater Sleeper sp. *Micropercops swinhonis* A Günther, 1873
Chinese Cyprinid sp. *Toxabramis swinhonis* A Günther, 1873

Robert Swinhoe (1836–1877) was born in Calcutta (Kolkata), India, but was sent to London (1852) to be educated. While at the University of London (1854) he was recruited into the China Consular Corps by the Foreign Office. Before he left for Hong Kong, he deposited a small collection of British birds' nests and eggs with the British Museum. His time in China as a diplomat gave him great opportunities as a naturalist; he explored a vast area, which had not been open previously to any other collector. As a result, he discovered new species at the rate of about one per month throughout the nearly two decades he was there. He was primarily an ornithologist, but his name is also associated with various Chinese mammals, reptiles and insects. He returned to London (1862) and brought part of his vast collection of specimens to meetings of the Zoological Society, as well as to their counterparts in France and Holland. He was somewhat taken aback by having to allow someone else to name the 200–plus new bird species which he had discovered, as he himself related: "…*I have been blamed by some naturalists for allowing Mr. Gould to reap the fruits of my labours, in having the privilege of describing most of my novelties. I must briefly state, in explanation, that I returned to England elated with the fine new species I had discovered, and was particularly anxious that they should comprise one entire part of Mr. Gould's fine work on the Birds of Asia, still in progress. On an interview with Mr. Gould, I found that the only way to achieve this was to consent to his describing the entire series to be figured, as he would include none in the part but novelties, which he should himself name and describe. I somewhat reluctantly complied; but as he has done me the honour to name the most important species after me, I suppose I have no right to complain.*" He began suffering from partial paralysis (c.1871) and his ill health forced him to leave China (1875). Four mammals, an amphibian and four reptiles are named after him, as well as twenty-seven birds.

Swire

Mariana Snailfish *Pseudoliparis swirei* ME Gerringer & TD Linley, 2017

Herbert Swire (d.1934), after whom Swire Deep in the Mariana Trench is named, was the First Navigating Sub-Lieutenant on HMS Challenger. He wrote a number of reports on that voyage, and the book: *The voyage of the Challenger : a personal narrative of the historic circumnavigation of the globe in the years 1872–1876 by Herbert Swire; illustrated with reproductions from paintings and drawings in his journals ; foreword by Roger Swire ; introduction by G. Herbert Fowler* (1938).

Swynnerton

Tilapia sp. *Astatotilapia swynnertoni* GA Boulenger, 1907

Charles Francis Massy Swynnerton (1877–1938) was an English naturalist and entomologist. He was born in Suffolk, spent his early years in India, and worked in Africa. He became manager of Gungunyana Farm close to the Chirinda Forest, Southern Rhodesia (now Zimbabwe) (1900). He used it as a base and worked on comprehensive collections of local plants, birds and insects. He was appointed (1919) the first game warden of Tanganyika (Tanzania). He later spent ten years (1929–1938) as head of tsetse fly research in East Africa. He published papers on many aspects of natural history, including: *On the birds of Gazaland, Southern Rhodesia* (1907). He obtained the tilapia type from his farm in Mozambique. He was killed in an air-crash on his way to Dar-es-Salaam, where he was to have been invested with the order of St Michael and St George. Four birds, an amphibian, a mammal and a reptile are also named after him.

Sychr

Sychr's Catfish *Corydoras sychri* SH Weitzman, 1960

Al Sychr was an aquarist who lived in Hayward, California. He provided the author with the holotype of this species.

Sycorax

Ruby Dragonet *Synchiropus sycorax* Y-K Tea & AC Gill, 2016

The Sycorax are a race of red-robed alien warriors in the UK science-fiction series 'Dr. Who'. The name was applied due to the fish's flamboyant coloration. (Presumably the word Sycorax was borrowed from Shakespeare's play *The Tempest*, where it is the name of Caliban's mother).

Sydney

Sydney's Pygmy Pipehorse *Idiotropiscis lumnitzeri* RH Kuiter, 2004

This is a toponym, referring to the city in Australia.

Sykes

Sisorid Catfish sp. *Glyptothorax sykesi* F Day, 1873

Colonel William Henry Sykes (1790–1872) was an English ornithologist and army officer. He saw plenty of action after he joined the Bombay Army aged fourteen (1804–1824) – this being a part of the armed forces of the Honourable East India Company. He was then appointed (1824) as a statistical reporter to the Bombay government, and later his statistical researches involved him in natural history. He wrote: *Catalogue of Birds of the Rapotorial and Incessorial Orders Observed in the Dukkan* (1832). He retired from active service (1833) and later became a Director of the East India Company, Rector of Aberdeen University, and Member of Parliament for Aberdeen. A mammal and six birds are named after him.

Sylvius

Neotropical Knifefish sp. *Gymnotus sylvius* JS Albert & FM Fernandes-Matioli, 1999

Silvio de Almeida Toledo Filho is a pioneering ichthyologist researching into the electro-biology of the genus *Gymnotus* from southeastern Brazil at the Department of Biology and Genetics, Biosciences Institute, University of São Paulo. There is also an allusion in the binomial to the Latin word *sylvi*, referring to the Atlantic rainforest where this species lives.

Symöns

Lake Kisale Cichlid sp. *Lamprologus symoensi* M Poll, 1976
African Rivuline sp. *Nothobranchius symoensi* RH Wildekamp, 1978

Professor Dr Jean-Jacques André Symöns (1927–2014) was a Belgian botanist-ecologist at the University of Lubumbashi, DR Congo. The Free University of Brussels awarded both his first degree in Chemical Sciences and his PhD in Botanical Science. He then became a researcher at the Lake Tanganyika station and afterwards Professor at l'Université Officielle du Congo Belge et du Ruanda-Urundi in Elisabethville (now Lumumbashi) where he taught botany and plant ecology and where his wife taught pharmaceutical chemistry. He returned to Belgium to become Professor at Vrije University, Brussels (1972), teaching botany and hydrobiology and also teaching the history of biological sciences at the University of Mons-Hainaut. His hobby was numismatics (the collection and study of coins). He continued to teach and write long after retirement age. His last book was: *Vegetation of Inland Waters*, part of the Handbook of Vegetation Science Series (2013). He discovered the rivuline and was honoured by Poll for his: 'interesting and fruitful' (translation) contribution to the study and faunal surveying of the Shaba (or Katanga) Plateau, DRC, where the cichlid occurs. Two odonates are also named after him.

Szabo

Upper Zambezi Mormyrid *Hippopotamyrus szaboi* B Kramer, H van der Bank & M Wink, 2004

Dr Thomas Szabo (1924–1993) was a Hungarian physician and neuroscientist. The St Emericus College, Budapest, awarded his bachelor's degree and the University of Budapest his MD (1947). For a short period (1946–1947) he was an Assistant at the Institute of Anatomy and Institute of Neurology there before gaining a fellowship from the Swiss government to work at the Institute for Brain Anatomy at the University of Zurich. While there he was summoned back to Hungary, refused to leave, and had his passport revoked – so becoming 'stateless'. He was not allowed to remain in Switzerland after his contract and was invited by his brother to go to Paris (1950). He worked at the CNRS (French National Centre for Scientific Research) looking at brain electro-sensitive pathways in various animals, including electric fishes. He completed his PhD (1956) (the year of the Hungarian Uprising) and decided to stay in France and was granted citizenship (1957). In the following years he underook field research in Chad and Belgian Congo (1958, 1959 & 1961) and French Guiana and Brazil (1959, 1964 & 1967). He also made research visits to the USA (1963–1964 & 1966). Within CNRS he was promoted from Attaché (1951–1956) and Chargé de Recherche (1957–1960) to Maître (1961–1973) and finally Directeur de Recherche (1974) of the newly formed Laboratoire de Neurophysiologie Sensorielle Comparée. He published over 300 scientific papers (1949–1994) right up until his death from cancer. He is considered one of the founding fathers of the field of electroreception, and was mentor and friend to the senior author who had been one of his post-doctoral students.

T

Taaf

Whiteleg Skate *Amblyraja taaf* EE Meissner, 1987

This is a form of toponym rather than an eponym: TAAF = Territoire des Terres australes et antarctiques françaises (French Southern and Antarctic Territories).

Taaning

(See under **Tåning**)

Taczanowski

Taczanowski's Gudgeon *Ladislavia taczanowskii* B Dybowski, 1869
White-edged Rockfish *Sebastes taczanowskii* F Steindachner, 1880
Armoured Catfish sp. *Chaetostoma taczanowskii* F Steindachner, 1882
Pencil Catfish sp. *Trichomycterus taczanowskii* F Steindachner, 1882
Suckermouth Catfish sp. *Astroblepus taczanowskii* GA Boulenger, 1890

Władysław Taczanowski (1819–1890) was a Polish zoologist. He was Conservator, then Curator, of the Royal University of Warsaw Zoological Cabinet (1862–1887), which he transformed from a teaching institution into a scientific centre. An outstanding zoologist and ornithologist who described many new species, he promoted the protection of birds of prey and (1866–1867) was a member of the Branicki expedition to Algeria. He wrote the 4-volume: *Ornithologie du Pérou* (1884–1886). He also wrote: *Faune Ornithologique de la Sibérie Orientale* (1891–1893), published posthumously in two parts, based on collections made by Dybowski, Godlewski and Jankowski, who were banished to Siberia by the Russians who occupied Poland. He was also an arachnologist, writing a list of spiders of Warsaw and describing spiders from French Guiana and Peru. Dybowski's description of the gudgeon does not identify the patronym, but the genus name is a Slavic spelling of the Polish 'Władysław'. Thus it is a rare case where the two parts of the scientific name both honour the same man. Twenty-one birds, two mammals and a reptile are also named after him. (See also **Ladislav**)

Taddy

Piranha genus *Taddyella* R Ihering, 1928 NCR
[Now in *Pygocentrus*]

This is an example of an unfortunate mistake by the author! When Ihering had to choose a replacement name for Eigenmann's preoccupied generic name *Rooseveltiella* he wanted to refer to Roosevelt's nickname 'Teddy', but erroneously wrote 'Taddy' instead. Perhaps it is as well that the name is not now regarded as valid. (See **Roosevelt**)

Taira

Basslet sp. *Pseudanthias taira* PJ Schmidt, 1931

Taira Aimori was a Japanese hero whose grave Schmidt said is on Amami-Oshima Island, Ryukyu Islands, Japan (type locality), near the village Urakami. A district is also named after him. He was presumably a member of the 12-century Taira samurai clan.

Tait

African Rivuline sp. *Nothobranchius taiti* B Nagy, 2019

Colin C Tait is a South African of Scottish parentage, who was the first to collect this species (1969) in Uganda while working as a fish ranger. He published field observations about *Nothobranchius* habitats in Zambia, as well as notes on their behaviour in captivity. He wrote: *Notes on the species Nothobranchius brieni Poll (Cyprinodontidae)* (1965). He was described by a colleague as "*...marvellous company to have in the long dark nights in the bush, with his love of jokes, and his fund of songs and stories.*"

Takagi

Goby sp. *Oxyurichthys takagi* F Pezold, 1998

Kazunori Takagi is an ichthyologist at the Tokyo University of Fisheries. He wrote: *A Study on the Scale of the Gobiid Fishes of Japan* (1951). His studies of gobioid oculoscapular canals revealed their significance to gobioid systematics, and led to him being honoured in the goby's binomial.

Takase

Coffee-bean Tetra *Hyphessobrycon takasei,* J Géry, 1964

Black-jacket Tetra *Moenkhausia takasei* J Géry, 1964

Roberto Takase was a Japanese immigrant to Brazil (1924). He was an aquarist described in one etymology as: "*...one of the fish-collection pioneers in the Brazilian Amazon*". He kept a retail store, now owned and managed by his son, in São Paulo. He collected the holotypes of both species.

Takeno

Japanese Loach sp. *Cobitis takenoi* J Nakajima, 2016

Makoto Takeno of the Graduate School of Agriculture, Kinki University, Osaka, Japan, discovered this species. Its existence was described in a paper Takeno co-wrote: *A distinctive allotetraploid spined loach population (genus Cobitis) from Tango District, Kyoto Prefecture, Japan* (2010). Although it was initially considered a form of *C. striata*, DNA sequencing proved otherwise.

Takeuchi

Takeuchi's Swallowtail Angelfish *Genicanthus takeuchii* RL Pyle, 1997

Hiroshi Takeuchi is a diver and marine photographer who first found and photographed this species.

Takita

Takita's Mudskipper *Periophthalmus takita* Z Jaafar & HK Larson, 2008

Toru Takita (1936–2014) was a Korean ichthyologist who spent much of his career at Nagasaki University, Japan (1963–2002). He was honoured for his contributions to our knowledge of mudskipper ecology. He was co-author of: *Two New Species of Periophthalmus (Teleostei: Gobiidae: Oxudercinae) from Northern Australia, and a re-diagnosis of Periophthalmus novaeguineaensis* (2004).

Talara

Dragonet sp. *Foetorepus talarae* SF Hildebrand & O Barton, 1949
Peruvian Mora *Physiculus talarae* SF Hildebrand & O Barton, 1949

This is a toponym: Talara, Peru, is where the holotypes were collected.

Talbot

Flame Cardinalfish *Apogon talboti* JLB Smith, 1961
Lesser Orange Brotula *Dermatopsoides talboti* DM Cohen, 1966
Talbot's Blenny *Stanulus talboti* VG Springer, 1968
Talbot's Damselfish *Chrysiptera talboti* GR Allen, 1975

Dr Frank Hamilton Talbot (b. 1930) is a fisheries scientist. marine researcher and museum director. The University of Witwatersrand, South Africa, awarded his bachelor's degree (1949) and the University of Cape Town his master's (1951) and doctorate (1959). He was a fisheries research scientist, British Colonial Service, Zanzibar (1954–1957); a marine biologist, South African Museum, Cape Town (1958–1959) and Assistant Director (1960–1963). He was at the Australian Museum, Sydney as Curator of Fishes (1964–1965) and Director (1965–1974). He was Professor of Environmental Studies, Macquarie University, Australia (1975–1981), Executive Director California Academy of Sciences, San Francisco, (1982–1988) and Director NMNH, Washington DC (1989–1994).

Taler

Zeta Trout *Salmo taleri* SL Karaman, 1933

Zdravko Taler (d.1954) was an amateur scientist and fishing expert who helped Karaman access remote sites to collect interesting trout populations in Montenegro, including the holotype of this species.

Taliev

Baikal Sculpin sp. *Batrachocottus talievi* VG Sideleva, 1999

Dmitrii Nikolaevich Taliev (1908–1952) was a Soviet Russian ichthyologist and limnologist, most notable for his work on Lake Baikal. While still at school he worked as a laboratory assistant at the Sverdlovsk Regional Museum. He attended the Leningrad Veterinary Institute (1925) then transferred (1926) to Leningrad State University. He was interested in both birds and fish. He became Senior Assistant, Pacific Research Institute of Fisheries and Oceanography, and then (1932–1947) – apart from a two-year stint at the Academy's Zoological Institute (1939–1941) – worked at the Baikal Limnological Research Station of the Academy of Sciences of the USSR (now Limnological Institute of the Siberian Division of the Russian Academy of Sciences) becoming Director (1947) until his death. He published 55 papers including: *A new genus Cottoidei from Lake Baikal* (1946) and *On the rate and the causes of divergent evolution in the Cottoidei from Lake Baikal* (1948). Taliev was married to Aleksanda Yakovlevna Bazikalova (q.v.), who also worked at the Limnological Research Station.

Tamang

Garra sp. *Garra tamangi* SD Gurumayum & L Kosygin, 2016

Lakpa Tamang is a Field Assistant at the Zoological Survey of India (2012–present). Previously he was a typist-cum-storekeeper for twelve years. During his work he has described nine fish new to science. He assisted the authors during their fieldwork in Arunachal Pradesh, where this species occurs.

Tamara

Moray sp. *Enchelycore tamarae* AM Prokofiev, 2005
[Junior Synonym of *Gymnothorax fimbriatus*]

Tamara Mikhailovna Ambrozhevich (1970) was a friend of the author Artem Prokofiev. She lived in Riga and was a Chief Accountant in a private company. They lost touch (c.2007) and Dr Prokofiev cannot remember any further biographical details.

Tamazula

Tuxpan Splitfin *Allodontichthys tamazulae* CL Turner, 1946
[Alt. Peppered Splitfin]

This is a toponym after the town of Tamazula de Giordano, Mexico.

Tan

White Cloud Mountain Minnow genus *Tanichthys* S-Y Lin, 1932

Tan Kam Fei was a Chinese boy-scout leader who discovered and collected the first specimens of the minnows (1932) and brought them to the attention of the author.

Tana

Azov Shad *Alosa tanaica* O von Grimm, 1901

This is a toponym referring to Tana or Tanais, the medieval trading city on the Sea of Azov, where it occurs.

Tanaka, H

Tanaka's Dottyback *Lubbockichthys tanakai* AC Gill & H Senou, 2002
Tanaka's Wrasse *Wetmorella tanakai* JE Randall & RH Kuiter, 2007

Dr. Hiroyuki Tanaka is a Japanese physician and aquarist. He has written numerous articles and several scientific papers as well as a book: *Angelfishes - a Comprehensive Guide to Pomacanthidae* (2003). The etymology for the wrasse states that the name is "*…in honor* *of Hiroyuki Tanaka, M.D., in appreciation of providing the paratypes of this species and their photographs.*"

Tanaka, S

Bitterling genus *Tanakia* DS Jordan & WF Thompson, 1914
Flounder genus *Tanakius* CL Hubbs, 1918
Tanaka's Snailfish *Liparis tanakae* CH Gilbert & CV Burke, 1912
Eelpout sp. *Lycodes tanakae* DS Jordan & WF Thompson, 1914
Eelpout sp. *Bothrocara tanakae* DS Jordan & CL Hubbs, 1925

Dr Shigeho Tanaka (1878–1974) was a Japanese ichthyologist who was Professor of Zoology at the Imperial University, Tokyo. He graduated (1903) from the University of Tokyo, which also awarded his doctor of science degree (1904). He went on to become a lecturer there, rising to Assistant Professor. He became Professor and Director of the Misaki Marine Research Department there (1938) before retiring (1940). After retirement, he taught biology at a high school and sometimes lectured at a local university. He wrote around 300 papers and 50 books, including co-writing a book on Japanese fish with DS Jordan, whom he met when he visited the USA, titled: *A catalogue of the fishes of Japan* (1913). He described more than 170 new species and also founded Japan's first Ichthyological Journal (Gyoyuku-zasshi). He was honoured in the bitterling genus as the "*…accomplished ichthyologist of the Imperial University of Tokyo, who described* T. shimazui *in 1908 and* Pseudorhodeus tanago *in 1909*."

Tanegasima

White-finned Blenny *Praealticus tanegasimae* DS Jordan & EC Starks, 1906
Goby sp. *Callogobius tanegasimae* JO Snyder, 1908

This is a toponym referring to Tanegashima, Osumi Islands, Japan, type locality [note spelling, without "h"].

Tang

Yellow-spotted Fanray *Platyrhina tangi* Y Iwatsuki, J Zhang & K Nakaya, 2011

D-S Tang was a Chinese ichthyologist at the University of Amoy, who was active in the 1930s. Iwatsuki showed that a species named by Tang as *Platyrhina limboonkengi* was in fact a junior synonym of *P sinesnsis*. Perhaps they named this ray for him as compensation. He wrote: *The Elasmobranchiate Fishes of Amoy* (1934).

Tangaroa

Tangaroa Shrimp-goby *Ctenogobiops tangaroai* HR Lubbock & NVC Polunin, 1977
Clingfish sp. *Modicus tangaroa* GS Hardy, 1983
Snailfish sp. *Paraliparis tangaroa* DL Stein, 2012

Not a true eponym but referring to Tangaroa, a powerful Maori/Polynesian deity, god of the sea.

Tangkahkei

Chinese Icefish *Neosalanx tangkahkeii* HW Wu, 1931

Grouper sp. *Epinephelus tankahkeei* H Wu, M Qu, H Lin, W Tang & S Ding, 2020

Tang Kah Kei (1874–1961) was described in the icefish etymology as the founder of Amoy University. His name is more commonly given as Tan Kah Kee, and he was also known as Chen Jiageng. He was a Chinese businessman, community leader and philanthropist who, at the age of 16 (1890), travelled from China to Singapore to help his father, who owned a rice trading business. In 1903, after his father's business collapsed, Tan started his own company and built a business empire from rubber plantations, manufacturing, sawmills, canneries, real estate, import and export brokerage, ocean transport and rice trading. He was one of the prominent overseas Chinese to provide financial support to China during the Second Sino-Japanese War (1937–1945).

Tanibe

Snaggletooth sp. *Astronesthes tanibe* NV Parin & OD Borodulina, 2001

Tat'yana Nikolaevna Belyanina is a specialist in oceanic fishes, working at the P P Shirshov Institute of Oceanology, Moscow. The binomial is formed by the first two letters of each of her names.

Tåning

Lanternfish genus *Taaningichthys* RL Bolin, 1959
Taaning's Lanternfish *Diaphus termophilus* AV Tåning, 1928
Slopewater Lanternfish *Diaphus taaningi* JR Norman, 1930
Onejaw sp. *Monognathus taningi* L Bertin, 1936
Striped Tubeshoulder *Mirorictus taningi* AE Parr, 1947
Lanternfish sp. *Hygophum taaningi* VE Becker, 1965

Dr Åge Vedel Tåning (1890–1958) was a Danish ichthyologist and an expert on lanternfish. He was a member of the company on a number of the voyages of the 'Dana' (q.v.), including the circumnavigation (1928–1930). He was also a director of the Carlsberg (q.v.) Laboratory and became Director of Danish Fisheries (1951) and at the time of his death was President, International Ocean Studies. He wrote: *Young Herring and Sprat in Faroese Waters* (1936).

Tanke

Tiger Pleco (catfish) sp. *Panaqolus tankei* CA Cramer & LM de Sousa, 2016

Andreas Tanke is a German aquarist and underwater photographger. He studied mathematics at Leibniz Universität, Hannover, and until recently (2017) worked for an Agricultural Engineering company. In the wording of the etymology, he is "*very dedicated to the genus Panaqolus, studying its behavior, reproduction, and differences between known forms, keeping these fishes in the aquarium, visiting their habitats, and publishing his findings.*"

Tanner

Eelpout sp. *Eulophias tanneri* HM Smith, 1902

Lieutenant-Commander Zera Luther Tanner (1835–1906) was an American naval officer, deep-sea explorer and oceanographer who developed an improved method of depth-sounding using instruments of his own design. He was involved in the design and construction of the Fish Commission steamers *Albatross* and *Fish Hawk* (1879–1894) then becoming their commander. He was described in Smith's etymology as "*...the foremost exponent of modern deep-sea exploration*". Two birds are also named after him.

Tansei Maru

False Brotula sp. *Parabrotula tanseimaru* M Miya & JG Nielsen, 1991
Goby sp. *Platygobiopsis tansei* M Okiyama, 2008

'Tansei Maru' is the name of a research vessel, owned by the Ocean Research Institute, University of Tokyo (now Japan Marine Science and Technology Center), which was responsible for collecting the holotypes of the goby and false brotula.

Taphorn

Armoured Catfish sp. *Hypostomus taphorni* CG Lilyestrom, 1984
Whiptail Catfish sp. *Farlowella taphorni* ME Retzer & LM Page, 1997
Characin sp. *Creagrutus taphorni* RP Vari & AS Harold, 2001
Pencilfish sp. *Lebiasina taphorni* CA Ardila Rodríguez, 2004
Characin sp. *Hemigrammus taphorni* RC Benine & GAM Lopes, 2007
Three-barbeled Catfish sp. *Phenacorhamdia taphorni* CL DoNascimiento & N Milani, 2008
Taphorn's Tube-snouted Ghost Knifefish *Sternarchorhynchus taphorni* CDC Machado de Santana & RP Vari, 2010
Tetra sp. *Hyphessobrycon taphorni* CA García-Alzate, C Román-Valencia & H Ortega, 2013
Serrasalmid characin sp. Myloplus taphorni MC Andrade, H López-Fernández & EA Liverpool, 2019

Dr Donald Charles Taphorn Baechle (b.1951) is an American zoologist and ichthyologist who lives in Venezuela and whose major interest is Venezuelan freshwater fish. The Southern Illinois University awarded his bachelor's degree (1971) and the University of Florida awarded his master's (1976) and doctorate (1990). He was founder and Director, Museum of Zoology, Biocentro-Universidad Nacional Experimental de Los Llanos, Venezuela (1978–2007) but still remains as curator of the fish collection. In retirement, he has moved back to the USA but is still active in ichthyology. A reptile is also named after him.

Tapu

Swallowtail sp. *Odontanthias tapui* JE Randall, AL Maugé & YB Plessis, 1979

Jean Tapu (1929–2018) worked in the Service des Peche (Papeete, Tahiti) and was also a world-champion spearfisher (1965 & 1967). He provided holotype, along with colour photographs of it. According to the authors he had provided other 'valuable' specimens of Tahitian fishes in the past.

Tarabini

Somali Giant Catfish *Pardiglanis tarabinii* M Poll, B Lanza & A Romoli Sassi, 1972

Dr Giovanni Tarabini Castellani (1910–1992) was Director of the leprosarium at Gelib, Somalia. He devoted forty years of his life to the study of leprosy. He qualified as a physician (1936) at the University of Modena. He wrote: *Relation between serum albumins and serum and leukemia immunoglobulins in leprosy* (1991) A local fisherman in Somalia caught the holotype and gave it to Tarabini who passed it to the authors.

Taranetz

Eelpout genus *Taranetzella* AP Andriashev, 1952
Goby subgenus *Taranetziola* SV Shedko & IA Chereshnev, 2005
Lumpfish sp. *Eumicrotremus taranetzi* GN Perminov, 1936
Taranetz Char *Salvelinus taranetzi* AG Kaganowsky, 1955
Goby sp. *Gymnogobius taranetzi* VI Pinchuk, 1978
Mud Skate *Bathyraja taranetzi* VN Dolganov, 1983
[Syn. *Rhinoraja taranetzi*]
Antarctic Sculpin sp. *Bathylutichthys taranetzi* AV Balushkin & OS Voskoboinikova, 1990
Sculpin sp. *Radulinopsis taranetzi* M Yabe & S Maruyama, 2001

Anatoly Yakovlevich Taranetz (1910–1941) was a Russian marine biologist and ichthyologist who was an eminent expert on Far Eastern fishes. He graduated from the Vladivostock Industrial College (1929) becoming an 'observer' in the raw materials sector of the Pacific Fisheries Research Centre; part of the Russian Academy of Sciences. He then became (1932) a Marine Researcher at TIRH Complex Pacific Expedition of the State Hydrological Institute before working (1933) at the Leningrad Zoological Institute where he presented his theses: *Freshwater fish of the North-Western basin in the Sea of Japan*. He conducted a number of expeditions on the Amur River (1930s) including to Sahalin (1934). He was a 'group leader' studying salmon (1939) and edited a guide to the fishing industry of the Far East: *Guide to the Fishes of the Soviet Far East and Adjacent Waters* (1941), one of around 30 papers he wrote. He was drafted into the army (November 1941) and was killed soon after (December) when his echelon was destroyed by enemy aircraft.

Tarasov

Prickleback sp. *Pseudalectrias tarasovi* AM Popov, 1933

Nikolay Ivanovich Tarasov (1905–1965) was a Soviet hydrobiologist. As an assistant observer (1922–1925), he worked on the Glavryba Azov-Black Sea fishing expedition led by N M Knipovich. He graduated from Leningrad State University (1927) and worked at the State Hydrological Institute. One of his interests was in marine bioluminescence, and he wrote: *The Living Light of the Sea* (1956). He is also known as a researcher in the biology of crayfish, as well being as the author of a large number of works on the hydrology and hydrobiology of the Far Eastern and Northern seas. However, he is probably best known for his work on 'technical marine biology', looking at organisms injurious to shipping and coastal industry.

Tariq

Asian Barbel genus *Tariqilabeo* MR Mirza & N Saboohi, 1990

Zafarullah Khan Tariq was the Deputy Director of the Department of Plant Protection, Government of Pakistan. He was honoured as he had collected specimens of *T. macmahoni* that were used in the authors' study. Among his published papers is: *Control of Sugarcane Pyrilla and Whitefly by Aerial Application of Emulsifiable Concentrates at Ultra-Low-Volume* (1970).

Tarleton Bean

Lanternfish genus *Tarletonbeania* CH Eigenmann & RS Eigenmann, 1890

Tarleton Hoffman Bean (1846–1916) was an American ichthyologist. Colombian (George Washington) University awarded his MD (1876) and Indian University his MS (1883) unexamined, on the basis of his accomplishments. He was a forester (his first interest was botany), fish culturist, conservationist, editor, administrator and exhibitor. He worked as a volunteer at the Fish Commission laboratory in Connecticut (1874) where he met Baird and George Brown Goode. He spent two decades (1875–1895) working in Washington at the USNM (1878–1888) and the Fish Commission (1888–1895). After resigning from the Fish Commission, he was Director of the New York Aquarium (1895–1898). He then worked on forestry or fishery exhibits at the World's Fairs in Paris (1900) and St Louis (1905). He was a fish culturist for New York State (1906) until his death in a motor accident. He wrote nearly 40 papers with Goode, culminating in: *Oceanic Ichthyology* (1896). A bird is named after him.

Taroba

Pencil Catfish sp. *Trichomycterus taroba* BW Wosiacki & JC Garavello, 2004

Eartheater Cichlid sp. *Gymnogeophagus taroba* J Casciotta, A Almirón, L Piálek & O Říčan, 2017

Tarobá was a warrior in a legend of the Kaingang people, who were the first inhabitants of the present-day province of Misiones in Argentina, particularly the Río Iguazú basin, above the falls, where these species occur. According to the legend, Tarobá and Naipí, a beautiful young maiden, angered Mboi, the guardian god of the río Iguazú, who created the falls to capture the lovers, transforming Naipí into one of the rocks of the falls, perpetually punished by its turbulent waters, and Tarobá into a palm tree on the bank, where, on sunny days, a rainbow overcomes the power of Mboi and serves as a bridge of love connecting Naipí and Tarobá.

Tarr

Tarr's Cardinalfish *Pseudamia tarri* JE Randall, EA Lachner & TH Fraser, 1985

A. Bradley Tarr is an ichthyologist and marine biologist whose bachelor's degree was awarded by the University of Hawaii, Manoa (1978). His master's degree was awarded by the University of Florida (2007). He was a curatorial assistant, Division of Fishes, Bishop Museum Honolulu, Hawaii (1978–1982). He was a research scientist at the King Fahd University of Petroleum and Minerals, Dhahran, Saudi Arabia (1983–1997) and a marine biologist, US Army Corps of Engineers, Jacksonville, Florida (2000–2015). He helped collect the holotype of this species.

Tarzoo

Green Discus *Symphysodon tarzoo* E Lyons, 1959

This is a name formed by the contraction of 'Tarpon Zoo', an ornamental fish export firm in Tarpon Springs, Florida, USA, with a fish collecting station in Leticia, Colombia, from which Earl Lyons' specimens originated.

Tate, R

Desert Rainbowfish *Melanotaenia splendida tatei* AHC Zietz, 1896

Professor Ralph Tate (1840–1901) was a British and Australian botanist and geologist. He became (1861) a teacher of natural science at the Philosophical Institution in Belfast, where he studied botany, publishing his *Flora Belfastiensis* (1863) while also investigating the Cretaceous and Triassic rocks of Antrim. He was appointed (1864) assistant at the museum of that society, writing severl papers, and *A Plain and Easy Account of the Land and Freshwater Mollusks of Great Britain* (1866). He then (1867) joined an expedition to Nicaragua and Venezuela. He was (1875) appointed Elder Professor of natural science at the University of Adelaide in South Australia, teaching botany, zoology and geology. He also became Vice President and then President (1878–1879) of what became the Royal Society of South Australia (1880). He travelled around Australia before joining the Horn Expedition (1894) when this species was first collected. He returned to England (1896).

Tate (Regan)

Driftwood Catfish genus *Tatia* A Miranda Ribeiro, 1911

Charles Tate Regan (1878–1943). (See **Regan**)

Tate Regan

Hillstream Loach sp. *Balitoropsis tatereganii* CML Popta, 1905
[Syn. *Pseudohomaloptera tatereganii*]

Charles Tate Regan (1878–1943) (See **Regan**)

Tate(urndina)

Peacock Gudgeon genus *Tateurndina* JT Nichols, 1955

The first part of the genus' name, Tate, here refers to the brothers George Henry Hamilton Tate (1894–1953), botanist and mammologist, and Geoffrey M Tate (1898–1964), expedition business manager and collector. They were Nichols' colleagues at the AMNH, and were honoured for their participation in the Archbold Expeditions to New Guinea during which the type was collected. The second part of the name *urndina*, presumably references the closely related genus *Mogurnda*.

Tatiana

River Stingray sp. *Potamotrygon tatianae* JPCB da Silva & MR de Carvalho, 2011

Tatiana Raso de Moraes Possato (1978–2006) was a biologist whose bachelor's degree was awarded by Universidade de São Paulo where she was studying for a master's degree when she died, tragically young. She had been an enthusiastic researcher of chondrichthyans, in particular freshwater stingrays.

Tatyana

Snaggletooth sp. *Astronesthes tatyanae* NV Parin & OD Borodulina, 1998

Tatyana Borisovna Agafonova works at the All-Russian Research Institute of Fisheries and Oceanography (VNIRO), Moscow. She collected the holotype of this species during the 1989 cruise of the Fishery Research Vessel 'Vozrozhdenie'.

Tauber

Lightheaded Dottyback *Pseudochromis tauberae* HR Lubbock, 1977

Ruth Tauber (no other information available) was helpful during Lubbock's visits to Kenya (the type locality).

Taunay

Three-barbeled Catfish genus *Taunayia* A Miranda Ribeiro, 1918
Armoured Catfish sp. *Ancistrus taunayi* A Miranda Ribeiro, 1918

Afonso d'Escragnolle Taunay (1876–1958) was the son of a Brazilian viscount when Brazil was still a monarchy. He graduated in Rio de Janeiro as a civil engineer (1900), and taught engineering in São Paulo (1904–1910). He succeeded **Ihering** (q.v.) as Director, Museum Paulista (1917–1939), and Professor, Faculty of Philosophy, Science, and Art, University of São Paulo (1934–1937). He was more interested in history than zoology, writing an 11-volume account of the coffee industry in Brazil (1929–1941). A reptile and a bird are also named after him.

Tavares

Brazilian Cichlid sp. *Australoheros tavaresi* FP Ottoni, 2012

Felipe Tavares Autran was a student when he first recognised this species as new, in his unpublished monograph (Autran, 1995) on the *Cichlasoma facetum* species complex. This study, although widely known among Brazilian ichthyologists, was never published. (See **Autran**)

Taverne

Kipepe Elephantfish *Paramormyrops tavernei* M Poll, 1972

Professor Louis P Taverne is a Belgian ichthyologist and paleo-ichthyologist. He was Professor both at the University of Brussels and the University of Bujumbura, Burundi. He has been a scientific collaborator at the Musée Royal de L'Afrìque Centrale, Tervuren, Belgium and Institut Royal des Sciences Naturelles de Belgique. Among his scientific papers (1960s through 2010s) was: *New insights on the osteology and taxonomy of tile osteoglossid fishes* Phareodus, Brychaetus *and* Musperia (*Teleostei, Osteoglossomorpha*) (2009). He described more than 100 new species, genera, families and even new orders. Poll (with whom Taverne wrote: *Description d'une espèce nouvelle de* Myomyrus *du Bas-Congo* (1967)) wrote of Taverne in his etymology that he was one "*…whose osteological studies have considerably advanced the classification of mormyrid fishes*."

Taylor, AD

Duckbill Eel sp. *Nettenchelys taylori* AW Alcock, 1898

Alfred Dundas Taylor (1825–1898) was a Commander in the Indian Navy. Alcock said of him that he was chiefly responsible for reviving the Marine Survey of India (1874). He wrote, among other books: *The India Directory for the Guidance of Steamers and Sailing Vessels* (1891).

Taylor, AS

Spotted Cusk-eel *Chilara taylori* CF Girard, 1858

Alexander Smith Taylor (1817–1876) was a collector, author and historian of California where he had first arrived (1848) and opened an apothecary's shop in Monterey (1849–1860) before moving to Santa Barbara and marrying. He had a number of diverse but large collections, much of which were lost in the San Francisco earthquake and fire (1906). His most noted writings include: *The Great Condor of California* (1859) and *Indianology of California* (1860–1863). He collected the holotype in Monterey, which, preserved in alcohol, he passed to Charles Girard.

Taylor, EH

Snake-eel sp. *Lamnostoma taylori* AW Herre, 1923

Dr Edward Harrison Taylor (1889–1978) was an American herpetologist. He went to teach in a Philippines village school (1912), returning briefly to his university, Kansas, to finish his master's. He was Chief, Division of Fisheries, Manila (1916–1920). He was Head of the Zoology Department of the Philippines (1923–1927) then worked at Kansas University (1927–1949), becoming full Professor (1934). He was an early broadcaster, giving a series of ten radio talks on herpetology (1932–1933). He took students on trips to Mexico (1937–1948), travelling in 'marginally reliable vehicles'; collecting took place at planned or unplanned stops. He collected after retirement in Costa Rica (1949), Thailand and Brazil, and wrote 19 papers on Philippine herpetology (1915–1928). In his etymology, Herre described him as: "*...a student of Philippine reptiles and amphibia*". Twenty-nine reptiles, twenty amphibians and a mammal are named after him.

Taylor, FH

Australian Sharpnose Shark *Rhizoprionodon taylori* JD Ogilby, 1915

Frank Henry Taylor (1886–1945) was an entomologist at the Institute of Tropical Medicine, Townsville, Queensland, Australia. He collected the holotype. He wrote a number of entomological papers including: *Medical Entomology in Australia* (1934). This was following his discovery of a mite that might have been a vector for Mossman fever.

Taylor, FHC

North Pacific Lanternfish *Tarletonbeania taylori* GW Mead, 1953

Dr Frederick Henry Carlyle Taylor (1919 –1987) was a Canadian oceanographer at Pacific Biological Station, Namaimo, British Columbia. The University of British Columbia awarded his bachelor's (1940) and master's (1947) degrees. The University of California awarded his doctorate (1956). He wrote: *Distribution and food habits of the fur seals of the North Pacific Ocean;*

report of cooperative investigations by the Governments of Canada, Japan, and the United States of America, February-July 1952 (1955).

Taylor, FW

Taylor's Sea Catfish *Cathorops taylori* SF Hildebrand, 1925

Frederic William Taylor (1876–1944) was an American agronomist and botanist and Director General of Agriculture for El Salvador (1923–1927). A bird is named after him.

Taylor, GW

Spinynose Sculpin *Asemichthys taylori* CH Gilbert, 1912
[Syn. *Radulinus taylori*]

Reverend George William Taylor (1854–1912) was the first Curator (1907–1912) of the Marine Biological Station at Departure Bay on the east coast of Vancouver Island. An Englishman by birth, he worked in a museum and also studied mining engineering in England before going to Canada (1882) where he took holy orders (1884), becoming a parish priest in Victoria, then Ottawa and eventually Wellington. There it was said of him that he was a 'tireless rambler, and collected without cessation'. He was elected a Fellow of the Royal Society of Canada and was appointed (1887) as Honorary Entomologist for British Columbia and became a founder member of the Natural History Society of British Columbia (1890). Interested in the salmon and halibut fisheries, he advocated (1898) the establishment of a Pacific coast biological station. While Curator he discovered the new sculpin subsequently named in his honour. He became seriously ill (2010) but was kept in the post although unable to continue his duties.

Taylor, LR

Yellow Cave Goby *Trimma taylori* PS Lobel, 1979
Perchlet sp. *Plectranthias taylori* JE Randall, 1980

Dr Leighton R Taylor Jr (b.1940) is a marine biologist who was Curator of Fishes (and later Director) at Waikiki Aquarium, Hawaii (1975–1986). He was also a professor of biology at the University of Hawaii. For a time, he combined his old calling with being a vintner in the Napa Valley, California, producing Cabernet Sauvignon and Merlot, but he and his wife, Linda, appear to have sold their winery (2008) and returned to live in Hawaii, where he became a member of the board of directors of a Hawaiian community group (2013). He described a number of fish including five sharks and among his published papers co-wrote: *Holacanthus griffisi, a new species of angelfish from the central Pacific Ocean* (1981). The goby etymology says he was honoured in the name "*…on the occasion of his appointment as director of the Waikiki Aquarium, continuing a tradition that each of the Aquarium's directors have a uniquely Hawaiian fish as a namesake*."

Taylor, RJ

Taylor's Garden Eel *Heteroconger taylori* PHJ Castle & JE Randall, 1995

Ron Josiah Taylor (1934–2012) was a renowned Australian professional diver, underwater photographer and shark expert as is his widow (Valerie May Taylor, née Heighes b.1935).

They were consulted on the *Jaws* (1975) films and others such as: *Orca* (1977) and *Sky Pirates* (1986). His first (1952) underwater interest was spearfishing and he became World Champion (1965, 1966, 1967 & 1968). They stopped killing sharks and became convinced conservationists. Their underwater documentary films (1962–1999) were acclaimed, and a TV series, *Blue Wilderness* followed (1992). They also worked on many feature films (1968–1995). He was honoured for his superb films and videos, and because his video of this eel prompted the second author to collect it. He died after a two-year battle with acute myeloid leukaemia.

Taylor, W Ralph

Caddo Madtom *Noturus taylori* NH Douglas, 1972
Taylor's Catfish *Neoarius taylori* A Roberts, 1978

Dr William Ralph Taylor (1919–2004) was an American ichthyologist whose university education was interrupted by WW2 service in the US Air Force, so that the University of Kansas only awarded his bachelor's degree (1947). The University of Michigan awarded both his master's (1951) and his doctorate. He worked as Curator of Fish, Smithsonian (1956–1979). He wrote: *A revision of the genus Noturus Rafinesque with a contribution to the classification of the North American catfishes* (1957).

Taylor, W Randolph

Taylor's Inflator Filefish *Brachaluteres taylori* LP Woods, 1966

William Randolph Taylor (1895–1990) was an American botanist, specialising in algae. He excelled as a collector of marine algae, in expeditions that took him throughout the Caribbean and northern South America, islands off western Mexico and the Galapagos Islands. He was appointed (1946) as senior biologist in 'Operation Crossroads' of the Department of Navy and conducted a botanical survey of Bikini and other Marshall Islands in the South Pacific prior to, and immediately after, the testing of atomic bombs. He collected the holotype in the Marshall Islands.

Tchang

Chinese Kissing Loach *Leptobotia tchangi* PW Fang, 1936
Perch-Barbel sp. *Percocypris tchangi* J Pellegrin & P Chevey, 1936
Sisorid Catfish sp. *Pseudecheneis tchangi* SL Hora, 1937
Sharpbelly (cyprinid) sp. *Hemiculter tchangi* PW Fang, 1942
Stone Loach sp. *Balitora tchangi* CY Zheng, 1982
Minnow sp. *Phoxinus tchangi* XY Chen, 1988
Rockling sp. *Ciliata tchangi* S Li, 1994
Snake-eel sp. *Ophichthus tchangi* W-Q Tang & C-G Zhang, 2002

Dr Tchung-Lin Tchang (1897–1963) was a Chinese zoologist and ichthyologist. He was Curator of Zoology at the Fan Memorial Institute of Biology and lecturer in zoology at the National University of Peking. One etymology described him as: "*...China's leading authority on ichthyology and fishery science.*" He published a number of papers such as: *Notes on a Fossil Fish from Shanxi* (1933).

Tchernavin

Titicaca Pupfish sp. *Orestias tchernavini* L Lauzanne, 1981

Vladimir Vyacheslavovich Tchernavin (1887–1949) was a Russian-born ichthyologist who became famous as one of the first and few prisoners of the Soviet Gulag system to escape abroad. After his father died (1902) he took part as a collector-zoologist in expeditions to the Altai region with the Russian explorer Vasili Sapozhnikov, and later led a series of scientific expeditions himself to the Altai Mountain and Sayanskii Mountain, Mongolia, the Tian Shan Mountains, the Amur River region, the Ussuriysk region on the Siberian-Manchurian border and to Lapland. He studied at St Petersburg University (1912–1917) interrupted by WW1 – being conscripted (1915), wounded and invalided out and then graduating (1916). He then taught at the Petrograd Agronomical Institute while studying for his master's. He became Professor of Ichthyology at the institute (1923), then moved (1926) to Murmansk as Director of the Northern Fisheries Trust giving up the production side (1928) to concentrate on research. He was arrested (1930) and imprisoned (1931), and sent to a gulag for five years. Initially doing hard labour, he was put to work as an ichthyologist at the camp fishery and later 'loaned out' as a lecturer and trainer of fish-farm managers. He used this relative 'freedom' to plan his escape (1932), taking his wife and son to Finland. He moved to England (1934) where he worked as an ichthyologist at the BMNH until his death. His account of his experiences: *I Speak for the Silent: Prisoners of the Soviets* (1935) and his wife's book *Escape From The Soviets* (1934) were among the first to give testimony of life under the Soviets, the GPU's operations and the Gulag.

Tchizh

Cod Icefish sp. *Nototheniops tchizh* AV Balushkin, 1976

Vladimir M Tcizh although he is not named in the etymology the fish honours Balushkin's childhood friend (according to R G Miller's *History and Atlas of Fishes of the Antarctic Ocean*, 1993).

Tchuvasov

Snaggletooth sp. *Astronesthes tchuvasovi* NV Parin & OD Borodulina, 1996

Vladimir Mikhailovich Chuvasov (See **Chuvasov**)

Teague

Driftwood Catfish sp. *Trachelyopterus teaguei* GJ Devincenzi, 1942
Long-ray Searobin *Prionotus teaguei* JC Briggs, 1956

Gerard Warden Teague (1885–1974) was a systematic ichthyologist and herpetologist who was British Vice-Consul for Paraguay. Born in Devon, he spent much of his working life in South America. He worked for the Midland Uruguay Railway Company, being the company's General Manager (1937), based in Paysandú where he was President of his local golf club. He was an honorary collaborator of Museo de Historia Natural, Montevideo. He lived in Lisbon (1960–1961), then in Quebec, before returning to live in Montevideo until his death. He wrote: *The armored sea-robins of America: a revision of the American species of the family Peristediidae* (1961). A reptile is also named after him.

Tecmin

Neotropical Rivuline sp. *Laimosemion tecminae* JE Thomerson, LG Nico & DC Taphorn, 1992

Tecmin honours a mining company rather than a person. Técnica Minera, or TECMIN, had been conducting geological and biological surveys in the Orinoco River basin of Venezuela, where this fish occurs. Their support made its collection possible.

Tecumseh

Shawnee Darter *Etheostoma tecumsehi* PA Ceas & LM Page, 1997

Tecumseh (1768–1813) is considered one of the greatest leaders of the Shawnee. His name means 'shooting star' or 'blazing comet'. He became leader of a multi-tribal confederacy in the early 19th century. When he heard of other Native Americans selling their lands to white settlers, he replied: "Why not sell the air, the great sea, as well as the earth? Did not the Great Spirit make them all for the use of his children?"

Teddy Roosevelt

Highland Darter *Etheostoma teddyroosevelt* SR Layman & RL Mayden, 2012

Theodore 'Teddy' Roosevelt (1858–1919), 26th President of the USA, needs no long biography here. He was honoured in the fish's binomial particularly for his enduring legacy in environmental conservation and stewardship, including the designation of vast areas as national forests, national monuments, and national parks.

Teegelaar

Lake Victoria Cichlid sp. *Haplochromis teegelaari* PH Greenwood & CDN Barel, 1978

Nico Teegelaar (1926–1976) was a Dutch scientific illustrator who illustrated a number of papers by Barel. The etymology says the cichlid is named "*…in honor of the late Nico Teegelaar, an outstanding Dutch biological artist whose work contributed much to the researches of the Zoology Department of Leiden University*", where the junior author worked.

Tee-Van

Prickly Brown Ray *Dipturus teevani* HB Bigelow & WC Schroeder, 1951
Flabby Whalefish sp. *Cetomimus teevani* RRR Harry, 1952
Flabby Whalefish sp. *Megalomycter teevani* GS Myers & WC Freihofer, 1966

John Tee-Van (1897–1967) joined the staff of the Bronx Zoo (1911) as an assistant keeper in the bird house. He famously brought two Giant Pandas from China to the Bronx Zoo (1941), and was at sea in the Pacific on his return trip during the attack on Pearl Harbour. He became William Beebe's assistant in the tropical research department and accompanied him on twenty-four expeditions. He retired having been Director of both the Bronx Zoo and the Coney Island Aquarium (1952–1962). He was honoured 'in appreciation of his helpful assistance' as editor-in-chief of the series: *Fishes of the Western North Atlantic*.

Teguh

Sulawesi Anchovy *Stolephorus teguhi* S Kimura, K Hori & K Shibukawa, 2009

Teguh Peristiwady is an Indonesian biologist. He is Senior Scientist at the Technical Implementation Unit for Natural Biota Conservation, Indonesian Institute of Sciences. He was formerly at CRDOA: Center for Research and Development of Oceanology, Indonesian Institute of Science, Ambon. He has published such scientific papers as: *First record of Odontanthias unimaculatus (Tanaka 1917) (Perciformes: Serranidae) from Indonesia* (2011). He was honoured "*...for giving the authors the opportunity to collect specimens.*"

Teijsmann

Airbreathing Catfish sp. *Clarias teijsmanni* P Bleeker, 1857

Johannes Elias Teijsmann (or Teysmann) (1808–1882) was a Dutch botanist and plant collector. He was Curator of the Buitenzorg Botanic Gardens in Java (1831–1869), which were subsequently renamed in his honour. He took part in numerous botanical expeditions in the Dutch East Indies (Indonesia). Four birds are named after him.

Teixeira

Damselfish genus *Teixeirichthys* JLB Smith, 1953

Gabriel Mauricio Teixeira (1897–1973) was a Madeiran who was (1946–1958) the Governor-General of Mozambique (the type locality of T. mossambicus, now regarded as a junior synonym of T. jordani). Smith wrote describing him as one "...who has for many years generously assisted [Smith's] researches in his territories, and whose administration has had a profound effect on the development of the country under his charge."

Tekomaji

Goby sp. *Amblygobius tekomaji* JLB Smith, 1959

This is a toponym referring to Tekomaji Island, Mozambique, western Indian Ocean, which is the type locality.

Telestes

Vairone (cyprinid) genus *Telestes* CL Bonaparte, 1837

The etymology is not explained by Bonaparte, and it may not be an eponym. However, Telestes was the name of the last king of ancient Corinth (748 BC) and of a poet of 5th-century Greece, so the name might refer to one of these.

Telfair

Fairy Mullet *Agonostomus telfairii* ET Bennett, 1832

Dr Charles Telfair (1778–1833) was an Irish physician, naval surgeon, botanist, sugar planter and, probably, rum smuggler. He lived in Mauritius but travelled widely in the Indian Ocean and further afield. He travelled in China (1826), where he acquired some banana plants that he sent to England. They were passed to the Duke of Devonshire, who successfully grew them in the glasshouses at Chatsworth. He imported Nile Crocodiles from Madagascar, where he spent time collecting, as Bennett reported in *Characters of a New Genus of Lemuridae, Presented by Mr. Telfair* (1832). A mammal, a reptile and some plants are also named after him.

Temminck

Kissing Gourami *Helostoma temminckii* G Cuvier, 1829
Broadfin Shark *Lamiopsis temminckii* JP Müller & FGJ Henle, 1839
Armoured Catfish sp. *Ancistrus temminckii* A Valenciennes, 1840
Dark Chub *Nipponocypris temminckii* CJ Temminck & H Schlegel, 1846
Threadfin Wrasse *Cirrhilabrus temminckii* P Bleeker, 1853
Surfperch sp. *Ditrema temminckii* P Bleeker, 1853
Samoan Silverside *Hypoatherina temminckii* P Bleeker, 1854
Goldribbon Soapfish *Aulacocephalus temminckii* P Bleeker, 1855
Long-snout Pipefish *Syngnathus temminckii* JJ Kaup, 1856

Coenraad Jacob Temminck (1778–1858) was a Dutch zoologist, illustrator and collector. He was the first Director of the National Museum of Natural History in Leiden (1820–1858). He was a wealthy man who had a very large collection of specimens and live birds. His chief interest was in ornithology, and his works include: *Manuel d'Ornithologie, ou Tableau Systematique des Oiseaux qui se Trouvent en Europe* (1815) and *Histoire naturelle générale des pigeons et des gallinacés* (1808–1815). Twenty-seven birds, thirteen mammals and two reptiles are also named after him.

NB Temminck did not name the chub species after himself; Schlegel wrote the description and coined the name, but the publication in which it appeared is credited to both authors.

Teng

Straight-tooth Weasel Shark *Paragaleus tengi* JSTF Chen, 1963
Loach sp. *Formosania tengi* M Watanabe, 1983

Dr Teng Huo-Tu (1911–1978) was a Taiwanese ichthyologist. His doctorate was awarded by Kyoto University, Japan (1962). He was, at one time, President of the Taiwan Fisheries Research Institute, Keelung, where he worked for many years. He was involved in the classification of chondrichthyes, especially sharks, so it is appropriate that a shark is named after him.

Tenison

Barbeled Dragonfish sp. *Eustomias tenisoni* CT Regan & E Trewavas, 1930
Tenison's Lanternfish *Protomyctophum tenisoni* JR Norman, 1930

Lieutenant-Colonel William Percival Cosnahan Tenison (1884–1983) was a British Army officer in the Royal Field Artillery. He was a Fellow of the LInnean Society and a talented painter and scientific illustrator, whose 'accurate and artistic drawings', according to the dragonfish etymology, are reproduced as plates in the authors' monograph.

Tennent

Doubleband Surgeonfish *Acanthurus tennentii* A Günther, 1861

Sir James Emerson Tennent (1804–1869) was a politician and traveller. Born in Ireland, he became (1832) Member of Parliament for Belfast. He was appointed (1845) Colonial Secretary of Ceylon (Sri Lanka), where he took an interest in the local flora and fauna, writing *Sketches of the Natural History of Ceylon* (1861). The surgeonfish holotype came from the collection of the Ceylonese physician and naturalist Edward Frederick Kelaart. Although Tennent is not mentioned in Günther's original text, there is little doubt as to who is intended. A reptile is also named after him.

TE-nox

Slaty Goby *Elacatinus tenox* JE Böhlke & CR Robins, 1968

This is an unusual case where the prefix *te* is based on the initials of the collectors of the holotype: James C Tyler (q.v.) and William N Eschmeyer (q.v.). it is combined with *nox* - meaning 'night', and referring to the goby's dark colour.

Terao

Eelpout sp. *Petroschmidtia teraoi* M Katayama, 1943
[Syn. *Lycodes teraoi*]

Dr. Arata Terao was Professor of Zoology and Director of the Zoological Laboratory, Imperial Fisheries Institute, Tokyo. 'Dr A Terao' was thanked by the author for his "*kindness extended to me in various ways.*" Although no further details are given by Katayama, we believe this refers to Arata Terao. Other taxa are named after him including a shrimp.

Teresa

Mountain Molly *Poecilia teresae* DW Greenfield, 1990
Terry's Dwarfgoby *Eviota teresae* DW Greenfield & JE Randall, 2016

Teresa 'Terry' Greenfield is the senior author's wife. She assisted in collecting the type material of the goby, and was thanked for providing field, editorial, and moral support to her husband over many years.

Teresinarum

Cuban Cavefish sp. *Lucifuga teresinarum* PA Díaz Pérez, 1988

Dr Maria Teresa del Valle Portilla, who is a physician and a full professor, and Maria Teresita de la Hoz Gonzalez, who is a biologist, are both members of the faculty at the Universidad de la Habana and colleagues of the author.

Ternetz

Black Widow Tetra *Gymnocorymbus ternetzi* GA Boulenger, 1895
Armoured Catfish sp. *Hypostomus ternetzi* GA Boulenger, 1895
Freshwater Croaker *Plagioscion ternetzi* GA Boulenger, 1895
Yellow King Piranha *Serrasalmus ternetzi* F Steindachner, 1908
Thorny Catfish sp. *Tenellus ternetzi* CH Eigenmann, 1925
Headstander sp. *Abramites ternetzi* JR Norman, 1926
[NCR: Junior Synonym of *Abramites hypselonotus*]
Pike-Cichlid sp. *Crenicichla ternetzi* JR Norman, 1926
Halftooth Characin sp. *Hemiodus ternetzi* GS Myers, 1927
Dwarf Sleeper sp. *Microphilypnus ternetzi* GS Myers, 1927

Red Hook Pacu *Myleus ternetzi* JR Norman, 1929
Pencil Catfish sp. *Typhlobelus ternetzi* GS Myers, 1944
Ternetz's Anostomus *Anostomus ternetzi* A Fernández-Yépez, 1949

Dr Carl Ternetz (1870–1928) was a Swiss-born ichthyologist and naturalist who first collected near Basel (1892) and subsequently collected in South America over a period of more than thirty years (from c.1895). He made extensive collections in French Guiana for the British Museum. A reptile and two amphibians are named after him.

Terofal

Characin sp. *Diapoma terofali* J Géry, 1964
African Rivuline sp. *Rhexipanchax terofali* HO Bergenkamp & V Etzel, 1981
[Junior Synonym of *Rhexipanchax schioetzi*]

Dr Fritz Terofal (1932–1988) was an ichthyologist who was Director of the Ichthyology Section, Bavarian State Collection of Zoology, Munich, Germany. He wrote: *Susswasserfische in europaischen Gewassern* (1984).

Terver

Terver's Lampeye *Plataplochilus terveri* JH Huber, 1981

Dr Denis Terver (b.1940) was a French ichthyologist, aquarist, writer and biologist who became the Deputy Director (1977) and then Director of Nancy Museum and Aquarium until retirement (2000). His BSc in biology was awarded by the University of Nancy as was his PhD (1975). Huber described him in his etymology as his good friend who had helped and heavily supported him for all the years while he prepared his thesis.

Tessmann

Robber Tetra sp. *Brycinus tessmanni* P Pappenheim, 1911
Squeaker Catfish sp. *Synodontis tessmanni* P Pappenheim, 1911

Dr Günter Theodor Tessmann (1884–1969), was a German explorer, ethnographer and plant collector. His honorary doctorate was awarded by the University of Rostock (1930). After education at the Reichskolonialschule, Witzenhausen (1902–1904) he went to Cameroon (1904) where he worked on a cocoa plantation and became an elephant hunter. He was put in charge of the Lübeck Pangwe Expedition (1907–1909) organised by the Lübeck Museum of Ethology. He was in Cameroon when WW1 put an end to another expedition, stared (1913). He had to leave the country, and went to Spanish Guinea where he was interned. After WW1, he turned his attention to South America and worked as an anthropologist in the Amazon region of Peru (1920–1926). He emigrated to Brazil (1936) and settled in Parana as a colonist, Eventually, he found a permanent post at Museu Paranaense (1947) and later at the institute of Biology in Curitiba, where he died. He collected the holotypes of both these species.

Tetha

Tetha's Dwarfgoby *Eviota tetha* DW Greenfield & MV Erdmann, 2014

Creusa 'Tetha' Hitipeuw (1969–2013) was an Indonesian marine scientist and conservationist. She took her first degree at Pattimura University in Ambon, Indonesia, while Vrije Universiteit in Belgium awarded her master's degree. After this she worked for various marine conservation projects in Kepulauan Aru in Maluku, Kepulauan Derawan in East Kalimantan, in the Bird's Head Seascape and the Teluk Cenderawasih of Papua, and in the Kei Kecil Island in Maluku. One of her focuses was on marine migration, having studied this in both leatherback turtles and whale sharks. She joined WWF (1996) for the rest of her career. The etymology says that, known to her colleagues as 'Tetha', she was a "... *passionate and highly-respected Indonesian marine conservationist who dedicated her career to saving the coral reefs and especially marine turtles of Indonesia, with a strong focus on Teluk Cenderawasih and the Bird's Head region of West Papua*" (where this goby occurs); she passed away at age 44 in 2013 shortly after the discovery of this species, after a brief and unexpected battle with cancer.

Tethys

Tethys Dragonet *Callionymus tethys* R Fricke, 1993

Tethys was one of the Titans of Greek mythology, a sea-goddess who was the wife of Oceanus (a personification of the great river thought to encircle the world).

Teugels

Claroteid Catfish sp. *Chrysichthys teugelsi* LM Risch, 1987
African Loach Catfish sp. *Paramphilius teugelsi* PH Skelton, 1989
Electric Catfish sp. *Malapterurus teugelsi* SM Norris, 2002
Congo Cichlid sp. *Lamprologus teugelsi* RC Schelly & MLJ Stiassny, 2004
Cross River Bichir *Polypterus teugelsi* R Britz, 2004
Congo Cichlid sp. *Nanochromis teugelsi* A Lamboj & RC Schelly, 2006
Eel Catfish sp. *Channallabes teugelsi* S Devaere, D Adriaens & W Verraes, 2007
Airbreathing Catfish sp. *Clariallabes teugelsi* CJ Ferraris, 2007
African Characin sp. *Distichodus teugelsi* V Mamonekene & EJ Vreven, 2008
African Rivuline sp. *Aphyosemion teugelsi* JR van der Zee & R Sonnenberg, 2010
African Barb sp. *Enteromius teugelsi* M Bamba, EJ Vreven & J Snoeks, 2011

Dr Guy G Teugels (1954–2003) was a Belgian ichthyologist. He took his master's in biology at the Katholieke Universiteit Leuven (KUL) (1977) and then started his PhD research on African clariid catfishes under the guidance of Professor Dr Dirk Thys van den Audenaerde at the Africa Museum in Tervuren, Belgium, presenting his doctoral thesis six years later (1983). He worked at the Laboratoire d'Ichtyologie Générale et Appliquée at the Muséum National d'Histoire Naturelle in Paris (1984) later returning (1988) to Musée Royale de l'Afrique Centrale to become Curator of Fishes, succeeding Poll and his old professor. He visited west and central Africa, collecting in Ivory Coast, Cameroon, Congo Brazzaville, Gabon and Benin. He co-authored over 100 publications in peer-reviewed journals, almost a third of them as a first author. He wrote numerous other scientific notes and reports. He then took a post as part-time lecturer at Leuven University. About a year and a half before his death he started chemotherapy to treat a tumour. During his treatment, he continued working and corresponding with his colleagues, mostly from home. Britz honoured him in the bichir's etymology for his "...*myriad influential contributions to the systematics of African freshwater fishes.*" In the barb etymology, he is described as "...*an outstanding ichthyologist who introduced the first and second author to fish taxonomy and greatly contributed to the knowledge of the African fishes over the last twenty years.*"

Teunis Ras

Lake Victoria Cichlid sp. *Haplochromis teunisrasi* F Witte & ELM Witte-Maas, 1981

Teunis Ras was a 'Dutch master fisherman' honoured for the help he gave the Haplochromis Ecology Survey Team in Mwanza Gulf of Lake Victoria, and "*especially for the pains he took in making the fishing gear and in teaching us how to use it.*"

Te Vega

Blue-striped Cave Goby *Trimma tevegae* DM Cohen & WP Davis, 1969

The 'Te Vega' is Stanford University's research vessel; a converted luxury yacht, from which the type was collected.

Thalassa

Gulper Eel sp. *Saccopharynx thalassa* JG Nielsen & E Bertelsen, 1985

The 'RV Thalassa' was a French research vessel from which the holotype was collected.

Thaman

Garden Eel sp. *Gorgasia thamani* DW Greenfield & S Niesz, 2004
Thaman's Dwarf-goby *Eviota thamani* DW Greenfield & JE Randall, 2016

Dr Randolph Robert Thaman (b.1943) is a Fiji-based American environmental scientist and conservationist who is Professor Emeritus of Pacific Islands Biogeography, University of the South Pacific, Fiji. The University of California, Berkeley awarded both his BA (1966) and MA (1968) and he completed his PhD at University of California, Los Angeles (1975). His career began as a lab assistant (1966–1967) at Forestry Remote Sensing Laboratory, University of California, Berkeley. He was then a Peace Corps Volunteer teaching at a school in Tonga (1967–1969 & 1971–1972). Back at the lab he was (1969–1970) Project Leader, NASA Earth Resource Inventory Project, followed by a period (1972–1974) as Research Coordinator Geography Remote Sensing Unit and part-time Lecturer in Cultural Geography, University of California, Santa Barbara. He joined (1974) the University of the South Pacific as Lecturer (1974–1978), became Senior Lecturer (1979–1985), Reader (1986–1991) and Professor (1991) (now Emeritus). Something of a linguist, he speaks Tongan, Fijian, French, Spanish and passable New Guinea Pidgin. He is the acknowledged resident Pacific Island authority on Pacific Islands biodiversity. He has authored a number of reports that have been key to the establishment of conservation projects in many parts of the Pacific. He has also been the recipient of many prestigious awards. He has published prolifically with many articles in magazines, working papers, project reports, scientific papers and books (1976–2014). He was honoured in the eel's binomial for "...*unending assistance*" to the authors in arranging their fieldwork and for promoting the conservation of Fiji's marine and terrestrial fauna; and in the goby's "...*for continuous support of the authors' work in Fiji over the years, without which their research would not have been possible*".

Tha Nho

Stone Loach sp. *Schistura thanho* JA Freyhof & DV Serov, 2001

Not an eponym but named for the 'friendly people' of the Tha Nho ethnic community, Binh Dinh Province, Việt Nam, the type locality.

Thavone

Stone Loach sp. *Schistura thavonei* M Kottelat, 2017

Thavone Phommavong works at LARReC (Living Aquatic Resources Research Center) and is responsible for database management and information dissemination of LARReC's technical and research publications, under the Data and Information Unit. Prior to studying at AIT in Bangkok, Thavone worked as a Technical Officer for the Capture Fisheries Unit, part of LARReC, where he assisted the Fisheries Programme with field surveys of fish migration studies in the Mekong River. He earned a degree in aquaculture at AIT (2004) then returned to LARReC. He was honoured for his "*...help and companionship during several, and sometimes difficult, fish surveys in Laos.*"

Thayer (Expedition)

Armoured Catfish sp. *Hisonotus thayeri FO* Martins & F Langeani, 2016

The Thayer Expedition (1865–1866) is: "*...considered one of the most important journeys performed in Brazil*" (see also next entry).

Thayer, N

Characin genus *Thayeria* CH Eigenmann, 1908

Dwarf Acara sp. *Laetacara thayeri* F Steindachner, 1875

Characin sp. *Hypomasticus thayeri* NA Borodin, 1929

Three-barbeled Catfish sp. *Brachyrhamdia thayeria* V Slobodian & FA Bockmann, 2013

Nathaniel Thayer Jr (1808–1883) was an American philanthropist and financier. He was a great benefactor of Harvard of which he was a Fellow (1868–1875), and he bore the costs of Agassiz's expedition (known as the Thayer Expedition – see above) to Brazil. Among their discoveries were the holotypes of *Laetacara thayeri* and *Hypomasticus thayeri.*

[Note that the catfish's binomial also refers to it having similar markings to *Thayeria* characins, and this resemblance is also reflected in the binomial of the Halftooth Characin sp. *Hemiodus thayeria*]

Thayer, S

False Black Tetra *Gymnocorymbus thayeri* CH Eigenmann, 1908

Stephen Van Rensselaer Thayer (1847–1871) was one of Nathaniel Thayer's sons (see above). He went on the Thayer Expedition as a volunteer. The holotype of this species was collected during that expedition.

Theager

Darter Characin sp. *Characidium theageri* H Travassos, 1952

[Junior Synonym of *C. rachovii*]

Gerard Warden Teague (1885–1974) – According to Travassos the type was collected by 'Theager' in Uruguay (1942). It has been suggested that *theager* was a typographical error and was intended to refer to Teague, who was collecting at the right place at the right time. (See **Teague**)

Theodora

Snake Catfish *Clarias theodorae* M Weber, 1897

Theodora Jacoba Sleeswijk née van Bosse (1874–1953) was a Dutch etcher and painter, mainly of landscapes. She went on an expedition to South Africa (1894) with her uncle and aunt. Her aunt was Ann Weber née van Bosse who was a botanist and her uncle was Max Weber, then Director Artis Zoological Museum, Amsterdam, and the describer of the above species. During and after WW1, she was socially active and campaigned for women's rights.

Theodore

Theodore's Threadfin Bream *Nemipterus theodorei* JD Ogilby, 1916

Edward Granville Theodore (1884–1950) was an Australian politician, nicknamed 'Red Ted', who served as Premier of Queensland (1919–1925). Ogilby honoured him "*in recognition of the formation by him of a Department of Fisheries.*"

Theodore Tissier

Dreamer sp. *Oneirodes theodoritissieri* G Belloc, 1938

The Président Théodore Tissier was the first research vessel commissioned (1933) by the French Merchant Navy for the Technical Office for Marine Fisheries (OSTPM). It had a diving turret that could descend to 800m. It was inaugurated (1933) at the Musée de la Mer de Biarritz and later transferred to the navy (1938) as an annex to its training school. During WW2 it took part in the evacuation of allied troops trapped on the beaches (1940). Seized by the British, it was immediately re-armed by the Free French Navy (FNFL). The vessel is named after the French jurist and politician Théodore Tissier (1866–1944), who was Vice President of the State Council (1928–1937). The dreamer (anglerfish) holotype was caught from the vessel off Portuguese Guinea (1936) by Gérard Belloc on behalf of the La Rochelle Natural History Museum.

Theophilus

Lesbos Stone Loach *Oxynoemacheilus theophilii* MT Stoumboudi, M Kottelat & R Barbieri, 2006

Theophilus Chatzimichael (1873–1934) was a prominent folk painter from the island of Lesbos, Greece, where the stone loach is endemic.

Thepass

Thepass' Sabretooth Blenny *Petroscirtes thepassii* P Bleeker, 1853

A H Thepass was a military surgeon in Ternate (Dutch East Indies) who supplied Bleeker with the holotype.

Therese (Hayes)

Therese's Sole *Aseraggodes therese* JE Randall, 1996

Therese Hayes was one of the collectors of the holotype (1991) at Midway Atoll, Hawaiian Islands. She has co-written various articles on Hawaiian fish, such as *Annotated Checklist of the Fishes of Midway Atoll, Northwestern Hawaiian Islands* (1993).

Therese (Princess)

Suckermouth Catfish sp. *Astroblepus theresiae* F Steindachner, 1907

Therese Charlotte Marianne Auguste von Bayern (Princess Theresa of Bavaria) (1850–1925) was a zoologist, botanist, anthropologist and explorer in Tunisia, Russia, the Arctic, Mexico, Brazil and western South America. Her father was Luitpold, Prince Regent of Bavaria. She was the first woman to receive an honorary degree from Ludwig Maximilian University, Munich (1897). She wrote books under the pseudonym 'Th. v. Bayer', such as: *Ausflug nach Tunis* (1880). Her South American anthropological collection is held by the State Museum of Ethnology, Munich. Two birds and a reptile are named after her.

Thérézien

Malagasy Goby sp. *Acentrogobius therezieni* A Kiener, 1963

Yves Thérézien (1925–2015) was a friend and colleague of André Kiener. He was an expert on freshwater algae and a hydrobiologist at MNHN, Paris. He was a fisheries researcher with Centre Technique Forestier Tropical, Madagascar (1958); inspector for the Water and Forestry Service, Madagascar (1960–1966) and designated collector of botanical specimens for Service Forestier de Madagascar. He worked for the Institut Nationale de la Recherche Agronomique, Guadeloupe (1979), and was on a Franco-Austrian hydrobiological expedition to the Lesser Antilles. After retiring (1980) he worked as a volunteer at MNHN, Paris. He wrote several scientific papers. He collected the type. Two reptiles are also named after him.

Thetis

Thorntail Stingray *Dasyatis thetidis* JD Ogilby, 1899

[Junior Synonym of *Bathytoshia lata*]

Thetis Fish *Neosebastes thetidis* ER Waite, 1899

HMCS 'Thetis' was on a trawling expedition off the coast of New South Wales (1898) during which the holotypes were collected. The vessel was named after a sea nymph in Greek mythology.

Thibaudeau

Laotian Shad *Tenualosa thibaudeaui* J Durand, 1940

Leon Emmanuel Thibaudeau (1883–1946) was a French Colonial administrator (1907–1942) who became Résident Supérieur, Cambodia. He began his career when posted to southern Annam (now Vietnam) (1907), earning a medal for his work during a typhoon and cholera outbreak there where he also quickly learned the local language. He was mobilised (1917) into the infantry until the end of WW1 (1919). He married and returned to Indo-China as deputy administrator and Resident for several provinces (1919–1928). He was appointed Director of Office and Inspector of Political and Administrative Affairs in Hue province (Vietnam) (1928–1931) then Acting Resident Superior (1933–1934). He was posted to Cambodia (1935) as Inspector of Political and Administrative Affairs and later (1936–1941) promoted to Résident Supérieur. He took an interest in the fishing industry there and worked with the Oceanographic Institute to enable local fishermen to gain more control over fish sales by grouping into co-operatives. He also became increasingly opposed to the Vichy regime there and retired from his post (1942). He was interned in a Japanese prison camp and, although released and repatriated to Bordeaux, the prison privations led to his death a few months later.

Thiele

Viviparous Brotula sp. *Grammonus thielei* JG Nielsen & DM Cohen, 2004

Werner Thiele (b.1966) is an Austrian self-taught underwater photographer. He was the first person to photograph and capture this species.

Thielle

Thielle's Anemonefish *Amphiprion thiellei* WE Burgess, 1981
[Probably a natural hybrid, perhaps between *A. sandaracinos* and *A. ocellaris*]

Mike Thielle was the owner of Reef Encounter in Hackensack, New Jersey (USA), an aquarium and pet-supply store. He donated two aquarium specimens to Burgess, which served as the types.

Thiemmedh

Thiemmedh's Horsefaced Loach *Acantopsis thiemmedhi* S Sontirat, 1999
[Alt Black-spotted Horseface Loach]

Dr Jinda(h) Thiemmedh (1902–1985) was a Thai zoologist and ichthyologist. He was awarded his degree in Biology and Diploma in Fish Culture from a Philippines University (1949) and his master's in Fisheries Management by Auburn University, Alabama, USA (1952). He was also later awarded honorary degrees. He worked in the Department of Fisheries and was Acting Dean of the Faculty of Fisheries, Kasetsart University (Thailand), and Sontirat's first teacher in ichthyology. He wrote a number of books and papers (1963–1982), such as: *Dolphin fish (Coryphaena hippurus, Linn). an addition to the list of fishes of Thailand* (1963), *Fishes of Thailand: Their English, Scientific and Thai Names* (1966) and *Ichthyology* (1982).

Thienemann

Mahseer sp. *Neolissochilus thienemanni* CGE Ahl, 1933

Professor Dr August Friedrich Thienemann (1882–1960) was a German limnologist, ecologist and zoologist. He studied (1901–1905) science and philosophy at the universities of Greifswald, Innsbruck and Heidelberg, culminating with his doctorate in Heidelberg with Robert Lauterborn. He was then (1907) head of the Biological Department of Fisheries and wastewater issues at the Zoological Institute of the Westfälische Wilhelms-University in Münster, where he conducted his famous research on the lakes of the Eifel. During his military service, he sustained a serious shrapnel wound at Reims (1914). When he recovered, he was transferred to military government service (1915). He was (1917) Professor of Zoology at the University of Kiel. He took part in the Deutschen Limnologischen Sunda-Expedition (1928–1929). He then became Director of the former Hydrobiologische Anstalt der Kaiser-Wilhelm-Gesellschaft (now the Max-Planck-Institut für Limnologie) at Plön until retirement (1960). He is best known for his work on the biology of Chironomidae (midges), and his contributions to the field of lake topology. He published a great many papers over half a century (1909–1959). He collected the mahseer type. A dragonfly is also named after him.

Thierry

Togo Killifish *Fundulosoma thierryi* CGE Ahl, 1924
[Syn. *Nothobranchius thierryi*]

Ahl did not identify the patronym, but the binomial presumably honours Gaston Thierry (1866–1904). Despite his French names, he was an Oberleutnant (First Lieutenant) in the Imperial German Army. He was one of those sent to Togoland (1896) to establish a series of bases, after the French and German governments had reached agreement on the border between German Togoland (now partly in Ghana and partly the state of Togo) and French Dahomey (now Benin), to enforce German control over the country. He left Togo (1899) and was killed (1904) in Cameroon (then a German colony) by a poisoned arrow. Thierry is known to have collected other natural history specimens and has a mammal, a reptile and two birds named after him.

Thiollière

Thiollière's Lanternfish *Diaphus thiollierei* HW Fowler, 1934

Victor Joseph de l'Isle Thiollière (1801–1859) was a French civil engineer, geologist and palaeo-ichthyologist, His main research was in fossils and his collection is now in the Muséum de Lyon. He was a member of both the Academy of Sciences, Belles-lettres and Arts de Lyon (1848–1859) and the Linnean Society of Lyon (1850–1858). He also reported on fishes collected by French priest and biologist Xavier Montrouzier (1820–1897) from the Woodlark Archipelago, Papua, New Guinea (1857).

Thiony

Thiony's Jawfish *Opistognathus thionyi* WF Smith-Vaniz, L Tornabene & RM Macieira, 2018

Thiony Simon (1985–2016) was an oceanographer, environmentalist and diver. The Universidade Federal do Espírito Santo awarded his BSc in Oceonography (2007), his MSc in Environmental oceanography (2010) and PhD in Animal Biology (2014). He was a post-doctoral fellow at UFES. In honouring him, the etymology described him as a 'colleague and dear friend' who passed away during preparation of the description. He had a diving accident while exploring mesophotic reefs. He collected most of the type material and dedicated his life to the study and conservation of Brazilian reef ecosystems.

Thisbe

Bream sp. *Acanthobrama thisbeae* JA Freyhof & M Özuluğ, 2014

This is not a true eponym but rather named for Thisbe, who was in love with Pyramus in Ovid's: *Metamorphoses*. Pyramus is also the ancient Greek and Latin name for the Ceyhan River, southern Anatolia, where this species is endemic so is actually a form of toponym.

Thoburn

Sucker genus *Thoburnia* DS Jordan & JO Snyder, 1917
Thoburn's Mullet *Xenomugil thoburni* DS Jordan & EC Starks, 1896
[Syn. *Mugil thoburni*]

Dr Wilbur Wilson Thoburn (1859–1899) was an American biologist. Allegheny College awarded his bachelor's and master's degrees, and finally his doctorate (1888). He was Professor of Geology and Botany in the Illinois Wesleyan University (1884–1888) and was also a lay preacher. He became Professor of Geology and Biology at the University of the Pacific (1888). He then taught bionomics (ecology) as an instructor in the zoology department at Stanford University, where he became a close friend of David Starr Jordan (q.v.). He died aged 39 of pneumonia.

Thollon

Tilapia sp. *Coptodon tholloni* HE Sauvage, 1884
African Characin sp. *Alestes tholloni* J Pellegrin, 1901

François-Romain Thollon (1855–1896) was a French botanist and collector who resigned (1882) from MNHN Paris to work in the Congo and as a member of the de Brazza mission in Gabon. He also collected in Nigeria and the Ivory Coast, but never returned to France being based in the Congo for the remainder of his life. Many taxa, including a bird, a reptile and a mammal, are named after him.

Thomas, HS

Western Ghats Glassy Perchlet *Parambassis thomassi* F Day, 1870
Red Canarese Barb *Hypselobarbus thomassi* F Day, 1874
Konti Barb *Osteochilichthys thomassi* F Day, 1877

Henry Sullivan Thomas FLS (b.1833) was born in Poona but educated at Haileybury, England (1853–1854). He worked for the Madras Civil Service (1855–1889), as Collector and Magistrate at South Kanara (1867–1873) and later as Collector at Tanjore (Thanjavur) (1874–1878). He was a member of the Revenue Board (1878–1884) and of the Governor-General's Legislative Council (1882–1883 & 1885). He was seconded on special assignment in relation to the pearl fisheries of Tuticorin and Ceylon (Sri Lanka) (1884). He was also a keen angler, hence his work: *The Rod in India: Being Hints How to Obtain Sport With Remarks on the Natural History of Fish, and Their Culture* (1873) and other books such as *Tank Angling in India* (1887). Day wrote that Thomas: "*...has paid great attention to the fishy inhabitants of his range*". (Note: it is unknown why Day consistently misspelt Thomas's surname, by adding an extra '*s*')

Thomas, J

French Congo Mormyrid *Mormyrus thomasi* J Pellegrin, 1938

Jean Thomas (1890–1932) was a French explorer and naturalist. He was awarded his CPN Certificate in Physics/Chemistry/Natural Sciences (1911) and later joined an infantry regiment being called up (1914) to fight in WW1. He was seriously wounded and his leg was barely saved from amputation after four operations. There was a long period of convalescence and he completed his BA (1917). He spent his time teaching and collecting animals, from which he created a local museum in his village: Saint-Paul-sur-Save. He entered the Ecole des Hautes Etudes as preparer of colonial animal production at the laboratory of the Muséum National d'Histoire Naturelle (Paris). He then collected natural history and anthropological specimens for the MNHN, undertaking seven exploratory missions in French Equatorial Africa (1922–1930), the main objective of which was to study the potential for fisheries development there. Before the first expedition to Morocco he was again seriously injured, the only survivor of four people involved in a motor vehicle accident. He also suffered from depression. He wrote two books about the expeditions: *A travers le sud Tunisien* (1930) and the posthumously published: *A travers l'Afrique équatorial sauvage* (1934). He collected the mormyrid type during his last expedition (1929–1930) and Pellegrin honoured him in the paper written some years after Thomas died.

Thomas, L

Thomas' Triplefin *Norfolkia thomasi* GP Whitley, 1964

Leonard R 'Len' Thomas was an Australian scientist who was Honorary Associate of the Australian Museum. He helped to plan, and accompanied, the Museum's (1962) Swain Reefs Expedition, during which paratypes of the triplefin were collected.

Thomas, NW

African Butterfly Cichlid *Anomalochromis thomasi* GA Boulenger, 1915
Claroteid Catfish sp. *Notoglanidium thomasi* GA Boulenger, 1916
Sherbo Mormyrid *Marcusenius thomasi* GA Boulenger, 1916
Goby sp. *Yongeichthys thomasi* GA Boulenger, 1916

Northcote Whitridge Thomas (1868–1936) was a British Government anthropologist who conducted field research in Nigeria (1909–1913) and Sierra Leone (1914–1915) particularly among the Edo, Yoruba, Kukuruku, Esa and Igbo people. He is noted for having made many early recordings on wax cylinders of local people and their music and songs. Trinity College, Cambridge University, awarded his MA and he collected many artefacts that he eventually donated to that University's Museum of Archaeology and Anthropology. He wrote a number of books, alone or as co-author, on folklore and anthropology, such as: *Bibliography of Folk-lore* (1905) and *Women of all Nations* (1908). A reptile is named after him.

Thomas, WS

Characin sp. *Piabina thomasi* HW Fowler, 1940
[Syn. *Bryconamericus thomasi*]

William Stephen Thomas (1909–2001) was a trained museologist who graduated from Harvard (1932) in history and American literature, with biology as a minor subject. He was a museum apprentice in Newark, New Jersey, an assistant registrar at the Metropolitan Museum of Art, New York, and took graduate courses in art history at New York University. He was Director of Education at the Academy of Natural Sciences of Philadelphia (1936–1942), after which he joined the US Navy for WW2 (and was later re-called during the Korean War). He was Director of the Rochester Museum and Science Centre, New York (1945–1973), retiring as Director Emeritus. He wrote: *The Amateur Scientist: Science as a hobby* (1942).

Thomasen

Thomasen's Snakelet *Halidesmus thomaseni* JG Nielsen, 1961

H B Thomasen collected the type in Karachi, Pakistan, while collecting live fishes for the Danmarks Akvarium (Charlottenlund); any fishes that did not survive were donated to the Zoological Museum (Copenhagen).

Thomass

Henry Sullivan Thomas (1833–fl1887), whose surname was inexplicably but consistently misspelt as Thomass by Francis Day. (See **Thomas, HS**)

Thomerson

Thorny Catfish sp. *Rhinodoras thomersoni* DC Taphorn & CG Lilyestrom, 1984

Dr Jamie Edward Thomerson (1935–2015) was Emeritus Professor of Biology at Southern Illinois University at Edwardsville, and Research Associate at the Field Museum. He was a National Aeronautics and Space Administration Pre-Doctoral Fellow at Tulane University, which awarded his PhD supervised by Royal D Suttkus. He published numerous articles on neotropical freshwater fish (1960–2005), and co-authored a book with David Greenfield: *Fishes of the Continental Waters of Belize* (1997). Donald Taphorn was one of the people who wrote Thomerson's obituary in *Copeia* when he died after a long battle with prostate cancer.

Thompson, AR

Rim-spine Searobin *Peristedion thompsoni* HW Fowler, 1952

Arthur R Thompson was the owner of a yacht, the 'Triton', which was used to dredge specimens – including the searobin holotype – off the coast of southern Florida.

Thompson, BA

Gumbo Darter *Etheostoma thompsoni* RD Suttkus, HL Bart Jr. & DA Etnier, 2012

Dr Bruce Alan Thompson (1946–2007) was an American fisheries biologist, ichthyologist and conservationist. He was interested in animals from an early age and studied entomology then ichthyology at Cornell University (1964–1968). He then attended Tulane where he studied under Suttkus (1968), later receiving his PhD (1977). He was not easy to overlook, dressing as loudly as he spoke, and one of his fellow students said of him "*...he learned* *to whisper in a sawmill!*" He then became Research Associate III in the Coastal Fisheries Institute at Louisiana State University, and after nearly 30 years at the institution, attained the rank of Associate Professor, Research. He devoted a great deal of time working with the World Health Organization on fisheries throughout the globe. He also gave presentations, many of them invited, in Greece, Cuba, South Africa, Mexico, France, UK and Australia. He was honoured "*…in recognition of his intense interest in the systematics and biology of darters.*" He died of kidney cancer just three weeks after diagnosis.

Thompson, DW

Poacher sp. *Freemanichthys thompsoni* DS Jordan & CH Gilbert, 1898

Sir D'Arcy Wentworth Thompson (1860–1948) was a Scottish mathematician and biologist. He was Professor of Natural History at University College, Dundee, noted for his translation (1910) of Aristotle's *History of Animals*. Jordan and Gilbert noted him as "*…the commissioner of Great Britain in the fur seal investigations in Bering Sea*" (1896–1897).

Thompson, JC

Goldspot Goby *Gnatholepis thompsoni* DS Jordan, 1904
Thompson's Snake-Eel *Muraenichthys thompsoni* DS Jordan & RE Richardson, 1908

Commander Joseph Cheesman 'Snake' Thompson (1874–1943) was a US Navy medical officer (retired 1929) and amateur herpetologist. His enemies knew him as 'Crazy Thompson', but to his friends he was 'Snake' because of his expertise in herpetology. A polymath, he had a depth of knowledge ranging from Asian religion to zoology. Columbia Medical School awarded his MD (1892) and after working as a physician, including a two-year stint in Yokohama Hospital, he joined the Navy (1897). As an assistant surgeon, he was posted to the US Marines China Relief Expedition (1900), which was sent to Peking to rescue foreigners and Chinese Christians from the Boxer Rebellion. He became full Ships Surgeon (1903). He spent two years (1909–1911) as a spy for the US army, travelling all over the Japanese Empire as he had grown up in Japan and spoke fluent Japanese. In the Philippines, he posed as a herpetologist studying coastal reptiles and amphibians but, in reality, charting invasion routes. He helped found the Zoological Society of San Diego and was Vice-President (1916) until called up (1917) as a neurosurgeon in a Navy Hospital. He became interested in Psychoanalysis and underwent it himself (1923), and was a member of the Washington Psychoanalytic Association (1924–1936). He became involved in archaeological explorations in Guam (1923). He also founded a cattery (1926), which he called 'Mau Tien', Cat Heaven, and at one time had 45 cats himself! On retirement from the Navy (1929) he moved to San Francisco where he later died of a heart attack. He collected the holotypes of the two fish named after him.

Thompson, JW

Thompson's Surgeonfish *Acanthurus thompsoni* HW Fowler, 1923
Thompson's Butterflyfish *Hemitaurichthys thompsoni* HW Fowler, 1923
Hawaiian Anthias *Pseudanthias thompsoni* HW Fowler, 1923

John W Thompson was an artist and modeller at the Bishop Museum, Honolulu (1901–1928). He developed a number of techniques which received approval from the US Fish

Commission, as his models were 'as good as any found in any museum'. At least one of these eponymous species was obtained in the market at Honolulu by him.

Thompson, WF

Rattail sp. *Coelorinchus thompsoni* CH Gilbert & CL Hubbs, 1920
Gold-eye Rockfish *Sebastes thompsoni* DS Jordan & CL Hubbs, 1925
Alaska Dreamer *Oneirodes thompsoni* LP Schultz, 1934
Bigeye Lanternfish *Protomyctophum thompsoni* WM Chapman, 1944
Notothen sp. *Patagonotothen thompsoni* AV Balushkin, 1993

William Francis Thompson (1888–1965) was a fishery biologist. He was the founding Director of the San Pedro laboratory of the California Fish and Game Commission (1917–1925). He was then founding Director of the International Fisheries Commission (1925–1937) (now International Pacific Halibut Commission) being recognised as an authority on ichthyology while at Stanford as DS Jordan's (q.v.) assistant. Stanford awarded his PhD (c.1930). He became Research Professor and Head of the Department of Fisheries (1930) at the University of Washington, Seattle, which is now the School of Aquatic & Fishery Sciences, but remained Director of the Commission. He was appointed Director of the newly formed International Pacific Salmon Fisheries Commission (1937–1943). Holding three posts, he resigned from the Halibut Commission. Later (1943) he resigned the Salmon Commission, becoming full-time Director of the School of Fisheries. He resigned his directorship of the School (1947) and was appointed Director and Research Professor of the newly created Fisheries Research Institute at the same University until retiring (1958).

[There is no etymology for the rockfish, but W.F. Thompson is the most likely candidate]

Thompson, Z

Deepwater Sculpin *Myoxocephalus thompsonii* CF Girard, 1851

Rev Zadock Thompson (1796–1856) was an Episcopalian priest, naturalist, geologist, geographer, historian and professor. He did not begin to study until, aged 12, he seriously wounded his foot with an axe and was unable to work on the family farm. For four years from the age of 17 he spent the winters teaching and the rest of the year studying, but a six-month illness led him to find another way of supporting himself by creating almanacs. He then (1820) enrolled in Vermont University, again teaching to support himself until graduating (1823), when he began writing his first history of Vermont. He began teaching at his alma mater (1825) where he helped create the natural history department and rose to professor. He taught school, edited magazines and was also ordained but did not receive recognition until late in life. He wrote a number of books, the most notable being: *History of the State of Vermont* (1833) and *History of the State of Vermont, Natural, Civil and Statistical* (1842). According to Charles Girard, the Rev. Thompson was an 'esteemed naturalist of Burlington, Vermont'. Thompson sent the undescribed sculpin species to Girard.

Thomson, CW

Scorpionfish sp. *Scorpaena thomsoni* A Günther, 1880
Pallid Sculpin *Cottunculus thomsonii* A Günther, 1882

Charles Wyville Thomson (1830–1882) was Professor of Natural History at Edinburgh University. He persuaded the Royal Society of London to ask the British Government to furnish a ship for a prolonged voyage of exploration across the globe. This resulted (December 1872) in the expedition by HMS 'Challenger', with Thomson serving as chief scientist, which made extensive collections. While Günther gives no etymology in either case, we believe the above is intended.

Thomson, K

Thomas' Armoured Catfish *Chaetostoma thomsoni* CT Regan, 1904

Kay Thomson, about whom we can find nothing beyond the fact that he collected the holotype in the Magdalena River basin, Colombia.

Thonner

Claroteid Catfish sp. *Chrysichthys thonneri* F Steindachner, 1912

Franz Thonner (1863–1928) was an Austrian taxonomist and botanist who was rich enough not to have to work for a living! He explored extensively in Europe and North Africa and (1896) made his first expedition to the Congo basin, collecting plants and making an ethnographic study of the region. He wrote: *Im afrikanischen Urwald. Meine Reise nach dem Kongo und der Mongalla im Jahre 1896* (1896).

Thor

Hatchetfish genus *Thorophos* AF Bruun, 1931
Goby genus *Thorogobius* PJ Miller, 1969
Thor's Scaldfish *Arnoglossus thori* HM Kyle, 1913
Silvery Pout *Gadiculus thori* EJ Schmidt, 1913

These genera and species are named after 'Thor', the first Danish research vessel specifically equipped for oceanic scientific research (named after the Norse god of thunder).

[NB. The cichlid genus *Thorichthys* has a totally different derivation, from the Greek θρώσκω (throsko) meaning 'to leap']

Thoreau

Dixie Chub *Semotilus thoreauianus* DS Jordan, 1877

Henry David Thoreau (1817–1862) was an author, poet, philosopher, abolitionist, surveyor, historian and naturalist. He is perhaps best known for his book *Walden* (1854), which reflects upon simple living in natural surroundings. His essay *Resistance to Civil Government* (1849) argues for civil disobedience to an unjust state. Jordan described him as "*…an excellent ichthyologist, one of the first to say a good word for the study of Cyprinidae*" as Thoreau wrote: "*…I am the wiser in respect to all knowledge, and the better qualified for all fortunes, for knowing that there is a minnow in the brook.*" As a pillar of American thinking, there is too much written on the man for us to elaborate further.

Thorpe

Thorpe's Unicornfish *Naso thorpei* MM Smith, 1966

Squaretail Kob *Argyrosomus thorpei* MM Smith, 1977
Bigeye Stumpnose *Rhabdosargus thorpei* MM Smith, 1979

Anthony R 'Tony' Thorpe was a South African lawyer and avid angler. He sent Smith three undescribed fish that he had discovered in KwaZulu Natal waters and supplied specimens and photographs.

Thosaporn

Palefin Threadfin Bream *Nemipterus thosaporni* BC Russell, 1991

Dr Thosaporn Wongratana is a Thai ichthyologist at the faculty of Science, Chulalongkorn University. He is an expert on engraulid fishes, and one of the authors of *Clupeoid Fishes of the World (suborder Clupeoidei): Chirocentridae* (1985). He was honoured in the name because he made his illustrations of *Nemipterus* species available to the author. (Also see **Wongratana**)

Thouin

Clubnose Guitarfish *Glaucostegus thouin* Anon. [Lacepède], 1798

André Thouin (1746–1824) was a French botanist and a colleague of Lacépède's at Le Jardin des Plantes (Muséum Nationale d'Histoire Naturelle), Paris. Thouin was honoured as he had helped secure a specimen of the guitarfish in Holland and transport it to France.

Thumberg

Brownspot Largemouth (cichlid) *Serranochromis thumbergi* FL de Castelnau, 1861

Carl Peter Thunberg (note spelling with an *n*, not an *m*) (1743–1828) was a Swedish naturalist who travelled through southern Africa (1772–1774) and is regarded as the founder of South African botany. The patronym is not identified by Castelnau, but probably honours Thunberg, despite the (apparent) spelling error.

Thurston

Bentfin Devil Ray *Mobula thurstoni* RE Lloyd, 1908
[Alt. Smoothtail Mobula]

Dr Edgar Thurston (1855–1935) was an ethnographer, natural historian and musicologist who qualified as a physician in England (1877). He was Superintendent of the Government Museum, Madras (Chennai), establishing the natural history and anthropology sections (1885–1910) during which time he gave access to Lloyd to study specimens. He returned to England (1910) and eventually settled in Cornwall, where he was a noted plant collector (1915–1926). He mainly published on ethnography but also wrote a book on the amphibians of southern India. A reptile is named after him.

Thys

West African Cichlid genus *Thysochromis* J Daget, 1988
Shellear sp. *Parakneria thysi* M Poll, 1965
Congo Cichlid sp. *Sargochromis thysi* M Poll, 1967
Squeaker Catfish sp. *Synodontis thysi* M Poll, 1971

African Barb sp. *Enteromius thysi* E Trewavas, 1974
African Rivuline sp. *Aphyosemion thysi* AC Radda & J-H Huber, 1978
Claroteid Catfish sp. *Chrysichthys thysi* LM Risch, 1985
Loach Catfish sp. *Doumea thysi* PH Skelton, 1989
Lake Bermin Cichlid sp. *Coptodon thysi* MLJ Stiassny, UK Schliewen & WJ Dominey, 1992
African lampeye sp. *Hylopanchax thysi* PHN Bragança, JR van der Zee, R Sonnenberg & E Vreven, 2020

Professor Dr Dirk Frans Elisabeth Thys van den Audenerde (b.1934) is a Belgian ichthyologist and Museum Director. He studied Agricultural Engineering (Applied Hydrobiology) at the University of Ghent (1951–1956) then did his National Service (1956–1958) as a Lieutenant in the Engineers Corps. He took his master's at the Free University of Brussels (1958–1959) during which time he had an assistant fellowship of NILKO (National Instituut voor Landbouwstudie van Kongo); undertaking specialised studies on African fishes in Brussels and Tervuren. A number of research jobs ensued and he joined Koninklijk Museum for Midden-Afrika (1962) as an attaché, rising to assistant (1963), and senior assistant (1965). He studied throughout this time and the University of Ghent awarded his PhD (1970). He became senior scientist (1970) and Head of the vertebrates section (1974), all under Poll's Directorship and studying tilapia systematics. He then became acting Director (1980) and applied for the Director's job as he thought it the only way to ensure decent funding for fish research. He then served as Director until retirement (1985–1999), and is now Director Emeritus. In his time, he more than quadrupled the African fish collection. He undertook many field expeditions to Senegal, Sierra Leone, Ivory Coast, Cameroon and Gabon collecting from a number of fish families and most often to Congo – much of this using an old and much repaired Land Rover which he bought with his own savings. During his time, he also lectured at the University as it ensured a steady stream of students at the museum. Poll, (q.v.) in naming the eponymous shellear, said that he collected in Katanga in: "*…difficult and even dangerous conditions*" (translation).

Tiantian

Burmese Bumblebee Barb *Pethia tiantian* SO Kullander & F Fang, 2005

Tiantian Kullander is the senior author's son. He, along with his brother 'Didi', "*…had to repeatedly suffer their parents' absence searching for these and other fish in faraway lands.*" (Also see **Kullander** & **Didi**)

Tiekoro

African Barb sp. *Enteromius tiekoroi* C Lévêque, G Teugels & DFE Thys van den Audenaerde, 1987

Tiekoro Sineogo is a local (West African) fisherman with whom the authors have worked since 1975.

Tiene

Tiene's Dwarfgoby *Eviota gunawanae* DW Greenfield, L Tonabene, MV Erdmann & DN Pada, 2019.

Dr Tiene Gunawan (See **Gunawan**)

Tihon

African Airbreathing Catfish sp. *Platyallabes tihoni* M Poll, 1944

L Tihon was Director, Laboratory of Industry and Commerce, Leopoldville (Kinshasa), Belgian Congo (Democratic Republic of the Congo). He supplied a photograph of a specimen that was discovered at a fish market and from which this species was described. With Poll he co-wrote: *Additional note on Stanley Pool fish* (1945).

Tikader

Stone Loach sp. *Aborichthys tikaderi* RP Barman, 1985

Dr Benoy Krishna Tikader (1928–1994) was an Indian (Bengali) zoologist and arachnologist; the leading expert of his time on Indian spiders. He held PhD and DSc degrees from Calcutta University. He worked for the Zoological Survey of India based in Kolkata. He wrote at least 48 papers and longer works, including: *Threatened Animals of India* (1983), *Handbook of Indian Spiders* (1987) and *Handbook, Indian Lizards* (1992). He was honoured for "…laboratory facilities and for his encouragement." At least seventeen spiders are also named after him.

Tilesius

Poacher genus *Tilesina* PJ Schmidt, 1903

Wilhelm Gottlieb Tilesius von Tilenau (1769–1857) was a German naturalist, physician and explorer. He was appointed (1803) professor at Moscow University, and participated as a ship's doctor, marine biologist and expedition artist on the frigate *Nadezhda* in the first Russian circumnavigation of the globe (1803–1806).

Tilston

Tilston's Whaler Shark *Carcharhinus tilstoni* GP Whitley, 1950
[Alt. Australian Blacktip Shark]

Richard Tilston was a naturalist who trained as a physician at Guy's Hospital, London (1841) and became a naval surgeon (1842). He was a Royal Navy Assistant Surgeon at Port Essington, Northern Territory, Australia. The type was collected near Port Essington during Tilston's time there.

Tilton

Plain Helmet Gurnard *Dactyloptena tiltoni* WN Eschmeyer, 1997

Thomas Tilton was a benefactor of ichthyological research undertaken by the Californian Academy of Sciences. The etymology says that the species was "*…named for Mr. Thomas Tilton, of San Francisco, in appreciation for the support of the Tilton family of the research activities at the California Academy of Sciences.*"

Timanoa

Sunrise Anthias *Pseudanthias timanoa* BC Victor, A Teitelbaum & JE Randall, 2020

Timothée, Maëlle & Noa Teitelbaum – 'Timanoa' is an amalgamation of parts of the names of the second author's three children: **Tim**othée, **Ma**ëlle, and **Noa**. Antoine Teitelbaum, is a

French agronomist. He obtained a diploma in tropical agronomics from ISTOM (France), doing most of his field research in small-scale fishery and rural-based aquaculture in Africa (Guinea, Malawi, Madagascar). He then graduated (MAppSc in Aquaculture and Marine Biology) from the James Cook University in Townsville, Australia, where he did research on pearl oysters, giant clams and mud crabs. He moved to Kiribati (2002–2004), back to France (2005) then the Marshall Islands (2005–2006). He worked (2006–2011) with SPC's Coastal Fisheries Programme as an aquaculture officer, specialised in mariculture and now works at the New Caledonia Aquarium. He collected the holotype in New Caledonia (2013).

Timle

Spotted Numbfish *Narcine timlei* ME Bloch & JG Schneider, 1801

The original description has no etymology but it seems that 'timle' derives from a Malay name for this fish, as the authors write: "*Hanc speciem Malaice* Pulli Timilei *i.e. rajam electricam maculatam vocari...*" ['This species is called by the Malays *pulli timilei*, i.e. spotted electric ray']

Tinant

Needle Fin-Eater *Belonophago tinanti* M Poll, 1939
Congo Cichlid sp. *Steatocranus tinanti* M Poll, 1939

Dr André Tinant (1901–1940), whose doctorate was awarded by the University of Louvain, was Secretary-General of the Palm-oil plantations of the Belgian Congo (Democratic Republic of the Congo) where he worked (1926–1930, 1932–1935 & 1935–1938). He also collected, including the holotypes of these species.

Ting

Golden-line Fish sp. *Sinocyclocheilus tingi* PW Fang, 1936

V K Ting (aka Ding Wenjiang) (1887–1936) was a renowned geologist, natural scientist, scholar, and the most eminent and competent organiser and administrator in China of his time. He was educated first in Japan (1902–1904) and then in the United Kingdom (1904–1911). He stayed in Tokyo learning Japanese and then went to Cambridge University (1904–1906) and Glasgow University (1907), studying zoology and geology. He received two bachelor's degrees from Glasgow (1911). He was Chief of the Section of Geology under the Ministry of Industry and Commerce of the Peking government of China (1913). He was then Director of the Geological Survey of China (1916–1921). He helped establish (1922) the Geological Society of China in Beijing, one of the earliest natural science organisations in China. He was its President (1923) and was re-elected (1929). He was Research Professor at Peking University (1931–1934). He travelled to the USA, Europe and the Soviet Union (1933). He was appointed (1934) Secretary General of the Central Academy of Sciences of China in Shanghai. The Ministry of Railways invited him (1935) to survey the Xiangtan coalmine in Hunan province to find coal for use by the Canton-Hankow Railway. He died while inspecting the Xiangtan coal mine; he lived in the coal mine and was poisoned by the old-fashioned coal stove in his bedroom. He was honoured "*...for his zeal in promoting the development of geological, paleontological and marine biological sciences in China.*"

Ting Ting

Moonstone Chromis *Chromis tingting* Y-K Tea, AC Gill & H Senou, 2019

Ting Ting is the mother of the first author, Yi-Kai Tea of the School of Life and Environmental Sciences, University of Sydney. This damselfish was named in her honour for her "*...unconditional love, support and encouragement.*"

Tinker

Tinker's Butterflyfish *Chaetodon tinkeri* LP Schultz, 1951
[Alt. Hawaiian Butterflyfish]

Spencer Wilkie Tinker (1909–1999) was a malacologist who became the second Director of the Waikiki Aquarium, Honolulu (1940–1973). He graduated from the University of Washington (1931) and went to the University of Hawaii (1932) as a teaching fellow in the zoology department. He served as a Captain in the US Army during WW2. He discovered the butterfish in Hawaii (1949). His major works include *Sharks and Rays of Hawaii* (1972) and *Fishes of Hawaii* (1978).

Tinkham, Captain ER

Darkspotted Scorpionfish *Sebastapistes tinkhami* HW Fowler, 1946
Wrasse sp. *Anapses tinkhami* HW Fowler, 1946
[Junior Synonym of *Anampses caeruleopunctatus*]

Captain Ernest R Tinkham of the United States Army Marines made a collection of 383 fish of 146 species or subspecies in the Ryukyu Islands off Japan (1945). He also collected other fish which were sent to Fowler from Saipan and the Marshall Islands. He did this, according to Fowler "*...in odd moments of relaxation from his duties with our armed forces.*" He is probably the same person who wrote *Iwo Jima* (1950).

Tinkham, Doctor ER

Loach sp. *Formosania tinkhami* AW Herre, 1934

Dr Ernest R Tinkham (1904–1987) was a Canadian entomologist. The University of Alberta awarded his BSc (1927), Montana State College his MSc in entomology (1928) and the University of Minnesota his PhD (1939). He began work (1928–1933) as an entomologist at the Pink Bollworm Lab, State Division of Entomology, Texas then went on to be an Instructor of Entomology at Lingnan University, Canton, China and Assistant Curator to its Natural History Museum (1933–1936). He was Science Artist at Bishop Museum, Hawaii (1936–1937), Head of the Department of Biology at Intermountain-Polytechnic Institute, Montana (1937–1938), Grasshopper Technician, Bureau of Entomology and Plant Quarantine, U.S. Department of Agriculture, North Dakota and Arizona (1938–1939), and Wildlife Biologist, State Game and Fishing Commission, Arizona (1941–1948). He thereafter (1948 until retiring) served as Manager and Entomologist at Coachella Valley Mosquito Abatement District. After his death, his insect collection was given to the California Academy of Sciences. He collected the loach type in China.

Tin Win

Spiny Eel sp. *Mastacembelus tinwini* R Britz, 2007
Gold-ring Danio *Brachydanio tinwini* SO Kullander & F Fang, 2009

U Tin Win (1944–2014) was a well-known Burmese aquarist who was Managing Director at the Hein Aquarium, Myanmar. He was previously a lecturer in the Chemistry Department of Yangon University and established the aquarium when he retired (1998). He had been a tropical fish hobbyist since childhood and kept and bred aquarium fish for three decades. After retiring from teaching he spent more than a decade collecting new species by exploring Myanmar's many rivers, lakes and streams and their varied habitats. He was honoured as a "*…dedicated aquarist, acknowledged collector, and exporter of aquarium fish from Myanmar.*"

Tirapare

San Borja Cichlid *Gymnogeophagus tiraparae* I González-Bergonzoni, M Loureiro & S Oviedo, 2009

María Luisa Tirapare was an 18th-century Guaraní woman who founded the now disappeared town of San Borja del Yí (close to the type locality), the last native town in Uruguayan land, where natives, fugitive African slaves, *gaúchos* (cowboys), and other outsiders lived together.

Tirbak

African Rivuline sp. *Aphyosemion tirbaki* J-H Huber, 1999

Peter J Tirbak (1932–2019) was a killifish hobbyist in Palo Alto, California. He was born in China to Russian refugees from the Bolshevik revolution. He was one of the discoverers of the species when in Gabon collecting fish for his aquariums.

Titan

Titan Pleco (catfish) *Panaque titan* NK Lujan, M Hidalgo & DJ Stewart, 2010

In Greek mythology, the Titans were a primeval race of immortal giants. Sometimes 'Titan' (singular) is used as a name for the sungod Helios.

Titcomb

Pencil Catfish sp. *Hatcheria titcombi* CH Eigenmann, 1917
[Junior Synonym of *Hatcheria macraei*]

John Wheelock Titcomb (1860–1932) was an American fish culturist. The Argentine government employed him to study possibilities of stocking fish for commercial or sports fisheries in Patagonia. The advice he gave was followed and the first salmonids were imported from the USA and introduced into Patagonian lakes. He wrote: *Aquatic Plants in Pond Culture* (1909).

Tittmann

Loosejaw (dragonfish) *Aristostomias tittmanni* WW Welsh, 1923

Otto Hilgard Tittmann (1850–1938) was a geographer, geodesist and astronomer. He joined the United States Coast and Geodetic Survey (1867) and was Superintendent of it (1900–1915) when he authorised use of the survey steamer 'Bache' for the South Atlantic expedition during which the holotype of this species was collected. He co-founded the National Geographic Society (1888) and was its president (1915–1919).

Tizard

Stone Loach sp. *Schistura tizardi* M Kottelat, 2000

Robert John Tizard (b.1971) is currently Senior Technical Advisor at the Wildlife Conservation Society (2005–2015) and also on the Board of Directors of Village Focus International (2001–2015). He was educated at Texas A&M University and graduated (1994) in geography and wildlife management. He previously worked on the Nam Ha Project at UNESCO (2004–2006), Sight Assessor at Fermata Inc (2003–2005) and as Lao Programme Coordinator at the World-Wide Fund for Nature (1997–2000). It was when at the latter that he helped pave the way for Kottelat's survey of Laotian fishes. The etymology of the loach says he was honoured: "*…in gratitude for his help at various stages.*" He has lived and worked in Asia since 1993. His main work is on the relationship between local people and protected areas in regard to land rights and natural resource management.

Tlaloc

Central American Killifish genus *Tlaloc* J Álvarez & J Carranza, 1951

Tlaloc was an Olmec rain god.

Tobije

Japanese Eagle Ray *Myliobatis tobijei* P Bleeker, 1854

This is not an eponym but based on a Japanese name; from *tobi*, the Japanese name for the Black Kite (*Milvus migrans*) and *jei* for ray, so literally translating as 'kite(-like) ray'.

Tobituka

Lead-hued Skate *Notoraja tobitukai* Y Hiyama, 1940

T Tobituka directed the trawling fishery survey that collected the type specimen.

Todd, A

African Rivuline sp. *Callopanchax toddi* HS Clausen, 1966

Alexander Todd was an entomologist who was Chief Technician at Fourah Bay College, University of Sierra Leone, Freetown. He discovered the type specimen on one of his numerous expeditions there on which he accompanied M Roloff of the Zoological Museum of the University of Copenhagen (which is where Clausen found it in their collection). A dragonfly is also named after him.

Todd, JL

Congo Cichlid sp. *Haplochromis toddi* GA Boulenger, 1905

Dr John Lancelot Todd (1876–1949) was a Canadian physician and parasitologist. McGill University awarded his BA (1898) and his MD (1900), after which he did lab work at the Royal Victoria Hospital. He then studied at the Liverpool School of Tropical Medicine in England. He joined their expeditions to The Gambia and Senegal (1902) where they studied diseases and treated patients, and the Congo Free State (1903–1905) studying sleeping sickness. He and his colleague JE Dutton both contracted the disease, with Dutton succumbing to it. He became (1907–1925) Associate Professor of Parasitology at McGill. He went on another LSTM expedition (1911) back to Gambia. He later worked for Canada's National Research Council then lived in England (1934–1939), returning when WW2 broke out. He helped collect the type on the Kasai River, on behalf of the Congo State Museum.

Togo

Tetra sp. *Hyphessobrycon togoi* AM Miquelarena & HL López, 2006

Carlos Togo was an Argentine ichthyologist, who, according to the etymology was: "*…a great expert and pioneer of ichthyofaunal research in pampasic lagoons.*" He co-wrote: *Piscicultura experimental* (1968)

Togoro

Armoured Catfish sp. *Pareiorhaphis togoroi* JC de Oliveira & OT Oyakawa, 2019

Eduado Shinji Togoro was an undergraduate student of biology at the Federal University of Juiz de Fora (1998–2001). He was honoured in recognition of: "*…his dedication and contribution to the knowledge of Serra da Mantiqueira fishes. Togoro collected, measured and studied hundreds of specimens of many species in the headwaters of four basins that originate at the Serra da Mantiqueira, in Minas Gerais state for his undergraduate dissertation.*"

Tokarev

Bigeye Notothen *Trematomus tokarevi* AP Andriashev, 1978

Dr Aleksey K Tokarev (1915–1957) served as ichthyologist on the First Soviet Antarctic Expedition. He died while returning from the Antarctic.

Tokranov

Tokranov's Lumpsucker *Microancathus tokranovi* OS Voskoboinikova, 2015

Alexei Mikhailovich Tokranov is an aquatic biologist at the Kamchatka Branch of Pacific Institute of Geography, Far East Branch of the Russian Academy of Sciences. Among his publications is: *Some biological features of rare and poorly-studied sculpins (Cottidae, Hemitripteridae, Psychrolutidae) in the Pacific Waters off the Northern Kuril Islands and Southeastern Kamchatka, Russian Federation* (2007). The etymology says: "*The species name was given in honor of the famous Russian ichthyologist, Far Eastern fish biology researcher Alexei Mikhailovich Tokranov.*"

Tokubee

Izu Catshark *Scyliorhinus tokubee* S Shirai, S Hagiwara & K Nakaya,1992

This species is named after the fishing boat and private lodge of Toshiyuki Iida, who captured the type specimens and is familiarly known as 'Tokubee-san' – 'Tokubee' being an old-fashioned male name in Japan.

Tolima

Armoured Catfish sp. *Ancistrus tolima* DC Taphorn Baechle, JW Ambruster, FA Villa-Navarro & CK Ray, 2013

Tolima refers firstly to Princess Yulima of the Pijao tribe, burned at the stake and martyred by the Spanish conquistadors, and secondly to Department of Tolima, Colombia, where the locality of the holotype is.

Tolmachov

Lake Yessey Char *Salvelinus tolmachoffi* LS Berg, 1926

Innokenty Petrovich Tolmachov (1872–1950) was a geographer, geologist and palaeontologist, who led an expedition to Khatanga, Russia, (1905–1906), and an expedition (1909–1910) by sledge along the coast from Kolyma to the Bering Strait. He wrote of the latter in: *Siberian passage : an explorer's search into the Russian Arctic* (1949). He collected the holotype of this species

Tolson

Sleeper sp. *Giuris tolsoni* P Bleeker, 1854

R P Tolson was a member of the trading house Anderson, Tolson & Co. in Batavia (now Jakarta). He discovered this species while exploring a coal mine near Meeuwenbaai (Seagulls Bay) in Java, Indonesia, and gave it to Bleeker.

Tom

Fighting Fish sp. *Betta tomi* PKL Ng & M Kottelat, 1994

Professor Dr 'Tom' Lam Toong Jin (b.1940), now Professor Emeritus, was head of the Department of Zoology, National University of Singapore where he spent his career (1969–present). He is currently (2018) Technical Advisor, Apollo Aquaculture Group, Singapore. The University of British Columbia awarded his BSc (1965) and PhD (1969). His reaearch foci are fish reproductive and larval physiology/endocrinology; fish larval nutrition & feed development; fish disease control and Aquaculture. Among his over 200 published papers are: *Application of endocrinology to fish culture* (1982) and *Hormones and egg/larval quality in fish* (1994). The etymology says the species was named after him because he: "*...has generously supported the authors' research over the years.*"

Tomas

Neotropical Rivuline sp. *Rivulus tomasi* FBM Vermeulen, S Valdesalici & JR García Gil, 2013

Joachim 'Charly' Tomas (1963–2012) was a German killifish hobbyist, aquarist, conservationist, adventurer and fish collector. He was also very interested in freshwater shrimps and poison-dart frogs. Well known in Germany and beyond as a hobbyist, but also an expert, he gave lectures and wrote articles and took photographs published in a number of magazines and scientific journals. He was the co-collector (with the senior author and others) of the species (2010) during an ichthyological survey of the middle drainage of the Orinoco at Tobogán da Selva, Venezuela. He died in a sporting accident two years later (25th February 2012). He also has a shrimp species, *Euryrhynchus tomasi*, named after him which he discovered (2006).

Tom Coon

Titicaca Pupfish sp. *Orestias tomcooni* LR Parenti, 1984

Dr Thomas G 'Tom' Coon is an ecologist, biologist, ichthyologist and limnologist who is Vice-Preident, Dean & Director of Oklahoma State University (2014–present). He was Director of Michigan State University Extension (2005–2014) and Professor in the Department of Fisheries and Wildlife there (1989–2014). He earned his BA in Biology from Luther College (1976), and his MSc (1978) and PhD in Ecology (1982) from the University of California-Davis. He was formerly Associate Dean for Graduate and International Programs in the College of Agriculture and Natural Resources and as Associate Department Chairperson and Acting Chairperson in the Department of Fisheries and Wildlife at MSU. He collected nearly 4,000 specimens (1979) of *Orestias* from the Titicaca basin. He sent a small sample of these to the Department of Ichthyology to the AMNH while carrying out a limnological study of the area under the auspices of the University of California, Davis.

Tomelleri

Gulf Chub *Macrhybopsis tomellerii* CR Gilbert & RL Mayden, 2017

Joseph Ralph Tomelleri (b.1958) is a biological illustrator, artist and scientist living in Leawood, Kansas. His BSc was awarded by Fort Hays State University (1980), as was his MSc in Biology (1984). He has illustrated many major publications such as *Fishes of the Central United States* (1990), *Peterson Field Guide to Freshwater Fishes of North America* (2011) and *Fishes of the Salish Sea: Puget Sound and the Straits of Georgia and Juan de Fuca (3-Volume Set)* (2019) among many others and many magazines. His principal research interest is native Mexican trout with the group Truchas Mexicanas. The authors said that his: "*…unsurpassed and meticulously rendered color illustrations of North American freshwater fishes have graced the pages of numerous scientific publications.*" He loves fly-fishing, and collecting and filming 16mm movie cameras and films.

Tominaga

African Silver Sweeper *Pempheris tominagai* K Koeda, T Yoshino, H Imai & K Tachinara, 2014

Yoshiaki Tominaga is a Japanese ichthyologist at the University Museum, University of Tokyo, who has made contributions to the taxonomy of sweepers (Pempheridae). Among his published papers is the co-written: *Posterior Extension of the Swimbladder in Percoid Fishes, with a Literature Survey of Other Teleosts* (1996).

Tomio

Snake-eel sp. *Ophichthus tomioi* JE McCosker, 2010

Dr Tomio Iwamoto (b.1939) is Curator Emeritus, Ichthyology, at the Institute for Biodiversity Science and Sustainability at California Academy of Sciences. UCLA awarded his BS (1961) and after spending six months in active duty as an Army reservist, he began working as a fishery biologist for the then US Bureau of Commercial Fisheries (now National Marine Fisheries Service) at the Exploratory Fishing & Gear Research Station in Pascagoula, Mississippi, and later at the field station on St Simons Island, Georgia. He then returned to graduate school at the University of Miami, Rosenstiel School of Marine and Atmospheric Science, which awarded his MSc and PhD. He spent a year at Oregon St University as a lecturer before starting employment with the California Academy of Sciences where he was Curator of Ichthyology for 39 years (1972–2011). He has been visiting researcher at a number of museums such as The Australian Museum (1993 & 2003). His principle research interest is the systematics of grenadiers, a group of more than 400 deep-sea fishes related to the cods. This began in the early 1960s while he was employed as a fishery biologist working in waters of the tropical western Atlantic. In pursuit of this interest, he has cruised on oceanographic and fishery vessels over most of the western tropical Atlantic, from Cape Hatteras throughout the Gulf of Mexico and Caribbean Sea to northern Brazil; the eastern tropical Atlantic off Angola; the Pacific coast of North America from the Bering Sea and the westernmost Aleutian Islands to southern California; to the Hawaiian Islands, the Philippines, New Caledonia, and the Norfolk Ridge and Lord Howe Rise north of New Zealand; He also undertook a study of the marine fishes of West Africa, collecting there aboard the Norwegian fishery research vessel 'Dr Fridtjof Nansen' (2005, 2007, 2010 & 2012). Among his most important publications is: *Grenadiers (families Bathygadidae and Macrouridae, Gadiformes, Pisces) of New South Wales, Australia* (2001). The etymology says that he is a friend and colleague of the author who captured the holotype.

Tomiyama

Rattail sp. *Nezumia tomiyamai* O Okamura, 1963

Ichiro Tomiyama of Tokyo University was honoured for his 'kindness' in letting Okamura examine his 'precious' collection. He was an adviser to the Japanese Emperor Akihito.

Tomoda

Japanese Catfish sp. *Silurus tomodai* Y Hibino & R Tabata, 2018

Dr Yoshio Tomoda works at the Department of Zoology at the National Science Museum, Tokyo. The etymology says: *The scientific name* tomodai *honors Dr. Yoshio Tomoda, the author of "Two New Catfishes of the Genus* Parasilurus *found in Lake Biwa-ko" (1961).*

Tompkins

Snailfish sp. *Paraliparis tompkinsae* AP Andriashev, 1992

Dr Linda S Tompkins is a molecular biologist who has been associated with the University of Arizona and Texas Tech University. Among her publications are the co-written: *New Species and New Records of Rare Antarctic Paraliparis Fishes (Scorpaeniformes: Liparididae)*

(1988) and *The Odonata of Washington County, Pennsylvania* (1979), along with many on bio-chemistry.

Tongareva

Tongareva Goby *Cabillus tongarevae* HW Fowler, 1927

This is a toponym referring to Tongareva (also called Penrhyn Island), northern Cook Islands, South Pacific, the type locality.

Tongiorgi

Stone Loach sp. *Oxynoemacheilus tongiorgii* TT Nalbant & PG Bianco, 1998

Paolo Tongiorgi (1936–2018) was an Italian marine biologist, taxonomist and zoologist. The University of Pisa awarded his degree in natural sciences (1959). He was a Fullbright scholar at the Museum of Comparative Zoology, Harvard University (1963–1964). He began his career as Assistant Lecturer at the Institute of Zoology, University of Bari (1960–1961), was then Assistant Professor of Zoology (1961) there, and took up the same post at Pisa University (1962–1965). He was Professor of Zoology at the University of Modena (1977–1999). Finally, he was Professor of Animal Biology Reggio Emilia Faculty of Agriculture (1999). He was also co-editor of the 'Italian Journal of Zoology'. He published at least 25 papers and longer works. He was honoured as the author's friend and for: "*…his help in the final editing of the special volume in which this description appeared*."

Tonozuka

Tono's Fairy Wrasse *Cirrhilabrus tonozukai* GR Allen & RH Kuiter, 1999
Spot-stripe Dottyback *Pseudochromis tonozukai* AC Gill & GR Allen, 2004

Takamasa 'Tono' Tonozuka was a Japanese underwater photographer who was a long-term resident of Bali and co-owner of 'Dive & Dive'. He co-wrote: *Pictorial Guide to Indonesian Reef Fishes* (2001), and wrote (illustrated with his sea-slug photographs) *Opisthobranchs of Bali and Indonesia* (2003). The dottyback etymology says that this species was named in his honour as it was Tonozuka "*…who discovered this species and brought it to the attention of the second author*." Tono and his wife Miki often assisted Allen collecting specimen fish on a number of occasions. (Also see **Miki**)

Toppin

East Coast Barb *Enteromius toppini* GA Boulenger, 1916

Lieutenant Fred Toppin (1878–1918) was an English professional collector of natural history specimens who was hired by the Natal Museum, South Africa (1905–1908) under the instruction to collect everything from snails to large mammals. He visited remote parts of Zululand, such as the junction of the two Umfolozi Rivers, Dukuduku Forest, and Kosi Bay. In correspondence with the museum at the time he wrote that he found the place 'dreadfully lonely' and also uncomfortable, saying: 'there are millions of ticks, you get nearly eaten to death in the bush'. He was also bitten by a snake, which objected to being 'collected' for the Museum. (Ironically, Toppin had already informed the museum Director that '*…this is a good locality for Reptilia*'.) The letters, particularly after the snakebite episode, convey the distinct impression that Toppin was eager to get out of the Umfolozi Game Reserve as quickly as possible, with or without his specimens. He applied (1908) to the Natal government for 500 acres (c.200 ha) of land anywhere between the Hluhluwe and Mzineni Rivers, in the Hlabisa district of Zululand. Five years later he asked for a site on which to erect a store on crown lands in the nearby Ubombo Division. He died of pneumonia and influenza whilst on active service with the South African Signal service during WW1. He collected the barb type.

Torgashev

African Killifish sp. *Nothobranchius torgashevi* S Valdesalici, 2015

Sergey Torgashev, Elektrostal, Moscow Oblast, Russia, is a collector and aquarist who discovered the species in Tanzania.

Tornier

African Cyprinid sp. *Labeobarbus tornieri* F Steindachner, 1906
Yellowtail Rasbora *Rasbora tornieri* E Ahl, 1922

Dr Gustav Tornier (1858–1938) was a German zoologist, anatomist, palaeontologist and taxonomist. He received his doctorate from Heidelberg University (1892). He worked at the zoological museum of the Friedrich-Wilhelms University, Berlin. He is best remembered (rather unfairly) for supporting the incorrect view that *Diplodocus* and other sauropod dinosaurs walked with a sprawling, lizard-like gait with widely splayed legs. Finds of sauropod footprints in the 1930s put Tornier's theory to rest. Three reptiles and six amphibians are also named after him.

Torre

Dwarf Catshark *Scyliorhinus torrei* L Howell-Rivero, 1936

Professor Carlos de la Torre y la Huerta (1858–1950) of Havana University was a malacologist regarded as the foremost Cuban naturalist of his generation. He was closely associated with the Smithsonian in Washington; DC before Castro took power. He was a leading figure in the Academia de Ciencias Medicas, Fisicas y Naturales de la Habana. It was he who first recognised this shark as a new taxon and granted Howell Rivero permission to study and describe it. An extinct Cuban mammal, a bird and a reptile are also named after him.

Torres

Characin sp. *Gephyrocharax torresi* JA Vanegas-Ríos, M de las Mercedes Azpelicueta, JM Mirande & MD Garcia Gonzàles, 2013

Mauricio Torres-Mejia is an ichthyologist and ecologist at Universidad Industrial de Santander, Colombia. He did graduate studies at the University of California, Riverside (2012). He co-wrote: *How does diet influence the reproductive seasonality of tropical freshwater fish? – A case study of a characin in a tropical mountain river* (2009). He collected the holotype of this species.

Tortonése

Tortonese's Goby *Pomatoschistus tortonesei* PJ Miller, 1969
Tortonese's Stingray *Dasyatis tortonesei* C Capapé, 1975

Dr Enrico Tortonése (1911–1987) was an Italian ichthyologist and specialist in echinoderms. He was a Professor at the Genoa Museum of Natural History. Capapé frequently cites Tortonése's work on Mediterranean sharks and rays.

Tosh

Brown Whipray *Maculabatis toshi* GP Whitley, 1939
[Alt. Black-spotted Whipray]

Dr James Ramsey Tosh (1872–1917) was a Scottish marine biologist and inland waterway engineer who was employed by the government of Queensland as a fisheries expert (1900–1903). The University of St Andrews, where he was employed as an assistant to the Professor of Natural History, awarded his master's degree (1894). He was working for the British Red Cross Society in Mesopotamia (Iraq) when he succumbed and died of heat stroke. He mentions the whipray in a report he wrote (1902–1903).

Toshiyuki

Toshiyuki's Dwarfgoby *Eviota toshiyuki* WD Greenfield & JE Randall, 2010

Toshiyuki Suzuki is a Japanese ichthyologist who teaches at Kawanishi-midoridai Senior High School. He is Chairman of the Investigation Committee of Rare Marine Fish, of the Ichthyological Society of Japan. His publications include: *Revision of the genus Acanthaphritis (Percophidae) with the description of a new species* (1996) and *Two new dwarfgobies (Teleostei: Gobiidae) from the Ryukyu Islands, Japan: Eviota flavipinnata and Eviota rubrimaculata* (2015). He collected the type and provided photographs of other *Eviota* species to the authors. They chose 'toshiyuki' over 'suzuki' because the latter is a common surname in Japan.

Tota

Characin sp. *Astyanax totae* C Ferreira Haluch & V Abilhoa, 2005

Adelinyr 'Tota' Azevedo de Moura Cordeira is a Brazilian biologist. She was honoured for her contributions to the fish collection at Museu de História Natural Capão, Parques de Curitiba, Paraná, Brazil. She co-wrote: *Catálogo dos peixes marinhos da coleçao dadivisão de zoologia e geologia da Prefeitura municipal de Curitaba* (1986).

Touré

Upside-down Catfish sp. *Synodontis tourei* J Daget, 1962

Moussa Touré was a forestry inspector in Mamou, Republic of Guinea, and later was National Inspector of Water and Forest, Conakry, Guinea.

Tournier

West African Cichlid sp. *Sarotherodon tournieri* J Daget, 1965

Jean-Luc Tournier (1907–1985) was a French pharmacist based in Saigon who later was an ethnographer during a mission to collect medicinal plants in French West Africa. He was the founder and Director of the Centre IFAN – *Institut Français d'Afrique Noire* – (1942) that became the Musée d'Abidjan, Côte d'Ivoire (1947) and another IFAN base in Ziéla. *He was also a notable collector of sculpture from the Ivory Coast.* His collection was bequeathed to the Natural History Museum of Besançon. An amphibian and a dragonfly are also named after him.

Tower

Bluetail Goodea *Ataeniobius toweri* SE Meek, 1904
[Alt. Striped Goodeid, Bluetail Goodeid]

Dr William L Tower (1872–1955) was an evolutionary biologist who was Professor of Zoology at the University of Chicago. He was educated at Harvard and Chicago. His studies on the evolution of chrysomelid beetles and the development of colour in insects are seminal. He discovered this species in San Luis Potosi, Mexico.

Townsend, CH

Dogtooth Lampfish *Ceratoscopelus townsendi* CH Eigenmann & RS Eigenmann, 1889
Belted Cardinalfish *Apogon townsendi* CM Breder Jr, 1927

Charles Haskins Townsend (1859–1944) was an American zoologist who worked for the US Fish Commission, and later became Director of New York Aquarium. He explored northern California (1883–1884) and the Kobuk River, Alaska (1885). He wrote *Field notes on the mammals, birds, and reptiles of northern California* (1887). He was the naturalist on at least one cruise of the US Fish Commission vessel 'Albatross', from which the holotype of the lampfish was collected. A mammal, ten birds and two reptiles are also named after him.

Townsend, FW

Townsend's Anthias *Pseudanthias townsendi* GA Boulenger, 1897
Scaly Dwarf Monocle Bream *Parascolopsis townsendi* GA Boulenger, 1901
Townsend's Fang-Blenny *Plagiotremus townsendi* CT Regan, 1905
Duncker's Pipehorse *Festucalex townsendi* PGE Duncker, 1915
[Alt. Red-and-gold Pipehorse]

Frederick William Townsend (1887–1948) was Captain of the Indian cable-ship 'Patrick Stewart', which he used for dredging as well as repairing telegraph cables in the Indian Ocean and Persian Gulf. He presented many fish specimens he found to the British Museum.

Townsend, JK

Townsend's Cusk Eel *Brotula townsendi* HW Fowler, 1900

John Kirk Townsend (1809–1851) was an American naturalist, ornithologist and collector. He trained as a physician and pharmacist but developed an interest in natural history. He was invited (1833) to join botanist Thomas Nuttall (q.v.) on an expedition across the Rocky Mountains to the Pacific Ocean. He later made two visits to the Hawaiian Islands (1835 and 1837). Audubon (q.v.), for his: *Birds of America* and *Viviparous Quadrupeds*, used

Townsend's bird and mammal specimens. Ironically Townsend died of arsenic poisoning, the 'secret' ingredient of the powder he formulated to use in taxidermy. Seven mammals are named after him. He collected the holotype of this species (1834).

Towoeti

Towoeti Halfbeak *Nomorhamphus towoetii* W Ladiges, 1972

This is a toponym referring to Lake Towoeti, Sulawesi, Indonesia.

Toyoshio

Jawfish sp. *Stalix toyoshio* G Shinohara, 1999

Named for the research vessel *Toyoshio-maru*, from which the type was collected.

Tracey

Tracey's Demoiselle *Chrysiptera traceyi* LP Woods & LP Schultz, 1960

Dr Joshua Irving Tracey Jr (1915–2004) was an ocean geologist at the US Geological Survey who specialised in the geological history of the Pacific Ocean. All three of his degrees were from Yale University; his BSc in physics & mathematics (1937), MSc in geology (1943) and PhD in geology (1950). He started his career at the outbreak of WW2 in Alabama, Georgia and Arkansas exploring for Bauxite. He worked at Bikini Atoll core drilling before and after the nuclear tests. He was then party chief in mapping the geology of Guam (1951–1954), on Midway during the 1960s, then at the National Science Foundation and Scripps Institution of Oceanography. In the 1970s he worked as the co-chief of the Glomar Challenger deep-sea drilling initiative in the Pacific. He retired (1985) but continued his research at the Smithsonian. Among his publications is: *Natural History of Ifaluk Atoll* (1961).

Traill

Sandpaper Fish *Paratrachichthys trailli* FW Hutton, 1875
[Alt. Common Roughy]

Charles Traill (1826–1891) was an amateur botanist and conchologist who was Postmaster of Stewart Island, New Zealand. He was born in Orkney, educated for two years at the University of Edinburgh, and was apprenticed to a lawyer. He emigrated to Australia (1849) to be a sheep farmer but left (1850) for the California gold fields. He eventually arrived in New Zealand (1856). During a visit to Stewart Island he discovered oyster beds and settled there (1871) to trade as a fish curer. The business was not a great success but he had land at Ulva and stayed on as a store keeper and postmaster. He also added to the charms of Ulva by importing song birds from England - nowadays not regarded as a good ecological move! He found the fish holotype "*...dead and floating on the surface of the water*" and presented it to the Otago Museum.

Traoré

African Barb sp. *Enteromius traorei* C Lévêque, GG Teugels & DFE Thys van den Audenaerde, 1987

Dr Kassoum El Hadj Traoré (d.2012) was a research hydrobiologist who was Secretary General of the Centre National de Recherche Agronomique, Ivory Coast. He was at the l'Institut d'Écologie tropicale d'Abidjan, Ivory Coast, where the species are endemic. He wrote: *Evaluation, après six années de monotoring [sic] (1980–1985), de l'impact des insecticides antisimulidiens sur l'ichtyofaune des cours d'eau ivoiriens traités dans le cadre de la lutte contre l'onchocercose* (1986). He is a friend and colleague of the author.

Trask

Tule Perch *Hysterocarpus traskii* WP Gibbons, 1854

John Boadman Trask (1824–1879) was a physician, amateur geologist, and founding member (1853) of the California Academy of Sciences, who was a friend of the author (who was also a founding member). He went to California (1850) as part of the US & Mexico Boundary Survey and afterwards became the first state geologist there. He wrote the report: Geology of the Sierra Nevada or California Range (1853) and Geology of the Sierra or California Ranges (1855) among others. He has been described as a 'pioneer of science on the west coast'. He obtained the types "*...through the kindness of Mr. Morris, from the fresh water lagoons of the Sacramento river, and from the river [itself], where they are found as high up as the fishermen have yet been.*"

Traude

African Rivuline sp. *Fundulopanchax traudeae* AC Radda, 1971

Traude Radda is the author's wife.

Trautman

Scioto Madtom *Noturus trautmani* WR Taylor, 1969
Gravel Chub ssp. *Erimystax x-punctatus trautmani* CL Hubbs & WR Crowe, 1956

Dr Milton Bernhard Trautman (1899–1991) was a self-taught ornithologist and ichthyologist from Ohio. He worked for the State of Ohio Department of Fish and Game (1926–1934) and was Assistant Curator of Fishes at the Museum of Zoology, University of Michigan and Assistant Director and Research Biologist for the Michigan Department of Conservation (1934–1939). He was a research biologist at the Franz Theodore Stone Laboratory of Ohio State University (1939–1955), Curator of Vertebrates at the Ohio State Museum of the Ohio Historical Society (1955–1970) and Curator of Birds at the Ohio State University Museum of Zoology (1970–1991). Lacking formal education, he received an honorary PhD from the College of Wooster (1951). He wrote: *The Fishes of Ohio* (1957). He was honoured in the name of the chub for: "*...his life-long thorough investigation of the fishes of Ohio.*"

Travassos, HP

Armoured Catfish sp. *Otothyris travassosi* JC Garavello, HA Britski & SA Schaefer, 1998
Characin sp. *Characidium travassosi* MRS Melo, PA Buckup & OT Oyakawa, 2016

Dr Haroldo Pereira Travassos (1922–1977) was an ichthyologist, a zoologist and a physician at the National Museum of Brazil (1942–1977), becoming Titular Professor of the Federal

University of Rio de Janeiro (1969). He graduated at the National School of Veterinary Medicine, Universidade Rural (1944), and as a physician at the Faculty of Medicine of the University of Brazil (1945). He was honoured for his contributions to the taxonomy of the genus *Characidium*. His father was Lauro Pereira Travassos (below).

Travassos, LP

Pencil Catfish sp. *Trichomycterus travassosi* P de Miranda Ribeiro, 1949
Characin sp. *Brycon travassosi* A Amaral Campos, 1950
[Junior Synonym of *Brycon orbignyanus*]

Dr Lauro Pereira Travassos Filho (1890–1970) was a Brazilian physician, zoologist, helminthologist and entomologist who travelled widely in Brazil. He studied medicine in Rio, graduating from the Faculty of Medicine (1913) although his doctoral thesis was entitled: *About the Brazilian species of the subfamily Heterakinoe*. He worked briefly at the Institute of Experimental Medicine but went on to take the Chair of Parasitology, Faculty of Medicine of São Paulo (1926). He lectured in Europe and took various academic positions in Brazil, establishing himself as a teacher but also as a researcher with many fieldtrips and describing hundreds of new species. He was Santos' teacher and introduced him to entomology. A school in São Paulo is also named after him. He was still dedicated to scientific endeavour and never stopped working until the day he died. He collected the holotypes of these species. His son was Dr Haroldo Pereira Travassos (above).

Travers, RA

Spiny Eel sp. *Mastacembelus traversi* EJ Vreven & GG Teugels, 1997

Dr Robert A Travers was employed in the Fish Section, BMNH. He wrote an obituary of Teugels published in *Cybium* in 2004. Among his many published papers are: *A review of the Mastacembeloidei, a suborder of synbranchiform telcostfishes* (1984) and *Diagnosis af a new African mastacembalid spiny eel genus Aethiomastacembalus gen. nov. (Mastacembeloidei: Synbranchiformes)* (1988). He was the first person to tentatively recognise part of the type series of the eel as belonging to a new species.

Travers, WT

King Dory *Cyttus traversi* FW Hutton, 1872

William Thomas Locke Travers (1819–1903) was a New Zealand explorer, lawyer, politician and naturalist. He was born in Ireland, educated in France, and joined the British Foreign Legion (1835). He then studied law in London, married, and emigrated to New Zealand (1849) where he served as resident magistrate in Nelson. He also became involved in local politics, as well as taking an interest in natural history. Among his successes was drafting the 1869 Act that established the Wellington Botanic Garden. He died as a result of an accident at Hutt railway station.

Trecul

Guadalupe Bass *Micropterus treculii* LL Vaillant & F Bocourt, 1874

Dr Auguste Adolphe Lucien Trécul (1818–1896) was a French botanist who visited North America (1848–1850) on a scientific mission. His unpublished reports are in the archives of the Museum of Natural History, Paris. We assume he was the collector of the fish.

Trefaut

Pencil Catfish sp. *Trichomycterus trefauti* WB Wosiacki, 2004

Dr Miguel Trefaut Urbano Rodrigues (b.1953). (See **Rodrigues**)

Treitl

Long-snout Corydoras *Corydoras treitlii* F Steindachner, 1906

Josef Treitl (1804–1895) was an Austrian bank and hospital director. He left a large sum of money to the Austrian Academy of Sciences, where Steindachner worked. This species is probably named after him, though Steindachner's text does not enlighten.

Trevelyan

Border Barb *Amatolacypris trevelyani* A Günther, 1877
Trevelyan's Char *Salvelinus trevelyani* CT Regan, 1908

Herbert Trevelyan (1847–1912) joined the army (1867) as a Lieutenant in the 32nd Regiment of Foot, which served in South Africa (1869–1877). He later served in the 16th Regiment of Foot as a Captain, transferred to the Royal Inniskilling Fusiliers (1881), served in Hong Kong and retired from the army as a major (1887). He regularly sent specimens, including the holotype of the barb, to the BMNH in London from South Africa and is honoured also in the names of a mammal and a reptile.

Trewavas

Deepsea Dragonfish *Eustomias trewavasae* JR Norman, 1930
Sisorid Catfish sp. *Glyptothorax trewavasae* SL Hora, 1938
Lake Tanganyika Threadfin Cichlid *Petrochromis trewavasae* M Poll, 1948
Trewavas' Tonguefish *Symphurus trewavasae* P Chabanaud, 1948
Garra sp. *Garra trewavasai* T Monod, 1950
Trewavas' Cichlid *Labeotropheus trewavasae* G Fryer, 1956
[Alt. Scrapermouth Mbuna]
African Characin sp. *Neolebias trewavasae* M Poll & J-P Gosse, 1963
Croaker sp. *Atrobucca trewavasae* PK Talwar & R Sathiarajan, 1975
New Grenada Drum *Protosciaena trewavasae* LN Chao & RV Miller, 1975
Leftvent sp. *Linophryne trewavasae* E Bertelsen, 1978
Priapiumfish sp. *Phenacostethus trewavasae* LR Parenti, 1986
Cameroon Cichlid sp. *Tylochromis trewavasae* MLJ Stiassny, 1989
Stone Loach sp. *Triplophysa trewavasae* MR Mirza & S Ahmad, 1990
Trewavas' Croaker *Johnius trewavasae* K Sasaki, 1992
Conger Eel sp. *Rhynchoconger trewavasae* A Ben-Tuvia, 1993
Lake Malawi Cichlid sp. *Copadichromis trewavasae* A Konings, 1999
Lake Malawi Cichlid sp. *Placidochromis trewavasae* M Hanssens, 2004

Dr Ethelwynn Trewavas (1900–1993) was an eminent British ichthyologist. She worked at the British Museum of Natural History where she was particularly known for her work on the *Cichlidae*. Reading University awarded her bachelor's degree and teaching certificate (1921). She worked as a teacher and then a demonstrator at King's College while continuing with her research. She began her employment at the museum (1935) as Assistant Keeper, becoming Deputy Keeper of Zoology (1958) until her retirement (1961). Her honorary doctorate was awarded by Stirling University (1986). Perhaps her most important work was the 583-page: *Tilapiine Fishes of the Genera "Sarotherodon", "Oreochromis" and "Danakilia"* (1983). She was honoured: "*...in appreciation of her kindness and outstanding contribution to systematics of fishes.*" (See also **E T** & **Ethelwynne**)

Tricia

Tricia's Garden Eel *Heteroconger tricia* PHJ Castle & JE Randall, 1999

Dr Patricia 'Tricia' J Kailola (See **Kailola** & also **Patricia (Kailoia)**)

Tricot

Lake Tanganyika Cichlid sp. *Benthochromis tricoti* M Poll, 1948

M (probably Monsieur) Tricot, Director of the Great Lakes Railroad Company (Albertville), was honoured for his interest in and concern for the Belgian Hydrobiological Mission to Lake Tanganyika (1946–1947), during which the type was collected.

Tristram

Israeli Cichlid genus *Tristramella* E Trewavas, 1942

The Reverend Henry Baker Tristram (1822–1906) was Canon of Durham Cathedral and a traveller, archaeologist, naturalist, antiquarian, and early supporter of Darwinism, who assembled a large collection of birds. He wrote a number of accounts of his explorations, including *The Great Sahara: Wanderings South of the Atlas Mountains* (1860), which he undertook in the company of his friend Upcher and *A Journal of Travels in Palestine* (1865) as he had collected cichlids in the Palestinian region for the British Museum. In his Sahara book he describes how he penetrated far into the desert and made an ornithological collection in the course of gathering material for his work. He writes interestingly on the indigenous peoples and their customs, as well as on the natural history of the region. He originally went there because of ill health. Despite his early penchant for collecting with a gun, Tristram went on to be a Vice-President of the RSPB (1904–1906). Fourteen birds, a mammal and a reptile are also named after him.

Trnski

Clingfish sp. *Dellichthys trnskii* KW Conway, AL Stewart & AP Summers, 2018

Dr Thomas 'Tom' Trnski (b.1963) is a German-born (to Croatian parents) marine biologist, now Head of Natural Sciences at Auckland Museum Tāmaki Paenga Hira, New Zealand. The University of Technology Sydney awarded his PhD (2003). He worked for the Australian Museum, Sydney (1985–2007), before moving to the Auckland Museum. He has led expeditions to document the biodiversity of New Zealand's outer islands and the wider South Pacific. Among his publications are the co-written *The Larvae of Indo-Pacific Shorefishes* (1989) and *Fishes from recent collections at the Kermadec Islands and new records for the region* (2015).

Troll

Pointy-nosed Blue Chimaera *Hydrolagus trolli* DA Didier & B Séret, 2002

Dr Raymond 'Ray' Michael Troll (b.1954) is an American artist (he calls himself a "fin artist"), musician, humourist and illustrator. He studied art at Bethany College, Lindsborg where he was awarded a BA in printmaking (1977). He did various day jobs financing his artwork, then (1981) went to Washington State University to do his master's degree (1981) where he focussed on his drawing skills. When he left, he took a teaching post back at Bethany College. He moved to Alaska (1983) and worked in his sister's retail fish market and also taught an art class and rented a studio. He began angling and drew the fish he caught, including a ratfish, which became his talisman. He started printing T-shirts (1984) with fishy themes and earns a good living from it. He put together a traveling show: *Dancing to the Fossil Record*, blending science and art (including the infamous 'Evolvo' art car). It ran from San Francisco (1995) and ended in Denver, Colorado (1999) and has since then been involved with other shows including acting as art director for the Miami Museum of Science's Amazon Voyage traveling exhibit. He has a music radio show in Alaska where he lives, and is a self-confessed rock & roll fanatic. He has had various bands including: 'The Squawking Fish', 'Zulu and the Robot Slave Boys' and the 'Rapping Ratfish Brothers', and now plays in 'The Ratfish Wranglers'. His *nom de soirée* is 'Ratfish Ray' and his personalised number plate 'RATFSH'. His pictures inspired Didier to name the chimaera after him, saying: "*It's kind of nice to name a species for someone, and I thought: Here's a chance to name one for someone who's really interested. It kind of looks like him, but with facial hair.*" He was awarded the Alaska Governor's award for the arts (2006), a gold medal for 'distinction in the natural history arts' by the Academy of Natural Sciences, Philadelphia (2007), and an honorary doctorate in fine arts by the University of Alaska Southeast (2008). Among his publications is the children's book: *Sharkabet, a Sea of Sharks from A to Z*.

Troschel

Troschel's Parrotfish *Chlorurus troschelii* P Bleeker, 1853
Glass Knifefish sp. *Rhabdolichops troscheli* JJ Kaup, 1856
Toothless Characin sp. *Pseudocurimata troschelii* A Günther, 1860
Panama Sergeant Major *Abudefduf troschelii* TN Gill, 1862
Chili Sea Catfish *Notarius troschelii* TN Gill, 1863
Mexican Cichlid sp. *Mayaheros troschelii* F Steindachner, 1867

Dr Franz Hermann 'Fritz' Troschel (1810–1882) was a German zoologist, malacologist, herpetologist, and ichthyologist. Universität Berlin awarded his doctorate (1834). He was Assistant to Lichtenstein at the Museum für Naturkunde, Humboldt-Universität, Berlin (1840–1849), and became Professor of Zoology, Friedrich-Wilhelms-Universität Bonn (1850). He wrote: *Über die Bedeutsamkeit of the naturgeschichtlichen Unterrichts* (1845) and co-wrote the three-volume: *Horae ichthyologicae* (1845–1849). A number of other taxa, including an amphibian and a reptile, are named after him.

Trow

Sea Catfish sp. *Galeichthys trowi* C Kulongowski, 2010
[Binomial sometimes amended to the plural *troworum*]

Eugene Trow Jr was a student studying the biology of *Galeichthys* in South Africa. He recognised the probable distinctiveness of this species. The etymology also remembers his father, Eugene Trow, Sr., who collected much of the type series.

Trunov

Snailfish sp. *Paraliparis trunovi* AP Andriashev, 1986
Trunov's Southern Cetomimid *Notocetichthys trunovi* AV Balushkin, VV Fedorov & JR Paxton, 1989
Rattail sp. *Coelorinchus trunovi* T Iwamoto & E Anderson, 1994
Barreleye sp. *Dolichopteryx trunovi* NV Parin, 2005
Eel Cod sp. *Muraenolepis trunovi* AV Balushkin & VP Prirodina, 2006
Deepwater Cardinalfish sp. *Epigonus trunovi* NV Parin & AM Prokofiev, 2012
Silver Roughy ssp. *Hoplostethus mediterraneus trunovi* AN Kotlyar, 1986
Dwarf Codling ssp. *Notophycis marginata trunovi* YI Sazonov, 2001

Dr Ivan Andreevich Trunov (1936–2005) was a Russian ichthyologist who for his entire career was at the Atlantic Research Institute of Fisheries and Oceanography and was finally a Senior Researcher there. Kharkov State University, Ukraine, awarded his bachelor's degree (1963) and his doctorate (1974). He took part in 15 major Atlantic Ocean expeditions over the period 1964–1994.

Trybom

Swedish Whitefish sp. *Coregonus trybomi* G Svärdson, 1979

Dr Arvid Filip Trybom (1850–1913) was a Swedish zoologist, entomologist and fisheries biologist who studied at Uppsala (1870), which awarded him an honorary doctorate (1907). He had administrative jobs in the fisheries industry, including being Fisheries Inspector of the Danish Agricultural Agency (1903), Secretary and Treasurer, Swedish Fisheries Association (1897–1901) and Fishery Affairs Director (1908). He took part in a number of major zoological research expeditions: as entomologist on Adolf Erik Nordenskiöld's expedition to Jenisej (1876), Lieutenant Herman Sandeberg's expedition to the Kolahal River (1877) and with Hjalmar Théel on a gunboat survey in Swedish waters. He wrote: *Handbook Fisheries and Fisheries* (1893). He was the first person to publish (1903) on the existence of this species and study its morphology.

Tryon

Tryon's Pipefish *Campichthys tryoni* JD Ogilby, 1890

Henry Tryon (1856–1943) was an English scientist who abandoned medicine in favour of natural science, particularly botany and entomology. He collected in Sweden and New Zealand before going to Queensland (1882), where he first became an Honorary Assistant at the Queensland Museum and then was officially employed there (1883–1893), becoming Assistant Curator (1885). His extracurricular activities for government departments, such as investigating the rabbit menace for the government of New South Wales (1888–1889), brought him into conflict with his Director, and he left, becoming a government entomologist (1893–1925). Ogilby named the pipefish after him "*...in remembrance of the very pleasant collecting trip which we enjoyed together there*" [Moreton Bay, Queensland; the type locality]. Two reptiles are also named after him.

Tschudi

Apron Ray *Discopyge tschudii* JJ Heckel, 1846
Titicaca Pupfish sp. *Orestias tschudii* FL Castelnau, 1855

Baron Dr Johann Jakob von Tschudi (1818–1889) was a Swiss explorer, physician, diplomat, naturalist, hunter, anthropologist, cultural historian, language researcher and statesman. He travelled to Peru (1838) where he spent five years exploring and collecting. He was appointed Swiss ambassador to Brazil (1860–1868). He wrote: *Untersuchungen uber die Fauna Peruana Ornithologie* (1844–1846). He also collected the type of the ray. Ten birds, six reptiles, five amphibians and five mammals are named after him.

Tshokwe

Dundo Elephantfish *Campylomormyrus tshokwe* M Poll, 1967

The etymology is not explained, but this probably refers to the Tschokwe (also spelled Chokwe) peoples of central Africa who inhabit the area where the species occurs.

Tsuchida

Snake-eel sp. *Ophichthus tsuchidae* DS Jordan & JO Snyder, 1901

[Sometimes included in *Ophichthus urolophus*]

Toyoza Tsuchida was assistant to zoologist Kakichi Mitsukuri who was the first Director (1898–1904) at the Misaki Marine Biological Station (the type locality), part of the Imperial University Tokyo. Jordan visited the Misaki Marine Biological Station (1900) and would have met Tsuchida; Snyder also visited the station (1906) well after the description was written.

Tsuchiga

Gudgeon ssp. *Squalidus gracilis tsuchigae* DS Jordan & Hubbs CL, 1925

Professor Yasukei Tsuchiga was a science teacher at the Yamada Middle School, Japan, who collected some of the fishes examined by Jordan and Hubbs.

Tsukawaki

Freshwater Dragonet sp. *Tonlesapia tsukawakii* H Motomura & T Mukai, 2006

Dr Shinji Tsukawaki is Professor of the Division of Terrestrial Environmental Studies at the Institute of Nature and Environmental Technology, Kanazawa University. His specialities include marine geology. Tohoku University awarded his DSc (1990), after which he was a post-doctoral researcher at the Japan Society of the Promotion of Science (1990–1992) and at the Graduate School of Science, Tohoku University (1992–1994). He was Associate Professor, Kanazawa University (1994–2010) prior to becoming full professor. He was thanked for "*his kind and invaluable assistance*" during the authors' ichthyological surveys in Cambodia.

Tsuruga

Silverside sp. *Hypoatherina tsurugae* DS Jordan & EC Starks, 1901

A toponym referring to the city of Tsuruga, Honshu, Japan.

Tuba

Stone Loach sp. *Seminemacheilus tubae* B Yoğurtçuoğlu, C Kaya, MF Geiger & J Freyhof, 2020

Tuğba (Tuba) Kaya is, according to the etymology, "*…the beloved wife of Cüneyt Kaya*" (the second author). She was honoured "*…for her endless patience and support with him and his work.*"

Tubb

Rudderfish genus *Tubbia* GP Whitley, 1943
King Rasbora *Rasbora tubbi* MR Brittan, 1954

Dr John Alan Tubb (1913–1985) was an Australian fisheries expert, academic and amateur naturalist. The University of Melbourne awarded his BSc and MSc (1936). He started his career as Fisheries Adviser to the Victorian Fresh Water Research Committee (1936–1938), then was Biologist with the CSIR Fisheries Division (1937–1940). He was Fisheries Survey Officer in the Colony of North Borneo (1947–1948) before being appointed Director of Fisheries there (1948–1953). It was here he collected the rasbora type specimen. He was Reader in Marine Biology and Director of the Fisheries Research Unit at the University of Hong Kong (1953–1955) and Technical Assistance Expert (FAO) on Inland Fisheries in Burma (1955–1956); then Regional Fisheries Officer for Asia and Far East (1956) and was Secretary (ex-officio) of the Indo-Pacific Fisheries Council (1956–1969) and Project Manager for the South Pacific Island Fisheries Development Agency (UN/SPC) in Noumea (1969–1971). He was Fisheries Officer (Policy and Institutions) for the United Nations Development Programme in Jakarta (1971–1972). He collected other marine organisms, e.g. near Melbourne, Australia (1930) & Tasmania (1941), which were deposited in the AMNH.

Tucano

Characin genus *Tucanoichthys* J Géry & U Römer, 1997
Corydoras (catfish) sp. *Corydoras tukano* MR Britto & FCT Lima, 2003

This genus and species are named after the Tucano Indians of the upper Rio Negro and Rio Uaupés area of Amazonas, Brazil.

Tucker, DW

Tucker's Frostfish *Benthodesmus tuckeri* NV Parin & VE Becker, 1970

Denys W Tucker (d. 2009) was a British zoologist who worked at the BMNH and was an expert in *Benthodesmus* and related fishes. His career at the museum began (1949) as a scientific officer, but started to falter after he claimed to have seen the Loch Ness Monster (1959). Dismissed from his post, he launched a legal campaign to be reinstated. The legal battle lasted seven years before the Court of Appeal finally dashed Tucker's hopes, and he never held an academic post again.

Tucker, G

Slender Filefish *Monacanthus tuckeri* TH Bean, 1906

George Tucker (1835–1908) was Archdeacon of Bermuda (the type locality) and was honoured "*...for his devotion to biological science in the colony.*"

Tucker, GV

Tucker's Pipefish *Mitotichthys tuckeri* EOG Scott, 1942

G V Tucker netted the holotype at Bridport, Tasmania. No further details of him are given.

Tudor Jones

Redback Dragonet *Synchiropus tudorjonesi* GR Allen & MV Erdmann, 2012

Paul Tudor Jones (b.1954) is an American philanthropist who was honoured for "*...his dedication and selfless service to the United States National Fish and Wildlife Foundation (NFWF).*"

Tugarina

Lower Amur Grayling *Thymallus tugarinae* IB Knizhin, AL Antonov, SN Safronov & SJ Weiss, 2007

Praskov'ya (Polina) Yakovlevna Tugarina (1928–2004) was a Russian ichthyologist as well as being a 'renowned' researcher of the graylings of Siberia and Russia's Far East, especially in the Lake Baikal Region. She studied at the biological-soil faculty, Irkutsk State University, where she spent her working life from being a post-graduate student to Professor of Vertebrate Zoology. She was elected dean of the biological-soil faculty (1970–1975) and was deputy director in science of the Research Institute of Biology at Irkutsk State University (1983–1990). She was Chief of Ichthyological Studies in Lake Khubsugul, Mongolia (1970–1995). She wrote: *Feeding and Food Relationships of Fish of the Baikal-Angara Basin* (1977) and *Graylings of Baikal* (1981). She died during an emergency heart operation.

Tuira

Armoured Catfish sp. *Spatuloricaria tuira* I Fichberg, OT Oyakawa & M de Pinna, 2014

Tuira was a Brazilian-Indian woman of Mebêngôkre/Kaiapó ethnicity who became a symbol (c.1989) of the resistance against construction of hydroelectric dams on the Rio Xingu, Brazil (where this catfish occurs).

Tukuna

Tetra sp. *Hyphessobrycon tukunai* J Géry 1965

This species is named after the Tukuna, an Indian tribe inhabiting the Upper Solimões River basin, Brazil.

Tuna

Tuna's Skate *Bathyraja tunae* MFW Stehmann, 2005

Dr María Cristina Oddone Franco, a post-doctoral researcher at the Instituto de Ciências Biológicas, Universidade Federal do Rio Grande, Brazil, was nicknamed 'Tuna'. Her bachelor's degree was awarded by Universidad de la República-UdelaR (2000), her master's by Universidade Federal do Rio Grande (2003) and her doctorate by Universidade Estadual Paulista (2007). She is one of the scientific editors for the Pan-American Journal of Aquatic Sciences. She has already written a number of papers such as: *Size at maturity of the smallnose fanskate Sympterygia bonapartii (Müller & Henle, 1841)(Pisces, Elasmobranchii, Rajidae) in the SW Atlantic* (2004).

Tung

Chinese Gudgeon sp. *Gobiobotia tungi* PW Fang, 1933

L M Tung was a zoologist at West Lake Museum, Hangzhou and Professor at the University of Zhejiang. He loaned the type specimen to Fang.

Tupi

Characin sp. *Astyanax tupi* M Azpelicueta, JM Mirande, AE Almirón & JR Casciotta, 2003

The Tupi are a tribe of people in the type locality of Chapada, Brazil.

Turan

Seyhan Scraper *Capoeta turani* M Özulug & J Freyhof, 2008
Aksu Goby *Ponticola turani* M Kovačić & S Engín, 2008
Turan's Minnow *Pseudophoxinus turani* F Küçük & SS Güçlü, 2014
Spirlin sp. *Alburnoides turani* C Kaya, 2020

Professor Dr Davut Turan is a Turkish ichthyologist. He is (since 2013) Dean of the Faculty of Fisheries at Erdogan University where he was Assistant Dean (2011–3013). He did his first degree at KTU (1992) and his masters (1997). Ege University awarded his PhD. He was honoured as the author of 'important papers on Anatolian *Capoeta*' and for his contributions to our knowledge of the fishes of Anatolia. He has written a great many papers including: *(Cyprinidae) Taxonomic revision of the species of the Barbus genus and their Distribution in Turkey* (2012) and *A new trout species described from the Alakır Stream in Antalya, Turkey* (2014).

Turbyne

Sole sp. *Solea turbynei* JDF Gilchrist, 1904

Captain Alexander Turbyne (d.1905) was in charge of the South African Government Fisheries steamer which obtained 'numerous specimens' of the sole. He had worked at the Marine Station, Granton, Edinburgh, and later at the Millport Marine Station (Scotland) before moving (c.1899) to South Africa. He died 'as the result of a gun accident' in East London, South Africa.

Turdakov

Stone Loach sp. *Dzihunia turdakovi* AM Prokofiev, 2003

Fedor Alekseevich Turdakov was a Soviet ichthyologist, a specialist in nemacheiline (loach) systematics and the fish fauna of Middle and Central Asia. He wrote: *Kyrgyzstan Fishes* (1952) and *Biological study of Lake Issyk-Kul* (1965).

Turkana

Lake Turkana Cichlid sp. *Haplochromis turkanae* HP Greenwood, 1974
Lake Turkana Catfish sp. *Chrysichthys turkana* M Hardman, 2008

These species are named after both Lake Turkana, Kenya, and the Turkana people, who are the predominant tribe in that area.

Türkay

Goby sp. *Obliquogobius turkayi* M Goren, 1992

Michael Türkay (1948–2015) of the Senckenberg Museum (Frankfurt, Germany) was a carcinologist, honoured for his contributions to the knowledge of the marine fauna of the Red Sea and Mediterranean. His obituary in the *Journal of Crustacean Biology described him as: "...a tower of strength in the world of crustaceans." He first contacted the museum when 16 – and a keen coleopterist with a vast collection – then went on to collect seashells. By the age of 19 he had published his first paper on brachyurans from South America's west coast. He studied* biology and chemistry at the Goethe-University, Frankfurt, aiming to become a teacher. He received a diploma in biology (1973) with a thesis on decapod crustaceans from various 'Meteor' expeditions in the eastern Atlantic. He had already published 16 scientific papers as a single author and became interim leader of the Crustacea Section of the museum (1974), at first as an assistant, and then (1983) as full curator and head of the section the same year that Goethe-University awarded his doctorate. He became head of the Department for Invertebrate Zoology at Senckenberg (1989) and Vice-Director (1995). He also taught at Goethe University, becoming extracurricular professor (2008). He made numerous cruises in the Atlantic and Pacific Oceans as well as the Red and Mediterranean Seas, and described and named 71 crustacean species, 26 genera, and one subfamily. He was honoured in the goby's name for his contributions to the knowledge of the marine fauna of the Red Sea and Mediterranean.

Turna

Nase (cyprinid) sp. *Chondrostoma turnai* SS Güçlü, F Küçük, D Turan, Y Çiftçi & AG Mutlu, 2018

Dr İsmail İbrahim Turna (1957–2016) was a Turkish ichthyologist who was Associate Professor in the Faculty of Fisheries at Süleyman Demirel University. The etymology says that the species is named "*...after İsmail İbrahim Turna (Süleyman Demirel University, Isparta, Turkey), who has made a great contribution to hydrobiology in Turkey.*"

Turner, CL

Highland Splitfin *Hubbsina turneri* F de Buen, 1940

Blackspotted Livebearer *Poeciliopsis turneri* RR Miller, 1975

Clarence Lester Turner (1890–1969) was an American ichthyologist who studied embryology and reproduction in fishes. Among his published works are: *The Crayfishes of Ohio* (1926) and *Collected Papers on Reproduction and Viviparity in Teleost Fishes* (1962).

Turner, G

Lake Malawi Cichlid sp. *Lethrinops turneri* BP Ngatunga & J Snoeks, 2003
Lake Malawi Cichlid sp. *Placidochromis turneri* M Hanssens, 2004

George Turner, Bangor University (Wales), has worked extensively on the fishes and fisheries of lakes Malawi and Lalombe. He was honoured for nearly 20 years of work on the ecology, ethology, taxonomy, and evolution of Lake Malawi cichlids; in addition, his (1996) book on the offshore cichlids of Lake Malawi was said by Hanssens to be an "*…important contribution to the taxonomic knowledge of the non-mbuna cichlids.*"

Turner, LM

Polar Eelpout *Lycodes turneri* TH Bean, 1879

Lucien McShan Turner (1848–1909) was an American naturalist and ethnologist who was a member of the Army Signal Corps and collected natural history and ethnological specimens for the USNM. He served as a meteorological observer for the Alaskan Signal Service at St Michael, Alaska (1874–1877), and then trained voluntary observers in the Aleutians (1878–1881) before being sent to Fort Chimo, Labrador, as an observer (1882–1884). He caught the eelpout holotype in Alaska (1876). Turner made extensive collections and also had a rapport with the local people, the Innu and Inuit, spending his free time studying and recording their culture, routines, language and stories. His pictures of them and their camps are among the earliest photographs of the Arctic. He wrote a number of books including *Contributions to the Natural History of Alaska* (1886) and *Ethnology of the Ungava District, Hudson Bay Territory* (1894). The US National Museum was said to be indebted to Turner "*…for large and valuable additions to its collections from Alaska.*" Two birds are also named after him.

Turner, PJ

Lanternfish sp. *Lampanyctus turneri* HW Fowler, 1934

Percy J Turner of Suva, Fiji, to whom Fowler was indebted for 'interesting' fishes from that island country. Fowler (q.v.) obviously had world-wide connections and not just his immediate circle in New England. Turner is recorded as having made a collection (1929) of 145 specimens representing 53 species.

Tursky

Čikola Riffle Dace *Telestes turskyi* JJ Heckel, 1843

Field Marshall Johann August Ritter von Tursky (1778–1856) was an Austrian who was Governor of Dalmatia (then part of the Austro-Hungarian Empire). The etymology says of him that his support allowed Heckel to study the little-known fishes of (present-day) Croatia.

Tussy

Betta sp. *Betta tussyae* D Schaller, 1985

Tussy Nagi is the wife of Austrain aquarist Peter Nagy. Together they brought the first examples of this species to Europe, collecting them in Malaysia (1979). (Also see **Nagy**)

Tutin

Titicaca Pupfish sp. *Orestias tutini* VV Tchernavin, 1944

Dr Thomas Gaskell 'Tom' Tutin (1908–1987) was Professor of Botany at the University of Leicester (1947–1973). He studied biological sciences at Downing College, Cambridge and, before he graduated, joined an expedition to Madeira and the Azores. After graduating (1930) he remained at Cambridge but took part in three further expeditions: Spain & Morocco (1931) and British Guiana (1933). He was a member of the Percy Sladen Trust Titicaca Expedition (1937). He was a demonstrator at Kings then became assistant lecturer at the University of Manchester. He joined the geographical section of Naval Intellegence (1942). He wrote *Flora of the British Isles* (1952) and edited the five-volume *Flora Europaea* (1964–1993).

Tuyuka

Characin sp. *Creagrutus tuyuka* RP Vari & FCT Lima, 2003

This species is named after the Tuyuka tribe of the Colombia-Brazil border region.

Tweddle, D & S

Lake Chilwa Cichlid sp. *Astatotilapia tweddlei* PBN Jackson, 1985
Dwarf Sanjika *Opsaridium tweddleorum* PH Skelton, 1996
Lizardfish sp. *Saurida tweddlei* BC Russell, 2015
Squeaker Catfish sp. *Chiloglanis tweddlei* RC Schmidt & JP Friel, 2017

Denis (b.1949) and Sharon Tweddle are husband and wife and together were honoured in the name of the dwarf sanjika for their contributions (e.g. study specimens, colour slides) to the study of Malawi fishes. He attended Bangor University, Wales (1967–1970), which awarded his BSc in Marine Biology/Zoology. He has worked in a number of fisheries posts in Uganda (1971–1972) and Malawi (1973–1995), rising to Senior Research Officer before becoming an independent adviser or consultant and working in much of southern Africa. He is currently (since 2013) Research Associate at the South African Institute for Aquatic Biodiversity, Grahamstown, South Africa and Project Coordinator at Namibia. He has published very widely including the paper: *The ecology of the catfish Clarias gariepinus and Clarias ngamensis un the Shire Valley, Malawi* (1978) and *First record of the river goby Glossogobius callidus (Smith, 1937) from Malawi* (2007). Sharon attended Rhodes University, Johannesburg, South Africa and is currently Senior Alumni Relations Officer there. She was also the Honorary Secretary of the Wildlife Society of Malawi (1990–1995). The lizardfish and the cichlid are named after Denis alone. Jackson honoured him for: "... *valuable contributions to the zoogeography of Malawi as well as his appreciation of the fact that the Lake Chilwa-Chiuta fauna contained a new haplochromine species.*"

Tweedie

Mahseer sp. *Neolissochilus tweediei* AW Herre & Myers, 1937
Alligator Hillstream Loach *Homalopteroides tweediei* AW Herre, 1940
Filefish sp. *Pseudomonacanthus tweediei* AF Fraser-Brunner, 1940
Kedah Danio *Devario tweediei* MR Brittan, 1956
Licorice Gourami sp. *Parosphromenus tweediei* M Kottelat & PKL Ng, 2005

Michael Wilmer Forbes Tweedie (1907–1993) was a herpetologist, ichthyologist, archaeologist and all-round naturalist who worked (1932–1971) at the Raffles Museum (now called the Singapore National Museum). He read Natural Science at Cambridge University, followed by a short period working as an oil geologist in Venezuela, before becoming assistant curator at Raffles Museum. After the end of WW2, he became Director of the museum (1946). Tweedie was involved in many biological and archaeological expeditions in South-East Asia and collected many specimens himself, including some of the fish species named after him. A reptile and an amphibian are also named after him.

Twist

Yellow-breasted Wrasse *Anampses twistii* P Bleeker, 1856
Redblotch Razorfish *Iniistius twistii* P Bleeker, 1856

Albertus Jacobus Duymaer van Twist (1809–1887) was Governor-General of the Dutch East Indies (1851–1856). He is known to have made a collection of zoological specimens during a journey to the Moluccas (1855).

Tydeman

Spikefish genus *Tydemania* M Weber, 1913
Rattail sp. *Coryphaenoides tydemani* M Weber, 1913

Vice-Admiral Gustaaf Frederik Tydeman (1858–1939) was a Dutch naval officer (1872–1915). As a Lieutenant he was commander, astronomer and hydrographer of the 'Siboga' Indonesian expedition (1898–1899), during which the rattail holotype was collected. In retirement, he undertook many duties and fulfilled many functions including member of the College for Fisheries until 1929, Member of the Defense Council (1915–1932) and Inspector of MIlitary Education in the navy (1916–1924). He wrote: *The birth of Earth and Moon* (1937).

Tyler

Pufferfish genus *Tylerius* GS Hardy, 1984
Bearded Puffer *Sphoeroides tyleri* RL Shipp, 1972
Tyler's Toby *Canthigaster tyleri* GR Allen & JE Randall, 1977
Deepwater Boxfish sp. *Polyplacapros tyleri* E Fujii & T Uyeno, 1979
Goldface Toby *Canthigaster jamestyleri* RL Moura & R Castro, 2002
Tyler's Coralbrotula *Ogilbia tyleri* PR Møller, W Schwarzhans & JG Nielsen, 2005

Dr James Chase 'Jim' Tyler (b.1935) is an ichthyologist and Senior Scientist Emeritus at the NMNH, Washington D.C., where he worked as Deputy Director/Acting Director (1985–1990). He graduated from George Washington University (1957) and was awarded his doctorate by Stanford University (1962). He was on the staff of the Academy of Natural Sciences of Philadelphia as Curator of Ichthyology and Herpetology (1962–1972). He was Director, Lerner Marine Laboratory, Bimini, Bahamas (1972–1975). He worked at the National Marine Fisheries Service, Washington, D.C., and Miami, Florida (1975–1980) and was Director, Biological Research Resources Program of the National Science Foundation in Washington, D.C (1980–1985). Among his publications is the co-authored (with Diane M Tyler (q.v.) to whom he is married): *Natural history of the sea fan blenny, Emblemariopsis pricei (Teleostei: Chaenopsidae), in the western Caribbean* (1999). He and his wife collected most of the materials for the study of the life history of this species near the Smithsonian laboratory on the island of Carrie Bow Cay, Belize. At least seven fossil fish are also named after him.

Tyon

Goby sp. *Rhinogobius tyoni* T Suzuki, S Kimura & K Shibukawa, 2019

Darsu Tyon (deceased) discovered the species and 'kindly informed' the authors of it for their study. He was a junior author of: *Redefinition and proposal of the new standard Japanese name for Rhinogobius sp. OR morphotype "Shimahire" (Perciformes, Gobiidae)* (2010).

Tyson

Arrow Wriggler genus *Tyson* VG Springer, 1983
Torrent Minnow sp. *Psilorhynchus tysoni* KW Conway & AK Pinion, 2016

Dr Tyson Royal Roberts (b.1940). (See **Roberts, TR**)

Tzanev

Riffle Minnow sp. *Alburnoides tzanevi* G Chichkoff, 1933

Panayot Tzanev was an assistant at the Zoological Institute of the Russian Academy of Sciences. He collected a large number of ichthyological samples, although there is no direct evidence that he collected the type of this species.

Tzinovski

Creamback Skate *Bathyraja tzinovskii* VN Dolganov, 1983

Dr Vladimir Diodorovich Tzinovskiy (b.1946) is a researcher at the P P Shirshov Institute of Oceanology, Moscow (part of the Russian Academy of Sciences), and an Arctic oceanographer who collected the holotype. During his career, he undertook many oceanic expeditions, mainly to the Arctic region. He co-wrote: *Hydrobiological studies in the Arctic Ocean at NP-23 (May-October 1977)* (1978).

U

Ubidia

Andean Catfish *Astroblepus ubidiai* J Pellegrin, 1931

Georges Ubidia was a student of parasitologist Otto Fuhrmann. He collected the holotype of this catfish, but we know nothing more about him.

Ubuirajara

Neotropical Rivuline sp. *Melanorivulus ubirajarai* WJEM Costa, 2012

Dr Ubirajara Ribeiro 'Bira' Martins de Souza (1932–2015) was a Brazilian entomologist who spent his career researching at the Museum of Zoology of the University of São Paulo (1959–2015) which had awarded his PhD (1975) and where he became a professor affectionately known by his colleagues and students as 'Bira'. The Universidade Federal de Viçosa awarded his bachelor's degree in agronomy (1954). He was President of the Brazilian Entomological Society (1983–1986) and was considered the most important coleopterist of his time, who named many new species. A frog species, *Bokermannohyla martinsi*, is also named after him. Costa said of him that he was a: "*...teacher and friend during my early formation in Systematics, providing great enthusiasm for studying taxonomy of killifishes about 30 years ago.*"

Uchida, K

Uchida's Eelpout *Zoarchias uchidai* K Matsubara, 1932
Goby sp. *Gymnogobius uchidai* K Takagi, 1957
[Syn. *Paleatogobius uchidai*]

Dr Keitarō Uchida (1896–1982) was Professor of Zoology (1942) at the fisheries department of Kyushu Imperial University, Japan. He was educated at the Imperial University of Tokyo. He was the person who recommended that Takagi study the gobies of Japan based on the sensory (lateral) line system as a taxonomic character, which is why he came to be honoured in the goby's binomial. Among his written works are the co-authored papers: *Marine ichthyological Distribution* (1935) and *Studies on the Larvae and Juveniles of Fishes accompanying Floating Algae-I* (1958).

Uchida, T

Stone Moroko ssp. *Pseudorasbora parva uchidai* Y Okada & SS Kubota, 1957
[Alt. Topmouth Gudgeon]

Dr Toru (Tohru) Uchida (1897–1981) was a Japanese zoologist and a professor at Hokkaido University, which was also the alma mater of the junior author. He edited several popular zoological encyclopedias and books on Japanese fauna, as well as writing more serious zoological works on systematics, and papers on a wide variety of fauna. He was honoured in celebration of his 60th birthday. A mammal is also named after him.

Udomritthiruj

Emperor Loach *Botia udomritthiruji* HH Ng, 2007
Stone Loach sp. *Schistura udomritthiruji* J Bohlen & VB Šlechtová, 2010

Kamphol Udomritthirug (b.1963) of Aquaricorp, Pathunthani, Thailand is an aquarist and ornamental-fish trader who provided the type specimens of these fish and associated data for them. He was also honoured for his continuous support of the authors' work on South-east Asian fishes and because he supplied material and data for other projects.

UENF

Damselfish sp. *Stegastes uenfi* R Novelli, GW Nunan & NRW Lima, 2000

The UENF, Unive Estadual do Norte Fluminense (Rio de Janeiro, Brazil) is a public university that encourages biological research (and where the first and third authors are affiliated).

Uejo

Guam Barred Butterflyfish *Roa uejoi* M Matsunuma & H Motomura, 2021

Takuya Uejo is a Japanese ichthyologist who was affiliated with the Graduate School of Fisheries, Kagoshima University. It was he who first recognised the uniqueness of this species whilst studying the taxonomy of the genus Roa. Among other works, he co-wrote, with Hiroyuki Motomura and others: Roa haraguchiae a new species of butterflyfish (Teleostei: Perciformes: Chaetodontidae) from Japan and the Philippines. Ichthyological Research (2020).

Ueno

Lumpfish sp. *Eumicrotremus uenoi* Y Kai, S Ikeguchi & T Nakabo, 2017

Dr Tatsuji Ueno, formerly Director of the Hokkaido Fisheries Experimental Station, was honoured for his contributions to the systematics of Cyclopteridae (lumpfish). He had served in the Japanese Air Force during WW2. He was also at the Faculty of Fisheries, Hokkaido University, which had awarded his master's degree and PhD. Among his published papers he wrote: *Studies on the Deep-water Fishes from off Hokkaido and Adjacent Regions* (1954) and co-wrote, with Koji Abe: *On Rare or Newly Found Fishes from the Water of Hokkaido* (1966). He also wrote the book *Cyclopteridae* (*Pisces*) (Fauna Japonica) (1970).

Ufermann

Cameroon Cichlid sp. *Benitochromis ufermanni* A Lamboj, 2001
Usumacinta Cichlid *Kihnichthys ufermanni* R Allgayer, 2002

Alfred Ufermann (d.2002) was a German aquarist, who played a significant role in getting Lamboj involved with the cichlids of West and Central Africa. He was also a friend of Robert Allgayer. He studied the nomenclature and systematics of cichlids and produced a catalogue of them with his friends Allgayer and Geerts (1987). Lamboj also says in his etymology of the cichlid that "*...one turns to him—mostly successfully—when it comes to unusual literature on cichlids.*" (translation)

Ugo

Dark Freckled Catshark *Scyliorhinus ugoi* KDA Soares, OFB Gadig & UL Gomes, 2015

Ugo de Luna Gomes is the son of the third author, the Brazilian ichthyologist Ulisses Leite Gomes (q.v.).

Uhler

Voodoo Whiff *Citharichthys uhleri* DS Jordan, 1889

Philip Reese Uhler (1835–1913) was an American librarian and entomologist specialising in Heteroptera (true bugs). He was educated at Harvard and taught by Louis Agassiz. He became Head of the Museum of Comparative Zoology Library & Insect Department there (1864–1867). He then became an Associate in Natural Sciences at Johns Hopkins University (1876). He was also Librarian at the Peabody Library in Baltimore. He made a number of collecting trips including to Haiti (where he obtained the holotype of this flatfish) and the Western USA, as well as to Europe (1888) to study insect collections in European museums. He was Secretary and then President (1873) of the Maryland Academy of Science. He wrote many papers on entomology and geology and also *Check-list of the Hemiptera Heteroptera of North America* (1886). A dragonfly is also named after him.

Ui

Giant Moray sp. *Strophidon ui* S Tanaka, 1918

Nuizo Ui (1878–1946) was an amateur naturalist who obtained or collected type. Tanaka named three other fishes after Ui, all now considered junior synonyms, all were obtained or collected by Ui at Tanaba, Wakayama Prefecture, Japan.

UIS

Pencil Catfish sp. *Trichomycterus uisae* CA Castellanos-Morales, 2008

UIS is an acronym of Universidad Industrial de Santander (Departamento de Santander, Colombia), near where this catfish occurs and where some of the paratypes are housed.

Ulbu Bunitj

Goby sp. *Egglestonichthys ulbubunitj* HK Larson, 2013

Not a true eponym as named for a people; the Ulbu Bunitj clan in Arnhem Land, Northern Territory, Australia, where the goby occurs.

Ulrey

Broken-line Tetra *Hemigrammus ulreyi* GA Boulenger, 1895

Albert Brennus Ulrey (1860–1932) was an American naturalist and marine biologist who was awarded both a bachelor's degree (1892) and a master's (1894) by Indiana University and was on the University staff as an instructor in zoology (1892–1894). He was Professor of Biology at Manchester College, Indiana (1894–1900) and the first marine biologist at the University of Southern California (1901) and finally Director of the University of Southern California's Marine Biological Station. Boulenger, in the etymology, states that Ulrey was the author of a very useful key [1895] to the determination of the species. He wrote: *The Marine Fishes of Southern California* (1922).

Uluguru

Shellear sp. *Kneria uluguru* L Seegers, 1995

This is a toponym referring to the Uluguru Mountains in Tanzania, where the species appears to be endemic.

Umali

Mindanao Barb sp. *Barbodes umalii* CE Wood, 1968

Agustin F Umali (1906–1996) was Senior Ichthyologist at the National Museum of the Philippines. The University of the Philippines awarded his bachelor's degree (1928). He taught in a school (1928–1929) before becoming an Assistant Ichthyologist at the Bureau of Science (1929–1936) and then a District Fisheries Officer (1936–1939). He was in government service as an ichthyologist or as an aquatic biologist and finally as Supervisor (Fisheries) Food Administration (1939–1944). With the end of WW2 approaching he went to the Philippine School of Fisheries (1944–1948). He was at the US Fish and Wildlife Service, Rehabilitation Office, Manila (1940–1950) and was Chief of the Geology Palaeontology Division, National Museum of the Philippines, Manila (1950–1950). Among his written works are: *Deep-sea Fishing in the Philippines* (1936) and *Reef Fishing in the Philippines* (1949). He was honoured in the name of the barb species because Umali encouraged Wood to study fishes in Mindanao, and was honoured for "*…his vast knowledge of pre-war ichthyology and long hours spent passing this information on to the writer.*"

Umeyoshi

Lizardfish sp. *Saurida umeyoshii* T Inoue & T Nakabo, 2006

Umeyoshi Yamada is a Japanese ichthyologist at the Seikai National Fisheries Research Institute, formerly known as Fisheries Agency. He was one of the authors of *Fishes of Japan* (2002) and co-wrote: *A new species of Zenopsis (Zeiformes: Zeidae) from the South China Sea, East China Sea and off Western Australia* (2006). He was the first person to recognise this species as new.

Umut

Clingfish sp. *Diplecogaster umutturali* M Bilecenoğlu, MB Yokeş & M Kovačić, 2017

Umut Tural (d.2012) was, according to the etymology, a "*promising conservation biologist passionate for marine life*". He was coordinator of the Kaş-*Kekova* Specially Protected Area Management Plan and Implementation Project on behalf of WWF-Türkiye, in which he actively participated by collecting clingfish samples. He passed away after a year-long battle with cancer.

Unam

East Pacific Nurse Shark *Ginglymostoma unami* L Del Moral-Flores, E Ramíz-Antonio, A Angulo and G Pérez-Ponce de León, 2015

This is an acronym of the Universidad Nacional Autónoma de México (UNAM).

Underwood

Central American Cichlid sp. *Talamancaheros underwoodi* CT Regan, 1906

Cecil Frank Underwood (1867–1943) was a British naturalist-taxidermist who collected in Costa Rica, including the type of this cichlid. He left London for Costa Rica (1889) to collect natural history specimens for a living, staying for the rest of his life. He was an all-round naturalist who collected for a number of overseas museums, and was a taxidermist at Costa Rica's National Museum. He described many new mammals from Central America, often with George Goodwin. Four mammals, two birds and two amphibians are also named after him.

UNIP

Unipa Chromis *Chromis unipa* GR Allen & M van N Erdmann, 2009

This is an acronym for Universitas Negeri Papua (State University of Papua, Monokwari), which invited the authors to teach an ichthyology field course, leading to the discovery of this species. They say: "*The University has played a critical role in training young Papuans in the natural sciences, and it is our sincere hope that this work will inspire students and lecturers at the University to actively explore the unique natural laboratory of Cenderawasih Bay [type locality] at their doorstep.*"

Uno

Duckbill sp. *Acanthaphritis unoorum* T Suzuki & T Nakabo, 1996

Masami Uno and Akira Uno collected holotype and some paratypes from off Hyogo Prefecture, the Sea of Japan.

Urado

Trumpetsnout *Macrorhamphosodes uradoi* T Kamohara, 1933

This is a toponym referring to Urado Market in Kochi Province, Japan, where the type was found. It had been taken at 100 fathoms off Urado.

Urashima

Searobin sp. *Pterygotrigla urashimai* WJ Richards, T Yato & PR Last, 2003

Urashima Taro is a hero in a Japanese folktale; a kindly young fisherman who rescues a turtle, which then carries him to the Dragon Palace under the sea.

Urbain

Sheatfish (catfish) sp. *Ompok urbaini* PW Fang & J Chaux, 1949

Dr Achille Joseph Urbain (1884–1957) was a French biologist, microbiologist and immunologist. He qualified as a veterinary surgeon at Lyon (1908); later he earned both a bachelor's degree (1912) and a doctorate (1920). He was a member of the French military and worked at a military veterinary laboratory as well as at the Pasteur Institute until resigning (1931). He was Director, Vincennes Zoo (1934–1942) and Director, MNHN, Paris (1942–1949). He described the Kouprey (Forest Ox, *Bos sauveli*), based on a young male individual captured in Cambodia and wrote: *Le Kou Prey ou bœuf sauvage cambodgien* (1937).

Urich

Three-barbeled Catfish ssp. *Rhamdia quelen urichi* JR Norman, 1926

Friederich William Urich (1872–1936) was a Trinidadian naturalist. He was a founder member of the Trinidad Field Naturalists Club (c.1891), most famous for discovering the cave-dwelling catfish which was named after him: originally as *Caecorhamdia urichi*, but now regarded as a troglomorphic form of *Rhamdia quelen*. He published his finding in the *Field Naturalists' Club Journal* (1895) as: *A visit to the Guacharo Cave of Oropuche.* He was described as Adjutant (1915) with the rank of Captain to the Military Department of the Government of Trinidad; presumably a temporary WW1 post. An amphibian, a bird and two mammals are also named after him.

Urraca

Dark-finned Sand Goby *Microgobius urraca* L Tornabene, JL Van Tassell & DR Robertson, 2012

Not an eponym but named for the Smithsonian Tropical Research Institute's research vessel 'Urraca', which contributed a wealth of information on fish diversity in the tropical Americas (1994–2007) and from which the type was collected (2003).

Urteaga

Urteaga's Dwarf Cichlid *Apistogramma urteagai* SO Kullander, 1986

Dr Jorge Andrés Urteaga Cavero is a Peruvian biologist. He was co-leader of the expedition during which the type was collected. He became Dean of the Faculty of Biological Science at the Universidad Nacional de la Amazonía Peruana (1992–1994). He was honoured for his 'considerable effort' in making the expedition a success.

Uruni

Titicaca Pupfish sp. *Orestias uruni* V Tchernavin, 1944

This is a toponym after Bahia de Uruni, Lake Titcaca

Urville

Congolli *Pseudaphritis urvillii* A Valenciennes, 1832
[Alt. Freshwater Flathead]
Urville's Longtom *Strongylura urvillii* A Valenciennes, 1846
[Alt. Urville's Needlefish]

and *Karyotypes and geographical distribution of ricefishes from Yunnan, southwestern China* (1988).

Uwisara

Myanmar Sea Bass *Lates uwisara* R Pethiyagoda & AC Gill, 2012

U Wisara (1889–1929) was a Burmese Buddhist monk who died in prison after a hunger strike against British rule in Burma. His death helped to galvanize the independence movement.

Uyeki

Korean Rose Bitterling *Rhodeus uyekii* T Mori, 1935

Dr Homiki Uyeki (1882–1976) was a Japanese botanist at Suigen Agricultural College, Korea – at that time controlled by Japan. The type locality is in Suigen, Korea. He wrote: *Woody plants and their distributions in Tyôsen* (1940). He graduated from Tokyo Imperial University (1907) and studied at Harvard (1918). He was awarded his doctorate by Tokyo Imperial University (1928). After WW2, he had to return to Japan and joined the Agricultural College, Ehime University, Matuyama (1945).

V

Vachell

Vachell's Glass Perchlet *Ambassis vachellii* J Richardson, 1846
[Alt. Telkara Perchlet]
Darkbarbel Catfish *Tachysurus vachellii* J Richardson, 1846
[Syn. *Pseudobagrus vachellii*]

Rev George Harvey Vachell (1799–1839) was chaplain to the British East India Company's factory at Macao. He was also a collector of botanical specimens for Kew Gardens. He graduated from Cambridge University (1821), arrived in Macao, China (1828), leaving finally (1836). A projected museum at the factory, with Vachell as curator, was never completed as the East India Company's trade monopoly ended (1834) and the factory was closed.

Vahl

Vahl's Eelpout *Lycodes vahlii* JCH Reinhardt, 1831

Jens Laurentius Moestue Vahl (1796–1854) was a Danish botanist and pharmacist. He graduated as a pharmacist (1819) then studied chemistry and botany. He took part in an expedition (1828–1829) to Greenland, searching, unsuccessfully for a lost settlement but he used the opportunity to collect many botanical specimens. This attracted royal sponsorship, enabling him to again (1830) visit Greenland before returning to Denmark (1836). He donated all he collected (including a number of fish) to Copenhagen University where Johannes Reinhardt was Professor of Zoology. He later took part in a French expedition to Spitsbergen (1838–1839). He was then an assistant at Copenhagen Botanical Garden (1840). He never completed his Greenlandic Flora, although he co-wrote *Flora Danica* (1861).

Vaillant, F Le

Laulao Catfish *Brachyplatystoma vaillantii* A Valenciennes, 1840

François Levaillant (1753–1824) was a French traveller, explorer, collector and naturalist. He was born in Dutch Guiana (Suriname), son of the French Consul, but later moved to France (1763). Birds attracted his early interest and he spent much time collecting specimens, and he thus became acquainted with many of Europe's private collectors. He went to the Cape Province of South Africa (1780–1784) in the employ of the Dutch East India Company, the first real ornithologist to visit the area. There he both explored and collected, eventually publishing a 6-volume book: *Historie Naturelle des Oiseaux d'Africa* (1801–1806), a classic of African ornithology. Published in Paris it contained 144 colour-printed engravings based on drawings by Barraband. Over 2,000 skins were sent to Jacob Temminck (q.v.), who financed this expedition A large collection of his specimens was lost when a Dutch ship was attacked and sunk by the English. Levaillant was opposed to the systematic nomenclature of Linnaeus and only gave the new species he discovered French names. Fourteen birds and an amphibian are also named after him.

Vaillant, LL

Loach genus *Vaillantella* HW Fowler, 1905
Vaillant's Grenadier *Bathygadus melanobranchus* LL Vaillant, 1888
Armoured Catfish sp. *Hypostomus vaillanti* F Steindachner, 1877
Chinese Gudgeon sp. *Pseudogobio vaillanti* HE Sauvage, 1878
Cusk-eel sp. *Dicrolene vaillanti* AW Alcock, 1890
Squeaker Catfish sp. *Synodontis vaillanti* GA Boulenger, 1897
Pike Cichlid sp. *Crenicichla vaillanti* J Pellegrin, 1903
Shortspine African Angler *Lophius vaillanti* CT Regan, 1903
Suckermouth Catfish sp. *Astroblepus vaillanti* CT Regan, 1904
Rasbora sp. *Rasbora vaillantii* CML Popta, 1905
Bagrid Catfish sp. *Bagrichthys vaillantii* CML Popta, 1906
Vaillant's Anchovy *Anchoviella vaillanti* F Steindachner, 1908
Barbeled Dragonfish sp. *Bathophilus vaillanti* EJG Zugmayer, 1911
Bagrid Catfish sp. *Pseudomystus vaillanti* CT Regan, 1913
Borneo Carp sp. *Thynnichthys vaillanti* MWC Weber & LF de Beaufort, 1916
Vaillant's Chocolate Gourami *Sphaerichthys vaillanti* J Pellegrin, 1930
Snailfish sp. *Paraliparis vaillanti* NV Chernova, 2004

Professor Léon Louis Vaillant (1834–1914) was a French herpetologist, ichthyologist and malacologist at the Muséum National d'Histoire Naturelle, Paris, well known for his researches in East Indian ichthyology. He was on four French naval expeditions aboard the 'Travailleur' (1880, 1881 & 1882) and the 'Talisman' (1883). Three reptiles and two amphibians are named after him. In honouring him in the name of the carp, the authors wrote that it was Vaillant who: "*...observed differences between this species and T. thynnoides in 1902.*"

Vaiula

Mottled Cardinalfish *Fowleria vaiulae* DS Jordan & A Seale, 1906

Vaiula was a Samoan fisherman at Apia, Upolu Island, Samoa Islands where the holotype of this species was collected.

Valade

Sleeper sp. *Eleotris valadei* P Keith, MI Mennesson & E Henriette, 2020

Pierre Valade is a hydrobiologist at Association Réunionnaise de Développement de l'Aquaculture (ARDA) and OCEA based on the island of Réunion (Indian Ocean). Among his published papers is the co-written: *Pelagic larval traits of the amphidromous goby Sicyopterus lagocephalus* display seasonal variations related to temperature in La Réunion Island. He "*...collected several specimens of this new species in Reunion Island and Mayotte.*"

Valdez

Large-tooth Flounder sp. *Citharichthys valdezi* F Cervigón Marcus, 1986

Julio Valdez was Professor at the Francisco de Miranda Experimental University. He co-wrote: *Los peces del Golfo de Venezuela* (1986). He was honoured as he: "*...generously made available to us all the material used in the description of the species.*"

Valdivia

Valdivia Black Dragonfish *Melanostomias valdiviae* AB Brauer, 1902
Topside Lampfish *Notolychnus valdiviae* AB Brauer, 1904

Both these species are named after the research vessel 'Valdivia' and the Valdivia expedition (1898–1899), Germany's first expedition to explore deep seas. The holotypes were collected during the expedition.

Valenciennes

Goby genus *Valenciennea* P Bleeker, 1856
Constellationfish genus *Valenciennellus* DS Jordan & BW Evermann, 1896
Valenciennes' Dragonet *Callionymus valenciennei* CJ Temminck & H Schlegel, 1845
Sumatran Silverside *Hypoatherina valenciennei* P Bleeker, 1854
Driftwood Catfish sp. *Ageneiosus valenciennesi* P Bleeker, 1864
Slender Tardoor *Opisthopterus valenciennesi* P Bleeker, 1872
Long-whiskered Catfish sp. *Parapimelodus valenciennis* CF Lütken, 1874
Greater Redhorse *Moxostoma valenciennesi* DS Jordan, 1885
Blue Morwong *Nemadactylus valenciennesi* GP Whitley, 1937
[Alt. Sea Carp, Queenfish]

Achille Valenciennes (1794–1865) was a French zoologist; primarily an ichthyologist and conchologist. In his early career, he classified much of Humboldt's neotropical collections and they became friends. He also made important contributions to parasitology. He worked as an assistant at the Muséum National d'Histoire Naturelle, Paris. He co-wrote with Cuvier (q.v.), under whom he had studied, the 22-volume *Histoire naturelle des poissons* (1828–1849), which he completed after Cuvier's death. In honouring him in the name of the Greater Redhorse the author wrote that he: "*...described this sucker in 1844 but used a preoccupied name (Catostomus [=Carpiodes] carpio Rafinesque 1820.*" Two reptiles are also named after him.

Valentin

Valentin's Sharpnose Puffer *Canthigaster valentini* P Bleeker, 1853

François Valentijn (1666–1727) was a Dutch minister and naturalist who spent some time in the Dutch East Indies. His massive work *Oud en Nieuw Oost-Indiën* ('Old and New East-India') describes the history of the Dutch East India Company. Bleeker refers to him as "*the well-known writer on the East Indies who first mentioned [the fish]*" (translation).

Valentin (Mbosi)

Lekoli Mormyrid *Petrocephalus valentini* S Lavoué, JP Sullivan & ME Arnegard, 2010

Valentin Mbossi is a Congolese field assistant. Sullivan described him as a *pinassier* [boatman] *extraordinaire* at Odzala National Park in the Republic of the Congo. They honoured him for field assistance, which they say is "*...as important as laboratory bench work and analysis when it comes to investigations of electric fish taxonomy, behavior and evolution*". His first name was selected to avoid confusion with a patronym which looks similar to his surname.

Valentina

Bristlemouth sp. *Margrethia valentinae* NV Parin, 1982
Snailfish sp. *Paraliparis valentinae* AP Andriashev & AV Neyelov, 1984

Valentina Aleksandrovna Mukhacheva was a specialist in gonostomatid systematics. She wrote: *Materials on the systematics, distribution, and biology of the species of the genus gonostoma (pisces, gonostomatidae)* (1987). She was the first to notice the bristlemouth species as distinct from *M. obtusirostra*

Van Bebber

Whitetail Reeffish *Chromis vanbebberae* EP McFarland, CC Baldwin, DR Robertson, LA Rocha & L Tornabene, 2020

Barbara Van Bebber is a professional underwater explorer (2010–2019) and the pilot of a submersible craft and supervisor (2020) of the Sub Centre Curaçao at U-Boat Worx. Previously she was a dolphin trainer (2006–2009), Media Manager at Curaçao Sea Aquarium (2006–2010) and a project manager (2007–2014) at Breathe-IT Bonaire. The etymology describes her as "*...one of the most accomplished submersible pilots in the Caribbean. Van Bebber was one of several skilled pilots of the 'Curasub' that assisted DROP with observations and collections of many new species, including this species.*"

Vance

Suckermouth Catfish sp. *Astroblepus vanceae* CH Eigenmann, 1913

Miss Lola Vance provided a small collection of fishes from Peru, including the holotype of this one. We can find no further information about her.

Van Cuong

Loach sp. *Parabotia vancuongi* VH Nguyen, 2005

This is a toponym; Văn Cuong is a village in Vietnam, near the type locality.

Vandelli

Parasitic Catfish genus *Vandellia* A Valenciennes, 1846
Parasitic Catfish genus *Paravandellia* A Miranda Ribeiro, 1912

Dr Domenico Agostino Vandelli (1735–1816) was an Italian naturalist who maintained a lengthy correspondence with Linnaeus. His doctorate in medicine and natural philosophy was awarded by the University of Padua (1758). He declined Catherine the Great's invitation to join the University of St Petersburg (1763) and worked in Portugal (1764–1816) where he was employed (1764) as a professor and lecturer at the University of Coimbra. He was also the first supervisor of the university's botanical garden and natural history museum. He was the first director of the botanical gardens at Palácio da Ajuda, Lisbon (1793) and said to be the founder of modern Portuguese natural history. He sent a collection of catfishes, including a number of holotypes, to Lacépède (1808) and it seems to have been a long time before the collection was processed until Valenciennes made a start (1846). A reptile is named after him.

Van den Berg

Neotropical Rivuline sp. *Austrolebias vandenbergi* J-H Huber, 1995

Leen Van den Berg is a Dutch aquarium enthusiast. With his son, Arjen, he collected some of the type specimens.

Van de Poll

Van de Poll's Molly *Poecilia vandepolli* TW van Lidth de Jeude, 1887

Jacob Rudolph Hendrick Neervoort van de Poll (1862–1925) was a Dutch entomologist and pioneer photographer who collected in the Netherlands Antilles. His specialty was the study of Coleoptera and he described several new beetle species.

Vander

Toothless Characin sp. *Cyphocharax vanderi* HA Britski, 1980

Vander M Britski, who helped to collect the holotype, is the author's brother.

Vanderbilt

Scarface Blenny *Cirripectes vanderbilti* HW Fowler, 1938
Vanderbilt's Cardinalfish *Gymnapogon vanderbilti* HW Fowler, 1938
Loach sp. *Homalopterula vanderbilti* HW Fowler, 1940
Vanderbilt's Chromis *Chromis vanderbilti* HW Fowler, 1941

George Washington Vanderbilt III (1914–1961) was a scientific explorer whose main interest was marine life. He owned several yachts, using them to combine exploration with the pleasure of sailing. He went on at least five major expeditions, including several to Africa. On board his schooner *Cressida* he sailed in the IndoPacific (1937–1939), most importantly to Sumatra to carry out a systematic study of more than 10,000 fish specimens including collecting the loach type. His fifth major expedition (1941) was on board his schooner 'Pioneer' to the Bahamas, Caribbean, Panama, Galapagos and islands off the Pacific coast of Mexico. He established the George Vanderbilt Foundation for scientific research. Fowler acknowledged Vanderbilt for his 'industry' and 'continued interest' in the development of the fish collection at the Academy of Natural Sciences of Philadelphia. He was found dead on the pavement in front of a skyscraper in San Francisco. The official verdict was that he committed suicide by jumping from his 101st floor apartment, but the truth remains a mystery. Two birds are named after him.

Van der Horst

Shrimp-Goby genus *Vanderhorstia* JLB Smith, 1949

Longfin Burrower *Apodocreedia vanderhorsti* LF de Beaufort, 1948
East African Cichlid sp. *Astatoreochromis vanderhorsti* H Greenwood, 1954

Professor Cornelius Jan van der Horst (1889–1951) was a Dutch zoologist who moved to South Africa (1928) and became Head of the Zoology Department, Witwatersrand University, Johannesburg. Among his published papers were: *The Burrow of an Enteropneust* (1934) and *The Placentation of Elephantulus* (1949).

Van der Loos

Mudslope Goby *Acentrogobius vanderloosi* GR Allen, 2015

Robert 'Rob' Van der Loos is the owner and operator of the live-aboard dive vessel 'Chertan', based at Alotau, Papua New Guinea. He has spent three and a half decades exploring the local waters. He was honoured for his generous assistance, which was instrumental in the discovery of this goby. (See also **Violaris**)

Van der Waal

Ocellated Spiny Eel *Mastacembelus vanderwaali* PH Skelton, 1976
Squeaker Catfish sp. *Synodontis vanderwaali* PH Skelton & PN White, 1990

Dr Ben C W Van der Waal was a biologist at the University of Venda for Science and Technology, Thohoyandou, South Africa, which he joined (1986) and was appointed professor (1990). He is now retired. He took part (2000) in a protest march to present a petition to the Department of Education in Thohoyandou. He was charged and convicted of taking part in an illegal strike and marching off campus during working hours! He was summarily dismissed (2003). He appealed, and the same year he was re-instated at the University, but on condition that he should 'keep quiet' for the next five years and that he would not be elected Head of Department of Biological Sciences. He co-wrote: *Epizootic ulcerative syndrome: Exotic fish disease threatens Africa's aquatic ecosystems* (2012). He collected the holotype of the spiny eel, with Skelton writing: "*The species is named after Mr B van der Waal, Senior Professional Officer in Charge of Fisheries in the Eastern Caprivi Government Service. Valuable systematic collections of fishes from this area, including this new species, have been submitted to the Queen Victoria an Albany Museum by Mr van der Waal.*"

Vandewalle

Pearlfish sp. *Encheliophis chardewalli* E Parmentier, 2004

Dr Pierre Vandewalle is a Belgian ichthyologist at the Laboratory of Functional Morphology, Institute of Zoology, University of Liège, Belgium where he is now a Professor and Director. He co-wrote: *Osteology and Myology of the Cephalic Region and Pectoral Girdle of* Bunocephalus Knerii, *and a Discussion on the Phylogenetic Relationships of the Aspredinidae (Teleostei: Siluriformes)* (2001). The species is named as a combination of the names of Dr M Chardon and Dr P Vandewalle of the University of Liège (Belgium), in recognition of their scientific accomplishments in ichthyology. (See also **Chardewall**)

Vanderyst

African Barb sp. *Enteromius vanderysti* M Poll, 1945

Father Hyacinthe Julien Robert Vanderyst R P (1860–1934) was a Belgian missionary, explorer, botanist, agronomist and entomologist in the Belgian Congo (1891) who collected the type under the auspices of the Musée Royal d'Afrique Centrale. He wrote: *Voyage d'etude de Leverville à la source de la Gobari. Missions belges de la Compagnie de jésus*, (1923). A number of plants are also named after him.

Van de Sande

Sculpin sp. *Gymnocanthus vandesandei* M Poll, 1949

Commandant Remi Van de Sande (1893–1969) was in charge of the Belgian training ship 'Mercator', from which the holotype was caught (1936).

Van Dragt

Armoured Catfish sp. *Micracanthicus vandragti* NK Lujan & JW Armbruster, 2011

Dr Randall 'Randy' Van Dragt is a biologist and ecologist who has a bachelor's degree awarded (1969) by Calvin College, Grand Rapids, Michigan, a master's from Cornell University (1971) and a doctorate (1985) awarded by the University of Rhode Island. He returned to Calvin College (1981) and is Professor of Biology there as well as currently directing the restoration of native prairie in Central Whidbey Island, Washington State. He co-wrote: *Creation care in horticulture: naturalizing the college campus* (2001). The etymology says Van Dragt's "*...patient introduction to tropical ecology and fish ecomorphology benefitted the first author immeasurably.*"

Vanessa

Butterfly Gurnard *Lepidotrigla vanessa* J Richardson, 1839

No etymology is given. *Vanessa* was used for a genus of butterflies (1807), which suggests a link to the fish's common name.

Van Heurn

Van Heurn's Rainbowfish *Melanotaenia vanheurni* MCW Weber & LF de Beaufort, 1922

Willem Cornelis van Heurn (1887–1972) was a Dutch taxonomist, civil engineer, botanist, educationalist, collector and biologist who worked for a period at the Natural History Museum, Leiden. He came from a wealthy family but chose to work all his life. He went to Suriname (1911), to Simeulue (off Sumatra) (1913), and to Dutch New Guinea (1920–1921). He then lived in the Dutch East Indies (mostly Java) (1924–1939), where he ran a laboratory for sea research; studied rat control on Java, Timor and Flores; was a schoolteacher; and served as head of the Botany Department at the Netherlands Indies Medical School before returning to Holland. Wherever he travelled or settled he collected natural history specimens which he meticulously prepared and labelled. Most he sent to the Leiden Museum, where he himself worked as an Assistant Curator for Fossil Mammals (1941–1945). He was a prolific writer, publishing c.100 articles on a wide range of topics, including such gems as 'The safety instinct in chickens' (1927), 'Cannibalism in frogs' (1928), 'Do tits lay eggs together as the result of a housing shortage?' (1955) and 'Wrinkled eggs' (1958). It was said of him in a memorial booklet published by the museum, 'He made natural history collections wherever he went and gave his attention to almost all animal groups. He was an excellent shot, and a competent preparator; his mammal and bird skins are exemplary.' Many different taxa, including a mammal, a reptile and five birds are named after him. He collected the rainbowfish type.

Van Heusden

East African Cichlid sp. *Haplochromis vanheusdeni* FDB Schedel, JP Friel & UK Schliewen, 2014

Hans van Heusden (b.1957) is a Dutch aquarist, angler and photographer (including underwater), who currently works for a construction company. He has collected in Africa and Latin America. He co-wrote the 2-volume: *Cichlids of Africa* (2016). According to the authors, he is: "*...one of the most dedicated cichlid naturalists, who has documented for the first time with underwater photographs and videos as well as with aquarium observations the behaviour and ecology of [this] new species and many other cichlids all over Africa.*"

Vanhöffen

Cosmopolitan Whipnose *Gigantactis vanhoeffeni* AB Brauer, 1902
Lanternfish sp. *Diaphus vanhoeffeni* AB Brauer, 1906

Dr Ernst Vanhöffen (1858–1918) was a German zoologist, geologist and botanist. He was a specialist in the study of medusa jellyfish. He studied at the Universities of Berlin and Königsberg, graduating (1888). He was at Stazione Zoologica Anton Dohrn, Naples (1889–1890). He took part in Drygalski's expedition to West Greenland (1892–1893) and in the Deutschen Tiefsee-Expedition on board the 'Valdivia' (1898–1899), during which the holotype of this species was collected. He also was a member of the Deutschen Südpolar-Expedition (1901–1903) on board the 'Gauss' under Drygalski's leadership. He taught at the University of Kiel, becoming a Professor (1901). He wrote: *Die Isopoden der Deutschen Südpolar-Expeditio 1901–1903* (1914). Vanhoffen Bluff in Antarctica is named after him.

Van Hyning

Golden Silverside *Labidesthes vanhyningi* BA Bean & ED Reid, 1930

Oather C Van Hyning (1901–1973) was an American naturalist, whose father, Thompson Van Hyning, was Director of the Florida State Museum. Although one source states that he graduated from Cornell University, another says he was forced to leave university prematurely, and never completed his degree, after he was caught drinking lab alcohol mixed with Coca-Cola by a campus policeman. In his youth, he seems to have been chiefly interested in the herpetology of Florida, made a large collection and published a number of papers. In later life his interests turned to orchids and bromeliads, and he made a series of fourteen expeditions to Mexico collecting a number of new species, and was at one time Director of the Bromeliad Society. He collected the silverside holotype. At least one bromeliad is also named after him.

Van Long

Loach sp. *Balitora vanlongi* HD Nguyen, 2005

This is a toponym referring to Văn Long, a pass near the Gâm River, Vietnam; the type locality.

Van Manen

Loach genus *Vanmanenia* SL Hora, 1932

Mari Albert Johan Van Manen (1877–1943) was a Dutch ethnographer, theosophist and orientalist who made a specialty of the study of Tibet and its texts. He was described as being the 'illustrious' General Secretary of the Asiatic Society of Bengal and a friend of the author.

Vanmelle

Vanmelle's Tonguefish *Symphurus vanmelleae* P Chabanaud, 1952

There is no etymology given. The holotype was taken during a voyage of 'Noordende III', Expedition Oceanographique Belge (1948–1849) on the Atlantic coast of Africa. The name may honour a (female) scientific illustrator named M. L. Van Melle of the Institut Royal des Sciences Naturelles de Belgique, which sponsored the expedition.

Vanneau

Goby genus *Vanneaugobius* CL Brownell, 1978

The 'Vanneau' was the name of the ship from which the type species, V. dollfusi, was collected.

van Oijen

Lake Victoria Cichlid sp. *Haplochromis vanoijeni* MP de Zeeuw & F Witte, 2010

Martien J P van Oijen is a Dutch ichthyologist at the Naturalis Biodiversity Centre (Leiden, Netherlands). He was honoured for his work on the taxonomy and ecology of haplochromine cichlids of Lake Victoria, and as one of the 'pioneers' of the Haplochromis Ecology Survey Team (HEST) of Leiden University that started its fieldwork (1977).

van Straelen

Sole genus *Vanstraelenia* P Chabanaud, 1950

Professor Victor Émile van Straelen (1889–1964) was a Belgian geologist and naturalist. He became Director, Belgian Royal Institute of Natural Science, Brussels (1925). He travelled widely, especially in Indonesia and Belgian Congo (DRC), and became (1933) President, Institute of National Parks of the Congo. He was the First President (1959–1964) of the Darwin Foundation. An amphibian is also named after him.

Van Tassell

Clementine Split-fin Goby *Psilotris vantasselli* L Tornabene & CC Baldwin, 2019

Dr James L Van Tassell is Research Associate-in-Residence in the Ichthyology Department at the AMNH (1976–present) and Adjunct Assistant Professor at the Department of Biology of Hofstra University where he was a research associate (2000–2010). Wagner College awarded his BSc (1969), Adelphi University his MSc (1975) and CUNY Graduate Center his PhD (1998). His research interests include the Systematics of fishes in the Gobiidae, with emphasis on the American seven-spined gobies and the ecology, distribution and systematics of fishes of the Canary Islands. He has undertaken several expeditions, such as a collecting trip to Venezuela (2006) and in Panama aboard the Smithsonian research vessel, RV Urraca (2007). He has published widely such as: *An Annotated Checklist of Shorefishes* of the Canary *Islands* (1985) and *The first record of Chromogobius britoi (Teleostei: Gobiidae) on the mainland European coast* (2005). He was honoured because his *"...work has contributed substantially to our understanding of the biology and systematics of the family Gobiidae, especially within the Gobiosomatini and other western Atlantic and eastern Pacific species."* Luke Tornabene was one of his students.

Vanzolini

Characin sp. *Leporinus vanzoi* HA Britski & JC Garavello, 2005

Tetra sp. *Hyphessobrycon vanzolinii* FCT Lima & N Flausino Jr, 2016

Professor Dr Paulo Emilio Vanzolini (1924–2013) was a Brazilian herpetologist, who worked at the AMNH, New York (late 1970s-1980s). His interest in herpetology started when he first cycled to secondary school, as he passed by snake pits which were maintained to 'milk' snakes to create antivenin. He was later allowed to help un-crate new snakes and help identify them too. He soon began a collection of his own at home, encouraged by his father and tolerated by his mother. He was invited when still a teenager to join a collecting expedition to Belém in Pará State. He was advised that medical school would be the best foundation for his zoological interests, so he spent two years in pre-med and six years at medical school, graduating MD from the University of Sao Paulo (1947) which was interrupted by two years in national service as a Corporal in the cavalry (1944–1946). He began work (1946) at the Zoological Museum, University of São Paulo, Brazil, but took a leave of absence (1947–1951) to study for his PhD at Harvard. He returned to the museum (1952) but worked part-time as his father was very ill with cancer. He took a part-time job producing a TV show which gave him financial stability, and he was able to take up a full-time post at the museum (1956). As scientific advisor to the Secretary of Agriculture (1959–1962) he re-organised the museum structure and then became Director there (1962–1993) and was responsible for assembling one of the largest collections of herpetofauna in the Americas. He (compulsorily) retired at 70, but still had a research affiliation with the Museum. He donated all his library to the museum (2008) and turned his attention to his second love, music, and is famous in Brazil as a composer of samba music and a poet. When he collected an award for his music on stage in his wheelchair (2013) he received a standing ovation. He wrote: *Elementary Statistical Methods in Zoological Systematics* (1993). Two mammals, seven amphibians and twelve reptiles are also named after him.

Varentsov

Riffle Minnow sp. *Alburnoides varentsovi* NG Bogutskaya & BW Coad, 2009

Petr Aleksandrovich Varentsov was a Russian naturalist who lived and travelled in the Transcaspian Province of the former Russian Empire. He took part in the 1892 survey of the Transcaspian region. He collected the type there (1896), and wrote a book on the geography and natural history of the area (1907).

Vari

Flannel-mouth Characin sp. *Semaprochilodus varii* RMC Castro, 1988
Toothless Characin sp. *Steindachnerina varii* J Géry, P Planquette & P-Y Le Bail, 1991
Toothless Characin sp. *Curimata vari* J Gaye-Siessegger & R Fricke, 1998
Characin sp. *Creagrutus varii* AC Ribeiro, RC Benine & CAA Figueiredo, 2004
Headstander sp. *Pseudanos varii* JLO Birindelli, FCT Lima & HA Britski, 2012
Characin sp. *Oligosarcus varii* NA de Menezes & AC Ribeiro, 2015
Driftwood Catfish sp. *Gelanoglanis varii* BB Calegari & RE dos Reis, 2016
Characin sp. *Xenurobrycon varii* MB Mendonça, LAW Peixoto, GM Dutra & AL Netto-Ferreira, 2016
Characin sp. *Nematocharax varii* SB Barreto, AT Silva, H Batalha-Filho,P Affonso & AM Zanata, 2018
Pencil Catfish sp. Trichomycterus varii L Fernández & J Andreoli Bize, 2018.
Whale Catfish sp. *Cetopsis varii* V Abrahão & MCC de Pinna, 2018
Bluntnose Knifefish sp. *Hypopygus varii* R Campos-Da-Paz, 2018
Angel Shark sp. *Squatina varii* FB Vaz & MR de Carvalho, 2018
Characin sp. *Astyanax varii* AM Zanata, R Burger, G Vita & P Camelier, 2019
Electric Eel sp. *Electrophorus varii* CD de Santana, WB Wosiacki, WGR Crampton, MH Sabaj, CB Dillman, RN Mendes-Júnior & N Castro e Castro, 2019
Armoured Catfish sp. Sturisomatichthys varii A Londoño-Burbano & RE Reis, 2019
Dwarf Tetra sp. *Odontocharacidium varii* EK de Queiroz Rodrigues & AL Netto-Ferreira, 2020
Characin sp. *Priocharax varii* GMT Mattox, CS Souza, m Toledo-Piza, R Britz & C Oliveira, 2020

Dr Richard Peter Vari (1949–2016) was a curator of fish at the Smithsonian. He was awarded his bachelor's degree (1971) by New York University and his doctorate (1977) by a joint graduate training programme run by the City University of New York and the AMNH. He was a post-doctoral fellow at the BMNH (1977–1978) and later in the Division of Fishes at the Smithsonian (1978), and then served on the curatorial staff (1980–2016) until his death from cancer. He revised most of the curimatids as well as many other groups of characiform and siluriform fish. He wrote more than 160 papers, naming over 200 new fish species. Several etymologies mention his outstanding contributions to Neotropical ichthyology. The one for the knifefish also praises his "*...enthusiasm and encouragement to many ichthyologists*", whilst Abrahão and de Pinna honoured his "*...inspiring role as a model of scientific and personal integrity to new generations of ichthyologists.*" (Also see **Irav**)

Varpachovski

Sharpbelly (cyprinid) sp. *Hemiculter varpachovskii* AM Nikolskii, 1903
Nikolai Arkadewich Warpachowski (see **Warpachowski**)

Varzea

Neotropical Rivuline sp. *Austrolebias varzeae* WJEM Costa, RE Reis & ER Behr, 2004
Characin sp. *Astyanax varzeae* V Abilhoa & LF Duboc, 2007

This is a toponym referring to the River Varzea, southern Brazil.

Vasil'yeva

Sakhalin Char *Salvelinus vasiljevae* SN Safronov & TV Zvezdov, 2005

Dr Ekaterina Denisovna Vasil'eva née Medvedeva (b.1952) is an ichthyologist who has both Russian and US nationality. Her bachelor's degree in zoology (1974) and her Doctorate of Science (1999) were both awarded by Moscow State University, where she has been a Research Scientist (1978–1985), Senior Research Scientist (1986–1990), Leading Research Scientist since 1990 and Chief of the ichthyological department of the Zoological Museum, Moscow, since 1986. She wrote: *Fishes of the basin of the Azov sea* (2013) This species was named after her "*...for her significant contribution to the study of fishes from the Russian Far East, including Salvelinus.*"

Vasnetzov

Viperfish sp. *Chauliodus vasnetzovi* NS Novikova, 1972

Vladimir Viktorovich Vasnetsov (1889–1953) was a Russian ichthyologist and ecologist. He wrote: *Fish growth as adaptation* (1947).

Vásquez

Armoured Catfish sp. *Chaetostoma vasquezi* CA Lasso Alcalá & F Provenzano Rizza, 1998

Dr Enrique Vásquez is a hydrobiologist at Fundación La Salle de Ciencias Naturales, Bolivar, Venezuela. He wrote: *The Orinoco river: A review of hydrobiological research* (1989). The etymology reads "*...for his 'pioneering' contributions to the limnology of Venezuela, where this catfish is endemic.*"

Vassali

Cusk-eel sp. *Parophidion vassali* GA Risso, 1810

Antonio Vassalli Eandi (1761–1825), also known as Anton Vassali Eandi, was a physicist, mathematician and an abbot and Professor of Physics at the University of Turin (1792–1825). He was ordained as a priest (1784) and in later life studied meteorology and became Director of the University's Observatory, Director of the Museum of Natural History and advised on the re-organisation of the Egyptian Museum.

Vauban

Searobin sp. *Lepidotrigla vaubani* L Del Cerro & D Lloris, 1997

The RV 'Vauban' was the vessel from which the holotype was taken (1985), off New Caledonia.

Vaughan

Spotback Scorpionfish *Pontinus vaughani* PS Barnhart & CL Hubbs, 1946

Dr Thomas Wayland Vaughan (1870–1952) was an American geologist and oceanographer. He graduated BSc from Tulane University (1888) then taught school (1889–1892), during which time he published a number of papers. He then studied biology at Harvard, receiving his BA (1893), MA (1894) and PhD (1903). He worked for the US Geological Survey as an assistant geologist (1894–1903) and participated in a number of surveys: West Indies (1901 & 1914), Panama (1911), Hispaniola (1919 & 1921), Virgin Islands (1919) and in the USA (1907–1923). He was then Director of the Scripps Institution of Oceanography until retirement (1924–1936). He became partially blind (1947) and died of a stroke.

Vayne

Righteye Flounder sp. *Poecilopsetta vaynei* J-C Quéro, DA Hensley & AL Maugé, 1988

Jean-Jacques Vayne is a French scientific illustrator. He has published with Quéro, including illustrating the books: *Les fruits de mer et les plantes des pêches Franmçaises* (1998) and *Guide des poissons de l'Atlantique européen* (2003) and co-authorship of new species: e.g. *Chromis durvillei : une nouvelle espèce de Pomacentridae de l'île de la Réunion (France, océan Indien) et premier signalement pour l'île de Chromis axillaris* (2009). He also illustrated the paper in which this species was described.

Vaz Ferreira

Neotropical Rivuline sp. *Austrolebias vazferreirai* HO Berkenkamp, V Etzel, JJ Reichert & H Salvia, 1994

Dr Raúl Vaz Ferreira Raimondi (1918–2006) was a Uruguayan biologist and ichthyologist. He was awarded a bachelor's degree by the National Museum of Natural History and Anthropology (1964) and a doctorate (1997). He was Professor and taught at the Faculty of Humanities and Sciences, University of the Republic, Montevideo (1950–1998) and was Director, Vertebrate Zoology Section (1962–1998). He was also in charge of Biology of the Scientific and Technical Department of the Oceanographic and Fisheries Service (1942–1960). He co-wrote: *Tres especies nuevas del género* Cynolebias *STEINDACHNER, 1876 (Teleostomi, Cyprinodontidae)* (1965).

Vazzoler

Vazzoler's Peacock Bass *Cichla vazzoleri* SO Kullander & EJG Ferreira, 2006

Gelso Vazzoler (1929–1987) was a Brazilian fisheries biologist and oceanographer. He was a former head of the Departamento de Biologia Acuática, Instituto Nacional de Pesquisas da Amazônia (Manaus, Brazil). He collected the holotype (1982).

Velain

Violet Warehou *Schedophilus velaini* H-E Sauvage, 1879
[Alt. African Barrelfish]

Charles Vélain (1845–1925) was a French geographer and geologist, and particularly an expert on volcanism. He took part in a mission to Saint Paul Island (1874) in the southern Indian Ocean, which is the type locality of this fish.

Velez

Velez Ray *Rostroraja velezi* F Chirichigno, 1973

Juan José Vélez Diéguez is an ichthyologist at Universidad Nacional del Callao, Peru where he is a professor. He was at Instituto del Mar del Perú. He wrote: *Peces Marinos Clave Artificial para Identifica los Peces Marinos Comunes en la Costa Central del Perú* (1980). He and the describer have often published together. He was honoured for his dedication to ichthyology in general and for his collaborations with the author.

Velioglu

Velioğlu's Chub *Alburnoides velioglui* D Turan, C Kaya, FG Atalay-Ekmekçi & ES Doğan, 2014

Dr Hasan Basri Velioğlu is a Turkish physician who is chief of staff at Rize State Hospital. He was honoured because he had 'eased' and contributed to the authors' studies through the use of radiography.

Venere

Characin sp. *Leporinus venerei* HA Britski & JC Birindelli, 2008
Characin sp. *Moenkhausia venerei* MG Petrolli, VM Azevedo-Santos & RC Benine, 2016

Dr Paulo César Venere is a Brazilian biologist and ichthyologist who graduated with a bachelor's degree (1986), a master's (1991) and defended his doctorate in genetics and evolution (1998), all from the Federal University of São Carlos. This was followed by post-doctoral training at Universidade Estadual de Botucatu, São Paulo. He became a full Professor in the Department of Biology and Zoology, Universidade Federal de Mato Grosso, Campus Universitário do Araguaia (2014). He was honoured for his: "*…contributions to our knowledge of the Rio Araguaia ichthyofauna of Brazil.*"

Veolia

Veolia Rainbowfish *Melanotaenia veoliae* Kadarusman, D Caruso & L Pouyard, 2012

Foundation Veolia Environment, based in France, supports non-profit activities related to sustainable development and the protection of the environment. It sponsored the Lengguru-Kaimana Expedition (October–November 2010) during which the rainbowfish type was collected.

Vera (Kasim)

Belitung Glass-Perch *Gymnochanda verae* HH Tan & KKP Lim, 2011

Vera Kasim is the wife of a good friend of the author, Gunawan Kasim (q.v.), who is a fish exporter. She was honoured for her generous assistance and logistic support. (Also see **Gunawan**)

Vera, J

Ecuadorian Deep-sea Scorpionfish *Trachyscorpia verai* P Béarez & H Motomura, 2009

Johnny Vera purchased the holotype (2006) at a fish market in Puerto López, Ecuador. He was apparently a local who kept an eye out for 'weird fish' on behalf of the senior author, Philippe Béarez.

Vera, JA

Suckermouth Catfish sp. *Astroblepus verai* CA Ardila Rodríguez, 2015

Jorge Augusto Vera Mantilla is a Colombian independent agronomist who helped capture the holotype.

Vérany

Large-scale Lanternfish *Symbolophorus veranyi* E Moreau, 1888

Chevalier Jean Baptiste Vérany (1800–1865) was a French naturalist and pharmacist. His specialty was the study of cephalopods. He co-founded the Muséum d›histoire Naturelle de Nice (1846), of which he was the Director. He wrote: *Catalogue des Mollusques cephalopodes, pteropodes, Gasteropodes nudibranches, etc. des environs de Nice* (1853).

Verco

Verco's Pipefish *Vanacampus vercoi* ER Waite & HM Hale, 1921
[Alt. Flinder's Pipefish]

Dr Sir Joseph Cooke Verco (1851–1933) was an Adelaide physician and conchologist. He qualified MRCS (1874), MB (1875) and DD & FRCS (1876), all at the University of London. He was a houseman at St Bart's (1876–1878) when he returned to Adelaide where, after a few years in general practice, he became honorary physician to Adelaide Hospital (1882–1912) and lecturer in medicine at the University of Adelaide (1887–1915) being associated with them until retirement (1919). Interested in shells as a boy, he took up the serious study of conchology (1887), dredging in the Great Australian Bight. His collection and library were donated to the South Australia Museum where he spent his time after retirement as honorary curator of molluscs.

Verdi

Bluntnose Knifefish sp. *Brachyhypopomus verdii* WGR Crampton, CDCM de Santana, JC Waddell & NR Lovejoy, 2017

Dr Absalon Lorgio Verdi Olivares is a Peruvian conservation biologist at National University of the Peruvian Amazon, Iquitos, where he was Director of the Graduate School. The national University of Trujillo awarded both his bachelor's degree and his doctorate, whilst Utah State University, Logan, USA, awarded his master's. He wrote: *Growth of Osteoglossum Bicirrhosum 'Silver Arahuana' fingerlings in controlled environments influenced by food frequency* (2014).

Verduyn

Lake Malawi Cichlid sp. *Copadichromis verduyni* A Konings, 1990

Dirk Verduyn, aka Verduijn (1942–2018), was a Dutch cichlid retailer and wholesaler. According to the etymology, he: "*...considerably advanced the aquarium hobby in Holland and across its borders.*"

Vereker

Dwarf Paradise Fish *Parapolynemus verekeri* W Saville-Kent, 1889

Captain the Honorary Foley Charles Prendergast Vereker (1850–1900) was a British naval officer on HMS *Myrmidon* during a surveying expedition in North-west Australia.

Verheyen E

Mormyrid sp. *Marcusenius verheyenorum* TM Baba, T Kisekelwa, CD Mizani, E Decru & E Vreven, 2020

Dr Erik K Verheyen (b.1958) is the son of Dr Walter Verheyen (below). Ghent University awarded his MSc (1982) and the University of Antwerp awarded his doctorate (1990). After post doctorial reasearch at the Royal Belgian Institute of Natural Sciences, he became staff scientist (2003) and Head of OD Taxonomy and Phlogeny of Recent Vertebrates there (2005 - present). He has published around 160 papers such as: *Origin of the Superflock of Cichlid Fishes from Lake Victoria, East Africa* (2003). He told us: "*The first part of my career was focussed on evolutionary work on African lacustrine cichlids. Subsequently, I became increasingly involved in molecular phylogenetic studies on rodents – initially as a support for my father's taxonomic work. I also coordinated a DNA barcoding project on the African bird collections my grandfather collected in the 1950's. Based on the collections of my father (and collections I gathered myself) I became involved in a series of molecular phylogeographic studies aimed at inferring the evolutionary histories of these small mammals on the African continent. Since the last 10 years our mammal work was increasingly carried out I the context of the study of zoonotic diseases.*" The etymology says the species is "*…dedicated to the late… …Walter Verheyen… …and his son Eric Verheyen… …for contributing… … to our knowledge of the vertebrate fauna of the Congo Basin.*"

Verheyen, W

Mormyrid sp. *Marcusenius verheyenorum* TM Baba, T Kisekelwa, CD Mizani, E Decru & E Vreven, 2020

Professor Dr Walter Norbert Verheyen (1932–2005), father of Dr Erik K Verheyen (above), was a Belgian zoologist who was (1966) a professor at the Biology Department of the University of Antwerp, having previously (1962) been a lecturer at the University of Ghent. After some studies on fishes, birds and primates, he became an expert mammologist who carried out fieldwork (mostly on rodents) from Sub Sahara Africa. His many publications include: *Une nouvelle sous espèce de Colobus polykomos (ZIMMERMANN) 1780* (1959) to: *The biology of Elephantulus brachyrhynchus in natural miombo woodland in Tanzania (1995).* Nicolas said in his etymology of the wood mouse "*The species is named in honor of our colleague the late Walter Verheyen, who initiated the taxonomic study on the genus* Hylomyscus, *and in recognition of his significant contribution to the systematics and biogeographic research on African small mammals.*"

Verigina

Spinyfin sp. *Diretmoides veriginae* AN Kotlyar, 1987

Inna Alexandrovna Verigina was curator of marine fishes, Zoological Museum, Moscow University. She helped Kotlyar over the course of many years.

Verloren

Verlorenvlei Redfin *Pseudobarbus verloreni* A Chakona, ER Swartz & PH Skelton, 2013

Not an eponym but a toponym, as the Verlorenvlei is a river in South Africa.

Vernay

Oriental Barb sp. *Hypsibarbus vernayi* JR Norman, 1925
Spottail Barb *Enteromius afrovernayi* JT Nichols & WR Boulton, 1927

Arthur Stannard Vernay (1877–1960) was an English-born art & antiques dealer, big game hunter, naturalist and philanthropist who lived in the USA. He developed a deep interest in natural history, and was a Trustee of the AMNH. Jointly with Colonel John Faunthorpe he financed six expeditions to Burma, India and Thailand (1922–1928) and financed a BMNH collecting trip to Tunisia (1925). He was a friend of Theodore and Kermit Roosevelt (q.v.) and, especially, of an American millionaire named Charles Suydam Cutting, with whom he travelled quite often; notably they journeyed to Lhasa (1935) and met the 13th Dalai Lama. Having sold his business and all his collections and antiques, he retired to the Bahamas (1940s) where he co-founded the Bahamas National Trust. Ten birds, a reptile and three mammals are also named after him.

Veronica

Charco Palma Pupfish *Cyprinodon veronicae* M de L Lozano-Vilano & S Contreras-Balderas, 1993

Verónica Contreras Arquieta is the junior author's daughter and the senior author's niece. She was honoured for her participation in the collecting trip (1984) during which the pupfish type was taken. (Also see **Cecilia**)

Verreaux

Southern Conger *Conger verreauxi* JJ Kaup, 1856

Jules Pierre Verreaux (1807–1873) was a French naturalist, collector, and dealer in natural history specimens; as was his brother Jean Baptiste Edouard Verreaux (1810–1868). Jules was an ornithologist and plant collector for the MNHN, Paris, which sent him to Australia (1842). He collected in Tasmania, New South Wales, and Queensland (1842–1850), returning to France with a reported 115,000 items (1851). Earlier he helped Andrew Smith in founding the South African National Museum, Cape Town. The Verreaux family traded at Maison Verreaux, a huge emporium for feathers and stuffed birds. They were ambitious taxidermists and gained notoriety for having once attended the funeral of a tribal chief whose body they then disinterred, took to Cape Town, and stuffed. In 1888 the Catalán veterinarian Francisco Darder, then Curator of the Barcelona Zoo, purchased the "specimen" from one of the brothers' sons, Edouard Verreaux. This controversial exhibit was on show in Barcelona until the end of the 20th century, when the man's descendants demanded that it be returned for a decent burial. Between them the brothers also have an amphibian, two mammals and twenty-one birds named after them.

Verrill

Wolf Eelpout *Lycenchelys verrillii* GB Goode & TH Bean, 1877
White-edged Moray *Gymnothorax verrilli* DS Jordan & CH Gilbert, 1883

Addison Emery Verrill (1839–1926) was an American marine biologist. He was a student of Louis Agassiz at Harvard, which awarded his degree (1862). He became the first Professor of Zoology at Yale, teaching there until he retired (1864–1907). He was also Professor of Comparative Anatomy at the University of Wisconsin (1868–1870. He was particularly interested in the invertebrate fauna of the Atlantic coast and became the expert of his day in cephalopods. He supplied the authors with type material from Yale University. One of his works: *Report upon the Invertebrate Animals of Vineyard Sound* (1874) is still the standard manual of the marine zoology of New England. In later life, he explored the Bermuda Islands with students, studying their geology as well as the marine fauna, and wrote: *The Bermuda Islands* (1903). In all he wrote over 350 papers, which included descriptions of over one thousand animal species.

Verrycken

Blenny sp. *Parablennius verryckeni* M Poll, 1959

C Verrycken was a radio-telegraph operator based in Banana (Atlantic coast of the DRC). He was also a sport fisherman who collected specimens for Poll.

Versluys

Versluys' Jawfish *Stalix versluysi* MCW Weber, 1913
African Barb sp. *Labeobarbus versluysii* M Holly, 1929

Dr Jan Versluys (1873–1939) was a Dutch zoologist and anatomist. The patronym is not identified in either of these species' descriptions, but we believe both very probably honour him. The University of Amsterdam awarded his bachelor's degree and the University of Giessen, Germany his doctorate. He lectured at the University of Amsterdam, was an Assistant Professor at Giessen and Professor of Zoology at Ghent, Belgium and finally at Vienna. He travelled widely in the Dutch East Indies (Indonesia), including serving as assistant to Max Weber on the 'Siboga' expedition.

Vespucci

Armoured Catfish sp. *Hisonotus vespuccii* FF Roxo, GSC Silva & C de Oliveira, 2015

Amerigo Vespucci (1454–1512) was an Italian navigator after whom the Americas are named. He is credited to have made four voyages to the 'New World', though there is doubt about the first and fourth of these. A letter published in 1504 purports to be an account by Vespucci of a first visit to the Americas (1497–1498), but some modern scholars doubt that the voyage took place and consider the letter a forgery. However, whoever wrote the letter made numerous authentic observations about native customs. Vespucci's second voyage (or possibly first 'real' one!) took place in 1499, when he joined an expedition in the service of Spain with Alonso de Ojeda as fleet commander. Vespucci and his backers financed two of the three ships in the small fleet. Their intention was to explore the coast of a new landmass found by Columbus on his third voyage. After reaching the coast of present-day Guyana, Vespucci and Ojeda separated, with Vespucci sailing south and discovering the mouth of the Amazon River.

Vettones

Alagón Spined Loach *Cobitis vettonica* I Doadrio Villarejo & A Perdices, 1997

This is not a true eponym but rather refers to the Vettones; the historical inhabitants of the sheep-raising area of west-central Spain, which coincides with the range of this species.

Vicente Spela

Glass Knifefish sp. *Eigenmannia vicentespelaea* ML Triques 1996

This is a toponym, referring to the Cave São Vicentie II, Tocantins River basin, Goiás, Brazil, which is he only known only known area where this species occurs.

Victor, BC

Galapagos Razorfish *Xyrichtys victori* GM Wellington, 1992

Dr Benjamin C Victor (b.1957) was born in South Africa, but educated in the USA, studying ichthyology at Cornell University which awarded his BA (1978). He was awarded his PhD (1986), MD (1992) and MBA (2000) by the University of California. While there (1993–1997) he was a Clinical Associate in the Department of Pathology. He worked briefly as a forensic pathologist (1997) in the Orange County Sheriff-Coroner's Office before taking up a position (1998) as Medical laboratory Director and CEO of the Ocean Science Foundation. He lists his interests as: fish biology, evolution, medicine, microbiology, mycology, geography, economics and his primary interest as being in the larval ecology of coral reef fishes and its role in determining biogeography and population dynamics. Apart from the USA, he has worked in Panama, Galapagos, Mexico, Brazil, Bali, Puerto Rico and the US Virgin Islands. He is also affiliated with the Guy Harvey Research Institute and works on shark tagging and genetic studies involving a variety of fish species. He has published c.40 papers. He was the co-discoverer of this species (1990) with the author.

Victor (Makushok)

Victory Whiptail *Asthenomacrurus victoris* YI Sazonov & YN Shcherbachev, 1982

Dr Viktor Markelovich Makushok (1924–1992) was an ichthyologist at the P P Shirshov Institute of Oceanology, Academy of Sciences of the USSR. He wrote: *The morphology and classification of the northern blennoid fishes (Stichaeoidae, Blennioidei, Pisces)* (1958). The whiptail's vernacular name seems to have been coined in a misunderstanding that the binomial is actually an eponym.

Victoria (Lake)

Lake Victoria Squeaker *Synodontis victoriae* GA Boulenger, 1906
Victoria Stonebasher *Marcusenius victoriae* EB Worthington, 1929
Lake Victoria Cichlid sp. *Paralabidochromis victoriae* PH Greenwood, 1956

This is a toponym referring to Lake Victoria, where these species occur.

Victoria

Characin sp. *Knodus victoriae* F Steindachner, 1907

This is a toponym referring to Victoria, a place near the mouth of the River Parnahuba in Brazil.

Victoria

Western Striped Cardinalfish *Ostorhinchus victoriae* A Günther, 1859

This is a toponym and refers to a large area between Perth and Carnarvon in Western Australia which in the 1840s was referred to as the Province of Victoria.

Victoria (Oliveros)

Neotropical Livebearer sp. *Phallotorynus victoriae* OB Oliveros, 1983

Maria Victoria Oliveros is probably related to the author, though we have not been able to find full details.

Vidal

Characin sp. *Characidium vidali* HP Travassos, 1967

Nei Vidal was a geologist and palaeontologist, a colleague of the author at the Federal University of Rio de Janeiro.

Viera

Neotropical Rivuline sp. *Anablepsoides vieirai* DTB Nielsen, 2016

Gilberto da Silva Vieira is a Brazilian biologist and environmentalist. He discovered and collected the type of the species with Maria da Conceição Pereira da Silva.

Viereck

Barbeled Dragonfish sp. *Melanostomias vierecki* HW Fowler, 1934

Henry Lorenz Viereck (1881–1931) was an American entomologist who specialized in Hymenoptera. He was known to be temperamental and often changed jobs at short notice. He studied medicine at Jefferson Medical College (1903–1905) but did not finish the course. He worked for the Connecticut Agricultural Station (1904–1905) as an assistant in a pathology laboratory (1905–1907), the Pennsylvania Department of Zoology (1907–1908), the Bureau of Entomology, USNM (1909–1913) as an entomological explorer for the California State Horticultural Commission (1914), the US Department of Agriculture Biological Survey (1916–1923), the Entomological Division, Canada Department of Agriculture (1923–1926) and at the Academy of Natural Sciences, Philadelphia from 1926. Fowler was 'indebted' for collections of fishes. He was killed by a hit-and-run motorist.

Vigna

Stone Loach sp. *Paracobitis vignai* TT Nalbant & PG Bianco, 1998

Dr Augusto Vigna Taglianti (b.1943) is an entomologist at the Department of Biology, La Sapienza University of Rome, which he joined as an assistant (1971) and where he taught

(1972–1982) vertebrate zoology and entomology. He became Professor of Entomology (1982). He originally graduated in natural sciences from the University of Rome (1966). He collected the holotype. He was President of the Italian Entomological Society (1996–2012) and wrote on Sardinian earwigs: *I Dermatteri di Sardegna* (Dermaptera) (2011).

Vigors

Snow Trout sp. *Osteobrama vigorsii* WH Sykes, 1839

Nicholas Aylward Vigors (1785–1840) was an Irish zoologist and politician. He served in the army during the Peninsular War (1809–1811), then returned to Oxford and graduated (1815). He was elected a Fellow of the Royal Society (1826), and was the first Secretary of the Zoological Society of London (1826–1833). He was a Member of Parliament at Westminster (1828–1835 & 1837–1840). He contributed a chapter on ornithology in *Zoology of Captain Beechey's Voyage* (1839). Ten birds are named after him, whilst he himself described over a hundred bird species.

Vilhena

Shellear sp. *Parakneria vilhenae* M Poll, 1965

Comandante Ernesto Jardim de Vilhena (1876–1967) was a Portuguese naval officer (1898–1920), during which time he was also a colonial administrator in Africa (1902–1912) and a politician, including being Minister for the Colonies (1917–1920). He combined his public career with being a businessman and was Director and Chairman of the Angola Diamond Company (1919–1955) and Chairman of the Portuguese Diamond Society (from 1958). Poll honoured him for his support of the study of Angolan zoology.

Villadolid

Priapiumfish sp. *Neostethus villadolidi* AW Herre, 1942

Dr Deogracias V Villadolid (1896–1976) was Director of the Bureau of Fisheries in the Philippines, including during the Japanese occupation.

Villar

Tilefish sp. *Lopholatilus villarii* A de Miranda Ribeiro, 1915

Frederico Otávio de Lemos Villar (1875–1964) was a Brazilian naval officer involved in scientific studies on the Brazilian coast, especially in relation to fisheries.

Villars

Armoured Catfish sp. *Isorineloricaria villarsi* CF Lütken, 1874
[Syn. *Squaliforma villarsi*]

Dr C-J-F-H Carron de Villars was a physician and zoologist. He provided the holotype of this species. We have been unable to find out anything about him except that he was for a time an oculist surgeon in the Sardinian Army.

Villasboas

Villasboas' Tube-snouted Ghost Knifefish *Sternarchorhynchus villasboasi* CDCM de

Santana & RP Vari, 2010
Armoured Catfish sp. *Harttia villasboas* OT Oyakawa, I Fichberg & L Rapp Py-Daniel, 2018

Orlando Villa Bôas (1914–2002), Cláudio (1916–1998) and Leonardo (1918–1961) Villas-Bôas were brothers. They were Brazilian explorers, anthropologists and authors who wrote: *Xingu; the Indians, their Myths* (1975) and were among the few non-missionaries to live permanently with indigenous tribes, treating them as equals and friends. They persuaded tribes to end their internecine feuds and unite to confront encroaching settlement. The etymology for the catfish says it honours the brothers: "*…Orlando, Claudio and Leonardo Villas Boas, three frontiersmen that led the Expedição Roncador-Xingu during the years of 1943 to 1949, with the mission to explore a wide and unknown territory of the Amazonian regions of Brazil. In 1949, the expedition reached the tablelands of the Serra do Cachimbo. One of the most important results of this expedition was the establishment of the Parque Indígena do Xingu in 1961, the first huge indigenous area in all South America.*" An amphibian is also named after them.

Villegas

Grey Grenadier *Trachyrincus villegai* G Pequeño Reyes, 1971
Luis Villegas was a marine biologist at the Department of Fisheries, University of Valparaiso, Peru.

Villwock

Panama Killifish sp. *Cynodonichthys villwocki* H-O Berkenkamp & VMF Etzel, 1997
Characin sp. *Astyanax villwocki* A Zarske & J Géry, 1999
Villwock's Killifish *Anatolichthys villwocki* T Hrbek & RH Wildekamp, 2003
[Syn. *Aphanius villwocki*]

Dr Wolfgang Villwock (1930–2014) was a Professor at the Zoological Institute and Zoological Museum, University of Hamburg. He wrote: *Die Titicaca-See-Region auf dem Altiplano von Peru und Bolivien und die Folgen eingeführter Fische für Wildarten und ihren Lebensraum* (1993). He collected some of the type series and made them available for the authors to study.

Vilma

Rainbow Copella *Copella vilmae* J Géry, 1963
Chocolate Neon Tetra *Hyphessobrycon vilmae* J Géry, 1966

Vilma Schultz was the wife of Harald Schultz (1909–1966) the ethnographer and fish collector.

Vincent, A

Pipefish sp. *Leptonotus vincentae* DC Luzzatto & ML Estalles, 2019

Dr Amanda Claire Jane Vincent is a marine conservationist who is the Director and Co-Founder of 'Project Seahorse'. The University of Western Ontario awarded her BSc (1981), the University of British Columbia her MSC (1985) and Corpus Christi College, Cambridge her PhD (1990). She has had academic posts from visiting scientist to Professor at universities in England, Sweden and Germany as well as several in Canada, culminating in being Professor at the Institute for the Oceans and Fisheries at the University of British Columbia (2002–present). She has written or co-written well over 100 papers and book chapters as well as the book *The International Trade in Seahorses* (1996). The authors describe her as someone "*...who does so much for the conservation of this group of fish*" and, in their etymology, say she was honoured because she "*...was the first person to study seahorses underwater, the first to document the extensive trade in these fishes and the first to initiate a seahorse conservation project.*"

Vincent, SG

Vincent's Sillago *Sillago vincenti* RJ McKay, 1980

S G Vincent was a Technical Officer at the Central Marine Fisheries Research Institute, Cochin, India, and collected the holotype (1980). He later worked at the Institute of Fish Eggs and Larvae Studies, Kuzhithurai, Tamil Nadu (c.2009). Among his published works are the co-written: *Invasion of Cliona margaritifera dendy and C. lobota hancock in the molluscan beds along the Indian coast* (1993) and *Perch fishery at Vizhinjam* (1994). Roland McKay thanked his friend (and colleague at the Institute during his time there) "*...for his valuable assistance in collecting specimens, obtaining information for this study recognising the two species in the field and assisting with the measurements.*"

Vincent (Gulf)

Cardinalfish genus *Vincentia* FL de Castelnau, 1872
Western Shovelnose Ray *Aptychotrema vincentiana* JW Haacke, 1885
St. Vincent's Gulf Dogfish *Asymbolus vincenti* AHC Zietz, 1908
[Alt. Gulf Catshark]

These species and genus are named after the location where the holotypes were caught – the Gulf of St Vincent, near Kangaroo Island, South Australia.

Vinciguerra

Lightfish genus *Vinciguerria* DS Jordan & BW Evermann, 1896
Sisorid Catfish sp. *Exostoma vinciguerrae* CT Regan, 1905
Stone Loach sp. *Schistura vinciguerrae* SL Hora, 1935

Dr Decio Vinciguerra (1856–1934) was an Italian physician, naturalist and ichthyologist at the Museo Civico di Storia Naturale di Genova (1883–1931). He studied at the University of Genoa and (1878) obtained a degree in Medicine and Surgery. Immediately after graduating he was appointed assistant to the Chair of Zoology and Comparative Anatomy in the University of Genoa. He participated in the Italian expedition to Tierra del Fuego (1882) as the official botanist and zoologist. He was (early 1890s) a Professor, Sapienza-Università di Roma, and Director of the Aquarium of Rome, which was also used as a fish hatchery. Two reptiles are also named after him.

Vinson

Vinson's Goby *Heterelеotris vinsoni* DF Hoese, 1986

Jean-Michel Vinson is a Mauritian zoologist and scientific illustrator, an expert on the fauna of the Mascarene Islands. His father, Joseph Lucien Jean Vinson (1906–1966), was also a zoologist (mainly an entomologist), and together they wrote: *The Saurian Fauna of the Mascarene Islands* (1969).

Vinton

Darter Characin sp. *Ammocryptocharax vintonae* CH Eigenmann, 1909
[Binomial originally given as *vintoni*]

Mrs C Vinton was, according to the author, "...*one of the few ladies who have visited the habitat of this species*" (in British Guiana). No further details are given, but this perhaps refers to Josephine Caroline Carstarphen, who married Lindley Vinton (1897) and moved with him to British Guiana.

Violaris

Alotau Goby *Acentrogobius violarisi* GR Allen, 2015

Julius Violaris is the owner of Nawe Constructions at Alotau, Papua New Guinea. He was honoured for "*allowing uninterrupted access to the survey site that yielded the new species, and for his generosity in continuing to provide an excellent home base for the Chertan, the live-aboard dive vessel that served as the logistic centre for the trip on which the new species was collected.*" (Also see **Van der Loos**)

Virginia

Sangama Cory (catfish) *Corydoras virginiae* WE Burgess, 1993

Mrs Virginia Schwartz was the wife of aquarium-fish exporter Adolfo Schwartz (q.v.), who collected the holotype of this species.

Vistonis (Lake)

Thracian Shad *Alosa vistonica* PS Economidis & AI Sinis, 1986

This is a toponym referring to Lake Vistonis, Greece, the shad's only known distribution.

Vital Brazil

Pencil Catfish sp. *Trichomycterus vitalbrazili* PJ Vilardo, AM Katz & WJEM Costa, 2020

Dr Vital Brazil Mineiro da Campanha (1865–1950) was a Brazilian physician, immunologist and biomedical scientist. He first discovered the polyvalent antiophidic serum, used to treat venomous snake bites. He founded both the Instituto Butantan (1899) and Instituto Vital Brazil (1919) and the catfish was first found in the campus grounds of the latter institute. He is regarded as one of the most important Brazilian scientists ever and has been honoured in many ways, including being portrayed on a banknote. Four species of snake are also named after him.

Vityaz

Flabby Whalefish genus *Vitiaziella* TS Rass, 1955
Eelpout sp. *Lycenchelys vitiazi* AP Andriashev, 1955
Vityaz Frostfish *Benthodesmus vityazi* NV Parin & VE Becker, 1970
Spearcheek Cusk *Monomitopus vitiazi* H Nielsen, 1971

Vitiaz Dragonfish *Eustomias vitiazi* NV Parin & Pokhil'skaya, 1974
Eelpout sp. *Melanostigma vitiazi* NV Parin, 1980
Cusk-eel sp. *Neobythites vityazi* H Nielsen, 1995
Rattail sp. *Coelorinchus vityazae* T Iwamoto, YN Shcherbachev & B Marquardt, 2004
Barreleye sp. *Dolichopteryx vityazi* NV Parin, TN Belyanina & SA Evseenko, 2009
Lizardfish sp. *Synodus vityazi* HC Ho, AM Prokofiev & KT Shao, 2010
Greeneye sp. *Chlorophthalmus vityazi* SG Kobyliansky, 2013

The Russian Research Vessel 'Vityaz' (also spelled Vitiaz) is famous for many cruises, during which the holotypes of several of the above fish were caught. NV Parin caught the holotype of *Dolichopteryx vityazi* during the vessel's 26th cruise (1959). The describers of the rattail use the feminine form *vityazae*; the only case reflecting the traditional female gender of ships.

Vivaldi

Vivaldi's Catshark *Bythaelurus vivaldii* S Weigmann & CJ Kaschner, 2017

Antonio Lucio Vivaldi (1678–1741) was an Italian composer, too well known to warrant a long biography here. The catshark was so named to reflect its relationship to *B. bachi* (see **Bach, JS**).

Vivien

Madagascar Wrasse *Macropharyngodon vivienae* JE Randall, 1978

Mireille Laurence Harmelin-Vivien (b.1947) is a French marine biologist and ichthyologist who is Directeur de Recherche CNRS Emeritus (2012), Institut Méditerranéen d'Océanologie, Marseille where she worked (1973–2012). She completed both her first (1972) and second (1979) PhD at Aix-Marseille University. She collected and illustrated the holotype (1972). She has written or co-written more than 260 papers (1971–2019), including: *The effects of storms and cyclones on coral reefs: a review* (1994) and *Patterns of trace metal bioaccumulation and trophic transfer in a phytoplankton-zooplankton-small pelagic fish marine food web* (2019).

Vlad

Toadfish genus *Vladichthys* DW Greenfield, 2006

Dr Vladimir 'Vlad' Walters (1927–1987) (See **Walters**)

Vladi

Scrapetooth Characin sp. *Apareiodon vladii* CS Pavanelli, 2006

Dr Vladimir 'Vladi' Pavan Margarido is a Brazilian biologist. The State University of Maringá awarded his bachelor's degree (1992), the Federal University of São Carlos both his master's (1995) and doctorate in genetics and evolution (2000). He is presently an Associate Professor at Universidade Estadual do Oeste do Paraná. According to the etymology, he noticed that this species had not been described based on cytogenetic features and "*...made hard efforts to collect good specimens*" for the type series.

Vladi Becker

Snailfish sp. *Careproctus vladibeckeri* AP Andriashev & DL Stein, 1998

Dr Vladimir Edwardovich Becker (1925–1995) was a Russian ichthyologist who specialised in the study of luminous anchovies. After graduating (1955) from the Ichthyology Faculty of Mosrybvtuz, the Moscow Technology Institute, he was a post-graduate student then an assistant (1957–1959) while he completed his dissertation (1959) *The influence of population density on the process of oogenesis in goldfish.* He moved on (1960) to the Institute of Oceanology at the Russian Academy of Sciences. He took part in 18 long expeditionary voyages to the Pacific, Atlantic, Indian and Southern Oceans on a number of research vessels: 'Vityaz, Baikal', 'Akademik Kurchatov', 'Akademik Vavilov', 'Akademik Keldysh', 'Akademik Ioffe' and 'Dmitry Mendeleev' the collections from which are at the Zoological Museum of Moscow State University.

Vladykov

Danube Whitefin Gudgeon *Romanogobio vladykovi* PW Fang, 1943
Vladykov's Lamprey *Eudontomyzon vladykovi* O Oliva & G Zanandrea, 1959
[Alt. Danubian Brook Lamprey]
Zagros Toothcarp *Esmaeilius vladykovi* BW Coad, 1988

Dr Vadim Dimitrievitch Vladykov (1898–1986) was a Ukrainian-born Canadian zoologist. Charles University, Prague, awarded his degree in zoology and anthropology. He took a post as assistant scientist at the Fisheries Research Board of Canada (1930) and became a Canadian citizen (1936). He then worked at the Laboratoire de Zoologie, Université de Montréal as Professor of Ichthyology. He then became a professor at Ottawa University (1958–1973) until retirement, thereafter being Professor Emeritus. He was also research associate at the National Museum of Natural Sciences. A prolific writer, he published more than 290 papers and contributed to longer works. He is also honoured in the name of a ship, 'Canadian Coast Guard Ship Vladykov'.

Vlamingh, CW

Blue-spotted Goatfish *Upeneichthys vlamingii* G Cuvier, 1829
Bignose Unicornfish *Naso vlamingii* A Valenciennes, 1835

Admiral Cornelis Willemsz de Vlamingh (<1678–1735) was a Dutch explorer and naval officer. He collected and, according to Cuvier, scientifically drew a 'great number' of fish. He was 'maitre d'equipage' of the Dutch East India Company in Bengal (1698) and Rear Admiral of the Fleet aboard *la Cour de Hollande* on the return journey (1715). He had earlier (1696–1698) been a Captain of a ship ('Wezeltje') that was part of a fleet commanded by his father (see next entry), which had sailed on a rescue mission for a previous ship that went missing (1694). They explored the coast of Australia and the Indian Ocean. His collection is in MNHN along with his drawings.

Vlamingh, W

Western Australian Gizzard Shad *Nematalosa vlaminghi* ISR Munro, 1956
[Alt. Perth Herring]

Munro supplied no etymology, but probably named for Willem Hesselsz de Vlamingh (1640–c.1698), a Dutch whaler and then sea captain of the Dutch East India Company who explored western Australia, where he is known to have landed (1697). It may, alternatively, be named for his son (above).

VNIRO

Roughy sp. *Hoplostethus vniro* AN Kotlyar, 1995

VNIRO stand for the All-Russian Research Institute of Fisheries and Oceanography. Kotlyar worked there for 20 years and was a member of the expedition that collected the holotype.

Vogt, D

Halfbeak sp. *Dermogenys vogti* M Brembach, 1982

Dieter Vogt (b.1933) is a German biologist and aquarist who co-wrote: *The freshwater fishes of Europe to the Ural and Caspian Sea* (1979).

Vogt, FX

Squeaker Catfish sp. *Atopochilus vogti* J Pellegrin, 1922

Monsignor Franz Xaver Vogt (1870–1943) was a German Catholic Missionary in Africa. He was ordained as a priest (1899) in the Congregation of the Holy Ghost and became Vicar Apostolic at Bagamoyo, German East Africa (Tanzania). He was declared persona non-grata and interned (1917). He transferred from Bagamoyo on being appointed Vicar Apostolic in Cameroon (1922). He sent the holotype of this species to the MNHN, Paris.

Volk

Volk's Sculpin *Cottus volki* AY Taranetz, 1933

Alexander M Volk (d.1943) was a friend of the author, and collected specimens of fish, amphibians and reptiles with him in the Russian Far East (1920s). Both of them worked at TINRO (The Pacific Scientific-Research Institute of Fisheries), and both died in action during World War II. We know he attended a technical school in Vladivostok and was a member of a young zoologist's club there (1927–1929). He probably entered college (1929). He was working as a Junior Researcher on the Aral Sea (1939) and in the same year travelled to the Kamyshlibashsky lakes (salt lakes in the Syrdarya river delta, located in the eastern part of the Aral region in the west of the Kyzylorda region of Kazakhstan). He conducted a survey (1940) of mountain lakes near the city of Alma-Ata in order to acclimatize economically valuable fish species there. He is known to have undertaken training at the Kharkov Infantry School (1942) and was sent to the front from there, probably as a junior sergeant commanding a squad. He died in action sometime the following year. We believe this to be the same person as Alexander Maximovich Volk (1910–1943) from Belarus, whose dates and whereabouts fit, but this remains speculation as records to confirm this no longer exist.

Vollmer

Glassy Perchlet sp. *Parambassis vollmeri* TR Roberts, 1995

Ernest Vollmer, Jr. (1924–1992) of San Luis Obispo, California, USA, was honoured for his "*...interest in ichthyological exploration*." He appears to have been a mine owner.

Volta

Electric Eel sp. *Electrophorus voltai* CD de Santana, WB Wosiacki, WGR Crampton, MH Sabaj, CB Dillman, N Castro e Castro, DA Bastos & RP Vari, 2019

Alessandro Giuseppe Antonio Anastasio Volta (1745–1827) was an Italian physicist, chemist and pioneer of electricity. He is credited with the invention of the electric battery and the discovery of methane. He is, of course, immortalised in the name given to a measurement of electrical current. He was honoured in the binomial because of the electrical discharge given off by this species, the most powerful of any known animal.

Volz

Rasbora sp. *Rasbora volzii* CML Popta, 1905

Dr Walter Volz (1875–1907) was a Swiss zoologist, traveller and ichthyologist at the Natural History Museum of Bern. He was in Indonesia and Thailand (1901–1902), and published three papers on the fishes of Sumatra (1903 & 1904). He travelled in Sierra Leone and Liberia (1906–1907) and was killed as 'collateral damage' by French troops conducting a punitive raid. The patronym is not identified in the etymology, but there is little doubt that Walter Volz is intended. A bird and an amphibian are also named after him.

Von Linné

Lake Victoria Cichlid sp. *Haplochromis vonlinnei* MJP van Oijen & MP de Zeeuw, 2008

Carl von Linné (1707–1788), rendered in Latin as Carolus Linnæus, was a Swedish naturalist known as the 'father of modern taxonomy'. He is too well known to require a lengthy entry here, but he was honoured in the cichlid's binomial on the occasion of the 250th anniversary of the official start of zoological nomenclature: taken to be the 10th revision of Linné's *Systema Naturæ*, published in 1758.

Vorderwinkler

Platinum Tetra *Hemigrammus vorderwinkleri* J Géry, 1963
Halftooth Characin sp. *Hemiodus vorderwinkleri* J Géry, 1964

William Vorderwinkler (1908–1970) was an internationally recognised aquarist and the publisher and editor of *Tropical Fish Hobbyist* magazine, in which several of Géry›s descriptions appeared. He wrote: *Breeding Livebearers* (1955) and, with HR Axelrod (q.v.), *Encyclopaedia of Tropical Fishes with Special Emphasis on Techniques of Breeding.*

Voronin

Snailfish sp. *Paraliparis voroninorum* DL Stein, 2012

Vladimir Voronin and Elena Voronina helped support the author's research on snailfish at the Zoological Institute of the Russian Academy of Sciences.

Vosmaer

Whitecheek Monocle Bream *Scolopsis vosmeri* ME Bloch, 1792

Arnout Vosmaer (1720–1799) was a Dutch naturalist and the Curator of the Menagerie and the Museum of the Stadtholder (Dutch Head of State). The French, on the orders of Napoleon Bonaparte, confiscated the collection (1795) and refused to give it back after the Netherlands regained its independence (1813). Vosmaer left an unpublished autobiography, *Memorie tot het leven van Arnout Vosmaer*. Bloch's original text has no etymology, but this seems the most likely candidate.

Vosseler

Pangani Nothobranch (killifish) *Nothobranchius vosseleri* CGE Ahl, 1924

Dr Julius Vosseler (1861–1933) was a German zoologist. He worked at the Amani Biological and Agricultural Institution, German East Africa (1903–1908), and discovered this species. Later he was Director of the Hamburg Zoo (1910). A bird and a reptile are also named after him.

Voulez

Three-barbeled Catfish sp. *Rhamdia voulezi* JD Haseman, 1911

Antonio Voulez was a Frenchman who lived in Serrinha Parana, Brazil. He assisted Haseman, including catching some of the type specimens, but we know nothing further about him.

Vrolik

Pearlscale Angelfish *Centropyge vrolikii* P Bleeker, 1853

Indian Ocean Pinstriped Wrasse *Halichoeres vrolikii* P Bleeker, 1855

Willem Vrolik (1801–1863) was a Dutch anatomist and a pioneer in vertebrate teratology. He was professor of anatomy, physiology and natural sciences at the Athenaeum Illustre (University of Amsterdam). As well as being a scientist he was a devout Christian and a deacon of the Lutheran Church.

Vua

Lake Malawi Cichlid ssp. *Protomelas marginatus vuae* E Trewavas, 1935

This is a toponym referring to Vua, near the north end of Lake Malawi, the type locality.

Vulcan

Volcano Rasbora *Rasbora vulcanus* HH Tan, 1999

Not a true eponym as it refers to Vulcan, the Roman god of fire, referring both to the fiery red colour of the fish and to the volcanic nature of Painan, West Sumatra, where the species occurs.

W

Waagen

Spotted Barb sp. *Puntius waageni* F Day, 1872

Dr Wilhelm Heinrich Waagen (1841–1900) was a German geologist and palaeontologist whose PhD was awarded by the University of Munich. He became an instructor in geology there (1866), also teaching Bavaria's young prince and princess. However, his Catholic faith meant he had little chance of becoming professor there. He took a post (1870) as an assistant in the geology section of the geological survey of India where he collected the type series of the barb. Unable to endure the climate he returned to Europe (1875), later (1877) becoming an instructor at Vienna University. He became Professor of Geology and Mineralogy at the German Polytechnic of Prague (1879) and then (1890) Professor of Palaeontology at the University of Vienna.

Waanders, HL

Waanders' Hard-lipped Barb *Osteochilus waandersii* P Bleeker, 1852
Waanders' Puffer *Tetraodon waandersii* P Bleeker, 1853

Hendrik Lodewijk van Bloemen Waanders (1796–1851) was a Dutch colonial administrator who was in Willemstad, Curacao (Dutch West Indies) (1831–1835), and at Semarang (Java) and Sumatra (1841–1845). His son, Henri Louis van Bloemen Waanders (1821–1883), was administrator of the tin mines of Banka (an island near Sumatra). It is uncertain whether the father or son is honoured in these binomials.

Waanders, JT

Shark Catfish sp. *Helicophagus waandersii* P Bleeker, 1858
Oriental Barb sp. *Puntioplites waandersi* P Bleeker, 1858

Jan Theodore van Bloemen Waanders (1820–1889) was a Captain, then Lieutenant-Colonel, of the Artillery, East Sumatra and an amateur naturalist. He was the son of Hendrik Lodewijk van Bloemen Waanders (above). He collected the types.

Waanders, PL

Tonguefish sp. *Cynoglossus waandersii* P Bleeker, 1854

Pieter Lodewijk van Bloemen Waanders (1823–1884) was a Dutch colonial civil servant and another son of Hendrik Lodewijk van Bloemen Waanders. He was Inspector on Sumatra's West Coast (1850), then at Palembang (c.1853–1859), and Assistant Resident in Bali being appointed as first Inspector in Bali (1856) then Assistant Resident of Banjoewangi and Resident of Palembang (1867). Bleeker says P **S** van Bloemen Waanders was an "… *ambtenaar en de binnenlanden van Palembang;*" a colonial official. However, we believe the initial 'S' may be a transcription error and actually refers to Pieter Lodewijk van Bloemen Waanders.

Wachters

Wachters' Killifish *Aphyosemion wachtersi* AC Radda & J-H Huber, 1978

Walter Wachters is an aquarist of Tremelo, Belgium. He was one of the collectors of the type in the Republic of Congo, along with Jan Buytaert. He was honoured in the binomial as a 'dear friend' and for his 'zealous' (translation) collecting efforts in West Africa. He is dedicated to the care and breeding of all *Aphyosemion* species, and soon after collecting this one he successfully bred it. (Also see **Mike**)

Wada

Blueberry Tetra *Hyphessobrycon wadai* MMF Marinho, FCP Dagosta, P Camelier & OT Oyakawa, 2016

Luiz Wada was trained as an engineer and is now a Brazilian aquarist and breeder of ornamental fish, who runs a business called 'Aquapet'. He was recognised for his: "*...help in many scientific researches with fishes.*"

Wadd

Blind Shark *Brachaelurus waddi* ME Bloch & JG Schneider, 1801

The original description does not explain the binomial. The shark was described on the basis of John Latham's drawing of it (now lost), and the name could be based on an aboriginal word 'waddi' or 'waddy' meaning a war club or type of tree, which, in turn may refer to the shark's shape.

Waddell

Tibetan Carp sp. *Gymnocypris waddellii* CT Regan, 1905

Lieutenant-Colonel Professor Dr Laurence Austine Waddell CB (1854–1938) was a noted British physician, army surgeon, orientalist, philologist, linguist, archaeologist, collector and explorer in Tibet. He accompanied Younghusband's (q.v.) expedition to Lhasa (1903–1904) as a cultural expert, as he had been stationed for years in Darjeeling and had repeatedly, in disguise, crossed the border and visited the forbidden land of Tibet to study its language, customs and culture. He was Professor of Chemistry and Pathology at the Calcutta Medical College (1881–1886). Among many articles, monographs and books he wrote: *The Birds of Sikkim* (1893) and *Lhasa and its Mysteries* (1905). He left his collection of c.700 books, notes and photographs etc. to Glasgow University. He preserved the type series in salt before presenting the specimens to the British Museum (Natural History). A bird is also named after him.

Wafic

Wafic's Eagle Ray *Aetomylaeus wafickii* RW Jabado, DA Ebert & S Al Dhaheri, 2022

Wafic Jabado is the father of the senior author, Lebanese marine biologist Rima Jabado. The ray was so named in recognition of his support for her work and on the occasion of his 73rd birthday.

Wagenaar

Congo Catfish sp. *Chrysichthys wagenaari* GA Boulenger, 1899

Lieutenant Jean-Clément-Frédéric Wagenaar (1869–1941) was a Belgian soldier who transferred to the service of the Independent State of the Congo (1886), He was in the Congo (1892–1895, 1896–1901, 1909–1912 & 1913–1914) in various positions including managing factories in Kasai (1909–1912) and being Director in Banana of the Society for the Study of Fisheries (1913–1914). In WW1, he fought with distinction in the Belgian Army. He collected Upper Congo fishes for Boulenger, including presumably the holotype of this species.

Wagner M

Scrapetooth Characin sp. *Saccodon wagneri* R Kner, 1863

Moritz Wagner (1813–1887) was a German traveller, naturalist, explorer and geographer. He explored in Algeria (1836–1839). He travelled through Persia (Iran), Georgia, and northern Iraq (1840s) and later through North and Central America and the Caribbean (1852–1855). He wrote: *Der kaukasus und das Land der Kosaken* (1847). He committed suicide at the age of 73.

Wagner, NP

Caspian Lamprey *Caspiomyzon wagneri* KF Kessler, 1870

Professor Dr Nicolai Petrovitch Wagner (1829–1907) was a Russian zoologist particularly interested in entomology. He was a professor (1871–1894) at St Petersburg University, where K.F. Kessler (q.v.) was a colleague. The University of Kazan awarded his first degree (1851) and PhD (1854), and he became Professor of Zoology there (1860). He is most famous for discovering paedogenesis (larval reproduction) in insects (1862). During his time at St Petersburg he mainly worked on zoological research at the White Sea, culminating in the publication: *The White Sea Invertebrates* (1855). Later in life he became President of the Russian Society of Experimental Psychology (1891). He was also a novelist and writer of children's stories such as: *Fairy Tales of Tomcat Murlyki* (1881).

Wahari

Armoured Catfish sp. *Lithogenes wahari* SA Schaefer & F Provenzano, 2008

Rúa-Wahari is the god of creation amongst the Piaroa people of Venezuela, where this species occurs.

Wahju

Sailfin Silverside sp. *Telmatherina wahjui* M Kottelat, 1991

Beni Nurtjahja Wahju (1934–2012) was a founding member of the Nature Conservancy's Indonesia Advisory Board and the founding chairman of YPAN, the Indonesian Natural

Heritage Foundation. Under his leadership, YPAN and the Conservancy collaborated on conservation efforts in Lore Lindu and Komodo National Parks. After graduating (1961) as a geologist at the Bandung Institute of Technology, he trained with the US Geological Survey, mapping the San Juan Mountains in Colorado (1963–1964). He then worked as a Senior Geologist for Inco, Canada; spent 25 years with PT Inco, where he rose to Vice President and Corporate Secretary; and served as President Director of PT Ingold Management for 17 years. He was Director of Vale Exploration (2008–2010), a PT Ingold subsidiary, and as its Advisory Commissioner until his death. The etymology acknowledges "*...his invaluable help and assistance, without which the ichthyological survey of the Malili Lakes would not have been possible.*"

Waiampi

Armoured Catfish sp. *Hypostomus waiampi* H Carvalho & C Weber, 2005

The Waiampi are an Amerindian ethnic group from north-eastern Brazil, living on an Indian reserve, western Cupixi River Basin, Amapâ, Brazil, where it appears this species is endemic.

Waikhom

Maharaja Barb genus *Waikhomia* U Katwate, P Kumkar, R Raghavan & N Dahanukar, 2020

Loach sp. *Aborichthys waikhomi* L Kosygin, 2012

Glassfish sp. *Parambassis waikhomi* K Geetakumari & C Basudha, 2012

Torrent Catfish sp. *Amblyceps waikhomi* A Darshan Singh, A Kachari, R Dtta, A Ganguly & DN Das, 2016

Dr Waikhom Vishwanath (b.1954) is an Indian ichthyologist who is at the Department of Life Sciences, Manipur University, Canchipur, where he is Professor of Life Sciences. Banaras Hindu University awarded his master's degree in zoology (1975) and Manipur University his doctorate (1984), immediately after which he joined the University's faculty. Additionally, he is a member of the Research Advisory Committee of the Central Inland Fisheries Research Institute, Barrackpore, India. He has a huge output of publications including: *On a collection of fishes of the genus* Garra *Hamilton from Manipur, India, with a description of a new species* (1993). He was honoured in the glassfish etymology "*...for his contributions to the ichthyology of freshwater fishes of northeastern India*".

Waimacu

Neotropical Rivuline sp. *Anablepsoides waimacui* CH Eigenmann, 1909

The name combines two different Amerindian tribes in Guyana, the Wai-Wai and the Macui.

Wai Si Han

Stone Loach sp. *Triplophysa waisihani* L Cao & E Zhang, 2008

Wai Si Han (Chinese spelling of Wais Khan) was the 10th-generation male offspring of the Mongolian emperor Genghis Khan, whose mausoleum is located in Dunmaza Town (Yining County, Xinjiang-Uighur, China); the type locality.

Waite

Ratfish sp. *Hydrolagus waitei* HW Fowler, 1907
[Junior Synonym of *Chimaera ogilbyi*]
Barracuda sp. *Sphyraena waitii* JD Ogilby, 1908
Southern Round Skate *Irolita waitii* AR McCulloch, 1911
Waite's Lefteye Flounder. *Arnoglossus waitei* JR Norman, 1926
Spotty-face Anchovy *Stolephorus waitei* DS Jordan & A Seale, 1926
Waite's Splitfin *Luzonichthys waitei* HW Fowler, 1931

Edgar Ravenswood Waite (1866–1928) was an English-born Australian zoologist and ichthyologist. After studying at Manchester University, he worked at the Leeds Museum (1888–1892). He went to work at the Australian Museum, Sydney (1892), where he was Curator of Ichthyology (1893–1905). He was in New Zealand (1906–1914) as Curator of the Canterbury Museum, Christchurch, but then returned to Australia, as General Director of the South Australian Public Library, Museum and Art Gallery (1914–1928). He took part in a number of expeditions including the (1907) Sub-Antarctic Islands Scientific Expedition. Among other works he wrote: *The Fishes of South Australia* (1923). He had malaria, contracted in New Guinea, which compromised his health, and he died of enteric fever while attending a meeting in Hobart, Tasmania. A reptile and a bird are also named after him.

Waitt

Barbeled Dragonfish sp. *Photonectes waitti* AJ Flynn & CI Klepadlo, 2012

Dr Theodore W 'Ted' Waitt (b.1963) is an American philanthropist and businessman who attended the University of Iowa (1982–1984) but did not graduate - his doctorate is honorary and awarded by the University of South Dakota. He made his money in the computer business. He is founder and chairman of the Waitt Foundation and of the Waitt Institute, which sponsored and directed the research vessel 'Seward Johnson' during her expedition (one of many that he has sponsored) to the equatorial western Pacific Ocean during which the holotype of this species was collected.

Waiwai

Glass Knifefish sp. *Eigenmannia waiwai* LAW Peixoto, GM Dutra & WB Wosiacki, 2015

This species is named after the Waiwai indigenous people whose home territory is near where the holotype was collected in Pará, Brazil.

Wajapi

Brazilian Cichlid sp. *Teleocichla wajapi* HR Varella & CLR Moreira, 2013

This is named for the Wajãpi people, also known as Waiapi or Oyampi, numbering less than 2000 people distributed in several tribes in Amapá, Brazil (where this cichlid occurs), and in French Guiana.

Wakiya

Seabass sp. *Malakichthys wakiyae* DS Jordan & CL Hubbs, 1925

Rockfish sp. *Sebastes wakiyai* K Matsubara, 1934
Thicklip Gudgeon ssp. *Sarcocheilichthys variegatus wakiyae* T Mori, 1927

Dr Yohiro (also spelt Yojiro) Wakiya was superintendent of the Korean Government Fisheries Experiment Station. He was honoured in the gudgeon's trinomial for his help in the preparation of Mori's paper. He collected the seabass holotype.

Walailak

Yellow-spotted Mudskipper *Periophthalmus walailakae* U Darumas & P Tantichodok, 2002

Both authors work at Walailak University Thammarat, Thailand and they named this species in honour of the University's 10th anniversary.

Waldron

African Barb ssp. *Labeobarbus bynni waldroni* JR Norman, 1935

Miss Fanny Waldron was a collector for the BMNH. She accompanied Willoughby P Lowe (q.v.) on his expedition to the Gold Coast (Ghana) (1934–1935) when she was reportedly well over 60. Norman spelt the trinomial *waldroni*, but this is sometimes amended to the feminine *waldronae*. A bird and a subspecies of monkey are also named after her.

Walecki

Amur Ide *Leuciscus waleckii* BT Dybowski, 1869
Vistula Barb *Barbus waleckii* H Rolik, 1970

Antoni Bazyli Wałecki (1815–1897) was a Polish zoologist who was Dybowski's colleague, and who became Assistant Director of the Cabinet of the Warsaw Zoo (1860–1862). He was then (1862–1893) Curator of Mineralogy at the Imperial University there. He noted large specimens of *Barbus cyclolepis* (misidentified as *B. petenyi*) in the Vistula River near Warsaw (1864), which are now recognised as *B. waleckii*, when undertaking a national survey of fish for the Polish fishing industry. He also worked in Eastern Siberia and around Lake Baikal where he had been exiled. He also wrote papers trying to reconcile evolutionary theory with Christianity.

Walker, BW

San Quintin Blenny *Paraclinus walkeri* C Hubbs, 1952
Elusive Signal Blenny *Emblemaria walkeri* JS Stephens, 1963
Walker's Anchovy *Anchoa walkeri* WJ Baldwin & NHC Chang, 1970
Sand Stargazer sp. *Myxodagnus walkeri* CE Dawson, 1976
Walker's Toadfish *Batrachoides walkeri* BB Collette & JL Russo, 1981
Masked Prickleback *Ernogrammus walkeri* WI Follett & DC Powell, 1988
Professor Stardrum *Stellifer walkeri* NL Chao, 2001
Longjaw Jawfish *Opistognathus walkeri* WA Bussing & RJ Lavenberg, 2003

Professor Dr Boyd Wallace Walker (1917–2001) was an American fisheries biologist; ultimately Professor Emeritus of Zoology-Fisheries at the University of California, Los Angeles, where he worked for over three decades (1949–1980). He began his undergraduate studies in zoology at Davis, then transferred to UC Berkeley and from there to the University of Michigan which awarded his MSc (1942). He volunteered for the army (1942) as a private in the infantry. He was discharged (1946) with the rank of Major. He was a doctoral student in ichthyology at the Scripps Institution of Oceanography but UCLA awarded his PhD (1948). He was not a prolific writer of papers but published a number mostly on the taxonomy of various fish. His major work was *Ecology of the Salton Sea*. He collected the anchovy type and encouraged the writing of the species' description. He is also renowned for giving a common name to the jawfish *Opistognathus rhomaleus*. He bestowed the ignoble title upon this little fish with a disproportionately large mouth, but legendary Baja California author Ray Cannon is the one who made it stick. The two fished together several times on collecting expeditions in the Sea of Cortez, which resulted in the eventual identification of dozens of new species. After one such trip, Walker told Cannon that the first *rhomaleus* brought to the university had been misplaced by one of his students. When he asked his class where "*that strange-looking fish*" went, another student asked Walker which strange-looking fish he was talking about, whereon the professor replied, "*You know, that big-mouthed bastard.*" Cannon, in his: *The Sea of Cortez* (1966), later explained that it was he who convinced Walker that there was no reason to call this fish by any other name. "*Walker agreed with my suggestion that, whenever reasonable, a thing should be known by the first name ever applied to it,*" Cannon wrote. "*This seemed a most reasonable situation, so the common name 'bigmouth bastard' stuck.*" However, it is more generally now called the Giant Jawfish. (Also see **Boyd Walker**)

Walker, HJ

Hagfish sp. *Eptatretus walkeri* CB McMillan & RL Wisner, 2004
Walker's Cusk-eel *Bassogigas walkeri* JG Nielsen & PR Møller, 2011

Harold J Walker Jr is a Senior Museum Scientist and Collections Manager at Scripps Institution of Oceanography, University of California, San Diego. His research interests include the taxonomy, systematics and zoogeography of marine fish. He has described numerous new species and has written a number of papers such as: *The world's smallest vertebrate, Schindleria brevipinguis, a new paedomorphic species in the family Schindleriidae* (2004).

Walker, I

Muckfish *Tarumania walkerae* M de Pinna, J Zuanon, L Rapp Py-Daniel & P Petry, 2017

Ilse Walker is a limnologist and ecologist at Instituto Nacional de Pesquisas da Amazônia, Manaus, Brazil. She collected the first specimen (1999) and was honoured "*…not only for her lifelong contribution to the knowledge of Amazonian ecology but also for having collected the first (and for some years, only) known specimen.*"

Walker, RBN

Walker's Mormyrid *Stomatorhinus walkeri* A Günther, 1867
Claroteid Catfish sp. *Chrysichthys walkeri* A Günther, 1899
Claroteid Catfish sp. *Notoglanidium walkeri* A Günther, 1903
African Barb sp. *Enteromius walkeri* GA Boulenger, 1904

African Rivuline sp. *Fundulopanchax walkeri* GA Boulenger, 1911

Robert Bruce Napoleon 'Brucie' Walker FRGS, FAS, FGS, CMZS (1832–1901) was a West African trader, explorer, anthropologist and natural history and ethnography collector. He spent most of his time (from 1851–1864 & 1865) in Gabon, where he collected several of the types and where he became Local Secretary for Gabon of the Anthropological Society of London. He made periodic visits back to Britain but kept returning to West Africa. Much of what he collected was sent to the British Museum or the Liverpool Museum, but later he sold items through commercial agents to many other institutions.

Walker, T

Eastern Spotted Gummy Shark *Mustelus walkeri* WT White & PR Last, 2008

Dr Terence Ivan Walker is an English-born Australian ichthyologist and expert on sharks. He received a BSc in zoology from Monash University (1970) while already working as a technical officer in the marine fisheries section of the (then) Victorian Fisheries and Wildlife Division. He has made a long-term study of the Gummy Shark (*Mustelus antarcticus*). He was honoured for: "*…dedicating a lifetime to the ecology and fisheries management of Australian chondrichthyans*."

Wallace, AR

West African Cichlid genus *Wallaceochromis* A Lamboj, F Trummer & BD Metscher, 2016

Wallace's Pike-Cichlid *Crenicichla wallacii* CT Regan, 1905

Payara (characin) sp. *Hydrolycus wallacei* M de Toledo-Piza, NA de Menezes & GM dos Santos, 1999

Driftwood Catfish sp. *Tetranematichthys wallacei* RP Vari & CJ Ferraris Jr, 2006

Neotropical Rivuline sp. *Melanorivulus wallacei* WJEM Costa, 2016

Freshwater Stingray sp. *Potamotrygon wallacei* MR de Carvalho, RS Rosa & MLG de Araújo, 2016

Miniature Killifish sp. *Fluviphylax wallacei* PHN de Bragança, 2018

Alfred Russel Wallace (1823–1913) was an English naturalist, evolutionary scientist, geographer and anthropologist and one of the greatest men of science of his age. He was the father of zoogeography. He was also a social critic and theorist, a follower of the utopian socialist Robert Owen. His interest in natural history began whilst working as an apprentice surveyor, at which time he also attended public lectures. He went to Brazil (1848) on a self-sustaining natural history collecting expedition. Even on his first expedition he was very interested in how geography limited or facilitated the extension of species' ranges. He not only collected but also mapped, using his surveying skills. He led an expedition to the Rio Negro and Rio Uaupés region (1850–1852). His return to England (1852) was a near disaster; his ship, the brig 'Helen', caught fire and sank with all his specimens, and he was lucky to be rescued by a passing vessel. He spent the next two years writing and organising another collecting expedition to the Indonesian archipelago. He managed to get a grant to cover his passage to Singapore (1862) and had the benefit of letters of introduction and the like prepared for him by representatives of the British and Dutch governments. He spent nearly eight years there, during which he undertook about 70 different expeditions involving a total of around 14,000 miles of travel. He visited every important island in the archipelago at least once, some many times. He collected a remarkable 125,660 specimens, including more than 1,000 new species. He wrote *The Malay Archipelago* (1869), which is the most celebrated of all writings on Indonesia and ranks as one of the 19th-century's best scientific travel books. He also published *Contributions to the History of Natural Selection* (1870) and *Island Life* (1880). His essay: *On the law which has regulated the introduction of new species*, which encapsulated his most profound theories on evolution, was sent to Darwin. He later sent Darwin his essay: *On the tendency of varieties to depart indefinitely from the original type*, presenting the theory of 'survival of the fittest'. Darwin and Lyell presented this essay, together with Darwin's own work, to the Linnean Society. Wallace's thinking spurred Darwin to encapsulate these ideas in *The Origin of Species*; the rest is history. Wallace developed the theory of natural selection, based on the differential survival of variable individuals, halfway through his stay in Indonesia. He remained for four more years, during which he continued his systematic exploration and recording of the region's fauna, flora and people. For the rest of his life he was known as the greatest living authority on the region and its zoogeography, including his discovery and description of the faunal discontinuity that now bears his name: Wallace's Line. This natural boundary runs between the islands of Bali and Lombok in the south and Borneo and Sulawesi in the north, and separates the Oriental and Australasian faunal regions. Two mammals, two reptiles, an amphibian and at least twelve birds are named after him as are four dragonflies and many other taxa.

Wallace, JH

Longnose Conger *Bathycongrus wallacei* PHJ Castle, 1968
Yellow-spotted Skate *Leucoraja wallacei* PA Hulley, 1970

Dr John H Wallace (fl.1921–fl.1986) was head of research at the *Oceanographic Research Institute, Durban, South Africa, specialising in skates and rays and later Director of the Port Elizabeth Museum (1975–1986).* He wrote: *The batoid fishes of the east coast of southern Africa* (1967).

Wallace, JH

Western Hardyhead *Leptatherina wallacei* JD Prince, W Ivantsoff & IC Potter, 1982

John H Wallace was the research assistant who set up the fish sampling laboratory for Professor I C Potter at the newly created Murdoch University (1970s). According to co-author Jeremy Prince, John had a background in studying fish biology from his previous employment with the Western Australian Fisheries Department's estuarine sampling group, which he put to good effect for Professor Potter who then went on to build up around him an active group of undergraduate, Honours and PhD students, which Jeremy Prince joined (1978–1980). It was John who effectively supervised their work and career development up until that stage, and he was a thoroughly likeable human being. Late in Prince's honours study (1979) he realised that what he'd thought was a single species of hardyhead was actually two, one of which had not been previously recognised. As John had been a great support to him, and was an unsung hero of the laboratory, he decided to name the new species for him. Eventually John became disillusioned with fisheries science and left to work as a full-time wood turner, which he ended up teaching, before moving on to practice and teach photography - particularly bird photography.

Walsh, F

Walsh's Fairy Wrasse *Cirrhilabrus walshi* JE Randall & RL Pyle, 2001

Fenton Walsh is an Australian ichthyologist who collected the type specimens in American Samoa. He has co-described other species, such as in: *A pictorial review of the Indo-Pacific labrid fish genus Pseudocoris, with description of a new species from the Coral Sea* (2008) with Jack Randall, and *Tosanoides bennetti, a new species of anthiadiine fish (Pisces: Serranidae) from the Coral Sea, Australia* (2019) with Gerald Allen.

Walsh, G

African Barb sp. *Enteromius walshae* V Mamonekene, AI Zamba & M Stiassny, 2018

Gina Walsh is (2015>) an Independent Research Associate Freshwater Ecologist at the University of Witwatersrand, South Africa, and an independent consultant at Flora, Fauna & Man Ecological Services. She previously (2008–2015) occupied a similar position at Ecotone Freshwater Consultants. The University of Johannesburg awarded her BSc (2005) and MSc (2008) and she is pursuing a part-time PhD at the University of Witwatersrand (2016–2020). The etymology says the barb is named "*...for our colleague Gina Walsh (University of Witwatersrand, South Africa), whose ongoing research continues to enhance conservation efforts throughout the region.*"

Walter (Dieckhoff)

Lake Tanganyika Cichlid sp. *Neolamprologus walteri* P Verburg & IR Bills, 2007

Lake Malawi Cichlid ssp. *Otopharynx lithobates walteri* AF Konings, 1990

Horst Walter Dieckhoff is a German cichlid aquarist who discovered, photographed, and filmed many new cichlids in the Great Lakes of Africa.

Walter (Heiligenberg)

Bluntnose Knifefish sp. *Brachyhypopomus walteri* J Sullivan, J Zuanon & C Cox Fernandes, 2013

Dr Walter F Heiligenberg (1938–1994) was an ethologist who was (1976–1994) Professor of Behavioural Physiology, Scripps Institute of Oceanography, University of California, San Diego, having joined earlier (1973). Before emigrating from Germany (1972) he had been awarded his doctorate by the University of Munich (1964) and had worked at the Max Planck Institute for Behavioural Physiology, Seewiesen (1964–1972). He was killed in an aeroplane crash.

Walter (Verheyen)

Tilapia sp. *Coptodon walteri* DFE Thys van den Audenaerde, 1968

Professor Dr Walter Norbert Verheyen (1932–2005) was a Belgian zoologist and mammologist who was a professor at the Biology Department of the University of Antwerp. He published widely on rodents and other small mammals of Africa: e.g. Leirs, Verhagen,

& Verheyen, *The basis of reproductive seasonality in* Mastomys *rats. (Rodentia: Muridae) in Tanzania.* He was a member of the expedition that collected the type and was honoured for photographing the holotype and other help in the field. Seven mammals are also named after him.

Walters

Walters' Toadfish *Batrachoides waltersi* BB Collette & JL Russo, 1981

Dr Vladimir 'Vlad' Walters (1927–1987) was an American ichthyologist. Cornell University awarded his bachelor's degree (1947) and master's (1948), and New York University his doctorate (1954). He was, variously, Assistant Curator of Fishes, AMNH (1956), Professor of Zoology, University of California, Los Angeles (1962–1965) and a fisheries research biologist, National Marine Fisheries Service. He wrote: *Body Form and Swimming Performance in the Scombroid Fishes* (1962). (See also **Vlad**)

Waltervad

Snapper sp. *Paracaesio waltervadi* WD Anderson & BB Collette, 1992

This is a toponym referring to the type locality: Walters Shoals, south of Madagascar, western Indian Ocean [*vadum* = Latin for shoal]

Walton (Family)

Walton Flasher Wrasse *Paracheilinus walton* GR Allen & MV Erdmann, 2006

The Walton Family Foundation was founded by American philanthropist Samuel Moore Walton (1918–1992) and his wife Helen Robson Walton (1919–2007). Samuel is best known as the man behind the Walmart retail corporation. The Foundation aims to tackle social and environmental problems, and supports the Bird's Head Seascape Initiative in eastern Indonesia (where this wrasse occurs).

Walton, HJ

Snowtrout sp. *Schizothorax waltoni* CT Regan, 1905

Walton's Mudskipper *Periophthalmus waltoni* FP Koumans, 1941

Lieutenant-Colonel Herbert James Walton (1869–1938) was a physician who qualified in London (1893), then joined the Indian Medical Service (1896). He was part of the force that relieved Peking (Beijing) in the 1900 Boxer Rebellion, and part of the Lhasa Expedition (1903–1904). He was co-author of: *The Opening of Tibet; an Account of Lhasa and the Country and People of Central Tibet and of the Progress of the Mission sent there by the English Government in the Year 1903–4* (1905). During this expedition, he collected 500 bird skins, and the type of the snowtrout. He retired from the army after WW1.

Walton, I

Wolftrap Seadevil sp. *Lasiognathus waltoni* RS Nolan & RH Rosenblatt, 1975

Sir Izaac Walton (c.1593–1683) was honoured for writing *The Compleat Angler* (1653). He is so well known as to require no biography here. Like Izaak, the seadevil – a type of deep-sea anglerfish – is a 'compleat angler', in that it appears to have a fishing rod (the projecting basal bone or pteropterygium), a fishing line (the illicium, a modified dorsal fin ray) and a lure in the form of a bioluminescent esca.

Walton, J

Stone Loach sp. *Schistura waltoni* HW Fowler, 1937

Joseph Walton (1817–1898) was an American naturalist, botanist and a contributor to the fish collection at the Academy of Natural Sciences of Philadelphia in its early history. His interest in natural history was sparked (1836) when attending Westtown Boarding-school in Chester County, Pennsylvania, which occupied 600 acres of woodland, swamp, valley and stream. He studied at Haverford College (1833–1836) then taught at his old school (1836–1846), after which he went into business. When he retired to New Jersey his interest was rekindled and he published a number of papers.

WAM

Cusk-eel sp. *Sirembo wami* JG Nielsen, W Schwarzhans & F Uiblein, 2014

WAM stands for Western Australian Museum, where the holotype is housed.

Wamuini

Mormyrid sp. *Marcusenius wamuinii* E Decru, JP Sullivan & E Vreven, 2019

Professor Soleil Wamuini Lunkayilakio is an ichthyologist who is a research scientist at the Institut Supérieur Pédagogique de Mbanza-Ngungu, DRC, and associated with the University of Liège. The etymology says the fish was dedicated "*...to Professor Soleil Wamuini Lunkayilakio of the Institut Supérieur Pédagogique de Mbanza-Ngungu (ISP Mbanza-Ngungu), to acknowledge his great contribution to the sampling effort in the area, which resulted in the discovery of the new species, and to recognize him as one of the first Congolese ichthyologists.*" Among his published papers are the co-written: *'Haplochromis' snoeksi, a new species from the Inkisi River basin, Lower Congo (Perciformes: Cichlidae)* (2010) and *The complex origins of mouth polymorphism in the Labeobarbus (Cypriniformes: Cyprinidae) of the Inkisi River basin (Lower Congo, DRC, Africa): insights from an integrative approach* (2018).

Wanda

Molly sp. *Poecilia wandae* FN Poeser, 2003

Vanda Marisa Freitas de Leite is a friend of the author and likes to be called 'Wanda' – perhaps she saw the eponymous film?

Wang, C-C

Snowtrout sp. *Schizothorax wangchiachii* PW Fang, 1936

Wang Chia-chi (1897–1976) was a zoologist and biologist who was Director of the National Research Institute of Biology, Academia Sinica. He was honoured for allowing Fang to: "*...stay a rather longer period in the European museums and institutes for carrying on his ichthyological work.*" He wrote: *Study of the Protozoa of Nanking* (1925).

Wang, FT

Chinese Cyprinid sp. *Ancherythroculter wangi* TL Tchang, 1932

Mr F T Wang (no other information is given in the original text) was the collector of the type series.

Wang, JW

Goby sp. *Rhinogobius wangi* IS Chen & LS Fang, 2006

J W Wang was a postgraduate student who gave the authors valuable assistance during their field trip (2002) in the Chinese provinces of Fujian and Guangdong.

Wangunu

Wangunu's Pygmygoby *Trimma wangunui* R Winterbottom & MV Erdmann, 2019

Noel Wangunu is a marine biologist; one of Papua New Guinea's foremost reef scientists and marine conservationists. Working with Conservation International he was instrumental in recruiting and training local islanders to monitor reefs (2009). He also assisted the junior author with collections and arranging local permits.

Wapisana/Wapixana

Dwarf Cichlid sp. *Apistogramma wapisana* U Römer, IJ Hahn & A Conrad, 2006
Pencil Catfish sp. *Potamoglanis wapixana* E Henschel, 2016

The Wapisâna (also given as Wapishana, Wapisiana or *Wapixana)* are an Amerindian tribe who inhabit the region in northern Brazil where these species occur. In recent decades, large parts of their tribal area have been devastated by excessive gold-mining and deforestation.

Ward, A

Ward's Tilefish *Branchiostegus wardi* GP Whitley, 1932

Alec Ward According to Whitley it was "*Mr. Alec Ward, who collected the specimen, and who has obtained many rare and interesting fishes on board the trawlers in recent years.*" Ward said (1831) that "*Most of the species added to the New South Wales fish-fauna in recent years have been caught in fairly deep water over the continental shelf. The continued interest of many friends aboard the trawlers and their help in securing specimens for the Australian Museum has been indispensable, and I desire especially to thank Mr. Alec Ward and Captain Knud Moller of the trawlers and Captain L. Comtesse of the dredge 'Triton' for the fishes they have submitted to me from time to time.*" *

Ward, CM

Ward's Damsel *Pomacentrus wardi* GP Whitley, 1927
Small-eyed Goby *Austrolethops wardi* GP Whitley, 1935
Northern Wobbegong *Orectolobus wardi* GP Whitley, 1939

Charles Melbourne Ward (1903–1966) was an actor, naturalist and marine collector. As his parents were entertainers, his childhood was peripatetic and his education divided between Sydney and New York. He became an actor and musician (1919) and played jazz saxophone and clarinet. He became a marine zoologist after a crab was named *Cleistostoma wardi* after him (1926). He collected all over the world from Australia to the Atlantic coast of the USA. He was an honorary zoologist at the Australian Museum Sydney (1929) where Whitley became a friend. He visited Papua New Guinea (1932) and collected ethnographic artefacts there. During WW2, he both entertained the troops and instructed on tropical hygiene. He opened a Gallery of Natural History and Native Art at Medlow Bath in the Blue Mountains (1943). It was described as combining "*...old curiosity shop and scientific exhibits.*" He suffered from diabetes and died of a coronary occlusion, leaving his scientific collections and library to the Australian Museum. He collected all the type specimens.

Ward, S

Ward's Sleeper Goby *Valenciennea wardii* L Playfair, 1867

Swinburne Ward (1830–1897) was a British diplomat, Her Majesty's Civil Commissioner in the Seychelles (1862–1868). He was also an amateur naturalist, and the extinct Seychelles Parakeet (*Psittacula wardi*) is also named after him.

Warming

Warming's Lanternfish *Ceratoscopelus warmingii* CF Lütken, 1892

Dr Johannes Eugenius Bülow Warming (1841–1924) was a Danish botanist and a main founding figure of the scientific discipline of ecology. He studied at the University of Copenhagen (1859–1863 & 1867–1869), having been in Lagoa Santa, Brazil (1863–1866) as secretary to the Danish palaeontologist Peter Wilhelm Lund. He then studied in Munich (1870) and at Bonn (1871), and finally was awarded his doctorate (1871) by the University of Copenhagen. He taught botany at the University of Copenhagen and the Pharmaceutical College (1873–1882). He was Professor of Botany, Stockholms högskola (later Stockholm University) (1882–1885). He was Professor of Botany, University of Copenhagen and Director, Copenhagen Botanical Garden (1885–1910). In addition to his stay in Brazil he took part in a number of expeditions including the Fylla expedition to Greenland (1884), Norway (1885 & 1887), Venezuela, Trinidad and the Danish West Indies (1891–1892) and the Faeroe Islands (1895). He wrote: *Om Grønlands Vegetation* (1887). He collected the holotype of this species.

Warpachowski

Sharpbelly (cyprinid) sp. *Hemiculter varpachovskii* AM Nikolskii, 1903

Nikolai Arkadevich Warpachowski was a Russian ichthyologist who named the species *Hemiculter bleekeri* and the related genus *Hemiculterella* in the Bulletin of the St Petersburg Academy of Sciences (1888).

Warren

Warren's Sixgill Sawshark *Pliotrema warreni* CT Regan, 1906

Professor Dr Ernest Warren (1871–1945) was an English zoologist who moved to Natal (1903) and became first Director of the Natal Museum, Pietermaritzburg (1903–1935). He championed the establishment of national parks in Natal (1920s-1930s). A reptile is also named after him.

Wartmann

Blaufelchen (Lake Constance whitefish) *Coregonus wartmanni* ME Bloch, 1784

Dr Bernhard Wartmann (1739–1815) was a Swiss physician and naturalist. He qualified as a physician at Montpellier and was a town councillor and public health doctor at St Gallen (1789). He described this species (1777) but did not use a Linnaean name.

Warzel

Dwarf Cichlid sp. *Dicrossus warzeli* U Römer, IJ Hahn & PM Vergara, 2010

Frank Martin Warzel (1960–2004) was a skilled German aquarist who, according to the authors: "*...dedicated most of his life*" to research on neotropical cichlids. He collected the type (1992) and was the first to import this species to Germany and observe its reproductive behaviour both in the field and in the aquarium.

Waser

Betta sp. *Betta waseri* R Krummenacher, 1986

Alfred Waser is a Swiss aquarist who discovered this species in Malaysia, which was described by the author from aquarium specimens. He breeds fish and is a leading light in an aquarist club.

Wass

Wass' Pipefish *Festucalex wassi* CE Dawson, 1977
Angler Snake-Eel *Glenoglossa wassi* JE McCosker, 1982
Fleckfin Dottyback *Pseudoplesiops wassi* A Gill & A Edwards, 2003
Anthias sp. *Odontanthias wassi* JE Randall & PC Heemstra, 2006

Dr Richard Charles Wass (b.1942) was the fisheries officer of the US Fish & Wildlife Service who collected many fishes in American Samoa in the 1970s and early 1980s. The University of Hawaii awarded his doctorate (1972). Until he retired, he was Refuge Manager, Hakalau National Wildlife Refuge in Hilo, Hawaii, and is now President of Friends of Hakalau Forest NWR. He has written several papers and longer works including: *An annotated checklist of the fishes of Samoa* (1984).

Wassink

Loach sp. *Homalopteroides wassinkii* P Bleeker, 1853
Kupang Cardinalfish *Ostorhinchus wassinki* P Bleeker, 1861

Major General Dr Geerlof Wassink (1811–1864) was a Dutch physician and military officer. He was a volunteer pupil teacher (1829) at the Utrecht National Military medical school and was appointed late that year as a Medical Officer, third class, in the Navy. He became (1853) Chief of the Medical Service in the Netherlands Indies when serving (1830–1864) as Surgeon-Major in a variety of posts and locations in the Netherland East Indies. He also suggested (1850) a Midwifery School should be established, but it did not occur until some years later (1861). Ill health forced him to return home for almost two years (1856–1857) but he returned to his duties once more. He received his final promotion to Major General (1862) two years before his death. The etymology honours him because through his kindness Bleeker received the type specimen.

Wassmann

Neotropical Rivuline sp. *Cynodonichthys wassmanni* H-O Berkenkamp & VMF Etzel, 1999

Klaus Wassmann of Munster, Germany, is an aquarist who was involved in the survey and planning of a study of killifish in Panama which Dr Etzel undertook. He has also travelled to Peru collecting fish and amphibians, specifically poison-dart frogs.

Watanabe

Black-edged Angelfish *Genicanthus watanabei* F Yasuda & Y Tominaga, 1970

Dr Masao Watanabe was a Japanese ichthyologist at the University of Tokyo. He was the first to report this species (1949), although he mistakenly thought it was the related species *G. caudovittatus*. Among his publications is the book: *Cottidae (Pisces) (Fauna Japonica)* (1960).

Watasé

Watase's Lanternfish *Diaphus watasei* DS Jordan & EC Starks, 1904
Slickhead sp. *Rouleina watasei* S Tanaka, 1909

Shozaburo Watasé (1862–1929) was a biologist who became (1901) Professor of Zoology at the Imperial University of Tokyo, his alma mater. He accepted (1886) a fellowship at Johns Hopkins University where he took his PhD (1889). He was then at Clark University in Massachusetts and the University of Chicago to become an instructor of cellular biology. He spent his summers at Woods Hole as a student and lecturer at the Marine Biological Laboratory (1888–1899). He was rare in his combination of physiological and morphological work, and of both plant and animal studies. He returned to Japan, and the Imperial University (1899). He wrote: *On the morphology of the compound eyes of arthropods* (1890). He presented the holotype of the lanternfish to Stanford University.

Waterhouse

Whiskered Prowfish *Neopataecus waterhousii* FL de Castelnau, 1872

Frederick George Waterhouse (1815–1898) was an English naturalist who made important contributions to the study of Australian fauna and flora. Soon after marrying (1852) he emigrated to Australia, first trying his hand in the goldfields then working as a surveyor. He became (1860) curator of the South Australian Institute Museum, which he had helped to found in 1856 with the donation of his own valuable collections. He made an expedition across Australia (1861), returning with a collection for the museum (1863). He provided Castelnau with fish specimens collected in a trawl off Gulf St Vincent, South Australia. A river in Northern Territory is also named after him.

Waterlot

Silverside sp. *Teramulus waterloti* J Pellegrin, 1932
Garra sp. *Garra waterloti* J Pellegrin, 1935
Squeaker Catfish sp. *Chiloglanis waterloti* J Daget, 1954
Squeaker Catfish sp. *Synodontis waterloti* J Daget, 1962

Emmanuel-Georges Waterlot (1877–1939) was a French collector of ethnological and natural history specimens (1913–1936) in French West Africa, both for the MNHN, Paris and for the BMNH, London. He also collected in French-speaking territories in southern and eastern Africa, including Madagascar, and wrote: *La sculpture sur bois à Madagascar* (1925).

Waterman

Whipnose Anglerfish sp. *Gigantactis watermani* E Bertelsen, TW Pietsch III & RJ Lavenberg, 1981

Talbot Howe Waterman (1914–2010) was Professor Emeritus of molecular, cellular and developmental biology at Yale. He is best remembered for his research on underwater optics, particularly on how aquatic animals use polarized light to navigate. He earned his BA, MA and PhD at Harvard University. He published his first scientific paper, on the behaviour of a water mite, just a year after finishing his undergraduate degree. He then joined the military as a scientific consultant and was stationed in the Pacific, where he gained experience in communications and radar navigation. He began research work (1946) at Yale and was a member of the faculty for the next four decades, but remained a very active emeritus professor thereafter.

Watermeyer

South African River Pipefish *Syngnathus watermeyeri* JLB Smith, 1963

F L E Watermeyer collected the type from the Bushmans River, South Africa, and sent it to JLB Smith. We have been unable to uncover any further biography.

Waterous

Goby sp. *Mangarinus waterousi* AW Herre, 1943

Willard Harry Waterous (1890–1964) was an American army physician and optometrist, a Major in the Medical Corps during WWII who was captured and interned in a POW camp for nearly three years during the Japanese invasion of the Philippines. He wrote the unpublished manuscript: Reminiscences of Dr W H Waterous Pertinent to World War II in the Philippines (1953). He stayed on in the Philippines for the rest of his life. Herre described him in the etymology as his "*…esteemed friend, army officer, and eminent physician*" who placed the resources of Hacienda Waterous, at Mangarin, Mindoro, Philippines (the type locality), at Herre's disposal

Waters

Armoured Catfish sp. *Neoplecostomus watersi* GSC Silva, L Reia, CH Zawadzki & FF Roxo, 2019

George Roger Waters (b.1943) is an English singer, songwriter and composer who is the bass player in one of the worlds' most famous rock bands – Pink Floyd. He was honoured – in somewhat awkward English – *"for his talent as musician and social awareness around world, specially his brave concerns to Brazilian economic, social and politic issues."*

Watson, HE

Indus Lotak *Cyprinion watsoni* F Day, 1872

H E Watson was Civil Officer at a 'station' in Sakkar, India (now Pakistan) and Deputy Collector of Sehwan. Dr Francis Day had been: *"…energetically assisted by the Civil Officer at the station, Mr. H. E. Watson,"* while collecting reptiles and other taxa in the Sind Hills.

Watson, RE

Goby sp. *Lentipes watsoni* GR Allen, 1997

Goby sp. *Stenogobius watsoni* GR Allen, 2004

Dr Ronald E Watson is an independent ichthyologist, underwater photographer and avid goby researcher at the Florida Museum of Natural History. He also has an association with the National University of Singapore. He was formerly in the military and while in Oahu, Hawaii (1978–1980), he met Jack Randall (q.v.) who got him interested in gobies. Stationed in Virginia he spent six years associated with the Smithsonian, then was posted to Germany where he worked with the ichthyology section of Forschungsinstitut Senckenberg in Frankfurt. He lived in Bavaria for a number of years (1992–1999) before returning to Florida. He was honoured for his 'fine contributions' to our knowledge of the then-recognised subfamily *Sicydiinae*. As an example, among his published papers are the descriptions: *Two new freshwater gobies from Halmahera, Maluku, Indonesia (Teleostei: Gobioidei: Sicydiinae)* (2006) and *A new species of Stiphodon from southern Sumatra (Pisces: Gobioidei: Sicydiinae)* (2008). He also has an interest in, and photographs, wild flowers.

Watters

African Rivuline sp. *Nothobranchius wattersi* E Ng'oma, S Valdesalici, K Reichwald & A Cellerino, 2013

Dr Brian R Watters is a South African Canadian geologist and killifish enthusiast who is an editor of the Journal of the American Killifish Association. He is the former Head of the Department of Geology, University of Regina, Saskatchewan, now Professor Emeritus. His BSc and Ph.D. were awarded by Cape Town, after which he was Geologist on the Ninth South African National Antarctic Expedition in Western Queen Maud Land (1967–1969). He was then Research Associate, Precambrian Research Unit, University of Cape Town (1969–1974) and Post-doctoral Research Fellow, Department of Geological Sciences, University of British Columbia (1975). He became Senior Geologist, Geological Survey of South Africa (1976). Among his published papers is: *Description and biogeography of Nothobranchius capriviensis, a new species of annual killifish from the Zambezi Region of Namibia* (2015) and he is co-author of *A World of Killies: Atlas of the Oviparous Cyprinodontiform Fishes of the World* (1996). Combining his disciplines, he examined more than a thousand *Nothobranchius* habitats, as well as pools that do not host *Nothobranchius* fishes. In all cases, the reasons for the presence or absence of *Nothobranchius* were primarily determined by the nature of the substrate.

Watts

Large-eye Bream genus *Wattsia* WLY Chan & RM Chilvers, 1974

Dr J C D Watts worked at the Tropical Fish Culture Research Institute, Malacca, Malaysia and at the Fisheries Development and Research Unit, Freetown, Sierra Leone. He was (1967–1968) the fourth Director of the East African Freshwater Fisheries Research Organisation. He wrote a number of articles in 1950s & 1960s such as: *The Saw-Fish of Sierra Leone* (1958) and *Seasonal Fluctuations and Distribution of Nitrite in a Tropical West African Estuary* (1961). He was honoured in the genus name for his "*...very significant contributions to the knowledge of the hydrography of the northern South China Sea made during his service from 1969 to 1972 with the Fisheries Research Station, Hong Kong.*"

Wauthion

Lake Tanganyika Cichlid sp. *Neolamprologus wauthioni* M Poll, 1949

René Wauthion was a Provincial Commissioner in the Belgian Congo. Poll honoured him for the 'encouragement and help' (translation) given to the Belgian Hydrobiological Mission to Lake Tanganyika (1946–1947), during which the type was collected. He wrote: *Le mouvement kitawala au Congo belge* (1950).

Wavrin

Orinoco Eartheater *Biotodoma wavrini* JP Gosse, 1963

Marquis Comte Robert Frédérick de Wavrin de Villers-au-Tertre (1888–1971) was a Belgian ethnologist, explorer and pioneer of cinema. He spent many years (1913–1937) exploring the least known areas of South America at his own expense, returning many times over twenty-five years, especially after his war service (1917–1919). During his travels in the Amazon and Orinoco basins he filmed tribal groups, documenting traditions and customs many of which have now disappeared. His most famous film was *In the Country of the Scalp* (1931). He also collected natural history specimens including the cichlid type (1935). WW2 found him stuck in Belgium (1940) where he wrote up his notes etc., and he never returned to South America. He entrusted his collections to museums in Belgium, London and Paris. His books include: *Mœurs et Coutumes des Indiens sauvages de l'Amérique du Sud* (1937), *Les bêtes sauvages de l'Amazonie* (1939) and *Mythology, Mythologie, Rites et Sorcellerie des Indiens de l'Amazonie* (1979). An amphibian is also named after him.

Wayampi

Characin sp. *Phenacogaster wayampi* P-Y Le Bail & MS de Lucena, 2010

The Wayampi are an indigenous people who live in the upper Rio Oiapoque system, French Guiana.

Wayana

Characin sp. *Phenacogaster wayana* P-Y Le Bail & MS de Lucena, 2010

The Wayana are an indigenous people who live in Rio Maroni system, French Guiana.

Wayuu

Hagfish sp. *Eptatretus wayuu* H-K Mok, LM Saavedra-Diaz & P Acero, 2001
Wayuu Sea Catfish *Cathorops wayuu* R Betancur-Rodríguez, P Acero & AP Marceniuk, 2012

These species are named after the Wayuu, an aboriginal people who live on the coastal region of the Guajira Peninsula (northern Colombia/north-west Venezuela), the type locality.

Weber, C

Armoured Catfish sp. *Pterygoplichthys weberi* JW Armbruster & LM Page, 2006
Armoured Catfish sp. *Hypostomus weberi* PH Carvalho, FCT Lima & CH Zawadzki, 2010

Dr Claude Weber works at the Department of Herpetology and Ichthyology, Muséum d'Histoire Naturelle de Génève. He co-wrote: *Identity of Hypostomus plecostomus (Linnaeus, 1758), with an overview of Hypostomus species from the Guianas (Teleostei: Siluriformes: Loricariidae)* (2012). He was honoured for his contributions to the knowledge of the genus *Hypostomus*.

Weber, D

Neotropical Rivuline sp. *Cynodonichthys weberi* JH Huber, 1992

Dale Weber (d.1997) of Novato, California, was a 'renowned' killifish hobbyist who made a number of collecting expeditions to Panama and Venezuela. He helped collect the type and other rivulines. He died in a car crash in Brazil whilst on a collecting trip. He contributed articles to aquarist hobby magazines, such as *Bluefin Killy* (1977).

Weber, MWC

Halfbeak sp. *Nomorhamphus weberi* GA Boulenger, 1897
Blacktip Tripodfish *Trixiphichthys weberi* BL Chaudhuri, 1910
Reef-flat Pipefish *Nannocampus weberi* G Duncker, 1915
Rattail sp. *Coelorinchus weberi* CH Gilbert & CL Hubbs, 1920
Bandfish sp. *Owstonia weberi* JDF Gilchrist, 1922
Weber's Chromis *Chromis weberi* HW Fowler & TH Bean, 1928
Loach sp. *Homalopteroides weberi* SL Hora, 1932
Searobin sp. *Peristedion weberi* JLB Smith, 1934
Weber's Mudskipper *Periophthalmus weberi* WB Eggert, 1935
Weber's Croaker *Johnius weberi* JDF Hardenberg, 1936
Borneo Catfish sp. *Ompok weberi* JDF Hardenberg, 1936

Max Wilhelm Carl Weber van Bosse (1852–1937) was a Germanborn Dutch physician and zoologist, and Director of the Zoological Museum in Amsterdam (1883) (when he became a naturalised Dutch citizen). He was educated in Germany at Bonn and Berlin. He did military service in the German army – half the time as a doctor and half as a hussar. He made a voyage in the small schooner *Willem Barents* (1881), appropriately to the Barents Sea. He combined the roles of watch-keeping officer, ship's doctor and naturalist. His wife was a skilled and learned botanist and after their marriage the Webers spent three summers in Norway where he could dissect whales and she could collect algae – her specialty. They made a number of other voyages to Sumatra, Java, Sulawesi and Flores (1888), and to South Africa (1894). He was co-author of the authoritative: *Fishes of the IndoAustralian Archipelago.* He also established 'Weber's Line', an important zoogeographical line between Sulawesi and the Moluccas, which is sometimes preferred over Wallace's Line (between Sulawesi and Borneo) as the dividing line between the Oriental and Australasian faunas. The *Siboga* expedition to Indonesian waters was carried out under Weber's personal leadership (1899–1900). He was honoured by Hora for the "*…valuable service rendered by him towards the study of Indo-Australian Fishes.*" A bird, two amphibians, three mammals and three reptiles are also named after him.

Webster

African Lampeye Killifish sp. *Procatopus websteri* JH Huber, 2007
[Syn. *Aapticheilichthys websteri*]

Kent Webster (b.1959) is an American aquarist at the Peninsula Hatchery, Gardena, California. He discovered this killifish and was honoured as he has "*…devoted much of his life to breeding aquarium fishes*", notably Australasian rainbowfish.

Weddell

Thorny Catfish sp. *Anadoras weddellii* FL de Castelnau, 1855

Hugh Algernon Weddell (1819–1877) was a botanist, born in England but raised in France. He explored in South America (1843–1847) as a member of the Castelnau (q.v.) Expedition to Brazil. Before leaving Paris, he had been particularly instructed to make a thorough investigation of the *Cinchona* plant, the source of quinine, in its native habitat. He wrote: *Voyage dans le Nord de la Bolivie et dans les Parties Voisines du Pérou* (1853) and the 2-volume *Chloris Andina – Essai d'une Flore de la Région Alpine des Cordillières de l'Amerique du Sud* (1855–1861). Three birds and a mammal are named after him. He was not related to the sealer Captain James Weddell (1787–1834), after whom the Weddell Sea and the Weddell Seal are named.

Wedl

Doublespine Seadevil sp. *Bufoceratias wedli* V Pietschmann, 1926

Anton Wedl (1864–1929) was an industrialist, wholesale merchant and philanthropist in New York who was born in Vienna and maintained links there being on the board of directors of the Austrian Commercial & Industrial Bank (1912). He was awarded the Medal of Honour by the University of Vienna, who was on their administrative board, because of the assistance he provided the students and teachers during the humanitarian emergency after WW1. He also supported the Vienna Museum of Natural History where Viktor Pietschmann was Curator of Fish. He was an important supporter of Austrian Jewish writers, despite the fact that he was an orthodox Catholic, and an intimate friend of the former Chief Rabbi Chajes.

Weeks

Mottled Ctenopoma *Ctenopoma weeksii* GA Boulenger, 1896

Mottled Bichir *Polypterus weeksii* GA Boulenger, 1898
African Cyprinid sp. *Leptocypris weeksii* GA Boulenger, 1899
Weeks' Mormyrid *Hippopotamyrus weeksii* GA Boulenger, 1902
Sicklefin Labeo *Labeo weeksii* GA Boulenger, 1909

Rev John Henry Weeks (1861–1924) was a pioneering Baptist missionary in Congo (1882–1912) who was also an ethnographer, explorer and diarist. He collected, including the labeo type, at his mission station in Monsembe, upper Congo River, Zaire (now Democratic Republic of the Congo). He was also a corresponding member of the Anthropological Institute and the Folklore Society. He wrote: *Among Congo Cannibals* (1913) and *Thirty Years on the Congo* (1914).

Wehrle

Goby sp. *Cryptocentrus wehrlei* HW Fowler,1937

Richard White Wehrle (1852–1937) was a jeweller and naturalist, who obtained many collections of fish and amphibians which he donated to the Academy of Natural Sciences, Philadelphia. He was described as "*...an indefatigable collector of cold-blooded vertebrates.*" An amphibian is also named after him.

Weigt

Weigt's Blenny *Starksia weigti* CC Baldwin & CI Castillo, 2011

Lee Alan Weigt (b.1960) is an American scientist who is the Director of the Smithsonian's Laboratories of Analytical Biology at the Natural History Museum. He was educated at Miami University, Ohio, which awarded his zoology BA (1982) and his MSc in herpetology and genetics (1985) and he was employed as a research associate there (1984–1988). He then went to the Smithsonian Tropical Research Institute where he started doing some marine biology and marine fish work (1988–1996). He became manager (1996–1998) at the Biochemistry Laboratories at the Field Museum and of DNA Sequencing Core at Virginia Tech (1998–2001) before joining the Smithsonian as lab manager (2001–2004). His more than fifty published papers include the co-written: *DNA Barcoding Fishes* (2012), *Biogeography of the tungara frog, Physalaemus pustulosus: a molecular prospective* (2005), and *Using DNA Barcoding to Assess Caribbean Reef Fish Biodiversity: Expanding Taxonomic and Geographic Coverage* (2012). He was honoured for his contributions to the DNA barcoding of fishes.

Weitzman, MJ

Tetra sp. *Hyphessobrycon weitzmanorum* FCT Lima & CLR Moreira, 2003

Marilyn Jean Weitzman née Sohner (b.1926) is a Research Associate on the fisheries staff at the Smithsonian. The tetra was named after Marilyn and her husband Stanley (below) for their: "*...life-long interest and extensive contributions*" to the knowledge of neotropical freshwater fish. They met in elementary school and were life-long partners and best friends. She graduated from high school a year earlier than her husband and started college while he was serving in the navy. She attended Marin Junior College and then went onto Berkeley, where she studied botany and landscape design. During their college years and later, Stan and Marilyn enjoyed traveling around California in search of plants and fishes. After coming to Washington, Marilyn moved from landscape design to ichthyology and has pursued her own studies of characiform fishes ever since. Her continuing studies of the Lebiasinidae include several ongoing projects. These include descriptions of new species of *Pyrrhulina* and *Copella* from Peru and Brazil. (Also see **Marilyn (Weitzman)**)

Weitzman, SH

Black Darter Tetra *Poecilocharax weitzmani* J Géry, 1965
Weitzman's Cory *Corydoras weitzmani* H Nijssen, 1971
[Alt. Twosaddle Cory]
Weitzman's Pearlside *Maurolicus weitzmani* NV Parin & SG Kobyliansky, 1993
Darter Characin sp. *Microcharacidium weitzmani* PA Buckup, 1993
One-sided Livebearer sp. *Jenynsia weitzmani* MJ Ghedotti, AD Meisner & PHF Lucinda, 2001
Tetra sp. *Hyphessobrycon weitzmanorum* FCT Lima & CLR Moreira, 2003
Characin sp. *Lophiobrycon weitzmani* RMC Castro, AC Ribeiro, RC Benine & ALA Melo, 2003
Characin sp. *Knodus weitzmani* NA de Menezes, AL Netto-Ferreira & KM Ferreira, 2009
Characin sp. *Odontostilbe weitzmani* J Chuctaya, CM Bührnheim & LR Malabarba 2018

Dr Stanley Howard Weitzman (1927–2017) was an American ichthyologist who was Research Scientist Emeritus, Division of Fishes, Department of Invertebrate Zoology, at the National Museum of Natural History, Smithsonian and a professor at George Washington University, Washington DC. The University of California Berkeley awarded both his bachelor's and master's degrees and Stanford University his doctorate. He has written at least 116 papers including: *A new species of characid fish of the genus Nematobrycon* (1944), *Phyletic studies of teleostean fishes, with a provisional classification of living forms* (1966) and *Two new species and a review of the inseminating freshwater fish genus Monotocheirdon from Peru and Bolivia* (2013). He was also an excellent artist and illustrated many of his works. The species *Hyphessobrycon weitzmanorum* was named after Stanley and his wife Marilyn (above).

Welander

Welander's Flathead *Rogadius welanderi* LP Schultz, 1966

Dr Arthur Donovan Welander (1908–1982) was a fisheries biologist at the School of Fisheries, University of Washington, his alma mater which had awarded his BS (1934), MS (1940) and PhD (1946). He succeeded Schultz (1936–1978) in teaching Ichthyology and, for much of that time, looking after its fish collection. He was an associate instructor in ichthyology (1937) while also working as a scientific assistant for the International Fisheries Commission, then Instructor (1943), ultimately rising to Professor (1958–1978). He was part of Operation Crossroads at Bikini during the summers (1946–1949) studying the radiation effects on living fishes. This work was funded by the U.S. Atomic Energy Commission and led to the establishment of the Laboratory of Radiation Biology in the School of Fisheries. Overall, he took part in around 25 expeditions. Among his c.40 published papers is: *Radioactivity in the reef fishes of Belle Island Eniwetok Atoll : April 1954 to November 1955 (1957).*

Welch, EJ

Welch's Grunter *Bidyanus welchi* AR McCulloch & ER Waite, 1917

Edwin James Welch (1838–1916) was an English naval cadet, surveyor, writer and journalist. After service in the Crimean War, he moved to Australia, where he took part in an expedition (1861) to discover the fate of the Burke and Wills expedition (an attempt to cross Australia from south to north). In later life he became a newspaper proprietor. He is mentioned in the description of the grunter as 'my old friend', though it is unclear which of the two authors he was friends with (perhaps both?).

Welch, TC

Robust Armoured-Gurnard *Satyrichthys welchi* AW Herre, 1925

Thomas Cary Welch Jr (1864–1928) was an attorney and civil servant in the Philippines. Herre says in his etymology: "*I take pleasure in naming this species for Thomas Cary Welch, as an appreciation of his interest in scientific matters and the assistance he has given my studies of Philippine fishes.*"

Welcomme

Lake Victoria Cichlid sp. *Allochromis welcommei* H Greenwood, 1966

Dr Robin Leon Welcomme (b.1938) is an English fisheries biologist and limnologist who formerly worked at the East African Freshwater Fisheries Research Organization (Jinja, Uganda). He is currently retired (1998) but is (from 2010) a Visiting Researcher at Imperial College Department of Life Sciences, London. He was employed in the late 1950s and into the 1960s as Assistant Scientific Officer at the Water Pollution Laboratory and later at the Salmon and Freshwater Fisheries Research Laboratory in the UK, then for East African Freshwater Fisheries. His PhD was awarded by Makerere College at the University of East Africa (1967). His c.100 published papers include: *Recent Changes in the Stocks of Tilapia in Lake Victoria* (1966), *Some general and theoretical considerations on the fish production of African rivers* (1974) and *Fisheries of the rivers of Southeast Asia* (2016) and his books include *Manual of Inland Fisheries: Ecology and Management* (2001). He collected the type and supplied ecological and other data to the author.

Wellington

Razorfish (wrasse) sp. *Xyrichtys wellingtoni* GR Allen & DR Robertson, 1995

Scorpionfish sp. *Scorpaena wellingtoni* BC Victor, 2013

Gerard M Wellington (d.2014) was Professor of Biology at the University of Houston. San José State University awarded his BA (1971) and the University of California, Santa Barbara, his PhD (1981). He began his teaching career at Houston as Assistant Professor of Biology (1982–1988), was then Associate Professor (1988–1994) and Professor (1994). He co-authored the book *Corals and Coral Reefs of the Galápagos Islands* (1983) and published numerous papers and articles. He was honoured in the name of the scorpionfish as "*...a pioneer in research on the marine biology and conservation in the Galápagos Archipelago. He developed the first plan for the Parque Nacional Galápagos as a member of the Peace Corps in the 1970s and in subsequent decades conducted many expeditions and surveys, especially* *on the effect of the El Nino-Southern Oscillation and climate change on the fragile marine ecosystem of the Galápagos Archipelago*."

Wellman

African Barb sp. *Enteromius wellmani* GA Boulenger, 1911

Dr Frederick Creighton Wellman (1871–1960) was an American medical missionary at Kamundongo - now Angola, but then in Portuguese West Africa (1896–1905) – during which period he took a break to take a formal qualification at the London School of Tropical Medicine. As a physician he specialised in tropical medicine. He wrote: *A criticism of some of the theories regarding the etiology of Goundou and Ainhum* (1906). He collected the holotype of the species named after him. After returning to the USA (1906) he started out on what turned out to be a memorable career in which he was married at least four times and divorced thrice and was variously a writer, an artist, a teacher, a linguist, a baggage porter, a bookkeeper, a farmer, an explorer, a mining engineer, a museum director but, firstly, he was the Founding Dean of the Tulane School of Hygiene and Tropical Medicine (1912–1914); which position he left abruptly (1914) and eloped to Brazil with a young woman, Elsie Dunn (1893–1963), whom he later married as his third wife, and who became well-known in her own right as the novelist Evelyn Scott. In Brazil, he successfully explored for diamonds and collected entomological specimens for the BMNH, London. He changed his named to Cyril Kay-Scott and after leaving Brazil (1919) settled in New York (1919–1923) and then in Europe (1923–1928) as a painter and writer, under pseudonyms including Cyril F C Kay-Scott and Richard Irving Carson. He eventually returned to the USA, set up and ran a successful art school in Santa Fé, New Mexico (1928–1941), and became the Dean of the College of Fine Arts, University of Denver (1931–1934). He published his autobiography, under the name Kay-Scott, *Life Is Too Short: An Autobiography* (1943). He has the distinction of appearing in 'Who's Who' under both his names.

Welwitsch

Angolan Happy (cichlid) *Chetia welwitschi* GA Boulenger, 1898

Dr Friedrich Martin Josef Welwitsch (1806–1872) was an Austrian botanist and explorer. He studied medicine and botany in Vienna before travelling to Portugal to escape the consequences of what was termed a 'youthful indiscretion'. He explored most of Portugal, forming a herbarium of 9,000 specimens. The Portuguese government sent him to explore and collect botanical specimens in Angola (1853–1861), where he accumulated c.5,000 specimens, many new to science. He caused an international quarrel by sending a large proportion of his collection to the BMNH, London, instead of to Lisbon. The Portuguese took the view that, as they had paid him, the collection belonged to them. The collection's duplicate specimens were split, so both museums got something out of it. Welwitsch moved to London (1863) and worked at Kew Gardens. A bird, two mammals and a reptile are also named after him, as is the plant genus *Welwitschia*.

Weng

Weng's Skate *Dipturus wengi* B Séret & PR Last, 2008

Dr Herman Ting-Chen Weng is a Queensland fisheries biologist. The University of Adelaide, South Australia, awarded his doctorate (1971). He wrote: *The Black Bream, Acanthopagrus butcheri (Munro): Its life history and fishery in South Australia* (1971). He established his own company in Brisbane, Fishing Weng Publications, through which he published and marketed a number of his own works including *Wonderful Fishes* (1982), a book for children. He was honoured as he "*...showed an enthusiastic interest in skates and collected the first validated Australian specimens of this species in 1983*".

Werestschagin

Crumbly Sculpin *Neocottus werestschagini* DN Taliev, 1935

Gleb Yurievich Werestschagin was a Russian limnologist and cladocerologist at the Leningrad Academy of Sciences. He led an expedition to Lake Baikal (1925–1927). Among his published papers are: *Methoden der hydrochemischen Analyse in ihren Anwendungen zu limnologischen Forschungen* (1929) and *Résultats d'une exploration scientifique du Baïkal en 1925–1927*(1929). At least one zooplankton is also named after him.

Werneke

Armoured Catfish sp. *Peckoltia wernekei* JW Armbruster & NK Lujan, 2016

David C Werneke is an American biologist and ichthyologist who is a long-serving Fish Collections Manager at the Auburn University Museum, Alabama, where he was awarded a bachelor's degree (2000) and a Graduare Research fellowship (2001–2002). He co-wrote: *Three new species of saddled loricariid catfishes, and a review of Hemiancistrus, Peckoltia, and allied genera (Siluriformes)* (2015). The etymology compliments him for: "*...diligence, camaraderie and humor during three expeditions to the upper Orinoco Basin.*"

Werner, A

Congo Cichlid sp. *Lamprologus werneri* M Poll, 1959
Werner's Killifish *Aplocheilus werneri* H Meinken, 1966
Featherfin Rainbowfish *Iriatherina werneri* H Meinken, 1974
Werner's Tetra *Hyphessobrycon werneri* J Géry & A Uj, 1987

Arthur Werner is a German aquarium fish importer who owns a company called 'Transfish'. He helped to collect the holotypes of these species.

Werner FJM

African Barb sp. *Enteromius werneri* GA Boulenger, 1905
Werner's Catfish *Clarias werneri* GA Boulenger, 1906
Labeo sp. *Labeo werneri* K Lohberger, 1929
African Barb sp. *Labeobarbus werneri* M Holly, 1929

Dr Franz Josef Maria Werner (1867–1939) was an Austrian zoologist and explorer. He taught at the Natural History Museum, Vienna. Here the Director, Steindachner, disliked him and forbade him access to the herpetological collection. He collected in North and East Africa, being in Egypt (1904) and the Sudan (1905), and made regular visits south to Uganda and west to Morocco until the outbreak of the First World War. His publications include: *Amphibien und Reptilien* (1910). He described dozens of amphibians, reptiles and a few fish, and has 28 reptiles, six amphibians and several arachnids named after him. The patronym is not identified in the descriptions of the labeo and the barb, but these binomials almost certainly honour Lohberger's and Holly's fellow Austrian and colleague.

Werner, G

Werner's Smooth-head *Conocara werneri* O Nybelin, 1947

Directeur G Werner was described as the "*donor of a projected Swedish expedition to explore the ocean depths*" (translation). No further information is available, and we have not been able to identify him further.

Werner (Schwarzhans)

Cusk-eel sp. *Bassozetus werneri* JG Nielsen & NR Merrett, 2000

Dr Werner Wilhelm Schwarzhans (See **Schwarzhans**)

Wertheimer

Thorny Catfish genus *Wertheimeria* F Steindachner, 1877
Armoured Catfish sp. *Pogonopoma wertheimeri* F Steindachner, 1867

There were two people called Wertheimer on the Thayer Expedition to Brazil (1865–1866), but Steindachner did not clarify which of them he was naming the fish after. The candidates are Louis Wertheimer and Achilles Wertheimer; the latter died during the expedition from a snakebite.

Weseth

Snubnose Blacksmelt *Bathylagoides wesethi* RL Bolin, 1938

Lars H Weseth (1895–1982) worked for the California Department of Fish and Game (1930–1960) as captain of a number of research vessels including 'Albacore', from which the holotype was collected. He later commanded 'N B Scofield' and 'Bonito'. He was noted for: "*...helpfulness and co-operation which he unfailingly extends to scientists working on board his vessel.*"

Wessel, C

Three-barbeled Catfish sp. *Pimelodella wesselii* F Steindachner, 1877

Carl Wessel was a Hamburg purveyor of natural history items to the Vienna Museum where Steindachner described several of them.

Wessel, R

Central American Cichlid sp. *Chortiheros wesseli* RR Miller, 1996

Richard 'Rusty' Wessel is a cichlid aquarist and, according to Miller, a "*...dedicated amateur naturalist devoted to gathering information on the ecology, behaviour, and identification of Middle American cichlids.*" He discovered this species and collected the type.

Wessels

Southern Churchill *Petrocephalus wesselsi* BJ Kramer & HF van der Bank, 2000

Pierre Wessels of Johannesburg, South Africa, participated in the authors' expedition to Caprivi (Namibia) and was described by them as a "*...nature conservationist and good friend.*"

Westermann

Long-whiskered Catfish sp. *Bergiaria westermanni* CF Lütken, 1874

Bernt Wilhelm Westermann (1781–1868) was a wealthy Danish businessman who collected insects. He travelled on business to India (where the holotype was collected) and Indonesia (via the Cape of Good Hope) and collected while there. He returned to Denmark (1817) where he became a ship-owner and sugar refiner. By the time of his death he had amassed over 45,000 species of insect, which are now housed in the collection of the University of Copenhagen.

Wetmore

Wrasse genus *Wetmorella* HW Fowler & BA Bean, 1928
Wetmore's Barb *Hypsibarbus wetmorei* HM Smith, 1931

Frank 'Alexander' Wetmore (1886–1978) was an American ornithologist and avian palaeontologist who conducted extensive fieldwork in Latin America. His first job was as a bird taxidermist at the Denver Museum of Natural History, Colorado (1909). He spent time in Puerto Rico studying its avifauna (1911). He travelled throughout South America for two years, investigating bird migration between continents, whilst working for the US Bureau of Biological Survey. He was Assistant Secretary of the USNM (1925–1946). He was President of the AOU (1926–1929), then became the USNM's sixth Secretary (1945–1952). Wetmore made a number of short trips to Haiti, the Dominican Republic, Guatemala, Mexico, Costa Rica and Colombia, and conducted a research programme in Panama (1946–1966), during which he made an exhaustive survey of the birds of the isthmus. He wrote: *A systematic classification for the birds of the world* (1930), which he revised twice (1951 & 1960). Therein he devised the Wetmore Order, a sequence of bird classification, which had widespread acceptance until recently (although still in use). His other publications included: *Birds of Haiti and the Dominican Republic* (1931) and *The Birds of the Republic of Panamá* (1965). Numerous taxa, comprising 56 new genera, species and subspecies of birds (both recent and fossil), mammals, amphibians, insects, molluscs and plants, are named after him. He wrote the first descriptions of 189 species and subspecies of living birds, mostly from Central and South America. Nineteen birds, an amphibian, a mammal and five reptiles are also named after him.

Weyns

African Cyprinid sp. *Leptocypris weynsii* GA Boulenger, 1899

Lieutenant Colonel Auguste François Guillaume Weyns (1854–1944) was a Belgian explorer who collected in Central Africa (1888–1903) including obtaining the type of this cyprinid. He was also the governor (1900–1903) of the semi-autonomous state of Katanga within the Congo, as 'representative' of the Comité Spécial du Katanga. A mammal and a bird are also named after him.

Weyrauch

Characin sp. *Microgenys weyrauchi* HW Fowler, 1945

Pencil Catfish sp. *Trichomycterus weyrauchi* HW Fowler, 1945

Dr Wolfgang Karl Weyrauch (1907–1970) was a German zoologist, malacologist, and ecologist who went to Peru (1938). He worked at the Museum of Universidad Nacional Mayor de San Marcos, Lima, and collected spiders and centipedes for the Department of Entomology (1940–1950). Thereafter, his function was taken over by Hans-Wilhelm and Maria Koepke. His private collection was deposited later in Instituto Miguel Lillo, Tucumán, Argentina, where he probably resided. A reptile is also named after him.

Whaler

Characin sp. *Gephyrocharax whaleri* SF Hildebrand, 1938

[Probably a Junior Synonym of *Gephyrocharax intermedius*]

Frederick Giles 'Fred' Whaler (1884–1951) of Balboa, Panama Canal Zone, was a Panama Canal employee who later became well known as a fishing guide. As a pilot and guide he accompanied President Franklin D Roosevelt on his first trip to Canal waters during his presidency. Fred Whaler was a friend of the author and "*...an ardent angler as well as a student of fishes, and who rendered valuable assistance when the specimens were taken.*"

Wheatland

Characin sp. *Henochilus wheatlandii* S Garman, 1890

Dr Henry Wheatland (1812–1893) first went to Harvard (1828) aged 16, and after graduating he studied medicine and received his medical degree from Harvard (1837). He was a founder member of the Essex County Natural History Society and its Secretary (1835–1848). The Society merged with the Essex Historical Society and became the Essex Institute, Salem, Massachusetts. Wheatland was the Institute's Secretary and Treasurer (1848–1868) and its President (1868–1893), despite being paralysed in his last years. He was also active in other organisations including being an original Trustee and Vice-President of the Peabody Museum of Archaeology and Ethnology. Garman named the characin after him for his "*...friendly interest and sympathy in favor of ichthyology and ichthyologists*" and the description was published by the Essex Institute.

Wheatland, A

Black-spotted Stickleback *Gasterosteus wheatlandi* FW Putnam, 1867

Dr Richard Henry Wheatland (1830–1863) of Nahant, Massachusetts collected the stickleback type (1859). He was a Harvard graduate. He was Cabinet Keeper of the Essex County Natural History Society and was most interested in fish and reptiles. He was the nephew of Henry Wheatland (above).

Wheeler, A

Goby genus *Wheelerigobius* PJ Miller, 1981

Jigsaw Moray *Uropterygius wheeleri* J Blache, 1967

Gorgeous Prawn-goby *Amblyeleotris wheeleri* NVC Polunin & HR Lubbock, 1977

Spotted Perchlet *Plectranthias wheeleri* JE Randall, 1980
Slender Armourhead *Pentaceros wheeleri* GS Hardy, 1983
Burma Hairfin Anchovy *Setipinna wheeleri* T Wongratana, 1983
Puffer sp. *Lagocephalus wheeleri* T Abe, O Tabeta & K Kitahama, 1984

Alwyne Cooper 'Wyn' Wheeler (1929–2005) was Curator of Fishes at the BMNH. He joined the London Natural History Society at the age of 13 and served his National Service as a radiographer and medical photographer in the Royal Army Medical Corps in both the United Kingdom and Jamaica, where he joined the Natural History Society of Jamaica. On leaving the army he applied to the BMNH for a post as an Assistant in the Department of Zoology. He was unusual in that his subsequent scientific career was achieved despite his never having obtained a university degree. He was Assistant in the Fish Section (1950–1954), Assistant Experimental Officer (1954–1958), Experimental Officer (1958–1987) and Curator of Fishes (1987–1989). He wrote more than 250 scientific papers and books (1955–2005), mostly about the taxonomy of historically important fish collections; identification and distribution of the British and European fish fauna; and the status of British fish in a changing environment. His more important works included: *Fishes of the World: an illustrated dictionary* (1975). He also wrote under the pen name Allan Cooper including: *Fishes of the World* (1969), and contributed columns to angling magazines. He was honoured in gratitude of "*many services*" by Blache and by Wongratana because his: "*...kind help during my time there was much appreciated.*" Polunin honoured him "*...for his help over the years, particularly with the authors' study of prawn-associated gobies of the Seychelles.*"

Wheeler, JFG

Blacktail Reef Shark *Carcharhinus wheeleri* JAF Garrick, 1982
[Sometimes regarded as a Junior Synonym of *C. amblyrhynchos*]

Dr John Francis George Wheeler (1900–1979) was a British zoologist who became (1932–1941) Director of the Bermuda Biological Station, and later of the East African Marine Fisheries Research Organization. Bristol University awarded his zoology degree (1922), after which he was an assistant lecturer there which included spending a number of years on board the research vessel 'Discovery II' on expeditions (1924–1927 & 1929–1930). He left BBSR (1942) when funding ran out and took a job in one of Bermuda's banks. He took a post (1943) as marine biologist in Mauritius. He was then in charge of a research unit in Zanzibar and at Lake Tanganyika (1951), becoming (by 1954) Director of the East African Marine Fisheries Research Organization. Among other works, he co-wrote *Southern Blue and Fin Whales* (1929) and wrote: *Sharks of the Western Ocean* (1959). His 1953 account of this species, as *C. amblyrhynchos*, is the first definitive record of it.

Whipple

Steelcolor Shiner *Cyprinella whipplei* CF Girard, 1856
Redfin Darter *Etheostoma whipplei* CF Girard, 1859

Lieutenant Amiel Weeks Whipple (1818–1863) was an American military engineer and surveyor. Fort Whipple (now Fort Myer) was named in his honour. He led (1853) the boundary survey team that collected the shiner type while exploring the route for the first transcontinental railway to the Pacific, after which he wrote: *Explorations and Surveys for a Railroad Route from the Mississippi River to the Pacific Ocean* (1855–1857) which contained the formal description of a cactus species. He served as a Brigadier General in the American Civil War and was mortally wounded by a sharpshooter at the Battle of Chancellorsville. On his deathbed he was promoted to Major General.

Whitaker

Whitaker's Sole *Aseraggodes whitakeri* LP Woods, 1966

Dr Douglas Merritt Whitaker (1904–1973) was the first Provost of Stanford University (1952–1955), having previously worked there as Professor of Biology. He served (1947) as a biologist with the US Army-Navy expedition in the Pacific to determine the effects of radiation fallout on marine life following the experimental atomic bomb explosions on Bikini Atoll.

White, EA

Balsa Splitfin *Ilyodon whitei* SE Meek, 1904

E A White lived in Mexico City and worked for the Interoceanic Railway of Mexico. He apparently took an interest in Meek's work and gave him assistance.

White, J

White's Seahorse *Hippocampus whitei* P Bleeker, 1855

John White (c.1756–1832) served as a surgeon in the Royal Navy (1781–1820), visiting India and the West Indies before 1787, when he was appointed Chief Naval Surgeon of the 'First Fleet' – the 11 ships which took the first 1,500 colonists to Australia, among whom were 778 convicts. Despite scurvy and dysentery only 34 people died on that voyage. He stayed in Australia and, as a keen naturalist, he accompanied Governor Phillip on two journeys of exploration. A colourful character, he fought a duel with his Third Assistant, William Balmain, which left them both slightly wounded. He had three legitimate children and also a son by a convict. The boy was brought up as part of his legitimate household. He kept a journal, which he sent to a friend in London, Thomas Wilson. This was published as: *Journal of a Voyage to New South Wales* (1790). White returned to London (1795) and was reluctant to return to New South Wales and (1796), faced with the alternatives of doing so immediately or resigning his appointment, he chose to resign. He served on various ships (1796–1799) and was Surgeon at the Sheerness Navy Yard (1799–1803) and at Chatham (1803–1820). The University of St Andrew conferred the degree of Doctor of Medicine on him (1797). An amphibian and a reptile are also named after him.

White, SRE

Little Kern Golden Trout *Oncorhynchus mykiss whitei* BW Evermann, 1906
[Alt. Little Kern River Rainbow Trout]

Stewart Edward White (1873–1946) was an American spiritualist who wrote novels about nature and adventure over the period 1900–1922, after which he and his wife wrote books they said had been channelled to them by spirits. The University of Michigan awarded both his bachelor 's degree (1895) and his master's (1903). He was a friend of President Theodore Roosevelt to whom he appealed to stop the overfishing of California's Golden Trout. Roosevelt complied and ordered the government to research their biology, propagation and distribution. Evermann was the leader of the team appointed to do the investigation. Among White's many books is the four-volume: *The Saga of Andy Burnett* (1930, 1932, 1933 & 1942.

White, TD

White's Pearlfish *Nematolebias whitei* GS Myers, 1942
[Alt. Rio Pearlfish]
Characin sp. *Brycon whitei* GS Myers & SH Weitzman, 1960

General Thomas Dresser White (1901–1965) was the Chief of Staff, United States Air Force, in which he served (1924–1961), having initially been an infantry officer (1920–1924). During his career, he was a military attaché in Russia (1934–1935), Italy and Greece (1935), and Brazil (1940–1942). During WW2, he held a number of senior commands in the Pacific theatre, particularly in the campaigns in the Philippines, Borneo and New Guinea. He was an ardent aquarist. He helped to collect the holotypes of these species. He died of leukaemia. His second wife became an ichthyologist and is also honoured in the name of a fish. (Also see **Constancia**)

White, WT

Philippine Guitarfish *Rhinobatos whitei* PR Last, S Corrigan & G Naylor, 2014

Dr William Toby White (b.1977) is an Australian ichthyologist whose particular interests are speciation and biodiversity of sharks and rays and fishery management in developing countries. Murdoch University, Perth, awarded his BSc (1997) and PhD (2003), and he was a post-doctoral researcher there (2004–2006). He is an ichthyologist at the Australian National Fish Collection which is part of the CSIRO Marine and Atmospheric Research facility in Hobart (2006–present). He has published c.150 papers and described 50 new sharks and rays as well as other fish. He was honoured in the guitarfish name for his "*… contributions to the taxonomic and biological knowledge of sharks and rays of the Western Central Pacific.*"

Whitehead, J

Borneo Loach sp. *Protomyzon whiteheadi* L Vaillant, 1894
Asian Perch sp. *Coreoperca whiteheadi* GA Boulenger, 1900

John Whitehead (1860–1899) was a British explorer, bird collector and naturalist who collected in Borneo (1885–1888), the Philippines (1893–1896) and Hainan (1899). He wrote *Exploration of Mount Kina Balu, North Borneo* (1893) and he might have been the first European to reach the summit of the mountain. He died of fever in Hainan, at the age of 38. He collected the loach type. Nineteen birds, an amphibian and five mammals are also named after him.

Whitehead, P

Indian Bandfish *Owstonia whiteheadi* PK Talwar, 1973

Whitehead's Sawtooth Pellonuline *Potamothrissa whiteheadi* M Poll, 1974
Whitehead's Deepwater Dragonet *Callionymus whiteheadi* R Fricke, 1981
Whitehead's Thryssa *Thryssa whiteheadi* T Wongratana, 1983
Whitehead's Round Herring *Etrumeus whiteheadi* T Wongratana, 1983
Burmese River Gizzard Shad *Gonialosa whiteheadi* T Wongratana, 1983
African Sprat sp. *Microthrissa whiteheadi* G Gourène & GG Teugels, 1989
Hatchet Herring sp. *Pristigaster whiteheadi* NA de Menezes & MCC de Pinna, 2000

Dr Peter James Palmer Whitehead (1930–1992) was a British biologist (BMNH), scholar, historian, and artist as well as a clupeoid specialist. After military service rising to Second Lieutenant in the Royal Artillery (1953) he graduated from Trinity Hall Cambridge and later was awarded his doctorate by them. He wrote the book: *Drawings of fishes from Captain Cook's Voyages* (1969). He wrote: *A Dutch 17th-Century Portrait of Brazil* (1989). According to one of the etymologies, he "*…contributed more than any other individual to the knowledge of clupeomorph fishes*". He was also honoured "*…for suggesting that there might be two different species of Pristigaster.*" Wongratana honoured him because Whitehead had encouraged him to make Indo-Pacific clupeoids the subject of his thesis. His (1963) revision of the genus *Etrumeus* formed a basis for Wongratana's study.

Whitehurst

Dusky Jawfish *Opistognathus whitehursti* WH Longley, 1927

Dr D D Whitehurst was a collector of specimens for the Smithsonian Institution, including the type of this species.

Whitelegge

Shadow Driftfish *Cubiceps whiteleggii* ER Waite, 1894
Banded Weedfish *Heteroclinus whiteleggii* JD Ogilby, 1894

Thomas Whitelegge (1850–1927) was an English naturalist who emigrated (1883) to Sydney, Australia, where he joined the Linnean Society of New South Wales. Perhaps his most important work was: *List of the marine and fresh-water Invertebrates of Port Jackson and neighbourhood* (1889).

Whitley

Whitley's Silverside *Hypoatherina tropicalis* GP Whitley, 1948
Whitley's Toadfish *Torquigener whitleyi* WEJ Paradice, 1927
Whitley's Boxfish *Ostracion whitleyi* HW Fowler, 1931
Lanternfish sp. *Diaphus whitleyi* HW Fowler, 1934
Melbourne Skate *Spiniraja whitleyi* T Iredale, 1938
Sole sp. *Liachirus whitleyi* P Chabanaud, 1950
Whitley's Splitfin *Luzonichthys whitleyi* JLB Smith, 1955
Whitley's Sergeant *Abudefduf whitleyi* GR Allen & DR Robertson, 1974
Whitley's Scorpionfish *Maxillicosta whitleyi* WN Eschmeyer & SG Poss, 1976
Bigcheek Snailfish *Psednos whitleyi* DL Stein, NV Chernova & AP Andriashev, 2001

Gilbert Percy Whitley (1903–1975) was a British-born Australian ichthyologist and malacologist. Born and educated in England, he migrated to Australia with his family

(1921) and joined the staff of the Australian Museum, Sydney (1922), while studying zoology at Sydney Technical College and the University of Sydney. He was appointed ichthyologist (1925) (later Curator of Fishes) at the Museum, a position he held until retirement (1964). During his term of office, he doubled the size of the ichthyological collection to 37,000 specimens through many collecting expeditions. He was also a major force in the Royal Zoological Society of New South Wales, of which he was made a fellow (1934) and which he served as President (1940–1941, 1959–1960 & 1973–1974) and edited its publications (1947–1971). He is commemorated by the Royal Zoological Society of New South Wales Whitley Awards for excellence in zoological publications relating to Australasian fauna.

Whitney

Humpback Smooth-Hound *Mustelus whitneyi* F Chirichigno, 1973

Dr Richard R Whitney (1927–2011) was a fisheries biologist. The University of Utah awarded his bachelor's degree (1949) and his master's (1951) and Iowa State University his doctorate (1955). He was at the University of California, Los Angeles, as a Research Biologist (1954–1957), at the Chesapeake Biological Laboratory (1957–1960), at the Bureau of Commercial Fisheries (1960–1967), Unit Leader at the Washington Cooperative Fishery Research Unit (1967–1983) and Professor, the School of Fisheries, University of Washington, Seattle (1983–1993) thereafter Professor Emeritus. He co-wrote: *Inland fishes of Washington* (2003). He was honoured for "*…teachings and guidance in the study of sharks.*"

Whitson

Whitson's Grenadier *Macrourus whitsoni* CT Regan, 1913

Sir Thomas Barnby Whitson (1869–1948) was a Scottish accountant who was Treasurer and Accountant of the Scotia Committee; 'Scotia' was the ship from which the type was collected during the Scottish National Antarctic Expedition (1902–1904). He was Lord Provost of Edinburgh (1929–1932). Edinburgh University awarded him an honorary doctorate (1931) in the same year that he was knighted by King George V. Cape Whitson in the South Orkneys is named after him.

Whitten

Goby sp. *Lentipes whittenorum* RE Watson & M Kottelat, 1994

Dr Anthony John 'Tony' Whitten (1953–2017) with his wife, Jane E J Whitten (b.1954), was zoologists and ecologists who work in Asia, specializing in Indonesian species. They were members of the Sub-department of Veterinary Anatomy, Cambridge (1981), which awarded Anthony his doctorate (1980). He was an adviser for the Center of Environmental Studies, University of North Sumatra (1981–1983). He was employed by Dalhousie University, Halifax, Nova Scotia (1983–1985), but spent 10 of those 12 years on assignment in Indonesia. From 1995 he was employed as a wildlife biologist and Senior Biodiversity Specialist, Asia Technical Department, World Bank. Athony died as the result of a car collision while cycling. They wrote: *Wild Indonesia: The Wildlife and Scenery of the Indonesian Archipelago* (1992). A reptile is also named after them.

Whymper

Suckermouth Catfish sp. *Astroblepus whymperi* GA Boulenger, 1890

Edward Whymper (1840–1911) was an English mountaineer who was also an illustrator and an explorer. He made the first ascent of the Matterhorn (1865). He led an expedition to Greenland (1867) during which he made an important collection of fossil plants. He returned to Greenland (1872) to survey its coastline. He was on an expedition to Ecuador (1880) to study the effects of altitude sickness and reduced pressure on the human body, and made the first ascent of Chimborazo (6,267 metres/20,561 feet). During this trip, he collected natural history specimens including the catfish type. He went to the Canadian Rockies (1901) and made the first ascent of Stanley Peak and of another mountain, now named Mount Whymper after him. He wrote: *Travels amongst the Great Andes of the Equator* (1892). An amphibian is also named after him.

Wichmann

Goby sp. *Sicyopterus wichmanni* M Weber, 1894

Wichmann's Mouth Almighty *Glossamia wichmanni* MCW Weber, 1907

Carl Ernst Arthur Wichmann (1851–1927) was a German geologist and mineralogist who graduated from Leipzig University (1874). He was Professor of Geology at Utrecht University (1879–1921) and founded the geological institute there. He took part in an expedition to the Dutch East Indies (Indonesia) to Celebes, Flores, Timor and Roti (1888–1889), and in the Dutch North New Guinea Expedition (1902–1903) during which the holotype of the mouth almighty was collected.

Wickler

Congo Cichlid sp. *Nanochromis wickleri* UK Schliewen & MLJ Stiassny, 2006

Dr Wolfgang Wickler (b.1931) is a German behavioural zoologist and author. He was a professor at the University of Munich (1969–1974) and again part-time (1976>). He was Director (1975–1999) and then Director Emeritus of the Max Planck Institut for Behavioral Physiology in Seewiesen, Germany, which he joined (1974) and where he had been a scientific assistant (1960–1969). He generously supported studies on central African cichlid fishes conducted by the senior author. His most well-known publication was *Das Prinzip Eigennutz* (The Principle of Self-Interest) (1981). He was honoured for his behavioural studies of cichlids in general and of benthic Congolese cichlids in particular.

Wickliff

Channel Shiner *Notropis wickliffi* MB Trautman, 1931

Edward Lawrence Wickliff (1893–1975) was an American biologist. With Trautman he wrote: *Water fowl of Ohio* (1928) and *Some food and game fishes of Ohio* (1941) as well as a number of other books and papers, such as: *A study of the food of the young small-mouth bass* (1920). Trautman honoured his: "*...loyal friend, who has done much in carrying on and furthering ichthyological research in Ohio.*"

Widdowson

Iraq Blind Barb *Typhlogarra widdowsoni* E Trewavas, 1955
[Syn. *Garra widdowsoni*]

A G Widdowson was an engineer at the Iraq Petroleum Company. He discovered this species (1954) but the specimens he collected died and did not get sent to researchers. He later asked some fellow cavers to collect more and so facilitated the collection of the type series which he sent to A C Hardy at Oxford who passed them on to the author. He wrote: *Explorers of subterranean by-ways. Interlude with the pipeline pot-holers* (1954).

Widegren

Valaam Whitefish *Coregonus widegreni* AJ Malmgren, 1863

Dr Hjalmar Abraham Teofil Widegren (1838–1878) was a Swedish ichthyologist and fisheries officer (1864). He went to Uppsala (1855) where he was awarded his doctorate (1863). The Agricultural Academy commissioned him to study the country's fisheries and fish fauna, during which investigation he made (1859) no less than 63 trips. He concentrated much of his research on salmon and wrote: *Contribution to the knowledge of Sweden's Salmonids* (1863) and *New Contributions* (1864).

Widodo

Whitefin Smooth-Hound *Mustelus widodoi* WT White & PR Last, 2006

Dr Johannes A O Widodo (b.1944) is a biologist at the Research Institute of Marine Fisheries, Jakarta, Indonesia, and has done much work on the shark and ray fisheries of Indonesia. He wrote: *A check-list of fishes collected by Multiara 4 from November 1974 to November 1975* (1976). White said that Widodo's "*...research on the shark and ray fisheries of Indonesia has provided important baseline data for this important faunal region.*"

Wiebel

Hong Kong Butterflyfish *Chaetodon wiebeli* JJ Kaup, 1863

Karl Maximilian Wiebel (1808–1888) was a German naturalist and a founding member of the Zoological Museum in Hamburg. After studying natural sciences in Bonn and Heidelberg and traveling through Germany, Belgium and Switzerland, he taught Physics and Chemistry in Frankfurt and elsewhere (1833–1836). He then took over the professorship at the Academic High School in Hamburg (1837–1881) and founded (1841–1877) a chemistry laboratory there, during which time (1870) he lost his sight.

Wiebrich

Neotropical Silverside sp. *Odontesthes wiebrichi* CH Eigenmann, 1928

Carlos Wiebrich was described by the author as being "*...liberal with help and information about the fishes at Valdivia [Chile]*". In his *Freshwater Fishes of Chile*, Eigenmann cites the source of fishes from Valdivia being supplied by Wiebrich many times. He seems to have been associated with the German School there. He also described him as: "*...a most ardent fisherman-naturalist*" who "*...made strenuous efforts to secure additional material*" for Eigenmann.

Wiener

Drum sp. *Robaloscion wieneri* H-E Sauvage, 1883
[Syn. *Sciaena wieneri*]
Pencil Catfish sp. *Plectrochilus wieneri* J Pellegrin, 1909

Dr Charles Wiener (1851–1913) was an Austrian-French explorer and linguist who travelled widely in Peru and came close to re-discovering Machu Picchu (1875). He wrote a book on his travels and Hiram Bingham, who is credited with Machu Picchu's re-discovery, is known to have read it and used the information that Wiener had gathered but not followed up. The University of Rostock awarded his doctorate for his: *Essai sur les institutions politiques, religieuses, économiques et sociales de l´Empire des Incas* (1874). He collected the holotype of this species.

Wiese

African Rivuline sp. *Scriptaphyosemion wieseae* R Sonnenberg & E Busch, 2012

Barbara Wiese is the partner of junior author Eckhard Busch. He collected with her in northern Sierra Leone (1993 & 2003), and during the second visit this species was collected. The etymology says: "*The new species is named after Barbara Wiese, who accompanied the second author on most of his research trips in West Africa and was the first who recognized it as a new Scriptaphyosemion species.*"

Wiggins

Opata Sucker *Catostomus wigginsi* AW Herre & VE Brock, 1936
Baja Blenny *Labrisomus wigginsi* C Hubbs, 1953

Dr Ira Loren Wiggins (1899–1987) was an American botanist and biologist. Stanford awarded him his doctorate, and he was on the staff there (1930–1964), rising from Assistant Professor of Botany to Emeritus Professor of Biology. He was also Director of the Stanford Natural History Museum and collected the sucker type and other rare Mexican fishes for it. He collected mainly botanical specimens, in the Galapagos Islands and Mexico as well as California and Alaska, where he spent time in the 1950s as Director of the Arctic Research Laboratory, Point Barrow. Two reptiles are also named after him.

Wilbur

Large-tooth Goby *Macrodontogobius wilburi* AW Herre, 1936

Dr Ray Lyman Wilbur (1875–1949) was an American physician who became the third President of Stanford University (1916–1943), which had awarded his BA (1896) and MA (1897), and the 31st US Secretary of the Interior (1929–1933) appointed by Hoover. Cooper Medical College awarded his MD (1899). He was an instructor in physiology at Stanford (1900) and became an assistant professor there while still practicing medicine full-time. He served as a chief of the conservation division of the US Food Administration during WW1. Hoover called him "*…my devoted friend and constant friend since boyhood*" and that "*During all his years, including his later chancellorship of Stanford, he has given a multitude of services to the people. Public health and education have been enriched over all these years from his sane statesmanship and rugged intellectual honesty. America is a better place for his* *having lived in it*." Herre honoured him as his support made it possible for Herre to visit the Palau Islands, the type locality. He died of heart disease.

Wildekamp

African Rivuline sp. *Aphyosemion wildekampi* HO Berkenkamp, 1973

Rudolf Hans 'Ruud' Wildekamp (1945–2019) was a Dutch aquarist and amateur ichthyologist who described at least twenty-seven fish species. He was also an authority on keeping and breeding killifish. Among his published works is the book: *A world of killies: Atlas of the oviparous cyprinodontiform fishes of the world* (1993). (Also see **Ruud Wildecamp**)

Wilder

Thorny Catfish sp. *Hassar wilderi* EM Kindle, 1895

Dr Burt Green Wilder (1841–1925) was an American comparative anatomist and naturalist. He was a student of Agassiz at Harvard, graduated as a physician (1862), served as a Surgeon in the Union army (1862–1865), and returned briefly to Harvard (1866). He was Professor of Neurology and Vertebrate Zoology at Cornell (1867–1910). He discovered (1862) that up to 140 meters (150 yards) of silk could be drawn from a living spider. This led him to investigate the habits of spiders and the qualities and usefulness of their silk. A reptile is also named after him.

Wiley

Fringed Cusk-eel *Lepophidium wileyi* CR Robins, RH Robins & ME Brown, 2012

Dr Edward Orlando Wiley III (b.1944) is a biologist who is Professor of Systematics and Evolutionary Biology and Curator Emeritus, Ichthyology, Natural History Museum, University of Kansas. Southwest Texas State University awarded his bachelor's degree (1966), Sam Houston State University his master's (1972), and City University of New York his doctorate (1976). He wrote: *Phylogenetics. The Theory and Practice of Phylogenetic Systematics* (1981), and has been described as '*...one of the most influential persons in the development of phylogenetic systematics and the use of the evolutionary species concept*.' An amphibian is also named after him.

Wilhelm

Wilhelm's Squirrelfish *Sargocentron wilhelmi* F de Buen, 1963

Wilhelm's Hawkfish *Itycirrhitus wilhelmi* RJ Lavenberg & LA Yañez, 1972

Dr Ottmar E Wilhelm Grob (1898–1974) was a Chilean physician and biologist. He graduated as a physician at the University of Chile (1923) then studied zoology at Heidelberg and Biology in Munich. He was founder of the School of Medicine, University of Concepcion (1924), where he was Professor of Medical Zoology and Professor of Pathology at the Dental School and eventually Director (1962–1968). He collected many fishes at Easter Island and provided good colour photographs of them. He wrote over 260 papers.

Wilk

Windward Triplefin *Enneanectes wilki* BC Victor, 2013

Les Wilk is a nuclear physicist, diver and underwater photographer. Among his papers is the co-written: *Epizoic Ophiothela brittle stars have invaded the Atlantic* (2012) and the book: *Reef Creature Identification: Florida Caribbean Bahamas* (2013). He was honoured for his contributions to the art of underwater photography of coral-reef animals and his development of the ReefNet underwater identification CDs.

Wilkes

Eelpout sp. *Lycenchelys wilkesi* ME Anderson, 1988

Captain Charles Wilkes (1798–1877) was a pioneering navy captain who led the US Exploring Expedition (1838–1842), which was a milestone in American science. It comprised four naval vessels, the flagship *Vincennes*, the *Peacock*, the *Porpoise* and the store ship *Relief*. Two New York pilot boats, the *Sea Gull* and the *Flying Fish*, were used as survey vessels close to shore. They visited Brazil, Tierra del Fuego, Antarctica, Australia, New Zealand, the west coast of North America, the Philippines and the East Indies. The two penetrations into Antarctic waters sighted land and provided the first proof of an Antarctic continent. Wilkes was a strict disciplinarian who was disliked by many of the crew. He took with him 82 officers, nine naturalists, scientists and artists, and 342 sailors. Of the latter, only 223 returned. During the voyage, 62 were discharged as unsuitable, 42 deserted, and 15 died of disease, injury or drowning. However, he brought back a wealth of geological, botanical, zoological, anthropological and other material, which was to be the foundation for much of American science. A bird is also named after him.

Will White

White's Goatfish *Upeneus willwhite* F Uiblein & H Motomura, 2021

Dr William Toby White (b.1977) (See **White, WT**)

Willdenow

Blunt-snouted Clingfish *Gouania willdenowi* A Risso, 1810

Professor Carl Ludwig Willdenow (1765–1812) was a German botanist who mainly undertook taxonomic studies based on collections sent to him by other naturalists. He studied pharmacy (1785) then medicine, gaining an MB (1789), after which he took over his father's apothecary business (1790–1798). During this time, he wrote his seminal work: *Principles of Botany* (1792). He then became (1798) Professor of Natural History at the Berlin Medical College, and later the Director of the Berlin Botanical Garden (1801), which holds his collections, and Professor of Botany at Berlin University (1810) until his death. His phytogeographical studies led to ideas that were later developed further by Humboldt and others.

Willert

Mnanzini Nothobranchius *Nothobranchius willerti* RH Wildekamp, 1992

Manfred Willert is a German aquarist. He was the co-discoverer of the species in Kenya, helping to collect the type and donating it to Wildekamp.

William

Kaalpens Goby *Coryogalops william* JLB Smith, 1948

The etymology is not explained by Smith, but believed to have been named after his son William (b. 1939) who is now a popular television science and mathematics teacher in South Africa.

Williamina

Oriental Cyprinid sp. *Parachela williaminae* HW Fowler, 1934

Williamina Wemyss Meyer de Schauensee née Wentz (b.1905) was the wife of Academy of Natural Sciences (Philadelphia) ornithologist Rudolphe Meyer de Schauensee. The Academy was 'greatly indebted' to her for many Siamese fishes. She accompanied her husband on many of his expeditions. Two birds are named after her.

Williams, FJ

Yellowbar Tilefish *Caulolatilus williamsi* JK Dooley & FH Berry, 1977

Frank Joseph Williams of Miami, Florida, was owner and captain of the commercial fishing vessel *Argo*. He caught the holotype (1975). The authors commented that Williams "*…has contributed many valuable deep-water fish specimens to us over the years.*"

Williams, JA

Lake Malawi Cichlid sp. *Pseudotropheus williamsi* A Günther, 1894

Joseph A Williams (d.1895) was an Anglican missionary. Williams drowned, along with Bishop Chauncy Maples, when their small boat capsized during a storm and sank in Lake Malawi. He collected several cichlids from Lake Malawi, including the type of this one (1891).

Williams, JD

Sickle Darter *Percina williamsi* LM Page & TJ Near, 2007

Dr James 'Jim' David Williams (b.1941) is a biologist and conservationist who worked in the Department of the Interior at the US Florida Fish & Wildlife Service and US Geological Survey (1974–2006), Florida Fish & Wildlife Conservation Commission (2007–2017), and is currently a research associate at the Florida Museum of Natural History. He previously was an Assistant Professor of Biology at three US universities (1969–1974). The University of Alabama awarded his BSc (1962), MA (1963), MSc (1965), and PhD (1969). His research foci are in systematics (taxonomy), and evolutionary biology of freshwater fishes and mussels. He is the co-author of *Vanishing Fishes of North America* (1983), *Freshwater Mussels of Alabama and the Mobile Basin in Georgia, Mississippi, and Tennessee* (2008), *Freshwater Mussels of Florida* (2014), and *Fishes in the Fresh Waters of Florida: an Identification Guide and Atlas* (2018). He also co-authored two editions of Audubon's field guide to fish, as well as other books on nonindigenous fishes and common and scientific names of fishes and molluscs. He was described by the authors as "*…a contemporary ichthyologist, malacologist, and natural historian extraordinaire.*" A freshwater mussel is also named after him.

Williams, JT

Splitfin sp. *Luzonichthys williamsi* JE Randall & JE McCosker, 1992
Williams' Triplefin *Enneapterygius williamsi* R Fricke, 1997
Williams' Rockskipper *Entomacrodus williamsi* VG Springer & R Fricke, 2000
Williams' Viviparous Brotula *Ungusurculus williamsi* W Schwarzhans & PR Møller, 2007
Williams' Blenny *Starksia williamsi* CC Baldwin & CI Castillo, 2011
Triplefin sp. *Helcogramma williamsi* M-C Chiang & L-S Chen, 2012
Williams' Goatfish *Parupeneus williamsi* S Shibuya & H Motomura, 2020

Dr Jeffrey Taylor 'Jeff' Williams is Collection Manager, Ichthyology, USNM. Florida State University awarded his bachelor's degree (1975), the University of South Alabama his master's (1979) and the University of Florida his doctorate (1986). He co-wrote: *When endemic coral-reef fish species serve as models: endemic mimicry patterns in the Marquesas Islands* (2016). He collected the holotype of the brotula. (See also **Jeff Williams**)

Williams, TM

Williams' Tonguefish *Symphurus williamsi* DS Jordan & GB Culver, 1895

Dr Thomas Marion Williams (1871–1947) took his BA at Stanford (1897) and MD at Columbia (1901) and practiced (1904) as a physician. DS Jordan's *The Fishes of Sinaloa* (1895) includes him in the authorship as 'assisted by'. This was a collecting expedition led by Jordan (1894) on which Williams is listed as one of three volunteer assistants. He is also listed as co-author with Jordan on two of the fish they described. He collected the eponymous species in sandy bottomed tide pools at Mazatlan, Mexico.

Williamson

Mountain Whitefish *Prosopium williamsoni* CF Girard, 1856

Lieutenant Robert Stockton Williamson (1825–1882) was an American soldier and engineer in the US Army Corps of Topographical Engineers. He was assigned to conduct surveys for proposed routes for the transcontinental railroad in California and Oregon, during which the holotype of this species was collected. Mount Williamson in California was perhaps also named after Robert S. Williamson (1864), or perhaps not: in an interview, Don McLain told John Robinson that he had named this peak for Will Williamson, a friend of his. When reminded of Lt. Robert Williamson, McLain added '*Well, yes, I named [it] for him too*'. Historian Don Hedly reported that McLain's widow told him this same story.

Willkomm

Iberian Cyprinid sp. *Pseudochondrostoma willkommii* F Steindachner, 1866

Heinrich Moritz Willkomm (1821–1895) was a German botanist and academic who collected (1844–1845 & 1850–1851) and studied the flora of Spain and Portugal, where this cyprinid occurs. He studied medicine in Leipzig and was later Professor of Natural History (1855–1868) in Tharandt. He was then (1868–1874) Professor of Botany and Director of the botanical garden at the University of Dorpat, followed (1874–1892) by similar positions at the University of Prague. In between appointments he went on a scientific expedition to the Balearic Islands. A grass species is named after him. The patronym is not identified by Steindachner, but probably honours this man.

Willughby

Triggerfish sp. *Balistes willughbeii* GT Lay & ET Bennett, 1839

Windermere Char *Salvelinus willoughbii* A Günther, 1862

Francis Willughby (sometimes spelled Willoughby) (1635–1672) was a British ichthyologist and ornithologist, described as "*...the first who with the practised eye of an ichthyologist examined the Charrs of England and Wales*," He was a student at Trinity College, Cambridge, where he was taught by the naturalist John Ray, who suggested they undertake scientific research together at Willughby's expense. Over the next decade they travelled all over the British Isles and in Western Europe, observing and collecting, Willughby concentrating on animals and Ray on plants. After they returned to England, they planned to publish the results of their studies but Willughby died from pleurisy whilst the work was in preparation. He left Ray enough money to proceed with the books, and Ray published Willughby's work on birds in Latin as *Ornithologiae libri tres* (1676) and in an English translation *The Ornithology of Francis Willughby* (1678). Ray also published Willughby's: *De Historia Piscium* (1686). Willughby was also interested in games but his scientific study of them was only published (as *Willughby's Book of Games*) in 2003. It contains descriptions of a number of 17th century games including a description of football, with descriptions of the pitch and the goals and how the teams should be selected and balanced, and how the football itself was made.

Will Watch

Seafarer's Ghost Shark *Chimaera willwatchi* PJ Clerkin, DA Ebert & JM Kemper, 2017

Not a true eponym as this is named after a ship's company – the 'hard-working' fishers on board the Sealord Corporation fishing vessel 'Will Watch', from which the type was collected.

Will White

White's Goatfish *Upeneus willwhite* F Uiblein & H Motomura, 2021

Dr William Toby White (b.1977) (See **White, WT**) He was honoured in the goatfish etymology "*...for collecting and photographing mullid specimens from S Indonesia and making them available for taxonomic research (including the types of Upeneus willwhite)*."

Wilson

Cape Conger *Conger wilsoni* ME Bloch & JG Schneider, 1801

The eel's patronym is not identified, nor can a very likely candidate be found. It was named from a specimen collected in Australia. Alexander Wilson (1766–1813) was a naturalist active at that time, but he was an American ornithologist, with no obvious connection to Australia or fish.

Wilson, B

Spotty Seaperch *Hypoplectrodes wilsoni* GR Allen & JT Moyer, 1980

Dr Barry Robert Wilson (1935–2017) was an Australian malacologist. The University of Western Australia awarded his PhD (1965), which he followed with post-doctoral studies in molluscan systematics at Harvard University. On his return to Australia he became Curator of Molluscs at the WA Museum, later becoming Head of Science there (1967–1979). He was then Director of the National Museum of Victoria (1979–1984) and Director of Nature Conservation in the Department of Conservation and Land Management in WA (1985–1999). He wrote a number of books on shells including the 2-volume *Australian Marine Shells* (1994). A mollusc is also named after him.

Wilson, C

Armoured Catfish sp. *Hypostomus wilsoni* CH Eigenmann, 1918

Charles Wilson was an Indianapolis businessman, who not only helped finance Eigenmann's trip to Brazil but went along with him and collected many specimens himself, including the holotype of this species.

Wilson, CB

Pygmy Rockfish *Sebastes wilsoni* CH Gilbert, 1915

Charles Branch Wilson (1861–1941) was an American marine biologist most well-known for his work on copepods (tiny crustaceans). Colby College awarded his bachelor's and master's degrees, and he was awarded his PhD by Johns Hopkins University (1910). He worked as a tutor at Colby and (1891) was a science professor at what is now the University of Southern Maine, then (1896) a professor of natural science at Westfield Teachers College. He was professor of biology and head of science there until retirement (1932). Throughout his professional life he was associated (from 1899) with the United States Fish Commission's Marine Biological Laboratory, which became United States Bureau of Fisheries (1902). He undertook many economic surveys of US rivers and lakes. He was also associated (from 1901) with NMNH, identifying and studying their copepods collection. He wrote three monographs of copepods. As well as studying copepods and teaching, he also wrote papers on a wide range of topics from the embryology of amphibians, nemertean worms, the economic importance of dragonflies, and more including the results of the various surveys of the Bureau of Fisheries.

Wilson, D

Wilson's Mangrove Goby *Mugilogobius wilsoni* HK Larson, 2001

Little Rainbowfish *Melanotaenia wilsoni* MP Hammer, GR Allen, KC Martin, M Adams & PJ Unmack, 2019

David Wilson works at Territory Wildlife Park, Berry Springs (Northern Territory, Australia). He was honoured for his: "*...continuing help and enthusiasm in collecting gobies and for promoting the aquarium care and appreciation of native Australian freshwater fishes*".

Wilson, EA

Spiny Icefish *Chaenodraco wilsoni* CT Regan, 1914

Dr Edward Adrian Wilson (1872–1912) was an English physician, polar explorer, painter and natural historian. He took part in two Antarctic expeditions, the 'Discovery' and the 'Terra Nova'. He died with Scott on the return journey from the South Pole. Scott wrote of his companion: "*Words must always fail me when I talk of Bill Wilson. I believe he really is the finest character I ever met.*"

Wilson, JB

Wilson's Weedfish *Heteroclinus wilsoni* AHS Lucas, 1891

John Bracebridge Wilson (1828–1895) was a headmaster and amateur naturalist, born in England who emigrated to Australia (1857). Chiefly interested in marine biology, he would spend school holidays dredging for shellfish, sponges and other fauna, some of which he kept in a home aquarium. He collected the type specimen of the weedfish.

Wilson, K

Fahal Sweeper *Pempheris wilsoni* JE Randall & BC Victor, 2015

Keith Duncan Peter Wilson (b.1953) is a British hydrobiologist, environmentalist and conservationist who is currently Technical Director, Al Fahad Environmental Services, UAE. The University of Reading awarded his BSc (1975) and King's College London his MSc in applied hydrobiology (1977). He has extensive knowledge and experience of establishing and managing marine protection areas and wetlands built up through a distinguished career in fisheries and marine environmental services. He was Assistant Biologist for the UK Southern Water Authority (1978–1983), then Fisheries and Conservation Officer there (1983–1989). Following this, he was Fisheries, Conservation and Recreation Manager, North West Region, National Rivers Authority (1989–1991). He moved to Hong Kong first as Fisheries Officer for their Agriculture and Fisheries Department (1991–1995), Marine Parks Officer (1995–1996) and Senior Fisheries Officer (1996–2003). He then became Ecological Consultant and Sole Proprietor of Dragonfly Ecological Services, Asia (2003–2008). He moved to UAE (2008), first taking up the post of Senior Manager (Environment), Nakheel, Waterfront and Palm Jebel Ali, Dubai (2008–2009) then becoming Director Marine Programme, Emirates Marine Environmental Group (2009–2013). He is recognised as an international expert on the Odonata of China. He has written or co-written more than eighty papers on fisheries management, wetland conservation and particularly odonata. He has authored several books such as *Hong Kong Dragonflies* (1995) and *Field Guide to the Dragonflies of Hong Kong* (2003). He has also on numerous occasions been called upon to serve as an expert witness in connection with planning disputes related to development proposals. He has described c.50 new species of Odonata from China and named four genera. He collected fishes in Oman for the authors' studies on *Pempheris*.

Wilson, L

Yellowfin Dottyback *Pseudochromis wilsoni* GP Whitley, 1929

Leonard Wilson of Darwin, Northern Territory of Australia, was said to have "*...made several collections of animals for the Australian Museum*". We do not have further details of him.

Wilson, RC

Wilson's Snailfish *Paraliparis copei wilsoni* WJ Richards, 1966

Robert C Wilson of the US Bureau of Commercial Fisheries, conceived and directed a bottom trawling survey aboard the RV 'Geronimo off Gabon', West Africa (1963). He was the author of several reports such as: *Marquesas Area Oceanographic and Fishery Data: January-March 1957* (1957). He was one of the principle officers involved in a tuna migration study in 1950 undertaking a number of voyages and co-ordinating the results from over 1000 fishing boats.

Wilson, WP

Cardinalfish sp. *Ostorhinchus wilsoni* HW Fowler, 1918

Dr William Powell Wilson (1844–1927) was a botanist who was President, Botanical Society of America (1895). After school he taught, then entered (1864) Michigan State Agricultural College working in the fields, orchards and gardens. He went to Harvard (1873) studying and being an instructor until graduating BSc (1878). He then spent a year in France followed by study at a number of German universities, finally being awarded his DS from the University of Gottingen (1880). After briefly returning to the USA and marrying, he went back to Germany for two years. He returned to Harvard but eventually settled in Philadelphia (1884–1896). He was Director, Philadelphia Commercial Museum (1893–1927), before which he was a professor at the University of Pennsylvania and Director of the School of Biology. He spent his following years establishing the Commercial Museum in Philadelphia, being its Director (1898–1927) until his death. He was honoured for having made his institution's collection of Philippine fishes available to Fowler.

Wilton

Notothen sp. *Patagonotothen wiltoni* CT Regan, 1913

David Walter Wilton (1873–1940) was an expert in skiing and sledging. He had extensive experience of Arctic life and work, having been born and lived for some years in the north of Russia, where he became an expert ski runner and adept at sledging. He joined (1896) the Jackson-Harmsworth Arctic Expedition (1894–1897) to Franz Josef Land. Sledging expeditions were undertaken during which much of the western archipelago was charted and some scientific programmes conducted. After this he returned to Edinburgh where he read zoology and botany at the University and Royal Colleges. He gained a sound knowledge of meteorology at the Observatory on the summit of Ben Nevis, and took part in an expedition to Turkestan and Western China. He was selected (1902) for the Scottish National Antarctic Expedition (1902–1904) as assistant zoologist on board the 'Scotia' and undertook the zoological work and also tested the specific gravities of water samples obtained throughout the expedition. On his return, Wilton contributed to the scientific reports of the expedition.

Wilverth

Wilverth's Mormyrid *Cyphomyrus wilverthi* GA Boulenger, 1898

[Syn. *Hippopotamyrus wilverthi*]

Captain Étienne Christophe Bernard Eugène Wilverth (1866–1916) was a Belgian army officer in the Congo. He collected numerous fishes from the area under the auspices of the Société d'Études Coloniales. As a lieutenant, he wrote: *Les Poissons du Congo* (1897) and later wrote the article: *Pêche et poissons au Congo belge* (1911).

Winch

Winch's Pygmygoby *Trimma winchi* R Winterbottom, 1984

Peter Winch was the owner and captain of the ketch 'Paille-en-Queue II'; it and he spent nine months ferrying equipment, supplies, and expedition members around the Chagos Archipelago, and, as the etymology points out "...without whom the expedition during which type was collected would not have been possible."

Winchell

Clear Chub *Hybopsis winchelli* CF Girard, 1856

Professor Alexander Winchell (1824–1891) was an American geologist. He graduated from Wesley University (1847), later receiving his LLD from there (1867). He taught school (1847–1853) and then was President of Masonic University in Selma, Alabama. He was Professor of Physics and Engineering (1853–1855) becoming Professor of Geology, Palaeontology, Zoology and Botany (1855–1872) at the University of Michigan. He was also Director of the Michigan Geological Survey (1859–1861 & 1869–1871). He was the first Chancellor of the University of Syracuse, New York (1872–1874). He was Professor of Geology and Zoology at Vanderbilt University (1875–1878), then returned to Michigan once again as Professor of Geology and Zoology. He wrote over 250 papers and books including *Sketches of Creation* (1870) and *The Reconciliation of Science and Religion* (1877), his collection of discussions, descriptions and essays: *Sparks From A Geologist's Hammer* (1887) and a comprehensive book on world history *World Life* (1889). He collected the chub type.

Winemiller

Stripetail Cichlid *Geophagus winemilleri* H López-Fernández & DC Taphorn Baechle, 2004

Dr Kirk O Winemiller runs the Winemiller Aquatic Ecology Lab at Texas A&M University (College Station, Texas, USA). Miami University awarded his BA (1978) and MSc (1981) and the University of Texas his PhD (1987). He is an aquatic ecologist who led the field expedition to the Río Casiquiare region of Venezuela, during which most of type series was collected. He has been a Research Associate of the Oak Ridge National Laboratory, a Lecturer at the University of Texas, Fulbright Research Scholar in Zambia and Ichthyology Curator, Texas Natural History Collection, Texas Memorial Museum. He was honoured for nearly two decades of contributions to ecology and tropical fish biology, many of which have been based on Venezuelan fishes.

Wingate

Nile Cichlid sp. *Thoracochromis wingatii* GA Boulenger, 1902

General Sir Francis Reginald Wingate (1861–1953) was an army officer and the first British Governor of Sudan (1899–1916). He was commissioned in the artillery (1860) and assigned to the Egyptian army (1883). He became Director of Egyptian Military Intelligence (1889) and was 'Sirdar' (commander-in-chief of the Egyptian Army) and Governor-General of the Anglo-Egyptian Sudan. He fought several battles against the forces of al-Madhi, the nationalist 'rebel'. He became British High Commissioner for Egypt (1917). He freed Father Joseph Ohrwalder and others held captive by Mahdi forces and translated his narrative into English: *Ten Years Captivity in the Mahdi's Camp 1882–1892 from the Original Manuscript of Father Joseph Ohrwalder* (1892). Boulanger dedicated the cichlid to Wingate: "*…to whose assistance, so kindly granted to Mr Loat [see* Haplochromis loati*] on his journey through the Soudan, the success of this part of the Nile Fish-Survey is in no small measure due.*"

Winge

Endler's Livebearer *Poecilia wingei* FN Poeser, M Kempkes & IJH Isbrücker, 2005

Dr Øjvind Winge (1886–1964) was a Danish biologist and pioneer of yeast genetics. The University of Copenhagen awarded his master's (1910) and his PhD. He was appointed (1910) chair of genetics at the Royal Veterinary and Agricultural University, Copenhagen, where amongst other works he wrote *The Textbook in Genetics* (1928). He was head of the Department of Physiology, Carlsberg Laboratory (1933–1956). He is described by the authors as the father of genetic engineering and he also described the many colour patterns and the genetics of sex-determination in the common guppy. (Also see **Endler**)

Winifred

African Rivuline ssp. *Aphyosemion celiae winifredae* AC Radda & JJ Scheel, 1975

Winifred Epie was one of the daughters of John Epie, Manager of the Meanje Rubber Estate, Cameroon. John Epie hosted Scheel (1973) on the latter's collecting trip. (Also see **Celia**)

Winter

Barreleye genus *Winteria* AB Brauer, 1901

Friedrich Wilhelm 'Fritz' Winter (1878–1917), a member of a family of high-quality printers, was scientific illustrator and photographer on the Valdivia Expedition (1898–1899) to sub-Antarctic waters. He was very interested in natural sciences and followed courses of lectures at the Senckenberg Society, Frankfurt-am-Main. During WW1, he served in the German army (1916–1917) and died from wounds received in action.

Winterbottom

Winterbottom's Pygmygoby *Trimma winterbottomi* JE Randall & N Downing, 1994
Pilbara Eelblenny *Congrogadus winterbottomi* AC Gill, RD Mooi & JB Hutchins, 2000
False Midnight Dottyback *Manonichthys winterbottomi* AC Gill, 2004
Characin sp. *Pseudanos winterbottomi* BL Sidlauskas & GM dos Santos, 2005
Negros Sole *Aseraggodes winterbottomi* JE Randall & M Desoutter-Meniger, 2007
Reefgoby sp. *Priolepis winterbottomi* Y Nogawa & H Endo, 2007
Viviparous Brotula sp. *Alionematichthys winterbottomi* PR Møller & W Schwarzhans, 2008

Winterbottom's Dwarfgoby *Eviota winterbottomi* DW Greenfield & JE Randall, 2010
Indian Dwarf Dartfish *Parioglossus winterbottomi* T Suzuki, T Yonezawa & J Sakaue, 2010
Coral Goby sp. *Gobiodon winterbottomi* T Suzuki, K Yanao & H Senou, 2012
Winterbottom's Flap-headed Goby *Callogobius winterbottomi* NR Delventhal & RD Mooi, 2013
Cardinalfish sp. *Foa winterbottomi* TH Fraser, 2020

Dr Richard 'Rick' Winterbottom (b.1944) is an evolutionary biologist and ichthyologist who was born in Zambia. He is Curator Emeritus of Ichthyology at the Royal Ontario Museum, Toronto, which he joined (1978) as Assistant Curator of Fishes. He is also Professor Emeritus in the Department of Ecology and Evolutionary Biology, University of Toronto. The University of Cape Town awarded his honours bachelor's degree and Queen's University, Kingston, Ontario his doctorate (1971), after which he did post-doctoral work at the Smithsonian and the Canadian Museum of Nature, Ottawa. He was a senior lecturer at Rhodes University, South Africa (1975–1978). He was honoured variously for 'significant' contributions to anostomine systematics and natural history; providing valuable advice on goby systematic; great contribution to the knowledge of the systematics of gobies, etc. He received (2006) the Gibbs Award from the American Society of Ichthyologists and Herpetologists for "*...an outstanding body of published work in systematic ichthyology*". He has described more than 120 new fish species, half with co-authors, and published more than 130 papers from *The familial phylogeny of the Tetraodontiformes (Pisces, Acanthopterygii), as evidenced by their comparative myology* (1974) to *Trimma tevegae and T. caudomaculatum revisited and re-described (Acanthopterygii, Gobiidae), with descriptions of three new similar species from the western Pacific* (2016). He is also honoured in the name of a fossil fish.

Wintersteen

Wintersteen Drum *Umbrina wintersteeni* HJ Walker & KW Radford, 1992
Amigo Stardrum *Stellifer wintersteenorum* LN Chao, 2001

John Wintersteen (d.1989) was an American ichthyologist at the University of California. He is described as being a "*...longtime researcher in the taxonomy of eastern Pacific sciaenids.*" The stardrum is named for both John and his mother, Bernice McIhenny Wintersteen. Ning Chao says: "*This publication is in memory of John Wintersteen and those he loved. All three species reported in this paper were first recognized by John Wintersteen in the 1960s while he was a graduate student at work on a taxonomic revision of* Stellifer *and related genera under the guidance of Dr. Boyd Walker at UCLA. In 1976, John passed the unfinished study and materials to me and we worked off and on across three countries and two continents, for over a decade. In 1989, when we were ready to submit the manuscript (near 300ms pages and 100 figures) to LACM, John had a fatal heart attack on his beloved sailboat on the high seas. The manuscript has since aestivated in the Amazon with me for another decade. I thank numerous colleagues, especially managers and curators at major fish collections, who have helped John and me during three decades.*"

Winton

Winton's Grunter *Hannia wintoni* JJ Shelley, A Delaval, MC Le Feuvre, T Dempster, TA

Raadik & SE Swearer, 2020

Timothy John Winton (b.1960) is an Australian author and environmentalist. He became (2017) patron of the newly established 'Native Australian Animals Trust' and a patron of the Australian Marine Conservation Society (AMCS) and is involved in many of their campaigns, notably their work in raising awareness about sustainable seafood consumption. He has always featured the environment and the Australian landscape in his writings, which include novels, plays and non-fiction for which he was declared a 'living treasure' (1997) by the National Trust. When he first heard about the fish to be named in his honour he says he thought "*I wonder what kind of fish it is and I wonder what its habits are and habitat is', and the redneck in me thought 'I wonder what it tastes like'.*"

Winz

Armoured Catfish sp. *Hypostomus winzi* HW Fowler, 1945

Carlos Winz may have been an optician in Colombia, as Fowler thanks "*Señor Carlos Winz of the Optica Moderna, Bogota.*" No further information is given, nor any reason why the catfish is being named after Winz.

Wirtz

Cameroon Goby *Wheelerigobius wirtzi* PJ Miller, 1988
Sao Tomé Clingfish *Apletodon wirtzi* R Fricke, 2007
Wirtz's Goby *Didogobius wirtzi* UK Schliewen & M Kovačić, 2008

Dr Peter Wirtz (b.1948) is an independent researcher of marine biology, a blenny taxonomist and an underwater photographer. He has taught Zoology, Behavioural Science and Marine Biology in Freiburg, Karlsruhe, Madeira and the Azores. He held the chair of Marine Biology at the University of Madeira, and subsequently the chair of Ecology & Evolution at the University of Freiburg. He is the author of seven books and more than 100 scientific articles, including *Underwater Guide to Madeira, Canary Islands & Azores Fish* (1994) and *Underwater Guide to Maldives Fish*. On several journeys along the coast of West Africa he discovered new fish and invertebrate species; 14 were named after him. His other interests include the evolution of social structures, the zoogeography of the marine fauna of the eastern Atlantic, and crustacean symbiosis. He collected the type specimens of all three eponymous fish species, as well as numerous additional gobies from the tropical and temperate eastern Atlantic Ocean: all of which are housed at the Bavarian State Collection of Zoology and Staatliches Museum für Naturkunde. The author of the Cameroon Goby said that "*…as on previous occasions,* (he) *kindly forwarded these non-blenniids to the author.*"

Wirz

Wirz's Goby *Apocryptodon wirzi* FP Koumans, 1937
[Syn. *Oxuderces wirzi*]

Dr Paul Wirz (1892–1955) was a Russian-born, Swiss anthropologist who studied the peoples of New Guinea. Whilst on an expedition he had a heart attack at the end of the year (1954) and died the next January. He had previously travelled in Java, then in the Indonesian part of New Guinea (1921–1922). He published a number of books about his studies there, such as: *Beiträge zur Ethnologie des Sentanier* (1928). The etymology only mentions 'Dr Wirz' who collected the type, but it is, in all likelihood this man who was honoured.

Wischmann

Neotropical Rivuline sp. *Moema wischmanni* L Seegers, 1983
[Syn. *Aphyolebias wischmanni*]

Hermann Josef Wischmann is a German aquarist and fish collector. A number of his photographs appear on FishBase. He was the co-discoverer (with H Fries) of this species in Peru (1981) and has collected with Seegers such as in Tanzania (1982–1983).

Wisner

Hagfish sp. *Paramyxine wisneri* CH Kuo, KF Huang & HK Mok, 1994
Lanternfish sp. *Diaphus wisneri* BG Nafpaktitis, DA Robertson & Paxton, 1995
Hagfish sp. *Eptatretus wisneri* CB McMillan, 1999
[Name replaced by *Eptatretus bobwisneri*]
Wisner's Lanternfish *Lampanyctus wisneri* BJ Zahuranec, 2000

Robert 'Bob' Lester Wisner (1912–2005) was an American ichthyologist. After military service, San Diego State University awarded his bachelor's degree (1947). He worked at Scripps Institution of Oceanography (1947) for his entire professional life. While there he took part in numerous collecting voyages including observing a nuclear detonation. He wrote many papers and longer works including: *The Taxonomy and Distribution of Lanternfishes (Family Myctophidae) of the Eastern Pacific Ocean* (1974). He was McMillan's colleague and she honoured him in the name of the hagfish for his invaluable assistance with her hagfish research and other contributions to ichthyology. He was also a keen fly-fisherman. (Also see **Bob Wisner** & **Fern**)

Witmer

Three-barbeled Catfish sp. *Pimelodella witmeri* HW Fowler, 1941

J S Witmer Jr. lived in Lancaster County, Pennsylvania (USA). Fowler appears to have had a very wide circle of friends and acquaintances who supplied him with specimens, from Pennsylvania and beyond, as his etymologies often use words similar to those he used in regard to Witmer to whom he was 'indebted for Pennsylvania fishes'. It is probable that he was part of the Witmer family that emigrated from Switzerland to Pennsylvania (1716). The family was successful in starting up a number of businesses as well as being prosperous farmers. J S Witmer Jr may have been Jacob S Witmer (b.1904).

Witte

African Characin sp. *Nannocharax wittei* M Poll, 1933
Upjaw Barb *Coptostomabarbus wittei* LR David & M Poll, 1937
Shellear sp. *Kneria wittei* M Poll, 1944
Banded Sole *Microchirus wittei* P Chabanaud, 1950
African Barb sp. *Labeobarbus wittei* J Banister & M Poll, 1973

Dr Gaston-François de Witte (1897–1980) was a Belgian herpetologist. He was in the Belgian Congo (1933–1935 & 1946–1949), where – though he concentrated on reptiles and amphibians – he also collected the types of several of the above fish. During his career, he was associated with both the Royal Museum for Central Africa in Tervuren and the Museum of Natural Sciences in Brussels. Five amphibians, four reptiles and two birds are also named after him.

Wolasi

Wolasi Ricefish *Oryzias wolasi* LR Parenti, RK Hadiaty, D Lumbantobing & F Herder, 2013

This is a toponym referring to the Wolasi District, Sulawesi Tenggara, Indonesia.

Wolf

Three-barbeled Catfish sp. *Pimelodella wolfi* HW Fowler, 1941

Herman Theodore Wolf of Philadelphia was an aquarist and horticulturist, who, according to Fowler "*...made several interesting collections of American fishes*" for the Academy of Natural Sciences of Philadelphia. He wrote: *Goldfish Breeds and Other Aquarium Fishes Their Care and Propagation, a Guide to Freshwater and Marine Aquaria, Their Fauna, Flora and Management* (1908).

Wolfe

Characin sp. *Leporinus wolfei* HW Fowler, 1940
Armoured Catfish sp. *Rineloricaria wolfei* HW Fowler, 1940

Thomas W Wolfe (sometimes spelt Wolf), along with Robert Hartwell and Robert T Petley, all from Cleveland, Ohio, assisted William Morrow (q.v.) in his collecting expedition (of freshwater fishes) to the Ucayali River basin of Peru (1937). We can find nothing further.

Wolff, G

Snailfish sp. *Paraliparis wolffi* G Duhamel & NJ King, 2007

Dr George Wolff is Professor of Earth, Ocean and Ecological Science at the University of Liverpool (1987–present). He trained as an organic geochemist, gaining his BSc (1979) and PhD (1983) at Bristol University where he was a Research Assistant (1983–1985). He was then a Research Fellow at the University of Strasbourg (1985–1987) before joining the Liverpool faculty. He has published many papers (1982–2018). The etymology states that the snailfish is: "*Named after Professor George Wolf, University of Liverpool, for his contributions to marine biogeochemistry and for always being happy on Mondays throughout an arduous cruise.*

Wolff, J

Duskyfin Glass Fish *Parambassis wolffii* P Bleeker, 1850
Bagrid Catfish sp. *Mystus wolffii* P Bleeker, 1851

J Wolff, who collected the holotypes of these species, was a military doctor and one of Bleeker's circle of friends who sent him fish specimens.

Wolli

Dwarf Cichlid sp. *Apistogramma wolli* U Römer, DP Soares, CR Garcia-Dávila, F Duponchelle, J-F Renno & IJ Hahn, 2015

Wolfgang 'Wolli' Friedrich is a German aquarist and tropical fish breeder. He was described by the authors as: "*...one of the most notable and skilled professional German fish breeders of recent decades until he finally closed down his facility in early 2014*". They also say: "*...his long-term work in breeding cichlid fishes has substantially contributed to our present knowledge of this family,*" and his "*...helpful and always constructive and mostly humorous comments on fish maintenance and breeding*" positively influenced the work of several ichthyologists.

Woltereck

Sailfin Silverside sp. *Paratherina wolterecki* HJ Aurich, 1935

Professor Dr Richard Woltereck (1877–1944) was a German zoologist and philosopher. He studied medicine and zoology at the University of Freiburg, receiving his PhD (1898). He joined the nine-month voyage (1898/1899) of the steam ship 'Valdivia', collecting plankton for embryological research. He spent much of WW1 working at the German embassy in Bern, Switzerland, where he met the writer and pacifist Hermann Hesse, and together they established an organisation to provide books to German prisoners of war. The two men also founded (1919) a progressive political magazine: *Vivos Voco*. Woltereck became "almost evangelically anti-materialistic". He spent 16 months (1931/1932) doing fieldwork in Southeast Asia, where he collected all four known species in the genus *Paratherina*, then moved to Turkey (1932) to take up a post at an agricultural college. He returned to Germany (1936) for treatment following an automobile accident.

Wolterstorff

Chinese Gudgeon sp. *Squalidus wolterstorffi* CT Regan, 1908
African Rivuline sp. *Austrolebias wolterstorffi* CGE Ahl, 1924

Dr Willy Georg Wolterstorff (1864–1943) was a German geologist and herpetologist who was a curator at Magdeburg Museum. An illness deprived him of his hearing and power of speech (1871) but he learned to lip-read. He was also very myopic and so had a lonely childhood, compensating by collecting and keeping amphibians; they remained his major lifelong interest. His death two years before spared him seeing the total destruction by the RAF of the museum and all his work, including 12,000 specimens in glass jars (1945). He received fishes from China collected by Martin Kreyenberg, which included the type of the gudgeon. A bird subspecies, a reptile and three amphibians are also named after him.

Wonder

Slickhead sp. *Narcetes wonderi* AWCT Herre, 1935
Niger Goby-Cichlid *Gobiocichla wonderi* RH Kanazawa, 1951

Frank C Wonder (1904–1963) was a taxidermist at the Field Museum of Natural History, Chicago, USA. He took part in a year-long scientific voyage on the 'Illyria' to the South Pacific (1928). He also took part in a voyage to Venezuela (1947). He collected the cichlid type.

Wongrat

Oriental Cyprinid sp. *Thryssocypris wongrati* C Grudpan & J Grudpan, 2012

Dr Prachit Wongrat of the Faculty of Fishes, Kasetsart University, Bangkok is a fish taxonomist. He was the authors' first teacher in ichthyology. He co-wrote: *Towards possible fishery management strategies in a newly impounded man-made lake in Thailand* (2007).

Wongratana

Round Herring sp. *Etrumeus wongratanai* JD DiBastista, JE Randall & BW Bowen, 2012

Dr Thosaporn Wongratana (see **Thosaporn**)

Woo Sui Ting

Snake-eel sp. *Ophichthus woosuitingi* JTF Chen, 1929

Sui-ting Woo was an assistant to Professor H N Fey at Sun Yat-sen University (Guangzhou, China), the finder of this species.

Wood, AT

Oldman Klipfish *Clinus woodi* JLB Smith, 1946

Alexander Thomas Wood (1872–1957) was a friend of the author who owned a cottage which Smith sometimes used as a base for fieldwork. Although the original text carries no etymology for the binomial, it seems likely that this is the person intended.

Wood, J

Kappie Blenny *Omobranchus woodi* JDF Gilchrist & WW Thompson, 1908

J Wood of Natal, South Africa, collected the blenny holotype along with "many specimens of marine animals". Unfortunately, the authors give no other information about him.

Wood, RC

Lake Malawi Cichlid sp. *Rhamphochromis woodi* CT Regan, 1922

Lake Malawi Cichlid sp. *Stigmatochromis woodi* CT Regan, 1922

Rodney Carrington Wood (1889–1962) travelled widely, particularly in Africa, farmed at times, worked as a game warden in Nyasaland (now Malawi) which he resigned from (1931), became a schoolteacher in Natal and was also employed by the railways in Nyasaland. His real love was natural history and he was an enthusiastic collector. Although he spent much of his life in Africa, he was in the Seychelles when he died. He presented to the BMNH a 'very fine collection' of Lake Malawi cichlids from hauls brought in by local fishermen, including the types of the above species. A mammal is also named after him.

Wood, TD

Wood's Razorfish *Novaculops woodi* OP Jenkins, 1901
[Alt. Hawaiian Sandy]

Dr Thomas Denison Wood (1865–1951) is sometimes known as the Father of Health Education. He was professor of hygiene and physical training at Stanford University. He was honoured in the binomial as he obtained two specimens of this wrasse in Honolulu.

Woodhead

Woodhead's Pygmy Angelfish *Centropyge woodheadi* RH Kuiter, 1988

Phillip Arthur 'Phil' Woodhead (b.1956) is an underwater photographer based in Cairns, Queensland, who brought the species to the author's attention. He has his own photography business: Wetimage Underwater Photography and is a photographic contributor to various IKAN publications; most recently *Cardinalfishes of the World* by R H Kutier and Toshikazu Kozawa. He was also a major contributor to BYO Guides *Coral Finder* (2012). He also edits the Nautilus Scuba Club newsletter. He recognised the angelfish as an undescribed species and sent photos to the author for identification.

Woodland

Rabbitfish sp. *Siganus woodlandi* JE Randall & M Kulbicki, 2005

Dr David John Woodland of the School of Environmental & Rural Science, University of New England, New South Wales, is an Australian ichthyologist and a specialist in the Siganidae (rabbitfishes). He wrote the book: *Revision of the fish family Siganidae with descriptions of two new species and comments on distribution and biology* (1990) and papers such as: *Description of a new species of rabbitfish (Perciformes: Siganidae) from southern India, Sri Lanka and the Maldives* (2014).

Wood-Mason

Tonguefish sp. *Symphurus woodmasoni* AW Alcock, 1889
Rattail sp. *Coryphaenoides woodmasoni* AW Alcock, 1890

Dr James Wood-Mason (1846–1893) was a British zoologist, lepidopterist and specialist in marine animals and made the first collection of molluscs from the Andaman and Nicobar Islands (1872). He worked at (1877) and then became (1887) Superintendent of the Indian Museum (Calcutta). In the same year he became vice-president of the Asiatic Society of Bengal. He co-wrote: *Natural History Notes from H.M. Indian Marine Survey Steamer "Investigator," Commander R.F. Hoskyn, R.N., Commanding* (1891). Suffering ill-health, he left India (1893) to return to England, but died at sea. Woodmason Bay in the Andaman Islands, along with two reptiles, are named after him. (Also see **Mason**).

Woods

Lightfish genus *Woodsia* SF Grey, 1959
Woods' Clingfish *Gobiesox woodsi* LP Schultz, 1944
Whitespot Soldierfish *Myripristis woodsi* DW Greenfield, 1974
Thorny Catfish sp. *Rhynchodoras woodsi* GS Glodek, 1976
Wood's Barbel-goby *Gobiopsis woodsi* EA Lachner & JF McKinney, 1978
Wood's Chromis *Chromis woodsi* JC Bruner & S Arnam, 1979
Swallowtail Bass *Anthias woodsi* WD Anderson & PC Heemstra, 1980

Loren P Woods (1914–1979) was Curator, Department of Ichthyology (1941–1978), Field Museum of Natural History, Chicago, where he worked (1938–1978), broken only by service in WW2 in the US Navy (1943–1946). He started as a guide-lecturer but soon (1941) transferred to the Division of Fishes as Curator. He participated in numerous collecting expeditions to the Indian Ocean, western Atlantic, southeastern Pacific, the Gulf of Mexico and the Caribbean. He co-wrote: *New species and new records of fishes from Bermuda* (1951), as well as the more populist: *Tropical Fish* (1971).

Woodward, AS

Woodward's Silverside *Doboatherina woodwardi* DS Jordan & EC Starks, 1901

Dr Arthur Smith Woodward (1864–1944) was an English palaeontologist and geologist, known as a world expert on fossil fish. He also described the Piltdown Man fossils, which were later determined to be fraudulent. He joined the staff of the Department of Geology at the BMNH (1882), became Assistant Keeper of Geology (1892), and Keeper (1901). He wrote: *Catalogue of the Fossil Fishes in the British Museum* (1889–1901). He was later Secretary of the Palaeontographical Society and became President of the Geological Society (1904). He is not related to Henry Woodward, whom he replaced as curator of the Geology Department of the BMNH. Woodward's contributions to palaeoichthyology resulted in him receiving many awards, but his reputation suffered from his involvement in the Piltdown Man hoax.

Woodward, BH

Knifejaw *Oplegnathus woodwardi* ER Waite, 1900
Woodward's Pomfret *Schuettea woodwardi* ER Waite, 1905
Woodward's Moray *Gymnothorax woodwardi* AR McCulloch, 1912

Bernard Henry Woodward (1846–1916) was a wine merchant in London (1875). He suffered from bronchial disease, and so moved to Western Australia for his health (1889). Here he worked as a government analyst (1889–1895), responsible for assaying and examining mineral oils. He also became Curator of the Geological Museum, Perth (1889). This expanded to become the Western Australian Museum and Art Gallery, and Woodward became Director (1897). He founded the Western Australia Natural History Society (1890). He wrote an article on: *National parks and the fauna and flora reserves in Australia* (1907). He was honoured in the moray eel's name for sending an '*interesting collection*' of '*new and little-known fishes*' (including this one) to McCulloch for study, and for "*...various kindnesses connected with the publication of this paper.*" Six birds and two mammals are also named after him.

Woollard

Barbeled Dragonfish sp. *Eustomias woollardi* TA Clarke, 1998

Dr George Prior 'Doc' Woollard (1908–1979) was an American geophysicist, a scientist and a visionary whose interests extended beyond his own discipline. Georgia Institute of Technology awarded his bachelor's degree (1932) and a master's degree (1934), after which he did another master's degree at Princeton (1935) where his doctorate was awarded (1937). He had a leading role in the creation of both the Polar Research Center at the University of Wisconsin and the

Hawaii Institute of Geophysics, University of Hawaii. He was Director of the latter (1963–1976), retiring as Professor Emeritus. He wrote: *International gravity measurements* (1963).

Woolman

Speckled Flounder *Paralichthys woolmani* DS Jordan & TM Williams, 1897

Dr Albert Jefferson Woolman (1861–1918) was a science teacher in Duluth, Minnesota. He collected fish in the USA and Mexico. He named at least one fish species after DS Jordan, whom he had studied under at Indiana University (graduating 1890) and collected in the Columbia River basin with CH Gilbert (1892–1893). His son founded Delta Airlines.

Woosnam

Upper Zambezi Squeaker (catfish) *Synodontis woosnami* GA Boulenger, 1911

Major Richard Bowen Woosnam (1880–1915) was a British army officer who had served in the Boer War in South Africa and was on the British Museum scientific expedition with R E Dent to Bechuanaland (Botswana) (1906–1907). He was a game ranger and collector in Kenya. He was killed in action during WW1 at Gallipoli (1915). Two mammals, a reptile and three birds are named after him.

Worthington

Lake Kyoga Cichlid sp. *Haplochromis worthingtoni* CT Regan, 1929
Labeo sp. *Labeo worthingtoni* HW Fowler, 1958
Lake Malawi Catfish sp. *Bathyclarias worthingtoni* PBN Jackson, 1959

Dr Edgar Barton Worthington (1905–2001) was a British zoologist with a particular interest in fisheries and waterways, and a pioneer explorer of African lakes and their fisheries. He was Director of the Freshwater Biological Association at Windermere (1937–1946). He lived and worked in East Africa (1946–1955) as Development Adviser to Uganda (1946–1950) and as Scientific Secretary to the Colonial Research Council and the East Africa Commission (1951–1955). He wrote: *A Development Plan for Uganda* (1949). A reptile is also named after him. (See also **Barton** & **Stella**)

Wosiacki

Tetra sp. *Hyphessobrycon wosiackii* CLR Moreira & FCT Lilma, 2017

Wolmar Benjamin Wosiacki is a Brazilian ichthyologist and a friend of the authors. He is a researcher and curator of the fish collection at the Museu Paraense Emílio Goeldi. The Federal University of Paraná awarded his BSc in Biological Sciences (1990) and his MSc in Zoology (1997) and the University of São Paulo his PhD (2002). Among his many papers is the co-authored: *The Electric Glass Knifefishes of the* Eigenmannia trilineata *species-group (Gymnotiformes: Sternopygidae): monophyly and description of seven new species.*

Wouter

Wouter's Pygmygoby *Trimma woutsi* R Winterbottom, 2002

Wouter 'Wouts' Holleman is an ichthyologist who is an Honorary Research Associate at the South African Institute for Aquatic Biodiversity. His particular interests are in the taxonomy of Indo-Pacific Tripterygiidae and the systematics of Southern African and other Indo-Pacific Clinidae. He has published at least ten papers, such as: *A review of the tripterygiid fish genus Enneapterygius (Blennioidei: Tripterygiinae) in the Western Indian Ocean, with descriptions of five new species* (2005). Winterbottom honoured him saying he was a "... friend, colleague, and indispensable field collaborator on many field trips."

Wowor

Daisy's Ricefish *Oryzias woworae* LR Parenti & RK Hadiaty, 2010

Sleeper Goby sp. *Eleotris woworae* P Keith, MI Mennesson, S Sauri & N Hubert, 2021

Dr Daisy Wowor (b.1956) is an evolutionary biologist and systematic carcinologist. The University of Maryland awarded her MSc (1980) and the University of Singapore her PhD (2005). She works at the Research Centre for Biology, Indonesian Institute of Sciences. She specialises in the freshwater decapods of Indonesia. Among her published papers are the co-written: *Foraging on mangrove pneumatophores by ocypodid crabs* (1989) and *Mitigating the impact of oil-palm monoculture on freshwater fishes in Southeast Asia* (2015). She collected the type series.

Wray

Wray's Gambusia *Gambusia wrayi* CT Regan, 1913

Charles Arthur Wray (c.1891–1936) worked at the BMNH. He was 'boy attendant' in geology (1891–1894) then joined the 'Index Museum' (1894–1936) as 'Higher Grade Clerk' until he retired. This was the part of the museum in the central hall which 'indexed' the major collections with small exhibitions. He collected the holotype in Jamaica.

Wright, JH

Skipjack Trevally *Pseudocaranx wrighti* GP Whitley, 1931

J H Wright was said to have "*presented many interesting specimens of Botany Bay fishes to the Australian Museum*". He is presumably the same person of this name recorded as being an Assistant Taxidermist at that Museum (1908–1916). He was still working at the museum in the 1920s as he accompanied the then Curator of Mammals and 'Zoologist' Ellis Troughton, as his assistant, collecting at several stations along the Trans-Australian Railway across the Nullarbor Plain.

Wright, P

Madagascar Rainbowfish sp. *Rheocles wrightae* MLJ Stiassny, 1990

Dr Patricia Chapple Wright (b.1944) is an American primatologist. She is Professor of Anthropology at the State University of New York at Stony Brook, having previously been an Assistant Professor at Duke University. As her dissertation for her doctorate, she wrote what has become a classic on the behaviour and ecology of the only nocturnal monkeys, the owl monkeys. She made her first visit to Madagascar (1986) and was in the party that discovered the Golden Bamboo Lemur (*Hapalemur aureus*). She was awarded Madagascar's national Medal of Honour (1995). The fish was named for her as she "*...has been unstinting in her efforts in the field of Madagascan rainforest management and conservation.*"

Wright, RR

Armoured Catfish sp. *Oxyropsis wrightiana* CH Eigenmann & RS Eigenmann, 1889

Robert Ramsay Wright (1852–1933) was a Scottish zoologist. The University of Edinburgh awarded both his bachelor's (1873) and master's degrees (1871). His career was spent in Canada as Professor of Natural History, University of Toronto (1874–1912) and when his title was changed, was the first Professor of Biology, University of Toronto (1887), the university's Dean of Arts (1901) and vice-president of the University (1902). He retired as Professor Emeritus (1912). He wrote: *An introduction to zoology: for the use of high schools* (1889). The etymology states that he "*...has contributed more than anyone else*" to knowledge of the anatomy of American catfishes.

Wu

Chinese Butterfly Loach *Sinogastromyzon wui* PW Fang, 1930
Chinese Cyprinid sp. *Sinibrama wui* H Rendahl, 1932
Prickleback sp. *Chirolophis wui* KF Wang & S-C Wang, 1935
Chinese Cyprinid sp. *Hemiculterella wui* KF Wang, 1935
Chinese Cyprinid sp. *Altigena wui* CY Zheng & YR Chen, 1983
[Syn. *Bangana wui*]
Chinese Barb sp. *Discocheilus wui* JX Chen & JH Lan, 1992
Golden-line Fish sp. *Sinocyclocheilus wui* WX Li & L An, 2013

Dr Wu Hsien-Wen, sometimes given as Wu Xian-Wen (1900–1985), was a Chinese zoologist and ichthyologist. He graduated in zoology at Xiamen (Amoy) University (1927) and the University of Paris awarded his doctorate in Science (1932). He taught biology at the Central University, Nanking (Nanjing) (1927–1934) and was Professor of vertebrate taxonomy and invertebrate zoology at the same establishment (1934–1936) and then Professor of comparative anatomy (1936–1937). During the Sino-Japanese war (1937–1945), he was senior scientist in the Institute of Zoology and Botany, Chungking (Chongqing) and concurrently (1941–1943) Professor of Zoology Fu Dan University. He was Professor of parasitology at Kiangsu (Jiangsu) Medical College (1947–1948) and was Deputy Director of the Institute of Hydrobiology, Academia Sinica, in charge of ichthyology (1950–1976) and was appointed Director (1977) becoming Honorary Director of the Institute of Hydrobiology (1983). He was a member of many learned societies including the Linnean Society, London (1983). He made many contributions to Chinese ichthyology and particularly to the systematics of Chinese fish, with his main work considered to be the 2-volume: *Monograph on Chinese Cyprinidae (Pisces)* (1997) covering 113 genera and 412 species.

Wu Han-Lin

Goby genus *Wuhanlinigobius* SP Huang, Z Jaafar & IS Chen, 2014
Wu's Skate *Dipturus wuhanlingi* C-H Jeong & T Nakabo, 2008
[Alt. China Skate]

Dr Wu Han-Lin is a Professor of Ichthyology at Shanghai Ocean University and Director of the Fisheries Laboratory (2012), having previously been at Shanghai Fisheries University

(2000). He co-wrote: *Fauna Sinica, Chinese fishes of the suborder Gobioidei* (2010). He was honoured in the skate's name for his 'great contributions' to Chinese ichthyology.

Wucherer

Armoured Catfish sp. *Hypostomus wuchereri* A Günther, 1864

Dr Otto Edward Henry Wucherer (1820–1874) was a Portuguese-born German physician and herpetologist. He qualified at Eberhard Karls Universität Tübingen and practiced at St. Bartholomew's Hospital, London, and in Lisbon. He discovered the cause of the tropical disease elephantiasis. He left Europe and settled in Salvador Bahia, Brazil (1843). He wrote: *Sobre a mordedura das cobras venenosas e seu tratamento* (1867). A bird, three reptiles and an amphibian are also named after him.

Wündsch

Wündsch's Killifish *Aphyosemion wuendschi* AC Radda & E Pürzl, 1985

Dr Leopold Wündsch was a professor at the Institute for General and Comparative Physiology at the University of Vienna until retiring. The fish was originally described as a sub-species: *Aphyosemion hanneloreae wuendschi* with the binomial being named after Hannelore Pürzl, the wife of the junior author Eduard Pürzl. (Also see **Hannelore**)

Wurtz

African Barb sp. *Labeobarbus wurtzi* J Pellegrin, 1908

According to the etymology, 'Monsieur le Docteur Wurtz' collected the type while traveling in French Guinea. Pellegrin asked Wurtz to collect for him and the MNHN, Paris, whilst he was in Senegal and French Guinea. Wurtz obliged and sent home 13 species, of which four were new to science. Pellegrin was impressed by how Dr Wurtz recorded life colours of his specimens before placing them in alcohol. We believe that this may be Professor Robert Wurtz (1858–1919), a bacteriologist, who studied under Louis Pasteur. He visited Ethiopia (1894 and 1897) on behalf of the French government to investigate epidemics there and was the first to take a smallpox vaccine to that country. On his return to France he was appointed to the Faculté de médecine de Paris, where he became a professor. He certainly travelled in French West Africa as evidenced by: *Mission en Afrique occidentale française pour le compte du ministère de l'Instruction publique et des Beaux-Arts du docteur Robert Wurtz, professeur agrégé de la faculté de Médecine de Paris « afin d'y poursuivre des recherches sur la prophylaxie des maladies coloniales » 1907/1908.*

Wyck

Crystal-eyed Catfish *Hemibagrus wyckii* P Bleeker, 1858

Dr Herman Constantijn Van der Wijck (1815–1889) was Regent of the Preanger Regencies, Java, Indonesia (which is the type locality) and invited Bleeker to a fishing party during which the river was poisoned with 'akar toeba' (the roots of which plant contain the ichthyocide, rotenone). Large numbers of stunned fishes were ladled from the water.

X

Xakriabá

Armoured Catfish sp. *Otocinclus xakriaba* SA Schaefer, 1997

The Xakriabá were an indigenous tribe which formerly inhabited the upper region of the São Francisco basin, Minas Gerais and Bahia states, Brazil.

Xantus

Limones Splitfin *Ilyodon xantusi* CL Hubbs & CL Turner, 1939
[Now included in *Ilyodon furcidens*]

Louis Janos (John) Xantus de Vesey (1825–1894) was a Hungarian who was on the staff of William Hammond, the collector. Whilst living in the USA (1855–1861), he sent the USNM 10,000 specimens! He is also renowned as a pathological liar. According to Schoenman and Benedek (1976), 'Xantus fled his native Hungary after taking part in the unsuccessful revolt against the Austrian Empire in 1848. A poor but educated and ambitious man, he wrote grandiose accounts of his American exploits. They were published in Hungary where he became famous. His letters make Private Xantus sound like he was in charge. Despite the fact that he plagiarized other travel accounts of the American West, lied about himself, and always claimed to be superior to those around him, Xantus did great work for Baird and the Smithsonian. Xantus once had a photo taken of himself as a US Navy captain, which was published in Hungary. Yet Xantus had never even served in the Navy'. The Austro-Hungarian Empire dispatched an expedition to Siam (Thailand), China and Japan (1868), with Xantus as one of the 18 members. They were charged with collecting botanical and zoological specimens and investigating local ethnography and arts. Xantus returned to Hungary (1870) with 155,644 specimens in 200 crates. Two reptiles and seven birds are also named after him.

Xavante

Neotropical Rivuline sp. *Plesiolebias xavantei* WJEM Costa, MTC de Lacerda & K Tanizaki-Fonseca, 1988

Darter Characin sp. *Characidium xavante* WJ da Graça, CS Pavanelli & PA Buckup, 2008

Characin sp. *Astyanax xavante* V Garutti & PC Venere, 2009

Armoured Catfish sp. *Hopliancistrus xavante* RR de Oliveira, J Zuanon, LH Rapp Py-Daniel, JLO Birindelli & LM Sousa, 2021

The Xavante are an indigenous ethnic group who live in the region between Rio das Mortes and Rio Culuene, Mato Grosso, Brazil. They are the largest Amerindian tribe in the area where the rivuline species was discovered.

Xaveriellus

Tetra sp. *Hemigrammus xaveriellus* FCT Lima, A Urbano-Bonilla & S Prada-Pedreros, 2020

Javier Alejandro Maldonado-Ocampo (1977–2019), who was killed when crossing a river in a small boat that overturned and was swept downstream, was honoured for his 'invaluable' contribution to the knowledge and conservation of Neotropical fishes. When Javier started as a student (1994), the third author, then his advisor, proposed that he study some diverse and poorly-known characin genera such as *Hemigrammus*. The name is a Latinization of *Javiercito* (diminutive of *Javier* in Spanish), an 'affectionate nickname' which the third author used to refer to him. (See **Maldonado**)

Xerente

Bumblebee Catfish sp. *Microglanis xerente* WBG Ruiz, 2016

The Xerente ethnic group are the native indigenous people who live along the Rio Tocantins basin, Brazil.

Xeruin

Characin sp. *Tyttobrycon xeruini* J Géry, 1973

This is a toponym referring to the Rio Xeruini system, middle Rio Negro basin, Brazil; the type locality.

Xingu

Thorny Catfish sp. Rhynchodoras xingui W Klausewitz & F Rössel, 1961

This is a toponym and refers to the Rio Xingu, Brazil.

Xinguano

Characin sp. *Cynopotamus xinguano* NA de Menezes, 2007

The Xinguano are an Amerindian tribe living in Parque Indígena do Xingu; a national park in the Rio Xingu basin, Brazil.

Xora

Orangehead Worm Eel *Scolecenchelys xorae* Smith JLB, 1958

This is a toponym referring to the Xora River (specifically, its mouth) in Transkei, South Africa; the type locality.

Xu

Spined Loach sp. *Cobitis xui* X-C Tan, P Li, T-J Wu & J Yang, 2019

Xu Xiake née Xu Hongzu, later known by the courtesy name of Zhenzhi (1587–1641), was a Chinese writer, traveller and geographer of the Ming Dynasty in China, best known for his famous geographical treatise. He travelled throughout China for more than 30 years, along with a servant called Gu Xing, recording his journeys throughout. He faced many hardships and was often dependent on the patronage of local scholars who would help him, including when he had been robbed. Local Buddhist abbots would pay him for recording the history of their monasteries. The records of his travels were compiled posthumously.

Y

Yabe

Eelpout sp. *Davidijordania yabei* ME Anderson & H Imamaura, 2008

Dr Mamoru Yabe (b.1952) is a Japanese ichthyologist who was at the Laboratory of Marine Zoology, Graduate School of Fisheries Sciences, Hokkaido University. Among other expeditions he collected fish aboard the RV 'Yakushi-Maru' in the Bering Sea (1979). He published more than 80 papers such as the co-written: *Fishes new to the eastern Bering Sea* (1981) and *Evolutionary History of Anglerfishes (Teleostei: Lophiiformes): A Mitogenomic Perspective* (2010). He was honoured for his contributions to the systematics of North Pacific fishes.

Yaguajal

Yaguajal Limia *Limia yaguajali* LN Rivas, 1980

This is a toponym after the Rio Yaguajal, Dominican Republic

Yahaya

Sleeper sp. *Giuris yahayai* P Keith & M Mennesson, 2020

Ibrahim Yahayai is head of the Département Biodiversité terrestre, Centre National de Documentation et de Recherche Scientifique des Comores (CNDRS). He was honoured "in recognition of his work on the fauna of this archipelago" (i.e. the Comoro Islands).

Yahgan

Tubeshoulder sp. *Normichthys yahganorum* RJ Lavenberg, 1965

The Yahgan, also called Yaghan or Yagán, are an indigenous people of Tierra del Fuego who, according to the author, "*...practiced shellfish conservation and avoided exhausting their food supply.*"

Yahyaoui

Yahyaoui's Barbel *Luciobarbus yahyaouii* I Doadrio, M Casal-Lopez & S Perea, 2016

Dr Ahmed Yahyaoui is an oceanographer and marine ecologist who is Professor of Biology in the Faculty of Sciences at the Mohammed V-Agdal University, Rabat, Morocco (1984–present). The University of Perpignan awarded his PhD (1983) and Mohammed V University of Rabat a further doctorate (1991). He has well over 100 publications including: *L'ichtyofaune des eaux continentals du Maroc* (2000). He was honoured for his contributions to "*...the knowledge of the fish fauna of Morocco and North Africa.*"

Yaldwyn

Yaldwyn's Triplefin *Notoclinops yaldwyni* GS Hardy, 1987

Dr John Cameron Yaldwyn (1929–2005) was a New Zealand marine biologist and zoologist who specialised in crustaceans. He was Curator of Crustaceans (1962–1968) at the Australian Museum in Sydney, before moving to the National Museum of New Zealand where he founded an archaeozoology department and became Director (1980). He cowrote several books and papers, such as: *Australian Crustaceans in Colour* (1971) and edited the detailed report on the distribution of flora and fauna in: *Preliminary Results of the Auckland Island Expedition 1972–1973* (1975). This formed the basis for a programme to rid the islands of non-native mammals, completed in the early 1990s. Several marine invertebrates are also named after him, as is an extinct New Zealand bird.

Yamada

Goby sp. *Obliquogobius yamadai* K Shibukawa & Y Aonuma, 2007

Umeyoshi Yamada worked at the Fisheries Agency of Seikai National Fisheries Research Institute. He wrote: *Fish and fisheries of the East China and Yellow Seas* (2007). He was honoured for his great contribution to our knowledge of fishes in the East China Sea.

Yamakawa

Perchlet sp. *Plectranthias yamakawai* T Yoshino, 1972

Northern Yellow-spotted Chromis *Chromis yamakawai* H Iwatsubo & H Motomura, 2013

Snake-Eel sp. *Ophichthus yamakawai* Y Hibino, JE McCosker & F Tashiro, 2019

Takeshi Yamakawa (b.1942) is a fish collecting specialist who is an ichthyologist at Kochi University (Japan). It was he who first recognised the chromis species as new, and made specimens available to the authors. He was honoured in the snake-eel's binomial for his great contributions to taxonomy and the study of Japanese fishes. Among his published works is the paper: *Studies of the fish fauna around Nansei Islands* (1979).

Yamamoto

Amur Minnow ssp. *Rhynchocypris lagowskii yamamotis* DS Jordan & CL Hubbs, 1925

Dr Senzi Yamamoto was a professor at the Imperial University of Kyoto. He helped the senior author acquire specimens from Japanese fish markets.

Yamanaka

Lefteye Flounder sp. *Arnoglossus yamanakai* A Fukui, U Yamada & T Ozawa, 1988

Kanichi Yamanaka was the Captain of the RV 'Yoko-Maru' from which the lectotype was taken.

Yama-no-kam

Waspfish sp. *Snyderina yamanokami* DS Jordan & EC Starks, 1901

This is a name from Japanese mythology but taken from the local name for the fish. According to the authors' etymology: "*It is said... to bear the local name of Yama-no-kami, or Mountain Goddess, in local mythology a woman with wings, capable of starting a storm.*"

Yamanoue

Seabass sp. *Parascombrops yamanouei* WW Schwarzhaus & AM Prokofiev, 2017

Yusuke Yamanoue is a Japanese ichthyologist in the Department of Aquatic Bioscience, Graduate School of Agricultural and Life Sciences at the University of Tokyo. He has a particular interest in *Mola* (Ocean Sunfish). Among his c.40 published works are: *Phylogenetic relationship of two Mola sunfishes (Tetraodontiformes: Molidae) occurring around the coast of Japan, with notes on their geographical distribution and morphological characteristics* (2009) and *Revision of the genus Verilus* (Perciformes: *Acropomatidae) with a description of a new species* (2016).

Yamasaki, K

Dottybelly Dottyback *Pseudochromis yamasakii* AC Gill & H Senou, 2016

Kimihiro Yamasaki is a professional Japanese diver and 'excellent underwater photographer', who collected the type and provided photographs. In fact, the fish had been reported by Japanese divers for quite some time, but it was only thanks to Yamasaki's efforts, who had previously photographed it, that a specimen was collected and scientifically described.

Yamasaki, Y

Black-tailed Membranehead *Hymenocephalus yamasakiorum* N Nakayama, H Endo & W Schwarzhans, 2015

Yasuko Yamasaki and her family operate a fishing trawler in Tosa Bay, Japan and nearby waters. They provide a large number of fish specimens to Kochi University.

Yamashiro

Yamashiro's Wrasse *Pseudocoris yamashiroi* PJ Schmidt, 1931
[Alt. Redspot Wrasse, Pearlescent Wrasse]

A Yamashiro was an English teacher on Okinawa. On a visit to that island (the type locality) (1926) Schmidt was assisted by "*Mr. A. Yamashiro (professor of the English language in the secondary school of Naha)*" who acted as a guide and interpreter. Unfortunately, no more was recorded.

Yamato

Eelpout sp. *Lycodes yamatoi* M Toyoshima, 1985

This is a toponym referring to the Yamato Bank, Sea of Japan, where this species is numerous.

Yamatsuta

Korean Striped Bitterling *Acheilognathus yamatsutae* T Mori, 1928

K Yamatsuta was a Japanese naturalist who was a teacher at the Mukden (now Shenyang) Higher Girls School during the period of Japanese occupation of Manchuria. He was honoured as it was he who: "*...obtained... ...a fine type specimen.*" He also collected botanical specimens in the Mukden region (1920s) and a number of plants are named after him.

Yanagita

Brown Pygmy-goby *Trimma yanagitai* T Suzuki & H Senou, 2007

Mitsuhiko Yanagita is a professional diver who works at NASO Dive at Ito, Shizuoka, Japan. He used to be on the staff of Izu Oceanic Park, Diving Center. He helped the authors collect and provided the type.

Yang, H-C

Hagfish sp. *Eptatretus yangi* HT Teng, 1958
Goby sp. *Myersina yangii* TR Chen, 1960
Taiwanese Blind Electric Ray *Benthobatis yangi* MR de Carvalho, LJV Compagno & DA Ebert, 2003

Hung-Chia (Hung-Jia) Yang was an ichthyologist and highly talented illustrator working at the Taiwanese Fisheries Research Institute, Kaohsiung. He carried out extensive research into Taiwanese cartilaginous fishes and was honoured for this and his 'superb fish illustrations'. He collected the goby type.

Yang, J-X

Hillstream Loach sp. *Erromyzon yangi* DA Neely, KW Conway & RL Mayden, 2007
Stone Loach sp. *Mustura yangi* T Qin, M Kottelat, YM Kyaw & X Chen, 2022

Dr Yang Jun-Xing is Secretary of the Party Committee of the Kunming Institute of Zoology and was the Deputy Director of the Kunming Institute of Zoology, Yunnan, China, where he has been a professor since 2009. His doctorate was awarded by the Kunming Institute of Zoology, the Chinese Academy of Sciences. He and his colleagues have been successful in restoring native wetland habitat in Lake DIanchi, Yunnan. He wrote or co-wrote: *The fishes of Fuxian Lake, Yunnan, China, with description of two new species* (1991) and *The identity of Schizothorax griseus Pellegrin, 1931, with descriptions of three new species of schizothoracine fishes (Teleostei: Cyprinidae) from China* (2006) among others. He was honoured in the hillstream loach name as he allowed the authors to review his draft manuscript on Chinese *Erromyzon.*

Yang, L

Yang's Longnose Catshark *Apristurus yangi* WT White, RR Mana & GJP Naylor, 2017

Dr Lei Yang (b.1980) has carried out extensive molecular phylogenetic work on sharks and rays. Saint Louis University awarded his PhD (2010). He is laboratory manager and a post-doctoral researcher at the Florida Museum of Natural History where Gavin Naylor is Director of their shark research programme. He has done some work on the taxonomy and phylogenetics of cypriniform fishes and chondrichthyan fishes. mong his more than 30 published papers are: *Threatened fishes of the world: Cranoglanis bouderius* (Richardson, 1846) (*Cranoglanididae)* (2009) and *Phylogeny and polyploidy: Resolving the classification of cyprinine fishes (Teleostei: Cypriniformes)* (2015).

Yano, K

Flagtail Shrimp-goby *Amblyeleotris yanoi* Y Aonuma & T Yoshino, 1996

Yano's Pygmy-goby *Trimma yanoi* T Suzuki & H Senou, 2008

Yano's Cheek-hook Goby *Ancistrogobius yanoi* K Shibukawa, T Yoshino & GR Allen, 2010

Sleeper Goby sp. *Valenciennea yanoi* T Suzuki, H Senou & JE Randall, 2016

Korechika Yano is a diving instructor and underwater photographer at the Dive Service YANO, Iriomote Island, Japan. He collected the types of the shrimp-goby and the pygmy-goby. He is the co-author of: *Nihon no haze = A photographic guide to the gobioid fishes of Japan* (2004). (Also see **Korechika**)

Yano, S

Yano's Snipe Eel *Labichthys yanoi* GW Mead & I Rubinoff, 1966

Shigeru Yano is a long-line fisherman from Honolulu, Hawaii. The etymology describes him as a "*...friend and fellow fisherman,*" whose maintenance and operation of nets and associated equipment contributed greatly to the success of the authors' trawling expedition. He was aboard the 'Anton Bruun' oceanic research vessel (1963) and the 'Townsend Cromwell' (1990) and other expeditions during their marine surveys of Hawaiian waters. In one paper it is said of him and his fellows that "*...the success of any trawling expedition is dependent in no small measure on those responsible for the maintenance and operation of the nets and associated equipment.*" Yano and his workmates "*...devoted themselves to this equipment with ability, understanding, and scrupulous care; without these master fishermen these cruises would have fallen short of their goals.*"

Yao Peizhi

Stone Loach sp. *Triplophysa yaopeizhii* TQ Xu, CG Zhang & B Cai, 1995

Yao Peizhi was vice-chairman of the Agriculture and Forestry Committee of Tibet, where this loach is endemic. The authors honoured him for his support of their research.

Yaquina

Lanternfish sp. *Lampadena yaquinae* LR Coleman & BG Nafpaktitis, 1972

Rattail sp. *Coryphaenoides yaquinae* T Iwamoto & DL Stein, 1974

The 'Yaquina' was a research vessel owned by Oregon State University. The holotypes of both species were collected from her.

Yarrell

Lightfish genus *Yarrella* GB Goode & TH Bean, 1896

Morid Cod sp. *Laemonema yarrellii* RT Lowe, 1838

Giant Devil Catfish *Bagarius yarrelli* WH Sykes, 1839

William Yarrell (1784–1856) was an English naturalist and bookseller. He is best known as the author of *The History of British Fishes* (2 vols., 1836) and *A History of British Birds* (3 vols., first ed. 1843). Lowe wrote that it was Yarrell who "*...first drew my attention to this very distinct and pretty species.*" Three birds are named after him.

Yarrow

Zuni Bluehead Sucker *Catostomus discobolus jarrovii* ED Cope, 1874
Green River Speckled Dace *Rhinichthys osculus yarrowi* DS Jordan & BW Evermann, 1891

Dr Henry Crecy Yarrow (1840–1929) was a surgeon, ornithologist, herpetologist and naturalist. He studied in Pennsylvania and Geneva for his MD (1861). He was Assistant Surgeon with the cavalry (1861–1862) and served in Virginia, then worked as the executive officer at Philadelphia's Cherry Street Hospital. He was ordered to Atlanta, Georgia (1866) and served through a severe cholera outbreak there and in New York (1867). He joined the Wheeler Survey (1871–1876) as both surgeon and naturalist. Thereafter he held many posts (1886–1917) as Medical Office, Professor of Dermatology, Curator of Reptiles at the Smithsonian and as Lt Colonel in the medical corps. The etymology of the sucker states it was he "*...whose zoological explorations in various portions of the United States have been productive of many interesting results.*" This included helping collect the type of the sucker, with Cope Latinizing the spelling of Yarrow's name, since 'y' and 'w' are absent in classical Latin. He was also honoured for his work on the fishes of the Colorado River.

Yasuda

Orange Pipefish *Maroubra yasudai* CE Dawson, 1983

Dr Fujio Yasuda was Professor of Ichthyology at the Laboratory of Ichthyology, Tokyo University of Fisheries. He published widely, including papers with JE Randall such as: *Centropyge shepardi, a new angelfish from the Mariana and Ogasawara Islands* (1979) and a number of books such as the co-authored *Pacific Marine Fishes* (1972). He also appeared, as himself, in an episode of the TV series Arthur C. Clarke's *Monsters of the Deep* (1980).

Yasuhiko Taki

Mekong Loach genus *Yasuhikotakia* TT Nalbant, 2002

Dr Yasuhiko Taki (1931-2020) was an Japanese ichthyologist and researcher '...who contributed very much' to the study of loaches, according to the etymology. He was at the Research Institute of Evolutionary Biology, Tokyo (1977) and at the Tokyo University of Agriculture (1985), of which he is Professor Emeritus. He is now President, Japan Wildlife Research Centre, having previously (late 1990s) been with System Science Consultants, Tokyo. He wrote: *Fishes of the Lao Mekong basin* (1974).

Yata

Characin sp. *Heterocheirodon yatai* JR Casciotta, AM Miquelarena & LC Protogino, 1992

This may look like an eponym but instead is a word derived from the Guaraní for palm tree. It refers to *Butia yatay*, a palm tree dominant at the type locality which is in Entre Ríos, Argentina.

Yatabe

Yatabe Blenny *Parablennius yatabei* DS Jordan & JO Snyder, 1900

Ryōkichi Yatabe (1851–1899) was a Japanese botanist. He was also an 'old friend' of the authors, after becoming Cornell University's first Japanese graduate. He drowned during a summer vacation.

Yatsu

Shrimp-Goby sp. *Cryptocentrus yatsui* I Tomiyama, 1936

Dr Naohide Yatsu (1877–1947) was Professor of Zoology, Tokyo Imperial University. He had studied at Columbia University, New York, and had worked at the Biological Station in Naples. Among his written works is: *An Experimental Study on the Cleavage of the Ctenophore Egg* (1910). He also has an amphibian named after him.

Yekuana

Armoured Catfish sp. *Pseudancistrus yekuana* NK Lujan, JW Armbruster & MH Sabaj Pérez, 2007

The Ye-kuana are an indigenous people inhabiting the upper Río Ventauri and other areas of southern Venezuela and northern Brazil, whose 'generous cooperation' made the authors' research possible. They are also known by variant spellings such as: Ye'Kwana, Ye'Kuana, Yekuana, Yequana and Yecuana (see next entry).

Yekwana

Characin sp. *Aphyocharax yekwanae* PW Willink, B Chernoff & A Machado-Allison, 2003

This species is named after the Ye'Kwana people (also 'Ye-kuana' – see above). They live in and oversee most of the Río Caura basin, Brazil. They are well known for their "*...fervid desire to protect and manage their home territory and its environment.*"

Yen

Bagrid Catfish sp. *Tachysurus yeni* VH Nguyēn & TH Nguyēn, 2005

Dr Mai Đinh Yên (b.1933) is a retired Vietnamese ichthyologist, formerly a professor of biology at Hanoi University (1964–1986) where he started work (1956), rising very quickly to the Head of the Department of Vertebrate Zoology (1960). He was known to his students as 'the king of fish research'. Among his publications are the books: *Types of freshwater fish in the northern provinces of Vietnam* (1969) and *Identification of Southern freshwater fish species* (1992). He was honoured as he had identified this catfish as a distinct population.

Yepez

Maracaibo River Stingray *Potamotrygon yepezi* MN Castex & HP Castello, 1970

Yepez's Tube-snouted Ghost Knifefish *Sternarchorhynchus yepezi* CDCM de Santana & RP Vari, 2010

Pencilfish sp. *Lebiasina yepezi* AL Netto-Ferreira, OT Oyakawa, J Zuanon & JC Nolasco, 2011

Dr Augustín Antonio Fernández-Yépez (1916–1977) was a Venezuelan architect and ichthyologist who was a curator at the Museo de Ciencias Naturales de Caracas. His brother,

Dr Francisco José Fernández Yépez (1923–1986), was an agronomist and entomologist and has several odonata and an earwig named after him. (See also **Fernández-Yépez**)

Yerger

Speckled Puffer *Sphoeroides yergeri* RL Shipp, 1972

Dr Ralph William Yerger (1922–2003) was an ichthyologist who was Professor of Biology at Florida State University. Pennsylvania State University awarded his BSc (1943). He joined the services and became Second Lieutenant of Infantry and (1944) saw active service in France, was promoted to First Lieutenant (1945), and was part of the army of occupation in Germany until being demobilised (1946). Back at Penn State he was awarded his MSc in Wildlife Management (1946). He entered Cornell graduate school (1947), being awarded his PhD (1950) then taking an assistant professorship at Florida State University where he stayed for the rest of his career. He participated in a number of collecting trips in Florida with RD Suttkus (q.v.) and they wrote together. He had collaborated with Shipp on an earlier paper.

Yersin

Stone Loach sp. *Schistura yersini* JA Freyhof & DV Serov, 2001

Alexandre Émile Jean Yersin (1863–1943) was a Swiss bacteriologist and humanist. Shortly after graduating in Paris (c.1882) he joined Dr Louis Pasteur's team and took French citizenship (1889). Later (1919) he was the Director of the Pasteur Institute in French Indochina (Vietnam), which he established and where he spent much of his life – especially around Dalat, where this species was collected. He was responsible for introducing the rubber tree to Vietnam and for the first quinine plantations there. Yersin discovered the bubonic plague bacillus *Yersina pestis*, which is named after him, and developed an antiserum. He also had a hand in the development of a vaccine for diphtheria. He took an interest in everything around him, which led him to explore in the Vietnamese highlands. He died at Nha Trang, in Annam. A bird is also named after him.

Yoka

African Characin sp. *Rhabdalestes yokai* A Ibala Zamba & E Vreven, 2008

Paul Yoka is the Director of the Institut de Développement Rural at the University of Brazzaville, Republic of Congo. He gave a great deal of help during the Léfini Expeditions (2004–2008), during which the holotype was collected. A Paul Yoka was arrested in Kinshasa (2016) for protesting at President Kabila's attempt to hold on to power: we are unclear if this is the same person, but whoever he is, he was held without charge or legal help and listed as a political prisoner.

Yoki

Characin sp. *Bryconamericus yokiae* C Román-Valencia, 2003

Yoki Román-Valencia is, we assume, the author's wife – since, in the etymology, he addresses her thus: "*...my dream witch, for her pains and patience with a husband who loves little fishes*". Apparently, 'dream witch' is a term of endearment, similar to 'bewitching woman'.

Yoli

Characin sp. *Chrysobrycon yoliae* JA Vanegas-Ríos, M de las Mercedes Azpelicueta & H Ortega, 2014

Yolanda 'Yoli' Ríos Nossa is the senior author's mother who "*...patiently encouraged and supported his academic formation in all senses.*"

Yonezawa

Freshwater Goby sp. *Rhinogobius yonezawai* T Suzuki, N Oseko, S Kimura & K Shibukawa, 2020

Toshihiko Yonezawa is a Japanese ichthyologist who works for the Foundation of Kagoshima Environmental Research and Service. Among his published work are the co-written papers: *Three New Species of the Ptereleotrid Fish Genus Parioglossus (Perciformes: Gobioidei) from Japan, Palau and India (2010)* and *Freshwater fishes of Yaku-shima Island, Kagoshima Prefecture, Southern Japan* (2010). He was honoured in the name because he: "*...offered much information and specimens to us for our study.*"

Yonge

Goby genus *Yongeichthys* GP Whitley, 1932
Nzoia Barb *Enteromius yongei* PJP Whitehead, 1960
Whip Coral Goby *Bryaninops yongei* WP Davis & DM Cohen, 1969

Sir Charles Maurice Yonge CBE FRS (1899–1986) was an English marine zoologist. He was honoured by Whitehead for the "*...interest he has shown and the assistance he has given to many aspects of fishery research in East Africa.*" He accompanied Davis & Cohen on the cruise where the goby type was taken. He led an expedition to the Great Barrier Reef (1928–1929) during which the genus was first taken.

Yoshi

Damselfish sp. *Pomacentrus yoshii* GR Allen & JE Randall, 2004

Satoshi Yoshii is a divemaster in the Marshall Islands who provided logistic arrangements for the collection of many of the type specimens.

Yoshida

Morid Cod sp. *Physiculus yoshidae* O Okamura, 1982

Miss Kiyoko Yoshida helped Okamura prepare the book in which this species was described. We have no further information about her.

Yoshigou

Threadless Cheek-hook Goby *Ancistrogobius yoshigoui* K Shibukawa, T Yoshino & GR Allen, 2010

Hidenori Yoshigou is (1998–present) a research zoologist and systematist with the environmental company, Chugai Technos, Hiroshima, Japan. His current (2018) project is cataloguing the fish collection at Hiwa Museum. He studied at the University of Ryukyu,

Faculty of Science and Graduate School of Engineering and Science (1992–1998). He provided five paratypes to the authors and valuable information about this goby. He has published more than 100 papers including a number in the series: *Fishes of Hiroshima* (2016–2017).

Yoshino

Yoshino's Goby *Gnatholepis yoshinoi* T Suzuki & JE Randall, 2009
Pygmy-goby sp. *Trimma yoshinoi* T Suzuki, K Yano & H Senou, 2015

Tetsuo Yoshino is a Japanese ichthyologist who is a member of the Department of Marine Sciences at the University of the Ryukyus, Okinawa, having formerly been at Kyoto University. He has written or co-written around 90 scientific papers, such as: *Title Records of Four Labrid Fishes (Osteichthyes, Labridae) from Japan* (1980) and *Review of the genus* Limnichthys *(Perciformes: Creediidae) from Japan, with description of a new species* (1999). He was honoured for his 'extensive' taxonomic research on the fishes of Japan. He was Senou's supervisor when the latter studied at the former's laboratory.

Younger

Golden Char *Salvelinus youngeri* GF Friend, 1956

Henry John Younger was a member of the eponymous brewery company. He lived on the Benmore Estate House, Cowal, Argyllshire, Scotland, where he developed an arboretum, known as the Younger Botanic Garden. He gave permission for Loch Eck, the holotype locality, to be 'netted'.

Younghusband

Tibetan Cyprinid sp. *Schizopygopsis younghusbandi* CT Regan, 1905

Lieutenant-Colonel Sir Francis Edward Younghusband KCSI, KCIE (1863–1942) was a British army officer, explorer and spiritual and political writer. He went to Sandhurst (1881) and was commissioned (1882) in the 1st King's Dragoon Guards. He took leave to make an expedition across Asia (1886–1887), mostly exploring Manchuria. He mapped mountains and also travelled across the Gobi dessert. Promoted to Captain (1889), he was sent with a party of Gurkhas to investigate uncharted Ladakh. He transferred to the Indian Political Service (1890). He led (1903–1904) a British expedition to (and *de facto* invasion of) Tibet, during which the type was collected. He later (1919) became President of the Royal Geographical Society and (1921–1924) Chairman of the British Everest Expeditions. He also had great faith in the power of Cosmic Rays and claimed there are extra-terrestrials with translucent flesh living on the planet Altair. After a religious experience in Tibet he was profoundly spiritual and founded the World Congress of faiths (1936), but died of a stroke after addressing one of their meetings. He was also a prominent advocate of 'free love'.

Younholee

Bangladeshi Guitarfish *Glaucostegus younholeei* KA Habib & MJ Islam, 2021

Youn-Ho Lee was the Ph.D supervisor of the senior author. He is also Principal Research Scientist/Professor and Vice-President of Korea Institute of Ocean Science and Technology

(KIOST). The etymology notes his contributions to our knowledge of marine biodiversity, DNA barcoding, and population genetics of marine organisms in the Western Pacific.

Yseux

African Characin sp. *Bryconaethiops yseuxi* GA Boulenger, 1899

Dr Émile Ghislain Joseph Yseux (1835–1915) was a professor of zoology at the Université Libre de Bruxelles, Belgium. He was also a physician and a founding member of the Royal Society of Public Medicine and Medical Topography of Belgium.

Yu

Loach sp. *Paralepidocephalus yui* TL Tchang, 1935

Professor Shou-Chie (Chao-ch'i) Yu was a Chinese biologist with a particular interest in Crustacea. He worked at the Fan Memorial Institute of Biology in Peiping. He often published in the institute's bulletin, such as: *On some species of shrimp-shaped Anomura from North China* (1931).

Yuanding

Snake-eel sp. *Cirrhimuraena yuanding* WQ Tang & CG Zhang, 2003

Professor Zhu Yuang-Ding (1896–1986) was a Chinese ichthyologist and educator. His early interest was sparked whilst at Ningbo High School near China's largest fishery, Zhoushan, where he wandered the docks and riverside. He graduated in Biology from Soochow University (1920) then undertook studies at Cornell, culminating in his MSc (1934) in entomology, then undertook his PhD at the University of Michigan. He became Professor of Biology at St John's University, Shanghai (having been employed there as a graduate), rising to Dean of the faculty. He wrote numerous books and papers on all aspects of fish biology, taxonomy, systematics, etc. and described many new species. The authors regard Professor Yuang-Ding as "*China's leading authority on ichthyology and fishery science.*"

Yuba

Goby sp. *Luciogobius yubai* Y Ikeda, K Tamada & K Hirashima, 2019

Takeo Yuba was the first to find this species, collecting the type series in Wakayama Prefecture with the first two authors and taking photographs of it. We have found no further biographical data.

Yunokawa

Viviparous Brotula sp. *Grammonus yunokawai* JG Nielsen, 2007

Kyo Yunokawa was at the School of Fishery Sciences, Kitasato University, Japan and at Ie-shima Diving Center, Okinawa, Japan. He co-wrote: *Karyotypic Variation Found among Five Species of the Family Platycephalidae* (2010). He photographed and caught (by hand in the back of a cave in absolute darkness) the only known specimen of this species (1998).

Yuri (Rudjakov)

Halfbeak sp. *Hyporhamphus yuri* BB Collette & NV Parin, 1978

Yuri Alexandrovich Rudjakov (b.1938) (See **Rudjakov**)

Yuri (Shcherbachev)

Beardfish sp. *Polymixia yuri* AN Kotlyar, 1982
Rattail sp. *Coelorinchus yurii* TN Iwamoto, D Golani, A Baranes & M Goren, 2006
Skate sp. *Notoraja yurii* VN Dolganov, 2020

Yuri Nikolaevich Shcherbachev (See **Shcherbachev**)

Yutaje

Britslenose Catfish sp. *Ancistrus yutajae* L De Souza, DC Taphorn & JW Armbruster, 2019

This is, primarily, a toponym referring to the type locality in Venezuela; the Yutajé river. However, it also alludes to the legend after which the river is named. Yu and Taje were young lovers whose tribes were at war. Ordered to be captured, they were surrounded and launched themselves off a cliff, invoking the aid of the god of waters – who supplied it, after a fashion, by turning the lovers into twin waterfalls.

Yuwono

Borneo Loach sp. *Homalopteroides yuwonoi* M Kottelat, 1998

Richardus Digdo Yuwono was an aquarist who was one of the founders of and a leading member of the Indonesian Ornamental Fish Association. He set up his business as a tropical fish dealer (1968) and the business is still run by his son. He was honoured for his continuous support of Kottelat's work on Indonesian freshwater fishes.

Z

Zaca

Zaca Blenny *Malacoctenus zacae* VG Springer, 1959

The 'Zaca' was a yacht owned by movie star Errol Flynn. In 1946 the yacht was taken on a scientific expedition along the Pacific coast of Mexico, during which the holotype was collected.

Zaiser

Miyake Bigeye *Priacanthus zaiserae* WC Starnes & JT Moyer, 1988

Martha J Zaiser formerly worked at the Tanaka Memorial Biological Station, Miyake-jima, Japan. At times she assisted Jack Moyer with underwater observations. Among her publications, she co-wrote: *Early sex change: A possible mating strategy of Centropyge angelfishes (Pisces: Pomacanthidae)* (with Jack Moyer) (1984) and *Synchiropus* moyeri, *a New Species of Dragonet (Callionymidae) from Miyake-jima, Japan* (1985). The authors say she made "*valuable contributions to the knowledge of the marine biogeography of the Izu Islands*."

Zajonz

Socotra Sweeper *Pempheris zajonz*i JE Randall & BC Victor, 2015

Uwe Zajonz (b.1967) is a Research Associate and Project Manager at the Senckenberg Biodiversity and Climate Research Centre, Frankfurt. He graduated with a Biology Diploma (1995) from Technical University of Darmstadt, Germany, then with an MSc in zoology. He started his career (1995) as a scientific assistant and fish taxonomist at Senckenberg and the Natural History Museum, Frankfurt. He has professional experience since then in deep sea fish taxonomy, reef fish ecology and biogeography, marine and fisheries biology, conservation, ecosystem services, and in freelance and institutional project management. He collected the type (1999).

Zakaria-Ismail

Malay barb sp. *Barbodes zakariaismaili* M Kottelat & KP Lim, 2021

Mohd Zakaria-Ismail is a Malaysian biologist. After studying at the College of Agriculture Malaya in Serdang (now Universiti Putra Serdang), he took his PhD at the Colorado State University, Fort Collins, USA. He was a Professor of Biology at the University of Malaya, Kuala Lumpur, until his retirement (2011). He is co-author of *Fishes of the Freshwater Ecosystems of Peninsular Malaysia* (2019). He was honoured in the binomial "…in appreciation for his work on the fish fauna of Malaysia."

Zakon

Zakon's Mormyrid *Petrocephalus zakoni* S Lavoué, JP Sullivan & ME Arnegard, 2010

Dr Harold H Zakon is a Professor (since 1999) at the Department of Neuroscience, College of Natural Sciences at the University of Texas, as well as Chairman of Neurobiology. After graduating with a science degree (1972) from Marlboro College he was a research assistant at Harvard Medical School before taking a research and teaching position at Cornell (1974–1981). Cornell University awarded his PhD in Neurobiology & Behaviour (1981), following which he did postdoctoral research at Scripps Institute of Oceanography, University of California, San Diego (1981–1983). He went on to teach zoology at The University of Texas, Austin, as Assistant Professor (1983–1988) and Associate Professor (1988–1993) becoming full professor (1994–1998). He has published more than 100 papers, particularly on the chemistry of the brain, including: *Androgen modulates the kinetics of the delayed rectifying K+ current in the electric organ of a weakly electric fish* (2007). He was honoured here for his many contributions to neuroethology, inspiring "*...a new area of research on genes that underlie electrolocation and electrical communication in gymnotiform and mormyroid fishes*".

Zaluar

Blenny sp. *Malacoctenus zaluari* A Carvalho-Filho, JL Gasparini & I Sazima, 2020

Ricardo Zaluar Passos Guimarães of the Laboratório de Biodiversidade de Recursos Pesqueiros, Universidade Federal do Rio de Janeiro, Brazil, is the 'ichthyologist friend' of the authors. He has contributed to the descriptions of several reef fishes from the Brazilian coast.

Zammarano

Somali Cave Catfish sp. *Uegitglanis zammaranoi* L Gianferrari, 1923

Lieutenant-Colonel Vittorio Tedesco Zammarano né Tedesco (1890–1959) (Zammarano was his stepfather's name) was an Italian traveller, zoologist, cartographer, geographer and hunter, who was involved with the Museo Civico di Milano. He wrote: *Azanagò non pianse*, (Azanagò does not cry) (1934) and *Auhér, mio sogno* (Auhér, my dream) (1935), both of which were published in Milan. He also wrote books on big game hunting in Somalia such as: *Hic sunt leones. Un anno di esplorazione e di caccia in Somalia* (Here be lions - a year of exploration and hunting in Somalia) (1924). de Beaux wrote an article entitled *Mammiferi della Somalia Italiana racconta del Maggiore Vittorio Tedesco Zammarano nel Museo Civico de Milano* (1924) (Mammals of Italian Somalia described by Major Vittorio Tedesco Zammarano of the Civic Museum of Milan). Except for fighting in the Italian Army at the Battle of the Piave River (1918), he appears to have spent much of the decade (1915–1925) in Africa and a film he made on location of his time there still exists. Zammarano observed that Ethiopian lions were 'more wary and cowardly' than those in other parts of Africa, with the result that hunting in Ethiopia was difficult: hardly surprising as he had helped to significantly reduce their numbers! A mammal is named after him.

Zanandrea

Po Brook Lamprey *Lethenteron zanandreai* VD Vladykov, 1955
Spined Loach sp. *Cobitis zanandreai* G Cavicchioli, 1965

Dr Giuseppe Zanandrea (1907–1965) was a Jesuit priest who worked at Istituto di Zoologia e Anatomia comparata, Universita di Padova (1959) and Istituto di Anatomia Comparata della Università di Bologna. The author honoured him as an "...invaluable advisor and collaborator" (translation), who identified this loach as a distinct taxon (1964).

Zanata

Characin sp. *Serrapinnus zanatae* FC Jerep, P Camelier de Assis Cardosa & LR Malabarba, 2016

Dr Angela Maria Zanata is a zoologist and ichthyologist at the Federal University of Bahia, Brazil (2007). The University of São Paulo awarded all her degrees, bachelor's (1991), master's (1995) and doctorate (2000). She did post-doctoral studies at both the Smithsonian (2001–2002) and the University of São Paulo (2003–2004 & 2014–2015). She is now (2017) an Adjunct Professor at the University of Bahia and Curator of the fish collection, as well as a collaborating researcher at both the Zoology Museum of the University of São Paulo and Feira de Santana State University. She co-wrote: *A new remarkable and Critically Endangered species of Astyanax Baird & Girard (Characiformes: Characidae) from Chapada Diamantina, Bahia, Brazil, with a discussion on durophagy in the Characiformes* (2017). She helped collect the holotype of this characin.

Zander

Sculpin sp. *Argyrocottus zanderi* S Herzenstein, 1892

Dr Alexander Karlovich Zander (DNK) came from a Prussian family and studied medicine (1883), receiving his medical degree (1894). He was hired as a 'senior ship doctor' for a trip to Sakhalin Island on the clipper *Rider*, whose mission was to make topographical images of the coastline (1888–1889). He discovered the eponymous species at Sakhalin Island in the Sea of Okhotsk (1890). **He** received his degree after the Sakhalin voyage; his dissertation was on cholera in children and corresponding hygiene problems. He worked (1890s) at the Nikolaevsky Hospital, where a cholera department and a bacteriological laboratory were opened (1892). One of his patients was the composer Pyotr Ilyich Tchaikovsky, who died (1893), possibly of cholera (and possibly of suicide). The etymology translates as: "*This species I named in honor of Dr. Zander, who discovered it and donated it with other collections of the School of Reformed Churches of St. Petersburg. Due to the friendly congeniality of the head teacher R. Haage, the Zoological Museum had the opportunity to acquire from the collections of Dr. Zander scientifically valuable objects by exchange.*" Biographical research conducted by Natalia Chernova (Russian Academy of Sciences), Hans-J. Paepke (Berlin Museum of Natural History), Erwin Schraml (Welt der Fische/World of Fishes), and Christopher Scharpf (The ETYFish Project).*

*Footnote – Researching this chap was a great example of what it sometimes takes to identify eponyms. *Tracking down the specific name* zanderi *required an international effort! We approached Christopher Scharpf of the wonderful ETYFish Project, who, in turn asked his German friend and* *colleague Erwin Schraml if he could research Zander's identity. Erwin asked Hans-J. Paepke, retired Curator of Ichthyology at the Berlin Museum of Natural History and a scholar of ichthyological history. Dr. Paepke forwarded our query to Dr. Natalia Chernova, an ichthyologist at the Zoological Institute of the Russian Academy of Sciences in St. Petersburg. Dr. Chernova spent a considerable amount of time digging through library and museum archives to compile everything she could find about Zander and we are grateful to them all for the facts for this entry.*

Zantedeschi

Taillight Shark *Euprotomicroides zantedeschia* PA Hulley & MJ Penrith, 1966

Giovanni Zantedeschi (1773–1846) was an Italian physician and botanist. He studied in Verona and Padua where he graduated in medicine and later practised as a surgeon. He was a very keen botanist, discovering several new species. German botanist Kurt Sprengel named the southern African plant genus *Zantedeschia* after him. The botanical museum in Molina (*Museo Botanico della Lessinia di Molina)* is dedicated to him. However, the shark derives this name via a convoluted route. The South African Arum Lily is called *Zantedeschia aethiopica* after the Italian physician. The trawler that caught the first shark specimen was called 'Arum' after the lily and the shark was, in turn, named after the trawler!

Zappa

Mudskipper genus *Zappa* EO Murdy, 1989

Frank Vincent Zappa (1940–1993) was well-known as a musician and composer. The etymology states he was honoured "*...for his articulate and sagacious defence of the First Amendment of the US Constitution*". The 1st amendment, one of the 10 amendments that constitute the Bill of Rights, deals with basic freedoms e.g, peaceful assembly, of speech, of religion, etc.

Zaret

Glass Knifefish sp. *Rhabdolichops zareti* JG Lundberg & F Mago-Leccia, 1986

Thomas M Zaret (1945–1984) was an ichthyologist and zoologist at Zoology Department, University of Washington, Seattle. The etymology states that he was the authors' "*...close friend, who contributed much to our knowledge of Rhabdolichops, planktivorous fishes and fish ecology.*" He co-wrote: *Species Introduction in a Tropical Lake* (1973).

Zarske

Mexican livebearer sp. *Gambusia zarskei* MK Meyer, S Schories & M Schartl, 2010
Brazilian Cichlid sp. *Cichlasoma zarskei* FP Ottoni, 2011

Dr Axel Zarske (b.1952) is a German ichthyologist who is a section head in the ichthyology department of Senckenberg Natural History Museum. He has published more than 170 scientific papers such as: *A new African Catfish Eutropiellus vandeweyeri* (1970) and *The type material of the Characiformes of the Museum of Natural History Berlin* (2012). He is the Editor in Chief of *Vertebrate Zoology*; the journal in which the cichlid description was published. The authors of the livebearer's description said: "*We name this species after Dr. Axel Zarske in recognition of his valuable contributions to discussions on the conservation biology and problems of endangered fishes such as G. zarskei.*"

Zarudny

Snowtrout sp. *Schizothorax zarudnyi* AM Nikolskii, 1897

Nikolai Alekseyivich Zarudny (1859–1919) was a Russian (Ukrainian) zoologist, traveller and ornithologist. He taught at the Military High School in Orenburg (1879–1892), during which time he undertook five expeditions through the Trans-Caspian region (Turkmenistan). He taught natural history at the Pskov Military School (1892–1906), and in this period he undertook four journeys through various parts of Persia (Iran), and was awarded the Russian Geographical Society's Przhewalski medal. He then worked in Tashkent (1906), continuing his Middle Asia exploration. He collected extensively, and his specimens are held by the Zoological Museum of the Academy of Science. He wrote: *Les reptiles, amphibiens, et poissons de la Perse orientale* (1903), and *Third Excursion over Eastern Persia (Horassan, Seistan and Persian Baluchistan) in 1900–1901* (1916). He was working on a book about Turkestan birds when he died of accidental poisoning. Nine birds, two mammals, two reptiles and several insects are also named after him. He collected the snowtrout type specimen.

Zawadzki

Armoured Catfish sp. *Pseudancistrus zawadzkii* GSC Silva, FF Roxo, R Britzki & C de Oliveira, 2014

Corydoras (catfish) sp. *Corydoras zawadzkii* LFC Tencatt & WM Ohara, 2016

Armoured Catfish sp. *Rineloricaria zawadzkii* GJ Costa-Silva, G Costa e Silva & C Oliveira, 2022

Dr Cláudio Henrique Zawadzki is a zoologist, limnologist and ichthyologist who is Associate Professor of Zoology at Universidade Estadual de Maringá, Paraná, Brazil, where he also was awarded his bachelor's degree (1993), his master's (1996) and his doctorate (2001). He co-wrote: *A new species of Aphanotorulus (Siluriformes: Loricariidae) from the rio Aripuanã Basin, Brazil* (2017). He was recognised for his comprehensive contributions to the knowledge of neotropical fishes, especially of the Loricariidae; Also, he is a 'dear friend' who directly participated in the professional development of Tencatt, the senior author of *Corydoras zawadzkii*.

Zeehaan

Southern Dogfish *Centrophorus zeehaani* WT White, DA Ebert & L Compagno, 2008

Zeehaan is the name of an Australian trawler that caught a giant squid *Architeuthis dux* near Portland, Victoria (2008), and from which the first specimens of the dogfish were collected from Tasmanian waters (1979).

Zekay

Ceyhan Spring Minnow *Pseudophoxinus zekayi* NG Bogutskaya, F Küçük & MA Atalay, 2006

Zekay Atalay is doubtless a relation of the third author, but their relationship was not explained in the original text.

Zé Lima

Three-barbeled Catfish sp. *Rhamdella zelimai* RE dos Reis, LR Malabarba & CAS de Lucena, 2014

Dr José Lima de Figueiredo aka 'Ze Lima' is a Brazilian ichthyologist who is a former researcher and Curator of Fishes at the Zoological Museum of the University of São Paulo. He is a professor at the University of São Paulo, which awarded his bachelor's degree (1969) and doctorate (1981). He co-wrote: *The northernmost record of* Bassanago albescens *and comments on the occurrence of Rhynchoconger guppyi (Teleostei: Anguilliformes: Congridae) along the Brazilian coast* (2011).

Zelinda

Zelinda's Parrotfish *Scarus zelindae* RL Moura, JL de Figueiredo & I Sazima, 2001

Zelinda Margarida de Andrade Nery Leão is a Brazilian marine geologist who is a professor at the Universidade Federal da Bahia (UFBA) (1980–present) where she was also a post-graduate researcher (1972). She graduated from the Faculty of Philosophy, Sciences and Letters of UFBA (1958) and researched at the Laboratory of Micro-palaeontology at the University of Paris (1963). Her PhD was awarded by the University of Miami (1982). Her main focus is the geology of reefs and their conservation. She has numerous published papers and book chapters. According to the etymology she is an "*... enthusiastic conservationist of Brazilian reefs.*"

Zenker

African Rivuline sp. *Epiplatys zenkeri* CGE Ahl, 1928
[Syn. *Epiplatys infrafasciatus zenkeri*]

Georg August Zenker (1855–1922) was a German botanist and gardener who collected in Cameroon, formerly a German protectorate. He established the settlement which later became the capital, Yaoundé. He had significant land holdings around Bipindi and devoted much time to collecting plants, insects, fish - and apparently human bones, which he is recorded disinterring in the area. He made a particular study of 'pygmies' and other native peoples. Nine birds, three mammals and a reptile are named after him.

Zenkevich

Cusk-eel sp. *Bassozetus zenkevitchi* TS Rass, 1955

Lev Aleksandrovich Zenkevich (1889–1970) was a Russian oceanographer and zoologist. He originally qualified as a lawyer at Moscow University (1912) and then at the natural science division of the department of physics and mathematics (1916). He was an assistant professor in the department of invertebrate zoology, Moscow State University (1925), and professor and department head (1930). He also worked (1947) at the Institute of Oceanography, Academy of Sciences of the USSR and, after 1949, concentrated his research on deep-water ocean fauna. He led many expeditionary cruises including the R/V 'Vitiaz' cruise on which the holotype was collected.

Zeppelin

Loach sp. *Lepidocephalichthys zeppelini* JC Havird & W Tangitjaroen, 2010

Not an eponym in that it was named after Led Zeppelin (1968–1980), a world-famous English rock band. The species was named after the band in reference to the Gibson EDS-1275 double-neck guitar that was played by Led Zeppelin's front man, Jimmy Page, because it reminded the authors of the diagnostic double *lamina circularis* of this species.

[NB. *lamina circularis* = the bony process or plate at the base of the first and second rays of the pectoral fin in most male specimens of Cobitid loaches]

Zetek

Characin sp. *Bryconamericus zeteki* SF Hildebrand, 1938

Professor James Zetek (1886–1959) was an American entomologist. He graduated from the University of Illinois (1911) and then worked for the US Government in the Panama Canal Zone (1911–1953). He became Professor of Biology and Hygiene, National Institute of Panama (1916–1918), and the first Director of the Smithsonian Tropical Research Institute on Barro Colorado Island, Panama Canal (1923–1956). His primary research interest was the study of termites and termite control. He co-wrote: *The Black Fly of Citrus and Other Subtropical Plants* (1920). A reptile, two amphibians and a dragonfly are named after him.

Zetgibbs

Snaggletooth sp. *Astronesthes zetgibbsi* NV Parin & OD Borodulina, 1997

Robert Henry Gibbs Jr. (1929–1988) was an American ichthyologist and conservationist, a long-standing curator at the Smithsonian and an expert on deep-sea and pelagic fishes. He was a member of the American Society of Ichthyologists and Herpetologists which honoured him posthumously with the establishment of the Robert H Gibbs Jr Memorial Award for Excellence in Systematic Ichthyology. The prefix 'zet' is from the Greek zeta; Gibbs referred to this taxon in unpublished papers as 'species Z'.

Zetti

Characin sp. *Cynopotamus zettii* NR Iriart, 1979
[Junior Synonym of *Cynopotamus kincaidi*]

Jorge Zetti (1938–1974) was an Argentine palaeontologist. He was a Researcher at the Museo de La Plata and the Museo Municipal de Ciencias Naturales de Mercedes.

Zeus

John Dory genus *Zeus* C Linnaeus, 1758

Derived from zaeus, an ancient Greek name for this fish dating back to Pliny's Natural History. More recent scholars have confused zaeus with the Greek god Zeus, culminating in an oft-repeated claim (1902) that the fish had also been called 'Piscis Jovii' in classical texts (Zeus being the equivalent to the Roman god Jove or Jupiter). However, no evidence for this claim has yet been brought forward.

Zhang

Chinese Gudgeon sp. *Microphysogobio zhangi* S-P Huang, Y Zhao, I-S Chen & K-T Shao, 2017

Professor Chun-guang Zhang (b.1955) is a Chinese ichthyologist at the Institute of Zoology, Chinese Academy of Sciences. He was honoured in recognition of his great contribution to taxonomic studies of fishes in China.

Zhao, TQ

Stone Loach sp. *Triplophysa zhaoi* AM Prokofiev, 2006

Zhao Tie-Qiao is a Chinese ichthyologist at the Shaanxi Institute of Zoology, Xian. He co-wrote: *Characteristics of the Fish-fauna of Qinghai-Xizang plateau and its geographical distribution and formation* (1991). He was honoured for his 'great contribution' to the study of nemacheiline loaches from northwest China.

Zharov

Snaggletooth sp. *Astronesthes zharovi* NV Parin & OD Borodulina, 1998

Dr Viktor L Zharov (1932–1998) was a Russian ichthyologist at the Atlantic Research Institute of Marine Fisheries and Oceanography, Kaliningrad, and a specialist in scombroid fishes. He was an early Russian researcher of the epipelagic fishes of the World Ocean. He wrote: *Reproduction of the Yellowfin Tuna* (Thunnus Albacores Bonaterre) *in the Atlantic Ocean* (1966).

Zheng Bao-Shan

Zheng's Loach *Heminoemacheilus zhengbaoshani* SQ Zhu & WX Cao, 1987

Zheng Bao-Shan is a Chinese ichthyologist who worked at the Institute of Zoology, Chinese Academy of Sciences and at the Beijing Museum of Natural History. He wrote: *The Fishes of Tumenjiang River* (1981). He was honoured for his contributions to Chinese ichthyology.

Zhirmunski

Seven-lined Prickleback *Ernogrammus zhirmunskii* AI Markevich & VE Kharin, 2011

Alexei Viktorovich Zhirmunsky (1921–2000) of the Russian Academy of Sciences was the founder and Director of the Institute of Marine Biology (Far Eastern Branch of the Russian Academy of Sciences, Vladivostok).

Zhou

Zhou's Scarlet Goby *Rhinogobius zhoui* F Li & JS Zhong, 2009

Zhou Hang of Shenzen, Guandong Province, China, is an ichthyologist and aquarist as well as an illustrator and photographer. He supplied the holotype and photographs of it

Zhu

Chinese Cyprinid sp. *Altigena zhui* C-Y Zheng & YR Chen, 1989
[Syn. *Bangana zhui*]

Goosefish sp. *Sladenia zhui* Y Ni, H-L Wu & S Li, 2012

Professor Zhu Yuang-Ding (1896–1986) was Director of the Shanghai Fisheries Institute, Shanghai, China (1958). He wrote: *Contributions to the ichthyology of China* (1930) and also co-edited: *Fishes of East China Sea* (1963), He was honoured as the co-author of an unpublished (1963) manuscript in which the cyprinid species was first described.

Ziebell

Ziebell's Handfish *Brachiopsilus ziebelli* PR Last & DC Gledhill, 2009

Alan Ziebell is a professional diver and abalone gatherer in Tasmania. He collected some of the first specimens and kept them in a marine aquarium at his home.

Zill

Zill's Tilapia *Coptodon zillii* P Gervais, 1848
[Alt. Redbelly Tilapia, St Peter's Fish]

Monsieur Zill was referred to by Gervais as "*...the distinguished naturalist*" (translation) who collected the type and sent it to Gervais at the MNHN, Paris. He is also known to have sent reptile specimens to Gervais from Algeria (1840s); despite which, his first name does not seem to have been recorded.

ZIN

Eelpout sp. *Aiakas zinorum* ME Anderson & AE Gosztonyi, 1991

ZIN is explained by the authors as being a shortened form of ZIAN – the Russian abbreviation for Zoological Institute, Academy of Sciences, USSR. The plural form (*-orum*) is intended to "*...honor the ichthyologists there who study the fishes of the Southern Ocean, and who have helped the senior author with many problems during his two visits, and who graciously permitted us to describe this new species.*"

ZISP

Snailfish sp. *Careproctus zispi* A Andriashev & DL Stein, 1998
Loach sp. *Micronemacheilus zispi* AM Prokofiev, 2004

ZISP stands for the 'Zoological Institute of the Russian Academy of Sciences', St Petersburg.

Zizette

Mentawai Sole *Aseraggodes zizette* JE Randall & M Desoutter-Meniger, 2007

Zizette is the nickname of Marie-Louise Bauchot (see **Bauchot**).

Zoe

Tetra sp. *Hyphessobrycon zoe* TC Faria, FCT Lima & WB Wosiacki, 2020

This refers not to a person, but a people; it is named in honour of the Zo'é, a Tupi-speaking people from the state of Pará, northern Brazil, where the tetra is found.

Zollinger

Waspfish sp. *Vespicula zollingeri* P Bleeker, 1848
Hillstream Loach sp. *Balitoropsis zollingeri* P Bleeker, 1853

Heinrich Zollinger (1818–1859) was a Swiss botanist, naturalist and explorer. He moved to Java (1842) to work in a botanical garden, and on small government-financed scientific expeditions. He was honoured because he gave his collection of Macassar (Indonesia) fishes to Bleeker.

Zorro (Fiction)

Snaggle-toothed Snake-eel *Aplatophis zorro* JE McCosker & DR Robertson 2001

This is not, strictly speaking, an eponym as it is named after a fictitious person: 'Zorro' who was a character created (1919) by New York-based pulp writer Johnston McCulley. He was a dashing swordsman hero who has been featured in numerous books, films, television series and other media, who made a Z-shaped cut with his sword as a trademark. The fish was named for the "*...remarkable coloration of the pore pattern along the face, reminiscent of the slash mark of the swordsman Zorro.*" (See **Zorro, M** below)

Zorro, M

Characin sp. *Myloplus zorroi* MC Andrade, MLAMAF Jégu & T Giarrizzo, 2016

Dr Mauricio Camargo-Zorro is a researcher at the Instituto Federal de Educação, Ciência e Tecnologia, São Paulo. He holds a bachelor's degree in biology and master's degree in zoology in addition to his doctorate, also in zoology. He is a specialist in fisheries policies for Amazonian countries, teaches on the environment and fishing resources at the Federal Institute of Paraíba, and is an Associate Researcher at the Mamirauá Institute for Sustainable Development. He was honoured for his: "*invaluable*" contribution to the fish fauna inventory from the Marmelos Conservation Area (boundaries of Amazonas and Rondônia states, Brazil. The binomial also is an allusion to the well-known fictional character, Zorro (see above), the *nom de guerre* of Don Diego de la Vega, because its features had masked its true identity.

Zu

Ribbonfish genus *Zu* V Walters & JE Fitch, 1960

Zu was the storm god of Babylonian mythology. The etymology commented that it was: "*...a fitting name ... since, until recent years, these fishes were known mainly from individuals found cast ashore in the wakes of storms.*"

Zuanon

Thorny Catfish sp. *Doras zuanoni* MH Sabaj Pérez & JLO Birindelli, 2008
Ghost Knifefish sp. *Sternarchogiton zuanoni* CD de Santana & RP Vari, 2010
Armoured Catfish sp. *Spectracanthicus zuanoni* CC Chamon & LH Rapp Py-Daniel, 2014

Dr Jansen Alfredo Sampaio Zuanon is a Brazilian ichthyolgist and ecologist. His bachelor's degree was awarded by the Paulista Júlio de Mesquita Filho State University (1985), his master's by the National Institute of Amazonian Research (1990) and his doctorate by the State University of Campinas (1999). He is presently a researcher at the National Research Institute of the Amazon, Manaus. He discovered *Doras zuanoni* and was the first ichthyologist to collect and identify *Spectracanthicus zuanoni* as being a new species.

Zuge

Pale-edged Stingray *Telatrygon zugei* JP Müller & FGJ Henle, 1841

This species' binomial *zugei* is after *zugu-ei*, the Japanese name for this species.

Zugmayer

Baluchistan Torpedo *Torpedo zugmayeri* R Engelhardt, 1912
[Status unclear: may be a junior synonym of *T. sinuspersici*]
Arrowtail (pelagic cod) *Melanonus zugmayeri* JR Norman, 1930
Zugmayer's Slickhead *Bathytroctes zugmayeri* HW Fowler, 1934

Professor Dr Erich Johann Georg Zugmayer (1879–1938) was an Austrian explorer, herpetologist and ichthyologist at the Bavarian State Zoological Collection, Munich. He visited Iceland (1902) and also explored the area around Lake Urmia, Persia (Iran), and collected in Tibet, Ladakh and Baluchistan (then part of India, now of Pakistan). He wrote, or co-wrote a number of papers and longer works including: *Die Fische von Balutschistan* (1913). Two reptiles, an amphibian and two birds are named after him.

Zuloaga

Thorny Catfish sp. *Doraops zuloagai* LP Schultz, 1944

Dr Guillermo Zuloaga Ramirez (1904–1984) was a Venezuelan geologist and naturalist who was a friend of the Phelps (q.v.) family. His doctorate was awarded by the Massachusetts Institute of Technology. He taught geology in Caracas' Central University and established the Ministry of Mines & Petroleum. He joined Creole Petroleum (1939), then its board of directors (1948). Two birds are named after him.

Zuneve

Headstander Characin sp. *Chilodus zunevei* J Puyo, 1946

M. (probably Monsieur) Zunêve is stated to have been a service agent, Eaux et Forêts (Waters and Forests), French Guiana. He provided the holotype, but we have been unable to find anything more about him.

Zur

Marg's Pygmy-goby *Trimma zurae* R Winterbottom, M van N Erdmann & NKD Cahyani, 2014

Margaret Zur is an ichthyology technician at the Royal Ontario Museum. She has, for several decades, curated both their material and borrowed material of Trimma for Winterbottom's research program. As well as this, she has been taking and compiling reams of data on the specimens, plotting distributions, summarizing results, and various other tasks; her work has been indispensable in keeping Winterbottom's Trimma project viable.

Zurstrassen

Goby sp. *Stenogobius zurstrasseni* CML Popta,1911

Dr Otto Karl Ladislaus zur Strassen (1869–1961) was a German zoologist. His doctorate was awarded by the University of Leipzig (1892). He was a member of the 'Valdivia' expedition (1898–1899) and was appointed Associate Professor of Zoology at Leipzig (1901). He was Director of the Senckenberg Natural History Museum (1909–1934) and Professor of Zoology at the University of Frankfurt (1914–1937). A bird is named after him.

Zvezda

Greeneye sp. *Chlorophthalmus zvezdae* AN Kotlyar & NV Parin, 1986

'Zvezda' is the name of the trawler from which the holotype was collected.

Zvonimir

Zvonimir's Blenny *Parablennius zvonimiri* G Kolombatović, 1892

Demetrius Zvonimir (d.1089) was King of Croatia and Dalmatia. The blenny was first identified from the Dalmatian coast. However, as the original text does not give an etymology, this remains a 'best guess' identification.